Handbook of
Thin Film Materials

Volume 4

Semiconductor and Superconductor Thin Films

Edited by

Hari Singh Nalwa, M.Sc., Ph.D.
Stanford Scientific Corporation
Los Angeles, California, USA

Formerly at
Hitachi Research Laboratory
Hitachi Ltd., Ibaraki, Japan

ACADEMIC PRESS
A Division of Harcourt, Inc.

San Diego   San Francisco   New York   Boston   London   Sydney   Tokyo

ACADEMIC PRESS
A division of Harcourt, Inc.
525 B Street, Suite 1900, San Diego, CA 92101-4495, USA
http://www.academicpress.com

Academic Press
Harcourt Place, 32 Jamestown Road, London, NW1 7BY, UK
http://www.academicpress.com

Library of Congress Catalog Card Number: 00-2001090614
International Standard Book Number, Set: 0-12-512908-4
International Standard Book Number, Volume 4: 0-12-512912-2

Printed in the United States of America
01 02 03 04 05 06 07 MB 9 8 7 6 5 4 3 2 1

1002666992

To my children
Surya, Ravina, and Eric

Preface

Thin film materials are the key elements of continued technological advances made in the fields of electronic, photonic, and magnetic devices. The processing of materials into thin-films allows easy integration into various types of devices. The thin film materials discussed in this handbook include semiconductors, superconductors, ferroelectrics, nanostructured materials, magnetic materials, etc. Thin film materials have already been used in semiconductor devices, wireless communication, telecommunications, integrated circuits, solar cells, light-emitting diodes, liquid crystal displays, magneto-optic memories, audio and video systems, compact discs, electro-optic coatings, memories, multilayer capacitors, flat-panel displays, smart windows, computer chips, magneto-optic disks, lithography, microelectromechanical systems (MEMS) and multifunctional protective coatings, as well as other emerging cutting edge technologies. The vast variety of thin film materials, their deposition, processing and fabrication techniques, spectroscopic characterization, optical characterization probes, physical properties, and structure-property relationships compiled in this handbook are the key features of such devices and basis of thin film technology.

Many of these thin film applications have been covered in the five volumes of the *Handbook of Thin Film Devices* edited by M. H. Francombe (Academic Press, 2000). The *Handbook of Thin Film Materials* is complementary to that handbook on devices. The publication of these two handbooks, selectively focused on thin film materials and devices, covers almost every conceivable topic on thin films in the fields of science and engineering.

This is the first handbook ever published on thin film materials. The 5-volume set summarizes the advances in thin film materials made over past decades. This handbook is a unique source of the in-depth knowledge of deposition, processing, spectroscopy, physical properties, and structure–property relationship of thin film materials. This handbook contains 65 state-of-the-art review chapters written by more than 125 world-leading experts from 22 countries. The most renowned scientists write over 16,000 bibliographic citations and thousands of figures, tables, photographs, chemical structures, and equations. It has been divided into 5 parts based on thematic topics:

> Volume 1: Deposition and Processing of Thin Films
> Volume 2: Characterization and Spectroscopy of Thin Films
> Volume 3: Ferroelectric and Dielectric Thin Films
> Volume 4: Semiconductor and Superconductor Thin Films
> Volume 5: Nanomaterials and Magnetic Thin Films

Volume 1 has 14 chapters on different aspects of thin film deposition and processing techniques. Thin films and coatings are deposited with chemical vapor deposition (CVD), physical vapor deposition (PVD), plasma and ion beam techniques for developing materials for electronics, optics, microelectronic packaging, surface science, catalytic, and biomedical technological applications. The various chapters include: methods of deposition of hydrogenated amorphous silicon for device applications, atomic layer deposition, laser applications in transparent conducting oxide thin film processing, cold plasma processing in surface science and technology, electrochemical formation of thin films of binary III–V compounds, nucleation, growth and crystallization of thin films, ion implant doping and isolation of GaN and related materials, plasma etching of GaN and related materials, residual stresses in physically vapor deposited thin films, Langmuir–Blodgett films of biological molecules, structure formation during electrocrystallization of metal films, epitaxial thin films of intermetallic compounds, pulsed laser deposition of thin films: expectations and reality and b″-alumina single-crystal films. This volume is a good reference source of information for those individuals who are interested in the thin film deposition and processing techniques.

Volume 2 has 15 chapters focused on the spectroscopic characterization of thin films. The characterization of thin films using spectroscopic, optical, mechanical, X-ray, and electron microscopy techniques. The various topics in this volume include: classification of cluster morphologies, the band structure and orientations of molecular adsorbates on surfaces by angle-resolved electron spectroscopies, electronic states in GaAs-AlAs short-period superlattices: energy levels and symmetry, ion beam characterization in superlattices, *in situ* real time spectroscopic ellipsometry studies: carbon-based materials and metallic TiNx thin films growth, *in situ* Faraday-modulated fast-nulling single-wavelength ellipsometry of the growth of semiconductor, dielectric and metal thin films, photocurrent spectroscopy of thin passive films, low frequency noise spectroscopy for characterization of polycrystalline semiconducting thin films and polysilicon thin film transistors, electron energy loss spectroscopy for surface study, theory of low-energy electron diffraction and photoelectron spectroscopy from ultra-thin films, *in situ* synchrotron structural studies of the growth of oxides and metals, operator formalism in polarization nonlinear optics and spectroscopy of polarization inhomogeneous media, secondary ion mass spectrometry (SIMS) and its application to thin films characterization, and a solid state approach to Langmuir monolayers, their phases, phase transitions and design.

Volume 3 focuses on dielectric and ferroelectric thin films which have applications in microelectronics packaging, ferroelectric random access memories (FeRAMs), microelectromechanical systems (MEMS), metal–ferroelectric–semiconductor field-effect transistors (MFSFETs), broad band wireless communication, etc. For example, the ferroelectric materials such as barium strontium titanate discussed in this handbook have applications in a number of tunable circuits. On the other hand, high-permittivity thin film materials are used in capacitors and for integration with MEMS devices. Volume 5 of the *Handbook of Thin Film Devices* summarizes applications of ferroelectrics thin films in industrial devices. The 12 chapters on ferroelectrics thin films in this volume are complimentary to Volume 5 as they are the key components of such ferroelectrics devices. The various topics include electrical properties of high dielectric constant and ferroelectrics thin films for very large scale integration (VLSI) integrated circuits, high permittivity (Ba, Sr)TiO$_3$ thin films, ultrathin gate dielectric films for Si-based microelectronic devices, piezoelectric thin films: processing and properties, fabrication and characterization of ferroelectric oxide thin films, ferroelectric thin films of modified lead titanate, point defects in thin insulating films of lithium fluoride for optical microsystems, polarization switching of ferroelecric crystals, high temperature superconductor and ferroelectrics thin films for microwave applications, twinning in ferroelectrics thin films: theory and structural analysis, and ferroelectrics polymers Langmuir–Blodgett films.

Volume 4 has 13 chapters dealing with semiconductor and superconductor thin film materials. Volumes 1, 2, and 3 of the *Handbook of Thin Film Devices* summarize applications of semiconductor and superconductors thin films in various types of electronic, photonic and electro-optics devices such as infrared detectors, quantum well infrared photodetectors (QWIPs), semiconductor lasers, quantum cascade lasers, light emitting diodes, liquid crystal and plasma displays, solar cells, field effect transistors, integrated circuits, microwave devices, SQUID magnetometers, etc. The semiconductor and superconductor thin film materials discussed in this volume are the key components of such above mentioned devices fabricated by many industries around the world. Therefore this volume is in coordination to Volumes 1, 2, and 3 of the *Handbook of Thin Film Devices*. The various topics in this volume include; electrochemical passivation of Si and SiGe surfaces, optical properties of highly excited (Al, In)GaN epilayers and heterostructures, electical conduction properties of thin films of cadmium compounds, carbon containing heteroepitaxial silicon and silicon/germanium thin films on Si(001), germanium thin films on silicon for detection of near-infrared light, physical properties of amorphous gallium arsenide, amorphous carbon thin films, high-T_c superconducting thin films, electronic and optical properties of strained semiconductor films of group V and III-V materials, growth, structure and properties of plasma-deposited amorphous hydrogenated carbon–nitrogen films, conductive metal oxide thin films, and optical properties of dielectric and semiconductor thin films.

Volume 5 has 12 chapters on different aspects of nanostructured materials and magnetic thin films. Volume 5 of the *Handbook of Thin Film Devices* summarizes device applications of magnetic thin films in permanent magnets, magneto-optical recording, microwave, magnetic MEMS, etc. Volume 5 of this handbook on magnetic thin film materials is complimentary to Volume 5 as they are the key components of above-mentioned magnetic devices. The various topics covered in this volume are; nanoimprinting techniques, the energy gap of clusters, nanoparticles and quantum dots, spin waves in thin films, multi-layers and superlattices, quantum well interference in double quantum wells, electro-optical and transport properties of quasi-two-dimensional nanostrutured materials, magnetism of nanoscale composite films, thin magnetic films, magnetotransport effects in semiconductors, thin films for high density magnetic recording, nuclear resonance in magnetic thin films, and multilayers, and magnetic characterization of superconducting thin films.

I hope these volumes will be very useful for the libraries in universities and industrial institutions, governments and independent institutes, upper-level undergraduate and graduate students, individual research groups and scientists working in the field of thin films technology, materials science, solid-state physics, electrical and electronics engineering, spectroscopy, superconductivity, optical engineering, device engineering nanotechnology, and information technology, everyone who is involved in science and engineering of thin film materials.

I appreciate splendid cooperation of many distinguished experts who devoted their valuable time and effort to write excellent state-of-the-art review chapters for this handbook. Finally, I have great appreciation to my wife Dr. Beena Singh Nalwa for her wonderful cooperation and patience in enduring this work, great support of my parents Sri Kadam Singh and Srimati Sukh Devi and love of my children, Surya, Ravina and Eric in this exciting project.

Hari Singh Nalwa
Los Angeles, CA, USA

Contents

Chapter 1.   ELECTROCHEMICAL PASSIVATION OF Si AND SiGe SURFACES
J. Rappich, Th. Dittrich

Chapter 2.   EPITAXIAL GROWTH AND STRUCTURE OF III–V NITRIDE THIN FILMS
Dharanipal Doppalapudi, Theodore D. Moustakas

Chapter 3. OPTICAL PROPERTIES OF HIGHLY EXCITED (Al, In) GaN EPILAYERS AND HETEROSTRUCTURES

Sergiy Bidnyk, Theodore J. Schmidt, Jin-Joo Song

Chapter 4. ELECTRICAL CONDUCTION PROPERTIES OF THIN FILMS OF CADMIUM COMPOUNDS

R. D. Gould

Chapter 5. CARBON-CONTAINING HETEROEPITAXIAL SILICON AND SILICON/GERMANIUM THIN FILMS ON Si(001)

H. Jörg Osten

Chapter 6. LOW-FREQUENCY NOISE SPECTROSCOPY FOR CHARACTERIZATION OF POLYCRYSTALLINE SEMICONDUCTOR THIN FILMS AND POLYSILICON THIN FILM TRANSISTORS

Charalabos A. Dimitriadis, George Kamarinos

Chapter 7. GERMANIUM THIN FILMS ON SILICON FOR DETECTION OF NEAR-INFRARED LIGHT

G. Masini, L. Colace, G. Assanto

Chapter 8. PHYSICAL PROPERTIES OF AMORPHOUS GALLIUM ARSENIDE

Roberto Murri, Nicola Pinto

Chapter 9. AMORPHOUS CARBON THIN FILMS

S. R. P. Silva, J. D. Carey, R. U. A. Khan, E. G. Gerstner, J. V. Anguita

Chapter 10. HIGH-T_c SUPERCONDUCTOR THIN FILMS

B. R. Zhao

Chapter 11. ELECTRONIC AND OPTICAL PROPERTIES OF STRAINED SEMICONDUCTOR FILMS OF GROUPS IV AND III–V MATERIALS

George Theodorou

About the Editor

Dr. Hari Singh Nalwa is the Managing Director of the Stanford Scientific Corporation in Los Angeles, California. Previously, he was Head of Department and R&D Manager at the Ciba Specialty Chemicals Corporation in Los Angeles (1999–2000) and a staff scientist at the Hitachi Research Laboratory, Hitachi Ltd., Japan (1990–1999). He has authored over 150 scientific articles in journals and books. He has 18 patents, either issued or applied for, on electronic and photonic materials and devices based on them.

He has published 43 books including *Ferroelectric Polymers* (Marcel Dekker, 1995), *Nonlinear Optics of Organic Molecules and Polymers* (CRC Press, 1997), *Organic Electroluminescent Materials and Devices* (Gordon & Breach, 1997), *Handbook of Organic Conductive Molecules and Polymers*, Vols. 1–4 (John Wiley & Sons, 1997), *Handbook of Low and High Dielectric Constant Materials and Their Applications*, Vols. 1–2 (Academic Press, 1999), *Handbook of Nanostructured Materials and Nanotechnology*, Vols. 1–5 (Academic Press, 2000), *Handbook of Advanced Electronic and Photonic Materials and Devices*, Vols. 1–10 (Academic Press, 2001), *Advanced Functional Molecules and Polymers*, Vols. 1–4 (Gordon & Breach, 2001), *Photodetectors and Fiber Optics* (Academic Press, 2001), *Silicon-Based Materials and Devices*, Vols. 1–2 (Academic Press, 2001), *Supramolecular Photosensitive and Electroactive Materials* (Academic Press, 2001), *Nanostructured Materials and Nanotechnology*–Condensed Edition (Academic Press, 2001), and *Handbook of Thin Film Materials*, Vols. 1–5 (Academic Press, 2002). The *Handbook of Nanostructured Materials and Nanotechnology* edited by him received the 1999 Award of Excellence in Engineering Handbooks from the Association of American Publishers.

Dr. Nalwa is the founder and Editor-in-Chief of the *Journal of Nanoscience and Nanotechnology* (2001–). He also was the founder and Editor-in-Chief of the *Journal of Porphyrins and Phthalocyanines* published by John Wiley & Sons (1997–2000) and serves or has served on the editorial boards of *Journal of Macromolecular Science-Physics* (1994–), *Applied Organometallic Chemistry* (1993–1999), *International Journal of Photoenergy* (1998–) and *Photonics Science News* (1995–). He has been a referee for many international journals including *Journal of American Chemical Society, Journal of Physical Chemistry, Applied Physics Letters, Journal of Applied Physics, Chemistry of Materials, Journal of Materials Science, Coordination Chemistry Reviews, Applied Organometallic Chemistry, Journal of Porphyrins and Phthalocyanines, Journal of Macromolecular Science-Physics, Applied Physics, Materials Research Bulletin*, and *Optical Communications*.

Dr. Nalwa helped organize the First International Symposium on the Crystal Growth of Organic Materials (Tokyo, 1989) and the Second International Symposium on Phthalocyanines (Edinburgh, 1998) under the auspices of the Royal Society of Chemistry. He also proposed a conference on porphyrins and phthalocyanies to the scientific community that, in part, was intended to promote public awareness of the *Journal of Porphyrins and Phthalocyanines*, which he founded in 1996. As a member of the organizing committee, he helped effectuate the First International Conference on Porphyrins and Phthalocyanines, which was held in Dijon, France in 2000. Currently he is on the organizing committee of the BioMEMS and Smart Nanostructures, (December 17–19, 2001, Adelaide, Australia) and the World Congress on Biomimetics and Artificial Muscles (December 9–11, 2002, Albuquerque, USA).

Dr. Nalwa has been cited in the *Dictionary of International Biography, Who's Who in Science and Engineering, Who's Who in America,* and *Who's Who in the World.* He is a member of the American Chemical Society (ACS), the American Physical Society (APS), the Materials Research Society (MRS), the Electrochemical Society and the American Association for the Advancement of Science (AAAS). He has been awarded a number of prestigious fellowships including a National Merit Scholarship, an Indian Space Research Organization (ISRO) Fellowship, a Council of Scientific and Industrial Research (CSIR) Senior fellowship, a NEC fellowship, and Japanese Government Science & Technology Agency (STA) Fellowship. He was an Honorary Visiting Professor at the Indian Institute of Technology in New Delhi.

Dr. Nalwa received a B.Sc. degree in biosciences from Meerut University in 1974, a M.Sc. degree in organic chemistry from University of Roorkee in 1977, and a Ph.D. degree in polymer science from Indian Institute of Technology in New Delhi in 1983. His thesis research focused on the electrical properties of macromolecules. Since then, his research activities and professional career have been devoted to studies of electronic and photonic organic and polymeric materials. His endeavors include molecular design, chemical synthesis, spectroscopic characterization, structure-property relationships, and evaluation of novel high performance materials for electronic and photonic applications. He was a guest scientist at Hahn-Meitner Institute in Berlin, Germany (1983) and research associate at University of Southern California in Los Angeles (1984–1987) and State University of New York at Buffalo (1987–1988). In 1988 he moved to the Tokyo University of Agriculture and Technology, Japan as a lecturer (1988–1990), where he taught and conducted research on electronic and photonic materials. His research activities include studies of ferroelectric polymers, nonlinear optical materials for integrated optics, low and high dielectric constant materials for microelectronics packaging, electrically conducting polymers, electroluminescent materials, nanocrystalline and nanostructured materials, photocuring polymers, polymer electrets, organic semiconductors, Langmuir-Blodgett films, high temperature-resistant polymer composites, water-soluble polymers, rapid modeling, and stereolithography.

List of Contributors

Numbers in parenthesis indicate the pages on which the author's contribution begins.

J. V. ANGUITA (403)
Large Area Electronics Group, School of Electronics, Computing and Mathematics,
University of Surrey, Guildford, United Kingdom

G. ASSANTO (327)
Department of Electronic Engineering and National Institute for the Physics of Matter
INFM RM3, Terza University of Rome, Via della Vasca Navale 84, 00146 Rome, Italy

SERGIY BIDNYK (117)
Center for Laser and Photonics Research and Department of Physics,
Oklahoma State University, Stillwater, Oklahoma, USA

J. D. CAREY (403)
Large Area Electronics Group, School of Electronics, Computing and Mathematics,
University of Surrey, Guildford, United Kingdom

L. COLACE (327)
Department of Electronic Engineering and National Institute for the Physics of Matter
INFM RM3, Terza University of Rome, Via della Vasca Navale 84, 00146 Rome, Italy

CHARALABOS A. DIMITRIADIS (291)
Aristotle University of Thessaloniki, Department of Physics, Thessaloniki 54006, Greece

TH. DITTRICH (1)
Technische Universität, München, Physikdepartment E16, Garching 85748, Germany

DHARANIPAL DOPPALAPUDI (57)
Boston MicroSystems Inc., Woburn, Massachusetts, USA

D. F. FRANCESCHINI (649)
Instituto de Física, Universidade Federal Fluminense, Avenida Litorânea s/n, Niterói,
RJ, 24210-340, Brazil

E. G. GERSTNER (403)
Large Area Electronics Group, School of Electronics, Computing and Mathematics,
University of Surrey, Guildford, United Kingdom

R. D. GOULD (187)
Department of Physics, Thin Films Laboratory, School of Chemistry and Physics,
Keele University, Keele, Staffordshire ST5 5BG, United Kingdom

QUANXI JIA (677)
Los Alamos National Laboratory, Superconductivity Technology Center,
Los Alamos, New Mexico, USA

GEORGE KAMARINOS (291)
LPCS, ENSERG, 38016 Grenoble Cedex 1, France

R. U. A. KHAN (403)
Large Area Electronics Group, School of Electronics, Computing and Mathematics,
University of Surrey, Guildford, United Kingdom

G. MASINI (327)
Department of Electronic Engineering and National Institute for the Physics of Matter
INFM RM3, Terza University of Rome, Via della Vasca Navale 84, 00146 Rome, Italy

THEODORE D. MOUSTAKAS (57)
Department of Electrical and Computer Engineering, Boston University,
Boston, Massachusetts, USA

ROBERTO MURRI (369)
Department of Mathematics and Physics, INFM and University of Camerino,
Via Madonna delle Carceri, 62032 Camerino, Italy

H. JÖRG OSTEN (247)
Institute for Semiconductor Physics, (IHP), Frankfurt D-15236, Germany

NICOLA PINTO (369)
Department of Mathematics and Physics, INFM and University of Camerino,
Via Madonna delle Carceri, 62032 Camerino, Italy

J. RAPPICH (1)
Hahn-Meitner Institut, Abteilung Silizium-Photovoltaik, Berlin D-12489, Germany

THEODORE J. SCHMIDT (117)
Center for Laser and Photonics Research and Department of Physics,
Oklahoma State University, Stillwater, Oklahoma, USA

S. R. P. SILVA (403)
Large Area Electronics Group, School of Electronics, Computing and Mathematics,
University of Surrey, Guildford, United Kingdom

JIN-JOO SONG (117)
Center for Laser and Photonics Research and Department of Physics,
Oklahoma State University, Stillwater, Oklahoma, USA

GEORGE THEODOROU (625)
Department of Physics, Aristotle University of Thessaloniki, 540 06 Thessaloniki, Greece

B. R. ZHAO (507)
Institute of Physics, Chinese Academy of Sciences, Beijing, China

Handbook of Thin Film Materials

Edited by H.S. Nalwa

Volume 1. DEPOSITION AND PROCESSING OF THIN FILMS

Volume 2. CHARACTERIZATION AND SPECTROSCOPY OF THIN FILMS

Volume 3. FERROELECTRIC AND DIELECTRIC THIN FILMS

Volume 4. SEMICONDUCTOR AND SUPERCONDUCTING THIN FILMS

Volume 5. NANOMATERIALS AND MAGNETIC THIN FILMS

Chapter 1

ELECTROCHEMICAL PASSIVATION OF Si AND SiGe SURFACES

J. Rappich

Hahn-Meitner Institut, Abteilung Silizium-Photovoltaik, Berlin D-12489, Germany

Th. Dittrich

Technische Universität, München, Physikdepartment E16, Garching 85748, Germany

Contents

Handbook of Thin Film Materials, edited by H.S. Nalwa
Volume 4: Semiconductor and Superconductor Thin Films

ISBN 0-12-512912-2/$35.00

1. INTRODUCTION

The most important invention for the development of microelectronics was the passivation of a Si surface with a thermal oxide, which provides excellent chemical and electronic stability [1–11]. Very low densities of electronic states are reached at Si/SiO_2 interfaces, for which the chemical and electronic passivation is well investigated. However, conventional thermal oxidation is not practicable for a number of applications demanding, for example, low thermal budget oxidation or processing in wet ambient. In such cases, electrochemical passivation is a good alternative. The advantage of electrochemical passivation is that specific chemical reactions, which lead to the formation of the passivation layer, are locally activated at the Si surface by an applied electrical potential. Another point is that electrochemical reactions can be well controlled by using the measured current as a monitor for reactions and injection currents on a submonolayer scale [2, 3, 5, 12–27].

SiGe is of great interest for modern semiconductor devices because of the increased mobility of charge carriers with the incorporation of Ge into the Si lattice, as used in heterobipolar transistors [28, 29] or metal-oxide-semiconductor structures [30]. Passivation of modern devices based on SiGe need low thermal budget processing below 500°C [29], because thermal oxidation of SiGe alloys above 600°C leads to Ge segregation at the oxide/SiGe interface [31, 32] and relaxation of the strained SiGe lattice of epitaxially grown thin film of SiGe on c-Si [33]. These processes induce defects in the crystal lattice and at the interfaces, which strongly affect the electronic behavior of the device [32–34].

The structural and electronic properties of electrochemically passivated Si surfaces have not been well investigated until now. The aim of this work is to give an overview of the present state of this field. We start with a detailed description of the in situ techniques of Fourier transform infrared spectroscopy (FTIR), surface photovoltage (SPV), and pulsed photoluminescence (PL), which give information about surface chemical bonds, electronic trap states, and nonradiative recombination centers at the Si surface, respectively. The combination of these methods gives the opportunity to correlate changes in chemical bonds with changes in trapping sites or intrinsic dangling bonds ($\bullet Si \equiv Si_3$), which act as nonradiative recombination defects. Stroboscopic measurements of PL and SPV signals with laser pulses in the nanosecond range have been developed to minimize the influence of the exciting light on the electrochemical processes [35, 36]. Interface state distributions are obtained by ex situ SPV measurements [37, 38].

The second and third sections are devoted to the electronic states of hydrogenated Si surfaces and their dependence on morphology. The termination of Si surfaces by hydrogen in HF solution was extensively studied over the last 20 years [39–97]. Hydrogenation of Si surfaces plays a key role in ultraclean processing of Si [98–102] and is the initial step for following treatments, like epitaxial growth of semiconductor material on c-Si [103–108] or thin gate oxide formation [109–119]. The ideally hydrogenated Si surface is free of electronic states in the forbidden gap. Thus, electronic states at hydrogenated Si surfaces should be related to isolated defects. Chemically and/or electrically active surface sites at hydrogenated Si surfaces can serve as reaction sites for organic molecules. The surface morphology has great influence on the electronic states at hydrogenated Si surfaces. Four types of hydrogenated Si surfaces are distinguished by their basic structural properties: HF-dipped, buffered NH_4F treated, electrochemically hydrogenated, and electrochemically etched–porous silicon. The electronic surface states of hydrogenated Si surfaces are described in this section. The electrochemical hydrogenation that takes place during the so-called current transient is investigated in detail. The role of local surface reconstruction for passivation and stabilization of hydrogenated Si surfaces is discussed.

The interface between an anodic oxide and Si (or SiGe) usually has a high amount of nonradiative recombination defects. The formation of interface defects starts with the onset of oxidation reactions. The surface roughness is very important for the formation of thin gate oxides [120–123]. The influence of the initial stages of oxide deposition differs from that of "bulk oxide" formation. In the former, the initial Si surface mainly defines the electronic properties, i.e., damage to the surface can be observed during the beginning of the deposition process by the required plasma source. No change in the interface occurs if the oxide layer is thick enough to head off the energy that is incorporated by the plasma [124, 125]. The situation strongly differs when the oxide layer is grown into the silicon bulk, where the interface is permanently changed by the formation process, i.e., diffusion of atoms, ions, or molecules through the oxide layer [2, 12]. It has been shown that the concentration of nonradiative surface defects (N_S) can be influenced by electrochemical treatments. The value of N_S at the anodic oxide/Si interface can be strongly reduced by the injection of electrons at cathodic potential or by optimizing the oxidation process at anodic potentials, where N_S correlates with the oscillatory behavior of the current in diluted acidic fluoride solution. Such passivation could be of interest for in situ conditioning of sensing surface structures in liquids. The formation of thin anodic oxides in alkaline solutions is also briefly described in this section.

Thick anodic oxides have good application potential when low thermal budget processes are needed instead of thermal oxidation. For example, Si surfaces in detectors based on very pure Si were passivated with anodic oxides in the past. A well-passivated anodic oxide/Si interface is formed during thermal treatment at elevated temperatures for short times. Steps and trenches can be well passivated with an anodic oxide due to the liquid contact used. In addition, anodic oxidation is independent of the crystal orientation and leads to a rounding of convex and concave steps caused by the electric field. Anodic oxidation is advantageous for the passivation of SiGe surfaces because diffusion of Ge in the layer system is prevented.

The chapter is concluded with summary and the outlook for further processing of Si- and SiGe-based devices.

2. IN SITU CHARACTERIZATION OF SURFACE BOND CONFIGURATIONS AND ELECTRONIC SURFACE STATES

2.1. Fourier Transform Infrared Spectroscopy–Attenuated Total Reflection

Silicon surfaces that are passivated by hydrogen or oxides are characterized by different Si-H and Si-O chemical bond configurations. FTIR was used to investigate the species on silicon surfaces by their specific vibration modes (i.e., stretching, bending, or wagging) during and after electrochemical treatments. In situ infrared (IR) techniques at the electrochemical interface have been developed over the last few years [126–131].

In the case of low infrared absorption of Si-H stretching modes and strong IR absorption of water molecules, we apply internal reflection techniques in the attenuated total reflection (ATR) configuration as used by Harrick [126]. With the use of this configuration, the IR light is coupled into the sample at one side, traverses the sample, and is collected at the other side. This procedure can be performed by using semiconductors that have a broad region of transparency between absorption due to lattice vibrations and the band to band transition (usually about 500 cm^{-1} to 10,000 cm^{-1}). Total reflection occurs when the angle of incidence (α) is greater than the inverse sinus of the relation of the refractive indexes of the medium (n_1) and the sample (n_2), $\alpha > \arcsin(n_1/n_2)$. This requirement can easily be realized by semiconductors that have a great refractive index (e.g., $n_{Si} \approx 3.5$, $n_{air} = 1$, or $n_{H_2O} \approx 1.33$, hence $\alpha > 23°$). Such measurements are typically performed at an incident angle of 45°. This technique provides high sensitivity to surface species, which is typically multiplied by the number of reflections. Furthermore, the technique becomes insensitive to electrolyte absorption because there is no need for the IR light to cross the electrolyte layer. However, there is an "evanescent wave" that penetrates into the surrounding medium by about $\lambda/14$ (λ is the vacuum IR light wavelength). The energy of this wave is partially absorbed by species on the sample surface and in a thin electrolyte layer of about 0.5–1 μm thickness. The "evanescent wave" can, therefore, be used to probe the surface of the sample and the electrolyte layer.

The long optical path through the multiple internal reflection (MIR)-ATR crystal (about 70 mm) leads to an IR absorption due to the Si lattice vibrations; hence, the usable FTIR spectra have to exceed 1500 cm^{-1}. To measure strong IR absorption caused by the asymmetric Si-O-Si stretching mode at about 1100 cm^{-1}, we reduce the length of the optical path to 10 mm with the use of a semicylinder or a micro-ATR crystal (a monolayer Si-H is not detectable with these crystals). Figure 1 shows schematically the optical path of the IR radiation through different types of Si-ATR samples and the Si surface species detectable by this method. The MIR-ATR crystals are Si trapezoids (Fig. 1a, free carrier concentration $n \approx 10^{15}$ cm^{-3}, 0° to 4° miscut), which are 52 mm in length and either 1 or 2 mm in thickness, leading to about 24 or 12 useful reflections at the electrolyte side, respectively. The single internal

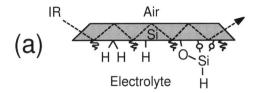

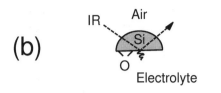

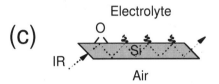

Fig. 1. Schematic drawing of the optical path of IR radiation through different types of Si-ATR samples and the Si surface species detectable. (a) MIR-ATR, trapezoid with 24 or 12 reflections per side. (b) SIR-ATR, semicylinder with one reflection. (c) Micro-ATR, parallelogram with four reflections per side.

reflection (SIR)-ATR crystal is an n-Si semicylinder (Fig. 1b, (111) oriented, $10 \times 10 \times 5$ mm^3, $n \approx 4 \times 10^{15}$ cm^{-3}), and the micro-ATR crystal is a Si parallelogram (Fig. 1c, (100) oriented, $7 \times 7 \times 0.5$ mm^3, $n \approx 2 \times 10^{15}$ cm^{-3}, four useful reflections). The incidence of the IR light was 45° for the MIR-ATR and micro-ATR and about 35° for the SIR-ATR. All crystals are cut from floatzone material. The internal reflection techniques are used to suppress the strong IR absorption of water in transmission/reflection experiments.

The electrochemical measurements are made in specially designed plastic cells, with the use of a platinum counter electrode and a 0.1 M KCl/AgCl/Ag reference electrode. Figure 2 shows the side and front views of the electrochemical cells for the respective ATR crystal ((a) MIR-ATR, (b) SIR-ATR, and (c) micro-ATR). The cells have quartz windows (W) to permit illumination for n-type Si. The reference electrode (RE) was close to the Si working electrode (WE), with one exception: the place was too small at the micro-ATR cell, so that RE had to be located outside the cell in a quartz tube. The silicon samples, mainly n-type, were contacted with InGa alloy. The sample surfaces exposed to the electrolyte were 7 cm^2 (MIR-ATR), 0.65 cm^2 (SIR-ATR), and 0.2 cm^2 (micro-ATR), respectively. A small pump and Teflon tubes were used to circulate or exchange the solution in the cells (volumes about 6 ml, 1 ml, or 0.1 ml). The electrolyte solutions were prepared from analytical grade purity reagents in triple-distilled water. The temperature of the solutions, kept under nitrogen, was $22 \pm 1°$C. The signal-to-noise ratio of a FTIR spectrum was improved by averaging 128 scans for hydrogen detection and 32–64 scans for oxide

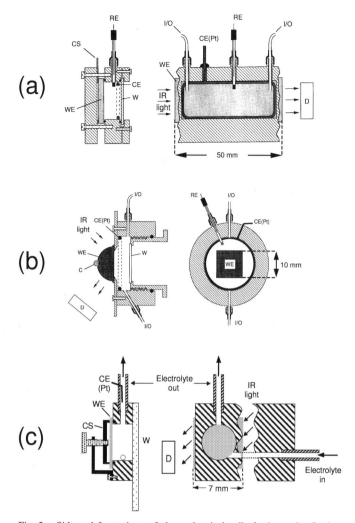

Fig. 2.   Side and front views of electrochemical cells for internal reflection configuration. (a) MIR-ATR. (b) SIR-ATR. (c) Micro-ATR. W, quartz windows; RE and WE, reference and working electrodes; C and CS, contact and spacer; I/O, inlet and outlet for the electrolyte; D, IR detector.

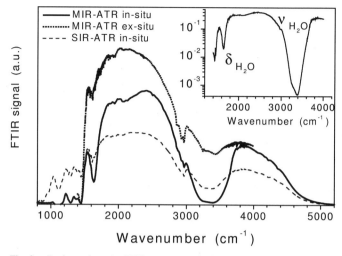

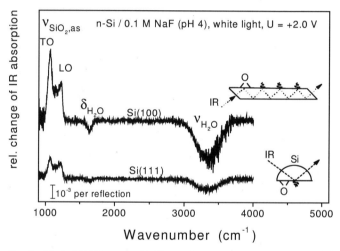

Fig. 3.   In situ and ex situ FTIR spectra as obtained with MIR- and SIR-ATR samples covered with a thin oxide layer in the respective configurations. The spectra are not corrected for the KBr beam splitter. Inset: In situ IR spectrum normalized to the ex situ IR spectrum.

Fig. 4.   Typical FTIR spectra of thin anodic oxides on n-Si(111) and (100) scaled with respect to the hydrogenated Si surfaces in the same solution, with the use of either the SIR-ATR or micro-ATR sample, as indicated. The anodic oxides were formed in 0.1 M NaF (pH 4) under white light illumination (about 10 mW/cm^2) at +2 V.

detection, with the use of a photovoltaic mercury-cadmium-telluride infrared detector with a processing time of 2 scans/s. The resolution of the IR spectra, which were recorded with a Bruker ifs 113v, was 4 cm^{-1} in aqueous solution and 1 cm^{-1} under nitrogen atmosphere.

Figure 3 shows FTIR spectra as obtained with the former presented ATR samples. The spectra are not corrected in any way and show the underlying intensity dependence due to the KBr beam splitter (400–8000 cm^{-1}, with a maximum around 2000 cm^{-1}). The solid and dotted lines denote the typical spectra for a MIR-ATR sample covered with a thin oxide layer in solution and under a nitrogen atmosphere, respectively. The inset shows the ratio of the two (please note the log scale here). The strong IR absorption due to the stretching mode, ν_{H_2O}, and bending mode, δ_{H_2O}, of liquid water around 3400 and 1650 cm^{-1} is well pronounced. In addition, the IR spectrum of the ex situ measured sample (dotted line in Fig. 3) exhibits the rotational fine structure of gaseous water, which is shifted to higher wavenumbers with respect to the liquid phase. The

IR absorption slightly below 3000 cm^{-1} is due to the plastic material (CH$_2$ groups) of the cells. The multiphonon absorption of the Si lattice vibrations leads to the strong decrease in the IR absorption below 1500 cm^{-1}. This influence can be reduced by taking a shorter optical path through the bulk Si as plotted for the SIR-ATR sample (dashed line), and the spectral region becomes available down to 900 cm^{-1}, where the asymmetric Si-O-Si stretching modes exist. Nevertheless, the sensitivity to the surface species is decreased with the reduction in the number of internal reflections. Table I summarizes the peak wavenumbers of some IR absorption due to Si surface species after [68, 117, 132–141].

Figure 4 shows typical IR spectra of thin anodic oxides on Si(111) and (100), for either the SIR-ATR sample (one reflec-

Table I.   Assignment of IR Absorption Due to Si Species

Species	Wavenumbers/cm^{-1}	Comments
Si$_3$-Si-H	2072,[a] 2071[b]	Coupled monohydride (ν_{as}) in rows at steps[a]
	2084,[a] 2083,[b] 2080[c]	Uncoupled monohydride (ν) on (111) terraces[a]
	2088.8,[a] 2088[b]	Coupled monohydride (ν_{ss}) in rows at steps[a]
Si$_2$-Si=H$_2$	2104.5,[a] 2100,[b] 2110[c]	Dihydride (ν_{ss})[a]
Si-Si≡H$_3$	ca. 2130,[a] 2134,[b] 2130[c]	Trihydride (ν_{ss} and ν_{as}) and constrained dihydride[a]
OSi-H	2100,[d] 2080[c] (ν)	
O$_2$Si-Si-H	2190,[d] 2200[c] (ν)	
O$_2$Si=H2	2200,[c] 2185[e] (ν), 975[e] (δ)	
O$_3$Si-H	2250,[c] 2245[d] (ν), 880[d] (δ)	
O$_2$Si	1060,[f] 1050,[g] 1065,[h] 1075[j]	TO3, ν_{as1}, strong
	1053 → 1060[i]	
	1170,[g] 1200[j]	TO4, ν_{as1}, weak
	1223,[g] 1252,[h] 1180 → 1217,[i] 1254[j]	LO3, ν_{as2}, strong
	1140,[g] 1165,[j]	LO4, ν_{as2}, weak
	800,[f] 820,[h]	TO2, δ

Vibration modes: ν, stretching (as, ss, asymmetric and symmetric); δ, bending.

Refs: [a][68], [b][132], [c][133], [d][134, 135], [e][136], [f][137, 138], [g][117], [h][141], [i][140] (blue-shift with increasing thickness of the oxide layer): [j][139].

tion, 64 scans) or the micro-ATR sample (four reflections, 32 scans). The anodic oxides are formed in 0.1 M NaF (pH 4) under white light illumination at +2 V. The spectra are normalized with reference to the IR spectrum of a hydrogenated Si surface in the same solution (up means an increase and down means a decrease in IR absorption of the respective surface species). The well-known splitting of the IR absorption at the asymmetric Si-O-Si stretching mode, $\nu_{SiO_2,as}$, into the parallel TO mode at 1050 cm^{-1} and the perpendicular component, LO mode at 1230 cm^{-1}, can be seen [142, 143]. On an ordered single crystal surface only the transverse mode (TO), parallel to the surface, would be expected for normal transmission. The fact that both bands are observed is due to the random orientation of the noncrystalline oxide layers on silicon and the nonnormal incidence of the IR light, which induces IR absorption by the longitudinal mode (LO) [144]. Extremely thin oxide films on silicon give rise to an LO–TO splitting, which can be ascribed either to an increase in the depolarization field or to dipolar interactions between Si-O oscillators as islands of native oxide growing and spreading across the silicon surface [142, 145]. Moreover, Figure 4 shows a reduction of the IR absorption at the stretching mode (about 3300 cm^{-1}) and the bending mode of water (around 1650 cm^{-1}). This finding is caused by the replacement of water by the oxide layer formed at the Si sample surface, which decreases the amount of water probed by the "evanescent wave." The changes of the integrated IR absorption in the range of the stretching modes of water and SiO$_2$ are about 4 times the magnitude for both samples which is equivalent to the increase in the amount of reflections for the micro-ATR. Furthermore, this finding shows that anodic oxides on (111) or (100) oriented Si surfaces have a similar thickness of oxides prepared at the

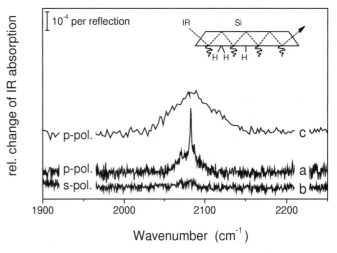

Fig. 5.   FTIR spectra of a hydrogenated Si surface under an argon atmosphere (a: p-polarized; b: s-polarized) and in contact with an electrolyte (c). The spectra are normalized to the oxidized surface in the corresponding environment.

same oxidation potential. The higher IR absorption of the TO mode in relation to the LO mode for the (100) sample points to a more ordered oxide layer on this type of surface than on the (111) oriented Si surface.

Figure 5 shows FTIR spectra of a hydrogenated Si surface under an argon atmosphere (a, b) and in contact with an electrolyte (c). The spectra are normalized with reference to the oxidized surface recorded in the same environment. The ex situ measured p-polarized spectrum (Fig. 5a) shows a sharp absorption peak of Si-H on the (111) oriented surface at 2082 cm^{-1} [59, 146] and a small amount of coupled monohydride on steps

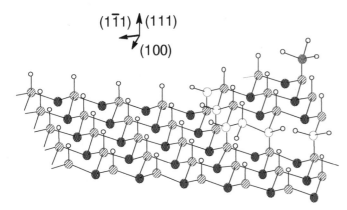

Fig. 6. Atomic arrangement of the Si(111) surface with a step on the right side (◨: Si atoms on (111) terraces—monohydrides, ○: Si atoms on steps—either monohydrides or dihydrides, ●: Si≡H₃ group).

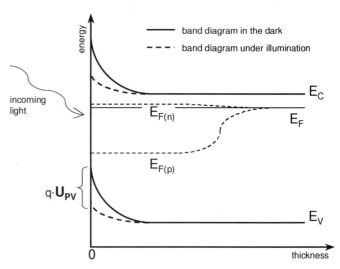

Fig. 7. Band scheme of an *n*-type semiconductor in contact with an electrolyte in the dark (solid line) and under illumination (dashed line).

at about 2070 cm^{-1} [59] (cf. Fig. 6), whereas the s-polarized spectrum (Fig. 5b) shows no pronounced features. The small absorption at the Si-H stretching mode results from the polarizer leakage and/or imperfect polarization at the Si surface [59]. The strong dependence on the polarization of the IR absorption of the Si-H species is due to the fact that the Si-H bonds are perpendicular to the (111) oriented Si surface (Fig. 6, left side). Therefore, the strongest interaction of Si-H bonds occurs with the p-polarized IR light (strongest absorption), where the electric field vector of the IR light is parallel to the Si-H bond. The line broadening of the SiH stretching mode in contact with water is presented in Figure 5c. This effect is attributed to the interaction of surface SiH groups with the water dipole of the electrolyte [147, 148].

Figure 6 shows a stick-and-ball model of the Si(111) surface with a step on the right side. The monohydrides on the (111) oriented terraces are perpendicular to the (111) surface and show a strong dependence on light polarization, which is not the case for the monohydride and dihydride species on the randomly distributed step facets.

2.2. Pulsed Surface Photovoltage

The spatial separation of excess electrons and holes leads to the development of a photovoltage. A comprehensive overview of SPV theory, experiment, and applications was recently made by Kronik and Shapira [149]. There are three basic mechanisms of charge separation: (i) built-in electric fields, (ii) different mobilities of excess electrons and holes (so-called Dember voltage), and (iii) preferential trapping of either positive or negative charges. The surface band bending (ϕ_S) can be measured by the SPV technique under in situ and ex situ conditions. In some parts of the article, the surface band bending is given in kT/q units (Y), where k, T, and q are the Boltzmann constant, temperature, and elementary charge, respectively. The SPV technique has been extended to ex situ measurement of the distribution of the surface state density (D_{it}) [37].

Figure 7 shows the band diagram of the surface region of *n*-type silicon in the dark and under illumination. In the dark, the surface Fermi-level position is usually in the midgap region due to surface states, whereas the bulk Fermi-level position (E_F) is close to the conduction band edge (E_C) for *n*-Si. The charge localized in the surface states is equal to the charge in the space charge region (Q_{SC}), with an opposite sign if the fixed charge is neglected. Therefore, the space charge region can be used to probe surface states.

Excess electrons and holes are generated under illumination, with photons having energy larger than the bandgap. The altered carrier concentrations are described by the Fermi-level splitting into separate quasi-Fermi levels for electrons and holes. The excess carriers of charge decrease the built-in electrical field. This light-induced drop of the band bending is determined as $q \cdot U_{PV}$, where U_{PV} is the surface photovoltage. It can easily be measured as the light-induced change in the contact potential difference with a Kelvin probe (for continuous wave (cw) or chopped excitation) or with an arrangement of a parallel plate capacitor (for pulsed excitation), as used in our experiments:

$$U_{PV} = \phi_S - \phi_S^0 \tag{2.1}$$

where ϕ_S and ϕ_S^0 are the surface band bending under illumination and in the dark, respectively. Please note that the reference potential of the PV measurement should not change as a result of the illumination. Because the back side of the sample is used as the reference electrode, the absorption length of the excitation light and the diffusion length of the light-induced excess carriers should be smaller than the thickness of the sample. The sign of U_{PV} is positive for *n*-type Si and negative for *p*-type Si in the case of depletion or weak inversion with respect to the illuminated front side electrode.

The concentrations of excess holes and electrons are equal in space ($\delta p = \delta n$). Assuming an unchanged space charge in the space charge region of the semiconductor in the dark and under illumination ($Q_{SC} - Q_{SC}^0 = 0$), a transcendent equation

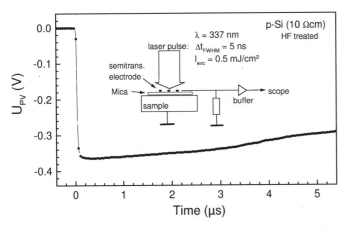

Fig. 8. Typical photovoltage transient and experimental set-up for transient PV measurements (laser pulse/semitransparent electrode/mica spacer/sample).

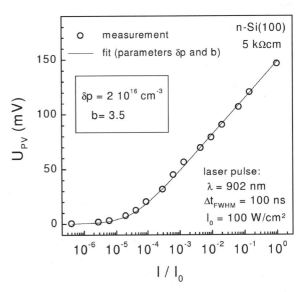

Fig. 9. Intensity dependence of the measured photovoltage amplitude for n-Si with high resistivity (open circles) and of the fitted Dember voltage (solid line).

for δp and Y_0 (in units of kt/q) was derived [150]:

$$\frac{\delta p}{n_i} = \frac{\lambda(e^{-Y_0} - e^{-Y}) + \lambda^{-1}(e^{Y_0} - e^{Y}) + (\lambda - \lambda^{-1})(Y_0 - Y)}{e^Y + e^{-Y} - 2} \quad (2.2)$$

where n_i is the intrinsic carrier concentration (1.4×10^{10} cm^{-3} for c-Si at room temperature), and λ is the doping factor ($\lambda = n_i/n$ and $\lambda = p/n_i$ for n- and p-type semiconductors, respectively).

The sign of the Dember voltage is given by the ratio of the electron and hole mobility ($b = \mu_n/\mu_p$). The Dember voltage is positive if the excess electrons are faster than the excess holes. This is the case for c-Si. The Dember voltage (in kT/q units) can be written as

$$U_D = \frac{b-1}{b+1} \cdot \ln\left(1 + \frac{\delta p}{n_i} \cdot \frac{b+1}{\lambda + b \cdot \lambda^{-1}}\right) \quad (2.3)$$

The surface photovoltage (in kT/q units) is then

$$U_{PV} = Y - Y_0 + U_D \quad (2.4)$$

The three equations (2.2), (2.3), and (2.4) give a set with four unknown values (Y_0, Y, U_D, and δp). Therefore, the excess carrier concentration should be obtained independently from the measurement of the band bending. For this purpose a new experimental parameter, the intensity of the exciting light, is introduced by the experiment. The application of well-defined laser pulses (with an intensity in the range of some W/cm^2 and a duration time on the order of nanoseconds) allows large signal SPV measurements with excellent reproducibility. A further advantage of application of nanosecond laser pulses is that trapping processes are less important. Time-dependent preferential trapping, for example, has been shown at the porous Si/Si interface [151].

Figure 8 shows a typical PV transient. The pulsed PV is measured in a parallel-plate capacitor arrangement (inset of Fig. 8). For ex situ measurements, the parallel-plate capacitor consists of a semitransparent front electrode, a thin mica spacer (thickness of the mica, some tens of micrometers), and the sample. The PV is measured with an oscilloscope via a resistance in the

GΩ range and a high impedance buffer. The maximum of the photovoltage is reached at the end of the laser pulse for n-type Si or a little bit later for p-type Si. The reason for this behavior is that the surface photovoltage and Dember voltage have opposite signs for p-type Si and therefore the PV amplitude can increase with decreasing concentration of excess charge carriers (δp) for high values of δp. The PV amplitude (U_{PV}) is measured just after the laser pulse has finished, and the band bending (Y_0) can be obtained from U_{PV} if δp is known.

Figure 9 shows the intensity dependence of the measured PV amplitude for n-Si with high resistivity (band bending is negligible because bulk and surface Fermi levels are both near midgap). The measured data are well fitted for the Dember voltage by Eq. (2.3), with $b = 3.5$ and $\delta p = 2 \times 10^{16}$ cm^{-3}. The obtained value of b is in good agreement with values published in the literature [152].

The experimentally obtained value of δp is characteristic for a given surface recombination velocity (S_0). The influence of S_0 on the maximum value of δp at the surface of n-Si ($\delta p^{(x=0)}$) is illustrated for different laser intensities (I_0) in Figure 10 (duration time of the laser pulse, 200 ns). These simulations are performed with a one-dimensional model for pulsed laser excitation (PC1D [153]). The values of $\delta p^{(x=0)}$ scale with I_0. The value of δp decreases significantly for $S_0 > 10^4$ cm/s. S_0 can be expressed by $\sigma \cdot \nu \cdot N_S$, where σ, ν, and N_S are the surface recombination cross section (on the order of 10^{-15} cm^{-3}), the thermal velocity of excess carriers, and the concentration of surface recombination defects, respectively. The surface recombination becomes important for the determination of the maximum excess carrier concentration if $N_S \geq 10^{12}$ cm^{-2}.

Figure 11 shows the simulated surface PV amplitude of n-Si as a function of the surface recombination velocity under pulsed laser excitation for different surface potentials (PC1D simulations [153]). δp is larger by one order of magnitude than the equilibrium carrier concentration, which is usually the case for

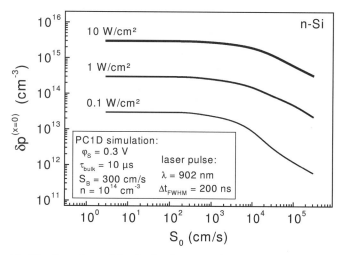

Fig. 10. Dependence of the simulated excess carrier concentration at the surface of n-Si on the surface recombination velocity under pulsed laser excitation for different intensities [153].

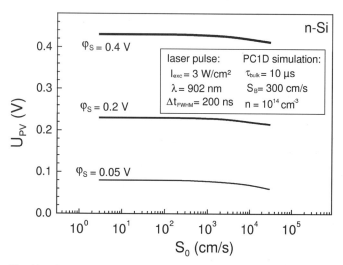

Fig. 11. Simulated photovoltage (PV) amplitude at the surface of n-Si as a function of the surface recombination velocity under pulsed laser excitation for different surface potentials [153].

our PV experiments. U_{PV} is larger than ϕ_S for low values in S_0 due to the Dember voltage, and U_{PV} starts to decrease remarkably for $S_0 > 10^4$ cm^{-3}. The decrease in U_{PV} at larger S_0 does not seem to be significant. However, it is a serious source of error in determining the distribution of the surface state density (D_{it}). This source of errors can be slightly reduced by increasing the equilibrium carrier concentration, which leads to a reduction of the Dember voltage. But, in this case, the accuracy of the determination of δp decreases, and the sensitivity of the SPV technique for measuring D_{it} decreases.

The ability to measure Y_0 with high accuracy by PV was used to obtain $D_{it}(E - E_i)$, where E_i is the bulk Fermi level of the intrinsic semiconductor [37, 154]. For such a measurement, a field voltage (U_F) is applied to the back contact of the Si sample for a certain period of time. The buffer is opened only during the PV measurement to avoid destruction of the buffer during switching of U_F. After the PV transient is recorded, a U_F pulse of identical time and amplitude but opposite sign is applied to discharge slow states [38].

The field voltage influences a charge at the semitransparent counter-electrode (Q_G). The value of Q_G is given by U_F and the thickness of the mica spacer (C_i, insulator capacitance). The condition of charge neutrality of the system is

$$Q_{SC} + Q_{it} + Q_G + Q_{fix} = 0 \qquad (2.5)$$

where $Q_{SC}(Y_0)$, Q_{it}, and Q_{fix} are the space charge, the charge in surface states, and the fixed charge, respectively. A variation in Q_G causes variations in Q_{SC} and Q_{it}. A change in Q_{SC} means that the surface Fermi level changes and, therefore, so does the charge in the surface states. The variation of Q_{it} with Y_0 determines the surface state distribution (D_{it}), which is given by

$$D_{it} = -\frac{1}{q}\frac{dQ_{it}}{dY_0} \qquad (2.6)$$

Using Eq. (2.5) as $dQ_{SC} + dQ_{it} + dQ_G = 0$, $dU_F = dU_i + dY_0$, and $dQ_G = C_i \cdot dU_i$, where U_i is the voltage drop across the mica spacer, the following expression for D_{it} can be derived [154]:

$$q \cdot D_{it} = C_i \cdot \left(\frac{dU_F}{dY_0} - 1\right) + \frac{dQ_{SC}(Y_0)}{dY_0} \qquad (2.7)$$

Equation (2.7) contains values that can be determined only experimentally. The SPV method works well when the surface is in depletion or weak inversion. The accuracy for determining δp and Y_0 can be increased by the so-called double-pulse method when the intensity of the exciting laser pulse is changed and $I_1/I_2 = \delta p_1/\delta p_2$ is considered [155]. It should be noted that the component of PV induced by preferential trapping cannot be separated in this kind of experiment, but its influence on the measurement can be minimized with the use of short laser pulses in the nanosecond range.

The minimal density of surface states (D_{it}^{min}) is reached near midgap for Si/SiO$_2$ interfaces. These so-called midgap states are fast traps with large capture cross section for both electrons and holes. The density of midgap states has to be minimized for electronic applications. The sensitivity of the SPV method to D_{it}^{min} (ΔD_{it}^{min}) is given by analysis of the turning point in the $U_{PV}(U_F)$ dependence. ΔD_{it}^{min} is limited by second term in Eq. (2.7). A serious source of experimental errors is the imperfect homogeneity of the mica spacer. Values for ΔD_{it}^{min} of about 10^{10} and 10^9 eV^{-1} cm^{-2} can be reached for dopant concentrations of 10^{15} and 10^{13} cm^{-3}, respectively.

The absolute values of fixed charges and of charged surface states cannot be obtained easily. The calculation of Q_{it} by integration over the acceptor and donor states demands a detailed knowledge of the distribution and character of involved defects. This renders an accurate determination of Q_{fix}. For most applications a knowledge of D_{it} is sufficient. Changes in Q_{fix} can be measured on the basis of shifts of the U_{PV} (U_F) characteristics along the U_F axis.

The experimental situation changes for in situ SPV measurements during electrochemical treatments. A typical exper-

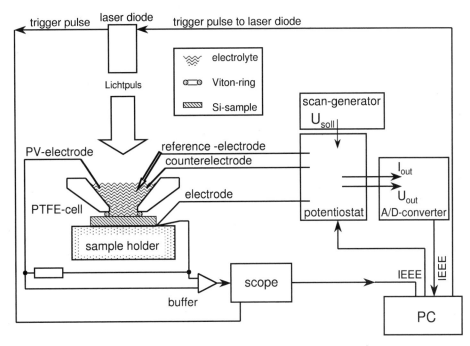

Fig. 12. Experimental setup for in situ SPV measurements during electrochemical treatment of semiconductor surface processing. Reprinted with the permission of the American Institute of Physics [36], copyright 2001.

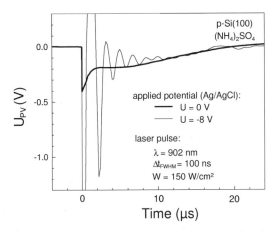

Fig. 13. Typical in situ measured PV transients during potentiostatic control of p-Si in $(NH_4)_2SO_4$ at 0 V (thick solid line) and −8 V (thin solid line). Reprinted with the permission of the American Institute of Physics [36], copyright 2001.

imental setup is shown in Figure 12 for PV measurements during electrochemical treatments. The band bending in the semiconductor can usually be obtained from the PV amplitude. The capacitance of the Helmholtz layer at the semiconductor/electrolyte interface is much larger than the space charge capacitance of the semiconductor. This makes a SPV analysis according to Eq. (2.7) impossible.

Figure 13 shows typical PV transients of p-Si in $(NH_4)_2SO_4$ at 0 V (thick solid line) and −8 V (thin solid line) during potentiostatic treatment plotted on a longer time scale. The PV transient at −8 V is very different from that measured at 0 V or the PV transients measured ex situ. First, the PV amplitude (not shown) is much higher than the band gap, and, second, the PV transient exhibit damped oscillations. The value of $q \cdot U_{PV}$ may be much larger than the band gap because the applied potential drops across the semiconductor. The oscillations in the PV transient at −8 V are caused by the time constant of the potentiostatic control. Therefore, the internal amplifier of the potentiostat should be slow enough if high U_{PV} is measured with high accuracy.

2.3. Pulsed Photoluminescence

Crystalline silicon (c-Si) is an indirect semiconductor. The forbidden band gap of c-Si is 1.1 eV. The rate of radiative interband recombination is very low for indirect semiconductors because phonons are involved during the transition process. Therefore, the radiative recombination lifetime is very large (more than 10 ms). Nonradiative recombination processes such as Shockley–Read–Hall (SRH) recombination are usually much faster. For this reason, the efficiency of the interband luminescence at room temperature is very low because of the high efficiency of nonradiative recombination processes. In other words, the recombination is dominated by nonradiative bulk and/or surface recombination processes.

The radiative interband recombination can be measured by PL techniques. The quenching of the PL signal contains information about nonradiative recombination. This circumstance has been used to monitor the change in nonradiative surface recombination during surface treatments and processing of c-Si by considering no change in the bulk lifetime.

The interband PL of c-Si can be excited with pulsed or cw lasers. The measured PL intensity is rather low for c-Si at room

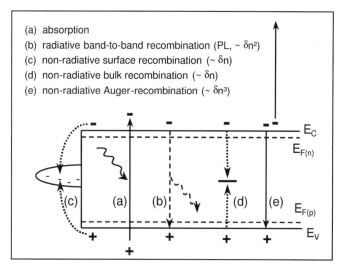

Fig. 14. Overview of elementary processes at a semiconductor surface under strong illumination.

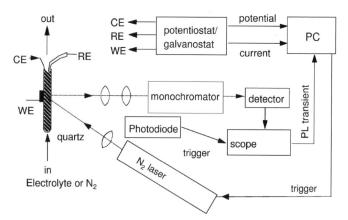

Fig. 15. Experimental set-up for in situ photoluminescence (PL) measurements during electrochemical treatment, after [35]. Reprinted with the permission of the Electrochemical Society, copyright 1997.

temperature. A high excitation intensity or extreme cooling of the Si sample is required to increase the PL intensity. The main disadvantages of PL excitation by cw lasers are (i) sample heating for high excitation levels, (ii) distortion of the electrochemical process by the high amount of excess carriers, and (iii) the unsuitability of cooling below about −10°C for electrochemical processing. These disadvantages are eliminated by excitation with laser pulses in the nanosecond range because of high excess carrier concentration for a very short period of time. The PL intensity can be measured by excitation with a single laser pulse, and it can be used very nicely for in situ stroboscopic probing of a c-Si surface during electrochemical processing.

Figure 14 gives an overview of elementary processes at a semiconductor surface under strong illumination ($\delta n \gg n, p$). The relevant processes are carrier diffusion, Auger recombination, nonradiative surface and bulk SRH recombination, and bimolecular radiative recombination. The efficiency of the radiative interband recombination is proportional to the product of the excess electron (δn) and hole concentrations (δp), whereas the efficiency of the SRH nonradiative recombination is proportional to the excess electron or hole concentration. Therefore, the PL intensity increases much more strongly with increasing excitation intensity (W) than the nonradiative SRH recombination. The efficiency of the Auger recombination is proportional to $\delta n^2 \delta p$ or $\delta p^2 \delta n$, and nonradiative Auger recombination limits the PL intensity at high values of W.

A typical setup for in situ PL measurements during electrochemical processing of Si surfaces is shown in Figure 15. The sample is placed in a quartz tube, and the electrolyte is pumped continuously through the tube. The Si sample serves as a working-electrode and a Pt wire as a counter-electrode. The reference-electrode is a Calomel electrode. The PL is excited with nitrogen lasers (wavelength, 337 nm; duration time of the laser pulse, $\Delta t_{FWHM} = 0.5$ or 5 ns; W up to 10 mJ/cm^2). The intensity of the N$_2$ laser is changed over several orders of

magnitude with glass plates as filters. The laser beam is slightly focused on the sample with a quartz lens (spot diameter about 3–4 mm). The light of the radiative interband recombination is collected using a lens with a short focal length and large diameter. A quartz prism monochromator is used to select the light at a wavelength of about 1.1 μm. Integrating InGaAs photodetectors with a high impedance preamplifier (EMM, integration time on the order of 10 ms) and Si avalanche photodiodes with fast amplifiers (EMM, time resolution 3 ns) are used for the detection of the integrated and transient PL signals, respectively. The oscilloscope is triggered with a photodiode by scattered light from the N$_2$ laser. Sometimes a filter made of silicon is used as an optical band-pass filter instead of the monochromator for the light of the radiative interband recombination of c-Si. The duty cycle of the stroboscopic measurements is, with respect to the lifetimes of excess carriers, on the order of 10^{-5}. Therefore, electrochemical processes at the anodic oxide/p-Si interface are not remarkably influenced by the PL and PV measurements.

The inset of Figure 16 presents the PL spectrum of c-Si at room temperature. PL transients are measured for photon energies at which the PL intensity has a maximum (about 1.1 eV). Figure 16 shows typical PL transients for c-Si at different excitation levels (N$_2$ laser pulses, wavelength 337 nm, $\Delta t_{FWHM} = 0.5$ ns). The absorption of ultraviolet light is very strong in c-Si, and excess carriers diffuse from the near-surface region into the bulk. The fast decay of the PL intensity at the shorter times is given by the fast reduction of the excess carrier concentration δn due to diffusion and Auger recombination. The decay of the PL intensity at the longer time is given by the so-called PL lifetime, which is a combined lifetime for surface and bulk recombination. The PL lifetime is about 25 μs for the PL transients (shown in Fig. 16), regardless of the excitation level. This fact is crucial to the calibration of the pulsed PL technique.

PL transients are simulated with a simple diffusion model in which band bending is not taken into account [156]. For high excitation levels, the values of excess holes and electrons can be considered as equal ($\delta n = \delta p$), and the bands are flat at the

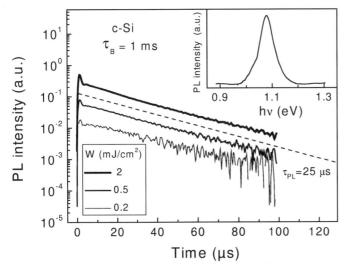

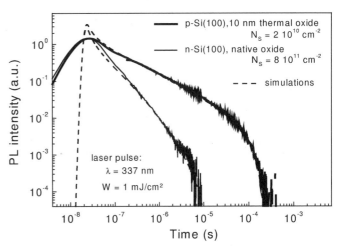

Fig. 16. Measured PL transients for c-Si at different excitation levels (N₂ laser pulses, wavelength 337 nm, $\Delta t_{FWHM} = 0.5$ ns). The inset presents the PL spectrum for c-Si at room temperature. The dashed line shows the decay of a PL transient for a PL lifetime of 25 μs. Reprinted with the permission of the American Institute of Physics [156], copyright 1999.

Fig. 17. Measured and simulated PL transients for Si surfaces covered with a thermal or native oxide.

semiconductor surface. The one-dimensional kinetic equation describing the excess carrier concentration can be written as [157]

$$\frac{\partial \delta n}{\partial t} = D \frac{\partial^2 \delta n}{\partial x^2} + G(x,t) - \frac{\delta n}{\tau_B} - \beta \delta n^2 - \gamma \delta n^3 \quad (2.8)$$

where D is the ambipolar diffusion coefficient (15 cm²/s for c-Si), $G(x,t)$ is the generation rate of nonequilibrium carriers, τ_B is the carrier lifetime in the bulk, β is the coefficient of interband radiative recombination (3×10^{-15} cm³/s for c-Si), and γ is the Auger recombination coefficient (2×10^{-30} cm⁶/s) [158].

The generation rate $G(x,t)$ can be expressed for pulsed laser excitation as

$$G(x,t) = \frac{W\alpha(1-R)}{h\nu_p(\Delta t_{FWHM}/2)\sqrt{\pi}}$$
$$\times \exp\left[-\alpha x - \frac{(t-t_0)^2}{(\Delta t_{FWHM}/2)^2}\right] \quad (2.9)$$

where W, $h\nu_p$, R (0.6 for $h\nu_p = 3.7$ eV), α (10^6 cm⁻¹ for $h\nu_p = 3.7$ eV), Δt_{FWHM}, and t_0 are the total energy density, the photon energy of the laser pulse, the reflection coefficient, the absorption coefficient, the laser pulse duration time (full width at half-maximum), and the time needed to reach the maximum light intensity, respectively.

The boundary conditions are

$$\left.\frac{\partial \delta n}{\partial x}\right|_{(x=0)} = \frac{S_f}{D}[\delta n(0,t) - \delta n_0] \quad (2.10a)$$

$$\left.\frac{\partial \delta n}{\partial x}\right|_{(x=d)} = \frac{S_b}{D}[\delta n(d,t) - \delta n_0] \quad (2.10b)$$

where d is the thickness of the sample (0.4 mm for the calculations), δn_0 is the equilibrium carrier concentration (10^{14} cm⁻³

for the calculations), and S_f and S_b are the surface recombination velocities at the front and back surfaces of the c-Si sample.

The surface recombination velocity depends on the concentration of surface nonradiative defects (N_s), their recombination cross section (σ), and the thermal velocity of the excess carriers (v, about 10^7 cm/s at room temperature):

$$S_f = S_b = \sigma v N_s \quad (2.11)$$

The recombination cross section (σ) may change over several orders of magnitude, but σ is on the order of 10^{-15} cm² for highly efficient recombination active centers [48]. Therefore, S is on the order of 100 cm/s for $N_s = 10^{10}$ cm⁻². For comparison, unusually low surface recombination velocities below 1 cm/s could be obtained for advanced c-Si (bulk lifetime larger than 100 ms) in HF [48]; i.e., N_s is below 10^8 cm⁻² in this case.

The transient and integrated PL intensities are given by

$$I_{PL}(t) = \beta \int_0^d \delta n^2(x,t)\,dx \quad (2.12a)$$

$$I_{PL}^{int} = \int_0^{t_m} I_{PL}(t)\,dt \quad (2.12b)$$

where $t_m \gg \tau_B$.

The quantum yield of PL can be calculated as

$$\eta_{PL} = \frac{I_{PL}^{int}}{(W/h\nu_p)(1-R)} \quad (2.13)$$

Measured and simulated PL transients on a log-log scale are shown in Figure 17 for Si surfaces covered with a thermal or native oxide. The bandwidth of the amplifier was 15 MHz for the measurements shown. The bulk lifetime was 0.9 ms for these samples. The first decrease in the PL intensity is mostly related to nonradiative Auger recombination. Diffusion of excess carriers dominates the decay of the PL intensity in the range between 100 ns and 20 μs, and nonradiative surface recombination limits the PL intensity at longer times for the well-passivated

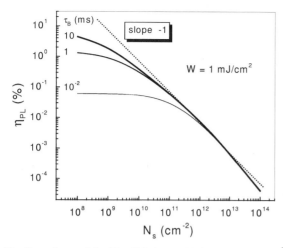

Fig. 18. Dependence of the PL efficiency on N_s for $W = 1$ mJ/cm^2 and different values of τ_B. Reprinted with the permission of the American Institute of Physics [156], copyright 1999.

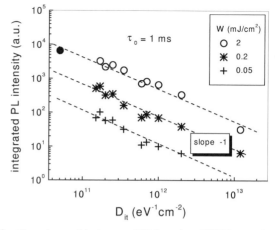

Fig. 19. Dependence of the integrated PL intensity of Si/SiO$_2$ samples on D_{it} for different excitation levels. The values of D_{it} were obtained by conventional capacitance/voltage measurements. Reprinted with the permission of the American Institute of Physics [156], copyright 1999.

sample (thermal oxide, 10 nm thick). The PL lifetime is about 40 μs for this sample. An excellent fit can be obtained for the sample covered with the thermal oxide ($N_S = 2 \times 10^{10}$ cm^{-2}). The regions where diffusion or nonradiative recombination of excess carriers dominates are not well distinguished for the sample covered with the native oxide. Nevertheless, the PL transient is well fitted for $N_S = 8 \times 10^{11}$ cm^{-2}. Thus, the absolute values of N_S can be obtained from PL transients if $\sigma = 10^{-15}$ cm^2.

The measurement of the integrated PL intensity is needed for stroboscopic in situ PL investigations of Si surfaces during electrochemical treatments. A theoretical analysis of the dependence of I_{PL}^{int} and η_{PL} on W was made with the use of Eqs. (2.12b) and (2.13) for different τ_B and N_s and for Δt_{FWHM} in the nanosecond range. The PL intensity is proportional to W^2 for W up to 1 mJ/cm^2 and saturates for higher W because of increasing influence of Auger recombination. The PL efficiency increases up to W of about 1–2 mJ/cm^2 and decreases for higher W. The excess carrier concentration is about 10^{17} to 10^{18} cm^{-3} for W of about 1 mJ/cm^2. The value of the excess carrier concentration at 1 mJ/cm^2 was also confirmed experimentally by PV measurements. The PL efficiency can reach values in the range of 1% for low N_s, high τ_B, and optimized conditions of excitation (W in the mJ/cm^2 range for Δt_{FWHM} in the nanosecond range).

The dependence of the PL efficiency on N_s is shown in Figure 18 for $W = 1$ mJ/cm^2 for different values of τ_B. The sensitivity of PL efficiency to changes in N_s (N_S sensitivity) is limited by the bulk carrier lifetime to lower values of N_S. The N_S sensitivity is in the range of 10^{11} cm^{-2} for $\tau_B = 10$ μs and can be improved to less than 10^8 cm^{-2} for $\tau_B = 10$ ms. From an experimental point of view, Si wafers with bulk lifetimes larger than 100 μs are needed to detect changes in N_S with a resolution better than 10^{10} cm^{-2}. If $\tau_B \geq 100$ μs, the PL efficiency is practically proportional to N_S^{-1} for $N_S > 10^{11}$ cm^{-2}.

The dependence of the integrated PL intensity of Si/SiO$_2$ samples on D_{it} is shown in Figure 19 for different excitation

levels. The values of D_{it} are obtained by conventional capacitance/voltage (CV) measurements. As remarked, for thermally oxidized c-Si the value of D_{it} in the minimum corresponds quite well to N_s, because intrinsically back bonded Si dangling bonds act as rechargeable and recombination centers, which dominate the interface state distribution in the range near midgap. The solid circle in Figure 19 denotes an oxidized Si sample for which no CV data but the PL intensity could be obtained. The integrated PL intensity is proportional to D_{it}^{-1} regardless of W. Therefore, the measurement of N_s with the use of stroboscopic PL excitation can be calibrated by only one set of PL and CV measurements of a Si sample (Figs. 18 and 19). This makes in situ PL measurements very manageable.

3. ELECTROCHEMICALLY HYDROGENATED Si SURFACES

Hydrogenation of Si surfaces takes place whenever an oxide layer on Si is etched back by an HF-containing solution. The formation of hydrogenated Si surfaces is one of the most important steps in device manufacturing. HF dip or buffered NH$_4$F treatments produce different kinds of surface morphology (i.e., rough or smooth), which is of interest for further processing (deposition, oxide growth, etc.). Four types of hydrogenated Si surfaces can be distinguished: (i) HF dip (a partial step in the RCA clean [99]), (ii) treatment in buffered fluoride solutions [59], (iii) electrochemical hydrogenation in diluted fluoride solutions [73, 85], and (iv) formation of porous Si (por-Si) in fluoride solutions [159]. The hydrogenation of these surfaces has been investigated by FTIR [48, 52, 53, 55, 59, 85, 147, 160] and high-resolution electron loss spectroscopy [51, 58, 67, 73, 161–164].

In this section, we show that electrochemically prepared, microscopically rough Si surfaces have very low defect concentrations, which will be interpreted by a special kind of

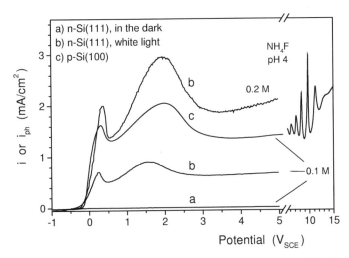

Fig. 20. Current–voltage curves of *n*-type Si(111) in the dark (a) and under white light illumination in 0.1 M or 0.2 M NH₄F (b) and of *p*-type Si(100) in 0.1 M NH₄F in the dark (c).

reconstruction of step facets. Furthermore, we show some results concerning the stability of such surfaces in the presence of oxidizing agents, acidic HF or alkaline solutions.

3.1. Electrochemical Hydrogenation in Diluted HF Solutions

The ideally hydrogenated Si(111) surface consists of atomically flat and unreconstructed facets [42, 62] and is covered by Si-H bonds [53]. Such surfaces can be prepared in concentrated buffered NH₄F solution (40%; pH 7.8) [42, 59]. The formation of flat surfaces on Si is important, for example, for heteroepitaxial deposition of other semiconductor material on Si or for the formation of thin gate oxides. Steps and kink sites open the pathway for leakage currents through an enhanced electric field at tips, which, at least, decreases the breakdown voltage. Furthermore, hydrogenated Si surfaces are free of surface states that act as recombination centers.

A well-controlled way to monitor the hydrogenation process can be performed by means of electrochemical processing. A current transient occurs at the end of the oxide dissolution in diluted HF solutions for *n*- and *p*-type material in the dark [85, 86, 164–168]. Figure 20 shows (photo)current–potential scans of *n*- and *p*-type silicon in 0.1 M and 0.2 M NH₄F (pH 4) under illumination and in the dark. At the first strong increase in the current, Si is oxidized to a divalent state and dissolves into the electrolyte, leading to a rough and at least porous structure. At higher anodic potentials, four positive charges (denoted by h⁺ (hole)), are consumed for the overall oxidation reaction [7] according to

$$Si + 2H_2O + 4\,h^+ \rightarrow SiO_2 + 4H^+ \qquad (3.1)$$

Therefore, *n*-type silicon needs illumination to ensure hole generation (electrons are majority carriers) by the incident light (curve b), whereas the hole concentration is high enough for *p*-Si (curve c), and no illumination is required (holes are ma-

jority carriers), as can be seen in Figure 20. There is no reaction (no current) of *n*-Si in the dark at anodic potentials (curve a). The Si surface is covered with oxide at potentials above the first current peak, and the thickness of the oxide layer depends on the applied potential [166, 168, 169]. Current oscillations occur when silicon is polarized around +6 to +12 V. These oscillations are damped at higher anodic potentials [168, 170].

While the oxide is being formed, it is simultaneously dissolved by small amounts of HF present in the electrolyte, leading to the well-known electropolishing behavior in such solutions [81, 171, 172]. Chemical etching of silicon oxide occurs according to the following reactions:

$$SiO_2 + 6HF \rightarrow SiF_6^{2-} + H_2O + 2H^+ \qquad (3.2)$$

$$SiO_2 + 3HF_2^- \rightarrow SiF_6^{2-} + H_2O + OH^- \qquad (3.3)$$

The etch rate for SiO₂ in fluoride solution is given by

$$k_e = a[HF] + b[HF_2^-] + c \qquad (3.4)$$

where *a* and *b* contain activation energy-type terms, $a = 2.5$, $b = 9.7$, and $c = -0.14$ [173]. It should be noted that etching by HF₂⁻ dominates at pH values above 2.5, because *b* is 4 times higher than *a*. The HF and HF₂⁻ concentrations can be calculated as follows. The solution contains F⁻ and H⁺ ions, which are coupled via dissociation reactions [174, 175],

$$k_1[HF] = H^+][F^-], \qquad k_1 = 1.3 \times 10^{-3} \quad \text{and} \quad (3.5)$$

$$k_2[HF_2^-] = [HF][F^-], \qquad k_2 = 0.104 \qquad (3.6)$$

The total concentration of fluor is given by

$$[F] = 2[HF_2^-] + [HF] + [F^-] \qquad (3.7)$$

The formation of an HF dimer, (HF)₂, is discussed for fluoride concentrations higher than 1 M [176], so that Eqs. (3.5), (3.6), and (3.7) are no longer valid at higher fluoride concentrations. Figure 21 shows the pH dependence of the concentrations of F⁻, HF, and HF₂⁻ for 0.1 M total concentration of fluoride as calculated from Eqs. (3.5) to (3.7). HF and F⁻ are the main components at lower and higher pH values, respectively. The concentration of HF₂⁻ has a maximum at a pH of about 3 and diminishes at pH values below 1 and above 5, where only HF and F⁻ exist, respectively. A solution consisting of 0.1 M NH₄F at pH 4 contains about 0.014% HF.

Figure 22 shows the behavior of the dark current transient during the etch-back process of an anodic oxide formed at +3 V. The dark current transients are monitored at different potentials after the light is switched off. At +0.5 V the well-known current transient with the typical current peak is obtained. At a potential near the flatband potential of about −0.6 V the dark current becomes slightly cathodic after a short period, and only a very small current peak can be observed. The dark current finally decreases to a constant negative value. The current peak disappears when a stronger cathodic potential (−0.9 V) is applied, and the dark current rapidly decreases to a constant value of about −50 μA/cm². The inset of Figure 22 shows the IR spectra obtained after the dark current

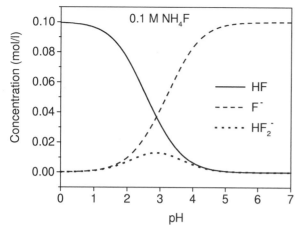

Fig. 21. Dependence of F^-, HF, and HF_2^- concentration on pH for 0.1 M total concentration of fluorine as calculated from Eqs. (3.5) to (3.7).

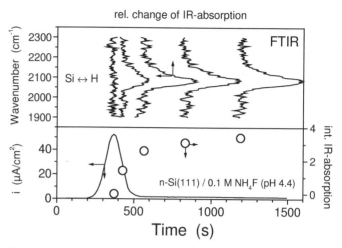

Fig. 23. Time dependence of the current (bottom) during the etch-back process of an anodic oxide in 0.1 M NH_4F (pH 4.4), and the relative change in the IR absorption in the Si-H stretching mode region measured at different times during the decay of the current (spectra are normalized to the oxidized surface). The open circles denote the integrated IR absorption as calculated from the FTIR spectra.

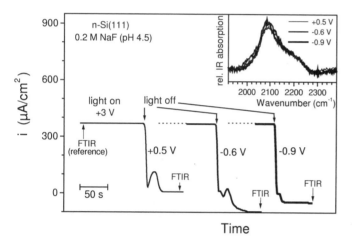

Fig. 22. Current during the etch-back process of an anodic oxide in 0.2 M NaF (pH 4.5) monitored at different potentials (a, b, and c: +0.5 V, −0.6 V, and −0.9 V) after switching off the light is switched off. The anodic oxide is formed at +3 V under illumination. The inset shows the respective IR spectra after the dark current transient has leveled out.

of Figure 23 (open circles). The H-termination is preserved over long etching periods. In addition, X-ray photoelectron measurements recorded when the current transient has decayed reveal an oxygen, fluorine, and carbon content below a tenth of a monolayer [86]. The current transient can be monitored in a high range of variation of anodic potentials at *n*-Si electrodes [166], and it was shown that the charge flow during the current transient increases slightly with increasing anodic potential at *n*-Si. The situation is different for *p*-type silicon, where holes are the majority carriers, and anodic oxidation occurs without illumination. Nevertheless, a narrow potential regime exists, located near the flatband potential between −0.4 and −0.6 V (see Fig. 20), to control the hydrogenation process and to protect the *p*-Si against oxidation reactions [35, 167, 168]. Therefore, the process of H-termination on *p*-Si surfaces can also be well monitored by measuring the time behavior of the current.

The anodic (or dark) current transient does not only depend on the potential, but also on the oxide thickness and the etch rate of the oxide, which is given by the pH of the solution used. Figure 24 shows the current transient in 0.2 M NH_4F for different pH values measured at +0.5 V. The anodic oxide is prepared in a 0.1 M solution of potassium hydrogenphthalate (pH 4) up to +10 V. The resulting thickness of the oxide layer is about 80 Å [164]. The transient occurs later in time with increasing pH (from 3 to 5.3), and the charge that passes the electrode increases. Surprisingly, the current did not decay at pH 5.3; moreover, it remains at a high level, and no hydrogenation takes place, as measured by FTIR spectroscopy [85].

The relative change in the IR absorption with respect to the Si-H-free surface is plotted in the inset of Figure 24 for different pH values of the electrolyte. The FTIR spectra are measured at a time that is two times larger than the transient width, to create comparable conditions for the experiments. The IR spectrum of the oxidized and 0.2 M NaF (pH 5.3) etched silicon

transients have reached constant values. The spectra are normalized with reference to the SiH-free and oxidized surface measured at +3 V under illumination. The IR spectra are quite similar to each other and show no distinguishing features. This means that the shape (slightly anodic or slightly cathodic) of the dark current transient has no influence on the hydrogenation process.

Figure 23 shows a series of IR spectra (top) recorded during the current transient (bottom) as a function of time. The spectra are normalized to the oxidized surface. There are no Si-H species detectable up to the maximum of the current transient (the detection limit is about 5–10% of 1 ML Si-H). The hydrogenation process starts after the maximum of the dark current transient. When the current begins to decay, the hydrogenation sets in, and the IR absorption in the range of the Si-H stretching modes increases and saturates when the dark current transient levels out. This is demonstrated by the integrated IR absorption (2000–2200 cm^{-1}), which is plotted at the bottom

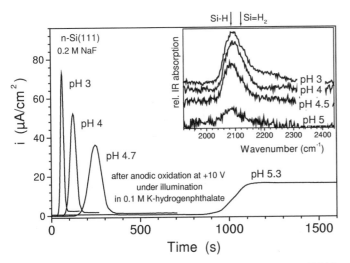

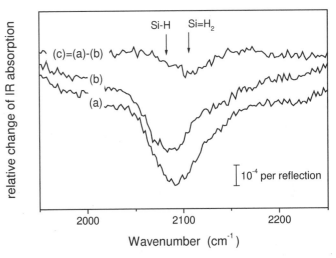

Fig. 24. Current during the etching process of an oxide-covered *n*-Si(111) surface in 0.2 M NaF at different pH values as a function of time ($U = +0.5$ V). The inset shows the respective IR spectra of the *n*-Si(111) surfaces (reference at pH 5.3—nonhydrogenated surface). The baselines of the spectra are shifted for better visualization.

Fig. 25. In situ FTIR spectra of an electrochemically hydrogenated Si surface in 0.1 M NaF (pH 4.0) obtained just after the anodic current transient has leveled out. The anodic oxidation is performed at $+1.5$ V (a) (no oscillations) and at $+6$ V (b) (with anodic current oscillation). The baselines are shifted for better visualization. Reprinted with permission of the Electrochem. Soc. Inc. [81], copyright 1994.

surface serves as our Si-H-free reference spectrum to eliminate the influence of the electrolyte on the IR absorption. A shoulder can be seen in the high-energy part of the spectrum at pH 3, which is attributed to the stretching mode of $Si=H_2$ species. With increasing pH this shoulder diminishes. This result is in good agreement with the fact that steps on a Si(111) surface have one or two dangling bonds, depending on the step orientation. The smallest amount of $Si=H_2$ was found after the two-step procedure. The treatment at pH 4 with a subsequent etching step at pH 4.9 leads to about 90–100% of a monolayer of hydrogen on the Si(111) surface [64, 76]. But there is still a slight asymmetry that points to a very small IR absorption due to $Si=H_2$. At pH 5, only a very small concentration of hydrogen silicon bonds (about 25% of spectrum d) exists on the surface.

The amount of hydrogen-silicon bonds at electrochemically hydrogenated Si surfaces depends very little on the way the oxide is formed before the etching back process. The anodic oxidation can be carried out with and without current oscillations. In the following, two different treatments are used: (i) anodic oxidation without current oscillation and subsequent etching back of the oxide layer (noOsc-surface) and (ii) anodic oxidation with current oscillation and subsequent etching back of the oxide layer (Osc-surface). The anodic oxidation and etching back of the oxide is performed in the same solution. The etch-back process is monitored by the current transient, and the hydrogenation is completed when the transient levels out.

Figure 25 compares in situ FTIR spectra of the different treatments, process (i) and (ii), applied with 0.1 M NH₄F (pH 4). The anodic oxidation is performed at $+1.5$ V for process (i) (Fig. 25a) (noOsc-surface) and at $+6$ V for process (ii) (Fig. 25b) (Osc-surface). The IR absorption due to hydrogen silicon bonds is a little bit stronger for the hydrogenated Si surface after process (i), the noOsc-surface. This stronger absorption is particularly prevalent in the region of the $Si=H_2$

bonds and, therefore, points to a higher microscopic roughness of the hydrogenated Si surface with prior electropolishing without oscillations. In addition, an increasing IR absorption due to $Si=H_2$ and $Si-H$ species has been obtained with increasing time of etching after the dark current transient has leveled out. At first, the amount of $Si=H_2$ increases and, finally, the total amount of $Si-H$ and $Si=H_2$ increases with increasing time. These processes are a result of roughening of the hydrogenated Si surface [177].

Figure 26 shows scanning-tunneling-microscopy (STM) micrographs of Si(111) surfaces hydrogenated electrochemically in 0.1 M NH₄F (pH 4) after anodic oxidation in the oscillating regime, process (ii), for 2 (a) and 30 (b) min [178]. For the STM studies *p*-type samples have been cut from a B-doped Si(111) wafer (resistivity 1 Ω cm) with a misalignment of 0.25° off the (111) orientation toward the (112) direction. The ultra-high-vacuum STM images are acquired at a constant current of 0.2 nA and a bias voltage of $+3$ V (for details see [178]). The overall surface morphology appears rough and is characterized by the formation of very flat hole-like structures with lateral dimensions of some 10 nm up to about 100 nm. The holes become more pronounced and tiny protrusions are dissolved with increasing time of anodic oxidation in the oscillating regime, resulting in a smoother surface structure at the microscopic scale. There is no indication of facet formation. Surface structures very similar to those observed in our experiments can also be produced during electrochemical oscillations in fluoride-free electrolytes [179]. It should be recalled that hydrogenated Si(111) surfaces are macroscopically flat but very rough on the microscopic scale after immersion in HF (40%). The NH₄F-treated (40%, pH 7.8) Si(111) surface is also macroscopically smooth; however, it exhibits a different microscopic structure. It consists of atomically flat terraces that are spaced

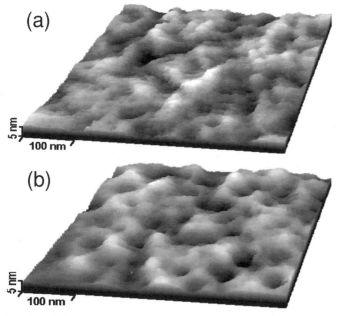

Fig. 26. STM micrographs of electrochemically hydrogenated Si(111) surfaces in 0.1 M NH₄F (pH 4.0) after anodic oxidation in the oscillating regime for 2 (a) and 30 (b) min. Reprinted with the permission of Elsevier Science B. V. [178], copyright 2000.

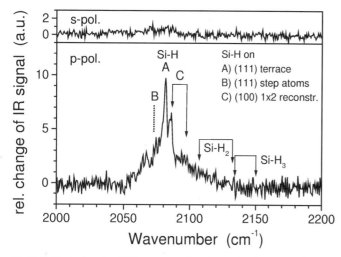

Fig. 27. Typical ex situ FTIR spectrum of an electrochemically hydrogenated Si(111) surface after anodic oxidation in the oscillating regime. The spectrum is normalized to the anodically oxidized surface. Reprinted with the permission of Elsevier Science B. V. [302], copyright 1999.

from one another by 0.31-nm-high bilayer steps. These terraces are only disturbed by point defects and triangular holes [62, 79, 178, 180].

Typical ex situ FTIR spectra of an electrochemically hydrogenated Si(111) surface, recorded after processing (ii) at +6 V, are presented in Figure 27 for s- and p-polarization of the IR light. These spectra are very different from the well-known FTIR spectra of flat hydrogenated Si(111) surfaces prepared in NH₄F (40%, pH 7.8) or HF-treated Si(111) surfaces [42, 52, 59]. The most striking feature of the FTIR spectra of electrochemically hydrogenated Si surfaces is their line broadening.

The broad IR absorption peak gives evidence of a high degree of disorder at the hydrogenated Si surface region. In addition, there is a small signal for s-polarized IR light. Furthermore, the p-polarized spectrum contains two narrow peaks in the range of the Si-H stretching modes (2081.4 and 2087 cm⁻¹), which are shifted with respect to the stretching mode of Si-H on terraces (2083 to 2084 cm⁻¹ [68, 132]). The IR absorption peaks at 2078 and 2081.4 cm⁻¹ may be due to Si-H on steps and Si-H groups on positions similar to Si-H on terraces, respectively. The peak at 2087 cm⁻¹ seems to be due to Si-H groups that are at positions similar to Si-H species on 1 × 2 reconstructed Si(100) surfaces [45]. The broad spectrum smears out in the range of the Si=H₂ stretching modes at higher wavenumbers. There is hardly any Si≡H₃ detectable on the surface. The existence of the two narrow Si-H peaks, the positions of which are different from that of the ideally hydrogenated Si(111) surface, indicates that the electrochemically hydrogenated Si(111) surface is free of well-oriented and ideally hydrogenated Si(111) facets.

3.2. Hydrogenated Si Surfaces in Alkaline Solutions

Alkaline solutions lead to a high etching rate of silicon [181–185], and the silicon remains hydrogenated [148]. Etching in alkaline solutions is of great interest for microstructuring of silicon devices (see, for example, [186]). (100) oriented Si surfaces become a pyramid-like structure in alkaline solutions that is used for light-trapping systems in solar cell devices. The reason for this behavior is that the ratios of the etch rate of dihydride on steps (SD) to monohydrides on steps (SM) and monohydride on (111) terraces (TM) in NaOH solution are about 40 and 5000, respectively [187, 188]. Therefore, (100) facets are etched much faster than (111) oriented step facets. The etching process can be stopped by applying an anodic current to the sample, which leads to a decrease in the amount of hydrogen silicon bonds [148, 184, 185]. Obviously the Si surface becomes oxidized and SiO₂ is etched much more slowly than silicon. This section shows some selected results concerning the stability of hydrogenated silicon surfaces in alkaline solutions inspected by FTIR spectroscopy.

The hydrogenation of a Si(111) surface is stable even at low anodic currents up to +3 μA/cm², as can be seen in Figure 28, where IR spectra, with the use of a MIR-ATR sample, are recorded at different current densities in 0.5 M NaOH. The spectra are scaled to the IR spectrum of the oxide-covered surface in the same solution. There is, however, only a very narrow anodic potential range in which the hydrogen-terminated surface (a) remains stable. At slightly increased anodic current above +4 μA/cm² (b and c) passivation occurs [148, 185], and the Si-H absorption gradually disappears until it vanishes (d). The electrode potential increases dramatically from −0.92 V (a) to −0.6 V (b), finally reaching +2 V (c), when passivation sets in at constant current. Obviously, a very sudden change in the mechanism of the corrosion occurs in this anodic potential range, which can be attributed to the formation of Si-OH surface bonds and their condensation to Si-O-Si bonds [148]. It

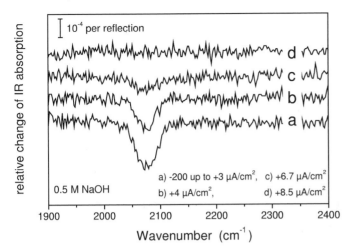

Fig. 28. Relative change in IR absorption of the hydrogenated n-Si(111) surface at different current densities in 0.5 M NaOH. (a) -200 μA/cm^2 up to $+3$ μA/cm^2. (b) $+6.7$ μA/cm^2. (c) $+8.5$ μA/cm^2. (d) As in (c), but 5 min later. The spectrum of the oxidized surface in the alkaline solution serves as the reference.

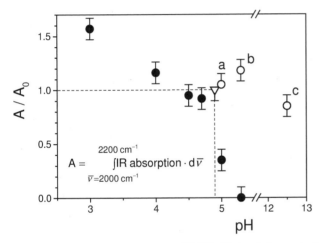

Fig. 29. Integrated IR absorption due to Si-H/Si=H$_2$ in relation to the integrated intensity at pH 4.9 with a pretreatment at pH 4.0 (A_0, open down triangle) as a function of pH (calculated from the spectra of Figs. 24 and 28). Solid circles: direct etching of the oxide in 0.2 M NH$_4$F; open circles: etching at pH 4.0 (a, c) or 4.5 (b) before etching at pH 5 (a), 5.3 (b), or 12.5 (c).

should be noted that Si-OH could not be detected by the ATR-FTIR techniques, as described in Section 2.

Figure 29 shows the integrated intensity of the IR absorption in the region of the Si-H stretching mode (2020 cm^{-1} to 2200 cm^{-1}) as a function of pH. The values are plotted in relation to the integral obtained at pH 4.9 with a pretreatment at pH 4.5 (open down triangle in Fig. 29, A_0), which corresponds to about 95% of a monolayer, as deduced from FTIR, UPS, and HREELS measurements [64, 73, 76, 85, 164]. The integral of the Si-H/Si=H$_2$ stretching mode region decreases with increasing pH. This decrease is attributed to the decrease in the Si=H$_2$ surface species due to a reduction in microscopic roughness of the silicon surface. At a pH above 4.7, the competition between the very slow etching process of the oxide layer and the etching or oxidation of the Si surface is reflected by a

strong suppression of the formation of hydrogen silicon bonds, which is completely suppressed at pH 5.3 in a 0.2 M NaF solution. A final etching step in a solution with pH 4.9 with a pretreatment at pH 4.5 leads to the smallest amount of Si=H$_2$ oscillators on the Si(111) surface. This result is in agreement with ex situ HREEL spectra, where Si=H$_2$ was present at pH 4.5, which disappeared after a subsequent dip in a solution with pH 4.9 [73]. In addition, Figure 29 reveals the stability of the hydrogenated Si surface with respect to alkaline etching processes when the acidic 0.1 M NaF (pH 4) solution is replaced with a solution of pH 5.3 or 12.5. Nevertheless, no H-terminated surfaces can be formed in solutions with such high pH values (the oxide is etched back, but no hydrogen-silicon bonds are detected).

3.3. Electronic States at Hydrogenated Si Surfaces

Another interesting point is the evaluation of defects on such H-terminated surfaces. The defect concentration, D_{it} and N_s, is related to the surface structure and surface morphology after the hydrogenated Si surface is formed. Recall that N_s is measured in situ by a pulsed PL technique and D_{it} is obtained from ex situ SPV experiments. Figure 30 shows N_s and PL intensity (top), D_{it} at midgap (middle), and the current (bottom) as a function of time during the dark current transient after the anodic oxide on n-Si(111) is etched back in 0.1 M NH$_4$F (pH 4) solution. The D_{it} value of wet anodic oxides measured with SPV is on the order of 10^{13} eV^{-1} cm^{-2}, which is one order of magnitude higher than the value obtained from PL measurements. This behavior is due to the different kind of defects measured by SPV (rechargeable defects) and PL (nonradiative recombination active defects). Both N_s and D_{it} start to decrease when the current tends to decay; i.e., hydrogenation of the surface sets in and saturates when the current transient levels out. Note that the PL intensity is limited by the lifetime of the excess charge carriers in the wafer material, so that the N_s of high-quality float zone Si could be 10^{10} cm^{-2} or less.

Anodic oxidation in the oscillating regime is important as the prior step for the formation of a hydrogenated Si surface with low density of surface states. As shown above, prolonged anodic oxidation in the oscillating regime leads to locally smoother surfaces. We note that the anodic current is lower and the maximum of the PL intensity is higher after the longer anodic oxidation in the oscillating regime. Therefore, the value of the anodic current transient, as well as the PL intensity, is a quantity describing the microscopic roughness of the Si surface under identical electrochemical conditions. It is important to note that the in situ PL intensity usually decreases in time after the hydrogenated Si surface is formed. This decrease in the PL intensity after the maximum is reached is caused by the onset of chemical etching at the Si surface in the electrolyte. The chemical etching is a dynamic process at the Si surface during which the well-passivated Si surface is disturbed and nonradiative recombination active surface defects are generated.

The maximum of the PL intensity can be strongly increased after repetition of the electrochemical treatment of anodic oxi-

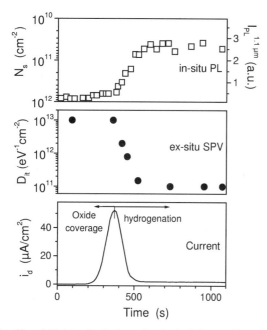

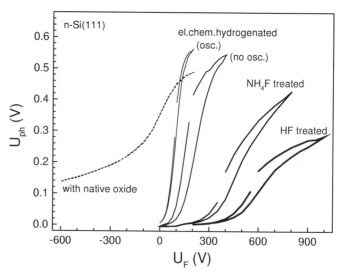

Fig. 30. N_S and PL intensity (top), ex situ obtained D_{it} at midgap (middle), and the current (bottom) as a function of time during the dark current transient after an anodic oxide was etched on n-Si(111) in 0.1 M NH$_4$F (pH 4) solution.

Fig. 31. Examples of SPV measurements of differently hydrogenated n-Si surfaces (solid lines) and of an n-Si surface covered with a native oxide (4 months of oxidation in air after treatment in NH$_4$F). Reprinted with the permission of Elsevier Science B. V. [302], copyright 1999.

dation with current oscillations and hydrogenation (process (ii), Osc-surface). Repeating this process (ii) leads only to a slight increase in the macroscopic surface roughness while tiny protrusions are dissolved or rounded. But N_S can be much more strongly reduced by this procedure than by prolonged anodic oxidation with only one oxide etch step [189]. The maximum of the PL intensity corresponds to a density of nonradiative surface defects on the order of 1×10^{10} cm^{-2}. Hydrogenation was also performed on Si surfaces after anodic oxidation when anodic current oscillations do not appear (process (i)). In this case, the microscopic roughness remains unchanged and the PL intensity does not depend on the repetition of process (i), and only a value of $N_S = 4 \times 10^{10}$ cm^{-2} could be reached [189].

The surface morphology of hydrogenated Si surfaces can be correlated with the density of surface states measured ex situ by SPV. Figure 31 shows examples of ex situ SPV measurements in an N$_2$ atmosphere of differently hydrogenated n-Si surfaces (solid lines) and of an n-Si surface covered with a native oxide (4 months of oxidation in air after treatment in NH$_4$F). For the oxidized surface, the neutral point (NP) ($U_F = 0$ V) is close to midgap, and the U_{ph} (U_F) characteristic is symmetric around NP. This is caused by the amphoteric character of the electronic states at the Si/SiO$_2$ interface, which are predominantly determined by Si dangling bonds [190, 191]. There is no hysteresis in the slope of the U_{ph} (U_F) characteristic of the oxidized Si surface. The U_{ph} (U_F) characteristic of hydrogenated Si surfaces generally shows a hysteresis, and NP is shifted to positive values of U_F. The latter is caused by an accumulation of fixed positive charges (Q_f) at the surface, so that the surface is in accumulation for p-type and in inversion for n-type Si (see also [192]). An n-type behavior of HF-treated Si surfaces has been found by Buck and McKim, who made surface

conductivity measurements [193]. The value of Q_f ranges between 10^{10} cm^{-2} in the case of electrochemical hydrogenation after anodic oxidation in the oscillating regime and 10^{12} cm^{-2} for HF-treated surfaces. The hysteresis tends to decrease with decreasing Q_f and with increasing slope of U_{ph} vs. U_F. It can be concluded that the microscopic rough hydrogenated Si surface (HF treated) has the largest concentration of surface defects, and the best passivation can be reached on the macroscopically relatively rough but microscopic smoothest surface after process (ii).

Slope, shift, and hysteresis of the U_{ph} (U_F) characteristics of hydrogenated Si surfaces sensitively depend on the regime of the SPV measurement. Figure 32 shows U_{ph} (U_F) characteristics obtained at different extensions of the U_F range for the same Si surface treated with process (ii), hydrogenation after anodic oxidation in the oscillating regime. The inset gives the distribution of surface states obtained from the SPV measurement with the lower extension of U_F. For the more extended U_F range, the value of U_F in the turning point is shifted to higher values, the hysteresis is increased, and the slope is reduced. Therefore, Q_f, hysteresis, and D_{it} are larger for larger extensions of U_F. Consequently, the distribution of D_{it} strongly depends on the condition of the measurement. To conclude, the surface states measured ex situ by SPV on hydrogenated Si surfaces are of the donor type, which have a broad distribution of trapping and detrapping times. Nevertheless, the lowest values of D_{it}^{min} can be obtained with high accuracy by turning point analysis from U_{ph} (U_F) characteristics, except when neither accumulation nor strong inversion is reached at the Si surface. The value of D_{it}^{min} is about 10^{10} eV^{-1} cm^{-2} for the n-Si(111) surface after process (ii) (see inset of Fig. 32).

For chemically hydrogenated Si(111) surfaces (preparation in buffered NH$_4$F), the lowest reported value of D_{it}^{min} is 2–5×10^{10} eV^{-1} cm^{-2} (ex situ measurements [194, 195]). Slightly

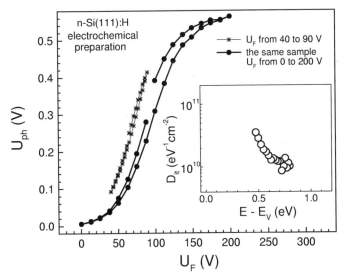

Fig. 32. U_{ph} (U_F) characteristics for different extensions of the U_F range for the same Si surface hydrogenated after anodic oxidation in the oscillating regime. The inset shows the distribution of surface states. Reprinted with the permission of Elsevier Science B. V. [302], copyright 1999.

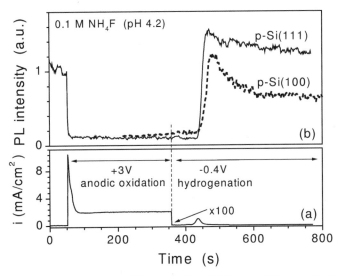

Fig. 33. Time dependence of the current (a) and PL intensity (b) for a cycle of anodic oxidation in 0.1 M NH$_4$F (pH 4.2) at +3 V followed by hydrogenation at −0.4 V for p-Si(111) and p-Si(100). The wafers are cut from the same Si ingot.

lower values of D_{it}^{min} are reached on electrochemically hydrogenated Si surfaces after process (ii). This is surprising for this surface morphology and shows that roughness on the microscopic scale is crucial. Efforts have been devoted to correlating D_{it}^{min} of chemically hydrogenated Si surfaces with surface roughness [195] determined by ellipsometry [78]. The obtained results have been interpreted in terms of a dangling bond model [196]. This model may be suitable for explaining the development of the distribution of D_{it} during the initial stages of oxidation [197], because the formation of dangling bonds at Si back bonds plays a major role [198]. However, this model is not applicable to hydrogenated Si surfaces, because the concentration of dangling bonds at a hydrogenated Si surface (in terms of dangling bond centers measured in porous silicon by electron paramagnetic resonance [199]) is much lower than D_{it} obtained by ex situ SPV.

The dangling bond concept can be applied to the in situ investigation of Si surfaces by PL during electrochemical processing. Dangling bonds are formed and passivated in the electrolyte during oxidation of Si surface atoms and etching depending on the chemical equilibrium. The surface recombination velocity is extremely low for hydrogenated Si surfaces in acidic solutions [48]. Therefore, the chemical equilibrium in acidic fluoride solution is shifted toward low oxidation rates and highly efficient passivation of surface states, probably by the reaction of protons with defects at the Si surface. Indeed, Trucks et al. [200] proposed a mechanism for the hydrogenation process on Si surfaces, which results in an extremely low etch rate for hydrogenated Si surfaces in fluoride-containing solutions. The concentration of dangling bonds on HF-treated Si surfaces is below the detection limit (less than 5×10^{10} cm^{-2}) of an electron spin resonance spectrometer and starts to increase during the initial oxidation process [201]. The development of dangling bonds, which act as nonradiative surface defects, can

also be probed in situ and ex situ by measurement of the PL quenching.

3.4. Role of the Etch Rate for Surface State Formation

The decrease in the PL intensity the maximum value is reached is correlated with a roughening of the surface on a microscopic scale during chemical etching in acidic fluoride solution. As outlined before, the IR absorption after electrochemical hydrogenation increases with increasing etch time in 0.1 M NH$_4$F (pH 4), and the increase is stronger for the Si=H$_2$ than for the Si-H bonds. The influence of surface orientation, oxidation rates, and temperature on the PL intensity measured in situ will be discussed in the following section.

The stability of the hydrogenated Si surface depends strongly on the surface orientation, as illustrated in Figure 33, which shows the current (a) and PL intensity (b) as a function of time for a cycle of anodic oxidation in 0.1 M NH$_4$F (pH 4.2) at +3 V followed by hydrogenation at −0.4 V for p-Si(111) and p-Si(100). The samples are cut from the same Si ingot to ensure the same bulk properties. Usually the maximum of the PL intensity is about 10–20% higher for Si(111) than for Si(100). The subsequent decrease in the PL intensity is much faster for the Si(100) than for the Si(111) surface. This behavior can be attributed to the higher etch rates in the (100) than in the (111) direction, as obtained in alkaline and buffered NH$_4$F solutions [188].

Figure 34 shows the influence of dissolved oxygen in the HF-containing electrolyte on the current transient and the PL intensity during and after hydrogenation of p-Si(111). The Si surface is oxidized at +8 V followed by the current transient recorded at −0.5 V in 0.1 M NH$_4$F (pH 3.5). The PL intensity increases drastically when hydrogenation occurs (decrease in N_s) and decreases with time in the HF-containing electrolyte.

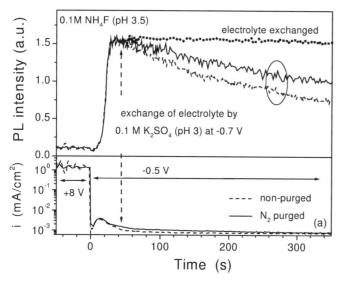

Fig. 34. Time dependence of the current (a) and PL intensity (b) for a cycle of anodic oxidation in 0.1 M NH₄F (pH 4.2) at +8 V followed by hydrogenation at −0.5 V. The electrolyte is either purged by N₂ (thin solid line) or is not (dashed line). The electrolyte is replaced with 0.1 M K₂SO₄ (pH 3) at −1 μA/cm² after the maximum of the PL intensity is reached (dotted line).

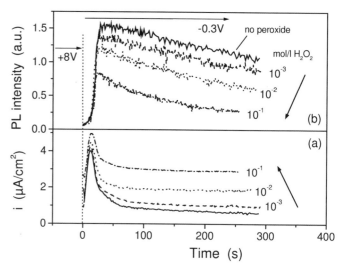

Fig. 35. Time dependence of the current (a) and PL intensity (b) for a cycle of anodic oxidation in 0.1 M NH₄F (pH 4.2) at +8 V followed by hydrogenation at −0.3 V (thick solid line). Different amounts of H₂O₂ are added to the electrolyte (dashed and dotted lines).

There is only a very narrow time window where the PL intensity is at a constant, high value directly after the hydrogenation is completed in the nonpurged solution (dashed line). This time region is somewhat prolonged when the solution is purged with N₂ before the experiment (solid line). Purging with N₂ leads to a strong reduction of oxygen in the solution. In both cases, the PL intensity also decreases with time. The time dependence of the current does not differ significantly for the purged and nonpurged solutions. The decay of the PL intensity in the diluted HF solution is due to etching of the surface (formation of nonradiative defects), which is enhanced by dissolved oxygen in the electrolyte. Oxygen in solution promotes the formation of Si-O bonds [88, 202], which are then dissolved by HF, leading to a roughening of the surface.

It is not only of interest from a practical point of view to preserve the low level of nonradiative surface defects for longer times. The decrease in the PL intensity after the maximum is reached is related to a partial destruction of the hydrogenated Si surface due to chemical etching of the Si surface. This can be avoided by replacing the acidic fluoride electrolyte with an acidic solution of, for example, 0.1 M K₂SO₄ (pH 3) at a small cathodic current density of −1 μA/cm² to protect the Si surface from oxide formation (dotted line in Fig. 34). Therefore, highly efficient etch stops can be integrated into electrochemical passivation procedures [203]. However, the cathodic current should not be too large, to avoid hydrogen evolution. The incorporation of hydrogen at cathodic potentials from the electrolyte into the top monolayers of the bulk Si leads to a partial disordering ("amorphization") [204] and to a decrease in the PL intensity [205]. Strong hydrogen evolution for very long times leads to a destruction of the p-Si surface and to the formation of etch pits [206].

The chemical oxidation rate at the Si surface in acidic fluoride solution can be strongly increased by the addition of small amounts of heavily oxidizing agents, like H₂O₂. Figure 35 shows the time dependence of the current (a) and PL intensity (b) for a cycle of anodic oxidation in 0.1 M NH₄F (pH 4.2) with different amounts of H₂O₂ at +8 V followed by hydrogenation at −0.3 V. The maximum of the PL intensity decreases, and the decrease in the PL intensity with time is faster and stronger with increasing H₂O₂ (from 10^{-3} to 10^{-1} mol/liter). This points to a faster etching of the hydrogenated Si surface.

Surface chemical reactions are thermally activated. They may also compete with surface electrochemical reactions, which are controlled by the potential. The time dependence of the PL intensity for anodic oxidation in 0.1 M NH₄F (pH 4.2) at +3 V followed by hydrogenation at −0.4 V at different temperatures is plotted in Figure 36. The maximum of the PL intensity decreases strongly with a slightly increased temperature of the solution. In contrast, the temperature dependence of the PL intensity of thermally oxidized Si is very weak in this narrow temperature range. Therefore, the strong change in the peak of the PL intensity is induced by a strong change in the rate of generation of surface defects, which act as nonradiative surface recombination defects. The PL intensity decays quite fast after it reaches its maximum for the lowest temperature (8°C), whereas the decay of the PL intensity is practically negligible for the highest temperature (44°C). The PL intensity depends only weakly on the temperature after longer etch times. This behavior shows that there is a great difference in the processes that lead either to the formation of hydrogenated Si surfaces or to chemical etching of hydrogenated Si surfaces by increasing the number of reactive surface sites

Figure 37 shows the Arrhenius plots of the inverse maximum PL intensity (open circles) and stationary PL intensity (solid circles), which were obtained during hydrogenation at

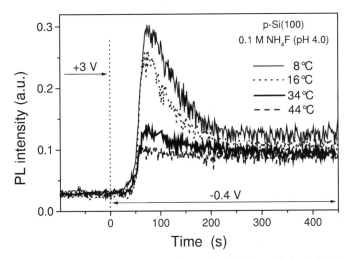

Fig. 36. Time-dependent PL intensity for a cycle of anodic oxidation in 0.1 M NH$_4$F (pH 4.2) at +3 V followed by hydrogenation at −0.4 V for different temperatures.

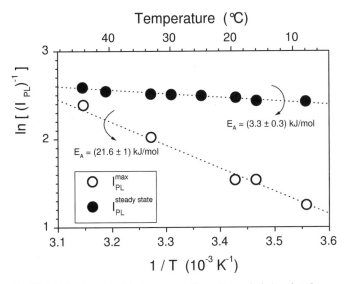

Fig. 37. Arrhenius plot of the inverse maximum (open circles) and stationary (filled circles) PL intensity obtained during hydrogenation at −0.4 V (Fig. 36). The PL intensity is corrected to the temperature dependence of the PL intensity of a well-passivated Si/SiO$_2$ sample.

−0.4 V (Fig. 36). The PL intensity is corrected to the temperature dependence of the PL intensity of a well-passivated Si/SiO$_2$ sample. The decrease in the maximum of the PL intensity is thermally activated with an activation energy of about 22 kJ/mol, whereas the activation energy at the steady-state PL intensity is only about 3.3 kJ/mol.

3.5. Local Reconstruction and Origin of Surface States

It was shown above that the local surface roughness plays an important role for the detection of surface states by SPV. The smoothest surface, i.e., the ideally hydrogenated Si(111) facet, should be free of surface states. However, a well-faceted hydrogenated Si(111) surface has a relatively large number of surface atoms at steps and corners. The bonds are weaker at such sites, and the probability of surface chemical reactions and adsorption of molecules is increased. Such defect sites cannot be avoided on a scale larger than the facets by simple chemical treatments. Furthermore, the size of a facet is thermodynamically limited.

The situation is different for electrochemical treatments. Under certain conditions, rounded shapes of Si surfaces can be created. Steps and corners are immediately smoothed by electrochemical reactions because of the strongly increased oxidation rate for the higher electric field at these sites. Therefore, all Si surface atoms are in a more or less identical position from the point of view of reactive surface sites after electrochemical hydrogenation. This excludes the existence of well-oriented facets and terraces on electrochemically hydrogenated Si surfaces. Figure 38 shows two possible atomic arrangements of Si-H and Si=H$_2$ at step facets (a) and (A), where the neighboring Si atoms on a (100) oriented facet (Si=H$_2$ groups) can be reassembled as shown in (b) and (B), respectively. The thick arrows denote the typical 1×2 reconstruction of a Si(100) surface [45, 57, 207]. Such a reconstruction of a step facet contains a relatively high degree of freedom in the bond angles and permits strain in the bonds, which can round off corners and steps. The driving force for such local reconstruction could be the electric field during the anodic oxidation in the oscillating regime. The broad FTIR spectra in the range of the Si-H$_x$ stretching modes of electrochemically hydrogenated Si surfaces show that there is a large variation in bond angles and/or bond lengths. In addition, there is only a very small fraction of Si-H surface bonds in well-defined positions. We believe that this observed "amorphization" of the Si surface concomitant with the formation of the rounded shapes is induced significantly by local reconstruction, which, in principal, is similar to well-known types of reconstruction of hydrogenated Si surfaces, like 1×2 and 1×3 [207, 208].

All Si surface atoms at the electrochemically hydrogenated Si surface with prior anodic oxidation are in more or less identical positions from the point of view of surface chemical reactions. One can suggest that the well-distinguished Si surface atoms at steps or corners at chemically hydrogenated Si surfaces are chemically more reactive than the Si surface atoms at the electrochemically hydrogenated Si surface. This is quite important when chemical or electrochemical treatments are interrupted by rinsing with water. The hydrogenated Si surfaces are hydrophobic and water cannot be adsorbed. However, when chemical or electrochemical surface treatments in fluoride solution are interrupted, there is a certain probability that surface dangling bonds, which may exist at the hydrogenated Si surface as intermediate states during the etch process, serve as sites for the adsorption of water molecules.

These adsorbed water molecules at the hydrogenated Si surface are the most probable candidates for the donor-type surface states measured by ex situ SPV. The water molecules can form [H$_3$O]$^+$ and related complexes [209]. The dependence of Q_f, hysteresis, and the slope of the U_{ph} (U_F) characteristics on the range of U_F showed that the surface states measured by SPV on hydrogenated Si surfaces have a broad distribution of trapping and detrapping time constants from the second or millisecond

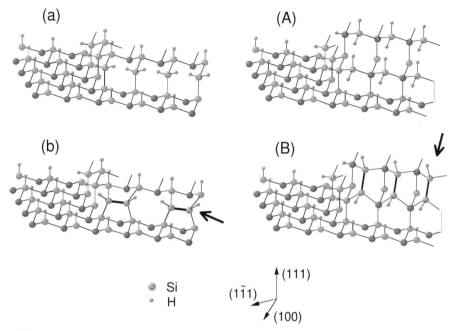

Fig. 38. Atomic arrangements of hydrogenated Si(111) surfaces with steps along the (100) direction (a) and (A), and possible types of respective reconstruction (b) and (B).

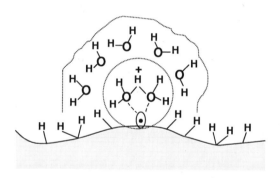

Fig. 39. Configuration for a Si dangling bond–$[H_5O_2]^+$ complex surrounded by physisorbed water molecules. Reprinted with the permission of Elsevier Science B. V. [178], copyright 2000.

range up to hours. Water molecules can be chemisorbed or physisorbed at surface sites that act as defect centers. Dangling bonds are the most probable defect centers for the chemisorption, and a complex of chemisorbed water molecules can serve as a defect center for physisorption. This behavior is schematically shown in Figure 39, where a possible configuration of a chemisorbed $[H_5O_2]^+$ and physisorbed water molecule at a dangling bond is sketched.

Poindexter [210, 211] postulated the presence of H_3O^+ for a model of chemical reactions of hydrogenous species in the Si/SiO_2 system to explain contradictions among experimental findings on the passivation of dangling bond centers and physisorption and chemisorption in the SiO_2 lattice. Dangling bonds are formed at the Si/SiO_2 interface during oxidation. Microscopic cavities exist around the dangling bonds and promote diffusion and adsorption of hydrogeneous species. Possible

chemical reactions are

$$H\text{-}Si \equiv +H_2O + h^+ \rightarrow \bullet Si \equiv +H_3O^+ \qquad \text{and}$$
$$\bullet Si \equiv +H_3O^+ + e^- \rightarrow H\text{-}Si \equiv +H_2O$$

for the capture of a hole or an electron, respectively. These reactions are controlled by the concentration of free holes or electrons at the Si surface, i.e., by the applied U_F. It is known from electron spin resonance experiments on hydrogenated por-Si surfaces that adsorption of water molecules can also decrease the concentration of dangling bonds by at least one order of magnitude [199]. A similar effect is known from experiments in ultra-high vacuum on Si(111) 7×7 surfaces [212] or Si(100) surfaces covered with a thin oxide [213], for which the concentration of nonradiative recombination active surface defects decreases strongly during the adsorption of water.

4. HYDROGENATED POROUS SILICON

The surface dissolution chemistry of silicon is still in question. It is generally accepted that holes are required in the initial oxidation steps for both pore formation and electropolishing (oxide formation) [171, 214, 215]. This means that hole generation mechanisms (i.e., illumination, high fields, etc.) are needed for significant dissolution of n-type material. Processes in which electron injection into the conduction band occurs have also been proposed by several authors [216, 217]. Current efficiencies have been measured, leading to approximately two and four electrons per dissolved Si atom during pore formation and electropolishing, respectively [218]. Independent of the type of anodic reaction, the final stable end product for Si in HF is H_2SiF_6 (or the SiF_6^{2-} ion). However, different kinds of kinetic processes are discussed in the literature during pore formation, involving

a. a continuous vacillation between hydrogenated and fluorinated Si surfaces [216],

$$=Si_2 - SiH_2 + 2F^- + h^+(Si) \rightarrow =Si_2 - SiF_2$$
$$+H_2(g) + e^-(Si)$$
$$=Si_2 - SiF_2 + 2HF \rightarrow SiF_4 + =Si_2H_2 \qquad (4.1)$$

b. the existence of a freely dissolved divalent state of Si, $SiF_2(aq)$, which undergoes a disproportionation reaction leading at least to a deposition of Si on the pore walls [219],

$$Si + 2HF + 2h^+ \rightarrow H_2(g) + SiF_2(aq)$$
$$2SiF_2(aq) + 2F^- \rightarrow Si + SiF_6^{2-} \qquad (4.2)$$

c. a competition between direct oxide formation via silanol followed by HF dissolution and tetravalent H_2SiF_6 formation [220, 221],

$$(1) \quad Si + 4OH^- + \lambda h^+ \rightarrow Si(OH)_4 + (4-\lambda)e^-$$
$$Si(OH)_4 \rightarrow SiO_2 + 2H_2O \qquad (4.3)$$
$$SiO_2 + 6HF \rightarrow H_2SiF_6(aq) + 2H_2(g)$$
$$(2) \quad Si + 2F^- + \lambda h^+ \rightarrow SiF_2 + (2-\lambda)e^-$$
$$SiF_2 + 2HF \rightarrow SiF_4(aq) + H_2(g) \qquad (4.4)$$
$$SiF_4 + 2HF \rightarrow H_2SiF_6(aq)$$

However, up to know spectroscopic data have not shown any kind of silicon oxide or silicon fluoride species during or after pore formation, leaving the validity of the reaction pathway (4.1 and 4.3) in question. The exact dissolution pathway is not known and is still under discussion. For more details on pore formation see, for example, [218, 222, 223].

4.1. pH Dependence of the Formation of Ultrathin Porous Si

Figure 40 shows a set of current voltage curves of n-Si(111) for different concentrations of NH_4F at pH 3.7 (scan rate 20 mV/s starting at -1 V, illumination intensity ~ 10 mW/cm^2). The current density shows a double peak structure at lower concentrations of NH_4F. The current density increases and the double peak structure diminishes with increasing NH_4F concentration until the first peak has vanished and a plateau is reached at higher anodic potentials (1 M NH_4F, pH 3.7). Simultaneously, the current becomes noisy because of heavy gas evolution according to reactions (4.1) to (4.4). Pore formation sets in before the first current peak maximum is reached, and with increasing potential the competing oxide formation overcomes the pore formation, leading, at least, to an oxide-covered surface that blocks the charge transfer into the electrolyte (passivation) and the current decreases.

It is well known that Si surfaces are hydrogenated during pore formation, as can be seen from Figure 41. This figure shows ex situ MIR-FTIR spectra at the Si-H stretching mode

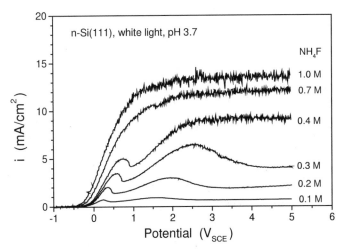

Fig. 40. Current–voltage curves of n-Si(111) in different concentrations of NH_4F at pH 3.7 (scan rate 20 mV/s starting at -1 V, illumination intensity ~ 10 mW/cm^2).

region of smooth hydrogenated n-Si(111) surfaces prepared in 0.1 M NH_4F (pH 4) after hydrogenation with prior anodic oxidation (a: s-polarization, b: p-polarization of the IR light) and after initial porous silicon formation at 0.15 mA/cm^2 for 5 s (c) and 50 s (d) during illumination with white light. The hydrogenated n-Si(111) surface exhibits a sharp IR absorption peak at about 2083 cm^{-1} due to Si-H groups perpendicular to (111) terraces, as expected for a flat (111) surface, and a small amount of coupled monohydride on steps on rows around 2070 cm^{-1} [132, 146] (see also Fig. 27). Spectra (c) and (d) are recorded after a positive electric charge of about 0.72 and 7.2 mC/cm^2, respectively, is passed. One can see that first, the IR absorption in the Si-H$_x$ stretching mode region is broadened, and, second, the IR absorption due to Si-H on (111) terraces is reduced and new IR peaks occur. These additional IR absorption peaks are a result of the roughening of the Si surface during the electrochemically induced etching process. The amount of coupled monohydride increases (ν_{as} and ν_{ss} are 2070 cm^{-1} and 2088 cm^{-1}, respectively), and a strong IR absorption appears at 2130 cm^{-1}, which is due to constrained dihydride or trihydride species. The peaks at 2092 cm^{-1} and 2114 cm^{-1} can be attributed to ν_{as} and ν_{ss} of dihydride species on (100) oriented facets (see Table I). It should be noted that the slight blue shift of these peak positions with respect to chemically prepared surfaces points to a special kind of strain on the porous surface. Furthermore, an IR absorption due to coupled monohydrides at steps at 2070 and 2088 cm^{-1} is superpositioned to the peak at 2092 cm^{-1}. Nevertheless, the amount of Si-H and Si=H$_2$ species (i.e., the IR absorption at this wavenumber range) increases with increasing charge flow (spectrum (d)) and overcomes at least the amount of Si-H groups on (111) terraces. On the other hand, the etching process involves no twice or threefold oxide back-bonded Si-H groups due to the missing IR absorption between 2200 cm^{-1} and 2300 cm^{-1}, whereas single oxide back-bonded Si-H species could not be excluded [69]. In addition, IR absorption due to the different Si-H species is no longer well resolved. The inset of

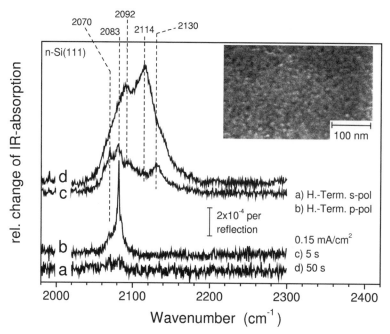

Fig. 41. Ex situ FTIR spectra of *n*-Si(111) surfaces in 0.1 M NH₄F (pH 4) after hydrogenation with prior anodic oxidation (a: s-polarization, b: p-polarization) and after initial porous silicon formation at 0.15 mA/cm² for 5 s (c) and 50 s (d). The inset shows a SEM graph (top view) of the porous surface layer formed at 0.15 mA/cm² for 100 s.

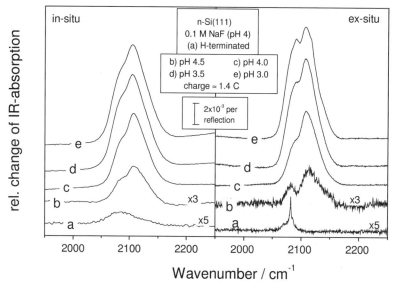

Fig. 42. In situ and ex situ FTIR spectra of *n*-Si(111) after electrochemical hydrogenation with prior anodic oxidation at +3 V in 0.1 M NaF (pH 4) (a) and after anodization at currents of one-third of the maximum in 0.1 M NaF with pH 4.5, 4.0, 3.5, and 3.0 (b to e, respectively, passed charge during anodization, 0.2 C/cm²).

Figure 41 shows a scanning electron microscope (SEM) image (top view) of an *n*-Si(111) surface after a charge of 18 mC/cm² is passed during illumination. One can see a random distribution of the starting hole formation, where the distances of the etch pits are on the order of 10 nm. These small sizes of the electrochemically induced structures lead to a drastic decrease in the amount of (111) terraces and a strong increase in facets of other types of orientation as reflected by the FTIR spectra.

The manipulation of nanoscaled structures at Si surfaces is interesting for possible applications and for getting a better understanding of the processes leading to the formation of porous silicon. The chemical equilibrium in the fluoride solution determines the etch rate of oxidized Si species and can easily be changed by the pH of the fluoride solution. Figure 42 compares in situ and ex situ FTIR spectra of *n*-Si(111) surfaces after electrochemical hydrogenation with prior anodic oxidation at +3 V

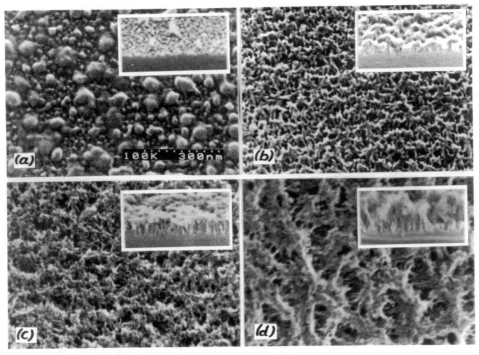

Fig. 43. SEM graphs of hydrogenated silicon surfaces after anodization in 0.1 M NH$_4$F with pH 4.5, 4.0, 3.5, and 3.0 (a to d, respectively). The amount of passed charge during anodization was 0.3 (a), 0.6 (b), and 1.2 (c, d) C/cm^2. Reprinted with permission from the American Institute of Physics [30], copyright 1995.

in 0.1 M NaF (pH 4) (a) and after anodization at currents of one-third of the first maximum in 0.1 M NaF with pH 4.5, 4.0, 3.5, and 3.0 (b to e, respectively; passed charge during anodization, 0.2 C/cm^2). The in situ measured FTIR spectrum of the H-terminated Si surface is broadened in comparison with the ex situ spectrum because of interaction of the surface species with the dipoles in the electrolyte [147, 148].

The ex situ recorded FTIR spectra are strongly broadened after the formation of porous silicon. Therefore, disorder dominates the Si-H bond configurations in porous silicon. The broadening of the in situ FTIR spectra is very similar to that of the ex situ spectra for porous silicon. The ex situ spectra are broadened by approximately only 5 cm^{-1} in the lower wavenumber region. This shows that the intrinsic broadening by disordered Si-H bonds is much more significant than the broadening due to dipole interaction with the water molecules. The shape of the FTIR spectra of (nano)porous Si is independent of the passed charge and pH (for pH <4.5).

The SEM graphs of the hydrogenated silicon surfaces after anodization in 0.1 M NH$_4$F with pH 4.5, 4.0, 3.5, and 3.0 (a to d, respectively) are presented in Figure 43. The amount of passed charge during anodization was 0.3 (a), 0.6 (b), and 1.2 (c, d) C/cm^2. The surface structures are very different for the different pH values. As shown by the dependence of the PL intensity on pH, the oxidation rate of Si is high for large pH because of the large amount of OH$^-$ ions [224]. In this case, a pebble-like surface structure is created during anodization, and the formation of porous Si is impossible. The formation of porous silicon layers can be seen for the lower values of pH (side views as

insets). There are two opposite tendencies for the formation of surface structures with decreasing pH. First, the dimensions of the tiniest parts of the Si skeleton are decreased, and, second, the amount of pores with larger dimensions is increased. As remarked, the shapes of the PL spectra did not depend on the pH. The missing correlation of the PL with the microstructure of porous Si indicates that the role of highly disordered Si surfaces for the PL of porous Si at room temperature is different from PL measurements at very low temperatures, where these measurements in the nanocrystallites are more important [223].

Figure 44 shows the pH dependence of the integrated IR absorption in the range of the Si-H modes for different amounts of passed charge during anodization in 0.1 M NaF (top) and of the calculated concentrations of HF, F$^-$, and HF$_2^-$ ions in the electrolyte (after Eqs. (3.4) to (3.7)) (bottom). The integrated IR absorption in the range of the Si-H modes is a measure of the internal surface area. The internal surface area increases with decreasing pH and tends to saturate at lower pH. The internal surface area does not scale with the amount of passed charge, which points to a competing process of chemical etching or oxidation during the formation of porous silicon. The thickness of nanoporous silicon layers is limited in diluted aqueous fluoride solutions, as shown by SEM [225]. The dependence of the integrated IR absorption does not correlate exactly with the concentration of HF or HF$_2^-$, and there is no real saturation of the integrated IR absorption at the lower pH values (pH up to 2). We speculate that porous silicon formation is possible by both HF and HF$_2^-$ species, whereas the efficiency of porous silicon formation is greater for the HF$_2^-$. As mentioned in Section 3,

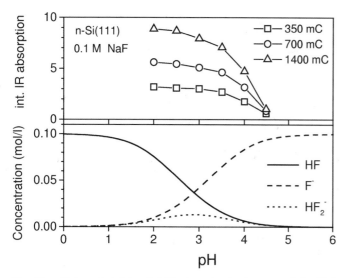

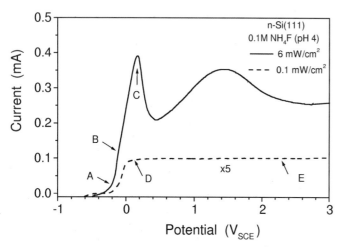

Fig. 44. pH-dependent integrated IR absorption in the range of the Si-H modes for different values of passed charge during anodization in 0.1 M NaF and pH-dependent concentrations of HF, F⁻, and HF₂⁻ in the electrolyte.

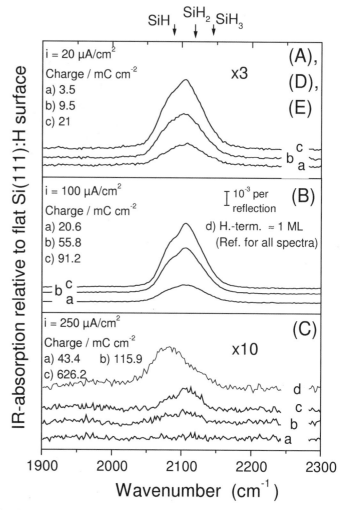

Fig. 46. In situ FTIR spectra of n-Si(111) obtained at the points marked in Figure 45 for different currents, illumination intensities, and anodization times.

Fig. 45. Current–voltage curves of n-Si(111) in 0.1 M NH₄F (pH 4) for different illumination intensities (scan rate 20 mV/s).

the etch rate of SiO₂ by HF₂⁻ is 4 times larger than for HF. In principle, the two chemical equations (4.1) and (4.2) can be reduced to one equation by setting HF₂⁻ in Eq. (4.1) instead of F⁻, which could make the formation of porous silicon faster:

$$=Si_2 - SiH_2 + 2HF_2^- + h^+(Si)$$
$$\rightarrow =Si_2H_2 + SiF_4 + H_2(g) + e^-(Si) \quad (4.5)$$

4.2. Competition between Hydrogenation and Electropolishing

The electrochemical equilibrium at the Si/diluted fluoride solution interface depends sensitively on the potential. Figures 45 and 46 reflect the potential dependence of the pore formation in 0.1 M NaF solution at pH 4. Figure 45 shows current–voltage curves of an n-Si(111) surface under strong (solid line) and

weak (dashed line) illumination intensities with white light. The photocurrent shows the typical behavior with a first strong increase and a second broad relative maximum at high illumination intensity. The potential dependence of the current differs strongly at low illumination intensity; the current shows no maximum at all. Moreover, the photocurrent is constant even at high anodic potentials. It is known that current multiplication occurs under such experimental conditions [226, 227]. The marked positions in Figure 45 denote the potentials where FTIR spectra are recorded. These spectra are plotted in Figure 46 with respect to the hydrogenated surface prepared after electropolishing in the same solution at +3 V (high-illumination condition) recorded after the current transient has decayed (spectrum d in Fig. 46C). Spectra taken at potential positions A, D, and E are very similar, so that only one set of spectra is shown for this low current density of 20 μA/cm² (position E). The IR absorption in the Si-Hₓ stretching mode region (Si-H and Si=H₂) increases with increasing charge flow. The increase in the IR absorption is much more pronounced at position B, where the current density is five times higher (100 μA/cm²;

Fig. 46, middle) than for A, D, or E. Surprisingly, the pore formation is strongly suppressed at position C, even for the high current densities observed at this potential. This is reflected by a very slight increase in the IR absorption with respect to the monolayer SiH covered surface (spectrum d), which is plotted at the bottom of Figure 46. There is no change in the IR absorption up to a charge flow of about 50 mC/cm². A charge of about 0.45 mC/cm² is needed to oxidize Si on a flat Si(111) from 0 to the +4 state. Therefore, more than 100 monolayers of Si have been etched without any change in the surface condition and morphology. A small increase in IR absorption due to Si=H₂ groups occurs after a charge flow above 100 mC/cm². The IR absorption after a charge flow of about 620 mC/cm² is equivalent to an increase in the surface roughness by about 40%, only in reference to a flat surface. From the charge that passes the electrode and the charge that is needed to oxidize Si to the +4 state, one can calculate the amount of Si dissolved during this etching process to a Si layer of a thickness of about 300 nm. This behavior shows that the competing oxide formation seems to overcome the pore formation at this potential region. Electropolishing leads to a smoothing of the surface. This behavior is reflected by the integrated IR absorption for the region of the Si-H$_x$ bonds. Figure 47 shows the first increasing part of the current voltage scan of n-Si(111) in 0.1 M NH₄F (pH 4) and the integrated IR absorption in the region of the Si-H stretching modes for a passed charge of 200 mC/cm² (right side). Hydrogen termination and porous silicon formation take place in the plotted part of the current voltage scan. Electropolishing starts at the maximum of the anodic current. The integrated IR absorption increases up to a current that is about 4 times lower than the current at the maximum. The integrated IR absorption decreases for currents higher than half of the current at the maximum. This behavior is typical for anodization in diluted aqueous fluoride solutions and reflects the competition between electrochemical reactions that lead to porous silicon or electropolishing.

In addition, the surface conditioning (pore formation versus electropolishing) is visualized in Figure 48, where SEM images of the Si surface are shown in a cross-sectional view at a tilt angle of 30° after different treatments. The SEM images are recorded after anodization at 12 μA/cm² (A), 100 μA/cm² (B), and 400 μA/cm² (C) in 0.2 M NH₄F (pH 3.2) after an electric charge of 100 mC/cm² has passed the electrode. The time positions are similar to positions A, B, and C in Figure 45. One can see the pore formation (Fig. 48A), which is enhanced in thickness at higher anodization currents (Fig. 48B), but the high current (just before the first maximum) leads to a very slight roughening of the surface (Fig. 48C), as was discussed with respect to FTIR experiments. The pores in Figure 48 are in the nanometer scale.

Figure 49 depicts the dependence of the PL intensity (open circles) and the respective current (solid line) of p-Si(100) in 0.2 M NH₄F (pH 3.2) on the applied potential. The potential has been corrected to the potential drop in the electrolyte. The scan rate of 20 mV/s is too fast for pore formation. Therefore, the dependence shown in Figure 49 gives information about the onsets

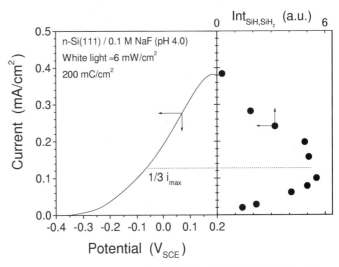

Fig. 47. Current–voltage scan of n-Si(111) in 0.1 M NH₄F (pH 4) and integrated IR intensity in the region of the Si-H stretching modes after a charge of 200 mC/cm² is passed. Reprinted with permission from the American Institute of Physics [229], copyright 1998.

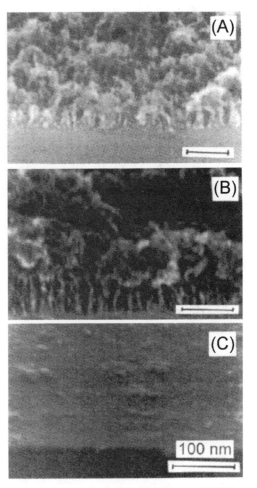

Fig. 48. SEM graphs (side views, tilt angle 30°) of p-Si(100) anodized in 0.2 M NH₄F (pH 3.2) at 12 (A), 100 (B), and 400 (C) μA/cm² after an electric charge of 100 mC/cm² has passed the electrode.

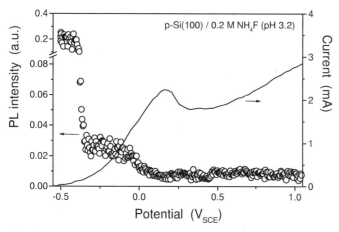

Fig. 49. Current–voltage scan and PL intensity of *p*-Si(100) in 0.2 M NH₄F (pH 3.2) during the voltage scan (scan velocity 20 mV/s). Reprinted with permission from the American Institute of Physics [229], copyright 1998.

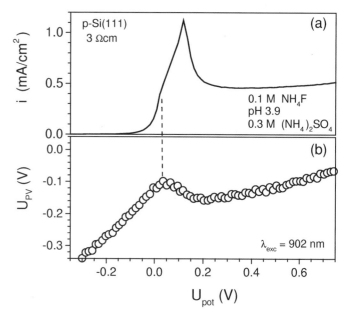

Fig. 50. Current–voltage scan (a) and corresponding PV scan (b) of *p*-Si(111) in 0.1 M NH₄F (pH 3.9).

of electrochemically correlated surface reactions with intermediates that act as nonradiative recombination surface defects.

The scan starts with an electrochemically hydrogenated Si surface with prior anodic oxidation at +3 V. The PL intensity drops sharply at about −0.4 V. The current is about 20 times lower at this potential than at the first current maximum. The sharp drop in the PL intensity is related to the onset of electrochemical reactions, which lead to the formation of porous silicon (etching). The rate of formation of nonradiative surface defects increases strongly at this potential, and the sharpness of the drop is evident for the sudden onset of porous silicon formation.

The PL intensity remains nearly constant with increasing potential up to about −0.1 V. Simultaneously, the current increases to a value that is about half of the current at the first maximum. The PL intensity decreases further with increasing potential. This decrease is slow, in contrast to the sharp drop at about −0.4 V, and the minimum of the PL intensity is observed when the first current maximum is reached. The PL intensity remains at a low level with a further increase in the potential. The range after the first current peak is characterized by electropolishing, and the silicon surface is covered by an oxide with a thickness in the monolayer range. The slow decrease in the PL intensity for potentials higher than −0.1 V marks the onset of electrochemical reactions leading to electropolishing. These chemical reactions compete with the reactions of porous silicon formation.

The potential dependence of the photovoltage, U_{PV}, gives information about the charge transfer at the Si surface. Figure 50 shows the current–voltage scan (a) and the corresponding U_{PV} scan (b) of *p*-Si(111) in 0.1 M NH₄F (pH 3.9). The anodic current changes the slope at a value that is about a third to a half of the anodic current at the first maximum. This is similar to the PL and current–voltage scan shown in Figure 49. The U_{PV} amplitude decreases linearly with increasing potential up to a potential of about 0 V. At this potential the anodic current changes the slope and the slow decrease in the PL intensity,

as shown in Figure 49, begins. There is no hint about any signature in the U_{PV} or the sharp onset of the process of porous silicon formation with increasing potential as observed by PL measurements (Fig. 49). Therefore, the charge transfer at the Si surface remains unchanged during H-termination or porous silicon formation.

The U_{PV} amplitude increases slightly with increasing potential between 0 and 0.2 V and decreases again toward accumulation with a further increase in the potential. The increase in the U_{PV} amplitude of *p*-Si(111) gives evidence for an increasing positive charge at the Si surface. The increase in positive charge at the Si surface is finished after the first maximum of the anodic current only. The onset of the increase in the U_{PV} amplitude with increasing potential coincides with the change in the slope of the anodic current and the onset of the slow decrease in the PL intensity. Therefore, the electrochemical reactions leading to electropolishing cause a dynamical storage of positive charge at the Si surface [228].

Smith and Collins [218] distinguished the current potential scan in diluted HF solutions in a region of hydrogenation, a transition region, and a region of electropolishing. The region of hydrogenation contains the two regimes of H-termination and porous silicon formation, which are characterized by two plateaus with the high and medium PL intensities (Fig. 49), respectively. The transition region is characterized by the slow decrease in the PL intensity and by a storage of positive charge at the Si surface. The PL intensity is at a low level during anodic oxidation, the region of electropolishing.

As known, holes are consumed at the Si surface during anodic oxidation. This positive charge is transferred to the electrolyte by a retarded process. Such a process should be mediated by a complex supporting an electron that is injected into the Si and recombines with the hole. The positive charge that re-

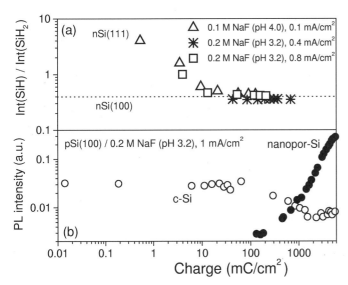

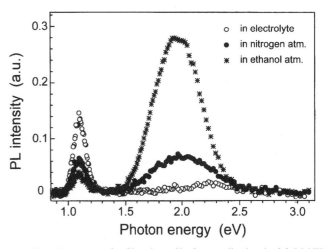

Fig. 51. Dependence of the SiH/SiH$_2$ ratio (a) and the PL intensities of c-Si (measured at 1.1 μm) and nanopor-Si (measured at 0.6 μm) (b) on the passed charge during anodization of Si(111) or Si(100) at different current densities and in different fluoride solutions. Reprinted with permission from the American Institute of Physics [229], copyright 1998.

Fig. 52. PL spectra of c-Si and por-Si after anodization in 0.2 M NH$_4$F (pH 3.2) in solution and after replacement of the electrolyte with nitrogen and ethanol atmospheres. Reprinted with permission from the American Institute of Physics [304], copyright 1997.

mains on the complex polarizes the Si surface within the inner Helmholtz layer. In contrast, there is no influence of charged surface complexes during porous silicon formation. Lehmann and Gösele [216] proposed a dissolution mechanism where a Si-H surface bond is weakened by the capture of a hole. Subsequently, fluoride ions react with the destabilized Si surface atom. One electron is transferred from a fluoride ion into the Si bulk during this reaction. The former surface hydrogen atoms lead to hydrogen gas evolution. The Si-Si back bonds are now strongly polarized and can be very quickly attacked by other polar molecules like HF, leading to a dissolution of a SiF$_4$ species, which further reacts with SiF$_6^{2-}$ ions, with F$^-$ ions present in solution (see reaction scheme (4.1)). Hence, the Si surface is again hydrogenated by a process similar to the chemical hydrogenation of Si surfaces as proposed by Trucks et al. [200].

Si surfaces can be roughened in a well-controlled manner. The surface area increases with increasing charge flow during anodization. Figure 51 shows the dependence of the ratio of SiH/SiH$_2$ on the passed charge during anodization of Si(111) or Si(100) at different current densities and in different fluoride solutions. A constant ratio of SiH/SiH$_2 = 0.4$ is reached after a charge flow of 10 mC/cm^2 (a). This ratio is independent of surface orientation and remains unchanged for the formation of porous silicon. In addition, Figure 51 shows the dependence of the PL intensity of c-Si (measured at 1.1 μm) and nanoporous Si (measured at 0.6 μm) (Fig. 51b) on the charge that passes the electrode [229]. The PL intensity of c-Si remains constant up to a passed charge of about 100 mC/cm^2, despite the strong increase in the surface area of the Si sample. This means that the parts of the surface that are not related to the anodization process do not contribute to the nonradiative surface recombination at all. In other words, the concentration of reactive surface sites remains constant during the surface roughening and porous sil-

icon formation. This is not surprising because the thickness of a porous layer is proportional to the passed charge.

The PL signal of por-Si arises after a passed charge of about 300 mC/cm^2 and increases strongly with further anodization. The PL intensity of c-Si starts to decrease at about 200 mC/cm^2, i.e., when the PL signal of por-Si occurs. The reason for this is that a certain amount of the exciting light is absorbed in the porous surface layer, where the excess carriers may recombine radiatively.

4.3. Electronic States at Internal Surfaces of Porous Si and Local Reconstruction

Transport of excess carriers of charge is important for the PL of por-Si. Usually, the PL intensity is negative correlated with the electric conductivity [230]. Figure 52 presents PL spectra of c-Si and por-Si after anodization in 0.2 M NH$_4$F (pH 3.2) solution and after replacing the electrolyte with nitrogen and ethanol atmospheres. The thickness of the por-Si layer is on the order of 70 nm, and the exciting light of the N$_2$ laser (wavelength 337 nm) is almost completely absorbed in the porous surface layer. The PL signal of por-Si increases strongly after the electrolyte is replaced with an ethanol atmosphere, whereas the PL signal of c-Si decreases. This negative correlation shows the influence of the ambient on the diffusion of charge carriers. The diffusion length of the excess carriers of charge is larger than about 50 nm in the electrolyte, whereas it is much shorter than 50 nm for por-Si in the ethanol atmosphere. For comparison, diffusion coefficients of excess carriers in mesoporous silicon measured by optical grating techniques amount to about 30–90 nm [231].

The interface state distribution is an important parameter for the characterization of the surface passivation from an electronic point of view. Ex situ SPV measurements can be used for ultrathin porous silicon (thickness 10–20 nm), for which the exchange of charge carriers with the c-Si bulk is not lim-

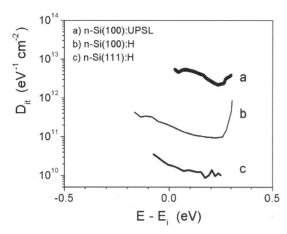

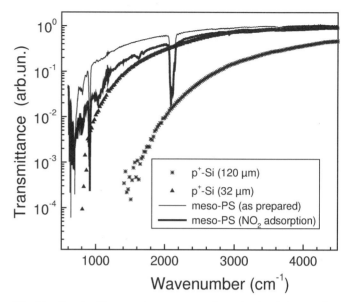

Fig. 53. Interface state distribution obtained from ex situ SPV measurements for electrochemically hydrogenated n-Si(111) and n-Si(100) surfaces and for a Si(111) surface covered with a 20-nm-thick por-Si layer.

Fig. 54. Ex situ IR transmittance spectra for as-prepared free-standing mesopor-Si film (thickness 75 μm), the same film of mesopor-Si with adsorbed NO_2 molecules, and the p^+-Si substrate for thicknesses of 120 and 32 μm. Reprinted with permission from *Physica Status Solidi* [235], copyright 2000.

ited by transport [232]. Figure 53 compares the interface state distributions for electrochemically hydrogenated n-Si(111) and n-Si(100) surfaces and for a Si(111) surface covered with a 20-nm-thick por-Si layer. The lowest value of D_{it} is reached on a Si(111) surface after electrochemical hydrogenation with prior anodic oxidation in the oscillating regime, as shown and discussed in Section 3. The density of surface states is higher for the n-Si(100) surface prepared under identical conditions than it is for the n-Si(111) surface. The overall surface state density of ultrathin por-Si is comparable to that of HF-treated Si surfaces, whereas the normalization to the internal surface area, which is about 600 m^2/cm^{-3} [233], leads to a value of D_{it} on an order similar to that of the n-Si(100) surface.

The excellent passivation of hydrogenated Si surfaces in por-Si can also be demonstrated for mesoporous Si prepared on highly doped p^+-Si substrates. An important question concerns the existence of free carriers of charge in the mesoporous Si. Usually, mesoporous Si has a large resistivity, and there is no evidence of free carriers of charge [234]. However, under certain preparation conditions absorption of infrared light by free carriers of charge is observed [235], as shown in Figure 54, where ex situ IR transmittance spectra are plotted for as-prepared free-standing mesopor-Si films (thickness 75 μm, without and with adsorbed NO_2 molecules) and for the p^+-Si substrate (thickness of 120 and 32 μm). The spectra show the typical absorption peaks of Si-H modes, and there is no evidence of oxide species at the internal surface of the mesoporous Si. The continuous underground of the IR spectra is characteristic of absorption by free carriers of charge. Our measurements give evidence of a high amount of free holes in the mesoporous silicon. One can conclude that the concentration of amphoteric or donor-type surface states is below 10^{11} eV^{-1} cm^{-2} with respect to the large internal surface area (about 600 m^2/cm^{-3} [233]). This shows that the concentration of reactive surface sites that may adsorb water molecules (donor type molecules) is very low for hydrogenated Si surfaces in mesoporous Si. The low concentration of compensating defects in mesoporous

silicon was also shown by intensity-dependent surface photovoltage measurements [236]. A space charge region is formed in the mesoporous silicon surface region, which depends on the concentration of free carriers. The porosity of the mesoporous Si modifies the dielectric constant, which allows us to handle this region in a manner similar to the method used for the space charge region of c-Si.

The number of electronic surface states (or reactive surface sites) at the inner hydrogenated Si surface of porous silicon is extremely low. This, together with the hindrance of carrier transport, is the reason for the highly efficient photoluminescence in nanoporous Si. The PL efficiency of nanoporous silicon can be strongly increased by initial oxidation in water [237]. Reactive surface sites are created during the initial oxidation. However, water molecules passivate effectively reactive surface sites. It is known from electron spin resonance experiments on hydrogenated por-Si surfaces that adsorption of water molecules can decrease the concentration of dangling bonds by at least one order of magnitude [199]. A similar effect is known from experiments in ultra-high vacuum on Si(111) 7 × 7 surfaces [212] or Si(100) surfaces covered with a thin oxide [213] for which the concentration of nonradiative recombination surface defects decreases strongly during the adsorption of water.

In Section 3, we relate the formation of disordered hydrogenated Si surfaces to electrochemically induced local reconstruction of the Si surface. The nanoparticles in porous silicon have round shapes; no faceting is observed. The argument of local reconstruction is supported by the fact that the ratio of SiH, SiH$_2$, and SiH$_3$ bonds saturates. Nevertheless, the ratio of dangling bonds at unreconstructed Si spheres is very different from the experimentally observed one. Figure 55 shows the dependence of the calculated relation of SiH$_2$/SiH and SiH$_3$/SiH bonds on the radius of a sphere of c-Si if reconstruction is ab-

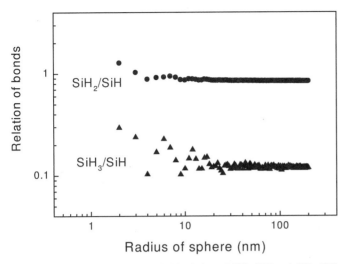

Fig. 55. Dependence of the calculated relation of SiH_2/SiH and SiH_3/SiH bonds on the radius of a sphere of c-Si when reconstruction is absent [238].

sent [238]. A similar approach of passivation of dangling bonds by hydrogen is used for theoretical calculations of the electronic structure of Si nanoparticles. The calculated ratio SiH/SiH_2 saturates at about 1, whereas the measured ratio SiH/SiH_2 saturates at about 0.4 for the as-prepared porous silicon. The difference means that the shapes of the Si nanoparticles are not really sphere-like, but have some preferential orientation in the (100) direction and/or that atomic steps at the surfaces of the spheres are smeared out by local reconstruction. In fact, porous Si is preferentially etched in the (100) direction. Obviously, more theoretical work is needed to finally solve the question of local reconstruction in porous silicon. Local reconstruction can be understood as a kind of amorphization of the hydrogenated Si surface, which keeps all surface atoms in more or less identical positions from the point of view of surface chemical reactions. A very thin amorphous Si surface layer with an extremely high amount of hydrogen would act as a passivation layer.

5. THIN ANODIC OXIDES ON Si

The most widely used method in Si device passivation is thermal oxidation, in a dry or wet oxygen atmosphere in a range of 700–1100°C for some minutes, depending on the thickness of the oxide [4, 9, 239]. Recently, low thermal budget processing like PECVD in any kind of variation [17, 117, 124, 240, 241], electrochemical oxidation procedures [2, 3, 5, 12–27], and other more or less exotic treatments like, for example, oxidation in ozone at 200°C [242] have been developed for device quality passivation of Si and even for SiGe epitaxial layers [32–34, 243–245].

The surface roughness is very important for the formation of thin gate oxides [120–123]. The influence of initial stages of oxide deposition differs from that of "bulk oxide" formation. In the former, the initial Si surface mainly defines the electronic properties; i.e., damage to the surface can be observed during

the beginning of the deposition process by the needed plasma source. No change in the interface occurs if the oxide layer is thick enough to head off the energy that is incorporated by the plasma [124, 125]. The situation is sharply different when the oxide layer is grown into the silicon bulk, where the interface is permanently changed by the formation process, i.e., diffusion of atoms, ions, or molecules through the oxide layer [2, 12].

The initial stages of formation of oxides are important for the electronic properties of the interface of thin oxide layers and have been widely investigated [133, 213, 246–255]. These properties include the space-resolved variation of tunneling current investigated by STM or atomic force microscopy (AFM) techniques [256], the homogeneity and morphology of the layer and the interface [257], the dependence of the current density on the number of steps at the Si/SiO_2 interface [258], and the strain at the interface during oxide growth [118, 125], which is correlated with the defect concentration [259]. Even effects of the initial electrode potential [260], amount of water and temperature [27, 261], and organic impurities in water [249] on the oxide growth conditions have been observed.

5.1. Initial States of Anodic Oxidation

X-ray photoelectron spectroscopy and high-resolution electron energy loss spectroscopy [198] revealed the formation of Si-OH groups on Si surfaces in contact with water. Infrared absorption experiments in the UHV chamber have recently been performed and show the formation of Si-H species with different types of oxide back bonds in the beginning of oxide formation [133]. The oxidation process of a hydrogenated Si surface starts with the breaking of Si-Si back bonds, leaving the Si-H bonds untouched at first [133]. Similar results have been obtained for electrochemical oxidation of H-terminated Si surfaces. Figure 56 shows FTIR spectra during anodic oxidation of Si(111) (A) and Si(100) (B) surfaces in 0.2 M Na_2SO_4 (pH 3) normalized to the hydrogenated state. The samples are cleaned and smoothed before oxidation by an electropolishing step at +3 V in 0.1 M NH_4F (pH 4) followed by the hydrogenation procedure described in Section 3. In the beginning, the Si-Si back bonds of the Si-H surface species are converted into Si-O-Si bonds, leading to a decrease in the IR absorption in the Si-H region (peak around 2090 cm^{-1}) and to an increase in Si-H species with an increasing amount of oxygen back bonds (OSi-H 2118, $O_2Si=H_2$ 2200, and O_3Si-H 2255 cm^{-1} [133–136]). Whereas the (111) oriented Si surface shows the subsequent formation of Si-H species with one, two, and three oxygen back bonds with increasing potential, the Si(100) surface shows the formation of Si-H species with two oxygen back bonds ($O_2Si=H_2$) only. This behavior points to a faster oxidation of the (100) surface, which is less stable than the Si(111) surface. The IR absorption due to Si-H completely disappears when the potential increases above +2 V. This relatively high potential is indicative of a preferred island formation of the oxide (3D growth) with a later onset of 2D growth, as also proposed by other authors [262]. In addition, Figure 57 shows FTIR spectra during oxide formation on Si(111) in the regime of the Si-O-Si asymmetric

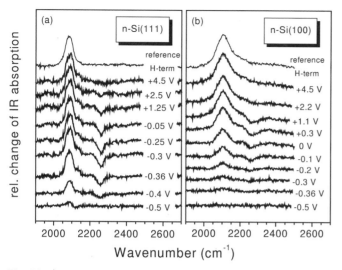

Fig. 56. IR absorption spectra in the range of Si-H stretching modes of *n*-Si(111) (a) and *n*-Si(100) (b) during anodic oxidation at different potentials. The spectra are normalized to the hydrogenated surface (reference).

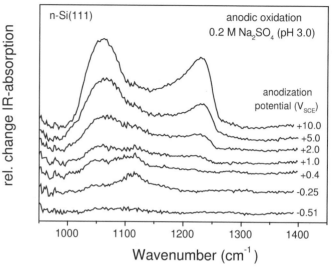

Fig. 57. Relative change in IR absorption in the range of Si-O-Si stretching modes during anodic oxidation of *n*-Si(111) at different potentials. The hydrogenated and oxide-free surface serves as the reference spectrum.

stretching mode, obtained with single internal reflection techniques. A broadened IR absorption with a maximum at about 1120 cm^{-1} occurs at small anodic currents at -0.25 V because of the formation of Si-O-Si groups [137–139]. This absorption peak decreases slightly in intensity with increasing anodic potential and splits into two peaks (easily seen at $+1$ V and above) that are centered at 1050 cm^{-1} and 1240 cm^{-1}, respectively. This indicates the formation of a thicker oxide layer (about 2–3 ML), where a disorder-induced vibrational effect is present [139], which leads to the LO–TO split [117, 139, 142, 144]. The former peak at about 1120 cm^{-1} (-0.25 V) seems to be due to a preferential order of the oxide formed during the beginning of the oxidation process (the LO4 mode is located at about 1140 cm^{-1} [117]); i.e., the bond angle of a Si-O-Si group at a Si surface bond is well defined by the crystalline lattice of the bulk material. The amount of this kind of ordered Si-O species, which exists only at the oxide/Si interface, decreases in relation to the growing thickness of oxide layer, and the IR absorption due to "normal" Si-O groups, which are highly disordered, predominates. From Figures 56 and 57 one can conclude that a mixture of fully (SiO$_2$) and partially ($_x$O-Si-H) oxidized parts coexists on the Si surface in this potential regime.

Figure 58 shows the relative change in IR absorption in reference to the hydrogenated and oxide free *n*-Si(111) surface at different anodic potentials under illumination in 0.1 M NaF (pH 4) with the application of a single internal reflection. The potential was stepped from -0.2 V to $+0.6$ V, $+0.9$ V, $+1.7$ V, or $+2.6$ V, respectively, as indicated by the thick arrows in the current–potential curve of the inset in Figure 58. Hydrogenated and porous Si is formed at the first strong increase in the current ($+0.6$ V, 130 mC/cm^2). The FTIR spectrum (a) exhibits typical IR peaks around 2100 cm^{-1}, which are broadened because of different kinds of Si-H$_x$ species of the porous layer ($x = 1, 2, 3$; mono-, di-, and trihydride, respectively). Nevertheless, no oxide species could be detected at this potential. The well-known split

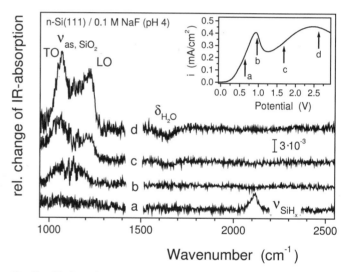

Fig. 58. IR absorption spectra during anodic oxidation of *n*-Si(111) in 0.1 M NaF (pH 4) at different potentials. The spectra are normalized to the hydrogenated Si surface obtained after etch-back of a thin anodic oxide layer (about one monolayer Si-H).

of the IR absorption at the asymmetric Si-O-Si stretching mode into parallel (TO, 1050 cm^{-1}) and perpendicular (LO, 1230 cm^{-1}) components can be seen at anodic potentials above the first current maximum. This split is not well resolved at $+0.9$ V (spectrum (b)). Moreover, the IR absorption of the asymmetric stretching mode of Si-O-Si is somewhat broadened, which is in contrast to the IR spectrum recorded at -0.25 V in fluoride-free solution (see Fig. 57), where a maximum is observed around 1120 cm^{-1}, which is attributed to a ordered Si-O-Si layer at the interface. The oxide/Si interface in fluoride-containing solution is no longer well defined. Moreover, the interface is permanently renewed because of the electropolishing behavior in such solutions [81, 171, 172], and, obviously, no well-ordered interface can be formed.

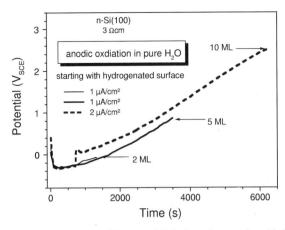

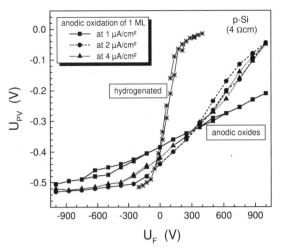

Fig. 59. Time dependence of the potential during galvanostatic oxidation of *n*-Si(100) in pure water. The anodization starts with hydrogenated surfaces.

Fig. 60. Dependence of the photovoltage amplitude on the field voltage for Si surfaces covered with thin anodic oxides (thickness about one monolayer) prepared at 1, 2, and 4 μA/cm^2. The dependence of the hydrogenated Si surface is shown for comparison.

Furthermore, Figure 58 shows a reduction of the IR absorption in the H-O-H bending mode region around 1650 cm^{-1}. This finding is correlated with the increasing amount of silicon oxide from spectra 58b to 58d (film growth), which replaces the surrounding water from the electrolyte. The amount of oxide at +2.6 V (Fig. 58d) is nearly three times greater than at +1.7 V (Fig. 58c).

The time dependence of the potential during galvanostatic oxidation of an *n*-type Si(100) wafer in pure water is plotted in Figure 59. The anodization starts with the hydrogenated surface. The potential decreases by about 0.7 V during the first 100 s of anodic oxidation. The increase in the potential after about 200 s gives evidence of the formation of a homogeneous anodic oxide layer, and the applied potential drops across the layer (the charge flow is equivalent for 1 ML). The potential increases by about 0.25 V after the formation of 2 ML. The anodic oxides, with thicknesses of 5 and 10 ML, are completed after the potential reaches +0.8 and +2.4 V, respectively. The linear increase in the potential with oxidation time is due to an increase in the potential drop across the oxide layer. The rate of oxide formation (dU/dt) for very thin anodic oxides cannot be increased significantly by increasing the anodization current (see thick solid and dashed lines). This behavior reveals that the formation of Si-O-Si bonds is the rate-limiting step. The PL intensity of the c-Si is constant during the growth of very thin anodic oxides regardless of the current density (i.e., for low current densities). As remarked, a similar behavior has been observed for initial porous silicon formation when the PL intensity, which is a measure of the rate of dangling bond formation, is constant with increasing current (see Section 4).

Figure 60 shows the ex situ measured dependence of the photovoltage amplitude, U_{PV}, on the field voltage, U_F, for Si surfaces covered with a very thin anodic oxide. The dependence of the hydrogenated Si surface is shown for comparison. The hydrogenated Si surface is in slight inversion because of positive charging of the surface. The slope of the U_{PV} (U_F) characteristics decreases strongly, and the band bending tends to stronger inversion after formation of the very thin anodic oxide layer. Therefore, interface states are generated, and the

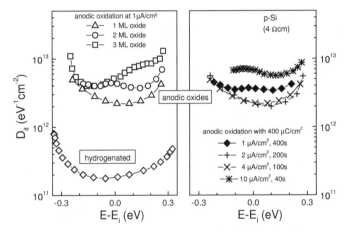

Fig. 61. Interface state density distribution (SPV analysis) for Si surfaces covered with anodic oxides prepared at an anodic current density of +1 μA/cm^2 (left, nominal thickness, 1, 2, and 3 ML) or as a function of the current density (right, nominal thickness, 1 ML). The dependence of the hydrogenated Si surface is shown for comparison.

density of slow hole traps increases because of initial anodic oxidation. Interestingly, the corresponding density of interface states has a minimum for a current of about +2 μA/cm^2 (highest slope of U_{PV} (U_F) characteristics among the very thin anodic oxides), whereas the inversion is the strongest. Positive charge may be partially compensated for by negatively charged acceptor states (for example, complexes associated with OH$^-$ ions). The probability of formation of negatively charged complexes increases with increasing surface state density, provided that the surface is covered by an oxide.

Figure 61 compares the distributions of surface states, D_{it}, for Si surfaces covered with anodic oxides of different thicknesses. D_{it} of the hydrogenated Si surface is shown for comparison. The value of D_{it}^{min} of the H-terminated surface is on the order of 10^{11} eV^{-1} cm^{-2} (analysis of the U_{PV}/U_F data from Fig. 60). D_{it} increases after initial anodic oxidation to values in the range of $2-5 \times 10^{12}$ eV^{-1} cm^{-2}. The lowest D_{it} has been observed after anodic oxidation with $+2$ μA/cm^2. The density of states in the minimum is practically independent of the thickness of the anodic oxide, whereas the density of states in the range toward the conduction band increases remarkably with increasing thickness of the very thin anodic oxide layer. This behavior is very similar to the formation of native oxides on hydrogenated Si surfaces in air or chemical solutions [197, 247]. The development of surface states in the range toward the conduction band can be interpreted as an increase of the number of acceptor-type surface states (exchange of charge with the conduction band) with increasing thickness of the very thin anodic oxide layer.

5.2. Passivation by Electron Injection at Cathodic Potentials

The density of electronic states at the anodic oxide/Si interface is usually quite high ($>10^{12}$ cm^{-2}), and thermal post-treatments are needed to improve the electronic passivation of the anodic oxide/silicon interface [27, 263]. The electronic states at the anodic oxide/Si interface depend strongly on the chemical equilibrium in which reactive surface sites such as dangling bonds are involved. It is shown in this section that the density of states at the anodic oxide/Si interface can be strongly reduced by optimizing the electrochemical equilibrium at the cathodic potential when electrons are injected into the thin anodic oxide layer. For this purpose, a thin anodic oxide (thickness about 6 nm) was formed on Si(100) samples of p-type doping (resistance 1 Ω cm) in 2 M (NH$_4$)$_2$SO$_4$ at $+10$ V, and the photovoltage and photoluminescence were probed stroboscopically during electron injection.

Figure 62 shows the current–potential (a) and photovoltage–potential (b) characteristics of a 6-nm-thick anodic oxide in 2 M (NH$_4$)$_2$SO$_4$ electrolyte for decreasing and increasing potential scan, U_{pot}. The PV transients were excited by single light pulses from a laser diode (wavelength 902 nm, duration time 100 ns, 10 μJ/cm^2). Electron injection starts at $U_{pot} = -2.1$ and -1.6 V for the branches of decreasing and increasing potential, respectively. The values of U_{PV} amount to -1.1 and -0.8 V at the potentials of -2.1 and -1.6 V. Breakdown fields of 1.8 and 1.3 MV/cm can be obtained for decreasing and increasing potentials if it is taken into account that U_{PV} corresponds to the potential drop across the p-Si sample. The flatband potential is shifted to lower U_{pot} for increasing potential by about 0.2 V in comparison with decreasing potential. This shows that the positive charge is decreased by about 3×10^{11} q/cm^2 after electron injection.

The specific role of the potential and current in the density of nonradiative recombination defects at the interface can be investigated by switching between anodic and cathodic directions of the current. Figure 63 shows an example of experiments with switching between anodic ($+8$ V), zero, and increasing cathodic potential. The anodic potential switch is used to create the same initial experimental conditions before the switch to the cathodic potential. The time dependence of the potential (a),

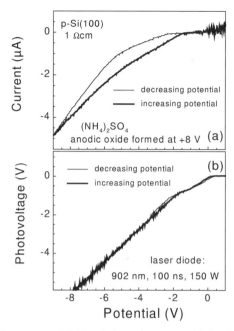

Fig. 62. Current–potential (a) and photovoltage–potential (b) characteristics of the anodic oxide/p-Si structure in (NH$_4$)$_2$SO$_4$ electrolyte. The photovoltage was excited with single pulses of a laser diode. Reprinted with the permission of the American Institute of Physics [36], copyright 2001.

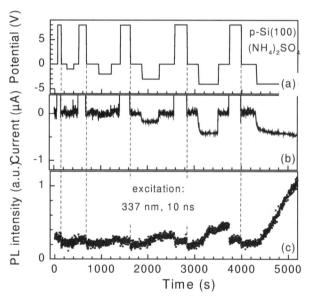

Fig. 63. Time dependence of the potential (a), current (b), and photoluminescence (c) of the anodic oxide/p-Si structure during switching experiments between $+8$ V and cathodic potentials of -1, -2, -3, and -4 V. The photoluminescence was excited with single pulses of a N$_2$ laser. Reprinted with the permission of the American Institute of Physics [36], copyright 2001.

current (b), and PL intensity (c) are plotted in Figure 63. The PL was excited with pulses from the N$_2$ laser (wavelength 337 nm, duration time 10 ns, 100 μJ/cm^2). The PL intensity is low during the application of the anodic potential and decreases further after switching to zero potential. The corresponding value of N_S is on the order of 10^{12} cm^{-2}.

The PL intensity remains almost unchanged during the application of low cathodic potentials ($U_{pot} \geq -2$ V). The Si surface is in strong inversion under these conditions as can be seen from Figure 62. Therefore, the influence of band bending on the recombination processes can be neglected and N_S is constant. In other words, the electrochemical equilibrium at the thin anodic oxide/silicon interface cannot be changed by a simple shift of the surface Fermi-level position from accumulation (anodic potential) to strong inversion (cathodic potential). The situation changes when electron injection becomes significant ($U_{pot} \leq -3$ V). The PL intensity starts to increase after switching from zero potential to a cathodic potential of $U_{pot} = -3$ V. Furthermore, the PL intensity increases with increasing time, i.e., with increasing injected charge. The PL intensity decreases again at a high anodic potential of $+8$ V. The increase in the PL intensity is much more significant at a potential of -4 V. Therefore, we can conclude that the electrochemical equilibrium at the thin anodic oxide/silicon interface is strongly affected by injected charge.

The PL intensity can be increased by up to more than one order of magnitude by electron injection. Nevertheless, the PL intensity saturates at high amounts of injected charge, as plotted in Figure 64. The PL intensity saturates faster at higher cathodic potentials. This is evident for the adjustment of a common chemical equilibrium that is characterized by a given value of N_S during the injection of electrons. We have to remark that the amount of injected charge that is needed for a certain increase in the PL intensity increases with increasing density of the cathodic current. The PL intensity at zero potential also strongly increases after electron injection but reaches only about half of the PL intensity at high cathodic potentials. This behavior reflects the influence of strong inversion on the surface recombination velocity. Surprisingly, the PL intensity can be further increased by switching between high cathodic and zero potential at shorter time intervals without the application of anodic potential (Fig. 64 time around 8000 s). This points out the importance of transport phenomena like drift or diffusion of ions through the oxide to establish the chemical equilibrium at the thin anodic oxide/Si interface.

The electron injection can be interrupted for short times during the ongoing passivation process. This allows a correlation between N_S and the projected charge at the thin anodic oxide/Si interface (Q_{ox}). Figure 65 presents an example of such a switching experiment where the applied potential (a), photovoltage (b), and PL intensity are plotted as functions of process time. The photovoltage is shown in the case where $U_{pot} = 0$ V. The p-Si surface is in inversion after switching from anodic to zero potential; i.e., there is a positive Q_{ox} of about 2×10^{11} cm^{-2} at the anodic oxide/Si interface. Despite the high band bending, the PL intensity is the lowest after switching from anodic to zero potential as mentioned above. U_{PV} at zero potential decreases from inversion to depletion during repetitive switching from cathodic to zero potential. The PL intensity increases at the same time. Hence, the electron injection neutralizes and/or anneals positive charges at the anodic oxide/p-Si interface and decreases N_S by more than one order of magni-

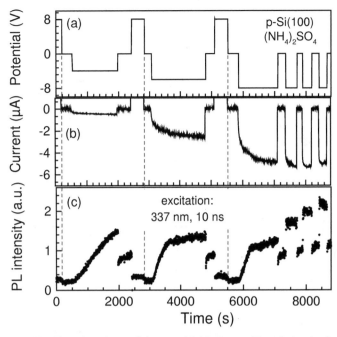

Fig. 64. Time dependence of the potential (a), current (b), and photoluminescence (c) of the anodic oxide/p-Si structure during switching experiments between $+8$ V and cathodic potentials of -4, -6, and -8 V. The photoluminescence was excited with single pulses of a N$_2$ laser. Reprinted with the permission of the American Institute of Physics [36], copyright 2001.

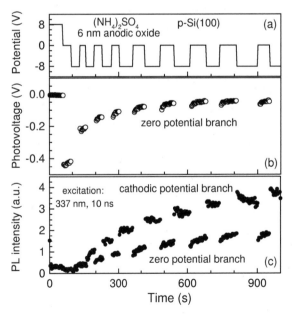

Fig. 65. Time dependence of the potential (a), photovoltage (b), and photoluminescence (c) of the anodic oxide/p-Si structure during switching between cathodic (-8 V) and zero potential. The experiment starts with anodic potential at $+8$ V. The photovoltage is shown only for the zero potential branch. The photovoltage and photoluminescence are excited with single pulses of a N$_2$ laser. Reprinted with the permission of the American Institute of Physics [36], copyright 2001.

tude. This correlation between Q_{ox} and N_S is given in Figure 66 for the zero potential branch (open circles). The data are obtained from Figure 65. N_S and Q_{ox} are reduced from 10^{12} to

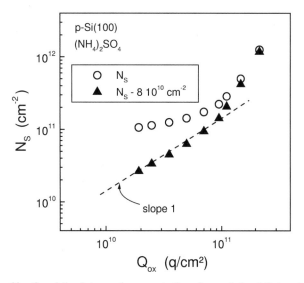

Fig. 66. Correlation between the concentration of nonradiative defect centers at the anodic oxide/p-Si interface, N_s, and the positive oxide charge, Q_{ox} (open circles). The triangles show N_S for N_{it}: 8×10^{10} cm^{-2}. The dependencies are obtained by analyzing the data of the zero potential branches in Figure 65. The slope 1 is indicated. Reprinted with the permission of the American Institute of Physics [36], copyright 2001.

10^{11} cm^{-2} and from 2×10^{11} to 2×10^{10} q/cm^2, respectively. N_S decreases monotoneously with decreasing Q_{ox} and tends to saturate at the lower values of Q_{ox} (N_S^{min} is about 8×10^{10} cm^{-2}). The triangles show the dependence of $N_S - N_S^{min}$ on Q_{ox}. The dashed line indicates the slope 1. It can be seen that $N_S - N_S^{min}$ is nearly proportional to Q_{ox} for values of Q_{ox} lower than 10^{11} q/cm^2.

The observed passivation of nonradiative recombination centers at the anodic oxide/p-Si interface is caused by injected electrons passing through the thin anodic oxide layer. The injected electrons change the chemical equilibrium at the anodic oxide/p-Si interface. There are two possible routes of chemical reaction that are induced by the injection of electrons. The first mechanism is connected with the drift of protons from the electrolyte/anodic oxide to the anodic oxide/Si interface. Injected electrons can react with protons near the anodic oxide/Si interface, and hydrogen atoms can diffuse back to the electrolyte/anodic oxide interface. Hydrogen could passivate Si dangling bonds, which act as nonradiative recombination defects. However, this mechanism does not explain the correlation between N_S and Q_{ox}. We favor a second mechanism that is related to the high amount of water in anodic oxides and its role of destabilization of H-Si $\equiv$ Si$_3$ bonds. The correlation between N_S and Q_{ox} can be well explained by this second mechanism as follows.

The correlation between N_S and Q_{ox} is very similar to the negative-bias-temperature instability (NBTI) of SiO$_2$/Si interfaces [264]. On the bases of many experimental findings, Poindexter proposed the following NBTI reaction [210]:

$$\text{H-Si} \equiv \text{Si}_3 + \text{H}_2\text{O} + \text{h}^+ \rightarrow \bullet\text{Si} \equiv \text{Si}_3 + \text{H}_3\text{O}^+ \quad (5.1)$$

In accordance with reaction (5.1) we propose the following reaction for passivation of Q_{ox} and N_S at anodic oxide/p-Si interfaces by electron injection:

$$\bullet\text{Si} \equiv \text{Si}_3 + [\text{H}_5\text{O}_2]^+ + \text{e}^- \rightarrow \text{H-Si} \equiv \text{Si}_3 + 2\text{H}_2\text{O} \quad (5.2)$$

Reaction (5.2) is, in principle, the inverse reaction of (5.1). Reaction (5.2) is controlled by the number of injected electrons and the amount of water present at the interface. The activation energy could be supported by the recombination energy of the electron. Reaction (5.1) dominates at anodic potentials, whereas reaction (5.2) is initiated by the injection of electrons at cathodic potentials.

N_S cannot be reduced below a certain level, whereas Q_{ox} decreases further. Therefore, a competitive reaction to (5.2) that neutralizes charged $[\text{H}_5\text{O}_2]^+$ complexes and creates new $\bullet\text{Si} \equiv$ Si$_3$ bonds should take part in the chemical equilibrium. Such a reaction could be

$$\text{H-Si} \equiv \text{Si}_3 + [\text{H}_5\text{O}_2]^+ + \text{e}^- \rightarrow \bullet\text{Si} \equiv \text{Si}_3 + 2\text{H}_2\text{O} + \text{H}_2\uparrow \quad (5.3)$$

The passivation of the anodic oxide/Si interface is limited by reaction (5.3). Reaction (5.3) can be partially suppressed by switching between cathodic and zero potentials as shown in Figure 64. We suggest that capture times of charge and/or polarization of Si-Si bonds are important for an inhibition of local reconstruction, which should be important for the activation of reaction (5.3).

5.3. Passivation by Process Optimization at Anodic Potentials

The density of nonradiative recombination centers at the anodic oxide/silicon interface is usually on the order of 10^{12} cm^{-2}. However, in this section we show that N_S can be decreased by more than one order of magnitude, even during anodic oxidation in diluted fluoride solution, with the use of a certain regime of anodic oxidation in the oscillating regime.

Figure 67 shows the maximum and minimum PL intensities during the anodic oxidation process of p-Si(100) in 0.1 M NH$_4$F (pH 4.2) as a function of the applied potential. The PL intensity has the lowest value at the lowest anodic potential (+3.5 V) during electropolishing and increases by more than one order of magnitude with increasing anodic potential in the oscillating regime. The current oscillations vanish for potentials higher than +10 V, and the PL intensity decreases for higher potentials. The strong increase in the PL intensity cannot be explained only by the increasing potential (p-type Si is in accumulation during anodic oxidation), because the PL intensity decreases for higher anodic potentials. The main reason for the strong increase in the PL intensity should be a structural change at the anodic oxide/p-Si(100) interface [178, 189].

As outlined in Section 3, an anodic current transient can be observed after complete removal of an oxide layer in a diluted fluoride solution at about -0.4 V. The dependence of the anodic oxide thicknesses that have been prepared on p-Si(100) at different potentials is monitored by anodic current transients in Figure 68. The potential is switched from the anodic oxidation

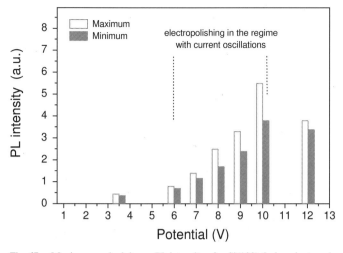

Fig. 67. Maximum and minimum PL intensity of p-Si(100) during electropolishing as a function of the applied potential (0.1 M NH$_4$F (pH 4.2)). Reprinted with the permission of Wiley VCH [168], copyright 1997.

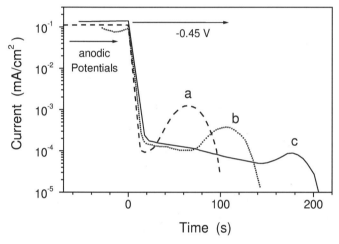

Fig. 68. Anodic current transients of p-Si(100) at -0.45 V after different anodic potentials (a, b, c: $+3.5$ V, $+9$ V, $+12$ V) in 0.1 M NH$_4$F (pH 4.2). Reprinted with the permission of Wiley VCH [168], copyright 1997.

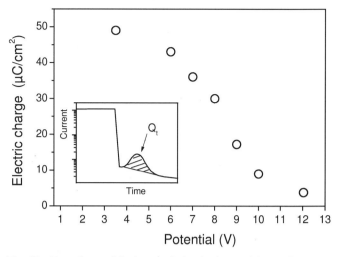

Fig. 69. Dependence of the integrated electric charge of the anodic current transient, Q_t, for p-Si(100) on the applied potential during the preceding electropolishing in 0.1 M NH$_4$F (pH 4.2). The inset illustrates the determination of Q_t. Reprinted with the permission of Wiley VCH [168], copyright 1997.

state (a, b, c: $+3.5$ V, $+9$ V, $+12$ V) to -0.45 V in the same solution, and the time base is scaled to the interruption of the current after the potential is switched. The onset of the anodic current transient appears at a later time with increasing oxidation potential because of the increased thickness of the oxide layer.

The time needed to reach the peak maximum of the anodic current transient correlates well with the period of the current oscillations for the potential range between $+4$ and $+10$ V, but it is slightly longer than the respective period [35, 168]. Therefore, the anodic current oscillations depend sensitively on the thickness of the anodic oxide layer. The oscillation period is given by the amount of oxide generated during one oscillation period (i.e., the oxidation potential) and by the etch rate of the electrolyte [35, 170, 179, 265, 266]. The frequency of the oscillation is proportional to the etch rate at fixed potential [170].

The anodic current peak that appears during the etch back

of the oxide layer is caused, in our opinion, by the oxidation of partially oxidized Si atoms at the Si surface during the hydrogenation process, as proposed by Gerischer and Lübke [166]. Figure 69 shows the dependence of the integrated electric charge of the anodic current transient at -0.45 V on the applied oxidation potential [168]. The integrated electric charge decreases with increasing oxidation potential-showing that the number of partially oxidized Si atoms at the interface decreases. It is important to note that the integrated electric charge of the anodic current transient is not correlated with the concentration of nonradiative recombination defects at the Si interface. It can be concluded that the quenching of the PL intensity is not related to the formation of partially oxidized Si surface atoms. Unfortunately, there is no direct experimental proof of this conclusion at the moment.

The PL intensity correlates with the anodic current oscillations during the anodic oxidation. An example is shown in Figure 70 for p-Si(100) at $+10$ V in 0.1 M NH$_4$F (pH 4.2). As usual, the maximum current corresponds to a minimum in the PL intensity and vice versa. The modulation of the PL intensity is related to a modulation of the concentration of dangling bonds at intrinsically back-bonded Si surface atoms. Breaking of Si-Si back bonds is one elementary step during the anodic oxidation of Si, and the concentration of dangling bonds at intrinsically back-bonded Si surface atoms is limited by the oxidation rate. As remarked, the anodic current is not necessarily correlated with the PL intensity and therefore with the oxidation rate [168].

The amount of oxide correlates with the oscillation period during anodic oxidation as shown by in situ ellipsometric measurements [267] or in situ FTIR spectroscopy [86, 92]. Figure 71 shows the time dependence of the current for n-Si(111) during two oscillation periods and the corresponding relative amounts of oxide. The inset shows the FTIR spectra normalized to the hydrogenated Si surface. The amount of oxide has a

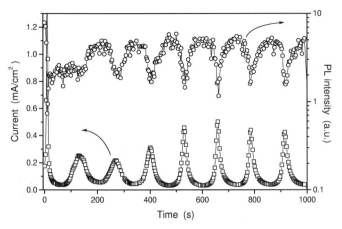

Fig. 70. Anodic current oscillations and oscillations of the PL intensity of
p-Si(100) during anodic oxidation at $+10$ V in 0.1 M NH$_4$F (pH 4.2). Reprinted
with the permission of Wiley VCH [168], copyright 1997.

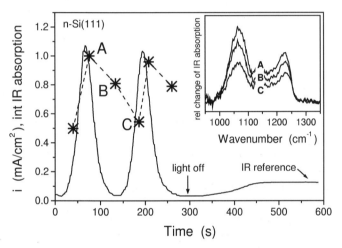

Fig. 71. Anodic current oscillations (solid line) and integrated IR absorption
of the Si-O-Si bonds for n-Si(111) in 0.1 M NaF (pH 4). The inset shows the
FTIR spectra normalized to the hydrogenated Si surface.

maximum (minimum) in the decreasing (increasing) part of the
oscillating current. The relative change in the amount of oxide
is about 50%. This finding reveals that the rate of formation of
anodic oxides changes strongly during the anodic current oscillations and that the maximum of the current is not simply
related to a pure injection current. As remarked, the FTIR spectra did not show any partially hydrogenated Si surface.

The homogeneity of the thin anodic oxide layer during the
current oscillations is important for a better basic understanding of the processes and for possible applications. The anodic
oxidation in the oscillating regime can be interrupted at the two
characteristic points when the oxidation rate is minimal (at the
minimum of oscillation) or maximal (at the maximum of oscillation). Figure 72 shows, as an example, the anodic current
transient at -0.4 V (a) and PL intensity (b) for p-Si (111) in 0.1
M NH$_4$F (pH 4) after anodic oxidation at $+8$ V and interruption
of the oscillations in the maximum (solid lines) or minimum
(dashed lines). The anodic current transient appears somewhat
earlier in time after interruption in the minimum than after in-

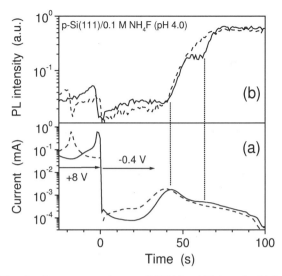

Fig. 72. Anodic current transient at -0.4 V (a) and PL intensity (b) for p-Si
(111) in 0.1 M NH$_4$F (pH 4) after anodic oxidation at $+8$ V and interruption of
the oscillations in the maximum (solid lines) or minimum (dashed lines).

terruption in the maximum of a current peak. This means that
the overall thickness of the anodic oxide is lower at the minimum of the current peak than at the maximum. The current
transient exhibits a shoulder after an etching time of about 60–
70 s, which gives evidence of a slight second maximum. This
behavior points to an inhomogeneous thickness of the anodic
oxide layer at the current maximum. The PL intensity decreases
just after interruption of the anodic oxidation, probably because
of the change in the electric field (Fig. 72b). The PL intensity increases within the first 20 s as a result of the ongoing chemical
reactions, which further passivate nonradiative recombination
centers at the anodic oxide/Si interface; i.e., these processes are
not interrupted after the potential is switched. The PL intensity
reaches a constant level with time and increases monotonously
after the maximum of the anodic current transient is reached,
after interruption of the anodic oxidation at the current minimum (dashed lines in Fig. 72). The PL intensity saturates when
the anodic current transient levels out. The situation is slightly
different for interruption of the anodic oxidation at the current
maximum (solid lines in Fig. 72). In this case, the PL intensity increases in two pronounced steps after the current transient
maximum is passed, which are related to the main and second
slight maximum of the anodic current transient (indicated by
the dotted lines). This behavior demonstrates that the thickness
of the anodic oxide at the current maximum of an oscillation
has two different mean values. The maximum of the PL intensity is practically independent of the time of interruption of the
oxidation process.

A new approach has been proposed by Föll et al. for the interpretation of the development of current oscillations during
anodic oscillations [268]. Föll postulated local current bursts
through a closed oxide. The current bursts occur for short times,
and they may under certain conditions interact space-resolved
to their synchronization in time (critical potential and etch rate).
Current bursts take place on the Å to nm scale [268], and their

synchronization should be responsible for the conditioning of the anodic oxide/silicon interface. The anodic oxidation rate decreases with decreasing current, and therefore the generation rate of nonradiative defects at the Si interface exhibits a minimum. The oscillation period and the oxidation rate increase linearly with the oxidation potential at constant etch rates of the solution [168]. This shows that the thickness of the oxide layer is important for damped or sustained current oscillations. To our understanding, the microscopic reason for triggering of the synchronized current bursts could also be related to the development of a space charge region inside the anodic oxide layer due to injected charge.

The condition for the development of current bursts can be changed by the structure at the anodic oxide/Si interface. The anodic oxidation in the oscillating regime should be interrupted by switching the potential to the hydrogenation state and restarting the oxidation process again to change the structure at the interface. The second cycle of anodic oxidation in the oscillating regime then starts with a microscopically smoother surface, and the probability of the development of current bursts due to structural nonuniformity on the microscopic level should be decreased. An example of anodic cycling between oxidation in the oscillating regime and hydrogenation is given in Figure 73 for p-Si(111) in 0.1 M NH$_4$F (pH 4). The current and the value of N_S are shown on the left and right sides, respectively. The potential is switched between $+8$ V and -0.4 V. As expected, the current at the maximum of the oscillations (starting from the second oscillation) decreases for repetitive cycling. The integrated charge of the anodic current transient decreased from the first to the second cycle, which is reflected by the reduced current density during the transient. The values of N_S at the maximum and minimum of the oscillations as well as for the hydrogenated surface decrease with each cycle. Figure 74 summarizes the results obtained from Figure 73: (i) decrease in the integrated charge of the anodic current transient (A), (ii) decreasing value of N_S at the hydrogenated Si surface down to 10^{10} cm^{-2} and lower (B, open square), and (iii) decrease in N_S at the oxidized Si surface to values below 10^{11} cm^{-2} (C) with increasing number of cycles. This demonstrates the great potential of anodic oxidation for low-temperature passivation of Si surfaces. Figure 74B also shows that repetitive cycling of electropolishing and hydrogenation does not lead to any improvement of N_S for anodic oxidation in the nonoscillating regime at $+3$ V (B, solid squares). We remark that the difference between the PL intensities at different potentials is not as strong for the p-Si(111) surface as it is for the p-Si(100) orientation.

The values of N_S at the anodic oxide/Si interface and the hydrogenated Si surface could be, to our opinion, further decreased by optimization of the electrolyte and the potential. There is a strong indication that interruption even after the first heavy current peak, before regular oscillations start, allows a further decrease in N_S. Obviously, any current minimum that follows a strong current maximum, which is induced by well-synchronized current bursts, is characterized by a very low concentration of reactive surface sites. This state of the anodic

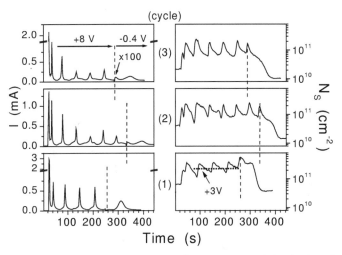

Fig. 73. Time dependence of the anodic current, I, and the density of nonradiative surface defect centers, N_S, as obtained from the PL intensity of p-Si(111) in 0.1 M NH$_4$F (pH 4) for three subsequent oxidation ($+8$ V)/hydrogenation (-0.4 V) cycles (1 to 3). The horizontal dotted line marks the value of N_S during anodic oxidation at $+3$ V (no oscillations).

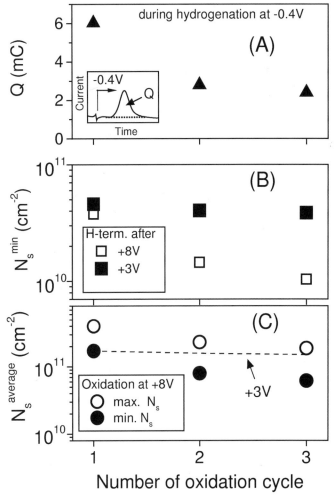

Fig. 74. The charge that passes the electrode during etch-back of the oxide formed at $+8$ V (A), the minimum values of N_S after hydrogenation at -0.4 V (B), and the average minimum and maximum values of N_S during anodic oxidation at $+8$ V and $+3$ V (C) (values of I and PL are taken from Fig. 73). Reprinted with the permission of Elsevier Science B. V. [189], copyright 2000.

oxide/silicon interface can be used to optimize the interface passivation. The current bursts occur preferentially at places where the electrical field is enhanced. A nonuniformity on the microscopic scale leads to the enhancement of the electrical field. The local input of energy during the current burst causes a fast local oxidation. This fast oxidation process is far from equilibrium and leads to a local reconstruction in such a way that the nonuniformity on the microscopic scale will be smoothed. The locally reconstructed interface is quite uniform and practically free of reactive surface sites, which would cause the development of new current bursts.

5.4. Formation of Oxides in Alkaline Solution

Figure 75 shows a typical treatment of the p-Si(111) surface (bottom: current density, i; top: PL intensity). At first, the oxide-free surface (HF dipped) is anodically oxidized in 0.1 M NH$_4$F (pH 4) at +3 V for about 15 min to obtain the same starting condition for each experiment. The PL intensity is low during oxidation, where the breaking of Si–Si bonds leads to a high amount of nonradiative recombination centers (i.e., to a high defect concentration, N_s, which is about 10^{13} cm^{-2}). The oxide is then etched back in the same solution by switching the potential to −0.4 V, and the well-known current peak occurs, which indicates the transformation from the oxidized to the hydrogenated state. Simultaneously, the PL intensity increases drastically when the current levels out. N_s on the Si(111) surface is then typically 10^{11} cm^{-2} or below [84, 189]. The PL intensity decreases slowly with time. This behavior is indicative of the formation of defects in the diluted HF solution (see Section 3). When an almost constant PL intensity is reached, the solution is replaced by an alkaline solution under galvanostatic conditions (−1 μA/cm^2) to avoid uncontrolled oxidation. The behavior of the PL intensity is shown for different alkaline solutions (pH 9.1, 11.2, and 13.5), which are made from sodium borate or KOH, respectively. The introduction of the al-

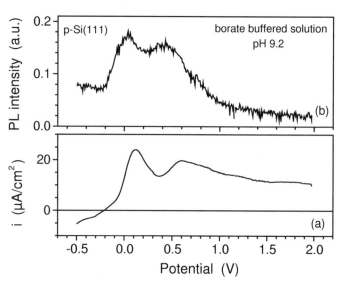

Fig. 76. Current–voltage scan (a) and PL intensity (b) of p-Si(111) in borate-buffered solution (pH 9.2).

kaline solution always leads to a lowering of the PL intensity. The higher the pH value, the higher is the etch rate of the H-terminated surface, and the higher is the defect concentration and, consequently, the lower is the PL intensity. After this procedure, a potential scan in the anodic direction is performed to investigate the electrochemically induced etch stop by hydroxide or oxide formation.

Figure 76 shows a potential scan of p-Si(111) in borate-buffered solution (pH 9.2) starting from the cathodic potential (bottom: current density, i; top: PL intensity). The current density increases with potential and exhibits two current peaks that are associated with the formation of a passivating oxide layer [16, 181, 186]. The PL intensity increases with potential when an anodic current starts to flow. The maximum of the PL intensity is reached when the first relative current peak starts to decay. The PL intensity increases again just before the second current maximum is reached. At higher anodic potentials (increasing oxide thickness), the PL intensity decreases to very low values (high defect concentration).

Figure 77 shows a potential step experiment. The hydrogenated Si surface is etched in borate-buffered solution (pH 9.2) at a potential of −0.24 V. Then the potential is switched to +0.07 V for 8 s, to the increasing part of the first current peak (see inset). The current increases and decays to a constant value of about 10 μA/cm^2. The PL intensity increases strongly because of the electrochemically induced etch stop [148, 269] and decays slowly to the initial value when the potential is switched back to −0.24 V. This behavior is a result of the etch back of the oxidized surface, leading obviously to a H-terminated Si surface again.

A logarithm plot of the two relative maxima of the PL intensity at the first (open circles) and second (solid circles) maxima during anodic current flow as a function of pH is presented in Figure 78. The PL intensity decreases linearly with increasing pH, with a slope of about −0.1 at the first relative maximum.

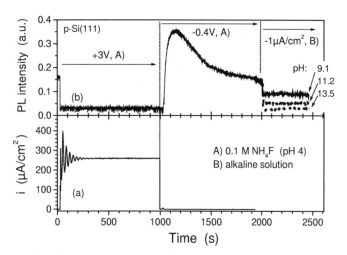

Fig. 75. Time dependence of the anodic current (a) and PL intensity (b) during anodic oxidation at +3 V and anodic current transient at −0.4 V in 0.1 M NH$_4$F (pH 4), followed by galvanostatic treatment in alkaline solutions (pH 9.1, 11.2, and 13.5) at a cathodic current of −1 μA/cm^2.

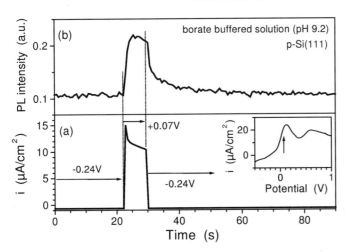

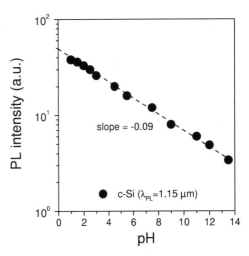

Fig. 77. Time dependence of the current (a) and PL intensity (b) of *p*-Si(111) in borate-buffered solution (pH 9.2) during the application of a potential step ($-0.24 \rightarrow +0.07 \rightarrow -0.24$ V). The inset shows the current potential scan of the sample, and the arrow marks the position of the anodic potential step.

Fig. 79. pH dependence of the PL intensity of the hydrogenated Si surface.

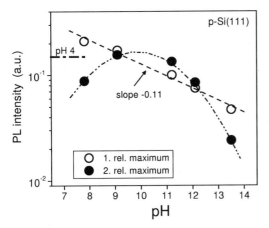

Fig. 78. pH dependence of PL intensity at the first and second maxima of the anodic current during a potential scan (open and solid circles; see also inset in Fig. 77). The horizontal dash-dotted line gives the value of the PL intensity at pH 4.

6. THICK ANODIC OXIDES ON Si

Low thermal budget processing of silicon-based semiconductors is of great interest for modern device passivation [28–30, 261, 270], where thermal oxidation is not suitable. Anodic oxidation is performed at room temperature [7, 271], and a post-anneal below 500°C follows [23, 27], which permits the passivation of devices with specific doping profiles or very high resistivity [23, 272], heat-sensitive materials like a-Si [15] or SiGe [32], without changing the morphology, crystallinity, or electronic properties. Furthermore, the liquid contact permits passivation and repassivation of etched grooves, trenches, etc., or of highly structured surfaces [27]. In addition, anodic oxidation of silicon (or SiGe) surfaces leads to a smoothing effect of the surface due to the electric field-enhanced growth process [27]. This behavior increases the breakdown voltage with respect to thermally grown oxides [270, 273]. The rate of oxide formation depends on the amount of water in the solution, the temperature of the solution, and the current density. The electronic properties (oxide charge and interface state density) depend only slightly on the temperature and current density, but they are strongly influenced by the amount of water [22]. The oxidation rate is strongly increased by the water content of the electrolyte, which then leads to a higher defect concentration at the interface. The final oxide thickness depends on the applied potential and the amount of water in the electrolyte.

6.1. Low Thermal Budget Processing

A solution of 0.04 M KNO_3 in ethylene glycol with different amounts of water is used for the preparation of thick anodic oxides ($d > 10$ nm) on Si or SiGe. A small amount of water (typically 0.35%) is necessary to reduce the rate of oxidation with respect to pure water, where the rate is too fast at higher anodic potentials. This leads to a worse passivation of the SiO_2/Si interface [261]. The total reactions in such a solution are as follows [7].

This behavior is a result of the higher etch rate of Si in higher concentrated alkaline solutions. In contrast, the PL intensity of the second relative maximum at higher anodic potentials shows a maximum around pH 9. The observed PL intensity points to N_s of about 5×10^{11} cm^{-2} eV^{-1} as deduced from the PL intensity of the hydrogenated state (typically 10^{11} cm^{-2} eV^{-1}) and the linear dependence of the PL intensity on N_s [156, 189]. Similar low values of N_s of SiO_2/Si interfaces have recently been observed in diluted HF solutions during electropolishing [189] (see also Section 4).

Figure 79 shows the dependence of the PL intensity of the hydrogenated Si surface on the pH. The linear behavior of the PL intensity is elongated into the direction of much lower pH values and shows the same dependence on pH as it does in alkaline solutions [224]. This points to a more universal mechanism of defect formation, where, for example, only OH$^-$ ions are the etching species.

Oxidation at the Si anode:

$$Si + 2H_2O + 4h^+ \rightarrow SiO_2 + 4H^+ \tag{6.1}$$

$$\begin{array}{c} CH_2 - CH_2 \\ | \qquad | \\ OH \quad OH \end{array} \rightarrow 2H_2CO + 2H^+ + 2e^- \tag{6.2}$$

$$2H_2O \qquad\qquad \rightarrow O_2\uparrow + 4H^+ + 4e^- \tag{6.3}$$

Reduction at the Pt cathode:

$$4H^+ + 4e^- \qquad\quad \rightarrow 2H_2\uparrow \tag{6.4}$$

$$NO_3^- + H_2O + 2e^- \rightarrow NO_2^- + 2OH^- \tag{6.5}$$

The efficiencies of reactions (6.1) to (6.3) are about 1%, 98%, and 1%, respectively [7]. This means that only 1% of the current leads to the oxidation reaction of Si with water to form silicon oxide.

The anodic oxidation of SiGe leads to an additional reaction,

$$Ge + 2H_2O + 4h^+ \rightarrow GeO_2 + 4H^+ \tag{6.6}$$

where it is known that GeO_2 is soluble in water [274–277].

Figure 80 shows a photograph of the front view of the quartz chamber for anodic oxidation for 3-inch (maximum) wafers (left) and a schematic drawing of the cell from a side view (right). The Si sample is sucked up onto an O-ring seal by a vacuum in the quartz tube and electrically contacted with a small Al plate. A Pt ring is used as a counter-electrode. The electrolyte in the chamber is stirred, and the temperature is controlled by a thermostat. It should be noted that the layout of an oxidation chamber is defined by process parameters so that, in principle,

several Si wafers can be oxidized at the same time with the use of another type of electrochemical cell.

After the anodic oxidation is completed, the oxidized sample is cleaned in $HCl:H_2O_2:H_2O$ (1:1:6), dried under nitrogen steam, and annealed in nitrogen or forming gas up to 800°C (maximum).

The oxidized samples are characterized by a great number of analytical and imaging techniques, and by capacitance–voltage (CV), surface photovoltage (SPV), and pulsed photoluminescence (PL) measurements.

Thermal oxidation in dry O_2 steam at 1000°C for Si and rapid thermal oxidation (RTO) at 800°C for SiGe epilayers are used for comparison.

6.2. Preparation of the Oxide Layer

Figure 81 shows the time dependencies of the potential and current on the temperature of the solution (A) and the initial current density at 40°C (B) during the oxidation process in ethylene glycol with 0.35% water. The oxidation starts with a galvanostatic process. In the beginning, the potential increases slowly. The increase in potential becomes faster with time because of the increasing thickness of the oxide layer. The thicker the oxide layer, the higher is the potential drop across the oxide. When the desired final potential is reached (in this case +120 V, oxide thickness about 54 nm), the process is continued potentiostatically until the current has decayed to an almost constant value, as can be seen in the bottom part of Figure 81. Increasing both the temperature of the solution and the current density

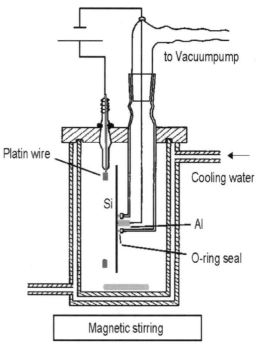

Fig. 80.   Front and side views of the electrochemical cell used for preparation of thick anodic oxides. Reprinted with the permission of Elsevier Science B. V. [27], copyright 2000.

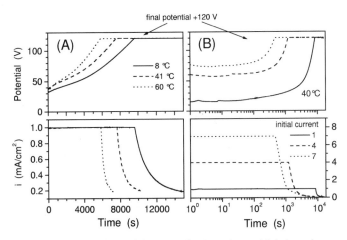

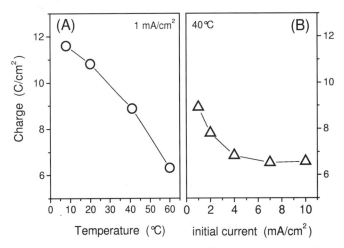

Fig. 81. Time dependence of the current density and potential during galvanostatic preparation of anodic oxides (A) at different temperatures and a constant current density of 1 mA/cm² and (B) at 40°C and different initial current densities. The final potential is 120 V (oxide thickness ~54 nm). Reprinted with the permission of Elsevier Science B. V. [27], copyright 2000.

Fig. 82. The electric charge that passes the Si anode as a function of the temperature of the electrolyte (A) and the initial current density (B) (values of current and time are taken from Fig. 81).

decreases the time of the oxidation process. The amount of electric charge that passes the electrode is plotted in Figure 82 as a function of the temperature of the electrolyte (A) and the initial current density (B) (values of current and time are taken from Fig. 81). The charge flow is reduced either by an increase in the temperature or by an increase in the current density. The charge flow saturates at a current density of about 7 mA/cm² at 40°C and does not change at higher current densities (Fig. 82B), whereas the charge decreases almost linearly with the temperature of the electrolyte and shows no tendency to saturate up to 60°C. The oxide thickness is independent of temperature and the current density; it depends only on the final potential and the amount of water in the electrolyte (see Fig. 83). Therefore, a constant charge at higher current densities, which is due to a reduction of the oxidation time, is a result of constant current efficiencies for reactions (6.1) to (6.3). The efficiency of oxide formation due to reaction (6.1) decreases with decreasing current densities, the competition of the other reactions overcomes this effect, and the net charge flow increases. The efficiency of oxide formation also shows a strong dependence on the temperature of the electrolyte; the efficiency at 60°C is twice that at 8°C. A reduction of the charge flow reduces the overall power consumption during anodic oxidation.

The oxide thickness depends linearly on the applied potential and on the amount of water in the supporting electrolyte, as can be seen from Figure 83 (4.5 and 11.5 Å/V for 0.35% and 0.6 % water, respectively).

6.3. Electronic Characterization

Figure 84 shows the interface state density distribution, D_{it}, as obtained from CV measurements. D_{it} is high for the as-prepared and cleaned oxide layers (about 5×10^{12} cm^{-2} eV^{-1} at midgap), regardless of the applied final potential during the growth process. D_{it} decreases by about one order of magnitude after drying in nitrogen at 400°C and annealing in forming gas

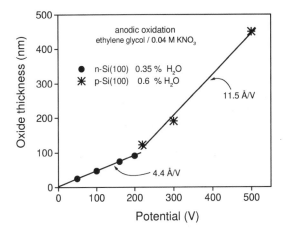

Fig. 83. Dependence of the thickness of an anodic oxide on the final potential for anodization in ethylene glycol with low contents of water.

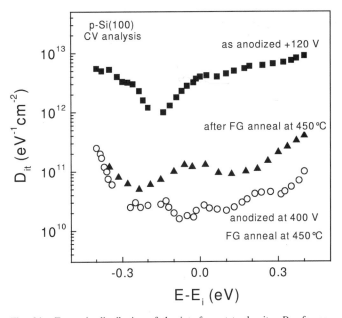

Fig. 84. Energetic distribution of the interface state density, D_{it}, for anodically oxidized p-Si(100) samples after different drying and annealing procedures.

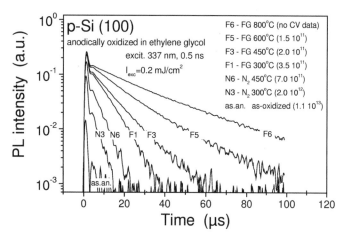

Fig. 85. PL transients of p-Si(100) covered with an anodic oxide processed in forming gas and N_2 at various temperatures. Reprinted with the permission of the American Institute of Physics [156], copyright 1999.

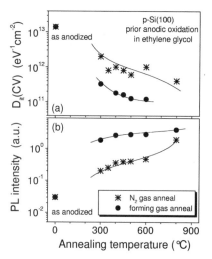

Fig. 86. Density of interface states, D_{it}, (a) and PL intensity (b) of anodically oxidized p-Si(100) as a function of the annealing condition.

at about 450°C. This procedure leads to a D_{it} value at midgap of about 1–2×10^{11} cm^{-2} eV^{-1}, whereas the D_{it} of thicker oxide layers (400 V, 180 nm, 3×10^{10} cm^{-2} eV^{-1} at midgap) is slightly below the value of the thinner layer prepared at 120 V (54 nm).

The dependence of D_{it} and of surface nonradiative defects (N_s) of anodically prepared oxide layers is investigated by CV and PL measurements, respectively. Figure 85 shows PL transients of differently annealed anodic oxide layers prepared on the same substrate, in the same solution, at the same current density and final potential to ensure comparative conditions. The inset shows the annealing condition and the D_{it} value at midgap as observed from CV measurement. No CV data have been obtained for sample F6. The lower the interface state density at midgap, the higher is the decay time of the time dependence of the PL. The integrated PL intensity scales with the reciprocal of D_{it} (see Section 2). Therefore, an increase in D_{it} by a factor of 10 decreases the PL intensity by the same factor. The as-anodized samples have high defect concentrations of about 10^{13} cm^{-2} eV^{-1} and a very low PL intensity. Annealing in nitrogen (or in a more pronounced way in forming gas) decreases the defect concentration at midgap to a typical value of about 10^{11} cm^{-2} eV^{-1} or slightly below. The oxide charges are in the range of 10^{12} cm^{-2}. Figure 86 shows D_{it} at midgap and PL intensity as a function of the annealing condition. There seems to be a plateau for the values of D_{it} and PL between 400°C and 600°C. The forming gas anneal decreases D_{it} by a factor of 10 with respect to the values obtained for the N_2 anneal. Note that the samples are dried in nitrogen at 400°C before the forming gas anneal.

6.4. Passivation of Steps and Trenches

Figure 87 shows SEM images of trenches on Si at a tilt angle of 30° (A: as prepared by CF$_4$-plasma etching; B: thermally oxidized at 1000°C; C: anodically oxidized). The as-prepared trench shows a sharp step that is also visible after thermal ox-

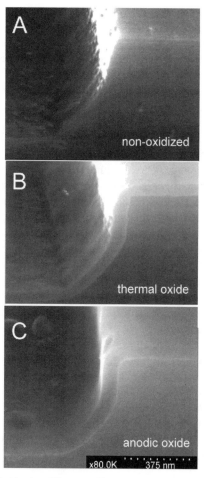

Fig. 87. SEM side view (tilt angle 30°) of concave edges of c-Si (A) before and after (B) thermal and (C) anodic oxidation.

idation. The electric field-enhanced anodic oxidation leads to a constant thickness of the oxide layer at the trench and to a rounding of the step where the oxide layer is thicker. The thickness of the thermal oxide (B) is somewhat smaller at the step.

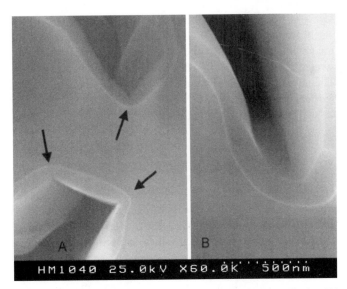

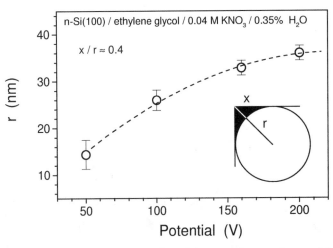

Fig. 88. SEM side view (tilt angle 30°) of convex edges of c-Si after (A) thermal and (B) anodic oxidation.

Fig. 89. Dependence of the radius of rounded concave shapes on the potential during anodic oxidation of Si edges (ratio x/r is constant, ~0.4).

A similar behavior can be observed by comparing thermal and anodic oxidation of a inner corner of a trench, as visualized by the SEM images in Figure 88. The arrows point to the narrowing of the thermal oxide at a corner (A), which is not the case for anodic oxides (B). The kinetics during thermal oxidation that lead to smaller oxide thickness at concave and convex steps is discussed intensively by Wolters and Duynhoven [110, 278].

The open circles in Figure 89 represent the radius of a circle, which follows the curvature of the rounding of the step (see inset), as a function of the applied potential (oxide thickness) as obtained from SEM images (like Fig. 87). The ratio of x/r is always constant and amounts to about 0.4. This means that the anodic potential at steps is enhanced by about 40% with respect to the applied potential, which leads to a 40% thicker oxide layer at the step. Recall that the thickness of anodic oxides depends linearly on potential (Fig. 83). This result is not surprising, because it is known that the potential is increased at steps and tips. In addition, such a constant oxide thickness and rounding of steps after anodic oxidation leads to a high breakdown voltage of about 12 MV/cm, whereas the breakdown voltage of thermal oxides is only about 8–10 MV/cm [270, 273]. The surface smoothing effect during anodic oxidation can easily be seen from AFM images presented in Figure 90. This figure shows microcrystalline silicon surfaces before and after anodization (A: as-deposited layer; B, C: after oxidation at +100 V and +160 V). The oxide layer (thickness about 50 nm (B) and 90 nm (C)) has been etched back by 2% HF before image recording. The surface of the microcrystalline silicon becomes smoother with increasing oxidation potential.

Figure 91 shows a highly structured Si surface (poly-Si) covered with a uniform anodic oxide layer, even in the very narrow tails and holes, which can be reached very well by the liquid electrolyte contact.

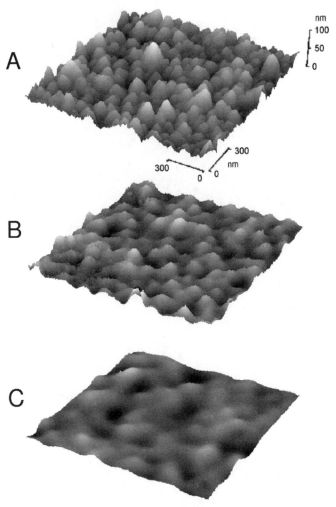

Fig. 90. AFM images of microcrystalline silicon surfaces (A: as-deposited layer, B, C: after oxidation at +100 V and +160 V, respectively). The oxide layer (thickness about 50 nm (B) and 90 nm (C)) was etched back by 2% HF before image recording.

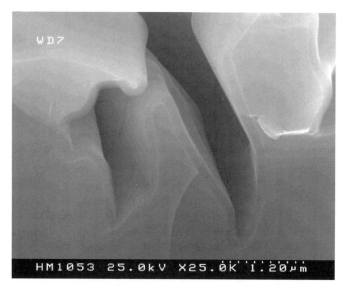

Fig. 91. SEM side view (tilt angle 30°) of a highly structured surface region of polycrystalline Si after anodic oxidation.

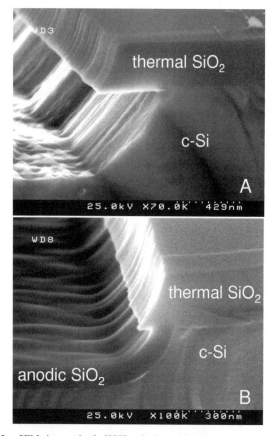

Fig. 92. SEM viewgraph of a KOH-etched trench in c-Si covered by a thermal oxide (A) and of the same trench after anodic oxidation (B).

This procedure is also applicable to repassivation, especially of etched structures on silicon, as presented in Figure 92. First, grooves are opened in a thermal oxide on Si down to the Si surface. These grooves are then etched by KOH, which leads to an underetching of the thermal oxide layer, whereas the etch rate

for Si is much faster than that for the oxide (Fig. 92A). The partly free-standing oxide layer can be seen at the top of the image. Such etched structures could then be completely repassivated by anodic oxidation due to the liquid contact (Fig. 92B). The anodic oxide layer connects very well to the underetched thermal oxide layer above.

7. ENHANCED PASSIVATION OF SiGe BY ANODIC OXIDATION

SiGe is of great interest for modern semiconductor devices because of the enhanced capability of bandgap engineering [279] and the increased mobility of charge carriers created by the incorporation of Ge into the Si lattice as used in heterobipolar transistors [28, 29] or metal-oxide-semiconductor (MOS) structures [30]. Thermal oxidation of SiGe alloys above 600°C leads to Ge segregation at the oxide/SiGe interface [31] and relaxation of the strained SiGe lattice of an epitaxially grown thin film of SiGe on c-Si [33]. These processes induce defects in the crystal lattice and at the interfaces, which strongly affect the electronic behavior of the device [32, 33]. Therefore, passivation of modern devices on the basis of SiGe requires methods of low thermal budget processing below 500°C [29]. Among these are plasma-assisted oxidation techniques [280–282], ion beam deposition [283], oxidation at high pressures [245], and anodic oxidation [32, 33]. Some of these techniques involve new problems. For example, plasma oxidation is usually performed in a UHV chamber where electrons and oxygen ions are excited and speeded up in the direction of the positively polarized SiGe samples. The deposition temperature is low (<200°C), but the ion bombardment may lead to local heating of the sample and, therefore, build-up of local irregularities (i.e., Ge cluster, Ge diffusion, etc.). High-pressure oxidation at 50–70 bar and 500°C defines high requirements for the chamber material and may be used for special devices only. Plasma-assisted oxidation or deposition techniques and anodic oxidation seem to be good candidates for low thermal budget passivation of SiGe alloys.

7.1. Defect Concentration at the Oxide/c-SiGe Interface

Figure 93 shows interface state densities, D_{it}, of oxidized CZ-grown c-Si and c-Si$_{1-x}$Ge$_x$ ($x = 0.038$, 0.057, and 0.096) as obtained from CV measurements as a function of energy with respect to the intrinsic level [224]. D_{it} at midgap of the anodic oxides on c-SiGe increases from 5×10^{11} to 18×10^{11} cm^{-2} eV^{-1} with increasing amounts of Ge (3.8 to 9.6 atomic %), which may be due to an increased stress by bond angle mismatch, in both the oxide layer and the SiGe lattice (Fig. 93b). The D_{it} values for thermal oxide on c-SiGe with 9.6 atomic % Ge is 3×10^{12} cm^{-2} eV^{-1} (Fig. 93a), which is half of the value observed by deep-level transient spectroscopy (6.9×10^{12} cm^{-2} eV^{-1} [284]). D_{it} of anodically oxidized CZ-grown c-Si is plotted in Figure 93a for comparison and shows the lowest interface state density. The increase in the amount of defect states with increasing Ge content of the bulk SiGe

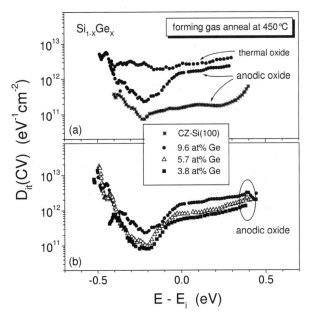

Fig. 93. Distributions of the interface state density, D_{it}, for $Si_{1-x}Ge_x$ after thermal and anodic oxidation followed by forming gas anneal at 450°C. (a) and (b) compare the roles thermal of anodic oxidation and of the Ge content, respectively.

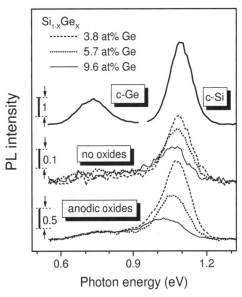

Fig. 94. PL spectra of c-Ge and c-Si (top), oxide-free c-SiGe (middle), and anodically and thermally oxidized c-SiGe (bottom) as a function of the Ge content.

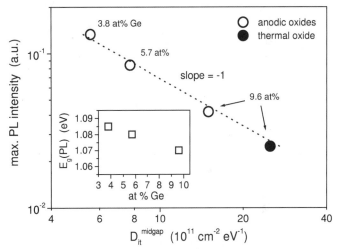

Fig. 95. Correlation of the PL intensity with the interface state density at midgap, D_{it}, for $c-Si_{1-x}Ge_x$ with different amounts of Ge. Inset: Bandgap of $c-Si_{1-x}Ge_x$ as a function of the Ge content as obtained from the maximum of the PL spectra of Figure 94. Reprinted with the permission of Elsevier Science B. V. [33], copyright 2000.

is also reflected by the decrease in the PL intensity measured at room temperature. Figure 94 shows PL spectra of c-Ge and c-Si (top), oxide-free c-SiGe (middle), and anodically and thermally grown oxides on c-SiGe (bottom) as a function of the Ge content. In general, the PL spectra due to the band-band recombination of c-SiGe shift to lower energy with respect to that of c-Si as a result of the reduced band gap energy (E_g) with increased Ge, regardless of the surface conditioning. This confirms a higher defect state density with increasing Ge in the crystals (dislocations, stress by Si-Ge-Si bond angle mismatch). However, there is a similar linear dependence of the maximum of the PL intensity on D_{it} at midgap (Fig. 95), as obtained for oxides on silicon (see Fig. 19). The inset shows E_g (taken from the maximum of the PL spectra, Fig. 94) as a function of the amount of Ge in the crystal. At least these bulk properties have an influence on the electronic behavior of the semiconductor/oxide interface.

The growth process of single crystals of SiGe is limited to a Ge content of about 10 and 20 atomic % in CZ and FZ material, respectively. Higher Ge content is needed for faster access of, for example, heterobipolar transistors (HBTs). To achieve higher mobility of the charge carriers, SiGe epitaxy is used on c-Si(100) substrates, which is compatible with the hole Si technology. The next paragraphs are devoted to anodic passivation of epitaxially grown SiGe layers on c-Si.

Figure 96 shows PL intensities of c-Si, c-Ge, and epi-SiGe layers (with ≈10 and ≈30 atomic % Ge) on c-Si after different oxidation treatments. The inset shows the PL spectra recorded at room temperature, which are shifted to longer wavelengths, as expected for the increasing content of Ge [285–287]. The largest PL intensity (lowest N_s value) of c-Si is reached af-

ter thermal oxidation (which includes a forming gas anneal), whereas the PL intensity is about one order of magnitude lower for the Si surface covered by a native oxide. The Si surface is also passivated very well after treatment in 1 M NH_4F solution (H-terminated). The PL intensity of the samples containing Ge decreases after the thermal oxidation in comparison with the untreated surface. The surface passivation of the SiGe epitaxial layer can be improved by the deposition of a Si cap, which protects the SiGe from dissolution in the fluoride solution. Similar results have been obtained by anodic oxidation of SiGe epitaxial layers [32].

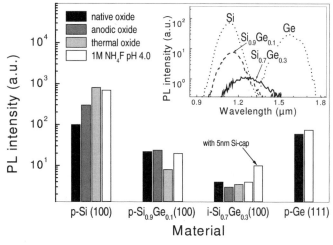

Fig. 96. PL intensities of c-Si, c-Ge, and epi-SiGe layers (with ~10 and ~30 atomic % Ge) on c-Si after different oxidation treatments. Inset: PL spectra of the samples recorded at room temperature. Reprinted with the permission of Elsevier Science B. V. [224], copyright 1998.

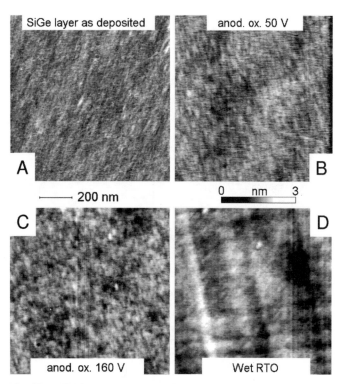

Fig. 97. AFM images (top view) of (A) as-deposited, (B, C) anodically oxidized (at 50 and 160 V, respectively), and (D) rapid thermally oxidized epi-Si$_{0.74}$Ge$_{0.26}$ layers.

7.2. Morphology of Oxidized epi-SiGe Samples

Figure 97 shows AFM images (top view) of (A) as deposited, (B, C) anodically oxidized at 50 and 160 V, respectively, and (D) rapid thermally oxidized epi-Si$_{0.74}$Ge$_{0.26}$ layers. There is no relaxation of the strained SiGe lattice visible for the low thermal budget processed oxides as observed for the RTO processed sample. The regular lines in the AFM image in Figure 97D reflect the relaxed steps of the strained epi-SiGe lattice after

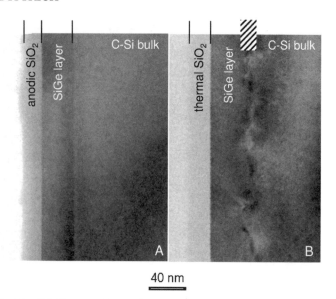

Fig. 98. TEM images of (A) anodically and (B) rapid thermally oxidized epi-Si$_{0.74}$Ge$_{0.26}$ layers on c-Si(100).

heating above 600°C at the RTO process. The anodically treated samples did not show such an effect.

Figure 98 shows transmission electron microscope (TEM) images of (A) anodically and (B) rapid thermally oxidized epi-SiGe layers on c-Si(100). From the left to the right of Figure 98A one can see the anodic oxide, with a thickness of ~18 nm; the SiGe layer (thickness ~34 nm); and the c-Si substrate. The image contrast of the epi-SiGe/Si interface is not high, but the interface can easily be seen. This is not the case for the thermally oxidized SiGe layer (Fig. 98B). One can see the slightly thicker oxide layer (~24 nm), but the epi-SiGe/Si interface cannot be well resolved. Most likely, the interface has many lattice distortions visualized by the strong irregular contrasts in the middle of the TEM image. Contrast differences in TEM could also be a result of the preparation method, but they are typically much greater in dimensions, as can be seen, for example, from the darker to the lighter top part in the left TEM image (Fig. 98A).

7.3. Oxide Composition

Figure 99 shows Auger depth profiles of oxygen (a), silicon (b), and germanium (c) of anodically (solid lines) and rapid thermally (dashed lines) oxidized epi-Si$_{1-x}$Ge$_x$ layer ($x = 0.26$). The Auger intensity of oxygen in SiO$_2$ is plotted in the top part of the figure and defines the oxide/SiGe interface. The anodic oxidation of the epi-SiGe layer shows Ge in the oxide layer and no Ge pile-up at the oxide/SiGe interface, which is not the case for thermal oxidation, where the oxide layer is free of Ge and the amount of Ge is enhanced at the oxide/SiGe interface. The Auger intensity of Ge is lower in the oxide layer than in bulk SiGe, which is due to dissolution of Ge by the oxygen incorporated into the SiGe lattice. Note that the Auger signal of Si ions in the oxide layer could not be measured; because of the experimental setup, the silicon Auger peak was

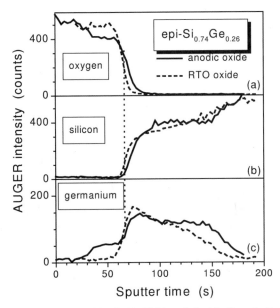

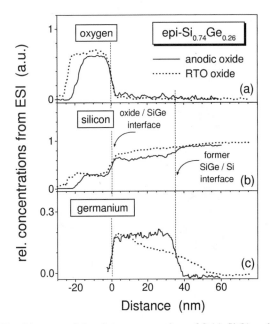

Fig. 99. Auger depth profiles O (a), Si (b), and Ge (c) of anodically (solid lines) and rapid thermally (dashed lines) oxidized epi-Si$_{0.74}$Ge$_{0.26}$ layers. The dotted line denotes the oxide/SiGe interface.

Fig. 100. Line scans of the relative concentrations of O (a), Si (b), and Ge (c) across the layer system as determined by ESI (the distance is scaled to zero at the oxide/SiGe interface). Reprinted with the permission of the Electrochemical Society [34], copyright 2001.

out of the detectable range. Nevertheless, a top surface layer of the anodic oxides of some nanometers in thickness is free of Ge. It is well known that oxidized Ge dissolves in aqueous solution [277]. Therefore, dissolution of Ge^{4+} ions occurs until a SiO$_2$ layer is formed, which acts as a diffusion barrier for the Ge^{4+} ions. This barrier has a thickness of about 5 nm in the case of the epi-SiGe layer with 26 atomic % Ge. The elemental distribution of the layer system as measured by Auger depth profiling is somewhat diffuse because of the sputtering process, which induces some inaccuracy. Therefore, we applied electron spectroscopy imaging (ESI) techniques to microtomed epi-SiGe samples. Line scans of the relative concentrations of O (a), Si (b), and Ge (c) across the layer system as determined by ESI are plotted in Figure 100. The distance is scaled to zero at the oxide/SiGe interface, which has been fixed by the position of the half-intensity of the oxygen-ESI signal at the interface. The concentration of Si is reduced in the oxide layer in comparison with the nonoxidized SiGe layer because of dilution by incorporated oxygen. The amount of Si and Ge across the SiGe layer is nearly constant for the anodically oxidized sample (solid lines), whereas the RTO sample (dashed lines) shows a slight increase in the concentration of Ge at the oxide/SiGe interface and a continuous decrease in the direction of the Si substrate. The former SiGe/Si interface is no longer well defined after RTO processing. The Ge atoms diffuse up to about 15 nm into the underlying Si substrate (as-deposited SiGe layer thickness, ~46 nm). This behavior was not well resolved by Auger depth profiling, where the sputtering process starting from the oxidized surface seems to lead to an increased inaccuracy with sputtering time, which is obviously due to the dimension of the sputtering club. Note that the ESI signal of Ge could not be measured (it is out of detector range). Therefore, no Ge is plotted in Figure 100 for the anodic or for the thermal

oxides. Ge is present only in the anodic oxide layer, as deduced from Auger (Fig. 99) and XP (Fig. 102) spectroscopy.

Elemental maps of these samples recorded by ESI are presented in Figure 101 to give an overall impression of the elemental distribution of the system. Each element is defined by a specific color for a better visualization (oxygen: blue, silicon: green, germanium: red; please note: computerized RGB mixing of the colors red and green leads to yellow). One can well distinguish between the oxide layers (blue) and the nonoxidized SiGe layer. The SiGe/Si interface is well defined only in the anodically oxidized sample (Fig. 101A). The RTO processed sample (Fig. 101B) exhibits a slight Ge enrichment at the oxide/SiGe interface and a diffusion of Ge into the Si substrate over the hole layer system. The black dotted line denotes the frontier of the Ge penetration into the Si substrate as measured in more detail by the ESI line scan (Fig. 100).

XP and FTIR spectra are recorded to gain more information about the oxidation states of Ge in the anodically oxidized SiGe samples (rapid thermal oxidized samples are free of Ge in the oxide layer). Figure 102 shows XP spectra of anodically and rapid thermally oxidized SiGe, which reveal the 4+ state for Ge in the oxide layer. Al$_{k\alpha}$ irradiation was used because Mg$_{k\alpha}$ excitation leads to an overlap of Ge-Auger and SiO$_x$ signals in the energy range between 102 and 110 eV. The thin top layer, which is free of Ge, has been etched back by diluted HF before the measurements. The thermal oxide on SiGe consists of pure SiO$_2$ with a signal at 105 eV only (99 eV for Si^{2p}), whereas the anodic oxide shows a peak at 35 eV, which is due to Ge^{4+} (29 eV for Ge^{3d}). In both cases there is a shift of 6 eV from the nonoxidized to the +4 state. From this measurement it can be

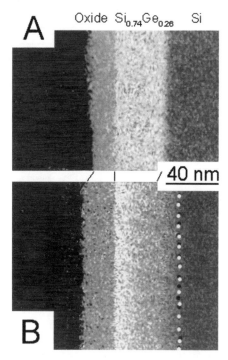

Fig. 101. Elemental maps of (A) anodically and (B) RTO processed Si$_{0.74}$Ge$_{0.26}$ layers recorded by ESI (oxygen: blue, silicon: green, germanium: red; note: computerized RGB mixing of the colors red and green leads to yellow). The black dotted line in (B) indicates the border of the Ge diffusion into the Si substrate after RTO processing. Reprinted with the permission of the Electrochemical Society [34], copyright 2001.

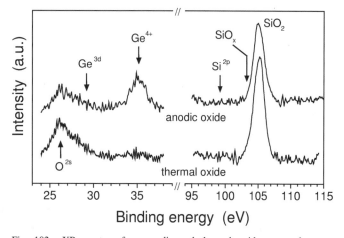

Fig. 102. XP spectra of an anodic and thermal oxide prepared on an epi-Si$_{0.74}$Ge$_{0.26}$/Si structure plotted in regions of the Si^{2p} and O^{2s} peaks. Reprinted with the permission of Elsevier Science B. V. [33], copyright 2000.

concluded that Ge is chemically bonded as Ge^{4+} ions into the anodic oxide network.

FTIR spectroscopy has been applied to clarify whether Ge is incorporated as GeO$_2$ and/or as mixed oxide species like Si-O-Ge into the oxide layer. Figure 103 shows FTIR spectra of anodically (A) and rapid thermally (B) oxidized epi-Si$_{0.74}$Ge$_{0.26}$ samples. Spectrum B exhibits no shoulder, is symmetric, and is due to pure SiO$_2$. The thick arrow in Figure 103 denotes the peak position of the Si-O-Ge asymmetric stretching mode as

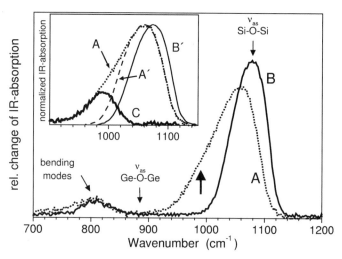

Fig. 103. FTIR spectra of anodically (A) and rapid thermally (B) oxidized epi-Si$_{0.74}$Ge$_{0.26}$ samples. Inset: On the maximum normalized IR absorption spectra of thermally oxidized Si (B′), anodically oxidized Si (A′), and the anodically oxidized SiGe layer (A).

obtained from the following analysis. The inset of Figure 103 shows the normalized IR absorption spectra (normalized to their maximum value) of thermally oxidized Si (B′), anodically oxidized Si (A′), and the anodically oxidized SiGe layer (A). The oxides on pure silicon have a nearly identical thickness of about 54 nm. The typical shift of ν_{as}(SiO$_2$) of anodic oxides (A′) to lower wavenumbers, which have a less compact network with respect to thermal oxides (B′), can easily be seen ($\Delta\nu$ about 20 cm^{-1}). Spectrum (C) is the difference in spectrum (A) and spectrum (A′), the anodic oxides on SiGe and Si, respectively. Therefore, spectrum (C) represents the part of the IR absorption of the anodically oxidized epi-SiGe (A), which is due to the incorporated oxidized Ge species. This peak position (centered at about 990 cm^{-1}) has recently been identified as the asymmetric stretching mode of Si-O-Ge (995 cm^{-1}) by dissolution experiments of plasma-oxidized Si$_{0.65}$Ge$_{0.35}$ layers [280]. However, no absorption peak has been detected in the region of the asymmetric Ge-O-Ge stretching mode (850 to 880 cm^{-1} [288]), which has been identified for plasma-oxidized SiGe layers with a similar amount of Ge (35 atomic %) [280]. It can be concluded that the Ge is homogeneously distributed in the SiGe layer during the growth process, so that each Ge atom has a neighboring Si atom (26 atomic % Ge in the layer). This distribution is not affected by the anodic oxidation process, and the Ge is chemically bonded in the network, leading to Si-O-Ge and not to GeO$_2$ clusters.

7.4. Photoluminescence Spectra of Oxidized SiGe Layers

All of these results (Ge segregation, lattice distortion, relaxation) affect the electronic properties (e.g., defect concentration) of the thin epilayer system, as can be seen from PL spectra plotted in Figure 104. The spectra are recorded at room temperature. Note that the PL intensities are corrected for the different reflectivity for the exciting N$_2$ laser beam (337 nm) of the re-

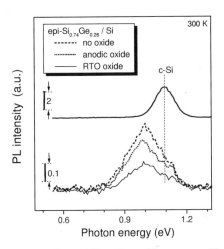

Fig. 104. PL spectra of an epi-Si$_{0.74}$Ge$_{0.26}$/Si structure without an oxide layer and covered by anodic or thermal oxide. The PL spectrum of c-Si is shown for comparison.

spective sample, which was measured by UV-Vis spectroscopy. The shift of the PL intensity is much more pronounced with respect to the c-SiGe samples (Fig. 94) because of the higher amount of Ge in the crystal lattice (26 atomic % Ge instead of 10%). The observed bandgap for the strained Si$_{0.74}$Ge$_{0.26}$ layer of about 0.98 eV at room temperature is in good agreement with previously reported values of 0.94 eV [286] (at 150 K) and 1.042 eV [287] (at 6 K) for strained Si$_{0.5}$Ge$_{0.5}$ and relaxed Si$_{0.7}$Ge$_{0.3}$ layers, respectively. A slight shoulder is visible in the higher energy part of the spectra, as a result of the diffusion of light-induced charge carriers excited in the SiGe into the Si substrate, where they recombine radiatively (see the dotted line). The highest PL intensity (and therefore the lowest N_s) has been observed for the as-deposited and Si-cap protected sample (dashed line). The PL intensity is only slightly reduced by the anodic oxide layer (dotted line), whereas the lowest PL intensity is measured for the rapid thermally oxidized sample (solid line). The PL intensity of the anodically oxidized SiGe layer is about a factor of 2 higher than of the RTO processed sample. This indicates that N_s is approximately half of the value of the RTO-processed sample because of a better passivation by anodic oxidation.

7.5. Conclusions and Outlook

An overview of electrochemical passivation of c-Si and SiGe surfaces is given in this work on the basis of the results obtained predominantly by our own in situ FTIR, PL, and PV investigations. Chemical and electronic surface passivation means that reactive surface sites and states in the forbidden gap are absent. Dangling bonds at Si atoms are free radicals, and, therefore, they serve as reactive surface sites. Dangling bonds at Si atoms can be well passivated by hydrogen, which forms Si-H$_x$ bonds, or by oxygen, because of the formation of a Si/SiO$_2$ interface.

There are some very significant differences between electrochemically and chemically passivated Si surfaces. The point is that the electrochemical reactions are driven by the potential and by the distribution of the electric field on the Si surface. The electrochemical reactions are faster at Si surface regions where the electric field is enhanced. This leads to an equipotential surface with similar overall morphology of the Si surface. As a consequence, the formation of atomically flat facets is impossible at electrochemically passivated Si surfaces, because there always exists an initial roughness at a Si surface, which is smoothed during the electrochemical treatment. The process of such local smoothing leads to a local reconstruction, which is characterized by a relatively high degree of disorder but with low densities of reactive surface sites and surface states. The lowest densities of surface states are reached on electrochemically hydrogenated Si surfaces. The ideal electrochemically passivated semiconductor surface consists of surface atoms that are all in the similar position from the point of view of chemical or electronic reactivity. This is the most striking difference from ideal chemically passivated Si surfaces that consist of surface atoms with very different positions, i.e., at facets and steps between facets.

The concept of dangling bonds as reactive surface sites and recombination active surface states is taken from experience with electronic properties of Si/SiO$_2$ interfaces and of a-Si:H bulk material. It is well known that the concentration of spin centers (dangling bonds at a Si atom with only Si back bonds) correlates with the density and distribution of interface states at the Si/SiO$_2$ interface [289] and in the bulk a-Si:H [290]. In the context of this work we assumed that the reactive surface sites are identical with the recombination active surface states. This assumption makes sense for the Si/electrolyte interface, because corrosive processes at the Si surface should go through localized intermediate states related to dangling bonds. For example, the importance of the etch rate on the PL intensity is shown in this work. However, it should be noted that the existence of dangling bonds at a clean Si surface does not mean, necessarily, that there are recombination active states in the forbidden gap. For example, the reconstructed Si(100) 2 × 1 surface contains a huge amount of dangling bonds (each Si-Si dimer contains two dangling bonds), but the nonradiative surface recombination is quite low. The dangling bonds are passivated during dissociative adsorption of water, whereas the PL intensity remains constant [213]. An opposite case is the reconstructed Si(111) 7 × 7 surface, which is characterized by a rather high nonradiative surface recombination [212]. The recombination active dangling bonds at Si(111) 7 × 7 surfaces belong to isolated Si ad-atoms. Dissociative adsorption of water molecules efficiently passivates such recombination active dangling bonds. Epoxide surface structures are also discussed as a possible origin for gap states at the Si surface [291].

The electrochemical reactions leading to hydrogenation and oxidation are discussed. However, it is practically impossible to develop satisfying models of the chemical and electrochemical reactions at Si surfaces, which include the formation and passivation of reactive surface sites as dangling bonds, an attempt has been made in [95].

The hydrogenated internal surface of porous silicon contains a low number of electronic states in the forbidden gap so that boron acceptors in meso-PS are not compensated for. The concentration of free charge carriers in meso-PS depends sensitively on the molecular or dielectric ambience [236]. This behavior is interesting for applications in sensors.

Thin anodic oxide p-Si interfaces can be passivated through the application of electron injection processes, which demonstrates the possibility of a real room temperature passivation of Si surfaces covered with thin oxides. Once more, water molecules at the anodic oxide/p-Si interface play a dominant role in the electrochemical reactions driving the passivation of gap states.

Thick anodic oxides are very homogeneous in their thickness because of the anisotropy of the process of anodic oxidation and the equipotential surface caused by the electric field. Passivation of trenches or steps can be well performed with the use of anodic oxidation, in contrast to the thermal oxidation process, which depends sensitively on the orientation of the crystal and the stress during thermal treatment.

Ge has the disadvantage that GeO_2 as a passivating layer is unstable in a wet atmosphere. Nevertheless, SiGe can be well passivated by anodic oxides regardless of the use of a thin Si cap layer, which is or is not oxidized. Low thermal budget processing, like anodic oxidation, is preferred for SiGe passivation to avoid diffusion processes.

The grafting of Si-H surface bonds by Si-O or Si-C surface bonds is intended to functionalize parts of a Si surface, for example, by attached organic molecules, which may serve as reactive sites for other organic molecules with specific chemical groups [292–296]. As remarked, electrochemical grafting is an alternative technique for exchanging surface chemical bonds that can also be performed by optical excitation with ultraviolet light of, for example, Si-H or Si-Cl surface bonds [297].

It was shown by adsorption of molecules of malic anhydride at reconstructed Si(100) 2×1 surfaces that the formation of Si-C bonds does not lead to the generation of new recombination active states in the forbidden gap [298]. Our first experiments with electrochemical grafting confirm the result from ultra-high-vacuum experiments. In the beginning, the electrochemical grafting of organic molecules leads to only a doubling of the concentration of surface nonradiative recombination defects, which decreases with further deposition time [299].

There are only a few inorganic materials such as silicon that are biocompatible. Chemical and biochemical microsystems can be fabricated on the basis of silicon, and, for example, microstructured porous silicon can be used as a biocatalytic surface [300]. Passivation and corrosion of Si surfaces are important in such systems and should be taken into consideration in medical applications [301].

Acknowledgments

The authors thank W. Füssel, J. Hersener, A. Klein, R. Knippelmeyer, A. Schöpke, I. Sieber, and V. Yu. Timoshenko for helpful and fruitful collaboration, and W. Fuhs and F. Koch for enlightening discussions. The authors are grateful to C. Murrell and O. Nast for critical reading of the manuscript. Daimler-Crysler AG and the Institute of Crystal Growth (IKZ) are acknowledged for providing us with the epi-SiGe and c-SiGe samples.

REFERENCES

1. J. Ligenza, *J. Phys. Chem.* 65, 2011 (1961).
2. W. A. Pliskin and H. S. Lehman, *J. Electrochem. Soc.* 112, 1013 (1964).
3. W. Waring and E. A. Benjamini, *J. Electrochem. Soc.* 111, 1256 (1964).
4. H. C. Evitts, H. W. Cooper, and S. S. Flaschen, *J. Electrochem. Soc.* 111, 688 (1964).
5. P. F. Schmidt and A. E. Owen, *J. Electrochem. Soc.* 111, 682 (1964).
6. R. M. Hurd and N. Hackerman, *Electrochim. Acta* 9, 1633 (1964).
7. E. Duffek, E. Benjamini, and C. Mylroie, *Electrochem. Technol.* 3, 75 (1965).
8. W. Kern and R. C. Heim, *J. Electrochem. Soc.* 117, 568 (1970).
9. E. A. Irene, *J. Electrochem. Soc.* 121, 1613 (1974).
10. S. T. Pantelides and M. Long, in "The Physics of SiO_2 and Its Interfaces" (S. T. Pantelides, Ed.), p. 339. Pergamon, New York, 1978.
11. O. L. Krivanek, D. C. Tsui, T. T. Shenh, and A. Kamgar, in "Physics of SiO_2 and Its Interfaces" (S. T. Pantelides, Ed.), p. 356. Pergamon, New York, 1978.
12. W. D. Mackintosh and H. H. Plattner, *J. Electrochem. Soc.* 124, 396 (1977).
13. A. Wolkenberg, *Solid-State Electron.* 24, 89 (1981).
14. J. J. Mercier, F. Fransen, F. Cardon, M. J. Madou, and W. P. Gomes, *Ber. Bunsenges. Phys. Chem.* 89, 117 (1985).
15. H. Hasegawa, S. Arimoto, J. Nanjo, H. Yamamoto, and H. Ohno, *J. Electrochem. Soc.* 135, 424 (1988).
16. R. L. Smith, B. Kloeck, and S. D. Collins, *J. Electrochem. Soc.* 135, 2001 (1988).
17. I. Montero, O. Sanchez, J. M. Albella, and J. C. Pivin, *Thin Solid Films* 175, 49 (1989).
18. I. Montero, L. Galan, E. De La Cal, and J. M. Albella, *Thin Solid Films* 193/194, 325 (1990).
19. J.-N. Chazalviel, *Electrochim. Acta* 37, 865 (1992).
20. H. J. Lewerenz, *Electrochim. Acta* 37, 847 (1992).
21. J. A. Bardwell, K. B. Clark, D. F. Mitchell, D. A. Bisaillion, G. I. Sproule, B. M. Dougall, and M. J. Graham, *J. Electrochem. Soc.* 140, 2135 (1993).
22. G. Mende, H. Flietner, and M. Deutscher, *J. Electrochem. Soc.* 140, 188 (1993).
23. G. Mende, H. Flietner, and M. Deutscher, *Nucl. Instrum. Methods Phys. Res., Sect. A* 326, 16 (1993).
24. P. Schmuki, H. Böhni, and J. A. Bardwell, *J. Electrochem. Soc.* 142, 1705 (1995).
25. M. J. Jeng and J. G. Hwu, *Appl. Phys. Lett.* 69, 3875 (1996).
26. D. Lapadatu and R. Puers, *Sens. Actuators A* 60, 191 (1997).
27. J. Rappich, *Microelectron. Reliab.* 40, 815 (2000).
28. K. Ismail, S. Rishton, J. O. Chu, K. Chan, and B. S. Meyerson, *IEEE Electron. Dev. Lett.* 14, 348 (1993).
29. U. König and J. Hersener, *Solid State Phenomena* 47/48, 17 (1996).
30. D. K. Nayak, J. C. S. Woo, J. S. Park, K. L. Wang, and K. P. MacWilliams, *IEEE Electron. Dev. Lett.* 12, 154 (1991).
31. F. K. LeGoues, R. Rosenberg, T. Nguyen, F. Himpsel, and B. S. Meyerson, *J. Appl. Phys.* 65, 172 (1989).
32. J. Rappich, I. Sieber, A. Schöpke, W. Füssel, M. Glück, and J. Hersener, *Mater. Res. Soc. Symp. Proc.* 451, 215 (1997).
33. J. Rappich and W. Füssel, *Microelectron. Reliab.* 40, 825 (2000).
34. J. Rappich, I. Sieber, and R. Knippelmeyer, *Electrochem. Solid State Lett.* 4(3), 1311–1313 (2001).
35. J. Rappich, V. Y. Timoshenko, and T. Dittrich, *J. Electrochem. Soc.* 144, 493 (1997).
36. T. Dittrich, T. Burke, F. Koch, and J. Rappich, *J. Appl. Phys.* 89(8), 4636–4642 (2001).

37. K. Heilig, H. Flietner, and J. Reineke, *J. Phys. D: Appl. Phys.* 12, 927 (1979).
38. K. Heilig, *Solid-State Electron.* 27, 394 (1984).
39. K. Fujiwara, *Phys. Rev. B* 24, 2240 (1981).
40. H. Wagner, R. Butz, U. Backes, and D. Bruchmann, *Solid State Commun.* 38, 1155 (1981).
41. F. Stucki, J. A. Schaefer, J. R. Anderson, G. J. Lapeyre, and W. Göpel, *Solid State Commun.* 47, 795 (1983).
42. Y. J. Chabal, *Phys. Rev. Lett.* 50, 1850 (1983).
43. H. Wagner and H. Ibach, *Festkörperprobleme* 23, 165 (1983).
44. J. E. Demuth and B. N. J. Persson, *J. Vac. Sci. Technol., B* 2, 384 (1984).
45. Y. J. Chabal and K. Rafghavachari, *Phys. Rev. Lett.* 53, 282 (1984).
46. S. Ciraci, R. Butz, E. M. Oellig, and H. Wagner, *Phys. Rev. B* 30, 711 (1984).
47. Y. J. Chabal, *Surf. Sci.* 168, 594 (1986).
48. E. Yablonovich, D. L. Allara, C. C. Chang, T. Gmitter, and T. B. Bright, *Phys. Rev. Lett.* 57, 249 (1986).
49. A. Venkateswara Rao and J.-N. Chazaviel, *J. Electrochem. Soc.* 134, 2777 (1987).
50. H. Gerischer and M. Lübke, *Ber. Bunsenges. Phys. Chem.* 91, 394 (1987).
51. M. Grundner and R. Schulz, *AIP Conf. Proc.* 167, 329 (1988).
52. Y. J. Chabal, G. S. Higashi, K. Raghavachari, and V. A. Burrows, *J. Vac. Sci. Technol., A* 7, 2104 (1989).
53. G. S. Higashi, Y. J. Chabal, G. W. Trucks, and K. Raghavachari, *Appl. Phys. Lett.* 56, 656 (1990).
54. D. B. Fenner, D. K. Biegelsen, and R. D. Bringans, *J. Appl. Phys.* 66, 419 (1989).
55. L. M. Peter, D. J. Blackwood, and S. Pons, *Phys. Rev. Lett.* 62, 308 (1989).
56. L. M. Peter, D. J. Blackwood, and S. Pons, *J. Electroanal. Chem.* 294, 111 (1990).
57. J. J. Boland, *Phys. Rev. Lett.* 65, 3325 (1990).
58. P. Dumas and Y. J. Chabal, *Chem. Phys. Lett.* 181, 537 (1991).
59. P. Jakob and Y. J. Chabal, *J. Phys. Chem.* 95, 2897 (1991).
60. U. Memmert and R. J. Behm, *Adv. Solid-State Phys.* 31, 189 (1991).
61. Y. J. Chabal, *Physica B* 170, 447 (1991).
62. K. Itaya, R. Suguwara, Y. Morita, and H. Tokumoto, *Appl. Phys. Lett.* 60, 2534 (1992).
63. P. Jakob, Y. J. Chabal, K. Raghavachari, R. S. Becker, and A. J. Becker, *Surf. Sci.* 275, 407 (1992).
64. T. Bitzer and H. J. Lewerenz, *Surf. Sci.* 269/270, 886 (1992).
65. S.-L. Yau, F.-R. Fan, and A. Bard, *J. Electrochem. Soc.* 139, 2825 (1992).
66. H. J. Lewerenz and T. Bitzer, *J. Electrochem. Soc.* 139, L21 (1992).
67. C. Stuhlmann, G. Bogdány, and H. Ibach, *Phys. Rev. B* 45, 6786 (1992).
68. P. Dumas, Y. J. Chabal, and P. Jakob, *Surf. Sci.* 269/270, 867 (1992).
69. M. Niwano, Y. Takeda, K. Kurita, and N. Miyamoto, *J. Appl. Phys.* 72, 2488 (1992).
70. M. B. Nardelli, F. Finocchi, M. Palummo, R. D. Felice, C. M. Bertoni, and S. Ossicini, *Surf. Sci.* 269/270, 879 (1992).
71. H. Yao and J. A. Woollam, *Appl. Phys. Lett.* 62, 3324 (1993).
72. P. Jakob, Y. J. Chabal, K. Raghavachari, P. Dumas, and S. B. Christman, *Surf. Sci.* 285, 251 (1993).
73. H. J. Lewerenz, T. Bitzer, M. Gruyters, and K. Jacobi, *J. Electrochem. Soc.* 140, L44 (1993).
74. U. Neuwald, H. E. Hessel, A. Feltz, U. Memmert, and R. J. Behm, *Surf. Sci. Lett.* 296, L8 (1993).
75. P. Dumas, Y. J. Chabal, and P. Yakob, *Appl. Surf. Sci.* 65/66, 580 (1993).
76. T. Bitzer, H. J. Lewerenz, M. Gruyters, and K. Jacobi, *J. Electroanal. Chem.* 359, 287 (1993).
77. P. Jakob, Y. J. Chabal, and K. Raghavachari, *J. Electron Spectrosc. Relat. Phenom.* 64/65, 59 (1993).
78. K. Utani, T. Suzuki, and S. Adachi, *J. Appl. Phys.* 73, 3467 (1993).
79. J. J. Boland, *Adv. Phys.* 42, 129 (1993).
80. T. Dittrich, H. Angermann, H. Flietner, T. Bitzer, and H. J. Lewerenz, *J. Electrochem. Soc.* 141, 3595 (1994).
81. J. Rappich, H. Jungblut, M. Aggour, and H. Lewerenz, *J. Electrochem. Soc.* 141, L99 (1994).
82. K. Usada, H. Kanaya, K. Yamada, T. Sato, T. Sueyoshi, and M. Iwatsuki, *Appl. Phys. Lett.* 64, 3240 (1994).
83. H. Bender, S. Verhaverbeke, and M. M. Heyns, *J. Electrochem. Soc.* 141, 3128 (1994).
84. S. Rauscher, T. Dittrich, M. Aggour, J. Rappich, H. Flietner, and H. J. Lewerenz, *Appl. Phys. Lett.* 66, 3018 (1995).
85. J. Rappich and H. J. Lewerenz, *J. Electrochem. Soc.* 142, 1233 (1995).
86. J. Rappich, M. Aggour, S. Rauscher, H. J. Lewerenz, and H. Jungblut, *Surf. Sci.* 335, 160 (1995).
87. H. N. Waltenburg and J. T. Yates, *Chem. Rev.* 95, 1589 (1995).
88. H. Ogawa, K. Ishikawa, M. T. Suzuki, Y. Hayami, and S. Fujimura, *Jpn. J. Appl. Phys.* 34, 732 (1995).
89. M. Matsumura and H. Fukidome, *J. Electrochem. Soc.* 143, 2683 (1996).
90. J. Rappich and H. J. Lewerenz, *Electrochim. Acta* 41, 675 (1996).
91. S. Miyazaki, J. Schafer, J. Ristein, and L. Ley, *Appl. Phys. Lett.* 68, 1247 (1996).
92. S. Cattarin, J.-N. Chazalviel, C. Da Fonseca, F. Ozanam, L. M. Peter, G. Schlichthorl, and J. Stumper, *J. Electrochem. Soc.* 145, 498 (1998).
93. A. Belaïdi, J.-N. Chazalviel, F. Ozanam, O. Gorochov, A. Chari, B. Fotouhi, and M. Etman, *J. Electroanal. Chem.* 444, 55 (1998).
94. H. Angermann, W. Henrion, M. Rebien, D. Fischer, J.-T. Zettler, and A. Röseler, *Thin Solid Films* 313/314, 52 (1998).
95. P. M. Hoffmann, I. E. Vermeir, and P. C. Searson, *J. Electrochem. Soc.* 147, 2999 (2000).
96. F. Bensliman, M. Aggour, A. Ennaoui, Y. Hirota, and M. Matsumura, *Electrochem. Solid-State Lett.* 3, 566 (2000).
97. J.-N. Chazalviel, A. Belaïdi, M. Safi, F. Maroun, B. H. Erné, and F. Ozanam, *Electrochim. Acta* 45, 3205 (2000).
98. T. Takahagi, I. Nagai, A. Ishitari, H. Kuroda, and Y. Nagasawa, *J. Appl. Phys.* 64, 3516 (1988).
99. W. Kern, *Semicond. Int.* 94 (1984).
100. L. A. Zazzera and J. F. Moulder, *J. Electrochem. Soc.* 136, 484 (1989).
101. J. Ruzyllo, A. M. Hoff, D. C. Frystak, and S. D. Hossain, *J. Electrochem. Soc.* 136, 1474 (1989).
102. W. Kern, *J. Electrochem. Soc.* 137, 1887 (1990).
103. J. Falta, D. Bahr, G. Materlik, B. H. Muller, and M. Horn von Hoegen, *Appl. Phys. Lett.* 68, 1394 (1996).
104. O. P. Karpenko, S. M. Yalisove, and D. J. Eaglesham, *J. Appl. Phys.* 82, 1157 (1997).
105. J. Falta, T. Schmidt, G. Materlik, J. Zeysing, G. Falkenberg, and R. L. Johnson, *Appl. Surf. Sci.* 162/163, 256 (2000).
106. T. Fujino, T. Fuse, J.-T. Ryu, K. Inudzuka, T. Nakano, K. Goto, Y. Yamazaki, M. Katayama, and K. Oura, *Thin Solid Films* 369, 25 (2000).
107. W.-X. Ni, A. Henry, M. I. Larsson, K. Joelsson, and G. V. Hansson, *Appl. Phys. Lett.* 65, 1772 (1994).
108. I. A. Buyanova, W. M. Chen, A. Henry, W. A. Ni, G. N. Hansson, and B. Monemar, *Appl. Surf. Sci.* 102, 293 (1996).
109. G. Hollinger, S. J. Sferco, and M. Lannoo, *Phys. Rev. B* 37, 7149 (1988).
110. D. R. Wolters and A. T. A. Z.-v. Duynhoven, *J. Appl. Phys.* 65, 5134 (1989).
111. S. P. Tay, A. Kalnitsky, G. Kelly, J. P. Ellul, P. DeLalio, and E. A. Irene, *J. Electrochem. Soc.* 137, 3579 (1990).
112. M. Hiroshima, T. Yasaka, S. Miyazaki, and M. Hirose, *Jpn. J. Appl. Phys.* 33, 395 (1994).
113. S. Hirofumi and I. Shuichi, *Mater. Trans., JIM* 36, 1271 (1995).
114. K. Kobayashi, A. Teramoto, Y. Matsui, M. Hirayama, A. Yasuoka, and T. Nakamura, *J. Electrochem. Soc.* 143, 3377 (1996).
115. D. J. DiMaria and J. H. Stathis, *Appl. Phys. Lett.* 70, 2708 (1997).
116. J.-H. Kim, J. J. Sanchez, T. A. DeMassa, M. T. Quddus, R. O. Grondin, and C. H. Liu, *Solid-State Electron.* 43, 57 (1999).
117. Y. Jia, Y. Liang, Y. Liu, Y. Liu, and D. Shen, *Thin Solid Films* 370, 199 (2000).
118. A. N. Itakura, T. Narushima, M. Kitajima, K. Teraishi, A. Yamada, and A. Miyamoto, *Appl. Surf. Sci.* 159/160, 62 (2000).
119. T. Watanabe and I. Ohdomari, *Appl. Surf. Sci.* 162, 116 (2000).

120. P. O. Hahn, M. Grundner, A. Schnegg, and H. Jacob, in "The Physics and Chemistry of SiO$_2$ and the Si/SiO$_2$ Interface" (D. Helms, Ed.), p. 401. Plenum, New York/London, 1988.

121. P. Jakob, P. Dumas, and Y. J. Chabal, *Appl. Phys. Lett.* 59, 2968 (1991).

122. L. Lai and E. A. Irene, *J. Appl. Phys.* 86, 1729 (1999).

123. G. J. Pietsch, U. Köhler, O. Jusko, M. Henzler, and P. O. Hahn, *Appl. Phys. Lett.* 60, 1321 (1992).

124. Y.-B. Park and S.-W. Rhee, *J. Appl. Phys.* 86, 1346 (1999).

125. T. Narushima, A. N. Itakura, T. Kurashina, T. Kawabe, and M. Kitajima, *Appl. Surf. Sci.* 159/160, 25 (2000).

126. N. J. Harrick, "Internal Reflection Spectroscopy." Wiley Interscience, New York, 1967

127. H. Neugebauer, G. Nauer, N. Brinda-Konopik, and G. Gidaly, *J. Electroanal. Chem.* 122, 381 (1981).

128. A. Bewick and S. Pons, "Advances in Infrared and Raman Spectroscopy." Wiley–Heyden, Chichester, 1985

129. F. Ozanam and J.-N. Chazalviel, *Rev. Sci. Instrum.* 59, 242 (1988).

130. Y. J. Chabal, *Surf. Sci. Rep.* 8, 211 (1988).

131. E. P. Boonekamp, J. J. Kelly, J. v. d. Ven, and A. H. M. Sondag, *J. Appl. Phys.* 75, 8121 (1994).

132. M. Nakamura, M.-B. Song, and M. Ito, *Electrochim. Acta* 41, 681 (1996).

133. M. Niwano, J. Kageyama, K. Kurita, K. Kinashi, I. Takahashi, and N. Miyamoto, *J. Appl. Phys.* 76, 2157 (1994).

134. J. C. Knights, R. A. Street, and G. Lucovsky, *J. Non-Cryst. Solids* 35/36, 279 (1980).

135. G. Lucovsky, *Sol. Energy Mater.* 8, 165 (1982).

136. G. Lucovsky, J. Yang, S. S. Chao, J. E. Tyler, and W. Czubatyj, *Phys. Rev. B* 28, 3225 (1983).

137. G. Lucovsky, P. D. Richard, D. V. Tsu, and R. J. Markunas, *J. Vac. Sci. Technol.* 4, 681 (1986).

138. G. Lucovsky, M. J. Manitini, J. K. Srivastava, and E. A. Irene, *J. Vac. Sci. Technol. B* 5, 530 (1987).

139. P. Lange, *J. Appl. Phys.* 66, 201 (1989).

140. J. E. Bateman, B. R. Horrocks, and A. Houlton, *J. Chem. Soc. Faraday Trans.* 93, 2427 (1997).

141. K. T. Queeney, M. K. Weldon, J. P. Chang, Y. J. Chabal, A. B. Gurevich, J. Sapjeta, and R. L. Opila, *J. Appl. Phys.* 87, 1322 (2000).

142. F. Ozanam and J.-N. Chazalviel, *J. Electroanal. Chem.* 269, 251 (1989).

143. H. Ogawa and T. Hattori, *Appl. Phys. Lett.* 61, 577 (1992).

144. D. W. Berreman, *Phys. Rev.* 130, 2193 (1963).

145. H. Shirai, *Jpn. J. Appl. Phys.* 33, L94 (1994).

146. P. Dumas, Y. J. Chabal, and G. S. Higashi, *Phys. Rev. Lett.* 65, 1124 (1990).

147. A. Venkateswara Rao, F. Ozanam, and J.-N. Chazalviel, *J. Electrochem. Soc.* 138, 153 (1991).

148. J. Rappich, H. J. Lewerenz, and H. Gerischer, *J. Electrochem. Soc.* 140, L187 (1993).

149. L. Kronik and Y. Shapira, *Surf. Sci. Rep.* 37, 1 (1999).

150. E. O. Johnson, *Phys. Rev.* 111, 153 (1958).

151. T. Dittrich and H. Flietner, *Mater. Res. Soc. Symp. Proc.* 358, 581 (1995).

152. S. M. Sze, "Physics of Semiconductors." Wiley, New York, 1981.

153. L. Elstner and T. Dittrich, unpublished observation, 1994.

154. Y. W. Lam, *J. Phys. D: Appl. Phys.* 4, 1370 (1971).

155. T. Dittrich, M. Brauer, and L. Elstner, *Phys. Status Solidi A* 137, K29 (1993).

156. V. Y. Timoshenko, A. B. Petrenko, M. N. Stolyarov, T. Dittrich, W. Fuessel, and J. Rappich, *J. Appl. Phys.* 85, 4171 (1999).

157. M. I. Galant and H. M. v. Driel, *Phys. Rev. B* 26, 2133 (1982).

158. E. Yablonovich and T. Gmitter, *Appl. Phys. Lett.* 49, 587 (1986).

159. A. G. Cullis, L. T. Canham, and P. D. J. Calcott, *J. Appl. Phys.* 82, 909 (1997).

160. Y. Morita and H. Tokumoto, *Appl. Phys. Lett.* 67, 2654 (1995).

161. M. Grundner and H. Jacob, *Appl. Phys. A* 39, 73 (1986).

162. S. Maruno, H. Iwasaki, K. Horioka, S.-T. Li, and S. Nakamura, *Phys. Rev. B* 27, 4110 (1983).

163. S. Zaima, J. Kojima, M. Hayashi, H. Ikeda, H. Iwano, and Y. Yasuda, *Jpn. J. Appl. Phys.* 34, 741 (1995).

164. T. Bitzer, M. Gruyters, H. J. Lewerenz, and K. Jacobi, *Appl. Phys. Lett.* 63, 397 (1993).

165. M. Matsumura and S. R. Morrison, *J. Electrochem. Soc.* 147, 157 (1983).

166. H. Gerischer and M. Lübke, *Ber. Bunsenges. Phys. Chem.* 92, 573 (1988).

167. F. Ozanam, J.-N. Chazalviel, A. Radi, and M. Etman, *Ber. Bunsenges. Phys. Chem.* 95, 98 (1991).

168. J. Rappich, V. Y. Timoshenko, and T. Dittrich, *Ber. Bunsenges. Phys. Chem.* 101, 139 (1997).

169. M. Aggour, PhD-Thesis, Tu-Berlin, Berlin, 1994.

170. H. J. Lewerenz and M. Aggour, *J. Electroanal. Chem.* 351, 159 (1993).

171. D. R. Turner, *J. Electrochem. Soc.* 105, 402 (1958).

172. F. Ozanam and J.-N. Chazalviel, *J. Electron Spectrosc. Relat. Phenom.* 64/65, 395 (1993).

173. J. S. Judge, *J. Electrochem. Soc.* 118, 1772 (1971).

174. R. E. Mesmer and C. F. Baes, *Inorg. Chem.* 8, 618 (1969).

175. D. Koschel, Ed., "Gmelin Handbook Fluor, Gmelin Handbook of Inorganic Chemistry," Vol. 3, Supplement. Springer-Verlag, Berlin, 1982.

176. S. Verhaverbeke, I. Teerlinck, C. Vinckier, G. Stevens, R. Cartuyvels, and M. M. Heyns, *J. Electrochem. Soc.* 141, 2852 (1994).

177. J. Rappich, unpublished observation.

178. T. Dittrich, V. Y. Timoshenko, M. Schwartzkopff, E. Hartmann, J. Rappich, P. K. Kashkarov, and F. Koch, *Microelectron. Eng.* 48, 75 (2000).

179. V. Lehmann, *J. Elektrochem. Soc.* 143, 1313 (1996).

180. G. J. Pietsch, U. Köhler, and M. Henzler, *J. Appl. Phys.* 73, 4797 (1993).

181. J. W. Faust and E. D. Palik, *J. Electrochem. Soc.* 130, 1413 (1983).

182. E. D. Palik, V. M. Bermudez, and O. J. Glembocki, *J. Electrochem. Soc.* 132, 871 (1985).

183. O. J. Glembocki, E. D. Palik, G. R. de Guel, and D. L. Kendall, *J. Electrochem. Soc.* 138, 1055 (1991).

184. P. Allongue, V. Costa-Kieling, and H. Gerischer, *J. Electrochem. Soc.* 140, 1009 (1993).

185. P. Allongue, V. Costa-Kieling, and H. Gerischer, *J. Electrochem. Soc.* 140, 1018 (1993).

186. S. A. Campbell, S. N. Port, and D. Schiffrin, *J. Micromechanics Microeng.* 2, 209 (1998).

187. P. Allongue, J. Kasparian, and M. Elwenspoek, *Surf. Sci.* 388, 50 (1997).

188. J. Kasparian and P. Allongue, in "ECS Meeting" (K. Rajeshwar, L. M. Peter, A. Fujishima, D. Meissner, and M. Tomkiewich, Eds.), Vol. 97-20, p. 220. Electrochem. Soc., Pennington, NJ/Paris, 1997.

189. J. Rappich, V. Y. Timoshenko, R. Würz, and T. Dittrich, *Electrochim. Acta* 45, 4629 (2000).

190. D. K. Biegelsen, N. M. Johnson, M. Stutzmann, E. H. Poindexter, and P. J. Caplan, *Appl. Surf. Sci.* 22/23, 879 (1985).

191. H. Flietner, *Surf. Sci.* 46, 251 (1974).

192. T. Dittrich, H. Angermann, W. Füssel, and H. Flietner, *Phys. Status Solidi A* 140, 463 (1993).

193. T. M. Buck and F. S. McKim, *J. Electrochem. Soc.* 105, 709 (1958).

194. H. Angermann, K. Kliefoth, and H. Flietner, *Appl. Surf. Sci.* 104/105, 107 (1996).

195. H. Angermann, W. Henrion, M. Rebien, J.-T. Zettler, and A. Röseler, *Surf. Sci.* 388, 15 (1997).

196. H. Flietner, in "7th Conf. Insulating Films on Semiconductors (INFOS)" (W. Eccleston and M. Uren, Eds.), p. 151. Hilger, Bristol, 1991.

197. H. Angermann, T. Dittrich, and H. Flietner, *Appl. Phys. A* 59, 193 (1994).

198. D. Gräf, M. Grundner, and R. Schulz, *J. Vac. Sci. Technol. A* 7, 808 (1989).

199. E. A. Konstantinova, V. Y. Timoshenko, P. K. Kashkarov, and T. Dittrich, *Thin Solid Films* 276, 265 (1996).

200. G. W. Trucks, K. Raghavachari, G. S. Higashi, and Y. J. Chabal, *Phys. Rev. Lett.* 65, 504 (1990).

201. K. Yokogawa and T. Mizutani, *Jpn. J. Appl. Phys.* 32, L635 (1993).

202. M. Morita, T. Ohmi, E. Hasegawa, M. Kawakami, and M. Ohwada, *J. Appl. Phys.* 68, 1272 (1990).

203. G. Schlichthörl and L. M. Peter, *J. Electroanal. Chem.* 381, 55 (1995).

204. K. C. Mandal, F. Ozanam, and J.-N. Chazalviel, *Appl. Phys. Lett.* 57, (1990).

205. J. Rappich, T. Dittrich, Y. Timoshenko, I. Beckers, and W. Fuhs, *Mater. Res. Soc. Symp. Proc.* 452, 797 (1997).

206. P. d. Mierry, A. Etcheberry, R. Rizk, P. Etchegoin, and M. Aucouturier, *J. Electrochem. Soc.* 141, 1539 (1994).

207. J. E. Northrup, *Phys. Rev. B* 44, 1419 (1991).

208. Y. J. Chabal and K. Raghavachari, *Phys. Rev. Lett.* 54, 1055 (1985).

209. V. F. Kiselev and O. V. Krylov, "Electronic Phenomena in Adsorption and Catalysis on Semiconductors and Dielectrics." Springer-Verlag, Berlin, 1987.

210. E. H. Poindexter, *J. Non-Crystal. Sol.* 187, 257 (1995).

211. E. H. Poindexter, *Z. Naturforsch.* 50a, 653 (1995).

212. T. Dittrich, T. Bitzer, V. Y. Timoshenko, and J. Rappich, submitted for publication.

213. T. Bitzer, T. Rada, N. V. Richardson, T. Dittrich, and F. Koch, *Appl. Phys. Lett.* 77, 3779 (2000).

214. J. B. Flynn, *J. Electrochem. Soc.* 105, 715 (1958).

215. R. Memming and G. Schwandt, *Surf. Sci.* 4, 109 (1966).

216. V. Lehmann and U. Gösele, *Appl. Phys. Lett.* 58, 856 (1991).

217. H. Föll, *Appl. Phys. A* 53, 8 (1991).

218. R. L. Smith and S. D. Collins, *J. Appl. Phys.* 71, R1 (1992).

219. M. J. Eddowes, *J. Electroanal. Chem.* 280, 297 (1990).

220. V. Labunov, V. Bondarenko, L. Glinenko, A. Dorofeev, and L. Tabulina, *Thin Solid Films* 137, 123 (1986).

221. X. G. Zhang, S. D. Collins, and R. L. Smith, *J. Electrochem. Soc.* 136, 1561 (1989).

222. K. H. Jung, S. Shih, and D. L. Kwong, *J. Electrochem. Soc.* 140, 3046 (1993).

223. D. Kovalev, H. Heckler, G. Polisski, and F. Koch, *Phys. Status Solidi B* 215, 871 (1999).

224. V. Y. Timoshenko, J. Rappich, and T. Dittrich, *Appl. Surf. Sci.* 123/124, 111 (1998).

225. T. Dittrich, I. Sieber, S. Rauscher, and J. Rappich, *Thin Solid Films* 276, 200 (1996).

226. J. Stumper, H. J. Lewerenz, and C. Pettenkofer, *Electrochim. Acta* 34, 1379 (1989).

227. J. Stumper, H. J. Lewerenz, and C. Pettenkofer, *Phys. Rev. B* 41, 1592 (1990).

228. J. Rappich, T. Burke, S. Lust, and T. Dittrich, to appear.

229. T. Dittrich, V. Y. Timoshenko, and J. Rappich, *Appl. Phys. Lett.* 72, 1635 (1998).

230. M. Ben-Chorin, A. Kux, and I. Schechter, *Appl. Phys. Lett.* 64, 481 (1994).

231. R. Schwarz, F. Wang, M. Ben-Chorin, S. Grebner, and A. Nikolov, *Thin Solid Films* 255, 23 (1995).

232. T. Dittrich, K. Kliefoth, I. Sieber, J. Rappich, S. Rauscher, and V. Y. Timoshenko, *Thin Solid Films* 276, 183 (1996).

233. G. Bomchil, R. Herino, K. Barla, and J. C. Pfister, *J. Electrochem. Soc.* 130, 1611 (1983).

234. G. Polisski, D. Kovalev, G. G. Dollinger, T. Sulima, and F. Koch, *Physica B* 273/274, 951 (1999).

235. V. Y. Timoshenko, T. Dittrich, and F. Koch, *Phys. Status Solidi B* 222, R1 (2000).

236. V. Y. Timoshenko, T. Dittrich, V. Lysenko, M. G. Lisachenko, and F. Koch, submitted for publication.

237. T. Dittrich and V. Y. Timoshenko, *J. Appl. Phys.* 75, 5436 (1994).

238. A. B. Petrenko, T. Dittrich, and V. Y. Timoshenko, unpublished observation, 1999.

239. D. E. Aspnes and J. B. Theeten, *J. Electrochem. Soc.* 127, 1359 (1980).

240. T. Yasuda, Y. Ma, Y. L. Chen, G. Lucovsky, and D. Maher, *J. Vac. Sci. Technol. A* 11, 945 (1993).

241. Y. Ma, T. Yasuda, and G. Lucovsky, *J. Vac. Sci. Technol. A* 11, 952 (1993).

242. S. Ichimura, K. Koike, A. Kurokawa, K. Nakamura, and H. Itoh, *Surf. Interface Anal.* 30, 497 (2000).

243. J. Xiang, N. Herbots, H. Jacobsson, P. Ye, S. Hearne, and S. Whaley, *J. Appl. Phys.* 80, 1857 (1996).

244. M. Mukhopadhyay, S. K. Ray, T. B. Ghosh, M. Sreemany, and C. K. Maiti, *Semicond. Sci. Technol.* 11, 360 (1996).

245. C. Caragianis, Y. Shigesato, and D. C. Paine, *J. Electron. Mater.* 23, 883 (1994).

246. T. Aiba, K. Yamauchi, Y. Shimizu, N. Tate, M. Katayama, and T. Hattori, *Jpn. J. Appl. Phys.* 34, 707 (1995).

247. H. Angermann, K. Kliefoth, W. Füssel, and H. Flietner, *Microelectron. Eng.* 28, 51 (1995).

248. H. Ibach, H. D. Bruchmann, and H. Wagner, *Appl. Phys. A* 29, 113 (1982).

249. N. Matsuo, N. Kawamoto, D. Aihara, and T. Miyoshi, *Appl. Surf. Sci.* 159/160, 41 (2000).

250. U. Neuwald, H. E. Hessel, A. Feltz, U. Memmert, and R. J. Behm, *Appl. Phys. Lett.* 60, 1307 (1992).

251. H. Ogawa, K. Ishikawa, C. Inomata, and S. Fujimura, *J. Appl. Phys.* 79, 472 (1996).

252. A. Shimizu, S. Abe, H. Nakayama, T. Nishino, and S. Iida, *Appl. Surf. Sci.* 159/160, 89 (2000).

253. M. Udagawa, M. Niwa, and I. Sumita, *Jpn. J. Appl. Phys.* 32, 282 (1993).

254. T. Umeda, S. Yamasaki, M. Nishizawa, T. Yasuda, and K. Tanaka, *Appl. Surf. Sci.* 162/163, 299 (2000).

255. M. Wasekura, M. Higashi, H. Ikeda, A. Sakai, S. Zaim, and Y. Yasuda, *Appl. Surf. Sci.* 159/160, 35 (2000).

256. R. Hasunuma, A. Ando, K. Miki, and Y. Nishioka, *Appl. Surf. Sci.* 162/163, 547 (2000).

257. M. Gotoh, K. Sudoh, and H. Iwasaki, *J. Vac. Sci. Technol. B: Microelectron. Nanometer Struct.* 18, 2165 (2000).

258. R. Hasunuma, A. Ando, K. Miki, and Y. Nishioka, *Appl. Surf. Sci.* 159/160, 83 (2000).

259. B. Nouwen and A. Stesmans, *Mater. Sci. Eng. A* 288, 239 (2000).

260. J. A. Bardwell, E. M. Allegretto, J. Phillips, M. Buchanan, and N. Draper, *J. Electrochem. Soc.* 143, 2931 (1996).

261. G. Mende, in "Semiconductor Micromachining—Techniques and Industrial Application" (S. A. Campbell and H. J. Lewerenz, Eds.), Vol. 2, p. 263. Wiley, New York, 1998.

262. F. Ozanam, A. Djebri, and J.-N. Chazalviel, *Electrochim. Acta* 41, 687 (1996).

263. G. Mende, E. Hensel, F. Fenske, and H. Flietner, *Thin Soild Films* 168, 51 (1989).

264. C. E. Blat, E. H. Nicollian, and E. H. Poindexter, *J. Appl. Phys.* 69, 1712 (1991).

265. H. J. Lewerenz and G. Schlichthörl, *J. Electroanal. Chem.* 327, 85 (1992).

266. H. J. Lewerenz, H. Jungblut, and S. Rauscher, *Electrochim. Acta* 45, 4615 (2000).

267. M. Aggour, M. Giersig, and H. J. Lewerenz, *J. Electroanal. Chem.* 383, 67 (1995).

268. H. Föll, J. Carstensen, M. Christophersen, and G. Hasse, *Phys. Status Solidi A* 182, 7 (2000).

269. E. D. Palik, J. W. Faust, H. F. Gray, and R. F. Greene, *J. Electrochem. Soc.* 129, 2051 (1982).

270. M. Glück, J. Hersener, H. G. Umbach, J. Rappich, and J. Stein, *Solid State Phenom.* 57/58, 413 (1997).

271. G. Mende and G. Küster, *Thin Solid Films* 35, 215 (1976).

272. J. v. Borany, G. Mende, and B. Schmidt, *Nucl. Instrum. Methods* 212, 65 (1983).

273. G. Mende and J. Wende, *Thin Solid Films* 142, 21 (1986).

274. M. V. Sulliva, D. L. Klein, R. M. Finne, L. A. Pompliano, and G. A. Kolb, *J. Electrochem. Soc.* 110, 412 (1963).

275. D. R. Turner, *J. Electrochem. Soc.* 107, 810 (1960).

276. A. Uhlir, *Bell Syst. Tech. J.* 35, 333 (1956).

277. H. Gerischer and W. Mindt, *Electrochim. Acta* 13, 1329 (1968).

278. D. R. Wolters and A. T. A. Z.-v. Duynhoven, *J. Appl. Phys.* 65, 5126 (1989).

279. R. People, *J. Quantum Electron.* 22, 1696 (1986).

280. M. Seck, R. Devine, C. Hernandez, Y. Campidelli, and J.-C. Dupuy, *Appl. Phys. Lett.* 72, 2748 (1998).

281. P. W. Li, H. K. Liou, E. S. Yang, S. S. Iyer, T. P. Smith, and Z. Lu, *Appl. Phys. Lett.* 60, 3265 (1992).

282. I. S. Goh, S. Hall, W. Eccleston, J. F. Zhang, and K. Warner, *Electron. Lett.* 30, 1988 (1994).

283. O. Vancuawenberghe, O. C. Hellman, N. Herbot, and W. J. Tan, *Appl. Phys. Lett.* 59, 2031 (1991).

284. C. G. Ahn, H. S. Kang, Y. K. Kwon, S. M. Lee, B. R. Ryum, and B. K. Kang, *J. Appl. Phys.* 86, 1542 (1999).

285. L. C. Kimerling, K. D. Kolenbrander, J. Michel, and J. Palm, in "Solid State Physics, Advances in Research and Applications" (H. Ehrenreich and F. Spaepen, Eds.), Vol. 50, p. 333. Academic Press, New York, 1997.

286. J. Olayos, J. Engvall, H. G. Grimmeiss, H. Kibbel, E. Kasper, and H. Presting, *Thin Solid Films* 222, 243 (1992).

287. L. P. Tilly, P. M. Mooney, J. O. Chu, and F. K. LeGoues, *Appl. Phys. Lett.* 67, 2488 (1995).

288. I. N. Chakraborty and R. A. Condrate, *J. Non-Cryst. Solids* 81, 271 (1986).

289. P. M. Lenahan, *Microelectron. Eng.* 22, 129 (1993).

290. J. Kocka, *J. Non-Cryst. Sol.* 90, 91 (1987).

291. M. Nishida, *Phys. Rev. B* 60, 8902 (1999).

292. M. Warntjes, C. Vieillard, F. Ozanam, and J. N. Chazalviel, *J. Electrochem. Soc.* 142, 4138 (1995).

293. C. H. Villeneuve, J. Pinson, F. Ozanam, J. N. Chazalviel, and P. Allongue, *Mater. Res. Soc. Symp.* 451, 185 (1997).

294. P. Wagner, S. Nock, J. A. Spudich, W. D. Volkmuth, S. Chu, R. L. Cicero, C. P. Wade, M. R. Linford, and C. E. D. Chidsey, *J. Struc. Biol.* 119, 189 (1997).

295. P. Allongue, C. H. d. Villeneuve, and J. Pinson, *Electrochim. Acta* 45, 3241 (2000).

296. M. P. Stewart, E. G. Robins, T. W. Geders, M. J. Allen, H. C. Choi, and J. M. Buriak, *Phys. Status Solidi A* 182, 109 (2000).

297. J. Terry, M. R. Linford, C. Wigren, R. Cao, P. Pianetta, and C. E. D. Chidsey, *Appl. Phys. Lett.* 71, 1056 (1997).

298. T. Bitzer, T. Dittrich, T. Rada, and N. V. Richardson, *Chem. Phys. Lett.* 331, 433 (2000).

299. P. Hartig, T. Dittrich, and J. Rappich, submitted.

300. M. Bengtsson, S. Ekström, J. Drott, A. Collins, E. Csöregi, G. Marko-Varga, and T. Laurell, *Phys. Status Solidi A* 182, 495 (2000).

301. L. T. Canham, M. P. Stewart, J. M. Buriak, C. L. Reeves, M. Anderson, E. K. Squire, P. Allcock, and P. A. Snow, *Phys. Status Solidi A* 182, 521 (2000).

302. T. Dittrich, M. Schwartzkopff, E. Hartmann, and J. Rappich, *Surf. Sci.* 437, 154 (1999).

303. T. Dittrich, S. Rauscher, V. Y. Timoshenko, J. Rappich, I. Sieber, H. Flietner, and H. J. Lewerenz, *Appl. Phys. Lett.* 67, 1134 (1995).

304. T. Dittrich, J. Rappich, and V. Y. Timoshenko, *Appl. Phys. Lett.* 70, 2705 (1997).

305. J. Rappich, Y. Timoshenko, and T. Dittrich, *Mater. Res. Soc. Symp. Proc.* 448, 51 (1997).

Chapter 2

EPITAXIAL GROWTH AND STRUCTURE OF III–V NITRIDE THIN FILMS

Dharanipal Doppalapudi
Boston MicroSystems Inc., Woburn, Massachusetts, USA

Theodore D. Moustakas
Department of Electrical and Computer Engineering, Boston University, Boston, Massachusetts, USA

Contents

1. INTRODUCTION

The family of III–V nitrides (the binaries InN, GaN, AlN, and their alloys) is one of the most important classes of semiconductor materials which has attracted significant attention for many years. All these materials are direct gap semiconductors with energy gaps ranging from 1.9 to 6.2 eV. Thus, they are ideal semiconductors for the fabrication of optical devices (light emitting diodes (LEDs), lasers, and detectors) operating in the visible and ultra violet (UV) parts of the electromagnetic spectrum, as well as electronic devices such as bipolar transistors and field effect transistors. Furthermore, due to the piezoelectric properties of these materials, they can also be used for the fabrication of electromechanical devices. The development of such devices is expected to affect a number of technologies such as information storage, full color displays, true color copying, lo-

Handbook of Thin Film Materials, edited by H.S. Nalwa
Volume 4: Semiconductor and Superconductor Thin Films
Copyright © 2002 by Academic Press

ISBN 0-12-512912-2/$35.00

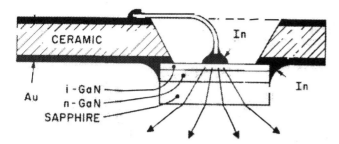

Fig. 1. Structure of GaN M-i-n LED. (From Pankove, Miller, and Berley-heiser [9] with permission from Elsevier Science.)

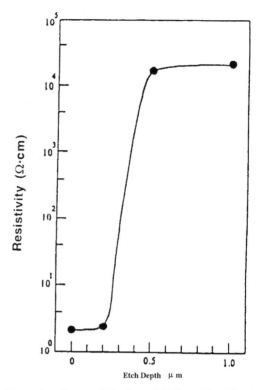

Fig. 2. Change in resistance of Mg-doped GaN film with etching depth from the surface. (From Amano et al. [11].) [Reproduced by permission of the Electrochemical Society, Inc.]

cal area networks, underwater communications, space-to-space communication, high temperature, and power electronics, as well as microwave electronics and sensors based on microelectromechanical systems (MEMS) technology. The development of optical devices, primarily LEDs and lasers has reached the stage of commercialization [1, 2]. A significant research effort is currently being devoted to the development of solar-blind UV detectors and various forms of electronic devices. A reference regarding the growth, structure, physical properties, and applications of gallium nitride has been published by Pankove and Moustakas [3, 4].

Gallium nitride was synthesized for the first time by Juza and Hahn [5] by the reaction of ammonia with hot gallium for the purpose of studying the crystal structure and the lattice constants of gallium nitride. This method produced small needles and platelets. Grimmeiss and Koelmans [6] used the same technique to produce small crystals of GaN to study their photoluminescence (PL) properties. Maruska and Tietjen [7] used the vapor phase epitaxy method to produce large area GaN films on sapphire substrates. The GaN films produced by these authors were always autodoped with n-type conductivity, a result attributed to the existence of nitrogen vacancies, due to the growth of such films at very high temperatures. This model was later questioned by Seifert and co-workers [8] who attributed the n-type conductivity to oxygen impurities due either to oxygen from the quartz tube employed during the growth or to atmospheric leaks. Films produced by this method became semi-insulating by doping them with zinc during the growth, which led to the first blue light emitting diode (LED) by Pankove, Miller, and Berkeyheiser [9]. Since this method was unable to produce p-type films, the fabricated device structure had the M-i-n configuration (M = metal) as shown in Figure 1. Such LED structures were found to emit in the blue, green, yellow, or red light, depending on the zinc concentration in the insulating layer [9].

During the 1970s, a large number of discoveries relating to physical properties and potential applications of GaN were made. However, the lack of sophistication of the epitaxial growth methods as well as the inability to dope the material p-type hindered progress in the full exploitation of GaN and its alloys with InN and AlN. In the late 1980s, Akasaki et al. [10] demonstrated p-type doping in GaN. Specifically, these authors were investigating the cathodoluminescence (CL) of GaN

doped with magnesium in a scanning electron microscope (SEM) apparatus. They observed that the CL becomes brighter when the film was exposed to the electron beam for longer periods of time. A systematic study of PL from such films before and after the low energy electron beam irradiation (LEEBI) showed that the luminescence increased by 2 orders of magnitude upon this treatment [10]. Hall effect measurements of the film subjected to the LEEBI treatment have shown that the sample was converted from semi-insulating to p-type. The LEEBI process was found to convert approximately the top 300 nm from semi-insulating to p-type, which is consistent with the penetration depth of the electrons. This was demonstrated by systematic etching studied by Amano and co-workers as illustrated in Figure 2 [11]. This phenomenon was accounted for by van Vechten, Zook, and Horning [12], who proposed that the shallow acceptor level of Mg was compensated by hydrogen atoms forming complexes with the Mg acceptors. Presumably the electron beam breaks the Mg–H bonds and releases hydrogen and thus Mg becomes a shallow acceptor with an ionization energy of 0.16 eV [13]. This conclusion was further confirmed by Nakamura and co-workers [14] who have shown that GaN doped with Mg converts from semi-insulating to p-type upon thermal annealing above 750°C in a nitrogen atmosphere. The role of hydrogen in the p-type doping of GaN was also confirmed by Moustakas and Molnar [15] who have doped GaN with Mg using molecular beam epitaxy (MBE). In this method which does not involve hydrogen during the growth, films with free hole concentration of 6×10^{18} cm^{-3} were produced without

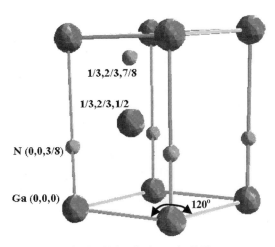

1/3,2/3,7/8

1/3,2/3,1/2

N (0,0,3/8)

Ga (0,0,0)

120°

Fig. 3.   Unit cell of wurtzite GaN.

Table I.   Properties of III-Nitrides in Wurtzite Structure

Property	AlN	GaN	InN
Lattice constants	$a = 3.11$ Å	$a = 3.19$ Å	$a = 3.55$ Å
	$c = 4.98$ Å	$c = 5.185$ Å	$c = 5.76$ Å
Bandgap	6.2 eV	3.4 eV	2.0 eV
CTE[a]	$\alpha_{//} - 4.2$	$\alpha_{//} - 5.59$ (300–900 K)	$\alpha_{//} - 2.7$–3.7
$\times 10^{-6}$ K^{-1}	$\alpha_{\perp} - 5.3$	$\alpha_{\perp} - 3.17$ (300–700 K)	$\alpha_{\perp} - 3.4$–5.7
		7.8 (700–900 K)	
Young's modulus[b]	$C_{11} = 398$	$C_{11} = 396$	$C_{11} = 271$
(Gpa)	$C_{12} = 140$	$C_{12} = 144$	$C_{12} = 124$
	$C_{13} = 127$	$C_{13} = 100$	$C_{13} = 94$
	$C_{33} = 382$	$C_{33} = 392$	$C_{33} = 200$
Refractive index	1.9–2.2	2.5–2.6	—
Bulk modulus[c]	199	176	122
(Gpa)			
Thermal conductivity	2.8	1.3	0.8
(W cm^{-1} K^{-1})			
Melting point[d] (K)	3487	2791	2146

[a]M. Leszczynski, T. Suski, P. Perlin, H. Tolsseyre, I. Grzegory, M. Bockowski, J. Jun, S. Porowski, and J. Major, *J. Phys. D: Appl. Phys.* 28, A149 (1995).

[b]H. Morkoc, "Gallium Nitride (GaN) I." Semiconductors and Semimetals (J. I. Pankove and T. D. Moustakas, Eds.), Vol. 50. Academic Press, New York, 1998.

[c]T. Azuhata, T. Sota, and K. Suzuki, *J. Phys.: Condens. Mater.* 8, 3111 (1996).

[d]J. van Vetchen, *Phys. Rev. B* 7, 1479 (1973).

requiring any postgrowth thermal annealing. In fact, postgrowth hydrogenation of the same GaN films by Brand et al. [16] have shown reduction in free hole concentration by orders of magnitude.

One major problem in the development of devices based on this class of materials is the lack of III-nitride substrates for homoepitaxial growth of devices. Thus, the great majority of work reported in the last few years dealt with the heteroepitaxial growth of GaN on various foreign substrates. Currently, there is a significant effort being devoted to the development of GaN substrates from the liquid phase [17] and by the development of quasi-GaN substrates by growing thick GaN films on a foreign substrate and then separating them from the substrate [18]. The development of *p*-type doping as well as the development of more sophisticated epitaxial methods have cleared the way for the fabrication and the commercialization of various optical and electronic devices discussed previously.

Gallium nitride and other III-nitrides have a wurtzite (hexagonal) structure, belonging to the P6₃Mc space group in which all the atoms are tetrahedrally coordinated. The wurtzite structure consists of two hexagonal close packed (hcp) lattices, which are displaced by $\pm 3c/8$, where c is the lattice parameter. The atomic positions in the GaN unit cell are shown in Figure 3. The equilibrium bulk lattice parameters for the wurtzite GaN are: $a = 3.189$ Å and $c = 5.185$ Å. Similarly, the lattice parameters for AlN and InN are (3.11, 4.98 Å) and (3.55, 5.76 Å), respectively. Some of the relevant properties of the wurtzite III-nitrides are listed in Table I.

Although GaN exists predominantly in the wurtzite structure, it can also be grown in zinc-blende (cubic) structure, depending on the substrate symmetry and the growth conditions [19–21]. The in-plane symmetry of (0001) wurtzite GaN (growth plane) is identical to the symmetry of (111) zinc-blende GaN. Furthermore, the cohesive energies of wurtzite and cubic phases are very similar [20], making conversion between the two phases to occur easily, mainly by creation of stacking faults of the close packed planes. Several innovative growth methods have been developed over the past few years to overcome these

difficulties and to obtain high-quality epitaxial films and to fabricate a wide variety of optoelectronic devices.

In Section 2, we review various growth methods commonly used for the growth of III–V nitrides as bulk crystals and epitaxial thin films. Section 3 is a review of the epitaxial growth of III-nitrides on various substrates and the techniques developed to improve the crystalline quality of the films. Although the main emphasis of this section is on the growth of wurtzite GaN, a brief review of GaN films with zinc-blende structure is given. Special focus is made on the nucleation and the early stages of growth. Section 4 reviews doping of the nitride films, which is essential in any device design. In Section 5, we review the microstructure of the GaN epitaxial films grown by various methods and we discuss the growth-microstructure-property correlations. Finally, Section 6 is a review of the issues related to ternary alloys—InGaN and AlGaN alloys.

2. GROWTH METHODS

In this section, we discuss the various methods used to grow binary nitrides and their alloys. GaN thin-film growth has been investigated by several growth techniques including hydride vapor phase epitaxy (HVPE), metal organic chemical vapor deposition (MOCVD), molecular beam epitaxy (MBE), pulsed laser deposition (PLD), and physical vapor deposition (PVD). Of these, the former three are the more popular methods and are discussed in detail in this section. First, we briefly discuss the growth of bulk GaN from liquid phase, since the development of this method will lead to substrates for homoepitaxial growth of thin films.

2.1. Growth of GaN and AlN Bulk Crystals

Growth of III-nitride crystals using standard methods (Czochralski, Bridgman) is extremely difficult because of the following factors: (1) the high melting temperature ($\sim$2800 K for GaN), (2) the relatively low sublimation–decomposition temperature compared to the melting temperature, (3) the very high equilibrium nitrogen vapor pressure (40 kbar at melting temperature), and (4) low solubility in acids, bases, and most other inorganic elements and compounds. Thus, novel bulk growth techniques must be employed for III-nitrides.

Growth of bulk AlN has been attempted using sublimation–recondensation methods [22, 23], or solution routes [24] in the 1970s. Slack and McNelly [22, 23] achieved the largest crystals (10-mm long and 3 mm in diameter) using the sublimation/recondensation method at 2250°C. Balkas and coworkers [25] reported seeded growth of bulk single crystals of AlN (0001) on 6H-SiC (0001) substrates in the range 1950–2250°C at 500 Torr of N_2.

Two major approaches have been used to grow bulk single crystals of GaN: vapor phase transport [26, 27] and solution techniques [17, 28–30]. Of these, the high-pressure synthesis from solution was primarily developed by the high pressure research center of the Polish Academy of Sciences [30, 31]. In this method, bulk GaN crystallization takes place from the solution in liquid Ga at high N_2 pressure and high temperature (2 Kbar, 1600°C). Hexagonal platelets 10 mm in size were produced from a solution in pure liquid Ga or Ga alloyed with 0.2–0.5% Mg [17]. Magnesium was found to increase the solubility of N in the Ga-Mg solution and allows crystallization to occur at lower values of N_2 overpressure and temperature. The crystals produced so far are generally of small dimensions ($<$1 in) with dislocation density between 10^3–10^5 cm^{-2}.

There has been considerable research to develop large area GaN and AlN wafers from gas phase reactions [32]. ATMI has developed free-standing high-purity GaN wafers of 2-in diameter, using hydride vapor phase epitaxy (HVPE). Similarly, Kyma Technologies have been developing a proprietary high transport rate physical vapor deposition (PVD) process to manufacture 2-in free-standing AlN wafers.

2.2. Hydride Vapor Phase Epitaxy

The hydride vapor phase epitaxy (HVPE) method has been used extensively for the development of arsenides and phosphides and for the commercialization of devices such as LEDs. Since this method is suitable for mass production, it has also been applied for III-nitrides in the 1970s [7] and has led to relatively high-quality GaN films at growth rates larger than 1 μm min^{-1}. Because of this high growth rate, the HVPE method has been successfully used for the quasi-bulk growth of GaP and GaN [18, 44]. An excellent review of GaN growth by this method was presented by Molnar [44].

The HVPE process is a chemical vapor deposition method, which is usually carried out at atmospheric pressure in a hot wall reactor. A novel aspect of this method is that the group III precursors are synthesized within the reactor vessel, which is typically done by the reaction of high-purity hydrogen chloride (HCl), with a group III metal at high temperature. In the case of gallium nitride (GaN), gallium monochloride (GaCl) is synthesized upstream in the reactor by reacting HCl gas with high-purity (better than 99.99999%) liquid gallium metal at 800–900°C. The GaCl is transported to the substrate where it reacts with ammonia (NH_3) at 1000–1100°C to form GaN, via the reaction,

$$GaCl + NH_3 \leftrightarrow GaN + HCl + H_2 \qquad (1)$$

This technique is often referred to as chloride-transport vapor phase epitaxy (VPE) because the group III element is transported to the substrate as a volatile chloride compound. Due to the relatively low vapor pressure of the metal chlorides, these molecules will tend to condense on the unheated surfaces. The use of a hot walled reactor and the *in situ* synthesis of the metal chlorides avoids the complicated vapor delivery systems and gas inlet heating which would be necessary if the metal chlorides were synthesized–stored externally. Moreover, the metal chlorides are hydroscopic and corrosive in nature. This would make maintaining source integrity during storage and processing difficult in the latter case. On the other hand, HCl is a highly corrosive gas, and care must be taken in its storage and transport to the reactor [44].

The nature of the chemistry involved in the GaN growth by HVPE differs substantially from other III–V semiconductors. For instance, in gallium arsenide (GaAs) growth by HVPE thermal disassociation of arsenic compounds (AsH_3) results in the formation of As_4 and As_2 molecules, which typically remain volatile and chemically reactive and thus participate in film growth. In GaN HVPE growth, the thermal disassociation of NH_3 results in the formation of N_2 molecules, which are extremely stable and unreactive in temperatures of interest. The viability of HVPE GaN growth lies in the relatively sluggish dissociation of NH_3, which enables the effective transport of reactive nitrogen to the growth surface [33]. Therefore, the growth of stoichiometric, homogeneous films over a large area requires efficient and uniform transport of NH_3 to the growth surface.

The majority of the studies on GaN films grown by HVPE have been carried out in reactors with a horizontal, coaxial configuration, as shown in Figure 4. This schematic shows HCl flowing over Ga melt in a pyrolitic boron nitride (PBN) boat in the inner tube and NH_3 flowing in the outer tube. Although the position of the two precursors can be reversed, it was found that the former method led to a higher Ga conversion efficiency [34]. Nitrogen is passed through the central "separator" tube to prevent premature reaction of the precursors before reaching the substrates. High-purity reactant and carrier gases are used. According to the report by Seifert et al., the use of hydrogen as

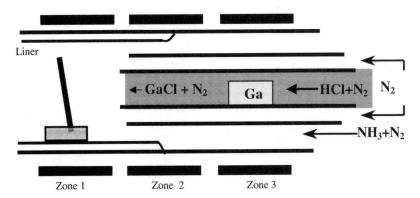

Fig. 4.   Schematic of the three-zone HVPE reactor and the gas flow system.

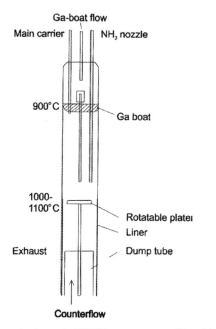

Fig. 5.   Schematic of a vertical HVPE reactor system. (From Molnar [44].)

the carrier gas improves the quality of the film in comparison to inert argon gas [8] as well as increases the growth rate of GaN. Many other researchers also use nitrogen as the carrier gas [35]. In this schematic, an additional liner covering the gas mixing zone starting at the end of the gallium tube downstream is used to prevent the GaN and other byproduct deposition on the main tube walls. Liners are cleaned after every growth in buffer oxide etch to remove the residual depositions. For the HVPE growth, the ideal furnace chamber material needs to be of high purity, excellent high-temperature properties, and high tolerance against abrupt temperature gradients. The reactor typically consists of a three-zone furnace to enable independent control of temperatures in different regions. The deposition rate depends strongly on the temperature of the substrate. In addition, the growth rate, uniformity, and crystalline quality vary with the growth parameters such as precursor flow rate. The growth of GaN by HVPE leads to the formation of gas phase adducts which in turn leads to undesirable gas phase reactions,

particle formation, and wall deposition problems (which induce reactor vessel cracking upon cool down). Due to the tendency for the formation of GaN, horizontal reactor designs suffer from gas phase depletion effects [36, 37].

Molnar et al. have also used HVPE systems in a vertical configuration [38, 39, 44]. The vertical design shown in Figure 5, facilitates substrate rotation during growth to improve uniformity. The substrate holder can be raised and lowered isothermally into a counterflow tube, in which a mixture of NH_3 and carrier gas is passed. The substrate can then be slowly lowered and cooled in an NH_3 environment to minimize decomposition.

2.2.1. GaN Growth Kinetics and Thermochemistry

Thermodynamics and the kinetics of GaN growth by HVPE have been investigated by several groups [34, 40, 41]. The HVPE process is inherently carbon free, making growth of high-purity films a little easier. In addition, the presence of efficient etchant halogens species helps remove excess metallic species from the growth surface. This self-stabilizing effect as well as the higher adatom mobility for chemisorbed Ga may account for the higher growth rate of the HVPE method. The self-stabilization allows the growth to occur under Ga-rich conditions of growth and this reduces the consumption of ammonia. Jacob, Boulou, and Bois [42] and Molnar et al. [48] reported that the introduction of addition HCl downstream of the Ga boat has a beneficial effect in improving the film properties. It was found that the growth rate decreased by more than a factor of 2 when HCl was introduced.

Ban studied the transport of Ga with HCl and Cl_2, the thermal decomposition of NH_3 and the reactions leading to deposition, using mass spectroscopy [33]. It was found that the monochloride (GaCl) was the only gallium chloride species resulting from halogen gallium interaction. Along with the transport of gallium, it is important to know the extent ammonia decomposes into nitrogen and hydrogen, because GaN cannot be deposited with the direct reaction between GaCl and N_2. Ammonia is a thermodynamically unstable gas at temperatures usually employed for growth of GaN crystals (over 1000°C), and over 99.5% of ammonia would decompose if equilibrium was achieved under the experimental conditions. However, it is known that the decomposition of ammonia is a sluggish reaction, which can be described by

$$NH_3(g) \rightarrow (1-x)NH_3(g) + x/2N_2(g) + 3/2xH_2(g) \quad (2)$$

where x is the mole fraction of NH_3 decomposing. When no catalyst was present, it was found that no more than about 4% of NH_3 decomposed at temperatures as high as 950°C.

Two thermodynamically feasible reactions have been postulated for the deposition of GaN,

$$GaCl(g) + NH_3(g) \rightarrow GaN(s) + HCl(g) + H_2(g) \quad (3)$$
$$3GaCl(g) + 2NH_3(g) \rightarrow 2GaN(s) + GaCl_3(g) + 3H_2(g) \quad (4)$$

Ban detected the presence of a $GaCl_3.NH_3$ complex, which suggested a reaction of $GaCl_3$ with NH_3:

$$GaCl_3(g) + NH_3(g) \rightarrow GaCl_3.NH_3(g) \quad (5)$$

This complex is thermodynamically unstable with respect to the following reaction,

$$GaCl_3.NH_3(g) \rightarrow GaN(s) + 3HCl(g) \quad (6)$$

Ban reported that the reactions (3) and (4) are fast. Reaction (5) is also fast as no $GaCl_3$ was detected in the experiment but reaction (6) is slow as the complex was detected in the system. The summation of reactions (4)–(6) yields the overall controlling equilibrium for deposition of GaN(s) as described in Eq. (3). Ban concluded that GaN deposition occurs by both routes.

At the temperatures employed in the typical HVPE process, both GaN and NH_3 are metastable, but their decomposition is much slower than the deposition rate of GaN, leading to a net accumulation of GaN.

Many researchers have reported the dependence of GaN growth rate on the growth parameters [40, 43]. Temperature of the substrate region is an important factor that controls the growth rate and the crystalline quality. At high growth temperatures, the chemical reaction is exothermic and the growth rate is limited by diffusive mass transport of reagents in the gas phase and is independent of substrate crystallographic orientation. At low temperatures, the growth rate is limited by the surface processes and mainly by the process of reagent adsorption. In this "kinetic" region, the growth rate is dependent on the substrate crystallographic orientation. In addition, in the low-temperature region the growth rate is improved when an inert gas carrier is used, whereas in the high-temperature region a higher growth rate is observed if hydrogen is used as the carrier gas [34].

It was found that the growth rate increases linearly with the gallium species [40, 43]. The flow rate of NH_3 does not influence the growth rate appreciably, since the growth is limited by the GaCl species [40]. However, the change in the III/V ratio in the system (which could result from the change in the flow rate NH_3) increases the etch pit density in the grown GaN [43]. Another parameter that alters the growth rate is the position of the substrate in relation to the gallium tube, since this determines the extent of mixing the precursor gases have gone through.

Assuming streamline flow, the gases mix by diffusion and they are more homogeneous away from the Ga tube. If the substrate is too close to the Ga tube, the gases will not be sufficiently mixed to obtain appreciable deposition. If the substrate is too far, gas phase nucleation can occur, resulting in polycrystalline deposition. The optimal position depends on the reactor design and is determined empirically. The orientation of the substrate with respect to the vertical (angle from the normal to the reactor flow) is also a factor that affects uniformity and the growth rate of GaN.

Although the use of high-purity gases and reactor materials has reduced the background concentration considerably, the growth of insulating GaN films has still remained a challenge with the HVPE method. The nitrogen vacancies in HVPE grown GaN were widely held responsible for high intrinsic doping levels observed in such films. This is supported by the thermal instability of GaN as well as impurity analysis which seem to indicate Si and O levels in these films are too low to account for the observed residual donor levels.

As mentioned before, GaN does not have a suitable lattice-matched substrate for epitaxial growth. Typically, GaN growth is carried out with C-plane sapphire and 6H-SiC, which have a lattice mismatch of 16 and 3.5%, respectively. Several nucleation mechanisms have been explored to improve the epitaxial quality of the films. The two prominent among them are the GaCl pretreatment [35, 38] and the use of RF-sputtered ZnO buffer layer [44, 48]. In addition, some researchers have employed a two-step growth method involving a low temperature GaN buffer layer followed by a higher temperature growth [43]. This technique has been used in MBE and MOCVD methods extensively and is discussed in detail in the following sections. Details of the substrate preparation processes are given in the next section.

2.3. Metal Organic Vapor Deposition

The metal organic chemical vapor deposition (MOCVD) or metal organic vapor phase epitaxy (MOVPE) is a nonequilibrium growth technique in which group III alkyls and group V hydrides are transported and react on a heated substrate. This method was pioneered by Manasevit [49] and Manasevit and Simpson [50] in the late 1960s and since then, it has been widely used for the production of semiconductor materials and devices. The method was proven as a successful manufacturing technique for both electronic and optoelectronic devices [51, 52]. The MOCVD method is also capable of producing layered structures, such as quantum wells and superlattices, and it is also used for atomic layer epitaxy (ALE) [53]. One area that MOCVD lags behind the molecular beam epitaxy (MBE) method is the development of in situ monitoring techniques. For example, the MBE method uses reflection high energy electron diffraction (RHEED) to monitor and to control the deposition process down to the atomic level. However, significant progress has been made over the past several decades in the understanding of the basic processes occurring during MOCVD growth, as

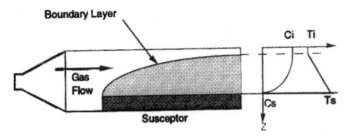

Fig. 6. Schematic of the MOCVD process (From [55].)

well as in developing the methods of purification of the various precursors.

It has generally been established [54] that during growth of III–V compounds such as GaAs, the growth rate varies with group III (trimethyl gallium), group V (arsine), and substrate temperature. The growth rate increases linearly with group III flow rate, is independent of group V flow rate, and shows three temperature regimes. At low temperatures, the growth rate increased linearly with T, but decreases with T at higher temperatures and is independent in the intermediate range. Optimum growth (best morphology and optoelectronic properties) generally occurs at this intermediate-temperature regime, in which it is believed that growth is controlled by diffusion of Ga-precursors across an intermediate (boundary layer) region [54, 55]. This boundary layer in a horizontal MOCVD reactor is schematically illustrated in Figure 6. Even though the flow rate of group V precursors does not influence the growth rate, its high concentration is critical in minimizing carbon incorporation into the growing crystal.

The MOCVD method was also successfully used for the growth of III-nitrides. The organometallic group III sources used in the growth of GaN include trimethyl gallium (TMGa) and trimethyl aluminum (TMAl) which are liquids and trimethyl indium, which is a solid. Ammonia is used as a group V source. Dopant materials include the metal organic precursor cyclopenta dienyl magnesium (Cp$_2$Mg) for p-type doping and silane (SiH$_4$) or disilane (Si$_2$H$_6$) for n-type doping. An excellent review of the MOCVD method for the growth of nitrides was published by DenBaars and Keller [55]. The basic MOCVD reaction describing the GaN deposition is

$$\mathrm{Ga(CH_3)_3 + NH_3 \rightarrow GaN + 3CH_4}$$

However, the details of the reaction are not well known and the intermediate reactions are thought to be complex. In analogy to the growth of GaAs [56, 57], it is likely [55] that elemental Ga is formed by homogeneous decomposition of TMGa. The ammonia on the other hand, is believed to decompose heterogeneously on the GaN surface to yield atomic nitrogen. Since the understanding of the growth processes is limited, optimization of MOCVD growth of GaN and its alloys with Al and In is done empirically by optimizing the growth temperature, V/III flux ratios, and substrate orientation. As in the case of GaAs, the growth of GaN occurs in the mass transport limited regime, which spans a wide temperature range (600–1100°C). In this regime, growth is limited by mass transport of the group III precursor to the growing surface.

2.3.1. MOCVD Reactor Design

Horizontal or vertical reactor designs are used for the growth of GaN. In the vertical design, gases are injected through the top. The substrate is held on a rotating SiC coated graphite susceptor that is perpendicular to the gas flow direction. Heating of the susceptor is accomplished by RF induction, resistance, or infrared lamp heating and the temperature is monitored with a pyrometer. In conventional MOCVD reactors, low gas velocities (2–4 cm s^{-1}) are employed leading to laminar flow conditions. In the horizontal design, the susceptor is situated parallel to the gas flow direction. Uniform growth is achieved by rotation or tilting of the susceptor to eliminate or to minimize reactant depletion along the flow direction. Several types of MOCVD reactor designs have been developed for mass production of III-nitride materials and devices. Both atmospheric and low-pressure reactors are produced by MOCVD equipment manufacturers (Aixtron GmbH, Emcore Corp., Nippon Sanso, and Thomas Swan Ltd.), that are capable of depositing up to seven wafers simultaneously.

Ammonia and metal organic precursors are very reactive and may react before reaching the substrate. Akasaki's group [58, 59] reported that the quality of GaN films improves upon modification of their MOCVD reactor to prevent the prereaction or the generation of adducts.

Nakamura introduced a modified two-flow MOCVD reactor for GaN growth [60], which is shown schematically in Figure 7. This reactor employs two different gas flows. The main flow carries the reactant gases (NH$_3$ and TMG) parallel to the substrate with high velocity using a quartz nozzle. The second flow carries inactive gases (N$_2$ and H$_2$) perpendicular to the substrate. The purpose of the second flow is to push the reactant gases toward the substrate, since they tend to rise by convection, as soon as they reach the hot substrate area. Using this two-flow MOCVD reactor, Nakamura reported that the morphology of GaN films on sapphire depends sensitively on the subflow rates [60]. Specifically, while the presence of subflow resulted in high-quality and uniform GaN films over 2-in sapphire substrates, its absence led to three-dimensional islands and discontinuous film.

Infrared radiation oscillations of the substrate heater transmitted through the growing film were used to monitor the GaN film growth using the two-flow reactor. Such a method was first implemented during the MBE growth of InGaAs and AlGaAs [61, 62]. This technique uses a narrow optical band pass IR radiation thermometer to measure the oscillations of the intensity of the transmitted IR radiation through the growing film. A similar approach was used in the MOCVD growth of ZnSe [63]. These authors used the same interference effect by observing the intensity oscillations of a reflected He–Ne laser beam. Nakamura has used the oscillation of the IR radiation to study the role of both the AlN and GaN buffers in the growth of GaN films [60].

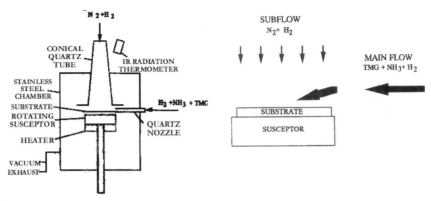

Fig. 7.   (a) Schematic of a modified MOCVD reactor for GaN growth. (b) A schematic of the two-flow MOCVD. (From Nakamura, Pearton, and Fsaol [60] with permission from Springer-Verlag.)

2.4. Molecular Beam Epitaxy

Molecular beam epitaxy (MBE) is a thin-film deposition process in which thermal beams of atoms or molecules react on the clean surface of a single-crystalline substrate, held at high temperatures and under ultrahigh vacuum conditions. Thus, contrary to the HVPE and MOCVD methods discussed previously, the MBE process is a physical method of thin film deposition. The vacuum requirements for the MBE process are typically better than 10^{-10} Torr. This makes it possible to grow epitaxial films with high purity and excellent crystalline quality at relatively low substrate temperatures. Additionally, the ultrahigh vacuum environment allows the study of the surface, interface, and bulk properties of the growing film in real time by employing a variety of structural and analytical probes such as reflection high energy electron diffraction (RHEED), Auger electron spectroscopy (AES), secondary ion mass spectroscopy (SIMS), X-ray photoelectron spectroscopy (XPS), and ultraviolet photoelectron spectroscopy (UPS).

Among the several analytical probes mentioned, reflection high energy electron diffraction (RHEED) is probably the most useful and most commonly used tool in the growth chamber to monitor and to control the growth process. RHEED consists of a well-collimated monoenergetic electron beam, which is directed at a grazing angle of about 1° toward the substrate. The primary electron beam has an energy of between 10 to 20 keV, resulting in an energy component perpendicular to the substrate of about 100 eV. Since the penetration depth of such a beam is only a few atomic layers, a smooth crystal surface acts as a two-dimensional grating and diffracts the electrons. The diffraction pattern is captured on a fluorescent screen. The applications of RHEED include the following. (1) Study thermal desorption or etching of oxides prior to growth: Removal of oxides is critical for epitaxy on substrates like Si and SiC, which tend to have a native oxide. (2) Control the initial stages of epitaxial growth: Surface of a smooth, single-crystal surface is expected to have a diffraction pattern in the form of a series of streaks perpendicular to the sample surface. A polycrystalline or a rough surface would result in a spotty diffraction pattern. Therefore, the effect of growth parameters on the epitaxy can be monitored

in real time. (3) Study surface reconstruction as a function of growth parameters. Since reconstruction introduces additional lattice periodicity, this would result in the appearance of superlattice diffraction reflections in the form of additional streaks. Polarity of the epitaxial films can be determined based on such reconstruction studies. (4) Estimate the growth rate from the time period of the RHEED oscillations.

Modern MBE deposition systems are designed to produce high-quality materials and devices at high throughput. The requirement of maintaining an ultrahigh vacuum environment while simultaneously improving the throughput was addressed through the design of MBE systems consisting of multiple chambers separated by gate valves. All commercially available equipment is constructed with at least three such chambers. The first chamber serves for sample introduction and is capable of a vacuum of approximately 10^{-8} Torr. The second chamber is capable of an ultrahigh vacuum (10^{-10} Torr) and acts principally as a buffer between the introduction and growth chambers. This chamber is also used for substrate preparation such as outgassing and sputter etching and for accommodation of surface analytical facilities. The third chamber, the growth chamber, is capable of an ultrahigh vacuum less than 10^{-10} Torr and its design criteria depend greatly on the nature of the material being deposited. Thus, the MBE method has emerged as a practical growth method for the growth of III–V compounds and devices and as an unique scientific tool for the study of thin-film growth phenomena *in situ*. In the following, we discuss in greater detail the use of this method for the growth of III-nitrides.

A typical MBE growth chamber used for the deposition of III-nitride films is schematically illustrated in Figure 8. Conventional Knudsen effusion cells are used for the evaporation of group III elements (Al, Ga, In) as well as the dopants (Si and Mg for n- and p-type doping, respectively). Since molecular nitrogen (N_2) does not chemisorb on Ga, active nitrogen is produced by a variety of methods. In the following, we describe three approaches and we point out the relative merits and problems of each.

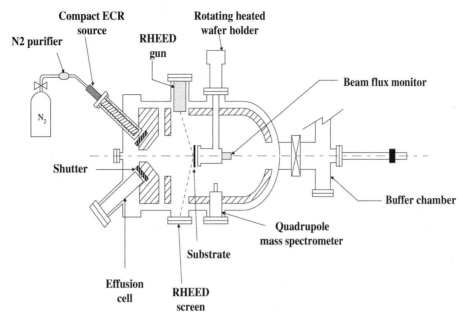

Fig. 8. A schematic of the MBE growth chamber.

2.4.1. Ammonia as a Nitrogen Source

In this method, ammonia is pyrolitically decomposed on the substrate surface at 700–900°C, a much lower temperature than that used in MOCVD or HVPE. Several groups have been developing the growth of GaN by this method [64–70] and the kinetics of ammonia-MBE GaN growth were presented by a number of researchers [68, 69]. According to these studies, the ammonia is thermally activated followed by surface reaction mitigated dissociation and finally it reacts with the metal (Ga, Al, or In) to form GaN.

When using ammonia as the nitrogen source, the growth rate increases in general with substrate temperature. Since the desorption of the group III element increases with temperature, the previous observation implies that the growth rate is limited by the density of thermally active ammonia on the substrate surface. The growth rates are generally as high as 1.2 μm h^{-1}, which is much higher than the plasma sources. Compared to MOCVD, this process consumes less ammonia. However, one problem with this method is that ammonia may react with the various filaments (effusion cell heaters, RHEED, mass spectrometer and ionization gauges). The films grown are generally semi-insulating with excellent optical properties [68–70] and this method demonstrated the fabrication of a high transconductance MODFET [71] and GaN/AlGaN separate confinement heterostructures [72].

2.4.2. Nitrogen Plasma Sources

Over the past few years, two methods have been developed for the activation of molecular nitrogen into a plasma. One is electron cyclotron resonance (ECR) assisted microwave plasma (∼2.45 GHz) while the second utilizes RF plasmas (13.5 MHz).

2.4.2.1. Microwave Plasma-Assisted ECR Sources

ECR plasma sources have been investigated and have been used by a number of groups for the growth of III-Nitrides by molecular beam epitaxy [73–84]. The design and the operation of compact ECR sources were discussed in detail by Molnar, Singh, and Moustakas [80] and Ohtani et al. [79]. In this section, design criteria and modifications that led to high-quality III-Nitride films and structures are summarized.

The compact-ECR source uses an axial electromagnet to stimulate a resonant coupling of the microwave energy (2.45 GHz) with the electrons, as shown in Figure 9 [85]. High-purity nitrogen gas is introduced through the gas feed inlet and is activated in the water cooled ECR zone. The cyclotron acceleration promotes electron–gas collisions and results in a high density plasma with gas ionization efficiencies as high as 10%. The efficient resonant coupling allows for the stable source operation at low growth pressures (∼10^{-5} Torr). This high degree of ionization and low process pressures make the ECR assisted plasma sources ideal for the excitation of inert gases like nitrogen.

At the low process pressures used, the mean free paths of the gaseous species as well as atoms from the effusion cells are very large (∼1 m at 10^{-4} Torr) and growth is carried out in a molecular flow regime where the transport occurs in a collisionless manner. However, the acceleration of the charged species in such a collisionless manner down the divergence of the magnetic field can lead to a plasma species with highly anisotropic energies [86]. The electrons in the plasma are well confined to magnetic field lines by their Larmor gyration around these field lines and are guided down the divergence of the magnetic field by a relaxation of this gyration [86]. On the other hand, the ions are poorly confined due to their comparatively large mass. However, the diffusion of the electrons induces an electric field

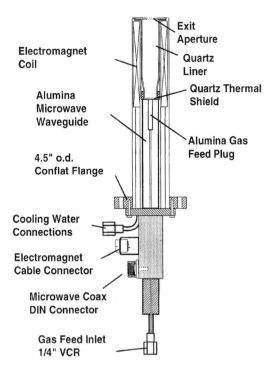

Fig. 9. Schematic of a compact ECR microwave plasma source. (From Moustakas [85].)

Langmuir probe and optical emission spectroscopy have also been used to characterize the ECR source and the effect of the exit aperture [80]. The plasma species were found to depend on the ECR source pressure, which is controlled by either varying the nitrogen flow rate or by changing the exit aperture diameter. Langmuir probe studies showed that the ion density is reduced by ~30% by the introduction a 1-cm exit aperture. Optical emission spectra for nitrogen plasma was collected and the intensity of emission peaks associated with nitrogen ions and activated neutral nitrogen molecules were measured. The ratio of the two peaks was used to characterize the plasma with different nitrogen flow rate and different exit apertures. It was found that the ratio of activated neutral nitrogen to the nitrogen ions increased marginally with an increase in the flow rate and more dramatically with a decrease in exit aperture diameter.

2.4.2.2. RF Plasma Sources

RF sources have been investigated by several groups [90–95] and have demonstrated the production of high-quality III-nitride films. A schematic of one commercially available RF plasma source is shown in Figure 10 [85]. The discharge tube and the beam exit plate are generally fabricated from quartz, boron nitride, or alumina. The RF coil around the discharge tube is cooled by either water or liquid nitrogen. The source is excited with inductively coupled 13.56-MHz power supplied by an RF generator capable of up to 600-W power. It was found that atomic nitrogen lines occurring at 700- to 900-nm spectral region dominate the emission spectra from these sources [92]. However, it has been demonstrated that the activated nitrogen produced is sufficient to lead to GaN growth rates close to $1 \mu m \, h^{-1}$ [95].

2.4.3. Nitrogen Ion Sources

The growth of GaN films by MBE using ionic N_2^+ species was reported by a few researchers [96–98]. Powell et al. [96] used a hot single tungsten grid to produce ionic N_2^+, and controlled the flux by varying the discharge current at a constant acceleration potential (~35 V). The flux contains primarily N_2^+ ions and the ions that are accelerated to greater than their molecular binding energy of 9.8 eV dissociate on the surface of the film. These authors reported that as the ion flux increases, the resistivity of the GaN films also increases.

Fu et al. [97] produced the nitrogen ion beam using a Kauffman-type ion source. They observed that the GaN films grown by this method show strong luminescence when the kinetic energy of the ions is about 10 eV, while the luminescence is not detectable when the kinetic energy of the ions exceeds 18 eV. They also found that the use of insulating substrates resulted in more non-uniform luminescence when compared to conductive SiC substrates, due to surface charging effects. Finally, Leung et al. [98] produced ionic nitrogen using a hollow ion source. In this source, a direct correct (dc) voltage generates a glow discharge, which is restricted to the area in the plasma chamber. The pressure difference between this chamber and the

that accelerates the ionic species along the magnetic field lines by Coulombic interaction, leading to the so-called ambipolar diffusion process [87]. The accelerated ions promote damage to the film through ion bombardment effects. This anisotropic kinetic energy of the ions has been shown to increase with distance from the source [88] and at the low pressure typically used in these sources it can reach even above 24 eV, the damage threshold for GaN [89]. Such high ion energies together with the high-power densities employed in the compact ECR plasma sources can also result in significant sputtering of impurities from the source and chamber walls. Replacing the quartz liner and the exit aperture with boron nitride reduces the active impurity concentration significantly.

The ECR source was modified by the incorporation of exit apertures to control the gas pressure within the source. The introduction of a 1-cm exit aperture results in a pressure increase in the ECR discharge area by at least a factor of 3 [80]. This increase in pressure affects the nitrogen activation mechanism. In the ECR source, the electrons accelerate rapidly while the heavier ions move slowly. At low pressure, electron mean free path is high and there is no redistribution of energy between the electrons and the ions. Thus, activated nitrogen is produced primarily by the relatively small concentration of high energy electrons. As the pressure inside the ECR source increases, the electron mean free path decreases, resulting in the redistribution of energy between electrons and molecules, leading to a higher concentration of activated nitrogen. Furthermore, the increase in pressure reduces the kinetic energy of charged species (which may sputter impurities off the walls) leading to an improvement in the film quality.

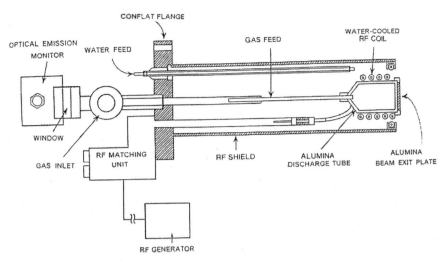

Fig. 10. Schematic of an RF plasma source. (From Moustakas [85].)

MBE growth chamber extracts the activated nitrogen with energies around 5 eV.

3. EPITAXIAL GROWTH

Gallium nitride and other III-nitrides have a wurtzite (hexagonal) structure, consisting of two hexagonal close packed (hcp) lattices, which are displaced by $\pm 3c/8$, where c is the lattice parameter. The equilibrium bulk lattice parameters for the wurtzite GaN are: $a = 3.189$ Å and $c = 5.185$ Å. Similarly, the lattice parameters for AlN and InN are (3.11, 4.98 Å) and (3.55, 5.76 Å), respectively. As mentioned before, GaN can also be grown in a zinc-blende (cubic) structure, depending on the substrate symmetry and the growth conditions [19, 20, 99]. Kurobe et al. have shown that good quality cubic GaN can be grown even on (0001) sapphire by MOCVD by controlling the Ga/N ratio and by using low-temperature buffers [99]. The unit cell of the cubic GaN has a lattice parameter of 4.50 Å [19]. The in-plane symmetry of (0001) wurtzite GaN (growth plane) is identical to the symmetry of (111) zinc-blende GaN. Furthermore, the cohesive energies of wurtzite and cubic phases are very similar [20], making conversion between the two phases occur easily, mainly by the creation of stacking faults of the close packed planes.

Good quality large area GaN substrates for the growth of GaN are currently unavailable. At present, bulk GaN crystals are limited in size (below 1 in, to date [100]), and are very expensive. Consequently, single crystal GaN has been predominantly grown by heteroepitaxial methods films on several foreign substrates including Si ((001) and (111) surfaces), GaAs, sapphire ((0001), (11$\bar{2}$0), (10$\bar{1}$0), and (10$\bar{1}$2) surfaces), 6H-SiC, ZnO, and LiGaO$_2$. In this section, studies on homoepitaxial growth of GaN are briefly reviewed, before going on to an in-depth discussion of the heteroepitaxial growth of GaN on various substrates.

3.1. Homoepitaxial Growth

Bulk GaN crystals are grown from Ga-melt under nitrogen over a pressure of about 15 kbars at temperatures ranging from 1300–1500°C [101]. The high melting point ($\sim$3000 K) and the high nitrogen overpressures of GaN make the growth of bulk GaN an extremely challenging task. Transmission electron microscopy (TEM) studies of the GaN crystals grown from liquid phase revealed that one side is atomically smooth, while the other side is rough with pyramidal shaped features. The difference in the surface quality is attributed to the polarity of the GaN ([0001] and [000$\bar{1}$]) at the two surfaces. The polarity is extremely important for the quality and the growth rate of GaN epitaxy. The [0001] direction, by convention, is chosen to be the direction of the Ga (cation) to the N (anion) bond in the c-axis. The lattice arrangement of a [0001] polar GaN crystal with the standard notation is shown in Figure 3. The [0001] polar surface is also referred to as Ga-polar or Ga-terminated. Convergent beam electron diffraction (CBED) studies to determine the polarity of GaN bulk crystals at the two surfaces have yielded contradictory results [102, 103]. Liliental-Weber et al. concluded that the long bonds along the c-axis in the direction from N to Ga point toward the smooth surface (therefore [000$\bar{1}$] or N polarity) and the rougher surface had Ga polarity [102]. Ponce et al., conducted CBED studies on GaN crystals grown by the same research group and concluded that the smoother surface had Ga polarity [103]. The reason for this discrepancy is not clear. However, both researchers observed that homoepitaxy on these substrates results in GaN epilayers that replicate the polarity of the substrate.

The polarity of the GaN surface was found to have a significant influence on the growth rate and the defect density of the epilayers. Liliental-Weber reported that the growth rate was approximately 60% higher on the smooth surface [104]. Homoepitaxy on the "smooth" surface requires a lower growth temperature than in the case of the "rough" surface. The epilayers on the smooth surface had threading dislocations and dislocation loops. As expected, the dislocation density is lower than

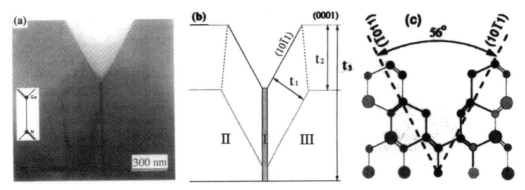

Fig. 11. Cross-sectional TEM micrograph of bulk GaN with [000$\bar{1}$] polarity. Two dislocations and an inversion domain are associated with the pinhole. The two polar directions have different growth rates, as evident from the thickness difference. The pinhole geometry is shown schematically in (b). (c) Atomic model of the GaN crystal in the [11$\bar{2}$0] projection, showing the atomic arrangement. (From Liliental-Weber et al. [106] copyright 1997, the American Physical Society.)

what is observed in heteroepitaxial layers (less than 10^8 cm^{-2}). The dislocation loops are formed due to the presence of threading dislocations in connected pairs, which is characteristic of a coherent interface [105]. Inversion boundaries (with opposite polarities across the boundary) originate from dislocation loops formed at the interface and their density was not larger than 5×10^5 cm^{-2}. These authors also observed pinholes associated with these dislocation loops. Figure 11a shows a TEM micrograph of an inversion boundary and the pinhole in an epilayer grown on the smooth surface (N-polarity, as defined by Liliental-Weber), which has a faster growth rate [106]. GaN in the inversion boundary has a lower growth rate, resulting in a pinhole. The formation of the pinhole is shown schematically in Figure 11b and the crystal structure at the surfaces is shown in Figure 11c. A more complete discussion of the polarity and the inversion domains is given in the subsequent sections. A much higher density of inversion domain boundaries and pinholes (in the range of 10^6–10^7 cm^{-2}) was seen in the epilayers grown on the rough crystal surface. Ponce et al. characterized the GaN films grown on the substrates with N-polarity and found that the dislocations with Burgers vectors [0001], $\langle 11\bar{2}0 \rangle$ and $\langle 11\bar{2}3 \rangle$ were present [107].

In addition to the bulk crystals, thick HVPE grown films also have been used for homoepitaxy. Thick HVPE films with considerably smaller defect densities have been grown on a foreign substrate and separated from the substrate using techniques such as laser ablation. Such free-standing GaN wafers provide a good surface for epitaxial growth of device structures.

3.2. Heteroepitaxial Growth

Epitaxial thin-film growth can be categorized into one of three modes: (i) Frank–Van der Merwe mode or layer-by-layer growth mode, (ii) Volmer–Weber mode or island growth mode, and (iii) Stranski–Krastanov mode or layer-by-layer followed by island growth method. The mode of nucleation and growth is determined by the interfacial free energies between the three surfaces: substrate (s), deposit (d), and vapor (v). Under equilibrium conditions, the deposit forms a hemispherical cap on the substrate surface with a contact angle θ given by [108],

$$\cos \theta = (\sigma_{sv} - \sigma_{sd})/\sigma_{dv}$$

where σ_{ij} is the interfacial free energy between the two surfaces i and j. When $\theta = 0$, the deposit wets the surface, resulting in growth in a two-dimensional layer-by-layer mode or a Frank–Van der Merwe mode.

In the case of homoepitaxy, since the substrate and the deposit have the same structure, $\sigma_{sd} = 0$ and $\sigma_{sv} = \sigma_{dv}$. Thus, the contact angle is zero and growth occurs in the layer-by-layer mode. In heteroepitaxy, the contact angle has a nonzero value, complicating the growth. In the presence of lattice misfit, layer-by-layer growth is never the equilibrium morphology, but is metastable with respect to cluster (island) formation. GaN epitaxy is particularly complicated since it can exist both in wurtzite and in zinc-blende structures. The structure of GaN films (wurtzite vs zinc-blende) is determined by the choice of substrate symmetry and by the choice of growth parameters. Several surface preparation and nucleation techniques have been developed, to accommodate the difference in free energy as well as the lattice mismatch between the substrates and the epitaxial layers, to obtain high-quality GaN films with the desired structure. In this section, growth of III–V nitrides on the various substrates listed earlier is described, keeping the early stages of growth in focus. Epitaxial growth of wurtzite GaN is discussed in detail followed by an overview of the research efforts to grow GaN in the zinc-blende structure.

3.2.1. GaN in Wurtzite Structure

Table II lists the various substrates used for the growth of wurtzite GaN and their lattice mismatch with GaN. However, the most widely used substrates for the growth of GaN are (0001) sapphire and 6H-SiC. In spite of a large mismatch with GaN, sapphire is the most commonly used substrate due to its low price, wide availability, and relatively simple pregrowth cleaning requirements. However, sapphire is an insulating material and cannot be used for all device applications. In contrast, SiC is a wide bandgap semiconductor with high thermal

Table II.   Growth Orientation Wurtzite GaN Epitaxy on Different Substrates

Substrate	Lattice Parameters	GaN Growth Direction	In-plane Orientation	Lattice Mismatch (%)
Sapphire (0001)	$a = 4.76$ $c = 12.991$	[0001]	$[11\text{–}20]_{GaN}//[10\text{–}10]_{sapphire}$ $[10\text{–}10]_{GaN}//[11\text{–}20]_{sapphire}$	13.8 13.8
Sapphire (11–20)	$a = 4.76$ $c = 12.991$	[0001]	$[10\text{–}10]_{GaN}//[10\text{–}10]_{sapphire}$ $[11\text{–}20]_{GaN}//[0001]_{sapphire}$	0.4 −1.9
Sapphire (10–10)	$a = 4.76$ $c = 12.991$	[01–13]	$[03\text{–}32]_{GaN}//[11\text{–}20]_{sapphire}$ $[11\text{–}20]_{GaN}//[0001]_{sapphire}$	+2.6 −1.9
Sapphire (10–12)	$a = 4.76$ $c = 12.991$	[11–20]	$[10\text{–}10]_{GaN}//[1\text{–}120]_{sapphire}$ $[0001]_{GaN}//[01\text{–}11]_{sapphire}$	13.8 1.1
6-H SiC (0001)	$a = 3.08$ $c = 15.12$	[0001]	$[10\text{–}10]_{GaN}//[10\text{–}10]_{SiC}$ $[11\text{–}20]_{GaN}//[11\text{–}20]_{SiC}$	3.45 3.45
ZnO (0001)	$a = 3.25$ $c = 5.21$	[0001]	$[10\text{–}10]_{GaN}//[10\text{–}10]_{ZnO}$ $[11\text{–}20]_{GaN}//[11\text{–}20]_{ZnO}$	−1.91 −1.91
Si (111)	$a = 5.43$	[0001]	$[11\text{–}20]_{GaN}//[11\text{–}10]_{Si}$	17

conductivity, and it can be used for the fabrication of high-power, high-frequency, and high-temperature integrated circuits for electronic applications coupled with GaN based optoelectronic devices.

3.2.1.1. Growth on Sapphire Substrates

Large area good quality single-crystal sapphire substrates are commercially available and require relatively simple pre-growth preparation. The substrates are degreased, etched in $H_3PO_4:H_2SO_4$ (1:3) for the removal of surface contaminants and mechanical damage due to polishing, followed by a deionized water rinse. The etching reveals atomic steps on the sapphire surface that facilitate nucleation and two-dimensional growth. Prior to epitaxial growth, the substrates are generally outgassed at elevated temperatures in the vacuum to rid the surface of moisture and any residual contaminants.

GaN films have been grown on several crystallographic planes of sapphire (Fig. 12) with varying success. Table II shows the growth direction, the orientation, and the lattice mismatch of GaN grown epitaxially on various faces of sapphire crystals. There have been very few studies done on the growth of GaN on M-planes [109, 110] and R-planes [111, 112] of sapphire and so far, these films show relatively inferior properties. On an R-plane, GaN grows in a [11$\bar{2}$0] direction, with a small tilt toward the sapphire [$\bar{1}$101] direction [77]. The films typically have ridgelike structures, with the side faces formed by {10–10} planes [109, 77]. Lei, Ludwig, and Moustakas, attributed this tilt to the large difference in the mismatch between GaN and sapphire in the two in-plane orientations [111]. A second possible origin of this surface morphology is the deviation of the substrate from the exact plane. GaN growth on an M-plane sapphire can occur in (11$\bar{2}$2), (10$\bar{1}$3), (1$\bar{2}$11), or (10$\bar{1}$1) directions, depending on the growth conditions. Therefore, it has been difficult to achieve high-quality epitaxial films

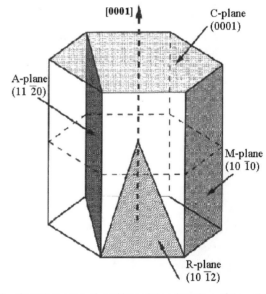

Fig. 12.   Crystal structure of sapphire and the orientations of the major planes used for GaN epitaxy.

on these two planes of sapphire. GaN growth on C-plane (0001) and A-plane (11$\bar{2}$0) of sapphire was more successful leading to the fabrication of several optoelectronic devices.

In spite of the largest lattice mismatch, (0001) sapphire is the most extensively used substrate, since this plane has the same hexagonal in-plane crystal symmetry as (0001) GaN. Several device structures have been successfully fabricated and tested on these substrates. One drawback with the C-plane sapphire is that there are no good cleavage planes perpendicular to the surface. This makes it difficult to form cleaved surfaces that are required in fabricating edge emitting laser diodes. A-plane sapphire, on the other hand, can be cleaved along the R-plane (10$\bar{1}$2) as well as along the M-plane (10$\bar{1}$0) (see Fig. 12). Thus,

III-nitride films grown on the $(11\bar{2}0)$ sapphire (A-plane) can be cleaved more easily to form edge emitting lasers. Nakamura et al. reported fabrication of a laser structure on A-plane sapphire to take advantage of the laser facet formation in the A-plane without the need for etching or polishing [113]. Such a device fabrication capability provides a strong motivation to understand the epitaxy on this substrate. In addition, since the lattice mismatch between GaN and A-plane sapphire is less than 2% [114], it is expected that better quality films may be achieved. Early research at Boston University has shown that GaN films grown on an A-plane sapphire with initial nitridation and low-temperature GaN buffer layers were smoother than those grown on a C-plane sapphire by a similar method [119, 120]. Doverspike et al. investigated the effect of GaN and AlN buffer layers on the growth of GaN on A-plane sapphire [122]. They too reported a better morphology and electron mobility in films grown on A-plane sapphire compared to C-plane sapphire. In the following, the discussion of the GaN films is limited to the epitaxy on the C- and A-planes of sapphire.

3.2.1.1.1. Epitaxial Orientation. As mentioned before, (0001) GaN has the same in-plane lattice symmetry as the C-plane sapphire. Figure 13a is a selective area diffraction (SAD) pattern from the GaN/sapphire interface near the $[11\bar{2}0]$ zone axis of GaN (and the $[11\bar{2}0]$ zone axis of sapphire). Interplanar spacing of the planes responsible for each diffraction spot was calculated from the distance of the diffraction spot from the transmitted spot (g), using the camera constant and the electron

wavelength values. Based on these calculations, lattice planes are assigned to each of the first-order diffraction spots, as shown in Figure 13b. The SAD pattern demonstrates excellent epitaxy and single-crystal structure of the GaN film, despite the large lattice mismatch. From the separation of GaN and sapphire diffraction spots, lattice mismatch of ∼14% was calculated, which is close to the lattice mismatch between GaN and Al_2O_3 bulk crystals. This mismatch and the sharpness of the diffraction spots indicate that the epitaxial GaN film (1.5-μm thick) is largely relaxed. SAD pattern from the other major zone axis ($[10\bar{1}0]$ of GaN and $[11\bar{2}0]$ of sapphire) is presented in Figure 13c and is indexed in Figure 13d. From the two electron diffraction patterns, it has been established that GaN grows with a 30° rotation with respect to the sapphire unit cell, with an orientation relationship:

$$[11\bar{2}0]_{GaN}//[10\bar{1}0]_{sapphire} \quad \text{and} \quad [10\bar{1}0]_{GaN}//[11\bar{2}0]_{sapphire}$$

X-ray diffraction studies confirm this orientation macroscopically (see, for example, Fig. 14). In Figure 14, the diffraction peaks at 34.52, 35.94, and 41.67° correspond to (0002) GaN, (0002) AlN buffer layer, and (0006) sapphire, respectively. The absence of any other peak indicates single-crystal quality of the epitaxial film. The width of the peak is representative of the crystal size (domain size in this case) and the inhomogeneous strain. Assuming that the contribution from the inhomogeneous strain is small, the domain size t can be estimated from the full width at half-maximum (FWHM) value B, using Scherrer's formula:

$$t_{domain} = \left[\frac{0.9\lambda}{B\cos\theta_B} \right]$$

For a measured B value of 180 arcsec, the domain size is calculated to be around 170 nm. The rocking curve at (0002) GaN diffraction is shown in the inset of Figure 14. The rocking curve has an FWHM of 370 arcsec, indicating good in-plane mosaicity of epitaxial films.

Several researchers have used crystallographic models to explain the epitaxial relation between the nitride epilayers and the (0001) sapphire surface [111, 115–117]. Lei, Ludwig, and Moustakas projected the (0001) basal planes of sapphire and GaN to show this 30° rotation (see Fig. 15) [111]. The lattice mismatch between sapphire and GaN as calculated from this projection is approximately 15% and is close to the experimentally measured value. It should be noted that without this rotation in the orientation (if $[10\bar{1}0]_{sapphire}//[10\bar{1}0]_{GaN}$), the mismatch is much higher, at 30%. More detailed crystallographic models have been developed by the researchers at Northwestern University to describe the epitaxy of (0001) AlN and GaN thin films on (0001) sapphire [115, 116]. Sapphire belongs to space group R-3c = D^6_3d (No. 167) and the crystal structure can be described as O^{2-} anions in a hexagonal close packed (hcp) arrangement, with Al^{3+} cations occupying two-thirds of the octahedral voids. A complete description of the sapphire crystal structure was given by Kronberg [118]. GaN and AlN with wurtzite structure (hexagonal symmetry) belong to space group $P6_3mc$ (No. 186). Sun et al. described the

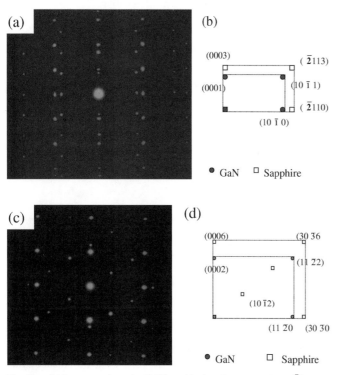

Fig. 13. SAD patterns from the GaN/sapphire interface, at (a) the $[10\bar{1}0]$ zone axis of sapphire and (c) the $[11\bar{2}0]$ zone axis of sapphire. Schematics of the indexed patterns for (a) and (c) are shown in (b) and (d), respectively.

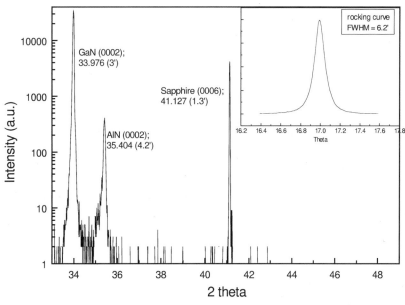

Fig. 14. XRD scan of a typical GaN film grown by MBE on sapphire, using an AlN buffer layer. Inset shows the (0002) rocking curve.

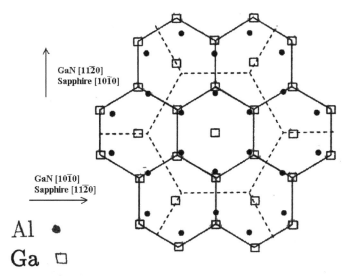

Fig. 15. Projection of bulk C-plane sapphire and GaN cation positions. The dots mark the aluminum atom positions and the dashed lines show the sapphire (0001) plane unit cells. The open squares mark the gallium atom positions and the solid lines show the GaN (0001) plane unit cells. (From Lei, Ludwig, and Moustakas [111].)

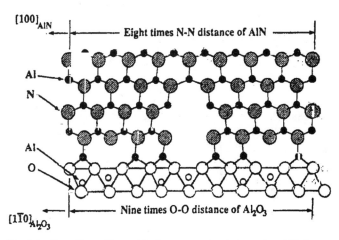

Fig. 16. Crystallographic model of a cross-sectional plane of an AlN film on (0001) sapphire. A close match is shown between eight times the N–N distance of AlN along [11$\bar{2}$0] and nine times the O–O distance of Al_2O_3 along [10$\bar{1}$0]. (From Sun et al. [115].)

atomic bonding between AlN and Al_2O_3 near the interface as shown in Figure 16 [115]. Here, eight times the N–N distance of AlN along [100] matches with nine times the O–O distance of Al_2O_3, which corresponds to a lattice mismatch of ∼11%. Grandjean et al. have shown this epitaxial relationship between GaN and (0001) sapphire using a lattice projection perpendicular to the interface (Fig. 17) [117]. It is evident from the figure that every sixth site of the GaN lattice closely coincides with every seventh site in the sapphire lattice along the interface, corresponding to a lattice mismatch of approximately 14.3%.

There are relatively few studies on the growth of GaN on A-plane (10$\bar{2}$0) sapphire [119–122]. In these studies, the preferred epitaxial orientation relationship was found to be $(0001)_{GaN}//(11\bar{2}0)_{sapphire}$. However, Kato et al. reported that the GaN films deposited on A-plane sapphire had two orientations with the (0001) and (10$\bar{1}$0) planes of GaN being parallel to the (11$\bar{2}$0) plane of sapphire [121]. Here, we show the crystallographic model of the interface for the A-plane sapphire/GaN system, assuming the former epitaxial relationship.

Figure 18a shows 20 × 20 Å crystallographic projections of (11$\bar{2}$0) and (0001) planes of sapphire and GaN, respectively, simulated using the Crystal Kit program [123]. The projections of the cleavage planes in the two lattices are marked in Fig-

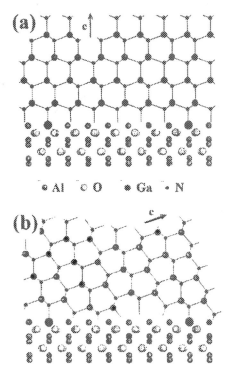

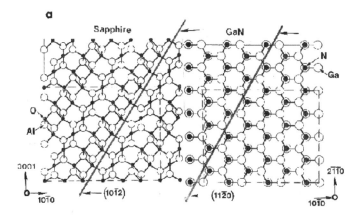

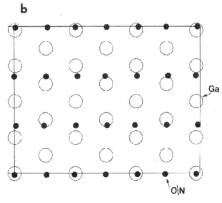

Fig. 17. Schematic of the epitaxial relationship between the GaN epilayers and the Al$_2$O$_3$ substrate. The Al$_2$O$_3$ and the GaN lattices are viewed in the [1$\bar{1}$00] and [11$\bar{2}$0] projections, respectively. In this projection, every sixth site in the GaN lattice matches up with the seventh site in the sapphire lattice implying a lattice mismatch of ~14.3%. (From Grandjean et al. [117].)

Fig. 18. Projections of sapphire and GaN lattices (a) (11$\bar{2}$0) and (0001) planes of sapphire and GaN, respectively, showing the cleavage planes. The rectangles depict the supercells of overlap. (b) A supercell of GaN (0001) epitaxy on sapphire (11$\bar{2}$0).

ure 18a. The line in sapphire is the projection of the (10$\bar{1}$2) plane (R-plane), which is shown to be approximately parallel to the projection of the (11$\bar{2}$0) plane in GaN. However, it should be noted that there is a 2.4° rotation between these two cleavage axes across the interface. The supercell of epitaxial overlap is outlined by the dotted rectangles (~13 × 16.5 Å) in the two projections. Figure 18b shows a superimposition of the two projections. The thickness in the z-direction (normal to paper) is restricted to a monolayer in each lattice, to include only the atoms closest to the interface that participate in the chemical bond. It is assumed here that the oxygen terminated top layer is completely replaced by N atoms during the nitridation process, after which a layer of Ga is attached, resulting in a [000$\bar{1}$] polarity.

Figure 19 is a schematic of the cross section of the sapphire/GaN interface showing the bonding across the interface. Figure 19a shows a cross section normal to the [10$\bar{1}$0] GaN zone axis (the [1$\bar{1}$00] zone axis in sapphire), while Figure 19b shows a cross section normal to the [11$\bar{2}$0] GaN zone axis ([0001] zone axis of sapphire). The cross section is taken from the line of the best match, i.e., the edges of the supercell outlined by a rectangle in the figure. Again, in these two models, a z-range (normal to the paper) of 2 Å is taken to show only the relevant atoms. In these studies, GaN is shown to be growing with a [000$\bar{1}$] polarity (Ga to N bond pointing down) in contrast to the [000$\bar{1}$] polarity seen in Figure 3. The supercell of epitaxial overlap is outlined by the rectangles in the two projections in

Figure 18a. With this supercell, the lattice match is as follows: six (10$\bar{1}$0) planes of GaN match with four planes of (10$\bar{1}$0) of sapphire, resulting in a lattice mismatch of −0.62%. In the perpendicular direction, eight planes of (11$\bar{2}$0) GaN match with one (0001) plane of sapphire, resulting in a mismatch of 1.8%. Based on the earlier projections, the expected distances between misfit dislocations are 179 and 443 Å along the [11$\bar{2}$0] and [1$\bar{1}$00] directions of GaN, respectively. This is in contrast to GaN growth on (0001) sapphire, where the lattice mismatch is 14% and the misfit dislocations are expected every 36 Å. Based on these models, we see that the lattice mismatch between the GaN and sapphire supercells for growth on (11$\bar{2}$0) sapphire is significantly smaller than the corresponding mismatch on (0001) sapphire. This raises the question whether the nitridation and low-temperature GaN buffer steps (that are used for growth on C-plane) are necessary for growth on A-plane sapphire. However, as can be seen in the projections, a large fraction of the atomic bonds in the supercell are highly distorted across the interface and do not match perfectly with the GaN structure, resulting in a strained GaN film. Therefore, a nucleation step may be required for growth on the A-plane as well. This has been verified experimentally, and is discussed later in this section.

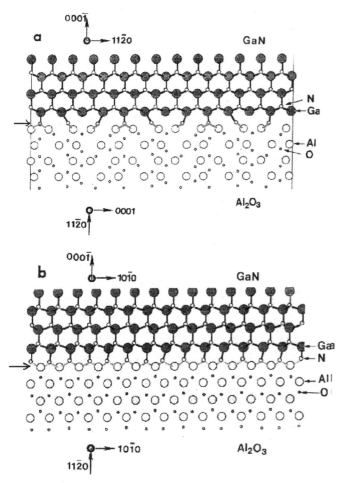

Fig. 19. Crystallographic model of the cross-sectional view of the interface, seen at (a) the [10$\bar{1}$0] GaN zone axis and at (b) the [11$\bar{2}$0] GaN zone axis.

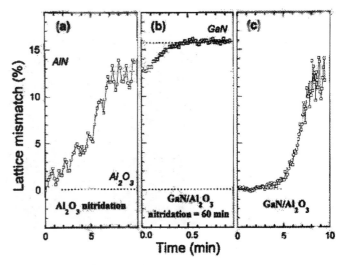

Fig. 20. Lattice-mismatch variation measured *in situ* by RHEED during (a) nitridation of C-plane sapphire, (b) growth of GaN on the sapphire surface after 60-min nitridation and (c) growth of GaN directly on the sapphire surface. Dashed lines indicate the calculated lattice mismatch with respect to the sapphire substrate. (From Grandjean, Massies, and Leroux [131].)

Traditional understanding is that epitaxial growth is possible only when the misfit between the substrate and the overgrowth is less than 15% [124]. Early attempts to grow GaN on sapphire substrates resulted in polycrystalline or very heavily defective epitaxial films. To accommodate the lattice mismatch, Amano et al. [125] and Akasaki et al. [126] developed a low-temperature AlN buffer layer followed by GaN epilayers at a higher temperature by the MOCVD method. This led to significant improvement in the GaN film quality. Subsequently, a low-temperature GaN buffer was developed by the Boston University [119, 74] group and the Nichia group [127] using MBE and MOCVD, methods, respectively. By depositing a 200-Å GaN buffer layer at 500–600°C prior to high-temperature growth, smooth and single-crystal GaN films with good transport characteristics have been achieved. For growth on sapphire, Moustakas et al. also demonstrated the importance of a "nitridation" step to further reduce the defect density and to improve epitaxy [119]. Subsequently, there have been numerous studies to understand the role of nitridation and buffer layers on C- and A-planes of sapphire.

3.2.1.1.2. Nitridation. Several researchers have shown that nitridation promotes two-dimensional and hexagonal growth

[110, 119, 129, 131–135, 137]. It is known that oxygen atoms terminate the sapphire crystal surface. This presence of oxygen species at the interface can complicate the epitaxy of GaN, by formation of a GaO layer at the interface. During the nitridation process, oxygen atoms near the surface are substituted by nitrogen atoms, thus forming a monolayer or a partial monolayer of strained AlN. It is believed that this facilitates the subsequent growth of GaN on a better (chemically) matched substrate. Note that in Figure 19, the top monolayer of O atoms in sapphire are shown to be replaced by N atoms. In addition, by bombarding the sapphire surface with ECR nitrogen plasma, any remaining impurities are removed and the surface atoms are provided with enough energy to recrystallize to a perfect lattice.

The formation of an AlN layer has been observed by several researchers, although the stoichiometry and the crystallinity of the AlN layer varied depending on the growth method. The formation of an AlN thin film upon exposure of the (0001) sapphire surface to nitrogen plasma has been documented by Moustakas et al., based on the observation of a streaky RHEED pattern [119]. The authors observed a broad diffraction line indicative of a thin and strained AlN layer. Although RHEED patterns are not accurate enough to determine the lattice constants, the lattice parameter of the AlN layer can be estimated compared to the known lattice parameters of the (relaxed) sapphire unit cell. It has been shown that the AlN grown by the plasma nitridation process is under compressive stress with an in-plane lattice constant of 3.05 Å after 5 min and 3.015 Å after 20 min, instead of a bulk crystal value of 3.11 Å [129, 130]. Thus, AlN film is strained by the (0001) Al$_2$O$_3$ lattice, which has a smaller lattice constant of 2.75 Å [111]. Grandjean, Massies, and Leroux studied the effect of nitridation of C-plane sapphire at 850°C by a gas source (NH$_3$) MBE [131]. Based on RHEED studies, these authors recorded the variation of the in-plane lattice constant as a function of nitridation time (see Fig. 20). After

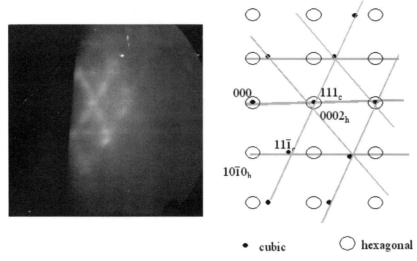

Fig. 21. (a) RHEED pattern from a GaN film grown directly on A-plane sapphire, viewed at the $[11\bar{2}0]_{GaN}$ zone axis, (b) schematic showing the two components of the RHEED pattern.

10 min of NH_3 exposure, the lattice mismatch with respect to the starting surface saturated around 13% (expected mismatch between AlN and Al_2O_3 is 12.8%), confirming the formation of a relaxed AlN layer. Vennegues and Beaumont have done detailed high-resolution TEM studies [132] to conclude that the AlN layer is about 10 atomic planes thick and is defective with Al vacancies and O atoms (incomplete conversion of Al_2O_3 to AlN). In spite of these significant differences in the observations, it is commonly agreed that a nitrided surface facilitates the growth of smooth and relaxed GaN films. Growth of the GaN film on a nitrided surface proceeds with the c-axis parallel to that of sapphire and $[11\bar{2}0]_{GaN}//[10\bar{1}0]_{sapphire}$ as seen in the electron micrographs shown in Figure 13. On the substrates without nitridation, a different orientation was observed: $[11\bar{2}0]_{GaN}//[10\bar{1}0]_{sapphire}$ and $[1\bar{1}03]_{GaN}//[11\bar{2}0]_{sapphire}$ [117]. In this orientation, the c-axis of the GaN layer is tilted 19° with respect to the Al_2O_3 (0001) plane.

In general, nitridation is highly efficient in MOVPE growth due to the much higher (>1000°C) temperatures employed. AlN thus might form rapidly, even as the GaN deposition is commenced. Uchida et al. investigated the influence of nitridation of C-plane sapphire on the properties of a GaN film grown by MOVPE [133]. *Ex situ* TEM studies show the formation of an amorphous AlO_xN_{1-x} layer, or a crystalline AlN layer surmounted by an amorphous AlO_xN_{1-x} layer. The authors made EDX measurements using a 1-nm diameter electron beam to confirm the presence of Al, O, and N in the nitrided layer. They hypothesize that NH_3 reacts with the surface layer, replacing oxygen with nitrogen and forming the amorphous layer which corresponds to five to six unit cells of sapphire. Excess nitridation leads to the accumulation of strain between the amorphous layer and the substrate causing roughness and three-dimensional growth, as confirmed by atomic force microscope (AFM) studies. Uchida et al. believe that the presence of an amorphous nitrided layer improves the surface morphology of the GaN films [133]. *In situ* TEM studies done by Yeadon et al.

indicate that nitridation reaction (under ammonia flux) starts with the nucleation of AlN islands, which proceed to form a continuous film [134]. The authors confirmed the formation of AlN by selective area diffraction studies and observed that the AlN/Al_2O_3 reaction interface moves progressively into the substrate.

In spite of the much better lattice match, nitridation was found to be necessary even for A-plane sapphire. Figure 21 shows a RHEED pattern observed during the early the stages of a GaN film grown directly on A-plane sapphire by the ECRMBE method [110, 135, 136]. The RHEED pattern shows the superimposition of diffraction from more than one crystal phase. Based on the RHEED patterns from thick GaN films, the camera length of the diffraction system was calculated and was used to estimate the interplanar spacing of the planes responsible for the pattern. It was determined that the pattern in Figure 21a is a superimposition of the RHEED pattern from the $[11\bar{2}0]$ zone axis of wurtzite GaN and the $[1\bar{1}0]$ zone axis of zinc-blende GaN. The analysis of the RHEED pattern distinguishing the two components is shown in Figure 21b. Although the films had a predominantly wurtzite structure after further growth, they were very defective. This was also observed in the X-ray diffraction (XRD) and TEM studies. XRD studies of the GaN layer grown without any nitridation showed a very weak peak at $2\theta = 34.5°$ corresponding to the (0002) plane of GaN, indicating poor epitaxial orientation of the film. In addition, there were other weak peaks suggesting that the GaN film is polycrystalline. Figure 22 is a cross-sectional TEM micrograph of such a GaN film, showing that the film is indeed polycrystalline and polymorphic. The inset electron diffraction pattern, taken from the region of the sapphire/GaN interface, shows an overlap of the diffraction patterns from sapphire and GaN. From the absence of any match between the two sets of diffraction spots, it is clear that there is no specific epitaxial relationship between the film and the substrate. Regions of the zinc-blende phase were clearly observed in electron diffraction

Fig. 22. Cross-sectional TEM micrographs of GaN grown on sapphire without nitridation. Inset diffraction shows random orientation. (From Doppalapudi et al. [135].)

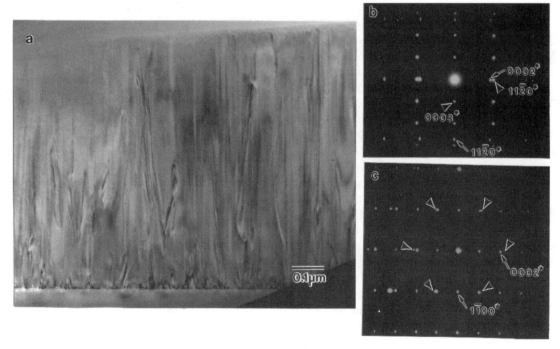

Fig. 23. (a) Cross-sectional TEM micrograph of a GaN film grown after nucleation by nitridation and a low-temperature buffer. The orientation relationship between the film and the substrate can be seen from the reciprocal lattice points of GaN (arrows) and sapphire (triangles) in the electron diffraction patterns from the film interface (b) at the $[10\bar{1}0]$ zone axis of GaN ($[10\bar{1}0]$ zone axis of sapphire), and (c) at the $[11\bar{2}0]$ zone axis of GaN ($[0001]$ zone axis of sapphire). The triangles in (c) represent the $\{11\bar{2}0\}$ family of planes. (From Doppalapudi et al. [135].)

analysis and are marked in the figure. Thus, optimization of the growth parameters, especially in the initial nucleation stages is very important to obtain GaN in the desired phase.

X-ray diffraction studies of the films grown on nitrided A-plane sapphire substrates showed that GaN grows with the (0002) plane parallel to the A-plane of sapphire. The intensity and the FWHM of the (0002) peak, as well as the surface mor-

phology improved significantly with nitridation for these samples. Figure 23a is the cross-sectional TEM micrograph of a GaN film grown after nitriding the A-plane sapphire substrate for 15 min. In contrast to the film shown in Figure 22, the film is single crystal GaN, with a defective buffer layer near the interface. The selective area diffraction (SAD) pattern near the $[10\bar{1}0]$ zone axis (of both lattices) from the interface is shown in Figure 23b. It shows that in reciprocal space, the $(11\bar{2}0)^*$ diffraction spots of sapphire line up with the $(0002)^*$ diffraction spots of GaN while the $(0003)^*$ diffraction spots of sapphire line up with the $(11\bar{2}0)^*$ diffraction spots of GaN, indicating that in real space, the planes responsible for the diffraction spots are parallel. The lattice mismatch can be estimated from the separation of these reciprocal lattice points. Figure 23c is the SAD pattern from the $[11\bar{2}0]$ zone axis of GaN ($[0001]$ zone axis of sapphire), also showing a good epitaxial orientation between the GaN film and sapphire. These results confirm that GaN epitaxy on A-plane sapphire occurs as

$$[11\bar{2}0]_{GaN}//[0003]_{sapphire} \quad \text{and} \quad [0002]_{GaN}//[11\bar{2}0]_{sapphire}$$

Figure 23 shows that the sample has domain boundaries and a high density of threading dislocations ($\sim10^{10}$ cm^{-2}). However, no cubic domains and very few stacking faults are present in the films. In the films with poor nucleation (insufficient nitridation), a mixture of two in-plane orientations was observed at the early stages of growth. The RHEED pattern showed GaN diffraction at both $[10\bar{1}0]$ and $[11\bar{2}0]$ zone axes, at the same orientation. These data clearly indicate the importance of the nitridation step of the A-plane sapphire for the epitaxial growth of GaN. This is due to the formation of AlN nuclei on the surface of Al$_2$O$_3$, which promotes wetting and facilitates two-dimensional growth, thereby improving the quality of the GaN film and suppressing the formation of the zinc-blende polytype.

The optimum nitridation time depends on the quality of the substrate surface and the misorientation with respect to the crystallographic axis. There is general agreement that prolonged nitridation leads to a rough surface [133, 137, 129], although this time varies widely depending on the particular nitridation method. Heinlein et al. conducted RF plasma nitridation studies at 400°C on c-plane sapphire substrates and reported that the substrate roughens drastically after 300 min [137]. The authors propose that nitridation initially leads to the formation of a smooth monolayer of AlN on the sapphire surface. Surface nitridation beyond this entails breaking of subsurface Al–O bonds, which causes a pronounced change in surface morphology and the onset of protrusion growth, similar to Stranski–Krastanov nucleation. Figure 24 shows the increase variation in surface roughness of the c-plane sapphire substrate with ECR

plasma nitridation carried out at 800°C [129, 130]. This roughness is presumably caused by the formation of AlN islands on the sapphire surface.

Formation of an AlN monolayer is also expected to establish the polarity of the subsequent GaN growth, since AlN and GaN have identical structural symmetry and very similar lattice parameters. Fuke et al. reported that nitridation processes result in an N-terminated surface whereas a low-temperature GaN buffer layer results in a Ga-terminated surface [138]. When nitridation is omitted, inversion domains which are domains of a polarity opposite to that of the matrix are observed [139]. Doppalapudi reported a decrease in IDB density [110] when the nitridation (with ECRMBE) time increased from 15 min to 35 min (see Fig. 25). Figure 25a is a TEM bright field micrograph showing a GaN film in which the nitridation of (0001) sapphire was carried out for 15 min (sample A). In contrast, Figure 25b shows a GaN film grown on a substrate that was nitrided for 35 min (sample B). Figure 25a shows the presence of a large number of inversion domain boundaries (IDBs), whereas these defects are absent in Figure 25b. We believe that this difference between the two films is due to a difference in nitridation time.

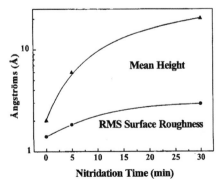

Fig. 24. Effect of nitridation on the roughness of the sapphire substrates, as measured by atomic force microscopy. (From Korakakis [129].)

Fig. 25. TEM bright-field cross-sectional images of GaN films (a) sample A with $\sim10^{10}$ cm^{-2} and (b) sample B with 2×10^9 cm^{-2} defect density.

As explained before, the polarity of the GaN film is reversed across an IDB. In the case of sample B, complete nitridation established a single polarity over the entire 2-in wafer, as evidenced by the lack of IDBs. Evidence from RHEED studies shows this polarity to be $[000\bar{1}]$ (N-polarity). In sample A, incomplete nitridation facilitated the formation of IDs of [0001] polarity (Ga-polarity), in a matrix of GaN with $[000\bar{1}]$ polarity. A more detailed discussion of polarity in GaN is given later in this chapter. In addition to the IDBs, sample A has a considerably higher density of threading defects ($>10^{10}$ cm^{-2}) closer to the film surface, as compared to sample B, which has 2×10^9 cm^{-2}.

Although the exact mechanism of surface modification caused by nitridation is still under discussion, the influence of this step on structure and properties of the resulting films grown by different methods has been well documented. Keller et al. investigated GaN films grown by MOCVD on C-plane sapphire with different nitridation time by controlling the NH$_3$ preflow time [140]. They observed lower dislocation density and better electrical and optical properties in the films with shorter nitridation. The authors attribute the better quality in these films to rough, faceted, and predominantly cubic GaN nucleation layers, which enhance dislocation interaction and annihilation. Grandjean, Massies, and Leroux report a similar improvement in optical properties (narrower band edge emission and reduced yellow PL) with optimized nitridation in the MBE grown GaN films [131]. Excess nitridation, however, results in the deterioration of the optical properties.

Influence of nitridation on the properties of GaN films grown on A-plane sapphire has also been investigated. One such study is summarized in Table III [84]. The table lists data on surface roughness and transport properties of GaN films grown with different nitridation times. A clear improvement of surface morphology was observed with an increase in nitridation time, as indicated by RMS roughness values measured by AFM. Figure 26 shows SEM micrographs of the three samples taken under the same magnification. It is clear from the figure that there is a dramatic improvement in surface morphology with nitridation time. Sample I in Figure 26a has a rough surface morphology, indicative of three-dimensional growth. Sample II (which was nitrided for 5 min) did not have a uniform surface morphology, with some portions of the surface being fairly smooth in Figure 26b and other regions being quite rough (not shown in this micrograph), indicating insufficient nucleation of AlN on the surface. Sample III, which was grown on sapphire nitrided

for 15 min has a very smooth, specular, surface. This improvement in the morphology is consistent with the hypothesis that nitridation promotes wetting and thus two-dimensional growth. The improvement in the crystalline quality is evident from the improvement in the value of electron mobility.

Photoluminescence measurements were carried out at room temperature and at 77 K to investigate the effect of the substrate nitridation on the optical properties of the films. Figure 27 shows the PL spectra of the three samples described before, both at 300 K (dotted lines) and at 77 K (solid lines). At 300 K, sample I had a broad peak at 3.3 eV with a low energy tail, while samples II and III show the excitonic lines of wurtzite GaN at 3.4 eV, indicating that nitridation is essential to improve film properties. To reveal the differences between the films more clearly, photoluminescence measurements were carried out at 77 K. Sample I, which was not nitrided, shows a very broad peak near 3.27 eV. Sample II, with 5-min nitridation, has the dominant peak at 3.27 eV with a weaker peak near 3.47 eV, associated with a neutral donor bound exciton of wurtzite GaN. At the low energy side of the 3.27 eV peak, two other peaks at 3.19 and 3.11 eV could be resolved, which are the 1 and 2 LO (longitudinal optical) phonon replicas of the 3.27-eV transition, respectively. Finally, for the case of the GaN film grown

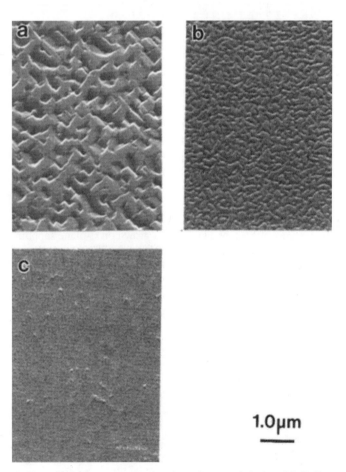

Fig. 26. SEM micrographs showing the surface morphologies of GaN films grown with (a) no nitridation, (b) 5 min nitridation, (c) 15-min nitridation. (From Doppalapudi et al. [135].)

Table III. The Effect of Nitridation on the Physical Properties of GaN Films on A-Plane Sapphire

Sample	Nitridation (mins)	RMS Roughness (nm)	n (cm^{-3})	μ (cm^2 V^{-1} sec^{-1})
I	0	40	1.1×10^{19}	1
II	5	4	9.8×10^{18}	89
III	15	1.5	7×10^{18}	119

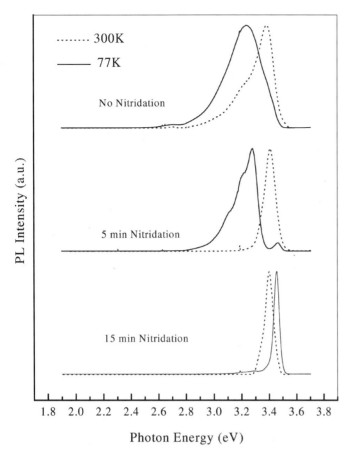

Fig. 27.   PL spectra from GaN films with different nitridation times taken at 300 K (dotted lines) and at 77 K (solid lines.) (From Doppalapudi et al. [135].)

3.2.1.1.3. Buffer Layers.

Epitaxial GaN films generally have a very high density of threading defects (10^8–10^{10} cm^{-2}) even after using a nitridation step. Most of the defects observed in GaN films are threading type, which originate in the nucleation layer of the GaN film, near the interface with sapphire. Island growth (Volmer–Weber mode) is the typical growth mode for nitrides on sapphire. During island coalescence, low angle tilt boundaries composed of edge dislocations are developed if the islands are slightly rotated or tilted with respect to each other. A majority of these dislocations are threading in the direction normal to the interface. Once formed, most of the threading edge dislocations are sessile due to the lack of a suitable glide plane. Therefore, optimization of the nucleation layer is essential to minimize the dislocation density in GaN films. As mentioned before, researchers have explored both GaN [74, 119, 127, 144–146, 122] and AlN [122, 147, 148] buffers.

The low-temperature GaN buffer layer as deposited, is generally polycrystalline and rough with a very high density of defects including stacking faults and grain boundaries [144, 145, 149]. In addition, the buffer layer consists of a mixture of hexagonal and cubic phases of GaN [110, 145, 149, 259]. This polycrystalline film accommodates the stress from the large lattice mismatch and reduces the propagation of defects generated at the interface by destructive interference and annihilation, thus reducing the dislocation density. As the temperature is increased to the growth temperature, this buffer layer is recrystallized and facilitates two-dimensional GaN epitaxy. Wu et al. have done TEM studies and reported that the grain size increases when the low-temperature buffer deposited at 600°C is heated to the growth temperature of 1080°C in an MOCVD reactor [145]. The authors also observed a partial conversion of zinc-blende GaN to wurtzite GaN. Thermal treatment of this low-temperature buffer layer influences the formation of defects such as dislocation and stacking faults. Keller et al., for example, noticed that the nucleation layers showed a tendency for island coarsening and increased roughness after heating to bulk GaN growth temperature [146]. It is important to optimize buffer deposition as well as the recrystallization by controlling the parameters such as ramp rate, annealing temperature and time, and the ambient (NH$_3$ or nitrogen, as the case may be) [150, 151]. The parameters have to be varied so that recrystallization would be facilitated without significant dissociation of GaN and formation of gallium-rich regions.

Sugiura et al. studied the effect of thermal treatment of GaN buffers grown by MOCVD at 550°C, on the quality of a subsequent GaN layer grown at 1100°C [150]. The authors reported that the best surface morphology and dislocation density were obtained for a ramp time of 12 min ($\sim$45°C min^{-1}). For a shorter ramp time, the buffer layer is not completely recrystallized and some amorphous regions remain. When the ramp time is much higher, the GaN buffer layer reevaporates or dissociates to form Ga droplets and a degradation of single-crystal nucleation sites occurs. In both cases, the resulting GaN films are rough and have stacking faults with bounding partial dislocations. The authors claim that the dislocation density in GaN films is determined by the density of single-crystal nuclei that

after 15 min of nitridation only one peak at 3.47 eV, with an FWHM of 45 meV can be seen. These data show the importance of optimizing the nitridation step to obtain good GaN epitaxial films. The FWHM values quoted for the completely nitridated sample at 300 K (60 meV) and 77 K (45 meV) are comparable to the state-of-the-art for GaN samples with similar carrier concentration ($\sim$10^{19} cm^{-3}). These values are also similar to the FWHM of Si-doped GaN films grown on C-plane sapphire with the same donor concentration [141]. This broadening is due to a combination of thermal broadening and potential fluctuations created by the random spatial distribution of impurity atoms in the host crystal and is consistent with the analysis of those fluctuations based on Morgan's impurity band broadening model [141].

Several researchers reported transitions near 3.27 eV as seen in this study [142, 143, 20]. While Lagerstedt and Monemar [142] and Ilegems and Dingle [143] attributed these peaks to donor–acceptor transitions in wurtzite GaN, Moustakas [20] proposed that this is a result of the zinc-blende domains or isolated stacking faults in the wurtzite structure. Such structural inhomogenities give rise to potential fluctuations in the conduction band of about 0.2 eV, which is the difference in the bandgap between wurtzite and zinc-blende GaN [20]. The direct correlation between the PL spectra and the cubic domains observed in the TEM studies strongly support the second hypothesis.

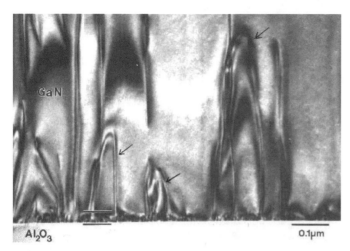

Fig. 28. TEM cross sectional image of a GaN film grown on sapphire by the MBE method. The low-temperature GaN buffer near the interface is indicated. (From Doppalapudi [110].)

form when the low-temperature buffer is ramped to the growth temperature, and is independent of the dislocation density in the buffer layer, which is due to the lattice mismatch between GaN and sapphire. Lin et al. have done a similar study on the dependence of GaN epilayer properties on the thermal treatment of the GaN buffer layer deposited at 525°C by low-pressure MOCVD [151]. Based on X-ray peak width, photoluminescence line width, and Hall mobilities, the authors concluded that a ramping rate of 20°C min^{-1} was the optimum. Deficiency of annealing time for thick buffer layers is also considered to increase the FWHM of the X-ray rocking curve due to poor crystallization [138].

The ability of the buffer to stop propagation of interfacial defects is shown clearly in Figure 28. The highly defective 20-nm thick GaN buffer is marked in the figure. It can be seen that a majority of defects stop at the buffer and only a few propagate through the thickness of the film. Some of these defects are dislocation half-loops (marked by arrows), which terminate within a short distance from the interface. However, some dislocations and IDBs propagate into the film. Although the density of these threading defects varies considerably with the nucleation and growth conditions, the density is generally high ($>10^8$ cm^{-2}) and is characteristic of all heteroepitaxial GaN films.

3.2.1.1.4. AlN Buffers. The defective GaN buffers are conductive and they may act as a leakage path for the lateral transport in the epilayer. AlN buffers on the other hand, are insulating in addition to minimizing the propagation of interfacial defects. The AlN buffers are typically grown at higher temperatures. It was observed that the AlN buffers improve the lattice mosaic of the epitaxial layer [110]. X-ray diffraction studies show that the FWHM of (0002) reflection of GaN films grown on an AlN buffer was narrower compared to the films of the same thickness grown using low-temperature GaN buffers. The studies also show that thick GaN films on AlN buffers are pseudomorphically relaxed. We believe that this is

due to the increased surface wetting of the AlN layer, facilitating two-dimensional growth.

Vaillant et al. deposited AlN buffers on (0001) sapphire by MOCVD at 800°C, and studied the effect of annealing at 1070°C for different time periods [148]. The authors report that the buffer layer is recrystallized by atomic migration through the matrix followed by a secondary recrystallization or Oswald ripening. During this secondary recrystallization, grain growth of AlN crystallites occurs, driven by lowering the interfacial energy. They also claim that strain plays an important role in determining the recrystallization process and the final morphology. Piquette et al. observed the formation of cubic domains even in AlN buffer layers deposited by MBE [147]. The authors report the growth of GaN in N-polarity (000$\bar{1}$) based on (3 × 3) and (6 × 6) reconstructions observed in RHEED patterns. However, most researchers report that AlN buffer leads to Ga-polarity [0001] in nitride films.

There have been reports of reduced dislocation density in the GaN films grown using AlN buffers, compared to those grown using GaN buffers [152]. However, the films on AlN buffers were more strained, probably due to the mismatch between the GaN epilayer and the AlN buffer. Films grown on AlN buffers had a higher density of pinholes or nanotubes [152]. This may be due to the difference in polarity of the two films. Pinholes are formed due to the slow-growing [000$\bar{1}$] domains in a [0001] polarity. Since AlN buffers lead to [0001] polarity, there is more of a chance of an N-polar domain. On GaN buffers, however, the matrix generally has N-polarity, making the pinholes less likely. Instead, pyramidal shaped surface roughness is often seen on these films, presumably as a result of faster growing Ga-domain in an N-polar matrix. It should be noted that optimization of nucleation and growth could lead to smooth films using either one of the buffer layers.

There have been several studies of GaN growth on AlN layers. There is a 2.5% lattice mismatch between the AlN and the GaN. Bykhovski, Gelmont, and Shur reported a critical thickness of 30 Å [153]. Widmann et al., however, observed GaN growth in the Stransky–Krastanov mode, with GaN three-dimensional islanding after two monolayers, resulting in elastic relaxation [154]. Figure 29 shows a TEM micrograph of a stack of 20 AlN/GaN layers grown by MBE [110, 155]. Different layers of AlN and GaN can easily be identified by the contrast difference. The layers with darker contrast are GaN (∼46 nm in thickness) and the layers with the lighter contrast are AlN (∼50 nm in thickness). A 26-nm thick AlN buffer layer is also seen near the interface. From the micrograph, it is evident that the AlN/GaN interface is always very smooth, whereas the GaN/AlN interface is relatively rough, especially after a few layers. It is conjectured that this results from AlN "wetting" the GaN surface better, and growing in a two-dimensional (Frank–Van der Merwe) mode, while GaN grows in a three-dimensional (Stranski–Krastanov or Volmer–Weber) mode. The superior wetting characteristics of AlN compared to GaN is evident from Figure 29. The three-dimensional growth mode of GaN may be a result of compressive stress arising from the growth on lattice-mismatched AlN. Similar roughening by the formation

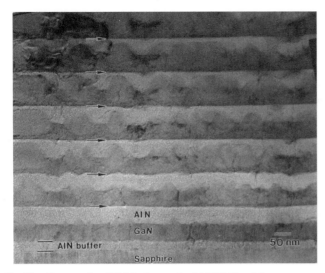

Fig. 29. Cross-sectional TEM micrograph of AlN/GaN multilayers grown on sapphire [110, 155].

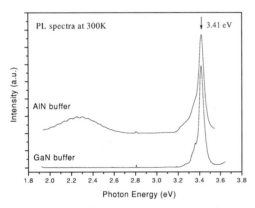

Fig. 30. PL spectra from GaN films grown on sapphire, using GaN and AlN buffers.

Table IV. The Effect of Low-temperature Buffer on the Properties of GaN Films Grown on A-Plane Sapphire

Sample	Buffer	Thickness (μm)	RMS Roughness (nm)	n (cm^{-3})	μ (cm^2 V^{-1} sec^{-1})
A	0	0.9	8	8.9×10^{18}	97
B	20 nm	1.0	2	7×10^{18}	118
C	0	1.5	9	2.7×10^{17}	135
D	20 nm	1.5	3	3.5×10^{17}	211

tributed the trap states to nitrogen vacancies, antisite defects, etc. The presence of an AlN buffer may facilitate the formation of such defects. This correlation between the yellow luminescence, the AlN buffers, and the film polarity is further discussed later in the chapter.

There have also been a few studies in using double buffer layers to improve the quality of the GaN films. Li et al. obtained improved film surface morphology by using an initial GaN buffer followed by an AlN buffer grown by MOCVD at 520–570°C [157].

The effect of low-temperature GaN and AlN buffers on the film properties was also studied on A-plane sapphire [110, 135, 136, 122]. Doverspike et al. reported that the GaN epilayers quality (based on X-ray and Hall measurements) is fairly independent of the GaN buffer layer thickness and the films are in general better than those obtained on C-plane sapphire [122]. This relative insensitivity of GaN film crystallinity to the buffer thickness is attributed to the lower lattice mismatch and better surface finish of A-plane sapphire, which facilitates more uniform buffer deposition. AlN buffer thickness, on the other hand, influenced the epilayers crystallinity and the best films were obtained for AlN buffer thickness of 160–175 Å. Doppalapudi et al. investigated the influence of the low-temperature GaN buffers deposited at 550°C on the structure and the properties of the epitaxial GaN films [135, 136]. The benefit of a low-temperature buffer was not obvious in heavily doped n-type GaN films, from *ex situ* TEM and Hall measurements. The results of some of the buffer studies are summarized in Table IV [135]. As seen in the table, 1-μm thick GaN films, GaN films with (sample A) and without buffer (sample B, with 20-nm thick buffer) layers had comparable dislocation densities ($\sim 8 \times 10^{10}$ cm^{-2}), and electron mobilities (~ 100 cm^2 V^{-1} sec^{-1}). However, XRD results show that the sample grown without a buffer step is more compressively strained, indicated by an increase in the c-lattice parameter. The c-lattice parameter was calculated to be 5.188 Å for the film grown on a low-temperature buffer, compared to 5.1896 Å for the film grown without the buffer. These observations were corroborated by *in situ* RHEED observations as well. It was observed that the epitaxial films were initially strained, and were relaxed as further deposition occurred. The films grown on low temperature buffers seem to relax very quickly, within the first 4–10 min of growth (equivalent of 10–25 nm). The films grown

of "V-defects" was observed previously, and was attributed to strain relief. Daudin et al. used this strain-relief mechanism to fabricate GaN quantum dots between smooth AlN layers [156]. They conducted RHEED studies to report that the GaN grown on AlN roughened after two monolayers and proceeded as island growth (Stranski–Krastanov mode).

There were significant differences in optical properties of the GaN films grown on the two types of buffers. Figure 30 compares the PL spectra from two GaN films, one grown with a GaN buffer and the other with an AlN buffer. Both films were undoped and were grown under the same growth conditions and to the same thickness. Although the PL peaks at 3.4 eV are very similar in both films, the films grown with the AlN buffer layer showed the presence of yellow emissions near 2.3 eV. Similar differences were seen in the films grown with higher doping. The correlation between the buffer type and yellow PL was made after observations from a large number of samples and we are certain that the yellow PL is not due to any other variations in the film growth. It is known that the yellow emission is due to a trap state in the middle of the GaN bandgap, although the origin of this state is not clear. Several researchers have at-

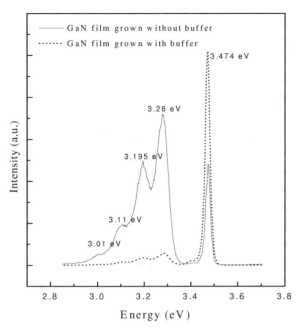

Fig. 31.   PL spectra at 77 K from samples A (solid line) and B (dotted line).

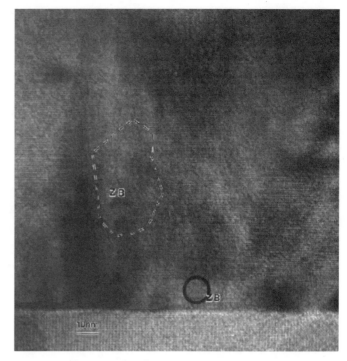

Fig. 32.   HREM image of sample A (without a buffer) from the interface region, showing cubic domains. (From Doppalapudi et al. [135].)

without a buffer, on the other hand, were found to be strained, sometimes even when the film thickness exceeded 1 μm.

Photoluminescence studies at 77 K of the heavily doped samples showed that the GaN film grown with a buffer (sample B) had a peak at 3.460 eV (corresponding to the donor bound excitons of wurtzite GaN) with an FWHM of 40 meV, as shown by the dotted line in Figure 31. The GaN film grown without a buffer (sample A) has a broader peak solid line in Fig. 31). This indicates that the buffer layer results a partial relief of stresses in the films. Furthermore, the PL spectra from the films grown without a buffer layer had a strong secondary peak at 3.27 eV, with LO phonon replicas at 3.19 and 3.11 eV. As discussed earlier, this peak was attributed to cubic domains and/or isolated stacking faults. Figure 32 shows a high-resolution electron micrograph of the sample A near the interface. Zinc-blende domains are clearly identified and are as large as 40 nm in some cases. We could not identify any such cubic domains in film B, which was grown on a low temperature buffer. This provides further support to our earlier correlation between the transition at 3.27 eV and the cubic features.

In the lightly doped films (samples C and D in Table IV), the benefit of the GaN buffer was more obvious. Specifically, the threading defects were predominantly inversion domains in the film without a buffer. The IDB density was observed to be a factor 5 higher in sample C (without buffer), compared to sample D (with 20-nm buffer). This difference in the defect character is clearly shown by TEM micrographs in Figure 33. Figure 33a is a bright-field micrograph from sample C, showing that the threading defects are predominantly domain boundaries, while Figure 33b shows a much smaller concentration of domain boundaries in sample D. This difference is also reflected in the transport properties of the films. The electron mobility in the film grown with a low-temperature buffer is a factor of 2 higher than in the film without a buffer, as shown in Table IV. This is believed to be due to the piezoelectric effect of the different polarities arising from the IDB in the sample C.

Figure 34 shows the PL spectra from samples C and D at 300 K (dotted line) and at 77 K (solid line). The PL spectra at 77 K from these samples also show peaks at 3.27 eV, which are not visible in the spectra taken at 300 K. The use of the buffer dramatically reduced the PL features attributed to cubic domains. The FWHM of the peaks associated with the neutral donor bound excitons for sample D are 38–40 and 22 meV at 300 and 77 K, respectively. These values are similar to the state-of-the-art in GaN films with $\sim 10^{17}$ cm^{-3} carrier concentration and are very close to the values expected by the cumulative effect of thermal broadening (39 meV at 300 K and 10 meV at 77 K), static disorder, and inhomogeneous strain [66, 141].

Figure 35a shows a θ–2θ XRD scan of a typical GaN film grown with nitridation and buffer on A-plane sapphire. The symmetric (0002) peak has an FWHM of 130 arcsec, which indicates a domain size of 230 nm, assuming that the contribution from the inhomogeneous strain is small. The rocking curve at (0002) diffraction is shown in Figure 35b. The rocking curve, an indicator of in-plane mosaicity of epitaxial films, has an FWHM of 600 arcsec, which is comparable to the GaN films grown on C-plane sapphire by MBE.

3.2.1.2. Growth on 6H-SiC Substrates

The use of sapphire substrates is limited by the poor heat dissipation in high-power GaN-based devices. SiC substrates find applications in such devices, due to their excellent thermal con-

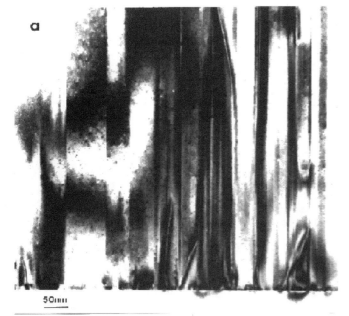

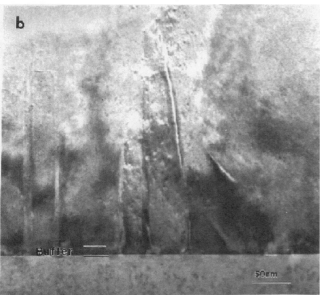

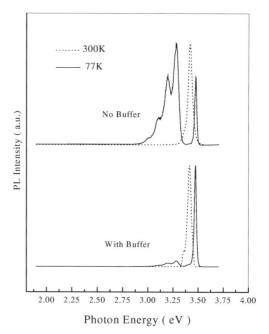

Fig. 34.   PL spectra from samples C and D taken at 300 K (dotted lines) and at 77 K (solid lines.) (From Doppalapudi et al. [135].)

Fig. 33.   TEM micrographs from samples C and D. The film grown without a buffer layer shows predominantly domain boundaries (a), whereas the film grown on a buffer has very few domain boundaries (b). (From Doppalapudi et al. [135].)

ductivity. In addition, SiC has a lower lattice and a thermal mismatch with GaN, compared to sapphire. Furthermore, SiC is a wide bandgap semiconductor and can be used as an active part of the device structure. A majority of the reported research has been conducted on Si-face (0001) 6H-SiC substrates. Commercially available SiC substrates typically have many fine scratches and rough surfaces with micrometer-sized holes called pinholes or micropipes. The peak-to-valley roughness on these substrates can be as high as 100 nm and the damage layer (from mechanical polishing) can be even deeper [158]. In addition, the SiC substrates have a thin film of native oxide (few atomic layers) that readily forms even in ambient conditions.

Hence, surface preparation of SiC substrates prior to the epitaxial growth is critical. In addition to standard cleaning, the substrates have to be immersed in HF solution to etch surface oxides and to passivate the surface with hydrogen. Several methods have been employed to etch or to anneal the SiC surface to remove the surface roughness and to expose the pristine crystal surface of (0001) SiC. Xie et al. annealed the substrates at 1450°C in H_2, C_2H_4/H_2, and HCl/H_2 for different time periods and found that the best results are obtained after annealing for 40 min in H_2 [159]. Brandt et al. used a combination of mechanochemical polishing and H_2 etching at 1600°C to improve the surface roughness to 2 nm [158]. Such a treatment leads to the disappearance of the scratches and gives rise to large atomically smooth terraces. AFM studies show that the terrace steps form a regular array and lie along the ⟨1–100⟩ directions [160, 161]. Another method of surface preparation was to anneal the substrates in an ultrahigh vacuum at 1000°C in Si flux to obtain an oxygen-free reconstructed surface [162]. To circumvent the problem of poor nucleation on the rough SiC substrates, researchers have also used initial SiC epilayers (p- and n-types) [163, 164]. Since the homoepitaxy on SiC is much easier, smooth surfaces are developed which facilitate the subsequent growth of epitaxial GaN. Use of such SiC epilayers led to high-quality GaN films on on-axis as well as on off-axis 6H-SiC and 4-H SiC substrates.

SiC and the III-nitrides have the same hexagonal symmetry and the epitaxial growth follows the substrate symmetry

$$(0001)_{GaN}//(0001)_{AlN}//(0001)_{SiC}$$
$$[11\bar{2}0]_{GaN}//[11\bar{2}0]_{AlN}//[11\bar{2}0]_{SiC}$$

Ponce et al. used high-resolution electron microscopy (HREM) and lattice imaging to explore the various bonding

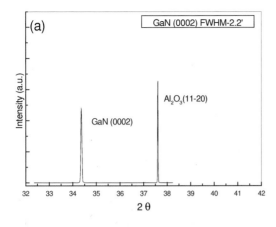

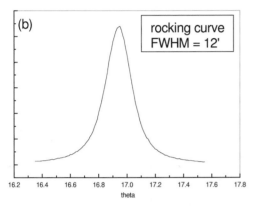

Fig. 35. (a) XRD pattern of a GaN film grown on A-plane sapphire, (b) rocking curve around the (0002) peak.

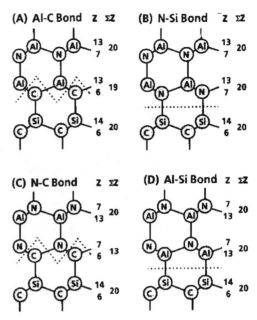

Fig. 36. Four possible atomic bonding configurations at the AlN/SiC interface. (From Ponce, VandeWalle, and Northrup [165], copyright 1996, the American Physical Society.)

possibilities at the SiC/AlN interface [165, 166]. The four possible bonding configurations for the atomically planar and abrupt interface are shown in Figure 36. The atomic numbers and the corresponding sum are shown at the right of each configuration. Based on these observations and the electronegativity values for Al, Si, C, and N (1.5, 1.8, 2.5, and 3.0, respectively), it was concluded that Si–N and Al–C bonds are preferred. Therefore, an Si-terminated SiC substrate generates epitaxy with Al in the top position of the basal plane (Al-polarity). The preferred bonding sequence is shown in Figure 36b.

Several nucleation and buffer layers have been investigated to improve the crystalline quality of the epitaxial films on 6H-SiC. The crystallinity of the epitaxial GaN is improved by initially flashing the SiC substrate with Al or Ga and exposing to the nitrogen precursor [129, 158]. This prevents the formation of amorphous silicon nitride, which may complicate epitaxy of good crystalline GaN. The process results in surface reconstruction and also establishes the polarity of the GaN epilayers.

It is possible to grow good quality GaN films directly on 6-H SiC without the use of any buffer layers, as observed by Korakakis [129] and Doppalapudi [110]. The substrates were dipped in HF solution prior to introduction into the MBE chamber. The wafers were heated to 900°C for a short time to desorb hydrogen prior to the GaN growth at 800–825°C. Growth was initiated by spraying a monolayer of gallium and exposing it to

nitrogen plasma to form GaN. The films showed a 2× RHEED reconstruction, indicating Ga-polarity. A cross-sectional TEM micrograph of a 160-nm thick GaN film grown without a buffer layer is shown in Figure 37. The micrograph shows a sharp interface and the high density of threading dislocations. Most of these dislocations are 1/3 [11$\bar{2}$0] edge dislocations. Selected area electron diffraction studies indicate that the GaN film is a single crystal with an epitaxial orientation relationship with SiC given as

$$(0001)_{GaN} \| (0001)_{SiC} \quad \text{and} \quad [11\bar{2}0]_{GaN} \| [11\bar{2}0]_{SiC}$$

The threading dislocation density in this film was measured to be 2–4×10^9 cm^{-2} near the top of the film. Typically this density diminishes with thickness due to dislocation interaction and annihilation. Most researchers report 10^{11}–10^{12} cm^{-2} dislocations near the interface and 10^8–10^9 cm^{-2} defects at a distance greater than 1 μm from the interface [173, 174]. The dislocation densities in the films grown in this study are comparable to these numbers. It is believed that the density would be considerably reduced in thicker films. In addition to threading defects, some dislocation loops are also seen in these films (marked by arrows), especially closer to the interface. There was no evidence of cubic phases and planar defects stacking mismatch boundary (SMB or pinholes) in this material. There are fewer stacking faults (none in this micrograph) than in the films reported previously [171]. The film described previously was also investigated by high-resolution electron microscopy (HREM). Figure 38 shows the presence of a sharp interface between SiC and GaN, and some steps are seen, as marked by arrows. In addition, certain amorphous regions are shown in the figure, suggesting imperfect surface preparation. A few stacking faults (SF) are also identified near the interface.

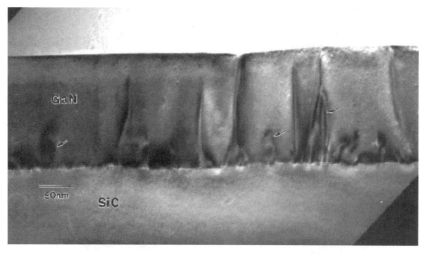

Fig. 37.   TEM micrograph of a GaN film on SiC, imaged at the $[1\bar{1}00]_{GaN}$ zone axis.

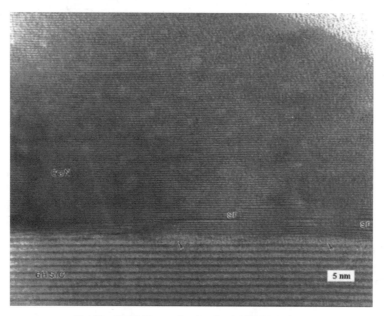

Fig. 38.   HREM image showing the GaN/SiC interface.

Although the preceding observations indicate that high-quality GaN films can be grown directly onto SiC substrates without using a buffer layer, it should be noted that a buffer layer may indeed be necessary to accommodate the thermal stresses.

Based on X-ray studies, Lantier et al. found that AlN buffers resulted in better quality GaN epilayers, than when GaN buffers were used [160]. The in-plane lattice mismatch between SiC and GaN is ~3.5%, where the mismatch between AlN and SiC is just 1%. The use of an AlN buffer, therefore, has obvious advantages. With optimized growth conditions, a very sharp SiC/AlN interface can be obtained, with no amorphous or polycrystalline features [174]. Although the lattice mismatch between GaN and AlN is ~2.5%, the epitaxy becomes considerably simpler, since the two lattices share identical crystal symmetry. The dislocation density near the GaN/AlN interface is high (~10^{12} cm^{-2}), but decreases rapidly on going away from the interface.

Use of AlGaN buffer layers should improve the epitaxy considerably, since they have a better thermal mismatch with SiC. Bremser et al. deposited AlGaN layers by organometallic vapor phase epitaxy (OMVPE) directly on both on-axis and off-axis 6H-SiC substrates and obtained smooth films [167]. Lin et al. experimented with several types of buffer layers including GaN, AlN, AlGaN to grow good quality GaN films by MOCVD on 6H-SiC substrates [168]. Based on Hall measurements and X-ray rocking curves, they concluded that a three-period GaN/Al$_{0.08}$Ga$_{0.92}$N (100 Å/100 Å) buffer stack facilitated the best GaN epilayers growth.

Smith et al. reported that high-quality GaN films can be grown on off-axis SiC substrates without the use of any buffer layers [171]. There is evidence that the substrates cut off a major crystallographic plane facilitate good epitaxial growth of compound semiconductors [169, 229], provided the substrates are sufficiently cleaned and etched. The surface of the off-cut substrates facilitates the nucleation of the epilayers at the closely spaced steps and promotes step-flow growth (Frank–van der Merwe mode). The steps also reduce the defect density by inhibiting the formation of antiphase domain boundaries (APBs). On a perfect substrate, any small tilt or roughness is in the form of monoatomic steps, which results in APBs. In an off-cut substrate, pairs of steps combine to form double height steps, thus preventing the formation of APBs.

Several groups have used SiC substrates for GaN growth, with and without using AlN buffers, and have analyzed the defect structure [170–174]. In addition to threading dislocation, pinholes [170], stacking mismatch boundaries [171, 173], stacking faults [172], and cubic phases [171] have been reported in these GaN films. In general, the epitaxial GaN films grown on SiC substrates are very defective, although the density of these defects is smaller than in the case of the films grown on sapphire and silicon. The general characteristics and effect of these defects are discussed in the next section.

3.2.1.3. Growth on Silicon (111) Substrates

Optimization of III-nitride growth on silicon substrates is essential to accomplish integration of Si electronics with III-nitride based devices such as LEDs and laser diodes (LDs). In addition, one could take advantage of the relatively inexpensive and well-established silicon processing technologies. GaN grown on (100) Si is predominantly cubic (zinc-blende structure), whereas the layers grown on (111) Si have a predominantly wurtzite structure. Although silicon has a cubic crystal structure, the (111) plane has a similar in-plane lattice symmetry to that of (0001) hexagonal GaN. Lei, Ludwig, and Moustakas reported that GaN films grown by MBE on (111) Si substrates have predominantly wurtzite structures, with a high density of stacking faults and cubic inclusions [111]. XRD and TEM studies of these samples confirmed that the zinc-blende phase constituted almost of 25% of the film [192]. Electron diffraction studies confirmed the alignment of the closest packed planes and directions,

$$(0001)_{w\text{-GaN}}//(111)_{z\text{-GaN}}//(111)_{\text{Si}}$$
$$[11\bar{2}0]_{w\text{-GaN}}//[\bar{1}10]_{z\text{-GaN}}//[\bar{1}10]_{\text{Si}}$$

Basu et al. [192] also observed the presence of wurtzite grains that are misoriented by a 30° twist along the [0001] axis with the rest of the wurtzite film. The authors predicted a lower interfacial energy than a random interface boundary, to explain the presence of these misoriented domains. Growth models of GaN thin films in hexagonal and cubic symmetry on (111) Si substrates is presented by Ohsato and Razeghi [175].

Commercially available (111) silicon substrates generally have excellent crystalline quality and surface finish. However, these wafers typically have a thin layer of native oxide on the surface. Therefore, the silicon wafers are etched in an HF solution before inserting them into the growth chamber. The high

Fig. 39. Plan-view bright-field TEM micrograph of a GaN film grown on (111) Si. Stacking faults, F, indicate the presence of cubic domains. The diffraction patterns indicate a highly oriented crystal symmetry. (From Basu, Lei, and Moustakas [192].)

lattice mismatch between silicon and GaN (17% as seen in Table II) necessitates the use of buffer layers to achieve good epitaxial layers. Low-temperature GaN buffers have been used to successfully minimize the zinc-blende inclusions. Nakada, Aksenov, and Okumura reported the growth of superior quality GaN films by MBE on nitrided Si (111) substrates compared to films grown directly on silicon substrates [176]. The nitride layer is about a monolayer thick in these experiments. PL and TEM studies show that good quality GaN films were obtained in spite of the amorphous silicon nitride layer. The epitaxial relation was found to be GaN [0001]//Si [111] and GaN (11$\bar{2}$0)//Si (1$\bar{1}$0). However, longer nitridation results in a thicker amorphous silicon nitride layer, which leads to poor epitaxy.

Figure 39 shows a plan-view bright-field micrograph of a GaN film grown on (111) silicon. Stacking faults (marked F) are seen, which lead to localized zinc-blende regions [192]. The microstructure consists of highly oriented domains, as seen by the inset electron diffraction. The presence of domains and the zinc-blende phase in these films was confirmed by dark-field TEM and cross-sectional studies. Ohtani, Stevens, and Beresford have grown GaN films on (111) silicon by ECRMBE using AlN buffers [177]. They performed X-ray diffraction studies to show the lattice mosaic and grain size improve with an increase in growth temperature.

3.2.1.4. Growth on Other Substrates

In addition to the foregoing text, several researchers have attempted GaN growth on ZnO, LiGaO$_2$, LiAlO$_2$ [178, 179], and GaAs. The (100) face of cubic LiAlO$_2$ has an excellent lattice match with (1$\bar{1}$00) GaN with only a 1.4% mismatch in the c-direction and 0.1% along the b-direction. However, the epitaxy is not straightforward since the lattice symmetry in the two crystals is very different. Hellman, Weber, and Buchanan observed that GaN grows in the (0001) direction with a tilt to accommodate the misfit [179]. Kryliouk et al. demonstrated high-quality GaN film growth with less than 10^7 cm^{-2} dislocation density on LiGaO$_2$ substrates by MOCVD [180]. The authors reported GaN growth in the {0001} orientation on the {001} substrate and the {1$\bar{1}$02} orientation on the {101} substrate. However, there is a high density of defects in the initial stages of the growth, probably due to the poor thermal stability of LiGaO$_2$ at the growth temperatures. Based on PL studies, Andrianov et al. reported the presence of cubic and wurtzite phases in the GaN grown on (001) LiGaO$_2$ by MBE [181].

ZnO has been investigated as a suitable substrate because of its close lattice match (1.9%) and its in-plane symmetry and stacking order match with the III-nitrides [110, 182–184]. In addition, by subsequent etching of the ZnO layer, free-standing GaN films can be obtained. Strained GaN layers can be pseudomorphically grown up to a thickness of ~100 Å on ZnO [184]. The importance of ZnO as a potential substrate for the growth of III–V nitrides is due to the possibility of developing lattice-matched heterostructures of In$_x$Ga$_y$Al$_{(1-x-y)}$N materials. Assuming Vegard's law, the composition of the alloy can be designed such that lattice parameter is matched to that of ZnO (3.25 Å), thereby reducing the defect density considerably. Such heterostructures span the spectral region from 2.8 to 4.5 eV. Since ZnO is polar like GaN, it is expected that films of a single polarity, determined by the polarity of the ZnO substrate, will be obtained. Since the steps on 2H-ZnO would be bilayer steps, stacking mismatch boundary (SMB) type of defects are less likely occur. Hamdani et al. reported that best quality GaN can be obtained on the oxygen face of the ZnO crystal [184]. In fact, an early investigation carried out by Sitar et al. (before the introduction of buffer layers) showed that the best quality GaN epilayers were obtained on ZnO substrates, when GaN was grown by MBE on different substrates including SiC, sapphire, Si, TiO$_2$, and ZnO [185].

Ideally, heteroepitaxial III–V nitrides should be grown on single-crystal ZnO substrates. However, such substrates are not available at reasonable sizes and cost. Several researchers have used polycrystalline ZnO layers on Si or sapphire crystals as the substrates for the GaN growth. However, such ZnO layers thermally dissociate at the GaN growth temperatures, as evidenced by RHEED observations [110]. The films grown with the AlN buffer were very rough, presumably due to the larger lattice and thermal mismatches between AlN and ZnO and/or due to poor crystallization of the AlN buffer. A low-temperature GaN or In-GaN buffer layer leads to smoother films with improved optical properties [110, 184]. However, it was observed that the epitaxial layers tend to delaminate during cool down and a postgrowth treatment with a slow ramp down from the growth temperature were necessary to prevent the spallation.

3.2.2. Growth of GaN in Zinc-Blende Structure

GaN films with a zinc-blende structure have been grown on a variety of cubic substrates including β-SiC [186, 187], GaAs [188, 189], MgO [190], and Si (100) [111, 174, 191, 192].

In the following, growth on Si (100) substrates is discussed, as these are the most commonly used substrates.

GaN films with a zinc-blende structure have been grown by the MBE method on both n- and p-type (001) silicon substrates [191]. As described for (111) silicon, (001) silicon substrates are cleaned and etched in HF solution to remove surface oxide prior to growth. As with the wurtzite films, single-crystal zinc-blende thin films were obtained using a two temperature-step growth process. Deposition of the buffer layer in the zinc-blende structure is required, to stabilize the epitaxial growth in the cubic phase. When the buffer layer is either amorphous or polycrystalline, the subsequent GaN film would grow in the wurtzite structure, which is thermodynamically a more stable phase. Lei et al. used RHEED studies to determine that an unreconstructed (001) silicon surface (i.e., 1×1) facilitates the formation of a single-crystalline GaN buffer in the zinc-blende phase [191]. Specifically, a GaN buffer layer of 300–900 Å was deposited at 400°C. The RHEED patterns from these buffers indicate that the layers have a zinc-blende structure with the [001] direction perpendicular to the surface. Although the buffer layers were defective and rough (indicated by broad spotty RHEED), the epitaxial GaN films grown at 600°C were smooth (indicated by streaky and sharp RHEED). The zinc-blende structure of the films was also confirmed by *ex situ* X-ray diffraction and TEM studies [191, 192]. When the buffer layers are deposited directly at the higher temperature (600°C), the GaN is polycrystalline with the wurtzite (or mixed) structure. Figure 40 compares the surface morphology of GaN films grown on p-type (001) silicon with and without the low-temperature buffer layer.

The two-step growth method promotes the layer-by-layer growth and improves the crystallinity. The lower temperature for the buffer layer is justified by the heteroepitaxial nucleation of GaN and the kinetics for formation of critical nuclei. At the low temperature employed, the sticking coefficient of the atomic species is high and a continuous film of GaN is quickly formed. At higher temperatures, the adatoms have a very short lifetime before they evaporate, leading to incomplete surface coverage and formation of three-dimensional clusters. However, if the temperature is too low, the surface mobility of the adatoms is limited, resulting in metallic clusters and/or amorphous regions. If a single-crystal zinc-blende buffer layer is successfully developed, growth at the second stage is homoepitaxial (which has no energy barrier) and a smooth GaN growth should follow in the layer-by-layer mode. Moustakas, Lei, and Molnar analyzed the SEM micrographs of the GaN films grown by the two-step method and estimated 2 orders higher growth rate in the lateral direction compared to the direction perpendicular to the surface [120]. This data clearly indicate that the two-step growth method leads to a quasi-layer-by-layer growth mode. The condensing material must reach the growth ledges by surface diffusion before new growth islands nucleate, for smooth step-flow growth. The second step growth should therefore be carried out at high enough temperatures to enable adatoms to diffuse to their equilibrium configuration and to improve the crystallinity of the film. Terraces on well-oriented

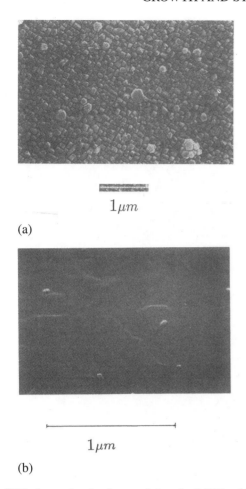

(a)

$1\mu m$

(b)

$1\mu m$

Fig. 40. SEM micrographs of surface morphology for GaN films (a) grown in one step at 600°C, and (b) grown in two steps with a low-temperature buffer at 400°C followed by a high temperature growth at 600°C. (From Lei et al. [191].)

surfaces are typically 100 to 1000-atoms wide; therefore step-flow growth at the rate of 0.1–1.0 monolayers sec^{-1} requires a diffusion coefficient of $D_s > 10^{-7}$ cm^2 s^{-1} [124]. However, too high temperatures are not desirable to epitaxy due to roughening transition. Lei et al. observed that growth at 800°C led to a very rough GaN film [191].

In addition to the nucleation process, the surface morphology of the GaN films are also influenced by the substrate (p- vs n-types) and the III/V (Ga/N) flux ratio. Lei et al. reported that under identical conditions, GaN films grown on n-type silicon substrates were smoother than those grown on p-type silicon and suggested that GaN wets n-type silicon better. This is in agreement with the observations of Morimoto, Uchiho, and Ushio, who reported that in vapor phase growth, GaN adheres to n-type silicon, but not to p-type silicon [193]. Lei et al. also reported the growth of smoother GaN films under low nitrogen pressure (or high Ga/N flux ratio).

Basu, Lei, and Moustakas investigated the GaN films grown on (001) silicon by TEM studies and found that the films were predominantly cubic (zinc blende) with (001) texturing [192]. However, a high density of defects including stacking faults, microtwins, IDBs, and hexagonal (wurtzite) domains were ob-

served. The majority of the stacking faults, microtwins, and hexagonal domains is formed along the {111} planes. Such faults are known to occur in face-centered cubic (fcc) materials with low stacking fault energy. In addition to the hexagonal regions formed due to the stacking faults on (111) zinc-blende GaN, domains of GaN that nucleated in the wurtzite phase were observed. Figure 41 is a high-resolution electron micrograph of a GaN/(100) Si interface showing hexagonal regions (H) and microtwins (T) and stacking faults [192]. The inset selective area diffraction (Fig. 41b) shows that there is a (001) texturing with a significant number of misoriented grains. Some of the twin defects are magnified in the inset, in Figure 41c. It is also clear from the figure that the defect density is considerably smaller away from the interface. Misfit dislocations were observed at the interface.

Several innovative methods have been explored to convert the silicon surface better suited for GaN epitaxy. For example, Hiroyama and Tamura carburized the Si (001) substrate to obtain a pseudo-SiC substrate, which has a better lattice match with cubic GaN [194]. The GaN films grown by gas source MBE on such substrates had a negligible wurtzite component and had a zinc-blende structure with (002) GaN parallel to the Si (001) surface. The films had PL spectra with a near-edge peak at 381 nm with an FWHM of 5.9 nm indicating good optical properties.

4. DOPING OF III-NITRIDES

The doping of wide bandgap semiconductors is thermodynamically difficult to achieve because the shift of the Fermi level toward the conduction or valence band upon doping results in the charging of native defects negatively or positively, which lowers their energy of formation and thus increases their concentration. Therefore, the Fermi level tends to stay in the middle of the gap [195]. Although this phenomenon is general to all semiconductors, it is particularly important to wide bandgap semiconductors because the reduction of the energy of the formation of the charged defects is equal to the shift of the Fermi level upon doping which may be several electron volts.

4.1. Unintentionally Doped Films

In the early stages of the field, the GaN films produced by the HVPE method were generally autodoped n-type with carrier concentration between 10^{18} and 10^{20} cm^{-3}. This n-type autodoping was initially attributed to nitrogen vacancies. The energy of formation of nitrogen vacancies is lower than that of Ga interstitials. Subsequent tight-binding approximation calculations [196] show that nitrogen vacancy is indeed a potential shallow donor in GaN. However, calculations of Neugebauer and Van de Walle [197] show that the energy of formation of nitrogen vacancies is 4 eV and therefore will not be present at a sufficiently high concentration at room temperature, to explain the large background electron concentration. *Ab initio* calculations show that the donor state contributed by nitrogen

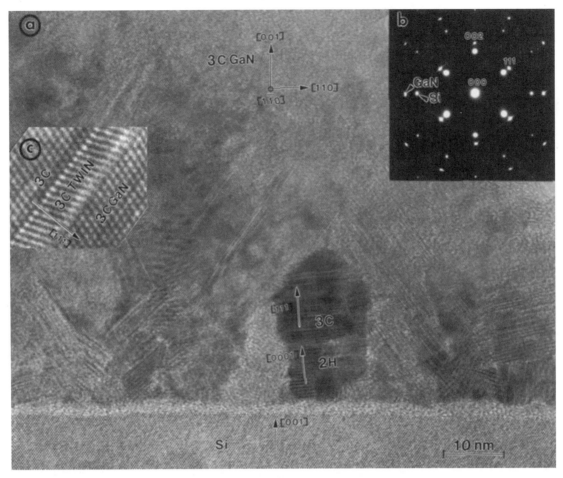

Fig. 41.   High-resolution TEM micrograph of the GaN buffer on (001) Si. The wurtzite regions and the microtwins are marked by *H* and *T*, respectively. The inset electron diffraction shows (001) texturing with a significant number of misoriented grains. A magnified region of the micrograph shows the presence of misfit dislocations at the interface. (From Basu, Lei, and Moustakas [192].)

vacancies is resonant with the conduction band and under high pressure the states move below the conduction band edge [198].

Some researchers also attribute the unintentional *n*-type doping to oxygen impurities entering into the nitrogen sublattice [199]. Oxygen during the HVPE growth can be the result of either atmospheric leaks or contamination from the quartz tube or from water impurity in ammonia. The role of oxygen as an *n*-type donor was demonstrated by Chung and Gershenzon [200]. These authors doped GaN with oxygen by introducing water vapor in the reaction chamber and observed a systematic increase in the electron concentration with water vapor pressure. High-purity ammonia has been developed by a number of vendors to address the issue of water impurities in ammonia used for HVPE, MOCVD, and MBE methods.

4.2. Films with *n*-Type Doping

Controlled *n*-type doping of GaN was accomplished by the incorporation of silicon during the growth. During MOCVD growth the precursor gas is silane, disilane, or tetraethyl silane.

In MBE growth, silicon is provided by evaporation using a standard Knudsen effusion cell. Figure 42 is an Arrhenius plot showing the carrier concentrations in a larger number of Si-doped GaN films. By varying the silicon cell temperature between 1000 and 1400°C, the carrier concentrations ranging from 10^{16} to 3×10^{19} cm^{-3} were deposited. From the plot, an activation energy of 4.5 eV is calculated, which is close to that of Si (4.2 eV), implying that the doping is primarily due to the incorporation of Si, and not due to impurities. Figure 43 is a plot of electron mobility vs electron concentration, estimated by Hall effect measurements for a large number of samples [201]. The electron mobility decreases with an increase in electron concentration, due to impurity scattering. Gaskill and co-workers [202] pointed out that, because the data for the unintentionally doped samples (open circles) are very similar to the data for intentionally doped samples (closed circles), the magnitude of compensation for doped and unintentionally doped samples is approximately the same. Temperature-dependent Hall effect measurements can be used to determine the activation energy of the Si-donor. Early work [203, 204], suggested activation energy of about 27 meV.

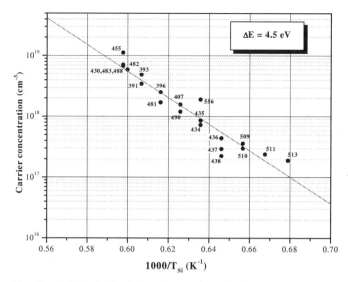

Fig. 42. Variation in the doping concentration with Si temperature in MBE grown GaN films.

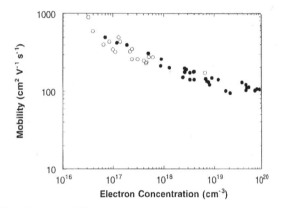

Fig. 43. Electron mobility vs carrier concentration in *n*-doped GaN films. (From Doverspike and Pankove [201].)

More recent investigations suggest that the real activation energy is between 12–17 meV [205, 289].

Silicon-doping of GaN films grown on sapphire appears to affect the stress in the films. Heavily Si-doped films develop cracks above certain critical thickness [206]. This critical thickness is larger for lightly doped films. In addition, Si-doping appears to affect the dislocation density. Ruvimov et al. [207] reported that doping GaN with silicon to the level of 3×10^{18} cm^{-3} decreases the dislocation density from 5×10^9 cm^{-2} (in undoped GaN films) to 7×10^8 cm^{-2}. The effect of dislocation on the mobility in *n*-doped GaN films is discussed in the next section.

Finally, it should be mentioned that Si-implantation with subsequent annealing resulted in *n*-GaN films with carrier concentration 4×10^{20} cm^{-3}. Burm et al. [208] reported that contact resistivity to such films is as low as 3.6×10^{-8} Ω cm^2. Nakamura, Mukai, and Senoh [209] reported *n*-type doping of GaN films with Germanium, to doping levels as high as 1×10^{19} cm^{-3}. This was done by flowing GeH$_4$ during MOCVD growth. However, it appears that the doping efficiency is about 1 order of magnitude smaller.

4.3. Films with *p*-Type Doping

GaN films can be doped *p*-type by incorporating group II elements in the sublattice. Such impurities include Zn, Cd, Mg, and Be, In the 1970s, GaN was doped *p*-type with Zn and the results indicate that Zn is a deep acceptor. Pankove provided evidence [210] that Zn incorporates in the Ga-sublattice giving rise to an acceptor level 0.21 eV above the valence band, and also incorporates in the N-sublattice, where it acts as a triple acceptor. There is one report of doping GaN *p*-type with carbon [211]. Carbon is expected to be an acceptor if it substitutes for N in GaN. Such films were reported to have hole concentration of 3×10^{17} cm^{-3} and mobility 103 cm^2 V^{-1} s^{-1}.

p-Type doping using magnesium has been routinely used over the past several years for the growth of GaN by either MOCVD [212, 213] or MBE [15]. As-grown Mg-doped GaN films by MOCVD are semi-insulating and some postgrowth treatment is required to activate the acceptors. This is due to hydrogen passivation of Mg-acceptors by the formation of Mg-H complexes. The phenomenon of *p*-type dopant passivation by hydrogen has generally been observed in many semiconductors [214]. The acceptors activation in Mg-doped GaN was reported first in 1989 by Amano et al. [212] who used low energy electron beam irradiation (LEEBI) treatment. Initially, it was assumed that local heating during LEEBI treatment breaks the H–Mg bonds. However, Li and Coleman [215] carried out the LEEBI treatment at low temperatures and concluded that the dissociation of Mg-H complexes by this method is not thermal. Subsequently, an alternate approach of Mg-activation by thermal annealing in a nitrogen environment was proposed by Nakamura et al. [216]. The annealing was done typically between 700 and 900°C.

GaN-doping with Mg by MBE was first reported by Moustakas and Molnar [15]. According to these authors, the as-grown Mg-doped GaN films were *p*-type without any postgrowth treatment. In fact, *p*-GaN films with free hole concentration of 6×10^{18} cm^{-3} were obtained. Figure 44 shows the dependence of the resistivity vs Mg temperature for two series of films grown at 750 and 700°C, respectively [130]. It is clear from the figure that both the substrate temperature and the Mg cell temperature influence the incorporation of Mg in the films. At lower Mg temperatures, the incorporated Mg is not sufficient to overcompensate the native defects or impurities, which results in a net *n*-type doping. At higher Mg-cell temperature, the incorporation of a high concentration of Mg may lead to Mg-clustering. There is also evidence that Mg-doping concentration can be increased by 2 orders of magnitude by expanding the C-lattice parameter from 5.16 to 5.22 Å [241]. This is achieved by engineering the residual strain by the control of growth parameters such as III/V flux ratio and buffer layer characteristics. Furthermore, high-pressure growth of GaN from Ga-Mg solutions have led to new compounds of Ga-Mg-N with different optoelectronic properties [217]. The effect of substrate temperature seen in Figure 44 can be accounted for by the reevaporation of Mg at higher growth temperatures. This conclusion is also consistent with the observations of Kamp et al. [218].

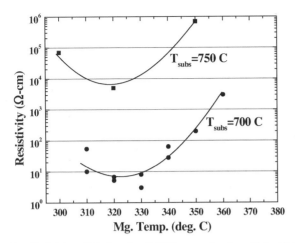

Fig. 44. Resistivity of Mg-doped p-GaN films as a function of Mg cell temperature. (From Ng et al. [130] with permission from Elsevier Science.)

It is understood that Mg in GaN films grown by MOCVD is facilitated by the formation of Mg-H complexes (which are later activated to generate acceptors). In the absence of hydrogen in plasma-assisted MBE growth of GaN, it has been proposed that Mg-incorporation is aided by the electron flux arriving at the substrate from the ECR source [130, 85]. Under such conditions of growth, the Mg dopants enter the lattice in their charged state, which require much less energy than the formation of neutral acceptors. To investigate the role of hydrogen in MBE grown films, Brandt et al. [219] conducted a study of postgrowth hydrogenation of Mg-doped GaN films The results show that posthydrogenation leads to a reduction in carrier concentration by more than an order of magnitude, which is attributed to the formation of Mg-H complexes. Such complexes were observed by IR and Raman scattering measurements [220].

The luminescence properties of Mg-doped GaN films appear to be different in films grown by the MOCVD and MBE methods. Details on these studies can be found in the articles by Doverspike and Pankove [201] and by Moustakas, respectively, [85].

4.4. Doping of InGaN and AlGaN Films

As-grown InGaN alloys are generally autodoped n-type. When such films are doped with silicon, their PL intensity increases by more than a factor of 30 [221]. In the early generation LEDs, the InGaN active region was codoped with Si and Zn to enhance donor–acceptor transitions. Yamasaki et al. also reported p-type doping of $In_{0.09}Ga_{0.91}N$ films [222].

Unlike GaN and InN, carrier concentration and mobility in AlN are very low, and undoped AlN is typically semi-insulating. Devices such as LEDs and laser diodes use cladding regions of AlGaN layers. Therefore, it is important to be able to dope $Al_xGa_{1-x}N$ n-type as well as p-type. In general, $Al_xGa_{1-x}N$ becomes more difficult to dope as the AlN mole fraction increases. It was found that as x increases from 0.4 to 0.6, the electron concentration and mobility decrease by up

to 5 orders of magnitude [223]. AlGaN films have been doped n-type with Si and have been doped p-type with Mg. The doping efficiency with both dopants was found to decrease with Al content [224–226]. The activation energy of Si dopant atoms in AlGaN is around 54 meV, compared to 17 meV in GaN samples.

p-Type doping of AlGaN samples was less successful. It was found that the activation energy of Mg in $Al_{0.075}Ga_{0.925}N$ was about 180 meV, compared to the value of 150 meV in GaN [225]. It has not been possible to dope $Al_xGa_{1-x}N$ films with Mg if $x > 0.13$. In $Al_xGa_{1-x}N$ films, formation energy of triply charged nitrogen vacancies is low. Such defects compensate the Mg-acceptors and decrease the hole concentration [223]. In addition, both experiments and theoretical calculations show that ionization energy of Mg increases as x in the $Al_xGa_{1-x}N$ layers increases [277].

5. STRUCTURE AND MICROSTRUCTURE OF EPITAXIAL GaN

The group III-nitrides (AlN, GaN, and InN) can crystallize in the wurtzite, zinc-blende, and rock salt structures. Among the three, the wurtzite structure is thermodynamically the most stable phase. The zinc-blende structure is metastable and may be stabilized by heteroepitaxial growth on substrates having cubic symmetry with appropriate growth conditions. It was found that GaN transitioned to a rock salt phase at high pressures. In this section, we concentrate on GaN films in the wurtzite structure, and therefore consider the other phases as defects. The crystal structure of the wurtzite GaN and the lattice orientation of the films near the interface with the various substrates have been discussed in the previous sections. In this section, the structure and the microstructure of the nitride epilayers and the various defects commonly observed are discussed.

Typically, GaN films grown on foreign substrates have a high density of numerous types of defects, which in other semiconductors, would make the material unsuitable for device fabrication. The defects observed in wurtzite GaN films include: (i) point defects—N and Ga vacancies, interstitials, and anitisite defects, (ii) line defects—misfit dislocations, threading dislocations (edge, screw, and mixed) and stacking faults (iii) three-dimensional defects—inversion domain boundaries (IDB), stacking mismatch boundaries (SMB), and zinc-blende domains. GaN based devices have been successfully developed in spite of an extended defect density that 5–6 orders higher than what would be an acceptable level in GaAs optoelectronics. This suggests the possibility that the majority of these defects are electronically inert and do not induce levels in the bandgap. We discuss the origin of the various defects observed, and we explore their influence on the electrical and optical properties of GaN films.

The major sources of the defects in nitride films are the high lattice and thermal mismatches with the substrates. The defects are generated to accommodate the lattice strain developed due to the mismatches. To minimize the defects in the nitride films, an understanding of the origin of the strain and the stress relief mechanisms is essential. Other sources of the defects in the epilayer are the presence of polytypes and the noncentrosymmetric nature of the GaN crystal structure which give rise to mixed phases and mixed polarities, respectively.

5.1. Residual Stresses

Epitaxial films may have two types of strain: (a) hydrostatic strain where the changes in the lattice parameters are from a high concentration of point defects and (b) biaxial strain due to lattice and thermal mismatches with the substrate. In this section, we discuss the biaxial strain (and stresses) in III-nitride films grown on sapphire and SiC, the origin of the strain and its implications to the properties of the films.

Epitaxy can be defined as the growth of a single-crystal A on the surface of a single-crystal B with a unique crystallographic orientation and with well-defined atomic bonding near the interface. The atoms near the interface have to accommodate the periodicity of both the substrate B and the epilayer A. In the case of heteroepitaxial growth, there is a misfit (f) between the two layers of similar structure, quantified from the difference in lattice parameters as

$$F = (a_A - a_B)/a_{avg}$$

In general, the epitaxial layer can be grown pseudomorphically, provided the misfit is sufficiently small. In a perfect crystal, this misfit is accommodated by homogeneous misfit strain, if the thickness of the epitaxial layer is smaller than a *"critical thickness"*. The epitaxial film in this case is said to be coherent or pseudomorphic with the substrate. Above this *critical thickness*, lattice strain is partially relaxed by the introduction of misfit dislocations near the interface. Stress relaxation by dislocation–defect generation and glide requires some minimum temperature. The brittle to ductile transition in crystals typically occurs around 0.4–0.6T_m, where T_m is the melting point of the crystal. However, the material can deform plastically at a lower temperature, particularly if the induced stresses are high. The melting point of GaN is ∼2800 K. MOCVD growth is carried out close to the transition temperature, while the MBE growth is typically below the transition range. Therefore, low-temperature growth techniques such as MBE can produce coherent films well beyond the critical thickness. Large lattice-mismatch epitaxy and the generation of misfit dislocations have been critically reviewed by Mahajan and Shahid [228] and van der Merwe [229].

In the case of large lattice-mismatch systems such as GaN/sapphire, it is not possible to elastically strain the film to the substrates even for a monolayer. In this case, there is approximate matching between m-planes of the film and n-planes of the substrate. Both sets of planes are perpendicular to the interface and $m = n + 1$ if the film has a smaller lattice parameter or $m = n - 1$ if the film has a larger lattice parameter than the substrate [281]. Such lattice matching can be seen in the atomic models of the interface shown in Figures 16, 17, and 19. In the case of GaN/C-sapphire, seven planes of $\{10\bar{1}0\}_{GaN}$ line

up with eight planes of $\{11\bar{2}0\}_{sapphire}$. Similarly, in the GaN/A-sapphire system, three planes of $\{10\bar{1}0\}_{GaN}$ match with two planes of $(10\bar{1}0)_{sapphire}$. The extra half-plane on one of the sides of the interface, a "geometrical misfit dislocation", accommodates only a part of the lattice mismatch. There is a "residual lattice strain" in the film, which builds up with the thickness. The residual strain in GaN/C-plane sapphire is ∼1% [281]. At a critical thickness, this strain may be relieved by the generation of more dislocations. However, if the temperature is not high enough, the residual strain and stresses remain in the films.

Thermal stresses are another important source of strain in the nitride epitaxial films. Sapphire has a coefficient of thermal expansion (CTE) of 8.5×10^{-6} K^{-1} in the c-direction and 7.5×10^{-6} K^{-1} in the (0001) plane. Similarly, 6H-SiC has a CTE of 4.68×10^{-6} K^{-1} in the c-direction and 4.2×10^{-6} K^{-1} in the (0001) plane. Thermal mismatches are expected based on the difference with the nitride CTE values listed in Table I. Obviously, thermal stresses vary significantly with the deposition temperature and the growth method and are generated as the films are cooled from the growth temperature (stress-free condition) to the room temperature. Therefore, these stresses are likely to be retained in the film after the growth unlike lattice-mismatch stresses.

There have been several studies of strain in GaN epilayers [230, 231], as well as strain in AlGaInN alloys grown on GaN [230, 232]. The residual stresses in GaN films have been calculated by X-ray diffraction studies [233], wafer bending measurements (for thick films) [234], as well as indirect methods including photoluminescence and Raman measurements [233]. An understanding of the origin of the residual strain and its control is essential to accurately design the bandgap of the optical devices. The strain in GaN samples varies significantly depending on the substrate used, as can be expected from the differences in the lattice and thermal mismatch between the epilayers and the substrates. Cheng et al. investigated strain in GaN epitaxial films grown on sapphire, SiC, and GaN substrates, using X-ray diffraction PL measurements [235]. They reported significant differences in the lattice parameters and the mosaic spread in the films. While the mosaic spread is a measure of inhomogeneous strain in the epilayer, the change in the lattice parameter is due to residual strain in the film resulting from lattice and thermal mismatches between the epilayer and the substrate. Cheng et al. reported an increase in the GaN c-lattice parameter in the case of the sapphire substrate and a decrease in the case of the SiC substrate. The data translate to compressive and tensile stresses in the GaN films grown on sapphire and SiC, respectively, consistent with the notion that thermal mismatch is the predominant cause for stresses. High tensile stresses in thin films generally cause cracking and therefore, it is critical to minimize the residual stresses in the films grown on SiC substrates. By using an AlN buffer, the net residual tensile stress can be considerably reduced, preventing the cracking.

Ning et al. concluded that the GaN films grown on sapphire substrates by MOCVD were under residual biaxial compression based on HREM studies [281]. The residual stresses in GaN films grown on sapphire are compressive in nature and can be as high as 1.5 GPa [241]. Lee et al. reported that such stresses induce a blueshift in the bound exciton peak energy with a linear coefficient of 42 meV GPa^{-1} [233]. The effect of stresses on the optical properties of the epitaxial films have been discussed in detail by Gil [236].

Amano et al. have shown that both AlGaN and InGaN ternary alloys can be grown pseudomorphically onto GaN layers based on X-ray diffraction studies around the $(20\bar{2}4)$ reciprocal lattice peak [232]. As expected from the lattice constants ($c_{InN} > c_{GaN} > c_{AlN}$), AlGaN is under tensile stress and InGaN is under compressive stress. In these studies, the thickness of $Al_xGa_{1-x}N$ ranges between 0.35–0.65 μm, while x was varied between 0 and 0.25. Korakakis, Ludwig, and Moustakas have carried out similar X-ray diffraction studies on 0.4 μm thick $Al_{0.25}Ga_{0.25}N$ grown on GaN by MBE and found that the AlGaN layer is relaxed [83]. This discrepancy between the results is surprising especially since the deposition temperature is considerably lower in the MBE method, compared to the MOVPE method. One possible reason for the difference is the much lower growth rate in the MBE method.

Apart from the mismatch with the substrate and the growth temperature, there are several other factors that affect the residual stresses. Some of the factors reported are: (i) buffer material, (ii) buffer thickness, (iii) buffer growth temperature, (iv) V/III flux ratio, and (v) doping concentration [241]. The buffer layer plays an important role in determining the residual stresses since the stress relaxation occurs mainly in the buffer region, which has a high density of dislocations and other defects. The epitaxial films grown with AlN buffers are expected to be more strained since stress relaxation is less likely in the AlN buffer, which is more refractory compared to the GaN buffer. A wide range of AlGaN buffers were investigated for the growth of GaN films on 6H-SiC and it was found that the residual stresses vary with the thickness and the composition of the buffer. The AlGaN buffer layer with 0.3-AlN mole fraction and 3000 Å thickness resulted in GaN films with the smallest residual stress. In general, the stresses decrease with an increase in GaN buffer thickness [241, 237]. It has also been reported that an increase in the III/V ratio reduces the strain relaxation; i.e., films are more strained [238, 239]. This is attributed to the reduction of strain-relieving defects with an increase in the III/V ratio. Lee et al. carried out systematic studies on the effect of Si-doping concentration on bound exciton peak energy and FWHM of X-ray $(10\bar{1}5)$ diffraction peak [233]. They concluded that Si-doping reduces the residual stresses by introducing defects. This is in contradiction with results by Ruvimov et al., who reported a reduction in dislocation density with Si-doping [240]. Other reports have shown an increase in the stresses at high silicon doping [241]. Lee et al. also observed a direct correlation between relative intensity of the yellow luminescence (near 2.2 eV) and Si-doping and speculated that the Si-induced defects originate this yellow luminescence. However, it should be noted that there are other propositions for the origin of this yellow luminescence. Liu et al. attribute the yellow luminescence to the electron-hole recombination at positively charged Ga interstitials local to dislocations [242].

Based on the previous discussion, it is possible to design the epitaxial film growth parameters to achieve a stress-free film. However, such stress relaxation may lead to a very rough film. There have been a few attempts to use porous substrates to grow strain-free GaN films. The porous substrates, being compliant, accommodate the stresses generated by lattice and thermal mismatches [243, 244]. Mynbaeva et al. formed porous GaN templates from GaN layers grown on SiC by anodizing the layers in an HF solution under ultraviolet light excitation [244]. GaN films grown by HVPE on these templates had improved properties and a stress reduction by an order of magnitude. It should be possible to grow strain-free GaN films directly onto porous single-crystal SiC templates.

There has been an effort to grow III-nitride thin films onto free-standing micromachined Si and SiC structures. By depositing functional nitride films onto MEMS (microelectromechanical structures), arrays of individually controlled devices can be fabricated, which may be used to make detectors, emitters, or sensors. When epitaxial films are grown on free-standing MEMS such as microcantilevers or membranes, stresses in the films are relieved by appropriate deflection of the MEMS structure. Furthermore, by appropriate control of growth parameters, these stresses can be minimized. An example of high-quality AlN films grown on SiC microcantilevers is shown in Figure 45 [245]. The SiC cantilevers shown in the micrograph were fabricated by Boston MicroSytems Inc., using a proprietary etch technology [245]. It is clear from the flat cantilever structures, that the stresses are minimal in these films. Such films are also expected to have a lower dislocation density, due to the conformal nature of the substrate.

5.2. Polarity

Due to the noncentro-symmetrical structure of (0001) wurtzite, GaN can be either nitrogen or gallium terminated. As seen from the GaN structure in Figure 3, each (0001) biplane of GaN consists of a monolayer of Ga and N atoms, respectively. The polarity of the GaN film is determined by the stacking sequence of these Ga and N monolayers. By convention, the film is said to have (0001) polarity when the vertical bond is from a Ga atom to the N atom above it, while the $(000\bar{1})$ polarity implies that this bond points vertically from an N atom to a Ga atom above it. Films with (0001) polarity are also termed "Ga-polar", "Ga-face," or "Ga-terminated." Based on this notation, the unit cell shown in Figure 3 has a Ga-polarity. However, it should be noted that polarity is not a surface property as these terms may suggest, but instead is a bulk property.

Knowledge and control of polarity in GaN films is essential for the design and the fabrication of many electronic devices. Various analytical tools have been used to determine the polarity of GaN films. Although results based on X-ray photoemission spectroscopy (XPS), Auger spectroscopy, X-ray photoelectron diffraction (XPD), and ion channeling techniques have been reported, convergent beam electron diffraction (CBED)

Fig. 45. SEM micrographs of AlGaN/GaN multiple quantum well structures grown on free-standing SiC MEMS fabricated by Boston MicroSystems' proprietary technology. The micrograph on the left shows an array of flat cantilevers and the micrograph on the right shows the tip of a U-channel cantilever [245].

study in a TEM is the most direct and reliable technique for such a study. A critical review of these studies has been published by Hellman [246]. There is still considerable disagreement among the researchers on the assignment of polarity in GaN films. While Ponce et al. [103] report the smooth face in GaN substrates made by UNIPRESS crystal growers to be a Ga-face, Liliental-Weber [104] indicated it to be an N-face. Depending on growth conditions, both polarities were found in epitaxial GaN films grown by MOCVD and MBE. Smooth films grown on a low-temperature buffer were characterized as Ga-face, while films grown directly on sapphire, with rougher surface were N-face.

An indirect method of polarity determination is by chemical etching using KOH or NaOH solution. It has been reported that a rough GaN layer with nitrogen termination can be polished to an atomically smooth surface using a KOH solution [247]. After further etching, the surfaces become faceted due to a preferential attack of the domain boundaries. The Ga-face films reportedly etch slower, due to their higher stability. The relative stability of the Ga-face film is supported by the observation of faster growth rate in this polarity compared to the N-face, by a factor of 10–20 [246]. The polarity of the GaN films is indicated by this difference in etching characteristics of the two surfaces.

The RHEED pattern also provides insight into the specific polarity. Smith et al. [248, 249] performed first principles total energy calculations and correlated the results with RHEED observations to determine the temperature dependence of the surface reconstructions for GaN with both polarities. They demonstrated 1×1 reconstruction for [000$\bar{1}$] polarity, which would change to 3×3 or higher order reconstruction when cooled below 300°C. Figure 46 shows an example of such reconstruction at the [11$\bar{2}$0] zone axis at growth temperature (a) and at 250°C (b) [110]. Although some inversion domains exist (as discussed later), the contribution from these defects is relatively small and is not reflected in the RHEED pattern. Similarly, [0001] polarity films showed 1×1 reconstruction at growth temperature and 2×2 reconstruction below 600°C.

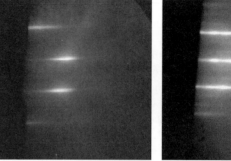

Fig. 46. RHEED pattern near [11$\bar{2}$0] from a typical GaN film at (a) growth temperature, 775°C and (b) after cooling down to 250°C.

It is possible to control the polarity of the epitaxial films grown by MBE by varying the growth conditions. GaN films grown on nitrided surfaces have a predominantly nitrogen terminated surface. These films tend to have a rough surface morphology. Since the nitridation step is likely to establish the polarity on the nonpolar sapphire surface, optimization of this step is essential. As shown in Figure 25b, the GaN film which was grown after nitridation for 35 min had virtually no inversion domains. In this case, the entire surface is terminated with N-atoms, causing GaN to form with the same polarity throughout. Steelmann-Eggebert et al. reported that [0001] polarity is obtained by growing low-temperature GaN directly (beginning with a monolayer deposition of gallium) on the sapphire surface without the nitridation step [250]. This is consistent with the observations of inversion domains of [0001] polarity in films with incomplete nitridation. It was also discovered that the presence of hydrogen leads to [0001] polarity in MBE grown GaN films [110]. In this case, instead of pure nitrogen, a mixture of nitrogen and hydrogen (9:1 ratio) was flown through the ECR source. Figure 47 shows a RHEED pattern from a GaN film taken at 500°C from such a sample, showing a clear 2×2 reconstruction. The reconstruction is stronger at a lower temperature and is observed down to 200°C. As described earlier,

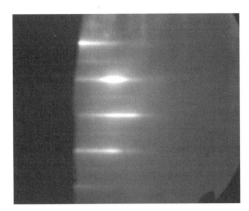

Fig. 47. RHEED pattern near [11$\bar{2}$0] from a GaN film grown in the presence of H$_2$, at 500°C.

such a reconstruction is indicative of GaN with [0001] polarity. Similar reconstructions were observed in homoepitaxial GaN films grown on HVPE GaN substrates, which are known to have [0001] polarity. It should be emphasized that this is an indirect method to determine polarity.

The polarity is expected to affect many of the material properties of GaN [110, 246, 255, 262]. Photoluminescence studies of [0001] polarity GaN films grown (in the presence of hydrogen) by MBE showed a set of broad peaks near 2.2 eV, in addition to the band edge emission near 3.4 eV. The emissions near 2.2 eV, called "yellow PL emission," are commonly attributed to a state in the middle of the energy gap, generated by point defects such as Ga-vacancies [251, 252], SMBs [34], antisite defects [253], etc. It is interesting to note that such "yellow PL" is commonly seen in GaN films grown by MOCVD and HVPE, which have [0001] polarity. On the other hand, such emissions are not commonly seen in MBE grown GaN films with [000$\bar{1}$] polarity. Although the exact nature of the defect responsible for yellow PL is not clear yet, it is hypothesized that these defects are facilitated by the growth in [0001] orientation. Similar yellow PL was also seen in GaN films grown with AlN buffers. We conjecture that these buffers lead to predominantly [0001] polarity in the GaN films. De Felice and Northrup [254] predicted [0001] polarity under Al-rich conditions, which may be the case in the AlN buffer layer growth conditions. Once the polarity is established near the interface, the GaN growth above would replicate the buffer polarity. This is supported by experimental evidence of Ga-polarity in such films, as evidenced by 2 × 2 reconstructions in RHEED patterns [110]. Detailed CBED studies need to be carried out to conclusively confirm the preceeding hypothesis.

The polarity in GaN films results in a piezoelectric effect, which has an implication in devices such as field effect transistors (FETs) and quantum well structures. Bykhovski, Gelmomt, and Shur predicted that this piezoelectric field at the coherently strained AlGaN/GaN heterojunctions can cause a large increase or decrease in the two-dimensional electron gas at the interface [255]. Therefore, depending on the polarity, an inversion of the epitaxial layer structure may be necessary to optimize the device performance.

5.2.1. Inversion Domains and Inversion Domain Boundaries

Inversion domains (ID) and their boundaries (IDBs) have been observed in GaN films grown by different methods [256–258]. These domains can be as large as 0.4 μm in some cases. The polarity of the GaN lattice switches from [0001] to [000$\bar{1}$] across an IDB. These IDs are commonly hexagonal in shape, with {10$\bar{1}$0} planes as side-walls [258]. However, Romano et al. observed IDs with {11$\bar{2}$0} facets as well [259]. Cheng et al. observe that the domain boundaries originate at the interface of the buffer and the epilayer and are formed from twin boundaries and 1/3 [111] atomic steps in the zinc-blende crystallites of the buffers layer [149]. They do not see such domains originating at the sapphire interface or the grain boundaries between hexagonal and cubic domains. Although the assignment of polarity is contentious, relative orientation can be detected easily in the case of films with mixed orientations. The inversion domains can be clearly identified in cross-sectional TEM studies, as filamental columns perpendicular to the interface, as marked by arrows and "ID" in Figure 48. Some researchers reported that in MOCVD grown films, the IDBs were sometimes terminated in the first 0.5 μm by forming "house-shaped" domains [277, 256]. The IDBs in this case were terminated by (1$\bar{1}$00) and (10$\bar{1}$2) planes. The mechanism of such a termination is not presently understood.

IDBs are formed when GaN domains nucleated in opposite polarities coalesce. By ensuring that the nucleation and the growth are initiated with only one of the species of the binary compound, these IDBs can be minimized. However, initiation with a single species still leads to IDBs if there are atomic steps on the substrate surface. For example, steps on the sapphire surface have a height of 2.16 Å ($c_{sapphire}/6$), which is significantly less than the bilayer spacing in GaN of 2.6 Å ($c_{GaN}/2$). Such a vertical mismatch between the epitaxial layer and the substrate can facilitate the nucleation of an inversion domain boundary. Figure 49 shows a schematic of an IDB formed at a surface step [256]. These IDs are formed by an exchange of the Ga and N sublattices and a relative translation of $c/2$ along [0001]. GaN grows with opposite polarity on either side of the IDBs. The IDBs are not commonly seen in SiC substrates since the polarity of the GaN films is fixed due to the substrate polarity. However, when the substrate is not properly cleaned, regions of amorphous layers may be present on the SiC surface, which could result in GaN films with mixed polarity and IDBs. The atomic structure of an inversion domain is shown in Figure 11. High-resolution electron microscopy (HREM) has been carried out to study these inversion domain boundaries [256].

In a simple inversion of sublattices across an IDB, there will be a wrong bond (cation–cation or anion–anion), which has a high energy of formation. An additional displacement along the c-axis by C/2 would avoid such bonding, keeping the tetrahedral coordination intact. In the case of GaN such translation is predicted based on lowest free energy considerations. Northrup et al. modeled IDBs with no wrong bonds and calculated that they have a low formation energy of 0.41 eV, which corresponds to a domain wall energy of only

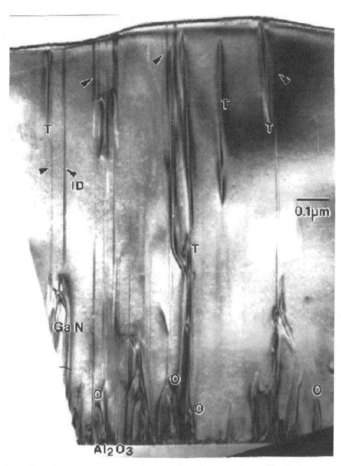

Fig. 48. Cross-sectional bright-field TEM micrograph of a GaN film grown on C-plane sapphire. "O" indicates dislocation loops, IDs are inversion domains and threading defects are labeled "T."

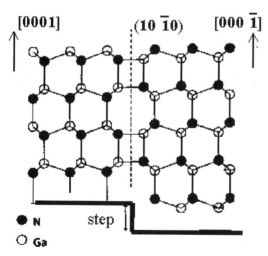

Fig. 49. Schematic of an IDB showing a GaN lattice with opposite polarities on either side. (From Romano, Northrup, and O'Keele [256] copyright 1996 by the American Physical Society.)

fects, once formed, are likely to propagate all the way to the surface of the film. Northrup et al. also concluded that properly coordinated IDs do not introduce any gap states, since the atomic coordination is not changed. However, vacancies or impurities may bind to these sites, which may be detrimental to the optoelectronic properties.

It has been reported that the polarity affects the growth rate and the surface morphology of the GaN films [260, 261]. GaN films with Ga-polarity have a higher growth rate than the N-polarity films. Figure 11a is a TEM micrograph of a pinhole in bulk GaN, showing such differences in the growth rate. The pinhole and the atomic model are shown schematically in Figure 11b and c. Under Ga-rich growth conditions, the growth rate of both polarities is approximately equal, resulting in very smooth GaN films. Under nitrogen-rich conditions, domains of Ga-polarity grow faster than the surrounding matrix of N-polarity, resulting in rough pyramidal morphology. Therefore, Ga-polarity samples are typically smooth with few IDBs, while N-polarity films are rough due to the presence of IDs. Several researchers have shown that these IDBs can be minimized by optimizing the nitridation and buffer layers [110, 258, 259, 262]. Romano et al. reported that MBE grown films had lower IDB density when they were grown without nitridation. The ratio of III/V fluxes can also effect the domain structure of the epitaxial films. The films grown under a high III/V flux ratio (Ga-rich conditions in the buffer as well as the high-temperature growths) revealed considerably higher inversion domains compared to the films grown with a lower III/V flux ratio [259]. This could be explained, assuming that the substrate was incompletely nitrided prior to growth, resulting in islands of thin AlN nuclei. Under N-rich conditions, the polarity established by the AlN nuclei is continued into the GaN film. Under Ga-rich conditions, GaN nucleates in the gap regions in between the AlN nuclei, with a polarity opposite to that on AlN, leading to a high density of inversion domains.

5.3. Polytype Defects

Wurtzite GaN is the stable phase on commonly used (0001) sapphire and SiC substrates, at growth temperatures above 600–700°C. However, due to comparable cohesive energies of wurtzite and zinc-blende GaN [42], the zinc-blende GaN phase is commonly seen near the interface, when a low-temperature buffer is used. In addition, the small difference in the formation energies of the two phases (of the order of 10 meV atom^{-1} [263]) promotes the introduction of stacking faults (SFs) in GaN films, especially close to the interface [192, 264]. Importance of the nucleation steps in avoiding such defects has been discussed in Section 3.

Stacking faults form by a simple change in the stacking sequence of the close packed planes. For example, an SF may be formed in wurtzite by changing the stacking sequence from ABABAB... to ABABACAC... or ABABCBC... (both intrinsic faults) or to ABABCABAB... (extrinsic fault). Stacking faults are equivalent to local transitions from the wurtzite (analogous to hcp) to the zinc-blende (analogous to fcc) phase. The

25 meV Å^{-2} [257]. Each sublattice (Ga and N) is shifted in the [0001] direction by $\pm c/8$. As seen in Figure 49, each atom is fourfold coordinated and the equilibrium bond lengths are close to the bulk bond lengths of 1.94 Å in GaN. Such low energy de-

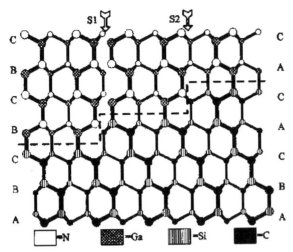

Fig. 50. Cross-section atomic model of wurtzite GaN on (0001) 6H-SiC. Steps on the SiC surface create SMBs sometimes as indicated by S1. There is no SMB created at step S2. (From Sverdlov et al. [173].)

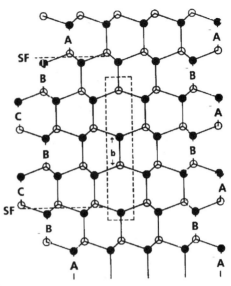

Fig. 51. Schematic of a stacking mismatch boundary, projected onto a (1$\bar{2}$10) plane. The filled circles denote N atoms and the open circles denote Ga atoms. The boundary lies in the [10$\bar{1}$0] plane and connects the two stacking faults on the left. The distance between the threefold coordinated Ga and N atoms in the boundary is 1.81 Å. (From Northrup, Neugenbauer, and Romano [257].)

SFs may also be stabilized by an impurity such as Mg, which is commonly used for *p*-type doping of GaN [263]. Bandic, McGill, and Ikonic applied local empirical pseudopotential theory and found that the SFs introduce electronic levels ~0.13 eV above the top of the valence band. They attribute the states to the heterocrystalline wurtzite–zinc-blende interface states. Researchers have observed Shockley as well as Frank partial dislocations, bounding these stacking faults [256, 266].

A second origin of these SMBs in GaN/SiC films are the surface steps [173]. If the wurtzite GaN films repeat the two topmost bilayers of the substrates, an atomic step can lead to two different GaN stacking sequences. In 6H-SiC with the stacking sequence ABCACB, there are three possible stacking sequences for wurtzite GaN depending on how the substrate stacking ends: substrate ABC leads to BCBC epilayers, BCA leads to CACA epilayers, and CBA leads to BABA epilayer sequence. An atomic step in the 6H-SiC surface, therefore, leads to GaN nucleation in two different stacking sequences. Suppose the termination of 6H-SiC is ABC and becomes ABCA after the step. The epitaxial GaN has the stacking sequences BCBC and CACA, respectively, in the two regions. The intersection of these two regions of the epilayer is an SMB. A cross-section atomic model of GaN grown on 6H-SiC with atomic steps is shown in Figure 50 [173]. Evidence for the origin of the SMBs at the substrate atomic steps is obtained from the high-resolution TEM studies. On the other hand, when the SiC surface termination changes from ABCA to ABCAC after the step, the GaN will have stacked as CACA and ACAC, respectively. This would not result in an SMB, as shown by S2 in the figure. From this discussion, it is clear that an atomic step in the surface of an isomorphic substrate (with a wurtzite structure), such as 2H-SiC or ZnO, will not result in SMBs in GaN epilayers. On sapphire, the stacking sequence is not as clear and the surface steps, in general, are expected to cause SMBs.

Stacking mismatch boundary has some threefold coordinated Ga and N atoms whose equilibrium bond lengths were found to be 1.81 Å [257], as shown schematically in Figure 51. Hence, there is significant lattice distortion compared to an IDB. Stacking fault energy is the highest for AlN and the lowest for GaN among the three nitrides [267]. The formation energy for the SMB in GaN was found to be 1.73 eV corresponding to a domain wall energy of 105 meV Å^{-2}. Such a boundary is expected to give rise to a state in the bandgap. Northrup, Neugenbauer, and Romano have shown that SMBs cause a transition state at 2.3 eV, which is very close to the region where yellow luminescence is observed in some films [257]. These defects can thus be detrimental to the optical properties of GaN films. However, since the formation energy of SMBs is quite high, these boundaries tend to terminate by the formation of stacking faults. Typically, SFs do not extend to very large distances. Stacking mismatch boundaries (SMB), also called double positioning boundaries (DPB), are boundaries between faulted and unfaulted regions. Since SFs have a low energy of formation, SMBs also terminate very easily and these boundaries are not seen beyond the buffer region in most films.

5.4. Dislocations

In any lattice-mismatched heteroepitaxy, misfit dislocations are expected since they accommodate the lattice mismatch between the two crystals. Misfit dislocations are observed at the interface, parallel to the interface in GaN films grown on sapphire and SiC [170, 172]. Ponce et al. measured the displacement between the misfit dislocations in GaN films grown on SiC using AlN buffers. The misfit dislocations at the SiC/AlN interface were separated by 30 nm, consistent with the ~1% lattice mismatch between the two crystals. Misfit dislocations were also documented at the AlN/SiC interface, separated by ~10 nm, close to the expected separation (11.3 nm) from the 2.5% mismatch. However, the focus of this discussion is on "threading dislocations," which constitute the majority of the dislocations observed in GaN films.

The threading dislocations or in GaN are aligned perpendicular to the interface and have pure edge, pure screw, or mixed characters. The threading defect density is generally higher than 10^8 cm^{-2}, and is often as high as 10^{10} cm^{-2} (see Figs. 25 and 48). Although most of the studies to date have been on C-plane sapphire and SiC, the types of defects observed are similar in films grown on different substrates. Many of these defects have dangling bonds that may act as scattering centers or as free carrier recombination centers, thus affecting electrical as well as optical properties. Since functional devices based on such defective GaN films have been successfully fabricated, it is possible that many of these defects are electronically inert, or may be passivated by some mechanism inherent to the growth method. Lester et al. argue that dislocations do not affect LED efficiency because they are not efficient nonradiative recombination centers [268]. The authors hypothesize that the relatively benign behavior of dislocations in GaN stems from the ionic character of bonding in these materials resulting in surfaces that do not cause Fermi level pinning. However, dislocations may influence the lifetime and may play an important role in many other devices such as laser diodes and high-temperature, high-frequency devices.

Many researchers have done detailed studies on various types of defects in GaN films using various transmission electron microscopy (TEM) tools like convergent beam electron diffraction (CBED), large area convergent beam electron diffraction (LACBED) and $g.b$ analysis etc. [107, 192, 259, 270–279]. Although the following discussion of defects in GaN films refers predominantly to epitaxy on C-plane sapphire, the issues involved are applicable to growth on other substrates as well. Ning et al. reported that the majority of the defects in the films they studied were pure edge dislocations with $b = 1/3$ [11$\bar{2}$0] [269], in agreement with observations made by other groups [259, 270–272]. The defect density in GaN thin films is also estimated by X-ray studies. The edge type threading dislocations do not contribute to the broadening of the symmetric (0002) X-ray diffraction rocking curve. Since the screw and mixed type threading dislocations constitute a small fraction of the total density, the FWHM of the (0002) rocking curve underestimates the dislocation density. As Heying et al. pointed out, rocking curves near an asymmetric reflection peak such as (10$\bar{1}$2) are better representative of the dislocation density in GaN films [271].

In addition to XRD and TEM studies, dislocation density can also be estimated by etching the thin film surface. The dislocations are preferentially etched using a suitable etchant and dislocation density can be estimated by counting the etch pit density. Molten KOH has been successfully used to reveal the hexagonal pyramid shaped etch pits, with the bases along $\langle 11\bar{2}0\rangle$ and (30$\bar{3}$2) facets [280]. However, this method may underestimate the dislocation density by more than an order of magnitude in very defective semiconductors like GaN.

The thermal and lattice mismatches between the film–substrate generate biaxial stresses that are parallel to the interface. Therefore, the resolved shear stresses on the (0001) basal plane and the {1$\bar{1}$00} prism planes are zero. Since most of the observed dislocations are observed on these planes, it is unlikely that they are caused by lattice–thermal stresses [281]. As discussed in the earlier sections, film growth occurs by island growth in large lattice-mismatched systems such as GaN on sapphire/SiC. In the case of (0001) oriented GaN, the nuclei are faceted with {1$\bar{1}$00} planes, which are the low energy planes in the GaN crystal. It is very likely that such nuclei are misoriented by a small angle, $\phi \sim 0.5°$, with respect to each other, which may result in a low-angle domain boundary when the adjacent nuclei coalesce. If the domains are rotated (tilted) with respect to each other along the c-axis, extra planes are needed to "fill" the gap. Therefore, edge dislocation parallel to the c-axis and the burgers vector $\mathbf{b} = 1/3$ [11$\bar{2}$0] are formed. These dislocations are separated by a distance $D = |\mathbf{b}|/\phi$ [281]. For example, a tilt angle ϕ of 0.5° can result in edge dislocations 30 nm apart, which accounts for the very high dislocation density seen in GaN epilayers. In the case of a twist boundary, i.e., when domains are rotated along an axis parallel to the interface, a combination of $\mathbf{b} = \langle 11\bar{2}0\rangle$ and $\mathbf{b} = \langle 0001\rangle$ screw dislocations are formed on the prism {1$\bar{1}$00} planes. Once formed, it is very difficult to annihilate these dislocations due to the lack of a good glide plane. The dislocation density decreases with thickness initially due to interaction between dislocations and annihilation. However, the density remains constant in films thicker than $\sim$0.5 μm. Ning et al. also observed basal dislocations with the same 1/3 [11–20] Burgers vector, in the first 150 nm of the film [269].

In addition to the edge dislocations with $\mathbf{b} = \langle 11\bar{2}0\rangle$, several researchers have observed other threading dislocations including screw dislocations with $\mathbf{b} = [0001]$, mixed dislocations with $\mathbf{b} = \langle 11\bar{2}3\rangle$, and dislocation half-loops (marked "O" in Fig. 48) in GaN [174, 269, 259, 270–274, 276]. During the island coalescence, certain domains are buried under the overgrowth of the other domains. The interface of these domains is accommodated by dislocations and appear as [0001] half-loops [174]. These half-loop dislocations lie on the {10$\bar{1}$0} planes and have a Bergers vector [0001]. The half-loops consist of two vertical screw dislocations connected by a horizontal edge dislocation. These [0001] dislocations have high energy of formation, and tend to end in the first half-micron of the film. Chien et al. believe that the $\langle 11\bar{2}3\rangle$ dislocations are formed by the reaction of the other two types of threading dislocations [174]:

$$1/3 \langle 11\bar{2}0\rangle + [0001] \rightarrow 1/3 \langle 11\bar{2}3\rangle$$

Some researchers associate nanotubes and pinholes to open-core threading screw dislocations [278]. Cherns et al. used large-angle convergent beam electron diffraction (LACBED) to investigate the hollow tubes of 5- to 25-nm diameter, seen in the GaN films grown on sapphire by MOCVD [282]. The authors found that these tubes contain screw dislocations with the Burgers vector [0001], with a magnitude of 5.19 Å. Qian et al.

have done scanning force microscopy (SFM) and TEM studies in the GaN films grown on sapphire and have observed nanotubes with 35- to 500-Å radii. These nanotubes are aligned along the growth direction [0001] with the internal surfaces of the open core formed by $\{1\bar{1}00\}$ prism planes. When the stored elastic energy of a dislocation is sufficiently large, the core will be empty with a radius proportional to the square of the Burgers vector. In GaN, an empty open core of the screw dislocation is formed to achieve an equilibrium between the elastic energy of the dislocation and the surface energy of the facets bounding the nanotube [278]. Lilental-Weber et al. report the observation of these pinholes at the grain boundaries [279]. Edge dislocations with $\mathbf{b} = 1/3 \langle 11\bar{2}0 \rangle$ have a much smaller Burgers vector and therefore have much smaller elastic energy, insufficient to form an open core.

Based on first principle calculations, Northrup predicted that these $[11\bar{2}0]$ type dislocations do not have gap states [283]. The surface state due to the dangling bond at the GaN surface has been investigated using photoemission studies [284, 285]. The surface states were found to be occupied and to be resonant with the valence band. Therefore, these states do not lie in the bandgap and do not act as recombination centers. This implies that any additional surfaces created by open-core dislocations should not affect the optical properties of GaN.

Elsner et al. explored the structure and the electronic properties of edge and screw threading dislocations using an *ab initio* local-density functional cluster method, AIMPRO, and a density functional based tight-binding method [286]. Comparison of line energies showed that screw dislocations are stabilized by an open core, while the threading edge dislocation prefer a filled core. Both types of dislocations are electrically inactive with a bandgap free from deep levels, although the stresses at the filled core of the edge dislocations may trap impurities that may contribute to a deep level in the bandgap. However, Xin et al. have done atomic resolution Z-contrast imaging studies to contradict the foregoing results and they concluded that the screw dislocations indeed have states in bandgap [287].

Doping of the GaN films makes a significant difference in the dislocation behavior. In undoped GaN, the spatial luminescence correlates well with dislocations, whereas this correlation is not evident in doped GaN films [107]. Figure 52 shows the variation of electron mobility with doping concentration in the various films. In single-crystal materials, the electron mobility is expected to increase with the lowering of carrier concentration, since there is reduced ionized impurity scattering at the lower doping levels. However, it can be seen that the data follow a set of bell-shaped curves. The dotted lines are guides to show this behavior. Weimann et al. [288] have, in collaboration with the Boston University group [110, 289], proposed a model to explain these experimental observations based on the interactions of dislocations with lateral transport of electrons. Edge dislocations have dangling bonds, which get negatively charged by the electrons present and act as scattering centers and impede the lateral movement of the electrons. Therefore, at high doping levels, lattice scattering dominates, while dislocation scattering dominates at lower doping concentrations.

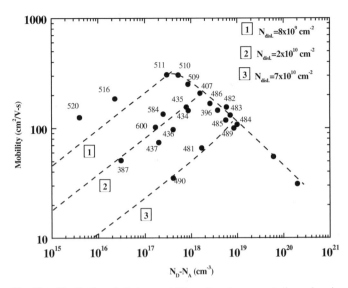

Fig. 52. Distribution of electron mobility and carrier concentration values in GaN films in a variety of GaN films. The dotted lines follow the theoretical predictions for films with different dislocation densities [288, 289].

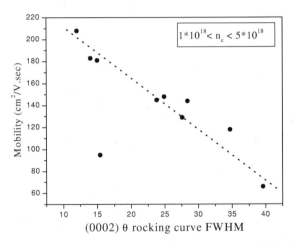

Fig. 53. Dependence of electron mobility on the FWHM of the (0002) rocking curve for GaN films with carrier concentration in the range of 1×10^{18}–5×10^{18} cm^{-3}.

In the films with higher dislocation density, this transition of a dominant mechanism is expected to occur at higher doping concentrations. The threading dislocation densities in some of the samples shown in the plot were measured by cross-sectional TEM studies and were found to agree qualitatively well with this model [110, 289].

The effect of the structure on electrical properties was also studied by correlating the FWHM of (0002) XRD rocking curves and electron mobilities. Since mobility is also expected to change with carrier concentration, films within a small range of carrier concentrations were compared. Figure 53 shows a plot of mobility vs FWHM for films with carrier concentrations between 1×10^{18} and 5×10^{18} cm^{-3}. It is clear from the plot that the mobility decreases with an increase in FWHM. It is believed that this is due to the scattering of the carriers near the boundaries of grains with a slightly different texture (orientation).

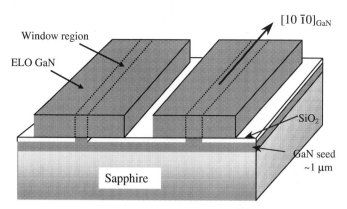

$[10\bar{1}0]_{GaN}$

Fig. 54. Schematic of ELO GaN growth structure.

Fig. 55. SEM micrographs of ELO GaN substrates grown with $\langle 1\bar{1}00\rangle$ oriented stripes at 1100°C with vertical side-walls. (From Doppalapudi et al. [293]. Reproduced by permission of the Electrochemical Society Inc.)

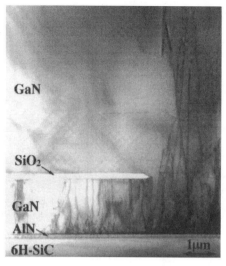

Fig. 56. Cross-section TEM micrograph of a laterally overgrown GaN layer using OMVPE. (From Nam et al. [291].)

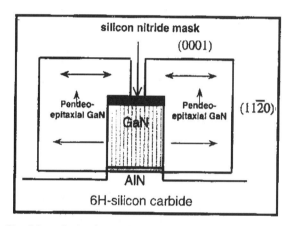

Fig. 57. Schematic showing pendeoepitaxial growth of GaN from an etched GaN seed with $(11\bar{2}0)$ side-walls. (From Linthicum et al. [296].)

Such a trend was not evident at lower carrier concentrations, possibly due to the more severe influence of threading dislocations described before.

Over the past few years, a number of groups have investigated the reduction of threading defects using the epitaxial lateral overgrowth technique [290–293]. Epitaxial lateral overgrowth (ELO) is a method of selective growth through an SiO$_2$ mask pattern formed on a pregrown semiconductor layer. ELO has been exploited for defect reduction in the 1960s in III–V compounds [294, 295]. ELO has mostly been applied using the MOCVD method, although a few groups are currently investigating this method using HVPE. A schematic of the epitaxial lateral overgrowth of GaN through a window in SiO$_2$ is shown in Figure 54. The ELO region shown in the schematic is expected to have a decreased dislocation density, since it grows off the GaN side-wall. A SEM micrograph of ELO GaN grown by the HVPE method with $\langle 1\bar{1}00\rangle$ oriented stripes is shown in Figure 55 [293].

Indeed, plan-view and cross-sectional TEM micrographs show a reduction in the threading dislocation density by a few orders of magnitude. For example, Figure 56 shows a cross-sectional TEM micrograph of ELO GaN grown using the OMVPE method [291]. It is evident from the micrograph that there is a dramatic improvement in the dislocation density in the ELO region. Using this ELO technique, the defect density was reduced to as low as 10^4–10^5 cm^{-2} in selected regions.

Although the ELO GaN is less defective, the window region and the region of coalescence between the overgrown regions still have a high defect density (see Fig. 56). Devices have been aligned and fabricated in the regions of low defect density. However, it is desirable to make large area defect free GaN films. Currently, there are two approaches to achieve this goal. One is to use a two-step ELO, where the overgrown GaN is patterned with SiO$_2$ a second time, to achieve selective growth nucleating from the defect free region of the first ELO. A second method is pendeoepitaxy which is a modification of the ELO process, where vertical propagation of threading defects is prevented by using a mask [296].

Figure 57 schematically shows the pendeoepitaxial growth on an SiC substrate [296]. First, a GaN seed layer with high dislocation density is grown by conventional methods, and is patterned as shown in the figure. The top of the seed is masked using silicon nitride and the epitaxial growth is initiated from

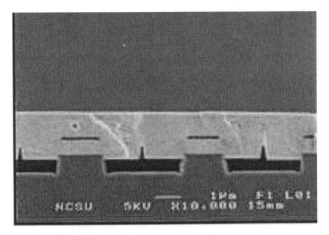

Fig. 58. SEM micrograph showing coalesced GaN grown by pendeoepitaxy. (From Linthicum et al. [297].)

the exposed side walls of the GaN seeds. The growth progresses as depicted by the arrows, eventually coalescing to form a continuous film. An example of such a continuous GaN film is shown in Figure 58 [297].

6. TERNARY ALLOYS

The ternary alloys AlGaN and InGaN are of great importance since they form an integral active part of many of the III–V nitride devices including emitters (LEDs and LDs) and detectors. As with the binary compounds, these ternary alloys have a direct bandgap and cover a wide range of energy spectrum. $In_xGa_{1-x}N$ alloys span the energy spectrum between 1.9 eV ($x = 1$) to 3.4 eV ($x = 0$), which includes the entire visible range, while $Al_xGa_{1-x}N$ compounds span the UV spectrum between, 3.4 eV ($x = 0$) and 6.2 eV ($x = 1$). The differences in atomic radii, bond strength, and the surface diffusivities between the cations have led to certain inhomogenieties in the ternary compounds. These include inhomogeneous stresses, phase separation, and long-range atomic ordering. In this section, growth of AlGaN and InGaN ternary families is discussed, with special focus on phase separation and atomic ordering.

6.1. InGaN Alloys

The InGaN alloy system has attracted special interest because of its potential for the formation of light emitting devices operating in a red to near UV region of the energy spectrum. Several designs of double heterostructures (DH), single quantum well (SQW), and multiquantum well (MQW) structures using InGaN active layers have been successfully grown and processed to commercialize light emitting diodes (LEDs) and laser diodes (LDs) [298–301]. In spite of all these developments, the growth and the properties of InGaN alloys have not been completely optimized or understood yet. Such an understanding and control of InGaN growth is essential to fabricate devices with predictable characteristics and longer lifetimes.

It has been commonly observed that the optical bandgap is seen at a lower energy than expected for the film composition [302, 303]. This has been attributed to a negative bowing parameter in the $In_xGa_{1-x}N$ bandgap, which can be expressed as

$$E_{InGaN}(x) = (1 - x)E_{GaN} + xE_{InN} + bx(1 - x)$$

where, b is the bowing parameter. McCluskey et al. predicted the bowing parameter to be temperature dependent, and reported a value of -3.8 eV at $x = 0.1$ [304]. In addition to bowing, the deeper level emissions can also be a result of phase separation.

6.1.1. Growth of Epitaxial InGaN

The two major problems hindering the use of InGaN alloys are the difficulties associated with indium incorporation into the epitaxial films and the phenomenon of phase separation in InGaN alloys. The current understanding of these problems in the research community is presented as follows.

The greatest difficulty in InGaN growth is that InN has a very high vapor pressure of nitrogen. The In–N bond is relatively weak compared to the Ga–N and Al–N bonds and dissociates easily under the growth conditions. At the standard MBE growth temperatures of GaN (750°C), the equilibrium vapor pressure (EVP) of nitrogen over InN is about 10 orders of magnitude greater than that over GaN [114]. This is especially problematic under MBE growth conditions, where operating chamber pressure during growth is between 10^{-4} and 10^{-5} Torr. Although the temperatures employed in MOCVD are much higher (>1000°C), the operating pressure is several orders higher, typically close to atmospheric pressure. At low pressures, the In–N bond dissociates and the indium atoms left behind are desorbed. Consequently, indium has a very low sticking coefficient at the typical growth temperatures for GaN, and the mole fraction of the indium in the films is considerably smaller than the indium fraction in the precursor.

One method of controlling InN dissociation is to reduce the growth temperature significantly. Increase of indium incorporation with reduction in growth temperature has been observed in films grown by MOCVD [305, 306], and MBE [110]. However, at too low temperatures, indium droplets may form, which act as sinks for InN, thereby preventing incorporation of high levels of indium in the film [110, 114, 305]. This segregation of indium as droplets results in films with poor optical and electrical properties, as evidenced by PL studies which show weaker and broader spectra. Films grown below 600°C show almost no photoluminescence [298]. A very good control of growth is required to achieve conditions under which InN dissociation is minimized in conjunction with the desorption of excessive indium. The problem is compounded when NH_3 is used as the nitrogen source, due to inefficient cracking at low temperatures [307]. In addition, regardless of the nitrogen source used, the crystalline quality of the films is inferior at lower temperatures due to poor surface mobility of the adatoms, resulting in highly columnar or even three-dimensional growth. Based on RHEED oscillation studies, Grandjean and Massies reported that the $In_xGa_{1-x}N$ growth mode undergoes a two-dimensional–three-dimensional transition after a few monolayers of growth when $x > 0.12$ [308], leading to films with very rough surfaces. A smaller degree of roughening was observed for films with $x < 0.12$.

A second method of improving the indium incorporation is by increasing the growth rate of the epitaxial film [110, 114, 306]. Increasing the growth rate by MBE, higher indium mole fractions were obtained using a lower fraction of the indium precursor. Furthermore, the necessity of low growth temperatures is relaxed, resulting in an improvement in the crystallinity of the films. At these higher growth rates, reevaporation of InN is suppressed due to the trapping of the molecules by the deposition of subsequent layers. Increasing the V/III flux ratio has also been used to achieve high indium mole fractions without the formation of indium droplets [114]. The overall concentration of indium in the film depends not only on the indium flux, but also on the gallium and nitrogen fluxes, V/III ratio, and In/(In + Ga) flux ratios. Optimizing these growth parameters, smooth InGaN films with an indium mole fraction up to 37% have been achieved [110, 309].

The lattice mismatch between InN and sapphire is greater than between GaN and sapphire, which makes the epitaxial growth even more challenging. There is also a significant lattice mismatch ($\sim$11%) between InN and GaN, which makes heteroepitaxy of InGaN with GaN or other III-N alloys difficult. This is another cause for failure to incorporate high indium mole fractions. The InGaN films are believed to be pseudomorphically strained to match the GaN lattice [304, 310]. Kawaguchi et al. [310] suggested that the indium atoms are excluded from the InGaN lattice to reduce the lattice mismatch during InGaN growth, a phenomenon they term as the "composition pulling effect." They observed that the defect density in the films increased with thickness. These defects accommodated the lattice mismatch, thereby reducing lattice deformation and enabling higher indium incorporation. This pulling effect was not observed in the films grown directly on sapphire with a low-temperature buffer, where the InGaN lattice is believed to be virtually strain-free due to the presence of the accommodating buffer layer [310, 325]. Another group reported the exact opposite observation; i.e., the indium incorporation in films grown on GaN was more than that on sapphire [326]. Since both studies were carried out on films grown by MOCVD at $\sim$800°C, the reason for the discrepancy was not clear.

Influence of lattice strain on the InGaN composition was observed in MBE grown films as well [110]. Two samples were deposited under identical conditions of growth temperature, In/(In + Ga) flux ratio, and V/III flux ratio. Sample A was grown on a thick GaN layer and sample B was grown directly on sapphire, with a low-temperature GaN buffer. In the case of direct growth on the low-temperature buffer (sample B), a high density of dislocations and defects was seen near the interface. Therefore, indium (and/or InN) can segregate more easily and can desorb from the surface. In the case of growth

on GaN (sample A), there is a much smaller defect density. The InGaN film grows with a fairly good epitaxial match to the GaN layer with less segregation. This resulted in a significantly higher growth rate for sample A ($\sim$3000 Å h^{-1}) compared to sample B ($\sim$1700 Å h^{-1}), which led to higher indium incorporation in the film A, as predicted by the growth rate dependence described before.

6.1.2. Phase Separation in InGaN Alloys

A major concern in the growth of InGaN alloys is phase separation at the growth temperatures, which originates from the 11% lattice mismatch between the two end components, InN and GaN. In such mixed compounds, the tetrahedral radii of the atomic species occupying a particular sublattice may be different from each other. The resulting strain in the layers leads to deviations from the homogeneity of the sublattice. Compositional inhomogeneities influence the bandgap and the optoelectronic properties of the InGaN films. An understanding of the nature of phase separation is essential not only to minimize its adverse effects on the film quality, but also to engineer this phenomenon in a controlled fashion when advantageous.

The majority of the III–V ternary and quaternary alloys are predicted to be thermodynamically unstable and to show a tendency toward clustering and phase separation [311–314]. Stringfellow [312] developed the delta lattice parameter (DLP) model for III–V compounds with the zinc-blende structure, to calculate the critical temperature (T_c) above which a particular ternary or quaternary system is completely miscible. The model is based on the correlation between the immiscibility and the difference in the lattice parameters of the components, and assumes that Vegard's law of solid solutions is applicable. However, there have been very few studies on the InN–GaN quasi-binary system and even fewer reports on phase separation in this system. Singh et al. [82] applied the DLP model using the lattice parameters of the zinc-blende structures of InN and GaN and found the critical temperature (T_c) above which the InN–GaN system is completely miscible to be 2457 K. Ho and Stringfellow used a modified valence-force-field (VFF) model and calculated a much lower T_c of 1473 K [315]. Since even this temperature is much higher than typical growth temperatures of InGaN films, phase separation is expected in these films based on thermodynamic considerations. However, there have been a few modifications to the phase diagram based on theoretical calculations. Zhang and Zunger predicted orders of magnitude increase in solubility due to dimerization driven by surface reconstruction [316]. Karpov reported that strain in the InGaN films could also increase the solubility, by skewing the equilibrium phase diagram to the higher indium side [317].

Zunger and Mahajan [311] reviewed several observations of traditional cubic III–V compounds mainly by SAD and dark-field imaging in TEM, which indicate that when the tetrahedral radii are different, two types of structural variations are observed; phase separation and atomic ordering [318–321]. Based on XRD and optical absorption studies, Singh and coworkers [82, 322, 323] provided the first evidence of phase separation in InGaN thick films grown by MBE based on XRD and optical absorption studies. Other researchers reported phase separation in thick InGaN films grown by MOCVD [324, 325].

Figure 59 shows an XRD pattern of an InGaN film with 37% (atomic) indium, grown at 725°C [309], showing a strong phase separated InN peak. The InGaN film in this case was grown on a thick (3000 Å) GaN film. In these data, a strong $In_{0.37}Ga_{0.63}N$ peak is also seen, as the phase separation was incomplete due to kinetic limitations. The phase separation in InGaN films has also been investigated using TEM techniques. To prevent dissociation of InN/InGaN, the samples have to be cooled using liquid nitrogen during the sample preparation (ion milling) as well as during the microscopy. Without the cooling, the InGaN foil disintegrated under the electron beam, leading to erroneous results [309]. Figure 60b is a dark-field image taken near the [0001] zone axis with $g = [10\bar{1}0]$, showing fine speckle contrast in the lattice, indicative of phase separation. El-Masry et al. reported a similar contrast in InGaN films grown by MOCVD with an In concentration of 49% [324], and attributed it to spinodal decomposition.

Figure 60a shows a plan-view SAD pattern of the $In_{0.37}Ga_{0.63}N$ film at the [0001] zone axis [309]. It is clear from the two sets of superimposed diffraction patterns in the image, that there is more than one phase in the material. For the wurtzite InGaN, six diffracted spots closest to the transmitted spot correspond to the $(10\bar{1}0)$ planes. The interplanar spacing, d was calculated to be 0.307 nm for the inner spot A and 0.288 nm for the outer spot B. These values match with the $(10\bar{1}0)$ plane spacings in InN and $In_{0.37}Ga_{0.63}N$, although the latter spot is more diffused and corresponds to a range of composition. Similar phase separation of InN (or InGaN with close to 100% In) was also seen from the diffraction studies from the cross section of the same film, taken at the $[11\bar{2}0]$ zone axis, as seen in Figure 61. This gives direct and conclusive evidence of phase separation in bulk InGaN alloys.

The phase separation in InGaN films is conjectured to be driven by strain due to the mixing of the two lattice-mismatched components of the InGaN alloy system. Indium atoms are excluded from the InGaN lattice to form an alloy of a different composition and to reduce the strain energy of the system. As discussed earlier, based on thermodynamic considerations, InN and GaN are immiscible at these growth temperatures. From the phase diagram calculated by Ho and Stringfellow based on a modified valence-force-field model, the solubility limit of InN in GaN is less than 5% at these growth temperatures (Fig. 62) [315]. The sample discussed before is identified as "A" in the figure, corresponding to 37% indium. The phase diagram predicts spinodal decomposition, resulting in $In_{0.93}Ga_{0.07}N$ and $In_{0.07}Ga_{0.93}N$ (corresponding to the solid lines). The data shown in Figures 60 and 61 are in agreement with the phase diagram. However, the $In_{0.37}Ga_{0.63}N$ peak with a range of lower indium compositions is seen, indicating that the phase separation was incomplete due to kinetic limitations.

The role of growth temperature on phase separation is investigated by growing several films at lower temperatures (650–675°C) on A-plane sapphire [309]. In this study, a number of thick InGaN films with indium content less than 35% were examined. Although XRD data did not show any InN related peaks, chemical inhomogeneity was observed in these films,

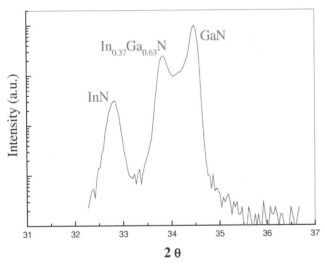

Fig. 59. X-ray diffraction pattern from an InGaN film with 37% indium. (From Doppalapudi et al. [309].)

Fig. 60. (a) Plan-view SAD pattern along the [0001] zone axis showing the superimposition of two sets of diffraction patterns from InN and InGaN. (b) Plan-view TEM micrograph of an InGaN film showing the contrast difference indicating chemical inhomogeneity.

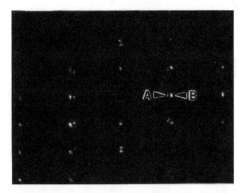

Fig. 61. Cross-sectional SAD pattern along the [11$\bar{2}$0] zone axis from the In$_{0.37}$Ga$_{0.63}$N film, showing two sets of superimposed diffraction patterns from InN and InGaN.

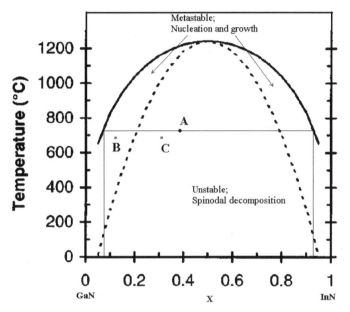

Fig. 62. Phase diagram of InN-GaN quasi-binary system [309, 315]. The data points A, B, and C describe the three samples.

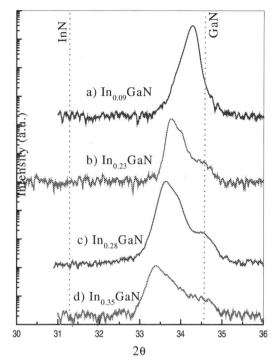

Fig. 63. X-ray diffraction patterns from as-grown InGaN films of different compositions. (From Doppalapudi et al. [309].)

especially in those with high indium concentrations. Figure 63 shows θ–2θ XRD patterns near the (0002) Bragg reflections of some of the InGaN films grown at these lower temperatures. Figure 63a shows a single peak corresponding to In$_{0.09}$Ga$_{0.91}$N. However, with an increase in indium composition, the (0002) Bragg reflection is broadened, indicating that the InGaN alloy is not chemically homogeneous. Such inhomogeneity could also be inferred from the SAD pattern shown in Figure 60a, where the diffraction spots corresponding to In$_{0.37}$Ga$_{0.63}$N (the outer set) are seen to be diffused and split, rather than being sharp. El-Masry et al. reported similar findings in the films grown by MOCVD [324].

Compositional inhomogeneity observed in the InGaN films may be a result of one of three possible mechanisms. One mechanism is the growth of InN precipitates, which may nucleate as a result of coalescence of indium droplets on the surface during growth. However, in our SEM analysis, we do not observe any droplets or InN related peaks to support this mechanism.

A second possibility is direct precipitation of InN from the bulk of InGaN lattice by a nucleation and growth mechanism. This involves relatively long-range diffusion and is strongly dependent on temperature and growth time. A third mechanism is spinodal decomposition, where phase separation occurs without nucleation, by "uphill" diffusion. Though this is also dependent on temperature and growth time, the initial diffusion lengths can be much shorter, and is therefore a more plausible mechanism at the relatively low temperatures used for growth of these films. At typical MBE growth temperatures (600–700°C), calculations by Ho and Stringfellow [315] predict that the alloy is metastable for compositions between 5 and 20% In (where phase separation can only occur by nucleation and growth), and unstable for In > 20% (where spinodal decomposition is expected) as seen from Figure 62.

Doppalapudi et al. investigated the phase separation mechanism by carrying out annealing studies on two InGaN films shown in Figure 62 [309]; sample B with 9% indium and sample C with 35% indium. To minimize film degradation, the annealing experiments were carried out at atmospheric pressure, in nitrogen ambient. Figure 64 shows the XRD scans near the (0002) peak of the sample B (In$_{0.09}$Ga$_{0.91}$N), which falls in the metastable region of the calculated phase diagram. There was no change in the film after annealing for 20 h at 725°C. In contrast, there were significant changes in the structure of sample C, whose composition is in the "unstable" region of the calculated phase diagram. Figure 65 shows the XRD scans near the (0002) peak of the film with 35% indium, annealed at various temperature–time conditions. Emergence of phase separated GaN can be clearly seen in the figure. After

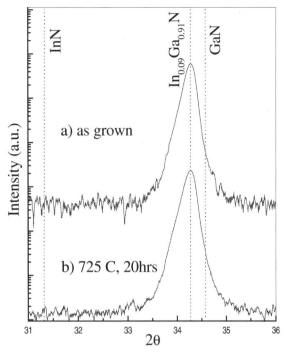

Fig. 64.   X-ray θ–2θ patterns from the In$_{0.09}$GaN film (a) as-grown and (b) after annealing at 725°C for 20 h. (From Doppalapudi et al. [309].)

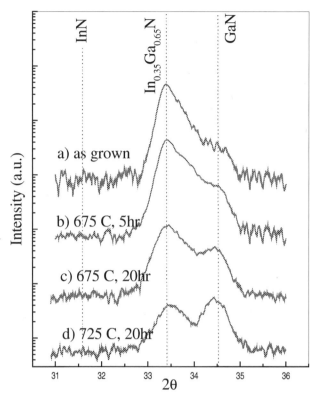

Fig. 65.   X-ray θ–2θ patterns from the In$_{0.35}$GaN film (a) as-grown and after annealing, (b) at 675°C for 5 h, (c) 675°C for 20 h, and (d) at 725°C for 20 h. (From Doppalapudi et al. [309].)

annealing for 20 h at 725°C, the intensity of the GaN peak exceeded even that of In$_{0.37}$Ga$_{0.63}$N, which is strong evidence of phase separation, possibly by spinodal decomposition. The absence of a corresponding InN peak at $2\theta = 31.35°$ is attributed to InN evaporation. This was confirmed by SEM data, which showed a roughened surface and a decrease in the film thickness. Spinodal decomposition occurs by "uphill" diffusion and can be initiated by small local fluctuations in composition. In the case of sample B, phase separation is predicted to occur by nucleation and growth, which usually requires diffusion over longer distances and hence takes longer times as compared to spinodal decomposition. These results are in excellent agreement with the phase diagram proposed by Ho and Stringfellow [315].

A correlation has been reported between dark spots observed in cathodoluminescence (CL) measurements and regions of compositional inhomogeneities such as phase separation [326]. Some researchers have reported that phase separation occurs in the form of quantum dots in InGaN based SQW and MQW structures [327]. These quantum dots, ranging from 5 to 20 nm in size, were seen in the InGaN quantum wells by TEM and were correlated to CL studies. Spontaneous emission was observed from these regions, as a result of quantum confinement in the three-dimensional indium-rich regions. Such confinement could also viewed to be beneficial in increasing the brightness of the emitters. McCluskey et al. performed annealing experiments on MQW structures and reported that significant phase separation was observed in the quantum wells only at temperatures higher than 950°C [328]. Thus, most MQW structures should

be free of indium segregation. It has been reported that the density of dislocations may affect the extent of phase separation [326, 329]. It is suggested that phase separation is aided by screw dislocations and that dislocations act as nonradiative recombination centers in InGaN MQW structures.

Phase separation also effects the transport properties in In-GaN films. The carrier mobility in general decreases due to the poorer crystallinity of the film and increases in the grain boundary area. However, since InGaN films are mostly used in the fabrication of light emitters, phase separation does not pose a major problem when it is appropriately engineered to emit at the desired wavelength.

6.1.3. Long-range Atomic Ordering

In addition to phase separation, the In$_x$Ga$_{1-x}$N films also showed atomic ordering. As mentioned before, the majority of the III–V ternary and quaternary alloys are predicted to be thermodynamically unstable at the low growth temperature and show a tendency toward clustering and phase separation. Thus, atomic ordering is usually not expected to occur. However, such a phenomenon was theoretically predicted [330] and was observed in many cubic III–V alloys [318, 320–322]. The Boston University group first reported long-range atomic ordering in AlGaN alloys [333] and then in InGaN alloys [309, 334].

The structure factor for a crystallographic plane (hkl) is given by the expression,

$$F_{hkl} = \sum_1^N f_N e^{2\pi i(hu_n + kv_n + lw_n)}$$

where the summation extends over all the N-lattice points (u_n, v_n, w_n) of the unit cell. The unit cell of the wurtzite structure (shown in Fig. 3) has two lattice points for the group III (Ga/In) atoms at $(0, 0, 0)$ and $(1/3, 2/3, 1/2)$ and two for nitrogen at $(0, 0, 3/8)$ and $(1/3, 2/3, 7/8)$. In this structure, the geometrical structure factor for an (hkl) plane is given as

$$F_{hkl} = f_1 + f_2 e^{2\pi i[(h+2k)/3 + (l/2)]}$$
$$+ f_N\left(e^{2\pi i(3l/8)} + e^{2\pi i(h/3 + 2k/3 + 7l/8)}\right)$$

where f_N is the structure factor of the nitrogen atom and f_1 and f_2 are scattering factors for the group III atoms occupying the lattice sites $(0, 0, 0)$ and $(1/3, 2/3, 1/2)$, respectively. From the preceding equation, $F_{hkl} = (f_1 - f_2)$, if l is odd and $h + 2k = 3n$, for any integer n. In a random InGaN alloy, indium and gallium atoms occupy the two lattice sites $(0, 0, 0)$ and $(1/3, 2/3, 1/2)$ with equal probability, resulting in equal values for f_1 and f_2. Therefore, the structure factor of $(000l)$ planes is zero if l is odd. In an ordered alloy, indium atoms occupy certain lattice sites preferentially, resulting in unequal values for atomic structure factors f_1 and f_2. As an example, for the case of a completely ordered lattice of an $In_{0.5}Ga_{0.5}N$ alloy, shown in Figure 66, the indium atoms preferentially occupy lattice sites in alternating basal planes. In this case, f_1 and f_2 become f_{Ga} and f_{In}, respectively, resulting in a nonzero structure factor for the (0001) plane.

Figure 67 is an XRD scan from the $In_{0.09}Ga_{0.91}N$ film, which clearly shows the presence of the (0001) peak, indicating ordering of the indium atoms in the alloy. Xiao et al. observed a (0001) XRD peak in pure GaN due to defect ordering. However, the ratio of intensities of the $(0001)/(0002)$ peaks in this study was only 10^{-6} [335], which is about 4 orders of magnitude less than what was observed in InGaN films, implying

that the ordering in the InGaN films is not defect related. Furthermore, such a superlattice peak was not observed in any of the GaN films grown in this study. The presence of superlattice diffraction peaks was observed in all InGaN films investigated.

Figure 68 is an SAD pattern obtained from an InGaN film with 35% indium, taken along the $[11\bar{2}0]$ zone-axis. Though the pattern also shows the presence of (0001) spots, this cannot be conclusively attributed to ordering since similar features could be observed even in pure GaN due to double diffraction. However, very prominent streaking of diffraction spots was observed in the $[0001]$ direction (Fig. 68). Such streaks in the SAD pattern were observed in AlGaAs alloys by Kuan et al. [332], who attributed them to partial ordering and the presence of antiphase boundaries in the growth direction. At the $[11\bar{2}0]$ zone axis, $\{0001\}$ spots result due to double diffraction of $\{10\bar{1}0\}$ and $\{10\bar{1}1\}$ spots, both of which have nonzero structure factors. To avoid double diffraction, the sample was tilted about the $[11\bar{2}0]$ SAD pattern, to illuminate the single row of $(000l)$ spots, where there is no double diffraction from the $(10\text{-}1l)$ row contributing

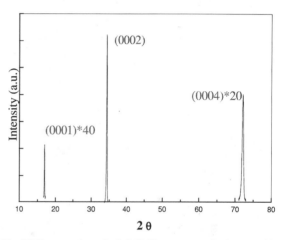

Fig. 67. XRD pattern from the InGaN film grown on A-plane sapphire. The presence of the (0001) peak indicates ordering in the lattice. (From Doppalapudi et al. [309].)

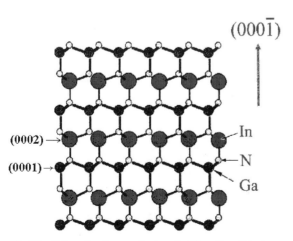

Fig. 66. Schematic of a completely orderd $In_{0.5}Ga_{0.5}N$ lattice.

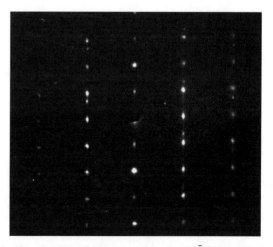

Fig. 68. Cross-sectional SAD pattern along the $[11\bar{2}0]$ zone axis from the $In_{0.35}Ga_{0.65}N$ film, showing the presence of the (0001) spots as well as streaking in the $[0001]$ direction. (From Doppalapudi et al. [309].)

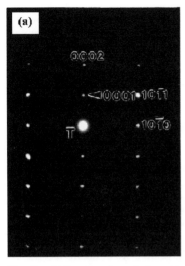

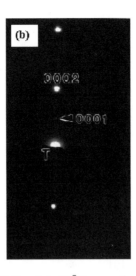

Fig. 69. SAD pattern from the $In_{0.09}Ga_{0.91}N$ film of the [11$\bar{2}$0] zone axis, tilted to illuminate only the (0001) spots. The arrows point to (0001) Bragg reflections.

Fig. 70. SAD pattern from the $In_{0.09}Ga_{0.91}N$ film at the [10$\bar{1}$0] zone axis. The arrows point to (0001) Bragg reflections.

to the presence of the (0001) spot. Figure 69a shows a similar diffraction pattern from the $In_{0.09}GaN$ film at the [11$\bar{2}$0] zone axis. In Figure 69b, the sample is tilted to illuminate the (000l) row of the SAD spots. The (0001) spots are still clearly seen, albeit with a weaker intensity indicating that the alloy is indeed ordered. Even more conclusively, SAD patterns of the InGaN film at the [10$\bar{1}$0] zone axis showed the presence of the superlattice (0001) diffraction spot, as seen by the arrows in Figure 70. In this case, the (0001) diffraction spots cannot be formed by double diffraction (since the {11$\bar{2}$1} planes have a zero structure factor), and can only be attributed to ordering.

To obtain an accurate estimate of the degree of ordering in these films, the intensity ratios of the (0001) to the (0004) Bragg reflections were compared. This normalization accounts for any variations in beam intensities, film thickness, and orientation effects. The XRD results showed that the ratio of (0001) to (0004)

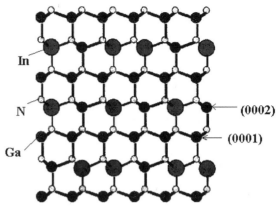

Fig. 71. Schematic of a completely ordered $In_xGa_{1-x}N$ lattice.

peak intensities increased with the composition of indium in the films. These experimental results were compared with the calculated intensity ratios for alloys ordered to different degrees. In general, the intensity of an ($hkil$) Bragg reflection is given by [336],

$$I_{(hkil)} = |F_{hkil}|^2 \left(\frac{1 + \cos^2 2\theta \cdot \cos^2 2\alpha}{\sin\theta \cdot \cos\theta} \right) \left(\frac{1}{\sin\theta} \right) A(\theta) \exp^{-2M}$$

where F_{hkil} is the structure factor for the ($hkil$) plane, the term in the brackets is the Lorentz-polarization parameter, $A(\theta)$ is the absorption factor, the exponential term is the Debye–Waller temperature factor and the $(1/\sin\theta)$ term accounts for the differences in the film volume exposed at different angles. In the equation, θ is the Bragg angle of the ($hkil$) reflection and α is the Bragg angle of the monochromator (13.63° in this case, for the Ge (111) plane we used). The structure factor F_{000l} is calculated from the atomic scattering factors of the constituent atoms. The group III atom positions in the unit cell are taken to be at (0, 0, 0) and (1/3, 2/3, 1/2) and nitrogen positions at (0, 0, 3/8) and (1/3, 2/3, 7/8), as shown in Figure 3. The long-range ordering parameter, S, is given by the relation,

$$S = \frac{r_A - x}{1 - x}$$

where r_A is the fraction of A sites occupied by the atoms of type A and x is the mole fraction of A in the compound. In a perfectly ordered ($S = 1$) $In_xGa_{1-x}N$ alloys, $r_A = 1$. In this case, all the indium atoms ($2x$) are assumed to be on the (0002) plane and the rest of the sites on the (0002) as well as all the sites on the (0001) plane to be taken up by Ga atoms. This is shown schematically in Figure 71.

The structure factor for the (000l) plane, F_{000l} is given by

$$F_{000l} = \left[2x(1 - r_{In})f_{In}^l + \{1 - 2x(1 - r_{In})\}f_{Ga}^l \right] \cdot \exp^{2\Pi i(0)}$$
$$+ \left[2r_{In}x \cdot f_{In}^l + (1 - 2r_{In}x) \cdot f_{Ga}^l \right] \cdot \exp^{2\Pi i(l/2)}$$
$$+ f_N^l \cdot \left(\exp^{2\Pi i(3l/8)} + \exp^{2\Pi i(7l/8)} \right)$$

where f^l is the atomic structure factor of the particular atom at θ corresponding to the (000l) diffraction peak.

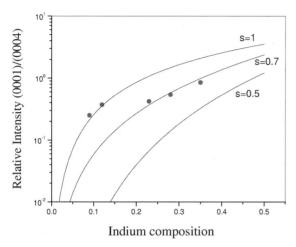

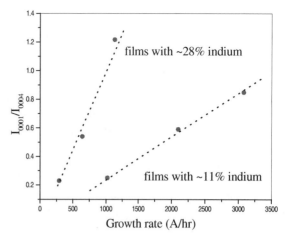

Fig. 72. Relative intensity of (0001) and (0004) peaks plotted as a function of indium composition, for different degrees of ordering, S. The solid circles indicate the experimental data. (From Doppalapudi et al. [309].)

Fig. 73. Variation in the long-range order parameter (S) with growth rate, in the InGaN films grown under same conditions and with same composition. (From Doppalapudi, Basu, and Moustakas [334].)

The ratio I_{0001}/I_{0004} is calculated from the previous expressions, using tabulated values for the various structure factors [336]. The absorption factor was calculated for the two θ-values, by estimating mass absorption coefficients for different indium compositions and taking the film thickness to be 0.5 μm. The ratio of Debye–Waller temperature factors for the two peaks was calculated to be 1.05, using the value of the Debye temperature of 770°C for GaN [337]. The curves in Figure 72 show the variation in the calculated relative intensity of (0001)/(0004), as a function of composition for different ordering parameters. The experimental data points shown are for samples grown under similar conditions of temperature, growth rate, and final thickness. The figure shows that at lower indium compositions, complete ordering occurs ($S = 1$). At higher indium concentrations, the films exhibit partial ordering ($S \sim 0.7$).

Long-range atomic ordering has been observed in several of the traditional III–V compounds, which have a zinc-blende type structure. Atomic ordering in these compounds is believed to be two-dimensional in nature, occurring on the growth surface and being subsequently trapped in the bulk due to kinetic limitations. Ordering in these compounds is caused by surface-reconstruction induced subsurface stresses which force preferential occupation of sites by atomic species of different radii [319]. In most of the atomic ordering observations, the epitaxial growth was carried out on [100] oriented substrates with zinc-blende structures.

Srivastava, Martins, and Zunger [330] performed first principle local density total minimization calculations for both ordered and disordered models for bulk GaInP and predicted that certain ordered phases could be thermodynamically stable at low temperatures. The stability of these ordered phases is due to the reduction of strain in the ordered vs disordered phases. It is conjectured that the driving force for ordering in the InGaN system is the lattice strain in the alloy due to differences in In–N and Ga–N bond lengths. Since the ordered phases can simultaneously accommodate two different bond lengths in the alloy

in a coherent fashion, they introduce less strain than would be present in a random alloy. This is consistent with the smaller degree of ordering observed in the samples with high indium composition, where strain is already partially relieved by phase separation. It is believed that this competition between ordering and phase separation is the reason why the data of Figure 72 do not follow the curves predicted by the kinematical scattering theory, for perfectly ordered material. In AlGaN alloys in which phase separation was not observed, the data are qualitatively consistent with the theory for perfect ordering [333]. The ordering observed in our films closely resembles the Cu–Pt-type ordering observed earlier in other III–V compounds with a zinc-blende structure [320, 321]. In these compounds, ordering is characterized by ½ {111} superlattice reflections, which are structurally similar to the (0001) superlattice reflections we observe in our wurtzite films. Total energy calculations done in other III–V compounds [338] suggest that the Cu–Pt-type ordered phase is thermodynamically unstable in bulk. This argument is supported by the observations of Suzuki, Gomyo, and Iijima [331], who showed that while $Ga_{0.5}In_{0.5}P$ grown on (001) GaAs is ordered, the same film grown on (111) B and (110) GaAs is disordered. However, once ordering is induced at the surface during growth, it is retained during further growth due to kinetic limitations.

It should be noted however, that Cu–Pt-type ordering has not been observed in the cubic alloys for growth along the [111] direction. The close packed symmetry of the (111) surface in these alloys is similar to the (0001) surface of the wurtzite InGaN alloys studied here. This indicates that the results for the cubic III–V alloys are not directly applicable to the wurtzite alloy system.

As discussed, the long-range ordering parameter, S, was a function of the film composition. It was found that S also depended strongly on the growth rate of the film. Figure 73 shows the variation of the degree of ordering with growth rate for two sets of films of same composition (28% In and 11% In) grown under similar conditions [334]. It is clear that the order parameter S increases monotonically with the growth rate in the films studied. This trend is in apparent contradiction with the observations in cubic III–V compounds [311], which show a decrease in ordering with growth rate.

This contrast in the growth rate dependence can be explained, if the ordering in InGaN alloys is in fact unstable at the surface and stable in the bulk. At faster growth rates, bulk thermodynamics would determine the structure and surface thermodynamics dominate at slower growth rates. A second possible explanation for the apparent contradiction may be in the large difference in growth rates of the two systems. The InGaN alloys in this study were deposited at considerably smaller growth rates (0.03 to 0.3 μm h^{-1}) compared to the traditional cubic alloys (1–3 μm h^{-1}). The slow arrival of atoms at the surface provides sufficient time for surface ordering to occur. Once the surface gets trapped in the bulk, it wants to disorder. However, disordering can occur effectively only in the near-surface region of the film. The absence of disordering in the bulk alloy is supported by the stability of the ordered phase for annealing up to temperatures of 725°C for 20 h. Thus, for faster growth rates, the time spent in the near-surface region is smaller, allowing for a smaller extent of disordering to occur (effectively "freezing" the ordered phase) and leading to a larger value of S. This strongly supports the contention that ordering is a surface phenomenon. At much higher growth rates (as in the case of cubic III–V compounds), there is not sufficient time for surface ordering to occur. Thus, in this regime, the degree of ordering decreases with growth rate.

Postgrowth annealing experiments were carried out in the films that do not show phase separation, to study the stability of these ordered phases. The InGaN films were first coated with SiO$_2$ to prevent any evaporation or dissociation of the film during annealing. The annealing studies were carried out in a tube furnace under a nitrogen atmosphere at 725 and 800°C for different time intervals. After each annealing experiment, the films were taken out and were studied by XRD, to get an estimate of the degree of ordering, by calculating the ratio of (0001) and (0004) Bragg reflection peaks. This ratio is plotted as a function of annealing time in Figure 74. The open triangles show the annealing results at 725°C, from the In$_{0.09}$Ga$_{0.91}$N film that was grown directly on sapphire. It is clear from the plot that the degree of ordering actually increases after heating the film for 60 h, implying that the ordered phase is actually thermodynamically stable. Annealing of the same film at 800°C for 10 more hours (solid triangle) did not result in any further changes. Based on this evidence, it is believed that the ordered phase in the InGaN system might be stable in the bulk. This is in contrast to other III–V compounds investigated, in which case, postgrowth annealing experiments indicated that the ordered phase is unstable in the bulk [321, 339]. Annealing studies were also carried out on another film with 12% indium, and grown on a thick GaN layer. In this case, heating at 800°C caused virtually no change in the degree of ordering. In fact, it appeared to decrease a little. It is possible, therefore, that epitaxial lattice-mismatch strain is playing a role in the ordering process in these films. These preliminary annealing studies suggest

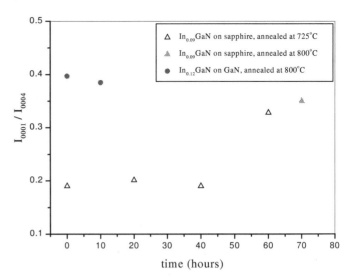

Fig. 74. Variation of degree of ordering with annealing. The film composition and the annealing conditions are indicated in the plot.

that the ordered phase may in fact be stable in the bulk, again in direct contrast with the observation of the cubic III–V alloys.

Another interesting observation can be seen in Figure 75a, which shows a bright-field cross-sectional TEM micrograph of a 1-μm thick InGaN film grown with 9% indium. A domain structure is clearly seen near the interface, with a domain size of 100–200 nm. Domains with two different contrasts can be seen and are designated as A and B. Selective area diffraction patterns obtained from the two regions are shown in Figure 75. Figure 75b is a [10$\bar{1}$0] diffraction pattern from region A, clearly showing the (0001) spot, which is evidence of long-range atomic ordering. The pattern from region B does not have the superlattice (0001) spot (Fig. 75c), indicating a disordered structure. From the foregoing observations, it is clear that In$_x$Ga$_{1-x}$N growth occurs with a domain structure with a mixture of ordered and disordered domains near the interface. The micrograph in Figure 75a also shows that this domain structure is not present near the surface. Electron diffraction studies reveal that the film is ordered near the surface across the entire film. We speculate that this is a result of heteroepitaxial stresses near the interface, which may promote the domain structure. The exact mechanism of such domain formation is not currently understood.

6.2. AlGaN Alloys

6.2.1. Growth of Epitaxial AlGaN

AlGaN alloys were grown both on (0001) sapphire and (0001) 6H-SiC substrates using MBE and MOCVD methods, using procedures similar to those described for the growth of GaN. However, the growth kinetics in the two methods are potentially different. Iliopoulos et al. [340] reported that in microwave plasma-assisted MBE growth, the composition of the AlGaN films varies with the ratio of groups III to V fluxes even though

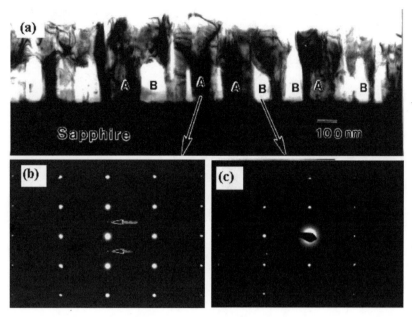

Fig. 75. (a) Cross-sectional TEM micrograph showing domain structure in the $In_{0.09}Ga_{0.91}N$ film, (b) SAD [10$\bar{1}$0] pattern from the region marked A, showing the presence of {0001} spots, (c) SAD [10$\bar{1}$0] pattern from the region marked "B," where the {0001} spots are absent. (From Doppalapudi, Basu, and Moustakas [334].)

the ratio of beam equivalent pressures of Al and Ga was chosen to be constant. In these experiments, films were grown at the same temperature and the same nitrogen plasma conditions (ECR power, flow rate) and with a constant Al/Ga flux ratio of 0.5. As the net (Al + Ga) flux was increased from a nitrogen-rich regime to group III-rich conditions of growth, the AlN mole fraction in the films increased from 50 to 89%. This suggests a significant change in the kinetics of growth with the III/V flux ratio.

Under group III-rich conditions, the available active nitrogen preferentially bonds with Al due to the higher bond strength of the Al–N bond compared to the Ga–N bond. This results in a higher sticking coefficient for Al and a smaller sticking coefficient for Ga. Thus, the films produced under these conditions have a higher AlN mole fraction than that predicted from the ratio of the beam equivalent pressures of Al and Ga. Under nitrogen-rich conditions, on the other hand, the sticking coefficient of Ga also increases, leading to AlGaN films with a composition close to what is expected from the ratio of beam equivalent pressures of Al and Ga. This change in growth kinetics also affects the surface roughness of the grown films. Films grown under group III-rich conditions are atomically smooth, while those produced under nitrogen-rich conditions have an RMS roughness of 3.31 nm [340].

The kinetics of growth of AlGaN films by MOCVD has not yet been reported. However, growth in this method always occurs under excess of ammonia, similar to the growth of AlGaAs. Thus, control of the AlGaN film composition is accomplished more directly by controlling the flows of TMAl and TMGa.

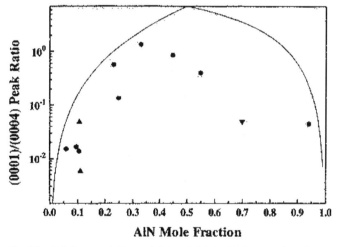

Fig. 76. Relative superlattice peak integrated intensity for a number of samples grown by MBE. The samples grown on sapphire are shown as circles and those grown on 6H-SiC are shown as triangles. The solid line is the theoretical prediction of the kinematical limit. (From Korakakis, Ludwig, and Moustakas [333].)

6.2.2. Long-Range Atomic Ordering

AlGaN films growth by both MBE [333, 340] and MOCVD [341] methods were found to undergo long range atomic order. Korakakis, Ludwig, and Moustakas [333] reported that XRD studies of films grown under nitrogen rich conditions show, in addition to the (0002) and (0004) diffraction peaks expected in wurtzite AlGaN, (0001), (0003), and (0005) diffraction peaks which are forbidden in random alloys. The presence of these (000l) peaks, where l is an odd number, indicate atomic order-

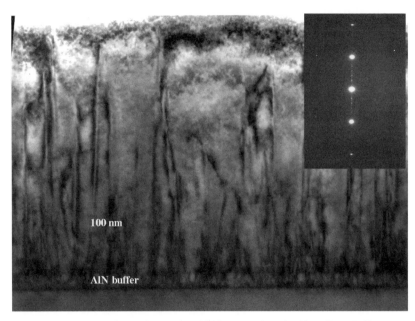

Fig. 77.   Cross-section TEM micrograph and SAD pattern from the $Al_{0.89}Ga_{0.11}N$ film grown under group III-rich conditions. (From Iliopoulos et al. [340].)

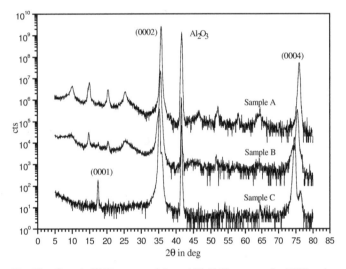

Fig. 78.   On-axis XRD spectra of three AlGaN films grown by MBE under different III/V flux ratios. (From Iliopoulos et al. [340].)

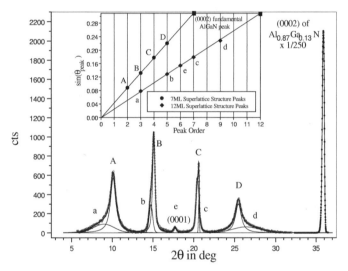

Fig. 79.   Extended on-axis XRD spectrum of the $Al_{0.89}Ga_{0.11}N$ film. The data are fitted to Voight functions. Sin θ vs integer number for the peaks A, B, C, D and a, b, c, d, e are plotted in the inset. (From Iliopoulos et al. [340].)

ing as in the case of InGaN alloys. The authors investigated a large number of AlGaN films and found that maximum ordering indeed occurs at $Al_{0.5}Ga_{0.5}N$ films, as expected (Fig. 76).

Iliopoulos et al. [340] observed a different type of superlattice structure under group III-rich conditions. Figure 77 shows a cross-sectional TEM micrograph of an $Al_{0.89}Ga_{0.11}N$ film grown under group III-rich conditions. The inset electron diffraction clearly indicates a superlattice structure that is more complex than the (0001) type of ordering observed before. XRD plots of AlGaN films grown under three different group III/V flux ratios is shown in Figure 78. The samples were grown keeping the other growth conditions very similar, and had different AlN mole fractions, as explained before.

Sample A ($Al_{0.89}Ga_{0.11}N$) was grown under a high III/V flux ratio and clearly has several superlattice peaks. Sample C ($Al_{0.55}Ga_{0.45}N$) was grown under nitrogen rich conditions and exhibits atomic ordering similar to that observed by Korakais, Ludwig, and Moustakas [333].

Iliopoulos et al. [340] analyzed the superlattice peaks observed in sample A by fitting the data to Voight functions (Fig. 79) and concluded that the sample has two intermixed superstructures; one with 7 monolayer periodicity and the other (minor) with 12 monolayer periodicity. The origin and the exact structure of these spontaneously formed superlattices have not yet been established.

REFERENCES

1. S. Nakamura, M. Senoh, N. Iwasa, and S. Nagahama, *Jpn. J. Appl. Phys.* 34, L797 (1995).
2. S. Nakmura, M. Senoh, S. Nagahama, N. Iwasa, T. Yamada, T. Matsushita, H. Kyoku, and Y. Sugimoto, *Jpn. J. Appl. Phys.* 35, L74 (1995).
3. "Gallium Nitride (GaN) I." Semiconductors and Semimetals (J. I. Pankove and T. D. Moustakas, Eds.), Vol. 50. Academic Press, New York, 1998.
4. "Gallium Nitride (GaN) II." Semiconductors and Semimetals (J. I. Pankove and T. D. Moustakas, Eds.), Vol. 57. Academic Press, New York, 1999.
5. R. Juza and H. Hahn, *Anorg. Allgem. Chem.* 234, 282 (1938).
6. H. Grimmeiss and Z. H. Kolemans, *Nature* 14a, 264 (1959).
7. H. P. Maruska and J. J. Tietjen, *Appl. Phys. Lett.* 15, 367 (1969).
8. W. Seifert, R. Franzheld, E. Buttler, H. Sobotta, and V. Riede, *Cryst. Res. Technol.* 18, 383 (1983).
9. J. I. Pankove, E. A. Miller, and J. E. Berkeyheiser, *J. Lumin.* 5, 84 (1972).
10. I. Akasaki, T. Kozowa, K. Hiramatsu, N. Sawak, K. Ikeda, and Y. Ishii, *J. Lumin.* 40–41, 121 (1988).
11. H. Amano, M. Kitoh, K. Hiramatsu, and I. Akasaki, *J. Electrochem. Soc.* 137, 1639 (1990).
12. J. A. van Vechten, J. D. Zook, and R. D. Horning, *Jpn. J. Appl. Phys.* 31, 3662 (1992).
13. I. Akasaki, H. Amano, M. Kito, and K. Hiramatsu, *J. Lumin.* 48-49, 666 (1991).
14. S. Nakamura, N. Iwasa, M. Senoh, and T. Mukai, *Jpn. J. Appl. Phys.* 31, 1258 (1992).
15. T. D. Moustakas and R. J. Molnar, *Mater. Res. Soc. Symp. Proc.* 281, 753 (1993).
16. M. S. Brandt, N. M. Johnson, R. J. Molnar, R. Singh, and T. D. Moustakas, *Appl. Phys. Lett.* 64, 2264 (1994).
17. S. Porowski, M. Bockowski, B. Lucznik, M. Wrobleski, S. Krukowski, I. Grzegory, M. Leszczynski, G. Nowak, K. Pakula, and J. Baranowski, *Mater. Res. Soc. Proc.* 449, 35 (1997).
18. R. P. Vaudo and V. M. Phanse, *Electrochem. Soc. Proc.* 98–18, 79 (1998).
19. T. Lei and T. D. Moustakas, *Mater. Res. Soc. Symp. Proc.* 242, 433 (1992).
20. T. D. Moustakas, *Mater. Res. Soc. Symp. Proc.* 395, 111 (1996).
21. T. Kurobe, Y. Sekiguchi, J. Suda, M. Yoshimoto, and H. Matsunami, *Appl. Phys. Lett.* 73, 2305 (1998).
22. G. A. Slack and T. F. McNelly, *J. Cryst. Growth* 34, 263 (1976).
23. G. A. Slack and T. F. McNelly, *J. Cryst. Growth* 42, 560 (1977).
24. C. O. Dugger, *Mater. Res. Bull* 9, 331 (1974).
25. C. M. Balkas, Z. Sitar, T. Zheleva, L. Bergman, I. K. Shmagin, J. F. Muth, R. Kolbas, R. Nemanich, and R. F. Davis, *Mater. Res. Proc.* 449 (1997).
26. E. Ejder, *J. Cryst. Growth* 22, 44 (1974).
27. D. Elwell, R. S. Fiegelson, M. M. Simkins, and W. A. Tiller, *J. Cryst. Growth* 66, 45 (1984).
28. C. J. Frosh, *J. Phys. Chem.* 66, 877 (1962).
29. R. A. Logan and C. D. Thurmond, *J. Electrochem. Soc.* 119, 1727 (1972).
30. J. Karpinski, J. Jun, and S. Porowski, *J. Cryst. Growth* 66, 11 (1984).
31. S. Porowski, *J. Cryst. Growth* 166, 583 (1996).
32. *Comp. Semicond.* 6, 13 (2000).
33. V. Ban, *J. Electrochem. Soc.* 119, 761 (1972).
34. W. Seifert, G. Fitzl, and E. Butter, *J. Cryst. Growth* 52 257 (1981).
35. K. Naniwae, S. Itoh, H. Amano, K. Itoh, K. Hiramatsu, and I. Akasaki, *J. Cryst. Growth* 99, 381 (1990).
36. M. J. Illegems, *J. Cryst. Growth* 14, 360 (1972).
37. S. A. Sufvi, N. Perkins, M. Horton, A. Thon, D. Zhi, and T. Kuech, *Mater. Res. Soc. Proc.* 423, 227 (1996).
38. R. Molnar, K. B. Nichlos, P. Maki, E. R. Brown, and I. Melngailis, *Mater. Res. Soc. Symp. Proc.* 378, 479 (1995).
39. R. Molnar, R. Aggarwal, Z. Liau, E. R. Brown, I. Melngailis, W. Gotz, L. T. Romano, and N. M. Johnson, *Mater. Res. Soc. Symp. Proc.* 395, 189 (1996).
40. A. Shintani and S. Minagawa, *J. Cryst. Growth* 22, 1 (1974).
41. J. Korec, *J. Cryst. Growth* 46, 655 (1979).
42. G. Jacob, M. Boulou, and D. Bois, *J. Lumin.* 17, 263 (1978).
43. K. Seth, M. S. Dissertation, Department of Electrical and Computer Engineering, Boston University, 2000.
44. R. Molnar, in "Gallium Nitride (GaN) II." Semiconductors and Semimetals (J. I. Pankove and T. D. Moustakas, Eds.), Vol. 57. Academic Press, New York, 1999.
45. T. Detchprohm, H. Amano, K. Hiramatsu, and I. Akasaki, *J. Cryst. Growth* 128, 384 (1993).
46. T. Detchprohm, K. Hiramatsu, H. Amano, and I. Akasaki, *Appl. Phys. Lett.* 61, 2688 (1992).
47. R. Molnar, P. Maki, R. Aggarwal, Z. Liau, E. R. Brown, I. Melngailis, W. Gotz, L. T. Romano, and N. M. Johnson, *Mater. Res. Soc. Symp. Proc.* 395, 221 (1996).
48. R. Molnar, W. Goetz, L. T. Romano, and N. M. Johnson, *J. Cryst. Growth* 178, 147 (1996).
49. H. M. Manasevit, *Appl. Phys. Lett.* 12, 136 (1968).
50. H. M. Manasevit and W. I. Simpson, *J. Electrochem. Soc.* 116, 1725 (1969).
51. P. D. Dapcus, *J. Cryst. Growth* 68, 345 (1988).
52. M. Razeghi and A. Hilger, "The MOCVD Challenge," Vol. 1. Adam Hilger, Bristol, U.K., 1989.
53. H. Sasaki, M. Tanaka, and J. Yoshino, *Jpn. J. Appl. Phys.* 24, 417 (1985).
54. J. O. Williams, "Growth and Characterization of Semiconductors" (R. A. Stradling and P. C. Klipstern, Eds.), Hilger, Bristol, New York, 1990.
55. S. P. DenBaars and S. Keller, in "Gallium Nitride (GaN) I." Semiconductors and Semimetals (J. I. Pankove, T. D. Moustakas, Eds.), Vol. 50. Academic Press, New York, 1998.
56. J. Nishijawa, H. Abe, and T. Kurabayashi, *J. Electrochem. Soc.* 132, 1197 (1985).
57. S. P. DenBaars, B. Y. Maa, D. P. Dapkus, and H. C. Lee, *J. Cryst. Growth* 77, 188 (1986).
58. H. Amano, N. Sawaki, and I. Akasaki, *J. Cryst. Growth* 68, 163 (1984).
59. I. Akasaki, H. Amano, Y. Koide, K. Hiramatsu, and N. Sawaki, *J. Cryst. Growth* 98, 209 (1989).
60. S. Nakamura, S. Pearton, and G. Fsaol, "The Blue Laser Diode" 2nd ed., Springer-Verlag, 2000.
61. S. L. Wright, T. N. Jackson, and R. F. Marks, *J. Vac. Sci. Technol. B* 8, 288 (1990).
62. A. J. Spring Thorpe and A. Majeed, *J. Vac. Sci. Technol. B* 8, 266 (1990).
63. T. Okamoto and A. Yoshikawa, *Jpn. J. Appl. Phys.* 30, L156 (1991).
64. S. Yoshida, S. Misawa, Y. Fuji, S. Tanaka, H. Hayakawa, S. Gonda, and A. Itoh, *J. Vac. Sci. Technol.* 16, 990 (1979).
65. S. Yoshida, S. Misawa, and S. Gonda, *Appl. Phys. Lett.* 42, 427 (1983).
66. Y. Moriyasu, H. Goto, N. Kuze, and M. Matsui, *J. Cryst. Growth* 150, 916 (1995).
67. R. C. Powell, N. E. Lee, and J. E. Greene, *Appl. Phys. Lett.* 60, 2505 (1992).
68. W. Kim, O. Aktas, A. E. Botchkarev, A. Salvador, S. N. Mohammad, and H. Morkoc, *J. Appl. Phys.* 79, 1 (1996).
69. M. Kamp, M. Mayer, A. Pelzmann, and K. J. Ebeling, *Mater. Res. Soc. Symp. Proc.* 449, 161 (1997).
70. N. Grandjean, J. Massies, P. Vennegues, and M. Leroux, *J. Appl. Phys.* 83, 1379 (1998).
71. O. Aktas, W. Kim, W. Fan, A. E. Botchkarev, A. Salvador, S. N. Mohammad, B. Sverdlov, and H. Morkoc, *Electron. Lett.* 31, 1389 (1995).
72. J. J. Schmidt, W. Shan, J. J. Song, A. Slavador, W. Kim, O. Atakas, A. Botchkarev, and H. Morkoc, *Appl. Phys. Lett.* 68, 1820 (1996).
73. M. Paisley, Z. Sitar, J. B. Posthil, and R. F. Davis, *J. Vac. Sci. Technol.* 7, 701 (1989).
74. T. Lei, M. Fanciulli, R. J. Molanr, T. D. Moustakas, R. J. Graham, and J. Scanlon, *Appl. Phys. Lett.* 58, 944 (1991).
75. S. Strite, J. Ruan, Z. Li, N. Manning, A. Salvador, H. Chen, D. J. Smith, W. J. Choyke, and H. Morkoc, *J. Vac. Sci. Technol. B* 9, 1924 (1991).
76. T. D. Moustakas, T. Lei, and R. J. Molnar, *Physica B* 185, 36 (1993).

77. C. R. Eddy, T. D. Moustakas, and J. Scanlon, *J. Appl. Phys.* 73, 448 (1993).

78. R. J. Molnar and T. D. Moustakas, *J. Appl. Phys.* 76, 4587 (1994).

79. A. Ohtani, K. S. Stevens, M. Kinniburg, and R. Beresford, *J. Cryst. Growth* 150, 902 (1995).

80. R. J. Molnar, R. Singh, and T. D. Moustakas, *J. Electron. Mater.* 24, 275 (1995).

81. R. Singh, D. Doppalapudi, and R. D. Moustakas, *Appl. Phys. Lett.* 69, 2388 (1996).

82. R. Singh, D. Doppalapudi, R. D. Moustakas, and L. T. Romano, *Appl. Phys. Lett.* 70, 1089 (1997).

83. D. Korakakis, K. F. Ludwig, and T. D. Moustakas, *Appl. Phys. Lett.* 72, 1004 (1998).

84. D. Doppalapudi, E. Iliopoulos, S. N. Basu, and T. D. Moustakas, *J. Appl. Phys.* 85, 3582 (1999).

85. T. D. Moustakas, "Gallium Nitride (GaN) II." Semiconductor and Semimetals (J. I. Pankove and T. D. Moustakas, Eds.), Vol. 57. Academic Press, New York, 1999.

86. F. F. Chen, "Introduction to Plasma Physics and Controlled Fusion," 2nd ed., Vol. 1. Plenum, Boston, 1984.

87. T. Matsuoka and K. Ono, *J. Vac. Sci. Technol. A* 6, 25 (1988).

88. W. E. Kohler, M. Romheld, R. J. Seebock, and S. Scaberna, *Appl. Phys. Lett.* 63, 2890 (1993).

89. K. W. Boer, "Survey of Semiconductor Physics," Vol. 1. Van Nostrand-Reinhold, New York, 1990.

90. R. M. Park, *J. Vac. Sci. Technol. A* 10, 701 (1992).

91. H. Liu, A. C. Frenkel, J. G. Kim, and R. M. Park, *J. Appl. Phys.* 74, 6124 (1993).

92. R. P. Vaudo, J. W. Cook, and J. F. Schetzina, *J. Vac. Sci. Technol. B* 12, 1242 (1994).

93. O. Brandt, H. Yang, B. Jenichen, Y. Suzuki, L. Daweritz, and K. H. Ploog, *Phys. Rev. B* 52, R2253 (1995).

94. S. Guha, N. A. Bojarczuk, and D. W. Kiser, *Appl. Phys. Lett.* 69, 2879 (1996).

95. H. Riechert, R. Averbeck, A. Graber, M. Schienle, V. Straub, and H. Tews, *Mater. Res. Soc. Proc.* 449, 149 (1997).

96. R. C. Powell, N. E. Lee, Y. W. Kim, and J. E. Greene, *J. Appl. Phys.* 73, 189 (1993).

97. T. C. Fu, N. Newman, E. Jones, J. C. Chan, X. Liu, M. D. Rubin, N. W. Cheung, and E. R. Weber, *Electron. Mater.* 24, 249 (1995).

98. M. S. Leung, R. Klockenbrink, C. Kisielowski, H. Fujii, J. Kruger, G. S. Sudhir, A. Anders, Z. Liliental-Weber, M. Rubin, and E. R. Weber, *Mater. Res. Soc. Proc.* 449, 221 (1997).

99. T. Kurobe, Y. Sekiguchi, J. Suda, M. Yoshimoto, and H. Matsunami, *Appl. Phys. Lett.* 73, 2305 (1998).

100. S. Porowski, Materials Research Society, Fall Meeting, Boston, 1998.

101. S. Porowski, B. Bockowski, B. Lucznik, M. Wroblewski, S. Korowski, I. Grzegory, M. Leszczynski, G. Novak, K. Pakula, and J. Baronowski, *Mater. Res. Soc. Symp. Proc.* 449, 35 (1997).

102. Z. Liliental-Weber, C. Kisielowski, S. Ruvimov, Y. Chen, J. Washburn, I. Grzegory, M. Bockowski, J. Jun, and S. Porowski, *J. Electron. Mater.* 25, 9 (1996).

103. F. A. Ponce, D. P. Bour, W. T. Young, M. Saunders, and J. W. Steeds, *Appl. Phys. Lett.* 69, 337 (1996).

104. Z. Liliental-Weber, in "Gallium Nitride (GaN) I." Semiconductor and Semimetals (J. I. Pankove and T. D. Moustakas, Eds.), Vol. 50. Academic Press, New York, 1998.

105. F. A. Ponce, D. P. Bour, W. Gotz, N. M. Johnson, H. I. Helava, I. Grzegory, J. Jun, and S. Porowski, *Appl. Phys. Lett.* 68, 917 (1996).

106. Z. Liliental-Weber, Y. Chen, S. Ruvimov, and J. Washburn, *Phys. Rev. Lett.* 79, 2835 (1997).

107. F. A. Ponce, D. Cherns, W. T. Young, and J. W. Steeds, *Appl. Phys. Lett.* 69, 3 (1996).

108. L. D. Landau and I. M. Lifshitz, "Statistical Physics," 3rd ed. Pergamon, Oxford, U.K., 1980.

109. J. Hwang, A. Kuznetsov, S. Lee, H. Kim, J. Choi, and P. Chong, *J. Cryst. Growth* 142, 5 (1994).

110. D. Doppalapudi, Ph.D. Dissertation, Department of Manufacturing Engineering, Boston University, 1999.

111. T. Lei, K. F. Ludwig, Jr., and T. D. Moustakas, *J. Appl. Phys.* 74, 4430 (1993).

112. C. Sun and M. Razeghi, *Appl. Phys. Lett.* 63, 973 (1993).

113. S. Nakamura, M. Senoh, S. Nagahama, N. Iwasa, T. Yamada, T. Matsushita, H. Kiyoku, and Y. Sigimoto, *Jpn. J. Appl. Phys.* 35, L217 (1996).

114. T. Matsuoka, T. Sakai, and A. Katsui, *Optoelectron. Dev. Technol.* 5, 53 (1990).

115. C. J. Sun, P. Kung, A. Saxler, H. Ohsato, K. Haritos, and M. Razeghi, *J. Appl. Phys.* 75, 3964 (1994).

116. P. Kung, C. J. Sun, A. Saxler, H. Ohsato, and M. Razeghi, *J. Appl. Phys.* 75, 4515 (1994).

117. N. Grandjean, J. Massies, P. Vennegues, M. Laugt, and M. Leroux, *Appl. Phys. Lett.* 70, 643 (1997).

118. M. L. Kronberg, *Acta Metall.* 5, 507 (1957).

119. T. D. Moustakas, R. J. Molnar, T. Lei, G. Menon, and C. R. Eddy, Jr., *Mater. Res. Soc. Symp. Proc.* 242, 427 (1992).

120. T. D. Moustakas, T. Lei, and R. J. Molnar, *Physica B* 185, 36 (1993).

121. T. Kato, H. Ohsato, T. Okuda, P. Kung, A. Saxler, C. Sun, and M. Razeghi, *J. Cryst. Growth* 173, 244 (1997).

122. K. Doverspike, L. B. Rowland,, D. K. Gaskill, and J. A. Freitas, Jr., *J. Electron. Mater.* 24, 269 (1994).

123. Distributed by Total Resolution, Berkeley, CA.

124. E. G. Baur, B. W. Dodson, D. J. Ehrlich, L. C. Feldman, C. P. Flynn, M. W. Geis, J. P. Harbison, R. J. Matyi, P. S. Peercy, P. M. Petroff, J. M. Philips, G. B. Stringfellow, and A. Zangwill, *J. Mater. Res.* 5, 852 (1990).

125. H. Amano, N. Sawaki, I. Akasaki, and Y. Toyoda, *Appl. Phys. Lett.* 48, 353 (1986).

126. I. Akasaki, H. Amano, Y. Kiode, H. Hiramatsu, and N. Sawaki, *J. Cryst. Growth* 89, 209 (1989).

127. S. Nakamura, *Jpn. J. Appl. Phys.* 30, 1705 (1991).

128. M. H. Kim, C. Sone, J. H. Yi, and E. Yoon, *Appl. Phys. Lett.* 71, 1228 (1997).

129. D. Korakakis, Ph.D. Dissertation, Electrical and Computer Engineering, Boston University, 1998.

130. H. M. Ng, D. Doppalapudi, D. Korakakis, R. Singh, and T. D. Moustakas, *J. Cryst. Growth* 189–190, 349 (1998).

131. N. Grandjean, J. Massies, and M. Leroux, *Appl. Phys. Lett.* 69, 2071, 1996.

132. P. Vennegues and B. Beaumont, *Appl. Phys. Lett.* 75, 4115, (1999).

133. K. Uchida, A. Watanabe, F. Yano, M. Kouguchi, T. Tanaka, and S. Minagawa, *J. Appl. Phys.* 79, 3487 (1996).

134. M. Yeadon, M. T. Marshall, F. Hamdani, S. Pekin, H. Morkoc, and J. M. Gibson, *Mater. Res. Soc. Symp. Proc.* 482, 99 (1998).

135. D. Doppalapudi, E. Iliopoulos, S. N. Basu, and T. D. Moustakas, *J. Appl. Phys.* 85, 3582 (1999).

136. D. Doppalapudi, E. Iliopoulos, S. N. Basu, and T. D. Moustakas, *Mater. Res. Soc. Sym. Proc.* 482, 51 (1998).

137. C. Heinlein, J. Grepstad, T. Berge, and H. Riechert, *Appl. Phys. Lett.* 71, 341 (1997).

138. S. Fuke, H. Teshigawara, K. Kuwahara, Y. Takano, T. Ito, M. Yanagihara, and K. Ohtsuka, *J. Appl. Phys.* 83, 764 (1998).

139. J. L. Rouviere, M. Arlery, R. Nie Bhur, K. H. Bachem, and O. Briot, *Mater. Sci. Eng. B* 43, 161 (1997).

140. S. Keller, B. P. Keller, Y. F. Wu, B. Heying, D. Kapolnek, J. S. Speck, U. K. Mishra, and S. P. DenBaars, *Appl. Phys. Lett.* 68, 1526 (1996).

141. E. Iliopoulos, D. Doppalapudi, H. M. Ng, and T. D. Moustakas, *Appl. Phys. Lett.* 73, 375 (1998).

142. O. Lagerstedt and B. Monemar, *J. Appl. Phys.* 45, 2266 (1979).

143. M. Ilegems and R. Dingle, *J. Appl. Phys.* 39, 4234 (1973).

144. D. Kapolnek, X. H. Wu, B.Heying, S. Keller, B. P. Keller, U. K. Mishra, S. P. DenBaars, and J. S. Speck, *Appl. Phys. Lett.* 67, 1541 (1995).

145. X. H. Wu, D. Kapolnek, E. J. Tarsa, B. Heying, S. Keller, B. P. Keller, U. K. Mishra, S. P. DenBaars, and J. S. Speck, *Appl. Phys. Lett.* 68, 1371 (1996).

146. S. Keller, D. Kapolnek, B. P. Keller, Y. Wu, B. Heying, J. S. Speck, U. K. Mishra, and S. P. DenBaars, *Jpn. J. Appl. Phys.* 35, L285 (1996).

147. E. C. Piquette, P. M. Bridger, Z. Z. Bandic, and T. C. McGill, *Mater. Res. Soc. Internet J. Semicond. Res.* 4S1, G3.77 (1999).

148. Y. L. Vaillant, R. Bisaro, J. Olivier, O. Durand, J. Duboz, S. Clur, O. Briot, B. Gil, and R. Aulombard, P2-45, Tokushima Meeting, 1998.

149. L. Cheng, Z. Zhang, G. Zhang, and D. Yu, *Appl. Phys. Lett.* 71, 3694 (1997).

150. L. Sugiura, K. Itaya, J. Nishio, H. Fujimoto, and Y. Kokubun, *J. Appl. Phys.* 82, 4877 (1997).

151. C. F. Lin, G. C. Chi, M. S. Feng, J. D. Guo, J. S. Tsang, and J. M. Hong, *Appl. Phys. Lett.* 68, 3758 (1996).

152. Z. Liliental-Weber, S. Ruvimov, T. Suski, J. W. Ager, W. Swider, Y. Chen, Ch. Kisielowski, J. Washburn, I. Akasaki, H. Amano, C. Kuo, and W. Imler, *Mater. Res. Soc. Symp. Proc.* 423, 487 (1996).

153. A. D. Bykhovski, B. L. Gelmont, and M. S. Shur, *J. Appl. Phys.* 78, 3691 (1995).

154. F. Widmann, B. Daudin, G. Feuillet, Y. Samson, M. Arlery, and J. L. Rouviere, *Mater. Res. Soc. Internet J. Nitride Semicond. Res.* 2, 20 (1997).

155. H. M. Ng, D. Doppalapudi, E. Iliopoulos, and T. D. Moustakas, *Appl. Phys. Lett.* 74, 1036 (1999).

156. B. Daudin, F. Widmann, J. Simon, G. Feuillet, J. L. Rouviere, N. Pelekanos, and G. Fishman, *Mater. Res. Soc. Internet J. Nitride Semicond. Res.* 4S1, G9.2 (1999).

157. X. Li, D. V. Forbes, S. Q. Gu, D. A. Turnbull, S. G. Bishop, and J. J. Coleman, *J. Electron. Mater.* 24, 1711 (1995).

158. O. Brandt, R, Muralidharan, P. Waltereit, A. Thamm, A. Trampert, H. von Kiedrowski, and H. H. Ploog, *Appl. Phys. Lett.* 75, 4019 (1999).

159. Z. Y. Xie, C. H. Wei, L. Y. Li, J. H. Edgar, J. Chaudhuri, and C. Ignatiev, *Mater. Res. Soc. Internet J. Nitride Semicond. Res.* 4S1, G3.39 (1999).

160. R. Lantier, A. Rizzi, D. Guggi, H, Luth, B. Neubauer, D. Gerthsen, S. Frabboni, G. Coli, and R. Cingolani, *Mater. Res. Soc. Internet J. Nitride Semicond. Res.* 4S1, G3.50 (1999).

161. V. M. Torres, J. L. Edwards, B. J. Wilkens, D. J. Smith, R. B. Doak, and I. S. T. Tong, *Appl. Phys. Lett.* 74, 985 (1999).

162. F. Boscherini, R. Lantier, A. Rizzi, F. D'Acapito, and S. Mobilio, *Appl. Phys. Lett.* 74, 3308 (1999).

163. C. M. Zetterling, M. Ostling, K. Wongchotigul, M. G. Spencer, X. Tang, C. I. Harris, N. Vordell, and S. S. Wong, *J. Appl. Phys.* 82, 2990 (1997).

164. S. Wilson, C. S. Dickens, J. Griffin, and M. G. Spencer, *Mater. Res. Soc. Internet J. Nitride Semicond. Res.* 4S1, G3.61 (1999).

165. F. A. Ponce, C. G. Van de Walle, and J. E. Northrup, *Phys. Rev. B* 53, 7473 (1996).

166. F. A. Ponce, M. A. O'Keefe, and E. C. Nelson, *Philos. Mag. A* 74, 777 (1996).

167. M. D. Bremser, W. G. Perry, T. Zheleva, N. V. Edwards, O. H. Nam, N. Parikh, D. E. Aspnes, and R. F. Davis, *Mater. Res. Soc. Internet J. Nitride Semicond. Res.* 1, 30 (1996).

168. C. F. Lin, H. C. Cheng, G. C. Chi, M. S. Feng, L. D. Guo, J. M. Hong, and C. Y. Chen, *J. Appl. Phys.* 82, 2378 (1997).

169. R. Hull, A. Fischer-Colbrie, S. Rosner, S. M. Koch, and J. S. Harris, *Appl. Phys. Lett.* 51, 1723 (1987).

170. Z. Liliental-Weber, H. Sohn, N. Newman, and J. Washburn, *J. Vac. Sci. Technol. B* 13, 1598 (1995).

171. D. J. Smith, D. Chandrasekhar, B. Sverdlov, A. Botchkarev, A. Salvador, and H. Morkoc, *Appl. Phys. Lett.* 67, 1830 (1995).

172. F. A. Ponce, B. S. Krusor, J. S. Major, Jr., W. E. Plano, and D. F. Welch, *Appl. Phys. Lett.* 67, 410 (1995).

173. B. N. Sverdlov, G. A. Martin, H. Morkoc, and D. J. Smith, *Appl. Phys. Lett.* 67, 2063 (1995).

174. F. R. Chien, X. J. Ning, S. Stemmer, P. Pirouz, M. D. Brember, and R. F. Davis, *Appl. Phys. Lett.* 68, 2678 (1996).

175. H. Ohsato and M. Razeghi, *SPIE* 2999, 288 (1997).

176. Y. Nakada, I. Aksenov, and H. Okumura, *Appl. Phys. Lett.* 73, 827 (1998).

177. A. Ohtani, K. S. Stevens, and R. Beresford, *Appl. Phys. Lett.* 65, 61 (1994).

178. X. Ke, X. Jun, D. Peizhen, Z. Yongzong, Z. Guoqing, Q. Rongsheng, and F. Zujie, *J. Cryst. Growth* 193, 127 (1998).

179. E. S. Hellman, Z. L. Weber, and D. N. E. Buchanan, *Mater. Res. Soc. Internet J. Nitride Semicond. Res.* 2, 30 (1997).

180. O. Kryliouk, T. Dann, T. Anderson, H. Maruska, L. Zhu, J. Daly, M. Lin, P. Norris, H. Chai, D. Kisker, J. Li, and K. Jones, *Mater. Res. Soc. Symp. Proc.* 449, 123 (1997).

181. A. V. Andrianov, D. E. Lacklison, J. W. Orton, T. S. Cheng, C. T. Foxon, K. P. O'Donnell, and J. Nicholls, *Semicond. Sci. Technol.* 12, 59 (1997).

182. T. Matsuoka, N. Yoshimoto, T. Sasaki, and A. Katsui, *J. Electron. Mater.* 21, 157 (1992).

183. M. J. Suscavage, D. F. Ryder, Jr., and P. W. Yip, *Mater. Res. Soc. Symp. Proc.* 449, 283 (1997).

184. F. Hamdani, M. Yeadon, D. Smith, H. Tang, W. Ki, A. Salvador, A. E. Bothkarev, J. M. Gibson, A. Y. Polyakov, M. Skowronski, and H. Morkoc, *J. Appl. Phys.* 83, 983 (1998).

185. Z. Sitar, M. Paisley, B. Yan, and R. F. Davis, *Mat Res. Soc. Symp. Proc.* 162, 537 (1990).

186. M. Paisley, Z. Sitar, J. B. Posthil, and R. F. Davis, *J. Vac. Sci. Technol.* 7, 701 (1989).

187. H. Liu, A. C. Frenkel, J. G. Kim, and R. M. Park, *J. Appl. Phys.* 74, 6124 (1993).

188. A. Kikuchi, H. Hoshi, and K. Kishino, *J. Cryst. Growth* 150, 897 (1995).

189. O. Brandt, H. Yang, A. Trampert, M. Wassermeier, and K. H. Ploog, *Appl. Phys. Lett.* 71, 473 (1997).

190. R. C. Powell, N. E. Lee, Y. W. Kim, and J. E. Greene, *J. Appl. Phys.* 73, 189 (1993).

191. T. Lei, T. D. Moustakas, R. J. Graham, Y. He, and S. J. Berkowitz, *J. Appl. Phys.* 71, 4933 (1992).

192. S. N. Basu, T. Lei, and T. D. Moustakas, *J. Mater. Res.* 9, 2370 (1994).

193. Y. Morimoto, K. Uchiho, and S. Ushio, *J. Electrochem. Soc. Solid State Technol.* 120, 1783 (1973).

194. Y. Hiroyama and M. Tamura, *Mater. Res. Soc. Internet J. Nitride Semicond. Res.* 4S1, G3.9 (1999).

195. Y. Bar-Youn and T. D. Moustakas, *Nature* 342, 786 (1989).

196. D. W. Jenkins, J. D. Dow, and M. Tsai, *J. Appl. Phys.* 72, 4130 (1992).

197. J. Neugebauer and C. G. Van de Walle, *Mater. Res. Soc. Proc.* 339, 687 (1994).

198. P. Perlin, T. Suzki, H. Teisseyre, M. Leszczynski, I. Grzegory, J. Jun, S. Porowski, P. Boguslawski, J. Berholc, J. C. Chervin, A. Polian, and T. D. Moustakas, *Phys. Rev. Lett.* 75, 296 (1995).

199. W. Seifert, R. Franzheld, E. Butter, H. Sobotta, and V. Riede, *Cryst. Res. Technol.* 18, 383 (1983).

200. B. C. Chung and M. Gershenzon, *J. Appl. Phys.* 72, 651 (1992).

201. K. Doverspike and J. I. Pankove, in "Gallium Nitride (GaN) I.", Semiconductors and Semimetals (J. I. Pankove and T. D. Moustakas, Eds.), Vol. 50. Academic Press, New York, 1998.

202. D. K. Gaskill, K. Doverspike, L. B. Rowland, and L. D. Rode, "International Symposium on the Compound Semiconductors" Vol. 425. IOP, Bristol, U.K., 1995.

203. R. J. Molnar, T. Lei, and T. D. Moustakas, *Appl. Phys. Lett.* 62, 72 (1993).

204. P. Hacke, A. Maekawa, N. Koide, K. Hiramatsu, and N. Sawaki, *Jpn. J. Appl. Phys.* 33, 6443 (1994).

205. W. Gotz, N. M. Johnson, C. Chen, H. Liu, C. Kuo, and W. Imler, *Appl. Phys. Lett.* 68, 3144 (1996).

206. H. Murakami, T. Asahi, H. Amano, K. Hiramatsu, N. Sawaki, and I. Akasaki, *J. Cryst. Gowth* 115, 648 (1991).

207. S. Ruvimov, Z. Liliental-Weber, T. Suski, J. W. Ager, J. Washburn, J. Krueger, C. Kisielowski, E. R. Weber, H. Amano, and I. Akasaki, *Appl. Phys. Lett.* 69, 990 (1996).

208. J. Burm, K. Chu, W. A. Davis, W. A. Schaff, L. F. Eastmann, and T. J. Eustis, *Appl. Phys. Lett.* 70, 464 (1997).

209. S. Nakamura, T. Mukai, and M. Senoh, *Jpn. J. Appl. Phys.* 31, 2882 (1992).

210. J. I. Pankove, *J. Lumin.* 7, 114 (1992).

211. C. R. Abernathy, J. D. Mackenzie, S. J. Pearton, and W. S. Hobson, *Appl. Phys. Lett.* 66, 1969 (1995).

212. H. Amano, M. Kito, K. Hiramtsu, and I. Akasaki, *Jpn. J. Appl. Phys.* 28, L2112 (1989).

213. S. Nakamura, M. Senoh, and T. Mukai, *Jpn. J. Appl. Phys.* 30, L1708 (1991).

214. J. I. Pankove, P. J. Zanzucchi, and C. Magee, *Appl. Phys. Lett.* 46, 421 (1985).

215. X. Li and J. J. Coleman, *Appl. Phys. Lett.* 69,1605 (1996).

216. S. Nakamura, T. Mukai, M. Senoh, and N. Isawa, *Jpn. J. Appl. Phys.* 31, L139 (1992).

217. T. Suski, P. Perlin, A. Pietraszki, M. Leszczynski, M. Bockowski, I. Grzegory, and S. Porowski, *J. Cryst. Growth* 207, 27 (1999).

218. M. Kamp, M. Mayer, A. Pelzmann, and K. J. Ebeling, *Mater. Res. Soc. Symp. Proc.* 449, 161 (1997).

219. M. S. Brandt, N. M. Johnson, R. J. Molnar, R. Singh, and T. D. Moustakas, *Appl. Phys. Lett.* 64, 2264 (1994).

220. M. S. Brandt, J. W. Ager III, W. Gotz, N. M. Johnson, J. S. Harris, Jr., R. J. Molnar, and T. D. Moustakas, *Phys. Rev. B Rapid Commun.* 49, 14,758 (1994).

221. S. Nakamura, T. Mukai, and M. Senoh, *Jpn. J. Appl. Phys.* 31, 2883 (1993).

222. S. Yamasaki, S. Asami, N. Shibita, M. Koike, K. Manabe, T. Tanaka, H. Amano, and I. Akasaki, *Appl. Phys. Lett.* 66, 1112 (1995).

223. S. C. Jain, M. Willander, J. Narayan, and R. Van Overstraeten, *J. Appl. Phys.* 87, 965 (2000).

224. H. Murakami, T. Asaki, H. Amano, K. Hiramatsu, N. Sawaki, and I. Akasaki, *J. Cryst. Growth* 115, 648 (1991).

225. T. Tanaka, A. Watanabe, H. Amano, Y. Kobayashi, I. Akasaki, S. Yamazaki, and M. Koike, *Appl. Phys. Lett.* 65, 593 (1994).

226. D. Korakakis, H. M. Ng, M. Misra, W. Grieshaber, and T. D. Moustakas, *Mater. Res. Soc. Internet J. Nitride Semicond.* 1, 10 (1996).

227. C. Stampfl and C. G. Van de Walle, *Appl. Phys. Lett.* 72, 459 (1995).

228. S. Mahajan and M. A. Shahid, *Mater. Res. Soc. Proc.* 144, 169 (1996).

229. J. H. van der Merwe, *Crit. Rev. Solid State Mater. Sci.* 17, 187 (1991).

230. S. C. Jain, M. Wilander, J. Narayan, and R. Van OverStraeten, *J. Appl. Phys.* 87, 965 (2000).

231. M. Nido, *Jpn. J. Appl. Phys.* 34, 1514 (1995).

232. H. Amano, T. Takeuchi, S. Sota, H. Sakai, and I. Akasaki, *Mater. Res. Soc. Symp. Proc.* 449, 1143 (1997).

233. I. Lee, I. Choi, C. R. Lee, and S. K. Noh, *Appl. Phys. Lett.* 71, 1359 (1997).

234. B. J. Skromme, H. Zhao, D. Wang, H. S. Kong, M. T. Leonard, G. E. Bulman, and R. J. Molnar, *Appl. Phys. Lett.* 71, 829 (1997).

235. T. S. Cheng, C. T. Foxon, G. B. Ren, J. W. Orton, Y. V. Melnik, I. P. Nikitina, A. E. Nikolaev, S. V. Novikov, and V. A. Dimitriev, *Semicond. Sci. Technol.* 12, 917 (1997).

236. B. Gil, in "Gallium Nitride (GaN) II." Semiconductor and Semimetals (J. I. Pankove and T. D. Moustakas, Eds.), Vol. 57. Academic Press, New York, 1999.

237. N. V. Edwards, M. D. Bremser, R. F. Davis, A. D. Batchelor, S. D. Yoo, C. F. Haran, and D. E. Aspnes, *Appl. Phys. Lett.* 73, 2808 (1998).

238. O. Briot, J. P. Alexis, B. Gil, and R. L. Aulombard, *Mater. Res. Soc. Symp. Proc.* 395, 411 (1996).

239. K. Funato, F. Nakamura, S. Hashimoto, and M. Ikeda, *Jpn. J. Appl. Phys.* 37, L1024 (1998).

240. S. Ruvimov, Z. Liliental-Weber, T. Suski, J. W. Ager, J. Washburn, J. Krueger, C. Kisielowski, E. R. Weber, H. Amano, and I. Akasaki, *Appl. Phys. Lett.* 69, 990 (1996).

241. C. Kisielowski, in "Gallium Nitride (GaN) II." Semiconductor and Semimetals (J. I. Pankove and T. D. Moustakas, Eds.), Vol. 57. Academic Press, New York, 1999.

242. H. Liu, J. G. Kim, M. H. Ludwig, and R. M. Park, *Appl. Phys. Lett.* 71, 347 (1997).

243. D. Zubia, S. H. Zaidi, S. R. J. Brueck, and S. D. Hersee, *Appl. Phys. Lett.* 76, 858 (2000).

244. M. Mynbaeva, A. Titkov, A. Kryzhanovski, I. Kotousova, A. S. Zubrilov, V. V. Ratnikov, V. Y. Davydov, N. I. Kuznetsov, K. Mynbaev,

D. V. Tsvetkov, S. Stepanov, A. Cherenkov, and V. A. Dimitriev, *Mater. Res. Soc. Internet J. Semicond. Res.* 4, 14 (1999).

245. D. Doppalapudi, R. Mlcak, A. Sampath, and T. D. Moustakas, to be published.

246. E. S. Hellman, *Mater. Res. Soc. Internet J. Nitride Semicond. Res.* 3, 11 (1998).

247. M. Seelmann-Eggebert, J. L. Weyher, H. Obloh, H. Zimmermann, A. Rar, and S. Porowski, *Appl. Phys. Lett.* 71, 2635 (1997).

248. A. R. Smith, R. M. Feenstra, D. W. Greve, M. S. Shin, M. Skowronski, J. Neugebauer, and J. Northrup, *Appl. Phys. Lett.* 72, 2114 (1998).

249. A. R. Smith, R. M. Feenstra, D. W. Greve, J. Neugebauer, and J. Northrup, *Phys. Rev. Lett.* 79, 3934 (1997).

250. M. Steelmann-Eggebert, J. L. Weyher, H. Obloh, H. Zimmermann, A. Rar, and S. Porowski, *Appl. Phys. Lett.* 71, 2635 (1997).

251. X. Zhang, P. Kung, D. Walker, A. Saxler, and M. Razeghi, *Mater. Res. Soc. Symp. Proc.* 395, 625 (1996).

252. J. Neugebauer, and C. G. Van de Walle, *Appl. Phys. Lett.* 69, 503 (1996).

253. T. Mattila, R. M. Nieminen, and A. P. Setsonen, *Electrochem. Soc. Proc.* 96, 205 (1996).

254. R. De Felice and J. E. Northrup, *Appl. Phys. Lett.* 73, 936 (1998).

255. A. Bykhovski, B. Gelmont, and M. Shur, *J. Appl. Phys.* 74, 6734 (1993).

256. L. T. Romano, J. E. Northrup, and M. A. O'Keele, *Appl. Phys. Lett.* 69, 2394 (1996).

257. J. E. Northrup, J. Neugenbauer, and L. T. Romano, *Phys. Rev. Lett.* 77, 103 (1996).

258. J. L. Rouviere, M. Arley, A. Bourret, R. Niebuhr, and K. H. Bachem, *Mater. Res. Soc. Symp. Proc.* 395, 393 (1996).

259. L. T. Romano, B. S. Krusor, R. Singh, and T. D. Moustakas, *J. Electron. Mater.* 26, 285 (1997).

260. B. Daudin, J. L. Rouviere, and M. Arlery, *Appl. Phys. Lett.* 69, 2480 (1996).

261. L. T. Romano and T. H. Myers, *Appl. Phys. Lett.* 71, 3486 (1997).

262. M. Sumiya, T. Ohnishi, M. Tanaka, A. Ohmoto, M. Kawasaki, M. Yoshimoto, H. Koinuma, K. Ohtsuka, and S. Fuke, *Mater. Res. Soc. Internet J. Nitride Semicond. Res.* G6.23 (1999).

263. Z. Z. Bandic, T. C. McGill, and Z. Ikonic, *Phys. Rev. B* 56, 3564, 1997.

264. M. J. Paisley and R. F. Davis, *J. Cryst. Growth* 127, 136 (1993).

265. R. C. Powell, N. E. Lee, Y. W. Kim, and J. E. Greene, *Appl. Phys. Lett.* 73, 189 (1993).

266. F. A. Ponce, J. S. Major, W. E. Plano, and D. F. Welch, *Appl. Phys. Lett.* 65, 2302 (1994).

267. A. F. Wright, *J. Appl. Phys.* 82, 5259 (1997).

268. S. D. Lester, F. A. Ponce, M. G. Craford, and D. A. Steigerwald, *Appl. Phys. Lett.* 66, 1249 (1995).

269. X. J. Ning, F. R. Chien, P. Pirouz, J. W. Yang, and M. Asif Khan, *J. Mater. Res.* 11, 580 (1996).

270. Z. Liliental-Weber, H. Sohn, N. Newman, and J. Washburn, *J. Vac. Sci. Technol. B* 13, 1598 (1995).

271. B. Heying, X. H. Wu, S. Keller, Y. Li, D. Kapolnek, B. P. Keller, S. P. DenBaars, and J. S. Speck, *Appl. Phys. Lett.* 68, 643 (1996).

272. L. T. Romano, J. E. Northrup, and M. A. O'Keele, *Appl. Phys. Lett.* 69, 2394 (1996).

273. J. E. Northrup, J. Neugenbauer, and L. T. Romano, *Phys. Rev. Lett.* 77, 103 (1996).

274. D. Kapolnek, X. H. Wu, B. Heying, S. Keller, U. K. Mishra, S. P. DenBaars, and J. S. Speck, *Appl. Phys. Lett.* 67, 1541 (1995).

275. M. J. Paisley and R. F. Davis, *J. Cryst. Growth* 127, 136 (1993).

276. W. Qian, M. Skowronski, M. DeGraef, K. Doverspike, L. B. Rowland, and D. K. Gaskill, *Appl. Phys. Lett.* 66, 1252 (1995).

277. X. H. Wu, L. M. Brown, D. Kapolmek, S. Keller, B. Keller, S. P. DenBaars, and J.S Speck, *J. Appl. Phys.* 30, 5 (1996).

278. W. Qian, G. S. Rohrer, M. Skowronski, K. Doverspike, L. B. Rowland, and D. K. Gaskill, *Appl. Phys. Lett.* 67, 2284 (1995).

279. Z. Liliental-Weber, S. Ruvimov, C. Kiesielowski, Y. Chen, W. Swider, J. Washburn, N. Newman, A. Gassmann, X. Liu, L. Schloss, E. R. Weber, I. Grzegory, M. Bockowski, J. Jun, T. Suski, K. Pakula, J. Baranowski,

S. Porowski, H. Amano, and I. Akasaki, *Mater. Res. Soc. Symp. Proc.* 395, 351 (1996).

280. T. Kozawa, T. Kachi, T. Ohwaki, Y. Taga, N. Koide, and M. Koike, *J. Electrochem. Soc.* 143, L17 (1996).

281. X. J. Ning, F. R. Chien, P. Pirouz, J. Yang, and M. A. Khan, *J. Mater. Res.* 11, 580 (1996).

282. D. Cherns, W. T. Young, J. W. Steeds, F. A. Ponce, and S. Nakamura, *J. Cryst. Growth* 178, 201 (1997).

283. J. E. Northrup, *Appl. Phys. Lett.* 72, 2316 (1998).

284. S. S. Dhesi, C. B. Stagarescu, K. E. Smith, D. Doppalapudi, R. Singh, and T. D. Moustakas, *Phys. Rev. B* 56, 10271 (1997).

285. T. Valla, P. D. Johnson, S. S. Dhesi, K. E. Smith, D. Doppalapudi, T. D. Moustakas, and E. L. Shirley, *Phys. Rev. B* 59, 5003 (1999).

286. J. Elsner, R. Jones, P. K. Sitch, V. D. Porezag, M. Elstner, Th. Frauenheim, M. I. Heggie, S. Oberg, and P. R. Briddon, *Phys. Rev. Lett.* 79, 3672 (1997).

287. Y. Xin, S. J. Pennycook, N. D. Browning, P. D. Nellist, S. Sivananthan, F. Omnes, B. Beaumont, J. P. Faurie, and P. Gibart, *Appl. Phys. Lett.* 72, 2680 (1998).

288. N. G. Weimann, L. F. Eastman, D. Doppalapudi, H. M. Ng, and T. D. Moustakas, *J. Appl. Phys.* 83, 3656 (1998).

289. H. M. Ng, D. Doppalapudi, T. D. Moustakas, N. G. Weimann, and L. F. Eastman, *Appl. Phys. Lett.* 73, 821 (1998).

290. A. Sakai, H. Sunakawa, and A. Usui, *Appl. Phys. Lett.* 71, 2259 (1997).

291. O. H. Nam, M. D. Bremser, T. S. Zheleva, and R. F. Davis, *Appl. Phys. Lett.* 71, 2638 (1997).

292. D. Kapolnek, S. Keller, R. Vetury, R. D. Underwood, P. Kozodoy, S. P. DenBaars, and U. K. Mishra, *Appl. Phys. Lett.* 71, 1204 (1997).

293. D. Doppalapudi, K. J. Nam, A. Sampath, R. Singh, H. M. Ng, S. N. Basu, and T. D. X. Moustakas, *Electrochem. Soc. Proc.* 98–18, 90 (1998).

294. F. W. Tausch, Jr. and A. G. Lapierre, III, *J. Electrochem. Soc.* 112, 706 (1965).

295. D. W. Shaw, *J. Electrochem. Soc.* 113, 904 (1966).

296. K. Linthicum, T. Gehrke, D. Thomson, E. Carlson, P. Rajgopal, T. Smith, D. Batchelor, and R. Davis, *Appl. Phys. Lett.* 75, 196 (1999).

297. K. Linthicum, T. Gehrke, D. Thomson, K. Tracy, E. Carlson, T. Smith, T. Zheleva, C. Zorman, M. Mehregany, and R. Davis, *Mater. Res. Soc. Internet J. Nitride Semicond. Res.* 4S1, G4.9 (1999).

298. S. Nakamura, T. Mukai, and M. Senoh, *Appl. Phys. Lett.* 84, 1687 (1994).

299. S. Nakamura, M. Senoh, S. Nagahama, N. Isawa, T. Yamada, T. Matsushita, H. Kiyoku, and Y. Sugimoto, *Jpn. J. Appl. Phys.* 35, L74 (1996).

300. I. Akasaki, S. Sota, H. Sakai, T. Tanaka, M. Koike, and H. Amano, *Electron. Lett.* 32, 1105 (1996).

301. S. Nakamura, M. Senoh, S. Nagahama, N. Isawa, T. Yamada, T. Matsushita, Y. Sugimoto, and H. Kiyoku, *LEOS Newsletter* 11, 16 (1997).

302. R. Singh, Ph.D. Dissertation, Electrical and Computer Engineering, Boston University, Boston, MA, 1997.

303. H. Riechert, R. Averbeck, A. Graber, M. Schienle, U. Straub, and H. Tews, *Mater. Res. Soc. Proc.* 449, 149 (1997).

304. M. D. McCluskey, C. G. Van de Walle, C. P. Master, L. T. Romano, and N. M. Johnson, *Appl. Phys. Lett.* 72, 2725 (1998).

305. E. L. Piner, F. G. McIntosh, J. C. Roberts, K. S. Boutros, M. E. Aumer, V. A. Joshkin, N. A. El-Masry, S. M. Bedair, S. X. Liu, *Mater. Res. Soc. Symp. Proc.* 449, 85 (1997).

306. S. Keller, B. P. Keller, D. Kapolnek, A. C. Abare, H. Masui, L. A. Coldren, U. K. Mishra, and S. P. DenBaars, *Appl. Phys. Lett.* 68, 3147 (1996).

307. K. S. Boutros, F. G. McIntosh, J. C. Roberts, S. M. Bedair, E. L. Piner, and N. A. El-Masry, *Appl. Phys. Lett.* 67, 1856 (1995).

308. N. Grandjean and J. Massies, *Appl. Phys. Lett.* 72, 1078 (1998).

309. D. Doppalapudi, S. N. Basu, K. F. Ludwig, Jr., and T. D. Moustakas, *J. Appl. Phys.* 84, 1389 (1998).

310. Y. Kawaguchi, M. Shimizu, K. Hiramatsu, and N. Sawaki, *Mater. Res. Soc. Proc.* 449, 89 (1997).

311. A. Zunger and S. Mahajan, in "Handbook on semiconductors" (S. Mahajan, Ed.), Vol. 3. North-Holland, Amsterdam, 1994.

312. G. B. Stringfellow, *J. Cryst. Growth* 58, 194 (1982).

313. M. B. Panish and M. Ilegems, "Progress in Solid State Chemistry" (M. Reiss and J. O. McCaldin, Eds.), p. 39. Pergamon, New York, 1972.

314. J. L. Martins and A. Zunger, *Phys. Rev. B* 30, 6217 (1984).

315. I. H. Ho and G. B. Stringfellow, *Appl. Phys. Lett.* 69, 2701 (1996).

316. S. B. Zhang and A. Zunger, *Appl. Phys. Lett.* 71, 677 (1997).

317. S. Y. Karpov, *Mater. Res. Soc. Nitride Internet J.* 3 (1998).

318. S. Mahajan, *Mater. Sci. Eng. B* 30, 187 (1995).

319. K. Lee, B. A. Philips, R. S. McFadden, and S. Mahajan, *Mater. Sci. Eng. B* 32, 231 (1995).

320. B. A. Philips, A. G. Norman, T. Y. Seong, S. Mahajan, G. R. Booker, M. Skowronski, J. P. Harbison, and V. G. Keramidas, *J. Cryst. Growth* 140, 249 (1994).

321. A. G. Norman, T.-Y. Seong, I. T. Ferguson, G. R. Booker, and B. A. Joyce, *Semicond. Sci. Technol.* 8, s9-s15 (1993).

322. R. Singh and T. D. Moustakas, *Mater. Res. Soc. Symp. Proc.* 395, 163 (1996).

323. R. Singh, W. D. Herzog, D. Doppalapudi, M. S. Unlu, B. B. Goldberg, and T. D. Moustakas, *Mater. Res. Soc. Symp. Proc.* 449, 185 (1997).

324. N. A. El-Masry, E. L. Piner, S. X. Liu, and S. M. Bedair, *Appl. Phys. Lett.* 72, 40 (1998).

325. A. Wakahara, T. Tokuda, X. Dang, S. Noda, and A. Sasaki, *Appl. Phys. Lett.* 71, 906 (1997).

326. H. Sato, T. Sugahara, Y. Naoi, and S. Sakai, M 2-7, Tokushima Meeting, 1998.

327. S. Chichibu, T. Azuhata, T. Sota, and S. Nakamura, *Appl. Phys. Lett.* 70, 2822 (1997).

328. M. D. McCluskey, L. T. Romano, B. S. Krusor, D. P. Bour, N. M. Johnson, and S. Brennan, *Appl. Phys. Lett.* 72, 1730 (1998).

329. T. Sugahara, M. Hao, T. Wang, D. Nakagawa, Y. Naoi, K. Nishino, and S. Sakai, *Jpn. J. Appl. Phys.* 37, L1195 (1998).

330. G. P. Srivastava, J. L. Martins, and A. Zunger, *Phys. Rev. B* 31, 2521 (1985).

331. T. Suzuki, A. Gomyo, and S. Iijima, *J. Cryst. Growth* 93, 396 (1988).

332. T. S. Kuan, T. F. Kuech, W. I. Wang, and E. L. Wilkie, *Phys. Rev. Lett.* 54, 201 (1985).

333. D. Korakakis, K. F. Ludwig, Jr., and T. D. Moustakas, *Appl. Phys. Lett.* 71, 72 (1997).

334. D. Doppalapudi, S. N. Basu, and T. D. Moustakas, *J. Appl. Phys.* 85, 883 (1999).

335. H. Z. Xiao, N. E. Lee, R. C. Powell, Z. Ma, L. J. Chou, L. H. Allen, J. E. Greene, and A. Rockett, *J. Appl. Phys.* 76, 8195 (1994).

336. B. D. Cullity, "Elements of X-ray Diffraction," Addison-Wesley, Reading, MA, 1978.

337. A. Dimitriev, *Mater. Res. Soc. Internet J. Nitride Semicond. Res.* 1, 46 (1996).

338. J. E. Bernard, R. G. Dandrea, L. G. Ferreira, S. Froyen, S.-H. Wei, and A. Zunger, *Appl. Phys. Lett.* 56, 731 (1990).

339. W. E. Plano, D. W. Nam, J. S. Major, Jr., K. C. Hsieh, and N. Holonyak, Jr., *Appl. Phys. Lett.* 53, 2537 (1988).

340. E. Iliopoulos, K. F. Ludwig, Jr., T. D. Moustakas, and S. N. G. Chu, *Appl. Phys. Lett.* 78, 463 (2001).

341. P. Ruterana, G. De Saint Jores, M. Laught, F. Omnes, and E. Bellet-Amalric, *Appl. Phys. Lett.* 78, 344 (2001).

Chapter 3

OPTICAL PROPERTIES OF HIGHLY EXCITED (Al, In) GaN EPILAYERS AND HETEROSTRUCTURES

Sergiy Bidnyk, Theodore J. Schmidt, Jin-Joo Song

Center for Laser and Photonics Research and Department of Physics, Oklahoma State University, Stillwater, Oklahoma, USA

Contents

Handbook of Thin Film Materials, edited by H.S. Nalwa
Volume 4: Semiconductor and Superconductor Thin Films
Copyright © 2002 by Academic Press

ISBN 0-12-512912-2/$35.00

1. INTRODUCTION

1.1. Historical Perspective and Economic Projections

At the turn of the new millennium, the group III–nitride semiconductor thin films, most notably AlN, GaN, InN, and their alloys, have experienced an unprecedented amount of attention in the research and industrial world. This attention has been sparked by the realization of high-brightness ultraviolet (UV), blue, green, and amber light-emitting diodes (LEDs) made of group III–nitride thin films. Yet by far the most anticipated event in nitride research has been the invention of the "blue laser diode" based on the ternary alloy InGaN.

The first observation of optically pumped lasing in GaN was made by Dingle et al. [1] in the early 1970s using needlelike GaN single crystals at a temperature of 2 K, but it took almost a quarter of a century before the first current injection lasing was observed. It all began with work done by Akasaki and Amano in 1989, who developed p-type doping and were the first to report the observation of room temperature coherent emission in an InGaN/GaN quantum well structure using pulsed current injection [2, 3]. In December 1995, the fabrication of laser diodes (LDs) using group III–nitrides was announced for the first time by Nakamura et al. [4] at Nichia Chemical, Inc. Since then, the performance of Nichia InGaN LDs has steadily improved, and the lifetime of LDs under continuous wave (cw) conditions at room temperature recently exceeded 10,000 hours, making them suitable for commercial applications. Inspired by this discovery, several industrial research laboratories and university research groups have also succeeded in the fabrication of InGaN/GaN/AlGaN-based blue LDs that operate at room temperature. Nowadays, almost all major multinational electronics companies, and many universities and research laboratories are involved in continuing studies and optimization of group III–nitride thin films for device applications.

To elucidate such a strong industrial interest in this field, we now give a brief economic analysis of how much of the compound semiconductor market could be captured in the future by devices based on nitride semiconductor thin films. In 1997, the total market for compound semiconductors (the group III–nitrides belong to this category) was $5.7 billion [5]. LEDs, LDs, and analog devices constitute about 77% of the compound semiconductor market. In the year 2002, the total market for compound semiconductors is estimated to increase and reach values as high as $9 billion a year.

The largest part of the compound semiconductor market belongs to LEDs. In 1998, the total market for LEDs was as much as 33% of the total market for compound semiconductors and was valued at approximately $2.5 billion. We can further divide this market into categories according to emission wavelength. Most LEDs are AlGaAs- or AlGaInP-based and are used for infrared applications, such as optical switches, position measurement devices, optical encoders, and fluid level indicators. The market for blue, green, and yellow InGaN LEDs is still emerging and represents a relatively small market share. With growth of 30–50% a year, this portion of the market will soon be a significant, if not the largest, portion of all the LED market. Currently, only a handful of high-tech companies possess the ability to produce InGaN-based LEDs in substantial quantities. Among the biggest suppliers are Nichia Corporation (Japan), which has shipments of over 30 million units per month, Cree Research Inc. (USA), which produces 20 million units per month, and Hewlett Packard (USA), which delivers approximately 10 million units per month. Among the smaller suppliers are EMCORE (USA), Toyoda Gosei (Japan), Infineon Technologies (Germany), and Samsung (Korea).

The current price of a blue LED varies from company to company within the range of $0.25–1.00 per piece, which is rather expensive compared to AlInGaP LEDs. However, the cost of InGaN LEDs is not considered to be an intrinsic problem. With the current rate of increase in production, the price of these LEDs is expected to drop to $0.10–0.50 per piece by the year 2003. From the technological standpoint, the cost of packaging InGaN LEDs is about the same as for any other LED. It has been reported that high optical quality GaN structures can also be grown on Si substrates [6]. By further developing this technology, the processing cost will drop substantially due to larger wafer area and the possibility for a backside contact (currently nitride-based LEDs require both contacts to be on the light-emitting side of the sample due to the insulating nature of the sapphire substrate). Recently developed white LEDs are also attracting attention. Current white LEDs are 10 times more efficient, last 1000 times longer, and have an emission spectrum closer to that of the sun than the spectrum of a regular incandescent bulb. The cost of a white LED array (with a power equivalent to that of an electric bulb), however, remains relatively high. Currently, several companies are attempting to commercialize white-LED-based products, capitalizing on low maintenance costs for such devices over their long lifetime (10–20 years).

It has been predicted that full-color displays and traffic lights will consume a considerable part of all nitride LEDs produced [7]. Current blue and green InGaN-based LEDs already have a higher external quantum efficiency and output power compared to AlInGaP-based amber LEDs. In fact, GaN-based LEDs are predicted to prevail in the market for all visible spectrum emitters. Perhaps the only exception is in the long-wavelength range (amber–red), where the efficiency of GaN-based LEDs drops dramatically.

Having examined the LED market, we could shift our attention to the second most important component of the compound semiconductor market, that is, LDs. The market for LDs is only slightly smaller than that for LEDs. It is also one of the fastest growing markets in the United States. In 1999 the total market for LDs was $2.15 billion. Most LDs are used for storage ($1 billion) and communications ($1 billion).

The capacity of optical storage is strongly limited by the spot size of the laser beam. The minimum spot size to which a laser can be focused is proportional to the wavelength of the laser light. Due to the shorter emission wavelength of lasers based on GaN thin films, storage and high-resolution printing are natural choices for GaN LD applications. The annual market in the world for printing is rather large ($100 billion). The historical trend in this field is to move to higher resolution and speed. To lower production costs, shorter wavelength and higher power laser diodes are necessary. There are some limitations, however, as to how short the wavelength needs to be. The organic photoreceptors used in printers should be sensitive to the laser wavelengths and expensive optics should not be required. A compromise is found somewhere around 400 nm. It has been estimated that through the utilization of a 5-mW InGaN-based laser with a high-quality beam profile, it would be possible to achieve printing speeds of up to 60 pages a minute.

In 1999 more than 300 million lasers were used for optical storage devices. By utilizing InGaN-based lasers, it will be possible to improve the resolution and increase the storage capacities of optical disks. Currently, digital video device/high definition (DVD/HD) technology is being developed that utilizes InGaN LDs with an emission wavelength of 410 nm. This will allow the laser beam to be focused to a spot approximately 0.24 μm in diameter, boosting the capacity of the laser disk to 30 GB. To read information from DVD/HD disks, a minimum laser output power of 5 mW (cw) is required; to record information, the output power should be at least 50 mW (pulsed). Current state-of-the-art nitride laser diodes meet these requirements.

Communication devices are dominated by InP and GaAs lasers that emit in the wavelength range of 0.8–1.6 μm. Some preliminary research to fabricate nitride-based LDs in this wavelength range has been done on GaInNAs. The desired spectral range of these lasers is dictated by the optical loss of modern optical fibers. Recently, there has been a lot of effort dedicated to the development of rare-earth-ion-doped GaN optical amplifiers and lasers [8]. The advantage of GaN over smaller gap semiconductors and glasses includes greater chemical stability, efficient carrier generation (by exciting the rare-earth ions), and physical stability over a wide temperature range. GaN also has a high level of optical activity even under conditions of high defect density, which would quench emission in other smaller gap III–V and in most II–VI compounds.

Finally, the high-power LD market for industry and medical applications has traditionally utilized InGaAsP/GaInP and AlGaAs/GaAs lasing structures. GaN-based LDs are expected to significantly contribute to this market once the development of high-power LDs is completed. Such a wide utilization of nitride thin films in light-emitting devices will inevitably amount to a several billion dollar a year market.

The uses of the III–nitride thin films in non-light-emitting applications are also attracting a considerable amount of attention. Due to their large direct bandgaps, GaN and its alloy with AlN (AlGaN) are promising materials for the development of solar-blind detectors. Sunlight in the spectral range of 250–300 nm is filtered out by ozone located 15 km above the surface of the earth. Thus any source of UV radiation within the atmosphere can be easily seen by a solar-blind detector even on a sunny day. This would allow for effective military applications, such as the tracking of missiles and war planes. A recently developed GaN-based linear array of UV detectors [9, 10] has an order of magnitude improvement in size, weight, and power consumption in comparison to photomultiplier tubes and microchannel plates now used to detect UV light. The UV photodetectors could also be used to detect UV light in hot environments and to detect UV emissions from flames. State-of-the-art nitride-based UV photodetectors have a remarkably good selectivity with a ratio of UV to visible responsivity that exceeds 8 orders of magnitude [11]. In 1999, the market for UV detectors was rather small, but it is expected to grow as much as $100 million a year.

In 1997, the first studies of optical properties of GaN and InGaN thin films at temperatures as high as 400°C under intense optical excitation were reported [12, 13]. Stimulated emission and a low temperature sensitivity of the stimulated emission threshold were observed over a large temperature range. This result suggested that GaN-based structures are excellent candidates for high-temperature optoelectronic devices. The market for high-temperature devices was estimated to be $400 million in 1999, but it has the potential to grow to as large as $12 billion. The reason this market remains small stems from the lack of semiconductors capable of high-temperature operation. Nitride devices are expected to revolutionize this field and to find many applications in automotive and aerospace industries.

The general increase in the worldwide communications market in recent years has resulted in a very fast rate of annual growth. In 1999, the number of cellular phones in use exceeded 400 million and continues to grow at a rate of 30% a year. The market for satellite communication is also growing at a rate of 25% a year and about 100 new satellites are launched every year. The market for radiofrequency (RF) semiconductors for satellite communication was estimated to be $200 million in 1998. The total market for RF semiconductors in 1999 was $4.5 billion. It has been overwhelmingly dominated by Si and GaAs, leaving less than 1% to other semiconductors. However, there are two newcomers to this market: SiC and GaN. The latter emerges for frequencies in the tens of gigahertz and RF powers up to 100 W. The combination of these two parameters leaves GaN as the only candidate for many military applications (Si devices are not operable at frequencies above 1 GHz and GaAs devices fail at RF powers exceeding 10 W). State-of-the-art field effect and heterojunction bipolar transistors based on GaN have already been shown to possess superior characteristics and are expected to extend the capabilities of RF electronics in the near future.

There are many other applications for group III–nitride thin films that utilize photovoltaic and photoconductive properties that were not included in the current discussion. The majority of these applications are presently unrealized. It is clear, however, that devices based on nitride semiconductor thin films are so versatile that in the near future they will inevitably become an essential part of our lives.

1.2. Challenges in Nitride Thin Film Research and Development

Having described some of the applications of the group III–nitrides, we now examine some of the challenges in research and development of nitride thin film devices. In spite of the substantial progress made by many industrial research laboratories and university research groups, there are many issues that merit further study. Even the best GaN-based LEDs and LDs have huge densities of dislocations, predominantly threading between the substrate and the surface. Whereas it appears that they have no effect whatsoever on the efficiency of light emission, there is great concern that these defects limit the life-time of the LDs. This great number of dislocations arises due to the lack of lattice-matched substrates. GaN growth onto a variety of substrates has been investigated. The most common substrate is sapphire, which is relatively cheap, but the lattice mismatch with GaN is rather high (16%). GaN can be grown on 6H-SiC, but the high cost of SiC wafers has not allowed the widespread use of this material. GaN growth has also been performed on spinel ($MgAl_2O_4$) substrates. None of these substrates, however, has resulted in the growth of high crystalline quality GaN epilayers (when compared to other technologically important semiconductors such as GaAs and Si).

A substantial decrease in dislocation density was observed in regions of lateral epitaxial overgrowth (LEO) compared to regions of conventional vertical growth. LEO occurs when GaN is grown from a mask pattern. Under optimized conditions, lateral to vertical growth rate ratios of up to 5 can be achieved. The LEO technique has resulted in a 4–5 order of magnitude reduction in dislocation density. This observation triggered research on similar experimental techniques to improve the crystalline quality of GaN epilayers, such as selective area growth and pendeo-epitaxy. These techniques allow the growth of high-quality GaN epilayers independent of the type of substrate. To facilitate the incorporation of GaN optoelectronic devices into Si-based electronics, there have been many successful attempts to grow GaN on Si substrates by LEO. However, there is still much to be done in this field to optimize the growth parameters and reduce the cost of processing.

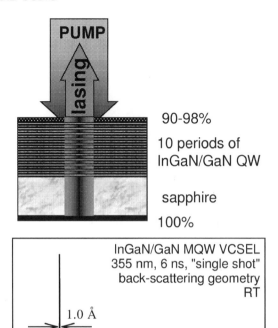

Fig. 1. One of the first demonstrations of room temperature lasing in an In-GaN/GaN VCSEL structure.

With the realization of edge-emitting InGaN-based lasers, the next milestone for group III–nitrides is the development of a vertical cavity surface-emitting laser (VCSEL). The output beam from the state-of-the-art edge-emitting "blue laser" is highly divergent due to diffraction from the narrow aperture formed by the edge of the active layer. Furthermore, the edge-emitting laser cannot be used to create two-dimensional arrays. Arrays can be used to create very large bandwidth (several terahertz) data links, while each individual device is only required to maintain a relatively low modulation rate. A two-dimensional array of VCSELs could also dramatically reduce the readout time in dense optical memory systems.

One of the first demonstrations of laser action at room temperature in InGaN/GaN multiple quantum well-based VCSELs is shown in Figure 1. To avoid lattice temperature heating effects due to large excitation density, the experiment was performed in "single shot" mode: the obtained spectra are the result of a single 6-ns-long pulse. Although optimization of the structure is still necessary, the feasibility of fabricating a nitride-based VCSEL has been clearly demonstrated. Currently, many research groups are pursuing the development of current-injection VCSELs. It might take several years before the first applications that utilize a blue vertical cavity surface-emitting laser emerge.

Finally, the nitride research community is putting a lot of effort into the realization of even shorter emission wavelength LDs for near- and deep-UV applications. It might be more challenging to manufacture LDs that emit photons with energies that exceed the GaN bandgap of 3.5 eV. In principle, these LDs could be realized by utilizing AlGaN compounds, but at this time it would be difficult to predict all the technological hurdles associated with short-wavelength LD development. The first step in this direction was taken by scientists at NTT's Basic Research Laboratory who fabricated a UV LED (346 nm) based on an $Al_{0.08}Ga_{0.92}N/Al_{0.12}Ga_{0.88}N$ multiple quantum well structure [14]. Also, room temperature stimulated emission was achieved in $Al_{0.26}Ga_{0.74}N$ epilayers with emission wavelength as short as 328 nm. This is the shortest wavelength stimulated emission observed in any semiconductor system.

Many aspects of the group III–nitrides have not been explored yet. This material system is very subtle from the viewpoints of optical characterization, growth, and device processing. Numerous exciting research studies are in progress on the group III–nitrides, and many more are anticipated.

1.3. Chapter Organization

This chapter intends to answer many questions related to the current state of knowledge regarding III–nitride thin films. The chapter is organized in the following way.

Section 2 deals with fundamental optical properties of GaN and its alloys. These include band structure, photoluminescence, pressure-dependent measurements, absorption, reflection, and photoreflectance.

Section 3 describes degenerate and nondegenerate nanosecond and femtosecond optical pump–probe experiments of GaN thin films. The experiments were performed for a zero and nonzero time delay between the pump and the probe beam. Pump–probe absorption experiments in InGaN thin films are also presented.

Section 4 gives a review of the general optical properties of semiconductors under high levels of optical excitation. A comprehensive picture of gain mechanisms in GaN and AlGaN epilayers as well as in GaN/AlGaN separate confinement heterostructures is presented. Conclusions are drawn based on numerous experiments performed in different time scales (from cw to femtosecond).

Section 5 is entirely devoted to studies of the optical properties of InGaN thin films and quantum well structures. The experimental results presented in this section strongly support the theory that carrier localization is the origin of spontaneous and stimulated emission in this material system. This section also summarizes temperature-dependent studies as well as studies related to various excitation power and wavelength conditions.

In Section 6, the optical characterization of GaN epilayers and InGaN/GaN multiple quantum wells at temperatures as high as 700 K is given. The studies in this section include high and low excitation density regimes, as well as damage mechanisms in GaN thin films.

Section 7 describes the research that has been done to study stimulated emission and lasing properties of GaN in the presence of self- and intentionally formed microcavities, scattering defects, and dislocations.

Section 8 discusses issues related to the near- and far-field stimulated emission patterns of GaN-based lasing structures. A novel technique for measuring optical confinement is introduced in this section.

A comprehensive summary of the optical properties of group III–nitride thin films concludes the chapter.

2. GENERAL OPTICAL PROPERTIES OF THE GROUP III–NITRIDES

2.1. Physical Properties and Band Structure

This section describes the emission properties of III–nitride thin films, with an emphasis on GaN, which is the most extensively studied member of the III–nitrides. The alloying of GaN with moderate ($x < 0.2$) molar concentrations of Al and In to form the ternary compounds $Al_xGa_{1-x}N$ and $In_xGa_{1-x}N$ allows the band-edge emission to be tuned over the entire near-UV to blue wavelength region. In particular, InGaN is well suited for generation of light in the blue–green spectral region, whereas AlGaN generates light in near and deep UV. Thus all three materials are of significant technological relevance. Because of the difficulties associated with indium and aluminum incorporation into GaN, InGaN thin films typically exhibit significantly broader band-edge transitions due to compositional fluctuations. Difficulties also exist in growing high optical quality AlGaN films. Because of this, this section concentrates primarily on the optical properties of the base material, GaN, where the band-edge transitions are significantly better behaved. (A detailed analysis of the optical properties of InGaN is given in Section 5.) We discuss a variety of experimental techniques such as photoluminescence, absorption, reflection, photoreflectance, pump–probe, and others. Excitonic recombination and the effects of strain are also included in this section.

All optical processes in semiconductors are related to their band structure. GaN crystallizes in either the cubic (zincblende), hexagonal (wurtzite), or rock salt structure. The wurtzite structure is by far the most common and is the subject of this section. In GaN, AlN, and InN, a hexagonal unit cell contains six atoms of each type and can be represented by two interpenetrating hexagonal close-packed sublattices, one of each type of atom, offset along the c axis by 5/8 of the cell height ($5c/8$). Table I summarizes some of the fundamental physical properties of GaN, AlN, and InN [15].

The conduction band minimum of wurtzite GaN has Γ_7 symmetry with quantum number $J_z = 1/2$. The maximum of the conduction band is also located at the Γ point, resulting in a direct fundamental bandgap. Crystal-field and spin–orbit coupling split the top of the valence band in GaN into three different subbands, denoted A, B, and C. The A band has Γ_9 symmetry, whereas the B and C bands have Γ_7 symmetry. The structure and symmetries of the bands are shown in Figure 2.

Table I.   Fundamental Properties of GaN, AlN, and InN

Property	GaN	AlN	InN
Bandgap E_g	3.39 eV (300 K)	6.2 eV (300 K)	1.89 eV (300 K)
	3.50 eV (1.6 K)	6.28 eV (5 K)	
Temperature coefficient dE_g/dT	-6.0×10^{-4} eV/K		-1.8×10^{-4} eV/K
Pressure coefficient dE_g/dP	4.2×10^{-3} eV/kbar		
Lattice constants a	3.189 Å	3.112 Å	3.548 Å
c	5.185 Å	4.982 Å	
Thermal expansion $\Delta a/a$	5.59×10^{-6} K^{-1}	4.2×10^{-6} K^{-1}	
$\Delta c/c$	3.17×10^{-6} K^{-1}	5.3×10^{-6} K^{-1}	
Index of refraction n	2.33 (1 eV)	2.15 (3 eV)	2.80–3.05
	2.67 (3.38 eV)		

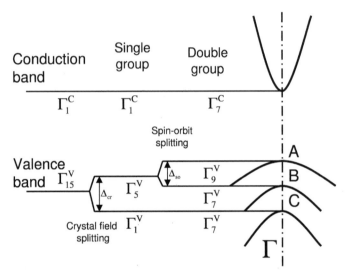

Fig. 2.   Structure and symmetries of the lowest conduction band and the uppermost valence bands in wurtzite GaN at the Γ point ($k \approx 0$).

At low excitation densities, free excitons (a bound state of an electron and hole) represent the lowest energy intrinsic excitation of electrons and holes in pure materials. For materials where the electron and hole are very tightly bound, the excitons are referred to as Frenkel excitons. Another type of exciton is a Wannier–Mott exciton, or simply a Wannier exciton. They are characterized as having relatively weak exciton binding energies, so the electron and hole are separated from each other by a comparatively large distance in the crystal. For GaN, excitons are well described by the Wannier formalism and for the rest of this chapter, the term "exciton" refers to a Wannier–Mott exciton.

Excitons associated with the Γ_9^V valence band (A band), the upper Γ_7^V valence band (B band), and the lower Γ_7^V valence band (C band) are often referred to as A, B, and C excitons. The ternary compounds InGaN and AlGaN typically do not exhibit excitonic features at room temperature, because these alloys contain high levels of compositional fluctuations and defects. Recent experimental results on higher quality AlGaN epilayers

have shown that at low temperatures, AlGaN has an excitonic resonance observable by absorption measurements. Once the quality of these ternary compounds is improved, we expect to see excitonic features in these materials over a wide temperature range.

2.2. Photoluminescence

At low temperatures, near-band-edge luminescence spectra observed from most GaN samples are dominated by strong, sharp emission lines that result from the radiative recombination of free and bound excitons, as shown in Figure 3. In addition to near-band-edge exciton emission, nominally undoped GaN samples often exhibit a series of emission structures in the energy range of approximately 2.95–3.27 eV, and a broad emission band in the yellow spectral region with a peak position around 2.2 eV, as shown in the inset of Figure 3. The intensity of these low emission bands relative to that of the exciton emissions varies from sample, to sample depending on the crystal quality. These two additional bands result from radiative recombination from impurity levels within the GaN bandgap.

The free exciton state of GaN can be described by the Wannier–Mott approximation, where the electrons and holes are treated as nearly independently interacting through their Coulomb fields. The Coulomb interaction reduces the total energy of the bound state relative to that of the unrelated free carrier states by an amount that corresponds to the exciton binding energy E_x^b. Free excitons exist in a series of excited states similar to the excited states of a hydrogen-like atomic system. Optical transitions can occur from discrete states below the bandgap E_g at the exciton energies:

$$E = E_g - \frac{E_x^b}{n^2} \qquad (2.1)$$

Bound exciton states involve both an exciton and an impurity. Excitons can be bound to neutral or ionized donors and acceptors. The energy of the photon produced through annihilation of an exciton is

$$h\nu = E_g - E_x^b - E_{BX} \qquad (2.2)$$

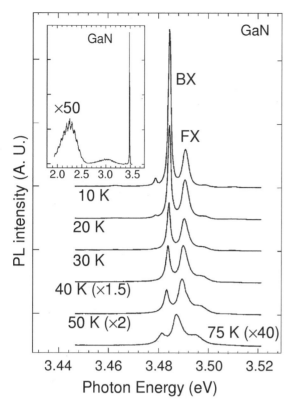

Fig. 3. Near-band-edge exciton luminescence spectra as a function of temperature taken from a 7.2-μm-thick GaN epilayer. The inset shows the spectrum of the same film over a wider range of photon energies taken at 10 K.

dislocations: however, the residual strain has a relatively strong influence on the optical properties of the sample. It is difficult to separate the effects of strain caused by lattice parameter mismatch from those that involve thermal-expansion mismatch to exactly determine their influence on the optical properties of GaN epilayers. We note, however, that the overall effect of residual strain generated in GaN on sapphire is compressive, which results in an increased bandgap, whereas the stress induced in GaN on SiC is tensile, which leads to a decrease in the measured exciton transition energies. The energy positions of the exciton resonances associated with A-, B-, and C-exciton transitions are also sample dependent.

The effects of strain become obvious when the lattice parameters of GaN are compared to those of virtually strain-free bulk GaN. In general, the introduction of strain changes the lattice parameters and generates variations in the electronic band structure. An example of the fit of experimentally observed exciton transition energies versus the A-exciton transition energy can be found in [16]. The fit gives an estimate of the coefficients of crystal-field splitting for the Γ_9 and Γ_7 orbital states and describes the spin-orbit coupling (see Fig. 2). The numerical values of these band-structure parameters can be used to estimate the deformation potentials of GaN.

At this time, however, there is no quantitative agreement between different research groups on the precise values of hydrostatic deformation potentials, which prompts further research in this field.

where E_{BX} is the exciton localization energy. In the Haynes approximation $E_{BX} \approx 0.1 E_i$, where E_i is the impurity binding energy.

Figure 3 shows that the bound exciton in GaN dominates the luminescence spectrum at low temperatures. Because as-grown GaN is always n type, bound excitons are expected to be bound to neutral donors. However, their intensity decreases with increasing temperature due to thermal dissociation and becomes unresolvable for temperatures in excess of 100 K. Free excitons were found to dominate the photoluminescence spectra at temperatures above 40 K. At temperatures above 200 K, excitons broaden and eventually band-to-band recombination of free carriers dominates the photoluminescence spectra.

By changing the composition of In and Al in InGaN and AlGaN alloys, it is possible to tailor the bandgap anywhere between 1.9 and 6.2 eV. The photoluminescence spectra of (Al, In) GaN alloys usually exhibit broad features and abnormal temperature behavior of the emission peak position, possibly due to fluctuations in alloy composition.

2.3. Strain Considerations

Due to the lattice mismatch and difference in thermal expansion coefficients between GaN epilayers and substrates, the effects of residual strain have to be taken into account when considering the excitonic energy transitions. Some degree of strain relaxation occurs through the formation of a large density of

2.4. Absorption

In high quality GaN epilayers, low-temperature absorption spectra are usually dominated by sharp excitonic resonances, as shown in Figure 4. Whereas the concentration of impurities in these samples is relatively small, bound excitons are not expected to contribute significantly to band-edge absorption. At 10 K, we can clearly observe three different absorption features associated with A, B, and C excitons. From theory, the C-exciton transition is only allowed for $E \parallel c$; thus, its intensity is significantly reduced in the transmission configuration (where $E \perp c$). However, if the experimental geometry is modified, an increase in the C−exciton absorption is expected [17].

Figure 4 also shows changes in absorption over the temperature range of 10–450 K. An excitonic resonance is clearly observed in the 300-K absorption data. In fact, excitonic resonances were observed at temperatures in excess of 100°C above room temperature. The absorption spectrum at each temperature was fitted to a double-Lorentzian functional form. The energy position of the A exciton was found to be well approximated by the Varshni equation

$$E(T) = E(0) - \frac{\alpha T^2}{\beta + T} \qquad (2.3)$$

with $\alpha = 11.8 \times 10^{-4}$ eV/K and $\beta = 1414$ K.

Similar experiments were performed on InGaN and AlGaN epilayers. The InGaN epilayers usually exhibit a very wide absorption edge (several hundred millielectronvolts) and have no

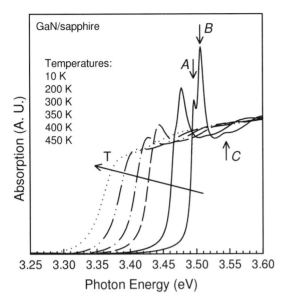

Fig. 4. Absorption spectra of a 0.38-μm-thick GaN epilayer in the vicinity of the fundamental bandgap. At 10 K, excitonic features associated with the A, B, and C excitons are observed.

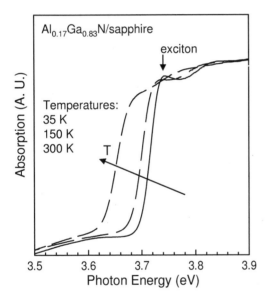

Fig. 5. Absorption spectra of a 0.44-μm-thick $Al_{0.17}Ga_{0.83}N$ epilayer in the vicinity of the fundamental bandgap. At low temperatures a weak excitonic feature is observed.

excitonic features in absorption. Excitons have been observed in AlGaN epilayers at low temperatures, as shown in Figure 5. The excitonic feature disappears for temperatures above 150 K. Difficulties in observing excitons in InGaN and AlGaN epilayers are related to material quality. We further note that the abnormal temperature behavior of the bandgap in these alloys cannot be adequately fitted by the Varshni equation.

2.5. Reflection and Photoreflectance

Recent results obtained from reflectance and photoreflectance measurements have clearly demonstrated the signatures of tran-

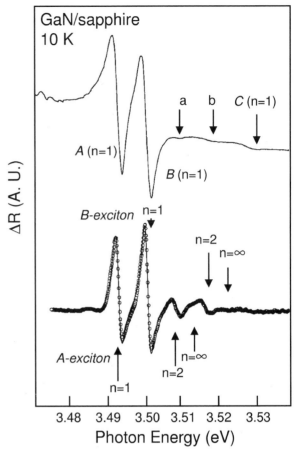

Fig. 6. Comparison of conventional reflectance (top) and photoreflectance (bottom) spectra taken from a 7.2-μm-thick GaN/sapphire sample at 10 K. Open circles denote experimental data and solid lines represent the best result of the least-squares fit to the photoreflectance data.

sitions related to the A, B, and C excitons, as well as the fundamental band-to-band ($\Gamma_9^V - \Gamma_7^V$) transition [39]. The unambiguous observation of these transitions allows a precise determination of their energy positions and the binding energy of excitons [using the hydrogenic model described by Eq. (2.1)]. Figure 6 shows reflectance and photoreflectance spectra from a GaN–sapphire sample at 10 K. As can be seen in the figure, photoreflectance is able to detect weak signals and contains more spectral features than typical reflectance spectra. This makes it easy to positively identify the nature of the transition.

Such identifications permit a direct estimate of the binding energy for the A and B excitons from the separation between the $n = 1$ and $n = 2$ states for excitons, assuming that the hydrogenic model based on the effective mass approximation is applicable. A binding energy of $E_x = 21$ meV for the A and B excitons was obtained [18]. GaN samples grown on SiC exhibit stronger C-exciton resonances compared to those grown on sapphire in photoreflectance spectra (see Fig. 7) and were used to obtain the binding energy of the C exciton. The best theoretical fit of the experimental data yielded a value of 23 meV for the binding energy of the C exciton.

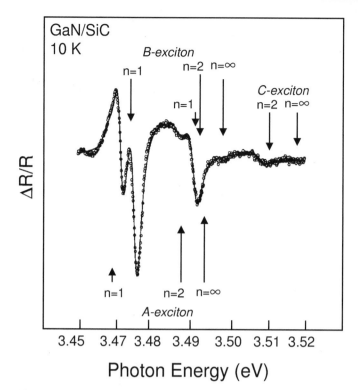

Fig. 7. Photoreflectance spectrum of a GaN epilayer grown on SiC. The sample exhibits strong transition signals associated with the C exciton.

Even though strong resonances associated with the formation of excitons often can be observed near the band edge of GaN by various spectroscopic methods, we found that the photoreflectance technique is the most reliable in determining exciton binding energies.

Thus far, only optical properties of nitride thin films at low excitation densities have been described, that is, the number of generated carriers has been $\ll 10^{17}$ cm^{-3}. At high excitation densities, a myriad of new and exciting optical properties occur. The next sections deals with optical properties of nitride thin films at pump carrier densities as high as 10^{20} cm^{-3}.

3. PUMP–PROBE SPECTROSCOPY OF HIGHLY EXCITED GROUP III-NITRIDES

3.1. Introduction to Pump–Probe Spectroscopy

In this section, the results of optical pump–probe experiments on III–nitride thin films are presented. These experiments provide a deeper understanding of the optical phenomena associated with high carrier concentrations in this material system than can be extracted from optical pumping experiments alone. In particular, the evolution of the near-band-edge transitions in GaN is examined as the number of free carriers is increased by direct band-to-band optical excitation. The band-edge evolution under nanosecond and femtosecond optical excitation is shown for excitation densities centered around typical stimulated emission (SE) thresholds of GaN epitaxial layers. The experiments in this section are broken down into five basic categories:

1. Nanosecond single-beam power-dependent absorption experiments.
2. Nanosecond nondegenerate optical pump–probe absorption experiments at zero time delay between the pump and probe pulses.
3. Nanosecond nondegenerate optical pump–probe absorption experiments with control of the time delay of the probe pulse with respect to the pump pulse.
4. Femtosecond nondegenerate optical pump–probe absorption experiments that utilize femtosecond optical pulses to study the ultrafast dynamics of the band-edge transitions in highly excited III–nitrides.
5. Nanosecond nondegenerate optical pump–probe reflection experiments performed on thick (on the order of several micrometers) GaN layers at zero time delay between the pump and probe pulses.

The first technique is the simplest in that it is a one beam experiment. The laser beam is passed through the GaN layer and spectrally tuned across the near-band-edge region of interest. The intensity of the laser is varied from scan to scan and the evolution of the band-edge transitions is monitored as a function of the laser intensity. This allows direct determination of the changes induced in the strength of a given transition by the excitation of that transition. The next experiment, nanosecond nondegenerate optical pump–probe absorption spectroscopy at zero time delay, allows the band-edge transitions to be monitored as the number of optically excited free carriers is increased. This experiment allows the excitation photon energy (and hence, the excess kinetic energy the photoexcited free carriers possess) to be freely varied as the band-edge transitions are monitored. The next nanosecond nondegenerate optical pump–probe experimental technique allows the probe beam pulse to be delayed with respect to the pump pulse. This experiment provides significant insight into the mechanisms responsible for the large nonlinearities observed in these experiments. The final technique, femtosecond nondegenerate optical pump–probe absorption spectroscopy, allows the ultrafast dynamics of the band-edge transitions in highly excited III–nitrides to be studied. Nondegenerate optical pump–probe reflection experiments on optically thick GaN layers are then presented to gauge the strength of the excitonic transitions in thick (on the order of several micrometers) GaN layers as the number of free carriers is increased beyond that required to achieve SE. Using reflection techniques provides a better estimate of the effects of the intense optical pump on the strength of the excitonic transitions by minimizing the effects of pump beam attenuation as it traverses the sample thickness. This experiment is valuable in that it allows pump–probe experiments to be performed on GaN layers of a quality that cannot be matched at this time by thin (on the order of several tenths of a micrometer) GaN layers.

The last section is devoted to nondegenerate optical pump–probe experiments performed on InGaN thin films. These films exhibit markedly different behavior than GaN thin films, helping to explain the differences in the SE characteristics of GaN and InGaN thin films.

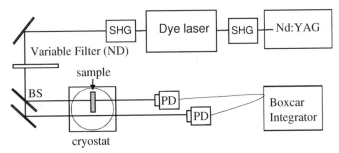

Fig. 8. Single-beam power-dependent absorption experimental setup. The phase matching condition in the second harmonic generating crystal is maintained by the autotracking unit as the dye laser emission wavelength is tuned. SHG, ND, BS, and PD refer to second harmonic generator, neutral density, beam splitter, and photodiode, respectively.

3.2. Single-Beam Power-Dependent Absorption Spectroscopy of GaN Thin Films

Single-beam power-dependent absorption experiments on GaN thin films provide a better understanding of how a transition being excited is affected by the excitation, that is, how an excitonic resonance decreases with increasing optical excitation of the resonance. The single-beam experiments provide complimentary information to nondegenerate pump–probe experiments, where a probe is used to monitor the band-edge transitions as the number of photoexcited free carriers is increased by a separate (e.g., above-gap) optical excitation source.

The single-beam power-dependent absorption experiments were performed using the second harmonic of the deep red radiation generated by a dye laser pumped by the second harmonic of an injection-seeded Nd:YAG laser with 4-ns full width at half-maximum (FWHM) pulse width and a 10-Hz repetition rate. The phase matching angle for second harmonic generation was maintained using an auto-tracking unit as the dye laser emission wavelength was scanned across the GaN band edge. The intensity of the transmitted beam through the sample was measured using a photodiode and a boxcar integrator. The transmitted signal was compared directly with a split-off portion of the beam that bypassed the sample to obtain the absorption spectra. The intensity of the beams was controlled using UV neutral density filters. The single-beam absorption experimental configuration is illustrated in Figure 8.

The results of single-beam power-dependent absorption experiments are shown in Figure 9 for a 0.38-μm-thick GaN layer. No noticeable change in the 10-K absorption spectrum was observed with increasing I_{exc} until I_{exc} exceeded ≈ 100 kW/cm^2. This provides an estimate of the I_{exc} required to reduce the oscillator strength of excitons in GaN. We note that this density is on the order of the SE threshold for GaN at 10 K.

3.3. Nondegenerate Optical Pump–Probe Absorption Spectroscopy of GaN Thin Films at Zero Time Delay

In this section, the results of nanosecond nondegenerate optical pump–probe experiments performed on metal–organic chemical vapor deposition (MOCVD)-grown GaN thin films are

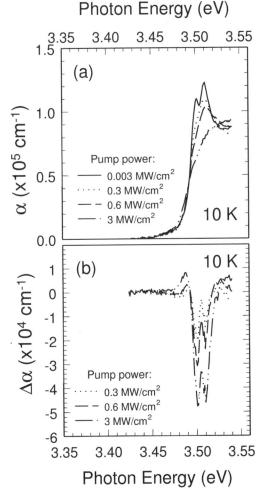

Fig. 9. (a) 10-K single-beam power-dependent absorption spectra for a 0.38-μm-thick GaN film grown by MOCVD on (0001) oriented sapphire. (b) Change in absorption, $\Delta\alpha(I_{exc}) = \alpha(I_{exc}) - \alpha(0)$, for the spectra in (a). The 1s A and B free exciton transitions are indicated in (a).

presented. Changes in the optical transitions due to excess photogenerated free carriers are considered as a function of excitation density and time delay at 10 K and room temperature using pump–probe spectroscopy with nanosecond and femtosecond optical pulses. At 10 K, strong, well-resolved features are present in the absorption and reflection spectra that correspond to the 1s A and B free exciton transitions. These features broaden and decrease in intensity in the nanosecond experiments due to high densities of photoexcited free carriers generated by the pump beam, resulting in extremely large values of induced transparency at 10 K and room temperature. In addition, large values of induced absorption are observed in the nanosecond experiments with increasing pump density in the below-gap region where gain is expected. The origin of the below-gap induced absorption is explored by monitoring the evolution of the band-edge transitions as a function of time delay after the nanosecond pump pulse. The results indicate a complex relationship exists between induced absorption and gain in GaN, and help to explain the relatively high

SE threshold of GaN compared to other wide (direct) bandgap semiconductors. This topic is further explored by comparing the experimental results to those obtained for InGaN thin films, the results of which are the subject of a subsequent section. Femtosecond experiments performed on highly excited GaN are also reviewed. These experiments give valuable insight into the recombination dynamics of highly excited GaN, including the evolution of exciton saturation and optical gain formation. The experiments in this section provide an explanation for the drastic reduction in SE threshold that results when indium is incorporated into GaN.

Nondegenerate optical pump–probe experiments have been performed on GaN thin films using nanosecond excitation sources. The excitation densities employed allowed the band-edge transitions to be monitored as the number of photoexcited free carriers was increased beyond that required to achieve SE, providing insight into the mechanisms involved in the SE process.

The experimental system consisted of an amplified dye laser pumped by the second harmonic of an injection-seeded Nd:YAG laser operating at 10 Hz. The deep red radiation from the dye laser was frequency doubled in a nonlinear crystal to produce near-UV wavelengths used to synchronously pump the GaN layers above their fundamental bandgap and a dye solution. The UV fluorescence from the dye solution was collected and focused onto the GaN layer. This broadband transmitted probe was then collected, coupled into a 1-m spectrometer, and spectrally analyzed using a UV enhanced, gated charge-coupled device (CCD). The pump wavelength was varied from 335 (3.700 eV) to 350 nm (3.541 eV), but no noticeable difference in experimental results was observed for the different pump wavelengths. A pump wavelength of 337 nm (3.678 eV) was used for the data presented in this section. The probe was kept many orders of magnitude lower (<200 W/cm^2) than the pump beam to avoid any nonlinearities due to the probe itself. Its spot size was kept at half that of the pump. Special care was taken to monitor the intensity of the probe and the pump (separately) from scan to scan to ensure consistent absorption coefficient values and luminescence compensation, when applicable. The sample temperature was varied between 10 K and room temperature through the use of a closed-cycle refrigerator. The pump–probe experimental configuration is illustrated in Figure 10. The experiments were also repeated with a scanned narrowband probe by employing a second, synchronously pumped dye laser. Consistent results were obtained.

Figure 11a shows the 10-K absorption spectra near the fundamental band edge of a 0.38-μm-thick GaN layer subjected to several different pump power densities. The unpumped absorption spectra agree very well with published cw absorption values for the same sample (see Fig. 4). Figure 11c shows the measured absorption changes with respect to the unpumped spectra, $\Delta\alpha(I_{exc}) = \alpha(I_{exc}) - \alpha(0)$, for the pump densities given in Figure 11a. Induced transparency ($\Delta\alpha$ negative) in the excitonic region and induced absorption ($\Delta\alpha$ positive) in the below-gap region are clearly seen with increasing pump density. Figure 11b shows room temperature absorption spectra for

several different pump densities. At room temperature, the A and B free exciton features seen in Figure 11a have broadened and merged into one, and the resulting induced transparency is about one-third that at 10 K. The below-gap induced absorption is seen in Figure 11d to be approximately half that at 10 K. The near symmetry between the induced transparency and induced absorption seen in Figure 11b and d appears to be coincidental. For pump densities greater than ≈ 3 MW/cm^2, the induced transparency changes very little, while the induced absorption continues to grow with increasing I_{exc}. The decrease in the free exciton absorption with increasing I_{exc} is attributed to many-body effects, such as exciton screening by free carriers, causing a diminution of their oscillator strength [19]. The origin of the below-gap induced absorption is still not well understood. It has been attributed in the literature to lattice heating [20–22]. Whereas the pump wavelength exceeds the bandgap energy, some lattice heating due to the transfer of energy from the photogenerated hot electrons to the lattice is expected. The amount of heating and its impact on the absorption properties of GaN thin films are not well understood. The presence of below-gap induced absorption has been confirmed in a variety of GaN layers of varying thickness and by several experimental groups, indicating that it is a basic property of state-of-the-art GaN thin films [20–25]. We note that these thin GaN layers are still optically thick (sample thickness $L = 0.38$ μm and $\alpha L \approx 4$), so the pump intensity is appreciably diminished as it traverses the sample thickness. Therefore, the resulting values of $\Delta\alpha$ presented here are lower limit values of the actual change with pump density. A better gauge of the change in the oscillator strength of the excitonic transitions for a given I_{exc} (e.g., the SE threshold) is provided by nondegenerate optical pump–probe reflection experiments on optically thick (on the order of several micrometers) GaN layers. Pump–probe reflection experiments are the subject of a subsequent section.

For cases of the relatively thin (<0.4-μm) layers used in the above study, there is the possibility that the GaN–sapphire interface of the sample contributes to the observed induced absorption as the pump density is increased due to the presence of a high density of defects and resultant electronic states that result from the large lattice mismatch between the GaN and the sapphire substrate. To clarify the role of the GaN–sapphire interface in these experiments, the experiments were repeated on high-quality thick (on the order of several micrometers) GaN layers. Although the thickness of the samples precluded the ability to observe excitonic features in transmission experiments, it did allow monitoring of the below-gap absorption changes. Due to the finite penetration depth of the pump beam, only a thin layer of the entire sample thickness was heavily pumped. This was used to study the role of the interface states on the induced absorption. First, the substrate side of the sample was pumped and probed. The resulting (below-gap) absorption changes were then compared to the thin sample case for the same experimental conditions. In this configuration, only a thin region (<0.4 μm) at the GaN–sapphire interface was heavily pumped, maximizing the role of interface states. The experiment was then repeated on the front surface of the thick sample.

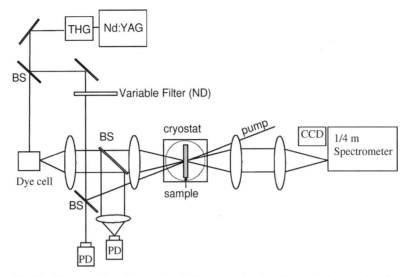

Fig. 10.   Nanosecond nondegenerate optical pump–probe absorption setup for zero time delay between the pump and probe pulses. SHG, ND, BS, and PD refer to second harmonic generator, neutral density, beam splitter, and photodiode, respectively.

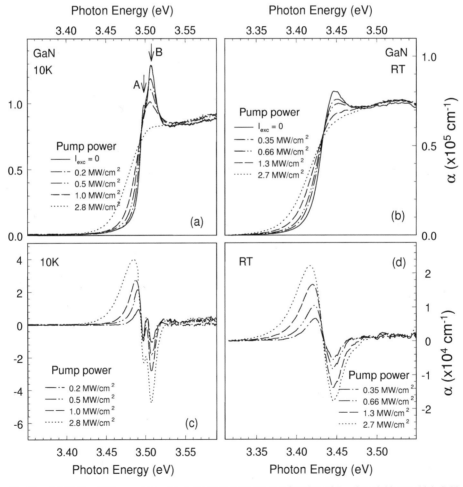

Fig. 11.   (a) 10-K and (b) room temperature absorption spectra as a function of I_{exc} for a 0.38-μm-thick GaN layer grown by MOCVD on (0001) oriented sapphire. (c), (d) Change in absorption spectra as a function of I_{exc}, $\Delta\alpha(I_{exc}) = \alpha(I_{exc}) - \alpha(0)$, for the spectra in (a) and (b), respectively. The 1s A and B free exciton transitions are clearly seen in the unpumped 10-K spectrum. The 1s A and B free exciton transitions have broadened and merged into one feature in the room temperature spectra of (b). Complete exciton saturation is seen for I_{exc} approaching 3 MW/cm^2 at both 10 K and room temperature. Induced transparency in the excitonic region and induced absorption in the below-gap region are clearly seen with increasing I_{exc}.

The results were again compared to the thin sample case for the same experimental conditions. This configuration ensured that a negligible amount of interface states was pumped. The results of both experiments consistently showed the same induced absorption, qualitatively and quantitatively (when the penetration depth of the pump beam was taken into consideration in the calculations of $\Delta\alpha$). Whereas there is a negligible amount of pump beam reaching the GaN-sapphire interface in the second experiment, the role of interface states in the below-gap induced absorption process was shown to be minimal.

Another possibility for the origin of the below-gap induced absorption lies in the present crystalline quality of GaN epitaxial film. It is well known that GaN epitaxial films contain a multitude of crystalline defects [26, 27]. Possibly these defects and resultant deep levels in the crystal are a contributing factor to the observed induced absorption. An other possibility is the presence of two-photon or excited state absorption. Due to the relatively narrow spectral width of the induced absorption, the possibility that excited state absorption causes the below-gap induced absorption seems remote.

The large values of induced transparency and induced absorption observed in this work, and the fact that excitons have been shown to persist in GaN epilayers well above room temperature [17] (see Fig. 4) indicate that the large optical nonlinearities will persist to well above room temperature. This suggests the III–nitrides may be well suited for optical switching applications. This switching behavior is illustrated in [24].

3.4. Nanosecond Experiments at Nonzero Time Delay

The previous pages have shown, in addition to the expected exciton screening with increasing free carrier concentration, that the emergence of significant below-gap induced absorption in the spectral region optical gain was expected. The below-gap induced absorption was shown to exceed 4×10^4 cm^{-1} at 10 K and 2×10^4 cm^{-1} at room temperature. Its presence is not consistent with two-photon or excited state absorption, and it was observed in optically thick samples as well as thin, eliminating the possible role of GaN–sapphire interface states in the process and illustrating that it is a common property of state-of-the-art GaN films. To better understand this behavior and shed light on its origin, the probe pulse was delayed with respect to the pump pulse in the previous experiments, allowing the time evolution (on nanosecond time scales) of the below-gap induced absorption to be studied.

The experimental system consisted of two independent dye laser systems pumped by the second harmonics of separate Q-switched Nd:YAG lasers. The deep red radiation from one of the tunable dye lasers was frequency doubled in a nonlinear crystal to provide the near- to deep-UV wavelengths needed to optically excite the GaN layers above their fundamental bandgap. The red–orange radiation from the other dye laser was also frequency doubled in a nonlinear crystal to generate the deep-UV wavelengths needed to pump the dye solution. The near-UV broadband fluorescence from the dye solution was collected and focused coincidental (spatially) with the pump beam

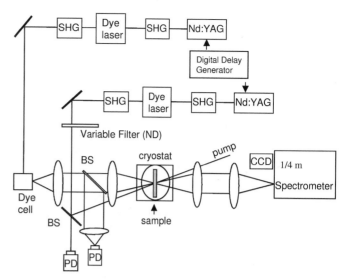

Fig. 12. Nanosecond nondegenerate optical pump–probe absorption setup with electronically controlled time delay between the pump and probe pulses. SHG, ND, BS, and PD refer to second harmonic generator, neutral density, beam splitter, and photodiode, respectively.

on the sample surface. The transmitted broadband probe beam was collected, coupled into a 1/4-m spectrometer, and spectrally analyzed using a UV-enhanced gated CCD. The two dye laser systems were temporally locked together electronically and a digital delay generator allowed the temporal delay between the two to be precisely varied. A temporal jitter better than 0.5 ns between the pump and probe pulses was achieved through creative use of various signal and delay generators in the process of slaving one Nd:YAG laser to the other. This experimental system provided significant advantages over the previous system in that the laser wavelengths used to pump the sample and the dye solution were independently variable, allowing optimization of the experimental conditions. The probe intensity was again kept many orders of magnitude lower than the pump intensity and its spot size was kept at approximately one-half that of the pump. A pump wavelength of 330 nm (3.756 eV) was used in the experiments described here. The temperature was varied between 10 K and room temperature through the use of a closed-cycle refrigerator. The experimental setup is illustrated in Figure 12. Both the pump and probe pulses had durations of approximately 4 ns FWHM. The zero of the delay was precisely set using a fast photodiode placed at the sample position and a 1-GHz digital signal analyzer. The delay was controlled by a digital delay generator with a better than 10-ps resolution for delays from 10 ps to over 1 s.

The first observation from this experiment is that the below-gap induced absorption is due to a shift in the GaN bandgap to lower energies. This is shown in Figure 13. The bandgap is seen to go through a transformation during the pump pulse duration, finally reaching a maximum change at a probe delay of about 5 ns with respect to the pump pulse. At this time, the excitonic features that disappeared during the pump pulse duration reappear at lower energy. The fact that the 1s A and B free excitonic transitions cannot be distinguished in the spectra of Figure 13 is a result of limited resolution afforded by the use of the 1/4-m spectrometer. The spectral positions of these transitions are marked in Figure 13 for completeness. The slight shift in their spectral position with respect to the data presented in the previous section is a result of different residual strain in different areas of this sample [28–31]. An interesting feature of the spectra shown in Figure 13 is the drastic reduction of the excitonic features in the −2-ns delay spectrum. This reduction is comparable to that observed in the zero time delay spectrum. The pump density at −2 ns is approximately one-half that at zero delay. This indicates that the pump density at which excitons are effectively screened by free carriers is significantly lower than was previously indicated in the zero time delay experiments. A considerable amount of below-gap induced absorption is also seen in the −2-ns delay spectrum. Another interesting feature of Figure 13 is that the below-gap induced absorption builds up with increasing delay after the peak of the pump pulse, reaching a maximum approximately 5 ns after the peak of the pump pulse. This was observed for all pump densities employed and at both 10 K and room temperature. This is significant in that it provides the following picture for gain development in GaN thin films: Stimulated emission and gain develop on the leading edge of the pump pulse. As the pump pulse continues in duration, gain continues to increase with increases in the pump pulse. At the same time, the below-gap induced absorption starts to grow. This leads to two competing effects in the below-gap region as the pump pulse continues: induced gain and induced absorption. The induced absorption quickly exceeds the gain. This results in the observation of only induced absorption in the probe spectra. As such, the results of this study indicate that the SE threshold of GaN measured in nanosecond experiments significantly overestimates the excitation density needed to achieve population inversion and optical gain. This overestimation is shown in a later section to partly explain the significant reduction of the SE threshold observed for InGaN thin films compared to GaN. The ultrafast dynamics of gain evolution in GaN will be explored in a subsequent section.

The evolution of the band edge for longer delays is shown in Figure 14. For delays longer than 5 ns, the band edge slowly returns to its prepumped position, finally recovering after approximately 100 ns. It is clear from the spectra in Figure 14 that as the band edge recovers, the excitonic resonances also decrease in linewidth. This behavior mimics the behavior observed in absorption measurements as the temperature is increased (see Fig. 4). The shift of the band edge to lower energy results in extremely large values of induced absorption in the below-gap spectral region, nearly doubling at 5 ns what was previously observed at zero time delay. The near symmetry between the induced transparency and absorption observed at zero time delay in the previous section is lost as the probe is delayed with respect to the pump. The same behavior was observed at room temperature. Of particular interest to optical switching applications is the enhanced induced absorption for 5-ns delay compared to the previously observed behavior at zero time delay. This results in a threefold increase in $\Delta\alpha$ and a threefold

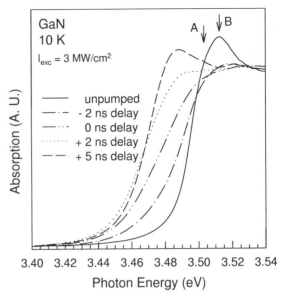

Fig. 13. 10-K absorption spectra of a 0.38-μm-thick GaN layer near the fundamental bandgap as a function of time delay across the pump pulse duration. The 1s A and B free exciton transition energies are indicated for completeness. The spectra show a clear shift in the bandgap to lower energy with increasing time delay of the probe across the pump pulse duration.

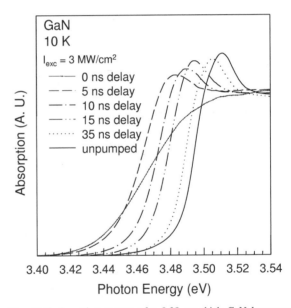

Fig. 14. 10-K absorption spectra of a 0.38-μm-thick GaN layer near the fundamental bandgap as a function of time delay after the pump pulse. The spectra show clear excitonic features that narrow as the bandgap returns to its prepumped position with increasing delay time. The bandgap recovered completely within ~85 ns after the pump pulse.

reduction in transmitted light in this spectral region (from approximately 56 to 18% at 3.41 eV for $I_{exc} \approx 3$ MW/cm^2). At room temperature, the band edge takes longer than 100 ns to completely shift back to its prepumped position.

To describe the shift in the band edge for various pump densities and probe delays, the effective lattice temperature rise ($T_{eff} - T_0$, where T_0 is the ambient temperature) required to shift the band edge the given amount was calculated. This was

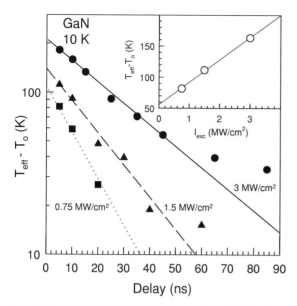

Fig. 15. Effective temperature rise in a 0.38-μm-thick GaN film ($T_{eff} - T_0$) at $T_0 = 10$ K induced by a strong optical pump as a function of time delay after the pump pulse. T_{eff} was calculated using the Varshni equation and the spectral position of the 1s B exciton as a function of delay for various I_{exc}. The Varshni equation fitting parameters (α, β) were obtained from cw absorption experiments performed on the same sample. The inset shows the maximum effective lattice temperature rise (at 5-ns delay) as a function of I_{exc}. The lines are given only to guide the eye.

done using the Varshni equation (see Eq. (2.3) and [32]) and the best fit values derived from temperature-dependent absorption measurements of the GaN film [17]. The calculated values for T_{eff} are shown in Figure 15 relative to the ambient temperature at 10 K for various excitation densities. At 10 K, T_{eff} drops off nearly exponentially for delays less than about 40 ns, with decay times on the order of 40 ns for $I_{exc} = 0.75$ MW/cm^2, up to approximately 90 ns for $I_{exc} = 3$ MW/cm^2. For delays longer than approximately 40 ns, T_{eff} exhibits nonexponential decay. At room temperature, T_{eff} decays nonexponentially for all delay ranges and excitation densities, with typical recovery times on the order of 140 ns. It is interesting to note that the maximum values for T_{eff} (at 5-ns delay) have a sublinear dependence on I_{exc} at both 10 K (see the inset of Fig. 15) and room temperature (not shown). In fact, if the 0 K rise in T_{eff} for $I_{exc} = 0$ is included in the plot, the rise in $T_{eff} - T_0$ with increasing I_{exc} exhibits saturation behavior. It should be noted that although T_{eff} is being used as a parameter to describe the shift in the band edge with increasing I_{exc} and for various delays, the behavior is not necessarily being ascribed to lattice heating alone. The 5-ns rise time, the recovery time (on the order of 50–100 ns), and the saturation behavior of $T_{eff} - T_0$ with increasing I_{exc} suggest that contributions from other mechanisms might be present and could influence the band-edge shift. Note that the Varshni equation, strictly speaking, describes variations in the bandgap energy under the condition that the lattice is in thermal equilibrium. Usually, this is not the case in nanosecond excitation experiments.

A well-known property of epitaxial GaN films is the presence of a large density of dislocations and other structural defects that result from the large lattice mismatch between the (sapphire) substrate and the GaN layer, as well as their differences in thermal expansion coefficients. In addition, a large concentration of deep levels often exists in epitaxially grown GaN films. These levels give rise to the so-called yellow-band luminescence (YBL) commonly observed when GaN is excited above its fundamental bandgap. The strength of the YBL relative to the band-edge luminescence was found to be a strong function of growth conditions. Layers with the strongest band-edge luminescence were typically found to exhibit the least YBL. We observed that a reduction in the YBL from a GaN layer typically was accompanied by a relative reduction in the SE threshold. The experiments of Taheri et al. [33] evidence the large concentration of these deep levels. In these experiments, strong laser-induced dynamic gratings were produced via two-step single photon absorption of 532-nm (2.33-eV) laser radiation. Electrons were promoted from the valence band to the conduction band of GaN via this two-step process with the deep levels providing the required intermediate states. The large magnitude of nonlinearities observed in the experiments of Taheri et al. suggests a considerable concentration of these mid-gap states. This YBL was found to be absent in the high-quality InGaN layers described throughout this chapter.

If these other properties of epitaxial GaN films are taken into account, another possible origin of the observed below-gap induced absorption exists in addition to lattice heating. The process is as follows: (1) During the pump pulse duration, the photogenerated free carriers quickly (on the order of several hundred picoseconds or less) relax to deep levels in the crystal. These deep levels are known to radiatively recombine (giving rise to YBL) with effective lifetimes on the order of tens to hundreds of nanoseconds. (2) These deep levels effectively screen the excitons and lead to the observed reduction in the bandgap, similar to what has been observed in indirect gap semiconductors while monitoring the direct gap excitonic transitions [34]. (3) This screening and bandgap reduction decreases as the carriers trapped at these deep levels recombine. This scenario, together with lattice heating effects, qualitatively explains the experimentally observed phenomena. The near zero time delay behavior is then explained as a competition between commonly observed changes in the bandgap of semiconductors with increasing optical excitation (such as exciton screening by free carriers, bandgap renormalization, and formation of optical gain), lattice heating, and screening by carriers trapped at deep levels.

3.5. Femtosecond Experiments

Femtosecond nondegenerate optical pump–probe absorption experiments have also been reported, characterizing the ultra-fast dynamics of band-edge transitions in highly excited GaN thin films [35, 36]. Carriers were optically excited using pulses from an amplified Ti:sapphire laser. The femtosecond pulses

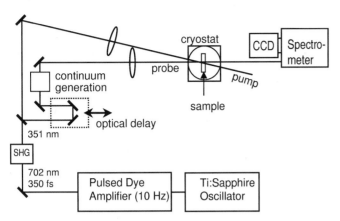

Fig. 16. Femtosecond nondegenerate optical pump–probe absorption experimental setup.

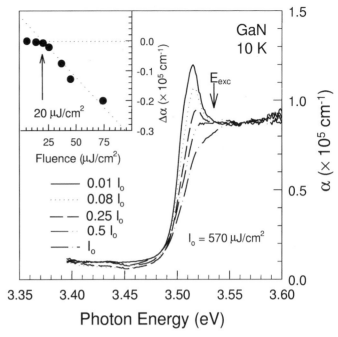

Fig. 17. 10-K absorption spectra of a 0.38-μm-thick GaN layer as a function of pump fluence. The inset shows the onset of exciton saturation at 20 μJ/cm^2.

were amplified to sufficient levels using a three-stage dye amplifier pumped by the second harmonic of a 10-Hz nanosecond Nd:YAG laser. The system produced 350-fs pulses at 702 nm (1.76 eV) with an energy of 300 μJ. The second harmonic (351 nm, 3.53 eV) of this radiation was generated in a nonlinear crystal and the output beam was split into two. One-half of the second harmonic output was used to excite carriers above the bandgap of GaN. The other half was focused onto a 3-mm-thick piece of quartz to create a broadband continuum probe source, a result of self-phase-modulation in the quartz. The probe beam was focused to a 150-μm diameter at the sample and the transmitted light was collected, coupled into a 3/4-m double spectrometer, and analyzed using a CCD detector. The pump beam was focused onto the sample with a larger spot diameter (250 μm) to reduce effects associated with the transverse beam profiles. The optical delay between pump and probe was accurately controlled using a computerized stepper motor delay stage. The experimental setup is illustrated in Figure 16.

Figure 17 shows 10-K band edge absorption spectra of a 0.38-μm-thick GaN layer as a function of pump fluence. Well-resolved A and B excitonic resonances were observed under low pumping conditions, but have merged in the data shown due to broadening with pump power and difficulties in obtaining clean, high-resolution data at 10 Hz. The spectra shown in Figure 17 were taken with a delay of 1 ps between pump and probe. At this time delay, initial transient effects (discussed subsequently) are gone and carriers have relaxed down to the band edge. The excitonic resonances were observed to decrease with increasing pump fluence due to plasma screening by free carriers, with the onset of saturation occurring at a pump fluence of 20 μJ/cm^2, as shown in the inset of Figure 17. At the highest pump densities, absorption above the excitonic resonances was also observed to saturate. For a GaN layer thickness of 0.38 μm and an absorption coefficient of 1.2 $\times$ 10^5 cm^{-1} at the excitation energy, the layers were optically thick. Therefore, the excited carrier density was not uniform throughout the 0.38-μm sample thickness. The large lattice mismatch between the sapphire substrate and the GaN epilayer makes it very difficult to grow high-quality thin GaN layers, and thinner samples were

observed to exhibit lower optical quality as judged by photoluminescence and absorption measurements. Although the carrier distribution was not constant throughout the sample thickness, the density of photoexcited carriers could still be estimated. At the onset of saturation, the average density of carriers in the first 0.1 μm was estimated to be about 1 $\times$ 10^{18} cm^{-3}, whereas the maximum density used in the study was about 1 $\times$ 10^{20} cm^{-3}.

Figure 18a shows the band-edge absorption as a function of time delay for a pump fluence of 160 μJ/cm^2. The excitonic resonance saturates within 1 ps and recovers slowly over the next 100 ps. Figure 18b shows the change in the absorption coefficient at the peak of the excitonic resonance as a function of time delay. The resonances recover with characteristic times of 17 and 23 ps for pump fluences of 160 and 730 μJ/cm^2, respectively. The longer recovery time at higher pump densities is explained by nonuniform photogenerated carrier densities throughout the sample thickness. For epilayers of GaAs, thin samples were found to exhibit fast absorption recovery ($<$50 ps) whereas the recovery in thicker samples was found to occur on much slower time scales ($\sim$1 ns) [37]. The thin GaAs samples recovered faster due to fast nonradiative surface recombination, whereas carriers in thicker samples decayed with the bulk radiative lifetime (on the order of nanoseconds). It is, therefore, expected that under low pumping conditions, carriers are primarily created in a thin layer near the GaN surface and can recombine via fast surface recombination. However, under very strong pumping conditions, absorption is saturated further into the sample thickness and higher carrier densities are created deeper into the GaN layer. Carriers located in the interior region of the sample recombine via slower, bulklike ra-

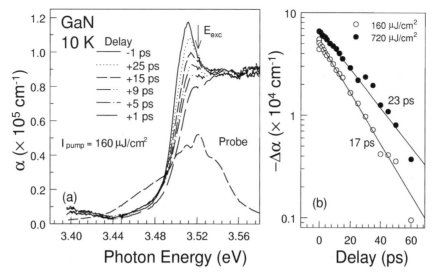

Fig. 18. (a) Absorption spectra as a function of delay between the laser pump and white-light continuum probe (dashed line) for a pump fluence of 160 μJ/cm^2. (b) Absorption saturation at the peak of the excitonic resonance as a function of time delay for pump fluences of 160 (full circles) and 730 μJ/cm^2 (open circles).

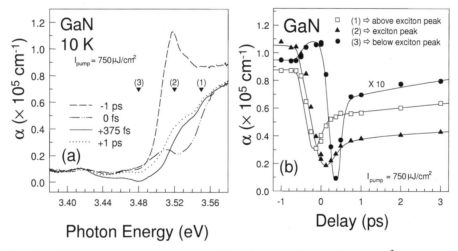

Fig. 19. (a) Absorption spectra as a function of delay for a pump fluence of 750 μJ/cm^2 showing the ultrafast near zero delay dynamics of a highly excited GaN film. Note the induced transparency at 3.48 eV. (b) Absorption as a function of delay for the three wavelengths denoted with arrows in (a). Note the induced absorption at zero delay followed by induced transparency in the below-gap region, reaching a maximum transparency at 400 fs. The solid lines are to guide the eye.

diative recombination and, thus, a slower recovery is observed in GaN films at higher pump fluences.

Figure 19a shows the absorption spectrum as a function of near zero delay for a pump fluence of 750 μJ/cm^2. The absorption spectrum at zero delay shows a dip in the absorption near 3.53 eV due to spectral hole-burning by the pump pulse. The carriers created at 3.53 eV quickly thermalize and relax toward the band edge, causing the hole to disappear. After 375 fs, transient-induced transparency attributed to the formation of an electron–hole plasma is observed below the bandgap at ~3.48 eV. The ultrafast dynamics near zero delay are more clearly illustrated in Figure 19b, where the absorption value is plotted as a function of delay. Curves are shown for detection at

the excitonic resonance (3.516 eV), as well as above (3.546 eV) and below (3.477 eV) the excitonic resonance, corresponding to the triangular markers in Figure 19a. The above-gap absorption is seen to saturate first, recovering quickly as carriers relax toward the band edge. The excitonic absorption saturates next and recovers after 1 ps to an intermediate value associated with excitonic phase-space filling. During the first few hundred femtoseconds, the below-gap data show induced absorption due to bandgap renormalization, which has a maximum magnitude of 1.5×10^3 cm^{-1} ($\pm 0.2 \times 10^3$ cm^{-1}). The induced absorption is caused by carriers injected into the sample above the gap that renormalize the bands before the carriers have time to relax down to the band edge [38]. The induced absorption changes to a strong induced transparency associated with the formation of a transient electron–hole plasma, which shows maximum transparency at 375 fs. This strong induced transparency recovers faster than the time resolution of the experiment to the intermediate level associated with band-filling. The induced transparency was observed over a range of pump fluences from ~400 to ~750 μJ/cm^2 and occurred just after the excitonic resonance was most strongly screened. The plasma state was observed to disappear quickly after a few hundred femtoseconds. The fast recovery of the induced transparency is most likely a result of fast nonradiative recombination, which acts to dissolve the plasma state. In addition, the presence of SE is also expected to quickly decrease the net measurable gain, because SE typically occurs on much faster time scales than spontaneous emission [39].

The observation of induced transparency and no net gain is most likely due to optical gain in a thin layer at the sample surface combined with absorption from band-tailing states further into the sample. For the relatively thick GaN layer used in the study, the highest carrier densities occur in a thin layer near the surface. In addition, no attempt was made to correct for reflection losses from the GaN surface. Although no net gain was observed, if reflection losses were accounted for and it was assumed that the change in absorption came from the first 50 nm of the epilayer, then a gain coefficient as high as 10^4 cm^{-1} was estimated. Such a large gain coefficient is consistent only with an electron–hole plasma. The gain was observed to disappear quickly due to depopulation that results from fast nonradiative recombination and/or stimulated emission in the plane of the sample. It is known from time-resolved photoluminescence measurements that the luminescence decay time of the exciton is dominated by nonradiative recombination channels [40]. Therefore, it is reasonable to expect that for samples with a lower defect density, the plasma state will persist longer and gain will be observed in transmission for GaN as it has been for GaAs-based samples.

3.6. Pump–Probe Reflection Spectroscopy of GaN Thin Films

In this section, the results of nanosecond nondegenerate optical pump–probe reflection experiments on GaN thin films are reviewed. These experiments were undertaken because GaN thin films used in pump–probe absorption experiments are typically optically thick, so the pump intensity is appreciably diminished as it traverses the sample thickness. Therefore, the resulting values measured for $\Delta\alpha$ are lower limit values of the actual change with pump density. A more precise measure of the influence of the pump density on the oscillator strength of the excitons can be achieved through optical pump–probe reflection measurements. In these measurements, only a thin layer (≈ 0.1 μm) at the sample surface is probed for wavelengths at and above the band edge, so the effect of pump beam attenuation is minimized. The effect of the pump on the band edge is, therefore, more easily compared to the results of optically pumped SE experiments.

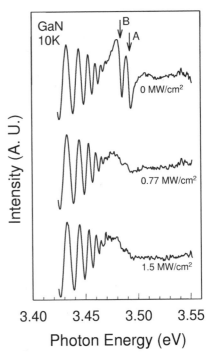

Fig. 20. Reflection spectra for a 4.2-μm-thick GaN layer grown by MOCVD on (0001) oriented sapphire as a function of I_{exc}. The 1s A and B free exciton transitions are clearly seen in the $I_{exc} = 0$ spectra (upper curve). The excitonic features are greatly diminished for $I_{exc} = 0.77$ MW/cm^2 and are completely indistinguishable from the system noise by 1.5 MW/cm^2. No clear red shift is observed in the excitonic positions as I_{exc} is increased from 0 MW/cm^2. The spectra have been displaced vertically for clarity. The oscillatory structures at longer wavelengths are the result of thin film interference.

The experimental configuration is identical to that of Figure 10 except the reflected instead of transmitted probe is collected, coupled into the spectrometer, and spectrally analyzed using the UV-enhanced gated CCD. An excitation photon energy of 3.676 eV (337 nm) was used in the reflection experiments.

Figure 20 shows the 10-K reflection spectra as a function of I_{exc} for a 4.2-μm-thick MOCVD-grown GaN layer whose optical properties were well characterized [41, 42]. The 1s A and B free exciton transitions are clearly seen in the $I_{exc} = 0$ spectra and are shown to be greatly diminished by the time I_{exc} reaches 0.77 MW/cm^2. We note that although the exciton resonances decrease with increasing I_{exc}, their positions remain relatively unchanged. This is because the decrease in the exciton binding energy is compensated by the red shift of the band edge due to bandgap renormalization and is consistent with observations in other materials [19]. The SE threshold for this sample was found to be $I_{th} \approx 0.3$ MW/cm^2 for the same excitation source [41]. The large decrease in the excitonic features at $I_{exc} = 0.77$ MW/cm^2 indicates that for pump densities that are a few times threshold, the gain in GaN is likely to arise from an electron–hole plasma rather than exciton-related mechanisms, even at 10 K.

3.7. Pump–Probe Absorption Spectroscopy of InGaN Thin Films

In this section, the results of nanosecond nondegenerate optical pump–probe absorption experiments on InGaN layers at zero time delay are presented. The study was initiated to help explain the drastic reduction in SE threshold that results when indium is incorporated into GaN to form InGaN. Considerable differences in the effects of a strong optical pump on the band-edge transitions are shown between InGaN and GaN thin films.

The $In_{0.18}Ga_{0.82}N$ layer used in this study was grown by MOCVD at 800°C on a 1.8-μm-thick GaN layer deposited at 1060°C on (0001) oriented sapphire. The $In_{0.18}Ga_{0.82}N$ layer was 0.1 μm thick and was capped by a 0.05-μm GaN layer. The average In composition was measured using high-resolution X-ray diffraction and assuming Vegard's law. We note that the actual InN fraction could be smaller due to systematic overestimation when assuming Vegard's law in this strained material system [43, 44]. The $In_{0.18}Ga_{0.82}N$ layer thickness given above is an estimation from the growth conditions.

The nanosecond nondegenerate optical pump–probe experimental system is similar to that illustrated in Figure 10 except the third harmonic (355 nm, 3.492 eV) of an injection-seeded Nd:YAG ($\approx$5-ns FWHM pulse width; 10-Hz repetition rate)

was used to synchronously pump the $In_{0.18}Ga_{0.82}N$ layer and a dye solution. The UV–blue fluorescence from the dye solution was collected and focused on the surface of the $In_{0.18}Ga_{0.82}N$ layer coincidental (spatially and temporally) with the pump beam. The broadband transmitted probe was collected, coupled into a 1/4-m spectrometer, and spectrally analyzed using a UV-enhanced gated CCD detector.

Figure 21 shows the absorption spectra of the $In_{0.18}Ga_{0.82}N$ subjected to various I_{exc} below and above I_{th}. The SE peak position is indicated for completeness. Note the absence of clear excitonic features in the unpumped absorption spectra. The absorption edge is seen to be considerably broader than those of the GaN films described in the previous sections ($\sim$50 meV compared to $\sim$30 meV for GaN). This difference is attributed to band tailing that results from difficulties in uniform indium incorporation. Usually this leads to a significant increase in the SE threshold. Instead, a reduction of over an order of magnitude in the SE threshold was observed for this film compared to high-quality GaN and AlGaN thin films. This is a typical result of the incorporation of indium into GaN, the origin of which is still not well understood. Clear bleaching of these band-tail states is observed with increasing optical excitation, as shown in Figure 21a. For clarity, the changes in the absorption spectra with respect to the unpumped spectrum, $\Delta\alpha(I_{exc}) = \alpha(I_{exc}) - \alpha(0)$,

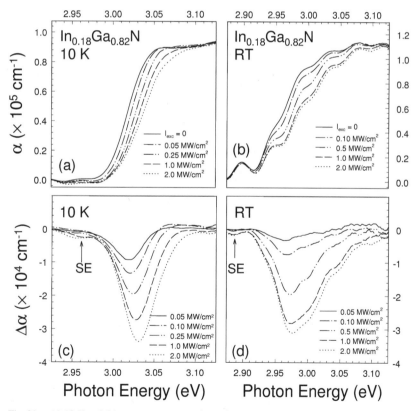

Fig. 21. (a) 10-K and (b) room temperature absorption spectra of a 0.1-μm-thick $In_{0.18}Ga_{0.82}N$ layer near the fundamental bandgap for several excitation densities above and below the SE threshold. (c) 10-K and (d) room temperature differential absorption spectra for the spectra shown in (a) and (b), respectively. The SE energy is indicated by an arrow in (c) and (d) for completeness. The differential absorption spectra exhibit a clear feature at the SE energy attributed to optical gain. No induced transparency is observed in the below-gap region.

are shown in Figure 21c for the spectra given in Figure 21a. The maximum of the observed absorption bleaching occurs on the high-energy side of the SE peak, originating approximately 65 meV above the SE peak position and blue-shifting slightly with increasing optical excitation. No significant increases in the bleaching spectra were observed for excitation densities that exceeded ~2 MW/cm^2. The feature peaked at ≈2.955 eV in the $\Delta\alpha$ spectra coincides spectrally with the SE peak and is attributed to optical gain, the magnitude of which (on the order of 10^3 cm^{-1}) is seen to be more than an order of magnitude smaller than the band-edge bleaching. The band-edge bleaching of the In$_{0.18}$Ga$_{0.82}$N film, which exceeds -3×10^4 cm^{-1} at $I_{exc} = 2$ MW/cm^2, is consistent with the bleaching behavior observed in GaN thin films. The In$_{0.18}$Ga$_{0.82}$N film thickness estimated from the growth conditions was used in the calculations of α and $\Delta\alpha$ presented in this section. Some error from this estimation is expected, but the calculated values of $\alpha(0)$ agree well with those of GaN films. Similar behavior was observed at room temperature, as illustrated in Figure 21b and d. Again, no significant increases in the bleaching spectra were observed for excitation densities in excess of ~2 MW/cm^2.

A striking difference between the absorption changes observed for the In$_{0.18}$Ga$_{0.82}$N film and those observed for GaN films is the absence of the below-gap induced absorption observed in GaN films with increasing optical excitation. This is significant in that it explains, at least in part, the drastic reduction in the SE threshold observed in InGaN films relative to GaN films. This is explained here as resulting from the lack of induced absorption and gain competition that is exhibited in GaN films. The incorporation of indium into GaN to form InGaN has been observed by us and other research groups typically to reduce or to eliminate the presence of yellow-band luminescence, and possibly reduce the number of deep levels that give rise to it. This adds further credence to the theory of their involvement in the formation of the below-gap induced absorption observed in GaN thin films. A detailed study is currently under way to better understand the effects of YBL on the bleaching behavior of highly excited GaN, InGaN, and AlGaN thin films.

Another difference in the bleaching behavior of the In$_{0.18}$Ga$_{0.82}$N film compared to GaN films is the modest reduction in the absorption bleaching maximum as the temperature is increased from 10 K to room temperature. As the temperature is increased from 10 K to room temperature, the maximum in the bleaching spectrum reduces only slightly, from $\Delta\alpha \approx -3.4 \times 10^4$ cm^{-1} at 10 K to $\approx -3.1 \times 10^4$ cm^{-1} at room temperature for $I_{exc} = 2$ MW/cm^2. This value differs greatly from the factor of almost 3 reduction observed in GaN films. The bleaching spectra of the In$_{0.18}$Ga$_{0.82}$N film at room temperature are significantly broader than those at 10 K, showing a 50% increase in FWHM over the 10-K spectra.

3.8. Summary

In this section, the results of nondegenerate optical pump–probe and single-beam absorption experiments on MOCVD-grown GaN films were presented. The evolution of the band edge was monitored in nondegenerate pump–probe experiments as the number of free carriers was increased by photoexcitation. Large optical nonlinearities in the region of the fundamental bandgap were observed in nanosecond experiments as the number of free carriers was increased by optical excitation to densities sufficient to observe SE. Exciton saturation was observed with increasing carrier concentration, with a resulting decrease in the absorption coefficient (induced transparency) approaching $\Delta\alpha = -4 \times 10^4$ cm^{-1} at 10 K and -2×10^4 cm^{-1} at room temperature. In addition, large below-gap induced absorption that exceeded $\Delta\alpha = 4 \times 10^4$ cm^{-1} at 10 K and 2×10^4 cm^{-1} at room temperature was observed as the pump density was increased to over 3 MW/cm^2. The exciton saturation is explained by many-body effects such as screening by excess free carriers. Nanosecond nondegenerate optical pump–probe transmission experiments with variable time delay between the pump and probe pulses showed the below-gap induced absorption to be a result of a large shift in the fundamental bandgap to lower energy. This shift was found to increase across the pump pulse duration, reaching a maximum within 1 ns after the end of the pump pulse. The band edge was shown to slowly (on the order of 100 ns) recover to its prepumped position. The maximum shift in the band edge (at 5-ns delay) was found to exhibit saturation behavior with increasing optical excitation. Lattice heating and screening due to carriers trapped at deep levels were suggested as a mechanism for the shift in the band edge. Single-beam power-dependent absorption experiments show enhanced absorption bleaching of the excitonic transitions for resonant excitation and a drastic reduction in the below-gap induced absorption. The experimental results indicate the states responsible for the induced absorption must first be created by the above-gap excitation of the pump beam and cannot be efficiently created by below-gap excitation.

The ultrafast dynamics of the band-edge transitions in highly excited GaN were also explored via femtosecond nondegenerate optical pump–probe transient absorption spectroscopy. These studies showed exciton saturation for pump fluences greater than 20 μJ/cm^2, which corresponds to an estimated carrier density of 1×10^{18} cm^{-3}. This exciton saturation was shown to occur within 1 ps and to recover with a characteristic time constant of ~20 ps. The recovery was shown to be slower at higher pump fluences as high carrier densities are created deeper inside the sample where slower bulklike recombination occurs. Induced absorption was also observed below the bandgap due to bandgap renormalization. This induced absorption was shown to change quickly to induced transparency due to a transient electron–hole plasma. The electron–hole plasma dissolved quickly after a few hundred femtoseconds due to depopulation by nonradiative emission and SE. The transient plasma state was shown to strongly screen the excitonic resonance during the first 1 ps, whereas exciton saturation at longer time delay was attributed to excitonic phase-space filling.

Nanosecond nondegenerate optical pump–probe absorption experiments on InGaN films were also detailed. Induced transparency in the band-edge region was observed and attributed to the filling of band-tail states with increasing optical excitation. This bleaching reached a maximum of approximately 3×10^4 cm^{-1} at 10 K and room temperature for an excitation density of 2 MW/cm^2. The InGaN films did not exhibit induced absorption in the below-gap region. Instead, features in the differential absorption spectra were clearly seen in the spectral region where SE was observed. These features were attributed to optical gain. The high SE threshold of GaN is due to induced absorption and gain competition with increasing free carrier concentration. The large values of induced transparency and absorption observed in these studies suggest the potential of new optoelectronic applications such as optical switching for III–nitride thin films.

4. GAIN MECHANISMS IN NITRIDE LASING STRUCTURES

4.1. Overview of Gain Mechanisms

To develop and optimize the lasing characteristics of group III–nitride-based laser diodes, it is important to understand the gain mechanisms in this material system. It has been found that the gain mechanisms in GaN, InGaN, and AlGaN thin films as well as InGaN/GaN multiple quantum wells and GaN/AlGaN separate confinement heterostructures vary greatly depending on the composition and geometry of the lasing medium. Gain in semiconductors occurs at a high level of optical excitation; consequently, this section concentrates on the optical properties of nitride thin films under high levels of optical pumping.

Theoretical discussion is limited to the optical properties of direct gap semiconductors, particularly the phenomena exhibited by and relating to highly excited III–nitrides. Typically, with increased excitation power, the sample emission undergoes dramatic changes in the band-edge region, which suggests the spectral features are due to interaction processes of excitons and/or free carriers with themselves or other quasiparticles. This leads to the basic "definition" of highly excited semiconductors used in this section: A highly excited semiconductor is one in which the density of electronic excitations has been increased (e.g., by optical pumping) to sufficiently high values that the interaction processes between the quasiparticles in the semiconductor that can be neglected at low densities become effective and give rise to new recombination processes.

The first type of quasiparticle interaction that is of interest to this section is that of exciton–exciton scattering. This process occurs in a regime of moderately high excitation intensity in which excitons have not been entirely dissociated due to screening by free carriers and scattering mechanisms, but are present at high enough densities that interactions between particles become appreciable and the following exciton–exciton scattering process results in optical gain. In this process, only part of the energy of the recombining exciton leaves the crystal as a photon. The remaining energy is transferred to another exciton.

Exciton–exciton recombination occurs when two excitons in the $n = 1$ hydrogen-like ground level scatter, promoting

one exciton to the $n = 2$ level (or higher). The other exciton recombines at a lower energy, conserving total energy. Inelastic exciton–exciton scattering manifests itself in the luminescence spectra of highly excited semiconductors by a new emission band, which is displaced from the free exciton position by approximately one exciton binding energy (E_x^b). These new emission bands, commonly referred to as P bands in the literature, are generally described to good approximation by

$$\hbar\omega_{\max} = E_g - E_x^b(1 - 1/n^2) \qquad n = 2, 3, 4, \ldots, \infty \quad (4.1)$$

where E_g is the bandgap of the semiconductor, E_x^b is the binding energy of the exciton, and $\hbar\omega_{\max}$ is the peak energy of the P_n band. The P_n lines are often broad (10–30 meV in GaN) compared to the exciton linewidths, even at low lattice temperatures, because the excitons have excess kinetic energy and, therefore, have an effective temperature that can be much higher than the lattice temperature. In addition, the exciton binding energy may change due to the presence of the dense excitonic gas created by the intense optical pump beam [45]. From Eq. (4.1) we see that the P bands extend from approximately $(1+3/4)E_x^b$ to $2E_x^b$ below the bandgap of the semiconductor (or, equivalently, from $3/4E_x^b$ to E_x^b below the excitonic resonance). In addition, there is the possibility that the P bands can extend to energies far below $E_g - 2E_x^b$ if one of the excitons is scattered high into the continuum. In this case, as the excitation increases, the bottom of the bands become filled and the unbound pairs created in the exciton–exciton scattering process must have higher and higher energies. As such, the kinetic energy $E_{e,h}^c$ can no longer be neglected. The shift of the P line toward lower energy can then be expressed as

$$\hbar\Delta\omega_{\max} = -E_{e,h}^c \quad (4.2)$$

If we assume that the bands are elliptical and consist of only one extremum, this band-filling gives a shift in the P-line maximum (with respect to the P_∞ line) of

$$\hbar\Delta\omega_{\max} = \left[\frac{1}{m_e^*} + \frac{1}{m_h^*}\right]\frac{h^2}{8}\left(\frac{3}{8\pi}\right)^{2/3}n^{2/3} \quad (4.3)$$

where n designates the number of free carriers per unit volume.

Other possible exciton-related optical gain processes in highly excited semiconductors, first introduced by Benoit à la Guillaume et al. [46], are exciton–electron recombination, exciton–hole recombination, exciton–LO phonon recombination, and excitonic molecule recombination.

If the concentration of free carriers is increased further (e.g., by band-to-band optical excitation), the density of electron–hole pairs can become high enough that the bound electron–hole pair states (excitons) are no longer stable. The electron–hole pairs can then form a collective state, described as an electron–hole plasma. This collective state is characterized by a phase transition at a critical temperature at which the excitons ionize as the density of electronic excitations is increased. This ionization is due to screening of the attractive Coulombic force between the electrons and the holes due to the presence of a high density of free carriers. The bound electron–hole pairs then become unstable and the system is said to undergo a Mott transition. The density at which this transition occurs can be estimated by the simple argument that bound states become impossible if the screening length becomes on the order of, or less than, the Bohr radius of the exciton ($k_D = 1/a_0$, where k_D is the inverse Debye–Hückel screening length and a_0 is the exciton Bohr radius in the ground state). The inverse Debye–Hückel screening length is given by

$$k_D = \left(\frac{8\pi n e^2}{\varepsilon_0 k_B T_p}\right)^{1/2} \tag{4.4}$$

where $k_B T_p$ is the thermal plasma energy and n is the density of free carriers. The density of free carriers at which the Mott transition occurs can then be estimated as

$$n = \frac{\varepsilon_0 k_B T_p}{8\pi e^2}\left(\frac{1}{a_0}\right)^2 \tag{4.5}$$

Optical gain that arises from an electron–hole plasma can be quite large, with gain coefficients approaching that of the above-gap absorption coefficient (10^3–10^4 cm^{-1} for traditional III–V semiconductors).

Having given this short introduction, an overview of the gain mechanisms in nitride thin films can now be presented. In the subsections that follow, it will be demonstrated that the dominant near-threshold gain mechanism in GaN thin films is inelastic exciton–exciton scattering for temperatures below ~150 K and an electron–hole plasma for temperatures above 150 K, whereas exciton–exciton scattering dominates the lasing spectrum of GaN/AlGaN separate confinement heterostructures from 10 K up to room temperature. Even though excitonic features were observed in AlGaN epilayers at low temperatures, it has been found that an electron–hole plasma is responsible for stimulated emission in AlGaN over the entire temperature range of 10–300 K. Finally an overview of critical role of localized carriers in stimulated emission of InGaN epilayers and InGaN/GaN multiple quantum wells will be given.

4.2. Origin of Stimulated Emission in GaN Epilayers

There have been many studies performed in an attempt to explain the origin of SE in GaN films at various temperatures. Amano and Akasaki [162] suggested that an electron–hole plasma is the most plausible origin of gain in GaN epilayers at room temperature. Catalano et al. [47] suggested exciton–exciton (ex-ex) scattering is the dominant gain mechanism at 80 K. Recently, Holst et al. performed gain spectroscopy on HVPE-grown GaN films and concluded that biexcitonic decay is responsible for the gain at 1.8 K at low excitation densities, whereas electron–hole plasma recombination dominates the spectra at higher excitation densities. In this section, a comprehensive review of the gain mechanisms in GaN epilayers using nanosecond optical pumping in the temperature range of 20–700 K is given.

The samples used in this work were GaN thin films of various thicknesses that were grown on sapphire and SiC. The samples were mounted on a copper heat sink attached to the custom-built wide temperature range heater system used in

conjunction with a cryostat cooled by a closed-cycle helium refrigerator. The SE part of this study was performed in an edge-emission geometry. A tunable dye laser pumped by a frequency-doubled, injection-seeded Nd:YAG laser was used as the primary optical pumping source. The deep red output of the dye laser was frequency doubled to achieve a near-UV pumping frequency. The emission was collected from one edge of the sample and coupled into a 1-m spectrometer, then spectrally analyzed with a UV enhanced multichannel analyzer. Low-power cw photoluminescence (PL) studies also were undertaken to measure the spontaneous emission as a function of temperature. A frequency-doubled Ar$^+$ laser (244 nm, 40 mW) was used as the excitation source. To avoid any spectral distortion of the spontaneous emission due to reabsorption, the laser beam was focused on the sample surface and spontaneous emission was collected from a direction near normal to the surface. Typical power-dependent emission spectra from a GaN epilayer are shown in Figure 22. At low excitation pump densities, the sample emission has a FWHM of about 15 nm, with the peak positioned at about 366 nm at room temperature. The peak appears to be red shifted in comparison to low excitation density cw PL due to reabsorption effects. As the excitation pump density is increased, a new peak emerges on the low-energy side of the spontaneous emission peak. The FWHM of this new peak is only 2 nm at room temperature. We note that this peak is strongly polarized and its intensity grows superlinearly with ex-

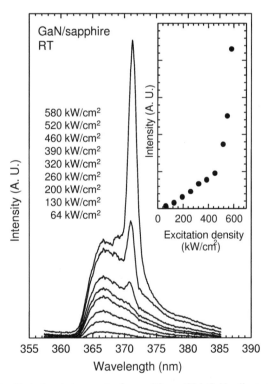

Fig. 22. Typical emission spectra from a 4.2-μm-thick GaN epilayer at room temperature for different pumping densities. The inset shows the dependence of the integrated emission intensity on the excitation power. From both spectral narrowing and the superlinear increase in emission intensity we determined the SE threshold to be 480 kW/cm^2.

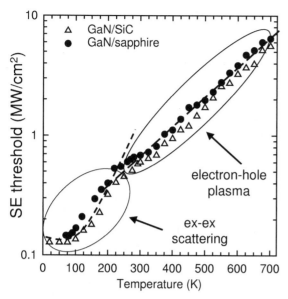

Fig. 23. SE threshold as a function of temperature for GaN thin films grown on SiC (open triangles) and sapphire (solid circles) in the temperature range 20–700 K. The SE threshold rises exponentially for temperatures that exceed 200 K, with a characteristic temperature of approximately 170 K. For temperatures below 200 K, the SE threshold is considerably reduced due to excitonic enhancement. The solid lines are given only to guide the eye.

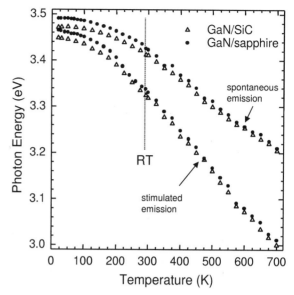

Fig. 24. The absolute energy positions of the spontaneous and SE peaks for GaN thin films grown on SiC (open triangles) and sapphire (solid circles). The difference in the energy positions (particularly at low temperature) between the two samples is the result of residual strain between the epilayers and the two different substrates.

citation power (as shown in the inset of Fig. 22). Based on the significant spectral narrowing, superlinear increase with excitation power, high degree of polarization, and directionality of the emission, we conclude that the new peak located at ~371 nm at room temperature is a result of SE.

Fischer et al. [17] convincingly demonstrated the presence of excitonic resonances in GaN thin films well above room temperature through optical absorption measurements. Excitons in GaN epilayers cannot be easily ionized due to their relatively large exciton binding energy. However, at near-SE-threshold (near-I_{th}) pump densities, the picture is not straightforward because screening of the Coulomb interaction weakens the binding of the exciton. In general, the existence of excitons depends on the strength of the Coulomb interaction, which in turn depends on the density and distribution of carriers among bound and unbound states [49]. Therefore, the observation of excitons at low excitation powers at room temperature does not assure their presence at SE pump densities.

To determine if the SE threshold density occurs above or below the Mott density (the critical carrier density beyond which no excitons can exist), it is informative to study the temperature behavior of the SE threshold in GaN epilayers grown on SiC (open triangles) and sapphire (solid circles), as shown in Figure 23 (note that the SE threshold is plotted on a logarithmic scale). A faster than exponential increase in the SE threshold occurs in the vicinity of 150 K. For temperatures above 200 K, the SE thresholds roughly follow an exponential dependence,

$$I_{th}(T) = I_0 \exp\left(\frac{T}{T_0}\right) \quad (4.6)$$

with $T_0 \cong 170$ K. This exponential behavior of the SE threshold is qualitatively similar to that observed in other material structures [50]. However, as the temperature is decreased below 200 K, a significant reduction in the SE threshold is typically observed. The decrease is due to a change in the dominant gain mechanism and manifests itself in a drastic increase in the SE efficiency at low temperatures. It has been predicted theoretically that in a material system with a relatively large exciton binding energy, inelastic ex-ex scattering can be expected to have the lowest SE threshold at low temperatures [49]. The presence of excitons at pump densities above the SE threshold in reflection spectra was also observed [23], as was described in Section 3.6.

The effects of excitons on SE can be better understood through examination of the power dependence of the SE peak position at near-I_{th} pump densities. The position of the spontaneous emission peak can be measured using low-power cw PL. The peak positions in the temperature range of 20–700 K for two samples grown on sapphire and SiC substrates are shown in Figure 24. The position of the spontaneous and SE peaks in the two GaN epilayers is influenced by residual strain that results from thermal-expansion mismatch between the epilayers and the substrates [160]. This difference in energy position for the two samples is largest at low temperature and gradually decreases as the temperature is increased.

To avoid strain-related complications, we should consider the relative energy shift between the spontaneous and SE peaks, $\Delta E = E_{spon} - E_{SE}$, as depicted in Figure 25. As the temperature is lowered below 150 K, ΔE asymptotically approaches the exciton binding energy ($E_x^b = 21$ meV) measured by photoreflectance [18]. However, at temperatures above 150 K, ΔE monotonically increases and reaches values as high as 200 meV

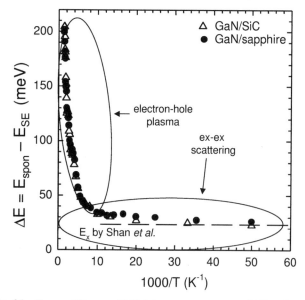

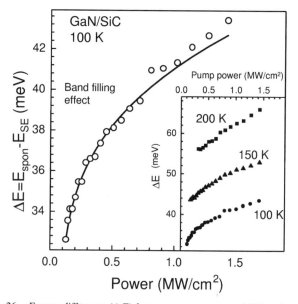

Fig. 25. Energy difference (ΔE) between spontaneous and SE peaks as a function of temperature for GaN thin films grown on SiC (open triangles) and sapphire (solid circles). At low temperatures, ΔE asymptotically approaches the exciton binding energy, indicating that the dominant near-I_{th} gain mechanism is inelastic ex-ex scattering. At elevated temperatures, the large value of ΔE indicates the increased presence of bandgap renormalization effects.

Fig. 26. Energy difference (ΔE) between spontaneous and SE peaks as a function of excitation density at 100 K. The solid line represents a theoretical fit of the experimental data (open circles) to Eq. (4.8). The inset shows the change in the behavior of ΔE at different temperatures. An abrupt near-I_{th} shift in ΔE at 100 K is related to band-filling effects associated with increased values of kinetic energy E_k^{e-h} during the ex-ex scattering recombination process. At elevated temperatures and/or high excitation powers, electron–hole plasma is the dominant gain mechanism.

at 700 K. The behavior of the energy difference between the spontaneous and SE peaks at low temperatures can be well explained by inelastic ex-ex scattering. In the case of ex-ex scattering, the energy difference between the two peaks can be estimated from [51]

$$\Delta E = E_{spon} - E_{SE}^{ex-ex} = (E_g - E_x^b) - (E_g - 2E_x^b - E_k^{e-h})$$
$$= E_x^b + E_k^{e-h}, \qquad (4.7)$$

where E_g is the bandgap energy and E_k^{e-h} is the kinetic energy of the unbound electron–hole pair created during the excitonic collision. At low excitation densities and low temperatures, the bands are considered to be empty. The unbound electron–hole pairs created during this process have a very small kinetic energy ($E_k^{e-h} \approx 0$); thus, ΔE approaches E_x^b as $T \rightarrow 0$ K, as shown in Figure 25.

For high temperatures ($T > 150$ K), the energy difference between spontaneous and SE peaks gradually increases from ~35 meV to a few hundred millielectronvolts. Both the large energy difference and the relatively high SE thresholds in this temperature range (Fig. 23) point to electron–hole plasma recombination as the dominant gain mechanism. In electron–hole plasma recombination, a large number of excited carriers cause bandgap renormalization effects that lead to a large value of ΔE. Under such high excitation conditions, excitons are dissociated by many-body interactions [162]. As further evidence to support the dominance of the electron–hole plasma gain mechanism in this temperature range, it should be pointed out that excitons have not been clearly observed in GaN at highly elevated temperatures ($T > 450$ K), even under extremely low excitation conditions [17]. Thus, electron–hole plasma recom-

bination is responsible for gain in GaN thin films at these elevated temperatures. Whereas no significant change in the behavior of the SE threshold or the SE peak position was observed for temperatures between 150 and 700 K (Figs. 23 and 25), it has been concluded that electron–hole plasma recombination is the dominant gain mechanism for all temperatures above 150 K.

At temperatures below 150 K, the effect of the kinetic energy E_k^{e-h} on the ex-ex scattering recombination process can be observed in the excitation density dependence of ΔE. As the excitation intensity or temperature is increased, the bottoms of the bands become filled. Thus, unbound electron–hole pairs created in the process of ex-ex collision must have higher energies, and the kinetic energy E_k^{e-h} can no longer be neglected. The inset in Figure 26 shows the power dependence of ΔE at three different temperatures near the point when the gain mechanism experiences a transition from inelastic ex-ex scattering to electron–hole plasma. For temperatures below 150 K, a rapid increase in ΔE at near-I_{th} pump densities is observed, as shown in the inset of Figure 26. This shift is associated with the band-filling effect, which causes increased values of E_k^{e-h}. For temperatures above 150 K, such a strong near-I_{th} shift in ΔE is not observed. At these temperatures, SE originates from electron–hole plasma recombination and the gradual increase in ΔE is caused by bandgap renormalization effects. For one-photon pumping and elliptical bands, the band-filling effect associated with ex-ex scattering gives a calculated line shift E_k^{e-h} that is proportional to $(I - I_{th})^{1/3}$ [51, 52]. Substitution of this expression into Eq. (4.7) yields

$$\Delta E = E_x^b + a(I - I_{\text{th}})^{1/3} \qquad (4.8)$$

A fit of the experimental data (open circles) taken at 100 K to Eq. (4.8) is shown in Figure 26 as the solid line. From the fit, an exciton binding energy of $E_x = 28$ meV and a SE threshold of $I_{\text{th}} = 100$ kW/cm^2 were obtained. These values are in reasonable agreement with experimental results and support the idea that ex-ex scattering is the dominant SE mechanism for temperatures below 150 K.

Whereas ex-ex scattering has a lower SE threshold than recombination from an electron–hole plasma (Fig. 23), it would be advantageous if SE were dominated by excitonic effects at room temperature and above [53]. One way this could be achieved is by introducing two-dimensional spatial confinement [54]. By tailoring the width of a GaN active layer located between AlGaN confinement layers, we can expect a significant increase in the exciton binding energy. For increased values of exciton binding energy, a strong reduction of the homogeneous broadening due to reduced Fröhlich interactions is expected [55] that could potentially extend the ex-ex scattering gain mechanism to room temperature. The SE threshold for such structures would be significantly reduced due to carrier confinement [56] and a shift in the dominant near-I_{th} gain mechanism from electron–hole plasma to ex-ex scattering. This reduction in the gain threshold density could substantially aid the development of GaN-based near-UV laser diodes.

4.3. Mechanism of Efficient Ultraviolet Lasing in GaN/AlGaN Separate Confinement Heterostructures

The successful fabrication of a blue laser diode was largely due to the realization that the incorporation of indium into GaN was concomitant with a dramatic lowering of the lasing threshold, an enhancement of the emission intensity [57], and improvements in the temperature characteristics of the material [13]. The small bandgap of InN (1.9 eV), however, is disadvantageous for the development of near- and deep-UV laser diodes. Lasing structures using a GaN-active medium are required to fabricate LDs with emission wavelengths that lie below 370 nm. An example of such a structure is described next.

The GaN-active medium structure under consideration was grown by MOCVD on 6H-SiC substrates with ~3-μm-thick GaN epilayers deposited prior to the growth of the active region. It contained a 150-Å-thick GaN active layer surrounded by 1000-Å-thick Al$_{0.05}$Ga$_{0.95}$N waveguide layers and 2500-Å-thick Al$_{0.10}$Ga$_{0.90}$N cladding layers symmetrically located on each side. This type of structure allows for separate optical and carrier confinements, and is generally referred to as a separate confinement heterostructure (SCH). The sample was mounted on a copper heat sink attached to a wide temperature range cryostat. Conventional PL spectra were measured in backscattering geometry using a frequency-doubled Ar$^+$ laser (244 nm) as the excitation source. To study the lasing phenomena, a tunable dye laser pumped by a frequency-doubled injection-seeded Nd:YAG laser was used as the primary optical pumping source.

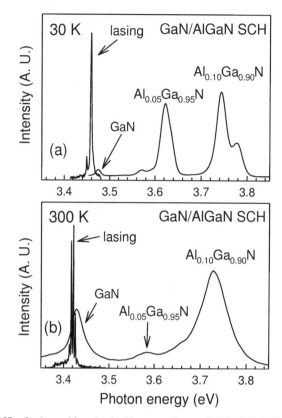

Fig. 27. Lasing and low-density PL spectra from a GaN/AlGaN SCH taken at (a) 30 and (b) 300 K. Lasing spectra were obtained at pump densities of $1.3 \times I_{\text{th}}$ for each temperature. The PL spectra have three characteristic peaks that are associated with the 150-Å-thick GaN-active region and the two AlGaN confinement layers with different aluminum concentrations.

Special precautions need to be taken to avoid fluctuations in sample position due to the thermal expansion of the mounting system. It is critical to "pin" the sample and obtain lasing modes from a single microcavity over the entire temperature range studied. Time-resolved photoluminescence measurements were performed using a frequency tunable pulsed laser (5-ps pulse duration, 76 MHz) as an excitation source and a streak camera (2-ps resolution) in conjunction with a 1/4-m monochromator as a detection system.

Typical low excitation emission spectra obtained in the surface-emission geometry at 30 (Fig. 27a) and 300 K (Fig. 27b) contain three distinct features associated with the active GaN layer, 5% aluminum waveguide layer, and 10% aluminum cladding layer. As seen in the figure, a doublet spectral feature can be seen at both of the AlGaN-related peaks. There are two possible explanations for this phenomenon. First, this feature could arise from a small alloy concentration difference between the two AlGaN cladding and two waveguide layers (it has been estimated that a 1% fluctuation in aluminum alloy concentration leads to a ~25-meV shift in the observed energy position [30]). The second explanation can be seen by examining the temperature-dependent photoluminescence spectra shown in Figure 28. Three distinct peaks associated with the 150-Å-thick GaN active region and the two AlGaN confinement

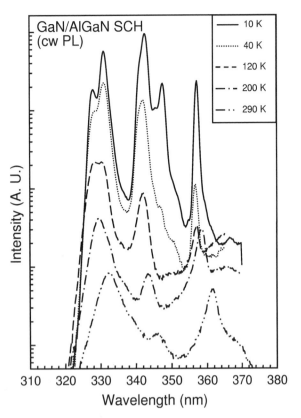

Fig. 28. Temperature-dependent spontaneous emission spectra. Each spectrum has three distinct peaks that are associated with the 150-Å-thick GaN active region and the two AlGaN confinement layers with different aluminum concentrations.

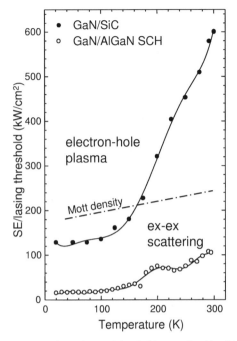

Fig. 29. Temperature dependence of threshold pump densities for a 4.2-μm-thick GaN epilayer (solid circles) and GaN/AlGaN SCH (open circles). The lasing threshold of the SCH was measured to be 15 kW/cm^2 at 10 K and 105 kW/cm^2 at room temperature. For the SCH, note that lasing occurs at pump densities much lower than those required for exciton dissociation over the entire temperature range studied. The solid lines are to guide the eye only.

layers with different aluminum concentrations persist over the entire temperature range studied, from 10 to 300 K. We note that the intensity of the "satellite" peaks (the doublet spectral features) quickly quenches with increasing temperature, suggesting that the levels associated with this transition have a low activation energy and could be impurity related.

At increased excitation powers, a series of equally spaced lasing modes, each with a FWHM of ∼3 Å, appears on the low-energy side of the GaN-active-region peak, as shown in Figure 27. Had the spontaneous emission been collected from the sample edge and not from the surface, the lasing modes would have appeared on the high-energy side of the spontaneous emission peak. This phenomenon is due to strong reabsorption that introduces a shift of several nanometers to the spontaneous emission peak. The lasing modes have a strong superlinear increase in intensity with excitation power. The laser emission has a narrow far-field pattern and is strongly polarized, with TE:TM ≥ 300:1. Cracks were observed on the sample surface along all three cleave planes associated with the hexagonal structure, with the majority running parallel to the length of the bar. The spacing between the lasing modes was correlated to the length of the microcavities formed by the cracks. Based on these observations, lasing was determined to be of microcavity origin [58]. Further refinements to the cleaving process have allowed samples to be cleaved in a manner that results in no

observable cracking, making them suitable for laser diode development.

A dramatic difference in the lasing threshold of the SCH structure in comparison to a thick GaN epilayer was observed over the entire temperature range studied, as shown in Figure 29. The SCH lasing threshold was estimated to be as low as 15 kW/cm^2 at 10 K and 105 kW/cm^2 at room temperature. Note that excitons in the GaN/AlGaN SCH cannot be easily ionized with temperature due to the relatively large exciton binding energy [17]. As was described earlier in this section, to screen the excitonic Coulomb interaction by free carriers (Mott transition) in GaN epilayers, a pump density of about 250 kW/cm^2 is required [59]. This pump density corresponds to a carrier density of ∼1.1×10^{18} cm^{-3}, calculated using a carrier lifetime of ∼35 ps and a penetration depth of 9×10^4 cm^{-1} (taken from [23] and [67]). This value of the Mott density is almost identical to that obtained from ultrafast studies described previously. At this carrier density, the gain mechanism in GaN epilayers switches from ex-ex scattering to electron–hole plasma, which is indicated in Figure 29 by a dotted line. Because the lasing threshold density of the SCH sample is considerably below the carrier density that corresponds to the Mott transition, excitons are expected to be present at pump densities above the lasing threshold and to make major contributions to the recombination dynamics (when comparing the lasing thresholds, it is necessary to take into account that the carrier lifetime of the GaN/AlGaN SCH active layer is not significantly different from that of GaN epilayers).

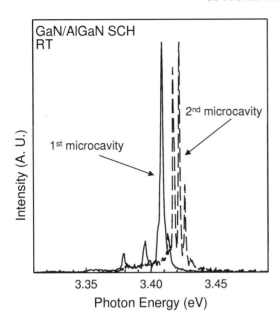

Fig. 30. Lasing modes obtained from two different microcavities. When studying the dependence of the energy positions of lasing modes, it is absolutely necessary that the modes are obtained from the same microcavity over the entire temperature range.

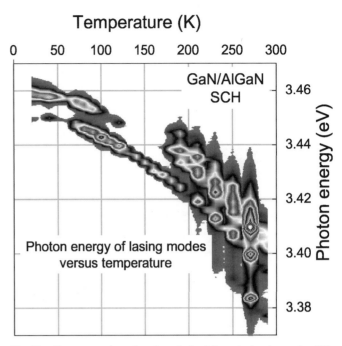

Fig. 31. Temperature-dependent data obtained for a single microcavity. This image was obtained by consistently taking lasing spectra with an optical multichannel analyzer at $1.3 \times I_{th}(T)$ while the temperature was gradually varied from 20 to 295 K.

The consequences of exciton dynamics on lasing in the GaN/AlGaN SCH can be better understood by studying the temperature dependence of the energy position of the emission peaks. However, it is an experimental challenge to obtain the temperature dependence of the lasing modes. Slight variations in the sample position inevitably result in different emission spectra. An example of two different sets of lasing modes from two adjacent cavities is depicted in Figure 30. When the cryostat temperature is varied, the length of the rod on which the sample holder is mounted also changes due to a nonzero temperature expansion coefficient. This results in spatial displacement of the sample. To avoid these complications, the cryostat should be mounted horizontally so that the rod expansions occur in the direction parallel to the direction of the laser beam and the temperature variations do not result in transverse displacement of the laser beam relative to the sample surface. This way, lasing can be obtained from the same microcavity over the entire temperature range.

An optical multichannel analyzer was used to consistently acquire spectra at $1.3 \times I_{th}$ for each temperature. Figure 31 shows a plot of mode energy position versus temperature. The color pallet corresponds to different emission intensities. As the temperature increases, the gain region broadens and additional modes can be observed.

The energy positions of the spontaneous emission peaks and lasing modes were extracted from Figures 28 and 31. The results of this analysis are summarized in Figure 32. For the purpose of comparison, the temperature evolution of the spontaneous and SE peaks from a thick GaN epilayer (solid lines) are also shown. The temperature behavior of the GaN thin film is distinctly different from that of the SCH. The temperature behavior of the energy peak positions of the SCH did not follow the Varshni equation. In fact, the spontaneous emission peaks blue-shift up to a temperature of 90 K and then red-shift thereafter. The position of the spontaneous emission from the 150-Å-thick GaN-active layer was located 0.15 eV below that of the $Al_{0.05}Ga_{0.95}N$ waveguide layer (open squares) and this difference remained temperature independent.

The behavior of the spontaneous emission and lasing from the two samples cannot be directly compared on this graph due to strain-related complications, such as different thermal expansion coefficients between AlGaN alloys and GaN. To avoid these complications, the discussion is once again restricted to an analysis of the relative energy difference ΔE between the lasing modes and the GaN-active-region peak, as illustrated in Figure 33 (open circles). The position of the lasing modes is one exciton binding energy below the GaN peak [18] and, most interestingly, remains temperature invariant from 10 to 300 K. This behavior is consistent with ex-ex scattering [59] being the dominant gain mechanism in the GaN/AlGaN SCH over the entire temperature range studied. To further corroborate this point, the energy difference (ΔE) between the spontaneous and SE peaks of a thick GaN epilayer (solid line) are plotted on the same graph. It has been shown previously that the dominant gain mechanism in GaN epilayers is ex-ex scattering for temperatures below 150 K. Similar to GaN epilayers, ΔE in the GaN/AlGaN SCH lies one exciton binding energy below the spontaneous emission peak. However, at elevated temperatures ($T > 150$ K), ΔE in GaN epilayers rapidly increases, in direct contrast to the SCH. This was previously attributed to the transition of the gain mechanism from ex-ex scattering to electron–hole plasma. Whereas this transition was not observed

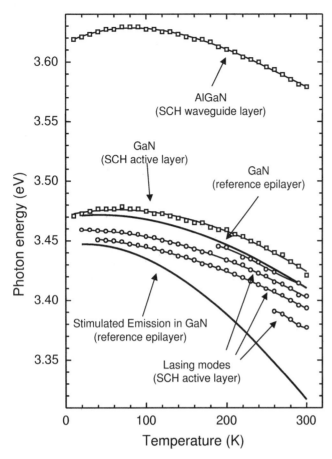

Fig. 32. Temperature dependence of low-density PL peak positions (open squares) and lasing modes (open circles) in a GaN/AlGaN SCH. For comparison, we also plotted the position of spontaneous and SE peaks from a 4.2-μm-thick GaN epilayer (solid lines). The two samples show distinctly different temperature evolutions associated with the different origins of SE/lasing at elevated temperatures.

in the SCH, ex-ex scattering is concluded to be the dominant gain mechanism in this lasing structure even at room temperature.

The carrier dynamics of the GaN/AlGaN SCH were studied through time-resolved photoluminescence and photoluminescence excitation (PLE) experiments. Figure 34 shows time-resolved data obtained at 10 K for the three major peaks depicted in Figure 28. The sample was excited using ~50-ps pulses at 302 nm. The mathematical modeling of the intensity decay of the peaks required fitting with several different exponentials. The short decay time of the GaN-active-layer peak indicates that the diffusion of carriers from carrier/waveguide regions into this layer is minimal.

This point is further supported by the PLE data depicted in Figure 35. To simplify the interpretation, photoluminescence data are plotted on the same scale. The scanning of the excitation source from shorter to longer wavelengths is always concomitant with the increase of the detector signal when it is set at the GaN-active-layer peak. This is consistent with the fact that the diffusion of carriers from outer layers into the GaN-active layer is rather weak.

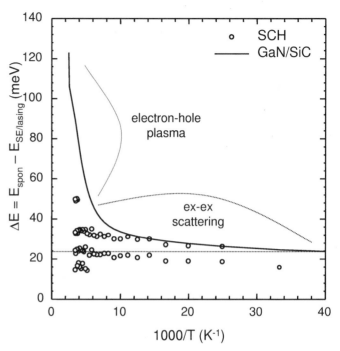

Fig. 33. Energy position of lasing modes relative to spontaneous emission from the GaN-active region in the SCH (open circles), and the energy difference between spontaneous and SE peaks in a 4.2-μm-thick GaN epilayer (solid line). The lasing modes in the SCH appear one exciton binding energy below the spontaneous emission peak over the entire temperature range studied, indicating that ex-ex scattering is the dominant lasing mechanism. On the contrary, the energy difference between the spontaneous and SE peaks in the GaN epilayer rapidly changes at ~150 K due to the gain mechanism transition from ex-ex scattering to electron–hole plasma.

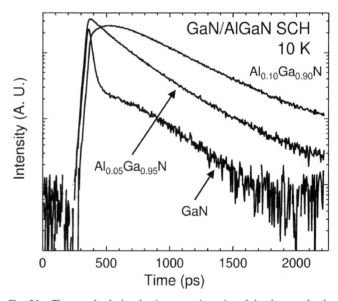

Fig. 34. Time-resolved photoluminescence intensity of the three peaks depicted in Figure 28 under picosecond optical excitation at 10 K. The emission from the $Al_{0.10}Ga_{0.90}N$ layer follows a single exponential decay pattern, whereas the emission from the $Al_{0.05}Ga_{0.95}N$ layer has two contributions to its decay, presumably due to both direct excitation and the diffusion of carriers from the $Al_{0.10}Ga_{0.90}N$ layer. The GaN-active layer has multiple contributions from different layers as well as a direct contribution from the laser excitation.

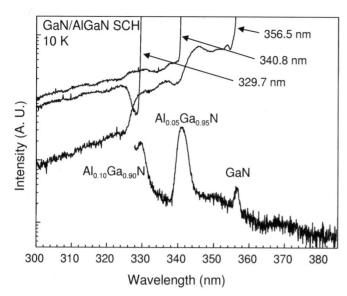

Fig. 35. PLE and PL spectra from a GaN/AlGaN SCH taken at 10 K. PLE spectra were obtained at different detection positions: 329.7, 349.8, and 356.5 nm.

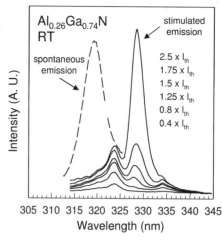

Fig. 36. Room temperature emission spectra for several excitation densities below and above the SE threshold for an Al$_{0.26}$Ga$_{0.74}$N thin film.

Note that enhancement of the exciton binding energy due to two-dimensional confinement is neither expected nor observed, because the thickness of the active region is a factor of 5 larger than the Bohr radius of excitons in GaN. In spite of this, very low values of the lasing threshold were measured. In fact, these lasing threshold values are comparable to those of state-of-the-art InGaN/GaN multiple quantum wells [13]. We believe that the low lasing threshold of the GaN/AlGaN SCH is due to improved carrier and optical confinement, as opposed to deeply localized states as in InGaN/GaN heterostructures (see Section 7.2). Such a low lasing threshold combined with recent improvements in the p-doping of AlGaN alloys indicate that realization of a GaN-active-medium UV laser diode is imminent.

4.4. Gain Mechanism in AlGaN Thin Films

The alloy Al$_x$Ga$_{1-x}$N has been a promising material for light-emitting devices and detectors, covering almost the entire deep-UV spectral range (3.39–6.20 eV at room temperature) with applications including satellite-to-satellite communication, remote chemical sensing, and compact tunable UV laser sources for medical purposes. AlGaN-based structures have been shown to possess superior waveguiding properties [60], resulting in the observation of low threshold and efficient lasing in GaN/AlGaN heterostructures [61]. The variable stripe gain measurements performed by Eckey et al. [62] suggested that under high optical excitation at 1.8 K, phonon-assisted processes play the dominant role in establishing gain in AlGaN epilayers. Recently, scientists reported observation of SE in AlGaN epilayers with Al concentrations as high as 26% and emission wavelengths as short as 328 nm, as shown in Figure 36. This is the shortest SE wavelength observed in any semiconductor system [63]. SE mechanisms in GaN epilayers and GaN/AlGaN separate confinement heterostructures have already been addressed. In this

section, a comprehensive temperature-dependence study of the optical properties of AlGaN at, below, and above the SE threshold from 30 to 300 K is presented.

A 2.5-μm-thick Al$_{0.17}$Ga$_{0.83}$N thin film used in this work was grown by low-pressure MOCVD on (0001) sapphire. Prior to the epilayer growth, a thin (~50-Å) AlN buffer layer was deposited on the sapphire. The high-density part of this work was performed in the traditional edge-emission geometry using a tunable dye laser pumped by a frequency-doubled injection-seeded Nd:YAG laser. A nonlinear crystal was used to double the frequency into the UV wavelength region. The resulting excitation pulse had a wavelength of 315 nm and was 4 ns in duration. Spontaneous emission experiments were performed in the traditional surface-emission geometry using a frequency-doubled Ar$^+$ cw laser (244 nm) as the excitation source. Absorption measurements were performed in the transmission geometry on pieces of the sample etched to a thickness of 0.4 μm using a high-energy plasma etching technique.

Emission spectra of the Al$_{0.17}$Ga$_{0.83}$N sample as a function of excitation pump density near the SE threshold are shown in Figure 37a and b for 30 and 300 K, respectively. At low excitation densities, only a broad spontaneous emission peak is observed at ~3.70 eV (~3.59 eV) with a FWHM of 38 meV (150 meV) at 30 K (300 K). As the excitation pump density increases to values that exceed the SE threshold, a spectrally narrower peak with a FWHM of 24 meV appears on the low-energy shoulder (~3.64 eV) of the spontaneous emission, as shown in Figure 37a. The intensity of the SE increases superlinearly with excitation power and appears to be strongly TE polarized. Similar behavior was observed at room temperature, as depicted in Figure 37b. When the temperature was raised from 30 to 300 K, the measured SE threshold increased from 0.3 to 1.1 MW/cm^2.

Absorption measurements were performed on pieces of the sample that were etched to 0.4 μm. Thinned AlGaN epilayers are necessary to obtain a high signal-to-noise ratio. The absorption edge appears to be blue-shifted by 217 meV (215 meV)

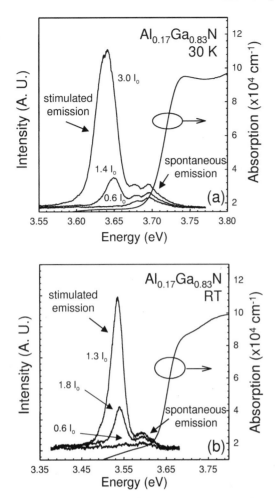

Fig. 37.   Stimulated emission spectra for the $Al_{0.17}Ga_{0.83}N$ sample at (a) 30 K and (b) room temperature as a function of excitation pump density. SE and SP label the stimulated and spontaneous emission peaks, respectively. The absorption spectra of the sample for both temperatures are also shown.

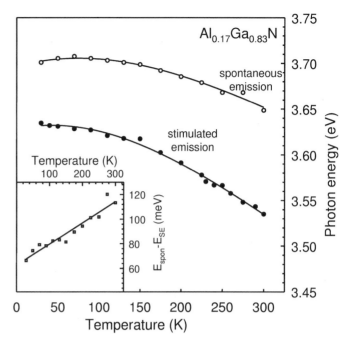

Fig. 38.   Photoluminescence and SE peak position as functions of temperature for the $Al_{0.17}Ga_{0.83}N$ sample. The inset shows the energy difference between positions of the PL (open circles) and SE (close circles) peaks versus temperature. The solid lines are to guide the eye only.

at 30 K (300 K) in comparison to the absorption edge of GaN epilayers [17]. The spectral width of the absorption edge of the sample (see Fig. 37) is very narrow (~40 meV) at 30 K, which suggests that potential fluctuations associated with aluminum incorporation are much smaller in AlGaN alloys than those due to In incorporation in InGaN alloys (for approximately the same molar fraction) [64]. From the absorption data, the penetration depth of the pump beam can be readily evaluated to be $d = 1/\alpha = 0.11$ μm.

To understand the origin of SE in AlGaN epilayers, the positions of the spontaneous and SE peaks in the $Al_{0.17}Ga_{0.83}N$ sample at various temperatures were measured (see Fig. 38). Interestingly, the spontaneous emission peak blue-shifts with increasing temperature up to 75 K and red-shifts thereafter. This type of anomalous temperature behavior is qualitatively different from that observed in GaN epilayers and cannot be adequately described by the Varshni equation [32]. There are two plausible explanations for this anomalous temperature behavior of the PL peak in AlGaN epilayers. Both piezoelectric fields [65] and potential fluctuations due to composition in-

homogeneity [66] have been found to substantially affect the energy position of the main emission peak in InGaN structures, sometimes causing anomalous behavior similar to that observed here. Extensive research on AlGaN thin films is currently underway to determine which mechanism is dominant in the AlGaN samples.

By comparing the energy difference between the spontaneous and stimulated emission peaks for the $Al_{0.17}Ga_{0.83}N$ sample over a broad temperature range (see the inset of Fig. 38), we can extract additional information as to the origin of gain in AlGaN alloys. As was shown previously for high-quality GaN thin films at low temperature, the energy difference between the spontaneous and stimulated emission peaks approaches the exciton binding energy [51], and the gain was determined to be due to exciton–exciton scattering. The SE peaks for the AlGaN sample are located at substantially lower energies than the spontaneous emission peaks. This energy difference varies from about 70 meV at 30 K to over 100 meV at room temperature. Whereas the exciton binding energy in this sample is expected to be in the range of 20–30 meV [17], the participation of excitons in SE can be excluded. Such a large energy difference between the two peaks is a consequence of bandgap renormalization effects associated with a high threshold carrier density, and SE is due to electron–hole plasma recombination.

To further corroborate the point that excitons do not take part in establishing gain in the AlGaN thin films, the behavior of the SE threshold and carrier lifetime with temperature should be examined, as shown in Figure 39. The carrier lifetime in AlGaN at 30 K is much larger (250 ps) than that observed in GaN epilayers (35 ps) [67]. The increase in carrier life-

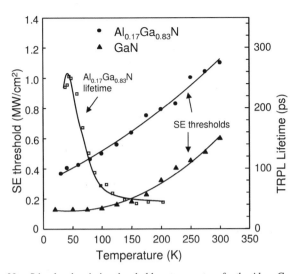

Fig. 39. Stimulated emission threshold vs. temperature for the $Al_{0.17}Ga_{0.83}N$ sample (closed circles). SE thresholds of a GaN epilayer (close triangles) are shown in the figure for the purpose of comparison. Carrier lifetimes (open squares) of the $Al_{0.17}Ga_{0.83}N$ sample as measured by TRPL are also shown as a function of temperature. The solid lines are to guide the eye only.

time can be explained by either the presence of a piezoelectric field or potential fluctuations due to compositional inhomogeneity. However, the increase of carrier lifetime did not result in a dramatic lowering of the SE threshold, contrary to the case of InGaN. At room temperature, the SE threshold of Al-GaN is about 1.8 times that of a high-quality GaN thin film. A monotonous increase in the SE threshold with increasing temperature was also observed. The SE threshold carrier density at 30 K can be readily evaluated if the recombination lifetime (250 ps) and penetration depth ($d = 0.11\ \mu m$) of the sample are taken into account. A pump density of $0.3\ MW/cm^2$ (the threshold value at 30 K) roughly corresponds to a carrier density of $1.2 \times 10^{19}\ cm^{-3}$. This value is 1 order of magnitude larger than the expected value for the Mott density of $1.1 \times 10^{18}\ cm^{-3}$ in GaN samples [61], at which point excitonic interaction is screened by free carriers generated by optical excitation. Excitons are likely to be screened out at these pump densities in AlGaN as well, leaving an electron–hole plasma as the only plausible gain mechanism. At elevated temperatures, reaching the SE threshold in AlGaN requires even higher levels of optical pumping. Thus, the dominant gain mechanism in the AlGaN thin film is recombination of an electron–hole plasma over the entire temperature range studied.

Contrary to observations in InGaN alloys [13, 68], the increase in carrier lifetime is not concurrent with a large reduction in the SE threshold in AlGaN thin films. Further improvements in sample quality are needed to reduce nonradiative recombination channels and to increase SE efficiency of AlGaN thin films. Despite all these difficulties, the AlGaN alloy is still a leading candidate for the development of UV laser diodes.

5. OPTICAL PROPERTIES OF InGaN-BASED HETEROSTRUCTURES

5.1. Physical Properties of InGaN Thin Films and Heterostructures

Understanding the physical mechanisms that give rise to spontaneous emission and stimulated emission in InGaN-based structures is crucial not only from the viewpoint of physical interest, but also for designing short-wavelength LEDs and LDs. To explain the spontaneous and SE characteristics of InGaN-based structures, several groups have proposed different mechanisms (alloy potential fluctuations, quantum-dot-like In phase separations, spontaneous polarization, strain-induced piezoelectric polarization, etc.). However, there has been much disagreement in the literature as to which of these mechanisms is dominant.

This section presents recent studies of the optical properties of InGaN-based structures with an emphasis on those aspects relevant to photonic applications. Experimental aspects of linear and nonlinear optical processes are discussed with special attention to InGaN thin films and various InGaN/GaN quantum structures. The experimental data presented were taken from various sample structures by a wide range of techniques to better understand both the spontaneous and SE properties. In this section, the results of luminescence, absorption, temporal evolution, and SE, as well as the influence of the In composition in the InGaN wells and the Si doping concentration in the GaN barriers on both the optical and structural properties are discussed. Also, the dependence of the emission spectra on temperature, excitation energy, excitation density, and excitation length is presented to clarify the critical role of the energy band-tail states in InGaN-based structures.

Several groups have stated that the recombination of carriers localized at band-tail states of potential fluctuations is an important spontaneous emission mechanism in InGaN/GaN quantum wells (QWs) [69–72]. The potential fluctuations can be induced by alloy disorder, impurities, interface irregularities, and/or self-formed quantum-dot-like regions in the QW active regions. It has been argued that the incorporation of indium atoms plays a crucial role in suppressing nonradiative recombination rates by the capture of carriers in localization centers that originate from quantum-dot-like and phase-separated In-rich regions [73–75]. Narukawa et al. [74] observed self-formed quantum dot-like features associated with In composition inhomogeneity in the well region of $In_{0.20}Ga_{0.80}N/In_{0.05}Ga_{0.95}N$ purple laser diode structures using cross-sectional transmission electron microscopy and energy-dispersive X-ray microanalysis. The main radiative recombination in these QWs was attributed to excitons highly localized at deep traps that probably originated from phase-separated In-rich regions acting as quantum dots [74, 75]. It was suggested that the self-formation of quantum dots may be a result of the intrinsic nature of InGaN ternary alloys to have In compositional modulation due to phase separation. It is likely that this carrier localization formed in the plane of the layers enhances the quantum efficiency by suppressing lateral carrier diffusion, thereby reducing the probability for carriers to enter nonradiative recombination centers. Moreover, the effective potential fluctuation due to the In composition modulation can be enhanced by the large bowing of the InGaN bandgap. The bandgap (E_g) of the $In_xGa_{1-x}N$ ternary alloy is given by

$$E_{g,InGaN}(x) = (1 - x)E_{g,GaN} + xE_{g,InN} - bx(1 - x) \quad (5.1)$$

where $E_{g,GaN}$ ($E_{g,InN}$) is E_g of GaN (InN) and b is a bowing parameter. The bowing parameter for the $In_xGa_{1-x}N$ ternary alloy was found to be much larger than generally assumed and is strongly composition dependent, as seen in experimental and theoretical studies of the bandgaps of strained $In_xGa_{1-x}N$ epilayers (e.g., $b \approx 3.2$ eV at $0 < x < 0.2$ [43, 76] and $b \approx 3.8$ eV at $x = 0.1$ [44]).

Understanding the origin and the detailed nature of the carrier localization in $In_xGa_{1-x}N$-based structures is of considerable scientific interest and technological importance. Recently, spatial variations in the optical properties of GaN- and $In_xGa_{1-x}N$-related materials were more directly investigated by a number of groups using submicrometer spatial resolution luminescence techniques, such as cathodoluminescence (CL), near-field scanning optical microscopy (NSOM), and scanning tunneling microscope-induced luminescence (STL). The optical information investigated by CL, NSOM, or STL complements high-resolution topographs simultaneously obtained by scanning electron microscopy, shear-force microscopy, or scanning tunneling microscopy, respectively. These local characterizations of optical properties from the spatially nonuniform samples provide spectral and spatial mapping of several chemical and physical quantities of interest.

The spatial variations of luminescence from InGaN/GaN QW structures [77, 78], as well as GaN films [79–81] have been studied by CL measurements. Chichibu et al. [77] obtained spatially resolved CL map images of MOCVD-grown InGaN/GaN single QWs at 10 K. These 3-nm-thick InGaN layers had In contents of 5, 20, and 50%. Both uncapped samples and samples capped by a 6-nm-thick undoped GaN layer were studied. Chichibu et al. estimated an In cluster size of less than 60 nm (the spatial resolution of the CL mapping technique, which is essentially limited by the diffusion length in the matrix). Zhang et al. [78] performed spatially resolved and time-resolved CL experiments for a MOCVD-grown InGaN/GaN single QW at 93 K to study the carrier relaxation and recombination dynamics. The InGaN layer had a 15% average In composition, was 4-nm thick, and was capped by a 70-nm-thick GaN layer. Local In alloy fluctuations on a scale of less than ~100 nm were observed for low excitation conditions, indicating strong carrier localization at In-rich regions. Time-resolved CL experiments revealed that carriers generated in the boundary regions diffuse toward and recombine at In-rich centers, with strong lateral excitonic localization prior to radiative recombination.

NSOM is an alternative approach to optical spectroscopy and mapping that is used to study the optical properties of semiconductors with spatial resolution well beyond the diffraction limit [82–86]. NSOM has various modes: (i) *illumination mode*, where the near-field probe acts as a tiny light source,

(ii) *collection mode*, where the near-field probe acts as a tiny light detector, and (iii) *illumination–collection mode*, where the near-field probe acts as both a tiny light source and a detector. In the case of the illumination mode, the spectrum can be influenced by carrier diffusion and relaxation, resulting in poorer spatial resolution than other modes. Crowell et al. [87] carried out a NSOM study of larger defects in an InGaN single QW, with illumination through a tapered silver-coated optical fiber with a ~100-nm-diameter aperture. They found no spectroscopic signature of localized carrier recombination at temperatures above 50 K, in striking contrast to the previously mentioned CL studies [77, 78]. A plausible explanation is the larger carrier diffusion lengths in the higher temperature and higher injection NSOM experiments. Vertikov et al. reported reflection NSOM results for InGaN epilayers and QWs in the illumination mode using an aluminum-coated fiber probe with a 100-nm aperture [88] and in the collection mode using an aluminum-coated fiber probe with a ~50-nm aperture [89]. By using the collection mode, they reported that the range of In alloy fluctuations reaches the 100-nm lateral scale [89].

STL is another alternative method that uses the localized filament current of a scanning tunneling microscope tip to generate photon emission locally [90]. STL studies of HVPE-grown GaN [91] and a MOCVD-grown InGaN/GaN multiple QW (MQW) [92] have been reported. Evoy et al. [92] observed that their STL images exhibited 30–100-nm scale fluctuations for a MOCVD-grown InGaN/GaN MQW. However, the STL images of the InGaN/GaN MQW exhibited a smooth variation of luminescence intensity, instead of the distinctively dark spots that may be expected in the case of local In composition modulation.

Some of the spatially resolved luminescence results are still controversial, depending on the samples, spatial resolution provided by the spectroscopic techniques, and other measurement conditions. Further work is needed to better understand the phase-separated and/or quantum-dot-like features, the correlation between local and macroscopic emission properties, and their roles in the recombination dynamics and lasing processes of InGaN-based structures.

It is equally important to understand that InGaN-based structures possess built-in spontaneous and strain-induced piezoelectric polarization fields. The built-in macroscopic polarization, which consists of (i) the spontaneous polarization due to interface charge accumulations between two constituent materials and (ii) the piezoelectric polarization due to lattice-mismatch-induced strain, plays a significant role in the wurtzite III–nitrides [93–103]. Spontaneous polarization has long been observed in ferroelectric materials. It was shown that wurtzite III–nitrides can have a nonzero macroscopic polarization even in equilibrium (with zero strain). From *ab initio* studies of the wurtzite III–nitrides, it was found that due to their low-symmetry crystal structure, they have very large spontaneous polarization fields (e.g., -0.081, -0.029, and -0.032 C/m^2 for AlN, GaN, and InN, respectively) [93]. The spontaneous polarization increases with increasing nonideality of the structure, going from GaN to InN to AlN, because of the sensitive dependence of the polarization on the structural parameters. In particular, the spontaneous polarization of AlN was found to be only about 3–5 times smaller than that of typical ferroelectric perovskites. On the other hand, electric polarization fields can be generated by lattice-mismatch-induced strain in strained-layer superlattices of III–V semiconductors [104]. Whether or not polarization fields are generated by the strain depends on the symmetry of the strain components, and the orientation of the polarization fields is determined by the superlattice (SL) growth axis. For the most commonly studied case of strained-layer superlattices made from zincblende structures grown along the [100] orientation, the piezoelectric effect does not occur, because only diagonal strain components are generated and diagonal strains do not induce an electric polarization vector in these zincblende structures. If any other growth axis is chosen, however, off-diagonal strain components are generated, resulting in electric polarization fields in zincblende superlattices. For instance, with a [111] growth axis in the zincblende structures, the polarization vectors point along the growth axis (perpendicular to the layers). Whereas one of the constituent materials is in biaxial tension while the other is in biaxial compression, the polarization polarity changes at the SL interfaces. These large alternating electric fields significantly change the SL electronic structure and optical properties. The wurtzite III–nitride structures grown along the [0001] orientation provide a similar interesting situation. It has been reported that if wurtzite nitride layers are under biaxial strain due to lattice mismatch, large piezoelectric fields can be generated because nitrides have very large piezoelectric constants (e.g., $e_{31} = -0.60, -0.49$, and -0.57 C/m^2 for AlN, GaN, and InN, respectively) [93].

An electric field applied to a QW structure changes the subband energy levels and bound state wave functions in the QW, and, hence, the optical transition energies and oscillator strengths. For bulk semiconductors, this is known as the Franz–Keldysh effect, which is associated with photon assisted tunneling. The effective energy gap is reduced because the energy bands are tilted by the applied electric field, and, therefore, the conduction and valence band wave functions have tails in the energy gap with some overlapping. Therefore, a small shift of the absorption edge to lower energies occurs in the presence of an electric field, resulting in a low-energy absorption tail below the fundamental energy gap. Whereas excitons are ionized by the electric field in bulk semiconductors, the excitons in bulk semiconductors play little part in this effect. The situation is quite different in QWs, since the quantum confinement is strong enough to prevent significant ionization for the applied field parallel to the growth direction [105]. The electric field produces a considerable red shift of the confined state energy levels, a phenomenon known as the quantum-confined Stark effect. In this case, the excitons are not dissociated, but only polarized under a large electric field. The quantum-confined Stark effect can be employed in a number of optical devices, including low-energy switches and high-speed electroabsorptive modulators.

Takeuchi et al. [98] proposed that piezoelectric fields due to lattice-mismatch-induced strain generate the quantum-confined Stark effect in InGaN/GaN QWs, resulting in a blue-shift behavior with increasing excitation intensity and a well-width dependence of the luminescence peak energy. They calculated a strain-induced electric field of 1.08 MV/cm for strained In$_{0.13}$Ga$_{0.87}$N grown on GaN, assuming $e_{31} = -0.22$ C/m^2. Because alloy composition fluctuations in the ternary InGaN could also cause the luminescence red shift, Im et al. [100] excluded this possibility in their time-resolved photoluminescence (TRPL) experiments by investigating the binary GaN in GaN/Al$_{0.15}$Ga$_{0.85}$N QWs. Based on their observations of (i) a decrease in the PL peak position and an increase in decay times with increasing well width from 1.3 to 10 nm and (ii) a PL peak shift to energies well below the GaN bandgap in the thicker layers (5 and 10 nm), they proposed that the piezoelectric field effect is significant. Similar experimental results were also observed for In$_{0.05}$Ga$_{0.95}$N/GaN QWs by the same group, so they concluded that the piezoelectric field effect is prominent in both GaN/AlGaN and InGaN/GaN QWs [101]. These polarization charges generate a built-in internal electric field directed along the growth direction and modify both electronic energy levels and wave functions. These internal polarization fields give rise to changes in optical matrix elements and can be screened by photogenerated electron–hole pairs. It has been argued that the spontaneous polarization and/or the strain-induced piezoelectric polarization play an important role in carrier recombination in both GaN/AlGaN and InGaN/GaN QWs [100, 101, 103]. However, such experimental results are not fully understood for InGaN/GaN QWs with the larger In composition used in state-of-the-art commercialized blue–violet laser diodes. To understand the influence of a rather large In composition on the characteristics of InGaN-based structures, a comparison study of structural and optical properties is often required, because any growth parameter changes (especially in InGaN-based structures) unexpectedly may affect other structural (e.g., quality and interface) properties, and hence, the optical and electrical properties.

5.2. Fundamental Optical Properties of InGaN-Based Structures

This section is devoted to the optical properties generally observed for InGaN epilayers and InGaN/GaN QWs. Recombination in InGaN-based materials is characterized by (i) a large Stokes shift between the emission peak and the absorption band edge, (ii) a blue shift of the emission peak with increasing excitation density, and (iii) a red shift of the emission peak with time (a rise in lifetime with decreasing emission energy). Interestingly, these phenomena have been attributed to carrier localization at potential fluctuations and/or built-in internal electric fields. The data presented in this section were taken by a wide range of techniques to better understand both the spontaneous and stimulated emission. The optical emission, absorption, excitation power emission dependence, and emission temporal dynamics were investigated by PL, PL excitation (PLE), optically pumped emission, and TRPL, respectively.

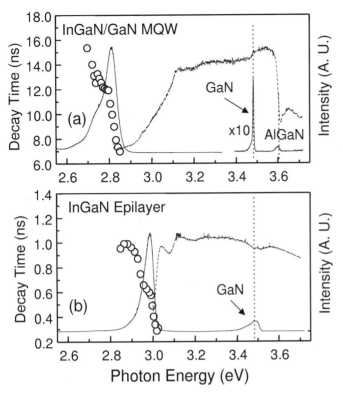

Fig. 40. 10-K PL (solid lines) and PLE (dashed lines) spectra for (a) an $In_{0.18}Ga_{0.82}N/GaN$ MQW and (b) an $In_{0.18}Ga_{0.82}N$ epilayer. Both were grown by MOCVD on c-plane sapphire. A large Stokes shift of the PL emission from the InGaN layers with respect to the band edge measured by PLE spectra is observed. The near-band-edge emission from the GaN and $Al_{0.07}Ga_{0.93}N$ layers was observed at 3.48 and 3.6 eV, respectively. The PLE contributions from the GaN layers [in (a) and (b)] and the $Al_{0.07}Ga_{0.93}N$ layer [in (a)] are clearly seen.

Finally, the possible explanations for the optical properties generally observed in InGaN-based structures are discussed.

Figure 40 shows typical 10-K PL and PLE spectra of the InGaN-related emission of (a) an $In_{0.18}Ga_{0.82}N/GaN$ MQW and (b) an $In_{0.18}Ga_{0.82}N$ epilayer, respectively. Both the MQW and epilayer were grown on c-plane sapphire films by MOCVD, following the deposition of a 1.8-μm-thick GaN buffer layer. The MQW sample consists of a 12-period SL with 3-nm-thick $In_{0.18}Ga_{0.82}N$ wells and 4.5-nm-thick GaN barriers, and a 100-nm-thick $Al_{0.07}Ga_{0.93}N$ capping layer. The InGaN epilayer is 100 nm thick and capped by a 50-nm-thick GaN layer. The average In composition was about 18% from X-ray diffraction analysis, assuming Vegard's law. It was observed that the MQW sample is fully pseudomorphic from symmetric and asymmetric reciprocal space mapping [106]. The PL spectra (solid lines) were measured using the 325-nm line of a cw He-Cd laser. The PLE spectra (dashed lines) were measured using quasimonochromatic light dispersed by a 1/2-m monochromator from a xenon lamp.

In Figure 40, the InGaN-related emission has a peak energy of ~2.8 and ~2.99 eV at 10 K for (a) the $In_{0.18}Ga_{0.82}N$ MQW and (b) the $In_{0.18}Ga_{0.82}N$ epilayers, respectively. The near-band-edge emission from the $Al_{0.07}Ga_{0.93}N$ cladding layer

[in (a)] and the GaN layers [in (a) and (b)] are also clearly seen at 3.60 and 3.48 eV, respectively. With the PLE detection energy set at the InGaN-related emission peak, the contributions from the GaN layers [in (a) and (b)] and from the $Al_{0.07}Ga_{0.93}N$ capping layer [in (a)] are clearly distinguishable, and the energy positions of the absorption edges are well matched to the PL peak positions. The absorption of the InGaN wells in the MQW increases monotonically, reaching a maximum at ~3.1 eV, and remains almost constant until absorption by the GaN barriers occurs at 3.48 eV. A large Stokes shift of the InGaN-related spontaneous emission peak with respect to the absorption edge measured by low-power PLE spectroscopy is clearly observed for both samples. The Stokes shift is much larger for the MQW than for the epilayer, indicating that the band-tail states responsible for the "soft" absorption edge are significantly larger in the MQW than in the epilayer, probably due to the influence of the MQW interfaces on the overall potential fluctuations.

The large Stokes shift between the emission and absorption edge is a common feature in InGaN epilayers and QWs. Recently, the Stokes shift and the broadening of the absorption edge from InGaN epilayers and diodes measured by different groups were plotted as a function of emission energy [107, 108]. Both the Stokes shift and the absorption broadening increase as the In composition increases, which causes the emission energy to decrease. These phenomena have been widely observed by several groups and attributed to either carrier localization, the piezoelectric field effect, or both.

Figure 41 shows the evolution of the 10-K emission spectra with increasing excitation pump density (I_{exc}) for (a) the $In_{0.18}Ga_{0.82}N/GaN$ MQW and (b) the $In_{0.18}Ga_{0.82}N$ epilayer. The spectra shown in Figure 41 were taken with an excitation energy of 3.49 eV and collected in a surface-emission geometry to minimize the effects of reabsorption on the emission spectra. (The SE peak is due to a leak of the in-plane SE at the sample edge. No SE was observed from the middle of the sample in this geometry, indicating the high quality of the sample structure [109].) The pump spot size was ~1 mm² and the excitation wavelength was the third harmonic (355 nm) of a Nd:YAG laser. As I_{exc} is increased, the spontaneous emission peak of the MQW blue-shifts until it reaches ~2.9 eV, and after this point it increases only in intensity until the SE threshold is reached. The SE, indicated by the arrows in Figure 41, develops on the low-energy side of the spontaneous emission peak. The blue shift of the spontaneous emission with increasing I_{exc} is attributed to band filling of localized states due to the intense optical pump. With increasing I_{exc}, the filling level increases and the PL maximum shifts to higher energies until sufficient population inversion is achieved and net optical gain results in the observed SE peak. The I_{exc}-induced blue shift is significantly larger for the MQW than for the epilayer (~80 meV for the MQW compared to ~14 meV for the epilayer), further indicating that the potential fluctuations are significantly larger in the MQW than in the epilayer.

There have also been a number of TRPL studies of InGaN epilayers and quantum wells. The measured recombination decay times for InGaN-related emission were generally in the

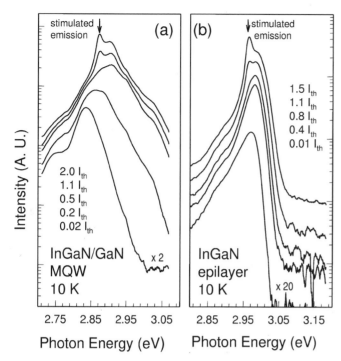

Fig. 41. Evolution of InGaN emission spectra from below to above the SE threshold, I_{th}, at 10 K for (a) an $In_{0.18}Ga_{0.82}N/GaN$ MQW and (b) an $In_{0.18}Ga_{0.82}N$ epilayer. The emission was collected in a surface-emission geometry. I_{th} for the MQW and epilayer was ~170 and 130 kW/cm², respectively, for the experimental conditions. A large blue shift of the emission is clearly seen with increasing excitation density for the MQW sample, showing band-filling of localized states.

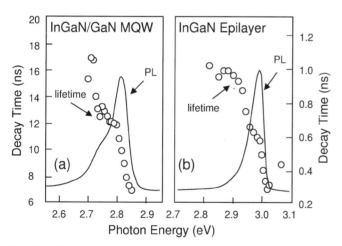

Fig. 42. 10-K decay time (open circles) as a function of detection energy across the PL spectrum for (a) an $In_{0.18}Ga_{0.82}N/GaN$ MQW and (b) an $In_{0.18}Ga_{0.82}N$ epilayer. A rise in lifetime with decreasing emission energy, resulting in a red-shift behavior of the emission with time, reflects that the InGaN-related emission is due to radiative recombination of carriers localized at potential fluctuations.

range of several hundreds of picoseconds to a few tens of nanoseconds, depending on the In composition of InGaN layer, the number of QWs, and/or other growth conditions [70, 110, 111].

Shown in Figure 42 is the effective recombination lifetime as a function of detection energy across the 10-K PL spectrum

of (a) an $In_{0.18}Ga_{0.82}N/GaN$ MQW and (b) an $In_{0.18}Ga_{0.82}N$ epilayer. The TRPL measurements were carried out using a tunable picosecond pulsed laser system consisting of a cavity-dumped dye laser synchronously pumped by a frequency-doubled mode-locked Nd:YAG laser as an excitation source and a streak camera for detection. The output laser pulses from the dye laser had a duration of less than 5 ps and were frequency doubled into the UV spectral region by a nonlinear crystal. The overall time resolution of the system was better than 15 ps.

Figure 42 shows a rise in the effective lifetime τ_{eff} with decreasing emission energy across the PL spectrum, resulting in a red shift of the emission peak energy as time progresses. This is evidence that the InGaN-related emission is due to radiative recombination of carriers localized at potential fluctuations. It is well known that the recombination of localized carriers is governed not only by radiative recombination, but also by the transfer to and trapping in the energy tail states [75]. The differences in τ_{eff} between the two samples indicate that the potential fluctuations that localize the carriers are significantly smaller in the epilayer than in the MQW. The longer lifetime for the MQWs in this work compared to those reported by other groups is probably due to a relatively larger degree of carrier localization caused by a larger number of QWs and/or different growth conditions used in this work [70, 110, 113].

An important parameter associated with the PL efficiency of photoexcited carriers is τ_{eff}, which is given by

$$\frac{1}{\tau_{eff}} = \frac{1}{\tau_r} + \frac{1}{\tau_{nr}} \qquad (5.2)$$

where τ_r is the radiative lifetime and τ_{nr} is the nonradiative lifetime. The nonradiative processes include multiphonon emission, capture and recombination at impurities and defects, the Auger recombination effect, and surface recombination, as well as diffusion of carriers away from the region of observation. Assuming a radiative efficiency η of 100% at $T = 10$ K and using the equation $\eta(T) = \tau_{eff}(T)/\tau_r(T) \approx I_{PL}(T)/I_{PL}(10 \text{ K})$, we can determine the radiative and nonradiative lifetimes as a function of temperature.

Figure 43 shows τ_{eff} as a function of temperature for (a) an $In_{0.18}Ga_{0.82}N/GaN$ MQW and (b) an $In_{0.18}Ga_{0.82}N$ epilayer. The rise in τ_{eff} with increasing temperature from 10 to ~70 (30) K for the MQW (epilayer) is indicative of recombination dominated by radiative recombination channels, whereas the decrease in τ_{eff} with increasing temperature for $T > 70$ (30) K indicates the increasing dominance of nonradiative recombination channels [114, 115]. The MQW has a significantly larger τ_{eff} than the epilayer for all temperatures studied, and the lifetimes of both are significantly larger than that of GaN epilayers and heterostructures [40, 110]. This is attributed to suppression of nonradiative recombination by the localization of carriers at statistical potential fluctuations that arise from the nonrandom nature of this alloy.

For GaN/$Al_{0.07}Ga_{0.93}N$ QWs, Lefebvre et al. [116] observed a significant spectral distribution of decay times from TRPL experiments that they interpreted in terms of localization of carriers by potential fluctuations due to alloy disorder and to well

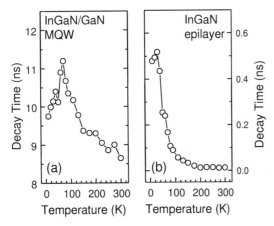

Fig. 43. Temperature dependence of the lifetimes for (a) an $In_{0.18}Ga_{0.82}N/GaN$ MQW and (b) an $In_{0.18}Ga_{0.82}N$ epilayer. The rise in effective recombination lifetime with increasing temperature observed from 10 to ~70 (30) K for the MQW (epilayer) is indicative of recombination dominated by radiative recombination channels, whereas the decrease in lifetime with increasing temperature for $T > 70$ (30) K indicates the increasing dominance of nonradiative recombination. We observed that the MQW has a significantly larger lifetime than the epilayer for all temperatures studied, and the lifetimes of both are significantly larger than that of GaN epilayers and heterostructures.

thickness variations. The inhomogeneous potential fluctuations can be caused by alloy composition fluctuation, well size irregularity, and/or other crystal imperfections such as point defects and dislocations, resulting in a spatial bandgap variation in the plane of the layers.

The influence of internal electric fields and carrier localization on optical properties in both GaN/AlGaN and InGaN/GaN QWs has been reported by several authors. In the case of In-GaN/GaN QWs, unfortunately, both internal electric field and carrier localization effects may be generally enhanced with increasing In alloy composition (at least up to 50%), and it is difficult to tell which effect is predominant. Although we may try to exclude (or reduce) the influence of alloy compositional fluctuation by introducing sample configurations with nonalloy- (or small-alloy-) active regions, such as GaN/AlGaN QWs, the results obtained from these samples may not necessarily be extended to the case of samples with alloy-active regions, such as the blue-light-emitting InGaN/GaN QWs.

Not only for InGaN/GaN, but also GaN/AlGaN [116] QW structures, most of the earlier works by several groups using optical properties to assign the responsible recombination mechanism relied mainly on the following experimental observations: (a) a large Stokes shift between the emission peak energy and absorption edge (Fig. 40), (b) a blue-shift emission behavior with increasing optical excitation power density (Fig. 41), and (c) a red-shift emission behavior with decay time (equivalently, a rise in decay time with decreasing emission energy; Fig. 42). Interestingly, although a number of research groups observed similar experimental findings for the QWs, their interpretations on the observations have been quite different. These phenomena have been interpreted in terms of either the built-in internal electric field effect or carrier localization,

so careful consideration is required to interpret these observations. This is because these observations may be—at least qualitatively—explained in terms of both the internal electric field and carrier localization models, as follows.

In the presence of an internal electric field, the electrons and holes are separated to opposite sides within the InGaN wells. An overlap between the wave functions of these spatially separated electrons and holes allows their radiative recombination within the wells at a lower energy than if there was no internal electric field [observation (a)]. When carriers are either optically or electrically generated, the generated carriers partially screen the internal electric field and, thus, separated electrons and holes come spatially closer together. Accordingly, the effective bandgap, which was reduced by the internal electric field, approaches the bandgap without the internal electric field [observation (b)]. As time proceeds after an excitation pulse, the generated carriers recombine and, hence, the internal electric field (that was screened by the generated carriers) recovers [observation (c)].

In the presence of carrier localization, on the other hand, large potential fluctuations result in the emission from localized carriers at potential minima, whereas the absorption mostly occurs at the average potential energy [observation (a)]. As the excitation power increases, the carriers can fill the band-tail states of the fluctuations and thus the average emission energy increases [observation (b)]. As time proceeds after an excitation pulse, the carriers go down to the lower band-tail states, resulting in the red-shift behavior with time [observation (c)]. Accordingly, the trend of experimental observations (a), (b), and (c) can be plausibly explained by the internal electric field *and/or* carrier localization models, especially for the case of In-GaN/GaN QWs. Because of this ambiguity, we cannot strongly argue which effect is predominant without more direct experimental evidence.

We note that a larger Stokes shift, a larger spontaneous emission blue shift with excitation density, and a longer lifetime for the MQW compared to those for the epilayer strongly reflect that the potential fluctuations are significantly larger in the MQW than in the epilayer. In general, the following effects can cause an emission shift in MQWs, but not in epilayers: (i) the quantum confinement (blue shift), (ii) the built-in internal electric field (red shift), and (iii) the potential fluctuations (red shift) related to the presence of MQW interfaces. As seen in Figure 40, the red-shifting influence is larger than the blue-shifting influence in the MQW. This indicates that the MQW emission properties are predominantly affected by the interfaces through (ii) and/or (iii). Although internal electric fields may be present in the MQWs, potential fluctuations are more likely to dominate the emission properties of the MQWs, because the 3-nm InGaN well widths are less than the exciton Bohr radius and too thin to cause substantial electron–hole separation. The large PL peak energy difference of ~190 meV between the MQW (~2.8 eV) and the epilayer (~2.99 eV) in Figure 40 shows that the potential fluctuations are larger in the MQW. Whereas the quality of upper layers is affected by that of lower layers or interface properties, the presence of interfaces may change the properties of the InGaN active region. In addition, the different growth conditions between the MQW and the epilayer may also affect the degree of potential fluctuations. In Section 5.3, we will see direct experimental evidence that carrier localization caused by potential fluctuations is the predominant effect for both spontaneous and stimulated emission in state-of-the-art InGaN QW devices.

5.3. Influence of In Composition in InGaN Layers

The emission characteristics of the III–nitride system are often influenced by the material combination (e.g., GaN/Al$_x$Ga$_{1-x}$N or In$_x$Ga$_{1-x}$N/GaN) and the alloy composition in the active region (e.g., $x < 0.1$ or $x > 0.15$ in In$_x$Ga$_{1-x}$N/GaN QWs), as well as other structural properties (e.g., well thickness, doping concentration, and number of QWs). There have been several studies on the dependence of structural, electrical, and optical properties on the growth conditions for the AlInGaN material system. Studies on the influence of In composition on the optical and structural characteristics of In$_x$Ga$_{1-x}$N-based structures are very important for the design of optical device applications and extending their wavelength range from UV to green–yellow. The effects of Si doping on both structural and optical properties are also crucial for both physical and practical aspects, because Si doping changes not only the carrier concentration, but also the growth mode, and thus influences optical performance. Therefore, to fully understand the influence of any growth parameter changes (e.g., In composition and Si doping) on the characteristics of InGaN-based structures, a comparison study of structural and optical properties is required because the structural quality may be very sensitively affected by changes in the growth parameters. In this subsection, we focus on the effects of In composition and Si doping on optical properties in conjunction with structural properties.

An increase in the In composition of InGaN results in an enhancement of both the strain and alloy potential fluctuations due to the increase in the degree of lattice mismatch and In phase segregation, respectively. The emission and absorption spectra of a range of commercial InGaN light-emitting diodes and high-quality epilayers have been summarized by Martin et al. [107], who demonstrated a linear dependence of the Stokes shift on the emission peak energy, using experimental spectra of both diode and epilayer samples, supplemented by data from the literature. The broadening of the absorption edge was shown to increase as the emission peak energy decreases. These results were discussed in terms of the localization of excitons at highly In-rich quantum dots within a phase-segregated alloy. More recently, Narukawa et al. [117] investigated the temperature dependence of radiative and nonradiative recombination lifetimes in undoped In$_{0.02}$Ga$_{0.98}$N UV-LEDs and observed that the incorporation of a small amount of In into the GaN layer improved the external quantum efficiency by suppression of nonradiative recombination processes. A recent study on the structural and optical properties of In$_x$Ga$_{1-x}$N/GaN MQWs with different In compositions is presented here [118]. Room temperature (RT) SE wavelengths of the samples used for this

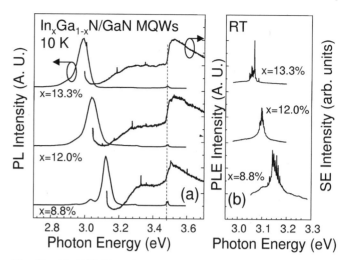

Fig. 44. (a) 10-K PL and PLE, and (b) RT SE spectra of five-period In$_x$Ga$_{1-x}$N/GaN MQWs with In compositions of 8.8, 12.0, and 13.3%.

work were between 395 and 405 nm, which is close to the operational wavelength of state-of-the-art violet current injection lasers.

Figure 44 shows 10-K PL and PLE, and RT SE spectra of MOCVD-grown In$_x$Ga$_{1-x}$N/GaN MQWs with different In compositions of 8.8, 12.0, and 13.3%. These samples consisted of (i) a 2.5-μm-thick GaN buffer layer doped with Si at 3×10^{18} cm^{-3}, (ii) a five-period SL consisting of 3-nm-thick undoped In$_x$Ga$_{1-x}$N wells and 7-nm-thick GaN barriers doped with Si at $\sim 5 \times 10^{18}$ cm^{-3}, and (iii) a 100-nm-thick GaN capping layer. To obtain samples with different In compositions in the In$_x$Ga$_{1-x}$N wells, trimethylindium fluxes of 13, 26, and 39 μmol/min were used for the different samples, while the In$_x$Ga$_{1-x}$N well growth time was kept constant. The PLE detection energy was set at the main In$_x$Ga$_{1-x}$N-related PL peak. With increasing In composition, the In$_x$Ga$_{1-x}$N-related PLE band-edge red-shifts and shows broadened features. By applying the sigmoidal formula

$$\alpha = \alpha_0 \left/ \left(1 + \exp\left(\frac{E_{\mathrm{eff}} - E}{\Delta E} \right) \right) \right. \tag{5.3}$$

to the PLE spectra [107], "effective bandgap" values of 3.256, 3.207, and 3.165 eV, and broadening parameter values of 23, 36, and 40 meV were obtained for the samples with In compositions of 8.8, 12.0, and 13.3%, respectively. This broadening of the PLE spectra with increasing In indicates that the absorption states are distributed over a wider energy range due to an increase in the degree of fluctuations in dot size and/or shape [107], or due to interface imperfection as observed in X-ray diffraction (XRD) patterns [118]. The Stokes shift increased from about 135 to 180 meV as the In composition increased from 8.8 to 13.3%. The large Stokes shifts and their increase with In composition can be explained by carrier localization [72, 75] or the piezoelectric effect [102], or a combined effect of both mechanisms [73, 107].

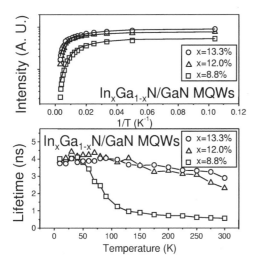

Fig. 45. (a) Integrated PL intensity as a function of $1/T$ and (b) carrier life-time as a function of T for InGaN-related emission in the $In_xGa_{1-x}N/GaN$ MQWs with In compositions of 8.8, 12.0, and 13.3%.

To check device applicability, SE experiments were performed on the $In_xGa_{1-x}N/GaN$ MQWs at RT (the normal device operating temperature). The SE spectra shown in Figure 44b were obtained for a pump density of $1.5 \times I_{th}$, where I_{th} represents the SE threshold for each sample. Below I_{th}, as the excitation power density is raised, the spontaneous emission peak blue-shifts due to band filling of localized states by the intense optical pump, as shown earlier. As the excitation power density is raised above I_{th}, a considerable spectral narrowing occurs. The emission spectra are composed of many narrow peaks of less than 0.1 nm FWHM, which is on the order of the instrument resolution. The SE thresholds were 150, 89, and 78 kW/cm^2 for the samples with In contents of 8.8, 12.0, and 13.3%, respectively. These thresholds are approximately an order of magnitude lower than that of a high-quality nominally undoped single-crystal GaN film measured under the same experimental conditions [41].

With increasing In composition from 8.8 to 13.3%, the SE threshold decreases while the FWHM of the high-resolution XRD SL-1 satellite peaks and the PLE band-edge broadening increase. The increase in the FWHM of the SL-1 peak with increasing In composition indicates the deterioration of interface quality due to nonuniform In incorporation into GaN layers. This interface imperfection or composition inhomogeneity is also reflected in the broadened band edge of PLE spectra. This interface fluctuation may be a source of scattering loss, and the absorption states distributed over a wider energy range can broaden the gain spectrum. Interestingly, although both these factors are disadvantageous to SE, a lower SE threshold density is observed for higher In composition. This is contrary to traditional III–V semiconductors such as GaAs and InP, for which the FWHM of the SL diffraction peaks is closely related to the optical quality of MQWs and the performance of devices using MQWs as an active layer [119, 120].

Figure 45a shows that as the temperature was increased from 10 to 300 K, the integrated PL intensities decreased by fac-

tors of 25, 6, and 5 for the $In_xGa_{1-x}N/GaN$ MQW samples with In compositions of 8.8, 12.0, and 13.3%, respectively. These results indicate that samples with a higher In composition are less sensitive to temperature change, possibly because of less thermally activated nonradiative recombination. To clarify this, temperature-dependent carrier lifetimes were measured by TRPL, as illustrated in Figure 45b. For $T < 50$ K, the lifetimes increase, indicating that radiative recombination dominates in these samples at low temperatures. With further increases in temperature, the lifetime decreases, reflecting that nonradiative processes predominantly influence the emission at higher temperatures. The lifetime starts to drop steeply at $\sim$50 K for the 8.8% In sample. The lifetime for the 12.0% In sample crosses over that for the 13.3% In sample at $\sim$150 K and reaches a lower value at 300 K. From an analysis of the temperature dependence of the integrated PL intensities and carrier lifetimes, we can extract 300-K nonradiative recombination lifetimes of 0.6, 2.7, and 3.6 ns for samples with 8.8, 12.0, and 13.3% In, respectively. These results are consistent with temperature-dependent PL data and indicate the suppression of nonradiative recombination for higher In composition samples.

The possible mechanism for these phenomena can be argued as follows. First, the effect of localization keeping carriers away from nonradiative pathways can be enhanced with the increase of In, as shown by the increase in the Stokes shift in Figure 44a [73, 74]. Second, the incorporation of more In into the $In_xGa_{1-x}N$ well layer can reduce the density of nonradiative recombination centers [117]. The RT SE threshold can be lowered by suppressing nonradiative recombination, because only the radiative recombination contributes to gain. In addition, a lower RT SE threshold for samples with higher In compositions indicates that this effect of suppressing nonradiative recombination overcomes the drawbacks associated with increasing interface imperfections. Note that this favorable effect of lowering the SE threshold seems to saturate with higher In composition, given that there is not much difference in carrier lifetimes, and therefore SE threshold densities, for the 12.0 and 13.3% In samples. This presents challenges in developing laser diodes with longer emission wavelengths.

As the In composition increases, the FWHM of SL diffraction peaks broadens due to the spatial fluctuation of interfaces. However, the RT SE threshold densities decrease with increasing In composition, and this is attributed to suppression of nonradiative recombination. The explanation for these phenomena may be the role of indium atoms in keeping carriers away from nonradiative pathways and/or in reducing the density of nonradiative recombination centers.

5.4. Influence of Silicon Doping in GaN Barriers

The effects of Si doping on both structural and optical properties are crucial, especially for designing practical devices based on III–nitride MQWs. The influence of Si doping on the optical properties of GaN epilayers [121, 122], $In_xGa_{1-x}N/GaN$ QWs [13, 72, 123–126], and GaN/Al$_x$Ga$_{1-x}$N QWs [127, 128] has been widely studied. Assuming an ideal case without structural

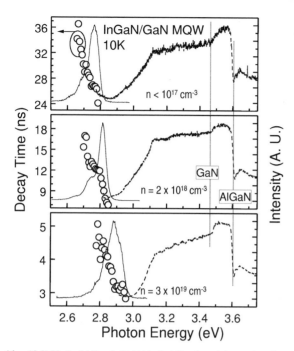

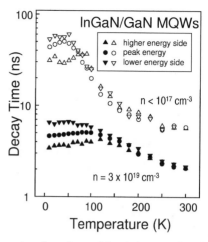

Fig. 47. Temperature dependence of decay times monitored above (up triangles), below (down triangles), and at (circles) the emission peak for the $In_{0.18}Ga_{0.82}N$/GaN MQWs with $n < 1 \times 10^{17}$ cm^{-3} (open symbols) and $n = 3 \times 10^{19}$ cm^{-3} (closed symbols) in the GaN barriers. The characteristic crossover temperature, T_c (which is determined by the radiative and nonradiative recombination rates), gradually increases as n increases: $T_c \sim 70$ K for $n < 1 \times 10^{17}$ cm^{-3}, $T_c \sim 100$ K for $n = 2 \times 10^{18}$ cm^{-3} (not shown), and $T_c \sim 140$ K for $n = 3 \times 10^{19}$ cm^{-3}.

Fig. 46. 10-K PL (solid lines), PLE (dashed lines), and decay time (open circles) of 12-period $In_{0.18}Ga_{0.82}N$/GaN MQWs with Si doping concentrations ranging from <1 × 10^{17} to 3 × 10^{19} cm^{-3} in the GaN barriers. The Stokes shift between the PL emission peak and the absorption band edge observed from PLE spectra decreases significantly as the Si doping concentration increases. Luminescence decay times measured by TRPL are shown as a function of emission energy. With increasing Si doping concentration, the 10-K lifetimes decrease from ~30 (for $n < 1 \times 10^{17}$ cm^{-3}) to ~4 ns (for $n = 3 \times 10^{19}$ cm^{-3}).

variations caused by Si doping, the Si doping dependence of the emission properties of QW structures provides useful information on the screening of the built-in internal electric field in QW structures. However, it has been reported that Si doping may not only dramatically screen the internal electric field, but also seriously change the structural and, hence, optical properties.

The influence of Si doping on both structural and optical properties (spontaneous and stimulated emission) in InGaN/GaN MQW structures has been intensively studied [72, 106, 124]. A series of InGaN/GaN MQW samples were grown on c-plane sapphire substrates by MOCVD specifically to study the influence of Si doping in the GaN barriers. The samples consisted of a 1.8-μm-thick GaN buffer layer and a 12-period MQW composed of 3-nm-thick $In_{0.18}Ga_{0.82}N$ wells and 4.5-nm-thick GaN barriers, followed by a 100-nm-thick $Al_{0.07}Ga_{0.93}N$ capping layer. To study the influence of Si doping in the GaN barriers, the disilane doping precursor flux was systematically varied from 0 to 4 nmol/min during GaN barrier growth. Accordingly, a doping concentration in the range of $n < 1 \times 10^{17}$–3×10^{19} cm^{-3} was achieved for the different samples, as determined by secondary ion mass spectroscopy and Hall measurements.

Figure 46 shows 10-K PL (solid lines), PLE (dashed lines), and decay times plotted as a function of emission energy (open circles) for the main InGaN-related PL peak. A large Stokes shift of the PL emission from the InGaN wells with respect to

the band edge measured by PLE is clearly observed. As the Si doping concentration increases, the Stokes shift decreases. The Stokes shift for the sample with $n = 3 \times 10^{19}$ cm^{-3} is ~120 meV smaller than that of the nominally undoped sample. The effect of Si doping on the decay time of the MQWs was also explored using TRPL measurements. Figure 46 (open circles) shows the 10-K decay time (τ_d) monitored at different emission energies. The measured lifetime becomes longer with decreasing emission energy and, hence, the peak energy of the emission shifts to the low-energy side as time proceeds. This behavior is characteristic of localized states, which in this case are most likely due to alloy fluctuations and/or interface irregularities in the MQWs. A decrease in τ_d with increasing Si doping, from ~30 (for $n < 1 \times 10^{17}$ cm^{-3}) to ~4 ns (for $n = 3 \times 10^{19}$ cm^{-3}), is clearly seen and can be attributed to a decrease in the potential fluctuations that lead to recombination.

Be very careful interpreting the low-temperature lifetime results, because nonradiative recombination processes (which can be caused by poor sample quality) may affect the measured lifetime. To clarify this point, the temperature dependence of the lifetime was investigated as shown in Figure 47. For the nominally undoped sample with $n < 1 \times 10^{17}$ cm^{-3} (open symbols), an increase of τ_d with temperature (up to around 40 ns for the higher energy side emission) was observed at temperatures below 70 K, in qualitative agreement with the temperature dependence of radiative recombination. As the temperature is further increased beyond a certain crossover temperature T_c, which is determined by the radiative and nonradiative recombination rates, the lifetime starts to decrease, because nonradiative processes predominantly influence the emission. Further evidence of this is given by the fact that the lifetimes become independent of emission energy at higher temperatures. Note that T_c gradually increases as n increases: $T_c \sim 70$ K for

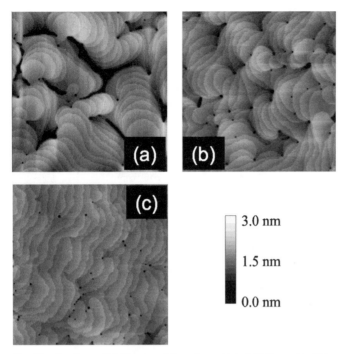

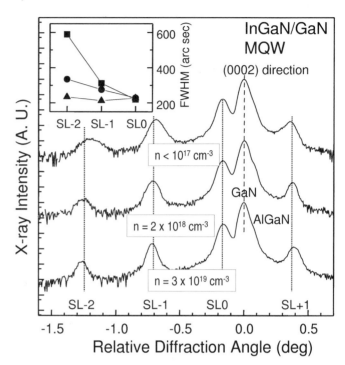

Fig. 48. 2 × 2-μm AFM images of three 7-nm-thick GaN films with different disilane flow rates during growth. The root-mean-square surface roughness estimated from AFM images was 0.49, 0.42, and 0.22 nm for GaN epilayers with the disilane flow rate of (a) 0, (b) 0.2, and (c) 2 nmol/min, respectively. A higher quality GaN surface morphology and a larger average terrace length were achieved by increasing Si incorporation.

Fig. 49. (0002) reflection high-resolution XRD curves of 12-period In$_{0.18}$Ga$_{0.82}$N/GaN MQWs with different Si doping levels in the GaN barrier layers. The variation in the FWHM of the higher order SL satellite peaks is shown in the inset as a function of the Si doping level (squares, $<1 \times 10^{17}$ cm^{-3}; circles, 2×10^{18} cm^{-3}; triangles, 3×10^{19} cm^{-3}).

$n < 1 \times 10^{17}$ cm^{-3}, $T_c \sim 100$ K for $n = 2 \times 10^{18}$ cm^{-3} (not shown here), and $T_c \sim 140$ K for $n = 3 \times 10^{19}$ cm^{-3}. This indicates that the decrease in lifetime with increasing n is due to a decrease of the radiative recombination lifetime itself rather than an increased influence of nonradiative recombination processes. Therefore, the decrease in lifetime with increasing n is mainly due to a decrease in potential localization, and hence, a decrease in the carrier migration time into the lower tail states in the MQW active regions.

To investigate the effect of Si doping on the GaN surface morphology, three reference 7-nm-thick GaN epilayers were also grown at 800°C with different disilane flux rates of 0, 0.2, and 2 nmol/min during growth [123, 124]. The surface morphology of these reference samples was investigated using atomic force microscopy (AFM) in the tapping mode. Figure 48 shows 2-×2-μm AFM images of the three reference GaN epilayers. The root-mean-square surface roughnesses estimated from the AFM images were 0.49, 0.42, and 0.22 nm for GaN epilayers with disilane flow rates of (a) 0, (b) 0.2, and (c) 2 nmol/min during growth, respectively. Thus, smoother GaN surfaces with a more homogeneous terrace length were achieved at higher Si doping levels.

High-resolution XRD experiments revealed that Si doping of GaN barriers improves the interface properties of the InGaN/GaN MQW samples [106], in good correlation with the optical properties. Figure 49 shows (0002) reflection high-resolution XRD ω–2Θ scans measured from 12-period In-

GaN/GaN MQW structures with different Si doping concentrations ranging from $<1 \times 10^{17}$ to 3×10^{19} cm^{-3} in the GaN barriers. The strongest peak is due to the GaN buffer layer, and the high-angle shoulder of the GaN peak is due to the AlGaN capping layer. All the spectra clearly show higher order SL diffraction peaks, indicating good layer periodicity. The SL period was determined from the positions of the SL satellite peaks. Note that the SL peaks for the nominally undoped sample ($n < 1 \times 10^{17}$ cm^{-3}) exhibit broadening with an asymmetric line shape, whereas the SL peaks for the Si-doped samples (e.g., $n = 3 \times 10^{19}$ cm^{-3}) show a more symmetric line shape. The broadening mechanism is partially due to spatial variation of the SL period, possibly caused by intermixing and/or interface roughness. The inset of Figure 49 depicts the variation in the FWHM of the higher order SL satellite peaks as a function of Si doping level. As n increases, the FWHM of the higher order SL satellite peaks narrows: the FWHMs of the second-order SL peaks were observed to be 589, 335, and 234 arcsec for $n < 1 \times 10^{17}$, $n = 2 \times 10^{18}$, and $n = 3 \times 10^{19}$ cm^{-3}, respectively. The ratio of integrated intensity between SL peaks is almost the same for all three samples, as expected for nominally identical SL structures that differ principally only in their degree of structural perfection. In addition, symmetric and asymmetric reciprocal space mapping scans were similar for the different samples, regardless of the Si doping concentration. Accordingly, it is concluded that Si doping in the GaN barriers significantly improves the structural and interface quality of the InGaN/GaN MQWs. Because incorporation of Si atoms can change the quality and/or surface free energy of the GaN barriers, it may affect the growth condition (or mode) of the subsequent InGaN wells and interfaces [121, 129, 130]. The observed (i) decrease in the Stokes shift and (ii) decrease in radiative recombination lifetime with increasing Si doping can be explained by a decrease in potential fluctuations and an enhancement of the interfacial structural qualities with the incorporation of Si atoms into the GaN layers. Therefore, although Si doping of the GaN barriers does not change the overall strain state (i.e., it does not cause relaxation of the lattice-mismatch strain), it does significantly affect the structural and interface quality of the InGaN/GaN MQWs. Similar effects were found for Si doping in InGaN barriers of $In_xGa_{1-x}N/In_yGa_{1-y}N$ QW structures [126, 131]. Spatially resolved CL studies showed that as the Si doping in InGaN barriers increases, the density of dotlike CL bright spots increases, in good agreement with the increased density of nanoscale islands observed by Uchida et al. [131].

5.5. Optical Transitions at Various Temperatures and Excitation Conditions

Ii is known that the temperature dependence of the fundamental energy gap is mainly caused by the changes of band structure induced by lattice thermal expansion and electron–phonon interactions. The temperature-induced change of the fundamental energy gap E_g can be generally given by the Varshni empirical equation [see Eq. (2.3)]. It has been reported that the PL emission from InGaN-based structures does not follow the typical temperature dependence of the energy gap shrinkage. An anomalous temperature-induced luminescence blue shift was observed in InGaN QW structures by several groups and attributed to the involvement of band-tail states caused by potential fluctuations [68, 132, 133]. Studies of the correlation between the temperature-induced anomalous emission behavior and its carrier dynamics for both InGaN epilayers and InGaN/GaN MQWs will be presented here [134]. A similar deviation from the typical temperature-induced energy gap shrinkage has been reported in ordered (Al)GaInP [135, 136] and disordered (Ga)AlAs/GaAs superlattices [137, 138].

Figure 50 shows the evolution of the InGaN-related PL spectra for (a) an $In_{0.18}Ga_{0.82}N$/GaN MQW and (b) an $In_{0.18}Ga_{0.82}N$ epilayer over a temperature range from 10 to 300 K. As the temperature increases from 10 K to T_I, where T_I is 70 (50) K for the MQW (epilayer), the peak energy position E_{PL} red-shifts 19 (10) meV. This value is about five times as large as the expected bandgap shrinkage of ~4 (2) meV for the MQW (epilayer) over this temperature range [139]. With a further increase in temperature, E_{PL} blue-shifts 14 (22.5) meV from T_I to T_{II}, where T_{II} is 150 (110) K for the MQW (epilayer). By considering the estimated temperature-induced bandgap shrinkage of ~13 (7) meV for the MQW (epilayer), the actual blue shift of the PL peak with respect to the band edge is about 27 (29.5) meV over this temperature range. When the temperature is further increased above T_{II}, the peak positions red-shift again. From the observed red shift of 16 (45) meV and the expected bandgap shrinkage

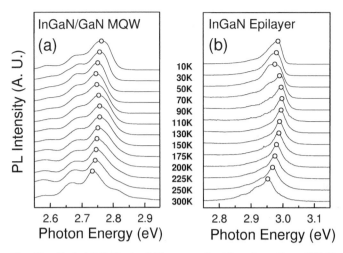

Fig. 50. Typical InGaN-related PL spectra for (a) an $In_{0.18}Ga_{0.82}N$/GaN MQW and (b) an $In_{0.18}Ga_{0.82}N$ epilayer in the temperature range from 10 to 300 K. The main emission peak of both samples shows an S-shaped shift with increasing temperature (open circles). All spectra are normalized and shifted in the vertical direction for clarity. Note that the turning temperature from red shift to blue shift is about 70 and 50 K for the $In_{0.18}Ga_{0.82}N$/GaN MQW and the $In_{0.18}Ga_{0.82}N$ epilayer, respectively.

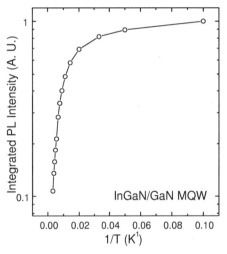

Fig. 51. Normalized integrated PL intensity as a function of $1/T$ for the InGaN-related emission in the $In_{0.18}Ga_{0.82}N$/GaN MQWs (open circles). An activation energy of ~35 meV is obtained from the Arrhenius plot.

of ~43 (51) meV from T_{II} to 300 K for the MQW (epilayer), an actual blue shift of the PL peak relative to the band edge is estimated to be about 27 (6) meV in this temperature range.

Figure 51 shows an Arrhenius plot of the normalized integrated PL intensity of the InGaN-related PL emission of the $In_{0.18}Ga_{0.82}N$/GaN MQW. The total luminescence intensity from this sample is reduced by only 1 order of magnitude from 10 to 300 K, indicating a high PL efficiency even at high temperatures. (For the MOCVD-grown GaN epilayer, the change in integrated PL intensity between 10 and 300 K is 2–3 orders of magnitude.) For $T > 70$ K, the integrated PL intensity of the InGaN-related luminescence is thermally activated with an activation energy of about 35 meV. In general, the quenching of the luminescence with temperature can be explained by thermal emission of carriers out of confining potentials with an activation energy correlated with the depth of the confining potentials. Whereas the observed activation energy is much less than the band offsets as well as the bandgap difference between the wells and the barriers, the thermal quenching of the InGaN-related emission is *not* due to the thermal activation of electrons and/or holes from the InGaN wells into the GaN barriers. Instead, the dominant mechanism for the quenching of the InGaN-related PL is thermionic emission of photogenerated carriers out of the potential minima caused by potential variations, such as alloy and interface fluctuations, as will be discussed later.

The carrier dynamics of the InGaN-related luminescence was studied by TRPL over the same temperature range. Figure 52 shows E_{PL}, the relative energy difference (ΔE) between E_{PL} and E_g at each temperature, and the decay times (τ_d) monitored at the peak energy, lower energy side, and higher energy side of the peak as a function of temperature. A comparison clearly shows that the temperature dependence of ΔE and E_{PL} is strongly correlated with the change in τ_d. For both the $In_{0.18}Ga_{0.82}N/GaN$ MQW and the $In_{0.18}Ga_{0.82}N$ epilayer, an overall increase of τ_d is observed with increasing temperature for $T < T_I$, in qualitative agreement with the temperature dependence of radiative recombination [114, 115]. As seen in the previous section, in this temperature range, τ_d becomes longer with decreasing emission energy and, hence, the peak energy of the emission shifts to the low-energy side as time proceeds. This behavior is characteristic of carrier localization, most likely due to alloy fluctuations (and/or interface roughness

in MQWs). As the temperature is further increased beyond T_I, the lifetime of the MQW (epilayer) quickly decreases to less than 10 (0.1) ns and remains almost constant between T_{II} and 300 K, indicating that nonradiative processes predominantly affect the emission. This is further evidenced by the fact that the difference between the lifetimes monitored above, below, and at the peak energy disappears for $T > T_I$, in contrast to the observations for $T < T_I$. This characteristic temperature T_I is also where the turnover occurs from red shift to blue shift for ΔE and E_{PL} with increasing temperature. Furthermore, in the temperature range between T_I and T_{II}, where a blue shift of E_{PL} is detected, τ_d dramatically decreases from 35 to 8 (0.4 to 0.05) ns for the MQW (epilayer). Above T_{II}, where a red shift of E_{PL} is observed, no sudden change in τ_d occurs for either the MQW or the epilayer.

From these results, the InGaN-related recombination mechanism for different temperature ranges can be explained as follows: (i) For $T < T_I$, whereas the radiative recombination process is dominant, the carrier lifetime increases, giving the carriers more opportunity to relax down into lower energy tail states caused by the inhomogeneous potential fluctuations before recombining. This reduces the higher energy side emission intensity and, thus, produces a red shift in the peak energy position with increasing temperature. (ii) For $T_I < T < T_{II}$, whereas the dissociation rate is increased and other nonradiative processes become dominant, the carrier lifetimes decrease greatly with increasing temperature and also become independent of emission energy. Due to the decreasing lifetime, the

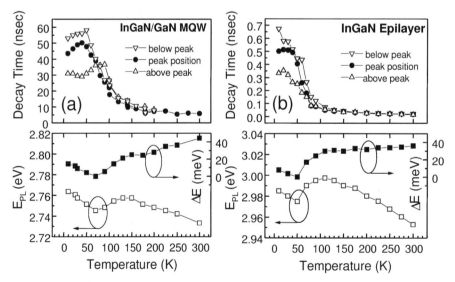

Fig. 52. InGaN-related PL spectral peak position E_{PL} (open squares) and decay time as a function of temperature for (a) an $In_{0.18}Ga_{0.82}N/GaN$ MQW and (b) an $In_{0.18}Ga_{0.82}N$ epilayer. ΔE (closed squares) is the relative energy difference between E_{PL} and E_g at each temperature. The minimum value of ΔE is designated as zero for simplicity. Note that the lower energy side of the PL peak has a much longer lifetime than the higher energy side below a certain temperature T_I, whereas there is little difference between lifetimes monitored above, below, and at the peak energy above T_I, where T_I is about 70 (50) K for the MQW (epilayer). This characteristic temperature T_I is also where the turnover occurs from red shift to blue shift of the InGaN PL peak energy with increasing temperature. A blue shift behavior of emission peak energy with increasing temperature is still seen at room temperature for the $In_{0.18}Ga_{0.82}N/GaN$ MQW, whereas this behavior is much less for the $In_{0.18}Ga_{0.82}N$ epilayer.

carriers recombine before reaching the lower energy tail states. This gives rise to an apparent broadening of the higher energy side emission and leads to a blue shift in the peak energy. (iii) For $T > T_{II}$, whereas nonradiative recombination processes are dominant and the lifetimes are almost constant [in contrast to case (ii)], the photogenerated carriers are less affected by the change in carrier lifetime, so the blue-shift behavior becomes smaller. Note that the slope of ΔE is very sensitive to the change in τ_d with temperature for both the InGaN/GaN MQW and the InGaN epilayer. Because this blue-shift behavior is smaller than the temperature-induced bandgap shrinkage in this temperature range, the peak position exhibits an overall red-shift behavior. Consequently, the change in carrier recombination mechanism with increasing temperature causes the S-shaped red-shift–blue-shift–red-shift behavior of the peak energy for the InGaN-related luminescence, and the anomalous temperature dependence of the emission is attributed to optical transitions from "localized" to "extended" energy tail states in the InGaN-based structures.

An interesting difference in the emission shift behavior of the InGaN/GaN MQW compared to the InGaN epilayer is the greater effective blue-shifting behavior even near RT, probably due to a different degree of carrier localization for the two structures. Except for this, a similar temperature-induced S-shaped emission behavior was observed for both the InGaN/GaN MQW and the InGaN epilayer, even though τ_d of the former is about 2 orders of magnitudes longer than that of the latter. This fact indicates that the anomalous temperature-induced emission shift mainly depends on the change in carrier recombination dynamics rather than the absolute value of τ_d.

5.6. Excitation Condition Dependence of Optical Transition

Energy selective spontaneous and stimulated emission studies of InGaN-based structures provide important information about the energy boundary between localized and extended (or delocalized) band-tail states, as well as the different carrier generation/transfer dynamics. We will review recent studies on the excitation energy dependence of the InGaN-related emission in InGaN/GaN MQWs investigated by energy-selective PL and PLE spectroscopy [140, 141].

Figure 53 shows 10-K $In_{0.18}Ga_{0.82}N$-related luminescence spectra measured with four different excitation photon energies E_{exc} of (A) 3.81, (B) 3.54, (C) 3.26, and (D) 2.99 eV. Each E_{exc} is indicated over the PLE spectrum for reference. The InGaN-related main and secondary peaks are shown at energies of 2.80 and ~2.25 eV, respectively. The oscillations on the main PL peak are due to Fabry–Perot interference fringes. When E_{exc} varies from above (curve A) to below (curve B) the near-band-edge emission energy E_g of the AlGaN capping layer ($E_{g,AlGaN}$), the relative intensity ratio of the main peak to the secondary peak noticeably changes. For $E_{exc} < E_{g,GaN}$, the secondary peak has nearly disappeared, whereas the main peak remains (curve C). When the excitation energy is further decreased to just above the InGaN main emission peak (curve D),

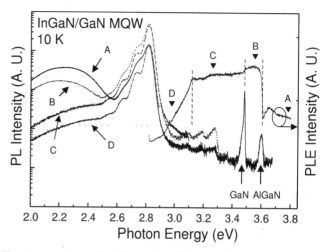

Fig. 53. Evolution of the PL emission of the $In_{0.18}Ga_{0.82}N$/GaN MQW for excitation energies of (A) 3.81, (B) 3.54, (C) 3.26, and (D) 2.99 eV. Note the change in the intensity ratio of the main peak to the secondary peak between spectra. The excitation photon energies are indicated over the PLE spectrum for reference.

no noticeable change is observed for the PL shape except for a decrease in the overall intensity of the emission. As the excitation energy is decreased (from curve A to curve D), the intensity of the lower energy side of the main peak is reduced, whereas that of the higher energy side is enhanced. This results in a ~7 meV blue shift of the peak energy and a narrower spectral width for the main peak with decreasing excitation energy. These facts strongly reflect that the recombination mechanism of the InGaN-related emission is significantly affected by the excitation (or carrier generation) conditions, as we will see later.

The upper part of Figure 54 shows the 10-K PLE spectra for the $In_{0.18}Ga_{0.82}N$-related main PL emission monitored at (a) 2.87, (b) 2.81, (c) 2.75, and (d) 2.68 eV. The PL spectrum for $E_{exc} = 3.81$ eV is also shown in this figure for reference. When the PLE detection energy is set below the peak energy of the main emission (curves c and d), the contributions from the InGaN wells, the GaN barriers, and the AlGaN capping layer are clearly distinguishable, whereas for the detection energy above and at the peak position of the main InGaN emission $E_{g,InGaN}$ (curves a and b), the PLE signal below $E_{g,InGaN}$ shows almost constant intensity across the $E_{g,GaN}$ region, indicating that carrier generation in the InGaN rather than in the GaN plays an important role. In both cases, the PLE signal above $E_{g,InGaN}$ is suddenly diminished, obviously due to the absorption of the $Al_{0.07}Ga_{0.93}N$ capping layer. When the detection energy is below $E_{g,InGaN}$ (curves c and d), the contributions of both the GaN and AlGaN regions are enhanced compared to the curves a and b: as the detection energy decreases, the PLE signal above $E_{g,GaN}$ is monotonically raised with respect to the almost flat region of the PLE signal between 3.15 and 3.4 eV. These facts imply that for $E_{exc} > E_{g,GaN}$, the lower energy side of the InGaN main emission peak is governed mainly by carrier generation in the GaN barriers and subsequent carrier transfer to the InGaN wells. From the different PLE contributions for the higher and lower energy sides of $E_{g,InGaN}$, one can expect

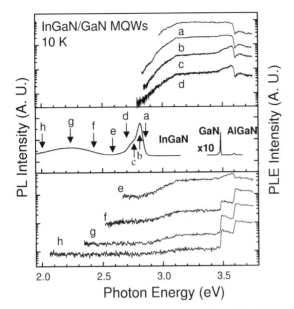

Fig. 54. 10-K PLE spectra taken at detection energies of (a) 2.87, (b) 2.81, (c) 2.75, (d) 2.68, (e) 2.57, (f) 2.41, (g) 2.24, and (h) 2.01 eV. The PL spectrum for an excitation energy of 3.81 eV is also shown for reference. The spectra are shifted in the vertical direction for clarity and the respective detection energies are indicated on the PL plot. As the detection energy decreases, the contribution of the $Al_{0.07}Ga_{0.93}N$ capping layer noticeably increases. When the detection energy is lower than 2.24 eV, the contribution of the $In_{0.18}Ga_{0.82}N$ wells is almost negligible.

different recombination mechanisms for various excitation energies, as will be described later.

The lower part of Figure 54 also shows the 10-K PLE spectra for the secondary peak taken for detection at (e) 2.57, (f) 2.41, (g) 2.24, and (h) 2.01 eV. When the detection energy is higher than 2.24 eV, the InGaN wells still partly contribute to the secondary peak emission (curves e and f), whereas for the detection energy below 2.24 eV, the contribution of the In-GaN wells almost disappears (curves g and h). Note that as the detection energy decreases, the contribution of the AlGaN capping layer is noticeably increased. These facts indicate that the main source of the secondary peak does not originate from the InGaN wells, but originates predominantly from the Al-GaN capping layer and partly from the GaN layers (consistent with the so-called yellow luminescence band). This observation was also confirmed by PL measurements using the 325-nm line of a He-Cd laser with varying excitation intensities. As the He-Cd laser excitation intensity increases, the relative emission intensity ratio of the InGaN main peak to the secondary peak increases. That is, the intensity of the main peak increases linearly, whereas that of the secondary peak saturates with increasing excitation intensity. This is another indication that the secondary peak is defect-related emission. Therefore, the main peak is due to the InGaN wells, whereas the secondary peak is mainly from the AlGaN capping layer and the GaN barriers rather than the InGaN wells. For $E_{exc} > E_{g,InGaN}$, most carriers are generated in the AlGaN capping layer and these photogenerated carriers partly migrate into the MQW region (corresponding to the main peak) and partly recombine via

defect-related luminescence in the AlGaN layer itself and the GaN barriers (corresponding to the secondary peak).

The PLE observations in the frequency domain are closely related to the carrier dynamics in the time domain. To clarify the temporal dynamics of the luminescence, TRPL measurements were performed at 10 K for different E_{exc}. Time-integrated PL spectra showed the same behavior as observed in the foregoing cw PL measurements for the InGaN-related main emission: a blue shift of the peak energy and a spectral narrowing of the lower energy side as E_{exc} decreases from above $E_{g,InGaN}$ to below $E_{g,GaN}$. The carrier recombination lifetime becomes longer with decreasing emission energy and, therefore, the peak energy of the emission shifts to the low-energy side as time progresses, as shown in Figure 42. No significant change in lifetime at and above the peak energy position ($\tau_d \sim 12$ ns at the peak position) was observed when E_{exc} was varied. However, the peak position reached the lower energy side faster for the $E_{exc} > E_{g,GaN}$ case than for the $E_{exc} < E_{g,GaN}$ case, and after ~ 20 ns, the peak position is almost the same for both cases. The starting peak position is lower for the $E_{exc} > E_{g,GaN}$ case than for the $E_{exc} < E_{g,GaN}$ case, so for the $E_{exc} > E_{g,GaN}$ case, the red-shifting behavior with time is smaller and most carriers recombine at relatively lower energies. The carriers generated from the GaN barriers (or AlGaN capping layer) migrate toward the InGaN wells, and the carrier transfer allows the photogenerated carriers to have a larger probability of reaching the lower energy states at the MQW interfaces. This may be due to more binding and scattering of carriers by interface defects and roughness near the MQW interface regions, because the carriers go through the MQW interfaces during the carrier transfer, enhancing trapping and recombination rates at the interface-related states. Therefore, the lower emission peak energy position and the spectral broadening to lower energies for $E_{exc} > E_{g,GaN}$ indicate that the lower (or deeper) energy tail states are more akin to the interface-related defects and roughness at the MQW interfaces than to the alloy fluctuations and impurities within the InGaN wells, whereas for $E_{exc} < E_{g,GaN}$, the carriers responsible for the emission are directly generated within the InGaN wells and thus recombine at relatively higher emission energies.

Another interesting feature in the PLE spectra of Figure 54 is different slopes for $E_{exc} < 3.0$ eV in curves b, c, and d. To investigate details of this phenomenon, energy-selective PL measurements were performed for $E_{exc} < E_{g,GaN}$. For the energy-selective PL study, the second harmonic of a mode-locked Ti:sapphire laser was used as a tunable excitation source to excite the sample normal to the sample surface. The emission was collected normal to the sample surface, coupled into a 1-m spectrometer, and spectrally analyzed using a CCD. The excitation photon energy from the frequency-doubled Ti:sapphire laser was tuned across the states responsible for the "soft" absorption edge of the InGaN layers. Spontaneous emission was observed for excitation photon energies over a wide spectral range above the spontaneous emission peak position (2.8 eV at 10 K). As the excitation photon energy was tuned from just below the bandgap of the GaN barrier layers to the high-energy

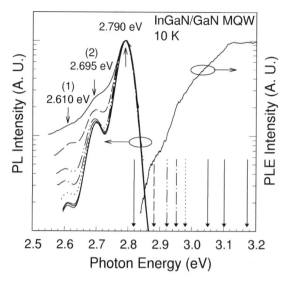

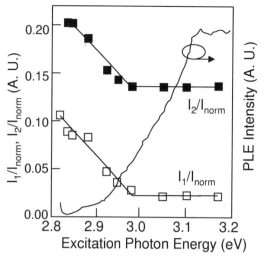

Fig. 55. Spontaneous emission spectra from an $In_{0.18}Ga_{0.82}N/GaN$ MQW with 2×10^{18} cm^{-3} Si doping in the GaN barriers as a function of excitation photon energy. The excitation photon energy for a given spectrum is indicated by the corresponding arrow on the x axis. The spectra have been normalized at 2.79 eV for clarity. The absorption edge measured by PLE is also shown.

Fig. 56. 10-K spontaneous emission intensity from an $In_{0.18}Ga_{0.82}N/GaN$ MQW at 2.61 (I_1) and 2.695 eV (I_2) relative to the peak emission intensity (at 2.79 eV) as a function of excitation photon energy. The solid lines are given only to guide the eye. A clear shift in the emission lineshape is seen for excitation photon energies below ~2.98 eV. The absorption edge measured by PLE is also given (solid line) for comparison.

side of the InGaN absorption edge, no significant changes in the spontaneous emission spectra were observed. The only change was a decrease in the emission intensity as the excitation photon energy was tuned below the higher energy side of the "soft" InGaN absorption edge, consistent with the reduction in the absorption coefficient with decreasing photon energy in this spectral region. However, as the excitation photon energy was tuned below approximately 2.98 eV, significant changes in the spontaneous emission spectra were observed. For excitation photon energies below ~2.98 eV, emission from the low-energy wing of the main InGaN PL peak became more and more pronounced with decreasing excitation photon energy. This is illustrated in Figures 55 and 56. The spectra in Figure 55 have been normalized at 2.79 eV for clarity. Figure 56 clearly shows the onset of this behavior for excitation photon energies below approximately 2.98 eV. This behavior indicates that the transition from localized to extended band-tail states is located at ~2.98 eV. As such, this energy defines the "mobility edge" of the band-tail states in this structure, where carriers with energy above this value are free to migrate and those of lesser energy are spatially localized by (large) potential fluctuations in the InGaN layers.

This red shift in the spontaneous emission is explained as follows: when E_{exc} is higher than the mobility edge, the photogenerated carriers can easily populate the tail states by their migration, but their lifetimes are relatively short due to the presence of nonradiative recombination channels. As E_{exc} is tuned below the mobility edge, the nonradiative recombination rate is significantly reduced due to the capture of the carriers in small volumes. This increase in lifetime with decreasing E_{exc} results in increased radiative recombination from lower energy states. The position of the mobility edge is seen to be ~180 meV above the spontaneous emission peak and ~130 meV below the start

of the InGaN absorption edge. Its spectral position indicates extremely large potential fluctuations are present in the InGaN-active layers of the MQW, leading to carrier confinement and resulting in efficient radiative recombination.

Energy-selective optically pumped SE experiments similar to the described energy-selective PL experiments were performed to elucidate whether the localized carrier recombination responsible for the spontaneous emission is also responsible for SE in these materials [142]. For energy-selective SE experiments, the second harmonic of an injection-seeded Q-switched Nd:YAG laser (532 nm) pumped an amplified dye laser. The deep red to near infrared radiation from the dye laser was frequency doubled in a nonlinear crystal to achieve the near UV to violet laser radiation needed to optically excite the InGaN/GaN MQWs in the spectral region of interest. The frequency-doubled radiation (~4-ns pulse width, 10-Hz repetition rate) was focused to a line on the sample surface. The excitation spot size was approximately 100×5000 μm. The emission from the sample was collected from one edge of the sample, coupled into a 1-m spectrometer, and spectrally analyzed using a UV-enhanced CCD.

The experiments were performed on $In_{0.18}Ga_{0.82}N/GaN$ MQW samples with undoped ($n < 1 \times 10^{17}$ cm^{-3}) and Si-doped ($n \sim 2 \times 10^{18}$ cm^{-3}) GaN barriers. Figure 57 illustrates the change in the SE spectra with decreasing E_{exc} for the InGaN/GaN MQW with undoped GaN barrier layers, whereas Figure 58 shows the behavior of the SE peak for the Si-doped MQW, which is presented in Figure 59. Figures 57 and 58 show the behavior of the SE peak as E_{exc} is tuned from above $E_{g,InGaN}$ to below the absorption edge of the InGaN-active layers. As E_{exc} is tuned to lower energies, no noticeable change is observed in the SE spectrum until E_{exc} crosses a certain value (~3.0 and ~2.95 eV for undoped and Si-doped MQWs, re-

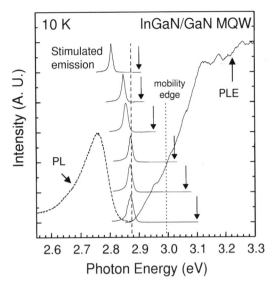

Fig. 57. SE spectra as a function of excitation photon energy, E_{exc}, for the In$_{0.18}$Ga$_{0.82}$N/GaN MQW with undoped ($n < 1 \times 10^{17}$ cm^{-3}) GaN barriers. "Mobility edge"-type behavior is clearly seen in the SE spectra with decreasing E_{exc}. The red shift of the SE peak is shown with decreasing E_{exc} as E_{exc} is tuned below the mobility edge (the dashed–dotted line). The excitation photon energies for the given SE spectra are represented by the arrows. The dashed line is given as a reference for the unshifted SE peak position. The SE spectra are normalized and displaced vertically for clarity. PL and PLE spectra are also given for comparison.

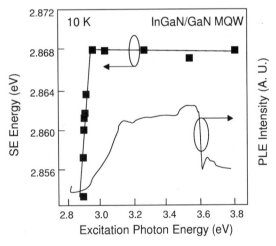

Fig. 58. SE peak position as a function of excitation photon energy, E_{exc}, for the In$_{0.18}$Ga$_{0.82}$N/GaN MQW with 2×10^{18} cm^{-3} Si doping in the GaN barriers. "Mobility edge"-type behavior is clearly seen in the SE spectra with decreasing E_{exc}. The solid lines are given only to guide the eye. The PLE spectrum is also given for comparison.

spectively), at which point the SE peak red-shifts quickly with decreasing E_{exc}. The red shift of the SE peak as E_{exc} is tuned below ~2.95 eV (for the Si-doped MQW) is consistent with the mobility edge behavior observed for the spontaneous emission, as already described. This behavior of the SE peak is due to enhanced population inversion at lower energies as the carriers are confined more efficiently with decreasing E_{exc}. The mobility edge measured in these experiments lies ~110 meV above the spontaneous emission peak, ~62 meV above the

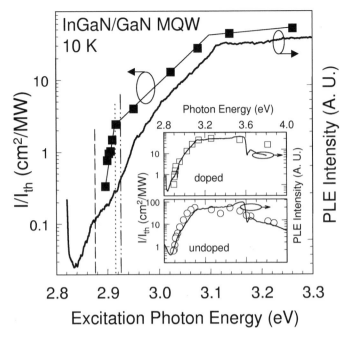

Fig. 59. Inverse SE threshold as a function of excitation photon energy E_{exc} (solid squares) shown in comparison with the results of low-power PLE experiments (solid line) for the Si-doped In$_{0.18}$Ga$_{0.82}$N/GaN MQW. The inset shows the same comparison over a wider energy range for both the undoped and Si-doped InGaN/GaN MQWs, illustrating the similarities over the entire energy range.

SE peak, and ~185 meV below the absorption edge of the InGaN well regions. The location of the mobility edge with respect to the spontaneous and stimulated emission peaks further indicates that large potential fluctuations are present in the InGaN-active regions, resulting in strong carrier localization. This explains the efficient radiative recombination (stimulated and spontaneous) observed from these structures as well as the small temperature sensitivity of the SE.

As a measure of the coupling efficiency of the exciting photons to the gain mechanism responsible for the SE peak, I_{th} was measured as a function of E_{exc} for the InGaN/GaN MQW with Si-doped GaN barrier layers. A comparison between this and the coupling efficiency obtained for the spontaneous emission peak measured by PLE is given in Figure 59, where $1/I_{th}$ is plotted as a function of E_{exc} to give a better measure of the coupling efficiency and afford an easier comparison to the PLE measurements. Four distinct slope changes are seen in the PLE spectrum. The first, at ~3.12 eV, marks the beginning of the "soft" absorption edge of the InGaN-active region, whereas the other three, located at ~2.96, 2.92, and 2.87 eV, suggest varying degrees of localization. The change in $1/I_{th}$ at ~3.1 eV is due to a decrease in the absorption coefficient below the absorption edge and is an expected result. The change in $1/I_{th}$ is coincident with a slope change in the PLE spectrum and is attributed to a significant decrease in the effective absorption cross section for excitation photon energies below the mobility edge. The inset of Figure 59 shows the same comparison over a wider energy range for both the Si-doped and undoped MQWs. A strong cor-

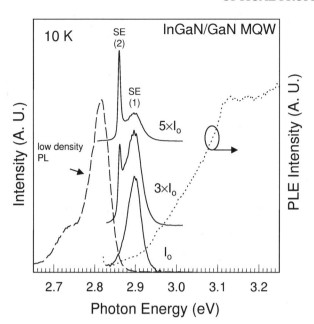

Fig. 60. 10-K SE spectra (solid lines) from an $In_{0.18}Ga_{0.82}N/GaN$ MQW sample subjected to several excitation densities, where $I_0 = 100$ kW/cm². The low-power PL (dashed line) and PLE (dotted line) spectra are also shown for comparison. The SE spectra are normalized and displaced vertically for clarity.

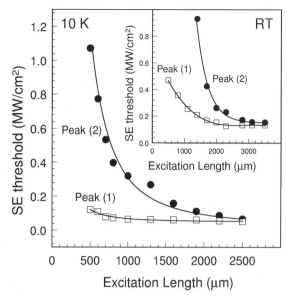

Fig. 61. SE threshold as a function of excitation length for SE peaks (1) and (2) at 10 K for the $In_{0.18}Ga_{0.82}N/GaN$ MQW. The solid lines are given only to guide the eye.

relation between the SE threshold and PLE measurements is clearly seen over the entire range for both samples. The correlation between the high carrier density behavior and the cw PLE results indicates that carrier localization plays a significant role in both the spontaneous and the stimulated emission processes.

5.7. Excitation Length Dependence of Stimulated Emission

Separate research groups have observed two different SE peaks from $In_xGa_{1-x}N/GaN$ MQWs grown under different growth conditions, illustrating dramatically different SE behavior in $In_xGa_{1-x}N$ MQWs for relatively small changes in the experimental conditions [143–145]. This also indicates that this two peak SE behavior is a general property of present state-of-the-art $In_xGa_{1-x}N$-based blue laser diodes. These results suggest that some of the varied results reported in the literature may be due to slightly different experimental conditions, resulting in significant changes in the SE behavior.

Typical power-dependent emission spectra at 10 K for the $In_{0.18}Ga_{0.82}N/GaN$ MQW sample with barrier Si doping of $n = 2 \times 10^{18}$ cm^{-3} are shown in Figure 60 for an excitation length L_{exc} of 1300 μm. At low I_{exc}, a broad spontaneous emission peak is observed at ~2.81 eV, consistent with low-power cw PL spectra. As I_{exc} is increased, a new peak emerges at ~2.90 eV [designated here as SE peak (1)] and grows super-linearly with increasing I_{exc}. If I_{exc} is further increased, another new peak is observed at ~2.86 eV [designated here as SE peak (2)], which also grows superlinearly with increasing I_{exc}. Both SE peaks (1) and (2) originate on the high-energy side of the low-power spontaneous emission peak (given by the dashed line in Fig. 60) and are red shifted by more than 0.2 eV below

the "soft" absorption edge. Both SE peaks are highly TE polarized, with a TE to TM ratio of ~200. SE peak (2) was studied in previous sections and attributed to stimulated recombination of localized states because energy-selective optically pumped SE studies showed "mobility edge"-type behavior in the SE spectra as the excitation photon energy was varied.

Figure 61 shows I_{th} of SE peaks (1) and (2) as a function of L_{exc} at 10 K and RT. Note that I_{th} for peak (2) is larger than that of peak (1) for all excitation lengths employed, but approaches that of peak (1) with increasing L_{exc} in an asymptotic fashion. The high I_{th} of peak (2) with respect to peak (1) and its increased presence for longer L_{exc} suggest that it results from a lower gain process than that of peak (1). Figure 62 shows the peak positions of SE peaks (1) and (2) as a function of L_{exc} at 10 K and RT. For $L_{exc} < 500$ μm, only SE peak (1) is observed and has a peak emission photon energy at 10 (300) K of ~2.92 (2.88) eV and an I_{th} of ~100 (475) kW/cm². As I_{exc} is increased and/or L_{exc} is increased, SE peak (2) emerges at 2.86 (2.83) eV at 10 (300) K. The peak positions were measured for I_{exc} fixed relative to the SE thresholds of the respective peaks, that is, $I_{exc} = 2 \times I_{th}$. As L_{exc} is increased, SE peak (1) shifts to lower energies due to a reabsorption process, whereas the SE peak (2) position is observed to be weakly dependent on L_{exc}. The apparent blue shift of SE peak (2) with increasing L_{exc} is a result of the experimental conditions. Whereas the I_{th} of SE peak (2) is a strong function of L_{exc}, the peak positions shown for small L_{exc} are for I_{exc} considerably higher than for large L_{exc}. The slight red shift of SE peak (2) with increasing I_{exc}, which is due to many-body effects and lattice heating, then manifests itself as the apparent blue shift with increasing L_{exc}. This phenomenon is also observed at 300 K, as shown in the inset of Figure 62. The red shift of SE peak (1) with increasing L_{exc} can be explained by gain and absorption competition in

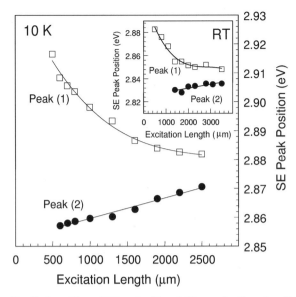

Fig. 62. Peak position of SE peaks (1) and (2) as a function of excitation length at 10 K for the $In_{0.18}Ga_{0.82}N$/GaN MQW. The inset shows the behavior observed at room temperature. The solid lines are given only to guide the eye.

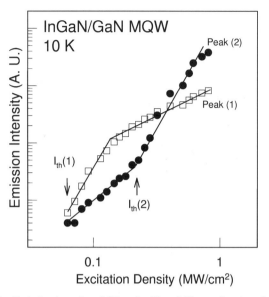

Fig. 63. Emission intensity of SE peaks (1) and (2) as a function of optical excitation density at 10 K, illustrating gain competition between SE peaks (1) and (2). The excitation length is 1300 μm. The respective SE thresholds I_{th} of SE peaks (1) and (2) are indicated for completeness. The solid lines are given only to guide the eye.

the "soft" absorption edge of the InGaN-active regions, where gain saturation with longer L_{exc} combined with the background absorption tail leads to the observed red shift. The fact that SE peak (2) does not experience a reabsorption-induced red shift with increasing L_{exc} is explained by the significant reduction of the absorption tail in this spectral region (see Fig. 60).

The gain saturation behavior of SE peak (1) is consistent with the observation by Kuball et al. [146] of a high-gain mechanism in the band-tail region of MQWs with similar active regions. The large spectral range that exhibits gain is explained by compositional fluctuations inside the active region. The red shift of SE peak (1) with increasing L_{exc} is consistent with observations of a red shift in the optical gain spectrum with increasing L_{exc} reported by Mohs et al. [147]. It is also consistent with the observation by Nakamura [148] that the external quantum efficiency of his cw blue laser diodes decreases with increasing cavity length. These similarities, combined with the relatively low I_{th} of SE peak (1) with respect to SE peak (2) and its similar spectral position with laser emission from diodes of similar structure [149], suggest that lasing in current state-of-the-art cw blue laser diodes originates from the gain mechanism responsible for SE peak (1). Its origin may lie in an entirely different degree of carrier localization than is responsible for SE peak (2). Further experiments are needed to clarify this issue.

The dependence of the emission intensity of peaks (1) and (2) on I_{exc} is shown in Figure 63 for $L_{exc} = 1300$ μm at 10 K. The emission of peak (1) increases in a strongly superlinear fashion ($\propto I_{exc}^{3.8}$) until the I_{th} of peak (2) is reached, at which point it turns linear, indicating that peak (2) competes for gain with peak (1). This is most likely a result of competition for carriers or reabsorption of the emitted photons. The presence of SE peak (2) is therefore seen to be deleterious to SE peak (1). The same process is observed at RT and for various excitation lengths. This gain competition may limit this material's

performance in high-power laser diode applications, where increased driving current and/or longer cavity lengths may result in a shift of the dominant gain mechanism and a drastic change in the emission behavior.

For the optical gain measurements, the variable stripe excitation length method was used [150, 151]. The samples were optically excited by the third harmonic (355 nm) of an injection-seeded Q-switched Nd:YAG laser ($\sim$5 ns FWHM, 10-Hz repetition rate). The excitation beam was focused to a line on the sample surface using a cylindrical lens and the excitation length was precisely varied using a mask connected to a computer-controlled stepper motor. The emission was collected from one edge of the samples, coupled into a 1/4-m spectrometer, and spectrally analyzed using a UV-enhanced CCD. The modal gain $g_{mod}(E)$ at energy E is extracted from

$$\frac{I_1(E, L_1)}{I_2(E, L_2)} = \frac{\exp[g_{mod}(E)L_1] - 1}{\exp[g_{mod}(E)L_2] - 1} \qquad (5.4)$$

where $L_{1,2}$ denote two different stripe excitation lengths and $I_{1,2}$ denote the respective measured emission intensities. Special care was taken to avoid gain saturation effects in the optical gain spectra by using stripe lengths shorter than those for which saturation effects occur.

The modal gain spectra at 10 K as a function of above-gap optical excitation density are shown in Figure 64a and b for an $In_{0.18}Ga_{0.82}N$/GaN MQW and an $In_{0.18}Ga_{0.82}N$ epilayer, respectively. The excitation densities in Figure 64 are given with respect to the SE threshold measured for long ($>$2000 μm) excitation lengths. A clear blue shift in the gain peak with increasing optical excitation is seen for the MQW. This blue shift was observed to stop for $E_{exc} > 12 \times I_{th}$. Further increases in I_{exc} result only in an increase in the modal gain maximum.

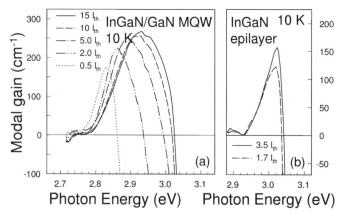

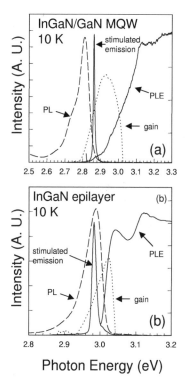

Fig. 64.   10-K modal gain spectra of (a) an $In_{0.18}Ga_{0.82}N$/GaN MQW and (b) an $In_{0.18}Ga_{0.82}N$ epilayer as a function of above-gap optical excitation density. The excitation densities are given with respect to the SE threshold I_{th} measured for long (>2000 μm) excitation lengths. A clear blue shift in the gain maximum and gain/absorption crossover point is seen with increasing excitation density for the $In_{0.18}Ga_{0.82}N$/GaN MQW sample. This trend is much less obvious for the $In_{0.18}Ga_{0.82}N$ epilayer.

Fig. 65.   PL (dashed lines), SE (solid lines), and modal gain (dotted lines) spectra taken at 10 K from (a) an $In_{0.18}Ga_{0.82}N$/GaN MQW and (b) an $In_{0.18}Ga_{0.82}N$ epilayer. The SE and gain spectra were measured for excitation lengths of >5000 and <200 μm, respectively. The maximum modal gain is 250 and 150 cm^{-1} for the $In_{0.18}Ga_{0.82}N$/GaN MQW and $In_{0.18}Ga_{0.82}N$ epilayer, respectively. The low-density PLE spectra are also shown for reference.

The maxima of the gain spectra in Figure 64a are red shifted by more than 160 meV with respect to the "soft" absorption edge of the InGaN well layers. The large shift in the gain maximum to higher energy with increasing I_{exc} is consistent with band filling of localized states in the InGaN-active layers. Similar behavior was observed at RT. The blue shift in the gain spectra of the InGaN epilayer is seen to be considerably smaller than that of the MQW and to stop at considerably lower excitation densities. The modal gain spectra of both samples correspond spectrally with the low-energy tail of the band-tail state absorption bleaching spectra: the crossover from absorption to gain corresponds approximately with the maximum in the observed bleaching.

For completeness, the relevant 10-K PL, PLE, SE, and modal gain spectra are shown together in Figure 65 for (a) the $In_{0.18}Ga_{0.82}N$/GaN MQW and (b) the $In_{0.18}Ga_{0.82}N$ epilayer. Note that the x axis of (b) covers half that of (a). Representative SE spectra are shown (solid line) for a pump density of $E_{exc} = 1.5 \times I_{th}$ (I_{th} denotes the SE threshold) and an excitation spot size of $\sim 100 \times 5000$ μm. Note that the SE peak is situated at the end of the absorption tail for both samples. The SE is seen to occur on the high-energy side of the low-power spontaneous emission peak for the MQW and slightly on the low-energy side for the epilayer. The modal gain spectra were taken with I_{exc} much greater than I_{th} and the excitation length less than 200 μm to minimize reabsorption-induced distortions in the spectra. The modal gain maxima are 250 and 150 cm^{-1} for the MQW and the epilayer, respectively. The SE peak for long excitation lengths (>5000 μm) is situated on the low-energy tail of the gain curve measured for small excitation lengths (<200 μm). This is explained by gain and absorption competition in the band-tail region of this alloy, where gain saturation with longer excitation lengths combined with the background absorption tail leads to a red shift of the SE peak with increasing excitation length. The modal gain spec-

trum for the MQW is seen to be significantly broader ($\sim$24 nm FWHM) than that of the epilayer ($\sim$7 nm FWHM), although both peak significantly below the onset of the "soft" absorption edge. It should also be noted that the optical gain maximum occurs only at photon energies below the mobility edge measured in energy-selective studies described in the previous section, giving further evidence that localized states are the origin of optical gain in the InGaN/GaN MQW structures.

5.8. Summary of Optical Properties of InGaN

As this section clearly shows, the InGaN-related spontaneous emission features are significantly affected by different carrier recombination dynamics that vary with temperature, because of band-tail states arising from large In alloy inhomogeneity, layer thickness variations, and/or defects in the MQWs. Increasing the temperature strongly affects the recombination features by means of thermal filling of band-tail states in the presence of carrier localization. In other words, thermally populated carriers at band-tail states recombine with different average emission peak energies, depending on the change in carrier lifetime and their thermal energy. Even at RT, the temperature-induced blue-shift behavior of the spontaneous emission peak of InGaN/GaN MQWs is still observable, and the emission peak energy is still lower than the mobility edge obtained by energy-selective PL and SE experiments. This suggests that carrier localization plays an important role in InGaN/GaN MQW structures, even at RT operation.

Excitation energy dependence studies of spontaneous and stimulated emission provide important information about the energy boundary between localized and extended band-tail states. An interesting change in PL spectra was observed with varying excitation photon energy above and below the GaN bandgap. The PL spectra were interpreted by noting the evolution of PLE spectra as a function of detection energy for the InGaN emission. From these results, the lower energy tail states are more related to the interface-related detects and roughness at the MQW interfaces rather than to alloy fluctuations and impurities within the InGaN wells. A mobility edge-type behavior was found in both PL and SE spectra as the excitation photon energy was tuned across the states responsible for the broadened absorption edge of the InGaN-active regions. The mobility edge is well above the PL and SE peak positions, but well below the absorption edge, indicating that both the spontaneous and stimulated emissions originate from carriers localized at low-energy band-tail states due to extremely large potential fluctuations in the MQW InGaN-active layers. Moreover, the dependence of the SE threshold density on excitation energy is very consistent with the PLE spectra, indicating that the coupling efficiency of the exciting photons to the gain mechanism responsible for the SE peak is correlated with that of the spontaneous emission mechanism. The correlation between the high-density behavior and the low-density cw PLE indicates that carrier localization plays a significant role in both the spontaneous and stimulated emission processes. The SE features in InGaN/GaN QW structures were further investigated through the dependence of the SE on excitation length and density. Two distinctly different SE peaks were observed with different dependencies on excitation length. The high-energy SE peak exhibits a strong red shift with increasing excitation length due to competition between an easily saturable gain mechanism and a background absorption tail, whereas the lower energy SE peak does not exhibit this reabsorption-induced red shift with increasing excitation length. The presence of the lower energy SE peak has been shown to be detrimental to the higher energy SE peak due to gain competition in the InGaN-active region. This competition may prove to be an obstacle in the design of InGaN-based high-power laser diodes, where high current densities and/or long cavity lengths lead to a shift in the dominant gain mechanism and a change in emission characteristics.

Many different experimental observations of the optical properties of InGaN-based heterostructures can be interpreted in terms of a built-in internal electric field *and/or* carrier localization. However, the results presented in this section for both low and high carrier densities are not explainable in the framework of an internal electric field or electron–hole plasma alone without consideration of carrier localization associated with potential fluctuation.

6. OPTICAL PROPERTIES OF NITRIDE THIN FILMS AT HIGH TEMPERATURES

6.1. High-Temperature Stimulated Emission and Damage Mechanisms in GaN Epilayers

New industrial and military applications require semiconductor devices that operate at high temperatures. Examples include the *in situ* monitoring of internal combustion engines for the automobile industry, electronics for ultrahigh-speed aircraft capable of withstanding large temperature variations, and electrical components for high-power electronic devices. It is likely that a myriad of potential applications have not even been thought of due to the unavailability of sturdy semiconductor-based optoelectronic systems. GaN-based structures are believed to have the potential to advance many current technologies dramatically by cutting costs and improving the performance of systems for wide-temperature applications.

Stimulated emission and laser action in GaN and related heterostructures at temperatures up to 300 K have been extensively studied in the literature (see [39], for example). Many scientists realized that the high efficiency of GaN-based LEDs has not been compromised in spite of an extremely high density of structural defects [26]. At the same time, GaN films were shown to possess a low-temperature sensitivity of the lasing threshold near RT [41]. This attribute suggested that further research was needed to explore the optical properties and suitability of GaN-based structures for high-temperature optoelectronic devices as well as to study the effects of sample quality on the performance of these devices. Some preliminary above RT stimulated emission experiments on GaN epilayers grown on sapphire and SiC [41, 152] as well as a theoretical prediction of the temperature sensitivity [153] have been published, but only recently have scientists reported on SE or lasing in GaN beyond 450 K [12, 13, 154].

High-temperature optical experiments on high-quality GaN thin films have resulted in some astonishing discoveries. For the first time in any semiconductor system, SE has been observed at temperatures as high as 700 K [12]. In this section, the temperature dependence of the energy positions for spontaneous and stimulated emission peaks at such high temperatures are explored. The temperature sensitivity of the SE threshold also is discussed and the values for the characteristic temperature are derived. Based on the observed unique properties of SE in GaN thin films at high temperatures, some possible device applications are discussed.

The GaN samples used in this study were all nominally undoped epitaxial films grown on (0001) sapphire and 6H-SiC substrates by MOCVD. Thin AlN buffers were deposited on the substrates at 775°C. The GaN layers were deposited at 1040°C on the AlN buffers. These conditions typically result in high-quality single-crystal GaN layers [155, 156]. The thickness of the layers ranged from 0.8 to 7.2 μm. Photoluminescence, absorption, reflection, and photoluminescence excitation spectra for these samples described in Section 2.

In this study, two different pumping configurations were applied. For high-temperature SE work, the samples were

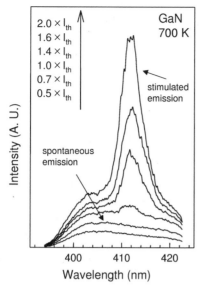

Fig. 66. Emission spectra at 700 K from a GaN film grown on a sapphire substrate. The FWHM of the SE peak is only 5 nm at 700 K.

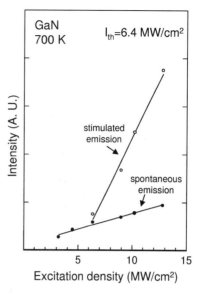

Fig. 67. Dependence of integrated emission intensity on the pump density for a GaN film at 700 K.

mounted on a copper heat sink attached to a custom-built wide temperature range cryostat/heater system. The third harmonic (355 nm) of an injection-seeded Nd:YAG laser with a pulse width of 6 ns and a repetition rate of 30 Hz was used as the pumping source. The laser beam was focused into a line on the sample surface using a cylindrical lens. The laser light intensity was continuously attenuated using a variable neutral density filter. The sample emission was collected from the sample edge.

To study spontaneous emission, a frequency-doubled cw Ar$^+$ laser (244 nm, 40 mW) was used as an excitation source. To avoid a spectral distortion of the spontaneous emission due to reabsorption, the laser beam was focused into an 80-μm-diameter spot on the sample surface and the spontaneous emission was collected from a direction near normal to the sample surface.

Stimulated emission was observed at temperatures as high as 700 K for thin GaN films grown on SiC and sapphire substrates. The emission spectra at 700 K for different excitation powers near the SE threshold are shown in Figure 66 for the GaN sample grown on sapphire. At excitation powers below the SE threshold, only broad reabsorbed spontaneous emission is present. As the pumping density increases and crosses the SE threshold, a significant spectral narrowing occurs and the peak intensity starts to increase superlinearly as shown in Figure 67, clearly indicating the onset of SE. The FWHM of the SE peaks for both samples is only 5 nm at 700 K. We note that the maximum of the SE peak at 700 K is located at approximately 412 nm, which is in the same wavelength range as the lasing from InGaN/GaN multiple quantum wells at RT reported in [157]. The polarization ratio TE:TM for the SE peak was measured to be greater than 20:1. The values of the SE threshold at 700 K were estimated to be 6.4 and 5.6 MW/cm^2 for GaN films grown on sapphire and SiC, respectively.

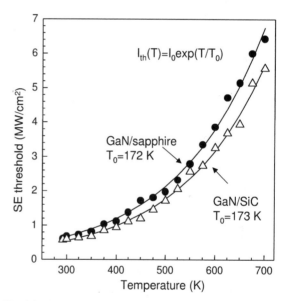

Fig. 68. Stimulated emission threshold as a function of temperature for GaN thin films grown on sapphire (solid circles) and SiC (open triangles). The solid lines represent the best least-squares fit to the experimental data. Characteristic temperatures of 172 and 173 K were obtained for samples grown on sapphire and SiC, respectively.

The temperature dependence of the SE threshold for the two different samples is shown in Figure 68. GaN epilayers grown on sapphire (solid circles) and SiC (open triangles) exhibited similar SE threshold trends and roughly followed an exponential dependence. The SE threshold for the sample grown on SiC was measured to be 0.57 MW/cm^2 at RT, 1.7 MW/cm^2 at 500 K, and 5.6 MW/cm^2 at 700 K. The solid lines in Figure 68 represent the results of the best least-squares fit of the experimental data to the empirical equation for the temperature dependence of the SE threshold [see Eq. (4.6)]. The characteristic temperature T_0 was estimated to be 172 and 173 K over the

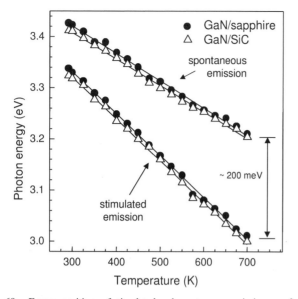

Fig. 69. Energy positions of stimulated and spontaneous emission as a function of temperature for GaN thin films grown on sapphire (solid circles) and SiC (open triangles). The solid lines represent a linear fit to the experimental data. The results of the fit are summarized in Table II. The energy separation between the peaks gradually increases from 90 meV at room temperature to 200 meV at 700 K, indicating that an electron–hole plasma is responsible for the SE mechanism in this temperature range.

Table II.   Results of a Least-Squares Linear Fit of the Energy Positions of Emission Peaks as a Function of Temperature[a]

Sample (thin films)	Spontaneous emission (eV)	Stimulated emission (eV)
3.7-μm GaN/SiC	$3.56–5.17 \times 10^{-4} \times T$	$3.56–8.15 \times 10^{-4} \times T$
7.2-μm GaN/sapphire	$3.58–5.35 \times 10^{-4} \times T$	$3.58–8.20 \times 10^{-4} \times T$

[a]For temperatures above 300 K. The small difference in coefficients is most likely due to different residual strain in the GaN epilayers grown on sapphire and SiC substrates.

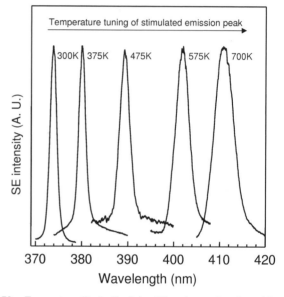

Fig. 70. Temperature "tuning" of the SE peak wavelength position. The broadening of the SE peak is considerably less than its shift in energy as the temperature is raised from 300 to 700 K for a GaN epilayer grown on SiC.

temperature range of 300–700 K for the GaN epilayers grown on sapphire and SiC, respectively, indicating the very low temperature sensitivity of the SE threshold. These obtained values of characteristic temperature are considerably larger than the near RT values reported for other material systems [158, 159], where small values of T_0 were found to be detrimental to above RT laser operation. Such a remarkably low temperature sensitivity of the SE threshold in GaN epilayers suggests GaN-based lasing mediums are well suited to high-temperature applications [13].

Figure 69 depicts the energy position of the spontaneous and SE peaks for a 7.2-μm-thick GaN film grown on sapphire (solid circles) and a 3.7-μm-thick GaN film on SiC (open triangles) versus temperature. We note that the energy position of the spontaneous emission peak was measured at very low excitation power to avoid any effects associated with bandgap renormalization and/or band filling. The position of the SE peak was measured at a pump density slightly above the SE threshold value at each temperature. For temperatures above 300 K, the energy positions of the stimulated and spontaneous emission peaks are well approximated by a linear fit (solid lines). The results of this fit are summarized in Table II. The behavior of the spontaneous and stimulated emission energy positions at elevated temperatures in GaN epilayers was found to be independent of the substrate. A small difference in the absolute energy position of spontaneous and stimulated emission peaks (and, subsequently, in the coefficients presented in Table II) can be explained by the different values of residual strain for GaN thin films grown on sapphire and SiC substrates [160]. The formulas presented in Table II constitute empirical values for the

energy positions of spontaneous and stimulated emission peaks for GaN thin films in the temperature range of 300–700 K.

Chow et al. [161] reported a many-body calculation of gain spectra in GaN and concluded that at RT the excitonic absorption decreases in amplitude with increasing carrier density and disappears with the appearance of gain. They predicted that the gain peak at high densities is shifted toward lower energies from the exciton resonance by several tens of millielectronvolts. As described here, the large energy separation between the spontaneous and stimulated emission peaks (it gradually increases from 90 meV at RT to approximately 200 meV at 700 K) was experimentally confirmed. Both this large energy difference and the relatively high values of SE thresholds in this temperature range effectively eliminate exciton-related effects from the consideration of SE mechanisms. Therefore, it has been concluded that free carrier recombination or an electron–hole plasma is the dominant SE mechanism in GaN thin films for temperatures above 300 K. We note that Amano and Akasaki [162] independently reached the same conclusion about the origin of SE.

The broadening of the SE peak was found to be considerably smaller than the temperature-induced shift in energy position,

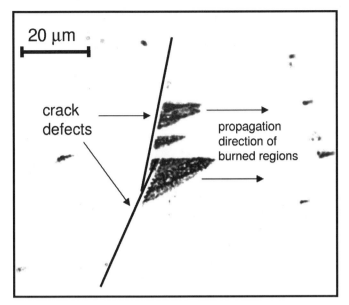

Fig. 71.   A microscope image of burn spots on the surface of a GaN film. Burn spots tend to originate at linear defects such as cracks near a sample edge and propagate toward the middle of the sample. Surface defects are shown to be a strong limiting factor for high-temperature SE in thin GaN films.

as shown in Figure 70. Raising the temperature from 300 to 700 K allows a "tuning" of the SE peak energy from the near-UV spectral region to blue (a range of more than 40 nm). No degradation of the GaN epilayers was observed in spite of the very high pumping densities required to reach the SE threshold at temperatures exceeding 700 K [154]. This unique property of SE in GaN thin films might result in new types of optoelectronic devices that utilize this temperature tuning. The low-temperature sensitivity of the SE threshold over such a large temperature range could potentially lead to the development of laser diodes that operate hundreds of degrees above RT.

The maximum temperature at which SE in GaN could be observed was directly related to the density of line defects (or cracks) on the sample surface. The SE threshold was previously shown (Fig. 68) to exponentially increase with temperature. This requires higher levels of optical pumping to reach the SE threshold. When pump densities on the order of 10 MW/cm^2 are reached, the sample surface starts to deteriorate rapidly. A microscope image of burn spots on the surface of a GaN film is shown in Figure 71. In spite of the uniform density of the excitation beam, most of the burn spots tend to originate at linear defects, such as cracks, in the vicinity of the sample edge. In the assumption of a single pass amplification (the sample facets were not specially prepared), the sample emission is amplified when the photon flux propagates along the longest path, that is, the SE intensity is strongest near the sample edges and weakest in the middle of the sample. This explains the fact that most of the burn spots were found near the sample edges. The burn spots always propagate toward the middle of the sample in the direction opposite to the amplification direction of the SE flux, as shown in Figure 71. Thus, the GaN epilayer quality becomes

a strong limiting factor for SE in bulk GaN at high temperatures.

To explain the burning mechanism in GaN epilayers, it is important to understand the effect of defects on SE. Under strong optical excitation, the defect spots dissipate a significant amount of heat that eventually leads to surface burning. In fact, once the damage threshold pumping density is exceeded, unpolarized surface-emitted SE is observed to increase and the burn spots tend to originate from linear surface defects, such as cracks. Therefore, surface quality is a strong limiting factor for the observation of SE at high temperatures. The high density of defects in GaN is associated with the film–substrate lattice mismatch and thermal-expansion coefficient mismatch. Even though significant progress in improving the epilayer quality has been made [163], a high density of structural defects might always be present in GaN-based working devices.

6.2. Stimulated Emission in InGaN/GaN Multiple Quantum Wells at Elevated Temperatures

The low-temperature sensitivity of the lasing threshold in ridge-geometry InGaN (MQW) structures near RT [164] in comparison to structures based on other materials, as well as the previously described success in the demonstration of high-temperature stimulated emission in GaN epilayers, prompted further research into potential high-temperature applications for III–V nitride MQWs. Until recently, there have been no reported studies on SE or lasing in InGaN/GaN/AlGaN-based heterostructures at temperatures above 80°C. In this section, we describe the results of an experimental study on SE in optically pumped InGaN/GaN MQW samples in the temperature range of 175–575 K. The effects of Si doping on the luminescence intensity and SE threshold density of InGaN/GaN MQWs also is discussed.

The InGaN/GaN MQW samples described in this section were grown on a 1.8-μm-thick GaN base layer by MOCVD. The GaN barriers were doped with Si at a concentration range of 1×10^{17} to 3×10^{19} cm^{-3} for the different samples studied. The MQWs were capped with a 100-nm-thick Al$_{0.07}$Ga$_{0.93}$N layer grown at 1040°C [165]. The number of periods in the MQW samples was 12. The nominal well and barrier layer thicknesses were 30 and 45 Å, respectively. To evaluate the interface quality and structural parameters such as the average In composition in the MQW and the period of the superlattice, the samples were analyzed with a four-crystal high-resolution X-ray diffractometer using Cu Kα_{I} radiation. The angular distances between the satellite superlattice diffraction peaks and GaN (0002) reflections were obtained by ω–2Θ scans. The spectra clearly showed higher order satellite peaks that indicate high interface quality and good layer uniformity.

The samples were mounted on a copper heat sink attached to a wide temperature range cryostat/heater system. This study was performed in the side-pumping geometry, where edge emission from the samples was collected into a 1-m spectrometer and recorded by an optical multichannel analyzer.

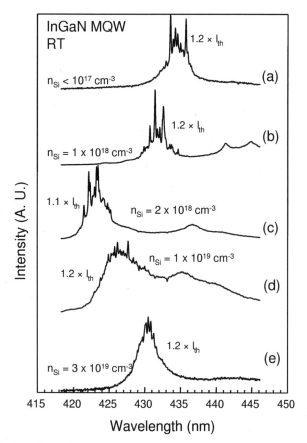

Fig. 72. Spectra taken at pump powers slightly above the SE threshold for samples with GaN barriers doped with silicon at concentration levels of (a) $<1 \times 10^{17}$ cm^{-3}, (b) 1×10^{18} cm^{-3}, (c) 2×10^{18} cm^{-3}, (d) 1×10^{19} cm^{-3}, and (e) 3×10^{19} cm^{-3}.

The effects of Si doping on the optical properties of bulk GaN [121, 122], InGaN/GaN MQWs [125, 165], and GaN/AlGaN [128] MQWs have been studied by many researchers. In this section, optical pumping on InGaN/GaN MQW samples with different Si doping is described. Typical spectra obtained at pump densities slightly above the SE threshold are depicted in Figure 72. All samples exhibited bright emission in the blue region of the electromagnetic spectrum. Only moderate concentrations of Si increased the luminescence intensity and reduced the SE threshold pump density in InGaN/GaN MQWs. Table III summarizes the Si concentrations as measured by secondary ion mass spectroscopy (SIMS) and the SE threshold densities at RT for the samples studied in this work. The maximum luminescence intensity and lowest SE threshold ($I_{th} = 55$ kW/cm^2 at RT) were observed for the sample with a Si concentration of 2×10^{18} cm^{-3} in the barrier layer. This threshold is 12 times lower than that of a high-quality nominally undoped single-crystal GaN film measured under the same experimental conditions.

Figure 73 shows the emission spectra for the InGaN/GaN MQW sample with a Si concentration of 2×10^{18} cm^{-3} for various temperatures and excitation powers. Dotted lines represent the broad spontaneous emission spectra taken at pump densities approximately half that of the SE threshold for each

Table III. Si Doping Concentrations and Stimulated Emission Threshold Densities for InGaN/GaN MQW Samples

Si concentration[a] in GaN barriers (cm^{-3})	Stimulated emission threshold density[b] (kW/cm^2)
$<1 \times 10^{17}$ (undoped)	58
1×10^{18}	58
2×10^{18}	55
1×10^{19}	92
3×10^{19}	165

[a] Measured by SIMS.

[b] At room temperature.

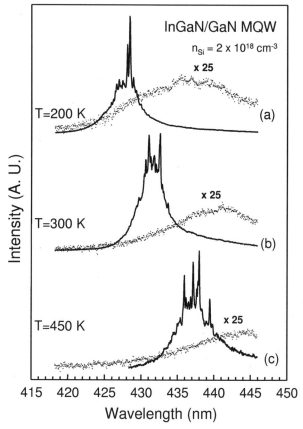

Fig. 73. Emission spectra for an InGaN/GaN MQW sample with a Si concentration of 2×10^{18} cm^{-3} in the barriers at (a) 200, (b) 300, and (c) 450 K. Spontaneous emission spectra (dotted lines) were taken under an excitation density of $0.5 \times I_{th}$ and SE spectra were obtained for a pump density of $2 \times I_{th}$, where I_{th} represents the SE threshold at the corresponding temperature.

temperature. Spontaneous emission in InGaN MQWs has been attributed to the recombination of excitons localized at certain potential minima in the quantum well [75, 166]. A more detailed discussion of the SE mechanisms is addressed later in this chapter. As the excitation power density is raised above the SE threshold, a considerable spectral narrowing occurs (solid lines in Fig. 73). The emission spectra comprise many narrow peaks

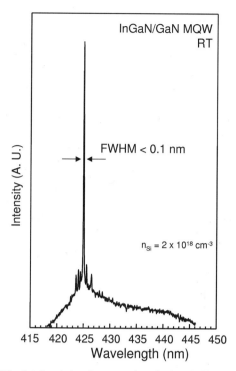

Fig. 74. Stimulated emission from a moderately doped (Si concentration of 2×10^{18} cm^{-3} in the barriers) InGaN/GaN MQW sample at room temperature. Stimulated emission spectra were strongly dependent on pumping and collecting configuration. Under certain conditions it is possible to generate single-peak SE even at room temperature.

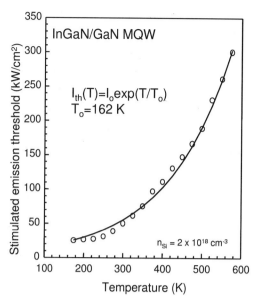

Fig. 75. Temperature dependence of the SE threshold in the temperature range of 175–575 K for the InGaN/GaN MQW sample. The line represents the best result of a least-squares fit to the experimental data (open circles). A characteristic temperature of 162 K is derived from the fit.

of less than 1 Å FWHM. The major effect of the temperature change from 200 (Fig. 73a) to 450 K (Fig. 73c) is a shift of the spontaneous emission and stimulated emission peaks toward lower energy. There was no noticeable broadening of the SE peaks when the temperature was varied over a range of 400 K.

The emission from the sample was found to be strongly dependent on the pumping and collecting configurations. It was possible to align the sample in such a way that even at RT only one narrow SE peak could be observed. An example of such a spectrum is shown in Figure 74. The FWHM of the peak (1 Å) was resolved only in the second order of a 1200 groove/mm grating in the 1-m spectrometer.

The temperature dependence of the SE threshold is shown in Figure 75 (open circles). Stimulated emission was observed throughout the entire temperature range studied, from 175 to 575 K. The SE threshold was measured to be ~25 kW/cm^2 at 175 K, ~55 kW/cm^2 at 300 K, and ~300 kW/cm^2 at 575 K, and roughly followed an exponential dependence. It is likely that such low SE threshold values are due to a large localization of carriers in MQWs. The solid line in Figure 75 represents the best result of a least-squares fit of the experimental data to the empirical form for the temperature dependence of the SE threshold [see Eq. (4.6)]. The characteristic temperature T_0 was estimated to be 162 K in the temperature range of 175–575 K for this sample. The derived value of characteristic temperature is considerably larger than the near RT values reported for laser structures based on other III–V and II–VI materials, where the

relatively small values of T_0 were a strong limiting factor for high-temperature laser operation. Such a low sensitivity of the SE threshold to temperature changes in InGaN/GaN MQWs opens up enormous opportunities for high-temperature applications using these materials. Laser diodes with InGaN/GaN lasing media can potentially operate at temperatures exceeding RT by a few hundred degrees kelvin.

It is commonly observed that an increase in temperature leads to a decrease in PL intensity. This change indicates the onset of efficient losses and a decrease in the quantum efficiency of the MQWs. At high temperatures, only a small fraction of excitons reach the conduction band minima, and most of them recombine nonradiatively. The modal gain depends only on radiatively recombining excitons. Therefore, the temperature increase efficiently decreases modal gain and leads to an increase in the SE threshold. To evaluate the number of electrons that recombine radiatively, the integrated photoluminescence intensity as a function of excitation power was studied for different temperatures, as shown in Figure 76. For the temperature range studied, it was found that under low excitation densities, the integrated intensity I_{integ} from the sample almost linearly increases with pump density I_p ($I_{\text{integ}} \propto I_p^\gamma$, where $\gamma = 0.8$–1.3), whereas at high excitation densities, this dependence becomes superlinear ($I_{\text{integ}} \propto I_p^\beta$, where $\beta = 2.2$–3.0). The excitation pump power at which the slope of I_{integ} changes corresponds to the SE threshold at a given temperature. Interestingly, the slopes of I_{integ} below and above the SE threshold do not significantly change over the temperature range involved in this study. This suggests that the mechanism of SE in InGaN/GaN MQWs at RT remains the same as we raise the temperature hundreds of degrees.

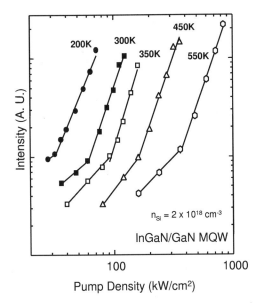

Fig. 76. Integrated intensity of InGaN/GaN MQW emission as a function of pump density for different temperatures. The slope change from 0.8–1.3 to 2.2–3.0 indicates the transition from spontaneous emission to stimulated emission.

The results described in this section are extremely intriguing because edge-emitted SE was observed at a record high temperature of 700 K for GaN thin films grown on SiC and sapphire substrates. Stimulated emission at such high temperatures has never been reported for any semiconductor-based system. The low SE threshold, as well as the weak temperature sensitivity, make GaN-based thin films an attractive material for development of laser diodes that can operate well above RT and create opportunities for the development of new optoelectronic devices capable of high-temperature operation.

7. MICROSTRUCTURE LASING

7.1. Origin of Surface-Emitted Stimulated Emission in GaN Epilayers

Gallium nitride (and its alloys) is a very intriguing material for optical applications, not only because of its large spectral emission range, but also due to the apparent controversy between structural and SE (lasing) properties. Due to the lack of ideal substrates for the growth of thin film nitrides, a large number of dislocations and cracks are naturally formed in the epitaxial layer to alleviate the lattice mismatch and the strain of postgrowth cooling. However, this does not always negatively affect lasing characteristics, but sometimes introduces interesting lasing properties due to self-formed high-finesse microcavities and efficient scattering centers.

Many experimental research groups have reported the observation of surface-emitted SE in GaN epilayers, but its origin was never fully understood and has led to much conflict of opinion among researchers. Understanding the origin of this surface-emitted SE is important to improve the lasing characteristics of edge-emitting lasers as well as to develop a vertical

cavity laser diode using this material. Vertical cavity SE from an MOCVD-grown GaN layer, as well as for an InGaN/GaN heterostructure, was reported by Khan et al. [167, 168]. The relatively small cavity lengths (1.5 and 4.1 μm) and low mirror reflectivities (<30%) indicate that SE within such a cavity requires extremely high gain in the active region. Bagnall and O'Donnell [169] theoretically estimated the value of the gain under similar conditions to be more than 10^5 cm^{-1}, which is a few orders of magnitude higher than the gain measured by the variable stripe length method reported by many authors [170, 171]. Yung et al. [172] also reported the observation of surface-emitted SE from a GaN film grown by ion-assisted molecular beam epitaxy. In contrast, Amano et al. [173, 174] studied SE in an MOVPE-grown GaN epilayer film and AlGaN/GaN heterostructures, and did not observe any surface-emitted SE for the epilayer film, even though SE was observed in a side-pumping (edge-emission) geometry. Interestingly, the same authors did observe surface-emitted SE from AlGaN/GaN heterostructures.

Lester et al. [26] found that GaN-based LEDs were highly efficient in spite of an extremely high density of structural defects ($\sim 10^{10}$ cm^{-2}). Similar GaAs LEDs have a dislocation density 6 orders of magnitude lower. A high density of structural defects might always be present in GaN-based working devices. Therefore, the effect of defects on lasing in GaN merits further investigation. Wiesmann et al. [171] compared the emission spectra from InGaN and AlGaN thin films with high-quality GaAs thin films and concluded that the observation of a SE peak in the direction perpendicular to the nitride layer plane was due to scattering of the in-plane SE. This conclusion was experimentally confirmed through spatially resolved surface-emitted SE measurements [109]. In this section, some of those results are reviewed and the present understanding of surface-emitted SE is given. Implications of this study on gain measurements in GaN and a possible explanation for the large discrepancies in reported gain values also are given [109, 167, 170].

The GaN samples used in this work were described in Section 6.1. The samples were mounted on a translation stage driven by a computer-controlled stepper motor. This setup allows the sample to be positioned with a precision of better than 1 μm. A tunable dye laser pumped by a frequency-doubled injection-seeded Nd:YAG laser was used as the primary optical pumping source. The visible output of the dye laser was frequency doubled to achieve a near-UV pumping frequency. The beam was focused to an $\sim$80-μm-diameter spot on the sample surface and the emission was collected in the backscattering geometry as shown in Figure 77. A 10× magnified image of the sample was projected onto the plane of the slits of a 1-m spectrometer. The signal from the sample was collected by an optical multichannel analyzer mounted on the spectrometer. This configuration ensured that the signal was collected in the direction normal to the sample surface. The overall spatial resolution of the system was better than 10 μm. All experiments were performed at RT.

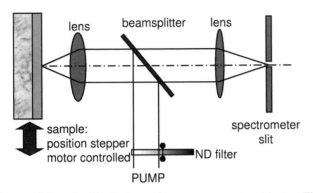

Fig. 77. Schematic of the backscattering geometry experimental setup. The sample is mounted on a translation stage driven by a stepper motor. A magnified image of the sample is projected onto the plane of the spectrometer slit.

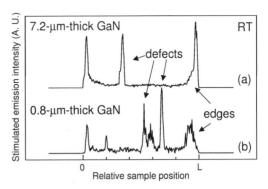

Fig. 79. Stimulated emission intensity versus position relative to the sample edges for GaN samples of thickness (a) 7.2 and (b) 0.8 μm. Stimulated emission appears to originate from the sample edges as well as from burned spots, cracks, and other imperfections. The widths of the samples on the graphs were normalized to line up the edges.

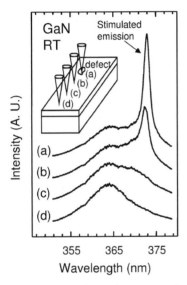

Fig. 78. Emission spectra in the backscattering geometry from a 7.2-μm-thick GaN epilayer as a function of the distance from a defect. The four spectra were taken at 50 μm intervals: spectrum (a) was taken at the defect, whereas spectrum (d) was taken 150 μm away from the defect. The picture shows the locations (relative to the defect) where the spectra were taken.

The emission spectra were found to vary strongly near defect regions. Stimulated emission spectra taken from the 7.2-μm GaN layer at different locations near the vicinity of a defect are shown in Figure 78. The sample was excited with a power density of 1.8 MW/cm^2 (approximately three times the SE threshold value measured in the side-pumping geometry). The broad emission feature at 363 nm corresponds to band-edge-related spontaneous emission from the GaN epilayer. The FWHM of the emission is about 10 nm. The narrow emission feature at 373 nm (with a FWHM of 2 nm) represents SE [41]. The SE peak appears as the excitation beam is moved close to the defect. In contrast to the SE collected from the sample's edge facets, the surface-emitted SE is not strongly polarized. The fact that only spontaneous emission was observed far from the defect and SE was observed only in proximity to the defect suggests that the defect acts as a scattering center, that is, the light is amplified inside the GaN layer while propagating paral-

lel to the sample surface until it is scattered out of the layer by the defect.

Figure 79 shows the dependence of the integrated SE intensity on the position on the sample surface for the two samples pumped with an excitation density of 4 MW/cm^2. Because the widths of the sample bars were different (3 and 4 mm), we normalized the x axis of the scan spectra to line up the samples' edges. A typical scan across the high quality 7.2-μm-thick GaN sample is shown in Figure 79a. Unpolarized SE is observed at the sample edges and at a burned spot on the surface. No SE from featureless parts of the sample was observed, even with pump power densities up to the damage point. From Figure 79b we observe that the 0.8-μm-thick GaN sample has many more points of origin for surface-emitted SE. In fact, once the damage threshold pumping density is exceeded, unpolarized SE coming from the surface can be observed in the backscattering geometry from any GaN sample. Thus the observed SE is not the result of a vertical amplification of light, but is due to a leak of the in-plane photon flux that propagates parallel to the surface. Cracks on the sample surface are the most efficient scattering centers for in-plane amplified SE.

The absorption coefficient of GaN is on the order of 10^5 cm^{-1} (see Section 2.4) in the energy region around the fundamental bandgap [17, 175]. This limits the penetration depth of light into the sample, so that even with pump power densities of a few megawatts per square centimeter, it is only possible to create a population inversion in a layer on the order of 0.5 μm in thickness, which constitutes the gain region in the vertical direction. On the other hand, the lateral spot size of the pump beam, even with tight focusing, is orders of magnitude larger. Thus, the gain region in the horizontal direction is considerably larger than that in the vertical direction. With poor optical feedback in the vertical direction, light is preferentially amplified along a direction parallel rather than perpendicular to the surface. This light amplified in the horizontal direction can then be scattered by surface imperfections and observed as surface-emitted SE.

When gain in an epilayer film is measured in the transmission geometry, we usually assumes that the beam is undergoing gain, that is, the probe beam is amplified only in the direction perpendicular to the surface. When optical feedback is poor, as in the case of GaN thin films, the amplification path is comparable to the thickness of the epilayer (a few micrometers). If the probe is scattered, though, the derivation of gain can be very complicated. Without making additional assumptions, the calculated gain could be orders of magnitude higher than when the gain is measured by the variable stripe technique, where the excitation length is on the order of a hundred micrometers and loss due to scattering is unaccounted for. Therefore, in GaN epitaxial layers with high levels of defects, caution is needed when the traditional techniques are used to measure gain.

The experimental results presented directly establish that the surface-emitted SE in GaN thin films originates from parts of the samples with poorer surface quality, such as cracks, burned spots, and other imperfections. For high-quality GaN films with nearly perfect surfaces, SE was not observed in backscattering geometry, even under excitation power densities close to the damage threshold. Based on these observations, it becomes obvious that the surface-emitted SE is due to scattering of the photon flux that propagates parallel to the surface by defects, rather than vertically amplified emission. These results show that the influence of imperfections in GaN epilayers cannot be ignored when gain values are determined from experiment, and may be one of the reasons there are large discrepancies in gain values measured in different experimental geometries.

7.2. Effects of Microcracks on the Lasing Characteristics of Nitride Thin Films

Microcracks in GaN-based structures can appear either as a result of cleaving or the natural relaxation of strain between the epilayer and the substrate during postgrowth cooling. Cracks formed in this manner have been found to introduce interesting effects on the emission properties of nitride thin films under high levels of optical excitation, such as high-finesse cavity modes.

Stimulated emission with Fabry–Pérot cavity modes in GaN grown on SiC was first observed by Zubrilov et al. [152]. The FWHM of the Fabry–Pérot modes was measured to be 0.2–0.5 nm and the distance between fringes was ~0.7 nm at RT. The cavity length was estimated to be 20 μm, which is much smaller than the sample size. Optical microscopy revealed that the sample had microcracks produced by cleaving. The cracks served as cavity mirrors and generated the Fabry–Pérot modes.

This work was extended by other scientists to systematically cover GaN/AlGaN separate confinement heterostructures. The sample description was given in Section 4.3. To evaluate the effects of cracks, the samples were cleaved into submillimeter-wide bars (note that when GaN is grown on SiC, it can be easily cleaved along the ($11\bar{2}0$) direction [176]). Before cleaving, the samples did not exhibit any noticeable defects on the surface. After the cleaving process, however, cracks were observed along all three cleavage planes associated with a hexagonal crystal structure, with the majority running parallel to the length of the bar as shown in Figure 80.

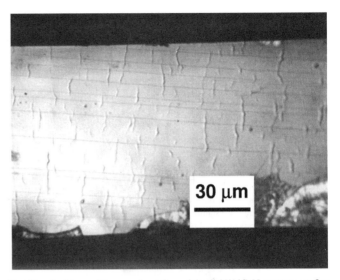

Fig. 80. A picture of the sample surface of the GaN/AlGaN separate confinement heterostructure after cleaving. Before cleaving, the sample exhibited no noticeable defects on the surface. After the cleaving process, however, cracks can be seen running parallel to the length of the bar.

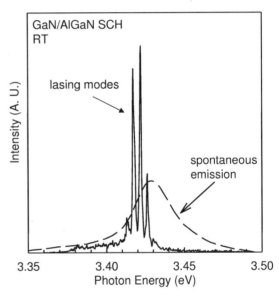

Fig. 81. Lasing and spontaneous emission in the GaN/AlGaN separate confinement heterostructure at room temperature. Lasing is believed to be of microcavity origin.

Samples were excited in the edge-emission geometry (342 nm, 6 ns) with an excitation spot in a shape of a line. As described in Section 4.3, an order of magnitude reduction in the lasing threshold was found in comparison to bulklike GaN epilayers. Such a low lasing threshold for the GaN/AlGaN separate confinement heterostructure is due to improved carrier and optical confinement as well as increased optical feedback introduced by cracks (see Section 8.2).

When the sample was excited above the lasing threshold, high-finesse cavity modes were observed. Typical emission spectra from a separate confinement heterostructure are shown in Figure 81. The spacing between the modes correlates well

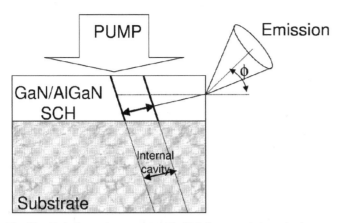

Fig. 82. Optical pumping geometry. The high-finesse emission exits the sample at a large angle ($\phi \sim 18°$). The cracks in the surface parallel to the cleaved edges ran through the active layer at an inclination consistent with the emission angle observed.

with the distance between the cracks depicted in Figure 80. Assuming that the unity round-trip condition is satisfied and there is no loss due to absorption in the GaN layer, the threshold gain can be estimated from [169]

$$g_{th} > \frac{\ln(R^{-2})}{2L} \qquad (7.1)$$

where R is the mirror reflectivity (20% for a GaN–air interface) and L is the distance between the cracks. A gain value on the order of 500 cm^{-1} was obtained, which is consistent with gain measured by the variable stripe technique.

It was found that by pumping different areas along the length of the bar, the emission exhibited varying degrees of cavity finesse. This is presumably due to the presence of cavities of varying quality formed by parallel cracks running through the active layer. Areas exhibiting high finesse consistently had a narrower far-field pattern than those exhibiting low finesse.

The far-field pattern also shows that the high-finesse emission exits the sample at an angle of $\phi \sim 18°$, as shown in Figure 82. To understand this phenomenon, the samples were examined in cross section using a scanning electron microscope. A large number of cracks lying at small angles ($\sim 7°$) to the c axis were detected. Taking into account the refractive index of GaN, the geometry of the cracks could be correlated to the emission angle through Snell's law.

Thus, the existence of microcracks in GaN-based structures (particularly those grown on SiC) results in many new features when the emission is collected from the sample edge. The directionality, spectral appearance, and lasing threshold are significantly affected. Consequently, we have to take into account the effects of these microstructures to interpret experimental data correctly.

7.3. Ring-Cavity Lasing in Laterally Overgrown GaN Pyramids

Microcavity lasers, in addition to vertical cavity surface-emitting lasers, offer many benefits that result from optical

confinement, including enhanced quantum efficiency and a greatly reduced lasing threshold. Modern growth techniques allow the formation of microcavities on nitride thin films with a variety of different geometries. These microcavities can also be arranged to form arrays. The arrays are a natural choice for applications that involve optical displays, imaging, scanning, optical parallel interconnects, and ultra-parallel optoelectronics. Furthermore, success in the fabrication of long lifetime cw InGaN edge-emitting laser diodes is largely due to a significant reduction in defects attained by using lateral overgrowth on sapphire substrates [177]. The selective area growth of wide-bandgap semiconductors is believed to be one of the most important methods to realize high performance laser diodes in the short wavelength region [178, 179]. Because of the large physical dimensions of their resonator cavities (several hundred micrometers), traditional edge-emitting lasers may be used only to construct one-dimensional arrays. On the other hand, surface-emitting or microstructure-based lasers [58] (with a typical cavity of a few micrometers) could potentially be used to develop two-dimensional laser arrays.

Some recent breakthrough results indicate that single and multimode RT laser action in GaN pyramids under strong optical pumping can be achieved [6]. The laser cavity in a pyramid was found to be of ring type, formed by total internal reflections of light off the pyramid surfaces. The mode spacing of the laser emission was correlated to the size of the pyramid.

The samples used in this study were grown by low-pressure MOCVD. First, a 0.10-μm-thick AlN buffer layer was deposited on a (111) Si wafer. A GaN thin film was subsequently grown, resulting in a thickness of about 0.15 μm. To prepare samples for selective growth, a 0.1-μm-thick Si$_3$N$_4$ masking layer was deposited on the wafer by plasma-enhanced chemical vapor deposition. Arrays of openings were created in this mask by photolithography and reactive ion etching. The openings were arranged in a hexagonal pattern with a spacing of 20 μm, and the average diameter of the openings ranged from 2 to 5 μm, depending on the size of the pyramid to be grown. GaN regrowth was then performed in the MOCVD reactor. Details on growth conditions are given in [180]. The result of the selective lateral overgrowth was a two-dimensional array of GaN pyramids. A scanning electron microscope (SEM) image of one of the samples is shown in Figure 83. The base diameters of the pyramids were estimated to be about 5 and 15 μm for the two different arrays of pyramids, which is considerably larger than the corresponding 2- and 5-μm openings in the mask, indicating substantial lateral growth of the pyramids. Transmission electron microscope pictures revealed a drastic reduction in defect densities.

The samples were mounted on a translation stage that allowed for a three-dimensional positioning of the sample with ~ 1-μm resolution. The third harmonic of an injection-seeded Q-switched Nd:YAG laser was used as the pumping source. The pulse width of the laser was varied from 5 to 25 ns by adjusting the Q-switch delay. The laser beam was focused to a spot with a diameter of 4 μm through a microscope objective. The laser light intensity was continuously attenuated using

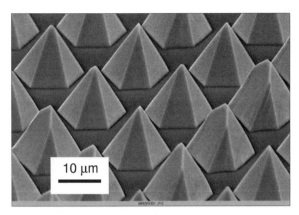

Fig. 83. SEM image of GaN pyramids with a 15-μm-wide hexagonal base, grown on a (111) Si substrate by selective lateral overgrowth.

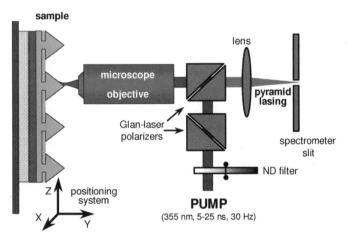

Fig. 84. Experimental geometry. The pyramids were individually pumped, imaged, and spectrally analyzed through a high-magnification telescope system. Two cross-polarized Glan-laser prisms were used to avoid a leak of the pump beam into the detection system.

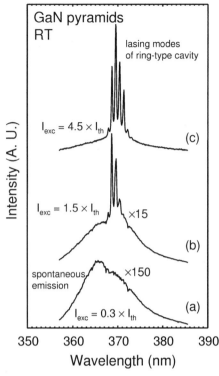

Fig. 85. Emission spectra of a 15-μm-wide GaN pyramid under different levels of optical pumping above and below the lasing threshold at room temperature. At pump densities below the lasing threshold, only a 14-nm-wide spontaneous emission peak is present, whereas at excitation levels above the lasing threshold, multimode laser action is observed.

a variable neutral density filter. This study was performed in a surface-emitting geometry where emission from the sample was collected through the same microscope objective in the direction normal to the sample surface as shown in Figure 84. Two cross-polarized Glan-laser prisms were used to avoid a leak of the pump beam into the detection system. A 30× magnified image of the sample was projected onto the slits of a Spex 1-m spectrometer and the emission spectra were recorded by an optical multichannel analyzer. The overall spatial resolution of the system was better than 5 μm. This configuration allowed us to pump, image, and spectrally analyze emission from separate pyramids.

Figure 85 shows the RT emission spectra for a 15-μm-wide GaN pyramid at several different pump densities near the lasing threshold. Note that these spectra correspond to emission from a single pyramid. At excitation densities below the lasing threshold, only a spontaneous emission peak with a FWHM of approximately 14 nm is present, as shown in Figure 85a. The energy position and spectral width of the peak are very similar to those observed from high-quality single-crystal GaN epilayers [41]. As the pump density is increased to values slightly

above the lasing threshold, several equally spaced narrow peaks with FWHMs of less than 0.3 nm appear, as illustrated in Figure 85b. The position of the peaks remains the same as the excitation density is raised. Due to the pump density dependence of the effective gain profile, the maximum intensity mode tends to hop to modes with lower energies as shown in Figure 85c.

The intensity dependence of the spontaneous and lasing peaks was studied as a function of excitation power. The integrated intensity of the spontaneous emission peak was observed to increase almost linearly with pump density ($I_{spon} = I_{pump}^{0.93}$) as illustrated in Figure 86 (open squares) for a 15-μm-wide GaN pyramid. On the other hand, the intensity of the lasing peaks experiences a strong superlinear increase ($I_{lasing} = I_{pump}^{2.8}$) with excitation power (open circles). The lasing threshold corresponds to an incident pump density of approximately 25 MW/cm². Such a high value of pump density does not accurately represent the real lasing threshold because most of the pump beam scatters off the surface of the pyramid. To estimate the coupling coefficient of the pump beam, we have to consider the surface roughness and the geometry of each specific pyramid.

Note that the spontaneous emission in Figure 85a does not exhibit a mode structure, indicating that it does not come from a laser cavity inside of the pyramid, but rather from the surface. On the other hand, the lasing peaks (as shown in Fig. 85b and c)

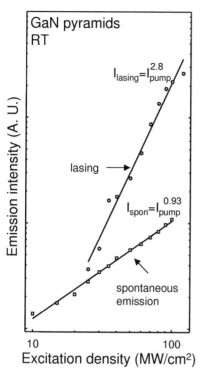

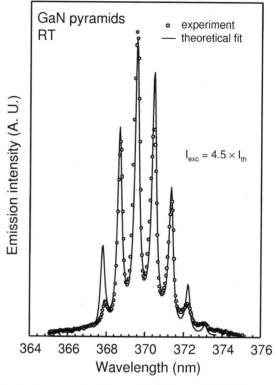

Fig. 86. Peak intensity as a function of excitation density. The intensity of the spontaneous emission peak (open squares) increases almost linearly with excitation power. The lasing peak intensity (open circles) exhibits a strong superlinear increase as the pump density is raised. The solid lines represent linear fits to the experimental data.

Fig. 87. Lasing modes in a 15-μm-wide GaN pyramid. High-finesse modes (open circles) were observed when the sample was excited above the lasing threshold. The solid line represents a fit to the experimental data with the assumption of a Gaussian gain profile. The perimeter of the cavity was estimated to be 58 μm.

appear to be modes from a high-finesse cavity. The perimeter of either a ring or standing-wave cavity that gives rise to these modes, taking into account the dispersion of the refractive index with wavelength, can be evaluated as [181]

$$p = \frac{\lambda^2}{(n - \lambda(dn/d\lambda))\Delta\lambda} = \frac{\lambda^2}{\bar{n}\Delta\lambda} = 58\ \mu\text{m} \qquad (7.2)$$

where $\lambda = 370$ nm is the center lasing wavelength, $\Delta\lambda = 0.89$ nm is the mode spacing, and $\bar{n} = n - \lambda(dn/d\lambda) = 2.65$ is the effective refractive index of GaN at 370 nm (taken from [182]). The high finesse lasing modes (open circles) were fitted by assuming a Gaussian profile. The result of the fit is shown Figure 87 as a solid line. The perimeter of the cavity appears to be several times larger than the diameter of the pyramid base. This result suggests that the photon flux propagating in the cavity undergoes several ($N \geq 4$) reflections inside of the pyramid prior to completing one round-trip.

For a semiconductor laser, the typical turn-on delay time is on the order of nanoseconds [183]. In addition, the electron and photon populations undergo oscillations before attaining their steady-state values. In practice these oscillations might require additional time (up to 15 ns) to become sufficiently dampened before mode intensities reach their steady-state values [184]. These initial transients are on the same order as our 6-ns excitation pulse. During the course of the electron and photon population oscillations, a considerable number of lasing modes experience temporary amplification. It is plausible that the observed multimode laser action in the GaN pyramids (as shown in Fig. 87c) is due to transient conditions associated with short-pulse excitation. Note that a large inhomogeneous broadening of the gain region (necessary to create a wide gain region) is not expected due to the high structural quality of GaN pyramids. This further corroborates the point that the width of the gain region (and thus multimode nature of lasing) may be the result of the transient conditions.

When the excitation pulse length was increased to ~20 ns, along with multimode laser action in the GaN pyramids, single-mode operation was also observed as shown in Figure 88 for a 15-μm-wide GaN pyramid. As the pump density was increased to values slightly above the lasing threshold, a very narrow peak with a FWHM of less than 0.3 nm appeared on the low-energy side of the spontaneous emission peak. Note that the FWHM of the SE peak in high-quality GaN epilayers is typically 2 nm at RT. As in the case of multimode operation, the intensity of the peak shown in Figure 88 increases superlinearly with excitation power.

A similar set of experiments was performed on the smaller pyramids. For the 5-μm-wide pyramids (Fig. 89b), the mode spacing increased to approximately 2.2 nm, which corresponds to a 23-μm cavity perimeter. The difference in mode spacing for the two different sized pyramids is inversely proportional to their physical dimensions (Fig. 89a and b).

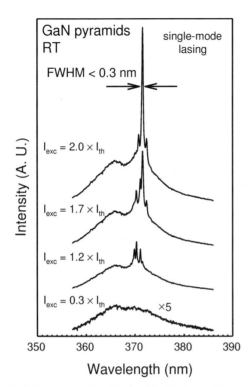

Fig. 88. Emission spectra of a 15-μm-wide GaN pyramid under different levels of optical pumping above and below the lasing threshold at room temperature (excitation pulse width is 25 ns). A single-mode laser peak of less than 0.3 nm FWHM was observed when the sample was pumped above the lasing threshold.

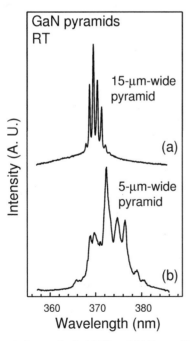

Fig. 89. Multimode laser action in (a) 15- and (b) 5-μm-wide pyramids. The mode spacings correspond to cavities of 58 and 23 μm, respectively.

To understand the geometry of the cavity formed inside the pyramids, we have to consider the limited penetration depth of the pump. It is expected that the gain region can lie only in

the vicinity of the surface rather than deep in the body of the pyramid. The existence of the highly modulated lasing spectra shown in Figure 87 with such a short gain region suggests that the losses associated with cavity mirror reflectivity and absorption at the lasing wavelength are very small. For multiple reflections, only highly reflective mirrors could provide a reasonable optical feedback. Under normal incidence, a GaN–air interface reflects approximately $R = 0.2$ of the incident signal. Having N near-normal reflections results only in $R^N \leq 10^{-3}$ (for $N \geq 4$) of the transmitted signal after one round trip in the cavity, which requires unrealistically high gain in a standing wave cavity with normally oriented end mirrors. Therefore, the cavity inside of the pyramid is most likely ring type formed by total internal reflections off the pyramid surfaces. This cavity could potentially lead to a large build-up of the electric field in the pyramid. The collected emission from the sample is believed to be only a scattered fraction of this field, as in the case of the scattered SE observed in GaN epilayers [109].

Picosecond time-resolved photoluminescence spectroscopy was used by Zeng et al. [185] to further investigate the optical properties of GaN pyramids. They found that (i) the release of the biaxial compressive strain in overgrown GaN pyramids on GaN/AlN/sapphire led to a 7-meV red shift of the spectral peak position with respect to a strained GaN epilayer grown under identical conditions; (ii) in the GaN pyramids on GaN/AlN/sapphire, strong band-edge transitions with much narrower linewidths than those in the GaN epilayer were observed, indicating the improved crystalline quality of the overgrown pyramids; (iii) PL spectra taken from different parts of the pyramids revealed that the top of the pyramid had the highest crystalline quality; and (iv) the presence of strong band-to-impurity transitions in the pyramids was primarily due to the incorporation of oxygen and silicon impurities from the SiO_2 mask. Studies of the optical modes in a ring-type three-dimensional microcavity formed inside the pyramids also have been reported in the literature [186].

Pyramids used in this study are much smaller than conventional edge-emitting LD cavities. Even though efficient carrier injection and emission coupling have yet to be developed, these GaN microstructures have the potential to be used as pixel elements and as high-density two-dimensional laser arrays [187, 188]. The Si substrate used to grow the pyramids might facilitate the integration of GaN microstructures into Si-based electronics.

8. IMAGING TECHNIQUES FOR WIDE-BANDGAP SEMICONDUCTORS

8.1. Transverse Lasing Modes in GaN-Based Lasing Structures

In a laser diode it is desirable to have a well-designed laser cavity that produces a narrow zero-order transverse mode pattern with smooth transverse amplitude, even phase profiles, and low diffraction losses. It is equally desirable to obtain a laser with the lowest lasing threshold possible. To achieve these objectives, the geometry of the laser cavity should be optimized, paying special attention to the waveguiding properties of the confinement layers. This section deals with issues related to near- and far-field SE patterns of GaN-based lasing structures and describes a novel experimental technique for evaluating optical confinement in wide-bandgap semiconductors.

The beam properties of a laser diode become critical when it is necessary to couple the laser output into another optical element. For example, a smaller beam divergence can increase the coupling efficiency of a laser diode to an optical fiber. In the DVD optical storage system (and even more so in DVD/HD), it is also necessary to reduce the perpendicular divergence angle to below 25° [189]. However, in the conventional separate confinement heterostructure laser, the tight optical confinement normally results in a large beam divergence in the direction perpendicular to the junction. Such a large transverse beam divergence leads to a low coupling efficiency and a highly asymmetric elliptical far-field pattern. To overcome this problem, different structures have been used to expand the size of the transverse electrical field in the optical cavities. This leads to improvements in both the near-field (Fresnel) and far-field (Fraunhofer) regions. However, the decrease of the far-field angle normally results in a higher threshold current. This higher current can cause power dissipation and heat generation problems that jeopardize the realization of the much longer lifetime that is essential for commercial applications of laser diodes. Only very limited attention has been paid to strict lateral carrier and photon confinement in InGaN-based laser diodes, because no reliable technique for measuring optical confinement exists.

Current state-of-the-art InGaN/GaN multiple quantum well laser diodes have many problems associated with poor optical confinement. The inferior optical confinement results in observation of higher order transverse modes, especially when the output power is increased to values substantially above the lasing threshold [190]. To suppress higher order modes, the transverse mode must be forced to be the fundamental mode over the entire range of operating currents by further narrowing the ridge width and improving optical confinement.

To understand emission profiles of blue LDs, intensity profiles of the near- and far-field patterns from InGaN/GaN multiple quantum well structures should be examined under high optical excitation. The samples in this study were pumped using the third harmonic of an Nd:YAG laser (355 nm, 6 ns). The emission from the sample edge was projected onto a Spiricon CCD camera using a $100\times$ microscope objective with NA = 0.9. The signal was averaged over 500 laser pulses and converted into a two-dimensional array of numbers for intensity analysis using a custom-programmed digital filter.

Typical emission intensity profiles when the sample was excited above the SE threshold are shown in Figure 90. The near-field pattern (Fig. 90a) resembles a line that lies along the sample surface. However, the far-field pattern bears the signature of a TEM$_{01}$ mode, consisting of two distinct bright spots. An example of such a pattern taken at a distance of 20 μm from the sample edge is shown in Figure 90b. The divergence of the

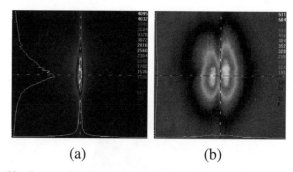

(a)   (b)

Fig. 90. Images of the beam propagation from (a) near-field to (b) 20 μm from the sample taken from an InGaN/GaN MQW sample at room temperature under pulsed excitation. Improvements in optical confinement are necessary to improve the far-field pattern of the laser emission.

beam in the direction perpendicular to the surface was estimated to be $\sim$35°.

To elucidate these experimental observations, we could perform simple computer simulations. Higher order transverse modes are usually a Hermite–Gaussian type in rectangular coordinates or Laguerre–Gaussian in cylindrical coordinates [191]. Preliminary analysis of the far-field pattern depicted in Figure 90b suggested that it did not represent the fundamental mode, but rather a higher order TEM mode.

The free-space Hermite–Gaussian TEM$_{nm}$ solutions can be written in the form [191]

$$\bar{u}_n(x, z) \propto H_n\left(\frac{\sqrt{2}x}{\omega(zt)}\right) \exp\left[-ikz - i\frac{kx^2}{2R(z)} - \frac{x^2}{\omega^2(z)}\right] \tag{8.1}$$

where $\omega(z)$ is the spot size and $R(z)$ is the radius of curvature. The first two (unnormalized) Hermitian polynomials are given by

$$H_0 = 1 \quad \text{and} \quad H_1(x) = 2x \tag{8.2}$$

The information given in Eqs. (8.1) and (8.2) can now be used to perform a computer simulation of the TEM$_{01}$ mode. The result of such a modeling is depicted in Figure 91. Note that the computer simulation depicted in Figure 91 closely resembles (at least qualitatively) the far-field pattern shown in Figure 90b.

Whereas the far-field pattern observed from the InGaN/GaN multiple quantum well sample is not in the shape of a Gaussian, it is reasonable to conclude that the fundamental mode (TEM$_{00}$) could not be observed due to poor optical confinement of the sample emission at excitation densities above the SE threshold. Thus, further improvement in waveguiding characteristics is needed. Unfortunately, the intensity profiles do not contain information about the depth of the optically confined region, and the "leak" of stimulated emission into the buffer–substrate region is rather difficult to estimate from this technique.

8.2. Novel Technique for Evaluating Optical Confinement in GaN-Based Lasing Structures

Although numerous theoretical models for evaluation of optical confinement have been given in the literature (see, e.g., [192]),

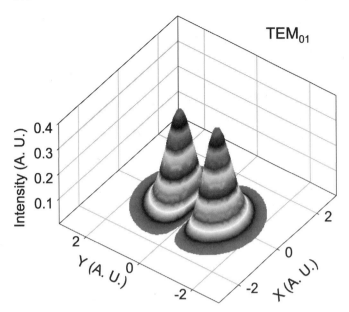

Fig. 91. Computer simulation of the TEM_{01} mode using the Hermite–Gaussian solutions given by Eq. (8.1).

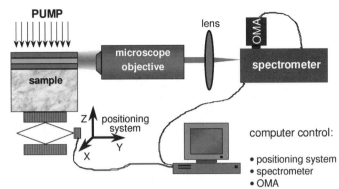

Fig. 92. Experimental configuration. A microscope assembly (NA = 0.9) was used to project the image of the sample onto the spectrometer slits. The microscope assembly was mounted on a translation stage with submicrometer resolution for scanning the sample perpendicular to the active layer.

experimental techniques have been limited mostly to measuring the beam divergence and extrapolating to obtain the thickness of the confinement layer under the assumption of Gaussian fields as was described previously. The limitations of this technique prompted one research group to develop an advanced measurement technique that can resolve the laser emission not only spatially, but also spectrally [60], by combining positioning and dispersion instruments in a single high-resolution experimental setup. An evaluation of the optical confinement in GaN-based lasing structures was performed by spectrally resolving the near-field pattern observed on the facets of the laser structures. It has been demonstrated that the confinement layers introduce a unique signature into the near-band-edge emission spectra that can be used to deduce the optical confinement characteristics. Information extracted through this technique can be applied toward lowering the internal optical losses and improving the transverse beam profile by eliminating optical leakage and the presence of higher order modes.

The configuration of the experimental setup is shown in Figure 92. A tunable dye laser pumped by a frequency-doubled injection-seeded Nd:YAG laser was used as the primary optical pumping source. The visible output of the dye laser was frequency doubled using a nonlinear crystal to achieve a near-UV pumping frequency. The laser beam was focused with a cylindrical lens into a line on the sample surface. The near-field pattern of the emission from the edge of the sample was imaged and spectrally resolved using a high-magnification optical assembly (NA = 0.9) in conjunction with a 1-m spectrometer and an optical multichannel analyzer. The optical assembly was mounted on a translation stage driven by a computer-controlled stepper motor that allowed the microscope objective to be positioned with submicrometer resolution relative to the sample. The spectral resolution of the system was better than 0.1 nm.

The samples used in this study were MOCVD-grown GaN films with thicknesses ranging from 1.7 to 7.2 μm as well as a GaN/AlGaN separate confinement heterostructure. The details of the sample growth were reported in Sections 4.3 and 5.1, respectively. Experiments were performed at RT with excitation power densities of two times the SE/lasing thresholds for each sample studied. The SCH exhibited low-threshold lasing at RT due to microcavity effects as described in Section 4.3.

Drastic changes in the emission spectra were observed for different positions of the sample relative to the microscope objective while exciting samples above the SE/lasing threshold. A three-dimensional plot of the spectrally and spatially resolved emission from a 7.2-μm-thick GaN epilayer is shown in Figure 93. The horizontal axes represent the emission energy and the position of the microscope objective relative to the active layer. Scanning of the near-field pattern under optical pumping was performed in the growth direction (normal to the sample active layer). The vertical axis indicates the intensity of the sample emission on a logarithmic scale. Utilizing this scanning technique enabled us to resolve fine spectral features that otherwise would be averaged out in spatially unresolved experiments. Three-dimensional plots obtained in this way contain a substantial amount of information about the optical confinement characteristics of the structure under investigation.

Several different cross sections of the three-dimensional plot from Figure 93 are shown in Figure 94. When the emission is collected at a significant distance above the surface of the epilayer (Fig. 94a), only nonreabsorbed surface-scattered spontaneous emission is observed. As the microscope objective is moved closer to the sample surface, a weak SE peak appears on the low-energy side of the spontaneous emission peak (Fig. 94b). In this case, the SE is amplified in-plane and scattered off the sample edge (see Section 7.1). When the optical axis of the objective is aligned with the GaN epilayer, only a strong SE peak is observed (Fig. 94c), because spontaneous emission is entirely reabsorbed in the direction parallel to the surface.

The sample emission becomes strongly modulated when it is collected from the sapphire substrate just below the GaN epi-

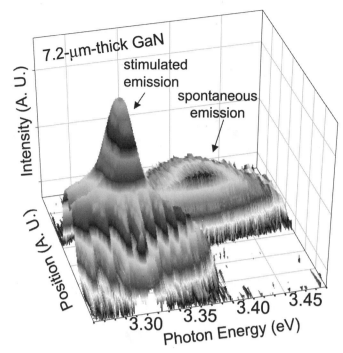

Fig. 93. Spatially resolved emission spectra from a 7.2-μm-thick GaN epilayer at excitation densities above the SE threshold. The horizontal axes indicate the emission photon energy and position of the microscope objective relative to the sample active layer. The vertical axis is the emission intensity plotted on a logarithmic scale.

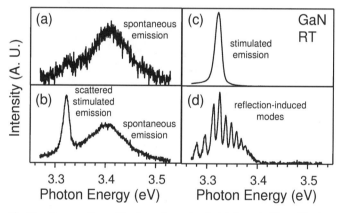

Fig. 94. Cross sections of the three-dimensional plot shown in Figure 93 taken at (a) substantially above the GaN epilayer, (b) slightly above the epilayer, (c) in line with the epilayer, and (d) slightly below the epilayer.

layer as shown in Figure 94d. This interference pattern results from multiple reflections within the active layer very close to the critical angle for total internal reflection between the epilayer and substrate interface, and the beams emerge at a grazing angle.

To understand this phenomenon the different refractive indices for the GaN epilayer, sapphire substrate, and air have to be considered. For simplicity, the refractive index of GaN is assumed to be $n \approx 2.4$. In general, however, we should account for the dispersion of the refractive index with wavelength, because SE is by its very nature close in photon energy to the

GaN bandgap. The photon flux that propagates inside the GaN epilayer is totally internally reflected off the GaN–air interface for angles larger than 25°. For the GaN–sapphire interface the incident angle has to be as steep as 49° for the beam to be totally reflected (assuming the refractive index of sapphire to be $n \approx 1.8$). Thus, for angles slightly below 49°, the photon flux is totally reflected off the air–GaN interface and is almost totally reflected off the sapphire–GaN interface as shown in Figure 95. Some of this light scatters off the sapphire–GaN interface and emerges at a grazing angle. This is the signal that is picked up by the apparatus and depicted in Figure 94d.

Therefore, this spatially resolved emission bears a spectral signature that can be used to deduce optical confinement characteristics. Combined with the near-band-edge dispersion of the GaN refractive index [193], this interference pattern can be used to obtain the thickness of the layer to which the emission is confined. In the simplest case (a GaN thin film), the fit of the experimental data into a tilted cavity Fabry–Pérot equation (see, e.g., [194]) yields the thickness of the epilayer. This has been confirmed to a high degree of accuracy in a variety of GaN samples as illustrated in Figure 96.

This work was extended to cover GaN/AlGaN SCH samples as shown in Figure 97. In addition to a strong lasing peak at 3.42 eV, weak modulated emission was observed in the low-energy spectral region of 3.31–3.40 eV. The observed mode spacing suggests that the modulated emission is due to the photon flux leaking from the waveguide region into the GaN buffer region as depicted in Figure 98.

A qualitative analysis of optical confinement can be given by further analyzing the data obtained by means of the technique described in this section. Whereas the edge emission is spectrally resolved, we can separately calculate the spectrally integrated intensity of the emission leaking into the buffer or substrate layers (I_{leak}) and the spectrally integrated intensity of SE/lasing ($I_{\text{SE/lasing}}$) for each position z of the microscope objective relative to the sample edge. The analysis can be extended by spatially integrating the SE/lasing and leak intensities, and evaluating the ratio of the integrals:

$$\Omega = \frac{\int I_{\text{SE/lasing}}(z)dz}{\int I_{\text{leak}}(z)dz} \qquad (8.3)$$

In general, Ω depends not only on the degree of optical confinement, but also on excitation power. Thus, it is necessary to maintain the power at the same level relative to the SE/lasing threshold for each of the samples. For the two spectra shown in Figures 93 and 97, the ratios were found to be $\Omega = 4$ and 10 for a GaN epilayer and a GaN/AlGaN SCH sample, respectively. The higher ratio corresponds to a higher degree of optical confinement.

Optical confinement has a significant effect on the lasing threshold and transverse beam profile of laser diodes. The described technique introduces a way to directly compare the quality of optical confinement in GaN-based lasing structures through the analysis of both spectrally and spatially resolved emission. By comparing the mode spacing as well as the inten-

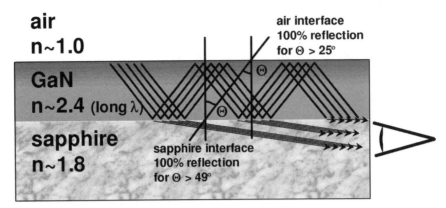

Fig. 95.   Formation of an interference pattern in a GaN epilayer. The difference in refractive indices for air, sapphire, and GaN creates a tilted Fabry–Pérot cavity.

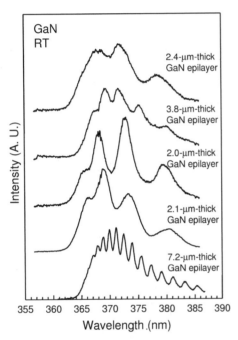

Fig. 96.   Edge emission from GaN epilayers with different thicknesses. Emission was taken with a microscope objective positioned slightly below the GaN epilayer. The spatially and spectrally resolved measurement technique gives precise information on the thicknesses of the epilayers.

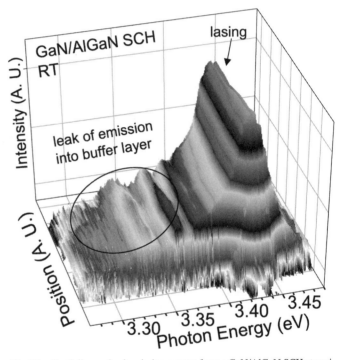

Fig. 97.   Spatially resolved emission spectra from a GaN/AlGaN SCH at excitation densities above the lasing threshold. Index guided modes that appear on the lower energy side of the lasing peak indicate a leak of emission from the waveguide region to the GaN substrate region.

sity ratio of the emissions, we gain insight into both the amount of emission leakage and the layer thicknesses.

9. SUMMARY

Many aspects of the III–nitrides have not been thoroughly explored yet. This chapter hopefully has answered many issues related to gain mechanisms, pump–probe experiments, microstructure lasing, imaging, and high-temperature optical properties of this material system, as well as provided some background information on the general properties of III–nitride thin films.

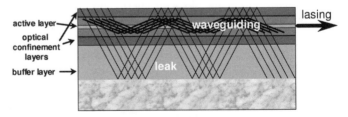

Fig. 98.   Mechanism of optical confinement and leak in a GaN/AlGaN separate confinement heterostructure. The observed mode spacing in Figure 97 suggests that the modulated emission is due to the photon flux leaking from the waveguide region into the GaN buffer region.

Nanosecond nondegenerate optical pump–probe absorption experiments on GaN and InGaN films were presented with particular emphasis on induced absorption and transparency in the band-edge region. Interestingly, the InGaN films did not exhibit induced absorption in the below-gap region as opposed to what was observed in GaN thin films. Instead, induced transparency in the differential absorption spectra was clearly seen in the spectral region where SE was observed. It was found that the high SE threshold of GaN is due to induced absorption and gain competition with increasing free carrier concentration. These issues have to be further studied before efficient optical switching devices based on the III–nitride thin films can be developed.

This chapter also covered the gain mechanisms in GaN thin films over the temperature range of 20–700 K. It was observed that for temperatures below 150 K, the dominant near-threshold gain mechanism is inelastic exciton–exciton scattering, characterized by a low SE threshold. For temperatures in excess of 150 K, the dominant gain mechanism was shown to be electron–hole plasma recombination, characterized by a relatively high SE threshold and a large separation between spontaneous and SE peaks.

The observation of SE in GaN at temperatures as high as 700 K was reported and issues pertinent to sample degradation were addressed. Stimulated emission at such high temperatures has never been observed in any other semiconductor system. By raising the temperature from 300 to 700 K, it is possible to achieve "tuning" of the SE peak in GaN from the near-UV spectral region to blue (a range of more than 40 nm). This unique property of SE might result in new types of optoelectronic devices that utilize this temperature tuning. Stimulated emission in InGaN/GaN multiple quantum wells was also described at elevated temperatures. The low sensitivity of the SE threshold creates opportunities for the development of light-emitting devices capable of high-temperature operation.

Achievements in efficient lasing in optically pumped GaN/AlGaN separate confinement heterostructures with the shortest lasing wavelength reported to date from a GaN/AlGaN-based structure also were given in this chapter. Remarkably low values of the lasing threshold were measured over a wide temperature range. Through an analysis of the relative shift between spontaneous and lasing peaks, combined with the temperature dependence of the lasing threshold, it was demonstrated that exciton–exciton scattering is the dominant gain mechanism that leads to low-threshold ultraviolet lasing over the entire temperature range studied.

Both the spontaneous and SE properties of InGaN/GaN multiple quantum well structures and InGaN epilayers were systematically investigated. The findings were explained in the context of the localization of photogenerated carriers associated with strong potential fluctuations in the InGaN-active region. The efficiency of state-of-the-art blue laser diodes and the origin of gain are believed to be due to the recombination of localized carriers from potential fluctuations of indium in the active layers.

Gain mechanisms in AlGaN alloys were also described in the temperature range of 10–300 K. Through a careful analysis of the carrier densities, it was concluded that the origin of gain is an electron–hole plasma recombination.

Study on the origin of surface-emitted SE in GaN epilayer films and a resolution to a longstanding controversy concerning GaN thin films also were given. Based on the experimental observations, it was clearly demonstrated that the surface-emitted SE reported by several groups is due to scattering of the photon flux that propagates parallel to the surface by defects. These results show that the influence of imperfections in GaN thin films cannot be ignored when determining gain values from experiment, and may be one of the reasons that there are large discrepancies in the gain values measured in different experimental geometries. The existence of microcracks in GaN-based structures (particularly those grown on SiC) affects the directionality, the spectral appearance, and the threshold of lasing. Consequently, one has to take into account the effects of these microstructures must be taken into account to interpret experimental data correctly.

The observation of single-mode and multimode laser action in GaN pyramids under strong optical pumping at RT also was presented. The cavity formed in a pyramid is three-dimensional ring type, introduced by total internal reflections of light off the pyramid surfaces. This study suggested that GaN microstructures have the potential to be used as pixel elements and high density two-dimensional laser arrays.

Finally, the intensity profiles for far-field emission from nitride-based structures were described. Good carrier and optical confinement were shown to be the critical parameters to maintain the fundamental Gaussian mode. A novel technique for evaluating the optical confinement in GaN-based lasing structures by studying their spectrally resolved near-field pattern under high optical excitation was introduced. This new technique may help to improve the transverse beam profile of GaN-based lasing structures.

Much of the research on the III–nitrides is still in progress. This chapter provides an up-to-date summary of research projects undertaken in this field at the time of this writing. The results reported in this chapter are directed toward future applications and aimed at explaining high-density carrier phenomena observed in this material system.

Acknowledgments

The authors are very grateful to Dr. Y. H. Cho for a comprehensive summary of his contributions to research on InGaN thin films and heterostructures. The authors also extend their gratitude to Dr. Y. H. Kwon, Dr. A. J. Fischer, J. B. Lam, and Dr. G. H. Gainer for their collaborative efforts. Finally, the authors thank Dr. B. D. Little for his editorial help.

REFERENCES

1. R. Dingle, K. L. Shaklee, R. F. Leheny, and R. B. Zetterstrom, *Appl. Phys. Lett.* 19, 5 (1971).

2. I. Akasaki, H. Amano, S. Sota, H. Sakai, T. Tanaka, and M. Koike, *Jpn. J. Appl. Phys.* 34, L1517 (1995).

3. H. Amano, M. Kito, K. Hiramatsu, and I. Akasaki, *Jpn. J. Appl. Phys.* 28, L2112 (1989).

4. S. Nakamura, M. Senoh, S. Nagahama, N. Iwasa, T. Yamada, T. Matsushita, H. Kiyoku, and Y. Sugimoto, *Jpn. J. Appl. Phys.* 35, L74 (1996).

5. J. Y. Duboz, Oral presentation, 3rd International Conference on Nitride Semiconductors, Montpellier, France, 1999, Mo_01.

6. S. Bidnyk, B. D. Little, Y. H. Cho, J. Krasinski, J. J. Song, W. Yang, and S. A. McPherson, *Appl. Phys. Lett.* 73, 2242 (1998).

7. S. Nakamura and G. Fasol, "The Blue Laser Diode," pp. 7–9. Springer-Verlag, New York, 1997.

8. A. J. Steckl and J. M. Zavada, *Mater. Res. Bull.* 24(9), 33 (1999).

9. Z. Huang, D. B. Mott, and P. K. Shu, *Tech. Briefs* 23(10), 50 (1999).

10. J. Schetzina, Oral presentation, The 6th Wide Bandgap III–Nitride Workshop, Richmond, VA, 2000, TP-2.0.

11. W. Yang, T. Nohava, S. Krishnankutty, R. Torreano, S. McPherson, and H. Marsh, *Appl. Phys. Lett.* 73, 978 (1998).

12. S. Bidnyk, B. D. Little, T. J. Schmidt, Y. H. Cho, J. Krasinski, J. J. Song, B. Goldenberg, W. G. Perry, M. D. Bremser, and R. F. Davis, *J. Appl. Phys.* 85, 1792 (1999).

13. S. Bidnyk, T. J. Schmidt, Y. H. Cho, G. H. Gainer, J. J. Song, S. Keller, U. K. Mishra, and S. P. DenBaars, *Appl. Phys. Lett.* 72, 1623 (1998).

14. T. Nishida and N. Kobayashi, "Proceedings of the 3rd International Conference on Nitride Semiconductors," Montpellier, France, 1999, p. 90.

15. K. P. O'Donnell, "Group III Nitride Semiconductor Compounds: Physics and Applications" (B. Gil, Ed.), pp. 1–18. Clarendon Press, Oxford, UK, 1998.

16. M. Tchounkeu, O. Briot, B. Gil, J. P. Alexis, and R. L. Aulombard, *J. Appl. Phys.* 80, 5352 (1996).

17. A. J. Fischer, W. Shan, J. J. Song, Y. C. Chang, R. Horning, and B. Goldenberg, *Appl. Phys. Lett.* 71, 1981 (1997).

18. W. Shan, B. D. Little, A. J. Fischer, J. J. Song, B. Goldenberg, W. G. Perry, M. D. Bremser, and R. F. Davis, *Phys. Rev. B* 54, 16,369 (1996).

19. C. Klingshirn and H. Haug, *Phys. Rep.* 70, 315 (1981).

20. H. Haag, P. Gilliot, D. Ohlmann, R. Levy, O. Briot, and R. L. Aulombard, *MRS Internet J. Nitride Semicond. Res.* 2, Art. 21 (1997).

21. H. Haag, P. Gilliot, R. Levy, B. Honerlage, O. Briot, S. Ruffenach-Clur, and R. L. Aulombard, *Phys. Rev. B* 59, 2254 (1999).

22. H. Haag, P. Gilliot, R. Levy, B. Honerlage, O. Briot, S. Ruffenach-Clur, and R. L. Aulombard, *Appl. Phys. Lett.* 74, 1436 (1999).

23. T. J. Schmidt, J. J. Song, Y. C. Chang, R. Horning, and B. Goldenberg, *Appl. Phys. Lett.* 72, 1504 (1998).

24. T. J. Schmidt, Y. C. Chang, and J. J. Song, *Proc. SPIE* 3419, 61 (1998).

25. T. J. Schmidt, J. J. Song, U. K. Mishra, S. P. DenBaars, and W. Yang, *Mater. Res. Soc. Symp. Proc.* 572, 433 (1999).

26. S. D. Lester, F. A. Ponce, M. G. Craford, and D. A. Steigerwald, *Appl. Phys. Lett.* 66, 1249 (1995).

27. S. D. Hersee, J. C. Ramer, and K. J. Malloy, *Mater. Res. Bull.* 22, 45 (1997).

28. B. Gil, O. Briot, and R. L. Aulombard, *Phys. Rev. B* 52, R17,028 (1995).

29. W. Reiger, T. Metzger, H. Angerer, R. Dimitrov, O. Ambacher, and M. Sturtzmann, *Appl. Phys. Lett.* 68, 970 (1996).

30. W. Shan, A. J. Fischer, J. J. Song, G. Bulman, H. S. Kong, M. T. Leonard, W. C. Perry, M. D. Bremser, and R. F. Davis, *Appl. Phys. Lett.* 69, 740 (1996).

31. D. Volm, K. Oettinger, T. Streibl, D. Kovakev, M. Ben-Chorin, J. Diener, B. K. Meyer, J. Majewski, L. Eckey, A. Koffman, H. Amano, I. Akasaki, K. Hiramatsu, and T. Detchprohm, *Phys. Rev. B* 53, 16,543 (1996).

32. Y. Varshni, *Physica* 34, 149 (1967).

33. B. Taheri, J. Hays, and J. J. Song, *Appl. Phys. Lett.* 68, 587 (1996).

34. H. Haug and S. Schmitt-Rink, *J. Opt. Soc. Am. B* 2, 1135 (1985).

35. A. J. Fischer, B. D. Little, T. J. Schmidt, J. J. Song, R. Horning, and B. Goldenberg, *Proc. SPIE* 3624, 179 (1999).

36. T. J. Schmidt, A. J. Fischer, and J. J. Song, *Phys. Status Solidi B* 215, 505 (1999).

37. G. W. Fehrenbach, W. Schaefer, and R. G. Ulbrich, *J. Lumin.* 30, 154 (1985).

38. C. V. Shank, R. L. Fork, R. F. Leheny, and J. Shah, *Phys. Rev. Lett.* 42, 112 (1979).

39. J. J. Song and W. Shan, "Group III Nitride Semiconductor Compounds: Physics and Applications" (B. Gil, Ed.), pp. 182–241. Clarendon Press, Oxford, UK, 1998.

40. W. Shan, X. C. Xie, J. J. Song, and B. Goldenberg, *Appl. Phys. Lett.* 67, 2512 (1995).

41. X. H. Yang, T. J. Schmidt, W. Shan, J. J. Song, and B. Goldenberg, *Appl. Phys. Lett.* 66, 1 (1995).

42. W. Shan, T. J. Schmidt, X. H. Yang, S. J. Hwang, J. J. Song, and B. Goldenberg, *Appl. Phys. Lett.* 66, 985 (1995).

43. T. Takeuchi, H. Takeuchi, S. Sota, H. Amano, and I. Akasaki, *Jpn. J. Appl. Phys.* 36, L177 (1997).

44. M. D. McCluskey, C. G. Van deWalle, C. P. Master, L. T. Romano, and N. M. Johnson, *Appl. Phys. Lett.* 72, 2725 (1998).

45. H. Buttner, *Phys. Status Solidi* 42, 775 (1970).

46. C. Benoit à la Guillaume, J. M. Debever, and F. Salvan, *Phys. Rev.* 177, 567 (1969).

47. I. M. Catalano, A. Cingolani, M. Ferrara, M. Lugarà, and A. Minafra, *Solid State Commun.* 25, 349 (1978).

48. J. Holst, L. Eckey, A. Hoffman, I. Broser, B. Schöttker, D. J. As, D. Schikora, and K. Lischka, *Appl. Phys. Lett.* 72, 1439 (1998).

49. I. Galbraith and S. W. Koch, *J. Crystal Growth* 159, 667 (1996).

50. See, for example, D. Wood, "Optoelectronic Semiconductor Devices," pp. 136–138. Prentice–Hall, New York, 1994.

51. R. Levy and J. B. Grun, *Phys. Status Solidi A* 22, 11 (1974).

52. X. H. Yang, J. M. Hays, W. Shan, J. J. Song, and E. Cantwell, *Appl. Phys. Lett.* 62, 1071 (1992).

53. T. Uenoyama, *Phys. Rev. B* 51, 10228 (1995).

54. R. Dingle, "Semiconductors and Semimetals," Vol. 24, pp. 18–23. Academic Press, New York, 1987.

55. H. Jeon, J. Ding, A. V. Nurmikko, H. Luo, N. Samarth, and J. K. Furdyna, *Appl. Phys. Lett.* 57, 2413 (1990).

56. T. J. Schmidt, X. H. Yang, W. Shan, J. J. Song, A. Salvador, W. Kim, Ö. Aktas, A. Botchkarev, and H. Morkoç, *Appl. Phys. Lett.* 68, 1820 (1996).

57. S. Nakamura, M. Senoh, S. Nagahama, N. Iwasa, T. Matushita, and T. Mukai, *MRS Internet J. Nitride Semicond. Res.* 4S1, G1.1 (1999).

58. J. J. Song, A. J. Fischer, T. J. Schmidt, S. Bidnyk, and W. Shan, *Nonlinear Optics* 18, 269 (1997).

59. S. Bidnyk, T. J. Schmidt, B. D. Little, and J. J. Song, *Appl. Phys. Lett.* 74, 1 (1999).

60. S. Bidnyk, B. D. Little, J. J. Song, and T. Schmidt, *Appl. Phys. Lett.* 75, 2163 (1999).

61. S. Bidnyk, J. B. Lam, B. D. Little, Y. H. Kwon, J. J. Song, G. E. Bulman, H. S. Kong, and T. J. Schmidt, *Appl. Phys. Lett.* 75, 3905 (1999).

62. L. Eckey, J. Holst, V. Kutzer, A. Hoffmann, I. Broser, O. Ambacher, M. Stutzmann, H. Amano, and I. Akasaki, *Mater. Res. Soc. Symp. Proc.* 468, 237 (1997).

63. T. J. Schmidt, Y. H. Cho, J. J. Song, and W. Yang, *Appl. Phys. Lett.* 74, 245 (1999).

64. T. J. Schmidt, Y. H. Cho, G. H. Gainer, J. J. Song, S. Keller, U. K. Mishra, and S. P. DenBaars, *Appl. Phys. Lett.* 73, 1892 (1998).

65. P. Riblet, H. Hirayama, A. Kinoshita, A. Hirata, and T. Sugano, *Appl. Phys. Lett.* 75, 2241 (1999).

66. H. S. Kim, R. H. Mair, J. Li, J. Y. Lin, and H. X. Jiang, *Appl. Phys. Lett.* 76, 1252 (2000).

67. J. J. Song and W. Shan, "Group III Nitride Semiconductors Compounds: Physics and Applications" (B. Gil, Ed.), pp. 186–191. Clarendon, Oxford, UK, 1998.

68. Y. H. Cho, G. H. Gainer, A. J. Fischer, J. J. Song, S. Keller, U. K. Mishra, and S. P. DenBaars, *Appl. Phys. Lett.* 73, 1370 (1998).

69. S. Chichibu, T. Azuhata, T. Sota, and S. Nakamura, *Appl. Phys. Lett.* 69, 4188 (1996); 70, 2822 (1997).

70. E. S. Jeon, V. Kozlov, Y. -K. Song, A. Vertikov, M. Kuball, A. V. Nurmikko, H. Liu, C. Chen, R. S. Kern, C. P. Kuo, and M. G. Craford, *Appl. Phys. Lett.* 69, 4194 (1996).

71. P. Perlin, V. Iota, B. A. Weinstein, P. Wisniewski, T. Suski, P. G. Eliseev, and M. Osinski, *Appl. Phys. Lett.* 70, 2993 (1997).

72. Y. H. Cho, J. J. Song, S. Keller, M. S. Minsky, E. Hu, U. K. Mishra, and S. P. Denbaars, *Appl. Phys. Lett.* 73, 1128 (1998).

73. S. F. Chichibu, A. C. Abare, M. S. Minsky, S. Keller, S. B. Fleisher, J. E. Bowers, E. Hu, U. K. Mishra, L. A. Coldren, S. P. Denbaars, and T. Sota, *Appl. Phys. Lett.* 73, 2006 (1998).

74. Y. Narukawa, Y. Kawakami, M. Funato, Sz. Fujita, Sg. Fujita, and S. Nakamura, *Appl. Phys. Lett.* 70, 981 (1997).

75. Y. Narukawa, Y. Kawakami, Sz. Fujita, Sg. Fujita, and S. Nakamura, *Phys. Rev. B* 55, R1938 (1997).

76. C. Wetzel, T. Takeuchi, S. Yamaguchi, H. Katoh, H. Amano, and I. Akasaki, *Appl. Phys. Lett.* 73, 1994 (1998).

77. S. Chichibu, K. Wada, and S. Nakamura, *Appl. Phys. Lett.* 71, 2346 (1997).

78. X. Zhang, D. H. Rich, J. T. Kobayashi, N. P. Kobayashi, and P. D. Dapkus, *Appl. Phys. Lett.* 73, 1430 (1998).

79. F. A. Ponce, D. P. Bour, W. Goltz, and P. J. Wright, *Appl. Phys. Lett.* 68, 57 (1996).

80. S. J. Rosner, E. C. Carr, M. J. Ludowise, G. Girolami, and H. I. Erikson, *Appl. Phys. Lett.* 70, 420 (1997).

81. S. J. Rosner, G. Girolami, H. Marchand, P. T. Fini, J. P. Ibbetson, L. Zhao, S. Keller, U. K. Mishra, S. P. DeaBaars, and J. S. Speck, *Appl. Phys. Lett.* 74, 2035 (1999).

82. D. W. Pohl, W. Denk, and M. Lanz, *Appl. Phys. Lett.* 44, 651 (1984).

83. U. Dürig, D. W. Pohl, and F. Rohner, *J. Appl. Phys.* 59, 3318 (1986).

84. E. Betzig, J. K. Trautman, T. D. Harris, J. S. Weiner, and R. L. Kostelak, *Science* 251, 1468 (1991).

85. E. Betzig and J. K. Trautman, *Science* 257, 189 (1992).

86. M. A. Paesler and P. Moyer, "Near-Field Optics: Theory, Instrumentation, and Applications," Wiley, New York, 1996.

87. P. A. Crowell, D. K. Young, S. Keller, E. L. Hu, and D. D. Awschalom, *Appl. Phys. Lett.* 72, 927 (1998).

88. A. Vertikov, M. Kuball, A. V. Nurmikko, Y. Chen, and S.-Y. Wang, *Phys. Lett.* 72, 2645 (1998).

89. A. Vertikov, A. V. Nurmikko, K. Doverspike, G. Bulman, and J. Edmond, *Appl. Phys. Lett.* 73, 493 (1998).

90. D. L. Abraham, A. Veider, Ch. Schonenberger, H. P. Meier, D. J. Arent, and S. F. Alvarado, *Appl. Phys. Lett.* 56, 1564 (1990).

91. B. Garni, J. Ma, N. Perkins, J. Liu, T. F. Kuech, and M. G. Lagally, *Appl. Phys. Lett.* 68, 1380 (1996).

92. S. Evoy, C. K. Harnett, H. G. Craighead, S. Keller, U. K. Mishra, and S. P. DenBaars, *Appl. Phys. Lett.* 74, 1457 (1999).

93. F. Bernardini, V. Fiorentini, and D. Vanderbilt, *Phys. Rev. B* 56, R10,024 (1997).

94. F. Bernardini and V. Fiorentini, *Phys. Rev. B* 57, R9427 (1998).

95. V. Fiorentini, F. Bernardini, F. D. Sala, A. D. Carlo, and P. Lugli, *Phys. Rev. B* 60, 8849 (1999).

96. M. Leroux, N. Grandjean, M. Laügt, J. Massies, B. Gil, P. Lefebvre, and P. Bigenwald, *Phys. Rev. B* 58, R13,371 (1998).

97. M. Leroux, N. Grandjean, J. Massies, B. Gil, P. Lefebvre, and P. Bigenwald, *Phys. Rev. B* 60, 1496 (1999).

98. T. Takeuchi, S. Sota, M. Katsuragawa, M. Komori, H. Takeuchi, H. Amano, and I. Akasaki, *Jpn. J. Appl. Phys., Part 2* 36, L382 (1997).

99. T. Takeuchi, C. Wetzel, S. Yamaguchi, H. Sakai, H. Amano, I. Akasaki, Y. Kaneko, S. Nakagawa, Y. Yamaoka, and N. Yamada, *Appl. Phys. Lett.* 73, 1691 (1998).

100. J. S. Im, H. Kollmer, J. Off, A. Sohmer, F. Scholz, and A. Hangleiter, *Phys. Rev. B* 57, R9435 (1998).

101. A. Hangleiter, J. S. Im, H. Kollmer, S. Heppel, J. Off, and F. Scholz, *MRS Internet J. Nitride Semicond. Res.* 3, 15 (1998).

102. H. Kollmer, J. S. Im, S. Heppel, J. Off, F. Scholz, and A. Hangleiter, *Appl. Phys. Lett.* 74, 82 (1999).

103. R. Langer, J. Simon, O. Konovalov, N. Pelekanos, A. Barski, and M. Leszczynski, *MRS Internet J. Nitride Semicond. Res.* 3, 46 (1998).

104. D. L. Smith and C. Mailhiot, *Phys. Rev. Lett.* 58, 1264 (1987).

105. D. A. B. Miller, D. S. Chemla, T. C. Damen, A. C. Gossard, W. Wiegmann, T. H. Wood, and C. A. Burrus, *Phys. Rev. B* 32, 1043 (1985).

106. Y. H. Cho, F. Fedler, R. J. Hauenstein, G. H. Park, J. J. Song, S. Keller, U. K. Mishra, and S. P. DenBaars, *J. Appl. Phys.* 85, 3006 (1999).

107. R. W. Martin, P. G. Middleton, K. P. O'Donnel, and W. Van der Stricht, *Appl. Phys. Lett.* 74, 263 (1999).

108. K. P. O'Donnell, R. W. Martin, and P. G. Middleton, *Phys. Rev. Lett.* 82, 237 (1999).

109. S. Bidnyk, T. J. Schmidt, G. H. Park, and J. J. Song, *Appl. Phys. Lett.* 71, 729 (1997).

110. C. I. Harris, B. Monemar, H. Amano, and I. Akasaki, *Appl. Phys. Lett.* 67, 840 (1995).

111. C. K. Sun, S. Keller, G. Wang, M. S. Minsky, J. E. Bowers, and S. P. DenBaars, *Appl. Phys. Lett.* 69, 1936 (1996).

112. C. K. Sun, T. L. Chiu, S. Keller, G. Wang, M. S. Minsky, S. P. Denbaars, and J. E. Bowers, *Appl. Phys. Lett.* 71, 425 (1997).

113. J. S. Im, V. Härle, F. Scholz, and A. Hangleiter, *MRS Internet J. Nitride Semicond. Res.* 1, 37 (1996).

114. B. K. Ridley, *Phys. Rev. B* 41, 12,190 (1990).

115. J. Feldmann, G. Peter, E. O. Göbel, P. Dawson, K. Moore, C. Foxon, and R. J. Elliott, *Phys. Rev. Lett.* 59, 2337 (1987).

116. P. Lefebvre, J. Allègre, B. Gil, A. Kavokine, H. Mathieu, W. Kim, A. Salvador, A. Botchkarev, and H. Morkoç, *Phys. Rev. B* 57, R9447 (1998).

117. Y. Narukawa, S. Saijou, Y. Kawakami, S. Fujita, T. Mukai, and S. Nakamura, *Appl. Phys. Lett.* 74, 558 (1999).

118. Y. H. Kwon, G. H. Gainer, S. Bidnyk, Y. H. Cho, J. J. Song, M. Hansen, and S. P. Denbaars, *Appl. Phys. Lett.* 75, 2545 (1999).

119. A. Krost, J. Böhrer, A. Dadgar, R. F. Schnabei, D. Bimberg, S. Hansmann, and H. Burkhard, *Appl. Phys. Lett.* 67, 3325 (1995).

120. H. Sugiura, M. Mitsuhara, H. Oohashi, T. Hirono, and K. Nakashima, *J. Cryst. Growth* 147, 1 (1995).

121. S. Ruvimov, Z. Liliental-Weber, T. Suski, J. W. Ager III, J. Washburn, J. Krueger, C. Kisielowski, E. R. Weber, H. Amano, and I. Akasaki, *Appl. Phys. Lett.* 69, 990 (1996).

122. E. F. Schubert, I. D. Goepfert, W. Grieshaber, and J. M. Redwing, *Appl. Phys. Lett.* 71, 921 (1997).

123. S. Keller, A. C. Abare, M. S. Minsky, X. H. Wu, M. P. Mack, J. S. Speck, E. Hu, L. A. Coldren, U. K. Mishra, and S. P. Denbaars, *Mat. Sci. Forum* 264-268, 1157 (1998).

124. Y. H. Cho, T. J. Schmidt, S. Bidnyk, J. J. Song, S. Keller, U. K. Mishra, and S. P. DenBaars, *MRS Internet J. Nitride Semicond. Res.* 4S1, G6.44 (1999).

125. P. A. Grudowski, C. J. Eiting, J. Park, B. S. Shelton, D. J. H. Lambert, and R. D. Dupuis, *Appl. Phys. Lett.* 71, 1537 (1997).

126. S. Chichibu, D. A. Cohen, M. P. Mack, A. C. Abare, P. Kozodoy, M. Minsky, S. Fleischer, S. Keller, J. E. Bowers, U. K. Mishra, L. A. Coldren, D. R. Clarke, and S. P. DenBaars, *Appl. Phys. Lett.* 73, 496 (1998).

127. A. Salvador, G. Liu, W. Kim, Ö. Aktas, A. Botchkarev, and H. Morkoç, *Appl. Phys. Lett.* 67, 3322 (1995).

128. K. C. Zeng, J. Y. Lin, H. X. Jiang, A. Salvador, G. Popovici, H. Tang, W. Kim, and H. Morkoç, *Appl. Phys. Lett.* 71, 1368 (1997).

129. H. J. Osten, J. Klatt, G. Lippert, B. Dietrich, and E. Bugiel, *Phys. Rev. Lett.* 69, 450 (1992).

130. D. J. Eaglesham, F. C. Unterwald, and D. C. Jacobson, *Phys. Rev. Lett.* 70, 966 (1993).

131. K. Uchida, T. Tang, S. Goto, T. Mishima, A. Niwa, and J. Gotoh, *Appl. Phys. Lett.* 74, 1153 (1999).

132. K. G. Zolina, V. E. Kudryashov, A. N. Turkin, and A. E. Yunovich, *MRS Internet J. Nitride Semicond. Res.* 1, 11 (1996).

133. P. G. Eliseev, P. Perlin, J. Lee, and M. Osinski, *Appl. Phys. Lett.* 71, 569 (1997).

134. Y. H. Cho, B. D. Little, G. H. Gainer, J. J. Song, S. Keller, U. K. Mishra, and S. P. DenBaars, *MRS Internet J. Nitride Semicond. Res.* 4S1, G2.4 (1999).

135. F. A. J. M. Driessen, G. J. Bauhuis, S. M. Olsthoorn, and L. J. Giling, *Phys. Rev. B* 48, 7889 (1993).

136. K. Yamashita, T. Kita, H. Nakayama, and T. Nishino, *Phys. Rev. B* 55, 4411 (1997).

137. A. Chomette, B. Deveaud, A. Regreny, and G. Bastard, *Phys. Rev. Lett.* 57, 1464 (1986).

138. T. Yamamoto, M. Kasu, S. Noda, and A. Sasaki, *J. Appl. Phys.* 68, 5318 (1990).

139. W. Shan, B. D. Little, J. J. Song, Z. C. Feng, M. Schurman, and R. A. Stall, *Appl. Phys. Lett.* 69, 3315 (1996).

140. Y. H. Cho, J. J. Song, S. Keller, U. K. Mishra, and S. P. Denbaars, *Appl. Phys. Lett.* 73, 3181 (1998).

141. T. J. Schmidt, Y. H. Cho, S. Bidnyk, J. J. Song, S. Keller, U. K. Mishra, and S. P. DenBaars, *SPIE Proc. Ultrafast Phenomena Semicond. III* 3625, 57 (1999).

142. T. J. Schmidt, Y. H. Cho, G. H. Gainer, J. J. Song, S. Keller, U. K. Mishra, and S. P. DenBaars, *Appl. Phys. Lett.* 73, 560 (1998).

143. T. Deguchi, T. Azuhata, T. Sota, S. Chichibu, M. Arita, H. Nakanishi, and S. Nakamura, *Semicond. Sci. Technol.* 13, 97 (1998).

144. T. J. Schmidt, S. Bidnyk, Y. H. Cho, A. J. Fischer, J. J. Song, S. Keller, U. K. Mishra, and S. P. Denbaars, *Appl. Phys. Lett.* 73, 3689 (1998).

145. T. J. Schmidt, S. Bidnyk, Y. H. Cho, A. J. Fischer, J. J. Song, S. Keller, U. K. Mishra, and S. P. DenBaars, *MRS Internet J. Nitride Semicond. Res.* 4S1, G6.54 (1999).

146. M. Kuball, E. S. Jeon, Y. K. Song, A. V. Nurmikko, P. Kozodoy, A. Abare, S. Keller, L. A. Coldren, U. K. Mishra, S. P. DenBaars, and D. A. Steiger-wald, *Appl. Phys. Lett.* 70, 2580 (1997).

147. G. Mohs, T. Aoki, M. Nagai. R. Shimano, M. Kuwata-Gonokami, and S. Nakamura, "Proceedings of the 2nd International Conference on Nitride Semiconductors," Tokushima, Japan, 1997, p. 234.

148. S. Nakamura, *MRS Internet J. Nitride Semicond. Res.* 2, 5 (1997).

149. M. P. Mack, A. Abare, M. Aizcorbe, P. Kozodoy, S. Keller, U. K. Mishra, L. Coldren, and S. P. DenBaars, *MRS Internet J. Nitride Semicond. Res.* 2, 41 (1997).

150. K. L. Shaklee and R. F. Leheny, *Appl. Phys. Lett.* 18, 475 (1971).

151. K. L. Shaklee, R. E. Nahory, and R. F. Lehny, *J. Lumin.* 7, 284 (1973).

152. A. S. Zubrilov, V. I. Nikolaev, D. V. Tsvetkov, V. A. Dmitriev, K. G. Irvine, J. A. Edmond, and C. H. Carter, Jr., *Appl. Phys. Lett.* 67, 533 (1995).

153. W. Fang and S. L. Chuang, *Appl. Phys. Lett.* 67, 751 (1995).

154. S. Bidnyk, B. D. Little, T. J. Schmidt, J. Krasinski, and J. J. Song, *SPIE Conf. Proc.* 3419, 35 (1998).

155. B. Goldenberg, J. D. Zook, and R. J. Ulmer, *Appl. Phys. Lett.* 62, 381 (1993).

156. T. W. Weeks, Jr., M. D. Bremser, K. S. Ailey, E. Carlson, W. G. Perry, and R. F. Davis, *Appl. Phys. Lett.* 67, 401 (1995).

157. S. Nakamura, M. Senoh, S. Nagahama, N. Iwasa, T. Yamada, T. Mat-sushita, Y. Sugiomoto, and H. Kiyoku, *Appl. Phys. Lett.* 70, 1417 (1997).

158. H. Shoji, Y. Nakata, K. Mukai, Y. Sugiyama, M. Sugawara, N. Yokoyama, and H. Ishikawa, *Appl. Phys. Lett.* 71, 193 (1997).

159. H. Jeon, J. Ding, A. V. Nurmikko, W. Xie, D. C. Grillo, M. Kobayashi, R. L. Gunshor, G. C. Hua, and N. Otsuka, *Appl. Phys. Lett.* 60, 2045 (1992).

160. W. Shan, A. J. Fischer, S. J. Hwang, B. D. Little, R. J. Hauenstein, X. C. Xie, J. J. Song, D. S. Kim, B. Goldenberg, R. Horning, S. Krishnankutty, W. G. Perry, M. D. Bremser, and R. F. Davis, *J. Appl. Phys.* 83, 455 (1998).

161. W. W. Chow, A. Knorr, and S. W. Koch, *Appl. Phys. Lett.* 67, 754 (1995).

162. H. Amano and I. Akasaki, "Proceedings of the Topical Workshop on III–V Nitrides," Nagoya, Japan, 1995, p. 193.

163. C. Kirchner, V. Schwegler, F. Eberhard, M. Kamp, K. J. Ebeling, K. Kornitzer, T. Ebner, K. Thonke, R. Sauer, P. Prystawko, M. Leszczynski, I. Grzegory, and S. Porowski, *Appl. Phys. Lett.* 75, 1098 (1999).

164. S. Nakamura, M. Senoh, S. Nagahama, N. Iwasa, T. Yamada, T. Mat-sushita, Y. Sugimoto, and H. Kiyoku, *Appl. Phys. Lett.* 69, 1477 (1996).

165. S. Keller, A. C. Abare, M. S. Minsky, X. H. Wu, M. P. Mack, J. S. Speck, E. Hu, L. A. Coldren, U. K. Mishra, and S. P. DenBaars, "Proceedings of the International Conference on Silicon Carbide, III–Nitrides and Related Materials," Stockholm, Sweden, 1997.

166. S. Chichibu, T. Azuhata, T. Sota, and S. Nakamura, *Appl. Phys. Lett.* 69, 4188 (1996).

167. M. A. Khan, D. T. Olson, J. M. Van Hove, and J. N. Kuznia, *Appl. Phys. Lett.* 58, 1515 (1991).

168. M. A. Khan, S. Krishnankutty, R. A. Skogman, J. N. Kuznia, D. T. Olson, and T. George, *Appl. Phys. Lett.* 65, 520 (1994).

169. D. M. Bagnall and K. P. O'Donnell, *Appl. Phys. Lett.* 68, 3197 (1996).

170. G. Frankowsky, F. Steuber, V. Härle, F. Scholz, and A. Hangleiter, *Appl. Phys. Lett.* 68, 3746 (1996).

171. D. Wiesmann, I. Brener, L. Pheiffer, M. A. Khan, and C. J. Sun, *Appl. Phys. Lett.* 69, 3384 (1996).

172. K. Yung, J. Yee, J. Koo, M. Rubin, N. Newman, and J. Ross, *Appl. Phys. Lett.* 64, 1135 (1994).

173. H. Amano, T. Asahi, M. Kito, and I. Akasaki, *J. Lumin.* 48 & 49, 889 (1991).

174. H. Amano, N. Watanabe, N. Koide, and I. Akasaki, *Jpn. J. Appl. Phys.* 32, L1000 (1993).

175. M. O. Manasreh, *Phys. Rev. B* 53, 16,425 (1996).

176. B. D. Cullity, "Elements of X-ray Diffraction," 2nd ed., p. 502. Addison–Wesley, Reading, MA, 1978.

177. S. Nakamura, M. Senoh, S. Nagahama, N. Iwasa, T. Yamada, T. Mat-sushita, H. Kiyoku, Y. Sugimoto, T. Kozaki, H. Umemoto, M. Sano, and K. Chocho, *Jpn. J. Appl. Phys.* 37, L309 (1998).

178. Y. Kato, S. Kitamura, K. Hiramatsu, and N. Sawaki, *J. Cryst. Growth* 144, 133 (1994).

179. S. Kitamura, K. Hiramatsu, and N. Sawaki, *Jpn. J. Appl. Phys.* 34, L1184 (1995).

180. W. Yang, S. A. McPherson, Z. Mao, S. McKernan, and C. B. Carter un-published; T. S. Zheleva, O. H. Nam, M. D. Bremser, and R. F. Davis, *Appl. Phys. Lett.* 71, 2472 (1997).

181. J. J. Song and W. C. Wang, *J. Appl. Phys.* 55, 660 (1984).

182. E. Ejder, *Phys. Status Solidi A* 6, 445 (1971).

183. D. Marcuse and T. P. Lee, *IEEE J. Quantum Electron.* QE-19, 1397 (1983).

184. G. H. B. Thompson, "Physics of Semiconductor Laser Devices," pp. 402–474. Wiley, New York, 1980.

185. K. C. Zeng, J. Y. Lin, H. X. Jiang, and W. Yang, *Appl. Phys. Lett.* 74, 1227 (1999).

186. H. X. Jiang, J. Y. Lin, K. C. Zeng, and W. Yang, *Appl. Phys. Lett.* 75, 763 (1999).

187. K. Iga, S. Kinoshita, and F. Koyama, *Electron. Lett.* 23, 134 (1987).

188. F. Koyama, K. Tomomatsu, and K. Iga, *Appl. Phys. Lett.* 52, 528 (1988).

189. W. D. Herzog, M. S. Ünlü, B. B. Goldberg, G. H. Rhodes, and C. Harder, *Appl. Phys. Lett.* 70, 688 (1997).

190. S. Nakamura, M. Senoh, S. Nagahama, N. Iwasa, T. Yamada, T. Mat-sushita, H. Kiyoku, Y. Sugimoto, T. Kozaki, H. Umemoto, M. Sano, and K. Chocho, *Appl. Phys. Lett.* 73, 832 (1998).

191. A. E. Siegman, "Lasers," pp. 642–648. University Science Books, Mill Valley, CA, 1986.

192. G. H. B. Thompson, "Physics of Semiconductor Laser Devices," pp. 132–162. Wiley, New York, 1980.

193. M. D. Bremser, T. W. Weeks, Jr., R. S. Kern, R. F. Davis, and D. E. Asp-nes, *Appl. Phys. Lett.* 69, 2065 (1996).

194. A. E. Siegman, "Lasers," pp. 398–447. University Science Books, Mill Valley, CA, 1986.

Chapter 4

ELECTRICAL CONDUCTION PROPERTIES OF THIN FILMS OF CADMIUM COMPOUNDS

R. D. Gould

Department of Physics, Thin Films Laboratory, School of Chemistry and Physics, Keele University, Keele, Staffordshire ST5 5BG, United Kingdom

Contents

1. INTRODUCTION

In this chapter, the electrical properties of thin films of some compounds based on cadmium are reviewed. While most of the emphasis is placed in the areas of lateral resistivity–conductivity, high field direct current (dc) conduction processes and alternating current (ac) conductivity in thin films, the more fundamental structural arrangements, and the band structures are also outlined. The materials covered are the II–VI cadmium compound semiconductors (or the *cadmium chalcogenides*) cadmium sulfide (CdS), cadmium telluride (CdTe) and cadmium selenide (CdSe), and the II–V semiconducting compound cadmium arsenide (Cd_3As_2). As seen later on, there are considerable similarities in the structure and the electrical properties of the II–VI chalcogenide materials, while the properties of Cd_3As_2 are distinctly different.

Figure 1 shows a portion of the right-hand side of the periodic table of the elements with the elements Cd (atomic number $Z = 48$), As ($Z = 33$), S ($Z = 16$), Se ($Z = 34$), and Te ($Z = 52$) indicated. Cd is a group II transition metal with electron configuration $[Kr]4d^{10}5s^2$, while As is a group V element with electron configuration $[Ar]3d^{10}4s^24p^3$. S, Se, and Te are group VI elements in the third, fourth, and fifth periods of the periodic table. Their electron configurations are, respectively, $[Ne]3s^23p^4$, $[Ar]3d^{10}4s^24p^4$, and $[Kr]4d^{10}5s^25p^4$. The properties of Cd_3As_2 include a body-centered tetragonal structure [1], a narrow direct energy bandgap, high mobility, and low resistivity, whereas the II–VI cadmium compounds crystallize in the zinc-blende (or sphalerite) and the wurtzite crystal structures [2], and are wide direct bandgap materials with low intrinsic conductivity.

Interest in the cadmium chalcogenides derives mainly from their uses in solar cell technology. As far back as the mid-1950s Loferski [3] pursued a theoretical investigation of the considerations governing the choice of optimum semiconductors for photovoltaic solar energy conversion. Among the many considerations contributing to the efficiency of *p–n* junction solar cells are the spectral distribution of the radiation reaching

Handbook of Thin Film Materials, edited by H.S. Nalwa
Volume 4: Semiconductor and Superconductor Thin Films
Copyright © 2002 by Academic Press
All rights of reproduction in any form reserved.

ISBN 0-12-512912-2/$35.00

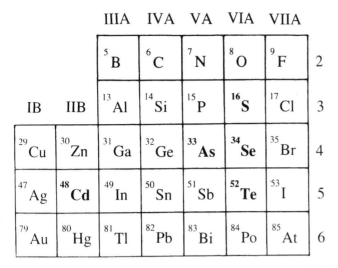

Fig. 1.   Region of the periodic table including the group IIB, VA, and VIA elements. Cd_3As_2 is a narrow direct gap II–V semiconductor, while CdS, CdSe, and CdTe are wide direct-gap II–VI semiconductors.

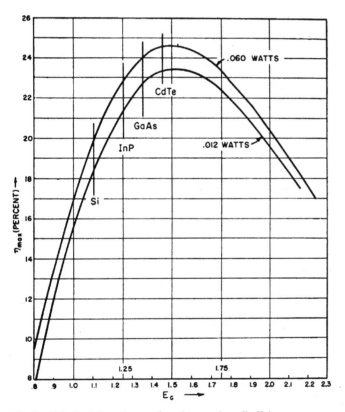

Fig. 2.   Calculated dependences of maximum solar cell efficiency η_{max} as a function of energy bandgap E_G on a cloudy day for power inputs of 0.060 (600 W m^{-2}) and 0.012 W cm^{-2} (120 W m^{-2}). Reproduced from [3]. Reprinted with permission from J. J. Loferski, *J. Appl. Phys.* 27, 777, copyright, American Institute of Physics, 1956.

the Earth's surface, including such effects as the absorption of radiation in the atmosphere, losses by reflection, and the electronic properties of the cell materials, all of which influence the maximum efficiency of conversion into electricity. Taking all these factors into account, it was concluded that solar cells

constructed from materials with energy gap E_G in the range 1.1 eV $< E_G <$ 1.6 eV would be likely to yield maximum efficiency η_{max} higher than that for Si. Figure 2 [3] shows the results of calculated maximum efficiency as a function of E_G for cloudy day conditions at two different power inputs. The figure shows that the efficiency does not decrease significantly with the incident power level. Furthermore, the energy gap of CdTe, which is indicated on the figure, is close to that required for the maximum theoretical efficiency. CdTe has a value of η_{max} of 26.6% under specified conditions, which exceeds that of the other semiconductors Si, InP, and GaAs which were considered by Loferski. The maximum efficiency of CdS solar cells would fall far short of this value. However, it is not normally possible to produce p-type CdS or CdSe films and thus their use is generally restricted to heterojunction solar cells [4]. These include n-CdS/p-CdTe and n-CdSe/p-CdTe heterojunctions which utilize the p-type doping capabilities of CdTe in conjunction with n-type layers of the other material. p–n homojunctions of CdTe are also utilized in solar cells.

Since it has a very narrow energy bandgap, Cd_3As_2 is certainly not suitable for use in solar energy conversion. However, its properties are considered ideal for use in thermal detectors based on the Nernst effect. The Nernst effect is the establishment of a voltage perpendicular to an applied magnetic field and to a temperature gradient, or conversely that of a temperature gradient perpendicular to an applied magnetic field and to a current. The basic requirements for the material comprising such a device are a high mobility and a high carrier concentration, a low energy bandgap (or band overlap as in a semimetal), and high electrical conductivity. Thermodetectors based on the Nernst effect have been successfully operated at room temperature using a Cd_3As_2–NiAs eutectic [5]. Another application for Cd_3As_2 is as a magnetoresistor for use in magnetic field measurements [6]. Below 200 K, both the mobility and the resistivity do not appear to be dependent on the temperature. Cd_3As_2 thin films are also well suited to use in Hall generators, devices which utilize the Hall effect to provide a voltage output in the presence of a magnetic field and an electric current [7].

In the course of this chapter, the basic structural features of the four materials are described in Section 2, both in terms of their bulk structures and their preferred orientations when deposited as thin films. The latter is largely determined by the substrate temperature during deposition, the film thickness, and certain other deposition variables. Following this, selected electrical properties are discussed in Section 3. Among those that are included are the formal band structures, of which examples are given for each material in Section 3.1. The lateral resistivity is also examined in Section 3.2, as this property is particularly important in solar cells where its value determines the internal resistance and therefore influences the maximum energy conversion efficiency, and in Cd_3As_2 films where low values are required for the reasons given earlier. Following this in Section 3.3, a major section in the work is concerned with the high field dc conduction processes observed in cadmium compound films. Before describing the properties of the materials, the various types of contact that may be applied to semiconductors and insulators are addressed in some detail, since the high field conduction properties are usually influenced, and sometimes almost completely determined by the contacts. The theory behind the various conduction properties, space-charge-limited conduction, Schottky and Poole–Frenkel field-lowering effects, tunneling, and hopping conduction are also discussed. Last the ac conduction behavior of the materials is addressed in Section 3.4. Although many models exist for ac conductivity and dielectric properties, some simple models which have previously been invoked to account for the ac behavior up to moderate frequencies are outlined, before an account of the observed behavior is given.

The scope of the chapter is limited to the topics referred to before. In particular, there is no detailed discussion of the use of cadmium compounds in solar cells, or the applications of Cd_3As_2. These issues are covered elsewhere in Refs. [4] and [8], respectively. There is no detailed discussion of the deposition methods currently used for each material. For comparability between different works, most quantities have been quoted in the SI system of units, with other conventional but convenient units such as the electron volt (eV) used sparingly. When the original work was presented using non-SI units these have normally been converted. Figures have been reproduced without alteration, but SI units have been used in the captions in addition to the original units where appropriate. The main emphasis in this chapter is placed on the physics of cadmium compound films and the processes which dominate their electrical conductivity. The account is not exhaustive, but is restricted to some of the more interesting and unusual structural and electronic properties of these films.

2. STRUCTURE

The cadmium chalcogenides can each occur in one of two different structural forms: the zinc-blende and wurtzite structures [2, 9]. Since these structures are common to all three materials, they are described at this point, rather than in the following sections for the individual materials. Cd_3As_2 crystallizes in the tetrahedral structure, and is covered in detail in Section 2.4.

On average, the number of valence electrons per atom in the II–VI compounds is four. Tetrahedral bonding then becomes probable. There are however two different ways in which tetrahedra formed from atoms of each of the constituent elements can penetrate one another (type A and B sites). The tetrahedra bases are parallel in both cases, but for type A sites the atoms of the first element are vertically above those of the second element, while for type B sites one tetrahedron is rotated by 60° with respect to the other. Type A sites are characteristic of the wurtzite lattice and type B sites are characteristic of the zinc-blende lattice, as shown in Figures 3 and 4, respectively, [10]. The illustrations of these two different structures are reproduced from the book by Sze [10], where they were presented in the context of the structures of many different semiconductor materials used in semiconductor devices.

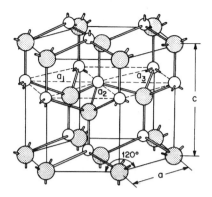

Fig. 3. Unit cell of the hexagonal wurtzite structure, where a and c represent the lattice constants of the unit cell. Reproduced from [10]. Reprinted by permission of John Wiley & Sons, Inc., from "Physics of Semiconductor Devices," S. M. Sze, 2nd ed., copyright 1981, John Wiley & Sons, Inc.

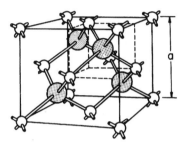

Fig. 4. Unit cell of the cubic zinc-blende structure, where a represents the lattice constant of the unit cell. Reproduced from [10]. Reprinted by permission of John Wiley & Sons, Inc., from "Physics of Semiconductor Devices," S. M. Sze, 2nd ed., copyright 1981, John Wiley & Sons, Inc.

In both the wurtzite and zinc-blende structures, each atom has four nearest neighbors. The wurtzite lattice can be described as two interpenetrating hexagonal close-packed (hcp) lattices, while the zinc-blende lattice is equivalent to two interpenetrating face-centered cubic (fcc) lattices. The planar spacings d_{hkl} in the hexagonal wurtzite structure are given by the expression,

$$\frac{1}{d_{hkl}^2} = \frac{4}{3}\left(\frac{h^2 + hk + k^2}{a^2}\right) + \frac{l^2}{c^2} \quad (1)$$

where h, k, and l are the Miller indices of the set of reflecting planes. a and c represent the dimensions of the unit cell shown in Figure 3, where a is the width of one face and c is the height of the hexagonal prism. In the cubic zinc-blende structure,

$$\frac{1}{d_{hkl}^2} = \frac{h^2 + k^2 + l^2}{a^2} \quad (2)$$

where a is the length of one side of the cubic unit cell shown in Figure 4. These equations are utilized extensively in determining the lattice constants and the preferential orientation in thin films using X-ray diffraction methods.

Before considering the individual structures in detail, it may be useful to consider the conditions for the growth of epitaxial films in the case of the cadmium chalcogenides. Substantial work has been performed in this area by Kalinkin et al. [11] and Muravjeva et al. [12]. This work has also been quoted in

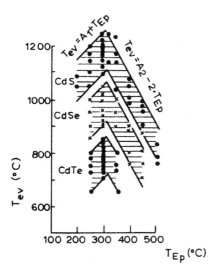

Fig. 5. Relationship between the evaporation temperature T_{ev} and the epitaxial temperature T_{Ep} in the growth of cadmium chalcogenide films on mica. Reproduced from [12]. Reprinted from *Thin Solid Films*, 5, K. K. Muravjeva et al. Growth and Electrophysical Properties of Monocrystalline Films of Cadmium and Zinc Chalcogenides, p. 7, copyright 1970, with permission from Elsevier Science.

previous reviews of the field [9, 13]. Extensive measurements were made of the evaporation of cadmium chalcogenides evaporated from Knudsen cells at a pressure of about 10^{-2}–10^{-3} Pa (10^{-4}–10^{-5} Torr). The substrates used were mica in all cases. Generally, it was found that for all the cadmium (and also for all the zinc) compounds the conditions for epitaxial growth were dependent on both the temperature of the substrate or epitaxial temperature T_{Ep} *and* the evaporator temperature T_{ev}. For each of the three materials, different combinations of T_{Ep} and T_{ev} were investigated, and it was found that below $T_{Ep} \approx$ 310–320°C, the evaporator temperature for epitaxy increased linearly with T_{Ep} according to $T_{ev} = A_1 + T_{Ep}$, where A_1 is a constant. Above $T_{Ep} \approx$ 310–320°C the evaporator temperature for epitaxy decreased linearly with T_{Ep} according to $T_{ev} = A_2 - 2T_{Ep}$ where A_2 is a second constant. These relationships between the epitaxial temperature and the evaporator temperature are illustrated in Figure 5 [12] for each of the cadmium chalcogenides. Values of the range of the constants A_1 and A_2 for which epitaxial behavior was observed are $A_1 = 800$–950°C and $A_2 = 1750$–1900°C for CdS, $A_1 = 600$–750°C and $A_2 = 1550$–1700°C for CdSe and $A_1 = 400$–550°C and $A_2 = 1350$–1500°C for CdTe [12]. It is evident from these values that for each substance the evaporator temperature T_{ev} lies within a range of about 150°C for values of T_{Ep} at which epitaxy is possible. In addition, for a given substrate temperature the evaporator temperature required for epitaxial CdS is highest, becoming progressively lower for CdSe and for CdTe. For example, for $T_{Ep} = 320$°C, the evaporator temperatures for epitaxy are 1120–1270°C for CdS, 920–1070°C for CdSe and 720–870°C for CdTe. It was also found that there were distinct relationships between the molecular weights of the compounds and the substrate and evaporator temperatures. At combinations of temperatures outside the limits given by the previous values

of A_1 and A_2 the films deposited were not epitaxial, being polycrystalline and only partially ordered.

The various structural features of the individual cadmium compounds are described in the following sections.

2.1. Cadmium Sulfide

CdS films are most commonly deposited by vacuum evaporation. Such films are not normally epitaxial, but polycrystalline. However, this is probably acceptable in the case of low cost solar cells. As mentioned earlier, the substrate and source temperatures influence the film structure. When the substrate is normal to the vapor stream, individual crystallites become aligned perpendicular to the substrate in a fiber texture. CdS films have also been prepared by radio frequency (rf) sputtering, by spray pyrolysis, and by silk screen printing. In addition, Williams et al. [14] deposited CdS films for solar cells by the method of electrophoresis. In this method, colloidal particles of CdS are deposited onto a conducting substrate by the application of an electric field. The CdS was precipitated from a reaction between H_2S and a compound of cadmium, such as cadmium acetate. Nevertheless, most of the structural work has been performed on evaporated films, and this is outlined in the following.

Evaporation is normally performed with the substrate held at an elevated temperature. Under these conditions, the hexagonal wurtzite structure is usually obtained. However, at lower substrate temperatures there are several reports of the cubic zinc-blende structure. When the substrate temperature was initially at room temperature, but rose to not greater than 100°C during deposition, the resulting films were of the cubic structure [15]. Films having this structure were also observed by Escoffery [16]. A polymorphic transition in epitaxial CdS films was described [17]. The behavior was somewhat complex in this case, in that at low substrate temperatures the hexagonal structure was obtained, with the cubic structure appearing at elevated substrate temperatures, but not as the result of annealing films deposited at a lower temperature. At still higher substrate temperatures the influence of the substrate became less, and the films reverted back to their stable hexagonal structure. Cubic modification films were also observed by Simov [18], but only under conditions where substrate temperature was in the range 270–300°C and the source temperature 600–700°C.

More typically hexagonal wurtzite films were deposited by Wilson and Woods [19]. All of their films had the same structure, irrespective of the substrate temperature, deposition rate, or film thickness. With increasing thickness all films showed an increasing degree of preferential alignment, with the c-axis directed perpendicular to the substrate plane. This has also been observed by Ashour et al. [20], and is illustrated by the diffraction traces in Figure 6 [20]. Figure 6a shows the X-ray reflections obtained from the initial CdS evaporant powder and Figure 6b shows those from the resulting film. This indicates a preferential orientation of the crystallites in the [002] direction of the hexagonal structure. The calculated values of the interplaner spacings of the (002) and the (004) planes observed in

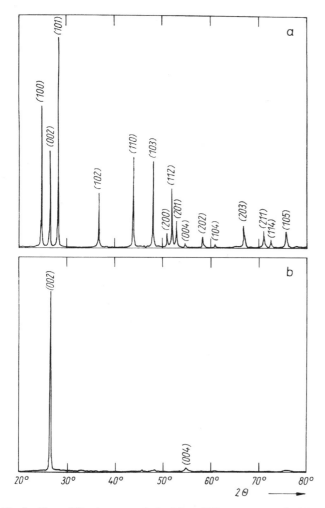

Fig. 6. X-ray diffraction traces obtained from CdS evaporant powder (a) and from an evaporated thin film (b). The film was preferentially oriented in the [002] direction of the hexagonal structure. Reproduced from [20]. Reprinted with permission from A. Ashour, R. D. Gould, and A. A. Ramadan, *Phys. Status Solidi A* 125, 541, copyright Wiley-VCH, 1991.

Figure 6b were 0.3398 and 0.1689 nm, respectively. The *c*-axis constant for the films was determined to be 0.6796 nm using Eq. (1). Careful measurements of the preferential orientation in CdS films were also performed by Hussain [21] in which thick (up to 100 μm) CdS films were deposited obliquely at vapor incidence angles of between 0 and 60° to the substrate normal. Up to a certain film thickness, the *c*-axis of the films was perpendicular to the direction of vapor incidence, but above this thickness the *c*-axis started to shift toward the substrate normal, eventually becoming aligned with it. Etching of a thin film deposited at oblique incidence confirmed that the top layers of the film had their *c*-axis normal to the substrate, whereas in the bottom layers it was parallel to the direction of vapor incidence. Similar orientation effects were also reported [19] in which films of a few tens of nanometers thickness were randomly oriented, but those of thickness about 20 μm were aligned with the normal to the substrate.

Measurements of mean grain size in CdS films normally show an increase with film thickness as shown in Figure 7 [19]

for films deposited at a substrate temperature of 220°C. Grain sizes were determined from the width of the X-ray diffraction peaks, using the method introduced by Scherrer [22]. The crystallite dimensions were found to increase almost exponentially with film thickness, and it can be seen from the figure that the apparent grain size is smaller when calculated using electron diffraction than by X-ray diffraction. Similar results have been obtained [20]. For film thicknesses up to about 2.5 μm, the mean grain size increased from about 10 up to 25 nm, and was thought to be related to the columnar growth of the films. It was also found that the largest crystallites were obtained for a substrate temperature of approximately 150°C, with clearly smaller crystallites at substrate temperatures of both 100 and 200°C. Concomitant with the increase in grain size with thickness was a *decrease* in the residual microstrain. The lowest microstrain values were obtained at a substrate temperature of 150°C, when the grain size was a maximum.

The dependence of grain size on deposition rate has also been investigated [19, 20]. Wilson and Woods [19] concluded that the crystallite grain size was not primarily a function of the deposition rate, since points on their grain size-film thickness plot shown in Figure 7 obtained from two series of films deposited at widely different rates were found to lie on the same curve. Conversely, in [20] there was a systematic decrease in the mean grain size from approximately 26 to 18 nm as the deposition rate increased from 0.2 to 0.9 nm s^{-1}. This effect was identified with a higher equilibrium concentration of nuclei during nucleation at higher deposition rates, which would naturally lead to a smaller mean grain size.

The morphologies of evaporated CdS films have been investigated by electron microscopy [20, 23]. The top surface of as-deposited films of thickness 25–30 nm were observed by scanning electron microscopy to consist of partially faceted hills of about 5-μm height and width about 15 μm at the base [23]. These workers also determined the grain size and the crystal defect structure using transmission electron microscopy. The grain size was about 2 μm at the top and at the center of the films, but was only of the order of 10–50 nm at the back surface adjacent to the substrate. Typical dislocation densities were of the order of 10^{13} m^{-2} (10^{9} cm^{-2}). Ultrafine-grain regions of 2–15 μm in diameter were observed. These were thought to be "spits" of material rapidly ejected from the source. The various morphological features observed in this electron microscopic study are shown schematically in Figure 8 [23]. Similar features have also been observed in this laboratory [20]. The partially faceted protrusions were reported, and cracks in the films were also present, as a result of differential thermal contraction between the CdS films and the glass substrates during cooling. Larger regions of CdS were seen, which were considered to have been ejected from the evaporation boat during deposition. This phenomenon has been referred to as "spattering" or "splattering" by Stanley [24], and its elimination is an important goal for deposition processes operating at higher temperatures.

Thus, CdS films are normally of the hexagonal wurtzite structure when deposited on heated substrates, and are preferentially oriented in the [002] direction. There is a fair measure of

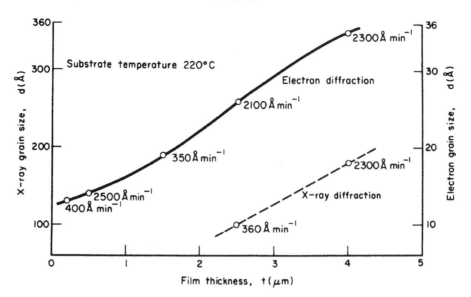

Fig. 7. Apparent grain sizes in evaporated CdS films deposited on glass as a function of thickness, measured using X-ray diffraction and electron diffraction. Reproduced from [19]. Reprinted from *J. Phys. Chem. Solids*, 34, J. I. B. Wilson and J. Woods, The Electrical Properties of Evaporated Films of Cadmium Sulphide, p. 171, copyright 1973, with permission from Elsevier Science.

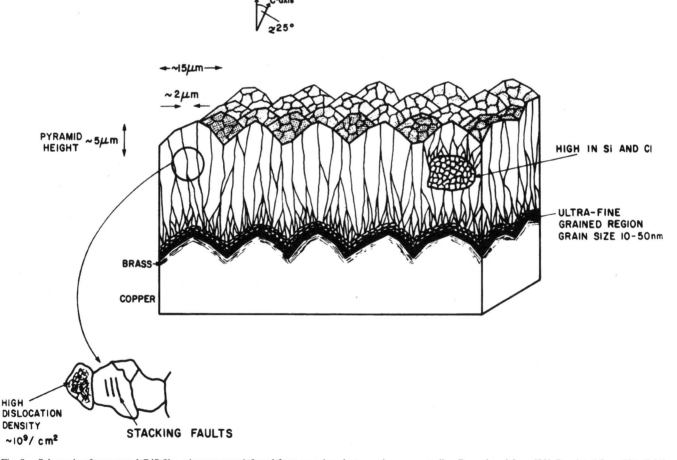

Fig. 8. Schematic of evaporated CdS film microstructure inferred from scanning electron microscopy studies. Reproduced from [23]. Reprinted from *Thin Solid Films*, 75, K. H. Norian and J. W. Edington, A Device-Oriented Materials Study of CdS and Cu$_2$S Films in Solar Cells, p. 53, copyright 1981, with permission from Elsevier Science.

agreement concerning the microcrystallite grain size and its dependence on film thickness and substrate temperature, although the dependence on deposition rate is less transparent. The morphology of CdS films is complex, as illustrated in Figure 8, and needs to be controlled to produce films of acceptable quality.

2.2. Cadmium Telluride

CdTe films have been deposited by several different deposition methods, including chemical deposition, sputtering, and evaporation, as well as by more novel methods such as screen printing [25] and the use of an epitaxial close-space technique [26]. Most work has concentrated on direct evaporation of the compound CdTe, rather than by coevaporation of the elements Cd and Te. The majority of workers report films which are of the cubic zinc-blende (or sphalerite) structure. Dharmadhikari [27] directly deposited CdTe films at a typical deposition rate of 0.5 nm s^{-1} and found that the films had a preferential (111) orientation, which was enhanced by a high substrate temperature. Deposition of CdTe onto a substrate held at room temperature gave material of the cubic structure [28], with a small amount of additional free Te. Higher temperature deposition at 150 and 250°C eliminated the free Te, but resulted in additional diffraction lines characteristic of the hexagonal form of CdTe. The lattice constant was $a = 0.0648$ nm, corresponding with that of the bulk material. Poelman and Vennik [29] deposited CdTe films at a higher rate of 4.5 nm s^{-1} onto glass substrates kept at temperatures of 240 or 280°C. They observed a strong diffraction peak corresponding to the (111) planes, with very faint lines corresponding to the (220) and (311) planes. The films were therefore preferentially oriented in the [111] direction. A similar preferential orientation has been observed many times [30–32] for CdTe films of the cubic structure.

In addition to the work of Thutupalli and Tomlin [28] mentioned before, there are additional occasional references in the literature to the hexagonal phase. It was observed that films with the wurtzite structure were obtained if CdTe and metallic Cd were coevaporated in an Ar atmosphere at a pressure of approximately 1 Pa [33]. Yezhovsky and Kalinkin [34] made detailed measurements of the structure of epitaxial CdTe films when condensed in a closed cell onto mica. They found that a combination of the rate of growth and the epitaxial temperature determined the type of films obtained. This was related to the relationship of these variables to the supersaturation coefficients and the condensation coefficients on the substrate. They found several different regions of film growth, i.e., region I cubic, region II mixed cubic and hexagonal, region III hexagonal, region IV quasi-single crystal of twinned cubic orientation, and region V textured. Region III corresponds approximately to an epitaxial temperature in the range 160–250°C, with the logarithm of deposition rate (Å s^{-1}) in the range 1.3–2.3 (approximately 2–20 nm s^{-1}). The orientation of the hexagonal phase was with the (0001) CdTe planes parallel to the (0001) mica substrate planes. This work showed that even at relatively low epitaxial temperatures of about 300°C a small excess of

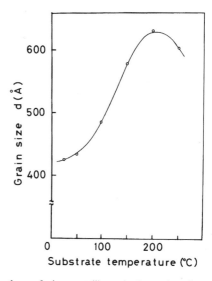

Fig. 9. Dependence of microcrystallite grain size on the substrate temperature during deposition in evaporated CdTe films deposited at a rate of 1.9 nm s^{-1}. Reproduced from [30]. Reprinted with permission from Y. Kawai, Y. Ema, and T. Hayashi, *Jpn. J. Appl. Phys.* 22, 803, copyright Japanese Institute of Pure and Applied Physics, 1983.

Cd yields films of the cubic modification. There was no evidence of wurtzite film structures resulting from an excess of Cd, as had been reported previously by Shalimova and Voronov [35]. These workers had observed a predominantly hexagonal structure when the Cd vapor pressure exceeded the Te vapor pressure. Yezhovsky and Kalinkin [34] also reported that excess Te resulted in films with the sphalerite structure, but with a reduced degree of perfection.

Measurements of microcrystallite grain size have been made by several workers [30, 32, 36] as functions of thickness, substrate temperature during deposition, and deposition rate. The mean grain sizes were determined by the Scherrer [22] method, as for the case of CdS films. The grain size was observed to increase with increasing substrate temperature from about 40–60 nm, as shown in Figure 9 [30] for films deposited at a rate of 1.9 nm s^{-1}. The maximum grain size occurred at a temperature of approximately 200°C. Very similar results have also been obtained by Ismail [36], where the maximum grain size was 50 nm, also at a substrate temperature of 200°C. In the latter work, the microstrain was also measured, and it was shown that this reduced by a factor of approximately 3 from 15×10^{-4} to 5×10^{-4} as the substrate temperature increased from room temperature to 200°C. Saha et al. [32] used the Scherrer method to determine the microcrystallite grain size and the microstrain as functions of the film thickness. In addition, the stacking fault probability, the dislocation density, and the photoyield were determined for films of thickness in the range 300–1500 nm. These results are shown in Figure 10 [32]. The microcrystallite grain size was observed to increase with film thickness from around 4 nm for a film of thickness 300 nm to a maximum of 20 nm for film thickness 700 nm. Above this thickness, the grain size decreased and the microstrain in-

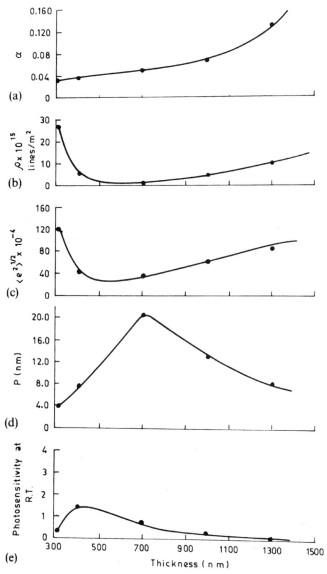

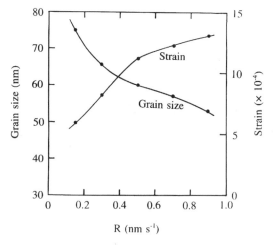

Fig. 11. Dependence of microcrystallite grain size and microstrain in evaporated CdTe films as a function of deposition rate R for films of thickness 350 nm deposited at a substrate temperature of 350°C. B. B. Ismail and R. D. Gould, unpublished [190].

Fig. 10. Dependence of stacking fault probability (a), dislocation density (b), microstrain (c), microcrystallite size (d), and photosensitivity (e) as functions of film thickness in evaporated CdTe films deposited at a rate of 3.5 nm s^{-1}. Reproduced from [32]. Reprinted from *Thin Solid Films*, 164, S. Saha et al., Structural Characterization of Thin Films of Cadmium Telluride, p. 85, copyright 1988, with permission from Elsevier Science.

creased. The stacking fault probability α increased steadily with the film thickness. According to these workers, the only difference between the zinc-blende and wurtzite structures is in the stacking of the atomic layers with the stacking pattern in the cubic type following an ABCABC... sequence, and that of the hexagonal type an ABABAB... sequence. Since the energy difference between the two forms is very small, faults in the stacking sequence of one type give rise to the other structural type at the fault. α represents the probability of layers undergoing stacking sequence faults. The increasing value of α with film thickness was ascribed to the formation of microcrystallites of the second structure, in which the growth of crystallites of the original phase was inhibited. Ismail [36] found larger

grain sizes in the range 45–55 nm, with grain size increasing with film thickness over the range 50–500 nm. However, the increase in grain size was most apparent when the film thickness was less than 70 nm, attaining a roughly constant value for films of greater thickness. There was a concomitant decrease in the microstrain over the range of approximately $(15–5) \times 10^{-4}$ as the film thickness increased. Measurements of the dependence of grain size on deposition rate were performed in the case of films of thickness 350 nm deposited at a substrate temperature of 150°C. The observed dependence is shown in Figure 11, together with that of the microstrain, for various deposition rates ranging from 0.15–0.9 nm s^{-1}. As the deposition rate increased, finer grain sizes were obtained, and at the same time the microstrain increased. The reduced grain size at higher deposition rates was interpreted according to the presence of a larger concentration of nuclei, from which the growth of finer grains is initiated. Similar behavior was also observed by Kawai, Ema, and Hayashi [30] in which films deposited at lower rates were polycrystalline, whereas those deposited at higher rates were amorphous. Specific values of the mean microcrystallite grain size given by these workers were 42.5 nm for a rate of 0.49 nm s^{-1}, 40 nm for 1.9 nm s^{-1}, and 30.6 nm for 6.7 nm s^{-1}. High quality films with a large grain size would therefore require deposition at a low rate. It was also noted [36] that films annealed at a temperature exceeding the substrate temperature during deposition resulted in an enhanced grain size. For example, a film of thickness 300 nm deposited at a substrate temperature of 150°C had a mean grain size of 54 nm; when annealed for 1 h at 300°C the mean grain size increased to 61 nm.

Early work on CdTe films [31] showed that photovoltages of up to 100 V cm^{-1} of film length could be generated in films that were deposited onto substrates at an oblique angle to the vapor stream. There was a general tendency for the photovoltage to decrease with the angle between the vapor stream and the normal to the substrate, with no photovoltage produced for

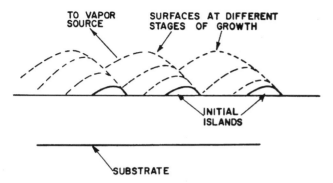

Fig. 12. Schematic of CdTe film buildup showing directional anisotropy in film growth due to deposition at an angle to the substrate normal. Reproduced from [31]. Reprinted with permission from B. Goldstein and L. Pensak, *J. Appl. Phys.* 30, 155, copyright, American Institute of Physics, 1959.

deposition at normal incidence. It was proposed that the photovoltage resulted from warping of the energy bands between individual crystallites. It was further suggested that this occurred when the film growth was directionally anisotropic, as shown in Figure 12 [31]. This type of structure was thought to arise only when the vapor stream is not normal to the substrate, and where the nucleating material depositing on a heated substrate has sufficient surface mobility to form islands rather than a uniform thin film. Additional material arrives at an angle, shadowing the far side of the island from the deposition source, and the shadowed side grows relatively slowly and is well ordered, while the exposed face grows toward the vapor source. When islands begin to touch each other there is expected to be a mismatch, leading to a disordered state in the region between the crystallites. Each interface will therefore have an ordered and a disordered side. At the interface between grains the band structure will be warped, leading to the establishment of photovoltages under illumination. Photomicrographs were produced showing the surfaces of epitaxial CdTe layers grown by the close-space technique [26]. It was found that smooth epitaxial layers were obtained when the deposition was on to (100), (110), and (111) Cd oriented substrates, but large hillocks tended to grow on the (111) Te substrate faces. When grown on mica substrates [34], it was observed that the morphology was significantly affected by the composition of the incident vapor phase. These were characterised by the ratios γ_{Cd} and γ_{Te} which depended on the ratio of the actual concentration to the stoichiometric concentration of the Cd atoms or the Te_2 molecules in the gas phase. When $\gamma_{Cd} = 53$ when CdTe and Cd were coevaporated, the films were of high quality consisting of large crystallites with degenerated (111) faces. When CdTe and Te were coevaporated and $\gamma_{Te} = 12$, formations resembling dendrites were observed. Macroscopic grain sizes of the order of 10 μm were observed in screen printed CdTe films [25]. These films were porous near the free surface. Scanning electron microscopy of CdTe films [29] revealed film structures which were of the order of 150-nm wide, in accordance with microcrystallite size determined by X-ray diffraction.

Thus, CdTe films are normally of the zinc-blende structure when deposited by vacuum evaporation. Films are normally preferentially oriented in the [111] direction. Deposition at higher temperatures can lead to a full or partial change to the wurtzite structure. There are also reports of the wurtzite phase when there is an excess of Cd or Te in the vapor stream during deposition. The microcrystallite grain size appears to increase with substrate temperature up to a maximum value at about 200°C, which corresponds with a minimum in the microstrain. There is also an increase with film thickness and a decrease with deposition rate. The surface morphology was affected by the direction of the incident vapor stream with respect to the substrate normal and by its composition.

2.3. Cadmium Selenide

CdSe films have been deposited by a variety of alternative deposition methods, including direct evaporation and condensation from the vapor phase [37–39], different varieties of sputtering including dc sputtering of Cd in an Se vapor [40], CdSe sputtered in a mixture of hydrogen selenide and argon [41] and rf sputtering [42], electrolytic codeposition [43], spray pyrolysis [44], and also by other *ad hoc* chemical methods. The type of crystal structure observed appears to depend to some extent on the substrate temperature during deposition, with films deposited at room temperature and up to about 200°C usually being of the wurtzite form, with a general tendency for films growing at higher temperatures to be of the cubic zincblende structure. This is not, however, a hard and fast rule, as many other factors including the stoichiometry of the vapor also influence the structure. Dhere, Parikh, and Ferreira [45] observed hexagonal films when deposited at room temperature. When Naguib, Nentwich, and Westwood [46] evaporated CdSe films onto glass substrates at 200°C, the wurtzite structure was observed by the use of electron diffraction. No preferential orientation effects were observed. At temperatures above 200°C, additional lines were seen in the electron diffraction patterns which were identified with the presence of the compound $Cd_3Se_4O_{11}$. This phase was also observed by Hamerský [38]. In earlier work, Shallcross [47] had deposited films onto substrates held at temperatures of 50 and 230°C, although films deposited at both temperatures were of mixed hexagonal and cubic structures. However, films deposited at the lower temperature had twice as much of the cubic as of the hexagonal phase, while those deposited at the higher temperature had three times as much cubic as hexagonal. Rentzsch and Berger [48] evaporated CdSe films at substrate temperatures between 200 and 500°C. In contrast with other workers, they found that the proportions of the two phases present remained roughly constant irrespective of the substrate temperature, at approximately 60% for the hexagonal phase and 40% for the cubic phase. With the vapor stream at normal incidence, the films were textured in the (0001) and (111) directions, respectively. Oduor [49] evaporated CdSe films at a rate of 0.5 nm s^{-1} onto substrates held at room temperature. The films were of the hexagonal structure, preferentially oriented in the [002] direction perpendicular to the substrate surface. Diffraction traces obtained from these films showed only the (002) peak. Films deposited at a substrate temperature of 200°C showed several peaks, which if indexed according to the hexagonal structure corresponded to the (100), (002), (101), (110), and (112) planes. Thus, the degree of preferential orientation was considerably reduced. It was, however, pointed out that the d-spacing corresponding to the (002) planes in the hexagonal structure, corresponds to that of the (111) planes in the cubic structure, and that it is possible that at the higher substrate temperature a mixture of the two structures is present. Däweritz and Dornics [50] stressed that in several Zn chalcogenides, as well as in CdTe, the hexagonal phase was formed when there was a nonstoichiometric vapor composition in which there was an excess of the metallic component, whereas an excess of the nonmetallic component led to the cubic structure. However, in this group of compounds the crystallographic axis ratios are greater than the ideal value of $c/a = \sqrt{(8/3)}$, whereas in CdS and CdSe $c/a < \sqrt{(8/3)}$. They argued that if the axial ratio was varied, either up or down, by the incorporation of an excess of either element, the ratio would approach the ideal value for one group but would have an increased deviation for the other group. This was tested by coevaporating CdSe with additional Se, the latter at various different rates. After deposition, the proportion of the cubic phase was determined as a function of the substrate temperature during deposition. The results of these experiments are shown in Figure 13 [50], where curve 1 represents CdSe only deposited at a rate of 1.4 nm s^{-1}, while curves 2 to 4 represent increasing proportions of coevaporated Se. It is clear that at low temperatures an increase in the Se evaporation rate decreased the proportion of the cubic phase, while at elevated substrate temperatures the proportion of cubic material increased with the Se evaporation rate. Owing to differences between the vapor pressures, an additional Se evaporation does not imply a Se excess at high temperatures, although at low substrate temperatures it does. Thus, an excess in Se (the nonmetallic element) was associated with the formation of the hexagonal modification, and supported their conjecture regarding the axial ratios. This model was later refined [2] to suggest that either the cubic or the hexagonal structure would be formed preferentially depending on the excess Cd concentration, since the vapor pressure of Cd differs from that of Se. A preferential occurrence of the cubic phase would be expected at higher temperatures and higher deposition rates. At a deposition rate of 8 nm s^{-1} and a substrate temperature of 25°C, the films were almost entirely hexagonal. An increase in the cubic fraction occurred when the substrate temperature was raised to 350°C and also when the substrate temperature remained constant, but the deposition rate was drastically reduced to 0.01 nm s^{-1}. Using this model, correlations were later found between the resistivity and the deposition conditions in terms of the vapor stoichiometry [51]. A hexagonal phase with a one-dimensional {001} texture orientation was obtained for films deposited onto unheated substrates. Hamerský [38] considered the effects of the oxygen partial pressure on evaporated films. Films deposited at room temperature and with lower oxygen partial pressure were of the hexagonal wurtzite structure, with the (002) planes par-

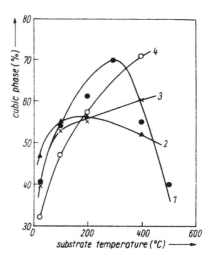

Fig. 13. Phase compositions in 1000-nm thick evaporated CdSe films with additional Se evaporation as a function of the substrate temperature during deposition. The CdSe deposition rate was 1.4 nm s^{-1} for each curve, with no Se deposition (1), and Se deposition rates of 0.1 nm s^{-1} (2), 0.3 nm s^{-1} (3), and 0.6 nm s^{-1} (4). Reproduced from [50]. Reprinted with permission from L. Däweritz and M. Dornics, *Phys. Status Solidi A* 20, K37, copyright Wiley-VCH, 1973.

allel to the substrate, while those deposited at a higher partial pressure had an approximately equal concentration of crystallites with (002) and (103) planes parallel to the substrate. There was also evidence of diffraction peaks corresponding to CdSeO$_3$.SeO$_2$ complexes.

Films sputtered using separate sputtered Cd and evaporated Se sources [40] and annealed at 650°C showed a polycrystalline hexagonal structure with the (002) planes parallel to the substrate, corresponding to the c-axis normal to the substrate. The hexagonal c-axis was also perpendicular to the substrate in rf sputtered films deposited at a substrate temperature below 400°C [42]. At temperatures above this value, an amorphous phase was observed for films deposited onto vitreous silica. In dc sputtered films deposited on indium–tin oxide substrates, the hexagonal phase was again produced, with the c-axis perpendicular to the substrate [41]. There was some evidence for the cubic phase in films deposited at 200°C, but deposition at 15°C resulted in films entirely of the hexagonal structure.

Measurements of grain size in CdSe films, performed using X-ray line broadening, are not as prevalent as for CdTe and the other Cd chalcogenides. Shallcross [47] reported that in evaporated films the average microcrystallite size was approximately 50 nm or larger in films of thickness 1–4 μm, and that the crystallite size increased with increasing substrate temperature. On glass substrates, the microcrystallite grain size increased from 5 to 40 nm as the substrate temperature was raised to 200°C [46]. An activation energy for the crystallite growth process was determined from an Arrhenius plot of grain size with respect to substrate temperature, and yielded a value of approximately 0.28 eV. It was also observed that there was no significant effect on the grain size of films deposited at 200°C by varying the deposition rate in the range 0.17–3.0 nm s^{-1}. For films deposited at 200°C substrate temperature at a rate of 1 nm s^{-1}, reflection

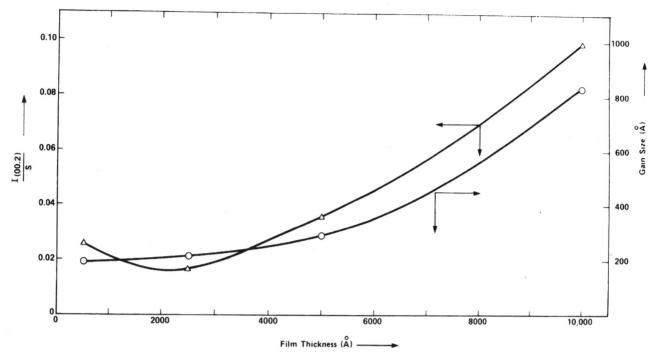

Fig. 14. X-ray diffraction intensity of the (002) reflection normalized to unit film thickness (I_{002}/s) as a function of film thickness for CdSe films evaporated onto glass (left-hand scale). Grain size normal to the substrate, as calculated from the (002) linewidth as a function of film thickness (right-hand scale). Reproduced from [46]. Reprinted with permission from H. M. Naguib, H. Nentwich, and W. D. Westwood, *J. Vac. Sci. Technol.* 16, 217, copyright, American Vacuum Society, 1979.

high energy electron diffraction (RHEED) patterns were obtained for films of varying thickness in the range 55 nm to 1 μm. An increase in the microcrystallite size and better ordering of the films was observed with increasing film thickness. Using X-ray diffraction methods, which sample the entire film thickness rather than only the surface as in RHEED, measurements of reflection intensity and linewidth of the (002) reflection were obtained. Figure 14 [46] shows the dependence of the normalized intensity of the (002) peak (ratio of the measured intensity to film thickness) and the grain size normal to the substrate, both as functions of film thickness. Although the normalized intensity of the (002) reflection increased with film thickness, for the (100) and (110) reflections there was a decrease with increasing film thickness and for the (101) and (112) reflections there was no change. Thus, there was an increase in the degree of preferential orientation in the [002] direction with increasing film thickness. This was attributed to the {002} oriented crystallites growing at the expense of those oriented in other directions shown by an increase in the (002) intensity which was accompanied by a decrease in the intensity of the other reflections. It was suggested that this resulted from recrystallization of the films as they became thicker, and it was argued that this was caused by compressive stresses in the films. A model was derived to account for the growth of the grains with increasing film thickness, which gave a correlation with increasing grain size for compressive stress and with decreasing grain size for tensile stress. In Figure 14 [46], it is shown that grain size normal to the substrate increased from about 20 to 80 nm as the film thickness increased from 100 nm to 1 μm. Polycrystalline

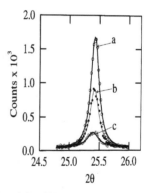

Fig. 15. Line profiles of the (002) reflection in evaporated hexagonal CdSe films deposited at a rate of 0.5 nm s^{-1} onto glass substrates at room temperature. Film thicknesses were 820 nm (a), 530 nm (b), and 250 nm (c). A. O. Oduor and R. D. Gould, unpublished [191].

films were also obtained in samples deposited for optical absorption studies [48]. Here, the mean crystallite size was 50 nm to 1 μm, depending on the film thickness and the substrate temperature. The work of [49] investigated the dependence of microcrystallite size on thickness and deposition rate. Figure 15 shows an expanded (002) hexagonal line profile for films of thickness 820 nm (a), 530 nm (b), and 250 nm (c), all deposited at 0.5 nm s^{-1} onto substrates held at room temperature. The peaks were fitted using a Lorenzian function, and the mean microcrystallite sizes were estimated to be 56, 51, and 39 nm for films a, b, and c, respectively. This is consistent with some earlier work in which the microcrystallite size increased with increasing film thickness. There was some evidence that films deposited at a rate of 0.5 nm s^{-1} had a larger grain size than those deposited at 0.1 nm s^{-1}, although as noted earlier Naguib, Nentwich, and Westwood [46] found no difference.

Film morphology has been observed by several workers. Tanaka [40] prepared films using Ar and H$_2$ sputtering gases; some films were subsequently annealed in Ar. Microscopic examination showed that sputtering in H$_2$ gas led to uniform grain growth of the films over the whole substrate surface, while films sputtered and annealed using Ar gave rise to geometric pillars and a textured structure. An investigation of the optimum conditions for obtaining CdSe epitaxial layers on (0001) oriented sapphire substrates was performed [52]. For substrate temperatures up to 460°C, the layers were polycrystalline, from 460 to 530°C a small degree of preferred orientation was observed, with a more significant textured structure for temperatures of 530 to 570°C. At higher temperatures of 570–620°C, a monocrystalline structure was obtained, varying in its degree of perfection depending on the actual temperature and the growth rate. Hexagonal pyramids with peaks, regular or irregular facets, and flat tops were also identified. Rf sputtered films [42] were porous when deposited at a substrate temperature of 550°C onto vitreous silica. When deposited onto substrates of single-crystal sapphire, the surface appeared shiny, and showed features of hexagonal flat tops and pyramids with edge lengths in the range 1–5 μm. Oduor [49] used a field-effect surface electron microscope (FESEM) to study the morphology of CdSe films. Films deposited at a substrate temperature of 206°C had bright uniform spots on their surfaces of diameter about 0.15 μm, which were thought to be "spattered" from the evaporation boat. After annealing, many of these disappeared, either as the result of coalescence or of detachment from the surface. Spattered particles were more prevalent on samples prepared at room temperature than for samples held at 206°C. Substrate heating therefore improved the film surface smoothness, in addition to its effects on the crystal structure.

Hence, in CdSe films both hexagonal and cubic structures are found, with the hexagonal form usually appearing at lower substrate temperatures. Films are usually aligned with the c-axis perpendicular to the substrate. Excesses in the Cd or Se vapor content during deposition can influence the film structure. Results concerning grain size are somewhat contradictory, with reports of grain size remaining constant [46] or increasing [49] with increasing deposition rate. Microcrystallite grain size tends to increase with film thickness and is usually of the order of a few tens of nanometers, although values up to 1 μm have been reported [48]. Pyramid-type structures, having pointed or flat tops, were reported when films were deposited onto sapphire substrates.

2.4. Cadmium Arsenide

The complex crystal structure of Cd$_3$As$_2$ has been the subject of some controversy in the past, and there are indications of several different phases. Its structure was first determined by von Stackelberg and Paulus in 1935 [53], who found it to be tetragonal with $a = 0.895$ nm and $c = 1.265$ nm. The As atoms were approximately in a cubic close-packed array and the Cd atoms were tetrahedrally coordinated. Each As atom was surrounded by Cd atoms at six of the eight corners of a distorted cube, with two vacant sites at diagonally opposite corners of the cube face. Steigmann and Goodyear [1] used improved crystallographic techniques to study Cd$_3$As$_2$ crystals. Weissenberg photographs indicated a body-centered tetragonal cell approximately four times the size of that determined previously [53] with $a = 1.267$ nm and $c = 2.548$ nm. These later values of a and c are, respectively, $\sqrt{2}$ and 2 times those reported in the earlier investigation. The ideal crystal lattice is illustrated in Figure 16 [1], which shows successive layers of atoms parallel to the (001) face of the unit cell. The ideal coordinates of the structure are tabulated, together with the coordinates of the final structure, in the article of Steigmann and Goodyear [1]. The Cd ions are displaced approximately 0.02 or 0.025 nm toward the neighboring Cd vacancies, while the As atoms are displaced only minimally from the positions of the ideal structure. In Figure 16, the circles represent Cd atoms in the plane of the diagram, while the crosses represent As atoms at $\Delta z = c/16$ below the plane of the diagram. V_1 and V_2 are the vacant Cd sites. In the tetragonal structure, the planar spacings d_{hkl} are given by

$$\frac{1}{d_{hkl}^2} = \frac{h^2 + k^2}{a^2} + \frac{l^2}{c^2} \qquad (3)$$

where the values of a and c are given previously.

At higher temperatures, there is evidence for a phase transition to other forms. A solid–solid phase transition has been reported at 578°C [54] or at 595°C [55]. Pietraszko and Łukaszewicz [55] reported three solid–solid phase transitions in all. The highest temperature β-form is a simple fluorite structure with $a_0 = b_0 = c_0 = 0.6403$ nm and exists above 600°C. The α''-phase exists from approximately 450–600°C and is a deformed fluorite structure with $a = \sqrt{2}c_0$, $b = \sqrt{2}b_0$, and $c = 2a_0$. The α'-phase exists between approximately 220 and 450°C and has $a = 2a_0$, $b = 2b_0$, and $c = 4c_0$. The α'-phase expressions also apply for the room-temperature α-phase, with very slight variations in the unit cell dimensions. The unit cells of the four different phases of Cd$_3$As$_2$ are illustrated in Figure 17 [55]. In the following, we are concerned only with thin films adopting the room-temperature α-phase.

Cd$_3$As$_2$ films are usually prepared by vacuum evaporation [6, 56, 57]. However, there is also a significant body of work on films prepared by pulsed laser evaporation (PLE), in particular with regard to the structure and morphology [58–60]. Sputtering has also occasionally been employed for deposition of Cd$_3$As$_2$ films. It was established in the early 1960s that Cd$_3$As$_2$ dissociates on evaporation according to the reaction [61, 62],

$$Cd_3As_2(s) \rightleftarrows 3Cd(g) + 0.5As_4(g)$$

Lyons and Silvestri [61] assumed that the compound exists as a monomer in the vapor phase, and calculated the vapor pressure from dew point measurements. These were compared with

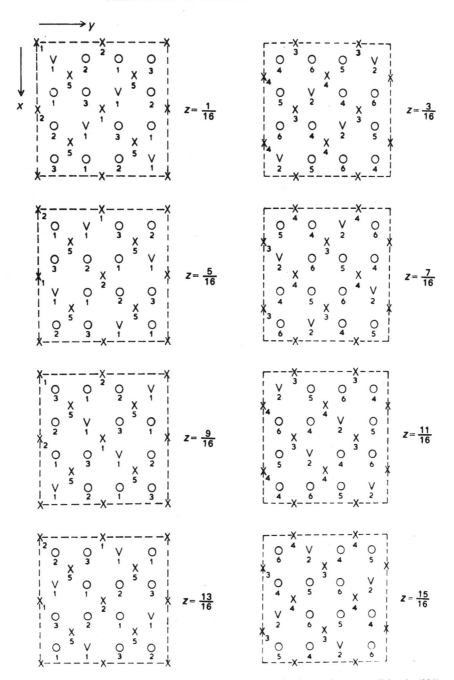

Fig. 16.   Idealized crystal structure of Cd_3As_2, showing successive layers of atoms parallel to the (001) face of the tetragonal unit cell. Circles represent Cd atoms in the plane of the diagram. Crosses represent As atoms at $\Delta z = c/16$ below the plane of the diagram. V_1 and V_2 are vacant Cd sites. Reproduced from [1]. Reprinted with permission from G. A. Steigmann and J. Goodyear, *Acta Crystallogr. B* 24, 1062, copyright Munksgaard International Publishers Ltd., Copenhagen, Denmark, 1968.

results obtained by direct pressure measurements using a Bourdon gauge, which demonstrated the vapor phase dissociation of the compound over the temperature range 434–695°C. Westmore, Mann, and Tickner [62] used a mass spectrometer to study the vapor above Cd_3As_2 over the temperature range 220–280°C. These workers pointed out that Cd_3As_2 has a much lower vapor pressure than either Cd or As. They measured the variation with temperature of the partial pressures of the various

species. Cd_3As_2 recondenses from the vapor with no apparent decomposition.

Żdanowicz and Miotkowska [63] performed a thorough investigation of the structure and the morphology of evaporated Cd_3As_2 films deposited onto mica substrates. The substrate temperatures during deposition were maintained at fixed values in the range 20–200°C. The deposition rate was controlled by varying the source temperature to cover a range of 2 orders of magnitude from 1 to 100 nm s^{-1}. Crystal structures and orientations were determined by detaching the films from the substrates and using both X-ray and electron diffraction analysis. Diffraction patterns indicated that the crystalline films were of the low-temperature tetragonal α-phase; amorphous films recrystallized following interaction with the electron beam in the electron microscope and also reverted to a tetragonal structure. The main findings of this work are illustrated in Figure 18 [63]. It was found that over the substrate temperature and deposition rate ranges investigated, the films could have an amorphous, polycrystalline or well-oriented crystal structure. Below approximately 125–145°C, the films were amorphous, with polycrystalline films resulting when this temperature was exceeded. The amorphous-to-crystalline transition was only weakly dependent on the deposition rate. However, the epitaxial temperature, which delineates polycrystalline and epitaxial films, increased with increasing deposition rate as shown in Figure 18 [63]. For deposition rates less than about

4 nm s^{-1} (40 Å s^{-1}), there was a direct transition from the amorphous to the well-oriented crystalline structure, whereas at higher rates there was a transitional polycrystalline region. The preferred orientation of the films was [001] with the c-axis of the tetragonal structure perpendicular to the plane of the substrate. Din [64] also reported that film structures were tetragonal, although there was no strong preferential orientation in this case; again, films were less crystalline at reduced substrate temperatures.

In films prepared by PLE, it was found that the structure was composed of amorphous agglomerates, typically 60–200 nm in size [58]. These workers also prepared films using the same deposition method, but with a decreased residual gas pressure, in an attempt to decrease the amorphous-crystalline transition temperature [59]. They found that films also crystallized in the tetragonal structure, but that the preferred orientation for films deposited above 140°C was with the a-axis perpendicular to the substrate plane, in contrast to films prepared by evaporation as noted earlier. Another feature noted in such films was that the d-values calculated from reflections obtained from the films deviated from the literature values for the bulk material. This was ascribed to stresses built up in the films. This was consistent with the thicker films having d-values closer to the bulk values than those of the thinner films.

There is little mention concerning the grain size of microcrystallites. Dubowski and Williams [59] observed microstructures of about 0.1-μm thickness and about 0.1–0.2 μm in diameter. There was an increase in their size up to 0.5 μm when the film thickness increased from 0.1 to 1.8 μm. Din [64] made some measurements of the linewidth of the (048/440) X-ray peak for Cd$_3$As$_2$ films of thickness 0.25, 0.30, and 0.35 μm.

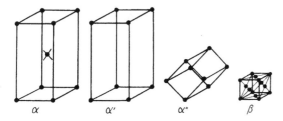

Fig. 17.   Unit cells of Cd$_3$As$_2$ crystals in different phases: α (room temperature), α' (220–450°C), α'' (450–600°C), and β (above 600°C). Reproduced from [55]. Reprinted with permission from A. Pietraszko and K. Łukaszewicz, *Phys. Status Solidi A* 18, 723, copyright Wiley-VCH, 1973.

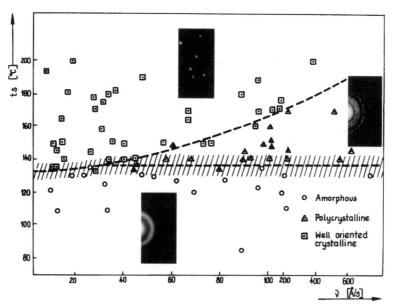

Fig. 18.   Effects of film growth rate v and substrate temperature t_s on the structure of Cd$_3$As$_2$ films vacuum evaporated onto mica substrates. Reproduced from [63]. Reprinted from *Thin Solid Films*, 29, L. Żdanowicz and S. Miotkowska, Effect of Deposition Parameters on the Structure of Vacuum-Evaporated Cadmium Arsenide Films, p. 177, copyright 1975, with permission from Elsevier Science.

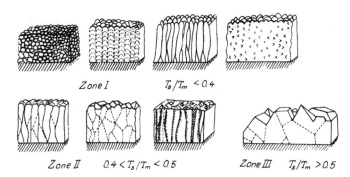

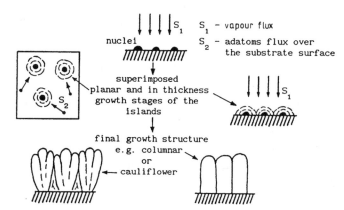

Fig. 19. Schematic three-zone model for the temperature dependence of the growth morphology in evaporated Cd_3As_2 films. T_s/T_m represents the ratio of the substrate temperature to the melting point of Cd_3As_2. Reproduced from [65]. Reprinted from *Thin Solid Films*, 67, J. Jurusik and L. Żdanowicz, Electron Microscope Investigations of the Growth Morphology of Cadmium Arsenide Films Vacuum Deposited at Various Substrate Temperatures, p. 285, copyright 1980, with permission from Elsevier Science.

Fig. 20. Postnucleation growth in evaporated Cd_3As_2 films considered as a consequence of two superimposed (planar and thickness) growth stages. Reproduced from [66]. Reprinted from *Thin Solid Films*, 214, J. Jurusik, Transmission Electron Microscopy Investigation of the Post-Nucleation Growth Structure of Amorphous Cadmium Arsenide Thin Films, p. 117, copyright 1992, with permission from Elsevier Science.

In each case, the substrate temperature was 180°C during deposition and the deposition rate was $0.5\ nm\ s^{-1}$. In all of the films, the microcrystallite grain size was of the order of 25 nm. There appeared to be a slight decrease in the grain size from 25.7 to 23.9 nm as the film thickness decreased from 0.35 to 0.25 μm. There also appeared to be a very minimal decrease in grain size from 24.5 to 23.9 nm when the deposition rate was reduced from 1 to $0.5\ nm\ s^{-1}$. A more significant increase in the mean grain size was observed after annealing at 200°C for 2 h. In this case, the grain size increased from 25.7 to 29.4 nm. However, these variations are unconfirmed as yet, and reflect only minor variations in the growth and crystallization processes.

The morphological characteristics of evaporated Cd_3As_2 films have been investigated by Jurusik and Żdanowicz using electron microscopy [65]. It was found that the microstructure of the films was strongly dependent on the substrate temperature T_s. These were interpreted in terms of a simple three-zone model in which the structure was correlated with the ratio T_s/T_m, where T_m is the material melting point, having a value of 994 K for Cd_3As_2 [65]. They found that for $T_s/T_m \leq 0.4$ the films were amorphous, consisting of spherical or elongated fibrous-like clusters, or were uniform with a large number of pores or cavities. For $0.4 < T_s/T_m < 0.5$ columnar growth was identified, while for $T_s/T_m \geq 0.5$ films with large crystallites and clear grain boundaries were observed. These structures are illustrated in Figure 19 [65]. A very detailed study was performed on amorphous Cd_3As_2 evaporated onto substrates maintained at 293 K [66, 67]. It was found that the size of postnucleation islands were functions of the substrate temperature, type of substrate, and the deposition rate. Furthermore, the shape and the type of growth of the islands varied with these parameters. At a low deposition rate ($1\ nm\ s^{-1}$), the initial nuclei grew until a certain size was reached, and new nuclei started to grow around the original islands. For a moderate deposition rate ($3.5\ nm\ s^{-1}$), the initial nuclei grew into islands which in turn increased in size until the islands had only narrow interfaces between them. Some islands formed poly-

gons. Films deposited at a high rate ($50\ nm\ s^{-1}$) consisted of small islands connected by relatively large low density areas. Films deposited at low rates were of approximately stoichiometric proportions (i.e., 40 at% As) while those deposited at higher rates had excess arsenic (typically 51.8 at% As). It was suggested that the growth process could be considered as a superposition of two growth stages as shown in Figure 20 [66]. The vapor flux onto the substrate is denoted by S_1 and the adatom flux over the substrate surface is denoted by S_2. In the first stage of growth, the initial nuclei enlarge owing to the adatom flux S_2. The second stage of growth incorporates atoms immediately from the incident flux S_1. This is usually very slow, but increases as the island surface area increases. At this stage, branched complex islands or only compact islands grow as the result of a shadowing mechanism. When the islands touch each other, the molecular flux S_2 over the substrate surface is eliminated, and the films grow in thickness as the result of the incident flux S_1 and the structure initiated in the earlier stage (i.e., large branched or smaller compact islands). The final growth structure gives rise to cauliflower-like or columnar structures. In later work [67], Cd_3As_2 films were evaporated from bulk crystallites of Cd_3As_2 or $CdAs_2$. For films grown from the Cd_3As_2 source, a pronounced columnar structure was observed, whereas those evaporated from the $CdAs_2$ source grew uniformly, and the columnar structure was absent. In the latter case, the initial island structure was not observed in the early stages of film growth. It was concluded that prerequisites for columnar growth were a limited surface mobility and an oblique component in the deposited flux, particularly where the films are deposited normally to the substrate surface.

Din [64] used scanning electron microscopy (SEM) to study the morphology of films at substrate deposition temperatures of 300, 418, 455, and 477 K. Films deposited at 300 K showed no significant surface structure, with amorphous regions. At 418 K, larger regions with typical dimensions 250–300 nm were observed, which were probably crystallites within an amorphous matrix. At 455 K, well-defined grains of size sometimes exceeding 300 nm covered the entire surface, whereas at 473 K larger grains were observed of size up to 600 nm. These grain sizes were considerably larger than typical microcrystallite grain sizes found by the same author [64] using the Scherrer method. This apparent discrepancy could be resolved if the larger grains observed by SEM are composed of many smaller microcrystallites. Annealing at 180°C for 2 h resulted in grains of larger size forming in amorphous films and an increase in grain size for films which are already polycrystalline.

We have seen that Cd_3As_2 films are deposited with a tetragonal structure (α-phase) at moderate temperatures, with amorphous films occurring in the lower temperature range. There are reports that microcrystalline films are deposited with the c-axis perpendicular to the substrate plane for films deposited by evaporation [63]; for films deposited by PLE the a-axis was perpendicular to the substrate plane [59]. There do not appear to be any substantial reports concerning microcrystallite grain size, but work indicates a value of typically 25 μm with slight variations depending on substrate temperature, film thickness, and deposition rate. A model of film growth has been proposed which appears to explain the structural features found in both amorphous and microcrystalline films reasonably satisfactorily [65–67]. Deposition at high substrate temperatures tends to increase the observed grain size, each of which is probably composed of many microcrystallites.

3. ELECTRICAL PROPERTIES

3.1. Band Structure

The band structure of a crystalline solid is a representation of its energy-wave vector (E–k) relationship. For a periodic potential within a crystal lattice, most texts on solid-state physics show, using the Bloch theorem, that the energy is periodic within the reciprocal lattice. Usually, the primitive cell of the reciprocal lattice, the first Brillouin zone, is considered and calculations of the E–k relationship are made along conventional symmetry directions. For the materials that are considered in this chapter, the relevant direct crystal lattices are the wurtzite, zinc-blende, and tetragonal structures. The unit cell of the reciprocal lattice is determined by that of the direct lattice, and so largely in turn is the energy band structure. Most optical and electronic properties are directly related to the band structure, thus justifying the major importance of this concept in solid-state physics. Earlier, it was noted that the cadmium chalcogenides can exist in both the wurtzite and the zinc-blende structures. In the following, only the band structure relevant to the most readily observed structures in thin films will be presented, although in principle a band structure for each crystal form will exist. For Cd_3As_2, the tetragonal and fluorite structures are relevant to band structure calculations.

Band structures in semiconductors may be determined by several different techniques, including expansion of the valence band and conduction band states as plane waves, pseudopotential methods, and the so-called **k.p** method to solve the Schrödinger equation. These methods are not discussed here, as they are beyond the scope of this chapter, although the methods used for the determination of each of the band structures included are noted. For further details, the reader is referred to the relevant reference.

3.1.1. Cadmium Sulfide

CdS is deposited mainly in the wurtzite structure [19]. In this case, the first Brillouin zone in the reciprocal lattice is as shown in Figure 21 [68]. The wurtzite band structures shown in some of the following diagrams start from the wave vector at point A in the reciprocal lattice, follow the line R to the point L, the line U to the point M, and the line Σ to the Γ point (0, 0, 0), before returning along the line Δ to point A. The diagrams then follow the line S to the point H, follow P to point K, and return via line T to the point Γ. The wurtzite structure calculations for CdS and CdSe were based on those for wurtzite ZnS, which in turn was based on the zinc-blende ZnS form factors reported previously [69]. The detailed electronic band structure for CdS calculated using the pseudopotential method is illustrated in

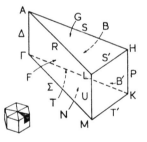

Fig. 21. Hexagonal first Brillouin zone of the wurtzite crystal structure showing lines and points of symmetry. Reproduced from [68]. Reprinted with permission from T. K. Bergstresser and M. L. Cohen, *Phys. Rev.* 164, 1069 (1967), copyright 1967 by the American Physical Society.

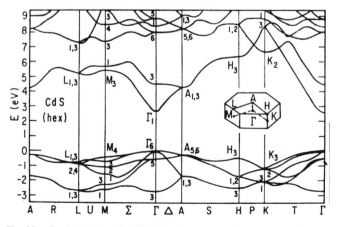

Fig. 22. Band structure of CdS of the hexagonal wurtzite structure. Reproduced from [68]. Reprinted with permission from T. K. Bergstresser and M. L. Cohen, *Phys. Rev.* 164, 1069 (1967), copyright 1967 by the American Physical Society.

Figure 22 [68]. In CdS, the fundamental energy bandgap corresponds to the $\Gamma_6 \to \Gamma_1$ transition. This transition is vertical on the energy band diagram since the lowest conduction band minimum is directly above the highest valence band maximum. This represents a transition in which there is no change in momentum, and CdS is therefore a direct-gap semiconductor with fundamental energy bandgap $E_g = 2.6$ eV. The relevant transitions and band splittings are tabulated and are discussed at length in the reference. It was also reported that there was no significant change in the form factors in going from the zinc-blende to the wurtzite crystal structure.

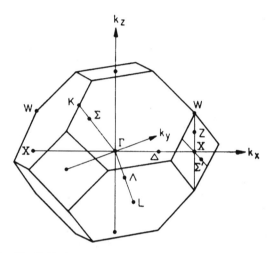

Fig. 23. First Brillouin zone for the face-centered cubic, diamond, and in particular the zinc-blende crystal structures showing lines and points of symmetry. The shape is that of a truncated octahedron. Reproduced from [71]. Reprinted with permission from G. Dresselhaus, *Phys. Rev.* 100, 580 (1955), copyright 1955 by the American Physical Society.

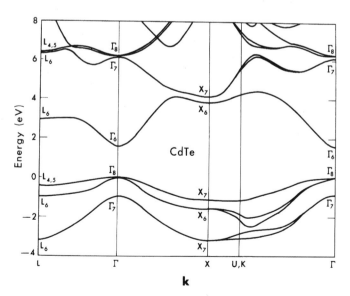

Fig. 24. Band structure of CdTe of the cubic zinc-blende structure. Reproduced from [73]. Reprinted with permission from D. J. Chadi et al., *Phys. Rev. B* 5, 3058 (1972), copyright 1972 by the American Physical Society.

3.1.2. Cadmium Telluride

For the case of CdTe, the zinc-blende structure is that most widely encountered. Parmenter [70] originally used group theory to investigate the symmetry properties of the one-electron energy bands of the zinc-blende crystal structure. These were studied at lines and points in the first Brillouin zone. The Brillouin zone for this structure is a truncated octahedron, as illustrated in Figure 23 [71]. The designation of points and directions shown in the figure was originated by Bouckaert, Smoluchowski, and Wigner [72]. The zinc-blende band structure for CdTe was originally determined by Cohen and Bergstresser [69] using the pseudopotential method. In this and the later band structure diagram given in the following, the diagram starts from the wave vector at the L point and follows the Λ direction to the Γ point. It then follows the Δ direction to the X point and then the line between the points X and K, the latter point designated U, K in Figure 24 [73]. Finally, the region along the line Σ between the points K and Γ is followed. Figure 24 [73] shows that CdTe has a direct energy bandgap of 1.59 eV, also located at the Γ point. This band structure was also calculated using the empirical pseudopotential method, and was correlated with the reflectivity spectrum of CdTe. A similar band structure and E_g value were obtained in the earlier work [69]. Again comprehensive data and discussion are given in the original publications [69, 73], including aspects of the spin–orbit splitting.

3.1.3. Cadmium Selenide

The band structure of CdSe in the wurtzite form was determined by Bergstresser and Cohen, along with that of CdS [68]. Having similar crystal structures, these two materials have the same Brillouin zone structure as shown in Figure 21. There is also a very considerable similarity between the energy band structures for the two materials determined in this work [68] by the pseudopotential method. However, the fundamental energy bandgap for CdSe was found to be 2 eV, less than the value of 2.6 eV for CdS, but associated with the same transition. The work on CdSe was presented in the form of energy band structure (E–k) diagrams similar to that of Figure 22 [68] for CdS, and spanning the k-vector symmetry directions between the points A, L, M, Γ, A, H, K, Γ. A similar, but slightly less detailed representation of the energy band structure of CdSe in the case of thin films rather than the bulk material was presented by Ludeke and Paul, and is shown in Figure 25 [74]. In this work, epitaxial thin films were grown by flash evaporation on cleaved BaF_2 substrates and optical transmission measurements were made for energies up to 6.5 eV in the ultraviolet spectral region. Since the epitaxial films were grown on the (111) BaF_2 face, the measurements were taken with an effective polarization of the electric field vector perpendicular to the c-axis of the wurtzite structure. The band structure diagram of Figure 25 [74] terminates at the L and K points, and is not plotted at energies above 7 eV, so some of the detail in the work of Bergstresser and Cohen [68] on the bulk crystal structure is absent. Nevertheless, both the band structures are similar, with a consistent

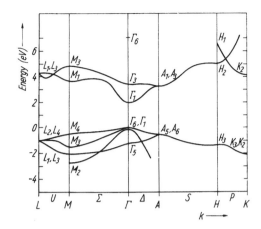

Fig. 25.   Band structure of CdSe thin films of the hexagonal wurtzite structure. Reproduced from [74]. Reprinted with permission from R. Ludeke and W. Paul, *Phys. Status Solidi* 23, 413, copyright Wiley-VCH, 1967.

direct energy bandgap $E_g \approx 2$ eV at the Γ point. The various peaks in the transmission spectrum of wurtzite CdSe were associated with various transitions in the conduction and valence bands.

3.1.4. Cadmium Arsenide

The band structure of Cd_3As_2 is extremely difficult to determine. This is related primarily to the exceptionally complicated crystal structure determined by Steigmann and Goodyear [1], which is shown in Figure 16. Lin-Chung [75] has pointed out that the body-centered tetragonal structure has 32 As ions and 48 Cd ions per unit cell, with 256 valence electrons per unit cell and a very involved structure factor. This would in principle entail solving secular determinants of the order of 1000, making a comprehensive theoretical study virtually impossible.

This worker used a pseudopotential method, and the results of this band structure calculation are given in Figure 28.

Band structure models which have been suggested include direct and indirect gaps, a single nonparabolic conduction band, and a conduction band with two separate valleys. Moreover, there is a wide variation in the value of the energy bandgap E_g proposed. Sexer [76, 77] proposed that the conduction band has two valleys and the carriers have two different effective masses. The simplified band structure is shown in Figure 26 [77]. Both of the bands are parabolic, and their minima are separated by $\varepsilon_d = 0.15$ eV. The effective mass in the lower valley $m_2 = 0.06m_0$, where m_0 is the free electron mass. In the higher valley, the effective mass $m_1 = 0.12m_0$. In the figure, (a) represents the situation when there is only a low concentration of carriers, and the Fermi level ξ_1 is above the bottom of the lower band, but below that of the higher heavier electron band. In (b), the situation is shown for a higher carrier concentration where the Fermi level has moved up to the bottom of the higher band and $\xi_1 = \varepsilon_d$. Finally, in (c) the situation is shown where there is a very high carrier concentration and the Fermi level has now risen above the minimum of the higher band. In this figure, ξ_2 represents the height of the Fermi level above the minimum of the higher valley.

Conversely, Armitage and Goldsmid [78] suggested that the conduction band was nonparabolic of the type proposed previously by Kane [79]. Using this assumption, it was possible to determine the effective mass from Hall effect measurements on Cd_3As_2, and set limits on the maximum values of E_g and the effective mass m^*. Values of $E_g < 0.1$ eV and $m^* < 0.01m_0$ were obtained. These effective mass values were in reasonable agreement with those of earlier work, and it was considered that the results provided strong evidence for a nonparabolic band of the Kane type. Additional evidence for a Kane-type single conduction band was provided by Blom and Schrama [80]. They measured the Hall and Seebeck coefficients, from which they

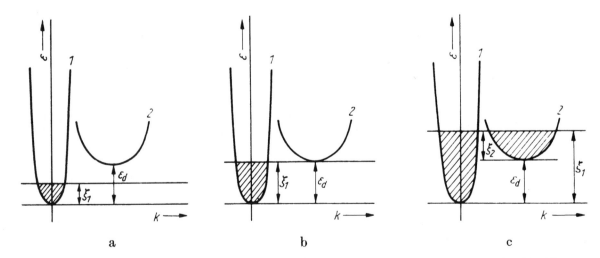

Fig. 26.   Band structure models for Cd_3As_2 for the case of a low carrier concentration (a), a higher carrier concentration (b) and for a very high carrier concentration (c). ε_d represents the energy difference between the bottom of the two conduction bands, and ξ_1 and ξ_2 represent, respectively, the positions of the Fermi level with respect to the bottom of the lower band and to that of the higher band. Reproduced from [77]. Reprinted with permission from N. Sexer, *Phys. Status Solidi* 21, 225, copyright Wiley-VCH, 1967.

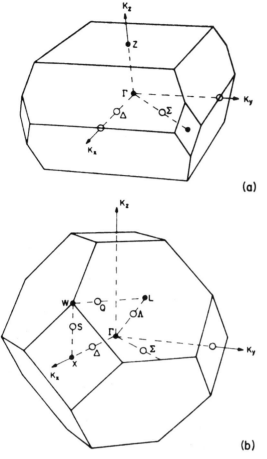

Fig. 27. First Brillouin zone of a body-centered tetragonal crystal structure (a) and the fluorite structure (b) showing lines and points of symmetry. (b) Is essentially the same as the face-centered cubic Brillouin zone illustrated in Figure 23. Reproduced from [75]. Reprinted with permission from P. J. Lin-Chung, *Phys. Rev.* 188, 1272 (1969), copyright 1969 by the American Physical Society.

were able to determine the effective mass for increasing electron concentrations in the range $(0.7–13.2) \times 10^{24}$ m^{-3}. The relationship between the effective mass and the carrier concentration was correlated with a Kane-type conduction band and also enabled values of $E_g = 0.15$ eV and $m^* = 0.012 m_0$ to be estimated at the bottom of the conduction band. Measurements of interband absorption and magnetoabsorption have been performed by Wagner, Palik, and Swiggard [81] and these were consistent with a Kane-type model for the light conduction and valence bands. They suggested an α-Sn-like zero bandgap model for Cd$_3$As$_2$.

Results obtained by Lovett [82] tended to support the two-conduction band model of Sexer, in that his results indicated that Cd$_3$As$_2$ has upper and lower conduction band minima, although both were probably not parabolic. These conclusions were reached since Hall coefficients were dependent on the magnetic field intensity and indicated the presence of two different types of electron carrier with different mobilities.

The Lin-Chung [75] model for the band structure of Cd$_3$As$_2$ was presented in 1969. Due to the immense complexity of the problem noted earlier, a pseudopotential was used instead of a

self-consistent crystal potential. Furthermore, the known crystalline structure [1], was simplified to render the problem more tractable. It was stressed that the Cd$_3$As$_2$ crystal structure is basically the simpler fluorite crystal structure with Cd vacancies distributed periodically through the crystal. The effect of vacancies was taken into account by a vacancy pseudopotential, which is the negative of the Cd atomic potential. The band structure calculation was then performed using the hypothetical fluorite structure, and the relationship between this and the real structure was explored. For the real structure, the Brillouin zone is that for the body-centered tetragonal lattice, and for the fluorite structure it is that for the face-centered cubic (fcc) structure. These are illustrated in Figure 27 [75], where (a) represents the first Brillouin zone of the real lattice and (b) represents that for the hypothetical fluorite structure. The volume of the Brillouin zone in the real structure is only 1/16th of that for the fluorite structure. As in the earlier discussions for Brillouin zones in the zinc-blende and wurtzite structures, the symmetry points and directions are indicated. The resulting overall calculated band structure for the hypothetical fluorite structure is shown in Figure 28 [75], where spin–orbit couplings and other relativistic effects were not included in the calculations. According to Lin-Chung [75] this resembles the band structures of InAs, GaAs, etc. [69]. Although a full description of the band structure is given by Lin-Chung, only a summary of the major points is appropriate here. The valence band maximum for Cd$_3$As$_2$ occurs along the $\langle 100 \rangle$ and $\langle 110 \rangle$ directions at the points $(0.065, 0, 0)$ and $(0.125, 0.125, 0)$, close to the center of the Brillouin zone. The direct gap at the Γ point ($\Gamma_{15} \rightarrow \Gamma_1$ transition) was 0.6 eV, somewhat higher than experimental values attributed to Cd$_3$As$_2$ previously. Having calculated the band structure of the hypothetical structure, the differences expected for the real crystal structure were examined. By considering the residual potential, the difference between the potential used in the calculations and the potential for the real crystal, it was argued that the band edge structure for the real crystal structure differed from that of the hypothetical structure at the Γ point. The effects of the small residual potential, which acts on electrons in the fluorite structure, would mix states having wave numbers differing by the reciprocal lattice vector of the real structure, and would only have a significant effect where the unperturbed states that were mixed have a very small energy difference. By considering X_1 and Γ_1, at the bottom of the conduction band, it was argued that since the energy difference is of the order of 0.5 eV, they would interact via the residual potential and push each other apart. The overall effect of these perturbations would be to reduce the fundamental energy gap E_g to approximately 0.2 eV or less, or even to a zero gap gray Sn-like structure as proposed earlier [81]. It was also pointed out that Sexer [77] had suggested two conduction bands with different effective masses, and that this was consistent with Lin-Chung's results [75].

A comprehensive investigation of the transport properties and optical features of the Cd$_{3-x}$Zn$_x$As$_2$ alloy system was performed by Rogers, Jenkins, and Crocker [83]. The optical measurements were performed both on the bulk alloy and on evaporated thin films. Since Cd$_3$As$_2$ and Zn$_3$As$_2$ are isomor-

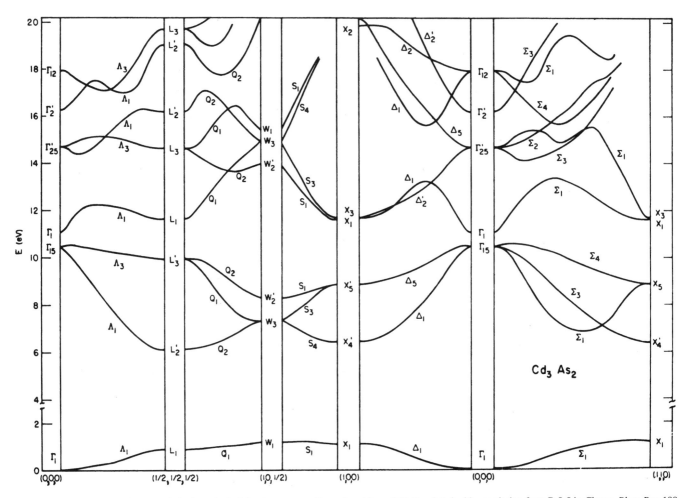

Fig. 28. Band structure of Cd$_3$As$_2$ in its hypothetical fluorite structure. Reproduced from [75]. Reprinted with permission from P. J. Lin-Chung, *Phys. Rev.* 188, 1272 (1969), copyright 1969 by the American Physical Society.

phous it was possible to prepare a range of compositions with x varying systematically in the range 0–3, where $x = 0$ represents pure Cd$_3$As$_2$ and $x = 3$ represents pure Zn$_3$As$_2$. Hall effect and resistivity measurements were made on the bulk samples, which were then ground down to thicknesses in the range 50–100 μm for measurements of infrared transmission and reflectivity spectra over the wavelength range 1–15 μm. Alloys with $x < 1.5$ were always n-type. The samples were intrinsic at temperatures above 300 K. For the degenerate n-type alloys and in particular Cd$_3$As$_2$, the Hall coefficient was independent of temperature. In the opinion of these workers, this would not occur if there were a heavy-mass conduction band near to the principal conduction band at the center of the zone, as proposed by Sexer [77]. From the transmission and reflectance data of bulk and thin film samples, it was found that the absorption coefficient had two distinct edges. In the bulk samples, this was characteristic of an indirect transition, and in the thin film samples a direct transition was observed. It was observed that there was a linear relationship between the square of the effective mass and $n^{2/3}$, where n is the carrier concentration, as had previously been reported by Armitage and Goldsmid [78], which implied a Kane-type conduction band. A full consideration of

these results led these workers to propose two alternative band structures for Cd$_3$As$_2$ and Cd$_{1.8}$Zn$_{1.2}$As$_2$. Those for Cd$_3$As$_2$ are shown in Figure 29 [83]. The left-hand figure shows the simpler band structure, which has two valence bands and one conduction band. There may be a significant overlap between the principal valence and conduction bands, making the material a semimetal. The principal valence band is positioned at π/a, with effective mass 0.5m_0. It was stressed that this differed somewhat from the energy band structure proposed by Lin-Chung [75], which included a second conduction band. The indirect gap structure proposed by Rogers, Jenkins, and Crocker [83] was justified primarily because of the form of the wavelength dependence of the absorption coefficient. The more complicated band structure shown in the right-hand figure was proposed when taking into account a discrepancy between $\varepsilon_G{}^*$ and $[(\varepsilon_G)_{\text{opt}}]_{\text{direct}}$ as shown. The heavy-mass conduction band is energetically too far from the light-mass band to influence free-carrier transport in Cd-rich alloys (in particular Cd$_3$As$_2$).

Measurements were made of thermomagnetic effects in Cd$_3$As$_2$ [84]. Values of carrier concentration and mobility were determined using measurements of carrier resistivity and Hall coefficient. The results were explained primarily by assuming

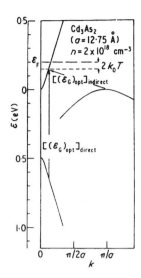

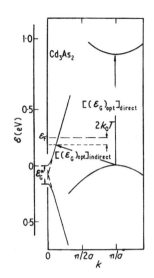

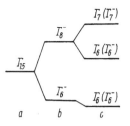

Fig. 29. Proposed band models for Cd_3As_2 having two valence bands and one conduction band (left-hand figure) and a more complicated structure devised to explain a discrepancy between the direct optical bandgap $[(\varepsilon_G)_{opt}]_{direct}$ and a smaller interaction energy ε_G^* (right-hand figure). Reproduced from [83]. Reprinted with permission from L. M. Rogers, R. M. Jenkins, and A. J. Crocker, *J. Phys. D: Appl. Phys.* 4, 793, copyright IOP Publishing Ltd., 1971.

Fig. 30. Valence band structure of arsenic and also phosphorus (bracketed) II_3V_2 compounds at the Γ point. This represents the basic Lin-Chung model (a), and the Lin-Chung model with the spin–orbit interaction included (b) and also with the tetragonal field distortion included (c). Reproduced from [91]. Reprinted with permission from B. Dowgiałło-Plenkiewicz, and P. Plenkiewicz, *Phys. Status Solidi B* 87, 309, copyright Wiley-VCH, 1978.

a band structure of the type previously proposed by Sexer [77]. For low carrier concentrations, only the lower band at $k = 0$ would be filled. The effective mass value derived was similar to that of Sexer in the lower band. A study of the Hall coefficient R was performed for Cd_3As_2 [85] samples having different carrier concentrations over the temperature range 4.2–300 K. A rise in R for concentrations of $(2–6) \times 10^{24}$ m^{-3} was identified with the existence of carriers of two different types in the conduction band, having different mobilities and effective masses. Using the data, it was possible to estimate the effective masses in the two conduction bands. In the lower band, the effective mass $m_1^* = 0.045m_0$ at 4.2 K and $0.054m_0$ at 300 K according to the work of Ballentyne and Lovett [84]. In the higher band, a temperature independent value of $m_2^* = 0.09m_0$ was adopted. Using these values, the dependence of R over the full temperature range explored was calculated and was compared with measured values of the authors and other workers. There was good agreement between this model and the experimental measurements, which was regarded as evidence for the band model of Sexer [77] shown in Figure 26.

Aubin, Brizard, and Messa [86] made measurements of the magneto-Seebeck effect and magnetoresistance measurements at 80 and 300 K. From these measurements, they also determined that there were two conduction bands. The effective mass of the bottom of the first conduction band was $0.012m_0$, in agreement with the earlier results of Blom and Schrama [80]. In heavily doped samples, the effects of the second conduction band were detected, with the energy difference between the two minima being 0.4 eV. Additionally, Iwami, Matsunami, and Tanaka [87] measured the electrical properties of Cd_3As_2 crystals cut along the [100], [001], or [110] axes. The magnetoresistance coefficient was found for carrier concentrations of

10^{24}, 3×10^{24}, and 10^{25} m^{-3}. The longitudinal magnetoresistance coefficient was not zero in several samples, giving support to the supposition of a nonsimple conduction band structure. Its value was close to zero for the sample with carrier concentration 3×10^{24} m^{-3}, but was nonzero for the other two concentrations. For the zero magnetoresistance coefficient sample, a single ellipsoidal Fermi surface elongated in the c-direction and located at the center of the Brillouin zone, as suggested by Rosenman [88], was consistent with the experimental results. However, the nonzero values of the magnetoresistance obtained for the other carrier concentrations could not be explained by this model. Furthermore, the fact that the magnetoresistance decreased with increasing carrier concentration was inconsistent with Sexer's model [76, 77]. It was suggested that to a first approximation, Cd_3As_2 could be considered as having cubic symmetry, with a germanium-type band structure for lower carrier concentrations. Infrared absorption studies on Cd_3As_2 thin films led Iwami, Yoshida, and Kawabe [89] to conclude that the material has a conduction band with two minima, the lower at $k \neq 0$ and the upper at $k = 0$. An energy bandgap value of $E_g = 0.11$ eV was obtained.

Bodnar [90] took into account the influence of the tetragonal field for a narrow-gap tetragonal semiconductor, and concluded that Cd_3As_3 has an inverted HgTe-type energy gap of value $E_g = -0.095$ eV. The conduction band and the first valence band come into contact only at two values of the wave vector, where the energy bandgap is effectively zero. The model was confirmed to some extent by some calculations of the spin–orbit and the crystal field splitting of various II_3V_2 compounds, including Cd_3As_2, at the Γ point [91]. The calculations were based on the Lin-Chung model [75] using the same assumptions and pseudopotential form factors. Owing to the great difficulties in performing calculations involving the symmetry of II_3As_2 compounds, the spin–orbit splitting in the fluorite structure, rather than in the real crystal structure was calculated. For Cd_3As_2, a splitting of 0.27 eV at the Γ point was determined. Second, the splitting due to the crystal field superimposed on the fluorite structure was investigated. The energy separation between the Γ_7 and the upper Γ_6 level was found to be 0.07 eV due to the tetragonal crystal field, while the energy difference between the Γ_7 and the lower Γ_6 level was equal to 0.31 eV corresponding to the spin–orbit splitting. These valence band splittings are illustrated in Figure 30 [91]. (a) shows the valence band at the Γ point for the Lin-Chung model, while (b) illustrates the spin–orbit splitting (0.27 eV), and (c) shows the effect of both the spin–orbit splitting and the tetragonal distortion (0.07 eV to the upper Γ_6 level and 0.31 eV to the lower Γ_6 level). This scheme gave values consistent with those determined earlier. Aubin, Caron, and Jay-Gerin [92] used a crystal field splitting of 0.04 eV and spin–orbit splitting of 0.30 eV, whereas Bodnar [90] used the **k.p** method to determine values of 0.05 and 0.27 eV, respectively.

Blom et al. [93] made measurements on the so-called electronic g^*-factor in Cd_3As_2 samples. This followed predictions of Wallace [94] that the effective g^*-factor would be considerably more anisotropic than the cyclotron effective mass, particularly at low carrier concentrations, if the model suggested by Bodnar [90] for the valence band in Cd_3As_2 was correct. Measurements were made on bulk samples in which the c-axis direction was perpendicular to the longest side of the sample, allowing measurements to be made at different angles. Electron concentrations of 3.5×10^{23} and 7.8×10^{23} m^{-3} were used. Anisotropy of Shubnikov–de Haas oscillations could be explained down to an electron concentration of 2×10^{23} m^{-3} using the Bodnar model [90]. Calculated variations of the anisotropy of g^* described the angular dependence well, but the absolute values differed from the theoretical curves by a factor of about 20%. A further calculation taking into account the oscillation periods and the masses of four samples resolved the anomaly. They concluded that the anisotropy could be correctly accounted for in terms of the Bodnar model, and that complete agreement between theory and experiment might be expected from extending the **k.p** theory with higher band and free-electron corrections.

The preceding discussion on the band structure of Cd_3As_2 has been fairly lengthy, and includes a wide variety of experimental and theoretical work. This level of coverage is justified by the fact that, unlike the cadmium chalcogenides, the full band structure of Cd_3As_2 remains undetermined. Another useful summary of the various work concerning the band structure of Cd_3As_2 was given by Lovett [8]. The experimental work has suggested correlations with Kane-type bands, with two conduction bands and with an inverted band structure. Detailed theoretical calculations appear to have been limited to the hypothetical fluorite structure, without spin–orbit splitting, and with attempts to correct this for a superimposed tetragonal crystal field. In the related material, Cd_3P_2, an attempt has been made to calculate the crystal structure for real tetragonal symmetry, including spin–orbit splitting [95]. The band structure determined was quite different to that determined using the hypothetical structure. A definitive band structure for Cd_3As_2 is unlikely to emerge before such calculations are applied to it.

3.2. Lateral Resistivity

Resistivity measurements on thin films are usually made using a simple Hall effect system, from which the Hall mobility may also be determined. Measurements have used the van der Pauw [96] technique, which is insensitive to the shape of the film samples. It was shown that both measurements of the resistivity and the Hall effect made on a flat sample of arbitrary shape, such as a thin film, can be performed without knowing the current flow patterns. The samples are required to have four small contacts on the circumference, to be of constant thickness, and not to contain isolated holes. The four contacts, A, B, C, and D are arranged in order around the circumference, and the transresistances $R_{AB,\,CD}$ and $R_{BC,\,DA}$ are measured. $R_{AB,\,CD}$ represents the potential difference between the contacts D and C divided by the current through the contacts A and B. $R_{BC,\,DA}$ is defined in a similar manner. It was shown by van der Pauw that the resistivity ρ is then given by

$$\rho = \frac{\pi d}{\ln 2}\left(\frac{R_{AB,\,CD} + R_{BC,\,DA}}{2}\right) f\left(\frac{R_{AB,\,CD}}{R_{BC,\,DA}}\right) \quad (4)$$

where d is the film thickness, f is a function of the ratio of $R_{AB,\,CD}$ to $R_{BC,\,DA}$ only, and is given graphically by van der Pauw [96] and in tabular form [97].

Resistivity depends on the concentration and the mobility of the electronic charge carriers. Depending on the details of the scattering mechanisms within the material, both of these are normally temperature dependent. Moreover activation energies for carrier excitation may be determined from temperature measurements. In the following, we are concerned with the dependence of resistivity (or its inverse, conductivity) on the film thickness, substrate temperature during deposition, and deposition rate. There is frequently a correlation with the grain size and the crystal structure, as described in Section 2. A selection of work performed on each of the various materials is presented as follows.

3.2.1. Cadmium Sulfide

CdS films are of particular interest in Cu_2S-CdS and CdS-CdTe solar cells. There are several studies concerning the variation of resistivity with various deposition parameters in CdS films aimed at providing data for such technology [19, 98–100], and some of the general features are outlined in the following.

The dependence of resistivity has been investigated by several workers, and there appears to be fairly general agreement that the resistivity can be explained on the basis of the Petritz grain boundary model [101]. Wilson and Woods [19] proposed the following expression for the mobility,

$$\mu = \mu_B \exp\left(-\frac{A}{kT}\right)\left(1 - \frac{2\lambda}{d}\right) \quad (5)$$

which includes both the mobility variation proposed by Petritz [101], i.e., $\mu_k = \mu_B \exp(-A/kT)$ and a simple model of diffuse scattering at the film surface, $\mu = \mu_k(1 - 2\lambda/d)$ proposed by Flietner [102]. In these expressions, μ_B is the single-crystal mobility, A is the barrier height at the intercrystalline grain boundaries, λ represents the effective mean free path between surface collisions, and d is the film thickness. μ_k is the bulk mobility, including the effects of grain boundaries and impurity

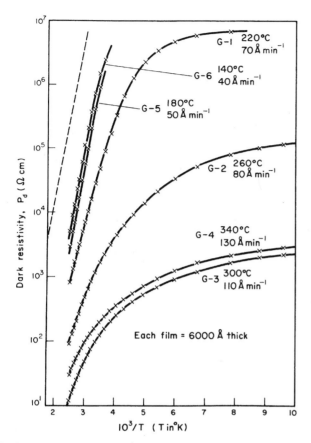

Fig. 31. Temperature dependence of the dark resistivity in evaporated CdS films deposited at different substrate temperatures and deposition rates. The broken line in the figure represents an activation energy of 0.50 eV. Reproduced from [19]. Reprinted from *J. Phys. Chem. Solids* 34, J. I. B. Wilson and J. Woods, The Electrical Properties of Evaporated Films of Cadmium Sulfide, p. 171, copyright 1973, with permission from Elsevier Science.

scattering. An essentially similar expression to Eq. (5) was earlier adopted by Kazmerski, Berry, and Allen [98] to account for mobility variations in CdS films. For a given film, the resistivity may be influenced not only by the temperature dependence of the mobility, but also by the carrier concentration. The overall resistivity ρ is then given by

$$\rho = \rho_0 \exp(E/kT) \qquad (6)$$

where ρ_0 is a constant and the activation energy may be obtained from the gradient of a $\log \rho - 1/T$ plot at high temperature.

Typical variations of resistivity as a function of inverse temperature are shown in Figure 31 [19]. The activation energy in the high-temperature region obtained from these data are in the range 0.29–0.50 eV, and includes a contribution from the activation energies of both the carrier density and the mobility. These values are reasonably consistent with those of Shallcross [103], who observed activation energies varying in the range 0.22–0.66 eV. Ray, Banerjee, and Barua [104] observed two distinct activation energies in CdS films which had not been subjected to heat treatment. Above 286 K, an activation energy of 0.26 eV was obtained for a film of thickness 0.41 μm, and an activation

energy of 0.16 eV for a thickness of 1.04 μm. At low temperatures, activation energies of 0.11 and 0.07 eV were determined, respectively, for these film thicknesses. A lower activation energy of 0.1 eV was also observed by Shallcross [103], which was identified with the carrier mobility. Ramadan, Gould, and Ashour [100] used van der Pauw measurements to determine the resistivity as a function of thickness in the range 0.32–1.0 μm deposited at a rate of 0.3 nm s^{-1} onto substrates held at 100 or 150°C. All of the higher temperature activation energies were of approximate value 0.5 eV. At low temperatures, activation energies of 0.02 eV were found, with a slightly higher value of 0.03 eV for a thicker 1.0-μm film. Shallcross [103] also observed a lower activation energy at low temperatures, and in this case the value of 0.1 eV was identified with the carrier mobility. In the case of the lower low-temperature activation energies described before [100], it was concluded that this indicated a hopping mechanism in the low-temperature region. Since in general it is accepted that the higher temperature activation energy represents a mean value of the intercrystalline barrier height, covering the effects of both the carrier concentration and the mobility, it is not unexpected that the values obtained by various workers differ, in the light of the different deposition methods and the conditions used. Nevertheless, the general behavior of resistivity with temperature is consistent, and indicates the existence of potential barriers between crystal grains which may be surmounted by thermal excitation. In Section 2.1, it was mentioned that microcrystallite grain sizes in CdS films are of the order of a few tens of nanometers, with macroscopic grains of the order of micrometers. Thus, grain boundary potential barriers in CdS films appear to be one of the major features determining the resistivity in the high-temperature range.

Most workers have also found that there is a decrease in resistivity with increasing film thickness, providing all other factors remain constant. Wilson and Woods [19] found that the dark resistivity at room temperature fell from approximately 10^5 to 10 Ω m in going from 0.1 to 4 μm in the case of films deposited at 250 nm min^{-1} (4.2 nm s^{-1}), or in going from 0.1 μm to 3 μm in the case of films deposited at 35–40 nm min^{-1}(0.6––0.7 nm s^{-1}). Hall effect measurements allowed these workers to differentiate between the effects of carrier concentration and mobility, and they concluded that variations in resistivity were mainly due to varying carrier concentrations, with surface scattering causing variations in mobility for low thicknesses. Equation (5) was proposed to account for the mobility variation. However, such an expression is unable to explain the very wide drop in resistivity associated with film thickness reported by Ramadan, Gould, and Ashour [100], where resistivity dropped from 10^8 to 10^3 Ω m for films deposited at a substrate temperature of 200°C. Kazmerski, Berry, and Allen [98] observed a slight increase in carrier concentration with increasing film thickness, and also that surface scattering affected the mobility of films of thickness up to 1 μm.

Buckley and Woods [99] performed a detailed study of the variation of resistivity with film thickness in CdS films, and found that there was a very strong correlation with the temperature of the source. There was a considerable difference in the

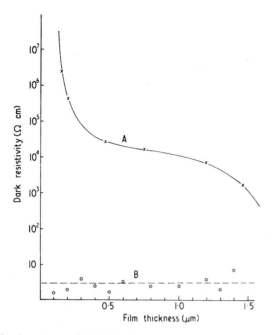

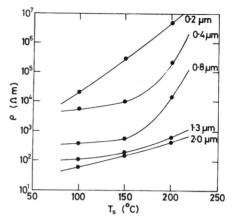

Fig. 33. Dependence of resistivity in evaporated CdS films on substrate temperature during deposition for different film thicknesses. Reproduced from [100]. Reprinted with permission from A. A. Ramadan, R. D. Gould, and A. Ashour, *Int. J. Electron.* 73, 717, copyright Taylor & Francis, 1992. Journal web site http://www.tandf.co.uk.

Fig. 32. Dependence of resistivity in CdS films on thickness for films evaporated in a closed system at temperatures of 800°C (A) and 850°C (B). Reproduced from [99]. Reprinted with permission from R. W. Buckley and J. Woods, *J. Phys. D: Appl. Phys.* 6, 1084, copyright IOP Publishing Ltd., 1973.

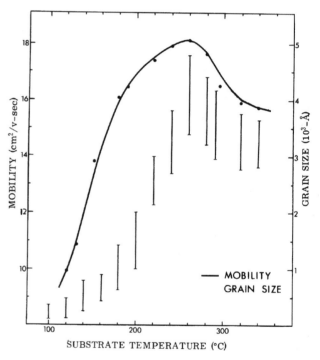

Fig. 34. Dependence of grain size and mobility in evaporated CdS films on substrate temperature during deposition, each indicating a maximum at 260°C. Reproduced from [98]. Reprinted with permission from L. L. Kazmerski, W. B. Berry, and C. W. Allen, *J. Appl. Phys.* 43, 3515, copyright, American Institute of Physics, 1972.

behavior of films evaporated at source temperatures of 800 and 850°C. Typical variations are shown in Figure 32 [99], where A corresponds to a temperature of 800°C and B to 850°C. For films deposited using the lower of these source temperatures, there was a substantial decrease in resistivity with increasing film thickness, but for films deposited using the higher source temperature the resistivity remained low and almost constant with respect to thickness within experimental error. They found that the carrier concentration for films deposited at the lower source temperature increased from 8×10^{17} to 8×10^{20} m^{-3} as film thickness increased to 1.5 μm, whereas for the films deposited at the higher source temperature the carrier concentration remained constant at about 5×10^{23} m^{-3}. They attributed these values to an increase in the Cd concentration in the deposited films with the deposition time. For the films deposited at 800°C, the Cd excess increased during the whole deposition time, while for the films deposited at 850°C, an equilibrium excess Cd concentration occurred early in the deposition of even the thinnest films. This careful work certainly confirms the increase in Cd concentration in the films with time, particularly at the lower source temperature, and seems the most plausible explanation for the decrease in resistivity with increasing film thickness generally observed.

There appears to be an increase in the film resistivity with increasing substrate temperature during deposition. The results of Ramadan, Gould, and Ashour [100] referred to earlier, were replotted in the form of resistivity as a function of substrate temperature, all other parameters remaining constant. The effect was considerably more pronounced for the thinner films, with an increase by a factor of 300 for a 0.2-μm film and an in-

crease only by a factor of 5 for a 2-μm film, as the substrate temperature increased from 100 to 200°C. These results are shown in Figure 33 [100]. Shallcross [103] observed a similar increase in resistivity for films deposited at substrate temperatures between 35 and 250°C. Kazmerski, Berry, and Allen [98] earlier reported that the carrier concentration increased for substrate deposition temperatures over the range 100–300°C, and also that the mobility increased at temperatures up to 260°C, above which it decreased. They also observed a similar maxi-

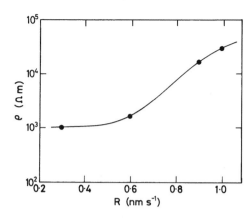

Fig. 35. Dependence of resistivity in evaporated CdS films on deposition rate for films of thickness 0.5 μm. Reproduced from [100]. Reprinted with permission from A. A. Ramadan, R. D. Gould, and A. Ashour, *Int. J. Electron.* 73, 717, copyright Taylor & Francis, 1992. Journal web site http://www.tandf.co.uk.

mum in grain size at this substrate temperature, which suggests that the mobility depends critically on the grain size. This correlation is shown in Figure 34 [98]. However, it has also been observed that the resistivity decreases with substrate temperature up to 150°C and increases at higher temperatures [19]. The increase in resistivity normally associated with increasing substrate temperature is probably related to the degree of microcrystallite preferential orientation. The degree of preferential orientation has been observed to decrease as the substrate temperature increases [19, 20]. It was already established that there is a positive correlation between the degree of crystalline orientation and the conductivity [105] and thus a decrease in the preferential orientation with increasing substrate temperature would result in an increased resistivity as observed. Furthermore, from the work of Wilson and Woods [19] it is apparent that some free Cd would be expected in films deposited at lower source temperatures, which would lower resistivity for lower substrate temperatures, becoming higher as the substrate temperature increases. The lower relative increase in resistivity with substrate temperature for the thicker films is probably a consequence of the increasingly Cd-rich vapor arriving at the substrate during the course of the deposition process.

Increasing the deposition rate normally increases the resistivity. A typical example of this behavior is shown in Figure 35 [100], for films of thickness 0.5 μm deposited with substrate temperature 100°C. Similar results had also been observed previously [19]. As the deposition rate increases, the mean microcrystallite size decreases [20], which would be expected to lead to a decrease in the mobility due to increased scattering, and also to a decrease in the free-carrier concentration due to trapping effects. Both these effects would lead to an increase in resistivity. Moreover, the degree of preferential orientation also decreases with increasing deposition rate [19, 20] which would also result in higher resistivity values.

Annealing of CdS thin films in both vacuum and air has been investigated by several workers. Early work showed that annealing in an oxygen-containing environment led to the appearance of CdO lines in the X-ray diffraction spectrum [106,

107]. Stanley [24] observed that annealing in air appears to have a greater effect on the resistivity than annealing in vacuum or in an inert gas. It was suggested [104] that the effect of heat treatment in air is to introduce oxygen atoms into the lattice of CdS films, thus introducing acceptor levels and lowering the Fermi level. This would explain an increase of 2 orders of magnitude in resistivity over that observed in films which were not heat treated. In contrast, other workers observed a decrease in resistivity after annealing. Yamaguchi et al. [108] found that in films deposited by vapor phase epitaxy and annealed in hydrogen, the resistivity decreased by 3 orders of magnitude. Akramov et al. [109] studied films deposited by vapor ion reaction and annealed in air at 400°C, and observed a resistivity decrease of 2 orders of magnitude. Ramadan, Gould, and Ashour [100] annealed films at different temperatures in the range 100–400°C for 1 h in either air or vacuum. For a film of thickness 0.5 μm deposited at a substrate temperature of 100°C, there was a decrease in resistivity from 1.7×10^3 to 1.0×10^3 Ω m when annealed in vacuum at 400°C, and to 0.7×10^3 Ω m when annealed in air at the same temperature, in agreement with earlier observations [24]. The decrease in resistivity after annealing is generally considered to be related to an increase in the microcrystallite grain size and a decrease in the defect density [24].

Resistivity variations in CdS films appear to exhibit different activation energies in different temperature ranges. At high temperatures, a higher activation energy of a few tenths of an electron-volt is observed, which is generally related to the thermally activated mobility variations and carrier excitation over intercrystalline potential barriers. At lower temperatures, a much lower activation energy is observed, which may result from carrier hopping behavior. Resistivity decreases with increasing film thickness. Buckley and Woods [99] associated this with the effect of the varying stoichiometry of the evaporation source, which becomes particularly important below a particular evaporation temperature. Additionally, surface scattering in thinner films is also thought to increase the resistivity. The substrate temperature during the deposition process also affects the film resistivity to a large extent. Although most investigators report that resistivity increases with the substrate temperature, attention has been drawn to both a maximum at 260°C and a minimum at 150°C. The generally observed increase in resistivity is thought to be related to the degree of preferred orientation of the microcrystallites. An increase in resistivity with increasing deposition rate is often associated with decreasing grain size, mobility, and free-carrier concentration, while the effects of annealing are to increase the mean grain size and thus to decrease the resistivity. Although these films are reasonably well understood, to obtain well-defined and predictable resistivity values, the deposition variables need to be very carefully controlled.

3.2.2. Cadmium Telluride

Although the bandgap of CdTe at 1.59 eV [22] is the lowest for the cadmium chalcogenides, the resistivity of the undoped

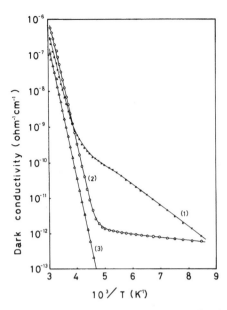

Fig. 36.   Dark conductivity of evaporated CdTe films as a function of reciprocal temperature for films deposited at 1.9 nm s^{-1} (1), 0.59 nm s^{-1} (2), and 1.7 nm s^{-1} (3). Reproduced from [30]. Reprinted with permission from Y. Kawai, Y. Ema, and T. Hayashi, *Jpn. J. Appl. Phys.* 22, 803, copyright Japanese Institute of Pure and Applied Physics, 1983.

material is relatively high, typically of the order of 10^4–10^6 Ω m [4]. For solar cell applications, low resistivity material is required, and doping with Cd or In can be employed to lower the resistivity value. Some of the published work on both undoped [30, 110] and doped [111, 112] films is examined in this section.

The dependence of conductivity on thickness in undoped CdTe films was investigated by Kawai, Ema, and Hayashi [30]. The films were prepared by conventional evaporation of CdTe. The dependence of conductivity (1/resistivity) determined by these workers is shown in Figure 36 [30] for deposition rates of 1.9 nm s^{-1} (1), 0.59 nm s^{-1} (2), and 1.7 nm s^{-1} (3). In each of curves (1) and (2), there were two different activation energies apparent, which were identified with impurity conductivity and intrinsic conductivity. The activation energy in the intrinsic region was 1.53 eV, close to the energy bandgap of crystalline CdTe. In the low-temperature regions of curves (1) and (2), different activation energies of 0.12 and 0.017 eV were estimated, respectively. The latter two values were close to values reported previously for the native acceptor of crystalline CdTe due to Cd vacancies, and due to the value of activation energy for native donors due to interstitial excess Cd. These workers observed a minimum in the conductivity (i.e., a maximum in the resistivity) at a deposition rate of 1.7 nm s^{-1}. This was identified with intrinsic conductivity, while the films deposited at rates of less than 1.7 nm s^{-1} were Cd-rich (*n*-type) and those deposited at higher rates were Te-rich (*p*-type). Ismail [36] observed a decrease in film conductivity from approximately 10^{-4} S m^{-1} at 325 K to 10^{-8} S m^{-1} at 150 K. At high temperatures, the activation energy was approximately 0.45 eV and at low temperatures a value of 0.046 eV was observed. There was no direct correlation between these values and known defects in CdTe, or with the energy bandgap.

The dependence of resistivity in CdTe films on thickness in the range 50–700 nm was investigated for films deposited at a substrate temperature of 150°C [110]. The resistivity decreased from 7.1×10^4 to 1.7×10^4 Ω m with the increase in thickness, with the most rapid decrease in resistivity occurring between 50- and 100-nm thickness. It was thought that this was related to the increasing degree of preferential orientation that takes place as the thickness increases, as had been observed by Wilson and Woods [19] in CdS films.

The dependence of conductivity on substrate temperature during deposition from room temperature up to 150°C in undoped CdTe films was investigated for films deposited at a rate of 1.9 nm s^{-1} [30]. These workers observed a decrease in conductivity up to a substrate temperature of 100°C where the minimum conductivity was 10^{-6} S m^{-1} (maximum resistivity of 10^6 Ω m). It was proposed that this corresponded to an improvement in the film stoichiometry. Ismail, Sakrani, and Gould [110] found a maximum resistivity value of 5×10^4 Ω m at a substrate temperature of 200°C for films deposited at a lower deposition rate of 0.5 nm s^{-1}. In this case, it was also considered that the maximum in resistivity corresponded with improved film stoichiometry. According to the suggestion of Kawai, Ema, and Hayashi [30], these films would be *n*-type owing to excess Cd, which is consistent with the lower resistivity values observed in this case.

Resistivity increased for increasing deposition rate up to a value of 5×10^5 Ω m at a rate of 1.9 nm s^{-1} [30], after which a decrease occurred. This was identified with the films going from *n*-type to intrinsic to *p*-type with increasing deposition rate. Resistivity was also found to increase over the range of 2×10^4 to 5.5×10^4 Ω m as the deposition rate increased in the range 0.2–1.0 nm s^{-1} [110]. These two sets of results are consistent, as the latter set did not cover deposition rates where the former set moved into the *p*-type region and the resistivity decreased once more. However, it was also argued [110] that the increase in resistivity with deposition rate is a consequence of a smaller microcrystallite size.

Kawai, Ema, and Hayashi [111] attempted to decrease the resistivity by coevaporating CdTe and Cd at a deposition rate of 0.8 nm s^{-1}. It was found that there was an exponential relationship linking the conductivity to the evaporation source temperature from 280–340°C, and that higher conductivity films could be obtained by increasing the source temperature. As the source temperature increased in this range, the films were initially intrinsic at a source temperature of 280°C, showed both intrinsic and extrinsic conductivity in the temperature range 300–340°C and became degenerate at a source temperature of 340°C. In this case, the conductivity was approximately 1.4 S m^{-1} and the corresponding resistivity was 0.7 Ω m. A different approach to lowering the resistivity was utilized by Huber and Lopez-Otero [112]. They used hot-wall epitaxy with two independent sources, one of CdTe and one of Cd, to control the stoichiometry. The films were in this case doped with In, so that they were *n*-type with carrier concentrations of up to 10^{23} m^{-3}. In

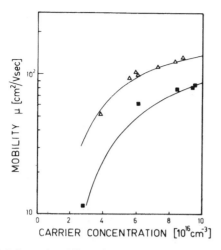

Fig. 37. Variations of mobility with carrier concentration at 300 K for two different In-doped CdTe thin film samples prepared by hot-wall epitaxy. The full curves are calculated from the grain boundary model proposed by Kamins [113], using μ_0 values of 0.22 m^2 V^{-1} s^{-1} (upper curve) and 0.24 m^2 V^{-1} s^{-1} (lower curve). Reproduced from [112]. Reprinted from *Thin Solid Films*, 58, W. Huber and A. Lopez-Otero, The Electrical Properties of CdTe Films Grown by Hot Wall Epitaxy, p. 21, copyright 1979, with permission from Elsevier Science.

many samples, the mobility was found to follow a relationship of the form $\mu = \mu_0 \exp(-E_B/kT)$, where E_B represents an energy barrier for electrons in the n-type films as proposed by Petritz [101]. The value of E_B was typically 20 meV. It was also observed that there was a tendency for the mobility to decrease with decreasing carrier concentration. Typical variations for two different samples are illustrated in Figure 37 [112]. This behavior was associated with the grain boundary model proposed by Kamins [113]. In this model, defects at the grain boundaries produce traps, which are initially neutral when empty. Free carriers captured by the traps establish a space-charge region, which results in an increase in barrier height with decreasing carrier concentration. The full curves in Figure 37 are calculated using this model with $\mu_0 = 0.22$ m^2 V^{-1} s^{-1} for the upper curve and $\mu_0 = 0.24$ m^2 V^{-1} s^{-1} for the lower curve. It was concluded that the mobilities for films deposited by this method were higher than those reported previously, reaching values of 0.01 m^2 V^{-1} s^{-1} at carrier concentrations of about 10^{22} m^{-3}.

Thus, the lateral resistivity in undoped CdTe films is normally too high for solar cell applications, and various doping strategies have been used to lower its value. In films evaporated solely from CdTe starting material, analysis of the measured activation energies suggested that the resistivity varies with deposition rate, reaching a maximum value at a rate of 1.7 nm s^{-1} where the material is intrinsic. At other values, the material appears to be extrinsic, n-type at lower rates and p-type at higher rates. Low-temperature activation energies were those corresponding either to interstitial Cd or to Cd vacancies, whereas in the intrinsic region the activation energy approached the value of the energy bandgap. An increase in resistivity with deposition rate at lower rates was also observed by other work-

ers. The resistivity decreased with increasing film thickness, and was related to an increased preferential orientation as the thickness increased. Maximum values of resistivity with respect to the substrate temperature during deposition were observed at 100°C [30] and at 200°C [110], having values of 10^6 and 5×10^4 Ω m, respectively. The resistivity found in the latter case was consistent with the lower deposition rate used, which would result in Cd-rich n-type films. Resistivity has been reduced by depositing the films using coevaporation with Cd and also by doping with In. In the coevaporated films, a resistivity value as low as 0.7 Ω m was attained. In In-doped films, the mobility decreased with decreasing carrier concentration, and such behavior was associated with trapping effects at grain boundaries. Higher mobilities of the order of 0.01 m^2 V^{-1} s^{-1} could be obtained. It is clear that for solar cell applications doped films are essential, and present work is aimed at improving the structural and electronic properties of such films.

3.2.3. Cadmium Selenide

As with CdS, CdSe films have been investigated quite extensively with regard to their electrical resistivity as functions of the temperature, thickness, and the deposition conditions. In addition to this work, there has been an interest in the varying resistivity of CdSe films in the presence of oxygen, and with the oxygen content of the films.

The variation of resistivity (or conductivity) with temperature has been studied by several workers. Shimizu [114] found dark current activation energies of 0.14 and 0.37 eV. The temperature range over which the lower activation energy was applicable was very wide for films deposited at a substrate temperature of 100°C. However, as the substrate temperature rose, this range became narrower, until for films deposited at 300°C only the higher activation energy was evident. These two activation energies were identified with trap levels which gave rise to high conductivity for films deposited at low temperatures (shallow traps) and to low conductivity for films deposited at high temperatures (deep traps). Berger, Jäniche, and Grachovskaya [115] investigated the conductivity of evaporated CdSe films. At low temperatures, the conductivity followed an exponential temperature dependence, with its value determined by the deposition conditions. In the higher temperature region, there was a rapid increase in conductivity. The conductivity below 400 K was identified with impurity conduction, whereas that at higher temperatures had an activation energy of about 1 eV, corresponding to approximately half the bandgap. Although the electron concentration in this range was independent of the film structure as in the case of intrinsic conductivity, it was concluded that the conductivity was neither intrinsic nor extrinsic, but was the result of a process in which intrinsic impurities acting as donors are thermally activated. Šnejdar, Berková, and Jerhot [116] measured the conductivity, carrier concentration and the mobility in evaporated CdSe films which had been annealed in Se vapor. The dependence of conductivity over the temperature range 200–360 K was also investigated. It was found that the conductivity decreased on annealing at

150–170°C from typically 10^4 to 1 S m^{-1} at room temperature. Unannealed low resistivity films were degenerate with scattering caused by a large number of Se vacancies. During annealing, the number of vacancies decreased, and the semiconductor became nondegenerate. Chan and Hill [37] pointed out that previous results obtained on CdSe films were contradictory, and probably due to insufficient control over deposition conditions and imprecise interpretation of the data. They measured the conductivity as a function of inverse temperature in evaporated films. The measurements usually showed two activation energies and there were linear portions in all of their plots of log (carrier concentration) against inverse temperature, which were interpreted as carrier excitation from compensated donor levels. Values of ionization energy E_D, donor concentration N_D, and acceptor concentration N_A were derived for all these plots and it was found that there was a good linear correlation between E_D and the logarithm of the compensation ratio $K_n = N_A/N_D$, particularly for lower values of K_n. These results are shown in Figure 38 [37], and were interpreted in terms of a lowering of the ionization energy caused by the potential energy of attraction between ionized donors and electrons. Dark conductivity measurements in pure CdSe films have exhibited two activation energies of 0.42 and 0.18 eV [117]. These were attributed to the presence of two donor levels in the bottom of the conduction band. Thermally stimulated current measurements gave activation energies of 0.44 and 0.17 eV, in good agreement with the electrical conductivity activation energies. By annealing CdSe films in vacuum, entirely reproducible resistivity-temperature characteristics could be obtained [118]. The curves each showed two activation energies, but the slopes varied, depending on the deposition conditions and the annealing temperature. However, it was considered that annealing of the films in vacuum after the resistivity reached a constant value allowed resistivity-temperature characteristics to be obtained, which were reproducible at the annealing temperature, providing that the annealing temperature was not exceeded. Furthermore, the large variations in resistivity of CdSe films deposited by other workers was attributed to nonstoichiometry of the films. In a comprehensive set of experiments on CdSe films [119], activation energies of 0.02 and 0.14 eV were found at low and high temperatures, respectively. The low-temperature value was associated with a hopping process, while the high-temperature value was thought to represent an intercrystalline potential barrier, of the type proposed by Petritz [101].

Several workers have measured the dependence of resistivity or conductivity on film thickness for both evaporated and sputtered films. Berger, Jäniche, and Grachovskaya [115] found that the room-temperature conductivity increased from about 500 to 820 S m^{-1} as the film thickness increased up to 1 μm. There was a further, much less rapid increase up to 900 S m^{-1} as the thickness further increased up to 4 μm. This variation was attributed almost entirely to variations in the mobility with thickness, since the shape of the mobility-thickness curve closely followed that of the conductivity-thickness curve; average mobility increased from about 3×10^{-3} to 6×10^{-3} m^2 V^{-1} s^{-1}

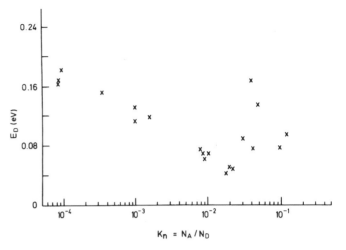

Fig. 38. Variation of donor ionization energy E_D with compensation ratio K_n in evaporated CdSe films. Reproduced from [37]. Reprinted from *Thin Solid Films*, 35, D. S. H. Chan and A. E. Hill, Conduction Mechanisms in Thin Vacuum-Deposited Cadmium Selenide Films, p. 337, copyright 1976, with permission from Elsevier Science.

at 1-μm thickness and then to 6.5×10^{-3} m^2 V^{-1} s^{-1} at 4-μm thickness. These results were also interpreted in terms of the intercrystalline potential barrier model [101]. It was explained that the mobility variations could not be accounted for by surface scattering of electrons, since the film thicknesses used were considerably larger than the mean free path of electrons. Following the results of Shallcross [103], in which the mean crystallite size increased with increasing film thickness, it was argued that the mean number of crystallite barriers *decreases* and results in an increase in the effective mobility. Above 1 μm, it was argued that only insignificant changes in the crystallite size occur with increasing film thickness. A similar decrease in resistivity was observed for films evaporated onto heated substrates [120]. For sputtered films [41], there was a maximum in film resistivity at a thickness of about 2 μm. The increase in resistivity with film thickness below 2 μm was quite different from the behavior observed in evaporated films. These workers attributed this to the fact that sputtered films are more likely to be stoichiometric than evaporated films, since in CdS films [99] increasing deviations from stoichiometry are observed with increasing film thickness. Work on evaporated CdSe films [119] has tended to support this view. Measurements were made at room temperature over the thickness range 0.07–1.30 μm on films deposited at a rate of 0.5 nm s^{-1}. The resistivity gradually increased with film thickness to a maximum of approximately 5 Ω m for a thickness of 1 μm, before falling below 10^{-3} Ω m between 1- and 1.3-μm thickness as shown in Figure 39 [119]. The initial increase in resistivity was consistent with improved film stoichiometry. The decrease at higher thicknesses was attributed not only to the effects of excess Cd atoms acting as donor centers, but also to an increase in crystallite size resulting in higher mobility values.

Substrate temperature during the deposition process appears to be important. Shimizu [114] noted that the ionization energies determined from temperature measurements vary, as

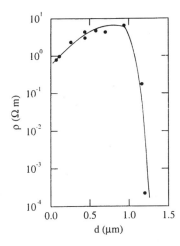

Fig. 39. Dependence of room-temperature resistivity in CdSe films on thickness for evaporated samples deposited at 0.5 nm s^{-1} onto substrates maintained at room temperature. Reproduced from [119]. Reprinted with permission from A. O. Oduor and R. D. Gould, Dependence of Resistivity in Evaporated Cadmium Selenide Thin Films on Preparation Conditions and Temperature, in Metal/Nonmetal Microsystems: Physics, Technology and Applications, *Proc. SPIE*, 2780, 80, copyright International Society for Optical Engineering, 1996.

discussed previously in the context of resistivity-temperature variations. Chan and Hill [37] found a very wide variation in resistivity from approximately 10^{-3}–10^2 Ω m as the substrate temperature increased from 20 to 100°C. They argued that as the substrate temperature increased the film composition changed from Cd-rich to Se-rich. Raoult, Fortin, and Colin [118] found that although the resistivity increased with increasing substrate temperature up to about 400 K, above this temperature there was a decrease in resistivity. Although this behavior is consistent with earlier work, a detailed description of the reasons was not given. Oduor and Gould [119] found a sharp increase in resistivity as the substrate temperature increased in the range 300–400 K, but remained approximately constant for substrate temperatures in the range 400–600 K, at a value of 6 Ω m. The initial increase in resistivity with substrate temperature in this case was identified with improved condensation kinetics (i.e., a higher condensation coefficient) of the atomic species onto the substrate. The fact that the resistivity did not decrease significantly for high substrate temperatures was attributed to the relatively low resistivity values obtained.

Deposition rate variations were observed by several workers [37, 115, 119]. A rapid increase in resistivity from 10^{-3} to 10^2 Ω m was observed as the deposition rate increased from 0.1 to 1.2 nm s^{-1}. Again this was explained in terms of a change from a Cd- to an Se-rich composition [37]. A similar increase in resistivity with deposition rate was observed by Berger, Jäniche, and Grachovskaya [115] (i.e., a decrease in conductivity with increasing deposition rate). Here at low temperatures, the conductivity followed an exponential (thermally activated) temperature dependence, whereas at higher temperatures all conductivity curves merged into a rapidly rising curve. Work from this laboratory [119] was performed using films of thickness 0.13 μm deposited onto substrates maintained at 373 K.

There was a rapid rise in resistivity for deposition rates in the range 0.05–0.30 nm s^{-1}, saturating to a maximum but constant value of 7 Ω m for deposition rates in the range 0.5–3.0 nm s^{-1}. The increase in rate was thought to decrease the microcrystallite grain size, causing both the mobility and the free-carrier concentration to decrease, as observed in CdS films [97]. This explanation differed from that of Chan and Hill [37], who proposed that the increase in resistivity was caused by a decrease in the concentration of Se vacancies.

The influence of oxygen in CdSe films has been studied quite intensively. Some typical results are discussed later. These include the interactions of oxygen on CdSe surfaces [121] and the instabilities in the electrical properties after exposure to oxygen [122]. Somorjai [121] studied the effects of oxygen exposure over the temperature range 0–360°C. It was clear that a drastic increase in the effective resistivity took place, since the film conductivity decreased. Freshly evaporated films had conductivities of the order of 0.01–0.10 S m^{-1}. When oxygen gas was allowed to interact with the surfaces, there was a very rapid decrease in the conductivity (increase in resistivity) by about 1 order of magnitude. The oxygen acted as an acceptor impurity. A longer term decrease in conductivity was also observed, in which the conductivity diminished by several orders of magnitude. For the first 5-min exposure, the conductivity change was extremely large (from 1 mA to 1 μA at a voltage of 1 V), after which further small decreases in conductivity continued to take place for up to 1 h. Both these effects were irreversible, in the sense that the oxygen could not be removed at the same temperature by continual evacuation under ultrahigh vacuum conditions. These films were then said to be "saturated" with oxygen; desorption of the adsorbed oxygen occurred rapidly if a temperature of 630°C was exceeded, with a rapid rise in the conductivity. Irreversible adsorption of oxygen was associated with chemisorption of oxygen in the CdSe lattice. Another effect was observed in addition to the irreversible adsorption. After the termination of irreversible oxygen desorption, evacuation of the chamber led to a further relatively small decrease in the conductivity. If the pressure was restored, the conductivity increased by the same amount. In this case, the oxygen molecules acted as donor impurities, and were only weakly held species adsorbed on the film surface. These two types of oxygen interaction could occur simultaneously, with oxygen acting both as an acceptor and as a donor impurity. Chan and Hill [122] noted that CdSe films show electrical instabilities when exposed to the atmosphere. Films deposited at a rate of 0.8 nm s^{-1} were unstable, while those deposited at 0.1–0.4 nm s^{-1} were usually stable. Similarly, films deposited at a substrate temperature above 55°C were unstable, while in those deposited at 55°C the conductivity decreased slowly with time in a stable manner. It was shown, by extending the analysis of Somorjai [121], that if the width of the depletion layer at the film surface is comparable with the film thickness, then for constant mobility, the conductivity would depend linearly on the logarithm of the time of exposure. Thus, although both the substrate temperature and the deposition rate require to be controlled to avoid unstable films, the model allowed calculations of the ionized donor concentration. Hamerský [123] also observed that an oxygen environment, and also oxygen incorporated into CdSe films during deposition, could affect the resistivity. A minimum in resistivity with respect to deposition rate occurred; their work showed that the position of the minimum depended on the ratio of the density of impinging molecules of CdSe to that of impinging oxygen molecules from the residual atmosphere during evaporation. Furthermore, the depth of the minimum in resistivity was dependent on the magnitude of the oxygen partial pressure during evaporation.

Thus, the resistivity variations in CdSe thin films have some features in common with those of the other cadmium chalcogenides, especially CdS. Temperature-dependence measurements frequently revealed two activation energies, which could be interpreted as evidence of trap levels, or with different conduction processes operating at lower and higher temperatures. The resistivity tended to decrease with increasing film thickness, which was associated with thickness dependent mobility variations and also with excitation of carriers over intercrystalline potential barriers, whose number decreased with increasing thickness. Comparison of the differences in the thickness dependence of resistivity between evaporated and sputtered films, led to the suggestion that, like evaporated CdS films, the variations in the resistivity were related to stoichiometry variations of the vapor stream during deposition. Resistivity increased with increasing substrate temperature during deposition, although some workers observed a maximum followed by a decrease at higher temperatures, while others observed a saturation value. Variations in stoichiometry and in condensation coefficients were suggested as possible causes of this effect. Resistivity increased with increasing deposition rate, which was variously interpreted as a change in stoichiometry from a Cd- to an Se-rich composition and to increases in the microcrystallite grain size. Oxygen appears to act as an acceptor (and sometimes also as a donor) impurity in CdSe. It is clear that although many of these effects are similar to those observed in the other cadmium chalcogenides, the effect of oxygen is somewhat unusual, particularly its role as both a donor and an acceptor impurity simultaneously, and thus future work in this area would be beneficial to our understanding of this type of film.

3.2.4. Cadmium Arsenide

The results of many experiments pertinent to the conductivity of Cd_3As_2, such as measurements of the energy bandgap, effective mass, and carrier concentration were more appropriately dealt with in Section 3.1.4. In addition to its very low (or even negative) bandgap, Cd_3As_2 possesses very high mobility values. For example, Iwami, Matsunami, and Tanaka [87] presented data showing Hall mobility values of $0.7-3.0$ m^2 V^{-1} s^{-1} in single crystals at 77 K. At 2.2 K, a value of 9 m^2 V^{-1} s^{-1} was observed, and this was applicable for temperatures up to about 15 K. Turner, Fischler, and Reese [124] had earlier reported Hall mobilities as high as 1.5 m^2 V^{-1} s^{-1} at 297 K. Thin film values are invariably lower than this, but are nevertheless high in comparison to most other semiconductors. Apart from some

unpublished work performed in this laboratory [64], there do not appear to be any systematic measurements of resistivity as functions of the deposition conditions and thickness of Cd_3As_2 thin films. However, some of the more important results concerning the individual variations in resistivity are discussed in the following.

The earliest measurement of the variation of resistivity with temperature in layers of Cd_3As_2 was reported by Moss [125] in 1950. Over the temperature range 90–390 K, the resistance R followed a relationship $R = R_0 \exp(\varepsilon/2kT)$, where R_0 is a constant and $\varepsilon = 0.14$ eV. Żdanowicz [7, 56] measured several electrical properties of evaporated Cd_3As_2 films. Some of these results are shown in Figure 40 [7] for 2-μm thick films deposited at a rate of 5 nm s^{-1} onto substrates held at a temperature of 165°C. From the figure, it is apparent that both the Hall coefficient R_H and the resistivity ρ are constant at temperatures below 300 K. Typical resistivity values for crystalline films were in the range $(1-7) \times 10^{-5}$ Ω m. The mobility R_H/ρ values are also shown in the figure, and are practically constant at low temperatures, having a value of approximately 0.25 m^2 V^{-1} s^{-1}. Sexer [77] measured the resistivity in the extrinsic (78–300 K) and in the intrinsic (>300 K) regions. In the extrinsic range, the material was degenerate, and had a resistivity value of the order of $(1-6) \times 10^{-6}$ Ω m, whereas in the intrinsic range a conductivity of 10^5-10^6 S m^{-1} (resistivity $10^{-5}-10^{-6}$ Ω m) was obtained. In the intrinsic region an energy bandgap of value 0.14 eV was determined, in agreement with the earlier measurements of Moss [125]. Goldsmid and Ertl [57] evaporated films of thickness 1 μm, after which they underwent heat treatment at temperatures of up to 600°C. Conductivity values of 1.14×10^5 and 2.92×10^5 S m^{-1} were obtained for two different samples. Dubowski and Williams [59] deposited films by pulsed laser evaporation (PLE) and measured Hall mobility and resistivity over the temperature range 77–300 K. They found that the temperature dependence of these parameters differed drastically depending on the film thickness. For films of thickness less than 0.4 μm, the resistivity decreased with increasing temperature, whereas for thicker films the resistivity increased with increasing temperature. A weak temperature dependence of mobility was observed for the thinner films, but for the thicker films the mobility increased with decreasing temperature by 20–30%, depending on the thickness, across the temperature range. Typical variations of resistivity and mobility are shown for a film of thickness 1.8 μm in Figure 41 [59]. Note that the temperature dependence of the resistivity is the inverse of that shown in Figure 40 [7], even though the thicknesses of the two films were similar. It was pointed out that in bulk Cd_3As_2 the resistivity also increased with increasing temperature. The distinct difference between the resistivity variations for thinner and thicker films was explained in terms of grain boundary scattering in the former case, which was much less important in the latter case. Din [64] found resistivity values of approximately 3×10^{-6} Ω m for a film deposited at 423 K, with a higher value of the order of 3×10^{-4} Ω m for a film deposited at room temperature. The dependence on substrate temperature is discussed in more detail later. Resistivity gen-

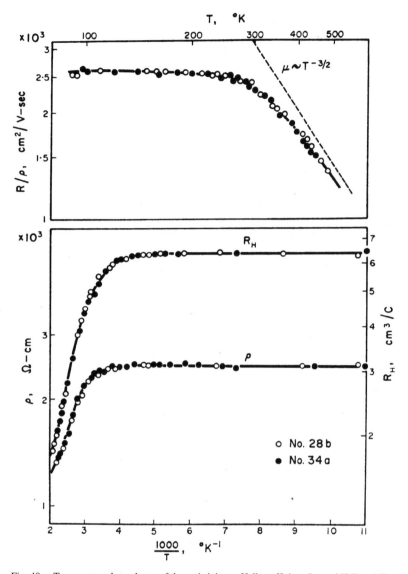

Fig. 40. Temperature dependence of the resistivity ρ, Hall coefficient R_H and Hall mobility R_H/ρ for two evaporated thin films of Cd_3As_2 obtained under the same deposition conditions (substrate temperature 165°C, deposition rate 5 nm s^{-1}, film thickness 2 μm). Reproduced from [7]. Reprinted from *Solid State Electron.* 11, L. Żdanowicz, Properties of Evaporated Hall Elements of Cadmium Arsenide, p. 429, copyright 1968, with permission from Elsevier Science.

erally decreased with increasing temperature, particularly at temperatures above 330 K. At lower temperatures, the change in resistivity was much less rapid, and the resistivity became almost constant at temperatures below about 200 K, as also observed previously [6, 7, 56]. The activation energies determined in the higher temperature regions were between 0.1 and 0.16 eV depending on the film thickness, while for lower temperatures values were very low and did not exceed 0.022 eV. This low value, obtained at low temperatures, is evidence of a hopping conduction process between closely spaced energy states within the forbidden energy gap. The higher temperature activation energies are similar to the value of 0.14 eV reported by Moss [125] and by Żdanowicz [56]. The origin of this activation energy is not definitively known. It may repre-

sent intrinsic excitation across the narrow energy bandgap of Cd_3As_2 or, as with the cadmium chalcogenides, it may arise from intercrystalline potential barriers of the type suggested in the model of Petritz [101], as described in Eq. (6). It should also be pointed out that the temperature dependence of resistivity in these thin films is different from that of single crystals. In this case Iwami, Matsunami, and Tanaka [87] reported a constant resistivity below 15 K, which increased slowly with increasing temperature, to reach a maximum value at 300 K. These workers estimated an intrinsic energy gap of 0.14 eV from the temperature dependence of $1/R_H T^{3/2}$, which is similar to the reported activation energies in thin films given before. It would be unwise to associate measured activation energies in thin films with the intrinsic energy gap, even though the mea-

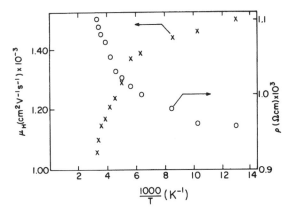

Fig. 41. Temperature dependence of resistivity and electron mobility for a Cd_3As_2 film of thickness 1.8 μm deposited by pulsed laser evaporation (PLE). Reproduced from [59]. Reprinted from *Thin Solid Films* 117, J. J. Dubowski and D. F. Williams, Growth of Polycrystalline Cd_3As_2 Films on Room Temperature Substrates by a Pulsed-Laser Evaporation Technique, p. 289, copyright 1984, with permission from Elsevier Science.

sured values are similar to those obtained in crystals. Additional work needs to be performed to determine whether the activation energy in these thin films is characteristic of the energy gap or of the height of internal potential barriers.

With the exception of the work of Żdanowicz, Żdanowicz, and Pocztowski [126] and some results of Din [64], measurements specifically designed to explore the variation of resistivity with thickness in Cd_3As_2 films do not appear to have been made. However, there are some thickness-related measurements which are mentioned in the literature. Matsunami et al. [127] mentioned that film thickness and the evaporation rate had little effect on the electrical properties, although the substrate temperature during deposition was very important. In later work [6], they stated that mobility in Cd_3As_2 films was not affected at thicknesses greater than 2 μm. In contrast, Żdanowicz, Żdanowicz, and Pocztowski [126], who made a thorough investigation of the variation in resistivity, Hall coefficient and Hall mobility in evaporated Cd_3As_2 films, detected quantum size effects in films of thickness 30–100 nm, where the film thickness is of the order of the de Broglie wavelength of the conduction electrons. For films of thickness up to 90 nm, deposited at a rate of approximately 4 nm s^{-1} at a substrate temperature of 160°C, oscillations in the resistivity (and the mobility) were observed, with a period of about 10 nm. This was in good agreement with theory, which predicted a period of about 8 nm. The theory also predicted that oscillations would occur only when the Hall mobility is greater than 0.1 m^2 V^{-1} s^{-1} at the 100-nm thickness. As mentioned previously, Dubowski and Williams [59] also found thickness-dependent resistivity and electron mobility in PLE-deposited films. Mobility typically increased linearly with film thickness from about 0.06 m^2 V^{-1} s^{-1} at 0.5 μm to 0.1 m^2 V^{-1} s^{-1} at 2 μm. This was attributed to a decreasing role of grain boundary scattering with increasing grain size (which is roughly proportional to film thickness). The thin film mobility for a thickness of 1.8 μm was typically 15 times smaller than that observed in single crystals with a similar elec-

tron concentration of 10^{24} m^{-3}. According to these workers, the effective surface scattering length in Cd_3As_2 films is less than 70 nm, whereas the thinnest sample investigated was of thickness 100 nm. Thus, surface scattering mechanisms were excluded in their consideration of the thickness dependence of mobility. Din [64] measured the resistivity of evaporated Cd_3As_2 films, deposited at a rate of 0.5 nm s^{-1} with a substrate temperature of 300 K. In general, the resistivity was found to decrease with increasing film thickness up to at least 1 μm. Over this range, the resistivity typically decreased from 10^{-2} to 10^{-5} Ω m. This range of thickness variation was in agreement with that described by Żdanowicz [128]. Din concurred with previous workers that the decrease in resistivity with increasing thickness was due to a combination of a reduction in intercrystalline and surface scattering as the mean grain size increased with film thickness. A reduction in scattering by lattice defects with increasing thickness was also mentioned as a possible cause of the resistivity-thickness variations observed. X-ray and scanning electron microscopy studies confirmed that the grain size increased with film thickness, and that scattering at grain boundaries would have a reduced effect for thicker films.

Early work on Cd_3As_2 films confirmed the influence of substrate temperature during deposition on the resistivity value [56]. Films deposited onto a cold support had a considerably higher resistivity than those deposited at temperatures above 150°C. At lower temperatures, the films were amorphous, whereas at higher temperatures X-ray diffraction measurements showed that the structure was very similar to the bulk material. Resistivity values for the lower substrate temperature films were in the range 10^{-4}–10^{-2} Ω m while those deposited at higher temperatures were in the range (2–7) × 10^{-5} Ω m; this was however, still higher than the bulk resistivity of (2–8) × 10^{-6} Ω m [128]. Matsunami et al. [127] found that the mobility increased with substrate temperature in the range 100–185°C, but fell thereafter as the substrate deposition temperature increased to 210°C. These workers also noted a change from amorphous to polycrystalline as the substrate temperature increased. Goldsmid and Ertl [57] similarly found increased mobility in films which were deposited at high temperature and/or heat treated at temperatures of up to 600°C without reevaporation under a layer of SiO of thickness 0.4 μm. Din [64] deposited films of thickness 0.5 μm at a rate of 0.5 nm s^{-1} onto substrates maintained in the temperature range 300–453 K. Resistivity values were typically 10^{-4} Ω m for films deposited at 300-K substrate temperature, decreasing to 10^{-6}–10^{-5} Ω m at 453 K. These results are similar to those of Żdanowicz [128], and the films also showed a transition from an amorphous to a crystalline state with increasing substrate temperature. Increasing grain size with increasing substrate temperature was consistent with the reduced resistivity. Jurusik [67] argued that higher substrate temperatures lead to faster surface diffusion, larger grain sizes, higher mobility, and thus lower resistivity. The resistivity-substrate temperature variations reported by most workers thus appear to be in reasonable agreement.

Żdanowicz [128] observed that for lower deposition rates the resistivity decreased with increasing deposition rate. A film of thickness 2.5 μm deposited at a rate of 3 nm s^{-1} and at a substrate temperature of 365 K had a relatively high resistivity of 2.39×10^{-3} Ω m, but a sample of thickness 2 μm deposited at 4.7 nm s^{-1} had a considerably lower resistivity of 2.3×10^{-5} Ω m. However, from this work it is difficult to reach a firm conclusion regarding the dependence of resistivity on deposition rate, since the substrate temperature and film thickness also varied. Żdanowicz et al. [129] reported that amorphous Cd_3As_2 films showed an increasing resistivity when the deposition rate was increased to a very high value of 10 nm s^{-1}. This was attributed to excess As, while films deposited at much lower rates showed excess Cd and thus a lower resistivity. Din [64] found that the resistivity decreased from 8×10^{-4} to 7×10^{-6} Ω m as the deposition rate increased from 0.2 to 1.3 nm s^{-1}. This was thought to be related to an increase in the Cd concentration with increasing deposition rate. However, at a rate exceeding 1.3 nm s^{-1} the resistivity increased slightly, and this may well have been due to an increase in the As content as proposed by Żdanowicz et al. [129]. It would therefore appear that the stoichiometry of the deposited material varies with the deposition rate, and that this effect is responsible for the resistivity variations observed.

Since it is fairly well established that Cd_3As_2 films move from an amorphous via a polycrystalline to a crystalline state as the substrate temperature during deposition is increased in the range 20–200°C [63], if high conductivity films are desired then it is simpler to deposit at the appropriate substrate temperature, rather than to use postdeposition annealing. In the mid-1970s, some work on annealing was performed, and this invariably resulted in decreased resistivity values. For example, Goldsmid and Ertl [57] prepared evaporated Cd_3As_2 films on substrates at deposition temperatures of up to 200°C, after which they were heat treated at temperatures of up to 600°C. They compared results on two samples, the first annealed at 475°C and the second annealed at 590°C. The first film had a Hall mobility of 0.34 m^2 V^{-1} s^{-1} and the second had a Hall mobility of 0.55 m^2 V^{-1} s^{-1}. The conductivity also increased from a value of 1.14×10^5 S m^{-1} for the first film to 2.42×10^5 S m^{-1} for the second (i.e., the resistivity decreased from 8.8×10^{-6} to 4.1×10^{-6} Ω m). There was also a moderate increase in the electron carrier concentration. Żdanowicz et al. [130] deposited amorphous Cd_3As_2 films on substrates cooled to temperatures as low as 90 K, after which they were annealed to room temperature. Film thicknesses were in the range 50 nm to 4 μm with deposition rates in the range 0.5–10.0 nm s^{-1}. They found that when the substrate temperature during deposition was less than a certain temperature $T_c \approx 220$ K that the resistivity was high, with a very small negative temperature coefficient of resistance up to a characteristic maximum. In this region, the curve was approximately reversible. At temperatures around T_c, there were strong irreversible changes in resistivity which decreased by 2–6 orders of magnitude, in some cases down to a value of around 10^{-5} Ω m, which is characteristic of crystalline films. In the final high temperature region above room tem-

perature, the resistivity remained low, with room-temperature values of the order of 10^{-5}–10^{-3} Ω m, much less than the values of 10^{-4}–10^{-1} Ω m for films deposited at room temperature. These workers suggested that since Cd_3As_2 dissociates in the vapor phase into Cd^+ and As_4^+ particles, low substrate temperatures freeze in the impinging particles at specific sites on the substrate. When the temperature is increased enough to allow surface migration, the compound is formed. An estimated activation energy for this process was 1.4 eV. Din [64] deposited films at a rate of 0.5 nm s^{-1} onto the substrates held at room temperature, or at 403 K. They were then annealed at 473 K in vacuum for 2 h. The initial resistivity was dependent on the thickness, as described earlier, varying from 1.2×10^{-3} Ω m for a thickness of 0.2 μm to 2.3×10^{-5} Ω m for a thickness of 1.51 μm, both using a substrate temperature of 300 K. After annealing, these values decreased to just over half of their original values. The decrease in resistivity for samples deposited at the higher substrate temperature was typically by only about 10%. For the moderately increased temperatures applied, the reduction in the resistivity was considered to result from annealing out of defects and possibly to the growth of small crystallites.

To summarize, the resistivities of Cd_3As_2 films are generally quite low, typically of the order of 10^{-5} Ω m, with the mobility usually just below 1 m^2 V^{-1} s^{-1}. High-temperature activation energies of 0.1–0.2 eV are observed, which may be identified either with a forbidden energy bandgap, or with intercrystalline potential barriers. Resistivity normally decreased with increasing temperature [7], but an increase has also been observed in the case of thick films deposited by PLE. Hopping conductivity is probably responsible for the conductivity at low voltages. Resistivity usually decreased with increasing thickness, although there was occasionally an increase for thicknesses greater than about 1 μm [64]. Explanations have been advanced in terms of varying composition, and reduced scattering due to increased grain size and lattice defects. Quantum size effects, manifested as an oscillating dependence of resistivity on film thickness, were observed for the first time in this material by Żdanowicz, Żdanowicz, and Pocztowski [126]. The influence of substrate temperature appeared to be related to the growth of amorphous films at low temperatures and crystalline films at higher temperatures, the latter leading to lower resistivities. Observed annealing effects were in accordance with this supposition. Resistivity tended to decrease with increasing deposition rate, although at very high rates there was a tendency for the resistivity to increase once more, possibly owing to an increasing As content in this case. The properties of this material are clearly complex, and it is capable of very low resistivity values and high mobilities when prepared under suitable deposition conditions. Some of the factors influencing the resistivity have been established with reasonable certainty, but deposition of films with a specific resistivity value still requires a certain amount of empirical optimization of the deposition parameters.

3.3. High Field Dc Conductivity

In this section, we consider the effects of dc electrical conductivity in which a voltage is applied across a cadmium compound thin film prepared in a sandwich configuration between metallic electrodes. It is clear that this situation differs considerably from that described in Section 3.2. Films can experience extremely high electric fields, when even low voltages are applied. Take, for example, a film of thickness 1 μm subjected to a voltage of only 10 V; the average electric field developed across the film is 10^7 V m^{-1}, which approaches the dielectric breakdown strength of many materials. Thinner films will obviously experience even higher electric fields for the same applied voltage. Simmons [131] has reviewed several high field conduction mechanisms that may occur in thin dielectric films. Many of these have been observed in cadmium compounds prepared as thin films. The origin of these processes will therefore be outlined later, together with the appropriate current density–voltage expressions which allow them to be identified. Before these are presented, however, it is first appropriate to discuss the different classes of electrical contact which may be applied to dielectric thin films. This is because the type of conduction process identified is frequently determined by the type of electrical contact made, and its interfacial properties. After discussing the various conduction mechanisms, those observed in each of the four materials will be discussed.

3.3.1. Electrical Contacts

Simmons [131] pointed out that the type of electrode used can greatly influence the conduction processes observed, and has classified the various conduction mechanisms as either *bulk-limited* or *electrode-limited*. In bulk-limited conductivity, the charge carriers are generated in the bulk of the material and the role of the electrodes is merely to apply a potential to generate a drift current. For electrode-limited conductivity, the interface between the electrode and the insulator or semiconductor presents a potential barrier to charge flow, which effectively limits the current. Contacts are generally *ohmic* or *blocking*, where the latter type of contact is sometimes known as a Schottky barrier. A transitional *neutral* contact also exists. The type of contact formed depends in principle on the relative work functions of the dielectric (semiconductor or insulator), ψ_i, and of the metal contact, ψ_m. In equilibrium, the vacuum and Fermi levels of both the electrode and the dielectric are continuous across the interface. With no voltage applied, the Fermi level must be flat, otherwise a current would flow in the absence of a voltage. In the bulk of the material, away from the interface, the energy difference between the vacuum and the Fermi levels will be equal to the work function ψ_i. The height of the potential barrier at the interface, ignoring the effects of surface states, is given by

$$\phi_0 = \psi_m - \chi \tag{7}$$

where χ is the dielectric electron affinity and ψ_m is the metal work function. However, as pointed out by Simmons, it is the

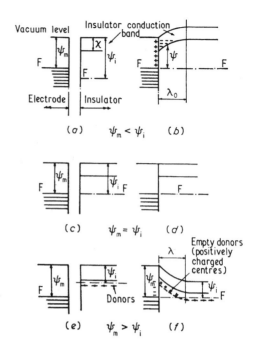

Fig. 42. Energy band diagrams showing the requirements and type of contacts for the case of an ohmic contact (a), (b), a neutral contact (c), (d) and a blocking contact (e), (f). F indicates the Fermi level. Reproduced from [131]. Reprinted with permission from J. G. Simmons, *J. Phys. D: Appl. Phys.* 4, 613, copyright IOP Publishing Ltd., 1971.

type of contact rather than its barrier height which determines the type of conductivity observed. In the following, the discussion is concerned mainly with electrons as the charge carriers. Simmons [131] argued that for insulators, with energy gap E_g greater than about 3 eV, conduction by holes could normally be ignored owing to relatively low hole mobilities and to immobilization of holes by trapping effects. In the present case, the cadmium chalcogenides are wide bandgap semiconductors, having values of $E_g \geq 1.59$ eV. Although this is not as wide as in insulators, these arguments are still applicable, n-type conductivity being the norm. For the rare cases of p-type conductivity, the various conditions regarding the relative values of the work functions are reversed [132]. For Cd_3As_2, in view of its very narrow energy bandgap, much of the discussion later is irrelevant. We see however, in Section 3.3.6 that there is, nevertheless, still some evidence for high field conduction processes in this material.

Ohmic contacts occur for electron injection in the case where $\psi_m < \psi_i$. To comply with the thermal equilibrium requirements, electrons are injected from the electrode into the dielectric (the insulator in Fig. 42). In Figure 42 [131], this situation is illustrated in (a) and (b). In a, are the energy band diagrams of the metal and the semiconductor or the insulator are shown, when they are not placed in contact. The vacuum level is constant in this case. F represents the Fermi level, which is lower in the case of the dielectric because of its higher work function. In (b), we see the situation when the metal electrode and the dielectric are in contact. Electrons are injected into the dielectric conduction band and penetrate a distance λ_0

beyond the interface. This region of negative space charge is termed an *accumulation region*. Since the Fermi level represents a constant occupation probability, it must remain constant for the two materials in the absence of an applied voltage or an electric field. Upward band bending of the conduction band occurs in the accumulation region, beyond which the dielectric is shielded from the electrode, and its band edges are similar to those in the absence of an electrode. The accumulation region acts as a reservoir of charge, which is capable of supplying carriers to the material as required by the bias conditions. With this type of contact, Simmons [131] pointed out that the conductivity is limited by the rate at which electrons can flow into the bulk of the material. This process is said to be bulk limited.

Blocking contacts (or Schottky barriers) occur when $\psi_m > \psi_i$. This is illustrated in Figure 42 (e) and (f). In this case, the flow of charge is reversed, with electrons flowing from the dielectric to the metal electrode on contact. This results in a positive space charge region near the interface in the dielectric, which is termed a depletion region, due to the absence of electrons. An equal and opposite negative charge is induced on the electrode, and the interaction between the two charges establishes a local electric field near the interface. In this case, the conduction band in the insulator bends downward as shown in (f), until the Fermi level in the bulk of the dielectric lies ψ_i below the vacuum level, as it also does in the absence of the contact as shown in (e). The free-electron concentration in the interfacial region in this case is much lower than in the bulk of the dielectric. The rate of flow of electrons is thus limited by the rate at which they can flow over the electrode–dielectric barrier, and the conductivity is then said to be electrode limited.

The case of a neutral contact is shown in Figure 42 (c) and (d). In this case $\psi_m = \psi_i$, and because of this equality there is no flow of charge across the interface, no band bending, and neither an accumulation nor a depletion region is set up. This situation is transitional between that for an ohmic and a blocking contact. At low applied biases, the conductivity is ohmic, but as the current level rises the contact becomes incapable of supplying sufficient current to balance that flowing in the dielectric and the process ceases to be ohmic.

We have seen that in principle the type of contact established depends on the relative values of ψ_m and ψ_i. In the real world, such considerations are rarely adequate to predict the type of contact formed, because of the frequent existence of *surface states*. These are usually due to the effects of unsaturated bonds and impurities at the interface [133] or to the departure from periodicity in the structure at the interface, and can drastically modify the shape and the height of the interfacial barrier. It is very difficult to accurately predict the presence of surface states, or their concentration. When they are present in significant quantities Eq. (7) does not apply. The type of contact, and thus the operative conduction process, can be determined almost entirely by the presence of surface states.

In Section 3.3.2, we consider first the bulk-limited conduction processes, where the conductivity does not depend on the barrier height at the interface. This is followed by a considera-tion of the electrode-limited processes, where the barrier height features prominently in the current density-voltage (J–V) equations.

3.3.2. Conduction Processes and Theory

One of the most important conduction processes in semiconducting thin films is space-charge-limited conductivity (SCLC). SCLC occurs only when the injecting electrode is an ohmic contact, and a reservoir of charge is therefore available in the accumulation region, without the need for carrier excitation over a potential barrier. At low voltages, the thermally generated carrier concentration exceeds the injected concentration, and the current density J, follows Ohm's law,

$$J = n_0 e \mu \frac{V}{d} \tag{8}$$

where n_0 is the thermally generated carrier concentration (normally electrons, but holes in the case of p-type materials), e is the electronic charge, μ is the mobility, V is the applied voltage, and d is the film thickness. When ohmic contacts are applied, majority carriers may be injected into the material, and when the injected carrier concentration exceeds that of the thermally generated concentration the SCLC current becomes dominant. Expressions for the SCLC current density are derived by considering both the drift and diffusion currents, and by solving Poisson's equation to obtain an expression for the electric field as a function of distance from the interface. Normally, traps will exist within the imperfect semiconducting film, and these have the effect of immobilizing a large proportion of the injected carriers. If the traps are shallow and are located at a discrete energy E_t below the conduction band edge, the SCLC current density is given by [131],

$$J = \frac{9}{8} \varepsilon_r \varepsilon_0 \theta \mu \frac{V^2}{d^3} \tag{9}$$

where ε_r is the relative permittivity of the material, ε_0 is the permittivity of free space, and θ is the ratio of free to trapped charge. In the absence of traps, θ does not appear in Eq. (9). The quantity θ is given by [134],

$$\theta = \frac{N_c}{N_{t(s)}} \exp\left(-\frac{E_t}{kT}\right) \tag{10}$$

where N_c is the effective density of states at the conduction band edge, $N_{t(s)}$ is the trap concentration residing at the discrete energy level, k is Boltzmann's constant, and T is the absolute temperature. There is a transition between the two conduction mechanisms described by Eqs. (8) and (9), which occurs at a transitional voltage V_t. This takes place when the injected carrier concentration first exceeds the thermally generated carrier concentration. An expression for the transition voltage is obtained by solving Eqs. (8) and (9) simultaneously to yield

$$V_t = \frac{8en_0d^2}{9\theta\varepsilon_r\varepsilon_0} \tag{11}$$

In many cases, particularly with amorphous and polycrystalline materials, a single shallow trap level is not present, and the combination of several poorly defined trap levels may lead to an overall exponential distribution of traps $N(E)$ [134] described by the relationship,

$$N(E) = N_0 \exp\left(-\frac{E}{kT_t}\right) \tag{12}$$

where $N(E)$ is the trap concentration per unit energy range at an energy E below the conduction band edge, N_0 is the value of $N(E)$ at the conduction band edge and $T_t > T$ is a temperature parameter which characterizes the distribution. It can be shown that the total concentration of traps comprising the distribution $N_{t(e)}$ is [135],

$$N_{t(e)} = N_0 k T_t \tag{13}$$

By assuming an exponential trap distribution as described by Eq. (12), Lampert [136] obtained the following expression for the current density:

$$J = e\mu N_c \left(\frac{\varepsilon_r \varepsilon_0}{e N_0 k T_t}\right)^\ell \frac{V^{\ell+1}}{d^{2\ell+1}} \tag{14}$$

This expression predicts a power-law dependence of J on V, with the exponent $n = \ell + 1$, where ℓ represents the ratio T_t/T. In a similar manner to the case for a single shallow trap level, there is also a transition voltage between ohmic conduction and SCLC dominated by an exponential trap distribution; the transition voltage is given in this case by [137],

$$V_t = \left(\frac{n_0}{N_c}\right)^{1/\ell} \frac{e N_0 k T_t d^2}{\varepsilon_r \varepsilon_0} \tag{15}$$

Starting from the preceding expressions for SCLC dominated by an exponential trap distribution, Gould [138] showed that measurements of J as a function of temperature at constant applied voltage in the SCLC region could be used to determine the mobility and the trap concentration. If the data are plotted in the form $\log_{10} J$ against $1/T$, the curves should be linear, and when extrapolated to negative values of $1/T$ they should all intersect at a common point irrespective of the applied voltage. The coordinates of this point are given by

$$\log_{10} J = \log_{10}\left(\frac{e^2 \mu d N_c N_{t(e)}}{\varepsilon_r \varepsilon_0}\right) \qquad \frac{1}{T} = -\frac{1}{T_t} \tag{16}$$

The gradients of the plots are

$$\frac{d(\log_{10} J)}{d(1/T)} = T_t \log_{10}\left(\frac{\varepsilon_r \varepsilon_0 V}{e d^2 N_{t(e)}}\right) \tag{17}$$

and the intercept $\log_{10} J_0$ on the $\log_{10} J$ axis is given by

$$\log_{10} J_0 = \log_{10}\left(\frac{e\mu N_c V}{d}\right) \tag{18}$$

This set of equations has been used to determine mobility and trap concentration in several cadmium chalcogenides showing this form of SCLC. These are discussed where appropriate in the following sections. We might also mention that Rose [134] explored the possibility of a uniform trap distribution, in addition to discrete energy levels and the exponential distribution. In the context of the present discussion, this has been observed

only in the case of CdTe films, and is therefore covered in Section 3.3.4. The previous account of SCLC has been concerned with single carrier injection only. The effects of double injection, when one electrode is ohmic for electrons and the other is ohmic for holes, are covered in detail in the standard text of Lampert and Mark [139] on current injection in solids, while a very useful description is also given by Lamb [133].

A second bulk-limited conduction process is the Poole–Frenkel effect. The effect is often referred to as the bulk analogue of the Schottky effect, a well-known *electrode*-limited process, which is itself covered later. Essentially, the Poole–Frenkel effect is the field-assisted lowering of the Coulombic potential barrier between carriers at impurity levels and the edge of the conduction band. A carrier which is trapped at a center is not able to contribute to the conductivity until it overcomes a potential barrier ϕ and is promoted into the conduction band. In the presence of a *high* electric field, the potential is reduced by an amount eFx where e is the electronic charge, F is the applied field, and x is the distance from the center. This situation is shown in Figure 43 [131] for the case of a donor level located an energy E_d below the bottom of the conduction band. The potential energy of the electron in the Coulombic field is $-e^2/4\pi\varepsilon_r\varepsilon_0 x$, where x is the distance from the center in the field direction. There is also another contribution to the potential energy, $-eFx$, arising from the applied field F. When both these contributions to the potential energy are taken into account, it is evident that there is a maximum in the potential energy for emission as shown in the figure. The effective potential barrier for Poole–Frenkel emission is lowered by an amount $\Delta\phi_{PF}$, where $\Delta\phi_{PF}$ ($\Delta\phi$ in the figure) depends on the electric field according to the relationship,

$$\Delta\phi_{PF} = \beta_{PF} F^{1/2} \tag{19}$$

where

$$\beta_{PF} = \left(\frac{e^3}{\pi \varepsilon_r \varepsilon_0}\right)^{1/2} \tag{20}$$

is termed the Poole–Frenkel field-lowering coefficient. The SI unit for the Poole–Frenkel coefficient is $J\,m^{1/2}\,V^{-1/2}$, although in this chapter and elsewhere it is normally quoted in a unit of $eV\,m^{1/2}\,V^{-1/2}$. Since the conductivity, and thus the current density, depends exponentially on the potential barrier for carrier excitation, and since this is *lower* by an amount $\beta_{PF} F^{1/2}$ (Eq. (19)), it follows that in the presence of a high field the current density is *increased* by an exponential factor. The current density then follows the relationship,

$$J = J_0 \exp\left(\frac{\beta_{PF} F^{1/2}}{kT}\right) \tag{21}$$

where $J_0 = \sigma_0 F$ is the low field current density. This is the form of the equation normally used in thin film work, owing to the almost universal presence of traps. However, if the low field conductivity σ_0 shows an intrinsic dependence and is proportional to the factor $\exp(-E_g/2kT)$ where E_g is the energy gap, then a factor of 2 would appear in the denominator of the exponential term of Eq. (21) [131]. Although J_0 in this equation

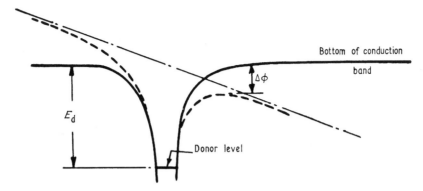

Fig. 43. Poole–Frenkel field-lowering effect at a donor center. Reproduced from [131]. Reprinted with permission from J. G. Simmons, *J. Phys. D: Appl. Phys.* 4, 613, copyright IOP Publishing Ltd., 1971.

depends on F, the variation is considerably less than that of the exponential term, and is frequently ignored in comparison. Since $F = V/d$, Eq. (21) may be rewritten,

$$J = J_0 \exp\left(\frac{\beta_{PF} V^{1/2}}{kT d^{1/2}}\right) \qquad (22)$$

If we therefore plot $\log J$ against $V^{1/2}$, a linear relationship should be followed, from which a value of the field-lowering coefficient can be obtained. This may then be compared with the theoretical value predicted by Eq. (20). For critical work, where the variation of the preexponential factor is also to be taken into account, then a linear plot of $\log(J/V)$ against $V^{1/2}$ should be obtained.

Several modified versions of Eq. (21) have been proposed to account for the current density variation in the presence of various combinations of centers, including donors and traps. Simmons [131] quoted an example of one of these for the case when the material contains donor levels below the Fermi level and shallow *neutral* traps. In this case, the coefficient of $F^{1/2}/kT$ is one half of the theoretical value given by Eq. (20). Gould and Bowler [140] have proposed the following expression for the dependence of Poole–Frenkel conduction, in which the electric field is nonuniform, and has a maximum value of $\alpha^2 F$, where F is the mean field:

$$J = \frac{2J_0 kT}{\alpha \beta_{PF} F^{1/2}} \exp\left(\frac{\alpha \beta_{PF} F^{1/2}}{kT}\right) \qquad (23)$$

This expression was originally proposed for the case of CdTe films, but interpretations of results in terms of this expression have also been made for other materials.

A third bulk-limited conduction process, which may in principle be observed in the cadmium compounds is known as *hopping*. This process is well known in noncrystalline materials, and is discussed at length in various books covering this area [141, 142]. In such materials, the lack of long-range order gives rise to a phenomenon known as *localization*, in which the energy levels do not merge into one another, especially in the region of the edges of the energy bands (band tails). Thus, for carriers to be transported through the material they progress by a series of jumps or "hops" from one localized energy level to another. Hopping occurs between the various localized energy levels when a low level of thermal energy is available. Since the localized levels are normally very closely spaced in energy, the thermal energy required is very small, and the process can take place at very low temperatures when other processes are excluded. Mott and Davis [142] argued that the conductivity σ exhibits different behavior in different temperature regions. At higher temperatures, thermal excitation of carriers to the band edges is possible and extended-state, or free band conductivity can take place, while at lower temperatures, where less thermal energy is available, hopping may occur. There are various different hopping regimes depending on the length of the hop. For example, *nearest neighbour* hopping is probably the simplest such process, while in *variable-range* hopping excitation takes place to a more distant neighbor, where the energy difference between the states is lower. For variable-range hopping, the conductivity follows a relationship of the type [141],

$$\sigma = \sigma_0 \exp\left(-\frac{A}{T}\right)^{1/4} \qquad (24)$$

where σ_0 and A are constants, and this relationship is commonly known as the Mott $T^{1/4}$ law, since a plot of $\log \sigma$ versus $T^{-1/4}$ should show a linear characteristic with negative slope.

Of the electrode-limited conduction processes, tunneling directly from the Fermi level of one electrode to the conduction band of the other is normally applicable only for very thin films of thickness less than about 10 nm when subjected to a high electric field. Tunneling is a quantum-mechanical effect, in which the wave function of the carrier is attenuated only moderately by the *thin* barrier, resulting in the carrier having a finite probability of existence on the opposite side of the barrier. Simmons [143] investigated the effects of tunneling between similar electrodes separated by a thin insulating film. The tunneling barrier presented by the film was assumed to be of arbitrary shape. For the general potential barrier, the probability that an electron could penetrate the barrier was assumed to be given by the Wentzel, Kramers, and Brillouin (WKB) approximation. It was found that different expressions for the current

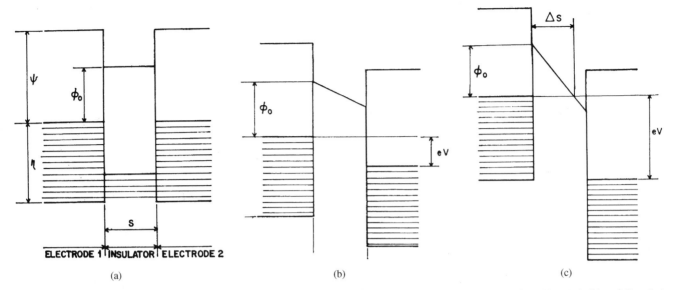

Fig. 44.   Rectangular potential barrier in an insulating film between metal electrodes for different applied voltages: $V = 0$ (a), $V < \phi_0/e$ (b), and $V > \phi_0/e$ (c). η represents the height of the Fermi level. Reproduced from [143]. Reprinted with permission from J. G. Simmons, *J. Appl. Phys.* 34, 1793, copyright, American Institute of Physics, 1963.

density were applicable, depending on whether the applied voltage V were very small, less than ϕ_0/e where ϕ_0 is the barrier height at the metal–dielectric interface, or greater than ϕ_0/e. Figure 44 [143] illustrates these three cases. s represents the film thickness, ψ represents the electrode work function, and η represents the height of the Fermi level. The derived expressions are too involved to be repeated here, but it should be evident from Figure 44, that for $V > \phi_0/e$ it is only necessary for electrons tunneling from the Fermi level of one electrode to penetrate a distance $\Delta s < s$ to reach unoccupied levels in the second electrode. Simmons [144] has further proposed that for $V > \phi_0/e$ a modified Fowler–Nordheim expression may be applicable for tunneling through an interfacial region. Here, the electric field at the barrier is sufficiently high to reduce its width, measured at the Fermi level, to about 5 nm. Under these circumstances, the current density is related to the voltage and the barrier thickness according to the expression,

$$ J = \frac{e^3 V^2}{8\pi h \phi_0 d_t^2} \exp\left(-\frac{8\pi (2m)^{1/2} \phi_0^{3/2} d_t}{3ehV} \right) \qquad (25) $$

where h is Planck's constant, ϕ_0 and d_t are the barrier height and effective thickness, respectively, of the tunneling barrier and m is the free-electron mass. Note that the barrier height ϕ_0 appears in this expression, both in the preexponential and in the exponential terms.

The Schottky effect is the field-assisted lowering of a potential barrier at the injecting electrode and is similar in origin to the Poole–Frenkel effect. This type of conduction process occurs when the film is too thick for direct tunneling to take place. The potential barrier at the metal-semiconductor interface is reduced by an amount $\Delta\phi_S$ which is given by

$$ \Delta\phi_S = \beta_S F^{1/2} \qquad (26) $$

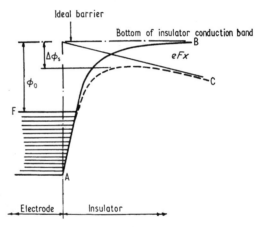

Fig. 45.   Schottky field-lowering effect at a neutral contact. Reproduced from [131]. Reprinted with permission from J. G. Simmons, *J. Phys. D: Appl. Phys.* 4, 613, copyright IOP Publishing Ltd., 1971.

where

$$ \beta_S = \left(\frac{e^3}{4\pi \varepsilon_r \varepsilon_0} \right)^{1/2} \qquad (27) $$

is the Schottky field-lowering coefficient. The barrier lowering process is similar to that for the Poole–Frenkel effect, and is illustrated in Figure 45 [131]. The difference between the Poole–Frenkel and Schottky coefficients given by Eqs. (20) and (27) is related to the different symmetry of the potential barriers; comparison of these two equations shows that $\beta_{PF} = 2\beta_S$. The basic thermionic emission equation of Richardson,

$$ J = AT^2 \exp\left(-\frac{\phi}{kT} \right) \qquad (28) $$

predicts the current density flowing *via* emission of carriers over a potential barrier of height ϕ at a temperature T, where A is the Richardson constant, whose theoretical value is 1.2×10^6 A m^{-2}. In this case, the current density depends only on the barrier height and the temperature, and does not require an applied voltage to have a nonzero value. If ϕ_0 is the zero-voltage barrier height, then the reduced barrier height ϕ is given by $(\phi_0 - \Delta\phi_S)$, so that Eq. (28) yields

$$J = AT^2 \exp\left(-\frac{\phi_0}{kT}\right) \exp\left(\frac{\beta_S F^{1/2}}{kT}\right) \qquad (29)$$

or

$$J = AT^2 \exp\left(-\frac{\phi_0}{kT}\right) \exp\left(\frac{\beta_S V^{1/2}}{kT d^{1/2}}\right) \qquad (30)$$

Thus, in principle the Schottky and Poole–Frenkel effects can be distinguished by the measured value of the field-lowering coefficient, which is twice as high in the Poole–Frenkel case as in the Schottky case.

This section has briefly reviewed most of the dc conduction processes which might be observed in cadmium compound films. Examples of those that have been identified are given in the following sections. In many structures, significant differences can be observed depending on the polarity, the magnitude of the applied voltage, and the temperature, not least because of the influence of surface states. It should also be noted that in many cases identification of the type of conduction process which takes place is very difficult, as the simple expressions of Eqs. (22) and (30) are only first approximations for simple cases, which in practice may be far more complex. It is often difficult to differentiate between SLC and the Poole–Frenkel effect on the sole basis of the J–V characteristics, as in many cases when the data are suitably plotted a linear correlation appears to be present in each case. It is wise therefore to use other information in assessing the category of conductivity.

3.3.3. Cadmium Sulfide

Early high field measurements on CdS single crystals were made at the same time as the development of the theory of space-charge-limited currents in the mid-1950s. Indeed, current density-voltage measurements were reported by Smith and Rose [145] directly preceding the classic article by Rose [134] on SCLC in materials containing traps. These experiments were performed on crystals of thickness approximately 50 μm supplied with In ohmic contacts as required for SCLC; in some experiments the samples were kept in the dark to eliminate photoelectric effects. In nonilluminated crystals, a $J \propto V^4$ relationship was observed, as would be expected for SCLC, and described by Eq. (14) as later formulated by Lampert [136]. Essentially, the same relationship was earlier derived by Rose [134], in which $I \propto V^{(T_c/T)+1}$, where T_c is a characteristic temperature describing the exponential distribution of trap levels in energy. The trap concentration, estimated from the SCLC measurements was 5×10^{18} m^{-3} within an energy range

kT below the conduction band edge. More comprehensive measurements were later performed by Bube [146] on CdS crystals of thickness 75 μm, grown from the vapor phase reaction of Cd and S, and also supplied with In electrodes. In this case, he found that, at each temperature investigated in the range 294–354 K, there was a region where the current was proportional to V^2 which extended over several orders of magnitude in current. This behavior was satisfactorily described by Eq. (9), and indicated that SCLC occurred, which was dominated by trap levels located at a discrete energy below the conduction band edge. Analysis of these results yielded a trap concentration of 10^{19} m^{-3} and a fairly deep trap level located at $E_t = 0.49$–0.53 eV below the conduction band edge. The ratio of free to trapped charge θ (see Eq. (10)) rose from 4.3×10^{-3} at 294 K to 2.9×10^{-2} at 354 K. However, analysis of the dependence of θ on temperature indicated a trap concentration of 5×10^{22} m^{-3} with $E_t = 0.28$ eV. It was suggested that the discrepancy between these two sets of values was the result of impact ionization emptying the traps, and that care must therefore be taken in interpreting an increase in current at the end of the SCLC region as complete filling of the traps.

Zuleeg [147] measured J–V characteristics on CdS thin film diode structures. For these measurements, the samples were fabricated with an ohmic In contact and an Au blocking contact, and the thickness of the films was 2.7 μm. At low applied voltages for both polarities, ohmic conduction was observed. Under forward bias (In injecting electrode) at higher voltages, SCLC was identified, as in the earlier single-crystal experiments. In general, Eq. (9) was obeyed, and the conductivity was dominated by a discrete trap level. Conversely, under reverse bias (Au injecting electrode) a field-lowering conduction process was observed. A typical set of results under reverse bias is shown in Figure 46 [147]. There is a clear region of the curve where the logarithm of the current is proportional to the square root of the applied voltage. In principle, this could result from either the Schottky or the Poole–Frenkel effects, and differential capacitance measurements were used to establish the presence of a Schottky barrier. Standard measurements of the capacitance C were made as a function of the applied voltage V, and plotted in the form $1/C^2$ against V. The resulting linear dependence enabled a trap concentration of 1.3×10^{22} m^{-3} and a Schottky barrier height of 0.83 eV to be determined. The value of the latter was close to that obtained for gold contacts to grown CdS single crystals. These measurements were later extended and confirmed by Zuleeg and Muller [148], who used Al or In for ohmic contacts and Au or Te for blocking contacts. Additional evidence for SCLC under forward bias was obtained by plotting the electrode spacing as a function of current for a fixed voltage, as shown in Figure 47 [148]. For a discrete trap level, Eq. (9) predicts that the slope of such a logarithmic plot should be -3, as obtained from the figure. Under reverse bias, $1/C^2$ versus V plots were linear, and indicated values of the electronic charge concentration of 1.3×10^{22} m^{-3} for a Au-CdS-In sample, 6.5×10^{21} m^{-3} for a Te-CdS-In sample, and 1.6×10^{22} m^{-3} for a second Te-CdS-In sample. Values of the

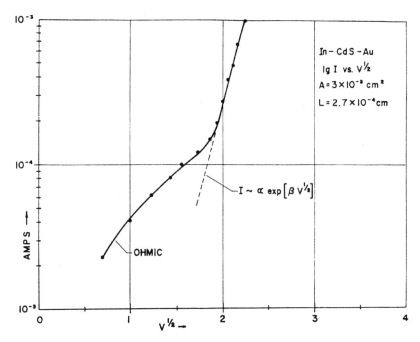

Fig. 46.   Reverse current measurements (Au negative) in an evaporated Au-CdS-In diode showing the Schottky effect above 4 V. Reproduced from [147]. Reprinted from *Solid State Electron.*
6, R. Zuleeg, Electrical Evaluation of Thin Film CdS Diodes and Transistors, p. 645, copyright
1963, with permission from Elsevier Science.

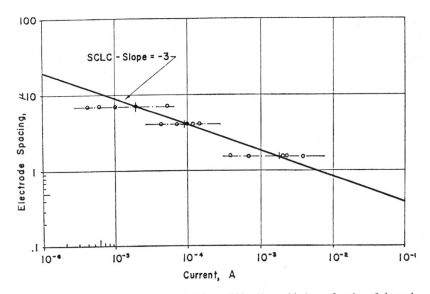

Fig. 47.   Logarithmic plot of current at 1 V forward bias (Au positive) as a function of electrode
spacing for three groups of five evaporated Au-CdS-In diodes. The slope of -3 indicates a reciprocal cube-law dependence on electrode spacing, as predicted by Eq. (9) for SCLC. Reproduced
from [148]. Reprinted from *Solid State Electron.* 7, R. Zuleeg and R. S. Muller, Space-Charge-
Limited Currents and Schottky-Emission Currents in Thin-Film CdS Diodes, p. 575, copyright
1964, with permission from Elsevier Science.

Schottky barrier height ϕ were found to be 0.78–0.90 eV for Au
blocking contacts and 1.1 eV for a Te blocking contact.

In coevaporated CdS films with In contacts, Pizzarello [149]
confirmed that the conductivity was ohmic at low voltages, followed by SCLC at higher voltages. The slope of a log I versus
log V plot was 2 at higher voltages. The trap concentration

was estimated from the onset of trap filling, yielding a value
of 2.5×10^{20} m^{-3}. This author also pointed out that in his
samples SCLC could only be observed using careful measurements undertaken under vacuum. Measurements in air did not
reveal this behavior, probably owing to the effects of adsorbed
gas on the surface. In this laboratory [150], similar conductiv-

ity behavior was observed in evaporated Al-CdS-In thin film structures. At low voltages, the ohmic carrier concentration was 10^{13}–1.2×10^{14} m^{-3}. In the higher voltage SCLC region, values of θ in the range 2.2×10^{-9}–2.7×10^{-8} were obtained. As expected, these are considerably lower than those obtained earlier in CdS crystals [146], owing no doubt to more severe trapping effects. Since activation energy measurements were not performed, calculations of the trap concentration were estimated using the single-crystal value of $E_t \sim 0.5$ eV [146], yielding values of the order of 2.9×10^{23}–3.6×10^{24} m^{-3}. This range of values is well in excess of the value of 10^{19} m^{-3} obtained on single-crystal CdS, but this is not unexpected in view of the imperfect structure of evaporated films. The low value of θ in comparison with those for single crystals is also in accordance with the enhanced trap concentration. SCLC was identified by investigating the dependence of $V_t \theta / n_0$ as a function of d^2, which from Eq. (11) should yield a linear relationship. This was indeed observed, and gave a satisfactory value of the relative permittivity. Furthermore, the slope of a $\log J$ versus $\log(\theta/d^3)$ plot at constant voltage was shown to have a value of approximately unity, as predicted by Eq. (9).

There have also been reports of a type of Poole–Frenkel conductivity in CdS films [151, 152]. Murray and Tosser [151] measured current-voltage characteristics in Al-CdS-Au structures, in which the Al and CdS layers were deposited by rf sputtering. A Cd target was used for the CdS layers, and sputtering was carried out in a mixture of H$_2$S and argon. The conductivity properties of these samples were interpreted in the same way as previous measurements by these workers on Al-ZnS-Au samples. There was a clear linear dependence of $\log I$ on $V^{1/2}$, which was symmetric as a function of the applied field, and did not therefore suggest Schottky-type effects at either of the electrodes. The Poole–Frenkel emission centers were associated with trapped particles from the discharge gases incorporated into the films. Measurements of the photocurrent, on exposure to light having photon energy less than the bandgap, showed Poole–Frenkel photoexcitation. In extensions to this work, Piel and Murray [152] noted that the samples initially showed asymmetric conductivity, but that after a stabilization process the characteristics became symmetric. They stressed that Poole–Frenkel conductivity is a bulk effect, and also the fact that there were no variations in capacitance, thus excluding the possibility of electric field penetration into the dielectric at the interface. Therefore, the Poole–Frenkel effect observed differs in detail from the classic behavior, described by Eqs. (21) and (22). Unfortunately, these workers did not present calculated values of β, whose theoretical value differs between the Poole–Frenkel and Schottky effects. As noted in the previous section, however, differentiating between Schottky and Poole–Frenkel behavior is unwise on the basis of the J–V characteristics only, and the fact that the curves were symmetric for different types of electrodes appears to rule out Schottky emission in this case.

The foregoing examples demonstrate that CdS films may exhibit several different conduction mechanisms. At low fields, ohmic conduction is observed, with thermally generated carriers responsible for the conductivity. Evaporated samples having ohmic injecting contacts such as In or Al generally show SCLC dominated by traps located at a discrete energy level [147–150]. Similar films having Au or Te injecting contacts, which are blocking for CdS, showed the Schottky effect [147] and the capacitance-voltage measurements have been used to determine the Schottky barrier height at the injecting electrode. In films prepared by rf sputtering, the samples did not show variations in capacitance with voltage, and the conduction was classified as bulk-limited. The linear $\log J - V^{1/2}$ dependence in this case was taken to be evidence of a modified Poole–Frenkel effect [151, 152], with the Coulombic centers provided by trapped sputtering gas atoms. The type of conductivity observed is directly dependent on the electrode material, and on the species of traps and/or Poole–Frenkel centers present.

3.3.4. Cadmium Telluride

SCLC was observed in semi-insulating etched CdTe crystals by Canali, Nicolet, and Mayer [153], who used this method to investigate the presence and the properties of traps in the material. Sample thickness was typically 160 μm, and the contacts were made using an aqueous suspension (aquadag). There was a clear dependence of the current on V^2 and on d^{-3}, as expected from Eq. (9). According to their calculations, the level of the current measured was consistent with trap-free conduction when single voltage pulses were applied; i.e., θ was absent from Eq. (9). Nevertheless, transient decay measurements of the current established that traps do play a role in the conductivity. From measurements of the detrapping time as a function of inverse temperature, it was established that traps are located at an energy $E_t \leq 0.65$ eV below the bottom of the conduction band. Steady dc current-voltage characteristics also showed a $J \propto V^2$ region with lower current levels, which was accounted for by the immobilization of charge carriers in the traps. The ratio of free to trapped charge θ was found to be of the order of 10^{-6}, whereas their theoretical estimate of the value was somewhat higher at about 10^{-5}. Typical trap concentrations were determined from the temperature measurements, yielding a value of $N_t \leq 8.5 \times 10^{17}$ m^{-3}.

In CdTe thin films, the earlier work suggested that although SCLC was present, it was dominated by a uniform distribution of trap levels within the energy bandgap [154, 155]. These workers made *in situ* current-voltage measurements on n-type CdTe thin films of thickness between 200 and 500 nm; however, the electrodes were applied on one surface, their type unspecified, and the typical separation was of the order of 200–400 μm [154]. In this work, it was found that the conductivity was ohmic at low voltages, but at higher voltages there was a region showing a linear dependence of $\log(I/V)$ on V, which indicated that $I \propto V \exp(\alpha V)$, where α is a constant. A similar situation had been explored theoretically by Rose [134], who predicted a current density-voltage characteristic of the form,

$$J = \frac{9}{8} \mu n_0 e \frac{V}{L} \exp(t V) \qquad (31)$$

where μ is the carrier mobility, n_0 is the thermally generated carrier concentration, e is the electronic charge, L is the electrode spacing, and $t = \varepsilon_r \varepsilon_0 e \eta_t L^2 kT$, where $\varepsilon_r \varepsilon_0$ is the permittivity, η_t is a constant trap concentration per unit energy range, k is Boltzmann's constant, and T is the absolute temperature. This expression was deliberately omitted from Section 3.3.2, since uniform trap distributions are rare in comparison with discrete trap levels and exponential trap distributions, and have not been observed in cadmium compound films, except in the present case. Based on the work of Rose [134], Lhermitte, Carles, and Vautier [154] modified the model for a uniform distribution of traps of width $\Delta E = E_2 - E_1$, where E_1 and E_2 are the limits of the distribution. They derived a modified expression for J [154] which is too complex to include here, but which produced an excellent description of their data in the temperature range 287–335 K. They concluded that the uniform trap distribution had a concentration in the range between 4.8×10^{19} and 2.4×10^{20} eV^{-1} m^{-3}, was of width $\Delta E = 0.107$ eV, and that the Fermi level was located 0.666 eV below the conduction band edge. The concentration and Fermi level position were in agreement with previous work. The model was extended to include the effects of photoconductivity [155], taking into account the electron–hole pair generation rate, the electron generation rate from the trap levels, and recombination. A carrier lifetime of 2 μs was determined using their data in the photoconductivity model.

The classic $J \propto V^2$ dependence indicated in Eq. (9) was unequivocally observed by Basol and Stafsudd [156] in CdTe films of thickness 0.5–1.0 μm. The films were electrochemically deposited onto evaporated Ni, and the counterelectrode was Au in all cases. When the Ni electrode was biased negative, clear SCLC was exhibited for a wide range of temperatures. Figure 48 [156] shows a set of such results spanning the temperature range 273–338 K. Activation energy measurements at 1-V applied voltage yielded a value of $E_t = 0.55$ eV and a trap concentration of 7×10^{21} m^{-3} was determined using Eq. (10). The conductivity was confirmed as SCLC by the linear dependence of current on d^{-3} at a fixed voltage level of 1 V. Under reverse bias (Au injecting electrode biased negative), SCLC was no longer observed as the Au contact is not ohmic for n-type CdTe; a barrier height of 0.75–0.85 eV was determined at the Au electrode. Dharmadhikari [27] made a very wide series of measurements on vacuum deposited CdTe films of thickness 100–250 nm having both Al base and counterelectrodes. At low voltages, ohmic conduction was observed, followed by the customary $J \propto V^2$ dependence. This was in turn followed by the so-called exponential region, where $J \propto V^n$, where $n = 6.0$–6.5, changing to 11–12. The variation of J with film thickness followed a $d^{-1.2}$ dependence in the ohmic region, a $d^{-3.12}$ dependence in the second region, and a $d^{-3.10}$ dependence in the exponential region. The first two dependences are within experimental error in accordance with those expected from Eqs. (8) and (9). The dependence of current on thickness in the third region does not unequivocally indicate SCLC of any particular type of trap distribution, although the voltage dependence suggests that an exponential

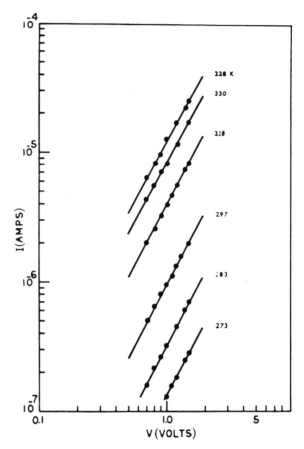

Fig. 48. V^2 dependence of current on voltage in an electrodeposited Ni-CdTe-Au sandwich structure of thickness approximately 1 μm showing SCLC for temperatures of 273–338 K. Reproduced from [156]. Reprinted from *Solid State Electron.* 24, B. M. Basol and O. M. Stafsudd, Observation of Electron Traps in Electrochemically Deposited CdTe Films, p. 121, copyright 1981, with permission from Elsevier Science.

trap distribution may be operative at higher voltages. In these films at very high fields above 5×10^7 V m^{-1}, negative resistance behavior was also seen which may be related to the very high n values determined in the exponential region. Nevertheless, in the V^2 region a trap level at typically 0.57–0.61 eV was determined, with trap concentrations of 5.95×10^{21} or 1.08×10^{22} m^{-3}. Values of these parameters were in reasonable agreement with those measured previously. Measurements on electrodeposited p-type CdTe films which formed a heterojunction with n-type CdS films have also been made by Ou, Stafsudd, and Basol [157]. The deposition method included the novel feature of heat treatment in air for 6 min at 350°C, which had the effect of converting the n-type films into p-type, which were of approximate thickness 1.6 μm. The usual $J \propto V^2$ dependence was observed, and activation energy measurements indicated that hole traps exist at an energy of 0.54 eV above the top of the valence band, with total concentration 10^{20} m^{-3}. In some cases, these workers found that the current increased more rapidly with voltage than predicted by a square law, and indicated that a uniform distribution of traps, essentially as described by Eq. (31), might be ap-

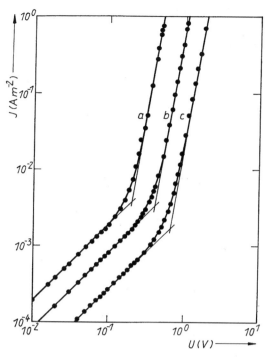

Fig. 49. Room-temperature current density-voltage characteristics for evaporated Al-CdTe-Al sandwich structures of thickness 0.31 μm (a), 0.45 μm (b), and 0.56 μm (c), showing ohmic conduction and SCLC. Reproduced from [158]. Reprinted with permission from B. B. Ismail and R. D. Gould, *Phys. Status Solidi A* 115, 237, copyright Wiley-VCH, 1989.

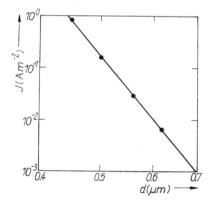

Fig. 50. Logarithmic plot of current density against CdTe thickness at 1-V applied voltage in evaporated Al-CdTe-Al sandwich structures. The derived slope of −15.3 implies a value of $\ell = 7.15$ in Eq. (14) for SCLC. Reproduced from [158]. Reprinted with permission from B. B. Ismail and R. D. Gould, *Phys. Status Solidi A* 115, 237, copyright Wiley-VCH, 1989.

plicable in this case. Processing of the data in this manner yielded an activation energy of 0.72 eV above the valence band edge with trap concentration per unit energy of approximately 10^{21} eV^{-1} m^{-3}.

Ismail and Gould [158] measured J–V characteristics in evaporated CdTe thin film sandwich structures of thickness 0.1–1.0 μm having ohmic Al electrodes. In this case, although the J–V characteristics were ohmic at low voltages, at higher voltages there was a $J \propto V^n$ dependence, with the exponent n typically 5.8–6.1. It is interesting to note that this is similar

to the value of 6.0–6.6 earlier determined by Dharmadhikari in his exponential region [27], although there was no hint of the extremely high values of around 12 which had also been reported previously. A typical set of J–V data is shown in Figure 49 [158], for three samples of thickness 0.31 μm (a), 0.45 μm (b), and 0.56 μm (c). The dependence is exactly as predicted by Eq. (14) for SCLC dominated by an exponential distribution of trap levels. Further evidence for this conduction mechanism was provided by a linear dependence of the transition voltage V_t on d^2, as predicted by Eq. (15). Equation (14) also predicts that $J \propto d^{-(2\ell+1)}$ at constant applied voltage. Figure 50 [158] illustrates this behavior for an applied voltage of 1 V for films of thickness in the range 0.45–0.62 μm. In this case, the slope was −15.3, implying a value of the parameter $\ell = 7.15$. Although this does not agree exactly with the more accurate values obtained from Figure 49 [158] for individual samples, it is clear evidence for SCLC dominated by an exponential trap distribution. The room temperature thermally generated electron concentration determined from these measurements was $(1.2–4.8) \times 10^{11}$ m^{-3} with total trap concentration $(3.25–4.03) \times 10^{23}$ m^{-3}. The temperature parameter T_t characterizing the trap distribution in Eq. (12) was generally in the range 1402–1493 K. Since this type of structure appeared to fulfil the conditions of an exponential trap distribution, it may in principle be used to determine the mobility and trap concentration as suggested previously [138] and discussed in Section 3.3.2. According to this method, plots of $\log_{10} J$ against $1/T$ performed at different applied voltages in the SCLC region should intersect at a common point when extrapolated to negative $1/T$ values, whose coordinates are given by Eq. (16). The gradients and the intercepts of the plots are given by Eqs. (17) and (18). A typical set of data is shown in Figure 51 [159] for a CdTe sample of thickness 0.25 μm between ohmic Al electrodes. Analysis of these results gave a mobility in the range $(4–8) \times 10^{-2}$ m^2 V^{-1} s^{-1} and a trap concentration of $(1.4–4.2) \times 10^{24}$ m^{-3}. Thus, from temperature measurements alone it was possible to determine both the mobility and the trap concentration, assuming only the relative permittivity and the effective density of states at the conduction band edge. This work confirmed that an exponential trap distribution dominated the SCLC, and illustrates the utility of this method for determining the mobility and trap parameters. These authors investigated the conduction properties of Al-CdTe-Al sandwich structures of thickness 260 and 400 nm using a material obtained from an alternative supplier [160]. In contrast with their earlier work [158], in this case the ohmic conductivity below an approximate voltage $V < 0.1$ V was followed first by a $J \propto V^2$ region (0.1 V $< V <$ 1 V) and then by a $J \propto V^n$ region ($V > 1$ V), in a similar manner to the results of Dharmadhikari [27]. In the ohmic region, the thermally generated electron concentration was similar to that obtained previously, and in the square-law region activation energy measurements indicated that a discrete trap level was present at $E_t \sim 0.5$ eV below the edge of the conduction band, in accordance with earlier work [155, 156]. The trap concentration $N_{t(s)}$ was estimated to be 8.8×10^{24} m^{-3}. In the power-law region, the behavior

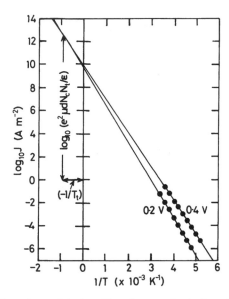

Fig. 51. Dependence of the logarithm of current density $\log_{10} J$ on inverse temperature $1/T$ for an evaporated Al-CdTe-Al sandwich structure of thickness 0.25 μm at two different applied voltages. The data are extrapolated to obtain intercepts on the $\log_{10} J$ axis and the coordinates of the intersection point, in accordance with Eqs. (16)–(18) for SCLC dominated by an exponential trap distribution. Reproduced from [159]. Reprinted with permission from R. D. Gould and B. B. Ismail, *Int. J. Electron.* 69, 19, copyright Taylor & Francis, 1990. Journal web site http://www.tandf.co.uk.

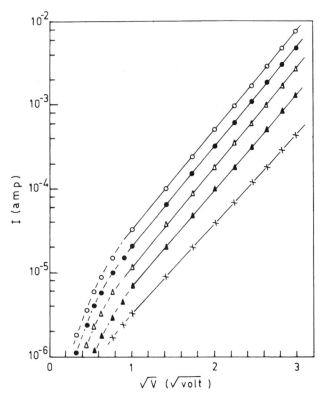

Fig. 52. Linear dependence of the logarithm of current on the square root of the applied voltage in an evaporated Al-CdTe-Al sandwich structure of thickness 722.5 nm at 0°C (bottom curve), 30, 55, 80, and 108°C (top curve). The behavior was identified with a form of the Poole–Frenkel effect. Reproduced from [162]. Reprinted from *Thin Solid Films*, 92, S. Gogoi and K. Barua, Dc Electrical Properties of Vacuum-Deposited CdTe Films, p. 227, copyright 1982, with permission from Elsevier Science.

was similar to that observed earlier [158], with trap concentration $N_{t(e)} \sim (7.04 - 9.39) \times 10^{23}$ m^{-3}. It was concluded that since the deposition processes were similar the appearance of the square-law region in the characteristic was related to the different suppliers of the materials. Nevertheless, the behavior in the ohmic and power-law regions were similar in both cases.

Poole–Frenkel conductivity was observed by Canali et al. [161], who made measurements of the carrier drift velocity as a function of the electric field in high resistivity compensated CdTe bulk samples. The mobility was found to vary linearly with the square root of the electric field and to yield a value of the Poole–Frenkel coefficient $\beta_{PF} = 2.5 \times 10^{-5}$ eV m$^{1/2}$ V$^{-1/2}$ in comparison with a theoretical value of 2.83×10^{-5} eV m$^{1/2}$ V$^{-1/2}$. The activation energy of the centers was estimated to be approximately 0.048 eV from temperature measurements, and their overall concentration was about 10^{23} m^{-3}. At electric fields greater than about 1.2×10^6 V m^{-1}, the Poole–Frenkel effect gave way to tunneling. In thin film samples, Gogoi and Barua [162] first observed field-lowering conductivity in Al-CdTe-Al samples. Figure 52 [162] illustrates some of their results for a sample of thickness 722.5 nm at temperatures of 0°C (bottom curve), 30, 55, 80, and 108°C (top curve); the dependence of $\log I$ on $V^{1/2}$ became linear at an electric field greater than 10^6 V m^{-1}, as predicted by Eq. (22). The measured value of β increased from 4.95×10^{-5} eV m$^{1/2}$ V$^{-1/2}$ at 0°C to 7.70×10^{-5} eV m$^{1/2}$ V$^{-1/2}$ at 108°C. These values were approximately twice the theoretical value for the Poole–Frenkel effect. To determine whether this behavior might be related to the electrode material (i.e., a type of Schottky ef-

fect), characteristics were measured using different electrode materials with different work functions (In and Ag). It was established that β did not depend on the electrode work function and thus Schottky emission was ruled out. The high values of β could not be explained by an alternative three-dimensional Poole–Frenkel process, but it was suggested that electrons produced by thermal ionization at donor-like centers, might hop between the sites as a result of thermal activation in a manner suggested by Jonscher and Ansari [163] in the context of photocurrents in silicon monoxide films. Reinterpretation of the data according to this model yielded good agreement with the predicted value of β and a satisfactory value of the barrier height between hopping sites of 0.154 eV, which was attributed to cadmium vacancies. Results similar to these were also obtained by Gould and Bowler [140] for evaporated Au-CdTe-Al film structures having a CdTe film thickness of 0.4–1.3 μm. The slopes m of linear $\ln J - V^{1/2}$ plots varied with temperature as would be expected from Eq. (22). Figure 53 [140] shows the dependence of this slope on the inverse temperature, and again there is a clear linear relationship, which allowed a value of $\beta = 4.4 \times 10^{-5}$ eV m$^{1/2}$ V$^{-1/2}$ to be derived. Other values of β calculated for different samples all yielded values greater than the theoretical Poole–Frenkel values. In this work, the value of

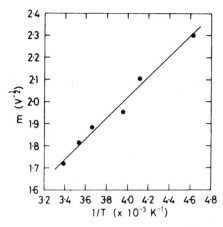

Fig. 53. Dependence of the slopes m of linear log $J - V^{1/2}$ curves on $1/T$ in an evaporated Al-CdTe-Al sandwich structure of thickness 1.2 μm, yielding a value of the field-lowering coefficient $\beta = 4.4 \times 10^{-5}$ eV m$^{1/2}$ V$^{-1/2}$. This behavior was also identified with a form of the Poole–Frenkel effect. Reproduced from [140]. Reprinted from *Thin Solid Films* 164, R. D. Gould and C. J. Bowler, D.c. Electrical Properties of Evaporated Thin Films of CdTe, p. 281, copyright 1988, with permission from Elsevier Science.

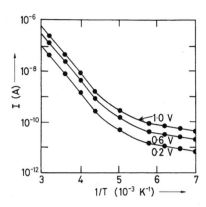

Fig. 54. Dependence of current on inverse temperature in an evaporated Al-CdTe-Al sandwich structure of thickness 267 nm at different applied voltages. Activation energies of approximately 0.08 eV at low temperatures indicate a hopping process. Reproduced from [165]. Reprinted with permission from R. D. Gould and B. B. Ismail, *Phys. Status Solidi A* 134, K65, copyright Wiley-VCH, 1992.

β was 1.5–2.75 times the theoretical value, whereas in the earlier work it was 1.75–2.68 times greater [162]. It was suggested that the similar behavior obtained in both cases may be related to individual centers experiencing an electric field greater than the mean value F. For a uniform distribution of local fields, Eq. (23) was derived. Interpretation of the results in terms of this expression suggested that localized electric fields are enhanced by up to approximately eight times the mean field. From the previous outline of both SCLC and Poole–Frenkel conductivity in CdTe films, it is unclear which mechanism may be observed in any particular sample. Indeed, in this laboratory both SCLC [158] and Poole–Frenkel conductivity [140] have been observed in films deposited using identical source materials and deposition conditions; furthermore electrode effects have been discounted by the work of Gogoi and Barua [162]. However, it was found that in p-type films which were deliberately doped with PbCl$_2$ the conductivity changed from SCLC to Poole–Frenkel [164]. These films yielded values of β approximately 2.6–2.8 times the theoretical Poole–Frenkel values and were therefore also interpreted in terms of Eq. (23), where the electric field was enhanced by a factor of 6.8–7.8. It was however unclear whether the Poole–Frenkel behavior resulted directly from the doping, or whether it was a consequence of the nonuniform electric field distribution.

On the basis of measurements of the frequency dependence of conductivity Dharmadhikari [27] had previously indicated that hopping may occur in Al-CdTe-Al structures at temperatures below 200 K. This behavior is discussed in Section 3.4.2. Following this work, Gould and Ismail [165] measured the dependence of dc current on temperature in similar structures at fixed voltages. Figure 54 [165] illustrates a typical set of results showing the dependence of current on inverse temperature for applied voltages of 0.2, 0.6, and 1.0 V over the temperature range 140–330 K. Each curve shows two linear ranges, with a

transitional region from approximately 175–423 K, suggesting that more than one conduction mechanism is involved. At the higher temperatures, the activation energy was 0.47 eV and at lower temperatures it was 0.08 eV. The higher temperature activation energy is consistent with SCLC, either controlled by a discrete trap level or by an exponential distribution of trap levels. However, the low-temperature variation followed an expression of the form [166],

$$I = I_0 \exp\left(-\frac{\Delta E}{kT}\right) \qquad (32)$$

where ΔE is the activation energy associated with a donor or trap band, and I_0 is the value of the current extrapolated to $1/T = 0$, and is given by

$$I_0 = e\mu_h N \frac{V}{d} A \qquad (33)$$

where μ_h is the mobility in the hopping regime, N is the concentration of centers, V is the bias voltage, and A is the active area. Assuming a previously measured value for the free band mobility μ_f of 4.6×10^{-2} m^2 V^{-1} s^{-1} [159], it was possible to estimate the density of the centers as $N \sim 9.5 \times 10^{12}$ m^{-3}, which yielded a value of the hopping mobility $\mu_h \sim 2.16 \times 10^{-5}$ m^2 V^{-1} s^{-1}. This value is clearly considerably lower than the free band mobility μ_f at higher temperatures and supports the existence of hopping at lower temperatures. It was also possible to estimate the extent of the electron wave functions $1/\alpha'$ using the model of Mott and Davis [142] in which

$$\alpha' = \left(\frac{2m^* \Delta E}{\hbar^2}\right)^{1/2} \qquad (34)$$

where m^* is the electron effective mass, $\hbar$ is Planck's constant divided by 2π, and ΔE now represents the bandwidth of the localized states. Following Dharmadhikari [27] in assuming m^* approximates to the free-electron mass and that ΔE is equal to the hopping activation energy yields a value of $1/\alpha' \sim 0.69$ nm, in comparison with a lattice parameter in our evaporated CdTe films of 0.648 nm [158]. This would tend to support hopping occurring between nearest neighbors, especially as there was no evidence of variable range hopping as described by Eq. (24).

It is perhaps unusual that Schottky field-lowering effects have not been claimed in CdTe films, whereas Schottky barriers to CdTe have been widely investigated. Ponpon [167] summarized the barrier heights between various metals and vacuum- and air-cleaved surfaces of n- and p-type CdTe; the barrier heights had been determined using several different methods, including photovoltaic measurements, current-voltage characteristics, capacitance-voltage measurements, and ultrasonic photoelectron spectroscopy. For Al contacts, Schottky barrier heights of less than 0.3 eV up to 1.0 eV were reported, with ohmic contacts in some cases. The contact characteristics appeared to be dependent on the nature of the CdTe surface, on the fabrication conditions, and on subsequent treatment. Ponpon [167] drew attention to the fact that TeO$_2$ and possibly CdO may occur after exposure to oxygen. Gould and Ismail [168] attempted to encourage oxidation by storing Al-CdTe structures in dry air for 2 weeks, before adding the top Al electrode. Far from showing an ohmic contact, these samples showed a differential capacitance C which varied with the applied voltage V according to $C^{-2} \propto V$, which allowed a value of the barrier height of 0.66–0.86 eV to be determined. Similar measurements had been performed previously at frequencies of 100 Hz to 100 kHz [169] in electrodeposited films having an Au electrode, and it was found that a $C \propto V^{-m}$ dependence was observed where $0 \leq m < 0.5$. This behavior is totally different to that observed in the oxidized films, but it was explained using a model which considered a large concentration of deep traps in the material. These workers assumed a barrier height of typically 0.85 eV [156] which gave a trap concentration of $(1.7–3.0) \times 10^{22}$ m^{-3}. In view of these and other accounts of Schottky barriers on CdTe films, the fact that the high field Schottky effect does not appear to have been observed is curious, even though experiments with electrodes of different work functions have established that field-lowering behavior in this material is normally a bulk rather than a surface effect [162].

In thin films of CdTe, tunneling does not appear to have been reported, although Canali et al. [161] reported this effect at high fields in bulk samples. A definite linear dependence of the logarithm of the drift mobility on inverse electric field E was observed for $E > 1.2 \times 10^6$ V m^{-1}. This is similar to the behavior predicted by Eq. (25) where $\ln J \propto E^{-1}$.

Thus, in CdTe thin films with ohmic electrodes (usually Al) SCLC has been observed which was controlled by traps located at a discrete energy level [156], or by uniform [154, 155], or exponential [158] trap distributions. In certain cases, both discrete trap levels and an exponential distribution of traps appeared to be present simultaneously [27, 158]. Mobilities in CdTe films were determined using measurements of current density as a function of inverse temperature at constant applied voltages [159] for films showing SCLC dominated by an exponential distribution of trap levels. Poole–Frenkel conductivity was also observed [140, 162] in Al-CdTe-Al structures,

although in both these works it was necessary to make additional assumptions to explain the results, either a field-lowering process for hopping conduction [162] or a nonuniform electric field distribution [140]. This type of conduction was also seen in CdTe films doped with PbCl$_2$ [164], which altered the conductivity from SCLC to Poole–Frenkel. Temperature measurements have indicated the existence of hopping conductivity at low temperatures [27, 165] which appears to be between nearest neighbors. The absence of the Schottky field-lowering effect in a material which also shows Schottky barriers is at present unexplained.

3.3.5. Cadmium Selenide

A disproportionate amount of the work performed on CdSe films has been limited to measurements on planar samples. Measurements of mobility, resistivity, and activation energies [51, 114–118] have been described in detail in Section 3.2.3. Trap concentrations and activation energies have been determined from thermally stimulated current measurements in evaporated films [170]. Trap concentrations of 10^{24}–10^{25} m^{-3} were found in as-deposited films. After baking, highly photosensitive films showed a decrease in the trap concentration, particularly for deep traps where the trap concentration was less than 10^{20} m^{-3}. A wide variety of trap levels were found for films deposited at different substrate temperatures in the range 100–300°C. Shallow traps were found at 0.12–0.19 eV and deep traps were found at 0.38–0.82 eV.

SCLC was first observed in dc sputtered CdSe films by Glew [41], who used one indium–tin oxide (ITO) contact and one In contact in a sandwich configuration. Film thicknesses were typically 0.6 μm. The I–V characteristics were asymmetric with respect to the electrode polarity. The dark current was ohmic at low voltages and showed SCLC with $I \propto V^2$ for higher voltages, as shown in Figure 55 [41] for a film of unspecified thickness in the range 0.4–4.0 μm. However, additional processing of these data and activation energy measurements were not performed. The light current showed ohmic conductivity up to a region of current saturation or dielectric breakdown, and current values were typically 3 orders of magnitude higher than the dark currents when illuminated with a tungsten lamp (color temperature 2800 K) at a power level of 30 W m^{-2}. Evaporated CdSe films of thickness 240–250 nm, sandwiched between Al electrodes have been investigated by Sharma and Barua [171]. Films were deposited at different substrate temperatures in the range 80–150°C and currents were generally higher in samples deposited at the lower substrate temperatures. For a film deposited at 135°C, the current was first ohmic and then followed a power-law dependence with exponent 2, giving way to a higher exponent of 3.6. On annealing at 130°C for 30 min the resistivity of such a film decreased from 1.8×10^6 to $7.9 \times 10^5 \Omega$ m. Annealed films showed $I \propto V^2$ and $I \propto V^{2.6}$ regions followed by trap filling. The results were interpreted in the usual fashion, with the square-law region identified with Eq. (9). This was confirmed by an $I \propto d^{-2.94}$

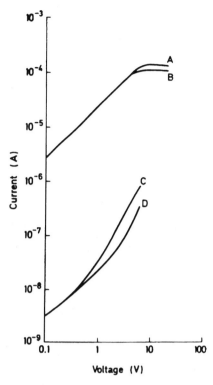

Fig. 55. Dependence of current on voltage in an In-CdSe-ITO sandwich structure prepared by dc sputtering of CdSe. The illumination and polarity conditions were: illuminated, In contact negative (A); illuminated, In contact positive (B); dark, In contact negative (C); dark, In contact positive (D). Reproduced from [41]. Reprinted from *Thin Solid Films* 46, R. W. Glew, Cadmium Selenide Sputtered Films, p. 59, copyright 1977, with permission from Elsevier Science.

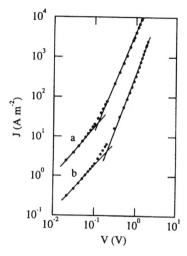

Fig. 56. Dependence of current density on voltage for evaporated Al-CdSe-Al sandwich structures showing SCLC for CdSe thicknesses of 0.62 μm (a) and 0.90 μm (b). Reproduced from [39]. Reprinted from *Thin Solid Films* 270, A. O. Oduor and R. D. Gould, Space-Charge-limited Conductivity in Evaporated Cadmium Selenide Thin Films, p. 387, copyright 1995, with permission from Elsevier Science.

dependence in the SCLC region, also consistent with this equation. It was also suggested that the higher power-law slopes apparent at higher voltages for films deposited at substrate temperatures of 120–135°C were consistent with an exponential distribution of traps. Typical data calculated from these measurements were: charge carrier concentration 1.85×10^{19} m^{-3}, trap concentration 1.55×10^{23} m^{-3}, and trap depth 0.21 eV below the conduction band edge. The electron mobility was 2.67×10^{-5} m^2 V^{-1} s^{-1} which reduced to an effective value of 7.4×10^{-8} m^2 V^{-1} s^{-1} when the effects of trapping were taken into account.

Pandey, Gore, and Rooz [172] used a modified electrodeposition process for growing n-type CdSe films for solar cell applications. They were deposited on Ni substrates in an aqueous acid electrolyte consisting of 0.3 mol^{-1} CdSO$_4$ and 9×10^{-3} mol l^{-1} SeO$_2$. Some films were prepared with the addition of 1.5×10^{-2} mol l^{-1} ethylenediaminetetraacetic acid (EDTA), to prevent the codeposition of metallic impurities present in the CdSO$_4$. After deposition, the films were annealed in air at 340°C. Front contacts to the films were made by evaporating Au dots of diameter 1 mm to form Ni-CdSe-Au structures. I–V characteristics for films of thickness 0.88 μm prepared both without (A) and in the presence of EDTA (B) exhibited an $I \propto V^2$ region above 2.4 V (A) and 2.5 V (B). The current level was slightly lower for the films which used EDTA in the deposi-

tion process. Both characteristics were consistent with Eq. (9). Temperature-dependence measurements of the SCLC current at a fixed applied voltage of 2.5 V followed an Arrhenius relationship allowing trap depths E_t of 0.2 and 0.17 eV to be determined using Eq. (10) for films deposited without and with EDTA, respectively. Using a value of $N_c = 1.17 \times 10^{24}$ m^{-3} in Eq. (10) trap concentrations of 5×10^{21} m^{-3} (without EDTA) and 1.7×10^{21} m^{-3} (with EDTA) were determined. Mobility values of 1.93×10^{-4} m^2 V^{-1} s^{-1} were calculated from the room-temperature resistivity and donor concentration of the films.

SCLC currents dominated by an exponential distribution of trap levels were observed in this laboratory [39] and the form of the J–V characteristics were similar to those of Sharma and Barua [171]. Figure 56 [39] shows the J–V characteristics for films of thickness 0.62 μm (a) and 0.90 μm (b) for sandwich structure films having two Al electrodes. The power-law exponents in the higher voltage regions were 2.48 (a) and 2.56 (b). These results were typical of ohmic conductivity followed by SCLC dominated by an exponential distribution of traps, whose characteristics are described by Eqs. (8) and (14), respectively. The transition voltage between the two processes V_t was shown to increase linearly with the square of the film thickness d^2, as predicted by Eq. (15). Furthermore, plots of measurements of J as a function of $1/T$ at constant applied voltage had voltage-varying slopes and the data were therefore analyzed according to Eqs. (16)–(18). Typical values of quantities derived from this analysis were mobility 3.82×10^{-5} m^2 V^{-1} s^{-1} and trap concentration 3.61×10^{24} m^{-3}. The mobility value was similar to the electron mobility of 2.67×10^{-5} m^2 V^{-1} s^{-1} measured by Sharma and Barua [171]. This work has been extended to the case of CdSe films having different electrode combinations of Al and Au [173]. Samples having symmetric Au

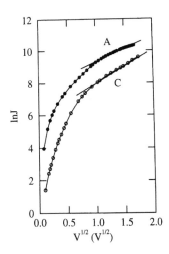

Fig. 57. Linear dependence of the logarithm of current density on the square root of the applied voltage in evaporated Au-CdSe-Au (A) and Al-CdSe-Au (C) sandwich structures of thickness 500 nm. The behavior was identified with a form of the Poole–Frenkel effect. Reproduced from [173]. Reprinted from *Thin Solid Films* 317, A. O. Oduor and R. D. Gould, A Comparison of the D.c. Conduction Properties in Evaporated Cadmium Selenide Thin Films using Gold and Aluminium Electrodes, p. 409, copyright 1998, with permission from Elsevier Science.

electrodes showed ohmic conductivity followed by SCLC with a thermally activated carrier concentration of 4.2×10^{20} m^{-3}, whereas samples having different electrodes had carrier concentrations of 4.2×10^{19} m^{-3} (Al on the substrate side) and 7.3×10^{19} m^{-3} (Au on the substrate side) and symmetric Al samples had a reduced carrier concentration of 2×10^{18} m^{-3}. Thus, the presence of an Au electrode appeared to increase the carrier concentration, with two Au electrodes enhancing this effect. It was concluded that this may result from diffusion of Au atoms within the CdSe films. SCLC dominated by discrete trap levels and by an exponential trap distribution were observed, depending on the details of the electrode combination. For those samples showing a trap level at a discrete energy level, an activation energy E_t of 0.193 eV was determined, in general agreement with that of 0.21 eV determined previously [171].

At higher voltages lying between 1 and 4 V, behavior consistent with field-lowering of a potential barrier was observed as shown in Figure 57 [173] for films with two Au electrodes (A) and with a bottom Al electrode and a top Au electrode (C). Experimental values of the field-lowering coefficient β obtained from these plots are 2.7×10^{-5} eV m$^{1/2}$ V$^{-1/2}$ (A) and 3.61×10^{-5} eV m$^{1/2}$ V$^{-1/2}$ (C) in comparison to a theoretical value for the Poole–Frenkel effect in CdSe of 2.55×10^{-5} eV m$^{1/2}$ V$^{-1/2}$. The measured values are somewhat higher than the theoretical value and probably represent the occurrence of a nonuniform electric field, as previously suggested for CdTe thin films [140], and whose J–V characteristic is described by Eq. (23). In addition, in agreement with the case for CdTe films, it does not appear that Schottky field-lowering effects have been observed, although Schottky barriers at the Al/CdSe interface of height 0.98 eV were determined from capacitance measurements [173].

Oduor [49] has also investigated high field conductivity in CdSe sandwich structures having both two Ag, Cu, or In electrodes or a combination of one of these metals and an Au electrode. The results of these studies mirrored most of the features observed for Al and Au electrodes discussed earlier, although definitive SCLC with an exponential trap distribution of total concentration 1.64×10^{23} m^{-3} was observed only in samples having two Cu electrodes. Field-lowering behavior was identified in many of the samples, with β values of typically $(2.23–3.17) \times 10^{-5}$ eV m$^{1/2}$ V$^{-1/2}$ for samples having at least one Ag electrode, $(2.85–3.47) \times 10^{-5}$ eV m$^{1/2}$ V$^{-1/2}$ (at least one Cu electrode) and approximately 2.2×10^{-5} eV m$^{1/2}$ V$^{-1/2}$ for In/Au electrode combinations; in general these were more appropriately identified with Poole–Frenkel emission than with the Schottky effect.

Chan and Hill [37] measured the lateral conductivity and the Hall coefficient at low temperatures in CdSe films. However, the applied electric field did not exceed 10^4 V m^{-1}, and this work cannot therefore be classified as high field measurements. Nevertheless, the measurements showed evidence of up to three different activation energies at majority carrier concentrations considerably below that given for the Mott metal–insulator transition [141]. In particular, the work was associated with impurity conduction *via* a hopping process as proposed by Mycielski [174]. Although this is consistent with the experimental observations, hopping behavior has not to the author's knowledge been reported under high field conditions in CdSe sandwich-type structures.

In CdSe films, SCLC has been observed by several workers [39, 41, 171–173]. Field-lowering behavior has also been identified at higher voltage levels [49, 173], although further work is necessary in this area. The observation of lateral hopping-type impurity conduction at low temperatures [37] is a clear indication that this type of conduction may occur at high fields, as in the case of CdTe films [27, 165].

3.3.6. Cadmium Arsenide

In Section 3.3.1, the different classes of contacts to semiconductors and insulators were discussed, and it was concluded that at least in principle, the type of contact depends on the relative values of the work functions of the metal electrodes and the dielectric. It was also indicated that such considerations were probably irrelevant for a very narrow (or possibly even overlapping [83] or negative [90]) bandgap such as that in Cd$_3$As$_2$. However, measurements on films of this material in this laboratory [175] have nevertheless shown evidence of field-lowering behavior. In this work, Cd$_3$As$_2$ films of thickness 0.1–1.0 μm were deposited by thermal evaporation at deposition rates of 0.5–5.0 nm s^{-1} onto glass substrates maintained at 20–120°C. Thin film sandwich structures were fabricated, with the bottom electrode of Ag and the top electrode of either Ag, Al, or Au. An investigation of the dependence of J on V showed no evidence of SCLC, but virtually all samples exhibited a $\log J \propto V^{1/2}$ dependence at high fields. When the applied voltage was increased sufficiently, instabilities characteristic of dielectric breakdown or electroforming were observed, which occurred at an electric field F_b value of up to 5×10^7 V m^{-1}. Using a relative permittivity value of 12 for Cd$_3$As$_2$ [176] theoretical values of the Poole–Frenkel field-lowering coefficient $\beta_{PF} = 2.19 \times 10^{-5}$ eV m$^{1/2}$ V$^{-1/2}$ and the Schottky field-lowering coefficient $\beta_S = 1.10 \times 10^{-5}$ eV m$^{1/2}$ V$^{-1/2}$ were calculated from Eqs. (20) and (27), respectively. For each sample, the ratio of the measured value of the field-lowering coefficient β and the theoretical value for the Poole–Frenkel effect β_{PF} was calculated. A value of 1 for the ratio β/β_{PF} would therefore indicate good agreement with Poole–Frenkel conductivity, while a value of 0.5 would represent good agreement with the Schottky effect since $\beta_{PF} = 2\beta_S$. Values of this ratio varied over the range 0.55–2.40 depending on the film thickness, deposition rate, and substrate temperature during deposition, and could not therefore be used, *per se*, to distinguish between the Schottky and Poole–Frenkel effects. For films with two Ag electrodes, of thickness 0.1–0.3 μm the ratio β/β_{PF} increased systematically from 0.55 to 0.84 for films deposited onto substrates maintained at room temperature, but were always between the theoretical values expected for the Schottky and Poole–Frenkel effects. However, for thicker films of the same configuration, albeit deposited at higher substrate temperatures, β/β_{PF} increased to values of greater than 2 for films of thickness 1.1 μm. It was suggested that this may represent a gradual transition from the Schottky to the Poole–Frenkel effect as the sample conductance becomes increasingly dominated by the bulk rather than by the interfaces with the electrodes. Confirmatory evidence for Schottky-type behavior for films of lower thickness was obtained in experiments on samples differing only in the metal used for one electrode, the second electrode being Ag in all cases. Typical results are shown in Figure 58 [175] for samples of thickness 200 nm deposited at a rate of 0.5 nm s^{-1} onto substrates maintained at room temperature. In one sample, the second electrode was Al and in the other sample it was Au. Although the measured β values were of the order of 1.5×10^{-5} eV m$^{1/2}$ V$^{-1/2}$ and comparable to that for samples with two Ag electrodes, it is clear from the figure that the current density is approximately an order of magnitude higher in the sample with the Au electrode than that with the Al electrode. This would tend to suggest different values of the barrier height ϕ_0 in Eq. (30), which may reflect differences in the work functions of the electrode materials and/or oxidation at the Al electrode.

Films of thickness 650 nm deposited at different rates of 1 and 6 nm s^{-1} showed similar log $J \propto V^{1/2}$ [175] behavior with a current density level only moderately higher in the case of the lower deposition rate, which may be attributed to variations in the resistivity of the Cd$_3$As$_2$ films. Values of β/β_{PF} in this case were in the range 1.1–1.3 suggesting Poole–Frenkel conductivity. It was suggested that values greater than 1 were evidence of a nonuniform electric field in the Cd$_3$As$_2$ films giving Poole–Frenkel emission as described by Eq. (23). Even higher values were obtained for films of thickness 1.1 μm deposited at a substrate temperature of 50–120°C. β/β_{PF} increased from 1.6 to 2.4 as the substrate temperature increased over this temperature

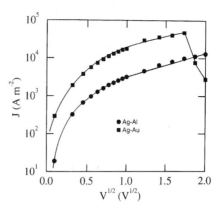

Fig. 58. Linear dependence of the logarithm of current density on the square root of the applied voltage in evaporated Cd$_3$As$_2$ sandwich structures of thickness 200 nm having one Ag electrode and a second electrode or either Al or Au. This behavior was attributed to the Schottky effect, with different values of the Schottky barrier height for the two samples. Reproduced from [175]. Reprinted from *Thin Solid Films* 340, M. Din and R. D. Gould, Field-Lowering Carrier Excitation in Cadmium Arsenide Thin Films, p. 28, copyright 1999, with permission from Elsevier Science.

range. In general, it was concluded that Poole–Frenkel conductivity occurred in the thicker films, and that increased local fields within the films were responsible for augmenting the β values observed.

Although not anticipated, Cd$_3$As$_2$ films exhibit characteristic field-lowering behavior. Among other features, values of the field-lowering coefficient β tend to increase with increasing thickness, and suggest a gradual transition from the Schottky effect to the Poole–Frenkel effect. Nonuniform electric fields appear to be present in the thicker films. Future work on this material should aim to verify the observed thickness dependence of β and to further investigate the effects of the electrode materials used.

3.4. Ac Conductivity

The ac properties of cadmium chalcogenide and Cd$_3$As$_2$ films have not been studied in great detail. Indeed, in some of the materials there is very little published work extant, with most of the effort being directed toward CdTe films. For this reason, the results on the cadmium chalcogenides are covered together in Section 3.4.2, rather than separately as in the earlier sections of this work. The ac conduction properties of Cd$_3$As$_2$ films have been investigated only in this laboratory [64], although owing to the somewhat different properties of this material, these are described separately in Section 3.4.3. Before discussing the experimental results, it is pertinent to first outline the theoretical background underpinning these studies; this is accomplished in Section 3.4.1. The approaches of different workers have fallen into two basic categories. Some workers have attempted to measure the real and imaginary parts of the permittivity, conductivity–conductance or capacitance [177, 178] and have analyzed their results in terms of the classical Cole–Cole description of absorption and dispersion in dielectrics [179]; a single relaxation process is then indicated by a semicircular locus in plots of the real versus the imaginary parts of the conductivity or permittivity. Other workers have considered the conductivity separately from the capacitance and loss tangent, and have attempted to explain the conductivity results in terms of existing models (typically involving a hopping process [180]) and the dielectric behavior using equivalent circuit models of the dielectrics having either Schottky barriers [181] or more usually ohmic contacts [182]. It is clear that for many of the results obtained models originally designed for different materials or circumstances have been adopted to explain the results. Nevertheless, there is a fair measure of coherence in the available results.

3.4.1. Theoretical Background

The Cole–Cole model for dispersion and absorption in dielectrics is based on an earlier work of Debye in which the differing values of the static and the high frequency relative permittivities are accounted for by dipole polarization. It was demonstrated by Cole and Cole [179] that under these circumstances plots of the real part of the relative permittivity as a function of the imaginary part would generate a circular locus. Rapoš et al. [177] in their work on CdTe sandwich structures, have summarized this and later work in the form of four equations, linking the real and imaginary parts of the complex permittivity (ε' and ε'') and of the complex conductivity (σ' and σ''). The relevant equations are

$$\varepsilon' = \varepsilon_\infty + \frac{\varepsilon_s - \varepsilon_\infty}{1 + \omega^2\tau^2} \tag{35}$$

$$\varepsilon'' = \frac{\sigma_s}{\omega} + \frac{\omega\tau(\varepsilon_s - \varepsilon_\infty)}{1 + \omega^2\tau^2} \tag{36}$$

$$\sigma' = \omega\varepsilon'' = \sigma_s + \frac{\omega^2\tau(\varepsilon_s - \varepsilon_\infty)}{1 + \omega^2\tau^2}$$

$$= \sigma_s + \frac{\omega^2\tau^2(\sigma_\infty - \sigma_s)}{1 + \omega^2\tau^2} \tag{37}$$

$$\sigma'' = \omega\varepsilon' = \omega\varepsilon_\infty + \frac{\omega(\varepsilon_s - \varepsilon_\infty)}{1 + \omega^2\tau^2}$$

$$= \omega\varepsilon_\infty + \frac{\omega\tau(\sigma_\infty - \sigma_s)}{1 + \omega^2\tau^2} \tag{38}$$

ε_s and σ_s and ε_∞ and σ_∞ are the static ($\omega \rightarrow 0$) and optical ($\omega \rightarrow \infty$) permittivity and conductivity and ω is the angular frequency. The relaxation time $\tau = (\varepsilon_s - \varepsilon_\infty)/(\sigma_\infty - \sigma_s)$ is the ratio of the relaxation part of the permittivity to that of the conductivity. The final terms in Eqs. (35)–(38) describe the relaxation process. The point was made that in samples for which there is more than one relaxation time, or a distribution of relaxation times, the final terms are replaced by sums or integral functions, respectively. Nevertheless, these expressions have been shown to be very useful in interpreting results on CdTe [177] and CdSe [178] samples having relaxation behavior requiring more than a single relaxation time for its explanation.

Conductivity measurements in cadmium chalcogenides have usually indicated a $\sigma(\omega) \propto \omega^s$ dependence, where $\sigma(\omega)$ represents the ac conductivity, ω is the angular frequency, and s is

an index whose value is dependent on the particular conduction mechanism involved. There are many alternative theoretical expressions for $\sigma(\omega)$ as a function of ω in the literature which follow the dependence mentioned before. A particularly useful expression was derived by Elliott [180] for chalcogenide glasses in which $\sigma(\omega)$ tends to increase with increasing frequency and decreases with increasing temperature. This conductivity expression was derived assuming a correlated barrier hopping process (hopping between pairs of closely spaced centers) at low temperatures and high frequencies, yielding

$$\sigma(\omega) = \frac{\pi^2 N^2 \varepsilon_r \varepsilon_0}{24}\left(\frac{8e^2}{\varepsilon_r \varepsilon_0 W_m}\right)^6 \frac{\omega^s}{\tau_0^\beta} \tag{39}$$

where N represents the density of localized states, $\varepsilon_r\varepsilon_0$ is the permittivity, e is the electronic charge, and τ_0 is the relaxation time. The power-law exponent in this model s is given by $1 - \beta$ at low temperatures, where $\beta = 6kT/W_m$ and W_m approximates the optical bandgap. Although originally derived for the case of chalcogenide glasses, in this laboratory Eq. (39) has also been found to satisfactorily account for the conductivity behavior in the cadmium chalcogenides: CdTe [183] and CdSe [49]. Notwithstanding the utility of this expression in the interpretation of some results, it should also be pointed out that Elliott [184] has reviewed several other ac conduction models in which the behavior of s as functions of both temperature and frequency are discussed. Furthermore, Böttger and Bryksin [185] argued that an increase in conductivity with increasing frequency is likely to be a consequence of disorder, rather than a direct indication of a hopping process. They argue that with increasing frequency charge carriers can move through clusters of decreasing size, so that well-conducting regions of finite size become increasingly effective. They also argue that if conduction is via hopping, a two-site hopping model is only appropriate at higher frequencies where the current is governed by transitions of electrons between pairs of sites, and at lower frequencies the current arises from transitions within highly conducting clusters of sites, this process being known as multiple hopping. In this hopping region, the value of s is also less than unity, and is given by $1 - [2/\ln(1/2)\pi\overline{\Omega}]$, where $\overline{\Omega}$ is a dimensionless frequency ($\overline{\Omega} \gg 1$). Thus, s approaches unity as the frequency is increased and also decreases with increasing temperature. Although it will be seen in the following section that variations of conductivity with frequency and temperature appear to suggest a hopping process as described by Eq. (39), it would therefore be unwise to rule out other hopping processes.

The dependences of the capacitance and loss tangent in sandwich structures was investigated theoretically by Simmons, Nadkarni, and Lancaster [181]. These workers assumed that the contacts to the dielectric film were Schottky barriers and that each of these two contacts possessed an equal Schottky barrier capacitance, C_s. The dielectric layer was modeled as a resistance R_b in parallel with a fixed capacitance C_b. The resistance

R_b was assumed to be thermally activated and given by

$$R_b = R_0 \exp\left(\frac{\phi}{kT}\right) \quad (40)$$

where compensating centers are present in the dielectric. A simple equivalent circuit model based on these assumptions of constant capacitance and a temperature-dependent resistance allowed expressions for the impedance Z, the series equivalent capacitance and resistance, and the parallel equivalent capacitance and conductance to be derived. They made the point that the quality factor Q, related to the loss tangent $\tan \delta$ by $Q = (\tan \delta)^{-1}$, had a minimum value which occurred at higher temperatures as the frequency increased.

Goswami and Goswami [182] developed a model, based on the earlier model of Simmons, Nadkarni, and Lancaster [181], for the ac dielectric properties of a zinc chalcogenide, zinc sulfide (ZnS). In their experimental work, they used ZnS thin film capacitors with Al electrodes, and in general the capacitance was found to decrease with increasing frequency and to increase with increasing temperature. The loss tangent showed a minimum at a given frequency, whose value varied with the temperature. This behavior was in direct contrast to that described by Simmons, Nadkarni, and Lancaster [181], in that the minimum in $\tan \delta$ would correspond to a maximum in Q, and not a minimum as predicted in the earlier model [181]. These workers proposed an equivalent circuit model in which there were no Schottky barriers at the dielectric interfaces, resulting in an equivalent circuit with only three elements. These were composed of a temperature-dependent resistance R as described by Eq. (40) for R_b, in parallel with a fixed capacitance C (both as in the model of Simmons, Nadkarni, and Lancaster), and a low value series resistance r accounted for by the leads or possibly also the contacts. Using a similar theoretical formulation to the earlier model, an expression for the equivalent series capacitance C_s and the loss tangent $\tan \delta$ were derived. (Note that, confusingly, in the model of Simmons, Nadkarni, and Lancaster C_s represents the Schottky barrier capacitance.) The relevant expressions for these quantities are

$$C_s = C + \frac{1}{\omega^2 R^2 C} \quad (41)$$

and

$$\tan \delta = \frac{1}{\omega R C} + \omega r C \quad (42)$$

respectively. The position of the minimum in the loss tangent is at an angular frequency $\omega_{\min}$ given by

$$\omega_{\min} = \frac{1}{\sqrt{r R} C} \quad (43)$$

Equation (41) predicts that the capacitance C_s will decrease with increasing frequency. In addition, since R decreases with increasing temperature (R_b in Eq. (40)), C_s is predicted to increase. Equation (42) predicts that $\tan \delta$ will first decrease with increasing frequency since the first term on the right-hand side of the equation is dominant at low frequencies, but will increase at higher frequencies where the second term becomes

dominant. Equation (43) predicts that the minimum in the loss tangent $\omega_{\min}$ will shift to higher frequencies with increasing temperature, as R decreases with increasing temperature from Eq. (40).

A somewhat more complex equivalent circuit model was suggested by Gonzalez-Diaz et al. [186] to explain their results on CdTe films. They suggested that their samples had a double-layer structure, each layer of which could be represented by a parallel resistance-capacitance (RC) network. The combined equivalent circuit consists of a series combination of these two networks. The interfacial layer, adjacent to the base electrode, was represented by a fixed resistance R_i in parallel with a thermally activated capacitance $C_i = C_{i0} \exp(-\phi_C/kT)$, where C_{i0} represents a constant capacitance and ϕ_C represents a barrier height. The second bulk layer, adjacent to the counterelectrode, was represented by a fixed capacitance C_b and a thermally activated resistance $R_b = R_{b0} \exp(\phi_R/kT)$, where R_{b0} is a constant resistance and ϕ_R is a barrier height. The results of these workers, which are discussed in Section 3.4.2, could be satisfactorily explained by their double-layer model.

Before the conclusion of this section, it is perhaps worth mentioning that dielectric effects may be evident in measurements of simple material properties, such as the dc or low-frequency relative permittivity ε_r. The value of ε_r for a given material frequently varies in results obtained by different workers, owing to differences in the deposition conditions or the supplier of the material. It is customary for a series of sandwich structures, of varying thickness, to be prepared and the relative permittivity obtained from the observed linear dependence of the capacitance C on the reciprocal thickness $1/d$. For example, Dharmadhikari [27] used this method in CdTe films to obtain a value of $\varepsilon_r = 8.25$ for films deposited at 27°C and 8.70 for films deposited at 150°C. Extrapolation of the data suggests that the linear plots intersect the C-axis, implying an additional small parallel capacitance. Similar measurements from this laboratory yielded similar results [187], although it was also noted that plots of the dependence of the reciprocal capacitance $1/C$ against thickness d were linear and intersected the $1/C$-axis, suggesting a series capacitance. These results suggested that there are elements of both parallel and series capacitance contributing to the overall capacitance, and two different equivalent circuits including these features were proposed. Analysis of these equivalent circuits showed that the measured low-frequency capacitance may be expressed in the form,

$$C = \frac{\hat{a} + \hat{b}d}{1 - \hat{c}d} \quad (44)$$

where $\hat{a}$, $\hat{b}$, and $\hat{c}$, are constants from whose values the series and parallel capacitance values and the relative permittivity may be obtained. Figure 59 [187] shows a sketch of this dependence. The function is clearly hyperbolic, and asymptotic at $C = -\hat{b}/\hat{c}$ and $d = 1/\hat{c}$, intersecting the axes at $C = \hat{a}$ and $d = -\hat{a}/\hat{b}$. In the figure, the full curve indicates the absolute limit of physical applicability of the model for $d > 0$, while the dashed curve represents the function for $d < 0$. It was shown

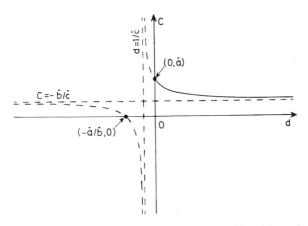

Fig. 59. Sketch of the dependence of capacitance C on film thickness d predicted by Eq. (44). The function is asymptotic at $C = -\hat{b}/\hat{c}$ and $d = 1/\hat{c}$, and intersects the axes at $C = \hat{a}$ and $d = -\hat{a}/\hat{b}$. Reproduced from [187]. Reprinted from *Thin Solid Films* 198, R. D. Gould, S. Gravano, and B. B. Ismail, A Model for Low Frequency Capacitance in Cadmium Telluride Thin Films, p. 93, copyright 1991, with permission from Elsevier Science.

that measurements of C as a function of d in CdTe films could be fitted to Eq. (44) using a least-squares three-parameter regression to obtain the values of $\hat{a}$, $\hat{b}$, and $\hat{c}$. Values of the relative permittivity of 9.55 and 9.62 approached those of the bulk material, and were clearly more precise than those determined using simpler approaches. The derived relative permittivity values were relatively insensitive to the series and parallel capacitance values determined for the equivalent circuits, showing changes of less than 1% for capacitance changes of 10%.

The conductivity and equivalent circuit models outlined before have been used to interpret data from measurements on cadmium chalcogenide and Cd$_3$As$_2$ thin films. At the present time, it is unclear whether a comprehensive Cole–Cole type analysis is more appropriate, or whether attempts to correlate the conductivity and dielectric behavior with more specific theories and equivalent circuits are more likely to prove fruitful. In the following sections, some of the few existing results are outlined and compared.

3.4.2. Cadmium Chalcogenides

As mentioned previously, the majority of the work on the cadmium chalcogenides has been on CdTe. Nevertheless, some of the early work on these materials was performed on CdS, although this was in the form of thin single crystals [188]. In this work, single-crystal platelets were used whose thickness was in the range 7–700 μm, although most experiments were performed on samples of thickness about 30 μm. Contacts were usually In or an In-Ga alloy, both of which gave ohmic contacts to CdS. For samples where a dc bias was not applied, liquid or Au contacts were used. Dc I–V characteristics measured on these samples revealed two different types of behavior. In group I crystals, the usual SCLC dependence was observed, with ohmic conduction at low voltages followed by SCLC behavior which appeared to be dominated by shallow traps as described by Eq. (9). In group II crystals, power-law behavior

was observed in the I–V characteristics at higher voltage levels, although the values of the power-law exponent were not quoted and the authors did not identify this behavior with SCLC dominated by an exponential distribution of trap levels as described by Eq. (14). In accordance with the classification of the dc behavior into the two groups, the frequency dependence of the capacitance was also found to differ. Group I samples showed a frequency-independent capacitance at frequencies from 200 Hz to 100 kHz. However, the values of the measured capacitance exceeded the calculated geometric capacitance values by a factor of 2 to 3, and depended on the voltage bias applied. Above 100 kHz, there was a decrease in capacitance with increasing frequency, with a frequency-independent lower value of capacitance obtained at frequencies above 1 MHz. Similar behavior was observed in the dependence of resistance on frequency, which was also constant up to 100 kHz, decreasing above this value according to a $1/f$ law before attaining a constant lower value at higher frequencies. The ac resistance depended strongly on the dc bias at lower frequencies, whereas at higher frequencies of the order of 10 MHz the influence of the dc bias was small. Group II crystals showed a capacitance–frequency variation in which the capacitance was proportional to the reciprocal of the frequency. There was no frequency-independent capacitance behavior even at very low frequencies. At high frequencies, the capacitance attained lower values, and the frequency at which this occurred increased with the applied dc voltage. Furthermore, the resistance of these samples was independent of frequency up to about 1 MHz, above which there was a decrease with increasing frequency. These workers suggested that in group I crystals, the increase in the capacitance above the geometric value was related to the dynamics of the injection of charge carriers. They derived an expression for the change in current ΔI corresponding to a voltage change of ΔV, in terms of trapping and emission frequencies. Although exact agreement with the experimental work was not obtained, there was a fair measure of agreement between the experimental results and the theory; it was concluded that simple shallow trap SCLC behavior was unlikely to be present, and that a distribution of trap times was likely to prevail. In group II crystals, where the capacitance measured was up to 1000 times the geometric capacitance, it was proposed that field emission from filled traps might occur, which was associated with a stable domain structure, in which a high field region is present near the cathode with a low field in the bulk of the crystal. They argued that this would lead to frequency-dependent capacitance behavior, although detailed calculations for this model were not carried out.

Rapoš et al. [177] made detailed measurements of the real and imaginary parts of the permittivity and conductivity in evaporated CdTe films having either Al or Au electrodes. Their analysis was based on the Cole–Cole model [179], as summarized by Eqs. (35)–(38). They found that it was better to express their results in terms of the capacitance C and the conductance G of the sample, rather than with the material properties ε and σ. They presented a plot of $C_p \tan \delta$ and $C_p \tan \delta - G_s/\omega$ as a function of C_p, and of $G'' - \omega C_{p\infty}$ as a function of G' with

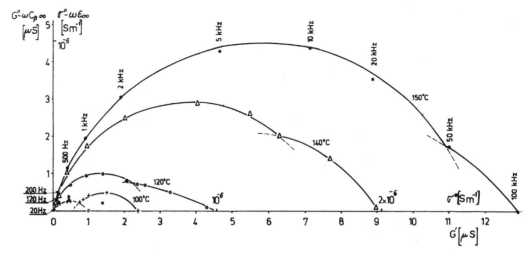

Fig. 60. A Cole–Cole conductance diagram for Al-CdTe-Al thin film sandwich structures. Reproduced from [177]. Reprinted from *Thin Solid Films* 36, M. Rapoš et al. Dielectric Properties of Me-CdTe-Me Thin Film Structures, p. 103, copyright 1976, with permission from Elsevier Science.

frequency as the free parameter. In these plots, C_p represents the equivalent parallel capacitance, whereas G_s, G' and G'' correspond to the conductivities σ_s, σ', and σ'', respectively, and $C_{p\infty}$ corresponds to ε_∞ in Eqs. (35)–(38). The shape of the capacitance plots differed from a semicircle centered on the horizontal axis which would correspond to a single relaxation time. It approximated to two semicircles, both with centers below the horizontal axis and suggesting a relaxation process with continuously distributed relaxation times, concentrated around two distinct values. These were identified with differences in the electrical properties under the electrodes as compared with the interior of the CdTe films. The conductivity plot is shown in Figure 60 [177]. With rising temperature the relaxation times varied, without changes in C_s (corresponding to ε_s) and C_∞ (corresponding to ε_∞) and was therefore due to changes in G_s and G_∞. From Figure 60, it is evident that only G_s changes significantly. This was in accordance with explanations for the relaxation processes caused by variations in the density of free charge carriers. The layers under the electrodes, of higher resistivity and lower free charge density than the interior, were identified as Schottky barriers.

As mentioned in Section 3.4.1, Gonzalez-Diaz et al. [186] presented a double-layer equivalent circuit model for ac conduction in sputtered CdTe films. They plotted both the reciprocal of the measured equivalent series capacitance C_s and the series resistance R_s as functions of the reciprocal absolute temperature $1/T$ at frequencies of 1 and 10 kHz in a sample of thickness 3 μm having a sputtered Al base electrode and a resistively evaporated Al counterelectrode. The somewhat complex plots were analyzed in terms of their model described in Section 3.4.1, which yielded values of the various parameters as follows: fixed capacitance in the bulk region $C_b = 245$ pF, constant resistance $R_{b0} = 5.6 \times 10^{-6}$ Ω, barrier height for thermal activation of resistance in the bulk region $\phi_R = 0.67$ eV, constant capacitance $C_{i0} = 2.5 \times 10^6$ pF (2.5 μF), and barrier height for thermal activation of capacitance in the interfacial

region $\phi_C = 0.15$ eV. The values of the fixed resistance in the interfacial region R_i was 13.4 kΩ at 1 kHz and 6.6 kΩ at 10 kHz. The values of these parameters varied considerably with the type of base electrode used, which were either evaporated or sputtered Al, sputtered Mo, or Mo sheet. Nevertheless, the predictions of the basic equivalent circuit model were followed in all cases. In particular, the variation in ϕ_C was very wide, spanning the range of 0.055 eV for a Mo sheet base electrode to 0.41 eV for a sputtered Mo electrode. The variations were correlated with energy levels of atoms which had diffused from the base electrode. The contribution from R_i was more important in thinner samples and for those grown at higher temperatures, also in agreement with an explanation based on a diffusion process. It was concluded that for an interpretation of the frequency dependence of the components representing the interfaces, experimental data over a wider frequency range were required.

Measurements of ac conductivity and variations in capacitance and loss tangent with frequency and temperature in CdTe films have been made both by Dharmadhikari [27] and by Ismail and Gould [183]. Both these sets of workers used Al electrodes and a sandwich configuration for their measurements, while the frequency range covered was either 100 Hz to 50 kHz [27] or 100 Hz to 20 kHz [183]. Measurements at frequencies in the megahertz range were not made by either group. The relationship $\sigma \propto \omega^s$ was found to hold quite well in one case [27], where the temperature range covered was 90–297 K. The value of the exponent s decreased with increasing temperature, and at 90 K it approached unity, having a value of 0.96. A range of higher temperatures of 203–339 K was explored by Ismail and Gould [183], and a typical set of their results is shown in Figure 61 [183]. In this case, s was also found to approach unity, with a typical value of 0.99 at 203 K in the frequency range 1–10 kHz. At higher temperatures, there was a tendency for the conductivity to become less frequency dependent, whereas at higher frequencies there was an increase

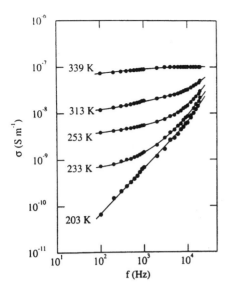

Fig. 61. Dependence of ac conductivity on frequency at different temperatures for an Al-CdTe-Al sandwich structure of thickness 663 nm. Reproduced from [183]. Reprinted with permission from B. B. Ismail and R. D. Gould, A.c. Conductivity and Capacitance Measurements on Evaporated Cadmium Telluride Thin Films, in Metal/Nonmetal Microsystems: Physics, Technology and Applications, *Proc. SPIE* 2780, 46, copyright International Society for Optical Engineering, 1996.

in the value of s to approximately 1.4, which was consistent with a limiting ω^2 dependence of the conductivity expected at higher frequencies [189]. The general features of the conductivity variations were consistent with the model of Elliott [180] for correlated barrier hopping at low temperatures and high frequencies. Using this model [183], it was estimated that the density of localized states involved in the conductivity was of the order of 9.9×10^{23} m^{-3}, in reasonable agreement with values calculated from dc SCLC measurements [158]. A typical set of data illustrating the dependence of conductivity on reciprocal temperature is shown in Figure 62 [27] for frequencies in the range 100 Hz to 50 kHz. In the higher temperature range, the activation energies were similar at all frequencies, having a value of 0.58 ± 0.02 eV, and similar to values determined from dc SCLC measurements. A value of 0.66 eV for the activation energy was also obtained from similar measurements [183], and was associated with a free band conduction process. Both sets of workers [27, 183] observed that at low temperatures the conductivity was very frequency dependent and yielded a low activation energy. A value of 0.013 eV was found in the temperature range 203–233 K [183]. It was concluded by both sets of workers that a hopping process was responsible for the low-temperature conductivity.

Capacitance measurements were also reported by these workers [27, 183]. Figure 63 [183] shows the dependence of capacitance on frequency for a sample of thickness 633 nm over the temperature range 190–298 K. At low temperatures, the capacitance was independent of frequency, whereas at higher temperatures and lower frequencies a dispersion in the capacitance was observed. Very similar results were also obtained by Dharmadhikari [27] who covered a wider temperature range of

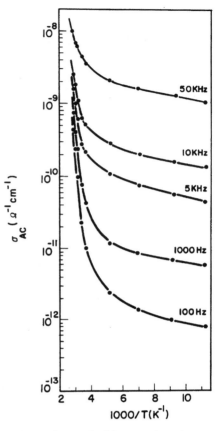

Fig. 62. Dependence of ac conductivity on reciprocal temperature for different fixed frequencies for an Al-CdTe-Al sandwich structure of thickness 151.5 nm. Reproduced from [27]. Reprinted with permission on payment of reproduction fee from V. S. Dharmadhikari, *Int. J. Electron.* 54, 787, copyright Taylor & Francis, 1983. Journal web site http://www.tandf.co.uk/journals.

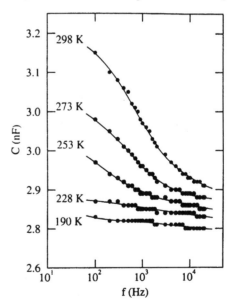

Fig. 63. Dependence of capacitance on frequency at different temperatures for an Al-CdTe-Al sandwich structure of thickness 356 nm. Reproduced from [183]. Reprinted with permission from B. B. Ismail and R. D. Gould, A.c. Conductivity and Capacitance Measurements on Evaporated Cadmium Telluride Thin Films, in Metal/Nonmetal Microsystems: Physics, Technology and Applications, *Proc. SPIE* 2780, 46, copyright International Society for Optical Engineering, 1996.

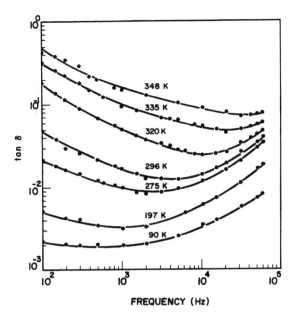

Fig. 64. Dependence of loss tangent on frequency at different temperatures for an Al-CdTe-Al sandwich structure of thickness 151.5 nm. Reproduced from [27]. Reprinted with permission on payment of reproduction fee from V. S. Dharmadhikari, *Int. J. Electron.* 54, 787, copyright Taylor & Francis, 1983. Journal web site http://www.tandf.co.uk/journals.

90–398 K. The results obtained by both sets of workers were essentially similar. It was noted by Dharmadhikari [27] that these results were similar to those obtained by Goswami and Goswami [182] for ZnS films and the capacitance equation derived by these workers (Eq. (41)) was directly applied to the results of Figure 63 [183]. These results are clearly as predicted by this expression, in that as the frequency increased the capacitance decreased to a constant value, and the capacitance increased as the temperature increased. The dependence of the loss tangent on frequency is shown in Figure 64 [27] for various temperatures over the frequency range up to 50 kHz. Although not directly compared with Eqs. (42) and (43) by the author, it is clear that the loss tangent follows the predictions of the model of Goswami and Goswami [182]. There is a minimum in tan δ, and tan δ is higher for higher temperatures. Furthermore, the minimum in the loss tangent shifted to higher frequencies for higher temperatures, as predicted by Eq. (43).

Finally, measurements of ac conductivity in CdSe films were made by Kaganovich, Maksimenko, and Svechnikov [178] and also by Oduor [49]. The capacitance was found to decrease with increasing frequency [178] from 200 Hz to 200 kHz, in accordance with Eq. (41). Furthermore, the loss tangent had a very broad minimum, which did not correspond to a distinct frequency value. These workers also plotted the series equivalent resistance R_s, the parallel conductance G_p, and the real and imaginary parts of the relative permittivity; these were similar to the expected characteristics of the dielectric dispersion in the presence of conductance and dielectric relaxation. Cole–Cole plots were also plotted of the real and imaginary parts of the relative permittivity, and these differed from the theoretical curve for a single relaxation time, a semicircle centered on

the horizontal axis. Each experimental curve could be approximated by two semicircles, with the centers below the horizontal axis. The results were similar to those of Rapoš et al. [177] for CdTe, as illustrated in Figure 60. It was concluded that the observed behavior of the dielectric dispersion in CdSe films was in good general agreement with the influence of inhomogeneities in the structure on the dielectric relaxation. Oduor [49] measured the dependence of the ac conductivity, capacitance, and loss tangent on frequency and temperature in CdSe films having Al electrodes. These results had some similarities with those of Dharmadhikari [27] and of Ismail and Gould [183] on CdTe films. The conductivity appeared to follow a hopping process at low temperatures and was thermally activated at higher temperatures with maximum activation energy of 0.36 eV, although the particular values of the power-law exponents s in the conductivity–frequency measurements differed from those expected for the Elliott [180] model. Capacitance variations were in accordance with the model of Goswami and Goswami [182], although the loss tangent did not exhibit a minimum in the frequency range covered. It was concluded that a minimum in tan δ might be apparent at a higher frequency than the maximum of 20 kHz investigated in this work.

In this section, the ac conductivity and the dielectric properties of cadmium chalcogenide films have been discussed. Use of Cole–Cole type plots have been used in the analysis of measurements on CdTe and CdSe films. For both materials, the experimental curves exhibit two semicircles, suggesting that relaxation could not be accounted for by a single process. Ac conductivity measurements in Al-CdTe-Al structures yielded reasonably consistent results between different groups. A $\sigma \propto \omega^s$ dependence was observed, which was consistent with hopping behavior at low temperatures and high frequencies. At higher temperatures, comparable activation energies of 0.58 and 0.66 eV were determined. In CdSe films, similar conductivity behavior was evident, with a maximum activation energy value of 0.36 eV for the free band conductivity. Capacitance invariably decreased with increasing frequency and in the case of CdS crystals, in which the capacitance greatly exceeded the geometric value, it was associated with dc SCLC and field-emission from traps. Results on CdTe films were generally consistent with a simple equivalent circuit model [182], although Gonzalez-Diaz et al. [186] observed more complex results in sputtered films, and proposed a double-layer model which included both constant and thermally activated resistance and capacitance to explain their results. In CdSe films, the capacitance variations were also consistent with the simpler model. Dielectric loss measurements in CdTe films displayed a minimum in tan δ, with a shift in the minimum value consistent with the simpler model [182]; in CdSe no minimum in tan δ was observed, probably as a result of the restricted frequency range explored. It is clear that both Cole–Cole analysis and measurements of conductivity and dielectric properties are useful tools in probing the ac behavior of cadmium chalcogenide films. A detailed study using both techniques, over a wide range of frequencies, temperatures, and materials is clearly required.

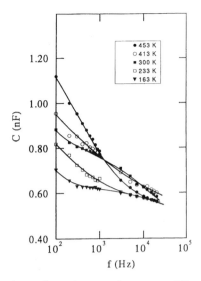

Fig. 65. Dependence of capacitance on frequency at different temperatures for an Ag-Cd$_3$As$_2$-Al sandwich structure of thickness 0.64 μm. M. Din and R. D. Gould, unpublished [192].

3.4.3. Cadmium Arsenide

Following the dc measurements on Cd$_3$As$_2$ films described in Section 3.3.6 in which high field conduction processes were observed, notwithstanding the very low resistivity and high mobility, investigations were made of the ac properties to ascertain whether these too show similarities with the results for the high resistivity cadmium chalcogenides [64]. Owing to the very low resistance (high conductance), it was not possible to make accurate measurements of the conductivity or loss tangent, but measurements of capacitance as functions of frequency, temperature, and thickness were performed. The effects of annealing and the use of different electrode materials were also investigated. Most of the measurements were performed on Ag-Cd$_3$As$_2$-Al samples, having a Cd$_3$As$_2$ layer of thickness in the range 0.12–1.31 μm. The basic capacitance variation as a function of frequency is shown in Figure 65 for a sample of thickness 0.64 μm. The temperatures corresponding to the different characteristics ranged from 165 to 453 K. The capacitance decreased with increasing frequency, and was particularly frequency dependent at higher temperatures and lower frequencies. At low temperatures, the capacitance decreased only slightly with increasing frequency, and at higher frequencies all the curves approached a constant capacitance value. This was similar to results obtained on CdTe [183] and CdSe [49]. At temperatures below about 220 K, the capacitance was insensitive to frequency, but increased significantly at higher temperatures for all frequencies. These results are qualitatively similar to those for CdTe and CdSe and follow the predictions of Eq. (41) for the simple equivalent circuit model. A geometric dependence of the capacitance on thickness was also observed, with films of thickness 1.31, 1.05, and 0.65 μm having progressively higher capacitance values. The major contribution to the capacitance appeared to derive from the bulk of the film, rather than from Schottky barriers at the interfaces. After annealing at 473 K for 2 h, the capacitance reduced much more quickly at higher frequencies in comparison with that for a fresh sample. This was attributed to a change in values of the quantities C and R in Eq. (41) after annealing. A significant variation in the capacitance was also observed over the entire range of frequency when the counterelectrode material varied. The capacitance values for Ag-Cd$_3$As$_2$-Au and Ag-Cd$_3$As$_2$-Ag samples were low in comparison with that for an Ag-Cd$_3$As$_2$-Al sample. It was suggested that in samples having an Al electrode, the overall capacitance may be affected by the contact impedance, where the existence of an oxide or other narrow region was considered possible.

The capacitance measurements described earlier are unusual, and require some confirmation. From both the dc results described in Section 3.3.6 and the ac results described before, it is apparent that Cd$_3$As$_2$ films have several features in common with those of the cadmium chalcogenides, notwithstanding differences in the electronic band structure, mobility, and dc conductivity values.

4. SUMMARY AND CONCLUSIONS

In this chapter, the basic structural features and some of the major electrical properties of cadmium chalcogenide and Cd$_3$As$_2$ thin films have been described. The structural properties frequently have a correlation with some of the electrical properties, such as the conductivity and the effects of potential barriers at the interfaces. The cadmium chalcogenides generally have two alternative structures: the zinc-blende and wurtzite structures. The specific type of structure which is displayed depends on several factors, particularly the substrate temperature during deposition and the evaporant vapor stoichiometry. The effects of many of these deposition variables are reasonably well understood, and films of a given structure can be deposited by controlling the deposition and substrate conditions. There is a good correlation between the evaporant temperature and the epitaxial temperature for each of the cadmium chalcogenides, which allows the best conditions for epitaxial growth to be provided. Cd$_3$As$_2$ films have a tetragonal structure when deposited at moderate substrate temperatures, although there are additional high-temperature phases having either fluorite or deformed fluorite structures. The effects of varying the deposition rate, substrate temperature, and film thickness on the microcrystallite grain size have been discussed for each of these materials, and how this in turn affects the electrical conductivity.

The fundamental band structures of each of the cadmium chalcogenides is well known, although that for Cd$_3$As$_2$ remains unclear. In their most prevalent thin film forms, CdS and CdSe crystallize in the wurtzite structure, and their band structures are very similar, with direct principal energy bandgaps of 2.6 and 2 eV, respectively. CdTe films are normally of the zinc-blende structure, with a direct energy bandgap of 1.59 eV. The band structure of Cd$_3$As$_2$ is complex, owing to its equally complex crystal structure. Various attempts have been made at suggesting or calculating a band structure, with varying degrees of success. These are in some ways contradictory, but what is clear is that this material has an energy bandgap of less than about 0.2 eV, although it is also probable that the principal bands overlap. Furthermore, the experimentally determined mobility is particularly large, and this factor has been addressed in some of the proposed band structures.

Lateral resistivity measurements have been made on each of the four materials considered, using either Hall effect or van der Pauw measurements. Most of the results concerning the variation of mobility or resistivity as a function of film thickness may be explained by assuming interfacial potential barriers between the individual crystalline grains comprising the films. The height of the potential barrier and its variation with temperature have a major influence on the value of the resistivity. Scattering at the film surfaces is also important when the film thickness is of the order of the mean free path between surface collisions. Variations from stoichiometry in the composition can have a significant effect on the resistivity in the cadmium chalcogenides. For example, in CdTe films n-type, intrinsic and p-type films may be obtained merely by varying the deposition rate; the maximum resistivity occurs for the deposition rate corresponding to the intrinsic material. In Cd_3As_2 films, contradictory results have been obtained regarding the temperature dependence of resistivity, with resistivity being reported as both decreasing and increasing (the latter in the case of films deposited by pulsed laser evaporation) as the temperature is increased. Quantum size effects have also been reported in films of this material, where the resistivity and the mobility were oscillatory with a period of approximately 10 nm.

High field dc conduction processes have been observed in all of the cadmium chalcogenides. Space-charge-limited conductivity (SCLC) was observed in all three materials, with the type of trapping regime identified varying between different workers. In CdTe in addition to the observation of discrete trap levels and exponential distributions of traps controlling the conductivity, there are also reports of a uniform distribution of trap levels. Schottky emission was reported in CdS films. Field-lowering behavior, most probably due to Poole–Frenkel emission, has been observed in both CdSe and CdTe films. For the latter material, a model was proposed in which the electric field is nonuniform, and which accounts for enhanced values of the Poole–Frenkel field-lowering coefficient. CdTe films have also been successfully doped with $PbCl_2$, which results in the dominant film conductivity changing from SCLC to Poole–Frenkel emission. Hopping conduction was identified in CdTe films and suggested in CdSe films. The absence of reports of the Schottky effect in CdTe and CdSe films is peculiar, in view of the fact that Schottky barriers to these materials have been reasonably well documented. Cd_3As_2 films have shown similarities in their high field dc conduction properties with those of the other cadmium compounds. In particular, field-lowering behavior was observed, and it was tentatively suggested that in view of the variation in the field-lowering coefficient with increasing film thickness there was a gradual transition from the Schottky to the Poole–Frenkel effect. It is still unclear why in these materials SCLC is observed in some circumstances and field-lowering behavior is observed in others. Some progress has been made in this area, where doping of CdTe films was found to change the dominant conduction process from SCLC to the Poole–Frenkel effect, although the exact criterion governing this remains largely unexplained.

Although only a limited quantity of work has been performed on the ac properties of cadmium compound thin films, the work that has been performed has tended to focus on two types of analysis. A Cole–Cole analysis has been performed in some studies, and has resulted in complex plane plots exhibiting two semicircles for both CdTe and CdSe films, implying that dielectric relaxation is not due to a single process. In other studies, the variation of the ac conductivity on frequency and temperature, as well as variations in capacitance and the dielectric loss factor have been investigated. Generally, a $\sigma \propto \omega^s$ dependence has been observed, particularly at low temperatures and high frequencies, which together with ac activation energy measurements, has indicated a hopping process. Free band conductivity was indicated at higher temperatures in both CdTe and CdSe films. Dielectric properties, such as capacitance and loss tangent, were reasonably consistent with a simple equivalent circuit model, although for sputtered CdTe films a more complex double-layer model was found to be necessary to account for the results. Capacitance variations were also observed in Cd_3As_2 films for the first time and were also consistent with the simple equivalent circuit model.

To conclude, it should be emphasized that much is known about cadmium chalcogenide films, including the effects of varying deposition parameters and electrode species on structural properties, lateral resistivity, and dc conductivity. Ac conductivity measurements are scarcer, and confirmation of existing results and new investigations need to be performed. Cd_3As_2 films have shown some interesting and unexplained properties, although these need to be corroborated. A major lesson from this work is the crucial role of control of the deposition properties and the processing of the materials. Nevertheless, inexpensive solar cell technology using the cadmium chalcogenides and the production of useful devices using Cd_3As_2 are clearly possible using existing relatively inexpensive technology. Significant improvements in the performance and the production of all the films considered in this chapter are likely to await further developments in the materials processing aspects of these materials.

Acknowledgments

The author gratefully acknowledges the experimental work and theoretical insights of all his collaborators in this field, in particular Dr. A. Ashour, C. J. Bowler, Dr. M. Din, Dr. S. Gravano, Dr. B. B. Ismail, Dr. A. O. Oduor, and Dr. A. A. Ramadan. In the course of this work, some of their theses have proved invaluable, and have greatly assisted in the search for reference materials.

REFERENCES

1. G. A. Steigmann and J. Goodyear, *Acta Crystallogr. B* 24, 1062 (1968).
2. L. Däweritz, *J. Cryst. Growth* 23, 307 (1974).
3. J. J. Loferski, *J. Appl. Phys.* 27, 777 (1956).
4. K. L. Chopra and S. R. Das, "Thin Film Solar Cells," Plenum, New York, 1983.
5. H. J. Goldsmid, N. Savvides, and C. Uher, *J. Phys. D: Appl. Phys.* 5, 1352 (1972).
6. H. Matsunami and T. Tanaka, *Jpn. J. Appl. Phys.* 10, 600 (1971).
7. L. Żdanowicz, *Solid State Electron.* 11, 429 (1968).
8. D. R. Lovett, "Semimetals and Narrow-Bandgap Semiconductors," Pion, London, 1977.
9. D. B. Holt, *Thin Solid Films* 24, 1 (1974).
10. S. M. Sze, "Physics of Semiconductor Devices," 2nd ed., Wiley, New York, 1981.
11. I. P. Kalinkin, K. K. Muravyeva, L. A. Sergeyewa, V. B. Aleskowsky, and N. S. Bogomolov, *Krist. Tech.* 5, 51 (1970).
12. K. K. Muravjeva, I. P. Kalinkin, V. B. Aleskovsky, and N. S. Bogomolov, *Thin Solid Films* 5, 7 (1970).
13. K. Zanio, "Cadmium Telluride." in "Semiconductors and Semimetals" (R. K. Willardson and A. C. Beer, Eds.), Vol. 13. Academic Press, New York, 1978.
14. E. W. Williams, K. Jones, A. J. Griffiths, D. J. Roughley, J. M. Bell, J. H. Steven, M. J. Huson, M. Rhodes, and T. Costich, *Solar Cells* 1, 357 (1979–1980).
15. J. F. Hall and W. F. C. Ferguson, *J. Opt. Soc. Am.* 45, 714 (1955).
16. C. A. Escoffery, *J. Appl. Phys.* 35, 2273 (1964).
17. K. L. Chopra and I. H. Khan, *Surf. Sci.* 6, 33 (1967).
18. S. Simov, *Thin Solid Films* 15, 79 (1973).
19. J. I. B. Wilson and J. Woods, *J. Phys. Chem. Solids* 34, 171 (1973).
20. A. Ashour, R. D. Gould, and A. A. Ramadan, *Phys. Status Solidi A* 125, 541 (1991).
21. S. B. Hussain, *Thin Solid Films* 22, S5 (1974).
22. P. Scherrer, *Gott. Nachr.* 2, 98 (1918).
23. K. H. Norian and J. W. Edington, *Thin Solid Films* 75, 53 (1981).
24. A. G. Stanley, in "Applied Solid State Science" (R. Wolfe, Ed.), Vol. 5, p. 251. Academic Press, New York, 1975.
25. N. Nakayama, H. Matsumoto, A. Nakano, S. Ickegami, H. Uda, and T. Yamashita, *Jpn. J. Appl. Phys.* 19, 703 (1980).
26. J. Saraie, M. Akiyama, and T. Tanaka, *Jpn. J. Appl. Phys.* 11, 1758 (1972).
27. V. S. Dharmadhikari, *Int. J. Electron.* 54, 787 (1983).
28. G. K. M. Thutupalli and S. G. Tomlin, *J. Phys. D: Appl. Phys.* 9, 1639 (1976).
29. D. Poelman and J. Vennik, *J. Phys. D: Appl. Phys.* 21, 1004 (1988).
30. Y. Kawai, Y. Ema, and T. Hayashi, *Jpn. J. Appl. Phys.* 22, 803 (1983).
31. B. Goldstein and L. Pensak, *J. Appl. Phys.* 30, 155 (1959).
32. S. Saha, U. Pal, B. K. Samantaray, A. K. Chaudhuri, and H. D. Banerjee, *Thin Solid Films* 164, 85 (1988).
33. I. Spînulescu-Carnaru, *Phys. Status Solidi* 15, 761 (1966).
34. Yu. K. Yezhovsky and I. P. Kalinkin, *Thin Solid Films* 18, 127 (1973).
35. K. V. Shalimova and É. N. Voronov, *Soviet Phys. Solid State* 9, 1169 (1967).
36. B. B. Ismail, Ph.D. Thesis, Keele University, U.K., 1990.
37. D. S. H. Chan and A. E. Hill, *Thin Solid Films* 35, 337 (1976).
38. J. Hamerský, *Thin Solid Films* 38, 101 (1976).
39. A. O. Oduor and R. D. Gould, *Thin Solid Films* 270, 387 (1995).
40. R. Tanaka, *Jpn. J. Appl. Phys.* 9, 1070 (1970).
41. R. W. Glew, *Thin Solid Films* 46, 59 (1977).
42. H. W. Lehmann and R. Widmer, *Thin Solid Films* 33, 301 (1976).
43. G. Hodes, J. Manassen, and D. Cahen, *Nature* 261, 403 (1976).
44. C.-H. J. Liu and J. H. Wang, *Appl. Phys. Lett.* 36, 852 (1980).
45. N. G. Dhere, N. R. Parikh, and A. Ferreira, *Thin Solid Films* 36, 133 (1976).
46. H. M. Naguib, H. Nentwich, and W. D. Westwood, *J. Vac. Sci. Technol.* 16, 217 (1979).
47. F. V. Shallcross, *RCA Rev.* 24, 676 (1963).
48. R. Rentzsch and H. Berger, *Thin Solid Films* 37, 235 (1976).
49. A. O. Oduor, Ph.D. Thesis, Keele University, U.K., 1997.
50. L. Däweritz and M. Dornics, *Phys. Status Solidi A* 20, K37 (1973).
51. N. G. Dhere, N. R. Parikh, and A. Ferreira, *Thin Solid Films* 44, 83 (1977).
52. T. M. Ratcheva-Stambolieva, Yu. D. Tchistyakov, G. A. Krasulin, A. V. Vanyukov, and D. H. Djoglev, *Phys. Status Solidi A* 16, 315 (1973).
53. M. von Stackelberg and R. Paulus, *Z. Phys. Chem.* 28B, 427 (1935).
54. A. Jayaraman, T. R. Anantharaman, and W. Klement, *J. Phys. Chem. Solids* 27, 1605 (1966).
55. A. Pietraszko and K. Łukaszewicz, *Phys. Status Solidi A* 18, 723 (1973).
56. L. Żdanowicz, *Phys. Status Solidi* 6, K153 (1964).
57. H. J. Goldsmid and M. E. Ertl, *Phys. Status Solidi A* 19, K19 (1973).
58. J. J. Dubowski and D. F. Williams, *Appl. Phys. Lett.* 44, 339 (1984).
59. J. J. Dubowski and D. F. Williams, *Thin Solid Films* 117, 289 (1984).
60. J. J. Dubowski, P. Norman, P. B. Sewell, D. F. Williams, F. Kròlicki, and M. Lewicki, *Thin Solid Films* 147, L51 (1987).
61. V. J. Lyons and V. J. Silvestri, *J. Phys. Chem.* 64, 266 (1960).
62. J. B. Westmore, K. H. Mann, and A. W. Tickner, *J. Phys. Chem.* 68, 606 (1964).
63. L. Żdanowicz and S. Miotkowska, *Thin Solid Films* 29, 177 (1975).
64. M. Din, Ph.D. Thesis, Keele University, U.K., 1998.
65. J. Jurusik and L. Żdanowicz, *Thin Solid Films* 67, 285 (1980).
66. J. Jurusik, *Thin Solid Films* 214, 117 (1992).
67. J. Jurusik, *Thin Solid Films* 248, 178 (1994).
68. T. K. Bergstresser and M. L. Cohen, *Phys. Rev.* 164, 1069 (1967).
69. M. L. Cohen and T. K. Bergstresser, *Phys. Rev.* 141, 789 (1966).
70. R. H. Parmenter, *Phys. Rev.* 100, 573 (1955).
71. G. Dresselhaus, *Phys. Rev.* 100, 580 (1955).
72. L. P. Bouckaert, R. Smoluchowski, and E. Wigner, *Phys. Rev.* 50, 58 (1936).
73. D. J. Chadi, J. P. Walter, M. L. Cohen, Y. Petroff, and M. Balkanski, *Phys. Rev. B* 5, 3058 (1972).
74. R. Ludeke and W. Paul, *Phys. Status Solidi* 23, 413 (1967).
75. P. J. Lin-Chung, *Phys. Rev.* 188, 1272 (1969).
76. N. Sexer, *Phys. Status Solidi* 14, K43 (1966).
77. N. Sexer, *Phys. Status Solidi* 21, 225 (1967).
78. D. Armitage and H. J. Goldsmid, *Phys. Lett.* 28A, 149 (1968).
79. E. O. Kane, *J. Phys. Chem. Solids* 1, 249 (1957).
80. F. A. P. Blom and J. Th. Schrama, *Phys. Lett.* 30A, 245 (1969).
81. R. J. Wagner, E. D. Palik, and E. M. Swiggard, *Phys. Lett.* 30A, 175 (1969).
82. D. R. Lovett, *Phys. Lett.* 30A, 90 (1969).
83. L. M. Rogers, R. M. Jenkins, and A. J. Crocker, *J. Phys. D: Appl. Phys.* 4, 793 (1971).
84. D. W. G. Ballentyne and D. R. Lovett, *Br. J. Appl. Phys. Ser. 2* 1, 585 (1968).
85. G. I. Goncharenko and V. Ya. Shevchenko, *Phys. Status Solidi* 41, K117 (1970).
86. M. Aubin, R. Brizard, and J. P. Messa, *Can. J. Phys.* 48, 2215 (1970).
87. M. Iwami, H. Matsunami, and T. Tanaka, *J. Phys. Soc. Jpn.* 31, 768 (1971).
88. I. Rosenman, *Phys. Lett.* 21, 148 (1966).
89. M. Iwami, M. Yoshida, and K. Kawabe, *Jpn. J. Appl. Phys.* 12, 1276 (1973).
90. J. Bodnar, "Proceedings of the III International Conference on Physics of Narrow-band Semiconductors," PWN, Warsaw, 1977.
91. B. Dowgiałło-Plenkiewicz and P. Plenkiewicz, *Phys. Status Solidi B* 87, 309 (1978).
92. M. J. Aubin, L. G. Caron, and J.-P. Jay-Gerin, *Phys. Rev. B* 15, 3872 (1977).
93. F. A. P. Blom, J. W. Cremers, J. J. Neve, and M. J. Gelten, *Solid State Commun.* 33, 69 (1980).
94. P. R. Wallace, *Phys. Status Solidi B* 92, 49 (1979).
95. P. Plenkiewicz and B. Dowgiałło-Plenkiewicz, *Phys. Status Solidi B* 92, 379 (1979).
96. L. J. van der Pauw, *Philips Res. Rep.* 13, 1 (1958).

97. A. A. Ramadan, R. D. Gould, and A. Ashour, *Thin Solid Films* 239, 272 (1994).

98. L. L. Kazmerski, W. B. Berry, and C. W. Allen, *J. Appl. Phys.* 43, 3515 (1972).

99. R. W. Buckley and J. Woods, *J. Phys. D: Appl. Phys.* 6, 1084 (1973).

100. A. A. Ramadan, R. D. Gould, and A. Ashour, *Int. J. Electron.* 73, 717 (1992).

101. R. L. Petritz, *Phys. Rev.* 104, 1508 (1956).

102. H. Flietner, *Phys. Status Solidi* 1, 483 (1961).

103. F. V. Shallcross, *Trans. Met. Soc. AIME* 236, 309 (1966).

104. S. Ray, R. Banerjee, and A. K. Barua, *Jpn. J. Appl. Phys.* 19, 1889 (1980).

105. F. I. Vergunas, T. A. Mingazin, E. M. Smirnova, and S. Abdiev, *Sov. Phys. Crystallogr.* 11, 420 (1966).

106. M. Bujatti and R. S. Muller, *J. Electrochem. Soc.* 112, 702 (1965).

107. W. Kahle and H. Berger, *Phys. Status Solidi* 14, K201 (1966).

108. K. Yamaguchi, N. Nakayama, H. Matsumoto, and S. Ikegami, *Jpn. J. Appl. Phys.* 16, 1203 (1977).

109. Kh. T. Akramov, A. Teshabaev, B. D. Yuldashev, and M. M. Khusanov, *Appl. Solar Energy* 3, 61 (1972).

110. B. Ismail, S. Sakrani, and R. D. Gould, *Solid State Sci. Technol.* 2, 287 (1994).

111. Y. Kawai, Y. Ema, and T. Hayashi, *Thin Solid Films* 147, 75 (1987).

112. W. Huber and A. Lopez-Otero, *Thin Solid Films* 58, 21 (1979).

113. T. I. Kamins, *J. Appl. Phys.* 42, 4357 (1971).

114. K. Shimizu, *Jpn. J. Appl. Phys.* 4, 627 (1965).

115. H. Berger, G. Jäniche, and N. Grachovskaya, *Phys. Status Solidi* 33, 417 (1969).

116. V. Šnejdar, D. Berková, and J. Jerhot, *Thin Solid Films* 9, 97 (1971).

117. K. C. Sathyalatha, S. Uthanna, and P. Jayarama Reddy, *Thin Solid Films* 174, 233 (1989).

118. F. Raoult, B. Fortin, and Y. Colin, *Thin Solid Films* 182, 1 (1989).

119. A. O. Oduor and R. D. Gould, *Proc. SPIE* 2780, 80 (1996).

120. M. K. Rao and S. R. Jawalekar, *Phys. Status Solidi A* 38, K93 (1976).

121. G. A. Somorjai, *J. Phys. Chem. Solids* 24, 175 (1963).

122. D. S. H. Chan and A. E. Hill, *Thin Solid Films* 38, 163 (1976).

123. J. Hamerský, *Thin Solid Films* 44, 277 (1977).

124. W. J. Turner, A. S. Fischler, and W. E. Reese, *Phys. Rev.* 121, 759 (1961).

125. T. S. Moss, *Proc. Phys. Soc. B* 63, 167 (1950).

126. L. Żdanowicz, W. Żdanowicz, and G. Pocztowski, *Thin Solid Films* 28, 345 (1975).

127. H. Matsunami, M. Iwami, K. Asano, and T. Tanaka, *Jpn. J. Appl. Phys.* 7, 444 (1968).

128. L. Żdanowicz, *Acta Phys. Pol.* 31, 1021 (1967).

129. L. Żdanowicz, W. Żdanowicz, Cz. Weçlewicz, and J. C. Portal, *J. Phys. (France)* 42, 1069.

130. L. Żdanowicz, G. Pocztowski, Cz. Weçlewicz, N. Niedźweidź, and T. Kwiecień, *Thin Solid Films* 34, 161 (1976).

131. J. G. Simmons, *J. Phys. D: Appl. Phys.* 4, 613 (1971).

132. E. H. Rhoderick, "Metal-Semiconductor Contacts," Clarendon, Oxford, U.K., 1978.

133. D. R. Lamb, "Electrical Conduction Mechanisms in Thin Insulating Films," Methuen, London, 1967.

134. A. Rose, *Phys. Rev.* 97, 1538 (1955).

135. R. D. Gould and M. S. Rahman, *J. Phys. D: Appl. Phys.* 14, 79 (1981).

136. M. A. Lampert, *Rep. Prog. Phys.* 27, 329 (1964).

137. R. D. Gould and B. A. Carter, *J. Phys. D: Appl. Phys.* 16, L201 (1983).

138. R. D. Gould, *J. Appl. Phys.* 53, 3353 (1982).

139. M. A. Lampert and P. Mark, "Current Injection in Solids," Academic Press, New York, 1970.

140. R. D. Gould and C. J. Bowler, *Thin Solid Films* 164, 281 (1988).

141. N. F. Mott, "Metal-Insulator Transitions," Taylor & Francis, London, 1974.

142. N. F. Mott and E. A. Davis, "Electronic Processes in Non-Crystalline Materials," 2nd ed., Oxford Univ. Press, Oxford, U.K., 1979.

143. J. G. Simmons, *J. Appl. Phys.* 34, 1793 (1963).

144. J. G. Simmons, *Phys. Rev.* 166, 912 (1968).

145. R. W. Smith and A. Rose, *Phys. Rev.* 97, 1531 (1955).

146. R. H. Bube, *J. Appl. Phys.* 33, 1733 (1962).

147. R. Zuleeg, *Solid State Electron.* 6, 645 (1963).

148. R. Zuleeg and R. S. Muller, *Solid State Electron.* 7, 575 (1964).

149. F. A. Pizzarello, *J. Appl. Phys.* 35, 2730 (1964).

150. A. Ashour Mohammed, Ph.D. Thesis, Minia University, Egypt, 1989.

151. H. Murray and A. Tosser, *Thin Solid Films* 36, 247 (1976).

152. A. Piel and H. Murray, *Thin Solid Films* 44, 65 (1977).

153. C. Canali, M.-A. Nicolet, and J. W. Mayer, *Solid State Electron.* 18, 871 (1975).

154. C. Lhermitte, D. Carles, and C. Vautier, *Thin Solid Films* 28, 269 (1975).

155. C. Lhermitte and C. Vautier, *Thin Solid Films* 58, 83 (1979).

156. B. M. Basol and O. M. Stafsudd, *Solid State Electron.* 24, 121 (1981).

157. S. S. Ou, O. M. Stafsudd, and B. M. Basol, *Thin Solid Films* 112, 301 (1984).

158. B. B. Ismail and R. D. Gould, *Phys. Status Solidi A* 115, 237 (1989).

159. R. D. Gould and B. B. Ismail, *Int. J. Electron.* 69, 19 (1990).

160. B. B. Ismail and R. D. Gould, *Int. J. Electron.* 78, 261 (1995).

161. C. Canali, F. Nava, G. Ottaviani, and K. Zanio, *Solid State Commun.* 13, 1255 (1973).

162. S. Gogoi and K. Barua, *Thin Solid Films* 92, 227 (1982).

163. A. K. Jonscher and A. A. Ansari, *Philos. Mag.* 23, 205 (1971).

164. R. D. Gould and B. B. Ismail, *Vacuum* 50, 99 (1998).

165. R. D. Gould and B. B. Ismail, *Phys. Status Solidi A* 134, K65 (1992).

166. Z. T. Al-Dhhan and C. A. Hogarth, *Int. J. Electron.* 63, 707 (1987).

167. J. P. Ponpon, *Solid State Electron.* 28, 689 (1985).

168. R. D. Gould and B. B. Ismail, *J. Mater. Sci. Lett.* 11, 313 (1992).

169. B. M. Basol and O. M. Staffsudd, *Thin Solid Films* 78, 217 (1981).

170. H. Okimura and Y. Sakai, *Jpn. J. Appl. Phys.* 7, 731 (1968).

171. K. N. Sharma and K. Barua, *J. Phys. D: Appl. Phys.* 12, 1729 (1979).

172. R. K. Pandey, R. B. Gore, and A. J. N. Rooz, *J. Phys. D: Appl. Phys.* 20, 1059 (1987).

173. A. O. Oduor and R. D. Gould, *Thin Solid Films* 317, 409 (1998).

174. J. Mycielski, *Phys. Rev.* 123, 99 (1961).

175. M. Din and R. D. Gould, *Thin Solid Films* 340, 28 (1999).

176. J.-P. Jay-Gerin, L. G. Caron, and M. J. Aubin, *Can. J. Phys.* 55, 956 (1977).

177. M. Rapoš, M. Ružinský, S. Luby, and J. Červenák, *Thin Solid Films* 36, 103 (1976).

178. E. B. Kaganovich, Yu. N. Maksimenko, and S. V. Svechnikov, *Thin Solid Films* 78, 103 (1981).

179. K. S. Cole and R. H. Cole, *J. Chem. Phys.* 9, 341 (1941).

180. S. R. Elliott, *Philos. Mag.* 36, 1291 (1977).

181. J. G. Simmons, G. S. Nadkarni, and M. C. Lancaster, *J. Appl. Phys.* 41, 538 (1970).

182. A. Goswami and A. P. Goswami, *Thin Solid Films* 16, 175 (1973).

183. B. B. Ismail and R. D. Gould, *Proc. SPIE* 2780, 46 (1996).

184. S. R. Elliott, *Adv. Phys.* 36, 135 (1987).

185. H. Böttger and V. V. Bryksin, "Hopping Conduction in Solids," VCH, Weinheim, Germany, 1985.

186. G. Gonzalez-Diaz, M. Rodriguez-Vidal, F. Sanchez-Quesada, and J. A. Valles-Abarca, *Thin Solid Films* 58, 67 (1979).

187. R. D. Gould, S. Gravano, and B. B. Ismail, *Thin Solid Films* 198, 93 (1991).

188. B. Binggeli and H. Kiess, *J. Appl. Phys.* 38, 4984 (1967).

189. A. K. Jonscher, *J. Vac. Sci. Technol.* 8, 135 (1971).

190. B. B. Ismail and R. D. Gould, unpublished work, Keele University, U.K.

191. A. O. Oduor and R. D. Gould, unpublished work, Keele University, U.K.

192. M. Din and R. D. Gould, unpublished work, Keele University, U.K.

Chapter 5

CARBON-CONTAINING HETEROEPITAXIAL SILICON AND SILICON/GERMANIUM THIN FILMS ON Si(001)

H. Jörg Osten

Institute for Semiconductor Physics, (IHP), Frankfurt D-15236, Germany

Contents

1. INTRODUCTION

The last decade has witnessed tremendous progress in the field of strained $Si_{1-x}Ge_x$/Si(001) heterostructures, from fundamental studies on growth and band structure engineering to a rapidly maturing technology. The accelerated development of the strained $Si_{1-x}Ge_x$/Si(001) arises from its fortuitous material properties [1]. For example, it is relatively easy to grow $Si_{1-x}Ge_x$ films because Si and Ge are completely miscible over the entire alloy range. Doping in the $Si_{1-x}Ge_x$ film is also no more difficult than that in Si. A completely miscible silicon/germanium alloy with a moderate lattice mismatch to Si

Handbook of Thin Film Materials, edited by H.S. Nalwa
Volume 4: Semiconductor and Superconductor Thin Films

ISBN 0-12-512912-2/$35.00

provides the additional advantages of heterostructure devices; a concept with tremendous success in group III/V optoelectronics and high-speed devices and circuits. SiGe alloys and heterojunctions will extend the performance of future Si-based devices [2]. At the forefront is the fastest Si-based device—the SiGe heterojunction bipolar transistor—which is now so sufficiently developed that some companies intend to put it on the market.

Because of a 4% larger atomic size of Ge than Si, a $Si_{1-x}Ge_x$ film is compressively strained when it is grown pseudomorphically (without misfit dislocations) on a Si substrate. The incorporation of Ge reduces the band gap of $Si_{1-x}Ge_x$, with most of the band-gap reduction exhibited as the valence band offset of the strained $Si_{1-x}Ge_x$/Si system.

The strain due to the incorporated Ge in the pseudomorphic $Si_{1-x}Ge_x$ limits the thickness of the film that can be grown without the generation of misfit dislocations, which are detrimental to device performance. Consequently, this poses a severe design constraint and limits its potential device applications. The $Si_{1-x}Ge_x$ on Si(001) system exhibits some severe limitations, like the existence of a critical thickness for perfect pseudomorphic growth that depends on the amount of germanium (the introduced strain). Above that thickness strain relief occurs by plastic flow, i.e., by injection and propagation of misfit dislocations. In addition, metastable SiGe layers (thicker than the critical thickness and grown at low temperatures) tend to relax during postgrowth thermal treatments, typical for device processing. The thickness has to be kept below that critical thickness—free tuning of crystallographic and therefore electronic structure is not possible. The main band offset between Si and strained SiGe is located in the valence band, that is, this system is much better suited to hole channel than to electron channel devices.

Over the past 5 years, researchers have been trying to add carbon to pseudomorphic $Si_{1-x}Ge_x$ on Si to relieve the strain [3]. Carbon is a good candidate because it has a much smaller atomic size and it is isoelectronic to both Si and Ge. Indeed, thicker films have been grown pseudomorphically with the addition of carbon. In the second section, it is shown that $Si_{1-y}C_y$ and $Si_{1-x-y}Ge_xC_y$ layers can also be grown pseudomorphically on Si(001), with molecular beam epitaxy (MBE) or different chemical vapor deposition (CVD) techniques. Although the bulk solubility of carbon in silicon is small, epitaxial layers that are more than 1% C can be fabricated. One of the most crucial questions is the relation between substitutional and interstitial carbon incorporation, which has a large impact on the electrical and optical properties of these layers. Substitutionally incorporated C atoms allow strain manipulation, including the growth of an inversely strained $Si_{1-x-y}Ge_xC_y$ layer. The mechanical and structural properties and the thermal stability and possible relaxation paths of those highly metastable systems are discussed in the third section.

The next topic is to see how C incorporation changes the electrical properties of the resulting $Si_{1-y}C_y$ or $Si_{1-x-y}Ge_xC_y$ materials. In the fourth section, we address the change in fundamental electronic properties due to the inclusion of C, primarily

in the band alignment of the tensile strained $Si_{1-y}C_y$/Si(001) and the still compressively strained $Si_{1-x-y}Ge_xC_y$/Si(001). After reviewing the experimental data of the effect of C on the band alignment of $Si_{1-x-y}Ge_xC_y$/Si, we propose a simple model to predict how C incorporation changes the band alignment of the epitaxial materials with the underlying Si substrate.

To use carbon-containing layers for device applications, one of the key parameters is the carrier mobility in the new, epitaxial materials. As a result, we study how C incorporation changes the hole and the electron mobility, through the fabrication of p- and n-type doped $Si_{1-x-y}Ge_xC_y$ and $Si_{1-y}C_y$ layers, respectively.

Furthermore, we investigate the possible existence of strain-stabilized, highly concentrated Si_nC layers embedded in silicon with up to 20% carbon (Section 5).

Then we go beyond the crystallographic and band structure engineering with C and demonstrate in Section 6 a potential application for $Si_{1-x-y}Ge_xC_y$ alloys in suppressing undesirable enhanced dopant diffusion, a fundamental problem in current Si and $Si_{1-x}Ge_x$/Si technologies. The incorporation of low concentrations of carbon ($<10^{20}$ cm^{-3}) into the SiGe region of a heterojunction bipolar transistor (HBT) can significantly suppress boron outdiffusion caused by subsequent processing steps. This effect can be described by coupled diffusion of carbon atoms and Si point defects. We discuss the increase in performance and process margins in SiGe heterojunction bipolar technology caused by the addition of carbon. SiGe:C HBTs demonstrate excellent static parameters, exceeding the performance of state-of-the-art SiGe HBTs. C also enhances high-frequency performance, because it allows one to use a high B doping level in a very thin SiGe base layer without outdiffusion from SiGe, even if postepitaxial implants and anneals are applied. Compared with SiGe technologies, the addition of carbon provides significantly greater flexibility to process designs. In conjunction with the improved AC performance, this will pave the way for the use of C in SiGe technology to widen process margins and boost static and dynamic performance.

Finally, we propose other possible device applications for this new semiconducting material.

2. GROWTH OF EPITAXIAL $Si_{1-y}C_y$ AND $Si_{1-x-y}Ge_xC_y$

2.1. Basic Considerations

Ge and Si are miscible in any composition, and solid solutions can be prepared in bulk or by epitaxial growth. However, although Ge and Si share a common crystal structure, the diamond cubic lattice, they do not have the same lattice parameter. The lattice parameter of $Si_{1-x}Ge_x$ alloys varies between 5.431 Å for Si ($x = 0$) and 5.657 Å for Ge ($x = 1$). Consequently, $Si_{1-x}Ge_x$ alloys cannot be grown epitaxially on Si or Ge substrates without introducing large amounts of strain. There is, however, an equilibrium critical thickness for perfect pseudomorphic growth that depends on the strain in the layer. Strain relief can occur by plastic flow, i.e., by injection and propagation of misfit dislocations, or elastically by the formation of islands [4]. It was shown that even for pure Ge a smooth film only a few monolayers thick can be grown pseudomorphically on a Si(100) substrate. When Ge is diluted with Si, the mismatch becomes smaller and thicker pseudomorphic layers can be grown. The main features of pseudomorphic growth of $Si_{1-x}Ge_x$ are fairly well understood. Epitaxial growth of carbon containing $Si_{1-y}C_y$ or $Si_{1-x-y}Ge_xC_y$ alloys is more complex, and several additional problems have to be addressed [5, 6]:

1. According to the binary Si–C phase diagram, stoichiometric SiC (silicon carbide) is the only stable compound. There are a number of silicon carbide phases and polytypes that, in equilibrium, form preferentially. Any alloy with a smaller C concentration is thermodynamically metastable. Such alloy layers can be achieved by kinetically dominated growth methods, like MBE [5, 6] or special kinds of CVD (e.g. rapid thermal CVD, RTCVD [7]) and relatively low growth temperatures. These methods generally work far from thermodynamic equilibrium conditions, allowing a kinetic stabilization of metastable phases. Once deposited as a random alloy at low temperature, $Si_{1-y}C_y$ appears to be able to withstand anneals up to significantly higher temperatures [8, 9].

2. In contrast to the Si/Ge alloy, carbon in silicon can easily form interstitial defect complexes. Thus, one of the most crucial questions is the relation between substitutional and interstitial carbon incorporation, which has a large impact on the electrical properties of these layers.

3. Carbon contamination on a silicon surface is known to interrupt epitaxy [10]. At temperatures needed to sublimate any kind of oxide layer (i.e., native oxide or protective layers grown by certain wet chemical precleaning steps) from the Si surface, adsorbed hydrocarbons transform partly into silicon carbide nanocrystallites. These nanocrystallites can cause stacking fault or twin formation [10]. For the growth of epitaxial $Si_{1-y}C_y$ layers the growth temperature has to be kept low enough to avoid the formation of carbide.

Another process that is known to allow the incorporation of elements into Si at concentrations far in excess of their bulk solubility limit is solid-phase epitaxy (SPE). SPE thus provides another possible synthesis route for the formation of metastable $Si_{1-y}C_y$ or $Si_{1-x-y}Ge_xC_y$ layers. These SPE approaches recrystallize amorphous mixtures of Si/Ge/C, obtained mainly by ion implantation with a well-defined temperature profile [11, 12]. Good quality over a relatively large range of compositions could be obtained. However, it is known that carbon inhibits the kinetics of Si SPE, so that higher temperatures would be required. SPE based on preamorphization by ion implantation presents the problem that implantation profiles are often not sharp enough to guarantee well-localized buried layers.

2.2. The Concept of Surface Solubility

Carbon has a low bulk solubility in Si (3×10^{17} atoms/cm^3 at the melting point [13]). Any alloy with a substitutional C concentration above the bulk solubility limit is thermodynamically

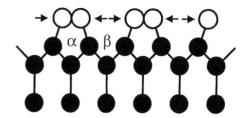

Fig. 1.   Cross section of a (2×1)-reconstructed Si(001) surface, showing two different sites for adatom incorporation.

metastable. In equilibrium, the solid solubility of a substitutional C impurity (i.e., the value of y for $Si_{1-y}C_y$) is given by

$$y = \exp(-(E - \mu)/kT \tag{2.1}$$

Here, E is the energy of a substitutional C atom in Si, μ is the chemical potential, and k is Boltzmann's constant. The concentration of C is limited by the formation of SiC precipitates, which prevents μ from rising higher than the value corresponding to a SiC reservoir [14].

However, during growth it is not the equilibrium bulk solubility that is important, but rather a "surface solubility." The increased solubility results from two factors [14]. First, the presence of a surface breaks the bulk symmetry and therefore creates sites for adatoms, which are energetically nonequivalent to bulk sites. The stress associated with the atomic size mismatch between C and Si is partially relieved, and the energy of the substitutional C decreases smoothly as it approaches the surface. Second, there is a stress field near the surface, associated with the atomic reconstruction there [15]. The coupling of the carbon stress to this spatially varying surface stress substantially lowers the carbon energy on certain sites. For a (2×1) reconstructed silicon (001) surface there are two nonequivalent sites (see Fig. 1). One site lies beneath the surface dimer (α site), and the other site lies between the dimers (β site). As discussed by Kelires and Tersoff [15], the α sites are under compressive stress. Thus they are favorable sites for a smaller atom such as C. The adatoms are rapidly buried on subsurface sites, which serve as a sink, and thus are immobilized. The high carbon density near the surface is frozen in as the crystal grows, permitting the growth of a highly supersaturated solid solution, e.g., high carbon concentrations in silicon or silicon/germanium. Recently, Tersoff used an empirical many-body potential to model the interaction between atoms in the system [14]. Including the surface reconstruction, he found an enhancement of 10^4 over equilibrium bulk solubility (Fig. 2). Thus, the surface solubility of carbon can easily be in the percentage range.

The lowest-energy position of carbon in Si is found to be the substitutional site. In addition, the observed activation energy for the diffusion of substitutional carbon (3.04 eV) is quite large [13]. This is the origin of the achieved stability of these metastable alloys. Carbon precipitation is a rare process that has been reported to occur only in the presence of the supersaturation of silicon self-interstitials or oxygen [16]. The mechanism described here has another possible consequence in addition

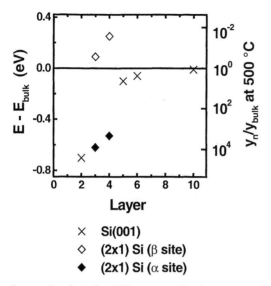

Fig. 2.   Energy of a substitutional C atom near a (2×1)-reconstructed Si(001) surface, relative to bulk energy. Filled and open diamonds correspond to α and β sites, respectively. The circles represent results for a nonreconstructed surface. Δy is the factor by which solubility is enhanced at each site. (Data are from [14].)

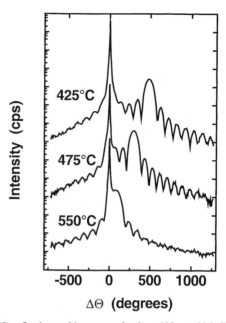

Fig. 3.   (400) reflection rocking curves for three 100-nm-thick $Si_{1-y}C_y$ layers on Si(001), grown at different growth temperatures with identical Si and C fluxes. Reproduced with permission from [19], copyright 1996, American Institute of Physics.

to enhanced solubility. If the C atoms have sufficient mobility within the first layers, then they will lie only on sites beneath the dimers, with negligible occupancy of the other sites. This ordering should not be lost as the layers are buried deeper, if the C is immobile in deeper layers. First indications of local ordering in $Si_{1-y}C_y$ layers have been observed by transmission electron microscopy (TEM) [17] and vibrational spectroscopy [18].

2.3. Substitutional versus Interstitial C Incorporation

The incorporation of impurities into epitaxially growing films is a classic problem in materials physics. An understanding of the factors limiting substitutional incorporation is helpful in the fabrication of nonequilibrium materials, such as strained $Si_{1-y}C_y$ layers grown heteroepitaxially on Si(001) substrates. MBE growth of C containing alloys pseudomorphically strained on Si(001) has been investigated as a function of growth conditions [19–21]. The carbon substitutionality is strongly influenced by the growth conditions, such as temperature and the Si growth rate. Similar trends were also observed with RTCVD [22]. All authors found a significant decrease in the substitutional carbon concentration with increasing growth temperature and/or decreasing growth rate. However, there is still a wide scatter in the obtained parameters (like energy barriers) needed to describe the experimental results. This is mainly due to the fact that the different authors did not take into account the different carbon concentrations used in their investigations. In this chapter, we show that the formation of interstitial carbon-containing defect complexes is a strain-driven process. Specifically, the energy barriers and hopping frequencies also depend on the carbon concentration.

2.3.1. Experimental Evidence

The amount of substitutional carbon can easily be determined from its influence on the lattice structure. X-ray diffraction is a useful tool for determining the tensile lattice distortion and for probing the structural quality by measurements of diffuse scattering. Figure 3 shows three typical rocking curves of $Si_{1-y}C_y$ layers grown at different temperatures with identical Si and C fluxes. The angular shift between the Si substrate peak at $\Theta = 0$ and the $Si_{1-y}C_y$ layer peak is directly related to the substitutional carbon content y_S. The intensity oscillations are related to the layer thickness. Measurements near the $Si_{1-y}C_y$ layer peak showed only a very weak diffuse scattering for all samples, even with high-intensity synchrotron radiation (regardless of the deposition temperature and the Si growth rate). This indicates the high structural perfection of the layers. Figure 4 shows substitutional carbon concentrations obtained from X-ray rocking curves for a series of samples grown at different temperatures with the same Si growth rate (0.3 Å/s) and carbon flux. There is a significant decrease in substitutional carbon concentration with increasing growth temperature.

The lattice constant of a $Si_{1-y}C_y$ layer is smaller than that of silicon; i.e., it will be tensile strained on Si(001) substrates. Thus, the same strain configuration used for high-mobility Si on relaxed $Si_{1-x}Ge_x$ buffers can be obtained directly on Si. $Si_{1-y}C_y$ channels embedded in Si could be an alternative to Si/SiGe, potentially obviating the need for a graded buffer and offering higher crystalline perfection [23]. An increase of in-plane electron mobility should be expected for low C contents in $Si_{1-y}C_y$ (see also Section 4). However, not all available experimental data support these predictions. In contrast to the results of Brunner et al. [24], which indicate a slight increase

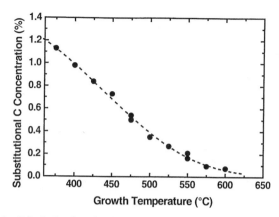

Fig. 4. Substitutional carbon concentration obtained from X-ray rocking curves as a function of growth temperature. All samples were grown with a silicon rate of 0.3 Å/s and identical C flux. The dashed line represents the best fit according to Eq. (2.4). Reproduced with permission from [19], copyright 1996, American Institute of Physics.

in mobility, our recent results on electron transport in $Si_{1-y}C_y$ alloys (with carbon concentrations in the percentage range) indicate that alloy, ionized, and neutral impurity scattering dominate over the expected gain due to strain [25]. Thus, one is led to assume that a significant fraction of the carbon atoms are not substitutionally incorporated, instead forming interstitial defect complexes. For example, there is a variety of interstitial Si–C complexes with energy levels within the $Si_{1-y}C_y$ gap [13], which can drastically influence transport properties or luminescence behavior.

One of the critical challenges for further development and application of strained, C-containing heterostructures is the ability to incorporate carbon atoms predominantly at substitutional positions. On the other hand, it has been shown that the addition of small amounts of C (below 10^{20} cm^{-3}) to SiGe provides a wider process margin and flexibility and substantially enhances the high-frequency performance of SiGe heterojunction bipolar transistors [26]. These transistors show almost ideal base current characteristics, even with carbon in the depletion regions, indicating the absence of interstitial defect complexes due to C incorporation. Thus, lower carbon concentrations appear to be predominantly incorporated substitutionally.

Recently, Kelires and Kaxiras reported atomistic calculations of the surface structure of Si with substitutional carbon atoms [27]. One of their important results is that the lowest energy configuration involves C atoms in the very top layer of the Si(001) surface, forming dimers in the (2 × 1) reconstruction. The authors suggest that the strong preference of C to remain on the surface is due to the bond between the two threefold coordinated C atoms (similar to the Si surface atoms forming dimers). The ability of C to form double bonds makes this a low-energy structure. This has important consequences. It implies that if C atoms have a chance to form C–C dimers upon arrival on the Si(001) surface, their subsequent incorporation into the growing film will be limited. The formation of such dimers depends on several factors, including the frequency of surface hops (i.e., the growth temperature), the time before further deposition im-

mobilizes the atoms (i.e., the growth rate), and the probability of finding another C atom (i.e., the total carbon concentration). In addition, some carbon evaporation sources deliver larger C fragments [28]. Therefore, we have to assume different incorporation kinetics for different C sources.

2.3.2. Model for C Incorporation into Growing Films

For a better understanding of C incorporation kinetics, we have to consider only the behavior of the atoms in the first monolayers. The carbon adatoms become rapidly buried at subsurface sites and are presumably immobilized. In [19] we proposed the following simple model: If the temperature is high enough for Si epitaxy, we can assume that all arriving C adatoms are rapidly incorporated at positions beneath the Si dimers. When another Si adatom arrives above a surface position filled with a C atom, either (i) it migrates further along the surface, or (ii) it binds to the carbon ("freezing in" of the C atom on lattice position), or (iii) it kicks the C atom out of a lattice position, forming a Si–C interstitial complex or the suggested C–C dimers on the surface. The substitutional versus interstitial carbon concentration is therefore primarily controlled by process (iii), i.e., by all parameters affecting the formation of the interstitial defect complexes. These parameters are the growth rate r (limiting the time available for defect formation), the growth temperature T, and an energy barrier E_C. An incorporated C atom has to overcome this barrier to form a defect complex (like a C–C dimer). Desorption processes are likely excluded for the used growth temperatures. Thus we can write a first-order rate equation for the above-discussed process,

$$dn_S(t)/dt = -kn_S(t) \tag{2.2}$$

where n_S is the number of substitutional carbon atoms. The frequency of defect formation k is given by

$$k = k_0 \exp(-E_C/kT) \tag{2.3}$$

If we assume that $n_S(0) = n_0$ is the initial substitutional carbon concentration in the first layer (which, of course, is a function of the Si growth rate r and the C flux density f, i.e., $n_0 = n_0(r, f)$), the above equations can be solved analytically:

$$n_S = n_0 \exp[-t_{pin}k] \tag{2.4}$$

The growth is continuous, which means there is a steady supply of new adatoms. After a certain time t_{pin} all carbon atoms are "pinned down"; i.e., they become immobile. We assume that $t_{pin} = a/(4r)$ (i.e., the time needed to deposit one monolayer, where a is the silicon lattice constant). The average C atom will be frozen in its position (interstitial or substitutional) after the time needed to deposit the next monolayer. Finally, we obtain the following concentration of substitutional carbon as a function of growth temperature T and growth rate r:

$$n_S(T, r) = n_0(r, f) \exp[-ak_0 \exp(-E_C/kT)/4r] \tag{2.5}$$

For a constant C flux f, the relation

$$n_0 \sim 1/r \tag{2.6}$$

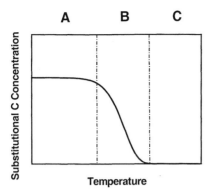

Fig. 5. Schematic representation of substitutional carbon incorporation in $Si_{1-y}C_y$ layers as a function of growth temperature for constant Si and C fluxes. Three distinct regions are marked. Reproduced with permission from [36], copyright 1999, American Institute of Physics.

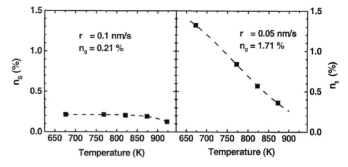

Fig. 6. Experimentally determined amount of substitutional carbon atoms for two sets of samples; each set was grown at a different Si and C flux. The growth rates are indicated. The dashed line represents the best fit according to Eq. (2.4). Reproduced with permission from [36], copyright 1999, American Institute of Physics.

where r is the silicon growth rate, holds for the initial coverage n_0.

Figure 5 illustrates schematically the behavior of Eq. (2.5). Generally, we can distinguish three different regions (labeled A–C in Fig. 5). In the first region, n_S is equal to n_0, i.e., we have complete substitutional incorporation, regardless of the growth temperature. For the highest temperatures in this region, we might find a slight decrease in n_S with increasing T. This behavior is similar to that reported by Zerlauth et al. [21]. In the middle region (B), we observe a relatively strong decrease in n_S with increasing growth temperature, which is similar to the results reported in [19] and [20]. For even higher growth temperatures (region C), almost no substitutional carbon incorporation should be possible. This agrees with earlier results obtained by Powell et al. [5]. These authors showed that there is only a narrow temperature window for good epitaxial growth of $Si_{1-y}C_y$ alloys with atom sources. If the substrate temperature is too high (above 600°C, according to [5]), a metastable regime no longer exists, and some deterioration in crystal quality (like a transition to 3D growth or a mixed alloy carbide phase) occurs. Fitting the data shown in Figure 4, we obtain $E_C = (0.5 \pm 0.1)$ eV for a total carbon concentration of $n_0 = 1.56\%$. Using the same approach, Dashiell et al. [20] obtained $E_C = (1.0 \pm 0.2)$ eV for $n_0 = 0.8\%$. It thus seems reasonable to assume that the energy barrier is lowered by the presence of surface strain, that is, E_C decreases with increasing n_0.

2.3.3. Discussion

To investigate a large range of n_0 values, we used different sets of samples [36]. Each set of experiments examined several 50-nm-thick $Si_{1-y}C_y$ alloy samples grown on Si(001) substrates with the same Si and C fluxes, but at different growth temperatures. The experimentally determined values for n_S as a function of growth temperature were fitted according to Eq. (2.5), with E_C and k_0 as free parameters. Figure 6 illustrates two examples, one set with a low carbon concentration (left figure) and one with a high carbon concentration (right figure).

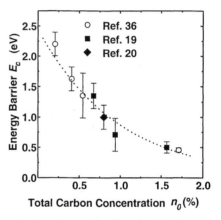

Fig. 7. Energy barriers E_C as a function of total carbon concentration obtained from fitting of our experimental data. In addition, a data point obtained recently by Dashiell et al. [20] is included. The dashed line represents a fit to a single exponential decay. Reproduced with permission from [36], copyright 1999, American Institute of Physics.

For the low total carbon concentration ($n_0 = 0.21\%$) combined with a high growth rate (0.1 nm/s), we find the A-type behavior as illustrated in Figure 5.

In contrast, the higher total C concentrations ($n_0 = 1.71\%$) and lower growth rates (0.05 nm/s) lead to a typical B-type behavior. The corresponding energy barriers obtained from the fitting procedure are (2.2 ± 0.2) eV and (0.45 ± 0.03) eV, respectively. The values for the total carbon concentration n_0 obtained from the fitting procedures agree well with those measured independently by secondary ion mass spectrometry (SIMS).

Figure 7 summarizes the different energy barriers as a function of total carbon concentration obtained from all experimental sets and data published by Dashiell et al. [20]. The energy barrier decreases with increasing carbon concentration. This behavior can be modeled by a first-order exponential decay (see Fig. 7). This leads to a value of E_C that approaches (2.5 ± 0.3) eV for $n_0 = 0$. This barrier is comparable with the barrier of 3.04 eV, obtained for the diffusion of substitutional carbon in bulk silicon with very low carbon concentrations [13]. For the low carbon concentrations used in typical HBT applications, the energy barrier is still larger then 2.2 eV. It is high

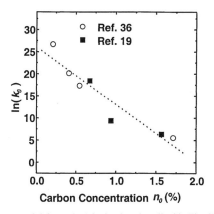

Fig. 8. Preexponential factor k_0 (obtained as described in Fig. 3) as a function of n_0. The dashed line represents a linear fit of $\ln(k_0)$ versus total carbon concentration. Reproduced with permission from [36], copyright 1999, American Institute of Physics.

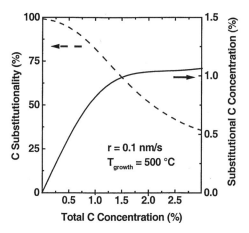

Fig. 9. Predictions of the ratio of substitutional to total carbon concentration (substitutionality, dashed line) and the resulting substitutional C concentration (solid line) as a function of the total carbon concentration. The assumed growth conditions are $r = 0.1$ nm/s, $T_{growth} = 500°C$. The calculation was based on Eq. (2.4), with dependencies obtained by fitting our experimental results (see Figs. 5 and 6). Reproduced with permission from [36], copyright 1999, American Institute of Physics.

enough to suppress any defect formation within the temperature window for epitaxial growth; that is, carbon will be predominantly incorporated substitutionally.

The preexponential frequency factor k_0 is shown in Figure 8 as a function of C concentration. k_0 decreases over several orders of magnitude with increasing n_0. This factor mainly describes the ability of adatoms to migrate on the surface. Thus, the presence of carbon on the surface appears to reduce surface diffusion significantly. Extrapolating to $n_0 = 0$, we obtain $k_0(0) = (2 \times 10^{10} \text{ to } 3 \times 10^{12})$ s^{-1}. This is a range similar to those typically assumed for surface migration of adatoms on an unperturbed Si surface.

There is another interesting consequence arising from the decrease in E_C and k_0 with increasing C concentration. Figure 9 shows the C substitutional fraction (n_S/n_0) as a function of total carbon concentration for $r = 0.1$ nm/s and $T_{growth} = 500°C$. The fraction of substitutionally incorporated carbon atoms de-

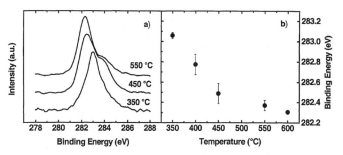

Fig. 10. C 1s photoelectron binding energy spectra for Si$_{1-y}$C$_y$ layers grown at different temperatures (a) and their peak binding energies versus growth temperature (b). Reproduced with permission from [29], copyright 1996, American Institute of Physics.

creases with increasing n_0. In Figure 9, we also depict the resulting amount of substitutionally incorporated carbon. It seems there is a saturation; that is, any further increase in the carbon flux leads mainly to an increase in the number of interstitial defects, with the amount of substitutionally incorporated carbon remaining constant. This behavior supports the often postulated existence of a limit for substitutional carbon incorporation. This limit depends, of course, on the specific growth conditions and the specific carbon source.

In summary, the carbon substitutionality in MBE-grown Si$_{1-y}$C$_y$ alloys on the initially (2×1) reconstructed Si(001) surface is strongly influenced by the growth conditions, such as growth temperature and Si growth rate, and by the amount of carbon itself. Reductions in growth temperature and increases in overall growth rate can lead to an increase in the amount of substitutionally incorporated carbon. This behavior is well described by a simple kinetic model, with the energy barriers as well as the preexponential frequency factor decreasing with increasing carbon concentration. Very low carbon concentrations can be predominantly incorporated substitutionally, regardless of growth temperature. For higher C concentrations, the substitutional carbon fraction is shifted to lower values and, in consequence, to an upper limit for substitutional C incorporation.

2.4. Segregation of Carbon-Containing Complexes

In vacuo X-ray photoelectron spectroscopy (XPS) was used to investigate Si$_{1-y}$C$_y$ layers grown pseudomorphically on Si(001) with constant carbon and Si fluxes but at different growth temperatures [29]. XPS is sensitive enough to trace changes in the carbon incorporation as a function of growth temperature. Figure 10a shows three measured C 1s spectra of Si$_{1-y}$C$_y$ alloy layers deposited at different growth temperature (350°C, 450°C, and 550°C, respectively). In Figure 10b we plot the C 1s binding energies versus growth temperature. All binding energies are between the values for silicon carbide (282 eV) and graphite (284.5 eV) [30]. The carbon 1s signals move toward lower binding energy as the growth temperature increases. The sample grown at the lowest temperature contains mainly substitutional carbon, whereas the sample grown at the highest

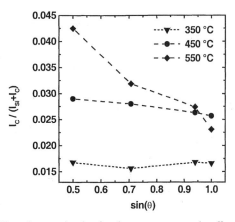

Fig. 11. Photoelectron signals of carbon atom versus take-off angle at different growth temperatures. The signal was normalized with respect to the Si 2p signal. Reproduced with permission from [29], copyright 1996, American Institute of Physics.

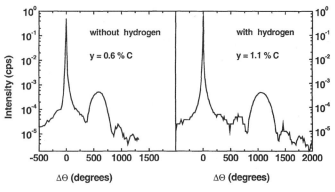

Fig. 12. X-ray rocking cirves of two 70-nm-thick $Si_{1-y}C_y$ layers on Si(001) grown at 475°C with identical Si and C fluxes but without and with hydrogen (rf source, $p = 2 \times 10^{-5}$ mbar). Reproduced with permission from [32], copyright 1997, Elsevier Science.

temperature shows nearly no substitutional carbon in the appropriate X-ray rocking curve.

Therefore, we assign the binding energies of 283.1 eV and 282.3 eV to substitutional and interstitial (nonsubstitutional) carbon atoms, respectively. This assignment agrees qualitatively with previously performed *ab initio* calculations, yielding a higher binding energy for substitutional carbon in silicon than for carbon bonded to silicon in a carbide structure [31].

To obtain information about the carbon depth distribution, we varied the take-off angle for the detected photoelectrons. Figure 11 shows the angle-dependent relative intensities of the C 1s core levels. As a measure of intensity we always used the integrated peak area. For increasing growth temperatures the normalized relative intensity of the carbon peak is strongly dependent on the take-off angle. Nearly 95% of the total information comes from a layer region within $3\lambda \sin(\theta)$ below the surface; thus XPS data do not deliver information from regions deeper than 3λ (~7 nm for the C 1s signal) from the sample surface.

If we assume that the concentration depth profile changes monotonically and slowly, then Figure 9 indicates an increase in carbon concentration toward the sample surface for increasing growth temperatures. This can only result from the segregation of nonsubstitutional carbon atoms during deposition. For a sample grown at 350°C, the relative intensity is nearly independent of take-off angle; therefore it is reasonable to assume that the concentration of substitutional carbon atoms is constant within 7 nm below the surface. For this case, we can estimate a C concentration of $(1.6 \pm 0.1)\%$. The relatively good agreement of our estimation (SIMS measurements yield $y = (1.8 \pm 0.2)\%$ for that sample) shows the capability of XPS, even for such small concentrations. We are not able to identify the segregating species at higher growth temperatures. They are probably some form of Si–C defect interstitial complexes and/or silicon carbide nanoparticles. The normalized relative intensity at $\theta = 15°$ corresponds to the averaged carbon concentration within $3\lambda \sin(15°) = 1.9$ nm. An Arrhenius-type behavior for

the surface carbon concentration could be fitted with an energy barrier of only $E_b = (0.25 \pm 0.02)$ eV, assuming that the concentration depth profile changes slowly within this depth. This barrier for the segregation of interstitial complexes containing C atoms is comparable to the Gibbsian heat of segregation found earlier for Ge in silicon (0.28 eV).

Strategies for changing the surface kinetics seem to be necessary to shift the window for a high substitutional-to-interstitial carbon ratio to process conditions suitable for good epitaxial quality. It is known that the presence of so-called surfactant atoms on the surface during growth can drastically alter the growth kinetics. Atomic hydrogen was supplied during MBE growth with the use of an *rf* source. The layers were grown at a hydrogen pressure above 10^{-5} mbar. Epitaxial layers could be grown under such conditions [32]. From X-ray rocking curves for layers grown under identical conditions (the same Si and C fluxes and the same growth temperature were used) with and without hydrogen we found a significant enhancement in the amount of substitutionally incorporated carbon atoms in hydrogen-mediated growth compared with commonly used MBE growth (Fig. 12).

2.5. Substitutional Carbon Incorporation during $Si_{1-x-y}Ge_xC_y$: Dependence on Germanium and Carbon

In the previous section we discussed only the incorporation of C into silicon. Here we extend the investigations to present results on the kinetics of substitutional C incorporation during MBE growth of $Si_{1-x-y}Ge_xC_y$ alloys [33]. We will show that the substitutional C incorporation can be dominated by the Ge or C concentration, depending on the total C concentration range. At low total C concentration, the Ge impact is insignificant, increasing total C concentration leads to less substitutional C incorporation efficiency. At high n_0, the impact of C concentration becomes marginal, and the presence of Ge inhibits the substitutional C incorporation.

Experiments were performed on 4″ Si(100) substrates in a multichamber Si MBE system (DCA Instrument) equipped

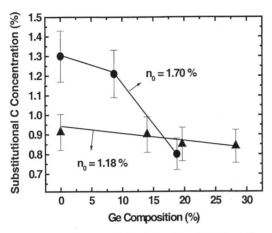

Fig. 13. The effect of Ge composition (x) on the substitutional C concentration (n_S) of the samples grown at 400°C with different C fluxes, i.e., different total C concentrations, n_0, but with the same total growth rate of 0.05 nm/s. Solid lines are guides for the eye. Reproduced with permission from [33], copyright 2000, American Institute of Physics.

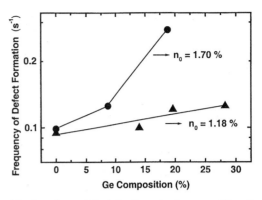

Fig. 14. The dependence of the defect formation frequency (k) on Ge composition (x). The defect formation frequency is calculated from the experimental results of Figure 13 and the kinetic model discussed in the text. Solid lines are guides for the eye. Reproduced with permission from [33], copyright 2000, American Institute of Physics.

with electron beam evaporators for Si and Ge, and a pyrolytic graphite filament sublimation source for C. Three sets of samples were grown at 400°C to ensure two-dimensional growth, monitored by reflection high-energy electron diffraction (RHEED). The total growth rate was 0.05 nm/s for all samples. In each set of samples, only the Ge composition was varied. The C flux was fixed to produce the same total C concentration in each set. Carbon desorption could be neglected. Samples were characterized by X-ray diffraction (XRD). The substitutional C concentration in samples without Ge was deduced from dynamical simulation of the measured rocking curves, assuming the nonlinear relationship given in [34]. To determine the substitutional C concentration in $Si_{1-x-y}Ge_xC_y$ samples, appropriate $Si_{1-x}Ge_x$ reference samples were grown and measured by XRD under identical conditions. XRD investigations yield the bulk concentration. We will treat the values also as surface Ge concentrations during growth (Ge segregation effects can be neglected at 400°C). The Ge composition in $Si_{1-x-y}Ge_xC_y$ was assumed to be the same as that in the corresponding SiGe reference samples. The error in the substitutional C concentration can be estimated to be less than 10%, mainly because of run-to-run variation in Ge composition.

Figure 13 shows the effect of the Ge content (x) on the substitutional C concentration (n_S) of the samples grown at 400°C with different C fluxes, i.e., different total C concentrations (n_0), but at the same total growth rate of 0.05 nm/s. The substitutional C concentrations in ternary $Si_{1-x-y}Ge_xC_y$ samples are always lower than those in the corresponding $Si_{1-y}C_y$ samples, indicating that the presence of Ge inhibits substitutional C incorporation. Mi et al. [35] did not find any correlation between the Ge and the substitutional C concentrations in $Si_{1-x-y}Ge_xC_y$ samples grown by RTCVD with the use of $SiH_4/GeH_4/SiCH_6$. The complex surface reactions during film growth with CVD, however, make it impossible to extract the effect of Ge composition because Ge content and total growth rate cannot be independently controlled. In the MBE ex-

periments, all other parameters could be kept constant, except for the Ge content, facilitating the extraction of the Ge compositional effect. We also note that the substitutional C concentration decreases with increasing Ge concentration for the samples grown with the same total C concentrations and total growth rates. This reduction is more significant in samples grown with higher C fluxes, leading to different substitutional C incorporation behavior during Si and SiGe growth. In $Si_{1-y}C_y$ alloy growth, the substitutional C concentration increases with the total C concentration. In contrast, in $Si_{1-x-y}Ge_xC_y$ samples with roughly the same Ge composition of 0.19, the substitutional C concentration decreases with the total C concentration (Fig. 13). The total C concentration at a certain C flux, also shown in Figure 13, was obtained from our previous experimental data and kinetic model of C incorporation into Si [19, 36] as discussed below.

In Section 2.3, we proposed a simple model for substitutional C incorporation into silicon. As there is no specific assumption on what other atoms, such as Si or Ge, are deposited simultaneously with C atoms, we use the same model for $Si_{1-x-y}Ge_xC_y$ growth.

For samples grown at the same temperature, we can reduce the number of free parameters by looking only at the defect formation frequency k (defined in Eq. (2.3)). Based on Eq. (2.5) and our previous data on $Si_{1-y}C_y$ growth, we have obtained the total C concentration in each set of samples with the same C flux and extracted the defect formation frequency for all samples. The dependence of k on Ge content is shown in Figure 14. There is a monotonic dependence on Ge concentration for C incorporation during MBE growth of $Si_{1-x-y}Ge_xC_y$ alloys. The defect formation frequency k increases with increasing Ge concentration for each set of samples with the same C flux, indicating that the presence of Ge atoms inhibits the substitutional C incorporation. A repulsive Ge–C interaction [37] may enhance the formation of interstitial C-containing defects and thus increase the defect formation frequency. Moreover, the effect of Ge on substitutional C incorporation is coupled with that of C. At low total C concentration, the effect of Ge is small but

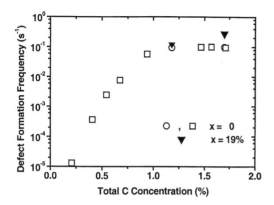

Fig. 15. The dependence of the defect formation frequency (k) on the total C concentration (n_0) for MBE growth of $Si_{1-y}C_y$ and $Si_{1-x-y}Ge_xC_y$ ($x \approx 0.19$) on Si(100). Reproduced with permission from [33], copyright 2000, American Institute of Physics.

becomes significant with increasing total C concentration. As a consequence, the effect of the total C concentration on substitutional C incorporation is mediated by the Ge content. As can be seen from Figure 14, there is only a small change in the defect formation frequency with increasing total C concentration for C incorporation into Si, whereas for C incorporation into SiGe with a Ge composition of about 19%, the defect formation frequency increases significantly with increasing total C concentration.

For a more complete picture of C incorporation, the dependence of k on n_0 is shown in Figure 15, in which previous data on $Si_{1-y}C_y$ growth are also displayed. In $Si_{1-y}C_y$ growth, k first increases with the increase in n_0 and then saturates with a further increase in n_0. However, we do not find a saturation of the defect formation frequency in $Si_{1-x-y}Ge_{1-x}C_y$ growth. Summarizing Figures 14 and 15, it can be concluded that the effect of Ge or C concentration can dominate the substitutional C incorporation, depending on the total C concentration range. At low n_0, the Ge impact is insignificant, and k increases with n_0, leading to lower substitutional C incorporation efficiency. At high n_0, the impact of C concentration becomes marginal, and the Ge concentration effect dominates the substitutional C incorporation.

The observed compositional dependence of k implies that the formation of interstitial C-containing defect complexes involves many coupled atomic processes, such as surface adsorption and diffusion and defect formation reactions. These atomic processes are poorly understood so far. The strain introduced by the incorporation of C and Ge makes these processes even more complicated. Nevertheless, the observed compositional dependence is already useful for materials growth. Following the compositional dependence of k, $Si_{1-x-y}Ge_xC_y$ alloys with high C substitutionality (n_S/n_0) can be obtained at small n_0 (similar to $Si_{1-y}C_y$ alloys). There should be a saturation of C substitutionality with increasing n_0, rather than the previously postulated saturation of n_S, in $Si_{1-y}C_y$ growth at a constant temperature and growth rate [36]. In $Si_{1-x-y}Ge_xC_y$ growth, however, there will be a saturation of n_S due to the continu-

ous increase of k with n_0. And the higher the Ge composition is, the smaller is the total C concentration at which the saturation starts to occur. The very low substitutional C incorporation during $Ge_{1-y}C_y$ growth, even at an extremely low growth temperature of 200°C (reported recently in [38]), is consistent with this prediction.

In summary, we showed that C incorporation into $Si_{1-x-y}Ge_xC_y$ is dominated by surface kinetics during MBE growth. The substitutional C concentration decreases with increasing Ge composition for the samples grown at 400°C with the same C flux and total growth rate. The reduction in the substitutional C concentration is small at low C flux, becoming increasingly significant at higher C fluxes. Following the compositional dependence of k, we predict a saturation of C substitutionality (n_S/n_0) with increasing n_0, rather than a saturation of n_S, in $Si_{1-y}C_y$ growth. In $Si_{1-x-y}Ge_xC_y$ growth, however, there will be a saturation of n_S due to the continuous increase in k with n_0.

3. MECHANICAL AND STRUCTURAL PROPERTIES

3.1. Strain Manipulation

The built-in strain and the composition of a pseudomorphic $Si_{1-x}Ge_x$ layer on a silicon substrate significantly affects the band structure and the energy gap as well as the band offsets. There is, however, an equilibrium critical thickness for perfect pseudomorphic growth that depends on the strain in the layer. For thicknesses larger than this critical thickness, a plastic strain-relief mechanism takes place, by injection and movement of misfit dislocations [39]. These dislocations can have a large influence on the electronic device properties. For a SiGe layer with 20% germanium ($x = 0.2$) the equilibrium critical thickness amounts to less than 20 nm [40]. It is possible to grow thicker pseudomorphic layers by reducing the growth temperature; these layers are metastable; i.e., any later temperature treatment (as is common in further device processing steps) can lead to the formation of misfit dislocations [41].

Adding a constituent with a covalent radius much smaller than that of silicon to a layer containing the larger Ge ($r_{Si} = 1.17$ Å, $r_{Ge} = 1.22$ Å, but $r_C = 0.77$ Å) opens the possibility of manipulating the strain. It was shown that it is possible to form an "inversely distorted" SiGe cell, i.e., a cell that is tetragonally distorted because of tensile stress instead of the usual compressive stress, by adding some carbon [17]. Using simple linear extrapolations between the different lattice constants and the appropriate relative concentration (Vegard's law for a ternary system; see Fig. 16), we can estimate the effective lattice constant in the $Si_{1-x-y}Ge_xC_y$ system as

$$a_{SiGeC} = (1 - x - y)a_{Si} + xa_{SiGe} + ya_{SiC} \qquad (3.1)$$

In the past, there has been some controversy about the validity of Vegard's law in ternary $Si_{1-x-y}Ge_xC_y$ alloys [42]. The linear approximation (between Si, Ge, and diamond) for the average lattice constants,

$$a(x, y) = (1 - x - y)a_{Si} + xa_{Ge} + ya_C \qquad (3.2)$$

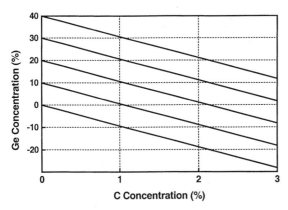

Fig. 16. Estimated strain reduction in SiGe layers due to the presence of carbon (ternary Vegard's law). Reproduced with permission from [17], copyright 1994, American Institute of Physics.

results in a Ge:C ratio of 8.2 for complete strain compensation. Using Vegard's approach one obtains 9.4 for extrapolation to silicon carbide (shown in Eq. (3.1)). *Ab initio* calculations predict values even larger than 10 (see, for example, [43]). These calculations also show a nonlinearity in the dependence of alloy lattice constants on the concentration ("bowing effects"). Recent experimental results also show a deviation from Vegard's law [34]. In our simple model considerations we will neglect this bowing and use Eq. (3.1) for an estimate of the average lattice constant. We define an "effective lattice mismatch" mf_{eff} as

$$mf_{eff} = [a(x, y) - a_S]/a_S \qquad (3.3)$$

This quantity is directly accessible by X-ray diffraction. A positive mf_{eff} stands for a compressively strained material, $mf_{eff} < 0$ indicates tensile strain, and $mf_{eff} = 0$ represents a fully strain-compensated $Si_{1-x-y}Ge_xC_y$ alloy [44]. For clarity, sometimes we will use the phrase "effective concentrations," i.e., the Ge or C concentration in the binary alloy that would yield the same lattice mismatch as mf_{eff}.

3.2. Microscopic Structure of $Si_{1-x-y}Ge_xC_y$ Alloys

Carbon atoms preferentially occupy substitutional lattice sites in silicon [13]. It has been demonstrated by *ab initio* total energy calculations that the substitutional site is the most stable position of C in Si [45]. The total energy of interstitial carbon (C_i) is about 3 eV higher than the energy of substitutional carbon (C_s) [46]. Isoelectronic substitutional carbon atoms represent neutral impurities in silicon and germanium crystals. They form strong covalent bonds with the four nearest-neighbor atoms of the host lattice and do not cause doping with extra electrons or holes. For crystalline $Si_{1-x-y}Ge_xC_y$ layers it is expected that the majority of C atoms are incorporated at lattice sites, as discussed in the previous section. This assumption is supported by the measured reduction of the lattice parameter when C is added to Si or $Si_{1-x}Ge_x$ crystals [17, 47]. However, C is well known to form a variety of defects in silicon [13]. These defects and defect complexes can be very

important for the electrical and optical properties of crystalline $Si_{1-x-y}Ge_xC_y$ alloys. For structural properties of the alloys, the much larger fraction of electrically inactive substitutional C atoms is of major interest. In the following, we focus on the discussion of substitutional $Si_{1-x-y}Ge_xC_y$ alloys.

3.2.1. Energetics of $Si_{1-x-y}Ge_xC_y$ Alloys

First, we discuss the energetics and stability of random $Si_{1-x-y}Ge_xC_y$ alloys. A substitutional alloy can be considered as a lattice gas with an average occupation of the lattice sites according to the stoichiometry. In general, the enthalpy of mixing,

$$\Delta H = \Delta E_{chem} + \Delta E_{strain} \qquad (3.4)$$

consists of two parts: a chemical term ΔE_{chem} deriving from the different strengths of the bonds and a strain term ΔE_{strain} deriving from the deformation of bonds.

In hypothetical, ordered zincblende structures of two group IV elements there is no strain. Consequently, the formation energy is purely chemical (ΔE_{chem}). Martins and Zunger [48] have studied the chemical energy contribution for Si–Ge and Si–C structures by first-principle total energy calculations. Formation energies are determined by the difference between the cohesive energies of the compounds and the cohesive energies (E_{coh}) of the elemental crystals. The formation energy of β-SiC (cubic silicon carbide) was found to be negative,

$$\Delta E_{chem}(\beta\text{-SiC}) = E_{coh}(\beta\text{-SiC}) - E_{coh}(Si) - E_{coh}(C)$$
$$= -0.33 \text{ eV/atom} \qquad (3.5)$$

whereas a slightly positive formation energy,

$$\Delta E_{chem}(SiGe) = E_{coh}(SiGe) - E_{coh}(Si) - E_{coh}(G)$$
$$= +0.009 \text{ eV/atom} \qquad (3.6)$$

was found for SiGe in a zincblende structure. The negative term ΔE_{chem} for SiC stabilizes the SiC phase with respect to the Si and C phases, whereas the positive energy for SiGe favors the phase separation into Si and Ge at low temperatures. Next, we use the energies (Eqs. (3.5) and (3.6)) for an estimation of the chemical contribution to formation energies of alloys. We assume that the A–A, B–B, and A–B bond strengths are transferable quantities. Then the chemical contribution to the energy of mixing can be assembled by counting the total number of bonds N_{AB} between different atoms A and B,

$$\Delta E_{chem} = 1/2 N_{AB} \Delta E_{chem}(AB) \qquad (3.7)$$

where $\Delta E_{chem}(AB)$ is the energy per atom as given in Eqs. (3.5) and (3.6).

Because of the considerable difference in the size of C, Si, and Ge atoms, crystalline $Si_{1-x-y}Ge_xC_y$ alloys show large microscopic strain. The contribution of microscopic strain ΔE_{strain} to the formation enthalpy results from the deformation of bond lengths and bond angles and can be described approximately by Keating's [49] valence force field model. The strain energy,

$$\Delta E_{\text{strain}} = \sum_{i,j} \alpha_{ij} \frac{3}{16r_{ij}^2} \left(x_{ij}^2 - r_{ij}^2\right)^2$$

$$+ \sum_{i,j,k} \beta_{ijk} \frac{3}{8r_{ij}r_{ik}} \left(x_{ij}x_{ik} + \frac{1}{3}r_{ij}r_{ik}\right)^2 \quad (3.8)$$

is expressed by bond stretching force constants α and bond bending force constants β for each kind of bond. The index i denotes the sum of all atoms with nearest neighbors $j, k = 1 \ldots 4$, and x_{ij} and r_{ij} are the strained and unstrained bond lengths, respectively. In contrast to the chemical contribution, the strain energy is always positive. Because of the large difference in the bond lengths occurring in $Si_{1-x-y}Ge_xC_y$ alloys, the bond distortions cannot be expected to be small, and anharmonic effects become important. Starting from *ab initio* density functional calculations, a bond length dependence of the bond stretching and bond bending force constants was introduced (see [50]), which can account for anharmonic effects of strongly distorted bonds.

Applying the anharmonic Keating model, we get strain contributions $\Delta E_{\text{strain}}(Si_{0.5}Ge_{0.5}) = 4.5$ meV/atom and $\Delta E_{\text{strain}}(Si_{0.99}C_{001}) = 16$ meV/atom to the formation enthalpy of random $Si_{0.5}Ge_{0.5}$ and $Si_{0.99}C_{001}$ alloys, respectively. If this strain energy for a $Si_{0.99}C_{001}$ alloy is divided into contributions of the individual C atoms, it corresponds to 1.6 eV per C atom. This energy is on the order of the formation energy of a substitutional C impurity in Si. *Ab initio* density functional calculations give a formation energy of 1.89 eV [45].

The stability of different phases at a given temperature (e.g., growth temperature) is controlled by Gibbs' free enthalpy $\Delta G = \Delta H - T \Delta S$, where H, S, and T are the enthalpy, entropy, and temperature, respectively. Consequently, the entropy S has to be included in the estimate of the stability of different structures at finite temperatures. The configurational entropy of a random substitutional $Si_{1-x-y}Ge_xC_y$ alloy is [51]

$$S_{\text{rand}} = -k_B N \big(x \ln(x) + y \ln(y) + (1 - x - y) \ln(1 - x - y)\big) \quad (3.9)$$

where N is the number of atoms and k_B is Boltzmann's constant. The entropy term favors the formation of disordered phases at high temperatures.

Using the energies given above, we obtain a formation enthalpy of 9 meV/atom for a random $Si_{0.5}Ge_{0.5}$ alloy. Note that the chemical contribution to the formation enthalpy of a random $Si_{0.5}Ge_{0.5}$ is half the value of zincblende SiGe (Eq. (3.6)), because only 50% of the bonds in the random alloy are heteroatomic. The entropy term compensates for the formation enthalpy for $T \sim 150$ K. Consequently, a miscibility temperature of about 150 K can be estimated for a $Si_{0.5}Ge_{0.5}$ alloy. The entropy term stabilizes the random alloy for $T > 150$ K. The structure and thermodynamics of $Si_{1-x}Ge_x$ alloys have been studied in detail by de Gironcoli et al. [52] with the use of Monte Carlo simulations with pair interactions determined from *ab initio* density functional calculations. $Si_{1-x}Ge_x$ was found to be a model random alloy with no chemical order; i.e.,

the different types of chemical bonds (Si–Si, Si–Ge, Ge–Ge) occur randomly according to the stoichiometry. A miscibility temperature of about 170 K was obtained for Ge contents of about 50%.

In contrast to $Si_{1-x}Ge_x$, the strain energy is very high in carbon-containing alloys ($Si_{1-y}C_y$ or $Si_{1-x-y}Ge_xC_y$). It forms the major part of the formation enthalpy of the alloy. $Si_{1-y}C_y$ alloys are metastable at moderate temperatures; i.e., the free enthalpy could be reduced by separation into Si and SiC phases. A rough estimate of the solubility of substitutional C in Si can be obtained from the energies given above. From the balance of the formation enthalpy H^f of substitutional C and the configurational enthalpy (Eq. (3.9)) one can estimate the solid solubility,

$$C_{\text{sol}} = C_0 \exp(-H^f / k_B T) \quad (3.10)$$

Identifying the formation enthalpy H^f with the calculated energy cost of 1.6 eV for transferring one C atom from a SiC reservoir into a substitutional configuration in Si, we obtain a solubility of $C_{\text{sol}} = 5 \times 10^{17}$ cm^{-3} at the melting point of Si. This value is surprisingly close to the experimental value $C_{\text{sol}} = 3 \times 10^{17}$ cm^{-3} [53], taking into account the approximations of our estimation. It should be noted, however, that the measured solubility of C in Si shows a stronger temperature dependence than that implied by Eq. (3.10). Bean and Newman [53] have measured an exponential temperature dependence of the solubility of C corresponding to $H^f = 2.3$ eV.

The occurrence of C–C bonds in $Si_{1-y}C_y$ and $Si_{1-x-y}Ge_xC_y$ alloys is energetically unfavorable. First, the negative chemical energy for SiC favors the formation of Si–C bonds, and, more importantly, the incorporation of C–C bonds into the lattice requires a large strain energy [46, 54]. For this reason almost all C atoms in substitutional $Si_{1-y}C_y$ alloys are bound to four Si neighbors. Those alloys with a preferential occurrence of certain types of bonds are called chemically ordered.

3.2.2. Local Strain Distribution

Next we discuss the local strains and nearest-neighbor bond lengths in substitutional $Si_{1-x-y}Ge_xC_y$ alloys. The distribution of the bond lengths is of fundamental importance when the properties of a substitutional semiconductor alloy are described from a microscopic viewpoint. We have applied the anharmonic Keating model to study the bond length mismatch problem in $Si_{1-x-y}Ge_xC_y$ alloys [44] to account for large bond distortions due to the incorporation of the much smaller C atoms. Model calculations for $Si_{1-x}Ge_x$ [55] and for $Si_{1-x-y}Ge_xC_y$ [44] as well as *ab initio* calculations for $Si_{1-x}Ge_x$ [52] have shown that the lattice parameter of the alloys is approximately given by Vegard's law.

Information on the microscopic strain situation can be obtained from the calculated bond length distribution as shown in Figure 17. The bond length distribution for $Si_{1-x}Ge_x$ plotted in the upper panel shows three distinct peaks corresponding to Si–Si, Si–Ge, and Ge–Ge bonds. Results for different compositions are given in Figure 17. The calculated bond lengths

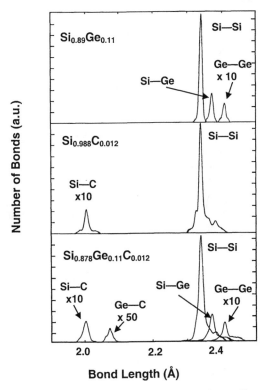

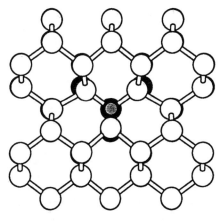

Fig. 19. Atomic structure in the vicinity of a substitutional C atom (gray) in Si (white) as calculated from the anharmonic Keating model. Atoms of a [110] plane are shown. Ideal Si lattice sites are indicated by black underlying circles.

Fig. 17. Calculated bond length distributions for $Si_{0.89}Ge_{0.11}$, $Si_{0.988}C_{0.012}$, and $Si_{0.878}Ge_{0.11}C_{0.012}$ layers pseudomorphically strained onto the Si(001) substrate. The Si–C, Ge–Ge, and Ge–C curves were multiplied by factors of 10 and 50, respectively. Reproduced with permission from [44], copyright 1994, American Physical Society.

The degree of relaxation from an average bond length toward a "chemical" value given by the sum of Pauling's covalent radii can be expressed by a dimensionless constant [48],

$$\varepsilon = \frac{R_{XY} - R_{av}}{R_{XY}^0 - R_{av}} \tag{3.11}$$

where R_{av} is the average bond length of the alloy as given by Vegard's rule, R_{XY} is the observed characteristic bond length in the alloy, and R_{XY}^0 is the X–Y bond length in the ordered (zincblende or diamond structure) XY compound. The limiting cases $\varepsilon = 1$ and $\varepsilon = 0$ correspond to bond lengths given by the sum of the covalent radii or by a common average bond length, respectively. The parameter ε and the analogous topological rigidity parameter introduced by Cai and Thorpe [56] are essentially determined by the relative strengths of bond bending and bond stretching force constants. *Ab initio* calculations [52] and model calculations [44, 55] for $Si_{1-x}Ge_x$ gave ε values of about 0.7 for Si–Si, Si–Ge, and Ge–Ge bonds. Figure 18 shows bond lengths in $Si_{1-x}Ge_x$ calculated with the anharmonic Keating model. X-ray absorption fine structure (EXAFS) measurements by Aldrich et al. [57] have confirmed this value. In contrast, it was concluded from previous EXAFS measurements [58] that Ge–Ge and Si–Ge bond lengths are almost independent of the alloy composition and are given as the sum of the covalent radii. This discrepancy has been discussed as originating from structural defects in the $Si_{1-x}Ge_x$ samples used [59] or from uncertainties in the interpretation of the EXAFS data [57].

Next we discuss the local atomic structure of $Si_{1-x-y}Ge_xC_y$ alloys (bottom panel in Fig. 17). Si–C bonds were found to be stretched by about 7% with respect to the bond length in cubic SiC. The distribution function of Si–Si bond lengths is asymmetric and shows distinct peaks and shoulders corresponding to Si–Si bonds with substitutional C atoms in the nearest neighboring positions. The local structure around a substitutional C atom is shown in Figure 19. A strong distortion is observed for the first and second nearest-neighbor shells of the C atom. In accordance with macroscopic elastic theory, the displacement decays as r^{-2} for larger distances r.

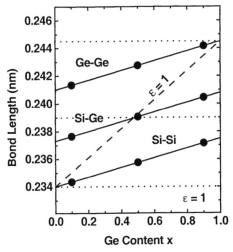

Fig. 18. Variation of the Si–Si, Si–Ge, and Ge–Ge bond lengths (solid lines) in a $Si_{1-x}Ge_x$ alloy as a function of the germanium concentration, as calculated with the valence-force field model (filled circles). The dashed line corresponds to the "Vegard" limit of common ideal lattice; the dotted lines correspond to Pauling's picture of transferable "chemical" bond lengths. Reproduced with permission from [44], copyright 1994, American Physical Society.

only approximately coincide with those of pure Si, pure Ge, and zincblende SiGe. At a given concentration, the bond lengths partially relax toward an average bond length, as given by the alloy lattice parameter at that concentration.

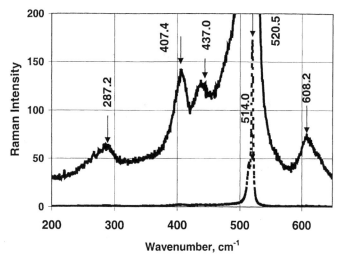

Fig. 20. Raman spectrum of a fully strain-compensated $Si_{1-x-y}Ge_xC_y$ layer. Reproduced with permission from [44], copyright 1994, American Physical Society.

Molecular dynamics calculations by Demkov and Sankey [60] indicated that the tetrahedral geometry of substitutional carbon atoms can be strongly distorted in regions of high carbon concentration. For a special arrangement of carbon atoms as second nearest neighbors, graphite-like carbon atoms in an sp^2 electronic configuration were found. Arrangements of substitutional carbon atoms as second nearest neighbors are energetically unfavorable (see also Section 5). A lower energy structure could be formed by rebinding some of the carbon atoms [60] or by moving the carbon atoms to more distant lattice sites. The kinetics of the growth process will determine which of these possible rearrangements will occur in practice.

3.3. Strain-Compensated Ternary Alloys

An interesting feature in Figure 16 is the line for zero misfit [44]. From X-ray rocking curves we found that for a Ge/C ratio of about 10 the lattice constant in the growth direction is indistinguishable from that of the substrate, indicating the absence of macroscopic strain. Such a layer exhibits the cubic silicon structure. The splitting in the energy gap in a twofold degenerated and a fourfold degenerated valley observed for strained SiGe layers should disappear. On the other hand, such a layer is not identical to a strain-free SiGe alloy [44].

An inside view of the bond distribution can be obtained from vibrational spectroscopy. Raman measurements were performed at room temperature in backscattering geometry with a Dilor xy triple-grating spectrometer. Supplementary Raman measurements were performed in other geometries to examine the SiGeC layer for a tetragonal deformation. The Si–Si mode would split in both scattering geometries under an existing biaxial strain. The Raman spectrum of the epitaxially grown $Si_{1-x-y}Ge_xC_y$ alloy corresponds well to a spectrum of a pseudomorphic $Si_{1-x}Ge_x$ alloy with the same Ge content (Fig. 20). An additional small peak at 608.2 cm^{-1} occurs that can be assigned to the localized C–Si mode. The Raman peak of the

Si–Si mode in the strain-compensated SiGeC layer is shifted to smaller wavenumbers from the bulk silicon. This shift is close to the value for the appropriate free-standing $Si_{1-x}Ge_x$ alloy and significantly larger than the value for a pseudomorphic $Si_{1-x}Ge_x$ layer on Si(001). Furthermore, there is no significant difference between the measured Raman shifts in backscattering from faces parallel and perpendicular to the layer surface. This proves the absence of a tetragonal distortion, in agreement with the X-ray measurements. The behavior of the Si–Si Raman frequency seems to be in contrast to the X-ray results that exhibit a compensation for the Ge-caused strain by C alloying. It becomes obvious if one considers that X-ray diffraction measures a mean value of the deformation over many unit cells, whereas the Raman scattering yields information about the short-range atomic arrangement. Let us assume a crystallographic structure $A_{1-x}B_x$, where the individual bond lengths R_{AB} are determined purely by the appropriate covalent radii (Pauling crystal),

$$R_{AB} = r_A + r_B \quad (3.12)$$

where r_A and r_B are the appropriate covalent radii for the atoms A and B. For a covalent crystal $A_{1-x}B_x$ where all atoms are fourfold coordinated, we obtain an average bond length R_{ave} of

$$R_{ave} = P_{AA}(x)R_{AA} + P_{AB}(x)R_{AB} + P_{BB}(x)R_{BB} \quad (3.13)$$

where

$$P_{AA}(x) = (1-x)^2$$
$$P_{AA}(x) = 2x(1-x)$$
$$P_{AA}(x) = x^2$$

are the probabilities that such bonds to occur in a perfectly stochastic distribution of atoms A and B. Substituting these probabilities into the expression for the average bond length, we obtain

$$R_{ave} = 2r_A + 2x(r_B - r_A) \quad (3.14)$$

or a linear dependence of the average bond length on x (Vegard's rule). X-ray methods average over the extension of the used X-rays; i.e., these methods always yield average bond lengths in agreement with Vegard's rule.

The microscopic structure of a fully strain-compensated $Si_{1-x-y}Ge_xC_y$ layer has been investigated in detail in [44] (see also Section 3.2). The Si–C bonds are stretched by about 7% with respect to the bond length in cubic silicon carbide, whereas the C atoms mainly strain the Si–Si bond in their neighborhood, and more distant Si–Si bonds are less affected. Although the average lattice constant of the strain-compensated $Si_{1-x-y}Ge_xC_y$ layer is the same as that of pure silicon, the alloy contains large internal (microscopic) strain. Calculations support the picture of only partial relaxation of the Si–Si and Ge–Ge bonds toward a common bond length [44]. The Si–C and Si–Ge bonds in strain-compensated $Si_{1-x-y}Ge_xC_y$ are strongly stretched. The calculated 7% stretching of Si–C in strain-compensated $Si_{1-x-y}Ge_xC_y$ corresponds to 35% relaxation toward the average bond length of the alloy (predicted by Vegard's law). The dependence of the Raman spectra on the strain and composition of $Si_{1-x-y}Ge_xC_y$ layers can be explained by the model calculations.

Through the combination of theoretical and experimental results for local Si–C phonon modes in dilute $Si_{1-y}C_y$ alloys, information concerning the short-range order was obtained [74]. Calculations using an anharmonic Keating model predict satellite peaks near the vibrational frequency of an isolated C impurity, associated with second-, third-, etc., nearest-neighbor C–C pairs. By comparing theoretical spectra with those obtained by Raman and infrared absorption spectroscopy, it was concluded that the probability of a third-nearest-neighbor coordination of the C atoms is considerably above that for a purely random alloy. This confirms earlier work predicting an attractive third-nearest-neighbor interaction of the C impurities due to elastic interactions [54].

3.4. Strain Relaxation in Tensile Strained $Si_{1-y}C_y$ Layers on Si(001)

$Si_{1-x}Ge_x$ alloys cannot be grown epitaxially on Si or Ge substrates without introducing large amounts of strain. Strain relief usually occurs by plastic flow, i.e., by injection and propagation of misfit dislocations. $Si_{1-x}Ge_x$ alloys can be grown to much larger thicknesses than predicted by mechanical equilibrium theory [61]. These layers become metastable. A later temperature treatment, typical for further Si technological steps, also causes the strain-relieving formation of extended misfit dislocations. Therefore, the inherent strain limits the compatibility of heteroepitaxial SiGe systems to common Si technologies because of its lower thermal stability. In addition, the formation of extended misfit dislocations always changes the in-plane lattice constant (loss of pseudomorphy), enabling the construction of virtual substrates with a lattice constant differing from that of silicon.

The temperature stability of tensile strained $Si_{1-y}C_y$ layers was investigated during and after postgrowth annealing [8, 9, 62–64]. Despite the tensile strain in a 100-nm-thick layer and the high carbon supersaturation, the samples were stable up to 800°C [9]. Beyond this temperature range, the substitutional carbon content started to decrease exponentially during isothermal annealing. However, the layers always remain pseudomorphical to the Si substrate; that is, they preserve the Si in-plane lattice constants. Although the number of possible nucleation centers for dislocation generation was quite high, we could not find any relaxation by the formation of extended misfit dislocations [9]. The observed relaxation effect can be explained by precipitation of carbon and/or nucleation and diffusion-limited growth of SiC nanocrystals. We conclude that in contrast to the mechanism of strain-relieved $Si_{1-x}Ge_x$, in comparably strained $Si_{1-y}C_y$ epilayers the main high-temperature process is precipitation, which does not alter the in-plane lattice constant. Thus, relaxed $Si_{1-y}C_y$ layers are not suitable as virtual substrates with an in-plane lattice constant smaller that of silicon. Similar results were obtained for the thermal stability of compressive strained $Si_{1-x-y}Ge_xC_y$ alloys [65].

In the following, we discuss in detail the strain relaxation behavior of a metastable tensile strained $Si_{1-y}C_y$ epilayer on Si(001) by comparing the layer before and after an annealing step with a variety of different diagnostic methods. We show that the main strain-relieving mechanism is the formation of carbon containing interstitial complexes and/or silicon carbide nanoparticles, which is similar to the behavior of carbon in silicon under thermodynamical equilibrium conditions (concentrations below the solid bulk solubility limit). Powell et al. [62] found silicon carbide precipitates that were about 5 nm in diameter after a 1050°C anneal by high-resolution transmission electron microscopy (TEM). The absence of any remarkable SiC precipitates in our sample only means that they are much smaller (if they exist). In the rest of the paper the term "carbon-containing interstitial complexes" also includes the possible existence of very small silicon carbide clusters (as a special form of an interstitial complex).

3.4.1. Experimental

The investigated layer was grown by solid-source MBE. The overall growth rate was 0.05 nm/s. First a Si buffer was deposited. Then we deposited a 230-nm-thick $Si_{1-y}C_y$ layer followed by a 30-nm Si cap layer. The whole growth sequence was performed without any growth interruption, and the growth temperature was kept constant at 400°C.

X-ray double-crystal diffractometer (DCD) measurements with the symmetrical (004)-reflection and CuK_α radiation were carried out to determine the lattice strain perpendicular to the sample surface in the $Si_{1-y}C_y$ layer. The depth profile of strain was obtained by computer simulations of DCD rocking curves based on a semikinematic algorithm [66].

The postgrowth annealing was performed in a nitrogen ambient in the high-temperature attachment of an X-ray diffractometer. A ramping speed of 80 K/min was used for heating to a final temperature of 920°C. This temperature was maintained for 270 min with an accuracy of about ±3°C. By conventional θ–2θ scanning (5 min per scan) with the symmetrical (004) reflection, the variation of the distance $\Delta(2\theta)$ between substrate and layer K_α doublet peaks could be studied in situ. We observe an exponential decrease in this distance in Figure 21. If the layer remains pseudomorphic and the in-plane lattice constant a_{II} of the layer does not change (controlled using asymmetrical reflection), $\Delta(2\theta)$ can be directly related to a decrease in the amount of substitutional carbon in the layer [9].

TEM investigations were performed on cross-sectional samples with a Philips CM30 microscope operating at 300 kV. Ion thinning equipment with a maximum energy of 12.5 keV of Ar^+ ions in a first step and about 3.5 keV in the final step was used for the cross section preparation.

The layers were analyzed by secondary ion mass spectrometry (SIMS) in an UHV-based Atomika 6500 instrument, with the use of mass-filtered Cs^+ primary beams. The focused beams were raster scanned over typical areas of 200×200 μm^2 at an angle of incidence of 65°. An energy of 5 keV has been chosen to obtain flat concentration profiles with moderate sputter rates

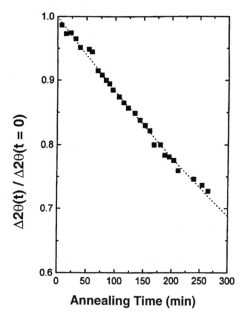

Fig. 21. Decrease in the normalized angular distance between substrate and layer X-ray (004) reflection during annealing at 920°C. The dotted line represents an exponential fit function. Reproduced with permission from [64], copyright 1996, Institute of Physics.

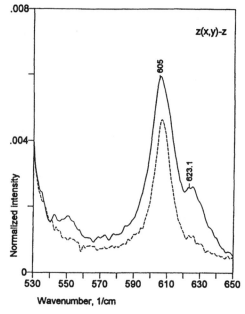

Fig. 22. Raman spectra of the $Si_{1-y}C_y$/Si heterostructure before and after annealing under identical experimental conditions. The difference spectra after subtraction of a Si spectrum are shown. At 605 cm^{-1} the localized mode of the carbon atoms in silicon occurs. The shoulder at 623 cm^{-1} is caused by a special C–Si–Si–C configuration. It disappears after annealing. Reproduced with permission from [64], copyright 1996, Institute of Physics.

for the films in a thickness range of several tens to hundreds of nanometers. MCs$^+$ cluster ions, where M is any element in the sample, have been detected for almost quantitative depth profiling of the C concentration in the layers. The quantification was performed with high dose implantation standards as references.

Backscattering analysis [67] using ion channeling techniques [68] was performed at the 1.7-MV tandem accelerator facility at the University of Western Ontario. A ^{4}He ion beam with energies of 1.5 and 2 MeV was used to measure the backscattering signal from silicon. For carbon analysis an ion beam energy of 4.3 MeV was used. At this energy the scattering cross section for carbon is 120 times higher than the Rutherford cross section [69], which allows the quantitative analysis of carbon concentrations below 1% in silicon. A detailed description of the experimental setup for the backscattering analysis and some of the measurements can be found in [70].

Raman measurements were performed with a μ-Raman spectrometer that was equipped with three identical gratings of 1800 grooves/mm, at a 500-mm focal distance. The first two gratings were used in a subtractive arrangement as a double monochromator. A spectral resolution of 3.2 cm^{-1} was achieved by the third, the spectrograph grating. In special cases a triple additive arrangement was used with a spectral resolution of 1.1 cm^{-1}. The sample was excited by the 514.5-nm line of an Ar$^+$ ion laser through a confocal optical system. The laser power was 20 mW. A microscope objective with a numerical aperture of 0.95 collected the scattered intensity. The inelastic scattered light was analyzed in the spectrometer and detected in a cooled charge-coupled device (CCD) matrix. From the localized carbon vibration a weak Raman intensity was expected because of the low carbon concentration and the thin $Si_{1-y}C_y$

layer compared with the penetration depth of the laser light in silicon. We estimated the contribution of the carbon scattering to be only 1/1000 of the whole intensity (dominated by the Si–Si mode). It was therefore important to carefully measure the scattering from the silicon substrate in the neighborhood of the carbon peak as well.

In the experiment a backscattering geometry was used, for which the selection rule prohibits two-phonon Raman scattering, with the exception of a weak peak at 615 cm^{-1} caused by a combination of optical and acoustic silicon phonons [71]. To separate the carbon contribution, a silicon spectrum was subtracted from the measured $Si_{1-y}C_y$/Si spectrum. The two spectra were measured successively under identical experimental conditions for equal measuring periods. They were normalized to the peak intensity of the Si–Si mode, and then the Si spectrum was subtracted. In Figure 22, difference spectra from the sample before and after annealing are shown.

3.4.2. Results

Figure 23 shows the rocking curve (RC) measured in the as-grown state. A pronounced layer peak can be seen at about $\Delta\theta = +1300''$, indicating a lattice contraction relative to the Si substrate. The reflectivity oscillations between substrate and layer peak are directly related to the thickness of the $Si_{1-y}C_y$ layer. The solid line in Figure 23 (right panel) shows the depth profile substitutional C content corresponding to the best fitted simulated curve (solid line in the left panels). The $Si_{1-y}C_y$ layer is about 230 nm thick. Assuming perfect pseudomorphy,

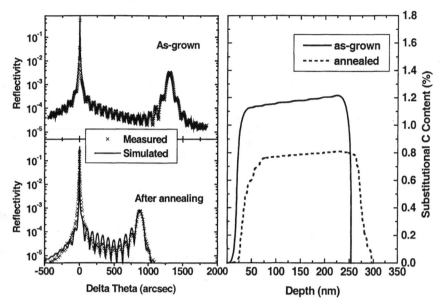

Fig. 23.   Measured (crosses) and calculated (solid line) double-crystal diffractometer rocking curve in the as-grown state and after annealing. The right panel represents depth profiles of the substitutional carbon concentration in the as-grown state (solid line) and after annealing (dashed line), used in the calculations shown in the left panels. Reproduced with permission from [64], copyright 1996, Institute of Physics.

the measured strain can be converted into a substitutional C concentration of 1.45%.

The observed exponential decrease in $\Delta(2\theta)$ during the annealing (Fig. 21) is typical for diffusion and precipitation of carbon [9]. The experimental and simulated rocking curves after annealing are also shown in Figure 23, as is the corresponding C concentration profile, as a dashed line (right panel). A significant shift of the layer peak to the substrate can be found, which is caused by a reduction of strain. The strain profile now shows a weaker decay in the near surface region than it did before annealing. X-ray diffraction measurements with asymmetrical reflections [72] confirm that the pseudomorphically grown $Si_{1-y}C_y$ layer remains pseudomorphic. Converting the measured strain into substitutional carbon concentration yields a reduction to only 0.97%.

The TEM results were obtained under various imaging conditions from the as-grown and the annealed sample. They demonstrate a good epitaxial relation between the Si substrate and the $Si_{1-y}C_y$ layer. In the upper part of the as-grown $Si_{1-y}C_y$ layer, we observe some crystallographic defects, opposite the "defect-free" lower part. These defects could act as sources for the generation of misfit dislocations during the later heat treatment. However, only small dislocation loops were generated during the anneal; but we did not find any elongated misfit dislocation network. The sample surface and the interface between the $Si_{1-y}C_y$ layer and the Si cap show some surface roughness with a characteristic length scale of about 50 nm.

Comparing the $Si_{1-y}C_y$ layer with the Si substrate (taken under strain-sensitive imaging conditions with $s = 0$), we conclude from the contrast fluctuations in the layer that there is a nonuniform strain distribution within the $Si_{1-y}C_y$ layer. We also found some contrast variations (striations) roughly paral-

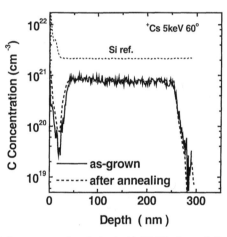

Fig. 24.   Carbon concentration obtained with SIMS before and after annealing. Reproduced with permission from [64], copyright 1996, Institute of Physics.

lel to the interface (most significant in the dark-field images). Perhaps we observe a phenomenon of "self-organization" similar to that observed by Claverie et al. [73] on their CVD-grown samples.

The contrast behavior before and after the annealing shows no remarkable differences. The observed fluctuation seems to be more pronounced in the annealed sample. Analyzing the diffraction pattern, no additional reflexes due to crystalline silicon carbide phases could be found. No larger SiC precipitates grown during the thermal treatment could be detected with TEM.

Figure 24 shows carbon SIMS depth profiles characterizing the total C amount in the layer before and after annealing. A relative sharp interface to the cap layer is formed, which is

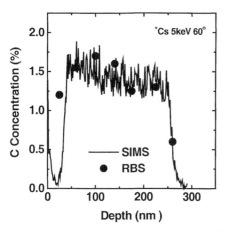

Fig. 25. SIMS depth profile in a linear scale. The observed slight increase in the C concentration toward the sample surface agrees with resonant RBS measurements, shown as solid circles. Reproduced with permission from [64], copyright 1996, Institute of Physics.

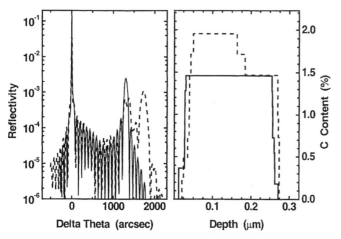

Fig. 26. Comparison of the DCD rocking curve (left) calculated with the strain profile (right) as shown in Figure 23 (solid line) and the rocking curve calculated with a strain profile related to carbon distribution measured with SIMS (dashed line) and shown in Figure 25. Reproduced with permission from [64], copyright 1996, Institute of Physics.

visible from the steep increase of the C profile. The broadened trailing edge of the profile toward the buffer layer can be related to SIMS artifacts. From Figure 24 we conclude that there is a nearly constant C concentration across the film thickness. The value of the concentration has been determined to be about 1.5% with the use of the MCs^+ technique and a high dose C implantation as a reference. This value is related to the sum of all carbon atoms. The agreement with the XRD results indicates that for the as-grown sample the vast majority of carbon atoms are substitutional. Figure 24 also shows a carbon SIMS depth profile of the same sample after annealing. No significant changes could be detected in comparison with the as-grown sample, indicating the absence of carbon out-diffusion.

In Figure 25 the SIMS profile of the as-grown sample is shown on a linear scale. We found a slight increase in the C concentration toward the surface, in good correlation with earlier resonant backscattering measurements [70] (shown as

solid squares in Fig. 21) and the TEM observations. Figure 26 demonstrates how such a variation in the strain profile would significantly change the X-ray rocking curves. Here a strain profile was used for the simulation that is related to the measured SIMS profile for the carbon distribution (Fig. 25). Especially in the region near the layer peak, the rocking curves should be completely different. In fact, the measured rocking curve (Fig. 19) is absolutely typical for a uniformly in-depth strained layer. Comparing all results, we conclude that the increase in the C concentration observed with SIMS and Rotherford backscattering is most probably related to the occurrence of carbon-containing complexes that do not affect the lattice strain but can create crystallographic defects.

For the ion channeling analysis the sample was aligned in the $\langle 001 \rangle$ direction. The minimum yield at an ion beam energy of 1.5 and 2 MeV was 6% to 8% for the silicon backscattering signal, compared with 3.5% measured on a reference area with no $Si_{1-y}C_y$ layer. One possible explanation for the increased yield of the $Si_{1-y}C_y$ layer is local lattice deformations around C atoms, as predicted in Section 3.2 based on theoretical structure investigations. Measurements of the annealed sample showed a thin oxide layer close to the surface. The channeling yield for silicon did not change because of the postgrowth annealing. The backscattering signal for carbon was analyzed with the use of a beam energy of 4.3 MeV. By measurements in random orientation, the carbon concentration was determined to be 1.5%, with a small increase toward the surface. This agrees with the SIMS results described earlier. For channeling in the $\langle 001 \rangle$ direction the minimum yield was 25%, which reveals that more than 75% of the carbon was located at substitutional sites. It should be noted that the channeling yield was higher close to the surface, where the increased carbon concentration had been found. This is an additional indication of the formation of carbon-containing interstitial complexes (including very small silicon carbide nanoparticles) as stated earlier. From an angular scan through the (110) plane, a strain distortion of 0.3° was found for the silicon and the carbon signal, which is in reasonable agreement with the value of 0.26° obtained by X-ray diffraction measurements. After the annealing the channeling yield for carbon had increased to 45%. This shows that at least 20% of the previously substitutional carbon had moved to interstitial lattice positions. In the (110) angular scan measured after annealing no sharp minimum for silicon and carbon could be found.

Raman difference spectra, before and after annealing, are shown in Figure 22. Both spectra show the peak of the localized C mode at 605 cm^{-1}. For the as-grown sample weak shoulders are visible at about 550 cm^{-1} and 623 cm^{-1} that are caused by C–Si–Si–C configurations [74]. These shoulders disappear in the spectrum from the annealed sample. The total integrated intensities in the region from 540 cm^{-1} to 650 cm^{-1} for the annealed sample are reduced by a factor of 0.64 ± 0.05 in comparison with the as-grown sample. The X-ray results yield a reduction in substitutional carbon by a factor of 0.67 due to the annealing. This fairly good agreement indicates that we observe only Raman bands due to substitutional carbon atoms in the region from 540 cm^{-1} to 650 cm^{-1}.

Next we searched for Raman lines due to silicon carbide precipitates in the annealed samples in the spectral region around 787 cm^{-1} and around 965 cm^{-1}. No indication of any such lines could be found. There are two possible causes for this: either such precipitates are not present or their size is so small that the Raman lines are strongly broadened [75] and are not visible. In good agreement with the TEM results, we can conclude from the Raman investigations that SiC precipitates are absent or, more likely, that they have diameters less than 3 nm.

3.4.3. Discussion

In Section 2 we discussed the relation between substitutional and interstitial carbon incorporation during growth. The interstitial carbons are not only highly mobile, but they can complex with many other impurities or defects (doping impurities, oxygen, silicon self-interstitial, or substitutional carbon) to produce deep centers. They are also associated with precipitate formation and nucleation for a variety of crystal defects. Carbon impurities (in the solid bulk solubility range) in silicon have been widely investigated recently (for a review see [13]). The lowest-energy position of carbon in Si is found to be the substitutional site. The observed activation energy for the diffusion of substitutional carbon (3.1 eV) is quite large. Carbon precipitation is a rare process that has been reported to occur only in the presence of supersaturation of silicon self-interstitials (I) or oxygen [16, 76]. Tersoff [46] calculated the energies of different carbon defects in silicon, using an empirical classical potential. Si self-interstitial should react with substitutional C in an exothermic "kick-out" process, forming interstitial C in the (100) split configuration. This interstitial can in turn bind to a second substitutional carbon, relieving stress, in different configurations with similar energies. The strong binding of interstitial C to substitutional C can easily be understood as arising from the relief of stress. Interstitial carbon in the split and in other configurations is under considerable compression, thus reducing the energy by binding to tensile stressed substitutional carbon.

Interaction of substitutional carbons with vacancies is very unlikely because the C dangling-bond energy is larger than that for Si [46], which would imply a repulsive interaction with the vacancy. In a silicon crystal that contains substitutional carbon atoms (C_s) and a severe amount of interstitial oxygen atoms (O_i) (e.g., CVD-grown layers), the following reaction will occur. The self-interstitial atoms will replace the C_s, forming interstitial carbon atoms C_i:

$$I + C_s \rightarrow C_i \tag{3.15}$$

The C_i atoms are mobile and will react with other impurities. One C_i can react with another C_s atom, forming the different dicarbon centers as discussed above. An interstitial oxygen O_i may also capture the C_i atom, forming the carbon-oxygen C center:

$$C_i + O_i \rightarrow CO \tag{3.16}$$

To what extent can the information obtained from silicon with carbon concentrations below the bulk solubility limit be extrapolated to our metastable strained alloys? The oxygen content in the investigated layers was around 10^{18} cm^{-3}, which is orders of magnitude smaller than the C concentration. Therefore, the formation of the carbon-oxygen C centers according to Eq. (3.16) cannot be the dominating mechanism. On the other hand, a primary reaction according to Eq. (3.15) is imaginable. After the anneals in an atmosphere containing oxygen we always find an oxide layer significantly thicker than native oxide. The number of self-interstitials created during the long annealing (and oxidation) could be sufficient for the formation of strain-relieving dicarbon centers. We have no measure for the number of created Si interstitials; values in the literature range over several orders of magnitude. Thus we cannot give a quantitative justification of this mechanism. The disappearance of the weak shoulders in the Raman spectra that can be assigned to substitutional carbon complexes like third-nearest neighbor configurations [74] after the annealing indicates that the formation of interstitial Si–C complexes will preferentially start in regions with higher local C concentrations and/or that the C interstitials are easily captured by another substitutional C atoms to release stress. In regions without third-nearest neighbor configurations the stress-relieving formation is limited by interstitial migration, whereas in the neighborhood of higher C concentration such a process can occur more easily.

To summarize, the absence of extended misfit dislocations after annealing combined with the absence of C out-diffusion leads to the conclusion that the observed strain relief is due to the formation of C-containing centers (like dicarbon centers) or very small and nondetectable silicon carbide precipitates [64]. The carbon atoms in strained $Si_{1-y}C_y$ layers pseudomorphically grown on Si(001) substrates behave similarly to carbon atoms with concentrations below the solid solubility limit in grown Si crystals.

3.5. Strain Relaxation of Ternary Alloys

In [9] we presented an extensive study of the behavior of strained $Si_{1-y}C_y$ layers during postgrowth annealing. From the temperature/time behavior of the changes in lattice spacing (measured *in situ* with an X-ray diffractometer) we were able to extract an activation energy for the strain-relieving mechanism of 3.3 eV. The same activation energy was obtained by Strane et al. [77] on heterostructures formed by solid-phase epitaxy from C implanted, preamorphized substrates with the use of rapid thermal annealing. The activation energy for bulk diffusion of substitutional carbon is 3.1 eV [13]. Warren et al. [65] performed thermal stability investigations on strained $Si_{1-y}C_y$ and $Si_{1-x-y}Ge_xC_y$ layers grown by rapid thermal chemical vapor deposition (RTCVD) on Si(001) substrates. They found a somewhat higher activation energy with the use of rapid thermal postgrowth anneals (RTAs). This higher overall activation energy could be explained by a lower self-interstitial concentration due to the short RTA process in comparison with our

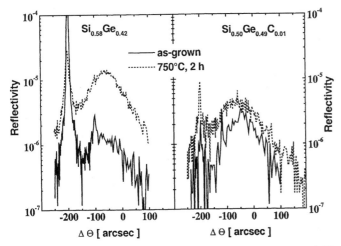

Fig. 27.   TDC measurement of the diffuse X-ray scattering for two buried, 30-nm-thick, strain-equivalent $Si_{1-x}Ge_x$ and $Si_{1-x-y}Ge_xC_y$ layers, respectively, in the as-grown state and after an anneal.

conventional long-time annealing under an oxidizing atmosphere. It is not clear whether the Ge in $Si_{1-x-y}Ge_xC_y$ layers has an additional influence on the activation energy. In any case, a carbon/germanium precipitation should be negligible.

In [78], the authors presented temperature stability measurements of MBE-grown $Si_{1-x-y}Ge_xC_y$/Si and $Si_{1-y}C_y$/Si heterostructures. The ternary $Si_{1-x-y}Ge_xC_y$ alloy with small $x \sim 0.1$ was found to have a higher activation energy for substitutional C loss than the binary $Si_{1-y}C_y$ alloy. The activation energy achieves a maximum (4.9 eV) in the strain-compensated layers and rapidly decreases as the Ge fraction is further increased. Thus, a ternary $Si_{1-x-y}Ge_xC_y$ strain-compensated layer grown on a Si substrate might hold the greatest potential for device processing at elevated temperatures for Si-based heterostructure technology.

To illustrate the stabilizing effect of C, we compare in Figure 27 triple-crystal measurements of diffuse X-ray scattering for two 30-nm-thick, buried $Si_{1-x}Ge_x$ and $Si_{1-x-y}Ge_xC_y$ layers in the as-grown state and after an anneal (750°C for 2 h). The two layers are strain-equivalent; that is, the C concentration is chosen to exactly compensate for the increase in Ge. In addition, both layers are well above the critical thickness limit—they are highly metastable. For the SiGe case, we find an increase in diffuse scattering of more than an order of magnitude, reflecting the formation of an extended network of dislocations. For the strain-equivalent SiGeC layer, the diffuse scattering intensity increases only slightly, indicating the suppression of plastic relation (formation of misfit dislocation) due to C.

3.6. Relaxed $Si_{1-x}Ge_x$/$Si_{1-x-y}Ge_xC_y$ Buffer Structures with Low Threading Dislocation Density

Heteroepitaxy of semiconductor alloys has opened up new possibilities for band structure engineering and novel devices, including strained-layer structures. However, to exploit these possibilities, one often needs a substrate that is lattice matched to the active layers. The usual approach is to grow a buffer layer of a material with the desired lattice constant, thick enough to relax the strain of mismatch. Because of the large lattice mismatch between Si and Ge, a relaxed SiGe buffer layer can be achieved only at the expense of creating misfit dislocations. A certain fraction of these misfit dislocations results in dislocations threading to the surface of the sample. Recently, the effect of threading dislocations (td's) on electron mobility was investigated [79]. It was found that the low-temperature mobility is sensitive to threading dislocations when their density exceeds 3×10^8 cm^{-2} and decreases by two orders of magnitude when the td density is 1×10^{11} cm^{-2}. The room temperature mobility is reduced under the same conditions by 10% and 50%, respectively. When a fully relaxed $Si_{0.7}Ge_{0.3}$ buffer is grown directly on Si, the resulting density of td's is typically in the range of 1×10^{10} and 1×10^{11} cm^{-2}. To reduce the td density, graded or stepped $Si_{1-x}Ge_x$ buffer layers have been investigated [80, 81]. The aim of the various buffer structures used has been to try to increase the length of the misfit dislocation per threading dislocation and thereby reduce the threading dislocation density. However, there are still several disadvantages in growing these thick buffer layers. First, the td density is still high (1×10^5 to 1×10^8 cm^{-2}). Second, the thickness of up to several microns of the structures means that the integration of mixed devices with the use of strained Si, unstrained Si, and $Si_{1-x}Ge_x$ materials is difficult because of problems with lithography.

Here we show a relaxed, only 1-μm-thick stepped buffer based on $Si_{1-x-y}Ge_xC_y$ with a low threading dislocation density ($<10^5$ cm^{-2}) [82]. A stepped $Si_{1-x}Ge_x$ buffer with the identical thickness and strain profile grown with the same temperature ramp yields a threading dislocation density above 10^7 cm^{-2}. This indicates that the addition of carbon is a promising method of discovering new relaxed buffers with low threading dislocation densities and flat surfaces. This buffer concept is based on the fact that the addition of carbon to a SiGe layer not only reduces the strain, but also stabilizes the layer. Because of the very strong local strain fields around the individual carbon atoms, dislocation glide requires a higher energy in $Si_{1-x-y}Ge_xC_y$ than in undisturbed, strain-equivalent $Si_{1-x}Ge_x$ on silicon [83, 84]. The ternary $Si_{1-x-y}Ge_xC_y$ system should be considered a new material with its own strain degree and relaxation behavior, rather than a $Si_{1-x}Ge_x$ film with an artificially reduced strain.

The growth experiments were performed on Si(001) wafers with the use of MBE. A typical overall growth rate was 0.1 nm/s. The samples were investigated with X-ray diffraction (XRD) and TEM. X-ray measurements were carried out with a double crystal diffractometer in a parallel (n,-n) setting, with 004 reflections and CuK$_\alpha$ radiation to obtain the lattice constant normal to the surface. To also measure the lattice constant parallel to the surface, an asymmetrical 422 reflection with steep radiation incidence and θ-scanning of the sample at fixed detector position were used (for details see [72]). The combination of the two measurements allows the determination of the strain and the degree of relaxation.

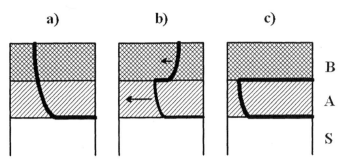

Fig. 28. Different effects due to dislocation gliding in a layered structure of two materials (A and B) with the same lattice mismatch to the underlying substrate (S). (a) The gliding of threading dislocations is the same in both layers. (b) The gliding is retarded in layer B relative to A, leading to the formation of elongated dislocations at the interface between layer A and B, shown schematically in (c) and observed experimentally. Reproduced with permission from [82], copyright 1997, American Institute of Physics.

To investigate the general relaxation behavior in a layered $Si_{1-x}Ge_x/Si_{1-x-y}Ge_xC_y$ system, we first grew the following structure. On a 60-nm-thick Si buffer we deposited (A) a 150-nm-thick $Si_{1-x}Ge_x$ layer (with 20% germanium) followed by (B) a 150-nm-thick $Si_{1-x-y}Ge_xC_y$ layer (with 25% germanium and ~0.5% carbon). The Ge and C concentrations were chosen such that the two layers have the same lattice mismatch to silicon. The thickness and concentration ensure that the individual layers far exceed the critical thickness for perfect pseudomorphic growth; i.e., relaxation should occur during growth. A sample grown at 500°C was investigated by cross-sectional XTEM. Elongated misfit dislocations are visible at the interface between the buffer layer and the $Si_{1-x}Ge_x$ layer. Figure 28 shows the observed mechanism schematically. At a certain point the dislocation is threading toward the surface. We notice that threading dislocation glide is retarded in layer B (shown schematically in Fig. 28b). It stays behind in comparison with layer A, resulting in a dislocation at the interface between layers A and B (Fig. 28c). The two layers are lattice matched. Even so, we find dislocations at their interface in TEM cross-sectional micrographs [82].

We used these results to design a 1-μm-thick, stepwise graded buffer. The basic idea is the following. The individual $Si_{1-x}Ge_x$ layers, grown and relaxed at relatively high temperatures, are covered by $Si_{1-x-y}Ge_xC_y$ with the same average lattice constant. Because of the lower dislocation mobilities in the ternary layers, the relaxation of the underlying $Si_{1-x}Ge_x$ layer is frozen in, also during further growth steps. The td's are also pinned and will generate new misfit dislocations if the strain increases again. In this way, the $Si_{1-x-y}Ge_xC_y$ layers decouple the relaxation in the individual $Si_{1-x}Ge_x$ layers and thereby suppress propagation of dislocation multiplication throughout the whole layered sequence.

Figure 29 shows schematically the complete growth sequence. After the growth of a 50-nm-thick Si buffer layer, we started with the growth of a 150-nm-thick $Si_{1-x}Ge_x$ ($x = 10\%$) layer followed by a 50-nm-thick lattice-matched $Si_{1-x-y}Ge_xC_y$ layer ($x = 15\%$, $y = 0.5\%$). This sequence was

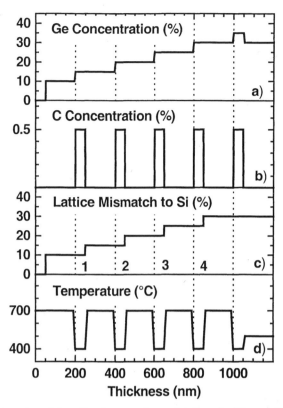

Fig. 29. Ge and C profiles for a stepwise graded $Si_{1-x}Ge_x/Si_{1-x-y}Ge_xC_y$ buffer on Si. The resulting effective lattice mismatch to silicon and the temperature ramping used are also shown. The individual strain steps are marked 1 to 4. Reproduced with permission from [82], copyright 1997, American Institute of Physics.

repeated four times, with the Ge concentration in the $Si_{1-x}Ge_x$ layers increased stepwise to 15%, 20%, 25%, and 30%, respectively. After the last $Si_{1-x-y}Ge_xC_y$ layer ($x = 35\%$, $y = 0.5\%$), again lattice matched to the previous $Si_{1-x}Ge_x$ layer, we finally deposited a 150-nm-thick $Si_{1-x}Ge_x$ layer with $x = 30\%$ at 500°C. To achieve a high degree of relaxation during the growth, the growth temperature of the $Si_{1-x}Ge_x$ layers should be high (we used 700°C). This temperature is outside the window for substitutional carbon incorporation (see Section 2). To incorporate carbon mainly in substitutional positions and therefore to influence the strain and relaxation, we had to ramp the temperature down to 400°C for the $Si_{1-x-y}Ge_xC_y$ layer deposition processes. The growth temperature ramping used is also shown in Figure 29d.

The resulting buffer layer was first investigated by X-ray methods. From the measured symmetric (400) and asymmetric (422) diffraction curves we obtain for the homogeneous $Si_{1-x}Ge_x$ top layer a germanium concentration of $(29.8 \pm 0.5)\%$ and a degree of relaxation of $(73 \pm 5)\%$ [85]. To estimate the density of threading dislocations we analyzed plan-view TEM micrographs. The occurrence of very few td's indicates that their density should be below 10^5 cm^{-2}.

A more detailed picture of the defect behavior can be obtained from cross-sectional TEM (see [82]). Surprisingly, we find significantly fewer dislocations at the last strain step (interface 4) compared with the other steps. At the other $Si_{1-x-y}Ge_xC_y$ layers, we see a preferential formation of misfit dislocations at all interfaces with a stepwise increase in the lattice constant (interfaces 1 to 3), i.e., between $Si_{1-x-y}Ge_xC_y$ and the following $Si_{1-x}Ge_x$. The cap layer is dislocation free in the XTEM micrographs.

To show the effect of carbon addition on the dislocation density, we compare the results to a stepwise graded $Si_{1-x}Ge_x$ buffer without carbon. This buffer was grown such that the strain profile was identical to the $Si_{1-x-y}Ge_xC_y$ layer; that is, the Ge concentration was varied according to the effective lattice mismatch shown in Figure 29c. We also used the same temperature ramp (Fig. 29d). Cross-sectional TEM micrographs show the expected high density of dislocation, including multiplication processes. Even in XTEM we see some dislocations threading to the surface. On the basis of plan-view TEM micrographs we estimate their density to be higher than 10^7 cm^{-2}, which is orders of magnitude higher than that of a $Si_{1-x-y}Ge_xC_y$ buffer that is equivalent in thickness and strain profile and grown with the same temperature ramp.

In summary, we have demonstrated the growth of a relaxed, only 1-μm-thick, stepwise graded buffer, based on a combination of $Si_{1-x}Ge_x$ and $Si_{1-x-y}Ge_xC_y$. This buffer concept relies on retardation of dislocation glide in $Si_{1-x-y}Ge_xC_y$ in comparison with strain-equivalent $Si_{1-x}Ge_x$ on silicon. The homogeneous $Si_{1-x}Ge_x$ layer with $x = 30\%$ on top of the buffer structure is $(73 \pm 5)\%$ relaxed. For the nonoptimized buffer growth we already estimate a threading dislocation density below 10^5 cm^{-2}. A stepped $Si_{1-x}Ge_x$ buffer with identical thickness and strain profile, grown at the same temperature ramping, yields a threading dislocation density above 10^7 cm^{-2}. This indicates that the addition of carbon is a promising method of discovering new relaxed buffers with low threading dislocation densities.

4. ELECTRICAL PROPERTIES OF C-CONTAINING ALLOYS ON Si(001)

4.1. Band Gap Changes and Band Offsets

The current knowledge of the band structure of tensile strained $Si_{1-y}C_y$ on Si(001) is still limited. Even less is known about strained ternary alloys. Assuming an average band structure for $Si_{1-x-y}Ge_xC_y$ alloys, Soref has suggested an empirical interpolation between Si, Ge, and diamond for the band gap [86]. This technique results in an increase in the fundamental gap of $Si_{1-y}C_y$ layers with increasing y. Alternatively, Demkov and Sankey have modeled $Si_{1-y}C_y$ alloys with supercells with random occupation of the lattice sites according to the alloy composition [60]. They found that the fundamental gap is reduced when a small percentage of carbon is added to the silicon lattice. Photoluminescence measurements of tensile strained $Si_{1-y}C_y$ layers yield a reduction in the fundamental gap, with the main offset occurring in the conduction band [87]. Other measurements of the dielectric function and critical points in $Si_{1-y}C_y$ and $Si_{1-x-y}Ge_xC_y$ strained alloys were

made with spectroscopic ellipsometry and electroreflectance [88–90]. From these experimental results, we can state that the virtual crystal approximation is not able to describe the changes in band structure for $Si_{1-y}C_y$ and $Si_{1-x-y}Ge_xC_y$ alloys correctly. Moreover, the carbon effects cannot be reduced to changes in strain only, as proposed earlier by Powell et al. [91]. To describe adequately the observed energy shifts for pseudomorphic carbon-containing layers, we have to consider at least strain-induced effects and effects due to alloying [89].

In this chapter we present an estimation for the band offsets and the fundamental band gap for $Si_{1-x-y}Ge_xC_y$ alloys strained on Si(001) [92, 93]. This estimation considers both bandlineup at the interface of two different materials and strain effects. Unknown material parameters have been adjusted to obtain the best agreement with experimental results for tensile strained, binary $Si_{1-y}C_y$ layers and the compressively strained binary $Si_{1-x}Ge_x$ layers, respectively. We can provide more reliable predictions for the ternary $Si_{1-x-y}Ge_xC_y$ alloy, including the strain-compensated case, from the interpolation between compressive strained $Si_{1-x}Ge_x$ and tensile strained $Si_{1-y}C_y$. The results obtained agree very well with the first experimental data for the effect of C on band structure properties in $Si_{1-x-y}Ge_xC_y$. Investigations of the direct transition and first electrical measurements indicate that the carbon effect on band structure is smaller than on the crystallographic structure [90, 94]. For a completely strain-compensated $Si_{1-x-y}Ge_xC_y$ layer (which should be cubic) we predict significant "Ge effects" (smaller gap than Si, valence band offset to Si), depending on the Ge content.

4.1.1. The Model

When two semiconductors are joined at a heterojunction, discontinuities can occur in the valence bands and in the conduction bands. Without strain (that is, for a lattice-matched interface), the bandlineup problems simply consist of determining how the band structures of the two materials line up at the interface. When the materials are strained, we have to consider

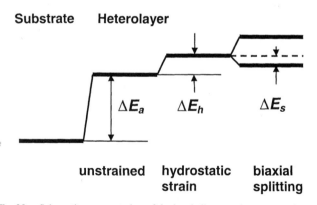

Fig. 30. Schematic representation of the band alignment between a substrate and a strained heterolayer. The three contributions shown are the "material effect" for the unstrained case ΔE_a, the shift due to hydrostatic strain ΔE_h, and the splitting of a degenerated band due to biaxial strain ΔE_S. Reproduced with permission from [92], copyright 1998, American Institute of Physics.

two additional effects on the band structure: hydrostatic strain will produce additional shifts, and uniaxial or biaxial strain splits degenerate bands. Figure 30 illustrates schematically the different contributions. Thus, the total change in a band can be expressed as

$$\Delta E = \Delta E_a + \Delta E_h + \Delta E_s \qquad (4.1)$$

where ΔE_a stands for the material differences (alloy effect for the unstrained case), ΔE_h is the shift due to hydrostatic strain, and ΔE_s is the splitting due to biaxial strain.

A similar approach has been used by Stein et al. [95] for the evaluation of experimental results for the conduction band offset. It should be noted that the individual contributions can have different signs, i.e., leading to partial compensation.

Considering ternary $Si_{1-x-y}Ge_xC_y$ on Si(001) interfaces, all terms in Eq. (4.1) depend on both x and y. The strain-induced contribution can be reduced to a dependency on the effective strain ε, i.e.,

$$\Delta E_{h,s}(x, y) = \Delta E_{h,s}(\varepsilon) \qquad (4.2)$$

where $\varepsilon = \varepsilon(x, y)$.

Perfect pseudomorphic growth leads to strain along the film plane (along both the x and y directions) of

$$\varepsilon_{||} = a_{Si}/a(x, y) - 1 \qquad (4.3)$$

The average lattice constant $a(x, y)$ was defined earlier in Eq. (3.2). With continuum elasticity theory, the perpendicular strain component (along z)

$$\varepsilon_\perp = -2(C_{12}/C_{11})\varepsilon_{||} \qquad (4.4)$$

can be calculated from the elastic constants.

Hydrostatic strain, corresponding to the fractional volume change, is given by the trace of the strain tensor, $\Delta V/V = \varepsilon_\perp + 2\varepsilon_{||}$. Thus, the hydrostatic contribution to the band shifts is given by

$$\Delta E_h = a_{v,c}(\varepsilon_\perp + 2\varepsilon_{||}) \qquad (4.5)$$

where $a_{v,c}$ stands for the appropriate hydrostatic deformation potential for the valence or conduction band, respectively.

All needed elasticity constants and deformation potentials where obtained in the following way. Because the values for Si and Ge are usually quite similar to begin with, a linear interpolation procedure is expected to give good results for different Ge concentrations. To get a parameter $p(x)$ for $Si_{1-x}Ge_x$ we use

$$p(x) = (1 - x)p_{Si} + xp_{Ge} \qquad (4.6)$$

The values used for Si and Ge are summarized in Table I. However, interpolating between silicon and diamond (or silicon carbide) does not seem to be the right way to get the dependency on y, because our investigations are restricted to only a few percent of carbon in silicon. A better assumption for this case is to use the silicon values, i.e., to neglect any influence of the low carbon concentration on deformation potentials and elasticity constants.

Table I. Material Constants for Si and Ge Used as Boundary Values for Interpolation to Obtain the Constants of Mixed Alloys. Reproduced with Permission from Springer-Verlag-Copyright 1982. Reproduced with permission from the American Physical Society Copyright 1989 and 1986

Parameter	Si	Ge	
Spin–orbit splitting Δ (eV)	0.04	0.30	(a)
Elasticity constants (GPa)			(a)
C_{11}	165.8	128.5	
C_{12}	63.9	48.3	
Hydrostatic deformation			(b)
potential (eV) for			(b)
Valence band a_v	2.46	1.24	
Conduction band a_c	4.18	2.55	
Biaxial deformation			
potential (eV) for			
Valence band b	−2.35	−2.55	(b)
Conduction band Ξ	9.16	9.42	(c)

(a) Experimentally obtained [96].

(b) Theoretically obtained [97].

(c) Theoretically obtained [98].

For the material-dependent term ΔE_a we have to assume additivity, i.e.,

$$\Delta E_a(x, y) = \Delta E_a(x) + \Delta E_a(y) \qquad (4.7)$$

where $\Delta E_a(x)$ and $\Delta E_a(y)$ can be extracted from the binary $Si_{1-x}Ge_x$/Si(001) and $Si_{1-y}C_y$/Si(001) cases, respectively.

Based on these general considerations, we are now ready to treat the valence and conduction band offsets separately. Finally, the changes in the fundamental band gap result as the sum of both.

4.1.2. Valence Band

First, we describe how the individual terms in Eq. (4.1) were obtained to estimate the valence band offsets at the $Si_{1-x-y}Ge_xC_y$/Si(001) interface. For simplicity we use the same notation as in the previous chapter, now always meaning shifts in the valence band.

The most direct way to predict values for $\Delta E_a(x)$ between Si and Ge is provided by the "model solid theory" [97]. This approach treats the band offsets as linear quantities, which can be obtained as differences between reference values that have been calculated for each semiconductor. Van der Walle obtained $\Delta E_a(1) = 0.68$ eV between Si and Ge [97]. Even in unstrained materials, spin-orbit interaction lifts the threefold degeneracy of the topmost valence band at the zone center in tetrahedral semiconductors: two bands are shifted up by $\Delta/3$, and one band is shifted down by $2\Delta/3$. This spin-orbit splitting value is quite different in Si (0.04 eV) and Ge (0.30 eV) [96]. Considering spin-orbit splitting, we obtain $\Delta E_a(1) = 0.77$ eV for the unstrained Si/Ge interface. Morar et al. [99] have derived an unstrained valence band offset between pure Si and Ge, based on spatially resolved electron-energy-loss spectroscopy in SiGe alloys. They found $\Delta E_a(1) = 0.78$ eV.

The hydrostatic contribution can easily be calculated with Eq. (4.5) and the appropriate values, interpolated from the data given in Table I:

$$\Delta E_h(x, y) = 2a_v(x)\big[a_{Si}/a(x, y) - 1\big]\big[1 - C_{12}(x)/C_{11}(x)\big]$$
(4.8)

To describe the splitting of the valence band due to perpendicular strain along [001] we followed [100]:

$$\Delta E_{s,1} = -\Delta(x)/6 + \delta E/4 + 0.5\big[\Delta^2 + \Delta\delta E + 9/4(\delta E)^2\big]^{1/2}$$
(4.9)

$$\Delta E_{s,2} = \Delta(x)/3 - \delta E/2$$
(4.10)

$$\Delta E_{s,3} = -\Delta(x)/6 + \delta E/4 - 0.5\big[\Delta^2 + \Delta\delta E + 9/4(\delta E)^2\big]^{1/2}$$
(4.11)

where δE is given by

$$\delta E = 2b(x)(\varepsilon_\perp - \varepsilon_\|)$$
(4.12)

and b is the potential for biaxial deformation (given in Table I). Note that these three values add up to zero; they only express the strain-induced splitting of the band with respect to the average value and do not introduce any shifts of the average.

The most crucial point for the calculation of the valence band offsets between Si and strained $Si_{1-x-y}Ge_xC_y$ is the estimation of $\Delta E_a(y)$, i.e., the alloy effect on the valence band due to the presence of carbon. No data for that contribution could be found in the literature. However, Kim and Osten [101] reported on the absence of a significant valence band offset caused by carbon incorporation in tensile strained $Si_{1-y}C_y$. These authors showed that carefully performed *in situ* XPS measurements allow the evaluation of valence band offsets in strained heteroepitaxial systems on Si(001). The results obtained for an $Si_{0.75}Ge_{0.25}$ alloy layer agrees very well with the known values, indicating the reliability of the method used. For an $Si_{0.977}C_{0.023}$ alloy layer tensile strained on Si(001), they could not find any significant valence band offset to Si. The strain-induced contributions $\Delta E_h + \Delta E_s$ alone lead to an offset of 31.4 meV for 1% carbon, whereas the dependency on y is nearly linear. This would result in a strain induced offset of 72 meV for the investigated layer with $y = 2.3\%$, that is, a value much larger than the uncertainty in the measurements. It seems to be logical to compensate for the strain-induced contribution to zero by assuming

$$\Delta E_a(y) \approx -3.14y \text{ eV}$$
(4.13)

Experience has shown that carbon concentrations above 3% start to destroy the epitaxial relation between the layer and the substrate. Therefore, we limit all calculations to carbon concentration between 0 and 3%. The Ge concentration considered did not exceed 30%.

Figures 31 and 32 summarize the results for the valence band offsets of strained $Si_{1-x-y}Ge_xC_y$ on Si(001). In Figure 31 we

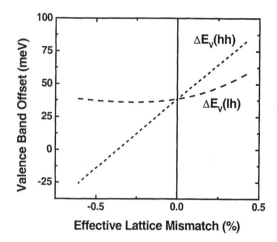

Fig. 31. Valence band offsets for a $Si_{1-x-y}Ge_xC_y$ layer on Si(001) with 10% germanium as a function of effective lattice mismatch. The C concentration y was varied between 0% (start point of the curves at the right side) and 3% (end point on the left side). Both the heavy hole band (highest band for compressive strained materials) and the light hole band (higher for tensile strained materials) are shown. Reproduced with permission from [92], copyright 1998, American Institute of Physics.

plotted ΔE_v versus the effective lattice mismatch for a layer with 10% germanium. The carbon concentration was varied between 0 and 3%, changing the strain from compressive to tensile. Several trends become visible. For compressively strained layers the highest valence band is the heavy hole (hh) band caused by the splitting $\Delta E_{s,2}$ according to Eq. (4.10). A light hole (lh) band ($\Delta E_{s,1}$, Eq. (4.9)) forms the highest valence band for tensile strained materials. The hh band decreases linearly with increasing C concentration (for a given Ge concentration). Opposite to that, the dependence of the lh band on C concentration is very weak, which means that ΔE_v is mainly dominated by the Ge content for tensile strained layers. Figure 32 summarizes results for only the highest valence band for different tensile and compressive strained $Si_{1-x-y}Ge_xC_y$ layers. This plot shows ΔE_v as a function of the effective Ge or C concentration for the compressive or tensile strained layers, respectively. The effective concentration corresponds to the concentration needed for identically strained binary layers.

The valence band offset between compressive strained layers and Si is generally much larger than that at the tensile strained layer/Si interface. Even so, we predict a measurable valence band offset for tensile strained $Si_{1-x-y}Ge_xC_y$ layers with high enough Ge concentration. For applications in hole confinement devices the compressively strained layers are of greater interest. For comparison we also added the known $\Delta E_v(x)$ dependence for strained $Si_{1-x}Ge_x$ layers. The curves for all ternary alloys are above that for the binary, showing that the decrease in the valence band offset caused by carbon addition is much smaller than that caused by an identical strain reduction due to lower Ge concentrations. Moreover, fully strain-compensated layers should still have a valence band offset to Si, whereas its value depends on the Ge and the C concentration needed for strain compensation. First experimental values for the valence band offset were extracted from capacitance–voltage characteristics

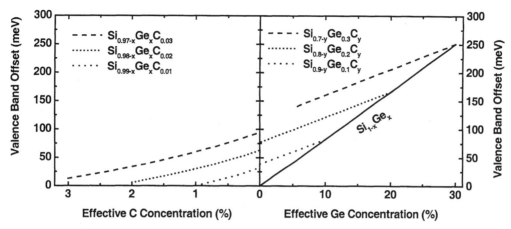

Fig. 32. Valence band offsets for compressive strained $Si_{1-x}Ge_x$ and $Si_{1-x-y}Ge_xC_y$ (with $x = 10\%$, 20%, 30%, whereas y varied between 0% and 3%) and tensile strained $Si_{1-y}C_y$ and $Si_{1-x-y}Ge_xC_y$ ($y = 1\%$, 2%, and 3%, x varied between 0% and 30%) plotted as a function of the effective lattice mismatch (expressed in "effective" Ge or C concentrations, respectively). Reproduced with permission from [92], copyright 1998, American Institute of Physics.

of p-Si/$Si_{1-x-y}Ge_xC_y$ MOS capacitors by Rim et al. [94]. For a compressive strained layer with 18% Ge and 0.8% C they found $\Delta E_V = (120\pm25)$ meV. Our model predicts for this concentration $\Delta E_V = 119.8$ meV. Using admittance spectroscopy, Stein et al. [95] obtained $\Delta E_V = (223\pm20)$ meV and (118 ± 12) meV for layers with $x = 39.4$, $y = 1.1$, and $x = 20.6$, $y = 0.4$, respectively. We calculate 291 meV and 156 meV for these layers.

4.1.3. Conduction Band

Now we investigate the effect on the conduction band. Analogously to the previous part, we will use the same notation, now always meaning shifts in the conduction band. Strain effects on the conduction band are similar to those discussed for the valence band in the previous chapter. Hydrostatic strain shifts the overall energetic position of the band according to Eq. (4.8), whereas the hydrostatic deformation potential has to be replaced by that for the conduction band $a_c(x)$. The degeneracy of the conduction bands in Si and Ge is different in nature from that of the valence bands. In Si, the conduction-band minima are found along the $\langle 100 \rangle$ directions (called Δ points) in reciprocal space; this implies that there are six minima, which occur at the same energy in unstrained material. Application of strain along [001] will affect the minimum oriented along [001] differently from those oriented along [100] and [010]. These shifts can be expressed as [102]

$$\Delta E_{s,1}(x, y) = 2/3 \Xi(x)(\varepsilon_\perp - \varepsilon_\parallel) \qquad (4.14)$$

$$\Delta E_{s,2}(x, y) = -1/3 \Xi(x)(\varepsilon_\perp - \varepsilon_\parallel) \qquad (4.15)$$

$\Xi(x)$ is the deformation potential for biaxial strain. $\Delta E_{s,1}$ that is along [001] is twofold degenerated (often called $\Delta(2)$ minimum), and $\Delta E_{s,2}$ (called $\Delta(4)$) is fourfold degenerated along [100] and [010]. Again, all six contributions add up to zero, and ΔE_s describes only the strain-induced splitting relative to the nonshifted average value. One additional problem arises from the fact that the conduction band minimum for germanium is on a different position in the reciprocal space (at the zero boundary

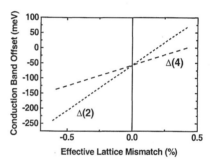

Fig. 33. Conduction band offsets for a $Si_{1-x-y}Ge_xC_y$ layer on Si(001) with 10% germanium as a function of effective lattice mismatch. The C concentration y was varied between 0% (start point of the curves at the right side) and 3% (end point on the left side). Both the $\Delta(4)$ band (lowest band for compressive strained materials) and the $\Delta(2)$ band (higher for tensile strained materials) are shown. Reproduced with permission from [92], copyright 1998, American Institute of Physics.

along the $\langle 111 \rangle$ directions, called the L point). Interpolations for the deformation potentials according to Eq. (4.6) have to be performed between the Δ points for Si and Ge; i.e., experimentally obtained values for Ge cannot be applied. To describe Ge-rich $Si_{1-x}Ge_x$ alloys ($x > 0.84$) we would have to interpolate between the L points for Si and Ge. In Table I, we included only the appropriate values for the Δ points.

Finally, we have to find the expression for the conduction band shifts at the unstrained interface $\Delta E_a(x)$ and $\Delta E_a(y)$. It is known that the conduction band offset in strained $Si_{1-x}Ge_x$ is very small, i.e.,

$$\Delta E_a(x) + \Delta E_h(x, 0) + \Delta E_s(x, 0) \approx 0 \qquad (4.16)$$

From the known relations for $\Delta E_h(x, 0)$ and $\Delta E_s(x, 0)$, we can estimate $\Delta E_a(x) \approx 0.44x$ [eV]. From their photoluminescence measurements on strained $Si_{1-y}C_y$ alloys, Brunner et al. [87] found that the band gap is decreasing roughly linearly with $\Delta E_g(y) \approx -6.8y$. As we already discussed, this change can be

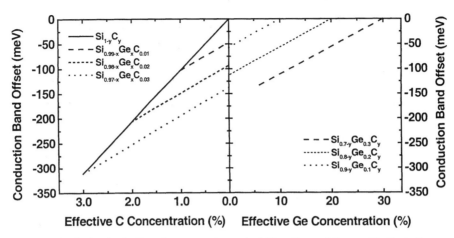

Fig. 34. Conduction band offsets for compressive strained $Si_{1-x}Ge_x$ and $Si_{1-x-y}Ge_xC_y$ (with $x = 10\%$, 20%, and 30%, y varied between 0% and 3%) and tensile strained $Si_{1-y}C_y$ and $Si_{1-x-y}Ge_xC_y$ ($y = 1\%$, 2%, and 3%, x varied between 0% and 30%) plotted as a function of the effective lattice mismatch (expressed in "effective" Ge or C concentrations, respectively). Reproduced with permission from [92], copyright 1998, American Institute of Physics.

mainly attributed to a change in the conduction bands, i.e.,

$$\Delta E_a(y) + \Delta E_h(0, y) + \Delta E_s(0, y) \approx -6.8y \text{ [eV]} \quad (4.17)$$

Calculating the strain-induced contributions $\Delta E_h(0, y)$ and $\Delta E_s(0, y)$ allows us to get an estimate for $\Delta E_a(y) \approx -4.85y$ [eV].

Figures 33 and 34 summarize the results for the conduction band offsets of strained $Si_{1-x-y}Ge_xC_y$ on Si(001). In Figure 33 we plotted ΔE_C versus the effective lattice mismatch for a layer with 10% germanium. Again, the carbon concentration was varied between 0 and 3%, changing the strain from compressive to tensile. Several trends become visible: for compressively strained layers the lowest conduction band is the fourfold degenerated $\Delta(4)$ band ($\Delta E_{s,2}(x, y)$ according to Eq. (4.15)). Opposite, the twofold degenerated $\Delta(2)$ band ($\Delta E_{s,2}(x, y)$ (Eq. (4.14)) forms the lowest conduction band for tensile strained materials. Both bands are linear functions of the lattice mismatch. Figure 34 summarizes results for only the lowest conduction band for different tensile and compressive strained $Si_{1-x-y}Ge_xC_y$ layers.

The conduction band offset between tensile strained layers and Si is generally larger than that at the compressive strained layer/Si interface, making tensile strained materials good candidates for electron-confined devices. Similar to the behavior of the valence band offsets, we also find that the decrease in the conduction band offset caused by germanium addition to $Si_{1-y}C_y$ alloys is much smaller than that caused by an identical strain reduction due to lower C concentrations. Furthermore, fully strain-compensated layers should still have a valence band offset to Si, whereas its value depends on the Ge and the C concentration needed for strain compensation.

4.1.4. Band Gap for the Strain-Compensated Alloy

Finally, we obtain the band gap narrowing by adding up both the valence and the conduction band offset. Figure 35 shows for

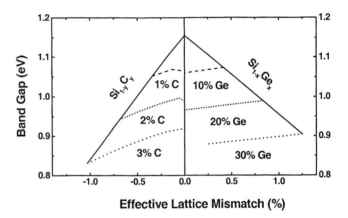

Fig. 35. Band gap narrowing for ternary $Si_{1-x-y}Ge_xC_y$ alloys, strained on Si(001) as a function of lattice mismatch. Both binary alloys (tensile strained $Si_{1-y}C_y$ and compressive strained $Si_{1-x}Ge_x$) are shown as references. Reproduced with permission from [92], copyright 1998, American Institute of Physics.

ternary $Si_{1-x-y}Ge_xC_y$ alloys strained on Si(001) the following trends: (i) The band gap for the ternary alloys is always smaller than that of silicon, regardless of strain state. (ii) The addition of C (Ge) to compressive strained $Si_{1-x}Ge_x$ (tensile strained $Si_{1-y}C_y$) leads to a smaller change in band gap narrowing than an equivalent strain reduction in the binary alloy (lower Ge or C content, respectively).

A case of particular interest is the fully strain-compensated $Si_{1-x-y}Ge_xC_y$ layer. For such a layer all strain-induced contributions to the band offsets vanish in Eq. (4.1); i.e., the hydrostatic term as well as the splitting is zero. That leads to the expression

$$\Delta E_{v,0}(x, y) = 0.68x - 3.14y + 1/3[\Delta(x) - \Delta(0)] \quad (4.18)$$

for the valence band offset, where the subscript 0 denotes the "zero strain" case, i.e., the fully strain-compensated layer.

The conduction band offset can be expressed as

$$\Delta E_{c,0}(x, y) = 0.44x - 4.85y \qquad (4.19)$$

If we define a strain compensation ratio $R = x/y$ (the value depends on the approximation used to obtain $a(x, y)$ and is 8.2 according to Eq. (3.2)) we get the following estimation for the band gap narrowing as a function of germanium concentration:

$$\Delta E_{g,0}(x) = x(0.33 + 1.71/R) \text{ [eV]} \qquad (4.20)$$

The band gap of a fully strain-compensated $Si_{1-x-y}Ge_xC_y$ is a linear function of Ge (or C) concentration. Those layers are strain free; i.e., there are no thickness limitations due to plastic relaxation. This opens a new way for band gap engineering of a material on Si(001) that is macroscopically strain free.

All band gap and alignment estimations made above were based only on substitutional carbon incorporation. However, the interstitial to substitutional carbon ratio is strongly influenced by the growth conditions as discussed earlier (Section 2).

4.2. Charge Transport

In this section, we present experimental results on Hall mobilities of electrons in tensile strained $Si_{1-y}C_y$ and holes in compressively strained $Si_{1-x-y}Ge_xC_y$ layers, respectively. For both cases, the measured charge carrier densities at room temperature are not affected substantially by the addition of a small concentration of carbon ($<1\%$), under identical growth conditions and with dopant fluxes. The measured charge carrier mobilities monotonically decrease with increasing carbon content, indicating the dominance of scattering on ionized centers [103].

N-type 200-nm-thick $Si_{1-y}C_y$ layers with different C contents ($0 \leq y \leq 0.85\%$) were grown on p-type substrates (50 Ω cm) by MBE at 400°C. The nominal Sb doping of about 1×10^{17} cm^{-3} was kept constant for all growth processes. Similarly, p-type 200-nm-thick $Si_{1-x-y}Ge_xC_y$ layers with 18% germanium and different C contents ($0 \leq y \leq 0.67\%$) were grown on n-type substrates. The nominal boron doping was kept constant for all layers at about 1×10^{17} cm^{-3}. Temperature dependencies of the resistivity ρ and of the Hall coefficient R_H were measured from 300 K down to 20 K. The Hall mobility $\mu_H = R_H/\rho$ and the charge carrier density $p, n = r_H/(eR_H)$ with the Hall factor r_H can be found from those parameters. The appropriate Hall factor is $r_H = 1$ for highly doped n-type Si and $r_H = 0.8$ for p-type $Si_{1-x}Ge_x$ [103].

The charge carrier mobility μ can be expressed as a function of the transport effective mass m^* and the transport scattering time τ_τ:

$$\mu = e/m^*\tau_\tau \qquad (4.21)$$

The scattering time τ_τ and thus μ are material parameters that represent all scattering mechanisms as carrier experiences. According to Mattiessen's rule, $1/\tau_\tau$ is the sum of all reciprocal scattering times associated with the respective scattering mechanism:

$$1/\tau_\tau = \Sigma 1/\tau_i \qquad (4.22)$$

Hence the mobility in a semiconductor is dominated by the scattering mechanism with the smallest time constant.

In the following we discuss the extent to which the addition of carbon affects the charge carrier mobility in strained $Si_{1-y}C_y$ and $Si_{1-x-y}Ge_xC_y$ layers on Si(001). Theoretical studies and Raman measurements show that the strain distribution in pseudomorphic layers containing C is highly nonuniform [104]. This could smear the band edges, cause band tails, and reduce the mobility of carriers. For tensile strained $Si_{1-y}C_y$, the band gap narrowing results mainly from the conduction band, making this material suitable for electron-confined devices [87]. Therefore, we concentrate on electron mobility in this material. In contrast, compressively strained layers of $Si_{1-x}Ge_x$ have a larger valence band offset to silicon, allowing hole confinement. For the ternary, compressively strained $Si_{1-x-y}Ge_xC_y$ materials, only the hole mobility is of interest [105].

4.2.1. Electron Transport in Tensile Strained $Si_{1-y}C_y$ Layers

Electron transport characteristics have been studied theoretically with Monte Carlo simulations [106]. The authors found that the dependence of mobility on C content depends critically on the (unknown) value for the alloy scattering potential. Varying that potential between 0 and 2.2 eV, they could deduce some general trends. The energy splitting in the strained alloys transfers electrons between nonequivalent valleys, from the upper to the lower valleys that have small effective mass in the plane. In addition, energy splitting suppresses the intervalley scattering between twofold and fourfold degenerate valleys. Both effects are responsible for a theoretical increase of in-plane electron mobility for low C contents in comparison with silicon. In contrast, increasing the carbon concentration leads to an enhancement of alloy scattering that decreases the charge carrier mobility. For small alloy scattering potentials the authors found the following behavior for the electron mobility as a function of C. With increasing carbon content the mobility first increases up to a maximum of approximately twice the value for silicon. A further increase in C content leads to a decrease in the mobility, which can be lower than the mobility in silicon, depending on the assumed scattering potential. Assuming a larger scattering potential, Monte Carlo simulations yield a monotonic decrease in electron mobility with increasing carbon content due to the dominance of alloy scattering over the strain-induced mobility gain. Brunner et al. reported that the mobility enhancement induced by strain is larger than the degradation caused by alloy scattering, leading to electron mobility values above the appropriate Si values, especially at low temperatures [107].

For device applications, higher charge carrier densities are usually required, when impurity scattering is the dominant mechanism and alloy scattering plays a minor role [108]. Therefore, we concentrate our investigations on C-containing layers with a charge carrier density of about 10^{17} cm^{-3}. Temperature dependencies of the measured electron densities for various C-containing $Si_{1-y}C_y$ samples are shown in Figure 36. They are

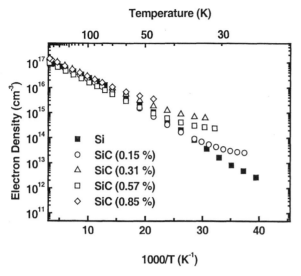

Fig. 36. Electron density vs. $1000/T$ in 200-nm-thick, Sb-doped layers of both Si and $Si_{1-y}C_y$, pseudomorphically strained on Si(001). Reproduced with permission from [103], copyright 1997, American Institute of Physics.

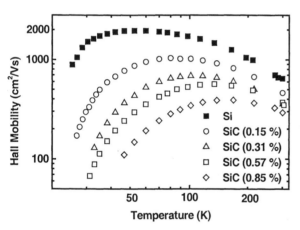

Fig. 37. Hall mobility of electrons as a function of temperature for Sb-doped Si and strained $Si_{1-y}C_y$ layers on Si(001). Reproduced with permission from [103], copyright 1997, American Institute of Physics.

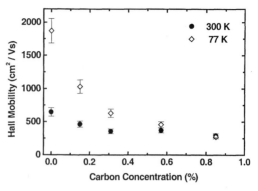

Fig. 38. Experimentally obtained electron mobilities at 300 K and 77 K as a function of carbon content in homogeneously Sb-doped $Si_{1-y}C_y$ layers on Si(001). Reproduced with permission from [103], copyright 1997, American Institute of Physics.

compared with the MBE-grown silicon reference layer (solid squares). At room temperature, the electron concentrations of all samples are the same within a factor of 2. With decreasing temperatures, the electron densities deviate nonsystematically with increasing C content. We have no explanation for this unusual behavior. The donor activation energy of Sb in the Si reference layer has been determined to be (27.2 ± 0.4) meV, corresponding well with the value for Sb in bulk silicon at the same donor level (26.4 meV according to [109]). The activation energy found from $n(T)$ analysis decreases with increasing C content in the $Si_{1-y}C_y$ layers and amounts to about 10 meV in case of the sample with the highest C content (0.85%, shown as open diamonds in Fig. 36).

Temperature dependencies of Hall mobilities of these samples are illustrated in Figure 37. The electron mobility in the pure Si layer at 300 K is about 650 cm²/Vs, close to the known value (around 700 cm²/Vs) for bulk Si [110] doped with Sb to

the same level. Within the investigated temperature range, the electron mobilities of $Si_{1-y}C_y$ alloys are always smaller than that of the Si reference sample.

In contrast to the results of Brunner et al. [107], these results on electron transport in $Si_{1-y}C_y$ alloys indicate that the effect of alloy, ionized, and neutral impurity scattering dominates over the expected gain due to strain. Adding C to Si reduces the electron mobility (more pronounced at lower temperatures; see Fig. 38).

In an earlier section we discussed the relation between substitutional and interstitial carbon incorporation. For the growth conditions chosen here, we can expect between 1 and 5×10^{18} cm^{-3} carbon atoms in interstitial positions (depending on the C content). C interstitials in silicon exist in neutral but also positive or negative charge states for concentrations below the solid solubility limit. Configurations of interstitial C-containing complexes in Si for high C content are unknown. Neutral impurity scattering has no or only a weak temperature dependence that decreases even with decreasing temperature [108]. It cannot describe the experimentally observed $\mu(T)$ dependencies. The decrease in the ionization energy, the results of the analysis of the $n(T)$ dependencies, and the observed $\mu(T)$ dependency itself point to the formation of both additional donors and compensating acceptors. The number of impurity scattering centers is characterized by the sum $N(D)^+ + N(A)^-$ of ionized donors and acceptors. Thus, the increase in the density of ionized impurity centers with increasing C content is the main reason for the observed temperature dependence of mobility at low temperatures. These centers could be formed in connection with the additional introduction of interstitial carbon or could be emitted from the used, extremely hot carbon evaporation source.

To evaluate the influence of interstitial carbon-containing defects on charge carrier transport in more detail, we now present experimental results on Hall mobilities of electrons in tensile strained $Si_{1-y}C_y$ layers with the same amount of substitutional carbon ($y_S = 0.4\%$ in all investigated samples) but a different concentration of interstitial carbon defects [111]. Based on modeling of the carbon incorporation as a function of growth conditions, we chose different growth temperatures and overall

Table II. Overview of the Investigated Samples Prepared for Mobility Measurements

	Growth condition		Target values		Measured values		
	T (°C)	R_{Si} (nm/s)	C_{subst} (%)	C_{inter} ($\times 10^{19}$ cm^{-3})	C_{subst} [a] (%)	C_{total} [b] ($\times 10^{20}$ cm^{-3})	Mobility [c] (cm^2/Vs)
A	519	0.078	0.4	1	0.41 ± 0.05	2.1 ± 0.3	854/2016
B	557	0.066	0.4	5	0.37 ± 0.05	2.6 ± 0.3	547/1221
C	562	0.041	0.4	10	0.41 ± 0.05	3.2 ± 0.4	Not measurable
Si	400	0.082					953/2859

[a] Obtained from X-ray diffraction.

[b] Measured with SIMS.

[c] Measured at room temperature/77 K.

growth rates to tune the interstitial carbon concentration between 1×10^{19} and 2×10^{20} cm^{-3}. Table II summarizes all investigated samples. All layers were identically doped with phosphorus. The lattice distortion due to misfit strain and therewith the band alignment is identical for all samples; therefore any contributions arising from alloy scattering should also be identical. However, we found differences in electron mobility of nearly a factor pf 2, correlating with the different concentration of interstitial carbon containing scattering centers. For the highest interstitial C concentration we could not even measure any reliable electrical data.

For comparison, we also include the results obtained from an identically grown and processed silicon layer.

The room temperature value of μ for the Si$_{1-y}$C$_y$ sample with the lowest amount of interstitial carbon differs by only about 10% from the Si value, whereas at low temperatures the mobilities of the Si$_{1-y}$C$_y$ sample are distinctly smaller than the ones for the pure Si sample. The expected gain in mobility due to the tetragonal distortion of the lattice is already more than compensated for by scattering effects due to the presence of carbon atoms. This tendency becomes even more pronounced for higher interstitial concentrations. For $y_I = 1 \times 10^{19}$ cm^{-3} (sample A) we obtain a weighted average over the profile $\langle \mu \rangle = 854$ cm^2/Vs at room temperature. When the interstitial carbon concentration increases to 5×10^{19} cm^{-3} (sample B) the room temperature mobility decreases to 547 cm^2/Vs. For a layer with an interstitial C concentration of 1×10^{20} cm^{-3} (sample C) or more, it was not even possible to obtain any reliable electrical measurements. The C interstitials themselves should not cause the observed decrease in mobility. The μ temperature dependence at low temperatures hints at a $T^{3/2}$ proportionality that is characteristic of scattering ions. Assuming that donor concentrations and concentrations of substitutional carbon are the same in all of the Si$_{1-y}$C$_y$ samples, the found $\mu(T)$ dependency leads to the conclusion that the increase in interstitial C concentration is accompanied by an increase in the concentration of compensating acceptors. These acceptors are probably electrically active C-containing clusters. Earlier in this section we reported on a systematic increase in low-temperature electron mobility with increasing substitutional C content. Now

we show that even for identically strained Si$_{1-y}$C$_y$ layers the mobility is decreasing with increasing concentration of interstitial carbon atoms. There is no contradiction between these two findings. As shown in Section 2, an increase in substitutional C content is always accompanied by an increase in interstitial carbon concentration (keeping the Si growth rate and growth temperature constant).

These results show that correlating transport properties with strain only (for example, by measuring lattice distortion with X-ray methods and extracting the substitutional C concentration) is not sufficient for a description of this material. Specific growth conditions can lead to very different electrical properties due to the different amounts of interstitial C atoms, even for pseudomorphically strained layers with the same lattice mismatch and band alignment.

4.2.2. Hole Transport in Compressively Strained Si$_{1-x-y}$Ge$_x$C$_y$ Layers

As in elemental semiconductors, the scattering mechanism for holes in Si$_{1-x}$Ge$_x$ alloys is more complex than for electrons, because of the different types of interacting holes [112]. In weakly doped layers, the lowering of the effective hole mass, concomitant with the lifting of the heavy-hole/light-hole degeneracy, exerts the major influence on mobility [113]. These effects become more pronounced in strained Si$_{1-x}$Ge$_x$ layers pseudomorphically grown on a Si(001) substrate.

First experimental results were presented by Levitas [114] and by von Busch and Vogt [115] in unstrained, undoped bulk SiGe alloys. The hole mobility is reduced considerably in comparison with silicon because of alloy scattering. More recently, Manku and Nathan [116] calculated hole mobilities in unstrained and strained alloys from a first-order perturbation solution of the Boltzmann transport equation. Scattering by nonpolar optical and acoustic phonons, as well as alloy scattering, was included in the calculations. These calculations showed that the differences in mobility between strained and unstrained Si$_{1-x}$Ge$_x$ materials are very small for low (<25%) Ge content. The reduction in strain due to the addition of less than 1% carbon should only have a marginal effect on the trans-

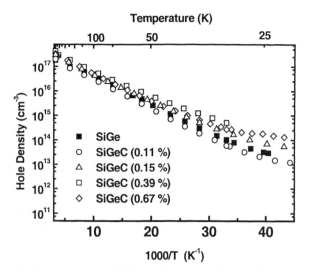

Fig. 39. Hole density vs. $1000/T$ in 200-nm-thick, B-doped $Si_{1-x}Ge_x$ and $Si_{1-x-y}Ge_xC_y$ layers with 18% germanium pseudomorphically strained on Si(001). Reproduced with permission from [103], copyright 1997, American Institute of Physics.

port behavior in strained $Si_{1-x-y}Ge_xC_y$ layers in comparison with strained $Si_{1-x}Ge_x$ layers.

In contrast to theoretically calculated values, the experimental results are much smaller [117, 118]. In a first article, People et al. [117] found for a strained SiGe layer with 20% germanium, homogeneously doped with 1×10^{18} cm^{-3} boron, a room temperature mobility of only 75 cm^2/Vs. Manku et al. [118] evaluated sheet resistances of strained $Si_{1-x}Ge_x$ layers (with up to 10% Ge) of heterojunction bipolar transistors in terms of doping-dependent hole mobilities. For all investigated boron concentrations (between 10^{16} and 10^{19} cm^{-3}), they found a monotonic decrease in μ with increasing x. The low mobility in $Si_{1-x}Ge_x$ shows already that scattering dominates over the strain effects.

We used homogeneously 10^{17} cm^{-3} boron-doped, strained $Si_{1-x}Ge_x$ layers with 18% germanium for our investigations of the influence of carbon on hole mobilities. Figure 39 shows the carrier density as a function of temperature for the investigated samples. Similar to the $Si_{1-y}C_y$ case, we observe an unusual and nonsystematic deviation from linear behavior at low temperatures for the sample containing carbon. The hole density is approximately the same at room temperature in these $Si_{1-x-y}Ge_xC_y$ layers as in the reference $Si_{1-x}Ge_x$ sample. This result is surprising in view of the known formation of B_i–C_s centers at temperatures above 150°C [119]. Analyzing the $p(T)$ dependencies gives an acceptor activation energy of boron in strained $Si_{1-x}Ge_x$ of 17.3 meV. The activation energy decreases to 13.0 meV for a $Si_{1-x-y}Ge_xC_y$ layer with 0.67% carbon.

The measured temperature dependencies of the Hall mobilities are shown in Figure 40. The hole mobility of the pure $Si_{1-x}Ge_x$ layer at 300 K is 97 cm^2/Vs. This value is even larger than those found by Manku et al. [118] for their highest Ge content (10%) and the same doping level. Over the investi-

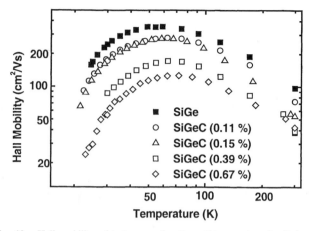

Fig. 40. Hall mobility of holes as a function of temperature for B-doped strained $Si_{1-x}Ge_x$ and $Si_{1-x-y}Ge_xC_y$ layers with 18% germanium on Si(001). Reproduced with permission from [103], copyright 1997, American Institute of Physics.

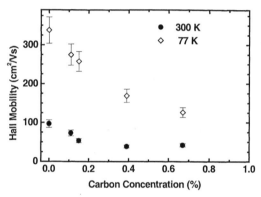

Fig. 41. Experimentally obtained hole mobilities at 300 K and 77 K as a function of carbon content in homogeneously B-doped $Si_{1-x-y}Ge_xC_y$ layers on Si(001). Reproduced with permission from [103], copyright 1997, American Institute of Physics.

gated temperature range, the hole mobilities of $Si_{1-x-y}Ge_xC_y$ alloys are always smaller than that of the $Si_{1-x}Ge_x$ reference sample. In analogy to the $Si_{1-y}C_y$ case, the increase in carbon results in remarkable changes in the $\mu(T)$ dependence at lower temperatures. Thus, the addition of carbon is accompanied by the formation of electrically active defects, the density of which increases with y, resulting in substantial decreases in mobility at low temperature. As for the growth of the $Si_{1-y}C_y$ layers, in addition to the constant density of dopant atoms, the measurements on $Si_{1-x-y}Ge_xC_y$ layers display an increase in the density of donor- and acceptor-like defects with increasing carbon concentration. The origin of this defect is not yet understood.

Figure 41 summarizes the mobility values versus carbon content at room temperature and at 77 K. The introduction of carbon leads to enhanced scattering and reduced mobility, with the effect being much larger at lower temperatures. As in the $Si_{1-y}C_y$ case, the increase in the density of ionized impurity centers with increasing C content is the main reason for the observed temperature dependence of mobility, especially at low temperatures.

Recently, p-type MOS field effect transistors (PMOSFETs) using tensile-strained Si C channel layers on Si substrates have been shown to provide mobility enhancement over Si epitaxial channel layers and Si bulk devices for the first time [120]. These layers did not exhibit the defects typically associated with a relaxed SiGe buffer structure. Furthermore, PMOSFET devices fabricated on these alloy layers demonstrate enhanced hole mobility over Si epi control and Si bulk. However, increased amounts of C may result in degraded device performance. Therefore, when such devices are fabricated, a balance must be struck to minimize C-induced interface charges and the alloy scattering rate.

5. HIGHLY CONCENTRATED PSEUDOMORPHIC $Si_{1-y}C_y$ LAYERS

So far, we have discussed pseudomorphically grown C-containing structures with carbon concentrations up to a few percent. In this section we present evidence that $Si_{1-y}C_y$ layers with $y \sim$ 20% can be grown pseudomorphically on a Si(001) substrate despite the large difference in the C and Si lattice constants. Such structures could only exist because of strain stabilization. It has been observed previously that semiconductor structures that are not found in the bulk phase diagram can be stabilized by the substrate-imposed strain in pseudomorphic epitaxial growth [48, 108]. The energy E_{epi} of an epitaxially confined structure can be decomposed as

$$E_{epi} = E_{equib} + E_{strain} \qquad (5.1)$$

where E_{equib} is the energy at the unconstrained equilibrium lattice constant and E_{strain} is the work needed to strain the material to match the Si substrate. An epitaxially stabilized structure X occurs when $E_{epi}(X)$ lies below that of the most stable bulk phase X_0 [in our case, $E_{epi}(\beta\text{-SiC})$]. Figure 42 illustrates that concept schematically.

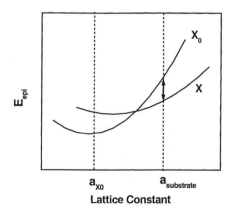

Fig. 42. Schematic presentation of the concept of strain stabilization. A structure X strained on a substrate can have a lower epitaxial energy E_{epi} than a structure X_0 with a lower equilibrium energy.

Recently, Rücker et al. [54] investigated the possible existence of such highly concentrated strain-stabilized Si_nC layers embedded in silicon. Calculations based on density-functional theory and a Keating model predict that embedded layers of certain structures with stoichiometry Si_nC where $n = 4, 5, \ldots$ are considerably more stable than isolated C impurities. The common feature of these structures is that the carbon atoms tend to arrange as third-nearest neighbors. Multilayer structures grown by molecular beam epitaxy strongly suggest that defect-free heterostructures with such high C concentrations can be fabricated.

5.1. Model Considerations

Very different "frozen-in" nonequilibrium structures can be obtained, depending on the growth conditions. A realistic simulation of the growth process is prohibitive, but we can study the energetics of the structures that might result. Focusing on the Si–C system, we need the total energy of a Si crystal in which some atoms have been replaced by carbon. This is essentially a lattice gas problem with a C–C interaction due almost exclusively to the interference of the strain fields. The energy depends on whether the Si–Si, Si–C, and C–C bond lengths can be accommodated without excessive bond bending. From these considerations, rather extreme arrangements of the C atoms could be competitive.

As an example, for a single embedded carbon (001) monolayer the Si–C bond length can relax freely by changing the interplanar spacing. The price paid is that some bond angles are distorted by a large amount. Another structure is obtained by distributing the C atoms over two adjacent planes without forming nearest-neighbor (nn) C–C pairs. Again, moving the lattice planes closer together can shorten the Si–C bonds, but here there is more freedom to minimize the bond-bending energy. Thus, it is possible that certain arrangements of the C atoms with a high local C concentration could have a low strain energy.

The specific question addressed here is whether there are stable structures that concentrate the carbon atoms in a narrow layer orthogonal to the (001) direction. Should such structures be found, they could possibly be fabricated by MBE under suitable conditions. This could open the way to quantum wells and superlattices involving $Si_{1-y}C_y$ layers with a high concentration of carbon atoms. For a thorough investigation, a fast but accurate method is needed to describe the bond-stretching and bond-bending forces in the mixed Si/C system. The method used was to tune a Keating model (see Section 3.2) to reproduce *ab initio* density-functional calculation energies for different trial structures [54]. The Keating model is fast enough to permit a reasonably exhaustive search through the possible arrangements of the C atoms on the Si lattice [49, 50]. In practice, the search included all periodic arrangements in which the unit cell contains up to two Si_nC formula units ($2 < n < 8$).

A nearest-neighbor C–C pair in the Si lattice leads to a large strain energy [46], and structures with such pairs were not considered. To simulate pseudomorphic growth, the x and y lattice

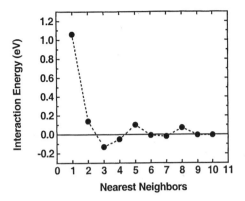

Fig. 43. Interaction energy for two substitutional C impurities as a function of distance obtained from the Keating model applied to a unit cell of 2000 atoms. The interaction is strongly repulsive for first- and second-nearest neighbors but shows an energy gain at the third-nearest neighbor distance. Reproduced with permission from [54], copyright 1994, American Physical Society.

constants were fixed at the bulk Si value. The energy was minimized with respect to the atomic positions and the unit cell height. For the low-energy structures found in this search, final *ab initio* calculations were then performed. For $n \geq 4$, the optimal energy lies below the impurity energy by about 0.4 eV. Thus a substantial energy gain is found, but only for specific arrangements of the C atoms. Inspection of the optimal geometries for the different values of n shows a common structural feature: the carbon atoms are arranged preferably as third-nearest neighbors. To understand this, consider the interaction energy between two substitutional carbon atoms in the Si lattice (Fig. 43).

The interesting feature is that the energy is negative by ~ 0.1 eV at the third-nearest neighbor distance. Here the carbon atoms lie opposite each other in a six-member ring. For an isolated ring of this type, the bond lengths can relax freely without bending the tetrahedral bond angles. The only other arrangement of C atoms on the ring with the same property is alternating Si and C atoms, which is the type of ring found in the ordered β-SiC structure.

Although the energy for more impurities is not simply a sum over C–C pair interaction energies, the results nevertheless suggest that the energy is reduced by maximizing the number of third-nn C–C pairs while avoiding first- and second-nn C–C pairs. A special role is taken by the case $n = 4$ with the stoichiometry Si$_4$C. This structure contains 20% carbon and is characterized by having the highest number of third-nn C–C pairs without first- or second-nn pairs.

5.2. Experimental Verification

The crucial question is now whether such highly concentrated embedded Si$_{1-y}$C$_y$ layers can be fabricated. It is generally assumed that Si–C alloys cannot be grown pseudomorphically on a Si substrate if the C content exceeds a few percent. The major uncertainty is whether growth is possible without the formation of silicon carbide microcrystallites; even in the optimized struc-

tures, the separation into pure silicon and carbide phases would still gain ~ 1.5 eV per carbon atom.

In [54] we described MBE growth and characterization of thin Si$_{1-y}$C$_y$ layers with $x \sim 0.2$. Here we summarize the main results. The crucial point of the growth procedure is to suppress the formation of stoichiometric silicon carbide by avoiding the availability of highly mobile Si and C at the same time. This is why the growth was interrupted between the Si and C deposition. The amount of carbon incorporated into the structure was determined by secondary ion mass spectroscopy (SIMS) as 1–1.5 monolayer C per individual C-rich layer. TEM investigations were made of cross-sectional samples (see [54]). From such images we estimate the layer to be about seven to eight monolayers thick, indicating a carbon concentration of 15–20%, in agreement with the theory. No crystallographic defects that usually appear when silicon carbide precipitates are formed were seen in any of the TEM images studied.

Recently, Ruvimov et al. [121] presented more detailed structural investigation of such layers, including superlattices with Si$_n$C layers. High-resolution 400-kV electron microscopy had been applied to study the microscopic structure of Si$_n$C layers grown by MBE on (001)-oriented silicon substrates. These authors found nearly perfect structures of the predicted Si$_n$C δ-layers embedded in Si. Local ordering of the carbon with a quasi-periodic variation of its distribution over the layer was observed on HREM micrographs under certain imaging conditions. The observed microroughness of the layer interfaces was mainly attributed to carbon diffusion during the formation of the Si$_n$C system. The growth of a silicon cap layer on top of a formed Si$_n$C layer starts with islands, which very soon coalescence to yield perfect crystalline silicon cap layers.

5.3. Formation of a Carbon-Rich Surface on Silicon

Understanding the microscopic processes that take place during crystal growth is desirable because, under suitable growth conditions far from equilibrium, materials can be trapped in a metastable state. The strain-stabilized Si$_n$C layers are the highest metastable C-containing system on Si. In this section we briefly discuss the formation of such layers (for more details, see [31]).

X-ray photoelectron spectroscopy (XPS) is a powerful tool for the investigation of local chemical environments. It is also a versatile technique with respect to investigation of growth kinetics [122]. In addition, XPS is a local probe with a probe depth below a few nanometers, compared with other techniques such as X-ray diffraction, with a probe depth on the order of microns. Using XPS, we show the existence of a carbon-rich Si$_n$C phase in silicon that differs from silicon carbide.

The first problem that has to be addressed is, when will these carbon-rich layers be formed? It is possible to assume a formation (i) during the deposition of carbon atoms on the silicon surface by a site exchange mechanism as well as (ii) during the deposition of silicon on an existing carbon layer by a similar mechanism. Using XPS measurements obtained for the individual stages of the growth sequence without leaving the vacuum, we were able to provide clear evidence for the first formation mechanism [31].

First, we measured the carbon 1s XPS spectra after the deposition of approximately 1 ML of carbon, keeping the substrate at the deposition temperature (600°C). The silicon 2p signal originating from the substrate was used as an internal reference. We obtained a binding energy of 282.8 eV for the C 1s line. This line is shifted by around 1.5 eV compared with the C 1s signal in graphite (284.3 eV). The measured signal position of 282.8 eV is close to that of silicon carbide (282.4 eV) on silicon surfaces formed during imperfect high-temperature *in situ* cleaning steps [123]. XPS signals are highly sensitive to the local binding environment of investigated atoms. The coincidence of the measured signal position only reflects the formation of Si–C bonds, but not of silicon carbide. A second interesting feature was the broadening of the C 1s signal. The measured full width at half-maximum (FWHM) of 1.8 eV indicated the formation of different chemical environments around the individual carbon atoms (we observe only an envelope signal).

After the XPS measurements a 1.5-nm-thick Si cap layer was deposited on top of the structure. The substrate temperature was kept at 600°C during the whole experimental cycle, including the XPS measurements. The C 1s XPS spectra obtained after the Si cap layer deposition do not differ significantly from those observed before. Besides a lower signal intensity due to the limited information depth of XPS, we found exactly the same line position and line width as before the Si deposition.

Summarizing the XPS results, we state that the formation of the Si_nC already takes place during the deposition of carbon on the silicon surface. This conclusion agrees well with recent results obtained by Kitabatake et al. [124]. These authors investigated the mechanism of SiC heteroepitaxial growth by carbonization of Si(100) surfaces at the atomic scale with molecular dynamics (MD) simulations and MBE experiments. Based on the MD simulations they predict the formation of SiC during the deposition of carbon atoms. The prediction of stoichiometric carbide instead of Si_nC structures is not in contradiction, because the simulations were performed for system temperatures up to 1300 K, where the formation of metastable Si_nC structures is not very likely.

In a second step we use XPS to obtain more detailed information about the formation kinetics. Approximately one monolayer of carbon was deposited on a silicon clean surface and cooled to room temperature. We measured the position of the C 1s signal with 283.6 eV and a FWHM of 1.89 eV. This measured binding energy is 0.8 eV, larger than that for the deposition at 600°C, indicating that the carbon signal is not dominated by Si–C bonds.

It is not clear what kind of carbon clusters will be evaporated from the sublimation source. Very likely we deposited a whole variety of small carbon fragments. This would explain the relatively large line width, because we measured only an envelope of different signals. The signal is also not identical to that of graphite for the same reason.

The layers were annealed *in vacuo* at 500°C for 45 min, followed by subsequent anneals at 550°C, 600°C, 650°C, 700°C,

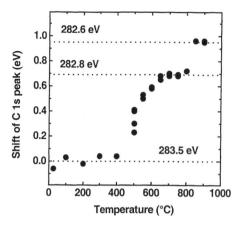

Fig. 44. Shift of the measured C 1s signal, depending on the annealing temperature for the last annealing step, relative to the position after deposition at room temperature. Three distinct plateaus (dotted lines) at the indicated core-level binding energies are observed. Reproduced with permission from [31], copyright 1995, American Physical Society.

and 750°C for 20 min at each temperature. At the end of each annealing step we measured the C 1s and the Si 2p (as a reference) XPS signal. We observe a significant shift in the C 1s signal shown in Figure 44. Starting after the 500°C anneal the signal shifts to lower binding energies (a Si–C bond is weaker than a C–C bond) and reaches a constant value of 282.8 eV at a temperature above 600°C. This line position is identical to that obtained earlier for the deposition at 600°C. We clearly see the formation of a Si–C phase. It appears that this phase has already formed completely at 600°C. This temperature would be surprisingly low for a silicon carbide formation, and the observed binding energy is still too large for carbide. Silicon carbide on Si(001) formation needs a minimum temperature above 900°C. To check whether we have formed silicon carbide or not, we used a additional high-temperature (950°C/15 min) anneal. After the anneal we measured an additional shift of 0.36 eV (see Fig. 44); the signal is now located as typically observed for silicon carbide on silicon.

Throughout the annealing the ratio of the integrated C and Si intensities remained constant within a mean deviation of 9.3%; i.e., there was no carbon loss due to desorption. The intensity of electrons (I) emitted from a depth (d) is given by the Beer–Lambert relationship,

$$I = I_0 \exp(-d/\lambda \sin\theta) \qquad (5.2)$$

where θ is the electron take-off angle relative to the sample surface. In our experimental setup we measured the photoelectrons normal to the surface, i.e., $\theta = 90°$. If we assume a constant amount of carbon distributed homogeneously over a thickness d, then we can express the XPS intensity as

$$I \sim (\lambda/d)\big(1 - \exp(-d/\lambda)\big) \qquad (5.3)$$

where λ is the inelastic mean free path of the photoelectrons, which is 23 Å for C 1s electrons under Mg-K$_\alpha$ excitation in Si [125].

For a known initial carbon coverage we can calculate the reduction in XPS intensity as a function of d. According to Eq. (5.2), a spread of the carbon monolayer over 100 monolayer of silicon (equivalent to the formation of a $Si_{1-y}C_y$ alloy with 1% carbon) would reduce the XPS intensity to only one-sixth of the initial value. For the formation of such an alloy we should observe a significant decrease in the relative carbon signal intensity. In contrast, throughout the annealing procedure, we have found that the ratio of the integrated C 1s and Si 2p intensities remained constant to within a mean deviation of less then 10%. Thus our measurements are compatible with a carbon concentration in the neighborhood of 15–20% but exclude a substantially lower concentration.

Based on all XPS results we summarize that we have observed the formation of a new Si–C phase that will be formed around 600°C. This phase is characterized by the dominance of carbon atoms bonded to silicon, but it is not identical to silicon carbide. It is further very localized within the topmost region of only some monolayers. The formation of this phase requires an activation energy of (0.6 ± 0.1) eV [31]. We believe that this phase corresponds to the strain-stabilized high-concentration Si_nC structures.

The existence of strain-stabilized Si_nC layers with carbon concentrations up to 20% has been demonstrated theoretically as well as experimentally. The structural predictions could be confirmed by different experiments. Up to now, nothing is known about the electrical properties of these layers.

6. DEVICE APPLICATION OF SiGe:C

6.1. Control of Dopant Diffusion by the Addition of Low Carbon Concentrations

In contrast to all of the other sections, the amount of carbon discussed in this section is on the order of dopant concentrations. To distinguish these materials from the $Si_{1-y}C_y$ or $Si_{1-x-y}Ge_xC_y$ systems discussed so far, we will use the same designation as for dopants, that is, Si:C or SiGe:C. We will show that such low carbon concentrations are sufficient to control dopant diffusion. On the other hand, the carbon concentration is too low to significantly affect the band alignment and strain of the epitaxially grown Si:C or SiGe:C layers. However, the carbon concentration used, in the range of 10^{20} cm^{-3}, is still orders of magnitude higher than the solid solubility limit; i.e., we have to consider metastable systems with carbon supersaturation.

A major challenge for the fabrication of ultrasmall devices for coming silicon technology generations is the control of steep dopant profiles. In this context, it is tempting to apply supersaturated carbon as a diffusion-controlling agent in silicon. The fundamental question for applications of C-rich Si in devices is whether electronic device parameters are affected by possible C-related defects.

Ban et al. [126] have reported suppressed diffusion of boron in the channel of a metal oxide semiconductor field effect transistor (MOSFET) due to the implantation of carbon. However, these authors have reported that carbon can also lead to poor activation of boron and degradation in MOSFET performance. In particular, increased off-stage leakage currents were measured for transistors with an implanted carbon dose above 10^{14} cm^{-2}. At present, it is an open question whether the observed degradation of MOSFET parameters is caused by the chosen preparation of the carbon-rich layer by ion implantation.

Recently, Gossmann et al. [127] announced the realization of MOSFET with a reduced reverse short channel effect due to suppressed transient diffusion in carbon-rich silicon. Little increase in leakage current was reported for devices with carbon.

First successful applications of C-rich layers were reported for SiGe heterojunction bipolar transistors (HBTs) [26, 128]. We [26, 129] have demonstrated that transistors with excellent static and dynamic parameters can be fabricated with epitaxial SiGe:C layers. The main result of this investigation was that carbon supersaturation can preserve steep doping profiles without degrading fundamental transistor parameters.

The present interest in the development of SiGe HBTs stems from their potential applications in integrated circuits operating at radio frequencies. Circuits operating in the range of several gigahertz are needed for wireless communication systems. Silicon bipolar transistors for high-frequency operation require short base transit times, low base resistance, and low extrinsic parasitics. Heterojunction devices with a larger band gap in the emitter region and a smaller band gap in the base region enhance high-frequency performance. A BiCMOS process based on modular integration of high-performance bipolar transistors into a state-of-the-art CMOS platform, with no change to the CMOS flow or parameters, is a promising candidate for cost-effective single-chip solutions for the wireless market [130].

One of the key problems in npn SiGe technology is to retain the narrow as-grown boron profile within the SiGe base layer during postepitaxial processing. Heat treatments and transient enhanced diffusion (TED) caused by annealing implantation damage can broaden the profile into the adjoining Si regions [131]. This can cause undesirable conduction band barriers and thus significantly degrade device performance. The most common solution to this problem is to grow undoped $Si_{1-x}Ge_x$ spacer layers between the p$^+$ $Si_{1-x}Ge_x$ and the emitter and collector layers. However, the thickness of the SiGe layer and therefore of these spacer layers must be minimized to avoid strain-induced defect formation. The incorporation of carbon into the epitaxial layer is a new way of suppressing boron outdiffusion [128]. Recently it was shown that low C concentrations ($<10^{20}$ cm^{-3}) in the SiGe base layer can significantly suppress TED of boron [26, 129].

To evaluate the potential of carbon for suppressing boron outdiffusion, we chose the process of depositing and annealing a heavily P-doped poly-Si emitter. Figure 45 shows SIMS boron profiles for buried, 30-nm-thick, box-shaped SiGe structures with and without a carbon background (within the SiGe layer), measured in the as-grown state and after processing. The layered structure without carbon exhibits strong boron outdiffusion. A background of substitutionally incorporated carbon

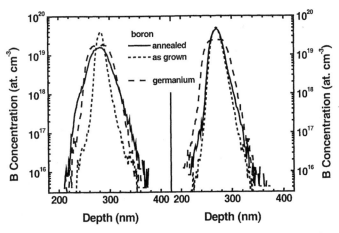

Fig. 45. SIMS profiles of boron before and after the formation of a phosphorous-doped poly-Si emitter annealed at 950°C for 30 s. (Left) Buried, box-shaped 30-nm-thick $Si_{0.8}Ge_{0.2}$ layer without and (right) with 1020 cm^{-3} carbon background.

atoms ($<10^{20}$ cm^{-3}) stops the boron outdiffusion nearly completely. Within the accuracy of SIMS, we could not find any differences in the boron profiles after phosphorous emitter formation and annealing (950°C/30 s) in comparison with the as-grown state (Fig. 45).

First we describe the physical mechanism for suppressed boron diffusion in carbon-rich Si and SiGe (below 0.2% carbon) based on an undersaturation of Si self-interstitials due to outdiffusion of carbon. Then we show a special realization of HBTs with highly doped, narrow base layers. This transistor construction is particularly vulnerable to performance degradation caused by enhanced boron diffusion. Therefore, it is an ideal candidate for demonstrating the advantages of carbon incorporation.

In the second half of this section, we show that heterojunction bipolar transistors with excellent static and dynamic parameters can be fabricated with epitaxial SiGe:C layers. The main result of this investigation is that carbon supersaturation can preserve steep doping profiles without degrading fundamental transistor parameters. As a result, the rf performance can be markedly improved. Compared with SiGe technologies, the addition of carbon provides significantly greater flexibility in process design and offers wider latitude in process margins.

We present performance data obtained from bipolar SiGe:C devices. In addition to comparable $1/f$ noise, the SiGe:C devices show (I_B-driven) Early voltage–current gain products greater than 20,000 V, exceeding values known for state-of-the-art SiGe HBTs. The high-frequency performance of SiGe:C HBTs also benefits from C incorporation into the SiGe base layer because it allows one to use a high B doping level in a very thin SiGe layer without outdiffusion from SiGe, even if postepitaxial implants are applied. Such implants were used for doping the external base regions of transistors and for selectively doping the HBT low-doped emitter (LDE) and collector (LDC). Finally, we present the first results on the modular integration in a 0.25-μm CMOS platform.

6.2. Effect of Carbon on Boron Diffusion

Diffusion of dopants as well as Si self-diffusion and diffusion of Ge and C in Si are mediated by point defects, i.e., vacancies and self-interstitials. It is generally accepted that both vacancy and interstitial mechanisms contribute to atomic transport in Si, whereas the direct exchange mechanism of nn atoms play a negligible role.

TED during the first annealing step after ion implantation accounts for a substantial fraction of dopant redistribution during device processing. The source of TED is an enhanced density of Si self-interstitials. Each implanted ion creates a trail of crystal defects through collisions with the lattice. These defects in turn can interact with substitutional impurities and enhance their mobilities during subsequent high-temperature steps. Substitutional C can stabilize steep B profiles during high-temperature process steps. The diffusion coefficient of B in Si is reduced by more than one order of magnitude when the concentration of substitutional C is elevated to about 10^{20} cm^{-3} [132]. Furthermore, TED of B is strongly suppressed in C-rich Si [133]. The effect of C on B diffusion can be explained by coupled diffusion of B and point defects in Si [134]. Diffusion of C out of C-rich regions causes an undersaturation of Si self-interstitials in the C-rich region, which in turn results in the suppressed diffusion of boron.

The basic mechanism for nonequilibrium point defect densities due to C outdiffusion is as follows [132]. Depending on growth conditions, carbon can be mainly substitutionally dissolved in Si [see Section 2]. Diffusion of C in Si is via a substitutional–interstitial mechanism. Mobile interstitial carbon atoms (C_i) are created through the reaction of Si self-interstitials (I) with immobile substitutional carbon (C_s):

$$C_s + I \leftrightarrow C_i \qquad (6.1)$$

and in the dissociative reaction,

$$C_s \leftrightarrow C_i + V \qquad (6.2)$$

where V is a vacancy. To conserve the total number of atoms, the flux of interstitial C atoms out of the C-rich region has to be balanced by a flux of Si self-interstitials into this region or a flux of vacancies outward. The individual fluxes are determined by the products of the diffusion coefficient D and the concentration C. For C concentrations $C_C > 10^{18}$ cm^{-3}, the transport coefficient of C may exceed the corresponding transport coefficients of Si self-interstitials and vacancies by

$$D_C C_C > D_I C_I^{eq} \qquad \text{and} \qquad D_C C_C > D_V C_V^{eq} \qquad (6.3)$$

where C_I^{eq} and C_V^{eq} are the equilibrium concentrations of Si self-interstitials and vacancies, respectively. Consequently, outdiffusion of supersaturated C from C-rich regions becomes limited by the compensating flux of Si point defects, which leads to an undersaturation of self-interstitials in the C-rich region.

Pronounced non-Fickian diffusion behavior was found experimentally for grown-in carbon at concentrations well above its solid solubility. In Figure 46 measured C profiles in Si are

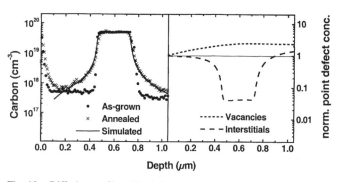

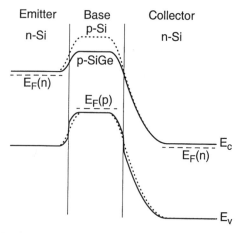

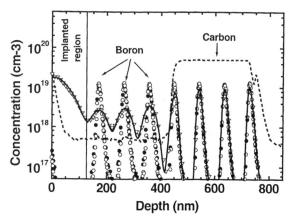

Fig. 46. Diffusion profiles of buried carbon layers annealed in a nitrogen atmosphere (900°C/45 min). SIMS profiles of as-grown (•) and annealed (×) samples are shown in the left panel together with results of model calculation. The corresponding normalized point defect densities are shown in the right panel.

Fig. 47. SIMS profiles of a boron doping superlattice implanted with 10^{14} cm^{-2}, 45 keV BF$_2$ ions and annealed in a N$_2$ atmosphere at 930°C for 30 s (triangles). Open circles are the as-grown B profile, and filled circles are the profile of an annealed sample (930°C, 30 s) without implantation. The solid line is the calculated B profile after implantation and annealing. Reproduced with permission from [136], copyright 1999, American Institute of Physics.

compared with model calculations [135]. A 50-nm-thick box-shaped C profile was grown by MBE and annealed in a pure nitrogen atmosphere. The model developed by Rücker et al. [132] can fit the carbon diffusion very nicely. The point defect distribution after annealing resulting from the calculation is also shown in Figure 46. In the carbon-rich region, there is a significant reduction in the interstitial density accompanied by an increase in the vacancy density. This has consequences for all dopant diffusion via point defects. Boron diffusion in Si is via an interstitial mechanism. The effective diffusion coefficient of B is proportional to the normalized concentration of self-interstitials. Accordingly, interstitial undersaturation results in suppressed diffusion of B. Moreover, the diffusion of phosphorous will be strongly reduced. In contrast, the diffusion of dopants that are diffusing by a vacancy mechanism (like As or Sb) will be enhanced by carbon incorporation [135].

The impact of supersaturated C on TED of B is illustrated in Figure 47. A buried layer, with a C concentration of about 5×10^{19} cm^{-3}, was located 450 nm below the surface. B doping spikes with doses of about 10^{13} cm^{-2} were incorporated into

Fig. 48. Band diagram of a SiGe HBT (full lines) for base-emitter voltage $V_{BE} = 0.5$ V and collector-emitter voltage $V_{CE} = 1.0$ V. The band edges of a Si bipolar transistor with identical doping profiles are shown for comparison (dotted line). Dashed lines indicate the Fermi levels for majority carriers.

the C-rich region and in the adjacent region without intentional C doping. Depth profiles of B were measured by SIMS for as-grown and annealed samples with and without implantation of BF$_2$ ions (Fig. 47). We found strongly enhanced diffusion of B spikes in the C-poor region of the implanted sample. In contrast, TED was suppressed almost completely for B diffusion markers located in the C-rich region. We have shown recently that the suppression of TED due to C can also be attributed to coupled diffusion of C and Si point defects [136]. Outdiffusion of supersaturated C restricts the penetration of excess interstitials into C-rich regions.

The quality of epitaxial layers for electronic device application can also be evaluated by studying the carrier generation lifetime in MOS-like structures [137]. For LPCVD-grown SiGe:C layer stacks similar to those used for HBT application, we typically measure lifetimes above 1 μs, i.e., values that are sufficient for successful application of those layers in high-performance devices with low noise. Recently we could also show that the electron mobility in the p-type base of npn bipolar transistors is not reduced by the incorporation of low C concentrations [138].

6.3. SiGe:C Heterojunction Bipolar Transistor

First, a short introduction to the mode of operation of HBTs will be given. After clarifying the advantage of narrow, highly doped base layers for the electronic performance of bipolar transistors, we address the problem of realizing such structures in a technological process. Then we show that transistors with excellent static and dynamic parameters can be fabricated with epitaxial SiGe:C layers. Compared with SiGe technologies, the addition of carbon provides significantly greater flexibility in process design and offers wider latitude in process margins. Finally, we demonstrate that the addition of carbon can suppress TED caused by a variety of subsequent processing steps. In particular, the limitation for lateral emitter scaling in a single-polysilicon technology, resulting from TED due to implantation

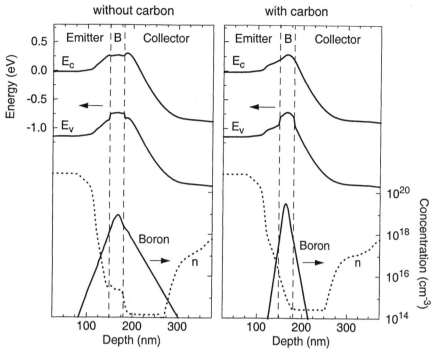

Fig. 49. Band diagram of SiGe (left) and SiGe:C (right) HBTs for a base voltage $V_{BE} = 0.7$ V and a collector voltage $V_{CE} = 1.0$ V. The boron profiles shown were used for the calculation of the band edges and electron densities n (dotted lines). Dashed lines indicate the $Si_{0.8}Ge_{0.2}$ layer.

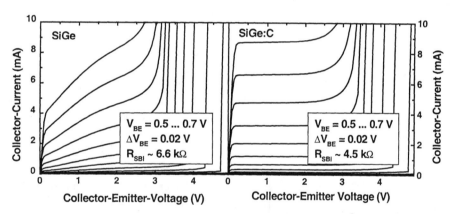

Fig. 50. Output characteristics of HBTs with an emitter area of 100×100 μm^2 and box-shaped, 30-nm-thick $Si_{0.8}Ge_{0.2}$ base profile. (Left) Base epitaxy without and (right) with carbon background. The devices were implanted after epitaxy with As and P.

of the external base regions can be eliminated. C incorporation also makes possible the use of selective collector doping, increasing the process and device flexibility.

6.3.1. Operation of Heterojunction Bipolar Transistors

The basic feature of a HBT is the use of materials with different band gaps for emitter and base. The emitter is formed of the material with the larger band gap. This is illustrated in Figure 48 by the band diagram of an npn HBT with a SiGe base and a Si emitter.

The consequence of the smaller band gap of the SiGe base layer is a larger current gain β in comparison with a Si bipolar transistor with the same doping profile. The barrier for the injection of electrons from the emitter into the base is lower for a SiGe base compared with a Si transistor.

This results in an increased electron injection current I_{nE} and an increased emitter efficiency,

$$\gamma = \delta I_{nE}/\delta I_E \qquad (6.4)$$

Here I_E is the total emitter current, which is the sum of the electron component I_{nE} and the hole component I_{pE}. The common-base current gain α increases linearly with increasing emitter efficiency γ and decreases approximately linearly with base doping. Consequently, in a heterojunction bipolar transistor, a higher base doping can be chosen for the same common-base current gain. The common-emitter current gain β is related to the common-base current gain by

$$\beta = \alpha/(1 - \alpha) \qquad (6.5)$$

Consequently, the increased electron injection at a heterojunction allows the realization of a required current gain with higher base doping. Higher base doping leads to reduced base resistance and improved frequency characteristics. A detailed analysis of high-frequency properties of SiGe HBTs was given by Heinemann [139] on the basis of device simulations within the drift-diffusion model.

SiGe HBTs with highly doped base layers were shown to be capable of maximum oscillation frequencies f_{max} as high as 160 GHz [140], cut-off frequencies f_T as high as 130 GHz [141], and current-mode-logic gate delays as low as 7.7 ps [142]. In large-volume production of integrated circuits, compromises have to be made between maximum peak frequencies, static transistor performance, and production cost. The additional degree of freedom in transistor design introduced by the use of a SiGe base has been exploited by the IHP group to develop a low-cost HBT module suitable for a range of applications in communication systems [129, 130].

Outdiffusion of boron from the SiGe layer has dramatic consequences for electronic transistor parameters. Boron diffusion into the Si regions causes the formation of parasitic conduction band barriers reducing the transistor's gain, speed, and Early voltage (see left panel in Fig. 49). The formation of a potential barrier for electrons at the base collector junction hinders the flux of electrons from the base to the collector. Consequently, electrons accumulate in the base region, as can be seen from the calculated electron densities shown in the left panel of Figure 49. As a result, transit frequencies are drastically reduced. The B profiles plotted in Figure 49 were obtained from SIMS measurements at SiGe and SiGe:C HBTs that have undergone identical post-epitaxial processing. The HBTs without C showed a strong broadening of the B profile caused by transient enhanced diffusion due to selective emitter and collector implantations. For SiGe:C HBTs, B outdiffusion from the heteroepitaxial layer is strongly suppressed.

Output characteristics as well as calculated transit frequencies of the corresponding SiGe and SiGe:C HBTs are shown in Figures 50 and 51, respectively. Outdiffusion of B causes degradation of the output characteristics and a strong reduction of peak frequencies in the case of the SiGe HBT.

6.3.2. Realization of Highly Doped, Narrow-Base-Layer HBTs

We turn now to the discussion of the effect of carbon doping on dopant profiles in a special HBT realization used at the IHP. A schematic cross section of the transistor is shown in Figure 52. The device structure and the technology used for HBT preparation were described elsewhere in great detail [129]. Therefore, we will only summarize the essential process steps. The HBTs were produced in a single-polysilicon technology [143]. Epi-free n-wells were produced by ion implantation following the oxide isolation formation. The layers were grown by LPCVD epitaxy (using a commercial tool) with low defect density necessary for high yield, and excellent profile control. A 30-nm-thick Si buffer, a 25-nm-thick SiGe base layer, and a Si cap layer were deposited without interruption. Half-graded Ge profiles were used. A substitutional C background of less than 10^{20} cm^{-3} was produced. The B doping was adjusted to obtain low internal base sheet resistivities (R_{SBi}) in the range of 1 to 4 kΩ. Other salient process details are as follows: self-aligned BF$_2$ implant after structuring of the poly emitter; short RTA to anneal the emitter; and Pt salicidation of emitter, base, and collector contacts. Postepitaxial processing was identical to our standard SiGe technology with one exception, namely for some wafers we introduced an additionally low-dose implantation to selectively dope the LDCs (SI-LDCs). Such implantation through the epilayer is only possible in the SiGe:C case, because of strong TED effects in SiGe.

Other than the addition of C during epitaxy, no changes were made to our standard SiGe technology. This means that the addition of carbon does not necessarily require any other changes in postepitaxial processing steps.

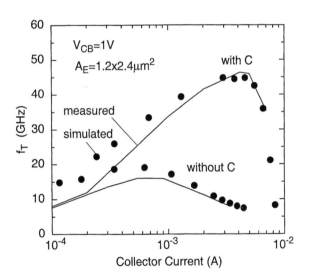

Fig. 51. Measured (lines) and simulated (dots) dependence of transit frequencies f_T on collector current at a voltage $V_{CB} = 1$ V for a SiGe:C HBT and a SiGe HBT.

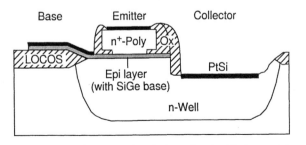

Fig. 52. Schematic cross section of a heterojunction bipolar transistor.

6.4. SiGe:C HBT Results

6.4.1. DC Characteristics

To highlight the impact of C addition on current gain, Early voltage, and LF low frequency noise, we compare the results with our SiGe-only devices. We note that the SiGe devices chosen for comparison represent or exceed the state of the art for the characteristics under discussion [144, 145].

We now describe our dc results, beginning with a discussion of the typical shortcomings in the dc performance of SiGe-only HBTs. Figures 53 shows Gummel plots, and Figure 54 shows the output characteristics for SiGe-only devices measured under I_B and V_{BE} drive conditions in the low-injection regime.

The strong change in the Early voltage with drive mode is a consequence of implantation-induced TED around the emitter perimeter. This is supported by the fact that the current gain and Early voltage increase from 70 to 150, and 10 to >200 V, respectively, when the overlap between the emitter poly and the active emitter region is increased and therefore the impact of TED is reduced (Fig. 55).

Turning to SiGe:C HBTs, Figure 53 shows the Gummel plot for these devices and Figures 56, 57, and 58 show the output characteristics at 240 K and 360 K. These data (and others not shown) demonstrate a gain of >200 and an I_B-

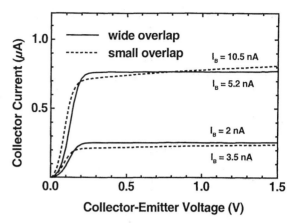

Fig. 55. I_B-driven output characteristics of two SiGe HBTs at 300 K with different emitter poly overlap ($A_E = 1 \times 2.7\ \mu m^2$). Larger overlap significantly reduces the impact of TED in the active emitter region and leads to high V_A.

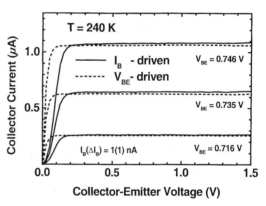

Fig. 56. I_B- and V_{BE}-driven output characteristics of a SiGe:C HBT at 240 K ($A_E = 1 \times 2.7\ \mu m^2$). Note that the transistor has only a small emitter poly overlap.

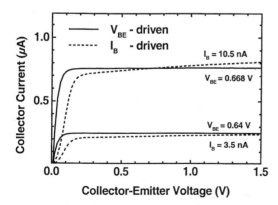

Fig. 53. Gummel plots of SiGe (dashed lines) and SiGe:C (solid lines) HBTs ($A_E = 1 \times 2.7\ \mu m^2$, $V_{CB} = 1$ V, $T = 300$ K). The difference in current gain is mainly due to B outdiffusion around the emitter perimeter.

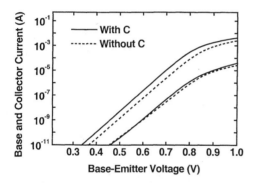

Fig. 54. I_B- and V_{BE}-driven output characteristics of SiGe HBT at 300 K ($A_E = 1 \times 2.7\ \mu m^2$). The V_A extracted from I_B-driven conditions (<10 V) is significantly lower than that obtained with constant V_{BE} (~200 V) due to TED around the emitter edge.

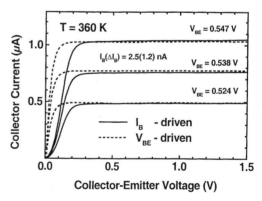

Fig. 57. I_B- and V_{BE}-driven output characteristics of a SiGe:C HBT at 360 K ($A_E = 1 \times 2.7\ \mu m^2$). The transistor has a small emitter poly overlap.

driven Early voltage exceeding 100 V over the temperature range 240–360 K. These values exceed the best data for SiGe in current-driven Early voltage times gain by a factor of 5 [146]. We emphasize that the SiGe:C HBTs have only a minimized emitter poly overlap.

A key question has concerned neutral base recombination (NBR) in SiGe:C HBTs. The excellent output characteristics of

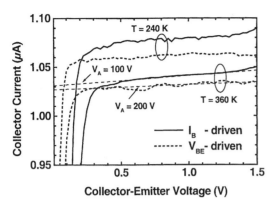

Fig. 58. High-resolution plot of I_B- and V_{BE}-driven output characteristics of a SiGe:C HBT at 240 K and 360 K. The dotted lines indicate the slope belonging to a V_A of 100 and 200 V, respectively. I_B (240 K, 360 K) = 3 nA, 4.8 nA and V_{BE} (240 K, 360 K) = 0.746 V, 0.547 V.

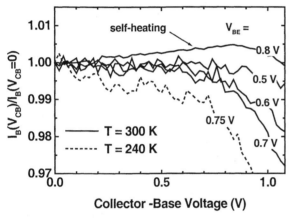

Fig. 59. Normalized I_B as a function of collector-base voltage for a SiGe:C HBT at 240 and 360 K. Only weak indication of NBR is visible at the lower temperature.

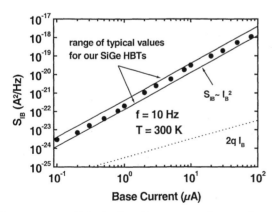

Fig. 60. Power spectral density S_{IB} of base current noise versus base current at $f = 10$ Hz for SiGe:C HBTs (circles, $A_E = 1 \times 2.7\ \mu m^2$, $T = 300$ K). No significant increase in the noise level for the SiGe:C devices compared with our SiGe HBTs can be observed.

SiGe:C HBTs constitute a strong indication that the addition of C has no deleterious effects on NBR. (This is particularly so because of current-gain feedback, which makes the current-driven Early voltage a sensitive indicator of NBR.) To make an addi-

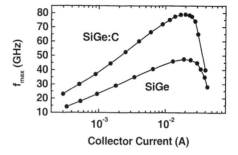

Fig. 61. Maximum oscillation frequency as a function of collector current for SiGe and SiGe:C multiemitter HBTs differing only in base doping (R_{SBi}(SiGe) = 4.0 Ω cm, R_{SBi}(SiGe:C) = 1.3 Ω cm).

tional direct estimate of NBR, we have measured the change in I_B with V_{CB} at constant V_{BE}. For values of $V_{CB} < 0.5$ V, where avalanche effects are negligible, there is essentially no change in the normalized I_B at 300 K, with only small changes at 240 K (Fig. 59). Whereas NBR has an impact on the design of precision analog circuits in the low-frequency regime, noise values at low frequencies must be low for use in high-quality rf circuits, like mixers or voltage-controlled oscillators. $1/f$ noise measurements of our SiGe:C HBTs (Fig. 60) yield a similar input current power spectral density S_{IB} compared with our SiGe HBTs [143].

6.4.2. High-Frequency Behavior

Suppressed boron outdiffusion enables the use of higher boron doses in the base layer. Figure 61 demonstrates the strong improvement in peak f_{max} from 50 to 80 GHz for the same transistor type, achieved by adding carbon and simultaneously increasing the base doping. Moreover, the SiGe:C HBTs show lower minimum noise figures with much higher associated gains at high frequencies and produce ring oscillator delays down to 12 ps.

In single-polysilicon HBT technology, a high-dose boron implantation is often used to dope the external base regions. Although the ions are not directly implanted in the active emitter region, they can nevertheless cause strong TED of boron. In [147] we investigated the influence of BF$_2$ implantation on narrow-base SiGe HBTs. The effective lateral ranges of the TED were extracted from electrical measurements on specially designed, laterally scaled devices. The estimated length of lateral boron outdiffusion depend on the shape of the Ge profile and on B concentration. It can easily reach several hundred nanometers and thus prevent scaling of the emitter width. The incorporation of small amounts of carbon into the SiGe base eliminates this constraint.

To demonstrate this, we increased the boron dose to obtain an internal base sheet resistance of 1 kΩ. The total SiGe width was only 25 nm. For a similarly processed SiGe HBT, we extracted a lateral boron outdiffusion of more than 400 nm on each side [147]. Figure 62 shows the cutoff frequency and the maximum oscillation frequency of SiGe:C HBTs for different emitter widths. Even for emitter widths as low as 0.5 μm, we do

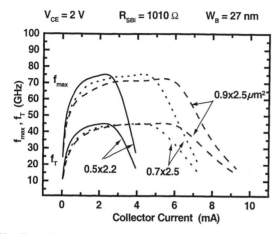

Fig. 62. Transit frequencies f_T and maximum oscillation frequencies f_{max} as functions of the collector current I_C for SiGe:C HBTs with $R_{SBi} = 1$ kΩ, SiGe layer thickness = 27 nm, and different emitter widths.

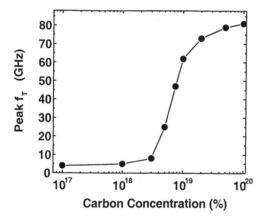

Fig. 63. Simulated TED effects on the maximum transit frequency as a function of C concentration located inside the SiGe layer (SI-LDC: 150 keV/5 × 10^{12} cm^{-2}).

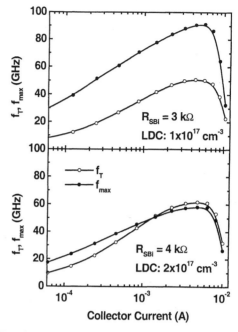

Fig. 64. Transit and maximum oscillation frequency versus collector current of HBTs differing in R_{SBi} and LDC doping level ($A_E = 4 \times (1 \times 1)$ μm^2, $V_{CE} = 2$ V).

Table III.   Minimum Single Stage Delay of ECL-Ring Oscillators with SiGe:C HBTs

R_{SBi} (kΩ)	LDC doping (10^{17} cm^{-3})	Delay (ps)
3.0	2	16–17
3.0	1	16–17
2.5	0.5	19–22
1.5	2	16–17
1.5	1	16–17
1.5	0.5	18–20

FI/FO = 1; emitter area = 1×1.4 μm^2.

not find any degradation in high-frequency performance. The overall reduction in f_T compared with the results shown earlier is caused by decreasing R_{SBi} without optimization of the transistor design, especially the boron position within the Ge profile. These results show that the addition of carbon allows us to scale a single-poly technology.

For some wafers we introduced an additionally low-dose implantation to selectively dope the LDCs (SI-LDCs). Such implantation through the epilayer is only possible in the SiGe:C case, because of strong TED effects in SiGe. Figure 63 shows simulated results for the TED effects on the maximum transit frequency f_T vs. C concentration located only inside the SiGe layer. As a SI-LDC implant, a low P dose was chosen (150 keV/5 × 10^{12} cm^{-2}). The calculation were based on the coupled diffusion model of C and Si point defects described above and on device simulations within the drift-diffusion model [139]. The observed drop of the peak transit frequency for C concentration below 10^{19} cm^{-3} is due to B outdiffusion from the SiGe layer for lower C concentrations. This clearly shows that SI-LDC implantation can only be used for HBTs with C.

Figure 64 shows f_T and f_{max} versus the collector current (I_C) for two HBTs differing in R_{SBi} and the LDC doping level. With the higher R_{SBi}, f_T peaks at 65 GHz. The maximum oscillation frequency can be increased up to 90 GHz by lowering R_{SBi} and the LDC doping. Ring oscillator data with minimum delays of 16 ps are shown in Table III. It also becomes evident from the ring oscillator data that the high-frequency performance can be further improved by optimizing the SI-LDCs. The thickness of the LDC was increased and its doping level reduced, compared with the LDC version implanted unselectively before base/emitter epitaxy. The extrinsic base-collector capacitance C_{BCex} is thus lowered, because the LDC is selectively doped. Additionally, R_{SBi} was lowered and the extrinsic base regions were optimized to reduce the base resistance R_B. Reducing R_B and C_{BCex} causes significant improvements in f_{max}, reducing the ring oscillator delays to 16 ps (Table III).

Substantial flexibility is available in this technology to trade off gain, V_{CE0}, f_T, and f_{max}. This can be achieved simply by varying the epitaxial base B dose and the conditions for the LDC implantation. (The latter is carried out through the emitter window without an additional mask step.)

6.4.3. *Modular Integration in a CMOS Platform*

The SiGe:C HBT module has been integrated in a $0.25\text{-}\mu m$, epi-free, dual-gate CMOS platform [130]. Our primary results can be summarized as follows: (1) A SiGe:C HBT module involving only four additional mask levels can be added to a $0.25\text{-}\mu m$ CMOS process, with no changes to the CMOS flow, to produce a BiCMOS technology with HBT peak f_T/f_{max} values of 55/90 GHz, at a B V_{CEO} of 3.3 V. This performance significantly surpasses the previous best in class for modularly integrated Si-BJTs and that of SiGe-HBTs in a bipolar-based BiCMOS process with epitaxial collectors. (2) The HBT integration does not significantly change the transistor parameters of the original CMOS platform, making possible the re-use of CMOS layouts and libraries. (3) The addition of C to HBTs is responsible for the high performance of the modular BiCMOS, because it prevents B outdiffusion from the highly doped, thin SiGe layer under the RTA conditions of the original CMOS flow. (4) The BiCMOS process produces a leakage-limited SiGe:C HBT wafer yield typically more than 70%, evaluated on the basis of an active HBT area of more than $10^4 \mu m^2$.

6.4.4. *Summary*

In conclusion, we have shown that SiGe:C HBTs containing small amounts of C offer excellent DC performance with high yield. Compared with SiGe technologies, the addition of carbon provides significantly greater flexibility in process design. In conjunction with the improved AC performance, this should pave the way for the use of C in SiGe technology to widen process margin and boost static and dynamic performance. Finally, we demonstrate the first modular integration of SiGe:C HBTs into a $0.25\text{-}\mu m$, epi-free, dual-gate CMOS platform.

7. SUMMARY AND OUTLOOK

We reviewed basic growth issues as well as some mechanical and electrical material properties of $Si_{1-y}C_y$ and $Si_{1-x-y}Ge_xC_y$ layers grown pseudomorphically on Si(001). This new material might overcome some of the constraints for strained $Si_{1-x}Ge_x$ and open new fields for device applications of heteroepitaxial Si-based systems. The incorporation of carbon can be used (i) to enhance SiGe layer properties, (ii) to obtain layers with new properties, or (ii) to control dopant diffusion in microelectronic devices. In detail, we would suggest the following applications for C-containing materials, based on the properties described in this book:

- Increase thickness, stability, Ge content of $Si_{1-x}Ge_x$ (advantageously for p-channel FET, npn HBT)

- Use strained $Si_{1-y}C_y$ on Si instead of Si on relaxed buffer (advantageously for n-channel FET, pnp HBT)
- Design new buffer concepts with $Si_{1-x-y}Ge_xC_y$ (as virtual substrates for hetero-FETs)
- Fabricate tensile/compressive strained superlattices on Si(001) for possible optoelectronic application
- Increase performance and process margins for SiGe HBTs by
  Suppressed transient enhanced diffusion of boron
  Reduced undoped SiGe spacer width

Research on C-containing materials heteroepitaxially grown on Si is still at an early stage. Different growth techniques have already been successfully demonstrated. However, one of the most crucial issues for all kinds of further applications is the ability to avoid interstitial carbon defect formation. Solutions to this problem will have a direct impact on the existing breakthrough potential of this material.

Acknowledgments

The author thanks D. Bolze, E. Bugiel, B. Dietrich, D. Endisch, G. G. Fischer, P. Gaworzewski, J. Griesche, B. Heinemann, M. C. Kim, W. Kissinger, D. Knoll, D. Krüger, G. Lippert, J. P. Liu, M. Methfessel, K. Pressel, H. Rücker, S. Scalese, R. Sorge, H. J. Thieme, M. Weidner, and P. Zaumseil for their contributions to the research reviewed in this chapter.

REFERENCES

1. S. C. Jain, "Germanium-Silicon Strained Layers and Heterostructures." Academic Press, Boston, 1994.
2. E. Kasper, *Curr. Opin. Solid State Mater. Sci.* 2, 48 (1997).
3. S. C. Jain, H. J. Osten, B. Dietrich, and H. Rücker, *Semicond. Sci. Technol.* 10, 1289 (1995).
4. H. J. Osten, H. P. Zeindl, and E. Bugiel, *J. Cryst. Growth* 143, 194 (1994).
5. A. R. Powell, K. Eberl, B. A. Ek, and S. S. Iyer, *J. Cryst. Growth* 127, 425 (1993).
6. H. J. Osten, *Mater. Sci. Eng., B* 36, 268 (1996).
7. J. L. Regolini, F. Gisbert, G. Dolino, and P. Boucaud, *Mater. Lett.* 18, 58 (1993).
8. M. S. Goorsky, S. S. Iyer, K. Eberl, F. K. LeGouech, J. Angiello, and F. Cardonne, *Appl. Phys. Lett.* 60, 2758 (1992).
9. G. G. Fischer, P. Zaumseil, E. Bugiel, and H. J. Osten, *J. Appl. Phys.* 77, 1934 (1995).
10. G. Lippert, "Contamination in Si Molecular Beam Epitaxy," PhD Thesis, University of the Armed Forces, Munich, 1994.
11. S. Im, J. Washburn, R. Gronsky, N. W. Cheung, K. M. Yu, and J. W. Ager, *Appl. Phys. Lett.* 63, 2682 (1993).
12. J. W. Strane, H. J. Stein, S. R. Lee, B. L. Doyle, S. T. Picraux, and J. W. Mayer, *Appl. Phys. Lett.* 63, 2786 (1993).
13. G. Davies and R. C. Newman, in "Handbook of Semiconductors" (T. S. Moss, Ed.), Vol. 3. Elsevier Science B. V., Amsterdam, 1994.
14. J. Tersoff, *Phys. Rev. Lett.* 74, 5080 (1995).
15. P. C. Kelires and J. Tersoff, *Phys. Rev. Lett.* 63, 1164 (1989).
16. W. J. Taylor, T. Y. Tan, and U. Goesele, *Appl. Phys. Lett.* 62, 3336 (1993).
17. H. J. Osten, E. Bugiel, and P. Zaumseil, *Appl. Phys. Lett.* 64, 3440 (1994).
18. K. Pressel, B. Dietrich, H. Rücker, M. Methfessel, and H. J. Osten, *Mater. Sci. Eng., B* 36, 167 (1996).

19. H. J. Osten, M. Kim, K. Pressel, and P. Zaumseil, *J. Appl. Phys.* 80, 6711 (1996).

20. M. W. Dashiell, L. V. Kulik, D. Hits, J. Kolodzey, and G. Watson, *Appl. Phys. Lett.* 72, 833 (1998).

21. S. Zerlauth, H. Seyringer, C. Penn, and F. Schäffler, *Appl. Phys. Lett.* 71, 3826 (1997).

22. T. O. Mitchell, J. L. Hoyt, and J. F. Gibbons, *Appl. Phys. Lett.* 71, 1688 (1997).

23. S. S Iyer, K. Eberl, A. R. Powell, and B. A. Ek, *Microelectron. Eng.* 19, 351 (1992).

24. K. Brunner, W. Winter, K. Eberl, N. Y. Jin-Phillipp, and F. Phillip, *J. Cryst. Growth* 175/176, 451 (1997).

25. H. J. Osten and P. Gaworzewski, *J. Appl. Phys.* 82, 4977 (1997).

26. H. J. Osten, G. Lippert, D. Knoll, R. Barth, B. Heinemann, H. Rücker, and P. Schley, *Techn. Digest IEDM* 1997, 803 (1997).

27. P. C. Kelires and E. Kaxiras, *J. Vac. Sci Technol., B* 16, 1687 (1998).

28. P. D. Zavitsanos and G. A. Carlson, *J. Chem Phys.* 59, 2966 (1972).

29. M. Kim, G. Lippert, and H. J. Osten, *J. Appl. Phys.* 80, 5748 (1996).

30. J. Chastain, Ed., "Handbook of X-ray Photoelectron Spectroscopy." Perkin-Elmer Corporation, Physical Electronics Division, Eden Prairie, 1992.

31. H. J. Osten, M. Methfessel, G. Lippert, and H. Rücker, *Phys. Rev. B* 52, 12179 (1995).

32. G. Lippert, P. Zaumseil, H. J. Osten, and M. Kim, *J. Cryst. Growth* 175/176, 473 (1997).

33. J. P. Liu and H. J. Osten, *Appl. Phys. Lett.* 76, 3546 (2000).

34. M. Berti, D. De Salvator, A. V. Drigo, F. Romanato, J. Stangl, S. Zerlauth, F. Schäffler, and G. Bauer, *Appl. Phys. Lett.* 72, 1602 (1998).

35. J. Mi, P. Warren, P. Letourneau, M. Judelewicz, M. Gailhanou, M. Dutoit, C. Dubois, and J. C. Dupuy, *Appl. Phys. Lett.* 67, 259 (1995).

36. H. J. Osten, J. Griesche, and S. Scalese, *Appl. Phys. Lett.* 74, 836 (1999).

37. P. C. Kelires, *Surf. Sci.* 418, L62 (1998).

38. R. Duschl, O. G. Schmidt, W. Winter, K. Eberl, M. W. Dashiell, J. Kolodzey, N. Y. Jin-Phillipp, and F. Phillipp, *Appl. Phys. Lett.* 74, 1150 (1999).

39. E. Kasper and H. Jorke, *J. Vac. Sci. Technol., A* 10, 1927 (1992).

40. D. C. Houghton, *J. Appl. Phys.* 70, 2136 (1991).

41. K. L. Wang and R. P. G. Karunasiri, *J. Vac. Sci. Technol., B* 11, 1159 (1993).

42. P. C. Kelires, *Appl. Surf. Sci.* 102, 12 (1996).

43. W. Windl, O. F. Sankey, and J. Menéndez, *Phys. Rev. B: Solid State* 57, 2431 (1998).

44. B. Dietrich, H. J. Osten, H. Rücker, M. Methfessel, and P. Zaumseil, *Phys. Rev. B: Solid State* 49, 17185 (1994).

45. A. Dal Pino, A. M. Rappe, and J. D. Joannopolous, *Phys. Rev. B: Solid State* 47, 12554 (1993).

46. J. Tersoff, *Phys. Rev. Lett.* 64, 1757 (1990).

47. J. A. Baker, T. N. Tucker, N. E. Moyer, and R. C. Buschert, *J. Appl. Phys.* 39, 436 (1968).

48. J. L. Martins and A. Zunger, *Phys. Rev. Lett.* 56, 1400 (1986).

49. P. N. Keating, *Phys. Rev.* 145, 637 (1966).

50. H. Rücker and M. Methfessel, *Phys. Rev. B: Solid State* 52, 11059 (1995).

51. R. J. Borg and G. J. Dienes, "The Physical Chemistry of Solids." Academic Press, Boston, 1992.

52. S. de Gironcoli, P. Giannozzi, and S. Baroni, *Phys. Rev. Lett.* 66, 2116 (1991).

53. A. R. Bean and R. C. Newman, *J. Phys. Chem. Solids* 32, 1211 (1971).

54. H. Rücker, M. Methfessel, E. Bugiel, and H. J. Osten, *Phys. Rev. Lett.* 72, 3578 (1994).

55. N. Mousseau and M. F. Thorpe, *Phys. Rev. B: Solid State* 46, 15887 (1992).

56. Y. Cai and M. F. Thorpe, *Phys. Rev. B* 46, 15872, 15879 (1992).

57. D. B. Aldrich, R. J. Nemanich, and D. E. Sayers, *Phys. Rev. B: Solid State* 50, 15026 (1994).

58. H. Kajiyama and Y. Nishino, *Phys. Rev. B: Solid State* 45, 14005 (1992).

59. N. Mosseau and M. F. Thorpe, *Phys. Rev. B: Solid State* 48, 5172 (1993).

60. A. A. Demkov and O. F. Sankey, *Phys. Rev. B: Solid State* 48, 2207 (1993).

61. R. People and J. C. Bean, *Appl. Phys. Lett.* 47, 322 (1985).

62. A. R. Powell, F. K. LeGoues, and S. S. Iyer, *Appl. Phys. Lett.* 64, 324 (1994).

63. A. R. Powell, K. Eberl, F. K. Legoues, and B. A. Ek, *J. Vac. Sci Technol., B* 11, 1064 (1993).

64. H. J. Osten, D. Endisch, E. Bugiel, B. Dietrich, G. G. Fischer, M. Kim, D. Krüger, and P. Zaumseil, *Semicond. Sci. Technol.* 11, 1678 (1996).

65. P. Warren, J. Mi, F. Overney, and M. Dutoit, *J. Cryst. Growth* 157, 414 (1995).

66. R. N. Kyutt, P. V. Petrashen, and L. M. Sorokin, *Phys. Status Solidi A* 60, 381 (1980).

67. W. K. Chu, J. W. Mayer, and M.-A. Nicolet, "Backscattering Spectrometry." Academic Press, New York, 1978.

68. L. C. Feldman, J. W. Mayer, and S. T. Picraux, "Material Analysis by Ion Channeling." Academic Press, New York, 1982.

69. J. A. Leavitt, L. C. McIntyre, Jr., P. Stoss, J. G. Oder, M. D. Ashbaugh, B. Dezfouly-Arjomandy, Z.-M. Yang, and Z. Lin, *Nucl. Instrum. Methods Phys. Res., Sect. B* 40/41, 776 (1989).

70. D. Endisch, H. J. Osten, P. Zaumseil, and M. Zinke-Allmang, *Nucl. Instrum. Methods Phys. Res., Sect. B* 100, 125 (1995).

71. P. A. Temple and C. A. Hathaway, *Phys. Rev. B: Solid State* 7, 3685 (1973).

72. P. Zaumseil, *Phys. Status Solidi A* 141, 155 (1994).

73. A. Claverie, J. Fauré, J. L. Balladore, L. Simon, A. Mesli, M. Diani, L. Kubler, and D. Aubel, *J. Cryst. Growth* 157, 420 (1995).

74. H. Rücker, M. Methfessel, B. Dietrich, K. Pressel, and H. J. Osten, *Phys. Rev. B: Solid State* 53, 1302 (1996).

75. P. M. Fauchet in "Light Scattering in Semiconductor Structures and Superlattices" (D. J. Lockwood and J. F. Young, Eds.), p. 229. Plenum, New York, 1991.

76. G. Davies, E. C. Lightowlers, R. C. Newman, and A. S. Oates, *Semicond. Sci. Technol.* 2, 524 (1987).

77. J. W. Strane, S. T. Picraux, H. J. Stein, S. R. Lee, J. Candelaria, D. Theodore, and J. W. Mayer, *Mater. Res. Soc. Proc.* 321, 467 (1994).

78. L. V. Kulik, D. A. Hits, M. W. Dashiell, and J. Kolodzey, *Appl. Phys. Lett.* 72, 1972 (1998).

79. K. Ismail, *J. Vac. Sci. Technol., B* 14, 2776 (1996).

80. E. A. Fitzgerald, Y. H. Xie, M. L. Green, D. Brasen, A. R. Kortan, J. Michel, Y. J. Mii, and B. E. Weir, *Appl. Phys. Lett.* 59, 811 (1991).

81. J. Tersoff, *Appl. Phys. Lett.* 62, 693 (1993).

82. H. J. Osten and E. Bugiel, *Appl. Phys. Lett.* 70, 2813 (1997).

83. E. Bugiel, S. Ruvimov, and H. J. Osten, *Int. Phys. Conf. Ser.* 146, 301 (1995).

84. H. J. Osten and J. Klatt, *Appl. Phys. Lett.* 65, 630 (1994).

85. E. Bugiel and P. Zaumseil, *Appl. Phys. Lett.* 62, 2051 (1993).

86. R. A. Soref, *J. Appl. Phys.* 70, 2470 (1991).

87. K. Brunner, K. Eberl, and W. Winter, *Phys. Rev. Lett.* 76, 303 (1996).

88. W. Kissinger, M. Weidner, H. J. Osten, and M. Eichler, *Appl. Phys. Lett.* 65, 3356 (1994).

89. S. Zollner, *J. Appl. Phys.* 78, 5209 (1995).

90. W. Kissinger, H. J. Osten, M. Weidner, and M. Eichler, *J. Appl. Phys.* 79, 3016 (1996).

91. A. R. Powell, K. Eberl, B. A. Ek, and S. S. Iyer, *J. Cryst. Growth* 127, 425 (1993).

92. H. J. Osten, *J. Appl. Phys.* 84, 2716 (1998).

93. S. Galdin, P. Dollfus, V. Aubry-Fortuna, P. Hesto, and H. J. Osten, *Semicond. Sci. Technol.* 15, 565 (2000).

94. K. Rim, S. Takagi, J. J. Welser, J. L. Hoyt, and J. F. Gibson, *Mater. Res. Soc. Symp. Proc.* 379, 327 (1995).

95. B. L. Stein, E. T. Yu, E. T. Croke, A. T. Hunter, T. Laursen, A. E. Bair, J. W. Mayer, and C. C. Ahn, *J. Vac. Sci. Technol., B* 15, 1108 (1997).

96. Landolt-Börnstein, "Numerical Data and Functional Relationships in Science and Technology," Vol. III/17a, O. Madelung, Springer-Verlag, New York, 1982.

97. C. G. Van der Walle, *Phys. Rev. B: Solid State* 39, 1871 (1989).

98. C. G. Van der Walle and R. M. Martin, *Phys. Rev. B: Solid State* 34, 5621 (1986).

99. J. F. Morar, P. E. Batson, and J. Tersoff, *Phys. Rev. B: Solid State* 47, 4107 (1993).

100. F. H. Pollack and N. Cardona, *Phys. Rev.* 172, 816 (1968).

101. M. Kim and H. J. Osten, *Appl. Phys. Lett.* 70, 2702 (1997).

102. C. G. Van der Walle, in "Properties of Strained and Relaxed Silicon Germanium," EMIS Datareview Series No. 12, Ed. E. Kasper, p. 99. Institution of Electrical Engineers, London 1995, pp. 99.

103. H. J. Osten and P. Gaworzewski, *J. Appl. Phys.* 82, 4977 (1997).

104. H. Rücker, M. Methfessel, B. Dietrich, H. J. Osten, and P. Zaumseil, *Superlattices Microstruct.* 16, 121 (1994).

105. S. S. Iyer, K. Eberl, A. R. Powell, and B. A. Ek, *Microelectron. Eng.* 19, 351 (1992).

106. M. Ershov and V. Ryzhii, *J. Appl. Phys.* 76, 1924 (1994).

107. K. Brunner, W. Winter, K. Eberl, N. Y. Jin-Phillip, and F. Phillip, *J. Cryst. Growth* 175/176, 451 (1997).

108. B. R. Nag, "Electron Transport in Compound Semiconductors." Springer-Verlag, Berlin, 1980.

109. G. F. Neumark and D. K. Schröder, *J. Appl. Phys.* 52, 885 (1981).

110. S. M. Sze, "Physics of Semiconductor Devices." Wiley, New York, 1981.

111. H. J. Osten, J. Griesche, P. Gaworzewski, and K. D. Boltze, *Appl. Phys. Lett.* 76, 200 (2000).

112. T. Manku and A. Nathan, *Phys. Rev. B: Solid State* 43, 12634 (1991).

113. K. Takeda, A. Taguchi, and M. Sakala, *J. Phys. C: Solid State Phys.* 16, 2237 (1983).

114. A. Levitas, *Phys. Rev.* 99, 1810 (1955).

115. G. von Busch and O. Vogt, *Helv. Phys. Acta* 33, 437 (1960).

116. T. Manku and A. Nathan, *IEEE Electron Device Lett.* 12, 704 (1991).

117. R. People, J. C. Bean, D. V. Lang, A. M. Sergent, H. L. Sörmer, K. W. Wecht, R. T. Lynch, and K. Baldwin, *Appl. Phys. Lett.* 45, 1231 (1984).

118. T. Manku, S. C. Jain, and A. Nathan, *J. Appl. Phys.* 71, 4618 (1992).

119. L. C. Kimerling, M. T. Asom, J. L. Benton, P. J. Drevinsky, and C. E. Caefer, *Mater. Sci. Forum* 38–41, 141 (1989).

120. E. Quinones, S. K. Ray, K. C. Liu, and S. Banerjee, *IEEE Electron Device Lett.* 20, 338 (1999).

121. S. Ruvimov, E. Bugiel, and H. J. Osten, *J. Appl. Phys.* 78, 2323 (1995).

122. E. D. Richmond, *Thin Solid Films* 252, 98 (1994).

123. G. Lippert, H. J. Thieme, and H. J. Osten, *J. Electrochem. Soc.* 142, 191 (1995).

124. M. Kitabatake, M. Deguchi, and T. Hirao, *J. Appl. Phys.* 74, 4438 (1993).

125. S. Tanuma, C. J. Powell, and D. R. Penn, *Surf. Interface Anal.* 17, 911 (1991).

126. I. Ban, M. C. Öztürk, and E. Kutlu, *IEEE Trans. Electron Dev.* 44, 1544 (1997).

127. H. J. Gossmann, C. Rafferty, G. Hobler, H. Vuong, D. Jacobson, and M. Frei, *Techn. Digest IEDM* 1998, 725 (1998).

128. L. D. Lanzarotti, J. C. Sturm, E. Stach, R. Hull, T. Buyuklimanli, and C. Magee, *Techn. Digest IEDM* 1996, 249 (1996).

129. D. Knoll, B. Heinemann, H. J. Osten, K. E. Ehwald, B. Tillack, P. Schley, R. Barth, M. Matthes, K. S. Park, Y. Kim, and W. Winkler, *Techn. Digest IEDM* 1998, 703 (1998).

130. K. E. Ehwald, D. Knoll, B. Heinemann, K. Chang, J. Kirchgessner, R. Mauntel, Ik-Sung Lim, J. Steele, B. Tillack, A. Wolff, K. Blum, W. Winkler, M. Pierschel, F. Herzel, U. Jagdhold, P. Schley, R. Barth, and H. J. Osten, *Techn. Digest IEDM* 1999, 561 (1999).

131. E. J. Prinz, P. M. Garone, P. V. Schwartz, X. Xiao, and J. C. Sturm, *IEEE Electron Device Lett.* 12, 42 (1991).

132. H. Rücker, B. Heinemann, W. Röpke, R. Kurps, D. Krüger, G. Lippert, and H. J. Osten, *Appl. Phys. Lett.* 73, 1682 (1998).

133. P. A. Stolk, D. J. Eaglesham, H. J. Gossmann, and J. M. Poate, *Appl. Phys. Lett.* 66, 1370 (1995).

134. R. Scholz, U. Gösele, J.-Y. Huh, and T. Y. Tan, *Appl. Phys. Lett.* 72, 200 (1998).

135. H. Rücker, B. Heinemann, D. Bolze, D. Knoll, D. Krüger, R. Kurps, H. J. Osten, P. Schley, B. Tillack, and P. Zaumseil, *Techn. Digest IEDM* 1999, 345 (1999).

136. H. Rücker, B. Heinemann, D. Bolze, R. Kurps, D. Krüger, G. Lippert, and H. J. Osten, *Appl. Phys. Lett.* 74, 3377 (1999).

137. H. J. Osten, G. Lippert, P. Gaworzewski, and R. Sorge, *Appl. Phys. Lett.* 71, 1522 (1997).

138. B. Heinemann, D. Knoll, G. G. Fischer, P. Schley, and H. J. Osten, *Thin Solid Films* 369, 347 (2000).

139. B. Heinemann, "2D Device Simulation of the Electrical Properties of SiGe HBTs," PhD Thesis, Culliviel Verlag, Göttingen, 1998.

140. A. Schüppen, U. Erben, A. Gruhle, H. Kibbel, H. Schumacher, and U. König, *Techn. Digest IEDM* 1995, 743 (1995).

141. K. Oda, E. Ohue, M. Tanabe, H. Shimamoto, T. Onai, and K. Washio, *Techn. Digest IEDM* 1997, 791 (1997).

142. K. Ohue, K. Oda, R. Hayami, and K. Washio, in "Proceedings of the 1998 Bipolar/BiCMOS Circuits and Technology Meeting" (IEEE, Piscatataway), p. 791, 1998.

143. D. Knoll, B. Heinemann, R. Barth, K. Blum, J. Drews, A. Wolff, P. Schley, D. Bolze, B. Tillack, G.Kissinger, W. Winkler, and H. J. Osten, in "Proceedings of the 28th European Solid-State Device Research Conference" (Editions Frontieres, Paris), p. 142, 1998.

144. H. J. Osten, D. Knoll, B. Heinemann, H. Rücker, and B. Tillack, in "Proceedings of the 1999 Bipolar/BiCMOS Circuits and Technology Meeting" (IEEE, Piscataway), p. 109, 1999.

145. D. Knoll, B. Heinemann, B. Tillack, P. Schley, and H. J. Osten, *Thin Solid Films* 369, 342 (2000).

146. A. J. Joseph, J. D. Cressler, D. M. Richey, R. C. Jaeger, and D. L. Harame, *IEEE Trans Electron Dev.* 44, 404 (1997).

147. B. Heinemann, D. Knoll, G. Fischer, D. Krüger, G. Lippert, H. J. Osten, H. Rücker, W. Röpke, P. Schley, and B. Tillack, in "Proceedings of the 27th European Solid-State Device Research Conference" (Editions Frontieres, Paris), p. 544, 1997.

Chapter 6

LOW-FREQUENCY NOISE SPECTROSCOPY FOR CHARACTERIZATION OF POLYCRYSTALLINE SEMICONDUCTOR THIN FILMS AND POLYSILICON THIN FILM TRANSISTORS

Charalabos A. Dimitriadis

Aristotle University of Thessaloniki, Department of Physics, Thessaloniki 54006, Greece

George Kamarinos

LPCS, ENSERG, 38016 Grenoble Cedex 1, France

Contents

1. INTRODUCTION

1.1. Definitions and Noise Sources

The voltage measured at the end of a device is not perfectly constant, but fluctuates around its average value. The time dependence of these fluctuations is referred to as "noise." Knowledge of the noise is important for the following reasons: (a) Concerning basic physics, noise can provide information on fundamental transport mechanisms, such as scattering mechanisms. (b) With regard to technology, noise can be used as a

Handbook of Thin Film Materials, edited by H.S. Nalwa
Volume 4: Semiconductor and Superconductor Thin Films
Copyright © 2002 by Academic Press
All rights of reproduction in any form reserved.

ISBN 0-12-512912-2/$35.00

powerful diagnostic tool for detecting defects in devices (bulk or interface states). (c) With regard to device physics, noise is a fundamental limitation to the sensitivity of a device. These mentioned reasons demonstrate the importance of noise studies in semiconductor devices.

Consider that the measured voltage $V(t)$ across a device fluctuates around a constant value V_o,

$$V(t) = V_o + \Delta V(t) \tag{1}$$

The random fluctuations $\Delta V(t)$ are characterized by their power spectral density (PSD), defined as the time averages of the Fourier transform of $\Delta V(t)$,

$$S_v(f) \sim \langle V(f)V^*(f) \rangle \tag{2}$$

In Eq. (2), $V(f)$ is the Fourier transform of $\Delta V(t)$ and f is the frequency. Based on the Wiener–Khinchin theorem [1], $S_v(f)$ can be written as

$$S_v(f) = 4 \int_0^\infty C(t) \cos(2\pi f t) dt \tag{3}$$

where $C(t) = \langle \Delta V(t) \Delta V(0) \rangle$ is the autocorrelation function of $\Delta V(t)$. The above definitions provide the method for noise measurements. First, the random signal $V(t)$ is sampled at times t_i, giving $V_i = V(t_i)$. Then the autocorrelation function is computed, and finally a fast Fourier transform procedure is used to complete $S_v(f)$ at the sampling frequencies f.

In semiconductors, five types of noise are of importance.

1. Thermal noise. This type of noise results from the thermal motion of the charge carriers in the sample. The thermal noise is generated in any physical resistor if a current is passed through it and cannot be reduced. For a resistance R, the thermal noise of the current or voltage fluctuations is characterized by the white spectrum without a current dependence,

$$S_v = 4kTR \tag{4}$$

$$S_I = 4kT/R \tag{5}$$

where k is Boltzmann's constant and T is the temperature.

2. Shot noise. Shot noise appears in a current, I, of discrete charge carriers crossing a potential barrier. The origin of the shot noise is due to thermal fluctuations of the charge carrier emission rate. The PSD of the shot noise is also characterized by the current-dependent white spectrum,

$$S_I(f) = 2qI \tag{6}$$

where q is the electron charge.

3. Diffusion noise. This type of noise is related to fluctuations in the number of charge carriers arising from the transport of carriers into and out of a given volume under consideration. In the case of one-dimensional diffusion, the PSD shows a $1/f^{1/2}$ behavior at low frequencies and $1/f^{3/2}$ at high frequencies [2]. The corner frequency, f_c, contains the information of the diffusion because $f_c = D/(\pi L^2)$, where L is the sample length and D is the diffusion coefficient.

4. Generation-recombination noise. Generation-recombination (G-R) noise is due to fluctuation in the number of free charge carriers N as a result of trapping and detrapping of the free charge carriers at traps in semiconductors. For a single trap of time constant τ, the PSD of G-R noise is characterized by the Lorentzian spectrum,

$$S_N(f) = \langle (\Delta N)^2 \rangle \frac{4\tau}{1 + (2\pi\tau)^2} \tag{7}$$

The Lorentzian noise shows a plateau at very low frequencies, followed by a $1/f^2$ decrease; the corner frequency f_c is directly related to the trap time constant ($\tau = 1/2\pi f_c$).

5. $1/f$ noise (Flicker noise). This type of noise is a fluctuation in the conductance with a PSD proportional to $f^{-\gamma}$, where the spectral exponent γ is nearly 1. Unlike the first four noise sources mentioned above, which are well understood, the origin of the $1/f$ noise is still under debate. The generally accepted models explaining the $1/f$ noise in semiconductor devices are based on carrier number or mobility fluctuation [3]. More details of the understanding of the $1/f$ noise in homogeneous semiconductors are given below.

1.2. $1/f$ Noise in Semiconductors

A simple way to approach the origin of $1/f$ noise is to assume that the $1/f$ noise spectrum is a summation of a large number of relaxation processes. For the relaxation process of a fluctuating quantity $\Delta X(t)$ with time constant τ, the Langevin equation is given by

$$\frac{d\Delta X(t)}{dt} = -\frac{\Delta X(t)}{\tau} + H(t) \tag{8}$$

where $H(t)$ is the Langevin random source, which has a white PSD. Making a Fourier transform of Eq. (8), one obtains the Lorentzian spectrum of $\Delta X(t)$,

$$S_{\Delta X}(t) = \langle (\Delta X)^2 \rangle \frac{4\tau}{1 + (2\pi f\tau)^2} \tag{9}$$

If there is a distribution $g(\tau)$ of time constants, and we integrate over a range of time constants between τ_1 and τ_2, the PSD is

$$S_{\Delta X}(t) = \langle (\Delta X)^2 \rangle \int_{\tau_1}^{\tau_2} \frac{\tau g(\tau)}{1 + (2\pi f\tau)^2} d\tau \tag{10}$$

The above equation will give a $1/f$ spectrum of the form

$$S_{\Delta X} = \frac{\langle (\Delta X)^2 \rangle}{\ln(\tau_2/\tau_1)} \frac{1}{f} \qquad \text{for } \frac{1}{2\pi\tau_2} < f < \frac{1}{2\pi\tau_1} \tag{11}$$

if $g(\tau)$ is inversely proportional to τ, i.e.,

$$g(\tau)d\tau = \frac{1}{\tau \ln(\tau_2/\tau_1)} d\tau \qquad \text{for } \tau_1 < \tau < \tau_2 \tag{12}$$

Thus, in identifying the origin of the $1/f$ noise, the problem is reduced to finding the correct physical processes that have a distribution in the relaxation time τ inversely proportional to τ. Below we will present two cases where the distribution of relaxation times satisfies Eq. (12).

1. The McWhorter model [4]. In the McWhorter model, the noise source is due to traps located in an oxide layer on a semiconductor. This model assumes that trapping and detrapping of charge carriers occurs during tunneling between the band states of the semiconductor and the trap states in the oxide layer. In this way, the total number of free charge carriers in the semiconductor is modulated and fluctuates in time. Trapping at a single trap with a time constant τ, which is inversely proportional to the tunneling probability, results in a Lorentzian spectrum. The tunneling probability varies exponentially with the distance x from the oxide/semiconductor interface, i.e.,

$$\tau = \tau_o \exp(x/\beta) \tag{13}$$

where β is the tunneling parameter ($\sim 10^8$ cm^{-1}). Assuming that the traps are homogeneously distributed in the oxide layer, it is obtained as

$$g(\tau) = \frac{dN}{d\tau} = \frac{dN}{dx}\frac{dx}{d\tau} \propto \frac{1}{\tau} \tag{14}$$

If the time constant of traps at the interface ($x = 0$) is $\tau_o \cong 10^{-12}$ s, then a distribution of traps in an oxide layer of thickness of 3 nm will yield a $1/f$ spectrum down to about 0.1 Hz.

2. The Dutta–Horn model [5]. In the Dutta–Horn model, it is assumed that there are some thermally activated processes with a relaxation time τ following the relationship

$$\tau = \tau_o \exp(E/kT) \tag{15}$$

where E is the energy of the trap from the bottom of the band. It has been shown that when the distribution of the activation energies is constant, $g(\tau) \propto 1/\tau$; hence $S_v \propto 1/f$. When the distribution of the activation energies is not constant but has a peak centered at E_o with a width much larger than kT, then a $1/f^\gamma$ spectrum is produced, with $\gamma = 1 \pm 0.3$. The width of the peak is determined by the disorder in the crystal. The slope S_v of the PSD is not constant over the whole frequency range, and the relationship between the temperature and the slope of S_v is

$$\gamma(\omega, T) = 1 - \frac{1}{\ln(\omega\tau_o)}\left[\frac{d \ln S(\omega, T)}{d \ln T} - 1\right] \tag{16}$$

where $\omega = 2\pi f$ is the angular frequency and f is the frequency. The Dutta–Horn thermal activation model has been used to study the interface trap distribution in Si N-channel n-metal-oxide-semiconductor field-effect transistors (N-MOSFETs) through noise measurements as a function of temperature [6–8]. When the Fermi level is located in a position where the trap states density is increasing toward the conduction band edge, γ is less than 1, and when the trap states density decreases γ is larger than 1. Thus, whenever γ becomes 1 (from less than 1 to more than 1) the Fermi level is located at the peak of the trap distribution.

Another way to approach the origin of the $1/f$ noise is with the mobility fluctuation model. Hooge [9] proposed that the relative PSD, S_R/R^2, can be normalized to the total number of the free charge carriers N, following the relationship

$$\frac{S_R}{R^2} = \frac{\alpha_H}{fN} \tag{17}$$

where α_H is the Hooge constant, varying in the range of 10^{-6} to 10^{-3}, depending on the material quality [10]. The empirical relationship of Eq. (17) was found to describe successfully the $1/f$ noise in many semiconductors and metals [11]. Thus, two general mechanisms can lead to $1/f$ fluctuations in semiconductors, namely fluctuations in the number of the charge carriers or fluctuations in their mobility. In some cases, for example, in the inversion layer of MOSFETs [12], number and mobility could fluctuate simultaneously.

1.3. $1/f$ Noise in MOSFETs

Two models have been invoked to explain the noise properties of MOS transistors [13]. One assumes that the drain current fluctuation is caused by fluctuations of the inversion charge near the semiconductor–oxide interface. If the effective mobility of the carriers is assumed to be independent of the insulator charge, then the normalized drain current spectral density is

$$\frac{S_{I_D}}{I_D^2} = \frac{g_m^2}{I_D^2} S_{V_{FB}} \tag{18}$$

where g_m is the device transconductance and $S_{V_{FB}}$ is the flat-band voltage spectral density [13]. Ghibaudo et al. [13] have shown that if the fluctuation of the insulator charge induces noticable changes in the effective carrier mobility μ_{eff} via Coulomb scattering, the normalized drain current spectral density S_{I_D}/I_D^2 is related to $S_{V_{FB}}$ by the relationship

$$\frac{S_{I_D}}{I_D^2} = \left(1 + \alpha'\mu_{eff}C_{ox}\frac{I_D}{g_m}\right)\frac{g_m^2}{I_D^2} S_{V_{FB}} \tag{19}$$

where α' is a constant correlated with the sensitivity of the mobility to the interface charge Coulomb scattering. If $\alpha' \approx 0$, i.e., if the mobility is almost independent of the interface charge, then Eq. (19) reduces to Eq. (18) and S_{I_D}/I_D^2 varies as $(g_m/I_D)^2$, provided $S_{V_{FB}}$ is weakly bias dependent. Following the conventional tunneling theory for charge carriers penetrating from the semiconductor into the oxide layer, the flat-band voltage spectral density is given by the expression [14, 15]

$$S_{V_{FB}} = \frac{q^2 kT \lambda N_{ox}}{WLC_{ox}^2 f} \tag{20}$$

where λ is the tunnel attenuation distance (about 0.1 nm), C_{ox} is the gate oxide capacitance per unit area, and N_{ox} is the slow oxide trap density (in cm^{-3} eV^{-1}).

The second model, originally introduced by Hooge [16], relies on the carrier mobility fluctuations. In this case, the normalized drain current spectral density of a uniform channel (i.e., in the linear region of operation) is [13]

$$\frac{S_{I_D}}{I_D^2} = \frac{q\alpha_H}{WLQ_i}\frac{1}{f} \tag{21}$$

where W is the channel width, L is the channel length, and Q_i is the inversion charge. In strong inversion, the inversion charge reduces to $Q_i = C_{ox}(V_G - V_T)$, where V_G is the gate voltage and V_T is the threshold voltage.

The mechanism responsible for the origin of the $1/f$ noise in MOS transistors can be identified from the S_{I_D}/I_D^2 versus I_D characteristics in a log-log scale [13]: if the normalized noise current varies as the inverse of the drain current in weak inversion, and if no obvious correlation with $(g_m/I_D)^2$ appears, one can conclude that Hooge mobility fluctuations may be the dominant noise source. If the carrier number fluctuation model idominates, S_{I_D}/I_D^2 starts from a plateau at weak inversion before decreasing as $1/I_D^2$ at strong inversion. In the last case, a direct comparison between the experimental data S_{I_D}/I_D^2 and $(g_m/I_D)^2$ with Eq. (18) or (19) will allow us to determine the parameter $S_{V_{FB}}$ and, thus, the trap density in the oxide N_{ox} from Eq. (20).

1.4. The Present Work

Regardless of the exact origin of the $1/f$ noise (McWhorter or Hooge theory), numerous studies have demonstrated the usefulness of low-frequency (LF) noise as a diagnostic tool for characterizing the quality and reliability of homogeneous semiconductor materials and devices (for a recent overview see [17]). However, thin films of polycrystalline semiconductors find widespread use in microelectronics. In particular, over the last two decades, polycrystalline silicon thin-film transistors (polysilicon TFTs) have attracted much research interest because of their applications in active matrix liquid crystal displays as pixel switching elements and static random access memories. In these applications, the noise property of TFTs is of major importance and must be taken into account in circuit design.

From the fundamental LF noise point of view, polysilicon TFTs present an interesting challenge. Indeed, besides the classical noise sources located at the front polysilicon/gate oxide interface, other sources may also contribute to the noise of a device. For very thin-film devices, the current transport and, therefore, the fluctuations will be affected by the back interface, giving rise to coupling effects. In addition, G-R centers in the bulk of the polysilicon layer (located at the grain boundaries and within the grains) will also contribute to the noise. These noise sources sometimes render the interpretation of the noise data for polysilicon TFTs quite difficult but challenging. In this chapter, we present systematic studies of the noise performance of polysilicon TFTs that have involved extensive research in recent years. More specifically, the LF noise spectroscopy technique as a diagnostic tool for the quality characterization of polycrystalline semiconductor thin films and of polysilicon TFTs is described, and the use of LF noise spectroscopy for studying aging effects in polysilicon TFTs is discussed.

2. NOISE OF POLYCRYSTALLINE SEMICONDUCTOR THIN FILMS

2.1. Introduction

In general, polycrystalline semiconductor devices exhibit poor electrical characteristics compared with single-crystal devices because of the presence of electrically active traps within the grains and at the grain boundaries. For practical applications, materials of low defect density are required. Low-frequency noise is becoming popular as a tool for characterizing qualitatively basic materials and semiconductor devices as well as determining the existence and location of trapping states. The basic reason for this is that the low-frequency fluctuations monitored during noise measurements are caused solely by the defects along the path of the current flow, explaining its high sensitivity to the defect properties of the materials [17]. Despite the extensive work done, noise characterization of polycrystalline semiconductor thin films has received little attention in the literature.

It is generally held that transport in polycrystalline thin films is dominated by hopping of carriers through defect states at the Fermi level at low temperatures, whereas at room temperature the conductivity is thermally activated, satisfying the Meyer–Neldel (MN) rule [18–20]. Recently, a room temperature low-frequency noise technique has been reported for defect characterization of polycrystalline iron disilicide (β-FeSi$_2$) semiconductor thin films [20]. Based on a thermally activated conduction mechanism satisfying the MN rule and taking into account mobility inhomogeneity across the thickness of the film, the power spectral density of the current fluctuations S_I was explained by fluctuations in the number of carriers via trapping–detrapping of holes of the valence band and the gap states. Through the combination of the noise with the Hall and conductivity experimental data, the trap density in polycrystalline thin films can be determined.

2.2. Noise Model

In polycrystalline semiconductor thin films, the origin of the $1/f$ noise of current fluctuations can be explained by fluctuations either in mobility (Hooge model) [11] or in the number of carriers (McWhorter model) [21]. When the current fluctuations are caused by mobility fluctuations, the power spectral density of the current fluctuations S_I obeys the Hooge relationship [11],

$$\frac{S_I}{I^2} = \frac{\alpha_H}{Pf} \tag{22}$$

where α_H is known as the Hooge parameter and P is the total number of carriers in the sample. The parameter α_H is a few times 10^{-3} when phonon scattering prevails and may strongly decrease if impurity scattering occurs [11].

To explain the current noise data by fluctuations in the number of carriers, the conduction mechanism in the polycrystalline semiconductor thin films has to be identified. As an example, the temperature dependence of the conductivity σ of a polycrystalline β-FeSi$_2$ thin film is shown in Figure 1. The plot of σ versus $10^3/T$ shows that at low temperatures ($T < 220$ K) the conductivity obeys Mott's relationship, $\sigma = \sigma_0 \exp[-(T_0/T)^{1/4}]$, for variable-range hopping [22] (inset of Fig. 1). At temperatures above 220 K, the observed continuous bending of the $\ln\sigma$ versus $10^3/T$ curve can be explained by

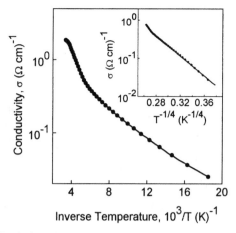

Fig. 1. Electrical conductivity σ as a function of the inverse temperature $10^3/T$ of the polycrystalline β-FeSi$_2$ thin film prepared by conventional furnace annealing at a temperature of 800°C for 1 h. The inset shows a plot of σ versus $T^{-1/4}$. (Reprinted from D. H. Tassis et al., *J. Appl. Phys.* 85, 4091, 1999, with permission.)

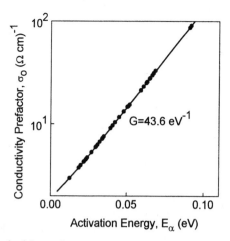

Fig. 2. Conductivity pre-factor σ_0 versus activation energy E_α obtained from the conductivity data of Figure 1. (Reprinted from D. H. Tassis et al., *J. Appl. Phys.* 85, 4091, 1999, with permission.)

the statistical shift of the Fermi level E_F with the temperature, using a proper density of states distribution [23].

As in any disordered semiconducting material exhibiting a large band tail of localized states, the low field conductivity of a p-type polycrystalline semiconductor is thermally activated [24],

$$\sigma = q\mu N_v \exp\left(-\frac{E_\alpha}{kT}\right) = \sigma_o \exp\left(-\frac{E_\alpha}{kT}\right) \quad (23)$$

where N_v is the effective density of states in the valence band and μ is the mobility of the carriers in the extended states. The activation energy E_α is the position of the Fermi level E_F with respect to the valence band edge E_v, i.e., $E_\alpha = E_F - E_v$. In determining σ_0 and E_α by differentiation of the $\ln\sigma$ versus $10^3/T$ curve of Figure 1, it is found that these parameters follow the MN rule, which is an empirical relationship and connects the different σ_0 and E_α values by [25]

$$\sigma_0 = \sigma_{00} \exp(GE_\alpha) \quad (24)$$

where σ_{00} and G are the Meyer–Neldel parameters as shown in Figure 2.

According to Eqs. (23) and (24), the fluctuation in the conductivity σ can be determined by

$$\delta\sigma = \left(\frac{\partial\sigma}{\partial E_F}\right)\delta E_F = \sigma\left(G - \frac{1}{kT}\right)\delta E_F \quad (25)$$

Therefore, the mean square fluctuation in σ is

$$\overline{\Delta\sigma^2} = \sigma^2\left(G - \frac{1}{kT}\right)^2\overline{\Delta E_F^2} \quad (26)$$

Using the relationship $p = N_v\exp[-(E_F - E_v)/kT]$, the mean square fluctuation in the Fermi level $\overline{\Delta E_F^2}$ can be written as a function of the mean square fluctuations in the number of free

holes $\overline{\Delta p^2}$ as

$$\overline{\Delta E_F^2} = \left(\frac{kT}{\Omega_{\text{eff}}p}\right)^2\overline{\Delta p^2} \quad (27)$$

where Ω_{eff} is the effective volume of the sample. Thus, Eq. (26) can be expressed in terms of $\overline{\Delta p^2}$ as

$$\overline{\Delta\sigma^2} = \sigma^2(GkT - 1)^2\frac{\overline{\Delta p^2}}{(\Omega_{\text{eff}}p)^2} \quad (28)$$

Because of trapping–detrapping processes of holes of the valence band and trapping states of continuous energy distribution characterized by a distribution of the emission time constants $g(\tau)$, the conductivity spectral density is given by the relationship

$$S_\sigma(\omega) = \sigma^2(GkT - 1)^2\frac{\overline{\Delta p^2}}{(\Omega_{\text{eff}}p)^2}$$
$$\times \int_{\tau_{\min}}^{\tau_{\max}} g(\tau)\frac{4\tau}{1 + (\omega\tau)^2}d\tau \quad (29)$$

where τ is the characteristic time constant of the emission process of the electronic state from which the emission takes place [26].

As in polycrystalline silicon [27, 28], the grain boundaries introduce a high density of tail states close to the conduction and valence bands [29]. For the exponential energy distribution of tail states from the valence-band edge,

$$N(E) = N_t\exp[-(E - E_v)/E_t] \quad (30)$$

where N_t and E_t are the characteristic distribution parameters, the resulting distribution of time constants for the trapping–detrapping of holes of the valence band in these states is [30]

$$g(\tau)d\tau = \frac{(kT/\tau)(\tau\nu)^{-kT/E_t}d\tau}{E_t[1 - e^{-(E_F - E_v)/E_t}]} \quad (31)$$

where $\nu = 10^{12}$ Hz is the frequency of the attempts to escape of the traps. Thus, Eq. (29) becomes

$$S_\sigma(\omega) = 4\sigma^2(GkT - 1)^2 \frac{\overline{\Delta p^2}}{(\Omega_{\text{eff}} p)^2}$$
$$\times \frac{kT}{E_t[1 - e^{-(E_F - E_v)/E_t}]\nu^{kT/E_t}}$$
$$\times \int_{\tau_{\text{min}}}^{\tau_{\text{max}}} \frac{d\tau}{\tau^{kT/E_t}(1 + \omega^2\tau^2)} \quad (32)$$

In the above integral it is reasonable to extend the integration from zero to infinity, as for any distribution of traps within the energy gap, the time constants of the trapping–detrapping process extend to a very large interval because these time constants depend on the energy distance between the Fermi level and the concerned trap. When $\tau_{\text{max}} \ll \tau \ll \tau_{\text{min}}$ and $-1 < kT/E_t < 1$, the analytical expression $[\pi/2 \cos(\pi kT/E_t)] \times \omega^{-1+(kT/E_t)}$ for the integral of Eq. (32) can be found in any integral table [31]. Thus, Eq. (32) is reduced to the relationship

$$S_\sigma(\omega) = 2\sigma^2(GkT - 1)^2 \frac{\overline{\Delta p^2}}{(\Omega_{\text{eff}} p)^2}$$
$$\times \frac{kT}{E_t[1 - e^{-(E_F - E_v)/E_t}]\nu^{kT/E_t}}$$
$$\times \frac{\pi}{\cos(\pi kT/2E_t)} \frac{1}{\omega^\gamma} \quad (33)$$

The frequency exponent γ is related to the thermal energy kT and the trap distribution parameter E_t by the simple relationship

$$\gamma = 1 - \frac{kT}{E_t} \quad (34)$$

When $\gamma < 1$, the trap distribution parameter E_t is positive, indicating that the trap density increases toward the valence band edge. When $\gamma > 1$, the trap distribution parameter E_t is negative, indicating that the trap density increases toward the mid-gap.

The fluctuation in the number of free holes of the valence band can be considered to be equal to the fluctuation in the total number of occupied trapping states at the energy range of kT around the Fermi level, i.e.,

$$\overline{\Delta p^2} = AkTN(E_F) \quad (35)$$

where A is the total interface area of the grain boundaries. Because $S_I/I^2 = S_\sigma/\sigma^2$, the power spectral density of the current fluctuations is given by the relationship

$$\frac{S_I(\omega)}{I^2} = \frac{2A(GkT - 1)^2}{(\Omega_{\text{eff}} p)^2} \frac{(kT)^2 N(E_F)}{E_t[1 - e^{-(E_F - E_v)/E_t}]\nu^{kT/E_t}}$$
$$\times \frac{\pi}{\cos(\pi kT/2E_t)} \frac{1}{\omega^\gamma} \quad (36)$$

From Eq. (36) it follows that the spectral current density shows a $1/f^\gamma$ behavior and S_I is proportional to I^2. From the measured value of the exponent γ, the distribution parameter E_t of the valence-band tails can be determined from Eq. (34). For a

uniform energy distribution of the tail states (i.e., for $E_t \to \infty$), the frequency exponent is $\gamma = 1$, and Eq. (36) becomes

$$\frac{S_I(\omega)}{I^2} = \frac{A(GkT - 1)^2}{(\Omega_{\text{eff}} p)^2} \frac{(kT)^2 N(E_F)}{(E_F - E_v)f} \quad (37)$$

From Eq. (36) or (37), the trap density at the Fermi level $N(E_F)$ can be determined from the hole concentration p, the MN parameter G, and the power spectral density of the current fluctuations S_I obtained from the experimental data of Hall, conductivity and noise measurements, respectively. By sweeping the Fermi level with temperature, the noise model can provide the trap distribution within the band gap of polycrystalline semiconductors.

2.3. Application in β-FeSi₂ Films

Polycrystalline β-FeSi$_2$ is a semiconductor material of growing interest because of its potential applications in optoelectronic devices [32]. The polycrystalline β-FeSi$_2$ film studied was grown on (100)-oriented single-crystal Si wafer, polished, and undoped with resistivity >3000 Ω cm. The substrates were chemically cleaned before they were loaded into the evaporation chamber of an ultra-high-vacuum system. At a base pressure of about 10^{-8} Torr, Si and Fe (99.999% purity) layers of 0.5 and 1.3 nm, respectively, were deposited sequentially by a dual electron-gun evaporation source, to form a total of 95 periods, giving an estimated silicide thickness of about 170 nm. The multilayer surface was coated with a 20-nm-thick amorphous Si layer to avoid contamination during the subsequent annealing at a temperature of 800°C for 1 h in an oil-free conventional furnace (base pressure about 10^{-6} Torr). Finally, the samples were etched in HF:HCl (1:20) solution for 30 s to remove the Si capping layer and any other Si that might have segregated to the surface by diffusion during the silicide formation.

The formation of the semiconducting phase β-FeSi$_2$ phase was identified by X-ray diffraction measurements. Hall measurements have shown that the silicide exhibits a p-type behavior with a hole concentration $p = 1.1 \times 10^{17}$ cm^{-3} and a mobility $\mu = 97$ cm^2/Vs. Transmission electron microscopy analysis has shown that the average grain size of the β-FeSi$_2$ film is about 50 nm. After the silicide formation in the shape of a rectangle with dimensions 3×7 mm, for noise measurements, two lateral electrical contacts were made by evaporating In on the film and subsequent annealing in a conventional furnace (base pressure about 10^{-6} Torr) at a temperature of 250°C for about 5 min. The Van der Paw technique was also used to measure the conductivity of the β-FeSi$_2$ material in the temperature range 50–300 K. The contacts were confirmed to be ohmic before the measurements were made. Figure 3 shows the current–voltage curve of the silicide film with the two lateral electrical contacts, and an ohmic behavior is observed for currents up to 0.01 mA.

The low-frequency noise measurements (0.1–100 Hz) were performed at room temperature with a HP35665 spectrum analyzer preceded by a low-noise voltage amplifier and a current–

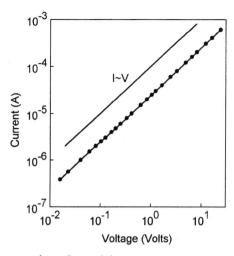

Fig. 3. Current–voltage characteristic across two lateral contacts of a polycrystalline β-FeSi$_2$ thin film in the shape of a rectangle with dimensions 3×7 mm. The silicide (170 nm thick) was prepared by conventional furnace annealing at a temperature of 800°C for 1 h. (Reprinted from D. H. Tassis et al., *J. Appl. Phys.* 85, 4091, 1999, with permission.)

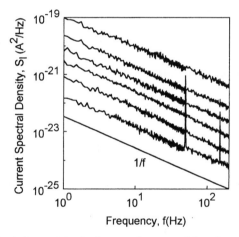

Fig. 4. Spectral current density S_I versus frequency of the polycrystalline β-FeSi$_2$ thin film of Figure 3 measured for the currents $I = 0.2, 0.4, 0.8, 1.5, 4, 6\ \mu$A. (Reprinted from D. H. Tassis et al., *J. Appl. Phys.* 85, 4091, 1999, with permission.)

voltage converter. The bias was supplied by CdNi batteries to avoid any external low-frequency noise. The noise measurements were performed in a frequency range where the excess noise is higher by one order of magnitude than the white noise observed at high frequencies and by two orders of magnitude than the total background noise observed by replacing the sample with a resistance of equivalent value. Figure 4 shows typical measurements of current noise spectra from 1 to 200 Hz for different values of the current I. At low frequencies, the current noise exhibits a $1/f$ behavior. Using the noise data of Figure 4, from Eq. (22) we found that the Hooge parameter is $\alpha_H \gg 1$, indicating that the mobility fluctuation model is rather unlikely.

The current noise spectra of Figure 4 have a frequency exponent $\gamma = 1$, indicating a uniform energy distribution of the tail states. Figure 5 shows the current noise spectral density S_I as a function of the current I derived from the noise data of Figure 4. The slope of the line indicates that S_I is proportional

to $I^{1.6}$, which is in disagreement with the theoretical prediction of Eq. (36) that $S_I \sim I^2$. Therefore, the use of Eq. (37) for estimation of the trap density $N(E_F)$ will lead to an erroneous result. A similar $I^{1.6}$ dependence of S_I was also observed in phosphorus-doped polycrystalline silicon films attributed to a layer distribution of the mobility across the thickness of the film [33]. Layers with small mobility can still contribute nonnegligible current, but negligible $1/f$ noise current because of the strong dependence of S_I on the mobility [33].

Next we consider the influence of the grain boundary barrier height inhomogeneity on S_I. The grain boundary area S is assumed to be divided into two areas: one area S_l, with a low barrier height ϕ_l, and another area S_h, with a higher barrier height ϕ_h. In a way similar to that of the Schottky barrier inhomogeneity caused by grain boundaries, the effective potential barrier height ϕ_b at the grain boundaries for the carrier transport across the polycrystalline material is expressed as [34, 35]

$$\exp\left(-\frac{q\phi_b}{kT}\right) = \left(1 - \frac{S_l}{S}\right)\exp\left(-\frac{q\phi_h}{kT}\right) + \frac{S_l}{S}\exp\left(-\frac{q\phi_l}{kT}\right) \tag{38}$$

Within the model of thermal emission, under bias voltage V_{gb} across a single grain boundary such that $qV_{gb} \ll kT$, the current I across the polycrystalline film is [36, 37]

$$I = I_o \exp\left(-\frac{q\phi_b}{kT}\right) \tag{39}$$

where $I_o = 2SA^*T^2\exp(-q\zeta/kT)$ and $q\zeta = E_F - E_v = kT\ln(N_v/p)$. In β-FeSi$_2$, the effective density of states in the valence band is $N_v = 2 \times 10^{19}$ cm^{-3}, and the effective Richardson's constant is $A^* = 102$ A^2/cm^2 K^2 [29]. Considering that only the sample area S_l with the lower potential barrier height ϕ_l at the grain boundaries contributes to the $1/f$ noise current, according to Eqs (38) and (39), the current I in Eqs. (36) and (37) can be written as

$$I = \frac{S_l}{S}I_o\exp\left(-\frac{q\phi_l}{kT}\right) = \frac{S_l}{S}I_o\exp\left(-\frac{\alpha q\phi_b}{kT}\right) \tag{40}$$

where the constant α is ≤ 1. The value of the constant α reflects the degree of the mobility inhomogeneity. When $\alpha = 1$, the carrier mobility is homogeneous across the thickness of the film. From Eqs. (38) and (39) it is obtained as

$$I^2 = \left(\frac{S_l}{S}\right)^2 I_o^{2(1-\alpha)}I^{2\alpha} = \left(\frac{\Omega_{\text{eff}}}{\Omega}\right)^2 I_o^{2(1-\alpha)}I^{2\alpha} \tag{41}$$

where Ω is the total sample volume. From Eq. (41), the power spectral density of the current fluctuations for exponential [Eq. (36)] and uniform tail states [Eq. (37)] distributions are written, respectively, as

$$\frac{S_I(\omega)}{I_o^{2-\beta}I^\beta} = \frac{2A(GkT-1)^2}{(\Omega p)^2}\frac{(kT)^2N(E_F)}{E_t[1-e^{-(E_F-E_v)/E_t}]_\nu kT/E_t} \times \frac{\pi}{\cos(\pi kT/2E_t)}\frac{1}{\omega^\gamma} \tag{42}$$

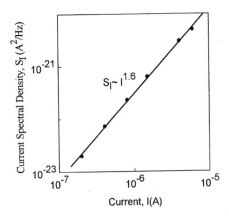

Fig. 5. Power spectral current density S_I at a frequency of $f = 10$ Hz as a function of the current, derived from the experimental data of Figure 3. (Reprinted from D. H. Tassis et al., *J. Appl. Phys.* 85, 4091, 1999, with permission.)

and

$$\frac{S_I(\omega)}{I_o^{2-\beta} I^\beta} = \frac{A(GkT - 1)^2}{(\Omega p)^2} \frac{(kT)^2 N(E_F)}{(E_F - E_v) f} \qquad (43)$$

where $\beta = 2\alpha$. From Eqs. (42) and (43) follows that $S_I \sim I^\beta$, where $\beta \leq 2$, in agreement with the experimental data of Figure 5. From the measured value of the exponent $\beta = 1.6$ and the data of Figure 5, using Eq. (43) we determined the trap density $N(E_F) = 2 \times 10^{11}$ cm^{-2} eV^{-1} at the Fermi level lying about 0.13 eV above the valence band edge. Thus, combined conductivity, Hall, and low-frequency noise measurements can be used to arrive at definitive conclusions regarding the experimental conditions for preparing polycrystalline semiconductors of higher quality.

3. NOISE OF THE DRAIN CURRENT IN POLYSILICON TFTs

3.1. Introduction

The electrical characteristics of polysilicon TFTs are strongly influenced by the presence of traps located mainly at the grain boundaries. Therefore, evaluation of the grain boundary trap properties is of major importance for the development of polysilicon TFT devices, as well as for accurate simulation and proper design of polysilicon TFT circuits. In previous works, to limit the complexity of the theoretical analysis, the polysilicon has been modeled with the "effective-medium" approach, where the grain boundary defects are assumed to be uniformly distributed over the entire volume of the film [28, 38–43]. This simplifying assumption can be justified in the case of small grain polycrystalline materials, where the grain size is much smaller than the Debye length. However, it is more realistic to assume that the defects are located principally at the grain boundaries, especially in polysilicon materials of large grain size and with few grain boundaries present within the channel of the transistor.

In polysilicon films, the grain boundary trap states are classified into two types: the acceptor-type traps that are located

in the upper half of the energy gap play an important role in n-channel TFTs. When the channel is formed, these trap states are negatively charged by trapping an electron. The donor-type traps that are located in the lower half of the energy gap play an important role in p-channel TFTs. When the channel is formed, these trap states are positively charged by emitting an electron.

The effective density of state distribution within the Si band gap can be modeled by the sum of a uniform distribution at the mid-gap of constant density and an exponential band tail distribution from the conduction or valence band edge. The deep states arise from silicon dangling bond states predominantly situated at the grain boundaries [44], whereas the tail states arise from distorted-bond defects [45]. Correlation between the TFT electrical characteristics and the corresponding gap-state density distribution has shown that the mobility is more closely associated with the trap states located near the band edges, whereas both subthreshold slope and threshold voltage are more closely associated with the trap states located near the mid-gap [44].

To characterize the grain boundary traps, techniques based on measurements of the drain current and the field effect conductance activation energy as a function of the gate voltage were mainly implemented [46–48]. With these techniques the energy distribution of the gap states has been derived, assuming the approach of uniform potential barrier distribution over the grain boundary plane. However, several studies of polysilicon films have shown spatial fluctuations of the potential barrier along the grain boundary plane [49–52]. Therefore, the significance of the gap states distributions obtained on the basis of uniform grain boundary characteristics is questionable. Barrier inhomogeneities over the grain boundary plane should be taken into account for accurate evaluation of the gap states in polysilicon TFTs.

Recently, LF noise was used as a diagnostic tool to characterize polysilicon TFTs [53–56]. Because the drain current is dominated by the potential barrier at the grain boundaries and the measured current fluctuations are due to the dynamic trapping and detrapping of carriers through the grain boundary traps, the LF noise technique is suitable for direct grain boundary trap evaluation. In this section, we present a LF noise technique based on the realistic model of traps localized at the grain boundary planes. Analysis of the drain current noise allows the quantitative estimation of the energy distribution of the grain boundary traps within the forbidden gap and the slow oxide traps density. The characteristics of the density of state distributions are correlated with the polysilicon growth conditions. Before the theoretical noise model in polysilicon TFTs is presented, the location of the noise sources and the grain boundary potential barrier inhomogeneities are discussed.

3.2. Empirical Relationship Between Noise and Grain Boundary Barrier Height

To investigate the location of the noise sources in polysilicon TFTs, the correlation between LF noise and potential barrier height V_b at the grain boundaries has been investigated [57].

It is demonstrated that a general empirical relationship exists between drain current spectral density S_I and potential barrier height V_b, indicating that the noise sources are located at the grain boundaries. In addition, this empirical expression facilitates the use of the LF noise as a practical tool to predict the value of V_b, a parameter that is directly related to the grain boundary trap states.

The studied devices were hydrogenated and nonhydrogenated high-temperature-processed polysilicon TFTs. First, polysilicon layers (220 nm thick) were deposited on thermally oxidized silicon wafers by low-pressure chemical vapor deposition (LPCVD), using SiH_4 source gas at 620°C and a pressure of about 0.4 mTorr. Then, the polysilicon layers were annealed at 1000°C for 3 h in dry N_2 ambient for grain enlargement. Transmission electron microscopy analysis has shown that the polysilicon layers have a columnar structure with an average grain size of about 100 nm. A standard self-aligned n-channel metal-oxide-semiconductor (MOS) process was used to fabricate TFTs with channel width $W = 50$ μm and length $L = 10$ μm. The gate dielectric was SiO_2 (45 nm thick) grown at 1000°C in a dry oxygen atmosphere. After the LPCVD polysilicon gates were deposited and patterned, phosphorous ions were implanted at a dose greater than 10^{15} cm^{-2} with an energy of 85 keV to form the source, drain, and gate contacts. The polysilicon films were undoped or boron or phosphorus doped to a concentration of about 6×10^{16} and 3×10^{17} cm^{-3}, respectively. The polysilicon films were hydrogenated by implanting hydrogen ions at a dose of 1×10^{16} cm^{-2} and an energy of 130 keV through the top passivating thick oxide, the polysilicon gate, and the gate oxide. Dopant activation was carried out at 450°C for 30 min in a H_2 ambient.

The transfer characteristics of the TFTs were measured with a HP 4155 semiconductor parameter analyzer. Noise measurements were performed at room temperature with a dynamic signal analyzer (HP 35665). The drain current noise was previously amplified by a low-noise EG&G model current–voltage converter and a low-noise NF5305 voltage amplifier. The gate and drain bias voltages were supplied by CdNi batteries to reduce any external low-frequency noise.

Considering that the conduction mechanism in the polysilicon film is described by thermionic emission over the potential barrier V_b, the drain current I_D is expressed as

$$I_D = I_o \exp\left(-\frac{qV_b}{kT}\right) \qquad (44)$$

where I_o is the current prefactor with weak temperature dependence. Thus, at each gate voltage V_G, the potential barrier V_b can be obtained from the slope of the Arrhenius plots of $\ln I_D$ versus $1/T$. Figure 6 shows the drain current as a function of the inverse temperature of a typical undoped hydrogenated polysilicon TFT, measured at different gate voltages and at a fixed drain voltage in the linear region of operation ($V_D = 0.1$ V).

Figure 7 shows the variation of the potential barrier height V_b with the gate bias V_G for undoped hydrogenated, boron-doped hydrogenated or nonhydrogenated, and phosphorus-doped nonhydrogenated polysilicon TFTs. As the gate bias voltage V_G

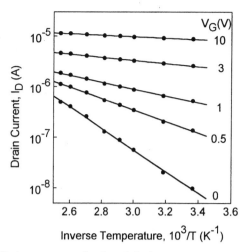

Fig. 6. Drain current as a function of the inverse temperature of a typical high-temperature polysilicon TFT, measured at drain voltage $V_D = 0.1$ V and various gate voltages V_G. (Reprinted from C. T. Angelis et al., *Appl. Phys. Lett.* 76, 118, 2000, with permission.)

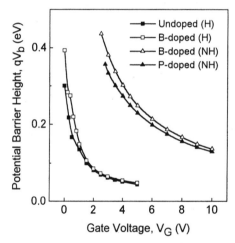

Fig. 7. Dependence of the potential barrier height at the grain boundaries on the gate voltage of high-temperature polysilicon TFTs. The polysilicon layers were hydrogenated (H) undoped and B-doped or non-hydrogenated (NH) B- and P-doped. (Reprinted from C. T. Angelis et al., *Appl. Phys. Lett.* 76, 118, 2000, with permission.)

increases, the potential barrier V_b decreases because of the increase in the induced carrier density in the channel, while the assumption of $V_b > kT/q$ for the thermionic emission theory is valid [58]. When the density of gap states is higher, the decrease in V_b with V_G is expected to be slower because the Fermi level is being pulled more slowly toward the conduction band. This occurs because most of the applied gate voltage is used to charge these states and not to induce free carriers in the channel.

Measurements of the drain current noise spectral density S_I have been performed on the above TFTs for different gate biases. The dependence of S_I on the gate voltage, measured at a frequency of $f = 10$ Hz, is illustrated in Figure 8. When the gate voltage increases, the drain current noise increases while the potential barrier V_b decreases, as shown in Figure 7. These results indicate the close relationship between S_I and $1/V_b$.

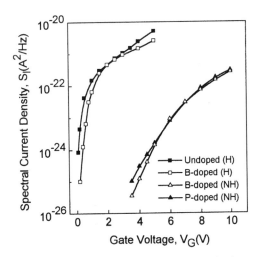

Fig. 8. Dependence of the drain current noise spectral density on the gate voltage of high-temperature polysilicon TFTs measured at a frequency of 10 Hz and in the linear region of operation ($V_D = 0.1$ V). The polysilicon layers were hydrogenated (H) undoped and B-doped or non-hydrogenated (NH) B- and P-doped. (Reprinted from C. T. Angelis et al., *Appl. Phys. Lett.* 76, 118, 2000, with permission.)

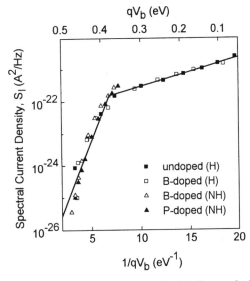

Fig. 9. The drain current noise spectral density plotted versus the inverse of the potential barrier height at the grain boundaries $1/qV_b$, formed from the data of Figures 7 and 8. (Reprinted from C. T. Angelis et al., *Appl. Phys. Lett.* 76, 118, 2000, with permission.)

Plots of S_I versus $1/V_b$, derived from the data of Figures 7 and 8, are shown in Figure 9. It is clearly seen that the data points of all devices converge, following a common dependence. In the logarithmic-linear plot of Figure 9, two linear regions can be distinguished. The following unique empirical analytical expressions can be written for the V_b dependence of S_I:

$$S_I = A_1 \exp(B_1/qV_b) \qquad (qV_b > 0.14 \text{ eV}) \qquad (45)$$

and

$$S_I = A_2 \exp(B_2/qV_b) \qquad (qV_b < 0.14 \text{ eV}) \qquad (46)$$

The corresponding coefficients are $A_1 = 7.4 \times 10^{-28}$ A^2/Hz, $B_1 = 1.78$ eV, $A_2 = 3.8 \times 10^{-23}$ A^2/Hz, and $B_2 = 0.205$ eV.

From the above results, the following conclusions can be obtained:

i. It can readily be seen that the drain current noise increases with the shrinking grain boundary potential barrier height V_b. In fact, a decrease in qV_b from about 0.4 to 0.1 eV results in an increase in S_I by four orders of magnitude. However, it has generally been recognized that the noise spectral density scales inversely with the effective size of elements containing noise sources [59]. Thus, the results of Figure 9 indicate that in polysilicon TFTs the noise sources are located at the grain boundaries.

ii. The empirical relationship between the drain current spectral density S_I and the potential barrier height at the grain boundaries V_b can be used to predict the value of V_b from noise measurements at room temperature.

iii. The drain current noise is a much stronger function of the grain boundary trap states than the potential barrier height V_b and will, therefore, be more sensitive to any kind of changes. Thus, the drain current noise can be considered a better monitor than a typical static device parameter for hot-carrier degradation studies [60].

3.3. Grain Boundary Barrier Height Inhomogeneities

In the linear region of operation of polysilicon TFTs, the drain current I_D is thermally activated, and, within the model of thermionic emission, I_D is described by Eq. (44). Considering uniform potential distribution over the grain boundary plane, this equation predicts that the Arrhenius plot of $\ln I_D$ versus $1/T$ should yield a straight line with a constant slope corresponding to the potential barrier height V_b. However, the curvature in the Arrhenius plots of conductivity in polysilicon TFTs indicates spatial potential fluctuations over the grain boundary plane [49–52].

To investigate the grain boundary potential barrier homogeneity, the drain current of undoped polysilicon TFTs was measured in the linear region of operation ($V_D = 0.5$ V) at different gate voltages V_G and temperatures ranging from 300 K to 420 K. The investigated TFTs were fabricated on polysilicon films (0.3 μm thick), deposited on oxidized silicon wafers by pyrolysis of silane at 630°C and pressures of 1, 10, and 40 mTorr. Transmission electron microscopy analysis has shown that the average grain size is about 117, 69, and 50 nm for the 1, 10, and 40 mTorr polysilicon layers, respectively [61]. A standard self-aligned n-channel MOS process was used to fabricate TFTs with channel width $W = 900$ μm and length $L = 10$ μm. A SiO$_2$ layer of thickness about 100 nm, deposited by atmospheric pressure vapor deposition, was used as a polysilicon gate. Source and drain contacts were formed by phosphorus ion implantation, followed by annealing at 600°C for 4 h for dopant activation. Details of the TFT fabrication processes are presented elsewhere [62].

Figure 10 shows plots of $\ln I_D$ versus $1/T$ of typical 1-, 10-, and 40-mTorr polysilicon TFTs at a gate voltage of $V_G = 10$ V.

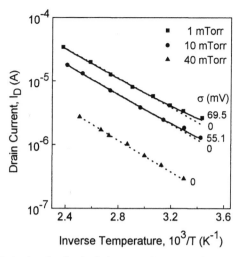

Fig. 10. Arrhenius plots for the drain current I_D at gate voltage $V_G = 10$ V of TFTs on LPCVD polysilicon films deposited at various pressures. The experimental data are fitted to Eqs. (44) and (48) for homogeneous (dotted lines) and inhomogeneous (solid lines) grain boundaries. (Reprinted from C. A. Dimitriadis, *IEEE Trans. Electron Devices* 44, 1563, 1997, with permission.)

It can readily be seen that in the 40-mTorr polysilicon TFT the plot is linear, whereas in the 1- and 10-mTorr polysilicon TFTs the plots are curved upward below 330 K. A detailed study of the polysilicon conductivity within a broad temperature range (100–300 K) arrived at the conclusion that the deviation from a straight Arrhenius curve is caused by potential barrier fluctuations over the grain boundary plane, rather than by a transition to a second conduction process [49, 50]. In addition, admittance frequency dispersion measurements verified that the grain boundary potential barrier inhomogeneity is described by a Gaussian distribution with a mean barrier height $\langle V_b \rangle$ a standard deviation σ [51]:

$$P(V_b) = \frac{1}{\sigma \sqrt{2\pi}} \exp \left[-\frac{V_b - \langle V_b \rangle}{2\sigma^2} \right] \quad (47)$$

Integration of Eq. (44) over the barrier distribution of Eq. (47) yields an effective barrier $V_{b,\text{eff}}$ for the drain current given by the relationship [49, 51]

$$V_{b,\text{eff}} = \langle V_b \rangle - \frac{\sigma^2}{2kT/q} \quad (48)$$

In Eq. (48), the second term results in a decrease in $V_{b,\text{eff}}$ with decreasing temperature, resulting in an upward curvature of the Arrhenius plot of the drain current.

In Figure 10, the best straight lines fitted to the data of the 1- and 10-mTorr polysilicon TFTs for $\sigma = 0$ are obtained for $V_{b,\text{eff}} = 0.235$ V and $V_{b,\text{eff}} = 0.23$ V, respectively (dotted lines). In this case, the theoretical lines deviate significantly from the experimental data. However, by considering σ as a varying parameter, excellent fitting of the experimental data with Eqs. (47) and (48) is obtained for values of σ shown in Figure 10. In the 40-mTorr polysilicon TFT only the grain boundaries are homogeneous as $\sigma = 0$, indicating no variation of the potential barrier over the grain boundary

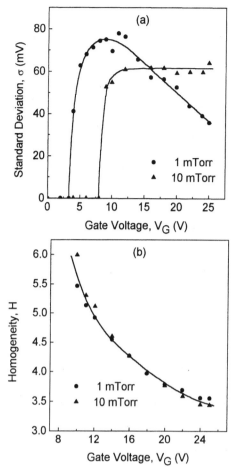

Fig. 11. Variation of (a) the standard deviation σ of the Gaussian grain boundary trap distribution and of (b) the grain boundary homogeneity H with the gate voltage V_G of typical 1- and 10-mTorr LPCVD polysilicon TFTs. (Reprinted from C. A. Dimitriadis, *IEEE Trans. Electron Devices* 44, 1563, 1997, with permission.)

planes. The calculated effective barrier is $V_{b,\text{eff}} = 0.286$ V and $V_{b,\text{eff}} = 0.271$ V for the 1- and 10-mTorr polysilicon TFTs, respectively. These values of $V_{b,\text{eff}}$ are overestimated compared with the values obtained for $\sigma = 0$. Thus, the assumption of a uniform grain boundary may lead to an erroneous estimation of the gap states distribution.

The observed grain boundary barrier properties can be related to the polysilicon structure. For deposition pressures of 1 and 10 mTorr, the polysilicon layers have a columnar structure with a (001) preferred orientation, whereas for a deposition pressure of 40 mTorr the structure changes to a striated one with a ⟨111⟩ twin texture [63]. The results of Figure 10 indicate that the grain boundary structure inhomogeneity is related to the columnar structure of the grains, i.e., to curved grain boundary surfaces. This conclusion is further supported by another work showing that spatial fluctuations of the potential barrier over the grain boundary can be caused by grain curvature [64].

The variation of σ with V_G of typical 1- and 10-mTorr polysilicon TFTs is illustrated in Figure 11a. The standard deviation σ increases abruptly from zero to a certain value when V_G

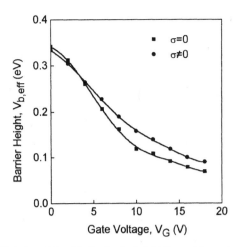

Fig. 12. Variation of the effective barrier height $V_{b,eff}$ at the grain boundaries with the gate voltage V_G of a typical 1-mTorr LPCVD polysilicon TFT for homogeneous ($\sigma = 0$) and inhomogeneous ($\sigma \neq 0$) grain boundaries. (Reprinted from C. A. Dimitriadis, *IEEE Trans. Electron Devices* 44, 1563, 1997, with permission.)

exceeds the threshold voltage V_T and then either remains constant (1-mTorr layer) or decreases linearly with increasing V_G (10-mTorr layer). The different behavior for the dependence of σ on V_G in polysilicon TFTs can be understood in terms of the structure of the layer: the polysilicon layer deposited at a pressure of 1 mTorr has a columnar tooth-shaped structure, whereas the polysilicon layer deposited at a pressure of 10 mTorr has a cylindrical structure [52]. Thus, with increasing depth below the film surface, it is reasonable to expect an increase in the curvature of the grains for a deposition pressure of 1 mTorr and constant curvature at a deposition pressure of 10 mTorr. Considering the above grain structure, the observed dependence of σ on V_G can be explained by the decrease in the channel depth with increasing gate voltage. The grain boundary homogeneity, defined as $H = \langle V_b \rangle / \sigma$, has been determined from the data for σ and $\langle V_b \rangle$. The larger value of H indicates more homogeneous grain boundaries. The variation of H with V_G of the 1- and 10-mTorr polysilicon TFTs is shown in Figure 11b. These results indicate that, in LPCVD polysilicon layers, H decreases with increasing gate voltage and is almost independent of the deposition pressure.

The experimental data of the effective barrier height $V_{b,eff}$ determined from Eq. (48) as a function of the gate voltage for the 1-mTorr polysilicon TFT, assuming homogeneous ($\sigma = 0$) and inhomogeneous ($\sigma \neq 0$) grain boundaries, are shown in Figure 12. It is clear that when the grain boundaries are assumed to be homogeneous, the measured effective barrier height $V_{b,eff}$ is underestimated, and this may lead to underestimation of the gap states evaluated by methods based on the experimental $V_{b,eff}$ versus V_G data [42, 48].

3.4. Noise Spectroscopy for Grain Boundary and Interface Trap Characterization

In polycrystalline semiconductors, crystallographic misfit between misoriented crystallites leads to electrically charged

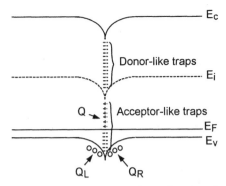

Fig. 13. Energy band diagram in the lateral direction along the channel of a p-channel polysilicon TFT. (Reprinted from C. T. Angelis et al., *IEEE Trans. Electron Devices* 46, 968, 1999, with permission.)

defect states at the boundary between the grains. Figure 13 illustrates the energy band diagram of the polysilicon in the neighborhood of a grain boundary of a p-channel polysilicon TFT along the channel at equilibrium (low drain voltage). A potential barrier V_b is formed at the grain boundaries because of trapping of gate-induced holes at the localized grain boundary donor-like states. Because the drain current is predominantly affected by the barrier height V_b, the current fluctuations can be considered to arise from changes in the potential barrier caused by fluctuations of the charge trapped at the grain boundary. Below we present a detailed analysis of the low-frequency noise in p-channel polysilicon TFTs. The current noise data are analyzed to yield quantitative information about the energy distribution of the grain boundary traps within the lower half of the forbidden band gap and the slow oxide traps density. A similar analysis holds for n-channel polysilicon TFTs in determining the energy distribution of the grain boundary traps within the upper half of the forbidden band gap.

Referring to Figure 13, because of charge neutrality conditions, the positive interface charge of areal density Q is compensated for by the negative charge of the depletion regions around the interface. For a polysilicon TFT operating in the linear region (small drain voltage V_D), the depletion regions around the interface are almost symmetrical, and therefore [65]

$$Q = Q_L + Q_R = 2Q_L = 2Q_R \quad (49)$$

where Q_R and Q_L are the areal charges of the negative acceptor ions within the depletion regions at the right-hand and left-hand sides of the boundary, respectively. If the concentration p of holes induced by the gate is large compared with the grain boundary trapped charge, the barrier height qV_b is given in terms of the depletion charge Q_R or Q_L by the relationship [58]

$$qV_b = \frac{Q_R^2}{2\varepsilon_{Si} p} = \frac{Q_L^2}{2\varepsilon_{Si} p} \quad (50)$$

or in terms of the interface charge Q by the relationship

$$qV_b = \frac{Q^2}{8\varepsilon_{Si} p} \quad (51)$$

where ε_{Si} is the permittivity of silicon.

The presence of the potential barrier V_b leads to a decrease in the drain current I_D of the polysilicon TFT because, according to the thermionic emission model, in the linear region I_D is described by [47]

$$I_D = \frac{W}{L} C_{ox}\mu_o \exp(-qV_b/kT)(V_G - V_{FB})V_D \quad (52)$$

where μ_o is the carrier mobility within the grain and V_{FB} is the flat-band voltage, which is approximately the gate voltage at which the drain currents of n- and p-channel TFTs are equal. The value of the gate oxide capacitance per unit area is $C_{ox} = \varepsilon_{ox}/t_{ox}$, where ε_{ox} is the permittivity of SiO_2 and t_{ox} is the oxide thickness. In polysilicon TFTs, correlation between drain current noise and potential barrier height at the grain boundaries V_b has shown that the noise sources are located at the grain boundaries (Section 3.2). Thus, fluctuation phenomena generally arise from trapping and detrapping processes in interface states within the boundary plane. Such processes result in fluctuations δQ of the interface charge Q determining the barrier height V_b through Eq. (51). The charge fluctuations δQ lead to fluctuations δV_b in the barrier height V_b, which in turn results in fluctuation δI_D of the drain current I_D. Therefore, the drain current spectral density is

$$S_I'(f) = N_g\left(\frac{\delta I_D}{\delta Q}\right)^2 \frac{S_Q(f)}{A_{gb}} \quad (53)$$

where N_g is the number of grain boundaries within the channel and $S_Q(f)$ is the grain boundary interface charge spectral density. The grain boundary area A_{gb} within the channel is $A_{gb} = Wt_{ch}$, where the channel depth t_{ch} varies with V_G according to the relationship [47]

$$t_{ch} = \frac{8kTt_{ox}\sqrt{\varepsilon_{Si}/\varepsilon_{ox}}}{q(V_G - V_{FB})} \quad (54)$$

However,

$$\frac{\delta I_D}{\delta Q} = \frac{\delta I_D}{\delta V_b}\frac{\delta V_b}{\delta Q} \quad (55)$$

Using Eqs. (51) and (52), Eq. (55) becomes

$$\frac{\delta I_D}{\delta Q} = \left(-\frac{q}{kT}\right)I_D\sqrt{\frac{V_b}{2q\varepsilon_{Si}p}} \quad (56)$$

The current spectral density is written as [65]

$$S_I'(f) = N_g\left(\frac{qI_D}{kT}\right)^2 \frac{V_b}{2q\varepsilon_{Si}p}\frac{S_Q}{A_{gb}} \quad (57)$$

From the continuously distributed donor-like states, we assume that only states close to the Fermi level E_F contribute to the noise spectra by interaction with the valence band via capture and emission of holes. Under this assumption, for a single donor-like trap located at the Fermi level with a time constant τ_p and concentration N_s, the charge fluctuation S_Q is described by the Lorentzian spectrum [66]

$$S_Q = \frac{q^2kT4\tau_pN_s}{1 + (2\pi f\tau_p)^2} \quad (58)$$

The time constant τ_p is given by the relationship $\tau_p = (\upsilon_{th}p_oS_p)^{-1}$, where $p_o = p\exp(-qV_b/kT)$ is the concentration of holes at the grain boundary, υ_{th} is the hole thermal velocity, and S_p is the capture cross section of the grain boundary interface states for holes. The concentration of holes p induced by the gate is $p = C_{ox}(V_G - V_{FB})/qt_{ch}$.

Equations (51) and (52), derived under the assumption of uniform potential barrier over the grain boundary plane, predict a Lorentzian spectrum for the frequency-dependent noise of the drain current. However, the spatial fluctuation of the potential barrier V_b along the grain boundary plane found in polysilicon layers (Section 3.3) leads to a modification of the drain current spectral density described by Eqs. (57) and (58). In a first approximation, the distribution $P(V_b)$ of the grain boundary potential barriers can be represented by the Gaussian distribution of Eq. (47) with mean barrier height $\langle V_b\rangle$ and standard deviation σ. As the charge carriers traverse the boundary, we assume that the contributions to the noise spectrum due to trapping and detrapping of the carriers at different barrier heights V_b are independent. The resulting drain current noise spectrum as a function of the frequency f is obtained by multiplying the spectrum of Eqs. (57) and (58) by the Gaussian distribution $P(V_b)$ and integrating over all barrier heights [65]:

$$S_I^G(f) = \frac{qN_g}{kT}\frac{2(qI_D)^2N_s}{A\varepsilon_{Si}p}$$
$$\times \int_0^\infty V_bP(V_b)\frac{\tau_p}{1 + (2\pi f\tau_p)^2}dV_b \quad (59)$$

The integral of Eq. (59) has been evaluated numerically with a computer minimization program [67]. This numerical evaluation can be conveniently performed if the measured spectrum is multiplied by $\omega = 2\pi f$ on a linear scale. Such a plot is more appropriate for the analysis of the noise spectra than the usual double-logarithmic representation. In such a case, the integral of the Lorentzian spectrum in Eq. (59) will appear as a broad symmetrical peak at a certain frequency. By fitting the experimental noise data with Eq. (59), the grain boundary trap parameters N_s and S_p and the potential barrier V_b distribution parameters $\langle V_b\rangle$ and σ can be determined. Because the Fermi level can be swept through the forbidden gap by variation of the gate voltage, the energy distribution $N_s(E)$ of the grain boundary trap states can be determined.

In addition to the Lorentzian spectrum, the linear representation of the measured noise spectra is particularly suitable for spectra that deviate even slightly from $1/f$ behavior. Pure $1/f$ noise will appear as a flat base line; thus small deviations from $1/f$ noise can easily be distinguished. According to an existing model for monocrystalline silicon MOSFET [68], the presence of a $1/f$ noise in a polysilicon TFT can be attributed to fluctuations of the inversion charge near the polysilicon–gate oxide interface. Analysis of the $1/f$ noise will enable us to evaluate the slow oxide traps density.

It is mentioned that the above noise analysis is valid when the transistor is operated in the linear regime (i.e., at low drain voltages). When a high drain voltage is applied, the potential barrier formed around the grain boundary is asymmetrical because of the forward bias of one side and the reverse bias on the other side of the boundary and Eqs. (49)–(51) are not valid. In addition, when the TFT is operated in the saturation regime, avalanche-induced excess noise is produced, originating from generation-recombination processes in the depletion region of the drain junction [69], which is superimposed on the drain current noise of Eq. (59).

3.5. Verification of the Noise Model

The noise model described in the above section has been verified in n- and p-channel high-temperature TFTs fabricated on (220-nm-thick) undoped hydrogenated polysilicon films. A SiO_2 layer of thickness 45 nm, grown at 1000°C in a dry oxygen atmosphere, was used as a gate dielectric. The physical channel width and length of the TFTs were 50 and 10 μm, respectively. The source and drain electrodes were phosphorus implanted for n-channel and boron implanted for p-channel devices. Details of the fabrication process of the polysilicon TFTs are presented in Section 3.2 and in [70]. The transfer characteristics of typical p-channel and n-channel polysilicon TFTs and their corresponding transconductance g_m, measured at drain voltages −0.1 and 0.1 V, are shown in Figure 14a and b, respectively.

Figure 15 shows typical current noise spectra of p-channel and n-channel polysilicon TFTs measured at fixed drain currents. In both cases, the spectra show a $1/f^\gamma$ ($\gamma \leq 1$) behavior over the measured frequency range. Using the noise data of Figure 15, the graphical representation of the product $2\pi f S_I$ as a function of the frequency is shown in Figure 16. As demonstrated in Figure 16a, we are just able to distinguish a broad maximum at a frequency of about 300 Hz for the p-channel TFT, whereas this maximum is shifted to a frequency well above 4 kHz for the n-channel TFT (Fig. 16b). These results indicate that, in both p- and n-channel undoped polysilicon TFTs, only the lower branch of the symmetrical broad peak is revealed in the measured frequency ranges. A broad maximum of the product $2\pi f S_I(f)$ was clearly observed in doped polysilicon TFTs. A typical example of B-doped polysilicon ($N_a = 3 \times 10^{17}$ cm^{-3}) is shown in Figure 17b, derived from the current noise data of Figure 17a. Analysis of the experimental noise data with Eq. (59) showed poor agreement with the measured curves. The experimental noise spectral density can be perfectly fitted to Eq. (58) only by adding a pure $1/f$ noise component, i.e.,

$$S_I(f) = S_I^G(f) + \frac{A}{f} \tag{60}$$

where A is a constant. In Figure 16 the solid lines were obtained by fitting the experimental data with Eqs. (59) and (60), using the values shown for the parameters N_s, S_p, $\langle V_b \rangle$, σ, and A. Similar fittings were also performed for various values of the gate voltage.

The distribution $P(V_b)$ of the grain boundary potential barriers is shown in Figure 18a and b for the p-channel and n-channel

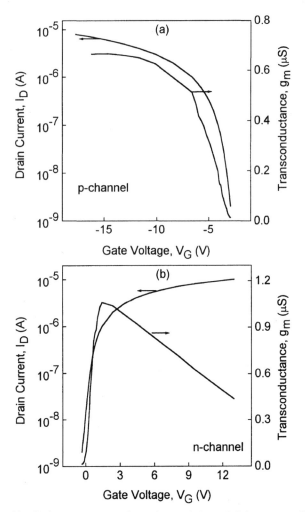

Fig. 14. Drain current–gate voltage characteristics and their corresponding transconductance for typical (a) p-channel and (b) n-channel polysilicon TFTs. (Reprinted from C. T. Angelis et al., *IEEE Trans. Electron Devices* 46, 968, 1999, with permission.)

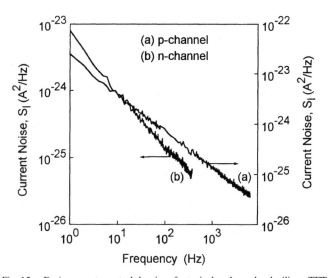

Fig. 15. Drain current spectral density of a typical p-channel polysilicon TFT at drain voltage $V_D = -0.1$ V and drain current $I_D = 10$ nA and of a typical n-channel polysilicon TFT at $V_D = 0.1$ V and $I_D = 30$ nA. (Reprinted from C. T. Angelis et al., *IEEE Trans. Electron Devices* 46, 968, 1999, with permission.)

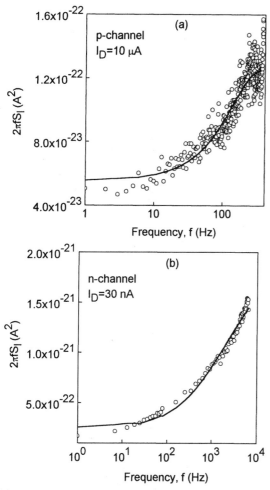

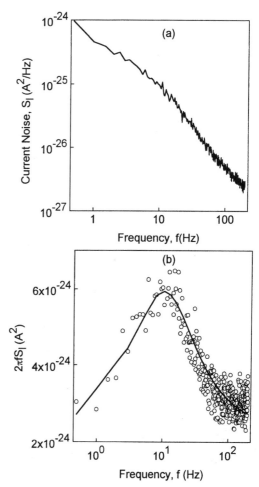

Fig. 16. The product of measured noise $S_I(f)$ and frequency $2\pi f$ obtained from the data of Figure 15. The solid lines were obtained from fitting with Eqs. (26) and (38), using the parameters (a) $\langle V_b \rangle = 0.25$ V, $\sigma = 60$ mV, $S_p = 4 \times 10^{-14}$ cm^2, $N_s = 8.5 \times 10^{11}$ cm^{-2}eV^{-1}, and $A = 5.5 \times 10^{-23}$ A^2 for the p-channel device and (b) $\langle V_b \rangle = 0.3$ V, $\sigma = 57$ mV, $S_n = 2 \times 10^{-14}$ cm^2, $N_s = 1.9 \times 10^9$ cm^{-2}eV^{-1}, and $A = 2.5 \times 10^{-22}$ A^2 for the n-channel device. (Reprinted from C. T. Angelis et al., *IEEE Trans. Electron Devices* 46, 968, 1999, with permission.)

Fig. 17. (a) Drain current spectral density of a typical n-channel polysilicon TFT at drain voltage $V_D = 0.1$ V and drain current $I_D = 30$ μA. The polysilicon layer was B-doped to a concentration $N_a = 3 \times 10^{17}$ cm^{-3}. (b) The product of measured noise $S_I(f)$ and frequency $2\pi f$ obtained from the data of Figure 17a. The solid line was obtained from fitting with Eqs. (26) and (38), using the parameters $\langle V_b \rangle = 0.114$ V, $\sigma = 64$ mV, $S_n = 2.6 \times 10^{-16}$ cm^2, $N_s = 6.2 \times 10^{15}$ cm^{-2}eV^{-1}, and $A = 2.3 \times 10^{-24}$ A^2. (Reprinted from C. T. Angelis et al., *IEEE Trans. Electron Devices* 46, 968, 1999, with permission.)

devices, respectively, with the drain current as a parameter. As expected, the whole distribution is shifted toward lower V_b with increasing hole density. According to the energy band diagram of Figure 13 for the p-channel device, the position of the Fermi level with respect to the valence band edge within the grain is $E_{F,G} = (q/kT)\ln(N_v/p)$, where N_v is the effective density of states in the valence band. Therefore, at the boundary the energy position of the Fermi level with respect to the valence band edge is $E_{F,B} = E_{F,G} + qV_{b,\text{eff}}$. The effective potential barrier height at the boundary $V_{b,\text{eff}}$ can be determined from Eq. (48). The grain boundary trap density can be determined from noise measurements in both p-channel and n-channel devices at various gate voltages while the Fermi level is sweeping the forbidden gap. The resultant distribution of localized acceptor/donor grain boundary states in the energy gap is plotted in Figure 19. It is clearly seen that the acceptor- and donor-type trap states are roughly divided into two groups: deep localized

states and tail localized states, which are increased as the energy level approaches the conduction and valence band edges.

The obtained $1/f$ component of the measured drain current spectral density $S_{1/f} = A/f$ can be explained by the two classical models developed for single crystalline silicon MOSFETs. The first model assumes that the current fluctuation is caused by fluctuations in the number of carriers. These fluctuations in the number of carriers are induced by fluctuations of the interfacial oxide charge due to dynamic trapping and detrapping of free carriers in slow oxide traps located close to the SiO$_2$–polysilicon interface Ghibaudo et al. [68] have shown that if the fluctuations give rise to a noticeable change in the mobility via Coulomb scattering, the normalized drain current density $S_{1/f}/I_D^2$ is related to the flat-band voltage spectral density $S_{V_{FB}}$ by Eq. (19).

The second model, originally introduced by Hooge [11], relies on the carrier mobility fluctuations. In this case, the nor-

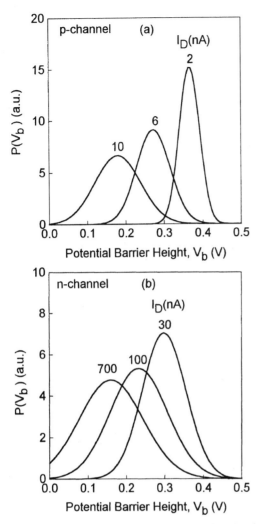

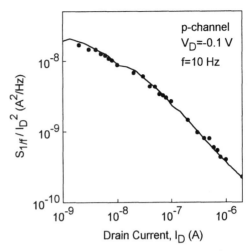

Fig. 20. Normalized drain current spectral density $S_{1/f}/I_D^2$ versus I_D of a typical p-channel high-temperature polysilicon TFT at drain voltage $V_D = -0.1$ V and a frequency of 10 Hz. The continuos line is the best fit to Eq. (39) with $S_{V_{FB}} = 4.7 \times 10^{-8}$ V^2/Hz. (Reprinted from C. T. Angelis et al., *IEEE Trans. Electron Devices* 46, 968, 1999, with permission.)

Fig. 18. The distributions of the grain boundary potential barriers for typical (a) p-channel and (b) n-channel polysilicon TFTs with the drain current as parameter. (Reprinted from C. T. Angelis et al., *IEEE Trans. Electron Devices* 46, 968, 1999, with permission.)

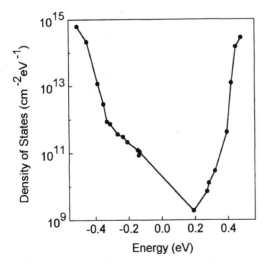

Fig. 19. Energy distribution of the grain boundary trap density within the forbidden gap of a polysilicon TFT. (Reprinted from C. T. Angelis et al., *IEEE Trans. Electron Devices* 46, 968, 1999, with permission.)

malized drain current spectral density of a uniform channel (i.e., in the linear region of operation) is

$$\frac{S_{1/f}}{I_D^2} = \frac{\alpha_H}{\overline{N}}\frac{1}{f} \qquad (61)$$

where α_H is the Hooge parameter and $\overline{N}$ is the total number of carriers in the device.

The mechanism responsible for the origin of the $1/f$ noise in polysilicon TFTs can be identified from analysis of the normalized drain current spectral density. In weak inversion, $S_{1/f}/I_D^2$ is expected to be inversely proportional to I_D for mobility fluctuations, whereas a plateau is usually observed for carrier number fluctuations. Figure 20 shows the plot of $S_{1/f}/I_D^2$ versus I_D for a typical p-channel undoped-hydrogenated polysilicon TFT, measured at a frequency of 10 Hz. From the shape of this curve, it is concluded that the $1/f$ noise can be ascribed to carrier number fluctuations.

Assuming that the effective carrier mobility is $\mu_{eff} = \mu_o \exp(-qV_{b,eff}/kT)$, Eq. (19) can be written as [65]

$$\frac{A^{1/2}}{g_m} = \left[1 + \alpha\mu_o C_{ox}\exp(-qV_{b,eff}/kT)(V_G - V_{FB})\right]$$
$$\times (2\pi f S_{V_{FB}})^{1/2} \qquad (62)$$

Based on Eq. (62), the plot of $A^{1/2}/g_m$ versus $(V_G - V_{FB})\exp(-qV_{b,eff}/kT)$ should give a straight line, which can be used to extract (a) the oxide trap density N_{ox} from the x axis intercept through Eq. (20) and (b) the Coulomb scattering coefficient α from the slope. Such plots are shown in Figure 21a and b for the p-channel and n-channel devices, respectively. In both cases, it is clear that the value of the parameter α is almost zero, indicating the validity of the trapping noise without correlated

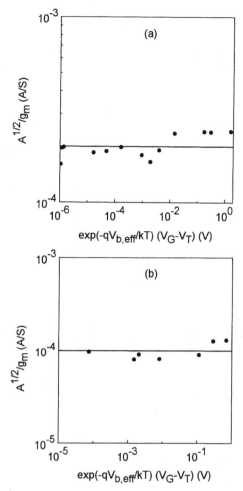

Fig. 21. Variation of the parameter $A^{1/2}/g_m$ with $\exp(-qV_b/kT)(V_G - V_T)$ for a typical (a) p-channel and (b) n-channel polysilicon TFT. (Reprinted from C. T. Angelis et al., *IEEE Trans. Electron Devices* 46, 968, 1999, with permission.)

mobility fluctuations; i.e., Eq. (62) reduces to [65]

$$\frac{A^{1/2}}{g_m} = (2\pi f S_{V_{FB}})^{1/2} \tag{63}$$

From the data of Figure 21a and b, the values obtained for the right-hand side of Eq. (63) are 2×10^{-4} and 1×10^{-4} A/S for the p-channel and n-channel devices, respectively. Using these values, from Eq. (20) we obtain the oxide trap state densities $N_{ox} = 2.7 \times 10^{19}$ cm^{-3}eV^{-1} and $N_{ox} = 6.7 \times 10^{18}$ cm^{-3}eV^{-1} for the p- and n-channel devices, respectively. Therefore, from analysis of the drain current noise data of polysilicon TFTs, both grain boundary and slow oxide trap properties can be evaluated.

3.6. Application in Excimer Laser-Annealed Polysilicon TFTs

Excimer laser annealing (ELA) has attracted much research activity recently as an alternative low-temperature processing method for the crystallization of amorphous silicon (α-Si) films or for the annealing of solid-phase crystallized (SPC) polysil-icon films [71–75]. The goal of these investigations was to optimize the ELA process to maximize the grain size of the polysilicon films with excellent structural quality and, therefore, to improve TFT performance.

Recently we have investigated the electrical properties of n-channel polysilicon TFTs, fabricated on fused quartz glass substrates with a standard low-temperature process [74, 75]. First, the glass substrate was covered by 200-nm-thick SiO$_2$, deposited by electron cyclotron resonance plasma-enhanced chemical vapor deposition (ECR-PECVD). On top of the oxide, intrinsic amorphous silicon (α-Si) films 50 nm thick were deposited by low-pressure chemical vapor deposition (LPCVD) at 425°C and a pressure of 1.1 Torr, using disilane (Si$_2$H$_6$) and helium as the reactant and dilution gas, respectively. Then some specimens were furnace-annealed in a nitrogen ambient at 600°C for 24 h, during which the amorphous films were transformed into polycrystalline state in solid phase. Both the α-Si and the SPC specimens were irradiated with various energy densities by KrF excimer laser ($\lambda = 248$ nm) at room temperature in air ambient. The pulse duration and repetition of the excimer laser light were 35 nm and 100 Hz, respectively. An optical homogenizer produced a uniform top-flat 0.2 mm × 300 mm line-shaped irradiated profile on the specimen surface. Following the patterning of the polysilicon film, a 120-nm-thick silicon dioxide (SiO$_2$) gate insulator film was formed by ECR-PECVD at 100°C. A tantalum film of thickness 750 nm was deposited by sputtering at 150°C and patterned to form the gate electrode. After the gate electrode formation, phosphine (PH$_3$) was implanted in the polysilicon film at 300°C, using the gate electrode as an implantation mask, to complete n-channel self-aligned TFTs. Annealing in a forming gas ambient (3% hydrogen and 97% argon) at 300°C for 3 h was performed to activate the implanted ions. The gate width W and length L of the investigated TFTs were $W = L = 10$ μm.

The structure of the polysilicon films, obtained after laser irradiation of α-Si and SPC films at various energy densities, was investigated by transmission electron microscopy analysis [73]. The most important structural parameters of polysilicon films prepared by laser crystallization of α-Si and SPC films subjected to ELA at different energy densities are summarized in Table I. Using α-Si as starting material, increase in the laser energy density from 210 to 280 mJ/cm^2 increases the mean grain size from about 67 to 145 nm, whereas the polysilicon films are characterized by a low in-grain defect density in all cases. Thus, an increase in the laser energy density is expected to have a beneficial effect on device performance. However, in contrast to the effect on the grain size, an increase in the laser energy density increases the interface roughness acting as a scattering center for the carriers, degrading the device performance.

The as-deposited SPC polysilicon film is characterized by a smooth surface and large grains (mean value about 2500 nm) with very high in-grain defect density. When the laser energy density increases from 260 to 320 mJ/cm^2, the quality of the SPC polysilicon material is significantly improved, showing a lower in-grain defect density, although the mean grain size remains about 2500 nm. However, an increase in the laser energy

Table I.   Structural Parameters of Polysilicon Films Prepared by Excimer Laser Annealing at
Various Energy Densities with a-Si and SPC Specimens as Starting Materials

Starting material	Laser energy density (mJ/cm²)	Mean grain size (nm)	% roughness of the film thickness	Intra-grain defect density
a-Si	210	67	—	Low
a-Si	260	125	33	Low
a-Si	270	143	—	Low
a-Si	280	145	75	Low
SPC	0	2500	0	Very High
SPC	260	2500	45	High
SPC	280	2500	45	High
SPC	320	2500	112	Low
SPC	340	100	—	Low

Reprinted from C. T. Angelis et al., *J. Appl. Phys.* 86, 4600, 1999, with permission.

density to 320 mJ/cm² results in an increase of the surface roughness to about 112% of the film thickness. A further increase in the laser energy density to 340 mJ/cm² results in 80% melting of the film area, and small grains with a mean diameter of about 100 nm coexist with the larger grains [74].

The TFT performance parameters of the effective electron mobility μ, threshold voltage V_T, and subthreshold swing voltage $S = \ln 10 \times [dV_G/d(\ln I_D)]$ were extracted from the I_D versus V_G characteristics in the linear region ($V_D = 0.1$ V). The parameters μ, V_T, and S of TFTs on polysilicon films, prepared by ELA of α-Si and SPC polysilicon films at various energy densities, are shown in Figure 22. It is evident that for the same laser energy density, the mobility and threshold voltage are superior in the case when SPC specimens are used as the initial Si material. In both cases of α-Si and SPC specimens as the starting materials, the TFT performance parameters are improved as the laser energy increases until the silicon film is completely melted. The energy density for complete melting of the 50-nm-thick SPC and α-Si specimens is 320 and 280 mJ/cm², respectively [74]. The improvement of the device performance parameters with increasing laser energy density is consistent with the measured structural properties presented in Table I.

It is has been established that the various TFT performance parameters are strongly correlated with the gap states distribution [76]. Therefore, to optimize the TFT fabrication process in relation to the starting material and the laser annealing conditions, measurement of the gap states distribution is essential. Based on the above described LF noise technique, the drain current noise data were analyzed to yield quantitative information about the oxide trap density located near the SiO$_2$/polysilicon interface and the energy distribution of the grain boundary trap states within the forbidden band gap.

The $1/f$ noise has been employed to determine the oxide trap density N_{ox} near the polysilicon/SiO$_2$ interface. Figure 23 shows the trap density N_{ox} as a function of the laser energy density of devices fabricated on polysilicon prepared by laser crystallization of α-Si films and a combined SPC and ELA

process. In both types of devices, an abrupt increase in N_{ox} is observed at high laser energy densities, related probably to the increased hillocks developed at the grain boundaries during laser irradiation (Table I). The higher values of N_{ox} found in SPC polysilicon TFTs, in comparison with the laser-annealed α-Si TFTs, can be attributed to contamination induced by the prolonged annealing of the SPC specimens and/or to the higher in-grain defect density inducing higher fluctuation of the insulator charge near the oxide/polysilicon interface.

The $1/f^{\gamma}$ noise data at various gate voltages have been employed to determine the energy distribution of the grain boundary trap states. Assuming that the traps at the grain boundaries are replaced by the same number of traps uniformly distributed throughout the film, the uniform distributed trap density N_t (in cm^{-3}eV^{-1}) is related to the areal trap density N_s (in cm^{-2}eV^{-1}) by the relationship $N_t = 6N_s/L_g$, where L_g is the average grain size [26]. The gap states distributions for differently processed polysilicon films are presented in Figure 24. It is apparent that the trap density is essentially lower in the SPC film than in the α-Si film processed under the same laser energy density. For both types of polysilicon films, the deep and tail state densities are reduced as the laser energy density increases. The higher mobility and lower threshold voltage of the laser-annealed SPC polysilicon TFTs seen in Figure 22 are correlated with the lower density of tail states.

3.7. Correlation Between Noise and Static Device Parameters

Low-frequency noise has been considered a very sensitive diagnostic tool for the quality and reliability assessment of electronic devices based on single-crystalline silicon [77]. An important issue is that when the technology parameters are changed, the device-to-device dispersion in the noise level may change by several decades, which is undesirable from the analog design point of view. Large efforts have been made to relate low-frequency noise to the static single-crystalline silicon MOSFET parameters, such as transconductance g_m, threshold

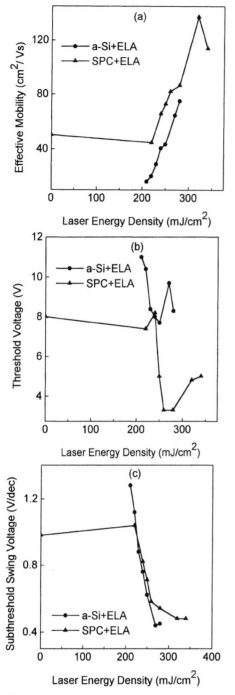

Fig. 22. (a) Field effect mobility, (b) threshold voltage, and (c) subthreshold swing voltage (measured at $V_g = -20$ V and $V_d = 0.1$ V) as a function of the laser energy density of excimer-annealed polysilicon TFTs. The devices were made from amorphous silicon crystallized by excimer laser annealing (a-Si + ELA) and a combined solid-phase crystallization and ELA process (SPC + ELA). (Reprinted from C. T. Angelis et al., *J. Appl. Phys.* 86, 4600, 1999, with permission.)

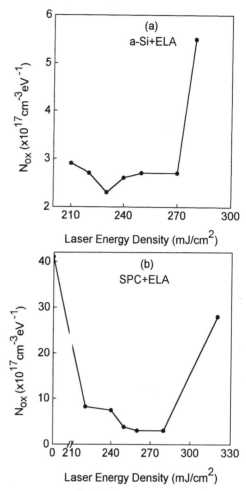

Fig. 23. Oxide trap density N_{ox} as a function of the laser energy density of excimer-annealed polysilicon TFTs. The devices were made (a) from a-Si crystallized by excimer laser annealing (a-Si + ELA) and (b) by a combined SPC and excimer laser annealing process (SPC + ELA). (Reprinted from C. T. Angelis et al., *J. Appl. Phys.* 86, 4600, 1999, with permission.)

regime. It is demonstrated that simple empirical relationships exist between the noise level and the parameters g_m, V_T, and S for the devices under study. The consequences of these findings from the fundamental and practical points of view are discussed.

The TFTs studied were fabricated on fused quartz glass substrates by processes described in Section 3.6. Two types of polysilicon films were used. In type A, amorphous silicon (a-Si) films of thickness 50 or 25 nm were first transformed into the polycrystalline phase by furnace annealing at 600° C for 24 h in nitrogen ambient (SPC). Then the 50- or 25-nm-thick SPC samples were irradiated at room temperature in air ambient by KrF excimer laser (248 nm) with energy densities varying up to 320 or 280 mJ/cm², respectively, before the polysilicon films melted. In type B, a-Si films (50 nm thick) were subjected to ELA with energy densities varying from 210 to 280 mJ/cm², before the polysilicon films melted. A standard self-aligned NMOS process was used to fabricate TFTs with a nominal area of $W \times L = 10\ \mu m \times 10\ \mu m$. As the gate oxide, a 120-nm-thick SiO₂ was formed by ECR-PECVD at a temperature of 100°C.

voltage V_T, or subthreshold slope S, because these parameters are closely related to the density of interface and/or oxide traps [78]. For the first time, the correlation between noise spectral density and static device parameters of a large number of n-channel polysilicon TFTs is reported for operation in the linear

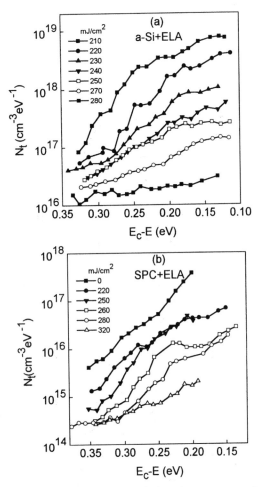

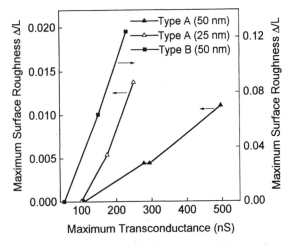

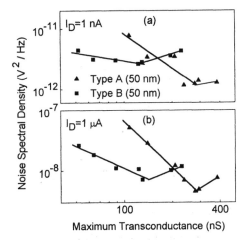

Fig. 24. Energy distribution of the grain boundary trap states of excimer-annealed polysilicon TFTs at different energy densities. The devices were made (a) from a-Si crystallized by excimer laser annealing (a-Si + ELA) and (b) by a combined SPC and excimer laser annealing process (SPC + ELA). (Reprinted from C. T. Angelis et al., *J. Appl. Phys.* 87, 1588, 2000, with permission.)

Fig. 25. Dependence of the parameter Δ/L for surface roughness with transconductance of polysilicon films prepared from SPC or a-Si films subjected to an ELA process at various laser energy densities.

Fig. 26. Input gate voltage spectral density versus maximum transconductance measured at a frequency of $f = 10$ Hz for $10\ \mu m \times 10\ \mu m$ polysilicon TFTs in linear operation ($V_D = 0.1$ V) for drain current (a) $I_D = 1$ nA and (b) $I_D = 1\ \mu A$. The polysilicon layers were prepared from SPC or a-Si films subjected to an ELA process at various laser energy densities.

The average grain size L_g and the surface roughness of the polysilicon films were investigated by transmission electron microscopy (TEM). Plane view and cross-sectional TEM images have shown that in 50-nm-thick samples of type A the average grain size remains about 2.5 μm for all specimens, and the polysilicon quality is improved, showing lower intragrain defect density as the laser energy density increases (Table I). A similar structural behavior was also observed for the 25-nm SPC polysilicon film with an average grain size of about 1.75 μm. In 50-nm-thick polysilicon films of type B, the average grain size increases from about 67 to 145 nm as the laser energy density increases. However, with increasing laser energy density, improvement of the crystallite quality results in a degradation of the surface roughness (Table I). As shown in Figure 25, in all samples the maximum surface roughness ratio Δ/L increases as the transconductance $g_m = \partial I_D/\partial V_G$ (i.e., the laser energy density) increases. In the above definition for surface roughness of polysilicon films, Δ is the maximum asperity height and L is the correlation length, which corresponds to the crystallite size near the surface of the film. However, with

a-Si as a starting material, the polysilicon layers show higher surface roughness in comparison with the polysilicon films prepared from SPC films as starting material (Fig. 25).

The data of the input gate voltage spectral density $S_{V_G} = S_{I_D}/g_m^2$ at a frequency of $f = 10$ Hz versus the maximum transconductance $g_{m,max}$ are plotted in Figure 26 for a set of 50-nm-thick polysilicon TFTs of types A and B. The noise spectral density S_{V_G} shows a power-dependent reduction with $g_{m,max}$, which can be represented by the empirical relationship

$$S_{V_G} = A' g_{m,max}^{-n} \tag{64}$$

where A' and n are empirical constants. For the devices of type A, the exponent n increases from about 1.8 for $I_D = 1$ nA to about 2.5 for $I_D = 1\ \mu A$, reflecting the change from device to device of the deep and tail states densities, respectively. The observed correlation between S_{V_G} and $g_{m,max}$ is expected because both physical quantities are affected by the in-grain and

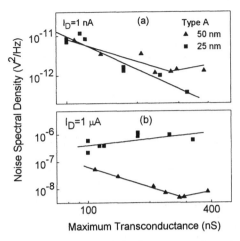

Fig. 27. Input gate voltage spectral density versus maximum transconductance measured at a frequency of $f = 10$ Hz for the polysilicon TFTs of Figure 25. The polysilicon layers (50 and 25 nm thick) were prepared from SPC films subjected to an ELA process at various laser energy densities.

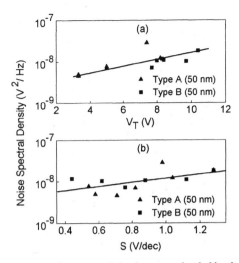

Fig. 28. Input gate voltage spectral density versus threshold voltage (a) and subthreshold swing voltage (b), measured at a frequency of $f = 10$ Hz for the polysilicon TFTs of Figure 25 in linear operation ($V_D = 0.1$ V and $I_D = 1$ μA). The polysilicon layers (50 nm thick) were prepared from SPC or a-Si films subjected to an ELA process at various laser energy densities.

grain boundary traps. Following the carrier number fluctuation model with correlated mobility fluctuations and considering carrier number fluctuations with dynamic trapping–detrapping at the interface, it has been demonstrated that there is a proportionality between S_{V_G} and the areal trap density in an interval of a few kT around the Fermi level. The observed qualitative trend between S_{V_G} and $g_{m,max}$ is correct because for degraded polysilicon quality (i.e., higher in-grain and grain boundary trap density), a reduced $g_{m,max}$ and an increased S_{V_G} are expected. However, in devices with transconductances above 2.5 μS, an increase in the noise spectral density with increasing $g_{m,max}$ is observed. This increase in the noise spectral density is probably related to the increased SiO_2/polysilicon interface roughness scattering (Fig. 25), causing an increased fluctuation of the scattering rate and, therefore, of the effective mobility in the channel of the transistor. A similar tendency is observed for devices of type B, with the influence of the surface roughness scattering on the noise spectral density being more pronounced because of the higher polysilicon surface roughness (Fig. 25).

The correlation between gate voltage noise spectral density and maximum transconductance for devices of type A with 50- and 25-nm-thick polysilicon layers is presented in Figure 27. For low drain current $I_D = 1$ nA, the noise spectral density follows the power-dependent reduction with $g_{m,max}$ law of Eq. (64), with the noise level reduced in devices with thinner polysilicon films because of the smaller total number of traps present in the polysilicon layer. However, for drain current $I_D = 1$ μA, the increased surface roughness in the thinner polysilicon films (Fig. 25) results in an increase in the noise spectral density with $g_{m,max}$, as shown in Figure 27b. These results indicate that surface roughness causing carrier scattering at the SiO_2/polysilicon interface is responsible for the increase in the low-frequency current fluctuations. Further work is needed to clarify the role of the surface roughness scattering in the device noise performance.

The correlation between S_{V_G} and V_T for polysilicon TFTs of types A and B is shown in Figure 28a for $I_D = 1$ μA. The

lower noise is found for the devices with the smallest V_T. A similar tendency was also observed in single-crystalline silicon n-MOSFETs [78]. From Figure 28a, an exponential reduction of S_{V_G} with V_T is derived from the empirical expression

$$S_{V_G} = A' \exp(-B' V_T) \qquad (65)$$

where A' and B' are empirical constants. The similar trend observed for both types of TFTs indicates that the relationship (65) is not restricted to one polysilicon TFT technology and may be more general. The correlation between the measured subthreshold swing voltage $S = \ln 10 \times [dV_G/d(\ln I_D)]$ and S_{V_G} is shown in Figure 28b for $I_D = 1$ μA. In a manner similar to that of the threshold voltage, the device with the smallest S yields the lowest noise, following the empirical relationship of Eq. (65).

In addition to modeling the dispersion in the low-frequency noise of a polysilicon TFT technology, an important practical consequence of Eq. (64) is that it can provide some predictive device properties. For example, extrapolation of the S_{V_G} versus $g_{m,max}$ straight line in Figure 26 can predict the expected device transconductance for a smooth polysilicon surface, because a one-to-one correlation between surface roughness and poor electrical characteristics of polysilicon TFTs has been reported [79]. Moreover, Eqs. (64), (65) show that the low-frequency noise is a much stronger function of the polysilicon film quality than the device parameters $g_{m,max}$, V_T, and S. Thus, the low-frequency noise is a better monitor than a typical static device parameter, for example, for hot-carrier degradation studies.

3.8. Noise of Very Thin Excimer Laser-Annealed Polysilicon TFTs

The use of very thin SPC amorphous silicon films has been shown to be beneficial for the TFT performance. The on/off current ratio increases drastically from about 10^5 for a film thickness of 50 nm (or more) to the order of 10^7 for very thin (25-nm) films [80]. Excimer laser annealing of SPC polysilicon films has been shown to be the method for producing high-quality TFTs [74]. Using electrical and low-frequency noise measurements, we have investigated the performance of TFTs fabricated on 50- and 25-nm-thick SPC polysilicon films prepared by a combined SPC and ELA process at different energy densities [81]. The device performance has been correlated with the trap properties of the polysilicon films.

The investigated n-channel polysilicon TFTs were fabricated on 50- or 25-nm-thick SPC polysilicon films, irradiated with various energy densities by KrF excimer laser at room temperature in an air ambient. TEM analysis of 50-nm-thick SPC polysilicon films has shown that the mean size is about 2.5 μm with a high density of in-grain defects (Table I). As the laser energy density increases up to 320 mJ/cm^2, there is a marked reduction in the in-grain defects, whereas the grain size remains unaffected. At an energy density of 340 mJ/cm^2, the 50-nm-thick SPC polysilicon film is completely melted. The reduction in the number of nucleation sites leads to the creation of small crystallites (of diameter 100 nm) through homogeneous nucleation of the molten layer. A similar behavior with the laser energy density was observed in the structure of the 25-nm-thick SPC polysilicon films characterized by a mean grain size of 1.75 μm and a laser energy density of 300 mJ/cm^2 for complete melting.

The TFT performance parameters of μ, V_T, and S were extracted from the I_D–V_G characteristics in the linear region ($V_D = 0.1$ V). The dependence of these parameters on the laser energy density, for devices with active layers of 50 and 25 nm, is shown in Figure 29. A degradation in the electron mobility and the threshold voltage and an improvement in the subthreshold swing voltage for the devices with an active layer thickness of 25 nm is observed. In both cases, the parameters μ, V_T, and S are improved as the laser energy density increases up to a critical value, above which the film is completely melted.

From analysis of the $1/f^\gamma$ noise data, the determined energy distributions of the grain boundary trap density N_t for 50- and 25-nm-thick SPC polysilicon films subjected to ELA at different energy densities, are shown in Figure 30. For both polysilicon films, the trap states density is reduced as the laser energy density increases. Although the grain size is smaller in the thinner polysilicon films, the 25-nm-thick polysilicon films present lower grain boundary trap density in comparison with the 50-nm-thick films when both are subjected to ELA at the same energy density. In devices with active layer thicknesses of 25 nm, there is probably insufficient thickness for complete band bending, resulting in less charge being trapped at the grain boundaries and therefore an apparent improvement in the density of states. The decreased number of traps in the active polysilicon volume improves the subthreshold swing voltage of the 25-nm-thick polysilicon TFTs seen in Figure 29c.

The normalized drain current spectral density $S_{1/f}/I_D^2$ of the $1/f$ noise fluctuation is shown in Figure 31a and b for the 50- and 25-nm-thick SPC polysilicon films, respectively. For the

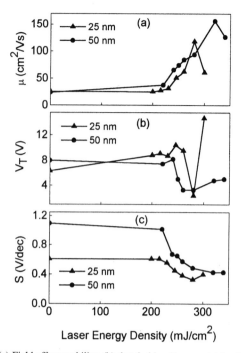

Fig. 29. (a) Field effect mobility, (b) threshold voltage, and (c) subthreshold swing voltage as a function of the laser energy density of TFTs fabricated on excimer-annealed SPC polysilicon films of thickness 50 and 25 nm. (Reprinted from C. T. Angelis et al., *Appl. Phys. Lett.* 74, 3684, 1999, with permission.)

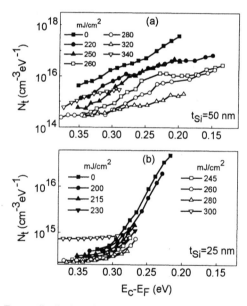

Fig. 30. Energy distribution of the grain boundary trap states of TFTs fabricated on SPC polysilicon films of thickness (a) 50 nm and (b) 25 nm, annealed by excimer laser at different energy densities. (Reprinted from C. T. Angelis et al., *Appl. Phys.Lett.* 74, 3684, 1999, with permission.)

devices with an active layer thickness of 50 nm, $S_{1/f}/I_D^2$ starts from a plateau at low drain currents before decreasing at higher drain currents. For the devices with an active layer thickness of 25 nm, $S_{1/f}/I_D^2$ starts almost from the same value as for the devices with an active layer thickness of 50 nm and passes through a maximum before decreasing at higher drain currents.

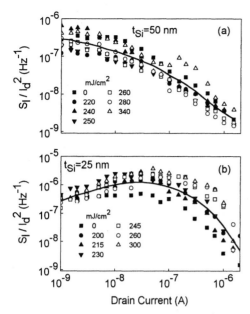

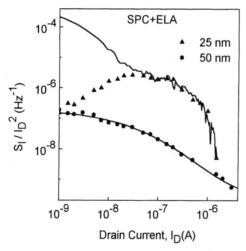

Fig. 31. Normalized drain current spectral density $S_{1/f}/I_D^2$ versus drain current of polysilicon TFTs at drain voltage $V_D = 0.1$ V and frequency 2 Hz. The devices were made on SPC polysilicon films of thickness (a) 50 nm and (b) 25 nm, annealed by excimer laser at different energy densities. (Reprinted from C. T. Angelis et al., *Appl. Phys. Lett.* 74, 3684, 1999, with permission.)

Fig. 32. Normalized drain current spectral density $S_{1/f}/I_D^2$ versus drain current of two typical polysilicon TFTs at drain voltage $V_D = 0.1$ V and a frequency of 2 Hz. The devices were made on SPC polysilicon films of thickness (a) 50 nm and (b) 25 nm, annealed by excimer laser at energy densities of 250 and 245 mJ/cm^2, respectively. The continuous lines are the best fit to Eq. (19) with $S_{V_{FB}} = 1.4 \times 10^{-8}$ V^2/Hz and $S_{V_{FB}} = 3 \times 10^{-6}$ V^2/Hz for the 50- and 25-nm-thick polysilicon TFTs, respectively. (Reprinted from C. T. Angelis et al., *Appl. Phys. Lett.* 74, 3684, 1999, with permission.)

Assuming the carrier number fluctuation model with correlated mobility fluctuations, $S_{1/f}/I_D^2$ is related to the flat-band voltage spectral density $S_{V_{FB}}$ by Eq. (19). A fit of the experimental $S_{1/f}/I_D^2$ data with Eq. (19) is shown in Figure 32 for some typical TFTs fabricated on 50- and 25-nm-thick films subjected to ELA irradiation. For the 50-nm-thick SPC polysilicon TFT, because $S_{1/f}/I_D^2$ varies with the drain current as $(g_m/I_D)^2$

according to Eq. (19), the $1/f$ noise in the present devices can be ascribed to the dynamic trapping of electrons by the gate oxide interface traps [68]. However, for the 25-nm-thick SPC polysilicon TFT, a significant deviation of the experimental $S_{1/f}/I_D^2$ data from Eq. (19) is observed in the low drain current region. This result indicates that the model based on the conventional mechanism of tunneling into the gate oxide traps is not applicable for the 25-nm-thick film. The observed peculiar behavior of $S_{1/f}/I_D^2$ with the drain current (Fig. 31b) and the higher values of the noise spectral density $S_{1/f}$ can be attributed to an additional noise arising from the dynamic trapping of electrons by the polysilicon/substrate oxide interface traps, because there is insufficient layer thickness for complete band bending. Interaction of the electrons with the gate and substrate oxide interface traps can explain the observed degradation in mobility and threshold voltage seen in Figure 29a and b. The above results indicate that the active polysilicon layer should be more than 25 nm thick to obtain the best device performance.

4. NOISE OF THE LEAKAGE CURRENT IN POLYSILICON TFTs

4.1. Introduction

Polysilicon TFTs present increasing interest because of their use as switching devices in active matrix liquid crystal displays. For such applications, the leakage current I_L has to be as low as 1 pA/μm to hold the signal levels for acceptable image quality [82]. However, polysilicon TFTs suffer from relatively high leakage currents, especially under high gate and drain voltages, generated by trap states present at the grain boundaries in the drain region. In view of this, extensive research was performed to identify the conduction mechanisms of the leakage current in polysilicon TFTs [83–87]. It has been shown that the conduction mechanism of I_L is not determined by a unique mechanism, but it depends on the gate bias V_G, drain bias V_D, device structure, and the polysilicon layer structure. All previous investigations were based on the analysis of the transfer characteristics at various temperatures.

Recently, low-frequency noise was used as a diagnostic tool for the quality assessment of polysilicon TFTs [53–56]. For the first time, we used combined measurements of the transfer characteristics at various temperatures and low-frequency noise measurements at room temperature to identify the conduction mechanisms and the origin of the leakage current in polysilicon TFTs [70]. Systematic investigations of the leakage current of high-temperature n-channel and p-channel TFTs, fabricated on undoped hydrogenated polysilicon films, are presented below. Details of the device fabrication processes are described in Section 3.2.

4.2. Conduction Measurements

Figure 33 shows typical transfer characteristics of n-channel and p-channel polysilicon TFTs, measured at room temperature and drain voltages $V_D = 5$ V and -5V, respectively. In the

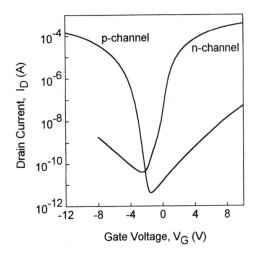

Fig. 33. Typical transfer characteristics of n-channel and p-channel polysilicon TFTs at drain voltages of 5 V and −5 V, respectively. The gate ratio is $W/L = 50/10$. (Reprinted from C. T. Angelis et al., *J. Appl. Phys.* 82, 4095, 1997, with permission.)

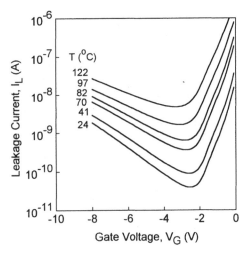

Fig. 35. Leakage current I_L versus gate voltage V_G of a typical n-channel polysilicon TFT measured at drain voltage $V_D = 5$ V and various temperatures. (Reprinted from C. T. Angelis et al., *J. Appl. Phys.* 82, 4095, 1997, with permission.)

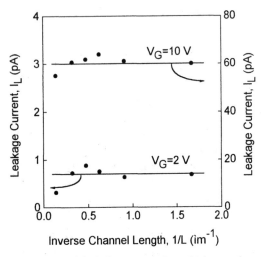

Fig. 34. Dependence of the leakage current I_L on the inverse channel length $1/L$ at drain voltage $V_D = -0.1$ V for typical p-channel polysilicon TFTs with channel width $W = 50$ μm. (Reprinted from C. T. Angelis et al., *J. Appl. Phys.* 82, 4095, 1997, with permission.)

off-state current region, the leakage current I_L increases considerably when the applied gate bias V_G increases. We have found that I_L is independent of L. The independence of I_L on $1/L$, as shown in Figure 34 for a typical TFT, indeed indicates that the leakage current originates from the reverse biased drain junction.

Generally, in polysilicon TFTs, the leakage current can arise from electron-hole pairs generated via grain boundary traps in the depletion region of the drain junction. The main mechanisms through which a trapped carrier at an energy level E_t can be generated in the conduction or valence band are (a) thermal generation due to thermal excitation of trapped carriers, (b) pure field emission due to field ionization of trapped carrier tunneling through the potential barrier into the conduction or valence band, and (c) thermionic field emission or Poole–Frenkel emis-

sion due to field-enhanced thermal excitation of trapped carriers in the conduction or valence band.

To clarify the leakage current mechanism, the temperature dependence of the drain current was measured at various drain voltages. The temperature dependence of the leakage current for a typical n-channel polysilicon TFT, measured at $V_D = 5$ V, is shown in Figure 35. The activation energies E_a of the drain current were derived from the Arrhenius plots of Figure 36a and b, corresponding to the n-channel and p-channel polysilicon TFTs, respectively. In the on-state current regime, according to Eq. (44), the activation energy of the drain current reflects the potential barrier height at the grain boundaries present within the TFT channel. In the off-state current regime, the activation energy of the minimum leakage current becomes about half the silicon band-gap energy at low drain biases, as shown in Figure 36. This result implies that the minimum leakage current is caused mainly by thermal excitation, because the thermal excitation current is proportional to the intrinsic carrier concentration (n_i), and n_i is proportional to $\exp(-E_g/kT)$, where E_g is the energy gap of silicon.

With increasing $|V_G|$ in the off-state regime, E_a decreases, indicating that the carrier generation rate is field enhanced. The temperature dependence of the leakage current depicted in Figure 35 rules out the mechanism of pure field emission. The observed field-enhanced carrier generation implies that the Poole–Frenkel effect is the most likely generation mechanism. According to this mechanism, the reversed-biased drain junction current is modeled as a enhanced generation current field of the form

$$I_L = I_{LO} \exp(\beta \sqrt{E_m}) \qquad (66)$$

where E_m is the peak electric field at the drain junction, β is the field enhancement factor, $I_{LO} = q A_j w_D U$, A_j is the junction area, w_D is the reverse-biased drain junction depletion region width, and U is the zero field carrier generation rate per unit

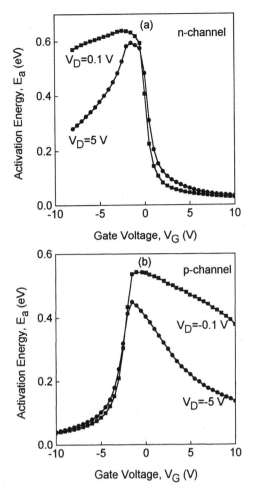

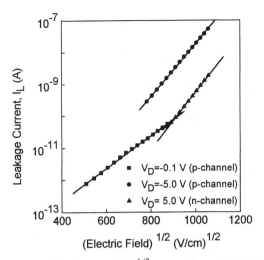

Fig. 37. Measured $\log I_L$ versus $(E_m)^{1/2}$ on a semilog scale of the junction at the drain end for n-channel and p-channel polysilicon TFTs at various drain voltages. (Reprinted from C. T. Angelis et al., *J. Appl. Phys.* 82, 4095, 1997, with permission.)

Fig. 36. Dependence of the drain current activation energy E_a on gate voltage V_G for (a) n-channel polysilicon TFT at drain bias 0.1 V and 5 V and (b) p-channel polysilicon TFT at drain bias −0.1 V and −5 V. The gate ratio is $W/L = 50/10$. (Reprinted from C. T. Angelis et al., *J. Appl. Phys.* 82, 4095, 1997, with permission.)

volume. The expression for the generation rate U due to grain boundary defects is [88]

$$U = \frac{\sigma_n \sigma_p \upsilon_{th} N_{gb} n_i}{\sigma_n \exp((E_t - E_i)/kT) + \sigma_p \exp((E_i - E_t)/kT)} \quad (67)$$

where σ_n and σ_p are the electron and hole capture cross sections, respectively; υ_{th} is the carrier thermal velocity; N_{gb} is the grain boundary trap density per unit area distributed throughout the grain; E_t is the trap level; and E_i is the intrinsic Fermi level. For simplicity, assuming that $\sigma_n = \sigma_p = \sigma$ and the carriers are mainly generated from mid-gap defect states N_d ($E_t = E_i$), the expression for the leakage current becomes [89]

$$I_L = q A_j w_D \sigma \upsilon_{th} N_d n_i \exp\left(\beta_{PF}\sqrt{E_m}\right) \exp\left(\beta_{TFE}\sqrt{E_m}\right) \quad (68)$$

In the above equation, β_{PF} and β_{TFE} are the current field-enhancement factors arising from the Poole–Frenkel effect and the thermal field emission, with expected values about 0.009 and 0.0011 $cm^{1/2}$ $V^{-1/2}$, respectively, at room temperature [89]. Assuming that the peak electric field E_m at the drain

junction is dominated by the vertical electric field at the interface [90],

$$E_m = \frac{|V_G - V_D - V_{FB}|}{t_{ox}(\varepsilon_{Si}/\varepsilon_{SiO_2})} \quad (69)$$

where ε_{Si} and ε_{SiO_2} are the permittivities of Si and SiO_2, respectively, and t_{ox} is the thickness of the gate oxide.

Figure 37 shows plots of $\log I_L$ versus $(E_m)^{1/2}$ obtained from the data of the n-channel and p-channel polysilicon TFTs at various drain voltages. In all cases, straight lines fit the data in the high electric field regions. This proves that I_L is dominated by Poole–Frenkel emission. With increasing drain voltage, the slope β of the straight lines increases. For a typical p-channel polysilicon TFT, β increases from about 0.01 $cm^{1/2}$ $V^{-1/2}$ at $V_D = -0.1$ V to 0.02 $cm^{1/2}$ $V^{-1/2}$ at high drain voltages, as shown in Figure 38. These results indicate that at the low value of $V_D = -0.1$ V the pure Poole–Frenkel emission is present ($\beta = \beta_{PF} \cong 0.009$ $cm^{1/2}$ $V^{-1/2}$), and at higher drain voltages the Poole–Frenkel effect is accompanied by thermal field emission ($\beta = \beta_{PF} + \beta_{TFE} \cong 0.02$ $cm^{1/2}$ $V^{-1/2}$) due to the increased electric field at the drain junction.

In the thermionic field emission conduction mechanism, the presence of trap states in the band gap assists the process by shortening the effective tunneling length of the carrier. In fact, polysilicon TFTs have many traps present in the band gap, and, hence, this situation occurs easily. The density of trap states in the silicon band gap consists of the grain boundary traps and the SiO_2/ polysilicon interface traps. Following the results of a recent work [42], the energy distribution of the grain boundary traps can be modeled by

$$N_{gb}(E) = N_d + N_t \exp\left(-\frac{E_c - E}{kT_t}\right) \quad (70)$$

where N_d is the density per unit area of deep levels with uniform energy distribution and N_t, T_t are the characteristic parameters

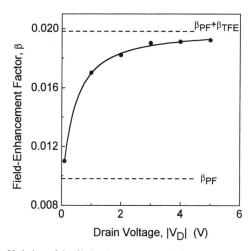

Fig. 38. Variation of the filed-enhancement coefficient β as a function of the drain voltage V_D of a typical p-channel polysilicon TFT. The dashed lines show the values of the field-enhancement coefficients $\beta = \beta_{PF} \cong 0.009$ cm$^{1/2}$ V$^{-1/2}$ and $\beta = \beta_{PF} + \beta_{TFE} \cong 0.02$ cm$^{1/2}$ V$^{-1/2}$. (Reprinted from C. T. Angelis et al., *J. Appl. Phys.* 82, 4095, 1997, with permission.)

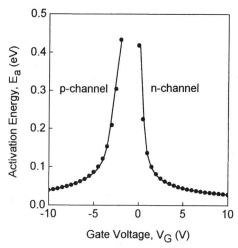

Fig. 39. On-state current activation energy E_a as a function of the gate voltage V_G of typical n-channel and p-channel polysilicon TFTs at drain voltages of 0.1 V and −0.1 V, respectively. The solid lines are the calculated curves from the method described in [42] for the trap distribution parameters (a) $N_d = 2 \times 10^{11}$ cm^{-2}, $N_s = 1.4 \times 10^{15}$ cm^{-2}, and $kT_s = 0.014$ eV for the n-channel TFT and (b) $N_d = 3 \times 10^{12}$ cm^{-2}, $N_s = 2.8 \times 10^{14}$ cm^{-2}, and $kT_s = 0.028$ eV for the p-channel TFT. (Reprinted from C. T. Angelis et al., *J. Appl. Phys.* 82, 4095, 1997, with permission.)

of the band tails with exponential energy distribution increasing toward the conduction band edge E_c. The deep levels are expected to give a larger capture emission cross section for carriers to tunnel through the band gap. The interface states distribution is considered to have the exponential distribution [42]

$$N_{is}(E) = N_s \exp\left(-\frac{E_c - E}{kT_s}\right) \tag{71}$$

where N_s and T_s are parameters characterizing the exponential distribution. The grain boundary and the interface states were determined with a method based on the variation of E_a with V_G in the on-state regime at low drain voltage (Fig. 39). The extracted values of the trap parameters are presented in the caption of Figure 39. It is found that, in both n-channel and p-channel TFTs, the total trap density consists of the exponential interface states N_{is} and the grain boundary deep states N_d with uniform energy distribution. Because the leakage current is proportional to the mid-gap trap density [Eq. (68)], the smaller leakage current of the n-channel polysilicon TFTs (Fig. 37) can be explained by the measured lower mid-gap trap concentration N_d (Fig. 39).

From the activation energy data in the off-state regime (Fig. 36), the following conclusions are obtained: (a) The activation energy decreases with increasing applied drain voltage. An increase in the electric field at the drain junction causes a decrease in the leakage current activation energy. This feature corresponds to the leakage current of the TFTs shown in Figure 33. That is, the increase in the leakage current in Figure 33 corresponds to the decrease in the activation energy shown in Figure 36b. In the n-channel polysilicon TFT, the decrease in the activation energy with increasing $|V_G|$ is less pronounced compared with the p-channel polysilicon TFT. This result implies that the field-induced process in the n-channel polysilicon TFT is suppressed.

One possible explanation for suppression of the electric field in the n-channel TFT is the development of localized fixed positive charges at the interface near the drain junction. If the existing interface states are assumed to be donor states, these are positively charged during the p-channel TFT operation because they are located above the Fermi level and neutral during the n-channel TFT operation because they are occupied by electrons. However, in the n-channel and p-channel polysilicon TFTs, positive phosphorus ions and negative boron ions, respectively, may diffuse fast through the grain boundaries from the drain contact toward the interface. In the case of p-channel polysilicon TFT, the negative charge of the boron ions at the interface near the drain junction is neutralized by the positive charge of the interface states. In contrast, in the n-channel polysilicon TFT the positive phosphorus ions lead to the development of positive fixed charges near the drain junction. Based on the above argument, we infer that the positive fixed charges suppress the electric field from the gate in the drain junction and finally reduce the leakage current in the n-channel polysilicon TFT. To strictly clarify the reasons for the superiority of the n-channel TFT, more systematic investigations are required.

4.3. Noise Measurements

The origin of the leakage current is further investigated by low-frequency noise measurements. Figure 40a and b shows typical measurements of noise spectra of n-channel and p-channel polysilicon TFTs, respectively, for different leakage currents. For the n-channel polysilicon TFT, the leakage current noise spectral density S_L exhibits an almost pure $1/f$ behavior. For the p-channel polysilicon TFT, at low leakage currents, one can notice that the $1/f$ dependence is turned into a $1/f^{1.5}$ depen-

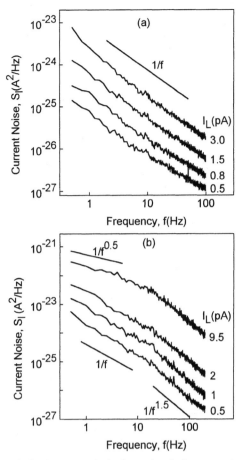

Fig. 40. Typical noise spectra obtained at various leakage currents from (a) n-channel and (b) p-channel polysilicon TFTs at drain voltages of 4 V and −5 V, respectively. (Reprinted from C. T. Angelis et al., *J. Appl. Phys.* 82, 4095, 1997, with permission.)

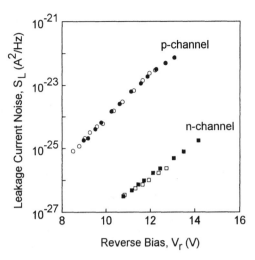

Fig. 41. Noise power spectral density of the leakage current S_L at a frequency of 10 Hz versus the reverse bias V_r of the drain junction at various drain voltages V_D of typical n-channel ($\square$, $V_D = 3$ V; $\blacksquare$, $V_D = 4$ V) and p-channel ($\bigcirc$, $V_D = -3$ V; $\bullet$, $V_D = -5$ V) polysilicon TFTs. (Reprinted from C. T. Angelis et al., *J. Appl. Phys.* 82, 4095, 1997, with permission.)

dence for frequencies above 10 Hz. However, at high leakage currents ($I_L > 2$ nA), the $1/f$ branch disappears in the low-frequency range; the noise spectra show a $1/f^{1.5}$ dependence at frequencies above 10 Hz with a changeover to a $1/f^{0.5}$ dependence. These features indicate that in the p-channel polysilicon TFT the noise source is probably related to diffusion noise [91, 92]. Such a noise source in a p-channel TFT reveals the presence of a stronger electric field at the drain junction in comparison with the n-channel TFT.

To clarify the origin of the leakage current in the n-channel and p-channel polysilicon TFTs, the leakage current noise spectral density S_L measured at a frequency of $f = 10$ Hz is plotted versus the reverse bias of the drain junction, $V_r = |V_G - V_D|$ (V_G and V_D are of opposite polarities), as shown in Figure 41. Clearly, in the two cases, different combinations of the drain and gate biases produce nearly the same off-state current noise for the same reverse bias. These results show that the leakage current is limited by the reverse biased drain junction, in agreement with results obtained from conduction measurements.

The $1/f$ frequency dependence of the leakage current noise can be explained by carrier fluctuation between the conduction or valence band and gap states, which occurs within the space charge region of the drain junction. Such a model was

also used in a previous work to explain the $1/f$ behavior in LPCVD polysilicon TFTs [55]. Because the leakage current is due to Poole–Frenkel emission, according to Eq. (66), the current fluctuation can be considered to arise from fluctuations of the peak electric field E_m at the drain junction. Thus, the fluctuation in I_L can be written as

$$\delta I_L = \frac{\partial I_L}{\partial E_m} \delta E_m \quad (72)$$

Assuming the voltage drop in the gate oxide and using Gauss' law at the polysilicon/oxide interface, an equation of the surface electrostatic potential (ψ_s) is

$$V_G = V_{FB} + \psi_s + \frac{\varepsilon_{Si} E(\psi_s)}{C_{ox}} \quad (73)$$

Assuming that the fluctuation of the flat-band voltage arises from the electrostatic potential fluctuation, from Eq. (73) the field fluctuation can be written as

$$\delta E_m = -\frac{2C_{ox}}{\varepsilon_{Si}} \delta V_{FB} \quad (74)$$

From Eqs. (72) and (74), the leakage current fluctuation δI_L is

$$\delta I_L = -\frac{2C_{ox}}{\varepsilon_{Si}} \frac{\partial I_L}{\partial E_m} \delta V_{FB} \quad (75)$$

Therefore, the leakage current noise spectral density is

$$S_L = 4 \left(\frac{C_{ox}}{\varepsilon_{Si}} \right)^2 \left(\frac{\partial I_L}{\partial E_m} \right)^2 S_{V_{FB}} \quad (76)$$

Because the flat-band voltage spectral density is $S_{V_{FB}} = S_{Q_{ss}}/(WLC_{ox}^2)$, where $S_{Q_{ss}}$ is the interface charge spectral density [68], Eq. (76) becomes

$$S_L = \frac{4}{\varepsilon_{Si}^2} \left(\frac{\partial I_L}{\partial E_m} \right)^2 \frac{S_{Q_{ss}}}{WL} \quad (77)$$

For a single trap with a time constant τ and concentration N_t, $S_{Q_{ss}}$ is given by the relationship [66]

$$S_{Q_{ss}} = \frac{q^2 kT 4\tau N_t}{1 + (2\pi\tau f)^2} \qquad (78)$$

For a continuous energy distribution of the trapping states with a distribution of time constants $g(\tau)$, the leakage current noise spectral density is given by

$$S_{Q_{ss}} = q^2 kT N_t \int_{\tau_1}^{\tau_2} \frac{4\tau g(\tau) d\tau}{1 + (2\pi\tau f)^2} \qquad (79)$$

For the continuously distributed gap states of constant density N_d, the distribution of time constants is

$$g(\tau) = A/\tau \qquad (80)$$

where $A = \ln^{-1}(\tau_2/\tau_1)$, which in practical cases has a value of 0.1 [26]. Substituting Eq. (80) into Eq. (79), and using Eqs. (66) and (77), we obtain the following expression for the leakage current noise spectral density:

$$S_L = \frac{\beta^2 N_d}{\varepsilon_{Si}^2 E_m WL} \frac{q^2 A}{f} I_L^2 \qquad (81)$$

Because the leakage current increases with increasing peak electric field E_m according to Eq. (66), from the above equation it follows that the power spectral density S_L should be proportional to I_L^γ with the exponent $\gamma < 2$; i.e., the product $S_L \times E_m$ should be proportional to I_L^2. Plots of $S_L \times E_m$ versus I_L for the n-channel and p-channel polysilicon TFTs are shown in Figure 42. The experimental data lie on straight lines proportional to I_L^2, showing the validity of the derived Eq. (81). From these straight lines and Eq. (81), we determine the trap densities $N_d = 3 \times 10^{12}$ cm^{-2} for the p-channel and $N_d = 2 \times 10^{11}$ cm^{-2} for the n-channel polysilicon TFTs, which are in excellent agreement with the values extracted from the on-state current activation energy data (Fig. 39). Thus, combined conduction measurements at various temperatures and low-frequency noise measurements at room temperature can provide thorough knowledge of the conduction mechanisms and the origin of the leakage current in polysilicon TFTs.

5. AVALANCHE-INDUCED EXCESS NOISE IN POLYSILICON TFTs

In polysilicon TFTs, an anomalous drain current increase in the output characteristics for high drain voltages is usually observed, which is known as the "kink effect" [38, 93]. Such an increase in the output conductance results in a degradation of the switching characteristics and an enhancement of the power dissipation in digital circuits. Numerical simulations have shown that this anomalous drain current increase in the output characteristics can be related to parasitic or pseudoparasitic bipolar effects [94, 95]. In particular, this effect is commonly ascribed to impact ionization occurring in the high

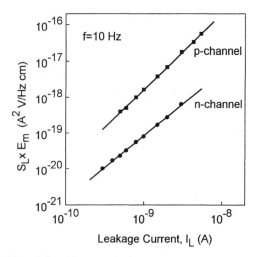

Fig. 42. Plots of $S_L \times E_m$ versus the leakage current I_L of typical n-channel and p-channel polysilicon TFTs. The noise power spectral density S_L was measured at a frequency of 10 Hz. (Reprinted from C. T. Angelis et al., *J. Appl. Phys.* 82, 4095, 1997, with permission.)

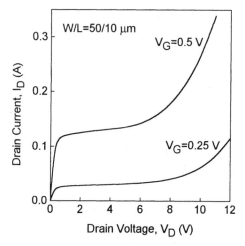

Fig. 43. Output characteristics for n-channel polysilicon TFT with channel width $W = 50\ \mu$m and length $L = 10\ \mu$m at different gate voltages. (Reprinted from C. A. Dimitriadis et al., *Appl. Phys. Lett.* 74, 108, 1999, with permission.)

electric field region close to the drain, with an avalanche generation rate related to recombination processes via gap states [93, 95]. Recently we investigated the noise of the avalanche-induced current in high-temperature polysilicon TFTs [69]. The origin of the noise was explained in terms of the carrier number fluctuation model. The experimental data were analyzed by a model developed on the basis of noise equations in silicon-on-insulator metal-oxide-semiconductor field-effect transistors (SOI MOSFETs).

Figure 43 shows typical output characteristics of n-channel polysilicon TFT with a channel width $W = 50\ \mu$m and a length $L = 10\ \mu$m. Above pinch-off, the drain current tends to saturate, and an anomalous current increase is observed at high drain voltages. Figure 44 shows typical measurements of drain current noise spectra obtained at a gate voltage $V_G = 0.25$ V and different drain currents I_D with the device operating in the linear, saturation, and avalanche regimes. The variation of

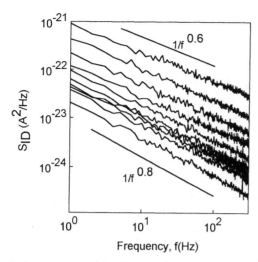

Fig. 44. Drain current spectral density S_{I_D} versus frequency, measured in the device of Figure 43 at gate voltage $V_G = 0.25$ V and drain currents $I_D = 20$, 26, 28.5, 30, 33, 40, 48, 61, 81, and 115 nA. (Reprinted from C. A. Dimitriadis et al., *Appl. Phys. Lett.* 74, 108, 1999, with permission.)

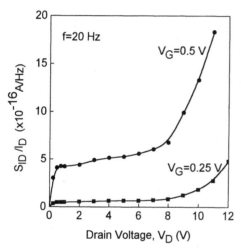

Fig. 45. Normalized S_{ID}/I_D drain current spectral density versus drain voltage V_D at a frequency $f = 20$ Hz and two different gate voltages, measured in the device of Figure 43. (Reprinted from C. A. Dimitriadis et al., *Appl. Phys. Lett.* 74, 108, 1999, with permission.)

the normalized drain current spectral density S_{ID}/I_D with I_D, shown in Figure 45, follows the drain current evolution with V_D of Figure 43. For all drain currents, the drain current noise shows a $1/f^\gamma$ (with $\gamma < 1$) behavior, indicating that the predominant mechanism for the origin of the drain current noise is similar in all operation regimes. However, in polysilicon TFTs operated in the linear region, the origin of the $1/f^\gamma$ noise has already been ascribed to carrier number fluctuations with dynamic trapping and detrapping at the grain boundary traps [96].

In polysilicon TFTs, quantitative analysis of the avalanche-induced current noise has been performed with a model for the excess noise of SOI transistors operated in the kink regime [69]. In SOI n-MOSFETs, fluctuation of the impact ionization current leads to fluctuation of the parasitic bipolar transistor, resulting in a peak-shaped noise overshoot [97]. The noise overshoot

phenomenon has been explained by generation-recombination (G-R) noise originating from traps in the depletion region of the Si film and interacting with the multiplication-generated holes in the high-field region near the drain. This leads to an expression for the drain current noise overshoot [98],

$$S_{I_D}(f) = g_m^2 \frac{q^2 N_T}{C_{ox}^2 W L} \frac{4}{(\tau_c + \tau_e)[(1/\tau_e + 1/\tau_c)^2 + \omega^2]} \quad (82)$$

where $N_T = N_{Tf} w_d$ is the effective trap surface density, N_{Tf} is the trap density in the film, w_d is the depletion region width, and τ_c, τ_e are the capture and emission time constants, respectively. In Eq. (82), considering that the noise maximum occurs for $\tau_c = \tau_e = \tau$ [98], the noise overshoot intensity can be approximated by the relationship

$$S_{I_D}(f) = g_m^2 \frac{q^2 N_T}{C_{ox}^2 W L} \frac{\tau/2}{[1 + \omega^2(\tau/2)^2]} \quad (83)$$

In polysilicon TFTs, as shown in Figure 45, the observed avalanche-induced excess noise is not represented by a noise peak but shows a continuous increase as the excess current increases. This behavior in S_{I_D} can be explained by considering an energy distribution of the gap states in the polysilicon film. For the exponential distribution of the band tails [96],

$$N_T(E) = N_t \exp\left(-\frac{E_c - E}{E_t}\right) \quad (84)$$

the distribution of the time constants $g(\tau)$ for the G-R process between free electrons in the conduction band and the gap states is [30]

$$g(\tau) = \frac{kT}{E_t \tau} (\tau \nu)^{-kT/E_t} \quad (85)$$

where N_t and E_t are the trap distribution parameters and $\nu = 10^{12}$ Hz is the attempt to escape frequency. Therefore, in polysilicon TFTs, the avalanche-induced current noise ΔS_{I_D} is given by the relationship

$$\Delta S_{I_D}(f) = g_m^2 \frac{q^2 N_t}{2 C_{ox}^2 W L} \frac{kT}{E_t(\nu)^{kT/E_t}} $$
$$\times \int_{\tau_{min}}^{\tau_{max}} \frac{1}{[1 + \omega^2(\tau/2)^2]} \frac{d\tau}{\tau^{kT/E_t}} \quad (86)$$

For $\tau_{min} \ll \tau \ll \tau_{max}$ and $-1 < kT/E_t < 1$, the integral in Eq. (86) reduces to [99]

$$\int_{\tau_{min}}^{\tau_{max}} \frac{1}{1 + \omega^2(\tau/2)^2} \frac{d\tau}{\tau^{kT/E_t}} = \frac{\pi}{2(\omega/2)^\gamma \sin(\gamma\pi/2)} \quad (87)$$

where $\gamma = 1 - kT/E_t$. From Eqs. (86) and (87), we finally obtain

$$\Delta S_{I_D}(f) = g_m^2 \frac{q^2 N_t}{4 C_{ox} W L} \frac{kT}{E_t(\nu)^{kT/E_t}} \frac{\pi^{1-\gamma}}{\sin(\gamma\pi/2)} \frac{1}{f^\gamma} \quad (88)$$

From Eq. (88) it follows that the avalanche-induced drain current noise spectral density exhibits a $1/f^\gamma$ behavior and

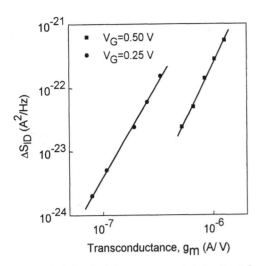

Fig. 46. Avalanche-induced drain current spectral density ΔS_{I_D} versus transconductance g_m at a frequency $f = 20$ Hz and two different gate voltages, measured in the device of Figure 43. (Reprinted from C. A. Dimitriadis et al., *Appl. Phys. Lett.* 74, 108, 1999, with permission.)

ΔS_{I_D} is proportional to g_m^2. Figure 46 shows the current noise spectral density ΔS_{I_D} as a function of g_m for two gate voltages. The transconductance was derived from the static characteristics of the device, and the excess noise ΔS_{I_D} was evaluated by subtracting the extrapolated saturation noise intensity from its measured values in Figure 45. The slope of the straight lines shows that ΔS_{I_D} is proportional to about g_m^3, which is not in agreement with the theoretical prediction of Eq. (88). In addition, in the linear and saturation regions, the power exponent of the frequency remains almost constant (about $\gamma = 0.8$). In the region above saturation, γ varies from about 0.8 to 0.6 as the avalanche-induced drain current increases (Fig. 44). These results indicate that when the device is operated in the avalanche regime, a redistribution of the gap states occurs, probably because of the generation of additional traps by hot-carrier effects in the high-field region of the drain junction. As a result, Eq. (88) underestimates the measured noise due to the additional noise introduced by the hot-carrier generated traps. Further work is required to separate the contribution of the impact ionization and hot-carrier effects on the anomalous drain current increase at high drain voltages.

6. HOT-CARRIER PHENOMENA IN POLYSILICON TFTs

In polysilicon TFT technology, it seems possible to integrate on the same substrate driving circuitry as well as switching devices when polysilicon TFTs are used for active matrix liquid crystal displays. However, at relatively high drain and gate voltages employed in driving circuits, the presence of grain boundary traps in the polysilicon significantly enhances the local electric field near the drain junction, and hot-carrier effects can be important [100–102]. Hot electron and/or hot hole injection into the gate oxide and interface states generation has

been reported during the application of various bias-stress conditions [103–105]. The oxide traps and interface states created by hot carriers degrade the device performance. This device degradation can be related to the parasitic bipolar transistor effect arising from the impact ionization in the high electric field region close to the drain [94]. It is therefore important to investigate the hot-carrier effects to determine the long-term reliability of polysilicon TFTs.

Low-frequency noise has already been used as a reliability tool in silicon devices and technologies because of the high sensitivity of fluctuations caused by the defects along the path of current flow in the device [105–109]. Recently we have shown that in polysilicon TFTs also the noise performance is closely related to the device reliability when hot carrier stress is applied [60]. In addition, polysilicon TFTs show a $1/f^\gamma$ noise ($0.5 \leq \gamma \leq 1$) behavior due to emission and trapping processes of carriers with exponential tail states [110]. Thus the evolution of the $1/f^\gamma$ noise of the drain current can give information about the model for the device aging during application of bias stress.

The mechanisms responsible for the device degradation under various bias stress conditions were investigated in n-channel TFTs, fabricated on 50-nm-thick SPC polysilicon films, subjected to excimer laser annealing, with energy density varying from 0 to 340 mJ/cm^2. All of the devices have a gate width of 20 μm, a length of 7 μm, and a gate oxide thickness of about 124 nm. Details of the device fabrication processes are presented in Section 3.6.

To predict the lifetime of polysilicon TFTs, it is necessary to study the time dependence of the device degradation at various gate and drain biases. In monocrystalline Si MOSFETs, the bias conditions are correlated with the maximum substrate current arising from impact ionization [111]. This is not possible in the present case, because polysilicon TFTs are fabricated on oxidized silicon wafers or glass substrates, and no contact for substrate current measurements is available. Hot-carrier-induced light emission along the channel width and near the drain junction can be used as an alternative tool to select the gate and drain bias voltages for maximum device degradation, without requiring an electrical contact on the substrate. In monocrystalline Si MOSFETs, comparative measurements of the emitted light intensity and the substrate current have shown that the photon emission phenomenon is unambiguously associated with hot-carrier effects [112, 113].

Photon emission monitoring has been used to determine the bias stress conditions for maximum degaradation of polysilicon TFTs. The photon emission characteristics of the devices were evaluated with a Hamamatsu photonics system, equipped with a coupled device camera (C4880). Figure 47 shows the light intensity as a function of the gate voltage for drain voltages $V_D = 12$ V and 14 V of typical polysilicon TFTs with a threshold voltage of about $V_T = 6$ V, determined from the transfer characteristics in the linear region. As the photon emission phenomenon is closely related to hot-carrier effects, these results show that the maximum device degradation occurs at a gate voltage close to the threshold voltage.

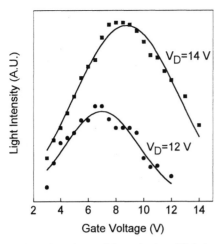

Fig. 47. Gate voltage dependence of the emitted total light intensity at drain voltages $V_D = 12$ and 14 V of a polysilicon TFT.

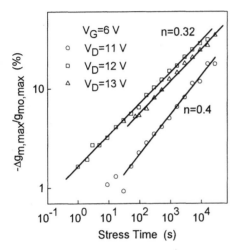

Fig. 48. Degradation of the maximum transconductance as a function of stress time during three different stress conditions. The slopes of the linear fits are also indicated.

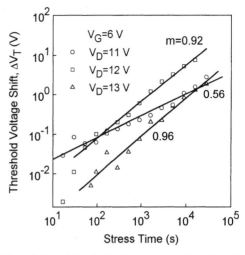

Fig. 49. Degradation of the threshold voltage as a function of stress time during three different stress conditions. The slopes of the linear fits are also indicated.

Three different biasstress regimes were applied in the above polysilicon TFTs. The gate bias stress voltage was selected to be close to the threshold voltage $V_G = 6$ V of the virgin devices and remained constant in all stress regimes. The drain voltage was $V_D = 11$ V for stress regime A ($V_D/2 < V_G$), $V_D = 12$ V for stress regime B ($V_D/2 = V_G$), and $V_D = 13$ V for stress regime C ($V_D/2 > V_G$).

For the bias stress regimes A, B, and C, Figure 48 shows the degradation (%) of the maximum transconductance, defined as $-\Delta g_{m,max}/g_{mo,max}$, where $\Delta g_{m,max} = g_{m,max} - g_{mo,max}$, $g_{mo,max}$ is the initial maximum transconductance, and $g_{m,max}$ is the maximum transconductance after stressing. However, during hot-carrier aging in crystalline silicon MOSFETs, it has been established that the degradation of the maximum transconductance follows a power-time-dependent law of the form At^n, with n varying from 0.2 to 0.8, depending on the applied bias stress conditions [114]. For the stress regime A, the time exponent is about $n = 0.4$, whereas for stress regimes B and C the exponent n decreases ($n = 0.32$). In addition, in polysili-

con TFTs, it has been shown that the degradation of $g_{m,max}$ is directly related to generation of states at the SiO_2/polysilicon interface near the drain [105].

To identify the mechanisms responsible for the device degradation under the different bias stress conditions, the threshold voltage shift ΔV_T from its initial value has been measured as a function of the stress duration (Fig. 49). In stress regime A, ΔV_T follows a power-time-dependent law with exponent $m = 0.56$, whereas in stress regimes B and C the shift of ΔV_T with stress time exhibits a stronger dependence ($m \cong 1$). The positive threshold voltage shift during stress indicates that more hot electrons than hot holes are injected into the gate oxide [115]. In stress regime A, we suggest that both interface states generation and hot-carrier (mainly electron) injection into the gate oxide occur. In stress regimes B and C, the experimental data indicate that electron injection is more pronounced than interface states generation.

The evolution with stress time of the static device parameters ($g_{m,max}$ and V_T) was correlated with noise measurements performed in the linear regime ($V_D = 0.1$ V) for constant drain current $I_D = 1\ \mu$A. Figure 50 illustrates the evolution of the power spectral drain current density $\Delta S_{ID} = S_{ID,S} - S_{ID,O}$ with stress time, where $S_{ID,O}$ and $S_{ID,S}$ are the power spectral densities before and after stressing, respectively. It is apparent that ΔS_{ID} obeys a power-time-dependent law of the form Kt^ρ, with ρ varying from about 0.9 to 1.4. Under the stress conditions A, where mainly interface states are generated, the exponent ρ has the higher value. By comparing Figures 48, 49, and 50, we conclude that the drain current noise is directly correlated with the $g_{m,max}$ degradation, indicating that the low-frequency noise in polysilicon TFTs is strongly interface states dependent. Moreover, hot-carrier injection into the gate oxide seems to play a less important role in the evolution of the low-frequency noise.

Under stress conditions B and C, the degradation of $g_{m,max}$ during stressing is almost identical, as shown in Figure 48, whereas the threshold voltage shift ΔV_T with stress time ex-

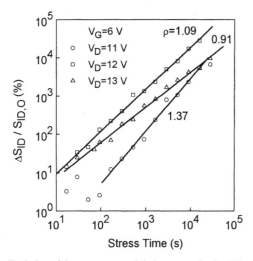

Fig. 50. Evolution of the power spectral drain current density (%) with stress duration. The noise was measured at $I_D = 1\ \mu A$ (in the linear regime) and $f = 100$ Hz. The slopes of the linear fits are also indicated.

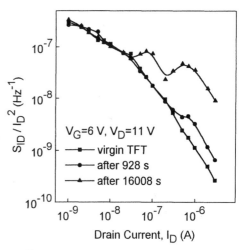

Fig. 51. S_{ID}/I_D^2 plot as a function of the drain current I_D at a frequency of $f = 10$ Hz. It is seen that aging does not influence the low drain current noise.

hibits different behavior with slightly different slopes ($m = 0.92$ and 0.96) and different magnitudes. The value of the time exponent m of the threshold voltage degradation with stress time is directly correlated with the time exponent ρ for the drain current noise evolution with stress time.

The origin of the noise generated after hot-carrier stressing in polysilicon TFTs can be investigated by measuring the normalized power spectral drain current density S_{ID}/I_D^2 as a function of the drain current I_D before and after stressing. Figure 51 shows typical plots of S_{ID}/I_D^2 versus I_D before and after stress A for different stress durations. It is clear that at the initial stage of stressing, the high drain current noise is only increased. In polysilicon TFTs, it is generally believed that the drain current in the subthreshold region is mainly associated with deep trap states predominantly situated at the grain boundaries, whereas the on-current is more associated with band tail states [116]. Therefore, we suggest that at the initial stages of stress A, the generated stress-induced interface states are mainly tail states, as indicated in Figure 50 by the noise peak observed at high drain currents. When the stressing proceeds further, a second noise peak is also observed at lower drain currents (Fig. 51), indicating that deeper states are generated in addition to the tail states.

In polysilicon TFTs, it is well known that the drain current noise exhibits a $1/f^{\gamma}$ (with $0.5 \leq \gamma \leq 1$) behavior in the low-frequency range [56]. This $1/f^{\gamma}$ noise can be explained by carrier emission and trapping processes between trapping states located within an energy range of kT around the Fermi level and the exponential tail states [30, 110]. For the exponential energy distribution of tail states described by Eq. (84), the frequency exponent γ is related to the characteristic energy E_t of the tail states distribution by the simple equation [110]

$$\gamma = 1 - \frac{kT}{E_t} \qquad (89)$$

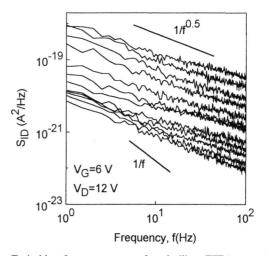

Fig. 52. Typical low-frequency spectra of a polysilicon TFT (measured in the linear regime for $I_D = 1\ \mu A$) after a stress duration of 0, 53, 95, 168, 297, 525, 928, 1640, 2900, 5125, 9057, and 16,008 s.

Therefore, during electrical stress, variation of the frequency exponent γ is directly related to variation of the generated exponential tail states.

Figure 52 shows the evolution of the drain current noise spectra during aging of polysilicon TFT under the stress regime B, measured in the linear regime ($V_D = 0.1$ V) for constant drain current $I_D = 1\ \mu A$. It is apparent that, apart from the increase in the power spectral density S_{ID} with stress duration, the frequency exponent γ also changes in the low-frequency domain. This result indicates that the density, as well as the profile, of the tail states is changing with stressing time. The evolution of the frequency exponent γ with stress duration for stress conditions A and B is shown in Figure 53. The frequency exponent γ decreases from about 0.9 to 0.6 after stressing for about 10^4 s. This result leads to the conclusion that the characteristic energy E_t of the tail states is also decreased, and, therefore, more tail states are generated. Thus, during aging of polysilicon TFTs, a combination of the evolution of the low-frequency noise and of

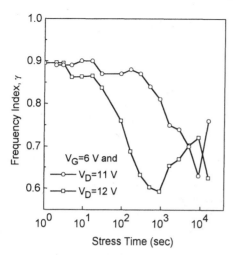

Fig. 53. Frequency index γ during stress time, calculated at low frequencies for $I_D = 1\ \mu$A in the linear region of operation.

the static device parameters with stress time can reveal (a) the aging process of the gate oxide and (b) the energy distribution of the stress-induced tail states.

7. CONCLUDING REMARKS

The low-frequency noise technique can be used as a diagnostic tool to characterize the electrical properties of polycrystalline semiconductor thin films and polysilicon TFTs, as well as to study aging effects in polysilicon TFTs. The overall results of the presented work are summarized as follows.

In polycrystalline semiconductor thin films, the power spectral density of the current fluctuations shows a $1/f$ noise behavior at low frequencies and, in general, is proportional to I^β ($\beta < 2$). An analytical model for the current noise spectral density has been developed, based on a thermally activated conduction mechanism satisfying the Meyer–Neldel rule and on a mobility inhomogeneity across the thickness of the film. By combining noise with Hall and conductivity experimental data, the proposed model provides the trap density at the Fermi level. A sweep of the Fermi level made by varying the temperature can yield the trap energy distribution within the forbidden band gap.

In polysilicon TFTs, a general empirical relationship between the drain current spectral density and the grain boundary potential barrier height has been found indicating that the noise sources are located at the grain boundaries. In addition, the occurrence of a curvature in Arrhenius plots of the conductivity in polysilicon TFTs suggests spatial potential fluctuations over the grain boundary plane described by a Gaussian-type distribution. Based on these results, a low-frequncy noise technique has been developed for a direct evaluation of the grain boundary trap properties and the slow oxide traps. Linear representation of the measured drain current noise spectrum multiplied by $2\pi f$ shows a transition from a pure $1/f$ behavior to a Lorentzian noise. An accurate analysis of the Lorentzian noise was performed, assuming potential fluctuations of the barrier height

over the grain boundary plane. The $1/f$ noise was analyzed with a trapping noise model developed for single crystalline silicon MOSFETs based on fluctuations of the inversion charge near the polysilicon oxide interface. Quantitative evaluation of the Lorentzian and the $1/f$ noise components yielded the energy distribution within the forbidden gap of the grain boundary traps and the slow oxide trap density. The grain boundary traps have an energy distribution with an increased density toward the conduction and valence band edges.

The effect of the laser energy density and the active layer thickness on the performance of excimer laser-annealed polysilicon TFTs was investigated by conduction and low-frequency noise measurements. The devices were made on polysilicon films prepared by excimer laser-induced crystallization of a-Si films or by a combined SPC and ELA process. The performance of the TFTs was correlated with the defect (grain boundary and oxide traps) properties of the corresponding polysilicon films derived from low-frequency noise measurements. Comparison of the various devices revealed that TFTs fabricated on SPC polysilicon processed by ELA are superior compared with the laser-crystallized a-Si polysilicon TFTs. An increase in the laser energy density has beneficial effects on the mobility, threshold voltage, and subthreshold slope until the silicon film has completely melted. However, the threshold voltage, the transconductance, and the leakage current were affected adversely by high laser energy densities because of the large hillocks that developed at the grain boundaries during the laser irradiation. The results indicate that the device performance can be improved further if a smooth polysilicon surface can be obtained with high laser energy density irradiation. When the active layer is increased from 50 to 25 nm, although the subthreshold characteristics are improved, the mobility and threshold voltage are degraded. The noise data indicate that this degradation is related to electron trapping in both gate and substrate oxide interface traps.

The leakage current in polysilicon TFTs can be investigated by combined conduction measurements at various temperatures and low-frequency noise measurements at room temperature. The results presented indicate that the leakage current is generated at the drain junction. The junction leakage current is due to two basic mechanisms, pure thermal generation at low electric fields and field-assisted Poole–Frenkel emission accompanied by thermal field emission at high electric fields. The leakage current is correlated with the grain boundary and the gate oxide/polysilicon interface traps determined from the on-state current activation energy data. In addition to the electric field strength in the drain junction, it is confirmed that the main factor determining the leakage current is the deep levels with uniform energy distribution within the silicon band gap known by a previous analysis. The origin of the leakage current was also identified by low-frequency noise measurements. It is shown that the analytical model developed for the leakage current noise spectral density provides a prediction for the grain boundary trap density.

Low-frequency noise measurements can also be used for investigation of avalanche-induced phenomena in polysilicon

TFTs. Based on a model for the excess noise in silicon-on-insulator MOSFETs operated in the kink regime, an analytical expression of this excess drain current noise in polysilicon TFTs has been derived, taking into account an exponential distribution of the gap states. Comparison of the noise experimental data with the theory indicates that, in addition to impact ionization, hot-carrier effects contribute to the anomalous increase in the drain current at high drain voltages.

Finally, hot-carrier phenomena in polysilicon TFTs can be investigated by low-frequency noise measurements, combined with measurements of the static device parameters. Correlation between low-frequency noise spectral density and static device parameters shows that the low-frequency noise is a much stronger function of the polysilicon film quality than the static device parameters, indicating that noise is a better monitor for hot-carrier degeradation studies. Different stress regimes were applied, and the evolution with stressing time of the maximum transconductance, threshold voltage, and noise were measured. The evolution of static device parameters with stressing time indicates that the possible degradation mechanisms are injection of hot electrons into the gate oxide and generation of interface states. The evolution of the low-frequency noise demonstrates that tail states are generated at the initial stages of the stress. As the stress proceeds, deeper states are generated.

REFERENCES

1. N. G. van Kampen, "Stochastic Processes in Physics and Chemistry," p. 61. North-Holland, Amsterdam, 1981.
2. K. M. van Vliet and G. R. Fassett, in "Fluctuation Phenomena in Solids" (R. E. Burgess, Ed.), Chap. 7, Academic Press, New York, 1965.
3. L. K. J. Vandamme, X. Li, and D. Rigaud, *IEEE Trans. Electron Devices* 41, 1936 (1994).
4. A. L. McWhorter, "$1/f$ noise and related surface effects in germanium" Ph.D. dissertation, Massachusetts Institute of Technology, 1955.
5. P. Dutta and P. M. Horn, *Rev. Mod. Phys.* 53, 497 (1981).
6. Z. Celic-Butler and T. Y. Hsiang, *Solid-State Electron.* 30, 419 (1987).
7. C. Surya and T. Y. Hsiang, *Solid-State Electron.* 31, 959 (1988).
8. H. Wong and Y. C. Cheng, *IEEE Trans. Electron Devices* 37, 1743 (1990).
9. F. N. Hooge, *Phys. Lett. A* 29, 139 (1969).
10. F. N. Hooge, *IEEE Trans. Electron Devices* 41, 1926 (1994).
11. F. N. Hooge, T. G. M. Kleinpenning, and L. K. J. Vandamme, *Rep. Prog. Phys.* 44, 479 (1981).
12. K. K. Hung, P. K. Ko, C. Hu, and Y. C. Cheng, *IEEE Trans. Electron Devices* 37, 654 (1990).
13. G. Ghibaudo, O. Roux, Ch. Nguyen-Duc, F. Balestra, and J. Brini, *Phys. Status Solidi A* 124, 571 (1991).
14. A. L. McWhorter, "Semiconductor Surface Physics," p. 207. Univ. of Pennsylvania Press, Philadelphia, 1957.
15. S. Christensson, I. Lundstrom, and C. Svensson, *Solid-State Electron.* 11, 797 (1968).
16. F. N. Hooge, T. G. M. Kleinpenning, and L. K. J. Vandamme, *Rep. Prog. Phys.* 44, 479 (1981).
17. Special issue, *IEEE Trans. Electron Devices* 41, no. 11 (1994).
18. C. A. Dimitriadis and P. A. Coxon, *J. Appl. Phys.* 64, 1601 (1988).
19. C. A. Dimitriadis, N. A. Economou, and P. A. Coxon, *Appl. Phys. Lett.* 59, 172 (1991).
20. D. H. Tassis, C. A. Dimitriadis, J. Brini, G. Kamarinos, and A. Birbas, *J. Appl. Phys.* 85, 4091 (1999).
21. A. Van der Ziel, "Noise in Solid State Devices and Circuits." Wiley, New York, 1986.
22. N. F. Mott, *Adv. Phys.* 16, 49 (1967).
23. K. Herz and M. Powalla, *Appl. Surf. Sci.* 91, 87 (1995).
24. W. Beyer and H. Overhof, "Semiconductors and Semimetals" (J. I. Pankove, Ed.). Academic Press, New York, 1984.
25. N. F. Mott and E. A. Davis, "Electron Processes in Non-Crystalline Materials." Clarendon, Oxford, 1979.
26. O. Jantsch, *IEEE Trans. Electron Devices* ED-34, 1100 (1987).
27. P. Migliorato and D. B. Meakin, *Appl. Surf. Sci.* 30, 353 (1987).
28. C. A. Dimitriadis, N. A. Economou, and P. A. Coxon, *Appl. Phys. Lett.* 59, 172 (1991).
29. D. H. Tassis, C. A. Dimitriadis, J. Brini, G. Kamarinos, and A. Birbas, *J. Appl. Phys.* 85, 4091 (1999).
30. F. Z. Bathaei and J. C. Anderson, *Philos. Mag. B* 55, 87 (1987).
31. I. S. Gradshtein and I. M. Ryzhik, "Table of Integrals, Series and Products," p. 292. Academic Press, San Diego, 1965.
32. M. C. Bost and J. E. Mahan, *J. Appl. Phys.* 64, 2034 (1988).
33. E. Loh, *J. Appl. Phys.* 56, 3022 (1984).
34. I. Ohdomari and K. N. Tu, *J. Appl. Phys.* 51, 3735 (1980).
35. I. Ohdomari and H. Aochi, *Phys. Rev. B: Condens. Matter.* 35, 682 (1987).
36. G. E. Pike and C. H. Seager, *J. Appl. Phys.* 50, 3414 (1979).
37. G. E. Pike and C. H. Seager, *Appl. Phys. Lett.* 35, 709 (1979).
38. M. Hack, J. G. Shaw, P. G. LeComber, and M. Willums, *Jpn. J. Appl. Phys.* 29, L2360 (1990).
39. T.-J. King, M. G. Hack, and I.-W. Wu, *J. Appl. Phys.* 75, 908 (1994).
40. G. Fortunato, D. B. Meakin, P. Migliorato, and P. G. Le Comber, *Philos. Mag.* 57, 573 (1988).
41. B. A. Khan and R. Pandya, *IEEE Trans. Electron Devices* 37, 1727 (1990).
42. C. A. Dimitriadis, D. H. Tassis, N. A. Economou, and A. J. Lowe, *J. Appl. Phys.* 74, 2919 (1993).
43. S. S. Chen and J. B. Kuo, *J. Appl. Phys.* 79, 1961 (1996).
44. W. B. Jackson, N. M. Johnson, and D. K. Biegelsen, *Appl. Phys. Lett.* 43, 195 (1983).
45. B. Faughnan, *Appl. Phys. Lett.* 50, 290 (1987).
46. J. Levinson, F. R. Sheperd, P. J. Scanlon, W. D. Westood, G. Este, and M. Rider, *J. Appl. Phys.* 53, 1193 (1982).
47. R. E. Proano and D. G. Ast, *J. Appl. Phys.* 66, 2189 (1989).
48. C. A. Dimitriadis, *J. Appl. Phys.* 73, 4086 (1993).
49. D. J. Thompson and H. C. Card, *J. Appl. Phys.* 54, 1976 (1983).
50. Y. Alpern and J. Shappir, *J. Appl. Phys.* 63, 2614 (1988).
51. J. H. Werner, "Polycrystalline Semiconductors III-Physics and Technology," Scitec Publications, Saint Malo, France, 1993, p. 213.
52. C. A. Dimitriadis, *IEEE Trans. Electron Devices* 44, 1563 (1997).
53. A. Corradeti, R. Leoni, R. Carluccio, G. Fortunato, F. Plais, and D. Pribat, *Appl. Phys. Lett.* 67, 1730 (1995).
54. C. A. Dimitriadis, J. Brini, and G. Kamarinos, in "14th International Conference on Noise in Physical Systems and $1/f$ Fluctuations," Leuven, Belgium, 1997, p. 434.
55. C. A. Dimitriadis, J. Brini, G. Kamarinos, and G. Ghibaudo, in "Proceedings of the 14th International Conference on Noise in Physical Systems and $1/f$ Fluctuations," Leuven, Belgium, 1997, p. 534.
56. C. A. Dimitriadis, J. Brini, G. Kamarinos, V. K. Gueorguiev, and Tz. E. Ivanov, *J. Appl. Phys.* 83, 1469 (1998).
57. C. T. Angelis, C. A. Dimitriadis, F. V. Farmakis, J. Brini, G. Kamarinos, V. K. Gueorguiev, and Tz. E. Ivanov, *Appl. Phys. Lett.* 76, 118 (2000).
58. J. W. Seto, *J. Appl. Phys.* 46, 5247 (1975).
59. P. Llinares, D. Celi, O. Roux-dit-Buisson, G. Ghibaudo, and J. A. Chroboczek, *J. Appl. Phys.* 82, 2671 (1997).
60. F. V. Farmakis, J. Brini, G. Kamarinos, C. T. Angelis, C. A. Dimitriadis, and M. Miyasaka, in "Proceedings of the 15th International Conference on Noise in Physical Systems and $1/f$ Fluctuations," Hong Kong, 1999, p. 457.
61. C. A. Dimitriadis, J. Stoemenos, P. A. Coxon, S. Friligkos, J. Antonopoulos, and N. A. Economou, *J. Appl. Phys.* 73, 8402 (1993).
62. C. A. Dimitriadis, P. A. Coxon, L. Dozsa, L. Papadimitriou, and N. A. Economou, *IEEE Trans. Electron Devices* 39, 2189 (1989).

63. D. Meakin, J. Stoemenos, P. Migliorato, and N. A. Economou, *J. Appl. Phys.* 61, 5031 (1987).

64. E. Scholl, *J. Appl. Phys.* 60, 1434 (1986).

65. C. T. Angelis, C. A. Dimitriadis, J. Brini, G. Kamarinos, V. K. Gueorguiev, and Tz. E. Ivanov, *IEEE Trans. Electron Devices* 46, 968 (1999).

66. F. N. Hooge, *IEEE Trans. Electron Devices* 41, 1926 (1994).

67. G. A. Evangelakis, J. P. Rizos, I. E. Lagaris, and I. N. Demetrakopoulos, *Comput. Phys. Commun.* 46, 401 (1987).

68. G. Ghibaudo, O. Roux, Ch. Nguyen-Duc, F. Balestra, and J. Brini, *Phys. Status Solidi A* 124, 571 (1991).

69. C. A. Dimitriadis, G. Kamarinos, J. Brini, E. Evangelou, and V. K. Gueorguiev, *Appl. Phys. Lett.* 74, 108 (1999).

70. C. T. Angelis, C. A. Dimitriadis, J. Brini, G. Kamarinos, V. K. Gueorguiev, and Tz. E. Ivanov, *J. Appl. Phys.* 82, 4095 (1997).

71. T. Sameshima and S. Usui, *Appl. Phys. Lett.* 59, 2724 (1991).

72. M. Cao, S. Talwar, K. J. Kramer, T. W. Sigmon, and K. C. Saraswat, *IEEE Trans. Electron Devices* 43, 56 (1990).

73. D. K. Fork, G. B. Anderson, J. B. Boyce, R. I. Johnson, and P. Mei, *Appl. Phys. Lett.* 68, 2138 (1996).

74. C. T. Angelis, C. A. Dimitriadis, M. Miyasaka, F. V. Farmakis, G. Kamarinos, J. Brini, and J. Stoemenos, *J. Appl. Phys.* 86, 4600 (1999).

74a. C. T. Angelis, C. A. Dimitriadis, M. Miyasaka, F. V. Farmakis, G. Kamarinos, J. Brini, and J. Stoemenos, *J. Appl. Phys.* 87, 1588 (2000).

75. M. Miyasaka and J. Stoemenos, *J. Appl. Phys.* 86, 5556 (1999).

76. B. Faughnan and A. C. Ipri, *IEEE Trans. Electron Devices* 36, 101 (1989).

77. L. K. J. Vandamme, *IEEE Trans. Electron Devices* 41, 2176 (1994).

78. E. Simoen and C. Claeys, *Appl. Phys. Lett.* 65, 1946 (1994).

79. A. B. Chan, C. T. Nguyen, P.K. Ko, S. T. H. Chan, and S. S. Wong, *IEEE Trans. Electron Devices* 44, 455 (1997).

80. T. W. Little, K.-I. Takahara, H. Koike, T. Nakazawa, I. Yudasaka, and H. Ohshima, *Jpn. J. Appl. Phys., Part 1* 30, 3724 (1991).

81. C. T. Angelis, C. A. Dimitriadis, F. V. Farmakis, G. Kamarinos, J. Brini, and M. Miyasaka, *Appl. Phys. Lett.* 74, 3684 (1999).

82. A. Chiang, I.-W. Wu, M. Hack, A. G. Lewis, T. Y. Hunag, and C.-C. Tsai, in "Extended Abstracts of the 1991 Conference on Solid State Devices and Materials," 1991, p. 586.

83. J. G. Fossum, A. Ortiz-Conde, H. Schichijo, and S. K. Banerjee, *IEEE Trans. Electron Devices* 32, 1878 (1985).

84. S. K. Madan and D. A. Antoniadis, *IEEE Trans. Electron Devices* 33, 1518 (1986).

85. M. Yazaki, S. Takenaka, and H. Ohshima, *Jpn. J. Appl. Phys., Part 1* 31, 206 (1992).

86. J. R. Ayres, S. D. Brotherton, and N. D. Young, *Optoelectron. Devices Technol.* 7, 301 (1992).

87. C. A. Dimitriadis, P. A. Coxon, and N. A. Economou, *IEEE Trans. Electron Devices* 42, 950 (1995).

88. S. M. Sze, "Physics of Semiconductor Devices," 2nd ed. Wiley, New York, 1981.

89. H. C. De Graaff and M. Huybers, *Solid-State Electron.* 25, 67 (1982).

90. S. S. Bhattacharya, S. K. Banerjee, B.-Y. Nguyen, and P. J. Tobin, *IEEE Trans. Electron Devices* 41, 221 (1994).

91. K. H. Duh, X. C. Zhu, and A. Van Der Ziel, *Solid-State Electron.* 27, 1003 (1984).

92. J. R. Hellums and L. M. Rucker, *Solid-State Electron.* 28, 549 (1985).

93. M. Hack and A. G. Lewis, *IEEE Electron Device Lett.* 12, 203 (1991).

94. G. A. Armstrong, S. D. Brotherton, and J. R. Ayres, *Solid-State Electron.* 39, 1337 (1996).

95. M. Valdinoci, L. Colalongo, G. Baccarani, A. Pecora, I. Policicchio, G. Fortunato, F. Plais, P. Legagneux, C. Reita, and D. Pribat, *Solid-State Electron.* 41, 1363 (1997).

96. C. A. Dimitriadis, J. Brini, G. Kamarinos, and G. Ghibaudo, *Jpn J. Appl. Phys.* 37, 72 (1998).

97. W. Fichtner and E. Hochmair, *Electron. Lett.* 13, 675 (1977).

98. E. Simoen, U. Magnusson, A. L. P. Rotondaro, and C. Clays, *IEEE Trans. Electron Devices* 41, 330 (1994).

99. J. I. Lee, J. Brini, A. Chovet, and C. A. Dimitriadis, *Solid-State Electron.* 43, 2181 (1999).

100. B. Doyle, M. Bourcerie, C. Berzgonzoni, R. Mariucci, A. Bravais, K. R. Mistri, and A. Boudou, *IEEE Trans. Electron Devices* 37, 1869 (1990).

101. N. D. Young, A. Gill, and M. J. Edwards, *Semicond. Sci. Technol.* 7, 1183 (1992).

102. L. Pichon, F. Raoult, T. Mahamed-Brahim, O. Bonnaud, and H. Sehil, *Solid-State Electron.* 39, 1065 (1996).

103. N. D. Young and A. Gill, *Semicond. Sci. Technol.* 5, 728 (1990).

104. F. V. Farmakis, C. A. Dimitriadis, J. Brini, G. Kamarinos, V. K. Gueorguiev, and Tz. E. Ivanov, *Electron. Lett.* 34, 2356 (1998).

105. F. V. Farmakis, C. A. Dimitriadis, J. Brini, G. Kamarinos, V. K. Gueorguiev, and Tz. E. Ivanov, *Solid-State Electron.* 43, 1259 (1999).

106. J. M. Pimbley and G. Gildenblat, *IEEE Electron Device Lett.* 5, 345 (1984).

107. Z. H. Fang, S. Cristoloveanu, and A. Chovet, *IEEE Electron Device Lett.* 7, 371 (1986).

108. P. Fang, K. K. Hung, P. K. Ko, and C. Hu, *IEEE Electron Device Lett.* 12, 273 (1991).

109. M.-H. Tsai and T.-P. Ma, *IEEE Electron Device Lett.* 14, 256 (1993).

110. C. A. Dimitriadis, J. Brini, J. I. Lee, F. V. Farmakis, and G. Kamarinos, *J. Appl. Phys.* 85, 3934 (1999).

111. C. M. Hu, S. C. Tam, F. C. Hsu, P. K. Ho, T. Y. Chan, and K. W. Terrin, *IEEE Trans. Electron Devices* 32, 357 (1985).

112. A. Toriumi, M. Yoshimi, M. Iwase, Y. Akiyama, and K. Taniguchi, *IEEE Trans. Electron Devices* 34, 1501 (1987).

113. B. Szelag and F. Balestra, *Electron. Lett.* 33, 1990 (1997).

114. P. Heremans, R. Bellens, G. Groeseneken, and H. E. Maes, *IEEE Trans. Electron Devices* 35, 2194 (1988).

115. T. Tsuchiya, *IEEE Trans. Electron Devices* 34, 2291 (1987).

116. T.-S. Li and P.-S. Lin, *IEEE Electron Device Lett.* 14, 240 (1993).

Chapter 7

GERMANIUM THIN FILMS ON SILICON FOR DETECTION OF NEAR-INFRARED LIGHT

G. Masini, L. Colace, G. Assanto

Department of Electronic Engineering and National Institute for the Physics of Matter INFM RM3, Terza University of Rome, Via della Vasca Navale 84, 00146 Rome, Italy

Contents

1. INTRODUCTION

1.1. The Scenario

In recent years, the growing need for greater than gigabit per second transmission rates and the spread of multimedia communications have determined a net transition toward optical signal processing, stimulating both the research and the development of high-performance and cost-effective optoelectronic components and systems. Optical communications are amenable to solution of the most critical problems related to electrical interconnection bottlenecks and speed, providing large bandwidths, immunity from noise, small power dissipation, and negligible cross-talk.

Handbook of Thin Film Materials, edited by H.S. Nalwa
Volume 4: Semiconductor and Superconductor Thin Films

ISBN 0-12-512912-2/$35.00

In the last decade, applied research in optoelectronics has addressed long-haul high-performance point-to-point transmission in the near infrared (NIR) at 1.3 and 1.55 micrometers, where silica optical fibers exhibit minimum losses and limited dispersion. This well established technology, along with the unprecedented demand for internet services, cable TV, video on demand, e-commerce, and so on, promotes a broader acceptance of fiber optic communications also in wide and local area networks (WAN, LAN), metropolitan area networks (MAN), fiber to the home (FTTH) connections, and even to the smaller distances characteristic of intrachip transmissions. In this framework, it has become essential to develop a suitable technology in the form of optoelectronic integrated circuits (OEIC). OEICs, which combine optical and electronic devices and functions, as opposed to state-of-the-art board- or card-assembled solutions, are expected to match the requirements for very high performance, while encompassing compactness and lightweight, reliability, low parasitics, mass-producibility and low cost. An ideal OEIC, such as the one depicted in Figure 1, incorporates light sources, waveguides, modulators, switches, and photodetectors along with analog electronics for high bandwidth amplification and filtering, and microprocessors for digital signal processing.

In spite of the potential advantages, however, to date OEIC have not replaced their discrete counterparts. This is because of existing technological challenges, which certainly are not softened by the incompatibility between the most common semiconductors for optoelectronics (III–V compounds) and "The Semiconductor" in electronics—silicon. To address this combination of materials, different approaches have been undertaken, resulting in two main integration streams: hybrid and monolithic.

Hybrid integration was developed along the mainstreams of silicon optical bench (SiOB) and wafer bonding (WB) technologies. SiOB stems from mature handling and processing capabilities of silicon substrates, including automated pick-and-place, and the realization of low-loss waveguides. Optoelectronic devices (based on III–V compounds) are grown, removed from their native substrates, and then aligned and bonded onto

Si [1]. This approach can integrate lasers, waveguides, splitters, wavelength filters, microlenses, mirrors, and photodetectors [2, 3]. Presently, silicon provides the substrate ("bench") for simple guided-wave networks with electrical interconnections and heatsink, but it could be employed to integrate signal processing and control electronics. WB is based on the fusion of a couple of mirror-polished, flat and clean wafers attracted to each other by van der Waals forces [4]. This yields a robust, atomically smooth and optically transparent bonding interface. Moreover, wafer bonding is not restricted to "homofusion," but can be extended to a variety of wafer combinations including III–V semiconductors on silicon [5].

Despite the outlined advantages and demonstration of several hybrid OEICs, however, many investigators believe that the benefits of integration and mass production are better pursued with monolihic integration. Monolithic integration is the most elegant solution, because it realizes embedded optoelectronic and electronic devices in a single fabrication process, and it is expected to avoid most of the alignment and interconnection problems, thus providing matchless yield and reliability. Conversely, monolithic integration requires heteroepitaxy, that is, the epitaxial growth of dissimilar semiconductor materials. Figure 2 is a display of the "playing field" for heteroepitaxy in optoelectronics: it shows the energy bandgap, the optical absorption edge, and the cubic lattice parameter of application relevant alloys and compounds. Commercially available substrates are also indicated. Whereas carrier transport through a heterointerface should not be affected by scattering, recombination, trapping, or other undesired effects, the material atomic structure has to be maintained across heterojunctions. This limits the choice of semiconductors to those with the same crystalline structure and very similar lattice constants, [i.e. to compounds which, in Figure 2, lie on vertical lines that correspond to commercial substrates (Si, GaAs, InP)]. Based on these considerations, the $Al_x Ga_{1-x}As$/GaAs system is the best

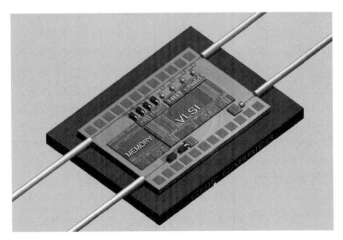

Fig. 1.   Picture of an optoelectronic integrated chip.

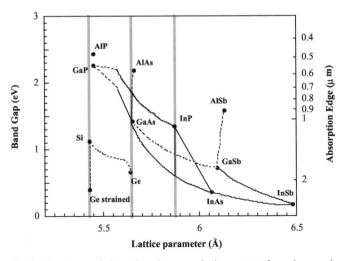

Fig. 2.   Bandgap and absorption edge versus lattice constant for various semiconductor materials. Solid and dashed lines correspond to direct and indirect bandgap semiconductor compounds, respectively. Data from T. P. Pearsall, *Crit. Rev. Solid State and Mater. Sci.* 15, 551 (1989).

candidate, because is spans the bandgap in the 1.4–2.0 eV range with a lattice mismatch within 0.1%. Other choices do imply a certain degree of mismatch, such as in the cases of SiGe alloys on Si and InGaAs on InP. After the pioneering work by Yariv and co-workers [6] on OEIC monoliths, many other examples were demonstrated, either on GaAs [7] or InP [8] substrates. On the one hand, although GaAs-based OEICs are quite advanced, their wavelength of operation (in terms of sources and detectors) is limited to the first fiber optics window around 850 nm, which is typical of LANs (see Fig. 2). On the other hand, InGaAs and InGaAsP on InP are the materials of choice in the 1.3–1.55 μm range, that is, second and third windows for optical communication systems, respectively.

Given the large number of applications and the excellent demonstrations of OEIC research prototypes, further development is expected in both GaAs and InP technologies. Nevertheless, an important obstacle is the huge financial and human investment required for a massive very large scale integration (VLSI) OEIC fabrication. To this extent, Si-based OEICs offer two viable alternatives: (i) integration of III–V compounds on Si and (ii) integration of group IV materials on Si. The first approach (i) relies on the epitaxy of InGaAs or InGaAsP on a silicon substrate with a substantial ($\approx$8%) lattice mismatch. Moreover, both growth and processing are made difficult because of the large mismatch in thermal expansion coefficients. Nevertheless, significant advances in the use of buffer layers have enabled demonstration of NIR lasers [9] and detectors epitaxially deposited on silicon [10]. In the second approach, as far as group IV epitaxy on silicon and regardless of the centrosymmetry and indirect bandgap of the electronic structure of Si, Ge, and their alloys, it has been demonstrated that near-infrared light can be emitted, guided, switched, and detected [11]. The most significant results are based on structurally related modifications (quantum confinement effects) of material properties to introduce a useful optoelectronic response while retaining the advantages of the mature Si technology. Group IV heterostructures, in fact, can be realized using nearly the same approaches employed by standard silicon technology. Silicon VLSI technology is unsurpassed: wafers are available in large sizes (30 cm diameter), with high crystal quality, a unique native oxide, and mechanical and thermal properties superior to III–V materials. The compatibility of Si-based optoelectronics with such an existing and well developed technology promises high yields and low-cost commercial chips for NIR optoelectronics (such as the one pictured in Fig. 3). Most Si-based OEIC could be fabricated today, with the exception of monolithically integrated laser sources. For the latter, promising candidates are erbium-doped Si [12], and SiGe strained multiple quantum wells (MQWs) [13] and superlattices [14]. Alternative solutions rely on external sources or modulated portions of the incoming light. Conversely, most of the technological problems related to the fabrication of low-loss (<1 dB/cm) waveguides have been successfully solved for silicon-on-insulator (SOI) [15] and SiGe [16]. Similarly, waveguide modulators and switches has been demonstrated using plasma dispersion (with extinction ratios in excess of 34 dB and switching times around

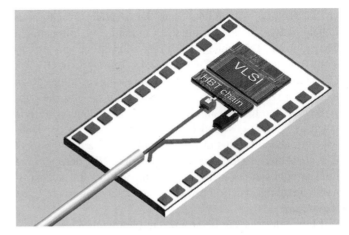

Fig. 3. Si-based OEIC.

200 ns [17]) and thermooptical effects [18]. Finally, normal incidence and guided-wave NIR photodetectors based on SiGe heterostructures and/or pure Ge on silicon have registered important advances.

Ge thin films on Si for NIR photodetectors are the main subject of this chapter. An introduction to near-infrared light detection is presented in the next subsection, along with a short overview of performances and costs of commercially available photodetectors. The main body of this chapter deals with material issues and the technology of SiGe on Si, including the heteroepitaxy of Ge-rich SiGe alloys on silicon and a discussion of the most common approaches to the realization of high quality epilayers for device applications. Then, we review SiGe-based NIR photodetectors, operating either at normal incidence or in guided-wave formats, and conclude with a short section on special detection systems. A brief appendix is devoted to the numerics useful in design and analysis. Throughout this chapter we discuss the most significant results obtained to date by us and other groups, point out material quality and overall device performance, and highlight cost, complexity, and compatibility with existing technologies and manufacturing processes.

1.2. Near-Infrared Detectors

Near infrared is commonly referred to as the spectral region with shortest wavelengths of infrared light, nominally from 750 to 2000 nm. In the NIR, most attention is focused on 850 nm (standard for LAN), 1300 and 1550 nm (second and third windows of fiber optics communications). As mentioned before, the use of 1.3 and 1.55 is rapidly expanding from long-haul point-to-point links toward shorter distances, involving a wider range of applications. For this reason, we restrict the following discussion to the NIR of second and third windows.

NIR photodetectors can be described and characterized with the usual figures of merit and definitions, including sensitivity, bandwidth, noise immunity, and reliability. Although these parameters should be large in every good light detector, in optical links, high speed, high sensitivity, and low noise can become crucial factors. With reference to the SONET (synchronous optical network), the Consultative Committee of International Telegraph and Telephone (CCITT) has allowed capacity expansions from the current OC-48 rate (2.48 Gbits/s) up to OC-768 (39.8 Gbits/s), with several links of decreasing capacities down to OC-1 (51.8 Mbits/s) between the backbone and the subscribers. Therefore, distinct tradeoffs and evaluation criteria (such as volume cost and life expectancy) have to be established to guarantee that optical technology has a broad impact in a consumer-oriented market. In addition, NIR detectors are used in applications such as environmental monitoring, astronomic imaging, spectroscopy, etc., where large areas are often required.

Light detectors can be grouped in photomultipliers, photoconductors, and photodiodes. The latter can be further distinguished as Schottky, p-n and p-i-n junctions, avalanche, and metal–semiconductor–metal (MSM). They are the devices of choice for optical communications, and we will focus on them. For a photodiode to work, the semiconductor energy gap E_g (cutoff wavelength) must be smaller (larger) than the energy of the photons (the wavelength of the light) to be detected. Therefore, from Figure 2 it is apparent that suitable materials for the NIR are Ge, InGaAs, and InGaAsP. Multiple heterostructures MQWs, and superlattices (SLs) can also be employed to take advantage of the possibility to tailor the absorption spectrum to specific wavelengths or applications through bandgap engineering [19].

Having selected the appropriate semiconductor or compound, an important figure to optimize is the external quantum efficiency η, which is defined as the number of photogenerated and collected electron–hole pairs per incident photon. A device-oriented parameter is the responsivity R at wavelength λ,

$$R = \frac{q\lambda}{hc}\eta_{\text{int}}(1 - \Theta_R)\left(1 - e^{-\alpha W}\right) \, A/W \qquad (1.1)$$

where q/hc is the wavelength that corresponds to 1 eV (q is the unit charge, h is the Planck constant, and c is the speed of light in vacuum), η_{int} is the internal quantum efficiency (i.e., the number of collected pairs per absorbed photon, close to unity for defect-free materials), Θ_R the reflectivity at the air–semiconductor interface, α is the optical absorption, and W is the thickness of the active layer. The terms $(1 - \Theta_R)$ and $(1 - e^{-\alpha W})$ account for the reflection losses at the surface and the absorption efficiency, respectively. Clearly, we need to keep $\Theta_R \approx 0$ and $\alpha W \gg 1$. The first requirement can be accomplished with antireflection coatings, whereas the second depends on both α and W, and ensures that most photons are captured. However, because an excessively large W could reduce the speed of response due to transit time limitations, high absorption materials are preferred. Typically, coefficients $\alpha > 10^4$ cm^{-1} keep absorption lengths in the 1–2 μm range and allow a sufficiently wide bandwidth.

The response speed depends on a combination of carrier diffusion, drift, and capacitance of the depletion layer. To minimize diffusion, as few carriers as possible have to be photogenerated outside the depletion region. The depletion ca-

pacitance, (which affects the RC time constant, R being the load) can be reduced by realizing small area detectors with large W. Finally, the drift time is minimized by ensuring that the carriers cross the depleted region at the saturation velocity and by reducing the width of the depletion layer. If the photodiode is well designed, the bandwidth is limited by the transit time effect. In terms of speed, the most common parameters are the 3 dB bandwidth $f_{3\,\text{dB}}$ and the full width of the pulse response at half maximum (FWHM). They are simply linked by $2f_{3\,\text{dB}} = 1/\text{FWHM}$.

NIR photodiodes are often operated in reverse bias to reduce both the transit time and the diffusion capacitance, and, sometimes, to improve their linearity. The reverse bias, however, increases the dark current, thus affecting the noise performances. The pertinent figure of merit in terms of noise is the noise equivalent power (NEP), which is defined as the minimum detectable root mean square (rms) optical power such that the signal-to-noise ratio (S/N) equals unity for a bandwidth of 1 Hz. The NEP is inversely proportional to the quantum efficiency and scales with the square root of the equivalent current I_{eq}, which is defined as the sum of background radiation current I_B, dark current I_D, and current associated to the thermal noise. The dark current, which increases with reverse bias, is one of the most important factors in the noise performance of NIR photodiodes. Table I summarizes characteristics, performances, and costs of fast, small, and large area Ge, InGaAs, and InGaAsP commercial devices.

Ultra-high performance is available in InGaAs-based photodiodes due to their unmatched noise performance. Although their cost is correspondingly high, specific applications in longhaul, high bit-rate links can accommodate it. At lower bit rates (≤ 2.5 Gbits/s) and for shorter distances, or in large area applications, Ge detectors can meet most of the requirements at a lower cost. In these applications, as previously mentioned, integrated Ge on silicon detectors can actually exploit the compatibility with Si technology in terms of functionality, compactness, reliability, production yield, and low cost, paving the way toward group IV OEICs.

2. SiGe TECHNOLOGY

2.1. Heteroepitaxy of SiGe

Heteroepitaxy differs from conventional epitaxy inasmuch as it is not limited to the growth of selectively doped layers on a substrate of similar material (homoepitaxy), but is able to combine electric and optical properties of different semiconductors. The enormous potentials of heterojunctions were identified at the very beginning of the "electronic era," soon after the invention of the transistor. In 1951, Shockley [20] proposed the heterojunction as an efficient emitter–base junction in a bipolar transistor and Kroemer [21] analyzed the benefits afforded in terms of current gain by an emitter with a wider gap than the base. Furthermore, heteroepitaxy allowed introduction of the concept of band-structure engineering, which led to quantum-confined structures such as quantum wells and superlattices

Table I. Model, Characteristics, Performances and Cost of a Variety of Commercial Ge, InGaAs, and InGaAsP NIR Photodetectors

Company	Model	Device type	Diameter (μm)	Responsivity 1300–1550 nm (A/W)	Dark current I_D	Cutoff frequency	NEP $pW/\sqrt{Hz}$	End-user cost (US$)
G.P.D. Corp.	GM3	Ge pin	100	0.65–0.85	1.0 μA @ 10 V	1.5 GHz	0.3	52
	GAV30	Ge APD	30	0.76–0.84	0.2 μA @ 1 V	2.5 GHz	—	200
	GM5	Ge pin	1,000	0.65–0.85	3 μA @ 10 V	55 MHz	1.5	52
	GM8	Ge pin	5,000	0.65–0.85	40 μA @ 3 V	1.6 MHz	1.6	150
Hamamatsu	B1720-02	Ge pin	1,000	0.8 @ 1550 nm	0.3 μA @ 10 mV	—	0.8	?
	B1919-01	Ge pin	5,000	0.8 @ 1550 nm	4 μA @ 1 0mV	—	2	?
G.P.D. Corp.	GAP5000	InGaAs	5,000	0.9–0.95	1 μA @ 0.3 V	1 MHz	0.28	940
	GAP300	InGaAs	300	0.9–0.95	1.0 nA @ 5 V	0.7 GHz	0.005	75
	GAP60	InGaAs	60	0.9–0.95	0.3 nA @ 5 V	2.4 GHz	0.001	100
Telcom Dev.	35PD10M	InGaAs	10,000	0.9–1.0	20 μA @ 1 V	—	—	?
	13PD75-F	InGaAs	75	0.85 @ 1300 nm	0.2 nA @ 5 V	1.5 GHz	—	?
	13PD55-TO	InGaAs	55	0.9 @ 1300 nm	0.55 nA @ 5 V	4 GHz	—	?
Fujitsu	FID3Z1LX	InGaAs	50	0.89 @ 1300 nm	0.05 nA @ 5 V	3 MHz	—	70
	FPD5W1KS	InGaAs	30	0.94 @ 1550 nm	20 nA @ 0.9 V	3 GHz	—	700
Hamamatsu	G3476-01	InGaAs	80	0.9–0.95	0.08 nA @ 5 V	3 GHz	0.002	?
	G5832-01	InGaAs	1,000	0.9–0.95	1 nA @ 5 V	35 MHz	0.02	100
	G5832-05	InGaAs	5,000	0.9–0.95	25 nA @ 1 V	0.6 MHz	0.1	1,400
	G7151-16	InGaAs	Array 16 80 × 200	0.9–0.95	0.2 nA @ 1 V	0.5 GHz	0.002	1,000
	G7230-256	InGaAs	Array 256 50 × 200	0.9–0.95	—	—	—	7,000
Discovery	DSC10ER	InGaAs	10	0.4 @ 1300	—	60 GHz	—	10,700
	DSC30S	InGaAs	30	0.7 @ 1300	—	16 GHz	—	2,500
	DSC50S	InGaAs	50	0.75 @ 1300	—	9 GHz	—	1,100
Alcatel	1951DMC	InGaAsP	—	0.9 @ 1300	5 nA @ 3 V	2.5 GHz	—	100
	1981 DMC	InGaAsP	—	0.9 @ 1550	1 nA @ 3 V	2.5 GHz	—	70

[19]. Full exploitation of heterostructures started in the 1960s, when epitaxial deposition became available.

Among others, the SiGe material system was initially investigated and considered to be very promising because of the mature and mutually compatible technologies related to silicon and germanium. The SiGe alloy, (i.e., $Si_{1-x}Ge_x$ with x the Ge molar fraction) is an easily miscible random alloy that enables a wide range of heterostructures by varying x between 0 and 1 (i.e., pure Si and pure Ge, respectively). The first successful growth of a SiGe heterostructure on silicon was performed by Kasper et al. [22] in 1975 via chemical vapor deposition (CVD). In the early 1980s, SiGe epitaxy received a boost due to the development of molecular beam epitaxy (MBE) and ultra-high vacuum chemical vapor deposition (UHV-CVD) [23]. Milestones of this progress are the first pseudomorphic MBE growth of SiGe on Si by Bean et al. [24] in 1984 and the first low-temperature UHV-CVD heteroepitaxy of SiGe on Si by Meyerson [25]. Subsequently, a large number of devices were proposed and demonstrated, such as the SiGe heterojunction bipolar transistor (HBT), the modulation-doped field effect transistor, resonant tunnel diodes, several near- and mid-

infrared photodiodes, light emitting devices, waveguides, and modulators as reviewed in [11, 26, 27].

A successful heteroepitaxy (in terms of crystal quality) can be accomplished if two materials have the same crystal structure and lattice parameters. Silicon and germanium have the same structure (diamond-like), but Ge has an atomic spacing about 4% larger than Si. This large lattice mismatch introduces strain, a crucial factor in the growth of SiGe epilayers on silicon. The first SiGe layers deposited on Si find it energetically convenient to adjust the lattice through compression in the growth plane and tensile strain along the normal, whereas the thick Si substrate remains substantially undistorted. As the growth proceeds, the large accumulated strain tends to let the Ge restore its own lattice spacing. This is referred to as *relaxation*, and the epitaxial film becomes *relaxed*. Relaxation takes place once the strain reaches a threshold level that corresponds to a critical thickness h_c and is always associated with generation of a large amount of line defects both in the growth plane (misfit dislocations) and perpendicular to the growth plane (threading dislocations). The first theoretical investigations of strained–relaxed transitions and defect formation date to the pioneering work by Van der Merve and co-workers [28, 29] and by Mattews et al. [30]. Extensive studies accompanied the technological advancement in the 1980s [31, 32].

Another important factor in the quality of SiGe heteroepitaxial layers and their applicability is the surface morphology. Three basic growth modes were initially identified by Bauer: (1) two-dimensional or layer-by-layer growth "Frank Van der Merwe" mode); (2) three-dimensional or island growth ("Volmer–Weber" mode); (3) hybrid growth, where an island follows a few two-dimensional monolayers ("Stransky–Krastanov" mode) [33]. Only the Frank Van der Merwe growth has practical applications.

In the absence of lattice mismatch the morphology is determined by the distribution of surface energies (substrate, epilayer, and interface). Parameters such as base pressure, substrate temperature, and flow of material need to be adjusted; however, the presence of large strain in SiGe makes the equilibrium more critical, because relaxation can also promote the formation of islands.

In summary, in SiGe heterostructures, the large lattice mismatch between Si and Ge causes accumulation of strain, which has to be properly managed. The strain can be retained by employing layers below the critical thickness or can be relieved through either generation of misfit and threading dislocations or island formation. These mechanisms are dealt with in the next subsections. At this point, however, it is useful to identify three critical figures that apply to SiGe heterostructures in the fabrication of devices: uniformity, quality, and epilayer thickness. The requirement for uniformity stems from the need for a flat surface, which allows standard planar processing; moreover, transport properties of granular structures are poor compared to single crystal epilayers. Quality and thickness are more application dependent and, therefore, amenable to trade-offs. For example, acceptable defect densities for minority carrier devices (such as bipolar transistors) are in the range 10^3–10^4 cm^{-2}, whereas for majority carrier devices [such as metal–oxide-semiconductor field-effect transistor (MOSFET)], dislocation densities of 10^5–10^6 cm^{-2} are still reasonable. The quality also depends on the amount of strain, which, in turn, depends on the Ge concentration in the SiGe alloy. Devices such as SiGe HBTs require a low Ge molar fraction (x in the 0.01–0.1 range), whereas modulation doped field effect transistors (MODFETs) and photodiodes require Ge-rich alloys, the former for deposition of a strained Si channel and the latter to increase the absorption efficiency. With regard to thickness, HBTs operate with very thin (and, therefore, strained) layers, typically on the order of the base width. MODFETs and photodiodes usually employ thicker (relaxed) epilayers that are well above the critical thickness.

The preceding excursion was meant only to sample the complexity and the variety of needs, problems, approaches, and applications somehow linked to the use of SiGe heterostructures. In the remainder of this chapter, we shall focus on SiGe photodiodes and in the next section, specifically, we concentrate on the fabrication of Ge-rich SiGe epilayers (including pure Ge) of suitable thickness for the NIR. In dealing with *relaxed* films,

we will discuss the most important techniques proposed for reducing the threading dislocation density.

2.2. SiGe for Near-Infrared Detection

One of the most important contributions to the responsivity [as evidenced in Eq. (1.1)] is represented by the factor $(1 - e^{-\alpha W})$, which is the absorption efficiency and tends to unity when both the optical absorption coefficient α and the thickness W of the absorbing layer are large. This efficiency is shown in Figure 4 as a function of W for some values of α between 1 and 10^5 cm^{-1}, which represents a typical range of values for semiconductor materials in the near-infrared spectral region.

Even with large absorption coefficients, it is apparent that W should be in the 0.1–1 μm range. This is often above the critical thickness h_c shown in Figure 5 (data from People [34]) for a SiGe strained alloy versus Ge fraction x. Strained epilayers of the mentioned thickness indeed can be grown using Ge concentrations below 20%. Unfortunately, the NIR optical absorption decreases with silicon content in the SiGe alloy and also de-

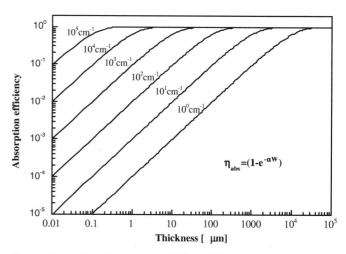

Fig. 4. Absorption efficiency versus thickness W for various absorption coefficients.

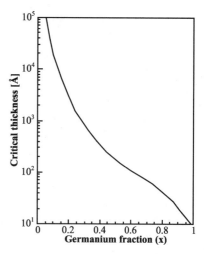

Fig. 5. Critical thickness of a Si$_{1-x}$Ge$_x$/Si epilayer versus x. After data from People [34].

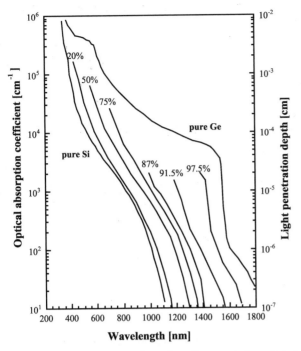

Fig. 6. Optical absorption spectra for various Ge concentrations. Data from Sze [35] (pure Si and Ge), Potter [36] (20, 50, and 75%), and Braunstein [37] (87, 91.5, and 97.5%).

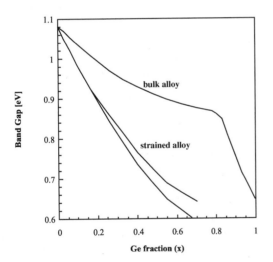

Fig. 7. SiGe bandgap versus Ge concentration in bulk (experimental data) and strained alloy (calculations). Data from People [34].

creases with wavelength, as clearly visible in Figure 6, where α is plotted versus wavelength for a set of compositions, including pure Si and pure Ge (data from Sze [35], and Potter [36] and Braunstein et al. [37]). Alloys with Ge concentrations larger than 50% are needed to cover the whole spectral range in the near infrared (1.3–1.55 μm); pure Ge exhibits the highest and most uniform absorption. The data reported in Figure 6 refer to bulk material and an increase in absorption is to be expected in strained epilayers due to strain induced bandgap reduction as shown in Figure 7 (data from People [34]), where the energy gap of bulk (experimental data) and strained alloy (calculations) is reported as a function of the Ge concentration. Unfortunately,

as far as we know, reliable data on the optical absorption of SiGe strained alloys are not available.

The calculated bandgap reduction due to strain is so pronounced that a bandgap similar to pure Ge (0.66 eV) can be observed in a $Si_{0.5}Ge_{0.5}$ strained alloy. However, even for a 50% alloy, the critical thickness is still too small for practical applications. Let us consider, for instance, the cases of $Si_{0.8}Ge_{0.2}$, $Si_{0.5}Ge_{0.5}$, and pure Ge. According to Figure 6, the α coefficients at 1.3 μm are 1, 10, and 8000 cm^{-1}, respectively. As can be seen from Figure 4, to achieve near-unity absorption efficiencies, layers of thickness W equal to 10 mm, 1 mm, and 1 μm have to be used, correspondingly. It is apparent from Figure 5, however, that such values are well above critical, even if we account for the strain effect on the bandgap.

When facing such material constraints, we have to undertake one of two possible approaches: (1) the growth of relaxed materials, with no thickness limitations (except technological ones) and (2) the growth of strained layers (below the critical thickness) to be employed in guided-wave geometries. The former is the subject of the next subsection, which is devoted to the deposition of high quality relaxed epilayers of Ge-rich SiGe alloys and pure Ge. As for the latter, although it ensures the absence of dislocations and defects, the unreleased strain limits the overall thickness and, therefore, the available absorption efficiency. In this case, only a guided-wave photodetector configuration can be adopted: the light is trapped and confined in the growth plane, where it is eventually absorbed versus propagation length rather than thickness (as for normal incidence). This approach is usually pursued in conjunction with strain symmetrization, which allows the fabrication of structures thick enough for light confinement: the growth of a SiGe layer is followed, before critical thickness, by a Si layer that partially releases the strain. These steps are repeated up to the desired thickness. Notice that the resulting overall critical thickness is that of a single strained layer with a Ge content equal to the average Ge over a single period of the multilayer [38]. Although the waveguide approach is well suited for OEICs, the absorption lengths can be quite large (and increase as the Ge diminishes) and may introduce speed limitations through the capacitance of the device. After treating relaxed films (Section 2.3), we go back to a comparison between devices based on strained and relaxed approaches in Section 3. Strained layer epitaxy and related material issue are not an issue in this work.

2.3. Relaxed SiGe Films: A Brief Overview

In relaxed films, high Ge concentrations can be used to maximize the absorption. For this reason, pure Ge seems to be the ideal candidate. However, even in the relaxed regime, the growth becomes more and more critical as the lattice mismatch increases, and pure-Ge epilayers are the most difficult to deposit. Ge-rich alloys with $x > 30\%$ (or mismatch $>1\%$) directly grown on silicon show threading dislocation densities on the order of 10^{11}–10^{12} cm^{-2} [39]. Nevertheless, their optical properties are so appealing that a large effort—both theoretical and experimental—has been devoted to reducing such values. A comprehensive theory is not yet available due to the wide range of mismatches, the large number of events that drive the dislocation dynamics (nucleation, annihilation, pinning), several experimental conditions (growth techniques, base pressures, flows, temperatures, substrate cleaning). Strain relief through relaxation is always associated with generation of a network of line defects that lie in the growth plane (at the interface) and are called *misfit dislocations*. A misfit dislocation cannot terminate within the bulk, but must either end upon itself, forming a closed loop, or terminate at the intersection with another defect. More generally, a misfit dislocation stops at the nearest free surface. An ideal picture, with defects entirely confined in the growth plane, requires the dislocations to sweep across the interface up to the end of the substrate. Dislocations can move, but their limited velocity (in terms of chip or substrate dimensions) and the proximity of other dislocations or the presence of pinning sites stop their propagation by bending them up. Defects that propagate across the epilayer are called *threading dislocations*. Hard to estimate, the density of threading dislocations is proportional to the lattice mismatch and, unless drastically reduced, is very detrimental to applications. Next, we review the pioneering work on the growth of pure-Ge relaxed films and then we outline the most successful approaches to lowering the threading dislocation density.

The growth of pure-Ge epilayers above critical thickness has been attempted by various groups. The deposition of pure Ge on Si, in fact, used to be of great interest as a way to provide a suitable substrate for GaAs deposition on Si. Because GaAs and Ge have similar lattice constants, a Ge buffer was employed to simplify the heteroepitaxy of a polar semiconductor on a nonpolar one with a relatively large lattice mismatch. One attempt consisted of growing an amorphous Ge film on silicon in high vacuum (10^{-7} torr) at room temperature, followed by crystallization annealing at $T = 650$–750°C and inspecting the crystal quality by reflection high energy electron diffraction (RHEED), transmission electron microscopy (TEM), and backscattering (channeling mode) [40]. Later, the same researchers were able to obtain similar results without recrystallization by a proper choice of substrate temperature [41]. Kuech and co-workers demonstrated the heteroepitaxial growth of Ge on Si by simple CVD [42], and performed a comparison between epilayers grown by CVD and physical vapor deposition (PVD) [43]. PVD was carried out at a baseline pressure of 6–8×10^{-10} torr on substrates heated up to 500°C, to fabricate Ge films from 500 Å up to 1 μm. CVD layers were grown from GeH_4 at atmospheric pressure and substrate temperatures in the 500–900°C range. The best films exhibited similar characteristics and an improvement in quality with thickness with a minimum dislocation density of 4×10^9 cm^{-2}. The resultant Ge films were highly p-doped in all cases, with concentrations of 2–4 $\times 10^{18}$ cm^{-3} and resistivities close to 10^{-2} Ω cm. The average Hall mobility increased with layer thickness from 100 to 300 cm^2/(V s), whereas the hole concentration remained constant. The authors inferred that both the carrier scattering and the acceptor-type impurities are associated with point or extended crystal defects incorporated into the Ge film [41].

Similar results [p-type concentration of 10^{19} cm^{-3}, resistivity about 10^{-3} Ω cm, and maximum mobility 150 cm^2/(V s)] were obtained by Garozzo et al. [44] by evaporation at 10^{-6} torr from a boron nitride crucible. They achieved deposition of single crystals at surprisingly low temperatures (375–425°C), but did not observe any significant improvement by lowering the base pressure to 10^{-9} torr. High p-type conductivity and improvement of crystal quality with increasing epilayer thickness have been consistently observed. Ohmachi et al. [45] reported a remarkably low residual doping of 10^{16} cm^{-3} and a carrier mobility as high as 1000 cm^2/(V s). Their excellent results, obtained with a technique (based on high vacuum evaporation) and growth parameters similar to those employed by others, suggest a high sensitivity to experimental conditions.

As PVD and atmospheric pressure CVD have been gradually replaced by molecular beam epitaxy and ultra-high vacuum CVD, Baribeau et al. [46] reported one of the first MBE grown pure-Ge epilayers, with threading dislocation density between 10^7 and 10^8 cm^{-2}. A similar defect density was obtained by Fujinaga [47] via low pressure CVD and thermal annealing in H_2. In 1991 at IBM, Cunningham et al. [48] reported for the first time the UHV-CVD epitaxial growth of pure Ge on Si.

It is evident that the threading dislocation density has been widely identified as the main limitation of Ge in device applications. Before summarizing and comparing the results of various groups over the years, it is worthwhile to briefly review the most used techniques for keeping such defect density to a minimum: (i) introduction of graded buffer layers, (ii) use of surfactants, (iii) carbon incorporation, and (iv) growth on processed substrates.

2.3.1. Graded Buffer Layers

Strain relaxation can be accomplished by the insertion of dislocations through a completely or partially relaxed buffer layer [49]. The equilibrium elastic strain during the growth of graded buffers can be calculated, predicting a reduction of threading dislocation (TD) density with the increase of buffer thickness [50]. Whereas the growth of relaxed films becomes more critical as the mismatch increases, it is intuitive to use an approach based on a series of low-mismatched heterointerfaces. To this extent, buffers with either stepwise or linearly graded compositions have been realized and tested. Early attempts to employ graded layering did not produce satisfactory results. A TD density of about 10^8 cm^{-2} has been obtained using a combination of step grading and thermal annealing, but the TD reduction has been attributed to the annealing process [51].

The effectiveness of linearly graded buffers grown by MBE was first demonstrated by Fitzgerald et al. [52, 53] at AT&T Bell Labs. They realized that previous attempts had failed due to the residual strain in the first layer of the buffer, which caused the mismatch in the next layer to be comparable to that between the second layer and the substrate, resulting in high dislocation density. In graded buffers, the strain is always low, regardless of the Ge concentration of the growing layer, provided dislocations nucleate continuously during the growth. Fitzgerald et al.

had evidence that, at low strain, dislocation nucleation is dominated by heterogeneous sources, the dislocation velocity is high, and misfits are long. Therefore, they raised the growth temperature (although normally avoided) to relax the first layer completely and chose the grading rate to maintain a low strain throughout the buffer. Various buffers were fabricated with a grading rate of 10% Ge for every micrometer ($10\%\ \mu\mathrm{m}^{-1}$) up to final compositions of 23, 32, and 50% and threading dislocation densities of 4.4×10^5, 1.7×10^6, and 3×10^6 cm^{-2}, respectively. A TEM cross section of a 32% sample is reproduced in Figure 8. The dislocation nucleation in the graded buffer and the high quality of the cap layer are apparent.

Using CVD and similar buffers, the same group bettered their results with TD densities between 10^5 and 10^6 cm^{-2} for grading rates between 10 and 40% Ge/μm [54]. The natural extension to the growth of pure Ge turned out to be more complicated because the larger lattice mismatch blocked the glide of dislocations and lead to their pileup. Following Fitzgerald's idea, Currie et al. [55] at MIT fabricated (by CVD) thick pure-Ge films on Si using compositionally graded buffers and introducing a chemical–mechanical polishing step (CMP) in the middle (at 50% Ge). The introduction of CMP is believed to halt the TD propagation and eliminate the effect of pileups (on the remaining high Ge concentration film) because of surface planarization. Currie et al. used grading rates of 10 and 5% Ge/μm (thus necessitating 10–20 μm thick buffers) and grew pure-Ge cap layers with thicknesses between 1.5 and 3 μm. The reported threading dislocation density is about 2×10^6 cm^{-2}, whereas samples without CMP exhibited TD density that was 1 order of magnitude larger. A schematic of the structure of the CMP sample with 10% Ge μm and its TEM cross section are reproduced in Figure 9.

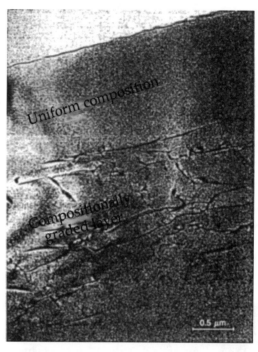

Fig. 8. Cross-sectional TEM of a sample with a 10% Ge/μm linear graded Si$_{1-x}$Ge$_x$ layer and a Si$_{0.7}$Ge$_{0.3}$ cap. Reproduced with permission from E. A. Fitzgerald et al., *Appl. Phys. Lett.* 59, 811 (1991). © 1991, American Institute of Physics.

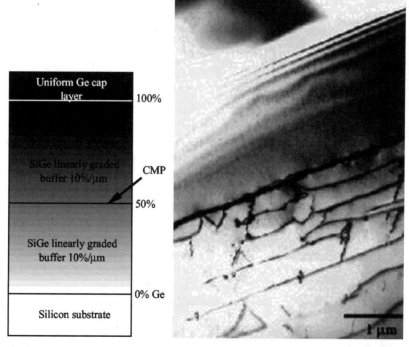

Fig. 9. Schematic of the structure and cross-sectional TEM of the sample with 10% Ge/μm linear graded buffer with CMP at half buffer and covered with a 2 μm pure-Ge layer. Reproduced with permission from M. T. Currie et al., *Appl. Phys. Lett.* 72, 1718 (1998). © 1998, American Institute of Physics.

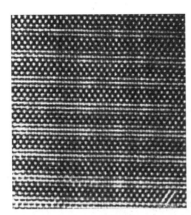

Fig. 10.   High resolution lattice image of a Si_5Ge_5 superlattice. Reproduced with permission from W. Jager et al., *Thin Solid Films* 222, 221 (1992). © 1992, Elsevier Science.

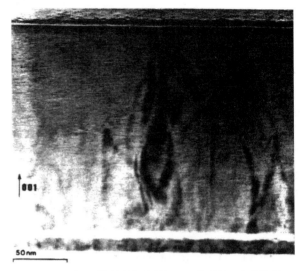

Fig. 11.   Cross-section TEM bright field image of a 145 period Si_5Ge_5 superlattice. Reproduced with permission from W. Jager et al., *Thin Solid Films* 222, 221 (1992). © 1991, Elsevier Science.

At the time as Fitzgerald et al. [52], LeGoues et al. [39] at IBM obtained similar results with step graded buffers, sometimes combined with linear grading. They showed that a low TD density could be achieved at a low growth temperature (450–550°C) and recognized the need for reducing it during growth, due to a melting point decrease associated with a rise in Ge content. Furthermore, they investigated the fundamental role of the initial surface (Si substrate or Si epilayer) in TD nucleation and dynamics. They demonstrated TD densities of about 10^4 cm^{-2} for low (20%) final Ge concentration using both linearly and step graded buffers, and observed TD densities of about 10^6 cm^{-2} in pure-Ge epilayers on a linearly graded buffer (20–>100%) deposited on a graded superlattice. The complications associated with the growth of Ge-rich relaxed alloys, either by CVD or by MBE, were further confirmed in later work at IBM, where TD densities of 10^6 cm^{-2} were obtained in $Si_{0.7}Ge_{0.3}$ on analogous step graded buffers [56].

To overcome the difficulties inherent in making a buffer layer graded up to 100%, Presting and Kibbel [57] proposed the use of Si_mGe_m short period superlattices and a buffer graded up to 50%. The alternation of Si and Ge ultrathin layers (of equal thickness and below the critical value for pure Ge on Si) provides strain symmetrization, yielding a critical thickness equivalent to a $Si_{0.5}Ge_{0.5}$ alloy (about 1000 Å). If such structure is deposited on a relaxed graded buffer ending with 50% Ge, it can be grown without thickness limitation. Unfortunately, although the TD density is substantially reduced with respect to the direct growth of such superlattices on Si, it was never below 10^8 cm^{-2} [58]. Figure 10 is a highly resolved lattice image of a Si_5Ge_5 superlattice: the abrupt interfaces demonstrate the actual feasibility of the superstructure. Figure 11 is a TEM cross-section image of a 145-period Si_5Ge_5 superlattice.

2.3.2. Use of Surfactants

The use of surface active species (surfactants) in heteroepitaxy was first introduced by Copel et al. [59], who proposed the use of a segregating surfactant to reduce the surface free energies of both substrate and epilayer, and, therefore, suppress island formation, as demonstrated in the growth of Si/Ge/Si(001)

structures with an As monolayer. Other group V elements such as Sb and Bi can also be employed as surfactants. The main effect of a surfactant seems to be the reduction of surface mobility, inhibition of island formation and reduction of segregation [59]. However, a site exchange mechanism also has been proposed [60]. Sakamoto et al. [61] demonstrated the growth of 12 monolayers of pure Ge on Si (above critical thickness) using an Sb monolayer. A remarkable result was obtained by Liu et al. [62] using MBE and a single Sb monolayer before epitaxy. They fabricated a graded buffer (25% Ge/μm) up to 50% Ge followed by a 0.3 μm thick $Si_{0.5}Ge_{0.5}$ cap layer, with a threading dislocation density of 1.5×10^4 cm^{-2}, which is 2 orders of magnitude lower than without Sb. Although surfactants have been successfully employed in MBE, they do not significantly affect the quality of CVD grown relaxed layers. This is probably due to the surfactant effect of H_2 from commonly used hydrogenated sources such as GeH_4 and SiH_4. H_2 was proposed as a surfactant by Copel et al. [63], and its beneficial effect was demonstrated in the growth of thick Ge films by CVD [64] and by MBE [65]. H_2 instead of group V elements can be convenient in MBE, because of its simpler dosage (via H_2 pressure) and the prevention of huge residual n-type doping ($\approx 10^{19}$ cm^{-3}). Only recently was a reduction in the background doping level from Sb obtained by a careful choice of growth temperature [66]. It is worth stressing, however, that the surfactants alone do not allow the growth of high quality Ge-rich epilayers; hence, they are commonly employed in conjunction with other techniques such as graded buffers.

2.3.3. Incorporation of Carbon

Incorporating carbon atoms into SiGe alloys to form SiGeC compounds is another promising approach to the growth of SiGe epilayers above their critical thickness. By varying x and y in $Si_{1-x-y}Ge_xC_y$, adjustable strain, bandgap and alignments can be tailored. The exchange of C for Ge atoms compensates

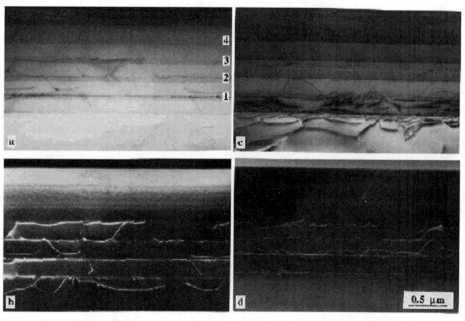

Fig. 12. Cross-sectional (a) bright and (b) dark field TEM micrographs of a $Si_{1-x}Ge_x/Si_{1-x-y}Ge_xC_y$ step-wise graded buffer structure grown in five steps. At the beginning of each step, a C concentration of 0.5% is added. For comparison, the TEM micrographs taken under the same conditions are shown for a $Si_{1-x}Ge_x$ buffer structure (c and d). Reproduced with permission from H. J. Osten and E. Bugiel, *Appl. Phys. Lett.* 70, 2813 (1997). © 1997, American Institute of Physics.

the strain, because of the smaller lattice parameter of C: the strain from compression can turn into tensile strain or even balance out. Assuming that Vegard's law is valid, that is, the $Si_{1-x-y}Ge_xC_y$ lattice constant changes linearly with composition, the equation

$$a_{Si_{1-x-y}Ge_xC_y} = a_{Si} + (a_{Ge} - a_{Si})x + (a_c - a_{Si})y$$

holds, where Si, Ge, and C lattice constants are 5.43, 5.64, and 3.56 Å, respectively. For example, for a Ge concentration of 20%, a C concentration of 2% can perfectly compensate for the strain. This technique was first implemented by Eberl et al. [67], who demonstrated its validity via MBE, growing different SiGeC compounds with 25% Ge and various C content. Their films exhibited good crystal quality (assessed by TEM) and a satisfactory strain compensation (by X-ray rocking curves). Similar results were obtained by Mi et al. [68] by rapid thermal chemical vapor deposition (RTCVD). Unfortunately, as C is added to a SiGe alloy (while the Ge molar fraction is held constant), the bandgap is expected to increase, partially thwarting the Ge effect. St. Amour et al. [69], however, observed that the bandgap increase is much less than it would be if the strain was reduced by lowering the Ge concentration without C. Otherwise stated, the incorporation of carbon allows a bandgap reduction (with the same strain) with respect to SiGe or, conversely, a strain reduction (with same bandgap). Characterization of thick (>1 μm) epilayers was performed by Osten and Bungiel [70], who grew a 1 μm thick relaxed step-graded buffer (up to 30% Ge) based on a combination of $Si_{1-x}Ge_x$ and $Si_{1-x-y}Ge_xC_y$ layers. The reported TD density was 10^5 cm^{-2}, which is 2 orders of magnitude lower than in a reference sample without C. Figure 12 shows cross-sectional bright (a) and dark

field (b) micrographs of the $Si_{1-x-y}Ge_xC_y$. The corresponding micrographs for $Si_{1-x}Ge_x$ are shown in (c) and (d) for comparison.

Despite the promising results, this approach is not suitable for growing epilayers with high Ge concentration. From the preceding equation, for 50% Ge, a C concentration in excess of 5% would be required, but the growth of random alloys with a high C dose is limited to about 4% by the low solubility of C in Si and by SiC precipitation [71].

2.3.4. Growth on Processed Substrates

As already pointed out, misfit dislocations tend to move along the heterointerface until they reach a free surface, unless they are stopped and pinned. The probability of being pinned, thus resulting in the generation of threading defects, depends on the sample size. A decrease in TD density can be expected either if the size is reduced or the substrate is processed to obtain selective epitaxial growth, or if it is miscut or a combination. In this way misfit dislocations are more likely to terminate on a free surface without affecting the film.

This approach was first proposed by Luryi and Suhir [72], who suggested patterning the substrate to reduce the stress energy in the epilayer by limiting the strained zone to a narrow region by the heterointerface. They calculated that good $Si_{1-x}Ge_x$ with $x < 0.5$ can be obtained by keeping the total strain energy (per unit area) below the threshold for generating dislocations, reducing the seeding window width to make the effective length of the stress zone comparable with the critical thickness, and performing a selective overgrowth [73]. Among other groups, Banhart and Gutjahr [74] employed liquid phase epitaxy to grow defect-free $Si_{0.9}Ge_{0.1}$ layers 1–3 μm thick on silicon oxide patterned Si substrates. For Ge concentration greater than 0.1, they observed confinement of dislocations close to the SiO_2 walls. Srikanth et al. [75] demonstrated the beneficial effect of miscut (6°) silicon on the quality of buffer layers graded from Si ($x = 0$) to pure Ge ($x = 1$), showing that samples grown on cutoff substrates exhibit reduced surface roughness and lower dislocation density. Although the combination of substrate patterning and epitaxial overgrowth strongly reduces dislocations, highly defective seeds are left, thus imposing iterations of the photolithographic and epitaxial processes. Recently, Langdo et al. [76] employed overgrowth on Si substrates patterned by interferometric lithography to obtain high quality films in a single patterning and epitaxial step.

2.4. Relaxed SiGe Films: Recent Approaches

High quality SiGe can be obtained for Ge fractions less than 30%, whereas for $x > 50\%$, we have to resort to involved techniques to significantly reduce the threading dislocations, but never below 10^6 cm^{-2} for pure Ge. The best growth of pure Ge, with a TD density of 2×10^6 cm^{-2}, was accomplished by Currie et al. [55], who employed a 10 μm thick graded buffer interrupted at 50% to perform CMP and then continued up to 100% to end with a 2 μm Ge cap. Such complex and thick structures are time consuming and expensive. If the growth has to be interrupted for CMP and then resumed, in fact, the high yield potential of parallel processing in CVD reactors vanishes. Moreover, thick structures reduce the compatibility with standard Si technology, because of large height mismatches between the Ge-based optoelectronics and Si-based electronics. For such reasons, two novel approaches have been investigated. The first is based on a simple thin, low-temperature Ge buffer for the subsequent growth of pure-Ge single crystals; the second employs polycrystalline Ge.

2.4.1. Epitaxial Pure Ge on Si

The idea of the low temperature (LT) Ge buffer layer comes from the need for a simple way to relax the large amount of strain in a very thin region close to the heterointerface, releasing the stress energy through the insertion of dislocations instead of island formation. In the absence of any strain, the growth of Ge on Si is expected to be layer by layer (two dimensional), but due to the lattice mismatch, "islanding" is produced after a few monolayers (Stranski–Krastanov growth). This growth also depends on the amount of mismatch. If the lattice mismatch is below a certain value, the stress tends to be released first by dislocation nucleation and then by island formation. Above this value, island formation is preferred and then dislocations are generated [77]. This explains why buffer layers up to low-concentrations can exhibit very low dislocation density. In all cases, the substrate temperature is crucial. For the growth of pure Ge on Si, a low-temperature regime can be identified where the rate is limited by the surface reaction and the mode is layer by layer. Above a certain temperature, however, the

growth turns into Stranski–Krastanov, and is controlled by diffusion and adsorption of GeH_4 from the gas phase [78]. When the surface is hydrogenated (as with CVD), it can be assumed that a substrate at T_{sub} below the desorption temperature T_{des} of H from Ge promotes a layer by layer growth because hydrogen significantly reduces the anisotropy of the surface energy, which is known to be responsible for island nucleation [77]. Therefore, a Ge LT buffer layer that is grown at $T_{sub} < T_{des}$ until the stress is released through the insertion of dislocation (a few tens of nanometers are enough) has been proposed [79]. At this low temperature the deposition rate is quite low (about 1 nm/min), but once the stress is released, the growth can proceed as in the homoepitaxial case at higher temperatures, thereby increasing the rate (above 10 nm/min) without promoting island formation.

Previously, only silicon LT buffers were considered. When SiGe is grown on LT Si, strain relaxation is promoted through the insertion of dislocations, because the relatively low growth temperature of the Si buffer reduces surface atom migration and a high density of defects is generated. Chen et al. [80] reported a TD density of about 10^6 cm^{-2} in a 500 nm thick $Si_{0.7}Ge_{0.3}$ alloy grown via MBE on a 200 nm LT Si buffer. A slightly lower TD density (10^5 cm^{-2}) was reported by Li et al. [81] who used a thinner (50 nm) LT Si buffer before the MBE deposition of $Si_{0.7}Ge_{0.3}$. Linder et al. [82] obtained TD densities on the order of 10^4 cm^{-2} in MBE $Si_{0.85}Ge_{0.15}$ on 100 nm thick LT

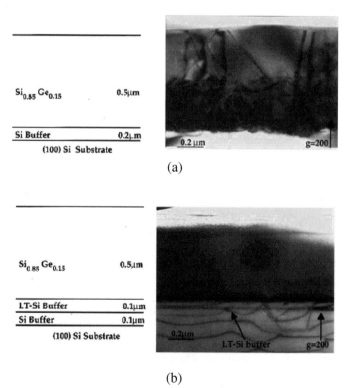

Fig. 13. Cross-section TEM images of 0.5 μm thick $Si_{0.85}Ge_{0.15}$ grown on Si (a) without any low-temperature Si (LT Si) buffer and (b) with 0.1 μm LT Si. Both samples have a 0.1 μm Si buffer layer grown at high temperature. Reproduced with permission from K. K. Linder et al., *Appl. Phys. Lett.* 70, 3224 (1997). © 1991, American Institute of Physics.

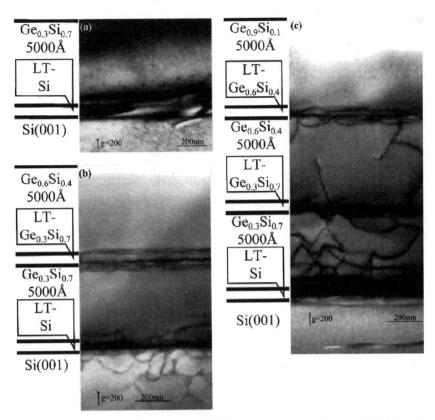

Fig. 14.   Layer structures and cross-section TEM images for (a) one, (b) two, and (c) three step LT Si/SiGe samples. The LT layers were grown at 400°C and the GeSi layers at 550°C. Reproduced with permission from C. S. Peng et al., *Appl. Phys. Lett.* 72, 3160 (1998). © 1988, American Institute of Physics.

Si buffers, with the quality appreciable from the TEM micrographs reproduced in Figure 13. They suggested that, because of the low growth temperature of the buffer, a large density of point defects is generated and as misfit dislocations propagate, they become trapped and annihilated.

Peng et al. [83] reported the growth of Ge-rich alloy on an LT Si buffer. They demonstrated the effectiveness of LT Si buffers also in multiple steps, measuring a TD density of 3×10^6 cm^{-2} in $Si_{0.1}Ge_{0.9}$ MBE grown on a three-step graded structure with a 90% final concentration. Sample layer structures are shown in Figure 14.

A new approach, based on an LT Ge buffer, permitted the epitaxy of pure Ge on silicon [84]. The (100) substrates were cleaned at 1100°C in a H$_2$ atmosphere in a UHV chamber, with basic pressure in the high 10^{-11} torr range. The surfaces were contaminant-free (C and O below 0.1%), 2×1 reconstructed, and flat, as confirmed X-ray photoelectron spectroscopy (XPS), RHEED, and atomic force microscopy (AFM). The Ge films were grown by low pressure CVD using high purity Ge without carrier gas and millitorricelli pressures. In the first deposition stage, the substrate was kept at $T = 330°C$ to grow a flat and relaxed epitaxial layer ~50 nm thick. Such temperature is lower than the the H desorption temperature from the Ge surface. Due to the H surfactant action (see foregoing text) at such temperature, nucleation of three-dimensional Ge islands is inhibited and the strain relaxes via the insertion of misfit dislocations.

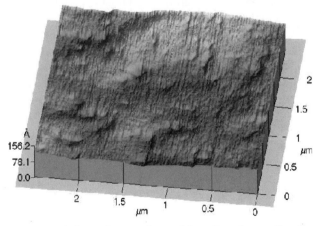

Fig. 15.   Atomic force microscopy image of the surface of a pure-Ge epilayer grown on Si through a LT Ge buffer layer.

Once rid of the excess elastic energy, the Ge growth can proceed as in the homoepitaxial case, at temperatures as high as 700°C to reduce the dislocations and the overall deposition time. The germanium is intrinsic (or unintentionally doped). AFM of the films, however, showed the presence of large Ge terraces (separated by 0.5 nm steps), which tend to widen with increased substrate temperature. The short-range roughness of about 0.5 nm did not vary significantly from sample to sample. A sample AFM picture is shown in Figure 15.

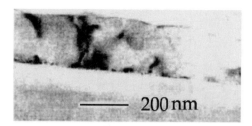

Fig. 16. Cross-sectional TEM micrograph of a Ge sample grown at a deposition temperature of 550°C on a Ge buffer grown at 330°C.

The RHEED spectra, although they confirm the film flatness, prove they are monocrystalline and 2×1 reconstructed. The defect distribution in the Ge films was evaluated by TEM, and a typical cross sectional image of a film grown at 550°C is displayed in Figure 16.

We observe that although most defects are confined in a narrow region near the Ge/Si interface, a few dislocations penetrate the Ge epilayer up to the free surface.

To investigate the optoelectronic quality of the Ge films for applications to light detection, a planar metal/Ge/metal interdigitated structure with 10 μm spacing was fabricated. This MSM configuration was lithographically defined on a silver layer evaporated onto the Ge. The metal contacts the semiconductor only in the interdigitated region through a properly windowed insulating layer meant to avoid the pad capacitance. No thermal treatments were carried out to prevent interdiffusion and alloying between Ge and Ag. Due to its good photocarrier collection, such a structure can be operated efficiently by coupling light from the top surface, even if half is reflected by the electrode fingers. The photocurrent was measured versus the applied electric field and the incident light power, as well as the free carrier transit time between the closely spaced electrodes, to quantify their mobility (free carriers photogenerated in the Ge layer can be collected only if their average lifetime exceeds that required to cover the distance between generation sites and electrodes). Assuming uniformity of both the applied field distribution and the generation across the sample, the fraction η_{col} of collected photocarriers versus field amplitude is given by the Hecht formula [85]

$$\eta_{\mathrm{col}} = \frac{L(E)}{d}\left[1 - e^{-d/L(E)}\right]$$

where d is the interelectrode spacing, and $L(E) = \mu\tau E$ is the drift length, and μ, τ, and E denote the carrier mobility, their lifetime, and the field amplitude, respectively. The collection efficiency saturates to unity for large values of E. Figure 17 is a typical graph of the measured photocurrent versus applied voltage. The measurement was carried out at a wavelength of 1.3 μm and submilliwat power. The data follow the Hecht formula and via a best-fit, a mobility-lifetime product of 10^{-6} cm^2/(V s) was estimated.

Time of flight experiments were carried out by illuminating the sample with 100 ps pulses at 1.32 μm and recording the outcome. The time response is visible in Figure 18 for two different voltages applied to the contacts.

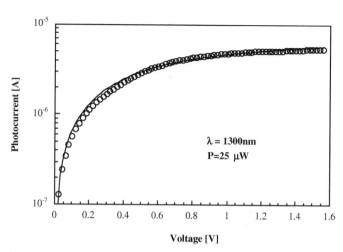

Fig. 17. Photocurrent as a function of the applied bias. Circles denote experimental data and the solid line is a fit from the Hecht model.

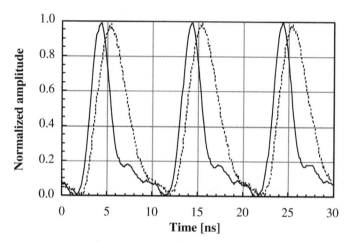

Fig. 18. Time response of the MSM test device excited by an optical pulse train (100 ps) from a Nd:YAG laser emitting at 1.32 μm. Solid line, bias voltage is 3 V; dotted line, bias is 0.4 V.

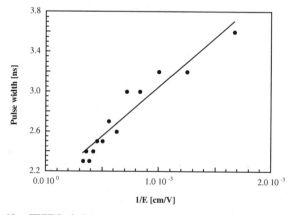

Fig. 19. FWHM of photocurrent response versus inverse electric field between the metal fingers. Points are experimental data; the solid line is a linear fit.

The broadening of the photocurrent FWHM can be attributed to the time required for the photocarriers to reach the electrodes (i.e., $\Delta\tau_{\mathrm{FWHM}} = d/\mu E$). Figure 19 graphs the measured pulse broadening versus the inverse electric field. The linear

Table II.   Annealing Parameters and Threading Dislocation Density of Samples A–E (A is as-grown)[a]

Sample ID	A	B	C	D
T_H (°C)/time (min)	NA	900/10	900/100	900/10
T_L (°C)/time (min)	NA	100/10	100/10	100/10
Number of cycles	NA	1	1	10
TD density (cm^{-2})	$9.5 \pm 0.4 \times 10^8$	$7.9 \pm 0.6 \times 10^7$	$7.8 \pm 0.5 \times 10^7$	$5.2 \pm 0.6 \times 10^7$

[a] All Ge epilayers are 1 μm thick and grown on LT Ge buffers. T_H and T_L are high and low annealing temperatures of the cyclic process. TD densities are measured by plan-view TEM. Data from Luan et al. [87].

slope gives a mobility $\mu = 1200 \pm 300$ cm^2/(V s) that is only three times lower than the value reported for bulk Ge crystals [3900 cm^2/(V s)].

By combining the data in Figure 19 with those in Figure 17, a lifetime of 1ns can be estimated, which is consistent with the presence of defects in the epitaxial layer (defects are expected to introduce states in the forbidden gap, thus reducing the carrier lifetime). Assuming that the lifetime τ is governed by recombination at defect centers, it can be related to the defect state density N_d and, for a recombination cross section $\sigma = 10^{-12}$ cm^2 (typical of bulk Ge [86]) and thermal speed $v_{th} = 10^7$ cm/s, $N_d = (\tau \sigma v_{th})^{-1} \approx 10^{14}$ cm^{-3} is obtained. Although no detailed information on the amount and spatial distribution of such defects is presently available, we expect a profile that peaks in proximity to the heterointerface. However, the high density of recombination centers suggests a high TD density in the epilayer, as well. Nevertheless, photodiodes based on this material have been fabricated and tested successfully, and their fabrication and performance are discussed in forthcoming sections.

Following the approach described above, Luan et al. [87] at MIT pursued the growth of pure Ge on Si with LT Ge buffer, but demonstrated that the introduction of postgrowth cyclic thermal annealing is effective in reducing the threading dislocation density. They reported TD densities down to 10^7 cm^{-2} in large wafers, and average TD densities of 2×10^6 cm^{-2} in Ge mesas (10 μm $\times$ 10 μm) on Si by combining cyclic annealing with selective area growth. To investigate the effects of this cyclic process between high (T_H) and low annealing temperatures (T_L), Luan and co-workers measured TD densities with plan-view TEM and etch-pit density counting (EPD). Table II summarizes annealing parameters and TD densities obtained in various Ge epilayers. Regardless of the actual parameters, more than 1 order of magnitude reduction in TD is always compared to the as-grown sample (A). The TD densities of samples B, C, and D show that increasing the number of cycles is more effective (in reducing TD) than extending the annealing time at T_H.

Because the Ge epilayers are first annealed at the high $T_H = 900$°C and the Ge melting point is 959°C, the whole Ge/Si structure is relaxed. Then the temperature drops at the low T_L, and the structure undergoes stress due to the thermal expansion mismatch between Ge and Si. The stress, in turn, induces faster dislocation glide and annihilation, which are responsible for TD density reduction at an enhanced rate (as the glide veloc-

Table III.   Annealing Parameters and TD Density of Samples D–F[a]

Sample ID	D	E	F
T_H (°C)/time (min)	900/10	900/10	900/100
T_L (°C)/time (min)	100/10	675/10	780/10
Number of cycles	10	10	10
TD density (cm^{-2})	$5.2 \pm 0.6 \times 10^7$	$4.2 \pm 0.1 \times 10^7$	$2.7 \pm 0.1 \times 10^7$

[a] Ge epilayers are 1 μm thick and grown on LT Ge buffers. The other parameters are as in Table I. Data from Luan et al. [89].

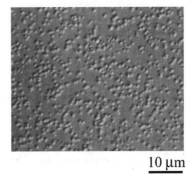

10 μm

Fig. 20.   AFM micrograph of sample E etched for EPD count.

ity goes up, TDs travel more and the probability of annihilation increases). Two factors control the dislocation velocity: thermal stress and dislocation glide energy barrier [88]: the former increases as T_L decreases; the latter prevents dislocation glide at low temperatures. Based on a calculation from Luan et al. [89], an optimum T_L of about 830°C for maximizing the dislocation velocity would be expected. To this extent, Luan et al. performed experiments with $T_H = 900$°C and $T_L = 100, 675,$ and 780°C, respectively ($T_L = 830$°C was not tested due to limitations of the three zone furnace), and the results are summarized in Table III. As predicted, the TD density decreases as TL approaches 830°C, validating the proposed understanding and optimization strategy.

Figure 20 is a micrograph of the etched sample E. Its average etch-pit density is $2.1 \pm 0.5 \times 10^7$ cm^2, in agreement with the value measured by plan-view TEM, $2.3 \pm 0.2 \times 10^7$ cm^2.

Figure 21 shows TEM cross sections of samples A (1 μm Ge on 30 nm LT Ge buffer without thermal annealing) and F (1 μm

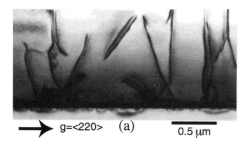

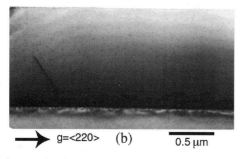

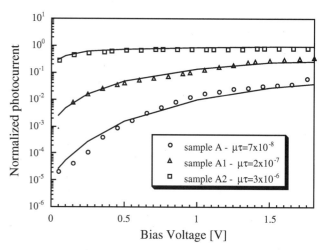

Fig. 22.   Measured MSM photocurrent response at 1.3 μm versus applied bias (symbols). The lines are fitted based on the modified Hecht model.

Fig. 21.   Cross-sectional TEM image of (a) sample A, Ge grown on LT GE buffer on Si, and (b) sample E, Ge grown on LT GE buffer on Si but followed by 10 cycles of thermal annealing between $T_H = 900°C$ and $T_L = 780°C$. The thickness of the Ge film is 1 μm, whereas the LT Ge buffer is 30 nm. Reproduced with permission from H.-C. Luan et al., *Appl. Phys. Lett.* 75, 2909 (1999). © 1999, American Institute of Physics.

of Ge grown on 30 nm LT Ge buffer with cyclic annealing as described in Table III). The effectiveness of the LT Ge buffer layer in confining most of the dislocations in a narrow region close to the heterointerface is apparent. Moreover, a reduction of the residual TD density in the Ge epilayer, associated to the thermal annealing, is also confirmed.

Based on these results, the group at MIT proposed a further reduction in TDs by gliding them to dislocation sinks such as mesa sidewalls. Small Ge mesas on Si were realized by selective growth on square windows etched in SiO$_2$ with sides from 10 to 100 μm. The resulting average TD density was about 2×10^6 cm^{-2}, with several TD-free sites among the 10 μm $\times$ 10 μm mesas, as also verified by AFM [87].

To assess the optoelectronic quality of Ge layers subjected to cyclic annealing, a number of samples were grown in optimized conditions (as evaluated by structural characterization) and processed in photodetector configurations. A hot wall UHV-CVD with a base pressure of 3×10^{-9} torr was used for Ge heteroepitaxy on Si p-type Si (100) wafers with resistivity in the range of 0.5–2 Ω cm. The Si substrates were precleaned in a Piranha solution (H$_2$SO$_4$:H$_2$O$_2$ = 3:1) for 10 min, and the native oxide was removed by dipping in HF solution (HF:H$_2$O = 1:5) for 15 s. Heteroepitaxy of Ge on Si was initiated at 350°C with a GeH$_4$ flow of 10 sccm (15% in Ar) and a total pressure of 15 mtorr. After depositing 30 nm of Ge, the furnace temperature was raised to 600°C and 1–4 μm of Ge (intrinsic or unintentionally doped) were grown on Si. Finally, the wafers were cyclically annealed between a high T_H and a low T_L. For EPD, a mixture of CH$_3$COOH (67 mL), HNO$_3$ (20 mL), HF (10 mL) and I$_2$ (30 mg) was employed. To investigate the im-

pact of Ge quality on the NIR optoelectronic characteristics of the devices, MSM planar interdigitated structures, with silver electrodes 10 μm apart on a 100 $\times$ 500 μm area and passivated by a polymeric insulating layer were fabricated.

The measured current density of MSMs was well above that in vertical structures, implying that most of the flow occurred in the Ge epilayer, despite a film thickness less than the contact spacing. A superlinear dependence of the dark current on applied bias suggests a space–charge limited current as the dominant transport mechanism. To evaluate the carrier mobility–lifetime product $\mu\tau$, the photocurrent versus voltage was measured, illuminating the MSM devices with laser light at 1.3 μm. The results are shown in Figure 22 for three samples: A (as-grown), A1 (after five annealing cycles between $T_H = 900°C$ and $T_L = 780°C$), and A2 (after 20 annealing cycles between $T_H = 900°C$ and $T_L = 780°C$).

After thermal annealing, the samples exhibit a dramatic improvement in collection efficiency at low bias and saturate more readily to maximum. A good fit to the data is a simple one-parameter (mobility–lifetime product) model based on the Hecht formula adapted to a nonuniform field distribution. The estimated mobility–lifetime products are quoted in Figure 22 and listed in Table IV. In the latter, calculated $\mu\tau$ products are well correlated to TD densities, demonstrating that both measurements are suitable for the optoelectronic characterization of the material.

A slight decrease of TD density in thicker samples (8×10^6 cm^{-2} in 4 μm thick Ge epilayers) was observed, that was related to either the larger volume for annihilation and/or thermal stress that promotes the motion of the dislocations. Time of flight measurements were also employed to evaluate the carrier mobility independently in the Ge epilayer. As described earlier, a train of 100 ps light pulses at 1.32 μm was shone on the MSM and the reduction in photocurrent pulse width (FWHM) was recorded versus applied bias. Figure 23 shows the results from sample A2 and Figure 24 shows the corresponding FWHM versus inverse average electric field: the linear fit yields a mobility

Table IV. Annealing Parameters and TD Densities of Samples A (as-grown), A1, and A2[a]

Sample ID	A	A1	A2
T_H (°C)/time (min)	NA	900/10	900/100
T_L (°C)/time (min)	NA	780/10	780/10
Number of cycles	0	5	20
$\mu\tau$ Product (cm²/V)	7×10^{-8}	2×10^{-7}	3×10^{-6}
TD density (cm⁻²)	$9.5 \pm 0.4 \times 10^8$	$2.7 \pm 0.1 \times 10^7$	$1.6 \pm 0.1 \times 10^7$

[a] All parameters are as in Table I.

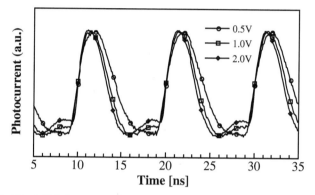

Fig. 23. MSM photocurrent time response of sample A2 illuminated by 100 ps pulses at 1.32 μm and for three bias voltages.

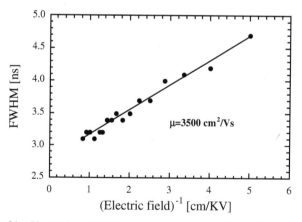

Fig. 24. Photocurrent FWHM dependence on inverse electric field in sample A2. Symbols denote data; the solid line is a linear fit. An electron mobility of 3500 cm²//(V s) can be estimated.

of 3500 cm²/(V s), the highest reported to date for pure-Ge epitaxially grown on silicon. This value is very close to that reported for undoped Ge with negligible scattering at impurity sites [86]. The carrier lifetime in sample A2, calculated from the measured mobility–lifetime product and the carrier mobility, is 850 ps. This is several orders of magnitude shorter than in electronic-grade bulk Ge, in the 10^{-3}–10^{-6} s range [86, 90]. This short lifetime can be related to recombination at threading dislocations that are still present in the epilayer. Whereas the link between carrier lifetime and dislocations in Ge has been investigated in bulk Ge crystal [91] and SiGe epilayers [92], us-

ing such data allows an epi-Ge carrier lifetime of about 40 ns in the presence of 1.6×10^7 cm⁻² TD density to be inferred. From this value we are led to believe that misfit dislocations close to the Ge/Si interface are responsible for the additional reduction in lifetime.

In summary, this simple two-step CVD technique followed by cyclic thermal annealing appears to be a successful approach to the epitaxial growth of pure Ge on Si, yielding TD densities below 10^7 cm⁻², mobility lifetime products in the 10^{-6} cm²/V range, and carrier mobilities of 3500 cm²/(V s), the latter presumably affected by additional misfit dislocations. Table V summarize the most significant achievements in Ge on Si relaxed epilayers by listing authors, deposition system, technique adopted for TD reduction, cap layer thickness, and TD density. The last column specifies whether operating devices actually have been realized.

In a subsequent section (Section 3.5) we report the fabrication and characterization of NIR photodiodes based on these materials.

2.4.2. Polycrystalline Ge on Si

A completely different approach to avoid threading dislocations relies on the granular structure of polycrystalline Ge film to prevent the accumulation of strain. Polycrystalline Ge can be easily deposited by standard evaporation systems at very low cost and on large areas. Moreover, the deposition temperature (300–400°C) is much lower than CVD heteroepitaxy, where substrate cleaning, deposition, and annealing require temperatures in the range of 600–1000°C. The resulting low thermal budget of poly-Ge evaporation ensures better compatibility with silicon technology and even the film deposition on electronic chips. Conversely, the electronic quality of the Ge cannot be compared to single crystal Ge epilayers. Nevertheless, this has been demonstrated as a suitable low-temperature and low-cost approach to the fabrication of NIR photodiodes.

In the past decade, polycrystalline SiGe alloys have received considerable attention as a viable alternative to polycrystalline Silicon in several integrated electronics applications such as the gate electrodes in complementary metal-oxide–semiconductor (CMOS) transistors [93] and the active layer of thin film transistors (TFT) for active matrix liquid crystal displays [94]. In addition to allowing deposition, crystallization, and dopant activation at lower temperatures and with a lower melting point, such SiGe films exhibit lower resistivity and larger grain size. However, to fulfill the required compatibility with VLSI technology, only SiGe with Ge concentrations <60% has been employed, because alloys richer in Ge have too low a melting point and unstable oxides. Polycrystalline Ge has received less attention, being mainly limited to applications as a suitable substrate for GaAs-based solar cells [95]. The small lattice mismatch between Ge and GaAs, in fact, and the simple poly-Ge deposition offer new ways to transfer high efficiency III–V solar cell technology from GaAs substrates to low-cost, large area substrates.

Table V. Comparison between the Most Significant Achievements in Ge on Si Relaxed Epilayers[a]

Reference	Deposition	Technique	Cap layer	TD (cm^{-2})	Device[b]
Maenpaa et al. 1981	Evaporation	Pure Ge layer 1 μm	—	4×10^{9c}	—
Baribeau et al. 1986	MBE	Step graded buffer to 100% + SL + annealing	1 μm	10^{7a}	—
Fitzgerald et al. 1991	MBE	Graded buffer 0–23%; grading rate 10% μm^{-1}	—	$4.4 \times 10^{5c,d}$	—
		Graded buffer 0–32%; grading rate 10% μm^{-1}	—	$1.7 \times 10^{6c,d}$	—
		Graded buffer 0–50%; grading rate 10% μm^{-1}	—	$3.0 \times 10^{6c,d}$	—
Presting et al. 1992	MBE	Fixed comp buffer 0–50% 50 nm thick	Si_5Ge_5 SL 200 nm	10^{10}–10^{9c}	Yes
		Graded buffer 0–40% (30% μm^{-1}) + const. comp. 40%	Si_6Ge_4 SL 200 nm	10^{8a}	Yes
LeGoues et al. 1994	MBE	Graded superlattice to 18% + step graded to 100%	0.5 μm	$10^{6a,b}$	—
Watson et al. 1994	RTCVD	Step graded buffer to 50% 5 step (10% μm^{-1})	1 μm	6.0×10^{5d}	—
		Step graded buffer to 50% 10 step (10% μm^{-1})	1 μm	3.7×10^{5d}	—
LeGoues et al. 1994	CVD	Step graded buffer to 30% 7–8 steps	1.4 μm	$10^{6d,e}$	—
Kissinger et al. 1995	APCVD	Step graded buffer 0–20% 4 steps	—	10^{6}	—
		Step graded buffer 0–20% 4 steps annealing	—	10^{4}–10^{3}	—
		Step graded buffer 0–20% 4 steps + optimized annealing	—	10^{3}–10^{2}	—
Li et al. 1997	MBE	LT Si buffer 30% Ge	0.8 μm	10^{5c}	—
Osten et al. 1997	MBE	$Si_{1-x-y}Ge_xC_y$ step graded buffer 0–30%	—	10^{5c}	—
Currie et al. 1998	UHVCVD	Graded buffer 0–100% grading rate 10% μm^{-1}	2 μm	$5.0 \times 10^{7c,e}$	Yes
		Graded buffer 0–100% grading 10% μm^{-1} + CMP	2 μm	$2.0 \times 10^{6a,b}$	Yes
Peng et al. 1998	MBE	LT-Si buffer + three step grading to 100% Ge	0.5 μm	3.0×10^{6a}	—
Liu et al. 1999	MBE	Graded buffer 0–50% 1 ML Sb surfactant	0.3 μm	1.5×10^{4b}	—
	UHVCVD	LT-Ge buffer + annealing	1 μm	$2.3 \times 10^{7a,b}$	Yes
Luan et al. 1999	UHVCVD	LT-Ge buffer + annealing + SE 100% Ge	1 μm	$<10^{6a,b}$	Yes
		LT-Ge buffer + annealing	1 μm	$1.6 \times 10^{7a,b}$	Yes
Our group 2000	UHVCVD	LT-Ge buffer + annealing 100% Ge	4 μm	$8.0 \times 10^{6a,b}$	Yes

[a] Pure Ge epilayers are denoted by starred references.

[b] This column specifies whether operating devices actually have been realized.

[c] Measured by cross on plan-view TEM.

[d] Measured by electron beam induced current.

[e] Measured by EPD.

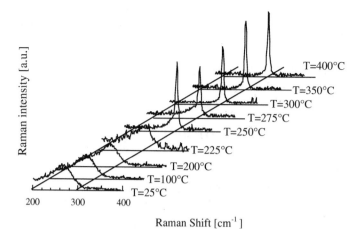

Fig. 25. Raman spectra of poly-Ge films grown on Si at different temperatures.

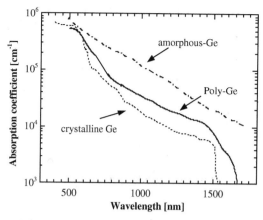

Fig. 26. Absorption spectra of poly-Ge and amorphous Ge compared to crystalline Ge.

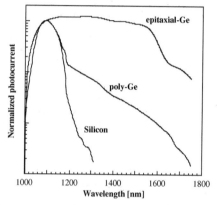

Fig. 27. NIR photocurrent versus wavelength of poly-Ge on n-type Si, compared with pure Si and epitaxial Ge on Si. The currents are normalized to their spectral maxima.

Poly-Ge was used for NIR photodiodes for the first time by Colace et al. [96]. Polycrystalline germanium was deposited by thermal evaporation in a vacuum chamber with a background pressure of 10^{-6} torr. During evaporation, the pressure was kept at 10^{-5}–5×10^{-6} torr. Different samples were grown to thicknesses between 0.12 and 1.8 μm on n- and p-type $\langle 100 \rangle$ silicon with resistivity of 2–3 Ω cm, at temperatures in the range 25–500°C and at a rate of 1.5 Å/s. The silicon substrates were chemically cleaned and the native oxide was removed in a buffered HF bath for 2 min, then rinsed in deionized water, and introduced into the vacuum chamber. The whole cleaning process was performed at room temperature and all the samples showed smooth specular surfaces, uniform over several square centimeters. The poly-Ge films were characterized via microprobe Raman spectroscopy using a He–Ne laser at 632.8 nm and intensities below 30 kW/cm² to prevent local recrystallization. A set of typical Raman spectra is reproduced in Figure 25: the displacement of the broad amorphous Ge–Ge transverse optical phonon band into the spectrally narrow mode around 300 cm⁻¹ indicates the transition from amorphous to polycrystalline Ge. The measurements, indeed, show the existence of a threshold temperature around 250°C that separates amorphous and polycrystalline phases, and is consistent with Raman analyses reported by other authors [97].

Absorption spectra were acquired by measuring transmitted and reflected light versus wavelength, and correcting the data for the presence of the substrate. Both thin (0.12 μm) and thick (1.8 μm) samples were used to operate with high signal dynamics in the visible and in the NIR, respectively. The absorption coefficient α is plotted in Figure 26 (solid line). The transition from amorphous to polycrystalline is emphasized by the neat appearance of absorption edges at 1.54 and 0.58 μm, related to direct transitions at Γ and Λ points in the crystalline Ge band structure, respectively [86].

To investigate the material photoresponse through photocurrent spectroscopy, a simple vertical device with metal contacts evaporated on poly-Ge and on Si (substrate) was fabricated. Figure 27 shows the spectrum of poly-Ge grown on n-type

silicon at 400°C compared with those from CVD-grown epitaxial Ge and Si. Polycrystalline Ge exhibits a significant (although much lower than epi-Ge) NIR photoresponse, which becomes vanishing small near the direct band edge of crystalline Ge (0.8 eV). Samples grown at temperatures below 250°C (amorphous Ge) do not respond in the NIR and their spectral features resemble silicon. Moreover, although poly-Ge on n-type Si exhibits NIR sensitivity (Fig. 27), poly-Ge grown on p-type Si does not show any appreciable photocurrent in this spectral range. This can be attributed to band alignment at the poly-Ge/p-Si heterointerface, which prevents photocarriers generated in Ge from being collected [96].

The electronic properties of poly-Ge films were studied through current–voltage (I–V) characteristics on MSM test structures at room temperature. As is visible in Figure 28, polycrystalline films exhibit a linear (ohmic) I–V behavior, with a resistivity of about 10^{-1} Ω cm that is nearly insensitive to temperature (see Fig. 29), as expected for an extrinsic semiconductor. For a mobility of 10 cm²/(V s) (as quoted by several authors [98, 99]) a free carrier concentration on the order of 10^{19} cm⁻³ can be inferred. Due to the low resistivity, the dark-current flow is confined in the poly-Ge film, and measurements

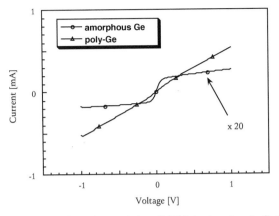

Fig. 28. Current–voltage characteristics of MSM structures in poly-Ge (triangles) and amorphous Ge (circles) at room temperature.

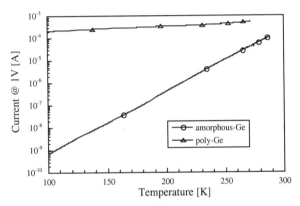

Fig. 29. MSM dark current at 1 V bias versus temperature in poly-Ge (triangles) and amorphous Ge (circles) structures.

on vertical poly-Ge/Si structures showed a rectifying behavior clearly that indicates the heterojunction barrier.

Amorphous Ge films, conversely, have a much higher resistivity (about 10^3 Ω cm) that forces the current through the Ge/Si heterojunction. As a consequence, the I–V characteristic of the a-Ge MSM resembles a pair of back-to-back diodes (see Fig. 28). From Figure 29, using a simple thermionic expression, a barrier height of 0.37 eV can be estimated, which is in good agreement with the valence band discontinuity at the Ge–Si heterojunction [44] and quite far from the height of metal–Ge Schottky barriers [100]. The latter suggests that the transport mechanism is the injection of holes from the Ge film into the Si substrate. The ohmic characteristics of the metal contacts on both amorphous and polycrystalline Ge were verified by acquiring I–V characteristics with closely spaced surface contacts.

Finally, the minority carrier lifetime was measured by Shockley-type experiments in the MSM structures: diodes were illuminated with 100 ps pulses at 1.32 μm (uniformly absorbed in the Ge layer) and the photocurrent recorded versus time. A 5 ns lifetime was estimated for the polycrystalline samples, with current (exponential) decay independent on the applied bias, as expected by lifetime-limited collection.

Despite the poor electronic properties (high unintentional doping, short carrier lifetime) of poly-Ge compared to single

crystal Ge epilayers, its optical absorption in the NIR make this approach amenable for photodetector applications. In subsequent sections (Sections 3.7, 4.2, and 4.3) we report the fabrication and characterization of NIR photodetectors based on polycrystalline Ge.

3. SiGe ON Si NIR PHOTODETECTORS: HISTORICAL OVERVIEW

In this section, we present a review of near-infrared photodiodes realized in Ge or SiGe thin layers on silicon substrates. A number of different structures, alloys, and geometries for light coupling have been attempted to improve responsivity and minimize dark current.

3.1. Early Devices

In the 1960s, the development of vapor growth techniques for epitaxial deposition of semiconductor layers stimulated a great deal of activity aimed at fabricating heterojunction devices. The first attempt to fabricate a Si/Ge heterojunction photodiode with response extended to the NIR was performed by Donnely and Milnes [101], who grew germanium onto cleaved silicon chips via CVI [102] and measured the photovoltaic response of diodes fabricated on substrates with doping ranging from 1.2×10^{14} to 5×10^{18} cm^{-3}; the Ge was n-doped 4–5 $\times 10^{16}$ cm^{-3}. With the exception of the sample on the lowest resistivity substrate, the diodes exhibited broad spectral responses up to 2.5 μm with two sign inversions in open circuit voltage (V_{oc}) as shown in Figure 30. This peculiarity was explained by presence of a high density of acceptor-like defect centers at the Si/Ge interface, resulting in a band diagram similar to that of two back-to-back connected Schottky junctions. Referring to the scheme in Figure 31, when short wavelength light is absorbed mainly in the high bandgap semiconductor (Si), V_{oc} is positive. As the photon energy decreases, Si becomes transparent and most photons are absorbed in the low gap semiconductor (Ge), until eventually the V_{oc} changes sign. Finally, when the incident photon energy is below the Ge bandgap, interface or Si gap state pumping takes place, extending the response to about 2.5 μm and the V_{oc} sign is positive again. The V_{oc} of diodes grown on the lowest resistivity substrate was appreciably different from zero only at wavelengths between the Si bandgap and the threshold for direct Ge transitions (1.55 μm). In this case, the photoresponse of the Si side of the heterojunction (both above bandgap and due to gap state pumping) is quenched because of the lack of built-in field, and the substrate acts as a passive filter for photon energies higher than the Si bandgap.

The main result of this first investigation was to establish that the optoelectronic properties of the Si/Ge heterojunction devices are controlled by the high density of states at the interface. Based on consideration of lattice mismatch, a defect density of about 6×10^{13} cm^{-2} was estimated.

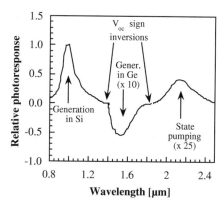

Fig. 30. Spectral response of Ge/Si heterojunction photodiodes. Data from [101].

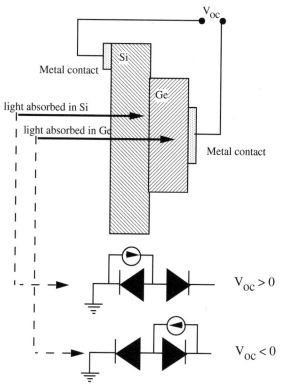

Fig. 31. The Ge/Si heterojunction diode reported in [101] (top) and equivalent circuit for short and long wavelength illumination (bottom). The resulting change in the sign of the open circuit voltage is evident.

3.2. The Work at Bell Labs

The initial report was followed by a long period of silence, which was broken in 1984 by the work of Luryi et al. [103]. At the end of the 1970s, the fast growing interest in long wavelength fiber optics telecommunication systems boosted research on sources and detectors working at photon energies well below the silicon bandgap. The development of very low-loss fiber working at 1.55 μm (NTT, 1978) and the experiments by NTT [32 Mbits/s through 53 km of graded-index fiber at 1.3 μm (1977)] and British Telecom [140 Mbits/s through 49 km of single-mode fiber at 1.3 μm (1981)] demonstrated the feasibility of long connections using the transparency windows cen-

tered at 1.3 and 1.55 μm. Note, however, that the rush toward reduction of optical losses in long-haul links was accompanied by the first attempts to bring the advantages of optical links directly to subscriber home. The Hi-OVIS experiment in Japan (announced in 1976 and working since 1978; 150 homes connected) and that of France Telecom (announced in 1978 and operative since 1986; 1500 home connected in Biarritz) were the first examples of the fiber to the home concept. Nevertheless, despite the lack of motivation in a possible cost reduction for FTTH transceivers, Luryi, Kastalsky, and Bean pointed to the performance improvements achievable by integrating the photodetector with a front-end amplifier on a single silicon chip.

3.2.1. The First Ge on Si p-i-n Photodiode

The Ge p-i-n detector described in [103] was grown on a stepwise 1.8 μm thick buffer on 0.5–3.0 Ω cm n-type (100) Silicon substrates. Such thick buffers separated the active intrinsic layer of the diode from the highly dislocated Si/Ge interface region. The diode exhibited good quantum efficiency (40% at 1.3 μm) in the photovoltaic mode, indicating the satisfactory transport properties of the Ge intrinsic layer. Moreover, the spectral response extended to 1.6 μm as shown in Figure 32. However, the still large dislocations density gave rise to a reverse dark current in excess of 50 mA/cm^2 at 1 V that does not compare well with a typical 0.1 mA/cm^2 in bulk Ge p-i-n photodiode. A year later, the same authors reported a strong improvement in both material quality and dark reverse current density by using a 10 period 10 nm Ge$_{0.7}$Si$_{0.3}$/50 nm Ge superlattice embedded in the buffer layer [104]. This additional buffer was positioned just before the intrinsic layer, and the corresponding strain was intended to trap and prevent dislocations from reaching the device active region. The TD density, measured by TEM, was indeed reduced by about 2 orders of magnitude from 10^9–10^{10} to 5 × 10^7 cm^{-2} and so was the dark reverse current density. Unfortunately, the quantum efficiency also dropped to 3% at 1.3 μm and even below at longer wavelengths.

Aiming to reduce the dark current, a new device structure was introduced by Temkin et al. [105]. Rather than pure Ge for the active layer, this device contained a coherently strained Ge$_x$Si$_{1-x}$/Si strained-layer superlattice (SSL) and employed a waveguide configuration to compensate for the reduced absorption. SSL were preferred to SiGe alloys of similar average composition because, as demonstrated in [106], they allowed the deposition of layers well above the critical thickness. The loss of responsivity at 1.55 μm, when compared to a pure-Ge detector, was not considered to be of capital importance given the attention devoted to 1.3 μm as the emerging wavelength for communications. Devices with various Ge content in the SSL were fabricated with thicknesses of the SiGe wells adjusted to preserve the commensurate growth [107]. As expected, the spectral responsivity extended to longer wavelengths as the Ge fraction increased (see Fig. 33). The maximum external quantum efficiency at 1.3 μm was 10.2% at a reverse bias of 10 V and for the sample with 60% Ge fraction in the wells. At the same bias, the dark current density was 7.1 mA/cm^2. Samples

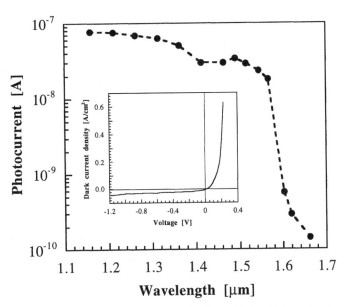

Fig. 32. Spectral response of the first p-i-n Ge on Si photodiode. In the inset, the current–voltage characteristic at room temperature is plotted. Data from [103].

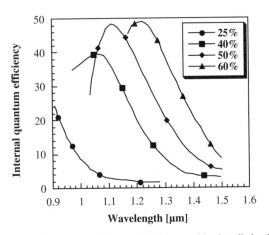

Fig. 33. Internal quantum efficiency of SSL waveguide photodiodes for different Ge percentages in the SiGe wells. Data from [105].

with lower Ge fraction exhibited lower dark current and responsivity. A fast response of 312 ps also was reported for a device excited by a 50 ps laser pulse at 1.3 μm.

3.2.2. The First SiGe on Si Avalanche Photodiode

Avalanche photodiodes (APD) are extremely appealing for optical communications, because their high gain can increase detectivity when either thermal or front-end amplifier noise is dominant over shot noise [108]. Whereas Si is an almost perfect material for APD fabrication due to the high ratio (5–500, depending on the electric field) of the ionization coefficient of electrons and holes, and the resulting low excess noise factor [109], Ge is cursed with a ratio below 2 and poor noise performance. These considerations, in conjunction with the successful demonstration of high-performance APDs with sep-

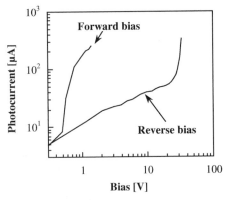

Fig. 34. Photocurrent of the avalanche photodiode described in [115] as a function of applied bias. The device exhibits photocurrent gain both in reverse bias, due to avalanche multiplication, as well as in forward bias, due to photoconductive behavior. Data from [115].

arate generation and multiplication sections (SAM) with III–V compounds [110] motivated the fabrication of NIR SAM APDs based on SiGe. The device was first designed [111], fabricated, and tested [112] at Bell Labs in 1986. Because APD requirements are very stringent in terms of material quality, the material selected for absorption of NIR photons was a SSL with light coupling in a waveguide geometry. The device was grown via MBE on a p$^+$-type silicon substrate, and consisted of a p$^+$pn$^+$ diode with an SSL and a buffer layer embedded between the substrate and the p-type layer. Among various SSLs with overall thickness ranging from 0.5 to 2 μm, one of the most effective was based on a 3.3 nm thick Ge$_{0.6}$Si$_{0.4}$ alloy with 29 nm Si spacers. The waveguide was 50 μm wide and 50–500 μm long. The authors reported a unity gain quantum efficiency of about 10% at 1.3 μm, yielding a maximum responsivity of 1.1 A/W once avalanche multiplication took place. It is worth noting that, as proven by a breakdown voltage close to 30 V, the avalanche occurred in the Si region rather than in the SSL, and had beneficial effects on the noise performances. Although the overall responsivity was lower than expected (probably due to additional coupling and scattering losses in the waveguide), the idea was exploited in a second device with an inverted doping structure (n$^+$pp$^+$ on a n$^+$-type substrate) [113]. This optimized structure exhibited more stable electrical characteristics and an external reponsivity as high as 4 A/W at 1.3 μm and 30 V reverse bias. Above this value, the formation of microplasmas, probably at dislocation sites, gave rise to a much higher (100–200) gain and noise, thus impeding signal discrimination. The APD response time of 100 ps prompted its use in a communication experiment performed with 1.275 μm light modulated at 800 Mbits/s by a pseudorandom pattern and a 45 km fiber with average 0.4 dB/km loss between source and (pigtailed) detector. The measured bit error rate (BER) reached 10^{-9} just below a 3 μA photocurrent, with no noise floor. The receiver sensitivity $\eta P = -29.4$ dB m was not far from that demonstrated in the same period with InGaAs pin detectors [114].

The described SAM APD, once it was forward biased, exhibited a large photoconductive gain (see Fig. 34 and [115]).

Its origin was suggested to be the preferential trapping and release of holes in the SiGe wells, and the resulting hopping-type transport across the SSL. In this configuration, the large dark current density, limited by the series resistence of the high resistivity buffer and cap layers employed in the structure, produced a much higher noise than in the avalanche mode. This resulted in a worsened receiver sensitivity of about -16 dB m at 180 Mbits/s and at a gain of 20.

3.2.3. Short Period Superlattice Detectors

In the following years, great expectations were generated by the reports of Pearsall et al. [116] and People and Jackson [117] on the optical properties of Si/Ge superlattices, with an emphasis on absorption enhancement in the 1.0–1.6 μm range induced by Brillouin zone folding. Based on these reports, Pearsall et al. [118] introduced a GeSi (4:4) superlattice detector built in a waveguide structure. The absorbing film consisted of four Ge monolayers followed by four Si monolayers, repeated five times (up to 5.2 nm) and embedded in a 20 nm Si buffer. The superlattice plus buffer structure was repeated 20 times, up to a total thickness of 500 nm. The waveguide, as in the previously described SLS detectors, was 50 μm wide and 400 μm long. The spectral response of this device, however, was peaked closer to the Si bandgap compared with an SSL detector of similar composition (Ge$_{0.6}$Si$_{0.4}$). This response could be ascribed to the Kronig–Penney superlattice potential, which shifted the band edge of the absorbing layer toward higher energies. The lack of responsivity improvements at the wavelength of interest (1.3 μm) and the relatively worse electrical characteristics of the diode, probably due to the very high number of heterointerfaces (80) traversed by the current flow, did not encourage further attempts in this direction.

3.3. European Efforts

In 1990, Canham [119] reported efficient photoluminescence from porous silicon. His findings, although limited to the visible spectrum, opened fascinating new perspectives on light generation and monolithic integration in silicon. In 1993, Soref [11] reviewed state-of-the-art Si-compatible light emission, modulation, guidance, and detection (very likely the most referenced work in this field).

It is in this framework, motivated by the promising results at the Bell Labs and by the continuous development of SiGe/Si growing techniques for the fabrication of HBTs, that a number of activites started on NIR photodetection integrated with silicon. Among them, in Europe, the research at the University of Berlin in collaboration with Daimler–Benz, and in the United States, the research at IBM and at the University of California Los Angeles (UCLA) in collaboration with AT&T. Once again, the focus was on 1.3 μm detectors based on strained or partially relaxed SiGe MQWs. The first important results date to the middle 1990s, which were preceded by few preliminary reports [120]. In 1994, Splett et al. [121] demonstrated a SiGe photodetector integrated with a waveguide for operation at 1.3 μm. The device structure was different from those previously outlined, because SiGe alloys of different compositions were used for the waveguide and for the detector. The latter was piled on the former (see Fig. 35), which consisted of p-type 2.5 μm thick Si$_{0.98}$Ge$_{0.02}$ alloy that had a slightly larger refractive index compared to Si and a 5 μm wide rib that confined the light in the transverse dimensions. To keep free carrier absorption losses below 1 dB/cm while allowing good p-type contact to the detector, the doping in the waveguide region was limited to 3×10^{17} cm^{-3}. The absorption (intrinsic) layer of the detector consisted of 20 periods of 5 nm thick Si$_{0.55}$Ge$_{0.45}$ quantum wells separated by 30 nm Si spacers. The detector was completed by n-type Si buffer and cap layers and aluminum contacts. Light was coupled from the waveguide to the detector in the overlapping region, employing various propagation lengths to maximize the responsivity. The maximum external efficiency measured at 1.282 μm was 11% at a reverse bias of about 7 V. The dark current density at the same bias and for a shorter device, which featured almost 1/3 of that efficiency, was

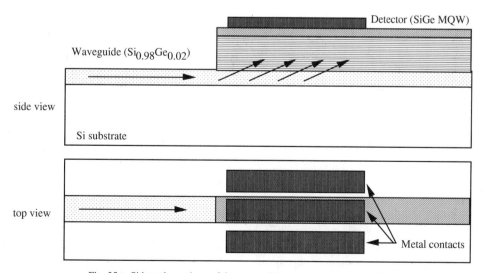

Fig. 35.   Side and top views of the waveguide detector described in [121].

about 1 mA/cm^2, whereas the saturation current in forward bias was almost 2 orders of magnitude lower. This fact and the lack of scaling of the reverse current with device area, underlined the presence of a large amount of leakage at the edges of the mesa, suggesting the possibility of large improvements. To single out the different contributions to the reverse bias dark current, in fact, it is sometime useful to fit the forward characteristic with the diode exponential law

$$J = J_s\left(e^{V/nV_t} - 1\right)$$

thus estimating the diode ideality factor (n) and the saturation current density (J_s). The latter is the sum of diffusion (Shockley) and a recombination term [122], and should coincide with the dark reverse current density J_d. The measured J_d is usually (if not always) larger than J_s due to surface and leakage effects, which can be minimized by a careful passivation of diode mesas. Therefore, in general, the J_d/J_s ratio can be regarded as a figure of the improvement feasible by simply optimizing the fabrication process.

The speed of the waveguide photodetector was tested in both the frequency and the time domains using a microwave network analyzer and recording the response to a light step function, respectively. Although both techniques gave a speed well in the gigahertz range, the latter evidenced the presence of a slow (50 ns) component in the step response, probably due to trapping of photogenerated holes in the quantum wells.

In the same year, the group at UCLA/AT&T reported a guided-wave detector with characteristics similar to those previously discussed [123]. In this device, however, guiding and absorbing layers were combined into a single SiGe MQW layer, as in the original design from Bell Labs [113]. The MQW itself was formed by thicker wells (14 nm) with higher Ge content ($x = 71\%$). Even Si spacers and cap layers where thicker, with an overall detector thickness slightly below 3 μm. The external quantum efficiency at 1.32 μm increased with the reverse bias peaking at 7% for 14 V, whereas the dark current density (at same bias) was 2.7 mA/cm^2. The need for such a high voltage to reach full carrier collection was justified in terms of MQW trapping of photogenerated holes at low bias. Detection of 0.5 Gbit/s (1.5 Gbits/s) pseudorandom NRZ (nonreturn to zero) optical signals with the waveguide detector connected to a Si bipolar (GaAs) preamplifier yielded well open eye diagrams, but no BER or sensitivity figures were reported.

A completely different approach was used in 1994 by Sutter et al. [124] who went back to the work of by Luryi et al. [103] with the aim of improving the normal incidence characteristics of a pure-Ge p-i-n detector. Whereas the deposition of a thick, complex buffer was one of the problems in growing pure Ge on Si, despite the unavoidable relatively large TD density, the Swiss group attempted to grow Ge on Si by direct MBE without gradual buffer. To prevent islanding, the growth was divided into two phases: First the temperature was kept low (420°C) while evaporating a 50 nm layer at a rate of 0.1 nm/s. Then the temperature was raised to 530°C for the remaining deposition time. The deposited Ge layer was 4 μm thick, whereas the first

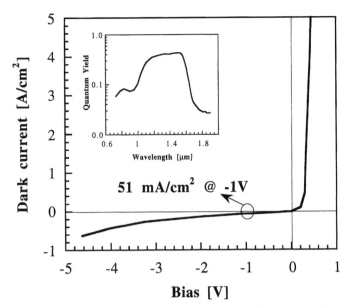

Fig. 36. Dark current voltage characteristic of the Ge p-i-n diode described in [124]. In the inset, the spectral responsivity of the same device is plotted. Data from [124].

2 μm was highly p-type doped ($N_A > 10^{17}$ cm^{-3}) and the following 2 μm was considerably less doped ($N_A < 10^{15}$ cm^{-3}). The p-i-n junction was formed by diffusion from a spin-on Sb/SiO$_2$ source at 640°C in a dry nitrogen atmosphere. Before dopant diffusion, the Ge layers were annealed at 700°C for various times. Mesa etching and metallization, as usual, concluded the fabrication process.

Extensive material quality assessment, performed by X-ray diffraction, Rutherford backscattering (RBS), and pit count after Schimmel etch, confirmed the good crystalline quality with a lower limit of 5×10^6 cm^{-2} to dislocation density.[1] It is interesting to note (and it will be worth recall later in the chapter) that annealing was effective in reducing the dislocation density. Despite the simpler fabrication process, electrical and optoelectronic characteristics of the device were similar to those in [103], with a spectral response extended to 1.6 μm and 43% peak external quantum efficiency at 1.55 μm in the photovoltaic mode (i.e., without external bias). Although the reverse dark current density was still large (51 mA/cm^2, see Fig. 36), an RC limited 530 ps risetime appears very promising.

The first report on a normal incidence p-i-n photodiode working at 1.3 μm and not based on pure Ge appeared in 1995 [126]. This device employed an SSL absorbing layer (similar to that used at Bell Labs for the waveguide photodetector) with overall thickness of 500 nm and single wells realized by 10 nm Si$_{0.5}$Ge$_{0.5}$, embedded into 40 nm Si spacers to prevent relaxation and proliferation of dislocations. The device had a p$^+$-i-n$^+$ structure and was antireflection coated with SiO$_2$ to optimize light coupling at 1.3 μm. The coating was

---

[1]Schimmel etch is a typical selective etch used to evidence defects in Ge and Si. See ASTM 10.05, F80, 1997 version (the current version does not contain this etchant). (For other selective etchants, see [125].)

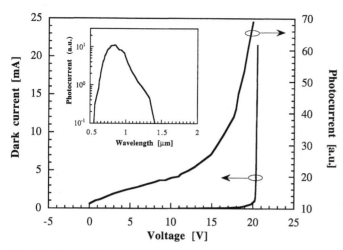

Fig. 37. Dark and photocurrent characteristics of the detector presented in [126]. In the inset, the spectral response is plotted. Data from [126].

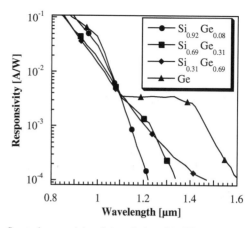

Fig. 38. Spectral responsivity of photodiodes with different amounts of Ge in the absorbing layer of the device. Data from [128].

deposited through a window opened on the aluminum top contact. A 1% external quantum efficiency was reported at 1.3 μm with a reverse bias of 4 V, whereas a much higher photocurrent was observed at voltage close to the breakdown (20 V). Once again, compared to pure-Ge detectors, both MQW and SSL SiGe devices needed a relatively large bias to collect all the photogenerated carriers. This need can be certainly ascribed to photocarrier trapping in the SiGe low bandgap wells. The good responsivity was accompanied by a dark current density of about 3 mA/cm^2 at 4 V, typical of SSL devices, and increased rapidly with bias up to 60 mA/cm^2 at 7 V (1/3 of the breakdown voltage) as shown in Figure 37.

3.4. The Introduction of SiGeC

Alhough a small amount of C helps to compensate for the Si/SiGe lattice mismatch (see Section 2), the precipitation of SiC severely limits the maximum quantity of C that can be incorporated into SiGeC films by quasiequilibrium techniques. This limit is far beyond the amount required for mismatch compensation at high Ge content. In addition, carbon raises the energy gap of the alloy, with detrimental effects on the absorption. Nevertheless, the use of nonequilibrium techniques like MBE or CVD with low substrate temperature can prevent precipitation. In 1996, the group at UCLA presented the first normal incidence p-i-n detector with a SiGeC active layer [127]. The device was grown on a p$^+$ silicon substrate and had an active layer 80 nm thick of Si$_{0.4-x}$Ge$_{0.6}$C$_x$ alloy with $x = 1.5\%$, as estimated by X-ray diffraction and RBS. Notice that 80 nm is far beyond the critical thickness of a 60% Ge alloy. The quantum efficiency, at normal incidence at 1.3 μm was about 1%, which is similar to the MQW device presented in [126]. However, the reverse bias dark characteristic was steeper, featuring 7 mA/cm^2 at 0.5 V and reaching breakdown at around 6 V. Although a SNR = 100 was reported at 1.3 μm and 0.1 V reverse bias, the light intensity actually employed was not specified.

In 1996, Battacharya's group at the University of Michigan reported impact ionization coefficients (α_p and α_n) in SiGe al-

loys of different composition [128]. The difference between the two ionization coefficients must be as large as possible to obtain good noise performance in APDs and, moreover, photocarrier multiplication has to be started by the carrier with the higher α [129]. The technique used to evaluate the ionization coefficients of electrons and holes independently of each other is based on the ability to obtain transport of a single type of carrier in the intrinsic layer of a p-i-n junction by shining highly absorbed light on the p or n side of the diode. Carrier pairs photogenerated close to the p (n) contact are separated by the electric field that drives the holes (electrons) toward the electrode, where they are collected, and accelerates electrons (holes) in the intrinsic layer. The investigated p-i-n (n-i-p) devices employed the silicon substrate as a p (n) contact and were grown via MBE using solid Ge and disilane (Si$_2$H$_6$) as sources. Alloys with $x = 8$–100% (pure Ge) were used for the 1 μm thick intrinsic layer, and either PH$_3$ or B doped the top layer in p-i-n and n-i-p diodes, respectively. As expected, increasing the germanium percentage progressively extended the spectral response to longer wavelengths, reaching 1.6 μm for the pure-Ge case (see Fig. 38). However, despite the relatively thick intrinsic layer, the responsivity at wavelength absorbed in the alloy never exceeded 10 mA/W. The measured α_n/α_p ratio (for an electric field intensity of 330 kV/cm) ranged from 10 to 0.75 as the Ge percentage in the alloy went from 0 to 100%. In particular, the value obtained for $x = 1$ is in good agreement with published results on bulk Ge. The alloy composition for which $\alpha_n = \alpha_p$ corresponded to x between 40 and 50%.

3.5. Optimized Waveguide Photodetectors

The optimized design of devices such as in [123] was carried out by Naval et al. in 1996 [130]. The authors numerically investigated a comprehensive range of parameters with the aim of maximizing the efficiency of waveguide photodetectors based on SSL at 1.3 μm. In an effort to comply with various constraints of physical and technological origin, they studied SSL with a maximum $x = 60\%$ to avoid extremely thin SSL layers. The waveguide width was set at 10 μm and its length was limited at 1 mm. The calculation took into account mode mismatch as well as Fresnel reflections at the fiber–waveguide interface, and the limited volume of SiGe absorbing layers with respect to the overall (guided) mode profile. The main outcome of the analysis is that by using the Dodson–Tsao model for the critical thickness h_c [131], a maximum efficiency of about 13% is expected with 0.4–1% SiGe composition, averaged over the SSL. This efficiency was obtained with different SSL structures, all characterized by a low number of periods and relatively high Ge content (48–55%) in the quantum wells. For higher average Ge content, the efficiency decreases due to a reduction in both coupling efficiency (the waveguide becomes thinner due to a reduced h_c) and mode confinement in the absorbing region. On the other hand, detectors with average Ge content lower than optimum suffer from a reduction in the relative volume occupied by the SiGe in the waveguide, corresponding to a longer absorption length, and a lower efficiency in a finite length device.

Better results can be obtained when the more optimistic model of People and Bean [132] for the critical thickness is used in the calculation. In particular, efficiencies as high as 60% were predicted for a Ge average percentage of 3.5% in the SSL. This is a direct consequence of the thicker SiGe strained layers allowed by this model, which, for the same alloy composition, permits better light confinement in the absorbing region. Although maximum efficiency is calculated for SSL thickness of about 10 μm (corresponding to the input fiber core diameter), a $\approx$1 μm limit imposed by realistic SSL deposition rates fixes a ceiling of about 16%. The reduction in efficiency, mainly due to fiber to waveguide light insertion losses, can be overcome with passive waveguide couplers, such as in Figure 35 and [120].

The hints derived from the preceding analysis were used in successive work, the results of which were published in [133], to fabricate a waveguide photodetector using SiGeC as the absorbing layer. The material was grown by rapid thermal CVD (RTCVD) using GeH_4 and C_2H_4 as sources for Ge and C, respectively. The substrate was kept at 625°C, and the final Ge and C content in the film was 55 and 1.5%, respectively, as obtained from X-ray diffraction and RBS measurements. The junction was formed by the p^+ Si substrate, an 80 nm thick unintentionally doped SiGeC layer, and a Si n^+ cap. The light confinement perpendicular to the wafer plane was due to the larger refractive index of SiGeC compared with silicon. The measured maximum external quantum efficiency at 0.3 V was 0.2 and 8% at 1.55 and 1.3 μm, respectively, with dark current density of 4 mA/cm^2.

In 1997, NEC Corporation (Japan) fabricated a waveguide detector on a SOI substrate, taking advantage of the large index step at the Si/SiO$_2$ interface for the light confinement, and with a trench receptacle for alignment of the fiber tip to the input facet of the photodiode [134]. The detector itself employed a 30 period SSL with 3 nm $Si_{0.9}Ge_{0.1}$ wells separated by 32 nm Si walls as the absorbing layer. The extremely low average Ge content imposed by the need to preserve the compatibility with the high temperature (950°C) steps used for bipolar and MOS transistors in Si technology resulted in a spectral re-

sponsivity only slightly different from Si. (Notice, however, that even detectors for the first window at 980 nm, where silicon is still absorbing, can benefit from the use of SiGe, because the indirect bandgap of silicon imposes an absorbing thickness of 100 μm, which is unacceptably large for high speed applications). For a reverse bias of 5 V, the device in [134] featured an external quantum efficiency of 25–29% at 980 nm that was independent of waveguide length down to 200 μm. A reverse dark current below 50 μA/cm^2 and a frequency response (-3 dB band) extending to 10.5 GHz indicated the good material quality. Of particular relevance is the original technique used to selectively grow the SiGe layers in the trench previously opened in the SOI overlayer. Because selective growth of SiGe (or Si) on Si (better than on SiO$_2$) is obtained in CVD at low Si_2H_6 and GeH_4 fluxes, and for limited layer thickness [135], an additional step to remove precursors on SiO$_2$ was introduced between each Si or SiGe growth step. This step was accomplished by a small amount of Cl$_2$ flow [136].

3.6. Toward Operation at 1.55 μm

After the mid 1990s, a large research effort in SiGe NIR detectors was devoted to the investigation of structures able to operate in the third spectral window for fiber communications, around 1.55 μm. Due to the vanishing absorption of SiGe alloys and MQWs at this wavelength, the material of choice was pure Ge subjected to treatments effective in reducing the dislocation density in the epitaxial film, as discussed in Section 2.3. A number of approaches were undertaken with different outcomes: the introduction of carbon, the use of graded or low-temperature buffers, and the growth of undulating SiGe layers. Even though the incorporation of carbon is still troublesome and has important drawbacks (mainly bandgap increase) the latter approach demonstrated effectiveness for waveguide detectors and the use of appropriate buffer layers yielded interesting results. In our opinion, this approach is the most promising. In the following discussion, we point out some noteworthy results with different methods.

3.6.1. GeC on Silicon

In 1998, Shao et al. [137] tried to fabricate normal incidence detectors by incorporating C into pure Ge films grown via MBE on Si. The device was a p-$Ge_{1-x}C_x$/n-Si heterojunction diode with x ranging from 0 to 2%. Recall that because nearly 10% C is required for a perfect lattice match between GeC and Si, the scarce solid solubility of C in Ge (10^8 atoms/cm^3) favors segregation, even when nonequilibrium techniques such as MBE or RTCVD [138] are adopted. The undercompensated GeC films in [137], in fact, were highly dislocated (10^{10} cm^2), as confirmed by cross-sectional TEM. Nevertheless, dark current in the milliampere per square centimeter range at 1 V reverse bias compared well with SiGeC devices reported in [127]. The responsivity extended to 1.55 μm (see Fig. 39) with a maximum value of 35 mA/W at 1.3 μm.

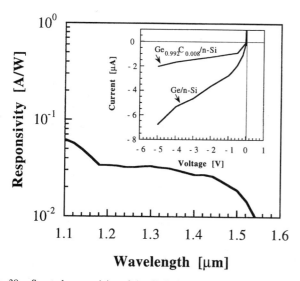

Fig. 39. Spectral responsivity of the GeC photodiode. In the inset, the dramatic reduction in the dark current obtained by alloying with C is shown. Data from [137].

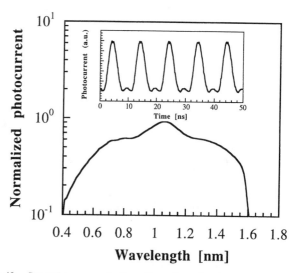

Fig. 40. Spectral response of a Ge on Si metal–semiconductor–metal detector. In the inset, the response to a train of pulses at 1.32 μm, each 100 ps width, is shown. Data from [139].

3.6.2. Low-Temperature Buffer

Planar MSM and vertical heterojunction photodetectors in pure epitaxial Ge on Si were first reported [139, 140] using a low-temperature Ge buffer layer. Although details on material growth and the beneficial effect of the hydrogen as a surfactant in UHV-CVD systems were summarized in Section 2.4, we want to point out that the response covered all three spectral windows (0.98, 1.3, and 1.55 μm) with a sharp drop that corresponded to the threshold for direct transitions at the Γ point of Ge band diagram (see Fig. 40). The responsivity at 1.3 μm was 240 mA/W at 1 V bias and at normal incidence. Taking into account Fresnel reflectivity losses and the Ge thickness (0.5 μm), an internal collection efficiency close to 90% was estimated. Thanks to the small parasitic capacitance typical of interdigited MSMs, the detector speed was limited by the photocarrier transit time, which, at low bias, decreased linearly with the inverse electric field. When excited by a train of optical pulses (100 ps width at 100 MHz repetition rate) generated by a Nd:YAG laser emitting at 1.32 μm, the MSM exhibited a photocurrent pulse width of 2.2 ns at a bias of 3 V.

After Kimerling and co-workers [141, 142] at MIT succeeded in fabricating low dislocation Ge on Si by a combination of low-temperature buffer and postgrowth annealing heterojunction, photodiodes with high speed and efficiency were demonstrated and reported. The devices consisted of a 1 μm unintentionally doped Ge layer on p-type (100) Si substrates. Mesas of different areas were realized to single out the effect of leakage on dark current. Ag top and bottom contacts were evaporated, patterned, and light coupled through the substrate. Due to the improved material quality, the photodetectors exhibited a high saturated responsivity of 550 and 250 mA/W at 1.3 and 1.55 μm, respectively, at reverse bias of a few hundred millivolts (see Fig. 41). The effect of the postgrowth annealing is quite evident when these values are compared with the much

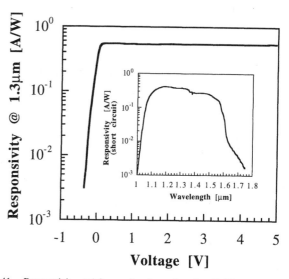

Fig. 41. Responsivity at 1.3 μm of an heterojunction Ge/Si photodiode as a function of the applied bias. In the inset, the spectral response of the same diode is plotted. Data from [142].

larger ones (a few volts) needed for full photocarrier collection in devices made with as-grown layers. Due to the small intrinsic layer thickness and large free carrier mobility, the response speed was RC limited, even in the smallest of 200×200 μm device, with a FWHM pulse response of 850 ps at 1.32 μm and 4 V (see Fig. 42). The dark characteristic put forward a rectifying behavior controlled by the p-Si/i-Ge heterojunction, with a saturated reverse current of 30 mA/cm^2, scaling with device area. We believe this is mainly due to generation in the well defected heterointerface region and that the incorporation of the latter into a doped layer in a fully optimized p-i-n device should lower this value. This hypothesis is supported by preliminary results obtained on p-i-n devices [143].

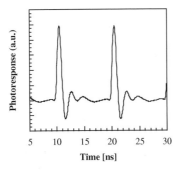

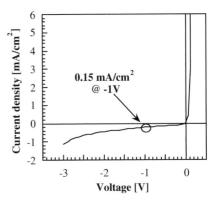

Fig. 42. Photocurrent response of a heterojunction photodetector made from Ge grown on Si followed by 20 annealing cycles. The photocurrent was excited by pulse train at 1.32 μm with 100 ps pulse width at 100 MHz repetition rate. The FWHM response time was 850 ps. Data from [142].

Fig. 43. Dark current voltage characteristic of a Ge on Si p-n diode grown using a gradual buffer layer and a CMP step. Note the record low current density at -1 V. Data from [144].

3.6.3. Optimized Graded Buffer

The lowest dark current density in Ge on Si p-n junctions was reported by Samavedam et al. [144] in 1998. They minimized TD density by the use of an optimized (patented) graded buffer complemented with a CMP step at an intermediate composition of $Si_{0.5}Ge_{0.5}$ on miscut Si wafers (see also Section 2.3). The p-n junction was laid entirely in the epitaxial Ge by introducing dopant sources during growth (PH_3 and B_2H_6 for n and p type, respectively). The square mesa photodiodes, of sizes ranging from 95 to 250 μm, exhibited a rectifying characteristic with quality factor 1.1 and dark current density lower than 0.2 mA/cm^2 (see Fig. 43). The latter was just three to five times larger than the theoretical value expected for a Ge diode with that doping profile. The responsivity at 1.3 μm was limited to 133 mA/W by the small (0.24 μm) thickness of the absorbing layer. Nevertheless, these diodes showed a nearly flat behavior with reverse bias from short circuit to 3 V, denoting good transport properties in the junction.

3.6.4. Undulating SiGe Layers

A completely different approach was used by Xu et al. [145]. Based on the observation that dislocation-free Si/SiGe/Si strained structures grown in the three-dimensional Stransky–Krastanov regime have an undulating aspect with preferential Ge segregation on thickness maxima, they fabricated a photodiode with a Ge-rich absorption region and a low TD density. The local increase of Ge content and thickness, while reducing the effect of quantum confinement, results in a significantly lower bandgap compared to flat QWs [146]. Undulating MQWs consisting of $Si_{0.5}Ge_{0.5}$ layers of nominal 5 nm thickness sandwiched in 12.5 nm thick silicon barriers were the absorption layer in a MSM waveguide detector fabricated on the overlayer of a SOI substrate, and had a thickness of 2 μm, a width of 65 μm, and a length of 240 μm. They exhibited high responsivity of 1.6 and 0.1 A/W at 1.3 and 1.55 μm, respectively. The very large quantum efficiency, which exceeded 100% at 1.3 μm, the observed sublinear dependence of the responsivity on light intensity, and the large dark current density imply the role of photoconductive gain.

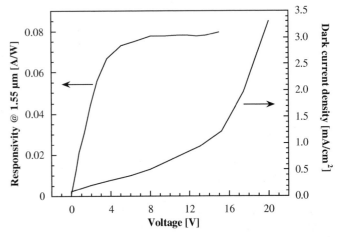

Fig. 44. Dark current density and responsivity at 1.55 μm of a waveguide detector integrated with a free carrier absorption modulator. Data from [147].

3.7. An Integrated Detector–Modulator

The first SiGe microsystem that integrated modulators and detectors on the same chip was demonstrated in 1998 by Li et al. [147]. The device consisted of a $Si_{0.96}Ge_{0.04}$ rib waveguide partially overlapping a $Si_{0.5}Ge_{0.5}$ MQW detector, similarly to that described in [121]. The waveguide section preceding the detector had a n^+-type cap to allow electron injection and light modulation by free carrier absorption. Measurements performed at 1.55 μm with a GaInAsP semiconductor laser demonstrated a maximum responsivity of 80 mA/W with an applied bias of 5 V and dark current of about 0.3 mA/cm^2 (see Fig. 44). This relatively high responsivity for a SiGe MQW-based detector was probably achieved thanks to the large device length (2 mm), despite the relevant insertion losses of the modulator (6.2 dB). Furthermore, by operating the optical switch, a modulation depth of 90% was achieved as measured directly by means of the integrated photodetector, with an injected current density of 6.6 kA/cm^2. The same group subsequently demonstrated a resonant cavity photodiode (RCP) optimized for normal incidence at 1.3 μm [148]. RCPs are detector embedded in an optical cavity resonant at a specific wavelength due to mirror selectivity. This increases the effective optical path length through repeated reflections (an example in silicon is reported in [149] for wavelengths below 1.1 μm). In the case of SiGe RCP's, the bottom mirror, which consisted of the Si/SiO$_2$ interface of a separation by implanted oxygen (SIMOX) substrate wafer, was not wavelength selective. Instead, the top mirror was fabricated by successive plasma enhanced chemical vapor deposition (PE CVD) of quarter wavelength Si/SiO$_2$ pairs of thicknesses 93 and 220 nm, respectively. The absorbing MQW region was grown via MBE at 600°C and consisted of 20 × 8 nm layers of Si$_{0.65}$Ge$_{0.35}$ each interleaved with 19 × 32 nm Si layers. The device responsivity was, expectedly, peaked at 1.28 μm with a value of 10.2 mA/W, which is about twice that obtained at wavelengths far from the peak. The dark current was ≈1 mA/cm^2.

3.8. Polycrystalline Ge Detectors

As pointed out in previous sections, the integration of NIR photodiodes with silicon electronics (e.g., VLSI) calls for harmonization of different and complex growth processes. In this scenario, low-temperature techniques are quite appealing because they do not impose modifications to the standard processing sequence. For instance, Ge-rich epitaxial alloys that require temperatures in the range 500–700°C, would force the Ge growth step to be placed after the MOS oxide realization and before the first metal deposition. Alternative techniques that allow the detector fabrication to be the last processing step, just before chip passivation, are, therefore, quite promising. In particular, polycrystalline Ge can be thermally evaporated at substrate temperatures below 300°C. Mesa poly-Ge/Si heterojunction diodes with photoresponse that extends to the NIR have been demonstrated and reported [150]. These devices exhibited a responsivity of 16 mA/W at 1.3 μm that was limited by the narrow diffusion length in such polycrystalline material and also demonstrated a fast (650 ps) pulse response at the same wavelength (Fig. 45). Diodes on n- and p-type

substrates demonstrated good rectifying characteristics: the heterojunctions were forward biased with Ge at positive (negative) voltages with respect to Si for n-type (p-type) substrates (see Fig. 46). Analyzing the reverse current dependence on bias and temperature allowed discrimination of two different current transport mechanisms in the diodes. In particular, although both currents evidenced a temperature activated process, heterojunctions grown on n-type substrates exhibited a flat reverse current independent of bias with an activation energy of 0.37 eV. On the contrary, the reverse current of diodes grown on p-type substrates slowly increased with the applied bias, following an exponential dependence on the fourth root of the voltage [151] (Fig. 47). Such dependence and the extrapolated barrier height (0.4 eV) are consistent with a model in which the current is limited by the barrier at the valence band discontinuity between Si and Ge [152]. Due to the presence of this barrier, no photocurrent was observed at photon energy below the silicon bandgap in devices grown on p-type substrates. The dark current density at room temperature for devices grown on n-type substrates was 2 mA/cm at 1 V. The current limitation of these devices (i.e., the relatively low responsivity) could be overcome in the future either by passivation of intergrain defects (by hydrogen treatments, for example) or by using very thin (100 nm) layers in

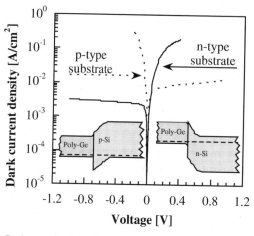

Fig. 46. Dark current voltage characteristics of poly-Ge/Si heterodiodes. The suggested band alignment at the heterointerface is reported for both cases. Data from [150].

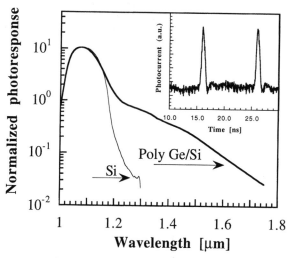

Fig. 45. Spectral response of a poly-Ge on Si detector compared with that of a Si device. The extension toward longer wavelength is evident. In the inset, the pulse response of the same poly-Ge on Si device is plotted. Data from [150].

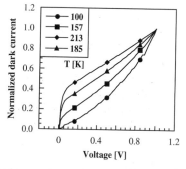

Fig. 47. Effect of temperature on the reverse bias characteristics of a poly-Ge/p-type Si diode. The curves are well fitted by a valence band barrier lowering model. Data from [150].

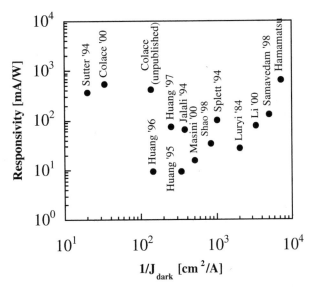

Fig. 48. Responsivity at 1.3 μm of a number of devices reported in the literature as a function of the inverse of the dark current density. A commercial (Hamamatsu) bulk Ge photodiode is reported for comparison. Data from various papers.

waveguide detectors. Preliminary results in this direction are reported in [153]. Finally, as described in the following section, the poly-Ge technology has proven to be very versatile in building functional devices.

3.9. Conclusions

This historical overview has shown the impressive progress in extension of the spectral coverage and the NIR sensitivity of SiGe-based photodiodes. In Figure 48, we have graphed the 1.3 μm responsivity versus inverse dark current density. Clearly, the quality of the detectors grown on silicon is approaching commercial (bulk) photodiodes (Hamamatsu in the top right corner) and, in our opinion, the first commercial devices soon will appear (for instance, in FTTHome systems). The opening of a market for such detectors presumably will boost widespread research in silicon-based optoelectronics.

4. FUNCTIONAL DEVICES

In this chapter we describe a number of SiGe on Si devices designed and fabricated to perform complex functions. In Section 4.1, we discuss a double SiGe SL diode that encompasses spectral tunability. Then we present the arrangement of several poly-Ge photodiodes in an array. Finally, in Section 4.3 we cover design, fabrication, and testing of a linear array with photodiodes of slightly differing spectral responsivities that is useful for wavelength identification of monochromatic light.

4.1. Voltage Tunable Detectors

Photodiodes with an adjustable spectral response have gained popularity recently for their inherent ability to simplify systems

for color image scanning [154] or environmental monitoring and imaging [155]. In optical communications, the extraction of a specific wavelength signal from a complex (broad or multichannel) spectrum is of great interest, in particular for long-distance wavelength division multiplexing (WDM) networks or secure links. As described in the following text, voltage tunable wavelength-selective photodetectors (VWP) can be used as basic components for encrypted transmissions or to recover information corrupted by noise.

Two back-to-back connected photodiodes that have different spectrally responses can form a device with voltage-controlled tunability. Such an example is the device presented in [156] and [157] that consists of a Si Schottky junction (sensitive to the visible) backconnected to a SiGe superlattice p-i-n diode (with sensitivity in the NIR). An intrinsic (semiinsulating) 400 μm silicon substrate was used to grow the SiGe sample. The buffer layer was a 650 nm thick undoped linearly graded layer of Ge_xSi_{1-x} followed by a 0.5 μm thick $x = 40\%$ SiGe undoped alloy. The absorbing region was a symmetrically strained superlattice consisting of 145 periods of n-doped Si_6Ge_4. Finally, a 2 nm n^+ Si layer capped the structure, as shown in Figure 49. Top and bottom electrodes were realized in aluminum, and the bottom electrode was windowed to let light into the structure. The two photodiodes, that is, the Al/Si Schottky barrier (top contact, Si diode) and the junction formed by intrinsic Si with the n-type SiGe superlattice (SiGe diode) resulted in the double backward diode sketched in Figure 49; the connecting resistor accounts for the i-Si bulk.

The spectral responsivity of the device can be evaluated as the weighted superposition of the two photodiode responses. Upon illumination, the photocurrent flow through the device is a combination of the two counterflowing contributions generated in the Si and SiGe diodes. Moreover, the different spectral responses and carrier-generation sites along the light path result in distinct photocurrent outputs in distinct wavelength regions. This is illustrated in Figure 50, where the light induced current is measured at various applied voltages using a lock-in technique. At $V_{bias} = 0.2$ V, the photocurrent is dominated by the contribution from the Si diode for photon energies (wavelengths) above (below) the Si bandgap, whereas it becomes negative when the Si contribution vanishes and the SiGe diode takes over. The figure also illustrates the dramatic effect of the bias on the spectral response: in particular, because a positive voltage corresponds to a forward (reverse) bias of the Si (SiGe) photodiode, the 1.2–1.3 μm response is greatly enhanced at bias greater than 0. On the contrary, a negative bias tends to suppress the SiGe contribution, resulting in a spectral response typical of a Si detector. As is apparent in Figure 50, in most cases, the outgoing photocurrent goes through a voltage-controlled zero.

4.1.1. Optical Decryption Based on the VWP

The tunability of the photoresponse and the presence of a sharp change of sign around the photocurrent null allow demonstration of deconvolution techniques for the NIR.

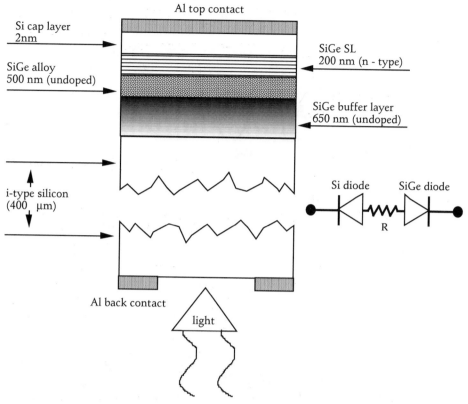

Fig. 49. Schematic of the double backward diode structure (left) and its electrical equivalent circuit (right).

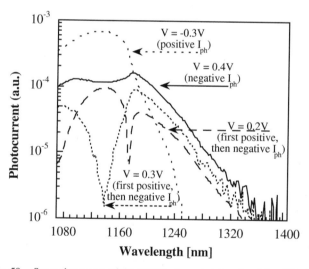

Fig. 50. Spectral response of the VWP in the near infrared and for different biases.

The decryption scheme that we describe in the following text, which was introduced in [157], permits optical communications in the presence of a level of signal-to-noise ratio that is too low for usual detectors. Let us assume that the transmitted signal S, modulating an optical carrier at wavelength λ_1, is intentionally jammed by an added disturbance $N(t)$, with a resulting $I_{out}(t) = S(t) + N(t)$ output at the receiver end. If the disturbance is known (i.e., it is purposely introduced

for encryption), signal recovery can be obtained by subtracting $N(t)$ from $I_{out}(t)$, making $N(t)$ the "key" to retrieve the information buried in $I_{out}(t)$. Such a key can be transmitted on a second carrier at λ_2, chosen in close proximity of λ_1 to disturb the heterodyne receiver and render filtering and separating the two wavelengths troublesome. However, the information can be recovered better by a detector with a photoresponse $I(\lambda_2)$ at λ_2 equal in amplitude, but opposite in sign to that at λ_1 (see Fig. 51). A suitable photodetector for such a task is the VWP described herein, where an exact balance $I(\lambda_2) = -I(\lambda_1)$ can be obtained at an appropriate bias. Although information recovery could—in principle—be obtained by using narrow-band filters with two single detectors and some additional electronics to linearly combine their outputs, there is no comparison in flexibility and compactness between these two schemes. Moreover, the voltage-controlled spectral response of the VWP can be exploited to precisely tune the decrypting receiver and/or in (most interesting) cases of time-varying encryption.

The proposed scheme was evaluated with the Si/SiGe VWP in the two-wavelength photocurrent experiment sketched in Figure 52. The light of wavelength $\lambda_1 = 1.3 \mu m$ (emitted by a semiconductor laser) was first modulated by an optical chopper at frequency f_1 (signal), and then spatially superimposed on and copropagated with a second beam at λ_2 (obtained by monochromator filtering of the light emitted by a tungsten lamp). A disturbance was applied to both light beams by chopping them at frequency f_2 close to f_1 before impinging the VWP. The photocurrent output from the latter and the noise am-

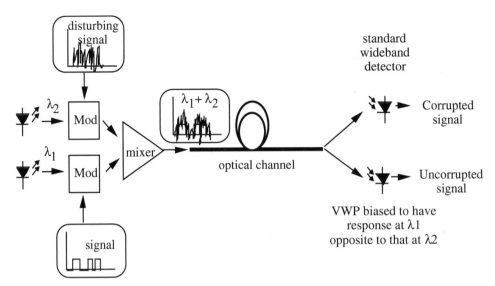

Fig. 51. Principle of operation of the encryption scheme using the VWP.

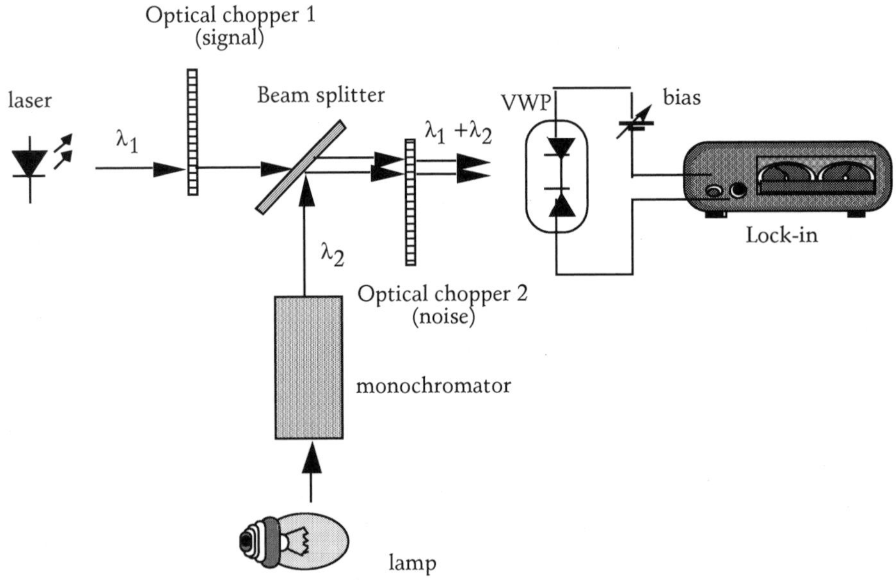

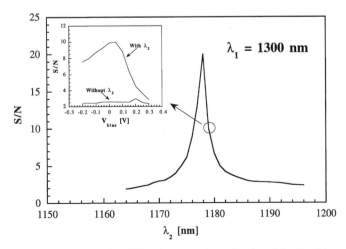

Fig. 52. Experimental setup used to validate the encryption scheme.

Fig. 53. Signal-to-noise (S/N) ratio measured as a function of the disturbing signal wavelength. In the inset, the effect of the bias applied to the VWP is evidenced.

plitude due to f_2 were measured by a lock-in amplifier locked in frequency and phase to f_1. Figure 53 shows the measured S/N versus wavelength λ_2. As predicted, the response strongly peaks where the photocurrents due to photons at λ_2 and λ_1, respectively, are equal in amplitude but opposite in sign. Far from that specific spectral location (or when light at λ_2 is suppressed as shown in the inset), the S/N reduces by a factor $\simeq 20$ with respect to its maximum. The effect of the bias applied to the VWP is evident in the inset.

4.2. Array of NIR Photodetectors

Linear and square arrays of NIR photodetectors are used in a wide range of applications: laser beam profiling, environmental monitoring, spectroscopic systems, and diagnosis of paintings [158] are just a few examples. In this framework, the

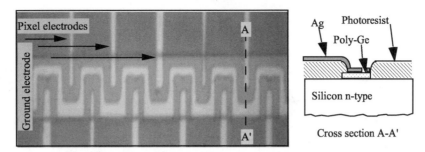

Fig. 54. Device schematic: (a) Detail of the 16-pixel photodetector array and (b) A–A' cross section.

possibility to integrate SiGe detector arrays with Si electronics is clearly quite appealing.

Here we describe a linear array of 16 photosensitive elements of polycrystalline Ge grown on Si using the low-temperature approach described in Section 3. The device layout and cross section are sketched in Figure 54. The equivalent circuit of each photoelement is a double backward Ge/Si heterojunction, despite the MSM geometry: this is due to the reduced thickness of the Ge layer compared to the interelectrode spacing. The multiple bends of the central ground electrode (visible in the micrograph in Fig. 54) define the 100 μm × 100 μm active pixel area. The whole array is fabricated on a mesa-etched Ge stripe and, outside the mesa, the metal lines are insulated from the substrate by a 2 μm thick photoresist layer, properly windowed on the sensitive Ge region. The device characteristics were acquired using a laser diode emitting at $\lambda = 1.3$ μm. At variance with the linear dependence of photoconductors, the photocurrent increased with applied bias, reaching saturation for voltages above 0.2–0.3 V and confirming the equivalent circuit previously mentioned. For incident powers in the 10^{-6}–10^{-3} W range, the photocurrent followed the light intensity with a linear trend and a maximum responsivity of 16 mA/W at $\lambda = 1.3$ μm (sample grown at 300°C).

The light beam induced current (LBIC) technique was employed to assess the quality of the process and to characterize the array in terms of photoresponse uniformity and pixel definition. The array was scanned by a 1.3 μm laser beam focused to a spot size of about 20 μm, whereas the generated photocurrent was recorded from the pixel under test, biased at 0.6 V. A typical LBIC map of a single pixel is reproduced in Figure 55. The spatial selectivity is apparent, with an abrupt response drop outside the perimeter of the sensitive pixel. A quantitative data analysis indicated a 10-fold signal reduction from the addressed pixel to the neighboring pixels. This guaranteed a low cross-talk between adjacent channels, which is of primary importance in most applications. By scanning all 16 pixels of the array, we could ascertain via LBIC a less than 2% fluctuation of the response from element to element.

The speed of the array was determined by recording the transient photocurrent due to a train of 100 ps laser pulses at $\lambda = 1.32$ μm. The device was biased at 3 V through a 50 Ω resistor and responded with a FWHM of about 2 ns, which is comparable to the reported value in MSM devices based on epitaxial Ge (see Section 3).

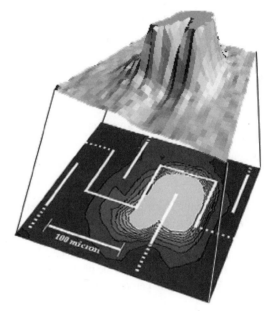

Fig. 55. LBIC mapping of one pixel of the array.

4.3. A NIR Wavemeter Integrated on Si

Determination of the wavelength of a monochromatic beam in the NIR is usually performed by complex systems like spectrum analyzers or wavemeters, which typical rely on the movement of mechanical parts (gratings, mirrors). Due to its intrinsic higher reliability, the monolithic integration of wavemeter functions on a single silicon chip is of great interest for applications such as the real time monitoring of the emission of tunable sources. One straightforward approach to realize this function is the use of an array of detectors with differently peaked spectral responses. The intrinsic responsivity that stems from optical and electrical properties of the photodetector semiconductor can be passively altered by narrow-band filtering of the incoming light with reflection gratings [159], waveguide phased array gratings [160], or Fabry–Perot cavities [161].

The detector array presented in [162] and more extensively in [163] employs a novel approach that is based on detecting a (wavelength dependent) interference pattern [164]. The operation of the device is illustrated in Figure 56. By placing a thin absorber within the standing-wave pattern generated by the light impinging the photodetector (forward propagating)

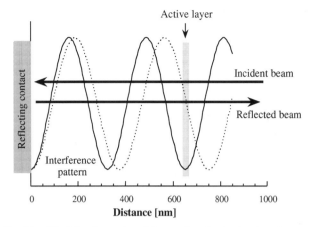

Fig. 56.   Principle of operation of the wavelength selective photodetector.

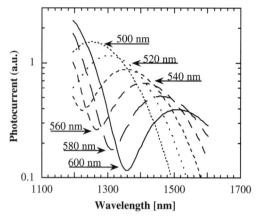

Fig. 57.   Calculated photocurrent spectra for different spacer layer thicknesses.

and reflected at the back metal contact (backward propagating), a strong spectral selectivity is obtained. Different wavelengths produce correspondingly different pattern amplitudes at the absorbing layer, which, in turn, gives rise to different current outputs. The light intensity at a distance x from the reflecting plane (left end in Fig. 56) is

$$I(x) = I_1 + I_2 - 2\sqrt{I_1 I_2} \cos(\varphi) \qquad (4.1)$$

where I_1 and I_2 are the (forward and backward) intensities and $\varphi = 2\pi x/\lambda$ is the relative phase between the interfering waves in x. The wavelength selectivity can then be maximized by careful design of the thickness of the spacer layer (i.e., the region between the sensitive layer and the backreflecting contact). From (4.1) the photocurrent can be easily cast in the form

$$J_{\text{ph}}(\lambda) \propto I(\lambda)\left(1 - e^{-\alpha t}\right) \qquad (4.2)$$

where I is the total light intensity, and α and t are the absorption coefficient and the thickness of the active layer, respectively. The second factor in (4.2) takes into account the absorption efficiency at wavelength λ. Notice that because I_1 is always greater than I_2 due to absorption in both the active and spacer layers, the photocurrent minima never reach zero. In Figure 57 we show the calculated spectral responses from (4.2) and (4.1) and various spacer thicknesses. A significant selectivity is predicted with spacer steps of 20 nm.

Based on the findings outlined in the preceding text, a six detector array was fabricated in poly-Ge on Si. The as-deposited poly-Ge film was thinned by successively dipping the wafer into a solution, yielding a 4 Å/s etch rate, to realize the various spacers (one per pixel). The final device is represented in Figure 58, whereas Figure 59 displays the measured photocurrent spectra for the six detectors: the agreement with Figure 57 is excellent.

To determine the wavelength of the impinging light, the discretized ($i = 1$–6) spectral response $M(i, \lambda)$ of the array elements can be taken as a vectorial base in the wavelength range 1200–1500 nm. Using this approach, the photocurrent J_{ph} from the ith detector is given by

$$J_{\text{ph}}^{i}(\lambda) = \sum_{\lambda} I(\lambda) M(i, \lambda) \qquad (4.3)$$

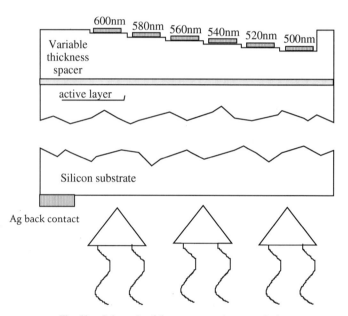

Fig. 58.   Schematic of the wavemeter (cross section).

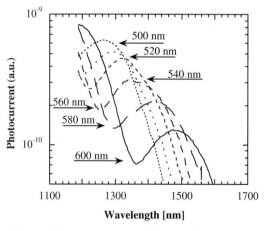

Fig. 59.   Measured photocurrent spectra of the photodiode array shown in Figure 58.

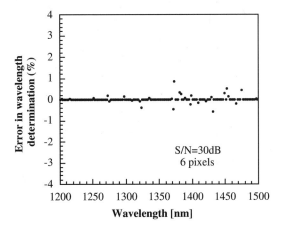

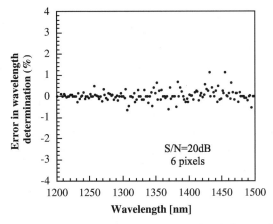

Fig. 60. Errors in determination of the input wavelength for two different signal-to-noise ratios: 30 dB (top) and 20 dB (bottom).

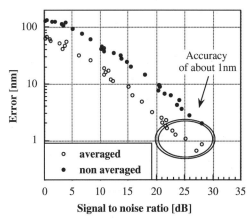

Fig. 61. Average error in determination of the wavelength as a function of the S/N.

over the whole spectral range, with larger S/N yielding more accurate results, although a few points are affected by the occasional occurrences of ill conditioning in the solution system. Finally, Figure 61 shows the error (in nanometers) versus the signal-to-noise ratio in the cases of nonaveraged (full circles) and averaged (open circles) data. The error is estimated as the mean value over the entire spectral range (1200–1500 nm). Remarkably, a good accuracy of 1 nm is reached in a wide interval of S/N values.

4.4. Conclusions

The examples illustrated in this section demonstrated the flexibility of the SiGe on Si technology for fabrication of functional devices capable of complex operation in the near infrared. These devices can find use in a wide range of applications from encryption of optical signals to wavelength detection.

APPENDIX: NUMERICAL SIMULATION OF RELAXED Ge/Si HETEROJUNCTIONS

In this Appendix, we deal with numerical simulations of Ge on Si heterojunctions, with particular emphasis on the effect of interface defects on the electrical properties of devices. As illustrated in Section 3, Ge on Si photodiodes are investigated for their high speed and efficiency over the entire NIR range. This motivated the search for a reliable tool that allows the determination of the photodiode performance based on structure.

A.1. Ge/Si Heterojunctions

Ge/Si heterojunction photodiodes can be numerically studied by solving the drift-diffusion equations

$$J_n = q\left(n\mu_n E + \frac{dn}{dx}D_n\right)$$

$$J_p = q\left(n\mu_p E - \frac{dp}{dx}D_p\right)$$

$$(A.1)$$

where $I(\lambda)$ is the light intensity at λ. Once a set of photocurrent values from each detector is known, the unknown spectrum $I(\lambda)$ can be resolved by inverting (4.3), However, due to the many involved unknowns, this simple procedure often yields incorrect results. If, however, an additional piece of information, such as the (nearly) monochromatic nature of the light, is introduced (i.e., the system is used as a wavemeter rather than as a spectrum analyzer), the results are more accurate. In addition, to minimize the effect of noise, a nonnegative least squares algorithm has to be used to invert (4.3). Whereas its outcome is not necessarily a single wavelength, the monochromaticity is only an aid to the precise wavelength determination by allowing, for example, a simple averaging of the results. The best attainable resolution is limited by the number of experimental points as well as by the signal-to-noise ratio in each measurement.

We tested the device in a series of experiments to evaluate its performance. Narrow-band light, generated through monochromator filtering of a tungsten lamp, was shone onto the array and the resulting photocurrent was collected from each pixel. The data, collected in 2 nm intervals, were the input to the nonnegative least squares algorithm. The results are plotted in Figure 60 (top and bottom) for two different S/N values. It can be seen that the wavelength determination is rather accurate

where the current densities ($J_{n,p}$) are given by superposition of the drift term, which is proportional to the electric field E and the free carrier density n (p) through the mobility. The diffusion term is linearly dependent on the free carrier gradient through the diffusion constant ($D_{n,p}$). When the Boltzmann approximation to the Fermi statistic functions holds (i.e., for nondegenerate semiconductors), the free carrier densities can be written in terms of the distance of the quasi-Fermi levels (E_{F_n} and E_{F_p}) from the conduction and valence band edges (E_c and E_v), and the effective density of states N_c and N_v:

$$n = N_c e_p \left(\frac{E_{F_n} - E_C}{kT} \right)$$

$$p = N_v e_p \left(\frac{E_V - E_{F_p}}{kT} \right) \tag{A.2}$$

Using the Einstein relationship $D/\mu = kT/q$, (A.1) can be recast in the compact form

$$J_n = \mu_n n \, \nabla E_{F_n}$$

$$J_p = \mu_p p \, \nabla E_{F_p} \tag{A.3}$$

(A.3) can be introduced in the continuity equations for electrons and holes, respectively,

$$\frac{\partial n}{\partial t} = \frac{\nabla \cdot J_n}{q} + G - R$$

$$\frac{\partial p}{\partial t} = \frac{\nabla \cdot J_p}{q} + G - R \tag{A.4}$$

and the latter can be solved in a finite difference scheme in conjunction with the Poisson equation and appropriate boundary conditions. This allows us to calculate the potential and free carrier distributions in the device.

The presence of a heterojunction is properly taken into account by (A.2) once suitable profiles for N_c, N_v, and the band edge positions are provided. The simulations then can be handled by software packages such as PC1D, which actually was used for the calculations in this chapter [165]. The results are satisfactory whenever quantum confinement effects at the heterojunctions are negligible.

A.2. Electric Equivalent of Relaxed Ge/Si Heterointerfaces

The results that follow were obtained by taking a uniform doping density of 10^{16} cm^{-3} both in the Si substrate and in the epitaxial Ge, either n- or p-type depending on the case. We conveniently name the diodes after their epi/substrate doping combination. Thus, for example, n-p is the diode formed by n-type Ge and p-type Si substrate. Two parameters represent the quality of the Ge epilayer and the Ge/Si interface: the minority carrier lifetime (τ) and the surface velocity (S). The role of S and τ on the reverse dark current density J_d of the diodes (both a sensitive parameter and a significant figure of

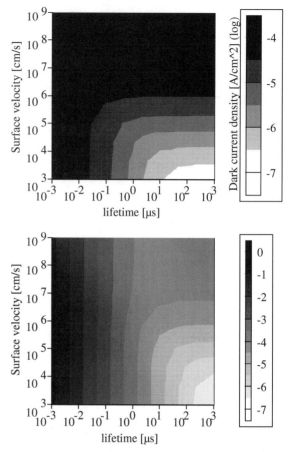

Fig. 62. Dark current density at 1 V reverse bias as a function of the free carrier lifetimes in epitaxial Ge film and the surface velocity at the Ge/Si heterojunction. Top, p-type Ge on n-type Si; bottom, n-type Ge on p-type Si.

quality) is illustrated in Figure 62. for p-n and n-p diodes, respectively. Clearly, in p-n diodes, J_d is nearly independent on both S and τ in a wide range, due to a current flow mainly controlled by the barriers at the heterointerface. Conversely, in n-p diodes J_d is strictly correlated to the generation in the Ge layer and at the interface. This correlation has important consequences on the choice of the best device structure. First, in p-n diodes, J_d cannot be reduced by improving material quality (increasing τ) as is the case for n-p diodes. Second, because the gathering of carriers photogenerated in the Ge epilayer occurs through the same mechanism as that of thermally generated carriers, n-p diodes are preferable over p-n for the absence of barriers at the heterointerface, suggesting the possibility of a complete photocarrier collection at low applied voltage.

In the simulation, we used minority carrier lifetimes of 1 ms and 1 μs in Si and Ge, respectively. These values were calculated from [72] by assuming a dislocation density of about 10^6 cm^{-2}, which is close to the best reported to date for an epitaxial Ge on Si [36]. The energy gap and electronic affinities of Si (Ge) were 1.12 (0.66) and 4.05 (4.13) eV, respectively. When dealing with illumination, light was supposed to shine on the Si side in order to properly study the contribution of the Ge layer at photon energies below the Si bandgap.

As underlined in Section 2, the heterointerface between a silicon substrate and a relaxed germanium layer epitaxially grown on it is rich in defects. These defects originate during epilayer growth and are caused by the 4% lattice mismatch between the two crystals. An estimate of the minimum defect density for a certain lattice mismatch can be performed on the basis of the difference ΔN_s in bond densities on the free semiconductor surfaces. For the Ge/Si (100) case, this is approximately 8% of the surface bond density; thus, $\Delta N_s \approx 5 \times 10^{13}$ cm^{-2}.

Defects from lattice mismatch often take the form of isolated dislocations, which, in turn, introduce deep electronic states in the bandgap [166]; in the case of germanium, they are found to be acceptor-like [167]. Moreover, due to their position close to mid gap, dislocation induced states are active also as recombination centers. The defect density can be assumed to be spatially peaked at the heterointerface because dislocations tend to annihilate as the epitaxial film thickens. The latter is especially true when the buffering techniques illustrated in Section 2 are employed.

On the basis of the foregoing considerations, the charge contribution of lattice mismatch defects was modeled as a distribution of acceptors, with density 5×10^{19} cm^{-3}, extending for 10 nm from the interface (i.e., total defects per unit area $= 5 \times 10^{13}$). The reduction of minority carrier lifetime (due to an increased probability of recombination at the defects) was taken into account by a surface (or interface) velocity $S = \sigma v_{\text{th}} \Delta N_s \approx 10^6$ cm/s.

A.3. Band Diagrams, Current–Voltage, and Wavelength Responses

Band diagrams of Ge/Si heterojunctions are shown in Figure 63 for various doping combinations. The role of defects at the interface is apparent. In particular, for n-type silicon, the defects give rise to a double backward junction regardless of the epilayer doping. Moreover, the heterojunction built-in potential mainly drops on the Ge side for n-p and n-n diodes. Therefore, these diodes are expected to behave better with photon energies below the Si bandgap. The band alignments presented in Figure 63 play an important role in the short circuit spectral responses shown in Figure 64. As described in Section 4 for voltage tunable wavelength-selective photodetectors, the n-n diode gives a photocurrent that changes sign in different spectral regions. In the p-n diode, instead, the smaller potential drop on the Ge side of the junction is not sufficient to change the photocurrent sign. In this case, the carriers photogenerated in Ge diffuse toward the substrate, where they are accelerated and collected by a much higher field. The spectral response is, thus, extended beyond that of silicon, but no sign inversion is visible. The most favorable situation concerns the n-p diode, where the band bending on the two sides of the heterojunction is concurrent and the large valence band discontinuity is partially

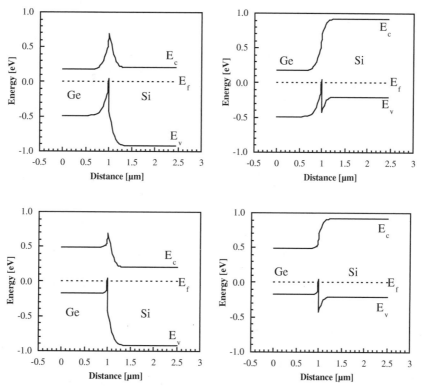

Fig. 63. Band diagrams of Ge/Si heterojunctions for different doping types of both Ge and Si. Note the effect of the delta-like acceptors used to model the defects at the interface.

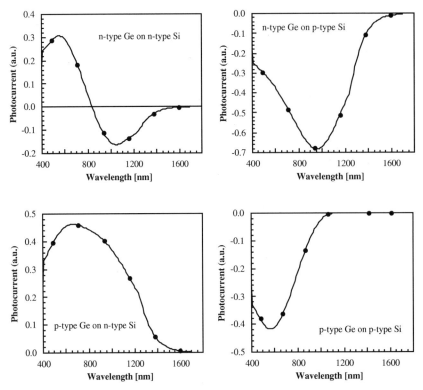

Fig. 64.   Spectral responses of the Ge/Si photodiodes.

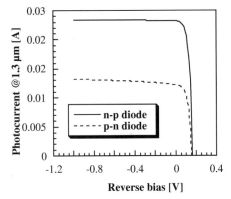

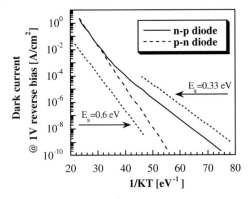

Fig. 65.   Effect of applied reverse bias on the collection efficiency. In the simulation, a light source with 100 mW/cm² intensity at 1.3 μm shining on the silicon side of the junction was used.

Fig. 66.   Dependence of the reverse dark current of n-p and p-n diodes on temperature. Thermal generation in the Ge epilayer ($E_a \approx 0.33$ eV) is clearly evidenced as the dominant mechanism at room temperature in n-p diodes.

compensated by the built-in drop in Ge. Finally, the presence of large barriers for both electrons (electrostatic) and holes (electrostatic plus the large valence band discontinuity) in the p-p diode prevents the carriers photogenerated in Ge from flowing through the heterojunction and contributing to the photocurrent. This mechanism explains the negligible response predicted at wavelengths where the thin (10 μm) silicon substrate becomes transparent.

Together with the 1.3 μm photocurrent dependence on reverse bias shown in Figure 65, these numerical findings indicate (and/or confirm) better performances of the n-p diode compared to other combinations.

Finally, in Figure 66 we plot the calculated dependence of dark reverse current on temperature. The activation energy of 0.33 eV (half of the Ge bandgap) evidenced by the n-p diode characteristic at room temperature and below demonstrates the importance of thermal generation in the epilayer to the dark current. At higher temperature, the slope gets much steeper due to the increasing generation contribution in the silicon substrate.

In conclusion, using a finite difference scheme based on drift-diffusion equations and including a model of the interface, the simulation of heterojunction diodes allows selection of the best substrate–epilayer doping to optimize the NIR collection efficiency.

REFERENCES

1. H. S. Hinton, *IEEE J. Selected Topics Quantum Electron.* 21, 14 (1996).
2. J. Gates, D. Muehlner, M. Cappuzzo, M. Fishteyn, L.Gomez, G. Henein, E. Laskowsky, I. Rayazansky, J. Simulovich, D. Syversten, and A. White, "Proceedings of the 48th IEEE Conference on Electronic Components and Technology," 1998, pp. 551–559.
3. F. Delpiano, B. Bostica, M.Burzio, P. Pellegrino, and L. Pesando, "Proceedings of the 49th IEEE Conference on Electronic Components and Technology," 1999, pp. 759–762.
4. U. Gosele, H. Stenzel, M. Reiche, T. Martini, H. Steinkirchner, and Q. Y. Tong, *Solid State Phenomena* 47–48, 33 (1996).
5. Z. H. Zhu, F. E. Ejeckam, Y. Qian, J. Zhang, Z. Zhang, G. L. Christenson, and Y. H. Lo, *IEEE J. Selected Topics Quantum Electron.* 3, 927 (1997).
6. C. P. Lee, S. Margalit, and A. Yariv, *Appl. Phys. Lett.* 32, 806 (1978).
7. M. Dagenais, R. F. Leheny, H. Temkin, and P. Bhattacharya, *J. Lightwave Technol.* 8, 846 (1990).
8. R. H. Walden, "Proceedings of the Symposium on Gallium Arsenide Integrated Circuits," 1996, pp. 255–257.
9. M. Sugo, H. Mori, Y. Sakai, and Y. Itoh, *Appl. Phys. Lett.* 60, 472 (1992).
10. A. M. Joshi, F. J. Effenberger, M. Grieco, G. H. Feng, W. Zhong, and J. Ott, *Proc. SPIE* 2290, 430 (1994).
11. R. A. Soref, *Proc. IEEE* 81, 1687 (1993).
12. S. Coffa, G. Franzó, and F. Priolo, *Mater. Res. Bull.* 23, 25 (1998).
13. Q. Mi, X. Xiao, J. C. Sturm, L. C. Lenchyshyn, and M. L. Thewalt, *Appl. Phys. Lett.* 60, 3177 (1992).
14. J. Engvall, J. Olajos, H. G. Grimmeis, H. Presting, H. Kibbel, and E. Kasper, *Appl. Phys. Lett.* 63, 491 (1993).
15. A. Rickman, G. T. Reed, B. L. Weiss, and F. Namavar, *IEEE Photon Technol. Lett.* 4, 633 (1992).
16. S. P. Pogossian, L. Vescan, and A. Vonsovici, *Appl. Phys. Lett.* 75, 1440 (1999).
17. B. Li, G. Li, E. Liu, Z. Jiang, C. Pei, and X. Wang, *Appl. Phys. Lett.* 75, 1 (1999).
18. G. Cocorullo, M. Iodice, and I. Rendina, *Opt. Lett.* 19, 420–422 (1994).
19. L. Esaki, *IEEE J. Quantum Electron.* QE-22, 1611–1624 (1986).
20. W. Shockley, U.S. Patent 2.569.347, 1951.
21. H. Kroemer, *Proc. IRE* 45, 1535 (1957).
22. E. Kasper, H. J. Herzog, and H. Kibbel, *Appl. Phys. Lett.* 7, 199 (1975).
23. E. Kasper and F. Schaffler, *Semicond. Semimetals* 33, 223 (1991).
24. J. C. Bean, T. T. Sheng, L. C. Feldman, A. T. Fiory, and R. T. Lynch, *Appl. Phys. Lett.* 44, 102 (1984).
25. B.S. Meyerson, *Appl. Phys. Lett.* 48, 797 (1986).
26. S. C. Jain and W. Hayes, *Semicond. Sci. Technol.* 6, 547 (1991).
27. T. P. Pearsall, *Prog. Quantum Electron.* 18, 97 (1994).
28. F. C. Frank and J. H. Van der Merwe, *Proc. R. Soc. London* 198A, 205 (1949).
29. J. H. Van der Merwe, *J. Appl. Phys.* 34, 117 (1963).
30. J. W. Mattews and A. E. Blakeslee, *J. Cryst. Growth* 27, 118 (1974).
31. S. C. Jain, J. R. Willis, and R. Bullough, *Adv. Phys.* 39, 127 (1990).
32. F. K. LeGoues, *Mater. Res. Bull.* 21, 38 (1998).
33. E. Bauer and H. Poppa, *Thin Solid Films* 12, 167 (1974).
34. R. People, *IEEE J. Quantum Electron.* 22, 1696 (1986).
35. S. M. Sze, "Physics of Semiconductor Devices," 2nd ed., p. 750. Wiley, New York, 1981.
36. R. Potter, "Handbook of Optical Constant of Solids," p. 465. Academic Press, New York, 1985.
37. R. Braunstein, A. R. Moore, and F. Herman, *Phys. Rev.* 109, 695 (1958).
38. E. Kasper and F. Schaffler, *Semicond. Semimetals* 33, 140 (1991).
39. F. K. LeGoues, B. S. Meyerson, J. F. Morar, and P. D. Kirchner, *J. Appl. Phys.* 71, 4230 (1992).
40. B. Y. Tsaur, J. C. C. Fan, and R. P. Gale, *Appl. Phys. Lett.* 38, 176 (1981).
41. B. Y. Tsaur, M. W. Geis, J. C. C. Fan, and R. P. Gale, *Appl. Phys. Lett.* 38, 779 (1981).
42. T. F. Kuech and M. Maenpaa, *Appl. Phys. Lett.* 39, 245 (1981).
43. M. Maenpaa, T. F. Kuech, M. A. Nicolet, and D. K. Sadana, *J. Appl. Phys.* 53, 1076 (1982).
44. M. Garozzo, G. Conte, F. Evangelisti, and G. Vitali, *Appl. Phys. Lett.* 41, 1070 (1982).
45. Y. Ohmachi, T. Nishioka, and Y. Shinoda, *J. Appl. Phys.* 54, 5466 (1983).
46. J. M. Baribeau, T. E. Jackman, D. C. Houghton, P. Maigne, and M. W. Denhoff, *J. Appl. Phys.* 63, 5738 (1988).
47. K. Fujinaga, *J. Vac. Sci. Technol. B* 9, 1511 (1991).
48. B. Cunningham, J. O. Chu, and S. Akbar, *Appl. Phys. Lett.* 59, 3574 (1991).
49. E. Kasper, *Surf. Sci.* 174, 630 (1986).
50. G. Heigl, G. Span, and E. Kasper, *Thin Solid Films* 222, 184 (1992).
51. J. M. Baribeau, T. E. Jackman, P. Maigne, D. C. Houghton, and M. W. Denhoff, *J. Vac. Sci. Technol. A* 5, 1898 (1987).
52. E. A. Fitzgerald, Y. H. Xie, M. L. Green, D. Brasen, A. R. Kortan, J. Michel, Y. J. Mii, and B. E. Weir, *Appl. Phys. Lett.* 59, 811 (1991).
53. Y. H. Xie, E. A. Fitzgerald, P. J. Silverman, A. R. Kortan, and B. E. Weir, *Mater. Sci. Eng.* B14, 332 (1992).
54. G. P. Watson, E. A. Fitzgerald, Y. H. Xie, and D. Monroe, *J. Appl. Phys.* 75, 263 (1994).
55. M. T. Currie, S. B. Samavedam, T. A. Langdo, C. W. Leitz, and E. A. Fitzgerald, *Appl. Phys. Lett.* 72, 1718 (1998).
56. P. M. Mooney, J. L. Jordan-Sweet, K. Ismail, J. O. Chu, R. M. Feenstra, and F. K. LeGoues, *Appl. Phys. Lett.* 67, 2373 (1995).
57. H. Presting and H. Kibbel, *Thin Solid Films* 222, 215 (1992).
58. W. Jager, D. Stenkamp, P. Ehrhart, K. Leifer, W. Sybertz, H. Kibbel, H. Presting, and E. Kasper, *Thin Solid Films* 222, 221 (1992).
59. M. Copel, M. C. Reuter, E. Kaxiras, and R. M. Tromp, *Phys. Rev. Lett.* 63, 632 (1989).
60. M. Jiang, X. Zhou, B. Li, and P. Cao, *Phys. Rev. B* 60, 8171 (1999).
61. K. Sakamoto, K. Miki, T. Sakamoto, H. Yamaguchi, H. Oyanagi, H. Matsuhata, and K. Kyoya, *Thin Solid Films* 222, 112 (1992).
62. J. L. Liu, C. D. Moore, G. D. U'Ren, Y. H. Luo, Y. Lu, G. Jin, S. G. Thomas, M. S. Goorsky, and K. L. Wang, *Appl. Phys. Lett.* 75, 1586 (1999).
63. M. Copel and R. M. Tromp, *Appl. Phys. Lett.* 58, 2648 (1991).
64. A. Sakai, T. Tatsumi, and K. Aoyama, *Appl. Phys. Lett.* 71, 3510 (1997).
65. A. Sakai and T. Tatsumi, *Appl. Phys. Lett.* 64, 52 (1994).
66. D. Reinking, M. Kammler, M. Horn-von-Hoegen, and K. R. Hofmann, *Appl. Phys. Lett.* 71, 924 (1997).
67. K. Eberl, S. S. Iyer, S. Zollner, J. C. Tsang, and F. K. LeGoues, *Appl. Phys. Lett.* 60, 3033 (1992).
68. J. Mi, P. Warren, P. Letourneau, M. Judelewicz, M. Gailhanou, M. Dutoit, C. Dubois, and J. C. Dupuy, *Appl. Phys. Lett.* 67, 259 (1995).
69. A. St. Amour, C. W. Liu, J. C. Sturm, Y. Lacroix, and M. L. W. Thewalt, *Appl. Phys. Lett.* 67, 3915 (1995).
70. H. J. Osten and E. Bugiel, *Appl. Phys. Lett.* 70, 2813 (1997).
71. R. A. Soref, *J. Vac. Sci. Technol. A* 14, 913 (1996).
72. S. Luryi and E. Suhir, *Appl. Phys. Lett.* 49, 140 (1986).
73. T. Bryskiewicz, *Appl. Phys. Lett.* 66, 1237 (1995).
74. F. Banhart and A. Gutjahr, *J. Appl. Phys.* 80, 6223 (1996).
75. B. Srikanth, B. Samavedam, and E. A. Fitzgerald, *J. Appl. Phys.* 81, 3108 (1997).
76. T. A. Langdo, C. W. Leitz, M. T. Currie, E. A. Fitzgerald, A. Lochtfeld, and D. A. Antoniadis, *Appl. Phys. Lett.* 76, 3700 (2000).
77. I. Daruka, J. Tersoff, and A. L. Barabasi, *Phys. Rev. Lett.* 82, 2753, 2756 (1999).
78. B. Cunningham, J. O. Chu, and S. Akbar, *Appl. Phys. Lett.* 59, 3574 (1991).
79. L. Di Gaspare, G. Capellini, E. Palange, F. Evangelisti, L. Colace, G. Masini, F. Galluzzi, and G. Assanto, 24th International Conference of Physics of Semiconductors, Jerusalem, 2–7 August 1998.
80. H. Chen, L. W. Guo, Q. Cui, Q. Hu, Q. Huang, and J. M. Zhou, *J. Appl. Phys.* 79, 1167 (1996).
81. J. H. Li, C. S. Peng, Y. Wu, D. Y. Dai, J. M. Zhou, and Z. H. Mai, *Appl. Phys. Lett.* 71, 3132 (1997).
82. K. K. Linder, F. C. Zhang, J. S. Rieh, P. Bhattacharya, and D. Houghton, *Appl. Phys. Lett.* 70, 3224 (1997).

83. C. S. Peng, Z. Y. Zhao, H. Chen, J. H. Li, Y. K. Li, L. L. W. Guo, D. Y. Dai, Q. Huang, J. M. Zhou, Y. H. Zhang, T. T. Sheng, and C. H. Tung, *Appl. Phys. Lett.* 72, 3160 (1998).

84. L. Colace, G. Masini, F. Galluzzi, G. Assanto, G. Capellini, L. Di Gaspare, and F. Evangelisti, *Solid State Phenomena* 54, 55 (1997).

85. V. Chu, J. P. Conde, D. S. Shen, and S. Wagner, *Appl. Phys. Lett.* 55, 262 (1989).

86. M. Neuberger, "Handbook of Electronic Materials." IFI/Plenum, New York, 1971.

87. H.-C. Luan, D. R. Lim, K. K. Lee, K. M. Chen, J. G. Sandland, K. Wada, and L. C. Kimerling, *Appl. Phys. Lett.* 75, 2909 (1999).

88. E. Kasper, Ed., "Properties of Strained and Relaxed Silicon Germanium," Vol. 12. INSPEC, The Institution of Electrical Engineers, London, 1995.

89. H.-C. Luan, D. R. Lim, L. Colace, G. Masini, G. Assanto, K. Wada, and L. C. Kimerling, "Proceedings of the Material Research Society Fall Meeting," Boston, 1999.

90. S. M. Sze, "Physics of Semiconductor Devices," p. 851. Wiley, New York, 1981.

91. G. K. Wertheim and G. L. Pearson, *Phys. Rev.* 107, 694 (1957).

92. L. M. Giovane, H. C. Luan, E. A. Fitzgerald, and L. C. Kimerling, "Proceedings of the Materials Research Society Fall Meeting," Boston, 1999.

93. T. J. King, J. R. Pfiester, J. D. Shott, J. P. Mc Vittie, and K. C. Saraswat, *Proc. IEDM* 253 (1990).

94. T. J. King and K. C. Saraswat, *Proc. IEDM* 567 (1991).

95. R. Venkatasubramanian, D. P. Malta, M. L. Timmons, J. B. Posthill, J. A. Hutchby, R. Ahrenkiel, B. Keyes, and T. Wangensteen, "Proceedings of 1st WCPEC," December 1994, p. 1692.

96. L. Colace, G. Masini, F. Galluzzi, and G. Assanto, "Proceedings of the Materials Research Society Fall Meeting," Boston, 1998.

97. F. Evangelisti, M. Garozzo, and G. Conte, *J. Appl. Phys.* 53, 7390 (1982).

98. H. Kobayashi, N. Inoue, T. Uchida, and Y. Yasuoka, *Thin Solid Films* 300, 138 (1997).

99. R. K. Ahrenkiel, S. P. Ahrenkiel, M. M. Al-Lassim, and R. Venkatasubramanian, "Proceedings of the 26th PVSC," Anaheim, CA, 1997, p. 527.

100. S. M. Sze, "Physics of Semiconductor Sevices," 2nd ed., p. 291. Wiley, New York, 1981.

101. J. P. Donnely and A. G. Milnes, *Solid-State Electron.* 9, 174 (1966).

102. W. G. Oldham, A. R. Riben, D. L. Feucht, and A. G. Milnes, *J. Electrochem. Soc.* 110, 53c (1963). This is only the abstract of a work presented at the Electrochemical Society Meeting held in Pittsburgh in 1963.

103. S. Luryi, A. Kastalsky, and J. C. Bean, *IEEE Trans. Electron. Dev.* ED-31, 1135 (1984).

104. A. Kastalsky, S. Luryi, J. C. Bean, and T. T. Sheng, "Proceedings of the Electrochemical Society," 1985, PV85-7, p. 406.

105. H. Temkin, T. P. Pearsall, J. C. Bean, R. A. Logan, and S. Luryi, *Appl. Phys. Lett.* 48, 963 (1986).

106. J. C. Bean, L. C. Feldman, A. T. Fiory, S. Nakahara, and I. K. Robinson, *J. Vac. Sci. Technol. A.* 2, 436 (1984).

107. J. C. Bean, "Proceedings of the 1st International Symposium on Si Molecular Beam Epitaxy" (J. C. Bean, Ed.), p. 339. Electrochemical Society, Pennington, NJ, 1985.

108. See, for example, A. Yariv, "Optical Electronics," p. 432. Saunders, Philadelphia, 1971.

109. F. Capasso, "Lightwave Communication Technology" (W. T. Tsang, Ed.). Semiconductors and Semimetal Series. Academic Press, New York, 1985.

110. J. C. Campbell, A. G. Dentai, W. S. Holden, and B. L. Kasper, *Electron. Lett.* 19, 818 (1983).

111. S. Luryi, T. P. Pearsall, H. Temkin, and J. C. Bean, *IEEE Electron Device Lett.* 7, 104 (1986).

112. T. P. Pearsall, H. Temkin, J. C. Bean, and S. Luryi, *IEEE Electron Device Lett.* 7, 330 (1986).

113. H. Temkin, A. Antreasyan, N. A. Olsson, T. P. Pearsall, and J. C. Bean, *Appl. Phys. Lett.* 49, 809 (1986).

114. S. Forrest, *J. Lightwave Technol.* LT-3, 347 (1985).

115. H. Temkin, J. C. Bean, T. P. Pearsall, N. A. Olsson, and D. V. Lang, *Appl. Phys. Lett.* 49, 155 (1986).

116. T. P. Pearsall, J. Bevk, L. C. Feldman, J. M. Bonar, and J. P. Mannaerts, *Phys. Rev. Lett.* 58, 729 (1987).

117. R. People and S. A. Jackson, *Phys. Rev. B* 36, 1310 (1987).

118. T. P. Pearsall, E. A. Beam, H. Temkin, and J. C. Bean, *Electron. Lett.* 24, 685 (1988).

119. L. T. Canham, *Appl. Phys. Lett.* 57, 1046 (1990).

120. A. Splett, B. Schuppert, K. Petermann, E. Kasper, H. Kibbel, and H. J. Herzog, *Dig. Conf. Integrated Photonic Res.* 10, 122 (1992); B. Jalali, L. Naval, A. F. J. Levi, and P. Watson, *SPIE Proc.* 1802, 94 (1992); V. P. Kesan, P. G. May, G. V. Treyz, E. Bassous, S. S. Iyer, and J.-M. Halbout, *Mater. Res. Soc. Symp. Proc.* 220, 483 (1991).

121. A. Splett, T. Zinke, K. Petermann, E. Kasper, H. Kibbel, H.-J. Herzog, and H. Presting, *IEEE Photon Techn. Lett.* 6, 59 (1994).

122. S. M. Sze, "Physics of Semiconductor Devices," p. 89. Wiley Interscience, New York, 1981.

123. B. Jalali, L. Naval, and A. F. J. Levi, *J. Lightwave Technol.* 12, 930 (1994).

124. P. Sutter, U. Kafader, and H. von Känel, *Solar Energy Mater. Solar Cells* 31, 541 (1994).

125. S. K. Ghandhi, "VLSI Fabrication Principles," p. 647. Wiley Interscience, New York, 1994.

126. F. Y. Huang, X. Zhu, M. O. Tanner, and K. L. Wang, *Appl. Phys. Lett.* 67, 566 (1995).

127. F. Y. Huang and K. L. Wang, *Appl. Phys. Lett.* 69, 2330 (1996).

128. J. Lee, A. L. Gutierrez-Aitken, S. H. Li, and P. K. Battacharya, *IEEE Trans. Electron Devices* 43, 977 (1996).

129. R. J. McIntyre, *IEEE Trans. Electron Devices* ED-13, 164 (1966).

130. L. Naval, B. Jalali, L. Gomelsky, and J. M. Liu, *J. Lightwave Technol.* 14, 787 (1996).

131. B. W. Dodson and J. Y. Tsao, "Proceedings of the Second International Symposium on Silicon Molecular Beam Epitaxy" (J. C. Bean and L. J. Scholwater, Eds.), pp. 105–113. Electrochemical Society, Pennington, NJ, 1988.

132. R. People and J. C. Bean, *Appl. Phys. Lett.* 47, 322 (1985).

133. F. Y. Huang, K. Sakamoto, K. L. Wang, P. Trinh, and B. Jalali, *IEEE Photon Technol. Lett.* 9, 229 (1997).

134. T. Tashiro, T. Tatsumi, M. Sugiyama, T. Hashimoto, and T. Morikawa, *IEEE Trans. Electron Devices* 44, 545 (1997).

135. K. Aketagawa, T. Tatsumi, and J. Sakai, *J. Cryst. Growth* 111, 860 (1991).

136. T. Tatsumi, K. Aketagawa, and J. Sakai, *J. Cryst. Growth* 120, 275 (1992).

137. X. Shao, S. L. Rommel, B. A. Orner, H. Feng, M. W. Dashiell, R. T. Troeger, J. Kolodzey, P. R. Berger, and T. Laursen, *Appl. Phys. Lett.* 72, 1860 (1998).

138. M. Todd, J. Kouvetakis, and D. Smith, *Appl. Phys. Lett.* 68, 2407 (1996).

139. L. Colace, G. Masini, F. Galluzzi, G. Assanto, G. Capellini, L. Di Gaspare, E. Palange, and F. Evangelisti, *Appl. Phys. Lett.* 72, 3175 (1998).

140. L. Colace, G. Masini, F. Galluzzi, G. Assanto, G. Capellini, L. Di Gaspare, E. Palange, and F. Evangelisti, *Mater. Res. Soc. Symp. Proc.* 486, 193 (1998).

141. G. Masini, L. Colace, G. Assanto, H.-C. Luan, K. Wada, and L. C. Kimerling, *Electron. Lett.* 35, 1 (1999).

142. L. Colace, G. Masini, G. Assanto, H.-C. Luan, K. Wada, and L. C. Kimerling, *Appl. Phys. Lett.* 76, 1231 (2000).

143. G. Masini, L. Colace, G. Assanto, H.-C. Luan, K. Wada, and L. C. Kimerling, unpublished data.

144. S. B. Samavedam, M. T. Currie, T. A. Langdo, and E. A. Fitzgerald, *Appl. Phys. Lett.* 73, 2125 (1998).

145. D.-X. Xu, S. Janz, H. Lafontaine, and M. R. T. Pearson, *Proc. SPIE* 3630, 50 (1999).

146. H. Lafontaine, N. L. Rowell, and S. Janz, *Appl. Phys. Lett.* 72, 2430 (1998).

147. B. Li, G. Li, E. Liu, Z. Jiang, J. Qin, and X. Wang, *Appl. Phys. Lett.* 73, 3504 (1998).

148. C. Li, Q. Yang, H. Wang, J. Zhu, L. Luo, J. J. Yu, Q. Wang, Y. Li, J. Zhou, and C. Lin, *Appl. Phys. Lett.* 77, 157 (2000).

149. V. S. Sinnis, M. Seto, G. W. Hooft, Y. Watanabe, A. P. Morrison, W. Hoekstra, and W. B. deBoer, *Appl. Phys. Lett.* 74, 1203 (1999).

150. G. Masini, L. Colace, and G. Assanto, *Mater. Sci. Eng. B* 69, 257 (2000).

151. R. S. Muller and T. I. Kamins, "Device Electronics for Integrated Circuits," p. 139. Wiley, New York, 1986.

152. L. Di Gasapare, G. Capellini, C. Chudoba, M. Sebastiani, and F. Evangelisti, *Appl. Surf. Sci.* 104–105, 595 (1996).

153. G. Masini, L. Colace, and G. Assanto, presented at the European Material Research Society 2000, Strasbourg, France.

154. D. Caputo, G. deCesare, F. Irrera, F. Lemmi, G. Masini, F. Palma, and M. Tucci, *Solid State Phenomena* 44–46, 943 (1995).

155. Y. Zhang, D. S. Jiang, and W. K. Ge, *Appl. Phys. Lett.* 68, 2114 (1996).

156. G. Masini, L. Colace, G. Assanto, and T. P. Pearsall, *J. Vac. Sci. Technol. B* 16, 2619 (1998).

157. G. Masini, L. Colace, F. Galluzzi, G. Assanto, T. P. Pearsall, and H. Presting, *Appl. Phys. Lett.* 70, 3194 (1997).

158. Available at http://www.sensorsinc.com/989001a.htm.

159. J. B. D. Soole, A. Scherer, H. P. Leblanc, N. C. Andreadakis, R. Bath, and M. A. Koza, *Appl. Phys. Lett.* 58, 1949 (1991).

160. C. Dragone, C. A. Edwards, and R. C. Kistler, *IEEE Photon. Technol. Lett.* 3, 896 (1991).

161. U. Prank, M. Mikulla, and W. Kowalsky, *Appl. Phys. Lett.* 62, 129 (1993).

162. G. Masini, L. Colace, and G. Assanto, *Electron. Lett.* 35, 1549 (1999).

163. G. Masini, L. Colace, and G. Assanto, Near Infrared Wavemeter Based on an Array of Polycrystalline Ge on Si Photodetectors, Photonics West, San Jose, CA, 2000.

164. D. A. B. Miller, *IEEE J. Quantum Electron.* 30, 732 (1994).

165. P. A. Basore and D. A. Clugston, "Proceedings of the 25th IEEE Photovoltaic Specialists Conference," Washington, DC, May 1996.

166. W. Bardsley, *Progr. Semiconductors* 4, 155 (1960).

167. W. T. Read, Jr., *Phylos. Mag.* 45, 775 (1954).

Chapter 8

PHYSICAL PROPERTIES OF AMORPHOUS GALLIUM ARSENIDE

Roberto Murri, Nicola Pinto

Department of Mathematics and Physics, INFM and University of Camerino, Via Madonna delle Carceri, 62032 Camerino, Italy

Contents

1. INTRODUCTION

The physical properties of crystalline III–V compounds with zinc blende structure are well known [1]. The situation of tetrahedral coordinated amorphous semiconductors is much more complicated. They can be easily investigated, but large structural variations can be found with the sample history. This makes the analysis of the experimental data difficult and the comparison among data collected on different samples not simple. The deposition conditions and the possible annealing processes influence physical properties of the material. The amorphous compounds contain defects, which alter the den-sity of states. They can be dangling bonds or bonds between like atoms, i.e., wrong bonds. When the atoms involved remain tetrahedral bonded, it is possible to consider two types of like bonds generated by an arsenic atom on a gallium site or vice versa. If the deposition conditions lead to isolated like bonds, we can consider each of them separately. Thus, arsenic like bonds act as deep donors and gallium like bonds act as deep acceptors. The bonding in III–V is largely ionic and thus chemical disorder produces large changes through the Coulomb interaction. The degree of tolerable chemical disorder depends on the compound ionicity, which determines the energy cost of a wrong bond compared to the normal heteropolar bond.

Handbook of Thin Film Materials, edited by H.S. Nalwa
Volume 4: Semiconductor and Superconductor Thin Films

ISBN 0-12-512912-2/$35.00

Both structural (dangling-bond type) and chemical (wrong-bond type) defects in amorphous III–V compounds introduce states in the mobility gap and/or at the band edges. Configurational disorder refers to variations in angle bond and bond lengths from values characteristic of the crystal.

Additional problems come from the necessity to control the film composition. Two desiderated properties of the film are in contradiction: amorphicity and stoichiometry.

At stoichiometry, a continuous random network (CRN) best describes III–V materials with even-membered rings only. The gap density of states is very similar to that of the crystalline compound, with an addition of states between the upper two peaks of the valence band. All this is a result of the loss of long-range order and not of the presence of As—As bonds. Wrong bonds seem not present and this can be due to the fact that Ga and As have almost equal atomic radii.

The chapter presented in Section 2 is the most general properties of material deposited with suitable growth parameters. Section 3 reviews composition, structural, and morphological properties, considering also theoretical results, while Section 4 is devoted to theoretical and experimental aspects of the density of states. Sections 5 and 6 present optical results and phonon spectra, respectively. Finally, Section 7 discusses the electrical transport properties and Section 8 discusses applications and devices.

We used a probably unusual structure of the chapter: in fact, we have chosen to present practically a "card" for each quoted article, giving the main experimental data, as an immediate reference.

2. DEPOSITION AND GROWTH PARAMETERS

2.1. Dc and Rf Sputtering

Alimoussa, Carchano, and Thomas [2] deposited a-GaAs:H films, controlling the partial pressure of hydrogen, p_{H_2}. The films were amorphous when substrate temperature, T_S, was lower than 30 °C, while crystallization started at about 350 °C. The hydrogen content and its depth profile were determined by nuclear reactions. The hydrogen content, C_{H_2}, starts from $\approx 40\%$ when p_{H_2} is $\approx 10\%$ of the total pressure, p_T, and saturates to an average value of $\approx 77\%$ when $p_{H_2} > 40\%$ of p_T. The introduction of hydrogen, reducing the number of ionized argon atoms, decreases the deposition rate, r_d, from 3 Å s^{-1} ($p_{H_2} = 0$) to about 1 Å s^{-1}, when $p_{H_2} = 40\%$ of p_T.

Thomas, Fallavier, Carchano, and Alimoussa [3] measured the composition and the hydrogen content in amorphous hydrogenated samples of GaAs. The substrate temperature was kept nearly constant to about 20 °C with a power density, W^*, of 0.8 W cm^{-2} and an argon partial pressure, p_{Ar}, of about 8 mTorr. The ratio H$_2$/(Ar + H$_2$) was finely adjusted to be 0, 1, 5, 10, 20, and 40%; the total thickness, d, was ranging between 600 and 700 nm. The nuclear reaction ^{1}H (^{15}N, $\alpha\gamma$) allowed measuring the hydrogen depth profile up to about 3.5 nm from the GaAs surface and a sensitivity better than 1 at.%. An artifact has been evidenced: the accumulation of hydrogen at the

substrate interface. By comparison, with results from Rutherford backscattering spectroscopy (RBS) of α particles at 8 MeV, they interpret this fact as due to a significant hydrogen distribution onto the glass substrate. The hydrogen content, C_{H_2}, can be expressed with respect to either the number of As or the number of Ga atom [4],

$$C_{H, As} = (M_{As}S_{As} + C^* M_{Ga} S_{Ga}) \frac{K_{St} N}{(6.023 \cdot 10^{17} - K_{St} N M_H S_H)}$$

where $C_{H, As}$ is the hydrogen content relatively to As; C^* is the at.% Ga/at.% As; M_{As}, M_{Ga} is the atomic mass of arsenic and gallium; S_{As}, S_{Ga}, S_H is the stopping power of arsenic, gallium, and hydrogen; N is the gamma-ray rate; $K_{St} = (C_{H_{St}}/N_{St}S_{St})$ is the normalization factor. All the quantities are thus referred to a standard of known hydrogen concentration, $C_{H_{St}}$.

The result evidences that hydrogen content increases rapidly as a function of the hydrogen partial pressure up to a saturation value, related to the argon pressure and the power density. The stoichiometry is largely shifted toward a high content of gallium (deficit of arsenic) for low p_{H_2}, the parity being only obtained at the highest values of C_{H_2}, i.e., when hydrogen content saturates. The stoichiometry ratio, Ga/As, shifts to 1.18–1.2 for $C_{H_2} \approx 5$–7 at.% and then it decreases to 1 when C_{H_2} saturates to about 45%. The density, ρ, of the films begins to increase from 4.8 g cm^{-3} ($C_{H_2} \approx 5$–7%) reaching a maximum, near the crystal value 5.2 g cm^{-3}, for $C_{H_2} \approx 12$–22.5% and then regularly decreases up to 4.6 g cm^{-3} when $C_{H_2} \approx 45\%$. All the most prominent features of the physical properties take place at low partial pressure of hydrogen, in agreement with results of [5]. All this happens before hydrogen saturation in the material and can be due to an initial loss of hydrogenated arsenic species.

Carchano, Lalande, and Loussier [6] studied the arsenic concentration in polycrystalline GaAs samples by electron beam excited X-rays spectroscopy. They were able to find a correlation between the arsenic concentration and the shift of the X-ray diffraction peak. Moreover, the minimum shift of the peak was found in samples with an arsenic concentration equal to that measured on electronic quality monocrystals of stoichiometric GaAs. They found the influence of each deposition parameter on the arsenic content, showing that the substrate temperature and the argon pressure are the most important parameters. The percentage shift of the X-ray diffraction peak (Fig. 1) tends to its minimum value of about 0–0.2% when the arsenic content, C_{As}, in the film is near 47 at.%. All the crystalline plane families show the same trend, even if with remarkable quantitative differences among the curves of different indices. By comparison with a measure on a monocrystalline stoichiometric sample, they were able to define that the resolution limit of the technique is affected by systematic error of (−3 at.%). In this way, the minimum shift of the peak is found in stoichiometric samples. The arsenic atomic concentration shows an excess of As for T_S lower than about 500 °C, when stoichiometry is reached. The atomic concentration of arsenic decreases from about 56% when $T_S \approx 300$ °C to about 49% when $T_S \approx 600$ °C. The decrease of the atomic C_{As} reflects the fact that As is more volatile

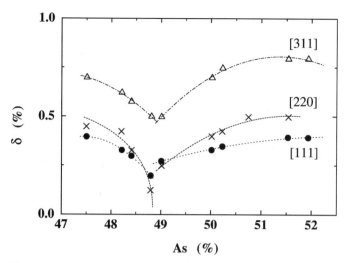

Fig. 1. Shift of the quoted X-ray diffraction peaks as a function of the As content in the films. (After [32], reprinted with permission by Editions de Physique.)

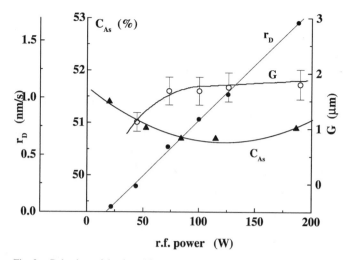

Fig. 3. Behaviors of the deposition rate, r_d; of the arsenic concentration, C_{As} and of the grain size, G, as a function of rf power. (After [7], reprinted with permission by Elsevier.)

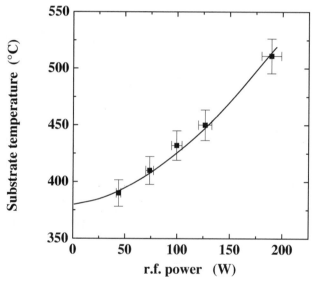

Fig. 2. Increase of the initial substrate temperature, $T_S = 380\ °C$, as a function of rf power. (After [7], reprinted with permission by Elsevier.)

than gallium. Dependencies were found for the atomic C_{As} as a function of the interelectrode distance, argon pressure, and sputtering power. In all cases, the dependence seems to be relatively fair.

Carchano, Lalande, and Loussier [7] studied the influence of some of the deposition parameters on the structure and the morphology of polycrystalline a-GaAs, grown on molybdenum foils. The substrate temperature, T_S, the radio frequency (rf) power, W, and the argon pressure, p_{Ar}, were kept all constant to 480 °C, 65 W, and 30 mTorr, respectively, to study the dependence on the interelectrode distance, D. In this way, they were able to evidence that the deposition rate, r_d, is constant (~ 0.8 nm s^{-1}) if D ranges from 20 to 30 mm, then decreases monotonically to 0.25 nm s^{-1} for $30 < D < 55$ mm. The As content shows a stoichiometric minimum for $D = 30$ mm, then

C_{As} tends to about 51.5% for the highest value of D. The dependence on the gas pressure was studied with $T_S = 400\ °C$, $W = 65$ W, and $D = 45$ mm. r_d increases rapidly while C_{As} decreases when $p_{Ar} < 40$ mTorr; after this value, r_d tends to saturate while the arsenic content continues its decrease, even if slower. It is interesting to consider the grain dimensions, G. It is nearly constant ($G \sim 1$ nm) for $p_{Ar} < 40$ mTorr, where r_d increases rapidly. Then, it grows slower, attaining a final value greater of more than 30%. The change of the argon pressure strongly influences the stoichiometry of the deposited films more the deposition rate. The dependencies on the third parameter, W, were considered, with $p_{Ar} = 30$ mTorr, $D = 35$ mm, and $T_S = 380\ °C$. In this case, it must be stressed the possible substrate temperature increase, due to energy transfer from the discharge to substrate itself. The authors show that an applied power of 200 W, could produce a temperature increase of more than 120–130 °C, with the initial substrate temperature equal to 380 °C (Fig. 2). In this case, the depositions rate, r_d, increases linearly with W from about 0.05 nm s^{-1} (20 W) to 1.7 nm s^{-1} (180 W). C_{As} varies around 51% for $48 < W < 190$ W, with a minimum between 100 and 150 W (Fig. 3). The grain size increases rapidly for $48 < W < 100$ W, then remains constant at about 2 nm. All these dependencies evidence the complexity of the deposition process and the difficulty to finely control the properties of the material.

Bandet et al. [8] and Aguir et al. [9] deposited a-GaAs films with a sputtering modified process: cathodic sputtering of gallium target and arsine (AsH_3) decomposition in rf glow discharge. The composition of the deposited films was determined by the ratio of the partial pressure of arsine to argon partial pressure in the gaseous phase. Arsine was first diluted in hydrogen, then argon was found to be more convenient. Microdiffractometry checked that as-deposited films were amorphous and secondary ion mass spectrometry (SIMS) that carbon and oxygen were present into the solid film. Weak photoluminescence signals were sometimes detected in gallium-rich samples. The optical constants were determined by classical spectrophoto-metric measurements in the range 450–2500 nm. The spectral absorption coefficient, α, shows the expected behavior for amorphous materials, being much higher in arsine-hydrogen than in arsine-argon samples. In near stoichiometric films, the Tauc's gap, E_g, tends to increase with the deposition temperature. It shows a nearly constant value around 1.4 eV if plotted as a function of the gallium content, up to $C_{Ga} \approx 40\%$. It then correctly decreases rapidly with C_{Ga}, assuming the value 1.1 eV for stoichiometric samples and decreasing up to 0.5 eV when $C_{Ga} \approx 70\%$. Aguir et al. [9] performed the plasma monitoring in the range 220–300 nm, to identify the intermediate chemical species with their evolution during the deposition process and then the film composition. The intrinsic difficulty of this deposition process is to keep the Ga/As ratio constant, by maintaining equilibrium between the gallium sputtering and the arsenic decomposition. The dc electrical conductivity was measured at room temperature after annealing at 373 K, as a function of the gallium content in the films. The conductivity increases rapidly for more than 9 orders of magnitude, going from As to Ga excess in a relatively narrow range (from 40 to 80% of Ga). The quality of alloys was improved by increasing the deposition temperature to 393 K: near stoichiometry, the conductivity decreases from about 10^{-5} $(\Omega\,cm)^{-1}$ to 10^{-7} $(\Omega\,cm)^{-1}$. The activation energy of the conductivity, E_σ, can be improved increasing the substrate temperature and reducing the pollution of the reactor. The statistical distribution of the conductivity prefactor, σ_0, shows a sharp minimum to 1 $(\Omega\,cm)^{-1}$ for nearly stoichiometric samples, with lateral values attaining 10^4–10^5 $(\Omega\,cm)^{-1}$. E_g remains constant for Ga content less than 40% and decreases before stoichiometry is reached, with a relatively strong influence of the substrate temperature. The optical absorption coefficient, α, was not modified by plasma hydrogenation for films containing gallium excess. However, for stoichiometric alloys and those with arsenic excess, α, decreases faster on the low-energy side and becomes weaker when argon is used as a diluent of arsine rather than hydrogen. The increase of the conduction, as well as the optical absorption, with the hydrogen partial pressure can be explained by the preferential fixation of hydrogen on gallium atoms. This fact induces defects in the band tails and a deficiency of gallium atoms in the alloy matrix. Consequently, defects (dangling bonds or wrong bonds) related to the arsenic species are created. The excess of As improves the electrical and optical properties of the films. An opposite result is obtained with an excess of gallium.

Casamassima et al. [10] deposited unhydrogenated a-GaAs films by rf sputtering at room temperature, only varying the power, W, in the range 100–400 W. Thickness of the samples varied from 0.5 to 1.2 μm and E_g ranges from 0.85 to 0.95 eV. They found a linear dependence of the measured deposition rate on W, from 1.4 Å s^{-1} to about 4.5 Å s^{-1}, in agreement with [2, 11]. RBS allowed measuring the composition of the films, using as a standard a stoichiometric sample grown by molecular beam epitaxy (MBE). The results were also compared to data obtained by the energy dispersive system (EDS). In samples deposited at $W = 400$ W, the ratio (As/Ga) = 1.04, while

it increases up to 1.11, when $W = 100$ W. The electrical conductivity, σ, as a function of T shows a typical hopping regime below 250 K. It is evidenced by the linear dependence of σ vs $T^{-1/4}$, in the range 140 K $< T <$ 250 K. In all samples, at $T > 250$ K, E_σ, varies between 0.33 and 0.40 eV. From space charge limited current (SCLC) measurements, the authors were able to compute the distribution of the density of states at the Fermi level as $N(E_F) = 6$–$7 \cdot 10^{16}$ eV^{-1} cm^{-3}. The same quantity is in the range 9×10^{17}–2×10^{18} eV^{-1} cm^{-3}, if computed from the Mott relations for hopping conduction, a well-known situation of overvaluation.

Carbone et al. [12] measured several physical properties of hydrogenated a-GaAs, deposited varying p_{H_2}, p_{Ar}, T_S with W constant to 200 W. The surface composition of the films was checked by RBS, (Fig. 4) while elastic recoil detection (ERD) allowed for an evaluation of the H content (Fig. 5). The samples generally show an As-deficient stoichiometry (Fig. 4) when deposited at the highest Ar partial pressure, that also limits the inclusion of H in the network. Classical optical absorption was

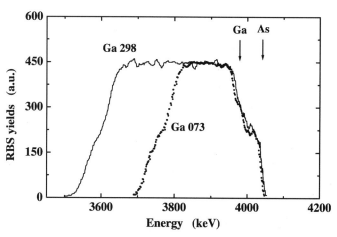

Fig. 4. RBS spectra of elemental composition for two a-GaAs samples: — Ga 298 ($T_S = 200$ °C, $p_{H_2} = 0.46$ Pa, $p_{Ar} = 0.80$ Pa, thickness = 0.90 nm); - - - - Ga 073 ($T_S = 134$ °C; $p_{H_2} = 0.075$ Pa, $p_{Ar} = 5$ Pa, thickness = 0.83 nm). (After [12], reprinted with permission by Società Italiana di Fisica.)

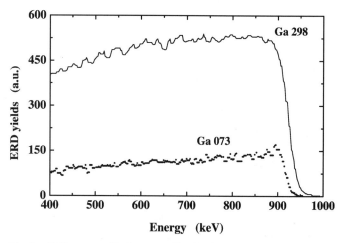

Fig. 5. ERD spectra of hydrogen profiles for the same samples of Figure 4. (After [12], reprinted with permission by Società Italiana di Fisica.)

performed to compute the optical constants. E_g varies with hydrogen, showing a rapid increase when H percentage inside the plasma is less than 5%, starting from 0.97 eV ($C_{H_2} = 0$) and growing up to a saturation value of 1.5–1.6 eV (5 at.% < C_{H_2} < 35 at.%). The index of refraction, n, at $\lambda = 1\ \mu$m is of the order of 3.2 for unhydrogenated samples, lower than the value 3.8 obtained with flash evaporated films [13]. It decreases to 2.4–2.5 for samples with the highest hydrogen content. In unhydrogenated samples, the energy gap was found to be 0.8–0.9 eV and RBS measurements gave Ga excess. For As-rich samples, [14] quotes an energy gap is about 1.1 eV: comparison of results is not straightforward, because As influences valence-band states while Ga influences conductance-band states.

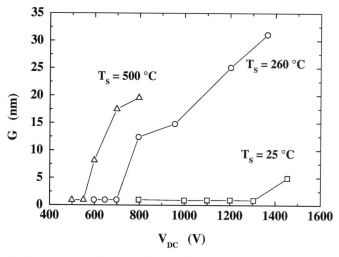

Fig. 6. Evolution of the crystallite size, G, vs the self-bias voltage V_{dc}, in GaAs films deposited on: glass substrates, $T_S = 25\ ^\circ$C ($\square$) and 260 $^\circ$C ($\bigcirc$), Mo substrates, $T_S = 500\ ^\circ$C ($\triangle$). The lines have been drawn only to guide the eye. (After [15], reprinted with permission by Elsevier.)

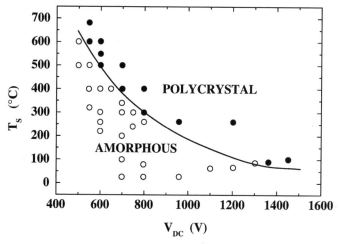

Fig. 7. Diagram showing the approximate frontier between the amorphous and crystalline area vs the self-bias voltage, V_{dc}, and the substrate temperature, T_S, used during the deposition of GaAs films on Mo substrates. Each circle represents the deposition of the GaAs film. (After [15], reprinted with permission by Elsevier.)

Seguin et al. [15] deposited on a crystalline or amorphous substrate at temperatures as high as 600 $^\circ$C by controlling the deposition parameters and, in particular, the dc self-bias voltage, V_{dc}. A strong dependence was found for the crystallite size, G, vs V_{dc}, when deposition is made on glass. At low temperature ($T_S = 25\ ^\circ$C), V_{dc} can attain $\approx$1300 V, (Fig. 6), before G starts to increase. When $T_S = 260\ ^\circ$C, V_{dc} reduces at $\approx$700 V, and reduces to 550 V if $T_S = 500\ ^\circ$C. Depositing on polycrystalline Mo substrates, amorphous films can be obtained with T_S as high as 600 $^\circ$C, but using a very low self-bias voltage ($V_{dc} \approx 500$ V). The relation between T_S and V_{dc} is now inverted. The phase diagram shows a roughly decreasing exponential line separating the amorphous and polycrystalline regions (Fig. 7): increasing T_S, V_{dc} should be decreased and vice versa. T_S and V_{dc} both influence the structure of the film: in fact, the deposition rate, r_d, increases rapidly with V_{dc} and decreases slightly with T_S. Measurements of the chemical composition of the films evidence that they contain an excess of As or Ga, according if T_S is lower or higher than 400 $^\circ$C, respectively. The possibility to grow at relatively high T_S, but with low self-bias voltage, can be attributed to the balance of high-energy sputtered atoms, which enhance the reevaporation of Ga and As adatoms and reflected neutrals. The possibility to deposit at high temperature can be interesting for device application of a-GaAs.

2.2. Flash Evaporation

Gheorghiu et al. [11] deposited amorphous films of GaAs, GaP, and GaSb and determined their composition. They found, according with other authors, that crucible temperature and the powder grain size are the crucial parameters for sample composition. Reproducible results under given evaporation conditions were obtained for GaSb and, to a lesser extent, for GaAs and, much more difficult, for GaP. Nearly stoichiometric amorphous films were deposited for GaAs and GaP with powder grain size in the range 160–250 μm, a crucible temperature of the order of 1700 $^\circ$C, and deposition rates in the range 0.1–0.2 nm s^{-1}. Increasing the grain powder size and decreasing the crucible temperature, an excess of the group V element was always found. No systematic dependence was found from substrate temperature.

Dixmier, Gheorghiu, and Theye [16] investigated the structure of thin (200–1000 Å) amorphous III–V compounds deposited at room temperature by electron diffraction. All the samples are nearly stoichiometric as determined by electron microprobe and α-particle backscattering. Microdensitometry was performed on photographs of diffraction patterns. They conclude that for nearly stoichiometric samples, the continuous random-network model of Connel–Temkin with even-membered rings only, that avoids the introduction of wrong bonds, describes the real structure of the amorphous materials, more adequately than other models. For nonstoichiometric films, in particular for a-GaSb samples, the excess atoms are incorporated at random with threefold coordination, into a tetra-coordinated amorphous GaSb little perturbed network. The typical fourfold configuration seems to exist only for samples very close to stoichiometry.

Dias da Silva et al. [17] studied the effect of deviation from stoichiometry in a-$Ga_{1-x}Sb_x$ films, preparing stoichiometric and Sb-rich samples. Many characterization techniques (optical spectroscopy, X-ray diffraction (XRD), electrical conductivity, Raman scattering, IR, electron energy loss (EEL), Auger) were used to study the physical properties of this material. Annealing of near stoichiometric films, evidences two different effects: an ordering of the amorphous matrix at T_A (lower than crystallization) and formation of GaSb and Sb crystallites at higher temperatures.

2.3. Other Methods of Deposition

Segui, Carrere, and Bui [18] used a glow-discharge process to deposit hydrogenated a-GaAs by decomposition of arsine (AsH_3) and trimethylgallium ($Ga(CH_3)_3$) or TMG, selected for its room temperature vapor pressure, allowing a relatively high growth rate.

Chen, Yang, and Wu [19] deposited films of a-GaAs by plasma enhanced chemical transport deposition (PECTD), starting from a mixture of arsine and trimethylgallium [(TMG) $Ga(CH_3)_3$]. They were able to obtain films with a smooth surface. The chlorine content in the material can change the optical energy gap. They found that PECTD grown samples are Garich as opposed to sputtered or evaporated films, which are As-rich. The optical gap ranges between 1.17 and 1.34 eV, following the increase of the chlorine contents from 4.2 to 9 at.%. The conductivity prefactor and the electron drift mobility have values $\sigma_0 = 1.5$–2×10^2 (Ω cm)$^{-1}$ and $\mu_e = 10^{-2}$–10^{-3} cm^2 V^{-1} s^{-1}, respectively.

Maury et al. [20] and Sahli, Segui, and Maury [21] using metal-organic chemical vapor deposition (MOCVD) grew $(GaAs)_{1-x}(SiC_2:H)_x$ thin films. They found that these materials have an inhomogeneous structure consisting of c-GaAs microcrystallites, amorphous GaAs, and amorphous Si_yC_{1-y}:H. The optical gap, E_g, of the films depends slightly on the film composition for $x < 0.32$ (being x the Si at.%), with values (0.87–1.09 eV) similar to those found in unhydrogenated a-GaAs. The amorphous GaAs phase is always present even for $x = 0$ and is responsible for the optical absorption edge. In the same x range, the refractive index at 3 nm is constant to $n \approx 3.0$, and then decreases abruptly to 2.2 when $x > 0.3$. The photoluminescence involves both the c-GaAs and the a-Si_yC_{1-y}:H phases. The electrical transport is controlled by the crystalline or amorphous part of GaAs, according to the volume fraction of a-Si_yC_{1-y}:H. The $x = 0.22$ value seems to be a transition point for the transport mechanisms. Below it, GaAs microcrystallites determine the transport paths, while, above it, the a-GaAs connective tissue limits the conductivity. This is sensitive to an increase in the defect density and/or relative volume of this disordered zone. The room temperature conductivity varies from 2.7×10^{-9} (Ω cm)$^{-1}$ when $x = 0.15$, to 1.3×10^{-6} (Ω cm)$^{-1}$, if $x = 0.29$. All the experimental results can be explained in the mainframe of a multiphase microstructure model. We can consider disordered and microcrystalline regions and their evolution changing the deposition parameters like the composition, x. The ac response of polycrystalline GaAs [21], deposited by using the same technique, allowed concluding that ac transport is controlled by grain boundaries, through the density of interface states.

Monnom et al. [22] obtained microcrystalline GaAs films in an MBE system at $T_S = 200$ °C under various flux ratios ($r = $ As/Ga varies from 0.3 to 15). X-ray diffraction, IR spectra, EDAX, and electrical conductivity were used to characterize the films. Mixed phases are generally obtained over the whole range of r, with As or Ga excess when $r > 1$ or $r < 1$, respectively. Around $r = 1$, near stoichiometric films are obtained and, film structure evolves from an amorphous to a microcrystalline structure. The room temperature resistivity changes by more than 5 orders of magnitude when r goes from about 0 (Ga-rich) to As-rich samples ($r = 2$). A plateau is found around stoichiometry ($r = 1$) and then conductivity saturates to about 5×10^4 (Ω cm). The activation energy of the resistivity resulted to be ≈ 0.52 eV.

Fujikoshi et al. [23] used different methods to fabricate doped a-GaP:H films for junction diodes. The hydrogen-reactive evaporation (RE) and sputtering (SP) in hydrogen are applied to deposit thin films, doped with zinc and sulfur. a-GaP:H is considered a useful material for optoelectronic devices working in the spectral region from green to blue. In fact, the room temperature energy gap of crystalline GaP is 2.26 eV and that of a-GaP:H should be higher. The conductivity of a-GaP:H films, made by RE and SP methods, increases drastically by doping. Rectification properties of the junction diodes present a complicated situation according to the various possible combinations of two active layers, their sequence and substrates used to deposit the diode. The same situation is found for photovoltaic effect and for the photoconductivity.

Yokoyama et al. [24] studied the solid-phase crystallization processes of MBE deposited amorphous GaAs at $T_S = 30$ °C. Different crystallization kinetics is active, according to different annealing temperature, T_A, range. When T_A is lower than 300 °C, the a-GaAs crystallization proceeds by solid-phase epitaxy; on the contrary, if T_A is greater than 400 °C, the a-GaAs changes to a single crystal through a polycrystal. The crystallization rate is strongly dependent on stoichiometry in the low T_A range, but less at higher temperatures.

Tsuji et al. [25] studied the pressure-induced amorphization in c-GaAs and c-GaP, from the quenched high-pressure phase below the transition pressure, p_t. X-ray diffraction of GaAs and GaP has been measured at high pressure (up to 27 GPa) and has been measured at low temperature (down 90 K). In GaAs, amorphization from the quenched high-pressure phase occurs with increasing temperature from 140 to 260 K at $p = 5$ GPa, and from 170 to 270 K if $p = 8$ GPa. The quenching of the orthorhombic (instead of zinc blende) high-pressure phase can be obtained releasing pressure at low temperature. The amorphous diffraction peak is four times wider than the typical diffraction peak of a crystalline phase. A pressure (p)–temperature (T)

phase diagram evidences that amorphization can take place in two different temperature regions: 160 and 220 K, in both cases decreasing p from 10 GPa. Similar results were obtained in GaP, even if with different temperatures and pressures: decompression from 27 GPa at constant temperature at 90 and 130 K. The transition pressure decreases with decreasing temperature. As the amorphization is the function both of the height of the potential barriers ΔU and on their pressure dependence, it is possible to discuss the results using a configuration-coordinate model. In GaAs and GaP, amorphization occurs when ΔU of the amorphous state is lower than that of the zinc blende phase. The ionic character of the bonding lowers ΔU and narrows the region of the amorphization.

Makadsi [26] used a combination of thermal and flash evaporation to deposit a-GaAs:H and a-Al$_x$Ga$_{1-x}$As:H. The films can be prepared with stoichiometric composition. The highest optical gap and the lowest Urbach's energy are attained for $x \approx 30\%$. This value of x seems to be the critical one for the physical properties of the material. In fact, the carrier concentration, n_C, attains its minimum value, while the Hall mobility, μ_H, and the activation energy of conductivity present their maximum. a-Al$_x$Ga$_{1-x}$As:H films with $x \leq 40$ and annealing temperature $T_A < 473$ K, are amorphous. The dc conductivity increases with Al content, while the activation energy increases from 0.6 to 0.8 eV when x goes from 0 to 30% and then decreases rapidly.

3. COMPOSITION, STRUCTURAL, AND MORPHOLOGICAL PROPERTIES

3.1. Theoretical Results

Molteni, Colombo, and Miglio [27] applied a tight-binding molecular-dynamics quenching of a liquid from 1600 to 300 K to calculate the structural, vibrational, and electronic properties of a-GaAs. They were able to calculate an average coordination number ≈ 4.09, in agreement with experimental results. The amount of threefold-coordinated sites (15–20%), larger than in a-Si (4–5%), is consistent with the structural model of [28], where this coordination is indicated as more stable in a-III–V than in a-Si. The fraction of wrong bonds is calculated to be $\approx 13\%$, a relatively low concentration. Electronic and vibrational density of state (DOS) were computed. The main features of the computed electronic DOS are in agreement with experimental X-ray photoelectron spectroscopy (XPS) data and vibrational spectrum. Both bond length and bond angles are locally distorted and the corresponding frequency of the stretching and bending modes are likely to be affected.

Tripathy and Sahu [29] computed the changes of the diamagnetic susceptibility, χ. This is an intrinsic electronic property of a material, related to the chemical bonding and their possible variations in the amorphous phase compared to the crystalline one. The authors assume that the continuous random network of Polk can adequately describe amorphous semiconductors. Disorder is due to deviation in the bond angles and an expression of χ can be written as a function of the distortion parameter, Δ, of the bond angle. Since the paramagnetic term, $\chi_p(\Delta)$, strongly depends on the overlapping integral, that decreases with an increase of disorder, $\chi_p(\Delta)$ also decreases. In III–V compounds, the particular feature is the ionicity. Raman and IR spectra, density of valence states of a-III–V compounds show a broadened version of their crystalline counterpart, suggesting short-range order, i.e., fourfold coordination. In compound semiconductors, the possibility of bonds between like atoms (wrong bonds) must be considered, destroying the chemical ordering. The authors construct a zinc blende structure where each atom of type A is surrounded by four nearest neighbor atoms of type B, and vice versa, calculating an expression for χ. Then, they introduce disorder essentially presented as bond-angle distortion, obtaining an expression for $\chi(\Delta)$, sum of three terms. They evidence that the Van Vleck-like paramagnetic term, $\chi_p(\Delta)$, reduces appreciably with an increase of Δ. However, in any case, this term is small compared to the Langevin-like diamagnetic term due to valence electrons, and so its variation cannot influence significantly the total susceptibility.

Mousseau and Lewis [30], using the so-called "activation-relaxation technique" and empirical potentials, built up two 216-atom tetrahedral CRN networks with different topological properties, further relaxed using tight-binding molecular dynamics. The first network corresponds to the well-known Polk-type network, randomly decorated with Ga and As atoms. The second one is made similar to the Connell–Temkin model introducing a minimum of wrong homopolar bonds, and therefore a minimum of odd-membered rings. The authors demonstrate that this last model is energetically favorable over the Polk model, allowing obtaining almost perfect fourfold coordination, realistic bond-angle distribution, and almost no wrong bonds. This result seems to correspond to the "real" a-GaAs. Moreover, most structural, electronic, vibrational properties are little affected by the differences in topology between the two models. Theoretical calculations evidence that wrong bonds, and their density, can be valuated only by direct measurements of the partial radial distribution function and not by indirect measurements, like XPS or Raman.

Ebbsjo et al. [31] used molecular-dynamics simulation with a new interatomic potential function, to compute structural correlation. They found an excellent agreement between the calculated structure factor, in particular the height and the width of the first peak, characteristic of the intermediate-range correlation, with X-ray diffraction experiments. The calculated energy difference between crystalline and amorphous systems is also in good agreement with electronic-structure calculations.

3.2. Experimental Results

3.2.1. EXAFS, X-ray, XPS

3.2.1.1. Rf and Dc Sputtering

Alimoussa et al. [32] studied the structural and compositional properties of thin films of GaAs by X-ray diffraction and nondispersive analysis of X-rays. According to the different substrate used for the deposition (glass, glass covered with

Mo, Ge without and with Mo) and to the different power, X-rays diffraction peaks evidence a polycrystalline structure preferentially with (111), (220), and sometimes (311) directions. Moreover, the dimensions of the grains tend to decrease with W. Another important result is the correlation between the composition and the shift of the X-ray diffraction peaks [6] (Fig. 1): the shift of the peaks shows a minimum when As content approaches 49% (i.e., stoichiometry, taking into account the experimental error). The shift is independent of the crystalline direction and that correlation can be used to control the stoichiometry of the deposited films.

Kärcher, Wang, and Ley [33] found that recrystallization of a-GaAs starts at about 280 °C and that the transverse dimensions of the crystallites increases from 30 to 120 Å, when T_A varies from 250 to 450 °C. During annealing, As—As bonds, in an As-rich surface, and Ga—Ga wrong bonds introduce two new structures inside the valence band, located at 15.0 and 9.0 eV, respectively.

Greenbaum et al. [34] studied configurational disorder and short-range bonding in films of a-GaAs by using ^{69}Ga and ^{75}As NMR. The ^{69}Ga, ^{71}Ga, and ^{75}As NMR line shapes were found to be second-order quadrupole broadened. Similar results obtained in c-GaAs, with a high concentration (10^{19} cm^{-3}) of charged defects exhibit a first-order quadrupole broadening. By comparison, it could be deduced that in a-GaAs the stronger quadrupole interaction is due to a higher configurational disorder degree rather than to electric fields associated with charged defects. The experimental quadrupole interaction can be simulated considering the electric field gradient generated by point charges located on specific nuclear sites, inside the angular distorted tetragonal structure. The distortion from the crystalline tetragonal symmetry can be evaluated to be about $\pm 9°$ from ^{75}As NMR results, neglecting the second and third nearest neighbors and assuming crystal-like bond lengths. Similar results are obtained from ^{69}Ga and ^{71}Ga data. However, quadrupole effects comparable to those observed and computed based on angular distortion, can be generated by a 5–10% variation in bond lengths.

Sedeek, Carchano, and Seguin [35] studied the hydrogen bonding in a-GaAs both with X-ray photoelectron spectroscopy (XPS) and infrared spectroscopy (IRS). XPS data evidence that hydrogenation induces a considerable modification of the Ga 3d-core level, leaving those of As unaffected: this can be interpreted with a preferentially linking of hydrogen to gallium. IRS measurements show that, independently of the hydrogen content, two strong bands are evidenced at 530–1430 cm^{-1}, together with several other less intense peaks. All the structures can be attributed to a possible vibration mode of the As—H or Ga—H bond. The 840 and 1000 cm^{-1} bonds can respectively be attributed to As—H bending mode and to As—H$_2$ bending mode. All these results agree with those of [36] and with the theoretical calculations of [28].

Baker et al. [37] prepared a-Ga$_{1-x}$As$_x$ ($x > 0.5$) films to investigate the local structure and bonding configurations by edge X-ray adsorption fine structure (EXAFS) measurements. The composition in the deposited films was determined by EDAX,

allowing to measure x in the range 0.5–0.85 and to verify a good homogeneity of the composition. The As-coordination changes from fourfold in near stoichiometric films to threefold for As-rich samples. The Ga coordination remains 4 in the studied x-range: 0.5–0.85. The amorphous network of As-rich samples consists of a mixture of tetrahedrally and trigonally bonded atoms. The EXAFS data are correlated to optical absorption curves, allowing to determine $E_g \approx 0.94$ eV, close to stoichiometry and a little increase to 1.0–1.05 eV, outside stoichiometry with $0.55 < x < 0.85$. IR absorption in stoichiometric films evidences the presence of the a-GaAs TO normal mode (≈ 250 cm^{-1}), broadened and shifted to a lower wave number from that found in c-GaAs (≈ 268 cm^{-1}). Increasing the As content, the intensity of the TO mode decreases and tends to disappear in the most As-rich samples.

Baker et al. [38] characterized In$_{1-x}$P$_x$ films deposited by rf sputtering with $0.4 < x < 0.9$. The samples are completely amorphous for $x > 0.5$, but In-rich material contains crystallites. EXAFS and XPS allowed evidencing that chemical order is predominant, with In—P bonds favored. Nevertheless, the degree of partial chemical order is consistent with thermodynamic models for amorphous covalent alloys.

3.2.1.2. Flash Evaporation

Gheorghiu and Theye [39] computed an average value of the distance of first neighbor with a good accuracy of 0.01 Å from an analysis of the EXAFS spectra collected in a-GaAs. The value is strictly corresponding to the crystalline one, and authors conclude that the density of wrong bond must be very low. A different situation is found for a-GaP, where both EXAFS data and optical properties evidence a much higher content of wrong bonds.

Dufour et al. [40] studied the disorder of a-GaP core levels, with Auger electron and X-ray induced photoelectron spectroscopies. The existence of partial chemical disorder is clearly demonstrated with a high proportion of Ga—Ga bonds, while the evidence for P—P bonds is less clear. Similar measurements on sputtered a-GaP layers did not evidence chemical disorder: a warning about the comparison among results from samples with different histories.

Gheorghiu et al. [11] in amorphous films of GaAs, GaP, and GaSb using electron diffraction, found that these compounds retain their tetrahedral coordination, similarly to a-Ge. For a-GaAs and a-GaP, the diffraction diagrams are insensitive to the film composition: the position and the width of the characteristic diffuse rings are independent of the excess of the group V element. On the contrary, for a-GaSb, the first main diffuse ring is gradually shifted toward a higher value of the scattering vector as the antimony concentration increases. Moreover, a new feature appearing for large excess of antimony modifies the whole shape of the diffraction pattern.

Senemaud et al. [41] performed photoelectron spectroscopy on a-GaAs films to study the effects of disorder and the defects on the core levels and valence-band distribution. They found that the chemical ordering is maintained, because any shift of the Ga and As 3d-core levels is not observed. For a-GaAs, the valence-band spectrum retains the characteristic three-peak structure of the crystal, with only a broadening of the peaks. Moreover, the same spectrum measured in stoichiometric and As-rich samples, evidence that in these last samples additional states appear at the top of the valence band. The most important effect is the splitting of the lowest peak at about 12 eV in two peaks located at 10.5 and 13 eV, respectively, corresponding to the presence of isolated As−As wrong bonds. Moreover, the absence of additional states below the bottom of the valence band suggests that the excess As atoms be distributed randomly in the GaAs network, without significant clustering.

Gheorghiu et al. [42] investigated the structure of nearly stoichiometric ($C_{Ga} \approx 50$ at.%) amorphous gallium arsenide films deposited at room temperature. The results obtained on these films were compared to those found in samples of a-Ge, deposited under similar conditions. The purpose was to discriminate between continuous random networks (CRN) with some proportion of odd-membered rings or only even-membered rings. The analysis of both the interference functions and the radial distribution functions has revealed significant differences between the atomic arrangements of the two materials beyond the second neighbors. A model containing only even-membered rings or a negligible proportion of odd-membered rings better describes the structure of the amorphous compound. On the contrary, that of the amorphous element is better accounted for by a model containing some proportion of odd-membered rings. Moreover, even if the average first-neighbor distance is larger than expected, the experimental data for a-GaAs do not show strong disorder effects linked to a less directional bonding and to a certain quantity of chemical disorder. Annealing of a-GaAs changes the short-range order very little but influences the dihedral angle distribution, important for the variations of the gap value.

Senemaud et al. [43] performed XPS measurements to compare the As and Ga core level spectra and the valence-band spectrum of stoichiometric a-GaAs to those of crystalline GaAs. The core level data suggest that the amorphous compound maintain chemical order (Fig. 8). The broadening of the core lines of the spectrum can be attributed to fluctuations of charges around the emitting atoms, only due to fluctuations of bond angles and lengths. Disorder provocates a broadening of the upper p-bonding peak in the valence band and its shift upward of the upper edge of the band. Moreover, a smoothing of the overall curve is evidenced, with crystalline peaks less pronounced.

Udron et al. [44, 45] performed a detailed investigation of the short-range order in a-GaAs samples, using EXAFS and anomalous X-ray scattering, for chemical disorder. From EXAFS oscillations, the authors were able to deduce that the average distance around both Ga and As atoms is the same (2.46 Å), close to crystalline one, 2.45 Å. The average coordination number, N, is 4.0–4.3 for Ga and is 3.7–4.0 for As: about 10% of As atoms are threefold coordinated. These undercoordinated As sites when inserted in a CRN type model force the increasing of the bond angles toward 120° at the defect site

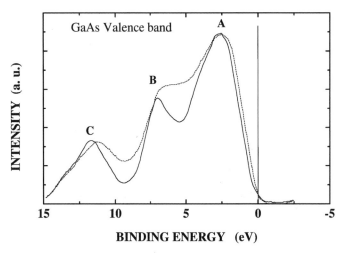

Fig. 8.   Valence-band photoelectron spectra from crystalline (continuous line) and amorphous (dashed line) GaAs. (After [43], reprinted with permission by Elsevier.)

and the structure presents four three-bond neighbors located at about 3.4 Å from the central defect. The three-coordinated As sites are expected to relax toward their natural bonding configuration (angle bond of 97°). The local environment of the defect As site should depend on the balance between the effects of electronic relaxation, which tends to decrease the bond angle and the tension of the atomic rings, which acts in the contrary way. The presence of these threefold coordinated As atoms plays an important role in the relaxation of the network, so giving support to the observation of unexpected bond distances. Moreover, Udron et al. [45] also compared the local atomic ordering in the III–V compounds: a-InP, a-GaP, and a-GaAs. a-InP, which exhibits the largest contrast between the two atomic species, remains four-coordinated with wide atomic distributions and is highly chemically disordered. In a-GaP, they also observed a high degree of chemical disorder, and dangling bonds are detected on the group V atomic sites. On the other hand, a-GaAs, which is less ionic, but is composed of atoms of similar size, tends to be mainly chemically ordered. A noticeable proportion of group V (As) undercoordinated sites is found in this last compound. They are accompanied by a local rearrangement in the topology of the amorphous network giving rise to a short heteropolar bond at 3.2 Å evidenced by anomalous X-ray diffraction.

3.2.1.3. Other Methods of Deposition

Matsumoto and Kumabe [46] found in MBE GaAs samples that amorphous or crystalline structure and the composition of amorphous films depend on the substrate temperature and on the ratio, m, between intensities of the molecular Ga and As beams. They found two well-separated regimes for $m > 1$ or $m < 1$. The composition ratio, $\gamma = As/Ga$, is independent of m and samples are spatially homogeneous, when $m > 1$ and $T_S > 90\,°C$. Moreover, stoichiometric films are grown depositing just below the crystallization temperature ($T_C = 240\,°C$):

for example, with $m = 10$ and $T_S = 190\,°C$, $\gamma = 1.03$. When $m < 1$, γ depends strongly on m and samples are spatially inhomogeneous.

Burrafato et al. [47] deposited InAs thin films (100–1500 Å) by vacuum evaporation at room temperature. Polycrystalline structure evidence an excess of In in the inner layers and the presence of In and As oxides at the surface of the films, where the atomic ratio of In to As was found to be 2.7($\pm$0.2):1.

Ridgway et al. [48] determined the structural parameters of amorphized GaAs from transmission EXAFS measurements. They started from a crystalline structure GaAs/AlAs/GaAs and through implantation with Ga and As ions, the first two layers were converted to amorphous phase. After this, the AlAs layer was dissolved with a selective chemical etching to separate the GaAs surface layer, 2.5-μm thick, from the GaAs substrate. The amorphous layers were then supported on adhesive Kapton film and stacked together to optimize the transmission coefficient. Relative to a crystalline sample, the nearest neighbor bond length and the Debye–Waller factor both increased for amorphous material. On the contrary, the coordination numbers both for Ga and As decrease from 4 to 3.85 atoms, in agreement with theoretical calculations [30, 49, 50].

3.2.2. Electron Microscopy

Murri et al. [51, 52] studied structure and morphology both of unhydrogenated and hydrogenated rf sputtered a-GaAs films. Transmission electron microscopy (TEM) and transmission high-energy electron diffraction (THEED) were made on three groups of samples. Only one parameter was allowed to change in each group. In the first one, the substrate temperature (20 °C $< T_S < 155\,°C$); in the second, the sputtering power (100W $< W < 400$ W); and in the last, the argon pressure (0.86 Pa $< p_{Ar} < 4.3$ Pa). The high magnification TEM and THEED images of samples deposited as a function of T_S show that the sample deposited at $T_S = 20\,°C$ is amorphous, homogeneous, and uniform. It contains small agglomerates having mean diameter, L, between 2.3 and 4.5 nm. At $T_S = 155\,°C$, L increases from 3.8 to 7.7 nm, TEM shows elongated structures and THEED pattern a third defined ring. Both the L and the THEED ring intensity decrease when W rises from 100 to 400 W. At these high values, the morphology tends to become irregular, the higher limit of L decreases to 5.2 nm, the structure of the films is amorphous but the quality is poor. A similar, even more evident trend is shown by samples deposited as a function of argon pressure, p_{Ar}. For samples deposited at $p_{Ar} = 0.86$ Pa, $L \approx 2.3$–9.0 nm, and five diffuse rings are present in the reflection high-energy electron diffraction (RHEED) pattern (Fig. 9). Increasing p_{Ar} up to 4.3 Pa, the morphology becomes irregular, a high density of agglomerates is present, and L ranges between 3.8 and 5.0 nm. These results of electron microscopy can be linked to the deposition parameters and their changes. The behavior of the deposition rate, r_d vs $(W/p)^{1/2}$, a measure of the energy of ions impinging on the film surface, is shown in Figure 10. A deposition process assisted by ionic bombardment can explain the rapid increase of r_d for $W \leq 200$ W. On

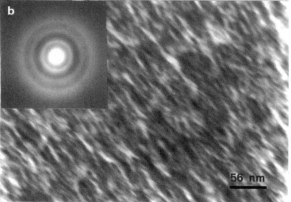
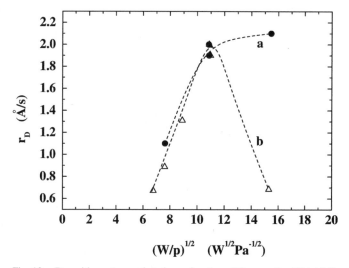

Fig. 9. High magnification images and corresponding THEED patterns (inset) for GaAs films deposited at: (a) $T_S = 20\,°C$ and (b) $T_S = 155\,°C$, with $p_{Ar} = 2.5$ Pa and $W = 200$ W constant. (After [51], reprinted with permission by Elsevier.)

Fig. 10. Deposition rate r_d, plotted as a function of the quantity $(W/p)1/2$. Curve (a) is for samples deposited as a function of the rf power, W; curve (b) for those grown varying argon pressure, p_{Ar}. The lines only indicate trends of the plotted quantities. (After [51], reprinted with permission by Elsevier.)

the contrary, the decrease of r_d when $(W/p)^{1/2} \approx 11$ ($W > 200$ W) can be explained invoking the concurrence of two processes: etching and deposition. Films deposited at $T_S = 20\,°C$,

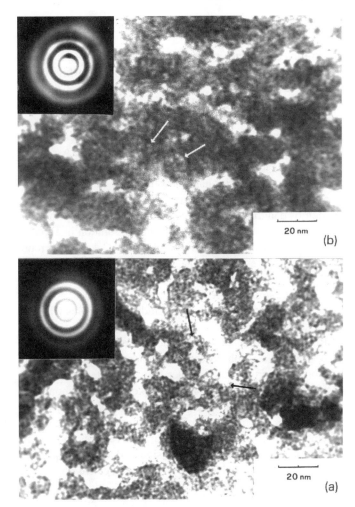

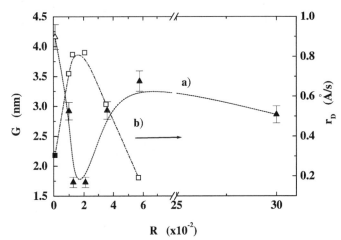

Fig. 12. Mean diameters of agglomerates, G, and deposition rate, r_d, as a function of $R = p_{H_2}/p_{Ar}$. Curves have been drawn to indicate trends of the plotted quantities. (After [52], reprinted with permission by Elsevier.)

Fig. 11. High magnification images and corresponding THEED patterns (inset) for unhydrogenated GaAs films deposited at: (a) $T_S = 20\,°C$, $p_{H_2} = 10.4 \times 10^{-2}$ Pa and (b) $T_S = 200\,°C$, $P_{H_2} = 7.3 \times 10^{-2}$ Pa, with $p_{Ar} = 4.9$ Pa always constant. (After [52], reprinted with permission by Elsevier.)

$W = 200$ W, and $p_{Ar} = 2.5$ Pa, have a very good quality, because they show a short-range order structure and a regular morphology, Figure 11. X-ray diffraction or THEED did not detect crystallization in our samples for annealing temperature up to 450 K. Murri et al. [52] made similar investigations on hydrogenated a-GaAs films. The deposition power was 200 W for all specimens. Samples deposited at the lowest $T_S = 20\,°C$, $p_{Ar} = 4.9$ Pa, and ($p_{H_2} = 10.4 \times 10^{-2}$ Pa) show a rather regular morphology with agglomerates (small region with a high degree of crystallographic order) uniformly distributed in the film, and mean diameter, L, of about 1.7 nm. At the highest $T_S = 200\,°C$, ($p_{Ar} = 4.9$ Pa and $p_{H_2} = 7.3 \times 10^{-2}$ Pa), the film is not homogeneous and $L \approx 3$ nm. The sample deposited at $T_S = 100\,°C$ has a columnar structure. Samples of the second group show an island structure, L is about 1.7 nm, separated by interstitial zones of low density. The effect of the addition of hydrogen in the gas phase (or argon dilution) is shown by a plot of the mean diameters of the agglomerates as a function of the ratio between hydrogen and argon pressure, $R = p_{H_2}/p_{Ar}$. It

can be seen (Fig. 12) that L decreases by a factor of 2.5 when R goes from 0 (unhydrogenated samples) to $R = 1 \times 10^{-2}$ and increases by a factor of 2 when $1 \times 10^{-2} < R \leq 6 \times 10^{-2}$ and then tends to remain constant for $R > 6 \times 10^{-2}$. The variation of p_{Ar} affects mainly r_d, while L is invariant to p_{Ar}.

Twigg, Fatemi, and Tadayon [53] show that solid-phase recrystallization of an amorphous GaAs film can form a monocrystalline film. They annealed at 775 or 850 °C amorphous films of GaAs, deposited by MBE at 190 °C. TEM observations indicate that annealing induces As precipitation, more in as-grown single crystals films than in the as-grown amorphous films. This fact insures a high concentration of As donors rather than a high concentration of precipitates that increase the resistivity. Electrical measurements evidence that annealed samples ($T_A = 775$ or 850 °C) have very high concentrations of n-type carrier (10^{18} cm^{-3}) and high mobility (800 cm^2 V^{-1} s^{-1}). The small volume fraction (0.02%) of As precipitates within the film, suggests that much of the excess As, remains as As$_{Ga}$ antisite defects which act as donors.

4. DENSITY OF STATES

4.1. Models, Calculations, and Theoretical Results

Kramer, Maschke, and Thomas [54] used the method of complex band structure (CBS) taking into account the short-range order of the amorphous structure by an approximate correlation function. The calculations were performed for c- and a-GaAs. GaP, GaSb, InP, InAs, and InSb were considered too. Results can be presented as a function of an adjustable parameter, α, a small number used to define the right Green's functions. The resulting CBS-band structures (Fig. 13) show, when compared with crystalline ones, that the real parts of energy are nearly unchanged and that the imaginary parts of the valence bands and of the first conduction band along the Γ-L axis are small.

O'Reilly and Robertson [55] calculated by the tight-binding recursion method the local electronic structure of like-atom

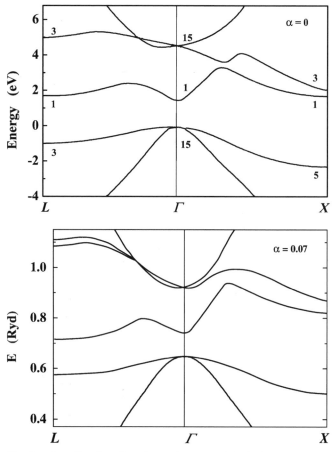

Fig. 13. Crystalline ($\alpha = 0$), and amorphous ($\alpha = 0.07$) CBS-band structures of III–V compounds. (After [54], reprinted with permission by Wiley–VCH.)

at this energy. The unrelaxed As dangling bond, $\theta = 109°$, presents an s-strong resonance at -9.9 eV and a resonance of an s/p-mixed character at the valence edge. If double occupied, θ relaxes to 97°, the s-state remains unchanged while the p-state moves to -0.8 eV inside the valence band. The main conclusion is that chemically ordered (high quality) a-GaAs containing only relaxed dangling bonds does not present intrinsic gap states. Like-atoms bonds, rather than dangling bonds could be considered as the dominant defect in a-GaAs, due to the role of the topological disorder, of the odd-membered rings, and because the antisite has a lower energy than the vacancy in c-GaAs. On the contrary, two arguments suggest a high proportion of dangling bonds in a-GaAs: (i) they satisfy the 8-N rule and they assume a more relaxed configuration in a-GaAs; (ii) networks with coordination higher than 2.4 are "overconstrained." A density of 1–10% of intrinsically broken bonds, can relax rather than reconstruct the a-GaAs network. Wrong bonds affect strain indirectly, via the ring statistics. All these defect configurations may be best studied in samples with different stoichiometry.

Robertson [56, 57] reviewed the electronic structure of defects in some amorphous semiconductors. Defects in random networks are structurally simpler than in crystals due to lack of periodicity. In a binary compound like GaAs, with bonds between unlike atoms, the simplest unit of chemical disorder is the like-atom bond (wrong bond). The most fundamental electronic parameter of a defect is its effective correlation energy, U. The strong electron–electron repulsion at doubly occupied localized states is assumed to be positive. Negative correlation energy expresses concisely the preference of electrons to pair into covalent bonds in the network, even at defects. a-GaAs is tetrahedral coordinated like c-GaAs, it obeys the 8-N rule, as N (average) $= 4$, which would require the coordination of both Ga and As to be 3. Ga—Ga and As—As wrong bonds are probably important defects in a-GaAs, but Ga_3 and As_3 sites should not be overlooked. The striking difference between a-GaAs and a-Si is that the various defects create gap states near E_c and E_v, or in the bands, rather than in the midgap as in Si. Robertson (Fig. 14) computed that the unrelaxed Ga_3 site (tetrahedral bond angle 109.5°) generates a gap state below E_c. Relaxed Ga_3 and As_3 sites give only resonances: chemically ordered a-GaAs does not possess deep gap states. For wrong bonds, the As—As bond does not give gap states, while the antibonding state of the Ga—Ga bond lies below E_c. The lower gap of a-GaAs (≈ 0.9 eV) compared to that (≈ 1.5 eV) of c-GaAs, is partially due to tailing and Ga—Ga bonds, with a possible contribution of defect clustering.

O'Reilly and Robertson [28] calculated the local electronic structure of bulk and defect sites using the tight-binding recursion method. The behavior of defects in amorphous-III–V semiconductors is strongly dependent on their deposition method (for example, quenching from the vapor). This intrinsically generates a high concentration of defects, according to deposition and annealing conditions. A second complication arises from the fact that their bulk bonding does not obey the (8-N) rule: being N the valence number, the atom's covalent coordination

bonds and threefold coordinated Ga and As sites in amorphous GaAs. In a-GaAs, dangling bonds and wrong bonds could occur as isolated defects, and can be considered as due to normal valence behavior and topological disorder, respectively. Amorphous semiconductors generally coordinate according to the 8-N rule, so the fourfold-coordinated sites obey the Mott's rule in pairs, while threefold-coordinated sites obey the same rule individually. This kind of defect is more stable in a-GaAs than dangling bonds in a-Si. The stable bond angle, θ, of a dangling bond depends on its occupation: it varies from $\theta = 109°$ for singly occupied levels to $\theta = 120°$ for empty states at Ga to $\theta \approx 97°$ for full states at As. An analysis of the different types of defects gives the following results for dangling and wrong bonds. As—As-like atoms ("wrong") bonds produce a well-defined splitting of the As s-band into a resonant bonding state, σ, at -12 eV and a bound antibonding state, σ^*, around -9.5 eV. The gap state of the As—As bond (σ^*) is at 1.4 eV, while its resonance is near -3 eV. Ga—Ga bonds introduce no gap states, producing several structures of different character. The unrelaxed Ga dangling bond ($\theta = 109°$) gives a gap state at 1.45 eV, just below the conduction-band edge, E_c. If the dangling bond is empty, the site relaxes to be planar, $\theta = 120°$. In this case, the lower state moves out of the gap to 1.8 eV, while the upper state falls to 3.8 eV: the true Ga p-dangling bond lies

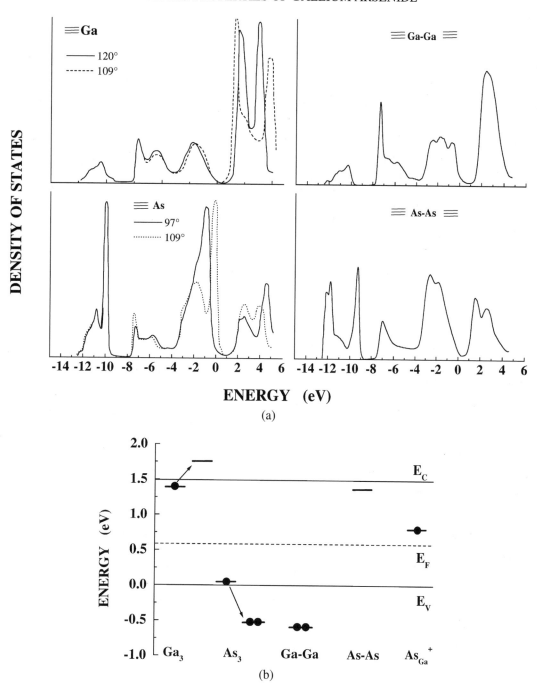

Fig. 14. Calculated defect levels for a-GaAs, in the bands, (a) and near the gap, (b). (After [57], reprinted with permission by Elsevier.)

is (8-N) or N, if N = 4. Individually, Ga and As have three and five valence electrons each, and so would both be expected to be trivalent, but the tetravalence of GaAs complex depends from the fact that the unit has eight electrons. The effect of the rule can be relevant in the alloys: in a-Ga$_x$As$_{1-x}$, an excess As creates both As—As bonds and trivalent As sites. The principal types of defects considered are undercoordinated atoms, dangling bonds, and like-atom bonds (wrong bonds, WB). The authors performed calculations on the following III–V amorphous compounds: AlP, AlAs, AlSb, GaP, GaAs, GaSb, InP,

InAs, and InSb. The bulk electronic structure computed for pure a-GaAs evidences that the gap of ≈1.55 eV is direct. Moreover, comparing with c-GaAs, its density of states (DOS) is much smoother. The leading edge of the valence band is much more shifted than the recession found in a-Si, and, finally, that the conduction-band edge does not move significantly. The optical gap close up, due to the presence of significant numbers of defect states at the band edges. As far as, defect configurations are concerned, they consider occupancy and defect bonding, trivalence, and relaxation effect of the bond angle and, finally, stoichiometry. The local DOS is computed for many types of defects: undercoordinated, relaxed and unrelaxed, anion and cation sites (dangling-bond configurations); anion and cation wrong bonds, and wrong-bond complexes; hydrogen related states. The main conclusions are that in all III–V compounds anion dangling bonds give rise to occupied acceptor-like states at or below the valence-band edge, E_v. Cation (Ga) dangling bonds produce empty donor-like states at or above the conduction-band edge, E_c. Isolated wrong bonds introduce gap states in some of the compounds. Usually, anion (As) wrong bonds introduce donor states near E_c, while cation wrong bonds introduce acceptor states near E_v. A much lower density of states at the Fermi level, E_F, is found in all these compounds, compared to a-Si, in agreement with experimental results [13, 58, 59]. Clusters of wrong bonds or isolated cation wrong bonds are suggested to be responsible for midgap states, in wide gap (as GaAs), or normal gap (as InP) compounds, respectively. The dangling-bond concentration is intrinsically high (1–5%) and predominates in annealed material. The effect of the stoichiometry deviations produces a combination of wrong bonds and trivalent sites of the excess species with an increased DOS at the midgap, where the Fermi level is locked.

Mosseri and Gaspard [60] analyzed the configurations of the wrong bonds in a tetracoordinated network in the attempt to go beyond the continuous random network (CRN) model. Experimentally, a-Si and a-Ge are well described by the Polk [61] model and III–V amorphous semiconductors seem better described by the Connell–Temkin model [16]. The two models differ because the latter has no odd-membered rings. Thus, the possible presence of like-atom bonds should lead to observable effects. The authors demonstrate that: (i) in an even two-dimensional or three-dimensional network there cannot be fully isolated wrong bonds; (ii) odd rings act as "sinks" for the wrong-bond strings. They compare the DOS for defect-free GaAs that shows filled valence band between -12.5 and 0 eV and the As and Ga s-bands separated by an ionic gap. Then, they distributed WB ordered arrangements to the structures and computed the local DOS on As and Ga sites with one WB or two WBs attached to them. The presence of WB, imposed by topology, induces a splitting of the As s-band, and new states in the ionic and bandgaps. Here, the states are predominantly on Ga sites in the lower part and on As sites in the higher part.

Agrawal et al. [62] performed a study of the electronic structure of defect complexes, in particular for the deep donor level EL2, which role is fundamental for technological applications of GaAs. On the other hand, this defect is of scientific interest because its properties show the behavior of metastability at $T < 100$ K. The authors employed a standard cluster Bethe lattice method (CBLM) and the linear combination of atomic orbital (LCAO) method for the description of electronic states. Their results are applicable to both the crystalline and amorphous phases. A right selection of the interaction parameters has to be made to reproduce the experimental data, in particular the density of states. Threefold coordinated Ga and As atoms (dangling bonds) are the simplest defects present in a-GaAs. The dangling bond at the cation Ga generates, Figure 15, a cal-

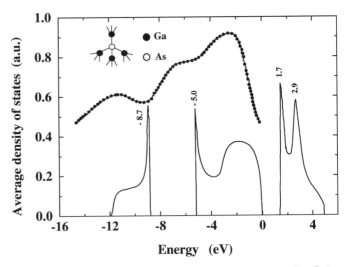

Fig. 15.   Calculated average electronic density of states (—) for bulk a-GaAs; (····), photoemission data. (After [62], reprinted with permission by Taylor & Francis.)

culated gap state of mixed s–p-character near the bottom of the conduction band, namely, at 1.31 eV. Whereas that at the anion As is mainly p-like and appears near the top of the valence band at 0.28 eV. The relaxation of the bond angles formed at the As dangling bond from 109.5 to 97° shifts the gap state to 0.04 eV. The calculation for the relaxation of the Ga dangling bonds from 109.5 to 120° gives a shift of the gap level from 1.31 to 1.49 eV. These shifts of the gap states toward the conduction- and valence-band edges are in agreement with the previous result [28]. Twofold coordinated atoms (double dangling-bond defects) produce defect states at 0.76 eV from the conduction band, while a two-fold coordinated atom at As site produces a state at -0.04 eV and the other at 0.98 eV. Ga vacancy introduces, at the immediate neighbors of the vacant site, a local density and a p-type state appears at 0.54 eV, while As vacancy generates gap states at 0.65 and 1.48 eV. The calculations evidence that As—As wrong bonds generate resonance states at 1.5 and 1.9 eV near the bottom of the conduction band, mainly of p-character and localized at the As atom. The peaks found in the calculated density of states are in reasonable agreement with experimental data of photoemission spectra [43]. Ga—Ga wrong bonds do not give rise to any state lying deep in the fundamental gap, but only to a state at 0.06 eV having a mixed Ga s–p-character. As far as As_{Ga} and Ga_{As} antisites (that is the occurrence of an As atom at a site normally occupied by a Ga atom and vice versa), the authors were able to reproduce the deep gap state at 0.75 eV above the valence-band edge. It was experimentally found, and can be generated by the As_{Ga} antisite, by adjusting only the value of the interaction parameters for the As—As bond. The Ga_{As} antisite generates gap states with mixed s–p-character and shifted at 0.19 eV toward the midgap. The last type of defects considered by the authors are As and Ga interstitial. The interstitial As bonded to four Ga atoms does not give gap states; on the contrary, the same interstitial atom when bonded to four As atoms generates states at 1.24 eV from the conduction band. The interstitial Ga bonded to four As atoms gives a gap state at 0.49 eV, localized at the Ga atom, and a second one just at the conduction band edge at 1.50 eV. The Ga interstitial bonded to four Ga atoms reveals a gap state lying at 1.41 eV. The authors review and discuss some of the proposed models for the EL2 defect in c-GaAs. However, by the comparison between the experimental and calculated density of states for different types of defects and their complexes, they suggest that the results for c-GaAs, can be easily applied to a-GaAs.

Xanthakis, Katsoulakos, and Georgiakos [63] theoretically investigated the electronic and atomistic structure of stoichiometric and As-rich, $Ga_{1-x}As_x$. They simulated the material as a ternary alloy of fourfold-coordinated As^{4+}, Ga^{3+}, and threefold-coordinated As^{3+}. They were able to compute, using the appropriate mathematical methods, the DOS in terms of the stoichiometry index, x and of the proportion of As^{3+} atoms and of the short-range order parameters. They deduced the structure of the material linking theoretical and experimental results. The number of As^{3+} atoms at which compensation occurs is a function of x only. Using this value and assuming no $As^{4+}-As^{4+}$ bonds, the authors are able to explain EXAFS data and optical gap as a function of x. Particular features of the valence band at $x = 0.6$ and the subsidiary shell of As neighbors of Ga. The simple model allows for compensation of the material with extra As forming As−As bonds and at least one As in the $\Theta = 97°$ configuration. The theoretical results are compared with experimental data [13, 37, 44, 64].

4.2. Experimental Results

Hauser, Di Salvo, and Hutton [65] measured the density of states (DOS) of sputtered a-GaAs both in as-deposited ($T_S = 77$ K) and annealed samples. They used resistivity as a function of the temperature, susceptibility, and ESR to compute the gap states at the Fermi level, $N(E_F)$, and the number of spin cm^{-3}, N_S. They found in as-deposited samples $N(E_F) = 9 \times 10^{18}$ states eV^{-1} cm^{-3}, and that conductivity is via a hopping mechanism. The disappearance of hopping conductivity can be obtained only by annealing at high temperature (at 300 °C: $N(E_F) < 1 \times 10^{17}$ states eV^{-1} cm^{-3}); after this, conductivity is an activated process with $E_\sigma = 0.64$ eV. The annealing has a similar effect on N_S (at RT: $N_S = 4$–6×10^{18} spins cm^{-3}, while at 600 °C, $N_S < 10^{16}$–2×10^{17} spins cm^{-3}). All these results and their dependence on temperature can be understood assuming that the overlap of donor and acceptor states in samples deposited at low temperatures can be removed by annealing and the structure evolves toward bulk equilibrium. Moreover, the more or less complete disappearance of dangling bonds upon annealing depends on how ionic or covalent are the bonds.

Kärcher, Wang, and Ley [33] by *in situ* photoemission spectra found that incorporation of hydrogen introduces a structure at 7.4 eV below the valence-band maximum. For $C_{H_2} \approx 50$ at.%, the shift is of about 0.5 eV, toward higher energies, in comparison to samples with $C_{H_2} \approx 0$ at.%. The Fermi level is located 0.57 eV above the top of the valence band in a-GaAs, while shifts to 0.63 eV in a-GaAs:H, independent of hydrogen content.

Ouchene et al. [66] used soft X-ray spectroscopy (SXS) and X-ray induced photoelectron spectroscopy (XPS) to study the electronic properties of flash-evaporated amorphous indium phosphide, InP. They found a broadening of the main peak at the top of the valence band (pure p-like states) leading to a $+0.6$-eV shift of the valence-band edge. Moreover, the peak corresponding to the mixed sp-states is smeared out and the valley between these two peaks is partially filled. A structure appearing at the bottom of the valence band in a-InP, reveals a disorder induced sd-hybridization. The presence of P−P and In−In wrong bonds is confirmed by these measurements, with a random distribution of P−P WB throughout the amorphous network.

Sedeek et al. [67, 68] deduced the density of gap states in vacuum evaporated amorphous gallium arsenide films using the space charge limited current (SCLC) method. The sandwich structure was Al/a-GaAs/Al with an active area of about 1 mm^2. The amorphous GaAs layer was deposited at 26 °C $\leq T_S \leq 40$ °C, and resulted to be very close to stoichiometry (Ga/As = 1.04) as measured by EDAX. SCLC measurements were performed at RT and the log I–log V characteristics consist of three different conduction regimes: (i) ohmic, I proportional to V, up to about 10^{-1} V (electric field strength between 3×10^2–8×10^2 V cm^{-1}). The conductivity activation energy ranges between 0.54 and 0.57 eV; (ii) a space charge trap limited current with I proportional to V^n, where $n > 2$ (electric field strength between 6×10^3–1.6×10^4 V cm^{-1}). The computed values of density of state, $N(E)$, are composed between 2×10^{15} and 2.5×10^{16} cm^{-3} eV^{-1}, in the energy range 0.44–0.56 eV below the conduction-band edge. A plot of log(I/V) vs $V^{1/2}$, in a high field regime evidences the absence of the Poole–Frenkel effect; (iii) a trap-free square law, where I is proportional to V^2, for $V > 1$ V. The trap density, computed using the Mark–Lampert theory, is about a factor 4 (from 6×10^{14} to 9×10^{15} cm^{-3}) below the values of the density of traps obtained from SCLC analysis. From all these data, the authors conclude that stoichiometric a-GaAs is essentially chemically ordered, with defects of the dangling-bond type only. The conductivity data for variable-range hopping were also studied. Moreover, using coplanar or sandwich configuration of contacts, the conductivity measured as a function of the temperature, evidences two completely different behaviors. It is thermally activated for coplanar structure with activation energy about to the half gap. On the contrary, three mechanisms are possible for sandwich structure. The values of trap density computed from the Mott equations for hopping conduction (4–6×10^{17} cm^{-3}) lie 2 or 3 orders of magnitude above those obtained from the SCLC method. They conclude that the midgap defect density in a-GaAs is less than that in a-Si prepared by glow discharge or sputtering. The result indicates that a-GaAs is essentially chemically ordered with defects of the dangling-bond type only. This result agrees with the prediction of Kuhl et al. [69] comparing studies of the relaxation of photoexcited carriers in a-GaAs and a-Si, using subpicosecond time-resolution experiments.

Greenbaum et al. [70] performed electron spin resonance (ESR) and ^{71}Ga and ^{75}As nuclear magnetic resonance (NMR) measurements on a-GaAs deposited on an SiO$_2$ substrate by molecular beam epitaxy. The a-GaAs film ($\sim$20-μm thick) has been removed from the substrate by immersion in a 50% aqueous HF solution to dissolve the SiO$_2$ layer; then the a-GaAs flakes were washed, dried, and packed into an ESR tube. The measurements were performed at $\sim$6 K. The ESR results evidence the four-line $S = 1/2$, $I = 3/2$ hyperfine features typical of the singly ionized arsenic antisite defect (As$_{Ga}$) observed in crystalline GaAs. It corresponds to the group formed by a singly ionized arsenic surrounded by four arsenic nearest neighbors, As^{4+}(As^{3-}). From intensity considerations, it is estimated that the As$_{Ga}$ defect density lies in the range 10^{17}–10^{18} cm^{-3}, 2–3 orders of magnitude greater than that found in as-grown crystalline specimens. The As$_{Ga}$ defect was suggested to be an important nonradiative recombination center in c-GaAs. Its presence, even if not clearly demonstrated, in a-GaAs also could represent a fundamental limitation to the efficiency of these materials for electronic device uses. ^{71}Ga and ^{75}As nuclear magnetic resonance linewidths exhibit substantial second-order quadrupole broadening consistent with slight deviations from crystalline bonding parameters, but with a complete preservation of the short-range tetrahedral order, even if strained.

Hoheisel et al. [71] performed ESR measurements in rf sputtered hydrogenated samples of a-GaAs, a-GaP, and a-InP. They used the X-band with a microwave power of 0.2 mW at a frequency of 100 kHz and the temperature in the range 20–300 K. Two resonances can be resolved. Even if g-values and linewidths are nearly constant in the range of H content investigated ($C_{H_2} = 0$–27%), the spin density, N_S, of a-GaAs computed for both lines 1 and 2, strongly depends on C_{H_2}. From the intensity of line 1, $N_S \approx 10^{18}$ cm^{-3} in unhydrogenated a-GaAs and attains about 7–8 $\times$ 10^{16} cm^{-3} as C_{H_2} increases up to $\approx$20 at.%. On the contrary, N_S values, computed from line 2, show an increasing behavior, starting from 7–8 $\times$ 10^{16} cm^{-3} at $C_{H_2} = 0$% up to 5 $\times$ 10^{17} cm^{-3} ($C_{H_2} \approx 27$ at.%). In parallel to the changes of N_S of the two resonances, the activation energy of the dark conductivity, E_σ, increases and the thermopower of samples is always positive. These two facts can be interpreted with a shift of the Fermi level from the valence band to midgap. The effects of the incorporated hydrogen on the ESR and the transport properties are completely reversible under thermal annealing due to its effusion. Two effusion maxima are evidenced at annealing temperatures of 480 and 610 °C, even if most of the incorporated hydrogen effuses at the highest temperature, for samples with $C_{H_2} = 18$ at.%. The ratio of the spin densities, $N_S(2)/N_S(1)$, calculated from lines 1 and 2, respectively, is about 1.6, at temperatures lower than 80 K. After these temperatures, the ratio shows a reversal of the relative intensities: at room temperature, $N_S(2)/N_S(1) = 0.5$. This reversal is even more pronounced for samples with higher C_{H_2}; however, it can be reversed again by decreasing the temperature. From all these experimental results, the authors argue that paramagnetic defects originating the line 1 are states located at group V atoms,

with two possibilities: charged As-antisite defects (As$_{Ga}$) or As dangling bonds. This last defect more probably originates the resonance. Line 2 relies on the defect related to group III atoms, like Ga dangling bonds. The increasing porosity with hydrogen content of the film could account for the inversion of the $N_S(2)/N_S(1)$ ratio around 80 K. An antiparallel spin pairing in high local density of As dangling bonds on internal surfaces (voids) or interfaces can account for this effect.

Von Bardeleben et al. [72] performed an electron paramagnetic resonance (EPR) study on stoichiometric and As-rich GaAs films prepared by flash evaporation, to discriminate between two different classes of elementary intrinsic defects: dangling bonds and like-atom bonds. The as-deposited stoichiometric material shows a paramagnetic spectrum, light insensitive, with a g factor of (2.06 $\pm$ 0.02) and a peak-to-peak linewidth $\approx$300 G: the spin ($S = 1/2$) concentration corresponds to 10^{18} cm^{-3}. On the contrary, the As-rich films show a highly asymmetric EPR spectrum with $g = 2.07$ and a linewidth $\approx$230 G. The spin ($S = 1/2$) concentration is essentially unchanged. After thermal annealing, the EPR spectrum does not change significantly, while the spin density remains constant up to $T_A = 150$ °C and then rapidly increases to a factor 2.5 times at $T_A = 250$ °C. The electrical resistance, constant up to 150 °C, decreases rapidly for higher temperatures due to sample crystallization. Similar measurements performed on As-rich a-GaAs films, showed a different behavior: neither the EPR spectrum nor the spin concentration changes up to $T_A = 250$ °C, but after an annealing at $T_A = 350$ °C, no EPR signal was detected any longer. The resistance drops abruptly at $T_A \geq 275$ °C, where crystallization rises. This low crystallization temperature was related to the presence of large internal strains, which induce the nucleation of crystallites at surfaces or cracks. The optical gap of crystallized stoichiometric films was found to be $\approx$1.4 eV.

Guita et al. [73] performed a comparative study of the photoabsorption spectra near the As and the Ga, L_3 edges in a- and c-GaAs. The crystalline samples were obtained by annealing of amorphous ones at 400 °C. By comparison, of AsL$_3$ spectra for a-As, a- and c-GaAs, three points are evident: (i) a shift to higher energies of the absorption edge for these two last materials; (ii) that it is steeper and, (iii) that it is located at the same energy, about 2 eV above E_F, in both cases. Similar characteristics are shown by GaL$_3$ photoabsorption spectra, even if with more pronounced effects. The absorption edge is shifted in a-GaAs from about 0.4 eV to higher energies and the whole curve is smoothed, when compared to that of c-GaAs. Only Ga s-states are present at the bottom of the conduction band and disorder does not induce a mixture of As and Ga states near the band edge.

Deville et al. [74] used ESR at 9.4 GHz and at 3.9 K to study the paramagnetic defects in amorphous GaAs deposited by sputtering. The measured curve, after subtraction of the spurious contributions, can be separated in three different signals. The first one gives a value of the Landè factor of $g = (2.06 \pm 0.01)$ and a width corresponding to (89.9 $\pm$ 0.5) mT. Assuming a Curie law, they found a defect density of 7 $\times$ 10^{17}

cm^{-3}. This signal can be attributed to the As$_{Ga}$ antisite defect. Signal 2 consists of a single broad line centered at $g_{eff} = (2.058 \pm 0.004)$, while signal 3 is of a single narrower line, centered at $g_{eff} = (1.925 \pm 0.002)$. Signal 1 is composed of four lines with equal intensities. The g and the hyperfine coupling, A, values of signal 1 are very similar to those found for the antisite defect in c-GaAs with its four hyperfine lines and $A \approx 90$ mT. In this defect, an arsenic atom replaces a Ga atom (antisite defect), and when the a-GaAs is not stoichiometric, like in this case, the antisite defect has a high probability to be generated. The defect concentration (3×10^{-5} defects/As atom) is lower than expected. Using the positions and widths of the ESR lines from the As$_{Ga}^+$ defect, the authors found that the paramagnetic electron is in an antibonding orbital. The concentration for the center corresponding to signal 2 was equal to that found for signal 1. It is natural to consider that signal 2 corresponds to the capture of the first electron, even if it was impossible to identify the defect responsible for its capture. The third less concentrated defect (4×10^{16} cm^{-3}) probably generates from some chemical impurity.

5. OPTICAL PROPERTIES

5.1. Theoretical Results

O'Leary [75] studied the role played by disorder in shaping the functional form of the optical absorption spectra of a-semiconductors. The author developed a semiclassical model for the density of states and the optical absorption spectrum of amorphous semiconductors. The formalism considers both the distributions of tail and band states, as well as the optical absorption spectra below and above the gap. The model assumes the optical absorption spectrum proportional to the joint density of states function, $J(\omega\hbar)$. Two principal parameters are used: the mean energy gap, E_{go}, and the energy gap variance, σ^{*2}, σ^* being a measure of the amount of disorder. The author can account for all changes in the optical absorption spectra, strictly by varying the disorder parameter, σ^*. The breadth of the absorption tail is a strong function of the disorder, while, E_{go}, is insensitive to the amount of disorder. As the disorder is decreased, the optical absorption spectra associated with these a-semiconductors, approach well-defined disorderless limits. At the same time, the energy gap associated with these limits results to be greater than the corresponding crystalline gaps. The independence of E_{go} from disorder can suggest assuming it as a fundamental gap, instead of as Tauc's gap. Comparison is made between theory and experimental absorption data on a-GaAs.

5.2. Experimental Results

5.2.1. Dc and Rf Sputtering

Paul et al. [76] measured the absorption coefficient as a function of photon energy, $h\nu$, in a-GaAs:H. They deduced the optical gap, E_{04}, as $h\nu$ when $\alpha = 10^4$ cm^{-1}, showing its increase with hydrogen partial pressure, p_{H_2}. This increase of E_{04}, implies that at least the cation–anion bond determines one of the two band-edge densities of states. No effects were found on the optical edge, due to the modification of the stoichiometry.

Alimoussa, Carchano, and Thomas [2] in a-GaAs:H found that the refractive index, n, shows the most important features when $p_{H_2} \leq 40\%$ of p_T. In fact, n increases with p_{H_2} up to a maximum value of about 4.37 ($h\nu = 2.17$ eV) for $p_{H_2} \approx 20\%$, then decreases and approaches the value of the unhydrogenated material when $p_{H_2} \geq 40\%$. The edge of the absorption coefficient shifts toward higher photon energies with hydrogen, and Tauc's gap increases from 1.06 eV ($p_{H_2} = 0\%$) to 1.45 eV ($p_{H_2} \approx 40\%$). In their article, the authors stress two points. The absorption coefficient at 2.5 eV (maximum of the solar spectrum) is about 25 times that of c-Si and twice that of a-Si:H. a-GaAs:H can have a gap lower or slightly greater than c-GaAs, while a-Si:H always has a gap greater than c-Si.

Kärcher, Wang, and Ley [33] found an increase of the optical Tauc's gap of about 0.3 eV, compared to that of unhydrogenated samples, with a corresponding recession of the conduction-band edge.

Murri et al. [77] performed a detailed study of the Urbach's tail both in hydrogenated and unhydrogenated a-GaAs films using photothermal deflection spectroscopy (PDS). Unhydrogenated samples deposited at variable W do not show any shift in E_g. On the contrary, a shift of the optical gap toward higher energies is evident in hydrogenated samples. The constancy of E_g for unhydrogenated samples can be attributed to the overlap between band-to-band excitation and other excitation processes active because of a high density of states in the "mobility gap." The Urbach's tail ($\alpha < 5 \times 10^3$ cm^{-1}) is due to transitions between valence tail states and extended conduction states and is present in all samples. The onset of a third region is observed for hydrogenated samples, which the authors attribute to defect extended state transitions. The integral of α on this low-energy range has been directly related [78] to the density of states, N_T, associated with dangling bonds and other defects, localized into the mobility gap. The Urbach's energy, E_U, for unhydrogenated specimens shows a dependence on the quantity $(W/p)^{1/2}$. This last quantity is a measure of the ions impinging on the growing film surface [79]. Starting from ≈ 260 meV, E_U decreases to ≈ 145 meV; then it tends to a constant value, with a rapid decrease from ≈ 220 to ≈ 180 meV for $(W/p)^{1/2} \approx 11$. These results can be related to the deposition rate, r_d, and the mean dimension of the agglomerates, L, both functions of $(W/p)^{1/2}$ [51]. $(W/p)^{1/2} \approx 11$ is a critical value for r_d: when $(W/p)^{1/2} < 11$, r_D increases rapidly; r_d is constant for $(W/p)^{1/2} > 11$. On the contrary, the mean size of the agglomerates, L, tends to remain constant for $(W/p)^{1/2} \leq 11$ and for larger values increases. The dependencies of r_d and L on $(W/p)^{1/2}$ were interpreted [51] in terms of a deposition process assisted by ionic bombardment when $(W/p)^{1/2} \leq 11$ and of the saturation of the sputtering yield at $(W/p)^{1/2} > 11$. In the case of hydrogenated films, E_U, E_g, and r_d decrease (Fig. 16) as the hydrogen pressure increases up to $p_{H_2} \approx 0.15$ Pa and then they tend to constant values.

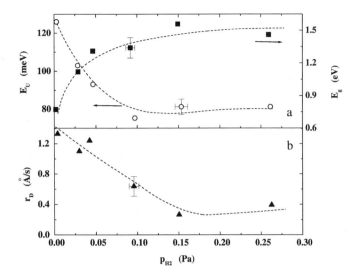

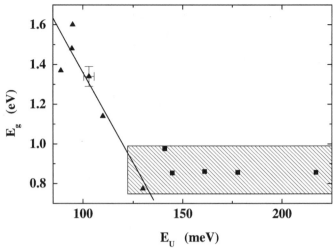

Fig. 16. (a) Plot of Urbach energy, E_U, and Tauc energy gap, E_g, vs hydrogen pressure. (b) Deposition rate, r_d, vs hydrogen pressure. The lines only indicate trends of the plotted quantities. (After [77], reprinted with permission by Elsevier.)

Fig. 17. Plot of E_g vs E_U for hydrogenated specimen (▲). The shaded area contains E_U values of unhydrogenated samples. The line indicates the best fit on the experimental points. (After [77], reprinted with permission by Elsevier.)

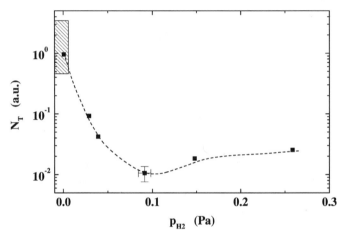

Fig. 18. Plot of the relative defect density, N_T, as a function of the hydrogen pressure. The shaded area contains N_T values of unhydrogenated samples. The line only indicates the trend of the plotted quantity. (After [77], reprinted with permission by Elsevier.)

E_U decreases up to ≈90–95 meV, much lower than the saturation value of the unhydrogenated samples. This is in agreement with results of structural characterization of the hydrogenated sample [52]. An evolution toward nanocrystalline structure is evidenced, with a reduction of r_d from ≈0.8 Å s^{-1} ($p_{H_2} \approx 1\%$ p_{Ar}) to ≈ 0.15 Å s^{-1} ($p_{H_2} \approx 6\%$ p_{Ar}). The influence of H on microcrystallinity of the a-GaAs films was also found by Carchano [14]. The influence of H in our samples is very similar to that found in [80] in sputtered a-Ge:H. In our case, we found an increase of the optical gap, E_g, of about 0.50 eV when p_{H_2} was increased from 0 to about 5% of p_{Ar}. In Figure 17, we present a plot of E_g vs E_U in unhydrogenated and hydrogenated specimens. E_g vs E_U does not show a Cody linear dependence [81] in unhydrogenated samples (shaded area of Fig. 17). The Cody relationship seems verified for the hydrogenated samples. PDS measurements are quoted by [14]. The authors computed the density of defects, N_T, from the part of $\alpha < 2 \times 10$ cm^{-1} for hydrogenated samples. The N_T values are given as relative quantities, assuming the value of $N_T = 1$ of an unhydrogenated sample. Figure 18 shows the plot of N_T vs p_{H_2}. A rapid decrease of N_T is evident when p_{H_2} increases up to 0.1 Pa. The shaded area contains the N_T values estimated in our unhydrogenated samples.

Sedeek [82] prepared a-GaAs films, whose C_{As}/C_{Ga} ratio was controlled by varying the deposition temperature in the range 40–250 °C. Sample composition was measured using microanalysis X-ray emission and their thickness was fixed at 400 nm. From classical optical absorption measurements, the author was able to compute, following Tauc's method, the energy gap, E_g, and the parameter B of the relation: $\alpha h\nu = B(h\nu - E_g)^2$. B^{-1} is the so-called edge width parameter representing the film quality. E_g is practically independent of T_S, going from 1.21 to 1.29 eV, when T_S increases from 40 to 250 °C. In the same temperature range, B^{-1} decreases from

2.9×10^{-6} to 1.8×10^{-6} eV cm and C_{As}/C_{Ga} decreases from 1.22 to 1.05, with the samples always As-rich. The author conclude that most of the defects associated with the As excess (dangling bonds and wrong bonds) are energetically located out of the fundamental absorption edge of a-GaAs.

5.2.2. Flash Evaporation

Stuke and Zimmerer [83] measured the optical properties over a broad range of photon energy of several III–V amorphous compounds (GaAs, GaP, GaSb, InP, InAs, and InSb) deposited in the substrate temperature range 210–300 K. The mass density, ρ, was estimated to be somewhat smaller (several percent) than the crystalline one: for example, c-GaAs: 5.31 g cm^{-3}, a-GaAs: (5.0 ± 0.5) g cm^{-3}. Room temperature reflectivity was measured in the range 1–12 eV at nearly normal incidence. Figure 19 shows the ε_1 and ε_2 spectra of amorphous and crystalline

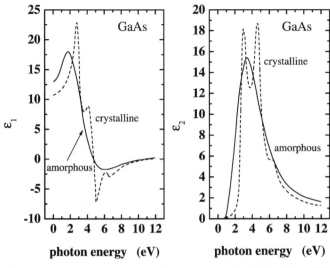

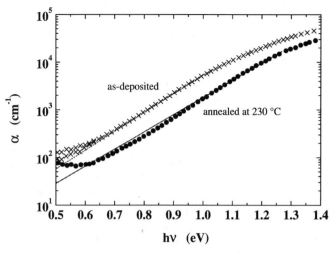

Fig. 19. ε_1 and ε_2 spectra of amorphous and crystalline GaAs. (After [83], reprinted with permission by Wiley–VCH.)

Fig. 21. Optical absorption edge for a-GaAs films, as-deposited ($\times$) and annealed at 230 °C ($\bullet$). (After [86], reprinted with permission by Elsevier.)

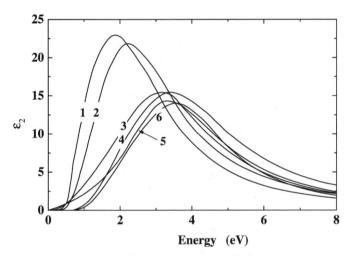

Fig. 20. Comparison of the ε_2 spectra of the amorphous III–V compounds: (1) InSb, (2) GaSb, (3) InAs, (4) GaAs, (5) InP, and (6) GaP. (After [83], reprinted with permission by Wiley–VCH.)

GaAs. General features of the spectra of the disordered material when compared to crystalline ones are:

(1) The smearing of the three crystalline peaks, E_1, E_2, and E_1' in one broad maximum. E_1 and E_1' are due to transitions between the spin-orbit split upper valence band near the L-point along the Λ-axis to the first and second conduction band, whereas E_2 is assigned to transitions near point X of the Brillouin zone.

(2) The shifts of the smeared spectra toward lower energies. It is relatively small for GaAs, where the ε_2 spectrum could be interpreted as an average of the ε_2-curve of the crystal. On the contrary, it is more relevant for GaP, with the amorphous curve well shifted toward lower energies. From a comparison of all the spectra, it can be deduced that the E_2 structure is the most affected by disorder. These ε_2 features are

reflected in the ε_1 curves, showing only one broad region of anomalous dispersion and a shift to lower energy. The static electronic dielectric constant $\varepsilon_1(0)$ for c-GaAs and a-GaAs are 10.8 and 13.0, respectively. Moreover, Stuke shows (Fig. 20) that the features of the measured spectra are common to all III–V materials (same crystal structure with tetrahedral short-range order) and that they can be interpreted in terms of long-range order loss. A modified Penn model can be applied to describe the optical properties of amorphous materials with tetrahedral short-range order. Moreover, the analysis of the data is difficult due to the sample history and many different models can be used in the attempt to reproduce the experimental data. In the case of the amorphous III–V compounds, the role of "wrong bonds" could be considered in computing $\varepsilon_1(0)$, systematically higher than in the equivalent crystal. This made it impossible to have a good fit on the ε_2 spectrum.

Gheorghiu and Theye [85] found an index of refraction of 3.75 for as deposited stoichiometric films ($E_g = 1.05$ eV). It evolves to 3.65 after annealing to 280 °C ($E_g = 1.17$ eV) and it evolves to 3.37 when crystallized to 360 °C. The absorption coefficient shows a rather sharp edge in as-deposited samples, and only a shift toward higher energies in annealed samples. The sharp edge could correspond to a presence of chemical order.

Theye et al. [86] measured the optical absorption edge of flash-evaporated stoichiometric a-GaAs films below photon energy of 0.5 eV, by spectrophotometric and photothermal deflection spectroscopy (PDS). The optical gap, E_g, evolves from (1.02 ± 0.02) to (1.15 ± 0.01) eV for as-deposited and 230 °C annealed samples. On the contrary, the Urbach's characteristic energy, E_U, does not change from its value (110 ± 3) eV. For nonstoichiometric (As excess of about 0.6) a-GaAs films, there is no (Fig. 21) change of the optical gap. The exponential behavior of the absorption edge suggests that the absorption be essentially related to band tail states both before and after annealing. Moreover, the fact that E_U retains the same value after annealing suggests a strong contribution from defect states located close to the band edge, superimposed to tail states, which are relatively insensitive to annealing.

5.2.3. Other Methods

Matsumoto and Kumabe [46] measured optical constants in MBE deposited films of controlled composition. They found a strong variation of the index of refraction with the composition: n increases from about 3.5 when $\gamma = $ As/Ga is very near to 1, and then increases up to about 4.8 when γ approaches 2. The absorption coefficient varies strongly with composition and tends to approach that of the crystalline film when the films are As-rich. The optical gap decreases slightly from about 1.0 eV when $\gamma = 1$, to about 0.8 eV if γ tends to 2. These results differ from the data of [39].

Segui, Carrere, and Bui [18] depositing by glow discharge from a mixture of arsine and trimethylgallium (TMG) found a strong dependence of the absorption coefficient from the hydrogen content that bonds in the solid film. The optical gap is around 1.34 eV when the ratio, P_{AsH_3}/P_{TMG}, of the partial pressures is equal to 10.

Flohr and Helbig [87] determined the optical constants of a-GaAs films by using photoacoustic absorption spectroscopy (PAS). The spectra of the optical absorption coefficient, α, and of the refractive index, n, were measured at room temperature, in the wavelength range 800–2000 nm, in samples with and without hydrogen. The spectral behavior of n decreases in unhydrogenated material monotonically from 5.5 at $\lambda = 1300$ nm to 4.2 when $\lambda = 1800$ nm. In samples with 5 or 40 at.% of hydrogen, n maintains the same behavior, but parallel shifted at shorter λ. The absorption edge shifts to shorter wavelengths with hydrogen, increasing its steepness. The inclusion of H in the film clearly saturates dangling bonds, but changes the packing density of the a-GaAs.

Feng and Zallen [88] performed reflectivity measurement, for photon energies from 2.0 to 5.6 eV in the electronic interband regime, in unannealed ion-implanted GaAs samples. The surface layer was amorphized by exposure to 45-keV Be$^+$ ions. The microstructure of the near-surface implantation-induced damage layer consists of a fine grain mixture of amorphous GaAs and GaAs microcrystals. By using a Lorentz-oscillator analysis, the optical dielectric function of the damaged layer was derived, while the optical constants were extracted using the effective-medium approximation. From the spectra, both the real and the imaginary part of the dielectric functions for the implanted samples as a function of the photon energy, it can be seen that the dielectric function is sensitive to the damage produced by ion implantation. In fact, the sharp spectral features are smeared out as ion fluence increases. These spectral changes are due to two separate effects: the increasing presence of the amorphous phase and the finite-size effects on the optical properties of the crystalline phase. With a right choice

of the photon energies (>2 eV), the optical penetration depth was less than the depth of the high-damaged layer (≈ 1500 Å). The measure is also made in a region with a volume uniform mixture of two components: microcrystals and amorphous material. Using the effective-medium approximation (EMA), the authors assume that the amorphous component has the same dielectric function as a pure amorphous material, but that the dielectric function of the microcrystalline components cannot be identified with that of crystal GaAs. In fact, elaboration of the experimental reflectivity data evidence that both real and imaginary parts of the dielectric function of the μc-GaAs differ appreciably from those of c-GaAs, especially in the vicinity of the sharp interband features and, in general, showing a much more smoothed behavior. For example, the E_1 and $(E_1 + \Delta E_1)$ doublet at near 3 eV, merge into a single peak as crystalline size becomes less than about 200 Å. The smoothing effect increases with the raise of the implanting dose: in fact, the average crystal size, L, decreases from about 500 to about 55 Å with the fluence increasing from 1×10^{13} to 5×10^{14} cm^{-3}. At the same time, the amorphous fraction increases from 3 to 20–25%. The evolution of the optical properties of μc-GaAs can be given in terms of the dependence of the E_1, $(E_1 + \Delta E_1)$, and E_2 (at about 5 eV) on decreasing L. These linewidths increase linearly and rapidly with L^{-1}, evidencing a finite-size effect. A possible simplified model introduces a short time for excited carriers to reach, and to be scattered, by the microcrystal boundary, with a severe limitation of the excited-state energy and a broadening of the excited-state energy. The lifetime is limited to about 2–5×10^{-14} s, when $L \approx 100$ Å.

Zammit et al. [89] studied subgap optical absorption in room temperature ion-implanted layers of GaAs using PDS. Implanting monocrystalline samples of GaAs, with different doses of 100-keV Si ions, the subgap absorption at 1 eV, for example, increases from $\approx 2 \times 10^{-1}$ cm^{-1} (unimplanted) to $\approx 2 \times 10^4$ cm^{-1} (10^{15} cm^{-2}). The absorption spectra show two well-defined region: a steep decrease in the fundamental band-edge region (Urbach's tail) and a much slower decrease at lower energies. An increase of E_U in amorphized samples is found compared to those measured in crystalline materials. In this last case, according to the growth method and possible treatments, E_U varies from (11 ± 1) to (33 ± 2) meV. In samples implanted with the lowest doses 1–5×10^{12} cm^{-2}, E_U raised to 18–34 (± 3) meV, and tends to increase rapidly for higher doses. The region of lower energies is characterized by energy E_1, variable in the range 0.175–0.52 eV, according to implantation dose, temperature, ...

6. PHONON SPECTRA

6.1. Theoretical Results

Ghosh and Agrawal [90] theoretically studied the vibrational excitations in different hydrogenated amorphous gallium-based alloys. They used the cluster Bethe lattice method introducing the angle-bending forces in a rotationally invariant valence force-field model, in which the noncentral interactions arise from the change in the bond angle formed by three atoms. With an appropriate choice of the values for the radial and the angle-bending force constants, the authors are able to demonstrate that their results are in good agreement with the experimental ones of [36]. Many frequencies were computed and assigned to localized vibrational and to various hydrogenated or deuterated modes, arising from the various stretching and angle-bending motions. Two broad bands appearing in the IR spectra at about 1400–1800 and 510–540 cm^{-1} have been assigned to the bridging positions of H atoms between two Ga atoms, in GaP, GaAs, and GaSb.

6.2. Experimental Results

6.2.1. Dc and Rf Sputtering

Stimets et al. [91] reported on phonon spectra in thick (typically, 40 μm) films. The analysis of the experimental curves was performed using interference fringes or Kramers–Krönig relations, where possible. $\varepsilon_1(\infty)$ measured in the spectral range 500–1000 cm^{-1} was found to be (12.25 ± 0.7). The phonon spectrum presents a strong absorption peak (4000 ± 400 cm^{-1}), located at 254 cm^{-1}, and three other less pronounced peaks to about 80, 140, and 530 cm^{-1}. These peaks are not found in c-GaAs, while the absorption peak is located at the TO phonon frequency of 273 cm^{-1} ($\mathbf{k} = 0$). The peak at 530 cm^{-1} (twice the frequency of the main peak) seems due to oxygen defects introduced by the sputtering target and/or by the limited base vacuum. Alternatively, a two-phonon process could be possible. From a comparison of IR absorption and Raman spectra, showing the principal IR peak, narrower and shifted to lower energy compared to correspondent Raman peak, Stimets concludes that the TO mode is IR active. However, both the transverse optical (TO) modes and longitudinal optical (LO) modes are Raman active, these last producing the broadening of the Raman peak at high energies.

Paul et al. [76] measured IR absorption in hydrogenated samples. The absorption appears to be between 1200 and 2000 cm^{-1} an unresolved superposition of a number of modes close in frequency.

Wang, Ley, and Cardona [36] performed a detailed IR absorption analysis of samples deposited at $T_S = 293$ K. Raman spectra show that samples are amorphous up to $T_S \approx$ 50 °C: TO and LO phonon peaks are evidenced in the spectra. Two well-defined, but broad, peaks are evidenced at 530 and 1460 cm^{-1}, respectively. Isochronal annealing shows that $T_A \approx$ 270–300 °C is both the starting crystallization temperature and the nearly completed hydrogen evolution. The two previously mentioned broad peaks at 530 and 1460 cm^{-1}, are assigned to Ga–H–Ga bridge wagging and stretching, respectively. The "bridging" position of an H atom between two Ga atoms tends to reduce the density of the wrong Ga–Ga bonds, which formation is favored by the high volatility of the arsenic. Taking into account the relations giving the vibration frequencies of the modes, Wang demonstrated that both bands at 530 and 1460 cm^{-1} correspond to a bridge position of hydrogen between two gallium atoms for the wagging and stretching mode,

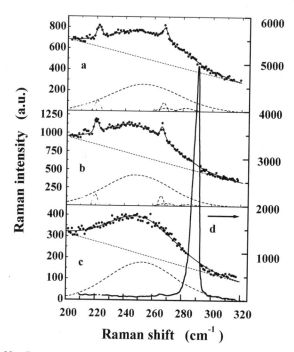

Fig. 22. Raman spectra of amorphous GaAs with corresponding decomposed spectra: film deposited at (a) $T_S = 20$ °C, $W = 200$ W, $p_{Ar} = 2.5$ Pa; (b) $T_S = 20$ °C, $W = 200$ W, $p_{Ar} = 3.4$ Pa; (c) initial crystalline sample which has been completely amorphized by 100 keV Si$^+$ implantation; (d) Raman spectrum from a (100) monocrystalline GaAs is shown for comparison. Narrow lines at 230 and 267 cm^{-1} are plasma lines. (After [92], reprinted with permission by Elsevier).

respectively. These two bands account for the most of the H induced IR absorption in a-GaAs:H.

Desnica et al. [92] studied the influence of deposition parameters on the structure of unhydrogenated gallium arsenide thin films by Raman spectroscopy. The study was based on the analysis of the first-order Raman spectra, allowing for a differentiation between the amorphous component and the crystallites of various sizes. The homogeneity and the uniformity of the samples were examined by TEM [51]. Raman spectra were collected at room temperature under low laser power (<0.5 W) to avoid heating effects using 488.0- or 514.5-nm light. For the 488-nm excitation line, the optical penetration depth of the incident light is $\alpha^{-1} \approx$ 80 nm in bulk, crystalline GaAs, whereas for the 514.5-nm line it is ≈100 nm. Since the samples were thicker than 0.3 μm, all signals were only dependent on the sample characteristic. Figure 22 depicts the Raman spectra of almost completely amorphous samples: the spectra (a) and (b) are from the films deposited at the lowest substrate temperature, 20 °C, deposition power 200 W, and argon pressure 2.5 or 3.4 Pa. In all cases, a peak is found at (250 ± 2) cm^{-1} with full width at half maximum (FWHM) ≈55 cm^{-1}. The fitting procedure gave (90 ± 3) and (98 ± 1)% amorphous phase, respectively; the deconvoluted spectra are presented at the bottom of the corresponding figures. Spectra (c) and (d) are presented for comparison and analysis. Spectrum (c) was measured in an amorphized GaAs sample by ion implantation of 100-keV Si$^+$ ions, while (d), with its Lorentzian shape,

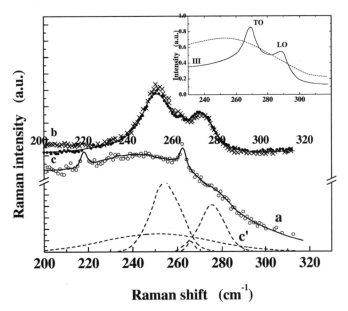

Fig. 23. Raman spectra of GaAs thin films deposited at various substrate temperatures: (a) $T_S = 20$ °C; (b) $T_S = 112$ °C; (c) $T_S = 155$ °C; (c') deconvoluted spectrum (c) depicting TO and LO phonon lines, symmetrically broadened and shifted toward lower frequencies, superimposed on the broad amorphous peak. Narrow lines at 220 and 267 cm^{-1} are plasma lines. Inset: Raman spectrum of ion-bombarded (111) crystalline GaAs (—) and amorphous material (- - - -). (After [92], reprinted with permission by Elsevier.)

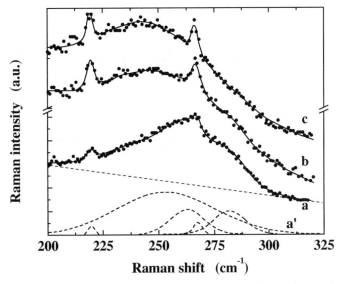

Fig. 24. The influence of the sputtering gas source on the Raman spectra of films deposited at 20 °C and 200 W deposition power: (a) $p_{Ar} = 1.7$ Pa; (a') deconvoluted spectrum (a); (b) $p_{Ar} = 2.5$ Pa; (c) $p_{Ar} = 3.4$ Pa. Narrow lines at 220 and 267 cm^{-1} are plasma lines. (After [92], reprinted with permission by Elsevier.)

was obtained on the (100) surface of the perfectly monocrystalline GaAs sample, before the ion implantation. A completely amorphous phase was obtained when the film was deposited at low substrate temperature (20 °C), relatively low deposition power (200 W), and high pressure of the sputtering gas (3.4 Pa). For the other combinations of deposition parameters

(Figs. 23 and 24), a mixture of polycrystalline and amorphous phases was obtained. For the samples with appreciable fractions of crystalline phase, large downshifts of both LO and TO lines were observed. From the analysis of the frequency shifts and the shapes of the Raman lines, the authors conclude that the films consist of a number of small, variously oriented microcrystallites embedded in an amorphous matrix. The large downshifts are attributed to a combination of defects formed during the film growth and the small sizes of the crystallites. The entire range from mostly microcrystalline to completely amorphous films can be obtained, as a function of the deposition parameters.

6.2.2. Flash Evaporation

Carles et al. [93] performed Raman measurements on flash-evaporated amorphous GaSb. They found that in the investigated range of composition most of the Sb atoms in excess of stoichiometry are incorporated into the amorphous GaSb network with their natural threefold coordination.

6.2.3. Other Methods of Deposition

Holtz et al. [94, 95] performed Raman spectra on 45-keV Be$^+$ ion-implanted GaAs (dose 5×10^{14} cm^{-2}), with a mixed amorphous-microcrystalline microstructure. They were able to evidence a new strong, low frequency peak at 47 cm^{-1} in the Raman spectra of amorphized samples alone: the peak has a nearly constant peak position and linewidth, and it is not present in either c-GaAs or pure a-GaAs, and increases with ion dose. It is strongly resonant near 1.7 eV, in the red, just above the value of the direct gap of c-GaAs, in contrast to the longitudinal-optic phonon line of microcrystals, which resonates in the violet, and the nonresonant broadbands of the amorphous component. The authors conclude that the peak is due to acoustic phonons made Raman active by the presence of microcrystal-amorphous regions, acting as extended defects. In [58], they performed a quantitative depth profile of the structural changes caused by a high dose of 45-keV beryllium implants in GaAs. The samples were chemically etched to remove the surface layers and were characterized by Raman spectra too, prior to any anneal. The structural model of the layer damaged by ion implantation seems to correspond to a mix of amorphous and crystalline phases of GaAs. The first 1500 Å result to be a high-damaged amorphized layer followed by a region, about 2500-Å thick, where the crystalline volume fraction and the crystallite size increase up to a depth of $\approx$4000 Å, where bulk is reached. For fluences ranging between 5×10^{13} and 5×10^{14} ions cm^{-2}, independent behaviors are presented by LO shift, linewidth, and intensity up to a depth of about 1500–2000 Å. For a fluence of 5×10^{14} ions cm^{-2}, the most superficial layers contain about 25% of amorphous volume fraction and crystallites which average size is about 60 Å. The LO Raman line of the crystalline fraction was studied as a function of excitation photon energy from 1.55 to 2.71 eV, showing that its intensity rises rapidly at low photon energy because of the increasing optical penetration depth. Finally, they found an increase in the Raman scattering efficiency above 2.5 eV, probably due to a resonance with the crystalline interband transition at 2.9 eV.

Zallen et al. [96, 97] compared the Raman scattering results in ion-implanted amorphized GaAs layers with those of true a-GaAs films deposited by sputtering and evaporation. They used $^{28}SiF_3^+$ ions at energy of 120 keV, to implant wafers of (100) oriented GaAs. Raman scattering measurements were carried out at room temperature, with laser lines ranging between 1.65 and 2.71 eV, and power less than 15 mW at the sample. In implanted samples, a sharp line at 292 cm^{-1} is well defined, but it tends to broaden if fluence of ion increases. Microcrystallinity of the implanted layer tends to increase, evidencing a broad-band continuum from 0 to 300 cm^{-1}. The authors compare the Raman spectra of three samples: the first one amorphized by ion implantation, the second one sputtered, and the last one evaporated. They evidence that allowing the light to be absorbed in a layer much smaller than the damage-layer thickness, no crystal line can be found in amorphized material, similar to other two real amorphous samples. Three broad overlapping bands are present at 70, 150, and 250 cm^{-1} for all samples and the depolarization ratio of 0.69 at 250 cm^{-1} is the same. A suggestion to use the nondestructive Raman technique is formulated by the authors. Zallen [97] presented an analysis of the optical properties of ion-implanted GaAs, with emphasis on the optical evidence that this material is a composite of amorphous GaAs and GaAs nanocrystals. a-GaAs can be formed by bombardment with heavy ions, but using lower doses or/and lighter ions implantation-induced damage can be studied. Nanocrystalline (or microcrystalline) materials display significant deviations (finite-size effects) from bulk-crystal behavior. The author underlines the role of the two-mode Raman-scattering behavior. It provides direct evidence for the heterogeneous microstructure with the coexistence of amorphous and microcrystalline phases. All the experimental results show the two-mode behavior. The LO-line shape changes (decrease in intensity, shift in peak position, asymmetric broadening) are finite-size effects revealing nanocrystallinity. For a nanocrystal of size L, due to a relaxation of the k-space selection rule, the smaller is L, the larger is the shift and the broadening of the Raman band: this provides a Raman method for estimating L. All these results provide support to the continuous-random network picture of a-solids: the amorphous phase is a distinct and separate phase and μc-GaAs is not a bridge between a-GaAs and c-GaAs. Finite-size effects have also been seen for electronic interband excitations in μc-GaAs, in the form of linewidth broadening linear in $(1/L)$. This dependence can be understood because a very small L implies a very short time to reach and to be scattered by the microcrystal boundary, thus limiting the excited-state lifetime and broadening the excited-state energy.

Ivanda, Hartmann, and Kiefer [98] observed a boson peak in GaAs amorphized by 100-keV Si$^+$ ion implantation. They measured and studied Raman spectra to evidence the common origin of the boson peaks observed in different tetrahedral a-semiconductors. A phase that corresponds to the boson peak is found, separated from the amorphous one and a fractal model can explain the properties of the boson peak. The presence of the boson peak in the form of a phase in tetrahedral a-semiconductors can give new ideas upon their structure and medium-range order.

7. ELECTRICAL TRANSPORT PROPERTIES

7.1. Dc and Rf Sputtering

Paul et al. [76] introduced H in the deposition chamber in parallel with similar procedures used for a-Si and a-Ge. The photoconductivity spectra are shifted to higher energies with hydrogen partial pressure, p_{H_2}, while dark conductivity, σ, is reduced by several orders of magnitude and reveals a regime of activated transport. Annealing decreases σ further, increasing the activation energy. Moreover, the room temperature conductivity decreases very rapidly with a small increase of p_{H_2}. The room temperature photoconductivity on as-deposited samples at $h\nu = 1.95$ eV shows an increase by a factor of 5. All these results allowed concluding about the reduction of the dangling bonds operated by the hydrogen, but no evidence about the question of the wrong bonds.

Alimoussa et al. [99] measured the electrical conductivity as a function of temperature, in a-GaAs samples, with substrate temperature in the range 30–400 °C. All samples are amorphous when deposited at temperatures lower than 400 °C. Stoichiometry is achieved for $T_S \approx 300$ °C: specimens are rich in As or Ga, depending if T_S is lower or greater than this value. The As or Ga excess could correspond to n- or p-doping. A plot of $\log \sigma$ vs $1000/T$ gives in all cases a linear behavior, according to the relation: $\sigma = \sigma_0 \exp(-E_\sigma/kT)$. Both σ_0 and E_σ when plotted as a function of the substrate temperature, show a maximum for $T_S \approx 300$ °C: 10^2 and 0.6 eV $(\Omega$ cm$)^{-1}$, respectively. At T_S greater or lower than 300 °C, both σ_0 and E_σ show a decreasing trend, supporting the hypothesis of a doping of the material from Ga or As excess, with conduction via extended states of valence or conduction band. The room temperature resistivity always is $\approx 10^8$ Ω cm or greater.

Najar et al. [100] studied the properties of a-GaAs doped by cosputtering of GaAs with pieces of molybdenum (Mo), placed on the target. The deposition was performed at 290 °C, optimized to obtain stoichiometric high resistivity amorphous films. The doping level was changed using Mo pieces of different area. According to the doping level in the films, the conductivity changes of many orders of magnitude. For example, at room temperature, σ varies for more than 5 orders: it increases from 10^{-8} $(\Omega$ cm$)^{-1}$ when Mo nominal concentration, $C_{Mo} \approx 0$ at.%, to 6×10^{-2} $(\Omega$ cm$)^{-1}$ if $C_{Mo} \approx 3$ at.%. In parallel, the activation energy of the conductivity varies from ≈ 0.6 to ≈ 0.15 eV, with a rapid decrease when $C_{Mo} \leq 0.5$ eV. The variation of the room temperature conductivity corresponds to a shift of 0.45 eV of the Fermi level toward the conduction band. The Tauc's optical gap evidences a downward shift of 0.1 eV for the doped sample with $C_{Mo} \approx 1.24$ at.%. The doping atoms introduce localized states within the mobility gap.

Mahavadi and Milne [101] studied the ac conductivity in a-GaAs:H in the frequency range of $10^{-2} \div 10^4$ Hz from 100 to 373 K. The analysis of the data allowed obtaining information about the loss type, activation energy of the loss process, and the phonon frequencies. The capacitance and conductance data were obtained on two kinds of samples: (i) an a-GaAs film sandwiched between two metallic electrodes (MSM structure): the bottom one always being Al, while the top was chosen from Au, Al, and In. The ohmic–nonohmic behavior of the contacts was checked by I–V characteristic; (ii) a metal–insulator–semiconductor (MIS) configuration, where silicon nitride or dioxide have been used. The total complex admittance dependent on the frequency can be expressed as

$$Y(\omega) = G(\omega) + i\omega C(\omega) = K[\sigma' + \sigma_1(\omega)] + i\omega K[\varepsilon_{hf} + \varepsilon_1(\omega)]$$

where K is a constant geometrical factor; σ' is the frequency-independent dc conductivity; ε_{hf} is the high frequency permittivity. The capacitance vs frequency shows different responses according to whether MSM or MIS structure is used, to hydrogenation of the material and to measurement temperature. The general shape of the loss curve is independent of the temperature, while its frequency dependence can be expressed as

$$X''(\omega, T) = X(0)F\big[\omega/\omega_p(T)\big]$$

where $X(0)$ is the temperature-dependent amplitude factor and $F[\omega/\omega_p(T)]$ is the spectral shape function. $\omega_p(T)$, the loss peak frequency at temperature T, is strongly temperature dependent and thermally activated

$$\omega_p(T) = \omega_0 \exp(-\Delta E/kT)$$

with: ω_0, phonon frequency and ΔE, activation energy (slope of the linear part, if any, in a plot of $\ln\omega$ vs $1/T$). Different results are found in MSM or MIS structures. According to the variation of the partial pressure of hydrogen from 0 to 10^3 μTorr in MSM, ω_0 increases from 3.5×10^{12} to 2.3×10^{13} Hz. The same trend is shown by the activation energy: from 0.50 to 0.67 eV. In MIS, the activation energy only increases from 0.50 to 0.59 eV, but the phonon frequency rises from 2.7×10^{11} to 2.0×10^{13} Hz with a narrower range of partial hydrogen pressure (0–300 μTorr). The phonon frequencies in unhydrogenated a-GaAs are 1 or 2 orders of magnitude smaller than those in a-GaAs:H and may be attributed to the larger number of dangling bonds in these films. The total conductivity measured as a function of the temperature in the frequency range 10^{-1}–10^4 Hz shows a knee around 220–250 K, separating the low temperature (nearly constant, true ac response) from the high temperature (thermally activated) behavior. The conductivity curves shift parallel each other toward higher values, increasing frequency. Hydrogenated samples tend toward the σ_{dc} value at high temperatures while unhydrogenated ones have no σ_{dc} term. The authors performed an analysis for the data at low temperatures comparing the quantum mechanical to the classical hopping model. The first one gives a frequency dependence of ac conductivity of the form $\sigma_1(\omega) = A\omega \ln^n(\nu_{ph}/\omega)$. $n = 4$ when hopping is over a pair of sites and $n = 3$ if hopping is over several sites and temperature dependent and $\nu_{ph} = \omega \exp(n/1 - s)$, s being the slope of the $\ln\sigma$ vs $\ln\omega$, at low temperatures. The factor A contains the density of states

at the Fermi level, $N(E_F)$, and α^{-1}, the extent of the localized wave function. Using the experimental values of s (close to 0.9), the right temperature dependence is found but phonon frequencies are unreasonably large, because they are exponentially dependent from T. Therefore, the quantum model cannot satisfactorily explain the experimental data on a-GaAs, with and without H. Different classical models consider the hopping of the charge carriers over barriers separating the localized sites. These models allow values of s greater than the maximum permissible 0.8 for the tunneling model. The temperature dependence of the conductivity can be expressed as: $\sigma(\omega) \propto (1/\omega\tau_0)^2$. The experimental data on a-GaAs and a-GaAs:H agree with these classical models, although the computed density of states is somewhat lower than the expected values. The ac conductivity in a-GaAs appears as due to carrier hopping over barriers randomly distributed in space and energy. The activation energy of the measured conductivity at higher temperatures for hydrogenated samples is 0.46 eV.

Carchano, Sedeek, and Seguin [14] performed characterization of high quality a-GaAs in view of the realization of solar cells. The films were prepared at T_S varying between 30 and 300 °C, 50 mTorr of pressure and the power applied to the target adjusted between 35 and 40 W. At $T_S = 290$ °C, the composition is stoichiometric and the conductivity activation energy $E_\sigma = 0.60$ eV, reaches its maximum value. Below this temperature, the material is As-rich and then n-doped, while at higher T_S it is Ga-rich and p-doped: the elevation of T_S leads to an effusion of As, more volatile than Ga, giving rise to a deficit of As. The electrical conductivity was measured as a function of temperature in three different samples. They were prepared at: (1) $T_S = 30$ °C; (2) $T_S = 30$ °C and $p_{H_2}/p_T = 20\%$; and (3) $T_S = 290$ °C. The second and third samples have the same linear behavior over many orders of magnitude, with $E_\sigma = 0.68$–0.70 eV, and a conductivity prefactor, σ_0, ranging between 700 and 1100 $(\Omega \text{ cm})^{-1}$. SCLC measurements allowed calculating a midgap density of states of 3×10^{16} eV^{-1} cm^{-3}, extending over 0.14 eV above the Fermi level. The first sample shows a continuously varying behavior in the whole temperature range, characteristic of a high density of states. The optical gap widens from 1.1 to 1.55 eV, in unhydrogenated stoichiometric samples and to 1.45 eV in $p_{H_2}/p_T = 20\%$ hydrogenated samples. The Urbach's characteristic energy decreases from 140 to 94 meV, when T_S increases from 35 to 290 °C. The decrease of the width of the band tails of 46 meV, can be explained in terms of a significant decrease in the number of wrong bonds and also in an improvement in the local short-range order. The H atoms bond preferentially to form Ga—H—Ga three center bonds. Doping with group VI elements, like Te and S, was demonstrated to be electrically ineffective. On the contrary, doping with Mo changed the room temperature conductivity by many orders of magnitude, leaving the optical gap unchanged.

Murri et al. [5] studied the electrical conductivity of a-GaAs as a function of the H content in the film. The total pressure, p_T, was maintained constant at about 2.5 Pa. The deposition temperature was $T_S = 20$ °C, and the rf power was $W = 200$ W. The H pressure, p_{H_2}, was varied and the H depth profile has been determined by elastic recoil detection (ERD) analysis. Quantitative evaluation of the H concentration has been obtained by comparison with: (1) a sample of silicon implanted with a known dose of H and (2) an amorphous hydrogenated silicon with a well-calibrated, known, and constant H content. The measurements of the conductivity were done in two different but, consecutive runs for all samples: the first run starting from room temperature and going up to 450 K (heating, run A) and the second one coming back to room temperature (RT) (cooling, run B). The measurements of the possible conductivity decay under isothermal conditions were performed on samples identical to those used for $\sigma(T)$ curves. The temperatures, T_A, used for isothermal annealing are always lower than T_K, the temperature above which $\sigma(T)$ shows a downward kink. The measured hydrogen content, C_{H_2}, in the film increases when p_{H_2} raises up to 6.5×10^{-2} Pa, then it saturates at 1.2–1.7×10^{22} at cm^{-3}. The increasing hydrogenation of the material leads to a decrease of the conductivity: the room temperature conductivity, σ_{RT} goes from 2×10^{-8} (Ω cm)$^{-1}$ when $p_{H_2} = 3 \times 10^{-2}$ Pa to 1.5×10^{-11} (Ω cm)$^{-1}$ with $p_{H_2} = 0.26$ Pa. The conductivity curves measured in run A show, for several samples, a not well-defined behavior from RT up to a variable value of T, after which the dependence of $\log \sigma$ vs $(kT)^{-1}$ tends to become linear and the dark conductivity is thermally activated. The value of the activation energy, $E_{\sigma A}$ tends to increase with p_{H_2}. The conductivity measured in run B shows in several cases a downward kink, which occurs to a T in the range 32 eV$^{-1} < (kT)^{-1} < 35$ eV^{-1}. We will define the temperature corresponding to the kink as T_K. The two conductivity curves A and B superimpose very well in the same sample at temperatures greater than T_K. At $T < T_K$, σ_A and σ_B tend to be different. The isothermal electrical conductivity was measured at different annealing temperatures, T_A as a function of the time. The experimental points were best fitted by curves whose equation is: $Y(t) = \exp[-(t/\tau)^\beta]$. The exponential law with long decay time can be attributed to dispersive diffusion of bonded hydrogen, when T_A is lower than the equilibrium temperature $T_{EQ} = T_K$. Great values of τ, like those found in this case, are characteristic of samples deposited under physical vapor deposition (PVD) conditions. Dispersion diffusion of bonded hydrogen is not the only way to induce the stretched exponential law. Crandall [102] also showed that a model of defect controlled relaxation gives stretched exponential time dependence for defect relaxation. The energy gap is strongly dependent on C_{H_2}: $E_g \approx 0.85$–0.9 eV for unhydrogenated samples with $E_U = 130$ meV, then it increases with the hydrogen content up to about 1.5 eV when $p_{H_2} \geq 0.15$ Pa and $E_U = 90$ meV. The low values of the conductivity prefactor, σ_{0A}, correspond to the low hydrogenation level of the films, that means a high density of states in the mobility gap. Conductivity data in the range 140 K $< T <$ 250 K can be fitted by the $T^{-1/4}$ law in our unhydrogenated samples. The density of localized states in the mobility gap is strongly reduced by hydrogenation. The activation energy of the conductivity varies with the hydrogen content: $E_{\sigma A} \approx 0.26$ eV for the unhydrogenated sample, then it increases rapidly with C_{H_2}, rises to about 0.70–0.80 eV

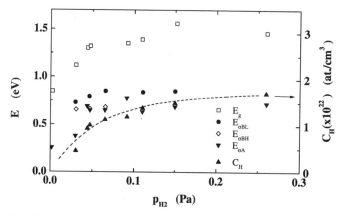

Fig. 25. Activation energies $E_{\sigma A}$, $E_{\sigma BL}$, $E_{\sigma BH}$ (see text), optical gap E_g, hydrogen content C_H as a function of the hydrogen pressure, p_{H_2}. (After [5], reprinted with permission by Elsevier.)

for $C_{H_2} \approx 10\%$, and then becomes independent of C_{H_2}. Figure 25 shows the behaviors of E_g, $E_{\sigma A}$, $E_{\sigma BL}$, $E_{\sigma BH}$, and C_{H_2} vs p_{H_2}: all these quantities show a rapid increase in the range $0 < p_{H_2} \leq 0.1$ Pa. The conclusion is that a high content of hydrogen in the sputtering chamber is not necessary: in fact, the main variation on physical properties are obtained when $P_{H_2} < 0.1$ Pa. The presence of a downward kink in the curve of the conductivity vs $(kT)^{-1}$ can be explained using the model of Spear et al. [103], which postulates a shift of the mobility edge of the conduction band, E_c, as a function of T.

Murri, Pinto, and Schiavulli [104] measured conductivity and Hall mobility as a function of temperature in a-GaAs films. The discharge power, W, or the argon pressure, p_{Ar}, was varied keeping the other deposition parameters constant. A van der Pauw configuration was used both for conductivity and for Hall-mobility measurements. The conductivity was measured using a dc method, while a double modulation technique was used to measure the Hall mobility, μ_H. The electric field was modulated at 13 Hz and the magnetic field was modulated at 20 Hz. An electronic chain allowed obtaining a modulated signal at 33 Hz, sent to the reference channel of a lock-in amplifier. Figures 26 and 27 show the dependence of the Hall mobility vs $1/kT$ for samples deposited varying W or p_{Ar}. At temperatures lower than 260 K (region 1), μ_H varies slowly with T and tends to saturate to 2×10^{-4} cm^2 V^{-1} s^{-1} for $T < 190$ K. In the second region (260 K $< T <$ 320 K), the two measured samples substantially show an undefined behavior for μ_H vs $1/kT$. In the third region, $T > 320$ K ($1/kT = 36$ eV^{-1}), the Hall mobility shows an exponential behavior as a function of $(1/kT)$, with $E_\mu \approx 0.25 \div 0.31$ eV. The conductivity can be expressed in the form $\sigma = \sigma_0 \exp[-E_{\sigma 3}/kT]$, where σ_0 and $E_{\sigma 3} \approx 0.33 \div 0.38$ eV, are the measured preexponential conductivity factor and the activation energy, respectively. Figure 28 shows μ_H vs $1/kT$ for samples of the second group, deposited varying p_{Ar}. μ_H is about 2 orders of magnitude lower than the previous ones and it becomes experimentally undetectable, $(1$–$2) \times 10^{-4}$ cm^2 V^{-1} s^{-1}, for $1/kT > 40$ eV^{-1} ($T \approx 290$ K). An activation energy can be computed for

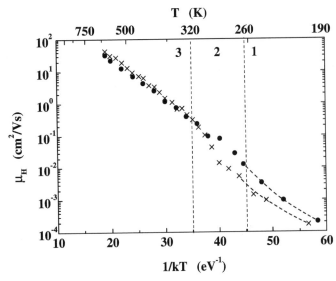

Fig. 26. Hall mobility, μ_H, as a function of $1/kT$ for samples deposited at: $T_S = 300\,°C$, $p_{Ar} = 1.7\,Pa$, $W = 200\,W$ ($\times$) and $300\,W$ ($\bullet$), respectively. The lines only indicate trends of the plotted quantities. (After [104], reprinted with permission by Società Italiana di Fisica.)

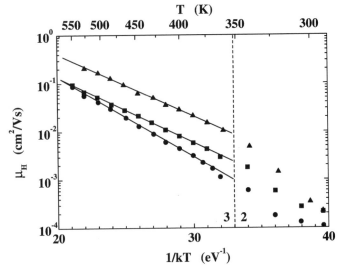

Fig. 27. Hall mobility, μ_H, as a function of $1/kT$ for samples deposited at: $W = 200\,W$, $T_S = 300\,°C$, and $p_{Ar} = 0.86\,Pa$ ($\blacksquare$), $1.1\,Pa$ ($\bullet$), and $4.3\,Pa$ ($\blacktriangle$), respectively. The lines only indicate trends of the plotted quantities. (After [104], reprinted with permission by Società Italiana di Fisica.)

$T > 350\,K$: $E_\mu \approx 0.28 \div 0.35\,eV$. The sign of the Hall mobility was always negative for all these samples. The conductivity as a function of $1/kT$ shows for $145\,K < T < 580\,K$ a downward kink at $T = T_K = 360\,K$ ($1/kT = 32\,eV^{-1}$) only for samples deposited at $p_{Ar} < 1.1\,Pa$. $E_{\sigma2}$, the conductivity activation energy for $250\,K < T < 360\,K$, varies in the range $0.27 \div 0.38\,eV$, whereas in the range $360\,K < T < 580\,K$, varies between 0.27 and 0.38 eV ($E_{\sigma3}$). At temperatures lower than 250 K, the $T^{-1/4}$ Mott law is verified, but not for all samples. μ_H and σ as functions of $1/kT$ behave as follows: (1) for $T \leq 250\,K$, the transport is controlled by variable range hopping; (2) when $250\,K < T < (350–360)\,K$, a transition happens from a predominant transport process to another; (3) for $T > (350–360)\,K$ the behaviors of μ_H and σ vs $1/kT$ are both thermally activated. The conductivity displays a well-defined linear dependence on $T^{-1/4}$ in the temperature range $140\,K < T < 260\,K$, with a slope of $0.019 \div 0.021\,eV$. μ_H also follows a $T^{-1/4}$ law in this region of temperatures with an exponential dependence: the slope is $0.022 \div 0.024\,eV$. The Hall mobility shows an activated behavior at $T > T_K \approx (350–360)\,K$ for all the measured samples. The transport takes place in localized states of the mobility gap, probably in the tail states of the conduction band, according to the values of the preexponential factor of the conductivity: $3 \times 10^{-2} < \sigma_0 < 60$ ($\Omega\,cm^{-1}$). The presence of a downward kink in several samples can be interpreted considering the possibility that samples are not stoichiometric. According to theoretical results [33], we can suppose the presence of a peak of electronic states, generated by gallium defect complexes, located between E_F and E_c. The energy distance of this peak from E_F agrees with the activation energy of the conductivity $E_{\sigma3}$.

Reuter and Schmitt [105] observed an anomalous photovoltaic effect (APVE) and a negative photoconductivity (NPC)

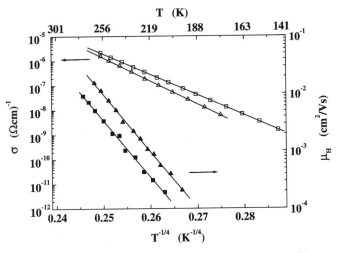

Fig. 28. Conductivity, σ (open symbols) and Hall mobility, μ_H (solid symbols) both as a function of $T^{-1/4}$ for the same specimens of Figure 26. Squares, sample 11 and triangles, sample 12. The lines only indicate trends of the plotted quantities. (After [104], reprinted with permission by Società Italiana di Fisica.)

in obliquely deposited thin, amorphous films of GaAs, Si, and GaAs-Si. Negative photoconductivity means that the conductivity of the illuminated sample is lower than in dark conditions. The authors assume that APVE and NPC are both caused by barriers with negative mobilities, moving in the extended states of valence and conduction bands of amorphous semiconductors. The negative mobilities are most likely caused by negative effective masses of the carriers. The authors demonstrate that very high photovoltages can be generated if carriers with positive mobilities in the localized states and carriers with negative mobilities in the extended states diffuse due to gradients of their densities. The best amorphous films were deposited as $T_S \leq 200\,°C$ in a dc sputtering system with substrates mounted obliquely to the normal of the target. All these films showed

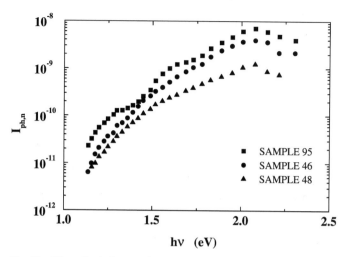

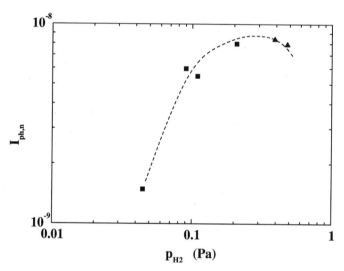

Fig. 29. Normalized photoconductivity as a function of the photon energy. Sample 95: $p_{H_2} = 0.21$ Pa, $p_{Ar} = 1.8$ Pa; sample 46: $p_{H_2} = 0.091$ Pa, $p_{Ar} = 2.4$ Pa; sample 48: $p_{H_2} = 0.045$ Pa, $p_{Ar} = 2.2$ Pa. $T_S = 20$ °C and $W = 200$ W, for all samples. Lines are drawn as a guide for the eye. (After [106], reprinted with permission by Elsevier.)

Fig. 30. Peak values ($h\nu = 2$ eV) of the normalized spectral photoconductivity (Fig. 29) as a function of the hydrogen pressure, p_{H_2}. Squares refer to samples deposited at $p_{Ar} > 1.1$ Pa and triangles refer to two samples deposited at $p_{Ar} < 1.0$ Pa. (After [106], reprinted with permission by Elsevier.)

APVE along the direction of the varying incidence angle of the atoms, while those deposited facing the target did not. The dependence of the two effects on temperature and photon energy shows that they are connected and that both can be explained if photocarriers with negative mobilities move in the extended states of amorphous semiconductors.

Coscia et al. [106] measured the spectral response of the photoconductivity in a-GaAs films with, and without, H, Figure 29. The peak value at $h\nu \approx 2$ eV of the normalized spectral photoconductivity, plotted as a function of hydrogen pressure, p_{H_2}, reaches the maximum value for $p_{H_2} \approx 0.21$ Pa and then tends to a constant value, $\approx 8 \times 10^{-9}$, Figure 30. The maximum value of the ratio ($\sigma_{ph}/\sigma_{dark}$) is 204 in samples deposited at $T_S = 20$ °C with about 25% hydrogen within the deposi-

tion chamber and an rf power of 200 W. The recombination lifetime of photogenerated carriers near the maximum of the photoresponse is $\approx 10^{-9}$ s, several orders of magnitude lower than in the best a-Si. These results are interpreted in terms of the possible effects that hydrogenation can induce in a compound semiconductor. That is, a less disordered network, a reduction of the band tail depth, a decrease of the bond angle fluctuations but an ineffectiveness of hydrogen on the density of wrong bonds.

Seguin et al. [107] performed a systematic study of the effects of the deposition parameters on the chemical composition of a-Ga_xAs_{1-x}. The Ar pressure was increased between 2.6 and 16.6 Pa, with the others parameters constant: the composition of the films did not change and only decreased the concentration of trapped Ar in the solid phase. All the films were slightly As-rich ($\approx 1\%$). Composition of the films was also independent of the argon flow rate, being always $x \approx 0.5$. The increase of the self-bias voltage, V_{dc}, leads to a slight increase of the Ga molar fraction, going from 0.48 ($V_{dc} = 500$ V) to less than 0.49 when $V_{dc} = 750$ V. Ga_xAs_{1-x} films were deposited on glass, Mo, and indium–tin oxide (ITO) coated glass substrates heated at temperatures from 25 to 600 °C. In all cases, the Ga molar fraction, x, shows an increase for $T_S \geq 200$ °C, going from 0.46–0.47 to 0.52–0.54. The slope of this increase is different. According to this, each substrate defines a different temperature for stoichiometric films: 300 °C for glass, 400 °C for molybdenum, and 500 °C for ITO coated glass. The authors conclude that the data are consistent with the hypothesis that the composition of the films is governed by the As reevaporation process at the surface of the growing film. The significant dependence found for x on T_S or V_{dc} support this hypothesis.

7.2. Flash Evaporation

Stuke [108] in one of the first articles on amorphous III–V semiconductors, reported on conductivity and thermoelectric power in samples deposited on a cold substrate at 140 K and then annealed for 15 min at different temperatures. Figure 31 shows the temperature dependence of the conductivity and of the thermoelectric power, S, for nine amorphous samples compared with a crystalline one. The thermoelectric power is positive and increases slightly with temperature: the conduction is extrinsic with an independent hole concentration. The contemporary strong variation of the conductivity has to be ascribed to the activated character of the mobility. The second step of the annealing ($T = 360$ K) leads to an appreciabe decrease of the conductivity and to a shift of the thermoelectric power to higher values. The disappearance of the acceptors is due to the connection of the dangling bonds, favored by the annealing. Increasing the temperature, the annealed, but still amorphous, samples show a transition from extrinsic to intrinsic conduction. It is evidenced by the sign reversal of S. In this region, can be evidenced the most striking characteristic of a-GaAs. The relative contribution of holes and electrons to conduction processes depends on the annealing state, whereas in c-GaAs

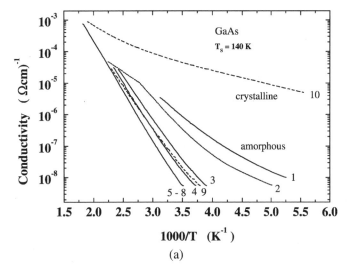

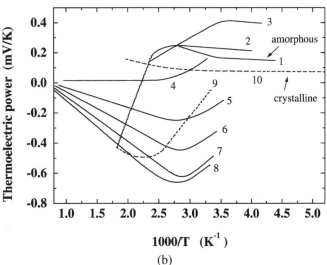

Fig. 31. Temperature dependence of conductivity, (a) and thermoelectric power, (b) of amorphous GaAs for different annealing states. $T_S = 140$ K, annealing time: 15 min; annealing temperature: (1) 300 K, (2) 360 K, (3) 425 K, (4) 460 K, (5) 490 K, (6) 520 K, (7) 540 K, (8) 570 K, and (9) 670 K. (After [108], reprinted with permission by Taylor & Francis, http://www.tandf.co.uk/journals.)

it is only determined by the band structure. The mobility ratio, $b = \mu_e/\mu_h$, can be plotted as a function of the annealing temperature. It starts even below 1 ($\mu_h > \mu_e$) and increases up to values near 10, similar to those of c-GaAs: electrons contribute to intrinsic conduction in annealed samples, more than holes. This fact can be connected with a decrease of the density of localized states near the valence-band edge, B_v, due to the annealing and the consequent increased hopping distance experienced by the holes. Stuke found similar results in a-GaSb too, but with hole mobility values higher than about 6 orders of magnitude than in a-GaAs: 5×10^{-1} cm^2 V^{-1} s^{-1} compared to 5×10^{-7} cm^2 V^{-1} s^{-1} at $T = 250$ K. Stuke examined three possible models (hopping process, trap-controlled drift mobility, and potential barriers within the amorphous semiconductor).

He concluded with the necessity to assume a combination of all three models to explain the electrical properties of these semiconductors.

Gheorghiu and Theye [85] deposited a-GaS films varying the stoichiometry according to the deposition conditions. As-deposited stoichiometric films present a relatively high RT resistivity (about $1-5 \times 10^4$ Ω cm). They remain amorphous when thermally treated up to 280 °C and afterward they show single resistivity activation energy of 0.55 eV, roughly corresponding to half of the optical gap (1.17 eV). Nonstoichiometric films show smaller RT resistivity (down to 10^3 Ω cm) when deposited, lower activation energy of the resistivity after annealing (0.35–0.40 eV) noncorrelated with the optical gap. Crystallized films show an RT resistivity around 10^2 Ω cm with common activation energy (0.10 eV). The analysis of the results leads the authors to conclude that the fundamental point is how excess As atoms are incorporated in the network.

Gheorghiu et al. [13, 40, 109] studied III–V amorphous samples (GaAs and GaP) to evidence the role of the wrong bonds according to the deposition conditions. The high temperature conductivity is activated with a single activation energy ($E_\sigma \approx 0.5$ eV) for a-GaAs, while annealed samples evidence a slightly higher $E_\sigma \approx 0.6$ eV. E_σ is of the order of half optical gap for a-GaAs, similar to a-Ge, and smaller than ($E_{opt}/2$) for GaP: $E_\sigma \approx 0.5$ eV, with $E_{opt} = 1.2-1.4$ eV in as-deposited samples. At low temperatures, the variable range hopping law $\sigma \propto (T_0/T)^{1/4}$ seems verified, with $T_0 \approx 10^9$ K, but variable according to the deposition conditions. This result disagrees with other data of literature, and evidences the strong dependence of the experimental results from the history of the samples: deposition conditions; thermal treatment, stoichiometry, and so on. Amorphous gallium arsenide seems to be chemically ordered, with a low density both of wrong bonds and of structural defects with the associated density of states. On the contrary, a-GaP contains a much higher density of wrong bonds. Reference [109] reports similar studies performed on a-GaSb samples.

Gheorghiu et al. [11], in amorphous films of GaAs, GaP, and GaSb, measured the *in situ* dc electrical conductivity immediately after the deposition. The conductivity is of the order of 3×10^{-6} (Ω cm)$^{-1}$ for nearly stoichiometric samples evaporated with a crucible temperature of ≈ 1700 °C. Small deviations from stoichiometry (0.85 < Ga/As < 1) have little influence on the value of σ. As the crucible temperature decreases to 1400–1500 °C with an increase of As content, σ increases to $10^{-5}-10^{-6}$ (Ω cm)$^{-1}$, according to other results of the literature [46]. Figure 32 shows the behavior of conductivity as a function of $1/T$ during two successive annealing cycles up to 580 and 635 K for an a-GaAs film with: Ga/As = 0.93, $T_S = 300$ K, previously annealed at $T_A = 540$ K. Crystallization is indicated by a steep increase in the film conductivity at about 56° K. Parallel to this, a decrease in the slope of the line σ vs ($T^{-1/4}$) at low temperature is evidenced, characteristic of the variable range hopping transport. It is localized in states near the Fermi level and their density increases with As excess content with respect to stoichiometry. Nevertheless,

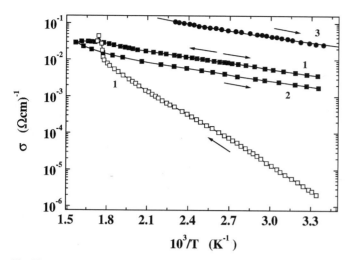

Fig. 32. Variations in the conductivity with the reciprocal temperature for an a-GaAs (Ga/As = 0.93) film during successive annealing cycles through crystallization: curve 1, stabilized amorphous film; curves 1′ and 2, crystallized film; curve 3, a polycrystalline film deposited directly at 430 K. (After [11], reprinted with permission by Elsevier.)

no clear correlation was found between deviation from stoichiometry and electrical conductivity, evidencing that also the distribution of excess As atoms throughout the film plays a fundamental role. Things are better for a-GaSb where a clear correlation between evaporation conditions and conductivity was evidenced. On the contrary, a much less clear situation was established for a-GaP. The conductivity of amorphous films decreases spontaneously as a function of time at the end of deposition, from its initial value to a near constant value. The atomic rearrangements eliminate some of the defects present in the as-deposited films. Thermal annealing can induce further rearrangements. The authors found that crystallization takes place at about 560 K, as evidenced by a steep increase in the film conductivity, independent of their composition. The first stage of crystallization is evidenced by the nucleation of isolated GaAs crystallites inside a still amorphous matrix. The increase of the conductivity corresponds to the coalescence of these crystallites. The crystallization process is strongly dependent on the annealing kinetics. In the diffraction figure, there is no evidence for the presence of segregated arsenic crystallites and the composition tends to become stoichiometric after the crystallization. The activation energy of the conductivity tends to evolve from 0.54 eV (about half the optical gap) in the amorphous films to 0.10 eV in crystallized samples, similar to that found in polycrystalline films.

7.3. Other Methods of Deposition

Guha and Narasimhan [110] vacuum deposited a-GaAs and measured the ac conductivity to evidence if the ω^s law is obeyed. They found a dependence of s on the temperature of measurement, on the thermal history, and on the dc conductivity (s decreases as the conductivity becomes smaller) in contrast

with the theoretical result, $s = 0.8$, constant. Hopping cannot account for all these dependencies. An effective-medium theory for inhomogeneous media can be used, even if the choice of the parameters to describe inhomogeneity is critical.

Matsumoto and Kumabe [46] prepared films of a-GaAs by molecular beam epitaxy (MBE), allowing a fine control of the compositional ratio, $\gamma = $ As/Ga, of the films, intrinsic to the deposition method. The RT resistivity is very high for nearly stoichiometric films ($\gamma \approx 1$, $\rho = 10^6$ Ω cm) and decreases by about 2 orders when γ approaches 2. On the contrary, the activation energy of the resistivity is independent of the film composition: $E_a \approx 0.4$–0.5 eV, about half the optical gap. The Fermi level exists near the middle of the mobility gap. Measurements of the thermoelectric power allowed to state that amorphous films are n-type where $\gamma \approx 1$–1.5 and are p-type where it is 1.5–2. The authors compare with their own results on sputtered films, normally Ga-rich, $\gamma \approx 0.9$–1 and p-type and conclude that in nearly stoichiometric films, the conduction type is determined by whether the film is Ga- or As-rich. The control of the conduction type can be achieved by controlling the composition ratio near the stoichiometry.

Burrafato et al. [47] measured resistivity and Hall coefficient as a function of temperature and magnetic field in InAs thin films (100–1500 Å) deposited by vacuum evaporation at room temperature. The Hall mobility always ranges from 10 to 100 cm^2 (V s)$^{-1}$. Intrinsic conductivity regime was not found in the investigated temperature range, avoiding the calculation of extrapolated energy gap. In the same way, it was not possible to evidence for impurity levels; in fact, temperature-dependent activation energy was always found, with values ranging from 1 to 10^2 meV, according to the measured sample. A surprising result was evidenced in the conductivity curves: an anomalous sharp variation separating two temperatures ranges with different conductivity slopes. The anomaly is observed in the range 135–150 K for all samples ($\Delta T \leq 5$ K), even if subjected to repeated thermal cycles. Only after an air exposure of several days, the discontinuity disappears. A numerical analysis best fit of the conductivity data allowed estimating the values of σ_0, T_0, and $(1/m)$ in the equation: $\sigma = \sigma_0 \exp(-T_0/T)^{1/m}$. σ_0 varies from ≈ 11 to $\approx 10^4$ (Ω cm)$^{-1}$ and T_0 from $\approx 2 \times 10^3$ to $\approx 1 \times 10^8$ K, respectively, in the low temperature region. In the high temperature range, σ_0 varies from $\approx 10^2$ to $\approx 10^4$ (Ω cm)$^{-1}$ and T_0 from $\approx 2 \times 10^3$ to $\approx 3 \times 10^5$ K. The exponent $(1/m)$ ranges from 1 to 1/2, at high temperatures and from 1/1.8 to 1/7.1 in the low temperature region. The disordered morphology of the films and its influence on transport mechanisms according to variable temperature can account for these variations and, through sharp reversible modification, probably, for conductivity kink.

Demishev et al. [111] studied ac and dc conductivities, magnetoresistance, hopping thermopower, and X-ray diffraction measurements in bulk a-GaSb, deposited by quenching under pressure. They found that activated temperature regime dominates for $T > 200$ K. A transition to variable range hopping takes place for 90 K $\leq T \leq 200$ K, while below 90 K the conductivity obeys the Mott law. The temperature dependence of the thermopower below $T \approx 25$ K is adequately described by a square-root law, $S(T) \approx T^{1/2}$, whereas at higher temperatures the dependence becomes linear, $S(T) \approx T$. High temperature ac conductivity and thermopower show appreciable deviations from the theoretical predictions.

8. APPLICATIONS, DEVICES

Eaglesham et al. [112] studied the ion beam damage in superlattices consisting of two materials with different damage rates. They formed a c-AlAs/a-GaAs superlattice by irradiation with 2-MeV Si^+ ions at 80–90 K of the structure c-AlAs-c-GaAs deposited by MBE on a GaAs substrate. With a right choice of the all irradiation conditions to enhance the selective damage of crystalline materials used in superlattice structures, the authors achieve the result to alternatively combine crystalline and amorphous layers. The irradiation was made at liquid nitrogen temperature with very high dose (5×10^{15} cm^{-2}); then the samples were allowed to return to room temperature, with a partial annealing of the damage. Cross-section TEM images show that GaAs layers are converted from crystalline to amorphous network, while AlAs layers remained undamaged: a-GaAs layers alternated to c-AlAs form the superlattice structure. A study of the built up damage allowed obtaining information on the amorphization mechanism. The bulk GaAs is almost completely amorphized by 1×10^{15} Si^+ cm^{-2}, and amorphous layers have formed at the center of each GaAs layer. With further increases in the dose, the amorphous-crystal interfaces move outward toward the AlAs layers, becoming flatter until at 5×10^{15} Si^+ cm^{-2} the interface is flat and lies within a few nanometers of the AlAs layer. These structures combine the possibility to use crystalline and amorphous materials and to tailor the band offsets using Fermi level pinning in the amorphous phase.

Fennouh et al. [113, 114] studied the electrical properties of the heterojunction formed sputtering a layer of undoped amorphous GaAs (500-nm thick). The layer was deposited on n-type crystalline Si wafers (impurity concentration of about 5×10^{15} cm^{-3}) with a backside n^+-diffused layer for ohmic contact. The a-GaAs layer was sputtered under the following conditions: Ar pressure, $P_{Ar} = 6.67$ Pa; rf power, $W = 25$ W, corresponding to a dc self-bias voltage, $V_{dc} = 600$ V and substrate temperature, $T_S = 150$ °C. The films presented a slight excess of As (51%). Ag contact was evaporated on the backside of an Si wafer while Au was evaporated on the a-GaAs layer with an area of 5×10^{-3} cm^2, to fabricate the heterojunction diode. Capacitance-voltage (C-V) characteristics, Figure 33, were measured at 1 MHz at room temperature. The C^{-2} vs V curve presents a linear behavior for reverse applied voltage (forward direction means positive bias applied to a-GaAs) from -5 to -1 V, being, after 0.5 V and up to 3 V, independent of V. The characteristic looks typical for a high frequency differential capacitance of this type of heterostructure. The total capacitance, C, of the junction is given by a series connection of C_a (amorphous layer) and C_c (crystalline substrate). Due to high frequency and the high resistivity of the a-GaAs film, the depletion layer extends in the c-Si (n) side only. From experimental considerations, the authors deduce $C_a \approx 23$ nF, as saturated capacitance with the forward bias. Under the assumption of abrupt heterojunction, they were able to calculate the mainband discontinuity, $\Delta E_v = 0.21$ eV, occurring in the valence band, Figure 34. Density of current, J, as a function of voltage, V, characteristics were measured at variable temperature in the range 213–402 K. The voltage interval was -1.5–$+1.5$ V. The diode exhibit a marked rectification characteristic for the measured temperature. Due to the quasi-continuous distribution of localized states in the mobility gap of a-GaAs it was necessary to use a multitunneling capture-emission process to interpret the J–V characteristics. The forward current assumes the form $J = J_O \exp(AV)$, for voltages less than 0.4 V, with J_O proportional to $\exp(-E_{af}/kT)$. Experimentally, the activation forward energy, E_{af}, was determined to be 0.34 eV, much lower than the activation energy of the dark conductivity (0.8 eV), due to electron emission process. The reverse current is proportional

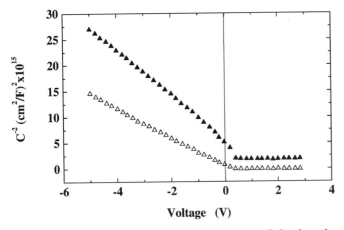

Fig. 33. Plot of inverse of square of capacitance vs applied voltage for a-GaAs-c-Si(n) heterojunction. Solid symbols represent the capacitance measured; open symbols represent the capacitance deduced from Eq. $(1/C) = (1/C_c) + (1/C_a)$, with C: total capacitance; C_c: depletion layer capacitance and C_a: GaAs amorphous layer capacitance. (After [113], reprinted with permission by Elsevier.)

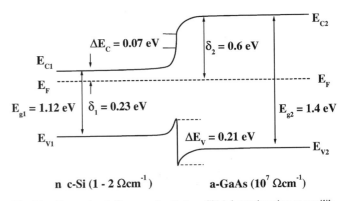

Fig. 34. Energy band diagram of a-GaAs-c-Si(n) heterojunction at equilibrium. (After [113], reprinted with permission by Elsevier.)

to $\exp(-E_{ar}/kT)(V)^{1/2}$, where the activation reverse energy, $E_{ar} = 0.55$ eV, is constant. The reverse current always exceeds the value of J_O, so indicating its possible limitation by a generation process in the depletion region. Similar results were obtained in a-GaAs/c-Si(p) heterojunctions.

Aguir et al. [115] studied two different types of structures: heterojunctions a-GaAs/c-GaAs(n) and the MIS structure a-GaAsN/c-GaAs(n). Nitrogen was introduced in gallium arsenide to increase the resistivity of the film. In both cases, GaAs substrates heavily doped ($N_D = 9 \times 10^{15}$ cm^{-3}) were used. a-GaAs films (500-nm thick) were deposited under the following optimized conditions: argon pressure, $p_{Ar} = 6.67$ Pa; rf power, $W = 25$–30 W, corresponding to a dc self-bias voltage, $V_{dc} = 600$–700 V and substrate temperature, $T_S = $ RT. The measured optical gap is 1.1 eV and the RT resistivity $\approx 10^5$ Ω cm. Metallic contacts were deposited using Au on a-GaAs ($\approx 0.8 \times 10^{-2}$ cm^2) and In on c-GaAs (n^+). C–V characteristics were measured at 1 MHz. A plot of C^{-2} vs V shows a nonlinear behavior giving information that the junction is not sharp, probably due to a high density of states at the interface amorphous crystal. The J–V curves show a strong rectification effect. In the attempt to increase the resistivity of a-GaAs, the authors introduced nitrogen at a given pressure, the total pressure is constant at 6.67 Pa. The substrate temperature was maintained to 493 K (instead of 293 K as in the previous case). The structure was made identical to the previous one. The curves J–V evidence a strong decrease in the intensity of the current, but without rectification effect. At forward voltage greater than 0.8 V, the J–V curve becomes nonlinear, evidencing injection of carriers. Room temperature C–V curves at three different frequencies (10, 100 kHz, and 1 MHz) are practically superimposed, evidencing a deep depletion layer. However, the presence of a fixed charge inside the a-GaAsN film is evidenced and the density of interface states is computed to be 10^{11} cm^{-2} eV^{-1}, lying at midgap.

9. LIST OF SYMBOLS

C_{As}	Arsenic content
C_{Ga}	Gallium content
C_{H_2}	Hydrogen content
d	Total thickness
D	Interelectrode distance
DOS	Density of state
E_c	Conduction-band edge
E_v	Valence-band edge
E_g	Tauc's optical energy gap
E_F	Fermi energy
E_{04}	Optical energy gap ($\alpha = 10^4$ cm^{-1})
E_{go}	Mean optical energy gap
E_U	Urbach's energy
E_μ	Activation energy of the Hall mobility
E_σ	Activation energy of the dark conductivity
g	Landè factor
G	Grain dimension
$J(\omega\hbar)$	Joint density of state function
L	Mean diameter of agglomerates
m	Ratio between intensities of two molecular beams
N	Coordination number
n	Index of refraction
$N(E_F)$	Gap state density at the Fermi level
N_S	Number of spin cm^{-3}
N_T	Density of defect state
n_C	Carrier concentration
p_{Ar}	Partial pressure of argon
p_{H_2}	Partial pressure of hydrogen
p_t	Transition pressure
p_T	Total pressure
R	Ratio between hydrogen and argon partial pressures
r	Ratio of two gas fluxes
r_d	Deposition rate
rf	Radio frequency
T_A	Annealing temperature
T_C	Crystallization temperature
T_S	Substrate temperature
U	Defect correlation energy
V_{dc}	Self-bias voltage
x	Stoichiometry index
W	Radio frequency power
W^*	Power density
WB	Wrong bond
Δ	Bond angle distortion parameter
α	Optical absorption coefficient
χ	Diamagnetic susceptibility
χ_p	Paramagnetic term of susceptibility
ε	Dielectric constant
γ	As/Ga, composition ratio
λ	Radiation wavelength
μ_e	Electron drift mobility
μ_h	Hole drift mobility
μ_{He}	Electron Hall mobility
μ_{Hh}	Hole Hall mobility
θ	Bond angle
ρ	Density of the material
σ_0	Conductivity prefactor
σ	Electrical conductivity
σ^{*2}	Energy gap variance
σ^*	Disorder parameter

REFERENCES

1. See, for example, "Properties of Gallium Arsenide." EMIS Datareviews Series no. 16 (M. R. Broze and G. E. Stillman, Eds.), INSPEC, IEE, London, 1996; S. Adachi, *J. Appl. Phys.* 58, R1 (1985).

2. L. Alimoussa, H. Carchano, and J. P. Thomas, *J. Phys. Paris* 42, C4-683 (1981).

3. J. P. Thomas, M. Fallavier, H. Carchano, and L. Alimoussa, *Nucl. Instrum. Methods* 218, 579 (1983).

4. R. E. Benenson, L. C. Feldman, and B. G. Bagley, *Nucl. Instrum. Methods* 168, 547 (1980).

5. R. Murri, N. Pinto, L. Schiavulli, R. Fukuhisa, and L. Mirenghi, *Mater. Chem. Phys.* 33, 150 (1993).

6. H. Carchano, F. Lalande, and R. Loussier, *Thin Solid Films* 120, 47 (1984).

7. H. Carchano, F. Lalande, and R. Loussier, *Thin Solid Films* 135, 107 (1986).

8. J. Bandet, J. Frandon, G. Bacquet, K. Aguir, B. Despax, and A. Hadidou, *Philos. Mag. B* 58, 645 (1988).

9. K. Aguir, M. Hadidou, P. Lauque, and B. Despax, *J. Non-Cryst. Solids* 113, 231 (1989).

10. G. Casamassima, T. Ligonzo, R. Murri, N. Pinto, L. Schiavulli, and A. Valentini, *Mater. Chem. Phys.* 21, 313 (1989).

11. A. Gheorghiu, T. Rappenau, S. Fisson, and M. L. Theye, *Thin Solid Films* 120, 191 (1984).

12. A. Carbone, F. Demichelis, G. Kaniadakis, F. Gozzo, R. Murri, N. Pinto, L. Schiavulli, G. Della Mea, A. Drigo, and A. Paccagnella, *Nuovo Cimento* 13, 571 (1991).

13. A. Gheorghiu and M. L. Theye, *Philos. Mag. B* 44, 285 (1981).

14. H. Carchano, K. Sedeek, and J. L. Seguin, "Proceedings of the Euroforum New Energies Congress," Vol. 3, p. 197. H. S. Stephens and Associates, Brussels, 1988.

15. J. L. Seguin, B. El Hadadi, H. Carchano, A. Fennouh, and K. Aguir, *J. Non-Cryst. Solids* 183, 175 (1995).

16. J. Dixmier, A. Gheorghiu, and M. L. Theye, *J. Phys. C: Solid State Phys.* 17, 2271 (1984).

17. J. H. Dias da Silva, J. I. Cisneros, M. M. Guraya, and G. Zampieri, *Phys. Rev.* 51, 6272 (1995).

18. Y. Segui, F. Carrere, and A. Bui, *Thin Solid Films* 92, 303 (1982).

19. K. Chen, Z. Yang, and R. Wu, *J. Non-Cryst. Solids* 77–78, 1281 (1985).

20. F. Maury, G. Constant, D. Jousse, and A. Deneuville, *Philos. Mag. B* 53, 445 (1986).

21. S. Sahli, Y. Segui, and F. Maury, *Thin Solid Films* 146, 241 (1987).

22. G. Monnom, C. Paparoditis, Ph. Gaucherel, S. Cavalieri, and A. Rideau, *J. Non-Cryst. Solids* 83, 91 (1986).

23. T. Fujiyoshi, M. Onuki, K. Honmyo, H. Kubota, and T. Matsumoto, *J. Non-Cryst. Solids* 137–138, 935 (1991).

24. S. Yokoyama, D. Yui, H. Tanigawa, H. Takasugi, and M. Kawabe, *J. Appl. Phys.* 62, 1808 (1987).

25. K. Tsuji, Y. Katayama, H. Kanda, and H. Nosaka, *J. Non-Cryst. Solids* 205–207, 518 (1996).

26. M. N. Makadsi, "Proceedings of the 4th International Conference on Physical Conductive Matter," 2000, April 18–21, p. 1. University of Amman, Jordan.

27. C. Molteni, L. Colombo, and L. Miglio, *Phys. Rev.* 50, 4371 (1994).

28. E. P. O'Reilly and J. Robertson, *Phys. Rev. B* 34, 8684 (1986).

29. P. C. Tripathy and T. Sahu, *Semicond. Sci. Technol.* 10, 447 (1995).

30. N. Mousseau and L. J. Lewis, *Phys. Rev.* 56, 9461 (1997).

31. I. Ebbsjo, R. K. Kalia, A. Nakano, J. P. Rino, and P. Vashishta, *J. Appl. Phys.* 87, 7708 (2000).

32. L. Alimoussa, H. Carchano, A. Fassi-Fihri, F. Lalande, and R. Loussier, *J. Phys. Paris* 43, C1-341 (1982).

33. R. Kärcher, Z. P. Wang, and L. Ley, *J. Non-Cryst. Solids* 59–60, 629 (1983).

34. S. G. Greebaum, R. A. Marino, K. J. Adamic, and C. Case, *J. Non-Cryst. Solids* 77–78, 1285 (1985).

35. K. Sedeek, H. Carchano, and J. L. Seguin, "Proceedings of the 8th International Photovoltaic Solar Energy Conference" (I. Solomon and B. Equer, Eds.), Vol. 1, p. 993. Kluwer Academic, Dordrecht/Norwell, MA, 1988.

36. Z. P. Wang, L. Ley, and M. Cardona, *Phys. Rev.* 26, 3249 (1982).

37. S. H. Baker, M. I. Manssor, S. J. Gurman, S. C. Bayliss, and E. A. Davis, *J. Non-Cryst. Solids* 144, 63 (1992).

38. S. H. Baker, S. C. Bayliss, S. J. Gurman, N. Elgun, B. T. Williams, and E. A. Davis, *J. Non-Cryst. Solids* 169, 111 (1994).

39. A. Gheorghiu and M. L. Theye, *J. Non-Cryst. Solids* 35–36, 397 (1980).

40. G. Dufour, E. Belin, C. Senemaud, A. Gheorghiu, and M. L. Theye, *J. Phys. Paris* 42, C4-877 (1981).

41. C. Senemaud, A. Belin, A. Gheorghiu, and M. L. Theye, *J. Non-Cryst. Solids* 77–78, 1289 (1985).

42. A. Gheorghiu, K. Driss-Khodja, S. Fisson, M. L. Theye, and J. Dixmier, *J. Phys. Paris* 46, C8-545 (1985).

43. C. Senemaud, E. Belin, A. Gheorghiu, and M. L. Theye, *Solid State Commun.* 55, 947 (1985).

44. D. Udron, M. L. Theye, D. Raoux, A. M. Flank, P. Lagarde, and J. P. Gaspard, *J. Non-Cryst. Solids* 137–138, 131 (1991).

45. D. Udron, A. M. Flank, P. Lagarde, D. Raoux, and M. L. Theye, *J. Non-Cryst. Solids* 150, 361 (1992).

46. N. Matsumoto and K. Kumabe, *Jpn. J. Appl. Phys.* 19, 1583 (1980).

47. G. Burrafato, N. A. Mancini, S. Santagati, S. O. Troja, A. Torrisi, and O. Puglisi, *Thin Solid Films* 121, 291 (1984).

48. M. C. Ridgway, C. J. Glover, G. J. Foran, and K. M. Yu, *Nucl. Instrum. Methods B* 147, 148 (1999).

49. E. Fois, A. Selloni, G. Pastore, Q. M. Zhang, and R. Carr, *Phys. Rev. B* 50, 4371 (1994).

50. H. Seong and L. J. Lewis, *Phys. Rev. B* 53, 4408 (1996).

51. R. Murri, F. Gozzo, N. Pinto, L. Schiavulli, C. De Blasi, and D. Manno, *J. Non-Cryst. Solids* 127, 12 (1991).

52. R. Murri, N. Pinto, L. Schiavulli, R. Fukuhisa, C. De Blasi, and D. Manno, *J. Non-Cryst. Solids* 151, 253 (1992).

53. M. E. Twigg, M. Fatemi, and B. Tadayon, *Appl. Phys. Lett.* 63, 320 (1993).

54. B. Kramer, K. Maschke, and P. Thomas, *Phys. Status Solidi B* 48, 635 (1971).

55. E. P. O'Reilly and J. Robertson, *Philos. Mag. B* 50, L9 (1984).

56. J. Robertson, *Philos. Mag. B* 51, 183 (1985).

57. J. Robertson, *J. Non-Cryst. Solids* 77–78, 37 (1985).

58. M. L. Theye and A. Gheorghiu, *Solar Energy Mater.* 8, 331 (1982).

59. A. Gheorghiu, M. Ouchene, T. Rappenau, and M. L. Theye, *J. Non-Cryst. Solids* 59, 621 (1983).

60. R. Mosseri and J. P. Gaspard, *J. Non-Cryst. Solids* 97–98, 415 (1987).

61. D. E. Polk, *J. Non-Cryst. Solids* 5, 365 (1971).

62. B. K. Agrawal, S. Agrawal, P. S. Yadav, J. S. Negi, and S. Kumar, *Philos. Mag. B* 63, 657 (1991).

63. J. P. Xanthakis, P. Katsoulakos, and D. Georgiakos, *J. Phys.: Condens. Matter* 5, 8677 (1993).

64. M. I. Manssor and E. A. Davis, *J. Phys.: Condens. Matter* 2, 8063 (1990).

65. J. J. Hauser, F. J. Di Salvo, Jr., and R. S. Hutton, *Philos. Mag.* 35, 1557 (1977).

66. M. Ouchene, C. Senemaud, E. Belin, A. Gheorghiu, and M. L. Theye, *J. Non-Cryst. Solids* 59–60, 625 (1983).

67. K. Sedeek, J. Monnom, C. Paparoditis, and Ph. Gaucherel, in Proceedings of the "Poly-Micro-crystalline and Amorphous Semiconductors" (P. Pinard and S. Kalbitzer, Eds.), p. 749. Les Editions de Physique, Paris, 1984.

68. K. Sedeek, *Thin Solid Films* 130, 47 (1985).

69. J. Kuhl, E. O. Göbel, Th. Pfeiffer, and A. Jonietz, *Appl. Phys. A* 34, 105 (1984).

70. S. G. Greenbaum, D. J. Treacy, B. V. Shanabrook, J. Comas, and S. G. Bishop, *J. Non-Cryst. Solids* 66, 133 (1984).

71. B. Hoheisel, J. Stuke, M. Stutzman, and W. Beyer, "Proceedings of the 17th International Conference on the Physics of Semiconductors" (J. D.

Chadi and W. A. Harrison, Eds.), p. 821. Springer-Verlag, New York, 1985.

72. H. J. von Bardeleben, P. Germain, S. Squelard, A. Gheorghiu, and M. L. Theye, *J. Non-Cryst. Solids* 77–78, 1297 (1985).

73. S. Guita, E. Belin, C. Senemaud, D. Udron, A. Gheorghiu, and M. L. Theye, *J. Phys. Paris* 47, C8-427 (1986).

74. A. Deville, B. Gaillard, K. Sedeek, H. Carchano, and K. W. Stevens, *J. Phys. Condens. Matter* 1, 9369 (1989).

75. S. K. O'Leary, *Appl. Phys. Lett.* 72, 1332 (1998).

76. W. Paul, T. D. Moustakas, D. A. Anderson, and E. Freeman, in "Amorphous and Liquid Semiconductors" (W. E. Spear, Ed.), p. 467. University of Edinburgh, 1977.

77. R. Murri, L. Schiavulli, N. Pinto, and T. Ligonzo, *J. Non-Cryst. Solids* 139, 60 (1992).

78. N. Amer and W. B. Jackson, in "Semiconductors and Semimetals" (J. Pankove, Ed.), Vol. 21, Part B, p. 83. Academic Press, Orlando, FL, 1984.

79. R. Murri, L. Schiavulli, G. Bruno, P. Capezzuto, and G. Grillo, *Thin Solid Films* 182, 105 (1989).

80. P. D. Persans, A. F. Ruppert, S. S. Chan, and G. D. Cody, *Solid State Commun.* 51, 203 (1984).

81. G. D. Cody, in "Semiconductors and Semimetals" (J. Pankove, Ed.), Vol. 21, Part B, p. 11. Academic Press, Orlando, FL, 1984.

82. K. Sedeek, *J. Phys. D: Appl. Phys.* 26, 130 (1993).

83. J. Stuke and G. Zimmerer, *Phys. Status Solidi B* 49, 513 (1972).

84. M. L. Theye, in "Amorphous and Liquid Semiconductors" (J. Stuke and W. Brenig, Eds.), p. 479. Taylor & Francis, London, 1974.

85. A. Gheorghiu and M. L. Theye, in "Amorphous and Liquid Semiconductors" (W. E. Spear, Ed.), p. 462. University of Edinburgh, 1977.

86. M. L. Theye, A. Gheorghiu, K. Driss-Khodja, and C. Boccara, *J. Non-Cryst. Solids* 77–78, 1293 (1985).

87. Th. Flohr and R. Helbig, *J. Non-Cryst. Solids* 88, 94 (1986).

88. G. F. Feng and R. Zallen, *Phys. Rev. B* 40, 1064 (1989).

89. U. Zammit, F. Gasparrini, M. Marinelli, R. Pizzoferrato, A. Agostini, and F. Mercuri, *J. Appl. Phys.* 70, 7060 (1991).

90. B. K. Ghosh and B. K. Agrawal, *J. Phys. C: Solid State Phys.* 19, 7157 (1986).

91. R. W. Stimets, J. Waldman, J. Lin, T. S. Chang, R. J. Temkin, and G. A. N. Connell, in "Amorphous and Liquid Semiconductors" (J. Stuke and W. Brenig, Eds.), p. 1239. Taylor & Francis, London, 1974.

92. I. D. Desnica, M. Ivanda, M. Kranjcec, R. Murri, and N. Pinto, *J. Non-Cryst. Solids* 170, 263 (1994).

93. R. Carles, J. B. Renucci, A. Gheorghiu, and M. L. Theye, *Philos. Mag. B* 49, 63 (1984).

94. M. Holtz, R. Zallen, and O. Brafman, *Phys. Rev. B* 37, 2737 (1988).

95. M. Holtz, R. Zallen, O. Brafman, and S. Matteson, *Phys. Rev. B* 37, 4609 (1988).

96. R. Zallen, M. Holtz, A. E. Geissberger, R. A. Sadler, W. Paul, and M. L. Theye, *J. Non-Cryst. Solids* 114, 795 (1989).

97. R. Zallen, *J. Non-Cryst. Solids* 141, 227 (1992).

98. M. Ivanda, I. Hartmann, and W. Kiefer, *Phys. Rev. B* 51, 1567 (1995).

99. L. Alimoussa, H. Carchano, F. Lalande, R. Loussier, and S. Najar, in Proceedings of the "Poly-Micro-crystalline and Amorphous Semiconductors" (P. Pinard and S. Kalbitzer, Eds.), p. 743. Les Editions de Physique, Paris, 1984.

100. S. Najar, K. Sedeek, F. Lalande, and H. Carchano, *Mater. Res. Symp. Proc.* 283 (1985).

101. K. K. Mahavadi and W. I. Milne, *J. Non-Cryst. Solids* 87, 30 (1986).

102. R. S. Crandall, *Phys. Rev. B* 43, 4057 (1991).

103. W. E. Spear, D. Allan, P. Le Comber, and A. Gaith, *Philos. Mag. B* 41, 419 (1980).

104. R. Murri, N. Pinto, and L. Schiavulli, *Nuovo Cimento D* 15, 785 (1993).

105. H. Reuter and H. Schmitt, *J. Appl. Phys.* 77, 3209 (1995).

106. U. Coscia, R. Murri, N. Pinto, and L. Trojani, *J. Non-Cryst. Solids* 194, 103 (1996).

107. J. L. Seguin, B. El Hadadi, H. Carchano, and K. Aguir, *J. Non-Cryst. Solids* 238, 253 (1998).

108. J. Stuke, in "Conduction in Low-mobility Materials" (N. Klein, D. S. Tannhauser, and M. Pollak, Eds.), p. 193. Taylor & Francis, London, 1971.

109. A. Gheorghiu, T. Rappenau, J. P. Dupin, and M. L. Theye, *J. Phys. Paris* 42, C4-881 (1981).

110. S. Guha and K. L. Narasimhan, *Phys. Rev.* 18, 2761 (1978).

111. S. V. Demishev, A. A. Pronin, M. V. Kondrin, N. E. Sluchanko, N. A. Samarin, T. V. Ischenko, G. Biskupski, and I. P. Zvyagin, *Phys. Status Solidi B* 218, 67 (2000).

112. D. J. Eaglesham, J. M. Poate, D. C. Jacobson, M. Cerullo, L. N. Pfeiffer, and K. West, *Appl. Phys. Lett.* 58, 523 (1991).

113. A. Fennouh, K. Aguir, H. Carchano, and J. L. Seguin, *Mater. Sci. Eng. B* 34, 27 (1995).

114. K. Aguir, A. Fennouh, H. Carchano, J. L. Seguin, B. Elhadadi, and F. Lalande, *Thin Solid Films* 257, 98 (1995).

115. K. Aguir, A. Fennouh, H. Carchano, and D. Lollman, *J. Phys III France* 5, 1573 (1995).

Chapter 9

AMORPHOUS CARBON THIN FILMS

S. R. P. Silva, J. D. Carey, R. U. A. Khan, E. G. Gerstner, J. V. Anguita

Large Area Electronics Group, School of Electronics, Computing and Mathematics, University of Surrey, Guildford, United Kingdom

Contents

Handbook of Thin Film Materials, edited by H.S. Nalwa
Volume 4: Semiconductor and Superconductor Thin Films
Copyright © 2002 by Academic Press

ISBN 0-12-512912-2/$35.00

1. INTRODUCTION

"Carbon is the King of mediocrity. Standing as it does in the midpoint of the periodic table, seeking neither other atom's electrons nor willing particularly to surrender its own, it is content to bond to itself and to form chains and rings of intricate form and of seemingly infinite variety." (P.W. Atkins, The Times Higher Education Supplement, February 1995)

Carbon is unique in its structure by being able to form one of the strongest materials known to man, diamond, or one that is soft, graphite, by virtue of the way in which each atom bonds to another. The material could have an bandgap as great as 5.5 eV or one that is negative. All of these variations have been made possible by the three different bond hybridizations that are available to carbon. These are tetrahedral diamondlike sp^3 bonds, trigonal graphitelike sp^2 bonds, and linear acetylenelike sp^1 bonds. See Figure 1. As in the crystalline case, each of these bond variations is available to its amorphous counterparts, and this enables the unique and large variation in the physical and material properties to be extended to amorphous carbon (*a*-C) thin films.

Due to the very many combinations of bond hybridizations available to amorphous carbon films, there is also a significant amount of confusion as to the terminology used to refer to a particular type of *a*-C thin film. It should be noted that the term diamondlike carbon (DLC) should generally be reserved for polycrystalline or nanocrystalline carbon films, whereas amorphous carbon films should generally fall into the categories of

(a) polymerlike amorphous carbon (PAC),
(b) graphitelike amorphous carbon (GAC),
(c) diamondlike amorphous carbon (DAC),
(d) tetrahedral amorphous carbon (TAC), and
(e) nanocomposite amorphous carbon (NAC).

We shall use the terms *a*-C and *a*-C:H as generic terms for amorphous carbon and its hydrogenated counterparts. The notation introduced above will be based more on the microstructure of the films as opposed to its properties which become more and more difficult to associate simply with a given structure due to the major advances being made in deposition techniques. As an example, we have seen the birth of a three-dimensional (3D) graphite structure based on a nanocomposite amorphous carbon whose nanoparticles are planar but still give a hardness that is comparable with a tetrahedral structure [1]. It is believed that the strength of this 3D planar structure arises from the closer in-plane bond lengths of the sp^2 rings. Two examples of NAC films are shown in Figure 2, where there is a mixed amorphous–nanocrystalline phase present, with the nanocrystals usually embedded in the amorphous carbon matrix giving it its unique properties. Due to the small dimensions of the nanocrystallites, the *a*-C films maintain their smooth mirrorlike surface finish with root-mean-square (RMS) values that are of the order of a fraction of a nanometer. It should be noted that in general TAC films are nonhydrogenated and are considered to have a higher sp^3 fraction and hardness when compared to DAC films. This is primarily due to the fact that DAC films have hydrogen in their structure, and the C–H bonds in sp^3 hybridizations contribute to the higher sp^3 ratio.

In this review chapter we will attempt to cover the evolution of *a*-C thin films and map out the significant contributions to the field by numerous research laboratories. We will examine the deposition and growth aspects as well as the growth models, which will bring us to the microstructure of these films. Then the impact of the microstructure and growth on the optical and electrical properties will be examined. The effect that defects play in the optoelectronic properties will then be mapped. We will then cover the electronic density of states of the material and evaluate the impact of defects on the measured electrical and optical properties. Subsequently, we will introduce some new results that show how ion implantation may allow a methodology to delocalize gap states within these films. It is indeed quite significant that it appears that this addition of disorder into the amorphous material helps to allow for better control of the conduction properties.

Finally, we will look toward the future uses of this material in the form of applications. The area which has generated the most amount of interest in the recent past has been electron field emission. We firmly believe that this interest will continue and develop and so we devote a full section to this subject. Other applications which include both passive and active thin *a*-C films will be covered separately. It should be noted that *a*-C has been used as a hard disc wear resistant coating commercially for many years and at present it is thought that all major hard disc drive manufacturers use an *a*-C protective film. All the information used in the applications section is from information

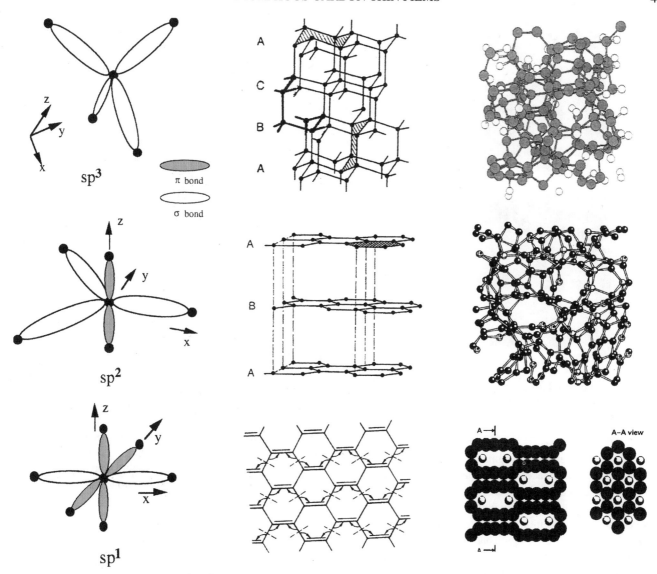

Fig. 1. A schematic representation of sp^1, sp^2, and sp^3 hybridized bonds and the resultant crystalline and model amorphous structures. The structures shown for sp^1 hybridized carbon are hypothetical structures showing the formation of triple bonded carbon from a graphite plane [36], and the packing of the carbon chain around Na atoms [37]. The model amorphous structure for sp^3 bonded carbon illustrates a TAC film that is hydrogenated (light circles) and has a density of 3.0 g cm^{-3} [38]. The model amorphous sp^2 structure is from [39] which is a-C that has 86% sp^2 bonding and has a density close to 2.0 g cm^{-3}.

available in the public domain. The authors do realize that a significant quantity of research is being conducted in companies on a number of commercially sensitive areas at present.

1.1. Carbon and Its Allotropes

Until the mid 1980s there were two main allotropic forms of carbon that could be readily identified and available, namely, diamond and graphite. Russian scientists in the mid 1970s attempted to introduce a third allotropic form of carbon called carbynes [2, 3] which was not accepted by the larger scientific community as a whole [4, 5]. In the mid 1980s C_{60} or buckminster-fullerene [6, 7] was added to the list of allotropic forms of carbon. The reason that each of these allotropic forms is of interest to us is their bond hybridization. Also each of these

crystalline forms has a unique structure that can give rise to properties that are mimicked in its amorphous counterparts.

1.1.1. Diamond

Diamond is the hardest substance known. In its crystalline form a completely tetrahedral structure gives it an optical bandgap of 5.5 eV due to the covalently bonded sp^3 C–C structure with only σ bond hybridizations. Diamond also has the distinction of having the largest known thermal conductivity due to the high speed of its acoustic phonon coupling. The first reported synthesis of diamond was in 1911 by von Bolton [8] using low temperature (100°C) and low pressure decomposition of acetylene. Systematic synthesis of diamond was then undertaken by Bundy [9] and Eversol [10] in the 1950s and 1960s. The main

Fig. 2. Examples of NAC films that show (a) dendritic growth of crystallites as discussed in [5] and (b) graphite nanoparticles in a TAC matrix (see [1]).

problem faced by the diamond community in using this material to its full capacity as an active device has been the suitability to confidently electrically dope the material *n*-type. Although N atoms which are found naturally in type IIb diamond can be introduced and sit substitutionally, they create a very deep donor state 1.7 eV below the conduction band edge. At present there is much excitement in the possible use of phosphorous as a shallow *n*-type donor in diamond. On the other hand, B can be used to dope this material *p*-type (0.37 eV above the valence band). Recently, a number of useful electronic devices have been fabricated using a surface conducting layer of hydrogen terminated diamond [11, 12].

1.1.2. Graphite

Graphite is a fully trigonal network of bonds that forms planar six-member aromatic rings of single and double bonded carbon. The sheets of planar sp^2 hybridized carbon are connected to planar bonds on either side by weak π bonds. The C–C distance of the planar double bonds is 1.42 Å in length when compared to the single bond length of 1.54 Å for sp^3 bonded carbon. The graphite sheets are 3.54 Å apart and due to the weak van der Vaals π bonds the sheets can slide relative to each other and this gives rise to good lubrication properties. In terms of bandgap, due to the highly delocalized π bonds between the sp^2 bonded graphite sheets there is an overlapping of the $\pi-\pi^*$ density of states (DOS) which makes the material behave as a metallic conductor.

1.1.3. Carbynes

Carbynes have a fully crystalline sp^1 bonded chainlike structure giving rise to a material whose bandgap and structure resemble that of an acetylenelike chain molecule with a mixture of single and triple bonded carbon forming a solid crystalline with a face-centered-cubic (fcc) structure [3]. As stated in the Introduction there is much debate as to the validity of these results [4]. To date, at least six forms of crosslinked linear carbon polytypes have been reported [13]. There are a number of reports that show carbon crystalline structures that do not conform to any of the above structures [5].

1.1.4. C_60 (Buckminster-Fullerenes)

This can be classified as an allotropic form by virtue of its structure being unique in that it consists of 60 carbon atoms formed into a large molecule which has a shape similar to that of a football. This form of carbon was initially observed in interstellar space [6] and now is being commercialized for various potential applications [14]. The 60 carbon atoms form into 20 hexagons and 12 pentagons. The pentagons give rise to the curvature in the otherwise fully sp^2 bonded network and thereby close the molecular cagelike structure. Although each carbon atom in C_{60} is threefold coordinated, due to the curvature involved in its bonding the films have a hybridization that can be regarded as an $sp^{2.5}$ hybridization [15]. Despite this fact, it can be used as a standard for a fully 100% π bonded network in electron energy loss analysis due to the intensities of the π to σ bond being compared against an unknown sample [16].

1.1.5. Amorphous Carbon (a-C)

Amorphous carbon is the generic term used to describe most disordered carbon films. Although there is no long range order present, both short and medium range order is preserved which gives rise to physical properties that can mimic those found in the crystalline material. In these materials by examining the nearest neighbor distance and the next nearest neighbor distance it is clear that the short/medium range order is very close to its crystalline counterpart. It is generally the bond angle that is different from that of the crystalline case which gives *a*-C no long range order. This angular or bond angle variation can also give rise to the narrowing of the bandgap by virtue of it creating highly localized states within the bandgap. Amorphous carbon films in general are broken down once more into two distinct categories; hydrogenated and nonhydrogenated films. (See Table I.)

Whenever a hydrocarbon source gas or plasma system is used to deposit the films some quantity of hydrogen gets incorporated within its microstructure. Since the hydrogen prefers to bond in an sp^3 hybridized manner, by terminating single bonds they tend to enhance the sp^3 or tetrahedral nature within the films. Using techniques such as the plasma beam source it has been reported that an sp^3 content as high as 80% has been recorded in the presence of 30% hydrogen within these highly

Table I. Typical Physical Properties for the Different Forms of Amorphous Carbon Thin Films

Category	Hardness (GPa)	sp^3 (%)	Optical bandgap (eV)	Density (g cm^{-3})	H (at.%)
PAC	soft	60–80	2.0–5.0	0.6–1.2	40–65
GAC	soft	0–30	0.0–0.6	1.2–2.0	0–40
DAC	20–40	40–60	0.8–4.0	1.5–3.0	20–40
TAC	40–65	65–90	1.6–2.6	2.5–3.5	0–30
NAC	20–40	30–80	0.8–2.6	2.0–3.2	0–30

tetrahedral hydrogenated a-C films [17]. This is in contrast to the more usual 40–60% sp^3 bonding found in the DAC films deposited using radiofrequency plasma enhanced chemical vapor deposition (RF PECVD), with ion bombardment (and H ~ 30–40 at.%). When no ion bombardment is used, as in the case of earthed electrode RF PECVD systems, highly polymeric a-C:H (PAC:H) films are deposited. These films have an sp^3 content between 60 and 80% (but mostly C–H bonds with little sp^3 C–C bonds) and high hydrogen contents (H 40–65 at.%). At the high end of the hydrogen contents, thin films deposited are more similar to hydrocarbon polymers such as polythene which possesses 66 at.% of H.

1.1.6. Nonhydrogenated a-C Films

Sputtering (with or without the use of a magnetron), evaporation, filtered cathodic vacuum arc (FCVA), mass selected ion beams (MSIB), and laser arcs are some of the more popular systems used for the deposition of nonhydrogenated or hydrogenated films with the deliberate addition of controlled amounts of hydrogen. In the case of hydrogenated carbon films, as the C–H bond energy (4.3 eV) is significantly higher than that of the Si–H bond (3.3 eV) it allows both the hydrogenated and nonhydrogenated films to be used in relatively high temperature environments. In fact these a-C(:H) films are extremely robust and are seldom attacked by concentrated acids or alkalis. Oxygen plasmas are one method in which the films can be etched, when required. In the case of sputtering it was thought that only highly sp^2 hybridized films could be obtained using this technique. Schwan et al. [18] clearly demonstrated that using a magnetron source which included an intense Ar ion plating environment, highly sp^3 bonded films could be created under the correct conditions. FCVA is a well known technique that has been used to deposit a-C films with high sp^3 contents that have ranged from 20 and 87%. Recently, Xu et al. [19] has shown that such a system can be produced commercially using an off axis energy filter to remove macroparticles generally transported as part of the plasma stream due to electrostatic considerations. Scheibe et al. [20] also have demonstrated the ability to deposit highly TAC films using a laser initiated plasma. Lifshitz et al. [21] have been using ion implantation to deposit a-C films using a model system, MSIB. They too get highly sp^3 films utilizing the ion bombardment effect which creates a metastable structure within the films that promotes sp^3 bond formation.

Laser initiated carbon arcs also have been reported to produce a-C films with high sp^3 contents. In a comparison of the two types of TAC films produced by laser vacuum arc and FCVA it was found that the physical properties of the films greatly differ despite the fact that both types of films have high sp^3 contents [22]. Therefore, the distribution of sp^2 sites as well as their atomic percentage must govern the material properties. Also, it has often been reported that the DC self-bias or the incident ion energy controls the sp^3 content. Xu et al. have produced TAC films under no bias, which are highly sp^3 hybridized, which suggests that the density of the plasma plays a key role in the deposition process. Early work by Davanloo et al. [23] on laser ablated a-C films reported high sp^3 films which have now been reclassified as moderately high sp^3 (closer to 60%). Preliminary results within our laboratory using plasma immersion ion implantation (PI^3) indicates that it is possible to grow high sp^3 films using the system in a CH$_4$ plasma under DC bias.

A recent addition to the nonhydrogenated family of samples has been the graphite nanoparticle containing TAC films [1]. In this case, the films are grown using a cathodic arc in a high He plasma regime that condenses some of the C ions into graphite nanoparticles before they impinge on the target (NAC).

Other nonhydrogenated films such as evaporated or sputtered carbon fall into the category of high sp^2 films as these a-C films are deposited at close to room (or low) temperature and there is no surface energy available for the bonds to rearrange themselves other than in their lowest energy configuration which is sp^2 or graphitelike (GAC).

1.1.7. Hydrogenated a-C Films

In the case of the hydrogenated films, hydrogen can be deliberately leaked into the deposition chamber in controlled quantities to change the properties of a-C films. More usually a hydrocarbon plasma is used during the deposition which means that H contents may vary between 20 and 60% depending on the type of growth system used. There are a number of different and varied configurations used and we will concentrate on the more popular systems covered in the literature. RF PECVD has been used to produce (good quality) films with varied properties and H contents between 20 and 60% with a bandgap E_g that ranges from 1 to 4 eV [24, 25]. Films can be deposited on the RF driven electrode to give a metastable structure that contains higher sp^3 diamondlike C–C contents. If films are to be grown on the earthed electrode, similar to the a-Si:H, the films generally contain a smaller number of defects as measured by electron paramagnetic resonance (EPR), higher hydrogen content, and higher bandgap [26]. Hydrocarbons containing gases such as CH_4, C_6H_6 and C_2H_2 are commonly used. Amorphous carbon films have also been grown using microwave plasmas as well as hybrid RF-microwave plasma systems.

The plasma beam source [17] provides a method of increasing the sp^3 content while decreasing the H content of DAC films by confining the plasma into a highly ionized dense beam. More recently, electron cyclotron wave resonance systems have achieved similar results [27]. This can be compared to the use of magnetically confined RF PECVD [28] which results in a significant variation in the film properties. The enhanced dissociation within the plasma helps electronically dope the deposited films [29]. Other methods to grow a-C:H films include the use of natural sources such as Camphor ($C_{10}H_{16}O$) [30] and $C_8Cl_4O_3$ [31]. Another interesting technique has been the helicon plasma system. In this case an antenna is used to excite the plasma at a resonance that gives rise to high dissociation. Inductively coupled plasmas too have been used to deposit films that are soft and have a high sp^2 content.

1.2. Historical Perspective

Since Aisenberg [32] in 1971 coined the term DLC (which more correctly should have been called DAC) for ion beam deposited carbon films that had some physical properties that could be associated with diamond, there has been much interest in the deposition, characterization, and utilization of a-C films. The field opened up in 1976 when Holland and Ojha used a RF PECVD system to produce hydrogen containing a-C thin films. Aksenov [33] then used a cathodic arc to produce TAC films. This allowed for the growth of a-C films that contained sp^3 contents that ranged from 20 to 80%, while the material still remains amorphous. Evaporation and the sputtering of carbon from solid sources has been performed for many years with little or no commercial applications as yet.

The drive in the area of thin film carbon deposition in the early days was based on the growth of polycrystalline diamond films [34]. But with time and a better appreciation of the properties of amorphous carbon films this area of material science has grown steadily with as many researchers or more working on this topic when compared to diamond films. The main difference associated with the two types of films was that in order to grow polycrystalline diamond films metastably, a highly ionized plasma such as a microwave dissociated plasma was required. In addition to the hydrocarbon source gas CH_4, a high percentage of H_2 dilution was required before the plasma impinged on a hot substrate of 800–1100°C. The high temperature made sure that there was enough surface energy associated with the process for the atomic H in the hydrocarbon to be dissociated. The hydrogen dilution gas also etched away the nondiamond component at this temperature to allow for the growth of diamond films. Other techniques used included hot filament CVD and oxyacetylene torch methods.

When initially proposed for industrial applications, a-C films were considered for hard transparent coatings that were infrared (IR) transparent. Since then with the knowledge gained in the study of these materials, the number of potential applications and uses has increased significantly.

1.3. Future Applications

At present, hard a-C coatings are used in hard disc drives to provide a wear resistant coating for the read/write media. It is also used to coat the read heads in CD and video head applications. The advantages of a-C coatings for these applications lie in the fact that by mixing sp^2 bonds with sp^3 bonds it is possible to add some lubricious qualities to the coating due to the stratified nature of the planar sp^2 bonds. The mixing of sp^2 with sp^3 also prevents the coating from becoming too brittle and prevents cracks propagating on impact. Similar wear resistant properties are used in a plethora of applications such as shaving blades, cutting edges, nails, pistons in car engines, decorative coatings such as jewelery and watches, etc. Other areas that utilize these wear resistant coating properties have been in the biomedical and optical lens areas. The properties of hard wear resistant DAC were predicted and demonstrated back in the early 1970s and since then they have been used as a proprietary technology by manufacturers. Another interesting and hitherto unutilized area has been in low fouling applications in heat exchangers. Hard a-C coatings have proved to be one of the best barrier layers that prevent the buildup of calcium based coatings that lower the efficiency of heat exchangers which is a multimillion dollar industry. These applications give a flavor as to the use of a-C as a mechanical coating. As the applications above have used a-C as a passive layer, its value-added component has not been as significant.

It is predicted that the impact a-C films will have in the future as an active layer material will be much more significant. Already prototype displays that utilize a-C cathodes have been demonstrated that utilize the property of low electric field threshold for electron emission. This property that was initially associated with negative electron affinity is now attributed to the electronic structure of the material and will be discussed more fully in a later chapter. As the cathodes used based on a-C are "nominally" flat, they will have a large number of advantageous properties for large area displays when compared to the use of conventional "Spindt" tips in field emission displays. This market is a multibillion dollar activity and has the potential of moving a-C to the status of one of the most value-added products available in advanced materials.

Other active applications in which the tuneable qualities of the bandgap are expected to portray a-C as a unique material are in electronic applications. Starting with the most basic component a blocking diode, the gap states in a-C have thus far prevented the fabrication of such a device. This is because any metal placed against a-C would allow carriers to leak into gap states and thus prevent the observation of a Schottky barrier. This has now been overcome by improvements to the material properties (see Section 6) as well as by the growth of super-lattice structures [35]. What this allows is the possibility of a plethora of electronic devices starting with the Schottky barrier and then moving to three-terminal MESFET and TFT applications. With this comes the possibility of providing tuneable electronic barriers. With the possibility of delocalized conduction proposed in Section 6 by the addition of disorder by ion implantation, the use of electronic components to exploit injection devices for electroluminescence applications becomes a distinct possibility. The use of *a*-C as a pseudo direct gap material for light emission also then becomes possible together with applications in sensors and photovoltaics. Some recent work also suggests the advantages of using *a*-C in electrochemical cells.

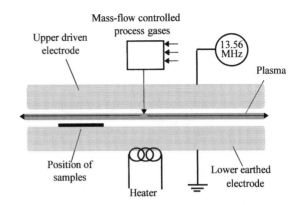

Fig. 3.   Schematic diagram for a RF PECVD system.

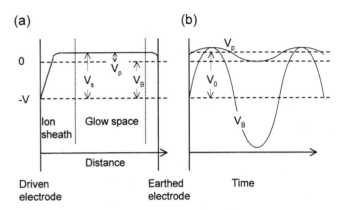

Fig. 4.   Variation of (a) the average voltage distribution in a RF PECVD system at the driven and earthed electrodes, and (b) the RF modulation potential with time together with the averaged bias potential experienced at the two electrodes.

2. DEPOSITION AND GROWTH

Amorphous carbon films can be deposited in a large number of ways. For example if only a sp^2 bonded material is required any simple evaporation technique will be capable of producing such a film. On the other hand, *a*-C films have been prepared using ion implanters where growth has been virtually layer-by-layer. Due to the vast variety of deposition equipment available this has meant that these films need to be characterized and carefully categorized according to the various classes of materials so that they can be compared with each other meaningfully. It will be seen that the growth process of the films will be intimately controlled by the deposition process and in particular by the plasma conditions that prevail during the growth.

2.1. Deposition Techniques

Due to the wide variety of deposition systems available for the growth of *a*-C films we will concentrate only on the more popular techniques that cover the full range of the material properties obtained in these films. A number of hybrid deposition techniques also exist and generally have a niche category of materials with specialist applications in mind. In general all techniques rely on positive C ions that are transported to the substrate from the source via a plasma, arc, ion beam, or evaporation.

2.1.1. Chemical Vapor Deposition

The research by Holland and Ojha [40] in 1976 can be cited as one of the most influential works that popularized *a*-C film deposition when they showed these films can be grown using a CVD process. They [40] replaced the SiH$_4$ used for *a*-Si:H deposition with CH$_4$ and obtained properties that were similar to those of Aisenberg [32]. PECVD is the most popular method and uses 13.56 MHz RF excitation of a hydrocarbon gas at a low pressure (<1 Torr). The growth temperature can be varied between 30 and 350°C, but in a large number of cases the deposition is performed at room temperature. In RF PECVD the gases are ionized between the two electrodes, one driven and

the other which is maintained at ground potential. See Figures 3 and 4.

The plasma developed between the plates depends critically on the chamber configuration but this would be fixed for each deposition system. The greater the asymmetry between the driven and the earthed plate (i.e., the driven electrode being smaller) the higher the self-bias that is developed. In general, due to the higher mobility of the electrons with respect to the ions a relatively large negative self-bias is observed across the plasma and the driven electrode. On the other hand some deposition systems are configured to have the driven electrode on top with the samples being placed on the bottom earthed electrode. In this case, although the DC self-bias developed across the sheath is negative, the value is very small and is comparable to the plasma potential, which can be monitored using a Langmuir probe. The substrate on either external electrode (driven or earthed) is bombarded by ions that acquire the plasma potential, which is critical in dictating the quality of the films. When films are grown on the earthed electrode they acquire a PAC structure that has a low defect density, wide bandgap, and high hydrogen content. Films grown on the driven electrode are generally DAC films with high defect densities, low bandgaps, and high hardness. It can be shown using the Child–Langmuir equation and the presence of a space charge limited current that the maximum energy of an ion striking a substrate placed on a driven electrode is

$$\frac{V_T}{V_P} = \left[\frac{A_T}{A_P}\right]^4 = [\text{Ratio}]^4 \qquad (2.1)$$

where V_T, V_P, A_P, and A_T are the DC self-bias plus plasma potential ($V_T = V_{DC} + V_P$), plasma potential, earthed electrode area, and the driven electrode area. Ions within the plasma sheath will not acquire a total energy equal to eV_T as there are likely to be many inelastic collisions within the mean free path of the ion. Assuming that the plasma chamber configuration were to be kept constant, Bubenzer et al. [41] developed an empirical relationship between the total energy E_T acquired by the ion before impacting on the substrate surface, V_T, and the pressure within the chamber. This was given as

$$E \sim \frac{V_T}{P^{0.5}} \qquad (2.2)$$

where P is the pressure. In most RF PECVD reactors the RF power can be equated to $(V_T)^2$, which then gives rise to the equation

$$E \sim \left[\frac{\text{Power}}{\text{Pressure}}\right]^{0.5} \qquad (2.3)$$

RF plasmas have an average electron energy of 2–3 eV with a low ionization efficiency of 10^{-4}–10^{-6}. Depending on the hydrocarbon gas used, the mixing gases (dilutant gases), pressure, and flow rate, the energy distribution of the ions that take part in the growth process can be very varied. Catherine [42] showed that by using a low frequency (25–125 kHz) discharge in place of the standard 13.56 MHz power supply, hard transparent crosslinked films too could be deposited. But no significant advantages were observed by using the lower frequency.

Koidl et al. [24] conducted a thorough study of the variations in the material properties as a function of negative DC self-bias (power) and showed that as the negative self-bias was increased from close to 0 to 1000 V, there was a steady decrease in the optical bandgap from 2.0 to 1.0 eV and hydrogen content from 40 to 25 at.%. Concomitant increases in the film density, refractive index, and microhardness were measured. These films were deposited at room temperature using only a hydrocarbon gas, C_6H_6, with the variation shown in Figure 5a. They further showed that if the process gas were to be changed from benzene (C_6H_6) to cyclo-hexane or n-hexane or methane (CH_4) this would have little effect on the refractive index, optical bandgap, hydrogen content, and sp^3 to sp^2 ratio at a DC bias of -400 V. They attributed this surprising fact to the efficient fragmentation of the energetic hydrocarbon on the growth surface. It was shown that for films that were deposited at low bias voltages and were polymerlike in nature, such a process gas independent phase was not observed, and that larger process gas molecules were incorporated in the growing films.

Catherine's [42] work showed the variation of the PECVD deposited a-C:H films when one diluted the hydrocarbon gas (CH_4) in a noble gas (Ar). As the ionization energies of noble gases such as Ar and He are significantly higher than the various radicals of the CH_4 molecule it is expected that the whole energy distribution of the RF discharge is shifted to higher energies. This also results in a higher dissociation rate as well as higher electron temperatures within the plasma. In the experiments of Catherine it was shown that the emission intensities of CH, H, and H_2 species all increased linearly with increasing Ar concentration within the plasma. More interestingly, when an 80% He dilution was used with CH_4 in their films, they clearly showed an energy window in the bias voltage at which films with an optimum density would be deposited. These data are shown in Figure 5b. Some additional points have been added from data calculated using Rotherford backscattering and elastic recoil detection analysis from the authors' group, which confirm the reproducibility of these experiments. The variation of a number of physical properties of hydrogenated a-C (a-C:H) when deposited using a hydrocarbon:noble gas mixture is shown in the Figure 5b. Amaratunga et al. [43] discussed the variation of the properties of hard DAC films giving rise to a deposition–etch process when a CH_4 gas is mixed with Ar during deposition of a-C films. In their data they showed that as the DC bias increased the growth rate decreased, with a concomitant increase in the refractive index. They showed that the refractive index is inversely related to the compressive stress, and that the compressive stress goes through a peak close to a negative self bias of 375 V during deposition (see Fig. 5b).

Silva et al. [25] followed this work by showing that a large controllable variation in the material properties can be obtained by using Ar dilution in PECVD plasmas, where once more the data pointed toward an optimum energy window at which a turning point to the material properties is observed. They indicated that as the DC self-bias is gradually increased the a-C:H films deposited go through a polymerlike (PAC) to diamondlike (DAC) transition which at very high bias energies converts back to a high sp^2 film due to extensive bombardment of the growing surface. They showed that a very small quantity, if any, of Ar is retained in the thin film with the large Ar atoms helping to compact the films during growth. The variation of the material properties as a function of DC self-bias for films deposited with a noble gas dilution is shown in Figure 5b. Two sets of EPR results are shown—(i) when Ar dilution is used and (ii) when He dilution is used. Clearly, all the properties shown do not peak at a particular energy but show that the behavior of the deposited film when diluted with a noble gas is different than when using only a hydrocarbon source gas. The low loss plasmon peak energy shown in the figure would give a measure of the valence electron density, which in theory should follow the density variation as shown in Catherine's work.

Typically, driven electrode deposited hard DAC films using this technique could have hydrogen contents of 20–40 at.%, while the soft PAC films deposited on the earthed electrode would generally have 40–60% of hydrogen. As hydrogen prefers to bond in a sp^3 hybridization, the PAC films could have an sp^3 content comparable to the DAC films. But most of these sp^3 bonds are in CH_3 and CH_2 configurations as observed by Fourier transform infrared, in comparison to the C–C sp^3 bond-

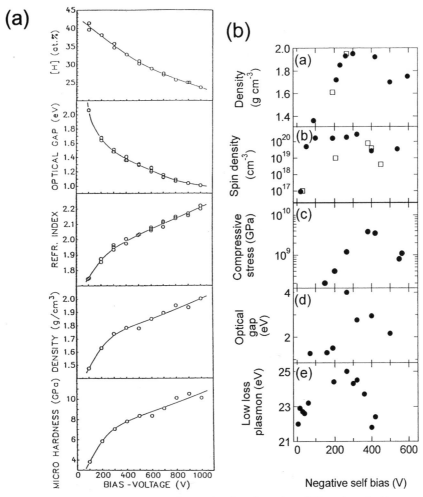

Fig. 5. Variation of the material properties as a function of negative self-bias voltage for (a) a pure methane plasma from Koidl et al. [24] and (b) for a methane/noble gas (He or Ar) mixture.

ing present in the DAC films. It is worth noting at this early stage that FTIR should never be used to estimate the hydrogen content in a-C:H films as up to 50% of the hydrogen has been shown by Grill and Patel [44] to be unbonded.

An interesting and potentially very useful variant to PECVD deposition was introduced by Silva et al. in 1994 [28]. They showed that by introducing magnetic confinement to the RF PECVD plasma they could get enhanced thin film properties due to the very high dissociation rates and higher electron temperatures within the plasma. For the first time they were able to move the Fermi level within the a-C:H films by up to 1 eV, which allowed them to demonstrate some evidence of electronic doping [29]. They were able to introduce up to 15 at.% of nitrogen to their films without distorting the optical bandgap significantly and study the variation in the material properties. Also, it was shown for the first time that changes in the joint density of states were occurring as a result of the nitrogen incorporation using electron energy loss spectroscopy (EELS) as shown in Figure 6 [29], and that these films allowed for the first demonstration of electron field emission at low threshold fields [45].

2.1.2. Sputtering

In CVD or sputtering the growth occurs atom-by-atom or molecule-by-molecule at temperatures much below the melting point of the material. The atom (or molecule) that arrives at the surface will tend to bind itself to the surface at a local minimum energy configuration by moving in a short range. As most growth is either at room temperature or close to that, there is little surface kinetic energy for the arriving atom (or molecule) to be able to find a global minimum either by atomic rearrangements or diffusion. This ensures that the deposited material possesses short range order but in an amorphous/disordered state. The above discussion applies to CVD or sputter processes where no DC bias has been applied. When either a self-bias or other energy source supplies the arriving species with superthermal energy, this would mean that the adatom could displace bulk or surface atoms and have enough energy to activate chemical reactions. This allows much more versatility to the growth techniques and therefore is widely utilized. Two main techniques are used for the growth of a-C by sputtering: RF bias sputtering and magnetron sputtering.

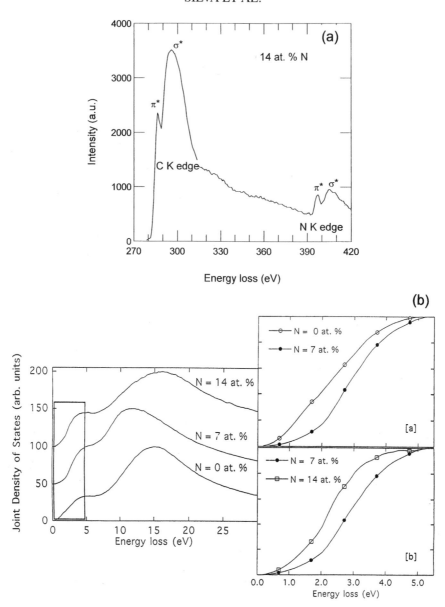

Fig. 6. (a) Typical EELS spectrum for a DAC:N film with assignment of the prominent bands and (b) the joint density of states as determined by EELS, from Silva et al. [29].

In the case of RF sputtering, an Ar plasma is struck below a graphite cathode which is connected to the driven electrode with respect to the system earth. The substrates are generally placed on the lower water cooled electrode which is negatively biased with respect to the system (and plasma). Typically the plasma potential is positive and varies between 10 and 20 V. Unhydrogenated films then can be deposited on the biased substrate as a function of negative bias, which varies the kinetic energy of the film forming species once it crosses the sheath space. In general, this technique is more used for metallization by the microelectronic industry, and in the case of a-C film deposition the magnetron sputtering technique is more widely used.

The main difference between the two sputter processes is that, in the case of magnetron sputtering, a transverse magnetic field with respect to the graphite cathode is applied such that the $\mathbf{E} \times \mathbf{B}$ field confines the drift of the electrons to the target surface. This means that a more intense plasma with a higher ionization efficiency is produced which allows for a higher deposition rate of the thin films. The higher carrier density within the plasma also allows for it to be operated at a lower pressure. Most often in the case of a-C films the unbalanced magnetron configuration is preferred over the balanced setup such that the plasma extends over both the target and the substrate. In this case, if Ar is used as the sputter gas it provides both the sputtering flux as well as the ion plating flux over the growing film. The unbalanced magnetron was developed by Savvides [46] and a schematic of the deposition system is shown in Figure 7. Using this technique Schwan et al. [18] have shown that it is possible to create a similar type of film developed by mass se-

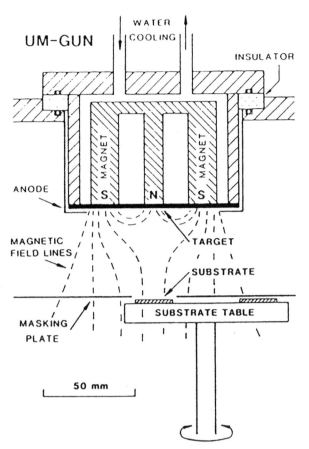

Fig. 7. Schematic of the unbalanced magnetron sputtering system from Savvides [48].

lected ion beams or filtered cathodic vacuum arc which have a high sp^3 content (87% by EELS). The density of these non-hydrogenated films was close to 3.1 g cm^{-3}. Interestingly, in these films the optical bandgap is close to zero which seems to indicate that all the sp^2 bonds must be clustered and percolating through the structure due to its high conductivity. In this case the high sp^3 content is attributed to a stress induced transformation from sp^2 to sp^3 states at compressive stresses above 14 GPa. The deposited films have quite unique properties when compared to other deposition processes for a-C thin films with evidence for highly ordered sp^2 graphitic planes oriented in the [002] direction with respect to the silicon (100) substrate [47].

Since the demonstration of the unbalanced magnetron to produce a-C films by Savvides [48] many other systems based on similar themes have been pursued by others. Cuomo et al. [49] produced highly TAC films using a dual ion beam sputtering process on cooled substrates. Andre et al. [50] produced a-C films with high densities with a different variant based on dual ion beam sputtering and ion assisted DC magnetron sputtering. Kleber et al. [51] produced a-C films in the balanced magnetron configuration with a film density of 2.0 g cm^{-3}. This is typical of sputtered films and a high sp^3 content is not expected.

2.1.3. Ion Beam Techniques

One of the major advantages of this type of system is that the process to produce the ion beam and the deposition process are independent of each other. This means that it is easier to optimize the various individual parameters or study a model system more easily using ion beam techniques. Any system whose process is based on ion beams would need to have the following components: an independent means of controlling the ion energy and flux, isolation of the ion generation from the thin film growth process, a narrow energy spread for the ion beam, and low pressure growth with control of the angle of incidence of the bombarding ions. Some of the more popular sources are broad beam Kaufman ion sources, ion beam accelerators, and ionized cluster beam sources. More recently some growth using plasma immersion ion implantation (PI^3) has also been reported.

Broad beam Kaufman ion sources were first developed in the 1960s for space propulsion applications as ion thrusters. Since then it has been used more so for thin film deposition with energy ranges typically 100–1000 eV. Deposition is carried out at very low pressures downstream of the source, with the process working best at high energies, resulting in high current densities such that potentially useful growth rates are achieved. The ion flux, ion energy, and direction of the flux can be fully characterized and can be independent of the deposition flux. Aisenberg and Chabot [32] produced the first a-C films using a C$^+$ ion source which was followed by Spencer et al. [52]. Weissmantel et al. [53] used a dual ion beam deposition system to produce similar DAC films where one beam provided the C$^+$ ions and the second beam provided Ar$^+$. A second ion beam may be an ideal way to get ionized nitrogen into the films and therefore could be examined either as a doping source or to produce high nitrogen containing carbon nitride films. Xu et al. [54] used a nitrogen ion source coupled with their cathodic arc system to introduce nitrogen very efficiently into the TAC films.

The use of ion beam accelerators to obtain MSIBs for better control and fundamental studies of the growth of high sp^3 TAC films was first proposed by Lifshitz et al. [55]. Subsequently, a number of research groups including Miyazawa [56], Hofsass [57], and Hirvonen [58] have examined films grown with MSIB. The deposition is carried out atom by atom using modified ion beam accelerators which offer isotropically pure C beams at very low pressures with excellent control over the flux and ion energy. Magnetic filtering is used to separate ion to mass ratios which are required for well collimated beams that are raster scanned over the sample area. The main limitations stem from the fact that ion beam self-sputtering and ion beam damage are present in the growth process and at low energies the growth rate is quite slow. Due to the excellent control offered by the process, beams of C$^+$, C$^-$, and C^{2-} have been used to deposit DAC thin films.

Cluster beams to deposit carbon films, especially for nanocluster deposition, have been pioneered by Milani et al. [59]. Cluster beam growth by macroaggregates having 500–2000 atoms loosely bound together are formed by an adiabatic expansion of evaporated material through a nozzle into a high vacuum chamber. The technique was developed by Takagi et al. [60] and is now widely used for intermetallic compounds and nanostructured materials. The advantages associated with the technique are due to the fact that any charge placed upon the cluster is equally distributed across each of the constituent atoms and thereby its use for ultrashallow implants cannot be overemphasized. Each of the atoms therefore imparts precisely the total ion energy given to the cluster divided by the size of the cluster upon impact with a surface. So the growth process possesses very low charge to mass ratios for the ionized cluster which prevents the buildup of surface space charge. The films grown using this technique have shown unique properties that avail themselves to electron field emission [61] and supercapacitors [62].

Plasma immersion ion processing (or deposition) has also been used to deposit DAC films. The work of Nastasi and co-workers [63, 64] shows that using a C_2H_2–Ar plasma immersion ion implantation process it is possible to produce films with varying sp^2/sp^3 ratios. In their work they clearly show a critical dependence of the microstructure of the film to negative pulse bias and the pulse bias duty factor. In our laboratory we have used a pure CH_4 plasma in DC pulse bias mode to produce DAC films with sp^3 contents as high as 80%. The advantage of using the PI^3 system is that it allows for high density plasmas to be used to produce non-line-of-sight coatings on nonplanar substrates with independent control of the ion energy.

2.1.4. Filtered Cathodic Vacuum Arc

The FCVA technique has become synonymous with the deposition of TAC films the world over. Since the first films were grown by Aksenov et al. [33] and then the technique was adopted by McKenzie et al. [65] a large number of groups have begun working with this system [66–70] and made many improvements to the growth process and fundamental understanding of the material properties. It is now considered to be a commercially viable system that is capable of producing macroparticle-free films to industry at close to room temperature [19]. The films produced are highly tetrahedral in nature (>80% sp^3) and extremely hard. The properties of the deposited films are crucially dependent on the ion energy of the carbon arc used for deposition. A schematic of the system typically used is shown in Figure 8. Xu et al. [19] have recently introduced an off plane double bend that further reduces the macroparticles in the thin films.

A carbon arc is struck between the cathode and a graphite striker (earthed anode), which is retracted once the arc has been struck. This produces a highly ionized beam of C^+ ions and neutrals which are further collimated by the use of an axial magnetic field. This curvilinear magnetic field is then used to steer the ionized C particles through a curved toroidal duct which makes use of the cross product of the electric and magnetic fields to filter out the neutrals. Therefore, the arc that is self-sustaining for a few minutes arrives at the off-line-of-sight target as a close to 100% ionized plasma, minus the neutrals and the macroparticles. The area of deposition may be increased by

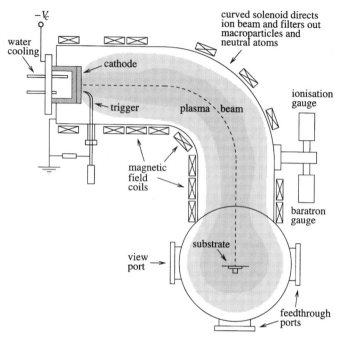

Fig. 8.   Schematic of the filtered cathodic vacuum arc system.

raster scanning the plasma beam over the substrate area. A negative bias is further applied to the substrate with respect to the anode (and system earth) which controls the growth properties. Films deposited using this technique are nanometer smooth (RMS roughness of less than an angstrom), with hardnesses approaching 60% of that of diamond, a friction coefficient below 0.10, and a Young's modulus close to 500 GPa. The ion energy distribution obtained shows the energy of this highly ionized beam without any applied bias to be close to 30 eV, with a full width at half maximum of 18 eV.

The versatility of the cathodic arc system in producing a range of materials with different properties has been well illustrated in the work of Amaratunga [1] and Coll [70]. It is shown that by having a high density gaseous environment or a gas jet next to the arc it is possible to rapidly condense the carbon ions in such a manner as to be able to produce nanoparticle containing thin films. These nanoparticles in most cases have a structure close to those of fullerenes and nanotubes embedded in the a-C thin films. Despite the high sp^3 content of the TAC films, with or without nanoparticles, close to 10% of sp^2 bonding within the films controls the optical bandgap and electronic conduction properties. The bandgap is generally below 2.5 eV, and the electronic properties are critically dependent on the impurities admitted into the system either in a gaseous form or by solid state incorporation in the graphite cathode. Fallon et al. [68] have conducted a detailed study of the microstructure of these films as a function of ion energy, with Chhowalla et al. [69] conducting a study on the behavior of the growth properties as a function of substrate temperature.

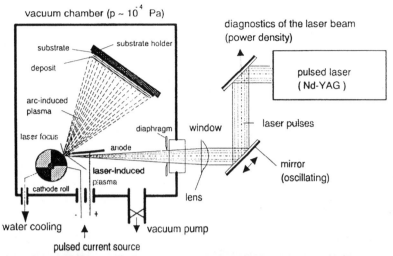

Fig. 9.   Schematic of the laser arc system used by Scheibe et al. [96].

2.1.5. Laser Techniques

In the mid 1980s it was shown that using laser ablation of graphite targets, or pulsed laser deposition (PLD) as it is better known, a-C films with properties varying from graphitelike (GAC) to diamondlike (DAC) can be deposited [71]. Recent work has taken this technique to the forefront of modern thin film growth by producing films with some of the highest sp^3 contents as well as hardness (60–80 GPa). The technique is popular due to its versatility in being able to deposit metal, ceramic, and polymer films with a high degree of control of its growth kinetics and, due to the rapid progress being made in the excimer laser field by providing high power pulsed lasers, increased growth rates. A very good review of the subject can be found in the paper by Voevodin and Zabinski [72].

An excimer laser focused through a transparent window impinges on a graphite target held in vacuum. This creates a high pressure carbon plasma that expands hypersonically in vacuum forming a high energy plume. The carbon ions thus created will

form a-C films with varying properties when they condense on the substrate depending on the laser, vacuum environment, and substrate condition. The laser fluence and surrounding environment need to be optimized in order to produce DAC thin films. Investigations of the plume dynamics have shown the leading edge (40 ns after laser initiation) of the plasma to have carbon atoms of energy close to 1500 eV, with a secondary peak (100 ns after initiation) at around 100–400 eV. The energy tail is shown to be composed of clusters of atoms. The leading edge has fast single atoms and ions with the subsequent peak being made up of slower ions, neutrals, and molecules. The mechanical properties of the films produced by PLD are second to none, but it is difficult to filter out macroparticles with this setup. Much work has also been conducted in producing nanocomposite coatings using this technique. By either having a composite sputtering target or a secondary sputter target for the mixing material, superhard amorphous composites can be produced. Most often due to the high kinetic energy associated with the plume the incorporated composite crystallizes in nanocrystals whose motions of dislocation are suppressed in favor of deformations at the grain boundaries which improve the toughness of the material [73].

A variant of PLD, namely laser controlled pulsed vacuum arc deposition or laser arc deposition, was introduced by Scheibe et al. [20]; see Figure 9. In this case a Q-switched Nd-YAG laser is used to ablate a graphite target which generates the plume of carbon ions. The laser in this case is only used to initiate a plasma that is sustained for a period of 100 ms with currents as high as 1000 A by a pulsed current source. The 100 ns laser pulse which is repeated at 1 kHz helps achieve a 10 mm/hour growth rate, with systematic erosion of the circular graphite drum. Variations of the film properties as a function of substrate temperature have been published by Silva et al. [22]. Sp^3 contents in excess of 80% with high (>700 GPa) Young's modulus have been demonstrated. The variations of the properties of the laser arc film with FCVA films have been compared as a function of growth temperature [22].

2.1.6. Other Techniques

There are a number of techniques being used today which are a hybrid of one or two of the above deposition systems. One of the more popular techniques is the dual electron cyclotron resonance (ECR)/RF plasma source where the surface wave couples microwaves at 2.45 GHz which are used to excite and dissociate the gas. A capacitively coupled RF supply is then used to transport ions to a downstream substrate holder at which the films are grown. The system was initially discussed by Martinu et al. [74] to grow DAC films. Subsequently, Turban et al. [75] and Godet et al. [76] have performed studies that have illuminated the physics of the plasma and growth process, including the material properties. Once more the films are generally kept at room temperature on a cooled electrode and the material properties vary critically as a function of the generated RF self-bias voltage. The uniqueness of the system lies in the fact that the plasma is ionized very efficiently by the ECR source, and by having a dual source involved the film growth is decoupled from the plasma generation stage. Therefore, much better control of the plasma can be afforded, together with the advantageous properties of RF thin film deposition.

One other technique that has produced some very interesting properties in the recent past has been the plasma beam source. A review of the technique can be found in the work of Weiler et al. [17]. As shown in Figure 10, RF power is applied via a capacitor and matching network to a moveable powered electrode. The magnets around the earthed and powered electrode help dissociate the plasma more efficiently as well as confine the plasma into a stream/beam. This beam is then directed toward the substrate on which the film growth takes place. Gases introduced into the plasma source ionize efficiently to produce a very well mixed plasma beam. Evidence for electronic doping of these a-C films has been published with the incorporation of nitrogen to the C_2H_2/Ar mixture which acts as a shallow dopant for the π^* states within the gap. Mobility values comparable to a-Si:H films have been obtained from Hall measurements [77].

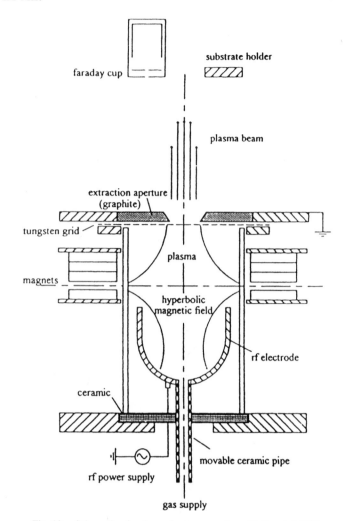

Fig. 10. Schematic of a plasma beam source from Weiler et al. [97].

2.2. Growth Models

2.2.1. Growth

Amorphous carbon films, especially the DAC and TAC films, are metastable in nature. Graphitelike sp^2 bonding is the energetically favorable hybridization in comparison to diamondlike sp^3 bonds. Metastable films are formed by raising the energy of the starting material either by thermal or chemical activation and then allowing the films to lose energy in a rapid manner (or quench). This de-energizing process is usually in the form of quenching and the films could move through a number of unstable and metastable phases before stabilizing to the final frozen-in configuration. Metastability could be compositional, structural, or morphological, where the free energy is lower than that which is required for atomic relaxation or rearrangement.

There have been many models put forward for the growth of a-C films. Of these, some of the more significant models are discussed below. It should be noted that in terms of the structure of a-C films there is still an ongoing debate as to the most energetically favorable configuration. Early work by Robertson [78] proposed a carbon structure that was composed mainly of

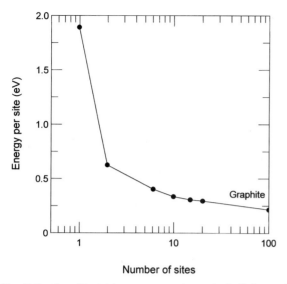

Fig. 11.   Estimation of the total energy versus cluster size by Robertson [79].

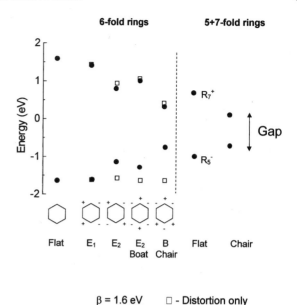

$\beta = 1.6$ eV   □ - Distortion only
● - Distortion + rotation

Fig. 12.   Band edge calculations using a Huckel approximation based cluster model by Robertson for 6-fold and 5–7-fold ring configurations from Robertson [79].

benzene like sixfold aromatic rings that fused together to form graphitic clusters. In the early models large sp^2 clusters were separated/segregated by the surrounding sp^3 matrix which controlled the material's mechanical properties. The reasons for the forming of the clusters were based on an energy gain argument where adjacent π bonds would pair up to form sp^2 sites and stabilize each other by forming parallel oriented pairs. These states would gain further energy if they were to form planar sixfold rings, which transformed themselves into a yet more favorable energy minima by clustering. See Figure 11 (from [79]) which shows the energy gain that can incurred by π states pairing up or clustering. As shown the isolated π state is quite unstable, with a energy gain of nearly 1.5 eV per site purely by pairing up with another π state. On the other hand, the energy gain beyond a six-member ring is minimal compared to the predicted disorder potential which is greater than 0.2 eV [80], and closer to 0.5 eV [38]. Therefore, with the current calculations, the driving force for clustering is not observed. In the original model [78] a relationship between the cluster size and the bandgap of

$$E_g = 6/M^{0.5} \qquad (2.4)$$

where M is the number of rings in the cluster, was derived. Yet, experimental values of DAC films prevented this model from being universally accepted due to the fact that a typical bandgap of 1.2 eV would have meant a 25-ring cluster with dimensions of the order of 1.2 nm. Neither high resolution transmission electron microscope [81] nor neutron diffraction studies [82, 83] support this view, although it is difficult to disprove such a model based on a handful of experimental results. Therefore, it was noticed that for the cluster model to be reconciled with the optical bandgap values there needs to be some modification to the theory.

Since then a number of simulation models have shown that the bonding within a-C is less structured, with a much higher probability of olefinic chainlike structures. Frauenheim et al. [84] computed the structure of a large number of a-C film den-

sities using local orbital simulations and found that in their work it was more likely for the sp^2 sites to be arranged in chains rather than rings. The bandgap values obtained by them matched those measured experimentally. In the case of TAC once more Drabold et al. [85] also found sp^2 sites to favor pairing up, which was expected due to the much smaller percentage of π bonds in their films. Recent experimental results by Anguita et al. [86] and Godet [76] in the PAC films clearly point toward a more olefinic chainlike structure with significant branching as opposed to clustering, but it should be noted that in the simulations by Galli et al. [80] and Wang et al. [87] there is evidence for clustering of sp^2 sites.

Robertson [79] has taken these contradictory findings and modified his original Huckel approximation based cluster model such that he incorporates the distortion of the sixfold (together with five- and seven-member rings) rings and shows that this would cause the mixing of σ and π states such that it tends to close the bandgap of a-C films. By this he reconciles the unusually large bandgap predicted in the previous work and also gives reasons as to why large areas of ordered regions cannot be observed in the microstructure of these films. In this later model the a-C films are proposed to be composed of roughly equal numbers of chains and ring structures. But, in order to account for the room temperature luminescence [88] and polarization memory effects observed in these films, the addition of negatively charged fivefold rings and positively charged sevenfold rings is included. These structures will give rise to germinate pair excitation of carriers but are distorted such that they pull down the bandgap of the material (see Fig. 12). Although Raman analysis in general supported the need for clusters in a-C films, results by Schwan et al. [18, 77] showed that there is still

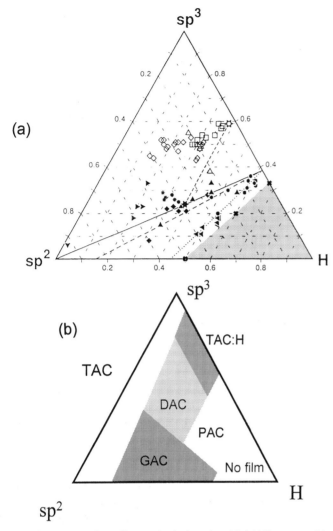

Fig. 13. Ternary phase diagram for hydrocarbon (a) initially proposed by Reinke et al. [89] and (b) the compositional space occupied by TAC, DAC, PAC, and GAC thin films.

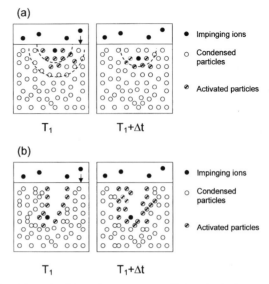

Fig. 14. Schematic of the surface atoms caused by ions impinging on a surface in (a) single thermal spike and (b) displacement spike, from Weissmantel et al. [53].

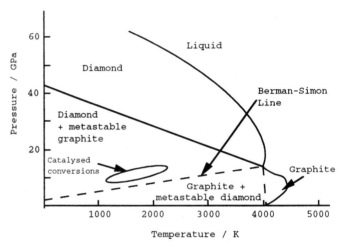

Fig. 15. The pressure–temperature phase diagram showing the Berman–Simon curve for carbon.

room for improvement of this model to universally cover all types of *a*-C films.

A ternary phase diagram for hydrocarbon films was introduced by Reinke et al. [89] for the analysis of *a*-C films in 1993. It has now been modified extensively and is used to categorize *a*-C films [90]. See Figure 13a and b. The diagram represents the allowed configurations for *a*-C films that have varying sp^2, sp^3, and hydrogen contents. Figure 13a shows the original diagram used by Reinke, whereas Figure 13b has been adapted to show the compositional space for the films defined in this review as TAC, DAC, PAC, and GAC. NAC cannot be represented on the diagram as it is generally restricted to TAC and DLC films that have diamond nanocrystallites or TAC and GAC films that have graphite nanocrystallites embedded in the matrix.

2.2.2. Thermal Spike

The proposal of a thermal spike for the formation of metastable phases of *a*-C was proposed by Weissmantel et al. [53] using the concepts introduced by Seitz and Koehler [91]. It was proposed that the temperature variation T at a distance r at a time t could be given by

$$T = \frac{E}{8[\pi K t]^{3/2}} [C\rho]^{1/2} \exp\left(\frac{C\rho r^2}{4Kt}\right) \quad (2.5)$$

where E is the energy provided by the impinging particle, K is the thermal conductivity of the film, C is the heat capacity, and ρ is the density. It was calculated that for an ion energy E of 100 eV, a region of radius 0.75 nm would be heated at least to the melting point, 3823 K, of diamond, with the thermal spike persisting over a time period of 7×10^{-11} s. This is long compared to the Debye temperature based vibrational

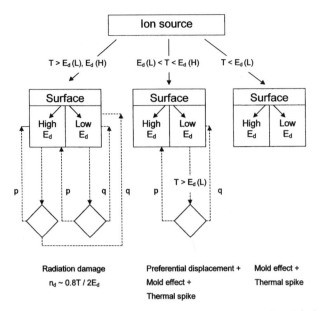

Fig. 16. A schematic illustration of the subplantation process where T is the energy transferred in a primary ion collision and E_d is the high and low displacement energies, from Lifshitz et al. [98].

period of 10^{-14} s. A schematic of the predicted variation of the surface of the thin film is shown in Figure 14, which has a predicted shock front giving rise to a pressure of 13 GPa. This pressure is well beyond the transformation stress of a sp^2 bond to a sp^3 according to the Bermon–Simon curve for carbon, Figure 15. Therefore, the thermal spike model of Weissmantel was one of the first to consolidate the concept of high quench rates (10–14 K/s) freezing in metastable phases of atomic arrangements in a-C films. The propagation of thermal spikes affecting a volume of the material and the vacancies and displacements left in the wake of a high energy collision are illustrated in Figure 14. One of the problems associated with this model was the concept of how effective a low mass ion such as carbon would be in creating energy cascades via thermal spikes at low ion energies within a material.

2.2.3. Subplantation

Subplantation as a concept was first proposed by Lifshitz et al. [55] for the growth of diamond from C^+ ions. In this model it was proposed that hyperthermal species with energies between 1 and 1000 eV could impinge on a substrate with enough energy to subimplant into the thin film and help form a continuous layer of thin film by subsurface growth. Although the initial model used incorrect displacement energies to show that one phase of carbon formed preferentially over another, the basic concept of subplantation has been universally accepted as a plausible route to the formation of highly metastable films such as a-C. In general the C^+ will lose energy by atomic displacements, phonon excitation, or electron excitation. If the energy of the carbon ion is greater than 25–30 eV, which is the displacement energy for graphite, there is a good chance that the displaced atom will be trapped interstitially or at a vacancy.

This would then constitute the first steps of a subsurface growth process. Ion penetrations of 0.3 to 30 nm have been calculated as the range with TRIM for C^+ energies of 10–10,000 eV. In this model it is proposed that three different time scales are present. At the outset a collisional stage gives rise to a thermal spike at a time step of 10^{-14} s to a small volume close to 1 nm. Energies per target atom could exceed 1 eV. These highly agitated atoms would then thermalize in a time step of 10^{-12} s while dissipating an average energy of below 1 meV per target atom. The final stage would be a long relaxation step that could last as long as 10^{-10} s, where chemical interaction, phase transformations, and diffusion could take place. Many experiments have been performed in developing this model and relating the growth process to material properties. Some of the variations with ion energy are shown in Figure 16, where three clear regions can be identified. First, at very low energies there is not enough energy for ions to penetrate the surface of the growing film and therefore there is surface growth. In this case most of the film is sp^2 bonded with low density and low bandgap. As the film growth only occurs via surface mobility there is much surface roughness associated with this phase. In the second stage the C^+ penetrates the subsurface and starts to densify this layer. An optimum high sp^3 content can be obtained during this period, with very smooth layers of film due to the smoothing out of any surface undulations by the energetic carbon deposition. The third stage of the energy regime is when enough energy is imparted into the film such that the carbon ions have increased mobility such that it suppresses the growth of the sp^3 rich metastable phase. C ions will tend to agglomerate into an energetically favorable sp^2 phase that will tend to increase the roughness of the film surface.

Variations of the material properties for MSIB [21], TAC [66, 92], and DAC [43] can be found in the literature. In the case of DAC films clear energy windows for optimum bandgap have been experimentally verified [25]. Robertson [90] and Davis [93] have independently analyzed the growth process that occurs during subplantation and introduced analytical expressions that describe the densificaton process as a function of energy. This can then be related to the sp^3 content, as there is a linear relationship between density and sp^3 content as shown in Figure 17 for magnetron sputtered films. A similar relationship can be found for most types of a-C films. The variation of the change in density arrived at by both Robertson and Davis is given by

$$\frac{\Delta f}{f_0} = \frac{f}{(1/\phi) - f + 0.016p(E_i/E_0)^{5/3}} \tag{2.6}$$

where Δf is the densification of the film at ion energy E_i, ϕ is the fraction of fast ions, f is the fraction of atoms that penetrate the surface, p is a constant close to unity, and E_0 is the excitation energy to escape from a trapped state to the surface. The $0.016p(E_i/E_0)^{5/3}$ is taken from the thermal spike model of Seitz and Koehler [91], and therefore this equation could be seen as a hybrid model. Using suitable fitting parameters the equation has been used to fit data from a number of groups.

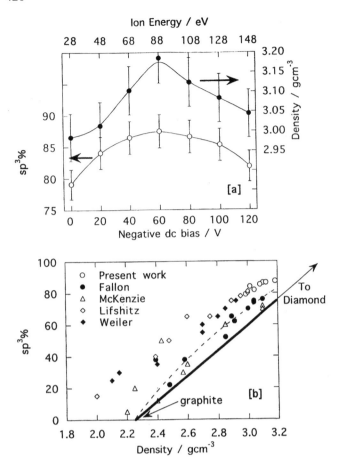

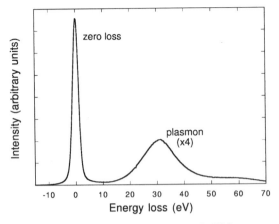

Fig. 18. Low-loss EELS spectrum for TAC.

Fig. 17. (a) The variation of carbon sp^3 bonding and density in TAC films as a fraction of ion energy and negative DC bias voltage. (b) Comparison of the sp^3 bonding in TAC films as a function of film density [17, 68, 99, 100]. The variation predicted [101] using a "rule of mixtures" is also shown (dashed lines).

2.2.4. Stress Induced Phase Transition

The importance of stress to the formation of sp^3 bonds was proposed and discussed by McKenzie et al. [94]. They were the first group to show evidence for the diamondlike bonding structure of TAC films, and they attributed the high sp^3 content to the compressive stress present in their films. When the compressive stress, as measured using a curvature method, is converted to the effective hydrostatic pressure within the film it was shown that the points lay above the Berman–Simon curve, which demarcates the graphite/diamond boundary on a temperature–pressure phase diagram, Figure 15. This allows for the nominally energetically less stable sp^3 bonds to form under stable phase conditions of high stress. The model proposes that it would be equally applicable to other material systems where more than one phase can be present by changing the pressure range via compressive stress. Recent results by the Sandia group have shown that stress-free, freestanding, highly tetrahedral films can be produced by annealing out any compressive stress. This appears to be feasible according to the models of Kelires [95] who using Monte Carlo simulations predicts highly inhomogeneous microscopic stresses of the order of GPa within

their material with the sp^3 bonds being compressive and sp^2 bonds being tensile. This implies that if the balance of bonding and the distribution of the sp^2 bonding is correct, there is a possibility of having stress stabilized sp^3 bonding with close to zero macroscopic stress.

3. MICROSTRUCTURE

3.1. Experimental Determination of Microstructure

3.1.1. Electron Energy Loss Spectroscopy

In a transmission electron microscope, "fast" electrons can be used to form magnified images of the microstructure of a specimen. In addition, because of the many different interactions that high energy electrons undergo as they travel through matter, electron microscopy allows a host of microanalytical techniques (such as cathodoluminescence, Auger electron spectroscopy, and energy dispersive X-ray spectroscopy). Among these perhaps the most valuable in ascertaining the structure and composition of amorphous carbon films is EELS.

Electron energy loss spectra are obtained by passing the electron beam transmitted through a sample through an energy dispersive magnetic prism onto a parallel electron detector array. By measuring the energy lost by electrons transmitted through the sample, information about thickness, electronic energy band structure, atomic density, chemical composition, and chemical bonding can be obtained.

3.1.1.1. Low-Loss Region

Figure 18 shows the low energy loss region of an EELS spectrum for TAC. The zero-loss peak corresponds to electrons that have undergone either "elastic" scattering in the forward direction or have passed through the specimen unscattered. The width of this peak reflects the energy resolution of the system. The integrated intensity of the zero-loss peak, I_0, is related to the integrated intensity over the entire spectrum, I_{total}, by

$$I_0 = I_{\text{total}} \exp(-t/\lambda) \qquad (3.1)$$

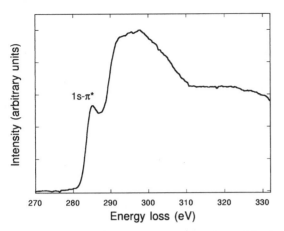

Fig. 19. EELS spectrum of a typical K-edge for highly graphitic carbon.

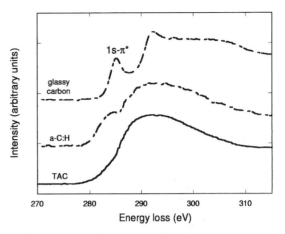

Fig. 20. Carbon K-edge EELS spectra for TAC deposited by filtered cathodic vacuum arc and a-C:H deposited by glow discharge from methane from Martin et al. [65], compared with that of glassy carbon.

where t is the thickness of the specimen and λ is the mean free path for all inelastic processes in the material. Equation (3.1) allows comparison of the thickness of samples of the same material, or if λ is measured from a sample of known thickness (or calculated from theoretical cross-sections) it can be used to calculate the absolute thickness for a sample. Additionally, the ratio t/λ can be used as a measure of the importance of multiple inelastic scattering and its effect on the extended EELS fine structure.

The broad peak in the low-loss region corresponds to energy loss to modes of collective excitation of valence electrons in the material known as *plasmons* (analogous to *phonons* in the vibrational modes of an atomic lattice). The plasmon energy E_p is given by

$$E_p = \hbar\omega_p \tag{3.2}$$

where ω_p is the plasmon frequency. Using a "quasi-free" electron model [102], ω_p is related to the valence electron density N_v by

$$\omega_p^2 = \frac{N_v q_e^2}{\epsilon_0 m} \tag{3.3}$$

where q_e is the charge of an electron, ϵ_0 is the permittivity of free space, and m is the *effective mass* of an electron (in this case approximately equal to the rest mass of a free electron). For many materials, including insulators, Eq. (3.3) can be used to estimate the atomic density (and hence physical density) of a sample, if the involvement of all electrons in the outer valence shells (four electrons per atom in the case of carbon) is assumed. Thus, given the marked difference between the density of graphitic carbon and diamond (1.3–1.55 g cm^{-3} for glassy carbon and 3.515 g cm^{-3} for diamond [102]), measurement of the plasmon energy loss gives an approximate gauge of the tetrahedral (or otherwise graphitic) nature of a given amorphous carbon film. Plasmon energies of around 31 eV are typical of highly tetrahedral, predominantly sp^3 bonded carbon films (cf. 33 eV for diamond [104] and <24 eV for graphitic amorphous carbon films) [105].

3.1.1.2. Inner-Shell Excitations

At higher energies in the EELS spectrum, features arising from the excitation of core electrons are observed. Such excitations generally involve the transition of inner-shell electrons to states in unoccupied levels above the Fermi energy of the solid. In materials with an electronic bandgap, the shape and structure of resulting EELS "edges" provide valuable information about the density of unoccupied states near the edge of the conduction band as well as the chemical makeup and bonding in a sample. Figure 19 shows the EELS K-edge (corresponding to transitions from K-shell states) for carbon with a high graphitic bonding component. The feature marked "$1s - \pi*$" is characteristic of states arising from sp^2 bonding and can be used to quantify the fraction of sp^2 bonded carbon atoms. This is done by comparing the integrated intensity of this feature as a fraction of the total main core-loss edge, to that measured from a standard of known sp^2 fraction—ideally glassy carbon with 100% sp^2 bonding but with no crystal structure—by the equation

$$f_{\text{sample}} = \frac{I_{\text{sample}\,\pi*}}{I_{\text{standard}\,\pi*}} \frac{I_{\text{standard}}(\Delta E)}{I_{\text{sample}}(\Delta E)} \frac{\exp(t_{\text{sample}}/\lambda)}{\exp(t_{\text{standard}}/\lambda)} f_{\text{standard}} \tag{3.4}$$

where $I_{\text{sample}\,\pi*}$ and $I_{\text{standard}\,\pi*}$ are the integrated intensities under the $\pi*$ peak for the sample and standard, respectively, $I_{\text{sample}}(\Delta E)$ and $I_{\text{standard}}(\Delta E)$ are the integrated intensities over an energy window of width ΔE, and $\exp(t_{\text{sample}}/\lambda)$ and $\exp(t_{\text{standard}}/\lambda)$ are thickness and material dependent factors which introduce a minor correction to account for the weakening of the $\pi*$ feature with multiple scattering (and can be determined using Eq. (3.1)).

3.1.1.3. EELS from Amorphous Carbon

Along with neutron diffraction (see Section 3.1.2) EELS provided some of the first evidence of the high atomic density and tetrahedral nature of TAC thin films deposited by filtered cathodic vacuum arc. Figure 20 shows the carbon K-edge from Martin et al. [65] showing the absence of the $1s - \pi*$ feature in

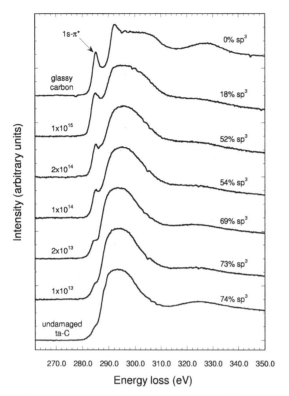

Fig. 21. Evolution of the carbon K-edge in ion damaged TAC as a function of 200 keV Xe^+ ion fluence, and a glassy carbon sample for comparison, from McCulloch et al. [107].

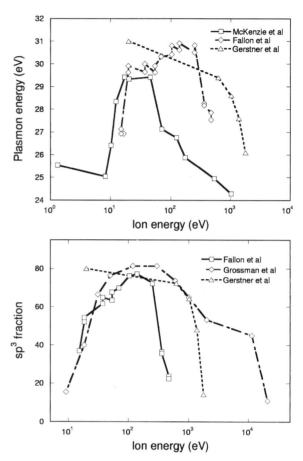

Fig. 22. Plasmon energy as a function of incident carbon ion energy from McKenzie et al. [94], Fallon et al. [68], and Gerstner et al. [110], and sp^3 fraction as incident carbon ion energy from Fallon et al. [68], Grossman et al. [108], and Gerstner et al. [110].

tetrahedral amorphous carbon deposited by FCVA indicating a low concentration of sp^2 bonding compared to a-C:H deposited by a methane glow discharge. Subsequent investigations calculated the total fraction of sp^3 bonded carbon atoms in these films from the K-edge EELS spectra to be in excess of 80% [106]. Figure 21 shows the evolution of the $1s - \pi^*$ feature as a result of 200 keV Xe^+ ion damage induced graphitization of TAC as a function of ion fluence (and the resultant sp^3 fraction calculated from the spectra) from McCulloch et al. [107].

EELS also provides an invaluable tool for determining the effect of different growth conditions on the bonding and density in carbon films. Of particular importance in the growth of tetrahedral carbon films is the role of the substrate bias and ion energy. Many groups have found maxima in both plasmon energy and sp^3 fraction as a function of incident ion energy [68, 94, 108, 109]. However, where this maximum occurs appears to vary from group to group (see Fig. 22), probably as a result of other factors such as ion fluence and substrate temperature. Indeed Gerstner et al. [110] found no maximum in sp^3 fraction above the ion energy of films deposited without applied substrate bias.

Another important factor is the effect on the density and bonding in films with the incorporation of impurities introduced to alter their electrical properties. Figure 23 shows the plasmon energy and sp^3 fraction as a function of hydrogen and nitrogen flow rate from Davis et al. [111, 112]. Hydrogen was introduced in an attempt to reduce the defect density in TAC and

nitrogen was introduced in order to dope it. The incorporation of hydrogen had little effect on both the sp^3 fraction, and only a small decrease in the density as inferred from the plasmon energy as higher flow rates, and was found to decrease the defect density by around an order of magnitude at flow rates up to 0.05 sccm (as inferred from space charge limited electrical characteristics; see Section 6) then to increase again at flows in excess of 0.1 sccm. In contrast and perhaps not surprisingly, the incorporation of nitrogen has a much greater influence on both the bonding and density, with sp^3 fraction and plasmon energy decreasing gradually with nitrogen flow rates up to around 4 sccm and decreasing rapidly above this. The effect of electron beam radiation damage while in the electron microscope was also found to be significantly greater for films containing as little as 1 at.% nitrogen, indicating a reduction in the structural stability of the carbon network with nitrogen incorporation.

3.1.2. Electron and Neutron Diffraction

While information about the chemical bonding in a film can be obtained from energy loss spectra, detailed structural information and the coordination and separation of atoms from one another in a sample can only be obtained by diffraction mea-

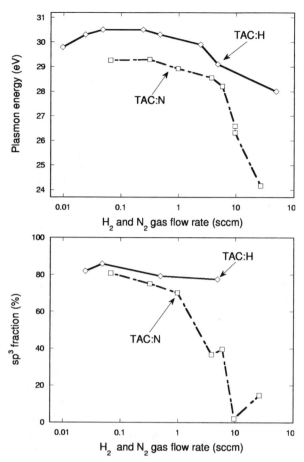

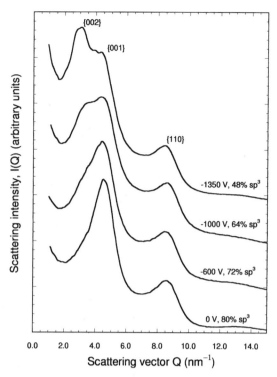

Fig. 24. Energy filtered electron diffraction intensities for TAC films deposited at various substrates biases from Gerstner et al. [110].

Fig. 23. Plasmon energy and sp^3 fraction as a function of hydrogen gas flow and nitrogen gas flow from Davis et al. [111, 112].

surement techniques. In the case of amorphous carbon films, electron diffraction and neutron diffraction have been particularly useful.

The principle advantage of electron diffraction over X-ray and neutron diffraction is that information can be obtained from extremely small regions of a specimen and significantly less material (particularly important in the case of thin films) is required. However, due to the difficulty in obtaining good qualitative reflection intensity data and the removal of the contribution of inelastically scattered electrons from the signal, it is only through the advent of computer interfaced electron energy loss spectrometers that both of these problems have been overcome and allowed electron diffraction to become a useful tool in the analysis of amorphous and polycrystalline materials.

By stepwise scanning the selected area electron diffraction pattern from a specimen across the entrance aperture to an electron energy loss spectrometer, a zero energy loss (elastic) diffraction intensity as a function of scattering angle can be obtained. The first practical use of this technique for the analysis of amorphous and polycrystalline films was demonstrated by Cockayne and McKenzie [113, 114].

Figure 24 shows the energy filtered diffraction pattern as a function of scattering vector Q, for TAC films deposited in the filtered cathodic vacuum arc at various substrate biases by

Gerstner et al. [110]. The appearance and evolution of a peak corresponding to the {002} reflection in graphite (corresponding to a lattice spacing of 0.335 nm) with increasing substrate bias indicate the emergence and clustering of sp^2 bonded atoms within the sp^3 network in the form of sheets, which are to some extent parallel and correlated in position.

In order to obtain more detailed quantitative information about structural parameters such as average bond length, bond angle, and coordination number in an amorphous material, both electron and neutron scattering intensity data can be use to obtain the reduced radial density function $G(r)$. The $G(r)$ is a measure of the spherical average of the deviation of the radial atomic density $\rho(r)$ from the average atomic density ρ_0 for the sample at a distance r from a typical or average atom center and is most commonly defined by the expression

$$G(r) = 4\pi r[\rho(r) - \rho_0] \quad (3.5)$$

which can be calculated from a Fourier-sine transformation of the diffraction data given by

$$G(r) = \sqrt{\frac{2}{\pi}} \int_0^{Q_{max}} Q[S(Q) - 1]D(Q)\sin(Qr)dQ \quad (3.6)$$

where Q is the scattering vector (given by $4\pi \sin(\theta)/\lambda$ where θ is the scattering angle and λ is the wavelength of the incident radiation), $D(Q)$ is a damping factor (commonly a sinc function in the case of electron diffraction [115]), and $S(Q)$ is the *structure factor* calculated from the scattering intensity $I(Q)$ which in the case of neutron diffraction takes into account attenuation, background, and container, multiple, and inelastic

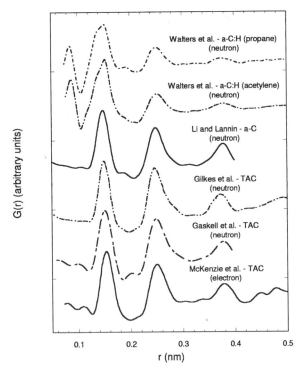

Fig. 25. Radial density functions for TAC from electron diffraction [94] and neutron diffraction [116, 119] for sputtered *a*-C from neutron diffraction (Li and Lannin [118] taken from Frauenheim et al. [84]), and for *a*-C:H from acetylene and propane precursors from neutron diffraction [83].

scattering (see Walters et al. [83] and Gaskell et al. [116]). In the case of inelastic electron scattering it is calculated by

$$S(Q) = I(Q)/Nf(Q)^2 \qquad (3.7)$$

where N is number of atoms in the sample (obtained by a least squares fitting (3.7) to $I(Q)$ at large scattering angles) and $f(Q)$ is the atomic scattering factor for electrons (see Cockayne and McKenzie [113, 114] and McKenzie et al. [94, 115]).

By fitting appropriate peaks (Gaussians convolved by an appropriate function to account for experimental parameters such as maximum collection angle) to the resulting $G(r)$ the average bond length and nearest neighbor distances can be determined, and the bond angle can be calculated by

$$\theta = 2\sin^{-1}\left(\frac{r_2}{2r_1}\right) \qquad (3.8)$$

where r_1 and r_2 are the first and second nearest neighbor distances, respectively. The average number of atoms bonded per atom, or coordination number, is then given by

$$C(r_1, r_2) = \int_{r_1}^{r_2} rG(r)dr + \frac{4\pi}{3}\rho_0[r_2^3 - r_1^3] \qquad (3.9)$$

Figure 25 shows the reduced radial density functions for TAC from electron diffraction [94] and neutron diffraction [116, 117], sputtered *a*-C from neutron diffraction [118], and *a*-C:H from neutron diffraction [83]. For TAC the $G(r)$ gives a first nearest neighbor distance, bond angle, and first nearest

neighbor coordination number of 0.153 nm, 110°, and 4.0 respectively from electron diffraction [94, 115] and 0.152 nm, 110°, and 3.8 respectively from neutron diffraction [116, 117]. These values are consistent with a predominantly tetrahedrally bonded carbon network, compared to 0.154 nm, 109.47°, and 4.0 respectively for diamond and 0.142 nm, 120°, and 3.0 for graphite.

However, less straightforward is the comparison of the second nearest neighbor coordination number calculated from the second peak in the $G(r)$. It is evident both in the study by Gaskell et al. [116] and in a subsequent higher resolution neutron diffraction study by Gilkes et al. [119] that this second peak is in fact made of two Gaussians at 0.248 and 0.275 nm. Gaskell et al. assume that both peaks are due to contributions by second nearest neighbors and arrive at a second nearest neighbor coordination number of 8.9. Gilkes et al., on the other hand, concede that the second peak at 0.275 nm may be made up of both second and third nearest neighbors and calculate a second nearest neighbor coordination of between 7.66 (in which all contributions to the second peak are third nearest neighbors) and 11.06 (in which the majority of contributions to the second peak are second nearest neighbors). It is important to note that if some number of atoms at 0.275 nm are indeed second nearest neighbors then the average bond angle will be somewhat higher than the tetrahedral angle 109.47°. This ambiguity underlines the fact that the structural information contained in $G(r)$ is not unique to any single structure, with many structures potentially resulting in the same $G(r)$. To some extent, therefore, any accurate discrimination between these structures must include reference to information obtained by other methods such as infrared and Raman spectroscopy (see Section 4.4.1). It is also helpful to compare these properties with those of computer models to form a better picture of the probable structure of these materials, and such models are reviewed in Section 3.2 with particular reference interpretation of the second coordination sphere in the $G(r)$.

Interpretation of the $G(r)$ for *a*-C:H with a high hydrogen content and mixture of sp^1, sp^2, and sp^3 bonding is similarly complex but perhaps less subtle than that of TAC. Peaks at nearest neighbor distances corresponding to hydrogen–hydrogen, carbon–hydrogen, and double and single carbon–carbon bonds are all evident [83]. While a determination of the relative proportion each of these bonding types is difficult, the accuracy in the determination of the peak positions does allow differentiation, in the case of sp^2 bonds, between chainlike olefinic bonding (with a first nearest neighbor distance of around 0.134 nm) and ringlike aromatic bonding (with a first nearest neighbor distance of around 0.142 nm). In the case of both samples analyzed by Walter et al. olefinic bonding was found to predominate, which is important with respect to the labels "polymeric" and "graphitelike" in the case of sp^2 rich films.

3.2. Computer Modelling of Growth and Structure

Computer simulation and modelling of amorphous materials range in sophistication from hand built structural models, to classical molecular dynamics using empirical interatomic potentials, to full blown *ab initio* quantum mechanical molecular dynamics. The advantage of less rigorous empirical approaches is their ability to model large systems, surfaces, and equilibrium processes, such as thermal annealing, without unreasonably long computation times. The main disadvantage, however, is their limited ability to model the complexity of large range of potential bonding environments of carbon atoms including mixtures of hybrid bonding states (sp^1, sp^2, and sp^3 bonding), and in particular π bonding.

3.2.1. Empirical Models

Owing to the fact that graphite is the stable allotrope of carbon at standard temperature and pressure, and that in most disordered forms of carbon sp^2 bonding is the favored phase, an important question regarding TAC is that of its formation mechanism (see Section 2). A key to this question may lie in the fact that when films of a high sp^3 bonded content are deposited, they always form with high compressive stresses, typically in the order of 5–6 GPa [120]. McKenzie et al. [94, 99, 115, 121] proposed that it is the formation of compressive stress due to the subplantation of carbon ions, either by direct or knock-on implantation into the growth region below the surface of the film, that creates an environment for the formation of carbon in which sp^3 or diamondlike bonding is the preferred phase—analogous to the way in which synthetic diamond is made industrially. Robertson [122], however, has proposed a different mechanism in which densification leads to preferential sp^3 bonding and compressive stress is simply a by-product of such densification. At best, it is contentious as to whether or not the formation of sp^3 carbon bonding is the result of compressive stresses or vice versa, or whether they just appear simultaneously in a related though not necessarily causal way. In simulations of two-dimensional systems analogous to carbon, Marks showed the evolution of compressive stresses induced during the growth of films with ions in a similar range to those produced in the cathodic arc, without any subplantation [123, 124]. Furthermore, a number of groups have shown that some of the stress in TAC can be relieved without a significant reduction in the sp^3 fraction [125, 126]. However, whether or not its formation is contingent on the presence of compressive stress during growth is as yet unresolved.

3.2.1.1. Hand Built Models

One of the first attempts to model the structure of amorphous carbon was conducted by Beeman et al. [127]. Carried out before the tetrahedral nature of filtered cathodic arc deposited amorphous carbon was confirmed, one of the aims of this study was to investigate the feasibility of tetrahedral bonding in *a*-C films proposed by Kakinoki et al. [128]. In this work hand built atomic models of random carbon networks containing various fractions of trigonal and tetrahedral bonding were constructed. These models were then relaxed by allowing the atoms to move in order to minimize their elastic strain energy as described by

a potential developed by Keating [129], and their radial density functions, vibrational density of states, and Raman spectra were calculated and compared to experiment. While it was found that the fraction of tetrahedral bonding was likely to be negligible in the majority of carbon films for which experimental data were available at the time, comparison of the experimental results from films analyzed by Kakinoki et al. was consistent with Beeman's model containing 9% fourfold coordinated carbon atoms.

3.2.1.2. Stillinger–Weber Potential

One of the first successful simulations of group IV semiconductors was applied to silicon in both solid and liquid phases by Stillinger and Weber [130]. The generalized potential energy function describing interactions between N identical particles can be expressed as the sum of all individual n-body contributions v_n by

$$U = \sum_i v_1(i) + \sum_{\substack{i,j \\ i<j}} v_2(i,j) + \sum_{\substack{i,j,k \\ i<j<k}} v_3(i,j,k)$$
$$+ \cdots + v_N(1,\ldots,N) \tag{3.10}$$

The Stillinger–Weber potential approximates this potential function by considering only two- and three-body contributions. The two-body term $v_2(i,j)$ depends only on the distance between atom pairs r_{ij} and is given by

$$v_2(r_{ij}) = \epsilon A \left[\frac{B}{r_{ij}^4} - 1 \right] \exp\left(\frac{1}{r_{ij} - a} \right) \quad r_{ij} < a$$
$$= 0 \qquad\qquad\qquad\qquad \text{otherwise} \tag{3.11}$$

and the three-body term $v_3(i,j,k)$ depending on the atomic separations r_{ij} and r_{jk} and the triplet bonding angle θ is given by

$$v_3(r_{ij}, r_{jk}, \theta) = \epsilon\lambda \left[\frac{\gamma}{r_{ij} - a} + \frac{\gamma}{r_{jk} - a} \right]$$
$$\times \left[\cos(\theta) - \cos(\theta_0) \right]^2, \quad r_{ij} < a \quad \text{and} \quad r_{jk} < a \tag{3.12}$$
$$= 0 \qquad\qquad\qquad\qquad \text{otherwise}$$

where θ_0 is the ideal bond angle for a triplet, ϵ is the bond strength, λ represents the resistance to change in bond angle, a is the maximum bond length, and A, B, and γ are "fitting" constants. The advantage of the Stillinger–Weber potential over those typified by the Keating potential used by Beeman et al. [127] is that it more accurately describes large atomic displacements from the ideal tetrahedral geometry of crystalline silicon thereby allowing both its liquid and amorphous phases to be modelled.

While originally intended for application to exclusively sp^3 bonded systems such as silicon and germanium, Bensan [131] and Mahon et al. [132] showed that the σ bond contribution of sp^2 and sp^3 carbon bonds had the same functional forms (when considered in reduced units) and could therefore both be implemented with the same potential. Using Hartree–Fock self-consistent field calculations of the energy of small carbon clusters, the Stillinger–Weber Eqs. (3.11) and (3.12) were thereby reparameterized to model trigonally [131] and tetra-hedrally [132] bonded carbon atoms. Gerstner and Pailthorpe [133] then integrated both these modified potentials into one molecular dynamics simulation to investigate the effects of many successive carbon ion impacts, simulating film growth with energies in the range of those used in the growth of TAC, into the surface of a 320-atom diamond lattice. The bonding hybridization of each individual atom in the lattice was dynamically determined using an algorithm developed by Marks [134] based on bond energy minimization, and modelling of the potential between sp^2 and sp^3 hybridized atoms was implemented using average values for ϵ and σ between those for diamond and graphite. Periodic boundary conditions were applied in the x- and y-directions to minimize the computation time required for the system to reach equilibrium. Thermal energy was extracted through the bottom-most layer using a random thermostating procedure [135]. The molecular dynamics of the system was carried out by solving Newton's equations of motion discretely using the Verlet algorithm [136, 137].

While the systems studied were too shallow to simulate the tetrahedral growth beyond the top graphitic layer observed experimentally on the surface of all TAC films, they did indicate, not surprisingly, that the rate of graphitization of the diamond surface increased with both incident energy and substrate temperature. The main deficiencies of the Stillinger–Weber potential with respect to carbon are its inability to effectively model π orbital interactions and the absence of a four-body term resulting in potentially unphysical dihedral angles. However, this is generally true of all empirical potentials for carbon.

3.2.1.3. Tersoff Potential

The Tersoff potential, also constructed to model silicon atomic networks which depart significantly from ideal tetrahedral geometries, attempts to solve the problem of simulating such systems with a more general approach than that adopted by Stillinger and Weber. Instead of considering interactions between atoms in the system in terms of one- to n-body interactions, Tersoff [138, 139] began from the observation that, from simple quantum mechanical arguments, the most important influence on the strength of bonding between pairs of atoms is the coordination number—the number of neighboring atoms which are close enough to form bonds—of both atoms. In general, the fewer atoms any given atom is bonded to, the stronger the bonds between those atoms—for example, the strength of individual bonds between trigonally bonded carbon atoms within a graphite plane is in fact greater than that between tetrahedrally bonded carbon atoms in diamond (the difference in bulk hardness being explained by the very weak π-bonding between planes). In the Tersoff potential then, the concept of bond order is introduced, and total potential energy of the system U is expressed only as the sum of pairwise potentials of each atom with every other atom in the system by

$$U = \frac{1}{2} \sum_{i,\, j \neq i} V_{ij} \qquad (3.13)$$

$$V_{ij} = f_{\text{cutoff}}(r_{ij})\left[a_{ij}\, f_{\text{repulsive}}(r_{ij}) + b_{ij}\, f_{\text{attractive}}(r_{ij})\right] \quad (3.14)$$

where $f_{\text{cutoff}}(r_{ij})$ is a smooth cutoff function to limit the range of the potential for computational convenience, $f_{\text{repulsive}}(r_{ij})$ represents the simple repulsive term between two atoms which includes the orthogonalization energy when atomic wavefunctions overlap (chosen to be a simple exponential of the form $f_{\text{repulsive}}(r) = A \exp(-\lambda_1 r)$), and $f_{\text{attractive}}(r_{ij})$ is the attractive term that represents bonding (also chosen to be a simple exponential of the form $f_{\text{attractive}}(r) = -B \exp(-\lambda_2 r)$), the coefficient a_{ij} is approximated in most cases to unity, and b_{ij} is an implicit measure of the *bond order*, which is dependent on the local environment and is a monotonically decreasing function of the coordination of atoms i and j, the strength of competing bonds, and the cosines of the angles with competing bonds. Further details of the form and nature of this potential with respect to amorphous silicon can be found in [138, 139].

The original intent in implementing the Tersoff potential was to attempt to make fewer assumptions about the specific nature of local bonding geometries and thereby be potentially more flexible in its ability to model undercoordinated forms of covalently bonded solids such as amorphous silicon and amorphous carbon. Specifically, it was felt that previous approaches artificially imposed a higher degree of coordination and arbitrarily excluded lower coordinated structures from the resulting models. The first attempts to use it to model amorphous carbon networks were made by Tersoff [140, 141] which yielded reasonable agreement with respect to experimentally determined elastic constants. However, they resulted in structures of significantly higher density for their sp^2 bonding content than experiment. In the latter paper [141]—in which the role of hydrogen in relieving strain in hydrogenated amorphous carbon was investigated—sp^3 bonding in pure carbon networks appears to be so unfavored by the Tersoff potential as to be unphysical, which suggests that in trying to allow for undercoordinated structures, in amorphous materials it has gone too far in the other direction. As in the case of the Stillinger–Weber potential, it seems that a neglect of π bonding and an inadequate account of modelling the dihedral angle in carbon are the main sources of difficulty. The principal effect of this, as pointed out by Stephan and Haase [142], is that in the absence of a significant repulsive term for π bonded atoms, an artificial excess of sp^2 bonding is allowed at high densities.

With this in mind, useful insights have been gained from simulations of amorphous carbon using the Tersoff potential. Ion beam growth of thin amorphous carbon films onto a crystalline diamond substrate at differing incident energies was simulated by Kaukonen and Nieminen [143]. While the density of their simulated films was higher and the sp^3 fraction lower than experimental results, they did reproduce the finding that maximums in density and sp^3 fraction occur with respect to ion energy at around 40–70 eV. They found that the formation of tetrahedral bonding results from competing effects of defects induced by thermal energy spikes caused by ion impacts, annealing of these defects at higher energies, and the high thermal conductivity of the diamond substrate preventing such annealing.

The largest body of work on the modelling of amorphous carbon using the Tersoff potential has been conducted by Kelires [144–146] and is reviewed in [95]. Amorphous carbon networks of various densities were generated by starting with between 216 and 550 atoms in a well equilibrated liquid state at approximately 9000 K which were then rapidly quenched to room temperature. While it is accepted that such a process does not directly mimic the thin film deposition process, it does produce stable structures which are believed to bear some resemblance to those in amorphous carbon films and so for its computational convenience is widely used (and used exclusively in tight-binding and *ab initio* approaches) throughout the literature. In investigating the structural stability of TAC, Kelires found his simulated networks were thermally stable up to a temperature of around 1200 K at which point they underwent a transition to a dense sp^2-rich phase, consistent with annealing experiments. Decomposition of the local bond stress found an inhomogeneous stress distribution between sp^2 and sp^3 bonded atoms, with sp^2 sites found mainly in tension and sp^3 sites found mainly in compression. This demonstrates the role that sp^2 sites may play in relieving the internal strain in the network, similar to the role of hydrogen found by Tersoff in *a*-C:H [141].

3.2.2. Tight-Binding Approaches

The next level of sophistication in modelling of amorphous materials employs an approach known as the *tight-binding* method and attempts to reach a compromise between the rigor of a computationally expensive *ab initio* quantum mechanical treatment and the limitations of a completely empirical approach. The term "tight-binding" covers a variety of approaches within fundamental electronic-structure theory which approximate a fully *ab initio* treatment by constraining the expansion of electron wavefunctions in the system to a basis set of atomiclike orbitals of the valence electrons only.

Following the treatment given by Frauenheim et al. [84] the initial step in the procedure is the quantum mechanical self-consistent determination of the effective one-electron potentials and corresponding wavefunctions using density-functional (DF) theory [147] within the local density approximation (LDA). These single-atom valence electron wavefunctions $\phi_\mu(\mathbf{r})$ are thereby used in the expansion of the electron wavefunction for all atoms at positions $\mathbf{R}_l$ by

$$\psi_i(\mathbf{r}) = \sum_\mu c_\mu^i \phi_\mu(\mathbf{r} - \mathbf{R}_l) \quad (3.15)$$

The coefficients c_μ^i are evaluated by diagonalization of the secular matrix

$$\sum_\mu^i (h_{\mu\nu} - \epsilon_i S_{\mu\nu}) = 0 \quad (3.16)$$

where $h_{\mu\nu}$ is the tight-binding Hamiltonian which introduces the contributions to the effective one-particle potential (which includes the electron–electron (Hartree) term and exchange–correlation term within the LDA), and $S_{\mu\nu}$ is the overlap matrix to correct for the nonorthogonality of the basis functions at different atom sites, to yield orbital energies ϵ_i and corresponding eigenfunctions. Then, in its simplest form, tight-binding expresses the total energy in the form

$$\begin{aligned} U_{\text{total}} &= U_{\text{binding}}(\{\mathbf{R}_l\}) + U_{\text{repulsive}}(\{\mathbf{R}_l - \mathbf{R}_k\}) \\ &= \sum_i n_i \epsilon_i(\{\mathbf{R}_l\}) + \sum_{l,\,k>l} V_{\text{repulsive}}(\{\mathbf{R}_l - \mathbf{R}_k\}) \end{aligned}$$
$$(3.17)$$

where n_i is the occupation number of orbital energy ϵ_i. The first term in (3.17) represents the binding or "band structure" energy, which includes the self-consistent LDA calculation of the single-atom valence electron orbitals. The second term in (3.17) represents the shortrange repulsive energy, which is generally modelled by an empirical short-range two-body potential $V_{\text{repulsive}}$, because of the computational expense of inclusion of the extended basis required to make an accurate quantum-mechanical calculation of this term. In sophisticated treatments $V_{\text{repulsive}}$ is chosen so that the potential energy curves of the bonded atomic pairs within the system reproduce those evaluated separately with a full LDA calculation.

Once an expression for the total energy of the system as a function of all atomic coordinates is found, the interatomic forces can be found from the gradients of this total energy at all atom sites. Frauenheim [84] uses the expression

$$\mathbf{F}_l = -\frac{\partial U_{\text{total}}}{\partial \mathbf{R}_l} = \sum_i n_i \sum_\mu \sum_\nu c_\mu^i c_\nu^i \left[-\frac{\partial h_{\mu\nu}}{\partial \mathbf{R}_l} + \epsilon_i \frac{\partial S_{\mu\nu}}{\partial \mathbf{R}_l} \right] - \frac{\partial U_{\text{repulsive}}}{\partial \mathbf{R}_l} \quad (3.18)$$

for calculating the force on atom l and position $\mathbf{R}_l$.

In the most extensive body of work of carbon using tight-binding Frauenheim et al. [38, 84, 148] have modelled a variety of pure and hydrogenated carbon networks with densities of between 2.0 and 3.5 g cm^{-3} by the rapid quenching of carbon atoms from a liquid state at approximately 8000 K. Systems sizes of 64 carbon atoms [84] and 128 carbon atoms, in addition to various numbers of hydrogen atoms, give structures of varying hydrogen content. As with many other simulation schemes, when compared with experimental results, the resulting films exhibited either lower sp^3 fractions at similar densities, or higher densities at similar sp^3 fractions with a discrepancy of around 1.2–1.5 times that of experimentally determined densities. It is probably that at least some of this may be due to an underestimation of the experimental density as a result of either the inclusion of lower density macroparticles during deposition, or other medium range density fluctuations larger than the size of the simulated supercell. The trends, however, are consistent with experiment demonstrating increasing sp^3 bonding with increasing density for networks of constant hydrogen content (including pure films) and increasing hydrogen content for networks of fixed density (with the exception of the highest density sample which showed a slight initial decrease hydrogen incorporation). The reduced radial distribution functions ($G(r)$)

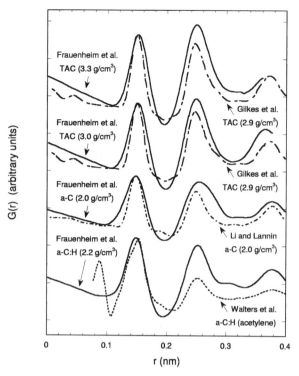

Fig. 26. Reduced radial density functions from various simulated carbon networks from Frauenheim et al. [84] compared with equivalent experimental results of Walters et al. [83], Li and Lannin [118] (taken from Frauenheim et al. [84]), and Gilkes et al. [119].

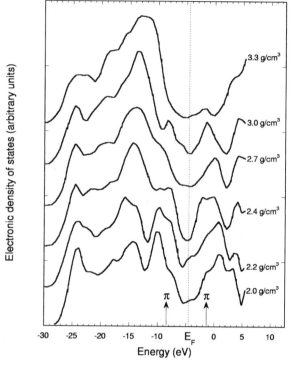

Fig. 27. Electronic density of states for hydrogen-free amorphous carbon structures at various macroscopic densities between 2.0 and 3.3 g cm^{-3} from Frauenheim et al. [84].

for various characteristic networks are shown in Figure 26 along with equivalent experimentally obtained $G(r)$'s from elsewhere in the literature for comparison.

In the highest density networks, most analogous to ion beam deposited TAC, analysis of the three-dimensional bonding topologies generally shows sp^2 bonded atoms in isolation or in pairs within an sp^3 bonded matrix with no extended aromatic ring structures present. At higher hydrogen contents and lower densities the structure increasingly becomes one of clusters of sp^3 bonded regions interconnected by sp^2 onded chainlike segments. In the lowest density networks containing intermediate concentrations of hydrogen (∼20 at.%), most analogous to plasma-enhanced CVD grown a-C:H, the network is dominated by a π bonded network of principally sp^2 hybridized atoms, with a small fraction of sp^1 and sp^3 bonding homogeneously distributed in chainlike segments. Consistent with the experimental neutron diffraction results of Walters et al. [83] (see Section 3.1.2), sp^2 bonding in these films also exhibits very little aromatic nature. Increasing the incorporation of hydrogen in these films removes all sp^1 bonding in favor of both sp^2 and sp^3 bonding.

It is generally accepted that the electronic properties of amorphous carbon films are dominated by contributions to the electronic density of states (EDOS) by π bonding states associated with sp^2 hybridized atoms within the material, which is confirmed in an analysis of the EDOS of all networks simulated by Frauenheim et al. [84]. An initial narrowing of the bandgap of their pure carbon networks occurred with increasing density

from 2.0 g cm^{-3}, which then widened again up to the maximum simulated density of 3.3 g cm^{-3} (see Fig. 27). In a subsequent work from the same group Stephan et al. [149] found that the EDOS is sensitive to the size and topology of π bonded clusters, with sp^2 atom pairs resulting in the widest bandgaps of around 3 eV. With the incorporation of hydrogen into these highest density networks an additional study found that reductions in the strain on π bonding resulted in the opening of an even greater bandgap of 3.3 eV [148].

A approach similar to that of Frauenheim et al. has been carried out by Wang et al. following a tight-binding treatment developed for carbon by Xu et al. [150]. Both unhydrogenated low density (2.2 g cm^{-3}) carbon networks analogous to sputtered a-C [39, 87] and high density (3.5 g cm^{-3}) carbon networks analogous to ion beam TAC [39, 151] were simulated.

The structure factor $S(Q)$ of the low density model gives a good fit to the experimental $S(Q)$ obtained by Li and Lannin [118] from sputtered a-C using neutron diffraction. The sp^1, sp^2, and sp^3 fractions in this model were 12%, 80.6%, and 7.4%, respectively, and its electronic properties were consistent with experimental observation of a small mobility gap of around 0.5 eV opening up between π states in a-C.

With respect to their high density carbon network [39, 151], as in the work by Frauenheim et al. [38, 84, 148], while a sp^3 fraction of around 80% is consistent with experiment, its density of 3.5 g cm^{-3} is again somewhat higher than expected. Similarly the peak corresponding to the second coordination sphere in the calculated reduced radial distribution function (see Fig. 28) is markedly higher than that determined by experiment.

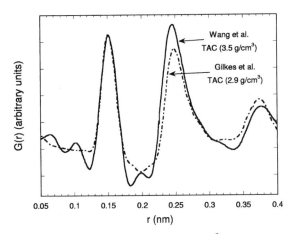

Fig. 28. Reduced radial density function of 3.5 g cm^{-3} simulated carbon network from Wang et al. [151] compared to that obtained for TAC by neutron diffraction from Gilkes et al. [119].

A more sophisticated technique developed by Sankey and Niklewski [152], which uses an extended atomic orbital basis set and avoids the need for an empirically fitted repulsive term ($V_{repulsive}(\{\mathbf{R}_l - \mathbf{R}_k\})$ in (3.17)), has been applied in a study of the structural, electronic, and vibration properties of TAC by Drabold et al. [85]. At a density of 3.0 g cm^{-3} (cf. 2.9 g cm^{-3} obtained experimentally) the 64-atom network they simulated had an sp^3 fraction of 91%, which sets it apart from most other models in that it is simultaneously the highest sp^3 fraction of any simulated amorphous carbon system and one of the lowest simulated densities for a predominantly tetrahedrally bonded network.

In this work they identify three types of defects of importance to the electronic structure of TAC—(i) π *defects* consisting of predominantly paired sp^3 bonded atoms, (ii) *strain defects* consisting of fourfold coordinated atom bond angles well in excess ($>150°$) of the ideal tetrahedral angle ($>109.47°$) which in amorphous silicon are known to contribute to localized electronic states within the mobility gap, and (iii) *stretched bond defects* consisting of fourfold coordinated atoms with bond lengths in excess of 0.173 nm. The most important of these is the π *defect* which is the principal structural difference between carbon and silicon bonding, and whose propensity to form pairs, identified throughout the literature, has significant implications with respect to the localization of associated electronic states, electronic conduction, and its contribution (or lack thereof) to the electron paramagnetic resonance signal of TAC. However, in simulations of systems in the order of 100 atoms it is perhaps not surprising that isolated sp^2 bonded atoms are rarely seen, as the presence of just one such atom in 100 would suggest defect densities well in excess of the $\sim 10^{20}$ spin cm^{-3} calculated from EPR measurements [109, 110] (see Section 5).

3.2.3. Ab Initio Molecular Dynamics

At the top of the spectrum of computer modelling approaches for both physical sophistication (in terms of interatomic potentials) and computational intensity is the *ab initio* quantum-mechanical molecular dynamics algorithm developed by Car and Parrinello [153]. Car–Parrinello molecular dynamics (CPMD) is based on the unification of classical molecular dynamics used to describe the motion of atoms under the influence of a given potential, with the DF theory of Kohn and Sham [147] which allows the accurate determination of electronic properties of an atomic network, with no prior assumptions about the interaction potentials and without constraining the description of the electron wavefunctions to a reduced basis set of atomlike orbitals (as in tight-binding approaches). As a consequence of the generality of this approach it is ideally suited (given enough computational power) to dealing with interactions between carbon atoms in different and intermediate hybridization states (such as sp^2–sp^2, sp^3–sp^3, and sp^2–sp^3 bonded atom interactions) and therefore to amorphous carbon systems.

Applied to carbon the Car–Parrinello method treats the atoms as positively charged cores (described by an appropriate pseudo-potential), each surrounded by four valence electrons. The total electron density is expressed in terms of occupied orthonormal Kohn–Sham (KS) orbitals by

$$n(\mathbf{r}) = \sum_i |\psi_i(\mathbf{r})|^2 \qquad (3.19)$$

whose wavefunctions $\psi_i(r)$ ($= \sum_k c_k^i \chi_k$) are expressed in terms of an expansion in terms of a large basis of plane waves $\chi_1, \ldots, \chi_m$. For a given static atomic configuration the ground electronic state is given by a point on the Born–Oppenheimer (BO) surface which is found by minimization of the energy functional

$$E(\{\psi_i\}, \{\mathbf{R}_l\}, \{\alpha_v\}) = \sum_i \int_\Omega \psi_i^*(\mathbf{r}) \big| -[\hbar^2/2m]\nabla^2 \big| \psi_i(\mathbf{r}) d\mathbf{r}$$
$$+ U(n(\mathbf{r}), \{\mathbf{R}_l\}, \{\alpha_v\}) \qquad (3.20)$$

where $\{\mathbf{R}_l\}$ is the set of atomic coordinates, $\{\alpha_v\}$ is the set of all possible external constraints such as volume Ω, strain, etc., and the potential $U(n(\mathbf{r}), \{\mathbf{R}_l\}, \{\alpha_v\})$ includes the Coulombic repulsion between ion cores, the electron–ion interaction, the Hartree electron–electron interaction, and the exchange and correlation interaction treated in the LDA.

Conventionally, minimization of Eq. (3.20) subject to orthonormality of the orbitals ψ_i leads to the self-consistent Kohn–Sham equations

$$\left[-\frac{\hbar^2}{2m}\nabla^2 + \frac{\partial U}{\partial n(\mathbf{r})} \right] \psi_i(\mathbf{r}) = \epsilon_i \psi_i(\mathbf{r}) \qquad (3.21)$$

whose solution requires repeated matrix diagonalization until self-consistency is achieved, the computational effort of which increases dramatically with system size and basis set. In principle this calculation must be executed for each and every new atomic configuration, i.e., at every time step if the ion cores are to be allowed to move toward a structural minimum, which makes any straightforward implementation of it into a conventional molecular dynamics scheme computationally prohibitive for any but the smallest systems. Furthermore, no such calculation can ever be completely self-consistent, which results in errors that give rise uncontrollable energy losses or gains within a system causing problems with stability [154]. However, the Car–Parrinello technique overcomes both these difficulties by approaching the specific problem of minimization of Eq. (3.20) by applying a variation on the concept of *simulated annealing* [155].

In conventional simulated annealing, a given functional, such as that in Eq. (3.20), is minimized with respect to a set of n free parameters which form an n-dimensional coordinate hyperspace by searching through that hyperspace randomly via a Monte Carlo algorithm, reducing the search radius as convergence is approached, in a manner analogous to the way in which the molecules of a liquid (their coordinates being analogous to the free parameters of the coordinate hyperspace) find their minimum (crystallize) as it is cooled. However, owing to the orthonormality constraint on the KS orbitals ψ_i, implementation of such minimization by random Monte Carlo means is not straightforward. Instead, the Car–Parrinello technique implements *dynamic simulated annealing* of the coordinate system by approaching it as a dynamic system and treating the variation of the free parameters of the system in the same way as motion of atomic coordinates in classical molecular dynamics. This is achieved by introducing the Lagrangian

$$L = \sum_i \frac{1}{2}\mu \int_\Omega |\dot{\psi}_i|^2 \, d\mathbf{r} + \sum_l \frac{1}{2} M_l \dot{\mathbf{R}}_l^2$$
$$+ \sum_\nu \frac{1}{2}\mu_\nu \dot{\alpha}_\nu^2 - E(\{\psi_i\}, \{\mathbf{R}_l\}, \{\alpha_\nu\}) \quad (3.22)$$

where $\dot{\psi}_i$, $\dot{\mathbf{R}}_l$, and $\dot{\alpha}_\nu$ are the time derivatives of the KS orbital, ionic position, and external constraint coordinates, respectively, M_l are the ionic masses, and μ and μ_ν are arbitrary "masslike" parameters of appropriate units. From the dynamics of the coordinate-space (i.e., the simulated annealing hyperspace), parameters $\{\psi_i\}$, $\{\mathbf{R}_l\}$, and $\{\alpha_\nu\}$ can then be described through the equations of motion

$$\mu \ddot{\psi}_i(\mathbf{r}, t) = \partial E / \partial \psi_i^*(\mathbf{r}, t) + \sum_k \Lambda_{ik} \psi_k(\mathbf{r}, t) \quad (3.23)$$

$$M_l \ddot{\mathbf{R}}_l = -\nabla_{\mathbf{R}_l} E \quad (3.24)$$

$$\mu_\nu \ddot{\alpha}_\nu = -(\partial E / \partial \alpha_\nu) \quad (3.25)$$

where Λ_{ik} are Lagrange multipliers introduced to ensure orthonormality of the KS orbitals. The physical significance of the ion dynamics described by Eq. (3.24) is real, relating the positions and velocities of ion centers, whereas the dynamics associated with Eqs. (3.23) and (3.25) are mathematical constructs introduced to perform the dynamical simulated annealing.

The important and innovative aspect of this approach is that the solution of Eqs. (3.23)–(3.25) allows the calculation and relaxation of the evolving electronic structure (embodied in the electron density distribution $n(\mathbf{r})$) and relaxation of the physical structure (through the molecular dynamics embodied in

the evolving positions of the ion centers $\{\mathbf{R}_l\}$) to be achieved simultaneously. Not only does Car and Parrinello's dynamic simulated annealing provide an efficient means of determining the electronic configuration of a system (i.e., its ground state on the BO surface) which then generates the forces which cause the ion cores to move, but it also tracks the electronic coefficients in such a way that it updates the total wavefunction to the BO surface corresponding to the new ionic configuration with each time step (provided that time step is sufficiently short). Furthermore, oscillation of the electronic coordinates within a narrow "envelope" in coordination-space (mediated by the small "fictitious temperature" associated with dynamics of the electronic orbitals embodied in Eq. (3.23)) close to the BO surface surprisingly results in negligible net energy accumulation or loss, to or from the system [154].

While CPMD implements the most accurate and general description of the molecular dynamics of small systems of around 100 atoms over real time periods of the order of picoseconds, it has been argued that the computational difficulty of ensuring the attainment of equilibrium by running for longer time periods, of achieving better statistics by repeating simulations more than a few times, limits the physical significance of any conclusions drawn from its results. However, with this caveat in mind, the generality of the method does allow for the potential of providing insights particularly into the applicability of various assumptions made by other simulation techniques.

The first application of the CPMD method to amorphous carbon was conducted by Galli et al. [80] by quenching a 54-atom cell at a constant density of 2.0 g cm^{-3} in a liquid state from a temperature of 5000 K down to 300 K over a period of 0.5 ps, which was then allowed to equilibrate for a further 0.3 ps. Two simulations where undertaken with different size basis sets for describing the KS orbitals with the greater using a set of 12,000 plane waves. A significant overestimation ($\sim$6%) of the bond lengths was observed in the structure modelled using the smaller basis, and a smaller though noticeable overestimation ($\sim$3%) in the $G(r)$ in comparison to both the experimental results of Li and Lannin and tight-binding results of Frauenheim et al. [84] when using the greater basis set was observed. Use of a larger basis is obviously limited by the available computational power which, it has been argued, in turn limits the applicability and potential generality of the Car–Parrinello approach to real systems. However, with the exponential increase in available computational power over the last decade, and with the obvious *caveats* in mind, this has become a less and less applicable objection to the technique.

Using CPMD code developed and maintained by the Parrinello laboratory at the Max Planck Institut fur Festkörperforschung, Stuttgart, this technique was applied by Marks et al. [156, 157], with 64 carbon atoms and a greater plane wave basis set than that used by Galli to the simulation of TAC. Similarly to Galli, the system was first heated to 5000 K, allowed to equilibrate for 0.36 ps, cooled to 300 K over a period of 0.5 ps, and allowed to equilibrate for a further 0.5 ps. The volume of the system was constrained to give a density of 3 g cm^{-3}. In contrast to the formation mechanisms of low density amorphous carbon films, this simulated formation regime may not be entirely removed from the physical mechanism by which TAC may be formed (apart from the important absence of the simulation of actual atomic impacts into a surface). In a separate paper Marks presents an analytical model based on the "thermal spike" theory of TAC formation (see Section 2.2.2) in addition to evidence from empirical molecular dynamics and CPMD simulations, along with the invariance of sp^3 fraction with various long quench times in tight-binding simulations, which suggest that subpicosecond cooling times from 5000 K to room temperature are not necessarily unphysical.

The sp^3 fraction in the resulting network was found to be 65%, lower than that found experimentally and similar to those found in other simulations carried out at the same density (with the exception of Drabold et al. [85]). The most surprising result of this work was the appearance of both three- and four-membered rings (three occurrences of each in the 64-atom simulation), long thought to be energetically unfeasible in bulk amorphous carbon films (it is an interesting historical point to note that TAC itself was also thought to be too energetically unfeasible to exist [158]). It has been argued [38] that the three- and four-membered rings present in these simulations are simply minor statistical artifacts of the individual simulation runs themselves, with little net contribution to the topology of TAC as a whole. However, a detailed comparison of the results with the unresolved issue of the surprisingly low coordination number calculated from neutron diffraction for a predominantly tetrahedrally bonded matrix suggests the occurrence and impact of these structures are perhaps more significant.

As presented in Section 3.1.2, analysis of the experimentally obtained reduced radial distribution function from TAC suggests a coordination number of around 8–9, significantly less than the expected value of just less than 12 for a material with up to 90% fourfold sp^3 bonding. In an attempt to resolve this discrepancy Gilkes et al. [119] suggested an asymmetric second neighbor distribution with the peak at 0.275 nm in the $G(r)$ consisting mainly of contributions by second neighbors. However, in their simulations Marks et al. found that by decomposing their results into pair distribution functions for first and second neighbor bond lengths the second neighbor peak is in fact symmetric, though broader than expected, and that therefore the asymmetry in the $G(r)$ is in fact entirely due to contributions by third nearest neighbors. Furthermore, the very presence of three- and four-membered rings in the network effectively reduces the total number of second nearest neighbors with respect to first nearest neighbors as the low second neighbor coordination indicates. In a three-member ring the second neighbor in one direction around the ring is in fact a first neighbor in the other, and in a four-membered ring the second neighbor in both directions contributes only one atom to the distribution instead of two. As further evidence of the presence of four-membered rings they point out that the peak at 0.215 nm identified in the $G(r)$ by Gilkes et al. can only conceivably be generated by the diagonal distance across a quadrilateral. The unexpected appearance of three- and four-membered rings underlines the importance of the generality of the CPMD approach in its po-

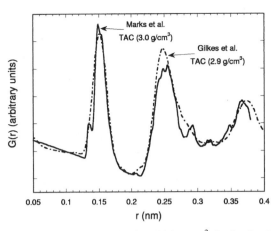

Fig. 29. Reduced radial density function of 3.0 g cm^{-3} simulated carbon network from Marks et al. [156] compared to that obtained for TAC by neutron diffraction from Gilkes et al. [119].

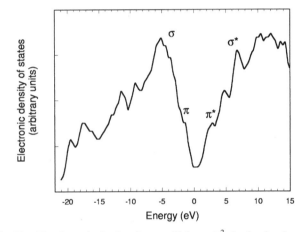

Fig. 30. The electronic density of states of 3.0 g cm^{-3} simulated carbon network from Marks et al. [156].

tential to provide insights that are less constrained by prior assumptions about the expected structure of a given atomic network.

Comparison of the $G(r)$ obtained in this work with that obtained by Gilkes et al. is shown in Figure 29. Apart from the agreement in both position and shape of the first two coordination peaks, the relative height of the two peaks is arguably better reproduced than any other simulated $G(r)$ in the literature at the time. Specifically, the tight-binding simulations of both Wang and Ho [151] and Frauenheim et al. [84] resulted in second neighbor peaks which were greater than the first neighbor peaks in contrast to that found from the experimental $G(r)$ for TAC and that found by Marks.

With respect to the nature of sp^2 sites, this work confirms the tendency for clustering found in other works including Frauenheim et al. [84, 148] and Drabold et al. [85], as well as their olefinic or chainlike nature. It also confirms the absence of isolated sp^2 sites, and their overwhelming tendency to form pairs. As pointed out by Drabold this pairing is consistent with EPR results which would make the occurrence of isolated sites unlikely in such small systems.

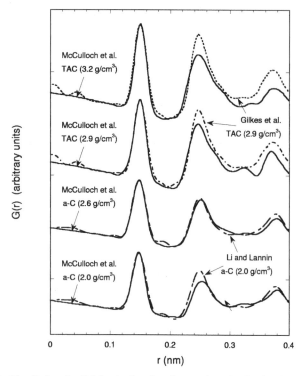

Fig. 31. Reduced radial density functions from various simulated carbon networks from McCulloch et al. [159], compared with equivalent experimental results of Li and Lannin [118] (taken from Frauenheim et al. [84]), and Gilkes et al. [119].

The electronic density of states of the structure is shown in Figure 30. It is difficult to infer too much from the electronic results of such a small system, or to determine the resultant mobility gap when the states at no point in an amorphous semiconductor go exactly to zero. Therefore, in order to relate the EDOS to a measurable quantity in TAC which reflects the EDOS, the Tauc gap (see Section 4.1) was calculated from a convolution of occupied and unoccupied states in this distribution, and using the assumption that the matrix elements for dipole transitions are independent of energy. This resulted in a Tauc gap of 0.5 eV, significantly lower than the measured value of around 2–2.5 eV in TAC, almost certainly due to the size of the system, and possibly also due somewhat to the over-representation of low-membered ring structures (as suggested by Frauenheim et al.).

Work by McCulloch et al. [159] has extended the work of Marks et al. with almost twice the number of atoms (125 carbon atoms), an even greater plane wave basis set, and the simulation of systems at a range of densities at 2.0, 2.6, 2.9, and 3.2 g cm^{-3}. Figure 31 shows the resultant reduced radial density functions at each of these densities compared to equivalent experimental results.

In the 2.0 g cm^{-3} network sp^2 bonding was found to predominate in layered regions bounded by sp^3 bonded atoms, indicated by the presence of a graphitic (002) reflection in the calculated reduced diffraction intensity. A significant amount of sp^1 bonding (15%) was also found, similar to the tight-binding results of Wang et al. [87] and Stephan et al. [149].

With respect to the high density networks, the best fit to the $G(r)$ data of Gilkes et al. is obtained from the 2.9 g cm^{-3} network with an sp^3 fraction of 57.6%, rather than the 3.2 g cm^{-3} network with an sp^3 fraction of 79.2%, closer to that found experimentally. This is attributed to the inclusion of significant amounts of sp^2-rich carbon in experimental samples (in part due to the insufficient filtering of macroparticles during deposition) during the collection of the large amount material required to carry out both an accurate measurement of density and the collection of neutron diffraction data. This argument could equally explain similar discrepancies found throughout the literature on the modelling of TAC and suggests that diffraction data will be best reproduced by simulated films with a density of up to 3.0 g cm^{-3}, but that high sp^3 fractions will be better reproduced at densities in excess of this.

In comparing the ring statistics of the 2.9 g cm^{-3} model to those of the TAC network simulated by Marks et al. the number of three-membered rings remains constant (at a total of three) even though the number of atoms in the system was doubled, while the number of four-membered rings was more consistent in that it exactly doubled (from three to six). This perhaps justifies the caution about the statistical significance of such structures raised by Frauenheim et al. [38] with respect to three-membered rings, though probably less so with respect to four-membered rings given the appearance (as mentioned above) of a peak corresponding to the quadrilateral diagonal distance at 0.215 nm in the neutron diffraction data.

4. OPTICAL PROPERTIES

4.1. Introduction

The optical properties of amorphous semiconductors are usually discussed in terms of a breakdown of the k conservation number with a loss of distinction between direct and indirect transitions. Band-to-band optical transitions are conventionally discussed in terms of a transition between the valence and conduction band, the strength of which is determined by the matrix elements. In this manner information with regard to the band tails and optical constants can be obtained. The refractive index n and absorption coefficient a can be related to the real, ϵ_1, and imaginary, ϵ_2, parts of the dielectric constant. In the one-electron approximation ϵ_2 takes the form [160]

$$\epsilon_2(\omega) = \frac{2}{V}\left[\frac{2\pi e}{\omega m}\right]^2 \sum_i \sum_f |\langle f|P|i\rangle|^2 \delta[E_f - E_i - \hbar\omega] \quad (4.1)$$

where V is the sample volume, P is the momentum operator, and the summations are over the initial states $|i\rangle$ and final states $|f\rangle$. For transitions between extended states (i.e., those that are delocalized), the matrix element evaluates to a^3/P_{CV}^2 with $P_{CV} \sim \hbar/a$. Equation (4.1) then can be expressed as

$$\epsilon_2(\omega) = 2\left[\frac{2\pi e}{\omega m}\right]^2 a^3 P_{CV}^2 \int_0^{\hbar\omega} g_{VB}(-E)g_{CB}(\hbar\omega - E)d\omega \quad (4.2)$$

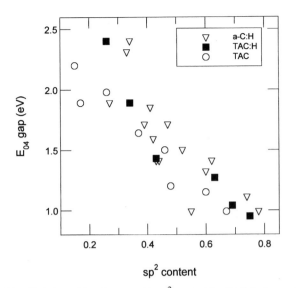

Fig. 32. Variation of the E_{04} gap with sp^2 content for DAC, hydrogenated, and nonhydrogenated TAC films. Adapted from Robertson [242] and references therein.

From Eq. (4.2) it is evident that ϵ_2 is a function of the joint density of states of the respective bands and will therefore depend on the shape of the bands. Furthermore, the absorption coefficient α is related to ϵ_2 via

$$\alpha(\omega) = \frac{\omega \epsilon_2(\omega)}{nc} \quad (4.3)$$

where n is the refractive index of the material. Equations (4.2) and (4.3) indicate that the absorption coefficient will depend upon the convolution of the density of states of the conduction and valence bands. To a first order the density of states can be expressed as a power law taking the general form

$$g_{VB}(-E) \propto E^p \quad \text{and} \quad g_{CB}(E) \propto (E - E_T)^q \quad (4.4)$$

where E_T can be regarded as a measure of the separation between the two bands. Inserting Eq. (4.4) into Eq. (4.2) results in

$$\omega^2 \epsilon_2(\omega) \propto (\hbar\omega - E_T)^{p+q+1} \quad (4.5)$$

For the case of parabolic bands $p = q = 1/2$, and Eq. (4.5) results in the well known Tauc energy gap relation

$$\alpha\hbar\omega = B(\hbar\omega - E_T)^2 \quad (4.6)$$

A plot of $(\alpha\hbar\omega)^1/2$ against $\hbar\omega$ has an intercept which gives E_T, the Tauc gap, and the parameter B, which has units of cm^{-1} eV^{-1}. The Tauc gap E_T is therefore an extrapolation of the density of states (for parabolic bands). In addition to the Tauc gap, the energy at which the absorption reaches certain values, most commonly 10^3 and 10^4 cm^{-1}, is often quoted. The energy gaps are usually referred to as the E_{03} and E_{04} gaps, respectively. The variation of the E_{04} gap on sp^2 content for DAC and hydrogenated and nonhydrogenated TAC is shown in Figure 32. The E_{04} gap clearly depends on the sp^2 content with a clear trend of a decrease in the E_{04} gap as the sp^2 fraction increases. This leads us to the conclusion that the optical gap is

ruled by the sp^2 content and was first examined in the cluster model.

The optical gap can be varied by changing the deposition conditions (e.g., self-bias in PECVD systems, C$^+$ ion energy in FCVA systems), by addition of N, or through annealing. Examples of the Tauc and E_{04} gaps as function of different deposition conditions will be presented later in this section and also in conjunction with the electrical properties of a-C films.

4.1.1. Optical Gap and Density of States

In 1997 Robertson and O'Reilly [78] proposed that the bandgap in amorphous carbon could be related to the size of the sp^2 clusters. They found that the most stable arrangement of sp^2 sites is in compact clusters of fused sixfold rings. For M sixfold rings the bandgap could be expressed as $E_g = 6/M^{1/2}$ eV. However, a number of deficiencies with this model became apparent. First, the size of the clusters required to explain the observed bandgaps would have to be so large that they should be readily observable by electron diffraction studies—which to date they have not been. For example, a bandgap of 1.2 eV in DAC would require a 25-ring cluster of approximately 1.2 nm in diameter. Such a well organized structure is unlikely to occur, especially in highly disordered DAC. In addition five- and seven-membered rings have been observed in the neutron diffraction of a-C [118]. In light of these experimental facts, Robertson revised his earlier model so that distorted clusters now play a more dominant role [79]. Various possible distortions of a sixfold ring were considered (cf. Fig. 12). In the case of the "chair" configuration a significant narrowing of the gap occurred. Furthermore, the gap was no longer symmetric about the midpoint as it was for planar sixfold rings. A fully distorted sixfold ring in the chair configuration gives a bandgap of 1.16 eV. On this basis it was proposed that the bandgap of 1.5 eV could be formed by the partial deformation of a planar sixfold ring. In the case of odd numbered rings, especially five- and sevenfold, it was noted in [78] that these would give rise to states near the Fermi level and would be singly occupied. In this way they would be paramagnetic. By considering negatively charged fivefold rings and positively charged sevenfold rings together Robertson proposed that such a combination would produce a bandgap of about 3 eV. However, if such a two-ring configuration could become distorted it would produce a bandgap of 0.8 eV.

In the original model of the DOS from Dasgupta et al. [160] the DOS consisted of a pair of symmetric bands, each with Gaussian shape and width. These distributions were symmetric and were separated by about 4 eV, corresponding to the separation of sixfold rings put forward by the cluster model together with another pair of bands 0.3 eV above and below E_F due to five- and sevenfold rings. Dasgupta et al. originally proposed that the former states were responsible for the optical absorption, whereas the latter were responsible for the observation of EPR signal.

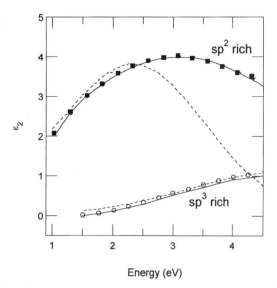

Fig. 33. Fits to the imaginary part ϵ_2 of the dielectric constant for sp^2 and sp^3-rich assuming a symmetric Gaussian model for the DOS as proposed in [160] (dashed) and the asymmetric band model (solid) of [161].

This model in which the shape of the bands could be modelled as a Gaussian with a particular energy position and width led to a number of consequences:

(i) There is no difference in the origin of the states sitting above or below the optical band edges.

(ii) There is no connection to the mobility gap and the degree of localization is related to the number of states at a particular energy.

(iii) Following from (ii), it is still possible for both the π and π^* states to be completely localized and still obtain a Tauc energy gap. This point is important since MD calculations reveal that even with sp^2 contents above 92% the π states are localized.

(iv) The value of the optical gap should be regarded purely as a method of characterizing the film and values such as E_{04}, etc., do not necessarily have any significance.

(v) Since the optical gap is controlled by the size of the sixfold rings then only an infinitely large sixfold cluster would produce states at the Fermi energy. No significant contribution from the tail states is expected in the EPR signal.

While this model has been used to explain the general feature of the DOS there have been problems applying it to the values of ϵ_2 obtained from spectroscopic ellipsometry [161]. The concept of asymmetric bands was employed by Fanchini et al. [161] in their explanation of the energy dependence of the imaginary part of the dielectric constant. Figure 33 shows the fits (dashed lines) to the imaginary part ϵ_2 for an sp^2- and for an sp^3-rich film assuming a symmetric Gaussian model for the DOS as proposed in [160]. It is evident that for the high sp^3 containing film that an adequate fit is obtained to the data only for $E > 2.5$ eV but not at lower energies. By contrast, for the sp^2-rich film a good fit is obtained at lower energies but not at higher ener-

gies above 2 eV. By considering many body effects within the Huckel model in which overlap between nearest neighbor atoms is non-negligible, Fanchini et al. obtained a much better fit to data as shown (solid lines).

Arena et al. [162] studied the DOS in TAC using conductivity spectra obtained from scanning tunnelling microscopy and observed E_F to be below the midgap (indicating an intrinsically p-type material). They also found that the valence band tail, associated with the π states, is steeper than the conduction band tail (π^* states), opposite to the case of a-Si [163]. They attributed the asymmetry to the effects of disorder. However, this result also shows that the disorder present is not simply due to the presence of different sp^2 configurations.

The optical properties (refractive index n and extinction coefficient k) as a function of wavelength of various a-C films have also be modelled in the confines of the Forouhi–Bloomer model (FB) [164] by Alterovitz et al. [165]. This early work showed that fits to n and k for both wide (2.3 eV) and narrow (1.2 eV) gap films were only possible over a limited energy range. It was also found in the fitting that the long wavelength refractive index decreased with increasing bandgap contrary to what should be observed as the amount of absorption from σ bonding becomes more dominant in the higher bandgap films. McGahan et al. modified the FB model by regarding the density of states to be of the form [166]

$$A_1(E - E_g)^2 + A_2(E - E_g)^{3/2} \qquad (4.7)$$

and obtained better fits to the data. They noted that assuming a Gaussian conduction and valence band density of states did not fit the measured values of n and k.

4.1.2. Urbach Tail Energy

In the vicinity of the band edge the optical absorption has an exponential energy dependence usually expressed in the form

$$\alpha(\hbar\omega) = \alpha_0 \exp([E - \hbar\omega]/E_U) \qquad (4.8)$$

where E_U is the Urbach energy. In a crystalline material E_U depends upon phonon induced structural disorder or the effects of charged impurities. In the case of an amorphous material E_U is governed by the static disorder (bond angle and bond length variation). In this way the Urbach absorption edge is a measure of the joint density of states and reflects the disorder broadening of the respective bands. In the case of low defect density a-Si:H the valence band tail has an exponential slope of about 45 meV which is larger than the slope of the conduction band (25 meV). In this way the Urbach energy of 55 meV for low defect density a-Si:H is mainly determined by the valence band. In the case of a-C based materials values of E_U tend to be higher and depend upon the optical gap. In PAC materials E_U tends to be about 520 meV for an E_{04} gap of 3.5 eV [88]. This reduces to about 250 meV for $E_{04} = 2.6$ eV. Addition of N affects both the optical gap and the Urbach energy. In the case of TAC films the optical gap remains approximately constant at about 2 eV for N concentrations of up to 0.2 at.% and then decreases to 0.75 eV (2 at.%) with subsequent increases in N closing the gap at about 10 at.% (see Fig. 34a). This is indicative of graphitization of the film and is mirrored by an increase in E_U from 250 to over 550 meV at the highest N content. The increase in E_U is a reflection of the broadening of the π and π^* bands which ultimately leads to closure of the gap.

A different behavior was reported by Silva et al. [29] in their DAC:N films. The optical gap first increases from about 1.6 eV, reaching a maximum about 7 at.% N, and then decreases. The Urbach tail width decreases from about 260 meV to reach its minimum around 7 at.% (175 meV) and then rises slightly to remain at around 200 meV. The increase in the optical gap and the reduction in E_U could be explained in terms of an initial reduction in the disorder in the films—mopping away some of the band tail states. The subsequent reduction of the optical gap is again due the inset of graphization. It is interesting to note that the electron paramagnetic spin density also reduces with the inclusion of N and then remains constant. Annealing of TAC films also shows the effects of graphization on the band tails. Conway et al. [167] showed that E_{04} gap is approximately constant to about 300°C and then begins to decrease. Over the same range of annealing E_U does not change significantly but increases above 350°C.

4.2. Infrared Absorption Studies

4.2.1. Introduction

The useful electro-optical and mechanical properties of a-C thin films and its alloys are attributed to their microstructure. There are numerous reports in the literature that offer valuable insight into the microstructure of these films using various characterization techniques.

The complications in determining the microstructure of the films arise from a combination of factors. First, a-C(:H) films can only be grown to a maximum thickness of a few hundred nanometers on silicon substrates. This value is reduced to a few hundred angstroms for the case of ultrahard DAC and TAC(:H) films. Attempts to grow thicker films result in the detrimental delamination of the films. Second, the metastable nature of the film's microstructure makes them extremely delicate to most characterization techniques. Even the small energy doses supplied by most characterization techniques, in the form of a particle beam or high energy electromagnetic radiation, will change the film's microstructure during the experimental measurement. This will lead to the determination of the microstructure of a different type of film than the one initially prepared.

The low energies involved in optical spectroscopic techniques make them particularly well suited for the investigation of such delicate microstructures. Infrared optical spectroscopy is an extremely powerful technique in characterizing the bonding found within the films. The high matrix element of the optically active stretching and bending modes of the C–H bonds give rise to a strong signal-to-noise ratio in modern Fourier transform spectrometers and yield strong absorption bands in the spectra. Also, the amorphous nature of the films causes

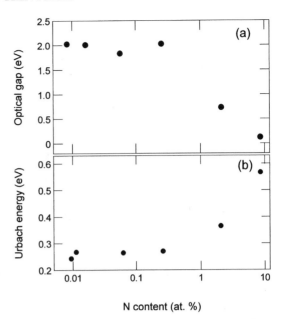

Fig. 34. (a) Optical gap and (b) Urbach energy for 100 eV C$^+$ FCVA deposited TAC as a function of N content. Adapted from [243].

symmetry breaking and the relaxation of the optical selection rules, allowing for the stretching modes of homonuclear covalent bonds to be observed. It is also a powerful and extremely sensitive technique used to determine the existence of alloying elements or impurities, covalently bonded to the structure of the films.

It should be noted that Fourier transform IR (FTIR) complements Raman scattering by providing detailed information on the nonsymmetric bonds in a-C thin films. It was shown by Kaufman et al. [168] that in the case of a-C films that contained aromatic sixfold benzenelike carbon rings that were IR-inactive, the addition of planar N bonds made them IR-active by breaking the symmetry. In the present study we will concentrate on a few key papers, each representative of a particular type of microstructure of a-C films. As FTIR has historically been a preferred tool for the characterization of a-C:H and a-C:N films, extensive literature in the field can be obtained by examining some of the references [17, 24, 25, 29, 44, 86, 168–177]. In our review we will examine in detail the FTIR spectra in PAC [86], DAC [25, 178, 172], and TAC [174] films. As there is considerable interest in examining nitrogenated a-C films, some of the results in addition of N into the microstructure of a-C will also be discussed [29, 86, 175].

4.2.2. Experimental Method

Most modern infrared spectrometers almost always use FT techniques for spectral detection. The FTIR spectrometer consists of a Michelson interferometer and this method offers greater sensitivity than moving diffraction grating spectrometers, because the detector monitors the entire spectrum simultaneously, rather than one frequency at a time. The most common way of obtaining the FTIR spectrum from films of a-C and its

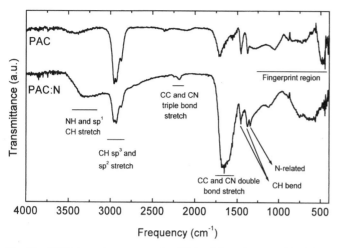

Fig. 35. FTIR absorption spectrum for a PAC and for a nitrogenated PAC film.

alloys is to grow a thick film of the material on a high resistivity silicon wafer substrate. It is important that the silicon is highly resistive, in order to obtain a clear spectra. The substrate is then mounted inside the spectrometer in transmission mode, so that the infrared beam trespasses the whole sample normally. The spectrum is then taken, using the spectrum from a plain substrate as the background signal.

A much enhanced signal-to-noise ratio is observed in the spectra when the measurement is obtained using diffuse reflectance infrared Fourier transform spectroscopy. In order to perform this type of measurement, the film must be deposited on a highly reflecting substrate (in the 4000 to 400 cm^{-1} region of the electromagnetic spectrum). Such a substrate can be either a highly doped silicon wafer or a metal-coated flat substrate. Suitable metals for this application are titanium (resistant to sputtering in the RF PECVD chamber) or aluminium (high reflectivity in the infrared). In this measurement, the infrared beam incident on the sample is transmitted through the film and reflected from the substrate back to the spectrometer, after traversing through the film a second time. The advantage of this method is that the infrared beam is not transmitted through the substrate and hence the noise level in the spectra is extremely low. This type of measurement is essential if accurate deconvolution of spectral absorption bands is required. Another technique to enhance the sensitivity of the FTIR measurement by avoiding the signal arising from the substrate is to dissolve the substrate in a liquid solvent and leave the freestanding film, which can be picked up using a silicon frame, as described by Silva et al. [25]. An often-used substrate for this procedure is single crystal NaCl, which is dissolved in water.

4.2.3. FTIR Studies of PAC Films

PAC is a particularly viable material to investigate by FTIR as films of PAC and PAC:N can be grown to up to 400 nm before delamination occurs. The thicker films often provide a clear FTIR spectrum. A typical spectrum from a PAC and PAC:N

films is shown in Figure 35. The figure also shows the assignment of the most important absorption bands.

Figure 35 shows how the hydrogen stretching vibration modes occur at the highest frequency, due to the small mass of the hydrogen atoms. The frequency region of the spectra where the stretching vibration modes occur is from 4000 to 1485 cm^{-1}. The frequency region between 1485 and 400 cm^{-1} is known as the fingerprint region and is where the absorption bands from bending modes occur. Bending modes in amorphous materials are difficult to examine due to the strained bond angles, and also due to the many different infrared active modes of oscillations that the solid network may have. The single valence of hydrogen gives rise to sharp bending modes in the spectra, which are useful for characterization.

4.2.3.1. Bending Modes

The symmetrical deformation mode of the methyl group (sp^3 CH_3) gives rise to the absorption band peaked at 1380 cm^{-1} [179]. The position of this band is relatively constant, and it is of great analytical importance, because this region does not contain other strong absorption bands. The asymmetric deformation vibration of the methyl group gives rise to a band centered around 1460 cm^{-1}. The proximity of this frequency to that of the scissoring vibration of the methylene group (sp^3 $-CH_2-$) gives rise to a convoluted absorption band peaked at 1457 cm^{-1}. This convoluted peak also contains the absorption band from the $=CH_2$ deformation, which occurs around 1420 cm^{-1}. The position of the deformation vibration of the methylene group can vary in the case of strained cycloalkanes, by up to 30 cm^{-1}. This may give rise to band broadening in PAC(:N). It may be possible that structures with sixfold rings exist in a-C:H. The bending modes of the sixfold rings are characteristic and show absorption bands at the lowest frequencies. As an example, both cyclohexane and benzene show structural C–C bending deformation vibration absorption bands centered around 522 and around 450 cm^{-1}, respectively. The strong and broad absorption band of a-C:H observed between 445 and 550 cm^{-1} may be due to bending modes of planes consisting of hexagonal structures, although such a band could also arise from the torsion of C=C double bonds [179]. Another C–C bending deformation mode of cyclohexane has an absorption band centered around 865 cm^{-1}. The spectra of a-C:H shows a complex structure around this frequency.

The in-plane rocking vibration of the CH bonds of the methylene group shows an absorption band centered around 720 cm^{-1} in straight chain alkanes. The infrared spectra of these films shows a very broad band around this region. The strong band centered around 890 cm^{-1} occurs in the spectra from all the films from the nitrogen series. It is attributed to a convolution of the absorption bands from the distortions of the ethyl group and of the C=C double bonds of the $R_1R_2C=CH_2$ group, where R_1 and R_2 are sp^3 carbon atoms. The absorption band from the $(CH_3)_2CH$-R group occurs at slightly higher frequencies, around 920 cm^{-1}. Also, cyclohexane shows a CH_2

bending frequency at 903 cm^{-1}. Again, the structure of this band is complicated.

A point of interest is that the nitrogenation of the PAC films gives rise to the absorption band centered at 1343 cm^{-1}. The assignment of this band is unclear, and it is attributed to the nitrogenation of the films. It probably arises from band splitting of the CH bending vibrations from 1375 cm^{-1} to lower wavenumbers. Films of PAC show a strong band peaked around 1050 cm^{-1}. As these films are nitrogenated, this band gets considerably weaker, and another band appears centered around 1140 cm^{-1}. The assignment of these bands is also not clear, but they may be related to the CH and C–C bending modes of sp^3 quaternary carbon atoms in structures such as R_1R_2CH-R_3 and a tetrahedrally bonded carbon to another four carbon atoms.

4.2.3.2. Stretching Modes

The absorption band found in the frequency range between 3570 and 3100 cm^{-1} is attributed to the convoluted absorption bands of the NH and sp^1 CH stretching vibrations. The band can be deconvolved into two Gaussians. The band for the a-C:H spectra arises mainly due to the sp^1 CH absorption frequency, due to the lack of nitrogen in the film. A small contribution from the NH stretching vibration was also detected in the PAC films, however, and it is speculated that this is due to the reaction between nitrogen and surface dangling bonds that may take place during the venting of the PECVD reaction chamber due the high porosity after growth.

The single valence of hydrogen makes the investigation of its bonding useful, since this will yield information about the atom that it is attached to. The fact that the films have a high hydrogen content, of the order of 55%, means that most of the carbon atoms within the films are attached to at least one hydrogen atom, and hence its hybridization state can be known (assuming all the hydrogen is bound). The sp^2 and sp^3 CH stretching band shows up clearly in the infrared spectra of all the samples, in the frequency region between 3070 and 2800 cm^{-1}. The band is composed of a convoluted series of different absorption bands that arise from the stretching vibrations of all the different types of CH bonds present in the sample, and also the individual bands from the different modes of the stretching vibrations for each type of CH bond containing group. The different groups containing sp^3 CH bonds are the CH_3, CH_2, and CH groups. The CH_3 group has symmetrical and asymmetrical stretch frequencies at 2872 and 2962 cm^{-1}, respectively. The sp^3 CH_2 group has a symmetrical and asymmetrical stretch centered at 2853 and 2926 cm^{-1}, respectively. These frequencies changed by one wavenumber for the case of cyclohexane. The sp^3 CH stretch occurs at 2890 cm^{-1}. These frequencies were determined from spectroscopy of alkanes.

Another type of CH bond occurs when the hydrogen is bonded to an sp^2 hybridized carbon atom. This bond has a higher force constant than that of the sp^3 CH bond, and so the infrared absorption bands will occur at higher frequencies, above 3010 cm^{-1}. The possible hydrogen containing groups that can be formed are the sp^2 CH_2 or sp^2 CH groups. For alkenes, the position of the absorption bands corresponding to the stretching vibrations of these groups can change, according to the position of the double bond in the alkene, but the bands typically occur at 3070 and 3015 cm^{-1}, respectively. Another important infrared absorption band is that of aromatic groups such as substituted benzene, which show an aromatic CH stretching vibration centered around 3060 to 3100 cm^{-1}.

The infrared spectra of PAC samples show no bands centered around 3070 cm^{-1}. The deconvolution of the CH stretching band into Lorentzian functions also shows no strong bands centered around 3015 cm^{-1}, but this does not rule out the possibility of a weak band existing at this frequency. It is generally accepted that in most PAC films, the number of sp^2 hybridized carbon atoms (including aromatic groups) that are bonded to hydrogen is very small, compared to the number of sp^3 hybridized carbon atoms bonded to hydrogen. Silva et al. [29] used a magnetically confined RF PECVD system to grow PAC:N films. Their analysis showed a much higher ratio of sp^2 hybridized carbon atoms, mainly in the form of aromatic rings.

Alkenes with conjugated double bonds (dienes) show characteristic frequency shifts when compared to alkenes with a single double bond. The interactions between the π electrons of the conjugated system decrease the C=C double bond order and increase the C–C single bond order. This means that the absorption bands of individual double bonds cannot be identified separately, and the entire carbon framework behaves like a single vibrational entity. Consequently, the C–C stretching vibrations occur in the 1600 cm^{-1} region. The strong absorption band from the C=N bond in the PAC:N samples makes difficult the observation of a possible band centered around 1600 cm^{-1}.

The C≡C and C≡N triple bonds have a very similar stretching resonant frequency, and they occur between 2270 and 2130 cm^{-1}. The band can hardly be observed in the PAC films, but it is clearly observable (although weak) in all the PAC:N samples. The weak signal-to-noise ratio for this absorption band made the deconvolution work of the band unfeasible. The amorphous nature of the a-C:H(:N) films ought to make the C≡C triple bond infrared active. This gave rise to the idea that there are few triple bonds in the a-C:H films but that this number was increased in the a-C:H:N films, where most of the triple bonds were C≡N bonds.

The stretching vibrations for the C=C and C=N double bonds are found in the frequency region between 1815 and 1490 cm^{-1}. Again, the similarity in the atomic masses of the C and N atoms and the similar spring constant of the C=C and C=N double bonds mean that the absorption bands occur at similar frequencies, and they appear as one convoluted absorption band. The strong signal-to-noise ratio of this absorption band made deconvolution possible, and the integrated intensities for the main components of the band were obtained. The spectra from the PAC:N films showed two bands. One strong band was centered around 1700 cm^{-1}, and another weak one was centered around 1555 cm^{-1}. The 1700 cm^{-1} band is attributed to the C=C double bond stretch (made optically active due to symmetry breakdown due to the amorphous nature of the material), and the 1555 cm^{-1} band is attributed to the

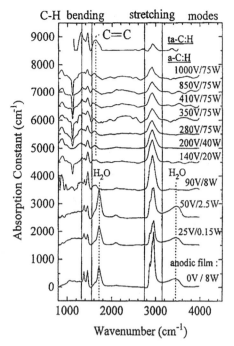

Fig. 36.   IR absorption bands for various *a*-C:H films. 0–90 V represents PAC films, and 200–1000 V resprents DAC films. A TAC:H film is also shown Reproduced with permission from [178], © 1996, Elsevier Science.

bands due to the nitrogenation of the samples is attributed to the electronegativity of the nitrogen atom, compared to the carbon atom.

An important point to stress is that the shape of absorption bands is described by a Lorentzian lineshape. In PAC, the C–C and C–N bonds found in the skeleton structure of the carbon network suffer from distortions due to the amorphous nature of the films. These distortions affect the position of the absorption band. The energy distortions are distributed according to Gaussian distribution functions, which are generally wider functions than Lorentzian functions. This means that the resulting absorption band shape is going to be determined mainly by the random distortions of the bonds, and their line shapes will be Gaussian distributed. However, the CH bonds are short and do not suffer from distortions or spatial confinements as much as the rest of the network bonds. This means that the effect of bond distortions on CH bonds is minimal, and the shape of the absorption bands is often well defined by Lorentzian functions.

A whole series of films deposited as a function of DC self-bias by Stief et al. [178] is shown in Figure 36. The PAC films are the ones deposited below a DC negative self-bias of 90 V, where the C–H stretching modes appearing between 2800 and 3100 cm^{-1} are well resolved. An expanded view of the modes visible is shown in Figure 37, for the film deposited at a negative bias of 25 V. Also shown in Figure 37 are the spectra for a DAC film (self-bias of 410 V) and a TAC:H film deposited using a plasma beam source for comparison. These will be discussed in detail later. A well resolved set of C–H bending modes is evident. For the PAC films, the vibrational modes are distinct and well defined due to the localized nature of the C–H vibrational

C=N double bonds. On nitrogenation of the films, the position of the C=C double bond band shifts to 1670 cm^{-1}, and the 1555 cm^{-1} band shifts to 1590 cm^{-1}. The position of the bands is the same for all the nitrogenated samples. The shifting of the

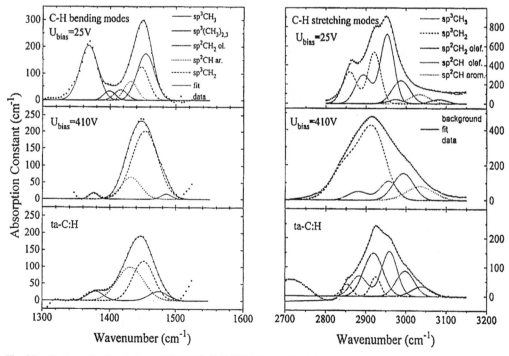

Fig. 37.   IR absorption bands for a PAC (25 V), DAC (410 V), and TAC film in the CH bending and stretching modes. Reproduced with permission from [178], © 1996, Elsevier Science.

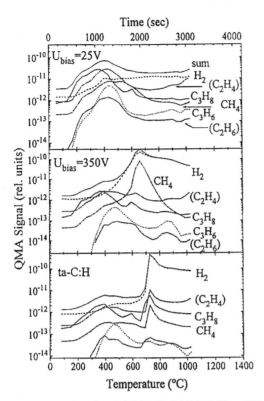

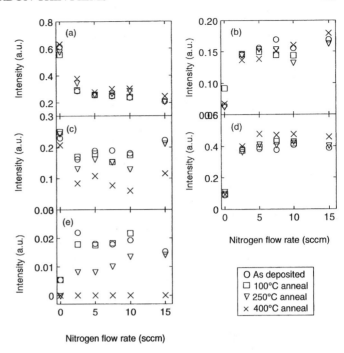

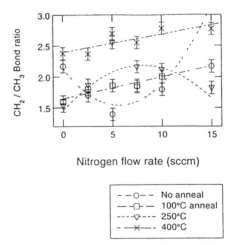

Fig. 38. Effusion transients for a PAC (25 V), DAC (350 V), and TAC films. Reproduced with permission from [178], © 1996, Elsevier Science.

Fig. 39. Normalized integrated absorption intensities of (a) sp^2 and sp^3 bonded CH stretch, (b) NH stretch and sp^1 bonded CH, (c) CC double bonds, (d) CN double bonds, and (e) CC and CN triple bonds as a function of nitrogen flow and anneal temperature in nitrogenated PAC film.

Fig. 40. The sp^3 CH$_2$/CH$_3$ ratio for the nitrogenated PAC films as a function of nitrogen flow and temperature.

modes. The variation of the films when subject to annealing is shown in the effusion transients in Figure 38. It is clear that for the PAC film deposited at −25 V, 70% of the hydrogen is released as C_3H_8, 17% as H_2, and 7% as C_3H_6 and CH_4 each. The films are stable up to 200°C, while the hydrogen transients were still continuing at 1050°C. Most of the hydrocarbon gases were completely exhausted after 600°C for the PAC film.

In the case of the nitrogenated PAC (PAC:N), the variation of the normalized integrated absorption bands is shown in Figure 39 as a function of anneal temperature. The sp^3 CH bands dominate the spectra, with the sp^2 CH bands contributing weakly to the signal [86]. As the fingerprint region of these spectra did not show any strong sharp absorption bands in the window 950–650 cm^{-1}, it was concluded that sixfold aromatic structures were not a major component of the polymeric films following the arguments of Williams and Fleming [180]. In their work [180] it was shown that nonaromatic films that contain C=C double bonds have a peak at 1600 cm^{-1}, but no features in the 3050 cm^{-1} or 950–650 cm^{-1} range. Anguita et al. [86] concluded that most of the PAC films were composed of polymer chains that had CH_2 repeat units that were terminated by CH_3 ends (all sp^3 hybridized). Using this argument they calculated the CH_2/CH_3 bond ratio as a function of nitrogen content and anneal temperature (see Fig. 40) and showed that on average a carbon alkanelike chain length of 1 nm is expected for these films. These chain structure values are much smaller than the cluster sizes predicted by Robertson [79]. In the analysis that followed, using bond formation energy calcu-

lations Anguita et al. discussed the thermal stability of the films and its implications on the deposition of good quality semiconducting PAC films.

4.2.4. FTIR Studies of DAC Films

The variation of the IR bands with increasing DC self-bias voltage is shown in Figs. 36 and 37. The DAC films are represented in the voltage range 200–1000 V. There is a major change in the structural properties when moving from PAC to DAC films, and this is represented in the IR spectra. It is clearly seen that both in the bending and stretching modes the intensity of the peaks

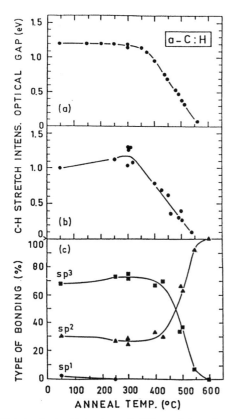

Fig. 41. Variation in the (a) bandgap, (b) CH stretch intensity, and (c) type of bonding for DAC films as a function of annealing temperature. Reproduced with permission from [24], © 1989, Trans Tech Publications Ltd.

decreases significantly. Also the sp^3 CH₃ vibrational peak at 1370 cm⁻¹ disappears. Most of the bonding that was sp^3 CH₃ is now being gradually replaced by sp^2 olefinic and aromatic bonds. A majority (40%) of the H is bonded as sp^3 CH₂. Significant percentages of olefinic (22%) and aromatic (23%) bonds in a sp^2 CH configuration are also found [178]. Bands usually associated with C–C single bonds below 1300 cm⁻¹ now become more prevalent, together with the C=C double bond associated with the aromatic clusters close to 1600 cm⁻¹. The porosity of the microstructure in the films is clearly decreasing by the reduced uptake of water vapor by the films presumably when exposed to the atmosphere. In the effusion experiments shown in Figure 38 a large percentage of the hydrogen (75%) is molecular, with a gradual effusion of methyl C₃H₈ groups at around 200°C. The main release of hydrogen occurs at 550°C, in the form of CH₄ and H₂. What remains beyond this is usually graphitic due to the collapse of the amorphous structure. A very similar picture is observed in the results of Koidl [24] shown in Figure 41, where variations in the bandgap, CH stretch intensity, and type of bonding are shown.

Essentially a similar variation can be observed in all DAC films independent of the deposition system used. This is beautifully highlighted when we consider DAC films deposited using a hybrid microwave-RF system constructed by Martinu et al. [181] shown in Figure 42 as a function of both Ar percentage and microwave power. Even the reduction of the 1375 sp^3 CH

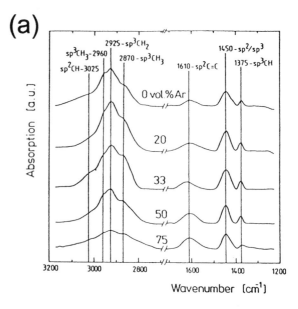

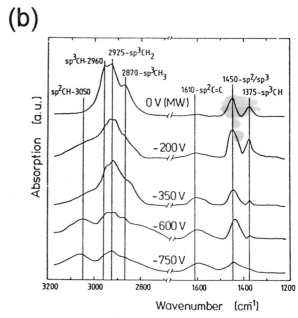

Fig. 42. Variation in the IR absorption spectra as a function of (a) Ar percentage of CH₄–Ar mixture for DAC films deposited at −350 V, and (b) microwave power (DC self-bias) for DAC films in a hybrid microwave–RF deposition system. Reproduced with permission from [181], © 1992, Elsevier Science.

peak is shown reproduced as the DC self-bias increases. At the highest bias voltages, it can be seen that there is a loss of hydrogen by the decrease in the intensity of the stretching and bending modes together with a spreading or diffusive effect of the peak shape. This is partly due to amorphization at the higher bias and partly due to graphitization.

A thorough analysis of the FTIR spectrum of DAC films deposited by RF magnetron sputtering has been reported by Clin et al. [182]. In their work, they sputtered a graphite target using an argon plasma containing 10% hydrogen as the reactant gas. Their analysis of the deconvolution of the CH stretching band showed a decrease of the sp^3 CH/sp^2 CH bond ratio as the negative target bias (i.e., RF power) increases. This was accompanied by an increase of the refractive index and a decrease of the E_{04} optical bandgap, suggesting a densification of the films with increasing RF power. The behavior was attributed to a decrease of the total H content, as well as a decrease in the porosity of the samples.

4.2.5. FTIR Studies of TAC Films

It should be noted that in order to produce TAC:H films using the plasma beam source whose properties in the IR are shown in Figures 36, 37, and 38, very specific conditions should be used [17]. When these optimized parameters are used extremely well defined IR bands are observed. The IR measurements suggest that the sp^2 C–H stretch modes are olefinic in nature as is shown to be the case in the neutron diffraction work of Walters et al. [83]. The sharpness of the stretching modes in the IR bands usually signifies polymeric soft films. In contrast, the effusion results clearly show the main threshold for hydrogen is at 700°C, which is much higher than the 200–400°C seen for the PAC films. This clearly shows that the structure of these films as well as the hydrogen uptake is significantly different. Chen et al. [174] observed the variation of the FTIR with increasing nitrogen content in their TAC films. They showed that in the case of the undoped films there were no IR bands visible, and in order to observe any bands nitrogen contents in excess of 15% were needed. When such a high alloying was used the nitrogen appeared to prefer to sit in a trigonal configuration and thus promote sp^2 bonding.

4.2.6. FTIR Studies of GAC Films

Some work on GAC films has been reported by Schwan et al. [77] when they used a plasma beam source to deposit a-C films. In the conditions used by them a sp^3 contents (by EELS) of less than 20% were obtained for all the bias voltages used. Stretching modes close to 3000 cm^{-1} were observed due to the use of C_2H_2 as the source gas for the plasma beam together with Ar, and an sp^3 content obtained from the ratio of sp^3 to sp^2 C–H bonds of 75% is predicted. This clearly illustrates the problems associated in using FTIR to obtain quantitative information on bond hybridizations. Effusion studies showed the hydrogen to be stable until above 500°C. Very interestingly, the bandgaps of these films were close to 2 eV, and so the microstructure of these films could not be explained using the simple cluster models developed by Robertson. It was thought that in these films the aliphatic (olefinic) bonding was more likely.

4.3. Photoluminescence Spectroscopy

4.3.1. Introduction

The classical description of photoluminescence (PL) is usually discussed in terms of the initial excitation of an electron from the valence band to the conduction band by the absorption of a photon of a particular wavelength. The resulting electron–hole (e–h) pair may form a quasi-particle, the exciton, but they

will eventually annihilate each other with a subsequent emission of light. During its lifetime, the exciton can move within the semiconductor (free exciton) or form a bound exciton. The energy levels of the exciton are given by hydrogenlike Rydberg states with an effective Bohr radius which often extends over several 10's of atomic sites. If light of energies significantly larger than the bandgap of the material is used for excitation, then the electron and hole are pumped high into the bands after which the carriers rapidly (typically femtoseconds) relax to the bottom (top) of the conduction (valence) band. Carriers lose energy to the lattice in the form of phonons. As an alternative to radiative recombination in which the emission of light results, nonradiative recombination energy may also be transferred to the lattice at specific sites, in the form lattice vibrations, thus not involving the creation of a photon. This type of recombination is termed nonradiative recombination (NRR). In a-Si:H paramagnetic defect sites are believed to act as sites for NRR. The role of paramagnetic sites as NRR sites in amorphous carbon films is discussed in Section 5.

Among all the useful properties of PAC thin films, that of strong, room temperature PL in the visible region of the electromagnetic spectrum is significant [183–186]. This makes amorphous carbon a potentially useful material for use in luminescent devices, being especially well suited as the active material in flat-panel displays, due to its ease of deposition over large areas using inexpensive RF PECVD techniques. There is a strong need for a thorough understanding of the PL process in amorphous carbon, to enable industrial exploitation of its full potential. The useful luminescence properties of other forms of carbon, such as carbon-fiber reinforced plastics, are already extensively used by material scientists to investigate the mechanical failure of these materials [187, 188]. In this case, luminescence is related to the relaxation that occurs after the destruction of an atomic/molecular single bond by the application of mechanical stress. The effect of field-emission-induced luminescence from carbon nanotubes has also been reported [189]. This part of the review, however, will place its emphasis on the PL properties of PAC and PAC:N thin films deposited by PECVD from hydrocarbon gas sources.

4.3.2. Photoluminescence in PAC Films

A search through the literature clearly shows that there is still a considerable amount of confusion about the origin of the PL in amorphous carbon. Most of the research in this area has been performed on the polymeric hydrogenated amorphous carbon (PAC) films. It is widely accepted that it is only the soft form of a-C that exhibits the strongest PL at room temperature [186, 190–192]. For this reason, this material has been the focus of many efforts in developing viable electroluminescent devices [193, 194]. PAC is also used as a dielectric in metal–insulator–metal switches in active matrix displays [195, 196]. The PL signal obtained from PAC films shows no sharp features like those observed in the PL spectra from crystalline materials. Rather, the PL spectra from PAC consist of broad bands, typical of amorphous semiconductors [163]. On excitation above

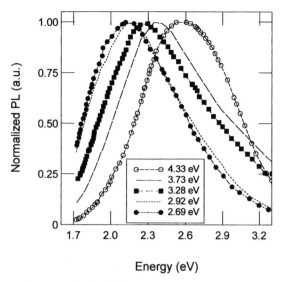

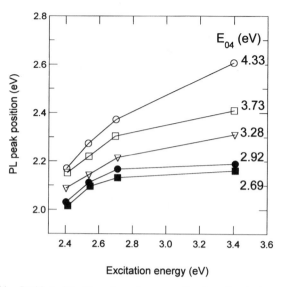

Fig. 43.   PL spectra obtained from a set of PAC films with different optical bandgap values, as indicated. Adapted from [201].

Fig. 44.   Position of the PL peak maximum as a function of excitation energy for films with different E_{04} bandgaps.

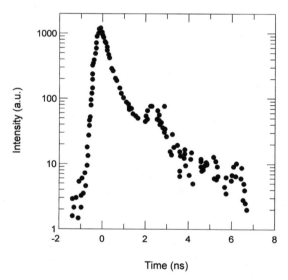

Fig. 45.   Time-resolved room temperature PL decay from a PAC film possessing an E_{04} gap of 3.4 eV. Reproduced with permission from [206], © 1998, Elsevier Science.

the optical bandgap, the broad band PL spectrum occurs over a broad region, several eV's wide. A similar broad band spectrum is observed from the PL signal obtained from polycrystalline diamond films grown by CVD [197], which suggests that PAC is present in this material, mainly at the grain boundaries.

Early workers [183, 198] investigated the effect of substrate DC self-bias during growth of PAC/DAC films on the PL prepared by RF PECVD using methane. They observed a significant decrease in the intensity of the PL signal with increasing substrate DC self-bias. This observation is consistent with the widely accepted model, since reducing the DC self-bias during growth leads to growth conditions that favor the formation of the highly luminescent PAC phase of a-C:H, due to the low defect density material being deposited Surprisingly, they also reported an invariance in the PL peak position with Tauc optical bandgap for a given excitation wavelength [198]. They used an excitation source of higher energy than the optical bandgap of their PL measurements. It is important to note that they obtained films with different optical bandgap values by using a different substrate DC self-biases to grow the different types of PAC films. Similar experiments repeated by other groups [88, 199–202] also show the very weak dependence of the peak position of the PL band with the optical bandgap for a given PL excitation wavelength. This invariance is shown as a small red-shift in the peak position of the PL spectra obtained from a set of films with different optical bandgap values, shown in Figure 43, by Rusli et al. [201, 202].

PL measurements performed by Silva [88] and later Rusli [201, 202] on similar films showed that the most influential parameter on the peak position of the PL signal was the excitation energy at which the PL measurement is performed. The position of the PL maximum as a function of excitation energy for films with different E_{04} optical bandgap values is shown in Figure 44.

Koós et al. [203] performed a very careful analysis of the PL spectra as a function of excitation energy for films grown by RF PECVD from pure methane. Their results were in good agreement with those reported in Figure 44. Both groups did not observe a significant change in the PL peak energy as a function of excitation energy in the range between 2.95 and 3.93 eV for a film with an E_{04} optical bandgap value of 3.1 eV. Koós et al. observed that excitation energies of 5.9 eV did not excite the characteristic PL band of PAC. A well defined PL spectrum was only when the excitation energy was decreased to a maximum value of 5.63 eV. Decreasing the excitation energy further lead to an increase in the PL efficiency, which peaked when the excitation energy was 3 eV.

4.3.3. Lifetime Studies of PAC Films

The PL decay processes in PAC films have been consistently reported to be about four orders of magnitude faster than in amorphous silicon [204–206]. The time-resolved room temperature PL decay of a PAC film (E_{04} optical bandgap of 3.4 eV) is shown in Figure 45, reproduced from [206]. The PL measurement was taken with a 1 ps pulsed laser at an excitation energy of 2.93 eV, monitoring the time evolution of the peak intensity at 2.36 eV. Figure 45 shows the initial rise of the signal, followed by a decay in intensity after $t = 0$, the time corresponding to the arrival of the laser pulse. The decay is not exponential, since it cannot be approximated by a linear function in the semilogarithmic plot. The peak at $t \approx 2$ ns was attributed to an artifact created by the electronics of the measuring system. Lormes et al. [206] noted that the decay of the PL intensity, $I(t)$, could be fitted to a stretched exponential decay according to

$$I(t) \propto \exp\left(-[t/\tau_{eff}]^{\beta}\right) \quad (4.9)$$

They obtained values of β of 0.44 and τ_{eff} of the order of 0.07 ± 0.02 ns. On this basis, a reasonable manner in which to define the average decay time constant (T_S) involving only one variable is given by [206]

$$T_s = \frac{1}{I_{total}} \int I(t)dt \quad (4.10)$$

where I_{total} is the total time integrated PL intensity. Using this expression, three important trends in the decay time of the PL signal were observed:

(1) It was observed that T_s increases almost linearly, from 0.5 to 1.5 ns for samples grown by RF PECVD with increasing substrate negative DC self-bias, from 0 to 1000 V.

(2) T_s is independent of the laser excitation power used to obtain the PL signal, but its value decreases by about a factor of 3 to 0.2 ns, after 50 minutes of laser illumination. It was also noted that increasing the illumination time has no further effect on T_s.

(3) T_s increases with increasing energy ΔE, where ΔE is the difference between the laser excitation energy, E_{exc}, and PL detection energy, E_{PL} ($\Delta E = E_{exc} - E_{PL}$). T_s increases by a factor of about 2, to 0.4 ns, as ΔE increases from 0.1 to 1.2 eV. The stretching to the exponential decay curve is a common feature in a-Si:H and a-C:H. This stretching was attributed to a decreasing decay rate which develops as recombination proceeds after pulsed excitation. In a-Si:H, this was explained by the increasing separation of e–h pairs as their density is reduced by the recombination process. This was used to explain how the PL decay became faster at larger e–h generation rates in a-Si:H. The constant decay time observed for a-C:H is in contrast with these observations. This fact, together with the short lifetime, suggest that the e–h

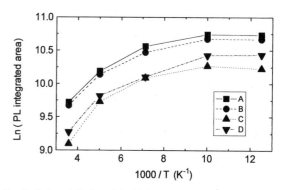

Fig. 46. Variation of the log of the PL intensity versus $1000/T$ for two samples, PL bands at 562 and 612 nm. A and B are integrated intensity of the 562 band from a PAC:N film with 2.5 and 15 sccm N_2, respectively. Points C and D refer to the integrated intensity of a band centered at 612 nm from PAC:N with 2.5 and 15 sccm N_2.

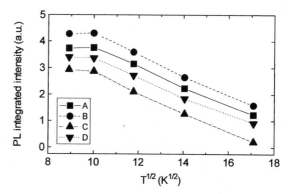

Fig. 47. PL integrated intensity as a function of the square root of the temperature. The bands are labelled as in Figure 46.

pairs are much more spatially localized in PAC than in a-Si:H.

4.3.4. Temperature Dependence of PL Intensity

The dependence of the PL intensity as a function of sample temperature is an area that has not attracted much attention in the literature and there are few reports. It is important to investigate the influence of temperature on the PL in PAC because this will give information about the quenching mechanisms in PAC films and the nature of the PL quenching sites. Experimental observations from most groups show that the temperature dependence of the integrated intensity for PL bands does not show an Arrhenius behavior. A typical example of this is shown in Figure 46, where a plot of the log of PL intensity versus $1000/T$ does not yield a straight line. The non-Arrhenius nature of the PL signal as a function of the temperature is widely accepted in the literature [190]. Points A and B are integrated intensity of the 562 nm band from PAC:N films with 2.5 and 15 sccm N_2, respectively. Points C and D refer to the integrated intensity of a band centered at 612 nm from PAC:N with 2.5 and 15 sccm N_2. It is also generally accepted that there is an increase in the PL intensity associated with decreasing the temperature of the sample during the PL measurement. Our experimental observations show that a good fit exists when the PL integrated intensity is plotted as a function of the temperature to the power $1/2$, as shown in Figure 47. The measurement taken at 80 K was made by exposing the sample to the excitation laser beam for more than 30 minutes for all the samples and hence caused a decrease in the PL intensity due to PL fatigue. The samples were grown by placing silicon substrates on the earthed electrode of a RF PECVD system using methane, nitrogen, and helium. The flow rates were 20 and 75 sccm for methane and helium, and the flow rate of nitrogen was varied from 0 to 15 sccm for different films.

4.3.5. PL Fatigue

PL fatigue has been reported in a-C based materials [173, 207–212]. It consists of a decrease in the PL intensity of the films after prolonged exposure to the excitation source. There is some doubt as to whether the decrease in the PL intensity is followed by a shift of the PL peak position. A reduction in the PL decay time by nearly a factor of 5 has been reported to follow the PL fatigue process for PAC films deposited by RF PECVD. Interestingly, a self-recovery of fatigued PL samples in the dark has also been observed [207]. PL fatigued samples placed in the dark for 10 minutes showed a recovery of the PL intensity of nearly 40%. The samples showed a 55% decrease in PL intensity due to initial fatigue, which lasted 2 minutes. Although most workers agree with the existence of PL fatigue in carbon films, some scientists report the observation of a significant enhancement in the PL signal intensity after exposure to the excitation source. This process was first reported by Koós et al. [211] for RF PECVD grown PAC films. The prolonged exposure to the excitation source was reported as "laser-soaking." A threefold increase in the PL intensity was observed after laser-soaking the samples for 1 hour, using the 488 nm line of an argon ion laser. A small blue-shifting of the PL intensity maximum was reported to follow the laser-soaking process. Similar results are reported for films grown by pulsed laser ablation, which result in a high quality form of a-C:H (DAC/TAC) film [210].

PL fatigue has been attributed to the creation of light-induced defects, created by the excitation source, which act as nonradiative recombination centers in the vicinity of a luminescent dipole. Such additional defects will both reduce the PL intensity and the PL decay rate, as confirmed by experimental results. The PL efficiency enhancement by laser-soaking has been attributed to the splitting of large clusters into many smaller ones, induced by the effect of the excitation source on the larger clusters. This explains the observed blue-shift of the PL peak observed after prolonged exposure. However, it must be stressed that sputtered a-C is a low bandgap material (hence high density of larger clusters), but it does not exhibit a PL efficiency enhancement by laser-soaking. Although PL fatigue and PL efficiency enhancement processes have both been reported to occur during prolonged exposure of films to the excitation light source, different reports on similar experiments discuss different combinations of these processes. For example, PL fatigue has been reported to occur immediately after exposure to the excitation source by some groups, while other groups suggest an initial increase in the PL efficiency followed by a decrease. The effect of prolonged exposure to the excitation source is thus still a matter of research, and a more rigorous analysis is needed.

It is well known that hydrocarbons exhibit photodecomposition by ultraviolet (UV) radiation [173], with the evolution of hydrogen containing molecules. It is likely that UV excitation sources used during PL measurements will have an effect on the microstructure of softer PAC films, which show the most promising PL characteristics. The decomposition of the polymerlike phase of PAC films is likely to reduce the PL efficiency, either by introducing defects or by increasing the size and/or density of the π-bonded clusters. This would result in a compromise, where the creation of defects would enhance the rate of nonradiative recombination, thus decreasing the PL efficiency, while the generation of clusters will increase the absorption coefficient of the films, thus increasing the e–h generation rate. One of the key areas of research focuses on determining the nature of the defects where nonradiative recombination takes place. It has been reported that these sites may not necessarily be paramagnetic defects. Experimental observations point to the fact that the creation of defects has a stronger effect.

Another macroscopic effect that decreases the PL efficiency is the mechanical application of high pressure to the films [212]. The investigations were carried out on reactively sputtered a-C:N films, which showed a very similar PL signal to RF PECVD grown films, when excited with 488 nm wavelength radiation. The results showed an exponential decrease in the integrated PL intensity with increasing pressure from 0 to 25 GPa. The PL integrated intensity was reduced to about 15% of its original value by the application of 8 GPa of mechanical pressure. It was observed that the PL intensity was recovered after decompression to ambient atmosphere.

4.3.6. Polarization Memory

The polarization memory that PECVD grown PAC exhibits was first reported by Koós et al. [206, 213] and has been observed and investigated by other groups [190, 202, 214–216]. The polarization memory consists of a considerable amount of the PL intensity having the same polarization as the excitation source. The intensity of the PL with polarization parallel (I_s) and perpendicular (I_p) is measured using a polarizer between the sample and the monochromator. The degree of polarization (DP) has been defined as [188]

$$\mathrm{DP} = \frac{I_s - I_p}{I_s + I_p} \quad (4.11)$$

It is noted that the intensity of the PL that has its polarization vector parallel to that of the excitation source is light that has the same polarization as the excitation source. The effect of PL polarization memory in an RF PECVD grown film with an E_{04} bandgap of 4.33 eV, when excited using a 2.41 eV source, is clearly observed in Figure 48a. In this figure, the solid and

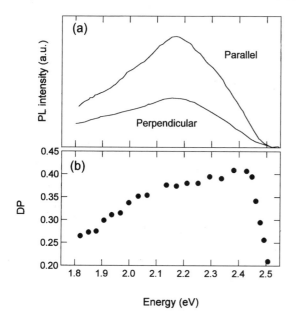

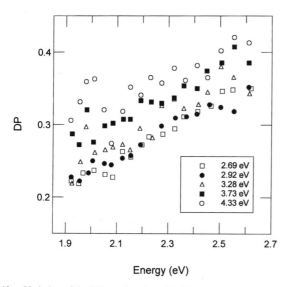

Fig. 48. (a) Effects of PL polarization memory on a PAC film (E_{04} gap of 4.33 eV) under 2.41 eV excitation. The polarizations refer to the polarization of the excitation source. (b) Degree of polarization (DP) for the PL spectra seen in (a). Reproduced with permission from [214], © 1996, American Physical Society.

Fig. 49. Variation of the DP as a function of the PL energy for a series of PAC films with different E_{04} optical bandgap values. Reproduced with permission from [214], © 1996, American Physical Society.

increasing PL energy above the excitation source. The dependence of the PL DP on the film's properties shows interesting results. The DP gives a measure of the degree of localization of the carriers. Figure 49 shows the variation of the DP as a function of the PL energy for a set of films with different E_{04} optical bandgap values reproduced from [214].

In Figure 49, it is observed that as the PL energy decreases from the value of the excitation source, the DP decreases almost linearly. The trend is observed consistently on films with different optical bandgap values. This result may be expected, since lower PL energies involve e–h pair thermalization for longer periods of time. This extended lifetime of the e–h pairs increases their probability of reorientation after their creation, thus losing their original orientation. Also, the e–h pair–lattice interactions involved during thermalization may contribute to changing the orientation of the e–h pairs. Another crucial trend that Figure 49 shows is that the DP is affected by the optical bandgap of the films, the general trend being that the DP increases with increasing optical bandgap of the films. The DP increases by about 0.1 for films with bandgap values ranging from 2.69 to 4.33 eV. This evidence suggests that carriers are more localized in high bandgap material.

4.3.7. Nitrogenated PAC Films

The better electronic and mechanical properties of PAC:N over PAC have led many scientists to diverge their research toward the nitrogenated material over the last decade. The PL properties of the nitrogenated material have been no exception in showing a significant improvement in performance. The exact role that nitrogen plays in PAC:N is yet not clear, but most agree that the nitrogenated alloy outperforms PAC in nearly all aspects.

The shape of the PL broad band observed from films of PAC:N is similar to ones obtained from PAC films. Only careful deconvolution of the PL signal into its constituent bands shows that nitrogenation gives rise to the appearance of a new PL band [86]. The reason it is difficult to distinguish PAC from PAC:N films from their PL band shapes is that the intensity of the contribution to the PL signal from the nitrogenation is small compared to the total PL signal from the material. Another aspect that clouds the issues on the difference in the PL band shape between these two materials is the fact that there is often a small amount of nitrogen found within the softer, more porous PAC films.

It is widely observed that the PL efficiency of PAC:N films increases as the concentration of nitrogen in the PECVD chamber increases. This trend is observed for nitrogen/methane gas mixture concentrations in the range between 0 and around 0.3. Increasing further the nitrogen concentration in the plasma has an adverse effect on the PL efficiency, and it starts to decrease once again. This trend is clearly observed in Figure 50 obtained from [244]. It is important to note from Figure 50, that there is not a significant change in the position of the PL band for the PAC:N films grown using different nitrogen concentrations in the gas mixture. However, there exists a significant change

broken lines correspond to polarization parallel and perpendicular to that of the excitation, respectively. Figure 48b shows the degree of polarization for the PL spectra seen in Figure 48a, from [214]. For PL events where the polarization is not conserved, $I_s = I_p$, and DP tends to 0. For a random distribution of dipoles, the theoretical limit for DP is 0.5 [216]. Figure 49 shows that DP for a-C:H can be as high as 0.4, which shows a strong PL polarization memory. Note that the peak in the polarization memory occurs for PL energies close to the excitation energy. Also note that the DP decreases very rapidly with

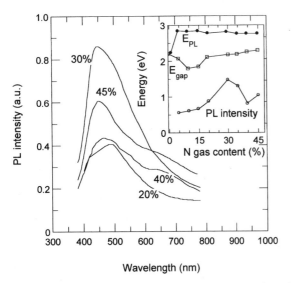

Fig. 50. PL spectra from PAC:N films for various N contents. The inset shows the dependencies of the main energy peak position (EPL), optical gap (E_{gap}), and the relative PL intensity with N content in the gas mixture. Reproduced with permission from [244], © 19??, ???.

in the PL peak position of the PAC and PAC:N films. Another point to state is that there is no apparent simple relation between the PL peak position and the optical bandgap. All these observations have been confirmed by various workers and are still not well understood. Early workers attributed the decrease in the PL intensity at the higher nitrogen concentrations to the formation of graphitelike bonding within the PAC:N films. Graphitelike bonding is most likely to occur when growing at high nitrogen concentrations.

Annealing PAC:N films at moderate temperatures (around 200–250°C) has a very similar effect on the PL signal to growing the films at high nitrogen concentrations. It is well known that annealing these films at such temperatures induces graphitelike bonding. The decrease in the PL efficiency with annealing has been attributed to the formation of this type of bonding. These two observations indicate that graphitelike bonding sites are nonradiative recombination centers.

Recently [217], EPR measurements performed on PAC:N films and discussed in Section 5 have shown that the density of paramagnetic defects is invariant as a function of nitrogen concentrations (ranging between 0 and 8 at.%). PL measurements performed on these films show, however, a dramatic increase in their PL efficiency upon nitrogenation [86]. These experimental observations put into question once more the nature of the nonradiative recombination sites in PAC and PAC:N materials and are further evidence against these sites being paramagnetic centers, as is the case for a-Si:H. To date there have been no reports that investigate the polarization memory of PAC:N films.

4.3.8. Possible PL Mechanisms

The polarization memory and the short lifetime of the PL process suggest that the e–h pair created during excitation remains closely correlated until luminescence and does not diffuse or

tunnel apart very easily. The close correlation between the electron and the hole renders their name as a geminate e–h pair.

An early mechanism for the PL process in PAC was suggested by Robertson et al. [190]. Their calculations explained the feasibility of a model based on the idea of geminate recombination within sp^2 carbon clusters embedded in an sp^3 amorphous matrix. This model was widely accepted, and most workers explained their observations in terms of this model. From the definition of the cluster model, the PL mechanism meant that the exciton created was confined to a particular cluster, which in most cases extends over many atomic sites. Subsequent discoveries, such as the very short lifetime of excitons in a-C:H and evidence for polarization memory, suggested that the degree of localization of the geminate pair was stronger than previously thought and was applicable to only a few atomic sites.

The strong localization of the e–h pair in wide bandgap PAC would mean that the Bohr radius of the exciton must be smaller than the dimensions of the cluster where the PL is believed to originate. The high possibility of a carbon sp^3 phase surrounding the cluster means that the amplitude of the exciton wavefunction decreases strongly in this material, and tunnelling to neighboring clusters is not expected to be a common process. This contradicts current understanding and questions the nature of the exciton that is being created.

Recently Piryatinskii et al. [218] investigated novel ways of postgrowth treatment of PAC and PAC:N films in order to enhance the intensity of their PL signal. The films were gown using a 13.56 MHz RF PECVD system using a $CH_4:H_2:N_2$ gas mixture. Their investigations were focused on the effect of H^+ ion implantation into PAC and PAC:N films. They concluded that in all cases implantation resulted in some decreasing of PL intensity because of ion beam induced film disordering and appearance of nonradiative recombination centers. However, strong UV irradiation of preliminary implanted films led to a marked increasing of PL intensity. In this case, the implanted hydrogen under the action of UV treatment can saturate dangling bonds, leading to PL intensity increasing. They also reported that the effect of strong UV irradiation alone, without implantation, has the effect of reducing the PL intensity of low nitrogen content a-C:H:N films but has no significant effect on the PL intensity of high nitrogen content a-C:H:N films. They attributed this to the strength of the carbon–nitrogen bonds [219].

4.4. Raman Spectroscopy

4.4.1. Introduction

There have been a large number of papers published on the use of Raman spectroscopy to obtain structural information about the different types of amorphous carbon. The study by Tamor and Vassell [220] published in 1994 is the one of the first comprehensive studies to examine the Raman spectra of both hydrogenated and unhydrogenated a-C thin films over a wide range of deposition conditions. In this part of the review the fundamentals required to understand the results from Raman measurements will be presented along with some representative results which demonstrate the current understanding of the topic.

Raman spectroscopy is a nondestructive optical technique that involves inelastic scattering from the vibrational modes in a sample. Raman scattering differs from infrared spectroscopy in that the former is concerned with the changes in the bond polarizability whereas the latter is concerned with the presence of a dipole in the vibration. For a single crystal, the intensity of a spectrum is determined by the particular geometry, polarization of the incident beam, and the k conservation rule. As the wavelength of the incident light is longer than the phonon wavelength only those phonons with wavevectors at the Brillouin zone center ($k \sim 0$) usually take part. In this way the Raman spectra from single crystals will display sharp peaks. This is referred to as first order scattering. Second order effects, in which two phonons of equal but opposite crystal momenta, may also effectively satisfy this condition and give Raman shifts that will have peaks at roughly twice those of the first order Raman shifts. The peaks tend to be broadened as a distribution of phonons can take part. In an amorphous material a greater number of phonons can take part in the Raman scattering process as the $k = 0$ conservation rule no longer holds and in this way the intensity of the Raman signal is reflected in the vibrational density of states (VDOS) of the material. The intensity is usually expressed in the Shuker–Gammon formula [221]

$$I(\omega) = \frac{n(\omega) + 1}{\omega} C(\omega) G(\omega) \qquad (4.12)$$

where $n(\omega) + 1$ is the boson occupation factor, $C(\omega)$ is the Raman coupling coefficient, and $C(\omega)$ is the VDOS of the amorphous material. The cross-section for Raman scattering also dictates the strength of a particular Raman peak. At visible wavelengths scattering from sp^2 C sites is much higher than that from sp^3 C sites. This is partly due to the fact that the local sp^2 C energy gap of (~2 eV) is comparable with the energy of the incident photons. In this way the contribution from the sp^3 sites is overshadowed by the contribution from the sp^2 sites. In order to examine the sp^3 sites it is necessary to use lower wavelength excitations (UV Raman scattering).

4.4.2. Raman Lineshapes

Raman lineshapes have been fitted using a combination of Gaussian lineshapes as well as Lorentzian lineshapes. One of the more common lineshapes is the Breit–Wigner–Fano (BWF) lineshape which has the form

$$I(\omega) = \frac{I_0[1 + 2(\omega - \omega_0)/Q\Gamma]^2}{1 + [2(\omega - \omega_0)/\Gamma]^2} \qquad (4.13)$$

where I_0 and ω_0 are the peak intensity and position respectively and Γ is the full width half maximum. $1/Q$ is the BWF coupling coefficient and in the limit of $1/Q \to 0$ a Lorentzian lineshape is recovered. The maximum of the BWF lineshape function is

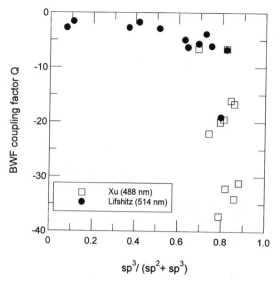

Fig. 51. Variation of the BWF parameter Q series of different prepared TAC films as a function of sp^3 content. Reproduced with permission from [222], © 2000, American Institute of Physics.

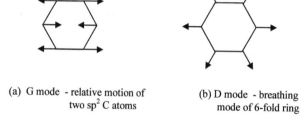

(a) G mode - relative motion of two sp^2 C atoms

(b) D mode - breathing mode of 6-fold ring

Fig. 52. Structures indicating the C sp^2 stretch vibrations (a) E_{2g} vibration of two sp^2 C atoms and (b) A_{1g} breathing mode found in a sixfold ring.

not at ω_0 but at $\omega_0 + \Gamma/2Q$. In this way Q measures the skewness of the peak and for the BWF curves the tail increases to lower frequencies for lower values of Q. Figure 51 shows the values of Q used for a series of TAC films as a function of sp^3 content [222]. Films with a high sp^3 content tend to have large negative values of Q and these tend to produce highly symmetrical Raman peaks. In such a situation Lorentzian lineshapes can also be used.

4.4.3. Fundamentals of Raman Spectroscopy in Graphite and Disordered Graphite

Many of the characteristics present in the visible Raman spectra of amorphous carbon thin films are discussed in terms of similar peaks observed in graphite and disordered graphite. It is therefore important to examine what is already known about the Raman spectra from these two materials. Graphite is composed of a series of stacked parallel layers of hexagonal planes with trigonal planar sp^2 bonding. Each sp^2 C atom bonds with three other sp^2 hybridized atoms to form σ bonds. The bond angle between the sp^2 hybridized orbitals is 120°, giving rise to threefold rotational symmetry, and this results in a continuous 2D arrangement of hexagons, as shown in Figure 1. The fourth valence electron is oriented perpendicular to this plane and is delocalized into a π orbital. The σ bond length is 0.142 nm, the spacing between the layers is 0.335 nm, and the layers are stacked in a hexagonal arrangement in which neighboring atoms in adjacent planes lie above and below (-ABABA-). The space group of graphite is therefore $D4_{6h}$ ($P6_3/mmc$). The decomposition of the zone center optic modes can be obtained using standard group theory and results in the following irreducible representations

$$\Gamma = A_{2u} + 2B_{2g} + E_{1u} + 2E_{2g} \qquad (4.14)$$

The A_{2u} and E_{1u} modes are infrared active and have been observed at 867 and 1588 cm^{-1}, respectively [223]. The B_{2g} mode is optically inactive but has been observed using neutron scattering at 127 and near 867 cm^{-1} [224]. Group theory analysis predicts that only the $2E_{2g}$ modes will be Raman active. The symmetry of the E_{2g} modes restricts the motion of the atoms to the plane and involves the in-plane bond stretching motion of pairs of sp^2 C atoms as shown in Figure 52a. Two different E_{2g} modes occur because adjacent planes can vibrate in phase or with the opposite phase; however, the energy separation between the two modes is small due to the weak interlayer coupling. The Raman spectrum measured from single crystal graphite by Tuinstra and Koenig (TK) [225] showed a single line at 1575 cm^{-1} and on the basis of group theory, as well as tentative calculations, this mode was assigned to the E_{2g} mode and is now universally referred to as the G peak. Second order Raman peaks, associated with two-phonon states, have also been observed in both single crystal graphite (SCG) as well as in highly oriented pyrolytic graphite (HOPG) by Nemanich and Solin [226] and by Wilhelm et al. [227]. A strong resonance is observed from HOPG at around 2725 cm^{-1} with weaker signals at $\sim$2450 and $\sim$3250 cm^{-1}. Wilhelm et al. showed that the strong second order resonance has a shoulder at 2697 cm^{-1} in SCG and at 2690 cm^{-1} in HOPG. The overall spectrum from 2300 to 3300 cm^{-1} was observed to be uniform and strongly polarized from which Nemanich and Solin concluded that the second order spectrum was due to overtones not to combinations of peaks.

In the case of microcrystalline graphite two peaks at 1575 cm^{-1} (G peak) and at 1355 cm^{-1} (later referred to as the D peak) were observed [225]. TK noted a linear behavior of the ratio of the intensities of the peaks I_D/I_G with the crystallite size, L_a, obtained from X-ray diffraction. This linear relationship indicates that the Raman intensity is proportional to the amount of boundary edge present and proposed the relationship

$$\frac{I_D}{I_G} = \frac{C}{L_a} \qquad (4.15)$$

where $C \sim 4.4$ nm measured at 514.5 nm. TK interpreted the presence of the D peak as being due to a relaxation of the selection rules for an A_{1g} breathing mode at the K zone edge (Figure 52b). This mode is normally forbidden but in the presence of disorder if can appear; it is for this reason that it is often referred to as the disorder peak. Some care must be exercised with this description, as the term "disorder" can be misleading.

Table II.   In-Plane Correlation Lengths as Measured by XRD and Ratio of I_D/I_G for Different Types of Disordered Graphite

Sample	L_a (μm)	I_D/I_G
SCG	>100	0
HOPG	$\sim$1	0
Flat micronic graphite	1–10	0.20
Ground natural graphite	<5	0.36
High temperature pyrocarbon	$\sim$0.02	0.16

Wilhelm et al. studied the ratio of the areas in the D and G peak and the measured values of L_a using X-ray diffraction (XRD) for a range of different graphite based materials. Table II shows that the ratio of I_D/I_G is similar for a flat micronic graphite sample and a pyrocarbon grown sample but the measured values of L_a differ by between 50 and 500. This implies that c-axis disorder is **not** the dominating factor and there must be smaller domains that are not observed by XRD and these play a significant role in the Raman process.

In addition to the D peak at 1355 cm^{-1}, a number of other peaks can appear depending on the level of imperfections in the graphite. In slightly damaged graphite (such as by layer-edge facets producing by fractures, etc.) the D peak can appear as a doublet at 1350 cm^{-1} (labelled D_1) and 1370 cm^{-1} (D_2). Vidano et al. [228] noted that the Raman shift in disordered graphite materials was dispersive—i.e., depends upon the wavelength of the excitation. They investigated samples of pyrolytic carbon deposited at 1800°C from methane, a glassy carbon (GC), a highly oriented large crystallite graphite prepared by compression annealing (CA), and a CA sample which had been subject to Ar ion beam damage (ion-CA). Using an excitation wavelength of 488 nm, they observed in the CA graphite sample peaks at 1580 cm^{-1} (G peak) as well as second order lines at 2695 cm^{-1} (G'_1) and 2735 cm^{-1} (G'_2). In the disordered samples additional lines appeared at 1360 cm^{-1} (D peak) and 1620 cm^{-1} (D' peak) as well as a second order peak at around $\sim$2950 cm^{-1} (D''). They further noted that the G'_1 and G'_2 appeared to merge into one band, labelled G', as the disorder increased.

As the wavelength of excitation increased, Vidano et al. noted that the D, G', and D'' bands red-shifted but that the G and D' bands did not. Figure 53 shows the wavelength dependence of the various peaks present. The D peak decreases from $\sim$1360 cm^{-1} (488 nm) to $\sim$1325 cm^{-1} (647.1 nm) whereas over the same range of excitation the G' peak shifts from 2725 cm^{-1} to 2645 cm^{-1}. This is almost twice the shift of the D peak and is consistent with the assignment that the G' peak is an overtone of the G peak. It is interesting to note that the D peak is absent from the highly oriented CA sample even though the G' peak is present and also moves. In the ion-CA sample the D peak is present and is observed to shift with excitation. This red-shift was observed to be independent of the orientation of the excitation. As noted by Vidano et al. the magnitudes of the band shifts are relatively small with a shift of the D band

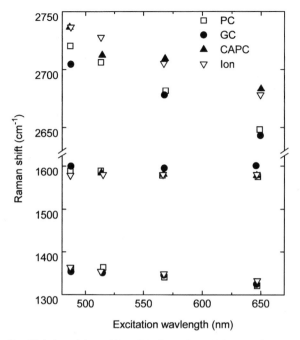

Fig. 53.   Variation of the position of (a) first order and (b) second order peaks as a function of excitation wavelength.

2697 and 2725 cm^{-1} though their position depends upon the incident wavelength. The G peak is associated with the in-plane movements of sp^2 C atoms. The G peak is not dispersive though its second order peaks are dispersive. In disordered graphitic materials the extent of disorder present will determine the intensity and position of the D peak. First the G peak will appear to increase from 1580 to above 1600 cm^{-1}. This is, in fact, due to the presence of an additional dispersive peak (D$'$) at around 1610 cm^{-1}. The D peak, which is dispersive, appears at about 1350 cm^{-1} and increases in intensity as boundary associated with the 2D graphite crystallite (the graphene layer). The Tu-instra and Koenig relationship (at a given wavelength) is often used to evaluate the in-plane correlation length, L_a, though this can differ from the size of crystallite domain as measured by XRD. The D peak is associated with the A_{1g} breathing mode and this is associated with the presence of sixfold rings. Second order peaks are also present. This analysis is applicable to graphite as well as to disordered graphite all which have sp^2 contents $\sim$100%. In the next section the transition from disordered graphite to amorphous carbon with sp^2 contents of less than 100% will be presented.

4.4.4. Visible Raman Spectroscopy of Amorphous Carbon

Tamor et al. [232] studied the visible Raman spectra and optical properties of a wide range of amorphous carbon films deposited using RF PECVD as a function of self-bias using feed gases of CH_4 and C_6H_6. The optical gap was observed to decrease with bias voltage ranging from about 1.8 eV (-200 V bias) to less than 0.8 eV (-1200 V). Above about -1200 V the optical gap begins to increase slightly. They used the original expression from Robertson and O'Reilly [78] relating the optical gap (in eV) to the cluster size. On this basis they concluded that cluster sizes varied from 1 nm at the lowest bias to 2 nm at -1200 V bias. Visible Raman spectra were performed and 459, 515, and 628 nm. Two bands were observed around 1500 cm^{-1} (G peak) and 1350 cm^{-1} (D peak). The Raman peak positions and the ratio of the intensities of the D peak to the G peak are shown in Figure 54 for the three different wavelengths. First, for a sample deposited at a particular bias the G peak shifts to lower frequencies as the excitation wavelength increases. This is in contrast to the situation found in disordered graphite where no dispersion of the G band is observed. Second, as the deposition bias is increased the G peak increases in frequency, eventually converging on 1582 cm^{-1}. Above a bias of about -1000 V the peak position remains approximately constant. The ratio the D peak to the G peak increases approximately linearly with bias, as shown in Figure 54b, and for a given sample deposited at a particular bias it is higher with longer wavelength excitation. At that time the conventional interpretation of the I_D/I_G ratio was based on the effects in disordered graphite. An increase in the I_D/I_G ratio implied that the cluster size was decreasing in contrast to what was inferred from the optical data. Tamor et al. concluded that "either the assumption of small graphitic domains and the associated optical gap calculations are inapplicable to amorphous carbon or else the I_D/I_G ratio is determined

of only 35 cm^{-1} and a shift of $\sim$75 cm^{-1} for the overtone G$'$ band compared to a change of over 5000 cm^{-1} in excitation frequency. From this the authors concluded that the red-shifts of the Raman spectra could not be attributed to the large longitudinal acoustic dispersion from the zone center to the zone edge.

A number of groups have proposed possible explanations for the dispersive nature of some of the Raman peaks. Baranov et al. [229] proposed that the dispersion was a result of selective coupling between phonon and electronic states that have the same magnitude of wavevectors ($\Delta k = \Delta q$) and as the photon energy rises (wavelength decreases) the number of phonons modes around the K point increases. Matthews et al. [230] have proposed that the peak shift is a result of the coupling between the high frequency optical modes near the K point with the transverse acoustic branch. They concluded that all the optic phonons on this branch that have the same magnitude Δq around the K point will contribute to the D and G$'$ bands. This is in contrast to Baranov et al. who proposed that only those phonons in the K-M direction will take part. Ferrari and Robertson [231] have addressed possible discrepancies between these different interpretations and have shown that the precise shape and symmetry of the dispersion curves around the K point are important. They noted that since the A_{1g} mode is singly degenerate that it is necessary to have a single degenerate band at the K point that is upward dispersing away from K. The branch chosen by Matthews et al. is doubly degenerate at the K point.

In summary, the first order Raman spectrum of ordered graphite consists a single peak (G peak) at about 1580 cm^{-1}. Second order doublet peaks (G$'_1$ and G$'_2$) also occur at about

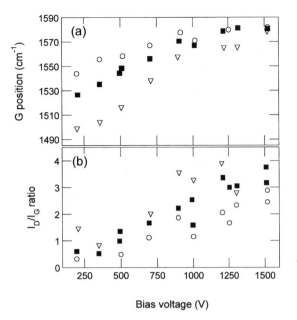

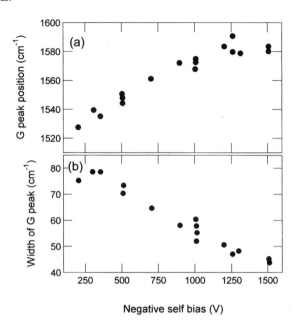

Fig. 54. (a) Position of the Raman G peak and (b) I_D/I_G ratio as a function of bias voltage at 459 nm (○), 515 nm (■), and 628 nm (▽).

Fig. 55. (a) Position of the Raman G peak and (b) G peak width as a function of bias voltage for DAC film deposited using RF PECVD using CH_4 as feed gas.

by factors other than the graphitic cluster size." They ruled out the former as the likely source of the discrepancy as the EELS spectra of their films indicated a strong graphitic structure present, and they concluded that the relative intensity of the D peak (to the G peak intensity) in amorphous carbon is not a useful indicator of graphitic cluster size. They interpreted the laser wavelength and bias voltage dependence of the G peak in terms of the amount of H termination in the clusters. For low bias samples the clusters are terminated with H and the vibrational modes associated with the reduction in the number of C edges and a weakening of the C–C bonds near the edge of a clusters. In a later comprehensive study Tamor and Vassell [220] studied a number of different DAC films and proposed that the Raman shift observed at 600 cm^{-1} is due to the presence of H. The incorporation of H leads to the greater presence of sp^3 sites so that a-C film can be viewed as an alloy mixture of the different sp^2-based and sp^3-based materials.

Tamor and Vassell also observed that for films deposited from CH_4 that the G peak moved up in frequency from 1520 to 1580 cm^{-1} as the self-bias increased to −1500 V and at the same time the G width decreased from about 80 to 50 cm^{-1}, as shown in Figure 55. The G peak linewidth is affected by cluster size, their distribution, as well as the stress in the film. Nemanich et al. [233] and Shroder et al. [234] found that the relative concentration and crystallite size are important in the analysis of spectra. They showed that the vibration modes of small crystallites can be described as phonons with an uncertainty in the wavevectors $\Delta k \sim 2\pi/d$, where d is the size of the crystallite. Schwan et al. [109] showed that for a range of films deposited by different methods that the ratio of the I_D/I_G peak depended on the G linewidth as reproduced in Figure 56. The films examined were PECVD grown DAC:H [220], sputtered GAC [220], TAC deposited using laser ablation [220],

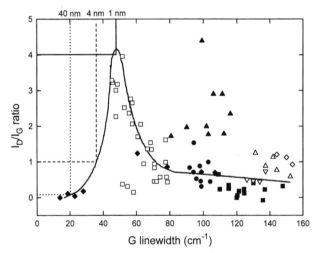

Fig. 56. I_D/I_G ratio vs G linewidth for DAC (□), MEPD deposited DAC (△), magnetically confined DAC films (▽), DAC films by Mariotto (◇), sputtered a-C (●), TAC films (■) and TAC films using magnetron sputtering with Ar ion plating (▲), and various graphites and other carbons by Knight [245] (◆). Reproduced with permission from [109], © 1996, American Institute of Physics.

TAC films deposited by magnetron sputtering [109], magnetically confined PECVD [88], as well as graphite and disordered graphite. All the films were examined using 514.5 nm excitation. In the case of microcrystalline graphite the small linewidth and low I_D/I_G ratio indicate high quality material. As the G linewidth becomes broader there is a maximum in the observed I_D/I_G ratio. Above a G linewidth of about 50 cm^{-1} the ratio begins to decrease and under the TK relationship given in Eq. (4.15) would indicate a crystallite size of less than 1 nm. Davis et al. [81] did not observe clusters bigger than 0.5 nm in RF plasma deposited DAC films. The low value of I_D/I_G and large linewidth for the TAC films indicate a very low concentration of sixfold rings. The deviation from the solid line in Figure 56 of the TAC films deposited by unbalanced magnetron sputtering was attributed to the high stress in these films. Finally it is worth noting that the position of the G peak also depends on the amount of H present in the films. The peak was observed to shift to lower frequencies as the H concentration increased. The loss of H helps to promote the formation of sp^2 bonds rather than sp^3 bonded C.

These results of the Raman measurements by Tamor et al. and Schwan et al. can be viewed in terms of an decrease in the sp^2 content (with an increase of the sp^3 content) as the bias voltage is reduced. In summary, for an increase in the sp^3 content,

  (i) the G peak becomes dispersive,
 (ii) the G peak position decreases, and
(iii) the TK relationship for the ratio of the D peak to the G peak breaks down and no longer reflects the in-plane correlation length. At this point cluster sizes are typically a couple of nm and I_D/I_G approaches 0.

Since the D peak is associated with the presence of sixfold clusters, at higher levels of disorder greater cluster distortion occurs and the number of such sixfold clusters will decrease. In this way Chhowalla et al. [235] and later Ferrari and Robertson [231] proposed that for small values of L_a, the intensity of the D peak is indicative of finding a six-fold ring in the cluster and is therefore proportional to the cluster area and therefore

$$\frac{I_D}{I_G} \propto L_a^2 \propto M \qquad (4.16)$$

Chhowalla et al. studied the effects of deposition temperature on the Raman spectra of FCVA deposited TAC. The G peak at ~1560 cm^{-1} was clearly observed in films deposited at 30°C but at higher temperatures a D peak at 1350 cm^{-1} was also observed and indicated the formation of sixfold graphitic rings. The variations of the G peak position, G peak width, and I_D/I_G ratio are shown in Figure 57. In conjunction with the Tauc gap, EELS measurements, and resistivity they concluded that films deposited at near room temperature consist of a small number of short olefinic chains which keep the bandgap wide and the resistivity low. As the deposition temperature rises the sp^2 sites begin to condense into clusters of increasing size (resulting in a reduction of the Tauc gap). Sixfold rings begin to form, resulting in an increase of the I_D/I_G ratio. According to the Robertson and O'Reilly model, discussed in the section (optical) for undistorted sixfold rings, the bandgap varies as the $1/M^{1/2}$, where M is the number of sixfold rings. In this way the ratio of I_D/I_G should be inversely proportional to the square of the bandgap E_g, if the gap is determined by sixfold rings. Figure 58 shows a linear relationship for the films studied by Chhowalla and also by using the data of Tamor et al. [232].

There have also been a number of studies using Raman spectroscopy of the effects of ion implantation in both TAC

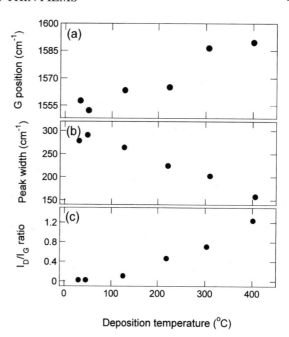

Fig. 57. Variation of (a) G peak position, (b) G peak width, and (c) I_D/I_G ratio as a function of deposition temperature for TAC films deposited using the FCVA technique.

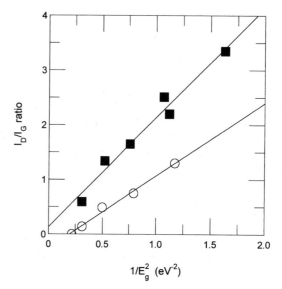

Fig. 58. The Raman I_D/I_G ratio against $1/E_g^2$ for a series of TAC films deposited at different substrate temperatures (o) and for series of films as a function of substrate bias (■).

[107] and GC [236, 237]. GC, produced by the decomposition of highly crosslinked polymers, is a nongraphitizing form of carbon as the tangled graphite ribbons prevent the formation of a proper 3D structure even at high temperatures [237]. The visible Raman scattering from a GC samples can been seen in Figure 59 [237]. It can be seen that the spectrum consists of two dominant peaks at 1360 cm^{-1} (D peak) and at 1588 cm^{-1} (G peak). This latter peak has a broad shoulder at about 1615 cm^{-1}, assigned to the D' peak. In the case of TAC films produced by the FCVA method only a broad band lying

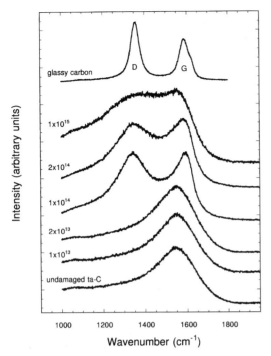

Fig. 59. The effects of 200 keV Xe ion implantation on the Raman spectrum of TAC implanted to the doses shown. Also shown for comparison is the Raman spectrum of glassy carbon.

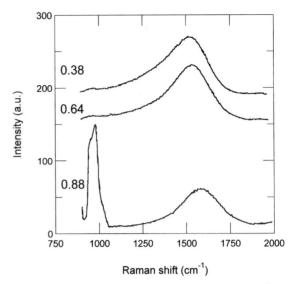

Fig. 60. Visible Raman spectra from TAC films with different sp^3 fractions. The spectrum from the sample with the highest sp^3 content was taken by Xu et al. using 488 nm excitation [54] and the remainder by Lifshitz et al. using 514 nm excitation [246]. The data have been fitted by a single BWF lineshape.

between the G and D peaks is observed. The Raman spectra of the films implanted with 200 keV Xe ions are also shown in Figure 59. It is evident that as the dose is increased above 10^{14} cm^{-1} the two broad bands become clearly visible and their appearance is attributed to the formation of a high concentration of sp^2 bonded C with good in-plane ordering. It is interesting to note that in the study of 320 keV Xe implantation into GC [236], the increased dose resulted in the initial broadening of the G and D peaks at a dose of up to 2×10^{14} cm^{-2} and then their subsequent merger into a single broad band at a dose of 6×10^{16} cm^{-2}.

4.4.5. Ultraviolet Raman Spectroscopy—probing the sp^3 Content

The cross-section for Raman scattering determines the strength of a particular Raman peak. As discussed in the introduction to Raman scattering, at visible wavelengths the contribution from the sp^3 sites is overshadowed by the contribution from the sp^2 sites. In order to examine the sp^3 sites it is necessary to use lower wavelength excitations. One of the first uses of UV Raman spectroscopy to examine the influence of sp^3 content was by Wagner et al. in their study of polycrystalline diamond [238]. They showed that the Raman spectra from polycrystalline diamond films depended upon the grain size and also the laser excitation energy. For a diamond film with grain sizes in the region of 50–100 μm, the Raman spectrum consisted of a single sharp line at 1332 cm^{-1} when excited with light of 2.33 eV energy. This line is the signature of the T$_{2g}$ mode in diamond and is associated with sp^3 C–C stretching modes.

Only very weak scattering was observed in the 1500 cm^{-1} region, indicating very little scattering from the sp^2 regions. However, as the laser energy was decreased a broad shoulder appeared at lower frequency. For a polycrystalline film with grains sizes of 15–100 nm, the vibration associated with the T$_{2g}$ modes was only clearly visible with an excitation of 4.82 eV, a broad but weaker peak around 1580 cm^{-1} was also observed at this wavelength. As the laser energy decreased to 3.53 eV the 1332 cm^{-1} was only just visible and the band at 1580 cm^{-1} had increased in size significantly. In addition a weaker band around 1150 cm^{-1} began to appear. For successively lower energies this 1150 cm^{-1} band increased in intensity whereas the two other bands began to merge. The band at 1150 cm^{-1} was attributed to the sp^3 content in the film.

Gilkes et al. studied the presence of the sp^3 bonding in FCVA grown TAC films using UV Raman (244 nm) spectroscopy [239]. In a more recent study the Raman spectra for several TAC films with known sp^3 contents have been studied [222]. Figures 60 and 61 show the visible (514/488 nm) and UV (244 nm) Raman spectra for three TAC films with different sp^3 contents. The visible spectra were all fitted with a BWF lineshape along with a linear background subtraction. The visible Raman spectra from each of the films consists of a broad peak centered around 1550 cm^{-1} associated with the G peak. (The square shaped peak in the sample with the highest sp^3 fraction is the second order mode from the underlying Si substrate. This sample has the highest optical gap and is the most transparent.) No evidence of the D peak is observed, indicating that no sixfold rings are present. The UV Raman spectra of the same samples shown in Figure 61 have two broad peaks at 1100 and 1600 cm^{-1}. It is evident from Figure 61 that as the sp^3 fraction increases (or conversely as the sp^2 content decreases) the position of the peak of the G band shifts to higher

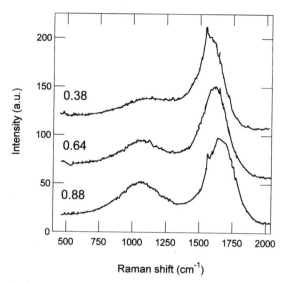

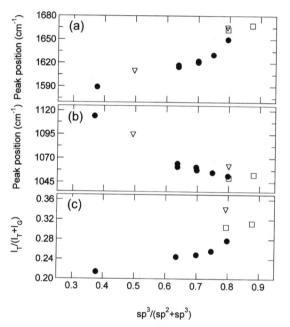

Fig. 61. Corresponding UV Raman (244 nm) spectra of the films shown in Figure 60. The data have been fitted by two Lorentzian lineshapes.

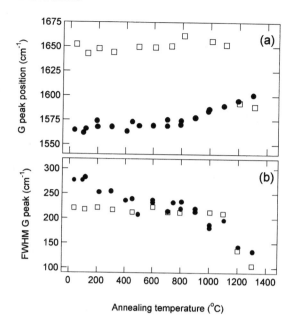

Fig. 63. (a) Position of the G peak and (b) width of the G peak as a function of annealing temperature measured under visible (●) and UV (□) excitation.

Fig. 62. Peak position of (a) the 1650 cm^{-1} band and (b) the 1100 cm^{-1} band; (c) ratio of the intensity of the 1100 cm^{-1} and 1650 cm^{-1} band as a function of sp^3 fraction. Adapted from [222].

frequency; the peak at 1100 cm^{-1} (T peak) decreases in frequency and finally the intensity of the T peak increases. This behavior is more clearly seen in Figure 62a–c for samples deposited by different groups using either FCVA or mass selected ion deposition [222]. Note that in Figure 62c the peak intensities rather than peak areas have been plotted. Gilkes et al. proposed that the peak at ~1600 cm^{-1} is most likely due to the presence of paired sp^2 sites which are enhanced by UV excitation. This peak position is larger than the usual sp^2 G peak at ~1550 cm^{-1} found using visible Raman spectroscopy, and the change in this band's position was attributed to two possible reasons. First, the relative proportion of sp^2 pairs to large

clusters as the sp^2 sites increase is possible. As the cluster size increases the bond order decreases with a resultant decrease in the Raman frequency as sp^2 content increases. The other possible explanation proposed by Gilkes et al. concerns whether an increase in the compressive strain will shift the Raman peak. The downward shift in the peak at 1100 cm^{-1} with increasing sp^3 fraction was attributed to the replacement of mixed sp^2–sp^3 bonds with more sp^3–sp^3 bonds. However, since these sp^3–sp^3 bonds are longer than the corresponding bond lengths in diamond, the Raman shift is lower than reported in the UV Raman spectrum of pure diamond—namely 1150 cm^{-1}. The tight-binding molecular dynamic calculations by Wang et al. [87] also suggest that in TAC in excess of 90% of the vibrational amplitude above 1600 cm^{-1} was localized on sp^2 sites. In addition the sp^3 stretching modes were found to be at 1100 cm^{-1}. Similar results were found by Drabold et al. [85].

Two more studies demonstrate the usefulness of UV Raman spectroscopy in studying the effect of annealing and also the addition of N. Ferrari et al. [125] used UV and visible Raman spectroscopy to study the effects of annealing of TAC films. Using UV Raman the as-deposited TAC films consisted of two broad peaks at 1050 cm^{-1} and a peak at 1660 cm^{-1}. They ascribed the band at 1660 cm^{-1} to sp^2 olefinic C=C stretching modes (G modes) and the band at 1100 cm^{-1} to sp^3 C–C bond vibrations (T peak). They proposed that the shorted sp^2 olefinic bands will produce Raman peaks at higher frequencies. By contrast the visible Raman spectra of the as-deposited film consisted of a single broad peak at 1567 cm^{-1}. Annealing of the films resulted in a shift of the G peak (1560 cm^{-1} in the visible spectrum) to higher frequencies as shown in Figure 63a. However, the 1650 cm^{-1} peak in the UV spectrum remains at about the same frequency up to an annealing temperature of 1000°C. There was also a change in the width of

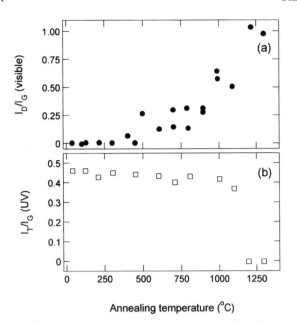

Fig. 64. (a) Ratio of I_D/I_G under visible excitation and (b) ratio of I_T/I_G under UV excitation.

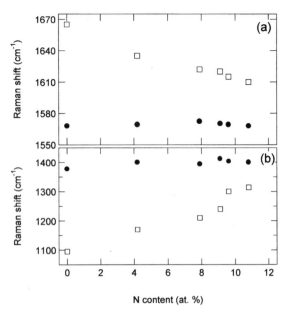

Fig. 65. Effect of nitrogenation on the positions of (a) G peaks and (b) D peak and T peak under visible (•) and UV (□) excitation as appropriate.

the G peak observed in the visible Raman spectrum, as shown in Figure 63b, from a width of ~260 to less than 175 cm^{-1} at 1000°C. In the study by Merkulov et al. of a-C films with sp^3 contents which ranged from 20 to 75%, an increase in the G position was also observed (from ~1580 to 1620 cm^{-1}). However, there was an increase in width of this peak. By contrast, in the study by Ferrari et al. the width of the 1650 cm^{-1} peak remains at approximately 220 cm^{-1} up to 1000°C from where it then falls. If the width of the G peak is taken as a measure of the size of the domain the reduction in the G width indicates an increase in the size of the sp^2 cluster. Changes in the relative peak intensities of the both pairs of peaks (I_D/I_G) in the visible and (I_T/I_G) in the UV are also accompanied by changes in the positions and linewidths as shown in Figure 64. Some care must be exercised with the interpretations of these peaks. However, the increase in the intensity of the D peak in the visible indicates the formation of sixfold rings. The ratio of the T band to the G band (1650 cm^{-1}) band is constant up to annealing temperatures of 1000°C, indicating that the sp^3 content in the films remained constant to 1000°C. This interpretation was also confirmed by EELS measurements and indicates the ratio of the I_D/I_G cannot be related to the sp^3 content. Finally, it should be noted that for annealing temperatures above 1000°C grains of β-SiC were formed. The results of the Raman measurements made using the two different wavelengths allowed Ferrari et al. to conclude that although sp^3 sites begin to convert to sp^2 sites at an annealing temperature of around 600–700°C these are not detected by UV Raman.

Shi et al. [240] used both UV and visible Raman to explore the effects of nitrogenation of FCVA grown TAC films. The N content was determined by Rutherford backscattering spectroscopy and by Auger electron spectroscopy and was 11.2 at.% for the maximum concentration. The visible (514 nm) Raman

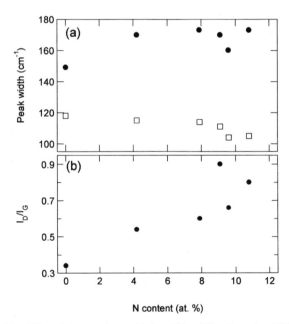

Fig. 66. Effect of nitrogenation on (a) the widths of G peaks under visible (•) and UV (□) excitation and (b) the ratio of I_D/I_G under visible excitation.

spectrum from a non-nitrogenated film had a peak at 1548 cm^{-1} (G peak) and a barely visible peak at 1377 cm^{-1} (D peak). The UV (244 nm) spectrum from the same non-nitrogenated sample consisted of two peaks at 1665 cm^{-1} and a peak at 1095 cm^{-1} (T peak). The effect of nitrogenation on the peak position for all these four peaks can be seen in Figure 65. As the N content increases the visible G peak only varies slightly in frequency and remains in the range 1568–1572 cm^{-1}, as shown in Figure 65a, whereas the peak at 1665 cm^{-1} decreases by 55 cm^{-1}, reaching 1610 cm^{-1} for the sample with 10.8 at.% N. The position of the D peak increases slowly from 1377 to 1412 cm^{-1}. The position of the T peak shifts upward by 219 cm^{-1} from 1095 to 1314 cm^{-1} (10.8 at.% N). Shi et al. also reported changes in the linewidths of the four peaks. The D peak increases from 149 (0 N) to 170 cm^{-1} at about 4.2 at.% and then remained approximately constant. The G peak under visible excitation decreased only from 113 (0 N) to 106 cm^{-1} (10.8 at.% N). A small reduction was also reported in the G peak under UV excitation from 110 to 105 cm^{-1}. The behaviors of both G peaks under visible and UV excitation are shown in Figure 66a. Shi et al. reported that no clear trend was evident for the width of the T peak with N. Finally it is evident in Figure 66b, that I_D/I_G increased with N content but that no trend was reported for the ratio of I_T/I_G. The increase in I_D/I_G increased and the reduction in width of the G peak with N was attributed to the generation of larger clusters. The lowering of the position of the G peak under visible and UV excitation could be attributed to sp^2 clusters with larger sp^2 aromatic rings as opposed to shorter sp^2 olefinic chains. The increase in the position of the T peak with N incorporation can be explained by the replacement of sp^3–sp^3 bonds with sp^3–sp^2 bonds, which are shorter and therefore appeared at higher energies.

Both Shi et al. [240] and Ferrari et al. [125] addressed the issue of strain and its possible implications for the position of the G peak. It was noted by Ager et al. [241] that the position of the G peak in TAC depends on the stress in the film. If a highly strained TAC film relaxes though delamination, Ager et al. showed that the position of the G peak changes by as much as 20 cm^{-1}. They found a shift of 1.9 cm^{-1} GPa^{-1} for biaxial-plane stress in TAC. Shi et al. noted that the difference in compressive stress between a non-nitrogenated film and one with 10.8 at.% N is 5.6 GPa which would manifest itself as a shift of 10.6 cm^{-1}. However, the G peak under visible excitation does not shift appreciably from which Shi et al. concluded that stress was not significant as the source of shift in the G peak. In the study by Ferrari et al. they showed near complete stress relaxation of their samples by annealing at 600°C. The stability of the G peak position and width in both UV and visible excitation indicate that only minor modifications in structure of the film have occurred. Finally it is worth noting that although the sp^2 content in both the films of Shi et al. and Ferrari et al. is increasing, the study by Shi et al. did not observe a major shift in the position in the G peak whereas in the study by Ferrari et al. the G peak increased by 33 cm^{-1}. The reason for this discrepancy is not known.

In summary using UV Raman allows the sp^3 content of amorphous carbon films to be examined. The sp^3 peaks are found at 1100 cm^{-1} and are referred to as the T peaks. The peak associated with sp^2 vibrations appears at higher wavenumbers under UV excitation than the corresponding peak under visible excitation. Visible Raman scattering allows probing of the sp^2 content as these are resonantly enhanced. The D peak is the signature of sixfold rings and the G peak is indicative of sp^2 bonding and can be seen in both in aromatic rings as well as in olefinic chains.

4.5. Conclusions

A considerable amount of information has been amassed concerning the study the optical properties of amorphous carbon. It is therefore useful to bring some of the main conclusion together. First, while the exact nature of the relationship between the energy gap and the size of clusters is not fully established, the general view persists that the larger the size of the cluster the small the energy gap. The role, if any, of distorted clusters, in particular, is not fully understood. Early models for the DOS assumed symmetric Gaussian bands but more recent experiments have cast into question whether this is the case. FTIR and Raman spectroscopy are powerful complimentary techniques and the development of UV Raman has helped to probe the role sp^3 sites play.

5. DEFECT STUDIES OF AMORPHOUS CARBON

5.1. Introduction

In order to gain further understanding of the role played by π states lying within the bandgap of a-C films, there has been a considerable body of research performed to look at the "defect" states especially those near the Fermi level. As noted previously sp^2 clusters may form into aromatic ringlike structures or olefinic chains which may or may not become distorted. The extent to which these clusters remain undistorted or become distorted is largely determined by the disorder within the a-C network with disorder tending to oppose clustering. Isolated sp^2 sites from an individual C atom may also form and give rise to a "dangling bond" (DB). The DB in a-Si:H has been characterised by a number of different techniques, one of which has been electron spin/paramagnetic resonance [163]. The DB in a-C is analogous to the CH$_3$ radical, and low temperature EPR measurements made by Fessender and Schuler [247] showed that the CH$_3$ radical can be considered as three C–H fragments oriented 120° apart in a plane. In this way the DB in amorphous carbon is planar rather than pyramidal-like as in a-Si.

From the discussion of Section 4, the density of states in the energy gap is principally determined by the π states made up of (i) isolated sp^2 sites, (ii) undistorted sp^2 clusters (in the form of chains or rings), and (iii) distorted sp^2 clusters. Whether any of these different possible structures will be paramagnetically active depends on whether there is a new unpaired electron spin available or not. Undistorted odd membered rings, such as five- and seven-membered rings, will produce a state which is near the Fermi level and will be EPR active. Doubly occupied states will not be EPR active.

The basic theory behind EPR is now well established and the interested reader is referred to the book by Wertz and Bolton [248] for further information and to the book by Poole [249] for more experimental information. In its most simplistic form, EPR occurs when there is a resonant absorption of microwave power, of frequency ν, between energy two levels. These energy levels induced by a magnetic field, of strength B, are known as the Zeeman levels [248]. The relative population of these energy levels is given by the standard Boltzman factor. For a defect with a net unpaired electron, giving a spin **S** of $\frac{1}{2}$, the resonance condition for absorption of electromagnetic radiation is

$$h\nu = g\mu_{\text{B}}B \qquad (5.1)$$

where h is Planck's constant and μ_{B} is the Bohr magnetron. The remaining parameter in Eq. (5.1) is known as the g value and is specific to each type of defect present. EPR experiments are carried out by placing the sample under investigation at the center of a resonant microwave cavity (typically X band $\sim$9.9 GHz) and monitoring the reflected microwave power. At resonance the absorption of microwaves alters the reflected microwave power. Usually, since the amount of reflected power is small, modulation techniques are used in which a slowing varying magnetic field (usually 100 kHz) is superimposed upon the Zeeman field. In this way the spectrum that is usually recorded is the first derivative of the absorbed microwave power. The g value for a particular resonance is calculated from Eq. (5.1) using the microwave frequency and the magnetic field at which the derivative of the microwave absorption is zero. This is often referred to as the zero crossing g value, g_0. For a free electron, the g value would be 2.0023 and any deviations from this value are determined by the extent of the spin–orbit interaction, which in turn is related to the localization of the electron. By performing a double integration of the EPR signal and by comparison with a known standard, a value of the concentration of the paramagnetically active species can be determined. Much of the early work on a-C based materials [250, 251] using EPR has concentrated on determining how the g value and spin density vary as a function of either deposition or postdeposition conditions but ignoring other important information such as the variation in the linewidth, lineshape, and relaxation times, all of which are now discussed in detail.

5.2. Lineshape Analysis and Relaxation Effects

Since EPR spectra are usually recorded in derivative mode, the linewidth is usually taken as the peak-to-peak linewidth (ΔB_{pp}) and is usually expressed in Gauss or mT. This can be related to the full width half maximum ($\Delta B_{1/2}$) of the absorption profile if the lineshape of the resonance is known. The lineshape is usually determined by whether the broadening of the signal is either homogeneous or inhomogeneous. Homogeneous line broadening processes are usually described as those processes that affect every defect identically, so that each defect center in the sample is capable of absorption of microwave energy over a (narrow) range of frequencies, but this range of frequencies is identical for each defect in the sample. Important sources of homogeneous broadening include the dipolar interaction between like spins and motional effects arising from fluctuating magnetic fields arising from exchange or hopping of spins; such broadening mechanisms will tend to produce a Lorentzian lineshape with $\Delta B_{1/2} = 1.7321\Delta B_{\text{pp}}$ [249]. Inhomogeneous line broadening is the overall effect of an ensemble

of defects, each of which has its own unique range of frequencies for absorption. Inhomogeneous broadening results from the defect experiencing different magnetic fields and the principal source in amorphous materials discussed here is from the hyperfine interaction with nuclear spins and high levels of atomic disorder. Inhomogeneous broadening tends to produce a more Gaussian lineshape with $\Delta B_{1/2} = 1.1776\Delta B_{\text{pp}}$ [249]. A more useful way to look at the different types of broadening, and one that reflects the dynamic nature of the spin system, is to regard homogenous broadening as arising from a particular broadening mechanism that is fluctuating rapidly compared with the time associated with a spin transition. If an EPR resonance is being broadened independently by two different mechanisms a convolution of the two lineshapes will result. A convolution of two Lorentzian lineshapes will produce a Lorentzian lineshape and the convolution of two Gaussian lineshapes will produce a Gaussian lineshape. In the case of a line that is independently broadened by both Lorentzian and Gaussian effects a Voigt profile will result.

Since EPR is based upon transitions between discrete energy levels it is important to understand how the absorbed microwave energy is dissipated. This is often discussed in terms of the spin–lattice relaxation time, T_1, and the spin–spin relaxation time T_2. The spin–lattice relaxation time is a measure of the how well energy can be dissipated from the spin system to the lattice; T_2 is a measure of the spin-to-spin effects. The relaxation times can be determined from the behavior of the peak-to-peak amplitude of the (derivative) spectrum h_{pp} and ΔB_{pp} with increasing microwave power as described in [249]. The spin–spin relaxation time can be determined directly from the spectrum before the onset of saturation from the equation

$$T_2 = \frac{2}{\sqrt{3}\gamma}\Delta B_{\text{pp}} \qquad (5.2)$$

where γ is the gyromagnetic ratio. This approach is only strictly correct for homogeneously broadened (Lorentzian lineshape) lines but has been generalized for other lineshapes [249]. Since T_1 and T_2 are sensitive to the presence of time averaged fields which can be produced by the motion of spins through exchange or hopping, their variation from sample to sample can give important information about the interaction of the spins within the defects.

In summary:

(i) g value: Each defect center has a particular g value and as such EPR is able to distinguish between different centers. In the absence of the spin–orbit interaction the g value would be that of the free electron 2.0023 and the deviation of the observed g value from this value is related to the degree of localization of the center.

(ii) Spin concentration: The concentration or density of a defect center can be determined if the spectrum is compared with that of a known standard such as Varian pitch in KCl or DPPH (1,1-dyphenil-1-2-picryl-hydrazil). Care must be taken to account for differences in the conductivity of the underlying substrate so that

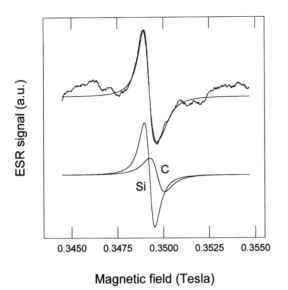

Fig. 67. Room temperature EPR spectrum from an PAC/Si film of 138 nm thickness. The spectrum has been fitted with two lines, one associated with C unpaired electrons and the other associated with defects at or near the film/Si interface. The individual lineshapes of the two defects are also shown.

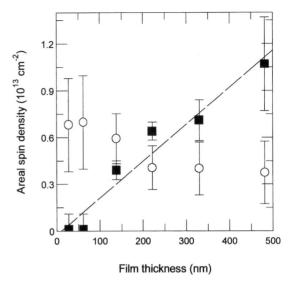

Fig. 68. Variation of the areal spin density with film thickness for the Si related defect (o) and for the carbon related center (■).

that Q-factor of the cavity is not significantly reduced. High resistivity substrates are usually used.

(iii) Linewidth and lineshape: The peak-to-peak linewidth (ΔB_{pp}) and the lineshape can give information about the dynamics of the spin system. Homogeneous broadening arises if the static plus time averaged magnetic fields are the same and will tend to lead to a Lorentzian lineshape. Inhomogeneous broadening will tend to produce a more Gaussian lineshape. A Voigt profile will result from the convolution of a Lorentzian with a Gaussian lineshape.

(iv) The spin–lattice relaxation time T_1 and the spin–spin relaxation time T_2 are sensitive to the fluctuations of fields which can arise from hopping or exchange. Long values of T_1 ($>10^{-4}$ s) indicate that the unpaired electron has a weak interaction with the lattice.

5.3. g Values

Table III shows the measured zero crossing g values, linewidths and shape, spin densities, and hydrogen content for a range of a-C samples. The first point to note is that the smallest defect concentrations ($<10^{18}$ cm^{-3}) are found in PAC films and it is from these samples that the greatest variation in g value from 2.0029 to 2.0045 is observed. In the samples measured by Sadki et al. [252], the large g value of 2.0045 was attributed to the interaction of the unpaired spins with nearby O atoms. The presence of O was confirmed by FTIR analysis of their films. By contrast no evidence of O contamination was observed in the films deposited by PECVD [86] and examined by Collins et al. [217]. They reported high zero crossing g values, g_0, and that g_0, ΔB_{pp}, and lineshape as well as the areal spin concentration, N_A, were dependent upon the film thickness.

Figure 67 shows the observed EPR spectrum from an PAC/Si film of 138 nm thickness. Collins et al. reported that as the film thickness increased the EPR signal became less symmetric and the linewidth larger. They observed that for films below 60 nm thickness g_0 remained constant at about 2.0050(3), ΔB_{pp} at about 0.56 mT, and N_A at about $(5 \pm 2) \times 10^{12}$ cm^{-2}. However, for films greater than about 60 nm thickness g_0 tended to decrease, reaching a value of 2.0046(2) for a film with a thickness of 481 nm. They also showed increases with thickness of ΔB_{pp} and N_A, but in the case of N_A the increase was slower than the increase in the film thickness.

This behavior of the spin density and g_0 on the thickness of film was explained on the assumption that the signal is the superposition of two lines of different g values. The first defect has $g \sim 2.0053$ and has a areal concentration independent of film thickness, and the second has g value ~ 2.0029 whose areal concentration increases linearly as the film thickness increases. The g value of the former center is typical of that found ion implanted Si which is usually attributed to Si related defects [253] and could originate from the film interfacial region and/or the silicon substrate. Spectra recorded from the Si substrate alone produced a weak, symmetric line with $g = 2.0053(4)$ and an average value of $N_A = 2 \times 10^{12}$ cm^{-2}. It was reported that all of the spectra from the PAC/Si samples could be fitted to the superposition of two Lorentzian lines: one with $g = 2.0053(3)$, $\Delta B_{pp} = 0.56$ mT and the second with $g = 2.0029(3)$, $\Delta B_{pp} = 0.83$ mT. The best fit to the measured spectrum using the above parameters for the PAC/Si film of 138 nm thickness is shown in Figure 67. Also shown there are the lineshapes of the two individual defect centers. The values of N_A for both lines as a function of film thickness are shown in Figure 68. The areal spin population of the $g \sim 2.0029$ line is proportional, within error, to the film thickness and implies that it is associated with C unpaired spins in the bulk of the film. From the slope of N_A versus thickness (dashed line) a volume

Table III. Zero Crossing g Values, Peak-to-Peak Linewidths, Spin Densities, and H Content for Different Amorphous Carbon Films

Type of carbon	Zero crossing g value (error)	ΔB_{pp} (mT),	lineshape	Spin density (cm^{-3})	H content (at.%)	Reference
PAC	2.0045	0.8	L+G	1.5×10^{17}	45–50	[252]
	2.0040	0.8	—	7.3×10^{17}	57	[255]
	2.0035	0.76	L	6.0×10^{18}	>22	[257]
	2.0029(3)	0.83	L	2.3×10^{17}	45–50	[217]
	2.0029(2)	—	L	$\sim 10^{18}$	64	[263]
	2.0029(1)	1.5	—	5×10^{17}	—	[277]
DAC	2.0031(2)	—	L	10^{21}	0.26	[263]
	2.003	0.3	L	2.9×10^{19}	—	[273]
	2.0028	0.4	L	4.5×10^{20}	0.25	[252]
	2.0027	0.27	L	2.0×10^{20}	—	[278]
	—	0.3	—	1.6×10^{20}	—	[279]
	2.0036	0.73	L	1.3×10^{20}	—	[255]
TAC	2.0024	0.44	L	10^{20}	—	[260]
TAC	2.0024	0.36	G	10^{20}	—	[260]
TAC	2.0023	0.34		1.4×10^{20} g^{-1}	—	[267]
TAC:H	—	0.18		0.8×10^{20}	0.23	[259]
TAC:H	—	0.25		1.5×10^{20}	0.25	[259]
TAC:H	—	0.40		3.0×10^{20}	0.28	[259]
GAC	2.0024(2)	0.31	L	10^{20}	—	[262]
	2.0027(2)	0.42	L	10^{19}	—	[263]

L = Lorentzian, G = Gaussian.

concentration, N_V, of $(2.3 \pm 0.3) \times 10^{17}$ cm^{-3} is obtained. Figure 68 also shows that the area concentration of the $g = 2.0053$ line is independent of the film thickness and the average value of N_A for this line is $(5 \pm 2) \times 10^{12}$ cm^{-2}. These results indicate that for low defect density polymeric PAC/Si films of thickness less than 200 nm, up to 50% of the paramagnetic defects present are not associated with the C related centers but with Si related centers which originate from the PAC/Si interface region and/or the underlying substrate.

In a study of RF sputtered a-C:H:Si films, DeMartino et al. [254] showed that the observed g value varied from 2.0057, for samples deposited at high SiH$_4$ flow rates, to 2.0027 for low SiH$_4$ flow rates. They also reported that a wide range of physical properties could be obtained from the films by changing the SiH$_4$ flow rates. Films that had graphiticlike characteristics in the absence of SiH$_4$ took more diamondlike properties (high Knoop hardness, low energy gap, and high spin density $\sim 3 \times 10^{20}$ cm^{-3}) with the addition of silane. They noted that the spin density was larger for films possessing the lower g value and attributed this to the fact that addition of silane will tend to form more sp^3 C due to the much larger H presence in the plasma.

5.4. Spin Densities

The volume spin densities, N_s, and ΔB_{pp} for three series of hydrogenated a-C films as a function of negative self-bias [255–257] are shown in Figure 69. In the case of the two studies by

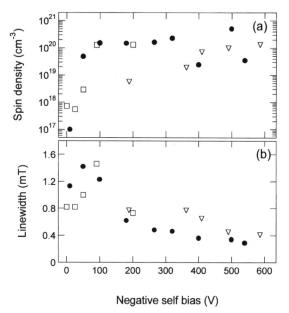

Fig. 69. Variation of the (a) volume spin density and (b) peak-to-peak linewidth with DC negative self-bias for three sets of films from Barklie et al. (●) [256], Ristein et al. (□) [255], and Zeinert et al. (▽) [257].

Ristein et al. [255] and Barklie et al. [256] both sets of films were deposited by RF plasma decomposition of either only CH$_4$ or a mix of CH$_4$ (10%) and Ar (90%), respectively. Films were deposited on the powered electrode in each case and the RF

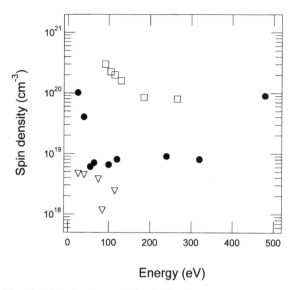

Fig. 70. Variation of volume spin density for three sets of TAC films grown by the FCVA method (●) [258], magnetron sputtering (MS) with Ar-ion plating (▽) [280] and plasma beam source (□) [259].

Gerstner et al. [110] have also used the FCVA technique to prepare TAC and deposited films with substrate biased in the range of 0 to −1750 V. Films deposited on Si substrates with biases between 0 and −500 V readily delaminated due to the high internal stress. Films deposited at higher biases had greater adhesion to the substrate. These films together with their substrates were ground into a power and EPR measurements were made on the mixture. Gerstner et al. reported that the EPR spectrum consisted of two Lorentzian lines, one with $g = 2.0028(1)$, while the other was associated with the Si dangling bonds which was attributed to the substrate being ground into a powder. The linewidth associated with C related defects varied from 0.38 to 0.69 mT, though no clear trend was observed as a function of deposition of bias. For samples which delaminated from the substrate a small reduction in the spin density with bias (5×10^{20} cm^{-3} at zero bias to 2×10^{20} cm^{-3} at −400 V bias) was reported. In contrast to the results of Amaratunga et al. [258] and Schwan et al. [109] no minimum in the EPR signal with bias was reported.

Fusco et al. [260] reported that the EPR signal from TAC grown by FCVA consisted of two centers both, with $g = 2.0024$, but one having a Lorentzian lineshape and the other having a Gaussian lineshape. They interpreted the Lorentzian line as being due to clustered C centers and the Gaussian line as being due to disorder. Since TAC is known to have a sp^2-rich surface layer it is possible that some of the observed EPR signal originates from this layer and not from the bulk of the film. Finally Palinginis et al. [261] have measured defect densities using drive level capacitance profiling. They found in undoped TAC a uniform defect density of $(3.5 \pm 1.0) \times 10^{17}$ cm^{-3} measured at room temperatures. From measurements at different temperatures they were able to account for thermally activate defects and the found that the total defect density is $(6 \pm 1.5) \times 10^{17}$ cm^{-3}. For TAC:N (n-type) they observed an order of magnitude higher defect concentration. Since the EPR density is typically 10^{20} cm^{-3} in TAC they accounted for the difference between the two defect density measurements by noting that EPR measures the neutral centers. Drive level capacitance profiling measurements examined transitions of the type $D^+ \rightarrow D^0 + h$ and $D^0 \rightarrow D^- + h$, though the latter transition is unlikely to occur due to the high correlation energy. Admittance measurements revealed an areal density greater than 2×10^{12} cm^{-2} at the film/Si interface.

Finally Demichelis et al. [262] reported that in sputtered a-C (GAC) that a g value of 2.0024 is obtained. This is much closer to the free electron g value and would imply a high degree of electron delocalization. Fancilulli et al. [263] reported that the best fit to the EPR spectrum came from a superposition of a single Lorentzian and a single Gaussian line. Clearly the different deposition conditions and whether H is present or absent have a major influence on the spin density. By suitable choice of deposition conditions, however, it is possible to reduce this value and yet retain the semiconducting properties of the films.

The results of the EPR measurements show that the lowest spin densities are found in PAC films which usually contain the highest concentration of H. As the deposition energy increases

power was adjusted which resulted in different values of the negative DC self-bias. In the study of Zeinert et al. [257] the films were produced by RF magnetron sputtering of a graphite target in an argon plasma containing 10% H. The results from both sets of PECVD grown films show a similar rapid increase in spin density with bias up to a bias of −100 V and the spin density remains constant at about a value of 10^{20} cm^{-3}. The films produced by magnetron sputtering also show an increase in the spin density but only reach a density of 10^{20} cm^{-3} at a negative self bias of 490 V, much higher than the PECVD grown material. The peak-to-peak linewidth initially increases, reaching a maximum also at about −100 V bias, and then decreases for the PECVD samples whereas the linewidth is initially smaller and then decreases. A detailed discussion of the behavior of the linewidth will be given in the next section.

The spin densities for three sets of TAC(:H) films deposited under different ion energies are shown in Figure 70. In the case of the TAC films grown from the FCVA method by Amaratunga et al. [258] the spin density decreases from about 10^{20} cm^{-3} for C ion energies below 25 eV to about 5×10^{18} cm^{-3} and remains below 10^{19} cm^{-3} for ion energies below 300 eV. The spin density then rises again at higher ion energies. It was reported in this study that N_s was lowest when the sp^3 content was about 60–70%. In the case of the films deposited using DC-biased unbalanced magnetron sputtering in the presence of Ar-ion plating described by Schwan et al. [109], spin densities of less than 5×10^{18} cm^{-3} were measured for a range of Ar ion plating energies. The lowest spin concentration of 1.2×10^{18} cm^{-3} was measured for energy of 85 eV. In the case of TAC:H films deposited using a plasma beam source derived from acetylene [259], the spin density decreased from a value of 3×10^{20} cm^{-3} at a C ion energy of 92 eV to 8×10^{19} cm^{-3} at 266 eV. In these films the H content varied only from 27 at.% (92 eV) to 23 at.% (266 eV).

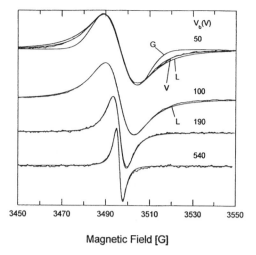

Fig. 71. Room temperature EPR spectra showing the effect of changing the negative self-bias from 50 to 540 V. The smooth lines are computer fits. Only Lorentzian fits are shown for $V_b = 100$, 190, and 540 V. For $V_b = 50$ V, Lorentzian (L) and Gaussian (G) fits are shown but the best fit has a Voigt (V) lineshape. The peak-to-peak linewidths have been normalized. Reproduced with permission from [256], © 2000, American Physical Society.

C–H bonds begin to break and are replaced with C–C and C=C bonds. The lower concentrations of H present and the higher ion energy result in unpaired electron and an increase in the spin density. It is also worth noting that the spin density from TAC films may mainly arise from the sp^2-rich surface layer. Even at these high spin densities ($\sim 10^{20}$ cm^{-3}) only a tiny fraction of the sp^2 states are paramagnetic. If a typical hydrogenated film has a density of 1.8 g cm^{-3}, 40 at.% H, and an sp^2 content of 0.2, then there are over 10^{22} π states cm^{-3}. In the PAC:H films the spin concentration is 1 spin per 10^5 π states. In this way the EPR active centers are the exception rather than the norm.

5.5. Linewidths, Lineshapes, and Relaxation Effects

5.5.1. Motional Effects

As discussed previously while the g values and spin densities arise from "static" properties of the defect centers, the linewidth and shape together with the relaxation times give important information about the "dynamic" response of the defects. Figure 71 shows the spectra from DAC:H/Si samples deposited at room temperature on the driven electrode of capacitively coupled PECVD system described previously [88]. Feed gases of CH$_4$ (10%) and Ar (90%) at a constant pressure of 300 mTorr were used and the RF power was varied such that the negative self-bias was varied from -50 to -540 V [256]. The variations of the spin densities and linewidths with bias have already been shown in Figure 69. As can be seen from Figure 71 increasing the self-bias results in a narrowing of ΔB_{pp} from 1.42 mT (-50 V) to 0.29 mT (-540 V). In addition it was reported that the lineshape changes to one that can be fitted to a Voigt profile for a self-bias of -50 V to one that is close to a Lorentzian lineshape at -100 V. At higher biases all of the lineshapes observed could be fitted to a Lorentzian lineshape. In order to

analyze the changes in the linewidth and lineshape with bias it is worth recalling that a Lorentzian lineshape will result from either dipolar broadening between like spins or from the effects of spin motion such as hopping or exchange. For dipolar broadening in magnetically dilute materials, Abragam [264] has shown that the half-width at half height of the integrated spectrum $\Delta B_{1/2}$ (measured in Gauss) can be related to spin density of N_s measured in cm^{-3} via

$$\frac{1}{2}\Delta B_{1/2} = \frac{2\pi^2}{3\sqrt{3}}g\mu_B N_s \tag{5.1}$$

Recalling that for a Lorentzian line $\Delta B_{pp} = \Delta B_{1/2}/\sqrt{3}$ gives

$$\Delta B_{pp} = 8.12 \times 10^{-20} N_s \tag{5.2}$$

For the three series of samples deposited at different biases V_b discussed in the previous section, Barklie et al., observed a Lorentzian lineshape was for biases of -100 V and greater [256]. At $V_b = -100$ V and the measured values of $\Delta B_{pp} = 1.22$ mT and $N_s = 1.5 \times 10^{20}$ cm^{-3}, Eq. (5.2) gives a predicted linewidth 1.22 mT and indicates that no other factor is required to explain either the linewidth or lineshape for this film. Ristein et al. [255] report that in their DAC:H films a Lorentzian lineshape is observed at a bias of -50 V and greater. At a bias of -100 V, the value of ΔB_{pp} is 1.46 mT and the spin density was measured to be only 1.3×10^{20} cm^{-3}. Equation (5.2) would predict a linewidth of 1.1 mT, slightly smaller than that measured.

For the films measured by Zeinert et al. [257], at a bias of -409 V, $N_s = 7.5 \times 10^{19}$ cm^{-3} and the measured value of ΔB_{pp} is 0.66 mT. Dipolar broadening alone would contribute 0.61 mT. At higher biases in both the sets of samples measured by Barklie et al. and Ristein et al. the spin density increases to about 1.5×10^{20} cm^{-3}. Since the value of N_s does not change significantly in the bias range from -100 to -300 V, Eq. (5.2) would predict that it should remain close to 1.12 mT. However, as seen in Figure 69 there is a continuous decrease in the linewidth in this range of biases and also in the RF sputtered samples at even higher biases. This is indicative that another mechanism is present which is determining the observed linewidth and so such reductions in linewidths are consistent with the presence of motional effects resulting from exchange or electron hopping.

In order to test the hypothesis that motional effects are occurring Barklie et al. measured the spin–lattice T_1 and spin–spin T_2 relaxation times for their samples as a function of bias using the power saturation method [249]. They observed T_1 to decrease from about 1.5×10^{-6} s at $V_b = -100$ V to 3.0×10^{-7} s at the highest bias they investigated (-540 V). Simultaneously there was an increase in T_2 from 4.5×10^{-9} s to 2.0×10^{-8} s over the same bias range, as shown in Figure 72. Using the approach described elsewhere [265], it can be shown that if the dipolar interaction is characterized by a frequency ω_p and a fluctuating magnetic field is characterized by a correlation time τ_c, then narrowing will occur if $\omega_p < 1/\tau_c$. In terms of τ_c, T_1 and T_2

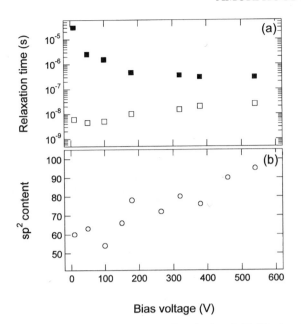

Fig. 72. Variation of the (a) spin lattice T_1 (■) and spin–spin T_2 (□) relaxation times and (b) sp^2 content measured by EELS as a function of negative self-bias.

Table IV. Variation of Spin–Lattice T_1 and Spin–Spin T_2 Relaxation Times for Different Forms of Amorphous Carbon

Type of carbon	T_1 (s)	T_2 (s)	Lorentzian component ΔB_{pp} (mT)	Temperature behavior of ΔB_{pp}	Reference
PAC	3×10^{-4}	6×10^{-9}	1.50	—	[256]
	2×10^{-4}	—	—	decreases	[263]
DAC	2×10^{-7}	—	—	decreases	[263]
	2.5×10^{-6}	5×10^{-9}	1.42	—	[256]
TAC	$<10^{-7}$	1.8×10^{-8}	—	—	[263]
DAC	2×10^{-7}	2.5×10^{-8}	—	—	[256]
	—	4.2×10^{-7}	—	—	

can be expressed as

$$\frac{1}{T_1} = \frac{2\gamma^2 \overline{B^2}}{1 + \omega_0^2 \tau_c^2} \quad \text{and} \quad \frac{1}{T_2} = \gamma^2 \overline{B^2} \tau_c + \frac{\gamma^2 \overline{B^2}}{1 + \omega_0^2 \tau_c^2} \quad (5.3)$$

where $\overline{B^2}$ is the magnitude of the magnetic field associated with fluctuation, g is the gyromagnetic ratio of the electron and $\omega_p = \gamma^2 \overline{B^2}$, and ω_0 is the Larmor frequency. In the case of $\omega_p < \omega_c < \omega_0$ Eq. (5.3) reduces to

$$\frac{1}{T_1} \approx \frac{2\omega_c \omega_p^2}{\omega_0^2} \quad \text{and} \quad \frac{1}{T_2} \approx \frac{\omega_p^2}{\omega_c} \quad (5.4)$$

so that the correlation frequency of the fluctuation can be determined from the relaxation times via

$$\omega_c \approx \omega_0 \left[\frac{T_2}{2T_1} \right]^{1/2} \quad (5.5)$$

For the sample deposited at −540 V the measured relaxation times are $T_1 = 3.0 \times 10^{-7}$ s and $T_2 = 2.0 \times 10^{-8}$ s, resulting in a correlation frequency of 1.1×10^{10} rad s^{-1}. The frequency of the dipolar field ω_p is given by $(\omega_c/T_2)^{1/2}$ and comes to 7.4×10^8 rad s^{-1}. The correlation frequency can thus be seen to be higher than the dipolar frequency with the result that the dipolar line is being averaged out by the effects of the fluctuating field.

Using Eq. (5.3) the magnitude of the fluctuating magnetic field $(\overline{B^2})^{1/2}$ is estimated to be about 4.2 mT. If no fluctuations were occurring then $\frac{1}{2}\Delta B_{pp}$ should be equal to $(\overline{B^2})^{1/2}$; for a value of 4.2 mT a spin density in excess of 3×10^{20} cm^{-3} would be required. Since the measured spin density is only 3.5×10^{19} cm^{-3} it was argued that this is evidence of clustering of spins at this bias and that the observed linewidth narrowing

was a result of the increasing strength of the exchange interaction, rather than electron hopping. This increase in the exchange in interaction is due to greater delocalization of the spin wave functions associated with the increase in the sp^2 content of the films as shown in Figure 72b. The sp^2 content, as measured by EELS, increased from about 60% to close to 100% at the highest biases.

Additional evidence for exchange interaction as the reason for the linewidth narrowing comes from the similar behavior of T_1 (becoming shorter) and T_2 (becoming longer) in solid solutions of DPPH dissolved to different concentrations in polystyrene [266]. This is due to the increasing strength of the exchange interaction though in this case it is as a result of the higher density of paramagnetic centers rather than greater delocalization of the spin [266]. The calculation just discussed can also be used to explain why the static dipolar field can be used for the sample deposited at −100 V. In this sample the measured relaxation times are $T_1 = 1.5 \times 10^{-6}$ s and $T_2 = 4.5 \times 10^{-9}$ s. This results in $\omega_c = 2.4 \times 10^9$ rad s^{-1} and $\omega_p = 7.3 \times 10^8$ rad s^{-1}. In this case ω_c is just fractionally larger than ω_p, so that the fluctuating frequency is not fast enough to average out the static linewidth. In this sample little exchange is occurring and the linewidth is determined by the static linewidth from dipolar broadening. Studies of relaxation times in types of film are one clear way to establish the presence of exchange effects. Table IV lists of the other studies of relaxation times and from the limited number of measurements there is some good agreement between different research groups for similar types of a-C.

5.5.2. Variation of Linewidth with Temperature

Additional information concerning dynamic effects can be found in the dependence of the linewidth with temperature. Figure 73 shows the variation of ΔB_{pp} with temperature for five series of different a-C related films [262, 267]. In the case of DAC, PAC, P doped TAC, and GAC:H, the linewidth decreases with increasing temperature. For example Demichelis et al. [262] showed that in highly sp^2 GAC films, ΔB_{pp} decreased from 0.54 mT at 10 K to 0.30 mT at 470 K. In P doped TAC, which is H free, Golzan et al. showed that ΔB_{pp}

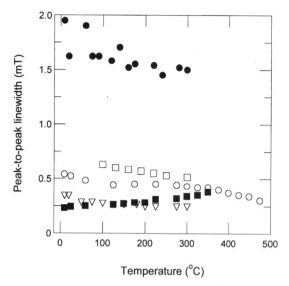

Fig. 73. Variation of the peak-to-peak linewidth ΔB_{pp} as a function of temperature for DAC ($\triangledown$), polymeric PAC ($\bullet$), P doped TAC ($\square$), graphiticlike carbon ($\circ$), and sputtered a-C ($\blacksquare$) [262, 267]. Only in the case sputtered films does ΔB_{pp} increase with increasing temperature.

decreased from 0.63 mT at 100 K to 0.52 mT at 300 K [267]. By contrast, in sputtered a-C films Demichelis et al. showed that ΔB_{pp} increased with temperature from 0.24 mT (10 K) to 0.35 mT (370 K). They observed that temperature dependence of the linewidth could be related to the conductivity $\sigma(T)$ via

$$\Delta B_{pp}(T) = \Delta B_{pp}(0) + C[\sigma(T)]^n \qquad (5.6)$$

where C is a constant and $\Delta B_{pp}(0)$ is the linewidth at 0 K. For the sputtered a-C films the best fit gave $\Delta B_{pp}(0) = 0.23$ mT and $n = 0.32$. Some care must be used in the use of Eq. (5.6). It was originally used by Voget-Grote et al. [268] to study the effects of temperature on the linewidth of the dangling bond center (D center, $g = 2.0055$) in evaporated a-Si and to relate it to the DC conductivity; they observed a decrease in the exponent n from 1 to 0.5 with increasing conductivity. Movagher et al. [269] interpreted these results as being due to effects of locally fluctuating magnetic fields which originated from the hopping motion of the electrons and are equivalent to field fluctuation with time at the site of the electrons.

By contrast, in the study of B and P doped a-Si:H, Dersch et al. [270] observed that the exponent n increased with doping. They interpreted their results as due to enhanced spin–lattice relaxation of the localized electrons due to exchange coupling with rapidly relaxing carriers in extended states. The values of T_1 have been measured at both room temperature (RT) and 77 K for GAC and for DAC:H. It was found that T_1 shortened from 1.6×10^{-6} s at 77 K to 4.2×10^{-7} s at RT for GAC [262]. In the case of the DAC:H film T_1 reduced from 2.0×10^{-5} s to 4.0×10^{-6} s at the two measuring temperatures. A reduction in T_1 in GAC as the temperature was increased and would result in the linewidth increasing in temperature, which was observed. However, T_1 also decreased in the DAC film and yet this was accompanied by a decrease of ΔB_{pp} with temperature. The rea-

son for this different behavior is not yet clear but may be a result of the fact that in DAC, T_1 is still significantly longer than T_2 such that the linewidth is still determined solely by T_2.

5.5.3. Variation of Linewidth for PAC Films

While a reduction in ΔB_{pp} is usually associated with motional effects, in particular the rapid exchange of spins, a reduction in the linewidth reported by Barklie et al. [256] and Ristein et al. [255] (Fig. 70) for films deposited at low biases was accompanied by a change in the lineshape. Barklie et al. reported that for the film deposited at -50 V bias, the lineshape could be fitted with either a Voigt profile or a superposition of a single Lorentzian and a single Gaussian. Ristein et al. reported that for biases greater than -50 V the lineshape is fully Lorentzian but did not report on the lineshapes for films they grew at lower biases.

Table V shows some of the measured parameters (spin density, linewidth, and H content) of two series of PAC films deposited with lower negative self-biases. It can be immediately seen that by raising the self-bias the spin density increases from 7.3×10^{17} cm^{-3} at zero bias to over 1.3×10^{20} cm^{-3} at -90 V bias. There is also a general increase in the measured value of ΔB_{pp} as the bias is raised. Table V also shows the predicted value of the linewidth, using Eq. (5.2), assuming that the only source of broadening was dipolar broadening. As can be seen the calculated values differ considerably from the measured values except at the higher biases of -90 V and above. Furthermore, dipolar broadening would produce a Lorentzian lineshape whereas both Voigt and/or Lorentzian and Gaussian lines have been reported for some of these low bias samples. The fact that non-Lorentzian lineshape may be present points to a source of inhomogeneous broadening. One of the principal sources of such broadening may come from the (unresolved) hyperfine interaction from the H^1 nucleus. In this way the lineshape is determined by the relative contribution from the (Gaussian) unresolved hyperfine interaction (UHI) to the (Lorentzian) dipolar contribution. The UHI will increase with H content and the dipolar contribution will reduce at lower spin concentrations. Barklie et al. have proposed that the Gaussian contribution from the UHI can be related to the H atomic concentration via the empirical relationship

$$\Delta B_{pp}^{UHI}(G) = (0.18 \pm 0.05) \times H \text{ (at.\%)} \qquad (5.7)$$

Values for linewidth predicted by Eq. (5.7) are shown in Table V. Some care must be taken in the use of this equation as it is based upon just three measurements so there may be a distribution of hyperfine coupling constants which would affect the hyperfine fields. However, Eq. (5.7) does show that it is quantitatively possible to discuss the measured linewidth at low biases in terms of the contribution from the hyperfine interaction from H.

The effects of the unresolved hyperfine interaction have also been attributed to the linewidth behavior with H content in the DC magnetron sputtered DAC:H film studied by Hoinkis et al. [271]. The EPR spectra, measured under vacuum, consisted of a

Table V. Measured Spin Density and Peak-to-Peak Linewidth, Predicted Linewidths Assuming Dipolar and Unresolved Hyperfine Interactions for Low Bias Films

V_{bias} (V)	Spin density (cm^{-3})	$\Delta B_{pp}^{measured}$ (mT)	$\Delta B_{pp}^{dipolar}$ (mT)	H content (at.%)	ΔB_{pp}^{UHI} (mT)	Reference
0	7.3×10^{17}	0.82	0.006	57	1.03	[255]
−10	3.0×10^{17}	1.13	0.0025	50	0.90	[256]
−25	5.5×10^{17}	0.82	0.0045	—	—	[255]
−50	2.9×10^{18}	1.00	0.024	—	—	[255]
	4.9×10^{19}	1.54	0.40	25	0.45	[256]
−90	1.3×10^{20}	1.46	1.06	—	—	[255]
−200	1.33×10^{20}	0.73	1.08	40	0.72	[255]

single Lorentzian line with $g = 2.002$ with no hyperfine structure present. They observed a decrease in ΔB_{pp} with H content from over 7 mT at H contents of less than 6 at.% to 0.28 mT at 24 at.%. Experiments performed on a sample with a 20 at.% H content at different microwave frequencies revealed that the linewidth is independent of frequency. This result indicates that the linewidth is not significantly affected by a distribution of g values or any g value anisotropy. At higher H contents the linewidth increased to 0.8 mT (36 at.%) and to 1.0 mT (39 at.%). The measured spin density decreased monotonically with H content and ranged from 8×10^{20} cm^{-3} at low H contents to 0.5×10^{20} cm^{-3} at 39 at.%. They interpreted the increase in ΔB_{pp} for a H content greater than 24 at.% as being due to UHI.

An alternative explanation has been proposed in which the observed linewidth of 0.28 mT at 24 at.% H is a result of exchange narrowing due to a high sp^2 content found in sputtered a-C [256]. The dipolar contribution to the linewidth for this sample, for which $N_s = 4.5 \times 10^{20}$ cm^{-3}, would be, via Eq. (5.2), 3.7 mT and the UHI would produce a linewidth of 0.43 mT, the combination of which, assuming a Voigt profile, far exceeds the observed value of 0.28 mT. Similar analysis for the film with 36 at.% H for which the measured $\Delta B_{pp} = 0.8$ mT and $N_s = 1.5 \times 10^{20}$ cm^{-3} reveals that $\Delta B_{pp}^{dipolar} = 1.22$ mT and $\Delta B_{pp}^{UHI} = 0.65$ mT and would result in a Voigt linewidth of 1.5 mT, which is still in excess of the observed value. Barklie et al. argue that the effects of exchange are still occurring but are reduced since the higher H content reduces the sp^2 content. At 39 at.% H, the measured values of ΔB_{pp} and N_s are 1.0 mT and 1.5×10^{20} cm^{-3} respectively. The predicted values of $\Delta B_{pp}^{dipolar}$ and ΔB_{pp}^{UHI} are 0.41 and 0.70 mT and the corresponding total Voigt linewidth would then be 0.9 ± 0.2 mT, which is close to the observed value. In this way the linewidth would be determined by a mixtures of exchange narrowing and unresolved hyperfine interactions and would produce a Voigt profile. As the H content is reduced to about 24 at.% contributions to the width and shape are dipolar and exchange interactions both of which will produce a Lorentzian lineshape. For lower H contents 2–20 at.% of the linewidth increases as the H content reduces. Hoinkis et al. argue that the simultaneous reduction in the film resistivity by about two orders of magnitude to a value under

10 Ω cm results in reductions in the spin relaxation times and consequent broadening of the lineshape.

Most of the measurements made by Hoinkis et al. were made under vacuum (10^{-6} Torr). Samples which were measured in air had slightly narrower linewidths and higher signal intensities. In order to examine the effects of the presence of O_2 gas, Hoinkis et al. backfilled their cavity with O_2. They observed an increase in linewidth but a decrease in the signal intensity such that the spin density was constant. This broadening effect was found to be reversible in that after the O_2 in the chamber was removed the lineshape narrowed to approximately the same linewidth that was observed in vacuum before the introduction of O_2. Other gases such as N_2, Ar, or a N_2/H_2 mixtures (all of which are diamagnetic) did not produce a significant change to when the sample was in vacuum. The reversible O_2-related linewidth broadening was attributed to O_2 permeating into film but without the formation of strong chemical bonds. Since the diamagnetic gases do not affect the linewidth, the line broadening was attributed the interaction of C unpaired electrons with the unpaired spins of O_2. No report of a variation in the g value was made and the EPR spectrum from the sample with 25 at.% H and exposed to 2 atmospheres of O_2 reveals a nearly Lorentzian lineshape. No change in g value was reported.

5.6. Effects of Nitrogenation

Figure 74 show the spin density for four different types of amorphous carbon film as a function of N content. The N content in each case was determined by either Rutherford backscattering spectrometry and/or elastic recoil detection analysis. The series of films examined included three sets of DAC:H films by Amir et al. [272], Silva et al. [29], and Bhattacharyya et al. [273] and one set of PAC films by Collins et al. [217]. In the case of the DAC:H films examined by Silva et al., the spin density decreases immediately with the inclusion of N from 4×10^{20} cm^{-3} (0 at.% N) to 9×10^{17} cm^{-3} (15 at.% N). They also reported a decrease in the linewidth from 0.50 to 0.15 mT over the same N concentration range. In the absence of exchange the dipolar contribution to the linewidth for a spin concentration of 4×10^{20} cm^{-3} would, via Eq. (5.2), predict a linewidth of 3.3 mT. Since this is considerably larger than the

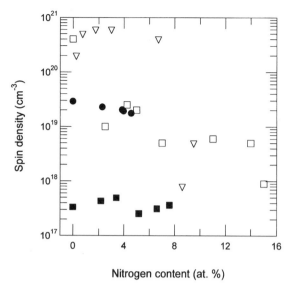

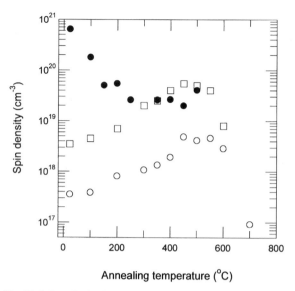

Fig. 74.   Variation of spin density with N content for DAC films by Amir et al. ($\triangledown$) [272], Silva et al. ($\square$) [29], and Bhattacharyya et al. ($\bullet$) [273], and one set of polymeric type by Collins et al. ($\blacksquare$) [217].

Fig. 75.   Variation of spin density with annealing temperature for TAC:H deposited using a plasma beam source using methane ($\bullet$), DAC films deposited at -60 V bias using a magnetically confined plasma containing 1 at.% N ($\square$), and a set of polymeric films grown by RF PECVD ($\circ$).

measured linewidth, some exchange narrowing is believed to be occurring in these samples. Since the addition of N in this film has also shown to increase the sp^2 content, which would result in greater wavefunction overlap, the reduction in the linewidth due to N is most likely an exchange effect.

In both the series of DAC samples measured by Amir et al. and by Bhattacharyya et al. reductions in the spin density also occurred but in the case of Amir et al., reduction in the defect density below 10^{20} cm^{-3} only occurred when the N content reached about 8.6 at.% N. The N content in the films studied by Bhattacharyya et al. never exceeded about 4.6 at.% but even in this limited range a reduction of spin density occurred from about 1.75×10^{20} cm^{-3} (0 at.% N) to 0.6×10^{20} cm^{-3} (4.6 at.% N). They recorded a single Lorentzian line for N-free and low N containing films with a g value of about 2.003 but from the film with 4.6 at.% N some hyperfine structure was observed. It was proposed that $\sim1\%$ of the spins are in the vicinity of the N atoms but are not localized upon them. They also observed that the addition of N resulted in an increase in ΔB_{pp} from 0.03 mT (0 at.% N) rising slowly to 0.06 mT (4.1 at.% N) then rising sharply to 0.16 mT at the highest N content. They attributed the increase in the linewidth as due to a reduction in the relaxation time. However, they did not indicate whether this the T_1 or T_2 relaxation time; nor did they give values for these two quantities. Finally, in the case of the PAC films, the addition of N appears to have no effect of the C related defect density which stays around 3.4×10^{17} cm^{-3}. This behavior is probably attributed to little defect passivation occurring in these low defect films. For GAC:N films produced by RF sputtering of a graphite target in an Ar/N$_2$ atmosphere, Demichelis et al. [274] reported that as the N partial pressure was increased from 1.0 to 4.7 Pa, the spin density decreased from 1.0×10^{19} to 1.1×10^{18} cm^{-3}. This was accompanied by an increase in the optical gap from 0.58 to 1.0 eV. (At a partial pressure of 4.7 Pa, the N content was

estimated to be about 30 at.%.) The measured g values were in the range of 2.0027–2.0032; no values for the linewidths were given. They argued that the reduction in the spin density is due to reduction in the disorder in the sp^2 phase of the material.

In addition to the incorporation of N as a n-type dopant, Golzan et al. [267] have studied the effects of P as a dopant source in FCVA grown TAC. They observed a single Lorentzian line with $g = 2.0023$ and $\Delta B_{pp} = 0.52$ mT at room temperature. The measured spin density was 3.3×10^{18} g^{-1}, nearly two orders of magnitude lower than undoped TAC. They interpreted the reduction of the spin density as being partly due to the pairing of the π orbital of the sp^2 C with the π orbital of P and due to passivation of dangling bonds.

The addition of N in most types of film results in a reduction of the spin density and would normally be interpreted in terms of the passivation of dangling bonds. This is almost certainly the case in the high defect density samples where the reduction of the spin density was accompanied by an initial increase in the Tauc gap (removal of gap states) followed by a reduction in the Tauc gap possibly due to graphization [29]. However, the effects of N may also be due to movement of the Fermi level with the result that that some paramagnetic states become diamagnetic. The fact that the spin concentration remains constant for the PAC films indicates that this may be the lower limit of the defect density.

5.7. Effects of Annealing

The effects of annealing on the spin density of three different series of films are shown in Figure 75. The four sets of films examined were TAC:H deposited using a plasma beam source with methane as feed gas, measured by Conway et al. [167], DAC:N deposited at -60 V bias using a magnetically confined

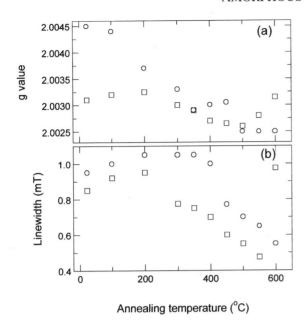

Fig. 76. Variation of (a) g value and (b) peak-to-peak linewidth for a DAC (□) and a polymeric (○) film with annealing temperature.

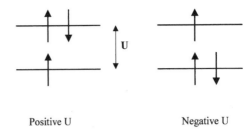

Fig. 77. The arrangement of electrons in orbitals for both positive and negatively correlation energies U for singly and doubly occupied orbitals.

plasma [88] which contained 1 at.% N, measured by Barklie et al. [275], and a set of PAC films grown by RF PECVD, measured by Collins et al. [217]. In the case of the TAC:H films the defect density reduces from 6.4×10^{20} cm^{-3} in as-deposited films to 2.6×10^{19} cm^{-3} for films annealed at 250°C for 30 min. Above this anneal temperature the spin density remains approximately constant. They did not give either g values or linewidths and whether there was any variation in these quantities with annealing. In the case of the DAC:N films, the samples were step annealed in flowing Ar at 10 min intervals. As can be seen from Figure 75, N_s increases continuously, reaching a maximum of 5×10^{19} cm^{-3} at around 450°C above which it drops sharply to 8×10^{18} cm^{-3} at 600°C. A similar trend was also observed in the PAC films which in which the spin density increased from 3.5×10^{17} cm^{-3} in as-deposited films to a value of 4×10^{18} cm^{-3} also at about 450°C. At higher anneal temperatures the spin density decreased and at 700°C was measured to be 9×10^{16} cm^{-3}.

Figure 76 shows that for both the DAC and PAC films the g value and the linewidth depend upon the anneal temperature. For both sets of films, a similar trend is observed in that the g value converges to around 2.0025 at anneal temperatures around 500–550°C. In addition the linewidth initially increases, reaching a maximum at about 200–300°C, and then continuously rises to an anneal temperature of 550°C. These reported trends of the spin density, g value, and linewidth can be explained by the loss of H and a subsequent increase in the amount of sp^2 C present. For temperatures above 400°C delocalization of the π electrons can occur with a subsequent shift of the g value to the free g value of 2.0023.

The rapid decrease in the spin density above 550–600°C is attributed to the removal of unpaired electrons through the formation of large graphitic sheets and a rise in the linewidth of

the DAC:H:N film at 600°C to a shortening of the spin lattice relaxation time due to interaction with the conduction electrons which are present if the graphitic sheets become parallel. Sadki et al. [252] reported a similar behavior in their annealing study of PAC films. As-deposited films had a spin density of 1.5×10^{17} cm^{-3}. Above an annealing temperature of 250°C they were unable to observe the EPR signal which implied that the spin density was below 10^{16} cm^{-3}. It was interpreted that the FTIR signal from these films annealed above 400°C shows the presence of aromatic rings and the C=C absorption band becomes very intense. They argued that the polymeric films consist of a small number of π delocalized clusters.

5.8. Correlation Energies and Photoyield Measurements

Apart from EPR, there are a number of other techniques that can give information about defects in a-C based materials but they may not be confined to just paramagnetically active states. EPR measurements give information concerning the density of singly occupied states. In the single electron approximation the density of states will be filled pairwise by electrons in order of the orbital energy. Within this approximation no EPR signals should be visible. In order to explain the presence of the EPR signal observed in many different firms of a-C, correlation effects which prevent the spin pairing from occurring need to be considered. The correlation energy U is regarded as the difference between the Coulomb repulsion energy of two electrons in the same orbital and the energy gained by lattice relaxation at the defect site when the second electron is added to the orbital. Positive and negative U centers have different properties. In the notation adopted from a-Si technology, consider a defect which is neutral when it contains a single electron, labelled D^0. The other two charge states occur when it is empty, D$^+$, and doubly occupied, D$^-$. Figure 77 shows the arrangement of electrons in orbitals for both positive and negatively U. If U is positive all the defect states will be single occupied and the sample will exhibit paramagnetism. By contrast no EPR will be visible from a negative U system and the Fermi level will be heavily pinned. Negative U centers are believed to be responsible for the absence of any EPR signal from amorphous chalcogenides [163]. Only the regime from $E_F - U$ to E_F will be singly occupied. Below $E_F - U$ the states are doubly occupied and above E_F the states will be empty.

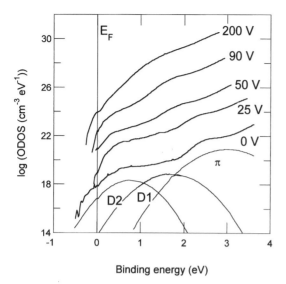

Fig. 78. ODOS with respect to the Fermi level for a series of *a*-C:H samples as a function of bias as determined from photoelectron yield spectroscopy. The curves are labelled with the self-bias which were prepared and are offset for clarity. For the polymerlike 0 V sample the fit of the three Gaussian bands D2, D1, and π to the density of states is also shown. The vertical bars in the spectra indicate which binding energy states have to be singly occupied to account for the measured spin densities. The labels at the bars are the corresponding lower limits for the correlation energies (in eV). Reproduced with permission from [255], © 1995, ???.

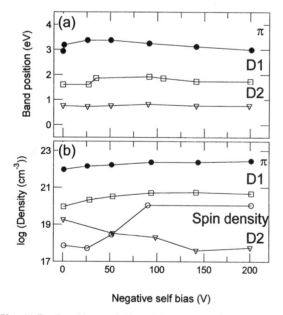

Fig. 79. (a) Band positions and (b) total integrated electron density for the three Gaussian bands fitted to the ODOS spectra in Figure 78. In addition, the EPR densities of the of samples are shown in (b). Reproduced with permission from [255], © 1995, ???.

One of the key parameters is the location of the DOS at the Fermi level E_F. This can be measured in a number of ways using photoelectron yield spectroscopy [255] and by electrical techniques such as by space charge limited current as discussed previously and by junction capacitance techniques such a drive level capacitance profiling [261]. Ristein et al. have

used photoelectron yield spectroscopy to determine the occupied density of states (ODOS) for a series of *a*-C:H samples as a function of bias as illustrated in Figure 78. In the case of the sample deposited at 0 V bias, which produced a PAC film, two defect bands labelled D1 and D2 and the π band centered at 3.0 eV have been fitted. All three bands were fitted using Gaussian functions. The peak DOS was measured to be 1.2×10^{22} cm^{-3} eV^{-1} with a full width at half maximum of 1.2 eV. According to Ristein et al. the two defect bands should be considered as a pure parameterization of the observed DOS. Extrapolating to the Fermi level the measured DOS, $D(E_F)$, is 1×10^{18} cm^{-3} eV^{-1}. As the bias voltage is increased the spectra can be seen to change and for the sample deposited at -200 V bias (a DAC film) the $D(E_F)$ is of the order of only 10^{16} cm^{-3} eV^{-1}. The variations of the band positions and electron densities for the three Gaussians fitted to the ODOS are shown in Figure 79. Also included are the EPR spin densities discussed previously. Recalling that states will remain singly occupied, and therefore paramagnetic, in the energy range from $E_F - U$ to E_F it is possible to estimate the correlation energy using the data in Figure 79 and the defect density. Integrating the DOS over an energy interval equivalent to U, to give the measured spin densities, will provide a lower limit on the value of U. Ristein et al. infer a value of 0.2 eV for the PAC sample. For the DAC sample deposited at -200 V bias the higher spin densities of 10^{20} cm^{-3}, coupled with the fairly low DOS near the Fermi level, mean that value of U must be at least 1.5 eV. Such a large correlation energy may either point to a high degree of localization of the gap states and/or to a strong suppression of lattice relaxation. Finally, it should be noted that surface band bending may also contribute to the value of U though its exact role has yet to be determined.

An attempt to estimate the correlation energy by examining the density of states inferred from optical absorption measurements and spin density measurements has been reported by Zeinert et al. [257]. Photothermal deflection spectroscopy (PDS), transmission spectroscopy in the visible and UV regions, and EPR measurements were made on RF magnetron sputtered films using a graphite target in an Ar plasma (90%) containing hydrogen (10%). The spin densities for these films are reported in Figure 79. By assuming that the DOS can be approximated by a Gaussian function of the form

$$G(E_F) = 2G_{\max} \exp\left(\frac{E_G^2}{2s^2}\right) \qquad (5.8)$$

values of E_G and s could be determined from the absorption spectra. The value of $G_{\max}$ was approximated from the number of π states via

$$N_\pi = s G_{\max} \qquad (5.9)$$

Zeinert et al. assumed that for a density of 1.5 g cm^{-3} and 50% sp^2 coordination that $N_\pi \sim 4 \times 10^{22}$ cm^{-3} and showed that as the optical gap increased from 1.6 to 2.5 eV the DOS decreased from 3×10^{20} to 3.5×10^{19} cm^{-3} eV^{-1}. These values are higher than those Ristein et al. quoted previously. Using the measured spin densities, values of U for the PAC films were found to 0.14 eV, similar to that reported by [255]. In the case of the DAC films deposited at higher bias film a value of U of 0.25 eV was reported. This is considerably less than that reported in the DAC films of Ristein et al. Zeinert et al., however, comment that the analysis of the optical absorption may overestimate the DOS.

5.9. Defects and Nonradiative Recombination

Many of the features of the photoluminescence from a-C have been discussed in Section 4.3. In this part of the review we shall concentrate solely on those factors which influence the PL efficiency from different types of films. In a-Si:H photon absorption above the optical gap results in an electron–hole pair that can fall into the tail states via thermalization. This may result in the carriers diffusing apart from one another. It is generally accepted that in a-Si:H the dangling bond (D-center) acts as a nonradiative center and if a carrier meets such a defect nonradiative recombination will occur. However, due to the more complicated defect structures that are present in the different types of amorphous carbon, in particular the different amount of sp^2 to sp^3 C as well as different sizes of the band tails in a-C:H, the PL mechanism may be more complicated. In the model proposed by Robertson [276] the PL is believed to originate from sp^2 clusters and the PL efficiency depends upon the defect density, so nonradiative recombination will occur if the e–h pair is generated within the tunnelling capture radius R_c of a nonradiative center. The relative quantum efficiency can be expressed as

$$\eta = \eta_0 \left[\frac{-4\pi}{3} R_c^3 N_s \right] \qquad (5.10)$$

where N_s is the spin density. This is the same expression used by Street for the case of a-Si:H in which the value of R_c ranged from 100 to 120 Å. In this description quenching occurs if a carrier tunnels out of the excited cluster to a defect center where it recombines. The decay of a tail state from any cluster will only occur in the sp^3 matrix, which is a fraction $(1 - z)$ of the total volume. Silva et al. [88] have modified Eq. (5.10) to take account of the relative amount of sp^2 bonding in the films such that

$$\eta = \eta_0 \left[\frac{-4\pi}{3} R_c^3 \frac{N_s}{1 - z} \right] \qquad (5.11)$$

The fraction of sp^2 bonding in the film is represented by z. Higher values of z indicate a greater sp^2 content which would result in an increase in the sp^2 cluster size. Figure 80 shows the fit of relative PL intensity from a number of films against the reduced defect density. From fitting the data to Eq. (5.11) capture radii of 10 and 15 Å: have been determined. This much reduced capture radius in comparison to a-Si:H was explained on the grounds that the reduced Bohr radius is a result of a higher Coulombic attraction of the e–h pairs.

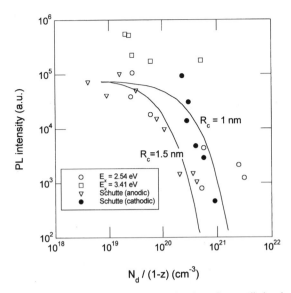

Fig. 80. Photoluminescence efficiency as a function of non radiative density. The solid lines are fits using Eq. (5.11) for different carrier capture radii as indicated. Reproduced with permission from [88], © 1996, ???.

5.10. Summary

In conclusion EPR has been used to examine a wide variety of amorphous carbon films. Measured values for C related centers are usually in 2.0026–2.0030 range, though in the case of GAC values closer to 2.0024 have also been reported. This is almost certainly due to a high degree of electron delocalization in a sp^2 graphitic phase. The spin density varies from 10^{17} cm^{-3} for soft PAC films up to 10^{20} cm^{-3} for harder DAC and TAC films. Addition of H or N can, for some films, reduce the spin density but usually by no more than one order of magnitude. Films deposited at higher bias or ion energy conditions tend to exhibit Lorentzian lineshapes which can be attributed to dipolar broadening, and a reduction in the peak-to-peak linewidth is indicative of exchange narrowing. Evidence for this comes from measured spin–lattice and spin–spin relaxation times. Films deposited at lower biases tend to have non-Lorentzian lineshapes and this is due to unresolved hyperfine interactions from H[1]. Annealing of samples does result in reduction of the spin density but at extreme temperatures (650°C) may result in the formation of large graphitic layers.

6. ELECTRICAL PROPERTIES OF AMORPHOUS CARBON

6.1. Conduction and Amorphous Semiconductors

Amorphous carbon possesses many attractive benefits as a large-area semiconducting material. The deposition is generally carried out at room temperature, which allows the use of plastic substrates. Furthermore, the films possess a controllable optical bandgap in the range of 1–4 eV [281], high electrical breakdown strength of the order of 10^6 V cm^{-1} [35], and low dielectric constant which makes a-C attractive as a high

ENERGY

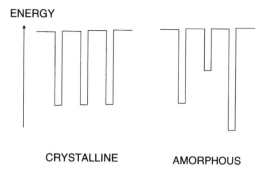

CRYSTALLINE   AMORPHOUS

Fig. 81.   Anderson model for the variation of the energy of potential wells in an amorphous material.

power semiconductor material. In this part of this review, first the generalized theory of conduction in amorphous materials will be described. Second, the electronic properties of the various types of a-C thin film will be discussed, with properties subdivided into two broad areas, the electronic structure of the intrinsic material, and n- and p-type "doping" using *in-situ* addition of precursors during deposition. Furthermore, these areas may be further subdivided into the various types of amorphous carbon which are commonly grown, namely PAC, DAC, TAC, and GAC. The electronic properties of each type of a-C will be commented upon and contrasted.

6.2. Electronic Band Structure of Amorphous Materials

Several texts describe the electronic structure and properties of amorphous semiconductors. The original models introduced by Mott and Davis [282] showed that amorphous materials exhibited short and medium range order but no long range order. Free electron theory shows that the bandgap of a material is caused by its crystalline order, as shown by the Anderson model of the potential wells for an amorphous semiconductor in Figure 81. In the case of a crystalline semiconductor there is a periodic potential of the ordered crystal. This leads to a constant phase relation between the wavefunction at different lattice sites. Therefore, the wavefunction has a well-defined k value defined by the $E - k$ relationship. In the case of an amorphous material this relationship breaks down, as the potential of the lattice is nonperiodic.

For a crystalline semiconductor, there exists a bandgap, caused by the broadening of the individual electronic states into well-defined bands, i.e., an occupied valence band that is separated from an empty conduction band by a gap that is usually less than 3 eV. This is due to the formation of standing waves being set up in the periodic lattice as a result of Bragg's diffraction theory, which results in the formation of allowed and forbidden bands in the $E - k$ diagram. However, in the case of an amorphous material, variations in the periodic potential caused by changes in bond angle and length result in the sharp band edges found in crystalline semiconductors to be replaced by band tails. Amorphous materials do possess a bandgap, though, as these bands are mostly influenced by short range order.

In amorphous films there are states within the crystalline bandgap, i.e., between E_V and E_C as shown in Figure 82. However, these states are a consequence of the amorphous nature of the solid and hence are not extended throughout the solid. Therefore, states in the crystalline valence and conduction bands are known as extended states and states found within E_C and E_V in the amorphous material are known as tail states that may or may not be localized. The mobility of carriers in localized states is much lower than in extended states. The E_C and E_V lines shown in Figure 82 are known as the mobility edges. Also, the existence of trapping centers and defects introduce states at the middle of the bandgap that tend to pin the Fermi level. The tail states are crucial in the conduction process as even a small concentration can have significant influence on the electronic properties. In general for a semiconductor the solution to Schrödinger's equation gives rise to a wavefunction for electronic states that is solved using the Bloch theorem. There is a constant phase difference between wavefunctions at different lattice sites due to the periodicity in crystalline semiconductors which have a k-space that extends throughout the crystal.

In an amorphous material the Bloch theorem can only be used with a varying nonperiodic potential function $V(r)$, whose disorder potential δV, though small, will tend to scatter electrons from one atom to another. Therefore, phase coherence is lost due to the disorder and as a result of the uncertainty principle there is a loss of momentum or k-conservation ($\Delta k \sim h/\Delta x \sim h/a_0$) in electronic transitions due to the uncertainty in the interatomic (Δx) or scattering length a_0. The loss of k-conservation due to disorder gives rise to a number of very important properties in amorphous materials and emphasizes the importance of the spatial location of the carriers which gives rise to the density of states. Electron and hole effective masses must be redefined as either tunnelling masses [283] obtained empirically or by simulation, as the usual curvature of the $E - k$ diagram cannot be used. Also, due to the disorder the carrier mobilities are decreased significantly due to frequent scattering. Most importantly due to the lack of momentum conservation there is no distinction between direct and indirect bandgaps in amorphous materials, with transitions allowed to occur whenever there is an overlap of states in real space.

6.3. Comparison with Amorphous Silicon

Amorphous silicon was first produced using evaporation or sputtering, and it was found that the DOS in the mobility gap was very high, resulting in the pinning of the Fermi level which prevented the doping of the material [163]. It was then found that amorphous silicon produced by the RF glow discharge of silane (SiH_4) had a reduced midgap DOS, allowing effective electronic doping by Group III or Group V elements. It was suggested that hydrogen had the ability to bond to the dangling bonds and hence reduce the density of states in the middle of the gap. However, the optimum atomic percentage of hydrogen was experimentally found to be 10%, which was at least two orders of magnitude higher than the amount required to satisfy all of the dangling bonds. Therefore, hydrogen also remained

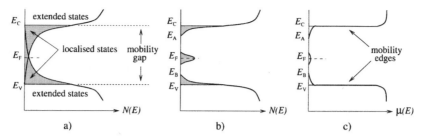

Fig. 82. Schematic representations of the density of states for an amorphous material.

in molecular form in a-Si:H and the material was in fact an alloy. Disadvantages included instability problems due to the presence of hydrogen.

Intrinsic a-Si:H was found to be slightly n-type [163]. Therefore, the addition of boron to the film had the effect of initially lowering and then increasing the conductivity as the Fermi level moved through the middle of the band gap. Activation energy measurements showed an increase followed by a decrease for the same reason. The doping mechanism was not thought to be able to be simply substitutional because of the 8-N rule [282], which states that all atoms (including dopants) in an amorphous network bond with their optimal valency. However, experimental evidence did show that substitutional doping was occurring for a-Si:H. Furthermore, the DOS was shown experimentally to increase with doping. This was not shown by EPR as the movement of the Fermi level by doping caused defects to become doubly occupied and hence nonparamagnetic, but it was shown in light-induced EPR.

6.4. Generalized Conduction in Amorphous Materials: Low Field Conduction

Most authors cite the measurements of conductivity and activation energy as being paramount in describing the electrical characteristics of amorphous carbon. These are commonly taken by measuring the resistance of the film by applying a voltage which results in a field strength of around 10^3 V cm^{-1}. Generally, at this field the current is shown to vary linearly with voltage. Also, this degree of electric field is applicable for planar (gap-cell) contacts onto a thin film, as amorphous carbon is typically grown in the thickness range of 50–300 nm, which results in a high electrical field if the electrical characteristics are measured in a sandwich structure with any appreciable voltage.

Due to the presence of localized states around the Fermi level and the band tails, electronic transport may occur in different regions of the DOS, as described below [282, 284]. These correspond to conduction in four different regimes.

(a) Transport by carriers excited beyond the mobility edges into extended states at E_C or E_V, leading to a straight-line dependence in the $\ln \sigma$ versus $1/T$ characteristic,

$$\sigma = \sigma_{\min} \exp\left[-\frac{E_C - E_F}{kT}\right] \qquad (6.1)$$

where $\sigma_{\min}$ is related to the extended state mobility and the density of states at the conduction band edge by $\sigma_{\min} = \mu_{\text{ext}} e N(E_C) kT$.

(b) Transport by carriers excited to localized states at the band edges and then hopping at the band tail, also leading to a straight-line $\ln \sigma$ versus $1/T$ dependence,

$$\sigma = \sigma_1 \exp\left[-\frac{E_A - E_F + w_1}{kT}\right] \qquad (6.2)$$

where w_1 is the hopping activation energy and E_A is the band-tail energy of conduction. σ_1 is found to be considerably lower than $\sigma_{\min}$, due to the reduced density of states and also the much reduced mobility. Although the mobility term for the hopping process has been shown by Mott [282] to possess a temperature dependence as shown in (d), the thermally activated temperature dependence of the excitation of carriers to the band tail will dominate this, resulting in the observed dependence.

(c) Transport by the hopping of carriers excited between localized sites at around E_F,

$$\sigma = \sigma_2 \exp\left[-\frac{w_2}{kT}\right] \qquad (6.3)$$

where w_2 is the hopping energy of activation. This also leads to a $\ln \sigma$ versus $1/T$ dependence.

(d) Finally, at lower temperatures, variable-range hopping (VRH) dominates for most materials with

$$\sigma = \sigma_2' \exp\left[-B/T^{1/4}\right] \qquad (6.4)$$

The constant $B = 2(\alpha^3/kN(E_F))^{1/4}$, where α is a measure of the decay of the wavefunction on a single potential well, and $\sigma_2' = \nu_{\text{ph}} e^2 N(E_F) \overline{R}^2$ where $\overline{R}$ is the average hopping distance and ν_{ph} depends on the phonon frequencies. The curved nature of the $\ln J$ versus $1/T$ characteristic in the case of most forms of a-C has led many authors to believe that VRH predominates up to room temperature and above [285]. However, others have refuted this by showing that the localization factor derived in the case of a-C is often unrealistic [286]. Finally, another form of hopping, known as nearest-neighbor hopping, is possible if the localization is very strong, such that the electron jumps to the state nearest in space because the term $\exp(-2\alpha R)$ falls off rapidly. This leads to an $\exp(-W/kT)$ dependence [282], where W is approximately $1/R_0^3 N(E_F)$, R_0 being the nearest neighbor hopping distance. See Figure 83.

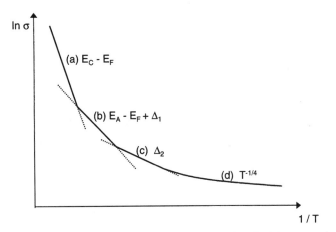

Fig. 83. Schematic of the temperature dependence of conductivity expected for an amorphous semiconductor.

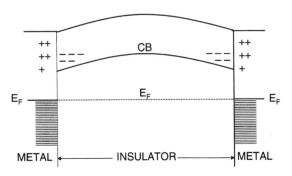

Fig. 84. Energy diagram showing two ohmic contacts on a trap-free semiconductor, demonstrating space-charge.

6.5. High Field Conduction

This applies in the electric field range of 10^5–10^6 V cm^{-1}. This would be achieved by performing electrical measurements through a thin film, using for example a metal–semiconductor–metal or metal–semiconductor–silicon structure. Normally, in this voltage region the current no longer varies linearly with voltage. The following mechanisms that describe these characteristics have been observed in a-C: space-charge-limited current (SCLC) [287, 288], Poole–Frenkel effect [289, 290], Schottky effect [291–293], and tunnelling [283, 294].

6.5.1. Space-Charge-Limited Current

This applies in MIM- and MIS-type structures where two ohmic contacts are separated by a space charge, as shown in Figure 84. This can first be analyzed assuming a trap-free undoped semiconductor [288]. The ohmic contacts are shown at the same potential. It can be seen that the accumulation regions are overlapping, and as a result the bottom of the conduction band is curved throughout its length. Therefore, space charge must extend through the length the semiconductor to allow for band bending. The result of this is that the conduction process is controlled by the space charge rather than the bulk properties of the material. In other words, the conduction is space charge limited. The derivations of the equations governing SCLC are well out-

lined by Lamb [295], and in the most simple case model the system as a parallel-plate capacitor with a transit time of carriers between the plates. It suffices to state that for a trap-free semiconductor with single carriers the J/E characteristic is

$$J \propto \frac{9\mu\epsilon_0\epsilon_r}{8d}E^2 \qquad (6.5)$$

where μ is the carrier mobility, ϵ_r is the relative permittivity, ϵ_0 is the permittivity of free space, and d is the film thickness. When the semiconductor contains shallow traps, a large fraction of injected space charge is condensed, resulting in a reduced density of free charge. The free charge to trapped charge ratio (Q) can be calculated by applying a simple Arrhenius dependence,

$$Q = \frac{N_c}{N_t} \exp\left(\frac{E_t}{kT}\right) \qquad (6.6)$$

where N_t is the number of shallow traps, and E_t is the energy distance between the trap and the conduction band edge [287]. For deep traps, Q is no longer a constant but depends on the depth and distribution of traps in the density of states. Therefore, it is possible, using a careful analysis, to extract some measure of the density of states at the Fermi level, from the current versus voltage data. This has been carried out in the case of DAC [296] and TAC [297]. The following scaling law has been predicted [298], assuming field-independent mobility:

$$\frac{J}{l} = f\left(\frac{V}{l^2}\right) \qquad (6.7)$$

6.5.2. Poole–Frenkel Effect

This mechanism is bulk-controlled, i.e., it occurs due to the intrinsic properties of the semiconductor and is not a contact-limited effect, and is therefore detrimental to device performance. Conduction in the bulk is governed by the manner in which the carrier leaves its originally occupied state at the Fermi level. As stated, these states are caused by defects due to the amorphous nature of the material. Some of these defects will be charged and hence contain carriers held by the electrostatic potential due to the charge. Hence, these trapping centers can be modelled by a coulombic potential well. The carrier can leave this well by a reduction of the well height caused by the application of an electric field, namely the Poole–Frenkel effect, which was first discussed in the 1930s [289] and modified in the 1970s by Hill [290]. Figure 85 shows the model diagrammatically.

The resulting J versus E characteristic, as shown by Hill [290], is given by

$$J = Ee\mu N_c \exp\left[\frac{E_g}{2kT}\right] \exp[\beta\sqrt{E}] \qquad (6.8)$$

where

$$\beta = \frac{1}{kT}\sqrt{\frac{e^3}{\pi\epsilon_0\epsilon_r}} \qquad (6.9)$$

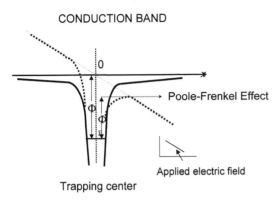

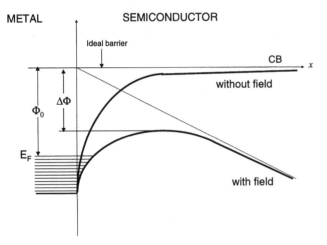

Fig. 85. Energy diagram showing the modification of the coulombic potential according to the single center Poole–Frenkel effect.

Fig. 86. Schematic diagram showing barrier lowering due to image charges, according to the Schottky effect.

and A^{**} is Richardson's constant, k is Boltzmann's constant, T is the absolute temperature, e is $1.6 \times 10^{-19} C$, ϵ_0 is the permittivity of free space, and ϵ_r is the relative permittivity. Therefore, by plotting $\ln J/E$ versus $E^{1/2}$, values of relative permittivity can be calculated, along with the trap density N_C.

6.5.3. Schottky Effect

At intermediate electric fields (10^4–10^5 V cm^{-1}), this is the dominant process controlling the current versus voltage characteristic in several materials, for example silicon nitride. This is known as the Schottky effect [291, 292] but must be distinguished from the standard Schottky barrier equations as it only occurs at fields at which the metal-to-semiconductor barrier height is reduced due to image forces attracting the electron to the surface of the metal, which is shown in Figure 86. The resulting current density is given by

$$ J = A^{**}T^2 \exp\left[\frac{\Phi_0}{kT} + \frac{\beta}{2}\sqrt{E}\right] \qquad (6.10) $$

where β has the same meaning as in Section 6.5.2. The dielectric constant and barrier height at zero voltage, Φ_0, can therefore be calculated from the above equation by plotting $\ln J$

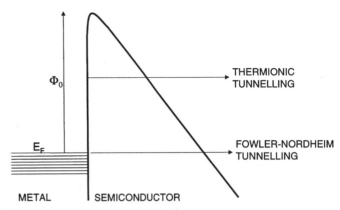

Fig. 87. Thermionic and Fowler–Nordheim tunnelling through a potential barrier.

versus $E^{1/2}$. In the case of a-C, this type of analysis is relatively rare, and there has been no proof as yet of a Schottky barrier on this material by applying to a fit to barrier lowering.

6.5.4. Tunnelling

At higher fields the electron can tunnel through the barrier, with or without thermionic assistance, as shown in Figure 87.

The derivation of thermionic tunnelling assuming thermionic field emission is outlined by Shannon [283]. For small carrier effective masses the total current is described by the equation

$$ J = A^{**}T^2 \exp\left[\frac{-(\Phi_0 + \alpha E_s)}{kT}\right] \qquad (6.11) $$

where E_s is the surface electric field and $\alpha = (kT)\delta \ln J/\delta E_s$ is the tunnelling constant. Generally, this type of tunnelling only occurs at electric fields of higher than 10^6 V cm^{-1} and is characterized by a straight line in the $\ln J$ versus E characteristic. The Fowler–Nordheim [294] process assumes an approximately triangular potential barrier and obeys the dependence

$$ J \propto \frac{V^2}{\Phi_0} \exp\left[\frac{-\Phi^{3/2}}{\beta V}\right] \qquad (6.12) $$

where Φ_0 is the barrier height and β is a constant which depends on the localized field enhancement. It must be noted that this model applies to tunnelling at the Fermi level. A straight-line $\ln J/V^2$ versus $1/V$ dependence is observed. In particular, though, this analysis has greater relevance in the case of surface electron field emission, as Fowler–Nordheim tunneling has often been cited as the process controlling the emission of electrons through the barrier to the vacuum level.

6.6. Electronic Structure and Properties of Amorphous Carbon

Amorphous carbon has been studied as an electronic material for about 30 years. However, the electronic structure of a-C, and hence the electronic properties, are complicated by the fact that both π and σ states exist within the bandgap, which contribute differently to the conduction properties. Therefore, the electronic properties of a-C are relatively poorly explained. In the first section, the electronic properties of amorphous carbon in its undoped state will be described. This will be subdivided into its two most commonly researched forms: a-C:H in its forms DAC, PAC, and GAC, and TAC.

As documented in Section 7, the gap states of a-C are highly localized. The eventual outcome of this is that in any form of amorphous carbon, conduction will always occur by some hopping process between localized states. Higher temperatures are therefore required for conduction in the extended states to be realized. In a later study, Robertson discusses the p-type conductivity of a-C as being due to the nature of the dangling bond defect. This is shown to be planar, by analogy to the methyl radical, and therefore contains a pure $p\,\pi$ orbital whose energy level lies at the p-orbital energy, which is at midgap. However, Robertson suggests that the dangling bond defect possesses a significant s content, due to s–p mixing. Therefore, the dangling bond energy level is lowered, as s states lie deeper than p states. This results in the p-type conductivity of the undoped forms of amorphous carbon [299].

The early experimental data for describing the electronic structure of a-C is summarized by Robertson [300]. Using current/temperature measurements, Robertson has reviewed work which has shown that for sputtered and glassy GAC, conduction follows variable-range hopping with an α^{-1} value of 1.2 nm and $N(E_F)$ of 10^{18} eV^{-1} cm^{-3}. Robertson also showed that the temperature dependence of conductivity was curved and did not fit any law particularly satisfactorily. Above 400 K it was ascribed to band-tail conduction. Chan et al. [35] showed that from a range of 300 to 500 K, it was not easy to ascribe any single conduction mechanism as the fits were similar with both band-tail and variable-range hopping. But with a fit to band-tail hopping made, an activation energy of 0.34 eV was derived, which was much smaller than half the bandgap, suggesting that the Fermi level was not at midgap.

6.7. Electronic Properties: GAC Films

The current versus temperature plots for GAC films were reported and investigated by Orzeszko et al. [285]. They evaporated GAC with or without a hydrogen atmosphere was compared to the resulting current versus temperature curves. Unhydrogenated films showed a reasonably close $T^{1/4}$ dependence, but the hydrogenated films showed a better T^{-1} relationship. This suggested that variable-range hopping played a greater role in the case of the nonhydrogenated film, as shown in Figure 88. Therefore, assuming a localization length α^{-1} as documented by Hill [290] of 7.5×10^6 cm^{-1}, the density of states $N(E_F)$ was carried out using the equations for variable-range hopping. It was found that $N(E_F)$ was two orders of magnitude lower in the case of nonhydrogenated films than hydrogenated films.

Similar measurements were carried out in more detail by Dasgupta et al. [286], who looked into the electronic proper-

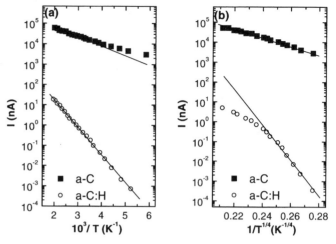

Fig. 88. Variation of current versus temperature for hydrogenated and unhydrogenated GAC films. Reproduced with permission from [285] © 1984, Wiley. (a) I versus $1000/T$, (b) I versus $1/T^{1/4}$.

ties of GAC:H, as compared to sputtered GAC. Both types of film were grown using RF sputtering. First, the conductivities of GAC and GAC:H were similar, implying that the optical gap and conductivity were ruled by the size and number of sp^2 islands rather than the hydrogen content. Second, in the case for films where the sp^2:sp^3 proportion was below its percolation threshold, the conduction at high temperatures appeared to follow some type of activated (band-tail) hopping process and some type of $T^{1/4}$ relationship at room temperature, implying variable-range hopping. This was in accordance the the behavior cited by Orzeszko et al. However, the calculations of α and $N(E_F)$ yielded unrealistic values ($\alpha^{-1} < 10^{-7}$ cm^{-1} and $N(E_F) > 10^{45}$ cm^{-3} eV^{-1} in one case). By fitting an Arrheniuslike relationship for the conduction at room temperature, and assuming a process of hopping between neighbouring sp^2 islands using a model from Mott [282], a good fit was obtained to the equation

$$\sigma = \left(2e^2R^2/kT\right)\nu_{ph}N_\pi T \exp(-w/kT - 2\alpha R) \quad (6.13)$$

The term R is the distance between adjacent sp^2 clusters and is crucial in calculating the conductivity variation of the material. For conduction at higher temperatures, variable-range hopping in the band tails was observed to fit the closest. For films with an sp^2:sp^3 proportion above its percolation threshold, hopping in the band tails appeared to be the only effective model. This has been contradicted by Helmbold et al. [301], a reason being that the values of thermopower generally found in the case of GAC (and DAC) are very small, in the order of 10–30 μV/K. Conduction in band tails should result in thermopowers several orders of magnitude higher due to an appreciable activation energy. Therefore, an alternative mechanism based on a multiphonon tunnelling model with weak electron–lattice coupling was proposed. A hopping process between sp^2 clusters was also assumed and the electron–phonon coupling was related to the mean lattice spacing and the localization length. Activation

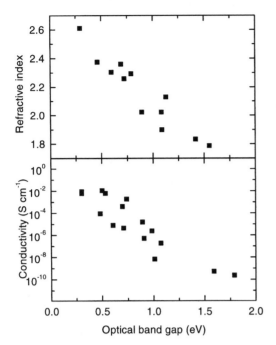

Fig. 89. Variation in optical band gap of DAC films as a function of DC bias voltage (see [201, 255, 220]).

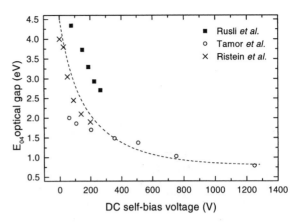

Fig. 90. Variation in refractive index and room temperature conductivity as a function of optical band gap. Reproduced with permission [302], © 1993, Wiley.

energies in the order of 0.015 eV were derived, which would explain the small thermopowers.

6.8. Electronic Properties: DAC Films

Robertson cites [300] that the optical absorption data suggests that most forms of a-C possess broad tails and hence a narrow bandgap (0.4–0.7 eV). However, DAC possesses a wider bandgap, due to the reduction of the band tail density of states as a result of the increase in sp^3 bonding. As already mentioned, the highly localized midgap states can dictate the conduction properties. These are formed as a result of the breaking of π bonds, as the π bonding is weaker than the σ bonding. If the π defect is on a conjugated system (e.g., an aromatic ring), it can delocalize with the ring. Any system with a half-filled π bonded site, including odd-membered rings, such as azulenes, as well as odd-numbered clusters, introduce states around E_F and are referred to as π defects. Dangling bonds are also classified as π defects, as the unpaired electron will preferentially occupy a π orbital.

6.8.1. Deposition Parameters

As already discussed, the DC self-bias voltage during deposition plays a critical role in determining the optoelectronic properties of the material. The variation of optical bandgap with DC bias voltage, according to Rusli et al. [201], Ristein et al. [255], and Tamor et al. [220], is shown in Figure 89. At bias voltages close to zero, the films possess a bandgap of approximately 4 eV and are polymeric in nature (PAC). As the DC bias voltage is increased, the bandgap shows a corresponding decrease as the film passes through a diamondlike (DAC) phase.

Eventually, above DC bias voltages of 600 V, the film becomes graphitic in nature (GAC) and the optical bandgap falls to below 1 eV.

However, relatively few studies have been carried out which refer to the effects on the electronic properties of DAC. Stenzel et al. [302] cites the effect of DC conductivity on the optical bandgap, as referred from earlier work. This variation is shown in Figure 90. This can be therefore be correlated to the effect of DC self-bias on optical bandgap, as documented by Silva et al. [281] and Ristein et al. [303]. It is commonly observed that an increase in the DC self-bias results in a decrease in optical bandgap. At low bias voltages, i.e., on the earthed substrate table of a PECVD process, the films are polymeric in nature and possess a Tauc optical bandgap of 2.5–3 eV [304]. At higher DC voltages the films become increasingly diamondlike and then graphitic in nature with the optical bandgap falling to eventual closure at the highest voltages. Stenzel et al. found that the room temperature conductivity showed an approximately exponential fall with the increase in optical bandgap, corresponding to values of 10^{-2} Ω^{-1} cm^{-1} with an optical bandgap of 0.3 eV (a GAC:H film grown with a DC bias of higher than 500 V), to 10^{-10} Ω^{-1} cm^{-1} with an optical bandgap of 1.8 eV (a DAC film grown with a DC bias of less than 100 V). Thus, the resistivity can be varied greatly by changing the optical bandgap of the material.

The effect of deposition temperature on the optoelectronic properties was documented by Jones and Stewart [305] who also grew films using an RF plasma. First, they found that both the optical bandgap and the activation energy of DAC decreased with increasing deposition temperature, from a value of 1.2 eV at a deposition temperature of 175°C to 0.7 eV at 350°C. This could be explained by the increase in sp^2 content resulting in a narrowing of the bandgap and hence activation energy. However, they found that the difference between optical bandgap and activation energy remained constant as the deposition temperature was changed, even though both were dependent on the deposition temperature. This indicated that the position of the Fermi level remained fixed relative to the valence band. How-

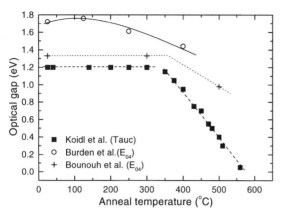

Fig. 91. Variation in optical band gap of DAC films as a function of annealing temperature (see [24, 307, 346]).

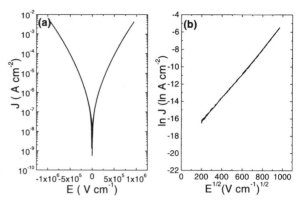

Fig. 92. Poole-Frenkel conduction in DAC:H, after Khan et al. (a) J versus E plot; (b) $\ln J$ versus $E^{1/2}$ plot.

ever, at room temperature the conduction appeared to move toward E_F, indicating hopping conduction.

The effect of thermal annealing on the optical bandgap of DAC is shown in Figure 91, which describes the variation according to Burden et al. [306] (in this case for DAC films grown using magnetically confined PECVD), Bounouh et al. [307] (direct vapor deposition), and Koidl et al. [24] (PECVD). What is generally found is an increase in sp^2 content above a graphitization threshold, commonly around 350°C, which has been correlated with the decrease in optical bandgap around this temperature, as well as a decrease in resistivity of around five orders of magnitude also around this threshold [304]. However, Burden et al. found that the optical bandgap increased at anneal temperatures of 125°C. It was suggested that the passivation of dangling bonds due to the movement of hydrogen during the anneal had sharpened the band tail and hence increased the bandgap.

6.8.2. I/V Characteristics

The current versus voltage (I/V) characteristic is essential in order to ascertain the dominant high field conduction mechanism which is taking place in these films. In the most simple case, DAC films grown on metal substrates show a symmetrical J versus E characteristic [308–310], suggesting that conduction is bulk-limited and not ruled by the differences in barrier heights at either contact. Egret et al. [308] showed that for DAC, the conduction was dominated by the Poole–Frenkel (PF) effect, which has been shown by others and is displayed in Figure 92. As a test of PF conduction, films were grown with a range of refractive indices of 1.8–2.3 by varying the self-bias voltage, which corresponded to an optical bandgap range of 1.2 to 2.2 eV. The dependence of the PF β factor with the refractive index was observed, and β was found to vary approximately with $1/n$ which suggested that the PF equation was applicable. Furthermore, it was found that as the optical bandgap increased over this range, the current density at zero voltage J_0 decreased from 10^{-7} to 10^{-12} A cm^{-2}, and the PF activation energy at high temperatures increased from 0.4 to 0.6

eV between bandgaps of 1.3 and 1.5 eV. At lower temperature the temperature dependence was nonlinear, as observed by others.

There is much useful information here which has not yet been analyzed to date. First, conduction in these films at high electric fields is bulk-limited and as the optical bandgap is increased, it is likely that the density of donor/acceptor trap states causing PF conduction decreases, causing a decrease in J_0. The PF (i.e., high field) activation energy follows the optical bandgap trend but always possesses a value just less than half of the bandgap, suggesting that these traps are in a defect band close to but not at midgap. It has been stated that p-type conductivity is likely in the case of DAC and this suggests that this PF activation energy is to the valence band via hole conduction. At high temperatures the carriers dictating PF conduction are likely to be excited to the valence band edge, resulting in a straight temperature dependence, but at room temperature and below they are retrapped around the defect band resulting in the curved temperature dependence akin to VRH or some alternative hopping process [286, 301] at lower electric fields.

A study to investigate the effect of contact material on the I/V characteristics was carried out by Konofaos et al. [311]. The films were grown on silicon substrates using a mixture of CH_4 and H_2 in a standard RF PECVD system, and substantial differences were observed in the I/V characteristics between using gold and aluminum top contacts. This was explained by the presence of a Schottky barrier, which possessed a different height for the two materials. Also, it was found that by growing films of different thicknesses, the J/E characteristics were not the same, suggesting contact limited conduction. However, Allon-Alaluf et al. [312] grew DAC films on aluminum substrates using a similar process and found that the I/V characteristics were symmetrical using gold or copper top contacts, indicating an ohmic contact. There appeared to be a low field asymmetry using aluminium top contacts, which was investigated using capacitance–voltage (C/V) measurements. This indicated an interfacial barrier between the film and the top contact. This was further justified using Auger emission spectroscopy which showed a substantial concentration (20%) of

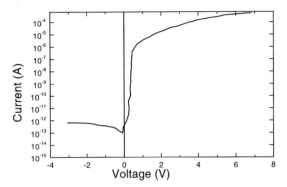

Fig. 93. Band gap-modulated DAC superlattice/Si heterojunction showing 8 orders of magnitude of rectification. Reproduced with permission from [35], © 1992, Elsevier Science. I versus V plot.

oxygen at the interface. It is possible that this result may explain those of Konofaos et al.

However, others, such as Silva et al. [296], have discussed these diamondlike films in terms of space-charge-limited current. From such measurements, using the analysis outlined previously and described in Lampert and Mark [298], an idea of the density of states at the Fermi level has been derived, the value being of the order of 10^{18} eV^{-1} cm^{-3}. This can be compared to a spin density as derived from EPR of the order of 1×10^{20} cm^{-3} [303]. Using this reference, an extracted correlation energy from photoelectron yield experiments is of the order of 1 eV, thus giving a density of states of 1×10^{20} eV^{-1} cm^{-3} (cf. Section 5). However, diamondlike films with a higher bandgap were also measured by Silva [313] and were found to obey Poole or Poole–Frenkel conduction. Therefore, it can be ascertained that most forms of DAC follow this type of conduction, and hence the density of defects dictates the conduction properties of the material.

6.8.3. DAC Heterojunctions

These has been much work which has provided evidence for a DAC:H/Si heterojunction. This was first shown by Amaratunga et al. [314] which showed that the I/V characteristic of the metal–carbon–silicon structure displayed rectifying behaviour. This suggested an a-C/silicon junction and has been shown by others [35, 315, 316]. Chan et al. [35] found that the forward to reverse rectification ratio was six orders of magnitude, by growing a bandgap-modulated DAC superlattice film, as is shown in Figure 93.

Capacitance-versus-voltage measurements are valuable when investigating the nature of the a-C/Si interface. The early evidence was obtained by Chan et al. [35, 317], following investigations carried out on sputtered a-C by Khan et al. [318, 319]. Chan showed that the observed frequency dependence of the C/V characteristic came about due to interface traps. This was modelled and the interface trap density was found to be of the order of 10^{10} cm^{-2} eV^{-1} and was comparable to that reported by Khan et al. [318]. This value was also far lower than the areal spin density as derived from EPR or deep level ca-

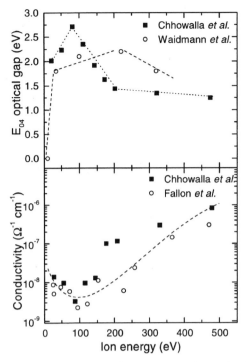

Fig. 94. Variation in optical bandgap and DC conductivity as a function of ion energy for TAC films.

pacitance profiling in TAC, which suggested that the dangling bonds in a-C films were not acting as trap centers. In comparison, Mandel et al. [320] carried out similar measurements and found that the gap state density of states was in the order of 10^{16} cm^{-3} eV^{-1}. However, Munindradasa et al. [315] correlated the C/V as a function of frequency with the field emission properties of the material. In this case, the amount of hysteresis was low and the flat band voltage was zero, suggesting an absence of fixed charges.

6.9. Electronic Properties: TAC Films

Even though TAC possesses a high sp^3 content of around 80%, there will always be a significant proportion of sp^2 bonds that will introduce π states into the bandgap. These will hence dictate the electronic properties of the material by being closer to the Fermi level.

6.9.1. Effect of sp^2/sp^3 Bonding

The change in the sp^2 content of TAC and its subsequent effect on the electronic properties of the material are of paramount importance in discussing the conduction within the material. Surprisingly, relatively few studies that concentrate on the electrical properties have been carried out. Fallon et al. [68] grew TAC films using FCVA and varied the bias voltage of deposition in order to see if there was an optimum ion energy for the promotion of sp^3 bonding. This was found at a bias voltage of 100 V, i.e., an ion energy of approximately 140 eV, and corresponded to the maximum in resistivity, along with other factors

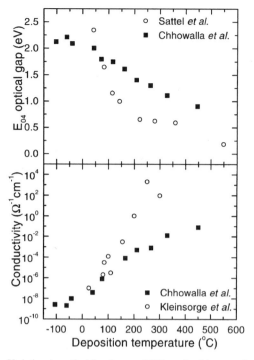

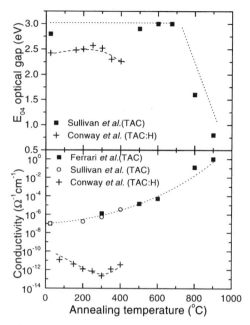

Fig. 95. Variation in optical bandgap and DC conductivity as a function of deposition temperature for TAC films.

Fig. 96. Variation in optical bandgap and DC conductivity as a function of annealing temperature for TAC and TAC:H films. Reproduced with permission from [125], © 1999, American Institute of Physics. Reproduced with permission from [126], © 1997, World Scientific. Reproduced with permission from [316], © 1998, American Institute of Physics.

such as compressive stress and plasmon energy. It was suggested that the bandgap is at its widest at this energy. Another explanation proposed is the decrease in conduction (variable-range or band-tail hopping) between sp^2 clusters, should the density of these clusters within an sp^3 matrix be reduced. Similar trends on optical and mechanical properties have been shown by Xu et al. [66], Chhowalla et al. [69], and Waidmann et al. [321]. Figure 94 shows the variation in optical bandgap and conductivity with ion energy, according to various authors [68, 69, 321]. It can be seen that Fallon et al. and Chhowalla et al. concur with the conductivity measurement, but the ion energy at which the optical bandgap is at its widest is 220 eV in the case of Waidmann et al. as opposed to 100 eV or less for the other authors. This illustrates the observation that different groups have measured different energies at which the peak in sp^3 content occurs. This could be due to differences in the degree of ionization or varying monochromacities of the ion beam. Another point to note is that most groups measure resistivity in TAC by the evaporation of planar (gap-cell) contacts which allow the measurement of surface resistance. This method is problematic as it has been shown [322] that TAC possesses a significant sp^2-rich surface layer, so reliable measurements of the bulk resistivity may not be obtained. In sone cases this layer is etched but one canot ascertain whether the surface layer is modified or terminated during this type of procedure. Therefore, the most reliable method for measuring bulk resistivity would be through the film using metal–carbon–metal structures.

The effect of varying the deposition temperature on the conductivity and optical band gap has also been measured and is illustrated in Figure 95 as derived by Chhowalla et al. [69], Sattel et al. [323], and Kleinsorge et al. [324]. It is observed that the optical bandgap shows a decrease from 2–2.5 eV to 0–1 eV as the deposition temperature is increased, and the conductivity shows an accompanying rise of 7–10 orders of magnitude over the temperature range. In the study by Chhowalla et al. [69], the sharp decrease in resistivity at higher deposition temperatures was explained by the increase in disorder of the film due to energetic ion bombardment leading to gap states. It could be explained similarly to the above result, though, i.e., an increase in the hopping probability between sp^2 clusters, as the sp^2 content is increased. Chhowalla et al. also measured the conductivity as a function of inverse temperature and found a linear dependence. This suggested band-tail conduction and as the activation energy was 0.45 eV, the Fermi level was not at midgap.

6.9.2. Annealing of TAC

The only significant studies into the effects of annealing on the electronic properties of TAC were carried out by Sullivan et al. [126] and Ferrari et al. [125]. This is shown in Figure 96. In addition, the annealing data in the case of TAC:H as derived by Conway et al. are shown and will be discussed later. In the case of TAC, a stress relaxation occurred around 600°C. Sullivan et al. modelled this stress relaxation as a small change in sp^2 content, of up to 6.5 at.%. Therefore, the variation in conductivity was compared with this change in sp^2 content. In particular, the conductivity appeared to be exponentially dependent on the sp^2 concentration of the film. It was postulated by the data that the enhancement of conductivity was governed by variable-range hopping with a term for the variable separation between sp^2 sites, assuming they occur in chains rather than rings. A typical chain length was estimated to consist of 13 carbon atoms. Therefore, the conduction was thought to be ruled by the length of sp^2 chains, which were lengthened through annealing (though only from 13 to ~14), coupled with chain-to-chain tunnelling.

Ferrari et al. also showed that the E_{04} optical gap stayed constant (at 2.7 to 3.0 eV) until 600°C, and therefore the change in sp^2 content was likely to be low. They also concluded that the decrease of resistivity of almost three orders of magnitude (initially ~10^7 Ω cm) over this temperature range could be accounted for by the slight increase in hopping centers. Above 600°C, though, the resistivity decrease was much greater and coincided with a fall in the optical bandgap, though the sp^3 percentage remained unchanged until 1200°C where it fell from 85% to 20%. This coincided with complete stress relief within the films. It was possible that this final transition was triggered by the formation of β-SiC. The Raman spectra associated with these films were discussed in Section 4.

6.9.3. TAC Heterojunctions

Tetrahedral amorphous carbon has been shown to form a heterojunction with silicon [120, 325], in spite of the fact that it was been shown that TAC forms a highly sp^2-rich surface layer as a consequence of the likely subplantation process which results in a densification below the film surface [81]. These heterojunctions possess strongly rectifying I/V characteristics and breakdown strengths above 100 V. From TAC/Si heterojunctions, it was shown that the current flow was space-charge-limited [297]. The turn-on voltages were 1.6 V for the TAC/n-Si heterojunction and 0.4 V for the TAC/p-Si heterojunction, suggesting that the material was p-type. Using the differential method, the density of states derived from the valence band was of the order of 10^{21} cm^{-3} eV^{-1}. Furthermore, Palinginis et al. [261] carried out junction capacitance measurements on TAC in order to derive the defect density. Using the drive level capacitance profiling technique, a defect density of 6×10^{17} cm^{-3} was determined. The value as derived from EPR was at least two orders of magnitude higher (cf. Section 5); this was explained by the fact that this technique used only would detect the D^+ or positively charged defect center.

6.9.4. Photoconductivity of TAC

Photoconductivity in the case of TAC was first reported by Amaratunga et al. [326]. It was found that the resistivity decreased from 10^7 to 10^5 Ω cm when the films were illuminated under AM1 light. The spectral response was then observed and it was found to show a peak at around 750–800 nm, corresponding to a photon energy of around 1.7 eV, which was close to the optical bandgap. The quantum efficiency of the film was

between 10 and 12.5% over the 450–800 nm range and then decreased. Also, it was found that the variation of photoconductivity was independent of temperature, indicating that the conductivity under light was dominated by carrier generation.

McKenzie et al. [327] found that photoconductivity decreases with increasing nitrogen content. Following this, Cheah et al. [328] fabricated nitrogenated TAC/p-Si heterojunctions and also observed that the optical absorption increases as a function of nitrogen content. Also TAC(:N) films were grown on p-Si substrates and the resulting reverse characteristic of the TAC:N/Si heterojunction was found to increase by three orders of magnitude. These factors also indicated promise for TAC:N as a material for solar cell applications

The latest and most detailed studies on the photoconductivity of TAC and TAC:H have been carried out by Ilie et al. [329, 330]. They first investigated the photoconductivity of the undoped forms and found first that above 200 K the photoconductivity decreased with temperature according to an activation energy, and below 200 K it was independent of temperature. Therefore, it was surmised that TAC was a low-mobility solid with $\mu\tau$ products in the order of 10^{-11}–10^{-12} cm^2 V^{-1} due to the high defect density. The main recombination centers were deep defects but tail states also caused recombination, and at 200 K there was a peak in conductivity corresponding to competitive recombination between two classes of centers. A later study showed that in the case of TAC:H, the photoconductivity is increased by nitrogen doping, as the material is doped n-type. The photosensitivity obtained was approximately 200 under 35 mW/cm^{-2}. The addition of nitrogen did not create extra charged defect recombination centers.

6.9.5. Hydrogenated TAC

A study was carried out by Davis et al. [112] which measured the properties of TAC deposited in the presence of varying flow rates of hydrogen, and it was found that there existed a certain threshold at approximately 0.05 sccm H$_2$. Below this threshold, the sp^3 fraction increased from 82 to 86% and the plasmon energy increased from 29.8 to 30.5 eV, suggesting in increase in carbon atom concentration with some hydrogen incorporation. Also, the optical bandgap increased from 1.9 to 2.1 eV and the trap density decreased, when using a SCLC analysis as in the case of Veerasamy et al. [297], the current injection was at its sharpest at 0.05 sccm. These measurements suggested that at these low hydrogen contents, there was a reduction in the band-tail density due to the increased sp^3 content, effectively as hydrogen would force an increase in the sp^3 content as it cannot bond to carbon in an sp^2 configuration. This would result in a reduction in defect states below the Fermi level and hence a movement of the Fermi level toward midband, causing increase in activation energy and resistivity. This was observed in the activation energy which increased from 0.22 eV in the undoped case to 0.36 eV at a flow rate of 5 sccm H$_2$ with a concurrent increase in resistivity from ~1×10^7 to ~3×10^7 Ω cm. However, above 0.05 sccm H$_2$, the trap density increased, and the sp^3 content decreased down from 83% to 77% at 5 sccm H$_2$.

It was postulated that this was due to an increase in the number of isolated sp^2 sites as the carbon–carbon coordination number decreased.

In the case of TAC:H grown using a plasma beam source, a variation in resistivity with ion energy similar to the case of TAC was found [331]. At an ion energy of 200 keV, the resistivity reached its maximum value of 2.5×10^7 Ω cm, indicating that this was also the energy at which the peak in the sp^3 content occurred. There has also been evidence of a reduction in the defect density by annealing [316]. This has been correlated to a decrease in conductivity and increase in both activation energy and bandgap, around an optimum annealing temperature of 573 K, as shown previously in Figure 96. This is similar to the trend described by Burden et al. [306] as shown in Figure 91, which had previously demonstrated that annealing at 250°C increased the optical bandgap of DAC. However, the study into TAC:H was carried out in more detail and showed that these films conducted via hopping at the band tails rather than at the Fermi level, as the $\ln J$ versus $1000/T$ plot showed a straight line from 250 to 500 K. Therefore, the decrease in conductivity and increase in activation energy as bought about by annealing were attributed to the sharpening of the π tails. This was later corroberated [332] by evaluating the reciprocal Urbach slope, which is proportional to the slope of the band tails in the energy range just below the E_{04} gap. This value was found to increase from 3.7 to 4.4 eV^{-1} from the unannealed sample to the sample annealed at 300°C, implying a reduction in the density of band-tail hopping sites. The postulated explanation for this was a redistribution of H atoms to passivate dangling bonds by moving along the atomic network. In this way a six-atom cluster would be converted, for example, into a four-atom cluster with a wider bandgap.

6.10. Electronic Modification of Amorphous Carbon

A significant proportion of the work in a-C is related to attempting to electronically modify or "dope" the material. The most commonly used method is to add gaseous nitrogen as an *in situ* dopant during deposition. This is performed for two reasons. First, nitrogen has been shown to be an n-type donor in the case of diamond, albeit a deep one, so one of the major objectives is to dope the material n-type. This is of interest because, as has already been mentioned, a-C is thought to be intrinsically p-type.

6.10.1. In Situ Doping of Hydrogenated a-C: DAC, GAC, and TAC:H

Several authors have reported the electronic doping of DAC by the addition of gaseous precursors into the deposition chamber during growth. The first reports of n- and p-type doping were shown by Meyerson and Smith [333]. Films were grown at using a DC glow discharge of acetylene at 150 and 250°C, and gaseous addition of PH$_3$ and B$_2$H$_6$ was investigated. It was found that in the case of boron addition for films grown at 250°C, the activation energy showed a small rise from 0.6

to 0.65 eV at a diborane flow of 0.1%, followed by a fall to 0.3 eV at diborane flow rates of up to 10%. The conductivity measurements showed the contrary, with a decrease from 10^{-11} to 10^{-12} Ω cm at 0.1% diborane, followed by an increase of up to 10^{-7} Ω cm. This indicated that the DAC films were intrinsically n-type and could be doped p-type by the addition of boron. Conversely, the films grown with increasing amounts of PH$_3$ showed only increases in conductivity (10^{-11} to 10^{-7} Ω cm) and decreases in activation energy (0.6 to 0.3 eV), further substantiating this claim. A later report [334] showed that the thermopower for B-doped films was positive and for P-doped films was negative, further indicating that doping was taking place. However, the magnitude of thermopower was small, and conductivity versus temperature measurements fitted a $T^{1/4}$ relationship, indicating a variable-range hopping mechanism. The most important point to note, though, was that there was no mention of bandgap in the study. It was later found that the incorporation of a range of dopants would have the effect of increasing the sp^2 content of the films, thus narrowing the bandgap, decreasing the activation energy, and increasing the conductivity though not necessarily as a result of electronic doping, but rather a reduction of E_g.

Jones and Stewart [305] then reported on the effects of nitrogen, phosphine, and diborane doping into DAC:H. Both diborane and phosphine were found to increase film conductivity. The effect of increasing the nitrogen content was to decrease the optical bandgap from the initial value of 1.2 to 1.0 eV at a nitrogen flow ratio of 0.1, increase the conductivity at 227 K from 10^{-7} to 10^{-4} Ω^{-1} cm^{-1}, and decrease the activation energy from 0.6 to 0.3 eV. However, the difference between optical bandgap and activation energy remained constant as the doping level was changed. This indicated that the position of the Fermi level remained fixed relative to the valence band irrespective of the doping level. Similar studies on the effects of nitrogen modification were carried out by Amir et al. [272], Stenzel et al. [302], and Khan et al. [304]. They reported similar trends in terms of conductivity, activation energy, and optical bandgap. Amir et al. performed thermopower measurements which showed a positive sign for films containing less than 1 at.% N, indicating conduction in the valence band tail. Increasing the nitrogen content of the films changed this to conduction around the Fermi level, as shown by the sharp reduction in the magnitude of the thermopower from 600 (0.2% N) to -4 μV K^{-1} (3% N). The change of the sign indicated n-type conduction; however, the doping effect was very weak and at still higher flow rates, the thermopower sign was once more positive. Stenzel et al. suggested that the enhancement of conductivity was due to a greater interconnectivity between existing clusters, resulting in enhanced hopping conduction.

In light of these contradictory findings, Helmbold et al. [301] investigated the addition of phosphine into the deposition chamber during DAC growth. They found that their current versus temperature measurements showed a $T^{1/4}$ dependence which was observed at and above room temperature. Therefore, a model based on variable-range hopping was postulated; however, conduction appeared to be dictated by a "characteristic temperature" at which all samples possessed the same conductivity. All other parameters could be explained by factors such as cluster diameter and lattice spacing. With this type of model, a shift in the Fermi level was excluded as changes in medium range order could account for the variations in conductivity.

Silva et al. [335, 336] grew films on the driven electrode of a magnetically confined RF PECVD source, with a low DC self-bias of around 60 V and varying flow rates of nitrogen. It was found that the optical bandgap increased from 1.65 to 2.1 eV, and the Tauc B parameter increased from 300 to 580 cm^{-1} eV^{-1} as the flow rate of nitrogen was increased from 0 to 7 at.%, which could be an effect of defect passivation. At higher levels of nitrogen (up to 13 at.%), the optical bandgap fell to less than 2.0 eV, due to sp^2 reordering. Activation energy measurements showed that there were two regimes of conduction, one at higher temperatures likely to be extended state conduction, and one at lower temperatures likely to be band-tail hopping. It was found that the activation energy of the high temperature mechanism appeared to increase from 0.5 to 0.92 eV and then decrease to 0.45 eV as the flow rate of nitrogen was increased, the peak coinciding at 7 at.% N once more. It was assumed that DAC is intrinsically p-type which agreed with the initial activation energy value and bandgap. This value appeared to increase to half of the optical bandgap, suggesting that the Fermi level had moved to the center of the bandgap, due to the compensation of defects which led to p-type properties. Following this, the activation energy decrease suggested n-type doping of the material. A later study [337] showed from EELS measurements that the point at which the Fermi level was suggested to be in the center of the bandgap coincided with the point at which the sp^3 fraction was at its highest. This point occurred at 7% atomic nitrogen, which appeared to be a turning point.

This work could be compared to other studies, for example by Schwan et al. [77] who deposited GAC films using a plasma beam source (PBS). The films formed possessed a low hydrogen content of around 10%, which increased on the addition of nitrogen, possibly due to the introduction of N–H bonds. Also observed was an increase followed by a decrease in the optical bandgap, agreeing with the findings by Silva et al. However, it was found that the electrical conductivity simply increased and the activation energy decreased slightly as the nitrogen flow rate was increased. This meant that the Fermi level was only moving very slightly and other effects rather than doping were occurring. For example, carriers may have been thermally activated to a π^* state, and the addition of nitrogen could have created N π^* states which were situated below the C π^* states, which the carriers were moving to. Therefore, the n-type dopant site could have manifested itself as a mixed C–N π^* state. In comparison to this, Conway et al. [331] also grew films using a PBS and measured the optical bandgap and conductivity as a function of nitrogen content. Here, it was found that the conductivity showed increases with nitrogen content from $\sim 10^{-8}$ to $\sim 10^{-5}$ Ω^{-1} cm^{-1}, and the optical bandgap showed a corresponding decrease from 2.0 to 1.0 eV (E_{04}). No turning point in the conductivity data as predicted by compensative doping was observed. The explanation for this was that the films

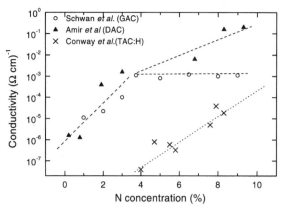

Fig. 97.   Variation of room temperature conductivity of nitrogen-doped DAC films (see [272, 280, 331]).

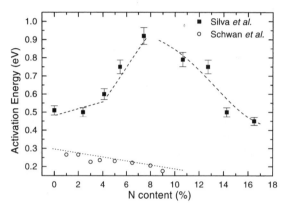

Fig. 98.   Variation of activation energy of nitrogen-doped DAC films (see [280, 335]).

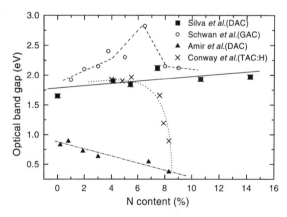

Fig. 99.   Variation of optical bandgap of nitrogen-doped DAC films (see [272, 280, 331, 335]).

were grown using acetylene as the source gas, which contained a nitrogen contamination corresponding to 4 at.% within the film. When methane was used, the polarity of the TAC:H/Si heterojunction was changed as the nitrogen concentration was increased, suggesting doping of the material through a compensative state.

Figures 97–99 show the variations in optical bandgap, room temperature conductivity, and activation energy as reported by

480 SILVA ET AL.

various groups [77, 272, 331, 335]. Only those who have related these parameters to the atomic nitrogen content within the films have been included. As can be seen, the greatest problem is ascertaining the variations in the undoped material, which depend on such factors as self-bias voltage and deposition temperature. Also, there are large differences in the magnitude of optical bandgap as reported by different groups. This could be related to different methods employed in measuring such factors. As has also been mentioned, activation energy measurements are problematic as some groups observe a $T^{1/4}$ relationship, some observe an Arrhenius relationship, and some observe two relationships depending on the measurement temperature.

However, the following may be surmised. First, nitrogen has the effect of increasing the conductivity. This could be either due to doping or an increase in the sp^2 content. Second, different groups have seen different trends in terms of optical bandgap and activation energy measurements. Some have observed a decrease in optical bandgap [272, 331] which equates to graphitization, and some have seen an increase in optical bandgap with nitrogen content [77, 337]. This could equate to a sharpening of the band tail (passivation) and hence a reduction in the defect density of the material. Meaurements made by Khan et al. [310] have shown that the Tauc bandgap increased from 0.7 to 1.0 eV as the N_2 flow rate was increased, and this was correlated to a reduction in Poole–Frenkel prefactor and decrease in paramagnetic defect density from 10^{20} to 10^{19} cm^{-3}. Although the latter can be explained by the pairing up of paramagnetic spins due to doping, this is still strong evidence for the passivation of midgap defects with the addition of nitrogen. Therefore, it suffices to say that there is evidence for electronic doping of DAC:H according to some groups, and also evidence for either defect passivation or graphitization according to others.

6.10.2. In Situ Doping of TAC

The first reported case of successful electronic doping of TAC was shown by Veerasamy et al. [338]. Phosphorus was incorporated into the films by use of a graphite/red phosphorus mixed cathode. It was found that with an increase in cathodic phosphorus percentage, the conductivity was found to rise and the activation energy fall. However, the evidence for n-type doping was weak as no initial increase in activation energy was reported. However, electrical measurements of undoped and P-doped TAC/n-Si heterojunctions showed a reversal in polarity, suggesting that an n-type heterojunction had been formed with the doped sample.

A later study [339, 340] documented successful n-type doping of TAC, this time by injecting nitrogen gas into the bend region of the filtered cathodic vacuum arc system. The evidence for n-type of doping of TAC was first a change in sign of thermopower from positive to negative. Second, the resistivity and activation energy of the films showed an increase followed by a decrease, which assuming intrinsically doped TAC was p-type, placed the undoped Fermi level at 0.3 V above the valence band edge. The compensative case at which the Fermi level

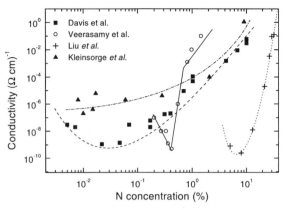

Fig. 100. Variation of room temperature conductivity of nitrogen-doped TAC films (see [112, 324, 339, 342]).

was found to be midgap corresponded to approximately 0.45 at.% nitrogen. Following this, the Fermi level was thought to move to within 0.2 eV of the conduction band edge on the addition of nitrogen. The optical bandgap was also observed to fall, indicating alloying of the material and sp^2 reordering. EELS analysis suggested that at low nitrogen levels there was little sp^2 reordering of the material, which was evident at higher nitrogen levels.

A similar study was reported soon afterward by Davis et al. [111, 327]. Similar trends in terms of variations of resistivity and activation energy were reported. However, in this case, a rectifying p–n junction consisting of nitrogen-doped TAC on undoped TAC was also found. These trends have been reproduced elsewhere [341, 342]. However, other studies [324, 343, 344] have seen only an increase in conductivity and a decrease in optical bandgap as the nitrogen content is increased, which is evidence only for sp^2 reordering.

In the case of Ronning et al. [343, 344], this could be as the films were grown using mass-selected ion beam deposition and nitrogen was added by alternating the ionic species layer-by-layer during deposition. This may have had an effect on the doping efficiency or controllability of activated dopant. However, it was found that conduction followed ohmic conduction at low fields, suggesting hopping conduction, and followed Poole–Frenkel conduction due to two trap states, one around E_F and one at the conduction band edge, at high fields. The addition of nitrogen or boron had the effect of modifying the I/V characteristics such that the density of one trap state increased far more than the density of the other state. This led to a kink in the I/V characteristic and suggested that the addition of the dopant had the effect of vastly increasing the density of states at the Fermi level. However, no change in activation energy values was observed, suggesting no shift in the Fermi level. Also, p–i–n diodes were fabricated and were not found to rectify, suggesting no doping effect, or a pinned Fermi level. With the MIS structures, heterojunction rectification was observed, but the polarity did not appear to change with respect to the doping of the film.

Figures 100–102 show the variations in conductivity, activation energy, and optical bandgap as a function of nitrogen

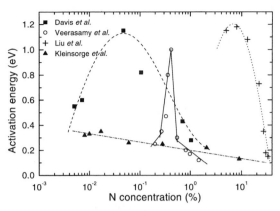

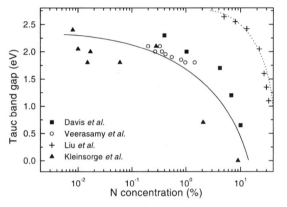

Fig. 101. Variation of activation energy of nitrogen-doped TAC films (see [112, 324, 339, 342]).

Fig. 102. Variation of optical bandgap of nitrogen-doped TAC films (see [112, 324, 339, 342]).

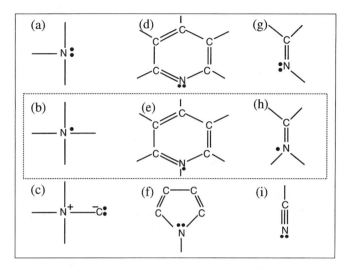

Fig. 103. Possible configurations of nitrogen in a-C:H:N. The lines represent bonds; dots are unpaired electrons. The items within the dotted box are the doping configurations (after Silva et al. [337]).

content, as reported by various groups [324, 327, 339, 342]. The atomic percentage of nitrogen content as a function of nitrogen flow rate is taken from other studies where necessary [66, 112]. What can first be noted is that at moderately high nitrogen contents (1%), the optical bandgap closes, presumably due to an increase in the sp^2 content. Second, the resulting variations differ significantly in terms of conductivity. This could be related to the relative fraction of sp^2 bonding which is also controlled by such factors as the incident ion energy, as mentioned earlier, or the monochromacity of the ion beam. Also, as has been mentioned earlier, these conductivity and activation energy measurements may be compromised as only the surface layer of the TAC film is probed, which may be sp^2-rich. Third, in terms of the evidence for doping, the compensated state varies considerably according to the different groups. This could be related to the method of nitrogen inclusion into the plasma, as Xu et al. injected nitrogen gas directly into the cathodic plume, resulting in a large degree of ionization. Liu et al. used an ion beam source, and the others added nitrogen simply into the deposition chamber or the torus. Finally, not all groups see a turning point in the values of conductivity and activation energy corresponding to the compensated state. This also suggests that the reactor configuration has a large role in determining whether doping is occurring or not.

Robertson [345] attempted to ascertain the role of N in the doping process. In the case of diamond, the donor level was about 1.7 eV from the conduction band edge. In the case of TAC, the substitutional N site was now shallow when compared to the π^* edge. However, the doping efficiency was likely to be low as most N would adopt its trivalent bonding configurations (see (a) in Figure 103), so the substitutional N_4^+ site (b) was less likely. Alternatively, the N_4 site could compensate existing midgap defects leading to the $N_4^+ C_3^-$ state (c), which would result in in a full C defect state and an empty N donor level, and hence a weak doping effect. However, in the case of a predominately aromatically bonded sp^2 material as in the case of higher nitrogen doping levels, N_4 could still substitute a C atom in an aromatic ring, contribute an electron into an $N\pi^*$ state, and weakly dope it (e). However, it would be far more likely in this case that nondoping pyridine- or pyrrolelike structures would be formed (f). For an olefinic sp^2 bonded region, once again a weak bonding effect can be brought about (h) in spite of other more likely configurations (g) or even sp^1 sites (i). Figure 103 shows these structures diagramatically, as shown by Silva et al. [337].

6.11. Summary

The electronic properties of a-C have been discussed with trends in terms of growth bias voltage/ion energy, deposition temperature, postdeposition annealing, and dopant addition. Variations in the electrical properties have been coupled with changes in factors such as paramagnetic defect density, sp^2 content, and optical bandgap. There is further evidence here for the point that conduction in a-C is governed by the density and distribution of the sp^2 hybridized states, whether these exist in chains or clusters.

It has been seen that there is evidence of nitrogen possessing the ability to dope a-C films n-type; however, it also appears that the doping process is complicated and a more careful analysis is required. In particular, electrical measurements have been shown to concentrate on conductivity, activation energy, and bandgap measurements, and useful information which may be derived from the high-field conduction properties has been largely ignored.

7. CONCEPTS OF LOCALIZATION AND DELOCALIZATION IN a-C

Here we attempt to discuss in greater detail the phenomenon of electron localization in a-C. It has already been mentioned that this property limits the electronic properties of a-C a great deal and results in hopping conduction, as explained by a variety of mechanisms, over a wide temperature range.

The disorder in an amorphous material first causes scattering and a reduction in the mobility of the carriers. If the disorder is increased, localization of the wavefunction may occur where no movement of carriers is possible at zero Kelvin. At finite temperatures tunnelling between localized states may occur. A localization length, R_0, of PAC has been measured to be 1.0–1.5 nm by Silva et al. [347], which is significantly larger than for a-Si (10–12 nm). Later calculations of the localization radius using photoconductivity measurements showed it to be as low as 0.2–0.3 nm for TAC and 0.9 nm for TAC:H [330]. Using the notation used by Anderson [348], it was shown that if

$$\delta V / B > 1 \qquad (7.1)$$

where $B(= 2zV)$ is the bandwidth of the conduction band due to the interaction between the atoms and z is the coordination number for σ states, we get

$$\delta V / B = \delta V / 2zV > 1 \qquad (7.2)$$

and

$$dV / V > 2z \qquad (7.3)$$

In a-Si which only permits σ bonds, localization only occurs at the tail states of both conduction and valence bands. In the case of a-C where both π and σ bonds are present, as well as defect states, the localization present can extend through the gap [149]. Chen et al. [349] performed atomistic calculations and showed that the entire π band is localized. As shown in Figure 104, the inverse participation ratio of the π states that control the conduction properties is very high. The simulation used a sp^2 bonding concentration of 80%, which is well over the percolation limit, and despite this the conduction within the material was poor. Interestingly in this case the disorder potential δV is expected to be quite small and so there must be another reason for the localization of the wavefunction. This effect was found to be the interaction between π states which is controlled by the projected dihedral angle ϕ of two bonds, as shown in Figure 105. In most cases the planes involving π bonds ended up being close to orthogonal and so there was little or no interaction as V, the potential well of each atom, tended

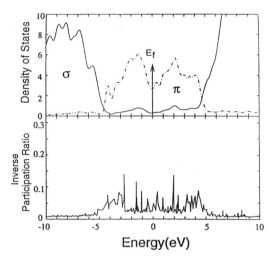

Fig. 104. Calculated partial density of states for an a-C network showing the high inverse participation ratio of the π states despite possessing 80% sp^2 bonds. Reproduced with permission from [299], © 1997, Elsevier Science.

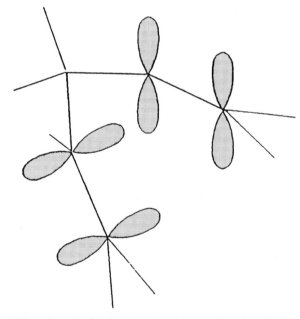

Fig. 105. Schematic of the low interaction between adjacent p-orbitals due to their random orientation.

to zero. This meant that $\delta V (= V_0 - V)$ was of a similar magnitude to V so $\delta V / V$ tended to be greater than 1, resulting in localization over the whole wavefunction.

This result is completely different from a-Si as in that case all the bonds are tetrahedral σ bonds. Therefore, the tail states in a-C will differ greatly from those of a-Si. It also means that in the case of PL, it is likely that the generated electron–hole pair will remain within a delocalized cluster, which is isolated from the remainder of the network and give rise to a germinate pair. This also means that the PL emission will be highly polarized as seen by experiments [201]. Another artifact of the localized tail states in a-C would be the intense room temperature PL as the carriers cannot diffuse apart easily and thereby get trapped into a nonradiative recombination path as seen by experiments [347].

7.1. Ion Implantation of *a*-C

Implantation of diamond has been carried out as a route to electronic doping of the material. Boron has successfully doped diamond *p*-type [350], and there has been evidence for the *n*-type doping of diamond using phosphorus [351]. The damage effects of ion implantation in diamond have also been widely reported. These include the onset of hopping conduction [352], and also the onset of sp^2 reordering [353]. There appears to be a threshold ion dose in both cases above which the film reorders into an sp^2-rich state and the electrical conductivity shows dramatic rises. This is related to the eventual overlap of damage cascades. However, at lower doses, it was found that diamond transforms into an amorphous sp^3 bonded phase, in a way similar to silicon.

One of the earliest reports of the effect of ion implantation into DAC films was conducted by Prawer et al. [354]. This study used a novel experimental technique where the resistivity of the film was measured *in situ* during ion irradiation. It was found that there was a threshold dose below which no changes were evident, and after which the resistivity collapsed by as much as six orders of magnitude. The bandgap showed a concomitant collapse, as did the hydrogen content. This suggested that the removal of hydrogen resulted in a depassivation of dangling bonds and a rise in the number of intergap states. This caused the increase in conductivity at approx. 10^{15} ions cm^{-2} (which was thought to be of the type of variable-range hopping in this case) and reduction in bandgap due to the smearing out of band tails. Graphitization was thought not to occur until higher doses, when the damage cascades were thought to overlap, as in the case of diamond, and the films underwent macroscopic graphitization.

This was similar to results found by Ingram et al. [355], Khan et al. [356], and Doll et al. [357], who also reported on *n*-type conductivity after DAC:H was implanted with nitrogen ions. Whether this was a doping effect, or a consequence of defect addition, was not ascertained. However, Khan et al. did not see a resistivity threshold that occurred at the same dose as for the optical bandgap collapse. This could be because the as-grown material was polymeric and hence intrinsically much more resistive. Therefore, the effect of small changes in resistivity could be measured. However, it was suggested that some degree of electron delocalization of the highly localized π states was occurring that accounted for the large changes in resistivity as a function of ion dose, before the onset of macroscopic reordering. Wang et al. [358] found an initial increase in resistivity with an implantation dose of 1×10^{15} ions cm^{-2}. IR measurements suggested that at this dose the C–H sp^3 concentration had increased, which would result in a decrease in the density of hopping centers and hence higher resistivity.

McCulloch et al. [107] implanted carbon and xenon ions into TAC and also performed *in situ* resistance measurements. In

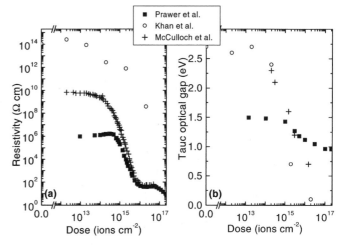

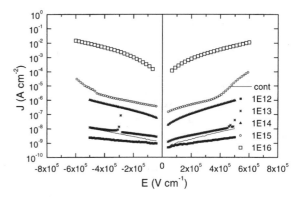

Fig. 106. Variation of room temperature conductivity and optical bandgap of PAC [356], DAC [354], and TAC [107] films as a function of carbon implant ion dose.

Fig. 107. Variation in *J* versus *E* characteristic of boron implanted PAC films as a function of boron ion dose.

this case, it was found that there was no threshold behavior in the resistivity trend, which was mirrored in the optical bandgap trend. This suggested that there was a regime at intermediate doses, in which a highly disordered sp^3-rich but nongraphitic material was being created. This was similar to the case in diamond and was corroborated using EELS measurements. For very high doses, it was thought that conduction was due to hopping between graphitic islands. It was also found, as by Doll et al. [357], that using Seebeck measurements undoped TAC was *p*-type and after carbon implantation the films became *n*-type. Therefore, it was suggested that sp^2 bonded material was being created, which contributed to the conductivity by acting as *n*-type dopants and also narrowed the bandgap. A later study [359] showed that the effect of irradiation at elevated temperatures was to increase the sp^2 fraction at higher doses and also the degree of reordering. Therefore, a graphitic material was favored at elevated temperatures, also as in the case of diamond. Figure 106 shows the variation in resistivity and optical gap with implantation dose for DAC, TAC, and PAC films, according to various researchers [107, 354, 356].

Researchers have attempted to improve the conduction properties in *a*-C with the addition of impurities such as hydrogen

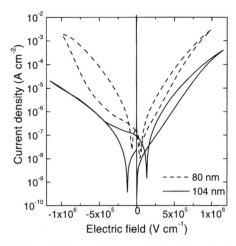

Fig. 108. Variation in J versus E characteristic of boron implanted a-C:H films at a dose of 2×10^{15} cm^{-2}, showing rectification in the case of the thicker film.

and nitrogen into the band structure. Some encouraging results have arisen in the case of FCVA [338], magnetron sputtering [77], and PECVD with very high nitrogen content (alloying) [337]. It is interesting to note that in the case of enhanced conduction with FCVA and magnetron sputtering, significant quantities of impurities have been needed to see any effects (when compared to traditional doping of crystalline semiconductors), but these effects were true and reproducible. Such results have not been seen for any other a-C films using other deposition techniques. One aspect that is common to both those techniques is the use of intense ion bombardment creating highly stressed and metastable structures. Extending this theme further we have used ion implantation in an attempt to delocalize electronic states within the a-C films that we have deposited with very low defect densities [310]. The variations of the electronic properties as a function of ion dose for B implantation are shown in Figure 107. It should be noted that the quantity of implant used is much smaller than those used in the previous studies, but the changes in conduction are greater. It can be seen that there is a current gain which we observe at a dose of 2×10^{15} cm^{-2} which is dependent on the applied voltage. A closer analysis reveals that this corresponds to a degree of hysteresis in the characteristic (see Fig. 108), which can be removed by implanting a thicker film (104 nm as opposed to 80 nm) resulting in the implant only going through part of the film. This suggests the formation of a Schottky junction, as true rectification has been demonstrated in spite of only metal contacts on either side of the film, for the first time in a-C. This suggests delocalization of the film into a material onto which Schottky contacts may be formed.

8. ELECTRON FIELD EMISSION

8.1. Introduction

Carbon based cathodes have been proposed as the emitter material for the next generation of flat panel displays based on field emission, due to their enviable property of emitting electrons from nominally flat surfaces at relatively low electric fields [360, 361]. This has attracted much publicity and research interest with some of the major multinationals producing prototype displays using this cathode material. With a global market for flat panel displays estimated at US$20 billion by the year 2002, this has given the amorphous carbon community much hope as a possible outlet for active devices. Initial interest in carbon films as cold cathodes was driven by the hope that these films possessed negative electron affinity [362]. This meant that as a cathode all that was needed was to ensure that there was a supply of electrons injected into the conduction band of the material, and that would allow for a suitable emission current at the front surface. Early experiments and calculations pointed toward negative or low electron affinity values whether the material was crystalline or amorphous, which allowed electrons to be extracted from the surface of the films at relatively low electric fields [362, 363]. Yet, more recently, it has been shown by models based on empirical results that the back contact to the carbon films can be as important to the emission of electrons at low electric fields [45, 364, 365] as the front surface [126, 366, 367]. This has meant that a thorough review of the emission process has been required in order for an acceptable model to be proposed. It is also still open to debate if there is an universally acceptable model, due to the plethora of different films available in the literature and the misunderstanding of trying to compare one set of films with another. A classic example is in the films deposited by Gröning et al. [368] which have been compared by a number of authors to films deposited by RF PECVD (for example, see [45]). A closer look at just the material properties shows that Groning's films are graphite-like (GAC) with no optical bandgap while the DAC films by Amaratunga and Silva have an optical bandgap close to 2 eV. Therefore, no parallels can be made in terms of field emission models and transfer of experimental results for films that have material properties that are dissimilar.

The primary reason for researching the field emission properties of amorphous carbon thin films has been the electron emission at low electric fields, which has allowed for its possible usage in flat field emission displays at voltages compatible with current complimentary metal oxide semiconductor (CMOS) technology driver circuitry. Flat cathodes have considerable advantages over their "Spindt" tip counterparts that are more expensive and more complicated to fabricate on a large area display. Yet, the initial designs for using a simple matrix addressed two-terminal flat carbon cathode [360] has not been forthcoming due to material property considerations as discussed later.

8.2. Theory of Field Emission

The electron emission that takes place from a metal surface is principally determined by the strength of the applied field in combination with the magnitude of the work function as shown in Figure 109. In general, as the electric field is increased first

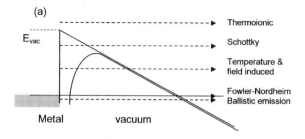

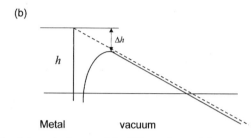

Fig. 109. Schematic diagram of the different forms of electron emission that can take place from a metal surface.

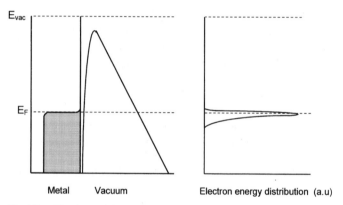

Fig. 110. The shape of the potential barrier when FN emission is taking place.

thermionic emission over the top of surface barrier (work function) will occur, followed by emission over the field reduced barrier (Schottky emission). Emission will then occur partly over the barrier if the temperature is high enough and partly through the barrier. As the temperature is lowered emission will come from electrons tunnelling through the barrier (Fowler–Nordheim tunnelling) from electron states near the Fermi level and finally at very high fields, emission from states below the Fermi level (ballistic electron emission) [369]. In the Fowler–Nordheim (FN) theory [370] of electron emission the shape of the potential barrier at the front surface is important. Figure 110 shows two possible potentials; the first potential is taken as a rectangular barrier of the form

$$V(z) = h - eEz \qquad (8.1)$$

where h is the surface barrier height, often regarded as the work function (ϕ) of the material. The second potential incorporates the image charge effects which in Systeme International units

can be expressed as

$$V(z) = h - eEz - \frac{e^2}{16\pi\epsilon_0 z} \qquad (8.2)$$

The latter potential results in a rounding of the potential barrier and a lowering of the barrier height Δh when compared with the triangular barrier. This results in a change in barrier height

$$\Delta h = e\left[\frac{eE}{4\pi\epsilon_0}\right]^{1/2} = 3.975 \times 10^{-5} E^{1/2} \qquad (8.3)$$

where the macroscopic electric field E is measured in V/m. For a field of 20 V/μm the barrier lowering due to image charge effects is approximately 0.17 eV. In addition to the assumption with regard to the shape of the potential barrier, the FN theory as applied to metals also assumes that [370]

(i) the metal has a free electron band structure,
(ii) the electrons are in thermodynamic equilibrium and obey Fermi–Dirac statistics,
(iii) the metal is at zero temperature,
(iv) the surface is smooth,
(v) the work function is uniform across the emitting surface and is independent of the applied field which is turn is assumed to be uniform above the surface,
(vi) the Jeffreys–Wantzel–Krames–Brillouin (JWKB) approximation may be invoked to evaluate the penetration coefficients.

Within these assumptions the emission current J can be expressed as

$$J = \frac{a E_{\text{local}}^2}{\phi} \exp\left(\frac{-b\phi^{3/2}}{E_{\text{local}}}\right) \qquad (8.4)$$

where a and b are constants with approximate values of 1.54×10^{-6} eV V^{-2} and 6.83×10^9 e$^{-3/2}$ V m^{-1}, respectively. The emission current density is determined from the emission current divided by the area of electron emission A, which will generally not be equal to the area of an anode as measured by experiment. The local electric field E_{local} is usually related to the macroscopically applied electric field E_{applied} by the equation

$$E_{\text{local}} = \beta E_{\text{applied}} \qquad (8.5)$$

where β is the field enhancement factor. Values of β for ellipsoidal tips or protrusions of radius r and height h can be given as [371]

$$\beta \approx 2 + \frac{h}{r} \qquad (8.6)$$

provided $h/r > 5$.

Fowler–Nordheim type emission of electrons would show that the peak of the electron energy distribution was centered at the Fermi level of the emitter with a sharp cutoff measured on the high energy side reflecting the thermal occupation of electron states. The low energy side is characterized by an

n⁺- Si   a-C(:H) layer   vacuum

Electron energy
distribution (a.u)

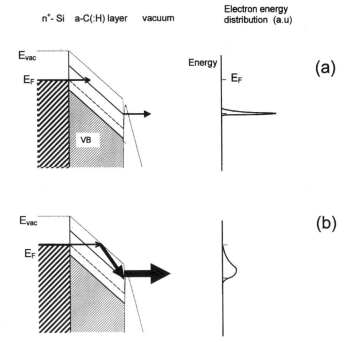

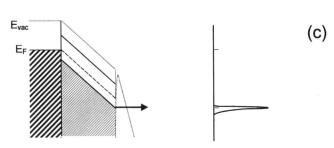

Fig. 111.   The predicted FEED as a function of the emission process due to (a) Fowler–Nordheim emission from the Fermi level at the back metal contact, (b) hot electron emission from the conduction band of a semiconductor, and (c) emission from the valence band of a semiconductor.

exponential decrease as the probability of electron emission diminishes and the barrier to tunnelling increases. In the case of a semiconductor, electron emission may also occur from the conduction band, valence band, defect states, and also surface states. Measurement of the energy distribution of the emitted electrons should be able to distinguish between the different mechanisms. If electron emission originates from the conduction band then the distribution of electron energy should be shifted to lower energies and possess a lower energy cutoff. This shift of the field electron energy distribution (FEED) is due to the potential drop between the surface and the back contact as shown in Figure 111a. In Figure 111b and c the predicted electron energy distribution for emission via hot relaxing carriers from the conduction band and from the valence band emission, respectively, are shown. Emission from the valence band maximum has been reported in p-type diamond and from the conduction band minimum from undoped diamond films. Furthermore, in low conductivity materials internal electric fields may be present, as observed by Geis et al. [364] in N doped diamond films.

Gröning et al. [368] have performed both FE measurements and also electron energy distribution measurements of N containing films. The films contained up to 17 at.% N and were found to be highly conducting. Using a 4 mm diameter stainless steel anode they observed threshold fields for 1 nA emission to be in the range of 20–25 V/μm. Plotting the I–V characteristic on a FN plot gave an apparent work function of 0.16 eV assuming $\beta = 1$. Energy resolved FE measurements were made by placing a grounded Cu TEM grid 50 μm above the samples, which was biased negatively at about −1000 V. The samples were illuminated with either ultraviolet radiation or X-rays, allowing ultraviolet photoelectron spectroscopy (UPS) and X-ray photoelectron spectroscopy analysis to be performed to allow measurement of the sample work function. They measured the onset of field emission at the Fermi level with a low energy tail and cutoff of UPS emission at 4.9 eV. Using this energy as the value of film work function they determined from the FN plot a field enhancement factor of 171 from which a local electric field of 6500 V/μm was calculated. Since emission was observed from the Fermi level with a high energy cutoff, characteristic of FN tunnelling, they concluded that no field penetration occurred. However, this is to be expected since the conductivity of the film is so high the emission would be predicted to be a "front" surface dominated process.

It is worth pointing out at an early stage an important experimental fact when comparing threshold fields between different research groups and with different types of samples. The two most common geometries used to measure FE characteristics are the sphere-to-plane geometry and the plane-to-plane geometry. In the former case an anode (either point or large area) is mounted between 10 and 100 μm away from the cathode material, a high voltage is applied to it, and the current is measured. In the case of the latter arrangement large area conductors such as indium tin oxide (ITO) coated glass are used. This latter configuration will tend to measure the turn on behavior of the lowest emitting areas (sites) first and will tend to give lower threshold fields when compared with the single point (probe) measurement tests.

8.3. Planar Emitter Structures Based on Carbon

It is worth categorising several types of amorphous carbon films for the analysis of their electron field emission process. Progressing with ascending energies required for the growth of the films, we will consider emission from PAC, DAC, TAC, GAC, and NAC films. It is also worth discussing that in the case of a-C films that do not possess large numbers of defect states within the films or graphitic or conducting regions in its microstructure, in general a conditioning process has to be encountered before the emission of electrons takes place. Subsequent to the initiation voltage required in the conditioning, emission takes place at successively lower threshold fields which reaches a minimum value after about four cycles. The hysteresis observed in this process of conditioning is shown clearly in Figure 112, and is nonreversible; i.e., once a cathode has been conditioned it will remain conditioned despite the surface being examined

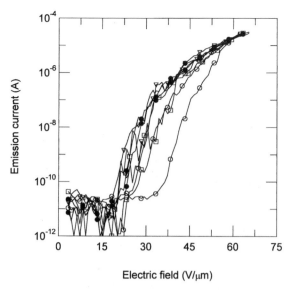

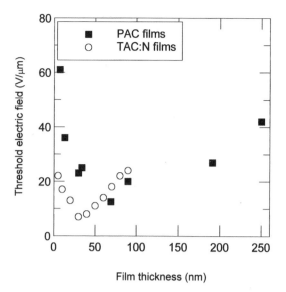

Fig. 112. Variation of the electron emission current as a function of electric field, for successive voltage cycles of a typical PAC film. Note the reduction in hysteresis for each successive cycle.

Fig. 114. Variation of threshold electric field as a function of film thickness for PAC/Si films.

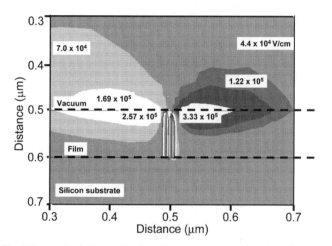

Fig. 113. A simulation study of conducting channels within the a-C matrix that give rise to local field enhancement.

days later even after exposure to atmospheric conditions. This points to the fact that surface termination or the front surface dipole effects may play less of a role than originally thought.

We will show that by examining the various models available in the literature, there are some hypotheses that are more suitable than others. According to our view, in most cases the crucial forming step of the field emitting process is the conditioning or first ramp up run in the experiment for a majority of the films. We show results based on simulation that incorporate inclusion of current channels or conducting regions.

The "worm-hole"-like current channels, if created during conditioning, are highly localized spatially and are well separated from each other. This means that around each of the conducting channels it is possible that high fields may be sustained by space charge that are necessary in order for a continuous electron emission process to be sustained. It should also be noted that in a dielectric such as a-C, which has a dielectric

constant ϵ_r of ~4, a conducting channel would represent a dielectric medium of 1 and thereby naturally give rise to field crowding effects that may be seen from the front surface as "internal" field enhancement. This local β factor would be further enhanced by inhomogenies within the amorphous films due to the localized nature of the films electronic structure as described in Section 7. Yet, it should be noted that for emission to take place the matrix surrounding this conducting channel must have its band structure suitably bent by the ionized carriers for electrons to be transported to the front surface of the films for emission. See Figure 113. Such a model could also be used to explain the very low thresholds reported in hydrocarbon polymers [372]. The highly localized channels could also act as conduits to remove any excess charge at the surface due to its relatively low resistivity compared to the bulk of the film. Recent simulation studies conducted show the channels to be positively charged in comparison to the surrounding matrix during electron emission, which would tend to indicate that the channels can effectively act as electron sinks as well as electron sources.

8.3.1. Polymerlike Amorphous Carbon Films

There are a number of reports on the electron field emission properties of these types of films [365, 373, 374]. These films have low defect densities and so generally require a high initiation electric field for emission to take place. Significant hysteresis is observed in the first cycle after initiation. Subsequent to the initiation process, the threshold field for electron emission has been shown to be controlled between 10 to 70 V/μm by varying the thickness of the undoped films by Forrest et al. [365] (Fig. 114). Despite some of these films being subject to very high electric fields of the order of 70 V/μm, no morphological change was observed with a scanning electron microscope after emission. The data show a very strong depen-

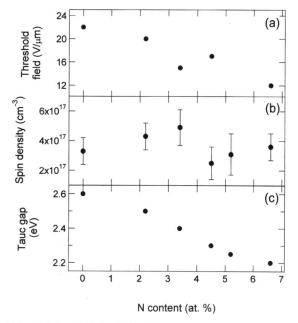

Fig. 115. Variation of (a) threshold field, (b) spin density, and (c) Tauc gap as a function of N content for PAC films.

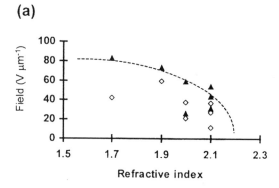

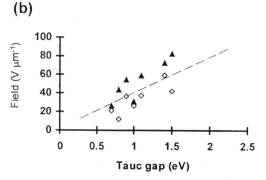

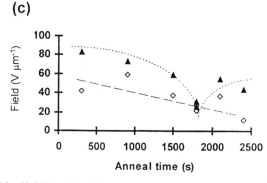

Fig. 116. Variation of the field emission as a function of annealing for PAC films: (a) versus refractive index, (b) versus Tauc optical gap, and (c) versus anneal time at 400°C.

dence of the conditioned threshold field on film thickness, with the required field showing a minimum turning point in the data at a film thickness close to 60 nm. The threshold field needed for emission increases on either side of this optimum thickness, with the surface of the films not exhibiting any morphological changes when examined by an atomic force microscope (AFM) as a function of film thickness. Therefore, it is difficult to attribute the emission readily to an external local field enhancement factor without further proof of its existence. Although sp^2 clustering has been proposed as being a factor in the controlling of the electron emission [366, 375], in these films the growth conditions have been kept constant throughout the deposition process and so no changes in the microstructure of the films are expected. The observed variations with film thickness in these films were explained using the interlayer model based on space charge induced band bending in thin films, as discussed later in this chapter.

In order to examine the other factors that control the emission properties of these films, the variation of the threshold field with nitrogen content for a series of films of thickness 60 nm were also examined. In addition to monitoring the threshold field, the EPR and Tauc optical gap were also examined, as shown in Figure 115. Addition of N has been shown to reduce the threshold field in DAC films [45] and also in TAC films [376]. However, both these types of films have had large initial paramagnetic defect concentrations and the nitrogen increase has generally ended up decreasing the spin density either by defect passivation or doping. In the case of PAC films the addition of nitrogen also decreases the threshold field, but there is little or no decrease in the defect density. Therefore, we are able to decouple the effects of defect passivation from field emission in this case. In the polymeric films the continued decrease in the Tauc gap without an increase in the spin density indicates that

the number of sp^2 clusters remains the same but that the cluster size is increasing. Therefore, the lowering of the threshold field with addition of N was attributed to an increase in the cluster size in the polymeric films. Yet, as discussed in the modelling of the emission process later in this chapter, for PAC films that have a large optical bandgap, low defect density, and a well defined heterojunction for the back contact, in most cases the limiting factor is controlled not by the front surface but by the back surface.

The variation of the emission properties as a function of annealing has also been reported [373]. It was first shown by Forrest et al. [374] that large changes in the refractive index and Tauc optical gap could be obtained by modest variations in the postannealing of these films. In this study an attempt was made to correlate the electron emission studies to the annealing. It was correctly concluded that no definite relationships could be obtained [374] due to the fact that there were also varia-

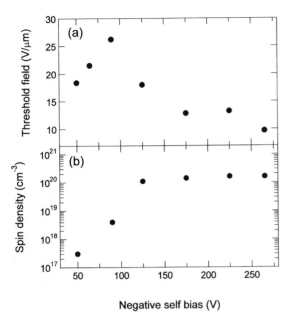

Fig. 117. Variation of (a) threshold electric field and (b) spin density as a function of negative self-bias during deposition for DAC films.

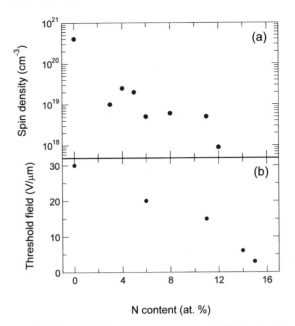

Fig. 118. Variation of (a) threshold electric field and (b) spin density as a function of nitrogen content at a fixed bias during deposition for DAC films.

tions in the film thickness as the films were subject to annealing. Therefore, Burden et al. [373] followed up the experiment making certain that the sample film thickness was kept constant by careful control of the deposition parameters in the study that followed. A summary of the variations observed by them is shown in Figure 116. The films discussed here were subject to 400°C anneals for times that varied from 0 to 2400 s. The dased lines in Figure 116a and b can be predicted using the space charge model described by Burden et al. In their analysis of electron field emission, as the film thickness was kept constant, they were able to use the refractive index as a microstructural parameter and empirically show that for Poisson's equation to be satisfied in a space charge induced band bending model the following equation could be derived:

$$E_{\text{th}} \approx A - \frac{Bn^2}{(C - Dn^2)} \tag{8.7}$$

n is the refractive index and A, B, C, and D are fitting parameters that were given values of 90, 10, 5, and 0.5 for the best fit for their films.

In the case of the threshold field variation with optical gap, they argued that if the valence band remained static with respect to the vacuum level when the bandgap varies [363], in the space charge induced band bending model the barrier that electrons need to overcome at the back n-Si/PAC heterojunction would be proportional to the bandgap (assuming fully depleted bands). Also discussed were the optimum anneal conditions for electron emission with annealing as shown in Figure 116c.

8.3.2. Diamondlike Amorphous Carbon Films

Good examples of DAC films are those deposited on the driven electrode of an RF PECVD system where during the

growth process there is also ion beam "subplantation" of growth species involved. In other deposition systems such as sputtering a similar result could be achieved by either having a second ion source that is used purely to ion bombard the growth surface or by the application of a negative self bias to the growth substrate. In the case of the samples used in this study for illustrative purposes, undoped a-C:H films were deposited using a standard capacitively coupled RF PECVD system with feed gases of CH_4 and He. The negative DC self-bias developed across the plasma sheath was varied in the range −50 to −265 V at a fixed pressure of 200 mT in the first set of samples shown in Figure 117. In this set care was once more taken in order to keep the film thickness constant around 60 nm. The variation of the EPR spin density is also shown in the figure in order to draw comparisons and to try and elucidate the mechanism that is responsible for the emission process. Similar studies have been performed where the threshold field has been plotted as a function of DC self-bias and a function of chamber pressure (100–1000 mT) by Silva et al. [377], but as the film thickness was not kept constant it is not possible to discuss definite trends.

The variation of the field emission properties as a function of nitrogen content for a fixed bias condition is shown in Figure 118. The variation in the EPR density as a function of nitrogen content is also shown for comparison. These doped films were deposited using a magnetically enhanced plasma deposition system described in detail elsewhere [88] using the feed gases CH_4, He, and N_2, with varying N_2 flow rates at a fixed DC self-bias and pressure. Changes due to N incorporation were attributed to electronic doping rather than graphitization. The doping effect observed in the films due to the nitrogen incorporation is weak, but clearly changes in the electronic properties such as DC conductivity and activation energy are observed [29]. All the films were deposited on water-cooled substrate holders, with the temperature at the surface of the films not exceeding 50°C during the deposition process. The DAC(:N) films were deposited on n^{++}-Si substrates, which were degreased and then Ar (He) plasma cleaned prior to the deposition of the DAC(:N) thin films.

The variation of the threshold field with DC self-bias is shown in Figure 117. As the self-bias is increased from -50 to -90 V, the threshold field increases from a value of 18 V/μm to a value close to 26 V/μm. For bias values above -90 V, the threshold field decreases to around 12 V/μm, where it remains approximately constant. Higher fields have been reported for undoped films by Lee et al. [378]. The measured spin density, N_s, of these films rises rapidly from 3×10^{17} to 1.1×10^{20} cm^{-3} at -125 V. At higher biases N_s increases only very gradually, finally reaching 1.6×10^{20} cm^{-3}. The Tauc gap falls rapidly from 2.6 eV (-50 V bias) to 1.3 eV (-125 V) and then falls gradually to about 1.1 eV at the highest bias voltages. In the case of ΔB_{pp} there is an initial increase in the linewidth, reaching a maximum of 1.4 mT at -125 V bias followed by a rapid decrease to 0.7 mT, which continues to decrease to 0.56 mT at -265 V bias.

These results show the existence of two regimes, above and below -125 V bias, and it is therefore convenient to discuss the results from the two regimes separately. For the sample deposited at -125 V bias the measured spin density is 1.1×10^{20} cm^{-3} and the observed linewidth is 1.4 mT. The Lorenzian lineshape can be explained using dipolar broadening. Above -125 V it is thought that the narrowing of the linewidth is due to rapid exchange of spins as the sp^2 clusters increase in size, as discussed in Section 5. This is consistent with the narrowing of the bandgap, and the reason for the converging of the threshold field to a limiting value close to 12 V/μm for these films. These films have a high sp^2 content ($>65\%$ [377]) as well as a high bandgap (>2 eV) and as a consequence overlapping of the clusters will result in electron delocalization and/or enhanced hopping between the clusters that will enhance the connectivity between the clusters. Application of an external electric field results in extraction of electrons from the film surface. Replenishment of the emitted electrons to the surface layer is easily accomplished due to the good connectivity between the clusters and low barrier at the back contact due to the low bandgap of the material. Emission from these high sp^2-rich films would be characterized as a "front surface"-type emission and since the sp^2 clusters are located at or about the Fermi level E_F they would be expected to yield a peak in the FEED at E_F. The high conductivity of these samples means the films would not be able to sustain any form of space charge. As the films would have a high percentage of defect states within the films it would not be possible to form a barrier (Schottky or heterojunction) at the back contact to the film. Therefore, as the back contact is likely to be ohmic in nature it would be able to readily supply electrons to the film and will not be controlled by the back contact process discussed by Amaratunga and Silva [45].

This was shown by Hart et al. [379] for $\sim$25 nm thick TAC films onto substrates of different metals (varying work functions). Several of these substrates (Ti, Mo, Si, and Cr) can

form interfacial carbides which could lower the barrier at the back contact and possibly lead to the formation of a "classical" ohmic contact. However, the authors noted that since there was a significant spread in the measured values of threshold field between the carbide and noncarbide forming substrates, the back contact does not play a significant role in determining the emission characteristics of TAC films. Indeed from Section 5 it was shown that the defect density of TAC is greater than 10^{20} cm^{-3}; such a defect density would tend to produce an electrically "leaky" junction whereby a space charge could not be sustained and is therefore not the determining factor for electron emission.

Reducing the self-bias to -90 V decreases the spin density by about two orders of magnitude and gives rise to an increase in the Tauc gap. The rise in the Tauc gap shows that major structural changes in the film are occurring at the low bias voltages with the films likely to have a higher sp^2 content due to C–H bonding. These films are often polymeric. At low self-biases a low concentration of small isolated sp^2 clusters forms. Initially, the number of C atoms in each cluster is small (as inferred from the large Tauc gap) but the greater energy available at higher biases allows the formation of a greater number of larger, but still isolated, clusters. This results in a reduction of the Tauc gap and an increasing spin density, though the electron delocalization in the cluster and/or hopping between clusters is kept to a minimum. This leads to poor connectivity between the clusters and results in films which need a higher applied electric field for the onset of emission and explains why the threshold field is higher at the lower bias voltages when compared with the films deposited the higher biases. The subsequent drop in E_T for the film at the lowest bias voltages indicates that another emission mechanism is also present in these low defect, polymeric, wide energy gap films. This additional mechanism is brought about by the presence of the high internal electric field (up to 20 V/μm) at the a-C:H/Si interface which produces "hot" electrons from the conduction band of the underlying Si substrate [45, 365]. The low concentration of the more conductive sp^2 clusters is unable to screen the field at the front surface. In this way emission from these low bias deposited films would be more correctly described as being determined by "the back contact" rather than being a "front surface" dominated process.

The variation of the threshold field and spin density with N content is shown in Figure 118. The threshold field in these nitrogenated DAC films drops monotonically from 20 to 3 V/μm as the atomic nitrogen content in the films is increased from zero to 15 at.%. The EPR spin density decreases by two orders of magnitude from 4×10^{20} to 4×10^{18} cm^{-3} when 7 at.% of nitrogen is introduced and remains close to 10^{18} cm^{-3} for subsequent increases in nitrogen. The linewidth at zero nitrogen decreases from 0.5 mT to close to 0.15 mT at the maximum nitrogen content. The intervening points between the minimum and maximum nitrogen content have a linewidth close to 0.21 ± 0.01 mT. These results indicate that the nitrogen in the films has a definite beneficial effect on the threshold field and the emission current density. The EPR results appear to indicate that other than in the first and last points of the data,

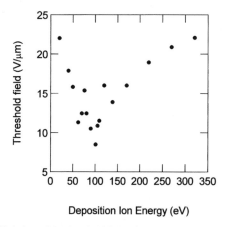

Fig. 119. Variation of the threshold field of TAC films as a function of deposition ion energy. Reproduced with permission from [382], © 1998, Elsevier Science.

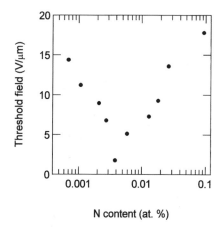

Fig. 121. Variation of the threshold field of TAC films as a function of nitrogen content. Reproduced with permission from [382], © 1998, Elsevier Science.

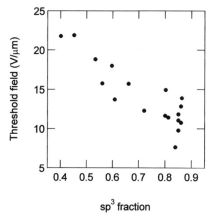

Fig. 120. Variation of the threshold field of TAC films as a function of sp^3 fraction. Reproduced with permission from [382], © 1998, Elsevier Science.

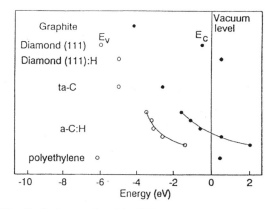

Fig. 122. Band edge energies of the C–H system. Reproduced with permission from [388], © 1997, Elsevier Science.

there is little change in the spin density and linewidth, both of which remain relatively low for these films. The films that were deposited using magnetic confinement had a bandgap close to 2 eV, with the controlling mechanism for field emission being attributed to band bending from space charge considerations. The invariance of the EPR properties appears to indicate that it is unlikely that the cluster sizes within the films are varying significantly during the increase in nitrogen content from 7 to 14 at.%.

8.3.3. Tetrahedral Amorphous Carbon Films

The variation of the threshold field for TAC films with ion energy is shown in Figure 119. It is well established that the sp^2 content in TAC films is primarily controlled by the deposition ion energy [92], which also controls many of the microstructural [68] and electronic [380] properties. It is very interesting to note that the threshold field for electron emission in TAC films can be so well correlated to the sp^2 content (Fig. 120). In the case of laser ablated TAC films an opposite trend to that shown in Figure 120 was observed with the films with the lowest threshold having the highest sp^2 content [381]. Recent

analysis of the same data by Ilie suggests that it is the clustering of sp^2 states that are more important in the field emission properties than the sp^2 content [366]. It should be noted at the outset that the TAC films discussed here have a large defect concentration that far exceeds the DAC or PAC films. A forming process or initiation field was needed before stable emission could be observed on some of the films [382], but not all [366]. Once more no surface damage was observed after forming. Field emission as a function of nitrogen content is shown in Figure 121.

The variation observed in the threshold field as a function of energy mirrors the variation of the sp^2 content, which shows a peak in the value close to 80% sp^2 at an ion energy of -100 V [68]. The results of Satyanarayana [382] were attributed to a change in the electron affinity which was controlled by the sp^2 content as shown in Figure 122. In the case of nitrogen doping there were clearly two regimes, with films deposited with high nitrogen content ($>0.4\%$) showing a increase of the threshold field which was attributed to increased graphitization of these films and thus reduced sp^2 content. On the other hand reasons for the lowering of the threshold fields at low nitrogen levels was not discussed in any detail. Cheah et al. [383], who obtained similar FE results, attributed the observed variations

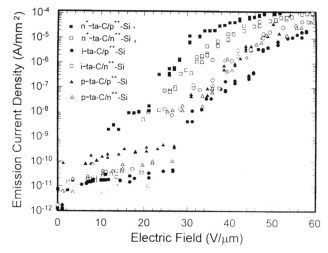

Fig. 123. Dependence of field emission on the substrate type and doping level for TAC films. Reproduced with permission from [383], © 1998, ???.

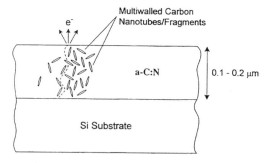

Carbon Nanoparticles Embedded in a Nitrogen-
Doped Amorphous Carbon Matrix

Fig. 124. Schematic diagram of multiwalled carbon nanotube fragments embedded in an *a*-C matrix. Reproduced with permission from [387], © 1999, MYU.

in their properties to space charge induced band bending that will be discussed in detail in the following section. It should be noted that TAC films deposited on different systems could have vastly differing properties.

Field emission studies on TAC films have also been performed by Cheah et al. [383], with similar threshold fields, though higher threshold fields were reported by Park et al. [384], Chuang et al. [385], and Missert et al. [381]. However, lower fields have also been observed by Talin et al. [367] and Xu et al. [386]. Results that show a dependence on the substrate on which the films have been deposited has also been reported by Xu et al. [383]. See Figure 123. In the figure it is clear that in order to explain the results a space charge induced band bending model must be invoked due to the dependence of the threshold field on substrate carrier type as well as the doping levels of the substrates.

8.3.4. Graphitelike Amorphous Carbon Films

There are a number of papers that have been published that show electron emission with very low threshold fields from polycrystalline graphite films. The example we will use in this study for a GAC film is a cluster assembled carbon thin film [61]. The films were deposited using a supersonic beam of carbon clusters in the laboratory of Milani [59], and the deposition process is described in detail under cluster beam sources. Raman spectroscopy was used to establish beyond reasonable doubt the graphitic nature of the deposited films with a prominent D band and a G peak at 1573 cm^{-1} which indicated ordered regions of sp^2 phase. The cluster size was around 2 nm, with a bandgap of close to 0.6 eV. Due to the graphitic nature of the film, the characteristics would be best described using a Fowler–Nordheim process, where the work function should be fixed at 5 eV. When this is performed a local enhancement factor of 350 is obtained. This is not an unreasonable number for a nanoclustered film, which is likely to have highly localized emission from the front surface.

8.3.5. Nanocomposite Amorphous Carbon Films

A typical NAC film that has nanoparticles embedded in an *a*-C matrix is shown schematically in Figure 124. The embedded nanoparticles in this case are fragments of multiwalled nanotubes that have been embedded in an *a*-C matrix using the cathodic vacuum arc technique for deposition. Simply by having a high pressure region of He or N_2 gas adjacent to the cathode striker Amaratunga et al. [387] showed that is was possible to rapidly condense the carbon arc into nanoparticles [1] such that composite films could be produced. Coll et al. [70] too showed similar microstructural results using the FCVA system to deposit cathodes. Typical results obtained from the NAC films are shown in Figure 125. Clearly, the threshold fields obtained are very low, with some scatter in the data as a result of various channels turning on and off. Each of these individual channels will have different emission characteristics, and the composite result is shown in Figure 125. The effect of nonuniform emission from the surface was shown by Amaratunga et al. when they etched the front surface of their films and showed that although there was an improvement in the emission threshold, this was at the expense of increased fluctuations in the emission characteristics. The results were explained using a Fowler–Nordheim tunnelling and hot electron emission composite model, in which the DAC matrix acts as an interlayer that moderates the emission and will be discussed later. When they compared the enhancement factors of their film assuming a work function of 5 eV, the β factor increased from 1200 to 1950 after the surface etch treatment. This further illustrated the nanocomposite nature of their films.

8.4. Modelling of the Electron Emission Process

The large variations available to the *a*-C microstructures make it difficult to postulate a single model that can explain all the observed characteristics within the films. But it is possible to formulate some basic rules that can be applied in the elucidation of the emission mechanism in the *a*-C films. The first attempts to explain the emission process in *a*-C (and polycrystalline diamond) based on negative electron affinity have been overtaken

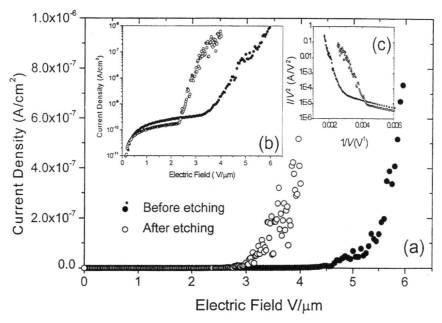

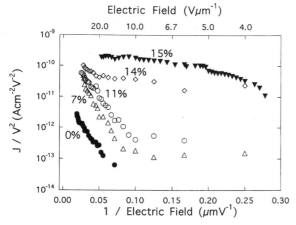

Fig. 125. Electron field emission characteristics of NAC films that show improvements to the threshold field after a surface etch treatment, at the expense of increased fluctuations. Reproduced with permission from [387], © 1999, MY.

Fig. 126. Fowler–Nordheim curve for nitrogenated DAC films. Reproduced with permission from [45], © 1996, American Institute of Physics.

by events that firmly point to this as being of secondary importance. In *a*-C, as shown by Robertson [388] in Figure 122, the low electron affinity of some of its configurations does help in expelling electrons from the films once it has entered the conduction band, but getting the electrons to the conduction band is not straightforward.

On an historic note, researchers first tried to explain the emission process at low electric fields using the FN equation. The FN curves for the nitrogen doped DAC films are shown in Figure 126. The threshold field is shown in Figure 118 and gives an increase in current density with increasing nitrogen content, with a concomitant reduction in the threshold field for electron emission. The FN curve shows linear regions that are expected for tunnelling currents. But the range of barrier values (assuming a β factor of 1 for mirror smooth films), 0.01–0.06 eV, and the range of emission areas, $\sim 10^{-24}$ cm², are nonrealistic when compared to experimental observations. Similar values were obtained for undoped DAC films [377], doped and undoped TAC films [382], and nearly all the forms of *a*-C films examined by researchers when a β factor of 1 was used for the mirror smooth films. According to FN theory, there are two parameters that control the emission process. First there is the barrier height between the electron source and vacuum at the surface, and second there is the β factor or local field enhancement. The barrier height is usually the work function for metals and the electron affinity for semiconductors both of which are invariant with doping. Therefore, the variation of the threshold field for emission with change in nitrogen content cannot be explained using classical FN theory. With the need to reconcile this apparent contradiction in the emission current as well as the barrier values and emission areas measured for *a*-C films in general, a thin film model for field emission was proposed that was based on a space charge induced band bending giving rise to hot electron processes [45]. In this model the *a*-C films acted as an interlayer with the true cathode being the underlying Si or metal substrate, with the crucial factor being the heterojunction or Schottky barrier that formed at the back contact being needed for the creation of hot electrons.

The need for space charge from the depleted bands of *a*-C can be viewed simply using basic semiconductor equations. If we were to take an ideal vacuum/semiconductor or insulator interface, where there is no accumulation of charge, and apply the Laplace theorem, from the conservation of electric flux as given by Gauss' law we find that the electric field within the thin film must be lower than the applied electric field by a factor given by the ratio of the material dielectric constant to that of vacuum ($\epsilon_{material}/\epsilon_{vacuum}$). As the dielectric constant of vacuum is 1, this means that in the case of a-C films, the maximum electric field that can exist within the material purely due to field penetration would be anything between 0.2 to 0.5 of the applied electric field outside the material. As we have seen emission of electrons below a macroscopic field of 10 V/μm regularly, it is difficult to see how a fraction of such an electric field will be able to allow for the emission of electrons from these cathodes. It should be noted that for there to be electron emission from a flat metal cathode electric fields as high as a 1000 V/μm are required.

On the other hand, if space charge based depletion is present in the a-C based cathodes, the solution to Poisson's equation gives rise to electric fields that are up to 5–10 times greater than the applied macroscopic electric field as shown by Amaratunga and Silva [45]. With fields of this magnitude, it is not difficult to envisage a hot electron process based on space charge induced bands as shown in Figure 126. As discussed previously, by the addition of inhomogenates in the thin films it is possible to increase the internal electric fields even further. The proposed model is applicable to any material system as long an the thin film can support space charge, and it has been used to explain emission from a-Si [389], a-SiC [390], and polycrystalline diamond [391] too. In the case of polycrystalline diamond, the space charge is supported across the grain boundaries of the diamond crystallites.

The proposed band diagram for emission from a-C films based on the space charge bending model is illustrated in Figure 127. Figure 127a shows the proposed band diagram at equilibrium for the heterojunction based hot electron model for electron emission. It is based on assuming that the a-C thin film is n-doped and has an electron affinity close to 2.5 eV (see Figure 122), a Fermi level of 0.5 eV based on the activation energy studies [29], and approximately 10^{16} cm^{-3} ionized nitrogen donors. It should be noted that the band bending at the n^{++}-Si/a-C interface is such that there is a collection of electrons at the surface which experience a barrier that oppose their movement into the conduction band of the a-C layer. When a bias is applied to the heterojunction and the anode voltage becomes positive, and the applied field is large enough to fully deplete the carbon thin film, it can be shown that the field observed by the electrons that tunnel through the heterojunction barrier into the conduction band is much greater than the applied electric field ($>$18 V/μm) due to the space charge considerations. This field decreases parabolically within the DAC:N thin film as it traverses toward the surface of the semiconductor (Fig. 127b). According to the band diagram shown in Figure 127b, the pool of electrons gathered at the n^{++}-Si/a-C interface has a built-in barrier that prevents the tunnelling/thermal excitation of electrons into the conduction band of the a-C layer before the interlayer band bending is large enough for the electrons to overcome the energy barrier for emission into the vacuum. The electrons can now lose up to 3 eV in phonon ionizations and still retain enough kinetic energy to enter the vacuum level in the a-C thin films and thus emit in to the vacuum. In the case of electrons that do not have enough energy to reach the vacuum,

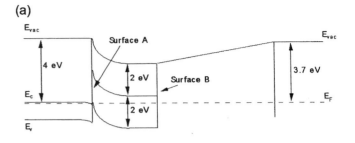

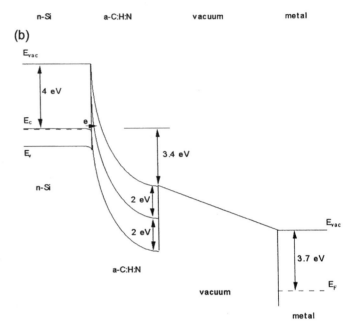

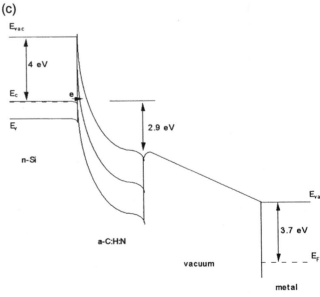

Fig. 127. The proposed field emission mechanism based on space-charge-induced band bending in a-C thin films.

they will collect at the front surface of the a-C thin film and decrease the band bending in the space charge layer as shown in Figure 127c. Yet, until the total band bending is lost, there will be some electrons emitted from the a-C thin films.

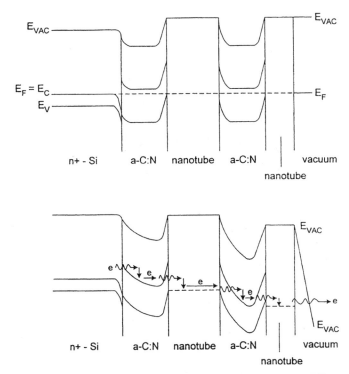

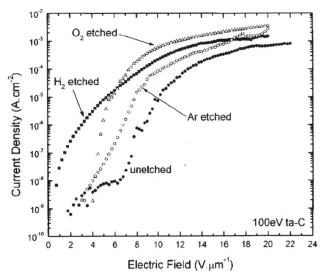

Fig. 129. Variation of the emission current density of as deposited TAC and after 2 nm etch by Ar, H_2, and O_2. Reproduced with permission from [379], © 1999, American Institute of Physics.

Fig. 128. The heterojunction based model for electron emission in NAC films. Reproduced with permission from [387], © 1999, MY.

Recent modifications to this model have now shown that it is possible for the collected negative charge at the surface to be removed from the surface via conducting dielectric channels. It has been shown using computer simulations that when a more conductive filament or channel is introduced into the thin film, there is a tendency for the electric field lines to emanate away from these "dielectric channels" such that on an E-field plot, as shown in Figure 113, electrons will be attracted toward them during electron emission. But it should be noted that having more than one channel within a space of 80 nm decreases the advantageous properties of the additions. The possibility of the existence of clustered sp^2 regions that could form such channels has been shown by Ilie et al. [366], using Raman analysis. The analysis performed is based on the modified analysis of Raman spectra introduced by Ferrari et al. [231] which show that the sp^2 clustering in TAC films varies as a function of deposition conditions. Such a mixed phase incorporated into an a-C film would be invisible to most surface analysis equipment available today. Also, it should be noted that the electric fields surrounding regions of these channels must be bent significantly due to space charge for the model to work.

A composite but serial "dielectric channel" space charge model was proposed by Ilie et al. [366], where they discussed a variant of the above discussion. In this case, the dielectric channel acted as a field enhancement conduit, whose β factor was used primarily for electron emission from the thin film surface at these localized regions. In the schematic diagram that was introduced the band bending shown was attributed to the filling up of amphoteric gap states close to the Fermi level due to negative space charge. In the discussion that followed, emis-

sion due to triple junction effects and/or the presence of space charge was not discounted. This model supersedes the earlier enhanced β factor due to the patch antenna surface termination model of Robertson [388, 392]. An emission process due to mixed sp^2/sp^2 phase materials was introduced by Xu et al. [393], where it was proposed that a step based space charge model could be used to explain the emission from thin TAC films. Also, the effects of surface termination on field emission properties were convincingly dismissed by their findings for different surface elements as will be discussed later.

Amaratunga et al. [387] also used the results based on the interlayer model [45] to successfully explain the variations of their NAC films. A band diagram explanation of the emission process of the NAC films is shown in Figure 128 with (a) no applied field and (b) applied field. In this case the internal junctions of the particles and the a-C matrix allow for heating of electrons within the thin film, with the emission process at the outer end of the particle being purely due to FN-like tunnelling (high β factors). Note that the band diagrams are symmetric about the y-axis due to the nature of the multiwalled nanotubes as well as the surrounding matrix. Variations in the threshold field with and without etching of the front surface was successfully explained by the surface to vacuum being either the host matrix or the nanotube.

Recently, Xu et al. [386] examined the electron field emission properties of a number of a-C thin films and came to the conclusion that the emission process cannot be described by any single mechanism. They proposed that for emission in carbon films there are two major regimes, with the intermediate region being a composite of the two. In their analysis all emission below 10 V/μm was attributed to thermionic emission, with emission for fields above 100 V/μm due to FN based field emission. This model was illustrated using numerical calculations for a-C films with varying barrier heights and fitting the results to the unified electron emission model of Murphy and Good. This analysis is totally consistent with the discussion above, with thermionic or hot electron emission virtually controlling all of the emission in most thin films. The only divergence is when high internal β factors or local enhancement comes into play due to mechanisms such as clustering.

Modinos and Xanthakis [394] have proposed an alternative mechanism for emission from the diamondlike carbon (DAC:N) films deposited and measured by Amaratunga and Silva [45]. The results of their calculations show the presence of a high density, $>10^{21}$ π states cm^{-3} eV^{-1}, in the mobility gap and it is from these π states near the Fermi energy that the electron emission originates. The addition of N results in an upward shift of the Fermi energy toward the conduction band. However, in order that the calculated emission current density matches the experimental values given in [45], Modinos and Xanthakis were forced to assume that the emission occurs at hemispherical protrusions at the surface. While they do not give dimensions of these protrusions, they have estimated that they at most produce an enhancement of the electric field by a factor of around 150. In this way the high density of π states and properties of the front surface determine the electron emission. No surface features or structures were observed in the DAC films by atomic force microscopy to support such an enhancement factor at the surface.

If we were to consider the rate limiting step in the electron emission process of a thin film semiconductor, it will either be at the back contact to the film, transport within the thin film, the emission from the front surface of the films, or a combination of two or more of the above. According to the hot electron model developed for the wide gap materials, the transport within the semiconductor and the emission at the front surface of the film is less of a limiting process than the back contact as long as the thin film can maintain the fully depleted space charge layer that supports the band bending. If the transport within the material is not hot electron based (highly defective or graphitic films), then bulk material properties become more and more significant, with this extending to the front surface due to clustering of the sp^2 states. The number and size of sp^2 clusters lying at or near the Fermi level then controls the emission and it is important that the clusters are large enough for efficient connectivity between clusters. In this way the a-C films deposited under different conditions will consist of a matrix of sp^2 C with regions of varying sp^2 cluster concentrations and sizes. Since these clusters will have different dielectric constants, the application of the external field will result in local field enhancements around the clusters and will aid in the emission of electrons. A similar conclusion concerning the importance of the size of the sp^2 cluster has also been reached by Ilie et al. based upon a correlation between the measured values of threshold fields for various types of amorphous carbon films. They determined that the optimum size of the clusters for emission, in-plane correlation length to be 1.5–2 nm depending on the specific type of film investigated. This model can also explain the nonuniformity of emission across the surface of the samples in terms of

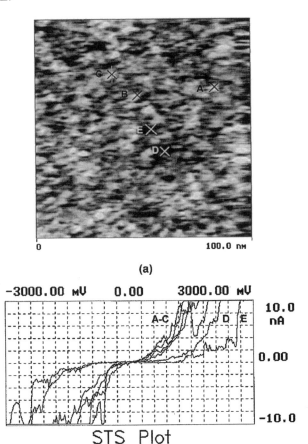

Fig. 130. CITS measurement of (a) STM image and (b) STS I–V curves for untreated TAC:N films. Reproduced with permission from [395], © 1999, American Vacuum Society.

different local concentrations of sp^2 clusters through the film extending from the surface.

Films deposited at high self-biases which have a high spin density $>10^{20}$ cm^{-3} (a high sp^2 content) and good connectivity between the clusters exhibit the lowest threshold fields. Such films would be characterized as "front surface emitters." Films with a comparable spin density but with poorer connectivity show higher threshold fields since the sp^2 cluster is not large enough. The films deposited at the low biases having low sp^2 content (low spin density) have lower threshold field by virtue of the good field penetration and the presence of a high internal electric field at the back contact.

8.5. Field Emission as a Function of Surface Modifications

Shi et al. [393] and Hart et al. [379] have studied the effects of postdeposition surface treatments using ion beam bombardment and ion plasmas, respectively. In the study performed by Shi et al. undoped and n^+ doped TAC deposited at 100 eV using the FCVA method was subjected to H, O, and Ar ion bombardment. Growth at 100 eV in undoped TAC results in the maximum sp^2 content and n-type doping was achieved by the addition of N. They noted that the threshold field in the undoped films decreased from 18 to 14 V/μm after treatment with each of the ions. For n^+-TAC ion beam treatment resulted in a decrease from 12 to 8 V/μm. The saturated emission current density of the post-treated films was noted to improve by about an order of magnitude and appeared to be independent of the ion used. Shi et al. also measured the effects on the threshold field of undoped TAC of varying the H ion energy from 50 to 800 eV. In this ion energy range the threshold field increased from 14 to 22 V/μm. Hart et al. studied the effects of reactive ion etching using H_2, O_2, and Ar gases as shown in Figure 129. Since TAC has an sp^2-rich surface layer extending about 1 nm from the surface, Hart et al. removed up to 2 nm of the surface, thereby allowing examination of the effects of this surface layer. As-deposited films had a threshold field of 8 V/μm. Etching in a H plasma reduced the threshold field to 3.0 V/μm if 0.5 nm was removed (incomplete sp^2 surface layer removal) and 3.9 V/μm if 2 nm (complete sp^2 surface layer removal). For O_2 etching the threshold field decreased to 5 V/μm for both a 0.5 and 2 nm etch. Ar plasma etching reduced the threshold field to 6 V/μm. Finally, emission maps using an ITO anode as a phosphor show that the emission site density measured at 20 V/μm increased from about 50–100 cm^{-2} in as-deposited films to 1000 cm^{-2} for O_2 etched films, 2000 cm^{-2} for Ar etched films, and 4000 cm^{-2} for H_2 etched films. Treatment with H_2 plasmas would be predicted to increase the probability of C–H surface termination. Therefore, within the confines of the negative electron affinity model of Robertson et al. discussed previously [388], it would be expected to reduce the emission barrier at the front TAC/vacuum interface. However, a reduction in the threshold field was also observed for O_2 plasma treatment which should raise the front surface barrier since the surface would be terminated with C–O bonds. The reasons for the reduction of the threshold field after O ion treatment are unclear. Since Hart et al. used a parallel plate geometry to measure FE characteristics the lowering of the threshold field is due to a higher emission site density and thus a greater distribution of sites with lower threshold fields. Atomic force microscopy showed that the RMS roughness of films did not change after plasma treatment and therefore surface features do not play a significant role.

AFM and also scanning tunnelling microscopy (STM) have also been employed by Cheah et al. [395]. AFM images showed that for n^+-TAC films the surface roughness did increase from 0.193 to 0.242 nm over a 500×500 nm^2 area after ion beam treatment. In addition nanoscale clusters were visible after treatment. Both STM and also current imaging tunnelling spectroscopy (CITS) have been used to attempt to correlate surface effects with tunnelling effects. STM images shown in Figures 130 and 131 reveal that H treated films contained clusters as well as untreated areas. CITS measurements of the clustered regions show that they require lower voltages for the onset of emission. UPS was used to examine both the untreated and post-treated films. In as deposited films UPS measurements indicate electron emission at about 4 eV with some emission attributed to defect levels at lower energies. After Ar ion treatment emission of photoelectrons occurs at 0.6 eV. Similar

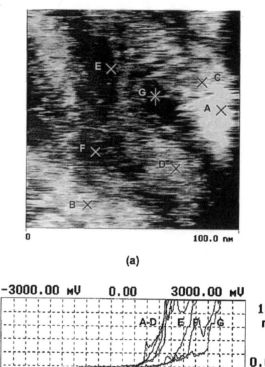

(a)

STS Plot

(b)

Fig. 131. CITS measurement of (a) STM image and (b) STS I–V curves for the post-treated TAC:N film. Reproduced with permission from [395], © 1999, American Vacuum Society.

emission is observed from diamond surfaces and Shi et al. attributed the emission at this level to a near 100% sp^2-rich phase. In this way Cheah et al. believe that after treatment nanoclusters of sp^2-rich material with sp^2-rich boundaries are formed. Since the sp^2 clusters have lower electron affinities than the sp^2-rich areas they provide an efficient way for electrons to be emitted into vacuum.

Further information about the role of surface treatment has been obtained from the implantation of C_{60} ions into PAC films by Carey et al. [396]. In this study films were implanted with 10 keV C_{60} ions to doses of 7.5×10^{13}, 1.25×10^{14}, and 2.65×10^{14} cm^{-2}. The ion source was a C_{60} sample evaporated in a heated crucible. C_{60} ions were also implanted at the highest dose into Si substrates and the average threshold field for virgin single crystal Si was 60 V/μm. The average threshold field for a conditioned nitrogenated and non-nitrogenated film as a function of C_{60} ion dose is shown in Figure 132a. It can be seen that after the initial dose of 7.5×10^{13} cm^{-2} the threshold field increased by nearly a factor of two for both films. However, at a higher dose of 1.25×10^{14} cm^{-2} the threshold fields

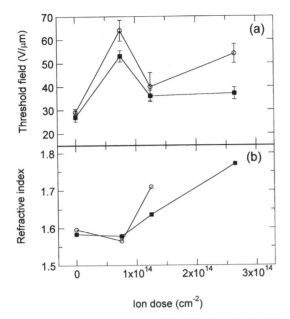

Fig. 132.   Variation of (a) the threshold field and (b) the refractive index with implant ion dose of C_{60}. Reproduced with permission from [396], © 2000, American Institute of Physics.

reduced to values 36 and 40 V/μm for the non-nitrogenated and nitrogenated films, respectively. The threshold field then remains constant for the non-nitrogenated film at the highest dose of 2.65×10^{14} cm^{-2}. For the nitrogenated film at the highest dose the threshold field again rises, approaching that measured for C_{60}^{+} implants into Si substrates ($\sim$60 V/μm). The refractive index of the films as a function of dose is shown in Figure 132b. No significant change is observed in the refractive index for either type of film after the lowest dose implantation. For higher doses the refractive index increases rapidly, reaching a value of 1.77 for the non-nitrogenated film, close to that of sp^2-rich graphite (1.8). In the case of the nitrogenated film, at the highest dose, no reliable refractive index or thickness data could be obtained. Finally, preliminary AFM has indicated that at the lowest implanted dose some nanostructural features 10–20 nm tall and 50–100 nm in diameter are observed. Even though surface features are present on the films after implanting to a dose of 7.5×10^{13} cm^{-2} and an increase in the threshold electric field compared with the unimplanted films is observed, we do not believe that the surface protrusions play a significant role in field emission. The field enhancement factor, β, can be related to the diameter of the surface r feature via

$$\beta = d/[kr(R - r)] \qquad (8.8)$$

where k is a constant ~ 5. For a vacuum separation d of 15 μm, this gives an enhancement factor of between 5 and 8. Silva et al. [397] showed that self-texturing of the surface of nitrogenated DAC films can occur leading to an enhancement of a factor of 8–10 due to changes in the surface morphology of these films. However, the current densities measured in field emission testing were over 100 times the current density from

a nontextured film from which they also concluded that the surface protrusions do not play a significant role in controlling the field emission properties. Molecular dynamic simulations indicate that a 10 keV C_{60} ion will on average penetrate 3.5 nm into a DAC film and will not have a significant effect on the bulk of the material. It is envisaged that the C_{60} fullerene on impact will break into sp^2-rich clusters that will get embedded into the surface of the DAC film. It was proposed that a surface sp^2-rich conductive layer will thus be formed, making it easier for charge collection at the surface to occur. This, in turn, alters the field experienced inside the semiconductor and, within the confines of the space charge induced band bending model described in Section 8.4, will make electron emission harder by increasing the energy barrier at the DAC/vacuum interface. Electrons that gain energy by traversing down the electric field due to the fully depleted interlayer will collect at the front surface of the cathode and thereby increase the surface potential (collection of negative charge) of the cathode. The subsequent reduction in the threshold field for the non-nitrogenated film at dose of 1.25×10^{14} cm^{-2} can be attributed to the reduction in thickness of the film discussed earlier (Section 8.3). It should be noted that in the case of the nitrogenated film, due to their lower mechanical strength and lower crosslinking, enhanced sputtering of the film will occur. Therefore, in the case of the highest ion dose for the nitrogenated film, it is believed that little or none of the original a-C:H:N film remains at the surface subsequent to the implant.

In summary, there are a plethora of different carbon films from which electron emission has been obtained at relatively low threshold fields. Amorphous carbon films appear to be unique in that most forms of carbon films give field emission at low fields. Each of these films has a unique structure with some common features with the catergorized a-C films discussed in this review. It is important that researchers first understand the basic microstructure of their a-C films before venturing into postulating a possible electron field emission mechanism in their films. There are a few more established emission models to which the electron emission from a number of these films can be attributed and the variations with the physical properties explained. It is envisaged that this topic will continue to grow for many years to come and it is imperative that a set nomenclature is established.

9. AMORPHOUS CARBON-BASED DEVICES

9.1. Electronic Devices

A number of devices have been fabricated based on amorphous carbon. One of the more straightforward electronic devices is probably the p-n junction. There has only been one reported case of this device as documented by McKenzie et al. [327]. A graded film of TAC/TAC:N was grown onto a plastic substrate and the resulting film was found to rectify by approximately one order of magnitude. However, it was reported that the defective surface layer may play a part in hindering the device performance.

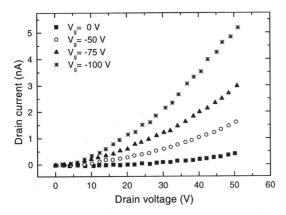

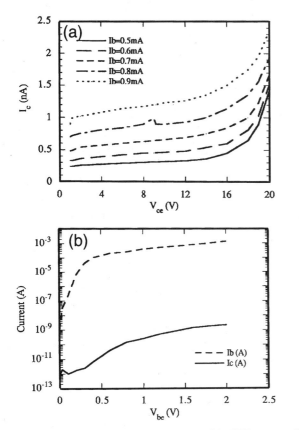

Fig. 133. TAC TFT drain characteristics as a function of gate bias. Reproduced with permission from [398], © 1996, IEEE.

Fig. 134. (a) Output and (b) Gummel characteristic of the HBT structure. Reproduced with permission from [317], © 1993, Elsevier.

Various groups have attempted to fabricate thin film transistors (TFTs). The only documented transistor based on *a*-C was documented by Clough et al. [398]. Here, a TFT was fabricated, and a coplanar design was chosen so that the TAC surface sp^2 layer could be etched using a nitrous oxide plasma. The channel length was varied between 10 and 30 μm and the gate insulator was silicon nitride. The drain characteristic for this transistor is shown in Figure 133. It was found that the TFT operated in the *p*-channel enhancement mode, as would be predicted if undoped TAC was *p*-type. The calculated field effect mobility

was only in the order of 10^{-6} cm^2 V^{-1} s^{-1}, and the characteristic did not saturate, suggesting band-tail rather than extended state conduction. The high field conduction showed a near V^2 dependence which was attributed to space-charge-limited current.

Chan et al. [399] has demonstrated a working *a*-C-based heterojunction bipolar transistor (HBT). A 50 nm DAC film was used as the emitter, which was grown onto *p*-implanted Si which was contacted as the base, and the underlying *n*-type Si substrate functioned as the collector. The collector current as a function of the collecter–emitter bias is shown in Figure 134. A Gummel plot of the characteristics obtained showing the variation of I_C versus I_B is also shown. It can be seen that the collector current increases with I_B and V_{CE}. This was shown to be due to carrier injection from the DAC emitter and not due the leakage current across the reverse-biased junction. No current gain was observed, though, and it was surmised that current injection took place from trap states in the DAC film, which suggested a pinned Fermi level in a deep acceptor band centered around 0.3–0.35 eV from the valence band edge.

NACs have been investigated for use in supercapacitors. These are electrochemical duoble-layer capacitors formed by two electrodes, a separator and electrolyte. The ions of the electrolyte are adsorbed onto the charged electrode, resulting in a "Helmholz layer." As the surface area must be maximized in order to increase the capacitance of the device, and an inert material is required, NAC has been used for the electrode. Furthermore, NAC may be grown over large areas on a variety of substrates which further enhances its potential in this role. A capacitor with series resistance $R = 1/13$ Ω, $C = 0.099$ F, and maximum power density of 506 kW/kg was fabricated, which suggested that NAC was very attractive for this particular application.

9.2. Novel Devices: Nonvolatile Memories, Antifuses, and MIMs

The other remaining applications of *a*-C films have been in a niche capacity. Gerstner et al. [400] suggested that TAC:N may have promise as a material for use in nonvolatile memories. It was observed that these films showed switching, in that that after a negative bias was applied across the film, a kink was observed in the forward bias, and the small signal resistance of the forward bias was found to fall from an "OFF" state to an "ON" state. Hence, the film possessed a "memory" of whether the last bias was negative or not. This is shown diagrammatically in Figure 135. This was explained by suggesting that the "OFF" state corresponded to conduction by hopping around the Fermi level, and the "ON" state corresponded to Poole–Frenkel conduction, according to I/V and $I/V/T$ data. Therefore, it was suggested that the negative switching bias resulted in electrons being promoted from deep acceptor traps to shallow donor traps, where they could conduct according to Pool–Frenkel conduction. The switch from "ON" to "OFF' was as a result of the accumulation of electrons in the conduction band enhancing the refilling of electrons in the acceptor traps. This memory effect

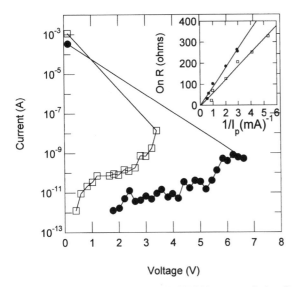

Fig. 135. *I/V* characteristic of nonvolatile TAC:N memory device. Reproduced with permission from [400], © 1998, American Institute of Physics.

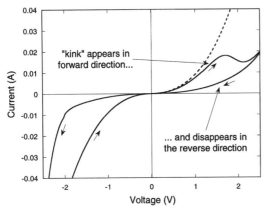

Fig. 136. *I/V* characteristic of antifuse based on DAC (□) and DAC:N:F (●). Reproduced with permission from [401], © 1998, IEEE.

bottom contact. The same would occur with excessive contact pressure resulting in punch-through of the film. Also, the short circuit may be only observed when a small voltage is applied. No information as to the reproducibility of these devices was documented and therefore there may be some conjecture as to whether this was a true antifuse effect or not.

It has already been stated that DAC is a promising material if used as a switching element in metal–insulator–metal based LCDs, as documented by Egret et al. [308]. This is due to the fact that in present TFT-based technology the function of the transistor is to allow charge to flow into or out of the pixel. Two terminal devices will perform a similar role at reduced cost, although the driver circuitry would be more complex. At high fields DAC exhibits PF conduction which allows a sharp turn-on voltage and hence suitability for this application. In particular, the dielectric constant of DAC is of the order of 5–8 which results in a high value for the PF β factor, resulting in a sharper turn-on voltage than other materials (Ta_2O_5 for example).

9.3. Solar Cells

It has also been mentioned that TAC:N shows promise as a photoconductive material [328]. This due to its benefits in terms of photoconductivity and high optical absorbance. However, others such as Lee et al. [402] have also attempted to fabricate *p–i–n* solar cells using boron-doped DAC (DAC:H:B) as the *p*-layer. As the DAC:H:B layer was too resistive for it to make up the whole contact, it was proposed to grow an ultrathin layer of it between the transparent conducting oxide (TCO) and *p*-SiC layer, in order for it to add to the SiC layer and enhance the built-in potential of the cell. It was found that the collection efficiency of the cell was enhanced with the DAC:H:B layer. This was thought to be due to the reduction in band bending experienced by the holes flowing toward the TCO front contact.

Another interesting potential application of *a*-C thin films is as an antireflection coating. Indeed, PAC films possess a wide Tauc bandgap (2.6 eV), low refractive index (1.5–1.7), and low optical absorbance. Furthermore, by varying the deposition conditions these parameters can be tuned so that multilayer antireflective (AR) coatings can be realized. A brief study of PAC-based AR coatings was carried out by Khan et al. [403]. It was found that it was possible to optimize film thicknesses to 70–85 nm and refractive indices to 1.5–1.7, which followed the required parameters for a double-layer AR coat onto a glass-fronted standard *a*-Si:H solar cell.

There has also been work carried out into the growth of boron-implanted C_{60} thin films, serving as the *p*-layer in heterojunction solar cells [404]. However, such films are beyond the scope of this review.

9.4. Summary

Some possible applications for *a*-C based devices have been demonstrated. It must be noted, though, that most of these roles are in a niche capacity and that the electronic properties of *a*-C

occurred with a write time down to 100 μs and the memory retention time was in the order of months.

An antifuse is a device which forms a state of low resistance when a threshold electric field is passed. *a*-C shows promise for such a device, such as low OFF-state leakage current, ON-state resistance, and dielectric constant. Most importantly, *a*-C antifuses do not show ON–OFF switching. Liu et al. [401] fabricated and reported upon DAC and DAC:N:F antifuses using a sandwich structure and Al top contacts, and the resulting *I/V* characteristic is shown in Figure 136. It was found that the breakdown voltage was about 1.4×10^6 V cm^{-1}. The breakdown was attributed to the temperature effect of current heating resulting in hydrogen loss and hence sp^2 reordering. This resulted in a graphitic phase being formed, which caused the observed low-resistance state. However, it must be noted that DAC sandwich structures are notoriously difficult to measure as any particulate matter underneath a top contact would render the device useless by causing a short circuit between the top and

have still not been sufficiently improved in order for the material to demonstrate the properties required for a large area semiconductor. There is some hope that the delocalization of gap states through ion implantation may serve to improve the electronic properties so that a range of semiconducting devices will be realized.

10. CONCLUSION

We have reviewed the state of the art in amorphous carbon thin films in the preceding sections. A new nomenclature based on the microstructure of the a-C films has been introduced to distinguish the various forms of amorphous carbon films available in the literature. This has helped us to review this field and it is envisaged that it will help in reducing the confusion that already exists in the field of carbon. Although the growth, microstructure, and properties of these films have been studied for over 30 years, we have yet to see the full potential of this material being exploited commercially. In terms of a robust and wear resistant protective coating a-C is used the world over for its enviable mechanical properties. In terms of an optical coating and a biocompatible thin film it is now entering a number of niche markets. As an electronic material, to date no applications have used a-C, but the time is near when the enviable electronic properties of this material will be utilized. There is still much to learn on the microstructural, electrical conduction, and optical properties.

REFERENCES

1. G. A. J. Amaratunga, M. Chhowalla, C. J. Kiely, I. Alexandrou, R. Aharonov, and R. M. Devenish, *Nature* 383, 321 (1996).
2. UCFMG and Group, in "Diamond Growth and Films" Elsevier, London/New York, 1989.
3. A. G. Whittaker, *Science* 200, 763 (1978).
4. P. P. K. Smith and P. R. Buseck, *Science* 216, 984 (1982).
5. S. R. P. Silva, K. M. Knowles, G. A. J. Amaratunga, and A. Putnis, *Diamond Relat. Mater.* 3, 1048 (1994).
6. H. Kroto, *Science* 242, 1139 (1988).
7. W. Kratschmer, D. L. Lowell, K. Fostiropopoulos, and D. R. Huffman, *Nature* 347, 354 (1990).
8. W. von Bolton, *Z. Elektrochem.* 17, 971 (1911).
9. F. P. Bundy, H. T. Hall, H. M. Strong, and R. J. Wentorf, *Nature* 176, 51 (1955).
10. W. G. Eversole, "Synthesis of Diamond." 1962.
11. Y. Mori, H. Kawarada, and A. Hiraki, *Appl. Phys. Lett.* 58, 940 (1991).
12. S. P. Lansley, H. Looi, W. D. Whitfield, and R. B. Jackman, *Diamond Relat. Mater.* 8, 946 (1999).
13. R. B. Heimann, J. Kleimann, and N. M. Salansky, *Nature* 306, 164 (1983).
14. J. Baggott, *New Scientist* 131, 34 (1991).
15. P. R. Surjan, *J. Molecular Structure* 338, 215 (1995).
16. A. P. Papworth, C. J. Kiely, G. A. J. Amaratunga, S. R. P. Silva, and A. P. Burden unpublished.
17. M. Weiler, S. Sattel, K. Jung, H. Erhardt, V. S. Veerasamy, and J. Robertson, *Appl. Phys. Lett.* 64, 2798 (1994).
18. J. Schwan, S. Ulrich, V. Batori, H. Ehrhardt, and S. R. P. Silva, *J. Appl. Phys.* 80, 440 (1996).
19. X. Shi, B. K. Tay, Z. Sun, L. K. Cheah, W. I. Milne, S. R. P. Silva, and Y. Lifshitz, in "Amorphous Carbon: State of the Art" (S. R. P. Silva,
J. Robertson, W. I. Milne, and G. A. J. Amaratunga, Eds.), pp. 262–271. World Scientific, Singapore, 1998.
20. H. J. Scheibe, D. Schneider, B. Schultrich, C. F. Meyer, and H. Ziegele, in "Amorphous Carbon: State of the Art" (S. R. P. Silva, J. Robertson, W. I. Milne, and G. A. J. Amaratunga, Eds.), pp. 252–261. World Scientific, Singapore, 1998.
21. Y. Lifshitz, in "Amorphous Carbon: State of the Art" (S. R. P. Silva, J. Robertson, W. I. Milne, and G. A. J. Amaratunga, Eds.), pp. 14–31. World Scientific, Singapore, 1998.
22. S. R. P. Silva, S. Xu, B. K. Tay, H. S. Tan, H. J. Scheibe, M. Chhowalla, and W. I. Milne, *Thin Solid Films* 317, 290 (1996).
23. C. B. Collins and F. Davanloo, *Surf. Coatings Technol.* 47, 754 (1991).
24. P. Koidl, C. Wild, B. Dischler, J. Wagner, and M. Ramsteiner, *Mater. Sci. Forum* 52&53, 41 (1989).
25. S. R. P. Silva, G. A. J. Amaratunga, and C. P. Constantinou, *J. Appl. Phys.* 72, 1149 (1992).
26. R. C. Barklie, M. Collins, and S. R. P. Silva, *Phys. Rev. B* 61, 3546 (2000).
27. N. A. Morrison, S. Muhl, S. E. Rodil, A. C. Ferrari, M. Nesladek, W. I. Milne, and J. Robertson, *Phys. Stat. Sol. A* 172, 79 (1999).
28. S. R. P. Silva, K. J. Clay, S. P. Speakman, and G. A. J. Amaratunga, *Diamond Relat. Mater.* 4, 977 (1995).
29. S. R. P. Silva, J. Robertson, G. A. J. Amaratunga, B. Rafferty, L. M. Brown, J. Schwan, D. F. Franceschini, and G. Mariotto, *J. Appl. Phys.* 81, 2626 (1997).
30. T. Soga, T. Jimbo, K. M. Krishna, and M. Umeno, in "Proceedings on the Specialist Meeting on Amorphous Carbon" (X. Shi, S. R. P. Silva, W. I. Milne, and G. A. J. Amaratunga, Eds.), pp. 206–217. World Scientific, Singapore, 2000.
31. V. Meenakshi and S. V. Subramanyam, in "Proceedings on the Specialist Meeting on Amorphous Carbon" (X. Shi, S. R. P. Silva, W. I. Milne, and G. A. J. Amaratunga, Eds.), pp. 224–229. World Scientific, Singapore, 2000.
32. S. Aisenberg and R. Chabot, *J. Appl. Phys.* 42, 2953 (1971).
33. I. I. Aksenov, V. A. Belous, V. G. Padalka, and V. M. Khoroshikh, *Sov. J. Plasma Phys.* 4, 425 (1978).
34. "Diamond 1999 Proceedings" (J. Robertson, H. Guttler, H. Karawada, and Z. Sitar, Eds.), Elsevier, Prague, 1999.
35. K. K. Chan, S. R. P. Silva, and G. A. J. Amaratunga, *Thin Solid Films* 212, 232 (1992).
36. A. M. Sladkov, *Sov. Sci. Rev. B* 3, 75 (1981).
37. Y. P. Kudryavtsev, R. B. Heimann, and S. E. Evsyukov, *J. Mater. Sci.* 31, 5557 (1996).
38. T. Frauenheim, J. Jungnickel, T. Koehler, P. Sitch, and P. Blaudeck, in "Amorphous Carbon: State of the Art" (S. R. P. Silva, J. Robertson, W. I. Milne, and G. A. J. Amaratunga, Eds.), pp. 59–72. World Scientific, Singapore, 1998.
39. C. Z. Wang and K. M. Ho, *Phys. Rev. B.* 50, 12429 (1994).
40. L. Holland and S. M. Ojha, *Thin Solid Films* 38, 17 (1976).
41. A. Bubenzer, B. Dischler, G. Brandt, and P. Koidl, *J. Appl. Phys.* 54, 4590 (1983).
42. Y. Catherine, *Mater. Sci. Forum* 52&53, 175 (1989).
43. G. A. J. Amaratunga, S. R. P. Silva, and D. R. McKenzie, *J. Appl. Phys.* 70, 5374 (1991).
44. A. Grill and V. Patel, *Appl. Phys. Lett.* 60, 2089 (1992).
45. G. A. J. Amaratunga and S. R. P. Silva, *Appl. Phys. Lett.* 68, 2529 (1996).
46. N. Savvides, *J. Appl. Phys.* 59, 4133 (1986).
47. J. Schwan, S. Ulrich, T. Theel, H. Roth, H. Ehrhardt, P. Becker, and S. R. P. Silva, *J. Appl. Phys.* 82, 6024 (1997).
48. N. Savvides, *Thin Solid Films* 163, 13 (1988).
49. J. J. Cuomo, J. P. Doyle, J. Bruley, and J. C. Liu, *Appl. Phys. Lett.* 58, 466 (1991).
50. B. Andre, F. Rossi, and H. Dunlop, *Diamond Relat. Mater.* 1, 307 (1992).
51. R. Kleber, M. Weiler, A. Kruger, S. Sattel, G. Kunz, K. Jung, and H. Ehrhardt, *Diamond Relat. Mater.* 2, 246 (1991).
52. E. G. Spencer, P. H. Schmidt, D. H. Joy, and F. J. Sansalone, *Appl. Phys. Lett.* 29, 118 (1976).

53. C. Weissmantel, K. Bewilogua, D. Dietrich, H. J. Erler, H. Hinnerberg, S. Klose, W. Nowick, and G. Reisse, *Thin Solid Films* 72, 19 (1980).

54. S. Xu, D. Flynn, B. K. Tay, S. Prawer, K. W. Nugent, S. R. P. Silva, Y. Lifshitz, and W. I. Milne, *Philos. Mag. B* 76, 351 (1997).

55. Y. Lifshitz, S. R. Kasi, and J. W. Rabalais, *Phys. Rev. Lett.* 62, 1290 (1989).

56. T. Miyazawa, S. Misawa, S. Yoshida, and S. Gonda, *J. Appl. Phys.* 55, 188 (1984).

57. H. Hofsäss and C. Ronning, in "Proceedings of the International Conference on Beam Processing of Advanced Materials," pp. 29–56. ASM International, Cleveland, 1996.

58. J. P. Hirvoden, J. Koskinen, R. Lappalainen, and A. Anttila, *Mater. Sci. Forum* 52&53, 197 (1989).

59. P. Milani, M. Ferretti, P. Piseri, C. E. Bottani, A. Ferrari, A. L. Bassi, G. Guizzetti, and M. Patrini, *J. Appl. Phys.* 82, 5793 (1997).

60. T. Takagi, I. Yamada, and H. Takabe, *J. Cryst. Growth* 45, 318 (1978).

61. A. C. Ferrari, B. S. Satyanarayana, J. Robertson, W. I. Milne, E. Barborini, P. Piseri, and P. Milani, *Europhys. Lett.* 46, 245 (1999).

62. L. Diederich, E. Barborini, P. Piseri, A. Podesta, P. Milani, A. Schneuwly, and R. Gally, *Appl. Phys. Lett.* 75, 2662 (1999).

63. D. H. Lee, K. C. Walter, and M. Nastasi, *J. Vac. Sci. Technol. B* 17, 818 (1999).

64. X. M. He, J. F. Bardeau, D. H. Lee, K. C. Walter, M. Tuszewski, and M. Nastasi, *J. Vac. Sci. Technol. B* 17, 822 (1999).

65. P. J. Martin, S. W. Filipczuk, R. P. Netterfield, J. S. Field, D. F. Whitnall, and D. R. Mckenzie, *J. Mater. Sci. Lett.* 7, 410 (1988).

66. S. Xu, B. K. Tay, H. S. Tan, L. Zhong, Y. Q. Tu, S. R. P. Silva, and W. I. Milne, *J. Appl. Phys.* 79, 7234 (1996).

67. D. M. Sanders, *J. Vac. Sci. Technol. A* 7, 2339 (1989).

68. P. J. Fallon, V. S. Veerasany, C. A. Davis, J. Robertson, G. A. J. Amaratunga, W. I. Milne, and J. Koskinen, *Phys. Rev. B* 48, 4777 (1993).

69. M. Chhowalla, J. Robertson, C. W. Chen, S. R. P. Silva, C. A. Davis, G. A. J. Amaratunga, and W. I. Milne, *J. Appl. Phys.* 81, 139 (1997).

70. B. F. Coll, J. Jaskie, J. Markham, E. Menu, and A. Talin, in "Amorphous Carbon: State of the Art" (S. R. P. Silva, J. Robertson, W. I. Milne, and G. A. J. Amaratunga, Eds.), pp. 91–116. World Scientific, Singapore, 1998.

71. C. L. Marquadt, R. T. Williams, and D. J. Nagel, *Mater. Res. Soc. Symp. Proc.* 38, 325 (1985).

72. A. A. Voevodin and J. S. Zabinski, in "Amorphous Carbon: State of the Art" (S. R. P. Silva, J. Robertson, W. I. Milne, and G. A. J. Amaratunga, Eds.), pp. 237–251. World Scientific, Singapore, 1998.

73. A. A. Voevodin, S. V. Prasad, and J. S. Zabinski, *J. Appl. Phys.* 82, 855 (1997).

74. L. Martinu, A. Raveh, D. Boutard, S. Houle, D. Poitras, N. Vella, and M. Wertheimer, *Diamond Relat. Mater.* 2, 673 (1993).

75. M. Zarrabian, G. Turban, and N. Fourches, in "Amorphous Carbon: State of the Art" (S. R. P. Silva, J. Robertson, W. I. Milne, and G. A. J. Amaratunga, Eds.), pp. 117–129. World Scientific, Singapore, 1998.

76. C. Godet, in "Amorphous Carbon: State of the Art" (S. R. P. Silva, J. Robertson, W. I. Milne, and G. A. J. Amaratunga, Eds.), pp. 186–198. World Scientific, Singapore, 1998.

77. J. Schwan, V. Batori, S. Ulrich, H. Ehrhardt, and S. R. P. Silva, *J. Appl. Phys.* 84, 2071 (1998).

78. J. Robertson and E. P. O'Reilly, *Phys. Rev. B* 35, 2946 (1987).

79. J. Robertson, *Diamond Relat. Mater.* 4, 297 (1995).

80. G. Galli, R. M. Martin, R. Car, and M. Parrinello, *Phys. Rev. Lett.* 62, 555 (1989).

81. C. A. Davis, S. R. P. Silva, R. E. Dunin-Borkowski, G. A. J. Amaratunga, K. M. Knowles, and W. M. Stobbs, *Phys. Rev. Lett.* 75, 4258 (1995).

82. T. M. Burke, *J. Non-Cryst. Solids* 164, 1139 (1993).

83. J. K. Walters, P. J. R. Honeybone, D. W. Huxley, R. J. Newport, and W. S. Howells, *Phys. Rev. B* 50, 831 (1994).

84. T. Frauenheim, P. Blaudeck, U. Stephan, and G. Jungnickel, *Phys. Rev. B* 48, 4823 (1993).

85. D. A. Drabold, P. A. Fedders, and P. Stumm, *Phys. Rev. B* 49, 16415 (1994).

86. J. V. Anguita, S. R. P. Silva, A. P. Burden, B. J. Sealy, S. Haq, M. Hebbron, I. Sturland, and A. Pritchard, *J. Appl. Phys.* 86, 6276 (1999).

87. C. Z. Wang, K. M. Ho, and C. T. Chan, *Phys. Rev. Lett.* 70, 611 (1993).

88. S. R. P. Silva, J. Robertson, Rusli, G. A. J. Amaratunga, and J. Schwan, *Philos. Mag. B* 74, 369 (1996).

89. P. Reinke, W. Jacob, and W. Moller, *J. Appl. Phys.* 74, 1354 (1993).

90. J. Robertson, in "Amorphous Carbon: State of the Art" (S. R. P. Silva, J. Robertson, W. I. Milne, and G. A. J. Amaratunga, Eds.), pp. 32–45. World Scientific, Singapore, 1998.

91. F. Seitz and J. S. Koehler, *Solid State Phys.* 3, 305 (1956).

92. S. R. P. Silva, S. Xu, B. K. Tay, H. S. Tan, and W. I. Milne, *Appl. Phys. Lett.* 69, 491 (1996).

93. C. A. Davis, *Thin Solid Films* 30, 276 (1993).

94. D. R. McKenzie, D. A. Muller, and B. A. Pailthorpe, *Phys. Rev. Lett.* 67, 773 (1991).

95. P. C. Kelires, in "Amorphous Carbon: State of the Art" (S. R. P. Silva, J. Robertson, W. I. Milne, and G. A. J. Amaratunga, Eds.), pp. 73–88. World Scientific, Singapore, 1998.

96. H. J. Scheibe and B. Schultrich, *Thin Solid Films* 246, 92 (1994).

97. M. Weiler, J. Robertson, S. Sattel, V. S. Veerasamy, K. Jung, and H. Ehrhardt, *Diamond Relat. Mater.* 4, 268 (1995).

98. Y. Lifshitz, S. R. Kasi, and J. W. Rabalais, *Phys. Rev. Lett.* 62, 1290 (1989).

99. D. R. McKenzie, D. A. Muller, B. A. Pailthorpe, Z. H. Wang, E. Kravtchinskaia, D. Segal, P. B. Lukins, P. D. Swift, P. J. Martin, G. A. J. Amaratunga, P. H. Gaskell, and A. Saeed, *Diamond Relat. Mater.* 1, 51 (1991).

100. Y. Lifshitz, G. D. Lempert, E. Grossman, I. Avigal, C. Uzan-Saguy, R. Kalish, J. Kulik, D. Marton, and J. W. Rabalais, *Diamond Relat. Mater.* 4, 318 (1995).

101. J. J. Cuomo, D. L. Pappas, J. Bruley, J. P. Doyle, and K. L. Saenger, *J. Appl. Phys.* 70, 1706 (1991).

102. R. F. Egerton, "Electron Energy Loss Spectroscopy in the Electron Microscopy." Plenum, New York, 1986.

103. J. Robertson, *Surface Coatings Technol.* 50, 185 (1992).

104. J. Daniels, C. V. Festenberg, H. Raether, and K. Zeppenfeld, *Springer Tracts Modern Phys.* 54, 77 (1970).

105. C. Weissmantel, K. Bewilogua, C. Schurer, K. Breuer, and H. Zscheile, *Thin Solid Films* 61, L1 (1979).

106. S. D. Berger, D. R. McKenzie, and P. J. Martin, *Philos. Mag. Lett.* 57, 285 (1988).

107. D. G. McCulloch, E. G. Gerstner, D. R. McKenzie, S. Prawer, and R. Kalish, *Phys. Rev. B* 52, 850 (1995).

108. E. Grossman, G. D. Lempert, J. Kulik, D. Marton, J. W. Rabalais, and Y. Lifshitz, *Appl. Phys. Lett.* 68, 1214 (1996).

109. J. Schwan, S. Ulrich, H. Roth, E. Ehrhardt, S. R. P. Silva, J. Robertson, R. Samlenski, and R. Brenn, *J. Appl. Phys.* 79, 1416 (1996).

110. E. G. Gerstner, P. B. Lukins, D. R. McKenzie, and D. G. McCulloch, *Phys. Rev. B* 54, 14504 (1996).

111. C. A. Davis, V. S. Veerasamy, G. A. J. Amaratunga, W. I. Milne, and D. R. McKenzie, *Philos. Mag. B* 69, 1121 (1994).

112. C. A. Davis, D. R. McKenzie, Y. Yin, E. Kravtchinskaia, G. A. J. Amaratunga, and V. S. Veerasamy, *Philos. Mag. B* 69, 1133 (1994).

113. D. J. H. Cockayne and D. R. McKenzie, *Acta. Cryst. A* 44, 870 (1988).

114. D. J. H. Cockayne, D. R. McKenzie, and D. A. Muller, *Microsc. Microanal. Microstruct.* 2, 359 (1991).

115. D. R. McKenzie, D. A. Muller, E. Kravtchinskaia, D. Segal, and D. J. H. Cockayne, *Thin Solid Films* 206, 198 (1991).

116. P. H. Gaskell, A. Saeed, P. Chieux, and D. R. McKenzie, *Philos. Mag. B* 66, 155 (1992).

117. P. H. Gaskell, A. Saeed, P. Chieux, and D. R. McKenzie, *Phys. Rev. Lett.* 67, 1286 (1991).

118. F. Li and J. Lannin, *Phys. Rev. Lett.* 65, 1905 (1990).

119. K. W. R. Gilkes, P. H. Gaskell, and J. Robertson, *Phys. Rev. B* 51, 12303 (1995).

120. G. A. J. Amaratunga, D. E. Segal, and D. R. McKenzie, *Appl. Phys. Lett.* 59, 69 (1991).

121. D. R. McKenzie, *J. Vac. Sci. Technol. B* 11, 1928 (1993).
122. J. Robertson, *Diamond Relat. Mater.* 2, 984 (1993).
123. N. A. Marks, Ph.D. Thesis, School of Physics, University of Sydney, 1996.
124. N. A. Marks, D. R. McKenzie, and B. A. Pailthorpe, *Phys. Rev. B*. 53, 4117 (1996).
125. A. C. Ferrari, B. Kleinsorge, N. A. Morrison, A. Hart, V. Stolojan, and J. Robertson, *J. Appl. Phys.* 85, 7191 (1999).
126. J. P. Sullivan and T. A. Friedmann, in "Proc. 1st Int. Spec. Meeting Amorphous Carbon" (S. R. P. Silva, J. Robertson, W. I. Milne, and G. A. J. Amaratunga, Eds.), pp. 281–295. World Scientific, London, 1997.
127. D. Beeman, J. Silverman, R. Lynds, and M. R. Anderson, *Phys. Rev. B* 30, 870 (1984).
128. J. Kakinoki, K. Katada, T. Hanawa, and T. Ino, *Acta Crystallogr.* 13, 171 (1960).
129. P. N. Keating, *Phys. Rev.* 145, 637 (1966).
130. F. H. Stillinger and T. A. Weber, *Phys. Rev. B* 31, 5262 (1985).
131. A. Bensan, Quantum Mechanical Bonding in Graphitic Structures, Honors Thesis, School of Physics, University of Sydney, 1990.
132. P. Mahon, B. A. Pailthorpe, and G. B. Bacskay, *Philos. Mag. B* 63, 1419 (1991).
133. E. G. Gerstner and B. A. Pailthorpe, *J. Non-cryst. Solids.* 189, 258 (1994).
134. N. A. Marks, Computational Study of Amorphous Carbon, Honors thesis, School of Physics, University of Sydney, 1991.
135. B. A. Pailthorpe, *J. Appl. Phys.* 70, 543 (1991).
136. L. Verlet, *Phys. Rev.* 159, 98 (1967).
137. L. Verlet, *Phys. Rev.* 165, 201 (1968).
138. J. Tersoff, *Phys. Rev. Lett.* 56, 632 (1986).
139. J. Tersoff, *Phys. Rev. B* 37, 6991 (1988).
140. J. Tersoff, *Phys. Rev. Lett.* 61, 2879 (1988).
141. J. Tersoff, *Phys. Rev. B* 44, 12039 (1991).
142. U. Stephan and M. Haase, *J. Phys. Condens. Matter* 5, 9157 (1993).
143. H. P. Kaukonen and R. M. Nieminen, *Phys. Rev. Lett.* 68, 620 (1992).
144. P. C. Kelires, *Phys. Rev. Lett.* 68, 1854 (1992).
145. P. C. Kelires, *Phys. Rev. B* 47, 1829 (1993).
146. P. C. Kelires, *Phys. Rev. Lett.* 73, 2460 (1994).
147. W. Kohn and L. J. Sham, *Phys. Rev. A* 140, 1133 (1965).
148. T. Frauenheim, G. Jungnickel, U. Stephan, P. Blaudeck, S. Deutschmann, M. Weiler, S. Sattel, K. Jung, and H. Ehrhardt, *Phys. Rev. B* 50, 7940 (1994).
149. U. Stephan, T. Frauenheim, P. Blaudeck, and G. Jungnickel, *Phys. Rev. B* 49, 1489 (1994).
150. C. H. Xu, C. Z. Wang, and K. M. Ho, *J. Phys. Condens. Matter* 4, 6047 (1992).
151. C. Z. Wang and K. Ho, *Phys. Rev. Lett.* 71, 1184 (1993).
152. O. F. Sankey and D. J. Niklewski, *Phys. Rev. B* 40, 3979 (1989).
153. R. Car and M. Parrinello, *Phys. Rev. Lett.* 55, 2471 (1985).
154. D. K. Remler and P. A. Madden, *Molecular Phys.* 70, 921 (1990).
155. S. Kirkpatrick, C. D. Gelatt, Jr., and M. P. Vecchi, *Science* 220, 671 (1983).
156. N. A. Marks, D. R. McKenzie, B. A. Pailthorpe, M. Bernasconi, and M. Parrinello, *Phys. Rev. Lett.* 76, 768 (1996).
157. N. A. Marks, D. R. McKenzie, B. A. Pailthorpe, M. Bernasconi, and M. Parrinello, *Phys. Rev. B* 54, 9703 (1996).
158. J. C. Angus and C. C. Hayman, *Science* 241, 913 (1988).
159. D. G. McCulloch, D. R. McKenzie, and C. M. Goringe, *Phys. Rev. B* 61, 2349 (2000).
160. D. Dasgupta, F. Demichelis, C. F. Pirri, and A. Tagliaferro, *Phys. Rev. B* 43, 2131 (1991).
161. C. Fanchini, A. Tagliaferro, D. P. Dowling, K. Donnelly, M. L. McConnell, R. Flood, and G. Lang, *Diamond Relat. Mater.* 9, 732 (2000).
162. C. Arena, B. Kleinsorge, J. Robertson, W. I. Milne, and M. E. Welland, *J. Appl. Phys.* 85, 1609 (1999).
163. R. A. Street, "Hydrogenated Amorphous Silicon." Cambridge Univ. Press, Cambridge, UK, 1991.
164. A. R. Forouhi and I. Bloomer, in "Handbook of Optical Constants of Solids II" (E. A. Palik, Ed.), p. 837. Academic Press, New York, 1991.
165. S. A. Alterovitz, N. Savvides, F. W. Smith, and J. A. Woollam, in "Handbook of Optical Constants of Solids II" (E. A. Palik, Ed.), p. 837. Academic Press, New York, 1991.
166. W. A. McGahan, T. Makovicka, J. Hale, and J. A. Woollam, *Thin Solid Films* 253, 57 (1994).
167. N. M. J. Conway, A. Ilie, J. Robertson, W. I. Milne, and A. Tagliaferro, *Appl. Phys. Lett.* 73, 2456 (1998).
168. H. Kaufman, S. Metin, and D. Saperstein, *Phys. Rev. B* 39, 13053 (1989).
169. B. Dischler, A. Bubenzer, and P. Koidl, *Appl. Phys. Lett.* 42, 636 (1983).
170. D. R. McKenzie, *Rep. Progr. Phys.* 59, 1611 (1996).
171. B. Dischler, A. Bubenzer, and P. Koidl, *Solid State Comm.* 48, 105 (1983).
172. J. Schwan, W. Dworschak, K. Jung, and H. Ehrhardt, *Diamond Relat. Mater.* 3, 1034 (1994).
173. M. Zhang and Y. Nakayama, *J. Appl. Phys.* 82, 4912 (1997).
174. D. Chen, A. W. S. P. Wong, and S. Peng, *Diamond Relat. Mater.* 8, 1130 (1999).
175. S. Bhattacharyya, A. Granier, and G. Turban, *J. Appl. Phys.* 86, 4668 (1999).
176. H. X. Han and B. J. Feldman, *Solid State Comm.* 65, 921 (1988).
177. N. Matsukura, S. Inoue, and Y. Machi, *J. Appl. Phys.* 72, 43 (1992).
178. R. Stief, J. Schäfer, J. Ristein, L. Ley, and W. Beyer, *J. Non-cryst. Solids* 198–200, 636 (1996).
179. L. H. D. N. B. Colthup and S. E. Wiberley, "Introduction to Infrared and Raman Spectroscopy." Academic Press, New York, 1975.
180. D. H. Williams and I. Flemming, in "Spectroscopic Methods in Organic Chemistry, p. 70. McGraw–Hill, UK, 1995.
181. L. Martinu, A. Raveh, A. Domingue, L. Bertrand, J. E. Klemberg, S. C. Gujrathi, and M. R. Wertheimer, *Thin Solid Films* 208, 42 (1992).
182. M. Clin, O. Durand-Drouhin, A. Zeinert, and J. C. Picot, *Diamond Relat. Mater.* 8, 527 (1999).
183. J. Wagner and P. Lautenschlager, *J. Appl. Phys.* 59, 2044 (1986).
184. S. Schutte, S. Will, H. Mell, and Fuhs, *Diamond Relat. Mater.* 2, 1360 (1993).
185. I. Watanabe, S. Hasegawa, and Y. Kurata, *Jpn. J. Appl. Phys.* 21, 856 (1981).
186. F. Demichelis, S. Schreiter, and A. Tagliaferro, *Phys. Rev. B* 51, 2143 (1995).
187. Y. Dekhtyar, Y. Kawaguchi, and A. Arnautov, *Internat. J. Adhesion Adhesives* 17, 75 (1997).
188. Y. Kawaguchi and S. Yamamoto, in "5th Proc. of the 11th Int. Symposium on Exoemission and Its Applications," p. 200. Gluholazy, Poland, 1994.
189. J. M. Bonard, T. Stöckli, F. Maier, W. A. de Heer, A. Châtelain, J. P. Salvetat, and L. Forró, *Phys. Rev. Lett.* 81, 1441 (1998).
190. J. Robertson, *Phys. Rev. B* 53, 16302 (1996).
191. Y. Hamakawa, T. Toyama, and H. Okamoto, *J. Non-Cryst. Solids* 115, 180 (1989).
192. F. Demichelis, Y. C. Liu, X. F. Rong, S. Schreiter, and A. Tagliaferro, *Solid State Comm.* 95, 475 (1998).
193. S. M. Kim and J. F. Wager, *Appl. Phys. Lett.* 53, 1880 (1988).
194. M. Yoshimi, H. Shimizu, K. Hattori, H. Okamoto, and Y. Hamakawa, *Optoelectronics* 7, 69 (1992).
195. T. Mandel, M. Frischholz, R. Helbig, and A. Hammerschidt, *Appl. Phys. Lett.* 64, 3637 (1994).
196. E. Leuder, in "Symposia Proceedings 377" (M. Hack, E. S. Schiff, A. Madan, M. Powell, and A. Matsuda, Eds.), p. 847. Materials Research Society, Pittsburgh, 1995.
197. M. S. Haque, H. A. Naseem, J. L. Shultz, W. D. Brown, S. Lai, and S. Gangopadhyay, *J. Appl. Phys.* 83, 4421 (1998).
198. J. Viehland, S. Lin, and B. J. Feldman, *Solid State Comm.* 82, 79 (1992).
199. D. A. I. Munindradasa, Ph.D. Thesis, University of Liverpool, 1999.
200. Rusli, G. A. J. Amaratunga, and S. R. P. Silva, *Opt. Mater.* 6, 93 (1996).
201. Rusli, J. Robertson, and G. A. J. Amaratunga, *J. Appl. Phys.* 80, 2998 (1996).
202. Rusli, J. Robertson, and G. A. J. Amaratunga, *Diamond Relat. Mater.* 6, 700 (1997).
203. M. Koós, I. Pócsik, J. Erostyák, and A. Buzádi, *J. Non-Cryst. Solids* 227–230, 579 (1998).

204. S. K. Chernyshov, E. I. Terukov, V. A. Vassilyev, and A. Volkov, *J. Non-Cryst. Solids* 134, 218 (1991).
205. Y. Masumoto, S. Shionoya, H. Munekata, and H. Kukimoto, *J. Phys. Soc. Jpn.* 52, 3985 (1983).
206. W. Lormes, M. Hundhausen, and L. Ley, *J. Non-Cryst. Solids* 227–230, 570 (1998).
207. G. H. Kim, J. H. Lee, and J. S. Chang, *Solid State Comm.* 89, 529 (1994).
208. C. L. Chen and J. T. Lue, *J. Non-Cryst. Solids* 194, 93 (1996).
209. N. M. Berberan-Santos, A. Fedorov, J. P. Conde, C. Godet, T. Heitz, and J. E. Bourée, *Chem. Phys. Lett.* 319, 113 (2000).
210. Z. F. Li, Z. Y. Yang, and R. F. Xiao, *Appl. Phys. A* 63, 243 (1996).
211. M. Koós, I. Pócsik, and L. Tóth, *Appl. Phys. Lett.* 65, 2245 (1994).
212. J. Zhao, R. Z. Che, J. R. Xu, and N. Kang, *Appl. Phys. Lett.* 70, 2781 (1997).
213. M. Koós, I. Pócsik, and L. Tóth, *J. Non-Cryst. Solids* 164–166, 1151 (1993).
214. Rusli, G. A. J. Amaratunga, and J. Robertson, *Phys. Rev. B* 53, 16306 (1996).
215. Y. Matsumoto, H. Kunitomo, S. Shinoya, H. Munekata, and H. Kukimoto, *Solid State Comm.* 51, 209 (1984).
216. K. Murayama and M. A. Bosch, *Phys. Rev. B* 25, 6542 (1982).
217. M. Collins, R. C. Barklie, J. V. Anguita, J. D. Carey, and S. R. P. Silva, *Diamond Relat. Mater.* 9, 781 (2000).
218. Y. P. Piryatinskii, N. I. Klyui, A. B. Romanyuk, and V. A. Semenovich, *Mol. Cryst. Liq. Cryst.* 324, 19 (1998).
219. A. Y. Liu and M. L. Cohen, *Phys. Rev. B* 41, 10727 (1990).
220. M. A. Tamor and W. C. Vassell, *J. Appl. Phys.* 76, 3823 (1994).
221. R. Shuker and R. W. Gammon, *Phys. Rev. Lett.* 25, 222 (1970).
222. K. W. R. Gilkes, S. Prawer, K. W. Nugent, J. Robertson, H. S. Sands, Y. Lifshitz, and X. Shi, *J. Appl. Phys.* 87, 7283 (2000).
223. R. J. Nemanich, G. Lucovsky, and S. A. Solin, *Solid State Electron.* 23, 117 (1977).
224. R. Nicklow, N. Wakabayashi, and H. G. Smith, *Phys. Rev. B* 5, 4951 (1972).
225. F. Tuinstra and J. K. Koenig, *J. Chem. Phys.* 53, 1126 (1970).
226. R. J. Nemanich and S. A. Solin, *Phys. Rev B* 20, 392 (1979).
227. H. Wilhelm, M. Lelaurin, E. McRae, and B. Humbert, *J. Appl. Phys.* 84, 6552 (1998).
228. R. P. Vidano, D. B. Fischbach, L. J. Willis, and T. M. Loehr, *Solid State Comm.* 39, 341 (1981).
229. A. V. Baranov, A. N. Bekherev, Y. S. Bobovich, and V. I. Petrov, *Opt. Spektrosk.* 62, 612 (1987).
230. M. J. Matthews, M. A. Pimenta, G. Dresselhaus, M. S. Dresselhaus, and M. Endo, *Phys. Rev. B* 59, R6585 (1999).
231. A. C. Ferrari and J. Robertson, *Phys. Rev. B* 61, 14095 (2000).
232. M. A. Tamor, J. A. Haire, C. H. Wu, and K. C. Hass, *Appl. Phys. Lett.* 54, 123 (1989).
233. R. J. Nemanich, J. T. Glass, G. Lucovsky, and R. E. Shroder, *J. Vac. Sci. Technol. A* 6, 1783 (1988).
234. R. Shroder, R. Nemanich, and J. Glass, *Phys. Rev. B* 41, 3738 (1990).
235. M. Chhowalla, A. C. Ferrari, J. Robertson, and G. A. J. Amaratunga, *Appl. Phys. Lett.* 76, 1419 (2000).
236. D. G. McCulloch, S. Prawer, and A. Hoffman, *Phys. Rev. B* 50, 5905 (1994).
237. D. G. McCulloch, D. R. McKenzie, and S. Prawer, *Philos. Mag. A* 72, 1031 (1995).
238. J. Wagner, C. Wild, and P. Koidl, *Appl. Phys. Lett.* 59, 779 (1991).
239. K. W. R. Gilkes, H. S. Sands, D. N. Batchelder, J. Robertson, and W. I. Milne, *Appl. Phys. Lett.* 70, 1980 (1997).
240. J. R. Shi, X. Shi, Z. Sun, E. Liu, B. K. Tay, and S. P. Lau, *Thin Solid Films* 366, 169 (2000).
241. J. W. Ager, S. Anders, A. Anders, and I. G. Brown, *Appl. Phys. Lett.* 66, 3444 (1995).
242. J. Robertson, *Philos. Mag. B* 76, 335 (1997).
243. B. Kleinsorge, A. C. Ferrari, J. Robertson, and W. I. Milne, *J. Appl. Phys.* 88, 1149 (2000).
244. N. I. Klyui, Y. P. Pityatinskii, and V. A. Semenovich, *Mater. Lett.* 35, 334 (1998).
245. D. Knight and U. White, *J. Mater. Res.* 4, 385 (1989).
246. Y. Lifshitz, *Diamond Relat. Mater.* 5, 388 (1996).
247. R. W. Fessender and R. H. Schuler, *J. Chem. Phys.* 39, 2147 (1963).
248. J. E. Wertz and J. R. Bolton, "Electron Spin Resonance, Elemental Theory and Practical Applications." Chapman and Hall, New York, 1986.
249. C. P. Poole, "Electron Spin Resonance," p. 476. Dover, New York, 1996.
250. R. J. Gambino and J. A. Thompson, *Solid State Comm.* 34, 15 (1980).
251. D. J. Miller and D. R. McKenzie, *Thin Solid Films* 108, 257 (1983).
252. A. Sadki, Y. Bounouh, M. L. Theye, J. von Bardeleben, J. Cernogora, and J. Fave, *Diamond Relat. Mater.* 5, 439 (1996).
253. M. N. Brodsky, R. S. Title, K. Weiser, and G. D. Pettit, *Phys. Rev. B* 1, 2632 (1970).
254. C. DeMartino, F. Demichelis, and A. Tagliaferro, *Diamond Relat. Mater.* 3, 547 (1994).
255. J. Ristein, J. Schäfer, and L. Ley, *Diamond Relat. Mater.* 4, 509 (1995).
256. R. C. Barklie, M. Collins, and S. R. P. Silva, *Phys. Rev. B* 61, 3546 (2000).
257. A. Zeinert, H. von Bardeleben, and R. Bouzerar, *Diamond Relat. Mater.* 9, 728 (2000).
258. G. A. J. Amaratunga, J. Robertson, V. S. Veerasamy, W. I. Milne, and D. R. McKenzie, *Diamond Relat. Mater.* 4, 637 (1995).
259. M. Weiler, S. Sattel, T. Giessen, K. Jung, H. Ehrhardt, V. S. Veerasamy, and J. Robertson, *Phys. Rev. B* 53, 1594 (1996).
260. G. Fusco, A. Tagliaferro, W. I. Milne, and J. Robertson, *Diamond Relat. Mater.* 6, 783 (1997).
261. K. C. Palinginis, Y. Lubianiker, J. D. Cohen, A. Ilie, B. Kleinsorge, and W. I. Milne, *Appl. Phys. Lett.* 74, 371 (1999).
262. F. Demichelis, C. DeMartino, A. Tagliaferro, and M. Fanciulli, *Diamond Relat. Mater.* 3, 844 (1994).
263. M. Fanciulli, G. Fusco, and A. Tagliaferro, *Diamond Relat. Mater.* 6, 725 (1997).
264. A. Abragam, in "Principles of Nuclear Magnetism," p. 126. Clarendon, Oxford, 1996.
265. C. P. Slichter, in "Principles of Magnetic Resonance," p. 198. Springer-Verlag, Berlin, 1990.
266. J. P. Goldsborough, M. Mandel, and G. E. Pake, *Phys. Rev. Lett.* 4, 13 (1960).
267. M. M. Golzan, D. R. McKenzie, D. J. Miller, S. J. Collocott, and G. A. J. Amaratunga, *Diamond Relat. Mater.* 4, 912 (1995).
268. U. Voget-Grove, J. Stuke, and H. Wagner, in "Proc. Int. Conf. on Structure and Excitations in Amorphous Solids," Vol. 31, Am. Inst. Phys., New York, 1976, p. 91.
269. B. Movagher, L. Schweitzer, and H. Overhof, *Philos. Mag. B* 37, 683 (1978).
270. H. Dersch, J. Stuke, and J. Beichler, *Phys. Stat. Sol. B* 107, 307 (1981).
271. M. Hoinkis, E. D. Tober, R. L. White, and M. S. Crowder, *Appl. Phys. Lett.* 61, 2653 (1992).
272. O. Amir and R. Kalish, *J. Appl. Phys.* 70, 4958 (1991).
273. S. Bhattacharyya, C. Vallee, C. Cardinaud, O. Chauvet, and G. Tuban, *J. Appl. Phys.* 85, 2162 (1999).
274. F. Demichelis, X. F. Rong, S. Schreiter, A. Tagliaferro, and C. DeMartino, *Diamond Relat. Mater.* 4, 361 (1995).
275. R. C. Barklie, M. Collins, J. Cunniffe, and S. R. P. Silva, *Diamond Relat. Mater.* 7, 864 (1998).
276. J. Robertson, *Phys. Rev. B* 53, 16302 (1996).
277. J. González-Hernández, R. Asomoza, and A. Reyes-Mena, *Solid State Comm.* 67, 1085 (1988).
278. S. Schutte, S. Will, H. Mell, and W. Fuhs, *Diamond Relat. Mater.* 2, 1360 (1993).
279. R. Kleber, K. Jung, H. Ehrardt, I. Muhling, K. Breuer, H. Metz, and F. Engelke, *Thin Solid Films* 205, 274 (1991).
280. J. Schwan, S. Ulrich, H. Ehrhardt, S. R. P. Silva, J. Robertson, R. Samlenski, and R. Brenn, *J. Appl. Phys.* 79, 1416 (1996).
281. S. R. P. Silva, G. A. J. Amaratunga, and C. P. Constantinue, *J. Appl. Phys.* 72, 1149 (1992).

282. N. F. Mott and E. A. Davis, in "Electronic Processes in Non-Crystalline Materials," Chap. 1. Oxford Univ. Press, Oxford, 1971.

283. J. M. Shannon and K. J. B. M. Nieuwesteeg, *Appl. Phys. Lett.* 62, 1815 (1993).

284. S. R. Elliott, "Physics of Amorphous Materials," 2nd ed. Longman, London, 1990.

285. R. Orzeszko, W. Bala, K. Fabisiak, and F. Rozploch, *Phys. Stat. Sol. B* 81, 579 (1984).

286. D. Dasgupta, F. Demichelis, and A. Tagiaferro, *Philos Mag. B* 63, 1255 (1991).

287. A. Rose, *Phys. Rev.* 97, 1538 (1955).

288. J. G. Simmons, *J. Phys. D* 4, 613 (1971).

289. J. Frenkel, *Phys. Rev.* 54, 647 (1938).

290. R. M. Hill, *Philos. Mag.* 23, 59 (1971).

291. P. Mark and T. E. Hartman, *J. Appl. Phys.* 39, 2163 (1968).

292. J. G. Simmons, *Phys. Rev.* 155, 657 (1967).

293. J. M. Shannon, J. N. Sandoe, I. D. French, and A. D. Annis, in "Amorphous Silicon Technology," Mater. Res. Soc., Pittsburgh, 1993, Vol. 297, Chap. 164, p. 987.

294. R. H. Fowler and L. Nordheim, in "Proceedings," Vol. A199, p. 173. Royal Society, London, 1928.

295. D. R. Lamb, "Electrical Conduction Mechanisms in Thin Insulating Films." Methuen, London, 1967.

296. S. R. P. Silva and G. A. J. Amaratunga, *Thin Solid Films* 253, 146 (1994).

297. V. S. Veerasamy, G. A. J. Amaratunga, C. A. Davis, W. I. Milne, P. Hewitt, and M. Weiler, *Solid State Electron.* 37, 319 (1994).

298. M. A. Lampert and P. Mark, in "Current Injection in Solids," p. 81. Academic Press, Cambridge, 1970.

299. J. Robertson, *Diamond Relat. Mater.* 6, 212 (1997).

300. J. Robertson, *Adv. Phys.* 35, 317 (1986).

301. A. Helmbold, P. Hammer, J. U. Thiele, K. Rohwer, and D. Meissner, *Philos. Mag. B* 72, 335 (1995).

302. O. Stenzel, M. Vogel, S. Pönitz, T. Wallendorf, C. V. Borczyskowski, F. Rozploch, Z. Krasilnik, and N. Kalugin, *Phys. Stat. Sol. A* 140, 179 (1993).

303. J. Ristein, J. Schäfer, and L. Ley, *Diamond Relat. Mater.* 3, 861 (1994).

304. R. U. A. Khan, A. P. Burden, S. R. P. Silva, J. M. Shannon, and B. J. Sealy, *Carbon* 37, 777 (1999).

305. D. I. Jones and A. D. Stewart, *Philos. Mag. B* 46, 423 (1982).

306. A. P. Burden, E. Mendoza, S. R. P. Silva, and G. A. J. Amaratunga, *Diamond Relat. Mater.* 7, 495 (1998).

307. Y. Bounouh, M. L. Theye, A. Dehbi-Alaoui, A. Matthews, and J. P. Stoquet, *Phys. Rev. B* 51, 9597 (1995).

308. S. Egret, J. Robertson, W. I. Milne, and F. J. Clough, *Diamond Relat. Mater.* 6, 879 (1997).

309. S. R. P. Silva, R. U. A. Khan, J. V. Anguita, A. P. Burden, J. M. Shannon, B. J. Sealy, A. P. Papworth, C. J. Keily, and G. A. J. Amaratunga, *Thin Solid Films* 332, 118 (1998).

310. R. U. A. Khan and S. R. P. Silva, *Internat. J. Mod. Phys. B* 14, 195 (2000).

311. N. Konofaos, E. Evangelou, and C. B. Thomas, *J. Appl. Phys.* 84, 4634 (1998).

312. M. Allon-Alauf, L. Klibanov, A. Seidman, and N. Croitoru, *Diamond Relat. Mater.* 5, 1275 (1996).

313. S. R. P. Silva, Ph.D. Thesis, Engineering Department, University of Cambridge, 1994.

314. G. A. J. Amaratunga, W. I. Milne, and A. Putnis, *IEEE Electron. Device. Lett.* 11, 33 (1990).

315. D. A. I. Munindradasa, M. Chhowalla, G. A. J. Amaratunga, and S. R. P. Silva, *J. Non-Cryst. Solids* 230, 1106 (1998).

316. N. M. J. Conway, A. Ilie, J. Robertson, W. I. Milne, and A. Tagliaferro, *Appl. Phys. Lett.* 73, 2456 (1998).

317. K. K. Chan, G. A. J. Amaratunga, S. P. Wong, and V. S. Veerasamy, *Solid State Electron.* 36, 345 (1993).

318. A. A. Khan, J. A. Woollam, Y. Chung, and B. Banks, *IEEE Electron Device Lett.* 4, 146 (1983).

319. A. A. Khan, J. A. Woollam, and Y. Chung, *Solid State Electron.* 27, 385 (1984).

320. T. Mandel, M. Frischholz, R. Helbig, and A. Hammerschmidt, *Appl. Phys. Lett.* 64, 3637 (1994).

321. S. Waidmann, M. Knupfer, J. Fink, B. Kleinsorge, and J. Robertson, *Diamond Relat. Mater.* 9, 722 (2000).

322. C. A. Davis, K. M. Knowles, and G. A. J. Amaratunga, *Surf. Coat. Technol.* 76–77, 316 (1995).

323. S. Sattel, J. Robertson, and H. Ehrhardt, *J. Appl. Phys.* 82, 4566 (1997).

324. B. Kleinsorge, A. Ferrari, J. Robertson, W. I. Milne, S. Waidmann, and S. Hearne, *Diamond Relat. Mater.* 9, 643 (2000).

325. N. Konofaos and C. B. Thomas, *Appl. Phys. Lett.* 61, 2805 (1992).

326. G. A. J. Amaratunga, V. S. Veerasamy, W. I. Milne, C. A. Davis, S. R. P. Silva, and H. S. MacKenzie, *Appl. Phys. Lett.* 63, 370 (1993).

327. D. R. McKenzie, Y. Yin, C. A. Davis, B. A. Pailthorpe, G. Amaratunga, and V. S. Veerasamy, *Diamond Relat. Mater.* 3, 353 (1994).

328. L. K. Cheah, X. Xu, E. Liu, and J. R. Shi, *Appl. Phys. Lett.* 73, 2473 (1998).

329. A. Ilie, N. M. J. Conway, B. Kleinsorge, J. Robertson, and W. I. Milne, *J. Appl. Phys.* 84, 5575 (1998).

330. A. Ilie, O. Harel, N. M. J. Conway, T. Yagi, J. Robertson, and W. I. Milne, *J. Appl. Phys.* 87, 789 (2000).

331. N. M. J. Conway, W. I. Milne, and J. Robertson, *Diamond Relat. Mater.* 7, 477 (1998).

332. N. M. J. Conway, A. Ferrari, A. J. Flewitt, J. Robertson, W. I. Milne, A. Tagliaferro, and W. Beyer, *Diamond Relat. Mater.* 9, 765 (2000).

333. B. Meyerson and F. W. Smith, *Solid State Comm.* 34, 531 (1980).

334. B. Meyerson and F. W. Smith, *Solid State Comm.* 41, 23 (1982).

335. S. R. P. Silva and G. A. J. Amaratunga, *Thin Solid Films* 270, 194 (1995).

336. S. R. P. Silva, B. Rafferty, G. A. J. Amaratunga, J. Schwan, D. F. Franceschini, and L. M. Brown, *Diamond Relat. Mater.* 5, 401 (1996).

337. S. R. P. Silva, J. Robertson, G. A. J. Amaratunga, B. Rafferty, L. M. Brown, J. Schwan, D. F. Franceschini, and G. Mariotto, *J. Appl. Phys.* 81, 2626 (1997).

338. V. S. Veerasamy, G. A. J. Amaratunga, C. A. Davis, A. E. Timbs, W. I. Milne, and D. R. McKenzie, *J. Phys. Condens. Matter* 5, 169 (1993).

339. V. S. Veerasamy, J. Yuan, G. A. J. Amaratunga, W. I. Milne, K. W. R. Gilkes, M. Weiler, and L. M. Brown, *Phys. Rev. B* 48, 17954 (1993).

340. G. A. J. Amaratunga, V. S. Veerasamy, W. I. Milne, D. R. McKenzie, C. A. Davis, M. Weiler, P. J. Fallon, S. R. P. Silva, J. Koskinen, and A. Payne, in "Proceedings, 2nd International Conference on the Application of Diamond Films and Related Materials," Saritaka, Japan, 1993, pp. 25–28.

341. X. Shi, H. Fu, J. R. Shi, L. K. Cheah, B. K. Tay, and P. Hui, *J. Phys. Condens. Matter* 10, 9293 (1998).

342. E. Liu, X. Shi, L. K. Cheah, Y. H. Hu, H. S. Tan, J. R. Shi, and B. K. Tay, *Solid State Electron.* 43, 427 (1999).

343. C. Ronning, U. Griesmeier, M. Gross, H. C. Hofsäss, R. G. Downing, and G. P. Lamaze, *Diamond Relat. Mater.* 4, 666 (1995).

344. H. Hofsäss, in "Proceedings, 1st International Specialist Meeting on Amorphous Carbon," p. 296. World Scientific, Cambridge, 1997.

345. J. Robertson and C. A. Davis, *Diamond Relat. Mater.* 4, 441 (1995).

346. A. P. Burden and S. R. P. Silva, *Appl. Phys. Lett.* 73, 3082 (1998).

347. S. R. P. Silva, J. Robertson, Rusli, G. A. J. Amaratunga, and J. Schwan, *Philos Mag. B* 74, 369 (1996).

348. P. W. Anderson, *Phys. Rev.* 109, 1492 (1958).

349. C. W. Chen and J. Robertson, *J. Non-Cryst. Solids* 227, 602 (1998).

350. J. F. Prins, *Phys. Rev. B* 38, 5576 (1985).

351. H. Hofsäss, M. Dalmer, M. Restle, and C. Ronning, *J. Appl. Phys.* 81, 2566 (1997).

352. J. F. Prins, *Phys. Rev. B* 31, 2472 (1985).

353. S. Prawer and R. Kalish, *Phys. Rev. B* 51, 15711 (1995).

354. S. Prawer, R. Kalish, M. E. Adel, and V. Richter, *J. Appl. Phys.* 61, 4492 (1987).

355. D. C. Ingram and A. W. McCormick, *Nucl. Instrum. Methods B* 34, 68 (1988).

356. R. U. A. Khan, D. Grambole, and S. R. P. Silva, *Diamond Relat. Mater.* 9, 645 (2000).

357. G. L. Doll, J. P. Heremans, T. A. Perry, and J. V. Mantese, *J. Mater. Res.* 9, 85 (1994).

358. W. J. Wang, T. M. Wang, and C. Jing, *Thin Solid Films* 280, 90 (1996).

359. D. G. McCulloch, D. R. McKenzie, S. Prawer, A. R. Merchant, E. G. Gerstner, and R. Kalish, *Diamond Relat. Mater.* 6, 1622 (1997).

360. N. Kumar, H. K. Schimidt, M. H. Clark, A. Ross, B. Lin, L. Fredin, B. Baker, D. Patterson, and W. Brookover, "SID '94 Digest," Society for Information Displays, 1994.

361. J. Jaskie, *Mater. Res. Soc. Bull.* 21, 59 (1996).

362. F. J. Himpsel, J. A. Knapp, J. A. van Vechten, and D. E. Eastman, *Phys. Rev. B* 20, 624 (1979).

363. J. Robertson, *Diamond Relat. Mater.* 5, 797 (1996).

364. M. W. Geis, J. C. Twichell, and T. M. Lyszszarz, *J. Vac. Sci. Technol. B* 14, 2060 (1996).

365. R. D. Forrest, A. P. Burden, S. R. P. Silva, L. K. Cheah, and X. Shi, *Appl. Phys. Lett.* 73, 3784 (1998).

366. A. Ilie, A. C. Ferrari, T. Yagi, and J. Robertson, *Appl. Phys. Lett.* 76, 2627 (2000).

367. A. A. Talin, L. S. Pan, K. F. McCarthy, H. J. Doerr, and R. F. Bunshah, *Appl. Phys. Lett.* 69, 3842 (1996).

368. O. Gröning, O. M. Küttel, P. Gröning, and L. Schlapbach, *Appl. Phys. Lett.* 71, 2253 (1997).

369. R. Gomer, "Field Emission and Field Ionisation." Oxford Univ. Press, Oxford, 1961.

370. R. H. Fowler and L. Nordheim, *Proc. Roy. Soc. London A* 199, 173 (1928).

371. G. E. Vibrans, Technical Report No. 353, Lincoln Laboratory, MIT (unpublished).

372. I. Musa, D. A. I. Munindrasasa, G. A. J. Amaratunga, and W. Eccleston, *Nature* 395, 362 (1998).

373. A. P. Burden, R. D. Forrest, and S. R. P. Silva, *Thin Solid Films* 337, 257 (1999).

374. R. D. Forrest, A. P. Burden, R. U. A. Khan, and S. R. P. Silva, *Surface Coat. Technol.* 577, 108 (1998).

375. A. R. Krauss, T. G. McCauley, D. M. Gruen, M. Ding, T. Corrigan, O. Auciello, R. P. H. Chang, M. Kordesch, R. Nemanich, S. English, A. Breskin, E. S. R. Chechyk, Y. Lifshitz, E. Grossman, D. Temple, G. McGuire, S. Pimenov, V. Konov, A. Karabutov, A. Rakhimov, and N. Suetin, in "Proc. IVMC '98," J. Vac. Sci. A, North Carolina, 1998, p. 162.

376. L. K. Cheah, X. Shi, E. Liu, and B. K. Tay, *J. Appl. Phys.* 85, 6816 (1999).

377. S. R. P. Silva, R. D. Forrest, D. A. Munindrasasa, and G. A. J. Amaratunga, *Diamond Relat. Mater.* 7, 645 (1998).

378. K. R. Lee, K. Y. Eun, S. Lee, and D. R. Jeon, *Thin Solid Films* 291, 171 (1996).

379. A. Hart, B. S. Satyanarayana, J. Robertson, and W. I. Milne, *Appl. Phys. Lett.* 74, 1594 (1999).

380. V. S. Veerasamy, J. Yuan, G. A. J. Amaratunga, W. I. Milne, K. W. R. Gilkes, M. Weiler, and L. M. Brown, *Phys. Rev. B* 48, 17954 (1993).

381. N. Missert, T. A. Friedmann, J. P. Sullivan, and R. G. Copeland, *Appl. Phys. Lett.* 70, 1995 (1997).

382. B. S. Satyanarayana, A. Hart, W. I. Milne, and J. Robertson, *Diamond Relat. Mater.* 7, 656 (1998).

383. L. K. Cheah, X. Shi, B. K. Tay, S. R. P. Silva, and Z. Sun, *Diamond Relat. Mater.* 7, 640 (1998).

384. K. C. Park, J. H. Moon, S. J. Chung, J. Jang, S. Oh, and W. I. Milne, *Appl. Phys. Lett.* 70, 1381 (1997).

385. F. Y. Chuang, C. Y. Sun, T. T. Chen, and I. N. Lin, *Appl. Phys. Lett.* 69, 3504 (1996).

386. N. S. Xu, J. Chen, and S. Z. Deng, *Appl. Phys. Lett.* 76, 2463 (2000).

387. G. A. J. Amaratunga, *New Diamond Frontier Carbon Technol.* 9, 31 (1999).

388. J. Robertson, *Diamond Relat. Mater.* 6, 212 (1997).

389. S. R. P. Silva, R. D. Forrest, J. M. Shannon, and B. J. Sealy, *J. Vac. Sci. Technol. B* 17, 569 (1999).

390. S. R. P. Silva, R. D. Forrest, A. P. Burden, J. Anguita, J. M, Shannon, B. J. Sealy, D. A. I. Munindradasa, G. A. J. Amaratunga, and K. Okano, in "Amorphous Carbon: State of the Art" (S. R. P. Silva, J. Robertson, W. I. Milne, and G. A. J. Amaratunga, Eds.), pp. 350–361. World Scientific Press, Singapore, 1998.

391. S. R. P. Silva, G. A. J. Amaratunga, and K. Okano, *J. Vac. Sci. Technol. B* 17, 57 (1999).

392. J. Robertson, *J. Vac. Sci. Technol. B* 17, 59 (1999).

393. X. Shi, L. K. Cheah, B. K. Tay, and S. R. P. Silva, *Appl. Phys. Lett.* 74, 833 (1999).

394. A. Modinos and J. P. Xanthakis, *Appl. Phys. Lett.* 73, 1874 (1998).

395. L. K. Cheah, X. Shi, and E. Liu, *Appl. Surf. Sci.* 143, 309 (1999).

396. J. D. Carey, C. H. Poa, R. D. Forrest, A. P. Burden, and S. R. P. Silva, *J. Vac. Sci. Technol. B* 18, 1051 (2000).

397. S. R. P. Silva, G. A. J. Amaratunga, and J. R. Barnes, *Appl. Phys. Lett.* 71, 1477 (1997).

398. F. J. Clough, W. I. Milne, B. Kleinsorge, J. Robertson, G. A. J. Amaratunga, and B. N. Roy, *Electron. Lett.* 32, 498 (1996).

399. K. K. Chan, Ph.D. Thesis, University of Cambridge, 1994.

400. E. G. Gerstner and D. R. McKenzie, *J. Appl. Phys.* 84, 5647 (1998).

401. S. Liu, D. Lamp, S. Gangopadhyay, G. Sreenivas, S. S. Ang, and H. A. Naseem, *IEEE Electron. Device. Lett.* 19, 317 (1998).

402. C. H. Lee and K. S. Lim, *Appl. Phys. Lett.* 72, 106 (1998).

403. R. U. A. Khan, R. A. C. M. M. van Swaaij, A. Vonsovici, and S. R. P. Silva, in "Proceedings, 16th European Photovoltaic Solar Energy Conference and Exhibition" (unpublished).

404. K. L. Narayan and M. Yamaguchi, *Appl. Phys. Lett.* 75, 2106 (1999).

Chapter 10

HIGH-T_c SUPERCONDUCTOR THIN FILMS

B. R. Zhao

Institute of Physics, Chinese Academy of Sciences, Beijing, China

Contents

1. INTRODUCTION

Following the discovery of high-T_c superconductors [1], the syntheses and investigations of high-T_c cuprate superconductor thin films were soon developed worldwide. The blasting fuse for such an upsurge is that people want imperatively to search the normal and superconducting properties of the cuprate perovskite compounds which are responsibe for the unexpectedly high-T_c superconductivity. While the ceramic feature of this kind of material is a big block to purpose, the experimental data based on such types of materials is actually extrinsic. The highly oriented thin films together with the single crystals are available since both of them can show intrinsic properties, such as sharp resistive transition; high critical current density had been obtained even in the early stage of investigation of high-T_c superconducting thin films. The intrinsic anisotropy associated with the lamellar structure is easily obtained for the well oriented thin films. The most distinctive feature of the thin film is

Handbook of Thin Film Materials, edited by H.S. Nalwa
Volume 4: Semiconductor and Superconductor Thin Films

ISBN 0-12-512912-2/$35.00

the large surface area (very large value of ratio between surface area to thickness), which is the reason the film is very useful for the investigation and application of high-T_c superconductors in the fields of surface (interface) science, optical science, and related devices.

By controlling the thickness of the thin film down to a value comparable with characteristic lengths, such as magnetic penetration depth and coherence length, the transport measurements performed with and without magnetic fields enable us to get intrinsic parameters of the materials. Also, by controlling the thickness, it is possible to grow the film with thickness down to one unit cell order of size, typically, for $YBa_2Cu_3O_{7-\delta}$, 1.2 nm, to understand the role of CuO_2 plane on superconductivity, i.e., on the preliminary stage of the occurrence of superconductivity, which is also a sufficient way to get information for the mechanism of high-T_c superconductivity.

It also should be specially indicated that based on the high quality thin film, the intrinsic and extrinsic pinning mechanisms can be investigated widely and deeply. The results of such investigation also can be used to develop the application of the thick films.

The thin film, due to its homogeneity and orientation in structure, the large ratio of length to width, and the thickness that can be obtained, may be designed as an ideal configuration for measurements to get definite superconductivity and transport properties, such as the determination of transition temperature, resistivity, Hall effect, and thermal power.

The thin film also is a special type which can almost avoid the weakness of the ceramic feature of high-T_c superconductors for microfabrication processes according to the design to realize the application of high-T_c superconductors, especially the superconducting electronic devices, such as superconducting quantum interference devices (SQUIDs), microwave devices, bolometers, and so on.

The investigation of high-T_c superconductor (HTSC) thin films associates with the following wide range and abundant contents: fabrication of thin films, including various fabrication methods, substrates, and buffer layers for growing the thin films on metal, semiconductor, and other kind of substrates; various types of thin films, including ultrathin films, large area thin films, and multilayers (heterostructures); physical properties; and device applications.

Since early stages of the investigation of high-T_c superconductors, efforts from many groups worldwide are focused on developing high-T_c superconducting thin films. Almost all high-T_c superconductors were fabricated into the thin films with excellent superconductivity as soon as they were discovered or synthesized in bulk type samples, or even for the materials that cannot be synthesized into the bulk sample with superconductivity. With the elapse of time, all possible techniques and lots of methods were unprecedentedly developed for preparation of high-T_c superconductor thin films. They are several types of sputtering methods (dc, rf and dc, rf magnetron sputteries) [2–5], ion beam sputtering [6–8], evaporation processing, including electron-beam evaporation [9, 10], molecular-beam epitaxy (MBE) [11], laser ablation [12], and laser-molecular-

beam epitaxy (laser-MBE) [13], chemical processing of the film synthesis, including metal–organic chemical vapor depositon [14–17], liquid phase epitaxy [18, 19], and sol–gel processing [20, 21].

In addition, other several novel chemical processings also have been developed. They are combustion chemical vapor deposition [22], flaming solvent spray [23], electrophoresis [24], dip-coating [25, 26], and so on.

So far, the high-T_c superconductor thin films are developed well for all main materials. They are:

$YBa_2Cu_3O_{7-\delta}$ ("123" phase) thin films,
BiSrCaCuO system thin films,
TlBaCaCuO system thin films,
HgBaCaCuO system thin films,
LaCuO ("214" phase) thin films,

and electron (n) type materials NdCeCuO system thin films, noncuprate high-T_c superconductor BaKBiO system thin films, and the thin films of infinite CuO_2 layer superconductors $CaCuO_2$, $SrCuO_2$, and $BaCuO_2$ with various dopants.

For various purposes several types of high-T_c superconductor thin films have been developed. The large area thin films were developed for microwave devices and transport applications. They also have strong application potential in industry. For example, they have application potential in electric power systems; the most promising application is resistive fault current limiting devices [27]. The low surface resistance ($<250~\mu\Omega$ at 77 K and 10 GHz order) is the most important feature for high-T_c superconductors to be used in microwave technology [28, 29].

As an extreme, the ultrathin films hold people's interest since the early stages of investigation of high-T_c superconductor thin films. First, $YBa_2Cu_3O_{7-\delta}$ thin films of $\sim$10 nm thickness were prepared [30]. In the early of 1990s, the YBaCuO thin films with thicknesses of unit cell (1.2 nm) were successfully grown, for which the zero resistance temperature, 30 K, has been obtained [31]. The superconducting transport in YBaCuO ultrathin films is likely to be two demensional in nature [30] and the Kosterlitz–Thouless (KT) resistive transition is considered to be the intrinsic feature in the CuO_2 conducting layers [32]. But the analysis of the conductivity near T_c in terms of the Alamazov–Larkin type fluctuation reveals three-dimensional (3D) behavior for such ultrathin films [33]. So the ultrathin films continuously receive much attention on superconductivity [34, 35], vortex dynamics [36], fabrication [37], and transport properties [38]. While the fabrication and investigation of ultrathin film are progressing well in many laboratories, the multilayers based on ultrathin film also received much attention and gradually became the most important branch in the field of high-T_c superconductor thin films. The first multilayer, $YBa_2Cu_3O_{7-\delta}/DyBa_2Cu_3O_{7-\delta}$, was prepared by Triscone et al. [39]; then lots of work was done experimentally and theoretically on $YBa_2Cu_3O_7/PrBa_2Cu_3O_{7-\delta}$ multilayers [40, 41]. The multilayers are associated with almost all superconductors (including CuO_2 infinite layers compound).

The main subjects are the role of the CuO_2 plane on the superconductivity [32], the coupling of CuO_2 planes [41], proximity effect [42], flux dynamics [43, 44], superconducting coupling and dimensionality (included KT phase transition) [45], and device applications.

For all the above works, various substrates and buffer layers are widely developed, including various kinds of perovskite structural substrates, semiconductors, and sapphire substrates. For developing the application of high-T_c superconductors on electric transport, various metallic substrates and buffer layers were developed for growing large area and thick high-T_c superconductor films.

The fabrication and investigation of high-T_c superconductor thin films also promote the development of other kinds of perovskite structural materials. Among them, the most important materials are ferroelectric and colossal magnetoresistance (CMR) materials which are the unique function materials receiving much attention. Typical materials for both of them are $Pb(Zr, Ti)O_3$ (ferroelectric) and $(La, Ca)MnO_3$ (CMR materials). The heterostructures of high-T_c superconductors and these nonsuperconducting perovskite materials were developed in the 1990s, prumpting a new interdisciplinary subject, oxide electronics, to arise in the field of condensed matter physics.

So far, lots of conferences and workshops have been held for high-T_c superconductor thin films, such as the series conferences of the science and technology of thin film superconductors, the branch conference of SPIE, the special session in M^{2}HTSC, and many other special international conferences and joint symposiums. Some reviews for techniques of thin film fabrication and properties have been made even in the early stages of investigation of high-T_c superconductors [46–49]. Then more reviews were made for wide investigations of HTSC thin films; various fabrication methods of high-T_c superconductor thin films and multilayers were reviewed by Stoessel et al. [50]. The fabrication and investigation of the superconducting $LnBa_2Cu_3O_x$ (Ln = Y, En) thin films and the multilayers of S/I/S were reviewed by Miyazawa et al. [51]. Norton have a rather complete review on the epitaxial growth and properties of several main kinds of high-T_c superconductor thin films and Josephson junction devices [52]. The thin film growth processing, the MBE for BiSrCaCuO systems, and the infinite CuO_2 layer, $(Sr, Ca)_mCu_nO_y$, were reviewed by Rogers et al. [53]. Hollmannt et al. reviewed the dielectric substrate materials that are suitable for preparation of $YBa_2Cu_3O_{7-\delta}$ thin film for microwave integrated circuits [54]. Cucolo and Prieto reviewed the different tunnel barriers for the c-axis oriented $Y(H_O)Cu_3O_{7-\delta}$ based on trilayer structures [55]. The thallium-based high-T_c superconducting thin films were reviewed by Bramley for their fabrications and applications to microwave communications [56]. Triscone and Fischer have given a complete review on $YBa_2Cu_3O_{7-\delta}/PrBa_2Cu_3O_{7-\delta}$ and infinite layer $(Sr, Ca)RuO_3/(Sr, Ca)CuO_2$ superlattices, including superconductivity and vortex dynamics [57]. The superconductivity in low-dimensional structures, such as nanopowders, thin films, and normal metal/superconductor interfaces have reviewed by Huhtinen et al. [58]. Gallop has reviewed the

wide applications of high-T_c superconductors and related functional materials [59]. The thick film is a special matter type of the high-T_c superconductors. It is different from both bulk sample and conventional thin films. It has been sythesized from some special methods, especially several kinds of chemical processings. The applications of thick films are mainly in magnetic shields and microwave devices. The large area thick films have been developed for HTSC electric power systems, for example, the resistive fault current limiting devices. Alford et al. gave a review of thick high-T_c superconducting films processing, properties, and development of devices, including major high-temperature superconductors, YBCO, BSCCO2212, BSCCO2223, TBCCO2212, TBCCO2223, and HgBaCaCuO1223 systems [60]. A tremendous number of papers associated with wide range studies on thick films have been published. This goes beyond the scope of the present review.

This review chapter makes every effort to try to give as complete an introduction as possible. But it cannot include all works on the high-T_c superconductor thin films. In fact, it is impossible as so many works have been published in the passed 14 years.

This chapter is arranged as follows: the fabrication of high-T_c superconducting thin films is arranged in Section 2, in which the main methods of fabrication of high-T_c superconducting thin films, substrates, and buffers are described. The high-temperature superconducting thin films are arranged in Section 3, in which various high-T_c superconducting materials thin films and various types of thin films (ultrathin thin films, multilayers, large area thin films, etc.) are included. Section 4 describes the transport properties of high-T_c superconducting thin films, in which resistivity, Hall effect, flux dynamics, and optical properties are included. The device applications are arranged in Section 5, in which SQUIDs, microwave devices, and other devices are included. In Section 6, the heterostructures of high-T_c superconductors and other related perovskite structural materials are described, including high-T_c superconductor/ferroelectric heterostructures and high-T_c superconductor/CMR materials heterostructures. In Section 7, a conclusion will be given.

2. FABRICATION OF HIGH-T_c SUPERCONDUCTOR THIN FILMS

2.1. Methods of Fabrication

Soon after the discovery of Bednorz and Muller [1], the preparation and investigation of high-T_c superconductor thin films became a interesting subject worldwide for developing fundamental research and applications of this new kind of superconducting material. It is well known that a vast number of techniques for depositing thin films were developed and successfully applied long before the discovery of high-T_c superconductivity. Therefore many of the existing techniques were available for deposition of high-T_c superconductor thin films. At the same time lots of new special processes were developed exclusively for preparing high-T_c superconducting thin films. Nevertheless, the deposition of high-T_c superconductor thin films encounters some problems caused by HTSC materials themselves. Most of them are complex, multicomponent materials with complicated factors to control the film composition and have significant influence on the physical properties of the films. Consequently, each deposition process is affected by those problems, and each process has its own benefits and drawbacks [50]. Based on the preparation conditions, these methods can be roughly divided into two categories: physical vapor deposition and chemical vapor deposition. Typical deposition methods of thin films are shown as follows:

- Physical process:

  - Sputtering

    - Dc/rf Diode sputtering
    - Magnetron sputtering
    - Ion beam sputtering

  - Evaporation

    - Electron beam evaporation
    - Molecular beam epitaxy
    - Laser ablation
    - Laser molecular beam epitaxy
    - Thermal plasma flash evaporation

- Chemical process:

  - Metal–organic deposition method
  - Chemical vapor deposition method
  - Metal–organic chemical deposition method
  - Combustion chemical vapor deposition method
  - Dip coating
  - Sol–gel method
  - Liquid phase epitaxy

2.1.1. Physical Process

2.1.1.1. dc/rf Diode Sputtering

Sputtering is one of the most common vacuum techniques currently applied for superconducting oxide thin film growth, mostly due to the simplicity of the physical processes and its peculiarities: versatility of the technique and flexibility for alteration and customization.

The sputtering deposition technique can withstand quite high gas pressures while still giving a controllable growth rate. Especially the target material can be used in the atmosphere of the oxygen gas. It therefore becomes a very attractive choice for growing the superconducting oxides that require a rather high oxygen partial pressure.

Among all sputtering equipment the simplest type is the dc diode sputtering system, shown as in Figure 1. The dc sputtering system is composed of a pair of planar electrodes. One of the electrodes is a cathode and the other is an anode. The front surface of the cathode is covered with target materials to be deposited. The substrates are placed on the anode. The sputtering chamber is filled with sputtering gas, typically argon

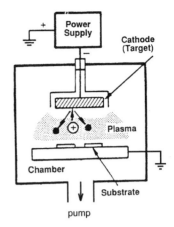

Fig. 1.   Schematic of a dc glow discharge sputtering system.

Table I.   Optimized Parameters for Growth of YBCO Film by Glow Discharge Deposition [61]

Target	YBa$_2$Cu$_3$O$_{7-x}$ pellet of 25 mm diameter and 3 mm thickness
Substrate	(100) MgO single crystal of 3 mm × 10 mm × 0.5 mm size
Substrate temperature	923–973 K
Sputtering gas pressure	3 Torr of Ar and 0.5 Torr of O$_2$
Sputtering rate	2 nm/min
Substrate–target distance	16 mm
Thickness of the film	500–1000 nm
Room-temperature resistance of the film	3–10 Ω

gas at 0.1 Torr. The glow discharge is maintained under the application of dc voltage between the electrodes. The Ar$^+$ ions generated by glow discharge are accelerated by the cathode fall and bombard the target resulting in the deposition of the thin films on the substrates. In the dc sputtering system the target is composed of metallic materials, since the glow discharge is maintained between the metallic electrodes. As an example, the optimized parameters for growth of YBCO film by glow discharge deposition are given in Table I [61].

In the dc diode sputtering system, if the targets are insulators, the sputtering discharge cannot be sustained because the accumulation of charge on the surface of target material without conductivity will lead to zero potential between cathode and anode, and the glow discharge become impossible. To sustain the glow discharge with the insulator target, a rf voltage is supplied to the target. This system is called rf-diode sputtering.

2.1.1.2. Magnetron Sputtering

The purpose of using a magnetic field in a sputtering system is to make more efficient use of the electrons and cause them to produce more ionization. In the magnetron system, the magnetic field can bound the electrons to move near the surface of the target; then bounded electrons can be accelerated by the ap-

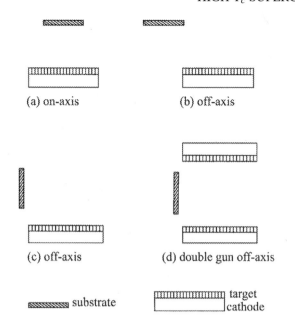

(a) on-axis   (b) off-axis

(c) off-axis   (d) double gun off-axis

████ substrate   target
                        cathode

Fig. 2.   Schematic drawings of typical sputtering configurations of the cathode and substrate.

plied voltage to ionize the Ar gas to increase the plasma density greatly and effectively increase the sputtering rate on the target surface. Due to the low working gas pressure, the sputtered particles traversing the discharge space encounter few collisions, which results in effectively a higher deposition rate than higher pressure deposition systems [62].

Single target magnetron sputtering has been proved to be one of the most powerful growth methods suitable for large-area deposition. The most serious issue in sputtering target is the negative ion bombardment effect [63]. This effect sometimes causes several serious problems, such as a large difference in composition between the target and the deposited film, interdiffusion with the substrates, and deterioration of the crystallization of the grown film. Solutions to that problem are: (1) the choice of special sputtering configurations, where a substrate is placed outside of the region of negative ion flux in an offset [64], off-axis [65], or double-gun off-axis configuration [66] where the substrates do not face the target and thus are not exposed to oxygen ions, shown in Figure 2; (2) placing the substrates facing the target but away from the sputter torch [67]; (3) adjusting the target composition to make up for the losses of certain elements [68]; (4) utilizing a multisource setup with adjusted sputtering rates [69]; (5) reduction of oxygen influences (usually by applying very low or no oxygen partial pressures in the process gas) [70]; or (6) employment of a separate gas feed (Ar to target, and O_2 to substrate) [71], as has been previously proven viable for the deposition of other oxide thin films. It also should be indicated that the negative oxygen ion bombardment is not too serious an influence an the growth and properties of the films even in the on-axis configuration of sputtering setup, especially in the case of higher pressure (80–100 Pa) of deposition gas with lower partial pressure of oxygen (the ratio of oxygen to argon <0.2).

The investigation indicates that the properties of the films are quite sensitive to the process parameters of magnetron sputtering, such as deposition mode (rf or dc) [72], gas composition (reactive/nonreactive) [73], configuration of substrate and target (on-axis, off-axis) [67], target alignment (single/multitarget) [69], and the common process parameters (power [74], substrate temperature [67], deposition gas pressure [75], etc.).

The unbalanced magnetron technique has also been employed to deposit superconducting YBCO films. In this sputtering system, the additional external magnetic field creates deposition conditions that avoid the shift in composition between the target and films.

The decision whether rf or dc voltage is applied to the sputtering cathode is in most cases connected to the state of conductivity for the individual experimental setup. As a rule of thumb, dc is more suitably used when sputtering a stoichiometric target (e.g., a high-T_c superconducting compound) with a significant oxygen partial pressure in process gas, because only in that case does the target may remain sufficiently conductive.

2.1.1.3. Ion Beam Sputtering

The ion beam sputtering deposition (IBSD) technique was first constituted one of the approaches to deposit thin superconductor films, mostly due to the long tradition and the good understanding of these processes, as well as the relatively easy experimental setup [7].

The principle of IBSD is based on the argon ion beam bombardment of the target. Usually the argon ions are created from a Kunfman ion gun and accelerated directly toward the target to bombard its surface, which is oriented at a 45 degree angle to the ion beam. The substrate is facing the target surface in a manner that the escaped particles from the target caused by bombardment can directly deposit on the substrates through the reaction process with gases (e.g., oxygen for growing high-T_c superconducting thin films) which is led to the area near the surface of the substrates. The reaction between gas and target particles is assisted by substrate heating to ensure the formation of the desired film structure and properties. Figure 3 is the schematic diagram of the ion beam sputtering apparatus [7].

With IBSD, film growth can be performed in a high-vacuum environment, in contrast to plasma sputtering deposition, and the following advantages can be listed:

(1) The ion-beam current and energy are determined by the structure of the ion source and operating conditions; they can be considered independently of the potential of the target and the substrate. Therefore, exact control of sputtering conditions is possible.

(2) By selecting the target potential arbitrarily, bombardment of the substrate by secondary electrons and secondary ions can be avoided. Thus the temperature increase of the substrate and damage of the deposited film can be avoided.

(3) The deposition rate of the film fabricated by the IBSD is lower compared with that for plasma sputtering

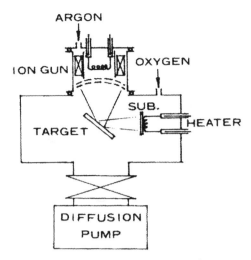

Fig. 3.   Schematic diagram of the ion-beam sputtering apparatus [7].

deposition. However, through periodic sputtering of different kinds of targets, formation of multilayer films can easily be accomplished. In such a case, by computer control, multilayered films with a complex structure can be reproduced with high accuracy.

(4) Deposition of oxide, nitride, and compound semiconductor films has been attempted experimentally by introducing a reactive gas or ion beam and creating a chemical reaction in the IBSD.

2.1.1.4. Electron Beam Evaporation

The electron beam heated evaporation source is a thermal evaporator like a resistance-heated source. It has three basic sections: the electron gun, the beam deflection magnetic lens, and the evaporant-containing hearth. The beam is formed in the gun, passes through the magnetic lens, and is focused upon the evaporant. Electron beam heated sources differ from resistance heated sources in two ways: the heating energy is supplied to the top of the evaporant by the kinetic energy of a high current electron beam, and the evaporant is contained in a water cooled cavity or hearth; heating by electron beam allows attainment of temperatures limited only by radiation and conduction to the hearth. Evaporants contained in a water-cooled hearth do not significantly react with the hearth, thus providing a nearly universal evaporant container.

Success in depositing films using electron beam heated sources is much more sensitive to application technology than in the case of sputtering. The evaporant quantity required by an electron beam heated source may be as little as 1 cm^3; yet the selection and conditioning of the evaporant are of the utmost importance for successful film deposition.

The deposition rate is strongly influenced by the variable characteristics of the electron beam gun and the evaporability of the material [76]. For melting materials, the rate increases with increasing power density (decreased spot size) in the melt, to the power dissipation limit. With semimelting materials, and subliming materials, which are not able to absorb the full power

of the beam, the power and rate can be increased only by increasing the spot area.

The following are four widely applicable guidelines for successful deposition of film from an electron beam heated evaporation source.

(1) Select a charge form with largest possible area to volume ratio. Avoid trying to evaporate powdered or granular materials.
(2) Use the largest hearth volume consistent with available evaporant charge and desired film.
(3) Use the largest beam spot area possible but still attain the required deposition rate.
(4) Increase the spot size if increasing the beam power causes misproportion or film pinholing.

Cui et al. [77] developed a reel-to-reel electron beam evaporation system to continuously deposit epitaxial CeO$_2$ and other oxide buffer layers for depositing YBCO films. Figure 4 is the schematic of reel-to-reel electron beam evaporation system [77]. It includes three interconnected high vacuum chambers—one for annealing of the as-rolled Ni tape, one for depositing buffer layers by electron beam evaporation, and one for depositing Y–BaF$_2$–Cu precursor films.

2.1.1.5. Molecular Beam Epitaxy

A distinct characteristic of MBE is the extremely low background pressure that allows the application of reflective high-energy electron diffraction (RHEED) analysis to track film characteristics on a monolayer basis during growth. The individual components that form the film are evaporated from separated sources. The evaporation sources may be electron-beam-heated or conventional resistance-heated effusion. In the most cases, activated oxygen or ozone is applied to the substrate to compensate for the low reactive gas pressure permitted. Figure 5 is sketch of a modified MBE setup [78]. It is equipped with a RHEED gun with a video recording system and a Ti sublimation pump used continuously during evaporation, which proved to be very effective in trapping oxygen and hydrogen [78]. The latter originates from the barium source.

A large number of investigations using this technique were undertaken to prepare high-T_c oxide thin film superconductors for fundmental and technological research. However, the oxide superconductors are ceramic materials with a layered structure and large cells with multiple cation species which make it considerably more difficult to grow film with atomically flat layers than that of III/V compounds. Their superconducting properties are dominated by the content and ordering of the oxygen atoms in the structure. On the other hand, the presence of oxygen in the MBE chamber is nearly detrimental for this type of process. Furthermore, some of these materials contain elements, like yttrium, which have very high melting points and low vapor pressures. This offers new challenges to the effusion cells. Therefore, one generally has to face many additional technical problems when using a standard MBE machine to prepare high-T_c oxide superconductor thin films.

Pay-out reel   In situ   Single E-beam,   Three E-beam,   Take-up reel
                  annealer   4-pocket, 6 cm³ ea, 3-pocket, 40cm³ ea,
                              buffer layer   Y-BaF₂-Cu
                              deposition chamber co-deposition chamber

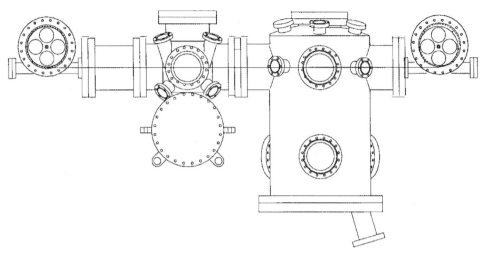

Fig. 4.   Schematic of reel-to-reel electron beam evaporation system [77]. Reprinted from *Physica C* 316, 27 (1999), with permission from Elsevier Science.

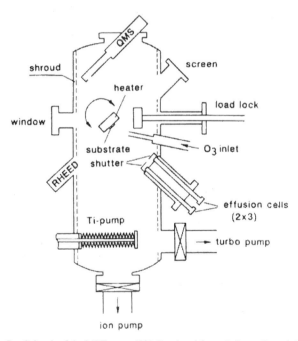

Fig. 5.   Schetch of the MBE setup [78]. Reprinted from *J. Cryst. Growth* 126, 565 (1993), with permission from Elsevier Science.

After the pioneering work of Kwo et al. [79] and Webb et al. [80], great progress has been made in the MBE study of high-T_c superconductor thin films. Kwo and co-workers [81] used atomic oxygen and electron beam sources and obtained high quality $YBa_2Cu_3O_{7-\delta}$ thin films. The creative use of pure ozone (O_3) by Berkley et al. [82] as the oxidizing agent was a great success in achieving full-effusion-cell evaporation and *in situ* growth of high-T_c oxide superconductors at sufficiently

low pressures [83]. Kawai et al. [84] developed a laser molecular beam epitaxy technique to grow Bi-based superconductor thin films. Wang et al. [85] have been succesful in using an Y_2O_3 crucible to evaporate yttrium in preparing $YBa_2Cu_3O_{7-\delta}$ thin film. Kawai et al. [86] achieved layer-by-layer growth of Bi-based compounds at very low substrate temperatures, and Klausmeier-Brown et al. [87] reported the first atomic length scale engineering of SIS planar junctions made of all high-T_c superconductor thin films by MBE growth.

It is the possibility to use MBE to grow films with precisely controlled atomic layers and without any macroparticles, so the MBE is a powerful tool for investigating problems of interface formation, nucleation, and film growth mode. The MBE has been proved to be very viable for the deposition of multilayer films with a period of only a few unit cell thickness, as required for an active device [88].

With the aid of MBE, complex superconductor thin films can be produced using the multilayer technique. In this technique, the individual components or suitable compounds of these are deposited separately and sequentially to form a structure of thin layers that form the desired homogeneous composition by diffusion in a postanneal. To promote a thorough interdiffusion during annealing to form the superconducting phase, it is necessary that the film thickness for each individual component is sufficiently small and the annealing temperature is enough high (above 800°C). To prevent undesired diffusion from the substrate due to the high annealing temperature, the entire superconductor multilayer system is usually separated from the substrate by a diffusion-resistant buffer layer, such as Ag [89] or ZrO_2 [90]. Low deposition rates are essential to ensure formation of a homogeneous and defect-free film. The film thickness ratios for the individual components determine the stoichiometric composition of the film after annealing.

2.1.1.6. Laser Ablation

Since KrF excimer laser ablation was first used for depositing superconducting YBCO thin films [30], pulsed laser deposition (PLD) has developed into a promising method. In this method, a pulsed laser beam is focused on a high-density stoichiometric target in an atomosphere at several tens Pa (40–50 Pa) so as to generate a plume of ejected atoms, ions, and molecules perpendicular to the target surface, regardless of the angle of incidence. The target is rotated slowly to prevent it from being pierced by the laser beam, which has a power density in the range 1–5 J/cm^2 on the target surface. The laser beam wavelengths used were in the range 190 to 350 nm, and it was found that films ablated at shorter wavelengths were smoother with fewer particulates and possessed lower normal-state resistivities and higher critical current densities. Figure 6 is the schematic diagram of a PLD apparatus.

One of the major benefits of PLD is the fact that it can deposit films with compositions exactly the same as that of target (except oxygen, which is added as a reactive gas). So there is no distortion in the composition between thin film and target, as occurs with some other deposition techniques. Another advantage of this method is in the controlling of the film thickness, which can easily be regulated by the number of laser pulses incident on the target. Furthermore, it has the ability to deposit stoichiometric superconductor films *in situ* without postannealing. However, the laser ablation method has some drawbacks: (1) The deposited films contain many fine particles; (2) the sharply pointed geometry ($\cos^n \theta$, with $8 < n < 12$) confines the emission plume to a small substrate area [91], thus complicating large-area deposition.

It was found [92] that the oxygen introduced into the plume would enhance the interaction and lead to a higher film quality. The process for growing thin films by this method is that keeping the substrate temperature at about 700°C and about 1 Pa oxygen partial pressure during deposition, followed by cooling

in flowing oxygen, resulted in high-quality *in situ*-growing *c*-axis-oriented epitaxial thin films [93]. This method is also used for YBCO thin film preparation on a variety of substrates, such as Si [94] and Al$_2$O$_3$ with or without a buffer layer.

Research results indicate that the laser power density is the most important parameter in superconductor thin film deposition by laser ablation techniques. However, the oxygen partial pressure and the temperature of substrate are also important. In fact, a proper combination of substrate temperature and oxygen partial pressure was found to be of crucial importance for obtaining the best superconducting properties. For YBCO deposited on MgO substrate [95], an optimum condition of around 10^6 W/cm^2 for an oxygen partial pressure of about 17 Pa or 3×10^8 W/cm^2 for 130 Pa was found, and the optimum substrate temperature was found to be around 750°C, independent of the laser power density and the oxygen partial pressure.

Olsan et al. [96] studied YBCO film resistance during deposition as a function of the number of laser shots. It was found that islands are formed during the first deposition shot and they start to coalesce only when a critical thickness is achieved. The critical thickness was found to increase with decreasing laser repetition rate. It was also found that the tetragonal-to-orthorhombic phase transition caused by oxygen filling at the end of deposition is very fast (about 30 ms).

2.1.1.7. Laser Molecular Beam Epitaxy

Figure 7 is a schematic diagram of the computer-controlled laser-MBE system equipped with a differentially pumped *in situ* RHEED system and an atomic oxygen source [13]. A similar laser-MBE system using ozone or other oxidants has been employed by Kawai [97] and Gupta [98]. Their works are mainly on growing various epitaxial cuprate thin films.

The RHEED system is used for *in situ* monitoring of film growth as well as film thickness. A 20 keV electron beam is focused to the substrate surface with an angle of incidence of 1.5°

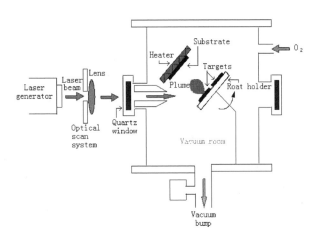

Fig. 6. Schematic diagram of PLD deposition apparatus.

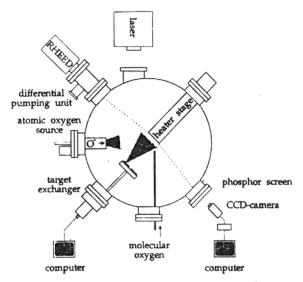

Fig. 7. Schematic diagram of the computer-controlled LMBE [13].

to 2.5°. The incidence angle was determined from the direct beam and the specular reflection image spot on the phosphor screen. For the ideal operation of RHEED during film growth, the basic chamber background pressure (P_0) should be kept below 10^{-4} Torr. For keeping the base chamber pressure and to satisfy the oxygen activity requirement for kinetic and thermodynamic stability of the high-T_c phase [99], the use of a differential pump and an atomic oxygen source is essential. To assure full oxidation of all ablated materials arriving at the substrate with each laser pulse, an additional continuous local spray of molecular oxygen is provided through a small nozzle (5 mm diameter) positioned as close as 1 cm in front of the substrate. The combined use of atomic and molecular oxygen results in an overall background pressure in the deposition chamber which is about 10^{-3} Torr. This pressure does not degrade the performance of the RHEED system.

As a template for epitaxial film growth, the substrate and its surface condition play a extremely important roles in determining the structure and surface morphology of the deposited film, especially during the initial stages of growth. To achieve layer-by-layer epitaxial film growth, it is of the utmost importance that the substrate is atomically smooth over a large area and that the lattice mismatch between the substrate and deposited film is as small as possible. The homoepitaxial growth of the SrTiO₃ film on (100) SrTiO₃ is a good example to illustrate this point [13].

The most impotant growth parameters for growing high-quality cuprate films are found to be the oxygen partial pressure, oxygen potency, substrate temperature (T_s), and growth rate [100]. In general, the substrate temperature should be high enough to ensure adequate atomic surface mobility at the film growth front and kinetic requirements for the high-T_c phase stability under the experimental oxygen ambient condition. However, the substrate temperature should be low enough to prevent interlayer diffusion of the film–substrate during deposition [101].

2.1.2. Chemical Processing

2.1.2.1. Metal–organic Deposition

The metal–organic deposition (MOD) method is a nonvacuum approach to form thin films and has the advantages of precise composition control and rapid deposition rates. Moreover, the MOD method is easily applicable to the formation of films with large areas and complicated shapes [102].

The MOD process involves dissolving metal–organic compounds of the individual elements in an appropriate solvent, mixing the solutions to achieve the desired stoichiometry, depositing these formulations on an appropriate substrate to make film, and then firing these wet films to obtain the desired inorganic film.

In MOD process, the formulation was spun onto substrates and heated to pyrolyze the film. It is determined that the temperature, heating rate, and pyrolysis temperature are all very important in affecting the superconducting properties of the

films. Films are currently being pyrolyzed by heating at 30°C per minute in flowing oxygen to 850°C. The thickness of typical single-layer fired film is 200 nm; the deposition and pyrolysis steps are repeated up to 10 times to give the desired film thickness. The films are then given a final anneal in oxygen at 920°C for 1 hr and furnace cooled [15].

Koike et al. [102] prepared the TlBa₂Ca₂Cu₃O_y thin films on SrTiO3 (100) substrates by the MOD method. In their works, the values of $T_{c\ zero} = 106$ K and $J_c = 6.3 \times 10^5$ A/cm² at 77 K and 0 T were obtained. In the magnetic field of 1 T perpendicular to the a–b plane, the J_c values remained at 2.0×10^4 and 6.0×10^5 A/cm² at 77 and 40 K, respectively.

2.1.2.2. Chemical Vapor Deposition (CVD)

Chemical vapor deposition techniques are vapor deposition techniques based on homogeneous and/or heterogeneous chemical reactions. These processes employ various gaseous, liquid, and solid chemicals as sources of the elements of which the film is to be made.

The CVD deposition results from a set of phenomena occurring in the gas phase and at the surface of the substrate. These phenomena can generally be divided into the following steps:

- Gas-phase phenomena:
  - homogeneous reactions
  - diffusion of reactions to the substrate surface (mass transport)

- Surface phenomena:
  - adsorption of reactants at the surface
  - heterogeneous chemical reactions
  - surface migration
  - lattice incorporation (deposition)

- Gas-phase phenomena:
  - desorption of reaction by-products from the surface
  - diffusion of products into the main gas stream

In the early stages of CVD-YBCO deposition, after deposition at rather low temperatures of 400°C, postannealing was applied [103], but soon deposition temperature was increased to 850–920°C. The in situ oxidation could be applied successfully [104]. Both films grown by postannealing and in situ [105] were all deposited with argon as a carrier gas for the thd-precursor vapors, and oxygen was separately introduced into the deposition chamber. The developments during that time contributed to lowering the deposition and in situ anneal temperatures, while the high-temperature annealing results in voids that affect film properties and are unsuitable for multilayer structures.

Further improvements were achieved when it was demonstrated that substituting O₂ with N₂O not only allowed lower deposition temperatures but also increased deposition rates. Tsuruoka et al. [106] investigated the influence of the reactive gases on film composition. They found that the flow ratio of the Y/Ba/Cu-thd vapors had to be readjusted for the new reactive gas. This indicates that for CVD techniques, the conditioning of the reactive gas affects film stoichiometry and thus the superconducting properties significantly. This is supported by the larger number of publications that discuss the topic of flux pinning centers affecting flux creep, and these centers are belived to be CuO platelets [107]. Another variation of the oxidizing gas was examined by using ozone [108].

2.1.2.3. Metal–organic Chemical Vapor Deposition

Metal–organic chemical vapor deposition (MOCVD) is another widely used process for preparing superconductor thin film oxides. A variety of oxides have been grown by MOCVD, including the important class of high-temperature superconducting Cu oxides. Particular attention has been given to the BiSrCaCuO and YBaCuO systems. Superconducting metal oxide films grown by MOCVD have been limited in performance largely by the novel precursors, the reactor designs compatible with the low vapor pressures, and the oxidizing nature of the growth ambient.

This growth process involves a preparation of organometallic precursors, such as metal alkyls, which are transported into the growth chamber where the formation of the desired compound takes place by its pyrolysis, and subsequent recombination of the atomic or molecular species in close proximity to the heated substrate. Most of the metal–organic precursors used belong to the β-diketonate family, with extensive use of Y(tmhd)$_3$, Ba(tmhd)$_2$, and Cu(tmhd)$_2$. The precursors for yttrium and copper have reasonable volatility and stability at moderate temperatures (around 100°C). Only Ba(tmhd)$_2$ has to be heated to teperatures higher than 200°C, which affects its long term vaporization stability. Oligomerization can occur which decreases volatility, leading to a compositional shift in the gas phase and in the YBa$_2$Cu$_3$O$_7$ film during deposition. The evaporation temperature for barium must therefore be very precisely controlled and kept relatively low, thus reducing the maximum available barium partial pressure into the deposition zone and limiting the growth rate by mass transport towards the substrate.

Oxygen (or wet O$_2$) necessary for the formation of oxide species is added to the carrier gas saturated with the precursors' vapors immediately before it enters the reaction chamber. In this sense, CVD is a chemical process which is not limited to line-of-sight deposition, thus allowing a uniform coating of large areas or complex three-dimensional structures. However, the main problem in high-T_c superconductor fabrication is to develop for each particular case, a suitable volatile precursor which will possess high purity and high vapor pressure and will stay at the operating temperature.

The main means of controlling growth rates, composition uniformity, and thickness in high-T_c superconductor thin films is by adjusting the evaporation rate and mass transfer of the precursors by the carrier gas into the MOCVD reaction chamber.

Early YBCO films prepared on MgO substrate by this method using Cu(acac)$_2$, Y(thd)$_3$, and Ba(thd)$_2$ at 170, 160, and 253°C, respectively, as precursor with nitrogen carrier gas

exhibited weak superconducting properties after annealing at about 900°C over a few hours [109]. It was found that the decomposition of Ba(thd)$_2$ at the vaporization temperature resulted in changing deposition rate and the film's composition. Aiming to improve stability and volatility, some groups have used fluorine-based precurors to deposit YBCO thin films on SrTiO$_3$ and MgO substrates, but with limited success [110].

In order to increase the stability of chemical vapor reactions and to improve the growth rate in the deposition process, alternative MOCVD techniques have been developed. The ultrasonic sensors, applied for the concentration monitor of binary gas mixture, are based on the principle that the velocity of sound is determined by the molecular mass and the ratio of the specific heats (Cp/Cv) of the species in the gas phase. Researchers in Superconductor Technologies Inc. used the ultrasonic sensors to control the composition of metal precursors and successfully prepared high quality YBCO films for microwave applications [111].

Real-time monitoring of thin film surfaces can support reproducible growth. An optical method is useful for this purpose in MOCVD. Optical reflectance measurements provide a method for in situ diagnostics of the crystal growth process in MOCVD of YBCO [112]. The thickness of the films and the mode of crystal growth are deduced from the reflectance measurements.

A new process was developed in which a pulverized mixture of the Y(tmhd)$_3$, Ba(tmhd)$_2$, and Cu(tmhd)$_2$ reagents is fed directly into the coating furnace by Ar carrier gas and vaporized before reaching the substrate [113]. The main advantage of this method over the conventional MOCVD is its high deposition rate (200 μm/hr versus 1–10 μm/hr), thus enabling better deposition control. High-density uniform films with zero resistance measuring around 82 K were thus obtained.

Figure 8 is a schematic of the atomic layer-by-layer MOCVD growth apparatus [114]. The main advantages of the atomic layer-by-layer MOCVD apparatus include the control of prop-

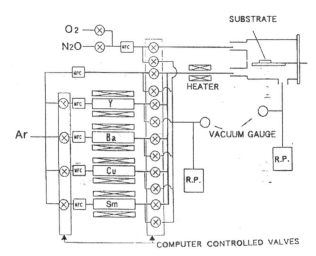

Fig. 8. A schematic diagram of the atomic layer-by-layer MOCVD apparatus [116]. Reprinted from *Appl. Surface Sci.* 112, 30 (1997), with permission from Elsevier Science.

erties at the atomic scale and the manufacturability at the large scale which is important for the application of high-temperature superconductors to future electron devices [114].

A complete novel approach to effectively eliminate the problems of instability of the Ba precursor was recently developed by Schulte et al. [115]. A modified, relatively simple, CVD technique which operates at high O_2 partial pressure (around 45 Pa) without any carrier gas was used in this method. Three temperature-controlled crucibles containing Y(thd)$_2$, Ba(thd)$_2$, and Cu(thd)$_2$ precursors and the substrates were placed in the same vacuum chamber, as shown in Figure 7. The vapors were mixed together and guided to the substrate's suface by means of a heated chimney without the use of any carrier gas. High-quality YBCO thin film with high c-axis orientation were thus obtained on SrTiO$_3$ substrate at 91.8 K and critical current density 10^6 A/cm^2 at 77 K.

2.1.2.4. Combustion Chemical Vapor Deposition Method

The combustion chemical vapor deposition (CCVD) process is an inexpensive, open-atmosphere deposition process that is not limited by the size of the deposition chamber and could be easily scaled and multiplexed to continuously fabricate buffer layer/superconductor and even passivation layer coating structure on kilometer lengths of textured substrate moving on a reel-to-reel system. It also can be used to deposit epitaxial buffers and high-temperature superconductors on oxide single-crystal substrates. CCVD does not use vacuum equipment or reaction chambers required by conventional techniques, while its coating quality rivals and even exceeds that of conventional methods. Compounds being studied with the CCVD process include the buffer layers cerium oxide (CeO$_2$), yttria stabilized zirconia (YSZ), strontium titanate (SrTiO$_3$), lanthanum alminate (LaAlO$_3$), yttria (Y$_2$O$_3$), and ytterbium oxide (Yb$_2$O$_3$), and two rare earth superconductors, YBa$_2$Cu$_3$O$_{7-\delta}$ (YBCO) and YbBa$_2$Cu$_3$O$_{7-\delta}$ (YbBCO) [116].

Figure 9 is a typical schematic representation for a CCVD system [116]. In CCVD, the film precursors are generally dis-solved in a solvent that also acts as the combustible fuel. This solution is atomized to form submicron droplets, which are then convected to and combusted in the flame. The heat from the flame provides the energy required to evaporate the droplets and to dissociate the precursors; the latter then react and vapor deposit onto the substrate to form the coating. Substrate temperature may be as low as 100°C, thus enabling deposition onto a wide variety of materials including plastics. Since CCVD processing occurs at ambient temperature and pressure, depositions within clean hoods equipped with ULPA filters are possible, thus minimizing the airborne contaminants that can be entrained in the deposited film. This ability is required for high-quality electronic coatings such as high-J_c superconductors.

2.1.2.5. Dip Coating

Among the different techniques used for the preparation of thin films of high-T_c superconducting materials, the dip-coating technique is relatively simpler where single-sided or double-sided films with thicknesses as low as 3 μm can be prepared even on curved surfaces. For the successful preparation of superconductor films, the selection of suitable substrate materials and the optimization of the processing conditions are critical [117].

For example, in the process of fabrication of Y-Ba-Cu-O thin films by dip coating, the concentrated mixed alkoxide solution used is as follows [118]:

Alkoxides are yttrium isopropoxide, barium ethoxide, and copper ethoxyethoxide. Y (O-iC$_3$H$_7$)$_3$ 0.4 M (mol/l) in toluene, Ba(O-C$_2$H$_5$)$_2$ 0.8 M, in ethanol, and Cu(OC$_2$H$_4$OC$_2$H$_5$)$_2$ 0.5 M in toluene are mixed with diethanolamine, so that the metal ratio of Ba, Y, and Cu becomes 2:1:3, and the solvent is exchanged for ethanol using a rotary vaporator in N$_2$. Thus a homogeneous and stable solution of the final concentration Y(O-iC$_3$H$_7$)$_3$ 0.1 mol/l, Ba(O-C$_2$H$_5$)$_2$ 0.2 mol/l, Cu(OC$_2$H$_4$C$_2$H$_5$)$_2$ 0.3 mol/l, HN(C$_2$H$_4$OH)$_2$ 0.6 mol/l is prepared. The solution exhibited appropriate viscosity to perform dip coating. The coating processes are repeated in air for the desired number of times after firing the dip-coated sample in O$_2$ at 550°C for 10 min each time. The final heat treatment is made in air by heating the specimen from 550 to 900°C in 2 h, holding for 1 h, then slowly cooling to 500°C in 6 h, holding again for 3 h, and finally cooling to room temperature in 5 h.

The dip-coating technique was found to be one of the easiest methods for obtaining good quality superconductor thin films, especially for YBa$_2$Cu$_3$O$_{7-\delta}$. Some investigations for YBa$_2$Cu$_3$O$_{7-\delta}$ by the dip-coating technique on several substrates, such as MgO (001), ZrO$_2$-Y$_2$O$_3$, and REBa$_2$NbO$_6$, were reported. The best film deposited on MgO by this method had a good properties with T_c (onset) = 84 K, T_c (R = 0) = 80 K, and zero-field critical current density of roughly 70 A/cm^2 at 77 K. The YBa$_2$Cu$_3$O$_{7-\delta}$ thin films fabricated on polycrystalline zirconia ZrO$_2$–Y$_2$O$_3$ showed $T_{c(onset)}$ = 94 K and $T_{c(zero)}$ = 60 K.

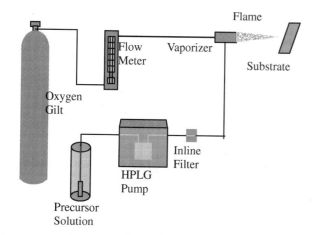

Fig. 9. Schematic representation of the system to deposit CCVD coatings [118] (*IEEE Trans. Appl. Supercond.* 9, 2426 (©1999 IEEE)).

2.1.2.6. Sol–Gel Method

Sol–gel techniques have been developed as viable methods for the fabrication of superconducting thin films [20, 119] for the facts that the sol–gel represents a low-cost, nonvacuum process. The sol–gel process involves synthesis of a polymerizable precursor solution by mixing or reacting metal alkoxides and metal–organic salts in a common solvent. The precursor solution can be spin-coated, dip-coated, sprayed, or painted on practically any substrate and calcined to obtain dense crystalline films. The major advantage of sol–gel processing over conventional solid-state reactions is that the polymeric network formation of the metal–organic complexes leads to atomic scale mixing, thus allowing a dramatic reduction in processing temperatures and time.

Shibata et al. [119] described the fabrication of *c*-axis oriented films of $RBa_2Cu_3O_{7-\delta}$ (R = Y, Nd, Y-Nd) on YSZ ceramic substrates by the sol–gel method as follows.

Alkoxides of R (Y or Nd), Ba, and Cu diluted in butyl alcohol were weighed to a certain mole ratio and were refluxed for 1 day. Then, water was added slowly to the solution with vigorous stirring. The solution was refluxed again for 10 hours and the solvents were evaporated at 120°C to make high viscous pastes involving the mixture of very fine oxide powders. The resultant pastes were painted on YSZ ceramic substrates by the screen printing technique. The samples were heated to 950°C for 0.5 h in an oxygen atmosphere and were cooled slowly to room temperature. Lower temperature were also used.

Sol–gel techniques are also used to fabricate the buffer layers for deposition of superconductor thin films, such as $LaAlO_3$ [120] and $BaCeO_3$ [121].

2.1.2.7. Liquid Phase Epitaxy

The liquid phase epitaxy (LPE) technique involves growth under thermoequilibrium conditions and is suitable for growing high-quality single-crystalline films with sufficient thickness, while the vapor deposition methods are suitable for the growth of oriented films of angstrom-order thickness but not single-crystalline films. As the process is performed under a near-equilibrium condition, homogeneous growth of high-quality film can be expected when the growth condition is properly controlled.

A large body of work on high-T_c superconductor thin films has been undertaken using LPE technique, such as $La_{2-x}Sr_xCuO_4$ [122], $PrBa_2Cu_3O_{7-\delta}$ [123], and $Bi_2Sr_2Ca-Cu_2O_x$ [124].

The process of the LPE method can be clarified through the preparation of $Bi_2Sr_2CaCu_2O_x$ (BSCCO) film as follows [124].

The LPE growth of the BSCCO film was performed by means of the flux method. KCl was chosen as the flux because it can be removed easily from the specimen after growth. The $Bi_2Sr_2CaCu_2O_x$ solute prepared by means of conventional solid-state reaction from powder reagents was mixed with the flux, charged in a high-purity alumina crucible, and then processed in the LPE furnace, as shown in Figure 10 [124]. The

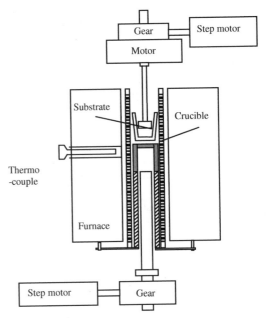

Fig. 10.   Illustration of LPE furnace [124].

temperature was first elevated to the melting temperature T_m and kept at that temperature for more than 1 h with stirring of the solution with an alumina bar in order to homogenize the solution. Then the temperature was slowly lowered to the supersaturation temperature T_1 at which point the substrate was dipped in the solution, and the growth of the film was performed. During growth the substrate was rotated at a constant rate to facilitate homogeneous growth of the film; then the substrate was removed from the solution and spun in the furnace for about 1 min to remove the residual flux.

Conditions related to the solute and solution had a considerable influence on the growth of the film. The melting temperature T_m and the growth temperature T_1 have considerable critical influence on the growth of the film. Lower melting temperature resulted in insufficient solubility of the solute. Higher melting temperature resulted in deterioration of the superconducting characteristics. This seems to be caused by the change of solute composition due to selective evaporation.

The lattice match between the film and the substrate is an important factor. It is desired that the lattice parameter and the thermal expansion coefficients of the substrates are similar to those of the films.

2.2. Substrate and Buffers

Due to the chemical composition, crystal structure, and anisotropic properties of high-T_c superconductors, substrates play a vital role in film properties and hence device characteristics. Since the discovery of high-T_c superconductors, a flood of research on substrates for depositing superconductor thin films including the influences of lattice mismatch, interface chemical stability, and substrate physical properties on crystalline, physical properties, and superconductivity has been carried out. The ideal substrate material is chemically compatible and has both good structure and thermal expansion matched to the superconductor thin films.

2.2.1. Perovskite Structural Substrates

Single-crystal $SrTiO_3$ has so far been found to be the best substrate for epitaxial growth of YBCO films on its (100) face [125]. Unfortunately, $SrTiO_3$ is a ferroelectric perovskite, characterized by large and temperature-dependent dielectric permittivity and very high microwave losses at low temperatures. So it is just an excellent substrate for YBCO films used for dc or low-frequency devices.

$LaGaO_3$ and $LaAlO_3$ were also successfully applied for growing high-quality epitaxial superconducting thin films. However, $LaGaO_3$ substrate produced serious twinning problems and the material itself is quite active chemically. $LaAlO_3$ has so far been the best choice for microwave application purposes [126]. Unfortunately, the $LaAlO_3$ crystals grown by the flux method are quite small and of poor quality; the fabrication of large single crystals by the Czochralski method is diffcult and leads to a high expense. Also the perovskites $NdGaO_3$, $PrGaO_3$, $YbFeO_3$, $YAlO_3$, and $LaSrGaO_4$ have similar properties and therefore are used as substrates.

Jose et al. [127] studied a new class of complex perovskites $REBa_2ZrO_{5.5}$ (where RE = La, Ce, Eu, and Yb). They proved that there is no detectable chemical reaction between $YBa_2Cu_3O_{7-\delta}$ and $REBa_2ZrO_{5.5}$ even under severe heat treatment at 950°C and that the addition of $REBa_2ZrO_{5.5}$ up to 20 vol% in $YBa_2Cu_3O_{7-\delta}$ shows no detrimental effect on the superconducting properties of $YBa_2Cu_3O_{7-\delta}$. Dielectric constants and loss factors are in the range suitable for their use as substrates for microwave applications. John et al. [128] proved that $Ba_2REHfO_{5.5}$ (where RE = La, Pr, Nd, and Eu) has the similar properties.

Another group of perovskite antimonates A_2MeSbO_6 (where A = Ba or Sr, Me = a rareearth, y, Sc, Ga, or In) and $A_4MeSb_3O_{12}$ (where A = Ba or Sr, and Me = Li, Na, K) were proved to be excellent candidates for use as a constituent in a high-T_c superconductor microwave device technology [129]. So far, more than 60 kinds of perovskite structural substrates have been developed and used in varying degrees. The lists in Table II are the perovskite structure substrates for deposition of high-T_c superconductor thin films.

2.2.2. Metallic Substrate

The major large-scale application of superconductivity has been the construction of high-field magnets. This application requires the conductors to carry high critical current densities ($> 10^4$ A/cm^2) at respectable magnetic fields greater than 1 T. Another important requirment is that the conductors possess high ductility, mechanical strength, and chemical stability. In particular, the conductors must be processed into long wires or tapes with a uniform microstructure. For stability, the superconductor tapes and wires must be sheathed with metallic materials that have a high electrical conductivity.

Ag, Al, and Ni are commonly used metallic substrates for high-T_c superconducting thin films due to their better electrical properties and heat conduction (seen in Table III).

2.2.3. Other Kinds of Substrates

The properties of some other kinds of substrates for deposition high-T_c superconductor thin films are listed in Table IV.

Highly oriented and epitaxial films have been grown on MgO (100) in spite of the large lattice mismatch [8]. MgO has a NaCl structure with a lattice constant of 4.2 Å. However, this material was found to be one of the least reactive substrates.

ZrO_2 stabilized by ~9 mol% YSZ is one kind of important substrate with a fluorite lattice structure of $a = 5.16$ Å for depositing high-T_c superconductor thin films, especially for YBCO. But some cases, the films deposited on the (100) face of stabilized ZrO_2 met relative poor quality [130] and some impurity phase, such as $BaZrO_3$, Y_2BaCuO_5, and $Y_2Cu_2O_5$, formed at the interface between substrate and film [131].

Films deposited on the most technologically important Si and Al_2O_3 substrates had very limited success due to the severe reaction between the high-T_c superconductor film and the substrate during the high-temperature annealing stage [8]. The reaction produces films with a graded compositional profile with insulating phases close to the film–substrate interface [132].

2.2.4. Various Kinds of Buffers

Buffer layers are commonly used to prevent interdiffusion or reaction of the films with the substrate [133]. Generally speaking, all buffer layers must be sufficiently perfect and have chemical stability. The annealing of the films at high temperature places stringent demands on the barrier material choice. A good lattice match and continuity of the buffer layer is one of the basic requirements, because grain boundaries provide an easy path for materials to diffuse. Any imperfections caused by poor buffer layer will greatly affect the junction quality and its performance because of the short coherence length of high-T_c cuprate superconductors. So far, several kinds of buffers were used; e.g., epitaxially grown MgO films on Si [134] and GaAs [135] were successfully used as buffer layers for the growth of in situ high-quality YBCO thin films. General requirements and classification of various buffer layers are well documented in the literature [136].

During the initial demonstration of the rolling-assisted biaxially textured substrates approach, high-J_c YBCO films were grown with a layer sequence of YBCO/YSZ/CeO$_2$/Ni substrates [137]. Here both buffer layers and YBCO superconductors were grown by pulsed laser deposition. Some scalable techniques such as electron beam evaporation and sputtering were used to produce high-quality CeO$_2$ and YSZ buffer layers [138]. The lists in Table V are the common buffer layers for deposition high-T_c superconductor thin films.

Table II.   Perovskite Structure Substrates for Deposition High-T_c Superconducting Thin Films

Substrate material	Crystal system	Space group	Lattice parameters (Å)			Density (g/cm³)	Dielectric constant ε (at 1 MHz)	Dielectric loss tan δ at 300 K
			a_0	b_0	c_0			
GdBa$_2$HfO$_{5.5}$	Cubic	Fm3m	8.364	8.364	8.364	7.925	20	5×10^{-3}
LaBa$_2$HfO$_{5.5}$	Cubic	Fm3m	8.312	8.312	8.312	7.862	25	6.6×10^{-3}
PrBa$_2$HfO$_{5.5}$	Cubic	Fm3m	8.541	8.541	8.541	7.268	24.5	5.2×10^{-3}
NdBa$_2$HfO$_{5.5}$	Cubic	Fm3m	8.352	8.352	8.352	7.812	26	6.0×10^{-3}
EuBa$_2$HfO$_{5.5}$	Cubic	Fm3m	8.349	8.349	8.349	7.908	21	2.0×10^{-3}
LaBa$_2$ZrO$_{5.5}$	Cubic	Fm3m	8.39	8.39	8.39	6.67	21	5.7×10^{-3}
CeBa$_2$ZrO$_{5.5}$	Cubic	Fm3m	8.55	8.55	8.55	6.30	25	6.1×10^{-3}
EuBa$_2$ZrO$_{5.5}$	Cubic	Fm3m	8.42	8.42	8.42	6.76	30	3.5×10^{-3}
YbBa$_2$ZrO$_{5.5}$	Cubic	Fm3m	8.39	8.39	8.39	7.04	33	4.1×10^{-3}
Ba$_2$DyZrO$_{5.5}$	Cubic	Fm3m	8.462	8.462	8.462	6.92	14	4×10^{-3}
PrBa$_2$SbO$_6$	Cubic	Fm3m	8.572	8.572	8.572	6.676	14.21	2.9×10^{-3}
SmBa$_2$SbO$_6$	Cubic	Fm3m	8.520	8.520	8.520	6.901	12.0	2.5×10^{-3}
GdBa$_2$SbO$_6$	Cubic	Fm3m	8.488	8.488	8.488	7.054	16.1	8.0×10^{-4}
YBa$_2$SnO$_{5.5}$							10	0.05
YBa$_2$NbO$_6$	Cubic	Fm3m	8.436	8.436	8.436	6.11	33	1.4×10^{-4}
PrBa$_2$NbO$_6$	Cubic	Fm3m	8.532	8.532	8.532	6.462	15	2×10^{-4}
NdBa$_2$NbO$_6$	Cubic	Fm3m	8.516	8.516	8.516	6.534	42	7.4×10^{-4}
SmBa$_2$NbO$_6$	Cubic	Fm3m	8.488	8.488	8.488	6.666	9	1×10^{-4}
EuBa$_2$NbO$_6$	Cubic	Fm3m	8.460	8.460	8.460	6.750	11	2×10^{-4}
BaZrO$_3$	Cubic		4.192	4.192	4.192	6.73	12	2.5×10^{-4}
BaHfO$_3$	Cubic		4.171	4.171	4.171			
LaAlO$_3$	Rhombohedral	R3m	5.364	5.364	13.11		21	4×10^{-4}
NdGaO$_3$	Orthorhombic		5.431	5.499	7.710		22	4×10^{-4}
SrTiO$_3$	Cubic	Pm3m	3.936	3.936	3.936	4.998	>300	3×10^{-2}
Ba$_2$GdNbO$_6$	Tetragonal		5.994	5.994	8.511			
KTaO$_3$	Cubic	Pm3m	3.989	3.989	3.989	7.012	243	1×10^{-3}
Ba$_2$DyNbO$_6$	Cubic	Fm3m	8.453	8.453	8.453	6.89	29	3×10^{-3}
Ba$_2$GdTaO$_6$	Tetragonal	F	5.997	5.997	8.518			
Ba$_2$EuSbO$_6$	Cubic	Fm3m	8.489	8.489	8.489	6.996	16.3	2.3×10^{-3}
Sr$_2$LaSbO$_6$	Cubic		8.325	8.325	8.325	5.91	13.6	
Sr$_2$PrSbO$_6$	Tetragonal		8.390	8.390	8.362	6.02	11.0	
Sr$_2$NdSbO$_6$	Tetragonal		8.365	8.365	8.320	6.13	11.0	
Sr$_2$SmSbO$_6$	Tetragonal		8.335	8.335	8.295	6.26	11.0	
Sr$_2$EuSbO$_6$	Tetragonal		8.320	8.320	8.300	6.30	10.8	2.2×10^{-3}
Sr$_2$GdSbO$_6$	Tetragonal		8.295	8.295	8.280	6.42	10.9	
Sr$_2$TbSbO$_6$	Tetragonal		8.280	8.280	8.248	6.48	11.0	
Sr$_2$DySbO$_6$	Tetragonal		8.248	8.248	8.224	6.64	11.1	
Sr$_2$YSbO$_6$	Tetragonal		8.231	8.231	8.216	6.56	10.9	
Sr$_2$HoSbO$_6$	Tetragonal		8.239	8.239	8.218	6.69	11.1	
Sr$_2$ErSbO$_6$	Tetragonal		8.222	8.222	8.204	6.77	11.1	
Sr$_2$TmSbO$_6$	Tetragonal		8.204	8.204	8.185	6.86	11.4	
Sr$_2$YbSbO$_6$	Tetragonal		8.190	8.190	8.176	5.87	11.2	
Sr$_2$LuSbO$_6$	Cubic		8.188	8.188	8.188	6.90	11.3	
Sr$_2$InSbO$_6$	Cubic		8.086	8.086	8.086	6.34	11.0	
Sr$_2$ScSbO$_6$	Tetragonal		8.019	8.019	8.063	5.49	12.5	
Sr$_2$GaSbO$_6$	Tetragonal		7.880	7.880	7.784	6.22	12.7	
Ba$_2$ScSbO$_6$	Tetragonal		8.172	8.172	8.196	6.54	24.3	
Ba$_2$InSbO$_6$	Tetragonal		4.174	4.174	4.134	7.33	15.2	

Table II. (Continued).

Substrate material	Crystal system	Space group	Lattice parameters (Å)			Density (g/cm^3)	Dielectric constant ε (at 1 MHz)	Dielectric loss tan δ at 300 K
			a_0	b_0	c_0			
Sr$_4$NaSb$_3$O$_{12}$	Cubic		8.180	8.180	8.180	5.52	8.36	
Ba$_4$LiSb$_3$O$_{12}$	Cubic		8.230	8.230	8.230	5.40	9.04	
Ba$_4$NaSb$_3$O$_{12}$	Cubic		8.221	8.221	8.221	6.65	14.8	
Sr$_4$KSb$_3$O$_{12}$	Cubic		8.275	8.275	8.275	6.62	14.1	
Ba$_2$DySbO$_6$	Cubic	Fm3m	8.431	8.431	8.431	7.259		
Ba$_2$LaNbO$_6$	Cubic	Fm3m	8.592	8.592	8.592			
Ba(Mg$_{1/3}$Ta$_{2/3}$)O$_3$	Trigonal	P3ml	5.782	5.782	7.067			
Y$_3$Ga$_2$(GaO$_4$)$_3$	Cubic	Ia3d	12.277	12.277	12.277	5.794		
Al$_5$Y$_3$O$_{12}$	Cubic	Ia3d	12.001	12.001	12.001	4.553		
Gd$_3$Ga$_5$O$_{12}$	Cubic	Ia3d	12.376	12.376	12.376	7.093		
PrGaO$_3$	Orthorhombic		5.458	5.490	7.733		24	
SrLaAlO$_4$	Tetragonal	I4/mmm	3.755	3.755	12.62	5.93	17	1.5×10^{-5}
YAlO$_3$	Orthorhombic	Pbnm	5.179	5.329	7.37		16	1×10^{-5}
Nd$_4$Ga$_2$O$_9$	Monoclinic	P2$_1$/c	7.733	11.032	11.456			
BaCuY$_2$O$_5$	Orthorhombic	Pbn	7.1319	12.180	5.6593	6.197		
NdAlO$_3$	Cubic		3.797	3.797	3.797		22.5	5×10^{-5}
LaGaO$_3$	Orthorhombic	Pbnm	5.487	5.520	7.752		25	7×10^{-3}
FeYbO$_3$	Orthorhombic	Pbnm	5.233	5.557	7.570	8.35	4–5	

Table III. Properties of Some Common Metallic Substrates for Deposition High-T_c Superconducting Thin Films

Substrate material	Crystal system	Space group	Lattice parameters (Å)			Density (g/cm^3)	Thermal expansion coefficient (K^{-1})	Resistivity (Ω/cm) (300 K)	Melting point (°C)
			a_0	b_0	c_0				
Al	Cubic	Fm3m	4.0494	4.0494	4.0494	2.697	23.2×10^{-6}	2.69×10^{-6}	660
Ni	Cubic	Fm3m	3.5238	3.5238	3.5238	7.372	12.7×10^{-6}	6.844×10^{-6}	1453
Ag	Cubic	Fm3m	4.0856	4.0856	4.0856	9.93	19.2×10^{-6}	1.6×10^{-6}	962
Cu–Al alloy									

Table IV. Properties of Some Other Kinds of Substrates for Deposition High-T_c Superconducting Thin Films

Substrate material	Crystal system	Space group	Structure	Lattice parameters (Å)			Density (g/cm^3)	expansion coefficient (K^{-1})	Dielectric constant ε (at 1 MHz)	loss tan δ 300 K	Thermal Melting point (°C)
				a_0	b_0	c_0					
YSZ	Cubic		Fluorite	5.16	5.16	5.16		10×10^{-6}	25	1×10^{-3}	2550
MgO	Cubic	Fm3m	NaCl	4.213	4.213	4.213	3.58	13.8×10^{-6}	10	9×10^{-4}	2800
TiO$_2$	Tetragonal	P4/mnm		4.584	4.584	2.953	4.26		100	1.4×10^{-4}	1840
ZrO$_2$	Orthorhombic	P2$_1$2$_1$2$_1$	Baddeleyite	5.016	5.016	5.230	6.220				2715
Si	Cubic	Fd3m	Diamond	5.43	5.43	5.43	2.330	2.6×10^{-6}	12		1410
α-Al$_2$O$_3$	Hexagonal	R3c	Corundum	5.57	5.57	8.64	3.96	8×10^{-6}	9.5	2.2×10^{-5}	2045
CaGdNbO$_6$	Tetragonal	I4/mmm	K$_2$NiF$_4$	3.663	3.663	12.01	5.94				1840
Sr$_2$RuO$_4$	Tetragonal	I4/mmm	K$_2$NiF$_4$	3.873	3.873	12.745	5.911				
SrLaGaO$_4$	Tetragonal	I4/mmm	K$_2$NiF$_4$	3.843	3.843	12.68	6.388				
CaNdAlO$_4$	Tetragonal	I4/mmm	K$_2$NiF$_4$	3.688	3.688	12.15	5.52		20	1.1×10^{-5}	
GaAs	Cubic	F43m	Sphalerite	5.6538	5.6538	5.6538	5.316	6×10^{-6}	13.18		1238

Table V. Common Buffer Layers for Deposition High-T_c Superconducting Thin Films

Buffer material	Crystal system	Space group	Structure	Lattice parameters (Å)			Density (g/cm^3)	Thermal expansion coefficient (K^{-1})	Melting point (°C)	Color
				a_0	b_0	c_0				
CeO$_2$	Cubic	Fm3m	Fluorite	5.4113	5.4113	5.4113	7.22		2600	Light gray
Gd$_2$O$_3$	Monoclinic	C2/m		14.095	3.5765	8.7692	8.298			White
Yb$_2$O$_3$	Cubic	Ia3		10.435	10.435	10.435	9.215		2227	Colorless
Y$_2$O$_3$	Cubic	Ia3		10.604	10.604	10.604	5.031	7.7×10^{-6}	2410	White
LaNiO$_3$	Rhombohedral	R		5.457		6.572	7.219	12.5×10^{-6}		
CuNd$_2$O$_4$	Tetragonal	I4/mmm		3.9437	3.9437	12,1693	7.300			Black
BaCeO$_3$	Tetragonal			6.212	6.212	8.804	6.363			
Ba$_2$PrCu$_3$O$_{7-x}$	Orthorhombic	Pmmm		3.823	3.877	11.793				Black
CoSi$_2$	Cubic	Fm3m		5.3640	5.3640	5.3640	4.954	9.5×10^{-6}		Dark gray
TiN	Cubic	Fm3m		4.2417	4.2417	4.2417	5.39	$8.0\text{--}9.0 \times 10^{-6}$	2930	Dark olive brown
SrRuO$_3$	Orthorhombic	Pnma		5.573	7.856	5.538	6.484			Dark blue–black
ZrO$_2$	Orthorhombic	P2$_1$2$_1$2$_1$		5.016	5.016	5.230	6.220		2715	White
BaTiO$_3$	Tetragonal	P4mm	Perovskite	3.99	3.99	4.03	5.85	19×10^{-6}	1620	
MgAl$_2$O$_4$	Orthorhombic			8.507	2.740	9.407	4.310	7.6×10^{-6}	2135	
AlAs	Cubic			5.662						
Ag	Cubic	Fm3m		4.0856	4.0856	4.0856	9.93	19.2×10^{-6}	962	
BaF$_2$	Orthorhombic	Pnam		6.7129	7.9245	4.0472	5.409		1280	White
Ti	Hexagonal	P6$_3$/mmc		2.950		4.686	4.503	8.5×10^{-6}	1668	
BaSO$_4$	Orthorhombic	Pbnm	Barite	7.1565	8.8811	5.4541	4.472		1350	Colorless
In$_2$O$_3$	Hexagonal	R3c		5.487		14.510	7.311		1565	Colorless
NiSi$_2$	Cubic	Fm3m		5.416	5.416	5.416	4.803			
MgO	Cubic	Fm3m	NaCl	4.213	4.213	4.213	3.58	13.8×10^{-6}	2800	
SrTiO$_3$	Cubic	Pm3m	Perovskite	3.936	3.936	3.936	4.998	8.63×10^{-6}	2060	
Al	Cubic	Fm3m		4.0494	4.0494	4.0494	2.697	23.2×10^{-6}	660	

3. HIGH-TEMPERATURE SUPERCONDUCTOR THIN FILMS

3.1. YBa$_2$Cu$_3$O$_{7-\delta}$ (Y-123) and Related Materials Thin Films

3.1.1. Introduction

Early in 1987, YBa$_2$Cu$_3$O$_{7-\delta}$ (YBCO), the first copper-oxide high-temperature (in the liquid nitrogen temperature region) superconductor, was discovered [176] which exhibits the highest superconducting transition temperature ($T_c = 90$ K) at normal pressure, and other interesting properties. This was followed by lots of research all over the world and many issues about superconductivity were understood and resolved [177, 181, 186, 189, 191, 193]. Even now, YBCO is still seen as the most typical high-T_c superconductor, and many new normal and superconducting properties and phenomena are observed based on it [183–185, 195, 196].

On the other hand, in the early stages of investigation of high-T_c superconductivity, YBCO thin film is naturally considered to be the ideal material for research on high-T_c superconductivity and development of applications for its remarkable stability [195, 196]. So following the preparation and investigation of (La,Sr)CuO (LSCO) thin films, the research on YBCO thin films rapidly developed worldwide [181, 183, 189, 191, 193]. YBCO thin films still play an important role in the field of high-T_c superconducting thin films on both fundamental research and thin film device application [193, 195].

In this section, fabrication and properties will be introduced. Some transport properties will be reviewed in another special section, Section 4.

3.1.2. Preparation and Characterization of YBCO Thin Films

There are several methods for growing YBCO thin films. In the early stages, magnetron sputtering (dc and rf) were used to fabricate YBCO thin films; then laser ablation was found to be a very useful tool for preparation of the YBCO films. At the same time the ion-beam sputtering method was also developed for preparing the YBCO thin films due to the advantage that its the composition of grown thin film is the same as that of the target. The chemical processing, such as CVD, MOCVD, the liquid phase epitaxy, and so on, is developed for fabrication of the thin films, especially for preparing the large-area thin film and depositing the film on substrates with complicated

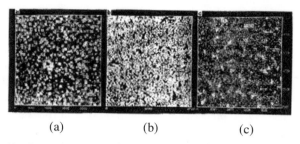

(a)   (b)   (c)

Fig. 11. Low-resolution AFM images of three films on (100) MgO substrates (a) 200 nm thick, (b) 20 nm thick, and (c) 10 nm thick. The scan size is 100 × 100 nm [140].

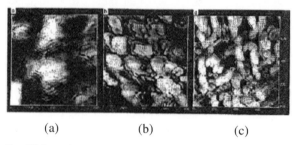

(a)   (b)   (c)

Fig. 12. High-resolution AFM images of three films on (100) MgO substrates (a) 200 nm thick, (b) 20 nm thick, and (c) 10 nm thick. The scan size is 800 × 800 nm [140].

shapes. During development of YBCO thin films, various kinds of substrates have also been developed, such as MgO, LaAlO$_3$, SrTiO$_3$, LaSrGaO$_4$, Nb-doped SrTiO$_3$, and NdGaO$_3$ single-crystal substrates. Of course, the above developed methods and substrates are also used to fabricate other kinds of high-T_c superconductor thin films.

Investigation of growth mechanism of YBCO thin films is most important for fabrication of YBCO thin films. The epitaxial growth of YBCO thin films on MgO(100) was first investigated by means of *in situ* RHEED [139], from which the layer by layer growth mode was observed. In 1991, Raistrick et al. investigated the microstructures of very thin sputtered films of YBCO deposited on (100) MgO substrate using scanning tunneling microscopy and atomic force microscopy (AFM) [140]. Figures 11 and 12 respectively show the low and high resolution AFM images of YBCO films with various thicknesses. It is shown that when the thickness reaches 10 nm, the substrate is completely covered by a fine-grained layer of the superconductor. The average grain size is about 100 nm. Many of these grains show evidence of spiral growth mode. In somewhat thicker films (20 nm), the grain size increases considerably to about 200 nm. Zheng et al. also studied the initial stages of epitaxial growth of c-axis-oriented YBCO thin films deposited by pulsed laser deposition on SrTiO$_3$ and MgO substrates by scanning electron microscopy (SEM) [141]. The images in Figures 13 and 14 showed that the films grown on SrTiO$_3$ have a sequential thickness variation, which reveals a transition from layerlike to island growth between 8 and 16 unit cells thickness; i.e., YBCO grows on SrTiO$_3$ by a Stranski–Krastanov mode. It also shows the existence of a slightly larger critical thickness for the introduction of screw dislocations into those films. In con-

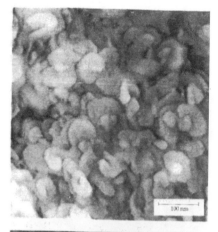

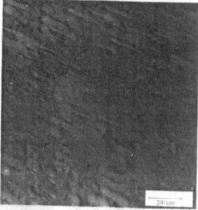

Fig. 13. STM images of $c_\perp$ YBCO films ~8 unit cells thick grown at 720°C on (001) MgO (top) and (001) SrTiO$_3$ (bottom) [141].

trast, the thinnest (~8 cell thick) YBCO films grown on MgO were found to have island growth mode with spiral growth features. For the case of SrTiO$_3$ substrate the spiral growth made was found in the film with thickness $\lesssim$6 unit cells.

3.1.2.1. YBCO Thin Film Fabricated by Magnetron Sputtering

Magnetron sputtering, as the earliest fabrication method, is the most important way to prepare YBCO thin films. In 1987. Burbidge et al. fabricated high-quality YBCO thin films by dc magnetron sputtering [142].

In the early stage the dc magnetron sputtering used for *in situ* deposition of YBCO thin films is performed by the following technique route: The sputtering target material of YBa$_{1.86}$Cu$_{2.86}$O$_y$ (change to 1:2:3 in films) was prepared by a solid reactive method. A mixture of Y$_2$O$_3$, BaO$_2$, and CuO in the stoichiometric ratio of Y:Ba:Cu = 1:1.86:2.86 was ground in an agate mortar. In order to achieve thorough mixing, the powder was sintered in air at 950–960 with 3–4 grindings during the sintered process. The sintered powders were then pressed into a disk with expected size (usually 3–7 cm in diameter with a thickness of 5 mm) and were sintered for more than 48 hours at 960°C in air. Finally, the target material were cooled down to room temperature with furnace. For the high-

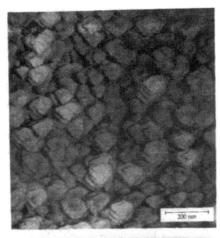

Fig. 14.   STM images of $c_\perp$ YBCO films ~16 unit cells thick grown at 720°C on (001) MgO (top) and (001) SrTiO$_3$ (bottom) [141].

quality target the most important parameter is not the large density, but the low resistivity and high superconducting transition temperature (above 91 K of zero resistance temperature). The high-quality target can be used for a long time regardless of the density. For sputtering the target material is bonded using silver epoxy to a water-cooled Cu backing plate of the magnetron target and inserted into a planar magnetron sputter source.

The base pressure of the vacuum system prior to deposition was in the range of $5-2 \times 10^{-6}$ Torr. During deposition, the diffusion or molecule pump (the molecule pump is encouraged to be used for keeping oil vapor free) was throttled and high-purity argon gas and oxygen gas was introduced. The sputtering gas pressure was 60–80 Pa with a ratio of oxygen to argon 1:3 or 1:4. A cathode voltage of 100–120 V was applied to the target with a dc current of 0.3–1.0 A according to the diameter of the target. Usually, the target was presputtered for cleaning the surface for 5 min before each deposition. The deposition temperature is in the range of 800–820°C. By this process the transition to zero resistance starts at 93 K but is not complete until about 30 K.

In 1987, Zhao et al. successfully prepared high-T_c YBCO thin films by the rf sputtering method [143]. The sputtering

voltage was in the range of 1.5–2 kV, the Ar background pressure was $1-5 \times 10^{-2}$ Torr, and the substrates were heated to about 200°C. After deposition, the films were annealed in oxygen atmosphere at 650°C for 1 hour followed by an annealing at 850°C for 1 hour. The T_c of the films deposited in this condition was 89 K.

Adachi et al. prepared YBCO thin films by a rf-magnetron sputtering technique [144]. Thin films were sputtered from $(Y_{0.4}Ba_{0.6})_3Cu_3O_7$ target onto substrates. In order to keep the films composition the same as the target's, pure argon was used for the sputtering gas and the substrates were not heated intentionally. The gas pressure was 0.4 Pa, the substrate temperature was 200°C, and the rf input power was 150 W. The as-sputtered films were nonconductive and colored metallic black, although the composition of the films was almost equal to that of the target within 10% accuracy. Then the sputtered thin films were annealed in air at 900°C for 1 hour. The heating and cooling rates were kept at about 400°C/h. After annealing, the films became conductive and showed superconducting transition with an onset temperature at 94 K, and the zero-ρ state was achieved below 70 K.

Bruyere et al. prepared the thin films by rf (13.56 MHz) diode reactive sputtering from an YBCO single target [145]. The distance between target and substrates is 3 cm. During the deposition the substrate temperature was kept at 390°C, the total pressure of the Ar/O$_2$(80/20) gas mixture was regulated at 8×10^{-3} Torr, and the total flow rate is 15 sccm. Before the deposition, the background vacuum pressure was ~10^{-6} Torr. In order to avoid the strong bombardment of the secondary particles on the substrate, a low rf power was used. The deposition rate was typically 1.8 nm/min. The thickness of the films is in the range of 0.5 to 1 μm. For testing deposition conditions, the different substrates have been used, but the study has been mostly focused on the properties of the films grown on the sintered and crystalline stabilized ZrO$_2$ (YSZ). The sintered ZrO$_2$ is stabilized with 10% of Y$_2$O$_3$ whereas the crystalline one is stabilized with 15% of Y$_2$O$_3$ which has a higher density and a lattice parameter of 5.15 Å + 0.005. The substrates are first cleaned with organic solvents and introduced into the vacuum system; then they are heated to 200°C for 10 hours before sputtering deposition. Targets of 50 mm diameter were sintered from mixed oxide powders of Y$_2$O$_3$, BaO, and CuO. After pressing under 5 tons/cm^2, sintering was performed at a temperature lower than 470°C to avoid a complete chemical reaction between the oxides. In this way the target has a good mechanical keeping and sputtering efficiency of the different elements. The zero-resistivity state is achieved at 87 K and the transition width is 3 K by using this kind of target.

Both rf and dc sputtering methods are almost identical for deposition of the YBCO thin films if the quality of target materials is good enough; i.e., the conductivity of the target materials is high.

Both on-axis and off-axis target to substrate configurations can be used to prepare YBCO thin films. For the above processes the on-axis configuration was used. Here, the off-axis configuration is introduced. As an example, Eom et al. fabricated YBCO thin films using an off-axis target to substrate configuration [65]. Stoichiometric targets were prepared from "freeze-dried" powders. The depositions were performed in a high oxygen partial pressure environment. Substrates were placed to face the side of the planar magnetron gun to avoid backsputtering damage from negative oxygen ions. The sputtering atmosphere consisted of 10 mTorr O_2 (or N_2O) and 40 mTorr Ar. The rf power (125 W) on the sputter gun generated a self-bias of -50 to -75 V and gave a deposition rate of about 0.5 Å/s. The block temperature was held at 600–700°C during film growth. After deposition, the chamber was immediately vented to 600 Torr of oxygen. In a typical run using a target of composition $Y_{17.7}Ba_{32.9}Cu_{49.4}$, the composition of the films at the extreme positions varied from $Y_{18.6}Ba_{32.8}Cu_{48.5}$ to $Y_{16.9}Ba_{33.1}Cu_{50.0}$, illustrating the 1:1 correspondence between target and film composition. The T_c of the film deposited by this substrate–target configuration is similar as that obtained in the on-axis case, i.e., $T_c \sim 75$–86 K.

Sofar, for magnetron sputtering stoichiometric (Y:Ba:Cu = 1:2:3) target is used it is easily to get YBCO thin films with high T_c, high J_c, and smooth surface. The deposition rate can be controlled precisely, so that the ultrathin films and multilayers can be grown according to the design. But deviation of the compositions between film and target may exist if the sputtering condition is not adjusted precisely. The magnetron sputtering method is still widely used to fabricate YBCO thin film today [146–148].

3.1.2.2. YBCO Thin Film Fabricated by Ion-Beam Sputtering

Not only magnetron sputtering but also IBS is also an important sputtering method to fabricate YBCO thin films. The ion-beam sputtering can be used with and without postannealing to prepare superconducting YBCO thin films. Klein et al. deposited oriented superconducting YBCO thin films on yttria-stabilized zirconia substrates by ion-beam sputtering of a nonstoichiometric target [149]. The deposition runs were performed in a Microscience IBEX-2000 deposition chamber evacuated by a diffusion pump fitted with a liquid-nitrogen trap and backed by a mechanical pump. The nonstoichiometric target of nominal composition $YBa_{2.10}Cu_{1.54}O_x$ was fabricated by a multiple-step reactive sintering process. Beam voltages of 750–1100 V were applied at beam currents of 15–50 mA to span a range of beam power from 11.25 to 37.5 W. A gas flow of 4 sccm of Ar to the ion source was employed for all sputtering conditions. A Cu plate positioned below the target was used to reduce the possibility of contamination arising from the chamber walls during sputtering. The stationary resistance heated substrate holder was oriented with its face parallel to the surface of the target at a center-to-center distance of 7.5 cm. Substrate temperatures above 600°C are commonly employed to obtain crystalline films. Higher substrate temperatures tend to promote the desired c-axis oriented film. A substrate temperature of 670°C was maintained during the majority of the depositions performed in this study. An oxygen partial pressure of 2×10^{-5} atm (2 Pa) is required to maintain the desired perovskite crystal

structure at the chosen 670°C substrate temperature. Since this pressure exceeds the usual 4×10^{-4} Torr of ion beam chamber pressure, a 4.5 sccm flow of oxygen was supplied by a flattened nozzles positioned immediately on either side of the substrate. Maximization of the film oxygen content during cooling is necessary to promote high T_c's in *in situ* grown films. The oxygen content of the films was further enhanced by backfilling the deposition chamber with 40 Torr of oxygen during the postdeposition cool-down period. The films deposited by this process exhibited zero-resistance temperatures as high as 80.5 K without postdeposition annealing.

Bagulya et al. also reported the deposition of YBCO thin films by IBS from a single nonstoichiometric target without postannealing [150]. In this system the target was bombarded with Ne^+ ions at an energy of up to 50 keV and a current of up to 10 mA. The substrate holder and target were held at ground potential. The beam was incident on the center of the target cooled by water at an angle of $\approx 45°$ to the target normal. To avoid the possibility of growing film contamination arising from target holder and chamber walls during sputtering, target sizes were substantially greater than the diameter of the focused ion beam. The substrate was mounted at a distance of about 30–35 mm from the target. Three nonstoichiometric targets with compositions $Y_1Ba_{2.4}Cu_{3.8}$, $Y_1Ba_{2.2}Cu_{3.8}$, and $Y_1Ba_{2.1}Cu_{3.6}$ were used. They were prepared by a standard solid reaction procedure from powders of yttrium, copper oxides, and barium carbonate. A relatively long presputter for about 20 h under normal operating conditions was required for a new target to stabilize the film composition. The films were deposited on (100) YSZ and (100) $SrTiO_3$ substrates. The depositions were carried out over the substrate temperature range 600–800°C. A pure heated oxygen gas flux was directed to the substrates by use of a specially designed substrate holder. This oxygen delivery system enables a uniform distribution of the gas flux across the substrates with a high pressure differential, the pressure at the substrates being up to 1–2 orders of magnitude higher than the pressure necessary to operate the ion gun in the sputtering chamber. The base vacuum pressure of the system was $\approx 10^{-6}$ Torr; the total O_2 pressure in the chamber during film growth was 10^{-3}–10^{-5} Torr. The local oxygen pressure close to the substrates was estimated to be 20 times higher than the pressure in the chamber. The maximum pressure that could be obtained close to the substrates was 20 mTorr. After deposition, the chamber was immediately filled with oxygen to ≈ 20 Torr and the film was cooled to $\approx 200°C$ in the oxygen atmosphere. The typical cooling time was approximately 20 min. The samples were taken from the chamber after the substrate temperature decreased to 200°C. Typical sputtering time was 0.5–3 h and films approximately 100–600 nm thick were deposited.

In latter years, the IBS method was also well used and developed. In 1996, a YBCO Josephson junction was prepared by a focused ion-beam technique [151]. In 1997, biaxially textured yttria-stabilized zirconia films for use as buffer layers of YBCO superconducting films have been deposited on polycrystalline Ni–Cr metallic substrates by using an ion-beam assisted deposition system with dual ion sources [152].

3.1.2.3. YBCO Thin Film Fabricated by Pulsed Laser Deposition

With the development of laser technology, PLD is more and more widely used to deposit YBCO thin films. This deposition method is relatively simple, very versatile, and does not require the use of ultrahigh vacuum techniques. The first successful preparation of thin films of YBCO superconductors using pulsed excimer laser evaporation of a single bulk material target in vacuum was achieved by Dijkkamp et al. [153]. Pellets with nominal composition $YBa_2Cu_3O_{7-\delta}$ were prepared in the usual way. This bulk material showed an onset temperature of 95 K and a transition width of 0.3 K. A pellet (1 cm in diameter, 2 mm thick) was mounted in a small vacuum system with a base pressure of 5×10^{-7} Torr and was irradiated through a quartz window with a KrF excimer laser (30 ns full width at half maximum, 1 J/shot) at 45° angle of incidence. A quartz lens was used to obtain an energy density of approximately 2 J/cm^2 on the target. The substrates were mounted at a distance of typically 3 cm from the pellet surface, close to the normal from the center of the laser spot. The laser was fired at a repetition rate of 3–6 Hz, for total of a few thousand shots. With each shot, a plume of intense white light emission could be observed normal to the sample surface. The deposited film thickness variation in the center of the substrate was about 20% within 0.25 cm^2. During deposition, the pressure in the system rose to about 1–2×10^{-6} Torr. To obtain a more constant deposition rate and avoid texturing, the pellet was slowly rotated. The films were deposited with the substrates heated to 450°C and annealed in an oxygen atmosphere for typically 1 h at 900°C followed by slow cooling to room temperature.

Shortly after Dijkkamp et al.'s work, Narayan et al. reported the formation of the YBCO thin films using a pulsed laser evaporation technique [154]. A pulsed excimer XeCl laser ($\lambda = 0.308 \ \mu m$, $\tau = 45 \times 10^{-9}$ s) with energy density varying from 2 to 4 J/cm^{-2} was used for deposition. The substrate temperature was maintained at 470°C during deposition. And high vacuum ($\sim 10^{-6}$ Torr) was also maintained by a turbomolecular pump. The oxygen contents of the thin films and microstructures were controlled by furnace annealing in an oxygen atmosphere at 860–900 and 650°C and then a slow cooling to room temperature.

Later, in 1989, Singh reported *in situ* processing of epitaxial YBCO superconducting films [155]. The excimer laser ($\lambda = 0.308 \ \mu m$, pulse duration = 45×10^{-9} s, energy density = 2–3 J/cm^{-2}) is used to ablate 1-2-3 (YBCO) targets and the deposition occurs on a substrate at a distance of about 5 cm from the target, which has been introduced in the fabrication method section. For the biased laser deposition, a ring is placed parallel and positively (300–400 V) with respect to the target, while the substrate is maintained at a floating potential. During the deposition process, the pressure was in the range of 180–200 mTorr

measured at a distance 25 cm away from the substrate–target assembly. It is found that a substantial improvement in the quality (epitaxial growth as well as the superconducting properties) can be obtained by biasing the ring with a voltage of +300–400 V dc. In the above pressure regime and the substrate–ring–target geometry, this bias voltage is not high enough to introduce a gaseous plasma in the chamber; it is in contrast to a discharge assisted deposition regime where high current may flow across the ring. It should be noted that oxygen is introduced near the substrate as opposed to the target in the plasma assisted deposition.

The YBCO thin films deposited by laser ablation have the same composition as that of the target material, easily getting T_c higher than 90 K and J_c higher than 10^6 A/cm^2, but it is not easy to overcome the phenomenon of outgrowth which is the problem of minimizing the surface resistivity R_s.

In the 1990s, the PLD method was more popular for fabricatation of YBCO thin films. Thivet et al. reported epitaxial growth by laser ablation of YBCO thin films on bare sapphire substrate in 1994 [156]. The optimum deposition temperature is 761°C. Norton et al. fabricated in-plane-aligned, c-axis-oriented YBCO films with superconducting critical current densities J_c as high as 700,000 amperes per square centimeter at 77 K [157]. In their work, the YBCO film was grown on thermomechanically rolled-textured nickel (001) tapes by PLD. List et al. also grew high-J_c YBCO thin films on biaxially textured Ni with an oxide buffer layer [158]. The buffer layer was deposited using conventional electron beam evaporation and rf sputtering, while the YBCO thin films were grown on these buffered substrates using pulsed laser deposition.

3.1.2.4. YBCO Thin Film Fabricated by Chemical Vapor Deposition

The CVD method can be used to deposit YBCO thin films with high T_c and high J_c. In 1988, Berry et al. first grew superconducting YBCO films by organometallic chemical vapor deposition [109]. Metal β-diketonates were decomposed thermally on MgO substrates in an oxygen-rich atmosphere to produce amorphous brown films. Subsequent annealing at 920°C in oxygen yielded dull gray films whose thickness corresponded to deposition rates of approximately 8 nm/min^{-1}. Theses films showed semiconductorlike behavior at higher temperatures, followed by a broad resistive transition from 80 to 36 K with the resistance becoming zero at $\sim$20 K.

Many other groups also reported the preparation of high-T_c YBCO films at 850–950°C by thermal CVD [159]. Ohnishi et al. reported the deposition of YBCO films with T_c $(R = 0) = 85$ K prepared at 715°C under a total gas pressure of 10 Torr, and oxygen partial pressures from 4.3 to 5.3 Torr by thermal CVD using a cold-wall-type reactor [160]. Kanehori et al. improved T_c $(R = 0)$ of the YBCO films deposited at 700°C from 66 to 83 K by reducing total gas pressure in a hot-wall-type reactor from 10 Torr (oxygen partial pressure: 4 Torr) to 1.5 Torr (oxygen partial pressure: 0.6 Torr) [161]. Then, Kanehori et al. were successful in the preparation of a YBCO film at 580°C by microwave plasma-enhanced CVD (PE-CVD) [162]. The $T_c (R = 0)$ and J_c (77.3 K, 0 T) of the film were 85 K and 10^5 A/cm^2, respectively. Zhao et al. obtained YBCO film with T_c $(R = 0) = 72$ K prepared at 570°C by introducing N_2O into a PE-CVD process [163]. The YBCO films they prepared at 670–730°C by PE-CVD with N_2O showed $T_c (R = 0) = 89$–90 K and J_c (77.3 K, 0 T) = 1.0–2.3 × 10^6 A/cm^2.

In 1991, Yamane et al. successfully prepared high-quality YBCO superconducting thin films at 700°C on a $SrTiO_3(100)$ single-crystal substrate without postannealing by thermal chemical vapor deposition under a low-oxygen partial pressure of 0.036 Torr in the hot-wall-type reactor (the total gas (oxygen + argon) pressure in this reactor is about 10 Torr) [164]. The sources used were 2,2,6,6-tetramethyl-3,5-hepanedionato(thd) of Y, Ba, and Cu. The evaporation temperatures of these sources were 124–129°C for Y(thd)$_3$, 240–242°C for Ba(thd)$_2$, and 106–118°C for Cu(thd)$_2$, which were set lower in order to decrease the concentration of these sources in the gas phase. The CVD reactor was a vertical hot-wall type. Each evaporated source was introduced into the reactor by Ar gas at a flow rate of 150 ml/min. Either pure oxygen gas ($O_2 > 99.9\%$) or 1% O_2 gas balanced with Ar gas was introduced into the reactor at a rate of 250 ml/min. The total flow rate of the gases was 750 ml/min. Total gas pressure was maintained at 10 Torr during film deposition. Oxygen partial pressures in the total gas introduced into the reactor were 3.6 and 0.036 Torr in the cases of the pure oxygen gas and the 1% oxygen gas, respectively. $SrTiO_3(100)$ single crystals (10 × 5 × 1 mm^3) with mirror surfaces were used as substrates. Deposition temperature was 700°C, which were measured with a chromel-alumel thermocouple installed under a substrate holder. The deposition process continued for a period of 30 min. After the deposition period, the films were cooled down to about 150°C at a rate of 10°C/min under 760 Torr of oxygen.

Because the CVD method has many advantages, it was also well developed and used in recent years. Ito et al. prepared YBCO thin films by MOCVD using liquid sources in 1996 [165]. Schmatz et al. fabricated thick YBCO films by MOCVD processes [166]. Watson et al. researched MOCVD methods of the high-T_c superconductoring YBCO [167]. In 1998, Abrutis et al. demonstrated the possibilities of a new single source injection metal–organic chemical vapor deposition technique for *in situ* growth of $YBa_2Cu_3O_7$ films with very high crystalline and transport quality [168].

3.1.2.5. *YBCO Thin Film Fabricated by Liquid-Phase Epitaxy*

LPE is a newer method to fabricate high-T_c superconductor YBCO thin films [169–171]. In 1997, Kitamura et al. prepared YBCO thin films by LPE on single-crystalline substrates and investigated flux pinning centers [172]. They found that the microstructure of the films is dominated by the misfit of the lattice constant between YBCO and substrate, while the misfit dislocations are introduced to release the stress caused by this kind of misfit. So the residual stress may be caused by other kinds of defects in the YBCO films. These defects introduced to the

YBCO films may act as the effective pinning centers to enhance the pinning force in the magnetic field.

In 1998, Klemenz et al. grew YBCO layers on (110) $NdGaO_3$ substrates by LPE and investigated the morphology formed by spiral growth mode [173]. For single dislocation with large Burgers vector b, the interstep distance increases linearly with the step height. For complex dislocations having $b > 1$, the interstep distance decreases as a function of the factor epsilon, as predicted by the BCS theory. The step density over this surface remains constant, which is an indication for diffusion-limited growth. Spirals with very small interstep distances are usually observed on vapor-grown layers. This can be explained by a much higher driving force, estimated to be (typically) $\Delta G = 37,000$ J/mol for laser-ablated films and $\Delta G((\Delta T = 2.5)) = 285$ J/mol for the film grown by LPE. At equilibrium, the hollow cores of the spirals should be equal to Frank's radius. The measurements of the cores agree fairly well with the theory, as do the available values of surface tension and shear modulus.

Relatively flat YBCO layers also could be grown by liquid-phase epitaxy, but the defect structure is currently introduced [174]. The first kind of defect is generated during the LPE growth, which depends not only on the system but also on the growth conditions (supersaturation, temperature, etc.), so it can be controlled by adjusting the growth conditions. The second type of defect occurs during the postgrowth phase, cooling and oxidation of the layers. The latter defects depend mainly on the inherent material properties of both the substrate and the film.

Depending on the growth conditions, macroscopic spirals can develop on LPE films. They are interesting because they provide a continuous source of steps, with large distances between the steps, which can propagate over the whole surface of the film. Flat areas of $\geq 10\ \mu$m between the monosteps could be used as a base for the fabrication of tunnel devices. The defects, generated during the growth, depend on the film thickness, misfit dislocations, and stacking faults which may compensate the misfit strain. After growth, during cooling/oxidation of the layers, the strain generated by the difference in the thermal expansion between the film and the substrate can be partially relaxed along the specific directions by the formation of twins, which results in surface roughening.

By LPE, depending on supersaturation, a- and c-axis, or mixed a/c-axis, oriented layers may be obtained on (110) $NdGaO_3$ substrate. Low supersaturation, typically $\sigma \approx 0.03$, is the experimental condition for c-axis growth. If the supersaturation is further decreased, the substrate is attacked with the formation of a (Y, Nd) BCO layer by etch-back and regrowth mechanisms, and an a-oriented film is obtained. If the supersaturation is increased ($\Delta T \geq 3$°C), mixed a/c-oriented or pure a-oriented YBCO layers are obtained. Such LPE films grown on low-misfit $NdGaO_3$ substrates show generally a good and homogeneous crystallinity, which results in the relatively flat surfaces, with macrosteps propagating over macroscopic dimensions. Depending on the growth conditions, LPE films grown on vertically mounted substrates show relatively strong thickness variations (sometimes up to 100%) between the center and the edge, whereas films grown on the horizontally mounted substrates show generally a homogeneous film thickness, except close to the edge. Therefore, measurements were done in the central part of LPE films showing negligible thickness variations over several mm.

3.1.2.6. YBCO Thin Film Fabricated by All-Iodide Precursors

In recent years, many new methods were performed to prepare YBCO thin films. For example, in 1999, YBCO films were grown on single-crystal substrates by an all-iodide precursors method [175]. First, the metal iodides were stored and weighed in an argon-filled glove box. Yttrium iodide (Alfa, 99.9%), barium iodide anhydrous (Alfa, 99.999%), copper (I) iodide (Alfa, 99.999%), ammonium iodide (Allied Chemical & Dye, 99.5%), 2-methoxyethanol (Alfa, spectrophotometric grade), dimethylformamide (Alfa), and polyoxyethylene isooctylphenyl ether $(4\text{-}(C_8H_{17})C_6H_{10}(OCH_2CH_2)_nOH,\ n = 10,\ \text{Aldrich})$ were used without further purification.

Yttrium iodide (0.393 g, 0.837 mmol) and barium iodide (0.655 g, 1.675 mmol) were dissolved together in 6 ml of 2-methoxyethanol to produce a clear pale yellow solution. Copper (I) iodide (0.478 g, 2.511 mmol) was suspended in 1 ml of dimethylformamide. Copper (I) iodide was dissolved by adding 0.35 ml of ammonium iodide/dimethylformamide solution (3.4 M). Although the solubility of copper (I) iodide is low, it was dissolved by the presence of relatively large concentrations of the counteranions to form the complex $CuI + nI^- \rightarrow [CuI_{n+1}]^{n-}$.

Ammonium iodide is soluble to 30–35% (w/v) in dimethylformamide, which leads to a combination solvent of 2-methosyethanol and dimethylformamide with ammonium iodide as a complexing agent to give a metal iodide solution. Mixing of the yttrium iodide/barium iodide/2-methoxyethanol solution and the copper (I) iodide/ammonium iodide/dimethylformamide solution at room temperature resulted in a clear pale yellow solution. The solution was concentrated under vacuum to obtain a total volume of 2 ml, giving a molarity of 2.5 based on total cation concentration. A solution suitable for spin coating was prepared by adding about 10 vol% of polyoxyethylene isooctylphenyl ether to increase the viscosity of the solution.

Then, the iodide solution was deposited onto rectangular (100) oriented $SrTiO_3$ single crystal substrates by spin coating with a photoresist spinner. Substrates were precleaned ultrasonically in acetone for 10 min, heated at 950°C for 30 min, and cooled to room temperature in the furnace in a stream of O_2 gas prior to coating. The spin coater was operated at 4000 rpm for 40 s after applying enough precursor solution to cover the substrate. Spin coating was performed at room temperature.

The heat treatment consisted of heating to a temperature of 820°C at a rate of 25–50°C/min, holding for 10–30 min at that temperature, and cooling at 5°C/min to 500°C. At that temperature, pure oxygen was changed to a low oxygen partial pressure of $P(O_2)$. Furnace atmosphere which was passed over the samples throughout the heating and the high temperature was hold. The low $P(O_2)$ furnace atmosphere was prepared by mixing

high-purity argon gas with an analyzed oxygen/argon gas mixture. At the end of the high-temperature annealing, a flow of 1 atm O_2 gas was restored.

As a processing solution for fabrication of oxide films, the all-iodide precursors method offers molecular level mixing of constituents leading to excellent chemical homogeneity and composition control. Shorter diffusion distances for reactants led to lower reaction temperatures for crystallization. Moreover, the process has the potential to form films on large size substrates in a short time without any high vacuum apparatus. This process also formed YBCO thin films, avoiding the formation of intermediate $BaCO_3$ and without the use of water. X-ray diffraction results from the films revealed a (100) cubic texture. The YBCO films had a T_c of 90.3 K, and the transport J_c was 1.3×10^5 A/cm^2 (77 K, 0 T) and 2.5×10^5 A/cm^2 (70 K, 0 T).

3.1.3. Properties of YBCO thin film

3.1.3.1. Superconducting Transition Temperature

The high superconducting transition temperature is a simple feature of YBCO thin films but it has the most importance. Soon after the bulk superconductivity in single-phase perovskite YBCO was found [176], the superconductivity of YBCO thin films received much attention.

Somekh et al. investigated on superconductivity of YBCO thin films in 1987 [177]. The samples they used were deposited onto R-plane sapphire substrates placed on a heater at 1050°C in a four-target ultrahigh vacuum (UHV) dc magnetron gettersputtering deposition system. The composition varied by up to 10% along the length of a single 12.5×3 mm^2 substrate. A transmission electron microscope (TEM) investigation of one film with an onset T_c of 75 K indicated the presence of at least two majority phases. The grain size was of the order of the film thickness. X-ray diffractometer traces from 10 samples, consistently show the same peaks and the main lines are similar to those observed for high-T_c bulk samples. Analysis of the variation of the resistive transition with composition leads to the conclusion that the high-T_c phase is $YBa_2Cu_3O_7$. This is inferred from the observation that the highest T_c in any given run always has a composition with a Y:Ba ratio of 1:2 and Cu in the range 55–60%. The structure and characteristics of this phase have been reported by Michel and Raveau [178]. In particular, at least two superconducting transitions associated with different phases exist as observed from the resistance measurements shown in Figure 15. The breadth of this transition is partly due to the two-dimensional grain structure of the film, in which at least 50% of randomly distributed material must be superconducting for a complete superconducting percolation path, compared with only 17% for three-dimensional distribution of grains in a bulk sample.

In 1987, Zhao et al. prepared YBCO thin films by a rf magnetron sputtering method [143]. The highest T_c onset is determined to be 89.5 K and the T_c midpoint is about 55 K as shown in Figure 16.

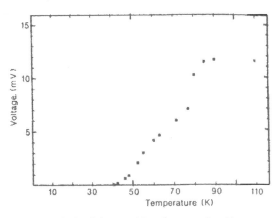

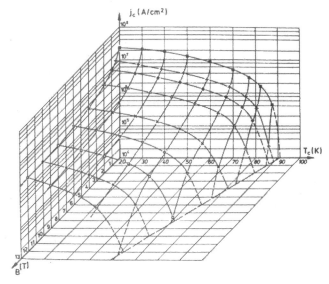

Fig. 15. Four terminal resistive transitions for a sample with mean composition $Y_{0.17}Ba_{0.34}Cu_{0.5}O_y$. The contact separation was 2 mm on a 3-mm-wide by 500-nm-thick film measured at 10 pA [177].

Fig. 17. Three-dimensional presentation of $J_c(B.T)$ for $B_{\parallel}c$-axis [181].

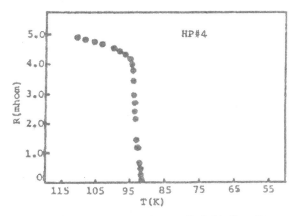

Fig. 16. Resistivity transition curve of Ba–Y–Cu–O thin film with zero resistance above 80 K [143].

After 1987, the superconducting transition temperature T_c of c-axis-oriented YBCO single phase thin films has been raised about 90 K in many laboratories worldwide.

3.1.3.2. Critical Current Density

Critical current density is another interesting property, especially for practical application. Soon after the YBCO was found in 1987, Chaudhari et al. grew epitaxial films of the YBCO compound on $SrTiO_3$ substrates and measured the critical current density [179]. They found that the superconducting critical current density in these films at 77 K is in excess of 10^5 A/cm^2 and at 4.2 K it is in excess of 10^6 A/cm^2. In the same year, Enomoto et al. epitaxially grew YBCO thin films on $SrTiO_3$ by magnetron sputtering and tested the critical current density in different directions [180]. The films exhibit zero resistance at 84 K with a transition width of 6 K. The $I_{c\parallel}$ and $I_{c\perp}$ are 1.8×10^6 and about 10^4 A/cm^2 at 77.3 K, respectively, where the symbols $\parallel$ and $\perp$ indicate parallel to and perpendicular to the basal plane, respectively. This shows the critical current density is two orders of magnitude larger in the direction along the basal plane than along the c-axis. These results strongly suggest a quasi-two-dimensional electronic structure in YBCO.

In 1991, Tome-Rosa et al. prepared oriented YBCO thin films with the c-axis perpendicular to the film plane by dc magnetron sputtering and measured the critical current densities on 5 to 10 μm wide strips as a function of magnetic field and temperature [181]. Typical values of the inductively measured superconducting transitions were about 90 K with a width less than 0.5 K. The carrier concentration per unit cell evaluated from Hall effect measurements was found to decrease linearly from 240 K with decreasing temperature extrapolating nearly through zero for $T = 0$. Highly resolved angular dependent measurements of the critical current density with magnetic field B perpendicular to the current but tilted from the c-axis show a very strong and sharp enhancement of J_c for the magnetic field parallel to the (CuO_2) layers ($B \perp c$). For the case of B parallel to the c-axis, the field and temperature dependence of the critical current density of a 100 nm thick film is measured and displayed three-dimensionally in Figure 17. At zero field and $T = 77$ K the J_c-value is 4×10^6 A/cm^2 and at $T = 20$ K, J_c reaches 3×10^7 A/cm^2. At low temperatures the field dependence of the critical current density is very weak. It is noted that the critical current density $J_c(B, T)$ can be described down to 30 K by a universal temperature function $J_c(T.B) = J_c^*(0, B) \times (1 - T/T_c(B))^{\alpha}$ with $\alpha = 1.5 \pm 0.1$.

Recent years, the critical current density is still researched widely. Schmehl et al. increased the critical current density of grain boundaries in YBCO films beyond the hitherto established limit by overdoping the superconductor [182]. Using Ca-doping, values for the critical current density were obtained that exceed the highest values for the undoped material by more than a factor of seven.

Dam et al. investigated the origin of high critical current density in $YBa_2Cu_3O_{7-\delta}$ superconducting thin films [183]. They got the large critical current density, which lies between 10^7 and 10^8 A/cm^2 at 4.2 K in zero magnetic field, and found its strong magnetic field dependence, i.e., the vortex pinning dependence of critical current density. But it is unclear which types of defects, dislocations, grain boundaries, surface corrugations, and anti-phase boundaries are responsible for vortex pinning. They make use of a sequential etching technique to address this question and found that both edge and screw dislocations, which can be mapped quantitatively by this technique, are the linear defects that provide strong pinning centers responsible for the high critical current density observed in these thin films. Moreover, they found that the superconducting current density is essentially independent of the density of linear defects at low magnetic fields. These natural linear defects, in contrast to artificially generated columnar defects, exhibit self-organized short-range order, suggesting that YBCO thin films offer an attractive system for investigating the properties of vortex matter in a superconductor with a tailored defect structure.

Trajanovic et al. measured the magnetic field angular and temperature dependences of the critical current density J_c in untwinned a-axis-oriented $YBa_2Cu_3O_{7-\delta}$ films [184]. The direction of the magnetic field with respect to the crystallographic axes was found to be the dominant factor to determine the behavior of J_c. Currents driven along the b direction yielded a peak of J_c at $0°$, corresponding to the magnetic field applied parallel to the b-axis in the CuO_2 planes. For supercurrents flowing along the c direction, a crossover from grain-boundary pinning to surface and interface pinning was observed as temperatures approached T_c.

Mezzetti et al. investigated the control of the critical current density in $YBa_2Cu_3O_{7-\delta}$ films by means of intergrain and intragrain correlated defects [185]. The aim of their work is to study the mechanisms controlling the current-carrying capability of YBCO thin films. A comparison of the magnetic properties between the film with intrinsic grain-boundary defects and that crossed by columnar defects is presented. Such properties have been studied by means of ac susceptibility measurements, resistivity measurements, and structural characterizations. The Clem and Sanchez model [186] is used to extract critical current values from the susceptibility data. In the virgin film, correlated grain-boundary defects were created among islands with homogeneous size, by means of the appropriate modifications in the growth process. Columnar defects were produced through 0.25-GeV Au-ion irradiation. The central issue concerns the investigation of the plateaulike features characterizing the log-log field dependence of the critical current density, the analysis of the J_c temperature dependence, and the analysis of the irreversibility line. An analytical expression of J_c vs B,

$$J_c(B) = J_c(0)\frac{k^2}{k^2 + B}\coth\left(\frac{\pi k}{2B^{1/2}}\right)$$

is given in order to compare the main issues with the experimental data. $J_c(0)$ and k are fitting parameters. This model suggests that the intergrain pinning acts the dominate role in the high-current/low-temperature regime through a network of frustrated Josephson junctions, while the intragrain pinning is effective near the irreversibility line.

3.1.3.3. Thermopower

In earlier days, the thermopower V_p was studied only on YBCO single crystal. In 1990, Mihailovic et al. observed a transient voltage between contacts along the c-axis of a YBCO single crystal which depends upon the application of a heat pulse [187]. They suggested this thermally induced voltage is a thermoelectric effect. In 1993, they reported the measurement results of this kind of thermoelectric effect in single crystals of YBCO as a function of doping, external electric field, and temperature [188].

Then, in 1996, thermopower was researched in thin films of YBCO by Grachev et al. [189]. The temperature behavior of V_p was found to exhibit such specific features as the peak just above T_c and the sign reversal at a higher temperature implying a sharp change in the spontaneous polarization P_s near T_c. Application of an external magnetic field with the aim of clarifying a possible role of the superconducting transition on the discontinuity of P_s showed a surprisingly strong effect on the thermoelectric peak involving a pronounced decrease in its height. At the same time, the expected shift of the peak position toward low temperatures is not observed. The temperature dependence of P_s considered simultaneously with data on behavior of the V_p peak in the magnetic field point out the preservation of P_s below T_c in the regions responsible for the thermoelectric effect. The origin of the $P_s(T)$ discontinuity can be attributed to a structural instability of the YBCO near T_c.

The experimental temperature dependence of the nonsteady state thermal voltage (NSTV) signal for the sample was shown in Figure 18, which demonstrates most clearly the specific features of $V_p(T)$ in the vicinity of the superconducting transition. The dependence may be divided into three parts: (i) a relatively sharp peak near T_c (region I), (ii) the region of a relatively weak increase and decay of the signal having the same sign as the peak (region II), and (iii) the region of the sign reversal of the signal (at $T = T_m$) and subsequent nearly linear growth (region III). Two experimentally established facts are very important for the interpretation of the signal. First, the

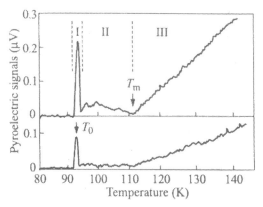

Fig. 18. Experimental curves for the temperature behavior of the for two different intensities of the incident light. There is the sign pyroelectric voltage in the YBCO thin film in the vicinity of T_c reversal of the voltage above T_m [189].

peak maximum was observed at a temperature higher than T_c (here T_c is the point at which the maximum slope in the curve of resistance vs temperature is obtained). Second, the temperature dependence of the signal was scaled with respect to the incident light intensity; i.e., only the magnitude of the signal linearly varied with the change of light intensity, while the temperature extents of the regions listed above remained unaltered.

The specific features revealed in the temperature behavior of NSTV can be simply explained by assuming that they have a thermoelectric origin. As is known, the magnitude and the sign of the thermopower are determined by the thermoelectric coefficient $P = dP_s/dT$. If the magnitude of P_s sharply changes, which is typically associated with a certain structural transition, a peak whose shape is determined by the nature of phase transition appears in the thermoelectric signal. In the case of small temperature variations dT ($dT < 10$ K), the peak shape, as well as the whole temperature behavior of $p(T)$, is determined by the $P_s(T)$ dependence alone. The magnitude of P_s typically falls with increasing temperature ($a < 0$). In this case different signs of the peak and voltages in region III imply a sharp growth of P_s with increasing temperature just above T_c. The presence of region II means a further increase in P_s up to the maximum value at $T = T_m$. It is apparent that the T_m position should not depend on the incident light power, consistent with experiments. Finally, region III corresponds to the onset of a decrease in P_s with increasing temperature. Thus the hypothesis about the thermoelectric origin of the NSTV signal explained the specific features of its temperature behavior without resorting to artificial considerations. It should be noted that a linear growth of V_p up to an anomaly in its behavior (chaotic changes in the sign or magnitude of the voltage) at $T \sim 200$ K was also observed in single crystals of YBCO.

3.1.3.4. C-Axis Transport Properties and Hall Effect

C-axis transport properties are very important for YBCO and were widely researched. Abrikosov presented a theory of the static c-axis conductivity of the YBCO with varying oxygen concentration in 1995 [190]. Prusseit et al. researched the c-axis transport properties [191]. They used oxygen deficient YBCO films for this purpose. Figure 19 shows the characteristic T_c vs δ variation of the films. Figure 20 exhibits the resulting c-axis resistivity as a function of temperature for four samples annealed under different conditions with transition temperatures indicated at the curves. The assigned anisotropy ratios γ have been deduced from the resistivity ratios in the ab- and c-direction $\gamma = (\rho_c/\rho_{ab})^{1/2}$ directly above the transition. Obviously, with decreasing oxygen content the c-axis transport changes from a metallic to a semiconducting behavior. Close to the metal–insulator transition the anisotropy increases drastically and γ-values were up to 80 for $\delta \approx 0.5$.

The Hall effect is another important transport property. The Hall resistivity of vortices that lie parallel to the CuO_2 layers in YBCO is shown to be negative, in contrast with the Hall effect observed with the field perpendicular to the layers [192]. In YBCO thin films, strong correlations between the Hall coefficient R_H, the transition temperature T_c, and the critical current density J_c were established in a series of epitaxial YBCO thin films as a function of oxygen deficiency δ [193]. The electrical transport properties of YBCO vary systematically with increasing oxygen deficiency δ. Both the resistivity $\rho(\delta)$ and the Hall coefficient $R_H(\delta)$ increased with δ at similar rates, and consequently the Hall angle, given by $\tan(\theta_H) \equiv R_H B/\rho$, changes only slightly (Fig. 21). In all films, the critical current densities $J_c(\delta, H = 0)$ measured in self-field extrapolated toward zero for compositions near the edge of the 90-K plateau (Fig. 22), while the temperature and field dependences of $J_c/J_c(H = 0)$ (Fig. 23) remained fixed for $\delta < 0.15$, both of which are suggestive of geometrical effects. For larger oxygen deficiencies ($\delta > 0.3$), the field dependences of $J_c(\delta, H)$ were similar to

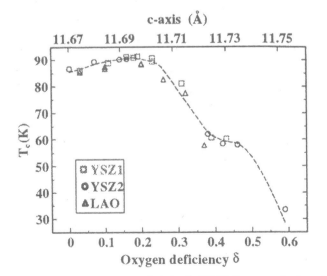

Fig. 19. T_c variation with oxygen depletion for YBCO films grown by thermal coevaporation on various substrates [191] (*IEEE Trans. Appl. Supercond.* 7, 2952 (©1997 IEEE)).

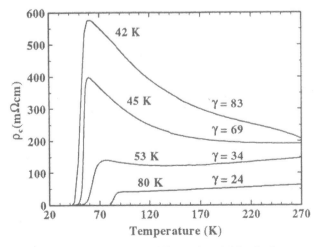

Fig. 20. Temperature dependence of the c-axis resistivity for four oxygen depleted YBCO mesas with transition temperatures and anisotropy ratios as indicated [191] (*IEEE Trans. Appl. Supercond.* 7, 2952 (©1997 IEEE)).

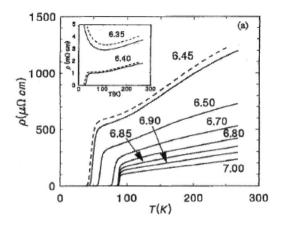

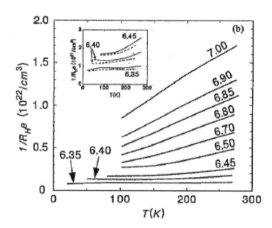

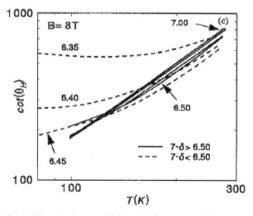

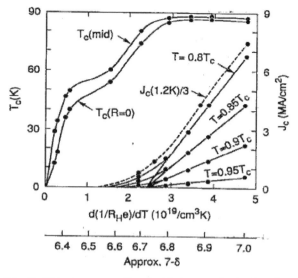

Fig. 22. Transition temperature T_c and critical current density J_c evaluated at fixed reduced temperatures, as a function of either the oxygen deficiency δ or the "apparent carrier density" from the Luttinger liquid theory. Notice that J_c extrapolates toward zero at the edge of the 90-K plateau, suggestive of phase separation. The solid (dashed) curves were obtained from a laser-ablated (BaF_2 processed) film. Extrapolations do not occur at exactly the same point due to uncertainties in the film thickness [193].

Fig. 21. Resistivity (a), inverse Hall coefficient (b), and inverse Hall angle (c) for a laser-ablated film of YBCO as a function of temperature and oxygen deficiency δ. Insert show the apparent changes in the orho-II phase with room-temperature annealing. The dashed curves in parts (a) and (b) represent the behavior immediately after quenching from 200°C whereas the corresponding solid curves represent the behavior after aging for 4 days. The inverse Hall angle is shown on a log–log plot to depict the relative insensitivity of the T^2 behavior to oxygen deficiency [193].

those observed in polycrystalline YBCO and the flux-flow Hall transitions exhibited systematic "noise" (Fig. 24), indicative of granularlike behavior. Pinning energies determined from both the field dependence of $J_c/J_c(H=0)$ and the resistive tran-

sitions in the field (Figs. 25 and 26) show plateaulike behavior near full oxygenation even though J_c decreases rapidly in the region. On the 90-K plateau, most films showed no broadening in the resistive transitions; however, all films showed some broadening in the transitions between the 90-K and 60-K plateaus. Films postannealed at low P_{O_2} usually showed "peaked" T_c behavior with δ unlike the high P_{O_2} postannealed films which typically show "flat" 90-K plateaus. However, the Hall coefficients (Fig. 27) were found not to rely on the processing conditions, suggesting that differences in the measurable carrier densities are not responsible for those different $T_c(\delta)$ patterns. The inverse Hall coefficients have linear temperature dependences, as predicted by the Luttinger liquid theory, and the implied carrier densities steadily diminish with increasing oxygen deficiency δ. Moreover, the relative insensitivity of the Hall angle to oxygen deficiency suggests that only one of the four predicted eletronic bands at the Fermi surface dominates the normal-state properties with filelds $H_{\parallel}c$. Critical currents extrapolate to zero as the oxygen content nears the edge of the 90-K plateau, suggestive of phase separation in which only the fully oxygenated phase has the high critical current density. On the 90-K plateau, no changes are seen in the pinning energies determined from the field dependence of J_c, in general agreement with the magnetic flux-creep studies of Ossandon et al. [194].

3.1.3.5. Microwave Property

Because of the excellent transport properties of the YBCO thin films, the microwave applications of YBCO thin films became more and more interesting and are widely researched.

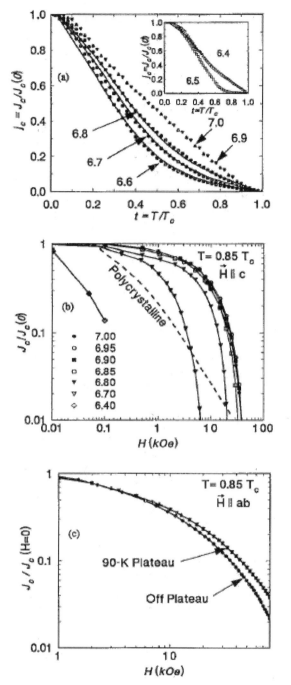

Fig. 23. Reduced critical current densities as a function of temperature (a), applied field for $H_{\parallel}c$ (b), and applied field for $H_{\parallel}ab$ (c) for various oxygen contents. No apparent changes in pinning energy occur while on the 90-K plateau even though $J_c(0)$ decreases rapidly. Interestingly, many of the curves in (a) behave as superconductor/normal metal/superconductor (SNS) proximity tunneling (solid curves) off the 90-K plateau. The dashed curve is a representative polycrystalline sample which indicates the granularlike behavior when $\delta > 0.2$ [193].

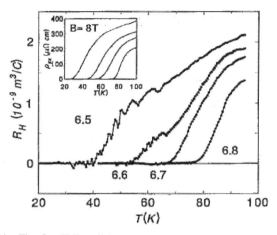

Fig. 24. Flux-flow Hall coefficients taken at 8 T for various oxygen contents. All Hall transitions with $\delta \geq 0.3$ show reproducible "onsets" indicative of a distribution of T_c's. However, these onsets were never observed on the 90-K plateau nor in any resistive transition as shown in the inset [193].

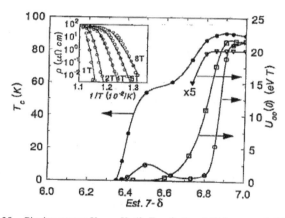

Fig. 25. Pinning energy $U_{00} = U_0$ (δ, $T = 0$, $B = 1$ T) (open symbols) and $T_c(\delta)$ (solid circles) as a function of oxygen content 7-δ for a film prepared by the BaF$_2$ process. The open circles represent $U_{00}(\delta)$ to the Arrhenius curves at various oxygen deficiencies. The inset depicts an actual fit at $\delta = 0$. The open squares (triangles) are $U_{00} \propto B^*$ for $H_{\parallel}c$ ($H_{\parallel}ab$). The values of $U_{00}(\delta)$ for $H_{\parallel}ab$ are scaled by a factor of 1/5 for clarity [193].

Microwave surface resistance R_s and microwave magnetic properties are two main properties of the high-T_c superconductor thin films to realize the microwave applications.

In 1990, Char et al. measured microwave surface resistance of YBCO thin films grown on Al_2O_3 substrates by a parallel-plate resonator technique [620]. In 1997, Findikoglu et al. measured the microwave surface resistance R_s of superconducting YBCO thin films grown on buffered polycrystalline alumina and Ni based alloy substrates using a parallel-plate resonator technique [195]. They observed a strong correlation between the low-power R_s and the in-plane mosaic spread of the films.

Magnetic property is another important microwave property. The magnetic property of HTSCs at microwave frequency has been of increasing interest since their magnetic behavior is related to microwave vortex dynamics. In 1997, Han et al. designed a cavity perturbation system (CPS) to simultaneously measure the resonant frequency (f_0) and the quality factor (Q) [196]. The magnetic susceptibilities ($\chi' - j\chi''$) of YBa$_2$Cu$_3$O$_{7-\delta}$/MgO thin films at microwave frequencies were investigated and could be analyzed by cavity perturbation theory. The transition temperature is 91 K and the temperature corresponding to the maximum imaginary part (χ'') is 85 K.

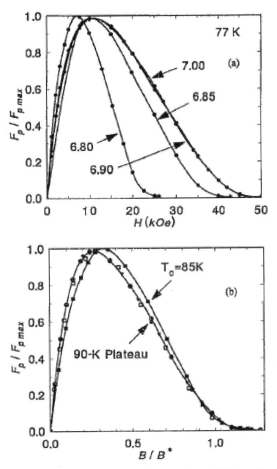

Fig. 26. Normalized pinning force density F_p taken at 77 K as a function of (a) B and (b) B/B^* both on and off the 90-K plateau. The universal F_p curve observed on the 90-K plateau suggests the field dependence of the pinning energy given by $U_{00} \propto 1/B$ is fixed across the plateau. However, off the plateau, the relative B dependence may change, as in the polycrystalline films, leading to the observed shift in $F_{p,\max}$ [193].

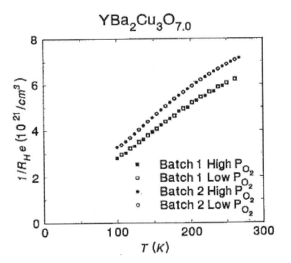

Fig. 27. Inverse Hall coefficients taken on two sets of film with identical precursors but postannealed under different conditions. The data suggest that growth conditions do not influence the overall carrier density. The slight differences between film batches may be attributed to small compositional differences in constituents or simply to errors in determining film thickness [193].

Table VI. Lattice Parameters of YBCO-Related-Material Thin Films

	a (Å)	b (Å)	c (Å)
$PrBa_2Cu_3O_{7-\delta}$	3.882	3.930	11.790
$HoBa_2Cu_3O_{7-\delta}$	3.8205	3.8851	11.682
$NdBa_2Cu_3O_{7-\delta}$	3.8590	3.9112	11.7412
$GdBa_2Cu_3O_{7-\delta}$	3.8350	3.8947	11.6992
$SmBa_2Cu_3O_{7-\delta}$	3.8458	3.9033	11.7272

However, the transition temperature and surface resistance (R_s) measured by a conventional transmission-line method were 86 K and 180 $\mu\Omega$ at 40 K, respectively. They found absolute differences between the cavity perturbation system and the transmission line systems and suggested that the CPS is one of the most promising techniques for measuring the magnetic susceptibility of HTSC thin film with a 0.1–1 μm thickness.

3.1.4. YBCO-Related-Material Thin Films

With the development of YBCO thin film research, YBCO-related-material thin films also became interesting in research for understanding the physics properties of high-temperature superconductivity and applications. Now, it has been proved that except on Pr, every rare-earth atom substitution for Y in YBCO can get high-temperature superconductivity. Some YBCO-related-material thin films even have more excellent properties and application potentials. Table VI shows the lattice parameters of some usual YBCO-related-material thin films.

3.1.4.1. $Y_{1-x}Pr_xBa_2Cu_3O_7$ thin film

$Y_{1-x}Pr_xBa_2Cu_3O_7$ (YPBCO) thin film is one of the most important YBCO-related-material thin films. Unlike other YBCO-related-material thin films, pure $PrBa_2Cu_3O_7$ does not show superconductivity. In $Y_{1-x}Pr_xBa_2Cu_3O_7$ thin film, the content of Pr can strongly change the carrier density of YPBCO and help us to understand the influence of Pr on superconductivity in the "123" phase. On the other hand, the controllable carrier density may lead to device applications of YPBCO thin films, e.g., one kind of field effect transistor [197, 198].

YPBCO thin films can be fabricated by many methods [199, 200]. Among these methods, laser ablation and magnetron sputtering are the efficient ones [201, 202]. In 1996, Dieckmann et al. prepared and investigated $Y_{1-x}Pr_xBa_2Cu_3O_7$ films with various Pr contents [202]. The $Y_{1-x}Pr_xBa_2Cu_3O_7$ films were deposited by a pulsed laser deposition process on $SrTiO_3$ substrates. A KrF excimer laser with a wavelength of 248 nm and a pulse length of 25 ns was used. For preparing Y-rich and Pr-rich films, the power density of the laser on the target was 1.67 and 2.29 J/cm^{-2}, respectively, the oxygen pressure was 8 and 3 Pa. For all films the substrate temperature during the deposition was 815°C and the target–substrate distance was 99.7 mm. The film thicknesses were about 300 nm to avoid Raman signals from the substrates. Various stoichiometric targets were prepared from

Y_2O_3 (99.999%), Pr_6O_{11} (99.99%), $BaCO_3$ (99.997%), and CuO (99.999%). All films were very smooth, appear shiny, and exhibit a low density of precipitates ($<10^5$ cm^{-2}).

Dieckmann et al. investigated phonon and magnon excitations in $Y_{1-x}P_x$BCO films by resonant and temperature dependent Raman spectroscopy [202]. The phonon spectra are used to examine the epitaxial growth quality and stoichiometry of the thin film samples. In part of the films ($0.5 \leq x < 1$) they found multiphase behavior and probably a substitution of Pr for Ba sites in the Pr-rich samples. Two-magnon excitations could be observed in both Heisenberg antiferromagnet (HAFM) and HTSC regimes [203]. The experimentally observed Raman lineshapes were explained in the framework of a damped HAFM. In this model the lineshapes are characterized by an exchange energy J and a phenomenologically introduced damping parameter Γ of the one-magnon states. With decreasing Pr content the exchange energy J and the damping parameter Γ increase almost linearly. The variation of J is roughly in conformity with the variation of the lattice parameter. The observed deviation is most likely due to band-structure effects. The increase of Γ is explained by scattering at the additional change carriers. There is no resonance enhancement in the pure Pr film and only a slight one in the mixture-type samples. A reason for this behavior might be a change of the band structure due to the unexpected substitution of Pr atoms on Ba sites.

The superconductor–insulator transition and the size effect on transport properties were also interesting and were observed in ultrathin $Y_{0.9}Pr_{0.1}Ba_2Cu_3O_y$ films by Kabasawa et al. [204]. Samples having planar $YBa_2Cu_3O_x$–ultrathin $Y_{0.9}Pr_{0.1}Ba_2Cu_3O_x$–$YBa_2Cu_3O_x$ structures were used. Transport properties of ultrathin $Y_{0.9}Pr_{0.1}Ba_2Cu_3O_y$ films were studied as a function of thickness and length scale. The $Y_{0.9}Pr_{0.1}Ba_2Cu_3O_y$ channel layer exhibited a superconductor–insulator (SI) transition when the film thickness was below ~ 8 nm. The critical sheet conductance of the SI transition was near $4e^2h$. The sheet conductance of insulator-phase channels shorter than 10 μm is higher than that of 100-μm-long channels with the same thickness due to a size effect. The results suggest that a thickness-dependent correlation length may exist in the insulator-phase $Y_{0.9}Pr_{0.1}Ba_2Cu_3O_y$.

3.1.4.2. $Y_xHo_{1-x}Ba_2Cu_3O_7$ thin films

$Y_xHo_{1-x}Ba_2Cu_3O_7$ ceramic samples were researched earlier [205, 206]. In 1994, Li et al. predicted a considerable increase of pinning force in solid solution $Y_xR_{1-x}Ba_2Cu_3O_7$, where R = rare earths, as opposed to corresponding unsubstituted Y and R-123 phases [207]. They consider that strain field induced by lattice mismatch between Y and R-123 unit cells results in enhanced pinning. In the case R = Ho the pinning force of such a strain field has been evaluated quantitatively to be comparable with that of screw dislocations and spherical inclusions in conventional superconductors.

$Y_xHo_{1-x}Ba_2Cu_3O_7$ thin films were also studied. C-oriented epitaxial films of $Y_xHo_{1-x}Ba_2Cu_3O_{7-\delta}$/$SrTiO_3$(100) ($x = 0$,

0.3 0.35 0.45, 1) were successfully grown by the MOCVD technique [208]. According to ac-magnetic susceptibility measurements, T_c of the films ranged from 90.0 to 91.7 K and J_c ($T = 77$ K, $H = 100$ Oe) was about 7×10^5 A/cm^{-2}. Superconducting properties of the films were found to be nearly independent of x value in solid solution formula. The absence of dependence of T_c and J_c on x may be due to the extremely close ionic radii of Y^{3+} and Ho^{3+}.

Among MOCVD methods [208, 209] a single source powder flash evaporation MOCVD setup with the vertical hot-wall reactor is one of the choices [208]. The main advantage of a single-source technique is reduction of experimental parameters which must be controlled precisely, since the individually controlled evaporator temperatures and gas flows result in stability of the deposition process. Substrates were cleaned by deionized water, acetone, and pentane. $Y(thd)_3$, $Ho(thd)_2$, and $Cu(thd)_2$ (thd = 2,2,6,6-tetramethylheptanedionate-3,5) were used as precursors. All the precursors were purified by vacuum sublimation and stored in a desiccator. Powder mixtures of the precursors with certain cation ratio (Y:Ho:Ba:Cu) were been placed in the single-source feeder and then a vibration device provided a gradual feeding of mixture microportions into a heated 270°C evaporator where nearly instant evaporation of the precursors takes place. Argon was used as a carrier gas and O_2 flow was supplied directly into the reactor. The oxygen partial pressure and the total pressure were 2.5 and 10 Torr, respectively. The substrate temperature was 800°C. In order to increase film uniformity, the substrate holder was rotated at the frequency 2 min^{-1}. After deposition the films were slowly cooled down to 400°C in the reactor; then 30 min annealing at 400°C in oxygen was performed.

Cucolo et al. report on the structural and electrical characterization of c-axis oriented $Y(Ho)Ba_2Ca_3O_7$-based trilayer structures with different insulating layers [55]. A comparative study of Y_2O_3, $SrTiO_3$, and $PrBa_2Cu_3O_7$ barriers is carried out. Structural characterization demonstrates the epitaxial growth of the heterostructures with sharp interfaces between the different layers. Resistivity and ac susceptibility dependences on temperature indicate no degradation of the superconducting properties of the bottom and top electrodes. The Josephson effect and quasi-particle tunneling are observed in the $YBa_2Cu_3O_7$/Y_2O_3/$YBa_2Cu_3O_7$ structures. The best conductance characteristics are obtained for $YBa_2Cu_3O_7$/$PrBa_2Cu_3O_7$/$HoBa_2Cu_3O_7$ trilayers that show quasi-ideal S–I–S behavior.

3.1.4.3. $NdBa_2Cu_3O_{7-\delta}$ thin films

$NdBa_2Cu_3O_{7-\delta}$ thin films are one kind of widely researched YBCO-related-material thin films. Cantoni et al. grew epitaxial $Nd_{1+x}Ba_{2-x}Cu_3O_{7-\delta}$ thin films on single-crystal $LaAlO_3$ and got $T_{c0} = 93$ K and J_c ($H = 0$, $T = 77$ K) 3×10^6 A/cm^2 [210]. Hakuraku et al. prepared $NdBa_2Cu_3O_{7-\delta}$ thin films on an MgO(100) substrate [211]. A highly c-axis-oriented thin film with T_c (zero resistance temperature) = 95.2 K was obtained.

Table VII.   Summary of Experimental Conditions

Ar gas flow rate (from A)	15 l/min
(from B)	3 l/min
(from C)	10 l/min
O_2 gas flow rate (from A)	~1.0 l/min
Ar carrier gas flow rate	0.3 l/min
rf input power	8 kW (4 MHz)
Pressure	240 Torr
Substrate temperature	700°C
Substrate distance	20–30 cm
Deposition time	60 min
Substrate	MgO (100)

The critical current density was 1×10^6 A/cm^{-2} at 77 K, and surface roughness was 35 nm.

NdBa$_2$Cu$_3$O$_{7-\delta}$ thin films can be prepared by other methods [210–213]. In 1998, Nagata prepared NdBa$_2$Cu$_3$O$_{7-\delta}$ thin films by mist inductively coupled plasma (ICP) evaporation and got a high superconducting transition temperature of 91–92 K [214]. Experimental conditions are listed in Table VII. The plasma torch consisted of a coaxial double tube of fused silica and a Cu coil wound outside of the tube. Cooling water was supplied between the inner and outer tube in order to protect plasma torch from high temperature Ar–O$_2$ plasma, which was operated at 8 kW and 4 MHz as the input power and frequency, respectively. Plasma becomes unstable by introducing organic solvents. Thus the selection of appropriate organic solvents was important to prevent the instability of plasma. 2-Methoxyethanol and organometallic sol diluted with this was used as a source material. Mist of a source material was generated by using an ultrasonic nebulizer, introduced into ICP with Ar carrier gas, and deposited onto single-crystalline MgO (100) substrates, which were placed on the substrate stage heated by a sheath heater. As the temperature of the tail flame of plasma was very high, substrates were heated not only by a sheath heater but also by plasma. Substrate temperature was measured by a thermocouple located at the back side of substrate stage. As substrates were heated by plasma, the temperature could be changed by adjusting the distance between the bottom of the plasma torch and the top of the substrate stage. Hereafter this distance was defined as substrate distance. The main factors which influence the preparation of the film would be substrate distance and oxygen partial pressure in the chamber.

The as-deposited NdBa$_2$Cu$_3$O$_{7-\delta}$ films were successfully obtained only under very narrow conditions; i.e., substrate distance was 25 cm and $\log(P_{O_2}/P_{Ar}) = 2.7$, and the films were grown not only oriented along the c-axis but also epitaxially. It was suggested that substrate distance influences both crystallization of the film and the thermal or physical states of the vapor of a source material in the tail flame region, and oxygen partial pressure in the chamber influences the crystallization of the film. It was necessary for the as-deposited film to anneal in an oxygen atmosphere for a long time for superconducting properties to emerge.

NdBa$_2$Cu$_3$O$_{7-\delta}$ thin films can also be deposited on STO substrates by an off-axis rf sputtering system [215]. The deposition temperature referred hereafter as the substrate temperature T_s was measured by a thermocouple inserted into the stainless steel substrate heater. The substrate was stuck on the heater by silver paste. The deposition gas was a mixture of argon and oxygen with different ratios. First, the deposition conditions were optimized. Different oxygen contents of the grown films were obtained by slightly changing the substrate temperature T_s, the ratio of P_{Ar}/P_{O_2}, and the annealing process.

The optimal deposition temperature and pressure for thin films with $T_{c0} \sim$ 89.5–90.5 K are 740°C $< T_s <$ 780°C and 1.2–1.8×10^{-1} mbar with the ratio of $P_{Ar}/P_{O_2} = 4/1$–$2/1$. Typical film thickness is ~2000 Å. In the X-ray diffraction patterns only (001) peaks for NdBa$_2$Cu$_3$O$_{7-\delta}$ and the STO substrate reflections are visible, indicating that the films are highly c-axis-oriented. The full width at half maximum of the (005) peak determined from the rocking curve is 0.2–0.3°, demonstrating a good crystallinity of the thin films.

A parabolic relation of T_{c0} to c-axis lattice parameter corresponding to a typical electronic phase diagram of T_c-n for high-T_c superconductors was observed. The oxygen out-diffusion was investigated by an *in situ* high-temperature X-ray diffraction. The result suggested that oxygen out-diffusion occurred below 770 K and resulted in an expansion of the c lattice parameter.

3.1.4.4. GdBa$_2$Cu$_3$O$_{7-\delta}$ thin films

GdBa$_2$Cu$_3$O$_{7-\delta}$ thin films are also one kind of widely investigated YBCO-related-material thin films. Li fabricated GdBa$_2$Cu$_3$O$_{7-\delta}$ thin films on LaAlO$_3$ single crystal substrates and got a T_{c0} of 92.5 K, a transition width of 0.57 K, and a critical current density of 3.6×10^6 A/cm^2 at 77 K [216].

High-temperature superconductor GdBa$_2$Cu$_3$O$_7$ thin films can be successfully grown *in situ* on LaAlO$_3$ single-crystal substrates by dc magnetron sputtering using a single planar target [216]. The disklike stoichiometric target of Gd–Ba–Cu–O with 40 mm diameter and 3–7 mm thickness was prepared from an adequate mixture of Gd$_2$O$_3$, BaCO$_3$, and CuO powders by the solid-state reaction method. A NdFeB magnet placed on the top of the target provided a plasma ring of 25 mm diameter. The sputtering gas is a mixture of 2×10^{-1} Torr oxygen and 4×10^{-1} Torr argon. The discharge was typically run at 120 V and 0.3 A. The (100) LAO substrate was placed 20–30 mm from the target on a Pt stripe which could be resistively heated up to 1000°C. The approximate angle between the target and the substrate was 60–80°C for minimizing the negative ion bombardment on the films. This geometry resulted in a deposition rate of typically 4–10 nm/min. After deposition, the chamber was vented with pure oxygen, and the temperature of the films was cooled down to 430°C. The film was held at this temperature for 10–20 min and was then allowed to cool down to room temperature.

Many other methods can been used to fabricate GdBa$_2$Cu$_3$O$_{7-\delta}$ thin films and more research work has been done on them. Stangl et al. reported pulsed-laser deposited GdBaSr-Cu$_3$O$_{7-\delta}$ and GdBa$_2$Cu$_3$O$_{7-\delta}$ films [217]. Li et al. grew GdBa$_2$Cu$_3$O$_{7-\delta}$ thin films on MgO substrates by magnetron sputtering deposition [218]. Dediu et al. investigated oxygen diffusion in GdBa$_2$Cu$_3$O$_{7-\delta}$ thin films by resistivity measurements in isobaric and isothermal regimes at an oxygen pressure of 1 mbar in the temperature interval 650–1050 K [219]. Cao et al. measured the resistive transition of an epitaxial GdBa$_2$Cu$_3$O$_{7-\delta}$ thin film in detail in a magnetic field up to 7.5 T and studied two-dimensional properties of GdBa$_2$Cu$_3$O$_{7-\delta}$ epitaxial thin films [220]. Fainstein et al. presented Raman scattering experiments on oxygen deficient GdBa$_2$Cu$_3$O$_x$ thin films ($x = 6.53$, 6.7, 6.8, and 6.93) as a function of photoexcitation and annealing induced oxygen disorder [221].

3.1.4.5. SmBa$_2$Cu$_3$O$_{7-\delta}$ thin films

SmBa$_2$Cu$_3$O$_{7-\delta}$ thin films are also important. The superconducting transitions of SmBa$_2$Cu$_3$O$_{7-\delta}$ thin films have an onset $T_c = 91$ K and widths of 7 and 5 K, respectively, for films grown on STO and MgO substrates. The critical current density at 60 K, estimated from the magnetization curve, is 3×10^4 A/cm^2 in a magnetic field of 2 T [222].

SmBa$_2$Cu$_3$O$_{7-\delta}$ thin films can also be prepared by many methods [223, 224]. Ditrolio et al. fabricated SmBa$_2$Cu$_3$O$_{7-\delta}$ thin films by PLD [222]. The ablation process was carried out by a Nd-YAG laser operating at $\lambda = 532$ nm, 10 ns of pulse width, 120 mJ/shot of energy in a spot area 8–10 mm^2. The thin films were grown on both SrTiO$_3$ (STO) and MgO substrates simultaneously. The ablation took place from a SmBa$_2$Cu$_3$O$_{7-\delta}$ high density target under different O$_2$ partial pressures. At an O$_2$ pressure of 3.5×10^{-4} mbar and $T_s = 735°$C a epitaxially grown single-phase films on both substrates were obtained. The films grown on STO show a higher degree of orientation with respect to the ones grown on MgO.

The properties of SmBa$_2$Cu$_3$O$_{7-\delta}$ thin films were also studied. Jiang et al. tested surface morphology of the in-plane epitaxially grown SmBa$_2$Cu$_3$O$_{7-\delta}$ films on SrTiO$_3$(001) substrates by scanning tunneling microscopy and grazing incidence X-ray diffraction [225]. Brunen et al. grew (110) and (103)/(013) oriented SmBa$_2$Cu$_3$O$_{7-\delta}$ thin films on SrTiO3 (110) by pulsed laser deposition and investigated the films with Raman spectroscopy [223].

3.1.4.6. LuBa$_2$Cu$_3$O$_{7-\delta}$ thin films

Compared with above YBCO-related materials, LuBa$_2$Cu$_3$O$_{7-\delta}$ have some differences. In general conditions, LuBa$_2$Cu$_3$O$_{7-\delta}$ is not a steady phase but a metastable phase. Pure LuBa$_2$Cu$_3$O$_{7-\delta}$ does not form in bulk and the nominal composition of Lu-1:2:3 contains Lu$_2$BaCuO$_5$, BaCuO$_2$, and some unreacted CuO and shows no superconductivity [226]. But the film can be seen as an extrem condition and the metastable LuBa$_2$Cu$_3$O$_{7-\delta}$ can be

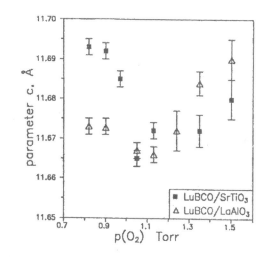

Fig. 28. Lattice parameter c of LuBa$_2$Cu$_3$O$_{7-\delta}$ in thin films vs deposition $P(O_2)$ ($T = 800°$C) after 1 h postannealing at 450°C in oxygen flow [227]. Reprinted from *J. Alloys Compounds* 251, 342, (1997), with permission from Elsevier Science.

really formed in thin film type. The LuBa$_2$Cu$_3$O$_{7-\delta}$ thin films show superconductivity, while the bulk LuBa$_2$Cu$_3$O$_{7-\delta}$ material does not [227]. Figure 28 shows the lattice parameter c of LuBa$_2$Cu$_3$O$_{7-\delta}$ in thin films vs deposition $p(O_2)$ ($T_s = 800°$C) after 1 h postannealing at 450°C in oxygen flow.

Doping Ca is an effective method to achieve LuBa$_2$Cu$_3$O$_{7-\delta}$ phase and get bulk superconductivity. Substitution of small amount of Ca ions in Lu$_{1-x}$Ca$_x$Ba$_2$Cu$_3$O$_{6+\delta}$ leads to the stabilization of the superconducting phase in this system although the sample contains a small amount of impurity phases. Pinto et al. presented a systematic investigation of superconductivity in argon annealed tetragonal samples of Lu$_{1-x}$Ca$_x$Ba$_2$Cu$_3$O$_{6+\delta}$ (δ less than or equal to 0.2) [228]. T_c increases with the concentration of Ca ions which indicates that Ca ions generate holes in the Cu–O layers. Powder neutron diffraction studies on argon annealed Lu$_{0.7}$Ca$_{0.3}$Ba$_2$Cu$_3$O$_{6+\delta}$ ($\delta = 0.25$, $T_c = 58$ K) show that there is a decrease in bond length of Cu(2)–O(3) and hence an increase in bond valence sum of Cu(2) consistent with oxidation of Cu–O layers whereas the other bond lengths that are associated with charge transfer from Cu–O chains remain close to the value of undoped YBa$_2$Cu$_3$O$_{6+\delta}$ (δ approximate to 0.2) tetragonal samples.

Among the fabrication methods, LuBa$_2$Cu$_3$O$_{7-\delta}$ films were mainly fabricated by MOCVD [229, 230]. In 1997, Samoylenkov et al. prepared epitaxial LuBa$_2$Cu$_3$O$_{7-\delta}$ films by flash evaporation MOCVD on LaAlO$_3$, SrTiO$_3$, and ZrO$_2$ (Y$_2$O$_3$) substrates [227]. The highest T_c and J_c (77 K, 100 Oe) values achieved were 89 K, 2.7×10^6 A/cm^{-2}, 88 K, 2.5×10^6 A/cm^{-2}, and 87 K, 1.0×10^6 A/cm^{-2}, respectively. An influence of $P(O_2)$ (at the deposition temperature $T = 800°$C) on structural and superconducting characteristics of LuBa$_2$Cu$_3$O$_{7-\delta}$ films was found to be similar to that in the case of YBa$_2$Cu$_3$O$_{7-\delta}$, in spite of the difference in morphology features and $J_c(T)$ dependencies. Occurrence of secondary phase inclusions in LuBa$_2$Cu$_3$O$_{7-\delta}$ films was analyzed by X-ray diffraction (XRD) and TEM. The "fully oxygenated" $LuBa_2Cu_3O_7$ was found to be overdoped; an increase of T_c value by 3 K was observed after an annealing of the films at reduced $P(O_2)$.

The properties of $LuBa_2Cu_3O_7$ thin films were investigated. Superconducting properties of oxygen deficient $LuBa_2Cu_3O_7$ thin films were researched by Samoylenkov et al. [231]. Srinivasu et al. report microwave surface resistance $R_s(T)$ measurements on $Lu_{1-x}Pr_xBa_2Cu_3O_{7-\delta}$ thin films *in situ* grown by the PLD technique, focusing on the effect of Pr doping in this system [232].

3.2. $Bi_2Sr_2Ca_{n-1}Cu_nO_y$ System Thin Films

3.2.1. Introduction

The $Bi_2Sr_2Ca_{n-1}Cu_nO_y$ (BSCCO) family of compounds presents a lot of features which have received much attention [53]. Especially, it is one kind of the main superconducting cuprates that present superconductivity up to a critical temperature (T_c) above the boiling point of liquid nitrogen, 77 K (when $n \geq 2$). Moreover, these materials are well suited together to create artificial superlattices since they have good accord between the *ab* plane lattice parameters, similar electronic structures, and reasonable matching of their charge carrier densities. The $Bi_2Sr_2Ca_{n-1}Cu_nO_y$ phases with $n = 1$ (2201), $n = 2$ (2212), and $n = 3$ (2223) are known to be the superconductors with zero resistance temperatures, T_{c0} of 10–20, 80, and 110 K, respectively, which are synthesized by a solid-state reaction method. Bi-2201 is of orthorhombic structure with lattice parameters $a = 5.362$ Å, $b = 5.3826$ Å, and $c = 24.384$ Å. Bi-2212 is also of orthorhombic structure and the corresponding lattice parameters are around 5.4, 5.4, and 30.8 Å, while Bi-2223 is characterized as tetragonal and the lattice parameters are $a = 3.82$ Å and $c = 37.1$ Å, respectively. The phases with $n = 4$ to 12, which have not been synthesized by any conventional solid-state reaction methods, were made by thin film growth methods. The phases with $n = 4$ and 5 were found to show T_{c0} at 84 and 30 K, respectively.

It is well known that BSCCO compounds are high-T_c cuprates that are less hazardous than those with $T_c > 77$ K, which contain toxic elements such as Hg, Tl. They also possess a unique structural feature of shearing along the micaceous Bi_2O_3 planes to produce platelike fragments which are then aligned to give well-oriented crystal structures with superconducting weak links. So this system has received considerable attention for use into wires and tapes for electric applications.

On the other hand, the *c*-axis conduction of BSCCO exhibits a Josephson effect behavior called the intrinsic Josephson coupling, which originates from the natural lamellar structure consisting of superconducting CuO_2 layers separated by nonsuperconducting barrier layers of Bi_2O_3 and SrO. In such configurations the collective motions of Josephson vortices have important application potentials for use in ultrafast millimeter-wave devices. So preparation and investigation of thin films are most important and are fundamental for applications to BSCCO systems and began soon after its discovery.

3.2.2. Fabrication and Characterization of BSCCO Film

3.2.2.1. Fabrication of BSCCO Film by MBE

MBE has several advantages for growing HTSC thin films [53]. (1) The base pressure is typically of the order of 10^{-10} Torr and the kinetic energy of evaporated species is relatively low, MBE has the potential to give very pure phases with atomically abrupt interfaces, smooth surfaces, and extremely low levels of contaminants and structural defects. (2) MBE permits control of epitaxial growth with a monolayer by monolayer mode and is therefore well suited to the synthesis of lamellar crystal structures such as the HTSC thin films. Furthermore, the freedom to choose the sequence of layers allows the realization of both periodic and aperiodic artificial superlattices. Such artificial configurations are of interest for device applications and investigations on the interaction between neighboring unit cells. (3) By means of heteroepitaxial growth, MBE can be used to synthesize phases that are even thermodynamically unstable. Therefore, MBE has been vastly accepted for preparing BSCCO film especially in recent years [37, 53, 233–238].

MBE growth of BSCCO can be considered as having several user-defined growth parameters.

Source (or Cation Flux) Temperature and Shutter Timing. In order to obtain 2D growth of BSCCO, incident fluxes for all the cation species tend to fall within the range 10^{14}–10^{15} atoms/cm^{-2}/s^{-1}. The usual three main growth modes, Volmer–Weber (3D growth), Stransky–Krastanov (3D–2D growth), and Frank–van der Merwe (2D growth), however, are different for each cation species and extremely dependent not only on the cation flux but also on various other parameters including substrate temperature (T_s), the oxidizing species, and pressure at substrates (P_s). RHEED is commonly adopted as an *in situ* characterization technique to fine tune the fluxes and manipulate the shutter timings.

Oxidant. MBE growth of HTSC thin films requires the introduction of an oxidant species reactive enough to fully oxidize the film *in situ*. Powerful oxidant species such as O_3, NO_2, or atomic oxygen must always be introduced. A fixed oxidant pressure between 1×10^{-5} and 2×10^{-5} Torr is typically maintained during growth.

The pumping power is of importance to keep the suitable background pressure near the substrates. This way, there is improved stability in the cation fluxes and enhanced mobility of adatoms at the substrate surface. This leads directly to improved growth characteristics and, ultimately, to better film quality.

After growth the films are usually cooled down to below 250°C before being removed from the growth chamber. Control of oxidant supply during cooling is necessary in order to determine the final oxygen content (carrier concentration) of the films and avoid decomposition of metastable phases. Eckstein et al. [239] report that, in order to maximize T_c, the reactivity of the oxidation environment during cooling should be varied according to the phase; for 2223 a more reactive oxidizing environment is needed than for 2212. It is mentioned that 2212 films cooled at the growth pressure of O_3 all the way down to 250°C are overdoped in oxygen and have both reduced T_c and increased normal-state resistivity. The oxidant pressure is hence reduced during cooling of 2212. 2223 grown under 2×10^{-5} Torr of 30% O_3-enriched O_2, however, requires a 20-fold increase in oxidant pressure during cooling in order to maximize T_c. Using a different approach Tsukada et al. [240] report maintaining the same O_3 pressure as during growth in order to avoid modification of the as-grown film structure but vary turn-off temperature T_{OC} (during cooling with oxidant supply) to control the oxygen content. The optimum T_{OC} is reported to be phase dependent at 300, 235, and 170°C for 2201, 2212, and 2223, respectively.

Substrate. (001)$SrTiO_3$ and MgO are commonly used as substrates for growing BSCCO films. Nonvicinal (001) Nd:$YAlO_3$ and Si(100) are also reported to be used as substrates however, unlike Nd:$YAlO_3$, successful use of Si as a substrate remains elusive because of interface reactions which produce remarkable diffusion tails in both materials.

Typically, the substrate is exposed *in situ* to the oxidizing gas at a high deposition temperature T_s (above 750°C) for half an hour prior to growth in order to clean the surface. Increasing T_s from 670 to 780°C is reported to improve the crystalline, decrease the normal-state resistivity, and increase T_c for BSCCO phases with $n = 1–4$. However, at higher T_s, for example, 820°C, even relatively stable phases such as 2212 become structurally unstable. It is found that, to obtain high T_c for the BSCCO phases, T_s should be strictly set between 650 and 800°C, since the absolute values depend greatly on how they are measured.

3.2.2.2. Fabrication of BSCCO Films by Laser Ablation

Pulsed laser ablation (PLA) was used much less frequently to grow epitaxial films in the BSCCO system [241–247] compared with YBCO-based film, multilayer, heterostructure, and superlattice structures. The reason partly lies in that several competing phases coexist during growth. To obtain phase purity, most epitaxial BSCCO films were often grown in two stages. The as-deposited films were annealed *in situ* or *ex situ* at different temperatures and oxygen pressures to enhance the growth of a particular phase. The c-axis of the films is always perpendicular to the substrate.

For PLA BSCCO films, the externally controlled growth conditions include the laser pulse repetition rate, laser energy density, substrate temperature, ambient oxidizer type and pressure, and the target composition. The PLA target was made from high-purity Bi_2O_3, $SrCO_3$, $CaCO_3$, and CuO powders by using a conventional solid-state reaction method. In brief, the KrF excimer laser ($\lambda = 248$ nm) beam with power density between 1 and 2 J/cm^2 and repetition rate from 0.5 to 4 Hz we used. The (001) MgO single crystal was usually used as the substrate. The substrate temperate was kept at 640–750°C, depending on the film composition and deposition conditions selected. The ambient oxygen pressure during

film deposition was varied according to the other controlled growth conditions. For instance, during the growth of Sr-free $Bi_y(Ca_{2-x}La_x)Ca_{n-1}Cu_nO_z$ ($n = 3, 4, 5, 6,$ and 7) films, as reported by Yoshida et al. [241], an oxygen partial pressure of 0.02 Torr was maintained. And annealing treatment should be made in O_2 or O_3 atmosphere after deposition. During the entire *in situ* process of growing Bi-2212 and Bi-2201 films, as reported by Zhu et al. [242], the ambient oxygen pressure during film deposition was set at 150 mTorr. It was found that the layer-stacking sequence, microstructure, and superconducting T_{c0} are sensitive to growth temperature and deposition rate. Low deposition rate improved both the T_c and microstructure of Bi-2212 films at a given growth temperature.

3.2.2.3. Fabrication of BSCCO Films by Sputtering

Several sputtering methods have been widely used on fabricating BSCCO films since 1989 [248–268]. Rf magnetron sputtering, as one layer-by-layer deposition technique, is also an effective method for controlling the number of CuO_2 planes in a unit cell of BSCCO [248–252].

Besides one target, improved sputtering apparatuses with two or three targets and/or off-axis configuration of target–substrate have been applied in order to obtain optimum preparation conditions, optimum film composition, and good film physical properties. Figure 29 (Figure 1 in [248]) shows an improved rf sputtering apparatus. Since the compositions of a pair of targets can be different, films with continuously varying compositions could be deposited on substrates placed between the targets. One of the targets (target A) had a fixed composition of $Bi_1Sr_1Ca_1Cu_1O_x$, while the other (target B) was selected from three compositions of $Bi_3Sr_5Ca_1Cu_6O_x$, $Bi_1Sr_2Ca_3Cu_4O_x$, and $Bi_1Sr_1Ca_1Cu_4O_x$. An improved rf magnetron sputtering apparatus, with three magnetron cathodes and off-axis substrate position being used, is schematically shown in Figure 30 (Figure 1 in [251]). The three targets are Bi, $Sr_5Cu_2O_7$, and $CaCuO_2$, respectively. The alternate sequence for a one-unit structure is $Bi \rightarrow Sr_5Cu_2O_7 \rightarrow CaCuO_2 \rightarrow Sr_5Cu_2O_7 \rightarrow Bi$, as shown in Figure 30. By varying the staying time at the $CaCuO_2$ target while keeping those at the Bi and $Sr_5Cu_2O_7$ target constant, the c dimension can be controlled.

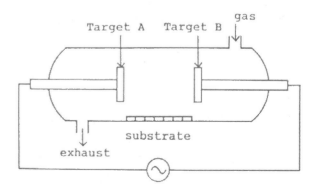

Fig. 29. Schematic representation of an improved rf sputtering apparatus [248].

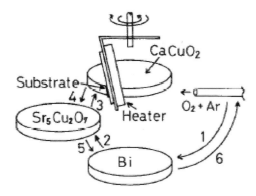

Fig. 30. A schematic diagram of the three targets and off-axis substrate geometry. The attached numbers show the deposition sequence [251].

Substrates that can be used for the growth of BSCCO films deposited by sputtering method include MgO, SrTiO₃, Al₂O₃, and YSZ. They are mounted, by coordinating with the other deposition conditions, to be either several centimeters above the target or off-axis like Figures 29 and 30 illustrate. During deposition, substrate temperature is held around 700–710°C. When at room temperature, however, a postannealing at about 800 to 890°C is necessary in order to obtain superconducting crystalline BSCCO films.

During sputtering, Ar or the mixture of Ar and O_2 atmosphere is selected depending on the desired composition and properties of BSCCO films. The pressure is varied according to the deposition apparatus and conditions selected. Using the rf sputtering apparatus shown in Figure 29, BSCCO thin films [248] were deposited in the atmosphere of an O_2/Ar (1/1) mixture at a pressure of 100 mTorr. Single-phase $(Bi_{0.8}Pb_{0.2})_2Sr_2Ca_2Cu_3O_6$ films synthesized by Marino et al. [250] were grown in Ar gas with pressure of 0.1 mbar. While using off-axis three-target magnetron sputtering, BSCCO thin films with n from 1 to 7 were obtained. The sputtering gas is the mixture of O_2/Ar (1/1) and the total pressure is set at 2 Pa [251].

3.2.2.4. Fabricating of BSCCO Films by the Chemical Process

The chemical process is also a good candidate for growing BSCCO thin film with simplicity, low cost, relative large-area film, and ease in shape formation. CVD, MOCVD, and LPE, which have been introduced previously, are also widely used for growing BSCCO films [269–282]. Here, some special chemical processes will be introduced.

Growing BiSrCaCu₂Oₓ Films by Pyrolysis of Organic Acid Salts. Pyrolysis process was selected for fabricating BSCCO films at the time around the discovery of BSCCO superconductors [283–286].

Nasu et al. [283] reported the fabrication of $BiSrCaCu_2O_x$ films, by pyrolysis of 2-ethylhexanoates, on Si(100) with epitaxially grown tetragonal ZrO_2 as buffer layer.

ZrO_2 films were deposited epitaxially on Si substrates heated at 800°C by electron-beam evaporation of monoclinic

ZrO_2 powder in oxygen atmosphere at pressure of about 5×10^{-5} Torr. Oxygen gas was introduced into the chamber during the deposition in order to oxidize evaporating ZrO or Zr dissociated by electron-beam bombardment. Si substrates used were p-type (100)-oriented single crystal Si wafers. The thickness of the ZrO_2 films was typically about 200 to 300 nm.

Bi-2-ethylhexanoate liquid and Sr-, Ca-, and Cu-2-ethylhexanoate powders were used as the starting materials for pyrolysis. Each 2-ethylhexanoate was dissolved into suitable solvents such as toluene. The resulting solution was mixed and stirred to contain Bi:Sr:Ca:Cu = 1:1:1:2 in atomic ratio and then dropped on the ZrO_2/Si(100) substrates and dried in air at room temperature. The samples were then heated at 500°C for 30 min in air. After repeating the above procedure 3 to 4 times, the samples were annealed at 950 to 980°C for 3 to 10 min. The surfaces of the samples were smooth and gray in color, their thickness was about 10 μm, and $T_{c(onset)}$ observed is 97 K while $T_{c(end)}$ is 50 K.

Growing Bi2212 Films Using Surface Diffusion Process. Using the surface diffusion process, Lee et al. [287] successfully synthesized $Bi_2Sr_2CaCu_2O_y$ films on alumina substrates with a $CuAl_2O_4$ buffer layer.

The substrates were prepared by a thermal deposition of the metallic Cu layer over alumina substrates in order to form a $CuAl_2O_4$ buffer layer and to react with a Bi:Sr:Ca = 2:2:2 (atomic ratio) coating powder prepared with starting powders of Bi_2O_3, $SrCO_3$, and $CaCO_3$, respectively. The mixture of these starting powders was ball-milled and calcined at 820°C for 48 h followed at 850°C for 48 h. The calcined powder was sieved to be less than 500 mesh and then was mixed with an organic binder (diethylene glycol mono-N-butylether acetate, ethylcelluose, terpinol, and olic acid) and screen printed on Cu-deposited alumina substrates. This composite sample was annealed at 860°C for 30 min under air atmosphere which led to the optimum superconducting properties of Bi2212 films ($T_{c(onset)}$ is 80 K and zero-resistance temperature is 72 K). Extending annealing time introduced a gradually broadening transition behavior and, eventually, typically semiconducting behavior. When the heat treatment was raised up to 890°C, the transition behavior gradually degraded. On the other hand, films annealed at temperatures below 860°C showed a semiconducting behavior in their resistivities, and a finite zero-resistance temperature cannot be obtained despite a long sintering time. Mean while the XRD patterns of the Cu-deposited alumina substrate, when annealed at 400°C, showed CuO, Cu_2O, Al_2O_3, and $CuAl_2O_4$ phases. As the annealing temperature increases up to 800°C, the peak of Cu_2O and Al_2O_3 phases decreases and the CuO and $CuAl_2O_4$ phases become dominant. During heating at temperatures between 600 and 800°C, the CuO and $CuAl_2O_4$ phases are unchanged because the $CuAl_2O_4$ layer has already formed between the alumina substrate and the unreacted CuO layer.

The $CuAl_2O_4$ buffer layer is produced by the diffusion of Cu into the alumina substrate. The microstructure of the $CuAl_2O_4$ buffer layer acts as a seed of a Bi-2212 grain growth and helps to grow a c-axis oriented microcrystal. The resulting optimum Bi-2212 film is highly c-axis-oriented with thickness of 20 μm and strongly aligns on the CuAl$_2$O$_4$ buffer layer of 1.5 μm thickness.

Growing Bi-2223 Films by Dip-Coating on Ba$_2$LaZrO$_{5.5}$. Jose et al. [288] synthesized Bi-2223 thick films by dip-coating on a new substrate, polycrystalline Ba$_2$LaZrO$_{5.5}$ (BLZO), whose dielectric constant and loss factor are in a range suitable for it to be a substrate material candidate for microwave applications.

BLZO was prepared, following the conventional solid-state reaction technique, from high-purity powders of BaCO$_3$, La$_2$O$_3$, and ZrO$_2$. Single-phase Bi-2223 powder was also prepared by the conventional solid-state reaction method from high-purity Bi2O$_3$, PbO, SrCO$_3$, CaCO$_3$, and CuO weighed in the precise stoichiometric ratio of (Bi$_{1.5}$Pb$_{0.5}$)Sr$_2$Ca$_2$Cu$_3$O$_x$.

The Bi-2223 suspension for dip-coating was prepared by thoroughly mixing fine superconducting Bi-2223 powder with isopropyl alcohol or *n*-butanol, and the viscosity of the suspension was controlled by the addition of fish oil. The polished and cleaned BLZO substrate was then dipped in the Bi-2223 suspension and dried. This procedure was repeated until a required thickness was attained. The dip coated films were then dried and heated at a rate of 5°C/min in air up to 880°C and kept at this temperature for 2 min and then annealed at 850°C for 6 h. Then the films were cooled at a rate of 1°C/min down to 800°C followed by furnace cooling to room temperature. Only single-phase Bi-2223 is detectable in the XRD patterns of resulting films. The thick Bi-2223 films has a $T_{c(0)}$ of 110 K and a current density of $\sim 4 \times 10^3$ A/cm^2 at 77 K. There is no detectable chemical reaction taking place between Bi-2223 and BLZO even after the severe heat treatment at 850°C.

3.2.2.5. Characterization of BSCCO Films

Several common methods have been applied to characterize the composition and structure of BSCCO films grown by MBE, laser ablation, sputtering, and chemical deposition processes, including SEM, energy dispersive X-ray analysis (EDX), Rutherford backscattering analysis (RBS), and XRD. The microstructure of the films is studied by SEM. Either EDX or RBS can be used to measure the accurate concentration of the films. XRD is suitable to investigate the impurities (if they exist), the orientation, and the crystallinity of the films.

The most commonly quoted parameter in the publications on growth of BSCCO films is T_c, though there are many important quality parameters which characterize an epitaxial superconducting thin film (i.e., T_c, J_c, average surface roughness). Therefore, films were often selected according to T_c for comparison. For many applications, T_c is only one of the important parameters.

Table VIII shows the highest values of T_c that can be established for each phase of the BSCCO films grown by MBE, sputtering, laser ablation, and chemical processes, together with some of the commonly quoted growth parameters.

3.2.3. Other Properties of BSCCO Films

3.2.3.1. Microwave Surface Impedance

Investigation on the temperature dependence of the penetration depth $\lambda(T)$ may provide information on the pair symmetry of high-temperature superconductors. Microwave surface impedance measurement is effective for this purpose.

Andreone et al. [289] measured the dependence of surface impedance of Bi-2212 film on temperature and rf magnetic field precisely. Magnetic field was found to always penetrate in grain boundaries. A $\lambda(0)$ with value of about 1 μm was extracted for the best measurement. The measured penetration depth showed a T^2 behavior, which is thought to be possibly related to extrinsic factors, in particular to the granularity of the films.

3.2.3.2. Transport Properties

Study on the peculiarities of the in-plane and out-of-plane electrical transport in high-temperature cuprates has drawn much attention since it might shed light on the mechanism of high-temperature superconductivity [290–292]. In the significant anisotropic BSCCO compound, the in-plane resistivity also has a metallic character like YBCO; however, the c-axis (out-of-plane) resistivity usually shows a peak, which increases in magnitude with decreasing oxygen concentration and an increasing external magnetic field parallel to the c-axis. Therefore, BSCCO is always treated as an ideal two-dimensional superconducting system.

Livanov et al. [290] reported the in-plane and out-of-plane transport properties of Bi2212 thin films. They found that the two-direction resistivities under magnetic fields up to 1 Tesla could be well described by the fluctuation theory in the Hartree approximation at temperatures above and below the zero-field critical temperature. The measured resistivities deviating from the theoretically predicted ones well below T_{c0} are interpreted as a pinning-induced phase transition from a strongly fluctuating normal phase into a pinned vortex phase.

3.2.3.3. Josephson Junctions (JJs) [53, 293]

Krasnov et al. [293] reported that the existence of the ac Josephson effect is an intrinsic HTSC Josephson effect which was directly evidenced by the observation of Fiske steps in Bi2212 intrinsic stacked Josephson junctions. This is supported by (a) observation of the step structure in the c-axis current–voltage (I–V) characteristics consisting of steps with nearly constant voltage and with the spacing between steps corresponding to the phase-locked geometric resonance at the lowest cavity mode branch, (b) the inverse proportionality of the step voltage to the length of the stack, and (c) the current amplitudes of the steps being periodic in applied magnetic field in the *ab* plane with the periodicity defined by the crystalline structure and the length of the mesa. This confirms that it is indeed the layered structure of Bi2212 which causes the appearance of atomic-scale intrinsic Josephson junctions.

Table VIII. Various Quality and Growth Parameters of BSCCO Films with $n = 1$–11

Phase n	T_c (R=0) (K)	T_c (onset) (K)	J_c (4.2 K) (A/cm^{-2})	Thickness (nm)	Substrate (001)	T_s (°C)	Buffer layer	Growth method	Atmosphere	P (Torr)	Growth rate (Å/s^{-1})	C (Å)	Reference
1	20	NG	NG	NG	STO	650,750	None	MBE	O_3	1×10^{-5}	~0.2	NG	[53]
2	>85	90	$>10^7$	50	STO	650,750	None	MBE	O_3	1×10^{-5}	NG	NG	[53]
3	86	>110	NG	30–100	STO	650,750	None	MBE	O_3	2×10^{-5}	0.15–0.3	27.2	[53]
4	75	~85	NG	130	STO	780,800	None	MBE	O_3	1×10^{-5}	NG	43.3	[53]
5	30	~80	NG	150	MgO	780,800	None	MBE	O_3	1×10^{-5}	NG	50	[53]
6	NA	NA	NA	30–80	STO + MgO	550,650	NG	MBE	NO_2	NG	NG	NG	[53]
7	NA	NA	NA	30–80	STO + MgO	550,650	NG	MBE	NO_2	NG	NG	NG	[53]
8	60	NA	NA	NA	NA	650,750	2212	MBE	O_3	NG	NG	NG	[53]
9	NA	NA	NA	30–80	STO + MgO	550,650	NG	MBE	NO_2	NG	NG	NG	[53]
10	NA	NA	NA	30–80	STO + MgO	550,650	NG	MBE	NO_2	NG	NG	NG	[53]
11	NA	NA	NA	30–80	STO + MgO	550,650	NG	MBE	NO_2	NG	NG	NG	[53]
2	71	93	5×10^6	400–600	MgO	740	None	PLA	O_2	0.15	~0.05	30.85	[242]
1	NA	NA	NG	NG	MgO	710	None	Sputtering	$Ar/O_2(1/1)$	~1.5	NG	24.61	[251]
2	60	NG	NG	NG	MgO	710	None	Sputtering	$Ar/O_2(1/1)$	~1.5	NG	30.98	[251]
3	67.8	NG	NG	NG	MgO	710	2201	Sputtering	$Ar/O_2(1/1)$	~1.5	NG	37.17	[251]
4	43.0	NG	NG	NG	MgO	710	None	Sputtering	$Ar/O_2(1/1)$	~1.5	NG	43.54	[251]
5	26.1	NG	NG	NG	MgO	710	None	Sputtering	$Ar/O_2(1/1)$	~1.5	NG	49.88	[251]
6	10.4	NG	NG	NG	MgO	710	None	Sputtering	$Ar/O_2(1/1)$	~1.5	NG	56.35	[251]
7	~4	NG	NG	NG	MgO	710	None	Sputtering	$Ar/O_2(1/1)$	~1.5	NG	62.66	[251]
3	102	105	NG	$\sim 1 \times 10^3$	MgO	600	None	Sputtering	Ar	0.076	~1.4	NG	[250]
5/1^a	41	NG	NG	78/39	MgO	700–710	None	Sputtering	$Ar/O_2(1/1)$	~1.5	NG	NG	[252]
1112^b	80	110	NG	$(1-2) \times 10^3$	MgO	RT	None	Sputtering	Ar	$(3-30) \times 10^{-3}$	4.5	15.38	[249]
1112^b	50	97	NG	$\sim 1 \times 10^4$	Si(100)	500	ZrO_2	Pyrolysis	air	–	–	30.646	[283]
2	72	80	NG	$\sim 2 \times 10^4$	Alumina	860	$CuAl_2O_4$	SDP	air	–	–	NG	[287]
3	110	~112	NG	$\sim 4 \times 10^4$	$Ba_2LaZrO_{5.5}$	880–850	None	Dip-coating	air	–	–	NG	[288]
2	83	NG	$>10^6$	250	$NdGaO_3$	800	None	LPE	air	–	–	NG	[289]

^{a}Bi-2245/2201 superlattice film.

^{b}Bi$_1$Sr$_1$Ca$_1$Cu$_2$O$_x$ film.

NG—not given; NA—not applicable; RT—room temperature; LPE—liquid phase epitaxy.

Note: Most films listed above have been postannealed in order to achieve optimum properties.

For the MBE grown trilayer the sandwich-type Josephson junction consisting of BSCCO phases has yielded a Josephson effect with both hysteretic (<15 K) and nonhysteretic (<65 K) I–V characteristics which show Shapiro steps under microwave radiation [53, 87, 294, 295].

SNS junctions were formed by the growth of a superconducting Bi2212 film into which unit cell thick barriers of BSCCO (B2201 or $Bi_2Sr_2(Ca,Sr,Bi,Dy)_{n-1}Cu_nO_y$ ($n = 4$ to 8)) were inserted [53]. The central Ca layers of the BSCCO barrier were doped with Bi, Sr, and Dy, to vary the conductivity of the barrier and allow the junction I_c to be tuned over four orders of magnitude, while maintaining a nearly constant I_cR_n product of about 0.5 mV [87, 294]. I–V characteristics of the junctions indicate that the barrier layers (thickness of between 25 and 44 Å) form continuous proximity effect junctions over areas of 30 μm × 30 μm. The use of Dy-doped 1278 barriers has yielded I_cR_n products in excess of 5 mV with ±5% uniformity in I_c [296–298]. Similar studies of SIS Josephson junction employing titanate barrier layers including $SrTiO_3$, $CaTiO_3$, $BaTiO_3$, $Bi_4Ti_3O_7$, and $DyTiO_3$ produced junctions with transport properties indicative of tunneling effects up to 50 K.

3.3. La$_2$CuO$_4$ System Thin Films

As for the simple and typical crystal structure and easily controlled doping, the La_2CuO_4 system is thought to be the prototype for the study of high-T_c superconductivity. For this system, a wide range Sr doping can be achieved from an antiferromagnetic insulator to a superconducting metal with $T_c \approx$ 40 K. However, high quality $La_{2-x}Sr_xCuO_4$ is not easy to grow, so the fabrication and properties of the single-crystalline thin films are important for the study of this system.

3.3.1. Fabrication

Below about 530 K, undoped La_2CuO_4 has a structure with orthorhombic symmetry and the room temperature lattice constants of this phase are $a = 5.354$ Å, $b = 5.401$ Å, and $c = 13.153$ Å. The tetragonal-to-orthorhombic (T–O) transition had been observed at about 530 K in the undoped La_2CuO_4. With the doping of Sr in the La_2CuO_4, the T–O transition temperature decreases to below the room temperature. Sputtering [299, 300], MBE, PLD, and co-evaporation [301] methods have been used to grow the $La_{2-x}Sr_xCuO_4$ thin films.

Sputtering includes dc sputtering and rf sputtering. The rf sputtering is always used to fabricate poor conductive materials such as La_2CuO_4 thin films, which cannot easily be deposited by dc sputtering due to the accumulation of the charge carriers on the surface of the target. In order to enhance the deposition rate, magnetron sputtering was uesd. A common process to fabricate $La_{2-x}Sr_xCuO_4$ thin films by magnetron sputtering is to deposit the materials on a substrate heated at about 800°C in sputtering gases. The sputtering gases were pure argon, mixtures of argon and 10–80% oxygen, or pure oxygen. In some cases, a following annealing of the as-deposited films is carried out in air or in oxygen at temperatures between 500 and

800°C [299]. Different Sr doped $La_{2-x}Sr_xCuO_4$ was fabricated at 780°C on $SrTiO_3$ substrate in sputtering gases with mixtures of argon and oxygen (argon:oxygen = 3:1). It was found that T_{c0} ($R = 0$) increases with the oxygen partial pressure of sputtering gases increasing from 15 to 40 Pa and deceases with the decreasing of the thickness [321]. The average of the content of the films fabricated by this method is always different from the content of the target. Off-axis magnetron sputtering was used to avoid this problem in some extent. Kwo et al. [302] fabricated the single domain, (103) oriented $La_{2-x}Sr_xCuO_4$ films of $0.04 \leq x \leq 0.34$ on vicinal (101) $SrTiO_3$ substrates using 90° off-axis sputtering.

MBE is another efficient method to deposit $La_{2-x}Sr_xCuO_4$ thin films. With this method, Locquet et al. [303–305] fabricated high-quality $La_{2-x}Sr_xCuO_4$ films on both $SrTiO_3$ and $LaSrAlO_3$ substrates. A double critical temperature (T_{c0} ($R = 0$) = 49.1 K) of $La_{1.9}Sr_{0.1}CuO_4$ thin film was deposited on $LaSrAlO_3$ substrate with a thickness of 15 nm [304]. The MBE-grown c-axis La_2CuO_4 films were also achieved with a high value of T_{c0} ($R = 0$) ~ 42 K on the $LaSrAlO_3$ substrates [305].

$La_{2-x}Sr_xCuO_4$ thin films can also be deposited by PLD on a substrate. The optimal substate temperature is about 780°C. The oxygen pressure during depostion is varied from 10 to 300 Torr, producing different growth rates. Recently, using the PLD method, Si et al. prepared $La_{1.85}Sr_{0.15}CuO_4$ thin films for studying the effect of strain and oxygenation on superconductivity of $La_{1.85}Sr_{0.15}CuO_4$ [306].

The selection of substrates is an important consideration to achieve high-quality $La_{2-x}Sr_xCuO_4$ thin films. Early in 1987, Nagata et al. had tried to deposit $(La_{1-x}M_x)_yCuO_{4-\delta}$ (M = Sr, Ba, Ca) on different substrates, such as quartz, alumina, sapphire, and YSZ, but the result was not quite good. In the same year, Suzuki fabricated c-axis-oriented epitaxial thin films on the $SrTiO_3$ substrates by sputtering. So far, $LaAlO_3$ and $LaSrAlO_3$ are well known to be the suitable substrates for preparing $La_{2-x}Sr_xCuO_4$ for their good lattice match.

3.3.2. Transport Properties

Suzuki and Hikita [307, 308] reported resistivity, Hall effect, magnetoresistance, and anisotropy in $La_{2-x}Sr_xCuO_4$ (LSCO) single-crystal thin films which were grown epitaxially on $SrTiO_3$(100) single-crystal substrates heated at about 800°C by rf single-target magnetron sputtering. In order to get the correct content of Cu, to the targets 20% more CuO than the nominal composition was added. The grown films are single crystalline with the c-axis perpendicular to the substrates.

3.3.2.1. Resistivity

The temperature dependence of resistivity is shown in Figure 31. A series of LSCO epitaxial films were prepared with x ranging from 0.04 to 0.36. The temperature dependence of resistivity showed metallic behavior below 150 K except for $x = 0.04$. For $x = 0.1$, no upturn was observed in the ρ–T curve above T_c, which was observed in bulk polycrystals

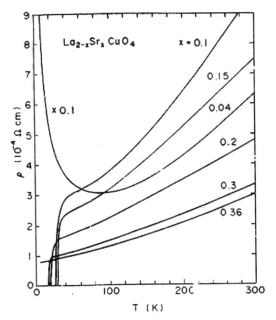

Fig. 31. Temperature dependence of resistivity ρ within the Cu–O planes for the $La_{2-x}Sr_xCuO_4$(001) epitaxial thin films with x from 0.04 to 0.36 [307].

[309, 310]. With the increase of x, the resistivity of the films decreases from 3.46×10^{-1} Ω/cm ($x = 0$) to 7.5×10^{-4} Ω/cm ($x = 0.15$) at $T = 300$ K, which is in agreement with the work on the polycrystal samples [309]. The temperature dependence of resistivity of the film with $x = 0.04$ shows incipient localization behavior below 150 K which can be scaled by the resistivity of variable range hopping in the 2D system. This fact indicates that the electronic conduction in the LSCO system is essentially 2D. Among these samples, the $x = 0.15$ one has the highest T_c ($R = 0$) ~ 28.6 K and the smallest transition width $\sim$2.7 K.

3.3.2.2. Magnetoresistance

As commonly observed in high-T_c superconductors, the magnetic field causes a profound broadening of the resistive transition, which is also observed in the LSCO system. Suzuki and Hikita [308] measured the resistive transition for the $La_{2-x}Sr_xCuO_4$ single-crystal thin films with x from 0.08 to 0.3 under the magnetic fields. A systematical change of the resistive transition with magnetic field was observed with the Sr concentration x as shown in Figure 32. In a lightly doped region, e.g., for the case of $x = 0.08$, the resistive transition exhibits a significant broadening under fields, especially when a field is parallel to the c-axis ($H//c$). As the Sr concentration is in the range from 0.1 to 0.15, the field-induced broadening of resistive curve becomes modest as x increases which is similar to what was reported by Kitazawa and co-workers [311, 312] in a $La_{2-x}Sr_xCuO_4$ bulk single crystal. For the samples with $x = 0.2$, 0.24, and 0.3, the applied magnetic field causes almost no broadening and, instead, it causes a parallel shift of the curves to lower temperatures. This behavior is similar to that of conventional type-II superconductors.

3.3.2.3. Hall Effect

The low-field Hall effect was studied by Suzuki [307] in $La_{2-x}Sr_xCuO_4$ single-crystal thin films with different Sr dopings. Figure 33 shows the temperature dependence of R_H and n_H for different Sr dopings. For $x = 0.04$, R_H is nearly constant over the temperature range measured. Carrier density for this composition derived from R_H is 4.8×10^{20} cm^{-3}, which nearly coincides with the Sr concentration. For $x = 0.36$, where the superconductivity is not observed till 4.2 K, R_H is pronouncedly reduced and exhibits complex behavior compared with other compositions. In this case, R_H decreases to 2.5×10^{-5} cm^3/C at 150 K and then abruptly increases to the highest value of 7.3×10^{-5} cm^3/C at 20 K. For the superconducting samples ($0.1 \leq x \leq 0.3$), R_H has a rather common temperature dependence. In this range of compositions, R_H increases with the decreasing of temperature to its maximum at $T \approx 10$–80 K, which is very similar to other Cu–O-based superconductors [313].

Except for $x = 0.04$, μ_H increases with decreasing temperature, which indicates that phonon scattering plays an important role on the transport properties. For $x = 0.04$, localization leads to μ_H decreasing.

3.3.3. Optical Properties

The reflectance (R) spectra and the transmittance (T) spectra for the LSCO epitaxial films were measured by Suzuki [307] (Fig. 34). The wavelength (λ) used in the measurement ranges from 0.4 to 4.5 μm. Except for $x = 0$, the plasma reflection was observed and become clearer as x increased from 0.04. R increased to about 75% at 4.5 μm when x increased to 0.36 and the plasma edge at $\lambda \approx 1.5$ μm changed only slightly with x. In addition to the plasma edge, there are small structures in the wavelength range shorter than 1.5 μm which reflects electronic band structure. The peak centered at $\lambda = 0.8$ μm shifts toward shorter λ with increasing x and disappears at $x = 0.1$, while the peak centered at $\lambda = 1.4$ μm also shifts to the shorter wavelength up to $x = 0.36$. The reflectance for the wavelength longer than 1.5 μm can be fit very well to the Drude model using the following equation:

$$\varepsilon(\omega) = \varepsilon_\infty \left[1 - \frac{\omega_p^2}{\omega^2 + i\omega\gamma} \right]$$

Here, ε_∞ is the optical dielectric constant, ω_p is the plasma frequency, and $\gamma = 1/\iota$ is the damping factor or the electron scattering rate.

The optical absorption is term determined primarily by the joint density of states (DOS), which reflects both DOSs above E_F and the valence-band structure. In the $La_{2-x}Sr_xCuO_4$ single-crystal films, the optical transmittance (T) in the infrared decreases dramatically with Sr doping due to the free-carrier absorption (Fig. 34). In the undoped La_2CuO_4 ($x = 0$), the rise in $T(\lambda)$ at $\lambda = 0.5$ μm and the large transmittance for $\lambda > 0.6$ μm indicate the existence of a bandgap of about 2 eV, which is consistent with what was reported by Ginder et al. [314] with

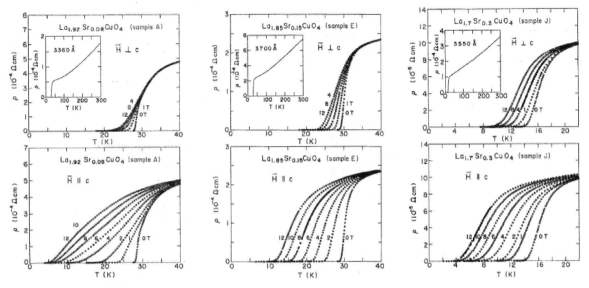

Fig. 32. Resistive curves for La$_{2-x}$Sr$_x$CuO$_4$ ($x = 0.08$, 0.15, 0.30) single-crystal thin films under various applied magnetic fields from 1 to 12 T with the field direction perpendicular to the c-axis (upper panel) and parallel to the c-axis (lower panel). The inset in the upper panel shows an extended view of $\rho(T)$ [308].

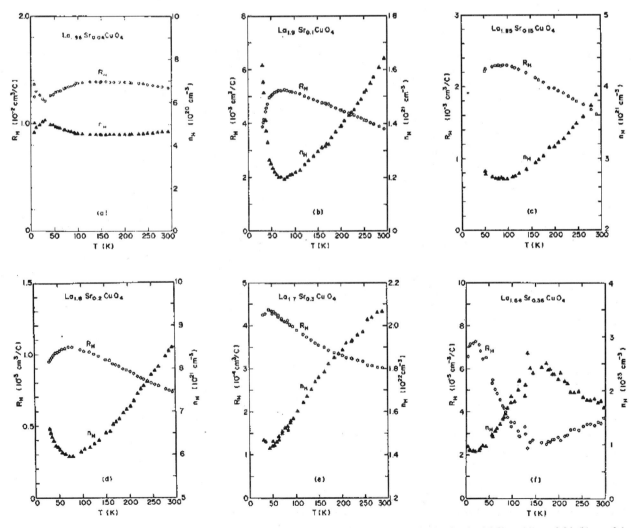

Fig. 33. Temperature dependence of Hall coefficient R_H and Hall carrier density n_H for the La–Sr–Cu–O epitaxial films. (a) $x = 0.04$, (b) $x = 0.1$, (c) $x = 0.15$, (d) $x = 0.2$, (e) $x = 0.3$, and (f) $x = 0.36$ [307].

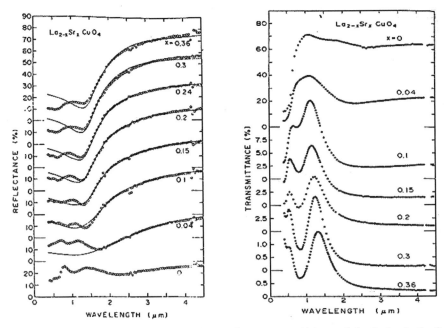

Fig. 34.   Reflectance spectra (left panel) and transmittance spectra (right panel) for the La–Sr–Cu–O (001) epitaxial films with x from 0 to 0.36. Scales are shifted by appropriate values. For the reflectance spectra, solid lines are Drude fits [307].

their photoemission measurements of La$_2$CuO$_4$. An additional absorption band centered at 0.85 μm emerged with increasing x. As x increases, the rise in $T(\lambda)$ shifts to shorter λ and, at the same time, the transmission peak splits into two, forming a broad absorption peak near 0.85 μm. With increasing x, the edge at longer λ shifts to lower energy, and the position of the absorption peak shifts slightly to lower energy too, while its strength increases systematically with x.

3.3.4. Strain and Oxygenation Effect

Although many attempts have been made to grow epitaxial films of LSCO, the superconducting transition temperatures of the films are significantly lower than that of the starting polycrystalline material (target) and have a strong dependence of thickness. Both the lower T_c and the strong thickness dependence of T_c (see Figure 35) [315] have been considered to be related to strain effects between the films and the substrates.

Sato et al. [301] reported the growth of (001) La$_{1.85}$Sr$_{0.15}$CuO$_4$ ultrathin films with thickness from 1 to 5 unit cells on LaSrAlO$_3$ substrates. Films with thickness of 4 unit cells exhibited almost bulklike transport properties and T_c ($R = 0$) of about 38 K. T_c decreased with decreasing thickness and disappeared in films less than 2 unit cells thick.

In 1997, Sato and Naito [316] grew (001) La$_{1.85}$Sr$_{0.15}$CuO$_4$ thin films on LaSrAlO$_3$ substrates with T_c ($R = 0$) as high as 44 K due to the plane stress produced by the lattice mismatch between the films and the substrates. In 1998, Locquet et al. [304] fabricated La$_{1.9}$Sr$_{0.1}$CuO$_4$ thin films with double critical temperature using epitaxial strain. The thin films were grown simultaneously on SLAO ($a = 3.754$ Å) and SrTiO$_3$ ($a = 3.905$ Å) with block-by-block molecular beam epitaxy at

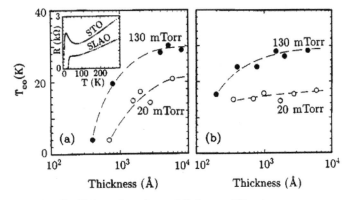

Fig. 35.   The thickness dependence of T_c for two different oxygen pressures for films deposited on STO (a) and SLAO (b) substrates. Inset: $R(T)$ for two similar films with thickness of 400 Å, deposited on STO and SLAO substrates [315].

750°C, where a sequence of two monolayers of (La,Sr)–O followed by one monolayer of Cu–O was repeated. The film grown on the LaSrAlO$_3$ substrate had a T_c of 49 K (double value of the bulk samples). To attain a higher T_c, a larger amount of compressive strain must be induced which can be described by the formula

$$T_c = T_c(0) + 2\frac{\delta T_c}{\delta \varepsilon_{ab}}\varepsilon_{ab} + \frac{\delta T_c}{\delta \varepsilon_c}\varepsilon_c \qquad (1)$$

where $\varepsilon = [d_{bulk} - d_{strained}]/d_{bulk}$, with d a lattice parameter, ε_{ab} indicates strain in plane, and ε_c indicates strain out of plane. The thickness is a compromise between avoiding the nucleation of misfit dislocations and minimizing the T_c reduction observed for ultrathin films [31, 499].

A model of the strain effect on the superconducting transition temperature T_c is presented with the basis of the extended $t - J$ model with d-wave symmetry [317]. The phenomenological Hamiltonian adopted in this model is

$$H_h = \sum_k \varepsilon_k c_k^+ c_k - V \sum_{\langle ij \rangle} n_i n_j \qquad (2)$$

where c_k^+ is the hole quasiparticle creation operator at momentum k, $n_I = c_i^+ c_i$ is the number operator of quasiparticles at site I, and V is the effective attractive correlation between nearest-neighboring pairs which is thought to be strain effect dependent. With a d-wave superconducting order parameter $\Delta_k = \Delta_0 f_k = \Delta_0(\cos k_x - \cos k_y)/2$, the superconducting transition temperature is determined by

$$1 = \frac{V}{N} \sum_k f_k^2 F_k \qquad (3)$$

where

$$F_k = \tanh(E_k/2T)/2E_k \qquad (4)$$

$$E_k = \sqrt{(\varepsilon_k - \mu)^2 + \Delta_k^2} \qquad (5)$$

along with the constraint condition for the density of holes n_H:

$$n_H = \frac{1}{2} - \frac{1}{N} \sum_k (\varepsilon_k - \mu) F_k \qquad (6)$$

Assuming the strain-induced change of pairing interaction V is responsible for the strain $\varepsilon = [d_{\text{bulk}} - d_{\text{strained}}]/d_{\text{bulk}}$, T_c will follow

$$\frac{dT_c}{d\varepsilon} = \frac{T_c^2}{W} \frac{d \ln V}{d\varepsilon} \qquad W = \frac{V}{2N} \sum_k f_k^2 \sec h^2 \left(\frac{\varepsilon_k - \mu}{2T_c} \right) \quad (7)$$

As the strain is very small, V can be expanded to follow $V(\varepsilon) = V(1 + d \ln V/d\varepsilon \varepsilon)$. Then,

$$\frac{1}{V(\varepsilon)} = \frac{1}{2N} \sum_k \frac{f_k^2}{\varepsilon_k - \mu} \tanh \left(\frac{\varepsilon_k - \mu}{2T_c} \right) \qquad (8)$$

and ε can follow the expression [318–320] $\varepsilon = \frac{2\mu b}{4\pi(1-\nu)M} \frac{\ln(h/b)}{h}$.

Similar to other cuprate superconductors, the oxygenation effect is important in LSCO systems. Oxygen vacancies are normally present in LSCO with the addition of strontium. An optimal oxygen partial pressure [315, 321] in the deposition process is investigated and some oxygenation methods are used to fill the vacancy in LSCO, such as high-pressure oxygenation [322] and ozone oxygenation [323]. It is intriguing that the film with a good oxygenation condition but tensile strain has a very similar superconductivity and normal-state properties to the film with a poor oxygenation condition and a compressive in-plane strain [306]. This indicates that oxygenation and strain are closely correlated. The compressive strain effect in a–b plane can induce a longer c-axis and thus make it easy to insert oxygen between the CuO_2 plane.

In contrast to the YBCO, BSCCO, and TBCCO thin films, the LSCO films were seldom studied for their lower potential for application. However, both LSCO bulk material and LSCO thin films are important for the fundamental study of superconductivity. All of the above is a brief presentation of the fabrication, superconducting, and normal-state properties of these kinds of materials.

3.4. TlBaCaCuO Thin Films

3.4.1. Introduction

Thin film is the most important material type for Tl-based HTSC for applications in passive microwave and active Josephson-effect devices and fundamental studies of normal and superconducting properties. So a high-quality Tl-based thin film is required. In the fabrication processing, the influence of thermodynamics and kinetics, the various potential thalliation techniques, and the selection of an appropriate substrate are important considerations, which were reviewed by Bramley et al. [56].

There are two Tl-based superconducting families, Tl–Ba–Ca–Cu–O (TBCCO) and Tl–Sr–Ca–Cu–O (TSCCO); both of them have a high T_c value. Most TBCCO materials have tetragonal crystal structures and may be described as consisting of alternating single or double Tl–O layers and perovskitelike Tl $Ba_2Ca_{n-1}Cu_nO_{2n+1}$ layers, while the phases in the TSCCO system are similarly described as $TlSr_2Ca_{n-1}Cu_nO_{2n+3}$ where $n = 1$–3.

3.4.2. Fabrication of Tl-based HTSC films

In the process of fabrication of Tl-based HTSC films, a delicate balance between the high temperature required to form the superconducting phases and the high volatility of thallium at these temperatures should be well considered. Both the partial pressures of oxygen ($P(O_2)$) and thallous oxide vapor ($P(Tl_2O)$) have an effect on the optimum processing temperature. Therefore, most routine TBCCO film fabrication is carried out by a two-step process involving the deposition of an amorphous precursor film (which may or may not contain thallium) using one of the standard thin film deposition techniques, followed by an *ex situ* thalliation anneal in the presence of thallous oxide vapor to form the crystalline superconducting phase.

3.4.2.1. Thermodynamic and Kinetic Factors Affecting Thalliation

Because of the volatility of Tl, the solid–vapor equilibrium existing above the thallous oxide source and the precursor film is a key point to be considered. The equilibrium reactions is presented as [324–326]

$$Tl_2O_3(c) \xleftrightarrow{k_1} T_2O(g) + O_2(g) \qquad (9)$$

$$2Tl_2Ba_2CaCu_2O_8(c) \xleftrightarrow{k_2} 4BaCuO_2(c) + 2CaO(c)$$
$$+ 2Tl_2O(g) + 2O_2(g) \quad (10)$$

where k_1, k_2 are equilibrium constants for reactions (1) and (2), respectively, and are given by

$$k_1 = P(Tl_2O)P(O_2) \qquad k_2 = P(Tl_2O)^2 P(O_2)^2$$

From the above equations, high values of $P(Tl_2O)$ and $P(O_2)$ can make for a reaction to the Tl(2212) formation. As the equilibrium constant k_2 increases with increasing temperature, higher oxygen and thallous oxide partial pressure are required to form Tl(2212). On the other hand, low values of $P(O_2)$ encourage increased evolution of Tl_2O vapor from a source powder.

Unfortunately, the fabrication of TBCCO thin films is more complicated than the above equilibrium process with the formation of other TBCCO compounds, such as

$$Tl(2201) \rightarrow Tl(2212) \rightarrow Tl(2223) \rightarrow Tl(1223)$$
$$\rightarrow Tl(1234) \rightarrow Tl(1245) \qquad \text{or}$$
$$Tl(2201) \rightarrow Tl(2212) \rightarrow Tl(2223) \rightarrow Tl1212)$$

according to the composition of the starting material [327–329].

Although single-phase Tl(2212) thin films were fabricated by several groups [330], it is difficult to prepare single-phase thin films of Tl(2223) and Tl(1223), because the lower temperature of formation of Tl(2212) results in the initial growth of Tl(2212) plus secondary phases at temperature below those required for Tl(2223) or Tl(1223) formation. In order to achieved single-phase Tl(2223), Tl(2212) must be transformed into Tl(2223) during the anneal [331–335]. The presence of a liquid phase is suggested to be necessary for the transformation [332, 336–339]. The transformation from Tl(2223) to Tl(1223) takes place by evaporation of Tl. As a result, it is common to find intergrowths of Tl(2212) or Tl(1223) phases in Tl(2223) thin films.

$P(Tl_2O)$ is one critical parameter in the processing of fabrication of Tl-based HTSC thin films. It has been found that with decreasing $P(Tl_2O)$ the order of phase stability was Tl(2212) $\rightarrow$ Tl(2223) $\rightarrow$ single Tl–O layer phases, which is consistent with the decreasing thallium content of the individual compounds [340–342]. Although the $P(Tl_2O)$ and $P(O_2)$ are not independent in a closed-crucible thalliation process, $P(Tl_2O)$ can be considered to be controlled by the Tl content of the thallous oxide source. According to this, the Tl_2O source is always a TBCCO compound of the same composition as the desired phase [332, 335, 343]. Since $P(Tl_2O)$ is much higher in equilibrium with Tl_2O_3 than with bulk TBCCO phases, initially the Tl_2O evolved from Tl_2O_3 rapidly; then the value of $P(Tl_2O)$ is controlled by the lower equilibrium $P(Tl_2O)$ over TBCCO. A mixture of Tl_2O_3 and TBCCO powder was used to achieve a higher $P(Tl_2O)$ at the start of thalliation [332, 344–346].

$P(O_2)$ is another important parameter in the thalliation process. Lower $P(O_2)$ can lower the thalliation temperature [332, 336, 337, 346–349] which can avoid film–substrate interactions, reduce Tl loss, and fabricate smoother films. On the other hand, a lower thalliation temperature requires longer annealing times to ensure the complete reactions. A partial phase diagram presented by Ahn et al. [337] shows the phases formed as a

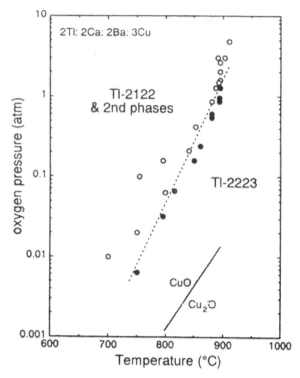

Fig. 36.   Phases observed in [Tl]:[Ca]:[Ba]:[Cu] = 2:2:2:3 samples as a function of annealing temperature and oxygen pressure: open circles = Tl-2212 and second phases; closed circles = Tl-2223 [337].

function of thalliation temperature and $P(O_2)$ (see Fig. 36). Lots of groups fabricated high-quality Tl(2223) [346, 348–352] and Tl(2212) [328, 353] films by thalliating at low temperatures in lower $P(O_2)$.

3.4.2.2. Ex Situ Film Fabrication Process

Ex situ film fabrication methods can divided into two categories due to the different geometries. One is crucible geometry which means the thallous oxide source and precursor film are held at the same temperature; another is two-zone furnace geometry in which the thallium source and precursor film are held at different temperatures. The crucible geometry is only well suited to small samples. Films with high values of T_c and J_c have been fabricated using this geometry. For the two-zone furnace geometry, the precursor film was placed in the high-temperature zone, while the thallous source is placed in the low-temperature zone. By controlling the temperature in the low-temperature zone, the $P(Tl_2O)$ in high-temperature zone which transports from the low-temperature zone can be controlled. The quality of films grown by these methods is inferior to that of those by crucible methods [331]. Recently, a new method named the hybrid two-zone-crucible process was devised by Siegal [354]. For this method, both the precursor film and the TBCCO source are placed in the high-temperature zone.

The disadvantage of the *ex situ* method is that it is only possible to grow highly textured films. However, a major advantage of this method is that it is possible to produce films on both sides of a substrate.

3.4.2.3. In Situ Film Fabrication Process

In situ film fabrication methods can control the film morphology well and allow a lower film deposition temperature. Using *in situ* film fabrication methods, it is possible to grow genuinely epitaxial films with the correct substrate. In contrast to the *ex situ* methods, it is more difficult to grow double-sided films by *in situ* methods, because the heating will destroy the first deposited side during the second side deposition. Betz et al. [355] fabricated TBCCO film with a two-stage process consisting of the deposition of a BCCO precursor film followed by an *in situ* heat treatment in the presence of Tl_2O vapor. Face and Nestlerode [356, 357] fabricated Tl(1212) thin films on $LaAlO_3$, $NdGaO_3$, and CeO_2-buffered sapphire substrates by off-axis magnetron sputtering in the presence of Tl_2O vapor. Further annealing in O_2 and Tl_2O was used to achieve a better superconducting properties ($T_c = 97$ K). Myers et al. [358–360] have fabricated thin films of $(Tl,Pb)SrCa_{1-x}Y_xCu_2O_7$ with T_c up to 83 K and thin films of $(Tl, Pb)SrCuO_5$ by the same technique.

Reschauer et al. [361–363] reported a successful truly *in situ* TBCCO film fabrication process that involves off-axis laser ablation from a BCCO target in the presence of Tl_2O vapor without any postdeposition process. The use of off-axis geometry allows a higher oxygen partial pressure during ablation, thus reducing the required partial pressure of Tl_2O. High-quality Bi-substituted Tl(1223) film with T_c up to 114 K, J_c (77 K) up to 8×10^5 A/cm^{-2}, and R_s (10 GHz, 80 K) less than 1 mΩ was fabricated by this method [364].

However, the *in situ* film deposition methods have only been used to grow phase-pure thin films of the single Tl–O layer phases. A possible reason is that it is too difficult to provide a high enough partial pressure of thallous oxide vapor to stabilize the double Tl–O layer phases.

3.4.2.4. Substrate Selection

The substrate plays a key role in the microstructural development of a growing film and hence in determining its superconducting properties [56]. Both the lattice matching and the surface topology can influence the growth of the epitaxial films and thus the properties of the films. Especially for fabrication of the Tl-based HTSC film by *ex situ* methods, the proper choice of substrate is needed to avoid film–substrate reactions during the high-temperature thalliation anneal.

$LaAlO_3$ is a rhombohedral perovskite with $a = 3.79$ Å, which has an excellent lattice match with the *a–b* plane of the TBCCO unit cells, so it is the most popular substrate for TBCCO thin film deposition. On $LaAlO_3$ substrate, epitaxial thin film can even grow in *ex situ* thalliation anneals because discrete islands of crystalline superconductor nucleate and grow with their *a*- and *b*-axes aligned with the *a*- and *b*-axes of the substrate [365]. Some very-low-angle grain boundaries are

formed eventually when these islands connect and the film has no well-defined grain structure. High-J_c and low-R_s TBCCO films [345, 366] must be grown on $LaAlO_3$ by avoiding high-angle grain boundaries.

$SrTiO_3$ is a cubic perovskite substrate which has an excellent lattice match with the *a–b* plane of both TBCCO and $YB_2Cu_3O_{7-\delta}$ ($a_{STO} = 3.91$ Å). However, the $SrTiO_3$ interacts with the film during the thalliation at high temperatures. It was identified that during the thalliation of Tl(2223) precursor films at temperatures around 860°C, Sr diffuses into the growing TBCCO film and substitutes for Ba, thus stabilizing the growth of Tl–O layer phase $Tl(Ba_{1-x}Sr_x)_2Ca_2Cu_3O_y$ [367, 368]. An interfacial Ba–Ca–Ti–O layer is also formed as a result of this interaction, but the superconductor film nevertheless has a strong relationship with the substrate lattice [369]. A similar chemical interaction was observed between Tl(2212) thin films and $SrTiO_3$ substrates [370]. The use of low thalliation temperature was found to avoid the chemical interaction between TBCCO thin films and $SrTiO_3$ substrates [371].

The most attractive property of MgO is its low isotropic dielectric constant and low loss tangent for microwave device design. Because of the large lattice mismatch ($a_{MgO} = 4.21$ Å), epitaxial films with biaxial alignment are difficult to achieve. Most films deposited on MgO show texture with their *c*-axes normal to the substrate but with no in-plane alignment [365, 369, 372]. The most common in-plane misalignments of the grains with the MgO substrate are 0° and 45°, but 27° misalignments are also found in some films [373]. The R_s (10 GHz, 80 K) measured on these Tl(2223) films was found to decrease monotonically with the degree of in-plane alignment increasing.

It is impossible to deposit TBCCO thin film on sapphire directly for severe film–substrate interactions during thalliation. Therefore, a buffer layer is needed to prevent film–substrate interactions and provide a suitable structure for the growth of epitaxial HTSC thin films.

3.4.2.5. Buffer Layer

The buffer layer material should be chemically stable with respect to both the film and the substrate and should have good lattice matching with the HTSC unit cell. CeO_2, Sr_2AlTaO_6 (SAT), $SrTiO_3$, and $LaAlO_3$ have been used as buffer layer during the deposition of TBCCO thin films.

CeO_2 has a cubic fluorite crystal structure with $a = 5.411$ and thus has a good lattice match with the HTSC unit cells when rotated through 45° with respect to the *a–b* plane. CeO_2 buffer layers with only 20-nm-thick on sapphire are sufficient to prevent the film–substrate interaction and allow the deposition of high-quality HTSC films [374, 375]. Holstein et al. fabricated Tl(2212) thin films onto CeO_2 buffered R-plane sapphire and with $T_c \sim 98$ K, J_c (75 K) $\sim 2.8 \times 10^5$ A/cm^{-2}, and R_s (10 GHz, 77 K) of 490 mΩ [376]. However, it was found that during thalliation at high temperatures around 840° a reaction occurred between the buffer layer CeO_2 and the Tl(2212) thin film to form a layer of BaCe(Tl)O$_3$ at the interface [377]. This problem can be avoided by lowering the required thalliation temperature by carrying out the thalliation annealing in Ar atmospheres.

SAT is a cubic perovskite material with a excellent lattice matching with the HTSC compounds. SAT can be grown epitaxially on (001) MgO substrates at temperatures around 780° [378]. High quality Tl(2212) thin film grown on SAT-buffered MgO was reported with T_c up to 103 K and J_c (77 K) up to 3×10^5 A/cm^{-2}.

SrTiO$_3$ has also been used as a buffer layer between TBCCO thin films and MgO substrates. Although film–substrate interactions has been also observed to be identical to those on bulk SrTiO$_3$ substrates [367, 368], buffer layer SrTiO$_3$ can improve the biaxial grain alignment compared with TBCCO films depostited directly on the MgO substrates [369]. Due to the technique problem of the deposition, LaAlO$_3$ was seldom used as a buffer layer. Especially, crystalline LaAlO$_3$ thin films have been deposited on a wide range of single-crystal substrates using slow cooling from the deposition temperature and/or a postdeposition anneal to induce crystalline film growth [369, 379–382]. The deposition of double Tl–O layered TBCCO thin films on LaAlO$_3$-buffered SrTiO$_3$ substrates is reported [369, 383].

The deposition of HTSC thin films onto metallic substrates is one possible route to the development of coated conductor technology and also for applications such as microwave cavities. Most of the metallic substrates are polycrystalline, which leads to a poor lattice matching to the HTSC unit cells. Some metals have severe high-temperature interactions with the thin films. For this reason, one or more oxide buffer layers must be employed to enable HTSC films to be grown on polycrystalline metallic substrates.

To achieve biaxial grain aligned superconductor films, two main approaches were adopted. An ion-beam-assisted (IBAD) technique has been used in the deposition of YBCO thin films onto the metallic substrates [384, 385]. Xiong et al. [386, 387] have deposited Tl(2212) thin films on YSZ-buffered polycrystalline alumina substrates with the best results of $T_c \sim 108$ K and J_c (77 K) $\sim 10^5$ A/cm^{-2}. Conventional thin film deposition techniques have been used to deposit aligned buffer layers onto an already textured substrate [388]. Ren and co-workers [389] have deposited Bi-substituted Tl(1223) thin films on rolling-assisted biaxially textured substrates (which describes the combination of a biaxially textured Ni substrates with aligned CeO$_2$ and YSZ buffer layers) using a postdeposition anneal in flowing Ar. The films were biaxially aligned with $T_c \sim 110$ K and J_c (77 K) $\sim 10^5$ A/cm^{-2}. A new technique called inclined substrate deposition has been used for the deposition of textured oxide films onto randomly oriented metallic substrates by laser ablation [390, 391] or by electron-beam evaporation [389]. This method simply involves inclining the metallic substrate with respect to the laser plume-evaporation source.

YSZ is always used as substrate or buffer layer with polycrystalline and metallic substrates. The biaxially YSZ layers can be easily grown on metallic substrates by the above methods. The best TBCCO films grown on YSZ single-crystal [392, 393] and YSZ-buffered [386, 387, 394] substrates have been fabricated in reduced $P(O_2)$ which can lower the thalliation temperature and thus avoid the film–substrate interaction. Moreover, a combination LaAlO$_3$/CeO$_2$/YSZ buffer layer was designed to avoid such interaction [395].

3.4.3. Microstructure

Microstructure has an important effect on the electronic properties of HTSC thin films. All the oxide films have a tendency for textured growth due to their strong directional bonding. Like all cuprate superconductors, the rate of growth in the c-axis direction is slower than that in a- and b-axis directions due to the strong anisotropy. Therefore, these materials will intrinsically adopt a platelike morphology. Good lattice match to the a–b plane will promote the growth with the c-axes normal to the substrates.

During the deposition of Tl(2212) and Tl(2223) films on LaAlO$_3$, the intergrowth of different numbers of Tl–O or Cu–O layers with such substrates was observed [396]. The films typically consisted of atomically flat terraces separated by growth steps of specific heights which were not always simply multiples or fractions of the unit cell height. Different from YBCO films, screw dislocations have not been observed on the surface of TBCCO. The Tl(2212) and Tl(2223) films were thought to grow in a simple layered manner. Figure 37 shows the microstructures of phase-pure Tl(2212) and Tl(2223) thin films on (001) MgO scaned by SEM [56]. For the Tl(2201) films in which some screwlike structures have been observed, the screwlike features were attributed to the disruption of simple layered growth by the roughness of the precursor film, inclusions of secondary phases, and an intrinsic feature.

3.4.4. Superconducting Properties

The superconducting properties can be mainly identified by the superconducting transition temperatures (T_c), the width of the transition (ΔT_c), the critical current density (J_c), and surface resistance (R_s). In the Tl-based HTSC thin films T_c and ΔT_c are related to the phase purity and the carrier density as other cuprate superconductors are. The carrier density is determined by doping and thallium loss. Apart from these, the intergrowths of secondary phases also have an effect on T_c and ΔT_c.

A high critical current density (J_c) is important for the application of these superconductors. In general, a high density of flux pinning and a low density of grain boundaries are needed to improve the critical current density (J_c). The epitaxial quality (in-plane alignment) of the film controls the grain boundaries and the lattice defects dominate the strength of flux pinning. Point defects, dislocations, and impurity phases are generally considered to contribute to flux pinning in HTSC materials and are necessary for very high values of J_c to be achieved [56]. It was found that J_c decreases with increasing film thickness due

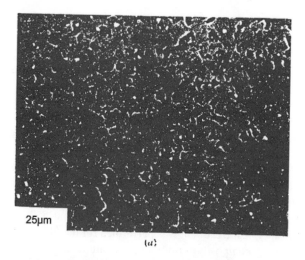

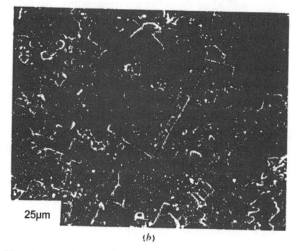

Fig. 37. Scanning electron micrographs showing the microstructure of phase-pure (a) Tl(2212) and (b) Tl(2223) thin films on (001) MgO [56].

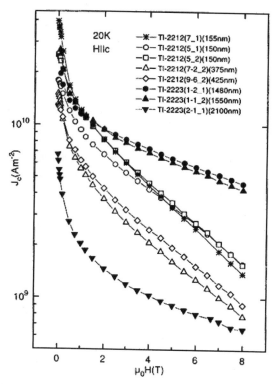

Fig. 38. Thickness dependence of the critical current density in Tl-2212 and Tl-2223 in magnetic fields up to 8 T at 20 K [396]. Reprinted from *Physica C* 268, 307 (1996), with permission from Elsevier Science.

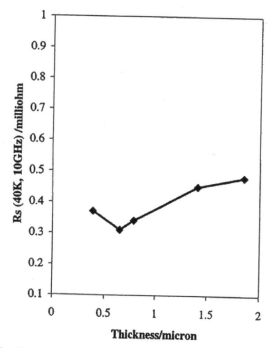

Fig. 39. The variation R_s (10 GHz, 80 K) with film thickness for Tl(2212) thin films on (001) LaAlO$_3$ substrates [56].

to the formation of stacking faults and low-angle grain boundaries as the film thickness increased [396], which was shown in Figure 38. For the TBCCO films, the *ex situ* thalliation process results in a highly faulted structure, which is beneficial to improve the flux pinning. Other methods were used to achieve high-density defects in TBCCO thin films, such as chemical substitution [397, 398], heavy-ion irradiation [399, 400], and thermal generation of stacking faults [401, 402]. However, the lattice defects will decrease the T_c and broad the transition width.

Surface resistance R_s is an important quantity for the application in microwave and other electronic applications. To achieve lower R_s, a film must consist of large, flat well-connected plates and a smooth surface with few nonsuperconducting secondary phases. The surface resistance R_s is a function of field penetration and the thickness (d) of the film (Fig. 39) [56, 403, 404]. When the field penetration depth is larger than the thickness, a leakage of field will exist; thus R_s will increase. For the HTSC thin films such as TBCCO, when the temperature is higher than T_c, the field penetration depth as far as the skin depth is larger than the ordinary thickness of the

films. While at $T < T_c$, the field penetration as far as the superconducting penetration depth (λ) is smaller than the typical thickness of the film. Therefore a lower R_s was achieved in the superconducting TBCCO films. Increases in λ for both small

(<0.3 μm) and large (>1 μm) thicknesses with a concomitant increase in the measured values of R_s were found [405].

3.4.5. Conclusion

Tl-based high-T_c thin films have a wide application potential in Josephson junction devices and passive microwave devices due to their high J_c and low R_s. For Tl(2201) thin film, due to its simple structure, a tetragonal crystal structure with one Cu–O plane between its two Tl–O layers, the fundamental studies of the superconductivity in this system are conducted [406–408]. However, the toxicity of Tl has limited the number of groups working actively on the fabrication of Tl-HTSC compound thin films.

3.5. Hg-Based Cuprate Thin Films

3.5.1. Introduction

Due to the highest T_c so far and the ability to sustain high current densities, the Hg-based high-T_c cuprate (HgBa$_2$Ca$_{n-1}$-Cu$_n$O$_{2n+2+\delta}$, HBCCO) has attracted considerable interest. Hg-1201 ($n = 1$), Hg-1212, and Hg-1223 are of tetragonal structure with space group being P4/mmm. Their critical temperatures T_c are around 30, 112, and 134 K, respectively. Despite the difficulties associated with the highly volatile nature of the Hg-based compounds, high-quality Hg-1212 and Hg-1223 thin films have been successfully fabricated. Current densities (J_c's) as high as $J_c \geq 10^7$ A/cm^{-2} at 5 K and $J_c \sim 10^5$ A/cm^{-2} at 110 K were achieved and the highest operational temperature of the SQUIDs made of the Hg-1212 thin films is 112 K [409–413]. The Hg-1223 film is the most promising for applications due to its high T_c of 134 K and high current carrying capability even at the high temperature of 120 K.

Many efforts have been paid on fabricating HBCCO films [20, 409–437]. HBCCO films were essentially made by two-step synthesis, i.e., annealing amorphous precursor films in Hg-contained atmosphere. The difficulties of synthesizing HBCCO films are threefold [413]. First of all, it is hardly possible to accurately control the processing parameters, such as Hg-vapor pressure, due to the highly volatile nature of the Hg-based compounds. This results in typically multiple superconducting phases plus a significant amount nonsuperconducting impurities in HBCCO samples, which substantially degrades the sample quality. Second, Hg vapor reacts with most metals as well as oxides, which prohibits epitaxial growth of HBCCO films on most technologically compatible substrates. Even on a few chemically stable substrates such as SrTiO$_3$, serious chemical diffusion in the film/substrate interface was observed. Consequently, most HBCCO films have to be made with large thickness ($\sim\mu$m) and most of them are c-axis-oriented uniaxial films with rough surface. Finally, the precursors are extremely sensitive to the air.

3.5.2. Fabrication of HBCCO Films

3.5.2.1. In Situ Preparation

Mizuno et al. [414, 415] reported the *in situ* preparation of HgBa$_2$CuO$_4$ (Hg1201) thin films as a preliminary work for *in situ* preparation of HBCCO films.

Hg1201 thin films were fabricated by rf magnetron sputtering which was carried out in a gas mixture of argon and oxygen of 0.6 Pa at various oxygen partial pressures. The Hg$_{2.5}$Ba$_2$Cu$_1$O$_{5.5}$ target was made by mixing HgO and Ba$_2$CuO$_3$ powders and pressing the mixture onto a copper plate. (100) SrTiO$_3$ single crystals were used as substrates. During the deposition, the substrate temperature was held at 550–600°C. No further heat treatment was performed for the films.

The Hg composition ratio and crystalline of the films was found to be related to the substrate temperature and the sputtering gas. Hg did not remain in the films when the substrates were heated at 600°C or higher. While when the substrate temperature was lower than 500°C, Hg was included in the film; however, the crystallinity of the films was bad. 550°C was nearly the optimum temperature of the substrates to obtain Hg contained films and good crystallinity. Although the Hg-1201 crystalline structure could be obtained despite a trace amount of Hg, the films were unstable and decomposed in the air within a few days. In addition, it was found that the Hg content in the deposited films was closely related to the sputtering gas, especially to the oxygen partial pressure. Hg remained in the film when the partial pressure of oxygen was lower. The optimum oxygen partial pressures to obtain good crystallized thin films were in the range from 0.1 to 0.01 Pa.

The lattice parameter c of in situ grown Hg-1201 films is ranged from 9.45 to 9.65 Å, which depends on the oxygen partial pressure. The more oxygen was introduced in the sputtering gas, the shorter the c-axis length was, finally resulting in more instably for the film. On the other hand, the resistivity in the normal state of the Hg-1201 films is fairly small and exhibits metallic behavior. But the superconducting transition is not so sharp and $T_{c(\text{onset})}$ ranged from 40 to 75 K. The maximum zero resistance temperature T_{c0} value was 40 K.

3.5.2.2. Two-Step Synthesis

A two-step synthesis is used to obtain HBCCO films: first, select and preparing precursor film and second, anneal the precursor film in Hg vapor. According to the type of precursor, this synthesis can be categorized into two classes, the amorphous type and the predesigned structure type. The former is named the conventional process while the latter is the so-called cation-exchange technique. The comparison of these two processes is schematically shown in Figure 40 (Fig. 1 in [413]). In a conventional process, object film is fabricated via reaction of simple-compound or amorphous precursor. In contrast, the film is formed by replacement of weakly bonded cation b in a precursor matrix by a volatile cation a in a cation-exchange process.

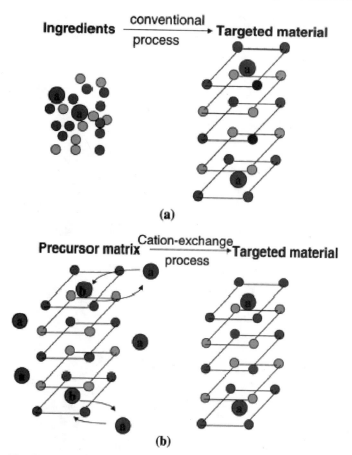

(a)

(b)

Fig. 40. Schematic description of syntheses of HBCO films of the conventional process (a) in which the object film is fabricated via reaction of simple-compound or amorphous precursor and the cation-exchange process (b) in which the film is formed by replacement of weakly bonded cation b in a precursor matrix by a volatile cation a [413].

Cation-Exchange Method. Wu et al. [413, 417–419] developed the cation-exchange technique on synthesis of high-quality c-axis-oriented HBCCO thin films. Tl-1212 was selected as a matrix for Hg-1212 since they have nearly the same crystalline structure while Tl1212 is much less volatile, insensitive to air, and thus easy to grow epitaxially on many single-crystal substrates.

Nonsuperconducting precursor Tl-1212 was sputtered by dc-magnetron sputtering from a pair of superconducting Tl-1212 targets onto (001)LaAlO$_3$ single-crystal substrates in a mixture of Ar and O$_2$ gases (Ar/O$_2$ = 4/1) at total pressure of 20 mTorr. The as-deposited film was amorphous and its thickness is typically controlled in the range 50–200 nm. The as-grown films were annealed in 1 atm O$_2$ at temperatures ranging from 800 to 850°C. The Tl-1212 films were then sealed in a precleaned and evacuated quartz tube with bulk pellets of Ba$_2$Ca$_2$Cu$_3$O$_x$ and HgBa$_2$Ca$_2$Cu$_3$O$_x$ and annealed in Hg vapor at typically 760–780°C for 3 to 4 hours to form Hg-1212 films. After the sintering, the films were further annealed at 300°C in a flowing O$_2$ atmosphere for 1 h to optimize the oxygen content of the films. Using this method, both thick Hg-1212 film (thickness > 200 nm) with T_c as high as 120 to 124 K and thin

Hg-1212 film (thickness < 100 nm) with T_c as high as 118 K are obtained.

Conventional Process. Kang et al. [420–422] reported the fabrication of high-quality c-axis-oriented Hg-1223 thin films using stable RE$_{0.1}$Ba$_2$Ca$_2$Cu$_3$O$_x$ (RE—rare earth) precursor. The precursor Ba$_2$Ca$_2$Cu$_3$O$_x$ thin films with thickness of ~1 μm were deposited on (100)SrTiO$_3$ substrates at a room temperature by PLD. After fabrication, the precursor film was put into a quartz tube together with a pressed RE-1223 precursor (fabricated by a conventional solid-state reaction method with RE:Ba:Ca:Cu = 0.1:2:2:3) and HgRE$_{0.1}$Ba$_2$Ca$_2$Cu$_3$O$_x$ precursor. The quartz tube was immediately evacuated to 10^{-6} Torr and sealed. In the typical sintering procedure the sample was fast heated to 750°C in 1 h, slow heated to 850°C in 30 min, held at this temperature for 1 h, and then furnace cooled to room temperature. Oxygen annealing was found to improve the superconductivity of Hg-1223 films. To optimize the oxygen concentration, the films were later annealed at 340°C for 12 h under flowing O$_2$ and T_c as high as 133 K for the Hg-1223 film was achieved.

Thin (Hg, Pb)-1212 films were also successfully grown on (001) MgO single crystal substrates by Plesch et al. [423]. First of all, Ba–Ca–Cu–O or Pb–Ba–Ca–Cu–O precursor films with thickness of 0.1 μm were deposited on MgO substrate by a sequential evaporation method. During the deposition, the substrate was kept either at 700°C or at room temperature. Then *ex situ* vacuum annealing of the precursor film was performed at 700°C at partial pressure of 10^{-6} Torr of dry oxygen. Compacted pellets of a mixture of HgO, PbO, BaO, CaO, and CuO in a stoichiometric ratio of 1.2:0.3:2:2:3 served as the Hg source. The precursor film was inserted between two such pellets in close contact and then put into a short alumina tube, which was then placed in a silica tube. The tube was purged with argon, evacuated, and sealed. All samples were annealed at 800°C for 60 min. For both the Pb-free and Pb-doped Hg-1212 films, T_c onset was about 132 K and the zero resistance temperature T_{c0} was 105 and 117 K, respectively.

C-axis-oriented (Hg, M)-12($n-1$)n (M = RE, Mo; $n = 2, 3$) thin films [424] were fabricated on (100) SrTiO$_3$ substrates. Amorphous HgO/M$_{0.1}$Ba$_2$CaCu$_2$O$_y$ multilayer precursor films were prepared by PLD. The typical thickness of each layer was approximately 2.5 and 3.5 nm, respectively. The precursor film with total multilayer thickness of 400 to 700 nm and a 50 nm thick HgO cap layer was sealed with two unreacted pellets in an evacuated quartz tube and heated at 650 to 775°C for 4 to 6 h. Some fabricated films were further postannealed in oxygen at 320°C for 24 h to adjust their oxygen content. The (Hg, RE)-1223 film with T_{c0} of 127.5 K and (Hg, RE)-1212 film with T_{c0} of 115 to 122 K were obtained.

Using a similar process, Gupta et al. [409, 410] deposited sequential layers of HgO and Ba$_2$CaCu$_2$O$_x$ on (100)SrTiO$_3$ substrates from two separated targets at room temperature and annealed the precursor films subsequently at ~800°C to produce the superconducting phase. The resulting c-axis-oriented

Hg-1212 film has a thickness of $\sim$300 nm and the highest T_{c0} achievable is 125.2 K.

Yun et al. [425, 426] reported the growth of a-axis-oriented Hg-1212 thin films on (100) LaAlO$_3$ substrate. The Hg-free precursor film (Ba$_2$CaCu$_2$O$_x$) was placed in a quartz tube together with a nonreacted stoichiometric HBCCO-1212 pellet as a mercury source and two stoichiometric Ba$_2$CaCu$_2$O$_x$ pellets; then an end of the quartz tube was closed and the other end narrowed to a small opening. The tube was put into a larger quartz tube which was evacuated. In this configuration, Hg vapor can be effectively confined around the precursor film, leading to a local high Hg-vapor pressure in a short annealing time. The samples had a heating rate of 50°C/min, avoiding the formation of CaHgO$_2$ impurities; then the samples were subsequently annealed for 5 min and cooled slowly to room temperature. The films with the predominantly a-axis-oriented grains were acquired from quenching at 700°C, exhibiting a $T_{c0} > 120$ K.

Table IX lists the synthesis condition, the precursor, and the physical parameters of HBCCO films mentioned in the text.

3.5.3. Surface Morphology of HBCCO Films

The physical properties of the film are also closely related to its surface morphology which is thus one of the factors for qualifying the films. SEM is useful for investigating the surface morphology of HBCCO films. Some typical SEM images of HBCCO films are presented here [436].

Figure 41 (Fig. 3 in [436]) shows the SEM images of $\sim$1 μm thick Hg-1212 films grown on (100) SrTiO$_3$ with T_{c0} of 121 K. Three distinct morphologies in the overlayer were recognized in the backscattered electron images: a granular structure adjacent to the substrate surface is observed as shown in Figure 41a, needlelike crystals are shown in Figure 41b, and a woolly object can be observed on the film surface (Fig. 41c). It is suggested that the composition of the grains close to the substrate surface consists of Hg-1212 superconductor, whereas the woolly object is probably a precipitated Hg-compound.

Figure 42 (Fig. 4 in [425]) is the SEM image of a-axis Hg-1212 film grown on (100) LaAlO$_3$ substrate with T_{c0} of 120 K. Heavy twinning observed by SEM is thought to result in the lengthening of the current paths and in an increase of the grain boundary resistances; hence such film displays less metallic behavior than the c-axis film in the normal state.

Figure 43 (Fig. 4 in [420]) is the SEM image of c-axis Hg-1223 film grown on (100) SrTiO$_3$ substrate with T_{c0} of 131 K. The well connected and irregular platelike shape grains in background can be observed. A small amount of bright grains which are scattered on or partially immersed on the surface can also been seen.

3.6. Infinite CuO$_2$ Layer Thin Films

3.6.1. Introduction

It is known that the well studied high-T_c superconducting cuprate compound has a finite number of CuO$_2$ layers which play a key

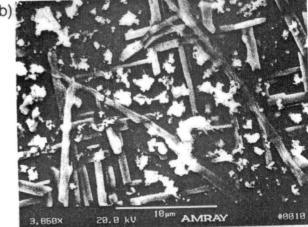

Fig. 41.   SEM images of the annealed Hg-1212 films: (a) granular structure of the layer close to the substrate surface, (b) needlelike crystals on the film, (c) woolly object on the film surface [436]. Reprinted from *Physica C* 276, 277 (1997), with permission from Elsevier Science.

role for the occurrence of high-T_c superconductivity. Therefore, the effect of both number and intercoupling of the CuO$_2$ layers on high-T_c superconductivity is a subject of wide interest.

In 1988, Roth and collaborators discovered the compound called the "all layer phase" or the "parent structure" of an all copper-oxide superconductor [438]. The first one is SrCuO$_2$

Table IX. Parameters on the Synthesis Condition, the Precursor, and Physical Properties of HBCCO Films Made by Several Groups

Phase	T_{c0} (K)	T_c (onset) (K)	J_c (110 K) (A/cm^{-2})	Thickness (nm)	Substrate (100)	T_a (°C)[a]	Precursor	Growth method — Two-step process: Method	T_s (°C)[b]	T_a (°C)[c]	C (Å)	Reference
c-axis 1201[e]	40	75	NG	400	STO	550–600	—	—	–	–	9.45–9.65	[414]
c-axis 1212	125.2	NG	10^5	300	STO	800	HgO + Ba$_2$CaCu$_2$O$_x$	PLD	RT	–	NG	[409, 412]
c-axis 1212	120–122[d]	NG	1 M	260	LAO	760–780	Tl-1212	Sputtering	NG	800–850	12.7	[413]
c-axis 1212	118[d]	NG	9.1×10^4	<100	LAO	760–780	Tl-1212	Sputtering	NG	800–850	NG	[417]
c-axis 1212	105–117	132	NG	200	MgO	800	Ba$_2$Ca$_2$Cu$_3$O$_x$	Evaporation	700 or RT	700	12.68–12.69	[423]
c-axis (Hg,Re) 1212	115–122	NG	5×10^6 (77 K)	NG	STO	650–775	HgO + M$_{0.1}$Ba$_2$CaCu$_2$O$_x$	PLD	NG	–	NG	[424]
c-axis (Hg,Re) 1223	127.5	NG	1.5×10^6 (77 K)	NG	STO	320 in O$_2$	HgO + M$_{0.1}$Ba$_2$CaCu$_2$O$_x$	PLD	NG	650–775	NG	[424]
c-axis 1223	131	133	2.6×10^5 (120 K)	NG	STO	340 in O$_2$	Ba$_2$Ca$_2$Cu$_3$O$_x$	PLD	RT	750–850	15.64–15.70	[420, 421]
a-axis 1212	>120	NG	NG	NG	LAO	700	Ba$_2$CaCu$_2$O$_x$	PLD	NG	–	a-axis: 3.86	[426]

[a] T_a—last annealing temperature for obtain HBCCO film.

[b] T_s—substrate temperature during depositing precursor.

[c] T_a—annealing temperature of as-prepared precursor film.

[d] From dc magnetization.

[e] In situ grown; RT—room temperature.

555

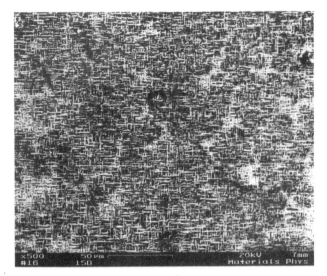

Fig. 42.   SEM image for an *a*-axis-oriented Hg-1212 thin film [425].

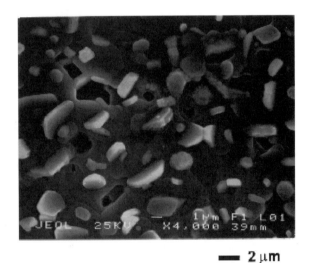

— 2 μm

Fig. 43.   SEM image of Hg-1223 films grown on STO [420].

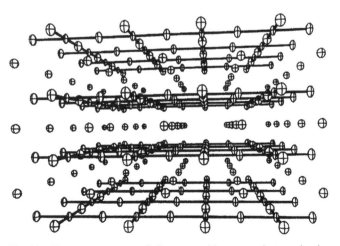

Fig. 44.   The parent structure of all copper oxide superconductors, showing infinite CuO_2 square-net layers separated by alkaline earth atoms [439].

which consists of alternating Sr^{2+} and $[CuO_2]^{-2}$ layers. Therefore, it developed compounds with common structural forms of CuO_2 planes separated by alkaline earth atoms, in the stacking sequence CuO_2–A–CuO_2–A···, as shown in Figure 44 [439]. After then, the $Sr_{1-x}Nd_x CuO_2$ and $(Ba, Sr) CuO_2$ have been synthesized, respectively [440, 441]. These oxide superconductors are unlike the well-known cuprate compounds. They must be synthesized through high-pressure (20–65 kbar) and high-temperature ($\geq 1000°C$) annealing. This is followed by a quenching process, and usually the multiphase samples are obtained by this route. Thin film growth processing is one kind of powerful technique for the synthesis of metastable phase material, such as A15 structural Nb_3Ge thin films [442] and B1 structural MoN_x thin films [443]. So thin film processing is very reasonably attempted to grow infinite CuO_2-layer phase compounds. Many groups have exerted effort in the preparation of infinite-CuO_2-layer thin films by using laser-ablation, laser molecular beam epitaxy [444–446], and rf-magnetron sputtering methods [447–449] and more recently the MBE technique [450].

The superconductivity was observed from resistivity or susceptibility measurements for $Sr_{1-x}Nd_xCuO_2$ thin films [445, 449, 451, 452]. Below, the growth processing for infinite layer compound thin films and related characterizations and physical properties are introduced.

3.6.2. Preparation Processing and Properties

3.6.2.1. $SrCuO_2$ Films and $SrCuO_2/Ca CuO_2$ Superlatices

The common preparation processing for infinite layer compound thin films is PLD. It is well known that PLD is a reliable technique for the growth of high-quality epitaxial $YBa_2Cu_3O_{7-\delta}$ thin films and other high-T_c superconductor thin films. It is also suggested that it be used to grow meatsable infinite layer phase thin films. The $SrCuO_2$ infinite layer phase thin films have been fabricated by PLD [444, 445, 451]. More recently, the $SrCuO_2/CaCuO_2$ infinite layer superlattices were prepared and characterized by Del Vecchio et al. [453]. For this process the target were made starting with a stoichiometric mixture of high-purity $SrCO_3$, $CaCO_3$, and CuO_2 powders. They were calcimined in air at 900°C for 24 hours, pressed form a disk, and finally heated at 920°C for 12 hours. (001) $SrTiO_3$ substrate is used for deposition. For depositing superlattices of $SrCuO_2$ and $CaCuO_2$, both targets were mounted and controlled by an external computer to realize layer-by-layer growth. The light source is XeCl excimer laser ($\lambda = 308$ nm) with a pulse length of 10 ns; the power density of the laser beam on target is several Joules per cm^2. During the growth process the oxygen pressure and substrate temperature were 0.2 mbar and 580°C, respectively. After deposition the films were cooled down to room temperature in about 10 minutes in an oxygen pressure of 300–400 mbar.

From the high-resolution rocking curve and the reciprocal space mapping, it is shown that the $SrCuO_2$ layers are unstrained while a tetragonal distortion in the strained $CaCuO_2$

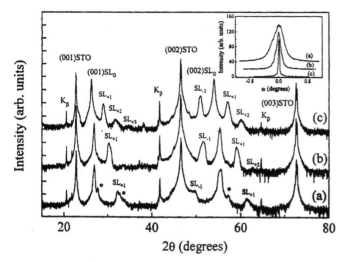

Fig. 45. Experimental θ–2θ X-ray scans of the $(3 \times 2)_{61}$ (a), $(5 \times 3)_{32}$ (b), and $(7 \times 3)_{21}$ (c) IL superlattice. The STO substrate and superstructure peaks (satellite peaks) are observed. The inset shows the X-ray rocking curve (ω-scan) of the (002) SL reflection. The FWHM result is $0.37°$, $0.12°$, and $0.09°$ for the $(3 \times 2)_{61}$ (a), $(5 \times 3)_{32}$ (b), and $(7 \times 3)_{21}$ (c) superlattices, respectively [453]. Reprinted from *Physica C* 288, 71 (1997), with permission from Elsevier Science.

compound occurs. The strain relaxation generates mismatch dislocations at the $SrTiO_3/Sr\,CuO_2$ interface, which causes a tilt of the whole superlattice.

The θ–2θ X-ray showing layer-by-layer growth of super-lattices and third grade satellite peaks can be observed in Figure 45.

$Ca_{1-x}\,Sr_xCuO_2$ films. $Ca_{1-x}\,Sr_xCuO_2$ has a relatively supple, layered structure consisting of CuO_2 planes separated by planes of alkaline earth elements. This material can be made superconducting either by electron doping [454] or by hole doping [441, 455]. Norton et al. [456] have prepared single-crystal $Ca_{1-x}Sr_xCuO_2$ thin films by pulsed laser depositions. The KrF exciter laser beam was focused to a horizontal line and scanned over a $\sim$25-mm-diam stoichiometric $Ca_{1-x}Sr_xCuO_2$ rotating target which was made by solid-state reaction of high-purity $CaCO_3$, $SrCO_3$, and CuO_2 powders at $1025°C$. (100) $SrTiO_3$ substrates were used. During deposition the oxygen pressure is 200 mTorr; the laser beam power density on target is 3–5 J/cm^2. The substrate temperature for deposition is $\sim$700°C. After deposition, the films were cooled down in 200 mTorr oxygen at rate of $10°C/min$.

X-ray diffraction shows only [00l] reflections correspond-ing to c-axis orientation for $Ca_{0.3}Sr_{0.7}CuO_2$ thin films. The impurity peaks are almost 10^{-3} as intense as the allowed $Ca_{0.3}Sr_{0.7}CuO_2$ reflections. The rocking curve through the (002) peak for the $Ca_{0.3}Sr_{0.7}CuO_2$ film indicates a full width at half maximum (FWHM) of $0.14°$. The rocking curve width does not change significantly as the Sr content is varied. A ϕ scan for the (202) peak shows that the film has a tetragonal structure with complete in-plane alignment with the substrate. The X-ray diffraction data of $Ca_{0.3}Sr_{0.7}CuO_2$ can be seen in

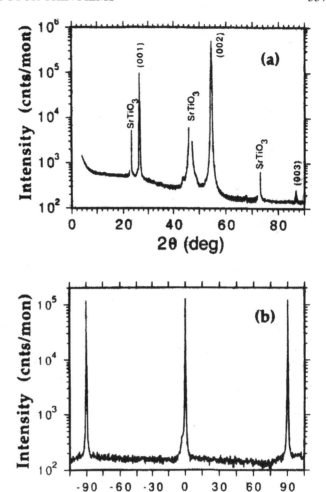

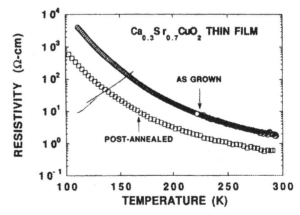

Fig. 46. X-ray diffractometry (Cu $K\alpha_1$ radiation) data for a $Ca_{0.3}Sr_{0.7}CuO_2$ thin film showing (a) the θ–2θ scan near the surface normal and (b) the Φ scan of the (202) peak [456].

Fig. 47. Resistivity as a function of temperature for a $Ca_{0.3}Sr_{0.7}CuO_2$ thin film grown at 700°C in 200 mTorr O_2 both before and after postannealing in 1 atm O_2 at 550°C for 1 h [456].

Figure 46. The temperature dependence of the resistively of such $Ca_{0.3}Sr_{0.7}CuO_2$ thin films shows semiconductorlike be-havior as shown in Figure 47. Films with other Sr contents grown at the same temperature (~700°C) showed similar characteristics. Such high resistivity can be thought to be consistent with the common feature that the flat and perfect CuO$_2$ planes in infinite layer film do not benefit from transport and superconductivity. In the case of postannealing of the film in oxygen at 550°C, a slight contraction of the c-axis was observed, and resistivity decreases. This implys that the enhancement of interply of CuO$_2$ planes under a decrease of c-axis length is an efficient way to decrease the resistivity even for creation of superconductivity.

In fact, the observed superconductivity in infinite layer compound thin films as well as in the bulk samples is generally associated with the presence of various defected layers acting as charge reservoirs [457]. Thus for providing necessary carriers the structural role of the CuO$_2$ planes is very important. It acts to accept the supplied charge carriers. The stress induced in a crystallographic lattice can either increase or decrease the possibility for the CuO$_2$ sheets to be optimally charged [458].

Vailionis et al. conducted a detailed investigation of the relation of charge carrier density of infinite layer films and the flatness of the CuO$_2$ plane. They found that the charge carrier density in the normal state is higher for films with the "buckled" CuO$_2$ planes compared to films exhibiting planar CuO$_2$ sheets [450].

The Ca$_{1-x}$Sr$_x$CuO$_2$ films used for their study were grown at temperatures of 550–600°C by oxide MBE on atomically flat (100) SrTiO$_3$ substrates to a thickness typically of 50 nm. The single-crystal substrates were prepared by extensive annealing in UHV at 1100°C, prior to film growth. They were characterized by RHEED and atomic force microscopy (AFM). Further RHEED patterns were taken during the Ca$_{1-x}$Sr$_x$CuO$_2$ film growth and they suggested that in the compositional range of $0 < x < 1$ highly ordered films were formed. Quantitatively analysis of the RHEED data gave cube-on-cube growth and the deduced a-axis lattice parameters of Ca$_{1-x}$Sr$_x$CuO$_2$ films were 0.39 nm in most cases. X-ray diffraction data show only (001) peaks resulting from a tetragonal infinite layer structure. Both RHEED and X-ray diffraction data indicate high-quality epitaxial growth of the Ca$_{1-x}$Sr$_x$CuO$_2$ films on atomically flat (100) SrTiO$_3$.

Due to the misfit between the substrate and the film they found that the tensile stress and compressive stress are induced for $x < 0.66$ and $x > 0.66$, respectively. For $x < 0.66$, the "shorter" Cu–O distance of 0.195 nm indicates elastical strain of film to match the substrate. For the film with $x > 0.66$, the observed "longer" Cu–O bonds of 0.198 nm suggest that the strain is possibly relieved by displacing the oxygen atoms from the CuO$_2$ plane by about 0.03 nm; the distorted CuO$_2$ sheet is shown in Figure 48.

Correspondingly, the behavior of normal-state resistance has been observed for the samples with "planar" and "buckled" CuO$_2$ planes. It is shown that the conductivity at 300 K is significantly higher for the "buckled" film. The observed lower conductivity, i.e., the lower hole density in the "planar" films, is possibly related to the tensile strain. This affects the Cu–O bonds but does not change the number of electrons in the

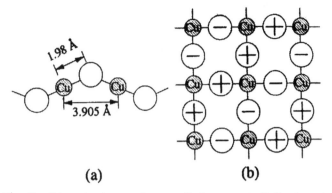

Fig. 48. Schematic pictures of oxygen displacements in CuO$_2$ planes in Ca$_{1-x}$Sr$_x$CuO$_2$ film: (a) side view of the CuO$_2$ sheet, (b) correlated displacement model reported by Toby et al. The (+) and (−) signs indicate up and down oxygen displacements from the CuO$_2$ plane [450].

bonds. A sufficient hole doping of such film requires shortening of the Cu–O bonds, but the tensile stress frustrates the shortening of the Cu–O bond. So the higher normal-state conductivity for samples with $x > 0.66$ indicates higher charge carrier density in infinite films with buckled CuO$_2$ plane. This may give the conclusion that the perfect and flat CuO$_2$ plane may be the main reason for poor normal-state conductivity and nonsuperconductivity for the infinite layer phase compound. Introducing the distortion in the CuO$_2$ plane possibly leads to high-T_c superconductivity.

3.6.2.2. Sr$_{1-x}$Nd$_x$CuO$_2$

Among the infinite layer phase compounds, Sr$_{1-x}$Nd$_x$CuO$_2$ is the one for which the c-axis length can be adjusted to be smaller by doping Nd (the ion radius of Nd^{3+} cation = 0.111 nm, radius of Sr^{2+} = 0.126 nm) to increase the carrier density in the CuO$_2$ plane. Niu and Lieber [445] prepared Sr$_{1-x}$Nd$_x$CuO$_2$ films by laser ablation. The condition of growth is similar to the other kinds of infinite layer phase compounds mentioned above. The Sr$_{1-x}$Nd$_x$CuO$_2$ targets were prepared by solid-state reaction of stoichiometric mixtures of SrCO$_3$, Nd$_2$O$_3$, and CuO at 950°C in air. The target material mainly consists of an orthorhombic structure, from which the deposited thin films are tetragonal c-axis-oriented infinite layer Sr$_{1-y}$Nd$_x$CuO$_2$ with the c-axis perpendicular to the (100) SrTiO$_3$ substrate surface. The values of the c-axis calculated from the X-ray diffraction data are 3.35, 3.34, 3.31, and 3.27 Å for the $x = 0$, 0.08, 0.16, and 0.24 materials, respectively. The decrease of c-axis lattice constant with increasing x is consistent with the substitution of the smaller Nd^{3+} cation for larger Sr^{2+}.

The electrical properties of the Sr$_{1-x}$Nd$_x$CuO$_2$ thin film show a strong x dependence. $x = 0$, SrCuO$_2$ is an insulator, and the thin films with $x = 0.08$, 0.16, and 0.24 show that the resistivity changes systematically with increasing x. For $x = 0.16$ and 0.24 the thin films show indications of superconductivity as shown in Figure 49, The systematic changes of electrical properties with increasing Nd^{3+} concentration are consistent with electron doping of the infinite layer material.

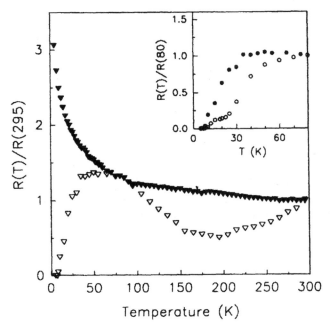

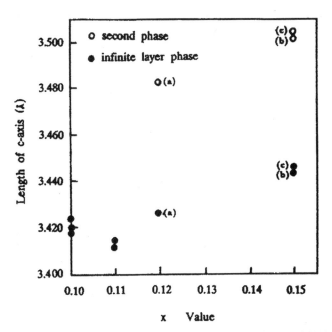

Fig. 49. Normalized resistance vs temperature curves recorded on $Sr_{0.92}Nd_{0.08}CuO_2$ (▼) and $Sr_{0.84}Nd_{0.16}CuO_2$ (▽) films. Conduction is activated up to 300 K in the $x = 0.08$ film but is metallic above 175 K in the $x = 0.16$ material. The inset shows the transition to a zero resistance state in $Sr_{0.84}Nd_{0.16}CuO_2$ (•) and $Sr_{0.76}Nd_{0.24}CuO_2$ (○) [445].

Fig. 51. The dependence of c-axis length on x value. (a)(a), (b)(b), and (c)(c) indicate the same thin film sample respectively [449].

Zhao et al. also have made investigation of $Sr_{1-x}Nd_xCuO_2$ thin films with various Nd contents [449]. The $Sr_{1-x}Nd_xCuO_2$ thin films for this investigation were prepared by rf-mangnetron sputtering. The $Sr_{1-x}Nd_xCuO_2$ targets were prepared by a technique similar to that described above. The rf-magnetron sputtering was performed under the following conditions. The argon pressure was 0.5–1 Pa; the (100) $SrTiO_3$ substrates were used under a deposition temperature of 650–700°C. X-ray diffraction analysis indicates that a single infinite layer phase $Sr_{1-x}Nd_xCuO_2$ with its characteristic (002) peak can be obtained for x values up to 0.11–0.115 as shown in Figure 50. This is different from that of [452] in which only a single phase can be obtained when the Nd concentration was less than 0.10. This difference may be attributed to the sputtering conditions. When $x = 0.12$, however, another *(002) peak appeared as shown in Figure 50c, with a further increase in x. The *(002) peak became stronger and even became a main one in the thin film with $x = 0.15$ as seen in Figure 50d. The *(002) peak comes from a long c-axis tetragonal phase which is induced by the extra doping of Nd. The lengths of the c-axis for the thin films with various Nd concentration are shown in Figure 51. This phenomenon mainly implies that when $x > 0.12$, the infinite layer phase tends to be unstable, and the thin film sample decomposes into two phases, the infinite layer phase and a second phase. So it can be suggested that the c-axis length has a minimum value in which the thin film is in a single infinite layer phase with a maximum concentration of Nd. It may be possible to find a way to grow single infinite layer compound $Sr_{1-x}Nd_xCuO_2$ thin film with smallest c-axis length and highest charge carrier concentration in the CuO_2 plane. Consequently, it is very favorable for the superconductivity of infinite layer CuO_2 phase thin films.

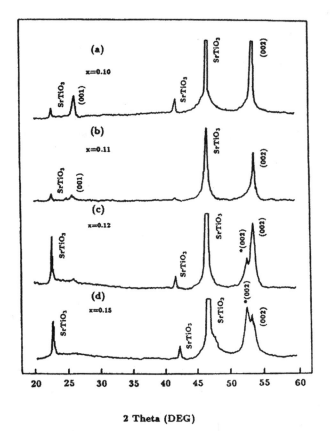

Fig. 50. The dependence of phase structure of $Sr_{1-x}Nd_xCuO_2$ thin films on the concentration of Nd determined by X-ray diffraction using Cu K radiation. (a) $x = 0.10$, (b) $x = 0.11$, (c) $x = 0.12$, and (d) $x = 0.15$ [449].

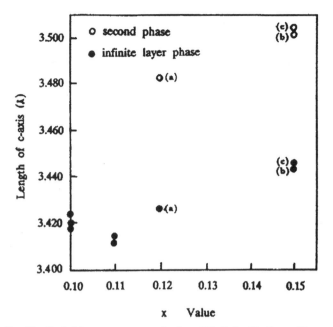

Fig. 52. Resistivity vs temperature for $Sr_{1-x}Nd_xCuO_2$ thin films with various concentrations x of Nd; the indication of superconducting transition was observed at 28 K. The inset shows the Meissner effect of the same sample [449].

The resistivity vs temperature for $Sr_{1-x}Nd_xCuO_2$ thin films with various concentrations x of Nd is shown in Figure 52. The drop of resistivity which comes from the superconducting transition is found at 28 K for $x = 0.11$. The inset shows the Meissner effect of the same sample obtained by magnetization measurements using the Quantum Design magnetization measurement system (MPMS-5) in a field of 50 Gauss.

So far, the investigations on infinite CuO_2 layer phase thin films may give the conclusion that the shorter c-axis length caused by substitution of a smaller radius ion for a larger one (e.g., in $Sr_{1-x}Nd_xCuO_2$) and "buckled" CuO_2 planes induced by oxygen displacement from CuO_2 plane along the c-axis (e.g., in $Ca_{1-x}Sr_xCuO_2$) are favorable for higher charge carrier density and consequently lead to higher conductivity and superconductivity.

3.7. $Ba_{1-x}K_xBiO_3$ (BKBO) System Thin Films

3.7.1. Introduction

Cubic bismuthates of $Ba_{1-x}K_xBiO_3$ (BKBO) exhibited the highest superconducting transition temperature ($T_c = 30$ K with $x \approx 0.4$) for copper-free superconductors before the discovery of MgB_2. Their unusual feature is the absence of two-dimensional metal–oxygen planes, which are widely believed to be an essential factor for the high transition temperature in cuprate systems. Many questions unresolved for the cuprates have clear answers for BKBO though BKBO exhibits some characteristics similar to high-T_c cuprate superconductors. Superconductivity in this kind of compound appears at the boundary of a metal–insulator transition, similar to that of cuprate superconductors. It will not become cubic and superconduct-

ing when potassium content $x < 0.37$. Apart from the high T_c, BKBO is a kind of conventional superconductor with a BCS-like gap. It also has a large isotope effect value for oxygen as well as a relatively long coherence (4 to 5 nm), which are advantageous for device fabrication.

For developing applications, BKBO film of high quality is first required. There are several methods for growing BKBO thin films, these are laser ablation, molecular beam epitaxy, coevaporation, sputtering, and liquid phase epitaxy. Various single-crystal substrates have been used to fabricate epitaxial BKBO thin film, such as MgO, $LaAlO_3$, $SrLaAlO_4$, $SrTiO_3$, Nb-doped $SrTiO_3$, $BaBiO_3$, etc. The as-grown films were usually oxygen deficient and required an additional oxidation anneal to achieve the maximum T_c.

3.7.2. Fabrication of BKBO Thin Film

3.7.2.1. BKBO Thin Film Fabricated by Laser Ablation

Laser ablation is a common method for growing BKBO thin films [459–461]. For this purpose a KrF excimer laser beam (350 mJ, 248 nm, 38 ns FWHM pulse duration) is one of the choices [459]. During the pulsed laser deposition of BKBO thin films, the focused laser power density was 1.5 to 3.0 J/cm^2.

Schweinfurth et al. and Moon et al. [460] reported single-phase BKBO films *in situ* grown in the (100) or (110) direction on MgO, $SrTiO_3$, $LaAlO_3$, and Al_2O_3 substrates. The transition temperature of BKBO films on (100) MgO substrate with zero resistance can be improved from 26 to above 28 K with a width of 0.3 K when raising the deposition temperature to 580–620°C in pure Ar at a pressure of 1 Torr, and oxygenating the films by soaking them for 12 min in 1 atm O_2 after the films are cooled to 410°C. The critical current densities can reach above 3×10^6 A/cm^2 at 4 K.

However, in order to avoid diffusion of potassium into the barrier layer, the deposition temperature should be as low as possible. Schweinfurth et al. [460] reported that when the substrates were heated in the temperature range of 350–450°C during the film deposition, the films grown on (100) $SrTiO_3$ exhibited T_c ($R = 0$) = 19.7 K with a transition width of 0.6 K, which is lower than that of the film grain at higher temperature ($\sim$600°C). Compared to the cuprate superconductor the physical property of the BKBO superconductor thin film strongly depends on the substrates. Although the 1.4% lattice mismatch for BKBO on (100) MgO is much smaller than the 10% mismatch for BKBO on (100) $SrTiO_3$, the transport properties for BKBO thin films grown directly on (100) MgO were poorer than those for films grown on (100) $SrTiO_3$. In most cases, the resistivity of BKBO films grown on (100) MgO is larger than that of BKBO film grown on (100) $SrTiO_3$; e.g., the ρ (30 K) of BKBO on MgO is at least a factor of 3 larger than for a film grown on (100) $SrTiO_3$ under the identical growth conditions. On the other hand, the oxidizer is also a very important for improving the quality of BKBO films. By introducing N_2O into the oxidation gases during *in situ* growth, Lin et al. [461] grew BKBO films on (100) MgO with substrate temperatures from

380 to 510°C with $T_{c(onset)}$ and $T_{c(zero)}$ in the ranges 30–21 and 28–16 K.

3.7.2.2. High-Oxygen-Pressure rf Sputtering

As mentioned before, the low-temperature growth of superconducting BKBO thin films is achievable by laser ablation, molecular beam epitaxy, coevaporation, and sputtering. However, postoxidation annealing in oxygen at temperature around 450°C is necessary for all these techniques to obtain the maximum T_c. By using a high-oxygen-pressure sputtering method, high-quality BKBO thin films on various substrates can be obtained without postannealing [462]. This can be seen from the following examples.

Using rf-magnetron sputtering under a gas pressure of 4–80 Pa with Ar:O$_2$ $\sim$ 1 (40–50%), the BKBO thin films deposited on (100) SrTiO$_3$ substrate at 400°C prepared by Iyori et al. showed a T_c of 16 K [462]. The BKBO films grown on (100) SrTiO$_3$ at 320°C by Sato et al. had a sharp T_c of 21 K (width $\sim$0.2 K) after oxygen annealing.

Under high oxygen pressure, Prieto et al. obtained BKBO thin films by *in situ* rf sputtering on different crystalline substrates. A T_c ($R = 0$) above 28 K is achieved, for which the transition width can be as small as 0.3 K, and resistivity ratios $\rho(300\ K)/\rho(30\ K)$ are in the range of 1.5 to 2.0. Critical current densities around 10^6 A/cm^2 at 4.2 K were also achievable. The brief preparing parameters are as follows. Rf sputtering at a frequency of about 7 MHz was used for depositing BKBO thin films on (001) SrTiO$_3$, (001) MgO, and (001) LaAlO$_3$ substrates. The sputtering gas was pure oxygen (99.998%) at unusually high pressures (3.0 to 3.5 mbar). The substrate temperature was varied from 400 to 440°C. The optimum target–substrate distance to produce smooth homogeneous films was 25 mm. The deposition rate was about 100 nm/h and typical film thicknesses ranged from 50 to 200 nm. After deposition, the chamber was flooded with pure oxygen at 1 bar, and the film was rapidly cooled down to room temperature within 10 min.

3.7.2.3. BKBO Thin Film Fabricated by Liquid Phase Epitaxy Technique

Liquid phase epitaxy [463, 464] is conducted in an electrochemical cell. BKBO films were grown at 240° for 5 h. On the stipulation of electrical conductivity of substrates and their small lattice mismatch with the deposited film, BaBiO$_3$ single-crystal-based substrates were chosen for electrochemical epitaxy of BKBO thin films. The substrate served as the anode. Two rods of 5 mm in diameter made of metallic bismuth served as cathode and reference electrode. The electrochemical deposition of the films was carried out in a potentiostatic regime at 0.7 V relative to the reference electrode. The initial current density was 3 to 5 mA/cm^2 and was reduced practically to a constant value of 0.4 to 1.7 mA/cm^2 depending on the composition of the substrate and that of the growing film. The average growth rate of films is of 1 to 4 μm/h. The primary formation of the film begins in the form of separated crystallites oriented

in the basal plane. Further growth process proceeds in the form of vicinal growth step formation.

Films deposited by this liquid phase epitaxy technique are all superconducting with T_c about 30 K when the potassium content x is between 0.33 and 0.50. The normal-state transport properties exhibit a metallic character of the resistivity temperature dependence with a $\rho(300\ K)/\rho(50\ K)$ ratio of more than 5 and $\rho(300\ K) = 1.5$ to 1.8 mΩ/cm. The as-grown BKBO films with thicknesses over 10 μm, prepared at the optimal anode current density, often exhibited superconductivity with fraction values close to 100%. The films 1 μm or thinner show initial growth stage character and the high film granularity caused the extremely low content of the superconducting phase.

Though not popular, MBE is also used for growing BKBO films. The AT&T Bell group [465] reported that (100) oriented BKBO films with good superconducting properties by MBE was achieved on (100) MgO substrate at 350°C followed by a postgrowth anneal in oxygen. The annealed films had superconducting transitions as narrow as 0.5 K at temperatures between 20 and 25 K. The critical current density could reach the range of 10^5 A/cm^2 at 4.2 K.

3.7.3. BKBO Film Related Junctions

3.7.3.1. BKBO/STNO (Nd-Doped SrTiO$_3$) All-Oxide-Type Schottky Junctions

BKBO is a suitable material for basic study of low-energy-type superconducting-base transistors for two primary reasons: (i) superconducting tunnel junctions with BKBO as an electrode have a clear superconducting-gap structure and act as an emitter/base junction where quasi-particles are injected into a base layer and (ii) epitaxial thin films can be grown at low temperatures (400 to 500°C).

BKBO/STNO heterojunctions were fabricated by depositing epitaxial BKBO thin films on (110) single-crystal STNO substrates [466]. The nominal Nb concentrations in raw materials were 0.01, 0.05, and 0.5 wt%. Before deposition, the substrates were treated with some methods in order to improve the surface conditions. After they were rinsed with acetone and ethanol in an ultrasonic bath, they were annealed in flowing O$_2$ at 1100°C for 1 h. The annealed substrates were transferred to a BKBO deposition chamber and thermally cleaned to eliminate the adsorbed contaminants by heating to 550°C for 40 min under an oxygen pressure of 3×10^{-3} Pa. After cleaning, the substrate was cooled to 400°C and BKBO epitaxial thin film was deposited using the rf-mangetron sputtering method. After a 100-nm Au film was deposited on the BKBO thin film, the Au/BKBO/STNO film was patterned into square junctions by conventional photolighography and ion milling. Aluminum wires were ultrasonically bonded to the STNO substrate to make ohmic contacts.

The growth and physical properties of BKBO films strongly depend on the surface feature of the substrates. Semilog plots of the current density–voltage (log $J - V$) characteristics of the BKBO/STNO (0.5 wt%) junctions at room temperature are

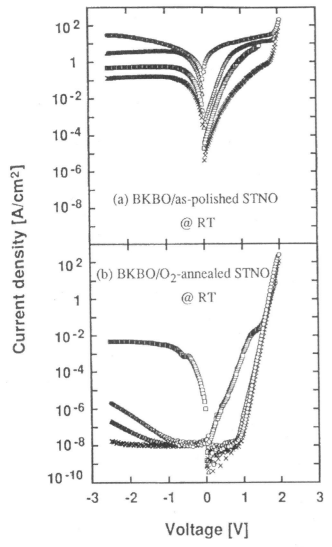

Fig. 53. Room-temperature log J–V characteristics of BKBO/STNO (0.5 wt%) Schottky junctions using (a) the as-polished STNO and (b) the O_2-annealed and thermally cleaned STNO. The junction sizes are 0.2 × 0.2 (circles), 0.3 × 0.3 (crosses), 0.5 × 0.5 (triangles), and 1.0 × 1.2 (squares) mm² [466].

lated to be 4.3×10^{17} cm⁻³. The $1/C^2$–V_a characteristics of the BKBO/STNO (0.05 and 0.5 wt%) junctions, on the other hand, exhibited nonlinear behavior. This anomalous behavior of $1/C^2$–V_a characteristics is suggested to originate from the distribution of either donor concentration or permittivity as a function of depth. The nonlinear $1/C^2$–V_a characteristics can be quantitatively explained by using a model that takes into account the electric-field-dependent permittivity and the presence of the interfacial layer. For heavily doped STNO, the permittivity was reduced by a built-in field larger than 100 kV/cm at room temperature.

3.7.3.2. BKBO/La₂₋ᵧSrᵧCuO₄ Superconducting Junctions

Tunneling junctions between different high-T_c superconductors are scientifically interesting. The BKBO possesses special benefit for the study on superconducting junctions. Based on BKBO layer, the junctions can be formed with other high-T_c superconductors, which not only can be used for device applications but also for fundamental research on intrinsic properties of other unknown superconductors.

BKBO/LSCO junctions introduced here were made by using MBE to grow BKBO on the LSCO films which were grown on various substrates by off-axis sputtering [467]. The substrates included (100) SrTiO₃, (110) SrTiO₃, and (100) SrLaAlO₄ (SLA). During growth of LSCO films, the sputtering gas consisted of a mixture of 10 mTorr O₂ and 70 mTorr of argon. The temperature of substrates was held at around 650°C. After growth, the LSCO films were cooled at 1 atm of O₂ over a 1.5 h period, followed by a slow cooling to room temperature. Before depositing BKBO, LSCO films were exposed to an oxygen plasma for half an hour at 305°C.

After BKBO growth, the multilayer films were patterned into junctions using conventional photolithography. BKBO/LSCO stripes were isolated by Ar ion milling, followed by SiO₂ evaporation to prevent shorting on the stripe edges. The junction areas were defined by milling away strips of the BKBO layer, which has a much faster mill rate than does for LSCO. Pairs of silver contacts to both top and bottom layers are then evaporated and patterned by lift-off, and the chip was annealed at 375°C for 2 h in oxygen. All processing was done in parallel to be able to compare the different types of samples. A final layer of gold is used to make wire bonding pads. The nominal junction areas were 80 × 140 and 60 × 90 μm.

Figure 54 (Fig. 3 in [467]) shows the conductance–voltage measured at 4.2 K for a junction fabricated on LSCO grown on (110) SrTiO₃. A current–voltage characteristic is shown in Figure 55 (Fig. 5 in [467]). Prominent in the conductance characteristics are the sharp quasi-particle peaks at the BKBO gap, the low conductance within the gap (except for the spike at $V = 0$), and the linearly increasing conductance appearing above the gap. The gap structure and linear background observed in Figure 54 are typical one of BKBO-native barrier-metal tunnel junctions. There are no features associated with any gap of LSCO.

shown in Figure 53 (Fig. 1 in [466]). For junctions using the as-polished substrate, and for junctions using O₂-annealed and thermal cleaned substrate. The log J–V characteristics for the former vary widely from junction to junction and exhibit nearly ohmic behavior. For the latter, in contrast, the I–V curves lie on a single line except for the large-area junction (1.0 × 1.2 mm²). This result indicates that oxygen annealing drastically improves the in-plane homogeneity of the I–V characteristics.

Capacitance–voltage measurements for such junctions were carried out at room temperature with a 15 mV rms test signal with frequency of 100 Hz; the BKBO/STNO (0.01 wt%) exhibited a linear dependence of $1/C^2$ on the applied reverse voltage V_a. Such dependence is predicted by the standard Schottky theory. The diffusion potential estimated by the intercept voltage of $1/C^2$–V_a characteristics was 1.73 V. The donor concentration N_D in the depletion region of a Schottky junction is calcu-

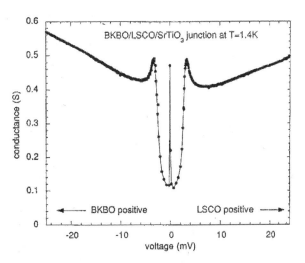

Fig. 54. Conductance–voltage characteristic for a $60 \times 100 \ \mu$m BKBO/LSCO junction fabricated on (110) STO, measured at 1.4 K. Note that the zero bias peak in the conductance corresponding to the supercurrent has been truncated [467]. Reprinted from *Physica C* 251, 133 (1995), with permission from Elsevier Science.

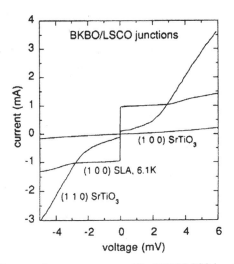

Fig. 55. Current–voltage curves measured for BKBO/LSCO junctions grown on (110) STO, (100) STO, and on (100) SrLaAlO$_4$ [467]. Reprinted from *Physica C* 251, 133 (1995), with permission from Elsevier Science.

BKBO/LSCO junctions fabricated on (100) SLA were qualitatively similar to those fabricated on SrTiO$_3$(110). The main difference was that the critical currents for the junctions on SLA were nearly 10 times larger. However, BKBO/LSCO junctions fabricated on (100) SrTiO$_3$ were qualitatively different from the former junctions. In Figure 56 (Fig. 6 in [467]), the current versus voltage to the 3/2 power for one of these junctions is plotted. A power-law fit to these data gives an exponent of 1.47 for positive bias, and 1.53 for negative bias.

3.8. Thin Films Related to C$_{60}$

3.8.1. Fabrication

In 1991, superconductivity was discovered in postassium (K)-doped C$_{60}$ (fullerides) [468]. After that, the thin films of those

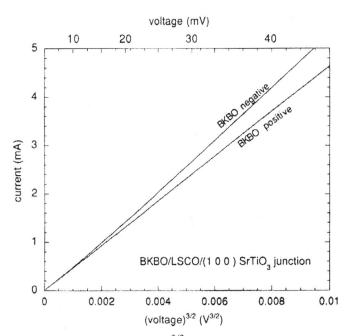

Fig. 56. Current plotted vs (voltage)$^{3/2}$, for a BKBO/LSCO junction grown on (100) STO [467]. Reprinted from *Physica C* 251, 133 (1995), with permission from Elsevier Science.

kinds of materials have attracted great attention because of their potential applications [469, 470].

Conventionally, the fabrication of this kind of film was performed by an evaporation method [471–473], that is, keeping C$_{60}$ film in the vapor atmosphere of the doped metal. For example, C$_{60}$ film is first deposited on substrates under $(1–2) \times 10^{-5}$ Torr, the temperature of the substrates is 135°C, and the evaporation temperature of C$_{60}$ is 450°C. The doping proceeds in a glass doping tube under 10^{-5} Torr. The resistance of C$_{60}$ is measured and the doping process should be stopped when the resistance of C$_{60}$ film reaches a minimum value and begins to increase. Whether the evaporation temperature of doped metal is lower or the temperature of C$_{60}$ film is higher will cause a longer doping time. In early experiments, the glass doping tube was put in a vertical furnace and usually it requires a long doping time. Wang et al. placed the tubular furnace and glass doping tube horizontally and kept the distance of the C$_{60}$ film from the tubular furnace at 3.5 cm and the distance between the doping source (potassium) and the C$_{60}$ film at 8 cm. An optimal temperature of the potassium is obtained at 160°C. Under these conditions, the temperature of C$_{60}$ film is 45°C. The doping process can be finished in 20 min [474].

3.8.2. Properties

Numerous experimental investigations of this kind of superconductor, both in granular and in crystalline structure, show unique properties and distinguish them from other superconductor materials. It was shown that superconductivity in alkali metal doped fullerences occurs in the face centered cubic crystal phase with the composition A$_x$B$_{3-x}$C$_{60}$ [475]. In the other family of fullerene superconductors, i.e., alkali-earth-doped

C_{60}, superconductivity occurs in simple cubic Ca_2C_{60} [476] and in body-centered-cubic Sr_6C_{60} [477] and Ba_6C_{60} [478]. Intercalation of fullerites with RE metals Yb, Sm, and Eu has led to superconductivity in $RE_{2.75}C_{60}$ compounds with $T_c \sim 6$ K [479].

3.8.2.1. Normal-State Properties

Doping C_{60} with alkali metals lowers the electrical resistivity from very high values (10^8–10^{10} Ω/cm for C_{60}) to a metalic-clike behavior in A_xC_{60}, because the doping results in a charge transfer to the C_{60} molecules and strongly increases the π–π overlap between them, which enhances the electrical conductivity. The minimum of the electrical resistivity occurs at $x = 3$, reaching a typical value of high resistivity metal [480]. The high resistivity can be explained by the relatively weak overlap of the electron wave functions between the adjacent C_{60}^{3-} ions and by the merohedral disorder in the alignment of adjacent C_{60}^{3-} ions.

3.8.2.2. Magnetization

Magnetization measurements show that for A_3C_{60}, the zero field cooling magnetization shows the flux exclusion from the sample, while the field cooling magnetization shows the flux expulsion. At fixed temperature a strong hysteresis appears in the magnetic field dependence of the magnetization [481]. That is good evidence that the flux lines penetrate the fullerence superconductors. Because of the three-dimensional structure of the fullerene superconductors, these flux lines should build up an Abrikovsov vortex lattice. Vortices are pinned by pinning centers and the pinning in fullerences is strong.

3.8.2.3. Electronic Transport Properties

Palstra et al. have reported the longitudinal resistivity and Hall-effect data of thin film K_3C_{60} in the magnetic fields up to 12.5 T. The resistivity is 2.5 mΩ/cm at room temperature and near the metal-to-insulator transition. The Hall coefficient is small as expected for a half-filled conduction band and changes sign at 220 K, showing that at half filling both electron and hole conduction are present. The resistivity is interpreted in terms of the granularity of the film which leads to zero-dimensional superconductivity in these systems with a length scale of 70 Å. A very large upper critical field of $\sim$47 T and a Ginzburg–Laudau coherent length of $\sim$26 Å were obtained [482]. The coherence length is very small and comparable to the short coherence length of high-T_c superconductors.

There are some further open questions on the superconducting properties of fullerenes. It will be worthwhile to search for the nature of superconductivity in the polymeric AC_{60}.

3.9. Ultrathin Films and Multilayers

3.9.1. Introduction

Based on the understanding of the role of the Cu–O plane and their coupling on high-T_c superconductivity, the ultrathin films

and multilayers mostly based on YBCO material are also developed in the early stage and more and more achievements have been obtained so far.

It is commonly understood that high-temperature superconductors are lamellar structural materials in which the CuO_2 planes take responsibility for conducting and superconductivity is weakly coupled and shows quasi-two-dimensional (2D) features for their physical properties. But people are puzzled as to what can be done to distinctly understand the role of the CuO_2 planes and their coupling on the superconductivity since the bulk samples (including single crystals) and conventional thin films all show body effects for both normal- and superconducting-state properties. Ultrathin films with thicknesses controlled at one or several unit cells are the ideal samples to directly provide the information on the role of the CuO_2 plane on superconductivity. This was performed mainly on $YBa_2Cu_3O_{7-\delta}$ ultrathin films by Venkatesan et al. [30], Xi et al. [33], and Terashima et al. [31]. This configuration, ultrathin superconducting films separated by an insulating layer (for which the thickness also is controlled to be one unit or several unit cells), i.e., sandwich or multilayers, is the ideal type of material for the investigation of coupling of CuO_2 planes. For example, the intercell couplings and intracell–interlayer couplings have been observed experimentally by Lowndes et al. [483] and Li et al. [484] and were theoretically proposed by Rajagopal et al. [41] for $YBa_2Cu_3O_{7-\delta}/PrBa_2Cu_3O_{7-\delta}$ superlattices. The c-axis direction Josephson coupling of the $YBa_2Cu_3O_7/Pr_{1-x}Y_xBa_2Cu_3O_7/YBa_2Cu_3O_7$ sandwich system was proposed by Umezawa et al. [485]. The proximity effect between superconducting and nonsuperconducting layers as another kind of medium to induce the coupling between superconducting layers in YBCO/PBCO/YBCO multilayers has been proposed theoretically [42, 486] and experimentally [45]. The superconductivity is also widely discussed based on the Koslerlitz–Thouless (KT) transition [483, 487, 570]. The ultrathin films and multilayers are also the typical types of materials for investigation of flux dynamics, since both of them are the ideal matter type for the formation of the vortex lattice with strong anisotropy, from which more new information about the vortex pinning and motion can be obtained. So far, most papers on the ultrathin films and multilayers are published in this field. In this chapter, the flux dynamics in high-T_c superconducting thin films has been arranged in a special section. The investigation of growth of ultrathin film is also an important subject to understand the growth mechanism of high-temperature superconducting thin films. The surface morphology of the initial growth stage of the high-T_c superconducting thin films has been observed and investigated by RHEED, X-ray diffraction, scanning tunneling microscopy (STM) [488], and atomic force microscopy [489]. In this chapter, the content will include the fabrication and physical properties of ultrathin films and multilayers.

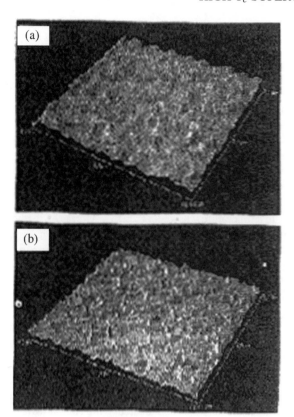

Fig. 57. (a) The AFM image of a smooth YSZ substrate; (b) AFM image of an ultrathin film of an average thickness of 4 nm. The outgrowths have already nucleated at such an initial growing stage [490]. Reprinted from *Physica C* 251, 330 (1995), with permission from Elsevier Science.

3.9.2. Fabrication of Ultrathin Films

Ultrathin films can be fabricated using many methods and on various substrates. Gao et al. studied the surface morphology of YBCO films with thickness down to 2 nm by scanning electron microscopy and atomic force microscopy [490]. The outgrowths were found to nucleate at the initial growing stage of the film. Thus the surface quality of the initially grown layer turned out to be of practical importance in controlling and improving the film-surface morphology. By employing a substrate with a very good surface and using a suitable deposition temperature, the nucleation of the outgrowths can be greatly suppressed and extremely smooth films up to hundreds several of nm can be obtained. Figures 57 and 58 show the surface morphology for the above cases. The YBCO ultrathin films were also studied by AFM. The AFM observation indicated that the film growth shows two modes, layer-by-layer and spiral growth mode. For the latter, the density of 10^6–10^7/cm^2 of screw dislocations was observed from 6-nm-thick film; with increasing thickness the density of dislocations increases. It was also found that within the thickness range from 2 unit cells to 12.5 nm, T_c and J_c are strongly dependent on the film thickness.

Sputtering and PLD are two main methods to fabricate YBCO ultrathin films. In 1989, Xi et al. grew ultrathin films of YBCO with thickness down to 2 nm on (100) SrTiO$_3$ and (100) MgO by magnetron sputtering [33]. A perfect substrate surface

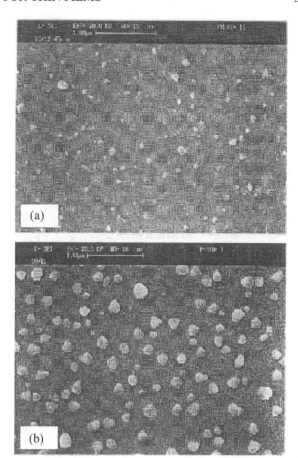

Fig. 58. Surface morphology of the ultrathin films at different growing stages: (a) for an ultrathin film with a thickness of 7 nm, (b) a 60-nm-thick film [490]. Reprinted from *Physica C* 251, 330 (1995), with permission from Elsevier Science.

achieved by annealing in oxygen atmosphere at 950–1000°C for 1 h was an essential parameter for the growth of ultrathin films. On the other hand, the optimized substrate temperatures (800°C for SrTiO$_3$ and 780°C for MgO) are also very important. An amorphous YBCO protection layer with a thickness of about 30–50 nm was deposited *in situ* at room temperature (RT) on the top of ultrathin films. The thickness of the films was determined by the deposition time at a deposition rate of 1 nm/min.

In 1990, Gao et al. prepared high-quality *c*-axis-oriented ultrathin YBCO films by off-axis rf-magnetron sputtering [491]. The sputter deposition itself took place in a vacuum chamber. A sintered YBa$_2$Cu$_3$O$_x$ target was mounted on a magnetron cathode (as in standard magnetron sputtering). The substrate heater is located off axis. By this configuration, a diffuse sputter process was achieved with a quite low deposition rate; on the other hand, no backsputtering occurs in this case and excellent stoichiometric growth can be obtained. An important improvement in this process was achieved by a reflection plate, which was mounted at about a 45° angle on the axis of the magnetron cathode. It yielded a better-directed partial stream to the substrate and further improved the quality of the films and the reproducibility of their properties. All films made this way have smooth surfaces. When the film is thinner as 7 nm, T_c (T_{c0} = 87–89.5 K, $T_{c,on}$ = 93 K) was obtained. The critical current density at 77.3 K was found to be strongly dependent on the film thickness. A maximum value was found to be 8×10^6 A/cm^2 at 77.3 K for a 10 nm film. Below 4 nm, no reasonable quality could be achieved on ZrO$_2$ and SrTiO$_3$ substrate. On atomically flat polished MgO single-crystal substrates, a transition was achieved with an onset of 85 K. When the film is 1.5 nm thick, the indication of the superconductivity was observed at 45.5 K. This means that the superconductivity can exist in an YBCO film with a thickness of just a unit cell ($\sim$12 Å).

PLD is also a widely used method to fabricate ultrathin YBCO films. In 1992, Guptasarma et al. prepared thin and ultrathin YBCO films on single-crystalline substrates of LiNbO$_3$ by PLD [492]. They used a KrF (λ = 248 nm) excimer laser (Lambda Physik EMG 200E with pulse width of 20 ns, power of 500 mJ/pulse) to deposit thin films with a 45° incident laser beam to the target surface. The ablation was performed with a power density of 2 J/cm^2 on the YBa$_2$Cu$_3$O$_{7-\delta}$ target with a zero resistance temperature of 92 K and transition width $\Delta T_c < 1$ K. The deposition chamber was in a background pressure of 5×10^{-7} Torr in which oxygen was incorporated to maintain a pressure of 100–200 mTorr during deposition. The target pellet was kept rotating during deposition to reduce cratering on it during ablation. With each pulse, a narrow plume of light was visible, pointing approximately normal to the pellet surface. The substrates were placed 6 cm away from the target. Deposition rates were found to be about 0.5 Å per pulse. Substrates were mechanically clamped onto a heated metal block whose temperature was maintained at 650°C.

Not only YBCO ultrathin films but also Ba–Sr–Ca–Cu oxide ultrathin films are important in the research of high-temperature superconductivity. Earlier in 1989, preferentially c-axis-oriented superconducting Bi–Sr–Ca–Cu oxide films as thin as 200 Å have where ben successfully grown on MgO (100) substrates by Yeh et al. [493].

For this processing the target was prepared by mixing an appropriate amount of Bi$_2$O$_3$, CuO, CaCO$_3$, and SrCO$_3$ powders, pressing them into a 5.0-cm disk, firing at 800–850°C for 6 h, regrinding, and pressing again at a pressure of 1.5 tons/cm^2. The disk was then sintered at 840°C in air and furnace cooled down to room temperature. During sputtering, the base pressure was 1.7×10^{-8} Torr, the substrate temperature was 25°C, and the sputtering gas was 30 mTorr Ar. The deposit rate of the film is $\sim$15 Å/min under a power of 25 W. At last, the film was annealed in oxygen at 870°C for 30 min. After the postoxygen anneal, the films exhibit a superconducting transition with an onset at 85 K and zero resistance at 80 K. The room-temperature resistivity of this film is 300 $\mu\Omega$/cm, which is comparable to that of single crystal. Using 2 μV/mm as the criterion, the critical current density as estimated by the transport measurement is on the order of 10^5 A/cm^2 at 50 K.

In the recent years, molecular beam epitaxy was also used to fabricate BSCCO ultrathin films [37]. The substrates used for ultrathin film growth were (001) SrTiO$_3$. The surfaces of the

substrates werte mechano-chemically polished and their edges were parallel to the ⟨100⟩ direction of the SrTiO3 crystal. In order to reduce the influences of substrate roughness on electric properties and surface morphology of the ultrathin films, these substrates were etched using an NH$_4$F–HF (BHF) solution (pH = 4.5–4.8). AFM images confirmed that the substrate surface after this procedure had atomically flat terraces of 100–200 nm in width and steps of approximately 0.4 nm in height corresponding to the unit cell size of SrTiO$_3$. After a 10-half-unit-cell (h.u.c.)-thick Bi-2201 was grown as a buffer layer, m-h.u.c.-thick Bi-2212 ultrathin films (m = 1, 2, 3, 4, 6) were grown and no cap layer was deposited. During growth, the substrate temperature was kept at 760°C. After growth, the substrate temperature was decreased and the ozone was supplied until the temperature was cooled down to 500°C. The superconducting transition temperatures of 2, 3, 4, 6-h.u.c.-thick Bi-2212 ultrathin films were 24, 60, 60, and 69 K, respectively.

EuBa$_2$Cu$_3$O$_{7-\delta}$ ultrathin films are were YBCO related films and can also be prepared by magnetron sputtering. Michikami et al. produced EuBa$_2$Cu$_3$O$_{7-\delta}$ ultrathin films on MgO (100) substrates at 580°C using planar-typde magnetron sputtering [494]. EuBa$_2$Cu$_3$O$_{7-\delta}$ films were prepared using dc magnetron sputtering from a single sintered 90 mm in diameter and 6 mm thick. A MgO (100) substrate was used which was washed ultrasonically in acetone and heated with Kantal wire radiation after washing. Before deposition, the substrate was heated at 700°C for 30 min in pure oxygen. Then, the film was sputter-deposited in Ar + 8.5%O$_2$ at T_s = 580°C and 8 Pa. The deposition rate was 38 Å/min. After deposition, the film was cooled down in the sputtering chamber. By this technique route, the c-axis-oriented films were epitaxially grown with a crystal orientation of [110]$_{EuBaCuO}$ ∥ [010]$_{MgO}$. Films more than 100 nm thick exhibited T_c endpoints of $\sim$90 K. As film thickness decreased, superconducting properties deteriorated. However, even for films 2-unit-cells thick, it is still superconducting.

3.9.3. Properties of Ultrathin Films and Multilayers

3.9.3.1. Superconductivity of YBCO Ultrathin Films

Ultrathin films have many special properties. High-temperature superconductivity in ultrathin YBCO films is one of the most important parameters for research. In 1989, Venkatesan et al. fabricated ultrathin films of YBa$_2$Cu$_3$O$_{7-\delta}$ in situ on (001) SrTiO$_3$ by pulsed laser deposition [30]. The zero resistance transition temperature (T_{c0}) is >90 K for films >30 nm thick. The critical current density (J_c at 77 K) is 0.8×10^6 A/cm^2 for a 30-nm film and $4–5 \times 10^6$ A/cm^2 for a 100-nm film. The T_{c0} and J_c deteriorate rapidly below 30 nm, reaching values of 82 K and 300 A/cm^2 at 77 K, respectively, for a 10-nm film. The films with thicknesses of 5 nm exhibit metallic behavior and possible evidence of superconductivity without showing zero resistance until 10 K. These results are understood on the basis of the defects formed at the film–substrate interface; the density of detects rapidly decreases over a thickness of 10 nm.

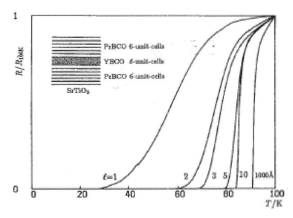

Fig. 59. Normalized resisitance [$R(T)/R(100$ K)] vs temperature for PrBCO (6 unit cells)/YBCO (l unit cells, $l = 1, 2, 3, 5, 10$)/PrBCO (6 unit cells)/SrTiO₃ trilayer films and a 1000-Å-thick film [31].

The defects were studied by ion channeling measurements and cross-section transmission electron microscopy. And the results suggest that the superconducting transport in these films is likely to be two dimensional in nature, consistent with the short coherence length along the c-axis of the crystals.

In 1991, Terashima et al. researched the superconductivity of 1-unit-cell-thick YBCO thin film [31]. They found for the first time that superconductivity can occur in a 1-unit-cell-thick layer of YBCO. The layer is grown, while monitoring with RHEED oscillations, on a nonsuperconducting PrBa₂Cu₃O₇ (PBCO) buffer layer and covered by a PBCO layer by reactive evaporation. Figure 59 shows the normalized resistance vs temperature. A reduction of the onset temperature is found to be mainly due to a decrease of the hole carrier density but not due to the absence of interlayer couplings. The KT transition (which will be described in detail below) is assumed to be one possible cause for the broad resistive transition. The resistive transition of 1-unit-cell-thick YBCO film is independent of the magnetic field which is applied parallel to the film (CuO₂) plane. The covering PBCO layer grown on the 1-unit-cell-thick YBCO layer has been found to be of significance in providing it with holes and thereby inducing superconductivity.

Norton et al. found that conductive Pr₀.₅Ca₀.₅Ba₂Cu₃O₇₋δ (PCBCO) buffer and cap layers significantly enhance the superconducting properties of ultrathin YBCO layers in trilayer structures with T_c ($R = 0$) $\sim$ 45 K for a 1-unit-cell-thick YBCO layer grown by PLD [34]. The PrCaBCO enhanced T_c is believed to be attributed to an increase of the carrier density and normal-state coherence length within the weak link kinks in the ultrathin YBCO layers. STM and STEM studies show that these kinks result from unit-cell steps on the growing surface of the buffer layer. The increased carrier density decreases the perturbation of the superconducting order parameter at the weak links, thus enhancing phase coherence across the macroscopic sample. Sun et al. reported the superconductivity of c-axis-oriented 2-unit-cell-thick YBCO film [495]. The KT transition model and fluctuation-enhanced conductivity theory can be used to fit the R-T behavior in the region of resistive

transition, and the intrinsic linear resistance behavior in the normal-state region of the YBCO ultrathin film is confirmed.

The origin of the T_c depression in ultrathin YBCO was also been investigated by Cieplak et al. from conductance measurements on YBCO between layers of Y₁₋ᵪPrᵪBa₂Cu₃O₇₋δ[(Y–Pr)BCO] [496]. The results show a transition from a bulk regime in the interior of the YBCO to a surface regime near the interfaces. The depression of the zero-resistance temperature in ultrathin YBCO is correlated with the depressed conductance in the surface layer. The results indicate that the changes are related to the presence of the interfaces, primarily to charge transfer between the layers, with little, if any, indication of a change in the intrinsic properties of the YBCO from bulk down to the thickness of a single unit cell. The results also show that the conductance is lower as a result of the conditions at the interfaces, such as charge transfer and imperfections. Conversely the results imply that except for the special features associated with the interfaces the properties of YBCO do not seem to change significantly as a function of thickness down to a single-unit-cell layer.

3.9.3.2. KT Transition in Ultrathin YBCO Films

Kosterlitz–Thouless transition is most important for ultrathin YBCO films. In 1992, Qiu et al. studied the KT transition in an YBCO/PBCO superlattice [570]. The thermally activated vortex–antivortex unbinding predicted by KT theory was observed. The resistance above the temperature T_{KT} showed an exponential inverse-square-root temperature dependence. Below the temperature T_{KT}, the current–voltage characteristics in the low-current limit obeyed a power-law relationship, and the universal jump of the exponent πK_R from 2 to 0 was observed around T_{KT}. All these phenomena are consistent with the Kosterlitz–Thouless theory. The main KT transition parameter $\tau_{KT} = (T_{c0} - T_{KT})/T_{KT}$ implies an enhancement of two-dimensionality in the superlattice. So KT transition is also a criterion for the two-dimensional superconductivity.

Matsuda et al. studied the resistive transition on ultrathin YBa₂Cu₃O₇₋δ films with the number of units n ranging from 1 to 10 based on KT transition [32]. The zero-resistance temperature, which is T_c = 30 K for n = 1, increases rapidly with increasing n. To interpret the results, topological excitation of vortices in thin films of a layered structure is discussed. A "vortex-string pair," in which vortices and antivortices piercing all n layers are pairwise bounded, is shown to be an important topological excitation in thin films. Dissociation of the vortex-string pair gives rise to the KT resistive transition in thin films of layered structure. The observed dependence of T_c on n is explained quantitatively by the increase of the projected two-dimensional carrier density with n in the framework of the KT theory. The result suggests that the KT transition at T_c = 30 K is intrinsic to the CuO₂ conducting planes in one YBCO layer and questions the earlier interpretation ascribing $T_c \approx$ 90 K in bulk YBCO to the KT transition. Figures 60 and 61 show the experimental result of the sheet resistance and the KT transition temperature.

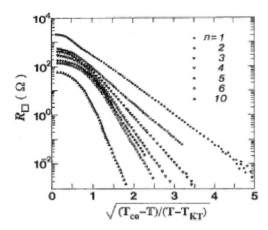

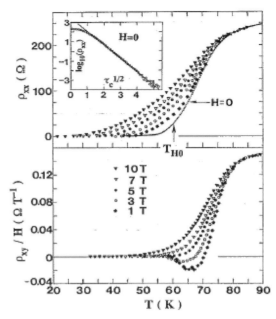

Fig. 60. The temperature dependence of the sheet resistance $R_\square$ of n-UCT YBCO films with $n = 1$–6 and 10 are shown by the $\log_{10} R_\square$ vs $[(T_{c0} - T)/(T - T_{KT})]^{-1/2}$ plot, where T_{KT} of $n = 1, 2, 3, 4, 5, 6,$ and 10 samples are 30.1, 58.0, 63.8, 72.4, 76.3, 79.9, and 80.8 K, respectively [32].

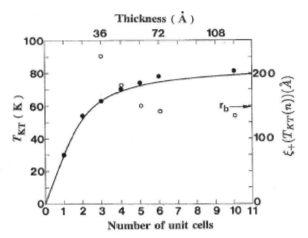

Fig. 61. The thickness dependence of the KT transition temperature of n-UCT YBCO films with $n = 1$–6 and 10 [32].

Fig. 62. ρ_{xx} and ρ_{xy}/H for [Y(1)/Pr(1)] $\times$ 10 in H perpendicular to the a–b plane. Inset: $\log_{10}(\rho_{xx})$ vs $\tau_c^{1/2} = [(T_{c0} - T)/(T - T_{KT})]^{1/2}$. The solid line in the inset shows theoretical values with $b = 2.5$, $T_{c0} = 82.0$ K, and $T_{KT} = 50.1$ K [497].

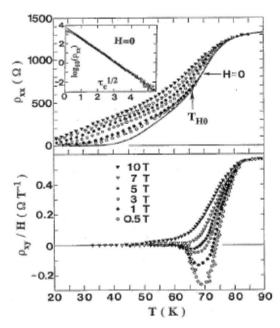

Fig. 63. ρ_{xx} and ρ_{xy}/H for Pr(6)/Y(1)/Pr(11) in H perpendicular to the a–b plane. Inset: $\log_{10}(\rho_{xx})$ vs $\tau_c^{1/2} = [(T_{c0} - T)/(T - T_{KT})]^{1/2}$. The solid line in the inset shows theoretical values with $b = 2.0$, $T_{c0} = 82.0$ K, and $T_{KT} = 30.1$ K [497].

As evidence of free vortex–antivortex excitation, Matsuda et al. also researched the disappearance of Hall resistance in 1-unit-cell thick YBCO [497]. Diagonal and Hall resistivities (ρ_{xx} and ρ_{xy}) in the normal and superconducting states are studied in 1-unit-cell-thick YBCO artificial materials. The transport properties in bulk YBCO are shown to be inherent properties of the individual conducting CuO_2 bilayer. Evidence of Kosterlitz–Thouless resistive transition as well as the excitation of free vortex and antivortex are presented. Figures 62 and 63 show T dependence of 2D sheet resistances ρ_{xx} and ρ_{xy} near the transition temperature for the $m = 1$ superlattice and the 1-UCT film, respectively, in magnetic fields perpendicular to the a–b plane.

In 1993, Norton et al. reported on the transport properties of ultrathin YBCO layers and compared the results with predictions of the Ginzburg–Landau Coulomb-gas (GLCG) model for 2D vortex fluctuations [498]. They found that the normalized flux-flow resistances for several ultrathin YBCO structures collapse onto a single universal curve, as predicted by the model. In addition, the values for the Kosterlitz–Thouless transition temperature T_{KT} and the Ginzburg–Landau temperature

T_{c0}, obtained by separate analyses of I–V and resistance data within the context of the GLCG model, are in general agreement. Finally, it is found that the properties of a series of YBCO/$Pr_{0.5}Ca_{0.5}Ba_2Cu_3O_{7-\delta}$ superlattice structures are consistent with the GLCG treatment for anisotropic 3D layered superconductors.

3.9.3.3. *Properties of Multilayers*

Multilayers, as an artificial construction of high-T_c superconductors, are also interesting and are investigated widely. In 1989, Triscone et al. for the first time epitaxially grew superlattices of Y–Ba–Cu–O/Dy–Ba–Cu–O by dc magnetron sputtering onto SrTiO$_3$ and MgO substrates [39]. The wavelengths range from 2.4 (twice the *c*-axis of YBCO) to 30 nm and X-ray diffraction shows satellite peaks characteristic of multilayers for each wavelength. The 2.4-nm-wavelength sample consists of, on the average, alternate planes of Y and Dy. The multilayers are superconducting with T_{c0}'s between 85 and 89 K, transition widths of 2 K, and resistivity ratios of about 3. These superlattices exhibit very large crystalline coherence in the growth direction and a modulation coherence as high as 36 nm.

In 1990, Triscone et al. artificially varied the coupling between ultrathin YBa$_2$Cu$_3$O$_7$ layers by interposing insulating planes of PrBa$_2$Cu$_3$O$_7$ in multilayers and studied the consequences of decoupling 12-Å YBa$_2$Cu$_3$O$_7$ unit cells on the superconducting properties [499]. They found that the T_c (midpoint) of a 12-Å/12-Å multilayer is 55 K, twice that of the corresponding layer. A linear reduction of T_c was observed with increasing thickness of the PBCO layer, suggesting that a minimum interlayer coupling is necessary for superconductivity to occur in the individual 12-Å YBCO layers. Figure 64 shows this linear reduction of T_c. The behavior seen in a magnetic field parallel to the layers demonstrates the increased decoupling of the layers as the PBCO thickness is increased.

In the same year, Lowndes et al. fabricated epitaxial, non-symmetric $M \times N$ superlattices in which YBCO layers either $M = 1, 2, 3, 4$, or 8 *c*-axis unit cells thick are separated by insulating PBCO layers N unit cells thick ($N = 1$ to $\sim$16) [483]. The T_{c0} for epitaxial YBCO/PrBCO superlattices initially decreases with increasing d_{pr}, but T_{c0} then saturates as the separation of the YBCO layers is increased further. The limiting T_{c0} values for 1-, 2-, 3-, 4-, and 8-cell YBCO layers separated by $\sim$190 nm in a PBCO matrix are $\sim$19, $\sim$54, $\sim$71, $\sim$81, and $\sim$87 K, respectively. Single-cell-thick YBCO layers, in particular, are superconducting in a PBCO matrix. The available evidence suggests that Josephson coupling of YBCO layers can occur through relatively thick PBCO layers, but PBCO also may severely modify the electronic structure of adjacent YBCO unit cells.

Rajagopal et al. developed a simple model of intralayer and interlayer couplings based on a generalized BCS pairing theory of superlattices of layered superconductors to correlate the observed trends in the T_c's of YBa$_2$Cu$_3$O$_7$/PrBa$_2$Cu$_3$O$_7$ superlattices [41]. These observed trends in the T_c's are obtained in this model if certain inequalitities are satisfied among the direct and that mediated by the PrBa$_2$Cu$_3$O$_7$ intercell–interlayer, intracell–interlayer couplings. These inequalities among the couplings may be reconciled in terms of the chemical structure of these systems.

Transport properties of *a*-axis-oriented YBCO/PBCO superlattices are also interesting. Eom et al. measured the resistive transition of *a*-axis-oriented YBa$_2$Cu$_3$O$_7$/PrBa$_2$Cu$_3$O$_7$ superlattices in magnetic fields [500]. They found for fields applied

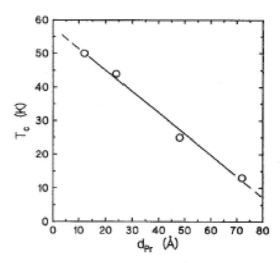

Fig. 64. T_c (10% of normal-state resistance) as a function of the PrBCO thickness for a series of multilayers with constant 12-Å YBCO layers [499].

parallel to the substrate (i.e., perpendicular to the CuO$_2$ planes) that 80% of the transition of a 48-Å multilayer was unaffected by magnetic fields up to 8 T. For a 24-Å/24-Å multilayer, it is observed that only the upper 30% part of the transition is insensitive to the magnetic field. Those results show that no substantial screening currents develop in the CuO$_2$ planes above the crossover, implying that each plane is divided into independent superconducting channels separated by insulating channels where the respective channels are determined by whether the rare-earth neighbor is Pr or Y.

Not only the T_c but also other properties of multilayers such as J_c were researched. Kwon et al. grew YBCO films with nominal thicknesses of 1–4 unit cells by pulsed laser deposition using (Pr$_x$Y$_{1-x}$)Ba$_2$Cu$_3$O$_{7-\delta}$ [(Pr$_x$Y$_{1-x}$)BCO] ($1 \geq x \geq 0$) as buffer layers and cap layers and researched the critical current density [501]. Films of 1 unit cell thickness were superconducting for all the x values while T_c increased when x was reduced. For adjacent layers of (Pr$_{0.6}$Y$_{0.4}$) BCO which is semiconducting, a T_c of 43 K and J_c of 2×10^6 A/cm^2 for $B\|ab$ and 4×10^5 A/cm^2 for $B\perp ab$ at $B = 7$ T and 4.2 K were obtained in a 1-unit-cell-thick YBCO layer. The J_c values of a few-unit-cell-thick YBCO layers nearly approached that of thick YBCO films. The results suggest the absence of significant weak-link effects in these films.

Many methods were performed in the ultrathin-layer YBCO/PBCO superlattice research. Li et al. presented Raman scattering studies of *c*-axis-oriented ultrathin-layer superconducting (YBa$_2$Cu$_3$O$_7$)$_m$/(PrBa$_2$Cu$_3$O$_7$)$_n$ superlattices in 1994 [502]. For the superlattice with a ($m = 2, n = 1$) sequence, Raman spectra revealed a new line in the spectral region around 320 cm^{-1}. It is interpreted as a mode represnting a combination of IR optical phonons of the Y-sublayers with an admixture of a B_{1g}-type Raman active vibration in the Pr sublayers. This new line, which is similar to those from the interior of the Brillouin zone of the original lattice, does not exhibit superconductivity-induced self-energy effects, although its counterpart in the pure substance does. No additional line is found in the ($m = 1$,

$n = 2$) superlattice in the same region, supporting this interpretation for the ($m = 2, n = 1$) sample.

Another interesting property of YBCO/PBCO multilayers is the vortex dynamics expressed as the critical scaling behavior of I–V curves in this system. Zhao et al. measured the I–V curves for YBa$_2$Cu$_3$O$_7$ (24 Å)/PrBa$_2$Cu$_3$O$_7$ (12 and 96 Å) multilayers [44]. The measurements were performed in the magnetic fields H up to 10 T with various orientations against the film surface and at temperatures well below the superconducting transition temperature. From the critical scaling analysis, the vortex–glass phase transition temperature T_g, and critical exponents z and v were extracted. It is found that these parameters depend on the thickness of PrBa$_2$Cu$_3$O$_7$ layers, magnetic fields, and their orientations. For YBa$_2$Cu$_3$O$_7$ (24 Å)/PrBa$_2$Cu$_3$O$_7$ (12 Å) multilayers, in the high field ($H_{\parallel}c$ and >1 T) the exponents ($z \sim 4.0$–6.0, $v \sim 1.1$–1.7) are basically consistent with the theoretical estimates of a three-dimensional vortex–glass transition, while for YBa$_2$Cu$_3$O$_7$ (24 Å)/PrBa$_2$Cu$_3$O$_7$ (96 Å) multilayers, when $H_{\parallel}c$, the critical scaling behavior is nearly consistent with that of a quasi-two-dimensional system at $H = 10$ T. Compared with single-layer YBa$_2$Cu$_3$O$_7$ thin films with thickness >1000 Å, it is found that a size effect in the vortex–glass transition of multilayers exists.

3.9.3.4. Proximity Effect in Multilayers

The proximity effect is a common effect but it is very interesting for multilayers. Wu et al. theoretically studied the superconducting transition temperature T_c of YBCO/PBCO superlattices [486]. They showed that the proximity effect for which the cooper pairs in the superconducting YBCO layer leak to the semiconducting PBCO layer plays a dominant role in explaining the depression of T_c in these superlattices with a minor modification from the charge-carrier-transfer effect across the Y and Pr interface.

Tachiki et al. also theoretically studied the proximity effect in SNS junctions composed of the oxides with isomorphic crystal structures by using a simple model in which the S and N layers have similar electronic structures, and the superconducting pairing interaction works even in the N layers [42]. The proximity effect is enhanced by the pairing interaction and the superconducting pair amplitude is connected between the S layers through the thick N layer which is primarily not superconducting due to a small number of holes. The proximity effect is almost unaffected by elastic impurity scatterings at low temperatures. The theoretical result suggests that the range of the proximity effect is possibly very long in the SNS junction of the superconducting oxides.

In 1993, Zhao et al. performed a study of dimensional crossover in YBCO/PBCO multilayers [45]. The superconductivity of YBa$_2$Cu$_3$O$_7$/PrBa$_2$Cu$_3$O$_7$ (YBCO/PBCO) multilayers has been studied under parallel and perpendicular magnetic fields H ranging from 0 to 7 T. The resistive transitions of the selected [YBCO/YBCO]s of 48 Å/100 Å and 48 Å/12 Å have been investigated. For both systems the linear behavior of $H_{c2}(T)$ versus T was found near T_c; when the temperature T decreases and the magnetic field increases, the $H_{c2}(T)$ versus T curve show nonlinear behavior. It is interesting to note that in the 48 Å/100 Å system when H ($\parallel$ ab plane) >5 T, the slope of the $H_{c2}(T)$ versus T curve increases rapidly and approaches -20 T/K. This behavior implies that the T_c of the 48 Å/100 Å system is almost independent of the magnetic field in the high field region. It may be considered that below T_c the dimensional crossover can be monitored by the magnetic field based on the Josephson effect enhanced proximity effect in YBCO/PBCO/YBCO multilayers. So when the magnetic field increases to a high enough value (5 T may be the case), such a proximity effect is weakened or even destroyed. Then the multilayers become two-dimensinal corresponding to superconducting YBCO layers, the applied magnetic field may almost pass through the PBCO layers, resulting in a large slope of dH/dT. Near T_c, however, the dimensional crossover must mainly be monitored by the behavior of the GL coherence length $\xi(T)$.

In recent years, much work associated with proximity effects has been done. Zhao et al. proposed a theory for the penetration depth of superconducting bilayers and multilayers [503]. The theory begins with an extension of the proximity-effect model developed by Golubov et al. [504], which is based on the Usadel equations.

Toyoda et al. investigated the temperature and voltage dependences of the conductance in a superconductor–semiconductor two-dimensional electron gas (2DEG) junction with a gate [505]. The systematically controlled reentrant behavior of the conductance was observed by using the gate to change the diffusion constant of the 2DEG. It is confirmed that the correlation energy of the proximity correction to the conductance is proportional to the diffusion constant of the normal part. They also examined the effect of the magnetic field on the reentrance and found that even a small magnetic field can change it drastically.

3.10. Large-Area Thin Films

3.10.1. Introduction

Due to the high critical current density and low microwave surface resistance (R_s), the large-area high-temperature superconducting thin films have increasingly received attention since the early 1990s [28, 506, 507]. The new and important application potentials of large-area high-T_c superconducting thin films are mainly in the following two aspects: (1) electric power systems, e.g., the fast responding fault-current limiter for which the area of 20 cm $\times$ 20 cm (thickness up to 5 μm) of the films is necessary for power of 100 kVA [27]; (2) the microwave device applications [28, 29, 508], the planar microwave devices which were prepared from large area high-T_c superconducting thin films with R_s value much lower than even high-purity normal metals.

For both electric power systems and microwave device applications, the high-quality high-T_c superconductor thin films are required which must be grown on large-area substrates with suitable buffer layers. Polycrystalline YSZ and sapphire with CeO_2 buffer are widely chosen. For microwave applications the excellent candidate of substrate is the R-plane cut (1102 oriented) sapphire, which is chemically stable and has a low microwave loss tangent ($\tan\delta < 10^{-6}$–10^{-8} around 10 GHz) [29, 54]. The deposition methods are normally chosen to be pulsed laser deposition [509, 510], magnetron sputtering [27–29, 506], and thermal co-evaporation [511], from which the YBCO films with excellent electronic properties, outstanding homogeneity in thickness, and composition up to 10 cm diameter have been obtained. One reasonable reactive co-evaporation system was designed [511] to be combined with a special substrate heater with a rotating disk holder separating oxidation and deposition zones, enabling a reactive high oxygen pressure zone in a surrounding high vacuum background. Plasma flash evaporation [512], is a relatively new method for the preparation of YBCO films. It uses a high-power inductively coupled thermal plasma to evaporate fine powders (the matter is completely decomposed into its atomic and ionic species) of YBCO which are carried by a gas stream to the plasma torch. Compared to PLD the mass of the evaporated material can be made much higher and can lead to higher deposition rates (the deposition rate reaches ~100 nm/min). Some other chemical processes were also used to prepare large-area films.

For depositing large-area high-T_c YBCO thin film polycrystalline substrates, e.g., the polycrystalline YSZ, were used. In this case, the textured YSZ was selected to be the buffer layer which is a rather good choice so far to achieve high J_c. The IBAD method is used in a dual ion-gun sputtering system for depositing YSZ [513, 514]. One of the ion guns operating at high voltage and beam current (1500 V, 200–250 mA) is used for sputter-etching of a YSZ target. The second ion gun operating at about 300 eV beam energy is used to bombard the substrate to influence the crystallographic orientation of the growing film (buffer layer). The angle between the impinging ion beam and the substrate normal has to be about 55°. The deposition rate for the YSZ layer can be controlled in the range of several to tens nm/min^{-1}.

Below the magnetron sputtering and laser ablation deposition methods and related applications for large-area high-T_c superconducting films will be introduced. We have a special section to introduce microwave applications in detail. So only applications to electric power system will be given in this section.

3.10.2. Sputtering of Large-Area HTSC Thin Films

The YBCO thin films can be deposited by dc magnetron sputtering method. The configuration of target and substrate can be on- or off-axis modes. Due to the possible secondary sputtering of negative oxygen ions to the deposited film and high-energy particles to bombard thin film surface, on-axis sputtering is avoided (in fact, in the case of high enough sputtering gas pressure, the mean free path of negative oxygen ions and high-energy particles becomes shorter and their energies become weaker when they arrive in the deposited film. Consequently,

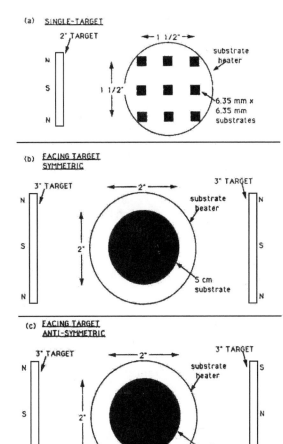

Fig. 65. This figure illustrates the geometry of the target and samples on the substrate holder for the single target geometry (a) and facing target geometry (b), (c). The targets, shown in cross-section, are located at a 90° angle from the heated substrate holder. This polarity of the sputter gun's magnets are shown for the two configurations which were used [506] (N. Newman, B. F. Cole, S. M. Garrison, K. Char, and R. C. Taber, *IEEE Trans. Magn.* 27, 1276 (©1991 IEEE)).

not much influence from them to the film surface needs to be considered), while an off-axis configuration is suggested. The typical configuration of targets and substrate is shown in Figure 65.

The substrates used for microwave devices are usually R-plane cut (1102) and M-plane cut (1010) sapphires *in situ* buffered with oxide buffer materials, for which both CeO_2 and MgO are the ideal candidates. For YBCO the lattice is not so well matched to MgO (the average mismatch is large as 9.0%, while for CeO_2 the average mismatch is small as 0.9%). YBCO grown on non-lattice-matched substrate may easily result in poor in-plane epitaxy and high-angle grain boundaries. So the CeO_2 buffered R-plane cut sapphire is a more attractive substrate for devices.

For sputtering deposition (also for laser ablation deposition), the large-area substrate is fixed on the heat plate located above a resistive heat element with silver paste. The heat elements can be made from doped Si plate, Pt wire, or Haynes alloy heater black. Cole et al. have tested the deposition of MgO

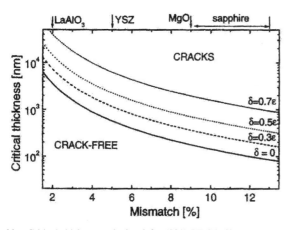

Fig. 66. Critical thickness calculated for (001) YBCO films vs lattice mismatch. The low limit is given by the curve $\delta = 0$ [29] (R. Wordenweber, J. Einfeld, R. Kutzner, A. G. Zaitsev, M. A. Hein, T. Kaiser, and G. Muller, *IEEE Trans. Appl. Supercond.* 9, 2486 (©1999 IEEE).

and CeO$_2$ buffer layers on sapphire and followed by deposition of YBCO [28]. The 5-cm-diam targets of MgO and CeO$_2$ were used to deposit buffer layers on 5-cm-diam M-plane cut and R-plane cut sapphire substrates with deposition temperatures of 450 and 700°C, respectively. The thickness for both buffer layers is ~50 nm. Then the YBCO films are deposited on MgO-buffered M-plane sapphire *in situ* by single-target off-axis sputtering and on CeO$_2$-buffered R-plane sapphire *ex situ* by dual-target off-axis sputtering at temperatures of 690 and 735°C, respectively. It was found from X-ray diffraction that for YBCO (103), MgO (200), YBCO (103), and CeO$_2$ (202) peaks, the amount of misoriented materials is below the detection limit (<0.03%) for the YBCO/MgO/M-plane sapphire system. The measured $T_c = 89$–90 K, $J_c > 1 \times 10^6$ A/cm^2 at 77 K and zero field, R_s at 94.1 GHz is 40–45 mΩ (450–510 $\mu\Omega$ when scaled as f^2 to 10 GHz in the center area of the wafer with ~3.0 cm diam).

For YBCO film on sapphire, the microcrack is always a block for getting a high-quality fit to the application. The microcrack is mainly considered to be caused by the substantial mismatch of the lattice parameters and coefficients of thermal expansion of sapphire with respect to YBCO. Wördenweber et al. [29] conducted a systematic investigation on the strain between YBCO and LaAlO$_3$, YSZ, MgO, and sapphire. The elastic strain (ε) related to the mismatch will lead to microcracks in thin film. When the thickness of the film increases and reaches a critical value, the elastic energy stored in the thin film becomes sufficient for the formation of structure defects, such as lattice misalignments, edge and screw dislocations, or cracks. The relation between critical thickness and mismatch can be concluded as shown in Figure 66. The critical thickness dc of YBCO must lie above the curve for strain $\delta > 0$. For sapphire substrate, the dc of YBCO is larger than 100 nm.

The release of strain can be obtained by some ways, for example, by introducing the structure imperfections. On the other hand, by changing the speed of the temperature decrease from the formation temperature of film to a temperature below

400°C, the degree of release of strain can be changed. When the temperature is slowly decreased, the strain may be released gradually; finally the formation of cracks may be avoided. This case has been observed in another kind of perovskite structural material, (La, Ca) MnO$_3$ films [515]. During the growth of (La$_{1-x}$Ca$_x$) MnO$_3$ ($x \geq 0.5$), the accumulated strain can be changed with temperature in the form

$$\sigma = \left| E_f \int_{T_V}^{T_1} (\alpha_F - \alpha_S)dT \right| \tag{11}$$

Here, E_f is the Young's modulus of thin film; α_f and α_s are the thermal expansion coefficients of thin film and the substrate, respectively (they are functions of temperature T). It was found that both annealing temperature and cooling down speed cause the thermal expansion difference between film and substrate which is the direct reason for the formation of the cracks in the film.

3.10.3. Pulsed Laser Deposition of Large-Area Films

Soon after the discovery of high-T_c superconductors, PLD emerged as a powerful and efficient technique for preparing high-T_c superconducting thin films of YBCO [153]. The PLD technique can provide an excellent and sufficient way to survey the film growth of various materials in a relatively short time. We have introduced of PLD in this chapter. In this section the preparation of large-area high-T_c superconducting thin film by PLD will be described and discussed.

Usually, by using PLD, thin films with areas of 0.5–3 cm can be fabricated easily. But the various microwave and digital electric device applications (e.g., microwave filters, resonators, and delay lines) need large-area films which opens a technique problem for the PLD method. So many groups have made every effort to improve the technique to fabricate large-area thin films. Errington et al. [516], Foltyn et al. [517], and Buhay et al. [518] have developed the PLD approaches used to deposit films on substrates with sizes over 5–10 cm in diameter. The ways for preparing such large-area films include utilizing two laser beams which are focused (or imaged) down onto small-diameter rotating targets along a fixed optical path as shown in Figure 67. These approaches are referred to as "offset" (OS) PLD, and "rotational/translational" (R/T) PLD, respectively. In OS PLD the target is positioned such that the center of the ablation plume (which nominally leaves normal to the target surface) impinges near the outer edge of two rotating substrates as shown in Figure 67a.

The OS PLD approach has been used to grown high-quality films of YBCO *in situ* on 5-cm-diam LaAlO$_3$ substrates [517] and deposit ferroelectric Bi$_4$Ti$_3$O$_{12}$ films over 10-cm-diam Si substrates for RAM applications [518]. In the R/T PLD approach, the substrate is both rotated and translated in a linear fashion as shown in Figure 67b. A computer controls both substrate motions such that uniform film thickness, compositions, and properties are obtained. This approach has been used to deposit YBCO film into 7.5-cm-diam LaAlO$_3$ substrate [519].

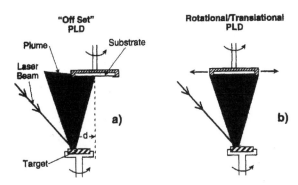

Fig. 67. (a) Schematic diagram of the "offset" large-area PLD approach. (b) Schematic diagram of the "rotation/translational" large-area PLD approach [509].

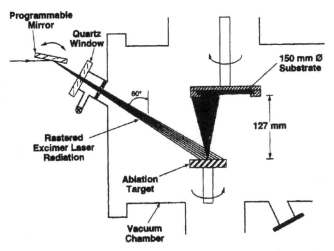

Fig. 68. Schematic diagram of large-area PLD utilizing laser beam rastering with target one-half the diameter of the substrate [509].

An alternative to these approaches uses computer-controlled larger-beam rastering over a large-diameter rotating target as shown in Figure 68. In this case a target at least one-half the diameter of the substrate is used, and its center is offset from the rotation axis of the substrate. This arrangement allows a large substrate to be coated with relatively small targets. In the other case, a large target for which the rotation axis is aligned with that of substrate can be used. A focused laser beam is rastered over the target using a mirror held in a programmable kinematics mount. In this case, the laser beam can be programmed to dwell longer at target locations where the center of the ablation plume impinges on the outer edge of the rotating substrate. With this type of approach the depositions can be performed on the substrate with an area of 12.5 cm [520] and developed further to deposit on substrates with diameters of 15 cm.

The uniformity of thickness and composition of large-area films is the most important factor for developing device applications. For the TlBaCaCuO HTSC precursor deposited by the R/T PLD technique on 5-cm-diam (100) LaAlO$_3$ substrate, the maximum variation in film thickness across the 5-cm-diam substrate was found to be 6% (±3%) [509]. For the Bi$_4$Ti$_3$O$_{12}$

deposited onto heated 10-cm-diam (100) Si substrate with 7-cm-diam target, the maximum variation in thin film thickness (between maximum and minimum values) was found to be 3% (±1.5%). The composition profile of YBCO can be mapped by energy dispersive X-ray analysis (EDXA). For 2-μm-thick YBCO film laser deposited onto a 15-cm-diam Si substrate using beam rastering over a 9-cm-diam stoichiometric target, the nominal film composition examined by EDXA showed that the difference between the maximum and minimum atomic concentration values for Y, Ba, and Cu actions were 1.4, 1.2, and 0.8 at.% (±0.7, ±0.6, and ±0.4 at.%), respectively, across the 15-cm-diam substrate. In comparison, the standard errors for the EDXA data were found to be ±1.48, ±0.17, and ±0.36 at.% for the Y, Ba, and Cu species, respectively. The R/T PLD approach has also demonstrated excellent composition uniformity after a high-temperature thallination step. The composition of thin film deposited onto a 5-cm-diam substrate held in a 12.5-cm-diam holder mapped by EDXA shows maximum variations between 1.0 and 2.8 at% for the Tl, Ba, Ca, and Cu elements.

For getting high-quality large-area high-T_c superconducting thin films, the uniformity heat for the substrate is also a key condition. As mentioned above the heater can be made from doped Si plate, Pt wire, and Haynes alloy heater elements. The substrates are usually attached with the silver paste to a heat plate placed above the heater element. But this kind of technique is just suitable for the substrate with a relatively small area; it is not suitable for large-area thin film deposition. Since the removal of thin and large-area substrates after deposition is often difficult and risks breakage, especially, it is impossible to make film deposition on both sides of a substrate. In order to avoid these problems, Auyeung et al. [510] constructed a substrate heater which simulates a blackbody radiation environment and employs mechanical means to secure the substrate. By using this heater the substrate is heated uniformity and films can be sequentially deposited onto both sides of a substrate. As an added feature for the large-area PLD system, this blackbody substrate heater is combined with a separate load-locked target chamber for quick and efficient target exchange for the deposition of large-area multilayer films. In order to scan the growth methods, the critical parameters, and other special features of large-area high-T_c superconducting thin films, a review table (Table X) is listed below.

3.10.4. Application of Large-Area High-T_c Superconducting Thin Films

As mentioned in the beginning of this section, large-area thin film has important application potentials in microwave devices and electric power systems due its low surface resistance and high critical current density. We devote a special section to introducing microwave applications, so we will give a description of applications in electric engineering power systems. Here, the application is not associated with high-field magnets or energy storage. For large-area thin (not too thin; the thickness is of micrometer order) films, the fault current limiter (FCL) is an

Table X.   Review of the Deposition Methods of Large-Area Thin Films

Material	Substrate	Method	Area	R_s	T_c (K)	J_c	University Institute
YBCO	r-Cu + sapphire with (001) CeO$_2$ of 20–30 nm	Magnetron sputtering	2 inch	R_s 10 GHz, R_s = 390 and 30 $\mu\Omega$ at 77 K and u.l. respectively	88–89	2–3.5 MA/cm^2	Institute fürschicht- and Ionentechnik IS2, Forschungszentrum Jûlich, Germany
YBCO	LaAlO$_3$	Magnetron sputtering	5 inch		~90		Conducts Inc. California
YBCO	LaAlO$_3$	PLD	2 inch		91	2–4 MA/cm^2	Research Laboratory Naval
YBCO	R.plane cut sapphire with suffer of CeO$_2$	Magnetron sputtering	5-cm-diam				Conducts Inc. California
YBCO	M-plane cut sapphire with suffer of MgO		5-cm-diam	40–45 $\mu\Omega$ (94.1 GHz) 450–510 $\mu\Omega$ (10 GHz)	85	1 MA/cm^2 (77 K)	Conducts Inc. California
YBCO (double-sided)	CeO$_2$/Al$_2$O$_3$/CeO$_2$	Magnetron sputtering	3 inch	≤25 $\mu\Omega$ (77 K, 2.5 MHz) ≤20–40 mΩ (77 K, 145 GHz)	≥88	≥3 MA/cm^2 (77 K)	Forschungszentrum karisruhe, INFP and IMF I, P.O. Box 3640, D-76021 Karisruhe, Germany
YBCO (double-sided)	MgO, LaAlO$_3$, YSZ	Thermal co-evaporation	4 inch	40–70 mΩ (77 K, 94 GHz)	≥88	>2 MA/cm^2	Techmsche Universität München, Physik Department E10, 85747 Garching, Germany
YBCO (double-sided)	YSZ/Si, GaAs/MgO O	Thermal co-evaporation	9 inch	500 $\mu\Omega$ (77 K, 10 GHz)	≥88.4	2 MA/cm^2	Techmsche Universität München, Physik Department E10, 85747 Garching, Germany
YBCO (double-sided)	LaAlO$_3$, CeO$_2$/sapphire, CeO$_2$/YSZ	Magnetron sputtering	3 inch	100 $\mu\Omega$ (77 K, 10 GHz)	≥88	4.6 MA/cm^2	Dupont centrol Research and development, P.O. Box 80304, Wilmington, DE USA
YBCO (double-sided)	LaAlO$_3$	Magnetron sputtering	2 inch	0.3–0.7 mΩ (77 K, 10 GHz)	≥90		Westinghouse Science and Technology Center 1310 Beulah Road Pittsburgh, PA 15235

Table X.   (Continued).

Material	Substrate	Method	Area	R_s	T_c (K)	J_c	University Institute
YBCO	LaAlO$_3$, YSZ	Magnetron sputtering	2 inch	<0.5 mΩ (77 K, 10 GHz)	>88	>2 MA/cm^2 (77 K)	General Research Institute for Nonferrous Metals, Beijing 100088, P. R. China
YBCO (double-sided)	LaAlO$_3$, YSZ	Magnetron sputtering	20 × 25	<0.9 mΩ (77 K, 10 GHz)	>90	>2 MA/cm^2 (77 K)	General Research Institute for Nonferrous Metals, Beijing 100088, P. R. China
Au/ YBCO/Y$_2$O$_3$	YSZ single	Thermal co-evaporation	Φ 10 cm			1300 kA/cm^2	Technical University of Munich, Munich, Germany
YBCO/ Y$_2$O$_3$/YSZ-biax	YSZ poly	Thermal co-evaporation	10 × 10			30 kA/cm^2	Technical University of Munich, Munich, Germany
YBCO	YSZ poly	PLD	1 × 12.5			19 kA/cm^2	Technical University of Munich, Munich, Germany
YBCO/ YSZ-biax	YSZ poly	PLD	5 × 5			45 kA/cm^2	Technical University of Munich, Munich, Germany
Au/YBCO/ CeO$_2$	Al$_2$O$_3$ single	Magnetron sputtering	Φ 5			3000 kA/cm^2	Technical University of Munich, Munich, Germany

important and possible application potential. At present, power stations are increasingly under demand to provide power that is safe and of high quality. The FCL becomes an attractive apparatus for this purpose. The rapid switching characteristic of the superconductor from the superconducting state to the resistive normal state under the large current over I_c (critical current), which is automatically reset after the large current lash, provides a basis for such a FCL. So the FCL has received more and more attention. Cave et al. [521] have discussed such an application.

Ries et al. [522] reported the first results of the performance of the FCL with YBCO switching elements. For the FCL with short samples of switching elements, the YBCO films were deposited on YSZ substrate by pulsed laser deposition, for which the configuration was designed as shown in Figure 69. The thickness and critical current density are 0.5 to 4.5 micrometers and 5 to 35 kA/cm^2, respectively. The contacts for current and voltage are made by sputtering silver pads onto the YBCO

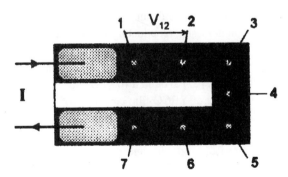

Fig. 69.   View of a U-shaped YBCO sample used for fault-current limiter. Substrate YSZ 20 × 10 × 0.2 mm^3 [522].

film and fixing wire by soldering, mechanical pressure, or chip-bonding to the pads.

The experimental setup is shown in Figure 70. All samples are mounted on a substrate holder, which is placed in a liquid

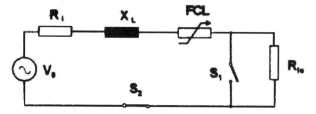

Fig. 70.   Limiter equivalent test circuit [522].

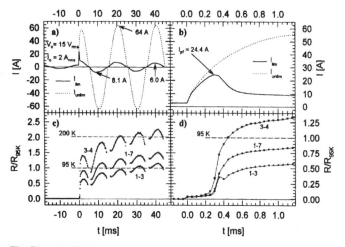

Fig. 71.   (a), (b) Limited and unlimited current as function of time. (c), (d) Normalized resistance between voltage pads as indicated in Figure 69 as a function of time. (b), (d) Expanded time scale [522].

nitrogen bath at atmosphere pressure. An ac voltage of 230 V and 50 Hz is transformed to the FCL test circuit. A typical experiment is on a sample with a 2.7-μm-thick YBCO layer with $I_C = 2$ A. The function of limitation of current is obvious; the experimental result is shown to compare with the unlimited case as in Figure 71. This indicates the application potential and prospect of high-T_c superconducting films on electric power systems.

The higher power FCLs have been designed and fabricated by Neumüller et al. [27] to build a 100 kVA power FCL. A critical current density of YBCO film of $\sim 3 \times 10^6$ A/cm^2 and an area of 400 cm^2 would be required, for which ceramic substrates of 4 and 8 inch sapphire are available.

For the physical properties and superconductivity, intrinsically, no difference exists between small-area and large-area thin films. But for the applications, the large-area thin films can be used to realize special purposes as mentioned above. The large-area thin film growth needs polycrystalline substrate, so a buffer layer is very necessary. For getting uniformity of thickness, composition, physical properties, etc., the technique and apparatus used to grow large-area thin films must be also special. So, in the future, it is most importat for this project to improve the substrate, buffer materials, and techniques.

4. TRANSPORT PROPERTIES IN HIGH-T_c SUPERCONDUCTOR THIN FILMS

4.1. Introduction

Ever since the discovery of HTSCs, much attention has been given to the studies of their transport properties such as resistivity, Hall effect, thermopower, and heat conductivity because they reveal the nature of electronic excitations near the Fermi surface and provide clues toward the ultimate understanding of high-T_c superconductivity mechanisms. HTSCs belong to strong correlated electron system where the electron–electron many-body effect plays an important role in determining the electronic properties. One of the important issues which remains unsolved in physics of the HTSCs is the nature of the characteristic excitations in the normal state. Many normal-state properties have been found to behave anomalously in contrast to the simple metallic or Fermi liquid behavior, possibly indicating the existence of quite exotic elementary excitations typical for strongly corrected electron systems. It is generally believed that in order to understand the mechanism of high-temperature superconductivity it is essential to understand their exotic normal-state properties.

Up to now, most of the reported works were performed on single crystals of LSCO, YBCO, BSCCO, etc., since they are available for many research purposes. But their sizes are very small and drag down research experiments, especially for nuclear magnetic resonance, far-infrared, and transport measurements for which the sample size is crucial. As a result researchers turn to epitaxially grown thin films in order to overcome the problem encountered in single crystals. There are many potential advantages to investigating the physical properties of HTSCs using thin films, compared with bulk samples. For example, the surface area of thin films is larger than single crystals and thus facilitates the measurements for infrared and Raman spectra for which a larger area can increase the signal strength and reduce the noise from the environment, resulting an improved signal to noise level. With a larger effective area it is also easy to make contacts with small resistance in the transport measurements. The configuration of the films can easily be fabricated to the desired pattern to facilitate the measurements. This is particularly true for resistivity, Hall effect, and critical current density measurements. The layered structure of HTSCs also makes thin film particularly useful to study the effect of dimensionality on the superconducting properties. The steadily advanced thin film process makes it possible to fabricate heterostructures based on HTSCs. As we know, HTSCs have a perovskite structure with large anisotropy, and artificial multilayers can be used to simulate the laminar structure of HTSCs and study the impact of interlayer coupling of CuO$_2$ planes on superconductivity and physical properties. By alternating the thickness of the insulating layer, it is possible for us to control the anisotropy and interlayer coupling to deeply study the HTSCs.

In this section we give a general review concerning what have been done and what is going on as for as the research of transport and optical properties of HTSCs using thin films. Because of the quality of thin films, we will mainly discuss the results obtained in those of YBCO, LSCO, and BSCCO. The landscape of this book limits us to research on thin films. This does not mean that work on single crystals is not important. However, to grow large-sized single crystals remains a task to be achieved in all the research laboratories. There are a lot of review articles concerning the electrical transport and magnetic properties of HTSCs in general; interested readers are referred to these articles [523, 524]. For the details of the general theory of superconductivity, readers are referred to the excellent book by Tinhkam [525].

4.2. Transport Properties

As conventional methods available in most research laboratories, transport property (resistivity, Hall effect, etc.) measurements have been widely used to evaluate HTSCs to obtain various parameters. It was found that for most of the cuprates such as LSCO and YBCO, the dominating carriers are holes residing in the CuO_2 plans. Transport measurements are also routinely utilized to determine the critical current densities and upper critical fields that are important parameters for the application of HTSCs as well as for fundamental research. Meanwhile, by measuring the resistivity in the mixed state under various magnetic fields, a lot of new physics associated with the high operation temperature, disorders, and anisotropy of oxide superconductors has been observed. In the following we will try to summarize some research work in this field performed on thin films.

4.2.1. Normal-State Resistivity

Among the remarkable features of HTSCs are the extraordinary transport properties in the normal state, which are inconsistent with the conventional electron–phonon scattering mechanism and have received much attention for the further consideration that the excitation interacting with carriers in the normal state might play an important role in the occurrence of high-temperature superconductivity. The dc resistivity yields information about single electronic states involving excitation of the electronic system where a transition takes place from an initial state to a different final state with the same energy. Reflecting the large anisotropy associated with their layered structure, the resistivity of HTSCs is characterized by large anisotropy for electronic transport in the a–b plane and along the c-axis. Usually the in-plane resistivity exhibits metallic behavior, and the out-of-plane resistivity has metallic behavior for optimal doping and a semiconducting behavior for underdoping. It is of great importance to understand the charge dynamics governing the electron scattering inside HTSCs. In some ways YBCO is a typical and relatively simple material for investigating the properties in the underdoped side of the phase diagram because the impurity potentials introduced by oxygen vacancies in the Cu–O chains can be expected to have less effect than that caused by Sr^{2+} ion substitution of La in the layer adjacent to the CuO_2

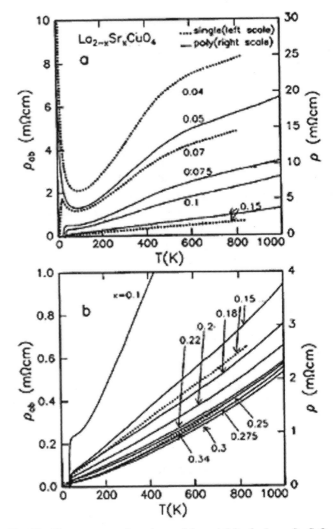

Fig. 72. The temperature dependence of the resistivity for $La_{2-x}Sr_xCuO_4$ single-crystal thin films [527].

planes in LSCO. In the mean time, from the quality point of view, YBCO thin film at present is the best and most reliable among all the HTSC thin films. Therefore most of the research has concentrated on YBCO thin films.

4.2.1.1. In-Plane Resistivity

A remarkable common feature of the normal-state in-plane resistivity of optimally doped HTSCs is the linear behavior over a large temperature range, i.e., $\rho_{ab}(T) = \alpha T + \beta$, α and β being two temperature independent parameters. The linear resistivity is observed without saturation up to 600 K for YBCO and above 1000 K for LSCO [526]. As we know, for a normal Fermi liquid, the resistivity caused by electron–electron scattering obeys a T^2 dependence. This linear dependence of $\rho_{ab}(T)$ has attracted much attention as a possible indication of exotic normal metallic state in HTSCs. The representative $\rho_{ab}(T)$ curves for $La_{2-x}Sr_xCuO_4$ films with different doping levels obtained by Takagi et al. are shown in Figure 72, from which it can be seen that a linear dependence of $\rho_{ab}(T)$ exists for the film with optimum doping at $x = 0.15$ [527]. Such linear behavior of $\rho_{ab}(T)$ is closely related to the $1/\omega$ decay of the free carrier conductivity at optical frequencies, which reflects an anomalous frequency dependent scattering rate proportional to ω instead of ω^2, as expected for a conventional Fermi liquid. For high-quality samples, the linear dependence of $\rho(T)$ can extrapolate to $T = 0$ K, giving a zero or even negative residual resistivity, in sharp contrast with a conventional metal in which a positive residual resistivity at $T = 0$ K is commonly observed and $\rho_{ab}(T)$ saturates at high temperature when the inelastic mean free path of the electrons approaches the lattice spacing. The absence of resistivity saturation at high temperatures is in sharp contrast with the behavior of the strong electron–phonon coupled conventional superconducting materials, possibly indicating a weak electron–phonon coupling in HTSCs. It was pointed out that the electron–phonon coupling constant estimated from the relation without $\rho_{ab}(T)$ saturation in HTSC is too small to account for the high T_c in terms of electron–phonon interaction [526]. No sign of saturation for $\rho_{ab}(T)$ of HTSCs also indicates that the mean free path l in HTSCs is much longer than the interatomic spacing. The long mean free path at high temperatures also excludes the existence of strong electron–phonon coupling.

For the resistivity dominated by electron–phonon scattering, the temperature dependence is given by the Bloch–Grüneisen formula,

$$\rho(T) \propto \left(\frac{T}{\Theta}\right)^5 \int_0^{T/\Theta} \frac{x^5\,dx}{(e^x - 1)(1 - e^{-x})} \tag{12}$$

where Θ is the effective Debye temperature. Equation (12) gives a nonlinear decrease of the resistivity for temperatures below a characteristic temperature T^* and linear resistivity for $T > T^*$, where $T^* = \eta\Theta$ with η a constant between 1 and 2 [528]. Martin et al. examined the $\rho_{ab}(T)$ of YBCO and BSCCO systems with the BG formula. They found that although the $\rho_{ab}(T)$ curves of YBCO and BSCCO can be fitted

with Eq. (12), an unreasonably low Debye temperature is obtained. For YBCO, the fitting value of Θ is about 200 K which is far smaller than the Θ value of 360 K determined from the specific heat measurements. For $Bi_2Sr_2CuO_{6+x}$ which has a T_c value of 10 K, the Debye temperature turns out to be 35 K which is out of the physically reasonable region. However, for $YBa_2Cu_4O_8$ (Y124) thin films the in-plane resistivity in can be fitted very well with Eq. (12) [528]. The result is shown in Figure 73. They argued that instead of T^* an effective Debye temperature $\Theta^* = (2k_F/G)\eta\Theta$ (k_F being the Fermi wave vector and G the reciprocal lattice vector) should be the temperature where the nonlinearty of resistivity at the low-temperature side appears. With their experimental data on Y124 film, they obtained a Θ^* value of 500 K.

The linear temperature dependence can also be obtained from band structure calculations in which the main contribution to transport current relaxation is supposed to come from the scattering of electrons by phonons. The transport relaxation time at high temperatures is estimated by

$$\frac{\eta}{\tau_{tr}} = 2\pi\lambda_{tr}kT\left(1 - \frac{\eta^2\langle\omega^2\rangle}{12k^2T^2}\right)$$

The electron–phonon coupling constant λ_{tr} is determined by the function $\alpha_{tr}^2 F(\omega)$ with the matrix element of electron–phonon interaction α_{tr}, which determines the relaxation of quasi-particle momentum. Using the experimental phonon density of states for LSCO and YBCO, a linear dependence of $\rho_{ab}(T)$ can be obtained over a wide temperature range [529].

For the parabolic energy bands $\varepsilon(k)$ of the conducting electrons, $\sigma \approx ne^2\tau/m^*$, where the hole density n calculated from the Hall density at 100 K is $\approx 6 \times 10^{21}$ cm^{-3}, the band effective mass m^* is about six times that of m_e, and scattering above T_c is due to the inelastic processes with a rate $\eta/\tau \approx 2kT$. The linear dependence on T of $\rho_{ab}(T)$ indicates that kT is much larger than the characteristic energy of the boson scattering and infers relatively weak coupling.

Both LSCO and YBCO are hole-type superconductors. There exists a series of oxides with electron-type conductivity among which $Nd_{2-x}Ce_xCuO_{4-\sigma}$ (NCCO) is a representative one. Its superconductivity is induced by substituting Nd^{3+} with Ce^{4+}, resulting in a CuO_2 plane with electron doping. It is of great interest to study the carrier dynamics in this material which may have similar superconducting mechanisms as those of YBCO and LSCO. The in-plane resistivity and magnetoresistance of $Nd_{1.85}Ce_{0.15}CuO_{4-\sigma}$ with different oxygen concentrations have been studied by Jiang et al. [530]. It was found that for the superconducting samples with superconductivity, the resistivity shows an upward curvature at low temperatures. The increasing oxygen concentration results in an enhanced residue resistivity with $d\rho/dT$ unaffected for $T > 80$ K, indicating an increasing impurity scattering by excess oxygen. The magnetoresistance in the upturn regime shows a crossover from positive to negative as the magnetic field is increased, which is a typical behavior induced by the competition between superconductivity and localization. The magnitude of

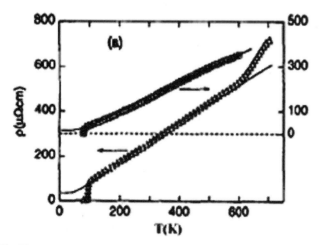

Fig. 73. Temperature dependence of ρ_{ab} of Y123 and Y124 thin films. The solid lines are BG fits with effective Debye temperatures $\Theta^* = 200$ K for Y123 and $\Theta^* = 500$ K for Y124 [528].

the Hall coefficient R_H decreases with decreasing oxygen content and R_H changes sign from negative to positive. This sign change could originate from the existence of two conducting bands inside NCCO. The holelike band could resulte from the buckling of the CuO_2 planes in NCCO. The removal of oxygen induces lattice distortion which affects the electronic states of the oxygen $2p$ bands. This could raise the energy of the oxygen p_z band through the Fermi level and result in lattice distortion induced hole state. Therefore, inside NCCO highly mobile impurity doped electrons coexist with a lattice distortion induced hole band on the CuO_2 planes. The occurrence of superconductivity in NCCO when oxygen is removed from the oxygenated sample is accompanied by the appearance of the hole band, which suggests that holes are also crucial for the occurrence of superconductivity in the electron-doped cuprates.

4.2.1.2. Doping Dependence

All the HTSCs have a common property in their crystalline structures; i.e., they are formed on the basis of a layered perovskite structure. Center to the structure are the CuO_2 planes separated by layers of other ions. The parent materials for all the HTSCs are Mott insulators showing long range antiferromagnetism. Charge carriers are introduced into the Mott insulator by doping. Hole carriers are introduced either by cationic substitution or by oxygen intercalation, the system undergoes an insulator to metal transition, and superconductivity emerges. The composition that gives the maximum T_c for a given system is sometimes called optimally doped as mentioned above. The lower carrier density side of the T_c maximum is referred to as the underdoped region and the opposite side as the overdoped region. The physical properties of the oxide superconductors are strongly influenced by the oxygen deficiency and the variation of the substitution of cations. The above mentioned linear temperature dependence of ρ_{ab} only exists in optimally doped samples. The overdoped samples with a high hole concentration showing a characteristic metallic conductivity with ρ_{ab} and ρ_c vary as T^2 is observed. In underdoped materials, an opening of a pseudogap in the normal-state has been observed. The presence of a normal-state pseudogap makes the carrier transport incoherent and results in a deviation of resistivity from the linear temperature dependence.

By measuring the high-quality single-crystal thin film of LSCO with different Sr substitutions, it is found that the linear behavior of $\rho(T)$ is confined to a narrow compositional range for optimum superconductivity [527]. The evolution of the in-plane resistivity upon Sr doping is given in Figure 72. It is evident that the T-linear reisistivity behavior holds true only for the Sr doping in which $x = 0.15$. Upon the decrease of Sr doping the resistivity increases. In the underdoped region below $x \sim 0.1$ the resistivity shows a pronounced S-shaped temperature dependence. At high temperatures, the slopes gradually decrease, indicating a saturation behavior which is in contrast with the behavior observed in optimally doped samples. It is widely accepted that the resistivity in a metal saturates at the high-temperature region where the mean free path l of the

charge carriers becomes comparable to the minimum scattering length, i.e., the interactomic spacing. An analysis of the magnitude of the resistivity with Boltzman transport theory yields a small Fermi surface rather than a large Fermi surface expected from a half-filled band. In the overdoped region, a power law $\rho(T) = \rho_0(T) + AT^n$, with $n = 1.5 \pm 0.2$, was observed. Such a power-law dependence is in contradiction to that of other HTSCs which show Fermi liquid behavior in the overdoped region for which the electron–electron scattering gives a T^2 dependence.

The evolution of ρ_{ab} upon the doping for the YBCO series shows similar behavior [531, 532]. The resistivity decreases with the increasing doping level. The resistivity at 290 K decreases from 4 mΩ cm for $\delta = 0.7$ to 300 $\mu\Omega$/cm for $\delta = 0.05$. Shown in Figure 74 is the temperature dependence of ρ_{ab} for YBCO with different oxygen concentrations. A reduction of oxygen concentration from the optimal doping level induces a systematic change from the linear temperature dependence; first a downward bending develops at low temperature. When $\delta > 0.5$ a upward bending develops at low temperature. The sample with $\delta = 0.7$ is not superconducting and shows a remarkable upward bending, indicating the onset of localization effects [531]. The residual resistivity obtained by extrapolating from high temperature does not increase appreciably as the doping level decreases, indicating that the main effect of introducing oxygen vacancies in the chain layer is to reduce the carrier concentration on the planes with no significant increase in scattering from random defect potentials [532]. Although up to now there is no definite explanation for the nonlinear behavior of ρ_{ab} in underdoped YBCO samples, it is proposed that such behavior is associated with the normal-state pseudogap as observed from far-infrared reflectivity and angular resolved photoemission spectroscopy [533, 534]. Levin and

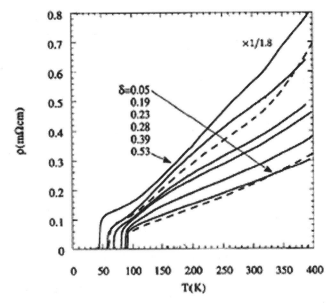

Fig. 74. Temperature dependence of ρ_{ab} for $YBa_2Cu_3O_{7-\delta}$ thin films with $0.53 < \delta < 0.05$. The dashed lines show $\rho_{ab}(T)$ data for single crystals of $YBa_2Cu_3O_{7-\delta}$ with $\delta = 0.0$ and 0.4 [531].

Quader proposed that the origin of the nonlinearity could be the coexistence of nondegenerate and degenerate carriers possessing different quasi-particle relaxation rates [535]. In the underdoped regime the change of slope $d\rho_{ab}/dT$ at T^* results from the thermal activation of nondegenerate carriers. In the overdoped regime, the nondegerate carriers tend to become degenerate and a second small Fermi energy emerges, resulting in a T^2 behavior in ρ_{ab}.

Superconducting YBCO has an orthorhombic crystalline structure which indicates that the resistivities along a- and b-axes are different. Measuring of the in-plane anisotropy is hindered by the difficulty in removing twinning. It is reported that at room temperature the anisotropy for the resistivity along the two main axes in the CuO_2 planes $\rho_b(T)/\rho_a(T)$ is roughly 2, while for thin films with unidirectional twinning, the value is about 6 [536]. The anisotropy of in-plane resistivity indicates a large contribution of the Cu–O chains along the b-axis to conductivity in YBCO.

4.2.1.3. c-Axis Resistivity

There is a large anisotropy for the resistivity in the a–b planes and along the c-axis. The anisotropy of the resistivity at room temperature for optimally doped samples attains a ρ_c/ρ_{ab} value of about 30 for YBCO, 200 for LSCO, and 10^5 for BSCCO [537]. Either a decrease of temperature or a reduction of hole concentration in CuO_2 planes results in a larger anisotropy due to a more rapid growth of ρ_c which is extremely sensitive to the oxygen stiochiometry. A linear dependence for $\rho_c(T)$ is sometimes observed in optimally doped samples; however, with a decreasing hole concentration, $\rho_c(T)$ tends to demonstrate a semiconducting behavior. More specifically, in samples with ρ_c below the Ioffe–Regel limit (for LSCO, it corresponds to $\rho_c \approx 10$ mΩ/cm) $d\rho_c/dT$ tends to be positive and above the Ioffe–Regel limit the sample generally shows semiconducting behavior with $d\rho_c/dT < 0$. Such a strong anisotropy and change of temperature dependence of $\rho_c(T)$ for a decreasing carrier concentration are in favor of the suggestion that $\rho_{ab}(T)$ and $\rho_c(T)$ have different natures. No definite correlation between ρ_c and T_c has been observed, which opens a question for the assumption that interlayer coupling is crucial for the enhancement of T_c.

Due to lack of high-quality a-axis oriented thin film for which the c-axis is accessible for transport measurements, few works have been done for the charge dynamics along the c-axis using thin films. A puzzling phenomenon in the transport properties of the HTSCs is the c-axis resistance peak near T_c often observed in single crystals as well as in thin films. It was found that near the onset of the transition the fluctuation is strong and the coupling between the CuO_2 layers is weak. This may be the main reason that the resistance measured across the superconducting layers shows a large peak near T_c. The position of the peak shifts to lower temperatures with increasing applied magnetic field from which $\rho_c(T)$ deviates from the extrapolation line and also decreases with increasing magnetic field. This is in contrast to the temperature dependence of $\rho_{ab}(T)$ for

which the temperature where $\rho_{ab}(T)$ deviates from the extrapolation and hardly changes with the magnetic field. Different mechanisms including interlayer tunneling, phase slippage, and thermal fluctuation have been proposed to explain such a peak. The magnetoresistance $\rho_c(B, T)$ was measured by Nygmatulin et al. on BSCCO thin films with a standard four-probe method up to a magnetic field of 8 T [538]. It was found that the behavior of ρ_c can be fitted well by using the fluctuation theory by including both the Maki–Thompson and Aslamazov–Larkin terms. The resistive peak originated from a competition between the paraconductivity dominating in the high-temperature regime and diaconductivity associated with the fluctuation decreases with the quasi-particle density of states in the temperature regime near T_c. The large anisotropy of HTSCs results in a reduction of the positive contribution to the hopping conductivity for a single particle propagating along the c-axis. On the other hand, the formation of nonequilibrium Cooper pairs gives rise to a suppressed single particle density of state at the Fermi level near T_c, resulting in a negative contribution to the fluctuation conductivity [539]. The relevant physical parameters obtained from the experimental fitting are $T_c\tau = 0.65 \pm 0.06$ and $T_c\tau_\phi = 9.5 \pm 1.5$, where $T_c = 89.8$ K and τ and τ_ϕ are the quasi-particle scattering time and phase pair-breaking lifetime, respectively.

4.2.2. Normal-State Hall Effect

The Hall effect plays an important role in determining the polarity and density of the charge carriers participating in the electronic transport. This is especially important for understanding the superconductivity mechanism. To measure the Hall effect, a transport current is flowing in the longitudinal direction while the voltage drops along both the longitudinal and transverse directions are taken. Due to the special geometry used in Hall effect measurements, thin films have an advantage over single-crystal samples they are since easy to make the required pattern on thin film sample photolithographically. From the Hall effect it is known that the dominating carriers in most of the HTSCs are holes and those holes form Cooper pairs upon lowering the temperature below T_c.

For the Hall measurements, the results are usually expressed in terms of Hall coefficient $R_H = -1/ne$, Hall mobility μ_H, and Hall angle $\tan\theta_H = \rho_{xy}/\rho_{xx} = \omega_c\tau_H$, where ρ_{xx} is the longitudinal resistivity, ρ_{xy} is the transverse resistivity, and τ_H is the scattering time. It is sometimes convenient to express R_H in terms of Hall density $n_H = 1/(R_He)$ which is generally normalized to unit cell volume. For HTSCs, the carrier density can be expressed as the number of mobile holes per Cu ion.

Like the in-plane resistivity, the Hall effect in the normal state of HTSCs has been of great interest due to the unusual temperature dependence of the Hall coefficient, e.g., $R_H \propto 1/T$ in YBCO [540]. This is in sharp contrast with what is usually observed in a normal metal whose R_H is a constant since it only depends on the carrier concentration which is independent of temperature. For general band structures, R_H involves

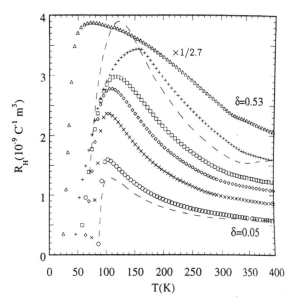

Fig. 75. Temperature dependence of R_H for YBa$_2$Cu$_3$O$_{7-\delta}$ with different oxygen concentrations. The measurement field was 7 T. The dashed line shows data for two single crystals of YBa$_2$Cu$_3$O$_{7-\delta}$ with $\delta = 0.0$ and 0.4 [531].

an average over the Fermi surface of derivatives of the dispersion $\varepsilon(K)$, weighted by the scattering time $\tau(K)$. It is difficult to explain the phenomenon $R_H \propto 1/T$ in the conventional band structure models. $R_H(T)$ of the high-T_c superconducting thin films has the same linear dependence on temperature as for crystals but its value is about two times lower than that in single crystal. Moreover, whenever the superconducting transition temperature is suppressed, the temperature dependence of R_H is also suppressed. In the mean time, numerous doping studies have shown that the temperature dependence of $1/R_H$ varies with the carrier density. The Hall coefficient for optimally doped samples varies as $1/T$, which results in the $1/T^2$ dependence of the Hall mobility μ_H. The $1/T^2$ dependence of μ_H is observed for HTSCs with different doping levels. The mechanism for such an anomalous behavior has been a controversy for a long time. Tremendous efforts have been devoted to understand the Hall effect in HTSCs in order to find clues for the mechanism of high-temperature superconductivity.

The doping dependence of the Hall effect has been studied by Carrington et al. and Wuyts et al. on YBCO thin films [531, 532]. Shown in Figure 75 is the result reported by Carrington et al. Upon decreasing doping, R_H increases, indicating a reduced carrier density. In the mean time, while the Hall coefficient deviates significantly from the $1/T$ behavior, the $1/T^2$ dependence of the Hall angle remains unchanged for all the doping levels. However, the $1/T^2$ dependence of the Hall mobility is not observed in Nd$_{0.1}$Sr$_{0.9}$CuO$_2$ thin films in the Hall effect measurements by Jones et al. [541]. Instead an anomalous T^3 inverse Hall mobility was observed, suggesting a mechanism other than electron–electron scattering.

4.2.3. Critical Current Density

Critical current density is determined as the highest current density that can pass through a superconductor without generating any resistance in the superconducting state. In actual measurements, critical current density is decided as the current density at which a certain voltage is generated. The application of a superconductor to a larger extent is determined by its ability to carry high dissipation-free current. One task for early work on thin film fabrication was to improve the T_c and J_c. Since the current passing through the electrodes will generate heat and thus affect the real temperature and destabilize the temperature at the film surface, to measure the critical current density, normally the film is patterned photolithographically to a microbridge with a typical dimensionality of 150×50 μm^2 in order to reduce the current passing through it. It is important to notice the difference between critical current density and pair breaking current density. The latter is referred to as the current density to breakdown the Cooper pair, which is generally more than one order higher than the critical current density.

The critical current density at zero temperature $J_c(0)$ can be estimated in the clean limit as $J_c(0) = ne^*\eta/\xi(0)m^*$. For YBCO, $J_c(0)$ is about 1.2×10^9 A/cm^2. In practice, the critical current density measurements give the dissipation-free current up to which the pinning force can stop the motion of the vortices. However, as we will mention below, when a finite transport current is flowing or an external magnetic field is applied, vortices will be generated inside the thin films. Those vortices will move under the influence of the Lorentz force. The motion of vortices will cause energy dissipation and give rise to a finite resistivity. The ability of a superconductor to carry dissipation-free current is due to the fact that pinning centers exist in the superconductor which provide the potential barrier to trap the vortices and block their motion. To improve the critical current density, it is important to study the pinning behaviors of the superconductor under magnetic field. Without the presence of pinning centers, the vortices will always move when they feels the Lorentz force and no absolutely zero-resistivity J_c can be obtained. However, at the presence of pinning centers which are generally some small scale defects inside the thin films, the vortices will be trapped in the pinning centers, resulting in zero resistivity. To date, for most YBCO thin films the J_c value is of the order of 10^6 A/cm^2. The highest reported value reached 5.0×10^7 A/cm^2 [542].

4.2.4. Vortex Dynamics and Dissipation Mechanisms

4.2.4.1. Kim–Anderson Theory

Because of the electromagnetic interaction and the elasticity of the vortices, in the presence of pinning centers, the vortices will prefer to form bundles in order to reduce the free energy of the vortex lattice. Each bundle is pinned by a pinning potential U. Under the action of thermal energy, the vortex bundle has the tendency to overcome the barrier to move with the escape probability proportional to $e^{-U/kt}$. As a result the jump rate for the vortex bundle to leave the barrier is

$$R = 2\pi \nu_0 e^{-U/kT} \tag{13}$$

where ν_0 is a characteristic frequency of the vortex bundle vibration.

When a transport current or flux density gradient is introduced, the potential barrier will tilt along the direction of current flow or flux density gradient, making the jump easier in the direction of decreasing flux density than in the opposite direction. The net jump rate will be the difference between the jump rate along the Lorentz force and opposite to it, i.e., $R = R_+ - R_- = 2\pi\nu_0 e^{-U/kT} \sinh JU/J_0 kT$, where J is the applied current density and J_0 is a characteristic current density.

4.2.4.2. Flux Pinning and Flux Creep

For a perfect type II superconductor, under magnetic field, any small current passing it will result in flux flow and in turn induce finite resistivity because of the Lorentz force exerted on the vortices. However, when there are defects inside a type II superconductor, those defects can act as pinning centers which would prevent the motion of vortices. When vortices are pinning by pinning centers, the vortices resemble the particles confined in a potential wall in quantum mechanics. At finite temperature, the thermal energy will assist the vortices in overcoming the potential barrier in that they have a tendency to move. With applied current, the potential barrier will be tilted and result in a diminishing energy barrier for the vortices to overcome; then flux creep occurs and finite resistivity appears caused by the motion of vortex core. When the transport current density is large enough, the potential barrier for the vortex motion will be smear out and the vortices will undergo free flow. Such a situation is called flux flow. The resistance ρ_{ff} from flux flow has been studied theoretically by Bardeen and Stephen who found that $\rho_{ff} \approx \rho_n H/H_{c2}$. It is well known that at H_{c2} the vortex cores of the Abrikosov lattice overlaps; therefore, the flux flow resistivity in this case equals to the normal state resistivity, consistent with that predicted by Bardeen–Stephen theory.

A unique phenomenon commonly observed in the HTSCs is the broadening of resistive transition under applied magnetic fields. For c-axis-oriented YBCO thin films, in case of the field parallel to the a–b plane, due to the strong intrinsic pinning, the transition width broadening is not obvious. When the field is perpendicular to the CuO_2 plane, the transition width is significantly enhanced although the onset temperature is hardly changed. Figure 76 shows the resistive transition broadening observed in an YBCO thin film under different magnetic fields [543]. The transition width (defined at the temperature interval between $0.1\rho_n$ and $0.9\rho_n$) increases considerably as the magnetic field increases. This is much different from that observed in conventional superconductors for which only a downshift of T_c rather than a broadening appears. Many explanations have been proposed to understand such an unusual behavior. Among them, thermally activated flux flow (TAFF) has been proved to be most convenient and physically meaningful especially for experimentists [544]. Due to the combination of large

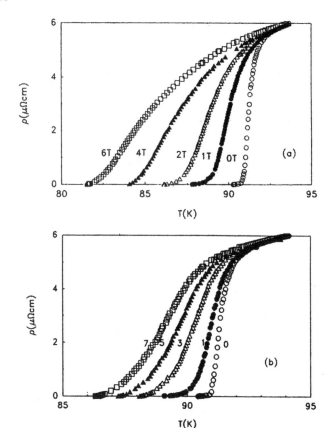

Fig. 76. Resistive transition at different applied magnetic fields for YBCO thin film. (a) $H//c$-axis; (b) $H \perp c$-axis [543]. Reprinted from *Physica C* 179, 176 (1991), with permission from Elsevier Science.

anisotropy, high transition temperature, and small coherence length, the pinning strength is weak and the potential barrier that traps the vortices is small, resulting in a high jumping probability for the vortices to overcome the barrier which gives rise to dissipation. This is the center point of the TAFF model. The TAFF model predicts a surprisingly simple form of resistivity, i.e.,

$$\rho = \rho_0 \exp(-U/kT) \tag{14}$$

where ρ_0 is a prefactor and U is the activation energy. It turns out that Eq. (14) can fit most experimental data quite well. To analyze the experimental data, the $\rho-T$ curves are usually plotted in an Arrhenius form $\ln \rho$ vs $1/T$. From the slopes in the Arrhenius plots, the activation energy for vortices can be obtained which can be used to obtain information regarding the mechanism of pinning and dissipation. Qiu et al. measured the field induced resistance broadening in oxygen deficient thin films as shown in Figure 77 [545]. Through a TAFF analysis with the help of Arrhenius plots, they observed two different TAFF regions which have different field dependent activation energies. The activation energy is found to be proportional to the inverse square root of B in the low-temperature region, while in the high-temperature region, the activation energy shows a logarithmic dependence of B. They proposed that the vortices are in the liquid state in both regions. In the low-

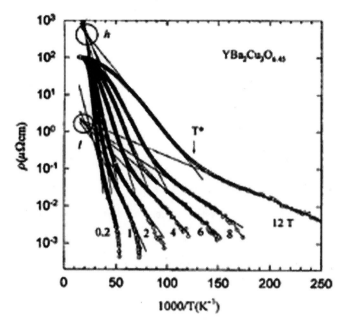

Fig. 77. Arrhenius plot of the resistive transition under different magnetic fields for an oxygen deficient YBCO thin film [545].

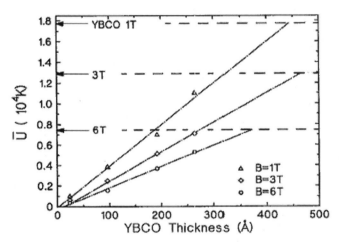

Fig. 78. Activation energy at 1, 3, and 6 T as a function of the YBCO thickness in multilayers with thick PBCO separation thickness [546].

temperature region, the $1/B^{0.5}$ dependence of $U(B)$ suggests that the vortex is in the 3D liquid state in which the vortices lose their transverse coherence while keeping the longitudinal coherence. The dissipation is controlled by the plastic deformation of the vortices through the generation of vortex double kinks. As the temperature increases to the high-temperature region, an interlayer coupling occurs which results in a quasi-2D behavior of the superconductor. The vortices are decoupled into 2D pancake vortices residing in each CuO_2 plane. The vortices in each CuO_2 plane become irrelevant to those in other CuO_2 planes. And the dissipation in this temperature region is governed by the nucleation and motion of edge dislocation pairs of the vortices.

The behavior of activation energy of vortices has also been used to analyze the interlayer coupling in YBCO/PBCO mul-

tilayers [546]. By measuring the $\rho(T)$ for YBCO/PBCO multilayers with different PBCO or YBCO layer thicknesses, the activation energies of the vortex are obtained as shown in Figure 78. It is found that the activation energy increases with increasing YBCO layer thickness d_{YBCO} while keeping the PBCO layer thickness at 96 Å. The linear dependence of $U(B)$ upon d_{YBCO} suggests that the YBCO layers are decoupled and that the vortices in each YBCO layer move independently. The $U(B)$ in these samples shows a logarithmic behavior, indicating a dissipation controlled by generation and subsequent motion of dislocation pairs as we expected in a two-dimensional sample.

The magnetoresistance of BSCCO thin films has been studied by Raffy et al. [547]. From the dependences of the resistance on applied magnetic field, temperature, and angle between field and film surface normal, they obtained a nice scaling behavior of the magnetoresistance and found that the magnetoresistance could be scaled with $H \sin \theta$, which means that only the field component along the surface normal contributed to the magnetoresistance. Such a result is a direct consequence of the two-dimensionality of the BSCCO. A similar result was obtained in YBCO/PBCO multilayers. However, the situation is different in the 90-K phase YBCO which has a smaller anisotropy than BSCCO; the magnetoresistance scale with $\cos^2 \theta + \sin^2 \theta$, which is the typical behavior for an anisotropic 3D superconductor.

4.2.4.3. Irreversibility Line and Upper Critical Field

The nature of the flux line structure and the phase transition between different vortex states have become central issues in the study of the physics of the vortex state in superconductors. In HTSCs, strong fluctuations greatly modify their H–T phase diagrams in which the presence of an irreversibility line separates the mixed state into two regimes. Below the irreversibility line the HTSCs are in a true superconducting state while above it a finite resistivity is always present due to the thermally activated flux flow. The nature of the irreversibility line is of great interest in the research of physics of the vortex state. It was found that the nature of the two regimes depends on the dimensionality and strength of the disorders inside the sample. For a sample with perfect crystalline structure in which pinning can be nearly neglected, the irreversibility line represents a first order transition line. While in samples with strong disorders, a second order transition sets in, replacing the first order transition.

The broadening resistive transition makes the determination of the upper critical field $H_{c2}(T)$ ambiguous. The small pinning potential also results in a high relaxation rate for the vortex motion and the magnetization of HTSCs remains reversible over a large temperature range below the onset temperature. The hysteretic magnetization associated with screening currents only occurs below the irreversibility line. It is important to notice that the irreversibility line is different from the $H_{c2}(T)$ line. The formation of the irreversibility line depends on the dimensionality and pinning strength of the sample. It is commonly believed that the irreversibility line corresponds to the first order melting line of the superconductor with weak pinning, the second order glass transition line caused in the superconductors with strong pinning.

4.2.4.4. Vortex Melting

The vortex dynamics in the mixed state of HTSCs remains a subject of intense research because of its fundamental importance for physics of the vortex matter. It was predicted theoretically and confirmed experimentally that the vortex system undergoes a second order transition from a high-temperature vortex liquid to a low-temperature vortex glass or Bose glass in superconductors with strong disorders. In a clean superconductor a first order transition from vortex liquid to a well ordered vortex lattice occurs.

For perfect HTSCs without pinning centers, an applied magnetic field will generate rigid and straight vortex lines with long range order in the transverse direction. The thermal vibration will disturb the equilibrium position of the vortex lines. It is expected that the long range order will be destroyed eventually via a first order melting transition as the temperature is raised. For a conventional superconductor, the melting transition can hardly be observed since it occurs near the $H_{c2}(T)$ line where the order parameter diminishes. However, for HTSCs, large thermal fluctuations enhance the lattice vibration and the melting transition line lies far below the mean field $H_{c2}(T)$ line, which makes it visible for experiment observation.

A flux line liquid is characterized by an absence of long range transverse order and a diminished shear modulus c_{66}. The melting transition temperature is defined as the temperature at which the magnitude of the displacement of the vortex line from its equilibrium position is larger than a critical portion of the vortex lattice spacing; i.e., $\Delta R = c_L a_0$, where $c_L \approx 0.1$–0.2 is the Linderman number. Due to the small entropy change associated with the melting transition, it is difficult to observe it thermodynamically by specific heat measurements. The melting transition demonstrates itself as a sharp resistive drop in the R–T transition. Because of the softening of the vortex lines, the resistance is found to be increased rapidly upon the melting transition from a vortex lattice state to a vortex liquid state. The first experimental evidence for the vortex melting was observed by Kwok et al. on YBCO single crystals who found a sharp resistive drop in the R–T transition curves, and there was a hysteresis for measurements upon warming and cooling samples [548].

For thin films in which the vortex mostly shows two-dimensional features, namely the longitudinal correlation length L_c is larger than the film thickness, it is preferred that the melting transition proceed via the Kosterlitz–Thouless transition through the small edge dislocation pairs. Due to the finite energy for the generation of small edge dislocation pairs, at a certain temperature there will be no highly ordered 2D vortex lattice up to some length scale. The 2D vortex lattice becomes unstable due to the unbinding of dislocation pairs thermally created. The shear modulus c_{66} drops sharply to zero when the vortex lattice melting criterion is fulfilled,

$$Ac_{66}a_0^2 d/kT = 4\pi \tag{15}$$

where a_0 is the vortex lattice spacing and A is a parameter representing the renormalization of c_{66} due to nonlinear lattice vibration and vortex lattice defects. Such a melting transition in 2D vortex lattice was observed by Berghuis et al. [549] in amorphous Nb_3Se thin films and by Qiu et al. [43] on YBCO/PBCO multilayers from an analysis of I–V characteristics. Their work suggests that a 2D vortex lattice can be melted locally within a scale of Larkin length through the unbinding of thermally generated dislocation pairs.

4.2.4.5. Vortex Liquid and Vortex Glass

A vortex lattice is a system which resembles the crystal lattice with elasticity. For HTSCs, due to the large anisotropy and thermal activation, the vortex lattice has a small shear modulus c_{66} and it would loose its rigidity along the c-axis and the vortices could be distorted along the c-axis and cause the system to behave like a 3D system. Under certain conditions the vortex lattice can be melted due to thermal fluctuations over a significant portion of the B–T phase diagram. Within this regime, crystalline long range order is absent; however, the superconducting order parameter is still preserved locally on a length scale smaller than the mean vortex spacing. As a result, a new configuration called a vortex liquid can be present in the H–T phase diagram. A vortex liquid is characterized by a small c_{66} and its resistivity generally follows that predicted by the TAFF theory.

As we mentioned previously, an Abrikosov lattice is not stable upon the flowing of a transport current. At a finite temperature, thermal energy will assist the flux to jump over the potential barrier. Therefore a theoretical zero-resistivity state cannot exist at an applied magnetic field when current passes through it. This could be a serious problem for the application of HTSCs that depend heavily on their ability to carry dissipation-free current. It was proposed by Fisher et al., by taking into account both pinning and collective effects of the vortex lines, that there should exist a state called vortex glass [550]. Due to the elasticity of the vortex system, in the presence of pinning centers they tend to form bundles with a characteristic scale of Larkin length L_c. At finite strength of disorders, upon the decreasing current density the size of the vortex bundle increases and the energy barrier for the vortex motion grows exponentially. Therefore a virtual state with zero dissipation should exist which Fisher called a vortex glass. A vortex glass phase is a true superconducting one with zero resistivity. An equilibrium phase boundary at a well defined temperature in the H–T phase diagram is expected.

The basic idea of the glassy behavior is that if there is a second order phase transition to this vortex glass similar to that in the theory of spin glasses, then both characteristic correlation length ξ and relaxation time τ_g of the fluctuations of the glassy order parameter diverge at the glass transition temperature T_g; i.e.,

$$\xi_g(T, B) = \xi_g(T, B) \left| 1 - T/T_g \right|^{-\nu}$$
$$\tau_g(T, B) \cong \tau_g(B) \left| 1 - T/T_g \right|^{-\nu z} \tag{16}$$

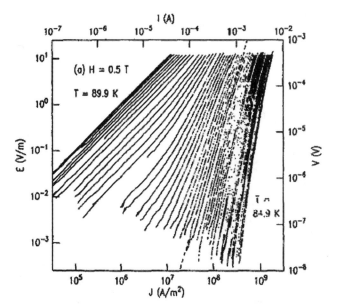

Fig. 79. $I-V$ curves measured on YBCO thin film at constant T for $H = 0.5$ T. All the $I-V$ curves can be scaled onto two universal curves [551].

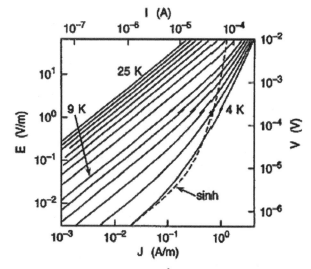

Fig. 80. $I-V$ curves measured in a 16-Å ultrathin YBCO film at a magnetic field of 0.5 T. Dashed line denotes $E = E_0 \sinh(J/J_0)$ [553].

The vortex glass picture predicts scaling laws; e.g., the electric field should scale as $E\xi_g^{z-1} = f_{\pm}(J\xi_g^{d-1})$ where d is the spatial dimension and $f_{\pm}(x)$ are scaling functions for the regions above and below T_g, respectively. For $x \to 0$ one has $f_+(x) = $ constant and $f_-(x) \to \exp(-x^{-\mu})$. At T_g, a power-law current–voltage curve is expected with $E \propto J^{(z+1)/(d-1)}$.

The existence of the vortex glass state was first confirmed by Koch et al. who measured the $I-V$ characteristics of YBCO thin films [551]. Based on the idea of a scaling law for the diverging coherence length ξ and coherence time ξ^z near a second order phase transition, they mapped the electric field E and current density J for a d-dimensional sample and obtained the scaling law

$$E(J) \approx J\xi^{d-2-z}\tilde{E}_{\pm}\left(J\xi^{d-1}\phi_0/k_B T\right) \quad (17)$$

J is scaled by a characteristic current density

$$J_0 = k_B T/\phi_0\xi^{d-1} \quad (18)$$

which vanishes as $T \to T_g$, where the second order transition occurs. Their experiment results measured at 0.5 and 4 T are shown in Figure 79. At T_g, the $I-V$ curves follow a power law which is required by a second order phase transition. For each field, the $I-V$ curve exhibits a power-law behavior with the power exponent equal to 3 at a characteristic temperature T_g. For temperatures above T_g, these $I-V$ characteristics show upward curvature, while those below T_g have downward curvature. Obviously for $T < T_g$, $\partial E/\partial j$ decreases rapidly at low current limit, indicating a zero linear resistivity of R_L and thus a true superconducting state. All the $I-V$ curves can be scaled to two curves by using Eq. (17) with T_g being determined as the temperature where the $I-V$ curves show a crossover from downward curvature to upward curvature. For $T < T_g$, the $I-V$ curve is seen to crossover from the critical power law at currents larger than J_0^- to an exponential behavior at low J, consistent

with expectations from the scaling theory. The extrapolated linear resistivities $R_L = \partial E/\partial j$ for $T > T_g$ are found to be scaled with $(T - T_g)^{\nu(z+2-d)}$. With $\nu = 1.7$, $z = 4.8$ is obtained in agreement with a scaling analysis of $I-V$ curves. This result was later obtained by various researchers on single crystals and thin films. All of these result point to the existence of a dissipation-free region in the $H-T$ phase diagram for HTSCs.

A crucial parameter for the existence of the vortex glass phase is the dimensionality. Theoretically it was predicted that the vortex glass phase is stable in 3D, the above prediction does not apply to a 2D superconductor. For a two-dimensional superconductor, as pointed out by Vinokur and Feigelman [552], the exponential growth of the energy barrier will be cut off due to the instability of vortices upon the generation of edge dislocations inside the 2D vortex lattice. The energy for the edge dislocations set up an upper limit for the energy barrier. As a result no 2D vortex glass phase is expected at finite temperature. However, vortex glass correlations will develop toward $T_g = 0$ K which means that the 2D vortex glass correlation length will diverge near zero temperature. The effect of dimensionality on the vortex glass behavior has studied by Dekker et al. using ultrathin films of YBCO [553]. Their results are shown in Figure 80. They found that for 2D superconductors, all the $I-V$ curves can be scaled nicely into one curve although the scale can only be accomplished with $T_g = 0$. This means that the vortex glass state can only exist at zero temperature. Their result was confirmed by Qiu et al. who measured the $I-V$ characteristics of YBCO/PBCO multilayers [550]. They found that when the applied current density was above some value, the $I-V$ curves could be well described by the exponential dependence between V and I. When the current density was decreased to below some characteristic value, the exponential decrease of voltage stopped and thermally activated behavior set in which demonstrated itself as a linear $I-V$ relationship. A similar conclusion was also obtained by Wen et al. on TlBaCaCuO thin films [554].

The picture of vortex glass arose from a situation in which the superconductor is filled with point disorders. The situation is different in the presence of strongly correlated disorders such as twin boundaries or column defects generated by irradiation with heavy ions. For such a system, by mapping the physics of flux lines onto the problem of localization of quantum mechanical bosons in two dimensions, Nelson and Vinokur [555] predicted the existence of new phase called Bose glass. The Bose glass phase lies below the flux liquid state in the H–T phase diagram for a superconductor with strong correlated disorders. The Bose glass phase is characterized by a resistivity which has a temperature dependence

$$\rho(T) \propto (T - T_{BG})^{\nu(z-2)} \quad (19)$$

where T_{BG} is the Bose glass transition temperature. The existence of the Bose glass transition was first observed with magnetometry by Krusin-Elbaum et al. [555] on YBCO single crystals and by Moshchalkov et al. [556] on TBCCO thin film irradiated by heavy ion implantation. Later with measurements of the resistivity, the relation predicted by Eq. (19) was observed in BSCCO thin films by Miu et al. [557].

4.2.4.6. Superconducting State Hall Effect

To understand the vortex dynamics of HTSC it is essential to study both the longitudinal as well as the transverse Hall resistivity in the superconducting state. The Hall effect in the mixed state is associated with flux flow which is generated on the moving vortices by a hydrodynamical force called Magnus force. In the superconducting state, due to the presence of a vortex core in which the carriers are in the normal state, it is expected that the Hall voltage will have the same sign as in the normal state, i.e., a positive sign from that of holes. However, it turns out that the Hall resistivity reverses its sign in the vicinity of T_c. This is in contrast to what is expected from a conventional theory for transverse resistivity in the mixed state. Several reasons such as two-carrier quasiparticle effects and unusual vortex motion were proposed in order to explain such an anomalous Hall effect in the superconducting state. A sign reversal of ρ_{xy} has also been observed in conventional superconductors, which would favor an explanation based on terms of vortex dynamics rather than exotic quasi-particle effects peculiar for HTSCs. It was found that sign reversal of ρ_{xy} is easily observed in samples with the mean free path l of the same order of ξ.

Hagen et al. studied the Hall resistivity ρ_{xy} and magnetoresistance ρ_{xx} of epitaxially grown c-axis YBCO thin films in magnetic fields up to 7 T [558]. Their results are shown in Figure 81. For temperatures above T_c (= 90 K), the ρ_{xy} is always positive at all magnetic fields, consistent with a hole-type conductivity in YBCO. However, when the temperature goes below T_c, ρ_{xy} shows a crossover from positive at high fields to negative at low fields. The magnitude of the field where the crossover occurs increases with the decreasing temperature. Based on an analog to the case of vortex motion in superfluid ^{4}He, they suggest that in addition to the drag force $-\eta V_L$ exerted on the vortex, there should be another term $-\eta' \hat{a} \times V_L$

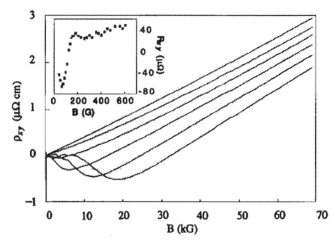

Fig. 81. Hall resistivity vs magnetic field in YBCO thin film near T_c. Notice that ρ_{xy} shows a sign reversal for temperatures below $T_c = 90$ K [558].

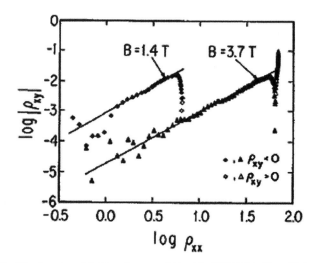

Fig. 82. Log–log plot of $|\rho_{xy}|$ vs ρ_{xx} obtained for YBCO thin film with temperature sweeps at two magnetic field values. A power-law dependence with $|\rho_{xy}| \propto \rho_{xx}^{\alpha}$ ($\alpha = 1.7 \pm 0.2$) is clearly observed [559].

along the tangent direction of the vortex motion added due to the Magnus force; here $\hat{a}$ is the unit vector along the field direction. Therefore the total Magnus force exerted on the moving vortex is $f = -\eta V_L - \eta' \hat{a} \times V_L$ which will give rise to an upstream vortex motion and reverse the sign of ρ_{xy}.

The correlation between ρ_{xy} and disorders in the superconducting state has been studied by Luo et al. on YBCO thin films [559]. They found that the magnitude of ρ_{xy} in the negative part scales very well with ρ_{xx} as shown in Figure 82, although a simple temperature dependence of either ρ_{xx} or ρ_{xy} was hard to determine. A power-law dependence $|\rho_{xy}| \propto \rho_{xx}^{\alpha}$ with $\alpha = 1.7 \pm 0.2$ was obtained, a relation that was later shown by Vinokur et al. to be a universal feature for any vortex system subject to strong disorders [560]. They argued that the scaling relationship arose from the vortex glass behavior associated with the large amount of point defects inside the thin film. As a result, pinning force could be an important factor for the occurrence of negative ρ_{xy}. However, the work done by Samoilov et al. on TlBaCaCuO thin films seems against the conclusion that the negative ρ_{xy} is related to pinning present in the sample [561]. They studied the mixed state Hall effect on TlBaCaCuO thin films before and after irradiation with high-energy Pb ions. It was found that ρ_{xy} is not altered by artificially introducing pinning centers. A sign reversal and power-law dependence of ρ_{xy} upon ρ_{xx} have also later been observed in YBCO/PBCO superlattices by Qiu et al. who argued in favor of a pinning induced sign reversal of ρ_{xy} [562]. They further discussed the explanation proposed by Dorsey based on time-dependent Ginzburg-Landau (TDGL) equations, incorporating an additional Hall term and assuming an imaginary component of the relaxation time for the order parameter [563]. In the TDGL model, ρ_{xy} is induced by both the motion of the vortices and that of quasi-particles in the regions surrounding the vortex cores. For the negative ρ_{xy} to occur, a quasi-particle spectrum with a positive energy derivative of the density of state average over the Fermi surface is required. This could happen in a superconductor with a complicated Fermi surface. Therefore, a sign reversal of ρ_{xy} will depend crucially on the shape of the Fermi surface and the position of the Fermi level.

4.2.5. Quantum Behavior in Ultrathin Films

Although the vortex glass state has a very high potential barrier for the vortex motion, there is another possibility for the motion of vortices, i.e., the quantum tunnelling of vortices. The quantum tunneling of vortices has a quantum origin which is different from the mechanism of thermal activation. When the temperature is low, quantum tunneling will replace thermal activation as the main source of dissipation. Quantum tunneling is independent of temperature. There have been a lot of reports on the existence of quantum dissipation. Non-Arrhenius behavior of the resistive transition at low temperatures has been observed in ultrathin films of Pb by Liu et al. [564] and amorphous $Mo_{43}Ge_{57}$ thin films by Ephron et al. [565]. For HTSCs, a large relaxation of the superconducting current at low temperature was observed, suggesting a quantum origin for the dissipation [566].

Liu et al. measured ultrathin films of Pb with a T_c value of 1.62 K [564]. The sheet resistance vs temperature has been measured under various applied magnetic fields. It was found that the resistance at low temperature appears to decrease exponentially with temperature in a non-Arrhenius fashion; i.e., $R \sim \exp(T/T_0)$. T_0 is nearly independent of the magnetic field. It is much lasier to observe the quantum effect in samples with high sheet resistance.

Similar results were obtained by Ephron et al. on amorphous $Mo_{43}Ge_{57}$ thin films which were highly disordered [565]. The sample had a T_c value of 500 mK and sheet resistance of 1350 Ω. Using low-frequency ac current and lock-in technique, they observed a crossover from an activation behavior at high temperature to a plateau at extremely low temperature which was thought to be evidence for the quantum tunneling of vortices. The activation energy extracted from the Arrhenius plot

followed a logarithmic behavior which was believed to originate from the dislocation–antidislocation nucleation process of the vortex lattice. They suggested that due to the quantum tunneling of the vortices at low temperature, no true zero-resistance state and thus no vortex glass state can exist in a 2D superconducting system at finite magnetic field.

Hoekstra et al. and Wen et al. [566] have measured the relaxation rate $Q(T)$ of superconducting currents on high-temperature superconducting samples in dirty and clean limits for temperatures down to 100 mK and magnetic fields up to 7 T. For YBCO/PBCO multilayers and underdoped YBCO thin films, which are in the dirty limit, they find that the relaxation rate is proportional to $\rho_n(0)/L_c(0)$, where $\rho_n(0)$ and $L_c(0)$ are the normal state resistivity and longitudinal Larkin correlation length at $T = 0$ K, respectively. However, for optimally doped YBCO and Y124 thin films which are in the clean limit, they observed a nearly sample independent relaxation rate $Q(0) \approx 0.022$ at $H = 1$ T. With a tunneling probability $P \propto \exp(-\pi n_s V)$, the relaxation rate is given by $Q(0) = 1/\pi n_s V(0)$, V being the volume enclosed by the tunneling trajectory of the vortex segment. A $Q(0)$ value of 0.022 implies that the number of superconducting charge carriers involved in the tunneling of a vortex segment is about 14.

4.2.6. Kosterlitz–Thouless Transition

The large anisotropy and short coherence length of HTSC result in the quasi-two-dimensional behaviors in superconductivity. One of the characteristic features of two-dimensional superconductivity is the KT transition associated with the thermal dissociation of vortex–antivortex pairs above a characteristic temperature T_{KT}. Below T_{KT}, when the separation distance r between the thermally excited vortex–antivortex pairs is smaller than the effective penetration depth λ_+, the vortex pairs tend to be bounded by an energy with a logarithmic dependence on r,

$$U(r) = 2E_c + \frac{\pi n_s^{2D} \eta^2}{2m^* \varepsilon} \ln(r/\xi) \tag{20}$$

where E_c is the vortex core energy, n_s^{2D} and m^* are the two-dimensional density and the effective mass of the superfluid electrons, and ε is the temperature dependent dielectric constant. The corresponding topological charge of a vortex is

$$q = \frac{\pi n_s^{2D} \eta^2}{2m^*} \tag{21}$$

As the temperature approaches T_{KT} from below, the average separation r of vortices of a pair diverges as $(T_{KT} - T)^{-1/2}$ and free vortices start to appear at T_{KT}. Above T_{KT}, vortex–antivortex pairs are present in thermodynamic equilibrium. The average distance between free vortices ξ_+ is proportional to $\exp((T_{c0} - T)/(T - T_{KT}))$. In the limit of zero magnetic field, the dissociation of vortex pairs gives rise to a finite resistivity with the characteristic temperature dependence

$$\frac{R}{R_0} = \alpha \exp\left(-2\left(b\frac{(T_{c0} - T)}{(T - T_{KT})}\right)^{1/2}\right) \tag{22}$$

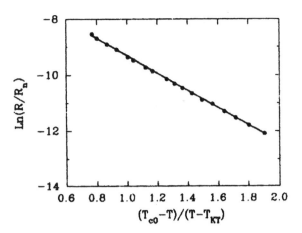

Fig. 83. The temperature dependence of resistance shown in $\ln(R/R_N)$ vs $[(T_{c0} - T)/(T - T_{KT})]^{1/2}$ plot. The full line is the fitting.

where R is the normal state resistivity, α and b are nonuniversal constants of order of unity, and T_{c0} is the mean field transition temperature.

For $T < T_{KT}$, an applied current will exert Lorentz force on the vortex pairs, and when the threshold current $I_{th} = (2e/h)kTK(l_m)$ ($K(l_m)$) being the KT renormalized stiffness constant and $l_m = \ln(\omega/\xi_{ab})$) is exceeded, the current induced pair breaking effect will be significant, resulting in a nonlinear resistivity which is proportional to $(j/j_0)^{\pi K_R}$, where $j_0 = eT_{KT}/\eta\xi_{ab}(T_{KT})$, j is the applied current density, and $K_R(T) = \eta^2 n_s^{2D}/m^* k_B T$ is the stiffness of the phase fluctuation. Therefore the I–V characteristic will show a power-law behavior $V = I^{N(T)}$ with $N(T) = 1 + \pi K_R(T)$. Since $K_R \to 2/\pi$ and n_s^{2D} is nonzero for $T \to T_{KT}$, and when $T > T_{KT}$, $K_R = 0$, therefore a universal jump of $N(T)$ from 3 to 1 is anticipated.

The Kosterlitz–Thouless transition has been observed in various HTSC thin films as well as superlattices [567–570]. Two main features associated with the KT transition were observed experimentally. The first feature is that in the limit of small current and zero applied magnetic field, thermal dissociation of vortex–antivortex pairs above T_{KT} into free vortices gives rise to a finite resistance with the characteristic temperature dependence given by Eq. (22). Indeed such a scaling of resistivity with T_{KT} has been observed from measurements of $\rho(T)$. Shown in Figure 83 is the work by Qiu et al. on a YBCO (48 Å)/PBCO (100 Å) superlattice [570]. The tail part of the R–T curve was plotted with the quantity $\ln(R/R_N)$ as a function of $(T_{c0} - T)/(T - T_{KT})$. The experimental data are found to follow a linear relationship as required by Eq. (22). A least-square fit to the experimental data gave a T_{KT} value of 62.50 K and $T_{c0} = 69.40$ K.

A second feature is that below T_{KT}, the I–V characteristics are nonlinear as a result of the current induced vortex pair breaking. Such a power-law behavior was observed by Qiu et al. for the I–V characteristics at the low excitation level. By fitting each individual I–V characteristic to a power-law relationship, the temperature dependence of the exponent $N(T)$ was observed. The values of $\pi K_R(T)$ extracted from $N(T)$ were

found to exhibit a jump around T_{KT}; this was consistent with what is expected from the KT theory. Due the possible finite size scaling and sample inhomogeneities, the jump is broadened.

The penetration depth λ can be estimated from the universal jump condition through the relation

$$\lambda_{ab}(T_{KT}) = \left(\frac{\Phi_0^2 l_c}{32\pi^2 \varepsilon k_B T_{KT}}\right)^{1/2} \quad (23)$$

where l_c is the effective film thickness and ε is the vortex dielectric constant which represents the screening of the vortex pair interaction by the existence of other intervening vortex pairs. The value of ε can be estimated from $K_0(T_{KT})$ which is the unrenormalized stiffness constant at T_{KT}. The relation between ε and $K_0(T_{KT})$ is given by $\varepsilon(T_{KT}) = \pi K_0(T_{KT})/2$. The value of ε at T_{KT} is estimated to be 1.1, which is several times higher than that for thin film. Inserting the value of ε into Eq. (23), λ_{ab} were calculated to be 0.87 μm. The zero temperature penetration depth can be obtained by using the empirical formula $\lambda_{ab}(0) = \lambda_{ab}(T)(1 - (T/T_{c0})^4)^{1/2}$. With $T_{c0} = 69.40$, $T_{KT} = 62.50$ K, and $\lambda_{ab}(T_{KT}) = 0.87$ μm, the value of $\lambda_{ab}(0)$ is calculated to be 0.5 μm, which is far larger than the typical value of 0.15 μm for $\lambda_{ab}(0)$.

If we know the correlation length l_c, then the effective mass of the superfluid electron m^* can be estimated from the unbinding energy of the paired vortices

$$m^* = \frac{\pi h}{4} \frac{n_s^{2D}}{k_B T_{KT}} \quad (24)$$

n_s^{2D} can be taken by the renormalization of the superfluid electron density over the effective thickness l_c as $n_s^{2D} = n_s^{3D} l_c$. For a typical value of $n_s^{3D}(0) \approx 10^{22}$ cm^{-3}, $n_s^{2D}(T_{KT}) = 1.9 \times 10^{14}$ cm^{-2} is obtained, which gives $m^* = 17m_e$. The large M^* value indicates a strong many-body interaction in the a–b plane.

The KT transition is concerned with the phase fluctuations and the establishment of quasi-two-dimensional long range order in the phase below T_{KT}. When the temperature is higher than T_{KT}, the fluctuation in the magnitude of the order parameter dominates the dissipation. The parameter $\tau_{KT} = (T_{c0} - T_{KT})/T_{c0}$ indicates for how large a temperature range the phase fluctuations dominate the dissipation. In the above mentioned work, $\tau_{KT} = 0.099$, which is far larger than the values obtained in YBCO thin films and single crystals [569]. The large τ_{KT} value means that in superlattices, the two-dimensionality and the phase fluctuations are greatly enhanced as compared with single crystals or single-layer thin films. As a result, the long range order in superlattices is much easier to break down. So, to increase the superconductivity, enhanced interlayer coupling in superlattices is necessary.

4.2.7. Superconducting Properties in Multilayers and Mesoscopic Samples

As we know, a superconductor has two important parameters λ and ξ, with λ characterizing the length scale to which the magnetic field will penetrate the superconductor and strength of interaction between vortices and ξ characterizing the length scale of the overlapping of superconducting order parameter. A natural question we will ask is what will happen if the size of a superconductor is reduced to the order of λ or ξ. The advanced nanotechnology nowadays enables us to fabricate thin films with size down to the micron or submicron scale. It turns out that when the size of a superconductor is comparable to λ, a strong confinement effect occurs due to the presence of the boundary which influences the distribution of circular superconducting currents. A peculiar effect arising from the presence of the geometry boundary is the paramagnetic Meissner effect observed in thin film quantum dots. Although the works described here are mainly performed on conventional superconducting thin films and few works have been done on high-temperature superconducting thin films, we believe similar physical phenomena should be observed in high-temperature superconducting thin films. Therefore we include these works here even though this book is specifically devoted to work on high-temperature superconducting thin films.

4.2.7.1. Paramagnetic Meissner Effect

It is well known that Meissner effect is one of the basic characteristics of superconductor, that is, the superconductor will expel flux upon entering the superconducting state, which differs from a perfect conductor. As a result of the flux expelling, the superconductor will exhibit a diamagnetic effect. The diamagnetic Meissner effect is one of the standard ways to determine the transition temperature of the superconductor. However, in some HTSCs, a puzzling paramagnetic Meissner effect (PME) was observed. It was first observed by Braunisch et al. in BSCCO single crystals [571]. It could only be observed in field cooling magnetization measurements when the applied magnetic field was very small, of the order of several Gauss. Different explanations have been proposed for the occurrence of the PME, including a spontaneous Josephson current in π junctions in d-wave superconductors, edge effects in strips, and size effect in mesoscopic superconductors. The idea that PME can occur in mesoscopic superconductor was first forward by Moshchalkov et al. by solving the coupled nonlinear GL equations [572]. It was found that under some conditions, the so-called giant vortex state has a lower energy than the mixed state with an Abrikosov lattice. This situation happens in type II superconductors with sizes compatible to the penetration depth λ. This theoretical prediction was confirmed by Geim et al.; their work was performed on mesoscopic Al films [573]. By using the ballistic Hall magnetometry, they were able to measure magnetic field down to small portion of ϕ_0 in a mesoscopic disk made from Al thin film with a thickness of 0.1 μm and a diameter less than 4 μm. They measured the H–T phase boundary with the help of a sensitive Hall array. The quantum state of flux entering and leaving has been directly observed and indeed they found at some portion of the H–T phase boundary, PME develops as the number of fluxiods changes. Their results are shown in Figure 84. It can be see clearly that at fields of

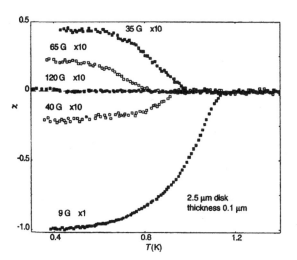

Fig. 84. Magnetic susceptibility of an aluminium thin film disk for various magnetic fields perpendicular to the film surface. A paramagnetic response is clearly observed [573].

35, 65, and 120 Gauss, the sample possesses paramagnetism. The paramagnetic state is found to be metastable which can be disturbed by external noise. The physical origin of the PME observed here is the occurrence of surface superconductivity in a mesoscopic thin film. Upon decreasing temperature, the flux can be captured as the third critical field inside the superconducting sheath compresses into a smaller volume, allowing extra flux to penetrate the superconductor and inducing a spatial redistribution of the superconducting circular current near the surface, resulting a circular current which generates a paramagnetic response.

4.2.7.2. Confinement Effect in Mesoscopic Superconducting Thin Film

For a mesoscopic superconductor with its size comparable to its penetration depth or coherence length, the nucleation of the superconducting state is strongly dependent upon the boundary conditions imposed by the sample configuration. Near the transition temperature, the problem of nucleation of the superconducting state can conveniently be solved using the linearized GL equations since the order parameter is small in magnitude. The linearized GL equations read

$$\frac{1}{2m}\left(-i\eta\nabla - \frac{e^*}{c}A\right)^2 \Psi = -\alpha\Psi \qquad (25)$$

where A is the vector potential defined by $H = \nabla \times A$, H being the applied magnetic field. By comparing Eq. (25) with the Schrödinger equation for a particle with mass m and charge e^*, we can find that $-\alpha$ corresponds to the energy term in the Schrödinger equation. With the suitable boundary condition,

$$\left(-i\eta\nabla - \frac{e^*}{c}\right)\Bigg|_n \Psi = 0 \qquad (26)$$

i.e., no supercurrent leaks out the superconductor along the direction normal to the surface. It is evident that when the sample geometry is changed, the corresponding boundary condition will be modified. As a result, the eigenenergy $(-\alpha)$ for the superconductor will be altered. The temperature dependence of $-\alpha$ can be expressed as

$$-\alpha = \frac{\eta}{2m\xi^2(T)} = \frac{\eta}{2m\xi^2(0)} \frac{T_c - T}{T_c} \qquad (27)$$

The onset temperature for the superconducting transition is decided by the lowest energy value for a certain field. By solving Eq. (27) for different applied magnetic fields, taking into account the flux quantization, it is possible to get the phase boundary for a mesoscopic superconductor. For a superconducting disk, an analytical solution can be obtained whose radian term is the hypergeometrical function. The boundary condition is solely determined by the radian term since the z-axis term is eliminated and the angular term just decides the phase of the order parameter and does not influence the magnitude of the order parameter. The onset of the superconductivity happens at a characteristic field H_{c3} which is higher than H_{c2}. This is called surface superconductivity.

The experiment was first done by Moshchalkov et al. who measured the H–T_c phase diagram for thin films of Al strips and disks to check the theoretical predictions about it [574]. The results are shown in Figure 85. They found that indeed when the sample size is compatible to λ and ξ, the sample geometry greatly affects the superconducting properties. For a superconducting disk (actually a square), the T_c determined from the resistivity measurements oscillates with H, resembling the Little–Park effect. They have a common cause, i.e., the quantization of flux quantum. Due to the quantum confinement of the flux, the superconductivity with lowest free energy will happen for the system with embedded flux quantum.

4.3. Optical Properties

Optical spectra of electronic states provide information on the collective electron–hole pair excitations within the energy of several eV near the Fermi surface. For semiconductors, visible light and ultraviolet light have been widely used to study the excitation states near the conduction bandgap. For superconductors the relevant energy scale is the superconducting energy gap 2Δ with an energy of the order of meV which lies in the infrared. An advantage of infrared measurement over other experimental techniques is that it is much less sensitive to the sample surface. Generally speaking, there are two experimental techniques for the infrared studies of superconductors, i.e., conventional Fourier transform infrared spectroscopy and time domain infrared spectroscopy. By means of these two infrared spectroscopy methods, several fundamental properties of HTSCs such as scattering rate, effective mass, and lifetime of quasi-particle excitations have been determined.

4.3.1. Infrared Properties

Infrared measurement is one of the techniques that provide the most fundamental and versatile probes of the superconducting

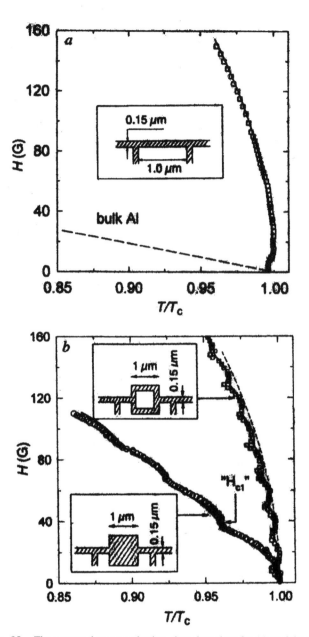

Fig. 85. The measured superconducting phase boundary for (a) straight Al line and (b) thin film Al disk, respectively [574].

energy gap. It measures the optical conductivity which has a gap as the threshold for pair excitations. Far-infrared reflectivity measurement has been widely used to study the interlayer electron transport in cuprate superconductors. Like resistivity, infrared properties are strongly influenced by the anisotropy of the crystalline structure and stoichiometry which controls the carrier density in the CuO_2 planes. The reflectivity of the electric field component parallel to the CuO_2 planes shows a metallic response, whereas reflectivity obtained in the polarization along the c-axis direction likes that of ionic insulators in the normal state. One peculiar feature for the carrier dynamics of the underdoped cuprate is the formation of a pseudogap well above T_c that persists in the superconducting state. The carrier transportation along the c-axis can hardly be understood within the framework of the Fermi liquid theory and recently it was proposed that interlayer Josephson tunneling is responsible for the occurrence of superconductivity in HTSCs.

4.3.1.1. a–b Plane Reflectivity

The interpretation of the optical conductivity data remains controversial. The normal-state optical conductivity cannot be described by a simple Drude model. Besides a strongly temperature dependent Drude absorption centered at $\omega = 0$, there exists a broad, nearly temperature independent mid-infrared absorption band. It has been observed that the mid-infrared absorption is absent in the undoped parent compounds such as La_2CuO_4 and $YBa_2Cu_3O_6$. For $La_{2-x}Sr_xCuO_4$, the mid-infrared absorption band develops with increasing dopant concentration and then exhibits saturation in the higher compositional range $0.1 \leq x \leq 0.25$. As a consequence of the redistribution of the O $2p$ and Cu $3d$ orbitals upon doping, spectral weight is rapidly transferred from the in-plane O $2p \rightarrow$ Cu $3d$ charge transfer excitations above 2 eV to the free carrier absorption and the low-energy excitations below 1.5 eV. Therefore, both the Drude and mid-infrared absorptions in HTSCs appear to be related to the introduction of holes on the CuO_2 layers by doping.

Infrared reflectivity has been measured by Gao et al. on $La_{2-x}Sr_xCuO_4$ thin films in the frequency region from 30 to 4000 cm^{-1} [575]. Show in Figure 86 are the representative spectra at different temperatures. A strongly damped plasma edge can be observed around 6000 cm^{-1}. The reflectivity increases with a decreasing temperature, in accordance with the metallic behavior of the resistivity. The reflectivity in the far-infrared region is rather high, near unity. Several optical phonons in the a–b plane are visible which are more obvious than in YBCO, perhaps due to a lower free-carrier concentration and a higher vibration oscillation length. The optical conductivity spectra $\sigma(\omega)$ obtained from the reflectivity spectra through the Kramers–Kronig transformation are shown in Figure 87. $\sigma(\omega)$ at the low-frequency limit is nearly equal to the dc conductivity and exhibits a Drude response. These spectra have common features for the in-plane optical conductivity of various cuprate materials near optimal doping, a sharp peak at $\omega = 0$ and a long tail extending to higher frequencies in the infrared region where $\sigma(\omega)$ falls as ω^{-1} rather than ω^{-2} decay as expected from a Drude model. The long tail part depends weakly on temperature in sharp contrast to the low-frequency part, $\omega < 300$ cm^{-1}. A loss of spectra weight can be seen in the superconducting state for $\omega < 150$ cm^{-1} due to a shift of weight to $\omega = 0$ accompanying the superconducting condensation, which is required in order to fulfill the frequency sum rule.

Two explanations have been proposed to understand the in-plane optical conductivity. In the two-component approach, the $\sigma(\omega)$ spectrum is decomposed into a Drude component which gives a sharp peak at $\omega = 0$ and a mid-infrared absorption band which is responsible for the long tail at high frequencies. The free carriers giving rise to the Drude component are also responsible for the dc conductivity above T_c and superconducting

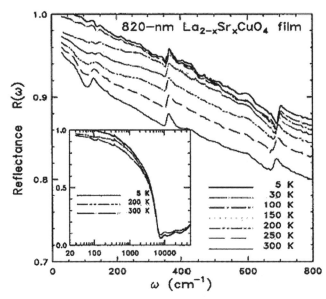

Fig. 86.   Measured reflectivity of a $La_{2-x}Sr_xCuO_4$ thin film at different temperatures. The inset shows the same data over a wider frequency region [575].

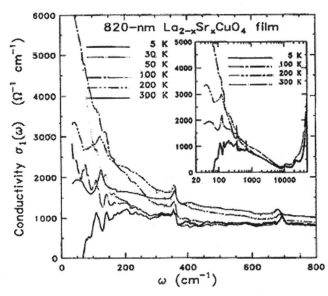

Fig. 87.   The real part of the a–b-plane conductivity of a $La_{2-x}Sr_xCuO_4$ thin film at different temperatures. The inset shows the same data over a wider frequency region [575].

condensation below T_c. The Drude carriers couple only weakly with phonons or other excitations which gives rise to a sharp $\sigma(\omega)$ peak at $\omega = 0$ as well as the linear behavior of $\rho(T)$. The bound carriers are responsible for the mid-infrared absorption which is weakly temperature dependent and shows slow frequency dependence. By subtracting the contribution of the bound carriers from the optical conductivity spectra, the plasma frequency ω_p and scattering rate $1/\tau$ can be obtained. The coupling constant γ was calculated to be 0.25, consistent with the weak coupling assumption and absence of resistivity saturation.

In the alternative one-component model, the sharp peak and long tail are due to the same carriers whose scattering rate and effective mass depend strongly on frequency. This approach leads to a broad range of optically inactive excitations in the mid-infrared region and has been described in the framework of marginal Fermi liquid and nested Fermi surface. The scattering rate $1/\tau$ is assumed to increase rapidly with ω, so that a small $1/\tau$ is responsible for the sharp peak at $\omega = 0$ in $\sigma(\omega)$ and a large $1/\tau$ for the long tail at higher frequency. The temperature dependent scattering rate $1/\tau(\omega)$ can be extracted from the experimental data with the help of the generalized Drude formula

$$\sigma(\omega) = \frac{\omega_p^2}{4\pi} \frac{m}{m*(\omega)} \frac{\tau(\omega)}{[1 - i\omega\tau(\omega)]} \quad (28)$$

where ω_p is the frequency of Josephson plasma. This approach gives linear dependence of $1/\tau(\omega)$ upon ω, in accordance with the linear relation behavior of $\rho(T)$.

Similar measurements have been performed by El Azrak et al. on YBCO and BSCCO thin films [576]. Their results can be successfully explained by a two-component model with a low-frequency Drude contribution and a mid-infrared absorption band. Both the scattering rate and effective mass of the carriers are found to increase linearly with frequency.

4.3.1.2. c-Axis Reflectivity and Josephson Plasma

The c-axis reflectivity exhibits a remarkable difference from the a–b plane reflectivity, especially in underdoped samples. Of particular interest is the development of a normal state pseudogap and buildup of the superconducting coherence for carrier transport along the c-axis [577]. A consequence of the superconductivity along the c-axis is the observation of Josephson plasma arising from the intrinsic Josephson junction associated with the layered structure of HTSCs. There have been a lot of reports on the observation of Josephson plasma in various high T_c cuprates; however, due to the difficulty in growing thin films with the c-axis lying in the direction along the substrate basal plane, nearly no work has been reported for infrared properties for HTSCs using thin films.

Using (110)-oriented YBCO thin films, Qiu et al. studied far-infrared reflectivity of underdoped YBCO films [578]. Polarized optical measurements were done using a rapidscan Fourier-type spectrometer in the far-infrared region (20–7000 cm^{-1}). Shown in Figure 88 is the reflectivity measured at five temperatures on one underdoped YBCO thin film with an oxygen deficiency $\delta = 0.35$ and a T_c value of 60 K. Above T_c, the contribution from optical phonons dominates the spectrum. A broadening reflectivity edge can be seen below 100 cm^{-1}, which could be a result of the high thermal energy and weak coherence in the transport of the carriers along the c-axis. As the temperature decreases to below T_c, the reflectivity edge sharpens and its position shifts to a slightly higher frequency than that in the normal state. It is attributed to the increasing coherence of carrier transport along the c-axis and the formation of the plasma excitations of the intrinsic Josephson junctions made of the CuO$_2$ planes, as suggested by Uchida et al. [579] and Tachiki et al. [580]. The very low c-axis plasma frequency

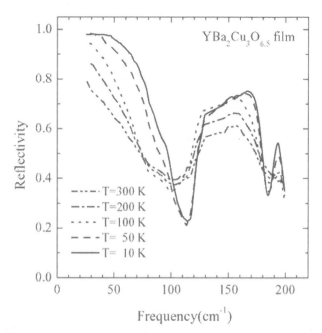

Fig. 88. c-axis far-infrared of an underdoped YBa$_2$Cu$_3$O$_{6.6}$ thin film at different temperatures.

corresponds to a carrier mass substantially larger than the c-axis optical mass obtained in band structure calculations. This plasma edge is associated with the presence of a zero frequency superconducting condensate in order to satisfy the conductivity sum rule $\int \sigma(\omega)d\omega = $ const. This frequency lies in the several tens of cm^{-1} order for La$_{2-x}$Sr$_x$CuO$_4$ and in the microwave region for Bi$_2$Si$_2$CaCu$_2$O$_8$, indicating a smaller anisotropy in this YBCO sample.

The Kramer–Kronig transformation is utilized to obtain the real part of the optical conductivity $\sigma(\omega)$ in order to gather information on the c-axis charge dynamics from the measured reflectivity. The real part of the conductivity is dominated by a series of optical phonons located at 130, 185, 232, 310, 380, and 625 cm^{-1}, respectively. As the temperature is cooled down, the contribution from the 185 and 232 cm^{-1} phonons is getting stronger and the position of the 310 cm^{-1} phonon shifts to lower frequency. The c-axis conductivity of YBCO shows a metallic background with a very large scattering rate and it is generally agreed that the transport is incoherent; i.e., the mean free path is less than one lattice spacing. The above facts suggest that in the normal state the interlayer coupling between the neighboring CuO$_2$ planes is very weak and that there is no coherence for the carrier transport along the c-axis. The carriers would move between the CuO$_2$ planes by tunneling which sometimes demonstrates itself as a resistive peak in the transport measurements along the c-axis. As the temperature decreases, the CuO$_2$ planes become superconducting and the phase coherence is built up for the carriers transporting along the c-axis.

The phonons were treated as separate channels of conductivity and subtract their contributions from the spectrum. A pseudogap can be seen to open up in the optical conductivity below 250 cm^{-1} even above T_c. In order to satisfy the conductivity sum rule, the spectral weight from the low-frequency part of the conductivity is transferred to higher energies above the upper limit of the frequency of their measurements. The frequency of the onset of the pseudogap is nearly independent of the temperature.

The influence of magnetic fields on the interlayer charge dynamics has also been studied by measuring the reflectivity under high magnetic fields up to 14 T which was applied perpendicular to the c-axis and parallel to the a–b plane. It is well known that a magnetic field can enhance the pair breaking effect which means that the superconducting carriers should decrease with the increase of applied magnetic field. This will in turn increase the penetration depth and therefore results in a shift of the Josephson plasma frequency considerably toward the lower frequency region. It was found that in the normal state, the magnetic field has little effect on the reflectivity. In the superconducting state, a noticeable effect can be observed below the plasma edge although it seems the magnetic fields hardly change the reflectivity at a frequency above that of Josephson plasma oscillation. The reflectivity decreases with increasing magnetic field, suggesting an enhanced pair-breaking effect.

4.3.2. Femtosecond Optical Properties

The transient optical response of HTSCs has been of great interest since it is directly correlated with the dynamics of pair breaking and recombination as well as having application to high-speed optical detectors with wide spectral range. Optical responses of HTSCs have been studied by exciting samples with chopped cw optical beams or pulsed lasers in which rather broad optical pulses with durations longer than 1 ps were used and transient voltage was measured. These experimental techniques limit the time resolution to the picosecond range. Later, ultrafast optical responses using femtosecond laser pulses have been reported and the subpicosecond response was interpreted in terms of relaxation of the photoexcited quasiparticles. The optical interaction in these materials determines the nonequilibrium superconducting properties which may reflect the pairing mechanisms. The optical response of HTSC has been done using the pump and probe technique with cw or pulsed lasers in time domain THz spectroscopy.

Quasi-particle dynamics in YBCO thin films has been studied by Han et al. using femtosecond time resolved spectroscopy by measuring the photoinduced changes in reflectivity [581]. Shown in Figure 89 are the transient photoinduced reflectivity $\Delta R/R$ for a superconducting YBCO thin film measured in the normal state and superconducting state, respectively. In the normal state, $\Delta R/R$ shows a step function response with a characteristic time scale of 3 ns, resembling the bolometric response in a metal. However, in the superconducting state, upon excitation, $\Delta R/R$ first decreases below zero with a characteristic time of 300 fs, followed by a longer recovery of several ps duration with $\Delta R/R < 0$, and finally it is restored to the positive value. The result obtained in the superconducting state

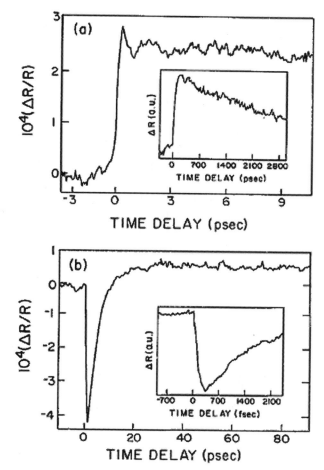

Fig. 89. Transient photoinduced reflectivity of a 3000-Å YBa$_2$Cu$_3$O$_{6.95}$ thin film at (a) 300 K and (b) 40 K [581].

suggests that there are two processes associated with the quasi-particle dynamics. The first is the avalanche multiplication of quasi-particles due to hot-carrier thermalization following photon absorption upon femtosecond beam excitation. This process is followed by a recombination of the photogenerated quasi-particle to form Cooper pairs via nonlinear kinetics on the order of 5 ps, which is dominated by the escape time of 2Δ phonons releases in the recombination process.

4.3.3. Terahertz Radiation from YBCO Thin Films

It is well known that ultrafast transient electromagnetic waves can be generated from the photoconducting antenna made of semiconductors by exciting femtosecond laser pulses. The ultrafast photocurrent transient is the source of THz radiation. It is also expected that the THz radiation can be emitted from superconductors if the supercurrent is modulated at a sufficiently high speed by femtosecond laser pulses.

Measurements of THz radiation generated by femtosecond laser pulses have been carried out by Hangyo et al. on YBCO thin films deposited on MgO substrate [582]. The films were c-axis oriented and had a typical thickness of 700 Å. A coplanar transmission line with a bridge structure at the center was patterned using a photolithographic technique. The sample is excited with a mode-locked Ti:sapphire laser operating at a repetition rate of 82 MHz with 80-fs pulses at a center wavelength of 794 nm. A time domain Fourier transformer experimental set was used for this experiment. The THz radiation emitted from the posited side through the MgO substrate was collimated by a paraboloidal mirror and focused by a second identical paraboloidal mirror onto a photoconductive detector with a dipole antenna. The receiving dipole antenna was made of two parallel 5-μm-wide Ni/Ge/Au alloy lines separated by 20 μm. A stardard pump-probe method was employed. The optical beam of the femtosecond laser pulses was divided equally with a beam splitter. One beam chopped at 2 kHz was used to excite the YBCO while the other was used to trigger the detector. The radiation signal was received with an antenna and measured with lock-in amplifier. Shown in Figure 90 are the electrical pulses of the freely propagating THz beam generated by the YBCO thin film measured at 11 K. A sharp pulse with a width of 1.5 ps is observed. This radiation is closely related to the photoinduced quasi-particles through pair breaking and their following recombination into Cooper pairs.

4.4. Conclusion

Thin films have contributed a lot to the understanding of physical properties of HTSC. However, despite the great effort

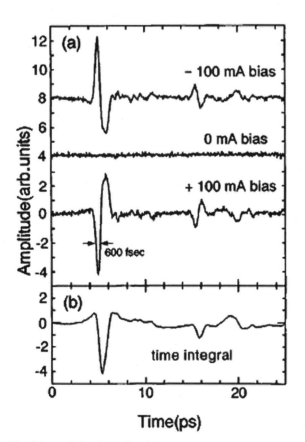

Fig. 90. Measured electrical pulse of the propagating THz radiation for various bias currents [582].

devoted to the research on HTSC, a lot of peculiar features in the physical properties of HTSC, such as the linear behavior of $\rho_{ab}(T)$ and its doping evolution, the temperature dependence of Hall coefficient, and the non-Drude behavior of optical conductivity, remain to be clarified. These films could also be important for the studies of the vortex dynamics in HTSCs which has emerged as a new field in condensed matter physics for a many-particle system subjected to strong disorders. To take advantage of thin films and to facilitate our further understanding of the physical properties of HTSCs, single-domained single-crystal thin films with different orientations are strongly demanded. With improved fabrication technology, various thin films with expected quality will be available, which will contribute to our final understanding of the mechanism of high-temperature superconductivity.

5. DEVICE APPLICATIONS

5.1. Josephson Junctions and Superconducting Quantum Interference Devices

5.1.1. Introduction

Josephson effect is one of the basic features for both conventional and high-T_c superconductors, so the Josephson junction is a powerful sensor to probe the physical properties of high-T_c superconductors and to develop devices for electronic applications, including SQUIDs, superconducting digital circuits, and superconducting transistors. The most exciting application is SQUIDs, which have a much higher sensitivity than traditional methods when used as flux sensors.

5.1.2. Josephson Effect

A configuration predicated by B. Josephson is that in two superconductors separated by a thin layer of insulator, the superconducting electrons (Cooper pairs) can pass through the insulator layer from one side to another as shown in Figure 91. This configuration is called a Josephson junction. In the case that nonvoltage is applied on the junction, a dc superconducting current will pass through the junction and follows the relation of $I = I_c \sin \phi$. Here, ϕ is the phasr difference between macroscopic superconducting tunneling current of junction. This phenomenon is termed the dc Josephson effect. When a voltage V is applied on the juntion, an ac current with frequency $f = (2e/h)V$ will flow through the junction which also can radiate electromagnetic waves with frequency of $f = (2e/hn)V$ with n the interger. This is termed the ac Josephson effect. The presence of the magnetic field will cause the Josephson current to oscillate with the well-known Fraunhofer pattern as shown in Figure 92.

The Josephson effect is a macroscopic quantum phenomenon. So Josephson junctions can be used to detect the flux quantum. For this purpose a dc SQUID is prepared and used. A dc SQUID is a superconducting loop consisting of two similar Josephson junctions working under a dc bias current (see Figure 93) [585].

The current–voltage curve of a SQUID looks like that of a single junction with the effective critical current equal to the sum of two component junctions. However, such a critical current is a periodic function of the magnetic flux applied to the superconducting loop with the maximum value and the minimum value corresponding to the integer and half-integer values of applied flux in unit of Φ_0, 2.07×10^{-7} gauss respectively. So the SQUID can be used as a probe with very high resolution to detect the weak magnetic field. In principle, it is convenient to make SQUID with high sensitivity by increasing the area of the superconducting loop. The most important event for the use of the Josephson effect in high-T_c superconductors is that the developed Josephson junction (tricrystal junction) was used to test the symmetry of superconducting order parameters and obtained the distinc d_{x2-y2} pairing symmetry (d-wave symmetry) [583, 584], which is different from that of s-wave superconductors.

5.1.3. Fabrication of Josephson Junctions

In the last decade, a tremendous effort has been made to fabricate HTSC Josephson junctions. Independent of the type, Josephson junctions are often made from YBCO and related materials, with artificial or natural weak links or barriers. The main categories of Josephson junctions devices are: (1) grain boundary junctions artificially nucleated in epitaxial films with c-axis normal to film plane (c-axis-oriented films) at defined ar-

eas, (2) junctions with artificial barriers, and (3) microbridges with a weakened/normalized superconductor link. Below we will discuss in detial the methods of fabrication of these kinds of Josephson junctions.

5.1.3.1. Controlled Grain Boundary Junctions

The earliest Josephson junctions made by high-T_c superconductors were naturally based on the grain boundaries in YBCO [586, 587]. Nowadays, better controlled and more reproducible grain boundary junctions are still preferred. Junctions that have extensively been investigated are bicrystal junctions [588–590], biepitaxial junctions [591], and thin film step-edge junctions [592, 593]. They all rely on various kinds of boundaries that occur in the film at specified positions defined by modified substrate surface.

Bicrystal Junctions. These junctions represent a very successful approach to nucleate a grain boundary only at a precisely defined location of an epitaxial film grown on bicrystal. The

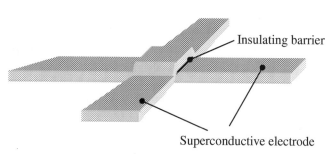

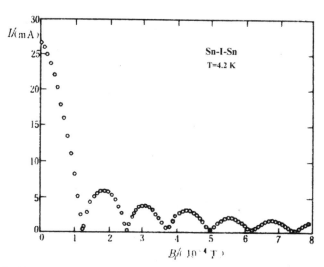

Fig. 91. Schematic of Josephson junction with superconductor/insulator/superconductor trilayer structure.

Fig. 92. Josephson critical current $I_c(B)$ as a function of the magnetic field B for the Sn/I/Sn junction.

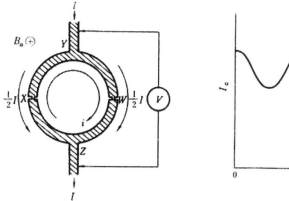

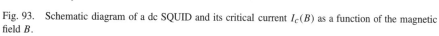

Fig. 93. Schematic diagram of a dc SQUID and its critical current $I_c(B)$ as a function of the magnetic field B.

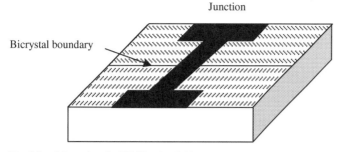

Fig. 94. Schematic of a YBCO microbridge on a bicrystal substrate with a certain misorientation angle. A bicrystal consists of two crystals that are pressed and sintered together with an angle. The lines indicate the grain boundary in the bicrystal. The superconducting thin films with a–b plane oriented along the lattice directions grow on the substrate and the grain boundary is formed.

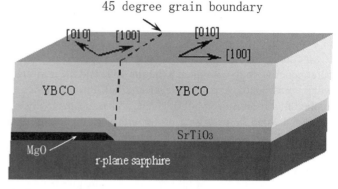

Fig. 95. Schematic view of the device structure with 45° degree grain boundary of YBCO based on R-plane sapphire substrate through SrTiO$_3$ buffer layer.

HTSC film, i.e., YBCO film, grows epitaxially on the bicrystal and replicates the boundary of the substrate. The critical current density across the grain boundary, J_{cGB}, is defined by the crystallographic tilt or twist angle θ between the two crystals. The junction is fabricated by patterning a narrow microbridge across the boundary (see Fig. 94).

Bicrystals can be formed by a solid-state intergrowing technique [594]. Two crystals are prepared with different crystallographic orientations at the conjugating surfaces and are joined together in high pressure and annealed at high temperature. A grain boundary is thus formed artificially between two pieces and is characterized by the misorientation angle θ given by the difference between the lattice directions in the crystals on either sides of the boundary (see Fig. 94). The bicrystal is a template for an epitaxially grown film. The common substrates used are SrTiO$_3$ and Y–ZrO$_2$. However, it has been shown that the grain boundary in the film wiggles back and forth over the bicrystal grain boundary on a scale of 15 nm due to the growth mechanism of the YBCO film [595].

This bicrystal technique has a major drawback: it cannot readily be extended to integrated circuits since all junctions are aligned along the bicrystal boundary. On the other hand, when used as SQUID devices, which are plagued by low yield and the presence of grain boundaries in the SQUID loop itself. These lead to excessive flux noise and hysteresis in the voltage-flux response of the SQUID.

Biepitaxial Junctions. This technique is estimated to be useful in integrating superconducting and semiconducting components, and it avoids the problems associated with the devices made from granular thin films and on bicrystal substrate. The method is quite general and can be used on a wide variety of substrates.

This approach to the artificial nucleation of grain boundaties is best explained with reference to Figure 95. The key technique is to fabricate 45° grain boundaries in YBCO films by controlling their in-plane epitaxy using seed and buffer layers deposited in R-plane-cut sapphire substrates. A prerequisite is the availability of a template or seed compound, which grows

epitaxially on a substrate but with a defined angle of rotation of major symmetry axis, and itself serves as a substrate for epitaxial growth of HTSC material or a buffer layer but without rotation. The substrate itself must also be suitable for HTSC or buffer growth without rotation. An opposite situation, i.e., the template which grows without rotation but causes HTSC or buffer rotation, can be equally useful. Typically the suitable seed layer is CeO$_2$ or MgO. Here take MgO as an example.

First deposit 3–30 nm of epitaxial MgO as a seed layer on a R-cut plane sapphire substrate; then mask the MgO with conventional photoresist and remove it from, for example, half of the substrate by either Ar ion milling or chemical etching with dilute phosphoric acid. SrTiO$_3$ buffer layer is deposited on both exposed sapphire surface and patterned MgO seed layer; thus SrTiO$_3$ film grows in two different orientations separated by a 45° grain boundary. Then immediately deposit YBCO, which grows epitaxially on the whole surface of the substrate and thereby reproduces the grain boundary. Finally the YBCO is patterned into an appropriate geometry [591].

The biepitaxial method attracts attention because it is useful for fabricating Josephson junctions on Si. This may be used in hybrid circuits integrating active superconducting elements and it interconnects with semiconductor devices. However, extra buffer layers have to be introduced in the biepitaxial system to solve the encountered problems, such as chemical interaction between YBCO and Si substrate, a large mismatch for the (001) Si and the (001) planes of YBCO, and a large difference of the thermal expansion coefficients of YBCO and Si. Commonly CeO$_2$/YSZ can be used as buffer layer between YBCO and Si; see Figure 96 [596].

Most processes give a 45° rotation, but misorientation angles of 18° and 27° have also been reported [597]. When the junction is used to make a dc SQUID device, the 45° grain boundary limits the $I_c R_n$ to rather low values: 0.1 to 1 mV, at 4.2 K, since the J_c of biepitaxial Josephson juctions with angle misorientations of 45° does not exceed 10^2–10^3 A/cm^2 at 77 K. This is rather insufficient for a low-thermal-noise operation and thus results in limitations of the practical application of SQUIDs consisting of such Josephson junctions.

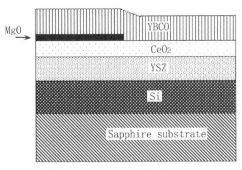

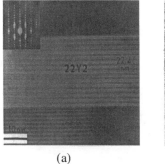

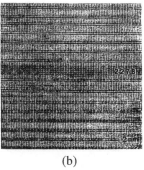

(a)   (b)

Fig. 96. Biepitaxial junction of YBCO based on the multilayer structure biexpitaxially growing on designed template of CeO₂/YSZ and MgO.

Fig. 98. High-resolution TEM cross-section micrograph of (a) Bi₂Sr₂-CaCu₂O₈₋δ/Bi₂Sr₂Cu₂O₈₋δ/Bi₂Sr₂CaCu₂O₈₋δ trilayer showing the interfaces between the superconducting 2212 films and the 22Y2 barrier. The inset shows the selected-area diffraction pattern at an interface [601]. (b) The 2212 electrodes are separated by a single 2278 molecular layer barrier, about 25 Å thick [602]. Reprinted from *J. Alloys Compounds* 251, 201 (1997), with permission from Elsevier Science.

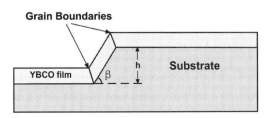

Fig. 97. Schematic of typical step-edge Josephon junction.

Thin Film Step-Edge Junctions. The process of fabrication of step-edge junctions is comparatively simple; however, subtly controlling the process for fabricating this kind of junction is very necessary. For this technique the heights of the steps comparable to the film thickness are made in the substrate SrTiO₃ or LaAlO₃ by etching. The step angle is larger than 45°. When the YBCO film is grown across the edge, the *c*-axis is tilted and at least two grain boundaries are formed as shown in Figure 97. It has been possible to adjust the junction parameters by adjusting the step height and angle, but the junction reproducibility has been low.

The behavior of step-edge Josephson junctions is somewhat smiliar to that of junctions on natural grain boundaries and bicrystal substrates. But in contrast to biepitaxial junctions, the J_c of step-edge junctions can be varied by three orders of magnitude, depending upon the step angle and height, and the ratio of film thickness to step height. J_c's of 10^4–10^5 A/cm² at 77 K can be readily obtained. The transmission electron microscopy shows that the geometrical variables identified above determine the orientations of the two grain boundaries which should, in turn, control the J_c of the junction [598].

5.1.3.2. SNS or SIS Josephson Junctions with Artificial Barriers

This technique is to create artificial barriers between two superconductors. Both normal metal and insulator barrier materials are used to fabricate SNS or SIS junctions. Categorizing by structure, there are two main type junctions, sandwich junctions and edge junctions.

In contrast to the grain boundary junction the barrier materials for making sandwich Josephson junctions between YBCO superconductors are much more formidable. It requires the reliable preparation of high-quality epitaxial YBCO thin films. The smooth surface and sharp interface are of primary concern. Of the electrical properties both high critical temperature and high critical current density are essential. In addition, deposition of a single superconductiong layer is not sufficient. It is necessary to make heteoepitaxial structures that involve the successive deposition of several thin epitaxial layers which must be superconducting, insulating, or metallic, respectively. The difficulty is to make a barrier that (i) survives after the high-temperature deposition process for the top electrode, (ii) is sufficiently thin, i.e., a few nm, but without pinholes to have a usefully high Josephson tunnel current density, and (iii) can allow the epitaxial growth of the top YBCO layer.

In spite of the difficulties in the PBCO sandwiched between two YBCO layers, YBCO/PBCO/YBCO structures have been successfully deposited *in situ* by a modified off-axis rf-magnetron sputtering technique [491], and Bi₂Sr₂CaCu₂O₈₋δ/Bi₂Sr₂Cu₂O₈₋δ/Bi₂Sr₂CaCu₂O₈₋δ structures were deposited *in situ* using a high-pressure dc-sputtering process. The thickness of the barrier is 22.4 nm as shown in the high-resolution TEM cross-section micrograph [599–601] (see Fig. 98a). Atomic layer-by-layer molecular beam epitaxy has been utilized to synthesize single-crystal thin films and heterostructures of cuprate superconductors and other complex oxides with atomically sharp interfaces, so it is an ideal method to fabricate planery tunnel junction. Bozovic et al. [602] fabricated BiSr₂Ca₇₋δ Dxₓ Cu₈O₁₉₊δ, in which only the Ca atom in the center layer is replaced by Dy. For this material the single unit cell consists of the bottom superconducting electrode, an insulating barrier layer (only a few Å thick), and the top superconducting electrode, thus constituting an artificial intracell Josephson junction; see Figure 98b).

As for as edge SNS Josephson junctions, the first result reported in the literature of a working SNS-like device was obtained by a group at Bellcore using laser deposited multilayers of YBCO/PBCO/YBCO/Au [603]. So far, YBCO Joseph-

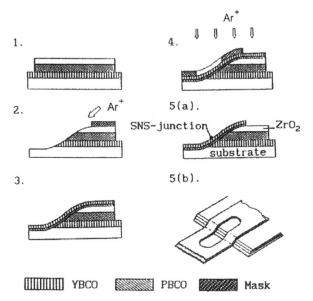

Fig. 99. Fabrication sequence for the edge SQUIDs: (1) *in situ* growth of the YBCO bottom layer, PBCO, and insulating ZrO$_2$. (2) Fabrication of the base electrode edge by the Ar ion-beam etching. (3) Deposition of the PBCO barrier and YBCO top electrode. (4) The sample is lithographically patterned and ion milled through to make the loop. Finally four silver electrodes are deposited by sputtering. (5) Schematic of a finished edge dc SQUID. (a) The cross-section view and (b) the perspective review [604] (*IEEE Trans. Magn.* 27, 3062 (© 1991 IEEE)).

son junctions with various barrier materials such as PBCO [604, 605] and Y$_{1-x}$Co$_x$Ba$_2$Cu$_3$O$_{7-z}$ (Co-YBCO) [606] have been fabricated. Some semiconductorlike materials, such as La$_{1-x}$Ca$_x$MnO$_{3-y}$, La$_{1-x}$Sr$_x$MnO$_{3-y}$, and Pr$_{1-x}$Ca$_x$MnO$_{3-y}$ have also been attempted for use as barrier materials for YBCO junctions [607–609]. However, the fabrication sequence for these edge Josephson junctions is almost the same. Here we introduce the YBCO/PBCO/YBCO heterostructural junction as an example.

There are two steps in the process to fabricate YBCO/PBCO/YBCO junctions in the *a–b* plane. In the first step, a YBCO layer with thickness of 30–200 nm and a PBCO top layer were grown *in situ* on (100) SrTiO$_3$ substrate. This PBCO layer is relatively thick (around 150–250 nm) in order to avoid Josephson coupling between two YBCO layers along the *c*-axis direction. This, however, gives rise to resistive shunting between the two superconducting electrodes. In some samples an extra ZrO$_2$ layer was used as insulator. Then a photoresist stencil covers a part of the multilayer and an argon ion beam with an incident angle of 45–65° is used to etch the unprotected area to fabricate an edge. After removing the photoresist stencil a thin PBCO barrier layer and YBCO top layer are deposited. It should be noted that the edge might be contaminated in between the two steps, so it is important to clean the edge with the Ar ion beam before depositing the barrier layer. Next a junction or a dc SQUID is formed across the edge using photolithographic patterning and ion beam etching. The whole process is shown in Figure 99 [604].

5.1.3.3. Weak Link Microbridges

A local weakening of coupling between two banks through a patterned epitaxial microbridge could be a simple and elegant way to obtain a functional weak link, of either flux-flow or Josephson device type. As early as 1987, Zimminerman reported break junctions made by YBCO sticks [610]. Nowadays, several methods such as "strip poisoning" [611], ion inplantation [612], electron-beam irradiation [613], or locally implanted with oxygen ions [614] are used to fabricate weak links with a highly localized damage ($Ln \ll 1 \mu$m) in the constriction area. However, junctions made by these methods are not reproducible and the junction pattern is too simple for device making.

5.2. Microwave Devices

5.2.1. Introduction

Due to the low micowave resistance (R_s) and high superconducting transition temperature, HTSC thin films have received much attention on microwave applications, and lots of work has been done on the HTSC microwave devices, such as high-Q cavities, low noise microwave oscillators, resonators, delay lines, filters, phase shifters, antenna arrays, etc. [615–618]. The low R_s of the HTSC thin films provides intrinsically narrow bandwidths and high quality factor in these devices. As the conductor loss, which is proportional to the R_s of the electrode, increases strongly with decreasing conductor resistance, the very low R_s of superconducting electrodes ($\sim \mu\Omega$ below about 80 K at 1 GHz for YBCO) permits design of compact devices with practically low operation voltage levels [619]. These are desirable to the researchers and have attracted considerable attention.

5.2.2. Microwave Properties of HTSC

Microwave surface resistance R_s and microwave magnetic properties are the main properties of the HTSC thin films to realize the microwave applications.

At microwave frequencies, current is confined to the surface of a metal and is essentially two-dimensional. Loss can therefore described in terms of a surface resistance R_s. According to Maxwell equation and the two-fluid model, we get

$$R_s^2 = \frac{\omega\mu_0\sigma_s}{2(\sigma_n^2 + \sigma_s^2)}\left[\sqrt{1 + \left(\frac{\sigma_n}{\sigma_s}\right)^2} - 1\right]$$

$$X_s^2 = \frac{\omega\mu_0\sigma_s}{2(\sigma_n^2 + \sigma_s^2)}\left[\sqrt{1 + \left(\frac{\sigma_n}{\sigma_s}\right)^2} + 1\right] \quad (29)$$

For normal metal, $\sigma_s = 0$, $\sigma = \sigma_n$; thus

$$R_s = X_s = \sqrt{\frac{\omega\mu_0}{2\sigma}} \quad (30)$$

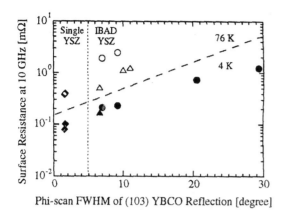

Fig. 100. Surface resistance at 10 GHz vs phi-scan FWHM of (103) YBCO reflection forYBCO films on single-crystal YSZ at 76 K (empty diamond) and 4 K (full diamond); IBAD YSZ buffered polycrystalline alumina at 76 K (empty circle) and 4 K (full circle); and IBAD YSZ buffered polycrystalline Ni-based alloy at 76 K (empty triangle) and 4 K (full triangle) [621] (*IEEE Trans. Appl. Supercond.* 7, 1232 (© 1997 IEEE)).

For conventional superconductors and the single crystals or high quality thin films of high-T_c superconductors,

$$\sigma_s \gg \sigma_n, \quad R_s = \tfrac{1}{2}\omega^2\mu_0^2\sigma_n\lambda^3, \quad X_s = \omega\mu_0\lambda \qquad (31)$$

This means that R_s is in proportion to $\omega^{1/2}$ for normal metal and ω^2 for superconductors. In the BCS theory, the densities of the normal-state electrons are decided by the Boltzman factor $\exp(-\Delta/kT)$, taking into account the linear temperature dependence of the resistivity of HTSC. This results in

$$R_s \propto \frac{\omega^2}{T}\exp\left\{-\frac{\Delta_0}{kT}\right\} \qquad (32)$$

at low temperatures.

Additional loss can arise from a number of mechanisms. The measured surface microwave impedance is an average of sample surface and will be affected by any inhomogeneity in surface properties.

In 1990, Char et al. measured microwave surface resistance on YBCO thin films grown on Al_2O_3 substrates by a parallel-plate resonator technique [620]. In 1997, Findikoglu et al. measured the microwave surface resistance R_s of superconducting YBCO thin films on buffered polycrystalline alumina and Ni-based alloy substrates using a parallel-plate resonator technique [621]. They observed a strong correlation between R_s and the in-plane mosaic spread of the YBCO films. R_s at 10 GHz vs phi-scan FWHM of (103) YBCO reflection for several YBCO films (on single-crystal YSZ, buffered Ni-based alloy, and buffered alumina substrates) was measured and is shown in Figure 100. In general, samples with smaller in-plane mosaic spreads, quantified by the phi-scan FWHM of the (103) YBCO reflections, show lower R_s values. At 76 K, the R_s of samples on buffered Ni-based alloy templates is, for a given FWHM value, about two times smaller than on buffered alumina templates. The best YBCO film with a FWHM of 6.6°(7°) on buffered Ni-based alloy (buffered alumina) shows an R_s of 0.51 mΩ (1.89 mΩ). At 4 K, all samples tend to show a

common and approximately exponential increase of R_s with increasing in-plane mosaic spread. The R_s increases monotonically from about 0.08 mΩ for YBCO with an in-plane mosaic spread of 1.6° on single-crystal YSZ to about 1.3 mΩ for YBCO with an in-plane mosaic spread of 29.3° on buffered alumina. The most textured YBCO film on buffered Ni-based alloy (buffered alumina) template shows an R_s of 0.17 mΩ (0.21 mΩ).

Piel and Muller [622] have presented a review of R_s measurements on HTSC thin films and the following characteristics can be concluded:

(1) The value of R_s depends greatly on the quality of the samples.
(2) R_s of textured thick film or bulk samples is lower than samples without textural sturcture.
(3) R_s of epitaxial films or single crystals is lower than thick film or bulk samples.
(4) R_s of HTSC is lower than that of the usual metals in the microwave field. For YBCO, with the increase of the frequency, the ratio of resistance of Cu to that of YBCO increases, which are 2, 8, 100, and 220 corresponding to frequencies of 500, 100, 10, and 1 GHz, respectively.

If we define a crossover frequency f_c as the frequecy below which the high-T_c superconductors outperform copper, we can find f_c equal to 10, 50, and 500 GHz for bulk ceramics, thick films, and high-quality epitaxial thin film, respectively.

The magnetic property is another important microwave property. The magnetic property of HTSC at microwave frequency has been of increasing interest since their magnetic behavior is related to microwave vortex dynamics.

In 1997, Han et al. designed a CPS to simultaneously measure the resonant frequency (f_0) and the quality factor (Q) [196]. The magnetic susceptibilities ($\chi' - j\chi''$) of $YBa_2Cu_3O_{7-\delta}$/MgO thin films at microwave frequencies were investigated and could be analyzed by cavity perturbation theory. The transition temperature is 91 K and the temperature corresponding to the maximum imaginary part (χ'') is 85 K. However, the transition temperature and surface resistance (R_s) measured by a conventional transmission-line method were 86 K and 180 μΩ at 40 K, respectively. They found absolute differences between the cavity perturbation system and the transmission line ones and suggested that the CPS is one of the most promising techniques for measuring the magnetic susceptibility of HTSC thin film with a 0.1–1 μm thickness.

The frequency responses of insertion losses were measured by the cavity perturbation method (a) and a transmission line method (b) in the range of T_c (= 91 K) to 10 K for YBCO thin film, which is shown in Figure 101. The resonance behaviors show the absolute difference between the cavity perturbation method to measure susceptibility and the conventional method to measure surface resistance. In the case of the cavity perturbation method, the quality factor decreased coming near to $T = 85$ K and increased below $T = 85$ K, while for the transmission line method the quality factor (Q) increased as

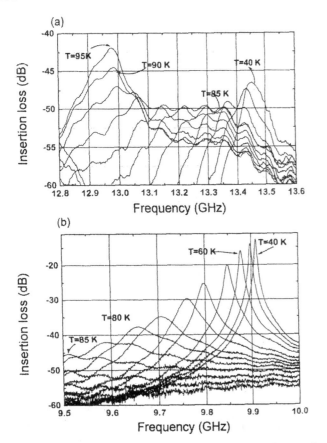

Fig. 101. Insertion loss as a function of frequency (a) using cavity perturbation method (b) using microstrip line method for YBCO/MgO thin film [196] (*IEEE Trans. Appl. Supercond.* 7, 1873 (© 1997 IEEE)).

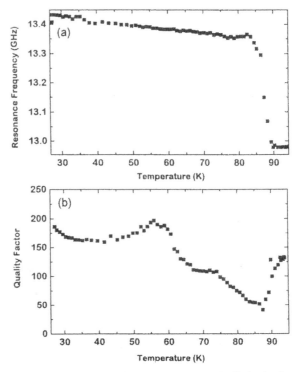

Fig. 102. (a) The shift resonance frequency and (b) the dissipation factor as function of temperature of YBCO/MgO thin film [196] (*IEEE Trans. Appl. Supercond.* 7, 1873 (© 1997 IEEE)).

a temperature decreased. The quality factor and the resonance frequency were found to be approximately 10^3 and 13 GHz near T_c, respectively. As predicted by simulation on the cavity, a dominant TE_{011} mode frequency was 13 GHz.

The resonant frequency shift and a quality factor related to temperature were measured and are shown in Figure 102. These resonance frequencies suddenly drop near T_c and then saturate at 80 K as the temperature is decreased. The temperature dependence of resonant frequency shift mainly depends on both the kinetic inductance and internal magnetic inductances; the former comes from the behavior of the superconducting Cooper pair carriers. When temperatures approach the critical temperature T_c, the kinetic inductance of the superconductor is decreased but the temperature dependence of phase velocity and resonant frequency are increased. The resonance frequency shift is attributed to the change of the inductance per unit length engaged with the change of the penetration depth. Since the phase velocity is inversely proportional to the square root of the inductance, the resonant frequency of the resonator is expressed by $f = V_p/2l$, where V_p is the phase velocity and l is the effective length of the resonator which is slightly longer than the physical length due to the coupling capacitance. Figure 102b shows the temperature dependence of quality factors proportioned to the ac losses and related to variation of $B(t)$

that are out-of phase with the ripple field. The maximum loss point was $T = 85$ K and agreed with the maximum insertion loss.

The surface resistance and microwave susceptibility were measured using the transmission line method and the cavity perturbation method, respectively, and are shown in Figure 103. The microstrip resonator was designed by microwave circuit theory, simulated by Supercompact™, and fabricated using high-temperature superconducting $YBa_2Cu_3O_{7-\delta}$ thin films. The characteristic impedance considering a 0.5 mm MgO substrate is 50 Ω and the fundamental resonance frequency is 9.2 GHz. The surface resistance of the YBCO is about 180 $\mu\Omega$ at 40 K. The result is fitted with the two-fluid model. The behavior of the surface resistance near T_c is well fitted to the theoretical model but showed significant deviations below 65 K for the residual resistance of the superconductor. The temperature dependence of the magnetic susceptibility at 13 GHz and $H_{ac} = 0.1$ Oe is shown in Figure 103b. The values were calculated from equations deduced from perturbation cavity theory which is based on measurement and analysis on the phase shift of the microwave signal envelope transmitted from the test cavity at resonance frequency. The onset transition temperature (T_c) was found to be 91 K. The measurements of the microwave complex susceptibility were carried out on the laser ablated YBCO thin film which have a area of 1.0×1.0 cm^2 and thickness of 300 nm. The absolute value of the experimentally determined χ'' was calculated under the assumption of a $\chi'(T = 10$ K, $H_{dc} = 0) = -1$. Corresponding to the superconducting critical transition, a steplike change in χ' and a peak in

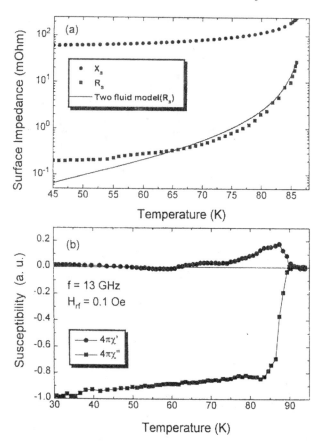

Fig. 103. (a) The temperature dependence of surface resistance for YBCO/MgO thin film. (b) The temperature dependence of the magnetic susceptibility at 13 GHz and $H_{ac} = 0.1$ Oe [196] (*IEEE Trans. Appl. Supercond.* 7, 1873 (© 1997 IEEE)).

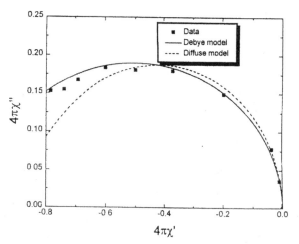

Fig. 104. χ' vs χ'' for YBCO/MgO thin film at $H_{dc} = 0$ Oe [196] (*IEEE Trans. Appl. Supercond.* 7, 1873 (© 1997 IEEE)).

χ'' were found. The χ' vs χ'' was measured and it is plotted in Figure 104, showing a symmetry shape with χ'' maximum at $\chi' = -0.45$. It was fitted with the Debye model and the diffuse model that are suggested in literature in various microscopic physical pictures for the rf susceptibility of nonideal superconductors [623].

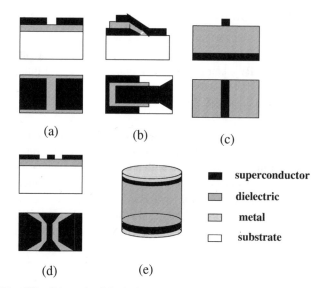

Fig. 105. Schematic of the basic components in the high-T_c superconductor microwave devices.

Thus, we can conclude that the microwave properties of high-quality epitaxial HTSC thin films is superb compared to that of usual metals. So the application of HTSC thin films in the microwave devices should be a great benefit to get low loss and good performance of the devices.

5.2.3. Configuration of the Components

Figure 105 shows the most widespread components of microwave devices based on the high-T_c superconducting thin films: conventional capacitors, planar capacitors, trilayer capacitors, coplanar lines, and disk resonators. To capitalize on the low loss of the high-T_c superconductor, it is important to select suitable dielectric materials as the substrate in these components, which should have low loss microwave properties and be structurally similar to high-T_c superconductors.

Planar Capacitor (Fig. 105a). In this component, the high-T_c superconducting thin film is deposited on a dielectric thin film, mainly a nonlinear dielectric material, such as ferroelectric thin film, to serve as the medium [624]. The typical thickness of the medium layer is $t = 0.3$–1.0 μm; the gap between two superconducting electrodes is $s = 2.0$–20 μm.

Trilayer Capacitor (Fig. 105b). The structure is also called the sandwich capacitor. The middle layer of the dielectric is usually less than 1 μm [625–629]. The thinner the dielectric layer, the smaller the operating voltage.

Microstrip Line (Fig. 105c). The substrate (usually, the thickness is $\sim$250 μm) is used as the dielectric layer with superconductor deposited on both sides of the substrate as the plates of the capacitor. It is widely used in the transmission line [630].

Coplanar Line (Fig. 105d). Many reports can be found concerning the coplanar line since it is widely used in superconducting microwave devices [629, 631, 632]. A coplanar waveguide (CPW) can be used as a phase shifter or a tunable half-wave length resonator. The multipole filter can be designed as a set of tunable CPW resonators.

Disk Resonator (Fig. 105e). The disk resonator is a kind of two-dimensional resonator. It is consists of a circular disk of dielectric film or single crystal with high-T_c superconductor deposited on both sides [633, 634]. As it can equalize the internal current distribution to avoid suffering from high peak current densities inside the resonators, it may improve the power handling characteristics of the devices. This kind of component is proposed to be used as the element in high-power microstrip filters for cellular applications.

5.2.4. Examples of High-T_c Superconducting Microwave Devices

5.2.4.1. Resonators

Wilker et al. reported the fabrication of 5-GHz resonators with coplanar line structure in 1991 [635]. The substrates are LaAlO$_3$ with the size of 10 mm $\times$ 5 mm $\times$ 0.5 mm. The central conductor was 0.5 mm in width and 7.8 mm in length. The gap between the center conductor and two ground conductors was as wide as 0.55 mm. Two kinds of superconducting films were used as the conductor layer: Tl$_2$Ba$_2$Ca$_2$Cu$_3$O$_8$ (TBCCO) and YBCO. For fabrication of TBCCO layers the Ba$_2$CaCu$_2$O$_5$ films were deposited first by rf-magnetron sputtering; then the films were annealed at 800–900°C with the powder of Tl$_2$Ba$_2$Ca$_2$Cu$_3$O$_8$ and Tl$_2$O$_3$ together to form the Tl–Ba–Ca–Cu–O films. YBCO films were deposited by the electron-beam evaporation method. The as-prepared films were patterned into the coplanar lines as described above by photolithography and ion-beam etching. Microwave properties was measured at 80 K. The quality factors Q of the resonators made by YBCO film and TBCCO film were 5900 and 11,800, which are respectively 25 times and 50 times higher than the resonator made by Au films.

Takemoto et al. have reported microstrip resonators using the conventional capacitor structure [636]. ErBa$_2$Cu$_3$O$_7$ (EBCO) films were grown *in situ* by MOCVD on both sides of polished LaAlO$_3$ substrate. Linear resonators with coupling gaps of 0.51 mm and meander line resonators with coupling gaps of 0.35 mm were fabricated by photolithography and chemical etching using dilute phosphoric acid solution. In all the processes, precautions should be taken in keeping away the high-T_c superconducting thin films from water or other harmful substances to avoid the degradation of the property. The films have a critical temperature of 92 K. The Er–Ba–Cu–O linear microstrip resonators demonstrated a quality factor of 3500 at 10 GHz and 4.2 K, which is 45 times better than silver resonators. The Er–Ba–Cu–O meander line resonators with a fundamen-

Fig. 106. Four-pole superconductor microstrip filter layout [638] (*IEEE Trans. Magn.* 27, 2549 (©1991 IEEE)).

tal frequency of 1.3 GHz have quality factor of 9600 at 4.2 K, about 50 times better than silver resonators.

Since the linear microstrip resonators suffer from high peak current density inside the resonators, as mentioned above, disk resonators have been designed to avoid this effect. Jenkins et al. have designed and tested such a type of resonators fabricated from Tl-based superconducting films [637]. TBCCO-2212 thin films were deposited on both sides of 20 $\times$ 20-mm LaAlO$_3$ substrate by rf-magnetron sputtering; then a disk resonator structure similar to that shown schematically in Figure 105d ($r = a$) was patterned. The R_s of such films measured at 24 GHz using a sapphire dielectric resonator is lower than 500 $\mu\Omega$, scaled to 10 GHz and at 80 K. The quality factor values of 3–12 GHz disk resonators have demonstrated considerable improvements when compared to both linear high-T_c superconductor microstrip resonators and comparable copper disk resonators. Additionally, the power handling of such resonators has been shown to be superior to that of conventional linear resonators fabricated from similar materials.

5.2.4.2. Filters

A microstrip bandpass filter is a large application area for high-T_c superconducting thin films. The qulity factor of normal metal microstrip lines is too low to support the low insertion loss and sharp filter skirts needed for channelizing communication bands on communication satellites, where nearly lossless filters are needed. So the superconducting filters have been investigated [631, 638–641].

In a joint effort of COMSAT Laboratory, AT&T Bell Laboratories, Holmdel, and Lincoln Laboratory, four-pole filters were fabricated successfully from YBCO thin films on LaAlO$_3$ dielectric substrate in 1991, as shown in Figure 106 [638]. The size of the substrate is 0.0425 $\times$ 13 $\times$ 23 mm. Single lines and ground planes were deposited on opposite sides of the substrate. A double-sided filter (with a YBCO single line and YBCO ground plane) and a single-sided filter (with a YBCO single line and a silver ground plane) were tested. The filter bandpass is slightly narrower than the nominal one for 3%. The double-sided device demonstrates an insertion loss of 0.3 $\pm$ 0.1 dB at 77 K and less than 0.1 dB at 13 K. The single-sided device has an insertion loss of 0.4 $\pm$ 0.1 dB at 77 K and less than 0.1 dB at 7 K. For both filters, the return loss is better than 10 dB over the

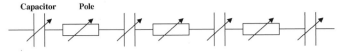

Fig. 107.   Designation of a three-pole half-wave bandpass filter [631].

Fig. 108.   The layout of the two-pole microstrip filter.

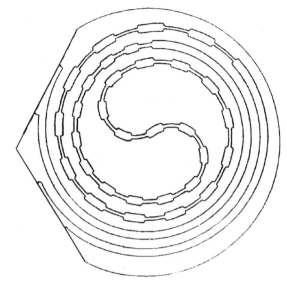

Fig. 109.   Layout of a chirp filter [640].

entire bandpass and better than 16 dB at the frequency of minimum insertion loss. The performances are much better than that of Au filters with the insertion loss of 2.8 dB at 77 K and Nb filters with 0.1 dB at 4.2 K.

In 1996, Findikoglu et al. reported the work of a three-pole half-wave bandpass coplanar waveguide filter incorporating a YBCO (0.4 μm thick)/SrTiO$_3$ (1.2 μm thick)/LaAlO$_3$ multilayer structure, designed as the configuration seen in Figure 107 [631]. The filter fits in an area of 1 cm^2 with a center frequency of 2.5 GHz and a bandwidth of 2%. Tuning is available by applying a separate bias voltage on each pole and also on each coupling capacitance. The broad passband shift of more than 15% is obtained with the applied voltage of 125 V (4×10^6 V/m). However, the insertation loss is quite high at 77 K and zero bias ($\sim$16 dB), although it is improved at 4 K and maximum bias ($\sim$2 dB).

A two-pole microstrip filter was designed by Subramanyam et al. [639]. A 300-nm-thick SrTiO$_3$ was deposited onto the 0.25-mm-thick LaAlO$_3$ substrate by laser ablation. Then, 350 nm of YBCO was deposited on SrTiO$_3$ and patterned into the shape shown in Figure 108. The opposite side of the substrate was covered by a gold layer to serve as the ground plane. Because of the nature of the construction, application of bias predominantly changes the coupling between the resonant sections. The center frequency shift is about 12% with an applied voltage of ±400 V at 30 K. The insertion loss is lower than 0.5 dB and the return loss is good (<-20 dB).

In an analog signal process, the high frequency and wide bandwidths require low conductor loss. So superconductor is proposed for use in the transversal filters which constitute the signal processing structures. Delay lines, in particular, chirp filters, which are the main components of the transversal filters, have been developed by some groups such as the Lincoln Laboratory. A stripline chirp filter was designed as shown in Figure 109, which uses 40-Ω, 120-μm-wide signal lines in the tapped portion of the device [640]. The dielectric is LaAlO$_3$ substrate and the conductor is YBCO film. The length of YBCO line is 0.7 m. The bandwidth of this chirp filter is 2.6 GHz centered on 4 GHz and the length of dispersive delay is 8 ns. The filter is flat-weighted, so that the response as a function of frequency is constant across the bandwidth.

5.2.4.3. Phase Shifters

The phase shifter simply consists of a transmission line with appropriate length. The transmission line must be matched to the external system and have a low loss; it must also have large phase shifts with low applied voltage.

Two types of phase shifters have been widely used: discrete phase shifters, in which switches introduce additional phase shifts into a circuit in discrete amounts, or continuous phase shifters in which a controlling parameter is adjusted to produce a change in the propagation coefficient in a fixed length of transmission line, thereby changing the phase by an amount proportional to the parameter change. The controlling parameter may be electric field, magnetic field, or light intensity.

An example of the discrete phase shifter made by Cole et al. used high-quality sputtered YBCO thin film on LaAlO$_3$ substrates, with R_s values of 20 $\mu\Omega$ at 4.2 K and 1 GHz rising to 300 $\mu\Omega$ at 77 K [642]. In the designation of the digital phase shifter, semiconductor PIN diodes serve as switches. The tested performance agreed reasonably well with simulation over 5% bandwidth. A 4-bit high-T_c superconductor phase shifter was designed with two different circuits; the maximum total insertion loss at 10 GHz is 1.6 dB for one and 1.07 dB for the other.

The widely used form of continuous phase shifter is the ferroelectric-superconductor phase shifter ultilizing the shift in permittivity of a ferroelectric insulator with applied electric field. Chrisey et al. have fabricated a microstrip transmission line based on a thin film heterostructure of Sr$_{0.5}$Ba$_{0.5}$TiO$_3$ (SBT)/YBCO. For temperatures between 100 to 11 K, frequencies less than about 0.5 GHz, and applied fields up to 100 kV/cm, a nearly linear change in field-induced phase shift was produced in the transimission line. A reasonable phase shift of about 80° was obtained at a voltage of about 8 V, which is much lower than the value of the applied voltage needed for the phase shifter in a ceramic form [643].

In a typical magnetically tuned phase shifter design, a superconductor microstrip line is coupled to a ferrite toroid, which contains the magnetic flux so the field in the superconductor is small enough and no extra loss is introduced [644, 645]. Dionne et al. have reported a phase shifter fabricated from a YBCO mi-

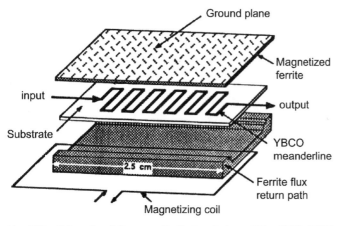

Fig. 110. Schematic of the magnetically tuned phase shifter [645] (*IEEE Trans. Appl. Supercond.* 5, 2083 (1995) (©1995 IEEE)).

Fig. 111. Layout of a optical modulated phase shifter [646] (*IEEE Trans. Appl. Supercond.* 3, 2899 (1993) (©1993 IEEE)).

crostrip line on LaAlO$_3$ coupled to a ferrite (Mn-doped YIG) toroid, as shown in Figure 110. The LaALO$_3$ is glued to the ferrite through the YBCO pressed against the ferrite. The ground plane is on the back of the ferrite and the fields from the microwave signal in the YBCO penetrate the ferrite to provide the interaction to generate the phase shift. Initial measurements at 77 K using a meander line structure providing nonreciprocal operation have demonstrated greater than 700° of differential phase shift at 10 GHz in a compact structure 2.5 cm long and 0.5 cm wide. The insertion loss is smaller than 0.7 dB with yields a figure of merit considerably greater than 1000 deg/dB [645].

An optically modulated phase shifter provided a mechanism for tuning their properties without disturbing the rf signal propagation. Microstrip delay lines of YBCO on 50.8-mm-diam and 0.25-mm-thick LaAlO$_3$ substrate have been fabricated as shown in Figure 111 and were investagted by Track et al. [646]. The width of the microstrip line is 80 μm and the characteristic impedance is 50 Ω. A linepitch of 2.54 mm has been chosen to minimize the coupling between adjacent lines. The total meander line length is 64 cm. The measured delay is 8 ns with operation from dc to 20 GHz. The measurements were performed in a Dewar bottle and in a closed cycle refrigerator, both with rf and optical access. Optical illumiantion of the lines re-

sulted in a phase shift of the transimitted signal. Using a 10 mW HeNe laser, phase shift was up to 360° at 20 GHz. The phase shift increases linearly with frequency and with optical intensity.

5.2.4.4. Antennas

A nonresonant antenna has a complex input impedance with a reactive component considerably greater than the much decreased radiation resistance, $R_{radiation}$. In a practical system, a matching network has to be used to "tune out" the antenna reactance to match the extremely low radiation resistance and finally drive the transmitting or receiving system. The reactance of the matching circuit cancels that of the antenna, $X_{radiation}$, and is in series with the transformed transmitter voltage and resistance R_{system}. For a lossless matching network, maximum power transfer requires $R_{system} = R_{radiation}$. However, ohmic losses in the antenna and matching network introduce an additional equivalent series resistance R_{ohmic} and decreasing the convention efficiency, η, by the factor $R_{radiation}/(R_{radiation} + R_{ohmic})$. Furthermore, the effective Q of the system at resonance is reduced to $X_{radiation}/(2R_{radiation} + R_{ohmic})$. The advantage of using superconductors in antenna systems is the virtual elimination of ohmic dissipation, enabling antenna systems to be devised with near 100% efficiency, even for antennas with dimensions much less than a wavelength.

With the development of the technique of thin film growth, HTSC thin film has also been employed in antenna systems [647–649]. Lancaster found that the "H" microstrip antenna is suitable for use as an efficient small antenna constructed of superconducting thin film materials [647]. An aperture feed and matching network are described which provide a convenient enhancement of the capabilities of the "H" antenna. It is likely that superconducting antennas will only have significant application when they are used in arrays. Three arrays are described demonstrating multiband self-diplexing multifrequency-enhanced bandwidth and multifrequency beam forming. Single-feed circularly polarized microstrip patch and patch array antennas for "direct-to-home" receiving systems at around 12 GHz, fabricated from YBCO, were fabricated and studied by Ali et al. [648, 649]. The antennas are found to show a very low axial ratio and a moderate bandwidth. The gain of the superconducting antennas showed a remarkable improvement over their gold counterparts. The receiving power of a four-element array fabricated from a single-side YBCO thin film on (100) MgO single crystal substrate is found to be 1.8 dB higher than that of a gold array with an identical configuration and with both measured at 77 K.

Besides those listed above, there are many other examples of the applications of HTSC in microwave devices. It is clear that microwave electronics is one of the earliest areas of application and commercialization of HTSC thin films. Many new devices based on the unique properties of HTSC can be expected to be designed and exploited in the near future.

6. HETEROSTRUCTURES

6.1. High-T_c Superconductor/Ferroelectric Heterostructures

6.1.1. Introduction

The perovskite structural ferroelectric has been studied for many years and was found to be particularly important for the application in piezoelectric, pyroelectric, electrostrictive, linear, and nonlinear optical devices. The ferroelectric material exhibits spontaneous polarization which has two or more optimal orientations and the orientation of the polarization can be shifted from one to another by application of an electric field.

Due to the similar crystal structure and different physical properties, the HTSC and perovskite ferroelectrics (FE) heterosturctures of HTSC/FE have been fabricated and investigated since early 1990s. The unique properties of HTSC and FE materials create the possibility for application of HTSC/FE heterostructures to several kinds of devices. For example, HTSC with low microwave loss in HTSC and FE with voltage controllable dielectric constant have been combined to implement voltage tunable low-loss microwave devices [650]. The modulations of the electric field on the free-carrier density of HTSC during the polarization of ferroelectric are relevant to the design of a nonvolatile superconductor field-effect transistor (SuFET) [651, 652]. The perfect interface feature of HTSC and FE makes HTSC a suitable electrode for ferroelectric memory devices since it can overcome the fatigue of ferroelectric [653, 654]. On the other hand, based on the features of superconducting and polarized states, the configuration of HTSC/FE heterostructures can be used to examine the mechanisms of high-T_c superconductivity and polarization. This is why the heterostructure of HTSC/FE has attracted wide interest in fundamental and application research.

The conventional ferroelectric materials used in electric devices are $PbTiO_3$, $BaTiO_3$, $Pb(Zr,Ti)O_3$ (PZT), $(Ba,Sr)TiO_3$ (BST), $(Bi,Sr)TaO$, etc., which may have high remnant polarization or high permittivity or good fatigue property. Most of the thin film deposition methods can be employed for the growth of HTSC/FE heterostructures, such as sputtering, laser ablation, sol–gel, MOCVD, etc. In this section, the syntheses and properties of heterostructures will be introduced.

6.1.2. Syntheses Methods of HTSC/FE Heterostructures

6.1.2.1. Sputtering

Magneron sputtering is used by many groups to grow HTSC/FE heterostructures [655–657]. For example, Hao et al. [657] fabricated the very smooth PZT/YBCO heterostructures on (100) STO substrate using magnetron sputteing. First, the a-axis YBCO film was deposited by dc sputtering at a temperature of 700°C which is $\sim$100°C lower than that for the growth of c-axis-oriented thin films; the sputtering gas pressure was 40 Pa atmosphere with 10 Pa patial pressure of oxygen. The deposition rate was kept as low as 50 nm per hour to ensure a good crystallization, and the thickness of YBCO films was between 40 and 50 nm. After deposition, high-purity oxygen of about 1 atm was poured into the chamber and the films were annealed at $\sim$400°C for 30 minutes and then cooled down to room temperature. Thereafter the PZT layer was deposited *in situ* on the YBCO film by rf magnetron sputtering in a gas pressure of 50–60 Pa with an oxygen–argon pressure ratio of 1:2.5 and at a temperature of 520 to 600°C. The rf power was 80 W; the deposition rate was kept below 80 nm/h. Each deposition process is extended 5–6 h. The PZT/YBCO heterostructure was slowly annealed in 600 Torr of oxygen and cooled down to room temperature at the same atmosphere. The XRD data show that the PZT layer is well c-axis-oriented and the bottom YBCO layer is a-axis-oriented. The very smooth surface of the heterostructure and the low density of defects in the film is proved by the AFM and SEM observations.

6.1.2.2. Laser Ablation

Laser ablation is most widely used due to its powerful advantages of reproducibility on film characteristics and high deposition rate [658, 659]. The short deposition time is a great benefit to resist the interdiffusion of two layers. An example of this method is given by Boikov et al. [658]. They have grown YBCO/PZT/YBCO multilayers on STO and Al_2O_3 (ALO) substrates using laser ablation. Substrates were heated to 400 and 700°C for deposition of PZT and YBCO, respectively, A flow of oxygen was directly against the substrate and the resulting pressure was 0.2 mbar during the deposition. A 300-nm-thick YBCO layer was first deposited on STO or ALO substrates. After annealing in 1 atm of oxygen, the film was cooled down to 400°C, and 300–500 nm of PZT film was deposited on the YBCO layer under the conditions described above. Then the bilayer was heated to 650°C in 1 atm of dry oxygen and annealed at this temperature for 15 min. After that, the temperature was raised to 700°C again for the growth of the top YBCO layer with a thickness of 300 nm. Finally, the multilayers were cooled down to room temperature in 1 atm of oxygen. The heterostructure on STO substrate was confirmed to be grown exptaxially with no lines other than the (00*l*) and (*h*00) observed in the XRD spectrum. For the heterostructure on ALO substrate, a weak impurity peak was found, which may be caused by the formation of a $Ba(Pb,Zr)O_3$ interface layer between YBCO and PZT. The case of creation of $Ba(Pb,Zr)O_3$ in the YBCO/PZT/YBCO/ALO heterostructure might be connected with the diffusion of Ba and Pb ions through the grain boundary due to the misorientation of the a–b plane of the bottom YBCO layer. This was also suggested to be the reason for the low T_c (46 K) of the bottom YBCO layer in the heterostructure on ALO (T_c = 91 K for the YBCO bottom layer on STO). PZT layers showed good ferroelectric properties with nice dielectric hysteresis loops.

Other deposition methods or combinations of different methods can also be used in the fabrication of HTSC/FE heterostructures, depending on the different materials and the various aims of the investigations. In selection of the growth method, at

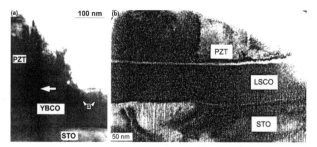

Fig. 112. TEM morphology of (a) PZT/YBCO/STO and (b) PZT/LSCO/STO [660, 661].

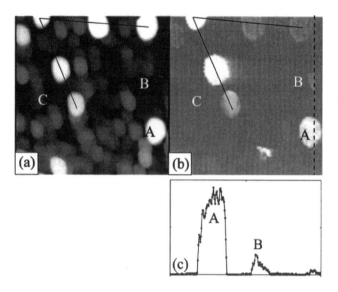

Fig. 113. (a) Surface morphology, (b) the leakage current image at a constant bias of 7.5 V, (c) the line scan along the marked straight line in (b). The image size was 5 μm × 5 μm. The solid lines are guidance for identifying the same position in Figures 114 and 115 [662].

least two factors should be taken into consideration: the method should be fit to grow high-quality films of the selected HTSC and FE materials; interdiffusion and interaction should be resisted during the growth process of the method.

6.1.3. Properties of the Heterostructure

Investigations on the properties of the heterostructures are mainly focused on the interface of two layers and the polarization-induced change of superconducting properties.

6.1.3.1. Surface and Interface Features

The surface and interface features of HTSC and FE layers are important because they are closely related to the properties of the heterostructures and are essential aspects to verify the quality of the heterostructures in many applications.

Hao et al. [657] investigated the surface feature of the PZT/YBCO heterostructure mentioned above by AFM and SEM. The smoothness of the bottom layer is suggested to be very important for the integration with the dielectric layer. In this report, average roughness of the bottom YBCO layer is about 1.7 nm in an arbitrarily selected 10 μm × 10 μm area, causing the very smooth surface of the whole heterostructure with the average roughness of about 1.7 nm also.

The ferroelectricity of PZT is found to connect with the surface feature. Cao et al. [660] also studied the surface feature of the PZT/YBCO heterostructure and suggested that the lattice mismatch will degenerate the smoothness of the surface. The interface of this heterostructure was also investigated by TEM and no interaction or interface layer between the PZT and YBCO films was detected, as shown in Figure 112a. However, TEM cross-section morphology of PZT/LSCO/STO heterostructures fabricated by a similar method shows an interface layer about 3 nm in thickness between the LSCO and PZT layers (Fig. 112b) [661]. As no evidence of interdiffusion and interaction between PZT and LSCO was seen from the SAED patterns, the lattice mismatch of PZT and LSCO has been suggested to be the main reason.

The leakage and ferroelectricity of the PZT/YBCO heterostructure were also studied by some groups [662–664]. Typical work is presented by Xie et al. [662] under the *in situ* testing conducting surface morphology by using a modified

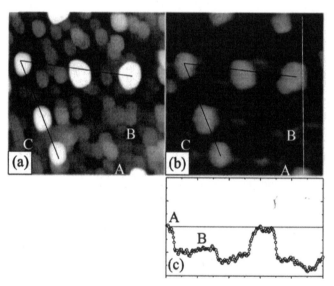

Fig. 114. (a) Surface morphology; (b) ferroelectric image. The area had been negatively polarized. (c) ID line scan in (a) and (b) [662].

conducting atomic force microscopy (C-AFM). The conducting tip of AFM works as the top electrode and the 50-nm-thick YBCO film works as the bottom electrode. Figure 113a, b shows the surface morphology and leakage current image obtained simultaneously at a constant tip–sample bias of 7.5 V in C-AFM mode. Figure 113c is the image line scan along the dashed line in Figure 113b. Figure 114a, b is the surface morphology and the ferroelectric image of the same area acquired by switching the AFM to the ferroelectric mode, while Figure 114c is the image line scan along the dashed line in Figure 114b. Before taking Figure 114, the area was polled *in situ* by applying a constant voltage of −5 V between the tip and

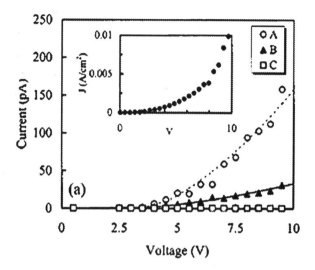

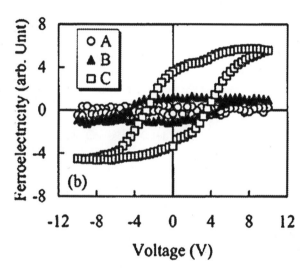

Fig. 115. (a) The I–V plots and (b) F–V loops on A, B, and C marked in Figures 114 and 115. Inset of (a), macroscopic leakage current density J–V plot [662].

the YBCO layer. Therefore, Figure 114b is the distribution of ferroelectric signal; the darker the area, the better the ferroelectricity. Compared with Figure 113b (in which the brighter the area, the larger the leakage current), it shows that the ferroelectric signal is directly related to the leakage current. In the region with a smaller leakage current, the ferroelectricity signal is strong and vice versa. This can be seen more clearly from the image cross section in Figures 113c and 114c. Figure 115 is the mbox$I$–V plots and ferroelectric hystersis (F–V) loops obtained in the typical regions marked by A, B, and C in Figures 113 and 114. It confirms the relationship of the leakage current and ferroelectricity mentioned above. The experiments indicate that the leakage occurs almost from particular grains as shown in Figure 113. However, it is not likely that these grains are from a nonferroelectric phase by examining the XRD data. The I–V characteristics of both point A and point B reflect a similar leakage mechanism, and the macroscopic measurements have revealed that the leakage in PZT is governed by

the model of space charge limited current (SCLC). Based on SCLC, the quantitative difference in points A and B is in support of the suggestion that the leaky grains are due to variations of the interface, which may result from either overgrowth of YBCO or interdiffusion between PZT and YBCO. This result, in addition to the previous studies the leakage current appeared at the grain boundaries [663, 664], would enrich our knowledge on the leakage mechanism.

Much work has been done and must be done further on the improvement of the surface and interface features for understanding the physical essence and matching the requisition of application.

6.1.3.2. Polarization-Induced Changes of Superconducting Properties

The spontaneous polarization of a ferroelectric would surely have some effect (the P-effect) on the electrical properties of superconductors. The discovery of high-T_c superconductors opened up new possibilities to study the P-effect due to the unusual properties of the new materials: the charge density, which is at lease an order of magnitude less than in conventional superconductors, and the anomalously low coherence length. Lemanov et al. reported their systematic study of polarization-induced changes of superconducting properties of high-T_c thin films on ferroelectric substrate [665]. YBCO thin films were deposited on BaTiO$_3$ and LiNbO$_3$ single-crystal substrates by magnetron sputtering. The thickness of the films was 100–300 nm with c-axis normal to film surface. Distinct changes of resistance with the reversal of spontaneous polarization of the ferroelectic substrate was observed. The superconducting transition temperature shifts about 0.5 K for YBCO/BaTiO$_3$ and 40 K for YBCO/LiNbO$_3$ heterostructures. The critical temperature with spontaneous polarization directed toward the film ($+P$) is lower than polarization directed away from the film ($-P$). The critical currents under different polarizations were also different. The nature of this change was discussed using the Thomas–Fermi model. The polarization-induced change of resistance in superconductor is caused by a change in free-carrier density in the superconductor in the polarization field. And since the shift at the beginning of the transition is much smaller and does not exceed 1 K, it was concluded that the shift of T_c is in fact the broadening of transition when polarization is direct toward the film. It may happen through the direct influence of the polarization on the parameters of weak links and/or the percolation in the network of weak links which determine the resistance in the transition region, thus broadening or narrowing the transition.

However, the polarization reversal of BaTiO$_3$ substrate in the heterostructure described above occurs at a high gate voltage of 100–250 V. In order to get obvious field effects at a small applied voltage, ferroelectric thin film must be made larger in the HTSC/FE heterostructure. Many groups have investigated the field effect of superconductor film modulated by a ferroelectric layer. Constantin et al. have fabricated the PZT/YBCO heterostructure on LaAlO$_3$(100) single-crystal substrate using off-axis rf-magnetron sputtering and investigated the polarization field effect on it [656]. The XRD on the 250-nm YBCO/250-nm PZT bilayer shows that YBCO is single 123 phase and has a good orthorhombic structure while the PZT exhibits high perovskite phase purity and (00l) orientation. The good ferroelectric properties of PZT were verified by $P-V$ loops with a high value of the remanent polarization of 61 μC/cm^2. The variation of T_{c0} as a function of gate voltage has been seen clearly even when the value of the applied voltage was lower than 15 V. Cao et al. observed that the resistance of the 12-nm-thick YBCO channel in YBCO/PZT heterostructure increased 7% when +1 V external voltage was applied across the 200-nm-thick PZT layer. By assuming a penetration depth of 1 nm applied to this sample within the framework of a parallel-conductor model, a resistance change of 8% was calculated by the authors, consistent with the experimental data [666].

The polarization-induced change in superconducting properties of superconductor is attractive and is utilized to produce a SuFET.

6.1.3.3. Dielectric and Ferroelectric Properties of HTSC/FE Heterostructures

The dielectric and ferroelectric properties, including the fatigue behavior, the dielectric loss, the polarization, and the optical readout, of a Au/PZT/YBCO heterostructure have been investigated in a temperature range from 20 to 300 K by Lin et al. [667–670]. The heterostructure was deposited on a MgO substrate by laser ablation. The dielectric loss has been found to drop a factor of 2 when the YBCO layer becomes superconducting. So when the YBCO layer becomes superconducting, the drastic change in the impedance of the heterostructure may be attributed to the drop of the dielectric loss, at the same time, the reduction of the interfacial barrier may attributed to the proximity effect. The remnant polarization and coercive field were also found to change with the temperature. A saturation polarization was found near 130 K and kept nearly constant below T_c of YBCO (70 K). This effect may be related to the combination of pinning of domain wall and the proximity effects near the PZT/YBCO interface. It has been proved in the early days that the fatigue property can be improved by using the oxide electrode, because the oxide electrode might reduce the oxygen vacancies in the ferroelectric layer. However, Lin et al. have found that the fatigue property of the heterostructure is much better below T_c. The measurement, using a signal of 5 MHz and 10 V peak to peak square wave, of the remnant polarization showed no fatigue at 10^{12} cycles below T_c, but a drop about 6% above T_c at the same cycles. From this phenomenon, it is also suggested that the most important benefit for HTSC/FE is especially at the temperature below T_c. The optical response for the heterostructure at different temperatures was obtained which showed a peak near T_c, probably ascribed to a potential barrier near PZT/YBCO interface. The ferroelectric switching and optical readout have been perfomed. The experiments have confirmed that the polariztion of a PZT/YBCO structure can be

switched at 5 V and the nondestructive optical readout can be made at wavelenth range of 250 to 500 nm.

These results have confirmed that the HTSC thin films are the ideal electrodes for future ferroelectric devices, especially when HTSC is in the superconducting state.

6.1.4. Application of HTSC/FE Heterostructures

HTSC/FE heterostrutures have be used for various applications, such as microwave devices, surface acoustic wave devices, optical detectors, ferroelectric dynamic random access memories, field-effect transistors, etc. The advantages of these applications have been described above. Here, a few examples are introduced.

6.1.4.1. Tunable Microwave Devices

The microwave devices fabricated from HTSC have many advantages over the conventional devices, under the postfabrication tuning. This can be efficiently achieved by intergrating ferroelectric materials into these devices and utilizing their dielectric nonlinearity, i.e., the strong electric-field dependence of the dielectric constant. A combination of HTSC and ferroelectric materials gives the possiblity for the realization of tunable narrow band filters, delay lines, and phase shifters for microwave frequencies up to several hundred GHz. These are mainly attributed to the low values of dielectric losses, the high dielectric nonlinearity of ferroelectric layer, and the low high-frequency losses of high-T_c superconductor at temperatures in an operating region well below T_c. Progress has been achieved during the last 10 years in this field and many devices have been reported. Some samples have been given in Section 5.2, such as phase shifters, tunable resonators, tunable filters, etc. For example, the disk made from bulk SrTiO$_3$ single crystal covered with double-sided YBCO films was used as a high-quality TM$_{010}$ mode tunable resonator [633]. The losses of the normal metal electrodes are much higher than those of the ferroelectric material. For the TM$_{010}$ mode which used superconductor as electrode, the losses become much lower. Experiments have been done on resonators formed by double-sided YBCO films on STO single crystal. The unloaded quality factor of the resonator was equal to 3000 under zero biasing voltage. The quality factor decreases drastically under increasing voltage. Experiments on YBCO grown on KTaO$_3$ single-crystal substrate have revealed approximately the same dependence of the loss on applied voltage as in the case of STO [671]. In the authors' opinion, the physical process on the interface between the ferroelectric crystal and the electrode could be responsible for the additional loss. So the attention should still be given to the growth of the HTSC film on the ferroelectic substrate. As a tunable component based on the planer capacitor, the YBCO/STO/LaAlO$_3$ multilayer has been investigated. STO with a thickness of 0.4 μm was deposited on LaAlO$_3$ by laser ablation; the gap $S = 5$ μm. YBCO was grown epitaxially on the STO layer. The quality factor is 7400 at 3 GHz and 78 K, which is sensitive to the temperature: below 78 K, the loss factor of the STO film increases; above 78 K, the dissipation in YBCO electrodes is switched on [672]. Chakalov et al. have fabricated YBCO/Ba$_{0.05}$Sr$_{0.95}$TiO$_3$ (BST) and investigated the properties of trilayer capacitor, planar interdigital capacitor, and coplanar waveguide based on this heterostructures. Efficient voltage tunability was demonstrated (up to 40%) at loss level tan $\delta = 0.01$–0.1. A compact YBCO/BST coplanar waveguide with a narrow gap of 18 μm was tested as an electrically tunable phase shifter and field-induced phase shifts of more than 180° were obtained by 35 V dc bias at 20 GHz. It was demonstrated the tuning in HTSC/FE devices can be significantly improved by proper choice of the ferroelectric material, accomplished epitaxial growth of the films, and decrease of the specific dimensions [629].

6.1.4.2. Field-Effect-Transistor

As we have described above, the transport properties of superconductors would be influenced by the polarization field. Using ferroelectrics to serve as a dielectric gate paves the way to boost the performance of superconducting FET. Ahn et al. have fabricated a FET based on a Au/PZT/DyBa$_2$Cu$_3$O$_7$ structure and obtained a resistivity change of 1% [673]. For the ultrathin GeBa$_2$Cu$_3$O$_{7-\delta}$ films, the modulation of superconductivity by PZT was also obtained by Ahn et al. The T_c is depressed by 7 K [673] as shown in Figure 116.

Figure 117 shows the configuration of a typical superconducting FET fabricated by Liu et al. [659]. A KrF excimer laser with the wavelength of 248 nm and a pulse duration of 25 ns was used to fabricate the FE/HTSC heterostructure. The power density of the laser beam on the target was about 2 J/cm^2. A 45-nm-thick YBCO film was first deposited on (100) SrTiO$_3$ substrate at 800–820°C under an oxygen pressure of 400 mTorr with a deposition rate of 2 nm/s, and it was annealed *in situ* in 1 atm of oxygen at 420°C for 30 min. The as-prepared YBCO film was patterned into a bridge of 20 μm width and 40 μm length by photolithography and chemical etching. Then a ~550-nm-thick PZT layer was deposited on the bridge through mechanical masking at 300°C in an oxygen pressure of 400 mTorr with a deposition rate of 7 nm/s. Finally, the whole heterostructure was annealed in at 640°C in 1 atm of oxygen for 10 min and cooled down slowly to room temperature. The XRD spectrum reflects the good epitaxial growth of PZT on YBCO as well as YBCO on STO with c-axis orientation. A 100-nm-thick Ag layer was deposited on the surface of the PZT/YBCO heterostructure by using the magnetron sputtering method. Six Ag contact electrodes were patterned by photolithography: four of them for four-probe method transport measurement of the YBCO channel and the other two to supply the gate voltage for field-effect measurements. With a very small gate area of ~6×10^{-6} cm^2, very good ferroelectric properties with a large saturation of ~60 μC/cm^2, large remnant polarization of 41 μC/cm^2, and small coercive field of 37 KV/cm were obtained. The maximum modulation of the channel resistance was about 3% under the gate voltage of ± 9 V at 64 K. In a similar structure with a larger gate area

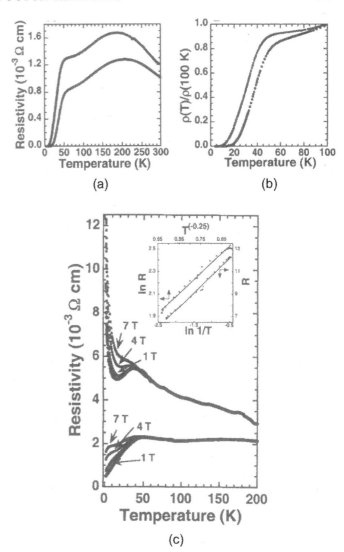

(a)　　　　　　(b)

(c)

Fig. 116. Resistivity versus temperature of PZT/20 Å GBCO/72 Å PBCO heterostructure, (a) (b) for the two polarization states of the PZT layer. The upper curve corresponds to a removal of holes from the p-type GBCO, resulting in an increase the normal-state resistivity normalized at 100 K and a depression of T_c by 7 K. (b) Expanded view of the date presented in (a). (c) Resistivity in magnetic fields. The upper curves show the case for which the holes have been depleted from the system. Inset in (c): Fits of the low-temperature insulating behavior VRH and ln(1/T) behavior [637].

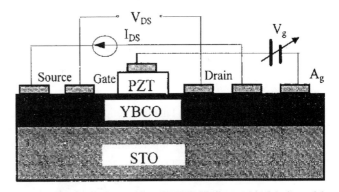

Fig. 117. Schematic diagram of Ag/PZT/YBCO three-terminal device and the circuit for ferroelectric field-effect measurement [659].

($150 \, \mu\text{m} \times 500 \, \mu\text{m}$), a stronger modulation with a value of 7% was obtained at a low gate voltage of 1 V [666]. This fact implies that the field effect was determined by not only the value of the polarization, but also some other factors. A small electrode area can minimize the defects of the PZT layer and get as high as possible polarization and breakdown field, but for the YBCO channel, the possibility of the existence of defects, such as a grain boundary where a strong field effect can be obtained due to lower carrier density, should be also minimized in the small gate area especially for high-quality YBCO layers. In order to get a strong field effect, the combination of both weak links and small electrodes may be the most sufficient way.

In summary, the investigation on the HTSC/FE heterostructures is very important for understanding the physical phenomena in this area and for the promising applications of these structures.

6.2. High-T_c Superconductor/CMR Material Heterostructures

6.2.1. Introduction

The research of manganites has received wide attention for many years. In the early 1950s neutron diffraction and other experiments have been performed on $La_{1-x}Ca_xMnO_3$ (LCMO) and revealed the complex structure and magnetic and electric properties which change closely with the doping level of Ca [674, 675]. Recently the so-called CMR effect has been discovered in these compounds [676, 677]. Such a function with great application potential promoted investigation on this kind of compound not only on the fundamental research, but also on device applications.

The CMR materials are formulated as the $RE_xA_{1-x}MnO_3$, where RE is trivalent rare earth (La, Pr, Nd, ...) and A is a divalent alkali earth (Ca, Sr, Ba, ...). Due to the strong Hund coupling and Jahn–Teller effect, carriers in CMR materials are almost 100% spin polarized. Thus, by controlling the x value, the $RE_xA_{1-x}MnO_3$ can be located in the doped ferromagnetic (FM) and metal region which are typically half-metal.

The fabrication of CMR manganites and high-T_c superconductor heterostructures is possible for the compatibility of their crystal structures and unique different physical features. So the expected new devices based on the heterostructures are proposed and developed. In this section, several kinds of heterostructures and devices will be introduced.

6.2.2. [FM/S]$_n$ Multilayers

If the metal manganites (being ferromagnetic phase in the phase diagram of manganites) are classified into the FM materials, then the CMR material/high-T_c superconductor multilayers can be termed as [FM/S] multilayers.

Usually, there is an intrinsic nonzero phase difference between two superconductors when they are in the Josephson coupling state. In case there is a ferromagnetic barrier layer the interaction of the spins in the two superconductors should be considered. The local exchange field in the ferromagnetic layer may produce a phase shift on both sides of superconducting wave functions as the Cooper pairs tunnel through the FM layers. For a certain thickness of magnetic layer the phase difference of ϕ in both sides of superconducting layers will be found which induces a change of T_c compared with the ordinary state with phase difference of 0. As a consequence there may be oscillations in T_c with the change of magnetic layer thickness [678]. It has been observed in Nb/Gd multilayers in which the superconducting transition temperature T_c oscillates when the thickness of Gd is changed in the range of 10–40 Å [679].

In [$La_{1-x}Ba_xMnO_3$/$YBa_2Cu_3O_7$ (LBMO/YBCO)]$_n$ [682] multilayers the CMR effects of the LBMO layer and superconductivity of the YBCO superconductor layer coexist below T_c of the YBCO [680, 681]. The measurement of the resistance shows anisotropy with respect to the orientation of the applied magnetic field as the usual case of YBCO in the magnetic field, in which it is understood that the YBCO layers are decoupled by the LBMO layers. Superconductivity can only be present in the YBCO layers. Such a result is verified by the following study in [$La_{1-x}Ca_xMnO_3$ (LCMO)/YBCO]$_n$ [683]. The resistivity can be scaled by the component of the magnetic field normal to the film surface. Thus the YBCO layers in such multilayers show a quasi-two-dimensional superconductivity. The electronic structure of a LCMO/$La_{2-1/6}Sr_{1/6}CuO_4$ (LSCO) multilayer system has been calculated using the linear muffin-tin orbital method under the atomic sphere approximation [684]. The result shows that the interaction between LSCO and LCMO in LCMO/LSCO multilayers depresses the MR peak and T_c, respectively.

6.2.3. S/FM/S—Josephson Junctions

It has been mentioned previously that the Josephson effect may be observed when two superconducting layers are coupled through a weak link. So a metal can also be seen as the barrier layer of the Josephson junction in some situations. In that case the metal layer can be treated within the proximity effect. The Cooper pair will enter the metal layer and its density will decay exponentially from the interface to the interior and a weak link forms between two superconducting layers. The Josephson current can pass through the metal layer as long as its thickness is smaller than the normal state coherence length ζ. In the conventional Josephson junction the critical current oscillates with the applied magnetic field due to the spatial modulation of the phase difference ° in the junction. When making substitution of FM layer for the insulator layer the exchange interaction (field) of the FM layer can modulate the ° in the junction. Due to the exchange field in the FM layer changing with the temperature, the oscillation of the Josephson critical current I_c will be changed with the temperature [685]. For a similar reason I_c should also oscillate with the thickness of the FM layer. Up to now some experimental evidence for such phenomena has been

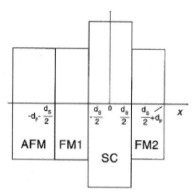

Fig. 118. Schematic picture of the superconducting spin-switch structure [692].

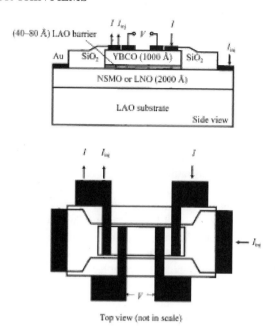

Fig. 119. Schematic of a quasi-particle injection device [695].

presented for [690] and the theoretical discussion have about these been made [685–689].

6.2.4. FM/S/FM—Spin Superconducting Valve

In a sandwich structure of FM/S/FM, there may be two distinctive different states for both sides of FM layers; i.e., the magnetizations are aligned parallel or antiparallel. Experimentally, to reach such a magnetic structure, several technologies can be adopted. The most commonly used method is antiferromagnetic layer pinning for which the magnetic structure is configured as the magnetic spin valve as shown in Figure 118. It arranges the spins such that the magnetization of the top FM layer will be flipped in an external magnetic field while that of the bottom FM layer is fixed. When the thickness of the superconductor layer is near the superconducting coherence length ξ, its critical temperature should be dominated by the proximity effect in such a sandwich configuration [691, 692]. There is still no clear support for above discussion in high-T_c superconductor/CMR manganites heterostructures. But LCMO/YBCO/LCMO trilayers have been fabricated and the aim of the research is focused on the magnetoresistance effect [693].

Another discussion associated with the structure is the magnetic exchange coupling through the superconductor. In FM/N/FM structures there is an indirect exchange coupling between two FM layers via the normal metal. The oscillatory coupling as a function of the thickness of normal metal can be observed experimentally. Similarly, it has been proved in theory that such an oscillation exists in FM/S/FM structures [694]. But the existence of a superconducting gap may influence indirectly exchange coupling between FM layers. The temperature dependence of such effect should be is weaker near the critical temperature and stronger at zero temperature.

6.2.5. Spin-Polarized Quasiparticle Injection Devices

Nonequilibrium superconductivity can be induced by the spin-polarized electron injection from a metallic ferromagnet, where the ferromagnet is used as the source of the spin-polarized electrons. Under the injection the superconducting order parameter will be suppressed and the critical current density and energy gap will be depressed. Thus a small injection current can suppress a large supercurrent. Such a device, called a spin-polarized electron injection device, is a new type of superconducting transistor. Several device patterns have been proposed. In a FM/I/S junction the spin-polarized electrons injected into the superconductor from the ferromagnet through the barrier layer will keep their spin state [695–697]. Actually, the insulating barrier layer inhibits the proximity effect between the superconductor and the ferromagnet [697, 698]. A special case may exist in which running a current parallel to the superconductor strip and partially through the ferromagnetic layer will effectively cause the superconductor to short out part of the ferromagnetic layer, resulting in some of the current being injected into the superconductor from the ferromagnet. All these device patterns will be discussed in detail below [102, 701].

A typical FM/I/S junction consists of $Nd_{1-x}Sr_xMnO_3$ (NSMO)/LAO/YBCO; the NSMO is the M layer, LAO is the insulator layer, and YBCO is S layer as shown in Figure 119 [695]. A $LaNiO_3$ (LNO)/LAO/YBCO trilayer is also fabricated to understand the difference in the injection effect of spin-polarized and unpolarized quasi-particles, where LNO is a nonmagnetic metal and the structure can be denoted as an N/I/S junction. For N/I/S junctions the critical current I_c decreases slightly when the injection current I_{inj} increases as shown in Figure 120a. Figure 120b shows a family of current–voltage curves for FM/I/S junctions with six different gate currents of negative polarity. The gain for FM/I/S is significantly larger than that for N/I/S junctions, which suggests that the spin-polarized quasi-particles generate a much larger nonequilibrium population. In YBCO/STO/LSMO junctions with a barrier layer thickness of 400 Å, negative current gains as large as 35 also have been observed [696]. In such kinds of junction no energy gap of the superconductor or other characteristic of the tunneling process in the junction has been observed and the spin-polarized current may be injected into the superconductor through the surface or interface states with one kind of barrier.

In thin film of the high-T_c superconductor $DyBa_2Cu_3O_7$ (DBCO) which is incorporated in a FM/I/S trilayers structure with I an undoped ultrathin (24 Å) layer of La_2CuO_4 (LCO) and FM a ferromagnetic underlayer of $La_{2/3}Sr_{1/3}MnO_3$(LSMO), it is found that its critical current density is strongly suppressed by current flowing in the FM layer as shown in Figure 121 [699]. Part of the current flows from the ferromagnetic layer into the superconductor and causes the shift of the current–voltage of the DBCO layer. The pulsed current technique is used to measure the effect of injection current as in the case of similar trilayers [698]. A slight increase in the critical current I_c is observed under small injection current I_{inj}, followed by a strong suppression of I_c under large I_{inj}. A "self-injection" mechanism is proposed to explain the abnormality under the small I_{inj}.

A third type of sample pattern has been devised to uncover the phenomena taking place at the interface between the high-T_c superconductor and CMR manganites as shown in Figure 122 [700]. Bilayer FM/S is grown on STO (001) substrates and is patterned to five S/FM/S junctions. The resistances of all S/FM/S junctions with different ferromagnet spacers (3–20 μm) are measured and the resistance of the FM/S interface is calculated by extrapolation of the results to zero length of the ferromagnet spacer. The I–V characteristics of the S/FM/S junctions are shown in Figure 122 (bottom). Below 60 K there is a dip in the curve that cannot be explained as the result of heating effect and may be the result of the injection of the spin-polarized quasi-particles. This may be correct since by replacing the FM layer with Au or paramagnetic LNO, such a conductance dip cannot be observed. The transport across the interface excludes the tunneling process since there is no barrier layer. The low-bias conductance of FM/S system has been discussed in detail by Zutic and Valls [701].

Scanning tunneling spectroscopy has been performed on LCMO (1000 Å)/YSZ (20 Å)/YBCO (1000 Å) trilayers [702]. The spin-polarized quasi-particles are injected into the superconductor by passing a current through the ferromagnetic LCMO layer. Two types of tunneling conductance dI/dV spectra have been observed on the YBCO layer at 4.2 K. One shows a distinct gap structure (Fig. 123). The left spectrum suggests that the YBCO layer has a d-wave pairing symmetry. The other type of spectrum shows a pronounced zero-bias conductance peak and can be attributed to tunneling in the a–b plane of a d-wave superconductor.

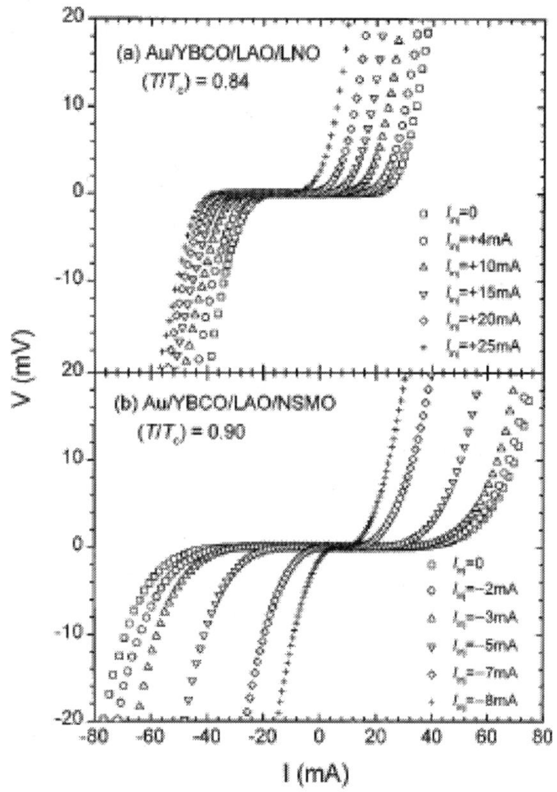

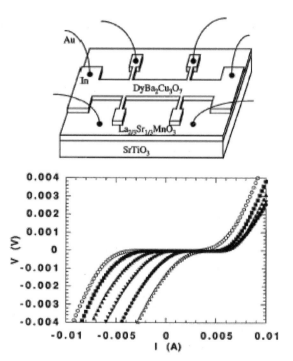

Fig. 120. I–V curves of two comparable devices at different injection currents, each with 80-Å-thick LAO barrier. (a) An S/I/N quasi-particle injection device (QPID) at 67 K ($t = 50.84$). (b) An S/I/F spin-polarized quasi-particle injection device (SP-QPID) at 80 K ($t = 50.90$) [695].

Fig. 121. Top: Geometry of the ferromagnet-superconductor sample used for measurements. The width of the $DyBa_2Cu_3O_7$ bridge was 300 μm and the distance between the voltage leads was 3 μm. The substrate was 6×6 μm in area. Bottom: Voltage–current characteristics of $DyBa_2Cu_3O_7$ layer of $La_{2/3}Sr_{1/3}MnO_3/La_2CuO_4/D_yBa_2Cu_3O_7$ heterostructure at $T = 50$ K when the parallel current in the ferromagnet equals 0 mA (open circles), 2 mA (squares), 4 mA (solid triangles), 6 mA (inverted triangles), 8 mA (solid circles), and 10 mA (open triangles) [699].

7. CONCLUSION

So far, the investigations of high-temperature superconductor thin films have obtained great achievements. Thin films have contributed a lot to fundamental research and device applications of high-T_c superconductivity.

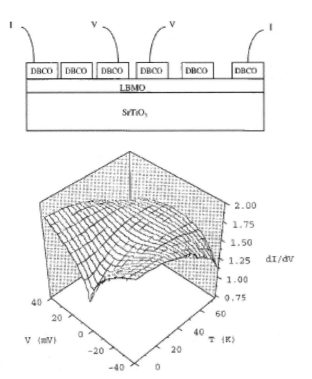

Fig. 122. Top: Geometry of the samples. The width of the DyBa$_2$Cu$_3$O$_7$ bridge was 400 μm, and the lengths of the La$_{2/3}$Ba$_{1/3}$MnO$_3$ spacers were 3, 5, 12, 32, and 102 μm, respectively. The substrate area was $6 \times 6 \ \mu$m^2. Bottom: Temperature and voltage bias dependence of the differential conductance of the interface between the ferromagnet La$_{2/3}$Ba$_{1/3}$MnO$_3$ and the superconductor DyBa$_2$Cu$_3$O$_7$ [701].

For fabrication of high-T_c superconductor thin films, many advanced and practical methods have been developed, and a great variety of substrates and buffers have been created. These are rich materials not only for preparing high-T_c superconductor thin films but also for developing many other kinds of material thin films. On the other hand, such methods, substrates, and buffer materials are also practically useful for developing thick, high-T_c superconductor films which have important application potentials for electronic device and electric power system.

The various kinds of material thin films and lots of types of high-T_c superconductor thin films have been developed. Nobody expected in 1987 that such great achievements on high-T_c superconductor thin films would be obtained after 14 years. To date the high-T_c superconductor thin films play an important role on all fundamental research and device applications. All discovered high-T_c superconducting materials have been made into superconducting thin films, even including those materials that can not be made into the bulk samples with superconductivity.

Ultrathin films and multilayers are special films with thickness sized in unit cell order or characteristic lengths. They show strong surface and interface effects, so they play an unusual role in fundamental research not only on electric and magnetic properties of high-T_c superconductors but also on the mechanism of high-T_c superconductivity. They also have special functions to develop various kinds of devices. The large-area thin films possess important special application potentials compared to conventional size thin films; it also should be noted that the substrates and buffer layers, as well as the technique and apparatus used for growing large-area thin films, still must be improved and developed.

Especially, thin films are the indispensable important basic materials for investigations on transport properties of high-T_c superconductors, such, as resistivity, Hall effect, flux dynamics, optical properties, and so on. These all have been done on high-

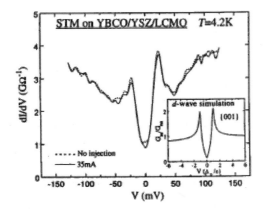

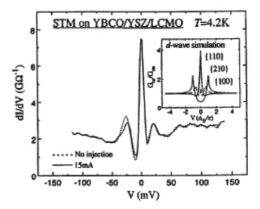

Fig. 123. Left: STM tunneling data taken on an YBCO/YSZ/LCMO heterostructure at 4.2 K, showing a distinct gap structure. The dashed curve is for 0 mA injection and the solid curve is for 35 mA (7×10^3 A/cm^2) injection. The inset shows the spectral simulation from [8] for c-axis quasi-particle tunneling on a d-wave superconductor. Right: STM tunneling data taken on a YBCO/YSZ/LCMO heterostructure at 4.2 K, showing a zero-bias conductance peak structure. The dashed curve is for 0 mA injection and the solid curve is for 15 mA (3×10^3 A/cm^2) injection. The inset shows the spectral simulations from [11] for various a–b-plane tunnelings on a d-wave superconductor [702].

T_c superconductor thin films and important new results have been obtained. These results have exposed some anomalous phenomena for high-T_c superconductors, such as the unusual temperature dependence of Hall effect and resistivity which open a way to probe the intrinsic property of high-T_c superconductors and even to get information about the mechanisms of high-T_c superconductivity. Understanding such unusual phenomena is not only necessary to improve the quality of the present thin films, but it also needed to make special types of thin films.

Investigation of heterostructures of high-T_c superconductors and other kinds of perovskite structural materials (ferroelectric and CMR materials, etc.) is an efficient way to understand and expose interactions of the charge and spins of these materials. Up to now, some new effects, such as field and spin-polarized electron tunneling effects, have been observed and studied.

So far, high-quality thin films were almost made from the superconductors in the optimal doped region. For understanding the mechanisms of high-T_c superconductivity, the underdoped and overdoped regions are very important, so it is very necessary to develop the thin films in both of regions in the next step.

The investigation of high-T_c superconductor thin films may develop new materials and devices and observe new effects. More great progress in this field is expected in the future.

Acknowledgments

I thank Drs. X. L. Dong, X. G. Qiu, Y. Lin, H. Wang, K. Tao, C. Cai, W. Z. Gong, and P. S. Luo for their help in preparing this chapter. I also acknowledge the assistance of Ms. B. Xu, Ms. X. P. Wang, Mr. B. Chen, Dr. B. Y. Zhu, Mr. H. L. Liu, Mr. L. Zhao, and Mr. H. W. Zhu.

Very useful discussions were conducted with Professor L. Lin, Professor Z. X. Zhao, Professor H. H. Wen, and other colleagues.

This work is supported by the National Natural Science Foundation Grant for State Key Program for Basic Research of China and the "Zhi-Shi-Chuang-Xin" Program of Chinese Academy of Sciences.

REFERENCES

1. Bedmorz and Müller, *Z. Phys. B* 64, 189 (1986).
2. H. Adachi, K. Setsune, and K. Wase, *Phys. Rev. B* 35, 8824 (1987).
3. A. S. Edelstein, S. B. Qadri, R. L. Haltz, P. R. Broussard, J. H. Claassen, T. L. Francavilla, D. U. Gubser, P. Lubitz, E. F. Skelton, and S. A. Wolf, *J. Cryst. Growth* 85, 619 (1987).
4. L. Li, B. R. Zhao, Y. Lu, H. S. Wang, Y. Y. Zhao, and Y. H. Shi, *Chinese Phys. Lett.* 4, 233 (1987).
5. T. Aida, T. Fukazawa, K. Takagi, and K. Miyauchi, *Jpn. J. Appl. Phys.* 26, L1489 (1987).
6. J. Yotsuya, Y. Suzuki, S. Ogawo, H. Kuwahara, K. Otani, T. Emoto, and J. Yamamoto, "5th International Workshop on Future Electron Devices," June 2–4, 1988, Miyagi-Zoa, Japan.
7. J. Gao, Y. Z. Zhang, B. R. Zhao, P. Out, C. W. Yuan, and L. Li, *Appl. Phys. Lett.* 52, 2675 (1988).

8. P. Madakson, J. J. Cuomo, D. S. Yee, R. A. Roy, and G. Scilla, *J. Appl. Phys.* 63, 2046 (1988).
9. M. Naito, R. H. Hammond, B. Oh, M. R. Hahn, J. W. P. Hsu, P. Rosenthal, A. F. Marshall, M. R. Beasley, T. H. Geballe, and A. Kapitulnik, *J. Mater. Res.* 2, 713 (1987).
10. J. Hudner, M. Östling, H. Ohlsen, L. Stolt, P. Nordblad, M. Ottosson, J.-C. Villegier, H. Morieeau, T. Weiss, and O. Thomas, *J. Appl. Phys.* 73, 3096 (1993).
11. F. Arrouy, J. P. Locquet, E. J. Williams, E. Mächler, R. Berger, C. Gerber, C. Monroux, J. C. Grenier, and A. Wattiaux, *Phys. Rev. B* 54, 1 (1996).
12. X. D. Wu, D. Dijkkamp, S. B. Ogale, A. Inam, E. W. Chase, P. F. Miceli, C. C. Chang, J. M. Tarascon, and T. Venkatesan, *Appl. Phys. Lett.* 51, 861 (1987).
13. T. Frey, C. C. Chi, C. C. Tsuei, T. Shaw, and F. Bozso, *Phys. Rev. B* 49, 3483 (1994).
14. K. Zhang, B. S. Kwak, E. P. Boyd, A. C. Wright, and A. Erbil, "Proceedings of the Conference on the Science and Technology of Thin Film Superconductors," November 14–18, 1988, Colorado Springs, pp. 271–280.
15. R. W. Vest, T. J. Fitzsimmons, J. Xu, A. Shaikh, G. L. Liedl, A. I. Schindler, and J. M. Honig, *J. Solid State Chem.* 73, 283 (1988).
16. R. Singh, S. Sinha, N. J. Hsu, J. T. C. Ng, P. Chou, H. S. Ullal, A. J. Nelson, and A. B. Swartzlander, "Proceedings of the Second Conference on the Science and Technology of Thin Film Superconductors," April 30–May 4, 1990, Denver, CO, pp. 303–310.
17. J. Hudner, O. Thomas, E. Mossang, P. Chaudouet, F. Weiss, D. Boursier, and J. P. Senateur, *J. Appl. Phys.* 74, 4631 (1993).
18. J. Takenoto, T. Inoue, H. Komatsu, S. Hayashi, S. Miyashita, and M. Shimizu, *Jpn. J. Appl. Phys. Lett.* 32, L403 (1993).
19. T. Kitamura, S. Taniguchi, I. Hirabayashi, S. Tanaka, Y. Sugawara, and Y. Ikuhara, *IEEE Trans. Appl. Supercond.* 7, 1392 (1997).
20. S. Reich and Y. Tsabba, *Adv. Mater.* 9, 329 (1997).
21. Z. C. Li, G. H. Cao, M. H. Fang, and Z. K. Jiao, *J. Mater. Sci. Lett.* 18, 91 (1999).
22. G. W. Book, W. B. Carter, T. A. Polley, and K. J. Kozaczek, *Thin Solid Films* 287, 32 (1996).
23. J. McHale, R. W. Schaeffer, A. Kebede, J. Macho, and E. Solomon, *J. Supercond.* 5, 511 (1992).
24. M. Ochsenkühn-Petropulu, I. Vottea, T. Arggropulu, L. Mendrinos, and K.-M. Ochsenkühn, *Mikrochim. Acta* 130, 289 (1999).
25. T. Nakada, M. Kazusawa, A. Kunioka, and T. Fujimori, *Jpn. J. Appl. Phys. Lett.* 27, L873 (1988).
26. J. Kurian and J. Koshy, *J. Supercond.* 12, 445 (1999).
27. H.-W. Neumüller, W. Schmidt, H. Kinder, H. C. Freghardt, B. Stritzker, R. Wördenweber, and V. Kirchhoff, *J. Alloys Compounds* 251, 366 (1997).
28. B. F. Cole, G. C. Liang, N. Newman, K. Char, and Zahar Chuk, *Appl. Phys. Lett.* 61, 1727 (1992).
29. R. Wördenweber, J. Einfeld, and K. Kutzuner, *IEEE Trans. Appl. Supercond.* 9, 2486 (1999).
30. T. Venkatesan, X. D. Wu, B. Dutta, A. Inam, M. S. Hedge, D. M. Hwang, C. C. Chang, L. Nazar, and B. Wilkens, *Appl. Phys. Lett.* 54, 581 (1989).
31. T. Terashima, K. Shimura, Y. Bando, Y. Matsuda, A. Fujiyama, and S. Komiyama, *Phys. Rev. Lett.* 67, 1362 (1991).
32. Y. Matsuda, S. Komiyama, T. Onogi, T. Terashima, K. Shimura and Y. Bando, *Phys. Rev. B* 48, 10498 (1993).
33. X. X. Xi, J. Geerk, G. Linker, and Q. Meyer, *Appl. Phys. Lett.* 54, 2367 (1989).
34. D. P. Norton and D. H. Lowndes, *Appl. Phys. Lett.* 63, 1432 (1993).
35. L. Z. Zheng, J. J. Sun, B. R. Zhao, C. F. Zhu, and C. L. Bai, *Chinese Phys. Lett.* 12, 625 (1995).
36. J. A. Chervenak and J. M. Valles, *Phys. Rev. B* 54, R15649 (1996).
37. H. Ota, S. Migita, Y. Kasai, H. Matsuhata, and S. Sakai, *Phys. C* 311, 42 (1999).
38. S. G. Zybtsev, *J. Low. Temp. Phys.* 107, 479 (1997).
39. J.-M. Triscone, M. G. Karkut, L. Antognazza, O. Brunner, and F. Fischer, *Phys. Rev. Lett.* 28, 1016 (1989).

40. D. H. Lowndes, D. P. Norton, and J. D. Budai, *Phys. Rev. Lett.* 65, 1160 (1990).

41. A. K. Rajagopal and S. D. Muhanti, *Phys. Rev. B* 44, 10210 (1991).

42. M. Tachiki and S. Takahashi, *Phys. C* 191, 363 (1992).

43. X. G. Qiu, B. R. Zhao, S. Q, Guo, J. L. Zhang, L. Li, and M. Tachiki, *Phys. Rev. B* 48, 16180 (1992).

44. B. R. Zhao, F. Ichikawa, M. Tachiki, T. Aomine, J. J. Sun, B. Xu, and L. Li, *Phys. Rev. B* 55, 1247 (1997).

45. B. R. Zhao, X. Q. Qiu, S. Q. Guo, J. L. Zhang, B. Xu, Y. Z. Zhang, Y. Y. Zhao, and L. Li, *Phys. C* 204, 341 (1993).

46. R. G. Humphreys, J. S. Satchell, N. G. Chew, J. A. Edwards, S. W. Goodyear, S. E. Blenkinsop, O. D. Dosser, and A. G. Cullis, *Supercond. Sci. Technol.* 3, 38 (1990); T. R. Lemberger, in "Physical Properties of High Temerature Susperconductors III" (D. M. Ginsberg, Ed.), pp. 471–523. World Scientific, Singapore, 1992.

47. M. Schieber, *J. Cryst. Growth* 109, 401 (1991).

48. J. A. Alarco, G. Brorson, T. Claeson, M. Danerud, U. Engström, Z. G. Ivanov, P.-Å. Nilsson, H. Olin, and D. Winkler, *Phys. Scripta* 44, 95 (1991).

49. R. K. Singh and J. Narayan, *J. Mater. Sci.* 26, 13 (1991).

50. C. H. Stoessel, R. F. Bunshah, S. Prakash, and H. R. Fetterman, *J. Supercond.* 6, 1 (1993).

51. S. Miyazawa, Y. Tazoh, H. Asano, Yasuhiro, Nagai O. Michikami, and M. Suzuki, *Adv. Mater.* 5, 179 (1993).

52. D. P. Norton, *Annu. Rev. Mater. Sci.* 28, 299 (1998).

53. D. J. Rogers, P. Bove, and F. H. Tcherani, *Supercond. Sci. Technol.* 12, R75 (1999).

54. E. K. Hollmannt, O. G. Vendik, A. G. Zaitsev, and B. T. Melekh, *Supercond. Sci. Technol.* 7, 609 (1994).

55. A. M. Cucolo and P. Prieto, *Internat. J. Modern Phys. B* 11, 3833 (1997).

56. A. P. Bramley, J. D. Oconnor, and C. R. M. Grovenor, *Supercond. Sci. Technol.* 12, R57 (1999).

57. J.-M. Triscone and F. Fischer, *Rep. Progr. Phys.* 60, 1673 (1997).

58. H. Huhtinen, R. Laiho, and P. Paturi, *Phys. Low-Dim. Struct.* 11/12, 93 (1998).

59. J. Gallop, *Supercond. Sci. Technol.* 10, A120 (1997).

60. N. M. Alford, S. J. Penn, and T. W. Button, *Supercond. Sci. Technol.* 10, 169 (1997).

61. P. Kuppusami, V. S. Raghunathan, A. G. Vedheswar, and Y. Hariharan, *Phys. C* 215, 213 (1993).

62. K. Wasa and S. Hayakawa, "Handbook of Sputter Deposition Technology," p. 99. Noyes, Park Ridge, NJ.

63. L. R. Gilbert, R. Messier, and R. Loy, *Thin Films* 54, 129 (1978).

64. H. Asano, M. Sahi, and O. Michikami, *Jpn. J. Appl. Phys.* 28, L981 (1989).

65. C. B. Eom, J. Z. Sun, K. Yamamoto, A. F. Marchall, K. Eluther, and T. H. Geballe, *Appl. Phys. Lett.* 55, 595 (1989).

66. O. Michikami and M. Asano, *Jpn. J. Appl. Phys.* 30, 939 (1989).

67. T. I. Selinder, Glarsson, U. Helmersson, and S. Rudner, *Supercond. Sci. Technol.* 4, 379 (1991).

68. K. Setsune, T. Kamada, H. Adachi, and K. Wasa, *J. Appl. Phys.* 65, 1318 (1988).

69. A. R. Krauss, O. Auciello, A. I. Kingon, M. S. Ameen, Y. L. Liu, T. Barr, T. M. Graettinger, S. H. Rou, C. S. Soble, and D. M. Gruen, *Appl. Surf. Sci.* 46, 67 (1990).

70. M. Migliuolo, A. K. Stamper, D. W. Greve, and T. E. Schlesinger, *Appl. Phys. Lett.* 54, 859 (1989).

71. G. C. Xiong and S. Z. Wang, *Appl. Phys. Lett.* 55, 902 (1989).

72. M. Migliuolo, R. M. Belan, and J. A. Brewer, "Proc. 3rd Annual Conf. on High-Temperature Superconductivity," Genoa, 1990, World Scientific, Singapore, 1990.

73. Z. Qi, W. Shi, J. Sun, H. Ke, and L. Liu, *J. Phys. D* 21, 1040 (1988).

74. J. D. Klein, A. Yen, and S. L. Clauson, *J. Appl. Phys.* 67, 6389 (1990).

75. V. Matijasevic, R. H. Hammond, P. Rosenthal, K. Shinohara, A. F. Marshall, and M. R. Beasley, *Supercond. Sci. Technol.* 4, 376 (1991).

76. E. B. Graper, *J. Vac. Sci. Technol.* 8, 333 (1971).

77. X. Cui, F. A. List, D. M. Kroeger, A. Goyal, D. F. Lee, J. E. Mathis, E. D. Specht, P. M. Martin, R. Feenstra, D. T. Verebelyi, D. K. Christen, and Mparanthaman, *Phys. C* 316, 27 (1999).

78. H. S. Wang, D. Eissler, W. Dietsche, A. Fischer, and K. Ploog, *J. Cryst. Growth* 126, 565 (1993).

79. J. Kwo, T. C. Hsieh, R. M. Fleming, M. Hong, S. H. Liou, B. A. Davidson, and L. C. Feldman, *Phys. Rev. B* 36, 4039 (1987).

80. C. Webb, S. L. Weng, J. N. Eckstein, N. Missert, K. Char, D. G. Schlom, E. S. Hellman, M. R. Beasley, A. Kapitulnik, and J. S. Harris, *Appl. Phys. Lett.* 51, 1191 (1987).

81. J. Kwo, *J.Cryst. Growth* 111, 965 (1991).

82. D. D. Berkley, A. M. Goldman, B. R. Johnson, J. Morton, and T. Wang, *Rev. Sci. Instrum.* 60, 3769 (1989).

83. B. R. Johnson, K. M. Beauchamp, T. Wang, J.-X. Liu, K. A. MacGreer, J.-C. Wan, M. Tuominen, Y.-J. Zhang, M. L. McCartney, and A. M. Goldman, *Appl. Phys. Lett.* 51, 1911 (1990).

84. T. Kawai, M. Kanai, T. Matsumoto, H. Tabata, K. Horiuchi, and S. Kawai, in "High T_c Superconductor Thin Films" (L. Correra, Ed.), p. 287. Elsevier, Amsterdam, 1992.

85. H. S. Wang, D. Eissler, Y. Kershaw, W. Dietsche, A. Fischer, and K. Ploog, *Appl. Phys. Lett.* 60, 778 (1992).

86. M. Kawai, S. Watanabe, and T. Hanada, *J. Cryst. Growth* 112, 745 (1991).

87. M. E. Klausmeier-Brown, G. F. Virshup, I. Bozovic, J. N. Eckstein, and K. S. Ralls, *Appl. Phys. Lett.* 60, 2806 (1992).

88. R. H. Ono, *Bull. Mater. Res. Soc.* 8, 34 (1992).

89. C. A. Chang, C. C. Tsuei, T. R. M. Guire, D. S. Yee, J. P. Boresh, H. R. Lilienthal, and C. E. Farrell, *Appl. Phys. Lett.* 53, 916 (1988).

90. N. Hass, U. Dai, and G. Deutscher, *Supercond. Sci. Technol.* 4, 262 (1991).

91. J. Cheung and J. Horwitz, *Bull. Mater. Res. Soc.* 17, 30 (1992).

92. R. K. Singh, D. Bhattacharya, P. Tiwari, J. Narayan, and C. B. Lee, *Appl. Phys. Lett.* 60, 255 (1992).

93. B. Raos, L. Schultz, and G. Endres, *Appl. Phys. Lett.* 53, 1557 (1988).

94. G. Koren, E. Polturak, B. Fisher, D. Cohen, and G. Kimel, *Appl. Phys. Lett.* 53, 2330 (1988).

95. P. Vase, S. Yueqiang, and T. Freltoft, *Appl. Surf. Sci.* 46, 61 (1990).

96. V. Olsan and M. Jelinek, *Phys. C* 207, 391 (1993).

97. M. Kanai, T. Tomoji, and S. Kawai, *Jpn. J. Appl. Phys.* 31, 331 (1992).

98. M. Y. Chern, A. Gupta, and B. W. Hussey, *Appl. Phys. Lett.* 60, 3045 (1992).

99. S. J. Pennycook, M. F. Chisholm, D. E. Jesson, R. Feenstra, S. Zhu, X. Y. Zheng, and D. J. Lowndes, *Phys. C* 202, 1 (1992).

100. T. Frey, C. C. Chi, C. C. Tsuei, T. Shaw, and G. Trafas, in "Layered Superconductors: Fabrication Properties and Applications," MRS Symposia Proceedings, Vol. 275, p. 61. Materials Research Society, Pittsburgh, 1992.

101. T. Frey, C. C. Chi, C. C. Tsuei, T. Shaw, and F. Bozso, *Phys. Rev. B* 49, 3483 (1994).

102. S. Koike, T. Nabatame, and I. Hirabayashi, *Jpn. J. Appl. Phys.* 32, L828 (1993).

103. H. Yamane, H. Masumoto, T. Hirai, H. Iwasaki, K. Watanabe, N. Kobayashi, and Y. Muto, *Appl. Phys. Lett.* 53, 1548 (1988).

104. K. Watanabe, H. Yamane, H. Kurosawa, T. Hirai, N. Kobayashi, H. Iwasaki, K. Noto, and Y. Muto, *Appl. Phys. Lett.* 54, 575 (1989).

105. T. Yamaguchi, S. Aoki, N. Sadakata, O. Kohno, and H. Osanai, *Appl. Phys. Lett.* 55, 1581 (1989).

106. T. Tsuruoka, R. Kawasaki, and H. Abe, *Jpn. J. Appl. Phys.* 28, L1800 (1989).

107. K. Watanabe, T. Matsushita, N. Kobayashi, H. Kawabe, E. Aoyagi, K. Hiraga, H. Yamane, H. Kurosawa, T. Hirai, and Y. Muto, *Appl. Phys. Lett.* 56, 1490 (1990).

108. H. Onishi, H. Harima, Y. Kusakabe, M. Kobayashi, S. Hoshinouchi, and K. Tachibana, *Jpn. J. Appl. Phys.* 29, L2041 (1990).

109. A. D. Berry, D. K. Gaskill, R. T. Holm, E. J. Cukauskas, R. Kaplan, and R. L. Henry, *Appl. Phys. Lett.* 52, 1743 (1988).

110. K. Shinohara, F. Munakata, and M. Yamanaka, *Jpn. J. Appl. Phys.* 27, L1683 (1988).

111. J. Musolf, *J. Alloys Compounds* 251, 292 (1997).

112. K. Higashiyama, I. Hirabayashi, and S. Tanaka, *Jpn. J. Appl. Phys.* 31, L835 (1992).

113. W. J. Lackey, W. B. Carter, J. A. Hanigofsky, D. N. Hill, E. K. Barefield, G. Neumeier, D. F. O'Brien, M. J. Shapiro, J. R. Thompson, A. J. Green, T. S. Moss, R. A. Jake, and K. R. Efferson, *Appl. Phys. Lett.* 56, 1175 (1990).

114. S. Yamamoto, A. Kawaguchi, K. Nagata, T. Hattori, and S. Oda, *Appl. Surface Sci.* 112, 30 (1997).

115. B. Schulte, M. Maul, W. Becker, E. G. Schlosser, S. Elschner, P. Haussler, and H. Adrian, *Appl. Phys. Lett.* 59, 869 (1991).

116. S. S. Shoup, S. Shanmugham, D. Cousins, A. T. Hunt, M. Paranthaman, A. Goyal, P. Martin, and D. M. Kroeger, *IEEE Trans. Appl. Supercond.* 9, 2426 (1999).

117. N. P. Bansal, R. N. Simon, and D. E. Farrell, *Appl. Phys. Lett.* 53, 603 (1988).

118. T. Nonaka, K. Kaneko, T. Hasegawa, K. Kishio, Y. Takahashi, K. Kobayashi, K. Kitazawa, and K. Fueki, *Jpn. J. Appl. Phys.* 27, L867 (1988).

119. S. Shibata, T. Kitagawa, H. Okazaki, and T. Kimura, *Jpn. J. Appl. Phys.* 27, L646 (1988).

120. S. S. Shoup, M. Paranthaman, and D. B. Beach, *J. Mater. Res.* 12, 1017 (1997).

121. G. N. Glavee, R. D. Hunt, and M. Paranthaman, *Mater. Res. Bull.* 34, 817 (1999).

122. H. Tanabe, I. Tanaka, S. Watauchi, and H. Kojima, *Phys. C* 315, 154 (1999).

123. M. Tagami, M. Nakamura, Y. Sugawara, Y. Ikuhara, and Y. Shiohara, *Phys. C* 298, 185 (1998).

124. T. Yamaguchi, F. Ueda, and M. Imamura, *Jpn. J. Appl. Phys.* 32, 1634 (1993).

125. C. M. Varma, S. Schmitt-Rink, and E. Abraham, *Solid State Commun.* 62, 681 (1987).

126. R. W. Simon, C. E. Platt, A. E. Lee, K. P. Daly, M. S. Wire, J. A. Luine, and M. Urbanik, *Appl. Phys. Lett.* 53, 2677 (1988).

127. R. Jose, Asha M. John, J. Kurian, P. K. Sajith, and J. Koshy, *J. Mater. Res.* 12, 2976 (1997).

128. A. M. John, R. Jose, J. Kurian, P. K. Sajith, and J. Koshy, *J. Amer. Ceram. Soc.* 82, 1421 (1999).

129. S. C. Tidrow, A. Tauber, W. D. Wilber, R. D. Finnegan, D. W. Eckart, and W. C. Drach, *IEEE Trans. Appl. Supercond.* 7, 1769 (1997).

130. Z. L. Bao, F. R. Wang, Q. D. Jiang, S. Z. Wang, Z. Y. Ye, K. Wu, C. Y. Li, and D. L. Yin, *Appl. Phys. Lett.* 51, 946 (1987).

131. M. J. Cima, J. S. Schneider, S. C. Peterson, and W. Coblenz, *Appl. Phys. Lett.* 53, 710 (1988).

132. A. Mogro-Campero, B. D. Hunt, L. G. Turner, M. C. Burrell, and W. E. Balz, *Appl. Phys. Lett.* 52, 584 (1988).

133. A. Mogro-Campero and L. G. Turner, *Appl. Phys. Lett.* 52, 1185 (1988).

134. D. K. Fork, F. A. Ponce, J. C. Tramontana, and Geballe, T. H., *Appl. Phys. Lett.* 58, 2294 (1991).

135. L. D. Chang, M. Z. Tseng, E. L. Hu, and D. K. Fork, *Appl. Phys. Lett.* 60, 1753 (1992).

136. R. S. Nowicki and M. A. Nicolet, *Thin Solid Films* 96, 317 (1982).

137. A. Goyal, D. P. Norton, J. D. Budai, M. Paranthaman, E. D. Specht, D. M. Kroeger, D. K. Christen, Q. He, B. Saffian, F. A. List, D. F. Lee, P. M. Martin, C. E. Klabunde, E. Hartfield, and V. K. Sikka, *Appl. Phys. Lett.* 69, 1795 (1996).

138. M. Paranthaman, A. Goyal, F. A. List, E. D. Specht, D. F. Lee, P. M. Martin, Q. He, D. K. Christen, D. P. Norton, J. D. Budai, and D. M. Kroeger, *Phys. C* 275, 266 (1997).

139. T. Terashima, Y. Bando, K. Iijima, K. Yamamoto, K. Hirata, K. Kamigake, and K. Terauchi, *Phys. C* 162, 615 (1989).

140. I. D. Raistrick, M. Hawley, J. G. Beery, F. H. Garzon, and R.J. Houlton, *Appl. Phys. Lett.* 59, 3177 (1991).

141. X.-Y. Zheng, D. H. Lowndes, S. Zhu, J. D. Budai, and R. J. Warmack, *Phys. Rev. B* 45, 7584 (1992).

142. D. S. Burbidge, S. K. Dew, B. T. Sullivan, N. Fortier, R. R. Parsons, P. J. Mulhern, J. F. Carolan, and A. Chaklader, *Solid State Commun.* 64, 749 (1987).

143. B. Zhao, H. Wang, Y. Lu, Y. Shi, X. Huang, Y. Zhao, and L. Li, *Chinese Phys. Lett.* 4, 280 (1987).

144. H. Adachi, K. Setsune, T. Mrrsuyu, K. Hirochi, Y. Ichikawa, T. Kamada, and K. Wasa, *Japan. J. Appl. Phys.* 26, L709 (1987).

145. J. C. Bruyere, J. Marcus, P. L. Reydet, C. Escribe-Filippini, and C. Schlenker, *Mater. Res. Bull.* 23, 429 (1988).

146. V. A. Marchenko, A. G. Znamenskii, and U. Helmersson, *J. Appl. Phys.* 82, 1882 (1997).

147. W. C. Tsai and T. Y. Tseng, *Mater. Chem. Phys.* 49, 229 (1997).

148. W. C. Tsai and T. Y. Tseng, *Japan. J. Appl. Phys. 1* 36, 76 (1997).

149. J. D. Klein, A. Yen, and S. L. Clauson, *Appl. Phys. Lett.* 56, 394 (1990).

150. A. V. Bagulya, I. P. Kazakov, M. A. Negodaev, and V. I. Tsekhosh, *Mater. Sci. Eng. B* 21, 5 (1993).

151. M. Fujimoto, K. Hayashi, K. Suzuki, and Y. Enomoto, *Japan. J. Appl. Phys. 1* 35, 90 (1996).

152. Y. J. Mao, C. X. Ren, J. Yuan, F. Zhang, X. H. Liu, S. C. Zou, E. Zhou, and H. Zhang, *Phys. C* 282, 611 (1997).

153. D. Dijkkamp and T. Venkatesan, *Appl. Phys. Lett.* 51, 619 (1987); H. B. Lu, S. F Xu, Y. J. Tian, D. F. Cui, Z. H. Chen, Y. Z. Zhang, L. Li, and G. Z. Yang, *J. Supercond.* 6, 335 (1993).

154. J. Narayan, N. Biunno, R. Singh, O. W. Holland, and O. Auciello, *Appl. Phys. Lett.* 51, 1845 (1987).

155. R. K. Singh, J. Narayan, A. K. Singh, and J. Krishnaswamy, *Appl. Phys. Lett.* 54, 2271 (1989).

156. C. Thivet, M. Guilloux-Viry, J. Padiou, A. Perrin, G. Dousselin, Y. Pellan, and M. Sergent, *Phys. C* 235-240, 665 (1994).

157. D. P. Norton, A. Goyal, J. D. Budai, D. K. Christen, D. M. Kroeger, E. D. Specht, Q. He, B. Saffian, M. Paranthaman, C. E. Klabunde, D. F. Lee, B. C. Sales, and F. A. List, *Science* 274, 755 (1996).

158. F. A. List, A. Goyal, M. Paranthaman, D. P. Norton, E. D. Specht, D. F. Lee, and D. M. Kroeger, *Phys. C* 302, 87 (1998).

159. F. Schmaderer and G. Wahl, *J. Phys. (Paris) Colloq.* 50, C5-119 (1989).

160. H. Ohnishi, Y. Kusakabe, M. Kobayashi, S. Hoshinouchi, H. Harima, and K. Tachibana, *Jpn. J. Appl. Phys.* 29, 1070 (1990).

161. K. Kanehori, N. Sughii, T. Fukazawa, and K. Miyauchi, *Thin Solid Films* 182, 265 (1989).

162. K. Kanehori, N. Sughii, and K. Miyauchi, "Proc. MRS Fall Meeting," M7-172, 1989.

163. J. Zhao, D. W. Noh, C. Chern, Y. Q. Li, P. Norris, B. Gallois, and B. Kear, *Appl. Phys. Lett.* 56, 2342 (1990).

164. H. Yamane, T. Hirai, K. Watanabe, N. Kobayashi, Y. Muto, M. Hasei, and H. Kurosawa, *J. Appl. Phys.* 69, 7948 (1991).

165. Y. Ito, Y. Yoshida, Y. Mizushima, I. Hirabayashi, H. Nagai, and Y. Takai, *Japan. J. Appl. Phys. 2* 35, L825 (1996).

166. U. Schmatz, C. Dubourdieu, O. Lebedev, G. Delabouglise, F. R. Weiss, and J. P. Senateur, *J. Low Temperature Phys.* 105, 1301 (1996).

167. I. M. Watson, *Chem. Vapor Deposition* 3, 9 (1997).

168. A. Abrutis, J. P. Senateur, F. Weiss, V. Bigelyte, A. Teiserskis, V. Kubilius, V. Galindo, and S. Balevicius, *J. Cryst. Growth* 191, 79 (1998).

169. H. J. Scheel, M. Berkowski, and B. Chabot, *Phys. C* 185-189, 2095 (1991).

170. C. Dubs, K. Fischer, and P. Goernet, *J. Cryst. Growth* 123, 611 (1992).

171. C. Klemenz and H. J. Scheel, *J. Cryst. Growth* 129, 421 (1993).

172. T. Kitamura, S. Taniguchi, I. Hirabayashi, S. Tanaka, Y. Sugawara, and Y. Ikuhara, *IEEE Trans. Appl. Supercond.* 7, 1392 (1997).

173. C. Klemenz, *J. Cryst. Growth* 187, 221 (1998).

174. C. Klemenz, I. Utke, and H. J. Scheel, *J. Cryst. Growth* 207, 62 (1999).

175. I. Matsebara, M. Paranthaman, A. Singhal, C. Vallet, D. F. Lee, P. M. Martin, R. D. Hunt, R. Feenstra, C.-Y. Yang, and S. E. Babcock, *Phys. C* 319, 127 (1999).

176. R. J. Cava, B. Batlogg, R. B. van Dover, D. W. Murphy, S. Sunshine, T. Siegrist, J. P. Remeika, E. A. Rietman, S. Zahurak, and G. P. Espinosa, *Phys. Rev. Lett.* 58, 1676 (1987).

177. R. E. Somekh, M. G. Blamire, Z. H. Barber, K. Butler, J. H. James, G. W. Morris, E. J. Tomlinson, A. P. Schwarzedberger, W. M. Stobbs, and J. E. Evetts, *Nature* 326, 857 (1987).

178. C. Michel, and B. Raveau, *Rev. Chim. Miner.* 21, 407 (1984).

179. P. Chaudhari, R. H. Koch, R. B. Laibowitz. T. R. McGuire, and R. J. Gambino, *Phys. Rev. Lett.* 58, 2684 (1987).

180. Y. Enomoto, T. Murakami, M. Suzuki, and K. Moriwaki, *Japan. J. Appl. Phys.* 26, L1248 (1987).

181. C. Tome-Rosa, G. Jakob, A. Walkenhorst, M. Maul, M. Schmitt, M. Paulson, and H. Adrian, *Z. Phys. B* 83, 221 (1991).

182. A. Schmehl, B. Goetz, R. R. Schulz. C. W. Schneider, H. Bielefeldt, H. Hilgenkamp, and J. Mannhart, *Europhys. Lett.* 47, 110 (1999).

183. B. Dam, J. M. Huijbregtse, F. C. Klaassen, R. C. F. vanderGeest, G. Doornbos, J. H. Rector, A. M. Testa, S. Freisem, J. C. Martinez, B. StaublePumpin, and R. Griessen, *Nature* 399, 439 (1999).

184. Z. Trajanovic, C. J. Lobb, M. Rajeswari, I. Takeuchi, C. Kwon, and T. Venkatesan, *Phys. Rev. B* 56, 925 (1997).

185. E. Mezzetti, R. Gerbaldo, G. Ghigo, L. Gozzelino, B. Minetti, C. Camerlingo, A. Monaco, G. Cuttone, and A. Rovelli, *Phys. Rev. B* 60, 7623 (1999).

186. J. R. Clem and A. Sanchez, *Phys. Rev. B* 50, 9355 (1994).

187. D. Mihailovic and A. J. Heeger, *Solid State Commun.* 75, 319 (1990).

188. D. Mihailovic, I. Poberaj, and A. Mertelj, *Phys. Rev. B* 48, 16634 (1993).

189. A. I. Grachev and I. V. Pleshakov, *Solid State Commun.* 101, 507 (1997).

190. A. A. Abrikosov, *Phys. C* 258, 53 (1996).

191. W. Prusseit, M. Rapp, and R. Semerad, *IEEE Trans. Appl. Supercond.* 7, 2952 (1997).

192. J. M. Harris, N. P. Ong, and Y. F. Yan, *Phys. Rev. Lett.* 71, 1455 (1993).

193. E. C. Jones, D. K. Christen, J. R. Thompson, R. Feenstra, S. Zhu, D. H. Lowndes, Julia M. Phillips, M. P. Siegal, and J. D. Budai, *Phys. Rev. B* 47, 8986 (1993).

194. J. G. Ossandon, J. R. Thompson, D. K. Christen, B. C. Sales, Y. Sun, and K. W. Lay, *Phys. Rev. B* 46, 3050 (1992).

195. A. T. Findikoglu, P. N. Arendt, J. R. Groves, S. R. Foltyn, E. J. Peterson, D. W. Reagor, and Q. X. Jia, *IEEE Trans. Appl. Supercond.* 7, 1232 (1997).

196. S. K. Han, J. Kim, K.-Y. Kang, and Y.-S. Ha, *IEEE Trans. Appl. Supercond.* 7, 1873 (1997).

197. S. Hontsu, H. Tabata, M. Nakamori, J. Ishii, and T. Kawai, *Japan. J. Appl. Phys. 2* 35, L774 (1996).

198. U. Kabasawa, H. Hasegawa, T. Fukazawa, Y. Tarutani, and K. Takagi, *J. Appl. Phys.* 81, 2302 (1997).

199. H. U. Habermeier, N. Jisrawi, and G. JagerWaldau, *Appl. Surface Sci.* 96-8, 689 (1996).

200. J. G. Wen, T. Usagawa, H. Zama, Y. Enomoto, T. Morishita, and N. Koshizuka, *Appl. Supercond.* 6, 199 (1998).

201. K. I. Gnanasekar, M. Sharon, R. Pinto, S. P. Pai, M. S. R. Rao, P. R. Apte, A. S. Tamhane, S. C. Purandare, L. C. Gupta, and R. Vijayaraghavan, *J. Appl. Phys.* 79, 1082 (1996).

202. N. Dieckmann, M. Rubhausen, A. Bock, M. Schilling, K.-O. Subke, U. Merkt, and E. Holzinger-Schweiger, *Phys. C* 272, 269 (1996).

203. M. Rubhausen, N. Dieckmann, A. Bock, U. Merkt, W. Widder, and H. F. Braun, *Phys. Rev. B* 53, 8619 (1996).

204. U. Kabasawa, T. Fukazawa, II. IIasegawa, Y. Tarutani, and K. Takagi, *Phys. Rev. B* 55, R716 (1997).

205. Y. Fang, L. Zhou, P. X. Zhang, P. Ji, X. Z. Wu, C. X. Luo, and M. R. Xing, *J. Supercond.* 5, 95 (1992).

206. V. K. Yanovsky, V. I. Voronkova, et al., *Cryogenics* 29, 648 (1989).

207. Y. Li, N. Chen, and Z. Zhao, *Phys. C* 224, 391 (1994).

208. A. A. Molodyk, O. Yu. Gorbenko, and A. R. Kaul, *J. Alloys Compounds* 251, 303 (1997).

209. I. S. Chuprakov and A. R. Kaul, *J. Chem. Vapor Deposition* 2, 123 (1993).

210. C. Cantoni, D. P. Norton, D. K. Christen, A. Goyal, D. M. Kroeger, D. T. Verebelyi, and M. Paranthaman, *Phys. C* 324, 177 (1999).

211. Y. Hakuraku, Z. Mori, S. Koba, N. Yokoyama, T. Doi, and T. Inoue, *Supercond. Sci. Technol.* 12, 481 (1999).

212. A. Takagi and U. Mizutani, *IEEE Trans. Appl. Supercond.* 7, 1388 (1997).

213. C. Cantoni, D. P. Norton, D. M. Kroeger, M. Paranthaman, D. K. Christen, D. Verebelyi, R. Feenstra, D. F. Lee, and E. D. Specht, *Appl. Phys. Lett.* 74, 96 (1999).

214. S. Nagata, N. Wakiya, Y. Masuda, S. Adachi, K. Tanabe, O. Sakurai, H. Funakubo, K. Shinozaki, and N. Mizutani, *Thin Solid Films* 334, 87 (1998).

215. W. H. Tang and J. Gao, *IEEE Trans. Appl. Supercond.* 9, 1590 (1999).

216. H. C. Li, H. R. Yi, R. L. Wang, Y. Chen, B. Yin, X. S. Ron and L. Li, *Appl. Phys. Lett.* 56, 2454 (1990).

217. E. Stangl, S. Proyer, M. Borz, B. Hellebrand, and D. Bauerle, *Phys. C* 256, 245 (1996).

218. Y. H. Li, A. E. StatonBevan, and J. A. Kilner, *Phys. C* 257, 382 (1996).

219. V. Dediu and F. C. Matacotta, *Phys. Rev. B* 54, 16259 (1996).

220. X. W. Cao, J. Fang, Z. H. Wang, X. J. Xu, R. L. Wang, and H. C. Li, *Phys. Rev. B* 56, 8341 (1997).

221. A. Fainstein, P. Etchegoin, and J. Guimpel, *Phys. Rev. B* 58, 9433 (1998).

222. A. Di Trolio, A. Morone, S. Orlando, and U. Gambardella, *Internat. J. Modern Phys. B* 13, 1055 (1999).

223. J. Brunen, T. Strach, J. Zegenhagen, and M. Cardona, *Phys. C* 282, 599 (1997).

224. A. Kazimirov, T. Haage, L. Ortega, A. Stierle, F. Comin, and J. Zegenhagen, *Solid State Commun.* 104, 347 (1997).

225. Q. D. Jiang, D. M. Smilgies, R. Feidenhansl, M. Cardona, and J. Zegenhagen, *Solid State Commun.* 98, 157 (1996).

226. E. Hodorowicz, S. A. Hodorowicz, and H. A. Elick, *J. Alloys Compounds* 181, 442 (1992).

227. S. V. Samoylenkov, O. Yu. Gorbenko, A. R. Kaul, Ya. A. Rebane, V. L. Svetchnikov, and H. W. Zandbergen, *J. Alloys Compounds* 251, 342 (1997).

228. R. Pinto, L. C. Gupta, Rajni Sharma, A. Sequiera, and K. I. Gnanasekar, *Phys. C* 289, 280 (1997).

229. S. V. Samoylenkov, O. Y. Gorbenko, I. E. Graboy, A. R. Kaul, V. L. Svetchnikov, and H. W. Zandbergen, *Acta Phys. Polon. A* 92, 243 (1997).

230. S. V. Samoylenkov, O. Y. Gorbenko, I. E. Graboy, A. R. Kaul, and Y. D. Tretyakov, *J. Mater. Chem.* 6, 623 (1996).

231. S. V. Samoylenkov, O. Y. Gorbenko, and A. R. Kaul, *Phys. C* 267, 74 (1996).

232. V. V. Srinivasu, P. Raychaudhuri, C. P. DSouza, R. Pinto, and R. Vijayaraghavan, *Solid State Commun.* 102, 409 (1997).

233. K. Yamano, K. Shimaoka, K. Takahashi, T. Usuki, Y. Yoshisato, and S. Nakano, *Phys. C* 185-189, 2549 (1991).

234. I. Yoshida, H. Furukawa, T. Hirosawa, and M. Nakao, *Appl. Surf. Sci.* 82–83, 501 (1994).

235. Vailionis, A. Brazdeikis, and A. S. Flodstrom, *Phys. Rev. B* 54, 15457 (1996).

236. S. Migita, Y. Kasai, H. Ota, and S. Sakai, *Appl. Phys. Lett.* 71, 3712 (1997).

237. M. Salvato, C. Attanasio, G. Carbone, T. DiLuccio, S. L. Prischepa, R. Russo, and L. Maritato, *IEEE Trans. Appl. Supercond.* 9, 2006 (1999).

238. M. Salvato, M. Salluzzo, T. DiLuccio, C. Attanasio, S. L. Prischepa, and L. Maritato, *Thin Solid Films* 353, 227 (1999).

239. J. N. Eckstein, I. Bozovic, K. E. von Dessonneck, D. G. Schlom, J. S. Harris, and S. M. Baumann, *Appl. Phys. Lett.* 57, 931 (1990).

240. I. Tsukada, H. Watanabe, I. Terasaki, and K. Uchinokura, "Studies of High Temperature Superconductors," Vol. 12 (A. V. Narlikar, Ed.). Nova, New York, 1994.

241. K. Yoshida, H. Sasakura, S. Tsukui, and Y. Mizokawa, *Phys. C* 322, 25 (1999).

242. S. Zhu, X.-Y. Zheng, E. Jones, and B. Warmack, *Appl. Phys. Lett.* 63, 409 (1993).

243. A. A. A. Youssef, T. Fukami, T. Yamamoto, and S. Mase, *Jpn. J. Appl. Phys.* 29, L60 (1990).

244. J. Lee, E. Naurumi, C. Li, S. Patel, and D. T. Shaw, *Phys. C* 200, 235 (1992).

245. W.-T. Lin, Y.-F. Cheng, C.-C. Kao, and K.-C. Wu, *J. Appl. Phys.* 74, 6767 (1993).

246. C. Marechal, R. M. Defourneau, I. Rosenman, J. Perriere, and C. Simon, *Phys. Rev. B* 57, 13811 (1998).

247. J. Perriere, R. M. Defourneau, A. Laurent, M. Morcrette, and W. Seiler, *Phys. C* 311, 231 (1999).

248. H. Koinuma, M. Kawasaki, S. Nagata, K. Takeuchi, and K. Fueki, *Jpn. J. Appl. Phys.* 27, L376 (1988).

249. M. Nakao, H. Kuwahara, R. Yuasa, H. Mukaida, and A. Mizukami, *Jpn. J. Appl. Phys.* 27, L378 (1988).

250. A. Marino, T. Yasuda, E. Holguin, and L. Rinderer, *Phys. C* 210, 16 (1993).

251. H. Narita, T. Hatano, and K. Nakamura, *J. Appl. Phys.* 72, 5778 (1992).

252. T. Hatano, A. Ishii, and K. Nakamura, *Phys. C* 273, 342 (1997).

253. M. Ohkubo, E. Brecht, G. Linker, J. Geerk, and O. Meyer, *Appl. Phys. Lett.* 69, 574 (1996).

254. T. Yotsuya, Y. Susuki, S. Ogawa, H. Imokawa, M. Yoshikawa, M. Naito, R. Takahata, and K. Otani, *Jpn. J. Appl. Phys.* 28, 972 (1989).

255. T. Matsushima, Y. Ichikawa, H. Adachi, K. Setsune, and K. Wasa, *Solid State Commun.* 76, 1201 (1990).

256. K. Setsune, K. Mizuno, T. Matsushima, S. Hatta, Y. Ichikawa, H. Adachi, and K. Wasa, *Supercond. Sci. Technol.* 4, 641 (1991).

257. Y. F. Yang, J. U. Lee, and J. E. Nordman, *J. Vacuum Sci. Technol. A* 10, 3288 (1992).

258. S. Kishida, H. Tokutaka, H. Kinoshita, and K. Fujimura, *Jpn. J. Appl. Phys.* 32, 700 (1993).

259. K. Harada, S. Kishida, T. Matsuoka, T. Maruyama, H. Tokutaka, K. Fujimura, and T. Koyanagi, *Jpn. J. Appl. Phys.* 35, 3590 (1996).

260. K. Harada, S. Kishida, T. Matsuoka, T. Maruyama, H. Tokutaka, and K. Fujimura, *Jpn. J. Appl. Phys.* 35, 4297 (1996).

261. J. H. Lu, H. C. Yang, and H. E. Horng, *Phys. B* 194–196, 2335 (1994).

262. K. Saito and M. Kaise, *Phys. Rev. B* 57, 11786 (1998).

263. Z. Mori, H. Ota, S. Migita, K. Sakai, and R. Aoki, *Thin Solid Films* 281–282, 517 (1996).

264. S. Migita, K. Sakai, H. Ota, Z. Mori, and R. Aoki, *Thin Solid Films* 281–282, 510 (1996).

265. T. Kobayashi, Y. Fukumoto, H. Hidaka, and M. Tonouchi, *IEEE Trans. Magn.* 25, 2455 (1989).

266. M. Manzel, L. Illgen, and R. Hergt, *Phys. Status Solidi A* 117, K119 (1990).

267. E. Brecht, G. Linker, T. Kroner, R. Schneider, J. Geerk, O. Meyer, and C. Traeholt, *Thin Solid Films* 304, 212 (1997).

268. Y. Z. Zhang, D. G. Yang, L. Li, B. R. Zhao, S. L. Jia, H. Chen, F. Wu, G. C. Che, and Z. X. Zhao, *Chinese Phys. Lett.* 15, 373 (1998).

269. H. Yamane, H. Kurosawa, H. Iwasaki, T. Hirai, N. Kobayashi, and Y. Muto, *Jpn. J. Appl. Phys.* 28, L827 (1989).

270. T. Kimura, H. Nakao, H. Yamawaki, M. Ihara, and M. Ozeki, *IEEE Trans. Magn.* 27, 1211 (1991).

271. N. Takahashi, A. Koukitu, H. Seki, and Y. Kamioka, *J. Crys. Growth* 144, 48 (1994).

272. K. Endo, H. Yamasaki, S. Misawa, S. Yoshida, and K. Kajimura, *Nature* 355, 327 (1992).

273. K. Endo, H. Yamasaki, S. Misawa, S. Yoshida, and K. Kajimura, *Phys. C* 185–189, 1949 (1991).

274. K. Endo, H. Shimizu, H. Matsuhata, F. H. Teherani, S. Yoshida, H. Tokumoto, and K. Kajimura, *IEEE Trans. Appl. Supercond.* 5, 1675 (1995).

275. V. N. Fuflyigin, A. R. Kaul, S. A. Pozigun, L. Klippe, and G. Wahl, *IEEE Trans. Appl. Supercond.* 5, 1335 (1995).

276. J. M. Zhang, H. O. Marcy, L. M. Tonge, B. W. Wessels, T. J. Marks, and C. R. Kannewurf, *Appl. Phys. Lett.* 55, 1906 (1989).

277. V. N. Fuflyigin, A. R. Kaul, S. A. Pozigun, L. Klippe, and G. Wahl, *J. Mater. Sci.* 30, 4431 (1995).

278. M. Matsubara and I. Hirabayashi, *Appl. Surf. Sci.* 82–83, 494 (1994).

279. G. Balestrino, V. Fogliette, M. Marinelli, E. Milani, A. Paolette, and P. Paroli, *IEEE Trans. Magn.* 27, 1589 (1991).

280. G. Balestrino, M. Marinelli, E. Milani, A. Paoletti, and P. Paroli, *Phys. C* 180, 46 (1991).

281. G. Balestrino, V. Foglietti, M. Marinelli, A. Paoletti, and P. Paroli, *Solid State Commun.* 79, 839 (1991).

282. K. K. Raina and R. K. Pandey, *J. Mater. Res.* 12, 636 (1997).

283. H. Nasu, H. Myoren, Y. Ibara, S. Makida, Y. Nishiyama, T. Kato, T. Imura, and Y. Osaka, *Jpn. J. Appl. Phys.* 27, L634 (1988).

284. T. Ohnishi, T. Hasegawa, K. Kishio, K. Kitazawa, and K. Fueki, *Annual Rep. Eng. Res. Inst. Faculty Eng. Univ. Tokyo* 47, 97 (1988).

285. M. Schieber, T. Tsach, M. Maharizi, M. Levinsky, B. L. Zhou, M. Golosovsky, and D. Davidov, *Cryogenics* 30, 451 (1990).

286. T. Manabe, T. Tsunoda, W. Kondo, Y. Shindo, S. Mizuta, and T. Kumagai, *Jpn. J. Appl. Phys.* 31, 1020 (1992).

287. K. Lee, I. Song, and G. Park, *J. Appl. Phys.* 74, 1459 (1993).

288. R. Jose, A. M. John, J. James, K. V. O. Nair, K. V. Kurian, and J. Koshy, *Mater. Lett.* 41, 112 (1999).

289. A. Andreone, C. Aruta, A. Cassinese, F. Palomba, R. Vaglio, G. Balestrino, and E. Milani, *Phys. C* 289, 275 (1997).

290. D. V. Livanov, E. Milani, G. Balestrino, and C. Aruta, *Phys. Rev. B* 55, R8701 (1997).

291. E. Silva, S. Sarti, M. Giura, R. Fastampa, and R. Marcon, *Phys. Rev. B* 55, 11115 (1997).

292. A. Odagawa, K. Setsune, T. Matsushima, and T. Fujita, *Phys. Rev. B* 48, 12985 (1993).

293. V. M. Krasnov, N. Mros, A. Yurgens, and D. Winkler, *Phys. Rev. B* 59, 8463 (1999).

294. G. F. Virshup, M. E. Klausmeier-Brown, I. Bozovic, and J. N. Eckstein, *Appl. Phys. Lett.* 60, 2288 (1992).

295. J. N. Eckstein, I. Bozovic, and G. F. Virshup, *Mater. Res. Soc. Bull.* 19, 44 (1994).

296. J. N. Eckstein, I. Bozovic, M. E. Klausmeier-Brown, G. F. Virshup, and K. Ralls, *Mater. Res. Soc. Bull.* 17, 27 (1992).

297. I. Bozovic, J. N. Eckstein, G. F. Virshup, A. Chaiken, M. Wall, R. Howell, and M. Fluss, *J. Supercond.* 7, 187 (1994).

298. J. N. Eckstein, I. Bozovic, and G. F. Virshup, *IEEE Trans. Appl. Supercond.* 5, 1680 (1995).

299. M. Kawasaki, M. Funabashi, S. Nagata, K. Fueki, and H. Koinuma, *Jpn. J. Appl. Phys.* 26, L288 (1987).

300. M. Suzuki and T. Murakami, *Jpn. J. Appl. Phys.* 26, L524 (1987).

301. H. Sato, H. Yamamoto, and M. Naito, *Phys. C* 274, 227 (1997).

302. J. Kwo, R. M. Fleming, H. L. Kao, D. J. Werder, and C. H. Chen, *Appl. Phys. Lett.* 60, 1905 (1992).

303. J.-P. Locquet, Y. Jaccard, A. Cretton, E. J. Williams, F. Arrouy, E. Mächler, T. Schneider, O. Fischer, and P. Martinoli, *Phys. Rev. B* 54, 7481 (1996).

304. J.-P. Locquet, J. Perret, J. Fompeyrine, E. Machler, J. W. Seo, and G. Van Tendeloo, *Nature* 394, 453 (1998).

305. A. Daridon, H. Siegenthaler, F. Arrouy, E. J. Williams, E. Machler, and J.-P. Locquet, *J. Alloys Compounds* 251, 118 (1997).

306. W. Si, H.-C. Li, and X. X. Xi, *Appl. Phys. Lett.* 74, 2839 (1999).

307. M. Suzuki, *Phys. Rev. B* 39, 2312 (1989).

308. M. Suzuki and M. Hikita, *Phys. Rev. B* 44, 249 (1991).

309. J. M. Tarascon, L. H. Greene, W. R. McKinnon, G. W. Hull, and T. H. Geballe, *Nature (London)* 235, 1373 (1987).

310. S. Uchida, *Phys. B* 148, 185 (1987).

311. K. Kitazawa, S. Kambe, M. Naito, I. Tanaka, and H. Kojima, *Jpn. J. Appl. Phys.* 28, L555 (1989).

312. S. Kambe, M. Naito, K. Kitazawa, I. Tanaka, and H. Kojima, *Phys. C* 160, 243 (1989).

313. P. Chaudhari, R. T. Collins, P. Freitas, R. J. Gambino, J. R. Kirtley, R. H. Koch, R. B. Laibowitz, F. K. LeGoues, T. R. McGuire, T. Penney, Z. Schlesinger, A. P. Segmüller, S. Foner, and E. J. McNiff, Jr., *Phys. Rev. B* 36, 8903 (1988).

314. J. M. Ginder, M. G. Roe, Y. Song, R. P. McCall, J. R. Gaines, and E. Ehrenfreunt, *Phys. Rev. B* 37, 7506 (1988).

315. M. Z. Cieplak, M. Berkowski, S. Guha, E. Cheng, A. S. Vagelos, D. J. Rabinowitz, B. Wu, I. E. Trofimov, and P. Lindenfeld, *Appl. Phys. Lett.* 65, 3383 (1994).

316. H. Sato and M. Naito, *Phys. C* 274, 221 (1997).

317. X. J. Chen, H. Q. Lin, and C. D. Gong, *Phys. Rev. B* 61, 9782 (2000).

318. J. H. van der Merwe, *J. Appl. Phys.* 41, 425 (1970).

319. J. H. Van der Merwe and W. A. Jesser, *J. Appl. Phys.* 64, 4968 (1988).

320. E. Olsson and S. L. Shinde, in "Interfaces in High-T_c Superconducting Systems" (S. L. Shinde and D. A. Rudman, Eds.), p. 116. Springer-Verlag, New York, 1994.

321. W. W. Huang, B. T. Liu, F. Wu, S. L. Jia, B. Xu, and B. R. Zhao, *Supercond, Sci. Technol.* 12, 529 (1999).

322. I. E. Trofimov, L. A. Johnson, K. V. Ramanujachary, S. Guha, M. G. Harrison, M. Greenblatt, M. Z. Cieplak, and P. Lindenfeld, *Appl. Phys. Lett.* 65, 2481 (1994).

323. H. Sato, M. Naito, and H. Yamamoto, *Phys. C* 280, 178 (1997).

324. W. L. Holsetein, *Appl. Supercond.* 2, 345 (1994).

325. W. L. Holstein, *J. Phys. Chem.* 97, 4224 (1993).

326. D. Cubicciotti and F. J. Keneshea, *J. Phys. Chem.* 71, 808 (1967).

327. R. Sugiese, M. Hirabayashi, N. Terada, M. Jo, T. Shimomura, and H. Ihara, *Phys. C* 157, 131 (1989).

328. R. Sugiese, M. Hirabayashi, N. Terada, M. Jo, T. Shimomura, and H. Ihara, *Japan. J. Appl. Phys.* 27, L2310 (1988).

329. E. Ruckenstein and C. T. Cheung, *J. Mater. Res.* 4, 1116 (1989).

330. S. L. Yan, L. Fang, M. S. Si, H. L. Cao, Q. X. Song, J. Yan, X. D. Zhou, and J. M. Hao, *Supercond. Sci. Technol.* 7, 681–684 (1994).

331. M. P. Siegal, E. L. Venturini, D. L. Overmyer, and P. P. Newcomer, *J. Supercond.* 11, 135 (1998).

332. W. L. Holstein and L. A. Parisi, *J. Mater. Res.* 11, 1349 (1996).

333. M. P. Siegal, E. L. Venturini, P. P. Newcomer, D. L. Overmyer, F. Dominguez, and R. Dunn, *J. Appl. Phys.* 78, 7186 (1995).

334. S. Narain and E. Ruckenstein, *Supercond. Sci. Technol.* 2, 236 (1989).

335. M. P. Sieagal, D. L. Overmyer, E. L. Venturini, P. P. Newcomer, R. Dunn, F. Dominguez, R. R. Padilla, and S. S. Sokolowski, *IEEE. Trans. Appl. Supercond.* 7, 1881 (1997).

336. J. Chrzanowski, X. M. Burany, A. E. Curzon, J. C. Irwin, B. Heinrich, N. Fortier, and A. Cragg, *Phys. C* 207, 25 (1993).

337. B. T. Ahn, W. Y. Lee, and R. Beyers, *Appl. Phys. Lett.* 60, 2150 (1992).

338. C. T. Cheung and E. Ruckenstein, *J. Mater. Res.* 5, 1860 (1990).

339. D. S. Ginley, J. S. Martens, E. L. Venturini, C. P. Tigges, C. Ashby, and S. Volk, *IEEE Trans. Appl. Supercond.* 3, 1201 (1993).

340. T. L. Aselage, E. L. Venturini, and S. B. Van Deusen, *J. Appl. Phys.* 75, 1023 (1994).

341. T. L. Aselage, E. L. Venturini, S. B. Van Deusen, T. J. Headley, M. O. Eatough, and J. A. Voigt, *Phys. C* 203, 25 (1992).

342. T. L. Aselage, E. L. Venturini, J. A. Voigt, and D. J. Miller, *J. Mater. Res.* 11, 1635 (1996).

343. M. Nemoto, S. Yoshikawa, R. Yussa, I. Yoshida, Y. Yoshisato and S. Maekawa, *Japan. J. Appl. Phys.* 35, 1720 (1996).

344. W. L. Holstein, L. A. Parisi, C. R. Fincher, and P. L. Gai, *Phys. C* 212, 110 (1993).

345. W. L. Holstein, L. A. Parisi, C. Wilker, and R. B. Flippen, *Appl. Phys. Lett.* 60, 2014 (1992).

346. W. L. Holstein, L. A. Parisi, C. Wilker, and R. B. Flippen, *IEEE Trans. Appl. Supercond.* 3, 1197 (1993).

347. J. Chrzanowski, S. MengBurany, W. B. Xing, A. E. Curzon, J. C. Irwin, B. Heinrich, R. A. Cragg, N. Fortier, F. Habib, V. Angus, G. Anderson, and A. A. Fife, *Supercond. Sci. Technol.* 9, 113 (1996).

348. A. E. Lee, C. E. Platt, J. F. Burch, R. W. Simon, J. P. Goral, and M. M. Al-Jassim, *Appl. Phys. Lett.* 57, 2019 (1990).

349. C. Y. Wu, F. Foong, S. H. Liou, and J. C. Ho, *IEEE Trans. Appl. Supercond.* 3, 1205 (1993).

350. B. J. Hinds, D. L. Schulz, D. A. Neumayer, B. Han, T. J. Marks, Y. Y. Wang, V. P. Dravid, J. L. Schindler, T. P. Hogan, and C. R. Kannewurf, *Appl. Phys. Lett.* 65, 231 (1994).

351. Y. Q. Tang Z. Z. Sheng, W. A. Lou, Z. Y. Chen, Y. N. Chan, Y. F. Li, F. T. Chan, G. J. Salamo, D. O. Pederson, T. H. Dhayagude, S. S. Ang, and W. D. Brown, *Supercond. Sci. Technol.* 6, 173 (1993).

352. T. Nabatame, Y. Saito, K. Aihara, T. Kamo, and S. P. Matsuda, *Japan. J. Appl. Phys.* 29, L1813 (1990).

353. S. L. Yan, L. Fang, Q. X. Song, J. Yan, Y. P. Zhu, J. H. Chen, and S. B. Zhang, *Appl. Phys. Lett.* 63, 1845 (1993).

354. M. P. Siegal, D. L. Overmyer, E. L. Venturini, F. Dominguez, and R. R. Padilla, *J. Mater. Res.* 13, 3349 (1998).

355. J. Betz, A. Piehler, E. V. Pechen, and K. F. Renk, *J. Appl. Phys.* 71, 2478 (1992).

356. D. W. Face and J. P. Nestlerode, *IEEE Trans. Appl. Supercond.* 3, 1516 (1993).

357. D. W. Face and J. P. Nestlerode, *Appl. Phys. Lett.* 61, 1838 (1992).

358. K. E. Myers, D. W. Face, D. J. Kountz, and J. P. Nestlerode, *Appl. Phys. Lett.* 65, 490 (1994).

359. K. E. Myers and L. Bao, *J. Supercond.* 11, 129 (1998).

360. K. E. Myers, D. W. Face, D. J. Kountz, J. P. Nestlerode, and C. F. Carter, *IEEE Trans. Appl. Supercond.* 5, 1684 (1995).

361. N. Reschauer, U. Speitzer, W. Brozio, A. Piehler, K. F. Renk, R. Berger, and G. Saemann-Ischenko, "1996 High Temperature Superconductors: Synthesis, Processing and Large-Scale Applications" (U. Balachandran, P. J. McGinn, and J. S. Abell, Eds.), p. 339. The Minerals, Metals and Materials Society, Warrendale, PA.

362. N. Reschauer, U. Speitzer, W. Brozio, A. Piehler, and K. F. Renk, *Appl. Phys. Lett.* 68, 1000 (1996).

363. N. Reschauer, U. Speitzer, W. Brozio, A. Piehler, K. F. Renk, R. Berger, and G. Saemann-Ischenko, "Applied Superconductivity 1995" (D. Dew-Hughes, Ed.), Inst. Phys. Conf. Ser. Vol. 148), p. 793. Institute of Physics, Bristol, 1995.

364. J. D. O'Connor, D. Dew-Hughes, N. Reschauer, W. Brozio, H. H. Wagner, K. F. Renk, M. J. Goringe, C. R. M. Grovenor, and T. Kaiser, *Phys. C* 302, 277 (1998).

365. A. P. Bramley, A. J. Wilkinson, A. P. Jenkins, and C. R. M. Grovenor, *J. Supercond.* 11, 71 (1998).

366. S. Huber, M. Manzel, H. Bruchlos, S. Hensen, and G. Muller, *Phys. C* 244, 337 (1995).

367. A. P. Bramley, B. J. Glassey, C. R. M. Grovenor, M. J. Goringe, J. D. O'Connor, A. P. Jenkins, K. S. Kale, K. L. Jim, D. Dew-Hughes, and D. J. Edwards, *IEEE Trans. Appl. Supercond.* 7, 1249 (1997).

368. A. P. Bramley, C. R. M. Grovenor, M. J. Goringe, J. D. OConnor, A. P. Jenkins, D. DewHughes, N. Reschauer, H. H. Wagner, W. Brozio, U. Spreitzer, and K. F. Renk, *J. Mater. Res.* 13, 2057 (1998).

369. A. P. Bramley, A. P. Jenkins, A. J. Wilkinson, C. R. M. Grovenor, and D. Dew-Hughes, "Applied Superconductivity 1997" (H. Rogalla and D. H. A. Blank, Eds.), Inst. Phys. Conf. Ser., Vol. 158, p. 201. Institute of Physics, Bristol, 1997.

370. W. L. Holstein, L. A. Parisi, R. B. Flippen, and D. G. Swartzfager, *J. Mater. Res.* 8, 962 (1993).

371. Y. Q. Tang, Z. Z. Sheng, W. A. Luo, I. N. Chan, Z. Y. Chen, Y. F. Li, and D. O. Pederson, *Phys. C* 214, 190 (1993).

372. S. Koike, T. Nabatame, T. Takenaka, B. Rai, K. Suzuki, Y. Enomoto, and I. Hirabayashi, *Trans. Mater. Res. Soc. Japan* 19A, 584 (1994).

373. S. H. Liou and C. Y. Wu, *Appl. Phys. Lett.* 60, 2803 (1992).

374. X. D. Wu, S. R. Foltyn, R. E. Muenchausen, D. W. Cooke, A. Pique, D. Kalokitis, V. Pendrick, and E. Belohoubeck, *J. Supercond.* 5, 353 (1992).

375. F. Wang and R. Wordenweber, *Thin Solid Films* 227, 200 (1993).

376. W. L. Holstein, L. A. Parisi, D. W. Face, X. D. Wu, S. R. Foltyn, and R. E. Muenchausen, *Appl. Phys. Lett.*, 61, 982 (1992); W. L. Holstein, Z. Y. Shen, C. Wilker, D. W. Face, and D. J. Kountz, "Advances in Superconductivity V" (Y. Bando and H. Yamamuchi, Eds.), Springer-Verlag, Berlin, 1993.

377. A. P. Bramley, S. M. Morley, C. R. M. Grovenor, and B. Pecz, *Appl. Phys. Lett.* 66, 517 (1995).

378. K. Y. Chen, S. Alfonso, R. C. Wand, Y. Q. Tang, G. Salamo, F. T. Chan, R. Guo, and A. S. Bhalla, *J. Appl. Phys.* 78, 2138 (1995).

379. X. F. Meng, F. S. Pierce, K. M. Wong, R. S. Amos, C. H. Xu, B. S. Deaver, Jr., and S. J. Poon, *IEEE Trans. Magn.* 27, 1638 (1991).

380. E. Sader, H. Schmidt, H. Kradil, and W. Wersing, *Supercond. Sci. Technol.* 4, 371 (1991).

381. E. Sader, *Supercond. Sci. Technol.* 6, 547 (1993).

382. V. Sandu, J. Jajlovsky, D. Miu, D. Draguinescu, C. Grigoriu, and M. C. Bunescu, *J. Mater. Sci. Lett.* 13, 1222 (1994).

383. G. Malandrino, A. Frassica, G. G. Condorelli, G. Lanza, and I. L. Fragala, *J. Alloys Compound* 251, 314 (1997).

384. Y. Iijima, K. Onabe, N. Futaki, N. Tanabe, N. Sadakata, O. Kohno, and Y. Ikeno, *J. Appl. Phys.* 74, 1905 (1993).

385. X. D. Wu, S. R. Foltyn, P. N. Arendt, W. R. Blumenthal, I. H. Campbell, J. D. Cotton, J. Y. Coulter, W. L. Hults, M. P. Maley, H. F. Safar, and J. L. Smith, *Appl. Phys. Lett.* 67, 2397 (1995).

386. Q. Xiong, S. Alfonso, K. Y. Chen, G. Salamo, and F. T. Chan, *Phys. C* 280, 17 (1997).

387. Q. Xiong, S. Alfonso, K. Y. Chen, G. Salamo, and F. T. Chan, *Phys. C* 282, 641 (1997).

388. A. Goyal, D. P. Norton, D. M. Kroeger, D. K. Christen, M. Paranthaman, E. D. Specht, J. D. Budai, Q. He, B. Saffian, F. A. List, D. F. Lee, E. Hatfield, P. M. Martin, C. E. Klabunde, J. Mathis, and C. Park, *J. Mater. Res.* 12, 2924 (1997).

389. M. Bauer, J. Schwachulla, S. Furtner, P. Berberich, and H. Kinder, "Applied Superconductivity 1997" (H. Rogalla and D. H. A. Blank, Eds.), Inst. Phys. Conf. Ser., Vol. 158, p. 1077. Bristol, Institute of Physics, 1997.

390. K. Hasegawa, N. Yoshida, K. Fujino, H. Mukai, K. Hayashi, K. Sato, T. Okhuma, S. Honjyou, H. Ishii, and T. Hara, "Proc. 9th Int. Symp. on Superconductivity," Sapporo, 1996.

391. W. A. Quinton and F. Baudenbacher, *Phys. C* 292, 243 (1997).

392. L. P. Guo, Z. F. Ren, J. Y. Lao, J. H. Wang, D. K. Christen, C. E. Klabunde, and J. D. Budai, *Phys. C* 277, 13 (1997).

393. C. A. Wang, Z. F. Ren, J. H. Wang, and D. J. Miller, *Appl. Supercond.* 3, 153 (1995).

394. Z. F. Ren, J. Y. Lao, L. P. Guo, J. H. Wang, J. D. Budai, D. K. Christen, A. Goyal, M. Paranthaman, E. D. Specht, and J. R. Thompson, *J. Supercond.* 11, 159 (1998).

395. P. A. Parilla, J. M. McGraw, D. L. Schulz, J. Wendelin, R. N. Bhattacharya, R. D. Blaugher, D. S. Ginley, J. A. Voigt, and E. P. Roth, *IEEE Trans. Appl. Supercond.* 7, 1969 (1997).

396. G. Samadi-Hosseinali, W. Straif, B. Starchl, K. Kundzins, H. W. Weber, S. L. Yan, M. Manzel, E. Stangl, S. Proyer, D. Bauerle, and E. Mezzetti, *Phys. C* 268, 307 (1996).

397. T. Kamo, T. Doi, A. Soeta, T. Yuasa, N. Inoue, K. Aihara, and S.-P. Matsuda, *Appl. Phys. Lett.* 59, 3186 (1991).

398. M. Paranthaman, M. Foldeaki, D. Balzar, H. Ledbetter, A. J. Nelson, and A. M. Herman, *Supercond. Sci. Technol.* 7, 227 (1994).

399. J. E. Tkaczyk, J. A. DeLuca, P. L. Karas, P. J. Bednarczyk, D. K. Christen, C. E. Klabunde, and H. R. Kerchner, *Appl. Phys. Lett.* 62, 3031 (1993).

400. V. Hardy, D. Groult, J. Provost, M. Hervieu, and B. Raveau, *Phys. C* 178, 255 (1991).

401. E. L. Venturini, P. P. Newcomer, M. P. Siegal, and D. L. Overmyer, *IEEE Trans. Appl. Supercond.* 7, 1592 (1997).

402. M. P. Siegal, E. L. Venturini, P. P. Newcomer, B. Morosin, D. L. Overmyer, F. Dominguez, and R. Dunn, *Appl. Phys. Lett.* 67, 3966 (1995).

403. W. L. Holstein, L. A. Parisi, Z. Y. Chan, C. Wilker, M. S. Brenner, and J. S. Martens, *J. Supercond.* 6, 191 (1991).

404. N. Klein, H. Chaloupka, G. Muller, S. Orbach, H. Piel, B. Roas, L. Schulz, U. Klein, and M. Peiniger, *J. Appl. Phys.* 67, 6940 (1990).

405. J. S. Martens, V. M. Hietala, E. L. Venturini, and W. Y. Lee, *J. Appl. Phys.* 73, 7571 (1993).

406. Z. F. Ren, J. J. Wang, and D. J. Miller, *Appl. Phys. Lett.* 71, 1706 (1997).

407. C. C. Tsuei, J. R. Kirtley, Z. F. Ren, J. H. Wang, H. Raffy, and Z. Z. Li, *Nature* 387, 481 (1997).

408. R. P. Vasquez, Z. F. Ren, and J. H. Wang, *Phys. Rev. B* 54, 6115 (1996).

409. A. Gupta, J. Z. Sun, and C. C. Tsuei, *Science* 265, 1075 (1994).

410. M. Rupp, A. Gupta, and C. C. Tsuei, *Appl. Phys. Lett.* 67, 291 (1995).

411. C. C. Tsuei, A. Gupta, G. Trafas, and D. Mitzi, *Science* 263, 1259 (1994).

412. L. Krusin-Elbaum, C. C. Tsuei, and A. Gupta, *Nature* 373, 679 (1995).

413. J. Z. Wu, S. L. Yan, and Y. Y. Xie, *Appl. Phys. Lett.* 74, 1469 (1999).

414. K. Mizuno, H. Adachi, and K. Setsune, *J. Low Temp. Phys.* 105, 1571 (1996).

415. H. Adachi, K. Mizuno, T. Satoh, and K. Setsune, *Jpn J. Appl. Phys.* 32 (12B), 1798 (1993).

416. Y. Tsabba and S. Reich, *Phys. C* 269, 1 (1996).

417. L. Fang, S. L. Yan, T. Aytug, A. A. Gapud, B. W. Kang, Y. Y. Xie, and J. Z. Wu, *IEEE Trans. Appl. Supercond.* 9, 2387 (1999).

418. Y. Y. Xie, J. Z. Wu, S. L. Yan, Y. Yu, T. Aytug, L. Fang, and S. C. Tidrow, *Phys. C* 328, 241 (1999).

419. S. L. Yan, Y. Y. Xie, J. Z. Wu, T. Aytug, A. A. Gapud, B. W. Kang, L. Fang, M. He, S. C. Tidrow, K. W. Kirchner, J. R. Liu, and W. K. Chu, *Appl. Phys. Lett.* 73, 2989 (1998).

420. W. N. Kang, R. L. Meng, and C.W. Chu, *Appl. Phys. Lett.* 73, 381 (1998).

421. W. N. Kang, S. I. Lee, and C. W. Chu, *Phys. C* 315, 223 (1999).

422. W. N. Kang, B. W. Kang, Q. Y. Chen, J. Z. Wu, S. H. Yun, A. Gapud, J. Z. Qu, W. K. Chu, D. K. Christen, R. Kerchner, and C. W. Chu, *Phys. Rev. B* 59, R9031 (1999).

423. G. Plesch, S. Chromik, V. Strbik, M. Mair, G. Gritzner, S. Benacka, I. Sargankova, and A. Buckuliakova, *Phys. C* 307, 74 (1998).

424. Y. Moriwaki, T. Sugano, S. Adachi, and K. Tanabe, *IEEE Trans. Appl. Supercond.* 9, 2390 (1999).

425. S. H. Yun, U. O. Karlsson, B. J. Jonsson, K. V. Rao, and L. D. Madsen, *J. Mater. Res.* 14, 3181 (1999).

426. S. H. Yun and U. O. Karlsson, *Thin Solid Films* 343–344, 98 (1999).

427. S. H. Yun, J. Z. Wu, S. C. Tidrow, and D. W. Eckart, *Appl. Phys. Lett.* 68, 2565 (1996).

428. Y. Moriwaki, T. Sugano, C. Gasser, A. Fukuoka, K. Nakanishi, S. Adachi, and K. Tanabe, *Appl. Phys. Lett.* 69, 3423 (1996).

429. Y. Moriwaki, T. Sugano, A. Tsukamoto, C. Gasser, K. Nakanishi, S. Adachi, and K. Tanabe, *Phys. C* 303, 65 (1998).

430. S. Miyashita, H. Higuma, and F. Uchikawa, *Jpn. J. Appl. Phys.* 33 (7A), L931 (1994).

431. Y. Yu, H. M. Shao, Z. Y. Zheng, A. M. Sun, M. J. Qin, X. N. Xu, S. Y. Ding, X. Jin, X. X. Yao, J. Zhou, Z. M. Li, S. Z. Yang, and W. L. Zhang, *Phys. C* 289, 199 (1997).

432. J. Zhou, Z. M. Ji, S. Z. Yang, P. H. Wu, Y. Yu, H. M. Shao, X. X. Yao, M. Y. Lai, and W. L. Zhang, *J. Mater. Sci. Lett.* 16, 1936 (1997).

433. J. Zhou, Z. M. Ji, S. Z. Yang, P. H. Wu, Y. Yu, H. M. Shao, X. X. Yao, M. Y. Lai, and W. L. Zhang, *J. Mater. Sci. Lett.* 17, 271 (1998).

434. S. Kumari, A. K. Singh, and O. N. Srivastava, *Supercond. Sci. Technol.* 10, 235 (1997).

435. Y. Tsabba and S. Reich, *Phys. C* 254, 21 (1995).

436. J. D. Guo, G. C. Xiong, D. P. Yu, Q. R. Feng, X. L. Xu, G. J. Lian, and Z. H. Hu, *Phys. C* 276, 277 (1997).

437. J. D. Guo, G. C. Xiong, D. P. Yu, Q. R. Feng, X. L. Xu, G. J. Lian, K. Xiu, and Z. H. Hu, *Phys. C* 282, 645 (1997).

438. T. Siegrist, S. M. Zahurak, D. W. Murphy, and R. S. Roth, *Nature* 334, 231 (1988).

439. R. J. Cava, *Nature* 351, 518 (1991).

440. M. G. Smith, A. Manthiram, J. Zhou, J. B. GoodEnough, and J. T. Market, *Nature* 351, 549 (1991),.

441. M. Takano, A. Azuma, Z. Hiroi, and Y. Bando, *Phys. C* 176, 441 (1991).

442. J. R. Gavaler, *Appl. Phys. Lett.* 23, 480 (1973).

443. Y. H. Shi, B. R. Zhao, Y. Y. Zhao, L. Li, and J. R. Liu, *Phys. Rev. B* 38, 4488 (1988).

444. M. Kanai, T. Kawai, and S. Kawai, *Appl. Phys. Lett.* 58, 771 (1991).

445. C. Niu and C. M. Lieber, *Appl. Phys. Lett.* 61, 1712 (1992).

446. Y. Terashima, R. Sato, S. Takeno, S. L. Nakamura, and T. Miura, *Jpn. J. Appl. Phys.* 32, L48 (1993).

447. I. Yazawa, N. Terada, Satoh, K. Matsutani, R. Sugise, M. Jo, and H. Ihara, *Jpn. J. Appl. Phys.* 29, L566 (1990).

448. N. Sugii, M. Ichikawa, K. Kudo, T. Sakurai, K. Yamamoto, and H. Yamauchi, *Phys. C* 196, 129 (1992).

449. B. R. Zhao, X. J. Zhou, J. Li, B. Xu, S. L. Jia, F. Wu, C. Dong, Y. S. Yao, L. Li, Y. Y. Zhao, and Z. X. Zhao, *Modern Phys. Lett. B* 7, 1585 (1993).

450. A. Vailionis, A. Brazdeikis, and A. S. Flodstron, *Phys. Rev. B* 55, R6152 (1997).

451. X. Li, T. Kawai, and S. Kawai, *Jpn. J. Appl. Phys.* 31, L934 (1992).

452. H. Adachi, T. Satoh, Y. Ichikawa, K. Setsume, and K. Wasa, *Phys. C* 196, 14 (1992).

453. A. Del Vecchio, L. Mirenghi, L. Tapfer, C. Aruta, G. Petrocelli, and G. Balestrino, *Phys. C* 288, 71 (1997).

454. G. Er, Y. Miyamoto, F. Kanamaru, and S. Kikkawa, *Phys. C* 181, 206 (1991).

455. M. Azuma, Z. Hiroi, M. Takano, Y. Bando, and Y. Takeda, *Nature* 356, 775 (1992).

456. D. P. Norton, B. C. Chakoumakos, J. D. Budai, and D. H. Lowndes, *Appl. Phys. Lett.* 62, 1679 (1993).

457. X. Li and T. Kawai, *Phys. C* 229, 251 (1994).

458. J. B. Goodenough, *Supercond. Sci. Technol.* 3, 26 (1990).

459. D. P. Norton, J. D. Budai, B. C. Chakoumakos, and R. Feenstra, *Appl. Phys. Lett.* 62, 414 (1993).

460. R. A. Schweinfurth, C. E. Platt, M. R. Teepe, and D. J. Van Harlingen, *Appl. Phys. Lett.* 61, 480 (1992); B. M. Moon, C. E. Platt, R. A. Schweinfurth, and D. J. Van Harlingen, *Appl. Phys. Lett.* 59, 1905 (1991); C. Ciofi, C. E. Platt, J. A. Eades, J. Amano, and R. Hu, *J. Mater. Res.* 9, 305 (1994).

461. W.-T. Lin, S.-M. Pan, and Y.-F. Chen, *J. Appl. Phys.* 75. 1179 (1994); J. Zhou, Z. M. Ji, S. Z. Yang, P. H. Wu, Z. G. Liu, and S. Y. Zhang, *Chinese J. Low Temp. Phys.* 18, 120 (1996); Z. M. Ji, J. Zhou, S. Z. Yang, P. H. Wu, G. S. Yue, Z. G. Liu, J. M. Liu, Z. C. Wu, S. Y. Zhang, Z. R. Xu, H. C. Zhang, and Z. L. Mu, *Chinese Sci. Bull.* 40, 686 (1995).

462. P. Prieto, U. Poppe, W. Evers, R. Hojczyc, C. L. Jia, K. Urban, K. Schmidt, and H. Soltner, *Phys. C* 233, 361 (1994); H. Sato, H. Takagi, and S. Uchida, *Phys. C* 169, 391 (1990); M. Iyori, M. Kamino, K. Takahashi, Y. Yoshisato, and S. Nakano, *Phys. C* 185-189, 1965 (1991); K. Takahashi, M. Iyori, M. Kamino, T. Usuki, Y. Yoshisato, and S. Nakano, *Jpn. J. Appl. Phys.* 30, L1480 (1991).

463. S. V. Shiryaev, S. N. Barilo, D. I. Zhigunov, V. V. Fedotova, A. V. Pushkarev, L. A. Kurochkin, and A. G. Soldatov, *J. Cryst. Growth* 198/199, 631 (1999).

464. S. V. Shiryaev, S. N. Barilo, N. S. Orlova, D. I. Zhigunov, A. S. Shestac, V. T. Koyava, V. I. Gatalskaya, A. V. Pushkarev, V. M. Pan, and V. F. Solovjov, *J. Cryst. Growth* 172, 396 (1997).

465. E. S. Hellman, E. H. Hartford, and E. M. Gyorgy, *Appl. Phys. Lett.* 58, 1335 (1991); E. S. Hellman, B. Miller, J. M. Rosamilia, E. H. Hartford, and K. W. Baldwin, *Phys. Rev. B* 44, 9719 (1991); D. Miller, P. L. Richards, E. J. Nicol, E. S. Hellman, E. H. Hartford, Jr., C. E. Platt, R. A. Schweinfurth, D. J. VanHarlingen, and J. Amano, *J. Phys. Chem. Solids* 54, 1323 (1993).

466. S. Suzuki, T. Yamamoto, H. Suzuki, K. Kawaguchi, K. Takahashi, and Y. Yoshisato, *J. Appl. Phys.* 81, 6830 (1997).

467. E. S. Hellman, J. R. Kwo, A. Kussmaul, and E. H. Hartford, Jr., *Phys. C* 251, 133 (1995).

468. F. Hebard, M. J. Rosseinsky, R. C. Haddon, D. W. Murphy, S. H. Glarum, T. T. M. Palstra, A. P. Ramirez, and A. R. Kortan, *Nature* 350, 600 (1991).

469. P. J. Benning, J. L. Martins, J. H. Weaver, L. P. F. Chibante, and R. E. Smalley, *Science* 252, 1417 (1991).

470. G. Wertheim, J. E. Rowe, D. N. Buchanan, E. E. Chaban, A. F. Hebard, A. R. Kortan, A. V. Makhija, and R. C. Haddon, *Science* 252, 1419 (1991).

471. H. Ogata, T. Inabe, H. Hoshi, Y. Maruyama, Y. Achiba, S. Suzuki, K. Kikuchi, and I. Ikemoyo, *Jpn. J. Appl. Phys.* 31, L166 (1992).

472. N. Okuda, H. Kugai, T. Uemura, K. Okura, Y. Ueba, and K. Tada, *Jpn. J. Appl. Phys.* 33, 1851 (1994).

473. R. C. Haddon, A. F. Hebard, M. J. Rosseinsky, D. W. Murphy, S. J. Duclos, K. B. Lyons, B. Miller, J. M. Rosamilia, R. M. Fleming, A. R. Kortan, S. H. Glarum, A. V. Makhija, A. J. Muller, R. H. Eick, S. M. Zahurak, R. Tycko, G. Dabbagh, and F. A. Thiel, *Nature* 350, 320 (1991).

474. L. B. Wang, H. Sekine, X. J. Xu, and Y. H. Zhang, *Chem. Phys. Lett.* 305, 85 (1999).

475. P. W. Stephens, L. Mihaly, P. L. Lee, R. L. Whetten, S. M. Huang, R. Kaner, F. Deiderich, and K. Holczer, *Nature* 351, 632 (1991).

476. A. R. Kortan, N. Kopylov, S. Glarum, E. M. Gyorgy, A. P. Ramirez, R. M. Fleming, F. A. Thiel, and R. C. Haddon, *Nature* 355, 529 (1992).

477. K. Tanigaki, I. Hirosawa, T. W. Ebbesen, J.-I. Mizuki, and J.-S. Tsai, *J. Phys. Chem. Solids* 54, 1645 (1993).

478. A. R. Kortan, N. Kopylov, S. Glarum, E. M. Gyorgy, A. P. Ramirez, R. M. Fleming, O. Zhou, F. A. Thiel, P. L. Treror, and R. C. Haddon, *Nature* 360, 566 (1992).

479. P. H. Citrin, E. Ozdas, S. Schuppler, A. R. Kortan, and K. B. Lyons, *Phys. Rev. B* 56, 5213 (1997).

480. G. P. Kochanski, A. F. Hebard, R. C. Haddon, and A. T. Fiory, *Science* 255, 184 (1992).

481. V. Buntar, F. M. Sauerzopf, H. W. Weber, J. E. Fischer, H. Kuzmany, M. Halushka, and C. L. Lin, *Phys. Rev. B* 54, 14952 (1996).

482. T. T. M. Palstra, R. C. Haddon, A. F. Hebard, and J. Zaanen, *Phys. Rev. Lett.* 68, 1054 (1992).

483. D. H. Lowndes, D. P. Norton, and J. D. Budai, *Phys. Rev. Lett.* 65, 1160 (1990).

484. Q. Li, X. X. Xi, X. D. Wu, A. Inam, S. Vadlamannali, and W. L. Mclaean, *Phys. Rev. Lett.* 64, 3086 (1990).

485. T. Umezawa, D. J. Lew, S. K. Streiffer, and M. R. Beasley, *Appl. Phys. Lett.* 63, 3221 (1993).

486. J. Z. Wu, C. S. Ting, W. K. Chu, and X. X. Yao, *Phys. Rev. B* 44, 411 (1991).

487. M. Rasolf, T. Edis, and Z. Tesanovic, *Phys. Rev. Lett.* 66, 2927 (1991).

488. X. Y. Zheng, D. H. Lowndes, S. Zhu, J. D. Budai, and R. J. Warmack, *Phys. Rev. B* 45, R7584 (1992).

489. J. J. Sun, B. R. Zhao, L. Z. Zheng, B. Xu, L. Li, J. W. Li, B. Yin, S. L. Jia, and Z. X. Zhao, *Phys. C* 269, 343 (1996).

490. J. Gao and W. H. Wong, *Phys. C* 251, 330 (1995).

491. J. Gao, B. Hauser, and H. Rogalla, *J. Appl. Phys.* 67, 2512 (1990).

492. P. Guptasarma, S. T. Bendre, S. B. Ogale, M. S. Multani, and R. Vijayaraghavan, *Phys. C* 203, 129 (1992).

493. J. J. Yeh and M. Hong, *Appl. Phys. Lett.* 54, 769 (1989).

494. O. Michikami, M. Asahi, and H. Asano, *Japan. J. Appl. Phys.* 29, L298 (1990).

495. J. Sun, B. Zhao, L. Zheng, B. Xu, J. Li, B. Yin, F. Wu, S. Jia, L. Li, and Z. Zhao, *Chinese Sci. Bull.* 41, 1951 (1996).

496. M. Z. Cieplak, S. Guha, S. Vadlamannati, C. H. Nien, and P. Lindenfeld, *Proc. SPIE* 2157, 222.

497. Y. Matsuda, S. Komiyama, T. Terashima, K. Shimura, and Y. Bando, *Phys. Rev. Lett.* 69, 3228 (1992).

498. D. P. Norton and D. H. Lowndes, *Phys. Rev. B* 48, 6460 (1993).

499. J.-M. Triscone, O. Fischer, Ø. Brunner, L. Antognazza, A. D. Kent, and M. G. Karkut, *Phys. Rev. Lett.* 64, 804 (1990).

500. C. B. Eom, J. M. Trisone, Y. Suzuki, and T. H. Geballe, *Phys. C* 185, 2065 (1991).

501. C. Kwon, Qi Li, X. X. Xi, S. Bhattacharya, C. Doughty, T. Venkatesan, H. Zhang, N. D. Spencer, and K. Feldmen, *Appl. Phys. Lett.* 62, 1289 (1993).

502. R. Li, R. Feile, G. Jakob, Th. Hahn, and H. Adrian, *J. Supercond.* 7, 213 (1994).

503. S. P. Zhao and Q. S. Yang, *Phys. Rev. B* 59, 14630 (1999).

504. A. A. Golubov, E. P. Houwman, J. G. Gijsbertsen, V. M. Krasnov, J. Flokstra, and H. Rogalla, *Phys. Rev. B* 51, 1073 (1995).

505. E. Toyoda, H. Takayanagi, and H. Nakano, *Phys. Rev. B* 59, R11653 (1999).

506. N. Newman, B. F. Cole, S. M. Garrison, K. Char, and R. C. Taber, *IEEE Trans. Magn.* 27, 1276 (1991).

507. X. D. Wu, S. R. Foltyn, P. Arendt, J. Townsend, C. Adams, I. H. Campbell, P. Tiwari, Y. Coulter, and D. Peterson, *Appl. Phys. Lett.* 65, 1961 (1994).

508. M. A. Hein, *Supercond. Sci. Technol.* 10, 867 (1997).

509. J. A. Greer and M. D. Tabat, *J. Vac. Sci. Technol. A* 13, 1175 (1995).

510. R. C. Y. Anyeung, J. S. Horwitz, L. A. Knauss, and D. B. Christy, *Rev. Sci. Instrum.* 68, 3872 (1997).

511. P. Berberich, B. Utz, W. Prusseit, and H. Kinder, *Phys. C* 219, 497 (1994).

512. B. Michelt, G. Lins, and R. J. Seeböck, "Appl. Supercond. 1995" (D. Dew-Hughes, Ed.), Inst. Phys. Cont. Ser., Vol. 148, p. 915. IOP, Bristol/Philadephia, 1995.

513. P. Arendt, S. R. Foltyn, X. D. Wu, J. Townsend, C. Adams, M. Hawley, P. Tiwari, M. Maley, J. Willis, D. Moseley, and Y. Coulter, *Mater. Res. Soc. Proc.* 341, 209 (1994).

514. J. Wiesmann, J. Hoffmann, A. Usoskin, F. Garlia, K. Heinemann, and H. C. Freyhardt, "Appl. Supercond. 1995" (D. Dew-Hughes, Ed.), Inst. Phys. Cont. Ser., Vol. 148, p. 503. IOP, Bristol/Philadelphia, 1995.

515. H. B. Peng, B. R. Zhao, Z. Xie, Y. Lin, B. Y. Zhu, Z. Hao, H. J. Tao, B. Xu, C. Y. Way, H. Chen, and F. Wu, *Phys. Rev. Lett.* 82, 362 (1999).

516. K. B. Erington and N. J. Ianno, *Mater. Res. Soc. Symp. Proc.* 191, 115 (1990).

517. S. R. Foltyn, R. E. Muenchausen, R. C. Dye, X. D. Wu, L. Luo, D. W. Cook, and R. C. Taber, *Appl. Phys. Lett.* 59, 1374 (1991).

518. H. Buhay, S. Sinharov, M. H. Francombe, W. H. Kasner, J. Tavlacchio, B. K. Park, N. J. Doyle, D. R. Lame, and M. Polinsky, *Proc. Integrated Ferroelectrics* 1, 213 (1992).

519. M. Lorenz, H. Hochmuth, H. Borner, D. Natusch, and K. Krehor, *Mater. Res. Soc. Symp. Proc.* 341, 189 (1994).

520. J. A. Greer and M. D. Tabat, *Mater. Res. Soc. Symp. Proc.* 341, 87 (1994).

521. J. R. Cave, D. W. A. Willen, R. Nadi, W. Zhu, and Y. Brissette, "Applied Superconductivity," Edinburgh, 3–6 July 1995, Inst. Phys. Conf. Ser., Vol. 148, p. 623. IOP, Bristol, 1995.

522. G. Ries, B. Gromoll, H. W. Newmuller, W. Schmidt, H. P. Kramer, and S. Fischer, "Applied Superconductivity," Edinburgh, 3–6 July 1995, Inst. Phys. Conf. Ser., Vol. 148, p. 635. IOP, Bristol, 1995.

523. A. P. Malozemoff, in "Physical Properties of High Temperature Superconductors I," (D. M. Ginsberg, Ed.), pp. 71–150. World Scientific, Singapore, 1990; P. B. Allen, Z. Fisk, and A. Migliori, in "Physical Properties of High Temperature Superconductors II" (D. M. Ginsberg, Ed.), pp. 213–264. World Scientific, Singapore, 1990; Y. Ive, in "Physical Properties of High Temperature Superconductors III" (D. M. Ginsberg, Ed.), pp. 285–361. World Scientific, Singapore, 1992.

524. G. Blatter, M. V. Feigel'man, V. B. Geshkenbein, A. I. Larkin, and V. M. Vinokur, *Rev. Mod. Phys.* 66, 1125 (1994); E. H. Brandt, *Rep. Progr. Phys.* 58, 1465 (1995).

525. M. Tinkham, "Introduction to Superconductivity," 2nd ed. McGraw-Hill, New York, 1996.

526. M. Gurvitch and A. T. Fiory, *Phys. Rev. Lett.* 59, 1337 (1987).

527. H. Takagi, B. Batlogg, H. L. Kao, J. Kwo, R. J. Cava, J. J. Krajewski, and W. F. Peck, *Phys. Rev. Lett.* 69, 2975 (1992).

528. S. Martin, M. Gurvitch, C. E. Rice, A. F. Hebard, P. L. Gammel, R. M. Fleming, and A. T. Fiory, *Phys. Rev. B* 39, 9611 (1989); S. Martin, A. T. Fiory, R. M. Fleming, L. F. Schneemeyer, and J. V. Waszczak, *Phys. Rev. B* 41, 846 (1990).

529. W. E. Pickett, *Rev. Mod. Phys.* 61, 433 (1989).

530. W. Jiang, S. N. Mao, X. X. Xi, X. Jiang, J. L. Peng, T. Venkatesan, C. J. Lobb, and R. L. Greene, *Phys. Rev. Lett.* 73, 1291 (1994).

531. A. Carrington, D. Walker, A. P. Mackenzie, and J. R. Cooper, *Phys. Rev. B* 48, 13051 (1993).

532. B. Wuyts, V. V. Moshchalkov, and Y. Bruynseraede, *Phys. Rev. B* 53, 9418 (1996).

533. C. C. Homes, T. Timusk, R. Liang, D. A. Bonn, and W. N. Hardy, *Phys. Rev. Lett.* 71, 1645 (1993).

534. Z. X. Shen, and D. S. Dessau, *Phys. Rep.* 253, 1 (1993).

535. G. A. Levin and K. F. Quader, *Phys. Rev. B* 62, 11879 (2000).

536. C. Villard, G. Koren, D. Cohen, E. Polturak, and D. Chateignier, *Phys. Rev. Lett.* 77, 3913 (1996).

537. T. Ito, Y. Nakamura, H. Takagi, and S. Uchida, *Phys. C* 185–189, 1267 (1991).

538. A. S. Nygmatulin, A. A. Varlamov, D. V. Livanov, G. Balestrino, and E. Milani, *Phys. Rev. B* 53, 3557 (1996).

539. L. B. Ioffe, A. I. Larkin, A. A. Varlamov, and L. Yu, *Phys. Rev. B* 47, 8936 (1993).

540. N. P. Ong, in "Physical Properties of High Temperature Superconductors II" (D. M. Ginzburg, Ed.), p. 459. World Scientific, Singapore, 1991.

541. E. C. Jones, D. P. Norton, D. K. Christen, and D. H. Lowndes, *Phys. Rev. Lett.* 73, 166 (1993).

542. B. Roas, L. Schultz, and G. Saemann-Ischenko, *Phys. Rev. Lett.* 64, 479 (1990).

543. X. G. Qiu, C. G. Cui, S. L. Li, M. X. Liu, J. Li, Y. Z. Zhang, Y. Y. Zhao, P. Xu, L. Li, L. F. Chen, P. F. Chen, N. Li, and G. T. Liu, *Phys. C* 179, 176 (1991).

544. P. H. Kes, J. Aarts, J. van den Berg, C. J. van der Beek, and J. A. Mydosh, *Supercond. Sci. Technol.* 1, 242 (1989).

545. X. G. Qiu, B. Wuyts, M. Maenhoudt, V. V. Moshchalkov, and Y. Bruynseraede, *Phys. Rev. B* 52, 559 (1995).

546. O. Brunner, L. Antognazza, J. M. Triscone, L. Mieville, and O. Fischer, *Phys. Rev. Lett.* 67, 1354 (1991).

547. H. Raffy, S. Labdi, O. Laborde, and P. Monceau, *Phys. Rev. Lett.* 66, 2515 (1991).

548. K. W. Kwok, J. Fendrich, C. J. van der Beek, and G. W. Crabtree, *Phys. Rev. Lett.* 73, 2614 (1994).

549. P. Berghuis, A. L. F. van der Slot, and P. H. Kes, *Phys. Rev. Lett.* 65, 2583 (1990).

550. M. P. A. Fisher, *Phys. Rev. Lett.* 62, 1415 (1992).

551. R. H. Koch, V. Fotlietti, W. J. Gallagher, G. Koren, A. Gupta, and M. P. A. Fisher, *Phys. Rev. Lett.* 63, 1511 (1989).

552. V. M. Vinokur, P. H. Kes, and A. E. Koshelev, *Phys. C* 168, 29 (1990); V. Feigel'man, V. B. Geshkenbein, and A. L. Larkin, *Phys. C* 167, 177 (1990).

553. C. Dekker, P. Wöltgens, R. Koch, B. Hussey, and A. Gupta, *Phys. Rev. Lett.* 69, 2717 (1992).

554. H. H. Wen, H. Radovan, F. Kamm, P. Ziemann, S. L. Yan, L. Fang, and M. S. Si, *Phys. Rev. Lett.* 80, 3859 (1998).

555. D. R. Nelson and V. M. Vinodur, *Phys. Rev. Lett.* 68, 2398 (1992); L. Krusin-Elbaum, L. Civale, G. Blatter, A. D. Marwick, F. Holtzberg, and C. Field Krusin-Elbaum, *Phys. Rev. Lett.* 72, 1914 (1994).

556. V. V. Moshchalkov, V. V. Metlushko, G. Guumlntherodt, I. N. Goncharov, A. Yu. Didyk, and Y. Bruynseraede, *Phys. Rev. B* 50, 639 (1994).

557. L. Miu, P. Wagner, A. Hadish, F. Hillmer, H. Adrian, J. Wiesner, and G. Wirth, *Phys. Rev. B* 51, 3953 (1995).

558. S. J. Hagen, C. J. Lobb, R. L. Greene, M. G. Forrester, and J. H. Kang, *Phys. Rev. B* 41, 11630 (1990).

559. J. Luo, T. P. Orlando, J. M. Graybeal, X. D. Wu, and R. Muenchausen, *Phys. Rev. Lett.* 68, 690 (1992).

560. V. M. Vinokur, V. B. Geshkenbein, M. V. Feigel'man, and G. Blatter, *Phys. Rev. Lett.* 71, 1242 (1993).

561. A. V. Samoilov, A. Legris, F. Rullier-Albenque, P. Lejay, S. Bouffard, Z.G. Ivanov, and L. G. Johansson, *Phys. Rev. Lett.* 74, 2351 (1995).

562. X. G. Qiu, G. Jakob, V. V. Moshchalkov, Y. Bruynseraede, and H. Adrian, *Phys. Rev. B* 52, 12994 (1995).

563. A. Dorsey, *Phys. Rev. B* 46, 8376 (1992).

564. Y. Liu, D. B. Haviland, L. I. Glazman, and A. M. Goldman, *Phys. Rev. Lett.* 68, 2224 (1992).

565. D. Ephron, A. Yazdani, A. Kapitulnik, and M. R. Beasley, *Phys. Rev. Lett.* 76, 1529 (1996).

566. A. F. Th. Hoekstra, R. Griessen, A. M. Testa, J. el Fattahi, M. Brinkmann, K. Weserholt, W. K. Kwok, and G. W. Crabtree, *Phys. Rev. Lett.* 80, 4293 (1998); H. H. Wen, H. G. Schnack, R. Griessen, B. Dam, and J. Rector, *Phys. C* 241, 353 (1995).

567. A. T. Fiory, A. F. Hebard, P. M. Mankiewich, and R. E. Howard, *Phys. Rev. Lett.* 61, 1419 (1988).

568. L. C. Davix, M. R. Beasley, and D. J. Scalapino, *Phys. Rev. B* 42, 99 (1990).

569. S. Vadlamannati, Q. Li, T. Venkatesan, W. L. Mclean, and P. Lindenfeld, *Phys. Rev. B* 44, 7094 (1991).

570. X. G. Qiu, B. R. Zhao, S. Q. Guo, J. L. Zhang, and L. Li, *Phys. C* 197, 195 (1992).

571. W. Braunisch, N. Knauf, V. Kataev, S. Neuhausen, A. Grütz, A. Kock, B. Roden, D. Khomskii, and D. Wohlleben, *Phys. Rev. Lett.* 68, 1908 (1992).

572. V. V. Moshchalkov, X. G. Qiu, and V. Brondock, *Phys. Rev. B* 55, 11793 (1997).

573. A. K. Geim, S. V. Dubonos, J. G. S. Lok, M. Henini, and J. C. Maan, *Nature* 396, 144 (1998).

574. V. V. Moshchalkov, L. Geilen, C. Strunk, R. Jonckheere, X. Qiu, C. Van Haesendonck, and Y. Bruynseraede, *Nature* 373, 319 (1995).

575. F. Gao, D. Romero, D. Tanner, J. Talvacchio, and M. Forrester, *Phys. Rev. B* 47, 1036 (1993).

576. A. El Azrak, R. Nahoum, N. Bontemps, M. Guilloux-Viry, C. Thivet, A. Perrin, S. Labdi, Z. Z. Li, and H. Raffy, *Phys. Rev. B* 49, 9846 (1994).

577. T. Timusk and B. Statt, *Rep. Progr. Phys.* 62, 61 (1999).

578. X. G. Qiu, H. Koinuma, M. Iwasaki, T. Itoh, S. Kumar, M. Kawasaki, E. Saitoh, Y. Tokura, K. Takehana, G. Kido, and Y. Segawa, *Appl. Phys. Lett.* 78, 506 (2001).

579. K. Tamasaku, Y. Nakamura, and S. Uchida, *Phys. Rev. Lett.* 69, 1455 (1992).

580. M. Tachiki, T. Koyama, and S. Takahashi, *Phys. Rev. B* 50, 7065 (1994).

581. S. Han, Z. Vardeny, K. Wong, O. Symko, and G. Koren, *Phys. Rev. Lett.* 65, 2708 (1990).

582. M. Hangyo, S. Tomozawa, Y. Murakami, M. Tonouchi, M. Tani, Z. Wang, K. Sakai, and S. Nakashima, *Appl. Phys. Lett.* 69, 2122 (1996).

583. C. C. Tsui, J. R. Kirtley, Z. F. Ren, J. H. Wang, H. Raffy, and Z. Z. Li, *Nature* 387, 481 (1997).

584. X.-Z. Yan and C.-R. Hu, *Phys. Rev. Lett.* 83, 1656 (1999).

585. J. C. Gallop, "SQUIDS, the Josephson Effects and Superconducting Electronics," p. 71. Hilger, Bristol, 1990.

586. W. R. McGrath, H. K. Olsson, T. Claeson, S. Ericsson, and L. G. Johansson, *Europhys. Lett.* 4, 357 (1987).

587. J. W. Ekin, A. I. Braginski, A. J. Panson, M. A. Janocko, D. W. Capone II, N. Zaluzec, B. Flandermyer, O. F. deLima, M. Hong, J. Kwo, and S. H. Liou, *Appl. Phys. Lett.* 58, 4821 (1987).

588. P. A. Nilsson, Z. G. Ivanov, D. Winkler, H. K. Olsson, G. Brorsson, T. Claeson, E. A. Stepantsov, and A. Y. Tzalenchuk, *Phys. C* 185, 2597 (1991).

589. D. Dimos, P. Chaudhari, and J. Mannhart, *Phys. Rev. B* 41, 4038 (1990).

590. P. A. Nilsson, Z. G. Ivanov, H. K. Olsson, D. Winkler, T. Claeson, E. A. Stepantsov, and A. Ya. Tzalenchuk, *J. Appl. Phys.* 75, 7972 (1994).

591. K. Char, M. S. Colclough, S. M. Garrisonk, N. Newman, and G. Zaharchuk, *Appl. Phys. Lett.* 59, 733 (1991).

592. G. Cui, Y. Zhang, K. Herrmann, C. Buchal, J. Schubert, W. Zander, A. I. Braginski, and C. Heiden, *Supercond. Sci. Technol.* 4, 130 (1991).

593. K. P. Daly, W. D. Dozier, J. F. Burch, S. B. Coons, R. Hu, C. E. Platt, and R. W. Simon, *Appl. Phys. Lett.* 58, 733 (1991).

594. E.a. Stepnatsov and A. Y. Tzalenchuk, *Phys. Scr.* 44, 102 (1991).

595. J. A. Alarco, E. Olsson, Z. G. Ivanov, P. A. Nilsson, D. Winkler, E. A. Stepantsov, and A. Y. Tzalenchuk, *Ultramicroscopy* 51, 239 (1993).

596. Yu. A. Boikov, Z. G. Ivanov, A. L. Vasiliev, and T. Claeson, *J. Appl. Phys.* 77, 1654 (1995).

597. R. P. J. IJsselsteijn, J. W. M. Hilgenkamp, M. Eisenberg, D. Terpstra, J. Flokstra, and H. Rogalla, *IEEE Trans. Appl. Supercond.* 3, 2321 (1993).

598. C. L. Jia, B. Kabius, K. Urban, K. Herrmann, G. J. Cui, J. Schubert, W. Zander, A. I. Braginski, and C. Heiden, *Phys. C* 175, 545 (1991).

599. J. Gao, B. Hauser, and H. Rogalla, *J. Appl. Phys.* 67, 2512 (1990).

600. A. M. Cucolo, R. Di Leo, A. Nigro, P. Romano, E. Bacca, W. Lopera, M. E. Gomez, and P. Prieto, *IEEE Trans. Appl. Supercond.* 7, 2848 (1997).

601. E. Baca, M. Chacon, W. Lopera, M. E. Gomez, P.Prieto, J. Heiras, R. Di Leo, P. Romano, and A. M. Cucolo, *J. Appl. Phys.* 84, 2788 (1998).

602. I. Bozovic and J. N. Eckstein, *J. Alloys Compounds* 251, 201 (1997).

603. R. H. Koch, W. J. Gallagher, V. Foglietti, B. Oh, R. B. Laibowitz, G. Koren, A. Gupta, and W. Y. Lee, unpublished.

604. J. Gao, W. A. M. Aarnink, G. J. Gerritsma, D. Veldhuis, and H. Rogalla, *IEEE Trans. Magn.* 27, 3062 (1991).

605. J. B. Barner, C. T. Roger, A. Inam, R. Ramesh, and S. Bersey, *Appl. Phys. Lett.* 59, 742 (1991).

606. K. Char, L. Antognazza, and T. H. Geballe, *Appl. Phys. Lett.* 63, 2420 (1993).

607. M. Kasai, T. Ohgno, Y. Kanke, Y. Kozono, M. Hanazono, and Y. Sugita, *Jpn. J. Appl. Phys.* 29, L2219 (1990).

608. M. Kasai, Y. Kanke, T. Ohno, and Y. Kozomo, *J. Appl. Phys.* 72, 5344 (1991).

609. J. Sakai, J. Hioki, T. Ohnishi, T. Yamaguchi, and S. Imai, *Jpn. J. Appl. Phys.* 37, 3286 (1998).

610. J. E. Zimmerman, H. A. Beall, M. W. Cromar, and R. H. Ono, *Appl. Phys. Lett.* 51, 617 (1987).

611. R. W. Simon, J. B. Bulman, J. F. Burch, S. B. Cooms, G. P. Daly, W. D. Dozier, R. Hu, A. E. Lee, J. A. Luine, C. E. Platt, S. M. Schwarzbek, M. S. Wire, and M. J. Zani, *IEEE Trans. Magn.* 27, 3209 (1991).

612. A. E. White, K. T. Short, R. C. Dynes, A. F. J. Levi, M. Anzlowar, K. W. Baldwin, P. A. Polakos, T. A. Fulton, and L. N. Dunkleberger, *Appl. Phys. Lett.* 53, 1010 (1988).

613. M. Gurvich, J. Macaulay, Z. Bao, B. Bi, S. Han, J. Y. Lin, J. Lukens, B. Nadgorny, J. M. Phillips, M. P. Siegal, and E. S. Hellman, Jpn. FED-102 Extended Abstracts 162, 1991.

614. S. S. Tinchev, *Supercond. Sci. Technol.* 3, 500 (1990).

615. A. M. Hermann, J. C. Price, J. F. Scott, R. M. Yandrofski, A. Naziripour, D. Galt, H. M. Duan, M. Paranthaman, R. Tello, J. Cuchiaro, and R. K. Ahrenkiel, *Bull. Amer. Phys. Soc.* 38, 689 (1993).

616. J. A. Beall, R. H. Ono, D. Galt, and J. C. Price, *IEEE Internat. Microwave Symp. Dig.* 1421 (1993).

617. D. Galt, J. Price, J. A. Beall, and R. H. Ono, *Appl. Phys. Lett.* 63, 3078 (1992).

618. A. Deleniv, D. Kholodniak, A. Lapshin, I. Vendik, P. Yudin, B. C. Min, Y. H. Choi, and B. Oh, *Supercond. Sci. Technol.* 13, 1419 (2000).

619. A. T. Findikogu, D. W. Reagor, K. O. Rasmussen, A. R. Bishop, N. Gronbech-Jensen, Q. X. Jia, Y. Fan, C. Kwon, and L. A. Ostrovsky, *J. Appl. Phys.* 86, 1558 (1999).

620. K. Char, N. Newman, S. M. Garrison, R. W. Barton, R. C. Taber, S. S. Laderman, and R. D. Jacowitz, *Appl. Phys. Lett.* 57, 409 (1990).

621. A. T. Findikoglu, P. N. Arendt, J. R. Groves, S. R. Foltyn, E. J. Peterson, D. W. Reagor, and Q. X. Jia, *IEEE Trans. Appl. Supercond.* 7, 1232 (1997).

622. H. Piel and G. Muller, *IEEE Trans. Magn.* 27, 854 (1991).

623. E. H. Brandt, *Internat. J. Mod. Phys. B* 5 (1991).

624. A. B. Kozyrev, E. K. Hollmann, A. V. Ivanov, et al., *Integr. Ferroelect.* 17, 257 (1997).

625. J. J. Kingston, F. C. Wellstood, P. Lerche, A. H. Miklich, and J. Clarke, *Appl. Phys. Lett.* 56, 189 (1990).

626. G. Brorsson, P. A. Nilsson, E. Olsson, S. Z. Wang, and T. Claeson, *Appl. Phys. Lett.* 61, 486 (1992).

627. J. S. Horwitz, C. M. Cotell, D. B. Chrisey, J. M. Pond, K. S. Grabowski, K. R. Carroll, H. S. Newmann, and M. S. Osofsky, *J. Supercond.* 7, 965 (1994).

628. M. R. Rao, *Thin Solid Films* 306, 141 (1997).

629. R. A. Chakalov, Z. G. Ivanov, Y. A. Boikov, P. Larsson, E. Carlsson, S. Gevorgian, and T. Claeson, *Phys. C* 308, 279 (1998).

630. J. Choi, S. Hong, B.-H. Jun, T.-H. Sung, Y. Park, and K. No, *Jpn. J. Appl. Phys.* 38, 1941 (1999).

631. A. T. Findikoglu, Q. X. Jia, X. D. Wu, G. J. Chen, T. Venkatesan, and D. W. Reagor, *Appl. Phys. Lett.* 68, 1651 (1996).

632. A. T. Findikoglu, Q. X. Jia, D. W. Reagor, and X. D. Wu, *Microwave Opt. Tech. Lett.* 9, 306 (1995).

633. O. G. Vendik, E. Kollberg, S. S. Gevorgian, A. B. Kozyrev, and O. I. Soldatenkov, *Electron. Lett.* 31, 654 (1995).

634. S. Gevorgian, E. Carlsson, P. Linner, E. Kolberg, O. Vendik, and E. Wikborg, *IEEE Trans. Microwave Theory Tech.* 44, 1738 (1996).

635. C. Wilker, Z. Y. Shen, P. Pang, D. W. Face, W. L. Holstein, A. L. Matthews, and D. B. Laubacher, *IEEE Trans. Microwave Theory Tech.* 39, 1462 (1991).

636. J. H. Takemoto, C. M. Jackson, H. M. Manasevit, D. C. St. John, J. F. Burch, K. P. Daly, and R. W. Simon, *Appl. Phys. Lett.* 58, 1109 (1994).

637. A. P. Jenkins, A. P. Bramley, D. J. Edwards, D. Dew-Hughes, and C. R. Grovenor, *J. Supercond.* 11, 5 (1998).

638. W. G. Lyons, R. R. Bonetti, A. E. Williams, P. M. Mankiewich, M. L. O'Malley, J. M. Hamm, A. C. Anderson, R. S. Withers, A. Meulenberg, and R. E. Howard, *IEEE Trans. Magn.* 27, 2549 (1991).

639. G. Subramanyam, F. V. Keuls, and F. A. Miranda, *IEEE Microwave Theory Tech. Dig.* 1011 (1998).

640. W. G. Lyons and R. S. Withers, in "Superconductor Technology." World Scientific, Singapore, 1991.

641. C. S Kim, S. M. Kim, S. C. Song, S. Y. Lee, H. K. Yoon, and Y. J. Yoon, *Appl. Surface Sci.* 154, 492 (2000).

642. G. C. Liang, X. H. Dai, D. F. Hebert, T. V. Duzer, N. Newman, and B. F. Cole, *IEEE Trans. Appl. Supercond.* 1, 58 (1991).

643. D. B. Chrisey, J. S. Horwitz, J. M. Pond, K. R. Carrol, P. Lubitz, K. S. Grabowski, R. E. Leuchtner, C. A. Carosella, and C. V. Vottoria, *IEEE Trans. Appl. Supercond.* 3, 1528 (1993).

644. S. D. Silliman, H. M. Christen, L. A. Knauss, K. S. Harshavrdhan, M. M. A. ElSabbagh, and K. Zaki, *J. Electroceramics* 4, 305 (2000).

645. G. F. Dionne, D. E. Oates, and D. H. Temme, *IEEE Trans. Appl. Supercond.* 5, 2083 (1995).

646. E. K. Track, R. E. Drake, and G. K. G. Hohenwarter, *IEEE Trans. Appl. Supercond.* 3, 2899 (1993).

647. M. J. Lancaster, H. Y. Wang, and J. S. Hong, *IEEE Trans. Appl. Supercond.* 8, 168 (1998).

648. S. Ohshima, K. D. Develos, K. Ehata, M. I. Ali, and M. Mukaida, *Phys. C* 335, 207 (2000).

649. M. I. Ali, K. Ehata, and S. Ohshima, *Supercond. Sci. Technol.* 13, 1095 (2000).

650. G. G. Vendik, L. T. Ter-Martirosyan, A. I. Dedyk, S. F. Karmanenko, and R. A. Chakalov, *Ferroelectrics* 144, 33 (1993).

651. J. Mannhart, D. C. Schlom, J. G. Bednorz, and K. A. Muller, *Phys. Rev. Lett.* 67, 2099 (1991).

652. J. Mannhart, J. Srobel, J. G. Bednorz, and Ch. Gerber, *Appl. Phys. Lett.* 62, 630 (1993).

653. R. Ramesh, W. K. Chan, B. Wilkens, H. Gilchrist, T. Sands, J. M. Tarascon, V. J. Keramidas, D. K. Fork, J. Lee, and A. Safri, *Appl. Phys. Lett.* 61, 1537 (1992).

654. J. Lee, L. Johnson, A. Safri, R. Ramesh, T. Sands, H. Gilchrist, and V. G. Keramidas, *Appl. Phys. Lett.* 63, 27 (1993).

655. M. Akinaga and S. Suzuki, *J. Korean Phys. Soc.* 35, S418 (1999).

656. C. Constantin, R. Ramer, I. Matei, L. Trupina, and M. Stegarescu, *Ferroelectrics* 225, 279 (1999).

657. Z. Hao, B. T. Liu, Y. Lin, B. Xu, H. J. Tao, B. Y. Zhu, Z. X. Zhao, and B. R. Zhao, *Supercond. Sci. Technol.* 13, 612 (2000).

658. Yu. A. Boikov, S. K. Esayan, Z. G. Ivanov, G. Brorsson, T. Claeson, J. Lee, and A. Safari, *Appl. Phys. Lett.* 61, 528 (1992).

659. B. T. Liu, Z. Hao, Y. F. Chen, B. Xu, H. Chen, F. Wu, and B. R. Zhao, *Appl. Phys. Lett.* 74, 2044 (1999).

660. L. X. Cao, Y. Xu, B. R. Zhao, L. P. Guo, J. Z. Liu, B. Xu, F. Wu, L. Li, Z. X. Zhao, A. J. Zhu, Z. H. Mai, J. H. Zhao, Y. F. Fu, and X. J. Li, *Supercond. Sci. Technol.* 9, 310 (1996).

661. B. T. Liu, W. W. Huang, Y. L. Qin, Z. Hao, B. Xu, X. L. Dong, F. Wu, H. J. Tao, S. L. Jia, L. Li, and B. R. Zhao, *Supercond. Sci. Technol.* 12, 344 (1999).

662. Z. Xie, E. Z. Luo, J. B. Xu, I. H. Wilson, H. B. Peng, L. H. Zhao, and B. R. Zhao, *Appl. Phys. Lett.* 76, 1923 (2000).

663. H. Fujisawa, M. Shimizu, T. Horiuchi, T. Shiosaki, and K. Matsushige, *Appl. Phys. Lett.* 71, 416 (1997).

664. Z. Xie, E. Z. Luo, H. B. Peng G. D. Hu, J. B. Xu, I. H. Wilson, B. R. Zhao, and L. H. Zhao, *J. Non-Cryst. Solids* 254, 112 (1999).

665. V. V. Lemanov, A. L. Kholkin, and A. B. Sherman, *Supercond. Sci. Technol.* 6, 814 (1993).

666. L. X. Cao, B. R. Zhao, Y. L. Qin, L. Li, T. Yang, and Z. X. Zhao, *Phys. C* 303, 47 (1998).

667. N. J. Wu, H. Lin, K. Xie, and A. Ignatiev, *Ferroelectrics* 156, 73 (1994).

668. H. Lin, N. J. Wu, K. Xie, X. Y. Li, and A. Ignatiev, *Appl. Phys. Lett.* 65, 953 (1994).

669. N. J. Wu, H. Lin, T. Q. Huang, S. Endictor, D. Liu, and A. Ignatiev, *Proc. SPIE* 2697, 502 (1996).

670. H. Lin, N. J. Wu, F. Geiger, K. Xie, and A. Ignatiev, *J. Appl. Phys.* 80, 7130 (1996).

671. E. Carlsson, E. Wikborg, and S. Gevorgian, *Ferroelectrics Lett.* 25, 141, (1999).

672. O. G. Vendik, E. K. Hollmann, A. B. Kozyrev, and A. M. Prudan, *J. Supercond.* 12, 325 (1999).

673. C. H. Ahn, J. M. Triscone, N. Archibald, M. Decroux, R. H. Hammond, T. H. Geballe, O. Fischer, and M. R. Beasley, *Science* 269, 373 (1995); C. H. Ahn, T. Tybell, L. Antognazza, K. Char, R. H. Hammond, M. R. Beasley, Ø. Fischer, and J.-M. Triscone, *Science* 284, 1152 (1999).

674. E. O. Wollan and W. C. Koeheler, *Phys. Rev.* 100, 545 (1955).

675. G. H. Jonker and J. H. Van Santen, *Physica* 16, 337 (1950).

676. R. Von Helmolt, J. Wecker, B. Holzapfel, L. Schultz, and K. Samwer, *Phys. Rev. Lett.* 71, 2331 (1993).

677. S. Jin, Th. H. Tiefel, M. McCormack, R. A. Fastnacht, R. Ramesh, and L. H. Chen, *Science* 264, 413 (1994).

678. Z. Radovic, M. Ledvij, L. Dobrosavljevic-Grujic, A. I. Buzdin, and J. R. Clem, *Phys. Rev. B* 44, 759 (1991).

679. J. S. Jiang, D. Davidovic, Daniel H. Reich, and C. L. Chien, *Phys. Rev. Lett.* 74, 314 (1995).

680. G. C. Xiong, G. J. Lian, J. F. Kang, Y. F. Hu, Y. Zhang, and Z. Z. Gan, *Phys. C* 282–287, 693 (1997).

681. P. Przyslupski, T. Nishizaki, N. Kobayashi, S. Kolesnik, T. Skoskiewicz, and E. Dynowska, *Phys. B* 259–261, 820 (1999).

682. G. Jokob, V. C. Moshchalkov, and Y. Bruynseraede, *Appl. Phys. Lett.* 66, 2564 (1995).

683. M. Isa, T. Nishizaki, M. Fujiwara, T. Naito, and N. Kobayashi, *Phys. C* 282–287, 691 (1997).

684. W. LY. Hu and Q. Q. Zheng, *Phys. C* 282–287, 1627 (1997).

685. A. I. Buzdin, L. N. Bulaevskil, and S. V. Panyukov, *JETP Lett.* 35, 178 (1982).

686. M. Kasai, T. Ohono, and Y. Kozono, *Mod. Phys. Lett. B* 7, 1923 (1993).

687. F. A. Demler, G. B. Arnold, and M. R. Beasley, *Phys. Rev. B* 55, 15174 (1997).

688. M. Kasai, T. Ohno, Y. Kanke, Y. Konono, M. Hanazono, and Y. Sugita, *Jpn. J. Appl. Phys.* 29, L2219 (1990).

689. S. Kishino, H. Kuroda, T. Shibutani, and H. Niu, *Appl. Phys. Lett.* 65, 781 (1994).

690. M. A. Bari, O. Cabeza, L. Capogna, P. Woodall, C. M. Muirhead, and M. G. Nlamire, *IEEE Trans. Appl. Supercond.* 7, 2304 (1997).

691. A. I. Buzidin, A. V. Vedyayev, and N. V. Ryzhyanova, *Europhys. Lett.* 48, 686 (1999).

692. L. R. Tagirov, *Phys. Rev. Lett.* 83, 2058 (1999).

693. K. Chahara, T. Ohno, M. Kasai, Y. Kanke, and Y. Kozono, *Appl. Phys. Lett.* 62, 780 (1993).

694. C. A. R. Sa de Melo, *J. Appl. Phys.* 81, 5364 (1997).

695. Z. W. Dong, R. Ramesh, T. Venkatesan, M. Johnson, Z. Y. Chen, S. P. Pai, V. Talyansky, R. P. Sharma, R. Shreekala, C. J. Lobb, and R. L. Greene, *Appl. Phys. Lett.* 71, 1718 (1997).

696. R. M. Stroud, J. Kim, C. R. Eddy, D. B. Chrisey, J. S. Horwitz, D. Koller, M. S. Osofsky, R. J. Soulen, Jr., and R. C. Y. Auyeung, *J. Appl. Phys.* 83, 7189 (1998).

697. K. Lee and B. Friedman, *Supercond. Sci. Technol.* 12, 741 (1999).

698. N.-C. Yeh, B. P. Vasquez, C. C. Fu, A. V. Samoilov, Y. Li, and K. Vakili, *Phys. Rev. B* 60, 10522 (1999).

699. V. A. Vasko, V. A. Larkin, P. A. Kraus, K. R. Nikolaev, D. E. Grupp, C. A. Nodman, and A. M. Goldman, *Phys. Rev. Lett.* 78, 1134 (1997).

700. V. A. Vasko, K. R. Nikolaev, V. A. Larkin, P. A. Kraus, and A. M. Goldman, *Appl. Phys. Lett.* 73, 844 (1998).

701. I. Zutic and O. T. Valls, *Phys. Rev. B* 60, 6320 (1999).

702. J. Y. T. Wei, N.-C. Yeh, and C. C. Fu, *J. Appl. Phys.* 85, 5350 (1999).

Chapter 11

ELECTRONIC AND OPTICAL PROPERTIES OF STRAINED SEMICONDUCTOR FILMS OF GROUPS IV AND III–V MATERIALS

George Theodorou

Department of Physics, Aristotle University of Thessaloniki, 540 06 Thessaloniki, Greece

Contents

1. INTRODUCTION

Silicon is an abundant and cheap material that dominates the area of microelectronics [1–3]. However, creation of advanced integrated circuits based on Si revealed the existence of different efficiency problems, for example, the power drain and shear number of wires that link Si chips impose limits on their performance. The solution to these problems could come from the use of optical rather than electronic inter-connects. Development of Si-based photonic components and subsequent integration of optic and electronic components on the same substrate will create optoelectronic integrated circuits and "superchips" that will perform much better than optical or electronic circuits alone. In addition, Si-based optoelectronic technology could find applications in fiber-optic transmitters and receivers, optical computer integrations, optical controllers, information display panels, and numerous other devices. Unfortunately, bulk crystalline Si has some disadvan-

Handbook of Thin Film Materials, edited by H.S. Nalwa
Volume 4: Semiconductor and Superconductor Thin Films

ISBN 0-12-512912-2/$35.00

tages. A relatively small and indirect bandgap and low carrier mobility make it unsuitable for use in photonic devices. These deficiencies have channeled research effort toward creation of an effective optoelectronic device through development of Si-based structures that do not possess the handicaps of crystalline Si. A promising way to achieve this goal seems to be the use of strained-layer epitaxy, doped heterostructures, and band engineering of quantum-confined structures [4–9]. Groups III–V semiconductor compounds are also important materials for use in optoelectronics [10–12]. Modification of their properties by strain or the creation of quantum structures improves their fitness for device uses.

The development of epitaxial techniques [4, 13], such as molecular beam epitaxy and special kinds of chemical vapor phase epitaxy, has enabled us to control the growth of individual semiconductor layers on an atomic scale. As a result, a strong basis for designing semiconductor materials with tailored electronic and optical properties has been established. The combination of different composites in the form of heterostructures and superlattices has opened up exciting possibilities for manufacturing novel devices and has increased basic understanding of electronic behavior in reduced dimensions.

For a long period of time, attention was focussed on lattice-matched materials for the construction of heterolayers, and strain was considered to be a deficiency. This approach considerably limited the associated development of electronic and optoelectronic semiconductor device structure. Laser emission wavelengths, frequency response, and transistor output power all suffered limitations in lattice-matched systems. Osbourn [14, 15] then drew attention to the fact that the growth of thin layers relaxes the requirement of lattice matching between materials, as long as any lattice mismatch can be elastically accommodated by strain in the films. The epitaxy of lattice-mismatched materials results either in a strained-layer configuration for the case where the layers are sufficiently thin or in strain relaxation, basically by a misfit dislocation network [4, 13]. In the strained-layer configuration, the thin layer's lattice constant in the growth plane is adjusted to that of the substrate, with a simultaneous relaxation of the material's lattice constant along the growth direction. The stored strain energy is proportional to the thickness of the layer, and below a critical thickness, the elastically strained layers are thermodynamically stable and a high quality of growth can be achieved. This type of growth is called pseudomorphic or coherent. The availability of device quality strained-layer epitaxy has provided additional design flexibility for device structures, allowing the rapid development of improved electronic and optoelectronic devices, thus expanding the applications that electronic and optoelectronic semiconductor devices can address.

In pseudomorphic growth, the relative lattice mismatch $\delta a_0/a_0$ between the host materials of the heterolayers range from almost zero (e.g., between GaAs and AlAs) to values as large as several percent (e.g., 7% between InAs and GaAs). It is now possible to grow high-quality strained-layer structures in which, for example, a single layer is accommodated in a semiconductor with significantly different lattice constant. Elasticity

theory, essentially a linear response formalism, is usually used for the treatment of the strains [12]. This approach is quite reasonable for systems with small lattice mismatch [e.g., InAs–GaSb ($\delta a_0/a_0 = 0.62\%$), GaSb–AlSb ($\delta a_0/a_0 = 0.65\%$)], whereas it may pose difficulties in handling large mismatches such as those exhibited between InAs and GaAs ($\delta a_0/a_0 = 7\%$).

The growth of strained layer structures has several advantages. Strained-layer structures may display new electronic and optical properties not seen in the unstrained constituent materials [2, 5, 6]. In fact, strain could be used to modify semiconductor band structures in a predictable way. In addition, artificial modification of the periodicity produced by periodic alternation of thin layer films, like quantum wells and superlattices, could modify the electronic and optical properties of the materials and yield properties not seen in natural materials. With this bandgap engineering it is possible to access wavelengths not otherwise achievable. The strain produced by pseudomorphic growth is called biaxial. In addition, uniaxial strain produced by external pressure applied along a given direction also introduces new physical effects that significantly modify semiconductor electronic properties. From the symmetry point of view, uniaxial strain is equivalent to biaxial strain and has similar effects on the properties of the material.

Groups IV and III–V semiconducting materials play a significant role in device construction. For this reason, their electronic and optical properties have been the subject of numerous experimental and theoretical work over the last decades. The use of such materials in superlattices and optomechanical devices has intensified the effort in this field to provide the best possible understanding their electronic and optical behavior and thus advance their technological applications [5]. The behavior of these materials under strain is an important factor that has significant implications. The change in optical properties due to strain contains information about the electronic structure and the strain deformation potentials, and plays a role in Brillouin [16] as well as Raman scattering [17]. In addition, data on piezooptical properties are desirable when an optical technique is used to monitor semiconductor growth processes [18] and to design optical modulators [19].

This chapter reviews the effect of homogeneous strain on the electronic and optical properties of group IV materials and III–V compounds. In Section 2, deformation potential theory for diamond and zincblende structures under uniaxial and biaxial strain is reviewed. In particular, the influence of strain on the band extrema and the optical gaps E_1 and $E_1 + \Delta_1$ is considered. Deformation potential theory is a macroscopic theory, which is unable to describe the microscopic picture of the material. One of the many methods that is capable of achieving a microscopic description is the tight-binding model. This model is suitable for efficiently describing systems with a large number of atoms per unit cell and for unveiling the underlying physics of the problem. The tight-binding model is presented in brief in Section 3. The properties of strained Si and Ge are reviewed in Sections 4 and 5, whereas Sections 6 and 7 deal with the electronic and optical properties of strained $Si_{1-x}Ge_x$ and

$Si_{1-y}C_y$ alloys. The review of group IV materials is concluded in Section 8 with a presentation of the properties of the artificial structure of strain Si/Ge superlattices. With regard to group III–V compounds, the piezooptical properties of GaAs and InP are reviewed in Section 9 and the artificial superlattice structures InAs/AlSb are discussed in Section 10.

1.1. Strain Tensor

The strain in a homogeneous material is described by the strain tensor $[\varepsilon]$, which for biaxial strain produced by pseudomorphic growth on a (001) or a (111) plane (or a uniaxial strain produced by pressure along a [001] or a [111] direction) has the form [20–22]

$$[\varepsilon]_{(001)} = \begin{pmatrix} \varepsilon_{xx} & 0 & 0 \\ 0 & \varepsilon_{xx} & 0 \\ 0 & 0 & \varepsilon_{zz} \end{pmatrix}$$

$$[\varepsilon]_{(111)} = \begin{pmatrix} \varepsilon_{xx} & \varepsilon_{xy} & \varepsilon_{xy} \\ \varepsilon_{xy} & \varepsilon_{xx} & \varepsilon_{xy} \\ \varepsilon_{xy} & \varepsilon_{xy} & \varepsilon_{xx} \end{pmatrix} \quad (1)$$

where x, y, and z denote the conventional crystallographic directions for diamond or zincblende structures. In the elastic approximation, we can find relationships between matrix elements of the tensor that are given as follows [23]:

(a) For a biaxial (001) strain configuration,

$$\varepsilon_{xx} = \varepsilon_x$$
$$\varepsilon_{zz} = -2\frac{C_{12}}{C_{11}}\varepsilon_x \quad (2)$$

(b) For a biaxial (111) strain configuration,

$$\varepsilon_{xx} = \frac{4C_{44}}{4C_{44} + C_{11} + 2C_{12}}\varepsilon_x$$
$$\varepsilon_{xy} = \frac{-(C_{11} + 2C_{12})}{4C_{44} + C_{11} + 2C_{12}}\varepsilon_x \quad (3)$$

In Eqs. (2) and (3), ε_x is the strain tensor component in the growth plane, which for pseudomorphic growth is equal to $\varepsilon_x = (a_s - a_0)/a_0$, where a_s is the lattice constant of the substrate and a_0 is the lattice constant of the unstrained material. In addition, C_{11}, C_{12}, and C_{44} are the elastic constants of the material. The corresponding expressions for a uniaxial strain configuration are given in [20].

Strain can be decomposed into two parts: the hydrostatic part, which gives rise to volume changes without changing the crystal symmetry, and the shear part, which, in general, reduces the symmetry of the crystal, but does not produce changes in the total volume. The hydrostatic component of the strain tensor is

$$[\varepsilon]_{hydro} = (1/3) \operatorname{Tr}[\varepsilon]\mathbf{1} \quad (4)$$

where $\operatorname{Tr}[\varepsilon]$ is the trace of the tensor and $\mathbf{1}$ is the unit tensor. The shear component is

$$[\varepsilon]_{shear} = [\varepsilon] - (1/3) \operatorname{Tr}[\varepsilon]\mathbf{1} \quad (5)$$

Uniaxial and biaxial strain configurations do not possess the same percentage of hydrostatic and shear components, even though they are equivalent from the symmetry point of view [22]. The symmetry changes produced by uniaxial or biaxial strain configurations result in a reduction of the diamond or zincblende crystal structure to tetragonal for a biaxial (001) (or a uniaxial [001]) strain configuration and to rhombohedral for a biaxial (111) (or a uniaxial [111]) strain configuration [21].

2. DEFORMATION POTENTIALS

Electronic states that play a significant role in determination of the electronic and optical properties of a semiconductor are those in the highest valence and lowest conduction band, and in particular those near the band extrema. For this reason, we are going to focus our attention on the behavior of these states under strain. A detailed analysis of the influence of strain on the band structure of bulk semiconductors can be found in Pollak and Cardona [20], Bir and Pinks [21], and Kane [24]. We present here a review of the main points of the influence of a (001) or (111) strain configuration, biaxial or uniaxial, on diamond and zincblende band structures. Symmetry points and lines of the Brillouin zone (BZ) that are of particular interest for these structures are those at the zone center (Γ point), the L point, and the Δ line. These points determine the fundamental bandgap of the material as well as the optical gaps E_0, E_1, and E_2. The band structure for unstrained Ge and GaAs, calculated by the tight-binding model [25], is given in Figure 1. Also shown in this figure are the transitions responsible for the critical points E_0, E_1, and E_2.

2.1. Γ Point States

For diamond and zincblende materials, the top of the valence band is located at the Γ point of the BZ. In the absence of strain and spin–orbit coupling, the top of the valence band is a p-like sixfold-degenerate orbital multiplet with symmetry $\Gamma_{25'}$ (Γ_{15}) for diamond (zincblende) structures [20]. Spin–orbit coupling splits the top of the valence band into a fourfold-degenerate $P_{3/2}$ multiplet that has Γ_8^+ (Γ_8) symmetry for diamond (zincblende) structures, and is labeled V_1 for the ($J = 3/2$, $M_J = \pm 1/2$) angular momentum-like state and V_2 for the ($J = 3/2$, $M_J = \pm 3/2$) angular momentum-like state, which represent the heavy-hole and light-hole states, respectively, and a $P_{1/2}$ doublet that has Γ_7^+ (Γ_7) symmetry and is labeled V_3 for the ($J = 1/2$, $M_J = \pm 1/2$) angular momentum-like state, which represents the spin split-off state. States V_1 and V_2 move up by an amount $\Delta_0/3$, whereas V_3 moves down by an amount $2\Delta_0/3$. The combination of spin–orbit coupling and strain produces shifts in the V_1, V_2, and V_3 valence states that are given in terms of the deformation potentials (DPs) by [20, 21, 26]

$$\Delta E_{V_1} = \delta E_h - \frac{1}{6}\Delta_0 + \frac{1}{4}\delta E + \frac{1}{2}\sqrt{\Delta_0^2 + \Delta_0(\delta E) + \frac{9}{4}(\delta E)^2} \quad (6)$$

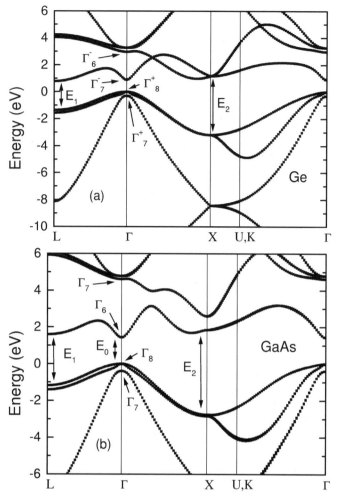

Fig. 1. Band structure of unstrained (a) Ge and (b) GaAs, obtained by a tight-binding method [25]. Also shown are the transitions responsible for the critical points E_0, E_1, and E_2.

$$\Delta E_{V_2} = \delta E_h + \frac{1}{3}\Delta_0 - \frac{1}{2}\delta E \qquad (7)$$

$$\Delta E_{V_3} = \delta E_h - \frac{1}{6}\Delta_0$$
$$+ \frac{1}{4}\delta E - \frac{1}{2}\sqrt{\Delta_0^2 + \Delta_0(\delta E) + \frac{9}{4}(\delta E)^2} \qquad (8)$$

δE_h represents the hydrostatic component of the shift, $\delta E_h = a_v \operatorname{Tr}[\varepsilon]$, where a_v is the hydrostatic valence band deformation potential, $\operatorname{Tr}[\varepsilon]$ is the trace of the strain tensor, and δE is the biaxial (or uniaxial) strain component. The component δE is given by

$$\delta E = \delta E_{001} = 2b(\varepsilon_{zz} - \varepsilon_{xx}) \qquad (9)$$

for a (001) strain and by

$$\delta E = \delta E_{111} = 2\sqrt{3}d\,\varepsilon_{xy} \qquad (10)$$

for a (111) strain, where b is the deformation potential for tetragonal strain symmetry [(001) strain configuration] and d is the corresponding deformation potential for rhombohedral strain symmetry [(111) strain configuration].

With the exception of semiconductors Si and α-Sn, the lowest conduction state at Γ point is an antibonding s-like state

with symmetry $\Gamma_{2'}$ (Γ_1) in the single group representation or Γ_7^- (Γ_6) in the double group representation for diamond (zincblende) structures. The lowest state is subject to hydrostatic strain shifts with no shear component and has an energy shift equal to

$$\Delta E_c = a_c \operatorname{Tr}[\varepsilon] \qquad (11)$$

where a_c is the hydrostatic conduction band deformation potential. For Si and α-Sn, the lowest conduction state at the Γ point is an antibonding p-like state that has Γ_{15} symmetry in the single group representation, for which the preceding relationship cannot be applied.

In the absence of strain, transitions between the $P_{3/2}$ valence states (V_1 and V_2 states) and the lowest s-like conduction state at the Γ point are labeled E_0 transitions; those from the $P_{1/2}$ state to the same conduction state are labeled $E_0 + \Delta_0$. The variation of E_0 with hydrostatic pressure, $\Delta E_0 = (a_v - a_c)\operatorname{Tr}[\varepsilon]$, leads to a direct measurement of the relative deformation potential $a_v - a_c$. However, in certain cases like heterojunctions, it is important to know not only the relative deformation potential, but also the deformation potential for the individual bands, because they are necessary to describe quantities like band discontinuity at the interface. To do so, the additional value for one of the band edge deformation potentials a_v or a_c is needed and it can be obtained from theoretical calculations.

2.2. Indirect Conduction-Band Minima

For certain semiconductors, the fundamental bandgap is indirect, and its variation under strain is influenced by variation of the conduction-band minimum that occurs either close to the X point along the [001] direction (referred to as Δ direction) or at the L point (along the [111] direction). The variation of these minima with strain depends on the type of strain as follows [27]:

2.2.1. (001) Strain Configuration

Under a (001) strain configuration, the six equivalent minima along the Δ lines split into two groups, one that includes four minima along the x and y directions and the other that has two minima along the z direction. Their energy shift is given by the expressions

$$\Delta E(\Delta_{xy}) = \left(\Xi_d^\Delta + \frac{1}{3}\Xi_u^\Delta\right)\operatorname{Tr}[\varepsilon] - \frac{1}{3}\Xi_u^\Delta(\varepsilon_{zz} - \varepsilon_{xx}) \qquad (12)$$

and

$$\Delta E(\Delta_z) = \left(\Xi_d^\Delta + \frac{1}{3}\Xi_u^\Delta\right)\operatorname{Tr}[\varepsilon] + \frac{2}{3}\Xi_u^\Delta(\varepsilon_{zz} - \varepsilon_{xx}) \qquad (13)$$

where Ξ_d^Δ and Ξ_u^Δ are the Δ line minimum deformation potentials, and $\Xi_d^\Delta + \frac{1}{3}\Xi_u^\Delta$ represents the hydrostatic component of the shift. For a semiconductor with a conduction-band minimum that occurs along the Δ line, variation of the fundamental gap under hydrostatic pressure is then given by

$$\Delta E_g^{i,\Delta} = \left(\Xi_d^\Delta + \frac{1}{3}\Xi_u^\Delta - a_v\right)\operatorname{Tr}[\varepsilon] \qquad (14)$$

For such a variation, the deformation potential $\Xi_d^\Delta + \frac{1}{3}\Xi_u^\Delta - a_v$ is directly measurable.

Under a (001) strain configuration, the conduction-band minima at the L point remain equivalent and their shift in energy is

$$\Delta E^L = \left(\Xi_d^L + \tfrac{1}{2}\Xi_u^L\right)\mathrm{Tr}[\varepsilon] \qquad (15)$$

where Ξ_d^L and Ξ_u^L are the L point deformation potentials. For a semiconductor with a conduction-band minimum that occurs at the L point, the variation of the fundamental gap under hydrostatic pressure is then given by

$$\Delta E_g^{i,L} = \left(\Xi_d^L + \tfrac{1}{3}\Xi_u^L - a_v\right)\mathrm{Tr}[\varepsilon] \qquad (16)$$

From such a variation, the relative deformation potential $\Xi_d^L + \tfrac{1}{3}\Xi_u^L - a_v$ is directly measurable.

2.2.2. (111) Strain Configuration

A (111) strain configuration splits the lowest conduction band at the L point into a singlet valley in the [111] direction and a triplet valley in the [$1\bar{1}\bar{1}$], [$\bar{1}1\bar{1}$], and [$\bar{1}\bar{1}1$] directions. Their energy shifts are given by

$$\Delta E(L^s) = \left(\Xi_d^L + \tfrac{1}{3}\Xi_u^L\right)\mathrm{Tr}[\varepsilon] + 2\Xi_u^L\varepsilon_{xy} \qquad (17)$$

$$\Delta E(L^t) = \left(\Xi_d^L + \tfrac{1}{3}\Xi_u^L\right)\mathrm{Tr}[\varepsilon] - \tfrac{2}{3}\Xi_u^L\varepsilon_{xy} \qquad (18)$$

for the singlet and triplet states, respectively. On the other hand, a (111) strain configuration leaves the conduction-band minima in the Δ directions equivalent, with a shift in energy equal to

$$\Delta E(\Delta) = \left(\Xi_d^\Delta + \tfrac{1}{3}\Xi_u^\Delta\right)\mathrm{Tr}[\varepsilon] \qquad (19)$$

2.3. Optical Gaps E_1 and $E_1 + \Delta_1$

Strain also influences the optical gaps E_1 and $E_1 + \Delta_1$. These gaps are connected to direct transitions between the higher valence and the lowest conduction bands along the Λ direction, near the L point. Electron–hole spin exchange interactions have a pronounced effect for a (001) strain, but such an effect has not been observed for a (111) strain. Changes in the optical gaps produced by strain are expressed in terms of the deformation potentials D_j^i. These changes depend on the strain configuration as follows [20, 26]:

2.3.1. (001) Strain Configuration

The energy for E_1 and $E_1 + \Delta_1$ optical gaps under a (001) strain are given by

$$E_\pm = E_1(0) + \delta_H + \tfrac{1}{2}\Delta_1 \pm \tfrac{1}{2}\sqrt{\Delta_1^2 + 4(\delta_J \pm \delta_S)^2} \qquad (20)$$

where $\delta_H = (3)^{-1/2}D_1^1\mathrm{Tr}[\varepsilon]$, δ_J is the exchange term, $\delta_S = \sqrt{2/3}D_3^3(\varepsilon_{zz} - \varepsilon_{xx})$, and Δ_1 is the spin–orbit splitting. The plus sign corresponds to the $E_1 + \Delta_1$ gap, whereas the minus corresponds to the E_1. In these relationships, D_1^1 describes the hydrostatic shift and D_3^3 describes the intraband splitting of the valence band [note that there is no L point valley splitting for a (001) strain]. Only the absolute value of D_3^3 can be obtained

from energy relationship (20), whereas its sign can be calculated from transition probabilities [20].

2.3.2. (111) Strain Configuration

A (111) strain causes splitting of the L point valleys into singlet [111] and triplet [$1\bar{1}\bar{1}$], [$\bar{1}1\bar{1}$], and [$\bar{1}\bar{1}1$] valleys. The valley splitting, together with the intraband splitting of the valence bands, gives rise to four different transitions, namely $E_\pm^s$ for the singlet valley, and $E_\pm^t$ for the triplet valleys. Their transition energies in terms of deformation potentials are given by

$$E_\pm^s = E_1(0) + \frac{\Delta_1}{2} + \delta_H + \frac{1}{2}\delta_{S'} \pm \frac{\Delta_1}{2} \qquad (21)$$

and

$$E_\pm^t = E_1(0) + \frac{\Delta_1}{2} + \delta_H - \frac{\delta_{S'}}{6} \pm \frac{1}{6}\sqrt{9\Delta_1^2 + 16\delta_{S''}^2} \qquad (22)$$

where $\delta_{S'} = 2\sqrt{3}D_1^5\varepsilon_{xy}$ and $\delta_{S''} = \sqrt{6}D_3^5\varepsilon_{xy}$, and the plus sign corresponds to a $E_1 + \Delta_1$ transition and the minus to a E_1. Deformation potential D_1^5 gives the interband contribution and D_3^5 gives the intraband component. The sign of D_3^5 once again cannot be determined from the energies; it must be calculated from the strain dependence of the transition intensities [20].

3. THE TIGHT-BINDING MODEL

Most of the theoretical results presented in this review were obtained by using a tight-binding (TB) model Hamiltonian with an sp^3 set of orbitals, including spin–orbit interaction [25]. Three-center representation and interactions up to the third neighbor were used in this model to obtain a good description of both the valence and the lower conduction bands. The tight-binding parameters were obtained by fitting existing bands for the unstrained materials. To calculate the electronic and optical properties of the strained materials, a modification of the tight-binding parameters, produced by a change in the distance between the atoms, was taken into account by the scaling formula [25]

$$H_{\alpha,\beta}(d) = H_{\alpha,\beta}(d_0)\cdot(d_0/d)^{\nu_{\alpha\beta}} \qquad (23)$$

where α, β represent atomic orbitals and d_0, d represent the unstrained and strained interatomic distances, respectively. In addition, uniaxial strain along the [001] direction lifts the degeneracy between p_x, p_y, and p_z orbitals. This modification is taken into account by the linear formula

$$\begin{aligned} E_p^{x,y} &= E_p + b_p(\varepsilon_\parallel - \varepsilon_\perp) \\ E_p^z &= E_p - 2b_p(\varepsilon_\parallel - \varepsilon_\perp) \end{aligned} \qquad (24)$$

where $E_p^{x,y}$ and E_p^z are the on-site p-orbital integrals, and $\varepsilon_\parallel$ and $\varepsilon_\perp$ denote the strain components parallel and perpendicular to the growth plane. The values for the scaling indices $\nu_{\alpha,\beta}$ and the orbital splitting parameter b_p were determined in such away as to obtain the best values for the Γ point deformation potentials [25].

The imaginary part, $\varepsilon_2(\omega)$, of the dielectric function was obtained by use of the relationship [28]

$$\varepsilon_2(\omega) = \frac{4\pi^2 e^2}{m^2 \omega^2} \sum_{c,v} \int \frac{2}{(2\pi)^3} \mid \langle \mathbf{k}, c \mid \mathbf{P} \cdot \mathbf{a} \mid \mathbf{k}, v \rangle \mid^2$$

$$\times \, \delta[E_{cv}(\mathbf{k}) - \hbar\omega] \, d\mathbf{k} \qquad (25)$$

where $|\mathbf{k}, c\rangle$ and $|\mathbf{k}, v\rangle$ stand for the wave functions of the conduction and the valence bands, respectively, and $E_{cv}(\mathbf{k})$ respresents the energy difference between the c conduction and the v valence band. $\mathbf{P}$ is the momentum operator and $\mathbf{a}$ is the polarization unit vector. In the present TB scheme, the momentum matrix elements were expressed in terms of the Hamiltonian matrix elements and distances between localized orbitals [29–31]. Integration in the BZ was performed within the linear analytic tetrahedron method [32, 33]. These calculations do not take into account many-body effects like electron–phonon and electron–electron interactions; therefore, the exciton contribution is not taken into account in the present model. The real part, $\varepsilon_1(\omega)$, is obtained from $\varepsilon_2(\omega)$ by use of the Kramers–Kronig relationships.

4. STRAINED Si

The influence of external strain on the electronic and optical properties of Si was extensively studied for a long time [20, 26, 34–48]. Values for band extrema deformation potentials for Si are given in Table I and values for optical gaps E_1 and $E_1 + \Delta_1$ DPs are given in Table II. As stated in Section 2, the sign for DPs D_3^3 and D_3^5 should be obtained from the strain variation of transition probabilities. Such an experimental investigation requires a relatively large value for spin–orbit splitting, a condition not applicable to Si. This fact, combined with a larger lifetime broadening for these critical energies of about 65 and 100 meV for low and room temperatures, respectively [50], makes the experimental determination of the signs extremely difficult. The latter were obtained from a theoretical investigation and the results are given in Table II. These signs are in agreement with established values for deformation potentials of other groups IV and III–V compounds shown in next sections. The agreement makes the theoretical prediction reliable.

4.1. Biaxial Strain

Si films under biaxial strain can be produced by pseudomorphic growth on a $Si_{1-x}Ge_x$ substrate. The lattice constant for $Si_{1-x}Ge_x$ alloys is well described by Vegard's law, that is, by a linear interpolation between the lattice constants of bulk Si and Ge (see Section 6).

Growth on a (001) substrate reduces the Si symmetry to a tetragonal D_{4h} body centered lattice. The extremum energies for the valence-band maxima (V_1, V_2, V_3), the conduction-band minima along the Δ directions, and the L point under biaxial strain were calculated using deformation potential theory [25],

Table I. Spin–Orbit Coupling Constant Δ_0 and Band Extrema Deformation Potentials for Si (all values are in eV)

Parameter	Experiment	Theory
Δ_0	0.04^a	
a_v	—	2.46^b
b	-2.10 ± 0.10^c	-2.35^d
d	-4.85 ± 0.15^c	-5.32^d
$\Xi_d^\Delta + \frac{1}{3}\Xi_u^\Delta - a_v$	1.50 ± 0.30^c	1.72^d
Ξ_u^Δ	8.6 ± 0.4^c	9.16^d, 8.86^e

^{a}Landolt–Börnstein [49].

^{b}Van de Walle [47].

^{c}Laude et al. [34].

^{d}Van de Walle and Martin [46].

^{e}Tserbak et al. [25].

Table II. Optical Gaps E_1 and $E_1 + \Delta_1$ Deformation Potentials D_i^j for Si

D_1^1	D_1^5	D_3^3	D_3^5	Ref.
-9.8 ± 1.3	6.5 ± 1.4	4.7 ± 0.5	3.0 ± 1.7	[37]
-8 ± 1	10 ± 2	5 ± 1	4 ± 1	[38]
-9.72 ± 0.56	7.83 ± 0.55	4.36 ± 0.62	5.04 ± 0.87	[45]
-8.5	12	-2.3	-6.5	[48]

Note: All values are in eV.

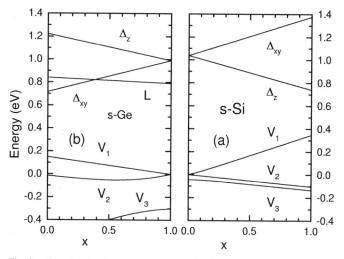

Fig. 2.   Energies for the three top valence bands at Γ (V_1, V_2, V_3), and the lowest conduction states along the Δ_{xy}, Δ_z lines and at the L point of the BZ for biaxially strained (a) Si and (b) Ge pseudomorphically grown on a $Si_{1-x}Ge_x(001)$ substrate, as a function of the Ge content, x, in the substrate.

and the results are shown in Figure 2a. The Δ line conduction-band minimum split into a doublet, Δ_z, and a quadruplet, Δ_{xy}. The doublet decreases in energy as the lattice constant of the substrate increases. The band structure of strained Si (s-Si) pseudomorphically grown on a Ge(001) substrate, calculated within a realistic TB model [25], is given in Figure 3. The top

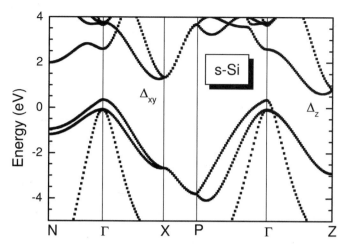

Fig. 3. Band structure of s-Si pseudomorphically grown on a Ge(001) substrate, calculated within a TB model [25].

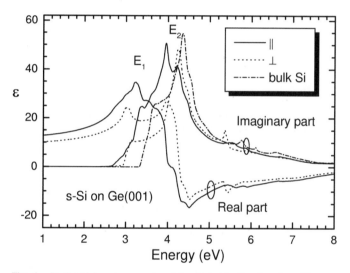

Fig. 4. Real and imaginary parts of the dielectric function for s-Si pseudomorphically grown on a Ge(001) substrate for polarizations parallel and perpendicular to the growth plane, calculated within a TB model [25]. The imaginary part of the dielectric function for bulk Si is also shown.

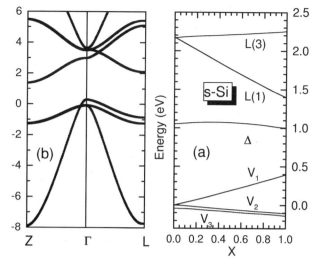

Fig. 5. (a) Energies for the three top valence states at Γ (V_1, V_2, V_3), and the lowest conduction states along the Δ_{xy}, Δ_z lines and at the L point of the BZ and (b) the band structure, for biaxially strained Si pseudomorphically grown on a $Si_{1-x}Ge_x(111)$ substrate, as a function of the Ge content, x, in the substrate. The results were obtained within a TB model [51].

Growth on a (111) substrate results in a rhombohedral D_{3d} structure. The band-edge energies calculated by deformation potential theory [51] are given in Figure 5. The Δ conduction-band minima remain equivalent, whereas the conduction-band minima at the L point split into a singlet $L(1)$ and a triplet $L(3)$ state: the singlet decreases in energy and the triplet increases as the lattice constant of the substrate increases. The calculated band structure [51] for s-Si pseudomorphically grown on a Ge(111) substrate is also given in Figure 5. The anisotropy between the bands along the ΓZ and ΓL directions, produced by the reduced symmetry, is clear. The top two valence bands along the ΓL direction exhibit an almost constant splitting that increases with Ge content in the substrate, x, and takes the value of about 0.4 eV for $x = 1$. On the contrary, parallel to the ΓZ direction, the splitting is quite small (~ 0.03 eV) and is independent of x.

Figure 6 shows the calculated dielectric function within the TB model [118] for s-Si grown on a Ge(111) substrate, and polarizations parallel and perpendicular to the growth plane. There is a strong anisotropy in the dielectric function for these two polarizations for energies up to about 5 eV. For larger energies, this anisotropy diminishes. The position of the E_2-like peak, as well as its width, differs significantly for the two polarizations: it is a sharp structure in perpendicular polarization and a broad structure in parallel polarization. The static dielectric constant $\varepsilon_1(0)$ for s-Si pseudomorphically grown on (111)-oriented $Si_{1-x}Ge_x$ substrates is given [51] in Figure 7 as a function of x. For polarization perpendicular to the growth plane, the static dielectric constant shows a slight variation with strain, whereas for polarization in the growth plane, $\varepsilon_1(0)$ changes more drastically, increasing almost linearly with strain.

of the valence band has a p_z character, and the higher valence bands have a splitting of about 0.3 eV along the Δ_{xy} direction and about 0.2 eV along the ΓN direction, whereas along the Δ_z direction, they are almost degenerate.

The dielectric function for s-Si pseudomorphically grown on a Ge(001) plane and obtained within a TB model [25] is shown in Figure 4, together with that of bulk Si. The present TB model describes well the available experimental data [50] in the entire energy spectrum except in a region close to the E_1 critical point, where exciton effects play an important role [25]. For s-Si, there is a strong anisotropy in the dielectric function for polarizations parallel and perpendicular to the growth plane for energies up to about 5 eV. For larger energies, this anisotropy diminishes. The most pronounced structure in ε_2 is the E_2-like peak. The position of this peak, as well as its width, differs significantly for the two polarizations: it is a sharp structure for perpendicular polarization and a broad structure for parallel polarization.

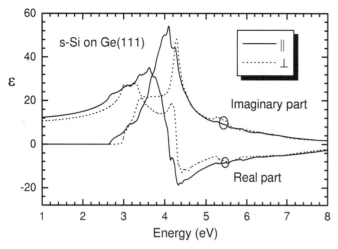

Fig. 6. Real and imaginary parts of the dielectric function for s-Si pseudomorphically grown on a Ge(111) substrate, calculated within a TB model [51], for polarization parallel and perpendicular to the growth plane.

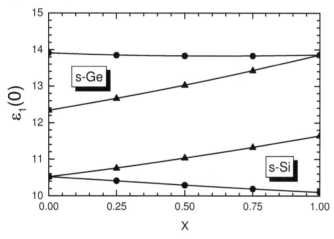

Fig. 7. Static dielectric constant, $\varepsilon_1(0)$, for s-Si and s-Ge pseudomorphically grown on a $Si_{1-x}Ge_x(111)$ substrate, as a function of the Ge content, x, in the substrate, for polarization parallel (▲) and perpendicular (●) to the growth plane, calculated within a TB model [51].

4.2. Uniaxial Strain

Uniaxial strain can be produced by an external pressure along the [001] and [111] directions. According to the usual convention, pressure is taken to be negative when the sample is compressed. The optical behavior of a material under uniaxial strain can be described by the piezooptical tensor $\{P_{ij}\}$. For cubic crystals that belong to the O, O_h, and T_d classes, there are three independent piezooptical tensor components [22, 48, 52] and they can be chosen in different ways: The most common choice is P_{11}, P_{12}, and P_{44}. Experimental results from Etchegoin et al. [45] and theoretical results from Theodorou and Tsegas [48] for the piezooptical tensor components P_{11} and P_{44} are shown in Figures 8 and 9, respectively. The main structures in P_{ij} appear in the regions around the critical points E'_0, E_1, and E_2. In the theoretical calculations, distinct structures at the critical points E'_0, E_1, and E_2 appear in P_1 and P_4; similar behavior is found in the results for the second deriva-

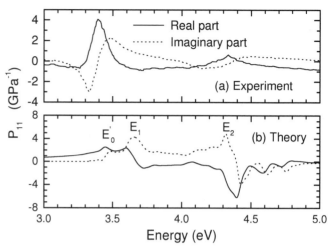

Fig. 8. Piezooptical tensor component P_{11} for Si. (a) Experimental results extracted from [45] and (b) theoretical results extracted from [48].

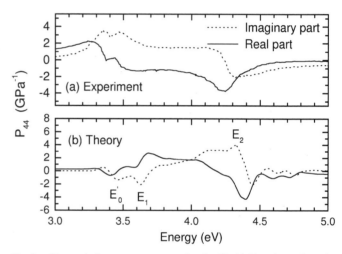

Fig. 9. Piezooptical tensor component P_{44} for Si. (a) Experimental results extracted from [45] and (b) theoretical results extracted from [48].

tive, $d^2\varepsilon/dE^2$, of the dielectric function. From the experimental point of view, Lautenschlager et al. [50] found distinct structures under pressure in $d^2\varepsilon/dE^2$ at the critical points E'_0, E_1, and E_2. As far as the piezooptical tensor components are concerned, Etchegoin et al. [45] did not find such distinct structures at critical points E'_0, E_1 for the component P_{11}, and found a broad structure at critical point E_2, whereas in P_{44}, they found small distinct structures in the range of critical points E'_0, E_1, and a well defined structure at critical point E_2. These differences require further investigation.

5. STRAINED Ge

The influence of external strain on the electronic and optical properties of Ge has been a subject of active research for a long time [20, 26, 35, 42, 43, 46, 47, 52–63]. Values for band extrema deformation potentials are given for Ge in Table III and for optical gaps E_1 and $E_1 + \Delta_1$ in Table IV. The experimental values for deformation potential D_1^5 refer to the triplet valley, because the experimental determination of D_1^5 was done with polarization parallel to the pressure direction, where only the triplet valley states contribute [52]. The theoretical results are given for both the triplet and the singlet valley, and the singlet valley deformation potential is larger than that for the triplet.

5.1. Biaxial Strain

Ge films with a biaxial (001) strain can be produced by pseudomorphic growth on a $Si_{1-x}Ge_x(001)$ substrate. The band-edge energies at the valence-band maxima (V_1, V_2, V_3), the conduction-band minima along the Δ direction, and the L point for strained Ge (s-Ge) were calculated as a function of the Ge content, x, in the substrate using deformation potential theory [25] and the results are shown in Figure 2b. Strain splits

Table III. Spin–Orbit Coupling Constant Δ_0 and Band Extrema Deformation Potentials for Ge

Parameter	Experiment	Theory
Δ_0	0.30^a	
a_v	—	1.24^b
b	-2.86 ± 0.15^c	-2.55^d
d	-4.85 ± 0.15^c	-5.50^d
$\Xi_d^L + \frac{1}{3}\Xi_u^L - a_v$	$-2.0 \pm 0.5^e, -3.8^f$	1.72^d
Ξ_u^L	16.2 ± 0.4^e	15.13^d

Note: All values are in eV.

^{a}Landolt–Börnstein [49].

^{b}Van de Walle [47].

^{c}Chandrasekhar and Pollak [26].

^{d}Van de Walle and Martin [46].

^{e}Balslev [53].

^{f}Paul and Warshauer [54].

the sixfold conduction-band minimum in the Δ direction into a doublet, Δ_z, and a quadruplet, Δ_{xy}. The quadruplet decreases in energy as the lattice constant of the substrate decreases. The conduction-band minimum changes position with strain, which is located in the Δ_{xy} direction for $x < 0.33$ and at the L point for $x > 0.33$. The band structure of s-Ge pseudomorphically grown on a Si(001) substrate, calculated within a realistic TB model [25], is given in Figure 10. The top of the valence band has a p_x, p_y character, and the higher valence bands along the ΓN direction split by strain by an almost constant amount of about 0.4 eV.

The dielectric function for s-Ge pseudomorphically grown on a Si(001) substrate, calculated within a TB model and calculated within a TB model [25], is shown in Figure 11, together with that of bulk (unstrained) Ge. The agreement between the present model and the available experimental data [64] for bulk Ge is very good in the entire energy spectrum [25]. For s-Ge, there is a strong anisotropy in the dielectric function for polarizations parallel and perpendicular to the growth plane for

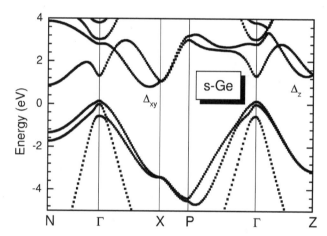

Fig. 10. Band structure of strained Ge pseudomorphically grown on a Si(001) substrate, calculated within a TB model [25].

Table IV. Optical Gaps E_1 and $E_1 + \Delta_1$ Deformation Potentials D_i^j for Ge

D_1^1	D_1^5	D_3^3	D_3^5	δ_{J_1}	δ_{J_2}	Ref.
$-9.6 \pm 0.8, -8.2 \pm 0.7$	11.3 ± 1.1	$5.8 \pm 0.6, 5.9 \pm 0.6$	6.1 ± 1.5	4.3 ± 0.4	2.0 ± 0.2	[26]
$-8.1 \pm 0.8, -9.9 \pm 0.5$	5.9 ± 1.2					[55]
-9.7 ± 1	7.5 ± 0.8	$2.2^{+1}_{-0.5}$	$1.5^{+0.6}_{-0.3}$			[56]
-7.8 ± 0.7	8.5 ± 0.8	2.6 (at $\mathbf{k} = 0$)	6.4 (at $\mathbf{k} = 0$)			[20]
-8.1 ± 0.8		5.9 ± 0.6		4.3 ± 0.4	2.0 ± 0.2	[57]
		3.0 ± 0.3	5.7 ± 0.6			[58]
-8.6	6.0					[59]
-9.5 ± 0.5						[60]
-10.7						[61]
$-10.4 \pm 0.5, -9.2 \pm 0.5$	8.5 ± 0.6	$3.4 \pm 0.3, 3.6 \pm 0.4$	$2.4 \pm 0.3, 2.5 \pm 0.4$			[62]
-8.6 ± 0.5	12.2 ± 0.5	$-5.6^{+0.6}_{-0.2}, -6.2^{+0.4}_{-0.1}$	-3.1 ± 0.9	6.5 ± 0.7	2.9 ± 0.3	[52]
-6.2	8.7 (triplet)	-2.8	-4.4 (triplet)			[63]
	9.7 (singlet)					[63]

Note: All values are in eV.

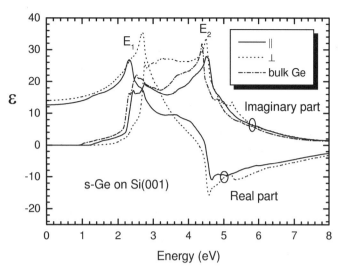

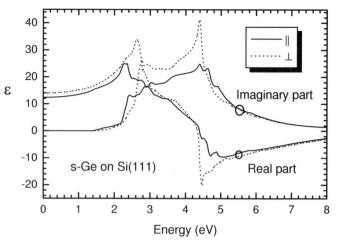

Fig. 11. Real and imaginary parts of the dielectric function for strained Ge pseudomorphically grown on a Si(001) substrate, for polarization parallel and perpendicular to the growth plane, calculated within a TB model [25]. The imaginary part of the dielectric function for bulk Ge is also shown.

Fig. 13. Real and imaginary parts of the dielectric function for strained Ge pseudomorphically grown on a Si(111) substrate, for polarization parallel and perpendicular to the growth plane, calculated within a TB model [51].

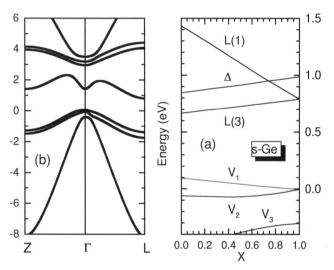

Fig. 12. (a) Variation in energy for the three top valence states at Γ (V_1, V_2, V_3), and the lowest conduction states along the Δ_{xy}, Δ_z lines and at the L point of the BZ and (b) the band structure, for biaxially strained Ge pseudomorphically grown on a $Si_{1-x}Ge_x(111)$ substrate, as a function of the Ge content, x, in the substrate, calculated within a TB model [51].

energies up to about 5 eV. For larger energies, this anisotropy diminishes. The most pronounced structure in ε_2 is the E_2-like peak around 4.2 eV. The position of this peak, as well as its width, differs significantly for the two polarizations: it is a sharper structure for the perpendicular polarization.

For s-Ge, pseudomorphically grown on a $Si_{1-x}Ge_x(111)$ substrate, the extremum energies as a function of the Ge content in the substrate, calculated by deformation potential theory [51], are given in Figure 12. The L point states split into a singlet $L(1)$ and a triplet $L(3)$ state, with the triplet lying at lower energies. The band structure for s-Ge pseudomorphically grown on a Si(111) substrate, calculated by a realistic TB model [51], is also given in Figure 12. The top valence bands

along the ΓL and ΓZ directions exhibit an almost constant splitting that increases with strain, and for $x = 0$ takes the values 0.2 and 0.36 eV for ΓZ and ΓL directions, respectively. The fourfold degeneracy at point X of the bulk Ge is lifted by strain (becoming the F point in the distorted BZ), and splits into two twofold-degenerate states with even and odd parity, respectively.

Figure 13 shows the calculated real and imaginary parts of the dielectric function in the TB model [51] for s-Ge grown on a Si(111) substrate, and polarizations parallel and perpendicular to the growth plane. There is a strong anisotropy in the dielectric function for polarizations parallel and perpendicular to the growth plane for energies up to about 4.5 eV. For larger energies, this anisotropy diminishes. The E_2-like peak appears around 4.5 eV. The peak for perpendicular polarization is quite strong and sharp, whereas for parallel polarization has almost half the strength with a considerably larger width and is split into two peaks. The static dielectric constant $\varepsilon_1(0)$ for s-Ge pseudomorphically grown on (111)-oriented $Si_{1-x}Ge_x$ substrates, as a function of x, is given [51] in Figure 7. For polarization perpendicular to the growth plane, the static dielectric constant shows a slight variation with strain, whereas for polarization in the growth plane, $\varepsilon_1(0)$ changes more drastically. In the latter case, it decreases almost linearly with strain.

5.2. Uniaxial Strain

Extensive investigations of the piezooptical properties of Ge have been performed by Etchegoin et al. [52] and Theodorou and Tsegas [63]. Figure 14 shows the experimental results of Etchegoin et al. [52] for P_{11}, together with the theoretical results of Theodorou and Tsegas [63]. There is some difference between theory and experiment. In particular, theoretical calculations predict a smaller strength for the E_1 structure and the E_0' structure located at a position of about 0.5 eV higher than in the experiment and with a larger strength. Apart from that, there is good agreement between theory and experiment. Figure 15

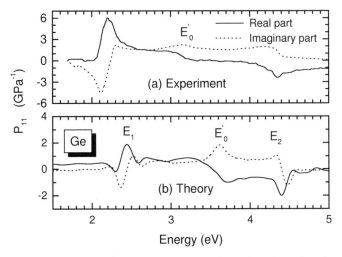

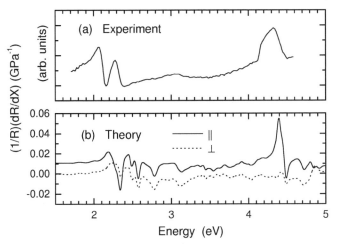

Fig. 14. Piezooptical tensor component P_{11} for Ge. (a) Experimental results extracted from [52] and (b) theoretical results extracted from [63].

Fig. 15. Differential reflectivity, $(1/R)(dR/dX)$, for polarization parallel and perpendicular to the pressure axis, for Ge under uniaxial pressure along the [111] direction. (a) Experimental results extracted from [52] and (b) theoretical results extracted from [63].

6. STRAINED $Si_{1-x}Ge_x$ ALLOYS

Binary group IV semiconductor alloys have attracted considerable attention because of their usefulness in electronic devices [2]. The quest to realize such materials is motivated by the desire to manipulate the bandgap of silicon, which can be achieved through alloying or by creating strained-layer superlattices (SLs). A prototypical and extensively studied system is silicon–germanium ($Si_{1-x}Ge_x$) alloys and SLs, usually grown on Si(100) substrates [2, 65–75].

Si and Ge crystallize in the diamond crystal structure with lattice constants equal to 5.431 and 5.657 Å for Si and Ge, re-

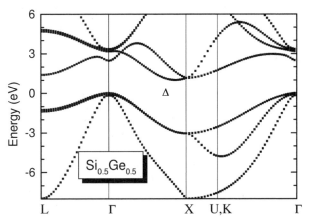

Fig. 16. Band structure of unstrained $Si_{0.5}Ge_{0.5}$ alloy, calculated in the virtual crystal approximation within a TB model [75].

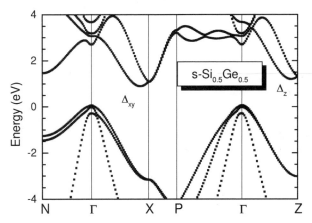

Fig. 17. Band structure of s-$Si_{0.5}Ge_{0.5}$ alloy pseudomorphically grown on a Si(001) substrate, calculated in the virtual crystal approximation within a TB model [75].

spectively, and form $Si_{1-x}Ge_x$ alloys with a lattice constant that is well described by Vegard's law, that is, by a linear interpolation between the lattice constants of bulk Si and Ge:

$$a(Si_{1-x}Ge_x) = (1 - x)a(Si) + xa(Ge) \qquad (26)$$

Elemental Si and Ge solids are indirect gap semiconductors that have a fundamental gap along the Δ direction with a value of 1.13 eV for Si and at the L point with a value of 0.76 eV for Ge. Figure 16 shows the band structure for unstrained $Si_{0.5}Ge_{0.5}$ alloy calculated by a TB model [75] in the virtual crystal approximation. The fundamental gap of that alloy is located along the Δ direction at about the position $k = 0.85(2\pi/a)$, where a is the lattice constant of the alloy, which has a calculated value equal to 1.0 eV.

For strained $Si_{1-x}Ge_x$ (s-$Si_{1-x}Ge_x$) alloys pseudomorphically grown on a Si(001) substrate, the sixfold degenerate conduction-band minimum along the Δ direction is split by strain into a twofold degenerate Δ_z minimum perpendicular to the growth plane and a fourfold degenerate Δ_{xy} minimum in the growth plane. Also heavy-hole (hh) and light-hole (lh) states at Γ are split by strain. For growth on a Si(001) substrate, the Δ_z minimum lies higher in energy than the Δ_{xy} minimum

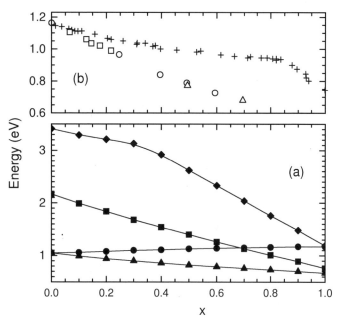

Fig. 18. (a) Variation of the direct gap (♦) and indirect gaps along the Δ_{xy} (▲) and Δ_z (●) directions as well as at the N (■) point, as a function of x, for s-Si$_{1-x}$Ge$_x$ alloys pseudomorphically grown on a Si(001) substrate. Theoretical results from [75]. (b) Experimental data for the fundamental gap of unstrained Si$_{1-x}$Ge$_x$ alloys (+) (from [71]) and s-Si$_{1-x}$Ge$_x$ alloys pseudomorphically grown on a Si(001) substrate (from [76] (□) and [67] (○ and △)).

and the hh state lies higher than the lh state. Figure 17 shows the band structure, calculated by a TB model [75], of s-Si$_{0.5}$Ge$_{0.5}$ alloy pseudomorphically grown on a Si(001) substrate, an indirect gap semiconductor with the conduction-band minimum at the point $k \cong 0.85 \times 2\pi/a$ along the Δ_{xy} direction, and a calculated gap equal to $E_g^i = 0.82$ eV. Figure 18 shows the variation of the calculated [75] direct and indirect gaps along the Δ_{xy} and Δ_z directions as well as at the N point, as a function of x, for s-Si$_{1-x}$Ge$_x$ alloys pseudomorphically grown on a Si(001) substrate. The same figure also shows experimental results for the fundamental gap of unstrained [71] as well as strained [67, 76] Si$_{1-x}$Ge$_x$ alloys pseudomorphically grown on a Si(001) substrate. For the unstrained alloys, the position of the absolute conduction minimum is located along the Δ direction for $x < 0.85$ and at the L point for $x > 0.85$ [65, 71, 75], whereas for strained alloys, the minimum of the conduction band is always along the Δ_{xy} direction. Also for strained alloys, the variation of the direct gap has a break at $x = 0.3$, which occurs because, for $x < 0.3$, the lowest conduction state at Γ is Γ_6^-, whereas for $x > 0.3$ it is Γ_7^-.

6.1. Optical Properties

Figure 19 shows theoretical results from a TB model [75] for the dielectric function for the s-Si$_{0.5}$Ge$_{0.5}$ alloy pseudomorphically grown on a Si(001) substrate. The same figure also shows the imaginary part of the dielectric function, ε_2, for unstrained (bulk) Si$_{0.5}$Ge$_{0.5}$ alloy. The consequence of the tetragonal crystal symmetry is an anisotropy between polarizations parallel

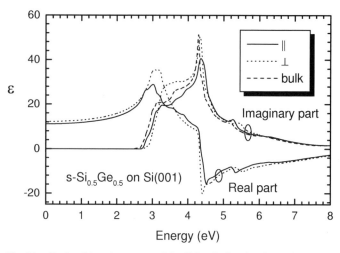

Fig. 19. Real and imaginary parts of the dielectric function for strained alloy Si$_{0.5}$Ge$_{0.5}$ pseudomorphically grown on a Si(001) substrate, for polarization parallel and perpendicular to the growth plane, as well as the imaginary parts of the dielectric function for bulk Si$_{0.5}$Ge$_{0.5}$ alloy. The results were obtained within a TB model [75].

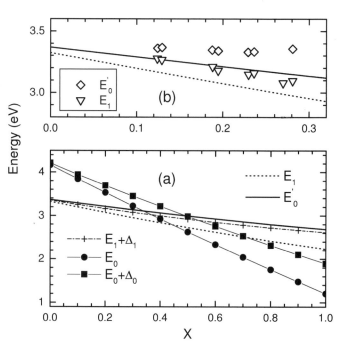

Fig. 20. (a) Calculated [75] critical energies E_0 (●), $E_0 + \Delta_0$ (■), E_1 (dashed line), $E_1 + \Delta_1$ (+), and E_0' (solid line) as a function of x for s-Si$_{1-x}$Ge$_x$ alloys pseudomorphically grown on a Si(001) substrate. (b) Experimental results [77] for E_1 (▽) and E_0' (◇) compared with the corresponding theoretical results [75].

and perpendicular to the growth plane. This anisotropy is significant for energies smaller than about 5 eV and diminishes for higher energies. Figure 20 shows the variation of the critical energies E_0, $E_0 + \Delta_0$, E_1, $E_1 + \Delta_1$, and E_0' as a function of x for s-Si$_{1-x}$Ge$_x$ alloys pseudomorphically grown on a Si(001) substrate. Critical point E_0 comes from vertical transitions at Γ between the highest valence band and the conduction state Γ_7^-, $E_0 + \Delta_0$ comes from transitions between the spin split-off band and the conduction state Γ_7^-, and E_0' comes from transitions between the highest valence band and the conduction Γ_6^- state. Critical point E_1 $(E_1 + \Delta_1)$ comes from vertical excitation between the highest (second highest) valence and the lowest conduction band near the N point. The same figure also shows experimental results [77] for the variation of E_1 and E_0'. The agreement between theory and experiment is very good. A similar variation with x is shown and the energy values differ by about 4%. The critical energies E_1 and $E_1 + \Delta_1$ show a variation similar to that of E_0'. The rapid variation of E_0 and $E_0 + \Delta_0$ with x comes from the corresponding rapid variation of the energy of the state Γ_7^-. The smaller variation with x of the state Γ_6^- implies a smaller variation of the critical energy E_0'.

7. STRAINED $Si_{1-y}C_y$ ALLOYS

The only stable solid compound that Si and C form in thermodynamic equilibrium is the equimolar strain-free zincblende 3C-SiC structure and the various stoichiometric polytypes [78, 79]. The solubility of carbon in silicon under equilibrium conditions is extremely low ($\leq 2 \times 10^{-3}$ at.% at its melting point [80]), because of the huge lattice constant mismatch and the cost in elastic energy (strained bonds) as carbon is incorporated into the lattice. Experimental efforts to overcome this obstacle have been based on nonequilibrium methods, such as growth of films by molecular beam epitaxy (MBE) [81, 82] and chemical vapor deposition (CVD) [83], which exploit the less constrained environment and the higher atomic mobility on surfaces.

For a long time, the electronic properties of $Si_{1-y}C_y$ alloys have been quite controversial. Theoretical arguments based on a linear interpolation scheme between the elemental bandgaps indicate that the incorporation of carbon, which in the diamond bulk phase has a larger bandgap than Si, yields a wider bandgap material [84, 85]. However, tight-binding-like quantum molecular dynamics calculations, based on localized atomic orbitals and the local density approximation, indicate that the $Si_{1-y}C_y$ alloys do not follow a mean-field-like (virtual crystal) behavior like $Si_{1-x}Ge_x$ alloys do and that the gap actually decreases with C content [86, 87]. Recent theoretical work by Theodorou et al. [88], based on a realistic TB model, presented detailed results for the electronic and optical properties of bulk as well as epitaxially strained $Si_{1-y}C_y$ alloys, and investigated the influence of the chemical effect, produced by the C content, and the epitaxial strain on the above-mentioned properties. In that investigation, the $Si_{0.984}C_{0.016}$ alloy was approximated by a $Si_{63}C_1$ supercell and the $Si_{0.968}C_{0.032}$ alloy was approximated by a $Si_{62}C_2$ supercell. The microscopic atomic structure of the alloy was determined using a semi-grand-canonical Monte Carlo method and empirical interatomic potentials [88–91].

Figure 21a presents the band structure [88] of the unstrained $Si_{0.968}C_{0.032}$ ($Si_{64}C_2$ supercell) alloy. The minimum of the conduction band is along the ΓX direction and the gap is equal to 0.826 eV. The same model applied to bulk Si, using a Si_{64} supercell, gave a value of 1.052 eV for the fundamental gap, implying that the fundamental gap for the unstrained $Si_{0.968}C_{0.032}$

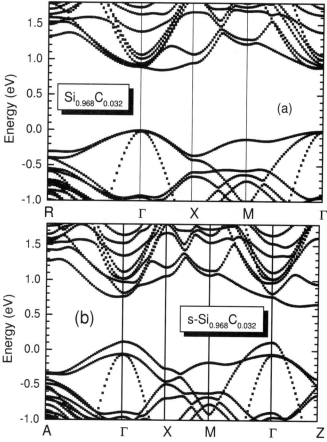

Fig. 21. Band structure for (a) unstrained $Si_{0.968}C_{0.032}$ ($Si_{62}C_2$ supercell) and (b) s-$Si_{0.968}C_{0.032}$ (s-$Si_{62}C_2$ supercell) alloy, calculated within a TB model [88], without taking into account spin–orbit coupling.

alloy is smaller than that of bulk Si, as suggested by experiments [92, 93] and other theoretical work [86, 94, 95]. The reduction is quantified by the coefficient $\gamma = -\Delta E_g(y)/y$, where y is the C content of the alloy. Theodorou et al. [88] gave a value $\gamma = 7$ eV for the unstrained alloys that represents the chemical contribution to the variation of the bandgap with C content in the alloy. The work of Demcov et al. [86] gave the result $\gamma = 28$ eV, a value that is exceedingly large.

Figure 21b shows the band structure [88] for strained $Si_{0.968}C_{0.032}$ (s-$Si_{0.968}C_{0.032}$) alloy, pseudomorphically grown on a Si(001) substrate. The absolute minimum of the conduction band is located along the ΓZ direction of the BZ for the tetragonal crystal structure. The strained alloy has a fundamental gap that is smaller than that of bulk Si by an amount $\Delta E_g = 0.53$ eV, giving a total value for the alloy $\gamma = 17$ eV, which contains both the chemical and the strain contribution. Whereas the chemical contribution to γ is 7 eV, the strain contribution is equal to 10 eV, giving a relative ratio of chemical to strain contribution of $7/10 = 0.7$.

Improvements [88] to the TB bandgaps produce a total value $\gamma = 13$ eV. An estimate for the strain contribution to γ can also be made by using deformation potential theory [88] and Si values for the deformation potentials. Such a calculation gives a strain contribution equal to 8 eV, implying a chemical contribution equal to 5 eV and a ratio of chemical to strain contribution equal to $5/8 = 0.62$, instead of 0.7. Experimental photoluminescence studies [92, 93] also revealed a strain contribution larger than the chemical one.

For a quantitative comparison with the experimental results, it must be taken into account that experiments were performed on a $Si_{1-y}C_y/Si(001)$ quantum well structure [92, 93], that is, a multiple quantum well structure with 52-Å-wide $Si_{1-y}C_y$ layers that represent the quantum well. In this case, a quantum well confinement effect should be taken into account, and with this correction, the value of γ for the quantum well structure becomes 9.6 eV. The reported measured value is $\gamma = 6.5$ eV. In addition, the carbon content for the alloy was estimated in the experiment by using Vegard's law. However, theoretical calculations [87, 89–91] and experimental measurements [96, 97] have shown that to obtain a reliable estimate of the carbon content in both bulk (unstrained) and epitaxially strained $Si_{1-y}C_y$ alloys, the significant deviations of lattice parameters from linearly interpolated values should be taken into account. Failure to do so results in an overestimation of the C concentration by about 30%. Correcting for this overestimation, the experimental value becomes $\gamma = 9.3$ eV, which is in excellent agreement with the theoretical results [88].

Band Offsets: Important parameters that are relevant to applications are the valence-band offset (VBO) and the conduction-band offset (CBO) at a $Si_{1-y}C_y/Si(001)$ interface. Experimental results [98] and theoretical calculations [88] predict a type I band alignment for the $Si_{1-y}C_y/Si(001)$ interface, where the holes and the electrons are localized in the $Si_{0.984}C_{0.016}$ layer, and a CBO of about 70% of the total band offset. For the $Si_{1-y}C_y/Si(001)$, the VBO and the CBO were found [88] to be 60 and 130 meV, respectively.

7.1. Optical Properties

Figure 22a shows the calculated [88] dielectric function for unstrained $Si_{0.984}C_{0.016}$ and $Si_{0.968}C_{0.032}$ alloys. The same figure also gives the calculated dielectric function for bulk Si. The consequences of the disorder are apparent: disorder leads to broader peaks and introduces a tail in ε_2 below the direct gap of Si at 3.4 eV. In addition, it was found that disorder shifts the E_2 peak, located at 4.3 eV, to lower energies, whereas the position of the E_1 gap, located at 3.5 eV, remains more or less unaffected.

For the s-$Si_{1-y}C_y$ alloys pseudomorphically grown on a Si(001) substrate, the dielectric function exhibits an anisotropy between polarizations parallel and perpendicular to the growth direction. The dielectric function for s-$Si_{0.984}C_{0.016}$ alloy is shown in Figure 22b for polarization parallel and perpendicular to the growth plane. The polarization anisotropy is relatively small, and the strongest part is between the critical points E_0' and E_2. The characteristics of the dielectric function curves are similar to those for the unstrained alloy.

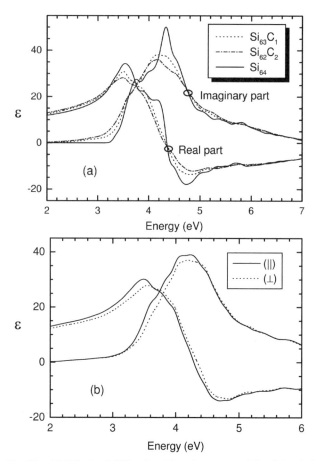

Fig. 22.   (a) Calculated [88] real and imaginary parts of the dielectric function for bulk Si_{64}, unstrained $Si_{0.984}C_{0.016}$ ($Si_{63}C_1$ supercell), and unstrained $Si_{0.968}C_{0.032}$ ($Si_{62}C_2$ supercell) alloys. (b) Calculated dielectric function for s-$Si_{0.984}C_{0.016}$ (s-$Si_{63}C_1$ supercell) alloy for polarizations parallel (solid line) and perpendicular (dotted line) to the growth plane.

8. Si/Ge SUPERLATTICES

In 1974, Gnutzman and Clausecker [99] predicted that the imposition of a proper new periodicity in an indirect gap semiconductor will fold the bands and may bring the minimum of the conduction band to the Γ point of the BZ, thus producing a direct gap material. Si is an indirect gap semiconductor with the minimum of the conduction band along the Δ direction and at about 80% of the distance between the Γ and the X point of the BZ. Imposition of a new periodicity that is five times the original could bring the conduction-band minimum to the Γ point. Modification of the lattice periodicity and construction of a direct gap material could be achieved by construction of Si/Ge SLs [25, 100–111]. Such a structure consists of a few atomic monolayers of Si and Ge that are stacked on top of each other in a periodic way. The relative band alignment of Si and Ge is such that the highest valence band is that of Ge and the lowest conduction band is that of Si [46]. This type of band alignment is called type II. A type II band alinement in a $(Si)_n/(Ge)_m$ SL implies that the highest valence band has mostly a Ge-like character and the lowest conduction band has mostly a Si-like character. Therefore, zone folding produced by a $(Si)_n/(Ge)_m$ SL grown along the [001] direction and having a period $n + m = 10$ is most likely to fold the Si-like conduction-band minimum to the Γ point. In addition to folding, to get a direct gap material, the conduction-band minimum in the growth plane must be shifted to higher energies than that along the growth axis. This shift can be achieved by straining the Si layers in the growth plane and thus splitting the Si-like sixfold-degenerate conduction-band minimum into a twofold-degenerate minimum, Δ_z, along the growth axis and a fourfold-degenerate minimum, Δ_{xy}, in the growth plane. Stretching (compressing) the Si layers in the growth plane reduces (increases) the energy of the Δ_z minimum relative to that of the Δ_{xy} minimum. Therefore, pseudomorphic growth of a $(Si)_n/(Ge)_m$ SL on a $(001)Si_{1-x}Ge_x$ substrate rich in Ge produces a direct gap material.

The large misfit of about 4.2% between bulk Si and Ge demands unconventional techniques to fabricate Si/Ge superlattices. Up to a critical thickness, the misfit can be accommodated by strain, whereas for larger thickness, the built-in strain is partly relaxed by the formation of misfit dislocations (for a review, see [4]). The critical thickness of Ge films pseudomorphically grown on a Si(001) substrate is about four to six monolayers. As a result, for $(Si)_n/(Ge)_m$ SLs pseudomorphically grown on a Si(001) substrate, the maximum value of m that can be used is 4. In addition, a second overall critical thickness exists that is equal to the total thickness of the superlattice, beyond which the material again partly relaxes by forming misfit dislocations. For $(Si)_6/(Ge)_4$ SLs pseudomorphically grown on a Si(001) substrate, the critical total thickness is about 20 nm [4]. To remove the second restriction of the critical total thickness, Kasper et al. [100] suggested strain-symmetrized or freestanding SLs. According to this idea, SLs are grown on a properly chosen substrate so that stresses in neighboring layers balance each other and the SL can stand alone after growth, because no external stress is applied. In particular, a strain-symmetrized $(Si)_n/(Ge)_m$ SL could be obtained by growth on an alloy $Si_{1-x}Ge_x(001)$ surface with $x \simeq m/(n+m)$ [46].

To calculate the electronic states of a $(Si)_n/(Ge)_m$ SL, the VBO between Si and Ge is needed. The VBO between strained Si and Ge pseudomorphically grown on a $Si_{1-x}Ge_x(001)$ alloy surface is given by [46]

$$\Delta E_v(x) = (1-x)\Delta E_v^{Si} + x\Delta E_v^{Ge} \tag{27}$$

where x is the concentration of germanium in the buffer alloy, $\Delta E_v^{Si} = 0.84$ eV is the VBO for growth on silicon, and $\Delta E_v^{Ge} = 0.31$ eV is the corresponding value for growth on germanium.

Figure 23 shows the band structure [25] of strain-symmetrized $(Si)_5/(Ge)_5$ SLs along symmetry lines of the BZ for the tetragonal D_{2d} crystal structure of the material that reveals a direct gap material with a gap equal to 0.75 eV. Figure 24 shows the variation for the direct and indirect gaps of $(Si)_5/(Ge)_5$ SLs pseudomorphically grown on a $Si_{1-x}Ge_x(001)$ buffer layer, as a function of Ge content, x, in the alloy [25]. As expected, the direct gap decreases with x, whereas the indirect gap increases.

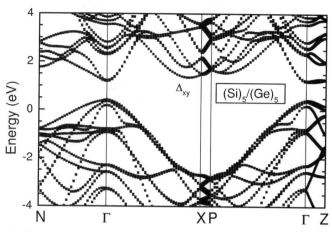

Fig. 23. Band structure of a strain-symmetrized $(Si)_5/(Ge)_5$ SL along symmetry lines of the BZ for the tetragonal D_{2d} crystal structure of the material [25].

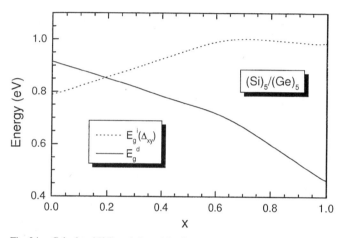

Fig. 24. Calculated [25] variation of the direct and indirect gaps of $(Si)_5/(Ge)_5$ SLs pseudomorphically grown on a $Si_{1-x}Ge_x(001)$ buffer layer, as a function of Ge content, x, in the alloy.

Figure 25 shows the calculated [31] dielectric function for strain-symmetrized $(Si)_5/(Ge)_5$ SLs together with that for unstrained (bulk) $Si_{0.5}Ge_{0.5}$ alloy [75]. As inset shows the imaginary part of the dielectric function for the SL close to the gap. From the figure it is evident that a strong polarization anisotropy exists for energies up to the critical point E_2 and diminishes at higher energies. At the low energy spectrum, between the energy gap of the SL (equal to 0.75 eV) and about ~ 2 eV, the imaginary part of the dielectric function for the SL is very low. This implies that transition probabilities from states near the top of the valence to those close to the bottom of the conduction band, at the Γ point, are very small. Calculations [112] showed that these transitions probabilities are 2–3 orders of magnitude smaller than typical values for GaAs. The reason is the fact that, in the SL, the lowest conduction states at Γ are produced by folding of the Si-like conduction-band minimum along the Δ_z direction. For bulk Si, transitions from the top of the valence to the bottom of the conduction band along the Δ line are forbidden (in the absence of electron–phonon interaction), because of momentum conservation. Folding brings the

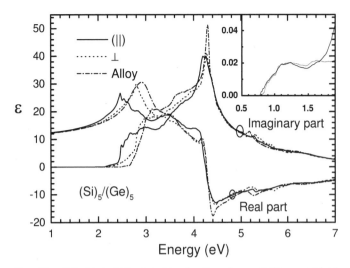

Fig. 25. Calculated [31] dielectric function for a strain-symmetrized $(Si)_5/(Ge)_5$ SL for polarization parallel and perpendicular to the growth plane. The calculated imaginary part of the dielectric function for unstrained $Si_{0.5}Ge_{0.5}$ alloy [75] is also shown. The inset shows the imaginary part of the dielectric function for the SL close to the gap.

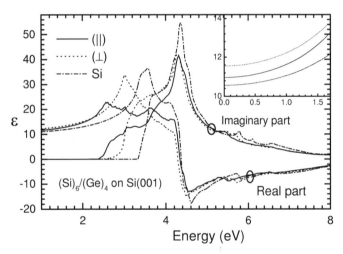

Fig. 26. Calculated [113, 114] dielectric function for $(Si)_6/(Ge)_4$ SLs pseudomorphically grown on a Si(001) surface, for polarization parallel and perpendicular to the growth plane, as well as that of bulk Si. The inset shows, for energies smaller than 1.5 eV, the real part of the calculated superlattice dielectric function for both polarizations, as well as that of Si.

conduction minimum from the Δ line to the Γ point and the superlattice potential makes the transitions allowed, but with a very small transition probability.

Figure 26 shows theoretical results [113, 114] for the dielectric function for $(Si)_6/(Ge)_4$ SLs pseudomorphically grown on a Si(001) substrate, as well as that of bulk Si. For energies smaller than 1.5 eV, the real part of the superlattice dielectric function is larger than that for bulk Si for both polarizations. In particular, the difference is approximately equal to 0.5 for polarization parallel to the growth plane and 1.0 for polarization perpendicular. This property could be utilized to produce electromagnetic confinement in a waveguide structure where the superlattice is the guiding material and Si is the cladding material.

8.1. Interface Intermixing

Photoluminescence (PL) measurements performed by Menczigar et al. [115] on SLs grown on a partly relaxed $Si_{1-x}Ge_x(001)$ alloy buffer showed strong luminescence in the infrared region. The luminescence signals were attributed to interband transitions of excitons localized at potential fluctuations in the superlattice. Menczigar et al. also observed a systematic shift of the bandgap energies to lower values with increasing period length. In the case of strain-symmetrized $(Si)_3/(Ge)_2$ SLs, the observed bandgap is identical to that of the corresponding $Si_{0.6}Ge_{0.4}$ alloy. The last result indicates the existence of interface intermixing. Schorer et al. [116], in Raman scattering experiments, found that phonon spectra of strain-symmetrized $(Si)_n/(Ge)_n$ SLs, with $n = 4, 5, 6, 8$, and 12, could not be explained with a model with abrupt-interface geometries. In addition, by taking into account interface intermixing, they recovered a general agreement between theory and experiment. Finally, measurements of the optical absorption coefficient, $\alpha(E)$, near the gap E_g of the strain-symmetrized $(Si)_n/(Ge)_n$ SLs gave a variation with energy proportional to $(E - E_g)^2$ that cannot be explained on the basis of ideal superlattices with sharp interfaces [117].

Theoretical studies [118] of the fundamental gaps for the ideal (abrupt interfaces) strain-symmetrized $(Si)_{10-n}/(Ge)_n$ SLs, as well as those of the corresponding bulk alloy $Si_{1-n/10}Ge_{n/10}$, showed that the ideal superlattices have a smaller gap than that of the corresponding alloys. This result can be explained as follows: Si/Ge SLs are type II superlattices, which means their valence band mostly is determined by their Ge layers and their conduction band is determined primarily by the Si layers. Because the valence band of Si is lower in energy than that of Ge, any diffusion of Si atoms into the Ge layers will modify the valence band of the superlattice in the direction approaching the valence band of Si, thus reducing its energy. In addition, the lowest conduction band of Ge is higher than that of Si and any diffusion of Ge atoms into the Si layers will modify the conduction band of the SL in the direction of the Ge conduction band, thus increasing its energy. It is, therefore, clear that the net result of interdiffusion of atoms across the interface is to widen the gap of the superlattice. Whereas the corresponding alloy of a given SL is the final stage of interface diffusion, it is clear that superlattices have a smaller gap than the corresponding alloys.

Figure 27 shows experimental results of Menczigar et al. [115] for the fundamental gap of strain-symmetrized $(Si)_{3n}/(Ge)_{2n}$ SLs. The $n = 0$ case represents the bulk alloy $Si_{0.6}Ge_{0.4}$. Theoretical results [118] for the ideal SLs, also included in the same figure, predict the correct variation of the fundamental gap with n, but with smaller values. The same figure also gives theoretical results for nonideal SLs, that is, SLs that have interface intermixing of the atoms. In particular, results were included for two sets of interface configurations: configuration A, for which the interface intermixing is extended into two atomic layers, one at each site of the interface, and configuration B, in which the intermixing is extended into four atomic layers, two at each site. In configuration A, the disorder

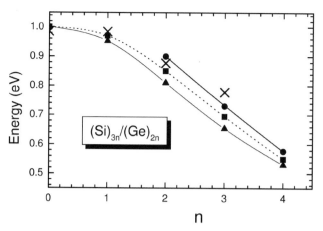

Fig. 27. Calculated [118] fundamental gap for strain-symmetrized $(Si)_{3n}/(Ge)_{2n}$ SLs vs. n. Results are given for ideal superlattices with abrupt interfaces (▲) and superlattices with disordered interfaces and interface configurations A (■) or B (●) as described in the text. In addition, the experimental data of Menczigar et al. [115] are shown (×). The $n = 0$ case refers to the bulk alloy $Si_{0.6}Ge_{0.4}$.

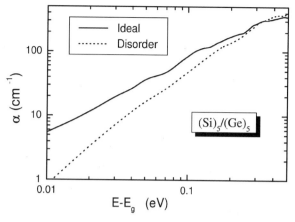

Fig. 28. Comparison between the calculated [120] absorption coefficients for strain-symmetrized $(Si)_5/(Ge)_5$ SLs with ideal interfaces to that of disordered interfaces of type B.

atomic planes at the interface are modeled so that 50% of the positions, chosen in a random way, are occupied by Ge atoms and 50% by Si atoms, whereas for configuration B, the positions in the first atomic layer from the interface are occupied by 50% Ge atoms and 50% Si atoms, and in the second layer, 75% of the positions are occupied by atoms of the host layer and 25% by atoms of the neighboring layer. As expected, the gap increases with the increase of interface intermixing. In particular, the experimental gap for $(Si)_3/(Ge)_2$ SLs is identical to that of the corresponding $Si_{0.6}Ge_{0.4}$ alloy, implying that the measured sample is practically an alloy. For the case of $(Si)_6/(Ge)_4$ SLs, the experimental gap is between the theoretical predictions for configuration A and configuration B SLs, being closer to the latter. For the case of strain-symmetrized $(Si)_9/(Ge)_6$ SLs, the experimental gap [115] is even larger than the theoretical predictions for configuration B superlattices, indicating that the sample used in the experiment is altered more drastically.

In addition to the gap, another characteristic quantity that describes the optical properties of a material is the absorption coefficient. Theoretical predictions [118] for the absorption coefficient for an ideal 5/5 SL, as well as a non-ideal 5/5 superlattice with type B disorder interface, are given in Figure 28. Near the band edge, the strength of the absorption for the disorder SL is lower than that of the ideal SL. In addition, intermixing also produces a smoother variation of the absorption coefficient. The ideal SL shows a near linear dependence, whereas for the disorder case, the dependence is nearly the square of the photon energy, in agreement with the experimental results [119, 120].

9. STRAINED GaAs AND InP

The influence of external strain on the electronic and optical properties of GaAs and InP has been extensively studied [20, 26, 43, 57, 121–130]. Values for band extrema deformation po-

tentials for GaAs and InP are given in Table V. Experimental and theoretical values for the deformation potentials for the optical gaps E_1 and $E_1 + \Delta_1$ are given in Tables VI and VII. There is a reasonable agreement between theoretical results and experimental data. The mentioned values for deformation potential D_1^5 refer to the triplet valley produced by strain along the [111] direction. Unlike Ge, Si, and GaAs, for which DP D_1^5 is positive [45, 48, 52, 63, 123, 124] and has values on the order of 10 eV, the corresponding experimental value for InP is negative, and has a small absolute value and a large error [125]. Theoretical calculations [124] predict a small positive value. On the other hand, if the experimentally measured value $D_1^1 = -9.1$ eV is used to analyze the tight-binding results, the value obtained [124] for the deformation potential D_1^5 is -2.4 eV, which is very close to the experimental result. These results imply that the absolute value of D_1^5 is small and its sign depends strongly on the value of D_1^1. The uncertainty in D_1^5 requires further investigation.

The piezooptical properties of GaAs and InP have been investigated experimentally in detail [123, 125] and a theoretical interpretation of the data was given [124] via the empirical pseudopotential and TB methods. The main points of these results follow.

9.1. GaAs

The calculated values [124], within a TB model, of the dielectric function for GaAs under uniaxial pressure of 1 GPa along the [001] direction, and polarization parallel and perpendicular to the pressure axis, are shown in Figure 29. For polarization parallel (perpendicular) to the pressure axis, the peak value at E_1 increases (decreases) with pressure, whereas the peak value at $E_1 + \Delta_1$ decreases (increases). This observation is in agreement with the experimental results [123]. Figure 30 shows experimental [123] and theoretical [124] results for the differential reflectivity with respect to pressure, $(1/R)(dR/dX)$, for pressure along [001], and for polarization parallel and perpendicular to the pressure axis. The results show a significant anisotropy between the two polarizations. In addition, several

Table V.   Spin–Orbit Coupling Constant Δ_0 and Band Extrema Deformation Potentials for GaAs and InP (all values are in eV)[a]

Compound	Δ_0	$a_c - a_v$	a_v	b	d
GaAs	0.34	−9.77, −6.70	1.16[b]	−1.7, −2.0	−4.55, −5.4, −5.3
InP	0.11	−6.35, −6.6	1.27[b]	−2.0, −1.55	−5.0, −4.2

[a] All values were taken from Landolt–Börnstein [49], unless otherwise indicated.

[b] From [47].

Table VI.   Deformation Potentials D_i^j for GaAs

D_1^1	D_1^5	D_3^3	D_3^5	δ_{J_1}	δ_{J_2}	Refs.
−7.6 ± 0.5	9.2 ± 0.9	3.4 ± 0.3	0.0 ± 0.5	10.2 ± 1.0	5.5 ± 0.8	(Exp.) [26, 57]
−7.9 ± 0.5		3.5 ± 0.3		9.9 ± 1.0	3.6 ± 0.5	(Exp.) [26, 57]
6.7 ± 0.5						(Exp.) [26, 57]
−6.9 ± 0.7	6.2 ± 0.6					(Exp.) [121]
−9.4 ± 0.9	8.5 ± 0.8	2.4 (at $\mathbf{k} = 0$)	8.5 (at $\mathbf{k} = 0$)	10	10	(Exp.) [20]
−8.0 ± 0.8		3.2 ± 0.3				(Exp.) [122]
−8.4 ± 0.8	12.0 ± 0.7	−5.4 ± 0.9, −4.3 ± 0.8	−6.4 ± 1.5	8.0 ± 1.5	6.1 ± 1.2	(Exp.) [123]
−6.4	4.0 (triplet)	−4.5	−10.7			(Theor.) [124]

Table VII.   Deformation Potentials D_i^j for InP

D_1^1	D_1^5	D_3^3	D_3^5	δ_{J_1}	δ_{J_2}	Refs.
−9.1 ± 1.4	−2.7 ± 6.4	−4.1 ± 0.3	−12.9 ± 2.5	3 ± 3	−25 ± 13	(Exp.) [125]
−2.06	5.84	−3.00	−6.59			(Theor.) [125]
−3.93	2.85	−3.05	−7.10			(Theor.) [125]
−6.9	0.4 (triplet)	−2.7	−7.3			(Theor.) [124]
	−2.4[a] (triplet)					(Theor.) [124]

[a] The value $D_1^1 = -9.1$ eV was used.

small peaks that were present in the calculations are smeared out in the experimental results, but overall there is reasonable agreement between theory and experiment. Finally, Figure 31 shows experimental [123] and theoretical [124] results for the piezooptical tensor component P_{11}. The important part of the main structure is located mainly between E_1 and E_2, with a very small value outside this region. The calculated spectrum is similar to that found in the experimental data. The main difference between the two spectra is in the peak intensities.

9.2. InP

The calculated values [124] for the imaginary part of the dielectric function for InP, within a TB model, under uniaxial pressure of 1 GPa, and polarization parallel and perpendicular to the pressure axis, are shown in Figure 32. The inset shows the calculated dielectric function at ambient pressure [124], together with the experimental results of Aspnes and Studna [131]. The main differences between theoretical and experimental results at ambient pressure are in the region of the E_2 structure. The

calculated E_2 peak is quite narrow and strong. In addition, at the low energy site of the E_2 peak, the calculated results show a shoulder, denoted by E_0', that was not present in the experiment. These two discrepancies appear in all optical properties calculated with the TB model under discussion. From the foregoing results, we obtain that, for polarization parallel (perpendicular) to the pressure axis, the peak value at E_1 increases (decreases) with pressure, whereas the peak value at $E_1 + \Delta_1$ decreases (increases). Figure 33 shows experimental [123] and theoretical results [124] for the piezooptical tensor component P_{11}. Again the calculated spectrum shows the structure E_0' that was not present in the experimental results. In addition, the calculated E_2 peak is too strong. Except for these two discrepancies, the agreement between theory and experiment is very good.

10. InAs/AlSb SUPERLATTICES

Heterostructure systems based on antimonide compounds and alloys have emerged as one of the most promising com-

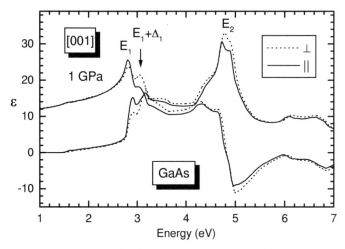

Fig. 29. Calculated [124] real and imaginary parts of the dielectric function for GaAs under 1 GPa pressure, and polarizations parallel and perpendicular to the pressure axis.

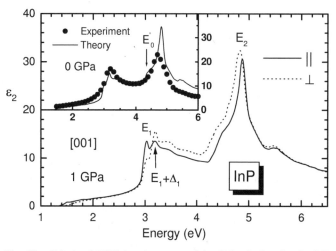

Fig. 32. Calculated [124] imaginary part of the dielectric function for InP under uniaxial pressure of 1 GPa, and polarization parallel and perpendicular to the pressure axis. The inset shows the calculated dielectric function at ambient pressure, together with the experimental results [131].

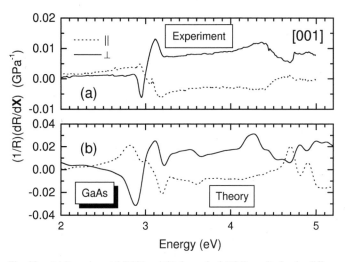

Fig. 30. (a) Experimental [123] and (b) theoretical [124] results for the differential reflectivity with respect to pressure, $(1/R)(dR/d\mathbf{X})$, for pressure along the [001] direction, and polarization parallel and perpendicular to the pressure axis, for GaAs.

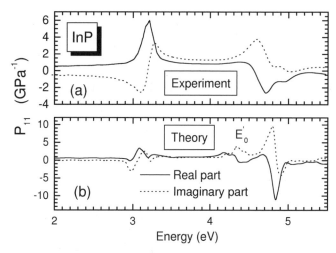

Fig. 33. Experimental [123] and theoretical [124] results for the piezooptical tensor component P_{11} for InP.

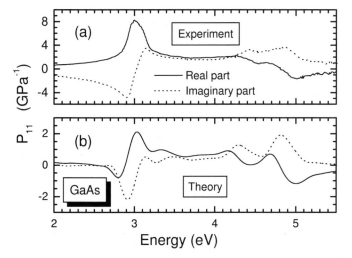

Fig. 31. (a) Experimental [123] and (b) theoretical [124] results for the piezooptical tensor component P_{11} for GaAs.

pounds for optical device applications. In particular, InAs/AlSb SL structures has been the subject of considerable attention, because of their interesting physical properties as well as their technological importance [18, 132–141]. The very large conduction-band offset [133] of 1.35 eV between InAs and AlSb, and the high electron mobility [8 × 10^5 cm^2/(V s) at 4.2 K] in the InAs compound make the materials potential candidates for use in the formation of key parts of high-speed electronic devices [142–144].

Bulk InAs and AlSb compounds crystallize in the zincblende structure with a small lattice constant difference of about 1.25% that makes the construction of InAs/AlSb SLs easy. The change of both anions and cations across an InAs/AlSb interface is a characteristic that makes the former interface different from those in GaAs/(Al, Ga)As and (Ga, In)As/(Al, In)As SLs. As a result, it is possible to obtain two different types of interfaces in the InAs/AlSb heterostructure [134]. The InSb-like interface, where the InAs layer terminates on an In monolayer and the adjoining AlSb layer starts with an Sb monolayer, and the AlAs-like interface, where the InAs layer terminates on an As monolayer and has an Al monolayer immediately after that. Therefore, InAs/AlSb SLs can be constructed to consist either of only one of the previously mentioned interface types or of alternating InSb/AlAs interfaces (IFs). InAs/AlSb SLs with InSb IFs only are denoted as type a SLs, those with only AlAs IFs are denoted as type b, and those with alternating InSb and AlAs IFs are denoted as type c. Ideal SLs that have sharp interfaces and an even number of atomic monolayers in the unit cell for each constituent material are type c [e.g., $(InAs)_6/(AlSb)_6$-c], and possess orthorhombic symmetry [140] (C_{2v} point-group symmetry), whereas ideal SLs with an odd number of atomic monolayers for both materials are either type a or b [e.g., $(InAs)_5In/Sb(AlSb)_6$-a or $As(InAs)_5/(AlSb)_6Al$-b] and possess tetragonal symmetry (D_{2d} point-group symmetry). In the former case SLs exhibit an optical anisotropy in the superlattice plane, whereas in the latter case, SLs should be isotropic in the layer plane. Experiments performed on InAs/AlSb SLs using spectroscopic ellipsometry and reflection difference spectroscopy [140] revealed the existence of an optical anisotropy, not only in structures with alternating AlAs/InSb interfaces, but also in those with only InSb or only AlAs interfaces. It has been proposed [141] that intermixing of the atoms across the interface is responsible for the observed anisotropy changes. However, the picture is not clear yet, especially for the case of SLs with only InSb interfaces. Further investigations, both theoretical and experimental, are necessary.

10.1. Electronic Properties

The band structure for $(InAs)_6/(AlSb)_6$-c SLs pseudomorphically grown on an AlSb substrate, calculated within a realistic TB model by Theodorou and Tsegas [145], is shown in Figure 34. The SL is a direct gap material, with a gap equal to 0.80 eV and orthorhombic symmetry that implies an anisotropy between the ΓX and ΓY directions that is evident in its band structure. In addition, the highest valence band in the ΓX direction for type-a SLs or in either the ΓX or the ΓY direction for type-c SLs is narrow. The states that belong to this band are localized at the InSb-like interfaces. In a primitive cell, there exist two InSb-like interface quantum wells for type-a SLs, one for type c, and none for type b. Figure 35 represents the probability amplitudes on the different atomic sites of the wave function for a state that belongs to the previously mentioned narrow valence band for the SL $(InAs)_5In/Sb(AlSb)_6$-a and that has a wave vector as indicated in the figure [145]. States that belong to the narrow valence band and have wave vectors along the [110] direction are localized in one of these InSb-like quantum wells, whereas states with **k** along the [$\bar{1}10$] direction are localized in the other. This can be understood as follows: In–Sb interface atoms form chains that are directed either along the [110] or the [$\bar{1}10$] direction, depending on the position of the interface. For type-c SLs, all In–Sb chains are directed along one of these directions, whereas for type-a SLs they are directed along

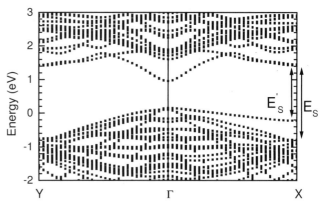

Fig. 34. Band structure [145] for $(InAs)_6/(AlSb)_6$-c SLs pseudomorphically grown on an AlSb substrate.

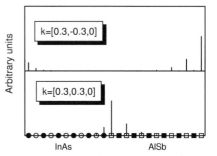

Fig. 35. Probability amplitude [145] (in arbitrary units) on different atomic sites in the unit cell for the upper valence state for $(InAs)_5In/Sb(AlSb)_6$-a SL with the **k** values as indicated in the figure. Circles denote InAs sites and squares denote AlSb sites; solid and hollow symbols denote cations and anions, respectively.

both. In the latter case, all chains that belong to equivalent interfaces are directed along the same direction, but chains that belong to neighboring, nonequivalent, interfaces are directed perpendicular to each other. Matrix elements that hop between neighboring In–Sb chains in the same interface are weak, because they are mediated through bonds to the InAs and AlSb layers, whereas elements that hop between neighboring atoms along the chain are strong. As a result, the band produced by the In–Sb interface chains is narrow in the direction perpendicular to the chain axis and wide along it. For instance, chains directed along [$\bar{1}10$] produce a narrow band along the [110] direction and a wide band along the [$\bar{1}10$] direction, and vice versa. The existence of localized interface states in InAs/AlSb SLs that are connected to InSb-like interface quantum wells was first proposed by Kroemer et al. [146] in response to a series of experimental studies [134] related to the observation of significant excess electron concentration. However, it is not clear that these localized states are, in fact, the source of the carriers [147–151].

Another interesting property, which initially was proposed by Dandrea and Duke [152], is that the energy gap for short-period InAs/AlSb SLs depends on the interfacial structure and can be varied by several hundreds of millielectronvolts by varying the interface structure, from type a to type b to type c SLs. The calculated [145] bandgaps of $(InAs)_nIn/Sb(AlSb)_{11-n}$-a,

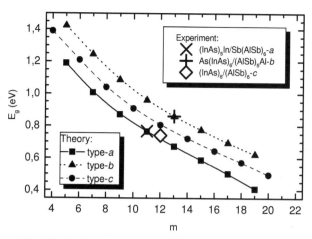

Fig. 36. Calculated [145] bandgap for $(InAs)_n In/Sb(AlSb)_{11-n}$-a, $As(InAs)_n/(AlSb)_{11-n}Al$-b, and $(InAs)_n/(AlSb)_{12-n}$-c SLs as a function of the InAs monolayers, m, in the primitive cell. The experimental results [141] for the energy gaps of SLs $(InAs)_5In/Sb(AlSb)_6$-a, $As(InAs)_6/(AlSb)_5Al$-b, and $(InAs)_6/(AlSb)_6$-c are also shown.

$As(InAs)_n/(AlSb)_{11-n}Al$-b, and $(InAs)_n/(AlSb)_{12-n}$-c SLs grown pseudomorphically on a AlSb(001) substrate, as a function of the number m of InAs monolayers in the unit cell (equal to $2n + 1$ for types a and b, and $2n$ for type c, are shown in Figure 36. These calculations [145] show that indeed the energy gap for type-b SLs is larger than that for the corresponding type-a SLs by about 0.2 eV, whereas type-c SLs have a variation in the middle. Figure 36 also shows the experimental results [141] for the energy gaps for $(InAs)_5In/Sb(AlSb)_6$-a, $As(InAs)_6/(AlSb)_5Al$-b, and $(InAs)_6/(AlSb)_6$-c SLs. There is good agreement between theory and experiment for the first two SLs, whereas the experimental value for the third SL almost lies on the curve that corresponds to type-a SLs. This discrepancy might indicate bad quality of the particular sample.

10.2. Optical Properties

The dielectric function, calculated within a TB model [145] and averaged over the three principal axes, is shown in Figure 37a for $(InAs)_6/(AlSb)_6$-c SLs. The critical points were obtained from the second derivative of the imaginary part of the dielectric function shown in Figure 37b and their energies are given in Table VIII. The same table also includes the critical point energies for the constituent bulk materials, as well as the corresponding average energy for each critical point. The positions of the SL critical points E_1, $E_1 + \Delta_1$, $E_2(X)$, and $E_2(\Sigma)$ are near the corresponding average energies for the InAs and AlSb critical points. The experimental energies for E_1 and $E_1 + \Delta_1$ are equal to 2.65 and 3.0 eV, respectively [141], in very good agreement with the calculated [145] values of 2.64 and 2.88 eV. Additional critical points, E_S and E_S', appear in the calculated spectrum and are connected to transitions from the valence to conduction bands along the ΓX and/or ΓY line as shown in Figure 34. The calculated [145] energy for E_S for $(InAs)_6/(AlSb)_6$-c SLs is 2.3 eV and the experimental value [141] is 2.45 eV. The transitions responsible for the E_S' critical point are from the

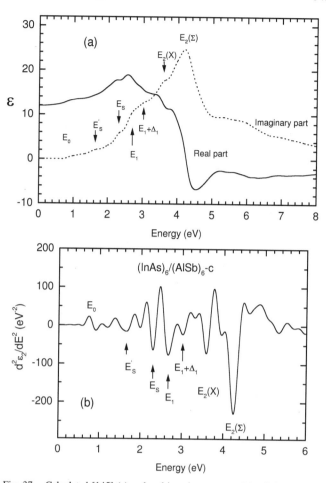

Fig. 37. Calculated [145] (a) real and imaginary parts of the dielectric function for $(InAs)_n/(AlSb)_{12-n}$-c SLs pseudomorphically grown on an AlSb(001) substrate and (b) second derivative of its imaginary part with respect to energy.

localized valence band to the lowest conduction band. The absence of the E_S' structure in the experimental results is probably caused by intermixing of the atoms at the interfaces that destroys, to a significant extent, the InSb interface quantum wells and the interface states produced by them. The absence could also be caused by relaxation effects, which are not included in the theoretical model. The calculated zero frequency dielectric constant, which corresponds to ε_∞, increases, as expected, with decreasing values of the energy gap, taking the values of 11.8, 11.95, and 12.1 for $As(InAs)_5/(AlSb)_6Al$-b, $(InAs)_6/(AlSb)_6$-c and $(InAs)_5In/Sb(AlSb)_6$-a SLs, respectively.

Finally, type-c SLs possess orthorhombic symmetry [140], with an optical anisotropy in the layer plane. The anisotropy differences between the principal axes [110] and [$\bar{1}10$] for ε_2 and the reflectivity are shown in Figure 38. The structure is similar in both ε_2 and reflectivity; Its strongest values are located mainly in the energy region between critical points E_0 and E_2.

11. SUMMARY

Thin strained films produced by pseudomorphic growth play a significant role in today's technology. In this review, an at-

Table VIII.   Critical Point Energies (in eV) for InAs, AlSb, and the SL $(InAs)_6/(AlSb)_6$-c Grown Pseudomorphically on a AlSb(001) Substrate

E_S	E_1	$E_1 + \Delta_1$	$E_2(X)$	$E_2(X) + \Delta_2$	$E_2(\Sigma)$	E_0'	E_2	Refs.
				InAs				
	2.40	2.66	4.14		4.47	4.76	5.19	(Theor.) [145]
	2.49	2.77			4.5 (E_0')	4.7 ($E_2(X)$)		(Exp.) [141]
				AlSb				
	2.68	3.15	3.37	3.64	4.18			(Theor.) [145]
	2.81	3.21		3.7 (E_0')	4.3			(Exp.) [141]
				$(InAs)_6/(AlSb)_6$-c				
2.3	2.65	3.0	3.6		4.25			(Theor.) [145]
2.45	2.64	2.88			4.15			(Exp.) [141]

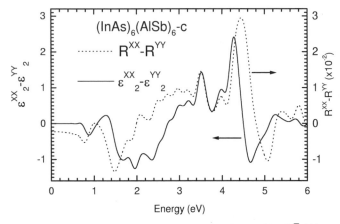

Fig. 38.   Calculated [145] anisotropy between directions [110] and [$\bar{1}$10] in ε_2 and reflectivity for $(InAs)_n/(AlSb)_{12-n}$-c SLs.

tempt was made to provide a flavor of the effects of strain on the electronic and optical properties of this vast class of materials. Groups IV materials and III–IV compounds are the constituents of the strained films that are described both macroscopically, using the deformation potential theory, and microscopically within the tight-binding model. The latter is appropriate to describe properties that are obtained from an integration in the entire Brillouin zone, like the optical properties, or systems with a large number of atoms per unit cell, like the superlattices and alloys. In addition, this model is unique in uncovering the physics of a complicated problem.

REFERENCES

1. T. S. Moss, Ed., "Handbook on Semiconductors," North-Holland, Amsterdam, 1980.
2. E. Kasper and K. Lyutovich, Eds., "Properties of Silicon, Germanium and SiGe:Carbon," EMIS Data Review Series, Vol. 24. INSPEC, London, 2000.
3. T. P. Pearsal, *Crit. Rev. Solid State Mater. Sci.* 15, 551 (1989).
4. E. Kasper and F. Schäffler, in "Semiconductors and Semimetals" (R. K. Willardson and A. C. Beer, Eds.), Vol. 33. Academic Press, New York, 1990.
5. R. K. Willardson and A. C. Beer, Eds., "Strained-Layer Superlattices: Physics of Semiconductors and Semimetals," Vol. 32. Academic Press, New York, 1990, and references therein.
6. H. Presting, H. Kibbel, M. Jaros, R. M. Turton, U. Menczigar, G. Abstreiter, and H. G. Grimmeiss, *Semicond. Sci. Technol.* 7, 1127 (1992).
7. J. C. Bean, *Proc. IEEE* 80, 571 (1992).
8. R. A. Soref, *Proc. IEEE* 81, 1687 (1993).
9. T. P. Pearsall, *Prog. Quant. Electron.* 18, 97 (1994).
10. W. J. Schaff, P. J. Tasker, M. C. Foisy, and L. F. Eastman, in "Semiconductors and Semimetals" (R. K. Willardson and A. C. Beer, Eds.), Vol. 33. Academic Press, New York, 1990.
11. H. Morkoc, B. Sverdlov, and G.-B. Gao, *Proc. IEEE* 81, 439 (1993).
12. S. Adachi, "Physical Properties of III–V Semiconductor Compounds," Wiley, New York, 1992.
13. R. Hull and J. C. Bean, in "Semiconductors and Semimetals" (R. K. Willardson and A. C. Beer, Eds.), Vol. 33. Academic Press, New York, 1990.
14. G. C. Osbourn, *J. Appl. Phys.* 53, 1586 (1982).
15. G. C. Osbourn, *IEEE J. Quantum Electron.* QE-22, 1677 (1986).
16. M. Cardona, in "Light Scattering in Solids II" (M. Cardona and G. Güntherodt, Eds.), p. 109. Springer-Verlag, Berlin, 1980.
17. B. J. Jusserand and M. Cardona, in "Light Scattering in Solids V: Superlattices and Other Microstructures" (M. Cordona and G. Güntherodt, Eds.), p. 49. Springer-Verlag, Berlin, 1989.
18. D. Toet, B. Koopman, P. V. Santos, R. B. Bergmann, and B. Richards, *Appl. Phys. Lett.* 69, 3719 (1996).
19. E. Hartfield and B. J. Thompson, in "Handbook of Optics" (W. G. Driscoll and W. Vaughan, Eds.), Chap. 17. McGraw–Hill, New York, 1978.
20. F. H. Pollak and M. Cardona, *Phys. Rev.* 172, 816 (1968).
21. G. L. Bir and G. E. Pikus, "Symmetry and Strain-Induced Effects in Semiconductors." Wiley, New York, 1974.
22. I. Balslev, in "Semiconductors and Semimetals" (R. K. Willardson and A. C. Beer, Eds.), Vol. 9. Academic Press, New York, 1972.
23. R. People and S. A. Jackson, "Strained-Layer Superlattices: Physics of Semiconductors and Semimetals" (R. K. Willardson and A. C. Beer, Eds.), Vol. 32. Academic Press, New York, 1990.
24. E. O. Kane, *Phys. Rev.* 178, 1368 (1969).
25. C. Tserbak, H. M. Polatoglou, and G. Theodorou, *Phys. Rev. B* 47, 7104 (1993).
26. M. Chandrasekhar and F. H. Pollak, *Phys. Rev. B* 15, 2127 (1977).
27. C. Herring and E. Vogt, *Phys. Rev.* 101, 944 (1956).
28. F. Wooten, "Optical Properties of Solids." Academic Press, New York, 1972.
29. N. V. Smith, *Phys. Rev. B* 19, 5019 (1979).
30. L. Brey and C. Tejedor, *Solid State Commun.* 48, 403 (1983).
31. C. Tserbak and G. Theodorou, *Phys. Rev. B* 50, 18, 179 (1994).

32. O. Jepsen and O. K. Andersen, *Solid State Commun.* 9, 1763 (1971).

33. G. Lehmann and M. Taut, *Phys. Status Solidi B* 54, 469 (1972).

34. L. D. Laude, F. H. Pollak, and M. Cardona, *Phys. Rev. B* 3, 2623 (1971).

35. G. W. Gobeli and E. O. Kane, *Phys. Rev. Lett.* 15, 142 (1965).

36. D. K. Biegelsen, *Phys. Rev. Lett.* 32, 1196 (1974).

37. K. Kondo and A. Moritani, *Phys. Rev. B* 14, 1577 (1976).

38. F. H. Pollak and G. W. Rubloff, *Phys. Rev. Lett.* 29, 789 (1972).

39. C. W. Higginbotha, M. Cardona, and F. H. Pollak, *Phys. Rev. B* 18, 4301 (1978).

40. M. H. Grimsditch, E. Kisela, and M. Cardona, *Phys. Status Solidi A* 60, 135 (1980).

41. D. K. Biegelsen, *Phys. Rev. B* 12, 2427 (1975).

42. M. Chandrasekhar, M. H. Grimsditch, and M. Cardona, *Phys. Rev. B* 18, 4301 (1978).

43. A. Blacha, H. Presting, and M. Cardona, *Phys. Status Solidi B* 126, 11 (1984).

44. Z. H. Levine, H. Zhong, S. Wei, D. C. Allan, and J. W. Wilkins, *Phys. Rev. B* 45, 4131 (1992).

45. P. Etchegoin, J. Kircher, and M. Cardona, *Phys. Rev. B* 47, 10,292 (1993).

46. C. G. Van de Walle and R. M. Martin, *Phys. Rev. B* 34, 5621 (1986).

47. C. Van de Walle, *Phys. Rev. B* 39, 1871 (1989).

48. G. Theodorou and G. Tsegas, *Phys. Status Solidi B* 207, 541 (1998).

49. O. Madelung, H. Schulz, and K. Weiss, Eds., "Landolt–Bornstein Numerical Data and Functional Relationships in Science and Technology," New Series, Vol. 17a. Springer-Verlag, Berlin, 1982.

50. P. Lautenschlager, M. Garriga, L. Viña, and M. Cardona, *Phys. Rev. B* 36, 4821 (1987).

51. C. Tserbak and G. Theodorou, *Phys. Rev. B* 52, 12,232 (1995).

52. P. Etchegoin, J. Kircher, M. Cardona, and C. Grein, *Phys. Rev. B* 45, 11,721 (1992).

53. I. Balslev, *Phys. Rev.* 143, 636 (1966).

54. W. Paul and D. M. Warschauer, in "Solids under Pressure" (W. Paul and D. M. Warschauer, Eds.). McGraw–Hill, New York, 1963.

55. U. Gerhardt, *Phys. Rev. Lett.* 15, 401 (1965).

56. D. D. Sell and E. O. Kane, *Phys. Rev.* 185, 1103 (1969).

57. M. Chandrapal and F. Pollak, *Solid State Commun.* 18, 1263 (1976).

58. I. Balslev, *Solid State Commun.* 5, 315 (1967).

59. R. L. Saravia and D. Brust, *Phys. Rev.* 178, 1240 (1969).

60. E. Scmidt and K. Vedam, *Solid State Commun.* 9, 1187 (1971).

61. Y. F. Tsay, S. S. Mitra, and B. Bendow, *Phys. Rev. B* 10, 1476 (1974).

62. J. Musilová, *Phys. Status Solidi B* 101, 85 (1980).

63. G. Theodorou and G. Tsegas, *Phys. Rev. B* 56, 9512 (1997).

64. L. Vina, S. Logothetidis, and M. Cardona, *Phys. Rev. B* 30, 1979 (1984).

65. R. Braunstein, A. R. Moor, and F. Herman, *Phys. Rev.* 109, 695 (1958).

66. J. P. Dismukes, L. Ekstrom, and R. J. Paff, *J. Phys. Chem.* 68, 3021 (1964).

67. D. V. Lang, R. People, J. C. Bean, and A. M. Serger, *Appl. Phys. Lett.* 47, 1333 (1985).

68. G. Abstreiter, H. Brugger, T. Wolf, H. Jorke, and H. J. Herzog, *Phys. Rev. Lett.* 54, 2441 (1985).

69. R. People, *Phys. Rev. B* 32, 1405 (1985).

70. Ch. Zeller and G. Abstreiter, *Z. Phys. B* 64, 137 (1986).

71. J. Weber and M. I. Alonso, *Phys. Rev. B* 40, 5683 (1989).

72. S. C. Jain, J. R. Willis, and R. Bullough, *Adv. Phys.* 39, 127 (1990).

73. M. M. Rieger and P. Vogl, *Phys. Rev. B* 48, 14,276 (1993).

74. Q. M. Ma, K. L. Wang, and J. N. Schulman, *Phys. Rev. B* 47, 1936 (1993).

75. G. Theodorou, P. C. Kelires, and C. Tserbak, *Phys. Rev. B* 50, 18,355 (1994).

76. D. Dutartre, G. Brémond, A. Souifi, and T. Benyattou, *Phys. Rev. B* 44, 11,525 (1991).

77. T. Ebner, K. Thonke, R. Sauer, F. Schaeffler, and H. J. Herzog, *Phys. Rev. B* 57, 15,448 (1998).

78. P. J. H. Denteneer and W. van Haeringen, *Phys. Rev. B* 33, 2831 (1986).

79. K. J. Chang and M. L. Cohen, *Phys. Rev. B* 35, 8196 (1987).

80. R. W. Olesinski and G. J. Abbaschian, *Bull. Alloy Phase Diagrams* 5, 485 (1984).

81. S. S. Iyer, K. Eberl, M. S. Goorsky, F. K. LeGoues, J. C. Tsang, and F. Cardone, *Appl. Phys. Lett.* 60, 357 (1992).

82. H. Rücker, M. Methfessel, E. Bugiel, and H. J. Osten, *Phys. Rev. Lett.* 72, 3578 (1994).

83. J. B. Posthill, R. A. Rudder, S. V. Hattangady, G. G. Fountain, and R. J. Markunas, *Appl. Phys. Lett.* 56, 734 (1990).

84. R. A. Soref, *J. Appl. Phys.* 70, 2470 (1991).

85. B. A. Orner and J. Kolodzey, *J. Appl. Phys.* 81, 6773 (1997).

86. A. A. Demkov and O. F. Sankey, *Phys. Rev. B* 48, 2207 (1993).

87. W. Windl, O. F. Sankey, and J. Menendez, *Phys. Rev. B* 57, 2431 (1998).

88. G. Theodorou, G. Tsegas, P. C. Kelires, and E. Kaxiras, *Phys. Rev. B* 60, 11,494 (1999).

89. P. C. Kelires, *Phys. Rev. Lett.* 75, 1114 (1995).

90. P. C. Kelires, *Appl. Surf. Sci.* 102, 12 (1996).

91. P. C. Kelires, *Phys. Rev. B* 55, 8784 (1997).

92. K. Brunner, K. Eberl, and W. Winter, *Phys. Rev. Lett.* 76, 303 (1996); K. Eberl, K. Bruner, and W. Winter, *Thin Solid Films* 294, 98 (1997).

93. O. G. Schmidt and K. Eberl, *Phys. Rev. Lett.* 80, 3396 (1998).

94. J. Xie, K. Zhang, and X. Xie, *J. Appl. Phys.* 77, 3868 (1995).

95. J. Gryko and O. F. Sankey, *Phys. Rev. B* 51, 7295 (1995).

96. M. Meléndez-Lira, J. Menéndez, W. Windl, O. F. Sankey, G. S. Spencer, S. Sego, R. B. Culbertson, A. E. Bair, and T. L. Alford, *Phys. Rev. B* 54, 12,866 (1996).

97. M. Berti, D. De Salvador, A. V. Drigo, F. Romanato, J. Stangl, S. Zerlauth, F. Schaffler, and G. Bauer, *Appl. Phys. Lett.* 72, 13 (1998).

98. R. L. Williams, G. C. Aers, N. L. Rowell, K. Brunner, W. Winter, and K. Eberl, *Appl. Phys. Lett.* 72, 1320 (1998).

99. U. Gnutzman and K. Clausecker, *Appl. Phys.* 3, 9 (1974).

100. E. Kasper, H. J. Herzog, H. Dambkes, and G. Abstreiter, *Mater. Res. Soc. Symp. Proc.* 56 (1986).

101. M. S. Hybertsen and M. Schluter, *Phys. Rev. B* 36, 9683 (1987).

102. S. Froyen, D. M. Wood, and A. Zunger, *Phys. Rev. B* 36, 4547 (1987); 37, 6893 (1988).

103. S. Satpathy, R. M. Martin, and C. G. Van de Walle, *Phys. Rev. B* 38, 13,237 (1988).

104. P. Friedel, M. S. Hybertsen, and M. Schluter, *Phys. Rev. B* 39, 7974 (1989).

105. T. P. Pearsall, *Crit. Rev. Solid State Mater. Sci.* 15, 551 (1989).

106. T. P. Pearsal, J. M. Vandenberg, R. Hull, and J. M. Bonar, *Phys. Rev. Lett.* 63, 2104 (1989).

107. T. P. Pearsall, J. Bevk, J. C. Bean, J. M. Bonar, J. P. Mannaerts, and A. Ourmazd, *Phys. Rev. B* 39, 3741 (1989).

108. R. Zachai, K. Eberl, G. Abstreiter, E. Kasper, and H. Kibbel, *Phys. Rev. Lett.* 64, 1055 (1990).

109. R. J. Turton and M. Jaros, *Mater. Sci. Eng. B* 7, 37 (1990).

110. U. Schmid, N. E. Cristensen, M. Alouani, and M. Cardona, *Phys. Rev. B* 43, 14,597 (1991).

111. J. Engvall, J. Olajos, H. G. Grimmeiss, H. Presting, H. Kibbel, and E. Kasper, *Appl. Phys. Lett.* 63, 491 (1993).

112. C. Tserbak and G. Theodorou, *J. Appl. Phys.* 76, 1062 (1994).

113. G. Theodorou, N. D. Vlachos, and C. Tserbak, *J. Appl. Phys.* 76, 5294 (1994).

114. G. Theodorou, C. Tserbak, and N. D. Vlachos, *J. Appl. Phys.* 78, 3600 (1994).

115. U. Menczigar, G. Abstreiter, J. Olajos, H. Grimmeiss, H. Kibbel, H. Presting, and E. Kasper, *Phys. Rev. B* 47, 4099 (1993).

116. R. Schorer, G. Abstreiter, S. de Gironcoli, E. Molinari, H. Kibbel, and H. Presting, *Phys. Rev. B* 49, 5406 (1994).

117. C. Tserbak and G. Theodorou, *Semicond. Sci. Technol.* 9, 1363 (1994).

118. G. Theodorou and C. Tserbak, *Phys. Rev. B* 51, 4723 (1995).

119. J. Olajos, J. Engvall, H. Grimmeiss, H. Kibbel, E. Kasper, and H. Presting, *Thin Solid Films* 222, 243 (1992).

120. T. P. Pearsall, L. Colace, A. DiVergilio, W. Jäger, D. Stenkamp, G. Theodorou, H. Presting, E. Kaspar, and K. Thonke, *Phys. Rev. B* 57, 9128 (1998).

121. D. D. Sell and S. E. Stokowski, in "Proceedings of the Tenth International Conference on the Physics of Semiconductors," p. 417. U.S. AEC, Oak Ridge, TN, 1970.

122. J. E. Rowe, F. H. Pollak, and M. Cardona, *Phys. Rev. Lett.* 22, 933 (1969).

123. P. Etchegoin, J. Kircher, M. Cardona, C. Grein, and E. Bustarret, *Phys. Rev. B* 46, 15,139 (1992).

124. G. Theodorou and G. Tsegas, *Phys. Status Solidi B* 211, 847 (1999).

125. D. Rönnow, P. Santos, M. Cardona, E. Anastassakis, and M. Kuball, *Phys. Rev. B* 57, 4432 (1998); erratum 59, 2452 (1999).

126. N. Suzuki and K. Tada, *Jpn. J. Appl. Phys.* 22, 441 (1983).

127. F. Canal, M. Grimsditch, and M. Cardona, *Solid State Commun.* 29, 523 (1979).

128. J. Camasel, P. Merle, L. Bayo, and H. Mathieu, *Phys. Rev. B* 22, 2020 (1980).

129. A. Gavini and M. Cardona, *Phys. Rev. B* 1, 672 (1970).

130. C. Priester, G. Allan, and M. Lannoo, *Phys. Rev. B* 37, 6519 (1988).

131. D. E. Aspnes and A. A. Studna, *Phys. Rev. B* 27, 985 (1983).

132. G. Tuttle, H. Kroemer, and J. H. English, *J. Appl. Phys.* 65, 5239 (1989).

133. A. Nakagawa, H. Kroemer, and J. H. English, *Appl. Phys. Lett.* 54, 1893 (1989).

134. G. Tuttle, H. Kroemer, and J. H. English, *J. Appl. Phys.* 67, 3032 (1990).

135. I. Sela, C. R. Bolognesi, L. A. Samoska, and H. Kroemer, *Appl. Phys. Lett.* 60, 3289 (1992).

136. C. R. Bolognesi, H. Kroemer, and J. H. English, *Appl. Phys. Lett.* 61, 213 (1992).

137. J. R. Waldrop, G. J. Sullivan, R. W. Grant, E. A. Kraut, and W. A. Harrison, *J. Vac. Sci. Technol. B* 10, 1773 (1992).

138. J. Spitzer, H. D. Fuchs, P. Etchegoin, M. Ilg, M. Cardona, B. Brar, and H. Kroemer, *Appl. Phys. Lett.* 62, 2274 (1993).

139. B. Brar, J. Ibbetson, H. Kroemer, and J. H. English, *Appl. Phys. Lett.* 64, 3392 (1994).

140. P. V. Santos, P. Etchegoin, M. Cardona, B. Brar, and H. Kroemer, *Phys. Rev. B* 50, 8746 (1994).

141. J. Spitzer, A. Höpner, M. Kuball, M. Cardona, B. Jenichen, H. Neuroth, B. Brar, and H. Kroemer, *J. Appl. Phys.* 77, 811 (1995).

142. G. Tuttle and H. Kroemer, *IEEE Trans. Electron. Devices* ED-34, 2358 (1987).

143. L. F. Luo, R. Beresford, and W. I. Wang, *Appl. Phys. Lett.* 53, 2320 (1988).

144. C. R. Bolognesi, M. W. Dvorak, and D. H. Chow, *J. Vac. Sci. Technol. A* 16, 843 (1998).

145. G. Theodorou and G. Tsegas, *Phys. Rev. B* 61, 10,782 (2000).

146. H. Kroemer, C. Nguyen, and B. Brar, *J. Vac. Sci. Technol. B* 10, 1769 (1992).

147. S. Ideshita, A. Furukawa, Y. Mochizuki, and M. Mizuta, *Appl. Phys. Lett.* 60, 2549 (1992).

148. D. J. Chadi, *Phys. Rev. B* 47, 13,478 (1993).

149. J. Shen, H. Goronkin, J. Dow, and S. Y. Ren, *J. Appl. Phys.* 77, 1576 (1995).

150. M. J. Shaw, P. R. Briddon, and M. Jaros, *Phys. Rev. B* 52, 16,341 (1995).

151. M. J. Shaw, G. Gopit, P. R. Briddon, and M. Jaros, *J. Vac. Sci. Technol. B* 16, 1794 (1998).

152. R. G. Dandrea and C. B. Duke, *Appl. Phys. Lett.* 63, 1795 (1993).

Chapter 12

GROWTH, STRUCTURE, AND PROPERTIES OF PLASMA-DEPOSITED AMORPHOUS HYDROGENATED CARBON–NITROGEN FILMS

D. F. Franceschini

Instituto de Física, Universidade Federal Fluminense, Avenida Litorânea s/n, Niterói, RJ, 24210-340, Brazil

Contents

1. INTRODUCTION

Since the suggestion by Liu and Cohen [1] of a hypothetical compound, carbon nitride or β-C$_3$N$_4$ (with the β-Si$_3$N$_4$ structure), whose mechanical properties would be similar to that of crystalline diamond, a strong research effort has been dedicated to the study of carbon–nitrogen-based solids [2, 3]. This effort constituted one of the first initiatives for a new way of doing research in Materials Science: the modeling of a new material with specific properties, for subsequent synthesis attempts.

Up to this moment, no successful synthesis of a single-phase crystalline solid with the β-carbon nitride structure has been reported. Nevertheless, this research effort led to the development of an entirely new class of materials, which have also found practical applications. This is case of the use of amorphous carbon nitride as a wear and corrosion protective coating for magnetic recording media [4].

The work on carbon nitride solids is strongly related to research on diamond-like carbon (DLC) materials [5, 6]. DLC materials are thin film amorphous metastable carbon-based

Handbook of Thin Film Materials, edited by H.S. Nalwa
Volume 4: Semiconductor and Superconductor Thin Films

ISBN 0-12-512912-2/$35.00

solids, pure or alloyed with hydrogen, which have properties similar to that of crystalline diamond (high hardness, low friction coefficient, high resistance to wear and chemical attack). This resemblance to diamond is due to the DLC structure, which is characterized by a high fraction of highly cross-linked sp^3-hybridized carbon atoms. To obtain this diamond-like structure at ambient conditions, DLC film deposition techniques always involve the bombarding of the film growing surface by fast particles with kinetic energy of the order of 100 eV. Since in β-C_3N_4 the carbon atoms should be sp^3 hybridized, one of the ways to search for carbon nitride was the study of nitrogen incorporation into DLC films.

One of the branches of this research has been the study of nitrogen incorporation in plasma-deposited hard amorphous hydrogenated carbon (a-C:H) films, the subject of this chapter. The research on plasma-deposited a-C(N):H films started even before Liu and Cohen's suggestion, with the study of electronic doping of a-C:H films reported by Jones and Stewart [7]. After that, followed the pioneering works of Han and Feldman [8], Amir and Kalish [9], and Kaufman, Metin, and Saperstein [10], which first reported important effects of nitrogen incorporation on the structure and the properties of a-C:H films. Further research on the field revealed some aspects of the nitrogen incorporation process in a-C:H films to place obstacles for obtaining an amorphous hydrogenated carbon nitride solid structurally analogous to β-carbon nitride. The first one is the limited nitrogen uptake observed in a-C(N):H films. No more than 30-at.% N incorporation could be achieved up to now. In fact, nitrogen contents lower than about 15-at.% N are the more frequent occurrence. The second one is that nitrogen incorporation results in a strong decrease in the sp^3-carbon atom fraction [11], which is the main responsibility for a-C:H film rigidity. Finally, one has the preferential bonding of hydrogen atoms to nitrogen atoms, which makes difficult the formation of a carbon–nitrogen extended network.

Despite such limitations, plasma-deposited a-C(N):H films were found to be used in a number of applications. The stress reduction induced by nitrogen incorporation [12] and consequent adhesion improvement, allowed the development of a-C(N):H antireflective coatings for Ge-based infrared detectors [13]. It was also found that N can electronically dope a-C:H films, and can strongly decrease the defect density, which gives prospects on its use as a semiconductor material [14]. Nitrogen incorporation was also found to decrease the threshold electric field in electron-field emission process [15], making possible the use of a-C(N):H films as an overcoat on emission tips in flat-panel display devices [16].

Research on plasma-deposited a-C(N):H films has been frequently included in the general discussion of carbon nitride solids [2, 3]. However, the presence of hydrogen in its composition, and the complexity of the deposition process, which introduces the nitrogen species in the already intricate hydrocarbon plasma-deposition mechanism, make a-C(N):H films deserve special consideration. This is the aim of the present work: to review and to discuss the main results on the growth, structure, and properties of plasma-deposited a-C(N):H films.

As this subject is closely related to a-C:H films, a summary of the main aspects relative to the plasma deposition of a-C:H films, their structure, and the relationship between the main process parameters governing film structure and properties is presented in Section 2. Section 2 is not intended to be a review, but only a guide to the main concepts and to the up-to-date key literature in the field.

Because they have been found to be strongly related, the nitrogen incorporation process and the growth kinetics are treated together in Section 3. The discussion of these themes includes the presentation of the main experimental results concerning nitrogen incorporation, the related effects of plasma chemistry and surface processes during film growth, and an attempt to model the growth kinetics.

In Section 4, the main aspects of the nitrogen-induced structural changes are presented, by the discussion of the most important characterization techniques. This presentation is complemented by an overview of a-C(N):H structure. Finally, in Sections 5 and 6, respectively, results concerning the mechanical properties, and the electrical and optical properties of a-C(N):H films are presented. As long as possible, they will be correlated with the observed structure changes.

2. AMORPHOUS HYDROGENATED CARBON FILMS

2.1. a-C:H Film Structure

The hard amorphous hydrogenated carbon films (a-C:H) belong to the class of materials called diamond-like carbon (DLC). These materials are designated as DLC due to their superlative properties, similar to that of crystalline diamond, such as high hardness, low friction coefficient, high chemical and wear resistance, and high electrical resistivity. These properties are closely related to the DLC film structure, which is formed mainly by sp^3- and sp^2-carbon atoms, with high sp^3 fraction (up to 80%), with a high degree of cross-linking of the amorphous network. Amorphous hydrogenated carbon films have a lower sp^3 fraction (up to 50%), in addition to the presence of a remarkable fraction of hydrogen atoms bonded to the network, in the 10- to 50-at.% range [5, 6].

The structure of the a-C:H films may be viewed, in a first approach, as a random covalent network [17, 18]. In this model structure, the fourfold sp^3 carbon, the threefold sp^2 carbon, and the singly coordinated hydrogen atoms, randomly bond with each other, forming a fully constrained network. To fulfil all of the constraints, the mean coordination number of the network should be equal to approximately 2.45 [17, 18]. Networks with higher mean coordination number are said to be overconstrained. The high sp^3-carbon atom fraction of hard a-C:H films leads to mean coordination numbers that are very often near 3 [19]. In this case, the formation of chemical bonds occurs with considerable distortion in chemical bond length and angle, giving rise to a strained amorphous network. This feature has been identified as the source of the high internal compressive stress [17], which is observed in all kinds of DLC. In DLC films, the internal stress can reach several GPa [5, 6], placing an obstacle to the production of thick films necessary for some applications. On the other hand, networks with mean coordination numbers smaller than 2.45 are said to be underconstrained, and are mechanically soft, like hydrocarbon polymers. The lower sp^2 fraction and the presence of a relatively high concentration of hydrogen atoms in their composition, make them less rigid and stressed than other kinds of DLC, such as highly tetrahedral, pure or hydrogenated amorphous carbons (ta-C and ta-C:H). These films can have hardness and density approaching that of crystalline diamond. In the C-H-sp^3-fraction ternary phase diagram, a-C:H films are situated near the rigidity border, at the side of hydrocarbon polymers [5].

Despite the diamond-like mechanical properties and electrical resistivity owned by a-C:H films, their optical properties are not comparable with that of crystalline diamond. Although a-C:H films were shown to be transparent in the infrared wavelength range (unless for the absorption bands due to, e.g., C—H groups), the transparency in the UV-vis wavelength range is remarkably smaller than that of diamond. The comparison of the bandgap as determined by optical absorption measurements—the optical gap—clearly shows the difference. The gap width of crystalline diamond is about 5.4 eV, whereas the gap width of hydrogenated diamond-like films are situated in the 1- to 4-eV [20] range. This relatively small gap observed in hydrogenated diamond-like (as also in other DLC materials) has been ascribed to the dominant role of the π electrons, arising from bond formation between sp^2-carbon atoms. Robertson and O'Reilly [21] and Robertson [22] proposed that the sp^2-carbon atoms in DLC films tend to cluster in the form of condensed sixfold aromatic rings, which also tend to cluster in the form of graphitic islands. Within this model, the bandgap width of DLC films is strongly correlated with the size of the π-bonded clusters, following an $M^{-1/2}$ dependence, where M is number of atoms in the cluster.

The structure of a-C:H films may be thus pictured as sp^2-carbon atoms in condensed aromatic clusters, dispersed in an sp^3-rich matrix, which confers to the network its characteristic rigidity. This situation can also be regarded as a random covalent network in which the sp^2 clusters of a defined size take part in the structure as an individual composed "atom" with its corresponding coordination number [17]. Such kinds of models have been successfully used to describe the dependence of a-C:H film mechanical properties on composition, hybridization, and sp^2 clustering [23].

2.2. a-C:H Film Deposition

It is well known that the stable crystalline form of carbon at ambient conditions is graphite, which is fully sp^2 hybidized. The synthesis of the fully sp^3-hybridized crystalline diamond is performed at high temperatures and pressures. So, the production of metastable carbon solids with a high fraction of sp^3-hybridized carbon at low temperatures and pressures is expected to require the use of out-of-equilibrium processes. In the case of DLC synthesis, this is achieved by submitting the growing film to the bombardment by energetic particles, as first reported by Eisenberg and Chabot [24]. This has been generally achieved by direct ion-beam deposition of C^+ ions accelerated with the required kinetic energy ($\sim$100 eV), by assisting the growth from a carbon vapor source with a noble gas ion beam, or by extracting and accelerating ions from a hydrocarbon plasma.

Roughly speaking, the condensation of the energy and momentum from the fast particle to the growing layer, locally generates conditions similar to the high temperatures and pressures needed for diamond synthesis. A more rigorous explanation is given by the subplantation model [25].

A version of the subplantation model [26] states different roles for the bombarding particles, as a function of their kinetic energies. Thinking on pure carbon deposition, for an intermediate energy range, at about 100 eV, the incoming fast particle penetrates into the subsurface, and stops at a lattice interstice, generating a local high-density state (sp^3 hybridization). For a lower energy range, the particle cannot penetrate into the surface, and is thus reflected or sticks to it, in the stable sp^2 configuration. For the higher energy range, the incoming ion generates a collision cascade, resulting in atom desorption and phonon activation, without generating a dense phase. This mechanism causes the structural parameters (sp^3 fraction, atomic density) and the mechanical properties (hardness, stress) to assume maximum values within a narrow ion-beam energy range around 100 eV.

Deposition of a-C:H films by plasma fundamentally retains the structure dependence on the bombarding ion energy. However, its mechanism is to some extent more complex, due to the presence in the deposition process species other than the atomic C^+ ions. In plasma deposition of a-C:H films, the diamond-like structure and the properties are determined by the hydrocarbon-derived molecular ions. However, film growth proceeds mainly by the attachment of neutral reactive radicals that result from hydrocarbon fragmentation [27]. So, in addition to the strong dependence of film structure on the bombarding ion energy, a strong dependence on the hydrocarbon used was also reported [28].

The reactions taking place at the growing surface of plasma-deposited a-C:H were reviewed by Jacob [29], and from this discussion emerged a framework to understand the a-C:H film deposition mechanism. This framework is to some extent equivalent to the subplantation model, since it emphasizes the role of energetic molecular ions. In addition, it takes into account the role of neutral radicals.

The previously mentioned picture of a-C:H deposition is based on some kinds of species coming from the plasma, and interacting with the film surface: Carbon-carrying neutral radicals and atomic hydrogen (slow fragments), plus carbon-carrying ions and hydrogen-derived ions. Each one of these species may play different roles in film growth, superimposing or competing with the other species. *Carbon-carrying ions* have low penetration depth and can stick to the surface, can displace hydrogen atoms (thus generating dangling bonds to which carbon-carrying radicals may stick), can recombine with dangling bonds, can activate lattice vibrations from direct energy transfer to film carbon atoms (consequently promoting

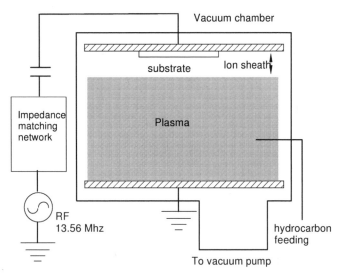

Fig. 1. Schematic of a-C:H plasma deposition by radio-frequency glow-discharge (RFPECVD).

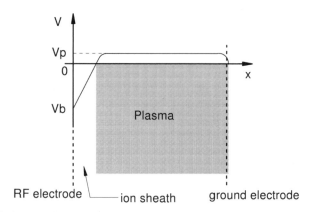

Fig. 2. Variation of the averaged in time electric potential across the interelectrode space.

carbon atom cross-linking). *Hydrogen ions* have a higher penetration depth than carbon (their range defines the thickness of the growth layer), can displace hydrogen atoms generating dangling bonds, and after stopping can hydrogenate sp^2-C atoms or can saturate dangling bonds. *Atomic hydrogen* can produce dangling bonds (by abstracting hydrogen atoms or by hydrogenating sp^2-carbon atoms), can erode carbon atoms at higher deposition temperatures, and its main role, can saturate dangling bonds. *Carbon-carrying radicals* react with surface dangling bonds contributing to the film growth, and thus saturate dangling bonds. Thus, the growth of a-C(H) films from hydrocarbon plasmas is a complex process and depends, in addition to the ion-beam energy, strongly on plasma chemistry details.

The radio-frequency glow-discharge method [30–34] has been the most used method in the study of a-C:H films. In this chapter, it is referred to as RFPECVD (radio frequency plasma enhanced chemical vapor deposition). Film deposition by RFPECVD is usually performed in a parallel-plate reactor, as shown in Figure 1. The plasma discharge is established between an RF-powered electrode and the other one, which is maintained at ground potential. The hydrocarbon gas or vapor is fed at a controlled flow to the reactor, which is previously evacuated to background pressures below 10^{-5} Torr. The RF power is fed to the substrate electrode through an impedance matching network, to assure maximum power transmission to the plasma. The coupling is made through a blocking capacitor, which is part of the matching network.

The presence of the blocking capacitor in the circuit gives rise to a great asymmetry on the electric potential within the interelectrode space. RF glow-discharge plasmas are divided into two regions, the plasma region (which emits light), and the ion sheath or cathode dark space. As referred to ground, the averaged-in-time electrical potential in the plasma region is small and positive, being zero at the grounded electrode. Across the ion sheath, there is a strong drop in the poten-

tial, which becomes negative, assuming the value V_B at the RF-powered electrode, as shown in Figure 2. The latter is the so-called self-bias potential V_B, which allows the extraction and the acceleration of the ions from the plasma toward the substrate electrode, giving rise to the ion bombardment necessary for diamond-like film growth. This effect is due to the much greater mobility of the plasma electrons, as compared to the ionic species. During an RF cycle, the electrons can move forth and back between the two electrodes, whereas the ions tend to be more concentrated near the RF electrode. This leads to an average-in-time positive net space charge, which polarizes the RF electrode negatively. The difference between the electron and ion motions also leads the traveling electrons to hit the RF electrode, neutralizing the positive charge built up at the RF electrode by the bombardment of the highly insulating a-C:H film. Otherwise, this positive charging would extinguish the discharge. These two effects—the self-bias potential and the charge neutralization—are essential to diamond-like a-C:H film growth, and are the reasons for the so widespread use of the RF-PECVD method.

The value of the self-bias potential was reported to strongly depend on the RF power and deposition pressure, following a $(W/p)^{1/2}$ dependence [30]. The reactor geometry also influences the value of V_B. As the area ratio between the RF electrode to the grounded electrode decreases, V_B is found to increase [30–34].

Although there are great advantages for the RFPECVD method, the direct current (dc) glow discharge has also been used for a-C:H film deposition [35]. By conveniently choosing a sufficiently high dc bias, a-C:H films as thick as 3 μm could be grown. This is possible because the electrical resistivity of a-C:H films greatly decreases with the applied electrical field, allowing the dc bias to reach the substrate surface. A variation of the RF flow-discharge method was reported, in which the substrate is placed in a negatively dc-biased electrode parallel and opposed to the RF-powered one [10].

There has been an increased interest in the use of denser plasmas, such as electron cyclotron resonance (ECR) plasmas for a-C:H film deposition [34, 36]. In ECR plasmas, the plasma discharge is sustained by microwave radiation (2.45 GHz),

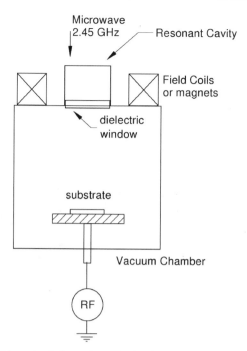

Microwave 2.45 GHz — Resonant Cavity

Field Coils or magnets

dielectric window

substrate

Vacuum Chamber

RF

Fig. 3. Schematic of plasma deposition by electron cyclotron resonance with an RF-biased substrate (ECRRF).

which is fed into the deposition chamber through a dielectric window placed at the end of a resonant cavity, as shown in Figure 3. In addition, a steady magnetic field is applied in the region near the window, to generate cyclotron resonance conditions for the electrons in the plasma. This greatly enhances the collisional process, thus increasing plasma ionization and fragmentation. This allows performing deposition at pressures lower than is possible in RFPECVD. The ion energies coming from ECR plasmas are far below 100 eV, the typical energy for the production of diamond-like films. To give the required kinetic energy to the plasma ions, the substrate is placed over a dc- or RF-biased electrode. In this way, the ions are extracted from the plasma and are accelerated toward the substrate. In addition to the increased plasma density, this method also allows the variation of bombarding ions energy, without introducing great changes in the energy fed to the plasma, as happens in RFPECVD. In this chapter, ECR plasmas with RF substrate biasing are referred to as ECRRF.

The use of the dense plasmas of a helical resonator plasma source [37] was reported, also together with RF biasing of the substrate electrode. A variation of the RF-glow-discharge method, in which a steady magnetic field is perpendicularly imposed to the RF-biased electrode surface, was used to deposit films with a gaseous mixture largely diluted in helium, to increase the plasma electron temperature [14]. Production of a-C:H films was also performed by using highly ionized plasma sources such as the plasma beam source [38, 39] or hydrocarbon and nitrogened ion-beam sources [40], to deposit or to assist the deposition process.

Since the main parameter influencing diamond-like carbon film structure is the energy of bombarding ions, it is expected

that the same happens with a-C:H films. In fact, it was found that in RFPECVD deposition of a-C:H films, the variation of substrate self-bias results in strong changes of film growth, composition, structure, and properties.

First, as the variation of self-bias in RFPECVD is performed by changing the RF power delivered to the plasma, in addition to changing the ion energy, it also changes the plasma fragmentation. So, when the self-bias increases, an increase on the production of ions as well as of reactive neutral radicals is also expected. As a consequence, the deposition rate is found to increase for increasing V_B [31–33, 33]. For higher values of V_B, a decrease in the deposition rate has been reported [33] and was attributed to the onset of film sputtering by ions with higher energy.

The hydrogen content of a-C:H films was found to continuously decrease with increasing self-bias, for films deposited by RFPECVD and ECRRF dual plasma, as a consequence of the increasing bombarding energy [31–33, 41]. As expected, the C-atom hybridization was found to strongly correlate with the self-bias, although in a more complex way than hydrogen-free DLC films.

Tamor, Vassel, and Carduner [19] measured by ^{13}C-nuclear magnetic resonance spectroscopy (^{13}C-NMR) the evolution of the C-atoms hybridization against the self-bias, in the up to −1000-V range, for a-C:H films deposited by RFPECVD in CH$_4$ atmospheres. As the self-bias increased, the overall carbon sp^3 fraction was found to continuously decrease. This in principle is contradictory with the subplantation model, from which an sp^3 fraction should pass through a maximum. However, the authors were able to extract from the NMR spectra, the contribution of each kind of carbon atoms present in the sample. They reported, upon increasing self-bias, a continuous increase on the sp^2 fraction of hydrogenated carbon atoms, together with the corresponding decrease of the sp^3 fraction of these atoms. On the other hand, the sp^3 fraction of nonhydrogenated carbon atoms was found to increase until reaching a maximum value of 20% at −200-V self-bias, and finally to decrease for higher self-bias. At the same time, the sp^2 fraction of nonhydrogenated carbon atoms was found to continuously increase. It must be thus noted that in plasma-deposited a-C:H films, the generation of quaternary carbon atom bonds occurs in parallel with the generation of sp^2 bonds that follows the dehydrogenation process.

The mechanical properties were generally found to follow the behavior described earlier. Amorphous hydrogenated carbon films deposited with very low self-bias, say below −100 V, are reported to be highly hydrogenated, mechanically soft, and nonstressed, being known as polymer-like a-C:H [31–33]. As the self-bias increases, the films become hard and stressed. Films deposited from methane show a clear maximum in mechanical hardness, in the −200- to −400-V self-bias range [32, 41]. As expected, the film density also follows the passing through a maximum behavior [41].

Films deposited from hydrocarbons with a higher number of carbon atoms, e.g., C$_6$H$_6$, show shifts in the position of the hardness and density maximums to higher self-bias [41]. Such behavior was ascribed to the fact the most abundant molecular ions that bombard the film growing surface are different for different hydrocarbons. In methane, the most abundant ions are expected to be the CH_n^+ ions, for acetylene, the most abundant ions are expected to be the $C_2H_n^+$ ions, for benzene, the most abundant ions are expected to be the $C_6H_6^+$ ions, and so on. Upon colliding with the surface, the molecular ions break up, being the carried energy shared between their constituent atoms. In this way, the energy delivered to the film surface is smaller as high is the number of carbon atoms, giving rise to a displacement in the optimum self-bias position to higher values [26].

The optical gap of a-C:H films was found to continuously decrease with increasing self-bias [42]. The gap shrinking was found to be strongly correlated to the variation of Raman spectra that is related to the increase of the graphitic clusters present in the a-C:H films. Accordingly, the electrical resistivity of C:H films was found to strongly decrease with substrate bias [43].

The deposition pressure also affects the structure and the properties of a-C:H, since it may significantly alter the ion energy. When accelerated across the ion sheath, the ions may collide with gas molecules and may lose energy, which no longer is equal to eV_B. Bubenzer et al. [31] suggested the ion energy to vary proportionally to $(V_B/p^{1/2})$. The main effect of the pressure increase in RFPECVD a-C:H deposition is to make the films more polymer-like, i.e., with increased hydrogen content and lower density, hardness and stress [32, 33]. To obtain hard a-C:H films, lower pressures are needed, assuring in this way a more effective bombardment.

3. NITROGEN INCORPORATION INTO a-C:H FILMS

3.1. Determination of Chemical Composition

The study of a-C(N):H films requires methods of chemical analysis capable of determining carbon, hydrogen, nitrogen, and major contaminants, and at the same time give information on the composition profile through the films. Although there have been reports on the use of conventional, bulk chemical analysis methods, like combustion analysis, surface and thin film analysis methods are preferred. This includes methods based on electron spectroscopy, like X-ray photoelectron spectroscopy (XPS) and Auger electron spectroscopy (AES) [44–46]; and nuclear techniques based on MeV-energy ion-beam scattering [44, 47]. The methods based on electron spectroscopy suffer from some limitations. First, they probe the very surface chemical composition, which may differ from film bulk composition. In addition, they cannot detect hydrogen. Their advantage is the additional ability to probe changes on the electronic structure, due to changes in the chemical bond environment.

The use of nuclear techniques allows the determination of C, N, H, O, and heavier contaminants relative fractions with great accuracy, and of the elements depth profile with moderate resolution (typically 10 nm). Rutherford backscattering spectroscopy (RBS) of light ions (like alpha particles) is used for

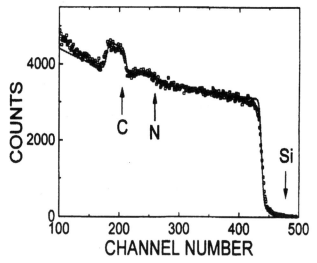

Fig. 4. RBS spectrum of a 12.5-at.% N a-C(N):H film. The continuum line is a computer simulation of the spectrum. (Reproduced from [59].)

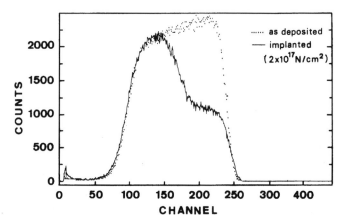

Fig. 5. ERDA spectra for as-deposited and N-implanted a-C(N):H films. (Reproduced from [47].)

the determination of carbon and heavier elements. Hydrogen contents are measured by forward scattering of protons by incident alpha particles (ERDA) elastic recoil detection analysis [44, 47].

Since the nuclear and electronic scattering cross sections for alpha particles are well known, the relative concentrations of the elements and their depth profiles can be easily obtained. The relative element concentrations are determined by the relative scattering intensities. The depth profile is obtained from the energy spread of the scattered particles, which lose energy before and after the nuclear collision, by inelastic scattering with electrons. The knowledge of the elements areal density and of the film thickness allows the determination of film density.

Figure 4 shows an RBS spectrum obtained from an a-C(N):H sample, together with spectrum simulation by a computer code, which allows the determination of the areal density of each element. To make this kind of analysis, previous knowledge of the hydrogen content is necessary [47]. Figure 5 shows ERDA spectra of a-C(N):H films, as deposited and after high-energy nitrogen ion implantation. In this figure, the hydrogen deple-

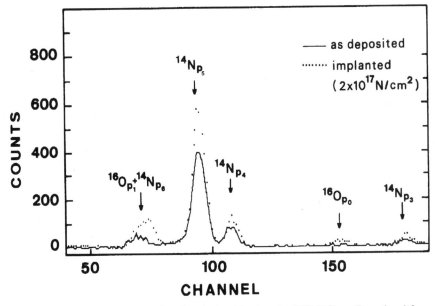

Fig. 6. Nuclear reaction spectra of as-deposited and implanted a-C(N):H films. (Reproduced from [47].)

tion profile is caused by the ion implantation process. Nitrogen contents of a-C(N):H films are very often rather low, making the nitrogen RBS signal small as compared to the carbon signal, and even smaller as compared to the substrate (usually Si single crystal wafers), as shown in Figure 4. To improve the accuracy in the determination of nitrogen, sometimes a nuclear reaction is used to obtain an independent measure of nitrogen content [46, 47]. This is the case for the $^{15}N(d, p)^{14}N$ reaction, which is exited by a deuteron beam with a 610-keV energy. Figure 6 shows the products of the nuclear reactions taking place upon deuteron irradiation, for the same samples shown in Figure 5. As occurs with hydrogen, nitrogen is also depleted by ion implantation treatment.

3.2. Nitrogen Incorporation and Growth Kinetics

RFPECVD (see Section 2.1) in atmospheres composed of mixtures of a hydrocarbon gas or vapor and a nitrogen-containing gas has been the most used deposition technique in the investigation of a-C(N):H films [8–10, 12, 14, 48–58]. Several hydrocarbon gases and vapors (methane, CH_4, acetylene, C_2H_2, butadiene, C_4H_{10}, cyclopentane, C_5H_{10}, benzene, C_6H_6) and nitrogen-containing gases and vapors (N_2, NH_3, CH_3NH_2) have been used in the precursor mixtures, within a wide range of deposition parameters. Due to the strong changes introduced on the deposition process and on film composition, structure and properties, the main focus of most works has been the study of the influence of nitrogen precursor partial pressure. Relatively few works have considered the effects of substrate bias or total deposition pressure.

In Table I are shown representative data concerning plasma deposition of a-C(N):H films, produced by different methods and under different gaseous mixtures and deposition parameters. In this table are displayed data on chemical composition

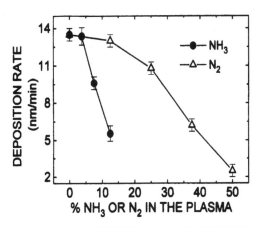

Fig. 7. Deposition rate of a-C(N):H films deposited by RFPECVD, as a function of NH_3 or N_2 partial pressure. (Reproduced from [56].)

(maximum N content, and range of variation of H content), deposition details (deposition pressure, self-bias, and atmosphere composition), and the method used for chemical composition determination.

In addition to the intended nitrogen incorporation, the addition of a nitrogen-containing gas to the deposition atmosphere also affects the film growth kinetics. It is always reported that a-C(N):H film deposition rates steeply decrease with increasing nitrogen-containing gas partial pressure [56], as shown in Figure 7. It must be noted that the observed decrease in the deposition rate is not a consequence of a pure dilution effect, as one could conclude from the fact that solid nitrogen does not exist at room temperature. As an example, the deposition rate decrease observed in a-C:H film deposition by RFPECVD in hydrocarbon-argon atmospheres occurs in a slower way than in deposition from methane-nitrogen or methane-ammonia atmo-

Table I. Chemical Composition and Deposition Parameters of Plasma-Deposited a-C(N):H Films

Reference	Deposition Method	Hydrocarbon	Nitrogen Precursor/Partial Pressure	Substrate Bias (V) or {Ion Energy (eV)}	Pressure (Pa)	Maximum N Content (at.%)	H Content (at.%)	Chemical Composition Determination Method
Grill and Patel [48]	DcPECVD	C_2H_2	N_2	500	20	14	19–37	RBS, ERDA
Amir and Kalish [9]	Dual chamber DcPECVD	CH_6H_6	N_2	150	2	9.5	—	NRA, AES
Chan et al. [49]	ECRRFPECVD	CH_4/Ar	N_2 (0–63%)	120	—	11 (C and N)	—	AES
Silva et al. [14]	Magnetic field enhanced RFPECVD	CH_4/He	N_2 (0–80%)	—	13	15	12–21	RBS, ERDA, NR
Rodil et al. [50]	Plasma wave ECR	C_2H_2	N_2	{80}	10^{-2}	30	32–26	RBS, ERDA
Kaufman, Metin, and Saperstein [10]	RFPECVDdc bias	C_5H_{10}/Ar	N_2 (0–95%)	380	3.5	11		RBS, XPS
Lacerda et al. [52]	RFPECVD	C_2H_2	CH_3NH_2 (0–40%)	350	3	12	12–16	RBS, ERDA, NR
Schwan et al. [53]	RFPECVD	C_2H_2	N_2 (0–70%)	—	7	20	30–38	Combustion analysis
Jacobsohn et al. [54]	RFPECVD	C_2H_2	N_2 (0–90%)	300	3	22	11–13	RBS, ERDA
Lee, Eun, and Rhee [55]	RFPECVD	C_6H_6	N_2 (0–80%)	500	13	10	22–27	Combustion analysis
			NH_3 (0–80%)	500	13	8	27	Combustion analysis
Franceschini, Achete, and Freire [12]	RFPECVD	CH_4	N_2 (0–50%)	370	8	11	12–17	RBS, ERDA, NR
Han and Feldman [8]	RFPECVD	CH_4	N_2	0 (Substrate in anode)	30	18	43–48	Combustion analysis
Freire and Franceschini [56]	RFPECVD	CH_4	NH_3 (0–25%)	370	8	13	16–20	RBS, ERDA, NR
Wood, Weydeven, and Tsuji [57]	RFPECVD	Not quoted	N_2 N/C ratio (~0–5)	1340	7	16	10–16	RBS, secondary ion mass spectroscopy (SIMS)

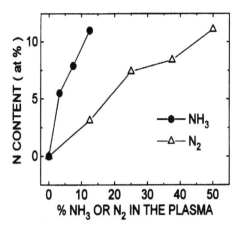

Fig. 8. Nitrogen content of a-C(N):H films deposited by RFPECVD, as a function of NH₃ or N₂ partial pressure. (Reproduced from [56].)

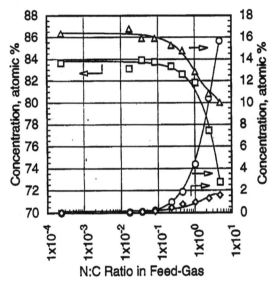

Fig. 9. Variation of C, N, H, and O contents as functions of an N/C ratio in the plasma. (Reproduced from [57].)

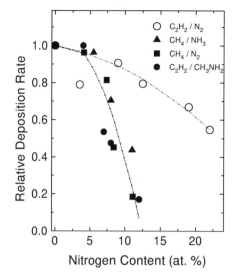

Fig. 10. Relative deposition rates of a-C(N):H deposited from several precursors as functions of the film nitrogen content. (Reproduced from [54].)

sphere [59]. This deposition rate drop is the responsible for the relatively low nitrogen content of a-C(N):H films, which was not reported to be higher than about 20 at.%, as shown in Table I.

Figure 8 shows the evolution of nitrogen content for the same samples shown in Figure 7, also as a function of the N-precursor partial pressure. Comparison of both N-content variations make clear that the nitrogen incorporation yield is quite higher for the NH₃-derived film, if we consider the nominal nitrogen content in the deposition

atmosphere. However, the clearly high reactivity of NH₃ plasma species does not lead to a higher maximum nitrogen incorporation, which is limited by the deposition rate.

Concerning the hydrogen content, some authors report a continuous decrease upon nitrogen incorporation [11, 48, 50, 53, 57], as depicted in Figure 9, which shows a typical evolution of the chemical composition of plasma-deposited a-C(N):H films. Other authors [12, 54] report only slight variations of the

hydrogen content, as shown in Table I. A more drastic effect on the hydrogen bonding in a-C(N):H films, the preferential bonding to nitrogen atoms, is discussed in Section 4.

Examination of Table I shows that in most cases the maximum N content is lower than 16 at.%. Nitrogen content in a higher range, like 20 at.% or more, were only observed for depositions performed with a C₂H₂-N₂ precursor mixture, suggesting a possible dependence of the nitrogen incorporation process on the hydrocarbon used. Comparison of the growth behaviors observed with use of different precursor mixtures confirms the suggestion. Figure 10 shows the deposition rates of RF plasma-deposited a-C(N):H films as a function of the nitrogen content in the film, normalized to the deposition rate obtained by the corresponding nitrogen-free film, taken from [54]. All the plots show a clear decrease of deposition rate upon nitrogen incorporation. The plots for CH₄-N₂, CH₄-NH₃, and C₂H₂-CH₃NH₂, seem to follow a very similar dependence, with the deposition rate having a trend to vanish at about 13 at.%. On the other hand, the deposition rate corresponding to C₂H₂-N₂ mixtures changes in a slower fashion as compared to the other ones. For about 20-at.% N incorporated into the film, the deposition rate is still half of the initial one. This behavior clearly shows that the higher nitrogen uptake of a-C(N):H films deposited by C₂H₂-N₂ plasmas is due to kinetic reasons.

Another point that one has to observe from analysis of Figure 10, is that despite the different precursor atmospheres, and consequently different N precursor partial pressures in the deposition, there is a coincidence of the deposition rate behavior upon nitrogen content (for mixtures other than C₂H-N₂). This points to a strong dependence of growth kinetics with nitrogen content.

In which, concerns about the quite different behavior shown in film deposition using acetylene as the hydrocarbon is not exclusive of a-C(N):H film deposition. In a-C:H film deposition, the use of acetylene has also shown to lead to higher deposition

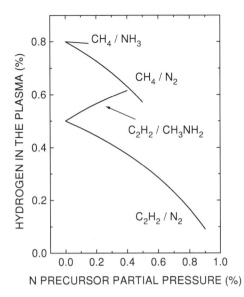

Fig. 11. Nominal hydrogen fraction in the deposition atmosphere against the nitrogen precursor partial pressure. (Reproduced from [54].)

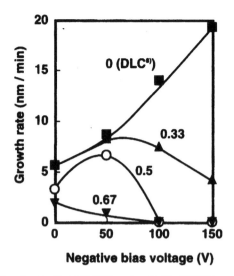

Fig. 12. Growth rate of a-C(N):H films deposited by ECRRF as a function of RF self-bias, for several N_2 partial pressures. (Reproduced from [62].)

rates as compared to methane [32, 33], and somewhat different film composition, structure, and properties as compared to several hydrocarbons [28]. One of the factors that leads to this situation is its relatively low hydrogen content, as compared to other hydrocarbons. Since the main role of the hydrogen species in a-C:H film deposition at room temperature is to saturate dangling bonds on the film surface, thus blocking surface sites initially available for the attachment of hydrocarbon reactive radicals [29]. Following this, we must expect that an increased hydrogen flux to the surface may lead to a decreased deposition rate. In fact, as Figure 11 shows, the nominal hydrogen content in the deposition atmosphere owned by the C_2H_2-N_2 mixture is much lower than that of other mixtures displayed in Figure 10.

The previously discussed dependence of the deposition rate on the hydrogen flux was also reported for a-C(N):H films deposited by an association of graphite-target N_2^+ ion-beam sputtering, and N_2^+/H_2^+ ion-beam assistance [60]. Although this deposition process may be not regarded as a typical plasma-deposition process, it retains some of its ingredients. In this work, increasing H_2 partial pressure in the deposition chamber resulted in a strong decrease in the deposition rate.

Table I includes results concerning a-C(N):H film deposition by ECRRF dual plasmas [49]. Despite the higher ionization and fragmentation of ECRRF plasmas, the results obtained are similar to those of glow-discharge plasmas. This is also the case for deposition performed by highly ionized methods such as plasma beam deposition [39].

Results from ECRRF plasma deposition of a-C(N):H films [61, 62] are useful to show the role of the bombarding ion energy on film growth. In the ECRRF plasmas, differently from RFPECVD, the energy of the bombarding ions can be varied—by the variation of the substrate RF self-bias—without great changes in the plasma. Inoue et al. [62] studied the deposition of a-C(N):H films by dual ECRRF plasmas in CH_4-N_2 atmospheres. In Figure 12 are plotted variations of the deposition

rate as functions of self-bias, for several values of N_2 partial pressure, being the ECR plasma power fixed. Beginning with the nitrogen-free film, an increase in the deposition rate is observed for increasing self-bias, reflecting ion-beam assisting of the growth process. For increasing nitrogen partial pressure, the deposition rate initially increases, passes through a maximum, and after falls down. As the N_2 fraction in the discharge increases, the deposition rate vanishing occurs for lower self-bias. For the higher N_2 partial pressure (0.67), only a decreasing behavior of deposition rate is observed. From the analysis of these plots, one can conclude that either nitrogen fraction (as in RFPECVD) or self-bias increase induces a deposition rate decrease. The observed deposition rate decrease with increasing self-bias was ascribed to film etching by N_2^+ fast ions.

3.3. Plasma and Surface Processes Affecting Film Growth

To understand the deposition mechanism of a-C(N):H films by plasma processes, one needs information on the plasma species that contribute to the growth process, and on the interaction of these species with the film growing surface.

Knowledge on the plasma species can be obtained by the use of plasma diagnostics techniques, such as optical emission spectroscopy (OES) and mass spectroscopy (MS). Both techniques are able to probe atomic and molecular, neutral or ionized species present in plasmas. OES is based on measuring the light emission spectrum that arises from the relaxation of plasma species in excited energy states. MS, on the other hand, is generally based on the measurement of mass spectra of ground state species.

A typical OES spectrum taken from CH_4-N_2 plasma used in a-C(N):H film deposition is shown in Figure 13 [63]. This figure also shows band assignment of the most relevant emitting spectroscopic systems for CH_4, N_2, and CH_4/N_2 plasmas. Figure 14, also taken from Ref. [63], shows the variation of the emission intensity of selected plasma species, as a func-

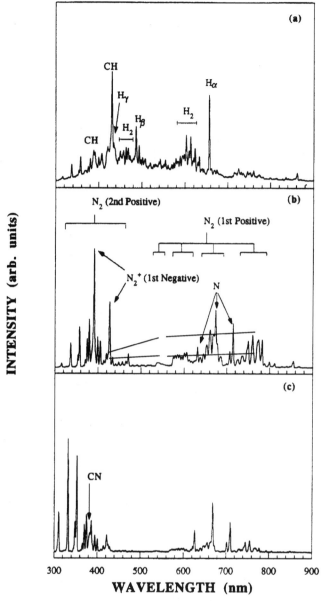

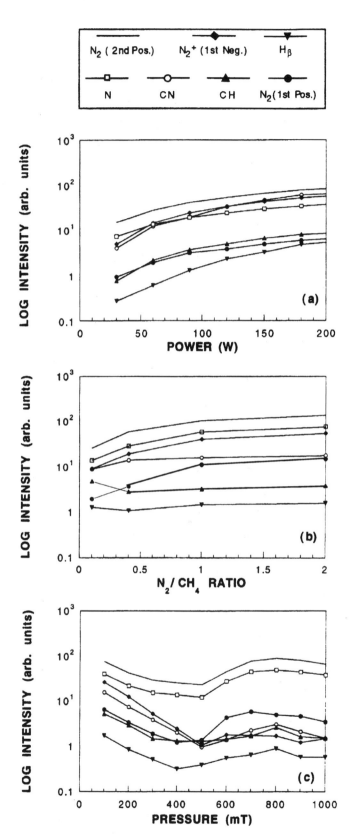

Fig. 13. Optical emission spectra taken from (a) CH$_4$, (b) N$_2$, and (c) CH$_4$/N$_2$ plasmas. (Reproduced from [63].)

Fig. 14. Emission intensity of the main spectroscopic systems for power, N$_2$/CH$_4$ ratio and deposition pressure variations. (Reproduced from [63].)

tion the of N$_2$ fraction in the discharge atmosphere, RF power, and deposition pressure. In Ref. [63], the analysis was focused on specific features of N$_2$, N$_2^+$, atomic N, CH, and CN species emission spectra, some of them designated in Figures 13 and 14 by the related spectroscopic nomenclature. Figure 14 shows that upon the increase of N$_2$ partial pressure, all the bands associated with nitrogen gas (N$_2$, N$_2^+$, and atomic N) were found to increase. In addition, a decrease of the C–H and an increase of the CN emission intensities were observed. This behavior has been identified as a possible source for the deposition rate decrease of a-C(N):H films, due to the decrease of C$_m$H$_n$ radical flux toward the growing surface. On the other hand, as the RF power increased, bands related to all species showed increased intensity, including the N$_2^+$ systems. This feature was associated to lack of film deposition observed above a critical power, by N_2^+ etching of the film.

The OES of CH_4-N_2 plasmas was also studied by Bhatasharyya, Granier, and Turban [64] (ECRRF) and Zhang et al. [65] (RFPECVD). The previously discussed decreasing behavior of CH emission intensity, and the increasing behavior of the CN emission intensity, for increasing N_2 partial pressure were also reported in both studies. In both works, the plasma was also probed by mass spectroscopy. In accordance with OES results, both works reported a decrease on the CH_3 fragment and an increase in the CN fragment mass signals. Bhatasharyya, Granier, and Turban [64] suggested that the increase of the CN abundance in the plasma would be responsible for the nitrogen incorporation, by $C\equiv N$ bonds formation in the film, as suggested by the transmission infrared spectra taken from the film (see Section 4.2). Bhatasharyya, Granier, and Turban also ascribed the deposition rate decrease to the diminishing of the CH radical abundance.

The correlation between the plasma diagnostics results with film growth process must be carefully discussed. Beginning with the reduction of CH radical abundance and the increase of CN radical abundance. Since nitrogen is being admixed in the deposition atmosphere as the expenses of hydrocarbon flow, the decrease of CH radical abundance should be expected of course, even in the absence of the formation of CN radicals. However, the deposition rate decrease may not be ascribed to this fact, at least as the only reason. First, the drop observed in the deposition rate of a-C(N):H films is too sharp to be attributed to a dilution effect. In addition, it was reported that the CH radicals are not the main source of a-C:H film growth [66], instead the C_2H radical was pointed as the most relevant one, even in a-C:H deposition from methane. From this point of view, the work of Bhatasharyya, Granier, and Turban [64] gives important additional information. Their neutral radicals MS results showed, in addition to the decrease of CH abundance, a twofold increase in the abundance of the C_2H radical upon an N_2 partial pressure increase to 40%. These facts show the need of a more detailed study on the role of carbon-carrying radicals in a-C(N):H deposition.

This is also the case of the CN radicals in the nitrogen incorporation process. Bhatasharyya, Granier, and Turban proposed, by comparing the film deposition rate with CN radical abundance, that CN radicals were responsible for the nitrogen incorporation, by reaction with the film surface [64]. In addition, a successful attempt to synthesize amorphous a-C(N):H and a-C(N) directly from CN radicals was reported [67]. On the other hand, failure was reported in the attempt to deposit a-C(N):H films from pure methylamine [52] CH_3NH_2, a precursor that may readily produce CN radicals. Under pressure and self-bias conditions that resulted in a-C(N):H growth from CH_4-N_2 plasma, no deposit was obtained using pure methylamine. Deposition performed under higher RF power and total pressure resulted in about a 30-at.% nitrogen content, but with a very low deposition rate.

It must be taken into account that parallel to the incidence of CN radicals, the growing surface is intensely bombarded by

N_2^+ fast ions. As pointed out by Zhang et al. [65], if CN radicals had a dominant role in the deposition process, no great differences would be found between film growth behavior on the RF powered and the grounded electrodes of a parallel plate RFPECVD deposition system. Since they are exposed to the same plasma, similar incidence of CN radicals is expected. In fact, Zhang et al. reported great differences in the growth behavior between the two electrodes [65]. At the RF-powered electrode, which is exposed to intense ion bombardment, the deposition rate strongly dropped with N_2 partial pressure. At the grounded electrode, where no bombardment is expected, an almost constant behavior of the deposition rate was found as the N_2 partial pressure increased.

To complement the preceding discussion, a more detailed discussion on the role of N_2^+ ion bombarding in the growth process is developed in the following paragraphs.

The relevance of N_2^+ ions bombarding was evidenced by the studies carried out by Hammer and Gissler [68] and Hammer et al. [69]. They reported the effects arising from a 150-eV N_2^+ ion bombardment of a pure carbon film. Hammer and Gissler [68] reported that N_2^+ ion bombardment results in the removal of carbon atoms from the films at rates as high as 0.5 C atoms per N_2^+ ion. Such a process was identified as a "chemical sputtering" since the sputtering yield was found to be substantially higher than that obtained [69] by noble ion bombardment with equivalent energies. This erosion process was found to result in the evolution of CN-containing molecules, such as CN and C_2N_2, as shown in Figure 15. This figure shows the mass spectrum of the evolved fragments in the N_2^+ ion bombardment process. In addition to the importance of the effect—high etch rates with relatively low energy beams—also the strong evolution of CN molecules, and even of the heavier C_2N_2 molecules must be noted.

One may observe that in plasma deposition of a-C(N):H films, as the nitrogen precursor gas partial pressure is increased, the bombardment of the film surface by N_2^+ ions will certainly occur. As this situation was shown to generate carbon erosion, and evolution of CN molecules, there is no reason to believe that CN fragments from the plasma would survive the bombardment, resting attached to the film.

A similar process was observed in the study of the etching process of a-C:H films by N_2 ECRRF plasmas [70]. It was found that RF-biased, previously grown a-C:H films were eroded by an N_2 plasma, with RF self-bias in the range of values of the results discussed in the previous session. As shown in Figure 16, the fragments evolved upon N_2^+ ion bombardment of a-C:H films are very similar to those presented by pure carbon films. This work also reports that ion bombardment also results in nitrogen incorporation. In another article, Hong and Turban reported that this process might indeed be more effective for nitrogen incorporation in amorphous carbon films. By alternating CH_4 and N_2 plasmas in a dual ECRRF plasma deposition system, they produced a-C(N):H films with a nitrogen content twice that could be obtained with CH_4-N_2 plasmas [71].

As a final comment on surface processes, the evaporation of N_2 molecules must not be neglected. Since one cannot have

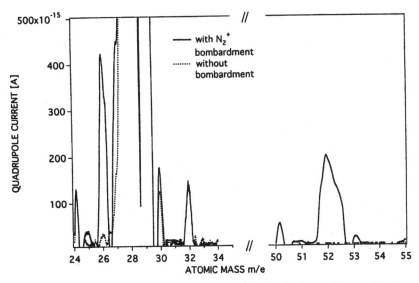

Fig. 15. Mass spectrum of molecules evolved from N_2^+ ion bombardment of a pure carbon film. (Reproduced from [68].)

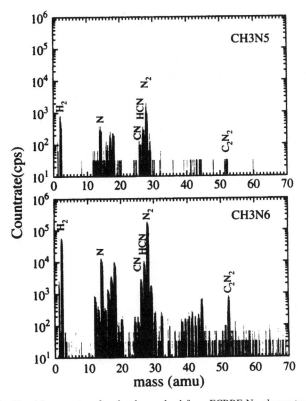

Fig. 16. Mass spectra of molecules evolved from ECRRF N_2 plasma treatment of an a-C:H film, for different treatment conditions. (Reproduced from [70].)

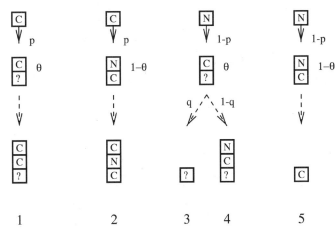

Fig. 17. Schematics of the process involved in the interaction of C and N species incident upon a carbon–nitrogen deposit. (Reproduced from [72].)

process. Stronger evidence of this suggestion may only be obtained by more detailed research on this theme, such as the measurement of the absolute sticking factors of the species involved.

3.4. Modeling of a-C(N):H Film Growth

A simple model for a-C(N):H film growth kinetics was proposed [72], and was based on some aspects of the discussion developed in the previous subsection. The model is based on the incidence of only two species over a deposit, one representing C-carrying radicals and ions, while the other represents energetic N_2^+ ions. It is also based on the fact that C-containing radicals are the main channel for film growth in plasma-deposited a-C:H films [27]. In addition, it is supposed that nitrogen incorporation occurs mainly by the aggregation of the nitrogen atoms that are products of the breaking up of N_2^+ fast ions after colliding and penetrating into the film subsur-

solid nitrogen at ambient temperature, as the nitrogen content increases, there will be an increased probability of N–N bonds formation. As pointed out by Marton, Boyd, and Rabalais [2], N–N bond formation should result in N_2 gas evaporation.

From the foregoing discussion, one may suggest that nitrogen ions play the dominant role in the a-C(N):H deposition

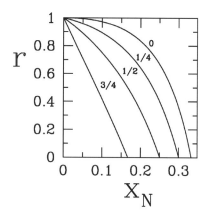

Fig. 18. Calculated relative deposition rate as a function of N content in film, for several values of the interaction parameter a. (Reproduced from [74].)

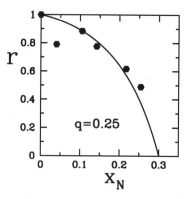

Fig. 19. Comparison of the two species model deposition rate as a function of N content with experimental results obtained from C_2H_2/N_2 plasmas. (Reproduced from [74].)

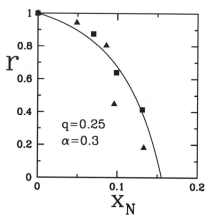

Fig. 20. Comparison of the two species model deposition rate as a function of N content, including the blocking of surface sites, with experimental results obtained from CH_4/N_2 plasmas. (Reproduced from [74].)

face. These two species were, respectively, C and N, which are impinged to the model deposit with complementary probabilities p and $(1 - p)$, being the variation of p analogous to the variation of N_2 partial pressures in hydrocarbon—N_2 plasmas. Upon falling over the surface of the deposit, these species may take part in five different processes at the surface, as shown in Figure 17, and described as follows:

(1) the aggregation of the C species over a surface C species, with probability equal to $p\theta$ (θ being the C surface coverage);

(2) the aggregation of a C species over a surface N species, with probability equal to $p(1 - \theta)$;

(3) the erosion of a surface C species by an impinging N species, with probability $(1 - p)q\theta$, q being an interaction parameter related to the carbon sputtering yield by fast N_2^+ ions;

(4) the aggregation of an N species over a surface C species, with probability $(1 - p)(1 - q)\theta$; and

(5) the evaporation of an N_2 molecule, as a consequence of an N species falling over a surface N species with probability $(1 - p)(1 - \theta)$.

Todorov et al. [73] made equivalent assumptions in the modeling of a-CN_x film growth within another context. They focused their attention on the calculation of the collision cascade details following subsurface penetration of C^+ and N^+ ions in direct ion-beam deposition of carbon–nitrogen films. Because of the presence of both carbon and nitrogen ionic species, they also included the erosion of nitrogen by fast C^+ ions, the process complementary to process (3).

Simple analytical calculations based on the previously discussed assumptions lead, among other results, to the determination of the film relative deposition rate as a function of the nitrogen content in the film, for several values of the interaction parameter q, as shown in Figure 18.

Several features of Figure 18 are related to the experimental results discussed in Section 3.2. For a range of the q interaction parameter ($q < 3/4$), the curves are very similar in shape with experimental a-C(N):H deposition rate vs N content plots.

As q is varied, a wide range of maximum nitrogen (or zero deposition rate) contents is spanned. In addition, it is interesting to note that in this model nitrogen evaporation seems to be the major factor limiting nitrogen incorporation. The figure shows that even for $q = 0$ (no carbon removal by nitrogen ions) the deposition rate shows a strong fall upon nitrogen incorporation, vanishing at about 33-at.% N.

Comparison of experimental deposition rate results from a-C(N):H deposition with C_2H_2-N_2 RFPECVD [54] showed a very good agreement (see Fig. 19) for a q parameter equal to 0.25. This q value is closely related to the 0.5 C atom per N_2^+ ion previously reported [68]. Within this model, no agreement was obtained for the deposition rate plots owning to other gas mixtures, which had deposition rate vanishing at about 13–15 at.% N (Fig. 10). As shown in Figure 18, for higher q values the deposition rate curves show a trend to change curvature, being qualitatively different from a-C(N):H experimental deposition rate plots.

Further improvement of the previously described model [74] allowed the inclusion of a sticking probability α superimposed to the aggregation of both carbon and nitrogen (C and N species in the model). This was done to represent the blocking of surface sites to the sticking of hydrocarbon neutral radicals by an increased hydrogen flux. In this way, it was possible to fit the deposition rate vs nitrogen content plots of a-C(N):H films deposited by RFPECVD in methane–nitrogen atmospheres, by supposing a sticking factor α equal to 0.3, whereas in C_2H_2-N_2 modeling α equals one. Figure 20 shows the comparison of CH_4-N_2 experimental and model results of the deposition rate.

The previously discussed model suggests how realistic is the assumption of N_2^+ ion bombardment as the main driving force of growth kinetics and nitrogen incorporation in a-C(N):H film growth.

4. CHARACTERIZATION OF a-C(N):H FILM STRUCTURE

4.1. Raman Spectroscopy

Raman scattering spectroscopy is used to probe the vibrational excitations of a sample, by measuring the wavelength change of a scattered monochromatic light beam. This is usually performed by impinging a monochromatic laser beam to the sample surface, and by recording the scattered beam spectrum.

Raman spectroscopy has been widely used in carbon materials characterization. The main features owing to the Raman spectra of carbon materials are the so-called D and G bands, located, respectively, at about 1350 and 1580 cm^{-1} wave numbers. The G band is the only feature in the first-order Raman spectrum of single crystalline graphite. It corresponds to the excitation of the in-plane E_{2g} and E_{1u} vibration modes of graphite, which are represented in Figure 21. The D band (D from disorder) is the additional feature that is present in the Raman spectrum of microcrystalle and disordered graphite. It arises from the breaking of translational symmetry within the graphene planes. In microcrystalline graphite, the ratio of the D to the G band intensities I_D/I_G was found to be inversely proportional to the crystal size across the basal plane.

In amorphous carbon films, the G and D bands are highly broadened, with the band intensity ratio I_D/I_G, maximum positions ω_G and ω_D, and line breadths Γ_G and Γ_D varying in a relatively broad range. The interpretation of the changes in the spectra parameters of DLC films is not straightforward as in crystalline graphite. Instead of direct relations with structural characteristics, empirical procedures have been used to extract useful information from Raman spectra of DLC materials. A usual approach is that coming from the investigation conducted by Dillon, Woolam, and Katkanant [75]. They performed Raman analysis in thermally treated (up to 1000 °C) a-C:H films deposited by RFPECVD and ion-beam deposition. To analyze the great changes introduced in the broad spectra, they were deconvoluted in their constituent D and G bands. The changes observed in the spectra parameters of the heat-treated samples were the following:

(a) The increase in the I_D/I_G integrated band intensity ratio up to about 600 °C, followed by a decrease;

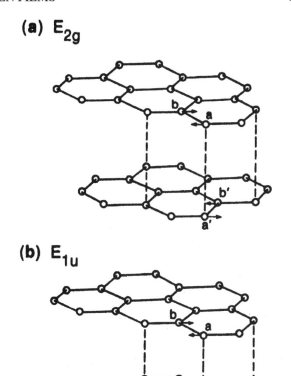

Fig. 21. E_{2g} and E_{1u} vibration modes of crystalline graphite. (Reproduced from [10].)

(b) The increase in the position of the G and D lines until stabilization for temperatures higher than 600 °C and;

(c) The narrowing of the D and G bands until stabilization for temperatures higher than 600 °C.

Such results were supposed to correspond to an increase in the size and/or number of the sp^2-carbon condensed ring domains present in the films upon increasing temperature, up to graphitization at 600 °C. After that, the expected behavior for graphitized carbon, the decrease of I_D/I_G upon the increase of crystal size is observed.

This picture was found to be consistent with the comparison of Raman spectra and optical gap of a-C:H films deposited by RFPECVD, with increasing self-bias [41]. It was found that both, the band intensity ratio I_D/I_G and the peak position ω_G increased upon increasing self-bias potential. At the same time, a decrease on the optical gap was observed. Within the cluster model for the electronic structure of amorphous carbon films, a decrease in the optical gap is expected for the increase of the sp^2-carbon clusters size. From this, one can admit that in a-C:H films, the modifications mentioned earlier in the Raman spectra really correspond to an increase in the graphitic clusters size.

In reality, several factors were mentioned as being responsible for this behavior, such as variations in bond angle distortion, in the internal stress or in the hydrogen content [40, 76], but all

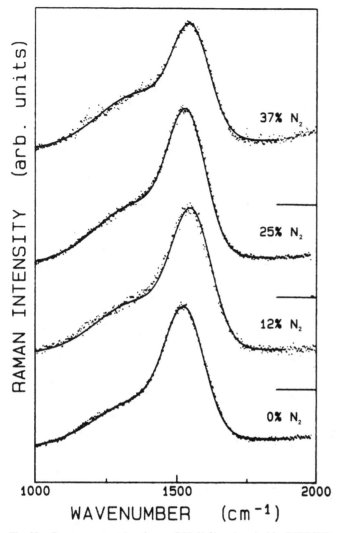

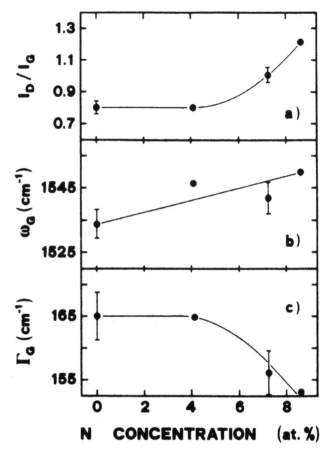

Fig. 23. Variation of the integrated intensity ratio I_D/I_G, the maximum position of the G band, ω_G, and its width Γ_G against film nitrogen content. (Reproduced from [78].)

Fig. 22. Raman spectra taken from a-C(N):H films deposited by RFPECVD in CH₄-N₂ atmospheres, for several values of the N₂ partial pressure. (Reproduced from [78].)

of them are also strongly correlated with the variation of optical gap width in amorphous carbon films. Theoretical work on Raman spectroscopy on DLC materials gave additional support for Dillon's interpretation [77].

Raman spectroscopy has been applied to probe the structural evolution of sp²-hybidized carbon aggregates in plasma-deposited a-C(N):H films [12, 14, 52, 54, 56]. Very often, due to the great breadth of D and G bands, the spectra deconvolution was done by fitting two Gaussian lines, as reported by Mariotto, Freire, and Achete [78]. They studied the Raman spectra of a-C(N):H films deposited by RFPECVD in CH₄-N₂ atmospheres. The obtained spectra are shown in Figure 22 and the corresponding results from the fitting procedure are shown in Figure 23. The analysis of the spectra showed an increase on the I_D/I_G band ratio from about 0.8 to about 1.3. At the same time, the position of the maximum of the G band was found to shift from 1530 to 1547 cm⁻¹, and to be narrowed from 165 to 154 cm⁻¹. All of these changes occurred within a 12.5-at.% N

incorporation, and were ascribed to an increase in the size or number of graphitic domains.

Other studies showed the same trends discussed earlier. Films obtained by RFPECVD, deposited at about the same conditions in other precursor atmospheres [79], showed the same behavior, with the I_D/I_G and ω_G following about the same dependence on nitrogen content as the one shown in Figure 23.

On the other hand, films deposited by RFPECVD in CH₂-N₂ atmospheres showed a different behavior [54]. In this study, in addition to the reported enhanced maximum nitrogen incorporation of 22 at.% (see Section 3.1), an almost constant behavior of the Raman parameters was observed for nitrogen content up to about 12 at.%. Above this limit, I_D/I_G and ω_G were shown to increase, and Γ_G was shown to decrease. This is evidence that the details related to the growth process, may give rise to differentiated sp²-clusters size evolution.

The Raman parameters reported for a-C(N):H films deposited by magnetic field enhanced RFPECVD in CH₄-N₂-He atmospheres also showed an intermediate N content range (about 7 at.%), with almost constant Raman parameters. In this case, the behavior was found to be associated with a nontypical variation of the C and N atom hybridization state, as discussed in Section 4.3.

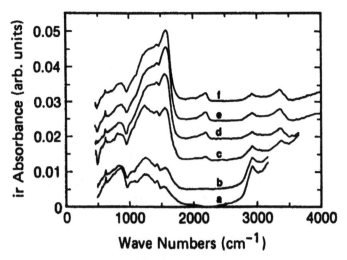

Fig. 24. Infrared absorption spectra taken from a-C(N):H films with several N contents. N increases from a to f. (Reproduced from [10].)

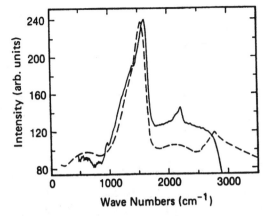

Fig. 25. Comparison of a Raman spectrum with an IR spectrum of an a-C(N):H film. (Reproduced from [10].)

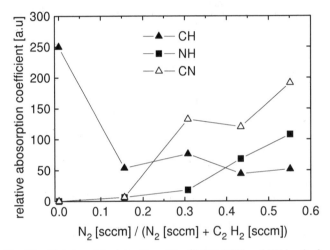

Fig. 26. Variation of the intensities of the IR CH, C≡N, and NH bands, for films obtained with several N_2 partial pressures in C_2H_2-N_2 plasmas. (Reproduced from [53].)

Results on Raman spectroscopy thus show that nitrogen incorporation, at least for a large enough N content, results in the increase of the graphitic clusters. This is contrary to the formation of an amorphous solid related to the β-C_3N_4 phase, which presumes sp^3-C hybridization and no clustering effects.

4.2. Infrared Spectroscopy

Infrared spectroscopy has been widely used to characterize amorphous hydrogenated carbon films [28, 31–33]. The main feature observed in the infrared spectra of diamond-like a-C:H films is a broadband in the 2900 cm^{-1} wave number range, which is due to the C−H stretching vibration of CH$_n$ groups containing carbon atoms in all the hybridization states. Although it has been sometimes used to determine the hybridization state of carbon atoms, by deconvoluting the CH broad band into its constituent bands [32, 33], this analysis is taught to give only qualitative information.

Nitrogen incorporation into a-C:H films was reported to give rise to dramatic changes in the infrared spectrum, as first reported by Kaufman, Metin, and Saperstein [10]. As shown in Figure 24 [10], as the nitrogen content increases, the spectra show the appearance, with increasing intensities, of absorption bands located at 2300 and 3500 cm^{-1}. These two bands have been respectively ascribed to the C≡N and N−H stretching bands. In addition, a continuous decrease of the CH stretching band is also reported. A broadband, located in the 1500 cm^{-1} wave number range also emerges with increasing intensity for increasing nitrogen content.

This broad band at 1500 cm^{-1} was ascribed by Kaufman, Metin, and Saperstein [10], to an IR observation of the amorphous carbon Raman D and G bands. This is forbidden by the selection rules, and has been attributed to the symmetry breaking introduced by the presence of CN bonds in the amorphous network. As carbon and nitrogen have different electronegativities, the formation of CN bonds gives the necessary charge

polarity to allow the IR observation of the collective C=C vibrations in the IR spectrum. This conclusion was stated by the comparison of spectra taken from films deposited from $^{14}N_2$ and $^{15}N_2$. In the $^{15}N_2$-film spectrum, no shift was observed for the 1500-cm^{-1} band, whereas all other bands shifted as expected from the mass difference of the isotopes. Figure 25 compares Raman and IR spectra taken from the same a-C(N):H film [10], showing an almost perfect overlapping of the two bands.

The variation of the IR band intensities upon nitrogen incorporation for RF plasma-deposited a-C(N):H films is shown in Figure 26, as reported by Schwan et al. [53]. As mentioned before, the intensity of the C−H stretching band decreases upon nitrogen incorporation, at the same time that an increase in the N−H stretching band intensity is observed. This suggests that hydrogen preferentially bonds to nitrogen in a-C(N):H films. Since the hydrogen content does not decrease at a rate higher than that of the nitrogen content (see Table I), it may be supposed that most of the nitrogen atoms would be hydrogenated, and thus at a terminal site of the amorphous network. The CN bonds shown in the IR spectra are also terminal bonds. As pointed out by Kaufman, Metin, and Saperstein [10], the IR spectra show that nitrogen atoms which are present in the IR-active groups may be located at the periphery of condensed aromatic rings.

Work on a-C(N):H films, deposited by Ar^+ ion-beam sputtering of graphite and ion-beam assistance by N_2^+/H_2^+ ions [80] lead basically to the same conclusions of Kaufman, Metin, and Saperstein [10]. In addition, these authors evidenced that nitrogen incorporation is not the only factor affecting the presence of the Raman band in the IR spectrum of a-C(N):H films. Hydrogen also plays a role, as evidenced by the theoretical simulation of the vibrational spectra of model organic molecules, and by the experimental comparison of Raman and infrared spectra of hydrogenated and nonhydrogenated carbon–nitrogen films. It was found that only for the hydrogenated films Raman and IR spectra were coincident, which was also corroborated by the model calculations.

4.3. Electron Energy Loss Spectroscopy

The hybridization of carbon atoms is the major structural parameter controlling DLC film properties. Electron energy loss spectroscopy (EELS) has been extensively used to probe this structural feature [5, 6]. In a transmission electron microscope, a monoenergetic electron beam is impinged in a very thin sample, being the transmitted electrons analyzed in energy. Figure 27 shows a typical EELS spectra from an a-C(N):H film [56]. The energy loss peaks shown in this figure correspond to the carbon and nitrogen K-edges, each one being composed of peaks corresponding to $1s-\sigma^*$ and $1s-\pi^*$ electronic transitions. The carbon and nitrogen sp^2-fraction estimates are obtained by comparing the π to σ peak area ratio to the one of a standard all-sp^2 sample, like highly oriented pyrolitic graphite HOPG [81].

The sp^2 C atom fraction variations against nitrogen content for a-C(N):H films deposited by RFPECVD from methane–nitrogen and methane ammonia atmospheres [11] are shown in Figure 28. In both cases, nitrogen incorporation results in a continuous increase of the sp^2-carbon atom fraction from about 50 to about 80%. Such a variation is in accordance with the previously discussed Raman spectroscopy results, which showed an increase in the size of graphitic domains present in the sample. The reported variations indicate a very strong effect, as compared to the relatively low nitrogen incorporation. Such kinds of effects are also reported for nitrogen incorporation in highly tetrahedral amorphous carbon [82], and seems to be inherent to the nitrogen incorporation process in DLC films.

However, different variations were also reported. A small decrease of the sp^2 fraction was observed for a 7-at.% nitrogen incorporation in films deposited by magnetically enhanced RF-PECVD from CH_4-N_2 mixtures highly diluted in He [15], as shown in Figure 29. At the same time, an N atom sp^2 fraction decrease from 75 to about 20% in the same nitrogen content range was observed. For higher nitrogen content, the C sp^2 fraction increased again, resulting in a variation from 75%, for the

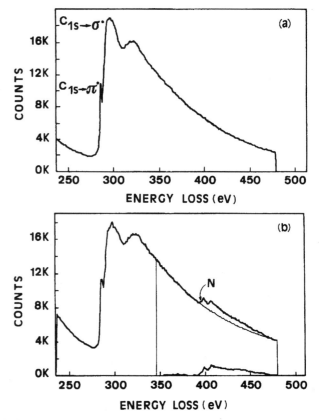

Fig. 27. Electron energy loss spectra taken from (a) an a-C:H film; and (b) an a-C(N):H film. (Reproduced from [56].)

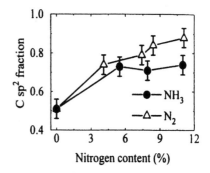

Fig. 28. Variation of the C sp^2 fraction as a function of N content in film, for films deposited from CH_4-N_2 and CH_4-NH_3. (Reproduced from [11].)

nitrogen-free film, to about 90%, for the 16-at.% N containing film.

It is important to stress that nitrogen incorporation in a-C:H films always result, at least above a certain level, in a strong decrease in the tetrahedrally bonded carbon atom fraction. Raman spectroscopy also gives support to this observation, because the increase in the size of graphitic clusters only can proceed with also increasing sp^2 fraction.

4.4. X-Ray Photoelectron Spectroscopy

X-ray photoelectron spectroscopy (XPS) is based on the measurement of the binding energy of core-level electrons ejected

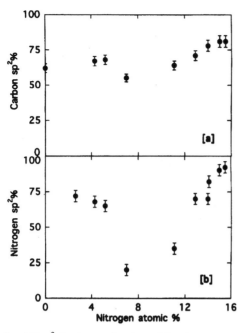

Fig. 29. C and N sp^2 fractions as functions of the nitrogen content in films deposited by magnetically enhanced RFPECVD in CH$_4$-He-N$_2$ atmospheres. (Reproduced from [14].)

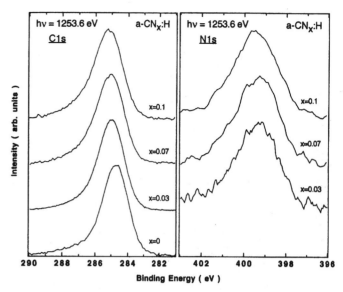

Fig. 30. C1s and N1s X-ray photoelectron spectra taken from a-C(N) films with several N contents, deposited by CH$_4$/N$_2$ ion beams. (Reproduced from [45].)

from the sample surface, as a consequence of monochromatic X-ray irradiation. The knowledge of the photoelectron production cross sections of the elements allows the estimation of sample chemical composition. This is done by comparing the corrected integral photoelectron intensities from each element present. In addition, it is expected that changes in the chemical environment of an atom may give rise to a shift in the photoelectron binding energy, due to charge transfer effects arising from chemical bonding [44]. However, the interpretation of the results is generally controversial [3], and shall be treated with care. The chemical shifts of XPS lines are small, strongly dependent on experimental factors such as surface charging effects or the presence of contaminants, and very often cannot be surely assigned to specific structural features, due to the lack of suitable model calculations.

In the following paragraphs, reports on XPS studies of a-C:N:H films are discussed. Most of the work that is discussed is not related to usual plasma deposition, but is related to ion-beam deposition, ion-beam assisted deposition, or other methods, because only a very few wide scope XPS studies on plasma-deposited a-C(N):H films were done up to this moment. Nevertheless, these results bring useful information on the role of hydrogen in the structure of a-C(N):H films.

Mansur and Ugolini [45] reported an XPS study on ion-beam deposited a-C(N):H films using a Kaufman-type ion source fed by CH$_4$-N$_2$ mixtures. They studied the changes in the C1s and N1s photoemission lines, for samples containing up to 10-at.% N (considering only C and N) (Fig. 30). Upon nitrogen incorporation, the C1s binding energy (BE) was found to shift from 284.8 to 285.2 eV. The corresponding linewidths (full width at half-maximum FWHM) were also found to increase, from 1.8

to 2.1 eV, being the broadening asymmetric toward higher BEs. At the same time, no changes were observed in the N1s line position or breadth, which remained fixed at 399 ± 0.1 and 2.4 eV, respectively. The shift of the C1s line to higher BEs was ascribed to the higher electronegativity of nitrogen atoms, as compared to carbon. Therefore, carbon–nitrogen bonding should result in charge transfer from carbon to nitrogen, and thus leading to the increase in the C1s binding energy.

Based on the IR spectroscopy, evidence that nitrogen incorporation into a-C:H films may occur in the periphery of condensed aromatic rings [10], Mansur and Ugolini compared the C1s spectra of the a-C(N):H films with that taken from benzene (C$_6$H$_6$) and pyridine (C$_5$H$_5$N). Data from benzene agreed well with that of pure a-C:H film (284.5-eV BE and 1.8-eV FWHM), but data from pyridine showed much higher BE (286.6 eV) than that shown by the 9-at.% N film. Since in pyridine, the nitrogen content (16.7%) is higher than that of the most nitrogenated film (9.0%), they made a linear extrapolation of the BEs to the actual film nitrogen content. The obtained shift, relative to the C1s BE of the a-C:H film, was still higher (2.5 times) than that observed for the 9-at.% N films, thus leading to discarding the nitrogen content as the only source of the observed difference. Taking into consideration the asymmetry of the line broadening, a two-phase model was proposed. Accordingly, the C1s line was decomposed by fitting two Gaussian lines. This procedure resulted for the carbon–nitrogen and pure carbon environments BEs of 286.2 and 285 eV, respectively, a 1.2-eV shift in a reasonably good agreement with the 1.4-eV shift observed for benzene and pyridine. The area ratio between the a-C:H and a-C(N):H features of the spectra was found to be 28%.

The bigger widths of the a-C(N):H N1s lines, as compared to that of pyridine, were also ascribed to a two-phase structure. As hydrogen is less electronegative than nitrogen, N—H bonding

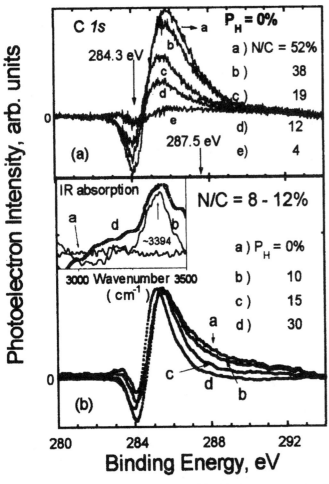

Fig. 31.   XPS C1s difference spectra, as explained in the text. (Reproduced from [83].)

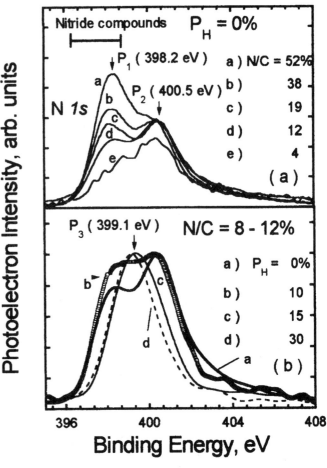

Fig. 32.   XPS N1s difference spectra, as explained in the text. (Reproduced from [83].)

should result in a decreased BE, as compared to that of nitrogen bonded only to carbon. The authors were also able to fit the N1s lines with two Gaussian lines, the one at lower BE being ascribed to N atoms bonded to C and H, whereas that at higher BE was ascribed to N bonded only to C.

Souto and Alvarez [83] closely studied the role of hydrogen in the XPS spectra of a-C(N):H films. They studied a-C(N):H films produced by reactive RF sputtering of a graphite target in Ar-N_2-H_2 atmospheres, onto unbiased substrates, maintained at 285 °C during deposition. Their study was done by the analysis of the core-level C1s and N1s spectra obtained from hydrogenated and nonhydrogenated CN_x films. The C1s line analysis was performed by subtracting the C1s spectrum of a pure carbon film from (a) the spectra obtained from hydrogen-free CN_x films, and (b) the spectra obtained from hydrogenated CN_x films (respectively, Fig. 31a and b). Figure 31a shows that nitrogen incorporation into a-C films results in the growing of a contribution at higher BEs, which was ascribed to charge transfer from carbon to nitrogen, as discussed before. On the other hand, hydrogenated films showed decreasing intensities in the higher energy tail of the difference spectra, for increasing hydrogen incorporation (Fig. 31b). This effect was attributed to

the preferential bonding of hydrogen to nitrogen atoms, which leads to the decrease of the number of CN bonds. Additional support for this hypothesis is given by the N1s photoelectron spectra, shown in Figure 32. Figure 32a shows an increased contribution of the peak located at 398.2 eV, named P1 in the figure, which was ascribed to nitrogen atoms sharing all of their valence electrons with carbon atoms. On the other hand, peak P2 at 400.5-eV binding energy was ascribed to N probably bonded to only one C atom. Figure 32b shows that hydrogen incorporation causes the collapse of P1 and P2 peaks, giving rise to the peak P3 at about 399 eV, which was identified to be probably due to a C−NH_2 bonding. This is additional evidence of preferential bonding of hydrogen to nitrogen atoms, even in a so different deposition method like reactive magnetron sputtering. The main effect of hydrogen incorporation is to avoid the formation of a C−N extended network.

A more detailed approach was used by Hammer, Victoria, and Alvarez [60], in the analysis of a-C(N):H films deposited by Ar^+ ion-beam sputtering of graphite targets, and N_2^+ and H_2^+ ion-beam assistance. They used four Gaussian lines to fit the N1s peaks of the hydrogen free CN_x film (398, 400.6 eV, and two low-intensity components at 402.6 and 404.6 eV). The assignment of these lines was performed by numerical simu-

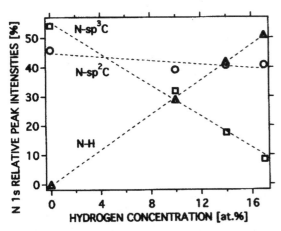

Fig. 33. Result of a-C(N):H films N1s peak decomposition. (Reproduced from [66].)

lations applied to representative N-containing molecules with several sp-, sp^2-, and sp^3-carbon atom arrangements, in agreement with previous extensive XPS work on a-CN$_x$ films. The 398-eV line was ascribed to N bonded to sp^3 C with isolated lone pair electrons. The peak at 400.6 eV was attributed to N atoms in graphite-like configurations, with the lone pair involved in p-bonding. The smaller peaks at higher BEs were not clearly identified, but were supposed to be due to N–N bonds. For the analysis of the hydrogenated films, the previously mentioned 399.1-eV peak due to N–H bonding also had to be included. The result of the fitting procedure applied to samples with different hydrogen contents is shown in Figure 33. This figure shows deep changes in film structure, with a strong decrease on the N-sp^3 C bonding, and the expected increase in the N–H bonding, all these occurring with an almost constant N-sp^2 fraction.

The previously discussed XPS results, despite being obtained from films deposited from deposition methods other than usual plasma deposition, give valuable information on a-C(N):H film structure. Results reported in the works of Mansur and Ugolini [45]; Souto and Alvarez [83]; and Hammer, Victoria, and Alvarez [60] are in total accordance with the picture that emerges from IR results: hydrogen bonds preferentially to nitrogen atoms, which are bonded mainly to sp^2-carbon atoms.

4.5. ^{13}C Nuclear Magnetic Resonance Spectroscopy

Solid-state nuclear magnetic resonance (NMR) on ^{13}C nuclei has been used for a-C:H characterization [19, 85]. It was shown that ^{13}C NMR is well suited for hybridization studies on amorphous hydrogenated carbon films, since it shows well-defined peaks for sp^3- and sp^2-carbon atoms. Its use, in conjunction with the knowledge on the sample chemical composition lead to the determination of the detailed carbon atom coordination in plasma-deposited a-C:H films with varying self-bias [19]. The combination of NMR on ^{13}C and ^{1}H nuclei, allowed the detailed determination of all the protonation states of sp^2- and sp^3-carbon atoms, in addition to the determination of the

bound-to-unbound hydrogen fraction in a-C:H films [84], in which concerns to hybridization determination, ^{13}C NMR spectroscopy is advantageous over other techniques, like electron energy loss spectroscopy and infrared spectroscopy. The high-energy electron beam used in EELS may damage the sample and may give rise to misleading results. Infrared spectroscopy can detect only IR-active bonds. Despite its power, NMR spectroscopy is scarcely used in a-C:H film characterization due to its low sensitivity. Samples as large as 70 mg in weight are needed [84], which may correspond to an about 350 cm^2 in area, 1-μm thick a-C:H film.

A ^{13}C-NMR study was on a-C(N):H RFPECVD deposited films in butadiene–nitrogen or butadiene–ammonia atmospheres was reported [85]. As the results, about a 65% sp^3 fraction was obtained for both the nitrogen-free film and the nitrogen-containing film deposited from nitrogen gas, while a 73% sp^3 fraction was obtained from the NH$_3$-derived film. The Auger electron spectroscopy estimates of the nitrogen content of the ammonia and nitrogen-derived films were about 5- and 2-at.% N, respectively. Such quite impressive sp^3-fraction enhancement was not confirmed by later studies on N$_2$/NH$_3$ comparison as nitrogen precursors [11]. The observed increase in the sp^3 fraction upon nitrogen incorporation with ammonia perhaps may be due to a higher hydrogen fraction, which may occur with the use of ammonia as the precursor gas [56].

Lamana et al. [86] also reported a ^{13}C-NMR study on a-C(N):H films. Despite being related to films deposited on the grounded electrode of an RFPECVD deposition reactor (no substrate bias at all), the obtained results are useful in the discussion of nitrogen modification of a-C:H film structure. For the nitrogen-free films, the authors observed the broad resonance bands related to the sp^2- and sp^3-hybridized carbon (Fig. 34). However, for the nitrogen-containing films sharp peaks were observed, superimposed on the broad ones, characterizing an ordering behavior. The authors fitted the spectrum to model structures composed of nitrogen-containing fused aromatic rings, terminated by NH$_2$ groups. No references were done by the authors on the evolution of the broad peaks in the spectra 3 and 4 (respectively, low nitrogen content and nitrogen-free film) shown in Figure 34. However, it must be noted that nitrogen incorporation leads to a decrease of the sp^3-carbon peak at 40 ppm, if we compare the spectrum 3 to that of the nitrogen-free sample, characterizing an increase in the sp^2-C atom fraction. For the higher nitrogen content samples, the model structures adopted for spectrum fitting are very similar to the previously discussed structure changes caused by nitrogen incorporation.

4.6. Overall Discussion

The results discussed earlier on the characterization of plasma deposited a-C(N):H films leads to a clear picture of the structure changes. Unless for a few exceptions, nitrogen incorporation leads to a strong increase of the C sp^2 fraction, as shown by electron energy loss spectroscopy. In parallel, Raman spectroscopy revealed that the graphitic clusters increase

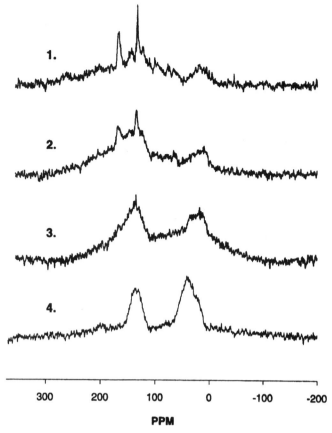

Fig. 34. NMR spectra taken from (1) a highly nitrogenated a-C(N):H film as deposited; (2) a highly nitrogenated a-C(N):H film annealed; (3) a low nitrogen content a-C(N):H film as deposited; (4) an a-C:H film as deposited. (Reproduced from [86].)

in size upon nitrogen incorporation. On the other hand, IR spectroscopy suggests, and XPS studies confirm, the preferential bonding of hydrogen to nitrogen, which bring two consequences. The first one is the location of N atoms in the border of the graphitic domains. The second one is to make N atoms as network terminators, impeding the formation of an extended C$-$N network. IR also shows the appearance of C$\equiv$N groups, which are network terminators as well.

At this point, we may suggest what would be the consequence of this microstrutural changes on the amorphous network as a whole. The foregoing modifications would lead to a lower cross-linking level, which is equivalent to a more open structure. This is in fact confirmed by other experimental techniques. Dopler-broadening observation of positron annihilation in a-C(N):H films [87] showed that nitrogen incorporation increases the open volume fraction in the film, what we would expect from the breaking of network interconnections. Hydrogen thermal effusion studies showed, for the same nitrogen-incorporated films, low temperature (about 200 °C) effusion peaks. As the compact a-C:H films usually show only high temperature (about 600 °C) hydrogen effusion peaks, the low temperature peaks present in a-C(N):H films

were assigned to the presence of open voids [88]. This may lead, as discussed in the next section, to great changes in the film mechanical properties, as a consequence at the loss of connectivity.

It will be interesting to discern if the preceding transformations are a consequence of electronic effects due solely to nitrogen incorporation, or if there is some influence of the growth process itself. Different theoretical approaches were used in the study of hydrogen-free carbon–nitrogen solids.

Weich, Widany, and Frauenheim [89] reported a molecular dynamics simulation study on the formation of carbon–nitrogen solids. They studied the nitrogen incorporation on amorphous carbon within several density ranges, and they found that the presence of nitrogen induces the continuous increase of the sp^2-C atom fraction in all the cases. Hu, Yang, and Lieber [90] studied the semiempirical methods of the nitrogen incorporation effects in carbon clusters formed by diamond cells, and they found that for nitrogen contents greater than about 12 at.% a transition to the sp^2 state is likely to occur. Both findings at least support the possibility that the sp^2-C atom increase observed in a-C(N):H films may be due to a chemical bonding effect. Both visions match with examples coming from our previous discussion: the more or less continuous hybridization changes observed in films deposited from CH$_4$-N$_2$ and CH$_4$-NH$_3$ (see Fig. 28), and the apparently sudden structural evolution, as revealed by Raman spectroscopy, for C$_2$H$_2$-N$_2$ deposited films [58].

In plasma deposition of a-C(N):H films, as the nitrogen precursor partial pressure increases, the bombardment by N$_2^+$ ions is strongly enhanced. On the other hand, it is known that ion bombardment governs the establishment of a highly cross-linked network. However, in addition to deposit energy upon incidence on film surface, nitrogen ions promotes chemical sputtering of carbon atoms. As pointed out by Marton et al. [91], this kind of damage by ion impact may act as a source for trigonal bond formation in carbon–nitrogen films, and thus may be playing a role in the development of a-C(N):H film structure.

5. MECHANICAL PROPERTIES

5.1. Hardness and Stress

Some of the first works on the mechanical properties of plasma-deposited a-C(N):H reported strong stress decrease upon nitrogen incorporation, without appreciable changes in the mechanical hardness [12, 13]. Such observation attracted attention to a-C(N)H films because high internal stress is a stringent limitation to the application of thin films as mechanical protective coatings. Table II shows results on hardness and stress variation upon nitrogen incorporation, into a-C:H films together with deposition details and hardness measurement method.

Analysis of Table II shows discrepancies in the hardness and stress behavior of a-C(N):H films. Although all the works

Table II. Mechanical Hardness and Internal Stress of a-C(N):H Films

Reference	Deposition method gas mixture	Maximum N content (at.%)	Hardness method	Stress range (GPa)	Hardness range (Gpa)
Wood, Wyedeven, and Tsuji [57]	RFPECVD/Hydrocarbon/N2	16	Vickers microhardness	0.4–0.8	10–20
Jacobsohn et al. [54]	RFPECVD C_2H_2-N_2	22	Nanoindentation	1.4–2.9	9.2–17.7
Schwan et al. [53]	RFPECVD/C_2H_2-N_2	20	Knoop microhardness	1.4–0.5	9–17
Franceschini, Achek, and Freire [12]	RFPECVD/CH_4-N_2	11	Vickers microhardness	1.5–2.4	19–21
Freire and Franceschini [56]	RFPECVD/CH_4-NH_3	11	Vickers microhardness	1.3–2.8	13–15
Houert et al. [92]	RFPECVD dc bias C_5H_{10}-N_2-Ar	8.2	Nanoindentation	—	13–23
Chan et al. [49]	ECRRF CH_4-N_2-Ar	12	Nanoindentation	—	9.5–4.5

reported a clear stress reduction upon nitrogen incorporation, the hardness sometimes is quoted as almost constant, or on the other hand clearly decreasing. In addition to the possible effect of different deposition methods and conditions, it can be easily seen that the differences in hardness testing methods are the major source for discrepancies. Constant hardness behavior is only reported with the use microindentation methods, like Vickers and Knoop microhardness. On the other hand, the use of low-load nanoindentation methods always led to a nitrogen-induced decrease in hardness. This is basically the consequence of two factors. The first one is the higher penetration depth owned by high-load microindentation, which allows the substrate mechanical properties to interfere in the measurement. The second one is the high elastic recovery of a-C(N):H films, which was found to be independent of nitrogen content [54]. In microindentation techniques, the penetration depth is determined by measuring the indentation dimensions after the load–unload process. On the other hand, nanoindentation methods measure the penetration depth during the load–unload process, allowing the suppression of elastic recovery effects from the hardness determination.

As mentioned earlier, nitrogen incorporation always leads to stress reduction. This is a straightforward consequence of the structural changes observed in a-C(N):H films. The combined increase in the sp^2-carbon atom fraction, the enlargement of the condensed aromatic ring clusters, the preferential bonding of H to N atoms, and the introduction of the terminal C≡N bonds may contribute to the decrease of network connectivity. This is equivalent to a decrease on the mean coordination number and hence to a reduction in the degree of overconstraining, thus leading to stress relief.

Figure 35 shows the internal stress variation for a-C(N):H films deposited by RFPECVD under similar conditions, from different precursor mixtures (CH_4-N_2 [12], CH_4-NH_3 [56], and C_2H_2-N_2 [54]). Despite all the nitrogen-free films showed about the same stress, the variations upon nitrogen incorporation are different. Even for the CH_4 derived samples, which almost showed the same variation of carbon atom sp^2 fraction (see Fig. 28), which was attributed to the differences in

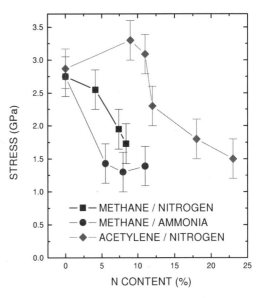

Fig. 35. Internal stress as a function of nitrogen content for several precursor mixtures. Data taken from [12, 54, 56].

the hydrogen content variation of the samples [11]. On the other hand, the C_2H_2 derived film shows a wide range of constant stress, followed by a strong decrease. This behavior closely follows the sp^2-carbon bonding variation upon nitrogen incorporations, as revealed by Raman spectroscopy on these films [54]. In this case, the Raman spectra parameters showed almost constant behavior for the region of constant stress.

This strong sensitivity to all of the parameters involved in the network connectivity, make internal stress measurement a useful diagnostic tool for the overall a-C(N):H film structure. This can be verified by the strong correlation showed between the internal stress and the atomic density, which is shown in Figure 36 [54]. The same structural evolution that led to the stress decrease, may also strongly affect the mechanical hardness. As pointed out by Robertson [23], structural features that lead to a decrease of network connectivity do not contribute to

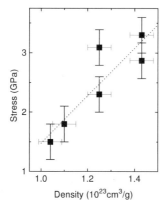

Fig. 36.   Internal stress as a function of film atomic density for films deposited from a C_2H_2-N_2 mixture. (Reproduced from [56].)

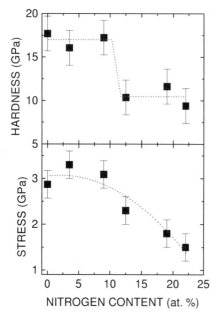

Fig. 37.   Internal stress and mechanical hardness as functions of nitrogen content, and for films deposited from a C_2H_2-N_2 mixture. (Reproduced from [56].)

network rigidity. Thus, the hardness variation must be strongly correlated with internal stress, as shown in Figure 37 [54].

5.2. Friction and Wear

Relatively few works considered the tribological properties of plasma-deposited a-C(N):H films. First, considering the coefficient of friction, nitrogen incorporation was not found to strongly modify it. Prioli et al. [93] studied the friction coefficient of a-C(N):H films against a Si_3N_4 probe in an atomic force microscope, in ambient air. Nitrogen incorporation up to about 10 at.% into the films (RFPECVD deposited in CH_4-NH_3 atmospheres) resulted in no appreciable changes of the friction coefficient, which remained in the 0.22–0.25 range. In addition, they also measured the film surface roughness, which was found to be very low, and to increase upon nitrogen incorporation, in the 0.11- to 0.24-nm range. Chan et al. [49] obtained similar be-

havior in a-C(N):H films deposited by ECRRF in CH_4-N_2-Ar atmospheres, with up to 12-at.% N incorporation. They essentially obtained constant friction coefficients in the 0.11 to 0.16 range, while film roughness increased from 0.2 to 0.7 nm. This roughness behavior was also reported by Silva, Amaratunga, and Barnes [94], and was ascribed to erosion effects arising from N_2^+ ion bombardment [74].

Considering the wear behavior of a-C:H films, it was reported that wear does not depend so strongly on film hardness as could be expected [95]. Wear behavior of a-C:H also depends on other factors, like the nature of the transfer layer—the layer modified by the wear process—and on the chemical reactions carried out by the ambient atmosphere. So, it could be expected that despite the strong drop in hardness, a-C(N):H films could have good wear properties. However, the few results reported on a-C(N):H films did not show an optimist wear performance for this material.

Dekempeneer et al. [96] studied the wear behavior of a-C(N):H films deposited by RFPECVD in CH_4-N_2 atmospheres, with up to 13-at.% N. The wear tests were done in a ball-on-disk tribometer, under air, at fixed 50% relative humidity. The initial friction coefficient was about 0.2 for all samples, and the wear life of the film was determined by the time necessary to achieve a friction coefficient equal to 0.5. Tests performed at 1- and 5-N normal loads showed wear lives greater than 24 h. On the other hand, tests performed with 10-N normal loads showed a strong decrease on the wear life, achieving a 12-h lifetime for the 13-at.% N sample. As should be expected, the studied samples also showed decreasing hardness and stress, and changes in the Raman spectra indicative of increased size or number of sp^2-C condensed ring domains. The same kind of behavior was reported by Wu et al. [97], in a study of friction and wear of ion-beam deposited a-C(N):H films. Despite the very low friction coefficient in ultrahigh vacuum conditions (about 0.02), the wear life was found to be very short, as compared with nitrogen-free a-C:H films.

6. OPTICAL AND ELECTRICAL PROPERTIES

As a semiconductor, the electrical and optical properties of a-C:H films are controlled by their electronic band structure. The electronic density of states of a-C:H films, as in a-Si:H and other amorphous semiconductors, does not show sharp edges and well-defined bandgaps as crystalline semiconductors. The disorder in the lattice potential gives rise to electronic localized states, which contribute to a tail in the density of states, leading it to expand into the gap [98]. The band is divided in two regions, one populated by the higher energy extended states like in crystalline semiconductors, and the tail localized states, which are separated from the extended states by a mobility edge. Since both conduction and valence bands have mobility edges, a mobility gap is thus defined. In addition to the tail states, disorder also gives rise to coordination defects, e.g., dangling bonds, thus introducing electronic defect states in the midgap.

In a-C:H, the tail states are dominated by π electrons, which results, as pointed out by Robertson [99, 100], in an enhanced localization as compared to a-Si:H, giving rise to higher band tail density of states and also to higher defect density in the midgap.

6.1. Optical Properties and Electron Spin Resonance

In the infrared (IR) wavelength range of the light spectrum, the a-C(N):H films are transparent, unless for the localized absorption bands due to CH, NH, C≡N and, if the films are nitrogenated enough, the IR Raman band. In the ultraviolet-visible (UV-vis) wavelength range, optical absorption is dominated by electronic transitions between valence band and conduction band states. So the parameter of most interest is the energy bandgap, or the optical bandgap. In amorphous semiconductors, because of the lack of sharp band edges, E_G is determined by the extrapolation of the bands. The usually adopted procedure is to assume parabolic band edges, which leads to the so-called Tauc relation,

$$\left[\hbar\omega\alpha(\hbar\omega)\right]^{1/2} = A(\hbar\omega - E_G)$$

which allows the determination of the optical gap width from measurements of the absorption coefficient α, which is in turn derived from optical transmittance and reflectance spectra.

Several optical gap variations upon nitrogen content in a-C(N):H films are displayed in Table III, together with deposition details and other optical and electrical parameters. With only one exception, all results quoted in Table III showed a clear reduction of the gap width upon nitrogen incorporation. The nitrogen-free film gaps spanned a relatively wide range of values, indicating different initial structures, which might be expected by the different deposition techniques and deposition parameters. As discussed in Section 2, nitrogen incorporation was found to more frequently lead to an increase in the C sp^2 fraction and in the size of the π clusters. Both modifications in a-C:H films are expected, as stated by Robertson and O'Reilly [21] and Robertson [22], to result in the gap shrinking, as observed. Unfortunately, data on the hybridization state of C atoms of these films are not always available.

The exception to the regularity discussed before is the work of Silva et al. [14], which reported a gap width passing through a maximum at 7-at.% N, with a remarkable variation in the 1.7- to 2.8-eV range. In accordance, a slight decrease of the sp^2 fraction of carbon atoms was reported around 7-at.% N, and also a strong decrease, of the sp^2 nitrogen atom fraction in the same nitrogen content range. The N sp^2 fraction varied from 75 to about 20%, and after increased to about 90%, for the higher nitrogen content.

In amorphous semiconductors, information about the width of the band tail states (or disorder) may also be extracted from the optical absorption spectra. For photon energies near bandgap energy, the optical absorption coefficient of amorphous semiconductors exhibit an exponential dependence on the photon energy, following the so-called Urbach relationship:

$$\alpha(\hbar\omega) = \alpha_o \exp\left[(E - \hbar\omega)/E_0\right]$$

It was found that in this energy range, α follows the same behavior of the joint (valence + conduction) density of states. Thus, E_0, may be interpreted as a measure of the structural disorder [98], as it represents the inverse of band tail sharpness.

Only three results of Urbach width variation are shown in Table III. Schwan et al. [53] reported a clear increase of the Urbach energy for the whole nitrogen incorporation range, typically from 100 to 250 meV, for several deposition pressures. Han and Feldman [8] reported a decrease of the Urbach energy in the 480- to 270-meV range. Silva et al. [14] reported a decrease of E_0 from 270 to 160 meV up to 7-at.% N, which was followed by a small increase to about 200 meV for the higher nitrogen content, indicating, as the results of Han and Feldman, a clear decrease in disorder.

This disorder decrease on alloying is in principle unexpected, because alloying introduces chemical disorder in addition to bond disorder. However, Silva et al. [14] reported results on the joint density of states as determined by EELS which showed a clear decrease in the density of tail states at 7-at.% N and showed a further increase for higher nitrogen content.

Table III. Optical (Tauc) Gap, Urbach Energy, and Spin Density of a-C(N):H Films

Reference	Maximum nitrogen content (at.%)	E_g (eV)	Urbach tail width (meV)	Spin density (spins/cm^3)
Han and Feldman [8] (optical)	18	2.5–2.0	480–270	2×10^{18}–6.4×10^{16} (spins/g)
Lin et al. [101] (ESR)				
Amir and Kalish [9]	9	0.8–0.4		5×10^{20}–$<10^{18}$
Silva et al. [14]	16	1.8–2.15–1.7	260–170–200	4×10^{20}–$<10^{18}$
Schwan et al. [53]	20	1.5–0.8	100–250	—
(several deposition pressures)		1.3–0.7	100–250	
		1.0–0.4	200–250	

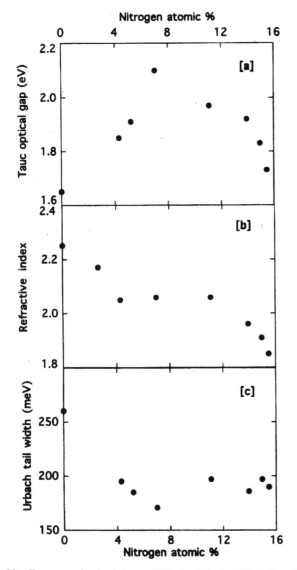

Fig. 38.   Tauc gap, refractive index, and Urbach width of a-C(N):H films. (Reproduced from [14].)

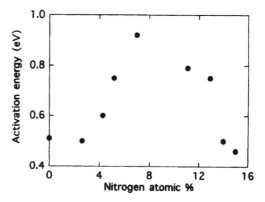

Fig. 39.   Electrical conductivity activation energy vs nitrogen content. (Reproduced from [14].)

6.2. Electrical Properties

Nitrogen incorporation is also found to increase the electrical conductivity of a-C(N):H films. Wood, Wyedeven, and Tsuji [57] reported a 4 orders of magnitude decrease in the room temperature volume resistivity, for a 16-at.% N incorporation. Equivalent results were reported by Schwan et al. [53]. What is argued is that it is caused by electronic doping or by the increase in the sp^2 fraction, in parallel with the increase in the size of graphitic domains.

Evidence on this question may be taken by the behavior of the electrical conductivity σ as a function of temperature. A thermally activated process (T^{-1} dependence on $\log(\sigma)$, Arrhenius plot) is expected if doping takes place, whereas a $T^{-1/4}$ dependence, characteristic of a variable range hopping at the Fermi level is expected for a nondoping situation.

Schwan et al. [53], Amir and Kalish [9], Lin et al. [101], and Stenzel et al. [102] reported a $T^{-1/4}$ behavior for the nitrogen-free as well as the a-C(N):H films conductivities. The only report of a thermally activated behavior in RFPECVD-deposited a-C(N):H films was from the work of Silva et al. [14]. In addition, they reported an N-content dependence on the activation energy of the thermally activated conductivity. Upon increasing the nitrogen content, the activation energy was found to increase up to 7-at.% N from 0.5 to more than 0.9 eV, and again reducing to about 0.5 when the nitrogen content increases to 16-at.% N, as shown in Figure 39.

This behavior led Silva et al. to suggest that nitrogen acts as a weak electron donor in these films. The variation of the activation energy was ascribed to the Fermi level moving up in the bandgap [14]. It must be noted that if this process is really identified as a doping process, its efficiency is very low (7-at.% N for the compensation point). In addition, one must take into consideration that the effect occurs after relatively strong nitrogen-induced structure changes.

In fact, an apparent doping effect was also reported by Schwan et al. [39] in a-C(N):H films deposited by the highly ionized plasma beam deposition technique in C_2H_2-N_2 atmospheres. Schwan et al. also observed thermally activated behavior for the conductivity. As reported by Silva et al. [14],

The variation of the Tauc gap and the Urbach energy obtained by Silva et al. are shown in Figure 38, together with the variation of the refraction index.

Information on the evolution of the defect states in the midgap complement the information on the tail states. Electron spin resonance is a useful tool to probe spin-carrying defects (such as dangling bonds) in amorphous semiconductors. All the reports on ESR spin density quoted in Table I show a remarkable decrease upon nitrogen incorporation. Lin et al. [101] reported the spin density only for two samples, nitrogen-free and 18-at.% N. Amir and Kalish [9] reported an almost constant spin density for a wide range of nitrogen content, being the decrease observed only for the 9.0-at.% sample. In all the cases, the spin signal decreased by more than 2 orders of magnitude. So, in fact, in two of the cases shown in Table III, nitrogen incorporation has led to reduction in the disorder and in the defect density of states.

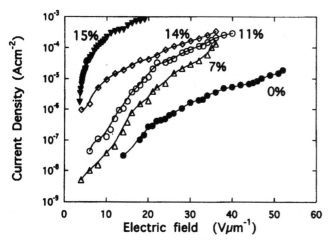

Fig. 40. Electron emission current density as a function of applied electric field, for a-C(N):H films with several N contents. (Reproduced from [15].

they also observed increasing optical gap, and decreasing ESR spin signal, but the Urbach energy was found to increase.

An interesting additional feature of the a-C(N):H films discussed in the previous paragraph [14] is their nitrogen-driven enhancement of the electric field-induced electron emission. Electron field emission is one of the alternatives considered for high-quality flat panel displays. Due to their presumed negative electron affinity, carbon-based materials are one of the possible choices for making emission cathodes. Nitrogen incorporation into a-C:H films was found to remarkably increase emitted electron current density, and to reduce the threshold field by electron emission [15], as shown in Figure 40. Such behavior was also reported by other workers [16], and was in fact used to improve the emission performance of Mo-tip emitters, by coating them with a-C(N):H films [16]. Further works emphasized the importance of surface plasma treatment [94], film thermal treatment after deposition [103], film thickness [104], and sp²-C clusters concentration and size [105] on the improvement of emission properties.

REFERENCES

1. A. Y. Liu and M. L. Cohen, *Science* 245, 841 (1989).
2. D. Marton, K. J. Boyd, and J. W. Rabalais, *Int. J. Mod. Phys. B* 9, 3527 (1995).
3. S. Muhl and J. M. Mendez, *Diamond Relat. Mater.* 8, 1809 (1999).
4. Patents: K. S. Goodson, S. Z. Gornicki, A. S. Lehil, C. K. Prabhakara, and W. T. Tang, Western Digital Corporation, US5855746-A; L. P. Franco, J. H. Kaufman, S. Metin, D. D. Palmer, D. D. Saperstein, and A. W. Wu, IBM Corporation, EP844603-A1.
5. J. Robertson, *Prog. Solid State. Chem.* 21, 199 (1991).
6. A. Grill, *Diamond Relat. Mater.* 8, 428 (1999).
7. D. I. Jones and A. D. Stewart, *Philos. Mag. B* 46, 423 (1982).
8. H. X. Han and B. J. Feldman, *Solid State Commun.* 65, 921 (1988).
9. O. Amir and R. Kalish, *J. Appl. Phys.* 70, 4958 (1991).
10. J. H. Kaufman, S. Metin, and D. D. Saperstein, *Phys. Rev. B* 39, 13053 (1989).
11. D. F. Franceschini, F. L. Freire Jr., and S. R. P Silva, *Appl. Phys. Lett.* 68, 2645 (1996).
12. D. F. Franceschini, C. A. Achete, and F. L. Freire Jr., *Appl. Phys. Lett.* 60, 3229 (1992).
13. S. Metin, J. H. Kaufman, D. D. Saperstein, J. C. Scott, J. Heyman, and E. Haller, *J. Mater. Res.* 9, 396 (1994).
14. S. R. P. Silva, J. Robertson, G. A. J. Amaratunga, B. Raferty, L. M. Brown, J. Schwan, D. F. Franceschini, and G. Mariotto, *J. Appl. Phys.* 81, 2026 (1997).
15. G. A. J. Amaratunga and S. R. P. Silva, *Appl. Phys. Lett.* 68, 2529 (1996).
16. E. J. Chi, J. Y. Shim, H. K. Baik, H. Y. Lee, S. M. Lee, and S. J. Lee, *J. Vac. Sci. Technol. B* 17, 731 (1999).
17. J. C. Angus and F. Jansen, *J. Vac. Sci. Technol. A* 6, 1778 (1988).
18. M. F. Thorpe, *J. Non-Cryst. Solids* 182, 135 (1995).
19. M. A. Tamor, W. C. Vassel, and K. R. Carduner, *Appl. Phys. Lett.* 58, 592 (1991).
20. A. Grill, *Thin Solid Films* 335–336, 189 (1999).
21. J. Robertson and E. P. O'Reilly, *Phys. Rev. B* 35, 2946 (1987).
22. J. Robertson, *Adv. Phys.* 35, 317 (1986).
23. J. Robertson, *Phys. Rev. Lett.* 68, 220 (1992).
24. S. Eisenberg and R. Chabot, *J. Appl. Phys.* 42, 2953 (1971).
25. Y. Lifshitz, S. R. Kasi, and J. W. Rabalais, *Phys. Rev. Lett.* 68, 620 (1989).
26. J. Robertson, *Diamond Relat. Mater.* 3, 361 (1994).
27. G. J. Vandentrop, M. Kawasaky, R. M. Nix, I. G. Brown, M. Salmeron, and G. A. Somorjay, *Phys. Rev. B* 41, 3200 (1990).
28. T. Schwarz-Selinger, A. Von Keudell, and W. Jacob, *J. Appl. Phys.* 86, 3988 (1999).
29. W. Jacob, *Thin Solid Films* 326, 1 (1998).
30. Y. Catheryne and C. Couderc, *Thin Solid Films* 144, 265 (1986).
31. A. Bubenzer, B. Dischler, G. Brandt, and P. Koidl, *J. Appl. Phys.* 54, 4590 (1983).
32. J. W. Zou, K. Reichelt, K. Schmidt, and B. Dischler, *J. Appl. Phys.* 65, 3914 (1989).
33. J. W. Zou, K. Schmidt, K. Reichelt, and B. Dischler, *J. Appl. Phys.* 67, 487 (1990).
34. A. Grill, "Cold Plasmas for Materials Fabrication," IEEE Press, New York, 1994.
35. A. Grill, *IBM J. Res. Dev.* 43, 147 (1999).
36. S. C. Kuo, E. E. Kunhardt, and A. R. Srivatsa, *Appl. Phys. Lett.* 59, 2532 (1991).
37. J. H. Kim, D. H. Ahn, Y. H. Kim, and H. K. Baik, *J. Appl. Phys.* 82, 658 (1997).
38. M. Weiler, S. Sattel, T. Giessin, K. Jung, V. Veerasamy, and J. Robertson, *Phys. Rev. B* 53, 1594 (1996).
39. J. Schwan, V. Batori, S. Ulrich, H. Ehrhardt, and S. R. P. Silva, *J. Appl. Phys.* 84, 2071 (1998).
40. H. W. Song, F. Z. Cui, X. M. He, and W. Z. Li, *J. Phys. Condens. Matter* 6, 6125 (1994).
41. M. A. Tamor and W. C. Vessel, *J. Appl. Phys.* 76, 3823 (1994).
42. M. A. Tamor, J. A. Haire, C. H. Wu, and K. C. Hass, *Appl. Phys. Lett.* 54, 123 (1989).
43. A. Grill, V. Patel, and S. Cohen, *Diamond Relat. Mater.* 3, 281 (1994).
44. L. C. Feldman and J. W. Mayer, "Fundamentals of Surface and Thin Film Analysis," North-Holland, Amsterdam, 1996.
45. A. Mansur and B. Ugolini, *Phys. Rev. B* 47, 10,201 (1993).
46. R. Kalish, O. Amir, R. Brener, A. Spits, and T. E. Derry, *Appl. Phys. A* 52, 48 (1991).
47. F. L. Freire Jr., D. F. Franceschini, and C. A. Achete, *Nucl. Instrum. Methods Phys. Res. B* 85, 268 (1994).
48. A. Grill and V. Patel, *Diamond Films Technol.* 2, 61 (1992).
49. W.-C. Chan, M.-K. Fung, K.-H. Lai, I. Bello, S.-T. Lee, and C.-S. Lee, *J. Non-Cryst. Solids* 124, 180 (1999).
50. S. Rodil, N. A. Morrinson, W. I. Milne, J. Robertson, V. Stolojan, and D. N. Jayavardane, *Diamond Relat. Mater.* 9, 524 (2000).
51. J. Seth, P. Padiyath, and S. V. Babu, *Diamond Relat. Mater.* 3, 221 (1994).
52. M. M. Lacerda, D. F. Franceschini, F. L. Freire Jr., and G. Mariotto, *Diamond Relat. Mater.* 6, 631 (1997).
53. J. Schwan, W. Dworschak, K. Jung, and H. Ehrardt, *Diamond Relat. Mater.* 3, 1034 (1994).

54. L. G. Jacobsohn, F. L. Freire Jr., M. M. Lacerda, and D. F. Franceschini, *J. Vac. Sci. Technol. A* 17, 545 (1999).

55. K.-R. Lee, K. Y. Eun, and J.-S. Rhee, *Mater. Res. Soc. Symp. Proc.* 356, 233 (1995).

56. F. L. Freire Jr. and D. F. Franceschini, *Thin Solid Films* 293, 236 (1997).

57. P. Wood, T. Wyedeven, and O. Tsuji, *Thin Solid Films* 258, 151 (1995).

58. J. Seth, P. Padiyath, and S. V. Babu, *Diamond Relat. Mater.* 3, 210 (1994).

59. D. F. Franceschini, F. L. Freire Jr., C. A. Achete, and G. Mariotto, *Diamond Relat. Mater.* 5, 471 (1996).

60. P. Hammer, N. M. Victoria, and F. Alvarez, *J. Vac. Sci. Technol. A* 16, 2941 (1998).

61. L. R. Shaginyan, F. Fendrych, L. Jastrabik, L. Soukup, V. Yu. Kulikovsky, and J. Musil, *Surf. Coat. Technol.* 116–119, 65 (1999).

62. T. Inoue, S. Ohshio, H Saitoh, and K. Kamata, *Appl. Phys. Lett.* 67, 353 (1995).

63. K. J. Clay, S. P. Speakman, G. A. J. Amaratunga, and S. R. P. Silva, *J. Appl. Phys.* 79, 7227 (1996).

64. S. Bhatasharyya, A. Granier, and G. Turban, *J. Appl. Phys.* 86, 4668 (1999).

65. M. Zhang, Y. Nakayama, T. Miyazaki, and M. Kume, *J. Appl. Phys.* 85, 2904 (1999).

66. C. Hopf, K. Letourneur, W. Jacob, T. Schwarz-Selinger, and A. von Keudell, *Appl. Phys. Lett.* 74, 3800 (1999).

67. H. Saitoh, H. Takamatsu, D. Tanaka, N. Ito, S. Ohshio, and H. Ito, *Jpn. J. Appl. Phys.* 39, 1258 (2000).

68. P. Hammer and W. Gissler, *Diamond Relat. Mater.* 5, 1152 (1996).

69. P. Hammer, M. A. Baker, C. Lenardi, and W. Gissler, *Thin Solid Films* 291, 107 (1996).

70. J. Hong and G. Turban, *Diamond Relat. Mater.* 8, 572 (1999).

71. J. Hong and G. Turban, *J. Vac. Sci. Technol. A* 17, 314 (1999).

72. F. D. A. Aarão Reis and D. F. Franceschini, *Appl. Phys. Lett.* 74, 209 (1999).

73. S. Todorov, D. Marton, K. J. Boyd, A. H. Al Bayati, and J. W. Rabalais, *J. Vac. Sci. Technol. A* 12, 3192 (1994).

74. F. D. A. Aarão Reis and D. F. Franceschini, *Phys. Rev. E* 61, 3417 (2000).

75. R. O. Dillon, J. A. Woolam, and V. Katkanant, *Phys. Rev. B* 29, 3482 (1984).

76. J. Schwan, S. Ulrich, V. Batori, H. Ehrhardt, and S. R. P Silva, *J. Appl. Phys.* 80, 440 (1996).

77. A. C. Ferrari and J. Robertson, *Phys. Rev. B* 61, 14,095 (2000).

78. G. Mariotto, F. L. Freire Jr., and C. A. Achete, *Thin Solid Films* 241, 255 (1994).

79. F. L. Freire Jr., *Jpn. J. Appl. Phys.* 136, 4886 (1997).

80. N. M. Victoria, P. Hammer, M. C. dos Santos, and F. Alvarez, *Phys. Rev. B* 61, 1083 (2000).

81. S. D. Berger, D. R. Mackenzie, and P. J. Martin, *Philos. Mag. Lett.* 57, 285 (1988).

82. V. S. Veerasamy, J. Yuan, G. A. J. Amaratunga, W. I. Milne, K. W. R. Gilkes, M. Weiller, and L. M. Brown, *Phys. Rev. B* 48, 17,954 (1993).

83. S. Souto and F. Alvarez, *Appl. Phys. Lett.* 70, 1539 (1997).

84. C. Donnet, J. Fontaine, F. Lefèbvre, A. Grill, V. Patel, and C. Jahnes, *J. Appl. Phys.* 85, 3264 (1999).

85. J. Seth, A. J. I. Ward, and V. Babu, *Appl. Phys. Lett.* 60, 1957 (1992).

86. J. LaManna, J. Bradok Wilking, S. H. Lin, and B. J. Feldman, *Solid State Commun.* 109, 573 (1999).

87. F. L. Freire Jr., D. F. Franceschini, and C. A. Achete, *Phys. Status Solidi B* 192, 493 (1995).

88. D. F. Franceschini, F. L. Freire Jr., W. Beyer, and G. Mariotto, *Diamond Relat. Mater.* 3 (1993).

89. F. Weich, J. Widany, and Th. Frauenheim, *Phys. Rev. Lett.* 78, 3226 (1997).

90. J. Hu, P. Yang, and C. M. Lieber, *Phys. Rev. B* 57, 3185 (1998).

91. D. Marton, K. J. Boyd, J. W. Rabalais, and Y. Lifishitz, *J. Vac. Sci. Technol. A* 16, 455 (1998).

92. R. Hauert, A. Gisenti, S. Metin, J. Goitia, J. H. Kaufman, P. H. M. Loosdrecht, A. J. Kellok, P. Hoffman, R. L. White, and B. D. Hermsmeier, *Thin Solid Films* 268, 22 (1995).

93. R. Prioli, S. I. Zanette, A. O. Caride, D. F. Franceschini, and F. L. Freire Jr., *J. Vac. Sci. Technol. A* 14, 2351 (1996).

94. S. R. P. Silva, G. A. J. Amaratunga, and J. R. Barnes, *Appl. Phys. Lett.* 71, 1477 (1997).

95. A. Grill, *Surf. Coat. Technol.* 94–95, 507 (1997).

96. E. H. A. Dekempeneer, J. Maneve, J. Smeets, S. Kuypers, L. Eersels, and R. Jacobs, *Surf. Coat. Technol.* 68–69, 621 (1994).

97. R. L. C. Wu, K. Myioshi, W. C. Lanter, J. D. Wrbanek, and C. A. De-Joseph, *Surf. Coat. Technol.* 120–121, 573 (1999).

98. R. A, Street, "Hydrogenated Amorphous Silicon," Cambridge Univ. Press, Cmabridge, U.K., 1991.

99. J. Robertson, *J. Non-Cryst. Solids* 198–200, 615 (1996).

100. J. Robertson, *Diamond Relat. Mater.* 6, 212 (1997).

101. S. Lin, K. Noonan, B. J. Feldman, D. Min, and M. T. Jones, *Solid State Commun.* 80, 101 (1991).

102. O. Stenzel, M. Vogel, S. Pönitz, R. Petrich, T. Wallendorf, C. V. Borczyskowski, F. Rozploch, Z. Krasolnik, and N. Kalugin, *Phys. Status Solidi A* 140, 179 (1993).

103. A. P. Burden, R. D. Forrest, and S. R. P. Silva, *Thin Solid Films* 337, 257 (1999).

104. R. D. Forrest, A. P. Burden, S. R. P. Silva, L. K. Chea, and X. Shi, *Appl. Phys. Lett.* 73, 3784 (1998).

105. J. D. Carey, R. D. Forrest, R. U. A. Khan, and S. R. P. Silva, *Appl. Phys. Lett.* 77, 2006 (2000).

Chapter 13

CONDUCTIVE METAL OXIDE THIN FILMS

Quanxi Jia

Los Alamos National Laboratory, Superconductivity Technology Center, Los Alamos, New Mexico, USA

Contents

1. TRANSPARENT CONDUCTING OXIDES

Common features of transparent conductive oxide (TCO) films are reasonably low resistivity values and good optical properties at a given optical wavelength. These unique properties of TCO films have made them very attractive for applications in photovoltaic devices, flat-panel displays, touch-panel controls, electromagnetic signal shielding, low-emissivity windows, defrosting windows, and electrochromic mirrors and windows.

1.1. Different TCO Materials

Many TCOs have been investigated in the past. The objective is to find TCO systems that show the highest conductivity and optical transparency for specific applications. Table I outlines some of the most commonly studied TCO materials [1–13]. Most research to develop highly transparent and conductive films has focused on binary and ternary compounds. However, multicomponent oxides composed of binary and/or ternary compounds with varied chemical compositions, such as $ZnO-SnO_2$, $Ga_2O_3-In_2O_3$, $ZnO-In_2O_3$, In_2O_3-

Handbook of Thin Film Materials, edited by H.S. Nalwa
Volume 4: Semiconductor and Superconductor Thin Films

ISBN 0-12-512912-2/$35.00

Table I.   Electrical Properties of Different Transparent Conducting Oxides

Material	Resistivity (m$\Omega\cdot$cm)	Carrier concentration (10^{20} cm^{-3})	Hall mobility [cm^2/(V·S)]	Reference
ZnO	9	0.23		[1]
ZnO:Al	0.14	>10	>40	[1]
ZnO:Ga	0.27	>20	>15	[1]
In$_2$O$_3$:Sn (ITO)	0.2–0.4	14.5	29	[2]
Zn$_2$SnO$_4$	17–50			[3]
ZnSnO$_3$	4	1.0	~15	[4]
CdSb$_2$O$_6$:Y	24.4	1.3	1.9	[5]
In$_4$Sn$_3$O$_{12}$	0.2	10	20	[6]
Zn$_2$In$_2$O$_5$	0.39	5	~30	[7]
GaInO$_3$	2.5	4	10	[8]
MgIn$_2$O$_4$	0.43	6.3	2.2	[9]
CuAlO$_2$	1,052	0.0013	10.4	[10]
CuGaO$_2$	15,873	0.017	0.23	[11]
SrCu$_2$O$_2$:K	20,704	0.0061	0.46	[12]
AgInO$_2$:Sn	167	0.27	0.47	[13]

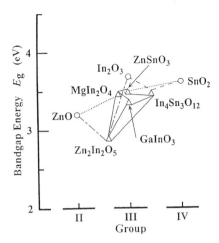

Fig. 1. Bandgap energy of transparent conductive oxide materials: circles, binary compounds; triangles, ternary compounds; lines, multicomponent oxides. Reproduced with permission from T. Minami, *MRS Bull.* 25, 38 (2000). © 2000, Materials Research Society.

SnO$_2$, In$_2$O$_3$-GaInO$_3$, In$_2$O$_3$-MgIn$_2$O$_4$, MgIn$_2$O$_4$-Zn$_2$In$_2$O$_5$, ZnSnO$_3$-Zn$_2$In$_2$O$_5$, GaInO$_3$-Zn$_2$In$_2$O$_5$, and In$_4$Sn$_3$O$_{12}$-Zn$_2$In$_2$O$_5$, have also been investigated [14]. Most of the TCO films investigated so far have n-type conductivity. Recently, p-type TCO films, such as CuAlO$_2$ [10], CuGaO$_2$ [11], and SrCu$_2$O$_2$ [12], have been developed. The growth of both n- and p-type TCO films makes it possible to fabricate p-n junctions based on transparent all-oxide materials [15].

The most important optical and electrical transport properties of TCO films are the visible absorption coefficient, electrical conductivity or resistivity, and carrier density and mobility. Other properties, such as bandgap energy, work function, thermal stability, chemical durability, and processing compatibility, should be also considered for specific applications of TCOs. Figure 1 shows the bandgap energy, and Figure 2 shows the relationship between the work function and the carrier density of TCO materials with different chemical compositions [16].

1.2. Figure of Merit for TCOs

The measure of the performance of TCOs (or so-called figure of merit) is the ratio of the electrical conductivity σ (measured per ohm per centimeter) to the visible absorption coefficient α (measured per centimeter) [17],

$$\sigma/\alpha = -\left\{ R_s \ln(T + R) \right\}^{-1} \qquad (1)$$

$$R_s = \rho/t \qquad (2)$$

where R_s is the sheet resistance, ρ is the resistivity in ohms centimeter, and t is the TCO film thickness in centimeters. The sheet resistance has units of ohms, but is conventionally specified in units of ohms per square ($\Omega/\square$). In Eq. (1), T is the

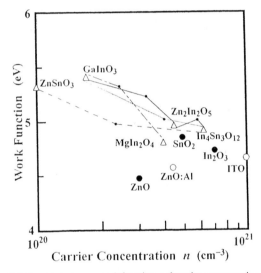

Fig. 2. Relationship between work function and carrier concentration of transparent conductive oxide films: circles, binary compounds; triangles, ternary compounds; dots, multicomponent oxides prepared with a composition of 50 wt.%. Reproduced with permission from T. Minami, *MRS Bull.* 25, 38 (2000). © 2000, Materials Research Society.

total visible transmission and R is the total visible reflectance. The larger is the value of σ/α the better is the performance of the TCO films [17]. From a materials point of view, a good understanding of the fundamental relationship between the conductivity and the transparency of the TCOs is the key to achieve the highest possible σ/α. The values of σ/α vary in a wide range from 0.1 to 7 Ω^{-1}, depending on the TCO materials [17].

An alternative measure of the performance of a transparent conductor, the maximum figure of merit occurring at 90% optical transmission, is expressed as [18]

$$\phi_{TC} = T^{10}/R_s = \sigma t \cdot \exp(-10\alpha t) \qquad (3)$$

where 90% optical transmission is T^{10} and the optical transmission T is given by the ratio of radiation I_0 entering the coating on one side to the radiation I leaving the sample on the opposite side so that

$$T = I/I_0 = \exp(-\alpha t) \qquad (4)$$

In this definition, the figure of merit of a coating with a specified σ and α is a function of its thickness, and it achieves a maximum value at $t_{max} = 1/10\alpha$ rather than at $1/\alpha$.

Note that increasing TCO film thickness to reduce the sheet resistance R_s for maximum σ/α is not necessarily the best approach, because a thicker TCO layer can sometimes lead to a large optical absorptance. Instead, it might be more effective to look for highly degenerated TCO materials with large carrier mobility, because this will enhance the electrical conductivity without degrading the optical properties of the TCO films.

1.3. Deposition Techniques

Thin TCO films with high transparency and low resistivity can be prepared by several different deposition techniques. These techniques include evaporation [19], magnetron sputtering [20], chemical vapor deposition [21], spray pyrolysis [22], screen printing [23], and pulsed laser deposition [24]. Sputtering is the most widely used deposition technique. Pulsed laser deposition recently has emerged as one of the most powerful techniques for growth of high performance TCO films. However, it should be noted that the electrical and optical properties of TCO films are very sensitive to processing parameters and deposition techniques. Experimental results also show that the electrical and optical properties of the TCO films can be changed by postdeposition heat treatment. In the following sections, both sputtering and pulsed laser deposition are described in detail to show how processing conditions and deposition techniques influence the properties of indium–tin-oxide (ITO) films. ITO is one of the most widely investigated TCO films. Table II outlines some typical properties of ITO films at room temperature.

1.3.1. Sputtering

The most widely used deposition technique for growth of ITO films is sputtering. Sputtering is accomplished by accelerating energetic particles onto a target surface with sufficient energy (50–1000 eV) to result in the ejection of one or more atoms from the target. A gas or a mixture of different gases with a pressure in the range of a few to several hundreds of millitorr is

introduced into the sputtering chamber through fine controlled valves to provide a medium in which a glow discharge can be initiated and maintained. The most common gas used for sputtering is Ar, due to its higher secondary emission coefficient for most materials to be sputtered. Figure 3 shows a schematic of the sputtering system used to deposit ITO films. The basic principles of the sputtering process (either dc or rf) can be found elsewhere [25]. The typical processing parameters for the growth of ITO films by sputtering are outlined in Table III.

To obtain the highest quality ITO films by sputtering, the target composition, sputtering power, substrate temperature during deposition, oxygen partial pressure need to be optimized. Note that a residual water-vapor partial pressure need to be optimized. Noted that a reduction in the resistivity of ITO films at fixed deposition temperature has been accomplished by introducing hydrogen into the deposition chamber during sputtering [26].

For a given chemical composition of the ITO material such as 90 wt.% In_2O_3 + 10 wt.% SnO_2, the substrate temperature during sputtering plays one of the most important roles in determining the resistivity of the ITO films. Figure 4 shows the resistivity of ITO films as a function of substrate temperature during deposition [27]. As can be seen from this figure, the resistivity decreases with increasing substrate temperature. The Hall mobility of the ITO film is believed to increase with increasing substrate temperature during deposition. This, in return, reduces the electrical resistivity of the ITO films.

Note that postannealing ITO films in air, vacuum, hydrogen, and/or forming gas leads to grain growth and changes in other physical properties of ITO films. For example, the effective bandgap energy increases from 3.638 eV for the as-deposited ITO film to 3.673 eV for film annealed at 550 °C. The refractive index of ITO films typically ranges from 1.83 to 2.15. In most cases, the refractive indices of annealed films are smaller than those of as-deposited films. Also, postannealing ITO films in air generally reduces the electrical resistivity up to a certain heat treatment temperature. The drop in resistivity after postannealing mostly has been attributed to reduction of oxygen in ITO, which increases the carrier concentration [28].

1.3.2. Pulsed Laser Deposition

Pulsed laser deposition (PLD) is based on physical vapor deposition processes, arising from the impact of high-powered short-pulsed laser radiation on solid targets and leading to the removal of material from the impact zone. PLD has been used to deposit high performance ITO films [24, 29]. The successful deposition of highly conductive and transparent ITO films at relatively low substrate temperature by PLD makes this technique very attractive, where a low deposition temperature is required because of the extremely sensitive nature of many substrates toward thermal budget.

Typical deposition conditions for the growth of ITO films by PLD are outlined in Table IV. Figure 5 is a schematic of a PLD system, which consists of a target holder and a substrate holder housed in a vacuum chamber. A high-powered laser is used as

Table II.   Typical Properties of Indium–Tin-Oxide Films at Room
Temperature

Resistivity (m$\Omega\cdot$cm)	0.1–0.5
Carrier density (cm^{-3})	10^{20}–10^{21}
Hall mobility [cm^2/(V·S)]	1–40
Energy bandgap (eV)	3.5–3.7
Work function (eV)	~4.8
Refractive index	1.8–2.2
Average transmittance ($\lambda = 0.4$–1.1 μm)	85–95%

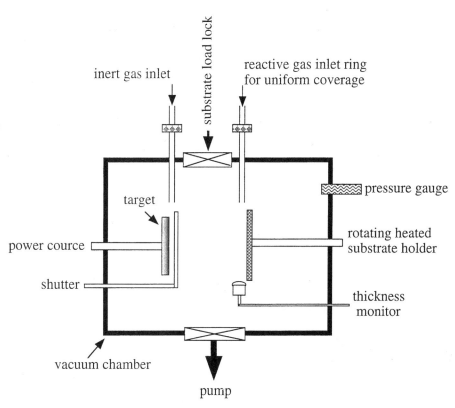

Fig. 3. Schematic of a magnetron sputtering system used for deposition of indium–tin-oxide films.

Table III. Typical Sputtering Deposition Conditions for Growth of Indium–Tin-Oxide Films

Target	90 wt.% In_2O_3 + 10 wt.% SnO_2
Substrate temperature	Room temperature – 400 °C
Total gas (Ar + O_2) pressure	2.0–15 mtorr
Oxygen partial pressure	0.5–1% O_2 at volume ratio
Power density	1–2 W/cm^2
Substrates	Glass, Si, GaAs, InP, etc.
Sputter configuration	rf or dc Magnetron

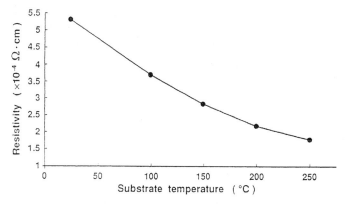

Fig. 4. Indium–tin-oxide film resistivity as a function of substrate temperature during deposition. Reproduced with permission from M. Higuchi et al., *J. Appl. Phys.* 74, 6710 (1993). © 1993, Institute of Physics.

an external energy source to vaporize materials and to deposit thin films. When the laser beam enters the chamber through the quartz window and hits the sintered target, it generates a plume that is directed perpendicularly to the surface of the target. The substrate is mounted on a heater block or substrate holder on the opposite side, facing the target. The basic principles of the PLD process and a complete overview of what is required to set up a PLD system can be found elsewhere [30].

PLD has some advantages compared to other deposition techniques. The film composition is quite close to that of the target, even for a multicomponent target. This is especially important when depositing complex multicomponent conducting oxide films. For ITO films deposited by PLD, the deposition rate has been found to decrease with increasing oxygen pressure [24]. The resistivity of the films, as shown in Figure 6, is not only a function of deposition temperature, but also of oxy-

Table IV. Typical Processing Conditions for Growth of Indium–Tin-Oxide Films by Pulsed Laser Deposition

Excimer laser	ArF (λ = 193 nm)
	KrF (λ = 248 nm)
	XeCl (λ = 308 nm)
Energy density	1–2 J/cm^2
Repetition rate	5–20 Hz
Target	90 wt.% In_2O_3 + 10 wt.% SnO_2
Oxygen pressure	1×10^{-1}–3×10^{-3} torr
Substrate temperature	Room temperature – 500 °C
Substrates	Glass, Si, GaAs, InP, etc.

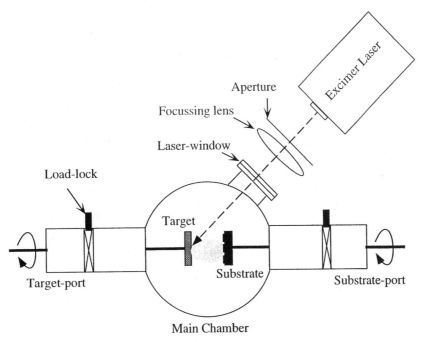

Fig. 5. Schematic of a pulsed laser system for deposition of transparent conductive oxide films.

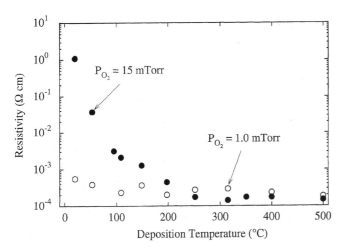

Fig. 6. Resistivity as a function of substrate temperature at oxygen pressures of 1 mtorr (open circles) and 15 mtorr (solid circles). Reproduced with permission from J. P. Zheng and H. S. Kwok, *Appl. Phys. Lett.* 63, 1 (1993). © 1993, American Institute of Physics.

gen pressure [24, 29]. The optical transmission increases with increasing oxygen pressure during film deposition. However, this increase is accompanied by an increase in film resistivity. ITO films with a transmission greater than 90% at wavelengths ranging from 600 to 900 nm have been demonstrated. At optimized oxygen pressures, films with resistivities of 0.56 and 0.14 mΩ·cm are realized on glass at deposition temperatures of room temperature and 310 °C, respectively [24]. ITO films (80 ± 5 nm) with a resistivities of 0.6 and 0.4 mΩ·cm are also obtained on InP at deposition temperatures of room temperature and 310 °C, respectively [29].

Note that PLD also has been used for to deposit of other TCO films such as GaInO$_3$ [8] and CuAlO$_2$ [31].

1.4. Epitaxial Growth of ITO Films

For many applications, it is fine to grow amorphous and/or polycrystalline transparent conducting oxides such as ITO films on specific substrates. However, for a better understanding of the conduction mechanisms and other potential applications of ITO films, it is necessary to fabricate high crystallinity ITO films. Recently, heteroepitaxial growth of ITO film on yttria-stabilized zirconia (YSZ) substrate has been conducted [32]. Figure 7 shows the standard X-ray diffraction 2θ scans for films grown on (a) glass and (b) YSZ substrates, respectively [32]. The heteroepitaxial relationships between ITO film and (100) oriented YSZ substrate, which are described as $(001)_{ITO}\|(001)_{YSZ}$ and $[100]_{ITO}\|[100]_{YSZ}$, can be deduced from the X-ray diffraction pole figure and Figure 7b.

In comparing the electrical properties of the epitaxial ITO film deposited by dc magnetron sputtering with those of poly-crystalline film grown on a glass substrate, experiments show that neither large angle grain boundaries nor crystalline orientation are revealed to be the dominant electron scattering factors in ITO films [32]. However, the resistivity of heteroepitaxial ITO film deposited by electron-beam evaporation on single-crystal YSZ substrate has been found to be significantly smaller than that of ITO film on glass. For example, the resistivity is 0.12 and 0.27 mΩ·cm for heteroepitaxial ITO on YSZ and ITO on glass, respectively [33]. Such a heteroepitaxial ITO film on YSZ has a Hall mobility of 38 cm^2/(V·s) and a carrier density of 14×10^{20} cm^{-3}. The much higher crystallinity of the ITO film deposited by the electron beam, compared to that deposited by dc magnetron sputtering, may explain the preceding conflicting results.

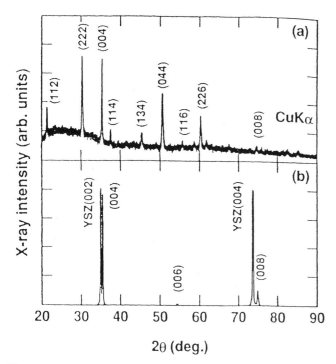

Fig. 7.   X-ray diffraction 2θ scan patterns for indium–tin-oxide films (a) grown on a glass substrate and (b) grown on a single crystalline (001) yttria-stabilized zirconia (YSZ) substrate. Reproduced with permission from M. Kamei et al., *Appl. Phys. Lett.* 64, 27/2 (1994). © 1994, American Institute of Physics.

1.5. Etching ITO Films

Chemical etchants, such as $HI:H_2O = 1:1$, $HCl:H_2O = 1:1-3$, and $H_3PO_4:H_2O = 1:1$, are commonly used to etch ITO films. Decreasing the concentration of the etching solution reduces the etching rate. For example, the etching rates at 60 °C for $HCl:H_2O$ ratios of 1:1 and 1:2 are 17.8 and 12.6 nm/s, respectively. The H_3PO_4-based etchant has an even lower etching rate compared to the other etchants mentioned [34].

2. RUTHENIUM OXIDE

Ruthenium oxide (RuO_2) crystallizes in the rutile structure, with lattice parameters of $a = b = 0.44902$ nm and $c = 0.31059$ nm. For single-crystal RuO_2, the room-temperature resistivity is approximately 35.2 $\mu\Omega\cdot$cm. The resistivity of RuO_2 is isotropic in the temperature range of 10–1000 K [35]. In other words, electrical resistivity along the [111], [011], [100], and [001] crystal directions are all the same with a value of around 35 $\mu\Omega\cdot$cm [36]. The residual resistivity ratio (RRR $= \rho_{300\,K}/\rho_{4.2\,K}$) of single-crystal RuO_2 can be as large as 20,000 [37].

2.1. Deposition Techniques

Several thin film growth techniques have been used to deposit RuO_2 thin films. These techniques include reactive sputtering [38–42], metallorganic chemical vapor deposition (MOCVD)

[43, 44], pulsed laser deposition [45–47], low-temperature chemical vapor deposition [48], and oxygen plasma-assisted molecular beam epitaxy [49]. The following subsections describe sputtering, MOCVD, and PLD for the deposition of RuO_2 thin films.

2.1.1. Sputtering

Sputtering, which includes dc sputtering, rf magnetron sputtering, and ion beam sputtering, has been used to deposit RuO_2 films. The target is high purity (99.95% or better) Ru. The deposition is commonly performed in a mixed O_2-Ar atmosphere. Stoichiometric RuO_2 films reportedly can be obtained over a wider range of sputtering conditions in the O_2-Ne mixture than in the O_2-Ar mixture [50].

The electrical resistivity of RuO_2 films formed by reactive sputtering can vary by an order of magnitude with the amount of oxygen in the chamber, even though all RuO_2 films are, within 8%, stoichiometric RuO_2 [51]. A minimum room-temperature electrical resistivity of about 40 $\mu\Omega\cdot$cm is observed at 50% O_2, although the RRR is monotonically decreasing with the percentage of O_2 and increasing with the grain size. A RRR of about 2 is a typical value for polycrystalline RuO_2 films. Electron scattering due to grain boundaries has been suggested as the dominant factor that influences the resistivity of polycrystalline RuO_2 thin films [52].

2.1.2. Metallorganic Chemical Vapor Deposition

Both cold and hot-wall MOCVD reactors have been used to deposit RuO_2 thin films [44, 53, 54]. Tris(2,2,6,6-tetramethyl-3,5-heptanedionato) ruthenium Ru(TMHD)$_3$, and Ru(C$_5$H$_5$)$_2$ are promising precursors for MOCVD RuO_2 thin films. The precursor vapor is commonly introduced into the reactor via a high-purity nitrogen carrier gas. Pure oxygen is used as the reactant gas. The total pressure in the reactor is in the range of 2–10 torr. The rate of film growth is controlled by the partial pressure of the precursor. This control can be achieved by adjusting the precursor source temperature, usually between 85 and 170 °C, at a fixed precursor carrier gas flow. Table V outlines MOCVD growth conditions for RuO_2 thin films [54].

The room-temperature electrical resistivity of RuO_2 films deposited by MOCVD widely varies. The high carbon content of earlier MOCVD RuO_2 films contributed to higher electrical resistivity values in the past [43]. Crack-free RuO_2 films with typical room-temperature resistivity of 35–40 $\mu\Omega\cdot$cm recently have been grown on SiO_2/Si in a temperature range of 275–425 °C by MOCVD [54]. Figure 8 shows the dependence of RuO_2 film resistivity on substrate temperature [55]. The lowest resistivity films are obtained in a temperature window between 275 and 450 °C. At temperatures above 500 °C, resistivity increases dramatically with increasing temperature and becomes immeasurable (>100 $\Omega\cdot$cm) above 700 °C. Deposition at high temperature (>450 °C) is thought to result in films of increasing surface roughness and, consequently, higher resistivity [55].

Table V.   Metallorganic Chemical Vapor Deposition Conditions for Growth of RuO$_2$ Thin Films.[a]

Substrate temperature	250–850 °C
Reactor pressure	4 torr
Ru(TMHD)$_3$ source temperature	110–118 °C
Flow rate of O$_2$ reactant gas	120 sccm
Flow rate of N$_2$ carrier gas	20 sccm
Flow rate of N$_2$ background gas	50 sccm
Film thickness	100–150 nm
Film growth rate	2.0–4.0 nm/min
Substrate materials	SiO$_2$/Si, Pt/Ti/SiO$_2$/Si

[a]The deposition is carried out from a horizontal, cold-wall quartz reactor (from [54]).

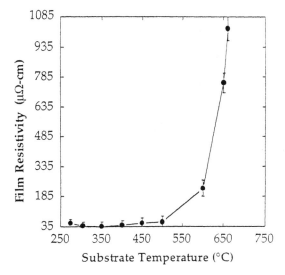

Fig. 8.   Substrate temperature effect on the resistivity of (110) oriented RuO$_2$ films grown by metallorganic chemical vapor deposition at a fixed oxygen concentration of 20%. Reproduced with permission from J. Vetrone et al., *J. Mater. Res.* 13, 2281 (1998). © 1998, Materials Research Society.

2.1.3. Pulsed Laser Deposition

Pulsed laser deposition has been established as one of the most powerful techniques to deposit many complex oxide materials that have superior properties compared to films deposited by other techniques. It recently was reported that RuO$_2$ thin films with high crystallinity were deposited by PLD [45, 56]. Both XeCl ($\lambda = 308$ nm) and ArF ($\lambda = 193$ nm) excimer lasers have been used to deposit RuO$_2$ thin films.

A ceramic RuO$_2$ pellet is usually used as the target for PLD growth of RuO$_2$ film. The RuO$_2$ pellet is made from RuO$_2 \cdot x$H$_2$O (55.85% Ru) powder. The powder is pressed into a disk and sintered at 500 °C in air for 48 h to form the target. Stoichiometric RuO$_2$ thin films can be deposited in a wide temperature range from room temperature to 700 °C. A wide range of oxygen pressures (0.5–200 mtorr) can be used if a ceramic RuO$_2$ pellet is used as a target. The oxygen pressure during deposition has less influence on both the structural and electrical properties of RuO$_2$ films deposited by PLD. The

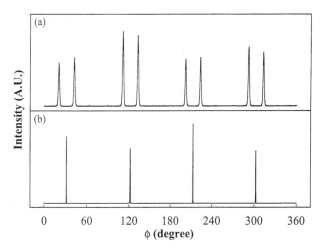

Fig. 9.   X-ray diffraction ϕ scans for a heterostructure of RuO$_2$ deposited at 700 °C on LaAlO$_3$ taken (a) on (101) reflection of RuO$_2$ and (b) (101) reflection of LaAlO$_3$. Reproduced with permission from Q. X. Jia et al., *Appl. Phys. Lett.* 67, 1677 (1997). © 1997, American Institute of Physics.

room-temperature resistivity of the films is mainly a function of substrate temperature during film deposition. The room-temperature resistivity is around 35 $\mu\Omega\cdot$cm for RuO$_2$ films deposited on several different substrates at 700 °C.

2.2. Epitaxial Growth of RuO$_2$ Films

Polycrystalline RuO$_2$ thin films can be deposited routinely on different substrates by sputtering and/or MOCVD. The residual resistivity ratio of the polycrystalline RuO$_2$ thin films is in the range of 1–2 compared with a value of over 20 for bulk single-crystal RuO$_2$ [36]. The room-temperature resistivity, which is one of the most important parameters for applications of RuO$_2$ thin film as an electrode material, is in the range of 100 $\mu\Omega\cdot$cm for most of the polycrystalline RuO$_2$ thin films. This value is three times higher than the 35 $\mu\Omega\cdot$cm of bulk single-crystal RuO$_2$. Epitaxial crystalline RuO$_2$ thin films are expected to have improved electrical conductivity due to the lack of grain-boundary related problems. The preparation of crystalline films is also very important for the study of fundamental properties of RuO$_2$, in particular, the surface structure of RuO$_2$.

2.2.1. RuO$_2$ on YSZ and LaAlO$_3$

Epitaxial RuO$_2$ films can be grown on single-crystal (100) LaAlO$_3$ ($a = 0.3793$ nm) and (100) YSZ ($a = 0.5139$ nm) substrates by PLD [45, 57]. The films deposited at temperatures from 400 to 700 °C on LaAlO$_3$ and/or YSZ are (h00) textured with respect to the normal of the substrate surface. The in-plane orientation of the RuO$_2$ film with respect to the major axes of the (100) LaAlO$_3$ substrate can be evaluated from X-ray diffraction ϕ scans. Figure 9 shows typical ϕ scans on (a) the (101) reflection of the RuO$_2$ film deposited at 700 °C and (b) the (101) reflection of the LaAlO$_3$ substrate [45]. Similar diffraction results are observed for RuO$_2$ on YSZ, except

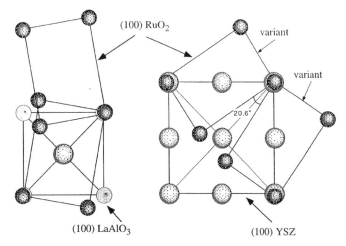

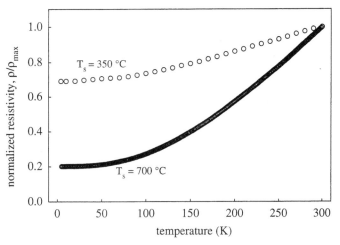

Fig. 10. Schematic of diagonal-type epitaxy of RuO₂ on LaAlO₃ and YSZ. This growth pattern results in the smallest lattice mismatches between the film and the substrate. Reproduced with permission from Q. X. Jia et al., *J. Vac. Sci. Technol. A* 14, 1107 (1996). © 1996, Materials Research Society.

Fig. 11. Dependence of normalized resistivity of RuO₂ thin films on the measurement temperature for films on LaAlO₃ deposited at 350 and 700 °C. Reproduced with permission from Q. X. Jia et al., *J. Vac. Sci. Technol. A* 14, 1107 (1996). © 1996, Materials Research Society.

that the film axes are rotated 45° with respect to the substrate major axes. The texturing relationship between the thin film and the substrate, based on X-ray diffraction results, is $(h00)_{RuO_2}||(h00)_{LaAlO_3}$ and $\langle 011 \rangle_{RuO2}||\langle 110 \rangle_{LaAlO_3}$ for RuO₂ on LaAlO₃, and $(h00)_{RuO_2}||(h00)_{YSZ}$ and $\langle 011 \rangle_{RuO_2}||\langle 001 \rangle_{YSZ}$ for RuO₂ on YSZ. One very interesting growth pattern that is evident in Figure 9 is that the diagonal of the rectangular basal plane lattice of RuO₂ is aligned with the cubic diagonal of the LaAlO₃ substrate ($\langle 011 \rangle_{RuO_2}||\langle 110 \rangle_{LaAlO_3}$). The lattice mismatch for the diagonal is only about 1.8%. This unique rectangle-on-cube arrangement gives rise to the two nearest RuO₂ ϕ-scan diffraction peaks neighboring the (101) reflection at an angle of about 21°, as shown in Figure 9a. Figure 10 is a schematic of the diagonal-type epitaxy of RuO₂ on LaAlO₃ and YSZ [58]. This growth pattern results in the smallest lattice mismatches between the film and the substrate. Note that the degeneracy of the RuO₂(101) peaks in the ϕ scans shown in Figure 9a comes from the different variants shown in Figure 10.

The epitaxial RuO₂ films are highly conductive, with a room-temperature resistivity of 35 ± 2 $\mu\Omega\cdot$cm, a value that is comparable to that of bulk RuO₂ single crystals. Figure 11 shows the typical normalized resistivity versus temperature characteristics of RuO₂ films on LaAlO₃ substrates deposited at two different temperatures [58]. RuO₂ films are metallic as long as the deposition temperature is above 100 °C. However, the RRR, which is a direct indication of the structural perfection or the impurity and defect content of the films, is a strong function of the deposition temperature. The higher is the deposition temperature, the larger is the RRR value. The RRR values of the crystalline RuO₂ thin films deposited on either LaAlO₃ or YSZ by PLD can be as large as 5, compared to values of 1–2 for polycrystalline RuO₂ thin films and a value of greater than 20 for bulk RuO₂ single crystals [45, 57].

2.2.2. RuO₂ on MgO

Epitaxial RuO₂ thin films have been grown on single-crystal (100) MgO ($a = 0.4213$ nm) substrates by different deposition techniques [44, 46, 49]. Films deposited at temperatures from 400 to 700 °C on MgO are $(hk0)$ textured with respect to the normal of the substrate surface. The in-plane orientation of the RuO₂ film with respect to the major axes of the (100) MgO substrate is deduced from X-ray diffraction (XRD) ϕ scans on the RuO₂(200) reflection and the selected-area electron diffraction patterns of the RuO₂ film. The overall orientation relationship between RuO₂ and MgO can be described as $(110)_{RuO_2}||(100)_{MgO}$ and $[001]_{RuO_2}||[011]_{MgO}$ in the case of the RuO₂ [001] direction being parallel to MgO [011]. Figure 12 shows the orientation relationship and domain structure of RuO₂ on MgO [59]. The (110) RuO₂ film on (100) MgO contains two variants that are related by 90° rotation about the substrate normal, as shown in Figure 12. The in-plane orientation relationship between the RuO₂ and MgO also can be expressed as $[\bar{1}10]_{RuO_2}||[0\bar{1}1]_{MgO}$.

The heteroepitaxial growth of RuO₂ on (100) MgO that adopts such an orientation relationship can be explained by minimization of the epitaxial misfit strain in the film. With the described orientation relationship, the lattice misfit along $[0\bar{1}1]$ MgO is

$$\frac{\sqrt{2}a_{MgO} - \sqrt{2}a_{RuO_2}}{\sqrt{2}a_{MgO}} \sim -6.6\% \qquad (5)$$

and the lattice misfit along [011] MgO is

$$\frac{(\sqrt{2}/2)a_{MgO} - c_{RuO_2}}{(\sqrt{2}/2)a_{MgO}} \sim -4.3\% \qquad (6)$$

In this case, the orientation relationship observed gives the smallest lattice misfit possible at the interface.

RuO₂ (110)   MgO (100)

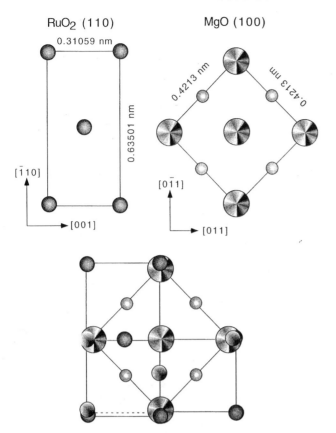

Fig. 12. Schematic that shows the orientation relationship between the RuO₂ film and the MgO substrate. The RuO₂ film contains two variants when the RuO₂(110) is superposed on MgO (100). Reproduced with permission from Q. X. Jia and P. Lu, *Philos. Mag. B* 8, 141 (2001). © 2001.

2.2.3. RuO₂ on Sapphire

Epitaxial RuO₂ films on both (0001) and R-cut sapphires have been reported [41, 60]. The films on (0001) sapphire deposited at elevated substrate temperatures possess a planar film-to-substrate relationship: $(h00)_{RuO_2}||(0001)_{Al_2O_3}$. The in-plane alignment is $[010]_{RuO_2}||\langle\bar{2}110\rangle_{Al_2O_3}$ and a threefold mosaic microstructure is observed. To obtain high crystallinity RuO₂ films on R-cut sapphire, a CeO₂ ($a = 0.5526$ nm) buffer layer can be used. In this case, the film-to-substrate relationship is $(h00)_{RuO_2}||(h00)_{CeO_2}||(01\bar{1}2)_{Al_2O_3}$. The ϕ scan of the (101) RuO₂ peak is somewhat similar to that of the RuO₂ on YSZ substrate [57]. The in-plane orientation relationship is $\langle 011\rangle_{RuO_2}||\langle 001\rangle_{CeO_2}||\langle 2\bar{2}01\rangle_{Al_2O_3}$.

2.2.4. RuO₂ on Si and SiO₂Si

Epitaxial growth of RuO₂ on Si has long been of interest for the development of novel electronic devices. For many electronic and optical devices, crystalline thin films are more attractive than polycrystalline because they exhibit greater stability, uniformity, and reproducibility.

To deposit crystalline RuO₂ thin films on (100) Si substrates, YSZ is used as an intermediate buffer layer [56]. The YSZ buffer layer is epitaxially grown on Si at a substrate tem-

perature of 800 °C. Following the YSZ layer deposition, the RuO₂ layer is deposited. X-ray diffraction 2θ scans of the RuO₂ thin films deposited above 400 °C show only $(h00)$ reflections of the RuO₂. The full width at half maximum (FWHM) of the ω-rocking curve, measured on the (200) peak of RuO₂, is 0.7° when the deposition temperature is 700 °C, which is comparable to the 0.6° value obtained on a single-crystal YSZ substrate. The in-plane alignment of each layer with respect to the major axes of Si is confirmed by X-ray diffraction ϕ scans on the RuO₂ (101), YSZ (202), and Si (101) reflections. The epitaxial relationship among RuO₂, YSZ, and Si can be described as $(h00)_{RuO_2}||(h00)_{YSZ}||(h00)_{Si}$ and $\langle 011\rangle_{RuO2}||\langle 001\rangle_{YSZ}||\langle 001\rangle_{Si}$.

Electrical measurements show that RuO₂ films on Si substrates deposited with optimized processing conditions are highly conductive, with a room-temperature resistivity of 37 ± 2 $\mu\Omega\cdot$cm. The RRR increases monotonically from 1.3 to 5.2 as the deposition temperature is increased from 300 to 700 °C.

The growth of well-textured RuO₂ on SiO₂/Si is more relevant in electronic devices because SiO₂ is used almost exclusively as a field oxide, as a passivation layer, or as an isolation material in Si-based circuitry. Highly conductive biaxially textured RuO₂ thin films can be deposited on technically important SiO₂/Si substrates by PLD, where YSZ produced by ion-beam-assisted deposition (IBAD) is used as a template to enhance the biaxial texture of RuO₂ on SiO₂/Si. The X-ray diffraction 2θ scan on RuO₂/IBAD-YSZ/SiO₂/Si shows that the RuO₂ is purely (200) oriented normal to the substrate surface. Diffraction peaks at 2θ angles of the FWHM of an ω-rocking curve on the RuO₂ (200) reflection ($2\theta = 39.8°$) are near 3.7°, compared with a value of less than 1.5° for RuO₂ on epitaxial YSZ/Si [56]. The in-plane orientation of the RuO₂ film is a strong function of the IBAD-YSZ quality; here the FWHM values of the ϕ scans from (220) YSZ are used as a measure of IBAD-YSZ quality. The better is the IBAD-YSZ biaxial texture, the better is the crystallinity of the RuO₂ layer.

The electrical resistivity of biaxially oriented RuO₂ on SiO₂/Si, which is in the range of 37 ± 2 $\mu\Omega\cdot$cm, is not a strong function of the FWHM value of IBAD-YSZ, where RuO₂ thin films are all deposited at 700 °C [61]. This value is comparable to a resistivity of 35 $\mu\Omega\cdot$cm for single-crystal bulk RuO₂. However, the RRR increases slightly with decreasing FWHM of IBAD-YSZ. A RRR of 2.5 for RuO₂ on IBAD-YSZ/SiO₂/Si can be compared with values of between 1 and 2 for polycrystalline RuO₂ thin films, above 5 from epitaxial RuO₂ on single-crystal substrates, and above 20 for bulk single-crystal RuO₂.

2.3. Etching RuO₂ Films

Due to the excellent chemical stability of RuO₂, it is difficult to use conventional wet etching techniques to etch RuO₂ films. Reactive ion etching that employs CF₄ or O₂ plasma is effective in etching RuO₂ films. The etching rate when CF₄ or O₂ plasma is employed is 2–5 times higher than that obtained by sputter etching [62]. RuO₂ also can be etched by reactive ion etching in

O_2/CF_3CFH_2 using SiO_2 as an etch mask. The etched profiles are both anisotropic and smooth. An etch rate of 160 nm/min has been achieved [63]. The possible etching mechanisms are

$$RuO_2 \text{ (solid)} + 1/2O_2 \rightarrow RuO_3 \text{ (gas)}$$
$$RuO_2 \text{ (solid)} + O_2 \rightarrow RuO_4 \text{ (gas)}$$

2.4. Applications of RuO_2 Thin Films

The unique structural, thermal, chemical, and electrical properties of RuO_2 make it valuable for a variety of potential applications. RuO_2 has very low resistivity, excellent diffusion barrier properties, good thermal stability, outstanding catalytic properties, and high chemical corrosion resistance. For example, RuO_2 is resistant to attack by strong acids, including aqua regia, and is thermally stable at temperatures as high as 800 °C [64]. However, the surfaces of RuO_2 are found to be easily reduced in vacuum, exhibiting different ordered surface structures depending upon O_2 partial pressure, temperature, and time of annealing [65]. The following text outlines the main applications of RuO_2 thin films.

RuO_2 thin films have been explored for applications in very large-scale integrated circuits. It can be used as a good diffusion barrier in silicon contact metallization with an Al overlayer [38, 42, 43] and as a conductor in an interconnection [66].

The fabrication of thermally stable Schottky contacts is very important for semiconductor devices and circuits. A metallic conduction mechanism and a work function of 4.80 eV [67] for RuO_2 make this material very attractive as a Schottky contact for semiconductors. For example, a Schottky barrier is formed at a RuO_2/CdS interface. The photovoltaic behavior of RuO_2/CdS has been observed [68]. The formation of a Schottky barrier at RuO_2/n-GaAs also has been reported [69, 70]. The good diffusion barrier property of RuO_2, which prevents the interdiffusion of Ga, As, and Ru atoms during the high-temperature rapid thermal annealing process, leads to the improved thermal stability of RuO_2 Schottky contacts to GaAs [70].

The use of RuO_2 as thin film resistors is another very attractive application in the microelectronics industry [39, 71–73]. RuO_2 thin film resistors on SiO_2/Si and/or ceramic alumina substrates with a temperature coefficient of resistance in the range of 0 ± 20 ppm/°C can be fabricated.

RuO_2 is very attractive as an electrode in terms of its electrical resistivity, thermal stability, processing compatibility, structural and chemical compatibility with high dielectric constant materials, and patterning capability [62, 63]. The electric and dielectric properties of $PbZr_xTi_{1-x}O_3$ (PZT) [74, 75], $BaTiO_3$ [76], $SrTiO_3$ [77], and $Ba_{1-x}Sr_xTiO_3$ [78, 79] thin films have been improved, in comparison to conventional Pt, by using RuO_2 as electrodes. The improvement of the electric and dielectric properties of thin film capacitors achieved by using RuO_2 as an electrode has been attributed to better structural and chemical compatibility, and a cleaner interface between the conductive oxides and the dielectric materials.

There are other applications of RuO_2 thin films. The potential applications of RuO_2 thin films as buffer layers or contact electrodes to high-temperature superconducting thin films have been investigated because of the good thermal stability and high conductivity of the RuO_2 thin films, and their structural and chemical compatibility with high-temperature superconductors [80]. The good catalytic properties of RuO_2 make it attractive in high-charge storage capacity devices also [81, 82]. Using RuO_2 as an electrode material for electrochemical capacitors has the advantages of a high capacitance of about 150–260 $\mu F/cm^2$ and low resistivity of the electrode.

3. IRIDIUM OXIDE

Iridium oxide (IrO_2) crystallizes in the rutile structure, with lattice parameters of $a = 0.451$ nm and $c = 0.315$ nm. For single-crystal IrO_2, the room temperature resistivities are 49.1 $\mu\Omega\cdot$cm and 34.9 $\mu\Omega\cdot$cm along the [001] and [011] directions, respectively [36]. The residual resistance ratio is 1400 [37].

3.1. Deposition Techniques

Thin films of IrO_2 have been deposited by electrochemical deposition [83], sol–gel processing [84], chemical vapor deposition [85], ion-beam mixing [86], rf magnetron reactive sputtering [87, 88], thermal oxidation [89], and pulsed laser deposition [90, 91]. The following subsections briefly reviews the processes of sputtering, pulsed laser deposition, and thermal oxidation to form IrO_2 films.

3.1.1. Sputtering

Reactive rf magnetron sputtering is a very powerful technique to deposit IrO_2 films. In reactive sputtering, the target is Ir with a purity of 99.99%. The oxygen partial pressure can be maintained at a level of around 2–5 mtorr. The total gas pressure (Ar + O_2) is usually kept at 5–10 mtorr. The crystalline nature and morphology of IrO_2 films depends strongly on the oxygen partial pressure, total pressure, and growth temperature [92]. The electrical properties of the IrO_2, on the other hand, are more closely related to the substrate temperature during deposition and/or postthermal treatment. Figure 13 shows the resistivity of (a) as-deposited IrO_2 films and (b) annealed IrO_2 films as a function of substrate temperature. In Figure 13, the postannealing is done at 700 °C in oxygen for 3 hr [88].

Note that the thermal stability of sputtered IrO_2 films depends on both the temperature and the environment. For example, IrO_2 films deposited by sputtering decompose at about 400 °C in air and 200 °C in vacuum [93].

3.1.2. Pulsed Laser Deposition

IrO_2 films can be deposited by ablating a pure polycrystalline Ir target under an ambient pressure of oxygen, where a KrF excimer laser ($\lambda = 248$ nm) is used [90, 91]. The IrO_2 growth rate is found to decrease from 0.5 to 0.2 μm/h when the substrate

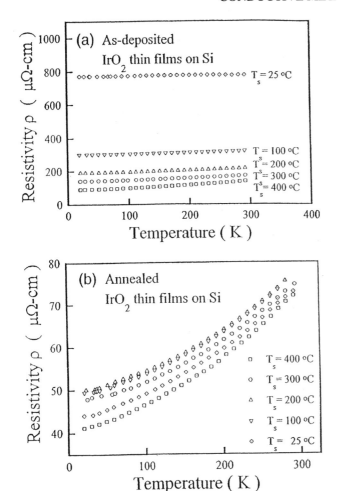

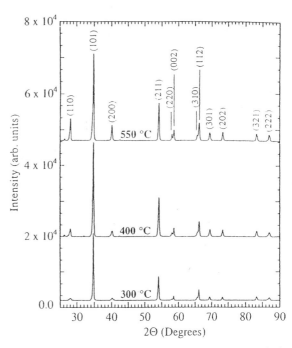

Fig. 14. Selected X-ray diffraction patterns of IrO_2 films deposited on Si (100) substrates at substrate temperatures ranging from 300 to 550 °C. Reproduced with permission from M. A. El Khakani et al., *Appl. Phys. Lett.* 69, 2027 (1996). © 1996, American Institute of Physics.

Fig. 13. (a) Resistivity of as-deposited IrO_2 films as a function of substrate temperature and (b) of annealed IrO_2 films as a function of substrate temperature. Reproduced with permission from P. C. Lias et al., *J. Mater. Res.* 13, 1318 (1998). © 1998, Materials Research Society.

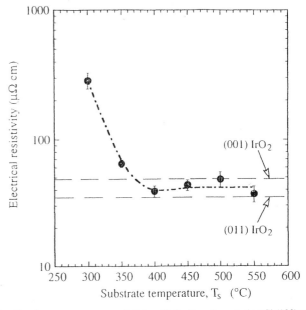

Fig. 15. Room-temperature resistivity of IrO_2 films deposited on Si (100) versus substrate temperature. For comparison, the two dashed lines indicate the resistivity values of the (011) and (001) oriented bulk single-crystal IrO_2. Reproduced with permission from M. El Khakani et al., *Appl. Phys. Lett.* 69, 2027 (1996). © 1996, American Institute of Physics.

temperature increases from 300 to 400 °C, and then to increase slightly to a value of 0.3 μm/h at a substrate temperature of 550 °C [90]. Both the structural and the electrical properties of the IrO_2 films are strongly influenced by the substrate temperature during deposition. The IrO_2 films deposited on Si (100) substrates at substrate temperatures in the range of 300–500 °C are polycrystalline, whereas those deposited at room temperature are amorphous. Figure 14 shows X-ray diffraction patterns of IrO_2 films deposited at different substrate temperatures [90].

Figure 15 shows the room-temperature resistivities of IrO_2 films deposited by PLD at different substrate temperatures [90]. The resistivity of the IrO_2 films drastically decreases with increasing deposition temperature from 300 to 400 °C, and then stabilizes around a mean value of about 42 $\mu\Omega\cdot$cm over the 440–550 °C deposition temperature range.

Note that IrO_2 films deposited at a substrate temperature less than 300 °C by pulsed laser deposition shows poor adhesion to the Si (100) substrates. On the other hand, IrO_2 films deposited at a substrate temperature above 350 °C exhibit very good adhesion to Si [90].

3.1.3. Thermal Oxidation

Thermal oxidation provides an easy way to form IrO_2. In this process, a thin film of Ir is first deposited by either sputtering or pulsed laser deposition on a substrate. The thermal oxidation of Ir is then carried out in oxygen at a certain temperature for a desired period of time. Oxidized Ir first forms on the surface and then extends into the bulk of the film. For example, after annealing Ir at 600 °C, the oxide layer (IrO_x) exists only near the surface region. At this annealing temperature, there is no evidence bulk oxidation. At an annealing temperature somewhere between 700 and 800 °C, oxygen starts to diffuse into the bulk of the film [94].

Note that thermally oxidized IrO_2 decomposes between 577 and 627 °C in ultra-high vacuum [95]. Films grown at 600 °C are only partially oxidized and completely decompose at temperatures above 515 °C. However, films grown at 700 and 800 °C are more stable and decompose at a temperature between 660 and 690 °C. Decomposition of the films grown at higher temperatures is believed to be from release of O_2 from the films [94].

3.2. Applications of IrO₂ Films

The large work function of 4.23 eV [96], low resistivity, and corrosion resistance makes IrO_2 very attractive in many applications. For example, it can be used as an electrode in a capacitor where high dielectric constant materials are used as dielectrics [87, 97]. It can be used further as a barrier material between Pt and Si due to its excellent barrier performance against oxygen diffusion [87]. Pt is the most commonly used electrode for capacitors on Si wafers because of its excellent oxidation resistance and high electrical conductivity. However, Pt also has the tendency to react with Si, which can affect the performance of the capacitors. By using an IrO_2 barrier layer between Pt and Si, this interdiffusion problem can be reduced. IrO_2 also can be used as an electrode for neural stimulation [98]. Other applications of IrO_2 films include oxygen and chlorine evolution anodes [99, 100], electrochromic displays [101], and pH sensors [102, 103].

4. STRONTIUM RUTHENATE

Strontium ruthenate ($SrRuO_3$) which crystallizes in the $GdFeO_3$-type orthorhombic distorted perovskite structure, has lattice parameters of $a = 0.5532$ nm, $b = 0.5572$ nm, and $c = 0.7850$ nm. $SrRuO_3$ also can be considered as a pseudocubic perovskite with a lattice parameter of $a \sim 0.3928$ nm. The Curie temperature (T_c) of single-crystal $SrRuO_3$ is around 160 K. Figure 16 shows the resistivity versus temperature characteristic of $SrRuO_3$ crystals [104]. The room-temperature resistivity is around 280 $\mu\Omega\cdot$cm. At $T > T_c$, the resistivity increases almost linearly with T up to more than 1000 K. The anisotropy of resistivity is less than 10% between the long axis and the thin axis [105]. The magnetic moments per ruthenium in bulk (films) above and below T_c are $1.6\mu_B$ ($1.4\mu_B$) and $2\mu_B$, respectively [106].

4.1. Deposition Techniques

$SrRuO_3$ films can be deposited by spray pyrolysis [107], reactive electron-beam coevaporation [108], metallorganic decom-

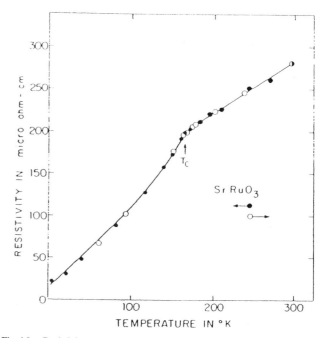

Fig. 16. Resistivity versus temperature for $SrRuO_3$ single crystals. Arrows indicate measurements taken on heating or cooling. Reproduced with permission from R. J. Bouchard and J. L. Gillson, *Mater. Res. Bull.* 7, 873 (1972). © 1972.

position [109], metallorganic chemical vapor deposition [110], laser molecular beam epitaxy [111], sputtering [112], and pulsed laser deposition [113]. The most commonly used deposition techniques are sputtering and pulsed laser deposition.

4.1.1. Sputtering

During the deposition of complex metal oxide materials such as $SrRuO_3$, $SrTiO_3$, and $YBa_2Cu_3O_7$ by sputtering, bombardment of the film surface by energetic ions can introduce serious problems such as damage to and resputtering of the growing film due to negative ion effects and radiation enhanced surface diffusion. To reduce the negative ion problem, two common approaches are in practice. One is to use relatively high gas pressure to reduce the energy of the negative ions [114]. Another is to place the substrate off position in which the substrate does not directly face the cathode [115]. The latter is called off-axis sputtering and has been used widely to deposit high-temperature superconducting thin films [116]. Figure 17 shows a typical arrangement of the substrate relative to the target configured for off-axis sputtering [115]. The disadvantage of this arrangement, however, is the low deposition rate. The difficulty with depositing uniform and large-area films may be another limitation of using this off-axis sputtering technique.

$SrRuO_3$ films deposited by off-axis sputtering exhibit low isotropic resistivity, excellent chemical and thermal stability, good surface smoothness, and high crystallinity quality. A substrate temperature of 680 °C, and a sputtering atmosphere that consists of 60 mtorr Ar and 40 mtorr O_2 are considered to be good conditions for producing high quality $SrRuO_3$ films [112].

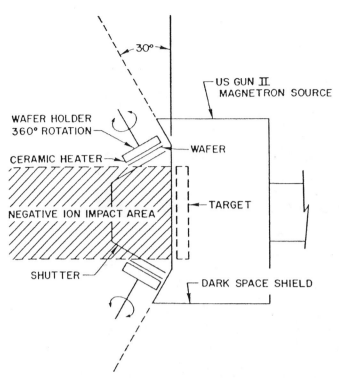

Fig. 17. Substrate arrangement relative to the target for an off-axis sputtering setup. Reproduced with permission from R. L. Sandstrom et al., *Appl. Phys. Lett.* 53, 444 (1998). © 1998, American Institute of Physics.

4.1.2. Pulsed Laser Deposition

Pulsed laser deposition is a very powerful deposition technique to deposit SrRuO$_3$ thin films. The SrRuO$_3$ target can be prepared by mixing appropriate molar ratios of SrCO$_3$ and Ru metal powder, grinding, heating in air at 1200 °C for about 12 h, and regrinding and reheating. The well-mixed fine powders are then pressed into disks that are used as laser ablation targets. Crystalline SrRuO$_3$ thin films have been grown successfully by PLD on many substrates, such as SrTiO$_3$ [117], LaAlO$_3$ [113], MgO with Pt as a buffer layer [118], and Si with YSZ as a buffer layer [119].

The surface morphology of SrRuO$_3$ is very different from that of high-temperature superconductor thin films, even when the same PLD processing conditions are used to grow the films. It is well known that particulates are found on YBa$_2$Cu$_3$O$_7$ film surfaces deposited by PLD. However, SrRuO$_3$ deposited by PLD shows no detectable or a very low density of particulates. Figure 18 shows the surface morphology of the SrRuO$_3$ thin films on LaAlO$_3$ deposited at different substrate temperatures [113]. The increase of the grain size at the film surface with increasing deposition temperature is quite obvious. It is apparent from the figure that the film surface is smooth. No particles are detected on the film surfaces, although crystalline films show some micropits due to incomplete grain coalescence. For example, epitaxial SrRuO$_3$ films (deposited at 650 °C) show a root mean square roughness of less than 1 nm on a test area of either 0.1×0.1 or 1.0×1.0 μm^2 [120]. Such smooth films deposited under optimized conditions are extremely im-

portant for use as electrodes in high dielectric constant thin film capacitors, because particles can generate problems in the devices, such as reduced breakdown voltage and enhanced leakage current density due to the decrease of the effective dielectric thickness. Large particles can render the devices useless if they short the top and bottom electrodes.

Electrical measurements show that SrRuO$_3$ films deposited by PLD with optimized microstructure are highly conductive with a room-temperature resistivity of around 280 $\mu\Omega\cdot$cm. This value is comparable to that of bulk single-crystal SrRuO$_3$ [104] and lower than the 340 $\mu\Omega\cdot$cm value of sputtered films [112]. As shown in Figure 19, the deposition temperature has an important influence on the room-temperature resistivity of the SrRuO$_3$ films [113]. The inset in Figure 19 shows the normalized resistivity ratio of SrRuO$_3$ films deposited at different substrate temperatures [120]. The room-temperature resistivity decreases by more than 3 orders of magnitude when the deposition temperature increases from 250 to 775 °C. The very weak dependence of film resistivity on oxygen pressure during film deposition in the range of 50–200 mtorr for a given deposition temperature indicates that the films are fully oxidized under optimized processing conditions. It has been argued that the resistivity is directly related to the crystallinity and the microstructure of the films. The SrRuO$_3$ thin films become crystalline and highly textured when deposited above 450 °C. Note that polycrystalline SrRuO$_3$ shows a much higher resistivity compared to crystalline SrRuO$_3$. For example, the room-temperature resistivity of polycrystalline SrRuO$_3$ films can be as high as 1130 $\mu\Omega\cdot$cm due to grain-boundary scattering [121].

SrRuO$_3$ thin films deposited above 600 °C show metallic resistivity characteristics with respect to temperature with a slight kink in the ρ versus T curves at ~160 K, corresponding to the Curie temperature of SrRuO$_3$, below which SrRuO$_3$ is ferromagnetic. However, films deposited at 250 °C exhibit semiconductor-like ρ–T behavior. The ρ–T curve is composed of three distinguishable regimes for SrRuO$_3$ films deposited at 450 °C (see Fig. 20). The resistivity of the film is almost constant above 160 K, metallic in the range of 93–160 K, and semiconductor-like below 93 K [113].

Residual resistivity ratios of the SrRuO$_3$ thin films are 3.4 and 8.4, respectively, for deposition temperatures of 650 and 775 °C [120]. RRR increases monotonically with increasing deposition temperature. RRR values are around 12 and less than 3 for bulk single-crystal SrRuO$_3$ [104] and thin film SrRuO$_3$ grown epitaxially by sputtering [112], respectively.

4.2. Applications of SrRuO$_3$ Films

The large work function of 5.21 eV [122], the excellent thermal and chemical stability, the relatively high electrical conductivity, and the structural compatibility with ferroelectric or high dielectric constant materials make SrRuO$_3$ very attractive as a bottom electrode for capacitors. As an electrode material for high frequency applications, it is essential that the SrRuO$_3$ be highly conductive. It should neither interact chemically with the

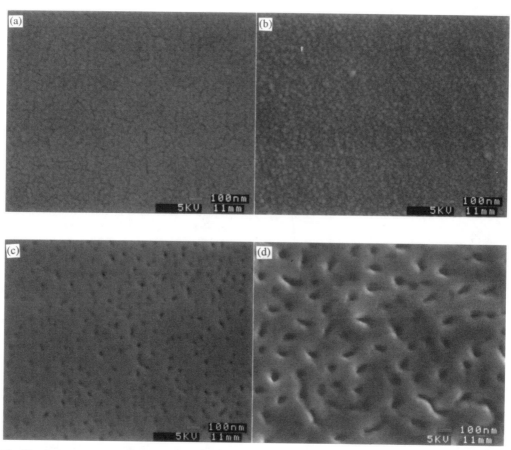

Fig. 18. Scanning electron microscopy graphs of SrRuO$_3$ thin films on LaAlO$_3$ deposited at (a) 250, (b) 450, (c) 650, and (d) 775 °C. Reproduced with permission from Q. X. Jia et al., *J. Mater. Res.* 11, 2263 (1996). © 1996, Materials Research Society.

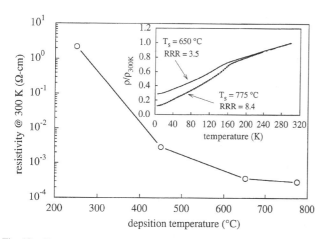

Fig. 19. Room-temperature resistivity of SrRuO$_3$ films as a function of deposition temperature. The inset shows the normalized resistivity of SrRuO$_3$ thin films deposited at 650 °C and 775 °C, respectively, as a function of testing temperature. Reproduced with permission from Q. X. Jia et al., *J. Mater. Res.* 11, 2263 (1996), and *J. Vac. Sci. Technol. A* 15, 1080 (1997). © 1997, Materials Research Society.

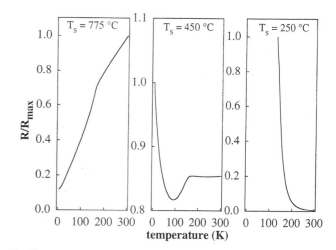

Fig. 20. Normalized resistivity of SrRuO$_3$ thin films deposited at different substrate temperatures (T_S) as a function of testing temperature. Reproduced with permission from Q. X. Jia et al., *J. Mater. Res.* 11, 2263 (1996). © 1996.

dielectric material nor form a low permittivity compound at the interface. In many applications, it also should not interact with the barrier layer, which is in contact with the electrode. It must be stable at elevated processing temperatures. Additionally,

considerations ought to be made for the following parameters when choosing electrode materials: the work function, the ability to be patterned either by conventional chemical wet etching or by dry etching methods, the stability of the surface, and processing compatibility and suitability. SrRuO$_3$ meets many of

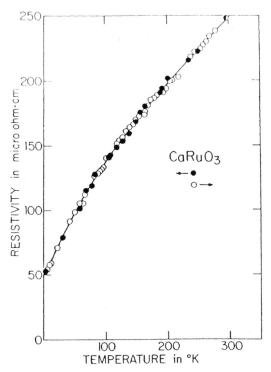

Fig. 21. Resistivity versus temperature for CaRuO₃ single crystals. Reproduced with permission from R. J. Bouchard and J. L. Gillson, *Mater. Res. Bull.* 7, 873 (1972). © 1972.

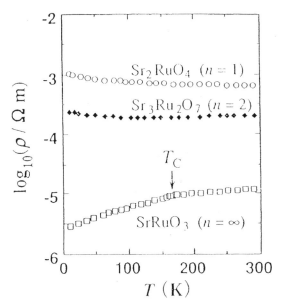

Fig. 22. Temperature dependence of the electrical resistivity for $Sr_{n+1}Ru_nO_{3n+1}$ ($n = 1, 2, \infty$). Reproduced with permission from M. Itoh et al., *Phys. Rev. B* 51, 16,432 (1995). © 1995, American Physics Society.

these requirements and has been used as an electrode for ferroelectric $Pb(Zr_{0.52}Ti_{0.48})O_3$ thin film capacitors [123]. These capacitors exhibit superior fatigue characteristics compared to those made with metallic Pt electrodes. By using $SrRuO_3$ as electrodes for $Ba_{0.5}Sr_{0.5}TiO_3$ [124] capacitors, the devices show markedly improved electric/dielectric properties, such as a high dielectric constant and low leakage current density. The extremely high thermal stability of the $SrRuO_3$ leaves it unaffected after dielectric thin film deposition. The $SrRuO_3$ surface has been shown that to keep the as-deposited microstructure, even after the deposition of $Ba_{0.5}Sr_{0.5}TiO_3$ thin film at 680 °C [125].

Other applications include the use of $SrRuO_3$ as a buffer layer for growth of high-temperature superconductor $YBa_2Cu_3O_7$ films on different substrates [126] and as a barrier for ramp-edge superconductor–normal-metal–superconductor ($YBa_2Cu_3O_7$-$SrRuO_3$-$YBa_2Cu_3O_7$) Josephson junctions [127]. $SrRuO_3$ also has been proposed as potentially useful in magnetooptic and electrooptic devices [128].

4.3. Other Conductive Ruthenates

The $Ca_{1-x}Sr_xRuO_3$ ($0 \leq x \leq 1$) system shows physical properties ranging from a metallic ferromagnetism in $SrRuO_3$ to a short range metallic antiferromagnetism in $CaRuO_3$ with the same cation valence. In other words, substitution of Ca in $SrRuO_3$ reduces the Curie temperature until the system finally becomes antiferromagnetic for the pure $CaRuO_3$ phase [129]. The lattice parameters of $CaRuO_3$ are $a = 0.5356$ nm, $b =$

0.5529 nm, and $c = 0.7660$ nm. Figure 21 shows the resistivity versus temperature plot for $CaRuO_3$ single crystals [104]. Epitaxial $CaRuO_3$ films [130] show a resistivity value similar to that of single crystals.

Similar to $SrRuO_3$, $Sr_{1-x}Ba_xRuO_3$ exhibits a simple perovskite structure in the whole region of the Ba/Sr ratio. The metallic conductivity is maintained in the whole region of finite Ba concentration. However, the room-temperature resistivity increases with increasing Ba concentration. The ferromagnetic ordering in $SrRuO_3$ is suppressed in $Sr_{1-x}Ba_xRuO_3$, as the tetragonal deformation increases and the Curie temperature decreases to 50 K in $BaRuO_3$ [131]. The good metallic conductivity and the structural/chemical compatibility with dielectric materials such as $Ba_{1-x}Sr_xTiO_3$ and PZT make $BaRuO_3$ very attractive as an electrode material [132, 133].

Another conductive ruthenate system is layered $Sr_{n+1}Ru_n$ O_{3n+1} ($n = 1, 2,$ and ∞). Figure 22 shows the temperature dependence of the resistivity for the layered perovskite system $Sr_{n+1}Ru_nO_{3n+1}$. The $SrRuO_3$ shows the slope change at T_c as described before. Sr_2RuO_4 and $Sr_3Ru_2O_7$ show an almost temperature-independent resistivity change [134]. The interesting point of this system is that Sr_2RuO_4 shows high metallic conductivity. It is nonmagnetic (unlike $SrRuO_3$). Importantly, it is the only known layered perovskite that is free of copper and yet superconducting at 0.93 K [135]. Interest in Sr_2RuO_4, from an applications point of view, is mainly due to its low resistivity and excellent lattice match with $YBa_2Cu_3O_7$. Sr_2RuO_4 has a tetragonal K_2NiF_4-type structure with lattice parameters $a = 0.387$ nm and $c = 1.274$ nm [136]. These properties make it an attractive candidate for use as conductive electrodes or a normal metal barrier in device applications of high-temperature superconducting films [137].

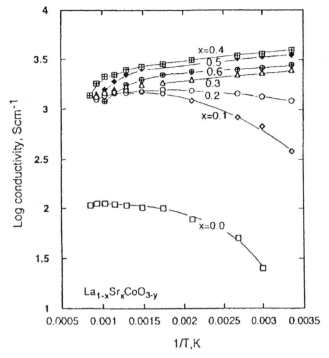

Fig. 23. Electrical conductivity of $La_{1-x}Sr_xCoO_3$ in air as a function of the reciprocal absolute temperature for different x values. Reproduced with permission from A. N. Petrov et al., *Solid State Ionics* 80, 189 (1995). © 1995, Elsevier Science.

5. STRONTIUM-DOPED LANTHANUM COBALTITE

Strontium-doped lanthanum cobaltite ($La_{0.5}Sr_{0.5}CoO_3$) has a pseudocubic lattice constant of 0.3835 nm. $La_{0.5}Sr_{0.5}CoO_3$ is a ferromagnet with a moment of approximately $1.5\mu_B$ per formula unit [138]. Decreasing the Sr content in $La_{1-x}Sr_xCoO_3$ leads to an increase in the electrical resistivity of the compound due to a reduction in the hole concentration. The crystal structure of $La_{1-x}Sr_xCoO_3$ changes from rhombohedral to cubic, depending on the value of x and the temperature. Figure 23 shows the electrical conductivity of $La_{1-x}Sr_xCoO_3$ (prepared using a standard ceramic technique) as a function of the reciprocal absolute temperature for different x values [139]. $La_{1-x}Sr_xCoO_3$ at compositions of $x = 0.4$–0.5 shows the lowest resistivity. A deficiency in oxygen is noted in $La_{1-x}Sr_xCoO_3$ that can lead to changes in the crystal and electronic structures, which in turn can affect the magnetic properties of the $La_{1-x}Sr_xCoO_3$. The room-temperature electrical resistivity of a sintered ceramic sample with $x = 0.5$ is around 90 $\mu\Omega$·cm [140]. For epitaxial $La_{0.5}Sr_{0.5}CoO_3$ films, there is evidence that the electrical resistivity of $La_{0.5}Sr_{0.5}CoO_3$ is isotropic [141]. $La_{0.5}Sr_{0.5}CoO_3$ is a p-type conductor, where holes are the carriers in this oxide. The work function of $La_{0.5}Sr_{0.5}CoO_3$ is in the range of 4.2–4.6 eV [142].

5.1. Deposition Techniques

$La_{0.5}Sr_{0.5}CoO_3$ films can be deposited by a variety of techniques, including the dipping–pyrolysis process [143], met-

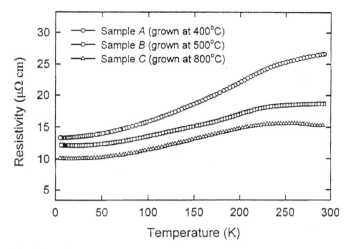

Fig. 24. Temperature dependence of resistivity for three samples of $La_{0.5}Sr_{0.5}CoO_3$ films. Reproduced with permission from G. P. Luo et al., *Appl. Phys. Lett.* 76, 1908 (2000). © 2000, American Institute of Physics.

allorganic chemical vapor deposition [144], sol–gel [145], metallorganic decomposition [146], freeze-drying [147], sputtering [148], and pulsed laser deposition [141]. Pulsed laser deposition is the most widely used technique for the growth of high quality $La_{0.5}Sr_{0.5}CoO_3$ films due to the easy control of stoichiometry. It is important to accurately control the stoichiometry of the film because the conductivity is strongly dependent on the chemical composition. An example of this dependency is shown by the La/Sr ratio (see Fig. 23).

5.1.1. Pulsed Laser Deposition

Figure 24 shows the temperature dependence of electrical resistivity for $La_{0.5}Sr_{0.5}CoO_3$ films deposited by pulsed laser deposition on $LaAlO_3$ grown at 400, 500, and 800 °C [149]. If the growth temperature is higher than 400 °C, it affects the electrical resistivity only slightly. Nevertheless, higher growth temperatures lead to relatively low electrical resistivity $La_{0.5}Sr_{0.5}CoO_3$ films.

The temperature dependence of resistivity measured for epitaxial $La_{0.5}Sr_{0.5}CoO_3$ films varies in quite a wide range, depending on the substrate materials and the processing conditions used to deposit the films. Figure 25 shows the resistivity as a function of measurement temperature for a $La_{0.5}Sr_{0.5}CoO_3$ film deposited at 650 °C on $LaAlO_3$ substrate [150]. For bulk $La_{1-x}Sr_xCoO_3$, the temperature dependence of resistivity can be fitted by the relationship of [151]

$$\rho = \rho_0 + AT^2 \tag{7}$$

For epitaxial $La_{0.5}Sr_{0.5}CoO_3$ film, the resistivity–temperature characteristics at higher temperature (150–300 K) can be fitted by [150]

$$\rho = \rho_0 + AT^{2.5} \tag{8}$$

as shown in the inset of Figure 25. At temperatures lower than 150 K, however, the relationship described by the Eq. (8) does not hold [150].

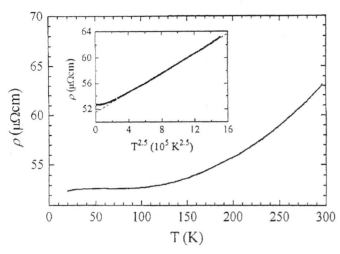

Fig. 25. The temperature dependence of resistivity measured for La$_{0.5}$Sr$_{0.5}$CoO$_3$ film grown at 650 °C. The inset shows the dependence of the resistivity on $T^{2.5}$. Reproduced with permission from W. B. Wu et al., *J. Appl. Phys.* 88, 700 (2000). © 2000, American Institute of Physics.

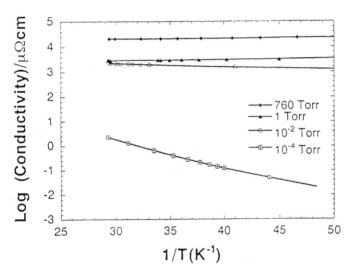

Fig. 26. Comparison of the dependence of electrical conductivity of La$_{0.5}$Sr$_{0.5}$CoO$_3$ films cooled in different oxygen partial pressures as a function of temperature. Reproduced with permission from S. Madhukar et al., *J. Appl. Phys.* 81, 3543 (1997). © 1997, American Institute of Physics.

The resistivity of La$_{0.5}$Sr$_{0.5}$CoO$_3$ films deposited by pulsed laser deposition is also a strong function of oxygen pressure during cooling down. Figure 26 is a plot of the logarithm of conductivity as a function of the inverse temperature of La$_{0.5}$Sr$_{0.5}$CoO$_3$ films on LaAlO$_3$ substrates [152]. When the film is cooled in 760 mtorr oxygen partial pressure, it shows a metallic resistivity versus temperature characteristic and a resistivity of ∼47 $\mu\Omega$·cm at room temperature. On the other hand, when the film is cooled in 10^{-4} torr oxygen pressure, it has a resistivity of 2 Ω·cm at room temperature. This film also shows a semiconducting behavior of resistivity with temperature [152].

Noted that the crystallinity of PLD-grown La$_{0.5}$Sr$_{0.5}$CoO$_3$ films is largely influenced by the lattice and the structure of the substrate. This, in turn, influences the electrical re-

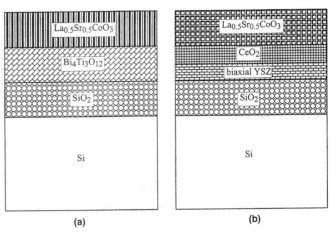

Fig. 27. Schematic of the multilayer structures used to produce highly oriented La$_{0.5}$Sr$_{0.5}$CoO$_3$ on SiO$_2$/Si (a) using Bi$_4$Ti$_3$O$_{12}$ as a template to produce La$_{0.5}$Sr$_{0.5}$CoO$_3$ with uniaxial normal alignment and (b) using biaxially oriented YSZ as a seed layer to produce La$_{0.5}$Sr$_{0.5}$CoO$_3$ with alignment both normal to and in the film plane. Reproduced with permission from R. Ramesh et al., *Appl. Phys. Lett.* 64, 2511 (1994) American Institute of Physics and Q. X. Jia et al. *J. Vac. Sci. Technol. A* 16, 1380 (1998). © 1998, Materials Research Society.

sistivity of epitaxial La$_{0.5}$Sr$_{0.5}$CoO$_3$ films. For example, the room-temperature resistivity of epitaxial La$_{0.5}$Sr$_{0.5}$CoO$_3$ film on MgO substrate is much higher than that of epitaxial La$_{0.5}$Sr$_{0.5}$CoO$_3$ film on LaAlO$_3$ substrate. The room-temperature resistivity of a (001) axis oriented La$_{0.5}$Sr$_{0.5}$CoO$_3$ film on MgO typically ranges from 130 to 200 $\mu\Omega$·cm [141], which is higher than the value of 90 $\mu\Omega$·cm for a bulk ceramic sample.

5.2. Oriented La$_{0.5}$Sr$_{0.5}$CoO$_3$ Films on SiO$_2$/Si

Epitaxial and/or well-textured La$_{0.5}$Sr$_{0.5}$CoO$_3$ films have been grown on SrTiO$_3$, MgO, LaAlO$_3$, and YSZ. The growth of well-textured La$_{0.5}$Sr$_{0.5}$CoO$_3$ on technically important SiO$_2$/Si is especially relevant in microelectronic devices, because SiO$_2$ is almost exclusively used as a field oxide, passivation layer, and/or isolation material in Si-based circuitry.

Figure 27 shows the generic structures used to construct highly oriented La$_{0.5}$Sr$_{0.5}$CoO$_3$ on SiO$_2$/Si. By using Bi$_4$Ti$_3$O$_{12}$ as a template as shown in Figure 27a, La$_{0.5}$Sr$_{0.5}$CoO$_3$ film with a uniaxial normal alignment is obtained [153]. By using a biaxially oriented YSZ seed layer and a structural template CeO$_2$, as shown in Figure 27b, highly oriented La$_{0.5}$Sr$_{0.5}$CoO$_3$ films with alignment both in plane and normal to the film plane can be obtained [154]. In the former case, ion-beam sputtering along with ion-beam assistance (IBAD) is used to deposit biaxially textured YSZ on the amorphous SiO$_2$ layer at room temperature. Pulsed laser deposition with a 308 nm XeCl excimer laser is employed to deposit both the structural template CeO$_2$ and the conductive La$_{0.5}$Sr$_{0.5}$CoO$_3$ thin films [154]. As can be seen from the drawing, both a biaxially oriented YSZ ($a = 0.514$ nm) seed layer and a structural template CeO$_2$ ($a = 0.541$ nm) are needed to accomplish the growth of highly oriented La$_{0.5}$Sr$_{0.5}$CoO$_3$.

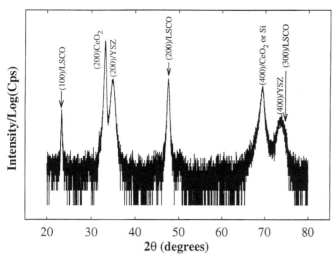

Fig. 28. X-ray diffraction 2θ scan for a $La_{0.5}Sr_{0.5}CoO_3$ (LSCO) film on SiO_2/Si. Both a biaxially oriented YSZ seed layer and a CeO_2 structural template are used to induce the growth of highly (100) oriented LSCO film on amorphous SiO_2. Reproduced with permission from Q. X. Jia et al., *J. Vac. Sci. Technol. A* 16, 1380 (1998). © 1998, Materials Research Society.

The $La_{0.5}Sr_{0.5}CoO_3$ deposited on biaxially oriented YSZ with a very thin intermediate CeO_2 layer on SiO_2/Si shows only (100) diffraction peaks from an X-ray diffraction θ–2θ scan, as shown in Figure 28 [154]. The growth of (100) oriented $La_{0.5}Sr_{0.5}CoO_3$ has great technical significance, because highly conductive $La_{0.5}Sr_{0.5}CoO_3$ with (100) orientation is preferred for applications where it is used in electrodes for ferroelectric thin film capacitors [155]. Note that the structural template of CeO_2 provides an optimum crystal structure for growth of thin films with desired textures. Efforts to grow $La_{0.5}Sr_{0.5}CoO_3$ thin films directly on single-crystal YSZ and/or biaxially oriented YSZ, on the other hand, resulted in (110) oriented films [154, 155].

The real significance of using the IBAD template is the accomplishment of biaxially oriented $La_{0.5}Sr_{0.5}CoO_3$ on SiO_2/Si. The degree of in-plane epitaxy of YSZ, characterized by a FWHM of the (220) YSZ peak, determines the degree of in-plane epitaxy of a $La_{0.5}Sr_{0.5}CoO_3$ layer. The relationship between the crystallinity of YSZ and that of $La_{0.5}Sr_{0.5}CoO_3$ is shown in Figure 29 [156]. The solid circles denote the FWHMs of the (220) $La_{0.5}Sr_{0.5}CoO_3$ peak obtained from an X-ray diffraction ϕ scan and the solid triangles denote the FWHMs of the X-ray diffraction rocking curve from the (200) $La_{0.5}Sr_{0.5}CoO_3$ peak. With enhancement in the in-plane epitaxy [i.e., decreasing FWHM of the (220) YSZ peak], the overall crystallinity of the $La_{0.5}Sr_{0.5}CoO_3$ layer improved, as evidenced by the reduction of the FWHM of the (220) and (200) peaks. As a reference, the FWHM value of the (220) peak of a $La_{0.5}Sr_{0.5}CoO_3$ film on a $LaAlO_3$ substrate deposited under the same conditions is also included in Figure 29.

The room-temperature electrical resistivity of the biaxially oriented $La_{0.5}Sr_{0.5}CoO_3$ thin films deposited at 700 °C on SiO_2/Si is found to be very close to epitaxial $La_{0.5}Sr_{0.5}CoO_3$ films on single-crystal substrates such as $LaAlO_3$, which has a

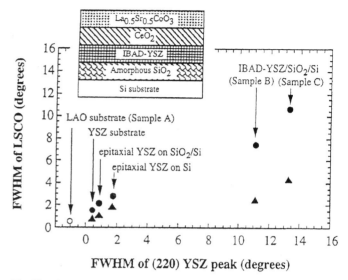

Fig. 29. Crystallinity of $La_{0.5}Sr_{0.5}CoO_3$ (LSCO) on various substrates plotted with respect to a FWHM of the X-ray diffraction ϕ scan of the (220) YSZ peak. Solid circles, FWHM obtained from X-ray diffraction ϕ scan of the (220) peak; solid triangles, FWHM from X-ray diffraction rocking curve of the (200) peak; open circle, FWHM of the epitaxial LSCO on a $LaAlO_3$ substrate. The inset is a schematic of the multilayer structure of LSCO on SiO_2/Si substrate. Reproduced with permission from C. Kwon et al., *Appl. Phys. Lett.* 73, 695 (1998). © 1998, American Institute of Physics.

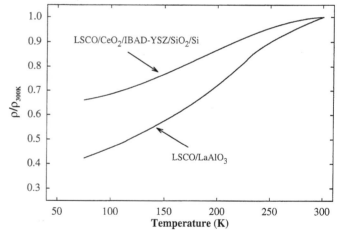

Fig. 30. Normalized resistivity versus temperature characteristics of an epitaxial $La_{0.5}Sr_{0.5}CoO_3$ (LSCO) on $LaAlO_3$ and a biaxially oriented LSCO film on SiO_2/Si. Both films are deposited at 700 °C and an oxygen pressure of 400 mtorr. Reproduced with permission from Q. X. Jia et al., *J. Vac. Sci. Technol. A* 16, 1380 (1998). © 1998, Materials Research Society.

very good lattice match with $La_{0.5}Sr_{0.5}CoO_3$. Figure 30 shows the resistivity versus temperature (normalized to the room-temperature resistivity) of $La_{0.5}Sr_{0.5}CoO_3$ films deposited at 700 °C on different substrates [154]. Both films show metallic behavior except for the difference in the temperature coefficient of resistivity (TCR) for the films. The room-temperature resistivity is basically the same with a value of around 110 $\mu\Omega\cdot$cm.

5.3. Applications of La$_{0.5}$Sr$_{0.5}$CoO$_3$ Films

La$_{0.5}$Sr$_{0.5}$CoO$_3$ is considered to be one of the good materials for high-temperature fuel cell electrodes and interconnections [157] due to its high electrical conductivity and oxygen diffusivity. Its high isotropic electrical conductivity and close lattice parameter matching to many ferroelectric materials of interest have made La$_{0.5}$Sr$_{0.5}$CoO$_3$ one of the premier electrode materials in memory devices. Experimental results have shown that ferroelectric thin film capacitors using La$_{0.5}$Sr$_{0.5}$CoO$_3$ electrodes exhibit superior fatigue and retention characteristics compared to capacitors using conventional Pt electrodes [155, 158].

For applications of La$_{0.5}$Sr$_{0.5}$CoO$_3$ films as electrodes for nonvolatile ferroelectric random access memories, epitaxial and/or well-textured La$_{0.5}$Sr$_{0.5}$CoO$_3$ films are preferable. The reduced grain-boundary scattering from an epitaxial La$_{0.5}$Sr$_{0.5}$CoO$_3$ film leads to a lower resistivity, which is a prerequisite for high frequency applications. As a bottom electrode and/or seed layer for ferroelectric thin film capacitors, well-textured La$_{0.5}$Sr$_{0.5}$CoO$_3$ also induces epitaxial or preferential oriented growth in subsequently deposited ferroelectric films. This is important because a highly oriented ferroelectric layer can produce a larger remnant polarization than a randomly oriented layer [153, 158].

Another consideration in selecting electrodes for ferroelectric capacitors is oxygen affinity. If the oxygen affinity is too low, oxygen in the electrode material will be depleted by the ferroelectric material, thus leaving an oxygen deficient layer of electrode material at the interface and thereby increasing the contact resistance [141]. There is evidence of a high kinetic barrier for oxygen incorporation and depletion in the La$_{0.5}$Sr$_{0.5}$CoO$_3$ lattice and thus better oxygen stability. Figure 31 compares the resistivity change over an oxygen pressure from 400 mtorr to 2×10^{-6} torr at 650 °C for both La$_{0.5}$Sr$_{0.5}$CoO$_3$ and YBa$_2$Cu$_3$O$_7$. The dependence of the electrical resistivity of La$_{0.5}$Sr$_{0.5}$CoO$_3$ on oxygen pressure is $R \sim P^{-0.057}$ for all pressures. Although this indicates the absence of a phase transition in this temperature range, it does show the same electrical property dependence on oxygen stoichiometry as the tetragonal phase YBa$_2$Cu$_3$O$_7$ [141].

Other applications of La$_{0.5}$Sr$_{0.5}$CoO$_3$ include its use as a buffer layer for the growth of high-temperature superconducting YBa$_2$Cu$_3$O$_7$ films on different substrates. The good chemical compatibility and structural match between YBa$_2$Cu$_3$O$_7$ and La$_{0.5}$Sr$_{0.5}$CoO$_3$ also make La$_{0.5}$Sr$_{0.5}$CoO$_3$ a candidate for being the normal-metal layer in the fabrication of ramp-edge superconductor–normal-metal–superconductor (YBa$_2$Cu$_3$O$_7$-La$_{0.5}$Sr$_{0.5}$CoO$_3$-YBa$_2$Cu$_3$O$_7$) Josephson junctions [159].

6. CONCLUDING REMARKS

Conducting metal oxide thin films such as TCO, RuO$_2$, IrO$_2$, SrRuO$_3$, and La$_{0.5}$Sr$_{0.5}$CoO$_3$, have been extensively investigated due to their potential applications in the field of microelectronic devices. The use of conventional conducting ma-

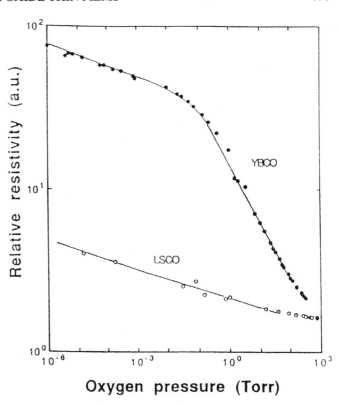

Fig. 31. Resistivity versus oxygen pressure for YBa$_2$Cu$_3$O$_{7-x}$ (YBCO) and La$_{0.5}$Sr$_{0.5}$CoO$_3$ (LSCO) films. The relative resistivity values are normalized at 745 torr. Reproduced with permission from J. T. Cheung et al., *Appl. Phys. Lett.* 62, 2045 (1993). © 1993, American Institute of Physics.

terials in microelectronic devices will remain. However, the exceptional spectrum of properties exhibited by conducting metal oxides makes these materials very unique compared to conventional metallic conductors. In addition to the applications that result from integrating conductive metal oxides with conventional semiconductor materials, the potential applications and advantages of using conducting metal oxides are tremendous in next generation electronic devices based on all-oxide materials.

For many applications, it is necessary to have superlative conductivity and optical transparency. From a materials point of view, it is very important to explore highly degenerated materials with large carrier mobility. Large carrier mobility enhances the electrical conductivity without degrading the optical properties. The development of high performance p-type transparent conductive oxides is also important for fabrication of transparent all-oxide-based active devices.

Preparation of nontransparent conductive metal oxides necessitates accurate control the chemical composition and oxygen content in the films, because the structural, electrical, and magnetic properties of these oxides are very sensitive to these parameters. Although many thin film deposition techniques for depositing conductive metal oxides have been explored only a few are very successful. It is necessary to optimize the processing conditions that not only can produce desired conductive metal oxide films, but also are compatible with the fabrication of conventional semiconductor devices.

REFERENCES

1. M. Hiramatsu, K. Imaeda, N. Horio, and M. Nawata, *J. Vac. Sci. Technol. A* 16, 669 (1998).

2. H. Kim, A. Pique, J. S. Horwitz, H. Mattoussi, H. Murata, Z. H. Kafafi, and D. B. Chrisey, *Appl. Phys. Lett.* 74, 3444 (1999).

3. H. Enoki, T. Nakayama, and J. Echigoya, *Phys. Status Solidi A* 129, 181 (1992).

4. T. Minami, H. Sonohara, S. Takata, and H. Sato, *Jpn. J. Appl. Phys.* 33, L1693 (1994).

5. K. Yanagawa, Y. Ohki, T. Omata, H. Hosono, N. Ueda, and H. Kawazoe, *Appl. Phys. Lett.* 65, 406 (1994).

6. T. Minami, Y. Takeda, S. Takata, and T. Kakumu, *Thin Solid Films* 308, 13 (1997).

7. T. Minami, H. Sonohara, T. Kakumu, and S. Takata, *Jpn. J. Appl. Phys.* 34, L971 (1995).

8. J. M. Phillips, J. Kwo, G. A. Thomas, S. A. Carter, R. J. Cava, S. Y. Hou, J. J. Krajewski, J. H. Marshall, W. F. Peck, D. H. Rapkine, and R. B. Vandover, *Appl. Phys. Lett.* 65, 115 (1994).

9. H. Unno, N. Hikuma, T. Omata, N. Ueda, T. Hashimoto, and H. Kawazoe, *Jpn. J. Appl. Phys.* 32, L1260 (1993).

10. H. Kawazoe, M. Yasukawa, H. Hyodo, M. Kurita, H. Yanagi, and H. Hosono, *Nature* 389, 939 (1997).

11. H. Yanagi, H. Kawazoe, A. Kudo, M. Yasukawa, and H. Hosono, *J. Electroceramics* 4, 407 (2000).

12. A. Kudo, H. Yanagi, H. Hosono, and H. Kawazoe, *Appl. Phys. Lett.* 73, 220 (1998).

13. S. Ibuki, H. Yanagi, K. Ueda, H. Kawazoe, and H. Hosono, *J. Appl. Phys.* 88, 3067 (2000).

14. T. Minami, *J. Vac. Sci. Technol. A* 17, 1765 (1999).

15. A. Kudo, H. Yanagi, K. Ueda, H. Hosono, H. Kawazoe, and Y. Yano, *Appl. Phys. Lett.* 75, 2851 (1999).

16. T. Minami, *MRS Bull.* 25, 38 (2000).

17. R. G. Gordon, *Mat. Res. Soc. Symp.* 426, 419 (1996).

18. G. Haacke, *J. Appl. Phys.* 9, 4086 (1976).

19. I. Hamberg and C. G. Granqvist, *J. Appl. Phys.* 60, R123 (1986).

20. M. A. Martinez, J. Herrero, and M. T. Gutierrez, *Solar Energy Mater. Solar Cells* 26, 309 (1992).

21. B. Mayer, *Thin Solid Films* 221, 166 (1992).

22. G. Frank and H. Kostlin, *Appl. Phys. A* 27, 197 (1982).

23. B. Bessais, H. Ezzaouia, and R. Bennaceur, *Semicond. Sci. Technol.* 8, 1671 (1993).

24. J. P. Zheng and H. S. Kwok, *Appl. Phys. Lett.* 63, 1 (1993).

25. J. L. Vossen and W. Kern, "Thin Film Processes II", Academic Press, New York, 1991.

26. S. Naseem and T. J. Coutts, *Thin Solid Films* 138, 65 (1986).

27. M. Higuchi, S. Uekusa, R. Nakano, and K. Yokogawa, *J. Appl. Phys.* 74, 6710 (1993).

28. W. F. Wu and B. S. Chiou, *Appl. Surf. Sci.* 68, 497 (1993).

29. Q. X. Jia, J. P. Zheng, H. S. Kwok, and W. A. Anderson, *Thin Solid Films* 258, 260 (1995).

30. D. B. Chrisey and G. K. Hubler, "Pulsed Laser Deposition of Thin Films", Wiley, New York, 1994.

31. H. Yanagi, S. Inoue, K. Ueda, H. Kawazoe, H. Hosono, and N. Hamada, *J. Appl. Phys.* 88, 4159 (2000).

32. M. Kamei, T. Yagami, S. Takaki, and Y. Shigesato, *Appl. Phys. Lett.* 64, 2712 (1994).

33. N. Taga, H. Odaka, Y. Shigesato, I. Yasui, M. Kamei, and T. E. Haynes, *J. Appl. Phys.* 80, 978 (1996).

34. B. S. Chiou and J. H. Lee, *J. Mater. Sci.* 7, 241 (1996).

35. K. M. Glassford and J. R. Chelikowsky, *Phys. Rev. B* 49, 7107 (1994).

36. W. D. Ryden, A. W. Lawson, and C. C. Sartain, *Phys. Rev. B* 1, 1494 (1970).

37. J. E. Graebner, E. S. Greiner, and W. D. Ryden, *Phys. Rev. B* 13, 2426 (1976).

38. L. Krusinelbaum, M. Wittmer, and D. S. Yee, *Appl. Phys. Lett.* 50, 1879 (1987).

39. Q. X. Jia, Z. Q. Shi, K. L. Jiao, W. A. Anderson, and F. M. Collins, *Thin Solid Films* 196, 29 (1991).

40. H. Kezuka, R. Egerton, M. Masui, T. Wada, T. Ikehata, H. Mase, and M. Takeuchi, *Appl. Surf. Sci.* 65–66, 293 (1993).

41. Q. Wang, D. Gilmer, Y. Fan, A. Franciosi, D. F. Evans, W. L. Gladfelter, and X. F. Zhang, *J. Mater. Res.* 12, 984 (1997).

42. E. Kolawa, F. C. T. So, E. T. S. Pan, and M. A. Nicolet, *Appl. Phys. Lett.* 50, 854 (1987).

43. M. L. Green, M. E. Gross, L. E. Papa, K. J. Schnoes, and D. Brasen, *J. Electrochem. Soc.* 132, 2677 (1985).

44. P. Lu, S. He, F. X. Li, and Q. X. Jia, *Thin Solid Films* 340, 140 (1999).

45. Q. X. Jia, X. D. Wu, S. R. Foltyn, A. T. Findikoglu, P. Tiwari, J. P. Zheng, and T. R. Jow, *Appl. Phys. Lett.* 67, 1677 (1995).

46. X. D. Fang, M. Tachiki, and T. Kobayashi, *Jpn. J. Appl. Phys.* 36, L511 (1997).

47. A. Iembo, F. Fuso, E. Arimondo, C. Ciofi, G. Pennelli, G. M. Curro, F. Neri, and M. Allegrini, *J. Mater. Res.* 12, 1433 (1997).

48. Z. Yuan, R. J. Puddephatt, and M. Sayer, *Chem. Mater.* 5, 908 (1993).

49. Y. Gao, G. Bai, Y. Liang, G. C. Dunham, and S. A. Chambers, *J. Mater. Res.* 12, 1844 (1997).

50. E. Kolawa, F. C. T. So, W. Flick, X. A. Zhao, E. T. S. Pan, and M. A. Nicolet, *Thin Solid Films* 173, 217 (1989).

51. L. Krusinelbaum, *Thin Solid Films* 169, 17 (1989).

52. S. Y. Mar, J. S. Liang, C. Y. Sun, and Y. S. Huang, *Thin Solid Films* 238, 158 (1994).

53. J. Si and S. B. Desu, *J. Mater. Res.* 8, 2644 (1993).

54. G. R. Bai, A. Wang, C. M. Foster, and J. Vetrone, *Thin Solid Films* 310, 75 (1997).

55. J. Vetrone, C. M. Foster, G. R. Bai, A. Wang, J. Patel, and X. Wu, *J. Mater. Res.* 13, 2281 (1998).

56. Q. X. Jia, S. G. Song, X. D. Wu, J. H. Cho, S. R. Foltyn, A. T. Findikoglu, and J. L. Smith, *Appl. Phys. Lett.* 68, 1069 (1996).

57. Q. X. Jia, S. G. Song, S. R. Foltyn, and X. D. Wu, *J. Mater. Res.* 10, 2401 (1995).

58. Q. X. Jia, X. D. Wu, G. Song, and S. R. Foltyn, *J. Vac. Sci. Technol. A* 14, 1107 (1996).

59. Q. X. Jia and P. Lu, *Philos. Mag. B* 81, 141 (2001).

60. C. L. Chen, Q. X. Jia, Y. C. Lu, J. L. Smith, and T. E. Mitchell, *J. Vac. Sci. Technol. A* 16, 2725 (1998).

61. Q. X. Jia, P. Arendt, J. R. Groves, Y. Fan, J. M. Roper, and S. R. Foltyn, *J. Mater. Res.* 13, 2461 (1998).

62. S. Saito and K. Kuramasu, *Jpn. J. Appl. Phys.* 31, 135 (1992).

63. W. Pan and S. B. Desu, *J. Vac. Sci. Technol. B* 12, 3208 (1994).

64. R. G. Vadimsky, R. P. Frankenthal, and D. E. Thompson, *J. Electrochem. Soc.* 126, 2017 (1979).

65. V. E. Henrich and P. A. Cox, "The Surface Science of Metal Oxides", Cambridge Univ. Press, Cambridge, UK, 1994.

66. J. A. Armstrong and M. Shafer, *IBM Tech. Disc. Bull.* 20, 4633 (1978).

67. M. Tomkiewicz, Y. S. Huang, and F. H. Pollak, *J. Electrochem. Soc.* 130, 1514 (1983).

68. A. J. McEvoy and W. Gissler, *J. Appl. Phys.* 53, 1251 (1982).

69. D. A. Vandenbroucke, R. L. Vanmeirhaeghe, W. H. Laflere, and F. Cardon, *J. Phys. D* 18, 731 (1985).

70. Y. T. Kim, C. W. Lee, and S. K. Kwak, *Appl. Phys. Lett.* 67, 807 (1995).

71. Q. X. Jia, K. L. Jiao, W. A. Anderson, and F. M. Collins, *Mater. Sci. Eng. B* 18, 220 (1993).

72. Q. X. Jia, K. L. Jiao, W. A. Anderson, and F. M. Collins, *Mater. Sci. Eng. B* 20, 301 (1993).

73. Q. X. Jia, K. L. Jiao, W. A. Anderson, and F. M. Collins, *J. Vac. Sci. Technol. A* 11, 1052 (1993).

74. D. P. Vijay and S. B. Desu, *J. Electrochem. Soc.* 140, 2640 (1993).

75. S. D. Bernstein, T. Y. Wong, Y. Kisler, and R. W. Tustison, *J. Mater. Res.* 8, 12 (1993).

76. Q. X. Jia, L. H. Chang, and W. A. Anderson, *J. Mater. Res.* 9, 2561 (1994).

77. K. Yoshikawa, T. Kimura, H. Noshiro, S. Otani, M. Yamada, and Y. Furumura, *Jpn. J. Appl. Phys.* 33, L867 (1994).

78. K. Takemura, T. Sakuma, and Y. Miyasaka, *Appl. Phys. Lett.* 64, 2967 (1994).

79. Q. X. Jia, A. T. Findikoglu, R. Zhou, S. R. Foltyn, X. D. Wu, J. L. Smith, Q. Wang, D. F. Evans, and W. L. Gladfelter, *Integrated Ferroelectrics* 19, 111 (1998).

80. Q. X. Jia and W. A. Anderson, *IEEE Trans. Components Hybrids Manufacturing Technol.* 15, 121 (1992).

81. J. P. Zheng, P. J. Cygan, and T. R. Jow, *J. Electrochem. Soc.* 142, 2699 (1995).

82. J. P. Zheng, T. R. Jow, Q. X. Jia, and X. D. Wu, *J. Electrochem. Soc.* 143, 1068 (1996).

83. R. O. Lezna, K. Kunimatsu, T. Ohtsuka, and N. Sato, *J. Electrochem. Soc.* 134, 3090 (1987).

84. A. Osaka, T. Takatsuna, and Y. Miura, *J. Non-Cryst. Solids* 178, 313 (1994).

85. C. Y. Xu and T. H. Baum, *Chem. Mater.* 10, 2329 (1998).

86. J. M. Williams, I. S. Lee, and R. A. Buchanan, *Surf. Coatings Technol.* 51, 385 (1992).

87. T. Nakamura, Y. Nakao, A. Kamisawa, and H. Takasu, *Appl. Phys. Lett.* 65, 1522 (1994).

88. P. C. Liao, W. S. Ho, Y. S. Huang, and K. K. Tiong, *J. Mater. Res.* 13, 1318 (1998).

89. Y. Sato, *Vacuum* 41, 1198 (1990).

90. M. A. El Khakani, M. Chaker, and E. Gat, *Appl. Phys. Lett.* 69, 2027 (1996).

91. M. A. El Khakani and M. Chaker, *Thin Solid Films* 335, 6 (1998).

92. H. J. Cho, H. Horii, C. S. Hwang, J. W. Kim, C. S. Kang, B. T. Lee, S. I. Lee, Y. B. Koh, and M. Y. Lee, *Jpn. J. Appl. Phys.* 36, 1722 (1997).

93. R. Sanjines, A. Aruchamy, and F. Levy, *J. Electrochem. Soc.* 136, 1740 (1989).

94. B. R. Chalamala, Y. Wei, R. H. Reuss, S. Aggarwal, S. R. Perusse, B. E. Gnade, and R. Ramesh, *J. Vac. Sci. Technol. B* 18, 1919 (2000).

95. M. Peuckert, *Surf. Sci.* 144, 451 (1984).

96. B. R. Chalamala, Y. Wei, R. H. Reuss, S. Aggarwal, B. E. Gnade, R. Ramesh, J. M. Bernhard, E. D. Sosa, and D. E. Golden, *Appl. Phys. Lett.* 74, 1394 (1999).

97. T. Nakamura, Y. Nakao, A. Kamisawa, and H. Takasu, *Jpn. J. Appl. Phys.* 33, 5207 (1994).

98. L. S. Robblee, J. L. Lefko, and S. B. Brummer, *J. Electrochem. Soc.* 130, 731 (1983).

99. S. Gottesfeld and S. Srinivasan, *J. Electroanal. Chem. Interfacial Electrochem.* 86, 89 (1978).

100. J. Mozota and B. E. Conway, *J. Electrochem. Soc.* 128, 2142 (1981).

101. S. Gottesfeld and J. D. E. McIntyre, *J. Electrochem. Soc.* 126, 742 (1979).

102. T. Katsube, I. Lauks, and J. N. Zemel, *Sens. Actuators* 2, 399 (1982).

103. K. Pasztor, A. Sekiguchi, N. Shimo, N. Kitamura, and H. Masuhara, *Sens. Actuators B* 12, 225 (1993).

104. R. J. Bouchard and J. L. Gillson, *Mater. Res. Bull.* 7, 873 (1972).

105. P. B. Allen, H. Berger, O. Chauvet, L. Forro, T. Jarlborg, A. Junod, B. Revaz, and G. Santi, *Phys. Rev. B* 53, 4393 (1996).

106. L. Klein, J. S. Dodge, C. H. Ahn, G. J. Snyder, T. H. Geballe, M. R. Beasley, and A. Kapitulnik, *Phys. Rev. Lett.* 77, 2774 (1996).

107. Y. Senzaki, M. J. Hampdensmith, T. T. Kodas, and J. W. Hussler, *J. Am. Ceram. Soc.* 78, 2977 (1995).

108. S. J. Benerofe, C. H. Ahn, M. M. Wang, K. E. Kihlstrom, K. B. Do, S. B. Arnason, M. M. Fejer, T. H. Geballe, M. R. Beasley, and R. H. Hammond, *J. Vac. Sci. Technol. B* 12, 1217 (1994).

109. J. P. Mercurio, J. H. Yi, M. Manier, and P. Thomas, *J. Alloys Compounds* 308, 77 (2000).

110. N. Okuda, K. Saito, and H. Funakubo, *Jpn. J. Appl. Phys.* 39, 572 (2000).

111. S. Ohashi, M. Lippmaa, N. Nakagawa, H. Nagasawa, H. Koinuma, and M. Kawasaki, *Rev. Sci. Instrum.* 70, 178 (1999).

112. C. B. Eom, R. J. Cava, R. M. Fleming, J. M. Phillips, R. B. Vandover, J. H. Marshall, J. W. P. Hsu, J. J. Krajewski, and W. F. Peck, *Science* 258, 1766 (1992).

113. Q. X. Jia, F. Chu, C. D. Adams, X. D. Wu, M. Hawley, J. H. Cho, A. T. Findikoglu, S. R. Foltyn, J. L. Smith, and T. E. Mitchell, *J. Mater. Res.* 11, 2263 (1996).

114. K. Tanabe, D. K. Lathrop, S. E. Russek, and R. A. Buhrman, *J. Appl. Phys.* 66, 3148 (1989).

115. R. L. Sandstrom, W. J. Gallagher, T. R. Dinger, R. H. Koch, R. B. Laibowitz, A. W. Kleinsasser, R. J. Gambino, B. Bumble, and M. F. Chisholm, *Appl. Phys. Lett.* 53, 444 (1988).

116. C. B. Eom, J. Z. Sun, K. Yamamoto, A. F. Marshall, K. E. Luther, T. H. Geballe, and S. S. Laderman, *Appl. Phys. Lett.* 55, 595 (1989).

117. P. Lu, F. Chu, Q. X. Jia, and T. E. Mitchell, *J. Mater. Res.* 13, 2302 (1998).

118. P. Tiwari, X. D. Wu, S. R. Foltyn, I. H. Campbell, Q. X. Jia, R. E. Muenchausen, D. E. Peterson, and T. E. Mitchell, *J. Electron. Mater.* 25, 51 (1996).

119. Q. X. Jia, H. H. Kung, and X. D. Wu, *Thin Solid Films* 299, 115 (1997).

120. Q. X. Jia, S. R. Foltyn, M. Hawley, and X. D. Wu, *J. Vac. Sci. Technol. A* 15, 1080 (1997).

121. Y. Nora and S. Miyahara, *J. Phys. Soc. Jpn.* 27, 518 (1969).

122. C. Yoshida, A. Yoshida, and H. Tamura, *Appl. Phys. Lett.* 75, 1449 (1999).

123. C. B. Eom, R. B. Vandover, J. M. Phillips, D. J. Werder, J. H. Marshall, C. H. Chen, R. J. Cava, R. M. Fleming, and D. K. Fork, *Appl. Phys. Lett.* 63, 2570 (1993).

124. Q. X. Jia, X. D. Wu, S. R. Foltyn, and P. Tiwari, *Appl. Phys. Lett.* 66, 2197 (1995).

125. Q. X. Jia, D. S. Zhou, S. R. Foltyn, X. D. Wu, A. T. Findikoglu, and J. L. Smith, *Philos. Mag. B* 75, 261 (1997).

126. X. D. Wu, S. R. Foltyn, R. C. Dye, Y. Coulter, and R. E. Muenchausen, *Appl. Phys. Lett.* 62, 2434 (1993).

127. L. Antognazza, K. Char, T. H. Geballe, L. L. H. King, and A. W. Sleight, *Appl. Phys. Lett.* 63, 1005 (1993).

128. L. Klein, J. S. Dodge, T. H. Geballe, A. Kapitulnik, A. F. Marshall, L. Antognazza, and K. Char, *Appl. Phys. Lett.* 66, 2427 (1995).

129. A. Kanbayasi, *J. Phys. Soc. Jpn.* 44, 89 (1978).

130. R. A. Rao, Q. Gan, C. B. Eom, R. J. Cava, Y. Suzuki, J. J. Krajewski, S. C. Gausepohl, and M. Lee, *Appl. Phys. Lett.* 70, 3035 (1997).

131. N. Fukushima, K. Sano, T. Schimizu, K. Abe, and S. Komatsu, *Appl. Phys. Lett.* 73, 1200 (1998).

132. S. M. Koo, L. R. Zheng, and K. V. Rao, *J. Mater. Res.* 14, 3833 (1999).

133. C. M. Chu and P. Lin, *Appl. Phys. Lett.* 72, 1241 (1998).

134. M. Itoh, M. Shikano, and T. Shimura, *Phys. Rev. B* 51, 16432 (1995).

135. Y. Maeno, H. Hashimoto, K. Yoshida, S. Nishizaki, T. Fujita, J. G. Bednorz, and F. Lichtenberg, *Nature* 372, 532 (1994).

136. J. Randall and R. Ward, *J. Am. Chem. Soc.* 81, 2629 (1959).

137. S. Madhavan, J. A. Mitchell, T. Nemoto, S. Wozniak, Y. Liu, D. G. Schlom, A. Dabkowski, and H. A. Dabkowska, *J. Crystal Growth* 174, 417 (1997).

138. M. A. Senarisrodriguez and J. B. Goodenough, *J. Solid State Chem.* 118, 323 (1995).

139. A. N. Petrov, O. F. Kononchuk, A. V. Andreev, V. A. Cherepanov, and P. Kofstad, *Solid State Ionics* 80, 189 (1995).

140. P. M. Raccah and J. B. Goodenough, *J. Appl. Phys.* 39, 1209 (1968).

141. J. T. Cheung, P. E. D. Morgan, D. H. Lowndes, X. Y. Zheng, and J. Breen, *Appl. Phys. Lett.* 62, 2045 (1993).

142. B. Nagaraj, S. Aggarwal, T. K. Song, T. Sawhney, and R. Ramesh, *Phys. Rev. B* 59, 16022 (1999).

143. K. Hwang, H. Lee, H. Ryu, Y. Lim, I. Yamaguchi, T. Manabe, T. Kumagai, and S. Mizuta, *Jpn. J. Appl. Phys.* 38, 6489 (1999).

144. Z. L. Wang and J. M. Zhang, *Philos. Mag. A* 72, 1513 (1995).

145. F. Wang, A. Uusimaki, and S. Leppavuori, *Appl. Phys. Lett.* 67, 1692 (1995).

146. J. V. Mantese, A. L. Micheli, A. B. Catalan, and N. W. Schubring, *Appl. Phys. Lett.* 64, 3509 (1994).

147. J. Kirchnerova and D. B. Hibbert, *J. Mater. Sci.* 28, 5800 (1993).

148. S. Sadashivan, S. Aggarwal, T. K. Song, R. Ramesh, J. T. Evans, B. A. Tuttle, W. L. Warren, and D. Dimos, *J. Appl. Phys.* 83, 2165 (1998).

149. G. P. Luo, Y. S. Wang, S. Y. Chen, A. K. Heilman, C. L. Chen, C. W. Chu, Y. Liou, and N. B. Ming, *Appl. Phys. Lett.* 76, 1908 (2000).

150. W. B. Wu, F. Lu, K. H. Wong, G. Pang, C. L. Choy, and Y. H. Zhang, *J. Appl. Phys.* 88, 700 (2000).

151. Y. Tokura, Y. Taguchi, Y. Okada, Y. Fujishima, T. Arima, K. Kumagai, and Y. Iye, *Phys. Rev. Lett.* 70, 2126 (1993).

152. S. Madhukar, S. Aggarwal, A. M. Dhote, R. Ramesh, A. Krishnan, D. Keeble, and E. Poindexter, *J. Appl. Phys.* 81, 3543 (1997).

153. R. Ramesh, J. Lee, T. Sands, V. G. Keramidas, and O. Auciello, *Appl. Phys. Lett.* 64, 2511 (1994).

154. Q. X. Jia, P. N. Arendt, C. Kwon, J. M. Roper, Y. Fan, J. R. Groves, and S. R. Foltyn, *J. Vac. Sci. Technol. A* 16, 1380 (1998).

155. R. Ramesh, H. Gilchrist, T. Sands, V. G. Keramidas, R. Haakenaasen, and D. K. Fork, *Appl. Phys. Lett.* 63, 3592 (1993).

156. C. Kwon, Y. Gim, Y. Fan, M. F. Hundley, J. M. Roper, P. N. Arendt, and Q. X. Jia, *Appl. Phys. Lett.* 73, 695 (1998).

157. J. B. Goodenough and R. C. Raccah, *J. Appl. Phys.* 36, 1031 (1963).

158. R. Dat, D. J. Lichtenwalner, O. Auciello, and A. I. Kingon, *Appl. Phys. Lett.* 64, 2673 (1994).

159. K. Char, L. Antognazza, and T. H. Geballe, *Appl. Phys. Lett.* 63, 2420 (1993).

Index